THE LIBRARY
ST. MARY'S COLLEGE OF MARYLAND
ST. MARY'S CITY, MARYLAND 20686

D1518818

ELSEVIER'S DICTIONARY OF PLANT NAMES AND THEIR ORIGIN

ELSEVIER'S DICTIONARY OF PLANT NAMES AND THEIR ORIGIN

by

DONALD WATTS
Bath, United Kingdom

2000
ELSEVIER
Amsterdam – Lausanne – New York – Oxford – Shannon – Singapore – Tokyo

ELSEVIER SCIENCE B.V.
Sara Burgerhartstraat 25
P.O. Box 211, 1000 AE Amsterdam, The Netherlands

© 2000 Elsevier Science B.V. All rights reserved.

This dictionary is protected under copyright by Elsevier Science, and the following terms and conditions apply to its use:

Photocopying
Single photocopies may be made for personal use as allowed by national copyright laws. Permission of the publisher and payment of a fee is required for all other photocopying, including multiple or systematic copying, copying for advertising or promotional purposes, resale, and all forms of document delivery. Special rates are available for educational institutions that wish to make photocopies for non-profit educational classroom use.

Permissions may be sought directly from Elsevier Science Global RightsDepartment, PO Box 800, Oxford OX5 1DX, UK; phone: (+44) 1865 843830, fax: (+44) 1865 853333, e-mail: permissions@elsevier.co.uk. You may also contact Global Right directly through Elsevier's home page (http://www.elsevier.nl), by selecting 'Obtaining Permissions'.

In the USA, users may clear permissions and make payments through the Copyright Clearance Center, Inc., 222 Rosewood Drive, Danvers, MA 01923, USA; phone: (978) 7508400, fax: (978) 7504744, and in the UK through the Copyright Licensing Agency Rapid Clearance Service (CLARCS), 90 Tottenham Court Road, London W1P 0LP, UK; phone: (+44) 171 436 5931; fax: (+44) 171 436 3986. Other countries may have a local reprographic rights agency for payments.

Derivative Works
Permission of the publisher is required for all derivative works, including compilations and translations.

Electronic Storage or Usage
Permission of the publisher is required to store or use electronically any material contained in this work. Contact the publisher at the address indicated.

Except as outlined above, no part of this work may be reproduced, stored in a retrieval system or transmitted in any form or by any means, electronic, mechanical, photocopying, recording or otherwise, without prior written permission of the publisher.
Address permissions requests to: Elsevier Science Global RightsDepartment, at the mail, fax and e-mail addresses noted above.

Notice
No responsibility is assumed by the Publisher for any injury and/or damage to persons or property as a matter of products liability, negligence or otherwise, or from any use or operation of any methods, products, instructions or ideas contained in the material herein. Because of rapid advances in the medical sciences, in particular, independent verification of diagnoses and drug dosages should be made.

First edition 2000

Library of Congress Cataloging in Publication Data
A catalog record from the Library of Congress has been applied for.

ISBN: 0-444-50356-0

∞ The paper used in this publication meets the requirements of ANSI/NISO Z39.48-1992 (Permanence of Paper).
Printed in The Netherlands.

Preface

The dictionary was conceived as part of the author's wider interest in plant and tree lore, and ethnobotanical studies. Knowledge of plant names can give insight into largely forgotten beliefs. It takes a folklorist to understand why the common red poppy is, or was, known as Blind Man. An old superstition has it that if the poppy was put to the eyes, it would cause blindness. Such names were probably the result of some taboo against picking the plant, for other names for the poppy show all kinds of results, thunder and lightning, headache, etc., if the plant was gathered. Similarly, other names surely must be the result of a country mother's admonition to her children, for a warning against eating berries that are poisonous, or even suspected of being poisonous, carries more weight when the name given to them reinforces the warning. Therefore, many such plants or fruits may be ascribed to the devil (Devil's Berries for Deadly Nightshade is an example. But the same name serves for Tutsan, one of the St John's Wort genus, and that is an example of the suspicion with which its berries are viewed).

Names can be purely descriptive, spectacularly so in many from the West country, and can also serve to explain the meaning of the botanical name. Beauty-Berry is an example: it is the name given to the American shrub that belongs to the genus Callicarpa, which is made up of two Greek words that mean beauty and berry.

Even though a surprising number of the names listed are still current, most of them belong to the past. Better education has ensured that there is only one name for a dandelion, and the advent of migration of labour has obscured local dialect. When the same name for a particular plant is recorded in Devonshire and Yorkshire, it is not an odd coincidence, but almost certainly the influence of migrant personnel.

Literary, or "book", names as they are known, have been included in this dictionary, as being a very important part of the whole. Many of them provide links in the transmission of words through the ages. Thor's Beard, for example, is a book name for houseleek, and has never been used in the vernacular. But it highlights the legend that houseleek is a lightning plant, and by reverse logic is a preserver from fire.

Abbreviations

(a) Textual

B & H	BRITTEN, James and Robert Holland - A dictionary of English plant names. English Dialect Society, 1886
D & G	DARTNELL, George Edward and Edward Hungerford Goddard - Wiltshire words: a glossary of words used in the County of Wiltshire. English Dialect Society, 1893 (reprinted 1991 by Wiltshire Life Society under title Wiltshire words)
D & G.1899	DARTNELL, George Edward and Edward Hungerford Goddard - Contributions towards a Wiltshire glossary [final part]; Wiltshire Archaeological Magazine, 1899
H & P	HYAM, Roger and Richard Pankhurst - Plants and their names: a concise dictionary. Oxford University Press, 1995
ODEE	Oxford dictionary of English etymology, edited by C T Onions Oxford, Clarendon Press, 1966
Oxf.Ann.Rec.	Oxfordshire and District Folklore Society. Annual Record
PLNN	Plant-lore notes and news no 1, May 1988 - no 35, 1994
RHS	ROYAL HORTICULTURAL SOCIETY - Gardeners' encyclopedia of plants and flowers. Dorling Kindersley, 1989

(b) Topographical

Beds.	Bedfordshire
Berks.	Berkshire
Bucks.	Buckinghamshire
Camb.	Cambridgeshire
Ches.	Cheshire
Corn.	Cornwall
Cumb.	Cumbria
Derb.	Derbyshire
Dev.	Devonshire
Dor.	Dorset
Dur.	co Durham
E Ang.	East Anglia
Ess.	Essex
Glam.	Glamorgan
Glos.	Gloucestershire
Hants.	Hampshire
Heref.	Herefordshire
Herts.	Hertfordshire

IOM	Isle of Man
IOW	Isle of Wight
Lancs.	Lancashire
Leics.	Leicestershire
Lincs.	Lincolnshire
Norf.	Norfolk
Northants.	Northamptonshire
Notts.	Nottinghamshire
N'thum.	Northumberland
Oxf.	Oxfordshire
Pemb.	Pembrokeshire
Som.	Somerset
Shrop.	Shropshire
Staffs.	Staffordshire
Suff.	Suffolk
Surr.	Surrey
Suss.	Sussex
War.	Warwickshire
Wilts.	Wiltshire
Worcs.	Worcestershire
Yks.	Yorkshire

Bibliography

ABLETT, William H - *English trees and tree-planting*. Smith Elder, 1880

ACKERMANN, A S E - *Popular fallacies explained and corrected*, 3rd ed. Old Westminster Press, 1923

ADDY, Sidney Oldall - *A glossary of words used in the neighbourhood of Sheffield*. Trübner for English Dialect Society, 1888

AITKEN, Hannah - *A forgotten heritage: original folk tales of Lowland Scotland*. Edinburgh, Scottish Academic Press, 1973

AKERMAN, John Yonge - *A glossary of provincial words and phrases in use in Wiltshire*. John Russell Smith, 1842

ALBERTUS MAGNUS - [Liber aggregationis] *The book of secrets of Albertus Magnus of the virtues of herbs, stones and certain beasts; also A book of the marvels of the world* edited by Michael R Best and Frank H Brightman. Oxford, Clarendon Press, 1973

ALEXANDER, J and D G Coursey - "The origins of yam cultivation" in UCKO, Peter J and G W Dimbleby (editors) - *The domestication and exploitation of plants and animals*. Duckworth, 1969

ALLAN, Mea - *The gardener's book of weeds*. Macdonald & Jane's, 1978

ALLIES, Jabez - *Antiquities and folklore of Worcestershire*. John Russell Smith, 1896

AMSDEN, Charles A - *Navajo weaving*. Santa Ana, Southwest Museum, 1934

ANDERSEN, Johannes C - *Myths & legends of the Polynesians*. Harrap, 1928

ANDERSON, Edgar - *Plants, man and life*. Berkeley, Univ. California Press, 1952

ARGENTI, Philip P and H J Rose - *The folklore of Chios*, 2 vols. Cambridge Univ. Press, 1949

ATKINSON, J C - *A glossary of the Cleveland dialect: explanatory, derivative, and critical*. John Russell Smith, 1868

ATTENBOROUGH, David - *The private life of plants: a natural history of plant behaviour*. BBC Books, 1995

AUBREY, John - *The natural history of Wiltshire*, edited by James Britten Devizes. Wiltshire Topographical Society, 1847

AWBERY, G M - "Plant names of religious origin in Welsh oral tradition" in VICKERY, Roy (editor) - *Plant-lore studies*. Folklore Society, 1984

BAKER, Anne Elizabeth - *Glossary of Northamptonshire words and phrases*, 2 vols. John Russell Smith, 1854

BAKER, H G - *Plants and civilization*. Macmillan, 1964

BAKER, Margaret - *Discovering the folklore of plants*, 2nd ed. Aylesbury, Shire Publications 1980 (Discovering series; no 74)

BAKER, Margaret - *Folklore and customs of rural England*. Newton Abbot, David & Charles, 1974

BAKER, Margaret - *The gardener's folklore*. Newton Abbot, David & Charles, 1977

BAKER, Margaret - *Wedding customs and folklore*. Newton Abbot, David & Charles, 1977a

BALCH, E E - In a Wiltshire village: some old songs and customs. *Antiquary*, vol 44, 1908, 379-382

BANKS, Mrs M Macleod - *British calendar customs: Scotland,* 3 vols. Folklore Society, 1937-41

BANKS, Mrs M Macleod - *British calendar customs: Orkney and Shetland.* Folklore Society, 1946

BARBER, Charles - *Early modern English.* Deutsch, 1976 (The language library series)

BARBER, Peter and C E Lucas Phillips - *The trees around us.* Weidenfeld & Nicolson, 1975

BARBOUR, John H - Some country remedies and their uses. *Folklore*, vol 8, 1897, 386-390

BARDSWELL, Frances A - *The herb garden.* A & C Black, 1911

BARKER, S G - Some indigenous dyestuffs of Travancore. *Government of Travancore, Department of Industries. Bulletin*, no 11, 1921

BARNES, William - *A glossary of the Dorset dialect with a grammar of its word shapening and wording.* Dorchester, M & E Case; London, Trübner, 1886

BARRETT, S A and E W Gifford - Miwok material culture. *Public Museum of Milwaukee. Bulletin*, vol 2, no 4, 1933

BARROWS, David Prescott - *The ethno-botany of the Coahuilla Indians of southern California.* Chicago, Univ. Chicago Press, 1900

BARTON, Benjamin H and Thomas Castle - *The British flora medica*, 2 vols. E Cox, 1837

BAYLEY, Harold - *Archaic England.* Chapman & Hall, 1919

BEAN, Lowell John - *Mukat's people: the Cahuilla Indians of southern California.* Berkeley, Univ. California Press, 1972

BECKWITH, Martha Warren - *Black roadways: a study of Jamaican folk life.* Chapel Hill, Univ. North Carolina Press, 1929

BELL, Hesketh J - *Obeah: witchcraft in the West Indies*, 2nd ed. Sampson Low, Marston & Co, 1893

BENNETT, Wendell C and Junius B Bird - *Andean cultural history.* Hale, 1965

BERKSHIRE FEDERATION OF WOMEN'S INSTITUTES - *The Berkshire book.* Reading, 1951

BERDOE, Edward - *The origin and growth of the healing art: a popular history of medicine in all ages and countries.* Swan Sonnenschein, 1893

BERGEN, Fanny D - Animal and plant lore. *American Folklore Society. Memoirs*, vol 7, 1899

BERGEN, Fanny D - *Current superstitions collected from the oral tradition of English speaking folk.* Boston, Houghton Mifflin, 1896 (American Folklore Society. Memoirs, vol 4, 1896)

BERNEY, Julia - Home thoughts from Oz. *The Countryman*, vol 94, no 3, Autumn 1989, 22-27

BIANCHINI, Francesco and Francesco Corbetta - *The fruits of the earth.* Cassell, 1975

BLACK, William G - *Folk medicine.* Folklore Society, 1883

BLACKWOOD, Beatrice - Use of plants among the Kukukuku of southeast-central New Guinea. *Pacific Science Congress, 6th, 1940. Proceedings.* vol 4, 111-126

BLAMEY, Marjorie, Richard Fitter and Alastair Fitter - *Wild flowers: the wild flowers of Britain and northern Europe*. Collins, 1974

BLAMEY, Philip and Marjorie Blamey - *Marjorie Blamey's flowers of the countryside*. Collins, 1980

BLOOM, J Harvey - *Folklore, old customs and superstitions in Shakespeareland*. Mitchell Hughes & Clark, 1930

BLOUNT, Margaret - The building of caravans for gypsies and showmen. *Journal of the Gypsy Lore Society*, 3rd series, vol 36, 1957, 121-125

BLUNT, Wilfred - *The art of botanical illustration*. Collins, 1950 (New naturalist series)

BLUNT, Wilfred - *Flowers drawn from nature*, edited by Gerard von Spoendanck. Leslie Urquhart Press, 1957

BOASE, Wendy - *The folklore of Hampshire and the Isle of Wight*. Batsford, 1976 (The folklore of the British Isles series)

BOLAND, Bridget - *Gardeners' magic and other old wives' lore*. Bodley Head, 1977

BOLTON, Eileen M - *Lichens for vegetable dyeing*. Studio, 1960

BONAR, Ann - *The complete guide to conservatory plants*. Collins & Brown, 1992

BONSER, Wilfrid - *The medical background of Anglo-Saxon England: a study in history, psychology and folklore*. Wellcome Hist. Med. Libr., 1963

BOTTRELL, William - *Traditions and hearth-side stories of west Cornwall*. Penzance, 1870

BRIGGS, Katharine Mary - *Nine lives: cats in folklore*. Routledge, 1980

BRIGHTMAN, Frank H and B E Nicholson (illustrator) - *The Oxford book of flowerless plants: ferns, fungi, mosses, liverworts, lichens, and seaweeds*. Oxford, Univ. Press, 1966

BRILL, Edith - *Life and tradition on the Cotswolds*. Dent, 1973

BRIMBLE, L J F - *Trees in Britain: wild, ornamental and economic, and some relatives in other lands*. Macmillan, 1948

BRIMBLE, L J F - *Flowers in Britain*. Macmillan, 1944

BRITTEN, James - The shamrock. *The Month*, vol 137, January-June 1921, 193-205

BRITTEN, James and Robert Holland - *A dictionary of English plant names*. English Dialect Society, 1896

BROADWOOD, Lucy E - A Swiss charm. *Folklore*, vol 16, 1905, 465-7

BROCKETT, John Trotter - *A glossary of North Country words, with their etymology, and affinity to other languages*, 3rd edition, corrected and enlarged by W E Brockett. Newcastle upon Tyne, Emerson Charnley, Bigg Market; London, Simpkin, Marshall 1846

BROUK, B - *Plants consumed by man*. Academic Press, 1975

BROWN, O Phelps - *The complete herbalist: or, The people their own physicians, by the use of nature's remedies*. The author, 1872

BROWN, P W F - Notes on names of the thorn. *Folklore*, vol 70, 1959, 416-8

BROWN, Raymond Lamont - *A book of superstitions*. Newton Abbot, David & Charles, 1970

BROWN, Theo - Tales of a Dartmoor village: some preliminary notes on the folklore of Postbridge. *Devonshire Association. Report and Transactions*, vol 93, 1961, 194-227

BROWNE, Sir Thomas - *Vulgar errors* (1646), edited by S Wilkin Bell, 1884

BROWNLOW, Margaret E - *Herbs and the fragrant garden*. Herb Farm, 1957
BUHLER, A - Primitive dyeing materials. *CIBA Review*, no 68, 1948
BUNYARD, Edward A - *Old garden roses*. Country Life, 1936
BURNE, Charlotte S - *Shropshire folklore*. Trübner, 1883
BURNE, Charlotte S - *The handbook of folklore*. Folklore Society, 1914
BURTON, Alfred - *Rush-bearing*. Manchester, Brook & Chrystal, 1891
CAMDEN, William - *Britannia*. Newly translated into English by Edmund Gibson, 1695
CAMPBELL, John Gregorson - *Superstitions of the Highlands and Islands of Scotland*. Glasgow, Maclehose, 1900
CANDOLLE, Alphonse de - *Origin of cultivated plants*. Kegan Paul, Trench & Co, 1884
CAREW, Richard - *Survey of Cornwall* (1602), Francis Lord de Dunstanville's edition. J Faulder, 1811
CARMICHAEL, Alexander - *Carmina Gadelica*, 2nd ed. Edinburgh, Oliver & Boyd, 1928
CARR, William - *The dialect of Craven, in the West Riding of the county of York*, 2 vols, 2nd ed. Wm Crofts, 1828
CHAMBERLIN, Ralph V - The ethno-botany of the Gosiute Indians of Utah. *American Anthropological Association. Memoirs*, vol 2, 1911, 331-405
CHAPLIN, Mary - *Riverside gardening*. Collingridge, 1964
CHAPMAN, Crisfield - Dog rose. *The Countryman*, vol 94, no 2, Summer 1989, 150
CHEVIOT, Andrew - *Proverbs, proverbial expressions and popular rhymes of Scotland*. Paisley, Gardner, 1896
CHOAPE, R Pearse - *The dialect of Hartland, Devonshire*. Kegan Paul, Trench, Trübner for English Dialect Society, 1891
CLAIR, Colin - *Of herbs and spices*. Abelard-Schumann, 1961
CLAPHAM, A R, T G Tutin and E F Warburg - *Flora of the British Isles*. Cambridge, Univ. Press, 1952
CLARE, John - *The natural history prose writings of John Clare*, edited by Margaret Grainger. Oxford, 1983
COATS, Alice M - *Flowers and their histories*. Black, 1956
COATS, Alice M - *The treasury of flowers*. Phaidon, 1975
COCKAYNE, Oswald - *Leechdoms, wort-cunning and star-craft of early England*, 3 vols. Longman, 1864
CODRINGTON, K de B - The use of counter-irritants in the Deccan. *Royal Anthropological Institute of Great Britain and Ireland. Journal*, vol 66, 1936, 369-377
COLES, William - *Adam in Eden: or, Nature's paradise*. 1657
CONDRY, W M - *The Snowdonia National Park*. Collins, 1966 (New naturalist series)
CONNOR, H E - The poisonous plants in New Zealand, 2nd rev ed. *Wellington. Department of Scientific and Industrial Research. Bulletin*, 1977
CONWAY, David - *The magic of herbs*. Cape, 1973
COOPER, Doris Muriel - personal communication
COOPER, William Durrance - *A glossary of the provincialisms in use in the county of Sussex*, 1834, 2nd ed. John Russell Smith, 1853
COPE, Sir William H - *A glossary of Hampshire words and phrases*. Trübner for English Dialect Society, 1883

COURTNEY, M A - *Glossary of words in use in Cornwall; 1. West Cornwall*. Trübner for English Dialect Society, 1880
COURTNEY, M A - *Cornish feasts and folklore*. Penzance, Beare & Son, 1890
CRAWFORD, I M - *The art of the wandjina: aboriginal cave paintings in Kimberley, Western Australia*. Melbourne, Oxford Univ. Press, 1968
CROMEK, R H - *Remains of Nithsdale and Galloway song*. Cadell & Davies, 1810
CROOKE, W - The folklore of Irish plants and animals. *Folklore*, vol 25; 1914, 258-9
CROSSING, William - *Folk rhymes of Devon*. Exeter, Commin, 1911
CULLUM, Elizabeth - *A cottage herbal*. Newton Abbot, David & Charles, 1975
CULPEPPER, Nicholas - *The English physician enlarged*. 1775
CUMMING, C F Gordon - *In the Hebrides*. Chatto & Windus, 1883
CUNNINGHAM, John J and Rosalie J Côté - *Common plants: botanical and colloquial nomenclature*. New York and London, Garland, 1977
CURTIS, William - *Flora londinensis: or, plates and descriptions of such plants as grow wild in the environs of London*, 3 vols, 1777-1798
DACOMBE, Marianne R (editor) - *Dorset up along and down along: a collection of history, tradition, folklore, flower names, and herbal lore, gathered by members of Women's Institutes*. Dorchester, DFWI, 1935
DALE, Ivan R and P J Greenway - *Kenya trees and shrubs*. Nairobi, Buchanan's Kenya Estates, 1961
DALE-GREEN, Patricia - *Cult of the cat*. Heinemann, 1963
DALYELL, John Graham - *The darker superstitions of Scotland*. Glasgow, Richard Griffin, 1835
DALZIEL, J M - *The useful plants of west tropical Africa*. Crown Agents, 1937 (appendix to Hutchinson, J and J M Dalziel, *Flora of west tropical Africa*)
DARLING, F Fraser and J Morton Boyd - *The Highlands and Islands*. Collins, 1964 (New naturalist series)
DARTNELL, George Edward and Edward Hungerford Goddard - *Wiltshire words: a glossary of words used in the County of Wiltshire*. English Dialect Society, 1893 (reprinted 1991 by Wiltshire Life Society)
DARTNELL, George Edward and Edward Hungerford Goddard - Contributions towards a Wiltshire glossary (final part). *Wiltshire Archaeological Magazine*, vol 29, 1899, 233-270
DAVEY, Norman - *A history of building materials*. Phoenix House, 1961
DAVIES, J C - *Folklore of west and mid-Wales*. Aberystwyth, 1911
DAWSON, Warren M - *A leechbook, or collection of medical recipes, of the 15th century*. Macmillan, 1934
DAWSON, Warren M - *The bridle of Pegasus: studies in magic, mythology and folklore*. Methuen, 1930
DEANE, Tony and Tony Shaw - *The folklore of Cornwall*. Batsford, 1975 (The folklore of the British Isles series)
DENHAM, M A - *A collection of proverbs and popular sayings relating to the seasons,etc*. Percy Society, 1846
DENHAM, M A - *The Denham tracts*, edited by James Hardy. Folklore Society, 1892
DENSMORE, Frances - Uses of plants by the Chippewa Indians. *Washington. Bureau of*

American Ethnology. Annual Report, 44th, 1926-7, 1928

DEVLIN, Terence - A posy of Euro-flowers. *The Countryman*, vol 94, no 1, Spring 1989, 114-115

DICKINS, Bruce - Yorkshire hobs. *Yorkshire Dialect Society. Transactions*, vol 2, 1942, 9-23

DICKSON, Isabel A - The Burry-man. *Folklore*, vol 19, 1908, 379-387

DIMBLEBY, G W - *Plants and archaeology*. John Baker, 1967

DINSDALE, F T - *A glossary of provincial words used in Teesdale in the county of Durham*. John Russell Smith, 1849

DIOSCORIDES [Herbal] - *The Greek herbal of Dioscorides*, illustrated by a Byzantine AD 512, Englished by John Goodyer AD 1655, edited and first printed by Robert T Gunther AD 1933. Oxford, Univ. Press, 1934

DODSON, Harry and Jennifer Davies - *The Victorian kitchen garden companion*. BBC, 1988

DORSON, Richard M - *Buying the wind: regional folklore in the United States*. Chicago Univ. Press, 1964

DOUGLAS, James Sholto - *Alternative foods: a world guide to lesser-known edible plants*. Pelham, 1978

DRAYTON, Michael - "Nymphidia" in HALLIWELL, J O - *Illustrations of the fairy mythology of Midsummer Night's Dream*. Shakespeare Society, 1845

DRIVER, Harold E - *Indians of North America*. Chicago Univ. Press, 1961

DRURY, Susan M - Plants and wart cures in England from the seventeenth to the nineteenth century: some examples. *Folklore*, vol 102, part 1, 1991, 97-100

DRURY, Susan M - Plants and pest control in England circa 1400-1700: a preliminary study. *Folklore*, vol 103, part 1, 1992, 103-6

DUNCALF, William G - *The Guinness book of plant facts and feats*. Enfield, Guinness Superlatives, 1976

DUNCAN, Angus - *Hebridean island: memories of Scarp*. East Linton, Tuckwell Press, 1995

DURAN-REYNALS, M L - *The fever-bark tree: the pageant of quinine*. W H Allen, 1947

DYER, T F Thiselton - *The folklore of plants*. Chatto & Windus, 1889

DYER, T F Thiselton - *Folklore of Shakespeare*. Griffith & Farrar, 1887

EARLE, John - *English plant names from the 10th to the 15th century*. Oxford, Clarendon Press, 1880

EDLIN, Herbert L - *The natural history of trees*. Weidenfeld & Nicolson, 1976 (World naturalist series)

EGAN, F W - Irish folklore: medical plants. *Folklore Journal*, vol 5, 1887, 11-13

ELLACOMBE, Henry N - *The plant-lore and garden-craft of Shakespeare*. 2nd ed W Satchell 1884

ELMORE, Francis H - Ethnobotany of the Navajo. *Albuquerque, Univ. of New Mexico, School of American Research. Monograph series*, vol 1. no 7, 1943

ELWORTHY, Frederic Thomas - *The West Somerset word-book: a glossary of dialectal and archaic words and phrases used in the west of Somerset and east Devon*. Trübner for English Dialect Society, 1888

EMBODEN, William A - *Narcotic plants*, new ed. Studio Vista, 1979

EMBODEN, William A - *Bizarre plants: magical, monstrous, mythical*. Studio Vista, 1974

EVANS, Arthur B - *Leicestershire words, phrases and proverbs*. William Pickering, 1848

EVANS, E Estyn - *Irish heritage: the landscape, the people and their work*. Dundalk, W Tempest, 1942

EVANS, E Estyn - *Irish folk ways*. Routledge, 1957

EVANS, George Ewart - *Ask the fellows who cut the hay*. Faber, 1956

EVANS, George Ewart - *The pattern under the plough: aspects of folk-life of East Anglia*. Faber, 1966

EVANS, George Ewart - Aspects of oral tradition. *Folk Life*, vol 7, 1969, 5-14

EVANS, George Ewart and David Thomson - *The leaping hare*. Faber, 1972

EVERETT, Thomas H - *Living trees of the world*. Thames & Hudson, 1969

FAIRWEATHER, Barbara - *Highland plant lore*. Glencoe and North Lorn Folk Museum

FERNIE, W T - *Herbal simples approved for modern uses of cure*, 2nd ed. Bristol, John Wright, 1897

FIELD, John Edward - *The myth of the pent cuckoo: a study in folklore*. Elliot Stock, 1913

FITZGERALD, David - Popular tales of Ireland. *Revue Celtique*, vol 4, 1879-1880

FLOWER, Thomas Bruges - Note on "Carduus tuberosus" - Linn. *Wiltshire Archaeological Magazine*, vol 3, 1857, 249-251

FLOWER, Thomas Bruges - The flora of Wiltshire. *Wiltshire Archaeological Magazine*, vol 5, 1859-..

FLÜCK, Hans - *Medicinal plants and their uses*. Foulsham, 1976

FLUCKIGER, Friedrich and Daniel Hanbury - *Pharmacographia*. Macmillan, 1873

FOGEL, Edwin Miller - *Beliefs and superstitions of the Pennsylvania Germans*. Philadelphia, American Germanica Press, 1915

FOLKARD, Richard - *Plant lore, legends, and lyrics, embracing the myths, traditions, superstitions, and folklore of the plant kingdom*, 2nd ed. Sampson, Low & Marston, 1892

FORBY, Robert - *The vocabulary of East Anglia*, 2 vols. J B Nichols & Son, 1830

FORSYTH, A A - *British poisonous plants*, 2nd ed. HMSO, 1968 (Ministry of Agriculture, Fisheries and Food. Reference book no 161)

FOSTER, Gertrude B - *Herbs for every garden*, rev ed. Dent, 1973

FOSTER, J J - Dorset folklore. *Folklore Journal*, vol 6, 1888, 115-119

FREETHY, Ron - *From agar to zenry: a book of plant uses, names and folklore*. Ramsbury, Crowood Press, 1985

FRIEND, Hilderic - *A glossary of Devonshire plant names*. English Dialect Society, 1882

FRIEND, Hilderic - *Flowers and flower lore*, 2 vols. Allen, 1883

GANNON, Ruth - *Decorating with house plants*. New York, Studio, 1952

GARRAD, Larch S - "Some Manx plant-lore" in VICKERY, Roy (editor) - *Plant-lore studies*. Folklore Society, 1984

GATES, Frank C - Weeds in Kansas. *Kansas State Board of Agriculture. Report*, no 243, 1941

GELDART, Ernest - *The art of garnishing churches at Christmas and other times: a manual of directions*, 1882

GENDERS, Roy - *The scented wild flowers of Britain*. Collins, 1971

GENDERS, Roy - *A history of scent*. Hamilton, 1972

GENG JUNYING et al. - *Practical traditional Chinese medicine and pharmacology: medicinal herbs*. Beijing, New World Press, 1991

GENTLEMAN'S MAGAZINE - *The Gentleman's Magazine Library*, being a classified collection of the chief contents of the Magazine from 1731 to 1868: popular superstitions, edited by George Laurence Gomme. Elliot Stock, 1884

GEPP, Edward - *An Essex dialect dictionary*, 2nd ed. Routledge, 1923 (reprint Wakefield, SR Publishers, 1969)

GERARD, John - *Herball* (Thomas Johnson's edition, 1636)

GIBBINGS, Robert - *Sweet Thames, run softly*. Dent, 1940

GIBSON, Alexander Craig - *The folk-speech of Cumberland and some districts adjacent*. John Russell Smith, 1869

GILL, William Walter - *A Manx scrapbook*. Arrowsmith, 1929

GILL, William Walter - *A second Manx scrapbook*. Arrowsmith, 1932

GILL, William Walter - *A third Manx scrapbook*. Arrowsmith, 1963

GILLILAND, H B - *Common Malayan plants*, 2nd ed. Kuala Lumpur, Univ. Malaya Press, 1962

GILMORE, Melvin Randolph - Uses of plants by the Indians of the Missouri River region. *Bureau of American Ethnology. Annual Report*, 33rd, 1919

GILMOUR, John and Max Walters - *Wild flowers: botanising in Britain*, 4th ed. Collins, 1969 (New naturalist series)

GIMLETTE, John D - *Malay poisons and charm cures*. Oxford, Univ. Press, 1915

GODDARD, C V and E H Goddard - Wiltshire words: addenda. *Wiltshire Archaeological Magazine*, vol 46, no 160, June 1934, 478-519

GODWIN, H - *The history of the British flora: a factual basis for phytogeography*. Cambridge, Univ. Press, 1956

GOLDIE, W H - Maori medical lore. *New Zealand Institute. Transactions*, vol 37, 1904, 1-120

GOODING, E G B, A R Loveless and G R Proctor - *Flora of Barbados*. HMSO, 1965 (Ministry of Overseas Development. Overseas Research Publication. 7)

GOOLD-ADAMS, Deenagh - *A conservatory manual*. Century Hutchinson, 1987

GORDON, Huntly - *The minister's wife*. Routledge, 1978

GORDON, Lesley - *Green magic: flowers, plants & herbs in lore and legend*. Ebury Press, 1977

GORDON, Lesley - *The mystery and magic of trees and flowers*. Exeter, Webb & Bower, 1985

GORER, Richard - *The flower garden in England*. Batsford, 1975

GRAHAM, Henry Gray - *The social life of Scotland in the eighteenth century*, 5th ed. Black, 1969

GRANT WHITE, J E - *Designing a garden today*. Studio Vista, 1966

GREENOAK, Francesca - *All the birds of the air: the names, lore and literature of British birds*. Deutsch, 1979

GREENOAK, Francesca - *God's acre: the flowers and animals of the parish churchyard*. Orbis, 1985

GREGOR, Walter - *Folklore of the north-east of Scotland*. Folklore Society, 1881

GREGOR, Walter - Some folklore from Achterneed: *Folklore Journal*, vol 6, 1888, 262-265
GREY-WILSON, Christopher - *The alpine flowers of Britain and Europe*. Collins, 1979
GRIEVE, Maud - *Culinary herbs and condiments*. Heinemann, 1933
GRIEVE, Maud - *A modern herbal*. Cape, 1931
GRIGSON, Geoffrey - *An Englishman's flora*. Phoenix House, 1955
GRIGSON, Geoffrey - *A herbal of all sorts*. Phoenix House, 1959
GRIGSON, Geoffrey - *A dictionary of English plant names (and some products of plants)*. Allen Lane, 1974
GRIGSON, Geoffrey - *The goddess of love: the birth, triumph, death and return of Aphrodite*. Constable, 1976
GROOME, Francis Hindes - *Gypsy folk-tales*. Hurst & Blackett, 1899
GROSE, Donald - *The flora of Wiltshire*. Devizes, WAHNS, 1967
GROSE, Francis - *A glossary of provincial and local words used in England*. John Russell Smith, 1839
GUNTHER, Robert T - See under L APULEIUS MADAURENSIS also DIOSCORIDES
GUNTHER, Robert T - The cimaruta: its structure and development. *Folklore*, vol 16, 1905, 132-161
GUTCH, M and M Peacock - *County folklore, Lincolnshire*. Folklore Society, 1908
GUTHRIE, E J - *Old Scottish customs: local and general*. Hamilton, Adams, 1885
HACKWOOD, Frederick William - *Staffordshire customs, superstitions & folklore*. Lichfield, Mercury Press, 1924 (reprint EP Publishing, 1974)
HAIG, Elizabeth - *The floral symbolism of the great masters*. Kegan Paul, Trench, Trübner, 1913
HAINING, Peter - *The warlock's book: secrets of black magic from the ancient grimoires*. W H Allen, 1972
HALLIWELL, James Orchard - *A dictionary of archaic and provincial words, obsolete phrases, proverbs, and ancient customs, from the fourteenth century*, 10th ed. John Russell Smith, 1881
HARDY, James - The popular history of the cuckoo. *Folklore Record*, vol II, 1879, 47-91
HARE, C E - *Bird lore*. Country Life, 1952
HARLAND, John - *A glossary of words used in Swaledale, Yorkshire*. Trübner for English Dialect Society, 1873
HARLEY, George Way - *Native African medicine, with special reference to its practice in the Mano tribe of Liberia*. Cambridge, Mass., Harvard Univ. Press, 1941
HARMAN, H - *Buckinghamshire dialect*. Hazell, Watson and Viney, 1929
HARPER, Roland E - Economic botany of Alabama, part 2. *Geological Survey of Alabama. Monograph 9*
HART, Cyril - *British trees in colour*, illustrations by Charles Raymond. Joseph, 1973
HARTLAND, Edwin S - *English fairy and other folk tales*. Walter Scott nd
HARTLEY, Marie and Joan Ingilby - *Life and tradition in the Yorkshire Dales*. Dent, 1968
HATFIELD, Audrey Wynne - *A herb for every ill*. Dentt, 1973
HATFIELD, V Gabrielle - *Country remedies: traditional East Anglian plant remedies in the twentieth century*. Woodbridge, Boydell Press, 1994
HAVERGAL, Francis T - *Herefordshire words and phrases*. Walsall, W Henry Robinson,

1887

HAWTHORN, Audrey - *Art of the Kwakiutl Indians, and other northwest coast tribes* Seattle, Univ. Washington Press, 1967

HAY, Roy and Patrick Synge - *The dictionary of garden plants in colour, with house and greenhouse plants*. Ebury Press/Joseph, 1969

HAYES, Elizabeth S - *Herbs, flavours and spices*, edited by F Newman Turner. Faber, 1963

HAYWARD, Bob - *Where the Ladbrook flows: memories of a village boyhood in Gastard, Wiltshire*. Corsham, Chris J Hall, 1983

HAZLITT, W Carew - *English proverbs and proverbial phrases*. Reeves and Turner, 1882

HEATHER, P J - Folklore from Naphill, Bucks. *Folklore*, vol 43, 1932, 104-110

HEMPHILL, Rosemary - *Herbs for all seasons*. Sydney, Angus & Robertson, 1972

HENDERSON, William - *Notes on the folk-lore of the northern counties of England and the borders*, new ed. W Satchell, Peyton for Folk-lore Society, 1879

HENKEL, Alice - Weeds used in medicine. *US Department of Agriculture. Farmers' Bulletin*, 188, 1904

HENSLOW, G - *Medical works of the fourteenth century*. Chapman & Hall, 1899

HEPBURN, Ian - *Flowers of the coast*. Collins, 1952 (New naturalist series)

HEWETT, Sarah - *Nummits and crummits: Devonshire customs, characteristics and folk-lore*. Thos Burleigh, 1900 (reprint Wakefield, EP Publishing, 1976)

HEWETT, Sarah - *The peasant speech of Devon*. Elliot Stock, 1892

HEWITT, J C B - The Iroquoian concept of the soul. *Journal of American Folklore*, vol 8, 1895, 107-116

HIBBERD, Shirley - *New and rare beautiful-leaved plants*, 2nd ed. Nimmo, 1891

HILL, Jason - *Wild foods of Britain*. A & C Black, 1939

HILL, Sir John - *The family herbal*. Bungay, C Brightley and T Kinnersley, 1820 (?)

HILL, Thomas - *The gardener's labyrinth* (1577), edited by Richard Mabey. Oxford, Univ. Press, 1987

HOGARTH, Peter with Val Cleery - *Dragons*. Allen Lane, 1979

HOLE, Christina - Notes on some folklore survivals in English domestic life. *Folklore*, vol 68, 1957, 411-419

HOLE, Christina - *Saints in folklore*. Bell, 1966

HOLLAND, Robert - *A glossary of words used in the county of Chester*. Trübner for English Dialect Society, 1886

HOUSE, Homer D - Wild flowers of New York. *State Museum of New York. Memoir*, 15, 1918 (72nd report of museum)

HUDSON, W H - *The book of a naturalist*. Nelson

HULME, F Edward - *Familiar wild flowers*, 2nd series. Cassell

HUNTER, Joseph - *The Hallamshire glossary*. Pickering, 1829 (reprint Sheffield University Centre for English Cultural Tradition and Language, 1983)

HURRY, Jamieson Boyd - *The woad plant and its dye*. Oxford, Univ. Press, 1930

HURSTON, Zora Neale - *Mules and men*. Philadelphia, Lippincott, 1935 (reprint New York, Harper & Row, 1990)

HUTCHINSON, John - *British wild flowers*. Penguin, 1955

HUTCHINSON, John and Ronald Melville - *The story of plants and their uses to man*.

Gawthorn, 1948

HUXLEY, Anthony - *Plants and planet*. Allen Lane, 1974

HYATT, Richard - *Chinese herbal medicine: an ancient art and modern healing science* (1978). New York, Thorsons, 1984

JACOB, Dorothy - *A witch's guide to gardening*. Elek, 1964

JAEGER, Edmund C - *Desert wild flowers*, rev ed. Stanford, Univ. Press, 1941

JAGO, Frederick W P - *The ancient language, and the dialect of Cornwall*. Truro, Netherton & Worth, 1882

JAMIESON, John - *An etymological dictionary of the Scottish language*, 2 vols. Edinburgh, Univ. Press, 1808

JEFFERIES, Richard - *Round about a great estate* (1880). Bradford-on-Avon, Ex Libris Press, 1987

JEFFREY, Percy Shaw - *Whitby lore and legend*, 2nd ed. Whitby, Horne & Son, 1923

JENKINS, J Geraint - *Life and tradition in rural Wales*. Dent, 1976

JENNESS, Diamond - The Ojibwa Indians of Parry Island. *National Museum of Canada. Bulletin*, 78, 1935

JENNINGS, James - *The dialect of the west of England, particularly Somersetshire*, 2nd ed. John Russell Smith, 1849

JENYNS, Roger Soane and William Watson - *Chinese art: the minor arts: gold, silver, bronze, cloisonné, Cantonese enamel, lacquer, furniture, wood*. New York, Universe Books, 1963

JOHNSON, C Pierpoint - *The useful plants of Great Britain*. Hardwicke, 1862

JOHNSON, Walter - *Folk memory*. Oxford, Clarendon Press, 1908

JOLLIE - *Sketch of Cumberland manners and customs*. Carlisle, F Jollie & Sons, 1811

JONES, David E - *Sinapia: Comanche medicine woman*. New York, Holt, Rinehart & Winston, 1972 (Case studies in cultural anthropology series)

JONES, Ida B - Popular medical knowledge in xivth century England. *Institute of the History of Medicine. Bulletin*, vol 5, 1937, 405-451, 538-588

JONES, Malcolm and Patrick Dillon - *Dialect in Wiltshire, and its historical, topographical and natural science contexts*. Trowbridge, Wiltshire County Council. Library and Museum Service, 1987

JONES, T Gwynn - *Welsh folklore and folk customs*. Methuen, 1930

JONES-BAKER, Doris - *Old Hertfordshire calendar*. Chichester, Phillimore, 1974

JORDAN, Michael - *A guide to wild plants: the edible and poisonous species*. Millington, 1976

KEARNEY, Michael - "Spiritualist healing in Mexico" in MORLEY, Peter and Roy Wallis (editors) - *Culture and curing: anthropological perspectives on traditional medical beliefs and practices*. Peter Owen, 1978

KEARNEY, Thomas H, Robert H Peebles et al. - *Arizona flora*, 2nd ed. Berkeley, Univ of California Press, 1960

KEBLE MARTIN, W - *The concise British flora in colour*, 3rd ed. Ebury Pr/M Joseph, 1974

KELLY, Isabel and Angel Palerm - *The Tajín Totonac*. Washington, Smithsonian Institution, 1952

KELLY, Walter K - *Curiosities of Indo-European tradition and folklore*. Chapman & Hall,

1863

KILLIP, Margaret - *Folklore of the Isle of Man*. Batsford, 1975 (The folklore of the British Isles series)

KINAHAN, G H - Irish plant-lore notes. *Folk-lore Journal*, vol 6, 1888, 265-267

KING, Margaret H - A partnership of equals: women in Scottish east coast fishing communities. *Folk Life*, vol 31, 1992-93, 18-35

KINGSBURY, John M - *Poisonous plants of the United States and Canada*. Englewood Cliffs, Prentice-Hall, 1964

KINGSBURY, John M - *Deadly harvest: a guide to common poisonous plants*. Allen & Unwin, 1967

KLUCKHOHN, Clyde, W W Hill and Lucy Wales Kluckhohn - *Navaho material culture*. Cambridge, Mass., Belknap Press of Harvard Univ. Press, 1971

KOURENNOFF, Paul M - *Russian folk medicine*. W H Allen, 1970

KRAPPE, Alexander Haggerty - *The science of folk-lore*. Methuen, 1930

KYEREMATEN, A A Y - *Panoply of Ghana*. Longmans, 1964

LA BARRE, Weston - *The peyote cult*. New Haven, Yale Univ. Press, 1938 (Yale University publications in anthropology no 19)

LANCASHIRE FEDERATION OF WOMEN'S INSTITUTES - *Lancashire lore: a miscellany of country customs, sayings, dialect words, rhymes, games, village memories and recipes*. Preston, 1971

LANGHAM, William - *The garden of health*. 1578

LAWRENCE, Berta - *Somerset legends*. Newton Abbot, David & Charles, 1973

LEASK, J T Smith - *A peculiar people, and other Orkney tales*. Kirkwall, W R Mackintosh, 1931

LEATHART, Scott - *Trees of the world*. Hamlyn, 1977

LEATHER, Ella M - *Folk-lore of Herefordshire*. Sidgwick & Jackson, 1911

LEHNER, Ernst and Johanna Lehner - *Folklore and odyssey of food and medicinal plants*. New York, 1962

LEIGH, Egerton - *Ballads and legends of Cheshire*. Longmans, 1867

LEIGHTON, Ann - *Early English gardens in New England*. Cassell, 1970

LELAND, Charles Godfrey - *Gypsy sorcery and fortune telling*. Unwin, 1891

LE STRANGE, Richard - *A history of herbal plants*. Angus & Robertson, 1977

LEWIN, Louis - *Phantastica: narcotic and stimulating drugs*. Kegan Paul, Trench, Trübner, 1931

LEWIS, George Cornewall - *A glossary of provincial words used in Herefordshire and some of the adjoining counties*. John Murray, 1839

LEYEL, Mrs C F - *The magic of herbs*. Cape, 1926

LEYEL, Mrs C F - *Herbal delights: tisanes, syrups, confections, electuaries, robs, juleps, vinegars and conserves*. Faber, 1937

LEYEL, Mrs C F - *Elixirs of life*. Faber, 1948

LINCOLN, Bruce - *Emerging from the chrysalis: studies in rituals of women's initiation*. Cambridge, Mass., Harvard Univ. Press, 1981

LINDLEY, John - *Medical and oeconomical botany*. Bradbury & Evan, 1849

LLOYD, John Uri - *Origin and history of all the pharmacological vegetable drugs, etc*, vol 1. Washington, American Drug Manuf. Asscn., 1921

LLOYD, Theodosia - Notes from Maidstone district, Kent. *Folk-lore*, vol 62, 1951, 328-9

LOCKWOOD, W B - *Languages of the British Isles past and present*. Deutsch, 1975 (The language library series)

LOEWENFELD, Claire and Philippa Back, with Patience Bosanquet - *Britain's wild larder*. Newton Abbot, David & Charles

LONG, Harold C - *Common weeds of the farm and garden*. Smith, Elder, 1910

LONG, Harold C - *Plants poisonous to live stock*, 2nd ed. Cambridge, Univ. Press, 1924 (Cambridge agricultural monographs series)

LONG, W H - *A dictionary of the Isle of Wight dialect, and of provincialisms used in the island*, 2nd ed. Portsmouth, W H Barrell and London, Simpkin Marshall, 1931

LOWSLEY, B - *A glossary of Berkshire words and phrases*. Trübner for English Dialect Society, 1888

MABEY, Richard - *Food for free*. Collins, 1972

MABEY, Richard - *Plants with a purpose: a guide to the everyday uses of wild plants*. Collins, 1977

MABEY, Richard and Tony Evans - *The flowering of Britain*, new ed. Chatto & Windus, 1989

McCLINTOCK, David and R S R Fitter - *The pocket guide to wild flowers*. Collins, 1956

McCLINTOCK, David - *The wild flowers of Guernsey*. Collins, 1975

MacCOLL, Ewan and Peggy Seeger - *Till doomsday in the afternoon: the folklore of a family of Scots travellers, the Stewarts of Blairgowrie*. Manchester, Univ. Press, 1986

MacCULLOCH, Sir Edgar - *Guernsey folklore*. Elliot Stock, 1903

MacCULLOCH, J A - *The misty isle of Skye*. Edinburgh, Oliphant, Anderson & Ferrier, 1905

MacDONALD, Christina - *Medicines of the Maori, from their trees, shrubs and other plants, together with foods from the same sources*. Auckland, Collins, 1974

McDONALD, Donald - *Fragrant leaves and flowers: interesting associations gathered from many sources, with notes on their history and utility*. Warne

McKAY, John G - *More West Highland tales*, vol 1. Edinburgh, Oliver & Boyd for the Scottish Anthropological and Folklore Society, 1940

MACMILLAN, A S (compiler) - *Popular names of flowers, fruits, etc., as used in the county of Somerset and the adjacent parts of Devon, Dorset and Wiltshire*. Yeovil, Western Gazette, 1922

McNEILL, Florence Marion - *The silver bough*, vol 1: Scottish folk-customs and folk-belief Glasgow, Maclehose, 1957; vol 2: A calendar of Scottish national festivals, Candlemas to harvest home. Glasgow, Maclehose, 1959

Mac NEILL, Maire - *The festival of Lughnasa: a study of the survival of the Celtic festival of the beginning of harvest*. Oxford, Univ. Press, 1962

McNEILL, Murdoch - *Colonsay: one of the Hebrides*. Edinburgh, Douglas, 1910

MacQUOID, Percy - *A history of English furniture*, vol 4: The age of satinwood. Lawrence & Bullen, 1908

MANDEVILLE, Sir John - *Travels*. Dent, 1928

MANKER, Ernst - *People of eight seasons*. Watts, 1965

MARROW, Norman - A bit of Black Country. *Lore and language*, vol 10, no 2, 1992, 11-

23
MARSHALL, Sybil - *Fenland chronicle: recollections of William Henry and Kate Mary Edwards collected and edited by their daughter*. Cambridge, Univ. Press, 1967

MARTIN, Martin - *A description of the western isles of Scotland* (1703), edited by Eneas Mackay Stirling, 1934

MARWICK, Ernest W - *The folklore of Orkney and Shetland*. Batsford, 1975 (The folklore of the British Isles series)

MASON, Otis T - Aboriginal American basketry. *US National Museum. Report*, 1901/2

MASON, Violet - Scraps of English folklore. XIX: Oxfordshire [Northleigh]. *Folk-lore*, vol 40, 1929, 374-384

MATTOCK, Mark - *Roses: a popular guide*. Poole, Blandford Press, 1980

MEGGITT, M J - *Desert people: a study of the Walbiri aborigines of Central Australia*. Sydney, Angus & Robertson, 1962

MILLER, William - *A dictionary of English names of plants applied in England and among English-speaking people to cultivated and wild plants,trees and shrubs*. John Murray, 1884

MITCHELL, Alan F - *Conifers in the British Isles: a descriptive handbook*. HMSO, 1972 (Forestry Commission Booklet no 33)

MITCHELL, Alan F - *A field guide to the trees of Britain and northern Europe*. Collins, 1974

MITTON, F and V Mitton - *Mitton's practical modern herbal*, rev. ed. Foulsham, 1982

MOLDENKE, Harold N and Alma L Moldenke - *Plants of the Bible*. Waltham, Mass., Chronica Botanica, 1952

MONTGOMERIE, Norah and William Montgomerie - *The well at the world's end: folk tales of Scotland*. Bodley Head, 1975

MOONEY, James - The mescal plant and ceremony. *Therapeutic Gazette*, vol 20, 1896, 7-11

MOOR, Edward - *Suffolk words and phrases: or, An attempt to collect the lingual localisms of that county*. R Hunter, 1823

MOORE, A W - *Folklore of the Isle of Man*. David Nutt, 1891

MOORE, A W, Sophia Morrison and Edmund Goodwin - *A vocabulary of the Anglo-Manx dialect*. Oxford, Univ. Press, 1924

MORLEY, George - *Shakespeare's greenwood: the customs of the country*. David Nutt, 1900

MORLEY, Peter and Roy Wallis (editors) - *Culture and curing: anthropological perspectives on traditional medical beliefs and practices*. Peter Owen, 1978

MORRIS, M C F - *Yorkshire folk-talk,with characteristics of those who speak it in the North of that county*. Henry Frowde, 1892

MORRISON, Jean - Unlucky flowers in Wiltshire. *Wiltshire Folklife*, vol 3, no 10, Spring 1983, 26-27

MORRISON, Sophia - *Manx fairy tales*. David Nutt, 1911

MUNRO, Neil Gordon - *Ainu creed and cult*. Routledge, 1962

NALL, John Greaves - *Great Yarmouth and Lowestoft... and an etymological and comparative glossary of the dialect of East Anglia*. Longmans, Green, Reader & Dyer, 1866

NARVAEZ, Peter - Newfoundland berry pickers "in the fairies": the maintenance of spatial and temporal boundaries through legendry. *Lore and Language*, vol 6, no 1, Jan. 1987, 15-49

NASH, Harry and Judy Nash - Swans and spear at Abbotsbury. *The Countryman*, vol 96, no 6, Dec/Jan 1991/92, 47-52

NASH, John - *English garden flowers*. Duckworth, 1948

NEWALL, Venetia - "The Jew as witch figure" in NEWALL, Venetia (editor) - *The witch figure: folklore essays by a group of scholars in England honouring the 75th birthday of Katherine M Briggs*. Routledge, 1973

NICHOLSON, John - *Folk lore of east Yorkshire*. Simpkin Marshall, 1890

NICKERSON, David - *English furniture of the eighteenth century*. Weidenfeld, 1963

NICOLAISEN, Age - *The pocket encyclopedia of indoor plants in colour*. Blandford Press, 1970

NICOLSON, James R - *Traditional life in Shetland*. Hale, 1978

NICOLSON, John - *Some folk-tales and legends of Shetland*. Edinburgh, Thomas Allan & Sons, 1920

NIELSEN, Margaret Steentoft - *Introduction to the flowering plants of West Africa*. ULP, 1965

NODAL, John Howard and George Milner - *A glossary of the Lancashire dialect*. Manchester, Alexander Ireland & Co, 1875 (Manchester Literary Club Publications)

NORBECK, Edward - *Religion in primitive society*. New York, Harper, 1961

NORTH, Pamela - *Poisonous plants and fungi in colour*. Blandford Press, 1967

NORTHALL, G F - *English folk-rhymes: a collection of traditional verses relating to places and persons, customs, superstitions, etc*. Kegan Paul, 1892

NORTHCOTE, Lady Rosalind - *The book of herbs*. Bodley Head, 1912

NORTHOVER, Douglas - The language of Old Burton, Burton Bradstock, Dorset. *Lore and Language*, vol 8, no 2, July 1989, 3-31

O'FARRELL, Padraic - *Superstitions of the Irish country people*, rev. ed. Cork, Mercier Press, 1982

OPIE, Iona and Moira Tatem (editors) - *A dictionary of superstitions*. Oxford, Univ. Press, 1989

O SUILLEABHAIN, Sean - *A handbook of Irish folklore*. Dublin, Folklore of Ireland Society, 1942

OXFORDSHIRE AND DISTRICT FOLKLORE SOCIETY - *Annual Record*, no 2-12, 1950-1960

PAGE, Robin - *Cures and remedies, the country way*. Davis-Poynter, 1978

PAINTER, Gilian and Elaine Power (illustrator) - *The herb garden displayed*. Auckland, Hodder & Stoughton, 1978

PALAISEUL, Jean - *Grandmother's secrets: her green guide to health from plants*, translated by Pamela Swinglehurst. Barrie & Jenkins, 1973

PALGRAVE, Olive H Coates (illustrator) and Keith Coates Palgrave - *Trees of Central Africa*. Salisbury, National Publications Trust, Rhodesia and Nyasaland, 1956

PALMER, A Smythe - *Folk-etymology: a dictionary of verbal corruptions or words perverted in form or meaning, by false derivation or mistaken analogy*. Bell, 1882

PALMER, Roy - *The folklore of Warwickshire*. Batsford, 1976 (The folklore of the British Isles series)

PARISH, Desmond and Marjorie Parish - *Wild flowers: a photographic guide*. Poole, Blandford Press, 1979
PARISH, William Douglas - *A dictionary of the Sussex dialect, and collection of provincialisms in use in the County of Sussex* (orig. publ. Lewes, Farncombe & Co, 1875). Expanded, augmented and illustrated by Helena Hall, together with some Sussex sayings and crafts priv. publ., 1957
PARKER, Angelina - Oxfordshire village folklore (1840-1900). *Folklore*, vol 24, 1913, 74-91
PARKINSON, John - *Paradisi in soli: paradisus terrestris*, reprinted from the 1629 edition. Methuen, 1904
PARKINSON, John - *Theatrum botanicum: the theatre of plants*. London, 1640
PATON, C I - *Manx calendar customs*. Glaisher for Folklore Society, 1939
PAYNE, Joseph Frank - *English medicine in the Anglo-Saxon times: two lectures delivered before the Royal College of Physicians of London, June 23 and 25, 1903*. Oxford, Clarendon Press, 1904 (The Fitz-Patrick lectures for 1903)
PEACOCK, Edward - *A glossary of words used in the Wapentakes of Manley and Corringham, Lincolnshire*. Trübner for English Dialect Society, 1877
PENNANT, Thomas - *A tour in Scotland and voyage to the Hebrides*, 2 vols, 2nd ed. Benjamin White, 1776
PERRING, Franklyn H, P D Sell and S M Walters - *A flora of Cambridgeshire*. Cambridge, Univ. Press, 1964
PERRY, Frances - *Flowers of the world*. Hamlyn, 1972
PERRY, Frances - *Beautiful leaved plants*. Scolar Press, 1979
PHILLIPS, Reginald W - "The first flora of Breconshire" in PHILLIPS, Thomas Richard (editor) - *The Breconshire border*. Talgarth, D J Morgan, 1926
PHILLIPS, Ted - Marigolds and the second law of gardening. *The Countryman*, vol 96, no 1, Feb/Mar 1991, 53-57
PHYSICIANS OF MYDDFAI - *Physicians of Myddfai: Meddygon Myddfai*, translated by John Pughe and edited by John Williams ab Ithal Llandovery, Roderic for the Welsh MSS Society, 1861
PICTON, John and John Mack - *African textiles: looms, weaving and design*. British Museum Publications, 1979
PLANT-LORE NOTES AND NEWS, no 1, May 1988-..
PLOWDON, C C - The hybrid pernettiana rose. *Rose Annual*, 1947
POLLOCK, Linda - *With faith and physic: the life of a Tudor gentlewoman: Lady Grace Mildmay, 1552-1620*. Collins & Brown, 1993
POLUNIN, Oleg - *Flowers of Europe: a field guide*. Oxford Univ. Press, 1969
POLUNIN, Oleg - *Trees and bushes of Europe*. Oxford Univ. Press, 1976
POMET, Pierre - *A compleat history of druggs, done into English from the originals*, 3rd ed. J & J Bonwicke, 1737
POOLE, Charles Henry - *The customs, superstitions and legends of the county of Stafford*. Rowney, 1875
PORTEOUS, Crichton - *The beauty and mystery of well-dressing*. Derby, Pilgrim Press, 1949
PORTER, Enid - *Cambridgeshire customs and folklore*. Routledge, 1969

PORTER, Enid - *The folklore of East Anglia*. Batsford, 1974 (The folklore of the British Isles series)
POTTER, Stephen and Laurens Sargent - *Pedigree: essays on the etymology of words from nature*. Collins, 1973 (New Naturalist series)
POWELL, John U - Folklore notes from South-west Wilts. *Folklore*, vol 12, 1901, 71-83
POWERS, Stephen - Aboriginal botany. *California Academy of Sciences. Proceedings*, vol 5, 1873-1875, 373-9
PRANCE, Ghillean Tolmie and Anne E Prance - *Bark: the formation, characteristics, and uses of bark around the world*. Portland, Or., Timber Press, 1993
PRIOR, R C A - *On the popular names of British plants*. Norgate, 1879
PUCKETT, Newbell Niles - *Folk beliefs of the southern negro*. Chapel Hill, Univ. North Carolina Press, 1926
PUTNAM, Clare - *Flowers and trees of Tudor England*. Hugh Evelyn, 1972
QUILLER-COUCH, Thomas - *Glossary of words in use in Cornwall 2. East Cornwall*. Trübner for English Dialect Society, 1880
RACKHAM, Oliver - *The history of the countryside*. Dent, 1986
RADFORD, E and M A Radford - *Encyclopedia of superstitions*. Rider, 1946
RAMBOSSON, J - *Histoire et légendes des plantes utiles et curieuses*. Paris, Firmin Didot, 1868
RAPPAPORT, Roy A - *Pigs for the ancestors: ritual in the ecology of a New Guinea people*. New Haven, Yale Univ. Press, 1968
RAVEN, Jon - *The folklore of Staffordshire*. Batsford, 1978 (The folklore of the British Isles series)
RAYMOND, Walter - *Under the spreading chestnut-tree: a volume of rural lore and anecdote*. Folk Press, 1920
READ, D H Moutray - Hampshire folklore. *Folklore*, vol 22, 1911, 292-329
REDFIELD, Robert and Alfonso Villa - Chan Kom: a Maya village. *Washington.Carnegie Institute. Publications*, no 448, 1934
REYNOLDS, Barrie - *Magic, divination and witchcraft among the Barotse of Northern Rhodesia*. Chatto & Windus, 1963
RIMMEL, Eugene - *The book of perfumes*. Chapman & Hall, 1865
RITCHIE, James T R - *The singing street*. Edinburgh, Oliver & Boyd, 1964
ROBBINS, Wilfred William, John Peabody Harrington, and Barbara Freire-Marreco - Ethnobotany of the Tewa Indians. *Washington. Smithsonian Institution. Bureau of American Ethnology. Bulletin*, 55, 1916
ROBERTSON, J Drummond - *A glossary of dialect and archaic words used in the County of Gloucester*. Kegan Paul, Trench, Trübner & Co for English Dialect Society, 1890
ROBERTSON, Ronald Macdonald - *More Highland folktales*. Edinburgh, Oliver & Boyd, 1964
ROBERTSON, Seonaid M - *Dyes from plants*. New York, Van Nostrand Reinhold, 1973
ROBINSON, F K - *A glossary of words used in the neighbourhood of Whitby*. Trübner for English Dialect Society, 1876
ROBINSON, W Clough - *The dialect of Leeds and its neighbourhood*. John Russell Smith, 1857
ROBSON, Eve and Norman Robson - *Botanical prints*. Collins & Brown 1991 (Classic

natural history prints series)

ROCHFORD, Thomas and Richard Gorer - *The Rochford book of house plants*. Faber, 1961

RODD, Rennell - *The customs and lore of modern Greece*. David Stott, 1892

RODWAY, James - Bush medicines in British Guiana. *Chemist and Druggist*, 1, Jan 1916, 54

ROGERS, Norman - *Wessex dialect*. Bradford-on-Avon, Moonraker Press, 1979

ROHDE, Eleanour Sinclair - *Herbs and herb gardening*. Medici Society, 1936

ROLLINSON, William - *Life and tradition in the Lake District*. Dent, 1974

RORIE, David - *Folk tradition and folk medicine in Scotland: the writings of David Rorie*, edited by David Buchan. Edinburgh, Canongate Academic, 1994

ROTH, H Ling - *The aborigines of Tasmania*. Halifax, F King, 1899

ROWLING, Marjorie - *The folklore of the Lake District*. Batsford, 1976 (The folklore of the British Isles series)

ROYAL HORTICULTURAL SOCIETY - *Gardeners' encyclopedia of plants and flowers*. Dorling Kindersley, 1989

ROYS, Ralph L - *The ethno-botany of the Maya*. New Orleans, Tulane Univ. of Louisiana. Middle American research series. Publications, no 2, 1931

RUDDOCK, Elizabeth - May-day songs and celebrations in Leicestershire and Rutland *Leicestershire and Rutland Archaeological Society. Transactions*, vol 40, 1964-5, 69-84

RUDKIN, Ethel L - *Lincolnshire folklore*. Gainsborough, Beltons, 1936

SACKETT, S J and William E Koch (editors) - *Kansas folklore*. Lincoln, Univ. Nebraska Press, 1961

SAFFORD, W E - An Aztec narcotic. *Journal of Heredity*, vol 6, pt 7, 1915, 291-311

SAFFORD, W E - Daturas of the Old World and New: an account of their narcotic properties and their use in oracular and in initiatory ceremonies. *Washington. Smithsonian Institution. Annual Report*, 1920

ST LEGER-GORDON, Ruth E - *The witchcraft and folklore of Dartmoor*, 2nd ed. Wakefield, EP Publishing, 1973

SALAMAN, Redcliffe N - *The history and social influence of the potato*. Cambridge, Univ. Press, 1949

SALISBURY, Sir Edward - *Weeds and aliens*, 2nd ed. Collins 1964 (New naturalist series)

SALISBURY, Sir Edward - *The living garden, or, The how and why of garden life*. Bell, 1936

SALISBURY, Jesse - *A glossary of words and phrases used in S E Worcestershire*. The author, 1893

SANECKI, Kay N - *The complete book of herbs*. Macdonald, 1974

SANFORD, S N F - New England herbs: their preparation and use. *New England Museum of Natural History. Special publication*, no 2, 1937

SAVAGE, Frederick G - *The flora and folk lore of Shakespeare*. Cheltenham, Burrow, 1923

SCHAUENBERG, Paul and Ferdinand Paris - *Guide to medicinal plants*. Guildford, Lutterworth, 1977

SCHENK, Sarah and E W Gifford - Karok ethnobotany. *Univ. California. Anthr. Rec.*, vol 13, no 6, 1952

SCHERY, Robert W - *Plants for man*. Allen & Unwin, 1954

SCHOFIELD, Eunice M - Working class food and cooking in 1900. *Folk Life*, vol 13, 1975, 13-23

SHARMAN, V Day - *Folk tales of Devon*. Nelson, 1952

SHAW, Margaret Fay - *Folksongs and folklore of South Uist*. Routledge, 1955 (3rd ed Aberdeen Univ. Press, 1986)

SIKES, Wirt - *British goblins: Welsh folklore, fairy mythology, legends, and traditions*. Boston, Osgood, 1881

SIMMONS, John editor - *Kew Gardens book of indoor plants*. Philip, 1988

SIMPSON, Eve - *Folklore in lowland Scotland*. Dent, 1908

SINGER, Charles - The herbal in antiquity and its transmission to later ages. *Journal of Hellenic studies*, vol 47, 1927, 1-52

SITWELL, Sacheverell and James Russell - *Old garden roses*, Part 1. Rainbird, 1955

SKEAT, Walter William - A group of ghost-words. *Philological Society. Transactions*, 1903-1906, 180-201

SKINNER, Charles M - *Myths and legends of flowers, trees, fruits and plants, in all ages and in all climes*. Philadelphia, Lippincott, 1925

SMITH, A W - *A gardener's dictionary of plant names: a handbook on the origin and meaning of some plant names*, rev ed (by William T Stearn). Cassell, 1972

SMITH, F Porter - *Chinese materia medica: vegetable kingdoms*, 2nd ed (revised by G A Stuart) (republished Taipei, Ku T'ing Book House), 1911

SMITH, Huron H - Ethnobotany of the Menomini Indians. *Milwaukee. Public Museum. Bulletin*, vol 4, no 1, 1923

SMITH, Huron H - Ethnobotany of the Meskwaki Indians. *Milwaukee. Public Museum. Bulletin*, vol 4, no 2, 1928

SMITH, J B - The name of the game: plant names as a key to some pastimes involving plants. *Folk Life*, vol 32, 1993-94, 92-100

SMITH, John - *A dictionary of popular names of the plants*. Macmillan, 1882

SMITH, Julia - *Fairs,feasts and frolics: customs and traditions in Yorkshire*. Otley, Smith Settle, 1989

SOUTHWICK, Charles H - *Ecology and the quality of the environment*. New York, Van Nostrand Reinhold, 1972

SPENCE, Lewis - *The fairy tradition in Britain*. Rider, 1949

SPIER, Leslie - Klamath ethnography. *Berkeley. University of California. Publications in American archaeology and ethnology*, vol 30, 1930

STARKIE, Walter - *Raggle-taggle*. 1933

STEELE, A B - Fungus folk-lore. *Edinburgh Field Naturalists and Microscopical Society. Transactions*, vol 2, 1887, 175-183

STERNBERG, Thomas - *The dialect and folklore of Northamptonshire*. John Russell Smith, 1851

STEVENSON, Matilda Coxe - Ethnobotany of the Zuñi Indians. *Washington. Smithsonian Institution. Bureau of American Ethnology. Annual Report*, 30th, 1908-9

STEWART, Alexander - *'Twixt Ben Nevis and Glencoe: the natural history, legends and folklore of the West Highlands*. Edinburgh, William Paterson, 1885

STORER, Bernard - *Sedgemoor: its history and natural history*. Newton Abbot, David &

Charles, 1972

SUMMERS, Robert - *Zimbabwe: a Rhodesian mystery*. Johannesburg, Nelson, 1963

SWAINSON, Charles - *The folklore and provincial names of British birds*. Folklore Society, 1886

TAMPION, John - *Dangerous plants*. Newton Abbot, David & Charles, 1977

TANNER, Heather and Robin Tanner - *Wiltshire village*, 3rd ed. Robin Garton, 1978

TAYLOR, A W - *Wild flowers of the Pyrenees*. Chatto & Windus, 1971

TAYLOR, Alice - *To school through the fields: an Irish country childhood*. Dingle, Brandon, 1988

TAYLOR, Joseph - *Arbores mirabiles*. W. Darton, 1812

TEIT, James - The Thompson Indians of British Columbia. *American Museum of Natural History. Memoir (Jessup North Pacific Expedition)*, vol 1, 1898-1900

THOMAS, Barry - *The evolution of plants and flowers*. Peter Lowe, 1981

THOMAS, Graham Stuart - *Shrub roses of today*. Phoenix House, 1962

THOMPSON, C J S - *Magic and healing*. Rider, 1947

THOMPSON, Flora - *Lark Rise to Candleford: a trilogy*. Oxford, 1945

THOMSON, William A R - *Herbs that heal*. A & C Black, 1976

THORNTON, Robert John - *A new family herbal: or, popular account of the nature and properties of the various plants used in medicine, diet, and the arts*. Phillips, 1810

THORPE, Benjamin - *Northern mythology*. Lumley, 1851

TOLSTEAD, William C - A flora in N.W. Iowa. *Iowa State College. Journal of Science*, vol 12, 1937-38

TONGUE, Ruth L - *Somerset folklore*. Folklore Society, 1965 (County folklore series, vol viii)

TONGUE, Ruth L - Folk-song and folklore. *Folklore*, vol 78, 1967, 293-303

TONGUE, Ruth L - *The chime child: or, Somerset singers*. Routledge, 1968

TOSCO, Uberto - *The flowering wilderness*. Orbis, 1972 (The world of nature series)

TRADESCANT, John - *Catalogue plantarum in horto Johannis Tradescanti, nascentium*, in ALLAN, Mea - The Tradescants: their plants, gardens and museum 1570-1622. Joseph, 1964

TREVELYAN, Marie - *Folklore and folk stories of Wales* (1853). Elliot Stock, 1909

TURNER, Nancy Chapman and Marcus A M Bell - The ethnobotany of the Coast Salish Indians of Vancouver Island. *Economic Botany*, 1969 (?), 63-104

TURNER, William - *The names of herbes* (1548), edited by James Britten. English Dialect Society, 1881

TURNER ETTLINGER, D M - *British and Irish orchids: a field guide*. Macmillan, 1976

TURRILL, William Bertram - *British plant life*, 3rd ed. Collins, 1962 (New naturalist series)

TUSSER, Thomas - *Thomas Tusser, 1557 floruit: his points of good husbandry*, edited by Dorothy Hartley. Country Life, 1931

TYNAN, Katharine and Frances Maitland - *The book of flowers*. Smith, Elder & Co, 1909

UDAL, John Symonds - *Dorsetshire folk-lore*. Hertford, 1922

UPHAM, Alan W - The flora of Windham County. *Connecticut. State Geological and Natural History Survey. Bulletin*, 91, 1959

USHER, George - *A dictionary of plants used by man*. Constable, 1974

VAUX, J Edward - *Church folklore: a record of some post-Reformation usages in the English Church, now mostly obsolete.* Griffith, Farran, 1894

VESEY-FITZGERALD, Brian - *Gypsies of Britain: an introduction to their history.* Chapman & Hall, 1944

VESTAL, Paul A and Richard Evans Schultes - *The economic botany of the Kiowa Indians as it relates to the history of the tribe.* Cambridge, Mass., Botanical Museum, 1939 (reprint New York, AMS Press, 1981)

VICKERY, Roy - *Plant-lore studies.* Folklore Society, 1984

VICKERY, Roy - *Unlucky plants.* Folklore Society, 1985 (Folklore survey series no 1)

VOIGHTS, Linda E and Robert P Hudson - "A drynke that men callen dwale to make a man to slepe whyle men kerven hem", in CAMPBELL, Sheila, Bert Hall and David Klausner - *Health, disease and healing in medieval culture.* Macmillan, 1992

WAKELIN, Martyn F - *English dialects: an introduction.* Athlone Press, 1972

WARING, Philippa - *A dictionary of omens and superstitions.* Souvenir Press, 1978

WARREN-WREN, S C - *Willows.* Newton Abbot, David & Charles, 1972

WASSON, R Gordon, Albert Hofmann and Carl A P Ruck - *The road to Eleusis: unveiling the secret of the Mysteries.* New York, Harcourt Brace Jovanovich, 1978

WATERS, Ivor - *Folklore and dialect of the lower Wye valley.* Chepstow, Moss Rose Press, 1982

WATT, John Mitchell and Maria Gerdina Breyer-Brandwijk - *The medicinal and poisonous plants of southern and eastern Africa*, 2nd ed. Edinburgh, Livingstone, 1962

WEBSTER, Helen N - *Herbs: how to grow them, and how to use them.* Massachusetts Horticultural Society, 1935

WEINER, Michael A - *Earth medicine - earth foods: plant remedies, drugs and natural foods of the North American Indians.* New York, Macmillan, 1972

WESLEY, John - *Primitive Physick: or, an easy and natural method of curing most diseases.* 1747 (9th ed 1761)

WESTERMARCK, Edward - *Ritual and belief in Morocco*, 2 vols. Macmillan, 1926

WESTROPP, Thos J - A folk-lore survey of county Clare. *Folk-lore*, vols 21-22, 1910-1911

WHITAKER, Thomas W and Glen N Davis - *Cucurbits: botany, cultivation and utilization.* New York, Interscience; London, Leonard Hill 1962 (World crops books series)

WHITING, Alfred F - Ethnobotany of the Hopi. *Flagstaff. Northern Arizona Society of Science and Art. Museum of Northern Arizona. Bulletin*, 15, 1939

WHITNEY, Annie Weston and Caroline Canfield Bullock - Folklore from Maryland. *New York, American Folklore Society. Memoirs*, vol 18, 1925

WHITTLE, Tyler and Christopher Cook - *Curtis's flower garden displayed: 120 plates from the years 1787-1807.* Oxford, Univ. Press, 1981

WILDE, Lady - *Ancient cures, charms and usages of Ireland: contributions to Irish lore.* Ward & Downey, 1890

WILKINSON, Gerald - *Trees in the wild, and other trees and shrubs.* Stephen Hope Books, 1973

WILKINSON, Gerald - *Epitaph for the elm.* Hutchinson, 1978

WILKINSON, Gerald - *A history of Britain's trees.* Hutchinson, 1981

WILKS, J H - *Trees of the British Isles in history and legend.* Muller, 1972

WILLIAMS, Alfred - *Folk-songs of the upper Thames*. Duckworth, 1923
WILLIAMS, Alfred - [selections] *In a Wiltshire village: scenes from rural Victorian life*, selected by Michael Justin, Davis Stroud, Alan Sutton (1981), 1992
WILLIS, John C - *A dictionary of the flowering plants & ferns*, 8th ed (revised by H K Airy Shaw). Cambridge, Univ. Press, 1973
WILLS, Norman T - *Woad in the Fens*, 3rd ed. Long Sutton, The author, 1979
WILSON, Edwin - Paradise lost: plant-lore on the far north coast of New South Wales. *Australian Folklore*, vol 9, 1994, 104-118
WILTSHIRE, Kathleen - *Wiltshire folklore*. Compton Chamberlayne, Compton Press, 1975
WIT, H C D de - *Plants of the world: the higher plants*. Thames & Hudson, 1966
WOOD-MARTIN, W G - *Traces of the elder faiths of Ireland*. Longman, 1902
WOODCOCK, Hubert B Drysdale and William Thomas Stearn - *Lilies of the world: their cultivation and classification*. Country Life, 1950
WRIGHT, Elizabeth - *Rustic speech and folklore*. Oxford, Univ. Press, 1913
WRIGHT, Thomas - *St Patrick's Purgatory*. John Russell Smith, 1844
YANOVSKY, Elias - Food plants of the North American Indians. *Washington. United States Department of Agriculture. Miscellaneous publication*, no 237, 1936
YARNELL, Richard Asa - Aboriginal relationships between culture and plant life in the Upper Great Lakes region. *Ann Arbor. Univ. of Michigan. Museum of Anthropology. Anthropological papers*, no 23, 1964
YEATS, William Butler - "The Celtic twilight", in *Mythologies*. Macmillan, 1959
YOUNG, Andrew - *A prospect of flowers: a book about wild flowers*. Cape, 1945
ZENKERT, Charles A - The flora of the Niagara frontier region. *Buffalo Society of Natural Sciences. Bulletin*, vol 16, 1934
ZOHARY, Michael - *Plants of the Bible*. Cambridge, Univ. Press, 1982

A

A-HUNDRED-FALD *Galium verum* (Lady's Bedstraw) Border Britten & Holland. Presumably because of the very large numbers of flowers produced by the average plant.

AAC *Quercus robur* (Oak) N Eng. Halliwell. This can only be a variant spelling; Morris, for Yorkshire, for instance, has Ak, and the pronunciation must be the same in both cases, though Ak may very well be pronounced Yak.

AAR *Alnus glutinosa* Scot. Grigson. One of the many dialectal forms of Alder, like Arl, Arn, Aul or Orl etc. The spelling may be idiosyncratic, but all these forms are there because the 'd' in alder seems to be intrusive. The OE was alor or aler.

AARON; HERB AARON; AARON'S ROOT *Arum maculatum* (Aaron) Prior (Herb Aaron) Henslow (Aaron's Root) Skinner. Whether it is spelled Aaron or Aron, it is simply a corruption of Arum.

AARON'S BEARD 1. *Aconitum napellus* (Monkshood) Som. Macmillan. 2. *Allium vineale* (Crow Garlic) Wilts. Macmillan. 3. *Hypericum calycinum* (Rose of Sharon) Dev. Friend.1882. Glos. J.D.Robertson. 4. *Cymbalaria muralis* (Ivy-leaved Toadflax) Scot. Grigson. 5. *Orchis mascula* (Early Purple Orchis) Berw. B & H. 6. *Saxifraga sarmentosa* (Mother-of-Thousands) Dev. Friend.1882. 7. *Spiraea salicifolia* (Bridewort) Lincs. Peacock. All probably descriptive - Crow Garlic and Rose of Sharon both have protruding bunches of stamens, while the thread-like shoots of Mother-of-Thousands, which is also sometimes called Old Man's Beard, account for the beard epithet. Ivy-leaved Toadflax hangs from a wall in such a way as to suggest a beard (and also has names like Mother-of-Thousands and Thread-of-Life). The only ones which are difficult to explain are the Orchis (it is probably a misnomer from a long time ago), and Monkshood.

AARON'S FLANNEL *Verbascum thapsus* (Great Mullein) Dor. Macmillan. Mullein has a great many 'flannel' names, all derived from the woolly leaves. It has been suggested that the word mullein itself is from French mol - soft, referring to the leaves, but not according to ODEE - see under MULLEIN. But we have Old Man's Flannel, Lady's Flannel, Our Lord's Flannel, Poor Man's Flannel, Adam's Flannel, and many others, as well as names like Duffle, Poor Man's Blanket, and Woollen. Even Fluffweed, from Somerset, probably means the same thing. The leaves were actually used to act like a piece of flannel when wrapped round the throat, and helped to relieve coughs and colds.

AARON'S PRIDE *Saxifraga umbrosa* Som. Macmillan. This is more probably *S spathularis x umbrosa* (London Pride).

AARON'S ROD 1. *Agrimonia eupatoria* (Agrimony) Som. Grigson. 2. *Arum maculatum* (Cuckoo-pint) Skinner. 3. *Kniphofia uvaria* (Red Hot Poker) Som. Macmillan. 4. *Solidago virgaurea* (Golden Rod) Corn, Som, War, Shrop. Vesey-Fitzgerald. 5. *Verbascum thapsus* (Great Mullein) Glos, Som, Lincs, Midl, Scot Grigson; USA Henkel. 6. *Verbascum virgatum* (Large-flowered Mullein) S Africa Watt. Descriptive names - the rod-like, straight and tall growth, except Cuckoo-Pint, where Aaron is a corruption of Arum. The biblical reference to Aaron's Rod is Numbers xvii 8. The word 'rod' in the Bible is often the equivalent of 'sceptre'.

ABACA *Musa textilis* (Manila Hemp) Schery. Abacá is in fact the Spanish name for Manila Hemp and its product, but it is known often in English-speaking communities by that name. The plant is a native of the Philippines, hence the Spanish language name.

ABASSIAN BOX *Buxus sempervirens* timber (Boxwood) Brimble.1948. Cf. Persian Box and Iranian Box, also given to boxwood.

ABBESPINE; AUBESPYNE; ALBASPYNE; ALBESPEINE *Crataegus monogyna* (Hawthorn) Mandeville (Abbespine); B & H (Aubespyne and Albaspyne); Ellacombe (Albespeine). All originally from Latin alba spina (Cf. the English common name Whitethorn), through French aubépine.

ABBEY; ABBY *Populus alba* Som. Macmillan (Abbey); N Eng. B & H; Som. Jennings (Abby). These are two of the variants of Abele, which comes from the Dutch abeel, and which in turn eventually comes from the dim. Latin albus, i.e. albellus. White Poplars were imported into this country from Holland in the 17th century.

ABEL *Populus alba* (White Poplar) Suff. Rackham.1976. This name appears in documents of the 13th and 14th centuries in Suffolk, apparently discounting the theory that the name, better known as Abele, is Dutch, and 17th century.

ABEL'S BLOOD, Drops of *Fuchsia sp* Dur.

ABELE; ABEEL *Populus alba* (White Poplar) Grigson. Gerard (Abeel). ("... in low Dutch abeel, of his horie or aged colour, and also Abeelbloome") Abele is from Dutch through O Fr abel, aubel, med. Latin albellus, diminutive of albus, white. White Poplars were imported from Holland into this country in the 17ty century. However, Rackham.1976 points out that the name Abel appears in Suffolk documents of the 13th and 14th centuries.

ABEOKUTA COFFEE *Coffea liberica* (Liberian Coffee) Dalziel.

ABLAMOTH *Dipsacus fullonum* (Teasel) Henslow.

ABLE-TREE *Populus alba* (White Poplar) Som. Macmillan. A variation on the more usual Abele, from the Dutch abeel, and so originally from Latin albus.

ABRAHAM, ISAAC AND JACOB 1. *Echium vulgare* (Viper's Bugloss) Lincs. Gutch. 2. *Pulmonaria officinalis* (Lungwort) Greenoak. 3. *Symphytum officinale* (Comfrey) Lincs. Gutch. 4. *Trachystemon orientale*. This is a typical name for flowers which have either variable, or three-shaded colours. Double names like Adam and Eve, or William and Mary, are very often applied to two-coloured flowers.

ABRAHAM, ISAAC AND JOSEPH 1. *Borago orientalis* Lincs. Gutch. 2. *Pulmonaria officinalis* Lincs. Gutch. From the three colours of flower on one stem. Cf. Abraham, Isaac and Jacob.

ABRAHAM'S BALM *Vitex agnus-castus* (Chaste Tree) Leyel.1937.

ABRAHAM'S BLANKET *Verbascum thapsus* (Mullein) Tynan & Maitland. 'Blanket', from the felty texture of the leaves; there are many similar names.

ABRAHAM'S OAK *Quercus pseudo-coccifera* See Genesis xviii.8. This is said to be the tree under which Abraham entertained the three angels.

ABRAMS CYPRESS *Cupressus abramsiana*.

ABRICOK *Prunus armeniaca* (Apricot) Turner. An early form of apricot, which comes either from Portuguese albricoque or Spanish alboricoque, and eventually through Arabic from Greek. Abricock was still being used ("nearly always so") in Somerset in Elworthy's day, i.e. late 19th century. The cok, or cock, ending shows its affinity with the Latin praecox - it flowers earlier than the peach (and fruits earlier too), hence the name Hasty Peach, and Precocious Tree.

ABSCESS-ROOT *Polemonium reptans*. Presumably because this was what it was used for. It is an American plant.

ABSINTH; ABSINTHE *Artemisia absinthium* (Wormwood) Clair (Absinth); (Absinthe) Ire. Wilde.1890. The flavouring for the drink absinthe comes from wormwood. The word itself, from absinthium, comes through Latin from Greek, and merely means wormwood, origin "alien".

ABYSSINIAN BANANA *Ensete ventricosum.*

ABYSSINIAN BASIL *Ocymum graveolens.*

ABYSSINIAN COFFEE *Coffea arabica* (Arabian Coffee) Dalziel.

ABYSSINIAN MILLET *Eleusine tocussa.*
ABYSSINIAN MUSTARD *Brassica carinata.*
ABYSSINIAN ROSE *Rosa sancta.* A form of Summer Damask from Ethiopia.
ABYSSINIAN TEA *Catha edulis* (Khat) Douglas.
ACACIA *Robinia pseudo-acacia.* Commonly called Acacia, but it should be False Acacia. The word itself is Greek, akakia, possibly from a word meaning point, an allusion to the thorns of the true Acacia, something that Robinia does not have.
ACACIA, Apple Ring *Acacia albida* (Ana Tree) Dale & Greenway.
ACACIA, Black-galled *Acacia drepanolobium.*
ACACIA, Bull's Horn *Acacia spadicigera.* Because of the swollen spines on the trunk and branches. The pulp attracts ants, which eat this and live inside the hollowed-out thorns, protecting the tree to some extent against other visitors.
ACACIA, Congo *Pentaclethra macrophylla* (Oil Bean) Dalziel.
ACACIA, Currawong *Acacia doratoxylon* (Brigalow) Usher.
ACACIA, False; ACACIA, Bastard *Robinia pseudo-acacia* (Acacia) USA Zenkert (False); Howes (Bastard). False Acacia = pseudo-acacia - and because it is not an acacia, but a Robinia.
ACACIA, Gum *Acacia senegal.*
ACACIA, Knife; ACACIA, Knife-leaf *Acacia cultriformis.*
ACACIA, Mescat *Acacia constricta.*
ACACIA, Mop-headed *Robinia pseudo-acacia var inermis.*
ACACIA, Mogador *Acacia gummifera.*
ACACIA, Mountain *Brachystegia tamarindoides.*
ACACIA, Persian *Albizia julibrissin.*
ACACIA, Prairie *Acacia angustissima.*
ACACIA, Raspberry *Acacia acuminata.*
ACACIA, Rose *Robinia hispida.* Pink flowers.
ACACIA, Rose, Large leafletted *Robinia hispida macrophylla.*
ACACIA, Sweet *Acacia farnesiana* (Cassie) Usher.
ACACIA, Thorny *Acacia nilotica.*
ACACIA, Three-thorned *Gleditschia triacanthos* Polunin. Three-thorned = triacanthos.
ACACIA, White-ball *Acacia angustissima* (Prairie Acacia) T H Kearney.
ACACIA NUT *Entada phaseoloides* (Nicker Bean) Hebr. Duncan.
ACANTHUS, Spiny *Acanthus spinosissimus.*
ACEROLA *Malpighia glabra* (Barbados Cherry) Brouk.
ACH *Apium graveolens* (Wild Celery) B & H. The same word as Ache, which is applied to several plants.
ACHE 1. *Apium graveolens* (Wild Celery) B & H. 2. *Bryonia dioica* (White Bryony) Corn. Grigson. 3. *Fraxinus excelsior* (Ash) Inv. B & H. 4. *Ranunculus sceleratus* (Celery-leaved Buttercup) B & H. Possibly from an O Fr word directly from Apium, and meaning parsley; this would explain celery, and celery-leaved buttercup. Ache for ash is a Scottishism. The Cornish use of the name for bryony is not explained.
ACHEWEED *Aegopodium podagraria* (Goutweed) B & H. Ache originally meant parsley, but there are other Ash-names for Goutweed.
ACHELEY; AKELEY *Aquilegia vulgaris* (Columbine). A corruption of columbine. The spurs are probably responsible for its use as an emblem of cuckoldom: "The blue cornuted Columbine, Like to the crooked horns of Acheley" (1599). The spur must give its derivation from Latin aquila, eagle.
ACHIRA *Canna edulis.*
ACHORN; AITCHORN *Quercus robur* - acorn. Achorn is from Ches. Halliwell. Aitchorn is a common variant on acorn, which is OE aecorn, from Latin acer, field. It seems to have meant the produce of the field, and was a general term for mast.
ACOM *Dioscorea bulbifera* (Potato Yam) W Indies Howes.

ACONITE *Aconitum napellus* (Monkshood) Brownlow. The derivation of aconite, and so of aconitum, has offered many permutations from the Greek. Rambosson derives it from the Greek meaning stone, "parce que cette plante croit dans les terrains pierreux" - this is probably akone, which Grieve.1931 quotes as a possible source, and says means cliffy or rocky. Barton & Castle derive the name from the same word, but now an actual place - Akone, a town of Bithynia, which was famous for poisonous herbs, and especially this one. Dyer.1887 derives it from the Greek meaning "without a struggle", because of the intensely poisonous nature. Grieve.1931 produces akontion, Greek for dart, because monkshood was used to poison arrows. Whatever the true derivation, old writers used the term aconite as synonymous with all that is deadly among plants.

ACONITE, Indian 1. *Aconitum ferox* (W Miller). 2. *Aconitum laciniatum*.

ACONITE, Nepal 1. *Aconitum ferox* (W Miller). 2. *Aconitum laciniatum* (Indian Aconite) Grieve.1931.

ACONITE, Winter *Eranthis hyemalis*. Presumably a misnomer which has stuck, although it actually is toxic, at any rate to animals. They won't eat it though, because of the burning taste. Parkinson has Yellow Aconite.

ACONITE, Yellow 1. *Aconitum vulparia* (Wolfsbane) Schauenberg. 2. *Eranthis hyemalis* (Winter Aconite) Parkinson.

ACORN *Quercus robur* fruit. Not 'corn of the oak', for this is directly from Latin acer, field, and the original meaning would be something similar to that conveyed by 'mast'. The soft 'c' of the Latin original is retained in some forms of acorn, such as Achorn or Atchern.

ACORN, Duck *Nelumbo lutea* (Yellow Lotus) Perry.

ACORN TREE *Quercus robur* (Oak) Som. Macmillan.

ACTION PLANT *Mimosa pudica* (Sensitive Plant) USA Cunningham.

ADAM AND EVE 1. *Aconitum napellus* (Monkshood) Som, Wilts. Macmillan Norf. B & H. 2. *Arum maculatum* (Cuckoo Pint) Som. Elworthy; Leics. Grigson; Lincs, Yks. B & H. 3. *Dactylorchis incarnata* (Early Marsh Orchid) B & H. 4. *Dactylorchis maculata* (Heath Spotted Orchid) N'thants. B & H; IOM Moore.1924; Yks. Carr. 5. *Lamium album* (White Deadnettle) Boland.1977. 6. *Orchis mascula* (Early Purple Orchis) Corn. B & H; Som. Elworthy; N'thants A E Baker. 7. *Pulmonaria longifolia* (Lungwort) Grigson. 8. *Pulmonaria officinalis* (Lungwort) Grigson. Monkshood - "when the hood (of the flower) is lifted up, there is an appearance of two little figures" - these would be the upper petals (similarly for White Deadnettle - see Adam-and-Eve-in-the-bower). Cuckoo Pint - B & H said that the dark spadices represent Adam, and the light ones Eve. But this is actually the archetypal male + female name suggested by the form of the plant, the spadix in the spathe, the emblem of copulation. As far as the various orchids are concerned, the name applies throughout to the tubers, which are in pairs, so these double names are often given. The pairs of tubers also give rise to rich dialect names meaning testicles. When applied to Lungwort, Adam and Eve, and many other double names like it, refers to the two (or even three) coloured flowers on the same stem. Joseph's Coat is another of these names.

ADAM-AND-EVE-IN-THE-BOWER *Lamium album* (White Deadnettle) Som. Macmillan. "because in the flower, if you hold it upside down, the stamens can be seen lying for all the world like two people side by side in a curtained bed" (Boland.1977).

ADAM'S APPLE 1. *Citrus limetta* (Lime) W Miller. 2. *Musa x paradisiaca* (Banana) Gerard. In the 16th and 17th centuries, banana was the favourite candidate for the tree of the knowledge of good and evil. Gerard reported that the Jews "suppose it to be the tree, of which Adam did taste".

ADAM'S FIG *Musa paradisiaca* (Cooking Banana) Bianchini.

ADAM'S FLANNEL 1. *Verbascum lychnitis* (White Mullein) Lincs. Peacock. 2. *Verbascum thapsus* (Great Mullein) Som. Friend; Ches. Holland; Yks. Carr; USA Henkel. From the texture of the leaves, like many others incorporating the word flannel, or blanket (Aaron's Flannel, Flannel-flower, Lady's Flannel, Blanket-leaf, Moses' Blanket, and many more). The leaves were actually used as if they were flannel; they were wrapped round the throat, to help relieve coughs and colds.

ADAM'S NEEDLE 1. *Scandix pecten-veneris* (Shepherd's Needle) Som. Macmillan; N'thum, Berw. Grigson. 2. *Yucca sp* J Smith. The "needle" names come from the appearance of the long beaked fruits. There are many of them, including Beggar's Needle, Old Wife's Darning Needle, Pook Needle, and, of course, considering the specific name, Venus' Needle. The yucca, though, gets the name from the sharp-pointed leaves.

ADAM'S NEEDLE-AND-THREAD *Yucca filamentosa* Som. Macmillan. Because the leaves have threadlike fibres on their margins, as well as sharp points, to give the "needle" epithet.

ADAM'S PLASTER *Polygonum persicaria* (Persicaria) Newfoundland Grigson. "It is reported that Dead Arsmart (i.e. Persicaria) is good against inflammations and hot swellings, being applied in the beginning" (Gerard).

ADDER-AND-SNAKE PLANT *Silene cucubalis* (Bladder Campion) Dev. Grigson.

ADDER'S COTTON *Cuscuta epithymum* (Dodder) Corn. Grigson. Cuscuta is from Greek kassuo, to sew together, and dodder is the plural of dodd, a bunch of threads (Fernie), hence, presumably, the reference to cotton.

ADDER'S EYES *Anagallis arvensis* (Scarlet Pimpernel) Herts. Grigson. A descriptive name; there are a number of 'eye' names for this, like Eyebright, Bird's Eye, Owl's Eye, etc.

ADDER'S FERN *Polypodium vulgare* (Common Polypody) Hants. Cope. 'Adder' is suggested by the rows of orange-coloured spores.

ADDER'S FLOWER 1. *Dactylorchis maculata* (Heath Spotted Orchis) Dev, Som. Macmillan. 2. *Endymion nonscriptus* (Bluebell) Dev, Som. Macmillan. 3. *Melandrium dioicum* (Red Campion) Dev, Som. Macmillan; Herts Grigson. 4. *Orchis mascula* (Early Purple Orchis) Dev, Som. Macmillan; Wilts D Grose.

ADDER'S FOOD *Arum maculatum* (Cuckoo Pint) Friend; Wilts. Goddard. Adder here is probably a corruption of OE attor, poison, though this is certainly not the derivation of the word. Adder's Meat occurs in Cornwall, too, and these, together with the Snake's Victuals, etc. group, were probably applied to any berries which looked a bit suspicious, whether they were actually poisonous or not.

ADDER'S GRASS 1. *Dactylorchis maculata* (Heath Spotted Orchis) Border B & H. 2. *Ophioglossum vulgatum* (Adder's Tongue) Tynan & Maitland. 3. *Orchis mascula* (Early Purple Orchis) B & H. 4. *Phalaris arundinacea* (Reed-grass) S Scot. Tynan & Maitland. Adder's Tongue and Snake Flower are recorded for the Early Purple Orchis, too. These names are almost certainly in some way descriptive, for the doctrine of signatures said they were to be used for snakebite.

ADDER'S MEAT 1. *Anthriscus sylvestris* (Cow Parsley) Som. Mabey. 2. *Arum maculatum* (Cuckoo Pint) Corn. B & H; Dev. Friend.1882; Som. Grigson; Glos. PLNN35. 3. *Mercurialis perennis* (Dog's Mercury) Herts. Grigson. 4. *Stellaria holostea* (Greater Stitchwort) Som. Macmillan. Suss. Parish. 5. *Tamus communis* (Black Bryony) Dev. Friend.1882. As far as Cuckoo Pint is concerned, Adder is probably a corruption of OE attor, poison. There are a dozen or so adder or snake names for Greater Stitchwort - it is an unlucky plant, protected by the fairies, and must not be gathered. Cornish children say they will be bitten by an adder if they pick it, and it certainly had some evil connotation, for names like Devil's Flower, Devil's Corn, etc., also occur. Dog's Mercury is easy to equate with Adder's Meat - it actually *is* poisonous, and so is Black Bryony. Cow Parsley's Somerset name Adder's Meat may very well be a misreading for Rabbit's Meat, which is quite common (Coney Parsley occurs, too).

ADDER'S MOUTH 1. *Iris foetidissima* (Foetid Iris) Som. Macmillan; Suss. Grigson. 2. *Orchis mascula* (Early Purple Orchis) Som. Macmillan. A general idea that Foetid Iris is poisonous probably related to the foul smell. The seeds are called Snake's Food in Devon, and Poison-berries also occurs. There are quite a few snake names associated with Early Purple Orchis; besides this, there are Adder's Flower, Tongue and Grass, and Snake Flower.

ADDER'S POISON *Tamus communis* (Black Bryony) Dev. Friend.1882.

ADDER'S SPIT 1. *Listera ovata* (Twayblade) Tynan & Maitland. 2. *Pteridium aquilina* (Bracken) Tynan & Maitland. 3. *Stellaria holostea* Corn. Grigson; Glos. PLNN35. Cf. Adder's Meat, Snake-flower, etc., for Stitchwort, and Adder's Tongue for Twayblade.

ADDER'S TONGUE 1. *Achillea ptarmica* (Sneezewort) Aber. Grigson. 2. *Arum maculatum* (Cuckoo Pint) Corn. B & H; Som. Elworthy. 3. *Geranium robertianum* (Herb Robert) IOW Grigson. 4. *Listera ovata* (Twayblade) Wilts Dartnell/Goddard. 5. *Ophioglossum vulgatum*. 6. *Orchis mascula* (Early Purple Orchis) Dev, Dor. Macmillan; Ches. B & H. 7. *Phyllitis scolopendrium* (Hartstongue Fern) Dev. Friend.1882. 8. *Sagittaria sagittifolia* (Water Archer) B & H. Presumably descriptive as far as Herb Robert (the seed vessels), and Water Archer are concerned. The shape of the leaf of Cuckoo-pint suggests the name, too. Twayblade gets it from the double leaf. Sneezewort is possibly in itself the descriptive derivation of this. Adder is OE attor, poison, when applied to cuckoo pint berries, and possibly the orchid, too.

ADDER'S TONGUE, American *Erythronium americanum* (Yellow Adder's Tongue) Grieve.1931.

ADDER'S TONGUE, Early *Ophioglossum lusitanicum*.

ADDER'S TONGUE, Giant *Erythronium oreganum*.

ADDER'S TONGUE, White *Erythronium albidum*.

ADDER'S TONGUE, Yellow *Erythronium americanum*.

ADDER'S TONGUE SPEARWORT *Ranunculus ophioglossifolius* (Serpent's Tongue Spearwort) McClintock.

ADDERS, Snakes-and- 1. *Anemone nemerosa* (Wood Anemone) Som. Macmillan. 2. *Ophrys apifera* (Bee Orchis) Som. Macmillan. Snake-flower is another name, from the same area, for Wood Anemone.

ADDERWORT 1. *Arum maculatum* (Cuckoo Pint) Cockayne. 2. *Polygonum bistorta* (Bistort) Cockayne. For Cuckoo Pint, Adderwort comes from OE athor (attor), poison (the shiny red berries, always suspicious). Bistort gets its adder- and snake-names from its writhed roots. By signature, it was used against snake-bite and poison, e.g. "Against snake-bite. Take waybread and agrimony and adderwort. Pound them in wine and give them to drink, and compound a salve of the same herbs. And take agrimony, make one ring round about the bite, it [the poison] will not pass any further; and bind the herbs again on the wound" (Storms). This is Anglo-Saxon, and Cockayne has a very similar remedy. But the cure lingered at least until Culpepper's day. He has "Both the leaves and Roots have a powerful faculty to resist all poison".

ADEN SENNA *Cassia acutifolia* (Alexandrian Senna) Dalziel.

ADHIB *Euphrasia officinalis* (Eyebright)

Halliwell. A name, Halliwell says, given in Dr Thomas More's ms additions to Ray.

ADONIS, Blood of; ADONIS FLOWER *Adonis annua* (Pheasant's Eye). It is Parkinson who calls this Adonis Flower.

ADRELWURT *Tanacetum parthenium* (Feverfew) Halliwell. He says it is in MS Harl. 978.

ADZUKI BEAN *Phaseolus angularis*.

AERIAL YAM *Dioscorea bulbifera* (Potato Yam) Alexander & Coursey. Because aerial tubers are produced in the leaf axils. Cf. Air Potato for this.

AEROPLANES *Acer pseudo-platanus* seeds (Sycamore keys) Som. Macmillan. Because they give the impression of flying as they spiral to the ground - helicopters would be more appropriate.

AFFADIL *Narcissus pseudo-narcissus* (Daffodil) W Miller. Affodil is closer to the original asphodelus than this.

AFFODILL *Asphodelus ramosus* (White Asphodel) Henslow.

AFFODILL, White *Asphodelus ramosus* (White Asphodel) Turner.

AFFON *Trecularia africana* (African Breadfruit) Wit.

AFFRODILE *Narcissus pseudo-narcissus* (Daffodil) Ches. B & H.

AFRICAN ARROWROOT 1. *Canna bidentata*. 2. *Tacca pinnatifida*.

AFRICAN BEECH *Faurea saligna*.

AFRICAN BLACKWOOD 1. *Dalbergia melanoxylon*. 2. *Peltophorum africanum* (Rhodesian Wattle) Palgrave.

AFRICAN BLACK WATTLE *Peltophorum africanum* (Rhodesian Wattle) Palgrave.

AFRICAN CEDAR 1. *Cedrus atlantica* (Atlas Cedar) Grieve.1931. 2. *Juniperus procera* (East African Cedar) Hora. Atlas Cedar is a native of North Africa. Cedar is Latin cedrus, from Greek kedros, which seems to have meant juniper as well as cedar, hence the frequent interchange of names.

AFRICAN CLOVER, Red *Trifolium africanum* (Cape Clover) Watt.

AFRICAN CORAL-WOOD *Pterocarpus soyauxii* (Barwood) Dalziel.

AFRICAN CORN LILY *Ixia sp.*

AFRICAN CUCUMBER *Momordica charantia* (Wild Balsam-apple) Dalziel.

AFRICAN DAISY 1. *Arctotis stoechadifolia*. 2. *Dimorphoteca aurantiaca* (Star of the Veldt) Hay & Synge.

AFRICAN EBONY 1. *Dalbergia melanoxylon* (African Blackwood) Dale & Greenway. 2. *Diospyros mespiliformis*.

AFRICAN FALSE ELM *Celtis integrifolia* (African Nettle Tree) Dalziel.

AFRICAN FOUNTAIN GRASS *Pennisetum setaceum.*

AFRICAN HARP *Sparrmannia africana* Rochford.

AFRICAN HEMP *Sanseviera guineensis*. Sansevieras are fibre plants, hence hemp.

AFRICAN KINO *Pterocarpus erinaceus* Dalziel. Kino is actually the coloured resin formed from the tree.

AFRICAN LABURNUM *Cassia afrofistula*.

AFRICAN LILY *Agapanthus africanus*.

AFRICAN LINDEN *Mitragyna stipulosa* Dalziel.

AFRICAN LOTUS *Zizyphus lotus* (Lotus Bush) Howes.

AFRICAN MAHOGANY *Khaya ivorensis*.

AFRICAN MAHOGANY, Smooth-barked *Khaya anthoteca* (White Mahogany) Dalziel.

AFRICAN MARIGOLD 1. *Dimorphoteca pluvialis* (Rain Daisy) Coats. 2. *Tagetes erecta*. Tagetes is a Mexican genus, but when names were being allotted, no-one seemed to know where they came from. Gerard called the same plant Turkey Gilliflower, and the smaller Tagetes patula is the French Marigold.

AFRICAN MILLET *Eleusine corocana* (Finger Millet) Brouk. Probably originally from India, and cultivated in India, Malaya and China, and also now in the wetter parts of Central Africa.

AFRICAN MYRRH *Commiphora africana*.

AFRICAN NETTLE TREE *Celtis integrifolia*.

AFRICAN NUTMEG *Monodora myristica* (Calabash Nutmeg) Dalziel.

AFRICAN OAK 1. *Chlorophora excelsa*. 2. *Oldfieldia africana*. Not oaks at all, of course; Oldfieldia belongs to the Spurge family, and Chlorophylla to the Moraceae.

AFRICAN PENCIL CEDAR *Juniperus procera* N Davey. J virginiana is the Pencil Cedar par excellence.

AFRICAN ROSEWOOD *Pterocarpus erinaceus* Dalziel.

AFRICAN RUE *Peganum harmala*.

AFRICAN SANDALWOOD *Spirostachys africana* (Tambootie) Summers.

AFRICAN STAR-APPLE *Chrysophyllum africanum*.

AFRICAN TEAK 1. *Oldfieldia africana* (African Oak) Sierra Leone Dalziel. 2. *Pterocarpus erinaceus* Dalziel.

AFRICAN TRAGACANTH *Sterculia tragacantha*.

AFRICAN TULIP TREE *Spathodea campanulata*.

AFRICAN WHITEWOOD *Enantia chlorantha* (African Yellow-wood) Dalziel. The timber is yellow, though. But it is as soft as most whitewoods.

AFRICAN WOOD OIL-NUT *Ricinodendron africanum*.

AFRICAN YELLOW-WOOD *Enantia chlorantha*.

AFTERNOON IRIS *Iris dichotoma*. The corolla never expands before midday. Vesper Iris is another name.

AG-LEAF; AG-PAPER *Verbascum thapsus* (Great Mullein) both Bucks. B & H. Ag is Hag, OE, modern hedge. There are a lot of variants of this, e.g. Hig, Higgis, High. Cf. also the names for Hawthorn, where the variants abound. Ag-paper, which becomes Rag-paper, also in Bucks, must be a corruption of Hag-taper, i.e. hedge taper. The whole plant was used as a torch at funerals, etc. The dried stalks were often used, smeared with grease or tar, as torches in the French fete de Brandons, on the 1st Sunday in Lent. Note the Welsh name tapr Mair, Mary's Taper.

AGALD; AGAR; AGARVE *Crataegus monogyna* fruit (Haws) Agald - Wilts. B & H; Agar - Bucks. Harman; Agarve - Suss. Parish. The first syllable is hag. Both haw and hag mean hedge. Hawthorn is the oldest of hedgerow trees, for it gets its name from OE haga, hedge or enclosure, and it was used from Saxon times onwards to make impenetrable fences.

AGASSE; AIGARCE *Crataegus monogyna* fruit (Haws) Hants, Suss. Grigson (Agasse); Suss. Parish (Aigarce). One of the large group which come eventually from OE haga, hedge (see Agald, Agar,etc). Hog-gosse also occurs in Sussex, and the Hog element is also OE haga.

AGATI *Sesbania grandiflora*.

AGAVE, South American *Agave americana* (Mescal) O P Brown. Agave is a name from Greek mythology - Agaue was the daughter of Cadmus and Harmonia. Agauos means illustrious.

AGGERMONY *Agrimonia eupatoria* Som. Elworthy.

AGGLE(BERRY); AGLEN; AGLON *Crataegus monogyna* fruit (Haws) Aggle - Corn. Courtney.1880; Dev. Choape; Aglen - Corn. Jago; Aglon - Wessex N.Rogers. All from OE haga, hedge.

AGLET; AGLET-TREE 1. *Crataegus monogyna* fruit (Haws) Dev. Friend.1882; Corn. Jago; Som. Macmillan. 2. *Corylus avellana* catkins (Hazel catkins) Gerard. Aglet-tree is applied to the tree itself in Cornwall. All are from OE haga, hedge. Hazel is as much a hedgerow tree as hawthorn. Gerard calls it Hedge-nut, for example.

AGOG *Crataegus monogyna* fruit (Haws) Berks. Lowsley. OE haga, hedge.

AGREEN *Senecio jacobaea* (Ragwort) Cumb. Grigson.

AGRIMONY 1. *Agrimonia eupatoria*. 2. *Agrimonia striata* USA House. Agrimony is based on agrimonia, a misreading for Latin argemonia, from Greek argemone, which is a type of poppy. This in turn comes from the

Greek argemon, meaning the white spot on the eye, which the plant (the poppy, presumably), was supposed to cure. Argemon derives from the word meaning white, or shining (Cf. argent). The eupatoria part of the botanical name denotes its use in liver complaints - "It openeth and cleanseth the liver, helpeth the Jaundice..." (Culpepper), etc. Or it has been suggested that it comes from Mithridates Eupator, King of Pontus, the first to use the herb.

AGRIMONY, Bastard 1. *Ageratum conyzoides* (Goatweed) W Miller. 2. *Aremonia agriminoides*. 3. *Bidens tripartita* Gerard. 4. *Eupatorium cannabinum* Gerard.

AGRIMONY, Dutch 1. *Bidens cernua* (Nodding Bur-Marigold) Gerard. 2. *Eupatorium cannabinum* (Hemp Agrimony) B & H.

AGRIMONY, Eupator's; AGRIMONY, European *Agrimonia eupatoria* (Agrimony) Le Strange.

AGRIMONY, Fragrant *Agrimonia odorata*.

AGRIMONY, Hemp *Eupatoriuum cannabinum*. 'Hemp' from the resemblance of the leaves to those of hemp.

AGRIMONY, Hemp, Trifid *Bidens tripartita* (Bur Marigold) Curtis.

AGRIMONY, Hemp, Water, Nodding *Bidens cernua* (Nodding Bur Marigold) Curtis.

AGRIMONY, Noble *Hepatica caerulea* B & H. Apparently coined by Lyte originally, probably because, like all Hepaticas, it was a sovereign remedy for liver complaints.

AGRIMONY, Three-leaved *Aremonia agriminoides* (Bastard Agrimony) W Miller.

AGRIMONY, Water 1. *Bidens cernua* (Nodding Bur Marigold) B & H. 2. *Bidens tripartita* (Bur Marigold) Gerard. 3. *Eupatorium cannabinum* (Hemp Agrimony) Ches. B & H.

AGRIMONY, Wild *Potentilla anserica* (Silverweed) B & H.

AGUE-ROOT *Aletris farinosa* (Unicornroot) Le Strange.

AGUE-TREE *Sassafras variifolium* (Sassafras Tree) Coles. Gerard had the name, too; he described it as Sassafras or the Ague-Tree, "of his vertue in healing the Ague". There is no other record of the use apparently.

AGUEWEED 1. *Gentiana quinquefolia* (Stiff Gentian) USA House. 2. *Eupatorium perfoliatum* (Thoroughwort) USA Henkel. Thoroughwort, among other things, is a sweat-inducer, and a tea made from the dried leaves used to be a common American domestic medicine for a cold. The leaves and flowering tops are still used for fever and ague. Feverwort also occurs in USA.

AGUEWEED, Indian *Eupatorium perfoliatum* (Thoroughwort) W Miller. See Agueweed.

AGWORM-FLOWER; HAGWORM-FLOWER *Stellaria holostea* Greater Stitchwort both Yks. (Agworm) B & H; (Hagworm) Grigson. Hagworm, or Agworm, is Yorkshire archaic dialect for a snake, and in this case, an adder. There are a number of other names, like Adder's Meat, Adder's Spit, Snake-flower, from the southern half of the country, as well as the Yorkshire examples. Reason for the names not known.

AIG; AIGIE-BERRIES *Crataegus monogyna* fruit Haws Lancs. Halliwell (Aig); Staffs. PLNN 17 (Aigie). This must be the same as Haig, or Hague.

AIGREEN; AYEGREEN *Sempervivum tectorum* (Houseleek) Lancs. Nodal (Aigreen); Barton & Castle (Ayegreen). Both mean evergreen.

AIK; AYK *Quercus robur* (Oak) both Scot. Jamieson. A Scots variant on the pronunciation, and thus the spelling, of Oak. Eike and Eke also occur in England. Wilks points out that Acton (London), Savernake (Wilts.), and Mallaig (Inv.) are all derived from this variant.

AIKRAN *Lobaria pulmonaria* (Lungwort Crottle) Bolton.

AILANTO *Ailanthus glandulosa* (Tree of Heaven) Chopra. Ailanthus, and thus Ailanto, is from an Amboyna name meaning tree of the gods.

AIL(E)S 1. *Avena sativa* (Oat) Dev. Friend.1882. 2. *Hordeum sativum* (Barley) Wilts. Dartnell & Goddard. Cf. Hoyles and Oils for oats, and Eyles, or Iles, for barley.

AILES, Pig's; ALES, Pig's *Crataegus monogyna* fruit (Haws) Som. Macmillan. "Pig's" in these cases is a misunderstanding. The immediate forebear of both these names is Peggaul, which has many variants.

AILIFF *Glechoma hederacea* (Ground Ivy) Suss. B & H. This is a variant of Alehoof, which, together with such as Allhoove, Allhose, Aller, etc., arose from its one-time use in brewing - it is a bitter herb, used to make the beer keep.

AINU ONION *Allium nipponicum*.

AIR PINE *Aechmea fasciata* (Urn Plant) Oplt.

AIR PLANT 1. *Kalanchoe pinnata* (Curtain Plant) USA Cunningham. 2. *Tillandsia cyanea*. In the case of the Tillandsia, given because of its capacity to absorb moisture from the atmosphere.

AIR POTATO *Dioscorea bulbifera* (Potato Yam) Schery. Cf. another name - Aerial Yam. Aerial tubers are produced in the leaf axils.

AIRACH, Dog's; AIRACH, Goat's; AIRACH, Stinking *Chenopodium vulvaria* (Stinking Orach) all Culpepper. Airach is Orach, which eventually comes from Atriplex, the botanical name for the Garden Orach.

AIRBELL *Campanula rotundifolia* (Harebell) A S Palmer. Another version of Hairbell.

AIRESS; ARESS *Galium aparine* (Goosegrass) both Yks. B & H. There are a number of names like this, mostly with initial H (Cf. Hairiff, Heiriff, Hairitch, etc.), all probably bearing some relation to old names descriptive of the bristly nature of the plant.

AIRPLANE PLANT *Chlorophytum comosum* (Spider Plant) USA Cunningham.

AIRSENS *Crataegus monogyna* fruit (Haws) Glos. J D Robertson.

AIRUP; AIRIF *Galium aparine* (Goose-greass) Yks. B & H (Airup); Lincs. B & H (Airif). Cf. Airess, etc., for the same plant.

AISE; AISEWEED; AISHWEED *Aegopodium podagraria* (Goutweed) B & H (Aise, Aiseweed); Gerard (Aishweed). These are variants of 'ash', and the weed, not a true native of Britain, but introduced as a potherb and medicinal herb in the Middle Ages, is widely known as Ground Ash, though better known as Ground Elder.

AISCHEN; AISH; AISHEN-TREE; ASHEN-TREE; ASHING-TREE *Fraxinus excelsior* (Ash) Inv. B & H (Aischen); Dor. Barnes (Aish); Dor. B & H (Aishen-tree); Glos. J D Robertson (Ashen-tree); Som. Tongue (Ashing-tree). All simple variants on Ash, made more obvious by the fact that ash derives from OE aesc.

AITEN; AITNAGH *Juniperus communis* (Juniper) both Scot. Grigson. Eatin-berries also occurs in Scotland, and the rather more complicated Etnagh-berries is recorded from Angus.

AIVERIN; AIVERN *Rubus chamaemorus* (Cloudberry) both Mor. B & H. There are other similar names in Scottish dialect - Cf. Averen, Averin, Evron, Everocks.

AK *Quercus robur* (Oak) Yks. Morris. This is probably the way to spell it, though it would be pronounced yak, and that is what Nicholson gives as the Yorkshire word.

AKEE *Blighia sapida*. Akee is an Ashanti word (the plant is a native of West Africa), while Blighia commemorates Captain Bligh, of HMS Bounty.

AKKARIN; ACKERN; AKKERN; AKRAN *Quercus robur* - acorn (Acorn) Yks. Addy (Akkarin); Heref. Havergal, Northants. Sternberg (Ackern, Akkern); Lancs. Nodal, Yks. Addy (Akran). Sternberg suggests OE akearn, Danish aaggern. The Yorkshire place name Ackroyd means a clearing among the oaks. O Norse akarn, Goth. akran, where it meant fruit in general; Nodal points out that it is only in the cognate tongues that the word applies solely to the fruit of the oak.

AKOKO *Newbouldia laevis*. A West African

name.

AKS, Bitter *Taraxacum officinale* (Dandelion) Shet. Grigson. Cf. Eksis-girse, also from Shetland.

AKYR *Quercus robur* - acorn (Acorn) Halliwell. "The bores fedyng is propreliche y-cleped akyr of ookys berynge and buck-mast" (MS Bodl. 546). Very close, at least in spelling, to its Latin original, acer.

ALAN *Nymphaea alba* (White Waterlily) Corn. B & H.

ALASKA BEAUTY *Claytonia acutifolia*.

ALASKA CURRANT *Ribes bracteosum*.

ALASKA CYPRESS *Chamaecyparis nootkatensis* (Alaska Yellow Cedar) Edlin.

ALASKA FLEABANE *Erigeron salsuginosus*.

ALASKA YELLOW CEDAR *Chamaecyparis nootkatensis*.

ALBASPYNE; ALBESPEINE *Crataegus monogyna* (Hawthorn) B & H (Albaspyne); Ellacombe (Albespeine). These, together with aubespyne and abbespine are all Latin alba spina, French aubepine, i.e. whitethorn, the other common name for the hawthorn.

ALBERTA WHITE SPRUCE *Picea canadensis var. albertiana*.

ALBIZIA, Plume *Albizia lophantha* (Brush Wattle) H & P. The specific name means 'with crested flowers', hence the 'plume' of this name.

ALCOCK SPRUCE *Picea bicolor*.

ALDER 1. *Alnus glutinosa*. 2. *Sambucus nigra* (Elder) co Clare Westropp. The use of alder for elder is a simple mispronunciation that has stuck. Alder itself is OE alor, aler (which accounts for many dialect forms like Aller, or Owler), related to MHGer. aller, alre, elre, else (note Halse or Alls in Devonshire). The OHGer. is elo, elawer - reddish-yellow, which is the colour of the fresh-cut timber. All these names derive ultimately from Latin alnus.

ALDER, Berry *Frangula alnus* (Alder Buckthorn) B & H. It is also known as Black Alder, or Black Aller. The berries as well as the bark and leaves are a violent purgative, and can be dangerous to children.

ALDER, Black. 1. *Alnus incana virescens* Amsden. 2. *Alnus glutinosa* (Alder) USA Zenkert. 3. *Frangula alnus* (Alder Buckthorn) Parkinson. 4. *Ilex verticillata* (Winterberry) USA Harper.

ALDER, Black, American *Alnus incana* (Hoary Alder) W Miller.

ALDER, Californian *Alnus rhombifolia* (White Alder) W Miller.

ALDER, Death *Euonymus europaeus* (Spindle Tree) Bucks. Grigson. It was thought unlucky to bring them into the house, but apart from this the leaves, bark, as well as the fruit, are violent purgatives and are dangerous to children and animals, perhaps not lethal, but may very well lead to loss of consciousness. "This shrub is hurtfull to all things, and namely to goats; he [Theophrastus] saith the fruit hereof killeth; so doth the leaves and fruit destroy goats especially..." (Gerard). On a lower level, Evelyn, in *Sylva*, said "... the powder of the berry, being bak'd, kills nits...".

ALDER, English *Alnus glutinosa* (Alder) Le Strange.

ALDER, European *Alnus glutinosa* (Alder) USA Zenkert. There are a number of American species, hence the occasional differentiation.

ALDER, Green *Alnus viridis*.

ALDER, Grey; ALDER, Grey, Norwegian *Alnus incana* (Hoary Alder) Clapham (Grey); Brimble.1948 (Norwegian Grey).

ALDER, Hazel *Alnus rugosa* (Smooth Alder) USA Youngken.

ALDER, Hoary *Alnus incana*.

ALDER, Italian *Alnus cordata*.

ALDER, Mountain 1. *Alnus tenuifolia*. 2. *Alnus viridis* (Green Alder) W Miller.

ALDER, Oregon *Alnus rubra* (Red Alder) Mitchell.1974. *Alnus oregana* used to be an alternative classification for this.

ALDER, Red 1. *Alnus rubra*. 2. *Alnus rugosa* (Smooth Alder) Grieve.1931.

ALDER, River *Alnus tenuifolia* (Mountain Alder) Johnston.

ALDER, Seaside *Alnus maritima*.
ALDER, Smooth 1. *Alnus rubra* (Red Alder) O P Brown. 2. *Alnus rugosa*.
ALDER, Speckled 1. *Alnus incana* (Hoary Alder) USA Tolstead. 2. *Alnus rugosa* (Smooth Alder) Wit.
ALDER, Spotted *Hamamelis virginica* (Witch Hazel) Grieve.1931.
ALDER, Tag 1. *Alnus rubra* (Red Alder) O P Brown. 2. *Alnus rugosa* (Smooth Alder) Grieve.1931.
ALDER, White 1. *Alnus incana* (Hoary Alder) Johnston. 2. *Alnus rhombifolia*. 3. *Clethra alnifolia* (Sweet Pepper Bush) USA Harper.
ALDER, Witch *Fothergilla gardeni*.
ALDER BUCKTHORN *Frangula alnus*. A buckthorn, though now taken out of the genus Rhamnus (it used to be called *Rhamnus frangula*), and as purging as the common Buckthorn. "... it is... a medicine more fit for clownes than for civil people, and rather for those that feed grossly, than for dainty people" (Gerard).
ALDER-LEAF MOUNTAIN MAHOGANY *Cercocarpus montanus* (Mountain Mahogany) USA T H Kearney.
ALDER-LEAVED BUCKTHORN *Rhamnus alnifolia*.
ALDER-LEAVED SERVICE-BERRY *Amelanchier alnifolia* (Western Service-berry) USA Elmore.
ALDERDRAUGHT; ALDERTROT *Heracleum sphondyllium* (Hogweed) both Som. Macmillan. These are variations on the common Eltrot, which becomes Eldertrot in Wiltshire, and so very close to Aldertrot.
ALDERN; ALDERNE *Sambucus nigra* (Elder) Wilts. Jones & Dillon (Aldern); B & H (Alderne).
ALDERNEY SEA LAVENDER *Limonium lychnidifolium*. It only grows on Alderney and Jersey.
ALE, Gill- *Glechoma hederacea* (Ground Ivy) Dev. Friend. Ale-gill was an infusion made of gill, used in brewing, and gill is from the French guiller, to ferment beer. Ground Ivy was the herb most commonly used in brewing beer before the advent of hops. There are many names, such as Alehoof and Allhoove, which are the remnants of the old practice, and the Gill names occur in all kinds of variations (Gilrumbithground, for instance, made more intelligible by the widespread Gill-run-by-the-ground).

ALECAMPANE; ALLICAMPANE; ALA-COMPANE *Inula helenium* (Elecampane). Alecampane is from Turner, Alacompane is in A S Palmer, and Allicampane is a Yorkshire name (Addy), in which the stress was put on the last syllable. A simple alternative to elecampane as the early form.

ALECOAST; ALECOST *Tanacetum balsamita* (Balsamint) (Alecoast) Gerard; (Alecost) B & H. Given, according to Parkinson, because it was used for flavouring ales. "Costus" may mean an aromatic plant used in making perfumes in the East, and according to ODEE, it is OE cost, from Latin costum, Greek kostos, Arabic qust, and eventually Sanskrit kusthas. Both Alecoast and Alecost are contracted to Coast and Cost, and Costmary is a common old name for this plant.

ALEHOOF *Glechoma hederacea* (Ground Ivy) Corn, Dev, Som, Suss, Shrop, Yks. Grigson; War. Bloom. From its one-time use in brewing. It means literally that which will cause ale to heave, or work. It has always been assumed that it was commonly used for this purpose, but the ODEE speaks of "its alleged use in brewing instead of hops". Gerard was sure enough, though apparently he didn't know why it was used. But there are some doubts of the derivation of the "hoof" part of the name. Friend ventured the derivation as from OE hufe, crown, alluding to the chaplet which crowned the ale-stake at an inn. Earle has another explanation - he says it is ael (el)-hofe, "another sort of hofe", which he says is the true name for the violet. Whatever the true derivation, Alehoof was the general and most common name for the Ground Ivy, and in some parts of the country, still is.

ALEPPO OAK *Quercus infectoria*. This oak is a native of west Asia and Cyprus.

ALEPPO PINE *Pinus halepensis*. A Middle East species, also called Jerusalem Pine.

ALERCE 1. *Fitzroya cupressoides*. 2. *Tetraclinis articulata* (Sandarac) Polunin.1976.

ALES, Pig's; AILES, Pig's *Crataegus monogyna* fruit (Haws) Som. Macmillan. These names started as Peggall, or something like it, for the word has many variants.

ALEXANDER'S FOOT *Anacyclus pyrethrum* (Pellitory-of-Spain) B & H. Gerard says this is a translation of the French pied d'Alexandre, "that is, pes Alexandrinus", without offering an explanation.

ALEXANDERS *Smyrnium olusatrum*. Like Alick, Allsanders, Alisanders, Alshinder, etc., this is a corruption of olusatrum, which is from Latin olus, or holus, a vegetable or potherb, and ate, black, probably from the dark colour of the fruits. Smyrnium is apparently from the Greek word for myrrh, possibly because the juice smells like it. Other attempts at derivation have been made. Plausible is the statement that it got the name because its homeland is the coastal region around Alexandria; not so plausible is the suggestion that the derivation should be looked for in the fact that an early name was Parsley of Macedon, Alexander's country.

ALEXANDRIA(N) LAUREL 1. *Calophyllum inophyllum*. 2. *Ruscus racemosus*.

ALEXANDRIAN PARSLEY *Smyrnium olusatrum* (Alexanders) Genders. Alexandrian is a corruption of olusatrum - olus, potherb, and atrum, black.

ALEXANDRIAN SENNA 1. *Cassia acutifolia*. 2. *Cassia angustifolia* (Senna) Thomson.1978.

ALFALFA 1. *Cuscuta epithymum* (Dodder) S Africa Watt. 2. *Medicago sativa* (Lucerne) Clapham. This is the Spanish word, formerly alfalfez, which is directly from Arabic, apparently meaning "the best sort of fodder" - applied, of course, to Lucerne. How Dodder came to get the name is not explained.

ALFILARIA *Erodium cicutarium* (Storksbill) Kansas Gates. Both this, and Filaree, another American name for the plant, come from Spanish alfiler, pin. These, the various needle names, like Pook-needle, Powk-needle, and Pink-needle, as well as the common name itself, with other names linking it with a bird's bill, derive from the shape of the seed-vessel with its sharpened end. The Karok Indian name for the plant means imitation pin, though this is probably a loan from the whites.

ALFOMBRILLA *Drymaria arenarioides*. A Mexican plant; the Spanish name means a rug or small carpet (alfombrilla can also mean measles).

ALGERIAN CEDAR *Cedrus atlantica* (Atlas Cedar) Wit. A native of the Atlas Mts., Algeria.

ALGERIAN FIR *Abies numidica*.

ALGERIAN IRIS *Iris stylosa*. The winter-flowering native of Algeria.

ALGERIAN OAK *Quercus canariensis*.

ALICE; SAUCY ALICE *Polygonum persicaria* (Persicaria) both Norf. Grigson. A name more usually associated with Alyssum, known as Sweet Alice. Alyssum is from the Greek a, not; lussa, madness, and both Friend for Devonshire, and Wood-Martin for Ireland, record its use for treating the bite of a mad dog. But I don't know of any such use for Persicaria, nor of any reason for the name.

ALICE, Sweet 1. *Lobularia maritima* (Sweet Alison). 2. *Arabis alpina* (Alpine Rock-cress) Dev. B & H. 3. *Pimpinella anisum* (Aniseed) A W Smith. Sweet Alison, or Alisson, is corrupted to Sweet Alice, and is simply the sound of the name Alyssum, which Folkard derives from a, not, and lussa, madness (Greek) - it was used for treating the bite of a mad dog. The change of l to n, and vice versa, is quite common in dialect, hence Anise becomes Alice, and Alison becomes Anise. The Arabis gets the name probably from an original simple error of recognition.

ALICK *Smyrnium olusatrum* (Alexanders) Kent Grigson. Like Alexanders, and all its various variants, this is a corruption of olusatrum, which in turn means a potherb (Lat. olus, or holus, potherb, and ater, atrum, black, probably from the dark colour of the fruits).

ALIGOPANE *Inula helenium* (Elecampane) Scot. B & H.

ALISANDER, Stinking *Senecio jacobaea* (Ragwort) N'thum, Stir. Grigson. Alisander is Alexanders (Smyrnium olusatrum), although Ell-shinders, or Yellow Elshinders occur for Ragwort only. But the 'stinking' epithet is wide-ranging; besides this, there are recorded Stinking Billy, Stinking Davies, Stinking Nancy, Stinking Willie and Stinking Weed. Perhaps Mare-fart can legitimately be included.

ALISANDERS; ALIZANDERS *Smyrnium olusatrum* (Alexanders). Alisanders - Parkinson; Alizanders - Coles. Corruptions of olusatrum, as Alexanders itself is.

ALISON, Hoary *Berteroa incana*.

ALISON, Small *Alyssum alyssoides*.

ALISON, Sweet *Lobularia maritima*. Alyssum is Gk a, not and lussa, madness. It is recorded both in Devonshire (Friend), and in Ireland (Wood-Martin), as being used for treating the bite of a mad dog. Madwort, Heal-dog, Heal-bite are all old names for the plant. Alison is merely a corruption of Alyssum, further shortened to Sweet Alice in places.

ALISSON *Lobularia maritima* (Sweet Alison).

ALISSON, White *Arabis alpina* (Alpine Rock-cress) B & H. Sweet Alice is also recorded from Devonshire for this, so it is probably a mistake in identification of garden flowers.

ALKAKENG; ALKEKENGY *Physalis alkekengi* (Winter Cherry) (Alkakeng) Turner; (Alkekengy) B & H. This comes through med. Latin from Arabic al-kakanj, or kenj. The same Persian word means a kind of medicinal resin, and also a nightshade. Physalis is of the same family (Solanaceae) as the nightshades, and in fact an old name for Winter Cherry was Red Nightshade.

ALKALI ASTER *Aster carnosus*. A plant of North American alkaline meadows.

ALKANET *Anchusa officinalis*. Alkanet is from French orcanette (sometimes spelt orcanète), by which name the plant is still known. This comes through the Italian alcana (as a diminutive) from the Arabic al-kanna, the name of another plant (Lawsonia inermis) - henna, which yields, like Anchusa, a red dye. "The Gentlewomen of France do paint their faces with these roots, as it is said" (Gerard). Anchusa is from the Greek meaning to paint or dye. It seems to be one of the most ancient of all face cosmetics. This use of the roots for making rouge led to the plant becoming known as an emblem of falsehood.

ALKANET, American *Lithospermum canescens* (Hoary Puccoon) W Miller.

ALKANET, Bastard *Buglossoides arvensis* (Corn Gromwell) Prior. "The girls in the northern parts of Europe paint their faces with the juice of this root on days of public festivity. The bark of the root tinges wax and oil of a beautiful red similar to that which is attained from the root of the foreign alkanet sold in the shops" (Taylor).

ALKANET, Cape *Anchusa capensis*.

ALKANET, Evergreen; ALKANET, Green *Pentaglottis sempervirens* Evergreen Alkanet is the usual name for this; McClintock calls it Green Alkanet. The root bark gives a red dye, like the true Alkanet.

ALKANET, Field *Lycopsis arvensis* (Small Bugloss) Clair.

ALKANET, Italian *Anchusa italica*.

ALKANET, Mountain *Arnica montana* (Arnica) W Miller.

ALKANET, Yellow *Anchusa ochroleuca*.

ALKEKENGI *Physalis pubescens* (Husk Tomato) Bianchini.

ALL SAINTS' WORT *Hypericum androsaemum* (Tutsan). A S Palmer. A mistaken rendering, according to Palmer, of the French toute-saine, the name which gave Tutsan.

ALL-BONE *Stellaria holostea* (Greater Stitchwort) Prior. Probably means Wholebone, from the skeleton-like stalks, perhaps. Gerard says "whereof I see no reason... for this is a tender herbe, having no such bony substance". But words like Breakbones, or Snap-jacks, etc., occur widely, simply because of the way the plant snaps off at the joints.

ALL-GOOD 1. *Chenopodium album* (Fat Hen) Hants. Grigson. 2. *Chenopodium bonus-henricus* (Good King Henry) Gerard. Good King Henry was known to Tudor botanists as tota bona - this is merely the English equivalent.

ALL-GOOD SMEARWORT *Chenopodium bonus-henricus* (Good King Henry) Leyel.1937. It may be a misprint, as All-good and Smearwort are separately recorded as names for this plant.

ALL-HEAL 1. *Artemisia vulgaris* (Mugwort) Krappe. 2. *Panax schinseng* (Ginseng) Halliwell. 2. *Prunella vulgaris* (Self-heal) Som, Ches, Yks. Grigson. 4. *Valeriana officinalis* (Valerian) Som. B & H; Wilts. Dartnell/Goddard. 5. *Viscum album* (Mistletoe) Scot. B & H. Mugwort was used in probably the greatest number of all folk remedies, so deserves the name. Self-heal bears the names Heal-all and Touch-and-heal, as well as All-heal, and was widely used as a wound herb. Similarly, Valerian too must have been regarded as a wound herb, for the names Cut-finger or Cut-finger Leaf are recorded from Wiltshire, and Cut-heal also occurs. The French name (or a French name) for Valerian is Guérit-tout, and the word valerian itself derives from Latin valere, to be well, be in good health. The use of All-heal for the mistletoe is strange, as the berries are toxic, causing gastro-enteritis, but the plant was used against epilepsy, and this is not altogether a fancy, for it contains an active principle that is anti-spasmodic and reduces blood pressure. But it is its association with the Druids that in all probability led to the name being given, because it is a sacred plant. Until quite recently, German people used to run about the villages at Christmas, knocking at doors and windows with hammers, and shouting Gut hyl, Gut hyl, which sounds like some archaic version of good health, and brings to mind the All-heal for the mistletoe, inevitably associated with Christmas time.

ALL-HEAL, Clown's 1. *Stachys palustris* (Marsh Woundwort) Curtis. 2. *Stachys sylvatica* (Hedge Woundwort) Fernie. The Woundwort names show the uses of this genus. One of them, betony, is "officinalis", i.e. in the official list of medicines (pharmocopeia), and indeed, all-heal ought to be applied to betony, for it has the longest list of ailments (longer even than mugwort), for which it was used. In this context, and similar ones, "clown's" simply means countryman's. Shakespeare uses the word in this sense, in Midsummer Night's Dream, for instance.

ALL-ROOT *Heracleum sphondyllium* (Hogweed) Som. Macmillan. All-root probably started existence as the quite common Eltrot.

ALLA *Oxalis acetosella* (Wood Sorrel) Dawson. An old contraction of Hallelujah, or Alleluia, both names for the plant, and there are other variants recorded. There is a legend that the colour of the flowers derived from drops of blood shed by Christ, and the plant was often put in the foreground of Italian pictures of the Crucifixion, especially by Fra Angelico. It is called Alleluia in Italy, too.

ALLAR; ALLER; ALLER-TREE *Alnus glutinosa* (Alder). Allar - Scot Jamieson; Aller - Dev. Choape; Som. Jennings; Dor. Barnes; Aller-tree Turner. OE alor, giving rise to many regional variations like Aul, Ellar, Arl, Arn, Orl, even to Howler in the North Country. Aller is also applied to the Ground Ivy, *Glechoma hederacea*, presumably as part of the group which includes Alehoof, etc.

ALLEGHENY BARBERRY *Berberis canadensis*.

ALLEGHENY BLACKBERRY *Rubus*

alleghiensis (Highbush Blackberry) Youngken.

ALLEGHENY VINE *Adluaria fungosa*.

ALLEMANDE, Pink *Cryptostegia grandiflora* (Rubber Vine) USA Kingsbury.

ALLEMANDE, Yellow *Allamanda cathartica*.

ALLELUIA 1. *Genista tinctoria* (Dyer's Greenweed) Shrop. Grigson. 2. *Oxalis acetosella* (Wood Sorrel) Som. Macmillan; USA House. There is a legend that the colour of the flowers of Wood Sorrel were from drops of blood shed by Christ, and the plant was often placed in the foreground of Italian pictures of the Crucifixion, especially by Fra Angelico. It is called Alleluia in Italy, too. There are a number of variants such as Allolida, Alla, Lujula, etc. Macmillan says Wood Sorrel is called Alleluia because it blooms between Easter and Whitsun, when the psalms sung are 113-117, all ending with Alleluia.

ALLEN-TREE; ELLEN-TREE *Alnus glutinosa* (Alder). Both are common variants of alder; these in turn vary to Ellar, or Ollar.

ALLER, Black 1. *Alnus glutinosa* (Alder) Som. Elworthy. 2. *Frangula alnus* (Alder Buckthorn) Corn, IOW Grigson; Dev. Choape; Som. Elworthy. Gerard, too, has this name for Alder Buckthorn.

ALLER, Whit *Sambucus nigra* (Elder) Som. Elworthy. The difference between alder and elder is so slight that the variants of each have become almost interchangeable. Thus alder itself appears in Ireland for elder, but so too do Eldern, Ellar, Ellen-tree, etc., and Alderne. Whit in this name is the distinguishing adjective - it means white (alder is usually a "black" tree).

ALLHOOVE; ALLHOSE *Glechoma hederacea* (Ground Ivy) both Suss. B & H. Variants of Alehoof, which means literally that which will cause ale to heave, or work. It has a tonic bitterness and was apparently once used in brewing.

ALLIENNE *Artemisia absinthium* (Wormwood) Guernsey MacCulloch.

ALLIGATOR APPLE 1. *Annona glabra* Roys. 2. *Annona palustris*. Presumably a thick-skinned fruit, to bear this name.

ALLIGATOR JUNIPER; ALLIGATOR-BARK JUNIPER 1. *Juniperus pachyphloea*. 2. *Juniperus deppeana*. Alligator Juniper is the usual name for these trees; Alligator-bark is used in USA by Elmore. The thick bark is divided into scaly squares, somewhat resembling an alligator's skin.

ALLIGATOR PEAR *Persea gratissima* (Avocado) Everett. One would assume that this was descriptive, from the feel of the outer skin, but apparently this is not so. The Aztec name for the fruit was ahuacatl, rendered into Spanish as aguacate, then further corrupted to avigato. Then, when the English name was coined, avigato became alligator.

ALLIGATOR PEPPER *Aframomum melegueta* (Melegueta Pepper) Dalziel. In this case, 'alligator' is a corruption of 'Melegueta'.

ALLOLIDA *Oxalis acetosella* (Wood Sorrel) B & H. A corruption of Alleluia, which see.

ALLOM-TREE *Ulmus procera* (English Elm) B & H. One of the many variants of the word elm. Most of them depend on how one would actually pronounce it.

ALLS-BUSH; HALSE-BUSH *Alnus glutinosa* (Alder) both Dev. B & H. Halse is more usually applied to hazel, when it has a number of variants.

ALLSANDERS *Smyrnium olusatrum* (Alexanders) Corn. Grigson. A corruption of Alexanders, itself a corruption of olusatrum.

ALLSCALE *Atriplex polycarpa* (Cattle Spinach) Jaeger.

ALLSEED 1. *Atriplex patula* (Orache) Halliwell. 2. *Chenopodium polyspermum* (Many-seeded Goosefoot) Curtis. 3. *Polygonum aviculare* (Knotgrass) B & H. 4. *Radiola linoides*.

ALLSEED, Four-leaved *Polycarpon tetraphyllum*.

ALLSPICE 1. *Calycanthus floridus*. 2. *Eugenia pimenta* fruit Rimmel. 3. *Pimento* *dioica*. 4. *Tanacetum balsamita* (Balsamint) Brownlow. Allspice got its name (at least for Pimento dioica) because the dried berries seemed to combine all the flavours of nutmeg, clove, cinnamon and juniper berries.

ALLSPICE, Californian *Calycanthus occidentalis*.

ALLSPICE, Carolina *Calycanthus fertilis*.

ALLSPICE, Jamaica *Eugenia pimenta* W Miller.

ALLSPICE, Japanese *Chimonanthus fragrans* (Wintersweet) Salisbury.1936.

ALMAN RADICE *Raphanus sativus* (Radish) Turner. Radice is an old spelling of radish, and nearer to the original, which is OE raedic. What is alman? (Halliwell has alman as a kind of hawk) Doesn't it just mean German?

ALM(E) *Ulmus procera* (English Elm) N'thants. B & H. One of many variations on the word elm.

ALMOND *Prunus amygdalus*. Almond is early English almande, O Fr alemande, almande (modern French amande), from med.Latin amandula (Cf. Spanish almendra, Portuguese amendoa, Italian mandola, mandorla, German Mandel, etc). Ultimately it is from Greek amygdalos, the name for the tree, itself probably from amysso, to lacerate, referring to the channels in the seed. Some say the name arose from the Hebrew word meaning vigilant, an allusion to the tree's early blooming, or perhaps the Hebrew was megealh el (sacred fruit).

ALMOND, Barbados; ALMOND, Bastard *Terminalia catappa* (Myrobalan) Howes (Barbados); Dale & Greenway (Bastard).

ALMOND, Desert Range *Prunus fasciculata*. Also known as the Wild Almond in its native N America.

ALMOND, Earth *Cyperus esculentus* (Tigernut) Schery.

ALMOND, Indian 1. *Arisaema triphyllum* (Jack-in-the-pulpit) Douglas. 2. *Terminalia catappa* (Myrobalan) Dale & Greenway. North American Indian for Jack-in-the-pulpit, but presumably Asiatic Indian for Myrobalan.

ALMOND, Jordan *Prunus amygdalus dulcis* (Sweet Almond) Hatfield. Jordan is from French jardin, in other words, a cultivated almond.

ALMOND, Turkey *Prunus amygdalus* Mitton. The almond tree probably is indigenous to Asia Minor and Persia.

ALMOND, Wild 1. *Prunus fasciculata* (Desert Range Almond) USA Jaeger. 2. *Terminalia catappa* (Myrobalan) Dawson.1930.

ALMOND, Wild, Nevada *Prunus andersonii*.

ALMOND-LEAVED PEAR *Pyrus amygdaliformis*.

ALMOND-LEAVED WILLOW; ALMOND WILLOW *Salix triandra* (French Willow) (Almond-leaved) Perring; (Almond) Grigson.

ALOE, American *Agave americana* (Mescal). Aloe is OE alwe, eventually from Greek Aloe, connoting both the plant and the drug obtained from it.

ALOE, Barbados *Aloë vera* (Mediterranean Aloe) Simmons. The plant was introduced into the West Indies, where the main production of aloine, better known as bitter aloes, is concentrated.

ALOE, Cape; ALOE, Zanzibar *Aloe vera* Mitton.

ALOE, Mauritius *Furcraea gigantea* (Mauritius Hemp) Schery. Indigenous to Brazil, but it is used commercially in Mauritius, hence the name. It is a fibre plant.

ALOE, Mediterranean *Aloe vera*.

ALOE, Partridge-breasted *Aloe variegata*.

ALOE, Tiger *Aloe variegata* (Partridge-breasted Aloe) USA Cunningham.

ALOE, Water *Stratiotes aloides* (Water Soldier) Grieve.1931.

ALOE, West Indian *Aloe vera* Mitton.

ALOE YUCCA *Yucca aloifolia*.

ALOES-WOOD *Aquilaria agalloche* (Eaglewood) Moldenke. The Old Testament references to "aloes" probably mean this wood.

ALOEWOOD *Cordia sebestena* timber Willis.

ALPENROSE *Rhododendron ferrugineum*.
ALPINE *Sedum telephium* (Orpine) Ches. B & H. In this case, Alpine is a corruption of Orpine.
ALPS, King of the *Eritrichium nanum*.
ALPS, Queen of the *Eryngium alpinum* (Alpine Eringo) Polunin.
ALRUNA *Mandragora officinalis* (Mandrake) Clair. An old English name which comes from German; it is a word that came originally from rune, but it is also a reference to Germanic spirit belief - it is difficult to separate the mandrake from the spirit associated with it. Alraun is still modern German for mandrake.
ALSHINDERS *Smyrnium olusatrum* (Alexanders) Scot. Grigson. A corruption of Alexanders, itself a corruption of olusatrum.
ALSIKE CLOVER *Trifolium hybridum*. Alsike is a place in Sweden, near Uppsala, recorded by Linnaeus as a place where this plant grew.
ALTILUDA *Oxalis acetosella* (Wood Sorrel) Henslow. An old version of Alleluia, which has a number of versions, applied to the Wood Sorrel. There is a legend that the colour of the flowers were from drops of blood shed by Christ, and the plant was often put in the foreground of Italian pictures of the Crucifixion, notably by Fra Angelico. It is called Alleluia in Italy, too.
ALTROT *Heracleum sphondyllium* (Hogweed) Som. Macmillan. Wilts. Jones & Dillon. One of the large number of variations of Eltrot.
ALUM *Symphytum officinale* (Comfrey) B & H. From the astringent properties of the plant.
ALUM-BLOOM *Geranium maculatum* (Spotted Cranesbill) Grieve.1931. An astringent, particularly the roots (it is called Alum-root, too), and so used for diarrhoea, dysentery, etc., and for piles and internal bleeding. It must be very mild, though, for it was used also for a tonic, said to be especially good for the delicate stomachs of infants and invalids, precisely because it is not irritating.

ALUM-ROOT 1. *Geranium maculatum* (Spotted Cranesbill) OP Brown. 2. *Heuchera americana*. The Cranesbill may be only mildly astringent, but not so the Heuchera. The root of this was used when a powerful astringent was needed, for example a wash for wounds or obstinate ulcers. The Indians used the leaves in a similar way for sores.
ALUMINIUM PLANT *Pilea cadieri*.
ALYSSON, Purple *Aubrietia purpurea* (Wall-cress) W Miller.
ALYSSUM, Desert *Lepidium fremontii*.
AMARANTH, Chinese *Amaranthus tricolor* (Joseph's Coat) A W Smith. Amaranth(us) is derived from the Greek meaning never-fading. It may possibly have been applied to the genus that now bears the name (certainly the flowers last a long time), but equally the Greeks may have had in mind a completely mythical plant. It was regarded as the emblem of immortality.
AMARANTH, Globe 1. *Amaranthus hypondriachus* (Prince's Feathers). 2. *Gomphrenia globosa*.
AMARANTH, Green 1. *Amaranthus hybridus* (Slender Pigweed) Blamey. 2. *Amaranthus oleraceus*. 3. *Amaranthus retroflexum* (Pigweed) USA Zenkert. 4. *Amaranthus viridis*.
AMARANTH, Prickly; AMARANTH, Spiny *Amaranthus spinosus* (Thorny Pigweed) Dalziel.
AMARANTH, Prostrate *Amaranthus blitoides* (Tumbleweed) USA Elmore.
AMARANTH, Purple *Amaranthus paniculatus*.
AMARANTH, Thorny *Amaranthus spinosus* (Thorny Pigweed) Watt.
AMARANTH PIGWEED *Amaranthus retroflexus* (Pigweed) USA Zenkert.
AMAZON LILY *Eucharis grandiflora*.
AMBARELLA *Spondias dulcis*.
AMBARI HEMP *Hibiscus cannabinus*.
AMBER 1. *Hypericum androsaemum* (Tutsan) Kent Grigson. 2. *Hypericum perforatum* (St John's Wort) Kent,USA Grigson. Should be strictly confined to Tutsan. The

name refers to the practice of gathering and pressing them in books. When dry, they have a very sweet smell, likened to ambergris. Other names, like Sweet Amber, Sweet Leaf, Bible Flower, Bible Leaf, or Book Leaf, have the same derivation.

AMBER, Liquid *Liquidambar styracifolia* (Sweet Gum) Grigson.1974. Liquidambar means exactly that - it is a reference to the gum.

AMBER, Sweet *Hypericum androsaemum* (Tutsan) Suss. B & H. See Amber.

AMBER PINE *Pinus succinifera*. It was one of the ancient sources of amber. The resin, over millions of years, loses its liquids and gases, and is left as a solid mixture of succinic acid and other resins and oils. This mixture is succinite amber.

AMBOINA-BERRY *Spondias dulcis* (Ambarella) Douglas.

AMBROISE; AMBROSE *Teucrium scorodonia* (Wood Sage) (Ambroise) Jersey B & H; (Ambrose) Dawson. Presumably from Ambrosia, itself used as a local name for other plants.

AMBROSIA 1. *Artemisia annua* (Sweet Wormwood) Webster. 2. *Chenopodium anbrosioides* (American Wormseed) USA. Henkel; S Africa Watt. 3. *Chenopodium botrys* (Jerusalem Oak) E Hayes. Ambrosia - the fabled food of the gods, from Greek meaning immortality, or the elixir of life.

AMEE; AMEOS *Ammi visnaga* (Bishop's Weed) Halliwell.

AMELIANCHIER, Red-branched *Amelanchier sanguinea* (Round-leaved Juneberry) W Miller.

AMELIANCHIER, Tree *Amelanchier arborea*.

AMERICAN ADDER'S TONGUE *Erythronium americanum* (Yellow Adder's Tongue) Grieve.1931.

AMERICAN ALKANET *Lithospermum canescens* (Hoary Puccoon) W Miller.

AMERICAN ALOE *Agave americana* (Mescal). Agave for mescal is a name from Greek mythology. She was the daughter of Cadmus and Harmonia.

AMERICAN ANGELICA *Angelica atropurpurea* (Purple-stemmed Angelica) Grieve.1931. A North American species, used there in much the same way as the European kind is here.

AMERICAN APPLE MINT *Mentha x gentilis* (Bushy Mint) GB Foster.

AMERICAN ARBOR-VITAE *Thuja plicata* (Giant Cedar) Brimble.1948.

AMERICAN ARROWHEAD 1. *Sagittaria latifolia* (Broad-leaved Arrowhead) RHS. 2. *Sagittaria rigida*. All the Arrowheads are so called from the shape of the leaf.

AMERICAN ASPEN *Populus tremuloides*.

AMERICAN BALSAM POPLAR *Populus balsamifera*. It is said that the Balsam Poplars get their name from the peculiar scent of the leaves and catkins.

AMERICAN BARBERRY *Berberis canadensis* (Allegheny Barberry) Le Strange.

AMERICAN BASIL *Ocymum americanum*.

AMERICAN BEECH *Fagus grandifolia*.

AMERICAN BELLBINE *Calystegia sylvestris*. A large bindweed of North America. Bellbine is one of a large group of names given to the bindweeds, but I don't know why it seems to be the main name for this American variety, when for the British species, *C sepium*, it appears to be confined to Suffolk.

AMERICAN BIRD CHERRY; VIRGINIAN BIRD CHERRY *Prunus virginiana* (Choke Cherry) both Ablett.

AMERICAN BITTERSWEET *Celastrus scandens*.

AMERICAN BLACK ALDER *Alnus incana* (Hoary Alder) W Miller.

AMERICAN BLACK POPLAR *Populus deltoides* (Carolina Poplar) Polunin.1976.

AMERICAN BLACK WILLOW *Salix discolor*.

AMERICAN BLACKBERRY *Rubus villosus*.

AMERICAN BLUEBELL *Campanula americana*. Not a bluebell in the English sense, but

at least of the same genus as the bluebell of Scotland.
AMERICAN BOXWOOD *Cornus florida* (Flowering Dogwood) Grieve.1931. New England Boxwood is also recorded.
AMERICAN BRAMBLE *Rubus odoratus* (Purple-flowering Raspberry) Grieve.1931.
AMERICAN BROOKLIME *Veronica americana*. Brooklime itself is V beccabunga, O Norse bekh, brook; bung is the name of a plant, so the "lime" part of brooklime is the equivalent of 'bung'.
AMERICAN CHERRY LAUREL *Prunus caroliniana*.
AMERICAN CHESTNUT *Castanea dentata*.
AMERICAN COLUMBINE *Aquilegia canadensis* (Canadian Columbine) Howes.
AMERICAN COWBANE *Cicuta maculata*.
AMERICAN COWSLIP *Dodecatheon meadia* (Shooting Star) McDonald.
AMERICAN CRANBERRY Vaccinium macrocarpon.
AMERICAN CRANBERRY-BUSH *Viburnum trilobum*.
AMERICAN CREEPER *Tropaeolum peregrinum* (Canary Creeper) Friend.1882. Canary, from the bright yellow colour, and not Canary Isles. On. the other hand, *T canariensis* is a synonym for *T peregrinum*.
AMERICAN CRESS *Barbarea verna* (American Land Cress) B & H.
AMERICAN CURRANT *Ribes americanum*.
AMERICAN CRANESBILL Geranium maculatum (Spotted Cranesbill) Grieve.1931.
AMERICAN DATE PLUM *Diospyros virginiana* (Persimmon) J Smith.
AMERICAN DOG TOOTH VIOLET *Erythronium americanum* (Yellow Adder's Tongue) Grieve.1931.
AMERICAN DOG VIOLET *Viola conspersa*.
AMERICAN DOGWOOD *Cornus florida* (Flowering Dogwood) Grieve.1931. Virginian Dogwood is also recorded, as are American Boxwood and New England Boxwood.

AMERICAN ELDER *Sambucus canadensis*.
AMERICAN ELM *Ulmus americana*.
AMERICAN EVERLASTING *Anaphalis margaretacea* (Pearly Immortelle) Grieve.1931. It is an American plant, but was introduced into this country very early, and was one of the first New World plants to escape.
AMERICAN FLY HONEYSUCKLE *Lonicera canadensis* (Early Fly Honeysuckle) Youngken.
AMERICAN FUMITORY *Fumaria indica*. See Fumitory for the derivation.
AMERICAN GENTIAN *Gentiana catesbei*.
AMERICAN GERMANDER *Teucrium canadense*. Germander is med Latin germandra, from chamaedreos, the late Greek form, earlier khamaidrus, ground oak (khamai, on the ground, and drus, oak). Jaeger says, though, that it may be a corruption of Scamander, a Trojan river.
AMERICAN GLOBE-FLOWER *Trollius laxa*.
AMERICAN GOOSEBERRY *Ribes cynosbati* (Prickly Gooseberry) Douglas.
AMERICAN GROMWELL *Lithospermuum latifolium*.
AMERICAN GROUNDNUT *Apios americana* (Groundnut) Bianchini.
AMERICAN HACKBERRY *Celtis occidentalis* (Hackberry) W. Miller.
AMERICAN HAWTHORN *Crataegus mollis* (Woolly Thorn) Douglas.
AMERICAN HELLEBORE *Veratrum viride* (American White Hellebore) USA Lloyd.
AMERICAN HEMP *Apocynum cannabinum* (Indian Hemp) Grieve.1931.
AMERICAN HOLLY *Ilex opaca*.
AMERICAN HOP *Humulus americanus*.
AMERICAN HOP-HORNBEAM *Ostrya virginiana*.
AMERICAN HORNBEAM *Carpinus caroliniana* (Blue Beech) USA Zenkert.
AMERICAN HORSE CHESTNUT *Aesculus californica* (Californian Buckeye) Hatfield.
AMERICAN HORSE MINT *Monarda punc-*

tata.

AMERICAN IPECAC *Euphorbia ipecacuanhae*.

AMERICAN IPECACUANHA *Gillenia stipulacea*.

AMERICAN IRIS, Dwarf *Iris verna*.

AMERICAN IVY *Ampelopsis hederacea* (Virginia Creeper) Grieve.1931.

AMERICAN JACOB'S LADDER *Polemonium van-bruntiae*. Jacob's Ladder, because the leaflets are arranged in successive pairs.

AMERICAN JUDAS TREE *Cercis canadensis* (Redbud) Brimble.1948.

AMERICAN LADY'S TRESSES *Spiranthes romanzoffiana* (Hooded Lady's Tresses).

AMERICAN LAND CRESS *Barbarea verna*.

AMERICAN LARCH *Larix laricina*.

AMERICAN LAUREL, Great *Rhododendron maximum* (Rose Bay) Howes.

AMERICAN LILAC *Centranthus ruber* (Spur Valerian) Dev. Friend.1882. Cf. German Lilac and Ground Lilac for this.

AMERICAN LINDEN *Tilia americana* (Basswood) Everett.

AMERICAN LIQUORICE *Glycyrrhiza lepidota*.

AMERICAN LIVERWORT *Hepatica triloba*. Liverwort means the same as Hepatica. It is in fact a mild remedy in liver disorders.

AMERICAN LOTUS *Nelumbo lutea* (Yellow Lotus) USA Yarnell.

AMERICAN MANDRAKE *Podophyllum peltatum* (May-apple) E Hayes. Wild Mandrake, also. This is the modern mandrake root of the pharmacists.

AMERICAN MOUNTAIN ASH *Sorbus americana*.

AMERICAN MULBERRY *Morus rubra* (Red Mulberry) Hora.

AMERICAN NIGHTSHADE *Phytolacca decandra* (Poke-root) USA Henkel; S Africa Watt.

AMERICAN OSIER, Red *Cornus amomum* (Silky Cornel) Grieve.1931. Red Dogwood and Red Willow are other American names.

AMERICAN OX-EYE *Heliopsis laevis*.

AMERICAN PENNYROYAL *Hedeoma pulegioides*.

AMERICAN PLANE *Platanus occidentalis* (Western Plane) Everett.

AMERICAN PLUM; AMERICAN WILD PLUM *Prunus americana* (Canada Plum) USA Elmore (Plum); Usher (Wild Plum).

AMERICAN PUMPKIN *Cucurbita foetidissima* (Buffalo Gourd) Bianchini.

AMERICAN RASPBERRY *Rubus idaeus var. strigosus* (Red Raspberry) USA Elmore.

AMERICAN SANICLE *Heuchera americana* (Alum-root) USA W Miller. Sanicle means a healing herb.

AMERICAN SARSAPARILLA 1. *Aralia nudicaulis*. 2. *Menispermum canadense* (Moonseed) Le Strange. True sarsaparilla is Smilax sarsaparilla, so the Aralia also bears the names False, or Wild Sarsaparilla. The word itself is Spanish zarzaparilla (whence French salsepareilla, and Italian salsapariglia), which comes from zarza, bramble, and so any thorny plant, plus parra, a vine or any twining plant. Hence the name means literally a thorny vine. One spurious derivation gives a Dr Parillo as the origin.

AMERICAN SEA LAVENDER *Limonium carolinianum*.

AMERICAN SENNA *Cassia marilandica*.

AMERICAN SILVER FIR *Abies balsamea* (Balsam Fir) Grieve.1931.

AMERICAN SLOE *Viburnum prunifolium* (Black Haw) Mitton.

AMERICAN SPEEDWELL *Veronica peregrina*. Now settled in Ireland, too.

AMERICAN SPIKENARD *Aralia racemosa*. Spikenard is applied in America to some of the Aralias, while in this country it at once calls to mind the genus Inula. It actually means something quite different - the aromatic substance from an Eastern plant. Spike is the branch of flowers (applied particularly to the lavender), while the plant in question was the nard. So in med. Latin it became spica nardi, easily becoming in the vernacular spikenard.

AMERICAN SPINACH *Phytolacca decandra* (Poke-root) Le Strange. Poke greens are a common potherb in eastern USA.

AMERICAN SQUIRREL-CORN *Dicentra canadensis* (Turkey Corn) McDonald.

AMERICAN SYCAMORE *Platanus occidentalis* (Western Plane) USA Barber/Phillips.

AMERICAN TARE *Vicia americana* (American Vetch) USA Elmore.

AMERICAN TURK'S CAP LILY *Lilium superbum* (Turk's Cap Lily) Yanovsky.

AMERICAN VALERIAN *Cypripedium parviflorum* (Small Yellow Lady's Slipper) USA Sanford. Its use as a nervine has given this name. All the Cypripediums either have this effect, or are supposed to have it, though it is probably doctrine of signatures (the likeness of the flowers to the female genitalia).

AMERICAN VERVAIN *Verbena hastata* (Blue Vervain) Howes.

AMERICAN VETCH *Vicia americana*.

AMERICAN WAKE ROBIN *Arisaema triphyllum* (Jack-in-the-pulpit) Le Strange.

AMERICAN WALNUT *Juglans nigra*.

AMERICAN WATERLILY *Nymphaea odorata* (White Pond Lily) Gorer.

AMERICAN WATERWEED *Elodea canadensis* (Canadian Pondweed) Howes.

AMERICAN WAYFARING TREE *Viburnum alnifolium* (Hobble-bush) USA Zenkert.

AMERICAN WHITE HELLEBORE *Veratrum viride*. Or American Hellebore.

AMERICAN WILD MINT *Mentha arvensis canadensis* (Canadian Mint) USA House.

AMERICAN WILD OLIVE *Osmanthus americanus* (Devilwood) Howes.

AMERICAN WILD VANILLA *Liatris odoratissima* (Deer's Tongue) Mitton.

AMERICAN WILLOWHERB *Epilobium adenocaulon*.

AMERICAN WORMSEED *Chenopodium ambrosioides*. Wormseed because it is anthelmintic, i.e. has the property of expelling worms. It is known as Wormweed in S Africa, and Wormseed Weed in Barbados, where the word has become corrupted to Wormwood, too.

AMERICAN WORMWOOD *Ambrosia artemisifolia*.

AMERICAN YEW 1. *Taxus brevifolia* (Western Yew) Leathart. 2. *Taxus canadensis* (Canadian Yew) USA Tolstead.

AMETHYST FLOWER *Browallia speciosa* (Bush Violet) USA Cunningham.

AMMI *Carum copticum*.

AMOISE *Solanum mammosum*.

AMUR CORK TREE *Phellodendron amurense*.

AMUR MAPLE *Acer ginnala*.

AMY *Ammi majus* (Bishop's Weed) Turner. See also Amee, Ameos.

AMYROOT *Apoocynum cannabinum* (Indian Hemp) Grieve.1931.

ANA TREE *Acacia albida*.

ANABASIS *Equisetum hyemale* (Dutch Rush) R T Gunther. It is glossed as such on a C12 MS of Apuleius. The word appears to mean a military march, applied especially to that of Cyrus the younger into Asia, as written of by Xenophon, and has the idea of an ascent, which was what the original Greek word meant. The reason for its application here is obscure, the only military connection being the North American Indian use in polishing their guns, giving rise to the name Gunbright.

ANANBEAM *Euonymus europaeus* (Spindle Tree) Cockayne. Cockayne suggests that this is German Anisbaum, but Anis is simply Anise, and if Anisbaum ever existed, it seems to have passed out of the language. Storms also has Ananbeam for the Spindle. The 'beam' part of the word is OE for tree.

ANANCIA; ENANCIA *Geum urbanum* (Avens) both Fernie. Halliwell has the verb anansy, to advance, exalt, and says it was probably avaunce. This would explain its use for avens.

ANCHOR PLANT *Colletia armata*.

ANCHOVY PEA *Grias cauliflora*.

ANDAMAN CRAPEMYRTLE *Lagerstroemia hypoleuca.*

ANDAMAN MARBLE *Diospyros kurzii.* The timber is also known as Zebrawood, for the same reason that the name 'marble' is given to it.

ANDREAS BROOM *Cytisus scoparius andreanus.*

ANDREWS, Loving *Geranium pratense* (Meadow Cranesbill) Wilts. Grigson.

ANDREWS, Polly *Primula vulgaris var. elatior* (Polyanthus) Wilts. Dartnell & Goddard; Glos. J.D.Robertson. A corruption of Polyanthus.

ANDROMEDA, Free-flowering *Andromeda floribunda.*

ANDROMEDA, Japanese *Andromeda japonica.*

ANDROMEDA, Large-flowered *Andromeda speciosa.*

ANDROMEDA, Marsh *Andromeda polifolia* (Bog Rosemary) Hutchinson.

ANDROSAEMME *Hypericum perforatum* (St John's Wort) Thorpe. An old book name, probably originally referring to Tutsan (*H. androsaemum*). It is suggested that it alludes to the decollation of the Baptist; the plant contains a reddish fluid.

ANDURION *Eupatorium cannabinum* (Hemp Agrimony) Lancs.. B & H. Probably a corruption of Eupatorium.

ANEMONE, Apennine *Anemone apennina* (Blue Anemone) RHS.

ANEMONE, Bastard *Pulsatilla vulgaris* (Pasque Flower) Gerard.

ANEMONE, Blue *Anemone apennina.*

ANEMONE, Canadian *Anemone canadensis* (Round-leaved Anemone) USA House.

ANEMONE, Crown *Anemone coronaria* (Poppy Anemone) Polunin.

ANEMONE, False *Anemonopsis macrophylla.*

ANEMONE, Long-fruited *Anemone cylindrica* (Slender-fruited Anemone) Gilmore.

ANEMONE, Long-sheathed *Pulsatilla vulgaris* (Pasque Flower) Geldart.

ANEMONE, Meadow *Anemone canadensis* (Round-leaved Anemone) USA Youngken.

ANEMONE, Mountain *Anemone appenina* (Blue Anemone) Curtis.

ANEMONE, Narcissus-flowered *Anemone narcissiflora.*

ANEMONE, Palestine *Anemone coronaria* (Poppy Anemone) Moldenke.

ANEMONE, Peacock *Anemone pavonia.*

ANEMONE, Plush *Anemone coronaria* Blunt.1957. It was Sir Thomas Hanmer, *Garden Book*, 1659, who called it this.

ANEMONE, Poppy *Anemone coronaria.*

ANEMONE, Round-leaved *Anemone canadensis.*

ANEMONE, Rue *Anemone thalictroides.* An American plant, the specific name showing it to be like the Meadow Rue, not the Rue proper.

ANEMONE, Slender-fruited *Anemone cylindrica.*

ANEMONE, Snowdrop *Anemone sylvestris* (Snowdrop Windflower) A.W.Smith.

ANEMONE, Spreading *Anemone patens.*

ANEMONE, Spring *Pulsatilla vernalis.*

ANEMONE, Tall *Anemone virginiana.* An American species which can be up to 3ft tall.

ANEMONE, Water 1. *Ranunculus aquatilis* (Water Crowfoot) W Miller. 2. *Ranunculus hederaceus* (Ivy-leaved Water Buttercup) Wilts Dartnell & Goddard.

ANEMONE, Wood 1. *Anemone nemerosa.* 2. *Anemone quinquefolia* USA Tolstead.

ANEMONE, Wood, Yellow *Anemone ranunculoides.*

ANEMONY, Meadow *Pulsatilla vulgaris* (Pasque Flower) Thornton.

ANENEMY, ENEMY *Anemone nemerosa* (Wood Anemone) Dev. Friend.1882 (Anenemy); Som. Elworthy; Lincs. B & H (Enemy). The consonantal misplacement is very familiar in dialect.

ANET *Anethum graveolens* (Dill) Gerard. A simple contraction of the generic name, Anethum. It appears again in Halliwell, so it must have been quite common.

ANEYS *Pimpinella amisum* (Anise) Halliwell.

ANGEL *Geranium robertianum* (Herb Robert) Dor. Macmillan.

ANGEL FLOWER *Achillea millefolium* (Yarrow) Som. Macmillan. Possibly because, in one of its aspects, yarrow is a death token. It is actually called Death-flower in Wales.

ANGEL GABRIEL *Lilium tigrinum* (Tiger Lily) Som. Macmillan. The reference must be to trumpets.

ANGEL ORCHID, White *Platanthera chlorantha* (Butterfly Orchid) Som. Macmillan.

ANGEL WINGS *Caladium bicolor 'Splendens'* (Fancy-leaved Caladium) USA Cunningham.

ANGEL'S EYES *Veronica chamaedrys* (Germander Speedwell) Dev. Grigson; Som. Macmillan; Oxf. Thompson. There are many 'eye' names for the speedwell, which are all purely descriptive. They include Eyebright, Bright Eye, Blue Eyes, Bird's Eyes, Cat's Eyes, God's Eye, and many others.

ANGEL'S TEARS 1. *Billbergia x windii (Billbergia nutans)* (Queen's Tears). H & P. 2. *Datura sanguinea* USA Cunningham. 3. *Helxine soleirolii* (Mind-your-own-business) USA. Cunningham. 4. *Narcissus triandrus*. 5. *Ornithogalum umbellatum* (Star of Bethlehem) Som. Macmillan. 6. *Veronica chamaedrys* (Germander Speedwell) Dev. Friend. Angel's Tears for the speedwell belongs to the same group which includes the many 'eye' names. But according to Alice Coats, the name for the little daffodil is accounted for by stories of a guide, or perhaps the guide's son, named variously Angel, Anghel, or Angelo, who was reduced to tears by his inability to keep up with the plant collector Peter Barr on the mountain.

ANGEL'S TRUMPET 1. *Datura arborea* Australia Watt. 2. *Datura metel* (Hairy Thorn Apple). 3. *Datura sanguinea* Cunningham. 4. *Datura stramonium* (Thornapple) Som. Grigson. 5. *Datura suaveolens*. Almost a generic name for the Daturas, but given in particular to D suaveolens, presumably because it is a cultivated plant in the West Indies.

ANGEL'S TURNIP *Apocynum androsaemifolium* (Spreading Dogbane) S USA Puckett.

ANGEL'S WING BEGONIA *Begonia coccinea*.

ANGELICA *Ligusticum apiifolium* USA Youngken.

ANGELICA, American *Angelica atropurpurea* (Purple-stemmed Angelica) Grieve.1931.

ANGELICA, Great *Angelica atropurpurea* (Purple-stemmed Angelica) W Miller.

ANGELICA, Purple-stemmed *Angelica atropurpurea*.

ANGELICA, Sweet *Myrrhis odorata* (Sweet Cicely) B & H. That is probably what is meant in Heywood's *Marriage Triumph*, 1613: .. And as they walke, the virgins strow the way. With costmary and sweet angelica.

ANGELICA, Wild *Angelica sylvestris*. This is herba angelica, 'angelic plant', so named on account of its reputed efficacy against poisons and pestilence.

ANGELICA TREE; ANGELICA TREE, Virginian *Aralia spinosa* (Hercules' Club) USA Kingsbury; (Virginian) W Miller.

ANGELICA TREE, Chinese *Aralia chinensis*.

ANGELICA TREE, Japanese *Aralia elata*.

ANGELS *Acer pseudo-platanus* seeds Som. Macmillan. For the same reason that produced the name Aeroplanes, in the same county.

ANGELS, Devils-and- 1. *Arum maculatum* (Cuckoo Pint) Som. Grigson; Dor. Dacombe. 2. *Plantago lanceolata* (Ribwort) Som. Macmillan. Cuckoo Pint has a number of these double names, like Knights-and-Ladies, Bulls-and-cows, Soldiers-and-angels, etc. Ribwort has Lords-and-Ladies, which is more usually associated with Cuckoo Pint, as well as a few others of the double names.

ANGELS, Fly-; ANGELS, Flying *Acer pseudo-platanus* seeds (Sycamore keys) Som. Macmillan.

ANGELS, Soldiers-and- *Arum maculatum*

(Cuckoo Pint) Dev. Macmillan. Confusion, probably, with Devil-and-angels, and Soldiers-and-sailors.

ANGELS-AND-DEVILS *Arum maculatum* (Cuckoo Pint) Som. Macmillan. The light parts of the flowers are the angels, and the dark parts the devils.

ANGLE-BERRY *Lathyrus pratensis* (Yellow Pea) N Eng Grigson. Halliwell has "a sore, or kind of hang-nail under the claw or hoof of an animal" as the meaning of Angle-berry. As the plant doesn't bear berries anyway, this is probably the reason the name occurs, as the forgotten cure for the complaint. Brockett, though, explained it by a verb, to angle, i.e. to cling to, plants or shrubs stronger than itself.

ANGLER'S FLOWER *Scrophularia aquatica* (Water Betony) Som. Macmillan. Presumably because it grows beside streams.

ANGOLA PEA *Cajanus indicus* (Pigeon Pea) Dalziel.

ANISE 1. *Anethum graveolens* (Dill) Turner. 2. *Lobularia maritima* (Sweet Alison) Dev. Friend. 3. *Magnolia salicifolia* USA Barber & Phillips. 4. *Myrrhis odorata* (Sweet Cicely) Dur. Grigson. 5. *Pimpinella anisum*. Anise is O. French anis (and modern French), Latin anisum, eventually from Greek anison, a word which ODEE suggests came into Greek from some unknown foreign source. French anis probably meant dill as well as anise, in fact anything umbelliferous with aromatic seeds. This would include Sweet Cicely, which is also known as Wild Anise, or, in Yorkshire, Annseed. When cultivated in Germany Sweet Cicely is known as Anise Chervil. The magnolia has scented flowers and leaves, so presumably this is why it is given the name in America. The odd one out is Sweet Alison. Friend points out that the change of l to n, and vice versa, is a common dialectal form.

ANISE, Chinese *Illicium verum* (True Anise) Grieve.1933.

ANISE, Purple *Illicium floridanum* (Poisom Bay) RHS.

ANISE, Star; ANISEED STAR *Illicium verum* (True Anise) both Grieve.1933. The fruit grows in the form of a star.

ANISE, Sweet *Foeniculum vulgare* (Fennel) A.W. Smith. It was known botanically at one time as Anethum foeniculum, and Anethum, although it is dill, seems to have had a similar philological origin as anise (Fr. anis seems at one time to have meant both).

ANISE, True *Illicium verum*. This is a small evergreen tree from China. The fruit, which grows like a star, is sold dried in Asia as a spice (and as a remedy against a number of diseases). In the West, the fruit is used to aromatize cordials and liqueurs like anisette. Anise itself is used for this drink, too. Why this should be regarded as the true anise, with verum as the specific name, I don't know. Other names include Star-anise, Chinese Anise, and Aniseed Star.

ANISE, Wild 1. *Carum kelloggii* USA Havard. 2. *Myrrhis odorata* (Sweet Cicely) Cumb. Grigson.

ANISE CHERVIL *Myrrhis odorata* (Sweet Cicely) Grieve.1931. The leaves and roots have a mild aniseed flavour, and the plant is cultivated in Germany under the name Anise Chervil.

ANISE-HYSSOP *Agastache anethiodora*. The leaves are aromatic. Anethiodora would mean "smelling like anise".

ANISEED *Pimpinella anisum*. Aniseed is used far more frequently for the plant as a whole than the real name, anise.

ANNAM PLUM *Flacourtia sepiaria*.

ANNASEED *Myrrhis odorata* (Sweet Cicely) Yks. Grigson. A corruption of aniseed, applied not to anise, but to Sweet Cicely, whose leaves and roots have a taste not unlike aniseed, but milder.

ANNUNCIATION LILY *Lilium candidum* (Madonna Lily) Woodcock.

ANNY; ANNYLE *Pimpinella anisum* (Anise) Anny W Miller; Annyle Halliwell.

ANSON'S SALLOW *Salix ansoniana*.

ANTARCTIC BEECH *Nothofagus antarctica*.

ANTELOPE HORNS *Asclepiodora decumbens* (Spider Milkweed) USA Elmore.
ANTELOPE SAGE *Eriogonum jamesii*.
ANTHURIUM, King *Anthurium veitchii*.
ANTILLES PEAR *Chrysophyllum argenteum*.
ANTIMONY, Vegetable *Eupatorium perfoliatum* (Thoroughwort) USA Henkel.
ANTIRRHINUM, Ivy-leaved *Cymbalaria muralis* (Ivy-leaved Toadflax) Curtis. The flowers are very like antirrhinum on a miniature scale, and many of the names, like Rabbit's-mouth, are common to the toadflaxes and antirrhinums.
AÑU *Tropaeolum tuberosum* (Mashua) Schery.
APACHE-PLUME *Fallugia paradoxa*. Because of the fancied resemblance of the feathery seed clusters to the plumed war bonnets of the Apache Indians.
APALANCHE *Ilex verticillata* (Winterberry) USA Cunningham.
APAREJO GRASS *Muhlenbergia utilis*.
APENNINE ANEMONE *Anemone apennina* (Blue Anemone) RHS.
APIMPE *Eleusine indica* Jamaica Beckwith.
APOSTLE PLANT *Marica northiana*.
APOSTLES, (The) *Ornithogalum umbellatum* (Star-of-Bethlehem) Som. Macmillan.
APOSTLES, Twelve *Pulmonaria officinalis* (Lungwort) Som. Tongue.1968. Presumably of the same variety as Abraham, Isaac and Joseph, and others like it - the double (or triple) name refers to the two-coloured flowers. There is an old Somerset song called The Twelve Apostles, part of which runs: "The Twelve Apostles in the garden plot do grow, Some be blue, some be red and others white as snow, They cure the ill of every man, whatever ill it be, But Judas he was hanged on an elder tree." This is quoted by Ruth Tongue.
APOTHECARIES' ROSE *Rosa gallica* (French Rose) Rohde. This was the rose most commonly used for medicinal purposes, hence the name. This use in medicine dates from a very remote period, and this rose was cultivated in England at Mitcham and in Oxford and Derby for the purpose. An infusion of the petals, acidulated with sulphuric acid, and slightly sweetened, used to be a common vehicle for other medicines.
APPALACHIAN TEA 1. *Ilex glabra* (Gallberry) Le Strange. 2. *Viburnum cassinoides* (Withe-rod) New Engl. Sanford.
APPEL-LEAF *Viola odorata* (Violet) Halliwell. This, Halliwell says, is the gloss of Viola in MS Harl. 978, and is the Anglo-Saxon word.
APPLE *Malus domestica*. O.E. aeppel - the ODEE cites a number of examples, including Welsh afal, to show a N. European base abl-. The tree itself is O.E. apuldor, O. Norse apal-dr, apparently surviving in the place names Apperknowle, Apperley, Appledore, Appledram, Applegarth (which is apalgardr, apple orchard).
APPLE, Adam's 1. *Citrus limetta* (Lime) W Miller. 2. *Musa x paradisiaca* (Banana) Gerard. In the 16th and 17th centuries, banana was the favourite candidate for the tree of the knowledge of good and evil. Gerard reported that the Jews "suppose it to be that tree, of whose fruit Adam did taste".
APPLE, Alligator 1. *Annona glabra* Roys. 2. *Annona palustris*.
APPLE, Amorous; APPLE OF LOVE *Lycopersicon esculentum* (Tomato) Amorous Apple - Parkinson; Apple of Love - Gerard. Apparently, tomatoes were considered to be aphrodisiac because of a mix-up in derivations. See Gold Apple.
APPLE, Balsam *Lycopersicon esculentum* (Tomato) Turner.
APPLE, Balsam, Wild *Momordica charantia*. It is also known as the Balsam Pear in America.
APPLE, Berk *Pinus sylvestris* cones (Pine cones) Yks. B & H.
APPLE, Bitter 1. *Colocynthis vulgaris* (Bitter Gourd) Wit. 2. *Momordica charantia* (Wild Balsam-apple) Sofowora. 3. *Solanum aculeastrum* S.Africa Watt. 4. *Solanum panduraeforme* S.Africa Watt. These are either

bitter-fruited plants, or else they are toxic. The Bitter Gourd is known also as the Bitter Cucumber, and Hill described the fruit pulp as "a very violent purge, but it may be given with proper caution...". Both the Solanums are called either Poison Apples, or Devil's Apples, names which emphasise the toxic nature.

APPLE, Bull *Passiflora laurifolia* (Jamaica Honeysuckle) Howes.

APPLE, Cane *Arbutus unedo* (Strawberry Tree) B & H. Parkinson also had heard this name, but obviously did not have any idea of the meaning, but Threlkeld says it is the Irish word Caihne that is meant. It certainly seems that the tree was introduced into England from Ireland, but what does caihne mean?

APPLE, Carthaginian *Punica granatum* (Pomegranate) W Miller. A simple translation of the Latin mala punica. The fruit was very popular in Carthage, apparently, or was it that Carthage was an early exporting centre?

APPLE, Cashew 1. *Anacardium occidentale* (Cashew Nut) Chopra. 2. *Mangifera occidentale*. The fruits of Mangifera taste like apples, and the bean-shaped fruits attached to the base are the source of cashew nuts. The word cashew is applied to *Anacardium occidentale*, and is Portuguese caju, or acaju, from which acajou, the French for mahogany, comes.

APPLE, Chess *Sorbus aria* (Whitebeam) Ches,Lancs. Grigson. This is cheese-apple.

APPLE, Chinese *Pyrus prunifolia* (Siberian Crab) Douglas).

APPLE, Coddled *Epilobium hirsutum* (Great Willowherb) Lincs Grigson; N'thants A.E. Baker. Coddled apples means stewed apples. Sod-apple, another name for the plant, means the same thing. There are many other similar names, e.g. Apple-pie, Gooseberry-pudding, Codlins-and-cream, all of which make it look as if Gerard's Codded Willow-herbe, "the flower groweth at the top of the stalke, comming out of the end of a small long codde", is an invention of his.

APPLE, Cone-deal *Picea abies* (Spruce Fir) E.Ang. B & H.

APPLE, Crab *Malus sylvestris*. Crab here is probably of Scandinavian origin. There is a Swedish dialect name skrabba for the wild apple, and scrab is a fairly common North country and Scottish name for the tree.

APPLE, Curacao *Eugenia javanica*. The fruits are pear-shaped, otherwise no information.

APPLE, Custard 1. *Annona reticulata*. 2. *Annona squamosa* (Sweetsop) Candolle. The edible fruit is sometimes chilled and the icecream-like pulp eaten with a spoon, so this is probably the origin of the name.

APPLE, Custard, Netted *Annona reticulata* (Custard-Apple) Chopra.

APPLE, Custard, Scaly *Annona squamosa* (Sugar Apple) India Dalziel.

APPLE, Custard, Wild *Annona chrysophylla*.

APPLE, Dead Sea 1. *Quercus lusitanica* (Cyprus Oak) Howes. 2. *Solanum sodomeum* (Apple of Sodom) W Miller.

APPLE, Deal 1. *Picea abies* cones (Spruce Fir cones) B & H. 2. *Pinus sylvestris* cones (Pine cones) Suff. Grigson N'thants. B & H. Deal is the name of the wood of either fir or pine; the cones are called Dealies in Suffolk, or Dealseys, and Delseed in Cornwall.

APPLE, Devil's 1. *Datura stramonium* (Thorn-apple) USA Henkel. 2. *Solanum aculeastrum* S. Africa Watt. 3. *Mandragora officinalis* (Mandrake) W Miller. Poisonous fruits, so the name Devil's Apple is applied.

APPLE, Earth 1. *Cyclamen europaeum* (Sowbread) Henslow. 2. *Helianthus tuberosus* (Jerusalem Artichoke) USA Sanecki.

APPLE, Eve's *Ervatamia dichotoma*. The fruits look tempting enough, but they are very poisonous. Moslems call it "the forbidden fruit of Eden".

APPLE, Fir 1. *Picea abies* (Spruce Fir) Surr, Herts, Lincs, Cumb. B & H. 2. *Pinus sylvestris* (Pine cones) Hants. Cope; N'thants. A.E. Baker.

APPLE, Goat *Datura stramonium* (Thorn-apple) S. Africa Watt.

APPLE, Gold *Lycopersicon esculentum* (Tomato) Gerard. The original Italian name was pomo dei mori (apple of the Moors), and this later became pomo d'ore (golden aple). It was introduced to France as an aphrodisiac, and the French misspelled its name as pomme d'amour. So the tomato eventually reached England under the name pome amoris, which name went back to America with the colonists.

APPLE, Golden *Spondias dulcis* (Ambarella) Schery.

APPLE, Grindstone *Malus sylvestris* fruit (Crabapples) Wilts. Mabey. This may sound as if it were an apt description for anyone trying to eat one of them raw, but actually there is an almost literal reason for the name. As Richard Jefferies said, country lads used to sharpen their knives by "drawing the blade slowly to and fro through a crab-apple; the acid of the fruit eats the steel like aquafortis".

APPLE, Hen *Sorbus aria* (Whitebeam) Mor. Grigson.

APPLE, Hog *Podophyllum peltatum* (May-apple) Hatfield.

APPLE, Honeysuckle The fungus that commonly grows on *Rhododendron nudiflorum* is called by American children Honeysuckle Apples; they apparently eat it. USA Bergen.

APPLE, Indian *Datura meteloides* (Downy Thorn-apple) USA Elmore.

APPLE, Jew's *Solanum melongena* (Egg Plant) W Miller.

APPLE, Kangaroo 1. *Solanum laciniatum*. 2. *Solanum aviculare*. Both Australian species.

APPLE, Love 1. *Epilobiuum hirsutum* (Great Willowherb) War. Grigson. 2. *Lycopersicon esculentum* (Tomato) Gerard. 3. *Mandragora officinalis* (Mandrake) Hartland 4. The tomato, in Gerard's time, was looked upon as an aphrodisiac. But, of course, it only got the name (and the reputation) through a mix-up of names. See Gold Apple for the explanation. Mandrake, too, was thought an aphrodisiac, and it became the symbol of the golden apple of Aphrodite; she was sometimes called Mandragoritis ("she of the Mandrake"). Great Willowherb is difficult to fit into this section, for surely there is no aphrodisiac quality attached to it. The local name probably came about because the plant does bear other 'apple' names - Apple Pie, Codlins-and-cream, etc.

APPLE, Mad 1. *Datura stramonium* (Thorn-apple) USA Henkel. 2. *Solanum insanum* Gerard. Gerard's name is given to a plant which he calls Solanum insanum, but which I can't trace anywhere else. The Datura is so-called because of the toxic effect it has. The famous quotation, arising from soldiers' gathering the young plants to cook them as pot herbs, is "the effect of which was a very pleasant Comedy; for they turn'd natural Fools upon it for several days...".

APPLE, Malay *Eugenia malaccensis*. The edible fruits are known as pomme rouge throughout Polynesia.

APPLE, Mammee; APPLE, Mammey *Mammea americana* (Mamey) Mammee - Candolle; Mammey - Brouk. Edible fruit, so apple. Mammee is the same word as mamey, which is Spanish, and has Haitian origin. The Latin name comes from this, not the other way round.

APPLE, May 1. *Passiflora incarnata* (Maypops) Schauenberg. 2. *Podophyllum peltatum*.

APPLE, May, Indian *Podophyllum hexandrum*.

APPLE, Median; APPLE, Persian *Citrus medica bajana* (Citron) both Brouk. This is what Theophrastus called citrons, obviously because they reached Europe via Persia.

APPLE, Monkey 1. *Annona glabra* Barbados Gooding. 2. *Annona palustris* (Alligator Apple) W Indies Dalziel. 3. *Strychnos spinosa* (Kaffir Orange) Palgrave.

APPLE, Mountain *Eugenia malaccensis* (Malay Apple) Douglas.

APPLE, Ooze *Datura meteloides* (Downy Thorn-apple) USA Elmore.

APPLE, Otaheite *Eugenia malaccensis* (Malay Apple) Howes.

APPLE, Paradise 1. *Lycopersicon esculentum* (Tomato) S Africa Watt. 2. *Malus sylvestris paradisiaca*. Tomato was called Paradies Apfel in Germany very early. It was believed the Turks brought it from the Holy Land, where the original Paradise was thought to have been. This South African name must have come down from the German, through Dutch.

APPLE, Pear, Chinese *Pyrus sinensis* (Sandy Pear) Douglas.

APPLE, Pie; APPLE, Pur *Pinus sylvestris* (Pine cones) N'thants. both A.E. Baker.

APPLE, Pillar *Malus tschonoskii*. 'Pillar' is a reference to its growth habit.

APPLE, Pin *Pinus sylvestris* cones (Pine cones) N'thants. B & H.

APPLE, Pine 1. *Picea abies* cones (Spruce Fir cones) B & H. 2. *Pinus sylvestris* cones (Scots Pine cones) B & H.

APPLE, Pitch *Clusia rosea* RHS.

APPLE, Poison *Solanum panduraeforme* S.Africa Watt.

APPLE, Pond *Annona glabra* Roys.

APPLE, Prairie *Psoralea esculenta* (Breadroot) Howes.

APPLE, Rose 1. *Eugenia jambolana* (Jambul) Grieve.1931. 2. *Syzygium moorei* (Coolamon) Australia) Wilson.

APPLE, Rose, Mankil; APPLE, Rose, Samarang *Eugenia javanica* (Curacao Apple) Douglas.

APPLE, Sand *Parinari mobola*.

APPLE, Satan's *Mandragora officinalis* fruit (Mandrake) Clair. A poisonous fruit, hence the name.

APPLE, Scalded *Melandrium dioicum* (Red Campion) Shrop. Grigson. Origin unknown, but it may possibly be a confusion with the various apple names associated with Great Willowherb.

APPLE, Sea *Eugenia grandis*.

APPLE, Serpent *Annona palustris* (Alligator Apple). W Indies Dalziel. The fruits must have particularly scaly skins to warrant these two names.

APPLE, Sod *Epilobium hirsutum* (Great Willowherb) Som. Macmillan; Wilts. Jefferies. Both coddled, which appears as Codlins-and-cream, Codlins, Coddled Apples and Sugar-Codlins, and sod, mean the same thing - boiled.

APPLE, Soda *Solanum aculeastrum* USA Kingsbury. Apple of Sodom is another name for this, so Soda here may derive from Sodom. The fruit is poisonous, hence the name Devil's Apple, and Bitter Apple.

APPLE, Sorb 1. *Sorbus domestica* fruit (True Service Tree) Turner. 2. *Sorbus torminalis* fruit (Wild Service Tree) Gerard. Halliwell quotes Sorb(e)as the name of the fruit as early as 1530. It is, of course, a form of the Latin sorbus, through French sorbe.

APPLE, Star *Chrysophyllum cainito*. The fruits are like apples.

APPLE, Star, African *Chrysophyllum africanum*.

APPLE, Star, Monkey *Chrysophyllum perpulchrum*. The fruit is reckoned inferior to the other Star-apples, hence 'Monkey'.

APPLE, Star, White *Chrysophyllum albidum*.

APPLE, Stub *Malus sylvestris* (Crabapple) Halliwell.

APPLE, Sugar *Annona squamosa*. The edible fruit is eaten fresh. Sugar Apple really means the same as Sweetsop. Custard Apple is another name for the fruit.

APPLE, Thorn *Datura stramonium*.

APPLE, Thorn, Downy *Datura meteloides*.

APPLE, Thorn, Green *Datura stramonium* (Thorn-apple) Dalziel.

APPLE, Thorn, Hairy *Datura metel*.

APPLE, Thorn, Purple 1. *Datura stramonium* (Thorn-apple) S.Africa Watt. 2. *Datura stramonium var. tatula*. The latter has pale purple flowers.

APPLE, Wild 1. *Euphorbia corollata* (Flowering Spurge) Le Strange. 2. *Malus sylvestris* (Crabapple) Gerard.

APPLE-BEARING SAGE *Salvia pomifera* Because it is usually infested with an insect which produces excrescences like oak galls.

APPLE BERRY *Billardiera scandens* An Australian plant, with edible berries. They taste like roasted apples, so they say, though in cultivation the flavour is insipid.

APPLE BERRY, Purple *Billardiera longiflora*.

APPLE DUMPLING *Billardiera scandens* (Apple Berry) Perry. Edible berries, tasting like roasted apples.

APPLE-FRUITED GRANADILLA *Passiflora maliformis* (Sweet Calabash) Grieve.1931. The "maliformis" part of the botanical name presumably refers to the apple-shaped seed vessels. Granadilla is the diminutive of granada, the Spanish for pomegranate.

APPLE-FRUITED ROSE *Rosa pomifera* (Apple Rose) McDonald.

APPLE-LEAVED WILLOW *Salix malifolia*.

APPLE MINT 1. *Mentha arvensis* (Corn Mint) Ire. E E Evans.1942. 2. *Mentha rotundifolia*. There is a distinct apple scent to the leaves of M rotundifolia.

APPLE MINT, American *Mentha x gentilis* (Bushy Mint) G.B.Foster.

APPLE MINT, Golden *Mentha x gentilis*.

APPLE MINT, Large *Mentha alopacuroides*.

APPLE MINT, Silver Mentha rotundifolia variegata Webster.

APPLE OF JERUSALEM *Momordica charantia* (Wild Balsam-apple) W.Miller.

APPLE OF PERU 1. *Datura stramonium* (Thorn-apple) Gerard; USA Henkel. 2. *Nicandra physalodes* (Shoo-fly Plant) USA Zenkert.

APPLE OF SODOM 1. *Atropa belladonna* (Deadly Nightshade) Emboden. 2. *Solanum aculeastrum* S.Africa Watt. 3. *Solanum incanum* (Hoary Nightshade) Moldenke. 4. *Solanum panduraeforme* S.Africa Watt. 5. *Solanum sodomeum*. Probably all because of either actual or supposed poisonous fruit. *S panduraeforme* has Poison Apple, and Bitter Apple. *S aculeastrum* has Bitter Apple, too, and also Devil's Apple. But note that Soda Apple occurs also for the latter. Deadly Nightshade has plenty of "devil" appropriations, and there is also the name Doleful Bells in Dorset - often, for other plants, this has "... of Sodom" added.

APPLE OF THE EARTH *Chamaemelum nobile* (Camomile) E.Hayes. Virtually a direct translation from Greek khamaemelon, earth-apple, from the apple-like scent of the flower.

APPLE PIE 1. *Artemisia vulgaris* (Mugwort) Ches. B & H. 2. *Cardamina pratensis* (Lady's Smock) Yks. Grigson. 3. *Chamaenerion angustifolium* (Rosebay) Dev. Friend. 4. *Epilobium hirsutum* (Great Willowherb) Som, Herts, Ess, Suff, Yks, N'thum. Grigson; Dev. Friend.1882; Glos. J.D.Robertson; Hants. Cope; Suss. Parish; Ches. Holland. It is said that the young shoots of Great Willowherb smell like apples (Jefferies wrote that country people said the leaves crushed in their fingers have something of the scent of apple-pie), and there are many 'apple' names for this. Rosebay probably got the name by association. I don't know any reasons for the others.

APPLE RING ACACIA *Acacia albida* (Ana Tree) Dale & Greenway.

APPLE RINGIE *Artemisia abrotanum* (Southernwood) Scot. Simpson. Simpson's explanation is that apple is from the old aplen, a church, and ringie is Saint Rin's, or St Ninian's, wood (Ringan was the Scots form of Ninian. Aitken offers another derivation. Appelez Ringan - pray to Ringan, became first Appleringan, then Appleringie. But Jamieson says that it is from Fr. apile, strong, and auronne, southernwood, which derives from abrotanum. Is there a connection with apill remgeis, a Scots term for a bead necklace?

APPLE ROSE *Rosa pomifera*.

APPLE SHRUB *Weigelia florida* Som. Elworthy. Presumably because of the similarity of the flowers to apple blossom.

APPLE TREE, Devil's *Euphorbia helioscopia* (Sun Spurge) Clack. B & H.

APRECOCK; APRICOCK *Prunus armeniaca* (Apricot). Gerard spelt it aprecock, and Parkinson apricock.

APRICOT *Prunus armeniaca*. All the various

spellings of Apricot (see Abricok, Aprecock Apricock) derived from the Latin praecox - it was called the Precocious Tree because it flowers and fruits earlier than the peach - thus Hasty Peach (Turner).

APRICOT, Desert *Prunus eriogyna*. It grows in the desert mountains of N.America.

APRICOT, Japanese *Prunus mume*.

APRICOT, St Domingo; APRICOT, Wild *Mammea americana* (Mamey) (St Domingo) Wit; (Wild) W Miller. A West Indian plant, with an edible fruit.

APRICOT VINE *Passiflora incarnata* (Maypops) USA. J Cunningham.

APRIL, Darling of *Primula vulgaris* (Primrose) Som. Macmillan.

APRON, Carpenter's *Lapsana communis* (Nipplewort) War. Grigson.

APRON, Devil's *Urtica dioica* (Nettle) Ire. C.J.S. Thompson.1947. The 'devil's' part of the name derives from the stinging properties; any poisonous plant is appropriated to the devil. Nettle has Devil's Leaf and Devil's Plaything. Can the 'apron' part of the name refer back to the use of nettle fibres as textiles? They were certainly used in Scotland for cloth up to the 18th century, and nettle linen is referred to as late as 1917 in the Tyrol.

APRON, Dolly's *Geranium robertianum* (Herb Robert) Dev, Som. Macmillan. Presumably from the colour and shape of the flowers - Baby's Pinafore, Dolly's Pinafore, Print Pinafore and Pink Pinafore occur as well as Dolly's Apron. Presumably also the miniature quality of the plant's flowers give the 'baby' and 'dolly' part of the names. They occur again as Doll's Shoes, Dolly's Nightcap, as well as Cry Baby.

APRON, Tanner's *Primula auricula* (Auricula) Glos. B & H. Presumably because of the toughness of the leaves. But this name is only applied to the yellow kind.

APS 1. *Populus tremula* (Aspen) Corn, Dev, Hants, IOW, Herts, Kent, Surr, Suss, War. Grigson; Som. Jennings; Wilts. Dartnell & Goddard; Glos. J.D.Robertson. 2. *Populus tremuloides* (American Aspen) USA Bergen. Is this the older form? See Gimbutas: old Prussian abse, Lithuanian apuse. In any case, the transposition of letters like this is common in dialect.

APSE; APSEN-TREE *Populus tremula* (Aspen) Som. Elworthy Dev. Friend.1882 (Apse); Corn. Jago (Apsen-tree). More often, it appears just as aps. Apsen must be the old adjectival form, in which case the Somerset version is probably the same, with the final letter missing.

APUAN WILLOW *Salix crataegifolia*. 'Apuan' refers to the Apuan Alps, where it grows.

ARA-ROOT *Maranta arundinacea* (Arrowroot) Lindley. Ara-root, and arrow-root, are both corruptions of Aru-root, names for the Aruac or Arawak Indians of South America, in spite of its being used as an antidote to arrow poison.

ARABIAN COFFEE *Coffea arabica*. The plant is probably a native of Ethiopia, but it was in Arabia that domestication first took place. It was brought there in the 6th century AD.

ARABIAN JASMINE *Jasminum sambac*.

ARABIAN RUE *Ruta tuberculata*.

ARABIAN STAR FLOWER *Ornithogalum arabicum* Tradescant.

ARABIAN TEA *Catha edulis* (Khat) Douglas.

ARABIAN VIOLET *Exacum affine* (Persian Violet) Bonar.

ARABIAN WOLFBERRY *Lycium arabicum*.

ARABIC, Gum *Acacia senegal*.

ARABIS, Garden *Arabis caucasica*. Arabis is Greek, possibly from Arabian araps.

ARACH; ARAGE; ARECHE *Atriplex hortensis* (Garden Orache) all B & H. These seem nearer to the original than Orache, which is Anglo-Norman arasche, O. French arache, and eventually from Latin atriplex.

ARAG, Stinking *Chenopodium vulvaria* (Stinking Orach) Scot. Graham. Arag is the same word as Arach. Stinking Arrach, and

Stinking Airach are also recorded.
ARAGE *Atriplex patula* (Orache) Halliwell. Cf. Orege, in Turner.
ARB RABBIT *Geranium robertianum* (Herb Robert) Dev. B & H. This is merely a dialect form of Herb Robert.
ARBALE; ARBEAL *Populus alba* (White Poplar) Arbale is Som. Elworthy; Arbeal is in B & H. Variants of Abele, from the Dutch abeel, from diminutive Latin albus ("... in low Dutch abeel, of his horie or aged colour, and also Abeelbloome" (Gerard)). The trees were imported from Holland in the 17th century, and are not native to Britain.
ARBEL, Dutch *Populus alba* (White Poplar) Som. Grigson. See Abele for derivation.
ARBER BEAN *Phaseolus vulgaris* (Kidney Bean) Turner. "because they serve to cover an arber for the tyme of Summer".
ARBESET *Arbutus unedo* (Strawberry Tree) B & H.
ARBOR-VITAE *Thuja occidentalis*. A book name, originating in France in C15, but almost never used in common speech, say Barber & Phillips. So what do they call this tree? White Cedar? Presumably it is the American Indians' view of this tree that gave the name. To people like the Kwakiutl, it really is a tree of life. The colour of Red Cedar gives an immediate association with blood, and they really look upon the tree as a source of life (but this isn't *T occidentalis*, but Giant Cedar, *T plicata*).
ARBOR-VITAE, Chinese *Thuja orientalis*.
ARBOR-VITAE, Giant; ARBOR-VITAE, American *Thuja plicata* (Giant Cedar) Clapham (Giant); Brimble.1948 (American).
ARBOR-VITAE, Japanese *Thuja standishii*.
ARBOR-VITAE, Nootka Sound *Thuja plicata* (Giant Cedar) W.Miller. A tree of the Pacific northwest, of which Nootka Sound is a small area.
ARBUTE-TREE *Arbutus unedo* (Strawberry Tree) Turner.
ARBUTUS, Trailing *Arctostaphylos uva-ursi* (Bearberry) Thornton. It was at one time known botanically as Arbutus uva-ursi, and Jamieson calls it Trailing Strawberry-tree.
ARBY, ARBY-ROOT *Armeria maritima* (Thrift) Ork. Jamieson.
ARCEL *Palmeria omphalodes* Bolton.
ARCHANGEL 1. *Angelica archangelica*. 2. *Galeobdalon luteum* (Yellow Archangel) Som. Macmillan. 3. *Lamium album* (White Deadnettle) Henslow. 4. *Lamium purpureum* (Red Deadnettle) Som. Macmillan; Glos, Leics. Grigson. 5. *Stachys sylvatica* (Hedge Woundwort) W.Miller. This is the medieval herba angelica, angel's plant, the tradition being that an angel had revealed its efficacy against such epidemic diseases as cholera and the plague. The angel, according to a German belief, was the archangel Raphael. "Some call this an herb of the Holy Ghost; others more moderate call it Angelica, because of its angelic virtues, and that name it retains still, and all nations follow it" (Culpepper). Because of all this, it was used as a preservative against evil spirits and witchcraft. Wiltshire folklore knows it as "the strongest protection against witchcraft". White Deadnettle is known as White Archangel, too, and Red Deadnettle is similarly Red Archangel, or Sweet Archangel.
ARCHANGEL, Balm-leaved 1. *Lamium orvala* (Balm-leaved Deadnettle) Whittle & Cook. 2. *Melittis melissophyllum* (Bastard Balm) Camden/Gibson.
ARCHANGEL, Black *Ballota nigra* (Black Horehound) B & H.
ARCHANGEL, Red 1. *Lamium purpureum* (Red Deadnettle) Gerard. 2. *Stachys sylvatica* (Hedge Woundwort) Turner.
ARCHANGEL, Sweet *Lamium purpureum* (Red Deadnettle) B & H.
ARCHANGEL, White *Lamium album* (White Deadnettle) Gerard.
ARCHANGEL, Yellow *Galeobdalon luteum*.
ARCHANGEL FIR *Pinus sylvestris* (Scots Pine) Wilkinson.1981. The timber was called by a number of different names,according to the port of export - Cf. Danzig Fir,Riga Fir,etc.
ARCHARDE *Quercus robur* acorn (Acorn)

Halliwell.

ARCHER, Water *Sagittaria sagittifolia*. From the shape of the leaves, as the other name Arrow-head shows.

ARCTIC BIRCH *Betula nana* (Dwarf Birch) RHS.

ARCTIC HEATHER *Andromeda floribunda* McDonald.

ARCTIC MOUSE-EAR *Cerastium edmonstonii*.

ARCTIC POPPY *Papaver radicatum* (Yellow Poppy) Polunin. A flower of the Canadian Arctic.

ARCTIC RASPBERRY *Rubus arcticus*.

ARCTIC RHODODENDRON *Rhododendron lapponicum* (Lapland Rose Bay) Blamey.

ARCTIC SAGEBRUSH *Artemisia frigida* (Fringed Sagebrush) USA Elmore.

ARCTIC SAXIFRAGE *Saxifraga nivalis*.

ARCTIC WOODRUSH *Luzula arcuata*.

ARD-LOSSEREY *Glechoma hederacea* (Ground Ivy) IOM Moore.1924. Manx ardlodhera, 'chief herb', presumably a measure of its usefulness as a medicinal plant.

ARECA PALM *Areca catechu*.

ARGANS *Origanum vulgare* (Marjoram) Dev. B & H. Argans is merely the Devonshire version of Organs or Organies, etc., corruptions of Origanum, which is from the Greek meaning joy of the mountain.

ARGEMONE *Potentilla anserica* (Silverweed) Halliwell.

ARGENTILL *Aphanes arvensis* (Parsley Piert) B & H. An old name, from the pale colour.

ARGENTINA; ARGENTINE *Potentilla anserica* (Silverweed) Argentina is in Aubrey's Natural History of Wiltshire, and Argentine is given by Gerard. Both from the same reason as Silverweed - the underside of the leaves. But Gerard gives as his reason for Argentine "of the silver drops that are to bee seene in the distilled water when it is put into a glasse, which you shall easily see rolling and tumbling up and downe in the bottome".

ARGENTINE DOCK *Rumex cuneifolius*. A South American plant, once known as *R magellanicus*, which places it nicely in its geographical setting. It is naturalized in dune slacks in S.W.England and S.Wales.

ARIN; ARN *Alnus glutinosa* (Alder) both Scot. Arin B & H; Arn Jamieson.

ARISTOCRAT PLANT *Haworthia chalwini*.

ARIZONA ASH *Fraxinus velutina* (Velvet Ash) Barber & Phillips.

ARIZONA BRAMBLE *Rubus arizonicus* (Arizona Red Raspberry) USA Elmore.

ARIZONA CYPRESS, Rough-barked *Cupressus arizonica*.

ARIZONA CYPRESS, Smooth *Cupressus glabra*.

ARIZONA FESCUE *Festuca arizonica*.

ARIZONA NUT PINE *Pinus edulis*. The nuts are an important food stuff for the American Indians who live in the area.

ARIZONA POPPY *Kallstroemeria grandiflora*.

ARIZONA RED RASPBERRY *Rubus arizonicus*.

ARIZONA WHITE OAK *Quercus arizonica*.

ARIZONA WILD GRAPE *Vitis arizonica*.

ARK, Dobbin-in-the- *Aconitum napellus* (Monkshood) Som. Macmillan. Doves-in-the-ark is more likely.

ARK, Doves-in-the- *Aconitum napellus* (Monkshood) Dor. Macmillan. It is claimed that this and other similar names, such as Lady Lavinia's Dove Carriage, Venus'-chariot-drawn-by-doves, Lady Dove and her coach and pair, all derive from the "dove-like nectaries".

ARK, Noah's 1. *Aconitum anglicum* (Wild Monkshood) Glos. B & H. 2. *Aconitum napellus* (Monkshood) Glos. Grigson. 3. *Aquilegia vulgaris* (Columbine) Som. Macmillan. 4. *Cypripedium parviflorum pubescens* (Larger Yellow Lady's Slipper) O.P.Brown. 5. *Delphinium ajacis* (Larkspur) Dor. Macmillan. Probably from the same group as Doves-in-the-ark, at least for the monkshoods and columbine.

ARKANSAS ROSE *Rosa arkansana*.

ARL; ORL; AUL; AULNE *Alnus glutinosa*

(Alder) (Arl) Dev, Som, Glos, Heref, Worcs, Shrop, Rad. Grigson; (Orl) Heref. G C Lewis; Shrop, Worcs. B & H; Som. Macmillan; Glos. J.D.Robertson (Aul) Heref. B & H; (Aulne) B & H. Hence aulen - made of alder.

ARMAND PINE *Pinus armandii* (Chinese White Pine) RHS.

ARMENIAN OAK *Quercus pontica* (Pontine Oak) RHS. It is from the Caucasus.

ARMS, Witches' *Galeopsis tetrahit* (Hemp Nettle) Dev. Macmillan.

ARMSTRONG *Polygonum aviculare* (Knotgrass) Suss. B & H. "from the difficulty of pulling it up" (B & H).

ARNBERRY *Rubus idaeus* (Raspberry) Yks. F K Robinson. Probably the same as the Scottish hainberry, and North Country hineberry. Hindberry, with a much wider spread, sounds nevertheless to be a rationalisation after the true derivation, whatever that is, was lost.

ARNICA *Arnica montana*. Perhaps an alteration of med. Lat. ptarmica, from the Greek meaning sneezewort, i.e. a plant which causes sneezing (Cf. Achillea ptarmica). Certainly there is a name Mountain Tobacco, but that is the only tentative confirmation.

ARNICKS *Ranunculus bulbosus* (Bulbous Buttercup) Dur. F.M.T. Palgrave. The impression given is that the bulbs of any of the Ranunculaceae that have them, were known as Arnicks.

ARNOT, Lousy 1. *Bunium bulbocastanum* (Great Earth-nut) Scot. Jamieson. 2. *Conopodium majus* (Earth-nut) Scot. Jamieson. Arnot is Earth-nut, so called from the chestnut-like tuberous rootstock, which is edible, and is often roasted like chestnuts (thus bulbocastanum). Lousy - they say that if you eat too much of the Lousy Arnots, lice will crowd into your hair. Other similar names are Lousy Arnut, and Lucy Arnut.

ARNOTTO; ANNATTO *Bixa orellana* Dalziel (Annatto). The fruits yield an orange to red dyestuff known as anatto, and as an artist's pigment it is known as anattoine. Arnotto comes from South American native names for the pigment - Arnotta, Arnatto, Anoto (Tamanac).

ARNTREE *Sambucus nigra* (Elder) Scot. B & H. Possibly because of the confusion elder/alder, for Arntree looks to be a more likely alder name. Indeed, one finds Alder, or Alderne, for the elder.

ARNUT 1. *Bunium bulbocastanum* (Great Earth-nut) Prior. 2. *Conopodium majus* (Earth-nut) Prior. Arnut = earth-nut.The rootstock is edible, and is roasted like chestnuts, and tastes like them, hence bulbocastanum.

ARNUT, Lousy; ARNUT, Lucy 1. *Bunium bulbocastanum* (Great Earth-nut) Scot. Jamieson. 2. *Conopodium majus* (Earth-nut) Perth, Aber. Grigson. Lucy *Conopodium majus* (Earth-nut) Fife Grigson. Arnut = earth-nut. Lucy = lousy. It was thought that eating them would breed lice.

ARNUT, Swine *Stachys palustris* (Marsh Woundwort) Banff. Grigson. Arnut is earth-nut, from the tubers on the rhizome. Swine's Beads and Swine's Murrills also occur from the Northern Isles.

AROLLA PINE *Pinus cembra* (Swiss Stone Pine) Polunin.

ARON *Arum maculatum* (Cuckoo Pint) Gerard. Like the various Aaron names, this is merely a corruption of Arum.

ARPENT; ARPENT-WEED *Sedum telephium* (Orpine) both Hants, Herts B & H. These are from orpiment, from which Orpine is derived. The Latin is auripigmentum, pigment of gold, possibly applied to Yellow Stonecrop, and finding its way to this Sedum in error.

ARRACH *Atriplex patula* (Orache) Parkinson. Arrach is nearer the original than Orache, for it is derived from O. Fr. arasche, arroche, from Latin atriplicam, atriplex earlier Gk. atraphaxis. Although Grieve.1931 says orache is a corruption of aurum, gold, because the seeds, mixed with wine, were supposed to cure the yellow jaundice.

ARRACH, Dog's *Chenopodium vulvaria* (Stinking Orach) Culpepper.

ARRACH, Goat's *Chenopodium vulvaria* (Stinking Orach) Culpepper. Any ill-smelling plant would be assigned to the goat.
ARRACH, Stinking *Chenopodium vulvaria* (Stinking Orach) Hill.
ARRACH, Wild 1. *Chenopodium vulvaria* (Stinking Orach) Grieve.1931. 2. *Atriplex patula* (Orach) Grieve.1931.
ARRADAN *Heracleum sphondyllium* (Hogweed) IOM Moore.1924.
ARROW, Green *Achillea millefolium* (Yarrow) Suff. Grigson; USA Henkel. Arrow here is a corruption of yarrow.
ARROW-BEARING TREE 1. *Acanthopanax ricinifolium* (Thorny Catalpa) Munro. 2. *Alnus hirsuta* Munro. 3. *Sambucus latipinna* Munro. All these trees are in Ainu folklore abodes of evil spirits. So, inan (protection) is made from them. Japanese farmers used an inan, stuck in the ground, to protect their millet from harm. The inan are cut pieces attached to rods, and looked upon as arrows.
ARROW GRASS, Marsh *Triglochin palustris.*
ARROW-GRASS, Sea *Triglochin maritima.*
ARROW-LEAF *Sagittaria latifolia* (Broad-leaved Arrowhead) New Engl. Sanford.
ARROW-REED *Cynerium saccharoides.*
ARROW VINE *Polygonum sagitattum* (Tear-thumb) USA Cunningham.
ARROW-WOOD 1. *Frangula alnus* (Alder Buckthorn) B & H. 2. *Philadelphus gordonianus* Calif. O.T. Mason. 3. *Viburnum acerifolium* Zenkert. 4. *Viburnum dentatum.* *Viburnum dentatum* wood was certainly used by the American Indians for arrows, and the other two American species probably were. Alder Buckthorn probably owes the name to the use in making skewers, as the names Butcher's Prick Tree, and Butcher's Prickwood, testify.
ARROW-WOOD, Indian *Euonymus atropurpureus* (Wahoo) Howes.
ARROWHEAD *Sagittaria sagittifolia* (Water Archer) B & H. A descriptive name, from the shape of the leaves.

ARROWHEAD, American 1. *Sagittaria latifolia* (Broad-leaved Arrowhead) RHS. 2. *Sagittaria rigida.*
ARROWHEAD, Arum-leaved *Sagittaria arifolia.*
ARROWHEAD, Broad-leaved *Sagittaria latifolia.*
ARROWHEAD, Chinese *Sagittaria chinensis.*
ARROWROOT 1. *Achillea millefolium* (Yarrow) Suff. B & H. 2. *Arum maculatum* (Cuckoo Pint) Fernie. 3. *Balsamorrhiza sagittata.* 4. *Heracleum sphondyllium* (Hogweed) Som. Macmillan. 5. *Maranta arundinacea. Maranta arundinacea*, sometimes known as Bermuda Arrowroot, produces a large amount of starch in the rhizomes. This is exported as an ingredient for arrowroot biscuits, and for use in puddings (blancmange), and dietetic foods; some arrowroot starch is used as a base for face powders, and also in glues. The word itself is a corruption of Aru-root, named for the Aruac, or Arawak Indians of South America.The plant is a native of Brazil, northern South and Central America, though wide-spread in cultivation in the tropics. It was known as an antidote for poisoned arrow-wounds in the 17th century, but this is certainly not the derivation. The Suffolk name for *Achillea millefolium* is probably a simple corruption of yarrow (the name Green Arrow occurs in the same region). But the name as applied to Cuckoo Pint is a reference to yet another name, Portland Sage. The tubers, when cooked, yielded the substance of that name, a substitute for arrowroot. "A pure and white starch, but most hurtfull to the hands of the Laundresse that hath the handling of it, for it chappeth, blistereth, and maketh the hands rough and rugged, and withall smarting" (Gerard). It was also once used in the French cosmetic industry and called Cypress powder. Arrowroot for Hogweed is a simple corruption of the name Aldertrot, itself better known as Eltrot. *Balsamorrhiza*, an American genus, seems to

get the name solely on the basis of leaf shape (*sagittata*).

ARROWROOT, African 1. *Canna bidentata*. 2. *Tacca pinnatifida* Howes.

ARROWROOT, Bermuda *Maranta arundinacea* (Arrowroot) W Miller. Given here as the name of the plant, but surely more usual as the name of the starch product.

ARROWROOT, Bombay; ARROWROOT, Indian *Curcuma angustifolia* Howes.

ARROWROOT, Brazilian 1. *Canna glauca*. 2. *Ipomaea batatas* (Sweet Potato) Howes. 3. *Manihot esculenta* (Manioc) Brouk.

ARROWROOT, British *Arum maculatum* (Cuckoo Pint) Le Strange. See Portland Arrowroot.

ARROWROOT, East Indian 1. *Curcuma angustifolia* Howes. 2. *Tacca pinnatifida* Brouk (East Indian).

ARROWROOT, English *Solanum tuberosum* (Potato) Howes.

ARROWROOT, Florida *Zamia integrifolia*.

ARROWROOT, Guyana *Dioscorea alata* (Greater Yam) Howes.

ARROWROOT, Hawaiian *Tacca hawaiiensis*.

ARROWROOT, Portland *Arum maculatum* (Cuckoo Pint) W Miller. See under Arrowroot. Portland because that is where it was grown.

ARROWROOT, Rosy *Calathea roseo-picta*.

ARROWROOT, Tahiti; ARROWROOT, Fiji *Tacca pinnatifida* Malaya Gilliland (Tahiti); Douglas (Fiji).

ARROWSCALE *Atriplex phyllostegia*.

ARSENIC, Plant *Colchicum autumnale* (Meadow Saffron) Schauenberg. The action of colchicine is similar to that of arsenic.

ARSENICKE *Polygonum hydropiper* (Water Pepper) B & H. Of the same group as Arsmart, etc., deriving from the irritant effect of the leaves.

ARSESMART 1. *Polygonum hydropiper* (Water Pepper) Turner. 2. *Polygonum persicaria* (Persicaria) Wilts. Grigson. Lincs. Peacock. 3. *Tanacetum parthenium* (Feverfew) Yks. Grigson. Water Pepper is the plant to which the name really belongs, and it derives from the irritant effect of the leaves. Cockayne says of this name: "it derives its name from its use in that practical education of simple Cimons, which village jokers enjoy to impart". There are a number of variants, such as Smartarse, and the Wessex Yes Smart, Smartweed (in America), etc. Then there is Culrage, which is simply the French translation of Arsesmart, brought back into English, and this has developed a whole lot of its own variants, from Ciderage through half a dozen names to Cow Itch. Persicaria got the name from the similarity of the two plants, but it is not nearly as effective as Water Pepper; in fact, Gerard called it Dead Arsmart - "it doth not bite as the other doeth". Dead in plant names usually carries the meaning of ineffective - Cf. deadnettle.

ARSESMART, Codded *Impatiens noli-me-tangere* (Touch-me-not) Gerard. Gerard translated the old name *Persicaria siliquosa* to produce Codded Arsesmart.

ARSESMART, Dead; ARSESMART, Mild *Polygonum persicaria* (Dead) Gerard; (Mild) Culpepper. "Dead" because "it doth not bite as the other doeth", the other being *P hydropiper*.

ARSESMART, Hot *Polygonum hydropiper* (Water Pepper) Culpepper. "Hot" in this case because of the peppery nature, as against "mild" for *Polygonum persicaria*.

ARTICHOKE, Chinese *Stachys affinis*. Introduced from north China in 1888; it has edible tubers which are eaten like potatoes.

ARTICHOKE, French; ARTICHOKE, Green *Cynara cardunculus var. scolymus* (Globe Artichoke) both Schery.

ARTICHOKE, Globe *Cynara cardunculus var. scolymus*. The original of artichoke is Arabic al-kharshuf, which was transliterated into Spanish-Arabic as al-kharstrofa, and into Italian as arcicioffo. But this name by popular etymology suggests the Italian arci, chief, and cioffo, horse collar. One of the French versions is Latinized into arti-cactus, cactus being the Latin name of the cardoon. Early spellings of the modern English version, artichoke, from 1531 onwards, show the etymological doubt - archecokk, hortichock and artichaux. Later came hortichoke (choking the garden) and Heartychoke (choking the gullet, or choking over the last bit to be eaten). According to a correspondent of Notes and Queries in 1873, the normal pronunciation in 18th century English was hartichoke.

ARTICHOKE, Japanese *Stachys sieboldii*. Distinguished from Chinese Artichoke, which is *S affinis*, though both are probably known as *S tuberifera*.

ARTICHOKE, Jerusalem *Helianthus tuberosus*. A corruption of the Italian girasole articiocco, where girasole means turn-sun, which is why the genus *Helianthus* generally is called Sunflower. Artichoke is eventually from the Arabic, and is co called because of the edible tubers, which instead of containing starch like potatoes, has the allied substance inulin, and also fructose, useful for diabetics.

ARTICHOKE, Jerusalem, White
Alstroemeria edulis Coats. Nothing to do with Jerusalem Artichoke - this ia a lily from San Domingo, and a food plant there.

ARTICHOKE, Sunflower *Helianthus tuberosus* (Jerusalem Artichoke) Grieve.1931.

ARTILLERY PLANT *Pilea microphylla* Rochford. Gunpowder Plant, Pistol Plant and Artillery Plant are all names for this house plant, because when the plant is shaken when in flower, it will produce clouds of pollen.

ARTS; HEARTS; HERTS; HORTS; HURTS *Vaccinium myrtillus* (Whortleberry) (Arts) Som, Wilts. Macmillan; (Hearts) Hants. B & H; (Herts) W.Eng. Halliwell; (Horts) Dev. B & H; (Hurts) Corn. Bottrell, Dev. Crossing, E.Ang. Grigson). All regional variations of Whorts = Whortleberries.

ARUM, Bog *Calla palustris* (Water Arum) W.Miller.

ARUM, Dragon *Dracunculus vulgaris*. There seems to have been a number of attempts to equate this with dragons or snakes. Dragons, Dragonwort, Dragance and Serpentine are all old names, and Edderwort also occurs, edder being OE for adder. Gerard describes the plant as being "... with spots of divers colours, like those of the adder or snake...".

ARUM, Three-lobed *Typhonium roxburghii*.

ARUM, Titan *Amorphophallus titanum*.

ARUM, Water *Calla palustris*. An American plant, which also has the names Water Dragons, and Female Dragons.

ARUM, White *Zantedeschia aethiopica* (Arum Lily) Grigson.1974.

ARUM-LEAVED ARROWHEAD *Sagittaria arifolia*.

ARUM LILY 1. *Arum maculatum* (Cuckoo Pint) Cambs. Porter. 2. *Zantedeschia aethiopica*.

ARVA, ARVI *Stellaria media* (Chickweed) Shet. Grigson.

ARYM *Scrophularia nodosa* (Figwort) IOM Moore.1924.

ASAFOETIDA *Ferula foetida*.

ASARABACCA 1. *Asarum europaeum*. 2. *Geum urbanum* (Avens) Dawson. 3. *Inula conyza* (Ploughman's Spikenard) Prior. "It appears from Pliny that the Asarum was not uncommonly confounded with the Baccharis; an English name was accordingly bestowed upon it, which is a curious compromise of the question, for it is a compound of both, viz Asarabacca" (Barton & Castle). This refers to *Asarum europaeum*, but as if to confirm Barton & Castle's point, Ploughman's Spikenard is called Baccharis by Gerard, while Prior gives Asarabacca for the same plant. The reason for the name being given to avens is not clear at all.

ASCENSION *Senecio vulgaris* (Groundsel) Fernie. Senecio, which is presumably from Latin senex, old man, is here corrupted into Ascension, and is further corrupted into Senshon, Sension, Sention, Sinsion, Simson, Simpson, etc.

ASCYRION *Hypericum tetrapterum* (St Peter's Wort) Turner.

ASH 1. *Fraxinus excelsior*. 2. *Syringa vulgaris* (Lilac) Glos. J.D.Robertson. OE aesc, OHG ask, ON askr. These old forms are still to be found in dialect - Esh in the North

Country, Ache and Aischen in Scotland, as well as Esch.

ASH, Arizona *Fraxinus velutina* (Velvet Ash) Barber & Phillips.

ASH, Basket *Fraxinus nigra* (Black Ash) Grieve.1931. The inner bark and the wood are used by various North American Indian groups to make baskets. Hoop Ash must be a related name.

ASH, Black 1. *Fraxinus excelsior* (Ash) War. F Savage. 2. *Fraxinus nigra*. 3. *Fraxinus sambucifolia*. Black, for the common ash, because of the colour of the young buds, presumably.

ASH, Blue 1. *Fraxinus quadrangulare*. 2. *Fraxinus pennsylvanica var. lanceolata* (Green Ash) Grieve.1931. 3. *Syringa vulgaris* (Lilac) Glos. B & H. The leaves and inner bark of *F quadrangulare* are said to yield a blue dye - this may account for the name in this case. Ash names for the lilac seem to be confined to Gloucestershire - Spanish Ash is recorded there, too.

ASH, Blueberry *Elaeocarpus reticulata*. Deep blue fruits in autumn.

ASH, Brown *Fraxinus nigra* (Black Ash) Howes.

ASH, Caucasian *Fraxinus oxycarpa*.

ASH, Chaney *Laburnum anagyroides* (Laburnum) Ches. B & H. Holland actually spells this Chainy, whicb is helpful, as there are various 'chain' names for laburnum.

ASH, Field *Sorbus aucuparia* (Rowan) Lyte.

ASH, Flowering 1. *Fraxinus cuspidata*. 2. *Fraxinus ornus* (Manna Ash) Lindley.

ASH, Flowering, Chinese *Fraxinus mariesii*.

ASH, French *Laburnum anagyroides* (Laburnum) Derb. B & H.

ASH, Green *Fraxinus pennsylvanica var. lanceolata*.

ASH, Ground 1. *Aegopodium podagraria* (Goutweed) Corn, Som, War, Lincs. Grigson; Ches. Holland. 2. *Angelica sylvestris* (Wild Angelica) N.Eng, Berw. Grigson. Ash in these cases may very well be ach, or ache, old names for parsley (they are both umbellifers).

ASH, Grunty *Aegopodium podagraria* (Goutweed) Bucks. Harman. Probably a corruption of Ground Ash, quite common, though not as frequent as Ground Elder for this. Harman, though, says grunty means tough in Buckinghamshire dialect.

ASH, Hedge *Sorbus aucuparia* (Rowan) Blount.

ASH, Hoop *Fraxinus nigra* (Black Ash) Grieve.1931. Hoop Ash presumably for the same reason that it is sometimes called Basket Ash. The North American Indians used both the inner bark and the wood with which to make baskets.

ASH, Jerusalem 1. *Isatis tinctoria* (Woad) W.Miller. 2. *Reseda luteola* (Dyer's Rocket) W.Miller.

ASH, Leather-leaved *Fraxinus velutina coriacea*.

ASH, Manna *Fraxinus ornus*. Manna in this case is the saccharine exudation of this tree - this is the manna of European medicine. In the Highlands, at the birth of a child, the midwife puts a green ash stick (*F excelsior* of course) into the fire, and while it is burning, lets the sap drop into a spoon. This is given as the first spoonful of liquor to the newborn baby. It is said that it is given as a guard against witches, and is possibly a recollection of the honey-like juice of the Manna Ash. Yggdrasil, the world tree, is generally taken to be an ash, and according to the myth, the morning dew of Yggdrasil was a sweet and wonderful nourishment.

ASH, Manna, Himalayan *Fraxinus floribunda*.

ASH, Mountain 1. *Populus tremula* (Aspen) Inv. B & H. 2. *Pyrus americana* USA Bergen. 3. *Sorbus aucuparia*. 4. *Viburnum cassinoides* (Withe-rod) USA Bergen.

ASH, Mountain, American *Sorbus americana*.

ASH, Mountain, Australian *Eucalyptus regnans*.

ASH, Mountain, Japanese *Sorbus matsumurama*.

ASH, Narrow-leaved *Fraxinus angustifolia*.

ASH, Oregon *Fraxinus oregana*.

ASH, Poison *Rhus vernix* (Swamp Sumach) Le Strange.

ASH, Pop *Fraxinus caroliniana* USA Harper.

ASH, Pot *Aegopodium podagraria* (Goutweed) Dev. Macmillan. Goutweed is not a true native to Britain, but was introduced in the Middle Ages as a pot herb and medicinal plant. The young leaves can be boiled like spinach and eaten. Actually, this name often appears as one word - Potash.

ASH, Prickly 1. *Aralia spinosa* (Hercules' Club) Everett. 2. *Xanthoxylum americanum*. Prickly Elder also occurs for Hercules' Club.

ASH, Prickly, Northern *Xanthoxylum americanum* (Prickly Ash) W.Miller. *X clava-herculis* is the Southern Prickly Ash.

ASH, Prickly, Southern *Xanthoxylum clava-herculis*.

ASH, Quaking *Populus tremula* (Aspen) Scot. Jamieson. Ash, in this case, should be asp.

ASH, Red *Fraxinus pennsylvanica*.

ASH, Rhodesian *Burkea africana* (Wild Syringa) Palgrave.

ASH, Rim *Celtis occidentalis* (Hackberry) W.Miller.

ASH, Rowan *Sorbus aucuparia* (Mountain Ash) Leland.

ASH, Single-leaved *Fraxinus anomala*.

ASH, Spanish *Syringa vulgaris* (Lilac) Glos. B & H. It is called Blue Ash also in the same county; the reason is obscure.

ASH, Stinking *Ptelea trifoliata* (Hop Tree) Hora.

ASH, Swamp, Black *Fraxinus nigra* (Black Ash) Grieve.1931.

ASH, Sweet *Anthriscus sylvestris* (Cow Parsley) Glos. B & H. Ash must be ach here, for this is an umbellifer.

ASH, Velvet *Fraxinus velutina*.

ASH, Wafer *Ptelea trifoliata* (Hop Tree) USA Yarnell. Is this really a misprint for 'water'?

ASH, Water 1. *Fraxinus nigra* (Black Ash) Grieve.1931. 2. *Fraxinus sambucifolia* (Black Ash) Youngken.

ASH, White 1. *Aegopodium podagraria* (Goutweed) Som. Elworthy. 2. *Fraxinus americana*. 3. *Sorbus aucuparia* (Mountain Ash) B & H. Ash for Goutweed (it occurs as Ground Ash and Pot Ash as well) is probably ache, an old name for umbellifers, parsley in particular.

ASH, Wild *Sorbus aucuparia* (Mountain Ash) N.Eng. Denham/Hardy.

ASH, Wood *Oxalis acetosella* (Wood Sorrel) Som. Macmillan.

ASH BARBERRY *Mahonia aquifolium* (Oregon Grape) Gorer.

ASH-LEAVED MAPLE; ASH-LEAVED NEGUNDO *Acer negundo* (Box Elder) (Maple) Zenkert; (Negundo) Hibberd. An older botanical name for this was *Negundo fraxinifolia*.

ASH OF JERUSALEM *Isatis tinctoria* (Woad) Turner. Probably invented by Turner. The reverse, Jerusalem Ash, is listed in W Miller.

ASHANTI PEPPER *Piper guineense* (West African Black Pepper) Dalziel. Cf. Benin Pepper for this.

ASHEN TREE; ASHING-TREE *Fraxinus excelsior* (Ash) Glos. J D Robertson (Ashen); Som. Tongue (Ashing).

ASHTHROAT *Verbena officinalis* (Vervain) C.J.S.Thompson. This was the Anglo-Saxon name for the plant. The dried root was used as a charm against ulcerated throat, a use which survived well into the twentieth century.

ASHWEED; AISEWEED; AISHWEED *Aegopodium podagraria* (Goutweed) (Ashweed) Wilts Dartnell & Goddard; (Aiseweed) B & H; (Aishweed) Gerard. All probably from ache, the old name for parsley, and umbellifers in general.

ASHY POKER *Plantago media* (Lamb's Tongue Plantain) Wilts. Macmillan. Ashy possibly from the hoary look of the plant generally (it is known as Hoary Plantain).

ASIATIC BARBERRY *Berberis asiatica* (Indian Barberry) W.Miller.

ASIATIC CORN *Zea mays* (Maize) Lehner. Asiatic, because the 16th century herbalists believed the plant had been brought by the Turks from Asia. Cf. Turkish Corn for this.

ASIATIC GLOBE FLOWER *Trollius asiaticus* (Orange Globe Flower) Grieve.1931.

ASIATIC ORACH *Atriplex hortensis* (Garden Orach) Schery. It is actually a native of Asia, and was not introduced into this country until 1548.

ASIATIC PENNYWORT *Hydrocotyle asiatica* (Indian Pennywort) Chopra.

ASP; ASPE *Populus tremula* (Aspen) Som. Jennings; Wilts, War, Ches, Yks, Cumb, N'thum, Ire. Grigson; Heref. Halliwell. Aspe is in Ellacombe. Asp is the original form, the OE and OHG name for the tree.Aspen was originally the adjective from asp.

ASP, Quakin *Populus tremula* (Aspen) Scot. Jamieson.

ASP, Rattling *Populus tremula* (Aspen) Lyte.

ASP, Shaking *Populus tremula* (Aspen) Ches. Holland.

ASP, White *Populus alba* (White Poplar) Turner.

ASPARAGUS; ASPARAGUS FERN *Asparagus officinalis*. The Latin form is directly from Greek asparagos, earlier aspharagos.

ASPARAGUS, Bath *Ornithogalum pyrenaicum* (Spiked Star of Bethlehem) B & H. French Sparrow-grass was the name under which the sprouts of the plant were sold in the Bath markets, to be eaten as asparagus. It is still picked and sold around Bath and Bristol.

ASPARAGUS, Cape *Aponogetum distachyum*. A South African asparagus substitute.

ASPARAGUS, Cossack *Typha latifolia* (False Bulrush) Lindley. It has sometimes been used as a food under this name.

ASPARAGUS, French; ASPARAGUS, Prussian *Ornithogalum pyrenaicum* (Spiked Star of Bethlehem) (French) Wilts. D.Grose; (Prussian) B & H. See under Bath Asparagus.

ASPARAGUS, Lincolnshire *Chenopodium bonus-henricus* (Good King Henry) Sanecki. This and another name, Lincolnshire Spinach, are the result of its having been once cultivated in that area.

ASPARAGUS, New Guinea *Setaria palmifolia* (Palm Grass) Rappaport.

ASPARAGUS, Sea *Asparagus officinalis prostratus*.

ASPARAGUS, South African *Asparagus laricinus*.

ASPARAGUS, Wild 1. *Chenopodium bonus-henricus* (Good King Henry) Tynan & Maitland. 2. *Ornithogalum pyrenaicum* (Spiked Star of Bethlehem). Dartnell & Goddard. Spiked Star-of-Bethlehem is also known as Bath Asparagus, for it was sold in the markets of Bath to be eaten like asparagus.

ASPARAGUS BEAN 1. *Vigna sesquipedalis* (Asparagus Cowpea) Bianchini. 2. *Psophocarpus tetragonolobus* (Goa Bean) Brouk.

ASPARAGUS BUSH *Dracaena manni*. A tropical African bush, whose young shoots are eaten like asparagus.

ASPARAGUS COWPEA *Vigna sesquipedalus*.

ASPARAGUS PEA *Lotus tetragonolobus*. The seeds and pods are edible, and often used as a vegetable.

ASPEN 1. *Populus nigra* (Black Poplar) Lincs. Gutch & Peacock. 2. *Populus tremula*. OE aesp(e), OHG aspa, Gk.espe, which is from the adjective espen, so asp is the original form, the word aspen, which would be the adjective of asp, being now a noun, and taking the place of the original form. In the Lincolnshire example, it is the Black Poplar that is the Shiver-tree, not the aspen; it was used for charms against the ague, the shivering sickness, just as *P tremula* was elsewhere.

ASPEN, American *Populus tremuloides*.

ASPEN, Euphrates *Populus euphratica* (Euphrates Poplar) Moldenke.

ASPEN, Great *Populus alba* (White Poplar) B & H.

ASPEN, Large-toothed *Populus grandidentata*.

ASPEN, Quaking; ASPEN, Trembling *Populus tremuloides* (American Aspen) USA (Quaking) Wit; (Trembling). H.Smith.1928.

ASPEN, Rocky Mountain *Populus tremuloides* (American Aspen) USA Elmore.

ASPHODEL, Bog *Narthecium ossifragum*. Asphodel is eventually from Greek, and there is a medieval Latin variant affodilus, becoming affodil, i.e. daffodil. Ossifragum means bone-breaker, because as it grows on wet moors and mountains, sheep pasturing there often suffered from foot rot, and this was attributed to their browsing on the plants.

ASPHODEL, False *Tofieldia pusilla* (Scottish Asphodel) USA Grieve.1931.

ASPHODEL, German *Tofieldia calyculatum*.

ASPHODEL, Lancashire *Narthecium ossifragum* (Bog Asphodel) Prior. Why Lancashire?

ASPHODEL, Marsh *Narthecium ossifragum* (Bog Asphodel) Prior.

ASPHODEL, Onion-leaved *Asphodelus fistulosus*.

ASPHODEL, Pyrenean *Asphodelus albus pyrenaeus*.

ASPHODEL, Scottish *Tofieldia pusilla*. It is very rare, and occurs only on hills in the north, hence the name.

ASPHODEL, White *Asphodelus ramosus*.

ASPHODEL, Yellow *Asphodeline lutea*.

ASPIDISTRA *Aspidistra elatior*. Aspidistra is made up of two Greek words, aspis, meaning a round shield, and astron, star; the reference is to the mushroom-like stigma.

ASS-EAR *Symphytum officinale* (Comfrey) B & H. Presumably from the shape of the leaves, but Cf. the French oreille d'Ane.

ASS-PARSLEY 1. *Aethusa cynapium* (Fool's Parsley) Prior. 2. *Anthriscus sylvestris* (Cow Parsley) A S Palmer. In spite of the tendency to want to make ass = fool, especially for Fool's Parsley, it almost certainly is ache, which is an old word for parsley. So it makes a pleonasm.

ASS'S FOOT *Tussilago farfara* (Coltsfoot) Prior. Cf. French pied d'Ane. There are a number of different foot or hoof names for coltsfoot. One is Donnhove, apparently from donn, an old word for a horse (donkey = little horse). Donnhove became Dunnhove, and eventually corrupted into Tunhoof. All these are from the shape of the leaves.

ASSAM ONION *Allium rubellum*.

ASSEGAI WOOD *Terminalia sericea* (Mangwe) Palgrave. Presumably assegais were made from the wood.

ASSYRIAN PLUM *Cordia myxa* (Sapistan) Dalziel.

ASTER 1. *Aster sp*. 2. *Callistephus sp*. Aster is directly from the Latin, and thence Greek, for the word meaning star, hence the name Starwort for many Michaelmas Daisies, which are all Asters. Callistephus is the annual, or Chinese Aster (Greek kallistos, most beautiful, and stephos, crown).

ASTER, Alkali *Aster carnosus*. A North American plant, confined to alkaline meadows.

ASTER, Beach *Erigeron glaucus*.

ASTER, Calico *Aster laterifolius* (Starved Aster) USA House.

ASTER, Cape *Felicia amelloides* (Blue Marguerite).

ASTER, China *Callistephus chinensis*. Pierre d'Incarville, a French missionary in China, sent the first seeds to France about 1730. Most of the annual asters derive from this species.

ASTER, Cornflower *Stokesia cyanea*.

ASTER, Desert *Aster abatus* (Mohave Aster) USA Jaeger. The desert in this case is the Mohave.

ASTER, Golden, Hispid *Chrysopsis hispida*.

ASTER, Golden, Maryland *Chrysopsis mariana*.

ASTER, Hairy *Aster canescens*.

ASTER, Heart-leaved *Aster cordifolius*.

ASTER, Heath, White *Aster ericoides* (Many-flowered Aster) USA Zenkert. It is

also known simply as White Heath in parts of America.
ASTER, Italian *Aster amellus.*
ASTER, Large-leaved *Aster macrophyllum.*
ASTER, Many-flowered *Aster ericoides.*
ASTER, Mecca *Aster cognatus.* An American plant like most of the asters. Why Mecca?
ASTER, Mexican, Purple *Cosmos bipinnatus.*
ASTER, Mohave *Aster abatus.* It is a plant which only grows on the dry mountain slopes; Desert Aster is another name.
ASTER, Mountain *Aster acuminatus.*
ASTER, New England *Aster nova-angliae.*
ASTER, New York *Aster novi-belgae.*
ASTER, Purple *Machaeranthera alta* USA Elmore.
ASTER, Purple, Late *Aster patens.*
ASTER, Purple-stemmed *Aster puniceus* (Red-stalked Aster) USA House.
ASTER, Red-stalked *Aster puniceus.*
ASTER, Sea *Aster tripolium.* The only British member of the genus, growing on salt marshes and muddy estuaries.
ASTER, Seaside *Aster spectabilis.*
ASTER, Showy, Late *Aster spectabilis* (Seaside Aster) USA House. *Spectabilis*, hence "showy".
ASTER, Smooth *Aster laevis.*
ASTER, Spiny *Aster spinosus.*
ASTER, Starved *Aster laterifolius.*
ASTER, Stokes's *Stokesia laevis.*
ASTER, Tansy 1. *Machaeranthera alta* USA Elmore. 2. *Machaeranthera ramosa.* 3. *Machaeranthera tanacitifolia* (Tahotta Daisy) USA Gates.
ASTER, Viscid *Machaeranthera alta* USA Elmore.
ASTER, White *Aster leucelene.*
ASTER, White, Upland *Aster ptarmicoides.*
ASTER, Whorled *Aster acuminatus* (Mountain Aster) USA House.
ASTER, Wild *Knautia arvensis* (Field Scabious) Som. Macmillan.
ASTER, Wood *Aster cordifolius* (Heart-leaved Aster) USA Perry.
ASTERLARGIA *Asperula odorata* (Woodruff) R.T.Gunther. This appears as a gloss on a 12th century manuscript of Apuleius Barbarus.
ASTHMA PLANT *Euphorbia hirta* S.Africa Watt. It is still used in western practice for asthma and bronchitis. In Australia, it is known as Queensland Asthma-weed.
ASTHMA WEED *Lobelia inflata* (Indian Tobacco) USA Henkel. The dried leaves and tops are used as an expectorant. In large doses it is an acro-narcotic plant, but in small doses it is useful for, among other things, asthma,the effect depending mainly on the alkaloids lobeline and lobelidine.
ASTHMA-WEED, Australian; ASTHMA WEED, Queensland *Euphorbia hirta* Australia Dalziel (Australian); Watt (Queensland). The plant was still being used in this century in western practice, according to Gimlette. Note the S.African name Asthma-plant, too.
ASTROLOGIA *Polygonum bistorta* (Bistort) Turner. This is probably a corruption of aristolochia, quite a different plant, transferred in some obscure way to bistort. There is some confirmation for this, as the names Osterick and Oisterloit also were once used for Bistort, and both of these come from the same source.
ATAMASCO LILY *Zephyranthes atamasco.* Presumably this is an area in the place where the plant is native - Mississippi, Florida or Virginia.
ATCHERN, (Oak) *Quercus robur* - acorn (Acorn) Ches. B & H. Like the more common Aitchorn, and the Lancashire Hatchhorn, this is a variant on the word acorn, used from the Midlands northwards; they all derive from OE aecorn, which in turn comes from Latin acer, field. This an example of the soft 'c' of the Latin original being retained in English.
ATHERLUS *Glechoma hederacea* (Ground Ivy) Ire. Wilde.1890.
ATIS ROOT *Aconitum heterophyllum.*
ATLANTIC BLUE CEDAR *Cedrus atlantica glauca.*
ATLAS CEDAR *Cedrus atlantica.* In spite of

'atlantica', the home of this cedar is the Atlas Mountains, as the names Algerian Cedar and African Cedar confirm. But *var. glauca* is usually known as Atlantic Blue Cedar.

ATTERLOTHE *Atropa belladonna* (Deadly Nightshade) Halliwell. OE attor, poison.

AUBEL; AWBEL *Populus alba* (White Poplar) B & H. This is one of the many variants of Abele, which comes from the Dutch abeel, and ultimately from Latin albus.

AUBERGINE *Solanum melongena* (Egg Plant) Watt. ODEE describes this as the fruit of the Egg Plant, *Solanum esculentum*, which must be the same as *S melongena*. The name is French, coming through Catalan from Arabic albadinjan, and eventually from Sanskrit vatimgana.

AUBESPYNE; ALBASPYNE; ALBESPEINE; ABBESPINE *Crataegus monogyna* (Hawthorn) Aubespyne and Albaspyne are in B & H; Albespeine in Ellacombe and Abbespine is from Mandeville. All are Latin alba spina - whitethorn.

AUGERS *Salix viminalis* (Osier) N'thants. A.E.Baker. Auger is probably the local form of osier. An augerholm in the same area is an osier-loft.

AUGERS, Screw *Spiranthes cernua* (Nodding Lady's Tresses) USA Howes. In this case Augers is the tool of that name, and the whole name is a perfect description of the spiralling habit of the whole genus.

AUGUST PLUM *Prunus americana* (Canada Plum) USA Elmore.

AUL, Ellern *Sambucus nigra* (Elder) Heref. G.C. Lewis. Aul is alder, while ellern can be either alder or elder, so this name probably compounds the confusion between the names.

AULNE *Alnus glutinosa* (Alder) B & H. A variant of the more common Allen, or Ellentree. Aul occurs in Herefordshire, and the form varies to many names, like Ellar, Ollar, Aller, Arl, etc., according to which part of the country one is in.

AUM; AUM-TREE *Ulmus procera* (English Elm) Yks. Morris, who spells it Awm; Cumb, N'thum. F. Grose; Dur. Dinsdale. Elm is an OE borrowing from the Latin ulmus, and has many variants, including these, and, close to them, Ome in Cumberland and Owm in Yorkshire.

AUNT HANNAH *Arabis alpina* (Alpine Rock-cress) E.Ang. E.Wright. A very rare native, growing only in one place in Skye, but well known as a garden plant, and, to have a name like this, affectionately known.

AUNT MARY'S TREE *Ilex aquifolium* (Holly) Corn. Grigson. Aunt is a term of endearment in Cornwall, so very probably Aunt Mary is the Virgin. The association with Christmas makes this explanation likely. Bayley, however, suggests it is equal to Aunt Maura, or Saint Maura, who he says was a fairy. He does say that the Cornish Aunt was reserved for old women (according to the Golden Legend, Mary died at the age of 72).

AUNTIE POLLY *Primula vulgaris elatior* (Polyanthus) Som. Macmillan. A felicitous example of the misplacement of syllables.

AURICULA *Primula auricula*. Auricula means little ear. Cf. Bear's Ears, etc., all from the shape of the leaf.

AURICULA TREE *Calotropis procera* (Swallow-wort) Dalziel.

AUSTRALIAN ASTHMA-WEED *Euphorbia hirta* Dalziel. This plant has a special reputation for asthma wherever it grows.

AUSTRALIAN BANYAN *Ficus macrophylla* (Moreton Bay Fig) RHS.

AUSTRALIAN BAOBAB *Adansonia gregori*.

AUSTRALIAN BEECH 1. *Nothofagus moorei*. 2. *Tectona australis*.

AUSTRALIAN BLACKWOOD *Acacia melanoxylon* (Blackwood) Willis. Other trees, not from Australia, are called Blackwood, hence the need for identification.

AUSTRALIAN BLUEBELL *Sollya heterophylla* (Bluebell Creeper) Cunningham.

AUSTRALIAN CHESTNUT *Castanospermum australe* (Moreton Bay Chestnut) Howes.

AUSTRALIAN CINNAMON *Cinnamonium*

oliveri (Oliver Bark) Mitton.
AUSTRALIAN EARTH CHESTNUT *Eleocharis sphacelata.*
AUSTRALIAN FUCHSIA *Correa ssp.*
AUSTRALIAN GOOSEBERRY *Muehlenbeckia adpressa.*
AUSTRALIAN GRASS *Gynerium argenteum* (Pampas Grass) Dev. Friend; Som. Macmillan. In Sussex, they call it Indian Grass.
AUSTRALIAN HAREBELL *Wahlenbergia gracilis.*
AUSTRALIAN KINO *Eucalyptus amygdalina* (Tasmanian Peppermint) Howes.
AUSTRALIAN LILLY-PILLY *Eugenia smithii.*
AUSTRALIAN MOUNTAIN ASH *Eucalyptus regnans.*
AUSTRALIAN NUT; AUSTRALIAN HAZEL NUT *Macadamia ternifolia* (Queensland Nut) Schery (Nut); Brouk (Hazel Nut). There is some similarity in appearance to hazel nuts.
AUSTRALIAN OAK *Eucalyptus regnans* (Australian Ash) Howes.
AUSTRALIAN PEA *Dolichos lablab* (Lablab) RHS.
AUSTRALIAN PINE 1. *Araucaria heterophylla* (Norfolk Island Pine) USA. Cunningham. 2. *Casuarina stricta* (She Oak) A.Huxley.
AUSTRALIAN PITCHER PLANT *Cephalotus follicularis.*
AUSTRALIAN QUININE *Alstonia constricta* Mitton. Because the bark of this tree is used just like cinchona, to combat malaria bouts.
AUSTRALIAN RIVER OAK *Casuarina cunninghamia* (Beefwood) Leathart.
AUSTRALIAN SANDALWOOD *Santalum lanceolatum.*
AUSTRALIAN SARSAPARILLA *Hardenbergia violacea* (Blue Coral Pea) RHS.
AUSTRALIAN SASSAFRAS *Atherosperma moschatum.*
AUSTRALIAN SNOW GUM *Eucalyptus niphophila.*
AUSTRALIAN VIOLET *Viola hederacea.*
AUSTRIAN BRIAR *Rosa foetida* (Capucine Rose) Plowdon.
AUSTRIAN COPPER BRIAR *Rosa foetida bicolor.*
AUSTRIAN PEA *Pisum sativum arvense* (Field Pea) Kingsbury.
AUSTRIAN PINE *Pinus nigra.*
AUSTRIAN ROCKET *Sisymbrium austriacum.*
AUSTRIAN YELLOW CRESS *Rorippa austriaca.*
AUTOGRAPH TREE *Clusia rosea* RHS.
AUTUMN BELLFLOWER, AUTUMN BELLS *Gentiana pneumonanthe* (Marsh Gentian) Gerard (Bellflower); Prior (Bells). Strictly, not a bellflower at all, in the sense of belonging to the Campanula species, but the shape of the flower justifies the names. Autumn Bells is also applied to *Gentianella amarella* (Autumn Gentian).
AUTUMN CORAL-ROOT *Corallorhiza odontorhiza.*
AUTUMN CROCUS *Crocus nudiflorus.*
AUTUMN DAMASK *Rosa damascena bifera.* Damask because of the tradition that the Damask Rose was brought to Europe by the Crusaders from Damascus; but the word damask may once have referred to the colour, a blush-red ("damasked cheeks" has been used in literature).
AUTUMN-FLOWERING ELDER *Sambucus canadensis* (American Elder) W.Miller.
AUTUMN GENTIAN *Gentianella amarella.*
AUTUMN HAWKBIT *Leontodon autumnalis.*
AUTUMN LADY'S TRESSES *Spiranthes spiralis* (Lady's Tresses) Clapham. A synonymous specific name is autumnalis. *S aestivalis* is Summer Lady's Tresses.
AUTUMN SAGE *Salvia greggii.*
AUTUMN SQUASH *Cucurbita maxima* (Giant Pumpkin) USA Elmore.
AUTUMN VIOLET *Gentiana pneumonanthe* (Marsh Gentian) B & H. Cf. Autumn

Bellflower and Autumn Bells for this. Gerard also has Harvest Bells.

AUTUMN'S GOODBYE *Parnassia palustris* (Grass of Parnassus) Tynan & Maitland. A very odd name, particularly as Summer's Goodbye, more appropriate, is also recorded.

AUTUMN'S WELCOME *Aster novae-angliae* (Michaelmas Daisy) Dev. Tynan & Maitland.

AUTUMNAL FURZE, Dwarf *Ulex minor* (Small Furze) Flower.1859.

AUTUMNAL SQUILL *Scilla autumnalis.* Squill is Scilla, but more immediately from its variant, Latin squilla, Greek skilla.

AUTUMNAL STAR HYACINTH *Scilla autumnalis* (Autumnal Squill) Camden/Gibson.

AUTUMNAL STARWORT *Callitriche autumnalis* (Narrow Water Starwort) Clapham.

AVANCE *Geum urbanum* (Avens) B & H. See Avens for derivation.

AVANDRILL *Narcissus pseudo-narcissus* (Daffodil) Ches. Holland. Cf. Averill, or Haverdrills, from the same area.

AVE GRACE *Ruta graveolens* (Rue) Suss. Loudon. Loudon was writing in 1838 when he said that rue "is to this day called Ave Grace in Sussex". It is of course Herb Grace, or Herb of Grace, both of which occur, as does Herb of Repentance. Rue, as Shakespeare well knew, was the symbol of regret and repentance. "To rue" is to be sorry for. The word is possibly derived from the same root as Ruth, meaning sorrow or remorse. Holy water was once sprinkled from brushes made of rue at the ceremony usually preceding the Sunday celebration of High Mass. This may be another reason for the Herb Grace series of names.

AVENS *Geum urbanum.* Avens comes through Old French avence, from med. Latin avencia, meaning unknown, though Barton & Castle offer the derivation from aveo, to rejoice. ODEE sticks to the unknown meaning.

AVENS, Alpine *Sieversia montana.*

AVENS, City *Geum urbanum* (Avens) B & H. City = urbanum.

AVENS, Drooping; AVENS, Nodding *Geum rivale* (Water Avens) B & H (Drooping); Prior (Nodding). From the habit of the flower.

AVENS, Mountain *Dryas octopetala.* Not an Avens at all - in fact, the leaf resembles that of the oak, so it was named after the dryad, nymph of the oaks.

AVENS, Mountain, Purple *Geum rivale* (Water Avens) Camden/Gibson.

AVENS, Purple *Geum rivale* (Water Avens) USA House.

AVENS, Pyrenean *Geum pyrenaicum.*

AVENS, Scarlet *Geum quellyon.*

AVENS, Tree *Cratoxylum cochinenense.*

AVENS, Water *Geum rivale.*

AVENS, White 1. *Geum canadense.* 2. *Geum virginianum.*

AVENS, Wood *Geum urbanum* (Avens) B & H.

AVENS, Yellow *Geum strictum.*

AVEREN, AVERIN *Rubus chamaemorus* (Cloudberry) Scot. Jamieson (Averen); Grigson (Averin). Meaning not known, though averish was a North-country word meaning the stubble left after reaping, while an aver was the horse used for ploughing averland, the land ploughed by the tenants for the use of a monastery, or the lord of the manor. Averoyne, though, in spite of looking very like these, is in fact an old name for southernwood. Other obvious derivatives exist for Cloudberry, all from Scotland - Evron, Aivern, Aiverin, Everocks.

AVERILL; HAVERDRIL *Narcissus pseudo-narcissus* (Daffodil) Averill B & H; Haverdril Ches. B & H. Meaning unknown - Haver usually refers to wild oats.

AVEROYNE *Artemisia abrotanum* (Southernwood) B & H. Averoyne is probably the original of a well-known name for the plant - Maiden's Ruin. A French name, Ivrogne, also derives from it. Averoyne, like the French auronne, must come directly from abrotanum.

AVOCADO PEAR *Persea gratissima.*

Avocado is Spanish for advocate (the same word) (Cf. French avocat). It was a popular substitute for the Aztec name for the fruit - ahuacatl, testicle. Another name for the avocado, alligator pear, came through another corruption from the original Aztec word.

AWGLEN *Crataegus monogyna* fruit (Haw) Corn. Jago. This occurs as Aglen in Cornwall too, and Aglon throughout Wessex. There are many other versions such as Agald, Eglet, Hague, etc., all of which eventually derive from OE haga, hedge.

AWL-LEAVED PEARLWORT *Sagina subulata* (Heath Pearlwort) Clapham.

AWL-WORT *Subularia aquatica*. Latin subula means awl (so this is the plant with the awl-shaped leaves).

AWMS *Hordeum sativum* awns (Barley awns) Yks. Harland.

AWNED-LEAVED BARBERRY *Berberis aristata* (Nepal Barberry) W.Miller.

AWNTLINGS *Hordeum sativum* (Barley - the awns) Yks. F K Robinson.

AWNUT *Bunium bulbocastanum* (Great Earth-nut) N.Eng. Brockett. A local variant of Arnut, itself a variant of Earth-nut.

AXE, Flower of the *Lobelia urens* (Acrid Lobelia) Dev. Friend. It is a native of the Atlantic side of France, Spain and Portugal. Its only British station is on Kilmington Common, near Axminster, hence the name.

AXSEED; AXWORT *Coronilla varia* (Crown Vetch) USA both House.

AXWEED *Aegopodium podagraria* (Goutweed) B & H. This, together with Acheweed, Ashweed, Aiseweed and Aishweed is from ache, parsley.

AYE-NO BENT *Lolium perenne* (Rye-grass) Glos. Grigson.1959. This is a typical name taken from the Love-me, love-me-not kind of divination done by pulling off the alternating spikelets. Tinker-tailor Grass is the best known of these names, but Cf. the Somerset Yes-or-no.

AYECOTE *Phaseolus coccineus* (Scarlet Runner) Bennett & Bird. Presumably this derives from the South American name. The plant was cultivated in the Andean region in pre-Columbian times.

AY(E)GREEN; AIGREEN *Sempervivum tectorum* (Houseleek) Barton & Castle (Ayegreen); Lancs. Nodal (Aigreen). i.e. evergreen.

AYK; AIK *Quercus robur* (Oak) Scot. both Jamieson. Wilks suggests that this variant, together with others like Eike and Eke, are responsible for the place names Acton, Savernake and Mallaig.

AYRON *Sempervivum tectorum* (Houseleek) Grieve.1931. Presumably a variation of Aigreen, or Ayegreen, and meaning evergreen.

AYRSHIRE ROSE *Rosa arvensis* (Field Rose) Bunyard.

AYVER *Lolium perenne* (Rye-grass) Corn, Dev. B & H. See Eaver.

AZALEA, Flame *Azalia lutea* USA House.

AZALEA, Hoary *Azalea canescens* (Mountain Azalea) USA House.

AZALEA, Indian *Rhododendron simsii*.

AZALEA, Mock *Menziesia ferruginea*.

AZALEA, Mountain 1. *Azalea canescens*. 2. *Loiseleuria procumbens* (Wild Azalea) Darling & Boyd.

AZALEA, Pink *Azalea nudiflora* (Purple Azalea) USA House.

AZALEA, Purple *Azalea nudiflora*.

AZALEA, Trailing *Loiseleuria procumbens* (Wild Azalea). J A MacCulloch.1905.

AZALEA, White *Azalea viscosa*.

AZALEA, Wild *Loiseleuria procumbens*.

AZALEA, Yellow *Azalea lutea*.

AZAROLE *Crataegus azarolus*. This must be the same as Spanish acerola - haw.

AZOB; EZOB *Hyssopus officinalis* (Hyssop) Grieve.1933 (Azob); B & H (Ezob). The Hyssopus of Dioscorides was so named - a holy herb.

AZTEC LILY *Sprekelia formosisissima* (Jacobean Lily) Simmons.

AZTEC MARIGOLD *Tagetes erecta* (African Marigold) RHS. The name makes more sense than African Marigold; after all, this is a Mexican genus.

AZTEC TOBACCO *Nicotiana rustica*
 (Turkish Tobacco) Howes.
AZZY-TREE *Crataegus monogyna*
 (Hawthorn) Bucks. Grigson.

B

BAA-LAMBS 1. *Corylus avellana* catkins (Hazel catkins) Som. Macmillan. 2. *Trifolium repens* (White Clover) Som. Macmillan. A typically inventive Somerset descriptive name. Lamb-sucklings also occurs in the north country, for clover, and there sucklings has the same meaning as honeysuckle - that which attracts the bees. For hazel catkins, Baa-Lamb's Tails has more relevance.
BAA-LAMBS' TAILS *Corylus avellana* catkins (Hazel catkins) Som. Macmillan.
BAALEM'S SMITE *Stellaria holostea* (Greater Sitchwort) Suff. Grigson.
BABE-AND-CRADLE *Fumaria officinalis* (Fumitory) Som. Macmillan. Babe-in-the-cradle, rather.
BABE-IN-THE-CRADLE 1. *Arum maculatum* (Cuckoo Pint) Som. Macmillan. 2. *Fumaria officinalis* (Fumitory) Som. Macmillan. 3. *Scrophularia aquatica* (Water Betony) Som. Macmillan; Wilts Goddard. In each of the cases, the name is fancifully descriptive, though it needs a prolonged look at the flowers to see the reason.
BABEL, Tower of *Verbascum thapsus* (Great Mullein) Som. Notes and Queries. vol 168, 1935.
BABES-IN-THE-WOOD *Stellaria graminea* (Lesser Stitchwort) Dev. Macmillan.
BABIES, Water *Caltha palustris* (Marsh Marigold) Som. Macmillan.
BABIES' BELLS *Narcissus pseudo-narcissus* (Daffodil) Wales Trevelyan. The 'bell' part of the name is obvious, but people say in some parts of Wales that only infants and young children can hear them ringing.
BABINGTON'S ORACHE *Atriplex glabriuscula*.
BABINGTON'S POPPY *Papaver lecoqii*.
BABUL TREE *Acacia arabica*. Babul, or Bablah, is the Hindi or Persian word for the tree.

BABY, Cocky; BABY, Cuckoo *Arum maculatum* (Cuckoo Pint) IOW. Long (Cocky-baby); Grigson (Cuckoo-baby). Cocky, of course, is the same as cuckoo.
BABY, Five-spot *Nemophila maculata*.
BABY, Pretty *Centranthus ruber* (Spur Valerian) Donegal Grigson. Baby here should probably be Betsy, which is a common name for Valerian, in such forms as Pretty, and Sweet, Betsy.
BABY, Water *Caltha palustris* (Marsh Marigold) Som. Macmillan.
BABY BLUE EYES *Nemophila insignis* (Californian Bluebell) USA Schenk & Gifford.
BABY CAKES *Geranium lucidum* (Shining Cranesbill) Dev. Macmillan.
BABY'S BONNET *Lathyrus odoratus* (Sweet Pea) Som. Macmillan. A perfectly recognisable descriptive name.
BABY'S BREATH *Gypsophila paniculata* A.W.Smith. A fancifully descriptive name, like Maiden's Breath, and fitting into the same category as Cloud Plant and Gauze Flower.
BABY'S CRADLES *Onobrychis sativa* (Sainfoin) Dor. Macmillan.
BABY'S PET *Bellis perennis* (Daisy) Som. Macmillan.
BABY'S PINAFORE *Geranium robertianum* (Herb Robert) Dev. Macmillan. There are also Dolly's Pinafore, from the same county, and Dolly's Apron, Pink Pinafore and Print Pinafore, too.
BABY'S RATTLE 1. *Ajuga reptans* (Bugle) Som. Macmillan. 2. *Melampyrum pratense* (Common Cow-wheat) Som. Macmillan. 3. *Rhinanthus crista-galli* (Yellow Rattle) Som. Macmillan. At least the name is obvious for Yellow Rattle; for Bugle it probably has something to do with the shape of the flowering stalk. For Cow-wheat, who knows? - but there is probably confusion between Cow-wheat and Yellow Rattle in the district.
BABY'S SHOES 1. *Ajuga reptans* (Bugle) Wilts. Dartnell & Goddard. 2. *Aquilegia vulgaris* (Columbine) Som. Macmillan.

BABY'S TEARS 1. *Helxine soleirolii* (Mind-your-own-business) McClintock. 2. *Peperomia rotundifolia* USA Cunningham.

BABYLON, Cypress of *Curcuma longa* (Turmeric) Pomet.

BABYLON WILLOW *Salix alba var. vitellina "pendula"* (Weeping Willow) Barber & Phillips.

BACCA, Boy's 1. *Clematis vitalba* (Old Man's Beard) Hants, Suss. Grigson. 2. *Heracleum sphondyllium* (Hogweed) Dev. PLNN27. Boys used to smoke their porous stalks. There is Gypsy's Bacca, too, as well as Smoking Cane and Smokeweed for Old Man's Beard.

BACCA, Gypsy's 1. *Clematis vitalba* (Old Man's Beard) Suss. Grigson. 2. *Teucrium scorodonia* (Wood Sage) Dor. Macmillan. 3. *Rumex acetosa* (Wild Sorrel) Som. Mabey. Probably for different reasons. Boys used to smoke the porous stems of Old Man's Beard (see Boy's Bacca). Presumably the dried leaves of Wood Sage served as gypsies' tobacco. But sorrel?

BACCA, Tom *Clematis vitalba* (Old Man's Beard) Suss. Parish (Tom Bacca). Cf. Boy's Bacca.

BACCHAR; BACCHARIS *Inula conyza* (Ploughman's Spikenard) E.Ang. Gerard (Baccharis); Halliwell (Bacchar). Possibly Asarabacca is meant; Prior certainly gives Asarabacca for the same plant. Halliwell gives Bacchar for "the herb ladies' glove", presumably this, for that name is recorded.

BACCHARIS, Broom *Baccharis sarothroides*. Sarothroides - like a broom shrub.

BACCOBOLTS; BACCYBOLTS *Typha latifolia* (False Bulrush) W.Miller (Baccobolts); IOW Long (Baccybolts).

BACCY, Poor Man's; BACCY-PLANT *Tussilago farfara* (Coltsfoot) both Som. Macmillan. Bechion, the plant in Dioscorides taken to be coltsfoot, was smoked against a dry cough, and it is still smoked in all herbal tobacco, as it is also in Chinese medicine. Gypsies smoke the dried leaves as beneficial for asthma and bronchitis, and it is also drunk as a tea for coughs.

BACCY LAMBS *Corylus avellana* catkins (Hazel catkins) Som. Macmillan. Baalambs, presumably.

BACHANG MANGO *Mangifera foetida*.

BACHELOR'S BUTTONS 1. *Achillea ptarmica* (Sneezewort) N'thants. A.E.Baker. 2. *Agrostemma githago* (Corn Cockle) B & H. 3. *Aquilegia vulgaris* (Columbine) Wilts. Dartnell & Goddard. 4. *Arctium lappa* burrs (Burs) Dev. Friend.1882. 5. *Bellis perennis* (Daisy) B & H. 6. *Caltha palustris* (Marsh Marigold) Som. Macmillan; Dor. Dacombe; Staffs. Raven. 7. *Centaurea cyanus* (Cornflower) Som. Macmillan; Derb, Yks. Grigson; USA Watt. 8. *Centaurea nigra* (Knapweed) Ire. Grigson. 9. *Centaurea scabiosa* (Greater Knapweed) Glos. B & H. 10. *Cephalanthus occidentalis* (Buttonbrush) Dev. Friend.1882. 11. *Geranium lucidum* (Shining Cranesbill) Lancs. Grigson. 12. *Geranium robertianum* (Herb Robert) Dev. Friend.1882; Som. Grigson. 13. *Gomphrenia globosa* (Globe Amaranth) S.Africa Watt. 14. *Kerria japonica* (Jew's Mallow) Som. Macmillan. 15. *Knautia arvensis* (Field Scabious) Dev. Friend.1882; Som. Grigson; Wilts. Akerman; Glos. J.D.Robertson; Berks. Lowsley. 16. *Lychnis flos-cuculi* (Ragged Robin) Dev, Som, Suss. Grigson. 17. *Melandrium album* (White Campion) Glos. Grigson; Suss, Yks. B & H; Suff. Moor. 18. *Melandrium rubrum* (Red Campion) Som, Dev, Worcs, War, Kent, Ess, Suss, Suff, Lancs Grigson; N'thants. A.E. Baker; Yks. Addy. 19. *Ranunculus aconitifolius fl pl* Gerard. 20. *Ranunculus acris* (Meadow Buttercup) Som. Grigson. 21. *Ranunculus acris fl pl* Dev. Friend.1882. 22. *Ranunculus bulbosus* (Bulbous Buttercup) Som. Grigson. 23. *Ranunculus repens* (Creeping Buttercup) Som. Grigson. 24. *Scabiosa columbaria* (Small Scabious) Wilts. Dartnell & Goddard. 25. *Stellaria holostea* (Greater Stitchwort) Som. Macmillan. Bucks, Suss. Grigson. 26.

Succisa pratensis (Devil's-bit) Glos. B & H; Hants. Cope. 27. *Tanacetum parthenium* (Feverfew) Dev. Friend.1882. 28. *Tanacetum vulgare* (Tansy) Som. Macmillan. 29. *Umbilicus rupestris* (Wall Pennywort) Dev. Friend.1882. 30. *Vinca minor* (Lesser Periwinkle) Som. Macmillan. Are they some divination flowers? Or roughly button-shaped? Perhaps it is simply a case of young men wearing them in their buttonholes to signal their availability to the girls.

BACHELOR'S BUTTONS, Billy *Stellaria holostea* (Greater Stitchwort) War. Grigson.

BACHELOR'S BUTTONS, Little *Geranium robertianum* (Herb Robert) Dev. Friend; Suss. Tynan & Maitland.

BACHELOR'S BUTTONS, White 1. *Melandrium album* (White Campion) Suff. B & H. 2. *Ranunculus aconitifolius fl pl* Ayr. B & H.

BACHELOR'S BUTTONS, Wild *Polygala lutea* (Orange Milkwort) USA House.

BACHELOR'S BUTTONS, Yellow 1. *Polygala lutea* (Orange Milkweed) USA Howes. 2. *Ranunculus acris fl pl* Ayr B & H.

BACHELOR'S PEAR *Solanum mammosum* (Amoise) Barbados Gooding. No explanation given, but it is grown for "love-magic" in Haiti, and the connection may lie there.

BACK, White *Populus alba* (White Poplar) E.Ang. Grigson. It is the backs of the leaves that show up white, hence the name.

BACK-TO-BACK *Viola tricolor* (Pansy) Som. Macmillan.

BACKACHE ROOT *Liatris spicata* (Button Snakeroot) Le Strange. It is known as Colic Root, too. The infused roots are still being prescribed by herbalists as a diuretic.

BACKWORT *Symphytum officinale* (Comfrey) B & H. "The slimy substance of the root made in a posset of ale, and given to drinke against the paine in the backe gotten by any violent motion, as wrestling, or overmuch use of women, doth in foure or five daies presently cure the same..." (Gerard). In spite of Gerard's claim, this is assuredly not backwort, but blackwort. Both this and Blackroot are old names, Blackwort old enough to appear in Turner. And Cf. the German Schwarzwurz.

BACLIN *Bidens cernua* (Nodding Bur-Marigold) Hants. B & H.

BACON *Rosa canina* the young shoots (Dog Rose) Som. Macmillan. Yorkshire children pick the shoots, peel off the skin, and eat them. They called it Bread-and-Cheese Tree. Somerset children obviously used to do the same.

BACON, Eggs-and- 1. *Linaria vulgaris* (Toadflax) Dev. Friend.1882; Som. Tongue; Wilts. Dartnell & Goddard; Glos. J.D.Robertson. 2. *Lotus corniculatus* (Bird's-foot Trefoil) Som, Glos, Suss, Rut. Grigson; N'thants A.E.Baker. 3. *Ranunculus arvensis* (Corn Buttercup) Ches. Grigson. 4. *Ranunculus heterophyllus* (Various-leaved Buttercup) Som. Macmillan. 5. *Ranunculus bulbosus* (Bulbous Buttercup) Ches. Grigson. 6. *Rubus caesius* the stalks (Dewberry) Som. Macmillan. 7. *Solanum tuberosum* flowers (Potato) Som. Macmillan.

BACON-AND-EGGS 1. *Linaria vulgaris* (Toadflax) Wilts. Dartnell & Goddard. 2. *Narcissus odorus* (Jonquil) Som. Macmillan. 3. *Ranunculus aquatilis* (Water Crowfoot) Som. Macmillan. Double names like this usually refer to two-coloured flowers, and in this case is particularly apt for Toadflax. Water Crowfoot is not so obvious, and Jonquil may very well be a misquotation for Pheasant's Eye.

BACON-WEED; BEACONWEED *Chenopodium album* (Fat Hen) Dor. Macmillan (Baconweed); Dor. Barnes (Beaconweed). Pigweed is another name, one which has spread to Canada, where according to Grieve.1931, it is grown as food for pigs and sheep. Gerard though said "it is reported that it killeth swine if they do eat thereof...". Actually Sowbane does occcur as a name for *Chenopodium rubrum*. Grigson is probably right when he says that such names as these, and Dirtweed, Muckhill-weed etc., merely denote rich, fat land.

BAD; BOD; BAWD *Juglans regia* (Walnut) Wilts. Dartnell & Goddard (Bad, Bod); Som. Halliwell (Bawd). Applied only to the outer shell of the nuts.

BAD MAN'S BREAD *Conopodium majus* (Earthnut) Yks. Grigson. Why should this be the property of the devil? It is perfectly edible, and wholesome.

BAD MAN'S OATMEAL 1. *Anthriscus sylvestris* (Cow Parsley) Yks. Mabey. 2. *Capsella bursa-pastoris* (Shepherd's Purse) Dur. Grigson. 3. *Conium maculatum* (Hemlock) Yks. (Nicholson); Dur, N'thum. Grigson; USA Henkel. The 'bad man' of these names is the devil, so each of these are in some way connected with the devil. Hemlock, being as poison-ous as it is, is obvious, and Devil's Oat and Devil's Flower are recorded from the same area as Bad Man's Oatmeal. Cow Parsley looks something like hemlock, so the connection holds good, particularly as Satan's Bread, Devil's Meal and Devil's Parsley are other names for this. It is suggested that meal, or oatmeal, come from the powdery appearance of the flowers. But what about Shepherd's Purse? There seems no obvious reason for the connection with the devil, though Witches' Pouches occurs as a Scottish name.

BAD MAN'S PLAYTHING *Achillea millefolium* (Yarrow) Fernie. The 'bad man' is the devil, and there are other names for yarrow connecting it with the devil - Devil's Nettle, Rattle, and Plaything. Clair suggested that it got the name because it was used in divination and the casting of spells. There certainly were plenty of magical uses of the plant, but nothing particularly baneful about any of them, and the use was in love divination, very widespread and well-known.

BAD MAN'S POSIES *Lamium purpureum* (Red Deadnettle) Yks, N'thum. Grigson. Black Man's Posies, which means exactly the same as this, is recorded in Cumbria.

BADGER, Smell *Geranium robertianum* (Herb Robert) Tynan & Maitland. Cf. Smell Foxes for this.

BADGER'S BANE *Aconitum vulparia* (Wolfsbane) Howes.

BADGER'S FLOWER *Allium ursinum* (Ramsons) Wilts. Dartnell & Goddard.1899. Presumably this is a reference to the smell.

BADGEWORTH BUTTERCUP *Ranunculus ophioglossifolius* (Serpent-tongue Spearwort). It was referred to as such in a news item of The Times 6 June 1994. Badgeworth is a village a mile or so southwest of Cheltenham.

BADIANE *Illicium verum* (True Anise) Grieve.1933.

BADINJAN *Solanum melongena* (Egg Plant) W Indies Howes.

BADMINNIE *Meum athamanticum* (Baldmoney) Scot. Grigson. This is just Scots for baldmoney.

BADMONEY *Gentianella amarella* (Autumn Gentian) Roxb. B & H. Baldemoyne, Baldmoney, Baldemayne, Baldymoney, and Baldwein are all names applied to various gentians, whereas the plant usually known as Baldmoney is *Meum athamanticum*.

BAEL, BEL *Aegle marmelos*.

BAG, Judas's *Adansonia digitata* (Baobab) Howes.

BAG FLOWER *Clerodendron thomsonae* (Bleeding Heart Vine) USA Cunningham.

BAGGER'S NEEDLE *Erodium moschatum* (Musk Storksbill) Worcs. Allies. The 'needle' part of the name is easy enough, and occurs with various first elements. It comes from the shape of the seed vessels, and gave rise to the common name Storksbill. But what about 'bagger'?

BAGPOD *Sesbania vesicaria*. Bladderpod is another name for this American plant, and it explains better the use of the word.

BAGS, Bull's 1. *Orchis mascula* (Early Purple Orchis) Scot. B & H. 2. *Orchis morio* (Green-winged Orchis) Scot. B & H. The tubers of these orchids are in pairs, a new one growing by the old, and the general appearance is of a pair of testicles, hence Bull's Bags in Scotland, and various other names such as Foxstones, Dogstones, Goatstones, etc., around the country.

BAGS, Flapper *Arctium lappa* (Burdock) Scot. B & H.

BAGS, Money 1. *Capsella bursa-pastoris* (Shepherd's Purse) Som. Macmillan. 2. *Impatiens noli-me-tangere* (Touch-me-not) Tynan & Maitland.

BAGS, Shepherd's *Capsella bursa-pastoris* (Shepherd's Purse) Turner. A variation on the common name Shepherd's Purse. Purse becomes bag, pouch, or pounce, scrip, and pocket, all with the same idea, and all deriving from the shape of the seed pods. Who says that country people of a few centuries ago had no imagination?

BAHAMA HEMP *Agave sisalina* (Sisal) Roys. It was introduced into a number of suitable areas for commercial cultivation from Yucatán, where it is native (Sisal is actually the name of a place there).

BAHAMA TEA *Lantana camara* (Jamaica Mountain Sage) Dalziel.

BAHAMA WHITEWOOD *Canella alba* (White Cinnamon) Howes.

BAHERA NUT *Terminalia bellirica* (Bastard Myrobalan) Chopra.

BAHIA ROSEWOOD *Dalbergia nigra* (Rosewood) Howes.

BAILEY'S MIMOSA *Acacia baileyana* (Cootamundra Wattle) Hora.

BAILEYA, Desert *Baileya multiradiata*.

BAILLIE NICOL JARVIE'S POKER *Kniphofia uvaria* (Red Hot Poker) W.Scot. Coats. According to Alice Coats, after a well-known incident in *Rob Roy*.

BAIRNWORT; BAIRNWORT, Bessy *Bellis perennis* (Daisy) both Yks. Grigson. Presumably these are variations on Banwort. "The Northern men call this herbe a Banwort because it helpeth bones to knyt agayne" (Turner).

BAIYAN-FLOWER *Bellis perennis* (Daisy) Lancs. Grigson. Probably from the same idea as Bairnwort, i.e. Banwort, the 'ban' part of it meaning bone. Like comfrey, it is a con-sound, helping to knit broken bones.

BAJRI *Pennisetum glaucum* (Pearl Millet) India Howes.

BAKEAPPLES; BAKED-APPLE BERRY *Rubus chamaemorus* (Cloudberry) Newfoundland Narváez (Bakeapples); USA Howes (Baked-apple Berry).

BAKER CYPRESS *Cupressus bakeri*.

BAKER LILY *Lilium bakerianum*.

BALD *Meum athamanticum* (Baldmoney) Genders. Just a straight abbreviation of Baldmoney, seeming to accentuate the idea of a dedication to Baldur.

BALD CYPRESS 1. *Taxodium distichum* (Swamp Cypress) A.W.Smith. 2. *Taxodium distichum var. mucronatum* (Mexican Bald Cypress) Mitchell.

BALD CYPRESS, Mexican *Taxodium distichum var. mucronatum*.

BALDAR HERB *Amaranthus hypondriachus* (Prince's Feathers) B & H.

BALDARE *Amaranthus caudatus* (Love-lies-bleeding) Prior. Prince's Feathers is recorded as Baldar Herb. Did Baldare for Love-lies-bleeding originate with Turner? Anything to do with the god Baldur?

BALDEMAYNE *Gentianella amarella* (Autumn Gentian) Dawson. This, as well as Baldemoyne, Baldmoney, Baldwein, etc., are old book names for various gentians.

BALDEMOYNE 1. *Gentianella amarella* (Autumn Gentian) Gerard. 2. *Meum athamanticum* (Baldmoney) Prior. See Baldemayne for variants of old book names for various gentians. As far as *Meum athamanticum* is concerned, this is a simple variant of Baldmoney.

BALDERRY 1. *Dactylorchis incarnata* (Early Marsh Orchid) Scot. Jamieson. 2. *Dactylorchis maculata* (Heath Spotted Orchid) Scot. Jamieson. 3. *Orchis mascula* (Early Purple Orchis) W Scot. B & H.

BALDEYEBROW *Anthemis cotula* (Mayweed) N.Eng. B & H. This is a corruption of Baldur's Brow, which becomes Baldur's Brae in Northumberland - see this

name.

BALDMONEY 1. *Gentiana lutea* (Yellow Gentian). 2. *Gentianella amarella* (Autumn Gentian) Gerard. 3. *Gentianella campestris* (Field Gentian). 4. *Meum athamanticum*. 5. *Peucedanum officinale* (Hog's Fennel) R.T.Gunther. Prior derives baldmoney from Latin valde bona, but Folkard says it comes from the name of the Scandinavian god Baldur. The name as applied to Hog's Fennel is doubtful, for it seems to appear only as a gloss to a 12th century manuscript of Apuleius. Various names derived from Baldmoney, like Badmoney or Baldemoyne, Baldemayne, Baldymoney, Baldwein, are all old book names applying to various gentians.

BALDUR'S BRAE *Anthemis cotula* (Maydweed) N'thum. Thorpe. This is Baldur's Brow, corrupted to Baldeyebrow in other parts of the north country. The name comes from the Scandinavian god Baldur, who was so fair of aspect that light issued from him. Swedish and Danish dialects have the same name, while in Iceland, it is the closely related *Matricaria inodorata* that is known as Baldur's Brow.

BALDWEIN *Gentianella amarella* (Autumn Gentian) B & H. As for Baldmoney.

BALDYMONEY *Gentianella amarella* (Autumn Gentian) B & H. See Baldmoney.

BALEWORT *Papaver somniferum* (Opium Poppy) Cockayne. This was the Saxon name, from balu, or bealu,which meant evil, or mischief. The word exists now solely as baleful or its adverb.

BALKAN MAPLE *Acer hyrcanum*.

BALKAN PINE *Pinus heldreichii*.

BALKAN PLUM *Prunus cerasifera* Anderson. This is the source of slibowitz, the plum brandy made by Balkan peasants.

BALL, Bitter *Tagetes micrantha*.

BALL, Blow *Centaurea cyanus* (Cornflower) Halliwell. Blow means a flower.

BALL, Blue *Succisa pratensis* (Devil's Bit) Suss. B & H.

BALL, Briar *Rosa canina* galls Worcs. J Salisbury. Recorded as a Worcestershire name, but this was the name under which they were sold by apothecaries.

BALL, Canker *Rosa canina* galls Som. Elworthy.

BALL, Dane *Sambucus ebulus* (Dwarf Elder) Som. Grigson. Better known as Dane's Blood, with a number of origin legends of that kind.

BALL, Fir *Pinus sylvestris* cones (Scots Pine cones) Shrop. B & H.

BALL, Gold 1. *Ranunculus acris* (Meadow Buttercup) Som. Grigson. 2. *Ranunculus bulbosus* (Bulbous Buttercup) Som. Grigson. 3. *Ranunculus repens* (Creeping Buttercup) B & H.

BALL, Golden 1. *Buddleia globosa* (Orange Ball Tree) McClintock. 2. *Helianthemum chamaecistus* (Rock Rose) Wilts. Jones & Dillon. 3. *Trollius europaeus* (Globe-flower) Dor. Macmillan; Ches. Holland; Lancs. Grigson. 4. *Viburnum opulus* (Guelder Rose) Som. Elworthy. The first and third are obvious, the fourth seems spurious, and. the second one rests on an observation in Place Names of Wiltshire.

BALL, Honey *Buddleia globosa* (Orange-ball Tree) Som. Elworthy.

BALL, Key *Pinus sylvestris* cones (Pine cone) Dev. B & H.

BALL, Lady's *Centaurea nigra* (Knapweed) Som. Grigson; Wilts. Dartnell & Goddard. OE boll meant any circular body, and the name became applied to knapweed in a number of forms - Gerard has Bullweed, and there are Bullheads (which could be confused with Hardheads if the real derivation was not known), Ballweed, Belweed, Bollwood, etc.

BALL, May *Viburnum opulus* (Guelder Rose) Dor. Dacombe. There is May Rose somewhere, and also King's Crown because it is claimed the King of the May was crowned with it. The Tisty-tosty, or May-tosty names and games are also connected with Guelder Rose.

BALL, Monkey *Strychnos spinosa* fruit (Kaffir Orange) S Africa Dalziel.

BALL, Satin *Calluna vulgaris* (Heather)

Som. Macmillan.

BALL, Silver *Viburnum opulus* (Guelder Rose) Som. Macmillan.

BALL, Sticky 1. *Arctium lappa* burrs (Burrs) Som. Macmillan; Cumb. Grigson. 2. *Galium aparine* (Goose-grass) Som. Macmillan.

BALL, Tossy *Viburnum opulus* (Guelder Rose) Som. Macmillan. Cf. Tisty-tosty, and May-tosty.

BALL, Whitsun 1. *Paeonia mascula* (Peony) Som. Macmillan. 2. *Viburnum opulus* (Guelder Rose) Som. Macmillan. Both have an official connection with Whitsuntide. Guelder Rose is the ecclesiastical symbol for Whitsun, and Peony has a dedication to St.George, as well as the name in Germany of Pentecost Rose.

BALL CLOVER *Gomphrenia caespitosa* USA T H Kearney.

BALL MOSS *Tillandsia recurvata*.

BALL-OF-THE-EARTH *Centaureum erythraea* Vesey-Fitzgerald. Presumably a gypsy name, but a mis-hearing - it should be gall-of-the-earth, which is an ancient name. Cockayne has Earthgall. It is one of the 'bitter herbs', and was once used like ground ivy, to make the beer keep.

BALL-TREE, Orange *Buddleia globosa*.

BALLAGAN *Lapsana communis* (Nipplewort) Ayr. Grigson. This comes from the Scottish word bolga, meaning a swelling. Bolgan-leaves is another Scottish name for the plant.

BALLAM; BELLUM *Prunus domestica institia* fruit (Bullace) Som. Macmillan.

BALLFLOWER *Trollius europaeus* (Globe Flower) Scot. Tynan & Maitland.

BALLHEAD ONION *Allium sphaerocephalum* (Round-headed Leek) Douglas.

BALLNUT *Juglans regia* nut (Walnut) Fernie. Probably a variant of Bannut.

BALLOCK-GRASS, BALLOCKS *Orchis mascula* (Early Purple Orchis) both B & H. Testicles, of course, from the appearance of the root tubers. There are many other names reflecting this.

BALLOCKS, Fool's *Orchis morio* (Green-winged Orchis) Lyte. There is also Fool's Stones. 'Fool's', perhaps,because anyone who believed the dictum, quoted as late as Hill's herbal, "it is supposed to be a strengthener of the parts of generation, and a promoter of venereal desires" was a fool.

BALLOCKS, Hare's *Himantoglossum hircinum* (Lizard Orchid) Lyte. The *hircinum* part of the botanical name links this with goats, rather than lizards, or hares. In fact, Goat's Cullions and Goat Stones are recorded for the plant; both cullions and stones mean the same as ballocks - testicles (root tubers).

BALLOON FLOWER *Platycodon grandiflorum*.

BALLOON VINE *Cardiospermum halicacabum* (Blister Creeper) USA Cunningham.

BALLOXE *Orchis mascula* (Early Purple Orchis) B & H. A variation of the usual spelling, presumably archaic, for the word meaning testicles, so applied because of the shape of the root tubers.

BALLWEED *Centaurea nigra* (Knapweed) B & H. OE boll meant any circular object, and it is reflected here in anything from Ballweed to Bullweed.

BALM *Melissa officinalis*. Balm is an abbreviation of balsam.

BALM, Abraham's *Vitex agnus-castus* (Chaste Tree) Leyel.1937.

BALM, Basil 1. *Acinos arvensis* (Basil Thyme) B & H. 2. *Monarda clinopodia* USA Zenkert.

BALM, Bastard *Melittis melissophyllum*.

BALM, Bee 1. *Melissa officinalis*. 2. *Monarda didyma* (Red Bergamot) New Engl. Sanford. Melissa means bee, and Gerard has "The hives of Bees being rubbed with the leaves of Bawme, causeth the bees to keep together, and causeth others to come unto them...", a belief still current in East Anglia, where they say that if this grows in the garden, the bees will not leave the hive. In Wiltshire, too, beekeepers rub the inside of the skeps with it. They call the bergamot Bee

Balm in America because the bees are so fond of its blossoms.

BALM, Bee, Mintleaf *Monarda fistulosa* (Purple Bergamot) USA Elmore.

BALM, Bee, Pony *Monarda pectinata* (Horse Mint) USA Elmore.

BALM, Calamint *Calamintha ascendens* (Calamint) Leyel.1937.

BALM, Field 1. *Calamintha nepeta* (Lesser Catmint) B & H. 2. *Glechoma hederacea* (Ground Ivy) USA Sanecki.

BALM, Honey *Melittis melissophyllum* (Bastard Balm) McDonald.

BALM, Indian *Trillium erectum* (Bethroot) Grieve.1931. An American plant, which the Indians used for a number of medicinal purposes.

BALM, Lemon *Melissa officinalis* (Balm) New Engl. Sanford. The leaves have got a very marked lemon fragrance.

BALM, Moldavian *Dracocephalum moldavicum*. A native of Moldavia and Siberia, and a regional substitute for true balm.

BALM, Molucca *Molucella laevis*.

BALM, Mountain *Calamintha ascendens* (Calamint) H.N.Webster.

BALM, Ox *Collinsonia canadensis* (Stone Root) Le Strange.

BALM, Sweet *Melissa officinalis* (Balm) H.N.Webster.

BALM, Tea *Melissa officinalis* (Bee Balm) Howes. A mistake for Bee Balm, perhaps? Not necessarily - balm tea has always been a favourite beverage.

BALM GENTLE *Melissa officinalis* (Balm) Turner.

BALM-LEAVED ARCHANGEL 1. *Lamium orvala* (Balm-leaved Deadnettle) Whittle & Cook. 2. *Melittis melissophyllum* (Bastard Balm) Camden/Gibson.

BALM-LEAVED FIGWORT *Scrophularia scorodonia*.

BALM MINT 1. *Mentha longifolia* (Horse Mint) Turner. 2. *Mentha piperita* (Peppermint) Hatfield.

BALM OF GILEAD 1. *Cedronella triphylla* Moldenke. 2. *Commiphora opobalsamum*. 3. *Melittis melisophyllum* (Bastard Balm) Wilts. Dartnell & Goddard. 4. *Populus x candicans* (Balsam Poplar) USA Brownlow. 5. *Populus trichocarpa* (Western Balsam Poplar) USA Mason.

BALM OF GILEAD FIR *Abies balsamea* (Balsam Fir) Grieve.1931. The blisters on the bark are a source of Canada balsam, or the aromatic needles are often made into balsam pillows. The balsam itself was once popular in domestic medicine as a stimulant and diuretic, and as an application to sore nipples, but is better known because it has the same refractive index as glass, so is used for cementing glass in optical instruments.

BALM OF THE WARRIOR'S WOUND *Hypericum perforatum* (St John's Wort) Som. Macmillan. This is a result of the doctrine of signatures. The glandular dots on the leaves look like perforations when held up to the light; the "perforations" were like wounds, and the plant on the strength of this became a wound plant.

BALM TREE *Viburnum lantana* (Wayfaring Tree) Wilts. Goddard.

BALMONY *Chelone glabra* (Turtlehead) USA House. Is this the same word as as baldmoney, transferred to America?

BALSA *Ochroma lagopus*. Balsa is Spanish for a raft or float. The famous rafts of Lake Titicaca are made from balsa-wood, hence the name is transferred to the plant.

BALSAM; BALSAMINE *Impatiens noli-me-tangere* (Touch-me-not) both Prior. Balsam is Latin balsamum, Greek balsamon, possibly of some Semitic origin (the Arabic is balasan). In the European languages it is the same as balm when the word is used for a medicament. Otherwise, it is applied to any plant of the genus *Impatiens*.

BALSAM, Canadian *Abies balsamea* (Balsam Fir) Barber & Phillips.

BALSAM, Field, Old *Gnaphalium polycephalum* (Mouse-ear Everlasting) O.P.Brown.

BALSAM, Garden *Melilotus caerulea* (Swiss Melilot) B & H.

BALSAM, Himalayan; BALSAM, Indian *Impatiens glandulifera* Howes (Indian).
BALSAM, Jericho *Balanites aegyptiaca*. The original Balm of Gilead, it is claimed.
BALSAM, Jumbie *Ocymum micranthum* Trinidad Laguerre. Jumbie is a ghost, or a spirit of some kind.
BALSAM, Kentish *Mercurialis perennis* (Dog's Mercury) Kent Grigson. Presumably for the same reason as Bristol Weed in that city for the same plant, and Wiltshire Weed in that county for elm (when there used to be some) - i.e. ironic - there are far too many of them growing. The balsam part of the name is probably ironic too; it is a poison, hence the "dog".
BALSAM, Mecca *Commiphora opobalsamum* (Balm of Gilead) Howes.
BALSAM, Orange *Impatiens capensis*.
BALSAM, Peruvian *Myroxylon balsamum var. pereirae*. It is the resin which is the balsam, very useful, too.
BALSAM, Seaside *Croton eleutheria* (Sweet-wood) Howes.
BALSAM, She *Abies fraseri* (Fraser's Balsam Fir) Brimble.1948.
BALSAM, Small *Impatiens parviflora*.
BALSAM, Water *Impatiens glandulifera* (Himalayan Balsam) Genders.1976.
BALSAM, White *Gnaphalium polycephalum* (Mouse-ear Everlasting) O.P.Brown.
BALSAM, Wild 1. *Impatiens capensis* (Orange Balsam) Grieve.1931. 2. *Impatiens glandulifera* (Himalayan Balsam) Sanecki.
BALSAM, Yellow *Croton balsamifer*.
BALSAM APPLE *Lycopersicon esculentum* (Tomato) Turner. This happens to be Turner's name, but when they first appeared in England, they were called apples of all kinds, Gold, Love (by Gerard), Amorous (by Parkinson), etc. The last two mentioned were named because it was thought that they had aphrodisiac qualities.
BALSAM APPLE, Wild *Momordica charantia*. It is also called Balsam Pear in the USA. The fruits are toxic, but at least in Haiti the plant is used for fevers, malaria, headaches and worms, so perhaps balsam could just be applied.
BALSAM-BOG *Azorella caespitosa*.
BALSAM FIR *Abies balsamea*. Balsam here because this is the source of Canada Balsam which is apparently taken from blisters on the bark. At one time it was a very popular American domestic medicine, used as a stimulant and diuretic, and as an application to sore nipples. But it is best known because it has the same refractive index as glass, so is used for cementing glass in optical instruments.
BALSAM FIR, Fraser's *Abies fraseri*.
BALSAM FIR, Southern *Abies fraseri* (Fraser's Balsam Fir) Brimble.1948.
BALSAM HERB *Tanacetum balsamita* (Balsamint) Culpepper.
BALSAM PEAR *Momordica charantia* (Wild Balsam Apple) USA Kingsbury.
BALSAM POPLAR *Populus x candicans*. Although it has been said that it was called Balsam Poplar from the scent of the leaves and catkins, the American Indians recognized a true balsam, which they made from the resinous buds that are rich in salicin. The salve was used not only for dressing wounds, but for putting up the nostrils to cure a cold in the head. The salicin probably decomposes into salicylic acid in the human system, so the effect would be similar to that of aspirin.
BALSAM POPLAR, American *Populus balsamifera*. The buds are used in just the same way as those of *P candicans*. See Balsam Poplar.
BALSAM POPLAR, Eastern *Populus balsamifera* (American Balsam Poplar) Mabey.1977.
BALSAM POPLAR, Western *Populus trichocarpa*. As with *P candicans*, Balm of Gilead is a name applied in America to this.
BALSAM-ROOT *Balsamorrhiza incana*.
BALSAM SPURGE *Euphorbia balsamifera*.
BALSAM TREE *Copaifera mopane* (Mopane) Palgrave.
BALSAM TREE, Euphrates *Populus euphratica* (Euphrates Poplar) Moldenke.

BALSAMEA, Mexican *Zauschneria californica* (Californian Fuchsia) USA Barrett & Gifford.

BALSAMINT *Tanacetum balsamita*. Culpepper called this Balsam Herb, and certainly an ointment is made from it for burns, bruises and skin troubles.

BALSAMWEED *Impatiens capensis* (Orange Balsam) Grieve.1931. Wild Balsam is also applied to this. It is an American plant which has established itself in parts of Britain. It grows like a weed along the banks of the Thames at Kew, and on the banks of the Wey.

BALTIC YELLOW DEAL; BALTIC REDWOOD *Pinus sylvestris* (Scots Pine) Hutchinson & Melville. A lot of timber from this species had been imported from the Continent under names like this. Cf. Danzig Pine, etc.

BAMBOO *Arundinaria sp*. The name seems to derive from a Malay word mambu, which was used both in Portuguese and English in the 17th and 18th centuries. But early forms (from the 16th century) were Dutch bamboes, and thence German Bambus; these were taken wrongly as plural forms so a spurious singular was introduced, and it stuck. (French bambou, Spanish and Portuguese bambu, Italian bambù, and of course English bamboo).

BAMBOO, Japanese; BAMBOO, Mexican *Polygonum cuspidatum* (Japanese Knotweed) Gorer (Japanese); A.W.Smith (Mexican).

BAMBOO, Mountain *Arundinaria alpina*.

BAMBOO-LEAVED OAK *Quercus myrsinaefolia*.

BAME *Melissa officinalis* (Balm) Som. Macmillan; Berks. Lowsley. That is the way that balm is pronounced in the area.

BANADLE; BANATHAL *Cytisus scoparius* (Broom) Corn. Jago (Banathal); Wales Grigson (Banadle). Obviously related to Cornish Bannal and Bannell, but meaning unknown. It has been suggested that this is the same as bundle - the bundle of heather or broom twigs tied round a stick to serve as a broom.

BANANA *Musa x paradisiaca*. Given in the 16th century as the native name in the Congo, though some have said it is related to Arabic banan, fingers, and banana, finger or toe (this may be sheer coincidence, though). The specific name may owe its existence to the opinion current in the 16th and 17th century that the banana was the tree of the knowledge of good and evil in the garden of Eden. Gerard reported that the Jews "suppose it to be that tree, of whose fruit Adam did taste", and he named it Adam's Apple Tree in consequence.

BANANA, Abyssinian; BANANA, Ethiopian *Ensete ventricosum*.

BANANA, Bush *Marsdenia australis*.

BANANA, Canary *Musa nana*.

BANANA, Cavendish *Musa nana* (Canary Banana) Hutchinson & Melville.

BANANA, Chinese *Musa nana* (Canary Banana) Dalziel.

BANANA, Cooking *Musa x paradisiaca*.

BANANA, Dessert *Musa x paradisiaca* (Banana) H & P.

BANANA, Dwarf *Musa nana* (Canary Banana) Dalziel.

BANANA, Edible *Musa x paradisiaca* (Banana) H & P.

BANANA, False *Ensete edulis*. False, because the fruit is not edible. But the plant is a staple subsistence crop of such people as the Gurage of S.W.Ethiopia, which area is known as the Ensete culture complex. It is the root, leaf-stem and inner bark which provide the food substance, which by various processes is made into a flour-paste to bake into bread. But virtually all the plant is used in some way, as insulation, fuel, wrapping, as well as medicinally and ritually.

BANANA, Flowering *Musa coccinea*, or *Musa ornata*.

BANANA, Japanese *Musa basjoo*.

BANANA, Ornamental *Musa x paradisiaca* 'Vittata'.

BANANA, Scarlet *Musa coccinea* (Flowering Banana) RHS.

BANANA-PASSION FRUIT *Passiflora mollissima.*
BANANA SHRUB *Magnolia figo* Perry.
BANANA YUCCA *Yucca baccata* (Datil) USA Yanovsky. The fruits are rather like bananas, in appearance as well as in texture - another name is Fleshy-fruited Yucca.
BAND PLANT *Vinca major* (Greater Periwinkle) Wales B & H. Really a translation of the Latin name. Vinculum means a band, and Vinca is derived from this.
BANEBERRY *Actaea spicata* (Herb Christopher) B & H. Because the berries are poisonous. The effects are severe, but seldom fatal - quickening of the heart action, with gastro-enteritis and dizziness, lasting for about three hours. European names, which translate something like troll's berry, or witches' or devil's berry, are given for the same reason.
BANEBERRY, Red *Actaea rubra.*
BANEBERRY, Western *Actaea arguta.*
BANEBERRY, White *Actaea alba.*
BANEBIND *Convolvulus arvensis* (Field Bindweed) Prior. Bane is a word which survives in the name of poisonous plants (fleabane, wolf's bane, etc), so in one sense it is not appropriate here. In another sense, coupled with bind, it is entirely appropriate.
BANEWORT 1. *Atropa belladonna* (Deadly Nightshade) Som. Macmillan. 2. *Bellis perennis* (Daisy) Yks, Cumb, N'thum. B & H. 3. *Ranunculus flammula* (Lesser Spearwort) Gerard. 4. *Ranunculus lingua* (Greater Spearwort) B & H. 5. *Ranunculus sceleratus* (Celery-leaved Buttercup) B & H. Bane is poison, applicable to Deadly Nightshade and to all the Ranunculi, which are acrid and cause blistering. Prior says that Banewort was applied to the Spearwort because it was thought harmful to sheep, an accurate statement, for Salmon says the harm it does them is to ulcerate the entrails. Daisy is the odd one out. The older form of Banewort is in this case Banwort, and here it means bone - "The Northern men call this herbe a Banwort because it helpeth bones to knyt agayne" (Turner).
BANJO FIG *Ficus lyrata* (Fiddle-back Fig) Rochford. There seems to be a heap of difference between banjoes, fiddles and lyres. However, there must be something about this West African variety which binds them together.
BANK CRESS *Sisymbrium officinale* (Hedge Mustard) Prior.
BANK THISTLE 1. *Carduus nutans* (Musk Thistle) B & H. 2. *Cirsium vulgare* (Spear Plume Thistle) B & H. Both of these thistles have the name Buck Thistle in Yorkshire. Which came first?
BANK THYME *Thymus drucei* (Wild Thyme) Berks. Grigson.
BANNAL; BANNELL *Cytisus scoparius* (Broom) both Corn. (Bannal) B & H; (Bannell) Jago. See also Banadle, Banathal. Bannell would be better with one 'l', for the word is Cornish banal, broom flower or besom.
BANNER, Dog *Anthemis cotula* (Maydweed) Yks. Grigson. Banner is probably not the same as the application to the daisy (bone). Dog Binder comes from the same area. The 'dog' part of it refers to the bad smell.
BANNER, Golden *Thermopsis rhombifolia* (False Lupin) USA Kingsbury.
BANNERGOWAN *Bellis perennis* (Daisy) Dumf. Grigson. The gowan part of the name is typically North Country and Scottish, and is the same as gule, or gold (i.e. the gold in Marigold). In actual fact, it seems to be applied to any daisy-like flower, irrespective of colour. The 'banner' part of the name is probably related to the word ban, bone - see Banwort.
BANNET-TREE; BANNUT-TREE *Juglans regia* (Walnut) both Glos. Grose (Bannet); B & H (Bannut); Bannut also Ches. Holland. Bannuts for the nuts themselves are more widespread; they occur in Glos, Heref, Shrop. B & H; Wilts. Akerman; Som. Jennings. This seems to have been the usual word in the West Country, walnut being reserved for the timber.

BANNOCK, Bird's *Oxalis acetosella* (Wood Sorrel) Scot. Tynan & Maitland. Cf. Cuckoo's Bread, etc.

BANNOTT; BANNIT *Juglans regia* (Walnut) Mon. Waters (Bannott); Worcs. J Salisbury (Bannit). Simple variations on the more usual Bannut, or even the compilers' guess at the spelling.

BANQUET HERB *Curcuma zedoaria* (Zedoary) Mitton. This was a very old book name, and refers to the use of a root preparation to ease flatulence and to aid digestion.

BANWOOD; BANWORT; BARNWORT *Bellis perennis* (Daisy) Yks. Grigson (Banwood); N'thum. B & H (Banwort); Halliwell (Barnwort). Ban here means bone - daisy was a consound (consolida), a bone knitter like comfrey (which is actually recorded as a name for the daisy). "The Northern men call this herbe a Banwort because it helpeth bones to knyt agayne" (Turner).

BANWOOD, Bessie *Bellis perennis* (Daisy) Yks. Grigson.

BANYAN *Ficus benghalensis*.

BANYAN, Australian *Ficus macrophylla* (Moreton Bay Fig) RHS.

BANYAN, Benjamin *Ficus benjamina* (Weeping Fig) Hora.

BAOBAB *Adansonia digitata*. Derivation unknown, probably from some African dialect.

BAOBAB, Australian *Adansonia gregori*.

BAOBAB, Baby *Adenium multiflorum* (Star of the Sabi) Palgrave.

BAR-ROOM PLANT *Aspidistra elatior* (Aspidistra) H & P. The plant is robust enough even for this environment.

BARBADINE *Passiflora quadrangularis* (Grenadilla) Howes.

BARBADOS, Curse of *Lantana camara* Nielsen.

BARBADOS ALMOND *Terminalia catappa* (Myrobalan) Howes.

BARBADOS ALOE *Aloë vera* (Mediterranean Aloë) Simmons. The plant was introduced into the West Indies, where the main production of aloine, better known as bitter aloes, is concentrated.

BARBADOS CEDAR *Juniperus bermudiana* (Bermuda Cedar) Howes.

BARBADOS CHERRY 1. *Eugenia uniflora* (Surinam Cherry) Dalziel. 2. *Malpighia glabra*.

BARBADOS LILAC *Melia azedarach* (Chinaberry Tree) Howes.

BARBADOS LILY *Hippeastrum equestre*.

BARBADOS NUT *Jatropha curcas*.

BARBADOS PRIDE *Caesalpina pulcherrima*.

BARBADOS SNOWDROP *Zephyranthes tubispatha*.

BARBAREUS *Berberis vulgaris* (Barberry) Mandeville.

BARBARY; BARBARYN *Berberis vulgaris* (Barberry) Ire. Wilde.1890 (Barbary); Halliwell (Barbaryn).

BARBARY GUM *Acacia gummifera* (Mogador Acacia) Hora.

BARBARY SQUASH *Cucurbita moschata* (Crook-neck Squash) Wit.

BARBATON DAISY *Gerbera sp.*

BARBER'S BRUSHES *Dipsacus fullonum* (Teasel) Som, Dor. Macmillan; Wilts. Dartnell & Goddard; Ess. Grigson. A descriptive name from the shape of the flower heads.

BARBERRY *Berberis vulgaris*. Berberis became in medieval Latin barbaris, the last part of the name soon becoming berry.

BARBERRY, Allegheny; BARBERRY, American *Berberis canadensis* Le Strange (American).

BARBERRY, Ash *Mahonia aquifolium* (Oregon Grape) Gorer.

BARBERRY, Asiatic *Berberis asiatica* (Indian Barberry) W.Miller.

BARBERRY, Awned-leaved *Berberis aristata* (Nepal Barberry) W.Miller.

BARBERRY, Californian *Berberis pinnata*.

BARBERRY, Colorado *Berberis fendleri*.

BARBERRY, Darlahad *Berberis aristata* (Nepal Barberry) Le Strange.

BARBERRY, Darwin's *Berberis darwinii*.

BARBERRY, Desert *Berberis fremontii*.

BARBERRY, Golden *Berberis stenophylla*.
BARBERRY, Holly-leaved *Mahonia aquifolium* (Oregon Grape) Grieve.1931.
BARBERRY, Indian 1. *Berberis aristata* (Nepal Barberry) Chopra. 2. *Berberis asiatica*. Both Himalayan species.
BARBERRY, Japanese *Berberis thunbergii*.
BARBERRY, Narrow-leaved *Berberis stenophylla* (Golden Barberry) W.Miller.
BARBERRY, Nepal *Berberis aristata*. A Himalayan species.
BARBERRY, Ophthalmic *Berberis aristata* (Nepal Barberry) Le Strange.
BARBERRY, Red-fruited *Berberis haematocarpa*.
BARBERRY, Tanner's *Berberis aristata* (Nepal Barberry) Chopra.
BARBINE *Convolvulus arvensis* (Field Bindweed) Shrop. B & H. The 'bar-' in this case is probably the same element as that in barley (OE bere). The reason then becomes obvious.
BARBORANNE *Berberis vulgaris* (Barberry) B & H. An old book name, of the kind that produced Barbareus, too.
BARDANA; BARDANE *Arctium lappa* (Burdock) Clair (Bardana); USA Henkel (Bardane). Bardane is the French name for burdock.
BARDFIELD OXLIP *Primula elatior* (Oxlip) Genders.1976. Woods near that Essex village are one of the very few places where Oxlip is still found.
BARFOOT, BARFOOT, He- *Helleborus foetidus* (Stinking Hellebore) War. B & H. As distinct from She-Barfoot, which is Green Hellebore. It is of course Bear's Foot, which is an old name still in use in parts, and given from the shape of the leaf.
BARFOOT, She- *Helleborus viridis* (Green Hellebore) War. Grigson. As distinct from He-barfoot (*Helleborus foetidus*).
BARGEMAN'S CABBAGE *Brassica campestris* (Field Mustard) Bucks. B & H. For it was that part of Buckinghamshire by the river Thames where the name was used.
BARILLA PLANT *Salsola kali* (Saltwort) USA Cunningham.
BARK-CLOTH TREE 1. *Antiaris africana*. 2. *Ficus natalensis* (Natal Fig) Wilkinson.1978. Like most of the figs, this is a fibre tree; it is the inner bark of Antiaris africana that is made into bark-cloth in Ashanti.
BARLEY *Hordeum vulgare*.
BARLEY, Dutch *Hordeum vulgare* (Barley) Turner.
BARLEY, Meadow *Hordeum secalinum*.
BARLEY, Mouse *Hordeum murinum* (Wall Barley) A S Palmer. The result of confusion between specific names. *Murinum* means 'of a mouse', and *murale* means 'of a wall'.
BARLEY, Sea *Hordeum marinum*.
BARLEY, Tabor *Hordeum spontaneum*.
BARLEY, Two-rowed *Hordeum distichum*.
BARLEY, Wall 1. *Hordeum murinum*. 2. *Lolium perenne* (Rye-grass) Lyte. 3. *Lolium temulentum* (Darnel) Coles.
BARLEY, Wheat *Hordeum vulgare* (Barley) Turner. Because the variety he was talking about "hath no mo Huskes on it than Wheat".
BARLEY, Wild *Hordeum murinum* (Wall Barley) Howes.
BARM; BARM-LEAF *Melissa officinalis* (Balm) Wilts. Wiltshire; W.Miller (Barmleaf). The word is spoken with a pronounced "r", as well as being written this way.
BARNABAS *Centaurea solstitialis* (St.Barnaby's Thistle). B & H. Barnabas is given because this was supposed to bloom about St.Barnabas' Day (11 June), though the usual flowering time is about a month later. Old beliefs die hard, though, as the name Midsummer Thistle shows.
BARNET; BARNUT *Juglans regia* (Walnut) Mon. Waters. Bannut is more usual.
BARNYARD ASTER *Anthemis cotula* (Maydweed) USA Tolstead.
BARNYARD GRASS *Echinochloa crus-galli frumentacea* (Japanese Barnyard Millet) USA Cunningham.
BARNYARD MILLET, Japanese *Echinochloa crus-galli frumentacea*.
BARON'S LADY'S FLOWER *Myrrhis*

odorata (Sweet Cicely) Kinc. H Gordon. The story is that the wife of the 5th Baron of Kincardine asked to be buried in the soil of her native Lochaber. Accordingly, soil was brought from her home, and it must have contained seeds of Sweet Cicely. Even now, 400 years later, clumps of it grow there, and nowhere else in Kincardineshire.

BARON'S MERCURY *Mercurialis annua* (French Mercury) W.Miller. Given to the male plant only.

BARRA BEAN *Entada phaseoloides* (Nicker Bean) Hebr. Duncan. For the capsules are sometimes washed up on the Atlantic shore of the Hebrides.

BARREN HEMP *Cannabis sativa* (Hemp) Lyte. Applied only to the male plant.

BARREN-WORT *Epimedium alpinum*. The name was coined by Gerard - "I have thought good to call it Barrenwort in English; because... being drunke it is an enemie to conception". It was once thought to produce no flowers, and it seldom does in full sunlight - another reason for the Barrenwort. "The people in the north give milk in which the roots have been boiled, to the females of the domestic animals, when they are running after the males, and they say it has the certain effect of stopping the natural emotions" (Hill).

BARROW-ROSE; BURROW-ROSE *Rosa pimpinellifolia* (Burnet-rose) both Pemb. B & H (Barrow); Grigson (Burrow). Barrow or burrow here mean sand-dunes, where the plants would naturally grow.

BARTRAM *Anacyclus pyrethrum* (Pellitory-of-Spain) Halliwell. Cf. Gerard's Bertram. They are both corruptions of pyrethrum.

BARTRAM'S OAK *Quercus x heterophylla*.

BARTSIA, Red *Odontites verna*. Bartsia is the name given to the genus (now split up into Odontites and Parentucellia) by Linnaeus in honour of Johann Bartsch, physician and botanist.

BARTSIA, Viscid *Parentucellia viscosa* (Yellow Bartsia) H C Long.1910.

BARTSIA, Yellow *Parentucellia viscosa*.

BARUS CAMPHOR *Dryobalanops camphora* (Camphor of Borneo) Howes.

BARWEED *Convolvulus arvensis* (Field Bindweed) Som. B & H. 'Bar-', as the first element, is here almost certainly from the same source as the 'bar-' in barley, and occurs for the same plant as Barbine in Shropshire, and as Bearbind elsewhere.

BARWOOD 1. *Baphia nitida* (Camwood) Schery. 2. *Pterocarpus angolensis*. 3. *Pterocarpus soyauxii*. Probably from the strong veining in the timber.

BASAM 1. *Calluna vulgaris* (Heather) Dev. B & H. 2. *Cytisus scoparius* (Broom) Dev. B & H. Basam is besom, i.e. a broom. Both these plants were used for just this purpose.

BASE BROOM *Genista tinctoria* (Dyer's Greenweed) Prior. Prior says that it was so-called because it was a base for dyeing, but it could be a case where base means low. In fact, Low Broom is a Yorkshire name for this plant.

BASE ROCKET *Reseda luteola* (Dyer's Rocket) Prior. The notes for Base Broom apply equally well to this name.

BASFORD WILLOW *Salix basfordiana*.

BASHELS *Taraxacum officinale* (Dandelion) Corn. Tynan & Maitland.

BASIER, BAZIER *Primula auricula* (Auricula) both Lancs. Nodal (Basier); B & H (Bazier). These are both corruptions of the name Bear's Ears, given a very long time ago from the shape of the leaf (hence auricula, too). The names seem to be typically Lancashire - there is a South Lancashire May song whose refrain runs. "The baziers are sweet in the morning of May". But the name Bezors had been recorded in Gloucestershire.

BASIL 1. *Ocymum sp.* 2. *Satureia vulgaris* USA Upham. *Satureia vulgaris* is a species of savory. Basilikos means royal in Greek, and the plant is important in Greek folklore. It is regarded as a prince among plants by the peasants. The name is preserved in French, too - it is known as herbe royale.

BASIL, Abyssinian *Ocymum graveolens*.

BASIL, American *Ocymum americanum*.

BASIL, Bush 1. *Clinopodium vulgare* (Wild Basil) Genders. 2. *Ocymum minimum*.

BASIL, Camphor *Ocymum kilimandscharicum*.

BASIL, Cow 1. *Saponaria officinalis* (Soapwort) S.Africa Watt. 2. *Vaccaria pyramidata* (Cow-herb) B & H. Cow-herb was thought to be good cow-fodder, so there are one or two other cow-names recorded. Soapwort has half a dozen cow names in South Africa, for some reason.

BASIL, Cow, Red *Centranthus ruber* (Spur Valerian) B & H. Cow-feet is another old book name for this.

BASIL, Curled *Ocymum crispum* (Japanese Basil) Parkinson.1640.

BASIL, Duppy *Ocymum micranthum* Barbados Gooding. Duppy is mosquito, and the name Mosquito Bush is also given to this in Barbados. The essential oil is methyl cinnamate, which gives off a pleasant smell, but appears to be repulsive to mosquitoes.(But the word duppy is more usually applied to a corpse).

BASIL, East Indian *Ocymum gratissimum*.

BASIL, Field *Clinopodium vulgare* (Wild Basil) B & H.

BASIL, Garden *Ocymum basilicum* (Sweet Basil) Culpepper.

BASIL, Green, Sweet *Ocymum basilicum* (Sweet Basil) H.N.Webster.

BASIL, Hedge *Clinopodium vulgare* (Wild Basil) Grieve.1931.

BASIL, Hoary 1. *Ocymum americanum* (American Basil) Dalziel. 2. *Ocymum canum*. Are 'canum' and 'americanum' mixed up in this instance?

BASIL, Holy *Ocymum sanctum*. This is the sacred basil of India, a protector from all misfortunes and disease. For a very long time, it has been cultivated in India as tulsi, a plant sacred to Vishnu.

BASIL, Japanese *Ocymum crispum*.

BASIL, Lemon *Ocymum citriodorum*.

BASIL, Lettuce-leaf *Ocymum crispum* (Japanese Basil) Painter.

BASIL, Monk's *Ocymum sanctum* (Holy Basil) W.Miller.

BASIL, Purple-stalked *Ocymum sanctum* (Holy Basil) W.Miller.

BASIL, Rough *Acinos arvense* (Basil Thyme) Parkinson.1640.

BASIL, Shrubby *Ocymum gratissimum* (East Indian Basil) Howes.

BASIL, Stone 1. *Acinos arvense* (Basil Thyme) Gerard. 2. *Clinopodium vulgare* (Wild Basil) B & H.

BASIL, Sweet *Ocymum basilicum*.

BASIL, Wild 1. *Acinos arvense* (Basil Thyme) Parkinson.1640. 2. *Clinopodium vulgare*.

BASIL BALM 1. *Acinos arvense* (Basil Thyme) B & H. 2. *Monarda clinopodia* USA Zenkert.

BASIL-SMELLING MINT *Mentha piperita citrata* Tradescant. Better known as Bergamot Mint, and one of the Bergamots, *Monarda clinopodia*, is known as Basil Balm.

BASIL THYME 1. *Acinos arvense*. 2. *Calamintha ascendens* (Calamint) Grieve.1931. 3. *Calamintha nepeta* (Lesser Catmint) H.N.Webster. 4. *Clinopodium vulgare* (Wild Basil) Fernie.

BASILWEED *Clinopodium vulgare* (Wild Basil) B & H.

BASIN, Blue *Geranium pratense* (Meadow Cranesbill) Som. Grigson. It is also known as Bassinet, or Bassinet Geraniuum. Bassinet came to mean perambulator, but it must originally have been the diminutive of basin. Descriptive of the flowers, of course.

BASIN, Break- *Veronica chamaedrys* (Germander Speedwell) Dev, Som. Macmillan. From the quick falling of the leaves, Grigson says, but that seems doubtful.

BASIN, Fairy or Fairy's 1. *Primula veris* (Cowslip) Som. Macmillan. 2. *Ranunculus acris* (Meadow Buttercup) Dev. Grigson. 3. *Ranunculus bulbosus* (Bulbous Buttercup) Dev. Grigson. 4. *Ranunculus repens* (Creeping Buttercup) Dev. Grigson. Fairy Cups and Fairy Bells give the same idea as far as cowslips are comcerned.

BASIN, Our Lady's *Dipsacus fullonum*

Som. Macmillan. Cf. Venus' Basin.

BASIN, Sugar *Stellaria holostea* (Greater Stitchwort) Som. Macmillan.

BASIN, Venus' *Dipsacus fullonum* (Teasel) Gerard. The upper leaves grow in pairs, and have their bases so grown together as to form a cup, capable of holding dew and rain ("...so fastened that they hold dew and raine water in manner of a little bason" (Gerard)). This water had the reputation of removing freckles, hence this name, given at a time when freckles were considered a very definite impediment to beauty. The water was also supposed to cure warts on the hands.

BASIN SAGEBRUSH *Artemisia tridentata* (Sagebrush) USA Elmore.

BASKET *Plantago lanceolata* (Ribwort Plantain) Wilts. Dartnell & Goddard. For little baskets used to be woven from ribwort stems, much as the Irish used to make them from rushes on Blaeberry Sunday, to put the fruit in.

BASKET, Beggar's *Pulmonaria officinalis* (Lungwort) Som. Macmillan; Ches. B & H.

BASKET, Butter *Trollius europaeus* (Globe Flower) Yks. Grigson.

BASKET, Gold *Aurinia saxatile* (Gold Dust) W.Miller.

BASKET, London 1. *Geum rivale* (Water Avens) Yks. Grigson. 2. *Geum urbanum* (Avens) Yks. B & H. This may have something to do with markets. Samuel Johnson defined a basket-woman as one who sold goods at a market, carrying her wares in a basket. Certainly the roots are aromatic in both species - the names Clove-root and Clovewort attest to that. An American species used to be called Chocolate Root.

BASKET, Old Maid's *Aquilegia vulgaris* (Columbine) Som. Grigson. Possibly a misreading of 'bonnet', but 'basket' serves quite well as a descriptive name.

BASKET, Pedlar's 1. *Cymbalaria muralis* (Ivy-leaved Toadflax) Som. Elworthy. Derb, Lancs. Grigson; Ches. Holland; Yks. Carr. 2. *Saxifraga sarmentosa* (Mother-of-Thousands) Shrop, Ches, Lancs. B & H.

Possibly because Ivy-leaved Toadflax and Mother-of-Thousands are wanderers, like pedlars. In this case, the true Toadflax would be an intrusion.

BASKET, Rattle 1. *Briza media* (Quaking-grass) Som. Grigson.1959. 1. *Pedicularis sylvatica* (Lousewort) Som. Macmillan. The 'rattle' part of the Lousewort name refers to the seed pods; when the seeds become ripe, they can be "rattled" in the pod. There are a number of other names like this. In fact the common name for *P palustris* is Red Rattle. Quaking-grass gets the name not only because of the ever-trembling seed heads, but also because of a connection with coins. Wiltshire children used to be told that if ever the spikelets were still, they would see that they were actually silver sixpences. Yellow Rattle also has coin connections.

BASKET, Shackle; BOX, Shackle *Briza media* (Quaking-grass) Som. Grigson.1959.

BASKET, Shekel 1. *Briza media* (Quaking-grass) Som. Notes and Queries, vol 168, 1935. 2. *Rhinanthus crista-galli* (Yellow Rattle) Dor. Macmillan. A variation on Shackle-basket.

BASKET ASH *Fraxinus nigra* (Black Ash) Grieve.1931. One of the chief basket-making materials of some of the American Indian tribes. The Meskwaki used both the inner bark and the wood. The name Hoop Ash is also given to this tree, and for the same reason.

BASKET OAK 1. *Quercus michauxii*. 2. *Quercus prinus* (Chestnut Oak) Mitchell.1974. Another of the important materials of the American Indians. The Cherokee strip splints from sections of the green log, and then smooth and gauge them with a knife.

BASKET WILLOW *Salix viminalis* (Osier) A.W.Smith. This is the best-known of the basket-making willows.

BASOM; BASOM, Green; BESOM; BESOM, Green *Cytisus scoparius* (Broom) Corn, Dev, Som. Grigson (Basom); Som. Macmillan (Besom); Som. Grigson (Green

Basom); Som. Macmillan (Green Besom). Basom is a variant of the commoner besom, meaning broom, and in the literal sense, this is the derivation of the word.

BASSAM 1. *Calluna vulgaris* (Heather) Dev. Friend. 2. *Cytisus scoparius* (Broom) Dev. B & H. Bassam = besom = broom. Both heather and broom were used as brooms.

BASSEL; BAZELL *Ocymum basilicum* (Basil) Tusser.

BASSINET 1. *Ranunculus acris* (Meadow Buttercup) Som. Macmillan. 2. *Geranium pratense* (Meadow Cranesbill) Coats. Bassinet comes from the French as a diminutive, so means a small basin, or it could be a skull cap, as worn under a helmet. Presumably, in both cases, it is the shape of the flowers which suggested the name. The Cranesbill is also known as Bassinet Geranium.

BASSINET, Brave *Caltha palustris* (Marsh Marigold) Prior. Brave, possibly because the flowers are larger than the buttercups; otherwise the meaning is the same.

BASSINET GERANIUM *Geranium pratense* (Meadow Cranesbill) Coats. See Bassinet.

BASSOM *Calluna vulgaris* (Heather) Dev. Friend. Another spelling of besom.

BASSWOOD 1. *Tilia americana*. 2. *Tilia cordata* (Small-leaved Lime) Lincs. B & H. The bark fibre, or bast, is the ready cordage of most American Indians. Become bass, it is known to all gardeners as the material for tying in plants, and for packing goods. Bast is the original form (OE baest), and the later form came about by suppression of the 't'.

BASSWOOD, White *Tilia heterophylla*.

BAST *Tilia cordata* (Small-leaved Lime) Lincs. B & H. The original form of bass, still recorded in the 19th century in Lincolnshire dialect.

BASTARD ACACIA *Robinia pseudacacia* (Common Acacia) Howes.

BASTARD AGRIMONY 1. *Ageratum conyzoides* (Goatweed) W.Miller. 2. *Aremonia agriminoides*. 3. *Bidens tripartita* (Bur Marigold) Gerard. 4. *Eupatorium cannabinum* (Hemp Agrimony) Gerard.

BASTARD ALMOND *Terminalia catappa* (Myrobalan) Dale & Greenway.

BASTARD ANEMONE *Pulsatilla vulgaris* (Pasque Flower) Gerard.

BASTARD BOX *Polygala chamaebuxus*. Box-leaved Milkwort is another name for this.

BASTARD CEDAR *Melia azedarach* (Chinaberry Tree) Dalziel.

BASTARD CINNAMON *Cinnamonium cassia* (Cassia) Moldenke. Because this is usually looked upon as an inferior kind of cinnamon.

BASTARD CLOVER *Trifolium hybridum* (Alsike Clover) S.Africa Watt. Presumably because, unlike most other clovers, this one can be poisonous to livestock.

BASTARD CRESS *Thlaspi arvense* (Field Pennycress) USA Cunningham.

BASTARD DAFFODIL *Erythronium americanum* (Yellow Adder's-tongue) USA Leighton.

BASTARD DITTANY *Dictamnus fraxinella* (False Dittany) Gerard.

BASTARD ELM 1. *Celtis occidentalis* (Hackberry) USA Vestal & Schultes. 2. *Celtis reticulata* (Western Hackberry) USA Elmore.

BASTARD FLAG *Acorus calamus* (Sweet Flag) Culpepper.

BASTARD FLOWER DE LUCE *Iris pseudo-acarus* (Yellow Iris) Gerard.

BASTARD FOXGLOVE *Maurandya scandens*. It was Curtis who gave it this name; the flower does bear some resemblance to that of a foxglove.

BASTARD HELLEBORE *Helleborus viridis* (Green Hellebore) Lyte.

BASTARD HEMP *Galeopsis tetrahit* (Hemp Nettle) Gerard.

BASTARD INDIGO 1. *Amorpha fruticosa* (Indigo-bush) H & P. 2. *Tephrosia purpurea*.

BASTARD IPECACUANHA *Asclepias curassavica*.

BASTARD JESUIT'S BARK *Croton*

eleutheria (Sweet-wood) Leyel.1937.

BASTARD KILLER, KILL-BASTARD 1. *Juniperus communis* (Juniper) Som. Grigson. 2. *Juniperus sabina* (Savin) Som. Elworthy (Bastard Killer); Yks. B & H (Kill-bastard). The berries of savin are notoriously abortifacient, and it is to this shrub the name properly refers, not to juniper. A reversal of the name, i.e. KILL BASTARD, is also recorded (Yks. B & H) for Savin.

BASTARD MYROBALAN *Terminalia bellirica.*

BASTARD NARCISSUS *Fritillaria meleagris* (Snake's-head Fritillary) Hudson. There are other Daffodil names for this.

BASTARD NIGELLA *Agrostemma githago* (Corn Cockle) Watt. Cf. Gerard's Field, and Wild, Nigella.

BASTARD OLEANDER *Thevetia peruviana* (Yellow Oleander) Chopra.

BASTARD OLIVE *Buddleia saligna* S.Africa Watt.

BASTARD PELLITORY *Achillea ptarmica* (Sneezewort) Tradescant. It was used medicinally as a sort of native substitute for Pellitory-of-Spain. Gerard has "the juice mixed with vinegar and holden in the mouth, easeth much the pain of the tooth-ach.. The herb chewed and held in the mouth, bringeth mightily from the braine slimie flegme like Pellitorie-of-Spaine; and therefore from time to time it hath bin taken for a wilde kinde thereof..." Gerard called it Wild Pellitory, and Culpepper had False Pellitory.

BASTARD PELLITORY-OF-SPAIN *Peucedanum ostruthium* (Masterwort) Gerard. Gerard must have given it this name for the same reason as he called Sneezewort Wild Pellitory.

BASTARD RHUBARB *Thalictrum flavum* (Meadow Rue) Gerard. False Rhubarb, and English Rhubarb, were other names, "which names are taken of the colour, and taste of the roots".

BASTARD RHUBARB, Spanish, Great *Thalictrum aquilegifolium* Gerard.

BASTARD SAFFRON *Carthamus tinctorius* (Safflower) Pomet.

BASTARD SERVICE TREE *Sorbus x hybrida* (*x thuringiana*).

BASTARD TEAK 1. *Butea monosperma*. 2. *Pterocarpus marsupium*.

BASTARD'S FUMITORY *Fumaria bastardii*. Mons. T. Bastard lived 1784-1846.

BAST(E)-TREE *Tilia europaea* (Common Lime) Lincs. C P. Johnson. Another version of bass, orig. bast.

BASTHAG-BWEE *Chrysanthemum segetum* (Corn Marigold) IOM Moore.1924. Manx bastag-vuigh, literally 'yellow basket'.

BASTWORT *Peucedanum officinale* (Hog's Fennel) R.T.Gunther. This appears only on a gloss to a twelfth century manuscript of Apuleius Barbarus.

BAT FLOWER *Tacca chantrieri*.

BAT-IN-WATER *Mentha aquatica* (Water Mint) Halliwell.

BATH, Fairy's *Geum rivale* (Water Avens) Dor. Macmillan.

BATH, Lady-in-the- *Dicentra spectabilis* (Bleeding Heart) Perry.

BATH, Venus' *Dipsacus fullonum* (Teasel) B & H. Gerard calls it Venus' Basin. The upper leaves grow in pairs, and have their bases so grown together as to form a cup, capable of holding dew and rain. This water was said to cure warts on the hands. More to the point, in view of the name being considered, the water was said to be used as a remedy for freckles. Culpepper said that the distilled water of the leaves was often used by women to preserve their beauty.

BATH ASPARAGUS *Ornithogalum pyrenaicum* (Spiked Star of Bethlehem) B & H. It is quite common round Bath, and is apparently still picked in the area and sold to be eaten like asparagus. French Sparrow-grass was the name under which it was sold in the Bath markets.

BATHURST BUR *Xanthium spinosum* (Spiny Cocklebur) Watt. This is a North American plant, but has strayed to many other parts of the world. It is a casual in this country. But the Bathurst of this name is presumably the Australian one (there is a Canadian town of this name too). Certainly the plant has spread to Australia.

BATOKA PLUM *Flacourtia indica* (Kaffir Plum) Palgrave.

BATS-IN-THE-BELFRY *Campanula trachelium* (Nettle-leaved Bellflower) Clapham. Possibly because of the way the flowers are clustered on the branches up the stem.

BATTER-DOCK *Petasites hybridus* (Butterbur) Ches. B & H. Presumably butter-dock. "...because the countrey huswives were wont to wrap their butter in the large leaves thereof" (Coles).

BATTINAN CRAPEMYRTLE *Lagerstroemeria piriformus*.

BATTLER, Cock *Plantago lanceolata* (Ribwort Plantain) Corn. Courtney. The flower heads are used as "soldiers" or "fighting cocks" by children. One child holds out a stem, and his opponent tries to decapitate it with another. The winners are numbered up as with the game of conkers.

BAUHINIA, Orchid *Bauhinia petersiana*.

BAUM-LEAF *Melittis melissophyllum* (Bastard Balm) W.Miller.

BAUME *Mentha spicata* (Spearmint) USA Cunningham.

BAW-TREE *Sambucus nigra* (Elder) Lincs. Peacock. Better known as Bour-tree, but there are a dozen or so variants.

BAWD'S PENNY *Meum athamanticum* (Baldmoney) Jason Hill. A corruption of Baldmoney through the Scottish pronunciation of 'bald'.

BAWDMONEY 1. *Gentianella amarella* (Autumn Gentian) A S Palmer. 2. *Meum athamanticum* (Baldmoney) Prior.

BAWDRINGIE *Meum athamanticum* (Baldmoney) Perth Grigson. A Scottish version of Baldmoney.

BAWM *Melissa officinalis* (Bee Balm) Hatfield. A spelling of local pronunciation. It is often spelt Barm, too.

BAY *Laurus nobilis*. Bay is from Latin bacca, meaning berry. Occasionally, one hears the word Bayberry applied to the berries - a pleonasm - but regularly applied to members of the genus *Myrica*.

BAY, Bull *Magnolia grandiflora*. This is the evergreen American magnolia with the overpowering scent, also known in America as Laurel Magnolia. Whenever the animal names bull, horse, or dog are applied to a name, something big (or sometimes coarse, in a derogatory sense) is implied. The botanical name bears this out - it is *grandiflora*.

BAY, Cherry 1. *Prunus laurocerasus* (Cherry-Laurel) Parkinson.1629. 2. *Prunus lusitanicus* (Portuguese Laurel) Macmillan.

BAY, Dwarf 1. *Daphne laureola* (Spurge Laurel) Gerard. 2. *Daphne mezereon* (Lady Laurel) B & H.

BAY, French *Chamaenerion angustifolium* (Rosebay Willowherb) Sanecki. Apparently given by William Robinson, *The English flower garden*. There is also French Willow and French Saugh for this.

BAY, Laurel *Laurus nobilis* (Bay) Dawson.

BAY, Loblolly *Gordonia lasianthus*.

BAY, Poison *Illicium floridanum*.

BAY, Rose 1. *Nerium oleander*. 2. *Rhododendron maximum*.

BAY, Rose, Californian *Rhododendron macrophyllum*.

BAY, Rose, Lapland *Rhododendron lapponicum*.

BAY, Stainless *Laurus nobilis* (Bay) Som. Macmillan. Presumably it means unspotted, as against the Cherry Laurel which has blotches on the leaves.

BAY, Swamp *Magnolia virginiana* (Southern Sweet Bay) A.W.Smith.

BAY, Sweet 1. *Laurus nobilis* (Bay) W.Miller. 2. *Magnolia virginiana*. The magnolia is also called White Bay in America.

BAY, Sweet, Southern *Magnolia virginiana* (Sweet Bay).

BAY, White *Magnolia virginiana* (Sweet Bay) USA Harper.

BAY, Wild *Viburnum tinus* (Laurustine) Gerard. A name given by Gerard, presumably deriving it from the Laurus part of the name.

BAY, Willow *Salix pentandra* (Bay Willow)

Staffs. Grigson.
BAY CHERRY *Prunus laurocerasus* (Cherry-laurel) Parkinson.1629.
BAY-LAMBS female flowers of *Pinus sylvestris* Yks. B & H. Note that the cones are known as Sheep in the same area.
BAY LAUREL *Laurus nobilis* (Bay) W.Miller.
BAY-LEAVED WILLOW *Salix pentandra*.
BAY OAK *Quercus petraea* (Durmast Oak) Berks. B & H. It probably refers to a local variety only.
BAY RUM TREE *Pimento acris*. Bay Rum is not from the Bay Tree, but from the leaves of this.
BAY WILLOW 1. *Chamaenerion angustifolium* (Rosebay) Gerard. 2. *Salix pentandra* (Bay-leaved Willow) Clapham. For Rosebay, this must be a mixture of Rosebay and Willowherb.
BAY WILLOW, Sweet *Salix pentandra* (Bay-leaved Willow) Brimble.1948.
BAYBERRY 1. *Laurus nobilis* - berries (Bay) B & H. 2. *Myrica pennsylvanicum*. 3. *Myrica gale* (Bog Myrtle) Grieve.1931. 4. *Pimento acris* (Bay-Rum Tree) Prance. These are all berry-bearing, so Bayberry itself must be a pleonasm, for bay derives from the Latin bacca, berry.
BAYBERRY, Californian *Myrica californica*.
BAYBERRY, Northern *Myrica pennsylvanicum* (Bayberry) USA Zenkert.
BAYONET, Spanish 1. *Yucca aloifolia*. 2. *Yucca baccata* (Datil) USA Elmore. 3. *Yucca glauca* (Soapweed) USA Gilmore. The sharp-pointed leaves of all the Yuccas are commemorated in various 'dagger' and 'bayonet' names.
BAYONET PLANT *Aciphylla squarrosa* (Spaniard) RHS. Descriptive - this is a thorny plant.
BAZZIES *Arctium lappa* (Burdock) Kent B & H. The name is given to the flower-heads.
BAZZOCK; BRAZZOCK *Sinapis arvensis* (Charlock) both Yks F K Robinson. These arose from a confusion in medieval documents with brassica. Yorkshire has got Brassock and Brassick, too. Why should the confusion be oonfined to that area?
BE-STILL TREE *Thevetia peruviana* (Yellow Oleander) USA Kingsbury. Possibly because all parts of the tree are highly poisonous.
BEACH ASTER *Erigeron glaucus*.
BEACH CLOVER *Trifolium fimbriatum*.
BEACH HEATHER *Hudsonia tomentosa*.
BEACH PEA *Lathyrus maritimus* (Sea Pea) USA Zenkert.
BEACH PINE *Pinus contorta*.
BEACH PLUM *Prunus maritimus*.
BEACH WORMWOOD *Artemisia stelleriana*.
BEAD-BIND *Tamus communis* (Black Bryony) Hants. Cope. Clearly printed as such, but surely something like Bearbind is meant? Unless, of course, the berries are the 'beads'.
BEAD PLANT *Nertera depressa*. A tiny plant with coral-red berries, hence the name.
BEAD PLANT, Coral- *Abrus precatorius* (Precatory Bean) W.Miller.
BEAD-TREE *Melia azederach* (Chinaberry Tree) Barber & Phillips. Because the berries were often used as beads.
BEAD VINE *Abrus precatorius* (Precatory Bean) Roys. Cf. the names Rosary Pea, and Prayer Beads.
BEADS *Sagina procumbens* (Pearlwort) Wilts. Dartnell & Goddard. A descriptive name, as is Pearlwort.
BEADS, Cuckoo's *Crataegus monogyna* fruit (Haws) Shrop. Grigson.
BEADS, Ladybird *Abrus precatorius* (Precatory Bean) Howes.
BEADS, Prayer *Abrus precatorius* (Precatory Bean) Grieve.1931. The beans are a highly dangerous poison, yet this name, and a corresponding American one, Rosary Pea, shows that they were used for rosary beads.
BEADS, Swine's 1. *Potentilla anserina* (Silverweed) Ork. Grigson. 2. *Scilla verna* (Spring Squill) Ork. Grigson. 3. *Stachys palustris* (Marsh Woundwort) Ork. Grigson. These all refer to the roots; Marsh

Woundwort has small tubers on the rhizomes - these are the "beads". The roots of silverweed have been a marginal food in the islands for centuries, perhaps not so marginal either. The squill is different. Perhaps the bulb was eaten (if only by swine).

BEAGLE *Primula veris* (Cowslip) Cambs. Tynan & Maitland. One of the local variants of the old name, Paigle.

BEAKED HAZEL *Corylus rostrata*.

BEAKED PARSLEY 1. *Anthriscus caucalis* (Bur Chervil). 2. *Anthriscus cerefolium* (Chervil) Barton & Castle. 3. *Anthriscus sylvestris* (Cow Parsley). 4. *Myrrhis odorata* (Sweet Cicely) Leyel.1937.

BEAKED WILLOW *Salix bebbiana*.

BEAKS, Long *Scandix pecten-veneris* (Shepherd's Needle) Dor. Macmillan. Descriptive, as the common name confirms.

BEAM TREE *Sorbus aria* (Whitebeam) W.Miller. This is a pleonasm, for beam is OE beam, tree.

BEAN, Adzuki *Phaseolus angularis*.

BEAN, Arber *Phaseolus vulgaris* (Kidney Bean) Turner. "because they serve to cover an arber for the tyme of Summer".

BEAN, Asparagus 1. *Psophocarpus tetragonolobus* (Goa Bean) Brouk. 2. *Vigna sesquipedalis* (Asparagus Cowpea) Bianchini.

BEAN, Barra *Entada phaseoloides* (Nicker Bean) Hebr. Duncan. For the beans are sometimes washed up on the Atlantic shores of the Hebrides.

BEAN, Black-eyed *Vigna unguiculata* (Cowpea) Bianchini.

BEAN, Blue *Lupinus perennis* (Wild Lupin) Le Strange.

BEAN, Bog 1. *Menyanthes trifoliata* (Buckbean). War, Yks, IOM, Cumb, N'thum. Grigson; Ire. Wilde.1890. 2. *Primula farinosa* (Birdseye Primrose) Yks. Grigson.

BEAN, Broad 1. *Canavalia ensiformis* (Jack Bean) Chopra. 2. *Vicia faba*.

BEAN, Buffalo 1. *Astragalus caryocarpus*. 2. *Thermopsis rhombifolia* (False Lupin) Johnston.

BEAN, Burma *Phaseolus lunatus* (Lima Bean) Kingsbury. It is also known as Java Bean, and Sieva Bean.

BEAN, Bush *Phaseolus vulgaris* (Kidney Bean) S.Africa Watt; USA Weiner.

BEAN, Butter *Phaseolus lunatus* (Lima Bean) Schery.

BEAN, Butter, Indian *Dolichos lablab* (Lablab) Dalziel.

BEAN, Cajan *Cajanus indicus* (Pigeon Pea) Schery.

BEAN, Calabar *Physostigma venenosum*.

BEAN, Cape *Phaseolus lunatus* (Lima Bean) Bianchini.

BEAN, Carolina *Phaseolus lunatus* (Lima Bean) USA Bianchini.

BEAN, Castor *Ricinus communis* (Castor Oil Plant) USA Kingsbury.

BEAN, Castor-oil *Jatropha curcas* (Barbados Nut) W Indies Beckwith.

BEAN, Chocolate *Theobroma cacao* (Cacao) Bennett & Bird. Of course, the beans produce cocoa, or chocolate.

BEAN, Civet *Phaseolus lunatus* (Lima Bean) USA Yanovsky. Presumably an extension of another name, Sieva Bean.

BEAN, Cluster *Cyamopsis tetragonolobus*.

BEAN, Congo *Cajanus indicus* (Pigeon Pea) Schery. Huxley records the name pois congo in Haiti.

BEAN, Coral; BEAN, Coral, Evergreen *Sophora secundiflora* (Mescal Bean) USA Norbeck (Coral); USA LaBarre (Evergreen Coral). Colour descriptive.

BEAN, Coral, Western *Erythrina flabelliformis*.

BEAN, Curcas *Jatropha curcas* (Barbados Nut) USA Kingsbury.

BEAN, Cut-eye 1. *Canavalia ensiformis* (Jack Bean) W Indies Dalziel. 2. *Dolichos ensiformis* (Horse-eye Bean) Beckwith.

BEAN, Dew *Phaseolus aconitifolius* (Moth Bean) Douglas.

BEAN, Donkey-eye *Mucuna urens* (Florida Bean) Howes. Cf. Horse-eye Bean for this.

BEAN, Duffin *Phaseolus lunatus* (Lima Bean) Willis.

BEAN, Egyptian *Dolichos lablab* (Lablab) Brouk.

BEAN, Fava *Vicia faba* (Broad Bean) Kingsbury. Fava here is merely the same as faba, which is Latin for a broad bean, so the name is in fact a pleonasm.

BEAN, Field 1. *Phaseolus vulgaris* (Kidney Bean) Schery. 2. *Vicia faba* (Broad Bean) F Savage.

BEAN, Fig *Lupinus sp* (Lupin) Coles. Coles's own name for lupins. He called them Flat Beans, too.

BEAN, Flat *Lupinus sp* (Lupin) Coles.

BEAN, Florida *Mucuna urens*.

BEAN, French *Phaseolus vulgaris* (Kidney Bean) Wit. Probably a better known name than Kidney Bean itself. French, because they were introduced from France in the 16th century.

BEAN, Garden *Phaseolus vulgaris* (Kidney Bean) Schery.

BEAN, Gemsbuck *Bauhinia esculenta* (Tsin Bean) Howes.

BEAN, Goa *Psophocarpus tetragonolobus*.

BEAN, Goat's *Menyanthes trifoliata* (Buckbean) Fernie. Buckbean is apparently Lyte's translation of the Flemish books boonen (goat's bean). This may not be true, but it is obviously the origin of Goat's Bean as a name.

BEAN, Ground *Amphicarpa bracteata* (Hog Peanut) USA Yarnell.

BEAN, Haricot *Phaseolus vulgaris* (Kidney Bean) Schery. Haricot is the same word as ragout, indicating that the beans were an essential part of this dish.

BEAN, Hibbert *Phaseolus lunatus* (Lima Bean) Chopra.

BEAN, Hog's 1. *Aster tripolium* (Sea Aster) Ess. Gerard. 2. *Hyoscyamus niger* (Henbane) Som. Macmillan. For the Sea Aster, "about Harwich it is called Hogs Beans, for that the swine do greatly desire to feed thereon, as also for that the knobs about the roots do somewhat resemble the garden bean" (Gerard). Hogbean, or as here, Hog's Bean, is a translation of Hyoscyanum.

Dioscorides gave it this name, because although to man and many animals it is poisonous, pigs could apparently eat it without harm. The French call it fève à cochons.

BEAN, Horse *Canavalia ensiformis* (Jack Bean) Douglas.

BEAN, Horse, Little-leaved *Cercidium microphyllum*.

BEAN, Horse-eye 1. *Dolichos ensiformis*. 2. *Mucuna urens* (Florida Bean) Le Strange.

BEAN, House; BEAN, Household *Vicia faba* (Broad Bean) Suss. Parish.

BEAN, Hyacinth *Dolichos lablab* (Lablab) Schery.

BEAN, Ignatius *Strychnos ignatii*.

BEAN, Indian 1. *Catalpa bignonioides* (Locust Bean) USA Harper. 2. *Dolichos lablab* (Lablab) Brouk. 3. *Erythrina flabelliformis* (Western Coral Bean) USA. T H Kearney.

BEAN, Jack *Canavalia ensiformis*.

BEAN, Java *Phaseolus lunatus* (Lima Bean) Kingsbury. Is this really Java? Or was it originally Fava, from Latin faba, bean?

BEAN, Jequirity *Abrus precatorius* (Precatory Bean) USA Kingsbury. Presumably a corruption of Precatory, which means having to do with prayer. The names Rosary Pea, or Prayer Beads, show that the beans were threaded as rosary beads.

BEAN, Jupiter's *Hyoscyamus niger* (Henbane) Jordan.

BEAN, Kiawe *Prosopis juliflora* (Mesquite) USA Kingsbury.

BEAN, Kidney *Phaseolus vulgaris*. Whatever the origin, the pods, reduced to ashes, were once prescribed for kidney complaints and dropsy - they are actually highly diuretic, and lower the blood-sugar level, and are still used in France for kidney and bladder disorders.

BEAN, Kidney, Chinese *Wistaria sinensis* (Chinese Wistaria) Hay & Synge.

BEAN, Kidney, Egyptian *Dolichos lablab* (Lablab) Dalziel.

BEAN, Kidney, Scimitar-podded *Phaseolus lunatus* (Lima Bean) Candolle.

BEAN, Lima *Phaseolus lunatus*. The designation *P lunatus* replaced the older *P limensis*.

BEAN, Locust 1. *Catalpa bignonioides*. 2. *Ceratonia siliqua* fruit (Carob) Wit. The fruit is imported into Britain under this name.

BEAN, Long, Chinese *Vigna unguiculata* (Cowpea) Dalziel.

BEAN, Lucky *Abrus precatorius* (Precatory Bean) Reynolds. *A precatorius* is the African lucky bean. Why? They are extremely poisonous, and are always associated in southern Africa with dangerous magic. When they are found decorating an object, it may safely be identified as being used in sorcery or witchcraft, etc.

BEAN, Mackay *Entada phaseoloides* (Nicker Bean) Australia Chopra.

BEAN, Madagascar *Phaseolus lunatus* (Lima Bean) Bianchini.

BEAN, Madras *Dolichos uniflorus* (Horsegram) Douglas.

BEAN, Mahogany *Afzelia cuanzensis* (Pod Mahogany) Dale & Greenway.

BEAN, Manila *Psophocarpus tetragonolobus* (Goa Bean) Douglas.

BEAN, Manioc *Pachyrrhizus tuberosus* (Yam Bean) Schery.

BEAN, Mat *Phaseolus aconitifolius* (Moth Bean) Schery.

BEAN, Mescal *Sophora secundiflora*.

BEAN, Moth *Phaseolus aconitifolius*.

BEAN, Multiflora *Phaseolus coccineus* (Scarlet Runner) USA Brouk. *P multiflorus* is an alternative botanical name.

BEAN, Mung *Phaseolus aureus* (Green Grass) Anderson. I should have thought this would have been the name for *P mungo*, but it seems not.

BEAN, Navy *Phaseolus vulgaris* (Kidney Bean) Howes. Presumably because beans appeared to be a major part of navy rations.

BEAN, Nicker *Entada phaseoloides*. Nicker means a water spirit, or even a crocodile. Is there any connection with this tropical climber?

BEAN, Nicker, Yellow *Caesalpina bonduc* A.W.Smith.

BEAN, Oil *Pentaclethra macrophylla*.

BEAN, One-eye *Canavalia ensiformis* (Jack Bean) W Indies Dalziel. Cf. Cut-eye, or Overlook, Bean.

BEAN, Ordeal *Physostigma venenosum* (Calabar Bean) Le Strange. The seeds are very poisonous, and have been used by some West African people as an ordeal - the prisoner had to eat a certain amount of the seed, and if he vomited within a specified period, thus ridding himself of the poison, he was acquitted as innocent, and of course, guilty if the seeds were retained. Certain types of crime were also punishable by forcing the guilty to drink an infusion of the crushed seed, which usually resulted in death within the hour.

BEAN, Overlook *Canavalia ensiformis* (Jack Bean) W Indies Dalziel. Cf. Cut-eye, or One-eye, Bean.

BEAN, Paigya *Phaseolus lunatus* (Lima Bean) H C Long.1924.

BEAN, Parang *Canavalia gladiata*.

BEAN, Patagonian *Canavalia ensiformis* (Jack Bean) Chopra.

BEAN, Paternoster *Abrus precatorius* (Precatory Bean) Howes. Because of their use as prayer beads.

BEAN, Poison, Fish *Tephrosia vogelii*.

BEAN, Pole *Phaseolus lunatus* (Lima Bean) Howes.

BEAN, Potato *Apios americana* (Groundnut) Douglas.

BEAN, Powdered *Primula farinosa* (Birdseye Primrose) Caith. B & H. It is also known in Yorkshire as Bogbean. 'Powdered' from the appearance of the plant - an alternative name is Mealy Primrose. But why 'bean' for this?

BEAN, Precatory *Abrus precatorius*. Precatory means pertaining to prayer, and the beans are used as rosary beads. Rosary Pea and Prayer Beads are names given in America.

BEAN, Princess *Psophocarpus tetragonolobus* (Goa Bean) Douglas.

BEAN, Rangoon *Phaseolus lunatus* (Lima Bean) H C Long.1924. Cf. Burma Bean for this.

BEAN, Red *Sophora secundiflora* (Mescal Bean) USA Norbeck.

BEAN, Rice *Phaseolus calcaratus*.

BEAN, Sabre *Canavalia ensiformis* (Jack Bean) Willis.

BEAN, Scotch *Vicia faba* (Broad Bean) Schery.

BEAN, Scurvy *Menyanthes trifoliata* (Buckbean) Fernie. It was regarded as a valuable medicinal plant, chiefly in Scotland and Ireland, usually as a bitter tonic, but it is probably a good anti-scorbutic.

BEAN, Sea *Entada phaseoloides* (Nicker Bean) Chopra. The beans are sometimes found floating in the sea.

BEAN, Sieva *Phaseolus lunatus* (Lima Bean) Kingsbury.

BEAN, Snuffbox *Entada phaseoloides* (Nicker Bean) Chopra. The seeds are often hollowed out and made into boxes.

BEAN, Sour *Tamarindus indica* (Tamarind) Parkinson.1640.

BEAN, Spangled *Mesembryanthemum crystallinum* (Sea Fig) Coats.

BEAN, String *Phaseolus vulgaris* (Kidney Bean) Gooding.

BEAN, Sugar *Phaseolus lunatus* (Lima Bean) Candolle.

BEAN, Sweet *Ceratonia siliqua* (Carob) Parkinson.1640.

BEAN, Sword 1. *Canavalia ensiformis* (Jack Bean) Willis. 2. *Canavalia gladiata* (Parang Bean) Schery.

BEAN, Sword, Seaside *Canavalia obtusifolia*.

BEAN, Tepary *Phaseolus acutifolius*.

BEAN, Tonka *Dipteryx odorata*.

BEAN, Tsin *Bauhinia esculenta*.

BEAN, Twig *Sorbus aucuparia* (Rowan) Suss. Parish. Bean in this case is beam, tree, and the whole name is a corruption of quickbeam, where quick is OE cuic, alive.

BEAN, Urd *Phaseolus mungo* (Black Gram) Schery. It is probably a native of India.

BEAN, Velvet *Stizolobium deeringianum*.

BEAN, Velvet, Florida 1. *Mucuna pruriens* (Cowage) Willis. 2. *Stilozobium deeringianum* (Velvet Bean) Schery.

BEAN, White *Phaseolus lunatus* (Lima Bean) Barbados Gooding.

BEAN, Wild 1. *Apios americana* (Groundnut) USA Bianchini. 2. *Lupinus perennis* (Wild Lupin) Le Strange.

BEAN, Windsor *Vicia faba* (Broad Bean) F.P.Smith.

BEAN, Winter *Lathyrus latifolius* (Garden Everlasting Pea) Suss. Parish. Presumably because it is everlasting, which in this case means perennial.

BEAN, Yam 1. *Pachyrrhizus tuberosus*. 2. *Sphenostylis stenocarpa*.

BEAN, Yam, Wayaka *Pachyrrhizus angulatus*.

BEAN, Yard-long *Vigna sesquipedalis* (Asparagus Cowpea) Bianchini.

BEAN, Yellow *Thermopsis rhombifolia* (False Lupin) USA Kingsbury.

BEAN CAPER *Zygophyllum dumosum*.

BEAN HERB *Satureia hortensis* (Summer Savory) Painter. Because, as a herb, it goes well with beans. The name is really a European, rather than English, one - like German Bohmer-Kraut, for example.

BEAN TREE *Cercis siliquastrum* (Judas Tree) Steele.

BEAN TREE, Black *Castanospermum australe* (Moreton Bay Chestnut) RHS.

BEAN TREE, Lucky 1. *Afzelia cuanzensis* (Pod Mahogany) Dale & Greenway. 2. *Erythrina abyssinica* Palgrave.

BEAN TREE, Snake *Swartzia madagascariensis*.

BEAN TREFOIL 1. *Anagyris foetida*. 2. *Laburnum anagyroides* (Laburnum) Parkinson.1629. 3. *Menyanthes trifoliata* (Buckbean) B & H.

BEANWEED *Pinguicula vulgaris* (Butterwort) Herts. Grigson.

BEAR, BERE *Hordeum sativum* (Barley). These ought to be interchangeble with barley, but in practice they seem to be confined to the now little grown four-rowed (actually six-rowed) variety, and that in Scotland. They are both from OE baere, bere. Barley is simply the adjectival form, i.e. baere plus lic.

BEAR BILBERRY *Arctostaphylos uva-ursi* (Bearberry) Aber. B & H. Bearberry, and thence Bear Bilberry and Bear Whortleberry got the name, according to Prior, because its fruit was a favourite food for bears! It sounds ridiculous, but Lloyd confirms that the berries are eaten by the North American bear.

BEAR BRUSH *Garrya fremontii.*

BEAR GRASS 1. *Camassia esculenta* (Quamash) USA Coats. 2. *Yucca filamentosa* Florida J.Smith. 3. *Yucca glauca* (Soapweed) Schery. All these are North American plants. The bulbs of the Quamash were eaten, not only by the Indians, but apparently by grizzly bears too.

BEAR HUCKLEBERRY *Gaylussacia ursina.*

BEAR SKEITERS 1. *Angelica sylvestris* (Wild Angelica) Scot. Dyer. 2. *Heracleum sphondyllium* (Goutweed) Mor. B & H. Bear here means barley. Children shoot barley, among other things, through the hollow stems, like a pea-shooter. The Somerset name Water Squirt obviously refers to a similar game.

BEAR WHORTLEBERRY *Arctostaphylos uva-ursi* (Bearberry) Gerard. See also Bear Bilberry.

BEAR'S BEARD *Verbascum thapsus* (Mullein) A S Palmer.

BEAR'S BREECH 1. *Acanthus mollis.* 2. *Heracleum sphondyllium* (Hogweed) Prior. Presumably goutweed got this name from a certain similarity between its leaves and those of Acanthus. Bear's Breech for Acanthus is a difficult one. Halliwell quotes as an archaism a verb to breech, meaning to flog or whip, and this is made interesting by another old name for the plant, Brank Ursin. Branks, it seems, was a word for a kind of halter or bridle. Ursin of course is the bear.

BEAR'S EARS 1. *Primula auricula* (Auricula) B & H. 2. *Saxifraga sarmentosa* (Mother-of-thousands) J.Smith. It is the shape of auricula's leaves which produced this name, which got corrupted to Boar's Ears, Bores' Ears, and ultimately to Baziers and Bezors. Why should the saxifrage be so called?

BEAR'S FOOT 1. *Aconitum napellus* (Monkshood) Notts. B & H. 2. *Alchemilla vulgaris* (Lady's Mantle) Hants, N'thum. Grigson; Derb. Camden/Gibson. 3. *Helleborus foetidus* Gerard. 4. *Helleborus niger* (Christmas Rose) Hill. 5. *Helleborus viridis* (Green Hellebore) Dor. Dacombe; Glos. Grigson. In each case, it was the shape of the leaf which gave rise to the name.

BEAR'S GARLIC; URSINE GARLIC *Allium ursinum* (Ramsons). B & H (Bear's); Geldart (Ursine). Possibly from the offensive smell, but why *Allium ursinum* in the first place?

BEAR'S GRAPE 1. *Arctostaphylos uva-ursi* (Bearberry) H & P. 2. *Phytolacca decandra* (Poke-root) S.M.Robertson.

BEARBERRY *Arctostaphylos uva-ursi.* Prior says it got its name because its fruit was a favourite food for bears. Surprisingly, Lloyd confirms that the berries are eaten by the North American bear. Arctostaphylos is from Greek arktos, bear,and staphyle, a bunch of grapes. Uva-ursi means the same thing - bear's grapes.

BEARBERRY, Alpine *Arctostaphylos uva-ursi* (Bearberry) Blamey.

BEARBERRY, Black *Arctous alpinus.* Black berries.

BEARBERRY, Red *Arctostaphylos uva-ursi* (Bearberry) Hutchinson. Red berries.

BEARBERRY HONEYSUCKLE *Lonicera involucrata.* Possibly because this has edible berries.

BEARBIND 1. *Calystegia sepium* (Great Bindweed) Heref, Bucks, Herts, Surr, Middx, Kent Grigson. 2. *Clematis vitalba* (Old Man's Beard) Kent B & H. 3. *Convolvulus arvensis* (Field Bindweed) Prior. 4. *Lonicera periclymenum* (Honeysuckle) Ches. Holland. 5. *Polygonum convolvulus* (Black Bindweed) Staffs. Grigson. This must be another case where bear means barley. It is

OE bere. Cf. Cornbind, etc.

BEARBINE 1. *Calystegia sepium* (Great Bindweed) Bucks, Surr. B & H. 2. *Convolvulus arvensis* (Field Bindweed) Heref. Havergal.

BEARD, Aaron's 1. *Aconitum napellus* (Monkshood) Som. Macmillan. 2. *Allium vineale* (Crow Garlic) Wilts. Macmillan. 3. *Cymbalaria muralis* (Ivy-leaved Toadflax) Scot. Grigson. 4. *Hypericum calycinum* (Rose of Sharon) Dev. Friend.1882. Glos. J.D.Robertson. 5. *Orchis mascula* (Early Purple Orchis) Berw. B & H. 6. *Saxifraga sarmentosa* (Mother-of-thousands) Dev. Friend.1882. 7. *Spiraea salicifolia* (Bridewort) Lincs. Peacock. All probably descriptive - Crow Garlic and Rose of Sharon both have protruding bunches of stamens, while the thread-like roots of Mother-of-thousands, which is also sometimes called Old Man's Beard, account for the beard epithet. Ivy-leaved Toadflax hangs from a wall in such a way as to suggest a beard (and also has names like Mother-of-Thousands and Thread-of-life). The only ones which are difficult to explain are the Orchis - it is probably a misnomer from a long time ago - and Monkshood.

BEARD, Bear's *Verbascum thapsus* (Mullein) A S Palmer.

BEARD, Bishop's *Anthriscus sylvestris* (Cow Parsley) Australia Hemphill.

BEARD, Buck's *Tragopogon pratensis* (Goat's Beard) B & H.

BEARD, Bushy *Clematis vitalba* (Old Man's Beard) Som. Macmillan.

BEARD, Capuchin's *Plantago coronopus* (Buck's Horn Plantain) Bianchini.

BEARD, Daddy's; BEARD, Daddy Man's *Clematis vitalba* (Old Man's Beard) Som. Macmillan.

BEARD, Devil's 1. *Anemone alpina* Friend. 2. *Pulsatilla vulgaris* (Pasque Flower) Tynan & Maitland. 3. *Sempervivum tectorum* (Houseleek) Wales C.J.S.Thompson. Semi-descriptive as far as houseleek is concerned - it is Jupiter's Beard and Thor's Beard as well.

Being a lightning protector probably accounts for the devil's ownership. But the Anemone? It is probably a complete mistake - but it crops up again for Pasque Flower, where it is claimed either the hairy stalks or the bushy head of seeds might account for the name.

BEARD, Goat's 1. *Astilbe sp*. 2. *Cichorium endivia* (Endive) Som. Macmillan. 3. *Filipendula ulmaria* (Meadowsweet) Dev, Dor. Macmillan. 4. *Hypericum perforatum* (St John's Wort) USA Watt. 6. *Spiraea aruncus* Broadwood. 6. *Tragopogon porrifolium* (Salsify) Australia Watt. 7. *Tragopogon pratensis*. All descriptive, in one way or another. Are Astilbe and Spiraea aruncus identical?

BEARD, Goat's, Purple *Tragopogon porrifolium* (Salsify) Grieve.1931.

BEARD, Grandfather's 1. *Clematis vitalba* (Old Man's Beard) Som. Macmillan. 2. *Equisetum arvense* (Common Horsetail) Dev. Macmillan. 3. *Geum triflorum* (Prairie Smoke) USA T H Kearney.

BEARD, Grandfy's *Clematis vitalba* (Old Man's Beard) Som. Macmillan.

BEARD, Grey *Clematis vitalba* (Old Man's Beard) Wilts. Dartnell & Goddard; Hants. Grigson. The exact equivalent of Old Man's Beard - very aptly descriptive.

BEARD, Hair- *Luzula campestris* (Good Friday Grass) N'thants B & H.

BEARD, Hare's *Verbascum thapsus* (Mullein) Gerard (also in USA Henkel).

BEARD, Hawk's *Crepis sp*.

BEARD, Jew's *Sempervivum tectorum* (Houseleek) A S Palmer. The French joubarbe, which is a version of the next entry, was responsible for Jew's Beard.

BEARD, Jove's *Sempervivum tectorum* (Houseleek) Conway. This one of a group which includes Jupiter's Beard, Thor's Beard, and Thunder Plant, which came about because of the firm belief that houseleek protected whatever it grew upon from lightning - "as good as a fire insurance", according to Mrs Wiltshire. It is even said that Charlemagne ordered that every dwelling in his empire should have them on its roof.

BEARD, Jupiter's 1. *Anthyllis vulneraria* (Kidney Vetch) Friend. 2. *Centranthus ruber* (Spur Valerian) A.W.Smith. 3. *Pulsatilla vulgaris* (Pasque Flower) Tynan & Maitland. 4. *Sempervivum tectorum* (Houseleek) Gerard. The name applied to Kidney Vetch appears in German, too, so possibly the recorded name is a simple translation of the German. Similarly, houseleek is joubarbe in French, and Devil's Beard is a Welsh name for the plant. Devil's Beard is recorded for the Pasque Flower, too.

BEARD, Monk's *Cichorium intybus* (Chicory) Sanecki. Probably from the French name, barbe de capucin.

BEARD, Old Man's 1. *Artemisia abrotanum* (Southernwood) Dor, Wilts. Macmillan. 2. *Chionanthus virginicus* (Fringe Tree) USA Thomson.1978. 3. *Clematis brachiata* S.Africa Watt. 4. *Clematis vitalba*. 5. *Equisetum arvense* (Common Horsetail) Som. Elworthy. 6. *Filipendula ulmaria* (Meadowsweet) Som. Macmillan. 7. *Hippuris vulgaris* (Mare's Tail) Dor. Grigson. 8. *Hypericum calycinum* (Rose of Sharon) Dev. Friend.1882. 9. *Nigella damascena* (Love-in-a-mist) Coles. 10. *Parmelia saxatilis* (Grey Stone Crottle) Nicolson. 11. *Rosa canina* the galls that grow on them Dev. Friend.1882. 12. *Saxifraga sarmentosa* (Mother-of-thousands) Dev. Friend.1882.

BEARD, Thor's *Sempervivum tectorum* (Houseleek) Grieve.1931. It is also Jupiter's Beard and Devil's Beard, but Thor is very appropriate, for it is a lightning plant (Thunder Plant is another recorded name). When growing on a roof, it acts as a protection against lightning. "Preserves what it grows upon from Fire and Lightning", Culpepper said. Mrs Wiltshire said it was "as good as a fire insurance". According to Irish belief, houseleek growing in the thatch preserves all the people in the house from scalds, burns and the danger of fire, as long as it remains untouched. The leaves were used medicinally in Scotland as a plaster for burns. As an indication of the antiquity of the general belief, note the claim by Albertus Magnus that he who rubbed his hands with the juice of houseleek would be insensible to pain when taking red-hot iron in his hands.

BEARD LICHEN *Usnea barbata*.

BEARD-TREE *Corylus avellana* (Hazel) B & H. In this case, it is filbeard, or filberd, filbert, i.e. St Philibert's nut, so called because it ripens round about the saint's day - 22 August OS.

BEARDED BELLFLOWER *Campanula barbata*.

BEARDED DARNEL *Lolium temulentum* (Darnel) Barton & Castle.

BEARDED FIG *Ficus citrifolia*. It is suggested that the name of Barbados (which may mean the land of beards) is derived from the many Bearded Figs there.

BEARDED PINK *Dianthus barbatus* (Sweet William) Macmillan. For *barbatus* is the specific name.

BEARDTONGUE, Hairy *Pentstemon hirsutus*.

BEARDTONGUE, Large-flowered *Pentstemon grandiflorus*.

BEARDTONGUE, Sharp-leaved *Pentstemon acuminatus*.

BEARDTONGUE, Smooth *Pentstemon laevigatus*.

BEARGRASS *Yucca glauca* (Soapweed) Schery.

BEARWORT *Meum athamanticum* (Baldmoney) B & H. Apparently this is a straight translation from the German name.

BEAU, Spangled *Mesembryanthemum crystallinum* (Sea Fig) Coats.

BEAUTIFUL FIR *Abies amabilis* (Red Silver Fir) Leathart.

BEAUTY, Alaska *Claytonia acutifolia*.

BEAUTY, Blue-eyed *Vinca major* (Greater Periwinkle) Som. Macmillan.

BEAUTY, Pink *Dianthus barbatus* (Sweet William) Som. Macmillan.

BEAUTY, Sleeping *Oxalis acetosella* (Wood Sorrel) Dor. Macmillan. Because of the way the leaves fold back.

BEAUTY, Wayside *Prunus spinosa* (Blackthorn) Som. Macmillan.
BEAUTY BERRY *Callicarpa americana* (French Mulberry) Howes.
BEAUTY BUSH 1. *Fuchsia sp* Tynan & Maitland. 2. *Kolkwitzia amabilis*.
BEAUTY LEAF *Calophyllum inophyllum*.
BEAUTY OF THE NIGHT *Mirabilis jalapa* Whittle & Cook. The flowers do not open until about 4 in the afternoon (hence another name, Four-o'-clock), and stay open until the next morning.
BEAVER-DAM BREADROOT *Psoralea castorea*. Breadroot because the tuberous starchy taproots furnished food for the American Indians and the early settlers.
BEAVER POISON 1. *Cicuta maculata* (American Cowbane) USA Kingsbury.1967. 2. *Conium maculatum* (Hemlock) USA Skinner. Why just beaver? Were they ever used deliberately to poison these animals?
BEAVER-ROOT *Heracleum lanatum* USA Gilmore.
BEAVER TREE *Magnolia virginiana* (Sweet Bay) J.Taylor. "...Beaver Tree, because the root being dainty to these animals, are frequently caught by it...".
BEAVERBREAD SCURF PEA *Psoralea castorea* (Beaver-dam Breadroot) Douglas.
BEAVERWOOD 1. *Celtis occidentalis* (Hackberry) Douglas. 2. *Celtis reticulata* (Western Hackberry) USA Elmore.
BEAWEED *Senecio jacobaea* (Ragwort) Scot. Grigson.
BECH *Fagus sylvatica* (Beech) Turner. This version is close to the OE bece.
BECKBEAN *Menyanthes trifoliata* (Buckbean) B & H. Is this a mispronunciation of buckbean? Or is it genuinely beck, meaning in dialect a brook? Brookbean is recorded as a local name.
BECKY-LEAVES *Veronica beccabunga* (Brooklime) Dev. Friend.1882. Beccabunga is O. Nor. bekh, brook, and this is obviously the same word.
BED, Fairy's *Scrophularia nodosa* (Figwort) Dor. Macmillan.
BED, Lady's *Galium verum* (Lady's Bedstraw) Dev. Macmillan.
BED, Lousy *Melandrium dioicum* (Red Campion) Cumb. Grigson. An odd name, but note that Fleabites is a Cornish name for this plant.
BED-FLOWER *Galium verum* (Lady's Bedstraw) B & H. The medieval legend was that the Virgin lay on a bed of bracken and Lady's Bedstraw during the Nativity. Bracken refused to acknowledge the child, and lost its flowers. *G. verum* welcomed the child, and, blossoming at that moment, found its flowers had changed from white to gold. Sudetenland women put it in their beds to make childbirth easier and safer.
BED-FURZE 1. *Ulex gallii* (Dwarf Frurze) Hants. Grigson. 2. *Ulex minor* (Small Furze) Hants. Cope. O Suileabhain says that furze was actually used in Ireland for bedding! Surely, for animals?
BEDDA NUT *Terminalia bellirica* (Bastard Myrobalan) Chopra.
BEDDING, Hottentot *Helichrysum crispum* S.Africa Watt.
BEDDYWIND, BETTYWIND *Convolvulus arvensis* (Field Bindweed) Wessex Rogers. Bedwind and Bedwine are other names, applied to more than Field Bindweed, though. The second element of Beddywind is obvious for this kind of plant, but the first element is conjectural, simply because there are so many variations - bed-, bell-, bes-, beth-, betty-, bane-, bear-, etc.
BEDEWEN, BEDWEN *Betula verrucosa* (Birch) both West. B & H.
BEDFORD WILLOW *Salix viridis*.
BEDGERY; PETGERY; PITCHERY *Duboisia hopwoodii* (Pituri) Lewin. All anglicizations of the native name,Pituri.
BEDLAM COWSLIP 1. *Primula veris x vulgaris* (False Oxlip) N'thants. A.E. Baker. 2. *Pulmonaria officinalis* (Lungwort) Som. Grigson. Bedlam is Bethlehem, and Bethlehem Cowslip is the name under which Coles described Lungwort. Jerusalem Cowslip, Jerusalem Sage and Bethlehem

Sage are other old names. The association for lungwort is the legend that during the flight into Egypt, some of the Virgin's milk fell on the leaves while she was nursing the infant Jesus, causing the white spots, which are still there (the white spots are also the "signature" for lungs, hence the use for pulmonary complaints, and the name lungwort). But for False Oxlip, one could argue that this is indeed a cowslip that has gone mad.

BEDSFOOT FLOWER *Clinopodium vulgare* (Wild Basil) Parkinson.1640. "because the branches, say some.. doe resemble the feet of a bed".

BEDSTRAW, Carpet-weed *Galium mollugo* (White Bedstraw) USA Zenkert.

BEDSTRAW, Catchweed *Galium aparine* (Goose-grass) USA Allan.

BEDSTRAW, Corn, Rough *Galium tricorne* (Small Goose-grass) Polunin.

BEDSTRAW, Cross-leaved *Galium boreale* (Northern Bedstraw) J A MacCulloch.1905. Quoted as such by MacCulloch, but should it not be Galium cruciatum, Crosswort?

BEDSTRAW, Dye *Asperula tinctoria* Usher. The roots yield a red dye.

BEDSTRAW, Fen *Galium uliginosum*. In this country, it is confined to the Fen country, and then only locally.

BEDSTRAW, Fragrant *Galium triflorum*. A North American species, sometimes called Lady's Bouquet, for the same reason that it is given the epithet 'fragrant'.

BEDSTRAW, Golden, Lady's *Galium verum* (Lady's Bedstraw) Yks. Grigson.

BEDSTRAW, Goose-grass *Galium aparine* (Goose-grass) USA T H Keraney.

BEDSTRAW, Heath *Galium saxatile*.

BEDSTRAW, Hedge; BEDSTRAW, Hedge, Great *Galium mollugo* (White Bedstraw) McClintock (Hedge); Clapham (Great Hedge).

BEDSTRAW, Hedge, Upright *Galium erectum*.

BEDSTRAW, Lady's 1. *Galium cruciatum* (Cross-wort) Yks. Grigson. 2. *Galium verum*. The medieval legend was that the Virgin lay on a bed of bracken and bedstraw during the Nativity. Bracken refused to acknowledge the child, and lost its flowers. *Galium verum* welcomed the child, and,blossoming at that moment, found its flowers changed from white to gold. In the Sudentenland, women put it in their beds to make childbirth easier and safer, and since women who have just had a child are susceptible to attack from demons, they will not go out unless they have some Lady's Bedstraw with them in their shoes. Its astringency probably also brought it in the bed against fleas.

BEDSTRAW, Marsh *Galium palustre* (Water Bedstraw) Clapham.

BEDSTRAW, Marsh, Slender *Galium debile*.

BEDSTRAW, Northern *Galium boreale*.

BEDSTRAW, Reclining *Galium aparine* (Goose-grass) USA Jaeger.

BEDSTRAW, Rough 1. *Galium aparine* (Goose-grass) Jago. 2. *Galium asprellum*.

BEDSTRAW, Shepherd's *Asperula cynanchica* (Squinancywort) Glos. J.D.Robertson.

BEDSTRAW, Shining *Galium concinnum*.

BEDSTRAW, Slender *Galium pumilum*.

BEDSTRAW, Swamp *Galium uliginosum* (Fen Bedstraw) W.Miller.

BEDSTRAW, Sweet *Galium triflorum* (Fragrant Bedstraw).

BEDSTRAW, Wall *Galium parisiense anglicum*.

BEDSTRAW, Water *Galium palustre*.

BEDSTRAW, White *Galium mollugo*.

BEDSTRAW, Yellow 1. *Galium cruciatum* (Cross-wort) Mitton. 2. *Galium verum* (Lady's Bedstraw) Yks. Grigson.

BEDSTRAW BELLFLOWER *Campanula aparinoides* (Swamp Bluebell) USA House.

BEDVINE *Clematis vitalba* (Old Man's Beard) Wilts. Tynan & Maitland.

BEDWIND 1. *Calystegia sepium* (Great Bindweed) Glos. B & H. 2. *Clematis vitalba* (Old Man's Beard) Dor. Macmillan; Wilts. Dartnell & Goddard. 3. *Convolvulus arvensis* (Field Bindweed) Glos, War, Hants. Grigson. One of the series that starts with Bellbine, or Bellbind, through Bethwine,

Bearbine, to Bedwind or Bedwine.

BEDWINE 1. *Clematis vitalba* (Old Man's Beard) Dor. Macmillan; Wilts. Dartnell & Goddard; Glos, IOW Grigson; Hants. Cope; Berks. Lowsley. 2. *Convolvulus arvensis* (Field Bindweed) Wessex Rogers. 3. *Polygonum convolvulus* (Black Bindweed) Hants. Cope. As for Bedwind. 'Wine' is the dialect variation of wind, just as 'bine' is the same in dialect as bind.

BEE, Honey *Lamium album* (White Deadnettle) Dev. Macmillan. In line with the series of names, all deriving from the bee-attracting quality of the flowers, which gives Bee-nettle, Honeysuckle, etc.

BEE, Humble *Ophrys apifera* (Bee Orchis) B & H. Bee Orchis is descriptive, and so are the various derivative names, like Bee Flower, Honey-bee Flower, etc. Humble Bee is the earlier version of Bumble Bee (the "b" only occurs in ME, apparently).

BEE BALM 1. *Melissa officinalis*. 2. *Monarda didyma* (Red Bergamot) New Engl. Sanford. Melissa means bee, and Gerard has: The hives of Bees being rubbed with the leaves of Bawme, causeth the bees to keep together, and causeth others to come unto them..." Wiltshire beekeepers still rub the inside of the skeps with it, and the same belief is current in East Anglia, where they also say that if this grows in the garden, the bees will not leave the hive. As far as Bergamot is concerned, the name is probably given because it is such a good bee plant, and the bees continually visit the flowers.

BEE BALM, Mintleaf *Monarda fistulosa* (Purple Bergamot) USA Elmore.

BEE BALM, Pony *Monarda pectinata* (Horse Mint) USA Elmore.

BEE BREAD 1. *Borago officinalis* (Borage) B & H. 2. *Trifolium pratense* (Red Clover) Glos. PLNN35; Kent. B & H. 3. *Trifolium repens* (White Clover) Som. Macmillan. All favourite bee plants - in fact, borage was at one time grown especially to attract the bees.

BEE CATCHER *Digitalis purpurea* (Foxglove) Dor. Macmillan. Presumably from the shape of the flowers, as the name Beehives is recorded also. But there is a more immediate reason for the name, for boys once used to wait till the bee had entered the flower, and then would close the entrance, to listen to the bee's struggles.

BEE FEED *Eriogonum fasciculatum* (Wild Buckwheat) USA Barrows.

BEE FLOWER 1. *Cheiranthus cheiri* (Wallflower) Lincs. B & H. 2. *Ophrys apifera* (Bee Orchis) Wilts. Dartnell & Goddard; IOW B & H. Descriptive for Bee Orchis, but for wallflower the reference must be to the scent.

BEE-IN-A-BUSH *Aconitum napellus* (Monkshood) Heref. Leather.

BEE NETTLE 1. *Galeobdalum luteum* (Yellow Archangel) Ches, Notts. Grigson. 2. *Galeopsis tetrahit* (Hemp Nettle) B & H. 3. *Galeopsis versicolor* (Large-flowered Hemp Nettle) Ches, Scot. B & H. 4. *Lamium album* (White Deadnettle) Som. Macmillan; Bucks. Harman; Leics, Notts, Lincs. Grigson. 5. *Lamium purpureum* (Red Deadnettle) Som. Macmillan; Lincs, Notts. Grigson. Possibly because they are all attractive to bees, but also possibly this is not 'Bee' at all, but a variation on Dea, or Deaf, or Dead.

BEE NETTLE, Red *Lamium purpureum* (Red Deadnettle) War. Grigson.

BEE NETTLE, White *Lamium album* (White Deadnettle) War. Grigson.

BEE ORCHIS *Ophrys apifera*. Descriptive.

BEE ORCHIS, Brown *Ophrys fusca*.

BEE PLANT, Colorado, Blue *Cleome serrulata* (Rocky Mountain Bee Plant) USA Elmore.

BEE PLANT, Rocky Mountain *Cleome serrulata*.

BEE PLANT, Yellow *Cleome lutea* (Yellow Spiderwort) USA T H Keraney.

BEE SOOKIES *Pedicularis sylvatica* (Lousewort) Scot. Grigson. Sookies is the Scottish equivalent of Suckle, as in Honeysuckle (which is a Hampshire name for this plant). It therefore means anything which is attractive to bees.

BEE WORT *Acorus calamus* (Sweet Flag) Cockayne. "That bees may not fly off, take this wort, and hang it in the hive" (Apuleius Herbarium).

BEE'S NEST 1. *Daucus carota* (Wild Carrot) Gerard. 2. *Heracleum sphondyllium* (Hogweed) Som. Macmillan. Carrot is better known as Bird's Nest, but both names occur because when the seeds begin to ripen, the umbel contracts to form a dense concave body, looking just like a nest. Or, to quote Fernie, "a concave semi-circle, or nest, which bees, when belated from the hive, will use as a dormitory". The name applied to Hogweed is probably an incorrect observation.

BEE'S REST *Caltha palustris* (Marsh Marigold) Som. Grigson. Possibly the same as Bee's Nest, but Somerset plant names are extremely imaginative.

BEECH *Fagus sylvatica*. OE bece, and MLG boke, which also produced OE boc, which then survived with shortened vowel as buck, hence Buckmast, etc., as well as several place names like Bockhampton (home farm among beeches). Beckwith, in Yorkshire (Beech wood) came from the other OE form. The Latin name, Fagus, according to ODEE, is the same word as Greek phagos, edible oak, and possibly comes from Greek phagein, to eat, so making Fagus as the beech "the tree with edible fruit".

BEECH, African *Faurea saligna*.

BEECH, American *Fagus grandifolia*.

BEECH, Antarctic *Nothofagus antarctica*.

BEECH, Australian 1. *Nothofagus moorei*. 2. *Tectona australis*.

BEECH, Black *Nothofagus solandri*.

BEECH, Blue *Carpinus caroliniana*. Not a beech, but a hornbeam - this is an American species, and it is also called Water Beech there.

BEECH, Chinese *Fagus englerana*.

BEECH, Clinker *Nothofagus truncata* (Hard Beech) Brimble.1948.

BEECH, Copper *Fagus sylvatica var. cuprea*.

BEECH, Cut-leaf *Fagus sylvatica var. heterophylla*.

BEECH, Dawyck *Fagus sylvatica var. fastigiata*.

BEECH, Dutch *Populus alba* (White Poplar) B & H. Probably a misreading of Abele or one of its derivatives. Abeel is Dutch for the poplar, which is not a native British tree. They were imported from Holland in the seventeenth century. Rackham, though, says that about ten thousand White Poplars were imported and planted in East Anglia. They were completely unknown to the country people, who just called them Dutch Beech.

BEECH, Evergreen *Nothofagus sp.*

BEECH, Fern-leaf *Fagus sylvatica var. heterophylla* (Cut-leaf Beech) Brimble.1948).

BEECH,Hard *Nothofagus truncata*.

BEECH, Hay 1. *Fagus sylvatica* (Beech) B & H. 2. *Carpinus betula* (Hornbeam) Bucks. Harman. Hay here is probably OE haga, hege - hedge. Both were once popular ornamental hedge trees.

BEECH, Horn; BEECH, Horned *Carpinus betula* (Hornbeam). B & H (Horn); Bucks. Harman (Horned). Hornbeam itself means the tree whose wood is tough like horn.

BEECH, Horse *Carpinus betula* (Hornbeam) Hants, Kent Grigson, Evelyn; Suss. Parish. 'Horse' in a plant or tree name usually means a coarse or useless variety (Cf. Horse Chestnut); but in this case, it is probably a variation of Hurst Beech, which occurs in the same three counties. Hurst means a small wood.

BEECH, Hurst; BEECH, Horst *Carpinus betula* (Hornbeam) Hants, Suss, Kent, Norf. Grigson; W.Miller (Horst). Hurst is a small wood or copse, OE hyrst, and has survived in a large number of place names (Ashurst, Lyndhurst, etc).

BEECH, Indian *Pongamia pinnata*.

BEECH, Mast *Fagus sylvatica* (Beech) Culpepper. i.e. the beech which gives mast, which is the name given to the fruit, OE maest. Strictly speaking, it is the fruit of any forest tree, especially as food for swine, so that acorns could come into this category, for the word comes from a Germanic original meaning food (the same word as meat, possibly).

BEECH, Mountain *Nothofagus cliffortioides*.

BEECH, Myrtle *Nothofagus cunninghamii* (Myrtle Tree) Howes.

BEECH, New Zealand *Nothofagus menziesii*.

BEECH, Oriental *Fagus orientalis*.

BEECH, Paper *Betula verrucosa* (Birch) Wilts. Macmillan.

BEECH, Purple *Fagus sylvatica var. purpurea*.

BEECH, Red *Nothofagus fusca*.

BEECH, Robel *Nothofagus obliqua*.

BEECH, Rusty-leaved *Fagus grandifolia* (American Beech) J.Smith. An alternate specific name is *ferruginea*.

BEECH, Silver *Nothofagus menziesii* (New Zealand Beech) Leathart.

BEECH, Southern *Nothofagus procera* (Raoul) RHS.

BEECH, Southern, Oval-leaved *Nothofagus betuloides*.

BEECH, Tanglefoot *Nothofagus gunnii*.

BEECH, Water 1. *Carpinus caroliniana* (Blue Beech) USA Zenkert. 2. *Platanus occidentalis* (Western Plane) W.Miller.

BEECH, Weeping *Fagus sylvatica var. pendula*.

BEECH, White *Carpinus betula* (Hornbeam) Ches. Holland. The timber is almost white.

BEECH-DROPS *Orobanche virginiana* (Cancer-root) Le Strange. A reference to its habitat.

BEECH-DROPS, False *Monotropa hypopitys* (Yellow Bird's Nest) USA Upham. There is a mistaken belief that it is a parasite upon the roots of Scots Pine and Beech - a belief that is mirrored in another name, Fir-rape. Actually, the plant has its own roots invested by a fungus.

BEECHWHEAT *Fagopyrum esculentum* (Buckwheat) Grieve.1931. The seeds resemble beech mast, hence this name, and Buckwheat, Bockwheat, etc. Buckwheat is from the Dutch boekweit, meaning beech wheat. The German Buchweize has exactly the same meaning.

BEEDY'S EYES; BIDDY'S EYES; BEATY EYES *Viola tricolor* (Pansy) all Som. Jennings (Beedy); Tongue (Biddy); Macmillan (Beaty). Biddy is St Bridget, and Ruth Tongue recorded the Somerset practice of making "Biddy's Bed" of the Mountain Pansy (*V lutea*). This is a case of the February festival celebrations being transferred to May Day.

BEEF, Bull 1. *Rosa canina* (Dog Rose) Ches. Holland. 2. *Rubus fruticosus* (Bramble) Ches. Holland. The young shoots, in both cases. Cheshire children used to eat the shoots, and call them Bull Beef.

BEEF, Roast *Iris foetidissima* (Foetid Iris) Prior. It is claimed that the bruised leaf smells like roast beef, a claim which does not really correspond to the name Foetid Iris. Gerard said that the leaves, being rubbed, were "of a stinking smell very lothsome".

BEEF SUET TREE *Shepherdia argentata* (Buffaloberry) Howes.

BEEFSTEAK 1. *Begonia feasti* Gannon. 2. *Iresine herbstii* (Bloodleaf) RHS. 3. *Perilla frutescens* (Perilla) G.B.Foster. Because of the very dark red foliage in all cases.

BEEFWOOD 1. *Banksia integrifolia* (White Honeysuuckle) Everett. 2. *Casuarina equisetifolia*. 3. *Grevillea striata* (Turraie) Australia Meggitt. 4. *Stenocarpus salignus* (Red Silky-oak) Everett. As far as the Casuarina is concerned, and possibly the others too, beefwood is applied to the timber - it is hard and red.

BEEHIVE *Digitalis purpurea* (Foxglove) Som. Macmillan. Descriptive, from the shape of the flowers. Bee Catchers is another Somerset name.

BEEHIVE, Prickly *Dipsacus fullonum* (Teasel) Dev. Macmillan. Descriptive - the shape of the flowers remind one of the old straw skeps.

BEELZEBUB 1. *Geum rivale* (Water Avens) Hants. Hants FWI. 2. *Pulmonaria officinalis* (Lungwort) Hants. Hants FWI. Why Beelzebub for lungwort? Most of its names belong to the opposite camp. Water Avens, on the other hand, has names like Egyptian attached to it, and that may have influenced someone to coin this one.

BEER, Poor Man's *Humulus lupulus* (Wild Hop) Dor. Macmillan.

BEERSEEM CLOVER *Trifolium alexandrinum* (Egyptian Clover) Schery.

BEESOM *Cytisus scoparius* (Broom) Dev. Grigson. One of the many spellings of besom, i.e. broom. The original broom, whether for domestic or magical (witch) purposes, was a stalk of the broom plant with a tuft of leaves at the end.

BEET *Beta vulgaris var. maritima.* OE bete, straight from Latin beta.

BEET, Cattle *Beta vulgaris var. maritima* (Beet) E Angl. G E Evans.1956. Better known in this context as Mangold Wurzel.

BEET, Chilean *Beta vulgaris var. cicla* (Swiss Chard) Bianchini.

BEET, Field *Beta vulgaris var. maritima* (Beet) Notes & Queries. vol 7.

BEET, Leaf; BEET, Spinach *Beta vulgaris var. cicla* (Swiss Chard) both Bianchini.

BEET, Marsh, Wild Limonium vulgare (Sea Lavender) W.Miller.

BEET, Sea *Beta vulgaris var. maritima* (Beet) Mabey.

BEET, Seakale *Beta vulgaris var. cicla* (Swiss Chard) Brouk.

BEET, Sicilian *Beta vulgaris var. cicla* (Swiss Chard) Bianchini.

BEET, Sugar *Beta vulgaris var. altissima.*

BEET, White *Beta vulgaris var. cicla* (Swiss Chard) Grigson.1974.

BEET, Wild *Amaranthus retroflexus* (Pigweed) USA Cunningham.

BEET CHARD *Beta vulgaris var. cicla* (Swiss Chard) Grigson.1974.

BEETLE, Marsh; BEETLE, Marish *Typha latifolia* (False Bulrush) both W.Miller.

BEETLE-NUT PALM *Areca catechu* (Areca Palm) USA Cunningham. Beetle = betel.

BEETRAW; BEETRIE *Beta vulgaris var. maritima* (Beet) (Scot) Jamieson.

BEEWEED, Rocky Mountain *Cleome serrulata* (Rocky Mountain Bee Plant) USA Elmore.

BEEWEED, Sonora *Cleome sonorae.*

BEEWOOD *Aster cordifolius* (Heart-leaved Aster) USA Perry.

BEGGAR, Make- *Sagina procumbens* (Common Pearlwort) B & H. The Norfolk name Poverty sums it up. Too much of this in the grass will make sure that soon there will be nothing else left.

BEGGAR BRUSHES *Clematis vitalba* (Old Man's Beard) Bucks. B & H. Old Man's Beard became known as an emblem of cunning, because it was used by beggars to cause false ulcers, by running the juice on themselves (like all the Ranunculaceae, it is very acrid). This plant secretes an irritating juice that causes a superficial sore; so much of an irritant that it could be fatal if ingested. Beggar's Plant is another name.

BEGGAR-LICE 1. *Galium aparine* (Goosegrass) Northants. A E Baker. 2. *Lappula virginiana* USA Youngken. The "lice" are the burrs in each case.

BEGGAR-TICKS 1. *Bidens frondosa.* 2. *Bidens bipinnata* (Spanish Needles) Canada Watt. 3. *Bidens pilosa* Canada Watt. The name refers to the burrs (the more usual name for most of the genus is Bur-Marigold). An American name for *B frondosa* is Sticktight. Beggar-ticks becomes Beggarsticks in S.Africa when applied to *B pilosa*.

BEGGAR-TICKS, Devil's *Bidens frondosa* (Beggar-ticks) Allan.

BEGGAR-TICKS, Smooth *Bidens laevis.*

BEGGAR-TICKS, Swamp *Bidens connata.*

BEGGAR'S BASKET *Pulmonaria officinalis* (Lungwort) Som. Macmillan; Ches. B & H.

BEGGAR'S BLANKET *Verbascum thapsus* (Great Mullein) Som. Macmillan; Cumb. Grigson. The woolly texture of the leaves produced a great variety of descriptive names, of which this is one. Flannel-flower, Aaron's Flannel, Blanket-leaf, Moses' Blanket, Woollen are just a few.

BEGGAR'S BURR *Arctium lappa* (Burdock) Vesey-Fitzgerald. There is an old proverb "sticks like a burr to a beggar's rags" (Baker) - so this name must have been recorded in Northamptonshire.

BEGGAR'S BUTTONS *Arctium lappa* (Burdock) either applied to the plant as a whole, as Dev. B & H; USA Henkel or simply to the burrs, as Dev. Halliwell.

BEGGAR'S LICE 1. *Daucus pusillus* USA Elmore. 2. *Galium aparine* (Goose-grass) Wilts, Hants, Glos, Bucks, N'thants. B & H. Descriptive - it is the burrs that the name refers to, for Goose-grass.

BEGGAR'S NEEDLE 1. *Erodium moschatum* (Musk Storksbill) Worcs. Allies. 2. *Scandix pecten-veneris* (Shepherd's Needle) Som. Macmillan; Midl. Halliwell. It is the long, beaked fruits that give the 'needle' name to both these plants. Note that in Worcestershire, as well as Beggar's Needle, the name for Musk Storksbill has become Bagger's Needle.

BEGGAR'S PLANT *Clematis vitalba* (Old Man's Beard) Skinner. See Beggar Brushes for the explanation of this.

BEGGAR'S STALK *Verbascum thapsus* (Great Mullein) Cumb. Grigson. A tall flower, so the origin is probably the same as that which gave such names as Shepherd's Staff or Aaron's Rod.

BEGGARMAN, Wild *Lychnis flos-cuculi* (Ragged Robin) Tynan & Maitland. From the ragged appearance of the flower.

BEGGARMAN CAKES *Althaea officinalis* (Marsh Mallow) Ire. O'Farrell. Quoted as Marsh Mallow, but perhaps Common Mallow was meant, as the seed vessels of the latter were certainly known as 'cakes' of one kind or another, notably Cheese cakes.

BEGGARMAN'S OATMEAL *Alliaria petiolata* (Jack-by-the-hedge) Leics. Grigson. The leaves are edible, and can be used as a salad, or as a potherb, or as an ingredient in sauces. It was often eaten raw with bread and butter, and it can even be fried with herrings and bacon. There is a garlic flavour, so there are names like Hedge Garlic, Garlic Mustard, or Poor Man's Mustard.

BEGGARSTICKS *Bidens pilosa* S.Africa Watt. A corruption of Beggar-ticks, referring to the burrs, which is the name in Canada.

BEGGARWEED 1. *Cuscuta epithymum* (Common Dodder) Dor. Grigson; Wilts. Dartnell & Goddard. 2. *Cuscuta europaea* (Greater Dodder) Dor. B & H. 3. *Galium aparine* (Goose-grass) N'thants. A.E.Baker; Notts. Grigson. 4. *Heracleum sphondyllium* (Hogweed) Beds. B & H. 5. *Polygonum aviculare* (Knotgrass) Beds. Grigson. 6. *Spergula arvensis* (Corn Spurrey) Beds. Halliwell. Names like Beggarweed usually mean either that too much of it will make the soil fit for nothing else, and so beggar one, as Hogweed, etc. does, or that the presence of the plant is a sure indication that the soil is no good anyway (Knotgrass, etc), and will still beggar one. Dodder is a parasite, so the connection here is obvious.

BEGGARY *Fumaria officinalis* (Fumitory) Grieve.1931. It often grows on waste ground, and this may be the explanation.

BEGH *Betula verrucosa* (Birch) Donegal Grigson.

BEGONIA, Angel's-wing *Begonia coccinea*.

BEGONIA, Elephant's-ear *Begonia scharfii*. From the size of the leaves.

BEGONIA, Eyelash *Begonia bowerae*. Long, curling hairs grow from the leaf edges, hence the 'eyelash' of the name.

BEGONIA, Iron Cross *Begonia masoniana*. So called from the markings on the leaves.

BEGONIA, Mapleleaf *Begonia dregei*.

BEGONIA, Metal-leaf *Begonia metallica*.

BEGONIA, Strawberry *Saxifraga sarmentosa* (Mother-of-thousands) Gannon. There is Strawberry Plant in Devonshire, too - the name is given because of its method of reproduction by runners. Begonia, perhaps from the shape of the leaves?

BEGONIA, Wild *Rumex venosus* (Veined Dock) Kansas Gates. It is also called Wild Hydrangea there.

BEGONIA TREE-BINE *Cissus discolor*.

BEGOON *Solanum melongena* (Egg Plant) W.Miller.

BEKKABUNG *Veronica beccabunga* (Brooklime) Shet. Grigson. Old Norse bekkr, brook + bung, the name of a plant. This, then, is the exact equivalent of Brooklime, for the last element of the common name is a plant name like bung.

BEL-BUTTONS *Caltha palustris* (Marsh Marigold) Tynan & Maitland. 'Bill Buttons', presumably.

BELAH *Casuarina lepidophloia*.

BELDAIRY; BILDAIRY 1. *Orchis mascula* (Early Purple Orchis) Aber. B & H. 2. *Orchis morio* (Green-winged Orchis) Aber. B & H. Whatever the meaning, it seems to have been forgotten in Scotland, for it is corrupted into Bull-dairy in parts.

BELDER-ROOT; BELDRUM *Oenanthe crocata* (Hemlock Water Dropwort) IOW B & H (Belder-root); Pemb. Grigson (Beldrum).

BELENE *Hyoscyamus niger* (Henbane) B & H. Probably bellen, i.e. furnished with bells. One of the OE names was haennebelle, henne-belle in some early ME texts, still existing as Hen-bell. But the French is mort-aux-poules, which is exactly what henbane means, and it is certainly deadly to poultry.

BELGIAN CAMOMILE *Chamaemelon nobile* Mitton. Is it cultivated particularly in Belgium?

BELGIAN ENDIVE *Cichorium intybus* (Chicory) Hemphill. It is also called French Endive sometimes. Chicory is cultivated on the Continent for the root, which is used either as an adulterant to, or a substitute for, coffee.

BELGIAN RED WILLOW *Salix alba var. cardinalis*.

BELL, Bog *Andromeda polifolia* (Bog Rosemary) Som. Macmillan.

BELL, Death *Fritillaria meleagris* (Snake's Head Lily) Cumb. B & H. This becomes Deith Bell in Scotland. "...from the dingy, sad colour of the bell-shaped flowers" (B & H).

BELL, Gold *Narcissus pseudo-narcissus* (Daffodil) Wilts. Macmillan.

BELL, Heather 1. *Campanula rotundifolia* (Harebell) Dor. Macmillan. 2. *Erica tetralix* (Cross-leaved Heath) Scot. B & H. Descriptive - shape of the flowers.

BELL, Hen- *Hyoscyamus niger* (Henbane) B & H. One of the OE names was haen-nebelle, and in some early ME texts it is henne-belle. So this is a direct descendant. Bellen would mean furnished with bells, but this is probably a misreading, and it should mean the same as bane - it is certainly deadly to poultry.

BELL, Lazarus *Fritillaria meleagris* (Snake's Head Lily) Dev. Friend. This is leper's bell - the reference is to the bells that lepers were made to carry and ring when approaching any other people.

BELL, Old Man's *Endymion nonscriptus* (Bluebell) Grieve.1931. Descriptive - there are a number of 'bell' names for bluebell in English dialect. The 'old man' part of this name is also mirrored in a number of "Granfer" names; perhaps the bent or drooping head is suggestive of age.

BELL, Turban *Nigella damascena* (Love-in-a-mist) Dor. Macmillan.

BELL-BOTTLE *Endymion nonscriptus* (Bluebell) Bucks. B & H. Descriptive - there are a number of 'bell' names for bluebell in English dialect. 'Bottle' in a plant name usually implies a bee-attracting plant (a honeysuckle, in fact), but it would appear that bluebells flower too early to be significant in this respect.

BELL FLOWER, Chilean *Lapageria rosea*.

BELL HEATH *Erica tetralix* (Cross-leaved Heath) Som. Grigson; Hants. Cope.

BELL-HEATHER *Erica cinerea*. Bell Ling also occurs in Yorkshire.

BELL LING *Erica cinerea* (Bell Heather) Yks. Grigson.

BELL PEPPER 1. *Capsicum annuum* (Chile) USA Cunningham. 2. *Capsicum grossum*.

BELL-ROSE *Narcissus pseudo-narcissus* (Daffodil) B & H.

BELL THISTLE *Cirsium vulgare* (Spear Plume Thistle) War, Yks. Grigson. Probably a corruption of Bull Thistle, itself a corruption of Bur Thistle.

BELL TREE *Fuchsia magellanica* (Wild Fuchsia) Cork A Taylor.

BELL-WEED 1. *Centaurea nigra* (Knapweed) W.Miller. 2. *Nicandra physalodes* (Shoo-fly Plant) S.Africa Watt. "Bell" for knapweed is OE boll, which means any circular body. The word has become variously ballweed, boleweed, bow-weed, bullweed, and a lot of others.

BELL WOODBIND *Calystegia sepium* (Great Bindweed) B & H. Bellbind, Bellbine, Bellbinder, etc also occur.

BELLADONNA *Atropa belladonna* (Deadly Nightshade) B & H. Literally, fair lady, which is also a book name for the plant. It refers to an ancient belief that the nightshade is the form of a witch called Atropai, who in fact was the eldest of the Fates, whose duty it was to cut the thread of life. There is an old superstition that at certain times the plant takes the form of an exceedingly beautiful enchantress, at whom it is dangerous to look. The name also refers to the custom, common on the Continent at one time, for women to use it as a cosmetic to make their eyes sparkle. Atropine is still used by oculists to dilate the pupils of the eyes.

BELLADONNA LILY *Amaryllis belladonna*. Large doses from the bulb are very poisonous, having a cardiac action.

BELLBIND 1. *Calystegia sepium* (Great Bindweed) Som. Macmillan; Ess, Suff. Grigson. 2. *Convolvulus arvensis* (Field Bindweed) Som. Macmillan. Ess, Norf, Cambs. Grigson. Presumably from the bell-shaped flowers, but there is such a welter of names which are near misses to this (like Bearbind, Bedwind, etc.) that it is difficult to sort them out and decide which is a corruption of what.

BELLBINDER *Calystegia sepium* (Great Bindweed) B & H.

BELLBINE *Calystegia sepium* (Great Bindweed) Suff. B & H. As Bellbind, etc.

BELLBINE, American *Calystegia sylvestris*.

BELLBLOOM *Narcissus pseudo-narcissus* (Daffodil) Halliwell. Cf. Bellflower, and various other "bell" names.

BELLE ISLE CRESS *Barbarea verna* (American Land Cress) B & H. Belle Isle is a small island between Newfoundland and Labrador. This plant is common there.

BELLFLOWER 1. *Abutilon avicennae* (China Jute) McDonald. 2. *Calystegia sepium* (Great Bindweed) Tynan & Maitland. 3. *Campanula sp.* 4. *Cheiranthus cheiri* (Wallflower) Parkinson.1640. 5. *Narcissus pseudo-narcissus* (Daffodil) Som. Jennings. Descriptive for the campanula, which is late Latin campana. Descriptive, too, for the daffodil, for which there are a number of "bell" names, such as Babies' Bells, a Welsh name. People said that only infants and young children could hear them ringing. Bellflower for the wallflower seems to be Parkinson's own invention - or was it an error for Beeflower, which he uses elsewhere?

BELLFLOWER, Autumn *Gentiana pneumonanthe* (Marsh Gentian) Gerard. Descriptive - Harvest Bells and Autumn Bells are other names for this.

BELLFLOWER, Bearded *Campanula barbata*.

BELLFLOWER, Bedstraw *Campanula aparinoides* (Swamp Bluebell) USA House.

BELLFLOWER, Canary *Canaria canariensis*. Canary Isles,of course,not the bird or the associated colour.

BELLFLOWER, Chimney *Campanula pyramidalis*. A very tall species, hence this name, and one or two "steeple" names recorded. But the real reason seems to be that,being very popular in Victorian times,they were used to put in fireplaces during the summer.

BELLFLOWER, Chinese *Platycodon grandiflorum* (Balloon Flower) Coats.

BELLFLOWER, Clustered *Campanula glomerata*.

BELLFLOWER, Corn *Legousia hybrida* (Venus' Looking Glass).

BELLFLOWER, Creeping *Campanula rapunculoides*.

BELLFLOWER, European *Campanula rapunculoides* (Creeping Bellflower) USA House.
BELLFLOWER, Giant 1. *Campanula latifolia*. 2. *Ostrowskia magnifica* (Great Oriental Bellflower). A.W.Smith.
BELLFLOWER, Hanging *Enkianthus quinqueflorus*.
BELLFLOWER, Heath *Campanula rotundifolia* (Harebell) Curtis.
BELLFLOWER, Italian *Campanula isophylla*.
BELLFLOWER, Ivy-leaved *Wahlenbergia hederacea*.
BELLFLOWER, Marsh *Campanula aparinoides* (Swamp Bluebell) USA House.
BELLFLOWER, Marsh, Blue *Campanula uliginosa*.
BELLFLOWER, Milky *Campanula lactiflora*.
BELLFLOWER, Narrow-leaved *Campanula persicifolia* (Peach-leaved Bellflower) Polunin.
BELLFLOWER, Nettle-leaved *Campanula trachelium*.
BELLFLOWER, Olympic *Campanula grandis*. But its native habitat is Siberia!
BELLFLOWER, One-flowered *Campanula uniflora*.
BELLFLOWER, Oriental, Great *Ostrowskia magnifica*.
BELLFLOWER, Peach-leaved *Campanula persicifolia*.
BELLFLOWER, Pyrenean *Campanula speciosa*.
BELLFLOWER, Rampion *Campanula rapunculus*. Rampion is the name given to members of the genus *Phyteuma*.
BELLFLOWER, Round-leaved *Campanula rotundifolia* (Harebell) Jamieson.
BELLFLOWER, Soft *Campanula mollis*.
BELLFLOWER, Spreading *Campanula patula*.
BELLFLOWER, Steeple *Campanula pyramidalis* (Chimney Bellflower) Parkinson. Steeple Bells also occurs in Devonshire.
BELLFLOWER, Tall *Campanula americana* (American Bluebell) USA House.
BELLFLOWER, Taurian *Campanula rapunculoides* (Creeping Bellflower) Allan. It was found in the 18th century in Tauria, and seed sent to the Chelsea Physic Garden, where they were successfully raised. But John Tradescant was growing them in his garden two hundred years before that.
BELLFLOWER, White *Calystegia sepium* (Great Bindweed) Tynan & Maitland.
BELLFLOWER, Yellow *Campanula thyrsoides*.
BELLIES, Fat *Silene cucubalis* (Bladder Campion) Som. Macmillan. Obviously, from the inflated calyx.
BELLOWS, Dead Man's 1. *Ajuga reptans* (Bugle) Berw. Grigson. 2. *Digitalis purpurea* (Foxglove) N Eng. Grigson. 3. *Pedicularis palustris* (Red Rattle) Som. Macmillan; Berw. Grigson. Bellows here is pillies - male members. What is the connection, though?
BELLOWS-FLOWER *Dicentra spectabilis* (Bleeding Heart) Som. Macmillan. Presumably descriptive (shape of the flowers).
BELLRAGGES *Rorippa amphibia* (Greater Yellow Cress) Turner.
BELLROSE *Narcissus pseudo-narcissus* (Daffodil) Som. Macmillan. It is actually pronounced bulrose, but it is a descriptive name, and there are a number of other 'bell' names for it. The 'rose' part of the name occurs again in Lent Rose.
BELLS *Fuchsia sp* Ches. B & H. Descriptive.
BELLS, Autumn 1. *Gentiana pneumonanthe* (Marsh Gentian) Prior. 2. *Gentianella amarella* (Autumn Gentian) Grigson. Descriptive, both from the appearance and the time of blooming.
BELLS, Babies' *Narcissus pseudo-narcissus* (Daffodil) Wales Trevelyan. The 'bell' part of the name is obvious, but people say in some parts of Wales that only infants and young children can hear them ringing.
BELLS, Bloody *Digitalis purpurea*

(Foxglove) Lanark. Grigson.

BELLS, Bow *Anemone nemorosa* (Wood Anemone) Worcs. Grigson. Probably from no other reason than the fact that Bow Bells is a kind of national byword.

BELLS, Candlemas 1. *Anemone nemorosa* (Wood Anemone) Som. Mabey & Evans. 2. *Galanthus nivalis* (Snowdrop) Wilts. Macmillan; Glos. B & H; Scot. Banks. Purification Flower, Fair Maids of February, reduced to Fair Maids in Norfolk, are other names on the same theme. The snowdrop was dedicated to the Purification of the Virgin (Candlemas) on 2 February, when it was the custom for young women dressed in white to walk in procession at the feast. One stage from this would mean that it was sacred to virgins in general, and at one time the receipt of snowdrops from a lady meant to a man that his attentions were not wanted. Near Hereford Beacon, which is said to be the one place in Britain where snowdrops may possibly be really native, a bowl of the flowers brought into the house on Candlemas Day is thought to purify the house. The time of flowering also accounts for the name as applied to wood anemone. Cf. Candlemas Caps.

BELLS, Canterbury 1. *Campanula glomerata* (Clustered Bellflower) Lyte. 2. *Campanula medium*. 3. *Campanula trachelium* (Nettle-leaved Bellflower) Gerard. 4. *Cardamine pratensis* (Lady's Smock) Norf. Gerard. It is said that the name derives from the similarity of the flowers to the bells carried by the horses of pilgrims going to Becket's tomb in Canterbury. Genders(1976), though, said the name was taken from the bells of Canterbury Cathedral, and that the flower was used by Chaucer's pilgrims as their emblem. Coventry Bells is often substituted. The Nettle-leaved Bell-flower is the original of the garden Canterbury Bell. It is only Gerard who claims the name in Norfolk for the Lady's Smock.

BELLS, Church 1. *Campanula medium* (Canterbury Bells) Som. Macmillan. 2. *Cymbalaria muralis* (Ivy-leaved Toadflax) Dor. Macmillan. 3. *Symphytum officinale* (Comfrey) Som. Macmillan.

BELLS, Cockle *Arctium lappa* burrs (Burrs) Corn. Jago. Cockle here is probably the same as cuckold (both names occur for the plant and the burrs).

BELLS, Coral *Heuchera sanguinea*. A combination of colour- and form-descriptive.

BELLS, Coventry 1. *Campanula medium* (Canterbury Bells) Gerard. 2. *Campanula trachelium* (Nettle-leaved Bellflower) Prior. 3. *Digitalis purpurea* (Foxglove) Dor. Macmillan. 4. *Pulsatilla vulgaris* (Pasque Flower) Cambs. Gerard. The Campanulas seem to bear the name as an alternative to Canterbury Bells. The foxglove, with a bell-shaped flower, is just a copy of the Campanula name, while the Pasque Flower name rests solely with Gerard - "in Cambridge-shire where they grow, they are named Coventrie Bels".

BELLS, Cow *Silene cucubalis* (Bladder Campion) Som. Grigson. There are various 'cow' names for this; it is said that they arise from a belief that bladder campion made cows want the bull.

BELLS, Crow 1. *Endymion nonscriptus* (Bluebell) Prior. 2. *Narcissus pseudo-narcissus* (Daffodil) B & H. 3. *Ranunculus bulbosus* (Bulbous Buttercup) War. Grigson.

BELLS, Crow, Yellow *Narcissus pseudo-narcissus* (Daffodil) B & H. The inference from this name is that Crow Bells should be reserved for bluebell.

BELLS, Dangle; BELLS, Dangling *Convallaria maialis* (Lily-of-the-valley) Som. Macmillan.

BELLS, Dead Man's 1. *Digitalis purpurea* (Foxglove) Dor. Macmillan; Dur, N'thum. Grigson; Scot. Jameson. 2. *Fritillaria meleagris* (Snake's Head Lily) Friend. 3. *Silene maritima* (Sea Campion) Mor. PLNN22. The foxglove has a number of similar names, either starting with "Bloody", e.g. Bloody Fingers, or "Dead Man's", e.g. Dead Man's Fingers. All spring from the belief that where the plant grows someone has been buried. As far as Dead Man's Bells is concerned, there was a belief that to hear them ring meant you were going to die soon. For the fritillary, this and other similar names come from "the dingy, sad colour of the bell-shaped flowers".

BELLS, Doleful *Atropa belladonna* (Deadly Nightshade) Dor. Macmillan.

BELLS, Dong *Narcissus pseudo-narcissus* (Daffodil) Som. Macmillan.

BELLS, Drooping *Galanthus nivalis* (Snowdrop) Som. Macmillan.

BELLS, Easter *Stellaria holostea* (Greater Stitchwort) Dev. Friend.1882. It is Easter Flower in Sussex, and Whit Sunday, or White Sunday, in Devonshire,too. So the 'bells' part of this name is probably purely fortuitous; the flowers are not obviously bell-shaped.

BELLS, Fairy 1. *Campanula rotundifolia* (Harebell) Dor, Som. Grigson; IOM Moore.1924. 2. *Convallaria maialis* (Lily-of-the-valley) Som. Macmillan. 3. *Digitalis purpurea* (Foxglove) Som. Grigson; N'thum. Denham/Hardy; USA Henkel. 4. *Disporum sp.* 5. *Narcissus pseudo-narcissus* (Daffodil) Dor. Grigson. 6. *Oxalis acetosella* (Wood Sorrel) Wales Northcote. 7. *Primula veris* (Cowslip) Som. Grigson. Not only are they all bell-shaped (including Wood Sorrel, at a pinch), but all these plants have distinct fairy connections.

BELLS, Fairy's *Endymion nonscriptus* (Bluebell) Som. Macmillan.

BELLS, Gold *Narcissus pseudo-narcissus* (Daffodil) Wilts. Macmillan.

BELLS, Golden *Primula veris* (Cowslip) Som. Macmillan.

BELLS, Hare's, English *Endymion nonscriptus* (Bluebell) Parkinson.1629. Gerard has English Harebell - he decided to give it the name "for that it is thought to grow more plentifully in England than elsewhere." Bluebell is still called Harebell in Devonshire, and the name is recorded in Shakespeare's time.

BELLS, Harvest 1. *Campanula rotundifolia* (Harebell) Flower.1859. 2. *Gentiana pneumonanthe* (Marsh Gentian) Gerard. A late-flowering gentian, as the names Autumn Bells, Autumn Violet, etc., testify. Harebell is also late-flowering.

BELLS, Hedge 1. *Calystegia sepium* (Great Bindweed) Som. Macmillan; Hants, IOW Grigson. 2. *Convolvulus arvensis* (Field Bindweed) Gerard.

BELLS, Hen *Hyoscyamus niger* (Henbane) B & H. OE names were haennebelle and haennepol. See also Belene, which on the face of it would mean the same as this.

BELLS, Ivy *Caltha palustris* (Marsh Marigold) Som. Macmillan.

BELLS, Lint *Linum usitatissimum* (Flax) Scot. B & H. Lint, of course, is the other name for flax.

BELLS, London *Calystegia sepium* (Great Bindweed) Dev. Macmillan.

BELLS, Monkey *Caltha palustris* (Marsh Marigold) Som. Macmillan.

BELLS, Old Man's 1. *Campanula rotundifolia* (Harebell) Scot Tynan & Maitland. 2. *Endymion nonscriptus* (Bluebell) Scot. Tynan & Maitland. In spite of being the Blue Bells of Scotland, it is often an unlucky plant in Scotland. The 'old man' of the name is usually meant as the devil, and it is known as Witch Bells, or Witches' Thimble in Scotland, too.

BELLS, Peach *Campanula persicifolia* (Peach-leaved Bellflower) Friend.1882.

BELLS, Pop *Digitalis purpurea* (Foxglove) Som. Macmillan. 'Pop', because children used to love to squeeze the unopened buds, which burst with a satisfying pop.

BELLS, Ring o' *Endymion nonscriptus* Lancs. B & H.

BELLS, River *Phygelius capensis* (Cape Figwort) S.Africa Perry.

BELLS, Roanoke *Mertensia virginica* (Virginia Cowslip) Howes.

BELLS, Rock *Aquilegia canadensis* USA House.

BELLS, St George's *Endymion nonscriptus* (Bluebell) Tynan & Maitland.

BELLS, St Peter's *Narcissus pseudo-narcis-*

sus (Daffodil) Wales Grigson.
BELLS, School *Campanula rotundifolia* (Harebell) Wilts. Dartnell & Goddard.
BELLS, Sea *Calystegia soldanella* (Sea Bindweed) Parkinson.1640.
BELLS, Sheep *Campanula rotundifolia* (Harebell) Dor. Grigson.
BELLS, Silver 1. *Anemone nemerosa* (Wood Anemone) Som. Macmillan. 2. *Viburnum opulus* (Guelder Rose) Wilts. Dartnell & Goddard. It is the garden, double, variety of Guelder Rose that gets this name. Macmillan records it for Cherhill, Wiltshire.
BELLS, Spring *Sisyrinchium douglassi* W.Miller.
BELLS, Steeple *Campanula pyramidalis* (Chimney Bellflower) Dev. B & H. Parkinson has Steeple Bellflower, too.
BELLS, Velvet *Bartsia alpina* (Alpine Bartsia) H & P.
BELLS, Water *Nymphaea alba* (White Waterlily) N.Eng. Grigson.
BELLS, White 1. *Convallaria maialis* (Lily-of-the-valley) Som. Macmillan. 2. *Galanthus nivalis* (Snowdrop) Som. Macmillan. 3. *Stellaria holostea* (Greater Stitchwort) Som. Macmillan.
BELLS, White, Little *Convallaria maialis* (Lily-of-the-valley) Som. Macmillan.
BELLS, Witch 1. *Campanula rotundifolia* (Harebell) Scot. Jamieson. 2. *Centaurea cyanus* (Cornflower) N.Eng. B & H; Scot. Jamieson. 3. *Digitalis purpurea* (Foxglove) N.Eng. Henderson. Harebell is an unlucky plant in parts of Scotland, in spite of being the bluebell of Scotland. As well as Witch Bells, there are also Witches' Thimble, and Old Man's Bell (old man in cases like this usually means the devil). In Sweden, it is called 'bell of the nightmare'. Foxglove is also a good candidate for the Witches' Bell - it is a fairy and witch plant, an unlucky plant to have in the house, or so they say in Scotland, and unlucky to have on board a ship for any purpose. There are many witch and fairy names for this. Cornflower is not so obvious - it is not bell-shaped to start with, and there is no connection with witchcraft that I know. And yet as well as Witch Bells, Witches' Thimble is also recorded in the North Country. Bell is probably a misreading for boll in this case.
BELLS, Wood *Endymion nonscriptus* (Bluebell) Bucks. B & H.
BELLS, Yavering *Ramischia secunda*. There is a standing stone at Yevering, in Glendale, known as Yevering Bell. Denham wondered if there was any connection with this very rare wintergreen.
BELLS, Yellow 1. *Narcissus pseudo-narcissus* (Daffodil) Som. Macmillan. 2. *Tecoma stans* (Yellow Elder) Everett.
BELLS OF IRELAND *Molucella laevis* (Molucca Balm) Perry.
BELLS OF SODOM, Drooping; BELLS OF SODOM, Mournful; BELLS OF SODOM, Solemn *Fritillaria meleagris* (Snake's Head Lily) all Dor. Grigson (Drooping); Dacombe (Mournful); Macmillan (Solemn). Sodom is probably a corruption of Sorrow - there is a name, Doleful Bells of Sorrow, recorded from Oxfordshire.
BELLS OF SORROW, Doleful *Fritillaria meleagris* (Snake's Head Lily) Oxf. Grigson. See the Bells of Sodom entries.
BELLWIND 1. *Calystegia sepium* (Great Bindweed) Bucks, Surr. Grigson. 2. *Convolvulus arvensis* (Field Bindweed) Bucks. Grigson.
BELLWINDER *Calystegia sepium* (Great Bindweed) B & H.
BELLY, Cow *Heracleum sphondyllium* (Hogweed) Som. Macmillan. Cow Billers is probably the original of this.
BELLY, Fish 1. *Cirsium eriophyllum* (Woolly Thistle) Cumb. B & H. 2. *Cirsium heterophyllum* (Melancholy Thistle) Cumb. Grigson. Possibly from the white underside of the leaves.
BELLYACHE BUSH *Jatropha gossypifolia*. An expressive enough name - that is what happens if you eat any part of it - perhaps not as severe as the poisoning caused by its near relative *J multifida*.

BELLYWIND *Clematis vitalba* (Old Man's Beard) Hants. Grigson. Merely a variation in the series which includes Bedwind, Bethwind, etc.

BELOTE OAK *Quercus gramuntia*. A Spanish species.

BELT, My Lady's *Filipendula ulmaria* (Meadowsweet) Scot. B & H. This would be a translation from the Gaelic.

BELTAINE, Shrub of *Caltha palustris* (Marsh Marigold) Ire. Wilde.1925. Beltaine is the May festival (Cf. the name May-flower, recorded in the west of England and in Ireland). The plant was certainly connected with the May Day festivities; in Shropshire, it was hung stalk upwards on doorposts on May Day, and in Ireland it was used extensively to protect cattle, etc., in the same way as rowan was used elsewhere.

BELVEDERE *Kochia scoparia* (Burning Bush) Gorer. This is, Gorer claims, from Parkinson. Is he right? Possibly - it is difficult to tell from the description Parkinson gives to Scoparia sive Belvidere Italorum - Broome Tode Flaxe.

BELWEED *Centaurea nigra* (Knapweed) B & H. This is probably from OE boll - any circular body. A whole series of the names exist - Bullweed, Ballweed, Bollweed, Boleweed, Bow-weed, etc. All refer to the shape of the involucre.

BEN 1. *Moringa oleifera* Turner. 2. *Silene cucubalis* (Bladder Campion) B & H. Gerard calls the latter White Ben. Ben is a corruption of Behen, an old name for the plant being cucubalis Behen. And Gerard describes it under the name Behen album, hence his White Ben.

BEN, White 1. *Silene cucubalis* (Bladder Campion) Gerard. 2. *Silene latifolia* USA House. In the latter case, Gerard's name travelled to America - this plant is also called Bladder Campion there. See Ben for explanations.

BEN-NUT TREE *Moringa oleifera* (Ben) Notes & Queries.1941.

BEND; BENT *Calluna vulgaris* (Heather) Ches. B & H. Presumably the same as bent, as applied to many grasses.

BENDOCK *Oenanthe crocata* (Hemlock Water Dropwort) Kent Grigson. Perhaps a corruption of bane-dock - the plant is highly poisonous.

BENE OF EGYPT *Nelumbo nucifera* (Hindu Lotus) Turner.

BENET, Herb *Geum urbanum* (Avens) J A MacCulloch.1905. See Bennett, Herb.

BENEWITH *Lonicera periclymenum* (Honeysuckle) Halliwell.

BENGAL CARDAMOM *Amomum aromaticum.*

BENGAL CLOCKVINE *Thunbergia grandiflora* (Sky Vine) Simmons.

BENGAL FIG *Ficus benghalensis* (Banyan) Rochford.

BENGAL GAMBOGE *Garcinia venulosa*. Gamboge is from Cambodia, whence it was brought around 1600.

BENGAL GRAM *Cicer arietinum* (Chick Pea) Douglas.

BENGAL INDIA-RUBBER TREE *Ficus elastica* (India Rubber Plant) Lindley.

BENGAL QUINCE *Aegle marmelos* (Bael) Brouk.

BENIN CAMWOOD *Baphia pubescens.*

BENIN MAHOGANY, BENIN WOOD *Khaya grandifoliola*. Benin Wood is the trade name for the timber.

BENIN PEPPER *Piper guineense* (West African Black Pepper) Dalziel. Cf. Ashanti Pepper for this.

BENISEED, Black *Polygala butyracea* Sierra Leone Dalziel.

BENJAMIN, Stinking *Trillium erectum* (Bethroot) Howes.

BENJAMIN BANYAN; BENJAMIN TREE *Ficus benjaminae* (Weeping Fig) Hora (Banyan); Howes (Tree).

BENJAMIN BUSH *Lindera benzoin* (Spice Bush) USA Yarnell. 'Benjamin' is a recognized derivative of *benzoin*, and of the gum obtained.

BENNEL *Senecio jacobaea* (Ragwort) Ire. Grigson. Obviously related to Benweed and its variants.

BENNELS, Lint *Linum usitatissimum* (Flax) Scot. B & H.

BENNERGOWAN *Bellis perennis* (Daisy) W.Miller. Cf. Bannergowan. Both of these are allied to the name Banwort (bonewort). "The Northern men call this herbe a Banwort because it helpeth bones to knyt agayne" (Turner).

BENNERT *Bellis perennis* (Daisy) Cumb. Grigson. This is the same word as Banwort, where ban means bone - daisies had a reputation as a consound.

BENNET 1. *Pimpinella saxifraga* (Burnet Saxifrage) Shrop. Grigson. 2. *Plantago lanceolata* seed heads (Ribwort Plantain). Wilts. Dartnell & Goddard. 3. *Plantago major* seed heads (Great Plantain) Wilts. Dartnell & Goddard. In the first case, Bennet must be a corruption of Burnet, but in the others, it is the same word as Bents.

BENNET, Herbal *Geum urbanum* (Avens) Glos. Grigson. See Herb Bennett.

BENNETT, Herb 1. *Conium maculatum* (Hemlock) Gerard. 2. *Geum urbanum*. 3. *Prunella vulgaris* (Self-heal) Som. Macmillan. 4. *Valeriana officinalis* (Valerian) Som. Macmillan. Herb Bennett comes directly from herba benedicta - blessed herb. "Where the root is in the house the Devil can do nothing, and flies from it; wherefore it is blessed above all other herbs." In spite of being called St Benedict's Herb, it is nothing to do with the saint, though the legend of the connection grew up steadily. St Benedict was the founder of the Benedictine order, and he was a hard ruler. His monks plotted to murder him by poisoning his wine. St Benedict made the sign of the cross over the glass, and it flew into pieces - so by association of the name, Avens became known as an antidote to poison. Why, then, should hemlock bear the name? Gerard noted the name,and St Bennet's Herb travelled to America, too. Earle notes herba benedicta for this in a C13 vocabulary. The name as applied to valerian seems straightforward enough. It is an All-heal (Guérit-tout in French). Valerian itself derives from Latin valere, to be well, be in good health. The association in the same area with the similarly styled Self-heal, is obvious.

BENNETT, Way 1. *Geum urbanum* (Avens) Friend. 2. *Lolium temulentum* (Darnel) Coles. The Bennett part of these names seems straightforward - see under Herb Bennett. But 'Way' is difficult. One is at first tempted to see OE weg here, in the sense that avens, at least, can be seen as a roadside herb. But darnel is a rye-grass, and Barton & Castle have recorded Ray-grass, while Culpepper has an intermediate form, Wray. So it is interesting to find that Fernie has Wild Rye for Avens. Barton & Castle want to ignore the derivation accepted for Rye, and suggest that Ray is probably from the French ivraie, drunkenness. Then if Way is the same word as Rye, then Bennett is *not* the "benedicta" of Herb Bennett at all, but is more likely to be from OE beonet, a word still preserved as Bent, a stiff, rush-like grass.

BENNISEED *Sesamum indicum* (Sesame) Notes & Queries.1941. Sesame oil is known as benne in West Africa.

BENT 1. *Ammophila arenaria* (Marram-grass) Shet. Nicolson. 2. *Plantago lanceolata* seed heads (Ribwort Plantain). Wilts. Dartnell & Goddard. 3. *Plantago major* seed heads (Great Plantain) Wilts. Dartnell & Goddard.

BENT, Aye-no *Lolium perenne* (Rye-grass) Glos. Grigson.1959. Because of the love-me, love-me-not game children used to play with it.

BENT, Black 1. *Agrostis gigantea*. 2. *Alopecurus myosuroides* (Slender Foxtail) H C Long.1910. 3. *Plantago lanceolata* (Ribwort) Bucks. B & H. Bent is a grass, OE beonet. Black is colour descriptive.

BENT, Brittle *Agrostis setacea*.

BENT, Brown *Agrostis canina*.

BENT, Colonial *Agrostis tenuis* (Fine Bent) Howes.

BENT, Common *Agrostis tenuis* (Fine Bent) Howes.

BENT, Couchy *Agrostis stolonifera* (White Bent) Wilts. Dartnell & Goddard.
BENT, Creeping *Agrostis stolonifera* (White Bent) McClintock.
BENT, Fine *Agrostis tenuis*.
BENT, Rhode Island *Agrostis canina* (Brown Bent) H & P.
BENT, Sea *Ammophila arenaria* (Marram) A Duncan.
BENT, Spike *Agrostis exarata*.
BENT, Sweet *Luzula campestris* (Good Friday Grass) Scot. B & H.
BENT, Velvet *Agrostis canina* (Brown Bent) Howes.
BENT, Water *Agrostis semiverticillata*.
BENT, White *Agrostis stolonifera*.
BENTWOOD; BENWOOD *Hedera helix* (Ivy) Berw. Grigson (Bentwood); Scot. Jamieson (Benwood). Presumably OE ben-den, which is the same word as bind, producing all the bindweeds and bindwoods (the latter is recorded in Scotland for ivy), and refers to the plant's habit.
BENWEED *Senecio jacobaea* (Ragwort) Scot, Ire. Grigson. Probably the same as Bindweed, which is recorded, with other variations, for Ragwort. According to Salisbury, bindweed is given as a name because one of the symptoms of jacobine poisoning in animals is extreme constipation.
BERBER *Berberis vulgaris* (Barberry) Scot. Jamieson. A simple contraction of Berberis.
BERBINE *Verbena officinalis* (vervain) Kent B & H. An old dialect form of vervain.
BERE; BEAR *Hordeum sativum* (Barley). See Bear.
BERG CYPRESS *Widdringtonia cupressoides* (Cypress Pine) Hora.
BERGAMOT 1. *Monarda sp.* 2. *Citrus aurantium var. bergamia* (Bergamot Orange) Rimmel. It is suggested in ODEE that this is derived from the town of Bergamo, in Italy. If this is so, it probably refers to the aromatic oil distilled from both the citrus and the herb. But it is much more likely that the town is Bergama, in Turkey, and not the Italian one.There is a kind of pear called a bergamot (Fr. bergamotte, Italian bergamotta, from Turkish begarmudi). The armudi part of this means pear, and beg means prince.
BERGAMOT, Prairie *Monarda citriodora*.
BERGAMOT, Purple *Monarda fistulosa*.
BERGAMOT, Red *Monarda didyma*.
BERGAMOT, Wild *Monarda fistulosa* (Purple Bergamot) USA House.
BERGAMOT, Wild, Pale *Monarda mollis*.
BERGAMOT MINT *Mentha citrata*.
BERGAMOT ORANGE *Citrus aurantium var. bergamia*. This is the species from which the perfume, also called bergamot, is obtained. The name derives from the town of Bergama, in Turkey.
BERK-APPLE *Pinus sylvestris* cones (Pine cone) Yks. B & H. Berk? Is it birk, rather - an old name for birch? Apple is often applied to the cones, and called,for instance, Deal Apples.
BERLIN POPLAR *Populus x berolinensis*.
BERMUDA ARROWROOT *Maranta arundinacea* (Arrowroot) W.Miller. Given here as the name of the plant, but more usually applied to the starch product.
BERMUDA BUTTERCUP *Oxalis pes-caprae*.
BERMUDA CEDAR *Juniperus bermudiana*.
BERMUDA LILY *Lilium longiflorum var. eximium*. A Japanese plant, but much grown since last century in the Bermuda lily industry.
BERRY, Red *Rhamnus californica*.
BERRY-GIRSE *Empetrum nigrum* (Crowberry) Shet. Grigson.
BERRY-TREE; BERRY-BUSH 1. *Flacourtia flavescens* (Niger Plum) Dalziel. 2. *Ribes uva-crispa* (Gooseberry) Yks. Morris; Ches. Holland; Lincs. Peacock; Dur. Dinscombe. Gooseberries are "berries", par excellence, and gooseberry pies are berry-pies. So they are in co Durham, and in Cheshire, too.
BERTERY; BUTTERY *Sambucus nigra* (Elder) both Yks. B & H. Both these are variants of Bour-tree.
BERTOLONIA *Gravesia guttata*.

BERTRAM 1. *Anacyclus pyrethrum* (Pellitory-of-Spain) Gerard. 2. *Tanacetum parthenium* (Mayweed) Prior. "after the Dutch name", according to Gerard, but it is more likely that these are simple corruptions of Pyrethrum, as is Halliwell's Bartram.

BESOM 1. *Calluna vulgaris* (Heather) Corn. Jago. 2. *Cytisus scoparius* (Broom) Som. Elworthy. OE besema, besma, from Teutonic originals, meaning unknown, but certainly coming to mean a broom. Both the plant broom and heather twigs were used for making brooms ("...some use to make brushes of heath both in England and in Germany" (Turner)). There are many variations on the word besom in dialect names for heather.

BESOM, Green *Cytisus scoparius* (Broom) Som. Elworthy.

BESOM HEATH *Erica tetralix* (Cross-leaved Heath) Hulme. See BESOM.

BESS, Bouncing 1. *Centranthus ruber* (Spur Valerian) Dev. Wright; Dor. Macmillan. 2. *Valeriana celtica* Dev. Friend.1882.

BESS, Crazy *Caltha palustris* (Marsh Marigold) Dor, Wilts. D & G.

BESS, Delicate 1. *Centranthus albus* Dev. B & H. 2. *Valeriana celtica* Dev. Friend.1882. Is this the same plant?

BESWIND; BESWINE 1. *Calystegia sepium* (Great Bindweed) Hants. Cope. 2. *Convolvulus arvensis* (Field Bindweed) S.Eng. Wakelin. (Beswind); Wessex Rogers (Beswine). One of the seemingly unending series of Bethwind, Bedwind, and so on.

BET, Bouncing *Saponaria officinalis* (Soapwort) Dor. Macmillan; USA Tolstead. Parkinson called it this in *Paradisus in Sole*, 1629. "Bouncing" because the plant does seem to do just that, rooting at the leaf nodes. Bouncing Betty and Sweet Betty are other names, though in Britain only.

BET, Crazy 1. *Caltha palustris* (Marsh Marigold) Dor. Dacombe; Wilts. Dartnell & Goddard. 2. *Leucanthemum vulgare* (Ox-eye Daisy) Wilts. Dartnell & Goddard. 3. *Nuphar lutea* (Yellow Waterlily) Som. Macmillan. 4. *Ranunculus acris* (Meadow Buttercup) Wilts. Grigson. 5. *Ranunculus bulbosus* (Bulbous Buttercup) Wilts. Grigson. 6. *Ranunculus ficaria* (Lesser Celandine) Wilts. Macmillan. 7. *Ranunculus repens* (Creeping Buttercup) Wilts. Grigson.

BETAYNE; BETEYNE *Stachys officinalis* (Betony) Turner (Betayne); Clair (Beteyne). Both earlier versions of Betony, which becomes Bidny or Bitny in dialect.

BETEL *Piper betel*. From ODEE, apparently originally a Portuguese word, taken from a Malay language which produced vettila.

BETEL NUT *Areca catechu* seed .

BETEL PALM *Areca catechu* (Areca Palm) H & P.

BETES *Polygonum bistorta* (Bistort) Turner.

BETHLEHEM, Sage of *Mentha spicata* (Spearmint) Lincs. Gordon.1985.

BETHLEHEM, Star of 1. *Anagallis arvensis* (Scarlet Pimpernel) Som. Macmillan. 2. *Anemone nemerosa* (Wood Anemone) Som. Macmillan. 3. *Campanula isophylla* (Italian Bellflower) Gannon. 4. *Hypericum calycinum* (Rose of Sharon) B & H. 5. *Ornithogalum umbellatum*. 6. *Ornithogalum thyrsoides* (Chincherinchee) S.Africa Watt. 7. *Passiflora coerulea* (Blue Passion Flower) Som. Macmillan. 8. *Stellaria holostea* (Greater Sitchwort) Dev, Som. Macmillan; Wilts. Dartnell & Goddard.1899; E.Ang, N.Ire. Grigson. 9. *Tragopogon pratensis* (Goat's Beard) Tynan & Maitland. It is said of *O umbellatum* that the flowers resemble the picture of the star that indicated the birth of Jesus, hence the name. Because of this, it is appropriated to the Epiphany (6 January) This name applied to the bellflower is a fairly recent florists' name. Stitchwort and Wood Anemone have a number of 'star' names, from the shape of the flowers, which also gives pimpernel the name. Passion Flower has a number of names from Christian religion.

BETHLEHEM, Star of, Drooping *Ornithogalum nutans*.

BETHLEHEM, Star of, Guernsey *Allium*

neapolitanum (Daffodil Garlic) McClintock.

BETHLEHEM, Star of, Neapolitan *Ornithogalum nutans* (Drooping Star of Bethlehem) Whittle & Cook. Clusius named it O. neapolitanum simply because a consignment of bulbs reached him from Naples.

BETHLEHEM, Star of, Spiked *Ornithogalum pyrenaicum.*

BETHLEHEM COWSLIP; BETHLEHEM SAGE *Pulmonaria officinalis* (Lungwort) Coles (Cowslip); Gerard (Sage). Jerusalem Cowslip and Jerusalem Sage are also applied to the lungwort. The probable reason is the legend invented to account for the spotted leaves - that of the Virgin's milk falling on them during the flight into Egypt. The association between the story and Bethlehem is obvious.

BETHLEM STAR *Ornithogalum umbellatum* (Star of Bethlehem) B & H.

BETHROOT 1. *Trillium erectum.* 2. *Trillium grandiflorum* (Wood Lily) USA Weiner. This seems to be birthroot - it was believed that the Indians used the plant to induce labour in childbirth. The actual name Birthroot occurs for both these two.

BETHWIND 1. *Calystegia sepium* (Great Bindweed) Glos, Bucks, Middx, Herts Grigson. 2. *Clematis vitalba* (Old Man's Beard) Glos. Grigson. 3. *Convolvulus arvensis* (Field Bindweed) S.Eng. Wakelin. 4. *Polygonum convolvulus* (Black Bindweed) Hants. Grigson.

BETHWINE 1. *Clematis vitalba* (Old Man's Beard) Glos. J.D.Robertson. IOW Halliwell; Suss. Parish. 2. *Convolvulus arvensis* (Field Bindweed) Wessex Rogers. Cf. Bethwind, for Old Man's Beard, from Gloucestershire.

BETON; BETONY *Stachys officinalis* (Wood Betony) Turner. Betony came through French bétoine from Latin betonica, itself, as Pliny has it, from vettonica. It is the Gaulish name of a plant discovered by a Spanish tribe, the Vettones. So betony is a Celtic name.

BETONY, Brook *Scrophularia aquatica* (Water Betony) B & H. It is said that the leaves resemble those of true betony - *Stachys officinalis.*

BETONY, Field *Stachys arvensis* (Field Woundwort) W.Miller.

BETONY, Head *Pedicularis canadensis* USA House. Presumably this started out as Wood Betony, which does occur as an American name.

BETONY, Marsh *Stachys palustris* (Marsh Woundwort) W.Miller.

BETONY, Paul's 1. *Veronica officinalis* (Common Speedwell) Browne. 2. *Veronica serpyllifolia* (Thyme-leaved Speedwell) Curtis. Referring to the former, Browne says "...wherof the people have some conceit in reference to St. Paul; whereas, that name is derived from Paulus Aegineta, an ancient physician of Aegina...".

BETONY, Water *Scrophularia aquatica.*

BETONY, Wild *Dryas octopetala* (Mountain Avens) co Clare Grigson. The leaf bears some resemblance to those of true betony, *Stachys officinalis.*

BETONY, Wood 1. *Ajuga chamaepitys* (Yellow Bugle) Ire. B & H. 2. *Ajuga reptans* (Bugle) Ire. Grigson. 3. *Pedicularis canadensis* USA Tolstead. 4. *Stachys officinalis.*

BETSY, Blue 1. *Vinca major* (Greater Periwinkle) Hatfield. 2. *Vinca minor* (Lesser Periwinkle) Dev. Grigson; Som. Macmillan.

BETSY, Crazy *Caltha palustris* (Marsh Marigold) Dor. Grigson; Wilts. Dartnell & Goddard. There is Crazy Bets and Crazy Betty, too, as well as Crazy, Crazy Lilies and Yellow Crazies, all in the same general Wessex area. The names are probably connected with another West Country name - Drunkards; the belief was that if you pick them, or even look long at them, you will take to drink. In much the same way, picking or looking long enough at poppies would give a headache. So it is reasonable to assume that picking or looking long at Marsh Marigold flowers would drive you crazy.

BETSY, Pretty 1. *Centranthus ruber* (Spur Valerian) Hudson. 2. *Saxifraga spathularis x umbrosa* (London Pride) Suff. B & H. Spur Valerian has many names in this vein, like Bouncing Bess, Bouncing Betty, Saucy Bet, Sweet Betsy, etc. But London Pride seems to have only this one in the series.

BETSY, Sweet 1. *Centranthus ruber* (Spur Valerian) Som, Kent Grigson. 2. *Dicentra spectabilis* (Bleeding Heart) Som. Macmillan.

BETTY, Bouncing *Saponaria officinalis* (Soapwort) Dor. Grigson.

BETTY, Crazy *Caltha palustris* (Marsh Marigold) Dor. Grigson; Wilts. Dartnell & Goddard. See Crazy Betsy.

BETTY, Jumping *Impatiens noli-me-tangere* (Touch-me-not) Suss. Parish. Cf. Jumping Jack from the same area. It is the explosive habit of seed distribution that provides the name.

BETTY, Pretty *Saxifraga spathularis x umbrosa* (London Pride) Dor. Macmillan; Worcs. J Salisbury.

BETTY, Red *Lobelia cardinalis* (Cardinal Flower) USA Gilmore.

BETTY, Sweet 1. *Centranthus ruber* (Spur Valerian) Som, Dor. Macmillan. 2. *Saponaria officinalis* (Soapwort) Grieve.1931. Bouncing Bet or Bouncing Betty are also recorded for both these.

BETTY-GO-TO-BED-AT-NOON *Ornithogalum umbellatum* (Star of Bethlehem) Shrop. B & H. "These flowers open themselves at the rising of the Sunne, and shut againe at the Sun-setting; whereupon this plant hath been called by some, Bulbus Solsequius" (Gerard). A number of plants have this characteristic, and so have various names to record the fact, the most common of which is Jack-go-to-bed-at-noon (also recorded for this one, which has also Eleven o'clock Lady, Six o'Clock, Sunflower, etc.).

BETTYWIND; BEDDYWIND *Convolvulus arvensis* (Field Bindweed) Wessex Rogers.

BEZORS *Primula auricula* (Auricula) Glos. B & H. This is bear's ears, so called from the shape of the leaf; auricula also confirms the ear shape. Bear's Ears has various corruptions - it becomes Boar's Ears, or Bores' Ears in Scotland. Close to Bezors is the Lancashire Baziers.

BHANDI TREE *Thespesia populnea* (Portia Tree) Howes.

BHUTAN CYPRESS *Cupressus torulosa* (Himalayan Cypress) Mitchell.1972.

BHUTAN PINE *Pinus griffithii*.

BIBIRU-BARK TREE *Nectandra rodiaei* (Greenheart) W.Miller. Bibiru is the native name in Guyana for the tree, and the bark has occasionally been used as a bitter tonic and febrifuge, like quinine.

BIBLE FLOWER 1. *Hypericum androsaemum* (Tutsan) Corn. Grigson. 2. *Tanacetum balsamita* (Balsamint) USA Boland.1977. Bible Leaf and Book Leaf are other names given to Tutsan which refer to the practice of gathering the leaves and pressing them in books, especially a Bible. When dry, they have a very sweet smell, which has been likened to ambergris, hence other names, such as Amber, Sweet Amber, Sweet Leaf. Exactly the same comments apply to Balsamint.

BIBLE FRANKINCENSE *Boswellia carteri*.

BIBLE LEAF 1. *Euphorbia amygdaloides* (Wood Spurge) Dev. Grigson. 2. *Hypericum androsaemum* (Tutsan) Corn, Som. Grigson. See BIBLE FLOWER for the Tutsan. Presumably the Wood Spurge must have been used in Devonshire in the same way for it to bear this name.

BICHY TREE *Cola acuminata* Laguerre.

BIDDY *Lotus corniculatus* (Bird's-foot Trefoil) IOM Moore.1924.

BIDDY'S EYE 1. *Viola arvensis* (Field Pansy) Som. Macmillan. 2. *Viola tricolor* (Pansy) Som. Tongue. Biddy means chick.

BIDIBID *Acaena anserinifolia* (Pirri-pirri Burr) New Zealand C Macdonald. Bidibid is a corruption of the Maori word piripiri.

BIDNY, BITNY *Stachys officinalis* (Wood Betony) Kent Grigson (Bidny); Dev. Fernie (Bitny). Both are variants of betony.

BIFOIL; BIFOIL, Herb *Listera ovata* (Twayblade) Prior (Bifoil); Gerard (Herb Bifoil). This means exactly the same as Twayblade - two-leaved.

BIG-ROOT 1. *Bryonia dioica* (White Bryony) E.Ang. G.E.Evans.1960. 2. *Ipomaea leptophylla* (Bush Morning Glory) USA Vestal & Schultes. 3. *Tamus communis* (Black Bryony) Wilts. Wiltshire; E.Ang. G.E.Evans/Thomson. A literally descriptive name in all cases, which is why either of the bryonies is also known as the English mandrake, the root being looked on as anthropomorphic, and sometimes artificially worked in order to produce the human shape. The Morning Glory is a relative of the sweet potato, and the large root used to be cooked in times of need by many of the Plains Indian tribes. The name is either spelled like this, or as a single word.

BIG-TREE *Sequioa gigantea* (Wellingtonia) Clapham. Wellingtonia is the English name - it is known as Big-tree, or Mammoth-tree in its native North America. It is the largest tree in the world.

BIGARRADE *Citrus aurantium var. bigaradia* (Seville Orange) Rimmel.

BIGCONE PINE *Pinus coulteri* (Coulter Pine) Mitchell.1974.

BIGCONE SPRUCE *Pseudotsuga macrocarpa* (Large-coned Douglas Fir) Brimble.1948.

BIGFACE, Father *Carduus nutans* (Musk Thistle) Som. Macmillan.

BIGG *Hordeum sativum* (Barley) Yks. Dickins. Evidently a name for barley, for bigg-chaff is referred to, and barley chaff is meant. Bigg Market, Newcastle, is also mentioned.

BIGGOTY LADY *Impatiens noli-me-tangere* (Touch-me-not) Som. Macmillan. According to Grigson, biggoty is a dialect word meaning saucy. This is connected in some way with the habit of explosive seed distribution.

BIGOLD *Chrysanthemum segetum* (Corn Marigold) Prior. Explained as tinsel, or false gold.

BIGROOT LADY'S THUMB *Polygonum muhlenbergii* (Swamp Persicaria) USA Yanovsky.

BIGSTRING NETTLE *Urtica dioica* (Nettle) Douglas. If this really is Bigstring, then it must be a reference to the fibres obtainable from the plant.

BILBERRY 1. *Vaccinium myrtillus*. 2. *Vaccinium uliginosum* (Northern Bilberry) USA A.W.Smith. The first element is probably of Norse origin (the Danish is bollebaer), from bolle,the name of the plant. Then berry was unncesssarily added, though bollebaer presumably means ball berry.

BILBERRY, Bear *Arctostaphylos uva-ursi* (Bearberry) Aber. B & H. Gerard knew it as Bear Whortleberry, too. See BEARBERRY for the significance of this.

BILBERRY, Bog *Vaccinium uliginosum* (Northern Bilberry) Hohn.

BILBERRY, Large *Vaccinium uliginosum* (Northern Bilberry) Ellacombe.

BILBERRY, Northern *Vaccinium uliginosum*.

BILBERRY, Red 1. *Vaccinium parvifolium*. 2. *Vaccinium vitis-idaea* (Cowberry) Jamieson.

BILBERRY, Tall *Vaccinium membranaceum* (Blueberry) Johnston.

BILBERRY, Whortle- *Vaccinium myrtillus* (Whortleberry) B & H.

BILDERS 1. *Apium nodiflorum* (Fool's Watercress) Dev. Grigson. 2. *Heracleum sphondyllium* (Hogweed) Corn. Jago; Dev, Som. Grigson. 3. *Nasturtium officinale* (Watercress) E.Ang, Ire. Grigson. 4. *Oenanthe crocata* (Hemlock Water Dropwort) Corn. Jago; Dev, IOW B & H. Billers seems to be an alternative in most cases. Halliwell in 1847 knew the word as "a kind of watercress". The name was originally OE billere, preserved in several place names, such as Bilbrook, in Somerset.

BILIMBI *Averrhoa bilimbi*.

BILL, Duck's 1. *Dicentra spectabilis* (Bleeding Heart) Som. Elworthy. 2. *Iris pseudo-acarus* (Yellow Iris) Som.

Macmillan. 3. *Syringa vulgaris* (Lilac) Dev. Friend.1882.

BILL, Goose 1. *Galium aparine* (Goose-grass) Halliwell. 2. *Geranium robertianum* (Herb Robert) Som. Macmillan. Cranesbill is more usual, though Storksbill occurs for Herb Robert too.

BILL, Hen's *Onobrychis sativa* (Sainfoin) B & H.

BILL, Parrot's *Clianthus puniceus*. Descriptive,from the shape of the flowers, exactly like a parrot's bill.

BILL, Stork's *Geranium robertianum* (Herb Robert) Som. Grigson. All the geraniums owe the name, and their common one, Cranesbill, to the beak-shaped fruits (geranos is Greek for crane), so they are either Stork's Bill, or Crane's Bill, and more often spelled as one word.

BILL, Swan *Iris pseudo-acarus* (Yellow Iris) Som. Macmillan. Cf. Duck's Bill, from the same area.

BILL BUTTONS 1. *Geum rivale* (Water Avens) Wilts. Grigson. 2. *Vinca minor* (Lesser Periwinkle) Tynan & Maitland. These are two of the very many flowers that bear the name Billy Buttons, or Soldier's Buttons, too. The name is presumably descriptive, from the shape of the flower.

BILLERS 1. *Apium nodiflorum* (Fool's Watercress) Halliwell. 2. *Heracleum sphondyllium* (Hogweed) Dev. B & H. 3. *Nasturtium officinale* (Watercress) Ire. B & H. 4. *Oenanthe crocata* (Hemlock Water Dropwort) Dev. B & H. As for Bilders.

BILLERS, Sweet *Heracleum sphondyllium* (Hogweed) Dev. B & H.

BILLION-DOLLAR GRASS *Echinochlea crus-galli* (Japanese Barnyard Millet) Howes. Presumably because of its grassy, lush growth.

BILLY, Stinking *Senecio jacobaea* (Ragwort) Lincs. Grigson. In the North Country and Scotland, roughly Lincolnshire northwards, there are half a dozen epithets starting with 'stinking' for Ragwort.

BILLY BEATTIE *Parietaria diffusa* (Pellitory-of-the-wall) Ire. Grigson. This is an Irish schoolboy name. When the need arose, they used to grab hold of it, with:
Peniterry, peniterry, that grows by the wall,
Save me from a whipping, or I'll pull you, roots and all.

BILLY BRIGHT-EYE *Veronica chamaedrys* (Germander Speedwell) Ire. Grigson.

BILLY BUSTER *Silene cucubalis* (Bladder Campion) Som. Macmillan. Presumably because children "bust" or pop the inflated calyx.

BILLY BUTTONS 1. *Agrostemma githago* (Corn Cockle) B & H. 2. *Althaea rosea* (Hollyhock) Som. Macmillan. 3. *Arctium lappa* burrs (Burrs) Dev. Friend.1882. 4. *Bellis perennis* (Daisy) Som, Wilts. Macmillan; Shrop. Grigson. 5. *Caltha palustris* (Marsh Marigold) Som. Macmillan. 6. *Geranium robertianum* (Herb Robert) Bucks. Grigson. 7. *Geum rivale* (Water Avens) Wilts. Grigson. 8. *Knautia arvensis* (Field Scabious) Som. Macmillan; Yks. B & H. 9. *Leucanthemum vulgare* (Ox-eye Daisy) Shrop. Grigson. 10. *Lychnis flos-cuculi* (Ragged Robin) War. Grigson. 11. *Malva neglecta* (Dwarf Mallow) Surr. Tynan & Maitland. 12. *Malva sylvestris* (Common Mallow) Som. Macmillan. 13. *Melandrium album* (White Campion) Som, Glos, War. Grigson. 14. *Melandrium dioicum* (Red Campion) War, Ess. Grigson. 15. *Ranunculus acris* (Meadow Buttercup) Som. Macmillan. 16. *Saxifraga granulata fl pl* B & H. 17. *Stellaria holostea* (Greater Stitchwort) Som. Macmillan. War. Grigson. 18. *Vinca minor* (Lesser Periwinkle) Tynan & Maitland.

BILLY CLIPPE; BILLY CLIPPER *Convolvulus arvensis* (Field Bindweed) Kent Grigson (Clippe); Shrop. Grigson (Clipper). This is OE clyppan, to embrace.

BILLY-GOAT WEED *Ageratum conyzoides* (Goatweed) Australia Dalziel. Because of the peculiar smell.

BILLY-O'-BUTTONS *Caltha palustris* (Marsh Marigold) Som. Macmillan. Plain

Billy Buttons is more usual.
BILLY WHITE'S BUTTONS *Stellaria holostea* (Greater Stitchwort) War. Grigson.
BILLY'S BUTTONS *Geum rivale* (Water Avens) Wilts. Macmillan.
BILLYCOCK, Parson's *Arum maculatum* (Cuckoo Pint) Som. Grigson. In Yorkshire, this becomes Parson's Pillycods, so making the meaning more obvious. The "pint" of the usual name means penis, and pillycods, or, as in Somerset, billycocks, means the same thing.
BILSTED *Liquidambar styraciflua* (Sweet Gum) Brimble.1948.
BIND 1. *Calystegia sepium* (Great Bindweed) E.Ang, Lincs. Grigson. 2. *Lonicera periclymenum* (Honeysuckle) Yks. Grigson.
BIND, Hairy *Cuscuta epithymum* (Common Dodder) Hants. Grigson.
BIND, Lily *Convolvulus arvensis* (Field Bindweed) Yks. B & H.
BIND-CORN 1. *Convolvulus arvensis* (Field Bindweed) N'thants. A.E.Baker. 2. *Fagopyrum esculentum* (Buckwheat) Halliwell. 3. *Polygonum convolvulus* (Black Bindweed) Turner. Halliwell clearly says Buckwheat, but it is probably Black Bindweed to which he is referring.
BINDER *Clematis vitalba* (Old Man's Beard) Hants. B & H.
BINDER, Dog *Anthemis cotula* (Maydweed) Yks. Grigson. Dog Banner occurs in the same area for this plant.
BINDWEED 1. *Convolvulus sp.* 2. *Lonicera periclymenum* (Honeysuckle) Yks. Grigson. 3. *Polygonum convolvulus* (Black Bindweed) Ches, E.Ang, Cumb. Grigson. 4. *Senecio jacobaea* (Ragwort) Scot. Grigson. 5. *Vicia hirsuta* (Hairy Tare) Herts. Grigson. OE bindan, which virtually has not changed. The meaning is obvious for all these except ragwort, where bindweed may possibly be a corruption of Bundweed, a much commoner Scottish name for the plant. But Salisbury maintains the Bindweed name, and explains that one of the symptoms of jacobine poisoning (caused by Ragwort) in animals is extreme constipation.
BINDWEED, Black 1. *Polygonum convolvulus*. 2. *Tamus communis* (Black Bryony) Fernie.
BINDWEED, Blue *Solanum dulcamara* (Woody Nightshade) B & H.
BINDWEED, Corn *Polygonum convolvulus* (Black Bindweed) Yks. B & H.
BINDWEED, Copse *Polygonum dumetorum*. Latin dumetum means a copse or thicket.
BINDWEED, Deer's-foot *Convolvulus arvensis* (Field Bindweed) Chopra.
BINDWEED, Field *Convolvulus arvensis*.
BINDWEED, Fringed *Polygonum cilinode*.
BINDWEED, Great 1. *Calystegia sepium*. 2. *Calystegia sylvestris* (American Bellbine) Blamey.
BINDWEED, Hairy *Calystegia pulchra*.
BINDWEED, Hedge 1. *Calystegia sepium* (Great Bindweed) Blamey. 2. *Convolvulus repens* (Trailing Bindweed) USA House.
BINDWEED, Hooded *Calystegia sepium* (Great Bindweed) Murdoch McNeill.
BINDWEED, Indian *Convolvulus major* (Morning Glory) Grigson.1974.
BINDWEED, Ivy *Polygonum convolvulus* (Black Bindweed). B & H.
BINDWEED, Jalap *Ipomaea purga* (Jalapa Purge) Thornton. Ipomaea is one of the Convolvulaceae. Jalap, for Jalapa, is from Spanish jalapa, originally Xalapa, from Aztec Xalapan, which apparently means sand by water. Anyway, the plant's native habitat is the eastern side of the Mexican Andes.
BINDWEED, Larger *Calystegia sepium* (Great Bindweed) Clapham.
BINDWEED, Lesser *Convolvulus arvensis* (Field Bindweed) Hatfield.
BINDWEED, Low *Convolvulus spithamalis* (Upright Bindweed) USA House.
BINDWEED, Mallow-leaved *Convolvulus althaeoides*.
BINDWEED, Purple *Convolvulus major* (Morning Glory) Whittle & Cook.
BINDWEED, Scammony *Convolvulus scammonia* (Scammony) Thornton. Scammony is

really the gum resin obtained from this plant. The word comes through O.French escamonie, scamonee, from Latin scammonea, itself from Greek skammonia. This gum resin, actually the dried, milky juice, has been known as a medicine from very ancient times (Theophrastus, in the 3rd century BC, for instance). It is used as an active cathartic, often with colocynth and calomel.

BINDWEED, Sea *Calystegia soldanella*.
BINDWEED, Shrubby *Convolvulus cneorum*.
BINDWEED, Silvery *Convolvulus cneorum* (Shrubby Bindweed) W.Miller.
BINDWEED, Small *Convolvulus arvensis* (Field Bindweed) USA House.
BINDWEED, Syrian *Convolvulus scammonea* (Scammony) Grieve.1931. It is native in Syria, but is certainly not confined to that country - it actually grows in waste, bushy places from Greece to south Russia.
BINDWEED, Trailing *Convolvulus repens*.
BINDWEED, Upright *Convolvulus spithamalis*.
BINDWEED-NIGHTSHADE *Circaea lutetiana* (Enchanter's Nightshade) Gerard. There is no apparent reason why Gerard gave it this name; it does not act like a bindweed at all.
BINDWITH *Clematis vitalba* (Old Man's Beard) Prior. i.e. a with to bind up faggots. It was used for binding instead of withies.
BINDWOOD 1. *Hedera helix* (Ivy) Scot. Jamieson. 2. *Lonicera periclymenum* (Honeysuckle) Yks. Grigson.
BINE, Great *Calystegia sepium* (Great Bindweed) B & H.
BINE-LILY 1. *Calystegia sepium* (Great Bindweed) Dor. Macmillan. 2. *Convolvulus arvensis* (Field Bindweed) Dor. Macmillan.
BINJAI *Mangifera coesia*.
BINNWOOD *Lonicera periclymenum* (Honeysuckle) Yks. B & H. Seems to be an intermediate form between Benewith and Bindwood.
BINWEED *Senecio jacobaea* (Ragwort) Scot. Jamieson. Bindweed occurs as well in Scotland for ragwort, but most probably this has nothing to do with bindweed - it is more likely to be related to Bun(d)weed.
BIRCH *Betula verrucosa*. Birch is OE birce, bierce, sometimes berc, or beorc, Dutch berk. According to ODEE this is one of the few Indo-European tree names (root bherga), Sanskrit bhurjas. Putnam says that these OE names all meant bark, in both senses, i.e. as part of the tree, and as a boat (Cf. North American birch bark canoes). Notice how these old names still survive in northern England and Scotland as Birk or Burk, and further south, in place names, such as Birkenhead (birchen headland), Berkeley, in Gloucestershire (birch wood), Berkhamstead, in Herts. (homestead among the birches). Evelyn says that Berkshire was derived from birch, and Wilks that Barking was Berecingum in AD 725 (birch tree dwellers).
BIRCH, Arctic *Betula nana* (Dwarf Birch) RHS.
BIRCH, Black 1. *Betula fontinalis* (Streamside Birch) USA Elmore. 2. *Betula lenta* (Cherry Birch) USA Harper. 3. *Betula nigra* (River Birch) Lindley.
BIRCH, Bog *Betula pumila* (Low Birch) USA Zenkert.
BIRCH, Brown *Betula pubescens* (Downy Birch) Howes.
BIRCH, Canadian *Betula lutea* (Yellow Birch) Hutchinson & Melville.
BIRCH, Canoe *Betula papyrifera* (Paper Birch) USA H.Smith 2. Because of its use (the bark, that is) for canoes among the North American Indians; actually the bark was used for hundreds of other purposes, utensils, etc.
BIRCH, Canyon *Betula fontinalis* (Streamside Birch) USA Elmore.
BIRCH, Cherry 1. *Betula fontinalis* (Streamside Birch) USA Elmore. 2. *Betula lenta*.
BIRCH, Downy *Betula pubescens*.
BIRCH, Dwarf *Betula nana*.
BIRCH, European *Betula verrucosa* (Birch) Flück.
BIRCH, Grey *Betula populifolia*.

BIRCH, Hairy *Betula pubescens* (Downy Birch) Hart.

BIRCH, Himalayan *Betula utilis*.

BIRCH, Himalayan, White-barked *Betula jacquemontii*.

BIRCH, Horn *Ulmus glabra* (Wych Elm) Som. Grigson. Presumably for the same reason that Wych Elms are occasionally called Hornbeam. In both cases, the identification may be wrong.

BIRCH, Lady *Betula verrucosa* (Birch) B & H. It is also known as the Lady of the Woods, which seems to have been Coleridge's invention originally. In Somerset, it is traditionally the tree of death. "The one with the white hand" appears to be a birch. This was a spirit which haunted the moorland near Taunton. It would rise at twilight out of a scrub of birch and oak and come drifting across to lonely travellers so fast that they had no time to escape. She was deadly pale with clothes that rustled like dead leaves, and her hand looked like a blasted branch. Sometimes she pointed a finger at a man's head, and then he ran mad, but more often she put her hand over his heart, and he fell dead.

BIRCH, Low *Betula pumila*.

BIRCH, Mahogany *Betula lenta* (Cherry Birch) Grieve.1931. It has a good, richly-marked timber for furniture, and pianoforte makers like it for its markings. Mountain Mahogany is another name. Presumably the usual name, Cherry Birch, refers to the timber, too.

BIRCH, Monarch *Betula maximowicziana*.

BIRCH, Paper 1. *Betula papyrifera*. 2. *Betula verrucosa* (Birch) Wilts. Dartnell & Goddard. Paper Birch as a name comes from the use of the bark, not necessarily just for paper. The North American Indians used it for all kinds of utensils, and canoes. The bark can be unrolled only by exposing it to the heat of a fire - when heated it becomes pliable, and retains any form into which it is moulded when softened in this way. The common Birch gets the name, and that of Ribbon-tree (in Lincolnshire) from the way the bark peels.

BIRCH, Paper, Indian *Betula bhojpattra*.

BIRCH, Red 1. *Betula nigra* (River Birch) USA Harper. 2. *Betula pubescens* (Downy Birch) Hutchinson.

BIRCH, Red-barked, Chinese *Betula albo-sinensis*.

BIRCH, River *Betula nigra*.

BIRCH, Rocky Mountain *Betula fontinalis* (Streamside Birch) USA Elmore.

BIRCH, Silver *Betula verrucosa* (Birch). These trees are usually called Silver Birch rather than simply Birch.

BIRCH, Streamside *Betula fontinalis*.

BIRCH, Swamp *Betula pumila* (Low Birch) USA Yarnell.

BIRCH, Sweet 1. *Betula fontinalis* (Streamside Birch) USA Elmore. 2. *Betula lenta* (Cherry Birch) USA Zenkert.

BIRCH, Warty *Betula verrucosa* (Birch) Hart. Warty, i.e. verrucosa.

BIRCH, Water *Betula fontinalis* (Streamside Birch) USA Elmore.

BIRCH, West Indian *Bursera simaruba* Howes.

BIRCH, White 1. *Betula papyrifera* (Paper Birch) USA Zenkert. 2. *Betula populifolia* (Grey Birch) USA Mason. 3. *Betula pubescens* (Downy Birch) Barber & Phillips. 4. *Betula verrucosa* (Birch) Grieve.1931. 5. *Fagus solandri*.

BIRCH, Yellow 1. *Betula excelsa*. 2. *Betula lutea*.

BIRCH-LEAF BUCKTHORN *Rhamnus betulaefolia*.

BIRCH-LEAVED MOUNTAIN MAHOGANY *Cercocarpus betuloides*.

BIRCHET *Betula verrucosa* (Birch) Suss. Parish.

BIRD 1. *Aesculus hipppocastanum* outer casing of fruit (Horse. Chestnut) Dev. Friend.1882. 2. *Arctium lappa* burrs (Burdock) Dev. Friend.1882. 3. *Galium aparine* (Goose-grass) Dev. Friend.1882. 'Bird' for 'burr' seems to have been quite common once in Devon.

BIRD BRIAR *Rosa canina* (Dog Rose)

Ches. Grigson.
BIRD CHERRY 1. *Prunus padus*. 2. *Prunus pennsylvanica* (Pin Cherry) USA Kingsbury.
BIRD CHERRY, American; BIRD CHERRY, Virginian *Prunus virginiana* (Choke Cherry) Ablett.
BIRD-DRINKING-AT-A-FOUNTAIN *Nigella damascena* (Love-in-a-mist) Dev. Tynan & Maitland.
BIRD EAGLES *Crataegus monogyna* fruit (Haws) Ches. Holland. Cf. Bird's Eggles, again from Cheshire. Eagles, eggles or agles are diminutives of Hagues, the usual name for haws in Cheshire.
BIRD EEN; BONNY BIRD EEN *Primula farinosa* (Birdseye Primrose) Cumb. Wright (Bird Een); Cumb. Grigson (Bonny Bird Een).
BIRD EYE *Callicarpa americana* Alab. Harper.
BIRD-IN-A-BUSH; BIRD-ON-A-THORN *Corydalis bulbosa* (Bulbous Fumitory) both N'thants. B & H.
BIRD KNOTGRASS; BIRD'S KNOTGRASS *Polygonum aviculare* (Knotgrass) Som. Macmillan (Bird); B & H (Bird's). Prior says that these, as well as other bird names (Cf. aviculare), such as Bird's Tongue and Sparrow's Tongue, come from the shape of the leaf.
BIRD OF PARADISE 1. *Aconitum napellus* (Monkshood) Som. Macmillan. 2. *Caesalpina pulcherrima* (Barbados Pride) S.Africa Watt. 3. *Poinciana gilliesii* (Poinciana) USA Kingsbury. 4. *Strelitzia reginae*. It is claimed that the flowers of Monkshood have dove-like nectaries; certainly some very fancy names are recorded, like Doves-in-the-ark, Lady Lavinia's Dove Carriage, etc. Strelitzia gets the name from the extraordinary similarity of the profile view of the flower to the head of some exotic bird.
BIRD PEPPER 1. *Capsicum annuum var. longum* (Paprika) Hatfield. 2. *Capsicum baccatum*. Paprika seeds are used by bird fanciers to improve canaries' plumage.
BIRD RAPE *Brassica campestris* (Field Mustard) Usher.
BIRD THISTLE *Cirsium vulgare* (Spear Plume Thistle) Worcs. Grigson. Probably, like Boar Thistle, Bow Thistle, etc., a corruption of Bur Thistle.
BIRD'S BANNOCK *Oxalis acetosella* (Wood Sorrel) Scot. Tynan & Maitland. Cf. Cuckoo's Bread, etc.
BIRD'S BRANDY *Lantana rugosa* S.Africa Watt.
BIRD'S BREAD *Sedum acre* (Biting Stonecrop) Prior. Prior says it blooms about the time the young birds are hatched (but Cf. pain d'oiseau, one of the French names for this plant).
BIRD'S BREAD-AND-CHEESE *Oxalis acetosella* (Wood Sorrel) Dev. Friend.1882. There are a number of names in this vein, most of them to do with the cuckoo - Cuckoo's Bread-and-cheese, Cuckoo's Meat, Gowk's Meat, etc.
BIRD'S CHERRIES *Crataegus monogyna* fruit (Haws) Som. Macmillan.
BIRD'S CLAWS *Lotus corniculatus* (Bird's-foot Trefoil) Dev. Macmillan.
BIRD'S EGGLE *Crataegus monogyna* fruit (Haws) Ches. Grigson. Eagles, agles or eggles are diminutives of Hague, the common name for haws in Cheshire and Lancashire. In its many forms, varying from Agog to Eglet, this is OE haga, hedge, showing that hawthorn has always been the ideal hedging plant.
BIRD'S EGGS 1. *Silene cucubalis* (Bladder Campion) Shrop. Grigson. 2. *Crataegus monogyna* haws (Haws) Ches. Holland. From the shape of the bladder, as far as bladder campion is concerned. When applied to haws, it is an extension of Bird's Eggle, or Bird Eagle. Eagles, or eggles, are diminutives of hagues, the common name for haws in Cheshire.
BIRD'S EYE 1. *Anagallis arvensis* (Scarlet Pimpernel) Wilts. Dartnell & Goddard; Oxf, Bucks. Grigson. 2. *Cardamine pratensis*

(Lady's Smock) Shrop, Yks, Cumb. Grigson. 3. *Euphrasia officinalis* (Eyebright) Som. Macmillan. 4. *Geranium robertianum* (Herb Robert) Corn, Bucks, N'thum. Grigson; Dev. Macmillan. 5. *Gilia tricolor*. 6. *Glechoma hederacea* (Ground Ivy) Oxf, Bucks, N'thants. Grigson. 7. *Lotus corniculatus* (Bird's-foot Trefoil) Som. Macmillan. 8. *Melandrium rubrum* (Red Campion) Dev, Som. Macmillan. 9. *Myosotis arvensis* (Field Forget-me-not) Som. Macmillan. Bucks, N'thants. B & H. 10. *Myosotis palustris* (Water Forget-me-not) Som, Hants. Grigson; N'thants. A.E.Baker. 11. *Pentaglottis sempervirens* (Evergreen Alkanet) Som. Elworthy. 12. *Primula farinosa* (Bird's Eye Primrose) Prior. 13. *Sagina procumbens* (Pearlwort) Suss. Grigson; USA. Allan. 14. *Saxifraga spathularis x umbrosa* (London Pride) Dev. Friend.1882. 15. *Stellaria holostea* (Greater Stitchwort) Worcs, Derb, Yks, Berw. Grigson. 16. *Stellaria media* (Chickweed) Som. Macmillan. 17. *Veronica beccabunga* (Brooklime) Som, Dor. Macmillan. 18. *Veronica chamaedrys* (Germander Speedwell) common. Grigson (includes Wilts. Dartnell & Goddard). 19. *Veronica hederifolia* (Ivy-leaved Speedwell) Ess. B & H. 20. *Veronica officinalis* (Common Speedwell) Wilts. Dartnell & Goddard. 21. *Veronica persica* (Common Field Speedwell) Wilts. Dartnell & Goddard; Bucks. B & H. 22. *Viola tricolor* (Pansy) Som. Macmillan. Descriptive in most cases, but there is a belief recorded in Leicestershire that if you pick Germander Speedwell, birds would come and pick your eyes out - probably a case of reverse etymology.

BIRD'S EYE, Blue *Veronica chamaedrys* (Germander Speedwell) Oxf, Suss. Grigson; Bucks. B & H.

BIRD'S EYE, Bonny *Cardamine pratensis* (Lady's Smock) Cumb. Grigson.

BIRD'S EYE, Pink *Geranium robertianum* (Herb Robert) Som. Macmillan; Bucks. Grigson.

BIRD'S EYE, Red 1. *Anagallis arvensis* (Scarlet Pimpernel) Som. Macmillan. 2. *Geranium robertianum* (Herb Robert) Prior. 3. *Melandrium rubrum* (Red Campion) Rad. Grigson.

BIRD'S EYE, Water *Veronica beccabunga* (Brooklime) Dev. Macmillan.

BIRD'S EYE, White 1. *Stellaria holostea* (Greater Stitchwort) Rad. Grigson. 2. *Stellaria media* (Chickweed) Bucks. Grigson.

BIRD'S EYE PRIMROSE *Primula farinosa*. Better as one word, i.e. Birdseye.

BIRD'S FOOT 1. *Lotus corniculatus* (Bird's Foot Trefoil) Som. Grigson. 2. *Ornithopus perpusillus*. The latter's common name is often spelt as a single word (birdsfoot).

BIRD'S FOOT, Orange *Ornithopus pinnatus*.

BIRD'S FOOT CLOVER 1. *Lotus corniculatus* (Bird's Foot Trefoil) Prior. 2. *Trigonella ornithopodioides* (Fenugreek) Prior. The second half of the botanical name of Fenugreek means exactly like a bird's foot.

BIRD'S FOOT FENUGREEK *Trigonella ornithopodioides* (Fenugreek) Clapham. Ornithopodioides means like a bird's foot. Fenugreek is a corruption of foenum-graecum (Grecian hay), because it was apparently used to scent inferior hay.

BIRD'S FOOT MILLET *Eleusine corocana* (Finger Millet) Brouk.

BIRD'S FOOT TREFOIL 1. *Lotus corniculatus*. 2. *Trigonella ornithopodioides* (Fenugreek) Prior.

BIRD'S FOOT VIOLET *Viola pedata*.

BIRD'S KNOTGRASS *Polygonum aviculare* (Knotgrass) B & H. Aviculare, and the various 'bird' names, such as Bird's Tongue, Sparrow's Tongue, etc., are all, according to Prior, given from the shape of the leaf.

BIRD'S MEAT 1. *Crataegus monogyna* haws (Haws) Dev. Choape; Som. Grigson. 2. *Plantago major* (Great Plantain) Aber. Grigson. The latter was a favourite cage-bird food - names like Canary-seed, Birdseed and Larkseed are also recorded from various parts of the country.

BIRD'S NEST 1. *Daucus carota* (Wild Carrot) Gerard. 2. *Listera ovata* (Twayblade) Som. Macmillan. The name is associated with the carrot rather than Twayblade. "The whole tuft (of flowers) is drawn together when the seeds is ripe, resembling a birde's nest" (Gerard). Cf. German and Dutch Vogelsnest.

BIRD'S NEST, Yellow *Monotropa hypopitys*. "Birds nest hath many tangling roots platted or crossed one over another very intricately, which resembleth a Crows nest made of sticks" (Gerard).

BIRD'S NEST FERN *Asplenium nidus*.

BIRD'S NEST ORCHID *Neottia nidus-avis*. Because of the tangled fibrous roots.

BIRD'S NEST ORCHID, Violet *Limodorum abortivum* (Limodore) D Parish.

BIRD('S) PEARS 1. *Crataegus monogyna* fruit (Haws) Som. Macmillan; Dor. Barnes. 2. *Rosa canina* fruit (Hips) Som. Macmillan.

BIRD'S TONGUE 1. *Acer campestre* fruit (Maple keys) Evelyn. 2. *Anagallis arvensis* (Scarlet Pimpernel) Norf. Grigson. 3. *Fraxinus excelsior* fruit (Ash keys) B & H. 4. *Pastinaca sativa* (Wild Parsnip) B & H. 5. *Polygonum aviculare* (Knotgrass) Som, Norf. Prior; N.Eng. Gerard. 6. *Senecio paludosus* (Great Fen Ragwort) B & H. 7. *Stellaria holostea* (Greater Stitchwort) B & H. 8. *Strelitzia reginae* (Bird of Paradise Flower Whittle & Cook.

BIRD'S TREE *Hedera helix* (Ivy) English gypsy name Groome. The legend is that this was the tree brought back by the dove into the ark, and that is the reason that birds are fond of clustering round it.

BIRDCAGE PLANT *Oenothera deltoides*. A descriptive name. This is a plant of sand dunes in the American southwest deserts. When conditions bring about the death of the plant, the stems curl upwards to make a hollow spherical lattice, the Birdcage of the common name.

BIRDLIME MISTLETOE *Viscum album* (Mistletoe) Flück. "...white translucent berries..whereof the best Bird-lime is made, far exceeding that which is made of the Holm or Holly bark..." (Gerard). It was made by drying, pounding, soaking in water for 12 days, and pulverizing again. It was used up to medieval times for taking small birds on the branches of trees, and also for catching hawks, which were decoyed by a bird between an arched stick coated with bird-lime. The use accounts for the generic name, Viscum.

BIRDS; BIRDS' WINGS *Acer pseudo-platanus* seeds (Sycamore keys) Som. Macmillan.

BIRDS ON THE BOUGH; BIRDS ON THE BUSH *Fumaria officinalis* (Funitory) both Dor. Dacombe (bough); Grigson (bush). Dicky Birds is a Wiltshire name, too.

BIRDSEED 1. *Plantago major* (Great Plantain) Dev. Friend.1882; Som. Macmillan. 2. *Senecio vulgaris* (Groundsel) Som. Macmillan. 3. *Sinapis arvensis* (Charlock) B & H. The first two at least have always been favourite feed for cage birds. Charlock seems only rarely, and in Scotland, to have been given the name.

BIRK *Betula verrucosa* (Birch) Lancs. Nodal; Yks. Morris; N Eng, Scot. Turner. Like the North country Burk, this is closer to the original Teutonic forms than Birch, which has softened the last consonant. The form is retained in various place names, like Birkenhead (birchen headland), Berkeley (birch wood), Birkhamstead (homestead among the birches). Evelyn cites Berkshire, too.

BIRTHROOT 1. *Trillium erectum* (Bethroot) Grieve.1931. 2. *Trillium grandiflorum* (Wood Lily) USA Weiner. Presumably this is what Bethroot means. Weiner explains that the name was given because it was believed that the Indians used it to induce labour in childbirth. It is in fact a well-known uterine stimulant.

BIRTHWORT *Aristolochia clematitis*. The name Aristolochia is derived from the Greek meaning best birth. The greenish-yellow flower constricts into a tube that opens into a globular swelling at the base. The swelling was interpreted as the womb, the tube as the birth passage. By the doctrine of signatures, it was used to help delivery, to encourage conception, and to purge the womb. It is a fact that the plant does have an abortive effect.

BIRTHWORT, Climbing *Aristolochia clematitis* (Birthwort) Thornton.

BIRTHWORT, Long *Aristolochia longa*.

BIRTHWORT, Saracen *Aristolochia clematitis* (Birthwort) Pomet.

BIRTHWORT, Sweet-scented *Aristolochia odoratissima*.

BISCUIT 1. *Geranium robertianum* (Herb Robert) Dev. Grigson. 2. *Potentilla erecta* (Tormentil) Grieve.1931. Herb Robert also bears this name in Somerset, except that it becomes Biscuit-flower there.

BISCUIT-FLOWER *Geranium robertianum* (Herb Robert) Som. Macmillan.

BISCUIT-LEAVES *Fagus sylvatica* (Beech) Som. Macmillan. The name implies that they are eaten, probably very young, just as children eat young hawthorn shoots.

BISCUIT-ROOT *Peucedanum farinosum*. The reference is to the floury,edible roots.

BISHOP, Five-faced *Adoxa moschatellina* (Moschatel) Polunin. From the same characteristic that gives the name Town Hall Clock.

BISHOP PINE *Pinus muricata*. A Californian species, discovered in 1832 by Coulter, at San Luis Obispo, hence the name Bishop Pine.

BISHOP'S BEARD *Anthriscus sylvestris* (Cow Parsley) Australia Hemphill.

BISHOP'S ELDER *Aegopodium podagraria* (Goutweed) IOW Grigson. 'Bishop's' because it was so often found near old ecclesiastical ruins: it was said to have been introduced by the monks in the Middle Ages.

BISHOP'S FLOWER *Stachys officinalis* (Betony) Wales Trevelyan. Does the purple colour provide the connection? Cf. Bishopwort or Bishop's Wort, for this.

BISHOP'S HAT *Epimedium alpinum* (Barrenwort) W.Miller.

BISHOP'S LEAF *Scrophularia aquatica* (Water Betony) Som. Macmillan; Yks. Gerard. A S Palmer suggested that the name arose from a misunderstand-ing of the French name, l'herbe du siège, as if siège were used here in the ecclesiastical sense of a see, instead of the other kind of seat in the necessary house. For Figwort was used for the fig, i.e. piles.

BISHOP'S POSY *Leucanthemum vulgare* (Ox-eye Daisy) Donegal Grigson.

BISHOP'S WEED 1. *Aegopodium podagraria* (Goutweed) Border B & H. 2. *Ammi visnaga*. For goutweed, see Bishop's Elder for a possible explanation.

BISHOP'S WIG *Arabis alpina* (Alpine Rock-cress) Ches. B & H. From the resemblance of the tufts in full flower to the wigs once worn by bishops.

BISHOP'S WORT 1. *Nigella damascena* (Love-in-a-mist) Gerard. 2. *Stachys officinalis* (Betony) Som, Glos. Grigson. For Love-in-a-mist, Britten & Holland hopefully said the reason may perhaps be because the carpels look something like a mitre. As far as Betony is concerned, the purple colouring may possibly be the reason for a rash of 'bishop' names.

BISHOPWEED *Mentha aquatica* (Water Mint) Hants. Cope.

BISHOPWORT 1. *Mentha aquatica* (Water Mint) Wilts. Dartnell & Goddard. Hants. Cope. 2. *Mentha pulegium* (Pennyroyal) Gordon.1985. 3. *Stachys officinalis* (Betony) Cockayne. Purple colouring, perhaps? But Cf. Churchwort.

BISOM; BISSOM 1. *Calluna vulgaris* (Heather) Som, Dor, Bucks, Yks. Grigson (Bisom); Dev. B & H (Bissom). 2. *Cytisus scoparius* (Broom) Dev. B & H (Bisom); Dev. Friend.1882 (Bissom). Besom is the more usual spelling, but there are many variations in dialect. The word means a broom (the plant broom got its name from its use as a household implement). Heather, too, is still made into a broom by tieing a bundle of twigs round a stick. "Bundle" seems to have produced another in this series of names - Banadle.

BISSY NUT *Cola nitida* (Kola) Jamaica Howes.

BISTORT *Polygonum bistorta*. Bistort is Latin bis, twice, and torta, twisted, and refers to the twisted roots, something which has produced a number of snake names, like Adderwort, Snakeweed, Dragonwort, etc. The doctrine of signatures decreed that these twisted roots should be used for snakebite and poison.

BISTORT, Alpine *Polygonum viviparum*.

BISTORT, Amphibious *Polygonum amphibium* (Amphibious Persicaria) Clapham.

BISTORT, Great; BISTORT, Greater *Polygonum bistorta* (Bistort) Carr (Great);Curtis (Greater).

BITCHWOOD *Euonymus europaeus* (Spindle Tree) Worcs. Grigson. There seems to be a general confusion in names with those of Dogwood (*Cornus sanguinea*), for not only is there Bitchwood, but Dog Timber, Dog-tree, Dog-rose and Dogwood, all applied to Spindle Tree. As with Dogwood, it was said that a leaf decoction was used to wash dogs free from vermin. (It seems, too, that small boys' nits were treated in the same way).

BITHWIND; BITHWINE *Convolvulus arvensis* (Field Bindweed) S.Eng. Wakelin (Bithwind); Hants. Jones & Dillon (Bithwine).

BITHYNIAN VETCH *Vicia bithynica*. Bithynia is the part of Asia Minor opposite Istanbul, and this plant is native there. It probably isn't in this country.

BITHYWIND 1. *Clematis vitalba* (Old Man's Beard) Wilts. Jones & Dillon. 2. *Convolvulus arvensis* (Field Bindweed) S.Eng. Wakelin.

BITHYWINE *Clematis vitalba* (Old Man's Beard) Wilts. D.Grose.

BITING STONECROP *Sedum acre*.

BITNY *Stachys officinalis* (Betony) Dev. Friend.1882.

BITNY, Water *Scrophularia aquatica* (Water Betony) Hunts. Marshall.

BITTER AKS *Taraxacum officinale* (Dandelion) Shet. Grigson.

BITTER APPLE 1. *Colocynthis vulgaris* (Bitter Gourd) Wit. 2. *Momordica charantia* (Wild Balsam-apple) Sofowora. 3. *Solanum aculeastrum* S.Africa Watt. 4. *Solanum panduraeforme* S.Africa Watt.

BITTER-BERRY *Solanum racemosum* Barbados Gooding.

BITTER BUSH *Eupatorium villosum*.

BITTER BUTTONS *Tanacetum vulgare* (Tansy) Mor. Grigson; USA Henkel. Tansy is bitter, and used in medicine as "bitters".

BITTER CAMOMILE *Matricaria recutita* (Scented Mayweed) Fernie.

BITTER CANDYTUFT *Iberis amara* (Candytuft) H C Long.1910. The specific name is *amara*, bitter.

BITTER CASSAVA *Manihot esculenta* (Manioc) A W Smith. 'Bitter', to distinguish it from the 'sweet' variety. The bitter forms contain greater amounts of hydrocyanic acid, toxic, so it has to be eliminated.

BITTER CUCUMBER *Colocynthis vulgaris* (Bitter Gourd) Polunin.

BITTER FLAX *Linum catharticum* (Fairy Flax) H C Long.1910.

BITTER-FLOWER; BITTER-MEDICINE *Sambucus nigra* (Elder) Som. Macmillan.

BITTER GENTIAN *Gentiana lutea* (Yellow Gentian) Howes.

BITTER GOURD 1. *Colocynthis vulgaris*. 2. *Momordica charantia* (Wild Balsam-apple) USA Kingsbury. Colocynthus means bitter gourd - an apt name, for it is exceedingly bitter.

BITTER HERB 1. *Centaureum erythraea* (Centaury) B & H. 2. *Chelone glabra* (Turtlehead) Grieve.1931. Centaury was once used, like Ground Ivy, and other bitter herbs, to make beer keep.

BITTER KOLA *Garcinia kola*.

BITTER LADY'S SMOCK *Cardamine amara* (Large Bittercress) Curtis.

BITTER LAND CRESS *Barbarea vulgaris*

(Winter Cress) Tynan & Maitland.
BITTER LEAF *Vernonia amygdalina* W.Africa Watt. It is not only the leaves, but also the root and stem, which are very bitter.
BITTER LETTUCE *Lactuca virosa* (Great Prickly Lettuce) Usher.
BITTER MELON 1. *Citrullus naudinianus*. 2. *Momordica charantia* (Wild Balsam-apple) Sofowora.
BITTER NIGHTSHADE *Solanum dulcamara* (Woody Nightshade) USA House. Cf. Bittersweet for this - a translation of dulcamara - amara dulcis.
BITTER ORANGE *Citrus aurantium var. bigaradia* (Seville Orange) Rimmel.
BITTER ROOT 1. *Apocynum androsaemifolium* (Spreading Dogbane) O.P.Brown. 2. *Apocynum cannabinum* (Indian Hemp) Grieve.1931. 3. *Gentiana lutea* (Yellow Gentian) Grieve.1931. 4. *Gentianella campestris* (Field Gentian) Grieve.1931. 5. *Lewisia rediviva*. It is the bitter principle that the gentians are valued for - for tonic drugs and "bitters". The dogbanes were used by North American Indians in a similar way.
BITTER-ROOT, Canadian *Lewisia rediviva* (Bitter-root) J.Smith.
BITTER RUBBERWEED *Hymenoxis odorata* (Bitterweed) USA Kingsbury.
BITTER THISTLE *Carduus benedictus* (Holy Thistle) USA Henkel.
BITTER TOMATO *Solanum incanum* (Hoary Nightshade) Howes.
BITTER VETCHLING *Lathyrus montanus* (Bittervetch) Blamey.
BITTER WILLOW *Salix purpurea* (Purple Osier) J.Smith.
BITTER YAM *Dioscorea dumetorum* (Cluster Yam) Alexander & Coursey.
BITTERCRESS 1. *Cardamina sp.* 2. *Nasturtium officinale* (Watercress) Camden/Gibson.
BITTERCRESS, Daisy-leaved *Cardamine bellidifolia*.
BITTERCRESS, Greater *Cardamine flexuosa* (Wavy Bittecress) Allan.
BITTERCRESS, Hairy *Cardamine hirsuta*.

BITTERCRESS, Large *Cardamine amara*.
BITTERCRESS, Meadow *Cardamine pratensis* (Lady's Smock).
BITTERCRESS, Narrow-leaved *Cardamine impatiens*.
BITTERCRESS, Nine-leaved *Cardamine enneaphyllos*.
BITTERCRESS, Radish-leaved *Cardamine raphanifolia*.
BITTERCRESS, Wavy *Cardamine flexuosa*.
BITTERCRESS, Wood *Cardamine flexuosa* (Wavy Bittercress) Clapham.
BITTERSGALL *Malus sylvestris* (Crabapple) Som. Macmillan. This applies to the fruit. Bittersgall means bitter-as-gall.
BITTERNUT HICKORY *Carya cordiformis* (Swamp Hickory) USA Cunningham. Butternut is more likely; it is probably a misprint - there are plenty in Cunningham & Côté.
BITTERSWEET 1. *Filipendula ulmaria* (Meadowsweet) Yks. Grigson. 2. *Solanum dulcamara*. For *Solanum dulcamara*, Bittersweet is a translation of dulcamara - amara dulcis. Gerard said they have "faire berries.. of a sweet taste at first, but after very unpleasant." The Meadowsweet name may possibly have the same meaning as the Cumbrian name Courtship-and-Matrimony - they say there it comes from the scent of the flowers before and after bruising.
BITTERSWEET, American *Celastrus scandens*.
BITTERSWEET, Climbing *Celastrus scandens* (American Bittersweet) USA Zenkert.
BITTERSWEET, False *Celastrus scandens* (American Bittersweet) Alab. Harper.
BITTERSWEET, Oriental *Celastrus orbiculatus*.
BITTERSWEET, Sea *Solanum dulcamara var. marinum*.
BITTERVETCH 1. *Lathyrus montanus*. 2. *Vicia orobus*. Vicia orobus, at least, often commands the name into two parts.
BITTERVETCH, Wild *Vicia ervilia*.
BITTERWEED 1. *Ambrosia artemisiaefolia* (American Wormwood) USA Fogel. 2.

Erigeron canadensis (Canadian Fleabane) USA House. 3. *Helenium autumnale* (Sneezeweed) USA Kingsbury. 4. *Helenium tenuifolium* (Fine-leaved Sneezeweed) Kansas. Gates. 5. *Hymenoxis odorata*. 6. *Populus alba* (White Poplar) N.Eng. Northall. 7. *Senecio glabellus* USA Kingsbury. Poplar bark is very bitter, so would anything be which has the specific name *artemisiaefolia*.

BITTERWEED, Fragrant *Hymenoxis odorata* (Bitterweed) USA T H Kearney.

BITTERWORT 1. *Gentiana lutea* (Yellow Gentian) Schery. 2. *Gentianella amarella* (Autumn Gentian). 3. *Gentianella campestris* (Field Gentian). 4. *Taraxacum officinale* (Dandelion) B & H. The bitter principle is in the roots of the gentians - it has been used in brewing instead of hops, but they are best known for bitter tonics, and "bitters". This is not a name obviously connected with the dandelion, but a few sorrel names have been given to it, and the leaves and juice are certainly sour.

BITTONY *Stachys officinalis* (Wood Betony) T Hill.

BLAB *Ribes uva-crispa* (Gooseberry) Scot. B & H. Just a variant of Blob, given because of the roughly globular shape of the fruit.

BLACEBERGAN *Rubus fruticosus* (Blackberry) B & H.

BLACK-BUTT *Eucalyptus pilularis* Howes.

BLACK GRASS *Medicago lupulina* (Black Medick) Bucks. Grigson. Black Hay, Black Trefoil, Black Nonesuch and Blackseed are all recorded.

BLACK-HEADED GRASS *Luzula campestris* (Good Friday Grass) Ches. B & H. Cf. Blackcaps, and the various Chimneysweeper names for this.

BLACK MAN 1. *Plantago lanceolata* (Ribwort Plantain) Som, Dor. Grigson. 2. *Prunella vulgaris* (Self-heal) Fernie.

BLACK MAN'S FLOWER *Prunella vulgaris* (Self-heal) Yks. Grigson.

BLACK MARY'S HAND *Dactylorchis maculata* (Heath Spotted Orchid)

Emboden.1974. Black, because the tubers are dark in colour; hand, because the tubers are divided into finger-like lobes. There are a number of "finger" and "hand" names for this.

BLACK ROOT *Veronica virginica* USA Lloyd.

BLACK SEA LILY *Lilium ponticum*. 'Ponticum' refers to the ancient Black Sea kingdom of Pontus, of northeast Asia Minor.

BLACKAMOOR'S BEAUTY; BLACKYMOOR'S BEAUTY 1. *Knautia arvensis* (Field Scabious) Som, Suss. Friend (Blackamoor). 2. *Scabiosa atropurpurea* (Sweet Scabious) Halliwell (Blackamoor); Som. Jennings (Blackymoor). 3. *Succisa pratensis* (Devil's-bit) Som. Elworthy (Blackamoor). Blackamoor here means gypsy - Gypsy Flower, Gypsy Rose, and Egyptian Rose are other names for the scabious.

BLACKAMORE *Typha latifolia* (False Bulrush) Som. Macmillan. Blackymore is also recorded, as well as other descriptive names like Blacktop, Blackie Toppers, etc.

BLACKBARK *Terminalia ivorensis* (Yellow Terminalia) Dalziel.

BLACKBEAD ELDER *Sambucus melanocarpa*.

BLACKBEETLE POISON *Lamium album* (White Deadnettle) Som. Macmillan.

BLACKBERN *Rubus fruticosus* fruit Lancs. B & H.

BLACKBERRY 1. *Empetrum nigrum* (Crowberry) Caith. Grigson. 2. *Ribes nigrum* (Blackcurrant) Lincs, Cumb, Berw. B & H; Yks. Nicholson. 3. *Rubus alleghe-niensis* (Highbush Blackberry) USA. Tolstead. 4. *Rubus cuneifolius* USA Harper. 5. *Rubus floridus* USA Harper. 6. *Rubus fruticosus*. 7. *Rubus villosus* (American Blackberry) USA Lloyd. 8. *Vaccinium myrtillus* (Whortleberry) Yks. Grigson. Blackcurrants were always called blackberries in Yorkshire, a name not given to bramble berries.

BLACKBERRY, Allegheny *Rubus alleghe-niensis* (Highbush Blackberry) Youngken.

BLACKBERRY, American *Rubus villosus*.
BLACKBERRY, California *Rubus vitifolius*.
BLACKBERRY, Cut-leaf *Rubus laciniatus*.
BLACKBERRY, Highbush *Rubus alleghe-niensis*.
BLACKBERRY, Running, Low *Rubus procumbens*.
BLACKBERRY, South African *Rubus pinnatus*.
BLACKBERRY, Sow-teat *Rubus alleghe-niensis* (Highbush Blackberry) USA Yarnell.
BLACKBERRY, Token; BLACKBERRY TOKEN *Rubus caesius* (Dewberry) both Wilts. B & H (Token Blackberry); Grigson (Blackberry Token). Grigson explains the name by the habit the plant has of producing quite a lot of fruit one year, and then another year they will be "mean, scanty, and not worth the picking or eating".
BLACKBERRY MOUCHERS *Rubus fruticosus* fruit Wilts. Jones & Dillon. Moochers or mouchers by itself would do just as well. The word meant truants, originally applied to children who went mooching (or mouching) from school in the blackberry season, then applied to the berries themselves.
BLACKBIDE; BLACK BOYD *Rubus fruticosus* fruit (Blackberry) Kirk, Wigt. Grigson (Blackbide); W.Scot. Jamieson (Black Boyd).
BLACKBLEG *Rubus fruticosus* fruit (Blackberry) Yks. Denham. Sometimes they are called Blacklegs, obviously a corruption of Blackbleg, because the names blegg, or blagg, are used in Yorkshire for the fruit.
BLACKBOWWOWERS *Rubus fruticosus* fruit (Blackberry) N.Eng. Halliwell. B & H has Black Bowours, which sounds more likely.
BLACKBRUSH 1. *Coleogyne ramosissima*. 2. *Florensia cernua* (Tarbrush) USA Kingsbury.
BLACKCAP 1. *Luzula campestris* (Good Friday Grass) Scot. B & H. 2. *Rubus leucodermis* (Western Raspberry) USA E.Gunther. 3. *Rubus occidentalis* (Thimble-berry) USA Tolstead. 4. *Typha latifolia* (False Bulrush) Som. Macmillan. Cf. Blackhead, etc for Bulrush, and the various Chimney-sweep names for Good Friday Grass.
BLACKCURRANT 1. *Ribes floridum* USA H.Smith.1928. 2. *Ribes nigrum*. Currant seems originally to have been Corinth. In O. French the fruit was called raisins de Corinthe, Corinth being their original place of export.
BLACKCURRANT, Wild *Ribes americanum* (American Currant) USA Youngken.
BLACKEYE ROOT *Tamus communis* (Black Bryony) North. See Gerard - the roots "do very quickly waste away and consume away blacke and blew marks that come of bruises and dry-beatings".
BLACKFELLOWS 1. *Bidens bipinnata* (Spanish Needles) Australia Watt. 2. *Bidens pilosa* Australia Watt. In South Africa, they are both known as Black Jacks.
BLACKHAW *Prunus spinosa* (Blackthorn) Ire. Grigson. A name which seems to compound the confusion between blackthorn and whitethorn. Another name, Blackthorn May, does the same.
BLACKHEAD 1. *Centaurea nigra* (Knapweed) H C Long.1910. 2. *Typha latifolia* (False Bulrush) Dev. Friend.1882. Som. Macmillan. Cf. Blackcap, Black Boys, Blacky-more, etc for bulrush, and Black Knapweed, Black Soap for knapweed.
BLACKHEARTS *Vaccinium myrtillus* (Whortleberry) Hants. Grigson. "Hearts" is Hurts, or Horts, the same word as Whorts.
BLACKIE TOPPERS *Typha latifolia* (False Bulrush) Som. Macmillan.
BLACKIE TOPS *Plantago lanceolata* (Ribwort Plantain) Som. Macmillan. Descriptive.
BLACKING PLANT *Hibiscus rosa-sinensis* (Rose of China) J.Smith. The red of the flowers becomes black when bruised, and is then used for colouring the eyebrows, and blacking shoes. With this information, the other name, Shoeflower, makes some sense.

BLACKINGIRSE *Filipendula ulmaria* (Meadowsweet) Shet. Grigson.

BLACKITE; BLACK KITE *Rubus fruticosus* fruit (Blackberry) B & H (Blackite); Cumb. Grigson (Black Kite). Kites is presumably a North Country word for berries; it occurs several times in the names for blackberries, Brammelkites, and Bumblykites, for instance. But the older authors insisted that the word meant belly.

BLACKJACK OAK 1. *Quercus marilandica*. 2. *Quercus nigra* (Water Oak) USA Vestal & Schultes.

BLACKLEG *Rubus fruticosus* fruit (Blackberry) Yks. Hazlitt. The last syllable is a shortening of bleg, which occurs either as Blackbleg, or by itself as blegg, or blagg, all in Yorkshire.

BLACKROOT *Symphytum officinale* (Comfrey) Dur. B & H. Turner has Blackwort for this. Cf. German Schwarzwurz.

BLACKSEED *Medicago lupulina* (Black Medick) Bucks. Grigson.

BLACKSMITH *Plantago lanceolata* (Ribwort Plantain) Som. Grigson.

BLACKSTRAP *Polygonum aviculare* (Knotgrass) Hants. Cope.

BLACKTHORN *Prunus spinosa*. To distinguish from the whitethorn - hawthorn.

BLACKTHORN MAY *Prunus spinosa* (Blackthorn) Middx. B & H. This name links it further with the hawthorn, so tending to confuse.

BLACKWEED *Euphorbia prostrata* Barbados Gooding. Also known as Black Milkweed in Barbados.

BLACKWOOD 1. *Acacia melanoxylon*. 2. *Dalbergia latifolia*. 3. *Haematoxylon campechianum* (Logwood) S.M.Robertson.

BLACKWOOD, African 1. *Dalbergia melanoxylon*. 2. *Peltophorum africanum* (Rhodesian Wattle) Palgrave.

BLACKWOOD, Australian *Acacia melanoxylon* (Blackwood) Willis.

BLACKWORT 1. *Symphytum officinale* (Comfrey) Turner. 2. *Vaccinium myrtillus* (Whortleberry) Fernie. The comfrey name is probably blackroot, and the name as applied to whortleberry implies black whorts.

BLACKY-MORE *Typha latifolia* (False Bulrush) Som. Macmillan; Wilts. Dartnell & Goddard.1899.

BLADDER, Pop *Digitalis purpurea* (Foxglove) Dor. Macmillan.

BLADDER BOTTLE *Silene cucubalis* (Bladder Campion) Som. Macmillan.

BLADDER CAMPION *Silene cucubalis*. Campion is probably the same word as champion; the word was used originally to mean "used as a garland", hence the connection. Bladder - this is the inflated calyx of the plant.

BLADDER CHERRY 1. *Physalis alkekengi* (Winter Cherry) McClintock. 2. *Physalis heterophylla* (Ground Cherry) USA Gates.

BLADDER GENTIAN *Gentiana utriculosa*.

BLADDER-HERB *Physalis alkekengi* (Winter Cherry) W.Miller. Physalis is from a Greek word meaning bladder - this from the inflated calyx.

BLADDER KATMIA *Hibiscus trionum*.

BLADDER NUT *Staphylea pinnata*. From the inflated capsules, and thence, like most of these 'bladder' plants, by doctrine of signatures, used for bladder complaints.

BLADDER NUT, American *Staphylea trifolia*.

BLADDER OF LARD *Silene cucubalis* (Bladder Campion) Som. Macmillan.

BLADDER POD 1. *Lobelia inflata* (Indian Tobacco) USA Henkel. 2. *Sesbania vesicaria* (Bagpod) USA Kingsbury.

BLADDER SAGE *Salazaria mexicana* (Paper-bag Bush) USA T H Kearney.

BLADDER SEED *Physospermum cornubiense*. Physospermum means just that.

BLADDER SOAPWORT *Saponaria officinalis* (Soapwort) S.Africa Watt.

BLADDER-STEM *Eriogonum inflatum* (Desert Trumpet) USA T H Kearney.

BLADDERWEED *Silene cucubalis* (Bladder Campion) Dor. Grigson.

BLADDERWORT 1. *Physalis alkekengi* (Winter Cherry). 2. *Utricularia sp.* Utricularia is a genus of submerged water plants which grow in peaty pools. They have "bladder-traps" which engulf small water creatures and digest them.

BLADDERWORT, Common *Utricularia vulgaris* (Greater Bladderwort) Gilmour & Walters.

BLADDERWORT, Greater *Utricularia vulgaris*.

BLADDERWORT, Irish *Utricularia intermedia*.

BLADDERWORT, Lesser *Utricularia minor* (Small Bladderwort) Storer.

BLADDERWORT, Small *Utricularia minor*.

BLADE, Healing *Plantago major* (Great Plantain) Scot. Denham/Hardy. Or Healing Leaf, where healing is pronounced helin. The Gaelic for the plant is slanlus, meaning a healing herb. It certainly seems to have been used as a remedy for virtually everything, but probably the emphasis here would be its use to stop bleeding of wounds, a use which is to be found in both Dioscorides and Pliny.

BLADES, Wood *Luzula sylvatica* (Great Woodrush) B & H.

BLAEBERRY 1. *Vaccinium myrtillus* (Whortleberry) Shrop, Lancs, Yks. Grigson; Ire. E.E.Evans; Scot. Jamieson. 2. *Vaccinium uliginosum* (Northern Bilberry) Prior. It is said that Blaeberry is close to its Scandinavian original, in that it is from blaa, dark blue.

BLAEBERRY, Bog *Vaccinium uliginosum* (Northern Bilberry) Darling & Boyd.

BLAEBOW *Linum usitatissimum* (Flax) B & H.

BLAGG; BLEGG *Rubus fruticosus* fruit (Blackberry) both Yks. Grigson. Presumably these both mean berry, for we also have Blackblegs in Yorkshire for blackberries.

BLAINY EYE *Euphorbia lateriflora* Dalziel.

BLANKET; BLANKET MULLEIN *Verbascum thapsus* (Mullein) Fernie (Blanket); Ches. Grigson (Blanket Mullein). All the 'blanket' and 'flannel' names for mullein derive from the texture of the leaves.

BLANKET, Abraham's; BLANKET, Beggar's *Verbascum thapsus* (Mullein) Tynan & Maitland (Abraham); Som. Macmillan; Cumb. Grigson (Beggar).

BLANKET, Devil's *Verbascum thapsus* (Mullein) Wilts. Macmillan. The appropriation to the devil is out of character - the rest of the 'blanket' and 'flannel' names are just the opposite - for instance, Our Lord's Flannel, Lady's Flannel, Moses' Blanket, etc.

BLANKET, Moses' *Verbascum thapsus* (Mullein) Som. Macmillan.

BLANKET, Poor Man's *Verbascum thapsus* (Mullein) Donegal Grigson.

BLANKET, Saviour's *Stachys lanata* (Lamb's Ears) Suss. Friend. It has the same consistency leaves as the Mullein, hence this name. Blanket-leaf also occurs in Devonshire.

BLANKET, Tear- *Acacia greggii* (Cat's Claw) USA Jaeger. It is an extremely thorny shrub.

BLANKET FLOWER *Gaillardia sp* RHS. Presumably from the striped effect of the flower colouring.

BLANKET-LEAF 1. *Stachys lanata* (Lamb's Ears) Dev. Friend.1882. 2. *Verbascum thapsus* (Mullein) Dev, Som, War. Friend.1882. USA Henkel.

BLASPHEME VINE *Smilax laurifolia* (Laurel Greenbrier) USA Cunningham.

BLATHERWEED; BLETHERWEED *Silene cucubalis* (Bladder Campion) both Dor. Grigson (Blather); Macmillan (Blether). i.e. Bladderweed.

BLATTERDOCK *Petasites hybridus* (Butterbur) Grieve.1931. Batter-dock is also recorded for this plant, so presumably this is Butterdock.

BLAVER 1. *Campanula rotundifolia* (Harebell) Scot. Grigson. 2. *Centaurea cyanus* (Cornflower) Berw, Roxb. Grigson. 3. *Papaver dubium* (Smooth Long-headed Poppy) Mor. B & H. Blaver seems to mean "belonging to corn" (Cf. French blé). The derivation from blue is presumably ruled out by the inclusion of a poppy - in any case the harebell is also known as Blue Blaver in Scotland. Perhaps the poppy only got the name by association, if it ever was genuinely called this.

BLAVER, Blue *Campanula rotundifolia* (Harebell) Scot. Grigson.

BLAVEROLE *Centaurea cyanus* (Cornflower) B & H. As blaver seems to mean "concerned with corn", blaverole may very well mean exactly "cornflower".

BLAW, Blue *Centaurea cyanus* (Cornflower) Turner. Gerard has Blue-blow, which becomes Blur-blow, and also Blue-bow later. Blue Blawort still exists in Scotland. Blaw is probably from the same original as blaver (corn) - it certainly does not mean blue, for Blue Blaw would then be meaningless; anyway, there is the Scottish expression "as blae as a blawort" to make the point quite clearly. Note - Turner actually spelled this as one word, Blueblaw.

BLAW-WEARY *Prunella vulgaris* (Self-heal) Scot. H C Long.1910.

BLAWORT 1. *Campanula rotundifolia* (Harebell) Scot. Jamieson. 2. *Centaurea cyanus* (Cornflower) Scot. Jamieson. See Blaver.

BLAWORT, Blue 1. *Campanula rotundifolia* (Harebell) Scot. Grigson. 2. *Centaurea cyanus* (Cornflower) Aber. Grigson.

BLAYBERRY; BLEEBERRY *Vaccinium myrtillus* (Whortleberry) N.Eng. B & H (Blayberry); Dur. F.M.T. Palgrave (Bleeberry). More usually Blaeberry.

BLAZING STAR *Liatris scariosa*.

BLEABERRY 1. *Vaccinium myrtillus* (Whortleberry) Yks. Harland. 2. *Vaccinium uliginosum* (Northern Bilberry) Prior. Blaeberry, or Bog Blaeberry also for Northern Bilberry. Is this a simple spelling mistake? No, Brockett has Bleaberry for bilberry, probably the same as Blayberry.

BLEB, Water *Caltha palustris* (Marsh Marigold) Lincs. Grigson. Bleb is more usually blob, a common dialect word for a blister. It is probable that the Marigold of the common name for this plant comes from a confusion from the OE marsh meargella, from mearh, meaning horse, and gealla, a blister. All the Ranunculaceae are acrid and will raise blisters - some have even been used deliberately for the purpose.

BLEDEWORT *Papaver rhoeas* (Red Poppy) Halliwell. From a plant list in MS Sloane 5, F 3, he says.

BLEEDING GRASS *Plantago major* (Great Plantain) Ire. Egan. Cf. Cut-grass for this. Irish people used to put the leaves on cuts, to stop the bleeding, and to take the pain away. Actually, this is an ancient usage,occurring for instance in the Anglo-Saxon version of Apuleius.

BLEEDING HEART 1. *Cheiranthus cheiri* (Wallflower) Dev. Friend; Wilts. Akerman; Glos. J.D.Robertson; Berks. Lowsley. 2. *Dicentra spectabilis*. 3. *Viola tricolor* (Pansy) Hants. Cope. It is interesting that Turner knew the red wallflower as Heartsease. There are also Blood Wall, and Bloody Walls, corrupted to Bloody Warrior. It is said that both Heartsease and Wallflower had a reputation for curing heart disease (the Viola with some justification, apparently). The Dicentra got the name from the colour and shape of the flowers, even if imagination has to be used.

BLEEDING HEART, Fringed *Dicentra eximea*.

BLEEDING HEART, Western *Dicentra formosa*. It is western North America that we are talking about.

BLEEDING HEART VINE *Clerodendron thomsonae*. The flowers are cream and blood-red in colour.

BLEEDING NUN *Cyclamen europaeus* (Sowbread) Skinner. In Palestine it is dedicated to the Virgin, because the sword of sorrow that pierced her heart is symbolized in the blood drop at the heart of the flower. The name Bleeding Nun occurs for the same reason.

BLEEDING TREE *Casuarina stricta* (She Oak) A.Huxley. This, and Fountain Tree, are names given because of the tree's ability to produce quite large amounts of water when the trunk or branches are cut.

BLEEDING WARRIOR; BLEEDY WARRIOR *Cheiranthus cheiri* (Wallflower) Som. Macmillan (Bleeding); Dev. Macmillan (Bleedy). See Bloody Warrior.
BLEEDING WILLOW *Orchis morio* (Green-winged Orchid) Bucks. B & H.
BLEEDWOOD *Pterocarpus angolensis* (Barwood) Howes.
BLEETS *Chenopodium bonus-henricus* (Good King Henry) Tusser.
BLESSED HERB *Geum urbanum* (Avens) Leyel.1937. A literal translation of the Latin herba benedicta. But as some will have it, from St Benedict - hence the occurrence of names such as St Benedict's Herb, Herb Bennett, etc.
BLESSED MARY'S THISTLE *Silybum marianum* (Milk Thistle) RHS.
BLESSED THISTLE *Carduus benedictus* (Holy Thistle) Ellacombe. From its reputation as a "heal-all". In fact, any plant that bears the name 'blessed' or 'holy' was supposed to have the power of counteracting poison - this would apply to the previous entry (Blessed Herb for Avens), which certainly has this reputation.
BLESSING, Stepmother's *Anthriscus sylvestris* (Cow Parsley) Shrop. Vickery.1985. It carries the same meaning as the better known Mother Die. Never give cow parsley to your mother, or she will die.
BLETHARD *Rumex sanguineus* (Blood-veined Dock) Derb. Grigson.
BLEWART *Veronica chamaedrys* (Germander Speedwell) B & H.
BLIB-BLOB *Caltha palustris* (Marsh Marigold) War. F Savage. See Blob.
BLIND EYES *Papaver rhoeas* (Red Poppy) Bucks. Harman; N'thants. A.E.Baker. From a superstition that if it were put to the eyes, it would cause blindness. In a similar vein, headache and earache have given a variety of names to the poppy.
BLIND FLOWER *Veronica chamaedrys* (Germander Speedwell) Dur. B & H. In co Durham, they say that if you look steadily at the flower for an hour, you will become blind.
BLIND MAN *Papaver rhoeas* (Red Poppy) Wilts. Dartnell & Goddard. See Blind Eyes.
BLIND MAN'S BUFF *Clematis vitalba* (Old Man's Beard) Som. Macmillan.
BLIND MAN'S HAND *Ajuga reptans* (Bugle) Hants. Grigson. There is at least one name with "Dead Man" in it too, so there is probably a connection between the two ideas, but what?
BLIND NETTLE 1. *Galeopsis tetrahit* (Hemp Nettle) Corn. Jago; Dev. B & H. 2. *Lamium album* (White Deadnettle) Dev, Som. Macmillan. 3. *Stachys sylvatica* (Hedge Woundwort) B & H. Deadnettles, being of the non-stinging variety, are often labelled blind, deaf, as well as dead. The woundwort owes the name to its vaguely nettle-shaped leaf.
BLIND TONGUE *Galium aparine* (Goose-grass) Dyer. Blind Tongue is a children's game, apparently. It probably involves drawing the underside of the leaves, which are very rough, across the tongue - so the name is probably more accurately Blood Tongue, which does exist as a name for Goose-grass in N.England and Scotland.
BLIND WITHY *Polygonum persicaria* (Persicaria) Wilts. D.Grose.
BLINDWEED *Capsella bursa-pastoris* (Shepherd's Purse) Yks. Grigson. Is there any possible connection with the Irish belief that shepherd's purse hung under the necks of sheep made them invisible to dogs? (see Wood-Martin).
BLINDYBUFF *Papaver rhoeas* (Red Poppy) Yks. Addy. See Blind Eyes for the explanation of this.
BLINKING-CHICKWEED *Montia fontana* (Blinks) Prior.
BLINKS *Montia fontana*. "The English name of Blinks has perhaps been given to this plant from the blossoms usually appearing in a half opened state, but when the sun shines on them they are fully expanded" (Curtis). But the name was earlier than Curtis, for Christopher Merret used the expression (in

Latin) in 1667 - "with little blinking flowers".

BLINKS, Corn *Centaurea cyanus* (Cornflower) Dev. Grigson.

BLISTER BUSH *Cissus sicyoides*.

BLISTER BUTTERCUP *Ranunculus scleratus* (Celery-leaved Buttercup) Howes.

BLISTER CREEPER *Cardiospermum helicacabum*. This means exactly what it says - the sap is highly irritant.

BLISTER CUP; BLISTER PLANT *Ranunculus acris* (Meadow Buttercup) Lincs. Grigson (Cup); Som. Macmillan (Plant). As with nearly all the buttercups, the juice of this one is extremely acrid, and will raise blisters. But, on the very old principle that like cures like, they say in Lincolnshire that it can be used as a cure for blisters. It is more likely that the cure consisted in raising a blister, quite a common practice at one time.

BLISTERING BUSH *Peucedanum galbanum* S.Africa Watt. This plant will blister the skin on contact.

BLISTERWORT *Ranunculus scleratus* (Celery-leaved Buttercup) B & H. From the blister-raising acrid juice.

BLITE *Chenopodium bonus-henricus* (Good King Henry) Prior. Prior says Latin blitum comes from a Greek word meaning insipid, and Evelyn's comment was "it is well-named, being insipid enough".

BLITE, Coast *Chenopodium rubrum* (Red Goosefoot) Howes.

BLITE, Frost *Chenopodium album* (Fat Hen) B & H. 'Frost' presumably for the same reason that it is called *album*.

BLITE, Stinking *Chenopodium vulvaria* (Stinking Orach) Curtis.

BLITE, Strawberry *Chenopodium capitatum*.

BLITHRAN *Potentilla anserina* (Silverweed) Ire. Grigson. A form of the original Irish name, given as brisclan. It is brisgein in Scottish Gaelic.

BLOB 1. *Digitalis purpurea* (Foxglove) Som. Macmillan; Ess, Suff. Grigson. 2. *Nuphar lutea* (Yellow Waterlily) Dor, N'thants, Yks. Grigson; Wilts. Akerman. 3. *Ribes uva-crispa* (Gooseberry) Scot. B & H. Blob means a bubble or blister, and in some forms (see below) refers to the ability of the plant to raise a blister. Here, though, the name is probably applied to describe the actual form - the globular form of the gooseberry, the bottle-shaped seed container of the Waterlily. For the foxglove, blob seems to mean something quite different. Children burst the flowers, which make a noise like "blob". Everywhere else, the noise is described as "pop".

BLOB, Blib *Caltha palustris* (Marsh Marigold) War. F Savage.

BLOB, Blue *Centaurea cyanus* (Cornflower) Hants. Grigson.

BLOB, Butter 1. *Caltha palustris* (Marsh Marigold) Yks. Grigson. 2. *Trollius europaeus* (Globe Flower) Scot. Tynan & Maitland. Blob is certainly blister here, as it would be in all cases of the Buttercup family.

BLOB, Honey *Ribes uva-crispa* (Gooseberry) Scot. B & H.

BLOB, Horse 1. *Caltha palustris* (Marsh Marigold) Som, Worcs, Leics, Notts. Grigson; Wilts. D.Grose; N'thants. A.E.Baker. 2. *Leucanthemum vulgare* (Ox-eye Daisy) N'thants. Grigson. 'Horse' in a plant name usually means a large, or coarse, variety. Thus Horse Daisy, which does occur widely for the Ox-eye, aptly describes it when compared with the common daisy, and Marsh Marigold is a large kind of buttercup. Why Blob for the Ox-eye, though?

BLOB, Mare *Caltha palustris* (Marsh Marigold) Som. Macmillan; Glos, War, Derb. Grigson; N'thants. A.E.Baker. The name Marsh Marigold itself almost certainly arose from a confusion about the OE name, meargealla, from mearh, horse, and gealla, blister. So Marigold itself in this case is merely the sound of meargella. And the 'mare' of Mare Blob is the first element in the OE name. Cf. Horse Blob for the same plant.

BLOB, May 1. *Caltha palustris* (Marsh Marigold) Glos. Brill; Wilts. Grigson;

Leics. Ruddock. 2. *Cardamine pratensis* (Lady's Smock) N'thants. A.E.Baker. 3. *Ranunculus sceleratus* (Celery-leaved Buttercup) N'thants. A.E. Baker. 4. *Trollius europaeus* (Globe Flower) Leics. Grigson.

BLOB, Mire *Caltha palustris* (Marsh Marigold) N'thants. B & H. Mare Blob, presumably, but it could be a recognition that it grows in the mud.

BLOB, Moll 1. *Caltha palustris* (Marsh Marigold) Worcs. Grigson; N'thants. A.E.Baker. 2. *Cardamine pratensis* (Lady's Smock) N'thants. Grigson. In both cases, the name means cow blister. Cf. the German Mollesblume for Marsh Marigold, where Molle means cow.

BLOB, Molly *Caltha palustris* (Marsh Marigold) N'thants. Sternberg.

BLOB, Water 1. *Caltha palustris* (Marsh Marigold) War, Oxf, Leics, N'thants, Derb, Yks. Grigson. 2. *Nuphar lutea* (Yellow Waterlily) Wilts. Dartnell & Goddard; Dor, N'thants. Grigson; Yks. Addy. 3. *Nymphaea alba* (White Waterlily) N'thants, Yks. Grigson. 4. *Ranunculus sceleratus* (Celery-leaved Buttercup) N'thants. A.E. Baker.

BLOB, Yellow *Caltha palustris* (Marsh Marigold) Leics. Grigson.

BLOCKWHEAT; BOCKWHEAT *Fagopyrum esculentum* (Buckwheat) both B & H. These are, like Buckwheat, from Dutch bockweit, and German Buchweize, meaning beech-wheat. The seeds resemble beech mast.

BLODEWORT *Polygonum hydropiper* (Water Pepper) B & H. i.e. Bloodwort.

BLOGDA *Caltha palustris* (Marsh Marigold) Shet. Grigson. Bludda and Blugda are other Shetland names, all obviously meaning the same, but what?

BLOOD, Butcher's *Melandrium rubrum* (Red Campion) Som. Macmillan.

BLOOD, Dane's 1. *Campanula glomerata* (Clustered Bellflower) Cambs. Dyer. 2. *Pulsatilla vulgaris* (Pasque Flower) Hants. Boase; Herts. Jones-Baker.1974; Ess, Norf, Cambs. Grigson. 3. *Sambucus ebulus* (Dwarf Elder) Camden. All these are associated with a tradition that they sprang from the blood of Danes killed in battle. The Bellflower used to be prolific on mounds which tradition says were made by Danes as monuments of the battle fought between Canute and Edmund Ironside in 1006. The Pasque Flower, too, grows on old earth-works traditionally associated with the Danes, or it was said that it only grew where battles had been fought between the English and the Danes. The same Cambridgeshire legend applied to the Clustered Bellflower is given by Camden, *Britannia*, 1586, for the Dwarf Elder. But the latter is more generally said to grow where blood has been shed, i.e. not specifically the blood of the Danes. It is probably the fact that the stems turn red in September that gave rise to this tradition.

BLOOD, Dragon's *Geranium robertianum* (Herb Robert) Som. Lawrence; Shrop. B & H. Bloodwort and Bloody May are recorded in other parts of the country. The Somerset name is confined to the area where there actually was a dragon legend - North Petherton.

BLOOD, Drops of *Anagallis arvensis* (Scarlet Pimpernel) Wilts. Macmillan.

BLOOD, Drops of Abel's *Fuchsia sp.*

BLOOD, Flesh-and- *Potentilla erecta* (Tormentil) Berw. Grigson. Bloodroot is also recorded in the same general area, because the roots give a red dye. Is Flesh-and-blood related to this?

BLOOD, Poor Man's *Orchis mascula* (Early Purple Orchis) Kent B & H. Presumably from the dark red spots on the leaves.

BLOOD, Witches' *Genista tinctoria* (Dyer's Greenweed) Mass. Leighton. Apparently because it grew on Gallows Hill, Salem.

BLOOD CUP *Asperula odorata* (Woodruff) Dor. Macmillan. It may be that the rough leaves gave rise to the name - yarrow, for instance, has various names like Nosebleed for that reason. On the other hand, woodruff was recommended by Gerard as a vulnerary. It is in fact useful for anti-coagulant purposes in heart disease (the hay smell is due to the coumarin the plant contains - this is the anti-coagulant).

BLOOD DROP EMLET *Mimulus luteus* Scot. Grigson.

BLOOD FLOWER 1. *Asclepias curassivica*. 2. *Haemanthus moschatus* McDonald. Both from the red colour of the flowers.

BLOOD HILDER *Sambucus ebulus* (Dwarf Elder) Norf. Fluckiger. It is said to grow where blood has been shed, and in Wales is called "Plant of the blood of man". There are also English names like Deathwort, Bloodwort, etc. It is associated in England with the Danes - wherever their blood was shed in battle, this plant afterwards sprang up. The fact that the stems turn red in September presumably gave rise to these traditions.

BLOOD LILY *Haemanthus katharinae*. Red flowers.

BLOOD OF ADONIS; ADONIS FLOWER *Adonis annua* (Pheasant's Eye) Parkinson. The scarlet flowers.

BLOOD OF MERCURY; MERCURY'S MOIST BLOOD *Verbena officinalis* (Vervain) Clair, and Fernie respectively. These names presumably arise from the use of the plant in ritual magic.

BLOOD OF ST JOHN *Hieraceum pilosellum* (Mouse-ear Hawkweed). It is Johannisblut in German. When gathered on St John's Eve, it brings good luck (but only on that evening). But it had to be uprooted, and then it was put in stables to preserve them from lightning.

BLOOD ORANGE *Citrus aurantium var. melitensis.*

BLOOD ROOT *Potentilla erecta* (Tormentil) N'thum, Scot. Grigson. Being rich in tannin, the roots used to be recommended for dysentery. Pennant records its use on the Island of Rum - "if they are attacked with a dysentery they make use of a decoction of the roots... in milk". Gerard notes that the powdered roots cure diarrhoea, and the bloody flux, and gypsies use the root decoction to stop internal haemorrhage.

BLOOD STAUNCH *Erigeron canadensis* (Canadian Fleabane) USA Henkel. Among its many medicinal uses in North America was one for arresting haemorrhage.

BLOOD TONGUE *Galium aparine* (Goosegrass) Ches, N'thum, Scot. Grigson. This, with Tongue-bleed, or Tongue-bluiders (a North Country name) all seem to refer to a children's game called Blind-Tongue, also given to the plant, which is described by Dyer "children with the leaves practise phlebotomy upon the tongue of those playmates who are simple enough to endure it" - common, apparently, in Scotland once.

BLOOD(Y)-VEINED DOCK *Rumex sanguineus*. Bloody-veined Dock is in J A MacCulloch.1905.

BLOOD VINE *Chamaenerion angustifolium* (Rosebay Willowherb) Hants. Cope. Presumably the colour gave the name.

BLOOD WALL *Cheiranthus cheiri* (Wallflower) Wilts. Macmillan; N'thants. A.E.Baker. Bloody Walls, too. See under Bloody Warrior, which is probably a corruption of Wall-yer.

BLOODBERRY *Rivina humilis* (Pigeonberry) Perry. The berries are scarlet, and the source of a red dye.

BLOODLEAF *Iresine herbsti*. This is a foliage house plant.

BLOODROOT 1. *Lacnanthes tinctoria*. 2. *Sanguinaria canadensis*. 3. *Geum canadense* (White Avens) Lindley. All North American plants. *Lacnanthes tinctoria* is also known as Redroot. The best known of these three is the *Sanguinaria*. The rootstock produces a reddish juice used by the Indians as a dye for clothes or basketry decoration, or for decorating their own bodies (then it serves a double purpose, for it is an insect repellent).

BLOODSTRANGE *Myosurus minimus* (Mousetail) Parkinson.1640. Probably a corruption of Bloodstay; Parkinson speaks of its use as a styptic.

BLOODWEED 1. *Asclepias curassivica* (Blood Flower) Grieve.1931. 2. *Polygonum lapathifolium* (Pale Persicaria) Donegal. Grigson.

BLOODWOOD 1. *Eucalyptus terminalis*. 2.

Pterocarpus angolensis (Barwood) Palgrave. The sap of the *Pterocarpus* is red and sticky, hence the name. A number of African peoples compare it with blood, and use it (or the wood) in rituals in which blood flows.

BLOODWOOD, Red *Eucalyptus gummifera*.

BLOODWOOD, Victoria *Eucalyptus corymbosa*.

BLOODWORT 1. *Achillea millefolium* (Yarrow) USA Henkel. 2. *Geranium robertianum* (Herb Robert) Som, Cumb. Grigson. 3. *Polygonum aviculare* (Knotgrass) B & H. 4. *Polygonum hydropiper* (Water Pepper) B & H. 5. *Rumex hydrolapathum* (Great Water Dock) Ches. Holland. 6. *Rumex sanguineus* (Blood-veined Dock) Parkinson.1629. 7. *Sambucus ebulus* (Dwarf Elder) B & H. 8. *Sanguinaria canadensis* (Bloodroot) Thomson. 9. *Sanguisorba officinalis* (Great Burnet) Clair. Most of these owe the name to their colour, but some have the additional reason of being blood-staunchers (though one should not lose sight of the possibility of the doctrine of signatures coming into play, i.e. the colour of the plant itself supported the idea of its medical use). Gerard, for instance, recommended Great Burnet as a vulnerary and stauncher of blood (Sanguisorba itself comes from Latin sanguis, blood). Water Pepper, too, with its reddish stalk and flower was used as a sanguinary, "because it draweth blode in places that it is rubbed on" (Gerard). Arnold of Villanova recommended all soldiers to drink the decoction of Water Pepper, or to have a powder of the plant to sprinkle on a wound. Knotgrass, too, was used in the same way, and a Somerset remedy for nosebleed is to rub knotgrass into the nostrils. Yarrow, too, was used as a styptic, but in addition was actually put into the nose to cause bleeding, ostensibly to cure megrim, but there was also a sort of love divination involving the same process. In East Anglia, a girl would tickle the inside of the nostril with a leaf of yarrow, saying: Yarroway, yarroway, bear a white blow; If my love love me, my nose will bleed now. There are a number of near variants to this rhyme, and the superstition spread to America - Bergen quotes a very similar rhyme. Prior though, suggests that it got this reputation by a mistranslation - the plant actually referred to is the very abrasive horsetail.

BLOODWORT, Burnet *Sanguisorba officinalis* (Great Burnet) Prior. See Bloodwort.

BLOODY BELLS *Digitalis purpurea* (Foxglove) Lanark Grigson.

BLOODY BONES 1. *Endymion nonscriptus* (Bluebell) Dor. Macmillan. 2. *Iris foetidissima* (Foetid Iris) Dor. Dacombe. 3. *Orchis mascula* (Early Purple Orchis) Dor. Macmillan. There are a number of "Dead Man" names for Early Purple Orchis, as if it were growing from a corpse. The dark red blotches on the leaves are enough to account for the 'bloody'. Why bluebell, though? Perhaps the clue is that the orchis is called Bluebell in Dorset, so it may be an example of mistaken recording.

BLOODY BUTCHER 1. *Centranthus ruber* (Spur Valerian) Som. Macmillan. 2. *Orchis mascula* (Early Purple Orchis) Heref. Havergal; Oxf. Oxf.Annl.Rec. 1951; N'thants. B & H. The colour again, and, with the Orchis, the reinforcement of the dark red blotches on the leaves.

BLOODY CRANESBILL *Geranium sanguineum*. Colour - it is also called Blood-red Cranesbill.

BLOODY DOCK *Rumex sanguineus* (Blood-veined Dock) Som. Elworthy. Bloodwort, too, for this. As Gerard explains "... of the bloudy colour wherwith the whole plant is possest".

BLOODY FINGERS 1. *Arum maculatum* (Cuckoo-pint) Fernie. 2. *Digitalis purpurea* (Foxglove) Som. Elworthy; Heref. Havergal; Yks, Cumb. Grigson; Scot. Aitken. 3. *Orchis mascula* (Early Purple Orchis) Glos. J.D. Robertson. The foxglove has several similar names - Bloody Bells, Dead Man's Bells, Bloody Man's Fingers,

Dead Man's Fingers, etc., all from the belief that where it grows someone has been buried. When applied to the Orchis, the name is descriptive, from the dark red blotches on the leaves, looking as if someone with bloody fingers had touched them.

BLOODY HEATH *Erica cruenta*.

BLOODY MAN'S FINGERS 1. *Arum maculatum* (Cuckoo Pint) Friend. 2. *Digitalis purpurea* Heref. B & H. 3. *Orchis mascula* (Early Purple Orchis) Glos. J.D. Robertson; Ches. Holland. 4. *Orchis morio* (Green-winged Orchid) Ches. Holland. As far as the Arum and Orchis are concerned, the name probably arose from the dark blotches on the leaves, as if someone with bloody fingers had touched them. Foxglove presumably got the name along with such as Dead Man's Fingers - there was a belief that where it grows, someone had been buried.

BLOODY MAN'S HANDS *Orchis mascula* (Early Purple Orchis) Kent B & H. See Bloody Man's Fingers.

BLOODY MARY *Geranium robertianum* (Herb Robert) Yks. Grigson. Colour descriptive.

BLOODY ROD, BLOODY TWIG *Cornus sanguinea* (Dogwood) both. B & H. "that which the Italians call Virga Sanguinea" (Gerard) - the twigs are dark crimson in winter.

BLOODY THUMBS *Briza media* (Quaking-grass) Worcs. J Salisbury.

BLOODY TRIUMPH *Trifolium incarnatum* (Crimson Clover) Dor. Macmillan. The local tradition is that the name commemorated a battle in which the winners decorated themselves with these flowers.

BLOODY WALLIER; BLOODY WALLS *Cheiranthus cheiri* (Wallflower) Halliwell (Wallier); McDonald (Walls). Cf. the Northants name Blood Wall, and of course Bloody Warrior.

BLOODY WARRIOR 1. *Cheiranthus cheiri* (Wallflower) Corn. Courtney.1880. Dev. Friend.1882; Som. Jennings; Wilts. Akerman; Dor. Dacombe; Glos. J.D. Robertson; Berks. Lowsley; Hants. Cope. 2. *Fritillaria meleagris* (Snake's Head Lily) Berks. Grigson. Warrior is probably, as far as wallflower is concerned, a corruption of the common name, or perhaps Wall-yer; in fact, Halliwell has Bloody Wallier as an established name. Bloody is colour descriptive, though there is the famous legend of the Scottish girl who died while attempting to escape with her lover. It is a tale connected with Neidpath Castle, near Peebles. Herrick, in his poem, said that Love

Turn'd her to this plant we call
The scented flower upon the wall.

Snake's Head Lily was said to have grown from a drop of Dane's blood, and so joins a number of plants connected with the death of Danes in battle.

BLOODY WILLIAM *Lychnis coronaria* (Rose Campion) Coats.

BLOODY WIRES *Cheiranthus cheiri* (Wallflower) Som. Macmillan. i.e. Bloody Warrior.

BLOOM, Gold *Calendula officinalis* (Marigold) B & H.

BLOOM-FELL; FELL-BLOOM *Lotus corniculatus* (Bird's-foot Trefoil) Scot. Grigson. Descriptive of habitat; Fell-bloom makes more immediate sense than Bloom-fell.

BLOOMY-DOWN *Dianthus barbatus* (Sweet William) Som. Elworthy.

BLOSSOM *Crataegus monogyna* (Hawthorn) Som. Macmillan. Just as May is the immediate identification of these flowers among all the others of Maytime, so Blossom is enough to identify them.

BLOW, Blue *Centaurea cyanus* (Cornflower) Gerard.

BLOW, Cathaw *Heracleum sphondyllium* (Hogweed) Cumb. Grigson.

BLOW, White *Saxifraga tridactylites* (Rue-leaved Saxifrage) B & H.

BLOW-BALL *Centaurea cyanus* (Cornflower) Halliwell. Blow, of course,

means a flower.
BLOW-FLOWER *Centaurea cyanus* (Cornflower) Som. Macmillan. The meaning of 'blow' must have been forgotten in this case, for it is just another word for flower.
BLOW-PIPE TREE *Syringa vulgaris* (Lilac) Skinner. Parkinson had Pipe Tree, and Gerard Blue Pipe. Is Blow-Pipe actually Blue Pipe?
BLOWBALL *Taraxacum officinale* (Dandelion) B & H; USA Henkel. Referring, of course, to the seed heads.
BLOWER *Taraxacum officinale* (Dandelion) B & H. The seed-heads.
BLUB, May *Caltha palustris* (Marsh Marigold) Wilts. Grigson. This is May Blob, where blob means blister. Like most of the Ranunculaceae, Marsh Marigold is acrid.
BLUBBERS, Water *Caltha palustris* (Marsh Marigold) Glos. J.D.Robertson. i.e. blob = blister.
BLUDDA, BLUGGA *Caltha palustris* (Marsh Marigold) Shet. Grigson.
BLUE-EYED GRASS *Sisyrinchium angustifolium*.
BLUE-EYED GRASS, Michaux's *Sisyrinchium mucronatum*.
BLUE-EYED GRASS, Prairie *Sisyrinchium campestre*.
BLUE-EYED GRASS, Stout *Sisyrinchium gramineum*.
BLUE-GRASS, Annual *Poa annua* (Annual Meadow-grass) USA Allan.
BLUE-GRASS, Kentucky *Poa pratensis* (Smooth Meadow-grass) USA Allan.
BLUE-RUNNER *Glechoma hederacea* (Ground Ivy) Bucks. Grigson.
BLUEBEARD 1. *Nigella damascena* (Love-in-a-mist) Som. Macmillan. 2. *Salvia horminoides* (Clary) Som. Macmillan.
BLUEBELIA *Lobelia erinus* (Garden Lobelia) Sam Cooper. Sam Cooper was an old scrap metal dealer who used to give up this business every year in order to go around selling his "bedding-out stuff". They went under some extraordinary names, of which Bluebelia was the most aptly outrageous.
BLUEBELL 1. *Aquilegia vulgaris* (Columbine) USA Hutchinson. 2. *Campanula parryi* Stevenson. 3. *Campanula rotundifolia* (Harebell) Dev. Friend.1882; Som. Macmillan. 4. *Endymion nonscriptus*. 5. *Mertensia virginica* (Virginia Cowslip) USA House. 6. *Orchis mascula* (Early Purple Orchis) Dor. Macmillan. 7. *Polemonium reptans* (Abscess Root) USA House. 8. *Vinca major* (Greater Periwinkle) Dev. B & H. 9. *Vinca minor* (Lesser Periwinkle) Dev. Friend. Harebell is the Bluebell of Scotland. The Virginia Cowslip is probably better known in America as False Bluebell, rather than just Bluebell.
BLUEBELL, American *Campanula americana*.
BLUEBELL, Australian *Sollya heterophylla* (Bluebell Creeper) Cunningham.
BLUEBELL, Californian *Phacelia minor*.
BLUEBELL, False *Mertensia virginica* (Virginia Cowslip) USA Tolstead. Sometimes, just Bluebell.
BLUEBELL, Little *Viola canina* (Dog Violet) Border B & H.
BLUEBELL, Mountain *Mertensia ciliata*.
BLUEBELL, New Zealand *Wahlenbergia albomarginata*.
BLUEBELL, One-flowered *Campanula uniflora* (One-flowered Bellflower) USA Elmore.
BLUEBELL, Scottish *Campanula rotundifolia* (Harebell) Polunin. Better known as the Blue Bell of Scotland.
BLUEBELL, Spanish *Endymion hispanicum*.
BLUEBELL, Swamp *Campanula aparinoides*.
BLUEBELL, Virginia *Mertensia virginica* (Virginia Cowslip) Howes.
BLUEBELL, White *Allium triquetrum* (Three-cornered Leek) McClintock. Seems a contradiction in terms.
BLUEBELL CREEPER *Sollya heterophylla*.
BLUEBERRY 1. *Billardiera longifolia* (Purple Appleberry) Howes. 2. *Cornus amomum* (Silky Cornel) Grieve.1931. 3. *Dianella nigra*. 4. *Rubus caesius* (Dewberry) Scot. Fairweather. 5. *Vaccinium mem-*

branaceum. 6. *Vaccinium myrtillus* (Whortleberry) Yks, Cumb, Ire. Grigson. 7. *Vaccinium pennsylvanicum* (Blue Huckleberry) USA. H.Smith.1923. There seems to be a lot of confusion in the common names of the genus *Vaccinium* - a number of them seem to be quite inter-changeable.

BLUEBERRY, Box *Vaccinium ovatum* (California Huckleberry) Yanovsky. Presumably because the leaves are like those of box. Cowberry (*V.vitis-idaea*) is sometimes called Flowering Box, for that reason.

BLUEBERRY, Canada 1. *Vaccinium canadense*. 2. *Vaccinium myrtilloides* (Sour-top Blueberry) USA Yarnell.

BLUEBERRY, Dryland *Vaccinium vacilans*.

BLUEBERRY, High-bush *Vaccinium corymbosum* (High-bush Huckleberry) USA Yarnell.

BLUEBERRY, Low, Late *Vaccinium vacillans* (Dryland Blueberry) USA Yarnell.

BLUEBERRY, Sour-top *Vaccinium myrtilloides*.

BLUEBERRY, Swamp *Vaccinium corymbosum* (High-bush Huckleberry) Chaplin.

BLUEBERRY, Sweet, Low *Vaccinium angustifolium*.

BLUEBERRY ASH *Elaeocarpus reticulatus*. Deep blue fruits in autumn.

BLUEBERRY ELDER *Sambucus caerulea*.

BLUEBERRY ROOT *Caulophyllum thalictroides* (Blue Cohosh) Grieve.1931.

BLUEBLAW *Centaurea cyanus* (Cornflower) Turner.

BLUEBOTTLE 1. *Aconitum napellus* (Monkshood) Dor. Macmillan. 2. *Campanula rotundifolia* (Harebell) Bucks. Grigson. 3. *Centaurea cyanus* (Cornflower) Gerard. 4. *Centaurea nigra* (Knapweed) Som. Macmillan. 5. *Echium vulgare* (Viper's Bugloss) Norf. Grigson. 6. *Endymion nonscriptus* (Bluebell) Som, Dor. Grigson; Wilts. Dartnell & Goddard; Suss. Parish. 7. *Vinca minor* (Lesser Periwinkle) Som. Macmillan. With the exception of harebell, and possibly bluebell, where it is probably a mistake, 'bottle' is a reference back to an old meaning - a bundle (usually of hay or flax). The simile is there - these flowers actually do look like bundles of blue petals.

BLUEBOTTLE, Corn *Centaurea cyanus* (Cornflower) N'thants. A.E.Baker.

BLUEMONY; BLUE MONEY *Pulsatilla vulgaris* (Pasque Flower) B & H (Bluemony); W.Miller (Blue Money). i.e. Blue Anemone. The sequence runs Blue Emony, Bluemony, Blue Money.

BLUET 1. *Centaurea cyanus* (Cornflower) Prior. 2. *Houstonia caerulea*.

BLUET, Mountain *Centaurea montana* (Mountain Knapweed) Howes.

BLUEWEED 1. *Echium vulgare* (Viper's Bugloss) Som, Hants. Grigson; USA House. 2. *Helianthus ciliaris*.

BLUGHTYNS *Caltha palustris* (Marsh Marigold) IOM Moore.1924.

BLUNDERBUSS *Dianthus caryophyllus* (Carnation) Som. Macmillan.

BLUNT-LEAVED EVERLASTING *Gnaphalium polycephalum* (Mouse-ear Everlasting) Grieve.1931.

BLUNT-SICKLE, HURT-SICKLE *Centaurea cyanus* (Cornflower) Tynan & Maitland (Blunt-sickle); Grieve.1931 (Hurt-sickle). Self-explanatory.

BLUSH, Maiden's *Momordica charantia* (Wild Balsam-apple) Howes.

BLUSHING PHILODENDRON *Philodendron erubescens*.

BLY *Rubus fruticosus* (Blackberry) Grieve.1931.

BO-TREE *Ficus religiosa* (Peapul) Folkard. i.e. the tree of enlightenment - Sinhalese bogaha (bo, enlightenment, or perfect knowledge, and gaha, tree). This is the tree under which Gautama became Buddha, the enlightened.

BOAB *Adansonia gregori* (Australian Baobab) Crawford.

BOAR-THISTLE 1. *Cirsium arvense* (Creeping Thistle) B & H. 2. *Cirsium vulgare* (Spear Plume Thistle) Grigson. Boar here is a corruption of bur.

BOAR'S EARS *Primula auricula* (Auricula)

N.Scot. B & H. Boar's Ears is a corruption of Bear's Ears, given to the plant from the shape of the leaves. But Jamieson says that a bear is called a boar in northern Scotland. Boar's Ears is also spelt Bores' Ears in the same area.

BOAR'S FOOT *Helleborus viridis* (Green Hellebore) Bucks. Grigson. Again, a corruption of Bear's Foot, from the shape of the leaf.

BOAT LILY *Rhoeo spathacea*.

BOATS 1. *Acer campestre* seeds (Maple keys) Som. Macmillan. 2. *Lathyrus odoratus* (Sweet Pea) Som. Macmillan.

BOATS, Fairy *Nuphar lutea* (Yellow Waterlily) Som. Macmillan.

BOATS, Water *Caltha palustris* (Marsh Marigold) Pratt. 'Boat' here is 'bout', or 'boot', from French bouton, button.

BOB *Trifolium pratense* flower heads (Red Clover) Ches. Holland. So clover is said to be in bob when it is in flower.

BOB, Dusty *Senecio cineraria* (Silver Ragwort) Som. Macmillan. Dusty Miller, although an American name for this particular plant, is nevertheless more widespread for others.

BOB, Fir *Pinus sylvestris* cones (Scots Pine cones) Shrop, Ches, Leics. B & H.

BOB, London 1. *Dianthus barbatus* (Sweet William) Lancs. Nodal. 2. *Geum rivale* (Water Avens) Tynan & Maitland. This name for Sweet William is apparently localised in the Calder Valley, near Garstang. Both London Pride, and London Tuft, are very old names.

BOB, Stinker; BOB, Stinking *Geranium robertianum* (Herb Robert) Som. Macmillan (Stinker); Midl. Grigson (Stinking). There are quite a number of "stinking" names for this, all applied because of the distinctive smell of the leaf when bruised.

BOB GINGER *Polygonum hydropiper* (Water Pepper) Cumb. Grigson. The name emphasises the acrid, irritant quality of the leaves, and there are a great variety of local names to the same effect.

BOB-GRASS *Bromus mollis* (Soft Brome) Wilts. Dartnell & Goddard. The Somerset equivalent is Lob-grass.

BOB ROBERT *Geranium robertianum* (Herb Robert) Dor. Macmillan.

BOB ROBIN *Melandrium rubrum* (Red Campion) Corn. Grigson; Som, Wilts. Macmillan.

BOBBERTY *Veronica chamaedrys* (Germander Speedwell) Hants. Hants FWI. Probably from a version of Botherum, though that is a North country word meaning 'trembling'.

BOBBIES 1. *Geranium robertianum* (Herb Robert) Som. Macmillan. 2. *Plantago lanceolata* (Ribwort Plantain) Dor. Macmillan. Bobbies is obvious for Herb Robert, but not for Ribwort - the clue is probably another Dorset name, Bobbins, for this plant.

BOBBIN AND JOAN *Arum maculatum* (Cuckoo Pint) N'thants. A.E.Baker. It is said that the name comes from the belief that the spadices are like the lace bobbins used in Buckinghamshire and Northamptonshire. Bobbin-Joan is also the name of an old country dance, sometimes known too as Bobbin-Joe (Barnaschone too mentions the "well-known tune of Bob and Joan". There is also the possibility that Bobbin here is the same as Robin (Wake-Robin), where robin means penis (Fr. robinet, an old meaning). Cf. Cuckoo Pint, etc.

BOBBIN JOAN 1. *Arum maculatum* (Cuckoo pint) N'thants. Sternberg. 2. *Lychnis flos-cuculi* (Ragged Robin) Dev. Tynan & Maitland.

BOBBINS 1. *Arum maculatum* (Cuckoo Pint) Bucks. Grigson. 2. *Nuphar lutea* (Yellow Waterlily) Scot. Jamieson. 3. *Nymphaea alba* (White Waterlily) Bucks. Grigson. 4. *Plantago lanceolata* (Ribwort Plantain) Dor. Macmillan. The name, as given to the waterlilies is applied properly to the seed vessels.

BOBBINS, Dog; BOBBINS, Dog-and- *Arum maculatum* (Cuckoo Pint) both N'thants. B

& H (Dog); A.E.Baker (Dog-and-). See Bobbin and Joan.

BOBBY-AND-JOAN *Arum maculatum* (Cuckoo Pint) N'thants. B & H.

BOBBY HOOD *Melandrium rubrum* (Red Campion) Som. Macmillan.

BOBBY ROSE *Trifolium repens* (White Clover) Corn. Grigson.

BOBBY'S BUTTONS 1. *Arctium lappa* burrs (Burrs) Som. Macmillan. 2. *Caltha palustris* (Marsh Marigold) Dev, Som. Macmillan. 3. *Centaurea cyanus* (Cornflower) Dev, Som. Macmillan. 4. *Centaurea nigra* (Knapweed) Som. Macmillan. 5. *Galium aparine* (Goose-grass) Som. Macmillan.

BOBBY'S EYE *Veronica chamaedrys* (Germander Speedwell) Som. Grigson; Hants. Cope.

BOBBY'S EYE, Red *Geranium robertianum* (Herb Robert) Wilts. Dartnell & Goddard; Hants. Grigson.

BOBBY'S EYE, White *Stellaria holostea* (Greater Stitchwort) Hants. Grigson.

BOBBYNS *Betula verrucosa* seed vessels (Birch seeds) Lothian Jamieson.

BODARK *Maclura pomifera* (Osage Orange) Everett. This is a corruption of bois d'arc; another name for the tree is Bow-wood (the Indians used it for just this purpose).

BODDLE *Chrysanthemum segetum* (Corn Marigold) Tusser. See Boodle.

BODKIN, Shepherd's *Scandix pecten-veneris* (Shepherd's Needle) Halliwell. All the "needle" names for this come from the shape of the long, beaked fruits.

BOG BILBERRY *Vaccinium uliginosum* (Northern Bilberry) Hohn.

BOG DAISY *Caltha palustris* (Marsh Marigold) Yks. Grigson.

BOG-FLOWER *Cardamine pratensis* (Lady's Smock) Yks. Grigson.

BOG HOP *Menyanthes trifoliata* (Buckbean) N Eng. Grigson. The reference here is to the bitterness of the plant, and its use for that reason in beer-making, as hops would be.

BOG HYACINTH *Orchis mascula* (Early Purple Orchis) Scot. Grigson.

BOG NUT *Menyanthes trifoliata* (Buckbean) Scot. Grigson. Presumably referring to the roots - they were once pounded up to yield an edible meal.

BOG SAGE *Salvia uliginosa*.

BOG STARWORT; BOG STITCHWORT *Stellaria uliginosum* Murdoch McNeill (Starwort).

BOG THISTLE *Cirsium palustre* (Marsh Thistle) B & H.

BOG TREFOIL *Menyanthes trifoliata* (Buckbean) Yks. Grigson.

BOGBANE *Menyanthes trifoliata* (Buckbean) IOM Moore.1924. Bogbean, of course, but 'bane' would come quite naturally in this context, for it means white, and has no connection with 'bane' as an English word.

BOGBEAN 1. *Menyanthes trifoliata* (Buckbean) N'thants A E Baker; War, Yks, Cumb, IOM Grigson; Ire. Wilde. 2. *Primula farinosa* (Birdseye Primrose) Yks. Grigson. Presumably 'bogbean' for 'buckbean' is a transformation to suit its habitat. But why 'bean' at all for Birdseye Primrose? It isn't the only example, for Powdered Bean is recorded, too, and from as far away as the north of Scotland.

BOGBERRY 1. *Potentilla palustris* (Marsh Cinquefoil) Donegal. Grigson. 2. *Vaccinium oxycoccus* (Cranberry) Ire. B & H.

BOGGART-FLOWER; BOGGART POSY *Mercurialis perennis* (Dog's Mercury) both Yks. Grigson. Boggart is the Yorkshire dialect form of bogey - a spectre or some form of apparition. The name is probably given because the plant is a poison (hence the 'dog'), and it generally had a bad name.

BOGSPINKS *Cardamine pratensis* (Lady's Smock) B & H.

BOGWORT *Vaccinium oxycoccus* (Cranberry) B & H.

BOHEMIAN CRANESBILL *Geranium bohemicum*.

BOHOLAWN *Senecio jacobaea* (Ragwort) Ire. Grigson. This is the same word as that given by Yeats as Bucalaun, from the Irish

Gaelic buadhghallan.

BOKHARA CLOVER *Melilotus alba* (White Melilot) B & H. According to McClintock, it is under this name that it is sown for fodder.

BOLAS; BOLLAS *Prunus domestica var. institia* (Bullace) Grieve.1931 (Bolas); B & H (Bollas). The same word as bullace.

BOLBONAC *Lunaria annua* (Honesty) Prior. This is from the Arabic. Gerard also uses the name.

BOLDO *Peumus boldus*.

BOLEWEED; BOLLWOOD *Centaurea nigra* (Knapweed) B & H. There are a number of variations on this name, e.g. Ballweed, Belweed, Bollwood, Bow-weed, etc. Boll, or bole, refers to the rounded seed vessel. Knapweed itself means the same thing.

BOLGAN LEAVES *Lapsana communis* (Nipplewort) Scot. Jamieson. Bolga is a swelling, so presumably the plant was used to relieve swellings.

BOLIAUN *Senecia jacobaea* (Ragwort) W.Miller. Like Boholawn, Bucalaun, and Bowlocks, this must be Gaelic buadhghallan, or buaghallan.

BOLIMONGE *Fagopyrum esculentum* (Buckwheat) B & H. This from Lyte. Cf. also Bullimony, Bullimong.

BOLIVIAN BARK *Cinchona calisaya* Chopra.

BOLIVIAN LEAF *Erythroxylon coca* (coca) Thomson. To distinguish from Peruvian Leaf, which is *E. truxillense*.

BOLLANE-BANE *Artemisia vulgaris* (Mugwort) IOM Moore.1924. Literally, white wort.

BOLLOMS *Prunus spinosa* - flowers (Blackthorn) Dor. Macmillan.

BOLTS *Trollius europaeus* (Globe Flower) B & H. Parkinson uses this name for any buttercup.

BOMAVIST BEAN *Dolichos lablab* (Lablab) Schery.

BOMBAY ARROWROOT; INDIAN ARROWROOT *Curcuma angustifolia* Howes.

BOMBAY EBONY *Diospyros montana*.

BOMBAY HEMP *Crotolaria juncea* (Sunn Hemp) Howes.

BONE-FLOWER 1. *Bellis perennis* (Daisy) N.Eng. Grigson. 2. *Globularia vulgaris* (Globe-flower) Schauenberg. Daisy, like comfrey, was a consound - a bone-knitter; in fact, comfrey is recorded as a name for the daisy. All the names like Banwort, Banewort, Bairnwort, etc., all from the north of England, refer to this medicinal property. "The Northern men call this herbe a Banwort because it helpeth bones to knyt agayne" (Turner).

BONES, Bloody 1. *Iris foetidissima* (Foetid Iris) Dor. Dacombe. 2. *Endymion nonscriptus* (Bluebell) Dor. Macmillan. 3. *Orchis mascula* (Early Purple Orchis) Dor. Macmillan. There are a number of 'dead man' names for Early Purple Orchis, as if it were growing from a corpse. The dark red blotches on the leaves are enough to account for the 'bloody'. Why bluebell, though? Perhaps the clue is that the orchis is called Bluebell in Dorset, so it may be an example of mistaken recording.

BONES, Dead Man's 1. *Linaria vulgaris* (Toadflax) USA Coats. 2. *Stellaria holostea* (Greater Stitchwort) Berw. Grigson. There seems to be no English equivalent for toadflax, but the name for stitchwort is quite common in one form or another, because of the way the stems snap off at the joints.

BONES, Fish 1. *Aesculus hippocastanum* (Horse Chestnut) Som. Macmillan. 2. *Potentilla anserina* (Silverweed) Som. Macmillan. Descriptive, from the shape of the leaf. For horse chestnut, the name is reserved for the leaves when the green part has been pulled off.

BONESET 1. *Eupatorium perfoliatum* (Thoroughwort) USA Tolstead. 2. *Symphytum officinale* (Comfrey) B & H. Comfrey is a consound, and was known in all the old herbals as the prime agent for helping broken bones to knit again. The name comfrey means to heal, or grow together, and similarly Symnphytum is from Greek sumphuo, to grow together. Vesey-Fitzgerald, writing in 1944, said that the gypsies still used the plant for this purpose. The American Thoroughwort has a single large leaf surrounding the stem. Such a fused leaf was the signature, and it was used to fuse or mend broken bones.

BONESET, Purple *Eupatorium purpureum.*

BONESET, Wood *Eupatorium perfoliatum* (Thoroughwort) USA Henkel.

BONEWORT *Viola lutea* (Mountain Pansy) Cockayne. If in Cockayne, this must be directly from the OE name, so there either must have been a forgotten use as a consound, or, more likely, the name should properly be applied to another plant, the most likely candidate being the daisy.

BONFIRE *Melandrium rubrum* (Red Campion) E Ang. Tongue.1967.

BONGAY *Aesculus hippocastanum* (Horse Chestnut) Suff. B & H. B & H do not specify, but presumably the horse chestnuut had some significance in, or association with, the town of the same name (though usually spelt Bungay) in Suffolk.

BONIN ISLES JUNIPER *Juniperus procumbens.* The Bonin Isles lie to the south of Japan.

BONNACE TREE *Daphne tinifolia* W.Indies J.Smith. As there is a Burn-nose Tree for this, Bonnace is presumably merely a corruption of this. Or is it the other way round?

BONNET, Herb *Geum urbanum* (Avens) Tynan & Maitland. Obviously a misreading of Herb Bennet, but the authority assures us that it was so called.

BONNET-STRINGS *Agrostis stolonifera* (?) (White Bent) Som. Elworthy. Bonnet in this case is the same as 'bent', for they both come from OE beonet.

BONNETS *Aquilegia vulgaris* (Columbine) Som. Macmillan; War. Grigson. Better known in this area as Granny Bonnets, but there are many 'bonnet', 'cap', or 'nightcap' names, all descriptive.

BONNETS, Baby's *Lathyrus odoratus* (Sweet Pea) Som. Macmillan.

BONNETS, Blue 1. *Centaurea cyanus* (Cornflower) Corn. Grigson; Som. Macmillan; Scot. Jamieson. 2. *Endymion nonscriptus* (Bluebell) Som. Macmillan. 3. *Jasione montana* (Sheep's Bit) Dumf. Grigson. 4. *Succisa pratensis* (Devil's Bit) Lanark. B & H.

BONNETS, Dolly's *Aquilegia vulgaris* (Columbine) Som. Macmillan.

BONNETS, Grandmother's 1. *Aconitum napellus* (Monkshood) Yks. Grigson. 2. *Aquilegia vulgaris* (Columbine) Dor,Som. Grigson. Monkshood shares with Columbine various descriptive names of the 'bonnet' or 'hood' class.

BONNETS, Granny *Aquilegia vulgaris* (Columbine) Wilts. Arthur Mizen; Suss. Doris Cooper.

BONNETS, Granny's 1. *Aconitum napellus* (Monkshood) Som. Macmillan. 2. *Antirrhinum majus* (Snapdragon) Som. Macmillan. 3. *Aquilegia vulgaris* (Columbine) Corn. Grigson; Dor. Dacombe; Hants. Boase. 4. *Calystegia sepium* (Great Bindweed) Wilts. Dartnell & Goddard. 5. *Delphinium ajacis* (Larkspur) Som. Macmillan. 6. *Digitalis purpurea* (Foxglove) Som. Macmillan. 7. *Geranium pratense* (Meadow Cranesbill) Som. Macmillan. 8. *Geum rivale* (Water Avens) Dor. Macmillan. 9. *Physalis alkekengi* (Winter Cherry) Som. Macmillan.

BONNETS, Lady's *Aquilegia vulgaris* (Columbine) Som. Macmillan. We already have Old Lady's Bonnet, plus Old Woman's, Grandmother's, Granny's Bonnet, etc.

BONNETS, Night *Bryonia dioica* (White Bryony) Dor. Macmillan.

BONNETS, Old Lady's *Aquilegia vulgaris* (Columbine) Som. Grigson.

BONNETS, Old Woman's 1. *Aquilegia vulgaris* (Columbine) Dev, Som. Macmillan. 2. *Campanula medium* (Canterbury Bells) Som. Macmillan. 3. *Capsella bursa-pastoris* (Shepherd's Purse) Som. Macmillan. 4. *Geum rivale* (Water Avens) Wilts. Dartnell

& Goddard.
BONNY BIRD EEN *Primula farinosa* (Birdseye Primrose) Cumb. Grigson.
BONNY BIRD'S EYE *Cardamine pratensis* (Lady's Smock) Cumb. Grigson.
BONNY RABBITS *Antirrhinum majus* (Snapdragon) Dev. B & H. i.e. Bunny Rabbits, a very common alternative to Snapdragon.
BONWORT *Bellis perennis* (Daisy) B & H. Bonwort = Bonewort, because daisies were regarded as a consound, an agent for helping bones that are broken to knit together again. Consound actually is recorded as an old name for this, and Banwort (see above) is the same word as Bonwort.
BOOAG *Rosa canina* hips (Hips) IOM Moore.1924.
BOODLE *Chrysanthemum segetum* (Corn Marigold) Norf, Suff. B & H; N'thants. A.E.Baker. Boodle, or Buddle, another East Anglian name, seems to derive from Dutch buidel, or boedel, the whole of one's possessions. The idea seems to be that in the purse is the Gold which is another of the Corn Marigold's names, with very many variations. Boodle has transferred itself to American usage, best known as caboodle ("the whole caboodle"), i.e. the exact meaning of the Dutch original. Check if Tusser came from East Anglia. He seemed to know the plant under the name boodle:

> The brake and the cockle be noisome too much,
> Yet like unto boodle no weed there is such.

BOOIN; BOUIN; BOWEN *Senecio jacobaea* (Ragwort) all Cumb. B & H (Booin); Grigson (Bouin and Bowen). These must all be related to the Irish Boholawn (Gaelic buadhghallan), which Yeats called Bucalaun.
BOOK LEAF 1. *Bupleurum rotundifolium* (Hare's Ear) Som. Macmillan. 2. *Hypericum androsaemum* (Tutsan) Dor. Macmillan. It was once quite common to pick the leaves of Tutsan and to press them in books. When dry, they have a very sweet smell, likened to ambergris, hence the name Amber, or Sweet Amber. Bible Flower, or Bible Leaf, are West Country names which derive from the same custom.
BOON TREE; BOUN-TREE *Sambucus nigra* (Elder) both N'thum, Scot Grigson (Boon); Jamieson (Boun). Bourtree and its variants are more common.
BOOR-TREE *Sambucus nigra* (Elder) Lancs. Nodal; Ire. O Suileabhain. There is a large group of names ranging from Boon-tree to Bull-tree and Bottery. Chambers explains them away by claiming that the elder was much planted for forming garden bowers! But surely it is likely that the group are akin to various Dutch and German words like boer, a peasant or farmer, buur, neighbour and even bauer, builder.
BOOR'S MUSTARD *Thlaspi arvense* (Field Penny-cress) Turner. Boor is corrupted to bowyer in another name, but boor in this sense merely means countryman, or peasant, much in the sense of 'yokel'. Churl's Mustard is also recorded.
BOORCOLE; BORECOLE *Brassica oleracea* (Cabbage) Grigson.1974. Both from Dutch boerenkool, i.e. peasant's kale.
BOOT BUTTONS *Ligustrum vulgare - berries* (Privet) Som. Macmillan.
BOOTRY *Sambucus nigra* (Elder) Yks. B & H. This is one of the very many variants of Bour-tree.
BOOTS 1. *Caltha palustris* (Marsh Marigold) Ches. Gerard. 2. *Sarracenia purpurea* (Pitcher Plant) Perry. As far as Marsh Marigold is concerned, Boots is less common than Bouts and other connected names like Marybouts and Meadow-bouts. All derive from the French bouton (d'or). Charlotte Burne, though, derived Bouts from the country pronunciation of bolt, meaning an arrow. The word is still extant,particularly in thunderbolt (she included Marsh Marigold in the list of lightning plants). The grotesque shape of Pitcher Plant must be responsible for the name in that case.

BOOTS, Cuckoo; BOOTS, Cuckoo's *Endymion nonscriptus* (Bluebell) Dor. Macmillan (Cuckoo); Som. Macmillan (Cuckoo's). It is also the cuckoo's stockings, and its hose.

BOOTS, Devil's *Sarracenia purpurea* (Pitcher Plant) Howes.

BOOTS, Lady's 1. *Cypripedium calceolus* (Lady's Slipper) Dev. Friend.1882. 2. *Lotus corniculatus* (Bird's-foot Trefoil) Dev. B & H. There are dozens of names linking these plants with boots, shoes, feet, toes, etc., all apparently vaguely descriptive.

BOOTS, Yellow *Caltha palustris* (Marsh Marigold) Ches. Holland. See Boots.

BOOTS-AND-SHOES 1. *Aconitum napellus* (Monkshood) Corn. Jago; Dev. Macmillan. 2. *Antirrhinum majus* (Snapdragon) Dor. Macmillan. 3. *Aquilegia vulgaris* (Columbine) Corn. Jago; Som. Macmillan. 4. *Cypripedium calceolus* (Lady's Slipper) Dev. Friend.1882. 5. *Fraxinus excelsior* seeds (Ash keys) Som. Macmillan. 6. *Lotus corniculatus* (Bird's-foot Trefoil) Dev. Friend.1882; Som. Macmillan; Hants. Boase. Columbine and Bird's-foot Trefoil have very many 'shoe' and 'foot' names. Monkshood probably got the name by a confusion with columbine, but the other two are perfectly good descriptive names.

BOOTS-AND-STOCKINGS *Plantago media* (Lamb's Tongue Plantain) Som. Macmillan.

BOR-TREE, BORTREE *Sambucus nigra* (Elder) Lancs. Nodal.

BORAGE *Borago officinalis*. Borage is straight from med. Latin borago, which itself may have come from an Arabic word meaning "father of sweat". The Arabian physicians used the plant as a diaphoretic, and in this country it was used at least till Tudor times for "hot fevers", and much later in France for fevers and lung complaints.

BORAGE, Eastern *Trachystemon orientalis* (Abraham, Isaac and Joseph) Polunin.

BORAGE, Indian *Coleus amboinicus*.

BORAGE, Wild *Echium vulgare* (Viper's Bugloss) Surr. Grigson. Same family.

BORDERING *Lobularia maritima* (Sweet Alison) Dev. Friend.1882. Presumably used too for any other plant serving the same purpose in the garden, but Sweet Alison is Bordering par excellence.

BORE-TREE 1. *Sambucus nigra* (Elder) Scot. Jamieson. 2. *Scrophularia nodosa* (Figwort) Cumb. B & H. The elder has very many variants on this name, but the Figwort seems out of place. However, *Scrophularia aquatica* has a Cornish name, Scaw-daver, which means water elder.

BORE'S EARS *Primula auricula* (Auricula) N.Scot. B & H. This is boar's ears, itself a corruption of bear's ears, from the shape of the leaf. But Jamieson says that a bear is called a boar in northern Scotland.

BOREAU'S FUMITORY *Fumaria muralis var. boraei.*

BORECOLE *Brassica oleracea var. acephala* (Kale) Howes. Dutch boerenkool, peasant's cabbage.

BORES *Arctium lappa* burrs (Burrs) Som. A S Palmer. i.e. burrs.

BORNEAN ROSEWOOD *Melanorrhoea curtisii* Gimlette. It is the red timber, streaked with black resinous lines, that is known to the trade as this, or Singapore Mahogany.

BORNEO, Camphor of *Dryobalanops camphora.*

BORNEO MAHOGANY *Calophyllum inophyllum* (Alexandrian Laurel) Howes.

BORR-TREE *Sambucus ebulus* (Dwarf Elder) Lancs, Scot, Ire B & H.

BORRAL *Sambucus nigra* (Elder) N'thum, Scot. Grigson.

BORRER'S SALTMARSH GRASS *Puccinellia fasciculata.*

BORSE *Acacia pendula* (Myrtle-wood) Australia Howes.

BORTREE *Sambucus nigra* (Elder) Lancs. Nodal.

BOSNIAN PINE *Pinus leucodermis.*

BOSS, Whitsun *Viburnum opulus* (Guelder Rose) Glos. J.D.Robertson; Heref. Leather.

BOSSELL; BOZZEL; BOZEN

Chrysanthemum segetum (Corn Marigold) Wilts. Dartnell & Goddard (Bossell, Bozzel); Hants. Cope (Bozzel); Hants. Grigson (Bozen).

BOSTON HORSETAIL *Equisetum ramosissimum*. This is Boston in Lincolnshire. The only station in this country is a riverbank near there.

BOSTON IVY *Ampelopsis tricuspidata*.

BOSWELL *Chrysanthemum segetum* (Corn Marigold) Herts. B & H. This must belong to the same group that contains words like Bussell, and Bozzom.

BOTANY BAY *Hydrangea hortensis* Corn. Jago.

BOTHAM *Chrysanthemum segetum* (Corn Marigold) Dor, Hants. Grigson.

BOTHEN 1. *Chrysanthemum segetum* (Corn Marigold) Hants. Cope. 2. *Tanacetum parthenium* (Feverfew) Corn. Jago.

BOTHEN, White *Leucanthemum vulgare* (Ox-eye Daisy) B & H.

BOTHERUM 1. *Chrysanthemum segetum* (Corn Marigold) Dor. Macmillan. 2. *Veronica chamaedrys* (Germaander Speedwell) Ches. Grigson. 3. *Veronica hederifolia* (Ivy-leaved Speedwell) Ches. B & H. Sometimes spelt Botherem. For Germander Speedwell, Dotherum is recorded as well, so it is suggested that both versions mean "trembling".

BOTHERY-TREE *Sambucus nigra* (Elder) Yks. A S Palmer. Toy pop-guns made from the branches are known as bothery-guns. Presumably this is the same as bor-tree, so the word 'tree' in effect appears twice in the name.

BOTTERY *Sambucus nigra* (Elder) Yks. Addy.

BOTTLE 1. *Chrysanthemum segetum* (Corn Marigold) Dor. Grigson. 2. *Cucumis sativa* (Cucumber) Dev. Halliwell. Bottle for cucumber is quite apt, as there is little but liquid in it, but for Corn Marigold the name is part of the series Boodle, Botham, Buddle, Bozzel, etc.

BOTTLE, Bell *Endymion nonscriptus* (Bluebell) Bucks. B & H.

BOTTLE, Bladder *Silene cucubalis* (Bladder Campion) Som. Macmillan.

BOTTLE, Brandy 1. *Allium ursinum* (Ramsons) Dor. Macmillan. 2. *Nuphar lutea* seed vessels (Yellow Waterlily) Som, Berks, Suss, Norf. Grigson; Wilts. Dartnell & Goddard. 3. *Rosa canina* hips Som. Macmillan. As far as Yellow Waterlily is concerned, the name is given either because they are said to smell like the stale dregs of brandy, or perhaps from the shape of the fruits, which are like bottles. Genders prefers to liken the smell to that of ripe plums. It is due to a combination of acetic acid and ethyl alcohol.

BOTTLE, Corn *Centaurea cyanus* (Cornflower) Dev. Macmillan; N'thants. A.E.Baker. Bottle here is probably derived from O. Fr. botte, a bundle, and according to the Oxford Dictionary, the word is still used in this sense (particularly a bundle of hay).

BOTTLE, Knap *Silene cucubalis* (Bladder Campion) Prior. Prior derives knap from Dutch knappe, to snap, crack.

BOTTLE, London *Prunella vulgaris* (Selfheal) Ayr. Grigson.

BOTTLE, Scent *Plantago media* (Lamb's Tongue Plantain) Dor, Som. Macmillan.

BOTTLE, Suck 1. *Lamium album* (White Deadnettle) Som. Macmillan; N'thants. Sternberg. 2. *Trifolium pratense* (Red Clover) N'thants. Grigson. 'Suck' here is used in the same sense as in honeysuckle. White Deadnettle is a good bee plant, as the various local names show - Bee Nettle, Honey-bee, Honeysuckle, etc. Clover, of course, is well-known as a bee plant.

BOTTLE, Titty *Rosa canina* hips (Hips) Som. Macmillan.

BOTTLE, White *Silene cucubalis* (Bladder Campion) Som. Macmillan; Cambs. Grigson.

BOTTLE, Yellow *Chrysanthemum segetum* (Corn Marigold) Kent, Yks. B & H.

BOTTLE-BRUSH 1. *Equisetum arvense* (Common Horsetail) Palaiseul. 2. *Equisetum* *sylvaticum* (Wood Horsetail) W.Miller. 3. *Hippuris vulgaris* (Mare's Tail) Hants. Cope; Yks. Grigson. 4. *Melaleuca leucadendron* (Cajuput) USA Cunningham. 5. *Spergula arvensis* (Corn Spurrey) Yks. Grigson. Bottle-brush for Mare's Tail and Corn Spurrey are descriptive, in their different ways, while Horsetail was actually used as a scourer, as well as looking the part - in fact, Bottlebrush is more apt a descriptive name than horsetail.

BOTTLE-BRUSH, Albany *Callistemon speciosus.*

BOTTLE-BRUSH, Lemon *Callistemon citrinus*. Lemon-scented leaves.

BOTTLE-BRUSH BUCKEYE *Aesculus parviflora.*

BOTTLE GOURD *Lagenaria siceraria* (Calabash) Wit. Because the ripe fruit is hollowed out, cleared of pulp, and used as a bottle. It is called Dipper Gourd for the same reason.

BOTTLE GRASS 1. *Equisetum arvense* (Common Horsetail) Leyel.1937. 2. *Setaria viridis*. For Horsetail,see Bottle-brush.

BOTTLE-OF-ALL-SORTS 1. *Centaurea cyanus* (Cornflower) Yks. Grigson. 2. *Pulmonaria officinalis* (Lungwort) Cumb. B & H. The cornflower application is probably from the old meaning of bottle - a bundle, which is what it meant in Yorkshire dialect, when it usually meant a bundle of hay or straw. The lungwort gets this name, along with a lot of others like Adam and Eve, Joseph and Mary, etc., from the two-coloured flowers.

BOTTLE PALM *Beaucarnea recurvata* (Pony Tail Plant) USA Goold-Adams.

BOTTLE TREE 1. *Adansonia gregori* (Australian Baobab) Everett. 2. *Sterculia urens* (Karaya Gum) USA Cunningham. Presumably simply from the general shape of the baobab's trunk, but not necessarily so, for the trunks of these trees have been used for water storage.

BOTTLES-OF-WINE *Dicentra spectabilis* (Bleeding Heart) Som. Macmillan.

BOTTRY *Sambucus nigra* (Elder) Yks. Morris.

BOUGH ELM *Ulmus glabra* (Wych Elm) Yks. Grigson.

BOUNCING BESS 1. *Centranthus ruber* (Spur Valerian) Dev. Wright; Dor. Macmillan. 2. *Valeriana celtica* Dev. Friend.1882.

BOUNCING BET *Saponaria officinalis* (Soapwort) Dor. Macmillan; USA Tolstead. Parkinson called it this in *Paradisus in Sole*, 1629. "Bouncing" is given because the plant does seem to do just that, rooting at the leaf nodes.

BOUNCING BETTY 1. *Centranthus ruber* (Spur Valerian) Dev, Dor. Grigson. 2. *Saponaria officinalis* (Soapwort) Dor. Grigson.

BOUNDARY MARK *Cordyline terminalis* (Ti Plant) W Indies Howes. In keeping with its specific name, which must imply a celebration of boundaries. The plant was grown at the four corners of a house, to drive away ghosts and demons.

BOUNTREE *Sambucus nigra* (Elder) N'thum. Grigson; Scot. Jamieson.

BOUQUET, Lady's *Galium triflorum* (Fragrant Bedstraw) Gilmore.

BOURBON LILY *Lilium candidum* (Madonna Lily) Woodcock.

BOURBON ROSE *Rosa bourboniana*. The original plant came from the island of Bourbon, off East Africa, in 1817.

BOURHOLM *Arctium lappa* (Burdock) B & H.

BOURNEMOUTH PINE *Pinus pinaster* (Maritime Pine) Barber & Phillips. It has been growing on the south coast since the sixteenth century, and Bournemouth is particularly associated with it.

BOURSAULT ROSE *Rosa pendula x chinensis.*

BOURTREE *Sambucus nigra* (Elder) Lincs, Ches, Lancs, Yks, Dur, West, N'thum, Ire. Grigson; Scot. Cromek. Chambers explains this from its being used in forming bowers! (apparently without reference to the variants, ranging from Boun-tree to Bottery). It is only fair to say that elders *were* planted around cottages, not to form bowers, but to protect from witchcraft.

BOUT; MEADOW BOUT *Caltha palustris* (Marsh Marigold) Prior (Bout); Shrop, Ches, Lancs. Grigson (Meadow Bout). Boots, or Yellow Boots, occur too - all apparently from the French bouton d'or, though Burne (speaking for Shropshire) says Bouts derives from the country pronunciation of bolts, i.e. arrows, and in this case thunderbolts. She included the plant in her list of lightning plants.

BOUT, Black *Plantago lanceolata* (Ribwort Plantain) Bucks. Grigson.

BOUTREY *Sambucus nigra* (Elder) Cumb. Jollie.

BOVISAND SAILORS *Centranthus ruber* (Spur Valerian) Dev. Friend.1882. Friend said that Bovisand was a place where the plant grew in profusion.

BOW, Blue *Centaurea cyanus* (Cornflower) Skinner. Possibly OE boll, but there are names like Blueblow, or - blaw, which seem to derive from another name Blaver, apparently meaning 'belonging to corn' (Cf. French blé).

BOW BELLS *Anemone nemerosa* (Wood Anemone) Worcs. Grigson.

BOW-KAIL *Brassica oleracea var. capitata* (Cabbage) Scot. Jamieson.

BOW THISTLE; BO THISTLE *Cirsium vulgare* (Spear Plume Thistle) Ches. B & H (Bow); Ches. Holland (Bo). i.e. Bur Thistle.

BOW-WEED *Centaurea nigra* (Knapweed) B & H. With Bow-wood, probably OE boll - any circular body. Cf. Boleweed, Ballwood for the same plant.

BOW-WOOD 1. *Centaurea nigra* (Knapweed) B & H. 2. *Centaurea scabiosa* (Greater Knapweed) B & H. 3. *Cedrula odorata* (West Indian Cedar) Hora. 4. *Maclura pomifera* (Osage Orange) Wit. The first two are from OE boll, which meant a circular body, but the other two mean literally what they say - wood to make bows from.

They are N.American trees, and the Indians used the wood just for this purpose.

BOWEL-HIVE; BOWEL-HIVE GRASS *Aphanes arvensis* (Parsley Piert) both Border B & H. Bowel-hive was a name for a children's complaint, described as inflammation of the bowels. This plant was used to cure it. Cf. Colicwort for the same plant.

BOWER, Lady's; BOWER, Lady-in-the- 1. *Clematis vitalba* (Old Man's Beard) Som. Macmillan. 2. *Nigella damascena* (Love-in-a-mist) W.Miller (Lady-in-the-bower). Old Man's Beard is better known by Gerard's name, Virgin's Bower.

BOWER, Virgin's 1. *Clematis virginiana* USA House. 2. *Clematis vitalba* (Old Man's Beard) Gerard. It has been said that the name was coined as a tribute to Elizabeth 1, but a German legend links it with Mary, for the story has it that Mary and Jesus sheltered under it during the flight into Egypt, but then what didn't they shelter under?

BOWER, Virgin's, Sweet *Clematis flammula* (Upright Virgin's Bower) Lindley.

BOWER, Virgin's, Upright *Clematis flammula*.

BOWER PLANT; BOWER VINE *Pandorea jasminoides*.

BOWL, Washing, My Lady's *Saponaria officinalis* (Soapwort) Mabey.1977.

BOWLOCKS 1. *Artemisia vulgaris* (Mugwort) Scot. Grigson. 2. *Senecio jacobaea* (Ragwort) Scot. B & H. Presumably the same, in the case of ragwort, as Irish Bucalaun and its variants, from buaghallan.

BOWMAN ROOT *Veronica virginica* USA Lloyd.

BOWMAN'S ROOT 1. *Apocynum cannabinum* (Indian Hemp) Grieve.1931. 2. *Euphorbia corollata* (Flowering Spurge) O.P.Brown. 3. *Gillenia trifoliata* (Indian Physic) Perry.

BOWN-TREE *Sambucus nigra* (Elder) Dur. Brockie. Close to Boun-, or Boon-tree, from Northumberland and Scotland, and part of a series of which the Scottish Bour-tree is best known.

BOWOUR, Black *Rubus fruticosus* fruit (Blackberry) N. Eng. B & H. Presumably the same as the Scots Boyds.

BOWSTRING *Sanseviera guineensis* (African Hemp) Grieve.1931.

BOWSTRING HEMP *Sanseviera metallica*.

BOWWOWER, Black *Rubus fruticosus* fruit (Blackberry) N.Eng. Halliwell. Cf. Black Bowour.

BOWYER'S MUSTARD 1. *Lepidium ruderale* (Narrow-leaved Pepperwort) Prior. 2. *Thlaspi arvense* (Field Pennycress) B & H. This is probably the same as Boor's Mustard (which is recorded for Field Pennycress), in the sense of peasant, or countryman.

BOX *Buxus sempervirens*. Box is Latin buxus, in turn from Greek puxos. Box (the receptacle) is properly made from boxwood. Buxus itself means flute, and the wood has been used from ancient times for musical and mathematical instruments.

BOX, Abassian *Buxus sempervirens* timber (Boxwood) Brimble.1948. Cf. Persian, or Iranian, Box.

BOX, Bastard *Polygala chamaebuxus*.

BOX, Burying *Buxus sempervirens* (Box) Lancs. Vaux. Evergreen, so a symbol of life everlasting, hence the connection with funerals. In Lancashire, a basin full of box sprigs often used to be put at the door of a house before a funeral. Everyone who attended was expected to take a sprig, to carry it in the procession, and then to throw it into the grave.

BOX,Cape *Buxus macowanii*.

BOX, Desert *Eucalyptus microtheca* (Coolibah) Howes.

BOX, Dutch *Buxus sempervirens var. suffruticosa*.

BOX, Dwarf *Buxus sempervirens var. suffruticosa* (Dutch Box) Ablett.

BOX, East London *Buxus macowanii* (Cape Box) Brimble.1948.

BOX, European *Buxus sempervirens* timber (Boxwood) Brimble.1948.

BOX, Flowering *Vaccinium vitis-idaea* (Cowberry) Ches. Holland. There is a resemblance between box leaves and those of cowberry.

BOX, French *Buxus sempervirens var. suffruticosa* (Dutch Box) Genders.

BOX, Gum, Chilean *Escallonia macrantha* W.Miller.

BOX, Iranian; BOX, Persian *Buxus sempervirens* timber (Boxwood) Brimble.1948.

BOX, Jack-in-the- *Arum maculatum* (Cuckoo-pint) Som. Macmillan.

BOX, Korean *Buxus microphylla*.

BOX, Long-leaved *Buxus longifolia*.

BOX, Minorca *Buxus balearica*. Not confined to Minorca, though; it grows in southern Spain, throughout the Balearics, and in Sardinia.

BOX, Mountain *Arctostaphylos uva-ursi* (Bearberry) Hatfield.

BOX, Penny *Rhinanthus crista-galli* (Yellow Rattle) Tynan & Maitland. Cf. Penny Rattle, and a lot of others.

BOX, Pepper *Papaver rhoeas* (Red Poppy) Som. Macmillan. Presumably from the shape of the capsule, or more simply from the large number of seeds in them. You can shake them like a pepper-box.

BOX, Rose *Cotoneaster microphylla*. Leaf resemblance?

BOX, Shackle 1. *Briza media* (Quaking-grass) Som. Grigson.1959. 2. *Pedicularis sylvatica* (Lousewort) Dev. Macmillan. Shekel-box, possibly, but the name is in the "Rattle Basket" mould. Lousewort seeds can be "rattled" in their over-large cases, and the grass seed-heads never seem to be still.

BOX, Shekel *Rhinanthus crista-galli* (Yellow Rattle) Dor. Macmillan. Money-box, in this case.

BOX, Turkey *Buxus sempervirens* timber (Boxwood) Brimble.1948. Cf. Persian, or Iranian, Box.

BOX, Victorian *Pittosporum undulatum*.

BOX, Yellow *Eucalyptus melliodora*.

BOX BLUEBERRY *Vaccinium ovatum* (California Huckleberry) Yanovsky. Presumably because the leaves are like those of box. Cowberry (*V.vitis-idaea*) is sometimes

called Flowering Box for that reason.
BOX ELDER *Acer negundo.*
BOX HOLLY *Ruscus aculeatus* (Butcher's Broom) Bianchini. Box-like leaves terminating in a prickle like holly?
BOX HUCKLEBERRY *Gaylussacia brachycera.*
BOX-LEAVED MILKWORT *Polygala chamaebuxus* (Bastard Box) Polunin.
BOX-OF-MATCHES *Acer campestre* leaves (Maple leaves) Som. Macmillan.
BOXBERRY *Gaultheria procumbens* (Wintergreen) USA Sanford.
BOXING GLOVES *Lotus corniculatus* (Bird's-foot Trefoil) Som. Macmillan. A variation on the familiar "fingers-and-thumbs" theme.
BOXTHORN 1. *Lycium andersonii.* 2. *Lycium halimifolium* (Duke of Argyll's Tea Plant) J.Smith. 3. *Lycium pallidum* (Rabbit Thorn) USA Elmore. More usually spelt as two words.
BOXTHORN, European *Lycium europaeum.*
BOXWOOD, American; BOXWOOD, New England; BOXWOOD, Virginian *Cornus florida* (Flowering Dogwood) all Grieve.1931.
BOY, Butcher 1. *Anacamptis pyramidalis* (Pyramidal Orchis) Som. Macmillan. 2. *Orchis mascula* (Early Purple Orchis) Som. Macmillan.
BOY'S BACCA 1. *Clematis vitalba* (Old Man's Beard) Hants, Suss. Grigson. 2. *Heracleum sphondyllium* (Hogweed) Dev. PLNN27. Boys used to smoke their porous stalks. By the spread of the various local names, such as Smoking Cane, Gypsy's Bacca, etc., this seems to have been a south of England habit.
BOY'S LOVE *Artemisia abrotanum* (Southernwood) Corn. Jago; Som. Jennings; Wilts. Akerman; Dor. Dacombe; Hants. Cope. Lad's Love, too - probably better known. A few branches of southernwood were generally added to the nosegay which courting youths gave the girls.
BOY'S MERCURY *Mercurialis annua* (French Mercury) Lyte. Lyte explains the name by referring to an old belief that when "dronken (it) causeth to engender male children". But he also records Girl's Mercury and Maiden Mercury as names for this, so where does his argument get one? Anyway, it is known in America as Boys and Girls, to confuse it still further.
BOYD; BOYD, Black *Rubus fruticosus* fruit (Blackberry) W.Scot. B & H (Boyd); Jamieson (Black Boyd).
BOYD'S PEARLWORT *Sagina boydii.*
BOYS, Black 1. *Plantago lanceolata* (Ribwort lanceolata) Wilts. Dartnell & Goddard. 2. *Typha lanceolata* (False Bulrush) Wilts. Dartnell & Goddard. Descriptive - ribwort is sometimes known as Black Plantain, and Black Boys ties in with a number of other local names, like Black Man, Chimney Sweeper, etc. Bulrush has also got the names Blackcap and Blackhead.
BOYS, Drummer *Centaurea nigra* (Knapweed) Dor. Macmillan. Drummer Heads, and Drumsticks too, elsewhere - descriptive, from the shape of the heads in profile.
BOYS, Naked 1. *Colchicum autumnale* (Meadow Saffron) Som, Wilts. Aubrey (Nat.Hist.Wilts). 2. *Crocus nudiflorus* (Autumn Crocus) Ches. Holland. 3. *Ranunculus fluitans* (River Crowfoot) Dor. Macmillan. Meadow Saffron flowers appear before the leaves, and so do those of Autumn Crocus.The leaves of River Crowfoot remain submerged while the flowers are in bloom, and so create the same effect.
BOYS, Star-naked *Colchicum autumnale* (Meadow Saffron) Norf. B & H. "Stark"-naked, of course. See Naked Boys.
BOYS, Yellow *Senecio jacobaea* (Ragwort) Donegal Grigson.
BOYS-AND-GIRLS 1. *Mercurialis annua* (French Mercury) USA Kingsbury. 2. *Primula veris* (Cowslip) Dor. Macmillan. 3. *Primula vulgaris* (Primrose) Wilts. Dartnell & Goddard. For the Mercury, see under Boy's Mercury. The long-pistilled, or "pin-eyed"

primrose flowers are Boys. The short-pistilled, or "thrum-eyed" ones are Girls.

BOYTON THISTLE *Cirsium tuberosum* (Tuberous Thistle) Wilts. Flower.1857. One of its rare habitats was a spot in Wiltshire between Boyton House and Fonthill.

BOZZOM 1. *Chrysanthemum segetum* (Corn Marigold) IOW Grigson. 2. *Leucanthemum vulgare* (Ox-eye Daisy) IOW Long.

BOZZOM, Yellow *Chrysanthemum segetum* (Corn Marigold) IOW Grigson.

BRACELET HONEY MYRTLE *Melaleuca armillaris*.

BRACKEN; BRAKE FERN *Pteridium aquilina*. Bracken is actually the old plural in -en of brake (OE bracce).

BRACKEN, Sweet *Myrrhis odorata* (Sweet Cicely) Cumb. Grigson. From the fern-like aromatic leaves. Coles called in Sweet Fern.

BRACTSCALE *Atriplex serenana*.

BRAGGE; DRAGGE *Lolium temulentum* (Darnel) B & H. Possibly connected with the sequence that includes Drunk, and so a comment on the toxic nature of darnel.

BRAKE, Cretan *Pteris cretica* (Tanle Fern) RHS.

BRAKE, Hog 1. *Ambrosia artemisiaefolia* (American Wormwood) Howes. 2. *Pteridium aquilina* (Bracken) USA Cunningham.

BRAKE, Pasture *Pteridium aquilina* (Bracken) USA Cunningham.

BRAKE, Rock *Polytrichum commune*.

BRAKE, Silver *Pteris argyraea* (Striped Brake) Perry.1979.

BRAKE, Striped *Pteris argyraea*.

BRAKE-OF-THE-WALL *Polypodium vulgare* (Common Polypody) B & H. A habitat name. Cf. Wall Fern.

BRAMBLE *Rubus fruticosus*. OE braembel, earlier braemel, bremel, which appears to be the same as OE brom (broom), with the suffix -le. Ellacombe said the name originally meant anything thorny, and Chaucer applied it to the Dog Rose. Certainly, Cock Bramble and Horse Bramble are still local names for wild rose. The word is applied to the fruit rather than to the bush in Scotland.

BRAMBLE, American *Rubus odoratus* (Purple-flowering Raspberry) Grieve.1931.

BRAMBLE, Arizona *Rubus arizonicus* (Arizona Red Raspberry) USA Elmore.

BRAMBLE, Blue *Rubus caesius* (Dewberry) B & H.

BRAMBLE, Cock 1. *Rosa canina* (Dog Rose) Suff. Grigson. 2. *Rubus fruticosus* (Blackberry) Norf, Suff. Grigson.

BRAMBLE, Elmleaf *Rubus ulmifolius*.

BRAMBLE, Ewe *Rubus fruticosus* (Blackberry) Som. B & H. More likely to be pronounced Yoe-brimmel, and often written that way. Surely it must really be He-bramble?

BRAMBLE, Hawk's Bill *Rubus fruticosus* (Blackberry) E.Ang. Grigson. Presumably from the shape of the thorns.

BRAMBLE, Heath *Rubus caesius* (Dewberry) B & H.

BRAMBLE, Hip *Rosa canina* (Dog Rose) Young. Bramble crops up a number of times in one form or another for the wild rose. There are Horse, Cock, Ewe Bramble, etc.

BRAMBLE, Horse *Rosa canina* (Dog Rose) E.Ang. Halliwell.

BRAMBLE, Moluccan *Rubus moluccanus*. A Malaysian plant.

BRAMBLE, Mountain *Rubus chamaemorus* (Cloudberry) B & H.

BRAMBLE, Palestine *Rubus sanctus*.

BRAMBLE, Stone *Rubus saxatilis*. 'Stone' describes its habitat - it grows on rocks and ledges in the north and west.

BRAMBLE BREER *Rubus fruticosus* (Blackberry) Langham. A mixture of the two prickly elements - bramble and briar.

BRAMBLEBERRIES *Rubus fruticosus* fruit (Blackberries) N.Eng. Halliwell.

BRAMBLEKITE; BRAMMELKITE *Rubus fruticosus* fruit (Blackberry) all Dur. Hazlitt (Bramblekite); Grigson (Brammelkite). Kite, in north country dialect, means the stomach.

BRAMMLE *Rubus fruticosus* fruit (Blackberry) Yks. Nicholson.

BRANDY, Bird's *Lantana rugosa* S.Africa

Watt.

BRANDY BOTTLES 1. *Allium ursinum* (Ramsons) Dor. Macmillan. 2. *Nuphar lutea* seed vessels (Yellow Waterlily) Som, Berks, Suss, Norf. Grigson; Wilts. Dartnell & Goddard. 3. *Rosa canina* hips (Hips) Som. Macmillan. As far as Yellow Waterlily is concerned, the name is given either because they are said to smell like the stale dregs of brandy, or perhaps because of the shape of the fruits, which are like bottles. Genders prefers to liken the smell to that of ripe plums. It is due to a combination of acetic acid and ethyl alcohol.

BRANDY MAZZARD *Prunus avium* (Wild Cherry) Dev. Grigson. Mazzard seems to be a dialect word for a particular kind of cherry. "It is in good esteem for making cherry-brandy" (Halliwell).

BRANDY MINT 1. *Mentha piperita* (Peppermint) West. B & H. 2. *Mentha spicata* (Spearmint) War. F Savage. Probably no connection, but peppermint is the mint used for making Mint Julep, but in that it is whisky that is used, not brandy.

BRANDY SNAP 1. *Linaria vulgaris* (Toadflax) Suss. Grigson. 2. *Stellaria holostea* (Greater Stitchwort) Suss. Grigson. In both cases, it is the "snap", rather than the "brandy" which is important. With toadflax, for the same reason that antirrhinum is called snapdragon, and with stitchwort, because of the brittle stalks, which tend to snap at the joints.

BRANK *Fagopyrum esculentum* (Buckwheat) E.Ang. F.Grose. This is a name which Tusser knew, and which possibly came from French branche (the same as our word branch). But the Spanish branca (the same in Italian) would have had the secondary meaning of "claw".

BRANK-URSIN (or spelt as one word) (or with a final e on ursin) *Acanthus lusitanicus* (Bear's Breech) Turner. If (see previous entry - Brank) that element means claw, this, quite reasonably, would mean bear's claw.

BRASHLACH; BRASHLAGH *Sinapis arvensis* (Charlock) N.Eng. Potter & Sargent (Brashlach); IOM Moore.1924 (Brashlagh). Presumably this is Brassica. Brassick and Brassock are Yorkshire names.

BRASILETTO 1. *Caesalpina sappan* (Brazil-wood) Barker. 2. *Haematoxylon campechianum* (Logwood) Pomet.

BRASS BUTTONS *Cotula coronopifolia* (Buttonweed) Polunin.

BRASSICK; BRASSOCK *Sinapis arvensis* (Charlock) both Yks. B & H. Charlock was known botanically as *Brassica arvensis*, or *Brassica sinapis* before being put into the genus *Sinapis*.

BRAVE BASSINETS; BRAVE CELANDINE *Caltha palustris* (Marsh Marigold) Lyte. Bassinets refers to the cup shape of the flower. Lyte coined these names, and Brave seems to mean simply big.

BRAWLINS 1. *Arctostaphylos uva-ursi* (Bearberry) N.Scot. Jamieson. 2. *Vaccinium vitis-idaea* fruit (Cowberry) Scot. Jamieson.

BRAZIL CHERRY *Eugenia michelli*.

BRAZIL NUT *Bertholletia excelsa*.

BRAZIL TEA; BRAZIL TEA, Jesuit's *Ilex paraguayensis* (Maté) Le Strange.

BRAZILIAN ARROWROOT 1. *Canna glauca*. 2. *Ipomaea batatas* (Sweet Potato) Howes. 3. *Manihot esculenta* (Manioc) Brouk.

BRAZILIAN CHERRY *Eugenia uniflora* (Surinam Cherry) Bianchini.

BRAZILIAN COCOA *Paullinia cupana* (Guarana) Mitton.

BRAZILIAN FLOSS SILK TREE *Chorisia speciosa*.

BRAZILIAN PEPPER *Schinus terebinthifolius*.

BRAZILIAN PINE *Araucaria angustifolia*. *A angustifolia* is sometimes listed as *A brasiliana*.

BRAZILIAN REDWOOD *Caesalpina bonduc*.

BRAZILIAN RHATANY *Krameria argentea*.

BRAZILIAN ROSEWOOD *Dalbergia nigra*

(Rosewood) Howes.
BRAZILWOOD 1. *Caesalpina bonduc* (Brazilian Redwood) Pomet. 2. *Caesalpina echinata*. 3. *Caesalpina sappan*. *C sappan* had the name originally, but the American *C echinata* is usually known as Brazilwood these days. The wood is not named after the country - the other way round, in fact. The word comes from Arabic braza, flaming red. When the South American trees were seen to be similar to the Asiatic ones, the name was transferred, and the country in which they grew was called Terra de Brazil.
BREAD, Bad Man's *Conopodium majus* (Earth-nut) Yks. Grigson. In all similar examples, "Bad Man" is to be taken as a euphemism for the devil. Similarly, "bread", like "meat" is not to be taken literally, but would mean food in general. Usually such a name would be reserved for poisonous plants, or plants thought to be poisonous. But why give it to Earth-nut? The tubers are perfectly wholesome.
BREAD, Bee 1. *Borago officinalis* (Borage) B & H. 2. *Trifolium pratense* (Red Clover) Glos. PLNN35; Kent. B & H. 3. *Trifolium repens* (White Clover) Som. Macmillan. These are all good bee plants, in fact there was a time when borage was grown especially for bees.
BREAD, Bird's *Sedum acre* (Biting Stonecrop) Prior. Prior explains this by saying that it blooms about the time the young birds are hatched. Cf. French pain d'oiseau.
BREAD, Cheese-and- 1. *Crataegus monogyna* (Hawthorn) Dur. F.M.T.Palgrave. 2. *Oxalis acetosella* (Wood Sorrel) Cumb. Grigson. In reverse order, there is Bread-and-cheese also in Lancashire. But there are a lot of these kinds of names scattered all over the country, and on the Continent, too.
BREAD, Cuckoo('s) 1. *Cardamine pratensis* (Lady's Smock) Dev, Som. Grigson. 2. *Oxalis acetosella* (Wood Sorrel) Som. Macmillan. Dor. Barnes. 3. *Plantago major* (Great Plantain) Le Strange. Lady's Smock is almost universally connected with the cuckoo, possibly because it is in flower when the cuckoo is about; the same applies to the Wood Sorrel.
BREAD, Daily *Rosa canina* (Dog Rose) Yks. B & H. Yorkshire children peel the young shoots and eat them in spring. This is one of the names they call what they eat.
BREAD, Devil's *Conopodium majus* (Earth-nut) Yks. Grigson. Cf. Devil's Oatmeal for this, from the same area.
BREAD, Hottentot *Dioscorea elephantipes*. This is one of the yams, growing in South Africa, so the description is apt.
BREAD, Loaves-of- 1. *Hyoscyanum niger* (Henbane) N'thants. Dyer. 2. *Malva sylvestris* (Common Mallow) Dor, Som. Macmillan. In both cases, it is the fruits which bear the name. The mallow seed case is like a cheese, and there are a number of 'cheese' names, too. Clare knew the henbane name in his county: see Shepherd's Calendar:

Hunting from the stack-yard sod
The stinking Henbane's belted pod
By youth's warm fancies sweetly led
To christen them his loaves of bread.

BREAD, Monkey 1. *Adansonia digitata* (Baobab) Everett. 2. *Adansonia madagascariensis*.
BREAD, St.John's *Ceratonia siliqua* - pods (Carob) Wit. The pods may have been the "locusts" eaten by St John the Baptist. The tree is known as Locust Tree, and the fruits are imported into Britain under the name of locust bean. Cf. the name St John's Sweetbread.
BREAD, Satan's *Anthriscus sylvestris* (Cow Parsley) Lincs. Gutch. The plant is connected with the devil in folklore, and many of its names are applied also to the hemlock, suggesting that the harmless Cow Parsley got the names and the connection by mistake.
BREAD, Sour, Ethiopian *Adansonia digitata* (Baobab) Douglas.
BREAD, Sow *Sonchus oleraceus* (Sow Thistle) Kent Grigson. When spelt as one

word, this is the usual name for Cyclamen.

BREAD, Wayside *Plantago major* (Great Plantain) Wilts. B & H. Probably arising from a misunderstanding of an old name for plantain - Waybread (OE weg-breade, waybreadth, i.e. a broad-leaved plant growing by the waysides). All the variety of names deriving from Waybread seem to belong to the North Country and Scotland, so perhaps the misunderstanding when in some way it reached Wiltshire is natural.

BREAD-AND-BUTTER 1. *Cardamine pratensis* (Lady's Smock) Tynan & Maitland. 2. *Linaria vulgaris* (Toadflax) Som. Macmillan. 3. *Potentilla anserina* (Silverweed) Som. Macmillan. For Toadflax, the name is probably a reflection of the two colours in the flower. Double names like this (Bacon-and-eggs, Egg-and-butter, etc) usually have this derivation. But that can't be the reason for the application to silverweed. The roots, though, are edible, and up in the Highlands and Islands, were quite important before the introduction of the potato. They were cultivated, so they grew quite large. It was said that a man could sustain himself quite well with silverweed roots on a square of ground of his own length. It was sometimes boiled in pots, or roasted on the stove, and sometimes dried and ground into meal for bread or porridge (see Carmichael, vol 4). But the Highlands are a long way from Somerset.

BREAD-AND-CHEESE 1. *Linaria vulgaris* (Toadflax) Wilts. Dartnell & Goddard. 2. *Lotus corniculatus* (Bird's-foot Trefoil) Som. Macmillan. 3. *Malva sylvestris* (Common Mallow) Som. Macmillan; Dor. Dacombe; Wilts. Dartnell & Goddard. 4. *Oxalis acetosella* (Wood Sorrel) Ches. Holland; Lancs. Nodal. 5. *Potentilla anserina* (Silverweed) Som. Macmillan. 6. *Rosa canina* (Dog Rose) Yks. Nicholson. 7. *Rumex acetosa* (Wild Sorrel) Dev. B & H. Toadflax, and possibly Bird's-foot Trefoil, owe the name to the two-coloured flowers (double names very often mean double colours). For Silverweed, see the explanation under Bread-and-butter. Mallow has a number of "cheese" names (Chucky-cheese, Lady's Cheese, Fairy Cheese, etc.) - the mature fruit is similar in shape to a cheese. Wood Sorrel and Wild Sorrel are both edible, in small quantities, and as for Dog Rose, it used to be quite a custom in Yorkshire for children to pick the shoots, peel off the skin, and eat them. None of these, or any other plant which bears a derived name, taste anything like bread and cheese, so it is probably a metaphor referring to the basic food value rather than the taste.

BREAD-AND-CHEESE, Bird's *Oxalis acetosella* (Wood Sorrel) Dev.Friend.1882. It is more usual for the bird to be specified - it is the cuckoo.

BREAD-AND-CHEESE, Cuckoo's *Oxalis acetosella* (Wood Sorrel) Som, Glos, Worcs, Shrop, Lancs, Cumb. Grigson; IOM Moore.1924. Possibly the time of flowering has something to do with the association with the cuckoo. It is certainly very widespread. It is pain de coucou in French, panis cuculi in medieval Latin, Kukuksbrot in German, and giorge-syre in Danish.

BREAD-AND-CHEESE, God Almighty's *Oxalis acetosella* (Wood Sorrel) Som. Elworthy.

BREAD-AND-CHEESE, Old Man's *Malva sylvestris* (Common Mallow) Som. Macmillan.

BREAD-AND-CHEESE-AND-CIDER 1. *Anemone nemerosa* (Wood Anemone) Dor. Macmillan. 2. *Malva sylvestris* (Common Mallow) Som. Macmillan. 3. *Oxalis acetosella* (Wood Sorrel) Som. Grigson. Why has Wood Anemone entered this series? It sounds like a mistake for Wood Sorrel.

BREAD-AND-CHEESE-AND-KISSES *Malva sylvestris* (Common Mallow) Som. Macmillan.

BREAD-AND-CHEESE BUSH, May *Crataegus monogyna* (Hawthorn) Som, Hants, E Ang. Grigson. The young leaves and shoots of hawthorn are regularly eaten by children in spring, and called bread-and-cheese.

BREAD-AND-CHEESE TREE 1. *Crataegus monogyna* (Hawthorn) Som. Macmillan; Wilts. Wiltshire; Mon. Waters; Suss. Fernie. 2. *Rosa canina* Yks. Nicholson.

BREAD-AND-CHEESE TREE, Cuckoo's *Crataegus monogyna* (Hawthorn) Wilts. Dartnell & Goddard; Suss. Parish Leics. Grigson.

BREAD-AND-CHEESE TREE, May *Crataegus monogyna* (Hawthorn) Norf, Lincs. Grigson. The young leaves and shoots of hawthorn are regularly eaten by children in spring, and called bread-and-cheese.

BREAD-AND-CIDER *Crataegus monogyna* (Hawthorn) Som. Macmillan.

BREAD-AND-MARMALADE *Sinapis arvensis* (Charlock) Som. Macmillan. Presumably relating to the mustard colour.

BREAD-AND-MILK 1. *Cardamine pratensis* (Lady's Smock) Fernie. 2. *Oxalis acetosella* (Wood Sorrel) Som. Macmillan.

BREAD TREE, Monkey *Adansonia digitata* (Baobab) Everett.

BREAD WHEAT *Triticum aestivum*.

BREADFRUIT 1. *Artocarpus communis*. 2. *Monstera deliciosa* (Ceriman) USA Cunningham. When roasted in their skins, breadfruit tastes, so it is claimed, exactly like freshly baked bread.

BREADFRUIT, African *Trecularia africana*.

BREADFRUIT, Highland *Ficus dammaropsis*.

BREADROOT *Psoralea esculenta*. The roots, rich in starch, can be ground into flour for bread making.

BREADROOT, Beaver-dam *Psoralea castorea*. A North American plant, which has tuberous, starchy taproots. These provided food for the Indians and the early settlers. But why "beaver-dam"?

BREADROOT, Indian *Psoralea esculenta* (Breadroot) Johnston.

BREADROOT, Iroquois *Arisaema triphyllum* (Jack-in-the-pulpit) USA Yanovsky. Both this and Indian Turnip are applied because the corms are eaten.

BREADROOT, Utah *Psoralea nephitica*.

BREAK-BASIN *Veronica chamaedrys* (Germander Speedwell) Dev, Som. Macmillan. From the quick falling of the petals, Grigson says.

BREAK-JACK 1. *Stellaria graminea* (Lesser Stitchwort) Dor. Macmillan. 2. *Stellaria holostea* (Greater Stitchwort) Dor. Grigson. Because the stems snap off at the joints. Cf. Breakbones, Snapjacks, etc.

BREAK-YOUR-MOTHER'S-HEART 1. *Anthriscus sylvestris* (Cow Parsley) PLNN 17. 2. *Conium maculatum* (Hemlock) Dor. Grigson. Because it is a poison? Cow Parsley gets many of hemlock's names, because of the similarity in appearance.

BREAKBONES *Stellaria holostea* (Greater Stitchwort) Ches. Prior. There are a number of names of this kind, most with 'break' or 'snap' in them. They refer to the method of growth, and the way they are likely to snap off at the joints.

BREAKSTONE 1. *Pimpinella saxifraga* (Burnet Saxifrage) B & H. 2. *Sagina procumbens* (Common Pearlwort) B & H). Saxifrage means literally break stone, and refers to the habitat - they will grow in crevices, and seem to have cracked the rock with their roots. The doctrine of signatures took this up. What could crack stone in nature should surely break up stones in the bladder, etc. So the old herbals gave recipes on these lines for any plant with a similar habit. For example: "The seed and root of Saxifrage drunk with wine, or the decoction thereof made with wine, causeth to pisse well, breaketh the stone in the kidnies and bladder, and is singular against the strangurie and stoppings of the kidnies and bladder" (Gerard). It is not very clear what Pearlwort is doing in this company.

BREAKSTONE, Parsley 1. *Alchemilla vulgaris* (Lady's Mantle) Suff, Scot. B & H. 2. *Aphanes arvensis* (Parsley Piert) Staffs. Grigson. Parsley Piert is from French perce-pierre, meaning break-stone (Prior lists

Percepier as an English name). So Parsley Breakstone is a strange mixture of French and English. The leaves may possibly show a tenuous likeness to those of parsley. Once again, the fact that it grows naturally in gravelly soil gave the doctrine of signatures the chance to claim the plant as a remedy for stones in the bladder. But it does actually have an effect on stone. Gypsies use an infusion of the dried herb for the gravel and other bladder troubles, and it was in demand in the 1939-45 war for bladder and kidney troubles; it is also valuable for jaundice. The name probably attached itself to Lady's Mantle by mistake, though *Aphanes arvensis* was once known as *Alchemilla arvensis*.

BREASTWORT *Herniaria ciliata* (Rupturewort) Le Strange. 'Burstwort',obviously.

BREATH, Baby's; BREATH, Maiden's *Gypsophila paniculata* A.W.Smith (Baby); Clapham (Maiden). Fancifully descriptive.

BREATH, Horse- *Ononis repens* (Rest Harrow) Som, Worcs. Grigson. Does this have the same sense as Rest Harrow? i.e. give the horses breath (breather).

BREATH-OF-HEAVEN *Diosma ericoides.*

BRECK SPEEDWELL *Veronica praecox.* A very rare British species, confined to arable fields in Breckland.

BRECKLAND CATCHFLY *Silene otites.*

BRECKLAND MUGWORT *Artemisia campestris.* Very rare, confined to Breckland grasslands.

BRECKLAND WILD THYME *Thymus serpyllum.*

BRECKON *Pteridium aquilinum* (Bracken) Yks. Harland.

BREECH, Bear's *Acanthus mollis.* There is an archaic verb to breech, meaning to flog or whip. Comparing this with another old name for the plant, Brank Ursin, one gets a picture of a bear being led, for branks was a word for a kind of halter or bridle. Ursin, of course, is the bear.

BREECHES, Chinaman's *Dicentra spectabilis* (Bleeding Heart) Coats. As with *D cucullaria*, which is called Dutchman's Breeches, it is the shape of the flowers which gave rise to the name - not at all a bad description when you look at them.

BREECHES, Dutchman's *Dicentra cucullaria.*

BREEM; BREAM *Cytisus scoparius* (Broom) Aber. B & H. Merely an approximation of the local pronunciation of broom.

BREER *Rosa canina* (Dog Rose) N.Eng. Halliwell.

BREER, Bramble *Rubus fruticosus* (Blackberry) Langham. Breer = briar, applied often to the blackberry, as well as to the wild rose.

BREER, Brid 1. *Rosa arvensis* (Field Rose) Ches. Holland. 2. *Rosa pimpinellifolia* (Burnet Rose) Ches. Holland. i.e. bird briar.

BREER, Buck *Rosa canina* (Dog Rose) N.Ire. Grigson. There are Buckie, or Bucky, too.

BREER, Dog *Rosa canina* (Dog Rose) Yks. Grigson.

BRENNET; BRUNNET; BROWN-NET *Scrophularia nodosa* (Figwort) Som. Grigson (Brennet); Dev. Grigson (Brunnet); Dev, Som. Grigson (Brown-net). These all mean the same. Brown-net means brown nettle.

BRERE *Rosa canina* (Dog Rose) Lancs. Nodal; Yks. Addy.

BRIAR 1. *Erica arborea* (Tree Heath) Grigson.1974. 2. *Rosa canina* (Dog Rose). Briar, whether spelt like this or as brier, seems to have meant any prickly bush, before standard English fixed it on to the Dog Rose. The earlier meaning is still apparent from the fact that locally the word (usually brier now) is used for bramble. When applied to the Tree Heath, the word is quite different, coming from French bruyère, heather, from Low Latin brucaria.

BRIAR, Austrian *Rosa foetida* (Capucine Rose) Plowdon.

BRIAR, Bird *Rosa canina* (Dog Rose) Ches. Grigson.

BRIAR, Bush *Rosa arvensis* (Field Rose) B & H.

BRIAR, Copper, Austrian *Rosa foetida var. bicolor.*
BRIAR, Dog's *Rosa canina* (Dog Rose) Hants. Grigson.
BRIAR, Green *Smilax rotundifolia* (Cat Brier) W.Miller. The more usual spelling is Greenbrier.
BRIAR, Hep; BRIAR, Hip *Rosa canina* (Dog Rose) Ches. B & H (Hep); Glos, Shrop. Grigson; Worcs. J Salisbury (Hip).
BRIAR, Persian *Rosa foetida* (Capucine Rose) Plowdon.
BRIAR, Roe *Rosa canina* (Dog Rose) Som. Grigson. Cf. the N.Ireland Buck Briar.
BRIAR, Sweet 1. *Acacia farnesiana* (Cassie) Barbados Gooding. 2. *Rosa micrantha* (Small-leaf Sweet Briar) Phillips. 3. *Rosa rubiginosa.*
BRIAR, Sweet, Narrow-leaved *Rosa agrestis.*
BRIAR, Sweet, Small-leaf *Rosa micrantha.*
BRIAR BALLS *Rosa canina* galls Worcs. J Salisbury. Recorded as a Worcestershire name, but this was the name under which they were sold by apothecaries.
BRIAR-BERRY *Rubus cuneifolius* USA Harper. Though it seems to be spelt 'brier' there.
BRIAR-BUSH; BRIAR-TREE *Rosa canina* (Dog Rose) Gerard (Bush); Turner (Tree).
BRIAR-ROSE *Rosa canina* (Dog Rose) N.Eng. Grigson.
BRID BREER 1. *Rosa arvensis* (Field Rose) Ches. Holland. 2. *Rosa pimpinellifolia* (Burnet Rose) Ches. Holland. i.e. bird briar.
BRID EEN *Melandrium rubrum* (Red Campion) Ches. Grigson. What is Bird's Eye in the south becomes Brid Een further north. The transposition of letters in dialect is very common.
BRID ROSE *Rosa pimpinellifolia* (Burnet Rose) Ches. Holland. i.e. Bird Rose.
BRIDAL FLOWER *Stephanotis floribunda* (Madagascar Jasmine) Simmons.
BRIDAL WREATH 1. *Campanula pyramidalis* (Chimney Bellflower) Som. Macmillan. 2. *Spiraea arguta.*

BRIDE, Mourning 1. *Knautia arvensis* (Field Scabious) Wilts. Macmillan. 2. *Scabiosa atropurpurea* (Sweet Scabious) Som. Tongue. It is difficult to know what brought this name and other similar ones, like Mournful Widow and Poor Widow on the Sweet Scabious, but there certainly was an association with death. Under the name of Sandade, it was much used once in Portugal and Brazil for funeral wreaths. There is no reason for the Field Scabious to have the name, and it is probably an error of observation.
BRIDE, Summer's; BRIDE, Sun's *Calendula officinalis* (Marigold) Coats (Summer's); T Hill (Sun's). Probably because it is said "to follow the sun".
BRIDEWEED *Linaria vulgaris* (Toadflax) Shrop, USA Grigson. Grigson says it was so named because it was used for "bride", a disease which pigs get.
BRIDEWORT 1. *Filipendula ulmaria* (Meadowsweet) Prior. 2. *Linaria vulgaris* (Toadflax) Shrop. Grigson. 3. *Spiraea salicifolia.* Meadowsweet - Bridewort, possibly from the resemblance of the flowers to the white feathers worn by brides, which is Prior's explanation, or perhaps because the plant was used for strewing the houses at wedding festivities. "The leaves and floures farre excell other strewing herbes for to decke up houses, to strew in chambers, halls, and banqueting houses in the summer time; for the smell thereof makes the heart merrie, delighteth the senses..." (Gerard). *Spiraea salicifolia* presumably for the same reason. Toadflax - Grigson says it was used for a disease which pigs get called "bride".
BRIDGET IN HER BRAVERY *Lychnis chalcedonica* (Jerusalem Cross) Wright.
BRIER *Rubus fruticosus* (Blackberry) Yks, N'thum, Donegal Grigson. An alternative spelling for briar, but it seems to be reserved for bramble.
BRIER, Cat *Smilax rotundifolia.*
BRIER, Horse *Smilax rotundifolia* (Cat Brier) USA Cunningham.

BRIER, Sand *Solanum carolinense* (Carolina Nightshade) USA Tolstead.

BRIER-BERRY *Rubus cuneifolius* USA Harper.

BRIGALOW 1. *Acacia doratoxylon*. 2. *Acacia harpophylla*.

BRIGHAM TEA; BRIGHAM YOUNG TEA 1. *Ephedra trifurca* (Long-leaved Joint-Fir) USA Elmore. 2. *Ephedra viridis* (Mountain Joint-Fir) USA Youngken. Also Mormon Tea, and other 'tea' names. Are these anything to do with the fact that a tea made from the dried plant was an Indian syphilis cure?

BRIGHT *Ranunculus ficaria* (Lesser Celandine) Halliwell. Is this a misprint in Halliwell? Grigson has Brighteye as a Devonshire name.

BRIGHT MEADOW *Caltha palustris* (Marsh Marigold) W.Miller. Probably a misreading for Meadowbright, which is recorded in Northants.

BRIGHTEYE; BRIGHT EYE 1. *Ranunculus ficaria* (Lesser Celandine) Dev. Grigson. 2. *Veronica chamaedrys* (Germander Speedwell) Som. Grigson.

BRILLIANT STAR *Kalanchoe blossfeldiana*.

BRIMBEL; BRIMBLE *Rubus fruticosus* (Blackberry) R.T.Gunther (Brimbel); Dor. Grigson; Cambs. Porter; Ches. Holland (Brimble). Grigson says that brimble is applied to the fruits only - very doubtful; Porter applies the name to the whole plant.

BRIMMEL, He-; BRIMMEL, Yoe- *Rubus fruticosus* (Blackberry) both Som. Grigson (He); Elworthy (Yoe). Is Ewe-bramble the same as Yoe brimmel? Seems probable, but then what is He-brimmel doing here? It may possibly be a still further corruption of Ewe. Halliwell, though, defined He-brimmle as a bramble of more than one year's growth.

BRIMMLE 1. *Rosa canina* (Dog Rose) Shrop. B & H. 2. *Rubus fruticosus* (Blackberry) Som. Jennings. 3. *Rubus fruticosus* fruit (Blackberry) Dev, Som, Shrop. B & H. Like Brimmel and Brimble, Brimmle is another West Country version of bramble.

BRIMMLE, Yew; BRIMMLE, Yoe *Rosa canina* (Dog Rose) Dev, Som. Grigson (Yoe); Som. Macmillan (Yew). Probably Ewe-bramble.

BRIMSTONE *Medicago lupulina* (Black Medick) Tynan & Maitland. Perhaps it is the colour of the flowers that gives the name.

BRIMSTONE-WOOD *Terminalia ivorensis* timber (Yellow Terminalia) Dalziel.

BRIMSTONEWORT 1. *Peucedanum officinale* (Hog's Fennel) Gerard. 2. *Peucedanum palustre* (Milk Parsley) B & H. Because their roots yield, as Coles says, "a yellow sap which waxeth quickly hard, and dry, and smelleth not unlike to brimstone".

BRINJAL; BRINJAUL; BRINGAL *Solanum melongena* (Egg Plant) Nielsen (Brinjal); C.Barber (Brinjaul); W.Miller (Bringal). The name came into English from Portuguese, who had it from Arabic, who in turn got it from Persian. It is in fact eventually from a Sanskrit word meaning good against the wind.

BRINTON ROOT *Veronica virginica* (Culver's Root) USA Lloyd.

BRISEWORT; BRISWORT 1. *Bellis perennis* (Daisy) B & H. 2. *Symphytum officinale* (Comfrey) B & H. Med. Bryswort, but it means bruisewort. Daisy "leaves stamped take away bruises and swellings proceeding of some stroke, if they be stamped and laid thereon; whereupon it was called in old time Bruisewort..." (Gerard). The same with comfrey; it is still a country remedy for cuts and bruises.

BRISKEN, Sheep's *Stachys palustris* (Marsh Woundwort) Ire. Grigson. Briosclan, or brisgean, is the Gaelic word for a type of edible root, probably silverweed (*Potentilla anserina*).

BRISTLE BENT *Agrostis setacea*.

BRISTLE FERN *Polystichum aculeatum* (Hard Shield Fern) USA Cunningham.

BRISTLE-GRASS, Green *Setaria viridis* (Bottle Grass) USA Allan.

BRISTLECONE PINE *Pinus aristata*.

BRISTLY FOXTAIL 1. *Setaria barbata*. 2. *Setaria viridis* (Bottle-grass) USA

Cunningham.

BRISTLY RHODODENDRON *Rhododendron barbatum.*

BRISTOL, Flower of; BRISTOW, Flower of *Lychnis chalcedonica* (Jerusalen Cross) Meager (Bristol); Parkinson (Bristow). Apparently this refers to the similarity of the colour to a popular bright red dye manufactured in Bristol in the 16th century.

BRISTOL ROCK-CRESS *Arabis stricta.* Its only location in Britain is on limestone in the Avon Gorge at Bristol.

BRISTOL WEED *Mercurialis perennis* (Dog's Mercury) Som. B & H. A Bristol name.

BRITISH ARROWROOT *Arum maculatum* (Cuckoo Pint) Le Strange. See rather Portland Arrowroot, or Portland Sago.

BRITISH COLUMBIAN PINE *Pseudotsuga menziesii* (Douglas Fir) N.Davey.

BRITISH MYRRH *Myrrhis odorata* (Sweet Cicely) Grieve.1931.

BRITTLE WILLOW *Salix fragilis* (Crack Willow) Warren-Wren. A slight pressure at the base of a twig will separate it from the branch with a cracking sound.

BRITTLEWORT *Salicornia europaea* (Glasswort) USA Chamberlin.

BRITTONS *Armeria maritima* (Thrift) Corn. Grigson. Cornish bryton.

BRIVET *Ligustrum vulgare* (Privet) Glos. B & H.

BROAD BEAN 1. *Canavalia ensiformis* (Jack Bean) Chopra. 2. *Vicia faba.*

BROAD CLOVER *Trifolium pratense* (Red Clover) IOW Grigson.

BROAD FIG *Ficus carica* (Common Fig) Dev. Friend.1882.

BROAD-GRASS 1. *Trifolium pratense* (Red Clover) Dor. Macmillan. 2. *Trifolium repens* (White Clover) Dor. Grigson.

BROAD-LEAF *Plantago major* (Great Plantain) Ches. Holland.

BROAD-LEAF, Canary *Plantago major* (Great Plantain) Som. Grigson. 'Canary' because it is one of the favourite cage birdseeds.

BROAD-LEAVED ELM *Ulmus glabra* (Wych Elm) B & H.

BROAD-LEAVED GARLIC *Allium ursinum* (Ramsons) Folkard. Ramsons is in fact the only broad-leaved garlic that is native to Britain.

BROAD-LEAVED GROUNDSEL; BROAD-LEAVED RAGWORT *Senecio fluviatilis* (Saracen's Woundwort) Murdoch McNeill (Groundsel); Clapham (Ragwort).

BROAD-LEAVED PLANTAIN *Plantago major* (Great Plantain) Hill.

BROAD-LEAVED THRIFT *Armeria arenaria.*

BROADWEED 1. *Heracleum sphondyllium* (Hogweed) Dor. Macmillan. 2. *Knautia arvensis* (Field Scabious) Dev. B & H.

BROCADE, Blue *Asperula azurea var. setosa.* Nurserymen's name for a variety of Blue Woodruff.

BROCCILO *Brassica oleracea var. italica* (Broccoli) Worcs. J Salisbury. A mispronunciation that must have become common enough for it to be listed as a local name.

BROCCOLI *Brassica oleracea var. italica.* Italian broccoli, the plural of broccolo, a cabbage sprout, or head, used in North Italian dialect as the name for cauliflower. Broccolo is the diminutive of brocco, a shoot.

BROCCOLI, Turnip, Italian *Brassica ruvo.*

BROCK *Brassica oleracea var. capitata* (Cabbage) N.Eng. Halliwell. Or at least, Halliwell says this is a North country word for a cabbage. The name must be Italian brocco, a shoot or sprout, preserved in English in the name broccoli.

BROKEN BONES PLANT *Oroxylum indicum* A Huxley. The name is given because of the tree's strange habit of shedding its huge leaves annually. The trunk becomes surrounded by a heap of pieces, hence the name.

BROKEN HEARTS *Clerodendron thomsoniae* (Bleeding Heart Vine) Howes.

BROME, Barren *Bromus sterilis.* Brome (and of course Bromus) comes from the Greek bromos, a kind of oats.

BROME, Compact *Bromus madritensis.*
BROME, Foxtail *Bromus rubens.*
BROME, Fringed *Bromus ciliatus.*
BROME, Great *Bromus diandrus.*
BROME, Hairy *Bromus ramosus.*
BROME, Meadow *Bromus commutatus.*
BROME, Rye *Bromus secalinus.*
BROME, Slender *Bromus lepidus.*
BROME, Smooth *Bromus racemosus.*
BROME, Soft *Bromus mollis.*
BROME, Upright *Bromus erectus.*
BROMPTON STOCK *Matthiola incana* (biennial form) (Stock). An 18th century name, from Brompton Park Nursery, near London.
BRONEY, Wood *Fraxinus excelsior* (Ash) B & H.
BROOK BETONY *Scrophularia aquatica* (Water Betony) B & H.
BROOK CRESS *Cardamine pennsylvanica.*
BROOK LOBELIA *Lobelia kalmii.*
BROOK SAXIFRAGE *Saxifraga rivularis* (Highland Saxifrage) McClintock.
BROOK-TONGUE *Cicuta virosa* (Cowbane) Cockayne.
BROOKBEAN *Menyanthes trifoliata* (Buckbean) B & H. 'Brook' here is probably a misreading of 'buck'. Not necessarily, though, for there is Beckbean, which means the same thing.
BROOKLEEK *Dracunculus vulgaris* (Dragon Arum) W Miller. A doubtful ascription.
BROOKLEM *Veronica beccabunga* (Brooklime) Gerard. This seems to be the earlier form. It appears as Broclempe in Cockayne, and Brokelemp in Dawson.
BROOKLIME 1. *Apium nodiflorum* (Fool's Watercress) Som. Macmillan; War. Grigson. 2. *Nasturtium officinale* (Watercress) Bucks. Grigson. 3. *Veronica beccabunga.* Beccabunga is O.Norse bekh, brook, and bung, the name of a plant. In Brooklime, 'lime' is equated with 'bung'. It was originally brokelemk (OE hleomoce, OHG lömeke). Brooklem (see previous entry) is the early form.

BROOKLIME, American *Veronica americana.*
BROOKMINT *Mentha aquatica* (Water Mint) Gerard. OE brocminte.
BROOKWEED 1. *Samolus floribundus* USA Zenkert. 2. *Samolus valerandi.*
BROOM 1. *Calluna vulgaris* (Heather) Bucks. B & H; Yks. F K Robinson. 2. *Cytisus scoparius.* 3. *Erica tetralix* (Cross-leaved Heath) Bucks. Grigson. 4. *Galium verum* (Lady's Bedstraw) Shrop. Grigson. The broom, i.e. the thing you sweep with, was originally a branch of the plant, whether the Cytisus or the heathers. The word itself (OE brom), when taken far enough back, is seen to be related to bramble. Lady's Bedstraw is probably given the name for a different reason, that of the applied meaning of broom as a sexual symbol (witches on a broomstick, broomstick marriages, etc).
BROOM, Andreas *Cytisus scoparius var. andreanus.*
BROOM, Base *Genista tinctoria* (Dyer's Greenweed) Prior. Prior says base for dyeing, but it could be base - low. There is a Yorkshire name Low Broom. The latter is probably right, but of course, it is a dyeplant.
BROOM, Black *Cytisus nigricans.*
BROOM, Butcher's *Ruscus aculeatus.* Because once butchers used to sweep their blocks with hard brooms made of the green shoots. In Italy, they were commonly used as household besoms.
BROOM, Church *Dipsacus fullonum* (Teasel) Som. Macmillan. From the resemblance of the flower heads in shape to the long "turk's head" brooms used for sweeping high places.
BROOM, Desert *Baccharis sarothroides* (Broom Baccharis) USA T H Kearney.
BROOM, Dyer's *Genista tinctoria* (Dyer's Greenweed) Prior.
BROOM, Fairy's *Dipsacus fullonum* (Teasel) Som. Macmillan. Cf. Church Broom for the same plant.
BROOM, Green *Cytisus scoparius* (Broom) Herts. B & H. Presumably because it was

actually the green twigs which were used to sweep with. Green Basom is a Somerset name, and Greenwood occurs there too.

BROOM, He *Laburnum anagyroides* (Laburnum) Fife B & H. Probably meaning High Broom, to distinguish it from the Low Broom, *Genista tinctoria*. Against this argument, though, is the fact that *Genista tinctoria* is actually called She Broom in the north of England.

BROOM, Heath; BROOM HEATH *Erica tetralix* (Cross-leaved Heath) B & H (Heath Broom); Hulme (Broom Heath). A heath that is used as a broom. See Broom.

BROOM, Hedgehog *Erinacea anthyllis*. It was Clusius who named it Erinacea, after the Latin for a hedgehog.

BROOM, Indigo *Baptisia tinctoria* (Wild Indigo) USA Sanecki.

BROOM, Irish *Cytisus scoparius* Mitton. Probably an American name for the plant - Scotch Broom certainly is.

BROOM, Low *Genista tinctoria* (Dyer's Greenweed) Yks. B & H. Cf. Base Broom.

BROOM, Madeira *Genista virgata*.

BROOM, Montpellier *Cytisus monspessulana*.

BROOM, Moonlight *Cytisus scoparius var. sulphureus*. A nurserymen's name, given because of the colour.

BROOM, Mount Etna *Genista aetnensis*.

BROOM, Portugal, White *Cytisus albus*.

BROOM, Prickly; BROOM, Thorn 1. *Genista anglica* (Needle Whin) both Turner. 2. *Ulex europaeus* (Gorse) both Gerard.

BROOM, Purple *Cytisus purpureus*.

BROOM, Pyrenean *Cytisus purgans*.

BROOM, Scotch 1. *Cytisus scoparius* (Broom) USA Zenkert. 2. *Genista canariensis*.

BROOM, She *Genista tinctoria* (Dyer's Greenweed) Ches, Yks. B & H. There is a He Broom - the laburnum.

BROOM, Spanish *Spartium junceum*.

BROOM, Spanish, White *Cytisus multiflorus*.

BROOM, Sweet *Ruscus aculeatus* (Butcher's Broom) Grieve.1931.

BROOM, Tenerife *Cytisus supranubens*.

BROOM, Warminster *Cytisus x praecox*.

BROOM, Weaver's *Spartium junceum* (Spanish Broom) Howes.

BROOM, White 1. *Borreria laevis* Barbados Laguerre. 2. *Retama raetam*.

BROOM, Woolly-podded *Cytisus grandiflorus*.

BROOM BACCHARIS *Baccharis sarothroides*.

BROOM CYPRESS *Kochia scoparia* (Burning Bush) Geng Junying.

BROOM DEER-WEED *Lotus scoparius*.

BROOM GROUNDSEL *Senecio spartioides*.

BROOM MILLET; BROOMCORN *Panicum miliaceum* (Millet) Brouk. (Broom Millet): Cunningham (Broomcorn).

BROOM SNAKEWEED *Gutierrezia microcephala* (Broomweed) USA Weiner.

BROOMBEERE *Rubus villosus* (American Blackberry) Grieve.1931.

BROOMCORN 1. *Sorghum saccharatum* (Sorgo) USA J.Smith. 2. *Sorghum vulgare* (Kaffir Corn) Schery. The inflorescences are used commercially for brooms. It was particularly grown in Ohio for this purpose.

BROOMDASHERS *Prunus cerasus* (Broom) Kent T Lloyd.

BROOMRAPE, Branched *Orobanche ramosa*. Parasitic, of course, but rape in this case means turnip, or in fact any tuberous plant.

BROOMRAPE, Carrot *Orobanche maritima*.

BROOMRAPE, Clove-scented *Orobanche caryophyllaceae*.

BROOMRAPE, Greater *Orobanche rapum-genistae*.

BROOMRAPE, Hemp *Orobanche ramosus* (Branching Broomrape). Rackham.1986.

BROOMRAPE, Ivy *Orobanche hederae*.

BROOMRAPE, Knapweed *Orobanche elatior* (Tall Broomrape).

BROOMRAPE, Lesser *Orobanche minor*.

BROOMRAPE, Picris *Orobanche picridis*.

BROOMRAPE, Purple *Orobanche purpurea*

(Yarrow Broomrape) Clapham.
BROOMRAPE, Red *Orobanche alba* (Thyme Broomrape).
BROOMRAPE, Thistle *Orobanche reticulata*.
BROOMRAPE, Tall *Orobanche elatior*.
BROOMRAPE, Thyme *Orobanche alba*.
BROOMRAPE, Yarrow *Orobanche purpurea*.
BROOMWEED 1. *Gutierrezia dracunculoides* D.E.Jones. 2. *Gutierrezia microcephala*. 3. *Malvastrum coromandelianum*.
BROOMWEED, White *Parthenium hysterophorus* Barbados Gooding.
BROOMWORT *Thlaspi arvense* (Field Penny-cress) B & H.
BROSEWORT *Hyoscyamus niger* (Henbane) Sanecki.
BROTHERWORT 1. *Mentha pulegium* (Penny-royal) Earle. 2. *Thymus drucei* (Wild Thyme) B & H. 3. *Thymus serpyllum* (Breckland Wild Thyme) Webster. Possibly because they may be grown in monks' gardens. The names of *M pulegium* and *T drucei* are connected in other ways (Puliall, for instance), and their uses may very well have been identical.
BROUSSA MULLEIN *Verbascum broussa*. It comes from Bursa, or Brusa, in Asia Minor.
BROWN-BACK *Ceterach officinarum* (Rusty-back) Dev. Friend.1882. Exactly descriptive.
BROWN-JOLLY *Solanum melongena* (Egg Plant) W.Indies Grigson.1974. A variant of the 17th century name Brinjal, itself descended from a Sanskrit word meaning good against the wind.
BROWN-NET *Scrophularia aquatica* (Water Betony) Dev, Som. Grigson. Probably Brownwort, which is a name Gerard gives it.
BROWN-SHILLERS *Corylus avellana* nut (Hazel nuts) Yks. Hunter. Only applied to the ripe nuts.
BROWNET 1. *Scrophularia aquatica* (Water Betony) Dev. Friend.1882. 2. *Scrophularia nodosa* (Figwort) Dev. Friend.1882.
BROWNWEED *Gutierrezia microcephala* (Broomweed) USA Elmore. Presumably this is just a variation on Broomweed.
BROWNWORT 1. *Ceterach officinarum* (Rusty-back) Payne. 2. *Prunella vulgaris* (Self-heal) Grigson. 3. *Scrophularia aquatica* (Water Betony) Gerard. 4. *Scrophularia nodosa* (Figwort) Gerard. As far as Self-heal is concerned, this is OE brunwyrt, perhaps from the brown colour of the stems and flowers. It was originally called Brunella, and it was supposed to cure a disease called in Germany die Braune, a kind of quinsey. This doesn't help with the two Scrophularias, though they are both called Brownet as well. For Rusty-back, the reason is obvious. Cf. Brown-back.
BRUISEROOT *Glaucium flavum* (Horned Poppy) S.Eng. Grigson. The orange-coloured sap was at one time used to put on bruises. Another name is Squat, or Squatmore, and squat means a bruise.
BRUISEWORT 1. *Bellis perennis* (Daisy) Gerard. 2. *Saponaria officinalis* (Soapwort) Som. Prior. 3. *Symphytum officinale* (Comfrey) Dawson. Daisy - "the leaves stamped take away bruises and swellings proceeding of some stroke, if they be stamped and laid thereon; whereupon it was called in old time Bruisewort..." (Gerard). Soapwort - gypsies still use the root decoction to apply to a bruise or black eye. Comfrey leaves are a well-known country remedy for cuts and bruises.
BRUM, French *Laburnum anagyroides* (Laburnum) Shrop. B & H. Brum is broom.
BRUMBLE; BRUMBLE, Cock *Rubus fruticosus* (Blackberry) Norf, Suff. B & H, Moor (Brumble); E.Ang. Forby (Cock Brumble).
BRUMLEYBERRY BUSH *Rubus fruticosus* (Blackberry) Border B & H. i.e. Brambleberry.
BRUMMEL(L) 1. *Genista tinctoria* (Dyer's Greenweed) Corn. Grigson. 2. *Rubus fruticosus* (Blackberry) Hants. Halliwell. 3. *Tamarix anglica* (Tamarisk) Corn. Grigson.
BRUMMELKITE; BUMMELTYKITE

Rubus fruticosus fruit (Blackberry) both Cumb. Grigson. Kite is Cumbrian and Northumbrian for belly (see Halliwell). Andrew Young said that children ate so many blackberries that their kites rumbled. Morris agreed with this, and he was a Yorkshireman.

BRUNELL; PRUNELL *Prunella vulgaris* (Self-heal) both Gerard. It is said the original name was Brunella, not Prunella, and Prior connects this with the German name for a kind of quinsy, for which the plant was used, called die Braune.

BRUNNET 1. *Scrophularia aquatica* (Water Betony) Dev. Grigson. 2. *Scrophularia nodosa* (Figwort) Dev. Grigson. This is probably Brown-net, which also occurs, and means brown nettle.

BRUSH 1. *Centaurea cyanus* (Cornflower) Stir. B & H. 2. *Centaurea nigra* (Knapweed) Dor. Macmillan. 3. *Cytisus scoparius* (Broom) Dor. Macmillan. 4. *Dactylorchis incarnata* (Early Marsh Orchid) Donegal. Grigson. 5. *Dipsacus fullonum* (Teasel) Som. Macmillan; Wilts. Dartnell & Goddard; Lincs. Grigson. 6. *Hippuris vulgaris* (Mare's Tail) Dor. Macmillan. Cornflower and Knapweed - descriptive (shape of the flowers) - imagine a chimney sweep's brush. Broom, because it was used literally as a brush. Teasel has other 'brush' names, like Barber's Brush, Clothes Brush, and so on. Mare's Tail has the name as a description also. The name Bottle Brush from Hampshire and Yorkshire is an accurate description. The orchid probably gets the name by description also, but that seems tenuous.

BRUSH, Barber's *Dipsacus fullonum* (Teasel) Som, Dor. Macmillan; Wilts. Dartnell & Goddard; Ess. Grigson.

BRUSH, Bear *Garrya fremontii*. Shape of the flowers?

BRUSH, Beggar's *Clematis vitalba* (Old Man's Beard) Bucks. B & H. There is also Beggar's Plant for this. The plant became known as an emblem of cunning, because it was used by beggars to gain pity. Like that of most of the Ranunculaceae, the juice is extremely acrid (Old Man's Beard is so much of an irritant that it can be fatal if ingested). Beggars used to run the juice on themselves to cause false ulcers.

BRUSH, Bottle 1. *Equisetum arvense* (Common Horsetail) Palaiseul. 2. *Equisetum sylvaticum* (Wood Horsetail) W Miller. 3. *Hippuris vulgaris* (Mare's Tail) Hants. Cope; Yks. Grigson. 4. *Melaleuca leucadendron* (Cajuput) USA Cunningham. 5. *Spergula arvensis* (Corn Spurrey) Yks. Grigson. All descriptive in various ways, and in the case of the horsetails, descriptive of usage as well.

BRUSH, Buck 1. *Ceanothus cuneatus*. 2. *Symphoricarpos occidentalis* (Western Wolfberry) Gilmore. 3. *Symphoricarpos orbiculatus* (Red Wolf-berry) USA Gates.

BRUSH, Buck, Mohave *Ceanothus vestitus*.

BRUSH, Burning *Euonymus atropurpureus* USA Gilmore. Burning Bush, too, which is what Burning Brush should be.

BRUSH, Chimney Sweeper's *Typha latifolia* (False Bulrush) Som. Macmillan.

BRUSH, Clothes'; BRUSH, Clothier's *Dipsacus fullonum* (Teasel) Som. Grigson; Wilts. Dartnell & Goddard. (Clothes'); Cumb. Grigson (Clothier's).

BRUSH, Coal Oil *Tetradymia glabrata* (Little-leaf Horsebrush) USA Kingsbury.

BRUSH, Comb-and- *Dipsacus fullonum* (Teasel) Wilts. Dartnell & Goddard. Or Brush-and-comb.

BRUSH, Deer *Ceanothus integerrimus*. Deer Brush apparently because the deer are fond of eating it. Miwok hunters used to look for deer particularly where this shrub grew.

BRUSH, Devil's *Pteridium aquilina* (Bracken) Waring.

BRUSH, Flue *Typha latifolia* (False Bulrush) Som. Macmillan. Descriptive, for the black top has also given such names as Chimney Sweeper, or Chimney Sweeper's Brush.

BRUSH, Fox's *Centranthus ruber* (Spur Valerian) Som. Macmillan; N.Ire. Grigson. Colour and shape.

BRUSH, Lady's *Dipsacus fullonum* (Teasel) Putnam. Cf. Lady's Brush-and-comb.
BRUSH, Little *Dipsacus fullonum* (Teasel) Som. Macmillan.
BRUSH, Poor Man's *Dipsacus fullonum* (Teasel) Som. Macmillan.
BRUSH, Rabbit 1. *Bigelovia graveolens* (Rabbit Wood). 2. *Chrysothamnus nauseosus var. bigelovii*. Are these the same plant?
BRUSH, Rabbit, Black-banded *Chrysothamnus paniculatus*. Black-banded, because a fungus very often causes blackish bands on the young twigs.
BRUSH, Rabbit, Douglas *Chrysothamnus confinis*.
BRUSH, Rabbit, Rubber *Chrysothamnus nauseosus*.
BRUSH, Red *Cornus stolonifera* (Red-osier Dogwood) Gilmore. It was used by various American Indian peoples to make a red dye.
BRUSH, Shaving *Centaurea nigra* (Knapweed) Shrop. Grigson. Descriptive - shape.
BRUSH, Sweep's 1. *Dipsacus fullonum* (Teasel) Dev. Macmillan. 2. *Luzula campestris* (Good Friday Grass) McClintock & Fitter. 3. *Plantago lanceolata* (Ribwort Plantain) Som. Macmillan. 4. *Plantago media* (Lamb's Tongue Plantain) Som. Macmillan. 5. *Tussilago farfara* (Coltsfoot) Som. Macmillan. All descriptive, either by shape or colour.
BRUSH, Tobacco *Ceanothus velutinus* USA Schenk & Gifford.
BRUSH-AND-COMB; BRUSH-AND-COMB, Lady's *Dipsacus fullonum* Som. Macmillan; Dor. Dacombe (Brush-and-comb); Som. Macmillan (Lady's Brush-and-comb).
BRUSH CHERRY *Eugenia luehmannii*.
BRUSHWATTLE *Albizia lophantha*.
BRUSSELS SPROUTS *Brassica oleracea var. gemmifera*. Apparently developed in the 15th century in that part of Europe which is now Belgium. Whether or not it originated in Brussels isn't known, but it was certainly cultivated there for centuries.

BRYLOCKS *Vaccinium myrtillus* (Whortleberry) Scot. Mabey.
BRYONY *Calystegia sepium* (Great Bindweed) Dev. Macmillan.
BRYONY, Black *Tamus communis*. ODEE merely quotes the derivation as a plant name - Latin bryonia, Greek bruonia (Dioscorides). Fernie, though, says it is from Greek bruein, to shoot forth rapidly.
BRYONY, English *Bryonia dioica* (White Bryony) Parkinson.1640. Possibly because it was also known as English Mandrake.
BRYONY, Red; BRYONY, Red-berried *Bryonia dioica* (White Bryony) B & H (Red); Barton & Castle (Red-berried).
BRYONY, White *Bryonia dioica*.
BRYSWORT *Bellis perennis* (Daisy) Gerard. Gerard calls it this, but it is really the medieval version of the later Bruisewort, through Briswort and Brisewort.
BUBBLE, May *Caltha palustris* (Marsh Marigold) Som. Macmillan; Wilts. Dartnell & Goddard. A reference to the blistering properties of the acrid juice of not only this, but most of the rest of the Ranunculaceae.
BUBBLES, Pig's *Heracleum sphondyllium* (Hogweed) Som. Elworthy. It used to be collected in the Taunton Deane area of Somerset as pig food. They were very fond of it, it seems.
BUBBLES, Water *Caltha palustris* (Marsh Marigold) Som. Macmillan; Oxf. PLNN 19; Bucks. Bucks F W I.
BUBBY WATER *Chrysophyllum cainito* (Star-apple) Sierra Leone Dalziel.
BUCALAUN *Senecio jacobaea* (Ragwort) Ire. Yeats.1959. This is Gaelic buaghallan, or buadhghallan, which also becomes Boholawn as an Irish local name.
BUCHU *Barosma betulina*.
BUCK 1. *Fagopyrum esculentum* (Buckwheat) Norf. Gerard. 2. *Fagus sylvatica* (Beech) Scot. B & H. A shortening of Buckwheat in the first case. For beech, it is the Scottish version of OE boece, or boc. Thus the fruit of the beech is Buckmast.
BUCK BREER *Rosa canina* (Dog Rose)

N.Ire. Grigson.

BUCK BRUSH 1. *Ceanothus cuneatus*. 2. *Symphoricarpos occidentalis* (Western Wolf-berry) Gilmore. 3. *Symphoricarpos orbiculatus* (Red Wolf-berry USA Gates.

BUCK BRUSH, Mohave *Ceanothus vestitus*.

BUCK THISTLE 1. *Carduus nutans* (Musk Thistle) Yks. Grigson. 2. *Cirsium vulgare* (Spear Plume Thistle) Yks. Grigson. Buck in these cases is probably a corruption of Bur.

BUCK'S BEARD *Tragopogon pratensis* (Goat's Beard) B & H. Buck is simply a variation of goat.

BUCK'S HORN; BUCK'S HORN PLANTAIN 1. *Plantago coronopus* (Buck's Horn) Parkinson.1640. 2. *Plantago maritima* (Sea Plantain) B & H.

BUCK'S HORN LICHEN *Evernia prunastri* (Stag's Horn Lichen) S.M.Robertson.

BUCK'S HORN TREE *Rhus typhina* (Stag's Horn Sumach).

BUCK'S HORN WELD *Catananche caerulea* (Cupid's Darts) Gerard.

BUCKBEAN *Menyanthes trifoliata*. Buckbean is apparently Lyte's translation of the Flemish books boonen (goat's beans). ODEE agrees, but Fernie says it may be a corruption of scorbutus, scurvy. It seems that the German name Scharboek *is* a corruption of scorbutus, and it certainly used to be of prime use in the treatment of scurvy.

BUCKBUSH *Salsola kali* (Saltwort) Australia Meggitt.

BUCKEYE, Bottlebrush *Aesculus parviflora*. Buckeye - because the fruit look like a deer's eye?

BUCKEYE, Californian *Aesculus californica*.

BUCKEYE, Dwarf; BUCKEYE, White *Aesculus parviflora* (Bottlebrush Buckeye) Chaplin (Dwarf); USA Harper (White).

BUCKEYE, Ohio *Aesculus glabra*.

BUCKEYE, Red *Aesculus pavia*.

BUCKEYE, Sweet *Aesculus octandra* (Yellow Buckeye) USA Kingsbury.

BUCKEYE, Yellow *Aesculus octandra*.

BUCKHORN *Plantago lanceolata* (Ribwort Plantain) USA H C Long.1910. In this country, it is *P coronopus* which is Buck's Horn rather than Ribwort.

BUCKIE; BUCKY *Rosa canina* (Dog Rose) N.Ire. Grigson; IOM Moore.1924. Buckbriar is recorded from Ireland, too. The hips are Buckie-berries, and the seeds Buckie-lice. In Orkney, buckies are snails.

BUCKIE-BERRIES *Rosa canina* hips (Hips) N.Ire. Grigson.

BUCKIE-FAALIE *Primula vulgaris* (Primrose) Caith. Grigson.

BUCKIE-LICE *Rosa canina* seeds (Dog Rose) S Scot, Ire. Grigson.

BUCKLE; BUCKLE, Horse *Primula veris* (Cowslip) Grieve.1931 (Buckle); Wilts. Macmillan; Kent Grigson (Horse Buckle). Possibly here a corruption of Paigle, a well-known though now archaic name for cowslip.

BUCKLER FERN *Dryopteris dilatata*.

BUCKLER FERN, Crested *Dryopteris cristata*.

BUCKLER FERN, Hay-scented *Dryopteris aemula*.

BUCKLER FERN, Narrow 1. *Dryopteris carthusiana* (Prickly Buckler Fern) RHS. 2. *Dryopteris spinulosa*.

BUCKLER FERN, Prickly *Dryopteris carthusiana*.

BUCKLER FERN, Rigid *Dryopteris villarsii*.

BUCKLER MUSTARD *Biscutella laevigata*.

BUCKLER-SHAPED SORREL *Rumex scutatus* (French Sorrel) Leyel.1937. 'Scutatus' means shaped like a buckler.

BUCKLER THORN *Rhamnus cathartica* (Buckthorn) Turner. Turner offers an explanation, talking about "fruite lyke a little buckeler", but this is too close to the common name to be a coincidence.

BUCKLEY CENTAURY *Centaureum calycosum*.

BUCKMAST *Fagus sylvatica* fruit (Beech mast) B & H. Mast (OE maest) means the fruit of any forest tree, not just beech, though it seems to have become confined to this one. Buck is simply the form of OE boc, for beech (with a long vowel) surviving with the shortened vowel. Buckwheat is another example.

BUCKRAMS 1. *Allium ursinum* (Ramsons) Gerard. 2. *Arum maculatum* (Cuckoo Pint) Suss. B & H. In the first case especially, the name may have been given because of the offensive smell, but it is just possible that both got the name from a supposed aphrodisiac connection.

BUCKTHORN 1. *Osmunda cinnamonea* (Cinnamon Fern) USA Yarnell. 2. *Prunus spinosa* (Blackthorn) Lincs. Peacock. 3. *Rhamnus cathartica*. Buckthorn appears to be a translation of the med. Latin cervi spina (stag's thorn). Or is it, as Prior suggested, the result of mistaking German Buxdon, box thorn, for Bocksdorn, buck's thorn? It is probably a mis-hearing when recorded for blackthorn, and the American application of the name to a fern is very strange.

BUCKTHORN, Alder *Frangula alnus*.

BUCKTHORN, Alder-leaved *Rhamnus alnifolia*.

BUCKTHORN, Californian 1. *Rhamnus californica* (Redberry) T H Kearney. 2. *Rhamnus purshiana*.

BUCKTHORN, Mediterranean; BUCK-THORN, Italian *Rhamnus alaternus* Italian Buckthorn is in RHS.

BUCKTHORN, Palestine *Rhamnus palaestina*.

BUCKTHORN, Purging *Rhamnus cathartica* (Buckthorn) North. The black berries are a powerful purgative, so much so that they are dangerous to children.

BUCKTHORN, Sea *Hippophae rhamnoides*.

BUCKWHEAT *Fagopyrum esculentum*. Perhaps from Dutch boekweit, meaning beech wheat (the seeds look rather like beech mast), but possibly it comes from an earlier form meaning goat wheat, i.e. a grain inferior to true wheat.

BUCKWHEAT, Black *Polygonum convolvulus* (Black Bindweed) War. Grigson.

BUCKWHEAT, Chinese *Polygonum cymosum*.

BUCKWHEAT, Climbing *Polygonum convolvulus* (Black Bindweed) C P Johnson.

BUCKWHEAT, Desert *Eriogonum deserticola*.

BUCKWHEAT, False *Polygonum sagitattum* (Tear-thumb) Grieve.1931.

BUCKWHEAT, False, Climbing *Polygonum scandens* (Climbing Knotweed) USA House.

BUCKWHEAT, Green *Fagopyrum tataricum*.

BUCKWHEAT, Kidney-leaved *Eriogonum reniforme*.

BUCKWHEAT, Mohave *Eriogonum mohavense*. Grows in the Mohave Desert.

BUCKWHEAT, Nodding *Eriogonum cernuum*.

BUCKWHEAT, Red-root *Eriogonum racemosum*.

BUCKWHEAT, Rough *Fagopyrum tataricum* (Green Buckwheat) Grieve.1931.

BUCKWHEAT, Running *Polygonum convolvulus* (Black Bindweed) Turner.

BUCKWHEAT, Tartary *Fagopyrum tataricum* (Green Buckwheat) Candolle. It is a native of Siberia.

BUCKWHEAT, Wild *Eriogonum fasciculatum*.

BUDDHA'S HEAD CITRON *Citrus medica*.

BUDDHIST PINE *Podocarpus sp* Goold-Adams.

BUDDLE; BUDLAND *Chrysanthemum segetum* (Corn Marigold) E.Ang. Grigson (Buddle); Norf. B & H (Budland). Better known as Boodle. It is just possible that is this from Dutch boedel, which means the whole of one's goods. Corn Marigold was a serious menace once in the wheatfield; Tusser wrote:

The brake and the cockle be noisome too much,
Yet like unto Boodle no weed there is such.

Farmers were fined, both in Scotland and in England, under an act of Henry II if Corn Marigold were found growing on their land. So if this derivation is right, it would mean the plant which was liable to take away all one's possessions by killing off the corn crop.

BUDDY-BUD; BUDDY-BUSS *Arctium lappa* (Burdock) both N.Eng. both B & H. Only applied to the flowers, it seems.

BUFFALO BEAN 1. *Astragalus caryocarpus*. 2. *Thermopsis rhombifolia* (False Lupin) Johnston.

BUFFALO BUR *Solanum rostratum* (Prickly Nightshade) USA Gates. Bur - prickly.

BUFFALO CLOVER *Trifolium reflexum*.

BUFFALO CURRANT *Ribes aureum*.

BUFFALO GOURD *Cucurbita foetidissima*.

BUFFALO THORN *Zizyphus mucronata*.

BUFFALO-WEED *Ambrosia trifida* (Great Ragweed) USA Allan.

BUFFALOBERRY *Shepherdia argentata*.

BUFFEL GRASS *Setaria chevalieri* S Africa Dalziel.

BUG ORCHID *Orchis coriophera*. It is said that the flowers smell of bed-bugs.

BUGBANE 1. *Actaea spicata* (Herb Christopher) Grieve.1931. 2. *Cimicifuga foetida*. 3. *Cimicifuga racemosa* (Black Snakeroot) Fluckiger. Herb Christopher got the name because of the offensive smell, which is said to drive away vermin. Black Snakeroot, which is also called Bugwort in the USA, got the name because of the power of the seeds "to drive away bugs" (Sanford).

BUGGIE FLOWER *Silene maritima* (Sea Campion) Shet. Grigson.

BUGLE 1. *Ajuga reptans*. 2. *Echium vulgare* (Vipcr's Bugloss) Hants. Grigson. 3. *Lycopsis arvensis* (Small Bugloss) R.T.Gunther. Bugle is Latin bugulla (it is still búgula in Spanish), and was still Bugull in medieval English. Barton & Castle suggest that bugula may be a diminutive of buglossum, hence the link with bugloss above.

BUGLE, Blue *Ajuga genevensis* (Erect Bugle) Polunin.

BUGLE, Brown *Ajuga reptans* (Bugle) Gerard.

BUGLE, Carpet *Ajuga reptans* (Bugle) USA Cunningham.

BUGLE, Creeping *Ajuga reptans* (Bugle) Thornton.

BUGLE, Erect *Ajuga genevensis*.

BUGLE, Jagged *Ajuga genevensis* (Erect Bugle) Turner.

BUGLE, Limestone *Ajuga pyramidalis*. It is confined to limestone rocks.

BUGLE, Nelson's *Ajuga reptans* (Bugle) Som. Macmillan.

BUGLE, Pyramidal *Ajuga pyramidalis* (Limestone Bugle) Polunin.

BUGLE, Sweet *Lycopus virginicus* (Virginia Bugleweed) Grieve.1931.

BUGLE, Tufted *Ajuga genevensis* (Erect Bugle) H.N.Webster.

BUGLE, Water *Lycopus virginicus* (Virginia Buglweed) Grieve.1931.

BUGLE, Yellow *Ajuga chamaepitys*.

BUGLE-BLOOM *Lonicera periclymenum* (Honeysuckle) Dor. Macmillan. This time bugle means the musical instrument. The name is in line with others like Trumpet-flower, and descriptive of the shape of the flowers.

BUGLER, Red, Royal *Aeschynanthus pulcher* (Lipstick Plant) RHS.

BUGLER, Scarlet *Pentstemon centrantifolius*.

BUGLEWEED 1. *Ajuga reptans* (Bugle) USA G.B.Foster. 2. *Lycopus asper*. 3. *Lycopus virginicus*.

BUGLEWEED, Northern *Lycopus virginicus*.

BUGLEWEED, Virginia *Lycopus virginicus*.

BUGLOSS 1. *Myosotis arvensis* (Field Forget-me-not) Dev. Grigson. 2. *Myosotis sylvatica* (Wood Forget-me-not) Dev. Friend.1882. Bugloss is the name given rather to plants of the geni Borago, Lycopsis, Anchusa and Echium. The word comes from Greek, and means literally 'ox-tongued', a reference to the characteristic shape and feel of the leaves. Why the name should be given to this Forget-me-not in Devon is not known.

BUGLOSS, Common *Borago officinalis* (Borage) Thornton.

BUGLOSS, Corn; BUGLOSS, Field *Lycopsis arvensis* (Small Bugloss) both B & H.

BUGLOSS, Dyer's *Anchusa tinctoria*. A famous dye plant, yielding, after treatment, a bright red dye. It was also used by cabinetmakers for staining wood, and, less honestly, for colouring and flavouring liquids spuriously sold as port.

BUGLOSS, Marsh, Sea *Limonium vulgare* (Sea Lavender) Parkinson.1640.

BUGLOSS, Officinal *Anchusa officinalis* (Alkanet) Thornton. Officinal, whether in the English form, or as the species name *officinalis*, means used in medicine, or made from a pharmacopeia recipe. It means official, a word which mostly superseded the older one.

BUGLOSS, Sea 1. *Anchusa officinalis* (Alkanet) Parkinson. 2. *Pulmonaria maritima*.

BUGLOSS, Siberian *Brunnera macrophylla*.

BUGLOSS, Small *Lycopsis arvensis*.

BUGLOSS, Snake's *Echium vulgare* (Viper's Bugloss) Gerard.

BUGLOSS, Spanish *Anchusa officinalis* (Alkanet) Culpepper. Possibly because of the Moorish origin of the name Alkanet.

BUGLOSS, Spotted *Pulmonaria officinalis* (Lungwort) Parkinson.

BUGLOSS, Viper's *Echium vulgare*. So called because the seeds are like the head of a viper. Echium is from a Greek plant name ekhion, from ekhis, viper.

BUGLOSS, Viper's, Purple *Echium plantagineum*.

BUGLOSS, Wall *Lycopsis arvensis* (Small Bugloss) Parkinson.1640.

BUGLOSS, Wild 1. *Anchusa officinalis* (Alkanet) Gerard. 2. *Echium vulgare* (Viper's Bugloss) Gerard. 3. *Lycopsis arvensis* (Small Bugloss) Turner.

BUGLOSS COWSLIP *Pulmonaria officinalis* (Lungwort) Prior. Bugloss-like leaves, and Primula-like flowers. There are a number of cowslip names for this.

BUGWORT *Cimicifuga racemosa* (Black Snakeroot) USA Weiner. In this case, Bugwort means rather Bugbane, because of the power of the seed "to drive away bugs" (Sanford).

BUKKUM-WOOD *Caesalpina sappan* (Brazil-wood) W.Miller.

BULKISHAWN, He *Senecio jacobaea* (Ragwort) Ire. Crooke. Bulkishawn is one of the versions of names deriving from Gaelic buaghallan. See also next entry.

BULKISHAWN, She *Tanacetum vulgare* (Tansy) Ire. Crooke.

BULL BAY *Magnolia grandiflora*. 'Bull' applied to a flower usually implies large, and because of that, coarse. 'Horse' has the same connotation, and to a lesser extent, 'dog'. In the case of this magnolia, the name is apt - its alternative specific name is *foetida*; these large flowers have an overpowering scent. It is said that the Indians would never sleep under magnolia in blossom.

BULL BEEF 1. *Rosa canina* (Dog Rose) Ches. Holland. 2. *Rubus fruticosus* (Bramble) Ches. Holland. The young shoots, in both cases. Cheshire children used to eat the shoots, and call them Bull Beef.

BULL BUTTERCUP; BULL CUP *Caltha palustris* (Marsh Marigold) Ess. Grigson (Buttercup); Som. Macmillan (Cup). A buttercup with large flowers, of course.

BULL DAIRY 1. *Dactylorchis incarnata* (Early Marsh Orchid) Scot. B & H. 2. *Orchis mascula* (Early Purple Orchis) Scot. B & H. This is a corruption of a Scottish name spelt Beldairy, or Bildairy.

BULL DAISY *Leucanthemum vulgare* (Ox-eye Daisy) Ches. (Holland); E.Ang, Yks, Cumb. Grigson. Bull = big, again.

BULL FACES; BULL PATES *Deschampsia cespitosa* (Tufted Hair-grass) H C Long.1910.

BULL FLOWER 1. *Althaea officinalis* (Marsh Mallow) Som. Fernie. 2. *Caltha palustris* (Marsh Marigold) Dev. Friend. Is this 'pool'-flower, rather?

BULL HAWS *Crataegus monogyna* haws (Haws) Yks. B & H. Only used for large haws, apparently.

BULL HEAD; BULLY HEAD *Centaurea nigra* (Knapweed) both Som. both Macmillan. In this case, 'bull' is OE bol, which meant any circular body.

BULL JUMPLING *Trollius europaeus* (Globe Flower) Kinross Grigson.

BULL NETTLE 1. *Jatropha stimuloca* (Spurge Nettle) Tampion. 2. *Solanum carolinense* (Carolina Nightshade) USA Gates. 3. *Solanum eleagnifolium* (Silverleaf Nightshade) USA. Stevenson. Spurge Nettle has spiny hairs like a nettle and cause similar, but more severe, skin irritation.

BULL PINE 1. *Pinus ponderosa* (Yellow Pine) Wilkinson.1981. 2. *Pinus sabiniana* (Digger Pine) Watt.

BULL RATTLE 1. *Melandrium album* (White Campion) Bucks. Harman. 2. *Melandrium rubrum* (Red Campion) Bucks. Grigson. 3. *Silene cucubalis* (Bladder Campion) Bucks, IOW B & H.

BULL RUSH *Caltha palustris* (Marsh Marigold) Dev, Som, Wilts. Grigson.

BULL SEG(G) 1. *Orchis mascula* (Early Purple Orchis) Scot. B & H. 2. *Orchis morio* (Green-winged Orchid) Scot. B & H. 3. *Typha latifolia* (False Bulrush) Scot. Jamieson. 'Seg' means 'sedge' as far as the bulrush is concerned, but that may not be the case for the orchids. It is pointed out that in the north of Scotland to segg means to castrate, so the reference may be to the testicle-like twin tubers.

BULL THISTLE 1. *Centaurea nigra* (Knapweed) Som. Grigson. 2. *Cirsium vulgare* (Spear Plume Thistle) Som. Macmillan; Dor, N.Ire. Grigson; USA Gates. 3. Silybum marianum (Milk Thistle) USA Kingsbury. For the knapweed, bull is OE bol, any circular body. The thistles are probably corruptions of Bur Thistle.

BULL TREE *Sambucus nigra* (Elder) Cumb. Grigson. This fits into the series whose best known example is Bour-tree.

BULL'S BAGS 1. *Orchis mascula* (Early Purple Orchis) Scot. B & H. 2. *Orchis morio* (Green-winged Orchid) Scot. B & H. Bags - testicles, from the shape of the tubers. There are many variations on this theme.

BULL'S COCKS *Arum maculatum* (Cuckoo-pint) Dor. Northover. Cf. Dog Cocks. 'Cocks' has exactly the same meaning as the 'pint' of Cuckoo-pint.

BULL'S EARS *Verbascum thapsus* (Mullein) Wales Grigson. From the texture of the leaves.

BULL'S EYES 1. *Aesculus hippocastanum* fruit (Horse Chestnut) Illinois. Dorson.1964. 2. *Althaea rosea* (Hollyhock) Fernie. 3. *Caltha palustris* (Marsh Marigold) Dev, Som. Macmillan Dor. Dacombe; Scot. Friend. 4. *Hypericum perforatum* (St John's Wort) Som. Macmillan. 5. *Leucanthemum vulgare* (Ox-eye Daisy) Corn, Ches. Grigson; Som. Macmillan. 6. *Malva sylvestris* (Common Mallow) Dev. Macmillan. 7. *Melandrium rubrum* (Red Campion) Dev. Friend.1882. 8. *Nuphar lutea* (Yellow Waterlily) Som, Dor. Macmillan. 9. *Papaver rhoeas* (Red Poppy) Som. Macmillan.

BULL'S FOOT; BULLFOOT *Tussilago farfara* (Coltsfoot) Turner (Bull's Foot); Gerard (Bullfoot). The '-foot' names are from the shape of the leaves.

BULL'S HEAD *Trapa bicornuta*. Unlike *Trapa natans*, whose seed has four "horns", this one has two - hence this name.

BULL'S HORN ACACIA *Acacia spadicigera*. The name is given because of the swollen spines on the trunk and branches. The pulp attracts ants, which eat this and then live inside the hollowed-out thorns, protecting the tree to some extent against other visitants.

BULL'S PEASE *Rhinanthus crista-galli* (Yellow Rattle) Donegal Grigson.

BULLACE; BULLERS; BULLESSE *Prunus domestica var. institia* Ches. Holland (Bullers); Gerard (Bullesse). Bullace seems to be O.French buloce, beloce - words applied to the sloe.

BULLBERRY 1. *Shepherdia argentata* (Buffaloberry) USA Havard. 2. *Vaccinium myrtillus* (Whortleberry) Gerard. In the second case, it is probably bilberry in another form, but it may have arrived there from blueberry.

BULLBINE *Clematis vitalba* (Old Man's Beard) Hants, Herts. Grigson.

BULLDOCK *Arctium lappa* (Burdock) IOM Moore.1924. Presumably just a rendering of 'burdock', but it could just be a case of 'bull' meaning 'large'.

BULLDOGS 1. *Antirrhinum majus* (Snapdragon) B & H. 2. *Caltha palustris* (Marsh Marigold) Dev. Grigson.

BULLEM *Prunus spinosa* (Blackthorn) Shrop. Grigson. Cf. Bullace for *Prunus domestica var. institia,* for bullace is from O.Fr. buloce, or beloce, which meant sloe. As well as being the Shropshire name for the tree, bullems are sloes in the same county.

BULLERS *Heracleum sphondyllium* (Hogweed) Som. Elworthy. Occasionally given to this as a plant, Elworthy says, but the word is usually applied to the flowers of any of the white umbellifers. Billers is more often quoted.

BULLIES 1. *Prunus domestica var. institia* (Bullace) Lincs. Fernie; N'thants. A.E.Baker. 2. *Prunus spinosa* fruit (Sloes) Lincs. B & H.

BULLIMONG; BULLY-MUNG *Fagopyrum esculentum* (Buckwheat) Gerard (Bullimong); E Ang. Forby (Bully-mung). Bolimonge is another early name. Bullimong is an East Anglian word given to a local dish of oats, peas and vetches. But earlier, buckwheat was the main ingredient.

BULLINS 1. *Prunus domestica var. institia* fruit (Bullace) Dev. Choape. 2. *Prunus spinosa* fruit (Sloes) Shrop. B & H.

BULLIONS *Prunus domestica var. institia* fruit (Bullace) B & H.

BULLISON 1. *Prunus domestica var. institia* (Bullace) Wilts. Macmillan. 2. *Prunus spinosa* (Blackthorn) Wilts. Macmillan.

BULLISTER *Prunus spinosa* Cumb, Scot, Ire. Grigson.

BULLOCK'S EYE 1. *Sedum acre* (Biting Stonecrop) N.Eng. Friend. 2. *Sempervivum tectorum* (Houseleek) N.Eng. Best & Brightman in Albertus Magnus.

BULLOCK'S HEART *Annona reticulata* (Custard Apple) A.W.Smith.

BULLOCK'S LUNGWORT *Verbascum thapsus* (Mullein) Gerard.

BULLOCKS *Arum maculatum* (Cuckoo Pint) Som. Macmillan. Bulls in Dorset, but there are a number of variations on this - Bulls-and-cows, Bulls-and-wheys, Stallions-and-mares, etc.

BULLOE *Prunus domestica var. institia* (Bullace) B & H.

BULLPOLL; BULLPULL *Deschampsia cespitosa* (Tufted Hair-grass) Wilts. Dartnell & Goddard. Cf. Bull Faces, and Bull Pates.

BULLRUSHES 1. *Caltha palustris* (Marsh Marigold) Dev, Som. Grigson. Wilts. Dartnell & Goddard. 2. *Dipsacus fullonum* (Teasel) Som. Macmillan. Marsh Marigold - "from some nursery legend that Moses was hidden among its large leaves" (Dartnell & Goddard).

BULLS 1. *Arum maculatum* (Cuckoo Pint) Dor. Grigson. 2. *Crataegus monogyna* (Hawthorn) Norf. B & H. Bulls-and-cows would be more usual for cuckoo-pint, on the analogy of Lords-and-ladies. The stems of hawthorn are known locally as bulls.

BULLS, Cows-and- *Arum maculatum* (Cuckoo Pint) Wilts. Goddard. Another variation on the "double" names, which start off with Lords-and-ladies.

BULLS-AND-COWS *Arum maculatum* (Cuckoo Pint) Som. Macmillan; War. Morley; N'thants. A.E.Baker; Lincs. B & H; Yks. Nicholson; Lancs. Grigson.

BULLS-AND-WHEYS *Arum maculatum* (Cuckoo Pint) Cumb, Yks. B & H. A whey is a heifer.

BULLS, COWS AND CALVES *Arum maculatum* (Cuckoo Pint) Dor. Dacombe.

BULLSLOP 1. *Primula veris x vulgaris* (Oxlip) Ches. Grigson. 2. *Primula vulgaris var. elatior* (Polyanthus) Ches. Holland. *Primula veris* is cowslip, originally cowslop; oxlip is a larger hybrid of cowslip, hence the prefix bull-. Is the ascription to polyanthus correct? Holland identifies the flower as *Primula variabilis*, which would indicate polyanthus.

BULLUMS 1. *Prunus domestica var. institia* (Bullace) Corn. Jago. 2. *Prunus spinosa* (Sloe) Dev. Hewett.1892; Dor. Dacombe.

BULLUM-TREE *Prunus domestica var. institia* (Bullace) Corn. Grigson.

BULLWEED *Centaurea nigra* (Knapweed) Gerard. Bull here is OE bol, and means a circular body.

BULLWORT 1. *Ammi visnaga* (Bishop's Weed) Coles. 2. *Scrophularia aquatica* (Water Betony) Prior. Is this pool-wort?

BULLY-BLOOMS *Prunus domestica var. institia* the flowers (Bullace) Grieve.1931.

BULROSE *Narcissus pseudo-narcissus* (Daffodil) W.Miller. This is the Somerset Bell-rose, which is pronounced bulrose. W.Miller obviously considered it spelt this way, too.

BULRUSH *Scirpus lacustris*. OE bolerush, i.e. the rush with a bole, or stem(A S Palmer), but B & H reckoned it was simply bull-rush, i.e. the large rush.

BULRUSH, False *Typha latifolia*. The true Bulrush is *Scirpus lacustris*. Bul- is presumably bull-, meaning large or coarse. But Elworthy, *A west Somerset word-book*, 1888, p 585, speaks of a tradition that the word was originally pool-rush.

BULRUSH, Lesser *Typha angustifolia*.

BULRUSH, Maori *Typha angustifolia* (Lesser Bulrush) Andersen.

BULRUSH MILLET *Pennisetum glaucum* (Pearl Millet) Brouk.

BULWAND 1. *Artemisia vulgaris* (Mugwort) Scot. Jamieson. 2. *Rumex obtusifolius* (Broad-leaved Dock) Shet. Grigson.

BULWAND, Grey *Artemisia vulgaris* (Mugwort) Shet, Ork. Jamieson.

BUM-PIPE *Taraxacum officinale* (Dandelion) Banff, Lanark Grigson.

BUMBLE, Cow *Heracleum sphondyllium* (Hogweed) Som. Macmillan.

BUMBLE-BEE 1. *Ophrys apifera* (Bee Orchis) Dev, Som. Macmillan. 2. *Prunella vulgaris* (Self-heal) Yks. Grigson.

BUMBLE-BEE FLOWER *Lamium purpureum* (Red Deadnettle) Som. Macmillan.

BUMBLE-BEE ORCHID *Ophrys bombyliflora*.

BUMBLEBERRY 1. *Rosa canina* fruit (Hips) Wilts. Dartnell & Goddard. 2. *Rubus fruticosus* fruit (Blackberry) Som. Macmillan.

BUMBLEKITE; BUMLYKITE; BUMMELTYKITE *Rubus fruticosus* fruit (Blackberry) Hants, N'thum. Grigson; Yks. Morris; Dur. Denham; Cumb. Grose (all Bumblekite); Cumb B & H (Bumlykite); Cumb. Grigson (Bummeltykite). Kite, in Yorkshire dialect, means stomach, and bumble means to hum, presumably saying that the fruit does not lie easily on the stomach.

BUMMEL *Rubus fruticosus* fruit (Blackberry) Cumb. Grigson. Bummle is an old North Country dialect word meaning to blunder, according to Halliwell. What connection with this? For blackberry, this is one of the Bumblekite series.

BUMMELBERRIES *Rubus fruticosus* fruit (Blackberries) Cumb. B & H.

BUMMELKITE 1. *Rubus fruticosus* fruit (Blackberry) Yks, Berw. Grigson; Cumb. Grose; Dur. Dinsdale. 2. *Rubus saxatilis* (Stone Bramble) Yks. B & H.

BUMMULL; BUMMALKYTE *Rubus fruticosus* (Blackberry) Hants. B & H (Bummull); Yks. Jeffrey (Bummalkyte).

BUMP, Butter *Trollius europaeus* (Globeflower) Yks. Grigson.

BUMS, Black *Rubus fruticosus* fruit (Blackberry) Cumb. Gibson. A hybrid - this is a mixture of Black-kites and Bummelkites, which in turn is Brummelkites, Brummel being a local variant of bramble.

BUN *Anthriscus sylvestris* (Cow Parsley) Yks. B & H.

BUNCH-O'-DAISIES *Achillea millefolium* (Yarrow) Dor. Macmillan. Descriptive.

BUNCH-OF-GRAPES 1. *Ampelopsis quinquefolia* the buds (Virginia Creeper). Som. Macmillan. 2. *Digitalis purpurea* (Foxglove)

Som. Macmillan.
BUNCH-OF-KEYS 1. *Fraxinus excelsior* fruit (Ash keys) Som. Macmillan. 2. *Primula veris* (Cowslip) Som. Macmillan. 3. *Ulex europaeus* (Furze) Som. Macmillan.
BUNCHBERRY 1. *Cornus canadensis.* 2. *Rubus saxatilis* (Stone Bramble) Yks, Cumb. Halliwell.
BUNCHBERRY ELDER *Sambucus microbotrys.*
BUNCHFLOWER *Melanthium virginicum.*
BUND 1. *Centaurea nigra* (Knapweed) E.Ang. Forby. 2. *Succisa pratensis* (Devil's-bit) E.Ang. B & H. See Bundweed also.
BUNDWEED 1. *Centaurea nigra* (Knapweed) E.Ang. Forby. 2. *Heracleum sphondyllium* (Hogweed) Suff. Moor. 3. *Senecio jacobaea* (Ragwoer) Suff, Scot. Grigson. 4. *Succisa pratensis* (Devil's-bit) E.Ang. B & H. Possibly the same as the Scottish Bunewand, which it is said means anything a witch flies on; so long-stalked plants, and this would include ragwort, growing in out of the way places, were all candidates for the name, hogweed more likely than most.
BUNGLEBERRY *Rubus saxatilis* (Stone Bramble) Cumb. Grigson. Bungle is the same word as Bumble, which appears in various guises, as Bummelkite, bummlekite, brummelkite, and in its own right as bumblekite. These are for both Stone bramble and blackberry.
BUNIAS, Crested *Bunias erucago.*
BUNIAS, Warted *Bunias orientale* (Warty Cabbage) Polunin.
BUNKS 1. *Calystegia sepium* (Great Bindweed) Norf. B & H. 2. *Cichorium intybus* (Chicory) Norf, Suff. Forby. 3. *Conium maculatum* (Hemlock) Norf. Grigson; USA. Henkel. Probably also used for any hollow-stalked umbellifer like hemlock.
BUNNEL 1. *Cannabis sativa* the dried stalks (Hemp) Cumb. B & H. 2. *Senecio jacobaea* (Ragwort) Scot. Jamieson.
BUNNEN; BUNNLE; BUNNERTS *Heracleum sphondyllium* (Hogweed) Yks, Scot. Grigson (Bunnen); Cumb, Lanark. Grigson (Bunnle); Scot. Jamieson (Bunnerts).
BUNNY MOUTH 1. *Antirrhinum majus* (Snapdragon) Suss. B & H. 2. *Misopates orontium* (Lesser Snapdragon).
BUNNY'S EARS 1. *Stachys lanata* (Lamb's Ears) RHS. 2. *Verbascum thapsus* (Mullein) Dor. Macmillan.
BUNNY RABBIT'S EARS *Lotus corniculatus* (Bird's-foot Trefoil) Som. Macmillan.
BUNNY RABBIT'S MOUTH 1. *Antirrhinum majus* (Snapdragon) Som. Macmillan. 2. *Cymbalaria muralis* (Ivy-leaved Toadflax) Som. Macmillan. 3. *Digitalis purpurea* (Foxglove) Som. Macmillan.
BUNNY RABBITS 1. *Antirrhinum majus.* 2. *Cymbalaria muralis* (Ivy-leaved Toadflax) Som. Macmillan. 3. *Digitalis purpurea* (Foxglove) Som. Macmillan. 4. *Linaria vulgaris* (Toadflax) Som. Macmillan. 5. *Lotus corniculatus* (Bird's-foot Trefoil) Som. Grigson.
BUNWAND; BUNEWAND; BUNWEED; BUNWORT *Heracleum sphondyllium* (Hogweed) Yks, Scot. B & H (Bunwand); Scot. Jamieson (Bunewand); Suff. Moor (Bunweed); Yks, Scot. Grigson (Bunwort). See Bundweed.
BUNWEDE 1. *Convolvulus arvensis* (Field Bindweed) Scot. Jamieson. 2. *Polygonum convolvulus* (Black Bindweed) Scot. Jamieson. 3. *Senecio jacobaea* (Ragwort) Scot. Jamieson.
BUNYA-BUNYA PINE *Araucaria bidwilli.* An Australian tree, and this name has an aboriginal origin. The seeds are known as bunya-bunya fruit, too.
BUR(R) 1. *Aesculus hippocastanum* outer casing of fruit (Horse Chestnut) Dev. Friend.1882. 2. *Arctium lappa* (Burdock) Turner. 3. *Cirsium vulgare* (Spear Plume Thistle) Yks, Scot. B & H. 4. *Humulus lupulus* (Hop) Kent, Suss. B & H. 5. *Pinus sylvestris* cones (Pine cones) Scot. Dickson. Whether with one or two r's, the word seems to be of Scandinavian origin.

ODEE compares it with Danish burre, bur, burdock, Swedish Kardborre, burdock. It may depend on something like bhrs, which would also be the base of bristle.

BUR, Bathurst *Xanthium spinosum* (Spiny Cocklebur) Watt.

BUR, Beggar's *Arctium lappa* (Burdock) Vesey-Fitzgerald.

BUR, Buffalo *Solanum rostratum* (Prickly Nightshade) USA Gates.

BUR, Clod *Arctium lappa* (Burdock) Yks, Cumb. B & H. The same as Clot Bur, a name it shares with agrimony.

BUR, Clot 1. *Agrimonia eupatoria* (Agrimony) Som. Grigson. 2. *Arctium lappa* (Burdock) Yks, Cumb. B & H. The fruits of burdock are called Clots, Cluts or Clouts. Agrimony seed vessels have hooked ends to their stiff hairs, so that they cling to anyone coming into contact with the plant, and so distribute themselves in exactly the same way as burdock.

BUR, Clote; BUR, Clote, Great *Arctium lappa* (Burdock) Gerard; B & H (Great..). Clote is usually applied to the Yellow Waterlily. Perhaps it is something to do with the shape of the leaves, but it is more likely, in this case, to be the same as Clot.

BUR, Cockle *Arctium lappa* (Burdock) Sanecki. When spelt as one word, the name is applied to members of the American genus *Xanthium*.

BUR, Cockly 1. *Agrimonia eupatoria* (Agrimony) Som. Grigson. 2. *Arctium lappa* (Burdock) Cumb. B & H.

BUR, Crockelty *Arctium lappa* (Burdock) Cumb. B & H.

BUR, Ditch 1. *Arctium lappa* (Burdock) Parkinson.1640. 2. *Xanthium canadense* USA Elmore. 3. *Xanthium strumarium* (Cocklebur) W Miller.

BUR, Great *Arctium lappa* (Burdock) Gerard. As against Small Bur, which is *Arctium minus*.

BUR, Hedge *Galium aparine* (Goose-grass) Stir. B & H.

BUR, Hurr *Arctium lappa* (Burdock) Shrop. B & H; Leics. A B Evans.

BUR, Khaki *Alternanthera repens* (Khaki Weed) S Africa Dalziel.

BUR, Lesser; BUR, Small *Arctium minus* (Lesser Burdock) both Parkinson.1640.

BUR, Louse *Arctium minus* (Lesser Burdock) Parkinson.1640.

BUR, New Zealand *Acaena sp.*

BUR, Pin *Galium aparine* (Goose-grass) Beds. Grigson.

BUR, Pirri-pirri *Acaena anserinifolia*.

BUR, Sand *Solanum rostratum* (Prickly Nightshade) USA House.

BUR, Scratch *Ranunculus arvensis* (Corn Buttercup) Beds. Grigson. Spiny seed vessels, hence other descriptive names like Hedgehogs and Prickleback.

BUR, Star *Centaurea calcitrapa* (Star Thistle) S Africa Watt.

BUR, Thorny *Arctium lappa* (Burdock) Grieve.1931.

BUR CHERVIL *Anthriscus caucalis*. Rough, bristly fruit.

BUR CLOVER 1. *Medicago arabica* (Spotted Medick) Schery. 2. *Medicago polymorpha* (Hairy Medick) USA Kingsbury.

BUR CROWFOOT *Ranunculus repens* (Creeping Buttercup) Yks. Grigson.

BUR FORGET-ME-NOT *Lappula myosotis*.

BUR GHERKIN *Cucumis anguria* (Gherkin) Whitaker & Davis.

BUR MARIGOLD 1. *Bidens bipinnata* (Spanish Needles) Dalziel. 2. *Bidens tripartita*.

BUR MARIGOLD, Nodding *Bidens cernua*.

BUR MARIGOLD, Small *Bidens cernua* (Nodding Bur Marigold) USA House.

BUR MARIGOLD, Trifid; BUR MARIGOLD, Tripartite *Bidens tripartita* (Bur Marigold) Le Strange.

BUR MEDICK 1. *Medicago minima*. 2. *Medicago polymorpha* (Hairy Medick) New Zealand Connor.

BUR-NUT *Tribulus terrestris* (Puncture Vine) Moldenke.

BUR OAK 1. *Quercus lobata* (Valley Oak) Powers. 2. *Quercus macrocarpa*.

BUR PARSLEY 1. *Caucalis daucoides*. 2. *Torilis japonica* (Hedge Parsley) Salisbury.1964.
BUR PARSLEY, Corn; BUR PARSLEY, Field *Torilis arvensis* (Spreading Bur Parsley) both Salisbury.1964.
BUR PARSLEY, Great *Caucalis latifolia*.
BUR PARSLEY, Spreading *Torilis arvensis*.
BUR-REED, Branched *Sparganium erectum*. Because the flowers look like burs.
BUR-REED, Floating *Sparganium angustifolium*.
BUR-REED, Least *Sparganium minimum*.
BUR-REED, Narrowleaf *Sparganium angustifolium* (Floating Bur-reed) USA Cunningham.
BUR-REED, Small *Sparganium simplex*.
BUR ROSE *Rosa roxburghii*.
BUR THISTLE 1. *Arctium lappa* (Burdock) Scot. Guthrie. 2. *Cirsium vulgare* (Spear Plume Thistle) B & H.
BUR THRISSIL *Cirsium vulgare* (Spear Plume Thistle) Scot. Jamieson. Thrissil - Scots dialect for thistle.
BUR TREE 1. *Arctium lappa* (Burdock) N Eng. Tynan & Maitland. 2. *Sambucus nigra* (Elder) N Eng. F Grose. Not 'bur' in the case of elder, but Bour-tree, which would be pronounced bottry.
BURDOCK 1. *Arctium lappa*. 2. *Xanthium canadense* USA Elmore.
BURDOCK, Lesser *Arctium minus*.
BURDOCK, Little *Cynoglossum officinale* (Houndstongue) E Ang. Grigson.
BURDOCK, Small *Xanthium strumarium*.
BURDOCK, Wood *Arctium nemerosum*.
BURGMOTT *Monarda didyma* (Bergamot) Corn. Jago.
BURGUNDY HAY *Medicago sativa var. sativa* (Lucerne) B & H.
BURHEAD *Galium aparine* (Goose-grass) N'thants. A E Baker; Notts. Grigson.
BURK *Betula verrucosa* (Birch) Yks, Cumb, N'thum. B & H. Birk is more usual, but they are both OE birce, MLG berke, etc.
BURLEEK *Veronica beccabunga* (Brooklime) IOM Moore.1924.

BURLER'S TEASEL *Dipsacus fullonum* (Teasel) Glos. Brill. A burl is a knot in wool or cloth, so to burl is to remove these knots. Teasel is OE taesel, taesan, to tease, in the sense of pluck or pull.
BURLEY *Nasturtium officinale* (Watercress) IOM Moore.1924.
BURMA BEAN *Phaseolus lunatus* (Lima Bean) Kingsbury.
BURMESE HONEYSUCKLE, Great *Lonicera hildebrandiana*.
BURN-NOSE TREE *Daphne tinifolia* W Indies J Smith. Cf. Bonnace Tree for this.
BURN PLANT 1. *Aloe chinensis* Simmons. 2. *Aloe vera* (Mediterranean Aloe) Simmons. The juice of the leaves quickly relieves the pain from burns.
BURNET *Poterium sanguisorba* (Salad Burnet) Turner. This ie ME burnet, meaning dark brown, the same word in fact as brunette, and refers to the brownish-red colour of the flowers.
BURNET, Canadian *Sanguisorba canadensis* (American Great Burnet) USA Upham.
BURNET, Evergreen *Sanguisorba officinalis* (Great Burnet) H N Webster.
BURNET, Fodder *Poterium polygamum*. Where this occurs in Britain, it can usually be regarded as a relic of cultivation.
BURNET, Garden 1. *Poterium sanguisorba* (Salad Burnet) Clair. 2. *Sanguisorba officinalis* (Great Burnet) H N Webster.
BURNET, Great *Sanguisorba officinalis*.
BURNET, Great, American *Sanguisorba canadensis*.
BURNET, Lesser 1. *Pimpinella saxifraga* (Burnet Saxifrage) Grieve.1931. 2. *Poterium sanguisorba* (Salad Burnet) Gerard.
BURNET, Salad *Poterium sanguisorba*. 'Salad', because of the cucumbery taste of the leaves, which were used in salads, or summer drinks. "...pleasant to be eaten in sallads, in which it is thought to make the heart merry and glad..." (Gerard).
BURNET, Salad, Greater *Sanguisorba officinalis* (Great Burnet) Hutchinson.
BURNET, Salad, Prickly *Poterium polyga-*

mum (Fodder Burnet) Genders.

BURNET, Upland *Poterium sanguisorba* (Salad Burnet Tynan & Maitland.

BURNET BLOODWORT *Sanguisorba officinalis* (Great Burnet) Prior. Cf. Bloodwort. Perhaps from the colour of the flowers, this was used as a vulnerary and stauncher of blood. "Burnet is a singular good herb for wounds.. it stauncheth bleeding and therefore it was named Sanguisorba..." (Gerard). *Sanguisorba* is from Latin sanguis, blood, and sorbere, to absorb.

BURNET ROSE *Rosa pimpinellifolia*. Probably so named because of the similarity of the leaves to those of Great Burnet.

BURNET SAXIFRAGE *Pimpinella saxifraga*.

BURNET SAXIFRAGE, Greater *Pimpinella major*.

BURNING BRUSH *Euonymus atropurpureus* USA Gilmore. This must be Burning Bush, which is another American name for this shrub. Brush is often used in America where bush would be favoured in Britain.

BURNING BUSH 1. *Clematis vitalba* (Old Man's Beard) Tynan & Maitland. 2. *Dictamnus fraxinella* (False Dittany) Brownlow. 3. *Euonymmus americanus* (Strawberry Bush) USA Moldenke. 4. *Euonymus atropurpureus* USA Zenkert Kochia sp. 3, 4, and 5 are descriptive, but the name for the *Dictamnus* is virtually literal. The volatile oil is so obvious around the plant that, so it is said, the atmosphere around it will often take fire if approached by a lighted candle, without damaging the plant. Old Man's Beard is the odd one out; perhaps the fact that the dry canes were at one time smoked like tobacco accounts for the name.

BURNING CLIMBER *Clematis vitalba* (Old Man's Beard) Tynan & Maitland. See Burning Bush.

BURNING FIRE *Taraxacum officinale* (Dandelion) Som. Macmillan. Probably connected with a sun dedication. Golden Sun is another name, but more important is its connection with St Bridget, as the spring goddess. Dandelion is her flower, as the sun. In Gaelic it is little flame of God.

BURRAGE, BURRIDGE *Borago officinalis* (Borage) Aubrey.1686 (Burrage); Dev. Friend (Burridge).

BURREN MYRTLE *Arctostaphylos uva-ursi* (Bearberry) Ire. Grigson. Because it grows on the Burren limestone.

BURRO-WEED STRANGLER *Orobanche ludoviciana var. cooperi*.

BURRO'S TAIL; BURROW'S TAIL *Sedum morganianum*. Burrow's Tail is probably a misprint. It is in Cunningham, and there are more than enough misprints there.

BURROW-WEED *Haplopappus heterophyllus* (Rayless Golden Rod) USA Kingsbury.

BURROWING CLOVER *Trifolium subterraneum*. Descriptive of its habit.

BURRY VERVAIN *Verbena lappulacea*.

BURST-BELLIES; BURST-BELLY PINK; BUSTERS *Dianthus plumarius* (Pink) Som. Macmillan. Presumably a reference to the way the flowers burst from the tight buds. The name is probably applied to carnations, too.

BURSTWORT *Herniaria ciliata* (Rupture-wort) Gerard. An obvious development from Rupture-wort, but it was probably an earlier name.

BURWEED 1. *Franseria acanthicarpa*. 2. *Galium aparine* (Goose-grass) Bucks, Herts, Notts. Grigson; N'thants. A.E.Baker. 3. *Xanthium spinosum* (Cocklebur) S.Africa Watt. 4. *Xanthium strumarium* (Cocklebur) Chopra.

BURWEED, Spiny; BURWEED, Thorny *Xanthium spinosum* (Cocklebur) Australia Watt (Spiny); S.Africa Watt (Thorny).

BURYING BOX *Buxus sempervirens* (Box) Lancs. Vaux. Evergreen, so a symbol of life everlasting, hence the connection with funerals. In Lancashire, a basin full of box sprigs was often put at the door of a house before a funeral. Everyone who attended was expected to take a sprig, to carry it in the procession and then to throw it into the grave.

BUSCH; BUSH TREE *Buxus sempervirens*

(Box) both Scot. Jamieson (Busch); B & H (Bush Tree). From the Latin buxus, like box.

BUSH, Sweet *Clethra alnifolia* (Sweet Pepper Bush) McDonald.

BUSH BEAN *Phaseolus vulgaris* (Kidney Bean) S.Africa Watt; USA Weiner.

BUSH COFFEE *Coffea stenophylla* (Narrow-leaved Coffee) Dalziel.

BUSH GREENS *Amaranthus caudatus* (Love-lies-bleeding) Sierra Leone Dalziel.

BUSH LAWYER 1. *Rubus australis* New Zealand Goldie. 2. *Rubus cissoides*. Like blackberry in England, where such names as Lawyers, County Lawyers, occur, it is given because once they have hold of you, they never let go.

BUSH OAK *Chlorophora excelsa* (African Oak) Dalziel. Not a comment on its stature, but on the kind of terrain in which it grows.

BUSH OKRA *Corchorus olitorius* (Long-fruited Jute) Dalziel.

BUSH PEPPER *Piper guineense* (West African Black Pepper) Dalziel.

BUSH YAM *Dioscorea praehensilis*.

BUSHMAN'S CANDLE *Sarcocaulon burmannii*.

BUSSELL, Yellow *Chrysanthemum segetum* (Corn Marigold) Berks. Grigson. This seems to be the local variant of a whole series, probably starting with Boodle, and going through Botham to Bozzom.

BUSY LIZZIE *Impatiens walleriana*.

BUTCHER 1. *Dactylorchis maculata* (Heath Spotted Orchid) Som. Macmillan. 2. *Melandrium album* (White Campion) Glos. Grigson. 3. *Orchis mascula* (Early Purple Orchis) Som. Macmillan; Heref. Leather. The orchids have dark red blotches on the leaves, looking as if someone with bloody fingers had touched them - hence this name, and also Bloody Man's Fingers, Bloody Butchers, etc. White Campion seems to have no connection with the idea of 'butcher' - one wonders if it is a mistake for Red Campion; the colour would then make sense. Apparently not, though, for Red Campion is specifically called Red Butcher.

BUTCHER, Bloody 1. *Centranthus ruber* (Spur Valerian) Som. Macmillan. 2. *Orchis mascula* (Early Purple Orchis) Heref. Havergal; Oxf. Oxf.Annl. Rec.1951; N'thants. B & H. See Butcher.

BUTCHER, Blue 1. *Ophrys apifera* (Bee Orchis) Som. Macmillan. 2. *Orchis mascula* (Early Purple Orchis) Clapham. Bee Orchis is probably an error.

BUTCHER, Red 1. *Melandrium rubrum* (Red Campion) Glos. Macmillan; Ches. Holland. 2. *Orchis mascula* (Early Purple Orchis) Kent B & H. See Butcher.

BUTCHER BOY 1. *Anacamptis pyramidalis* (Pyramidal Orchid) Som. Macmillan. 2. *Orchis mascula* (Early Purple Orchis) Som. Macmillan.

BUTCHER FLOWER *Orchis mascula* (Early Purple Orchis) Som. Macmillan.

BUTCHER'S BLOOD *Melandrium rubrum* (Red Campion) Som. Macmillan.

BUTCHER'S BROOM *Ruscus aculeatus*. It is said that the name comes from the fact that once butchers used to sweep their blocks with hard brooms made of the green shoots of this. In Italy, they were commonly used as household besoms.

BUTCHER'S PRICK-TREE 1. *Euonymus europaeus* (Spindle Tree) B & H. 2. *Frangula alnus* (Alder Buckthorn) B & H. In both cases, skewers were made from the wood, though Spindle was thought the best wood. "The butchers doe make skewers of it, because it doth not tainte the meate as other wood will doe..." (Aubrey, Nat. Hist. Wilts.) See other names for the Spindle, like Skiver, Skiverwood, Prick-timber, etc.

BUTCHER'S PRICKWOOD *Frangula alnus* (Alder Buckthorn) Prior. Again, because butchers' skewers were made from the wood.

BUTTER *Ranunculus ficaria* (Lesser Celandine) Som. Macmillan.

BUTTER, Bread-and- 1. *Cardamine pratensis* (Lady's Smock) Tynan & Maitland. 2. *Linaria vulgaris* (Toadflax) Som. Macmillan. 3. *Potentilla anserina*

(Silverweed) Som. Macmillan. Bread-and-cheese, too. Certainly, the roots of silverweed were used as a marginal or famine food in the Highlands and Ireland, roasted or boiled. But the names, and in Somerset, too, suggest the casual eating of the leaves, as with hawthorn.

BUTTER, Cheese-and- *Ranunculus ficaria* (Lesser Celandine) Som. Macmillan. This seems to be a mixture of the Somerset Cheesecups and the Devonshire Cream-and-butter.

BUTTER, Cream-and- *Ranunculus ficaria* (Lesser Celandine) Dev. Macmillan.

BUTTER, Eggs-and- 1. *Ranunculus acris* (Meadow Buttercup) Ches. B & H. 2. *Ranunculus bulbosus* (Bulbous Buttercup) Ches. B & H. 3. *Linaria vulgaris* (Toadflax) Dev, Som, Wilts. Macmillan. 4. *Narcissus maialis var. patellaris* (Pheasant's Eye) Som. Tongue.1968. It is a common children's game to hold buttercups under the chin to see if they like butter.

BUTTER, Golden *Ranunculus flammula* (Lesser Spearwort) Dor. Macmillan.

BUTTER, Midshipman's; BUTTER, Subaltern's *Persea gratissima* (Avocado Pear) Grigson.1974.

BUTTER-AND-BREAD *Crataegus monogyna* (Hawthorn) Yks. B & H. This is but one variation of the "Bread-and-cheese" name given by children when they eat the new leaves in spring.

BUTTER-AND-CHEESE 1. *Malva sylvestris* (Common Mallow) Dor, Som. Macmillan. 2. *Oxalis acetosella* (Wood Sorrel) Som. Macmillan. 3. *Ranunculus acris* (Meadow Buttercup) Som. Grigson. 4. *Ranunculus bulbosus* (Bulbous Buttercup) Som. Grigson. 5. *Ranunculus ficaria* (Lesser Celandine) Som. Macmillan. 6. *Ranunculus repens* (Creeping Buttercup) Som. Macmillan.

BUTTER-AND-EGGS 1. *Iris pseudo-acarus* (Yellow Iris) Oxf, Bucks, N'thants. B & H. 2. *Leucanthemum vulgare* (Ox-eye Daisy) Tynan & Maitland. 3. *Leucojum vernum* (Spring Snowflake) Dor. B & H. 4. *Linaria vulgaris* (Toadflax) Dev. Friend; Dor. Barnes; Wilts. Dartnell & Goddard; Glos. J.D. Robertson; Kansas Sackett & Koch. 5. *Lotus corniculatus* (Bird's-foot Trefoil) Som, Glos, War. Grigson; Suss. Parish. 6. *Narcissus incomparabilis* Wilts. Dartnell & Goddard; Dev, Surr, Lancs. B & H. 7. *Narcissus maialis var. patellaris* (Pheasant's Eye) Corn. Quiller-Couch; Dev. Friend. 8. *Narcissus pseudo-narcissus* (Daffodil) Som. Elworthy. 9. *Oxalis acetosella* (Wood Sorrel) Som. Macmillan.

BUTTER-AND-SUGAR *Linaria vulgaris* (Toadflax) Wilts. Macmillan.

BUTTER BASKET *Trollius europaeus* (Globe Flower) Yks. Grigson.

BUTTER BEAN *Phaseolus lunatus* (Lima Bean) Schery.

BUTTER BEAN, Indian *Dolichos lablab* (Lablab) Dalziel.

BUTTER-BLOB 1. *Caltha palustris* (Marsh Marigold) Yks. Grigson. 2. *Trollius europaeus* (Globe Flower) Scot. Tynan & Maitland.

BUTTER BUMP *Trollius europaeus* (Globe-flower) Yks. Grigson.

BUTTER BUSH *Cephalanthus occidentalis* (Butterbrush) USA Cunningham.

BUTTER CHURN 1. *Nuphar lutea* (Yellow Waterlily) Mor. Grigson. 2. *Ranunculus acris* (Meadow Buttercup) War. Grigson. 3. *Ranunculus bulbosus* (Bulbous Buttercup) War. Grigson. 4. *Ranunculus repens* (Creeping Buttercup) War. Grigson. Waterlily is Churn, too, in Oxfordshire.

BUTTER CRESS *Ranunculus acris* (Meadow Buttercup) Bucks. B & H.

BUTTER DAISY 1. *Leucanthemum vulgare* (Ox-eye Daisy) Dev, Dor. Macmillan. 2. *Ranunculus acris* (Meadow Buttercup) Bucks. B & H. 3. *Ranunculus bulbosus* (Bulbous Buttercup) Bucks. Grigson. 4. *Ranunculus repens* (Creeping Buttercup) Bucks. Grigson.

BUTTER DOCK 1. *Arctium lappa* (Burdock) Corn. Jago. 2. *Petasites hybridus* (Butterbur) Flück. 3. *Rumex alpinus*

(Monk's Rhubarb) Grigson. 4. *Rumex longifolius*. 5. *Rumex obtusifolius* (Broad-leaved Dock) Shrop. B & H. Ches. Holland. The docks got this name because the leaves were once used to wrap butter. The Scottish equivalent was Smair-dock. Butter Dock for *R.obtusifolius* became Bitter Dock when the name travelled to America. Presumably burdock leaves were used in a similar way, and certainly butterbur leaves were.

BUTTER DOCKEN; BUTTER LEAVES *Rumex alpinus* (Monk's Rhubarb) both Cumb. Grigson. As for the previous entry, the name must have been given because the leaves were used to wrap butter. Docken is merely the old plural of dock.

BUTTER FINGERS *Anthyllis vulneraria* (Kidney Vetch) Som. Grigson. A combination of colour- and shape-descriptive names. They are known as Lady's Fingers, or Fingers-and-thumbs, with a number of variations.

BUTTER FLOWER; BUTTERFLOWER 1. *Caltha palustris* (Marsh Marigold) Wilts. Dartnell & Goddard. 2. *Lotus corniculatus* (Bird's-foot Trefoil) Ork. Leask. 3. *Ranunculus acris* (Meadow Buttercup) Gerard. 4. *Ranunculus bulbosus* (Bulbous Buttercup) Dev. Friend. Som, Herts, Derb. Grigson. 5. *Ranunculus repens* (Creeping Buttercup) Gerard.

BUTTER HAW *Crataegus monogyna* fruit (Haws) Oxf, Norf. Grigson.

BUTTER JAGS *Lotus corniculatus* (Bird's-foot Trefoil) N.Eng. Prior. Prior suggests that this might once have been Buttered-eggs. But jags is an old North-country word for rags, and this is possible.

BUTTER LEAVES 1. *Atriplex hortensis* (Garden Orache) Glos. B & H. 2. *Rumex alpinus* (Monk's Rhubarb) Cumb. Grigson. Both because the leaves were used once to wrap butter.

BUTTER PAT *Viola canina* fruit (Heath Dog Violet) Lancs. B & H.

BUTTER PLANT; BUTTER ROOT *Pinguicula vulgaris* (Butterwort) Selk. Grigson (Plant); Gerard (Root). It is suggested that butterwort got the name because of the greasy feel of the leaves. The plant is carnivorous, and the upper surface of the leaves secretes a sticky fluid, which effectively catches flies, etc. Once caught, the leaves fold over to secure the prey. And then the leaf secretes an acid fluid which can dissolve and digest the prey.

BUTTER ROSE 1. *Primula veris* (Cowslip) Dev. Friend. 2. *Primula vulgaris* (Primrose) Dev. Fernie. 3. *Ranunculus acris* (Meadow Buttercup) Dev. Friend. 4. *Ranunculus bulbosus* (Bulbous Buttercup) Dev. Grigson. 5. *Ranunculus repens* (Creeping Buttercup) Dev. Grigson.

BUTTER TREE 1. *Bassia latifolia* (Mahua) Chopra. 2. *Crassula paniculata*. 3. *Pentadesma butyracea* Dalziel. In the last case, a reference to the fat that is extracted from the seeds.

BUTTER WINTER *Chimaphila umbellata* (Umbellate Wintergreen) Le Strange.

BUTTERBLEB *Caltha palustris* (Marsh Marigold) Yks. Grigson. Butter - the colour of the flowers; blob (bleb in this case) means blister (most of the Ranunculaceae have acrid juice, some of them have juice acrid enough to raise a blister on the skin of the unwary).

BUTTERBUMPS 1. *Ranunculus acris* (Meadow Buttercup) Yks. Grigson. 2. *Ranunculus bulbosus* (Bulbous Buttercup) Yks. Grigson. 3. *Ranunculus repens* (Creeping Buttercup) Yks. Grigson. 4. *Trollius europaeus* (Globe Flower) Yks. Grigson.

BUTTERBUR 1. *Petasites hybridus*. 2. *Xanthium canadense* USA Elmore. Butterbur, "because the countrey huswives were wont to wrap their butter in the large leaves" (Coles). It has no burs, but is named from its broad resemblance to burdock.

BUTTERBUR, Creamy *Petasites japonicus*.
BUTTERBUR, White *Petasites albus*.
BUTTERBUR COLTSFOOT *Petasites hybridus* (Butterbur) Prior. Presumably, like coltsfoot, the flowers appear before the

leaves.

BUTTERCHOPS *Ranunculus ficaria* (Lesser Celandine) Som. Macmillan. Is the "-chops" part of the name actually "-cups"?

BUTTERCREESE 1. *Ranunculus bulbosus* (Bulbous Buttercup) Bucks. Grigson. 2. *Ranunculus repens* (Creeping Buttercup) Bucks. Grigson.

BUTTERCUP 1. *Caltha palustris* (Marsh Marigold) Dev, Som. Grigson. 2. *Potentilla anserina* (Silverweed) Bucks. B & H. 3. *Ranunculus sp.* ODEE suggests buttercup is a fusion of Butter-flower with Goldcup, or Kingcup. Buttercup is the usual name for most of the Ranunculus species; even Lesser Celandine (*R.ficaria*) gets the name in Wiltshire, Gloucestershire, and in Yorkshire.

BUTTERCUP, Badgeworth *Ranunculus ophioglossifolius* (Serpent's-tongue Spearwort). It is referred to as such in a news item in The Times of 6 June 1994. Badgeworth is a village a mile or so southwest of Cheltenham.

BUTTERCUP, Bermuda *Oxalis pes-caprae*. Butter-yellow flowers.

BUTTERCUP, Big *Caltha palustris* (Marsh Marigold) Som. Grigson.

BUTTERCUP, Bulbous *Ranunculus bulbosus*.

BUTTERCUP, Bull *Caltha palustris* (Marsh Marigold) Ess. Grigson. 'Bull' usually means large, or even coarse, as does 'horse' or 'dog'.

BUTTERCUP, Californian *Ranunculus californicus*.

BUTTERCUP, Celery-leaved *Ranunculus sceleratus*.

BUTTERCUP, Common *Ranunculus acris* (Meadow Buttercup) North.

BUTTERCUP, Corn *Ranunculus arvensis*.

BUTTERCUP, Creeping *Ranunculus repens*.

BUTTERCUP, Devil's *Ranunculus sceleratus* (Celery-leaved Buttercup) Tynan & Maitland. Presumably because it is the most poisonous of all the buttercups.

BUTTERCUP, Fan-leaved *Ranunculus flabellatus* (Jersey Buttercup) Clapham.

BUTTERCUP, Fern *Potentilla anserina* (Silverweed) Wilts. Dartnell & Goddard. An apt descriptive name.

BUTTERCUP, Field, Tall *Ranunculus acris* (Meadow Buttercup) USA Kingsbury.

BUTTERCUP, Giant *Ranunculus acris* (Meadow Buttercup) New Zealand Connor.

BUTTERCUP, Golden *Caltha palustris* (Marsh Marigold) Som. Macmillan.

BUTTERCUP, Grass-leaved *Ranunculus gramineus*.

BUTTERCUP, Grassland *Ranunculus lappaceus*.

BUTTERCUP, Hairy *Ranunculus sardous*.

BUTTERCUP, Hispid *Ranunculus hispidus*.

BUTTERCUP, Horse *Caltha palustris* (Marsh Marigold) Dev, Som. Elworthy. Cf. Bull Buttercup. Both mean usually large, or even coarse.

BUTTERCUP, Jersey *Ranunculus flabellatus*. Confined to Jersey in GB.

BUTTERCUP, Marsh *Ranunculus septentrionalis* (Swamp Buttercup) USA House.

BUTTERCUP, Meadow *Ranunculus acris*.

BUTTERCUP, Pyrenean 1. *Ranunculus amplexicaulis*. 2. *Ranunculus pyrenaeus*. Are these the same plant?

BUTTERCUP, Scented *Potentilla anserina* (Silverweed) Som. Macmillan.

BUTTERCUP, Scilly *Ranunculus muricatus*. Naturalised, and a pest in bulb fields in Scilly and Cornwall.

BUTTERCUP, Sequier's *Ranunculus seguieri*.

BUTTERCUP, Small-flowered *Ranunculus parviflorus*.

BUTTERCUP, Snow *Ranunculus glacialis* (Glacier Crowfoot) Gilmour & Walters.

BUTTERCUP, Spiny-fruited *Ranunculus muricatus*.

BUTTERCUP, Swamp *Ranunculus septentrionalis*.

BUTTERCUP, Tall *Ranunculus acris* (Meadow Buttercup) USA Allan.

BUTTERCUP, Thora; BUTTERCUP, Thore's *Ranunculus thora*.

BUTTERCUP, Water 1. *Caltha palustris*

(Marsh Marigold) Dev. Friend.1882; Oxf, Suss. Grigson. 2. *Ranunculus flammula* (Lesser Spearwort) Wilts. Dartnell & Goddard.
BUTTERCUP, Water, Ivy-leaved *Ranunculus hederaceus.*
BUTTERCUP, White 1. *Parnassia palustris* (Grass of Parnassus) Border B & H. 2. *Ranunculus aconitifolius.*
BUTTERCUP, White, Large *Ranunculus platanifolius.*
BUTTERCUP, Wild *Ranunculus multifidus* Kenya Watt.
BUTTERCUP, Woolly *Ranunculus lanuginosus.*
BUTTERCUP CROWFOOT *Ranunculus acris* (Meadow Buttercup) Phillips.
BUTTERCUP TREE *Cochlospermum religiosum* (White Silk Cotton). Yellow flowers.
BUTTERED EGGS 1. *Chrysosplenium oppositifolium* (Opposite-leaved Golden Saxifrage) Wilts. Macmillan. 2. *Linaria vulgaris* (Toadflax) Dev. Friend; Wilts. Dartnell and Goddard; Kansas Sackett & Koch. 3. *Lotus corniculatus* (Bird's-foot Trefoil) Som. Macmillan; Cumb. Grigson. 4. *Narcissus incomparabilis* Wilts. Dartnell & Goddard; Dev, Surr, Lancs. B & H.
BUTTERED FINGERS *Anthyllis vulneraria* (Kidney Vetch) Som. Macmillan.
BUTTERED HAYCOCKS *Linaria vulgaris* (Toadflax) Yks. Grigson. A haycock is a conical heap of hay; buttered refers to the colour of the flowers. The whole is quite a good descriptive name.
BUTTERFLY 1. *Acer pseudo-platanus* seeds (Sycamore keys) Som. Macmillan. 2. *Lathyrus odoratus* (Sweet Pea) Som. Macmillan.
BUTTERFLY, Blue *Delphinium ajacis* (Larkspur) Som. Macmillan.
BUTTERFLY BUSH 1. *Buddleia davidii* A.W.Smith. 2. *Clerodendron myricoides.* Because the flowers attract butterflies.
BUTTERFLY BUSH, Blue *Clerodendron ugandense.*
BUTTERFLY FLOWER 1. *Schizanthus sp.* 2. *Viola tricolor* (Pansy) Conway.
BUTTERFLY IRIS *Iris spuria.*
BUTTERFLY LADIES *Papaver rhoeas* (Red Poppy) Dor. Macmillan.
BUTTERFLY MILKWEED; BUTTERFLY WEED *Asclepias tuberosa* (Orange Milkweed) Bergen (Milkweed); Tolstead (Weed).
BUTTERFLY ORCHID *Platanthera chlorantha.*
BUTTERFLY ORCHID, Pink *Orchis papilionacea.*
BUTTERFLY PEA *Clitoria ternatea.*
BUTTERFLY VIOLA *Viola cornuta.*
BUTTERNUT 1. *Carya cordiformis* (Swamp Hickory) USA Yarnell. 2. *Juglans cinerea* (White Walnut) USA Tolstead. Another name for the walnut, Oilnut, probably gives sufficient explanation.
BUTTERNUT, Shea *Butyrospermum parkii.*
BUTTERPLATE *Ranunculus flammula* (Lesser Spearwort) Berw. Grigson.
BUTTERPRINT *Abutilon avicennae* (China Jute) USA Bergen. Because the pods were used once to stamp butter, or pie-crusts - the China Jute pods were called Pie-print, or Pie-marker in the States, too.
BUTTERPUMP *Nuphar lutea* (Yellow Waterlily) Dor. Macmillan. Butter churn, perhaps?
BUTTERWEED *Erigeron canedensis* (Canada Fleabane) B & H; USA Henkel. Bitterweed also, so it seems likely that Butterweed is merely a version of that.
BUTTERWORT 1. *Pinguicula vulgaris.* 2. *Sanicula europaea* (Sanicle) Palaiseul. *Pinguicula vulgaris* is a carnivorous plant. The upper surface of the leaves secretes a sticky fluid. Insects, seeds, etc., becoming attached to the fluid, set up an irritation of the leaf, which slowly folds over and secures the objects. The leaf then secretes an acid fluid which has the power of rapidly dissolving and digesting the offending substances. It is the greasy feel of the leaves which gives the name Butterwort.
BUTTERWORT, Alpine *Pinguicula alpina.*

BUTTERWORT, Great *Pinguicula grandiflora.*
BUTTERWORT, Irish *Pinguicula grandiflora* (Great Butterwort) W.Miller.
BUTTERWORT, Pale *Pinguicula lusitanica* (Western Butterwort) McClintock.
BUTTERWORT, Western *Pinguicula lusitanica.*
BUTTERY EGGS *Narcissus odorus* (Jonquil) Wilts. Macmillan.
BUTTERY-ENTRY *Viola tricolor* (Pansy) Derb. B & H. A contraction of the extraordinary, but cumbersome, name Meet-her-in-the-entry-kiss-her-in-the-buttery, if that is the correct order.
BUTTON SNAKEROOT 1. *Eryngium aquaticum.* 2. *Eryngium yuccifolium.* 3. *Liatris spicata.*
BUTTON-SOOREE *Arctium lappa* seeds (Burs) IOM Moore.1924. Literally, 'courting button' - "holding to her like a button-sooree".
BUTTONBRUSH *Cephalanthus occidentalis.*
BUTTONHOLE FLOWER *Euphorbia marginata* (Mountain Snow) Coats.
BUTTONHOLE PLANT *Claytonia perfoliata* (Spring Beauty) McClintock.
BUTTONHOLE ROSE, D'Orsay's *Rosa virginiana pl* Sitwell.
BUTTONS 1. *Arctium lappa* burrs (Burrs) Som. Elworthy; Cumb. B & H. 2. *Dipsacus fullonum* (Teasel) Som. Macmillan. 3. *Galium aparine* burrs (Goose-grass) Wessex Rogers. 4. *Malva moschata* (Musk Mallow) Som. Macmillan. 5. *Tanacetum parthenium* (Feverfew) Som. Elworthy. 6. *Tanacetum vulgare* (Tansy) Yks. Vesey-Fitzgerald. Burrs, Feverfew and Tansy own Buttons as good descriptive names. Musk Mallow seeds presumably are the reason, but Buttons for Teasel is far from obvious.
BUTTONS, Bachelor's 1. *Achillea ptarmica* (Sneezewort) N'thants. A.E.Baker. 2. *Agrostemma githago* (Corn Cockle) B & H. 3. *Aquilegia vulgaris* (Columbine) Wilts. Dartnell & Goddard. 4. *Arctium lappa* burrs (Burrs) Dev. Friend.1882. 5. *Bellis perennis* (Daisy) B & H. 6. *Caltha palustris* (Marsh Marigold) Som. Macmillan; Dor. Dacombe; Staffs. Raven. 7. *Centaurea cyanus* (Cornflower) Som. Macmillan; Derb, Yks. Grigson; USA Watt. 8. *Centaurea nigra* (Knapweed) Ire. Grigson. 9. *Centaurea scabiosa* (Greater Knapweed) Glos. B & H. 10. *Cephalanthus occidentalis* (Buttonbrush) Dev. Friend.1882. 11. *Geranium lucidum* (Shining Cranesbill) Lancs. Grigson. 12. *Geranium robertianum* (Herb Robert) Dev. Friend.1882; Som. Grigson. 13. *Gomphrenia globosa* (Globe Amaranth) S.Africa Watt. 14. *Kerria japonica* (Jew's Mallow) Som. Macmillan. 15. *Knautia arvensis* (Field Scabious) Dev, Som. Grigson; Wilts. Akerman; Glos. J.D.Robertson; Berks. Lowsley. 16. *Lychnis flos-cuculi* (Ragged Robin) Dev, Som, Suss. Grigson. 17. *Melandrium album* (White Campion) Glos. Grigson; Suss, Yks. B & H; Suff. Moor. 18. *Melandrium rubrum* (Red Campion) Dev, Som, Worcs, War, Kent, Ess, Suss, Suff, Lancs. Grigson; N'thants. A.E. Baker; Yks. Addy. 19. *Ranunculus aconitifolius fl.pl.* Gerard. 20. *Ranunculus acris* (Meadow Buttercup) Som. Grigson. 21. *Ranunculus acris fl.pl.* Dev. Friend.1882. 22. *Ranunculus bulbosus* (Bulbous Buttercup) Som. Grigson. 23. *Ranunculus repens* (Creeping Buttercup) Som. Grigson. 24. *Scabiosa columbaria* (Small Scabious) Wilts. Dartnell & Goddard. 25. *Stellaria holostea* (Greater Stitchwort) Som. Macmillan. Bucks, Suss. Grigson. 26. *Succisa pratensis* (Devil's-bit) Glos. B & H; Hants. Cope. 27. *Tanacetum parthenium* (Feverfew) Dev. Friend.1882. 28. *Tanacetum vulgare* (Tansy) Som. Macmillan. 29. *Umbilicus rupestris* (Wall Pennywort) Dev. Friend.1882. 30. *Vinca minor* (Lesser Periwinkle) Som. Macmillan. Are they some divination flowers? Or roughly button-shaped?
BUTTONS, Bachelor's, Billy *Stellaria holostea* (Greater Stitchwort) War. Grigson.
BUTTONS, Bachelor's, Little *Geranium*

robertianum (Herb Robert) Dev. Friend; Suss. Tynan & Maitland.

BUTTONS, Bachelor's, White 1. *Melandrium album* (White Campion) Suff. B & H. 2. *Ranunculus aconitifolius fl pl* Ayr. B & H.

BUTTONS, Bachelor's, Wild *Polygala lutea* (Orange Milkwort) USA House.

BUTTONS, Bachelor's, Yellow 1. *Ranunculus acris fl pl* Ayr B & H. 2. *Polygala lutea* (Orange Milkweed) USA Howes.

BUTTONS, Beggar's *Arctium lappa* (Burdock) Dev. B & H; USA Henkel; or, confined to the burrs - Dev. Halliwell.

BUTTONS, Bel *Caltha palustris* (Marsh Marigold) Tynan & Maitland.

BUTTONS, Bill 1. *Geum rivale* (Water Avens) Wilts. Dartnell & Goddard. 2. *Vinca minor* (Lesser Periwinkle) Tynan & Maitland. Billy Buttons for Water Avens, too, in Wiltshire - but this applies to a lot of other plants as well.

BUTTONS, Billy 1. *Agrostemma githago* (Corn Cockle) B & H. 2. *Althaea rosea* (Hollyhock) Som. Macmillan. 3. *Arctium lappa* burrs (Burrs) Dev. Friend.1882. 4. *Bellis perennis* (Daisy) Som, Wilts. Macmillan; Shrop. Grigson. 5. *Caltha palustris* (Marsh Marigold) Som. Macmillan. 6. *Leucanthemum vulgare* (Ox-eye Daisy) Shrop. Grigson. 7. *Geranium robertianum* (Herb Robert) Bucks. Grigson. 8. *Geum rivale* (Water Avens) Wilts. Grigson. 9. *Knautia arvensis* (Field Scabious) Som. Macmillan; Yks. B & H. 10. *Lychnis flos-cuculi* (Ragged Robin) War. Grigson. 11. *Malva neglecta* (Dwarf Mallow) Surr. Tynan & Maitland. 12. *Malva sylvestris* (Common Mallow) Som. Macmillan. 13. *Melandrium album* (White Campion) Som, Glos, War. Grigson. 14. *Melandrium rubrum* (Red Campion) War, Ess. Grigson. 15. *Ranunculus acris* (Meadow Buttercup) Som. Macmillan. 16. *Saxifraga granulata fl pl* B & H. 17. *Stellaria holostea* (Greater Stitchwort) Som. Macmillan. War.

Grigson. 18. *Vinca minor* (Lesser Periwinkle) Tynan & Maitland.

BUTTONS, Billy's *Geum rivale* (Water Avens) Wilts. Macmillan.

BUTTONS, Billy-o'- *Caltha palustris* (Marsh Marigold) Som. Macmillan. Plain Billy Buttons is more usual.

BUTTONS, Billy White's *Stellaria holostea* (Greater Stitchwort) War. Grigson.

BUTTONS, Bitter *Tanacetum vulgare* (Tansy) Mor. Grigson; USA Henkel.

BUTTONS, Blue 1. *Centaurea cyanus* (Cornflower) Som. Macmillan. 2. *Geranium pratense* (Meadow Cranesbill) Som. Macmillan. 3. *Jasione montana* (Sheep's-bit) Som, Dor, Cumb. Grigson. 4. *Knautia arvensis* (Field Scabious) Dor. Dacombe; Wilts. Dartnell & Goddard. 5. *Scabiosa columbaria* (Small Scabious) Wilts. Dartnell & Goddard. 6. *Succisa pratensis* (Devil's-bit) Som. Macmillan; Ches. Holland. 7. *Vinca major* (Greater Periwinkle) Dev. B & H. 8. *Vinca minor* (Lesser Periwinkle) Dev. Friend.1882.

BUTTONS, Bobby's 1. *Arctium lappa* burrs (Burrs) Som. Macmillan. 2. *Caltha palustris* (Marsh Marigold) Dev, Som. Macmillan. 3. *Centaurea cyanus* (Cornflower) Dev, Som. Macmillan. 4. *Centaurea nigra* (Knapweed) Som. Macmillan. 5. *Galium aparine* (Goose-grass) Som. Macmillan.

BUTTONS, Boot *Ligustrum vulgare* - berries (Privet) Som. Macmillan.

BUTTONS, Brass *Cotula coronopifolia* (Buttonweed) Polunin.

BUTTONS, Clitch 1. *Arctium lappa* (Burdock) Dev. Friend. 2. *Galium aparine* (Goose-grass) Dev. B & H. Clitch is a Devonshire dialect word meaning to stick, or to adhere to. Cf. Clutch Buttons in the same county for goose-grass.

BUTTONS, Clutch *Galium aparine* (Goose-grass) Dev. Friend. Cf. Clitch-buttons.

BUTTONS, Coachman's *Knautia arvensis* (Field Scabious) Som. Macmillan.

BUTTONS, Cockle *Arctium lappa* (Burdock) Corn. Courtney.1880; Dev. Friend.1882;

USA Henkel. This is probably originally Latin coccum, the word used for the insect kermes. The burrs cling like insects. But the word cockle became cuckle, cuckold, and cuckoo.

BUTTONS, Cuckle; BUTTONS, Cuckold('s); BUTTONS, Cuckoldy *Arctium lappa* (Burdock) Dev. Friend.1882 (Cuckle); Dev. Choape (Cuckold); Cumb. B & H (Cuckoldy). Probably all starting with Cockle Buttons, which see. Cuckoo Buttons is the next stage, but this is shared with two other plants.

BUTTONS, Cuckoo 1. *Arctium lappa* burrs (Burdock) Som. Elworthy. 2. *Arctium minus* (Lesser Burdock) Dev. B & H. 3. *Cirsium vulgare* (Spear Plume Thistle) Som. Grigson. The original point of departure would have been Cockle Buttons, then through Cuckle, Cuckold's, Cuckoldy, to this.

BUTTONS, Devil's 1. *Geum rivale* (Water Avens) Tynan & Maitland. 2. *Knautia arvensis* (Field Scabious) Corn. Deane & Shaw. 3. *Succisa pratensis* (Devil's-bit) Corn Courtney.1880. Field Scabious is probably a mistake for the Devil's-bit.

BUTTONS, Fluffy *Salix caprea* (Goat Willow) Som. Macmillan.

BUTTONS, Gentleman's *Succisa pratensis* (Devil's-bit) Shrop. B & H.

BUTTONS, Golden *Tanacetum vulgare* (Tansy) Dev. Grigson.

BUTTONS, Gookoo *Cardamine pratensis* (Lady's Smock) Som. B & H. One of a number of 'cuckoo' names for this plant.

BUTTONS, Grandfather's *Caltha palustris* (Marsh Marigold) Som. Macmillan.

BUTTONS, Horse *Malva sylvestris* (Common Mallow) Donegal Grigson.

BUTTONS, Lady's *Stellaria holostea* (Greater Stitchwort) Som. Macmillan.

BUTTONS, Linen *Convallaria maialis* (Lily-of-the-valley) Som. Macmillan.

BUTTONS, Musk *Aster tripolium* (Sea Aster) Glos. Grigson.

BUTTONS, Old Man's 1. *Arctium lappa* burrs (Burrs) Dev. Macmillan. 2. *Caltha palustris* (Marsh Marigold) Som. Grigson. 3. *Ranunculus acris* (Meadow Buttercup) Som. Grigson. 4. *Ranunculus bulbosus* (Bulbous Buttercup) Som. Grigson. 5. *Ranunculus repens* (Creeping Buttercup) Som. Grigson.

BUTTONS, Policeman's *Caltha palustris* (Marsh Marigold) Som. Macmillan.

BUTTONS, Purple *Knautia arvensis* (Field Scabious) Som. Macmillan.

BUTTONS, Quaker *Strychnos nux-vomica* (Snakewood) Le Strange.

BUTTONS, Sailor; BUTTONS, Sailor's 1. *Knautia arvensis* (Field Scabious) Som. Macmillan. 2. *Melandrium rubrum* (Red Campion) Som. Tynan & Maitland. 3. *Succisa pratensis* (Devil's-bit) Dev. Macmillan. 4. *Stellaria holostea* (Greater Stitchwort) Hants. Grigson.

BUTTONS, Shimmy-and- *Calystegia sepium* (Large Bindweed) Dor. Grigson.

BUTTONS, Shirt 1. *Achillea ptarmica* (Sneezewort) Coats. 2. *Melandrium album* (White Campion) Som. Macmillan. 3. *Stellaria holostea* (Greater Stitchwort) Dor. Macmillan. Wilts. Jones & Dillon. The sneezewort attribution of this name is usually reserved for the double form.

BUTTONS, Shirt, Devil's *Stellaria holostea* (Greater Stitchwort) Som. Macmillan. There are a number of 'button' names for this, all from the button-like capsules. The 'devil' connection is widespread, too. It is the Devil's flower in Somerset, and his nightcap, his eyes in Wales, and his corn in Shropshire.

BUTTONS, Skirt 1. *Stellaria holostea* (Greater Stitchwort) Som. Grigson. 2. *Stellaria media* (Chickweed) Dor. Grigson.

BUTTONS, Soldier's; BUTTONS, Soldier 1. *Anemone nemorosa* (Wood Anemone) Som. Macmillan. 2. *Aquilegia vulgaris* (Columbine) Som, Wilts. Macmillan. 3. *Arctium lappa* (Burdock) Som. Macmillan; Wilts. Dartnell & Goddard. 4. *Caltha palustris* (Marsh Marigold) Som. Macmillan. 5. *Galium aparine* (Goose-grass) Cumb. Grigson. 6. *Geranium robertianum* (Herb

Robert) Bucks. Grigson. 7. *Geum rivale* (Water Avens) Wilts. Macmillan. 8. *Helianthemum chamaecistus* (Rock Rose) Som. Macmillan. 9. *Knautia arvensis* (Field Scabious) Som. Macmillan. 10. *Melandrium rubrum* (Red Campion) Yks. Grigson. 11. *Nuphar lutea* (Yellow Waterlily) Som. Grigson. 12. *Nymphaea alba* (White Waterlily) Som. Macmillan. 13. *Prunella vulgaris* (Self-heal) Som. Macmillan. 14. *Ranunculus acris* (Meadow Buttercup) Som. Macmillan. 15. *Ranunculus bulbosus* (Bulbous Buttercup) Som. Grigson. 16. *Ranunculus repens* (Creeping Buttercup) Som. Grigson. 17. *Rosa pimpinellifolia* (Burnet Rose) Kirk. Grigson. 18. *Stellaria holostea* (Greater Stitchwort) Bucks. Harman.

BUTTONS, Soldier's, Lousy *Melandrium rubrum* (Red Campion) Lancs. Grigson.

BUTTONS, Spanish *Centaurea nigra* (Knapweed) USA Cunningham.

BUTTONS, Stick 1. *Arctium lappa* (Burdock) Som. Macmillan; USA Henkel. 2. *Galium aparine* (Goose-grass) Som. Macmillan.

BUTTONS, Sticky 1. *Arctium lappa* (Burdock) Dev. B & H. 2. *Galium aparine* (Goose-grass) Wessex Rogers.

BUTTONS, Teddy *Knautia arvensis* (Field Scabious) Som. Macmillan.

BUTTONS, Yellow *Tanacetum vulgare* (Tansy) Som. Macmillan.

BUTTONWEED 1. *Centaurea nigra* (Knapweed) Suss. Grigson. 2. *Cotula coronopifolia*. 3. *Heracleum sphondyllium* (Hogweed) Suff. Moor.

BUTTONWOOD *Platanus occidentalis* (Western Plane) New Engl. Harper.

BUXBAUM'S SPEEDWELL *Veronica persica* (Common Field Speedwell) McClintock. Johann Christian Buxbaum (1683-1730) was a German botanist who first described the plant (in 1727) from fields around Constantinople.

BY-THE-WIND *Clematis vitalba* (Old Man's Beard) Wilts. Dartnell & Goddard. A mis-pronunciation, or a mis-hearing, of Bethwind, probably.

BYSMALE *Althaea officinalis* (Marsh Mallow) Henslow. The 'male' part of the word would be mallow - in fact, it was also known, according to Halliwell, as Bysmalow.

C

CABARET *Asarum europaeum* (Asarabacca) W.Miller. Presumably a corruption of Asarabacca.

CABBAGE *Brassica oleracea var. capitata*. The earlier form was cabache, which is Old French caboche, thence Latin caput, head (the specific name includes '*capitata*', 'with a head').

CABBAGE, Bargeman's *Brassica campestris* (Field Mustard) Bucks. B & H. It was that part of Buckinghamshire by the river Thames that used this name.

CABBAGE, Bastard *Rapistrum rugosum*.

CABBAGE, Black *Brassica oleracea var. acephala* (Kale) Bianchini.

CABBAGE, Celery *Brassica pekinensis* (Chinese Cabbage) Schery. There is a celery-like flavour to this, hence the name.

CABBAGE, Chinese 1. *Brassica sinensis* (Chinese Mustard) Usher. 2. *Brassica pekinensis*.

CABBAGE, Dan's *Senecio latifolius* S.Africa Watt.

CABBAGE, Dune *Rhyncosinapis monensis*. Confined to the British Isles, very local by the sea.

CABBAGE, Field *Brassica campestris* (Field Mustard) Jordan.

CABBAGE, Hare's Ear *Conringia orientalis*. Cf. Hare's Ear by itself, presumably from some descriptive similarity.

CABBAGE, Indian *Yucca glauca* (Soapweed) USA Elmore. The central part of the spike used to be eaten as "cabbage" by the Indians.

CABBAGE, Isle of Man *Rhyncosinapis monensis* (Dune Cabbage) Clapham.

CABBAGE, John's *Hydrophyllum virginianum* (Virginian Waterleaf) Perry.

CABBAGE, Lundy *Rhyncosinapis wrightii*. Confined to Lundy Island.

CABBAGE, Meadow *Symplocarpus foetidus* (Skunk Cabbage) Le Strange.

CABBAGE, Paiute *Stanleya pinnata* (Prince's Plume) USA Jaeger. An American species, much used as greens by the Indians, but surely not just by the Paiute?

CABBAGE, Poor Man's 1. *Polygonum bistorta* (Bistort) Lancs. Grigson. 2. *Rumex patientia* (Patience Dock) Lancs. Nodal.

CABBAGE, St Patrick's *Saxifraga spathularia*. Perhaps because it grows in the west of Ireland, where the saint lived. It was first recorded from co Kerry in 1697. But the name may be the result of a mis-translation. It was claimed in *Cybele Hibernica* (2nd ed 1898) that the genitive form of the Irish for 'Patrick' and 'fox' are similar enough to be confused in this name.

CABBAGE, Savoy *Brassica oleracea var. bullata*. Originally from Savoy?

CABBAGE, Sea 1. *Brassica oleracea* (Wild Cabbage) Grigson. 2. *Crambe maritima* (Sea Kale) B & H. 3. *Verbascum thapsus* Glam. Grigson.

CABBAGE, Skunk 1. *Symplocarpus foetidus*. 2. *Veratrum californicum* (False Hellebore) USA Kingsbury. From the smell, presumably; certainly, the former is labelled foetidus. The bad smell is apparent when the plant is disturbed.

CABBAGE, Strand *Crambe maritima* (Sea Kale) Donegal Grigson.

CABBAGE, Wallflower *Rhyncosinapis erucastrum*. The older botanical name was *Brassica cheiranthoides*, hence this name.

CABBAGE, Warrigal *Tetragonia tetragonioides* (New Zealand Spinach) C Macdonald.

CABBAGE, Warty *Bunias orientalis*.

CABBAGE, Wild *Brassica oleracea*. This, it seems, is the parent plant of all the cabbages. Brassica is of Celtic origin, from bresic, cabbage.

CABBAGE DAISY *Trollius europaeus* (Globe Flower) Scot. Grigson.

CABBAGE FLOWER; CABBAGE SEED *Sinapis arvensis* (Charlock) both Som. Macmillan. Cf. the various Kale and Rape names given to Charlock.

CABBAGE GUM *Eucalyptus pauciflora*.
CABBAGE LETTUCE *Lactuca sativa var. capitata*. Because it makes a head, like cabbage.
CABBAGE PALM *Dracaena indivisa* (Fountain Dracaena) McClintock. Cf. Cabbage Tree for this.
CABBAGE ROSE 1. *Paeonia mascula* (Peony) Som. Macmillan. 2. *Rosa centifolia*.
CABBAGE THISTLE *Cirsium oleraceum*.
CABBAGE TREE 1. *Cussonia kirkii*. 2. *Dracaena indivisa* (Fountain Dracaena) New Zealand. Everett. 3. *Vernonia conferta* Dalziel. The Cussonia is an African tree, but this Dracaena is a native of New Zealand, and got the name because the early settlers used the young shoots as vegetables.
CABBAGE TURNIP *Brassica oleracea var. caulo-rapa* (Kohlrabi) Schery. Presumably this arose simply as a translation of Kohlrabi, the German name, from Italian cavoli-rape, cabbage turnips (Latin caulis, rapa).
CABBISH *Brassica oleracea var. capitata* (Cabbage) Dur. Dinsdale.
CACAO *Theobroma cacao*. This is the Spanish word, which was derived from the Nahuatl cacauatl - the uatl tree. Cacao became altered in English to cocoa.
CACOMITE *Tigridia pavonia* (Tiger-flower) USA Driver. This is a Mexican name adopted in the USA.
CACTUS, Christmas *Schlumbergera x buckleyi*.
CADDELL; CADWEED *Heracleum sphondyllium* (Hogweed) both Dev. B & H.
CADLOCK 1. *Raphanus raphanistrum* (Wild Radish) War. B & H. 2. *Sinapis arvensis* (Charlock) Grigson. Wild Radish bears the name Charlock in many places, so it is natural that it should carry the variants as well. Cadlock, as well as Carlock, are very common corruptions of Charlock; these in turn vary enormously - Charnock, Chadlock, Kedlock, even Warlock, and there are very many more.
CADLOCK, Rough *Sinapis arvensis* (Charlock) N.Eng. Halliwell. Smooth Cadlock is Wild Rape, Halliwell says. But aren't these both the same plant - charlock?
CAGGE *Iris pseudo-acarus* (Yellow Flag) B & H. Cagge is the same as seg, or sedge, segg being OE for a small sword (Cf. the various dagger or sword names for this).
CAIN-AND-ABEL 1. *Aquilegia vulgaris* (Columbine) Wilts. Dartnell & Goddard. 2. *Conopodium majus* (Earth-nut) Dur. Grigson. 3. *Dactylorchis incarnata* (Early Marsh Orchid) Border B & H. 4. *Orchis mascula* (Early Purple Orchis) Fernie. Double names like this usually imply bicoloured flowers, but that doesn't seem to apply in this case. As far as the orchids go, the reason is probably that the tubers are double, and perhaps anthropomorphic - the tubers of the Early Marsh Orchid are called Adam-and-Eve in places. Earthnut must get the name for the same reason. Columbines are better recognized as Cains-and-Abels, i.e. in the plural rather than in the singular.
CAISE *Conium maculatum* (Hemlock) Yks. B & H. Kex is the more common name, and really should be applied to the dried stalks - "as dry as a kex" is one of Hazlitt's English proverbs. He also has "as hollow as a kex", "as light as a kex" and "as sapless as a kex".
CAJAN BEAN *Cajanus indicus* (Pigeon Pea) Schery.
CAJUPUT *Melaleuca leucadendron*.
CAKERS; CAXLIES *Heracleum sphondyllium* (Hogweed) both Som. Macmillan. These are more immediately recognisable under the widespread name of Kex.
CAKES, Baby *Geranium lucidum* (Shining Cranesbill) Dev. Macmillan.
CAKES, Beggarman *Althaea officinalis* (Marsh Mallow) Ire. O'Farrell. Quoted as Marsh Mallow, but perhaps Common Mallow was meant, as the seed vessels of the latter were certainly known as 'cakes' of one kind or another, notably Cheese cakes.
CAKES, Cow 1. *Heracleum sphondyllium* (Hogweed) Border Grigson. 2. *Pastinaca sativa* (Wild Parsnip) Roxb. Grigson. Cow-flops for these, too, but in Cornwall. 'Cakes'

is 'Keeks', and the same as Kex.

CAKES, Lady *Oxalis acetosella* (Wood Sorrel) Dumf. Grigson.

CAKES, Penny *Umbilicus rupestris* (Wall Pennywort) Corn. Jago; Dev, Pemb. Grigson.

CAKESEED 1. *Conium maculatum* (Hemlock) Som. Macmillan. 2. *Heracleum sphondyllium* (Hogweed) Som. Macmillan. Cf. Cakers and Caxlies for Hogweed, and Cakezie for Hemlock.

CAKEZIE *Conium maculatum* (Hemlock) Som. Macmillan. Cakeseed is the immediate forebear. It then goes directly to Caxes, and so to Kex.

CALABAR BEAN *Physostigma venenosum*.

CALABASH *Lagenaria siceraria*. French calebasse, calabasse, Spanish calabaza. The Catalan form with "r" (carabassa) is probably closer to the original, and there are other Romance forms with "r". The word is probably of oriental origin; Cf. Persian Kharbuza - water-melon.

CALABASH, Sweet *Passiflora maliformis*.

CALABASH CUCUMBER *Lagenaria siceraria* (Calabash) Willis.

CALABASH NUTMEG *Monodora myristica*. The seeds contain an oil that tastes of nutmeg, and, indeed, are sometimes used as nutmegs.

CALABASH TREE *Crescentia cujete*. They have gourd-like fruits which, when the pulp is removed, are made into drinking vessels (calabashes).

CALABRIAN PINE *Pinus brutia*.

CALADIUM, Fancy-leaved *Caladium bicolor* 'Splendens'.

CALADIUM, Painted *Caladium picturatum*.

CALAF OF PERSIA WILLOW *Salix aegyptica*.

CALAMANDER WOOD *Diospyros quaesita* timber Willis.

CALAMINT 1. *Calamintha ascendens*. 2. *Nepeta cataria* (Catmint) R.T.Gunther. Calamint is from O.Fr. calament, med.Lat. calamentum, Lat. calaminthe, Gk. kalaminthe, perhaps from kalos, beautiful, and minthe, mint.

CALAMINT, Bush *Calamintha ascendens* (Calamint) Turner.

CALAMINT, Cushion *Clinopodium vulgare* (Wild Basil) Mabey. 'Cushion', from the dome-shaped whorl of flowers.

CALAMINT, Hedge *Clinopodium vulgare* (Wild Basil) Grieve.1931.

CALAMINT, Hore *Calamintha ascendens* (Calamint) Turner.

CALAMINT, Lesser *Calamintha nepeta* (Lesser Catmint) Genders.

CALAMINT, Mountain *Calamintha ascendens* (Calamint) Gerard.

CALAMINT, Wood *Calamintha sylvatica*.

CALAMINT BALM *Calamintha ascendens* (Calamint) Leyel.1937.

CALAMONDIN *Citrus mitis*. Kalamunding is the name given to this in Tagalog, the national language of the Philippines.

CALATHIAN VIOLET *Gentiana pneumonanthe* (Marsh Gentian) Gerard. This is Latin calathus - basket. The name was given by Pliny to another plant, and misappropriated to this one.

CALCALARY; CALSCALARY *Cypripedium calceolus* (Lady's Slipper) Dev. Friend.1882.

CALE *Brassica oleracea var. capitata* (Cabbage) Wilts. Halliwell. From Aubrey's Natural History of Wiltshire, merely a local variant of Cole, if it isn't just Aubrey's spelling of Kale.

CALENDAR, Shepherd's *Anagallis arvensis* (Scarlet Pimpernel) Dev. Friend.1882. It makes a little sense when the other 'Shepherd's' names are considered. Shepherd's Weatherglass, Shepherd's Warning, etc., all relate to weather lore:

Pimpernel, pimpernel, tell me true,
Whether the weather be fine or no;
No heart can think, no tongue can tell,
The virtues of the pimpernel.

CALEY PEA *Lathyrus hirsutus* (Hairy Pea) USA Kingsbury.

CALF'S FOOT 1. *Allium vineale* (Crow Garlic) B & H. 2. *Arum maculatum* (Cuckoo Pint) Gerard. 3. *Sinapis arvensis* (Charlock) Glos. J.D.Robertson; War. Grigson. 4. *Tussilago farfara* (Coltsfoot) Som. Macmillan. Presumably all from the shape of the leaves? But that can hardly include Crow Garlic.

CALF'S SNOUT 1. *Antirrhinum majus* (Snapdragon) Tynan & Maitland. 2. *Misotapes orontium* (Lesser Snapdragon) Turner. Cf. also for this Weasel-snout, Toad's Mouth, Lion's Mouth, Bunny Mouth, etc.

CALF'S SNOUT, Broad *Antirrhinum majus* (Snapdragon) Turner.

CALFKILL *Kalmia angustifolia* (Sheep Laurel) USA Kingsbury. All species of Kalmia have the reputation of being poisonous, and particularly to cattle. The names of this species include Lambkill and Sheepkill as well as Calfkill. "It kills sheep and lesser animals, when they eat of it plentifully..." (Taylor).

CALICO ASTER *Aster laterifolius* (Starved Aster) USA House.

CALICO BUSH *Kalmia latifolia*.

CALICO PLANT *Aristolochia elegans*.

CALIFORNIA BAYBERRY *Myrica californica*.

CALIFORNIA BLACKBERRY *Rubus vitifolius*.

CALIFORNIA DODDER *Cuscuta californica*.

CALIFORNIA FERN *Conium maculatum* (Hemlock) USA Kingsbury. Cf. Nebraska Fern.

CALIFORNIA HOLLY *Photinia arbutifolia*.

CALIFORNIA HUCKLEBERRY *Vaccinium ovatum*.

CALIFORNIA JOINT FIR *Ephedra californica*.

CALIFORNIA JUNIPER *Juniperus californica*.

CALIFORNIA LILAC *Ceanothus integerrimus* (Deer-brush) Calif. Mason.

CALIFORNIA LIVE OAK *Quercus agrifolia*.

CALIFORNIA NUTMEG *Torreya californica*.

CALIFORNIA PEPPER TREE *Schinus molle* (Peruvian Mastic). Kearney. A Peruvian tree, but much planted as an ornamental street tree in California. 'Pepper', because, besides having pungent foliage, it was thought at one time that the seeds would actually produce pepper.

CALIFORNIA PITCHER PLANT *Darlingtonia californica* (Cobra Plant) USA Cunningham.

CALIFORNIA REDBUD *Cercis occidentalis* (Western Redbud) Brimble.1948.

CALIFORNIA ROSE BAY *Rhododendron macrophyllum*.

CALIFORNIAN ALDER *Alnus rhombifolia* (White Alder) W.Miller.

CALIFORNIAN ALLSPICE *Calycanthus occidentalis*.

CALIFORNIAN BARBERRY *Berberis pinnata*.

CALIFORNIAN BLACK OAK *Quercus kelloggii*.

CALIFORNIAN BLUEBELL *Phacelia minor*.

CALIFORNIAN BUCKEYE *Aesculus californica*.

CALIFORNIAN BUCKTHORN *Rhamnus purshiana*.

CALIFORNIAN BUTTERCUP *Ranunculus californicus*.

CALIFORNIAN CYPRESS *Cupressus goveniana*.

CALIFORNIAN ELDER *Sambucus glauca* (Blue Elder) W.Miller.

CALIFORNIAN EVERLASTING *Gnaphalium decurrens var. californicum*.

CALIFORNIAN FEVER-BUSH *Garrya fremontii* (Bear Brush) Grieve.1931. Or just Fever-bush.

CALIFORNIAN HAREBELL *Campanula prenanthoides*.

CALIFORNIAN HAZEL *Corylus rostrata var. californica*.

CALIFORNIAN HONEYSUCKLE *Lonicera hispidula var. californica*.

CALIFORNIAN HYACINTH *Brodiaea laxa*.
CALIFORNIAN LAUREL *Umbellularia californica*.
CALIFORNIAN MUGWORT *Artemisia vulgaris var. heterophylla*.
CALIFORNIAN PANTHER LILY *Lilium nevadense*.
CALIFORNIAN PLANE *Platanus racemosa*.
CALIFORNIAN POISON OAK *Rhus diversiloba* (Poison Oak) Le Strange.
CALIFORNIAN POPPY 1. *Escholtzia californica*. 2. *Romneya coulteri* (Californian Bush Poppy) Salisbury.1936.
CALIFORNIAN RED FIR *Abies magnifica*.
CALIFORNIAN SASSAFRAS *Umbellularia californica* (Californian Laurel) Le Strange.
CALIFORNIAN SLIPPERY ELM *Fremontia californica* (Flannel Bush) Le Strange.
CALIFORNIAN SOAPROOT *Chloragalum pomeridianum* (Soap Plant) Usher. The plant gives a soapy lather, like soapwort.
CALIFORNIAN SPEARMINT *Chenopodium ambrosioides* (American Wormseed) New Zealand Le Strange.
CALIFORNIAN WHITE CEDAR *Libocedrus decurrens* (Incense Cedar) Grieve.1931.
CALIFORNIAN YEW *Taxus brevifolia* (Western Yew) Dallimore.
CALISAYA *Cinchona calisaya*.
CALL, Yellow *Ranunculus arvensis* (Corn Buttercup) IOW Long. Yellow-cup, perhaps? No, it must be Yellow Caul.
CALL-ME-TO-YOU *Viola tricolor* (Pansy) Som. Macmillan. Gerard has Cull-me-to-you, and there is also a Cuddle-me-to-you, which sounds as if it may be the origin of the other two.
CALLA LILY *Zantedeschia aethiopica* (Arum Lily) Simmons.
CALLARDS *Brassica oleracea var. capitata* (Cabbage) IOW Halliwell. Applied to the leaves and shoots of cabbages. Better known as Collards.
CALLIANDRA, Hairy-leaved *Calliandra eriophylla*.
CALLOCK *Sinapis arvensis* (Charlock) N'thants. A.E.Baker. A corruption of charlock. Cadlock and Carlock are more common.
CALOMEL, Sweet; CALOMEL, Vegetable *Acorus calamus* (Sweet Flag) both Watt. Calomel is apparently mercurous chloride, said to be from Greek kalos, beautiful, and melas, black, because in its first preparation a black powder turned into a white one. The reason for calling the plant "Vegetable" calomel now becomes clear, but why calomel at all? What does the specific name mean, if not a reed?
CALTHROP *Tribulus terrestris* (Puncture Vine) Chopra.
CALTRAP 1. *Centaurea calcitrapa* (Star Thistle) Cockayne. 2. *Tribulus terrestris* (Puncture Vine) Chopra. Saxon calcetreppe, from calcem, heel, and the Latin form of trap. Caltrop also, and see this for a much more likely source.
CALTROP 1. *Caltha palustris* (Marsh Marigold) Culpepper. 2. *Centaurea calcitrapa* (Star Thistle) Prior. 3. *Ranunculus acris* (Meadow Buttercup) Som. Grigson. 4. *Ranunculus bulbosus* (Bulbous Buttercup) Som. Grigson. 5. *Ranunculus repens* (Creeping Buttercup) Som. Grigson. 6. *Trapa natans* (Water Chestnut) Brouk. 7. *Tribulus terrestris* (Puncture Vine) USA Kingsbury. According to Halliwell, a caltrop is "an instrument with four spikes, so contrived that one of the spikes always stands upwards, no matter in what direction it is thrown", apparently used in wolf hunting, but the main use was against cavalry horses. These things were thrown down to maim them. Caltrop is from the Latin calcitrapa. So the derivation in these cases is probably descriptive in some way - and certainly so in the case of Water Chestnut, for the nuts really do look like this.
CALTROP, Hairy *Kallstroemeria hirsutissima* (Carpet Weed) USA Kingsbury.
CALTROP, Water *Caltha palustris* (Marsh

Marigold) Fernie.

CALVARY CLOVER *Medicago arabica* (Spotted Medick) McClintock. The spots on the leaves account for the name. Like many other plants with spotted leaves with names like this (Gethsemane for instance) the legend that it grew on Calvary, and that the spots are Christ's blood preserved for ever on the plant, is responsible for the name. A German superstition taken to America said that Calvary Clover would only germinate if sown on Good Friday.

CALVES, Cows-and- *Arum maculatum* (Cuckoo Pint) Wilts. Goddard; Glos. J.D.Robertson; N'thants. A.E.Baker; War. Morley; Dev, Dor, Som, Bucks, Shrop, Notts, Lincs, Yks, Cumb. Grigson.

CALVES, Sucky *Arum maculatum* (Cuckoo Pint) Som. Grigson.

CAMAFINE *Chamaemelon nobilis* (Camomile) Ork. Leask. One of a number of variations on camomile.

CAMAS, Death *Zigadenus gramineus*. A poisonous lily from America.

CAMASS *Camassia esculenta* (Quamash) Hay & Synge.

CAMBIE-LEAF 1. *Nuphar lutea* (Yellow Waterlily) N.Scot. Jamieson. 2. *Nymphaea alba* (White Waterlily) N.Scot. Jamieson.

CAMBUCK *Conium maculatum* (Hemlock) E.Ang. Halliwell. Halliwell has this as the dry stalk of a plant, as of hemlock, just as the Kex and Kecksies group get applied to quite a number of different plants.

CAMEL 1. *Achillea millefolium* (Yarrow) Dev. Grigson. 2. *Chamaemelon nobilis* (Camomile) Corn. Courtney.1880; Som. Grigson. 3. *Pulicaria dysenterica* (Yellow Fleabane) Dev. Grigson. Camel, and Camil, are obviously the same as camomile.

CAMEL'S FOOT TREE *Bauhinia purpurea* (Orchid Tree) Usher.

CAMEL THORN *Acacia giraffae*.

CAMEROON CARDAMOM *Aframomum hanburyi*.

CAMIL *Chamaemelon nobilis* (Camomile) Corn, Som. Grigson. See Camel.

CAMLICK *Heracleum sphondyllium* the dried stalks (Hogweed) Suff. B & H.

CAMMANY *Chamaemelon nobile* (Camomile) Lancs. Grigson.

CAMMICK 1. *Hypericum perforatum* (St John's Wort) IOW Long. 2. *Ononis repens* (Rest Harrow) Dev, Som, Bucks. Grigson; Dor. Halliwell; IOW Long. 3. *Senecio jacobaea* (Ragwort) IOW Long. Cammick, or Cammock, seems to be given to almost any yellow-coloured flower in the Isle of Wight, but it should properly be reserved for Rest Harrow, which is not yellow-flowered.

CAMMIL *Achillea millefolium* (Yarrow) Dev. Friend.1882. See Camel.

CAMMOCK 1. *Achillea millefolium* (Yarrow) Dev. Macmillan. 2. *Hypericum perforatum* (St John's Wort) Hants. Cope; S.Africa Watt. 3. *Lotus corniculatus* (Bird's-foot Trefoil) Dev. Grigson. 4. *Ononis repens* (Rest Harrow) Gerard. 5. *Peucedanum officinale* (Hog's Fennel) Cockayne. 6. *Pulicaria dysenterica* (Yellow Fleabane) Hants. Cope. 7. *Scandix pecten-veneris* (Shepherd's Needle) Prior. 8. *Senecio jacobaea* (Ragwort) Hants. Cope. OE cammoc, commuc. It means a crooked beam or tree, or timber prepared that way - so the diverse selection above are all probably descriptive. The name is still common for Rest Harrow, and is included in the Wiltshire list. In Hampshire, almost any yellow flower is known as Cammock, so is the derivation different in this case?

CAMMOCK, Dumb; CAMMOCK, Whin *Ononis repens* (Rest Harrow) Som. Macmillan (Dumb); B & H (Whin).

C(H)AMOMILE *Chamaemelon nobile*. From French camomile, Latin chamomilla, Greek khamaemelon, meaning earth-apple, a name given from the apple-like scent of the flower.

CAMOMILE, Belgian *Chamaemelon nobile* Mitton. Presumably because it is cultivated in Belgium?

CAMOMILE, Bitter *Matricaria recutita* (Maydweed) Fernie.

CAMOMILE, Blue *Aster tripolium* (Sea

Aster) Gerard. Purple Camomile, too, but later, and shared with Pheasant's Eye.

CAMOMILE, Corn *Anthemis arvensis*.

CAMOMILE, Disc *Matricaria matricaroides*.

CAMOMILE, Dog's 1. *Anthemis cotula* (Maydweed) B & H. 2. *Matricaria recutita* (Scented Mayweed) Prior. 3. *Tripleurospermum maritimum* (Scentless Maydweed) B & H. Probably because of the bad smell (of the first named).

CAMOMILE, Double *Chamaemelon nobile* (Camomile) Flück. Single Camomile is *Matricaria chamomilla*, Scented Mayweed.

CAMOMILE, Dyer's *Anthemis tinctoria* (Yellow Camomile) Coats. The specific name *tinctoria* indicates its use as a dye plant. Lambert records its use in Finland and Russia for the production of a yellow dye.

CAMOMILE, False; CAMOMILE, Scentless *Tripleurospermum maritimum* (Scentless Maydweed) G.B.Foster.

CAMOMILE, False, Sweet *Matricaria recutita* (Scented Mayweed) H & P. 'Sweet', because this is scented.

CAMOMILE, Foetid; CAMOMILE, Stinking *Anthemis cotula* (Maydweed) Australia Watt (Foetid); Yks. B & H (Stinking). It has a bad smell. Dog's Camomile is another name given for the same reason.

CAMOMILE, German; CAMOMILE, Hungarian *Matricaria recutita* (Maydweed) Brownlow (German); Watt (Hungarian). It grows wild in great quantities in the Debrecson region of Hungary - there it is harvested for use as an ingredient of hair rinses.

CAMOMILE, Ox-eye *Anthemis tinctoria* (Yellow Camomile) Coats.

CAMOMILE, Purple 1. *Adonis annua* (Pheasant's Eye) Turner. 2. *Aster tripolium* (Sea Aster) B & H.

CAMOMILE, Red *Adonis annua* (Pheasant's Eye) Gerard.

CAMOMILE, Roman *Chamaemelon nobile* (Camomile) Clair. First so named by the 16th century German humanist Joachim Camerarius, who found it growing near Rome.

CAMOMILE, Scotch *Chamaemelon nobile* (Camomile) B & H. Apparently known as such in the 19th century pharmaceutical industry.

CAMOMILE, Single *Matricaria recutita* (Scented Mayweed) Flück). *Chamaemelon nobile* is Double Camomile**.**

CAMOMILE, Spanish *Anacyclus pyrethrum* (Pellitory-of-Spain) Lloyd.

CAMOMILE, White *Chamaemelon nobile* (Camomile) B & H.

CAMOMILE, Wild 1. *Anthemis cotula* (Maydweed) Gerard. 2. *Matricaria recutita* Bucks, Ches. B & H.

CAMOMILE, Yellow *Anthemis tinctoria*.

CAMOMILE FEVERFEW *Matricaria recutita* (Scented Mayweed) C P Johnson.

CAMOMILL; CAMOMINE; CAMOWYNE *Chamaemelon nobile* (Camomile) Parkinson (Camomill); Shrop, Scot. Grigson (Camomine); Scot. Jamieson (Camowyne). There are lots of variants of camomile - another close to these is Camovyne, but it is applied to *Anthemis cotula* as well.

CAMOROCHE *Potentilla anserina* (Silverweed) B & H. Halliwell says this is "the wild tansy", but B & H correct him. Anyway, silverweed has always been known as tansy, or wild tansy.

CAMOTE *Ipomaea batatas* (Sweet Potato) Kelly & Palerm. This is one of the Mexican names.

CAMOVYNE 1. *Anthemis cotula* (Maydweed) Scot. Grigson. 2. *Chamaemelon nobile* (Camomile) Scot. Jamieson. Gerard knew Maydweed as Wild Camomile.

CAMOVYNE, Dog's *Tripleurospermum maritimum* (Scentless Maydweed) Ross Jamieson. 'Dog' or 'Dog's' in front of a name usually means a large or coarse variety. 'Horse' means the same. Cf. Horse Daisy and Dog Daisy for this plant.

CAMPANA, La *Verbena officinalis* (Vervain)

Guernsey MacCulloch.

CAMPANELLE *Calystegia sepium* (Great Bindweed) Som. Macmillan. Bell-shaped flowers. Cf. Bellbind, Hedge Bells, etc.

CAMPEACHY WOOD *Haematoxylon campechianum* (Logwood) S.M.Robertson. This and the specific name are given because, although the tree grows in Central America and northern South America, it was from Campeachy on the Gulf of Mexico that it was taken to be propagated (successfully) in the West Indies.

CAMPERDOWN ELM *Ulmus glabra var. camperdownii*. The reference is to the house of that name near Dundee.

CAMPHIRE 1. *Crithmum maritimum* (Samphire) Cumb. Grigson. 2. *Lawsonia inermis* (Henna) Watt. A mixture of sources here. An initial letter change from Samphire has produced Camphire for *Crithmum maritimum*, but the source for *Lawsonia inermis* must be camphor. Camphire for camphor is actually the earlier form.

CAMPHIRE-TREE *Cinnamonium camphora* (Camphor) Hill. This is the earlier form of Camphor.

CAMPHOR 1. *Cinnamonium camphora*. 2. *Tanacetum balsamita* (Costmary) Painter. French camphre, O.French camphore, med. Latin camphora (probably through Spanish alcanfar), Arabic kafur, Sanskrit karpura. The fact that the word is derived from the Arabic is some indication that our knowledge of the drug is also derived from that source. Camphire is the earlier form in English. Given to costmary, the name seems to suggest a quite different smell than does Mint Geranium, for instance, also given to it.

CAMPHOR, Japanese *Cinnamonium camphora* (Camphor) S.Africa Watt.

CAMPHOR BASIL *Ocymum kilimandscharicum*.

CAMPHOR LAUREL *Cinnamonium camphora* (Camphor) Fluckiger.

CAMPHOR OF BORNEO *Dryobalanops camphora*.

CAMPHOR OF MALAYSIA *Dryobalanops*

camphora (Camphor of Borneo) Tosco.

CAMPHOR PLANT *Balsamita vulgaris* Brownlow.

CAMPHOR WORMWOOD *Artemisia camphorata*. This plant is camphor-scented.

CAMPHORWOOD, East African *Ocotea usambarensis*.

CAMPION, Bladder 1. *Silene cucubalis*. 2. *Silene latifolia* USA House. Bladder, from the inflated calyx. Campion is first recorded in the 16th century, from Lobel and Lyte, and seems to be from Greek lukhnis stephanomatike (used for garlands); on this the derivation of campion has been based, i.e. the same word as champion.

CAMPION, Corn *Agrostemma githago* (Corn Cockle) B & H.

CAMPION, Evening *Melandrium album* (White Campion) H C Long.1910. Cf. Evening Close, from Dorset, and Shades of Evening, from Somerset.

CAMPION, Meadow *Lychnis flos-cuculi* (Ragged Robin) B & H.

CAMPION, Mexican *Silene laciniata* (Mexican Catchfly) USA Elmore.

CAMPION, Moss *Silene acaulis*.

CAMPION, Mountain *Silene acaulis* (Moss Campion) Camden/Gibson.

CAMPION, Red *Melandrium rubrum*.

CAMPION, Rock *Silene rupestris*.

CAMPION, Rose 1. *Agrostemma githago* (Corn Cockle) S.Africa Watt. 2. *Lychnis chalcedonica* (Jerusalem Cross) Gerard. 3. *Lychnis coronaria*.

CAMPION, Rose, Wild *Melandrium rubrum* (Red Campion) Gerard.

CAMPION, Round *Silene cucubalis* (Bladder Campion) Vesey-Fitzgerald. 'Round', for the same reason as 'bladder'.

CAMPION, Sea *Silene maritima*.

CAMPION, Spanish *Silene otites* (Breckland Catchfly) W.Miller.

CAMPION, Starry *Silene stellata*.

CAMPION, White *Melandrium album*.

CAMPION OF CONSTANTINOPLE *Lychnis chalcedonica* B & H. This is a Crusader reference to the Maltese Cross shape of each flower. Cf. Maltese Cross, Knight Cross, Jerusalem Cross, etc.

CAMPROOT *Geum strictum* (Yellow Avens) USA House.

CAMWOOD 1. *Baphia nitida*. 2. *Pterocarpus angolensis* (Barwood) Picton & Mack. According to ODEE, said to be from an African dialect word, kambi. An erroneous aplication of the name in the case of Barwood.

CAMWOOD, Benin *Baphia pubescens*.

CAMWOOD, Walking-stick *Baphia polygalacea*.

CAN, Lily *Nuphar lutea* (Yellow Waterlily) Fife, Perth Grigson.

CAN, Water 1. *Nuphar lutea* (Yellow Waterlily) Som. Macmillan. 2. *Nymphaea alba* (White Waterlily) War, Lincs. B & H.

CAN DOCK 1. *Nuphar lutea* (Yellow Waterlily) Prior. 2. *Nymphaea alba* (White Waterlily) Som, War, Notts. Grigson.

CANACH; CANAVAN *Eriophorum angustifolium* (Cotton-grass) Scot. J.G.McKay (Canach); Ire. Lockwood (Canavan). Canavan is Irish ceannbhán, literally white head. Canach must be the same, with the 'white' element somehow left out.

CANADA BLUEBERRY 1. *Vaccinium canadense*. 2. *Vaccinium myrtilloides* (Sour-top Blueberry) USA Yarnell.

CANADA CROOK-NECK SQUASH *Cucurbita moschata* (Crook-neck Squash) Bianchini.

CANADA GARLIC *Allium canadense*.

CANADA HEMLOCK *Tsuga canadensis* (Eastern Hemlock) Yanovsky.

CANADA LILY *Lilium canadense*.

CANADA MAPLE, Scarlet *Acer rubrum* (Red Maple) Hay & Synge.

CANADA PLUM *Prunus americana*. 'Canada', because it is actually cultivated in that country, the fruit being very popular there.

CANADA POTATO *Helianthus tuberosus* (Jerusalem Artichoke) Parkinson. They call the tubers Namara Potatoes in S. Africa.

CANADA THISTLE *Cirsium arvense* (Creeping Thistle) USA H.Smith.1945.

CANADA VERVAIN *Verbena canadensis*.

CANADA VIOLET *Viola canadensis*.

CANADA WILD RICE *Zizania aquatica*.

CANADIAN ANEMONE *Anemone canadensis* (Round-leaved Anemone) USA House.

CANADIAN BALSAM *Abies balsamea* (Balsam Fir) Barber & Phillips.

CANADIAN BIRCH *Betula lutea* (Yellow Birch) Hutchinson & Melville.

CANADIAN BITTER-ROOT *Lewisia rediviva* (Bitter-root) J.Smith.

CANADIAN FLEABANE *Erigeron canadensis*.

CANADIAN GOLDEN ROD *Solidago altissima*.

CANADIAN HEMP *Apocynum cannabinum* (Indian Hemp) Grieve.1931. or American Hemp, Black Indian Hemp.

CANADIAN MILK VETCH *Astragalus carolinianus*.

CANADIAN MINT *Mentha arvensis var. canadensis*.

CANADIAN MOONSEED *Menispermum canadense* (Moonseed) Le Strange.

CANADIAN PONDWEED *Elodea canadensis*.

CANADIAN POPLAR 1. *Populus canadensis*. 2. *Populus monolifera* (Northern Cottonwood) Ablett.

CANADIAN PRIMROSE, Dwarf *Primula mistassini*.

CANADIAN ST JOHN'S WORT *Hypericum canadense*.

CANADIAN SNAKEROOT *Asarum canadense*.

CANADIAN SPRUCE *Picea canadensis* (White Spruce) N.Davey.

CANADIAN TEA *Gaultheria procumbens* (Wintergreen) USA Sanford. The leaves have been used as a substitute for tea, hence this name and such as Mountain Tea, Teaberry.

CANADIAN WATERWEED *Elodea canadensis* (Canadian Pondweed) Gilmour & Walters.

CANADIAN WORMWOOD *Artemisia*

canadensis.

CANADIAN YELLOW PINE *Pinus strobus* (White Pine) N.Davey.

CANADIAN YEW *Taxus canadensis*.

CANAIGRE *Rumex hymenosepalus* (Tanner's Dock) Schery.

CANARY BANANA *Musa nana*. That, at any rate, is the name under which the fruit is sold in Europe.

CANARY BELLFLOWER *Canaria canariensis*. Canary Isles, of course, and not the bird or the associated colour.

CANARY BIRD BUSH *Crotalaria agatiflora* (Lion's Claw) RHS. The flowers are a yellowy-green in colour, hence the name.

CANARY BIRD FLOWER *Tropaeolum peregrinum* (Canary Creeper) Macmillan. Canary here surely must refer to the Canary Isles, as *T canariensis* is a synonym of *T peregrinum*.

CANARY BROAD-LEAF; CANARY FLOWER *Plantago major* (Great Plantain) both Som. Grigson. A bird seed, as many other names testify.

CANARY CREEPER *Tropaeolum peregrinum*.

CANARY FOOD; CANARY SEED 1. *Plantago major* (Great Plantain) Som. Macmillan. 2. *Senecio vulgaris* (Groundsel) Som. Macmillan. i.e. a birdseed.

CANARY GRASS *Phalaris canariensis*.

CANARY GRASS, Lesser *Phalaris minor*.

CANARY GRASS, Reed *Phalaris arundinacea* (Reed-grass) Hepburn.

CANARY HOLLY; MADEIRA HOLLY *Ilex perado*.

CANARY ISLAND JUNIPER *Juniperus cedrus*.

CANARY ISLAND LAUREL *Laurus canariensis*.

CANARY PALM *Phoenix canariensis*.

CANARY PINE *Pinus canariensis*.

CANARY WHITEWOOD *Liriodendron tulipifera* (Tulip Tree) Willis.

CANCER *Melandrium rubrum* (Red Campion) Scot. Grigson.

CANCER JALAP *Phytolacca decandra* (Poke-root) USA Henkel. Jalap is a purgative - Spanish jalapa, from Aztec Xalapan, "sand by the water" (xalli - sand, atl - water).

CANCER-ROOT 1. *Orobanche fasciculata* (Piñon Strangleroot) USA Elmore. 2. *Orobanche virginiana*. 3. *Phytolacca decandra* (Poke-root) Watt.

CANCER-ROOT, One-flowered *Orobanche uniflora*.

CANCERWORT 1. *Kickxia elatine* (Sharp-leaved Fluella) B & H. 2. *Kickxia spuria* (Round-leaved Fluella) B & H. 3. *Veronica officinalis* (Common Speedwell) USA Cunningham. The Fluellas were once used in the treatment of cancerous ulcers, and speedwell was used for ulcers by Gerard's recommendation.

CANDELABRA CACTUS *Euphorbia lactea* USA Kingsbury.

CANDELABRA FLOWER *Brunsvigia orientalis*.

CANDELABRA TREE 1. *Araucaria angustifolia* (Brazilian Pine) Leathart. 2. *Euphorbia ingens*.

CANDIES *Knautia arvensis* (Field Scabious) Antrim B & H.

CANDLE-BARK GUM *Eucalyptus rubida*.

CANDLE-NUT TREE *Aleurites moluccana*. The oily kernels of the seeds were once used for making candles in its native Malaysia.

CANDLE PLANT 1. *Dictamnus fraxinella* (False Dittany) Willis. 2. *Senecio articulatus*. As with another name, Burning Bush, given to False Dittany because of the fact that the whole plant secretes a volatile oil which smells like lemon-peel. It is said that the atmosphere round it will often take fire if approached by someone with a lighted candle, without injuring the plant.

CANDLE TREE 1. *Parmentaria cerifera*. 2. *Pentadesma butyracea* Dalziel. The hanging fruits of Parmentaria cerifera look like yellow wax candles. In the case of Pentadesma butyracea the reference is to the fat that is extracted from the seeds, used for soap and candle making in Africa.

CANDLEBERRY 1. *Myrica gale* (Bog

Myrtle) Som. Macmillan. 2. *Myrica pennsylvanica* (Bayberry) USA Leighton. The berries were still used for candles in New England in Sanford's day. They are boiled, and the scum rising to the surface skimmed off and moulded into fragrant candles. The name was probably given by the early New England settlers, who made night lights from the white wax crust of the berries.

CANDLEGOSTES *Orchis mascula* (Early Purple Orchis) B & H. A variation of Gandergosses.

CANDLEMAS BELLS 1. *Anemone nemorosa* (Wood Anemone) Som. Mabey & Evans. 2. *Galanthus nivalis* (Snowdrop) Wilts. Macmillan; Glos. B & H; Scot. Banks. The snowdrop was dedicated to the Purification of the Virgin (Candlemas) (2nd Feb.), when it was the custom for young women dressed in white to walk in procession at the feast. A stage from this would mean that it was sacred to virgins in general, and at one time the receipt of snowdrops from a girl meant to a man that his attentions were not wanted. Near Hereford Beacon (the one place in England where it may possibly be native) a bowl of snowdrops brought into the house on Candlemas Day is thought to purify the house. Cf. Purification Flower for the snowdrop, and Candlemas Caps for the wood anemone.

CANDLEMAS CAPS *Anemone nemorosa* (Wood Anemone) Som. Macmillan. As with snowdrops (previous entry) it is the time of flowering which accounts for the name.

CANDLES 1. *Aesculus hippocastanum* flowers (Horse Chestnut) Som. Macmillan. 2. *Fraxinus excelsior* seed (Ash keys) Som. Macmillan; Dor. Halliwell. 3. *Sedum acre* (Biting Stonecrop) Som. Macmillan. Descriptive, especially for Horse Chestnut flowers. But why is there no name like Fairy's Candles for them? The story is that it was the Horse Chestnut that kept its candles burning to light the fairies home after their dance.

CANDLES, Bird's; CANDLES, Lady's; CANDLES, Our Lady's; CANDLES, Virgin Mary's *Verbascum thapsus* (Mullein) Grigson (Bird's); Som. Grigson (Lady's); Som. Macmillan (Our Lady's); Ire. Grigson (Virgin Mary's). Mulleins were used as torches at funerals, etc. The dried stalks were often used, smeared with grease or tar, as torches in the French fête des Brandons, on the first Sunday in Lent. Cf. the names Torches, Torchwort, Lady's Taper, King's Taper, Hag-taper, etc. Note that the Lady in Lady's Candle is Our Lady, the Virgin Mary. But Bird's Candle does not really exist in English - it is only a translation from the Welsh.

CANDLES, Bushman's *Sarcocaulon burmannii*.

CANDLES, Christmas *Aesculus hippocastanum* flowers (Horse Chestnut) Som. Macmillan. Just as the sight of the tree in full bloom seems to be a reminder of the Christmas Tree, so the individual flower spikes are the Christmas candles.

CANDLES, Desert *Eremurus sp* (Foxtail Lily) Perry.

CANDLES, Devil's 1. *Mandragora officinalis* (Mandrake) C.J.S.Thompson. 2. *Zantedeschia aethiopica* (Arum Lily) Corn. Deane & Shaw. Mandrake gets the name from the belief that the leaves would give out light at night-time. In the first century A.D., Josephus said that in the dungeons of the castle of Machaeras at Baras, there grew a root that was flame coloured and shone like lightning on persons who attempted to approach. There may be some basis in fact for the belief that mandrake shines at night - for some reason, its leaves are attractive to glow-worms. This name is not genuine in English, for it is actually a translation from the Arabic. Arum Lily gets the name, firstly as a description ('candle'), then, for 'devil's', because of its association with the dead.

CANDLES, Our Lord's *Yucca whipplei* USA Elmore.

CANDLES, Roman 1. *Aesculus hippocastanum* flowers (Horse Chestnut) Som.

Macmillan. 2. *Kniphofia uvaria* (Red Hot Poker) Som. Macmillan. Christmas Candles is recorded in the same county for Horse Chestnut flowers, and Robber's Lanterns is a name in the same vein, from Dorset. The name applied to Red Hot Poker is an obvious one.

CANDLES, Swamp *Lysimachia terrestris* (Bulb-bearing Loosestrife) USA House.

CANDLESTICK SENNA *Cassia alata*.

CANDLESTICKS 1. *Geranium robertianum* (Herb Robert) Dev, Dor. Macmillan. 2. *Orchis mascula* (Early Purple Orchis) Som. Macmillan. 3. *Sedum acre* (Biting Stonecrop) Som. Macmillan.

CANDLESTICKS, Devil's 1. *Glechoma hederacea* (Ground Ivy) Som, War. Dyer; Yks. Addy. 2. *Onobrychis sativa* (Sainfoin) Som. Macmillan. A strange name, for generally speaking ground ivy has been reputed a protector against witchcraft, fairies, etc. Equally strange for sainfoin.

CANDLESTICKS, Fried *Orchis mascula* (Early Purple Orchis) Som. Macmillan. Candlesticks here is probably Gander-gosses.

CANDLESTICKS, Great *Lychnis chalcedonica* (Jerusalem Cross) Tynan & Maitland. Cf. Lampflower for the same plant.

CANDLESTICKS, Lady's *Primula vulgaris var.elatior* (Polyanthus) N.Eng. B & H. Descriptive.

CANDLESTICKS, St Olaf's *Moneses uniflora*.

CANDLEWEEK FLOWER; CANDLEWICK FLOWER *Verbascum thapsus* (Mullein) both Cockayne. Mullein was once called Lucernaria, or wick-plant, as it was useful for wicks for lamps. It was the fluff from the leaves, collected by penitents, that was used for the wicks of altar candles. Cf., too, the various taper, candle and torch names.

CANDLEWICK STATICE *Limonium suworowii*.

CANDY, Corn Marigold of *Chrysanthemum coronarium* (Annual Chrysanthemum) Parkinson.1629. Candy = Crete. Perhaps a little too specific, but our familiar annual chrysanthemums had been introduced from southern Europe by Parkinson's time.

CANDY, Dittany of *Origanum dictamnus* (Dittany) Henslow. Gerard's Dittanie of Crete is the same name. Dittany itself is from Mount Dicte, in Crete.

CANDY, Sugar *Rosa canina* the shoots and the hips (Dog Rose) Wilts. Macmillan.

CANDY MUSTARD *Iberis amara* (Candytuft) Prior. Prior explains that the plant was brought from Crete, or Candy.

CANDYTUFT *Iberis amara*. Candy is Candia, another name for Crete. Tuft is descriptive of the "corymbose habit of the plant" (Grigson.1974).

CANDYTUFT, Bitter *Iberis amara* (Candytuft) H C Long.1910. The specific name is *amara*, bitter.

CANDYTUFT, Evergreen *Iberis sempervirens* (Perennial Candytuft) A.W.Taylor.

CANDYTUFT, Naked-stalked *Teesdalia nudicaulis* (Shepherd's Cress) Curtis.

CANDYTUFT, Perennial *Iberis sempervirens*.

CANDYTUFT, Persian *Aethionema sp.*

CANE, Indian *Canna indica* (Indian Shot) Tradescant. Cane in this case must surely be Canna?

CANE, Sweet *Acorus calamus* (Sweet Flag) Grieve.1931.

CANELLA, Red *Cinnamodendron corticosum* (False Winter's Bark) Le Strange. The reference is to *Canella alba*, White Cinnamon.

CAÑIGUA; CAÑIHUA *Chenopodium pallidicaule*. An Andean food plant.

CANKER 1. *Papaver rhoeas* (Red Poppy) Glos. PLNN35; Norf. Grigson; Suff. Moor. 2. *Rosa canina* fruit (Rose hips) Dor, Cambs, Ess, Norf. Grigson. 3. *Rosa canina* (Dog Rose) Dev. F.Grose; Dor, Som, Ess, Norf, Lincs, Cambs. Grigson. 4. *Taraxacum officinale* (Dandelion) Glos. J.D.Robertson. Possibly, in the case of the Red Poppy, because of its supposed detrimental effect on arable land.

CANKER-BALLS *Rosa canina* galls Som. Elworthy.
CANKER-BERRY *Rosa canina* fruit (Rose hips) Wilts. Dartnell & Goddard; Kent Grigson.
CANKER-FLOWER *Rosa canina* (Dog Rose) B & H.
CANKER LETTUCE *Pyrola rotundifolia* (Round-leaved Wintergreen) O.P.Brown.
CANKER ROOT 1. *Coptis trifolia* (Goldthread) USA Cunningham. 2. *Limonium carolinianum* (American Sea Lavender) USA House. Both the Indians and the early settlers used Goldthread root as a remedy for sore and ulcerated mouths, and for mouth cankers. Cf. the name Mouthroot.
CANKER ROSE 1. *Papaver rhoeas* (Red Poppy) Norf, Suff. B & H. 2. *Rosa canina* (Dog Rose) Dev. Friend.1882; Som, Kent Grigson; Wilts. Dartnell & Goddard; Herts. Jones-Baker.1974. The galls on the Dog Rose are known as cankers. So does Canker Rose actually mean the rose on which cankers grow?
CANKER WEED *Senecio jacobaea* (Ragwort) Halliwell.
CANKER WEED, White *Prenanthes alba* (White Lettuce) USA Cunningham.
CANKERWORT 1. *Senecio jacobaea* (Ragwort) Fernie. 2. *Taraxacum officinale* (Dandelion) USA Henkel.
CANNA-GRASS *Eriophorum angustifolium* (Cotton-grass) Scot. Cumming. An anglicization of Gaelic canach.
CANNELL *Cinnamonium zeylanicum* (Cinnamon) Mandeville. Often with one 'n' - see a 13th century poem on the land of Cockaigne - "Pudrid with gilofre and canel". Halliwell has canel. Cf. Italian cannella, and French cannelle, because the sticks of rolled bark look rather like pieces of reed, which is what cannella means.
CANNON-BALL PLANT *Aspidistra elatior* (Aspidistra) Perry. Called that by the Victorians on account of its tolerance of shade, fluctuating temperatures, dust, smoke and general neglect.

CANOE BIRCH *Betula papyrifera* (Paper Birch) USA H.Smith.1928. American Indian birch bark canoes.
CANOE CEDAR *Thuja plicata* (Giant Cedar) Wit. Presumably because the Northwest Coast Indians made canoes from this.
CANOE-WOOD *Liriodendron tulipifera* (Tulip Tree) W.Miller. "...affords excellent timber for many uses: particularly the trunk, which is frequently hollowed, and made into a canoe sufficient to carry many people..." (J.Taylor).
CANT-ROBIN *Rosa pimpinellifolia* (Burnet Rose) Fife Grigson.
CANTALOUP MELON *Cucumis melo var.cantalupensis*. Cantaloup comes through French from Italian Cantalupo, the name of a former summer residence of the popes near Rome. The melon was cultivated there after its introduction from Armenia.
CANTERBURY BELLS 1. *Campanula glomerata* (Clustered Bellflower) Lyte. 2. *Campanula medium*. 3. *Campanula trachelium* (Nettle-leaved Bellflower) Gerard. 4. *Cardamine pratensis* (Lady's Smock) Gerard. Coventry Bells is also common for the Campanulas. The flowers are likened to the small bells carried by pilgrims to the shrine of St Thomas at Canterbury; though Genders said the name was taken from the bells of Canterbury Cathedral. The flower was used as their emblem by Chaucer's pilgrims.
CANTERBURY BELLS, Cup-and-saucer *Campanula calycanthema*.
CANTERBURY JACK *Bryonia dioica* (White Bryony) Kent Grigson.
CANTON CASSIA *Cinnamonium cassia* (Cassia) Leyel.1937. Cf. Chinese Cinnamon.
CANYON BIRCH *Betula fontinalis* (Streamside Birch) USA Elmore.
CANYON GRAPE *Vitis arizonica* (Arizona Wild Grape) USA Elmore.
CANYON OAK *Quercus chrysopelis* (Maul Oak) USA Schenk & Gifford.
CAORTHANN *Sorbus aucuparia* (Rowan)

Ire O Suileabhain.

CAP, Blue 1. *Centaurea cyanus* (Cornflower) Som. Macmillan; Kent Grigson; N'thants. A.E.Baker. 2. *Knautia arvensis* (Field Scabious) Som. Macmillan. 3. *Succisa pratensis* (Devil's-bit) Prior.

CAP, Candlemas *Anemone nemerosa* (Wood Anemone) Som. Macmillan.

CAP, Cuckold's *Aconitum napellus* (Monkshood) Ess, Cambs, Norf. B & H. Cf. Cuckoo's Cap for *A anglicum*.

CAP, Cuckoo's 1. *Aconitum anglicum* (Wild Monkshood) Shrop. B & H. 2. *Aconitum napellus* (Monkshood) Ches. Holland. Cf. Cuckold's Cap for *A napellus*.

CAP, Dalmatian *Tulipa sp* Gerard. Parkinson has Turk's Cap for tulips.

CAP, Fairy *Digitalis purpurea* (Foxglove) Som. Grigson; Lancs. Harland & Wilkinson.1873; Ire. Starkie.1933; USA Henkel.

CAP, Fairy's *Campanula rotundifolia* (Harebell) Dev, Som. Grigson; Wilts. Macmillan.

CAP, Finger *Digitalis purpurea* (Foxglove) Som. Macmillan.

CAP, Fool's 1. *Aquilegia vulgaris* (Columbine) Yks. Grigson. 2. *Solanum dulcamara* (Woody Nightshade) Som. Macmillan.

CAP, Friar's *Aconitum napellus* (Monkshood) Dev. Friend.1882. Friar's Cap = Monk's hood.

CAP, Gentleman's *Tulipa sp* (Tulip) Som. Macmillan.

CAP, Granny's 1. *Aquilegia vulgaris* (Columbine) Dor. Macmillan. 2. *Geum rivale* (Water Avens) Wilts. Dartnell & Goddard.1899.

CAP, Granny's, Egyptian *Geum rivale* (Water Avens) Wilts. Dartnell & Goddard.1899. Egyptian is a Wiltshire name for this plant; so is Granny's Cap. Can there really be a genuine case of putting the two names together in order to make something completely unwieldy?

CAP, Honey *Pedicularis palustris* (Red Rattle) Donegal Grigson.

CAP, Huff *Agropyron repens* (Couch-grass) Shrop. Grigson.1959. The name apparently means bully.

CAP, Huntsman's *Scrophularia aquatica* (Water Betony) Corn. B & H.

CAP, Kettle; CASE, Kettle *Orchis mascula* (Early Purple Orchis) S.Eng. B & H (Case); IOW Long (Cap). Cf. Kettle-pad, from Hampshire. Probably the same as the Kentish Keatlegs, or Skeatlegs, where the OE sceat meant to swathe.

CAP, Penny *Umbilicis rupestris* (Wall Pennywort) Dev, Som. Grigson.

CAP, Silver *Impatiens capensis* (Orange Balsam) USA Sanecki. Can cap be capensis?

CAP, Soldier's 1. *Aconitum anglicum* (Wild Monkshood) N'thants. B & H. 2. *Aconitum napellus* (Monkshood) N'thants. A.E.Baker. 3. *Orchis mascula* (Early Purple Orchis) Som. Macmillan.

CAP, Turk's 1. *Aconitum anglicum* (Wild Monkshood) N'thants. B & H. 2. *Aconitum napellus* (Monkshood) N'thants. A.E.Baker. 3. *Lilium martagon*. 4. *Tulipa sp* (Tulip) Parkinson.1629. Mackay says the name Tulip is derived from a Turkish word, tulband, meaning turban - hence Turk's Cap. Gerard has Dalmatian Cap for the tulip.

CAP, Turkey *Fritillaria meleagris* (Snake's Head Lily) Hudson. 'Turkey' may be 'Turk's' - it is called Turk's Head in Warwickshire, but equally it may be the equivalent of guinea-hen. The petals are spotted like this bird, and in fact meleagris is from the Greek meaning a guinea-hen.

CAP, White *Spiraea tomentosa* (Hardhack) Grieve.1931.

CAP, Witches' *Helianthus annuus* (Sunflower) Dev. Macmillan.

CAP-AND-FRILLS, Gentleman's; CAP-AND-FRILLS, Golden *Ranunculus ficaria* (Lesser Celandine) both Som. Macmillan.

CAP-DOCKEN *Petasites hybridus* (Butterbur) Yks. B & H. "the leaves are very great, like to a round cap or hat, called

in Latine Petasus..." (Gerard).
CAPE ALKANET *Anchusa capensis*.
CAPE ALOE *Aloe vera* Mitton.
CAPE ASPARAGUS *Aponogetum distachyum*. A South African asparagus substitute.
CAPE ASTER *Felicia amelloides* (Blue Marguerite).
CAPE BEAN *Phaseolus lunatus* (Lima Bean) Bianchini.
CAPE BLUE WATERLILY *Nymphaea capensis*.
CAPE BOX *Buxus macowanii*.
CAPE CHESTNUT *Calodendrum capense*.
CAPE CLOVER *Trifolium africanum*.
CAPE COWSLIP *Lachenalia aloides*.
CAPE CUDWEED *Gnaphalium undulatum*.
CAPE FIG *Ficus capensis*.
CAPE FORGET-ME-NOT *Anchusa capensis* (Cape Alkanet) W.Miller.
CAPE GOOSEBERRY 1. *Physalis bunyardii*. 2. *Physalis pubescens* (Husk Tomato) Bianchini.
CAPE HOLLY *Ilex mitis*.
CAPE HONEY FLOWER *Melianthus major*.
CAPE HONEYSUCKLE *Tecomaria capensis*.
CAPE HYACINTH *Galtonia candicans* Coats.
CAPE IVY *Senecio mikanioides* (German Ivy) Australia Watt.
CAPE JASMINE *Gardenia jasminioides*.
CAPE LEADWORT *Plumbago capensis*.
CAPE LILY *Crinum sp*.
CAPE MAHOGANY *Trichilia emetica*.
CAPE MARIGOLD 1. *Dimorphoteca aurantiaca* (Star of the Veldt) Perry. 2. *Dimorphoteca pluvialis* (Rain Daisy) Friend.
CAPE MISTLETOE *Viscum capense*.
CAPE PEA *Phaseolus vulgaris* (Kidney Bean) S.Africa Watt.
CAPE PONDWEED *Aponogetum distachyanum* (Cape Asparagus) RHS.
CAPE TREASURE FLOWER *Gazania pavonia*.
CAPE TULIP *Homeria collina*.
CAPE WEED *Cryptostemma calendula*.
CAPE WILLOW *Salix mucrinata*.
CAPER *Capparis spinosa* flower buds. Greek kapparis; the final 's' was thought eventually to be a plural, and was dropped.
CAPER, Dog *Capparis canescens*.
CAPER, False; CAPER, Wild *Euphorbia lathyris* (Caper Spurge) J.Smith (False); Prior (Wild). The fruits are used green as a caper substitute, but this can be poisonous, and they have been known to be fatal.
CAPER, Mitchell's *Capparis mitchelli*.
CAPER, Mount *Dactylorchis incarnata* (Early Marsh Orchid) Ire. B & H.
CAPER BUSH; CAPER PLANT; CAPER TREE *Euphorbia lathyris* (Caper Spurge) IOW Grigson (Bush); E.Ang, Yks. Forby (Plant); USA Bergen (Tree).
CAPER SPURGE *Euphorbia lathyris*. See under False Caper.
CAPEROILES *Lathyrus montanus* (Bittervetch) Scot. Grigson. Obviously the same word as gave the Scottish Knapperts, and its variants.
CAPITATE MINT *Mentha aquatica* (Water Mint) H C Long.1910.
CAPON TREE *Viburnum lantana* (Wayfaring Tree) Cumb. Rowling. Presumably this was originally Coven-tree, still quite common, and Cobin Tree. The name was given to any tree which stood before a Scottish mansion, under which the laird went out to greet his visitors. So the word is covyne, a meeting or trysting place.
CAPON'S FEATHER 1. *Aquilegia vulgaris* (Columbine) Halliwell. 2. *Valeriana officinalis* (Valerian) B & H.
CAPON'S TAIL 1. *Aquilegia vulgaris* (Columbine) B & H; USA Hutchinson. 2. *Centranthus ruber* (Spur Valerian) Friend. 3. *Valeriana officinalis* (Valerian) Turner.
CAPMINT *Calamintha ascendens* (Calamint) Yks. B & H. As Catmint also occurs for this, Capmint must be the same word.
CAPPADOCIA, Oak of *Ambrosia elatior*.

CAPPADOCIAN MAPLE *Acer cappadocicum*.
CAPPER *Capparis spinosa* (Caper) Gerard. Merely Gerard's alternative spelling.
CAPRIFIG, Wild *Ficus carica var.sylvestris*.
CAPRIFOLY *Lonicera caprifolium* (Perfoliate Honeysuckle) Prior. Caprifoly is med. Latin caprifolium, goat leaf, a mistake for capparifolium, caper leaf (Prior).
CAPRIFOY *Lonicera periclymenum* (Honeysuckle) Dev. Grigson. Med. Latin caprifolium - goat leaf, a mistake,according to Prior (see Caprifoly). But Goat's Leaf is a recorded Devon and Somerset name.
CAPSICUM *Solanum capsicastrum* (Winter Cherry) Harvey. Perhaps from the Latin capsa, case.
CAPTAIN-OVER-THE-GARDEN *Aconitum anglicum* (Wild Monkshood) Yks. B & H. In accord with the military nature of the helmet-like flowers.
CAPUCHIN'S BEARD *Plantago coronopus* (Buck's Horn Plantain) Bianchini.
CAPUCINE ROSE *Rosa foetida*.
CAPULI *Prunus virginiana* (Choke Cherry) USA Driver. There is a Prunus capulí - the word seems to be Mexican, and is probably a native name.
CARAMAILE; CARMALE; CARMILE; CARMYLIE *Lathyrus montanus* (Bittervetch) all Scot. all B & H. From the Gaelic caermeal, the name of the plant, itself coming from corr, a crane, and meilg, a pod. It is Pys y garanod, Crane's peas, in Welsh, too.
CARAMBOLA *Averrhoa carambola*.
CARBERRY *Ribes uva-crispa* (Gooseberry) Yks, N.Eng. B & H. Carberry is, or was, the usual word for gooseberry in the north country. According to A S Palmer, it is from the AS gar, thorn.
CARD TEASEL; CARD THISTLE *Dipsacus fullonum* (Teasel) Gerard; USA Henkel (Card Teasel); B & H (Card Thistle). Originally, the implement consisting of teasel heads set in a frame, for raising the nap of cloth, was called the card, and the process carding.
CARDAMOM *Elettaria cardamomum*. The spice can be either the seed, or the dried fruit. The Latin cardamomum comes from the Greek kardamomom, from kardamon, cress plus amomom, Indian spice.
CARDAMOM, Bengal *Amomum aromaticum*.
CARDAMOM, Cameroon *Aframomum hanburyi*.
CARDAMOM, Round *Amomum cardamomum*.
CARDIACKE *Alliaria petiolata* (Jack-by-the-hedge) B & H.
CARDINAL, Red *Erythrina herbacea*.
CARDINAL CLIMBER *Ipomaea x multifida*. Crimson flowers,hence this name.
CARDINAL FLOWER *Lobelia cardinalis*. The colour of the flowers.
CARDINAL FLOWER, Blue *Lobelia syphilitica* (Great Lobelia) USA Kingsbury. Rather an anomaly as a name, as Cardinal Flower (*L. cardinalis*) is named for colour only.
CARDOON *Cynara cardunculus var. altilis*. This is French cardon, from carde, the edible part of the artichoke, and eventually from Latin carduus, thistle. Cardoon in English was first of all applied to the artichoke.
CARE; CARR *Sorbus aucuparia* fruit (Rowan berries) Corn. Courtney.1880; Dev. Friend.1882 (Care); Corn. Courtney.1887 (Carr). This is old Cornish caer, meaning a berry, becoming Keer in the same area, Cayer in Pembrokeshire, and Cuirn, or Keirn, in the Isle of Man (Caorthann in Ireland).
CARE-TREE *Sorbus aucuparia* (Rowan) Corn, Dev. Grigson.
CARELUCK *Sinapis arvensis* (Charlock) Glos. J.D. Robertson. Better known as Carlock; this is a rather extreme example in the series.
CARETT *Daucus carota* (Carrot) R.T.Gunther.
CARIB YAM *Rajania cordata*.
CARICATURE PLANT *Graptophyllum pictum*. The curiously-marked leaves sometimes

bear some resemblance to faces, hence this name. The generic name, *Graptophyllum*, is from Gk graptos, paint or write, and phyllon, leaf.

CARILLA *Momordica charantia* (Wild Balsam-apple) Chopra. Cf. Karela, obviously the same word.

CARL HEMP *Cannabis sativa* (Hemp) W.Miller. i.e. male hemp. Hemp was divided into male and female. Male hemp was for ropes, sacking, etc., i.e. coarse use; female hemp was for sheets and other domestic uses. Cf. Churl Hemp, which means the same.

CARLDOD *Plantago lanceolata* (Ribwort Plantain) Banff, Aber. Denham/Hardy. See Carldoddie, the more usual name.

CARLDODDIE 1. *Cirsium heterophyllum* (Melancholy Thistle) Ayr. Grigson; Forf. B & H. 2. *Plantago lanceolata* (Ribwort Plantain) Mor. Denham/Hardy. 3. *Plantago major* (Great Plantain) Forf. B & H. The flower heads of Ribwort are used as "soldiers", or "fighting cocks" by children. One child holds out a soldier, and the other tries to decapitate it with another, and so on, like Conkers. In parts of Scotland, the game of Soldiers is called Carldoddie. One derivation is that the name comes from the two names of the Chevalier and King George - Charles George, or that the Carls were the supporters of Charles Stuart, and the Doddies the supporters of King George. Doddy is a local (Aberdeen) name for George.

CARLICUPS *Caltha palustris* (Marsh Marigold) Som. Macmillan.

CARLIN-HEATH; CARLIN-HEATHER *Erica cinerea* (Bell Heather) Yks, Scot. Grigson (Heath); Scot. Jamieson (Heather). This must equate with the North Country She-heather, better applied to *E. tetralix*. This to distinguish from He-heather, which is *Calluna vulgaris*.

CARLIN-SPURS *Genista anglica* (Needle-whin) Scot. Jamieson. The spurs refer to the spines.

CARLINE THISTLE; CARLINE, Common *Carlina vulgaris* Carline Thistle is the usual name for this. Common Carline is in Murdoch McNeill. There is a tradition that the Emperor Charles V dreamt that the decoction of the root would cure the plague then rife in his army(Aubrey). This, then, would be to account for the name Carline; but surely the Charles is more likely to have been Charlemagne? Gubernatis offers the sequence Carduus sanctus - carolus sanctus - carline - erba carlina (in Italian), and that sounds about right.

CARLINE THISTLE, Dwarf *Cirsium acaulon* (Picnic Thistle) Gerard.

CARLINE THISTLE, Stemless *Carlina acaulis*.

CARLOCK *Sinapis arvensis* (Charlock) common. One of the many corruptions of Charlock.

CARLOCK CUPS *Caltha palustris* (Marsh Marigold) Som. B & H. Cf. Carlicups.

CARMEL *Lathyrus montanus* (Bittervetch) Yks, Scot. Grigson. From the Gaelic name of the plant, caermeal. It gives Caramaile, Carmale, Carmylie, and a number of others.

CARN-NOCK *Ononis repens* (Rest-harrow) Wilts. Tynan & Maitland. One would question the accuracy of this record; it looks very like a misreading of Cammock.

CARNADINE *Dianthus caryophyllus* (Carnation) B & H.

CARNATION *Dianthus caryophyllus*. Carnation is more properly coronation, from Latin corona, chaplet, presumably because the flowers were used in making chaplets. Gerard, though, speaks of the "pleasant Carnation colour, whereof it tooke his name". The colour seems to have a different derivation - from Latin, carn, caro - flesh, hence a flesh-coloured flower.

CARNATION, Spanish *Caesalpina pulcherrima* (Barbados Pride) Barbados Gooding.

CARNATION POPPY *Papaver somniferum* (Opium Poppy) Chopra.

CARNAUBA WAX PALM *Copernicia cerifera*. Carnauba wax, the leaf wax of this tree, is the usual base of shoe polishes.

CAROB-TREE 1. *Ceratonia siliqua*. 2.

Jacaranda procera. Arabic kharrub(a).
CAROBE-TREE *Cercis siliquastrum* (Judas Tree) Turner.
CAROLINA BEAN *Phaseolus lunatus* (Lima Bean) USA Bianchini.
CAROLINA BLACK POPLAR *Populus angulata* (Common Cottonwood) Brimble.1948.
CAROLINA HEMLOCK *Tsuga caroliniana*.
CAROLINA JESSAMINE *Gelsenium sempervirens*. A yellow climber, very similar to jasmine, as its other names, like False Jasmine, Wild Jessamine, etc., indicate.
CAROLINA LILY *Lilium michauxii*.
CAROLINA NETTLE *Solanum carolinense* (Carolina Nightshade) USA Allan.
CAROLINA NIGHTSHADE *Solanum carolinense*.
CAROLINA POPLAR *Populus deltoides*.
CAROLINA SPRING BEAUTY *Claytonia caroliniana* (Broad-leaved Spring Beauty) USA House.
CAROLINA TEA *Ilex vomitoria* (Cassine) Schery.
CAROSELLA *Foeniculum vulgare var. piperitum* (Sicilian Fennel) E.Hayes. Obviously an Italian name, but why should this be like a merry-go-round? Although, any umbellifer, by the nature of the way they display their flowers, deserve the name - Cf. Cart-wheels, etc.
CARPATHIAN BURL ELM *Ulmus procera* timber (English Elm) USA Wilkinson.1978.
CARPENTER, Herb 1. *Ajuga reptans* (Bugle) Gerard. 2. *Prunella vulgaris* (Self-heal) B & H. In both cases, the plant had the reputation of curing small wounds, hence the name Herb Carpenter, as being most likely to cause the condition.
CARPENTER, Proud *Prunella vulgaris* (Self-heal) Ches. Holland.
CARPENTER'S APRON *Lapsana communis* (Nipplewort) War. Grigson.
CARPENTER'S CHIPS *Nasturtium officinale* (Watercress) Glos. Grigson.
CARPENTER('S) GRASS 1. *Achillea millefolium* (Yarrow) B & H; USA Henkel. 2. *Prunella vulgaris* (Self-heal) Som, Ches. Grigson. In both cases, as the plants are styptic, one presumes that the name reflects the cure of small wounds such as carpenters experience.
CARPENTER'S HERB 1. *Achillea millefolium* (Yarrow) B & H. 2. *Ajuga reptans* (Bugle) Som. Macmillan. 3. *Prunella vulgaris* (Self-heal) Gerard. See Carpenter's Grass.
CARPENTER'S SQUARE *Scrophularia nodosa* (Figwort) Grieve.1931.
CARPENTER'S WEED *Achillea millefolium* (Yarrow) Fernie. See Carpenter's Grass.
CARPET, Golden *Sedum acre* (Biting Stonecrop) Som. Macmillan.
CARPET, Squaw's *Ceanothus prostratus*.
CARPET BUGLE *Ajuga reptans* (Bugle) USA Cunningham.
CARPET PLANT, Green *Herniaria glabra* (Smooth Rupture-wort) USA Cunningham.
CARPET WEED 1. *Kallstroemeria hirsutissima* 2. *Mollugo verticillata*.
CARPET WEED, Mountain *Euphorbia montana* (Mountain Spurge) USA Elmore.
CARPET WEED BEDSTRAW *Galium mollugo* (White Bedstraw) USA Henkel.
CARR-GOLD *Chrysanthemum segetum* (Corn Marigold) Lancs. Grigson. Carr is boggy ground, but that is not the environment that Corn Marigold would like.
CARRAG 1. *Rumex obtusifolius* (Broad-leaved Dock) IOM Moore.1924. 2. *Plantago major* (Great Plantain) IOM Moore.1924.
CARRAN *Spergula arvensis* (Corn Spurrey) IOM Moore.1924.
CAR(R)AWAY *Carum carvi*. Carraway seeds were known to the Arabians, who called them Karawya, which is the origin of carraway, as well as of Carvi and the Spanish alcarahueya. The word can be spelt with one or two r's.
CARRAWAY, Black 1. *Nigella sativa* (Fennel-flower) G B Foster. 2. *Pimpinella saxifraga* (Burnet Saxifrage) Thomson.1978.
CARRAWAY, Corn *Petroselinum segetum* (Corn Parsley) McClintock.

CARRAWAY, False *Carum oregonum.*
CARRAWAY, Russian *Nigella sativa* (Fennel-flower) G.B.Foster. Russian, because the seeds are used on Russian rye bread.
CARRAWAY, Whorled *Carum verticillatum.*
CARRAWAY, Wild *Anthriscus sylvestris* (Cow Parsley) Banff. B & H.
CARRAWAY THYME *Thymus herba-baroni.* The leaves used to be rubbed on the baron of beef, to give the distinctive carraway flavour.
CARRION-FLOWER *Smilax herbacea.* From the foul smell of the flowers, likened to that of decaying flesh.
CARRION-FLOWER, Warty *Stapelia verrucosa var. robusta.* All species of Stapelia attract pollinating flies by the yellow or livid colouring of their flowers, and by giving off a stench of rotting flesh or fish.
CARROT 1. *Conium maculatum* (Hemlock) Uist Banks. 2. *Daucus gingidum* (Sea Carrot) Uist Carmichael. Carrot is O.French carotte, Latin carota, from Greek Karoton.
CARROT, Moon *Seseli libanoticum.*
CARROT, Native *Geranium parviflorum* (Small-flowered Cranesbill) Tasmania H.L.Roth. The Tasmanians used to roast the large, fleshy roots for food, hence it was called there Native Carrot.
CARROT, Sea *Daucus gingidum.*
CARROT, Spring *Potentilla anserina* (Silverweed) Freethy. The roots are edible, and have always been a marginal or famine food.
CARROT, Wild 1. *Cicuta maculata* (American Cowbane) USA Kingsbury.1967. 2. *Daucus carota*. 3. *Daucus pusillus* USA Elmore.
CARROT BROOMRAPE *Orobanche maritima.*
CARROT PLANT *Escholtzia californica* (Californian Poppy) Som. Macmillan. Presumably because of the look of the root.
CARSLOPE *Primula veris* (Cowslip) Yks. Grigson.
CARSONS 1. *Cardamina pratensis* (Lady's Smock) S.W. Scot. Grigson. 2. *Nasturtium officinale* (Watercress) B & H. i.e., at any rate for Lady's Smock, growing on carse - low, rich, damp land.
CARTAPHILAGO *Gnaphalium sylvaticum* (Heath Cudweed) Turner.
CARTHAGENA BARK *Cinchona lancifolia* (Colombian Bark) Le Strange.
CARTHAGINIAN APPLE *Punica granatum* (Pomegranate) W.Miller. A simple translation of the Latin mala punica. The fruit was very popular in Carthage, apparently, or was it simply that Carthage was an early exporting centre?
CARTHAMINE *Carthamus tinctorius* (Safflower) Moldenke.
CARTHAMUS, Dyeing *Carthamus tinctorius* (Safflower) Buhler.
CARTHUSIAN PINK *Dianthus carthusianorum.*
CARTWHEEL *Conium maculatum* (Hemlock) Som. Macmillan. It is the shape of the umbel which gives this name.
CARTWHEEL-FLOWER 1. *Heracleum montegazzianum* (Giant Hogweed) Perry. 2. *Heracleum villosus* Chaplin. Descriptive.
CARUWAIE *Carum carvi* (Caraway) Gerard.
CARVER'S TREE *Tilia europaea* (Lime) Ablett. Because this is such a good, close-grained timber, not subject to warping, and therefore excellent for decorative carving. All the Grinling Gibbons carvings are in lime.
CARVI, Dog's *Anthriscus sylvestris* (Cow Parsley) Shet. Grigson. Presumably this means Dog's Carraway. Cow Parsley has been called Wild Carraway in Gloucestershire.
CARVIE; CARVY; CARVER *Carum carvi* (Carraway) Scot. F.M.McNeill (Carvie); Som. Halliwell (Carvy); Dor. Northover (Carver). All from the specific aname.
CASCA BARK *Erythrophlaeum guineense* (Ordeal Tree) Le Strange.
CASCADE FIR *Abies amabilis* (Red Silver Fir) Everett.
CASCARA SAGRADA *Rhamnus purshiana* (Californian Buckthorn). A mild laxative, marketed under various names, including

Sennapods. This name, meaning sacred bark, was given by the Spanish pioneers.

CASE, Water *Apium nodiflorum* (Fool's Watercress) Corn. Jago. i.e. watercress.

CASEWEED; CASEWORT *Capsella bursa-pastoris* (Shepherd's Purse) Gerard (Caseweed); Culpepper (Casewort). Case is from Latin capsa, box.

CASHES 1. *Anthriscus sylvestris* (Cow Parsley) Cambs. Turner. 2. *Conium maculatum* B & H; USA Henkel. This is the same word as the more usual Kex, or Kecksies.

CASHEW, Pear *Anacardium occidentale* (Cashew) Bianchini. Pear, from the general shape of the fruit.

CASHEW APPLE 1. *Anacardium occidentale* (Cashew Nut) Chopra. 2. *Mangifera occidentale*.

CASHEW NUT *Anacardium occidentale*. According to ODEE, this is Portuguese cajú, a variation of acajú (whence the French acajou, mahogany). The Portuguese word comes directly from the Tupi name for the tree.

CASKETS *Brassica oleracea var. capitata* (Cabbage) Dur. F.M.T. Palgrave. Cabbage stalks always used to be called Caskets in co Durham. Probably, from the sense of Brockett's entry, the word was used for any plant stem or stalk, not just cabbage stalks, though the name seems to be associated more with cabbages than with anything else.

CASPERE *Alliaria petiolata* (Jack-by-the-hedge) B & H.

CASPIAN LOCUST *Gleditsia caspica*.

CASPIAN WILLOW *Salix acutifolia*.

CASSABULLY *Barbarea vulgaris* (Winter Cress) S. Eng. B & H (including Corn. Courtney.1880). This is taken from Halliwell.

CASSADA, Wild *Jatropha gossypifolia* (Bellyache Bush) Dalziel.

CASSAVA; CASSAVA, Bitter *Manihot esculenta* (Manioc) Redfield & Villa (Cassava); A.W.Smith (Bitter Cassava). This is from a Haitian word originally, casavi; the present form is an alteration of this through French cassave. Bitter cassava, to differentiate from the Sweet Cassava (*Manihot dulcis*). Bitter or sweet according to the hydrocyanic acid content - low in the sweet kind, high in the bitter, and so it has to be eliminated.

CASSAVA, Sweet *Manihot dulcis*. See Bitter Cassava.

CASSE-WEED *Capsella bursa-pastoris* (Shepherd's Purse) B & H. This is Caseweed, case in this instance being Latin capsa, from which is derived Capsella.

CASSENA *Ilex vomitoria* (Cassine) Dallimore.

CASSIA *Cinnamonium cassia*. The Latin cassia comes ultimately from a Hebrew word qatsa, meaning to strip off, and refers to the bark.

CASSIA, Canton *Cinnamonium cassia* (Cassia) Leyel.1937. Cf. Chinese Cinnamon.

CASSIA, Desert *Cassia armata*.

CASSIA, Foetid *Cassia tora* (Sicklepod) Dalziel.

CASSIA, Horse *Cassia grandis*.

CASSIA, Long-pod *Cassia abbreviata* (Long-tail Cassia) Rhodesia Watt.

CASSIA, Long-tail *Cassia abbreviata*.

CASSIA, Purging *Cassia fistula* (Indian Laburnum) Willis. This is the source of the purgative Cassia pod.

CASSIA, Silver; CASSIA, Wormwood *Cassia artemisoides* RHS (Wormwood).

CASSIA BEAN *Cassia fistula* (Indian Laburnum) Watt.

CASSIA CINNAMON *Cinnamonium cassia* (Cassia) Watt.

CASSIASISTRE *Cassia fistula* (Indian Laburnum) Halliwell. Cassia tree?

CASSIDONY *Lavandula stoechas* Prior. This is an old book name, which according to Prior arose from the Latin stoechas sidonia, from Sidon, where the plant is indigenous.

CASSIDONY, Golden *Helichrysum stoechas*. Cassidony is an old book name for Spanish Lavender - *Lavandula stoechas*.

CASSIE *Acacia farnesiana*. The flowers are the cassie flowers of perfumery. They are the base of the various compounds for "violet" scents.

CASSINE *Ilex vomitoria*.

CASSOCKS *Agropyron repens* (Couch-grass) Grigson.1959; Wilts. Dartnell & Goddard.

CAST-IRON PLANT *Aspidistra elatior* (Aspidistra) Bonar. Like Cannon-ball plant, another Victorian name, it was given because the plant can withstand practically any adverse environmental conditions.

CAST-ME-DOWN *Lavandula stoechas* (Spanish Lavender) B & H. This is a corruption of Cassidony.

CAST-THE-SPEAR *Solidago virgaurea* (Golden Rod) Dor. Macmillan.

CASTINGS *Prunus spinosa* fruit (Sloes) Dev. Macmillan. This must be the same as the name, Keslings, given in the same county to bullace.

CASTLE, Single 1. *Orchis mascula* (Early Purple Orchis) Dor. B & H. 2. *Orchis morio* (Green-winged Orchid) Dor. Macmillan. Cf. the Devonshire Single Guss. It must be the same as Gandergosses.

CASTLE GILLIFLOWER *Matthiola incana* (Stock) Lyte.

CASTOCKS *Brassica oleracea var. capitata* stalks (Cabbage) Scot. C P Johnson. Kail-stalks, presumably. Cf. Caskets in N England.

CASTOR BEAN *Ricinus communis* (Castor Oil Plant) USA Kingsbury.

CASTOR OIL BEAN *Jatropha curcas* (Barbados Nut) W Indies Beckwith.

CASTOR OIL PLANT 1. *Fatsia japonica* (Rice Paper Plant) Rochford. 2. *Ricinus communis*. Castor oil must originally have been Casto oil, for the name came from Spanish agno casto. "Chaste", because of an early. reputation of being able "to soothe the passions". Given its undoubted laxative virtues, that may very well have been true.

CASTOR WOOD *Magnolia virginiana* (Sweet Magnolia) W.Miller.

CAT BED *Centranthus ruber* (Spur Valerian) Lincs. Grigson.

CAT BRIER *Smilax rotundifolia*.

CAT CHOOPS *Rosa canina* hips (Rose hips) Cumb. Grigson. Choops, or choups, are the North country words for hips. Sometimes they are called Chowps, too. Calling them Cat Choops is also north country, instead of the more usual Dog - hips, choops, berries, etc.

CAT HAWS *Crataegus monogyna* haws (Haws) N'thants. A.E.Baker; Lincs, Cumb. B & H; Yks. Morris; Dur. F.M.T. Palgrave. Brockett suggested that 'cat' here is the old word 'cates', meaning food. Haws are perfectly edible, of course, but what about all the other 'cat' names?

CAT HEATHER 1. *Erica cinerea* (Bell Heather) Scot. Grigson. 2. *Erica tetralix* (Cross-leaved Heath) Aber. B & H.

CAT HEP *Rosa pimpinellifolia* (Burnet Rose) N.Eng, Berw. Grigson. Hep for hip is common enough in the north. Cat-rose and Cat-whin are also recorded for this rose.

CAT HERB *Ballota africana* S.Africa Watt.

CAT HIPS *Rosa canina* (Dog Rose) B & H.

CAT JUGS *Rosa canina* hips (Rose hips) Yks, Dur. B & H. Jug is presumably the same word as choop.

CAT OAK *Acer campestre* (Field Maple) Yks. Grigson. Dog Oak, too, is recorded. Possibly both these names are to show it is not a true oak.

CAT O'NINE TAILS 1. *Corylus avellana* catkins (Hazel catkins) Dev. B & H. 2. *Typha latifolia* (False Bulrush) Som. Macmillan.

CAT PEA 1. *Lotus corniculatus* (Bird's-foot Trefoil) Corn. Grigson. 2. *Vicia cracca* (Tufted Vetch) Scot. Grigson.

CAT POSY *Bellis perennis* (Daisy) Som. Macmillan; Cumb. Grigson. The German has Katzenblume.

CAT PUDDISH *Lotus corniculatus* (Bird's-foot Trefoil) Cumb. Grigson. There are a lot of 'cat' names for Bird's-foot Trefoil. Cat's Clover, Cat-in-clover, Catten-clover, Cat's Claws, Cat Pea, etc.

CAT RASH; CAT RUSH *Euonymus europaeus* (Spindle Tree) both B & H. There is also Cat-tree, or Catty-tree - so almost certainly these are corruptions of Gatteridge. In this case, the 'cat' part means goat (OE gat).

CAT ROSE 1. *Rosa arvensis* (Field Rose) Ches. B & H. 2. *Rosa canina* (Dog Rose) Ches. Grigson. 3. *Rosa pimpinellifolia* (Burnet Rose) Yks. Grigson. *Rosa arvensis* gets the name, it is said, to distinguish it from Dog Rose, an explanation which falls down as soon as Dog Rose is called Cat Rose.

CAT TAIL, Broad-leaf; CAT TAIL, Common; CAT TAIL FLAG *Typha latifolia* (False Bulrush) USA E.Gunther (Broad-leaf); USA Kingsbury (Common); USA Elmore (Flag).

CAT THYME 1. *Teucrium marum*. 2. *Teucrium polium* J.Smith.

CAT-TRAIL *Valeriana officinalis* (Valerian) Yks. Grigson. Cf. Cat's Love, German Katzenkraut, and French herbe aux chats. Cats are said to be wildly fond of it. Tapsell's *Four-footed beasts*, 1658 - "The root of the herb Valerian, is very like to the eye of a cat and wheresoever it groweth, if cats come thereunto, they instantly dig it up for the love thereof,... for it smelleth moreover like a cat".

CAT-TREE 1. *Cornus sanguinea* (Dogwood) Shrop. B & H. 2. *Euonymus europaea* (Spindle Tree) Bucks. Grigson. In both cases, this is a corruption of Gatteridge, and so 'cat' here is goat (OE gat). Cf. Catwood and Catty-tree.

CAT WHIN 1. *Genista anglica* (Needle Whin) Yks, Cumb, Kirk, Wigt. Grigson. 2. *Ononis repens* (Rest Harrow) Som, Yks. Grigson. 3. *Rosa canina* (Dog Rose) Yks. Morris. 4. *Rosa pimpinellifolia* (Burnet Rose) Yks, N'thum. Brockett. 5. *Ulex gallii* (Dwarf Furze) Cumb. Grigson. 6. *Ulex minor* (Small Furze) Cumb. B & H. The Yorkshire name applied to the Dog Rose is pronounced catchin).

CAT'S CLAWS 1. *Acacia greggii*. 2. *Anthyllis vulneraria* (Kidney Vetch) Som. Macmillan; Mor. Grigson. 3. *Lotus corniculatus* (Bird's-foot Trefoil) Som. Macmillan; Bucks. Grigson. 4. *Ranunculus repens* (Creeping Buttercup) Lancs. Grigson. 5. *Robinia neomexicana* (New Mexican Locust)

USA Elmore. 6. *Rubus fruticosus* (Bramble) Som. Macmillan.

CAT'S CLOVER *Lotus corniculatus* (Bird's-foot Trefoil) N'thum. Grigson.

CAT'S EAR *Agrostemma githago* (Corn Cockle) Dor. Macmillan.

CAT'S EAR, Common *Hypochoeris radicata* (Long-rooted Cat's Ear) McClintock.

CAT'S EAR, Long-rooted *Hypochoeris radicata*.

CAT'S EAR, Smooth *Hypochoeris glabra*.

CAT'S EAR, Spotted *Hypochoeris maculata*.

CAT'S EYES 1. *Chamaenerion angustifolium* (Rosebay) Shrop. B & H. 2. *Geranium robertianum* (Herb Robert) Dev. Tynan & Maitland; Dor, Hants. Grigson. 3. *Hottonia palustris* (Water Violet) Som. Macmillan. 4. *Myosotis sylvatica* (Wood Forget-me-not) Herts. B & H. 5. *Veronica chamaedrys* (Germander Speedwell) Corn, Glos, Cumb. B & H; Dev. Friend.1882; Hants. Cope. 6. *Veronica persica* Ess. B & H.

CAT'S FACE *Viola tricolor* (Pansy) Som, Suss. Grigson.

CAT'S FOOT 1. *Antennaria dioica*. 2. *Antennaria neglecta* USA Zenkert. 3. *Glechoma hederacea* (Ground Ivy) N.Eng. Prior. 4. *Gnaphalium polycephalum* (Mouse-ear Everlasting) Grieve.1931.

CAT'S FOOT, Lesser *Antennaria neodioica*.

CAT'S HAIR 1. *Euphorbia hirta* (Hairy Spurge) Sofowora. 2. *Spergula arvensis* (Corn Spurrey) Corn. Grigson.

CAT'S HEAD *Salix sp.* catkins (Willow catkins) E.Ang. B & H.

CAT'S HOOF *Glechoma hederacea* (Ground Ivy) Tynan & Maitland.

CAT'S KEYS *Fraxinus excelsior* fruit (Ash keys) Norf. B & H.

CAT'S LOVE 1. *Centranthus ruber* (Spur Valerian) Wilts. Dartnell & Goddard. 2. *Valeriana officinalis* (Valerian) Wilts. Macmillan; Yks. Grigson. Cats do love it.

CAT'S LUGS *Primula auricula* (Auricula) Roxb. B & H. The leaves are more usually compared with bears' ears, rather than cats'.

CAT'S MILK 1. *Euphorbia amygdaloides*

(Wood Spurge) Glos. PLNN35. 2. *Euphorbia helioscopia* (Sun Spurge) Worcs. Prior.

CAT'S PAW 1. *Antennaria dioica* (Cat's Foot) Cumb. Grigson. 2. *Glechoma hederacea* (Ground Ivy) Folkard. 3. *Lotus corniculatus* (Bird's-foot Trefoil) Wilts. D.Grose.

CAT'S PEAS *Cytisus scoparius* (Broom) Corn. Grigson.

CAT'S TAIL 1. *Aconitum napellus* (Monkshood) Shrop. B & H. 2. *Amaranthus caudatus* (Love-lies-bleeding) Dev. Friend.1882. 3. *Corylus avellana* catkins (Hazel catkins) Dev. Friend.1882; Som. Tongue; Wilts. Dartnell & Goddard; Suss. Parish; N'thants. A.E.Baker. 4. *Echium vulgare* (Viper's Bugloss) E.Ang. B & H. 5. *Equisetum arvense* (Common Horsetail) Wilts. Dartnell & Goddard; Ches. Holland. 6. *Eriophorum angustifolium* (Cotton-grass) J.G.McKay. 7. *Hippuris vulgaris* (Mare's Tail) Dev, Som, Oxf. Grigson. Hants. Cope; E Ang. Forby. 8. *Juglans regia* catkins (Walnut) Lyte. 9. *Phleum nodosum*. 10. *Salix fragilis* catkins (Crack Willow catkins) Som. Macmillan. 11. *Typha latifolia* (False Bulrush) Som. Macmillan. Often as one word, i.e. Catstail.

CAT'S TAIL, Alpine *Phleum commutatum*.

CAT'S TAIL, Blue *Echium vulgare* (Viper's Bugloss) Herts. Grigson.

CAT'S TAIL, Greater; CAT'S TAIL GRASS, Great *Typha latifolia* (False Bulrush) Curtis (Cat's Tail); A.Stewart (Cat's Tail Grass).

CAT'S TAIL, Pointed *Phleum phleoides*.

CAT'S TAIL, Redhot *Acalypha hispida* (Chenille Plant) Hay & Synge.

CAT'S TAIL, Sand *Phleum arenarium*.

CAT'S TAIL, Smaller *Typha angustifolia* (Lesser Bulrush) Curtis.

CAT'S WHISKERS 1. *Gynandropsis gynandra* West Indies Dalziel. 2. *Tacca chantrieri* (Bat Flower) RHS.

CAT'S WORT *Nepeta cataria* (Catmint) USA Henkel.

CATALONIA SAFFRON *Carthamus tinctorius* (Safflower) Parkinson. Parkinson has Spanish Saffron for this, too.

CATALONIAN JASMINE *Jasminum grandiflorum* (Spanish Jasmine) Grieve.1931.

CATALPA *Catalpa bignonioides* (Locust Bean) Grigson.1974. Catalpa sometimes becomes Catawba in America. The name is from a Creek Indian name for the tree - Katuhlpa, meaning head without wings.

CATALPA, Thorny *Acanthopanax ricinifolium*.

CATALPA, Western *Catalpa speciosa*.

CATAPUCE *Euphorbia lathyris* (Caper Spurge) Chaucer. Used by Chaucer in the Nun's Priest's Tale, and identified as this species by Mabey & Evans.

CATAWBA *Catalpa bignonioides* (Locust Bean) USA Harper. i.e. Catalpa.

CATBERRY 1. *Nemopanthus mucronata* (Mountain Holly) USA Yarnell. 2. *Ribes uva-crispa* (Gooseberry) Cumb. Grigson. This name is confined to hedge gooseberries.

CATCH-TRAP *Drosera rotundifolia* (Sundew) Tynan & Maitland.

CATCHFLY 1. *Antirrhinum majus* (Snapdragon) Lincs. B & H. 2. *Silene armeria*. Catchfly was the name given to *Silene armeria* by Gerard, because of its glutinous stalks.

CATCHFLY, Alpine 1. *Silene alpestris*. 2. *Viscaria alpina*.

CATCHFLY, Alpine, Red *Viscaria alpina* (Alpine Catchfly) Clapham.

CATCHFLY, Berry *Cucubalis baccifer*.

CATCHFLY, Breckland *Silene otites*.

CATCHFLY, Drooping *Silene pendula*.

CATCHFLY, English *Silene anglica* (Small-flowered Catchfly) Hutchinson.

CATCHFLY, Flaxfield *Silene linicola*.

CATCHFLY, Forked *Silene dichotoma*.

CATCHFLY, German, Red *Viscaria vulgaris* (Red Catchfly) Clapham.

CATCHFLY, Hairy *Silene dichotoma* (Forked Catchfly) USA Zenkert.

CATCHFLY, Italian *Silene italica*.

CATCHFLY, Large-flowered *Silene elisabetha*.

CATCHFLY, Mexican *Silene laciniata.*
CATCHFLY, Mountain *Silene montana.*
CATCHFLY, Night-flowering *Melandrium noctiflorum.*
CATCHFLY, Nodding *Silene pendula.*
CATCHFLY, Northern *Silene wahlbergella.*
CATCHFLY, Nottingham *Silene nutans.* Nottingham, because it was common on the rocks upon which Nottingham Castle stands.
CATCHFLY, Red *Viscaria vulgaris.*
CATCHFLY, Rock *Silene rupestris* (Rock Campion) Blamey.
CATCHFLY, Sand *Silene conica.*
CATCHFLY, Sleepy *Silene antirrhina.*
CATCHFLY, Small-flowered *Silene anglica.*
CATCHFLY, Spanish *Silene otites* (Breckland Catchfly) Clapham.
CATCHFLY, Sticky *Viscaria vulgaris* (Red Catchfly) Blamey.
CATCHFLY, Sticky, White *Silene viscosa.*
CATCHFLY, Striated *Silene conica* (Sand Catchfly) Clapham.
CATCHFLY, Sweet William *Silene armeria* (Catchfly) Clapham.
CATCHFLY, Valais *Silene vallesia.*
CATCHGRASS; CATCHROGUE *Galium aparine* (Goose-grass) Ches. Holland (Catchgrass); Scot. Jamieson (Catchrogue). These, like many others, refer to the clinging powers of the burrs.
CATCHTHORN *Zizyphus abyssinica.*
CATCHWEED 1. *Galium aparine* (Goose-grass) Yks. Nicholson. 2. *Rubia peregrina* (Wild Madder) Som. Macmillan.
CATCHWEED BEDSTRAW *Galium aparine* (Goose-grass) USA Allan.
CATCLUKE; CATLUKE *Lotus corniculatus* (Bird's-foot Trefoil) both Scot. Jamieson. i.e. cat's claws.
CATECHU, Black *Acacia catechu* (Cutch Tree) Le Strange.
CATECHU, Pale *Uncaria gambier* (Gambier Cutch) Le Strange.
CATEN-AROES *Leucanthemum vulgare* (Ox-eye Daisy) Lancs. Grigson.
CATERPILLAR, Prickly *Amaranthus spinosus* (Thorny Pigweed) Barbados Gooding. *Amaranthus viridis* is called White Caterpillar on the same island.
CATERPILLAR, White *Amaranthus viridis* (Green Amaranth) Barbados Gooding. Cf. Prickly Caterpillar for *A spinosus* here.
CATERPILLAR GREASEWOOD *Sarcobatus vermicularis* (Greasewood) Jaeger.
CATESBY'S LILY *Lilium catesbaei.* This is Mark Catesby (c1680-1749), who first described this lily in the *Natural History of Carolina.*
CATFOOT POPLAR *Populus nigra* (Black Poplar) Lancs. Grigson. This is apparently a reference to the dark spots in the timber.
CATHARTIC FLAX *Linum catharticum* (Fairy Flax). J A MacCulloch.1905. Cf. Purging Flax. It is certainly effective as a purge, but thoroughly dangerous to use.
CATHAW-BLOW *Heracleum sphondyllium* (Hogweed) Cumb. Grigson.
CATHAY LILY *Cardiocrinum cathayanum.* Cathay is China.
CATMINT 1. *Calamintha ascendens* (Calamint) B & H. 2. *Glechoma hederacea* (Ground Ivy) Tynan & Maitland. 3. *Nepeta cataria.* 4. *Oreganum vulgare* (Marjoram) Tynan & Maitland. 5. *Teucrium scorodonia* (Wood Sage) Tynan & Maitland. Cats "are much delighted with catmint, for the smell of it is so pleasant unto them, that they rub themselves upon it, and swallow or tumble into it, and also feed on the branches very greedily" (Gerard).

If you set it
The cats will eat it
If you sow it
The cats will not know it.

Ground Ivy has much of the appearance of a dwarf catmint, hence the inclusion of the name.
CATMINT, Garden *Nepeta faasenii.*
CATMINT, Hairless *Nepeta nuda.*
CATMINT, Lesser *Calamintha nepeta.*
CATNEP; CATNIP; CATNAP *Nepeta*

cataria (Catmint) Brownlow (Catnep); USA Tolstead (Catnip); Tynan & Maitland (Catnap). Often found simply as Nep, or Nip, and their derivatives.

CATNUT *Conopodium majus* (Earth-nut) Yks. Grigson.

CATRUP *Nepeta cataria* (Catmint) USA Henkel.

CATRUSH (and spelt as separate words) 1. *Equisetum arvense* (Common Horsetail) Ches. Holland. 2. *Euonymus europaeus* (Spindle-Tree) B & H. For spindle-tree, there is Cat-rash also. Probably both these are merely corruptions of the name Gatteridge, common in the southern half of the country, a word which means goat-tree (OE gat).

CATS, Dogs-and- *Trifolium arvense* (Hare's-foot Trefoil) Mor. Grigson.

CATS, Pussy- *Corylus avellana* catkins (Hazel catkins) Wilts. Dartnell & Goddard. This is more usually Pussy Cat's Tails, or some other animal's tail.

CATS-AND-DOGS *Salix sp* catkins (Willow catkins) Corn. Jago; Wilts. Macmillan.

CATS-AND-EYES *Fraxinus excelsior* seeds (Ash keys) Yks. Morris. Probably an error for Cats-and-keys.

CATS-AND-KEYS 1. *Acer campestre* fruit (Maple keys) Dev. Friend.1882. 2. *Acer pseudo-platanus* fruit (Sycamore keys) Dev. B & H. 3. *Fraxinus excelsior* fruit (Ash keys) Dev. Friend.1882.

CATS-AND-KITTENS 1. *Corylus avellana* catkins (Hazel catkins) Worcs. B & H. 2. *Salix sp* catkins (Willow catkins) N'thants. A.E.Baker.

CATS-AND-KITLINGS *Salix sp* catkins (Willow catkins) N'thants. A.E.Baker.

CATSKIN 1. *Corylus avellana* catkins (Hazel catkins) Som. Elworthy. 2. *Salix sp* - catkins (Willow catkins) Som. Elworthy. Not as it appears, but possibly a contraction of Cats-and-kittens. After all, catkin does mean kitten, '-kin' being the diminutive. Cf. German Kätzchen and Dutch katteken.

CATTERIDGE TREE *Cornus sanguinea* (Dogwood) B & H. Gatteridge is more usual. Both 'cat' and 'gat' mean goat.

CATTIJUGS *Rosa canina* hips (Hips) Yks. Atkinson. A development of Cat's Jugs, i.e. Cat's Hips.

CATTIKEYNS *Fraxinus excelsior* fruit (Ash keys) Wilts. Dartnell & Goddard. An example of the old plural in '-n'.

CATTLE-BEET *Beta vulgaris* (Beet) E Angl. G E Evans.1956. Better known in this context as Mangold Wurzel.

CATTLE-KEYS *Fraxinus excelsior* fruit (Ash keys) Derb. Addy. A variant of Cats-and-keys.

CATWOOD 1. *Cornus sanguinea* (Dogwood) Shrop. Grigson. 2. *Euonymus europaeus* (Spindle Tree) Bucks. Grigson. Not cat, but OE gat - goat. Cf. Cat-tree.

CATTY-TREE *Euonymus europaeus* (Spindle Tree) Shrop. Grigson. See Catwood.

CAUCASIAN ASH *Fraxinus oxycarpa*.

CAUCASIAN COMFREY *Symphytum caucasicum* (Blue Comfrey) W.Miller.

CAUCASIAN ELM *Zelkova carpinifolia*.

CAUCASIAN FIR *Abies nordmanniana*.

CAUCASIAN LILY *Lilium monadelphum*.

CAUCASIAN LIME *Tilia euchlora* (Crimean Lime) Duncalf.

CAUCASIAN OAK *Quercus macranthera*.

CAUCASIAN SPIKENARD *Inula glandulosa*.

CAUCASIAN SPRUCE *Picea orientalis* (Oriental Spruce) Mitchell.1972. Grows from the Caucasus to Asia Minor.

CAUCASIAN STONECROP *Sedum spurium*.

CAUCASIAN WHORTLEBERRY *Vaccinium arctostaphylos*.

CAUCASIAN WINGNUT *Pterocarya fraxinifolia*.

CAUL *Brassica oleracea var. capitata* (Cabbage) Dev. Choape. The local variant of 'cole'.

CAUL, Yellow 1. *Ranunculus acris* (Meadow Buttercup) IOW B & H. 2. *Ranunculus bulbosus* (Bulbous Buttercup) IOW Grigson. 3. *Ranunculus repens* (Creeping Buttercup)

IOW Grigson. Like Bassinet, another name for the flower, this probably comes from the shape of the flower. Bassinet is French meaning a small basin, or a skull cap, such as worn under a helmet. Caul, being the membrane with which some babies' heads are covered at birth, must mirror the same idea.

CAULIFLOWER 1. *Brassica oleracea var. botrytis*. 2. *Sambucus nigra* flowers (Elder flowers) Som. Macmillan. Cauliflower is adapted from Italian cavolfiori - cabbage flowers.

CAVE *Brassica oleracea var.capitata* (Cabbage) N.Eng. Halliwell.

CAVENDISH BANANA *Musa nana* (Canary Banana) Hutchinson & Melville.

CAWLE, Sea *Calystegia soldanella* (Sea Bindweed) B & H. Cawle here is Cole - Sea Cole is also recorded for this.

CAXES 1. *Anthriscus sylvestris* (Cow Parsley) B & H. 2. *Conium maculatum* (Hemlock) B & H. 3. *Daucus carota* (Wild Carrot) Dor. Macmillan. Better known as Kex, and applied usually to the dried stalks of umbellifers.

CAYER *Sorbus aucuparia* (Rowan) Pemb. Grigson. The Welsh is cerdinen, cerddynen, and there are various other words from the Celtic languages. Old Cornish caer meant a berry, and Keer, Care are names from Cornwall. IOM has Keirn and Cuirn, while the Irish word is Caorthann.

CEARÁ RUBBER *Manihot glaziovii*.

CEDAR 1. *Cedrus sp.* 2. *Thuja occidentalis* (Arbor-vitae) USA H.Smith.1923. Cedar is ultimately from Greek kedros, which apparently could mean either cedar or juniper.

CEDAR, African; CEDAR, Algerian *Cedrus atlantica* (Atlas Cedar) Wit.

CEDAR, Atlantic, Blue *Cedrus atlantica var. glauca*.

CEDAR, Atlas *Cedrus atlantica*.

CEDAR, Barbados *Juniperus bermudiana* (Bermuda Cedar) Howes.

CEDAR, Bastard *Melia azedarach* (Chinaberry Tree) Dalziel.

CEDAR, Bermuda *Juniperus bermudiana*. Cf. Barbados Cedar.

CEDAR, Brown-berried *Juniperus oxycedrus* (Prickly Juniper) Moldenke.

CEDAR, Canoe *Thuja plicata* (Giant Cedar) Wit. The Indians made dugout canoes from this timber.

CEDAR, Chilean *Austrocedrus chilensis*.

CEDAR, Chinese *Cedrela sinensis* (Toon) Leathart.

CEDAR, Cigar-box *Cedrela odorata* (West Indian Cedar) N.Davey. It is used for making cigar boxes.

CEDAR, Clanwilliam *Widdringtonia juniperoides*.

CEDAR, Cyprus *Cedrus brevifolia*.

CEDAR, Deodar *Cedrus deodara* (Deodar) B Thomas. See Deodar.

CEDAR, Giant *Thuja plicata*.

CEDAR, Ground *Juniperus communis* (Juniper) Brimble.1948. Cf. Creeping, or Dwarf, Juniper.

CEDAR, Incense *Libocedrus decurrens*.

CEDAR, Incense, Chilean *Austrocedrus chilensis* (Chilean Cedar) Leathart.

CEDAR, Indian *Cedrus deodara* (Deodar) Dyer.

CEDAR, Japanese *Cryptomeria sp.*

CEDAR, Lebanon *Cedrus libanii*.

CEDAR, Mlanje *Widdringtonia whytei*. Mount Mlanje, in Malawi, is the best environment for these trees.

CEDAR, Mountain *Libocedrus bidwillii*.

CEDAR, Mountain, Red *Juniperus scopulorum* (Western Red Cedar) USA Elmore.

CEDAR, New Zealand *Libocedrus bidwillii* (Mountain Cedar) Grieve.1931.

CEDAR, Oregon *Chamaecyparis lawsoniana* (Lawson Cypress) Wit.

CEDAR, Pencil; CEDAR, Pencil, Virginian *Juniperus virginiana* Watt; N.Davey (Virginian). It is said that no other wood has been found which has just the right physical properties for the casing of lead pencils.

CEDAR, Pencil, African *Juniperus procera* (East African Juniper) N.Davey. *J.virginiana* is the Pencil Cedar par excellence.

CEDAR, Port Orford *Chamaecyparis lawso-*

niana (Lawson Cypress) USA Schenk & Gifford. Port Orford, on the Oregon coast, where it thrives particularly well.

CEDAR, Prickly; CEDAR, Sharp *Juniperus oxycedrus* (Prickly Juniper) Grieve.1931 (Prickly); Mitchell.1972 (Sharp).

CEDAR, Red 1. *Juniperus pinchotii*. 2. *Juniperus virginiana* (Virginian Juniper) USA Tolstead. 3. *Libocedrus decurrens* (Incense Cedar) Hay & Synge. 4. *Thuja plicata* (Giant Cedar) Hawthorn. 5. *Uapaca guineensis* Dalziel.

CEDAR, Red, Colorado *Juniperus scopulorum* (Western Red Cedar) Cunningham.

CEDAR, Red, Eastern *Juniperus virginiana* (Virginian Juniper) Wit.

CEDAR, Red, Jamaican *Cedrela odorata* (West IUndian Cedar) Brimble.1948.

CEDAR, Red, Japanese *Cryptomeria japonica* (Japanese Cedar) Polunin.1976.

CEDAR, Red, Pacific *Thuja plicata* (Giant Cedar) Wit.

CEDAR, Red, Southern *Juniperus silicicola*.

CEDAR, Red, Western 1. *Juniperus scopulorum*. 2. *Thuja plicata* (Giant Cedar) Clapham.

CEDAR, Rocky Mountain *Juniperus scopulorum* (Western Red Cedar) USA Elmore.

CEDAR, Salt *Tamarix glauca* USA Elmore.

CEDAR, Stinking 1. *Ailanthus glandulosa* (Tree of Heaven) Chopra. 2. *Chamaecyparis nootkatensis* (Alaska Yellow Cedar) Edlin. 3. *Torreya taxifolia*.

CEDAR, Summit *Athrotaxus laxifolia*.

CEDAR, Tasmanian *Athrotaxus selaginoides* Everett.

CEDAR, Tasmanian, Smooth *Athrotaxus cupressoides*.

CEDAR, Virginian *Juniperus virginiana* (Virginian Juniper) J.Smith.

CEDAR, West Indian *Cedrela odorata*.

CEDAR, White 1. *Chamaecyparis thyoides*. 2. *Melia azedanach* (Chinaberry Tree) USA Kingsbury. 3. *Pycnanthus kombo* (False Nutmeg) Dalziel. 4. *Thuja occidentalis* (Arbor-vitae) USA Zenkert. False Nutmeg timber looks rather like cedar, but it is softer and lighter, and not durable.

CEDAR, White, Californian *Libocedrus decurrens* (Incense Cedar) Grieve.1931.

CEDAR, White, Eastern; CEDAR, White, False; CEDAR, White, Northern *Thuja occidentalis* (Arbor-vitae) RHS (Eastern); Grieve.1931 (False); A.W.Smith (Northern).

CEDAR, Willowmore *Widdringtonia schwarzii*.

CEDAR, Yellow 1. *Chamaecyparis nootkatensis* (Alaska Yellow Cedar). Hawthorn. 2. *Thuja occidentalis* (Arbor-vitae) Grieve.1931.

CEDAR, Yellow, Alaska *Chamaecyparis nootkatensis*.

CEDAR GUM *Eucalyptus gunnii* (Cider Gum) Barber & Phillips. Is this genuine, or just an invention by the authors?

CEDAR OF GOA *Cupressus lusitanica* (Mexican Cypress) Mitchell.1972.

CEDAR PINE *Pinus glabra* (Spruce Pine) Everett.

CEDAR-WOOD *Artemisia abrotanum* (Southernwood) Ork. Leask. i.e. southernwood, though it is not obvious.

CEDRAT *Citrus cedrata*.

CELANDINE; CELANDINE, Wild *Impatiens capensis* (Orange Balsam) USA Grigson; USA Bergen.1896 (Wild). The name is applied to *Ranunculus ficaria*, or *Chelidonium*, and only incidentally, and in America, to this. The word is eventually Greek khelidonion, from khelidon, swallow - see Lesser Celandine for the legend.

CELANDINE, Brave *Caltha palustris* (Marsh Marigold) Lyte. Lyte must have been using the word brave in the sense of showy, grand.

CELANDINE, Greater *Chelidonium majus*.

CELANDINE, Lesser *Ranunculus ficaria*. Pliny says that swallows restore their sight by this plant. The name appears in Theophrastus, who says that the flower blooms when the swallow wind blows. There is a saying from Devonshire (re the restoration of sight legend):

Fennel, rose, vervain, celandine and rue
Do water make which will the sight renew.

CELANDINE, Major *Chelidonium majus* Folkard. i.e. Greater Celandine.

CELANDINE, Tree *Bocconia frutescens*.

CELANDINE CROCUS *Crocus korolkowi*.

CELANDINE POPPY *Chelidonium majus* (Greater Celandine) USA Kingsbury. Lesser Celandine is unrelated to Greater Celandine, *Chelidonium majus*, which is a member of the Poppy family.

CELERIAC *Apium graveolens var. rapaceum*. Same word as celery, with "arbitrary use of the suffix -ac" (ODEE).

CELERY, German *Apium graveolens var. rapaceum* (Celeriac) Bianchini.

CELERY, Green *Apium australe* (Maori Celery) New Zealand C. Macdonald.

CELERY, Italian, Upright *Apium graveolens* G.M.Taylor. Celery seems to be French celeri, from an Italian dialect word selleri, from the Latin selinon, selinum.

CELERY, Maori *Apium australe*.

CELERY, Prostrate *Apium australe* (Maori Celery) New Zealand C.Macdonald.

CELERY, Turnip-rooted *Apium graveolens var. rapaceum* (Celeriac) Bianchini.

CELERY, Water *Ranunculus sceleratus* (Celery-leaved Buttercup) W.Miller.

CELERY, Wild 1. *Apium graveolens*. 2. *Peucedanum galbanum* S.Africa Watt. 3. *Smyrnium olusatrum* (Alexanders) IOW Grigson. The name celery comes eventually from the Latin selinum, so linking it with parsley, with which it shares some other names. Gerard calls it Marsh Parsley, or Water Parsley, for celery as a name had not come in in Gerard's time (probably Evelyn.1664).

CELERY CABBAGE *Brassica pekinensis* (Chinese Cabbage) Schery. There is a celery-like flavour to this.

CELERY-LEAVED BUTTERCUP *Ranunculus sceleratus*.

CELERY PINE *Phylloclados trichomanoides* (Tanekaha) Duncalf.

CELERY-SEED *Rumex obtusifolius* (Broad-leaved Dock) Suss. B & H.

CELERY-TOP PINE *Phyllocladus rhomboidalis*.

CELLAR, Salt *Oxalis acetosella* (Wood Sorrel) Dor. Macmillan. Cellar must surely be a mere corruption of sorrel?

CELTIC NARD *Valeriana celtica* R T Gunther.1934.

CELTIS, Stink *Celtis cinnamonea*.

CEMBRIAN PINE *Pinus cembra* (Swiss Stone Pine) Ablett.

CENCLEFFE *Narcissus pseudo-narcissus* (Daffodil) Halliwell. There is a Welsh name for the daffodil - Cennin Pedr, Peter's Leeks. Is there a connection? And is Cencleffe some Celtic word?

CENTAURY *Centaureum eyrthraea*. The medicinal properties of the plant were said to have been discovered by Chiron the centaur.

CENTAURY, Buckley *Centaureum calycosum*.

CENTAURY, Channel *Centaureum tenuiflorum*. A rare plant from the Dorset and Isle of Wight coast.

CENTAURY, Corn *Centaurea cyanus* (Cornflower) B & H. Not a centaury, of course,though Centaurea must have the same derivation as Centaureum.

CENTAURY, Dumpy *Centaureum capitatum*.

CENTAURY, Greater 1. *Blackstonia perfoliata* (Yellow-wort) Clair. 2. *Centaurea scabiosa* (Greater Knapweed) Parkinson.1640.

CENTAURY, Guernsey *Exaculum pusillum*. Very rare, and confined to the northern part of Guernsey.

CENTAURY, Lesser 1. *Centaureum erythraea* (Centaury) Gerard. 2. *Centaureum pulchellum* (Slender Centaury) Blamey.

CENTAURY, Little *Centaureum eyrthraea* (Centaury) B & H.

CENTAURY, More *Blackstonia perfoliata* (Yellow-wort) Clair. More usually means root. Does it here?

CENTAURY, Narrow-leaved *Centaureum littorale* (Sea Centaury) Hepburn.

CENTAURY, Perennial *Centaureum scilloides.*
CENTAURY, Rock *Centaureum beyrichii.*
CENTAURY, Sea *Centaureum littorale.*
CENTAURY, Slender *Centaureum pulchellum.*
CENTAURY, Small *Centaureum eyrthraea* (Centaury) Gerard.
CENTAURY, Yellow 1. *Blackstonia perfoliata* (Yellow-wort) Ches, Yks Grigson. 2. *Cicendia filiformis* (Slender Cicendia) Blamey.
CENTAURY GENTIAN *Centaureum erythraea* (Centaury) Grieve.1931.
CENTINODE, CENTYNODY *Polygonum aviculare* (Knotgrass) both B & H. Cf. the less ambitious Ninety-knot, from Shropshire.
CENTORY; CENTURY *Centaureum erythraea* (Centaury) Camden/Gibson (Centory); Dev, Ches. Friend (Century).
CENTRE OF THE SUN *Centaureum erythraea* (Centaury) Worcs. Fernie. Centre of the sun is a simple corruption of Centaureum.
CENTRY *Anagallis tenella* (Bog Pimpernel) Wilts. Dartnell & Goddard. Probably a mistaken identity - it surely should be centaury? Though there is some slight resemblance between the two.
CENTURY PLANT 1. *Agave americana* (American Aloe) Safford.1915. 2. *Agave utahensis* Elmore. Because it was said to flower only once in a hundred years, something which is actually quite possible. It does only flower once, and the period of preparation during which it accumulates food material can be anything from five to a hundred years.
CERIMAN *Monstera deliciosa.*
CERNOYLE *Lonicera periclymenum* (Honeysuckle) B & H.
CERVISE TREE *Sorbus torminalis* (Wild Service) Aubrey (Nat.Hist.Wilts.).
CESTUS, White *Drimys winteri* (Winter's Bark) Pomet.
CETEWAYO *Solanum tuberosum* (Potato) S.Africa Watt.
CETYWALL *Valeriana officinalis* (Valerian) Halliwell. i.e. Setwall, which was what Gerard called it. It was also the name Chaucer used for the plant. Halliwell's version, he says, comes from Percy's Reliques.
CEYLON IRONWOOD *Mesua ferrea.*
CEYLON LEADWORT *Plumbago zeylanicum.*
CEYLON OAK *Schleichera oleosa.*
CEYLON ROSE *Nerium oleander* (Rose Bay) Watt. This becomes Selon's Rose sometimes.
CEYLON TEA TREE *Elaeodendron glaucum.*
CHADDOCK; CHADLOCK; CHEDLOCK *Sinapis arvensis* (Charlock) Tynan & Maitland (Chaddock); Gerard (Chadlock); Yks. B & H (Chedlock). These are but three of the very many corruptions of Charlock, which run from Cadlock to Warlock.
CHAFEWORT *Filago germanica* (Cudweed) Turner. Cf. Chaffweed, in Gerard, for the same plant.
CHAFFWEED 1. *Centunculus minimus.* 2. *Filago germanica* (Cudweed) Gerard. 3. *Gnaphalium sylvaticum* (Heath Cudweed) Turner.
CHAIN, Blue *Wistaria sinensis* (Chinese Wistaria) Som. Macmillan.
CHAIN, Gold *Sedum acre* (Biting Stonecrop) B & H.
CHAIN, Gold, White *Melilotus altissimus* (Tall Melilot) Som. Grigson.
CHAIN, Golden 1. *Laburnum anagryoides* (Laburnum) Wright (inc. Worcs. J Salisbury). 2. *Lathyrus pratensis* (Yellow Pea) Wilts. Macmillan.
CHAIN, Golden, Blue *Wistaria sinensis* (Chinese Wistaria) Som. Macmillan. In other words, the description of a blue laburnum.
CHAIN, Golden, Tall *Melilotus altissimus* (Tall Melilot) Som. Grigson.
CHAIN, Lady's *Laburnum anagryoides* (Laburnum) Dev, Dor. Macmillan. Golden Chain is common for this, and there is a Yorkshire name, Lady's Fingers.
CHAIN, Locket-and- *Dicentra spectabilis* (Bleeding Heart) Som. Macmillan.

CHAIN, Locket-and-, Gold; CHAIN, Watch-and-, Gold *Laburnum anagryoides* (Laburnum) Som. Macmillan.

CHAIN, Love-in-a- *Sedum reflexum* (Yellow Stonecrop) Cumb. Grigson.

CHAIN, Silver; CHAIN, Watch-and-, White *Robinia pseudacacia* (Common Acacia) B & H (Silver Chain); Som. Macmillan (White Watch-and-chain).

CHAIR ELM *Ulmus glabra* (Wych Elm) Wilkinson.1978. It must be used for chair-making somewhere.

CHAIRS, Tables-and-; CHAIRS-AND-TABLES *Buxus sempervirens* seed (Box) Som. Macmillan. Cf. the Wiltshire Pots-and-kettles for box seed.

CHALICE FLOWER, CHALICE LILY *Narcissus pseudo-narcissus* (Daffodil) B & H (Flower); Tynan & Maitland (Lily).

CHALICE VINE *Solandra maxima*.

CHALK MILKWORT *Polygala calcarea*.

CHALK PLANT *Gypsophila paniculata* Som. Macmillan.

CHALK THISTLE *Cirsium acaulon* (Picnic Thistle) H C Long.1910.

CHAMIZO *Atriplex canescens*. Chamizo in Spanish means a half-burnt log.

CHAMOCK *Ononis repens* (Rest-Harrow) B & H. A variation on the more usual Cammock.

CHAMOIS CRESS *Hornungia petraea* (Hutchinsia) Polunin.

CHAMPION, Corn *Agrostemma githago* (Corn Cockle) S.Africa Watt. Is champion here champaigne = open country? No, it can't be. Surely this is Corn Campion, which form is in B & H.

CHAMPION OAK *Quercus maxima* (Red Oak) J.Smith.

CHANEY ASH *Laburnum anagryoides* (Laburnum) Ches. B & H. Is Chaney a place name, or does the name refer back to the various 'chain' names? Cf. particularly the common Golden Chain. Helpfully, Holland spells it Chainy as if in confirmation.

CHANGE-OF-THE-WEATHER *Anagallis arvensis* (Scarlet Pimpernel) Som. Macmillan. Pimpernel is a famous weather forecaster.

CHANGEABLES *Hydrangea macrophylla* (Hydrangea) Dev. M.Baker.1977; Som. Macmillan. The flowers can be pink, or, if treated, blue. Actually, though, the flowers are colour variable year by year, without any artificial aid from the gardener.

CHANNEL CENTAURY *Centaureum tenuiflorum*. Very rare, and confined to the Channel coast, in Dorset and the Isle of Wight.

CHANNELLED HEATH *Erica canaliculata*.

CHANY OYSTER *Callistephus chinensis* (China Aster) Glos. J.D.Robertson; Oxf. A S Palmer. Just a way of saying China Aster, of course. Chiny Oyster, from nearby, is closer to the original.

CHAPARRAL, Ground *Eriogonum jamesii* (Antelope Sage) USA Vestal & Schultes.

CHAPARRAL LILY *Lilium rubescens*.

CHAPLET FLOWER *Stephanotis floribunda* (Madagascar Jasmine) Perry.

CHARD, Beet; CHARD, Swiss *Beta vulgaris var. cicla*. Swiss Chard is the common name; Beet Chard is in Grigson.1974. Chard is French carde, ultimately from Latin carduus, thistle. Evelyn uses card as well as chard. The name was first used in English for the edible part of an artichoke, and later transferred to this.

CHARE *Cheiranthus cheiri* (Wallflower) B & H. Halliwell has this also, and it seems to be but one of a variety of alternatives to the generic and specific names. Cf. Keiry, Cheiry, Chier, Churl. Cheiranthus seems to mean hand-flower.

CHARIOT, Jacob's; CHARIOT, Venus' *Aconitum napellus* (Monkshood) B & H (Jacob's); S.Africa Watt (Venus').

CHARIOT, Horses-and- *Briza media* (Quaking-grass) Som. Grigson.1959.

CHARIOT-AND-HORSES 1. *Aconitum anglicum* (Wild Monkshood) Herts. B & H. 2. *Aconitum napellus* (Monkshood) Le Strange.

CHARIOT TREE *Ougeinia dalbergoides*.

CHARITY *Polemonium caeruleum* (Jacob's Ladder) Dor. Macmillan; Cumb. Grigson.

CHARLICK *Sinapis arvensis* (Charlock) Hants. Cope.

CHARLIE, Creeping 1. *Glechoma hederacea* (Ground Ivy) USA Grigson. 2. *Lysimachia nummularia* (Creeping Jenny) A.W.Smith. 3. *Sedum acre* (Biting Stonecrop) Dev. Friend.1882.

CHARLOCK 1. *Raphanus raphanistrum* (Wild Radish) B & H. 2. *Sinapis arvensis*. OE cerlic, which according to ODEE was applied to the genus *Mercurialis*. But it must have attached itself to the Field Mustard very early, to judge from the variety of dialect forms. They go from Carlock, Chedlock, etc to such as Harlock and Warlock, and a number of forms in 'k', such as Kecklock and Kerleck, etc.

CHARLOCK, Jointed; CHARLOCK, White *Raphanus raphanistrum* (Wild Radish) B & H (Jointed); Berks, Ess. B & H (White).

CHARNOCK *Sinapis arvensis* (Charlock) Potter & Sargent.

CHASBOL; CHESBOL *Papaver rhoeas* (Red Poppy) both Scot. Jamieson. This is OE chespolle, chesboke, from AS ceosel, pebble (as in Chesil Bank). Cheesebowls, then (another name from the same source), would mean a ball of pebbly seeds.

CHASBOW *Papaver somniferum* (Opium Poppy) Scot. B & H. See Chasbol.

CHASE-THE-DEVIL *Nigella damascena* (Love-in-a-mist) B & H.

CHASSE *Papaver rhoeas* (Red Poppy) Halliwell. Presumably related to the Scottish Chasbol, and then Cheesebowl, all from OE chespolle, which comes from ceosel, a pebble (i.e. the stony seeds).

CHASTE NUT *Castanea sativa* (Sweet Chestnut) Ire. A S Palmer. That is probably how chestnut was pronounced.

CHASTE TREE; CHASTE WILLOW *Vitex agnus-castus* Pomet (Chaste Willow). Chaste here is the Latin castus, which also is responsible for the word castor (castor oil). Willow, because of the pliant branches.

CHASTEY *Castanea sativa* (Chestnut) Halliwell. Quoted as such in a plant list in ms Sloane 5,f.4. Cf. Chaste Nut.

CHATHAM ISLE FORGET-ME-NOT; CHATHAM ISLE LILY *Myosotidium hortensis* C.I. Forget-me-not is the usual name; C.I. Lily is from Perry (New Zealand).

CHATS 1. *Acer pseudo-platanus* seeds (Sycamore keys) Norf, Suff, Grigson; Yks. Carr. 2. *Fraxinus excelsior* (Ash) Norf, Suff. B & H; Yks. Addy. 3. *Pinus sylvestris* cones (Pine cones) Yks. Atkinson. Halliwell says that catkins (presumably of any tree) are called Chats in the West Country.

CHATTERBOX 1. *Antirrhinum majus* (Snapdragon) Brimble. 2. *Epipactis giganteum* (Stream Orchis) Usher. 3. *Geranium robertianum* (Herb Robert) Dor. Macmillan. For Snapdragon, at least, this is obviously of the same derivation as the various 'mouth' names.

CHATTS *Pinus sylvestris* cones (Pine cones) Yks. Morris.

CHAW-LEAF *Sorbus aria* (Whitebeam) IOM Gill.1963. Chaw, of course, is chew. Children do, or at least used to, chew the leaves.

CHAWS *Crataegus monogyna* fruit (Haws) Mor. Grigson.

CHAYOTE; CHOYOTE *Sechium edule*. This is a corruption of the Aztec word chayotl.

CHEADLE *Mercurialis perennis* (Dog's Mercury) Cockayne.

CHEADLE DOCK *Senecio jacobaea* (Ragwort) Ches. Grigson. There is also Cradle Dock for this, but more probably it forms part of the complex containing Kedlock, Kadle-dock, etc.

CHEAT *Lolium temulentum* (Darnel) Dor. Barnes; IOW Long. Because it is so like corn, Long says.

CHEBULE *Terminalia catappa* (Myrobalan) Copra.

CHECKERBERRY 1. *Gaultheria humifusa* (Teaberry) USA Elmore. 2. *Gaultheria procumbens* (Wintergreen) USA House. 3.

Mitchella repens (Partridge-berry) USA Le Strange.

CHECKERBLOOM *Sidalcea malviflora var. asprella* (Prairie Mallow) USA Schenk & Gifford.

CHEDDAR PINK *Dianthus gratianapolitanus.* Only known at Cheddar Gorge.

CHEDDIES *Solanum tuberosum* (Potatoes) Som. Macmillan.

CHEESE 1. *Malva neglecta* (Dwarf Mallow) Glos. J.D.Robertson. USA Tolstead. 2. *Malva sylvestris* (Common Mallow) Corn. Courtney.1880; Wilts. Dartnell & Goddard; Glos. J.D.Robertson; Hants. Cope; IOW Long; Worcs. J Salisbury. 3. *Paeonia mascula* (Peony) B & H. 4. *Tanacetum vulgare* (Tansy) Le Strange. All the 'cheese' names given to the mallows arise from the shape of the mature fruits, just like a cheese. As far as the peony is concerned, the name might have come in error from the mallow, or even the poppy (Cf. Cheesebowls), but there is an earlier form, Chesses.

CHEESE, Bread-and- 1. *Linaria vulgaris* (Toadflax) Wilts. Dartnell & Goddard. 2. *Lotus corniculatus* (Bird's-foot Trefoil) Som. Macmillan. 3. *Malva sylvestris* (Common Mallow) Som. Macmillan; Dor. Dacombe. 4. *Oxalis acetosella* (Wood Sorrel) Lancs. Nodal. 5. *Potentilla anserina* (Silverweed) Som. Macmillan. 6. *Rosa canina* (Dog Rose) Yks. Nicholson. 7. *Rumex acetosa* (Wild Sorrel) Dev. B & H.

CHEESE, Bread-and-, Bird's; CHEESE, Bread-and-, Cuckoo; CHEESE, Bread-and-, Cuckoo's *Oxalis acetosella* (Wild Sorrel) Friend (Bird's); Cumb. Grigson (Cuckoo); Som, Glos, Worcs, Shrop, Lancs, Cumb. Grigson (Cuckoo's).

CHEESE, Bread-and-, God Almighty's *Oxalis acetosella* (Wood Sorrel) Som. Grigson.

CHEESE, Bread-and-, Old Man's *Malva sylvestris* (Common Mallow) Som. Grigson.

CHEESE, Butter-and- 1. *Malva sylvestris* (Common Mallow) Dor, Som. Macmillan.

2. *Oxalis acetosella* (Wood Sorrel) Som. Macmillan. 3. *Ranunculus acris* (Meadow Buttercuup) Som. Grigson. 4. *Ranunculus bulbosus* (Bulbous Buttercup) Som. Grigson. 5. *Ranunculus ficaria* (Lesser Celandine) Som. Macmillan. 6. *Ranunculus repens* (Creeping Buttercup) Som. Macmillan.

CHEESE, Chakky; CHEESE, Chock *Malva sylvestris* Corn. Jago (Chakky); Dev. Friend (Chock). Probably from Truckles-of-cheese.

CHEESE, Chucky 1. *Crataegus monogyna* the young leaves (Hawthorn) Dev. B & H. 2. *Malva sylvestris* (Mallow) Som. Macmillan. In the first instance, the 'cheese' is the young leaves that children love to eat in spring. For mallow, the 'cheese' is a reference to the shape of the seed capsules.

CHEESE, Cuckoo *Oxalis acetosella* (Wood Sorrel) Dev. Grigson.

CHEESE, Custard *Malva sylvestris* (Common Mallow) Lincs. Grigson.

CHEESE, Doll *Malva neglecta* (Dwarf Mallow) Yks. B & H.

CHEESE, Dutch *Malva neglecta* (Dwarf Mallow) Ches. B & H.

CHEESE, Fairy 1. *Malva neglecta* (Dwarf Mallow) Yks. B & H. 2. *Malva sylvestris* seed (Common Mallow) Som, Yks. Friend.

CHEESE, Frog *Malva sylvestris* (Common Mallow) Oxf. B & H.

CHEESE, Jack; CHEESE, Jacky; CHEESE, Jacky's *Malva sylvestris* fruits (Common Mallow) all Som. Macmillan. Chucky cheese must have been the original, and that may possibly have come from Truckles-of-cheese.

CHEESE, Lady's *Malva sylvestris* (Common Mallow) Dor. Grigson.

CHEESE, Paddock *Asparagus officinalis* (Asparagus) Halliwell. Paddock is toad. This is an ancient name, according to Halliwell, but the reference is unclear.

CHEESE, Remedy, Joynson's *Tanacetum vulgare* (Tansy) Bucks. B & H. No comment in B & H.

CHEESE, Sheep's *Agropyron repens*

(Couch-grass) Grigson.1959.

CHEESE, Truckles-of- *Malva sylvestris* (Common Mallow) Som. Macmillan.

CHEESE-AND-BREAD 1. *Crataegus monogyna* (Hawthorn) Dur. F.M.T.Palgrave. 2. *Oxalis acetosella* (Wood Sorrel) Cumb. Grigson.

CHEESE-AND-BREAD, Cuckoo *Oxalis acetosella* (Wood Sorrel) Cumb. Grigson.

CHEESE-AND-BUTTER *Ranunculus ficaria* (Lesser Celandine) Som. Macmillan. This seems to be a mixture of the Somerset Cheesecups and the Devonshire Cream-and-butter.

CHEESE-CAKE 1. *Anthyllis vulneraria* (Kidney Vetch) Som. Macmillan. 2. *Lotus corniculatus* (Bird's-foot Trefoil) Som. Macmillan; Worcs, Yks. Grigson.

CHEESE-CAKE, Fairy *Medicago lupulina* (Black Medick) Wilts. Macmillan.

CHEESE-CAKE FLOWER *Malva sylvestris* (Common Mallow) Som, Suss. Grigson; Wilts. Dartnell & Goddard.

CHEESE-CAKE GRASS *Lotus corniculatus* (Bird's-foot Trefoil) Yks,N.Eng. Atkinson.

CHEESE-LOG *Malva sylvestris* (Common Mallow) Bucks. B & H.

CHEESE-RENNET; CHEESE-RENNING; CHEESE-RUNNING *Galium verum* (Lady's Bedstraw) Som, War, Cumb. Ire. Grigson (Rennet); Gerard (Renning); Ches. Prior (Running). The use of Lady's Bedstraw as a substitute for rennet is testified as least as far back as Dioscorides. "The people of Tuscanie or Hetruria doe use to turne their milke, that the Cheese which they make of Sheepes and Goats milke might be the sweeter and more pleasant in taste, and also more wholsome, especially to breake the stone, as it is reported", and "The people of Cheshire, especially about Namptwich, where the best Cheese is made, doe use it in their Rennet, esteeming greatly of that Cheese above other made without it". It is used in the Highlands, and in Wiltshire and Gloucestershire, for colouring cheese. In fact, Double Gloucester has nettle mixed with the Lady's Bedstraw.

CHEESEBOWLS 1. *Papaver rhoeas* (Red Poppy) Som. Macmillan. 2. *Papaver somniferum* (Opium Poppy) Prior. Cheesebowls is OE chespolle, chesboke, from AS ceosel, a pebble (as in Chesil Bank). Cheesebowls would mean, then, a ball of pebbly seeds. There are a number of variants, including Chesbow and Chasbol.

CHEESECUPS *Ranunculus ficaria* (Lesser Celandine) Som. Grigson.

CHEESEWEED *Malva parviflora* USA Barrett & Gifford. So the 'cheese' name travelled to the USA. This species is actually a Mediterranean one, introduced into North America.

CHEIPER *Iris pseudacorus* (Yellow Flag) Roxb. B & H. From the shrill noise children used to make with the leaves.

CHEIRY *Cheiranthus cheiri* (Wallflower) Turner. Turner's version of the specific name is closer to the original than a lot of others - Cf. Chare, Chier, and even Churl.

CHEMISE *Calystegia sepium* (Great Bindweed) Som. Macmillan.

CHEMISE, Lady's 1. *Anemone nemerosa* (Wood Anemone) Som. Macmillan. 2. *Calystegia sepium* (Great Bindweed) Som. Macmillan. 3. *Stellaria holostea* (Greater Stitchwort) Som. Macmillan. It is the white flowers, like washing hung out to dry, that give the name.

CHENILE *Hyoscyanum niger* (Henbane) B & H. Presumably because this plant has a lot of sticky hairs. The French chenille means a hairy caterpillar.

CHENILLE PLANT *Acalypha hispida*.

CHENOOK LIQUORICE *Lupinus littoralis*.

CHEQUER-TREE; CHEQUER-WOOD *Sorbus torminalis* (Wild Service Tree) Suss. W D Cooper; Kent. Grigson (Tree); Kent Grigson (Wood). A television programme by Richard Mabey suggested that the name is derived from the pub name "The Chequers". In effect, it means the tree belonging to The Chequers. Sounds rather doubtful, and it seems more likely to be descriptive in some way (bark?). Cf. Chequered Lily for Snake's

Head Lily.
CHEQUERED DAFFODIL; CHEQUERED LILY; CHEQUERED TULIP *Fritillaria meleagris* (Snake's Head Lily) Gerard (Daffodil) Parkinson (Lily); B & H (Tulip). From the markings on the petals. *Fritillaria* is from Latin fritillus, a dice-box, and the reference is extended to the chequer-board on which the dice were thrown.
CHEQUERS *Sorbus torminalis* fruit (Wild Service Tree - fruit) Evelyn. See Chequer-tree, or Chequer-wood, for the tree itself.
CHERIMOYA *Annona cherimola*.
CHERISAUNCE; CHEVISAUNCE *Cheiranthus cheiri* (Wallflower) both Coats. Both variations on the Latin name ("the comforter", apparently referring to the perfume). See Spenser, Lycidas: The pretty paunce and the chevisaunce. Shall match with the faire flower-de-luce. But Cherisaunce is clearly an error for chevisaunce.
CHERRY 1. *Erythroxilon ovatum* Barbados Gooding. 2. *Prunus cerasus,etc*. Cherry is Latin cerasus, Greek kerasos, from the name of the tree, though Kerasos may actually be the name of a place. The word comes through ME cherise by loss of the final -se, as if the original had been mistaken for a plural.
CHERRY, Barbados 1. *Eugenia uniflora* (Surinam Cherry) Dalziel. 2. *Malpighia glabra*.
CHERRY, Bay; CHERRY BAY *Prunus laurocerasus* (Cherry-laurel) both Parkinson. The confusion between what has become known simply as laurel, and the true laurel (bay) is shown in these names.
CHERRY, Bird 1. *Prunus padus*. 2. *Prunus pennsylvanica* (Pin Cherry) USA Kingsbury. *Prunus avium* seems the likeliest contender for this name, but this is known as the Wild Cherry.
CHERRY, Bird, American; CHERRY, Bird, Virginian *Prunus virginiana* (Choke Cherry) both Ablett.
CHERRY, Bird's *Crataegus monogyna* fruit (Haws) Som. Macmillan.
CHERRY, Black 1. *Atropa belladonna* (Deadly Nightshade) Grieve.1931. 2. *Prunus virginiana* (Choke Cherry) USA Driver. A dangerous name to give to Deadly Nightshade. A child may think them perfectly edible. The usual attribution is to the devil.
CHERRY, Black, Wild *Prunus virginiana* (Choke Cherry) USA Zenkert.
CHERRY, Bladder 1. *Physalis alkekengi* (Winter Cherry) McClintock. 2. *Physalis heterophylla* (Ground Cherry) Kansas Gates.
CHERRY, Brazil *Eugenia michelii*.
CHERRY, Brazilian *Eugenia uniflora* (Surinam Cherry) Bianchini.
CHERRY, Brush *Eugenia luehmannii*.
CHERRY, Choke *Prunus virginiana*. Presumably, 'choke' indicates that these cherries are not for eating.
CHERRY, Choke, Western *Prunus demissa*.
CHERRY, Christmas *Solanum capsicastrum* (Winter Cherry) Rochford & Gorer.
CHERRY, Cluster; CHERRY, Cluster, Wild *Prunus padus* fruit (Bird Cherry) B & H; Camden/Gibson (Wild).
CHERRY, Cornel; CHERRY, Cornelian *Cornus mas*. Cornel Cherry is the usual name; Cornelian Cherry is in A.W.Smith. Cornel seems to be derived from German - OHG kornulberi - but surely must come at some stage from the Latin cornus as the name of the tree. That Latin is from Greek kranos. Cornelian for cornel is quite common, but the two words have quite distinct derivations, and one is not an alternative for the other, the Latin original of cornelian being cornelius.
CHERRY, Crab *Prunus avium* (Wild Cherry) Bucks. Grigson. Presumably from the same idea that produced crabapple for wild apple, though in that case crab is really the name (without apple).
CHERRY, Devil's 1. *Atropa belladonna* (Deadly Nightshade) N'thants. Tynan & Maitland. 2. *Bryonia dioica* (White Bryony) Dyer.1889. 3. *Solanum dulcamara* (Woody Nightshade) Som. Macmillan. This would mean a poison cherry; anything noxious would be assigned to the devil. It doesn't matter whether the poisonous nature is actual,

as with the Deadly Nightshade, or just assumed, as with the other two.

CHERRY, Dog *Cornus sanguinea* (Dogwood) Prior. Gerard knew the fruits as Dogberries. The name probably comes from the traditional use of the wood for skewers - a "dog" means a sharp spike.

CHERRY, Dog's *Bryonia dioica* (White Bryony) Som. Macmillan. As with Devil's Cherries, the use of 'dog's' indicates the suspicious nature of the berries.

CHERRY, Fire *Prunus pennsylvanica* (Pin Cherry) USA Kingsbury. Is this from the colour of the cherries?

CHERRY, Fool's *Prunus padus* fruit (Bird Cherry) B & H.

CHERRY, Ground 1. *Physalis bunyardii* (Cape Gooseberry) USA Watt. 2. *Physalis fendleri*. 3. *Physalis heterophylla*.

CHERRY, Ground, Bladder, Large *Physalis viscosa*.

CHERRY, Ground, Clammy *Physalis heterophylla* (Ground Cherry) USA House.

CHERRY, Ground, Hairy *Physalis pubescens* (Husk Tomato) Jaeger.

CHERRY, Ground, Ivy-leaved *Physalis hederaefolia*.

CHERRY, Ground, Prairie *Physalis lanceolata*.

CHERRY, Ground, Smooth *Physalis subglabrata*.

CHERRY, Ground, Virginia *Physalis virginiana*.

CHERRY, Heart, Wild *Prunus avium* (Wild Cherry) Camden/Gibson.

CHERRY, Higan *Prunus cerasus var.subhirtella*.

CHERRY, Hog *Prunus padus* (Bird Cherry) W.Miller. Sounds reasonable for a fruit that has little to recommend it, but significantly, there are a lot of names like Hagberry for this, so the 'hog' of this name is probably OE haga, hedge.

CHERRY, Hollyleaf *Prunus ilicifolia*.

CHERRY, Hottentot *Maurocenia capensis*.

CHERRY, Indian 1. *Rhamnus caroliniana* USA Harper. 2. *Rhamnus crocea*.

CHERRY, Japanese *Prunus serrulata*.

CHERRY, Jerusalem *Solanum pseudocapsicum*. Probably merely a florists' name - it is a pot plant.

CHERRY, Jerusalem, False *Solanum diflorum*.

CHERRY, Laurel *Prunus caroliniana* (American Cherry Laurel) USA Harper.

CHERRY, Long *Cornus mas* (Cornel Cherry) Turner.

CHERRY, Manchurian *Prunus maackii*.

CHERRY, Morello *Prunus cerasus*. This must be Italian morello - nearly black, ultimately from med. Latin Maurus, Moor.

CHERRY, Mountain *Prunus angustifolia* (Chickasaw Plum) Usher.

CHERRY, Naughty Man's *Atropa belladonna* (Deadly Nightshade) Som,Bucks. Grigson. 'Naughty Man' is always a euphemism for the devil, and is applied to anything that is, or is deemed to be, poisonous. Naughty Man's Cherry is, then, a simple admonition to children to leave the berries alone.

CHERRY, Pie *Prunus cerasus* (Morello Cherry) USA Leighton. Because this is the species cherry pies are made from.

CHERRY, Pin *Prunus pennsylvanica*.

CHERRY, Pitanja *Eugenia uniflora* (Surinam Cherry) Dalziel.

CHERRY, Polstead *Prunus padus* (Bird Cherry) Suff. B & H. There is a village named Polstead in Suffolk, so perhaps the cherries were abundant there - they were sold in Suffolk under this name.

CHERRY, Red, Wild *Prunus pennsylvanica* (Pin Cherry) USA Kingsbury.

CHERRY, Rosebud *Prunus cerasus var. subhirtella* (Higan Cherry) Mitchell.1974.

CHERRY, Rum *Prunus virginiana* (Choke Cherry) USA Zenkert. Country people used to infuse the cherries in brandy as a flavouring, so presumably they did the same with rum. It is said that birds often get drunk eating these cherries.

CHERRY, St Lucie *Prunus mahaleb*.

CHERRY, Sand 1. *Prunus besseyi*. 2. *Prunus*

pumila.

CHERRY, Satan's *Atropa belladonna* (Deadly Nightshade) Yks. Grigson. As with Naughty Man's Cherry, this is a warning to children to leave the poisonous berries alone.

CHERRY, Snake's *Cornus sanguinea* (Dogwood) Som. Macmillan.

CHERRY, Sour 1. *Prunus cerasus* (Morello Cherry) Clapham. 2. *Eugenia crynantha*. The Eugenia is a variety growing in Australia, and the common name is applied there.

CHERRY, Sour, Chinese *Prunus cantabrigiensis*.

CHERRY, Surinam *Eugenia uniflora*. A South American species, known also as Brazilian, Barbados, Pitanja or Cayenne Cherry.

CHERRY, Tibetan *Prunus serrula*.

CHERRY, West Indian *Malpighia glabra* (Barbados Cherry) Brouk.

CHERRY, Wild *Prunus avium*.

CHERRY, Winter 1. *Cardiospermum halicacabum* (Blister Creeper) Chopra. 2. *Physalis alkekengi*. 3. *Physalis pubescens* (Husk Tomato) Bianchini. 4. *Prunus cerasus var. subhirtella autumnalis*. 5. *Solanum capsicastrum*.

CHERRY, Yoshino *Prunus x yedoensis*.

CHERRY BAY 1. *Prunus laurocerasus* (Cherry Laurel) Parkinson. 2. *Prunus lusitanicus* (Portuguese Laurel) Macmillan.

CHERRY BIRCH 1. *Betula lenta*. 2. *Betula fontinalis* (Streamside Birch) USA Elmore.

CHERRY LAUREL *Prunus laurocerasus*.

CHERRY LAUREL, American *Prunus caroliniana*.

CHERRY PEPPER *Capsicum cerasiforme*.

CHERRY-PIE 1. *Chamaenerion angustifolium* (Rosebay) Dor. B & H. 2. *Epilobium hirsutum* (Great Willowherb) Som, Dor, Worcs, Notts. Grigson. 3. *Heliotropum europaeum*. 4. *Valeriana dioica* (Marsh Valerian) Wilts. Macmillan. 5. *Valeriana officinalis* (Valerian) Wilts. Dartnell & Goddard. The willowherbs particularly have attracted a number of these names. Great Willowherb has Apple Pie, Gooseberry Pie, Plum Pudding, etc., as well as Codlins-and-cream and such like. Marsh Valerian has Gooseberry Pie, as well. Have these names something to do with the smell?

CHERRY-PLUM *Prunus cerasifera*.

CHERRY TOMATO *Lycopersicon cerasiforme*.

CHERRY-WOOD *Viburnum opulus* (Guelder Rose) B & H. This is an eighteenth century name, presumably referring to the timber.

CHERRY-WOODBINE *Lonicera alpigena* (Alpine Honeysuckle) Polunin.

CHERRYSTONE JUNIPER *Juniperus monosperma* (One-seedeed Juniper) USA Yanovsky. Because it is "one-seeded".

CHERVELL 1. *Anthriscus cerefolium* (Chervil) Gerard. 2. *Lonicera periclymenum* (Honeysuckle) B & H.

CHERVIL 1. *Anthriscus cerefolium*. 2. *Anthriscus sylvestris* (Cow Parsley) B & H. 3. *Chaerophyllum temulentum* (Rough Chervil) Yks. Grigson. Chervil is AS cerfille, Latin cerefolium.

CHERVIL, Anise; CHERVIL, Spanish *Myrrhis odorata* (Sweet Cicely) both Grieve.1933. When cultivated, these names are applied. 'Spanish' seems merely to be 'anise'. The 'anise' names are frequent too in the wild, especially in the north country.

CHERVIL, Bur *Anthriscus caucalis*. This plant has rough, bristly fruits.

CHERVIL, Corn *Scandix pecten-veneris* (Shepherd's Needle) Turner.

CHERVIL, Cow 1. *Anthriscus sylvestris* (Cow Parsley) Som. Mabey. 2. *Myrrhis odorata* (Sweet Cicely) Grieve.1933.

CHERVIL, Garden *Anthriscus cerefolium* (Chervil) Barton & Castle. Garden Chervil, or Salad Chervil, to distinguish it from the wild Chervil, better known as Cow Parsley.

CHERVIL, Giant; CHERVIL, Great *Myrrhis odorata* (Sweet Cicely) G.B.Foster (Giant); B & H (Great). Cf. Giant Sweet Chervil, Mock, Sweet, or Cow, Chervil, and others.

CHERVIL, Golden *Chaerophyllum aureum*.

CHERVIL, Hemlock *Torilis japonica* (Hedge Parsley) B & H.

CHERVIL, Mock 1. *Anthriscus sylvestris* (Cow Parsley) Turner. 2. *Myrrhis odorata* (Sweet Cicely) Langham. 3. *Scandix pecten-veneris* (Shepherd's Needle) B & H.

CHERVIL, Needle *Scandix pecten-veneris* (Shepherd's Needle) W.Miller. The 'needle' names are from the long, beaked fruits.

CHERVIL, Rough 1. *Anthriscus caucalis* (Bur Chervil) Barton & Castle. 2. *Chaerophyllum temulentum*. 3. *Torilis japonica* (Hedge Parsley) B & H.

CHERVIL, Salad *Anthriscus cerefolium* (Chervil) A.W.Smith. This and Garden Chervil are to distinguish it from Cow Parsley, *Anthriscus sylvestris*.

CHERVIL, Sweet; CHERVIL, Sweet, Giant 1. *Myrrhis odorata* (Sweet Cicely) B & H (Giant Sweet. Chervil is in Clair). 2. *Anthriscus cerefolium* (Chervil) Genders.

CHERVIL, Turnip-rooted *Chaerophyllum bulbosum*.

CHERVIL, Wild 1. *Anthriscus sylvestris* (Cow Parsley) B & H. 2. *Scandix pecten-veneris* (Shepherd's Needle) Gerard.

CHERVIN *Sium sisarum* (Skirret) Douglas.

CHESBOL *Papaver rhoeas* (Red Poppy) Scot. Jamieson.

CHESBOUL; CHESBOWL *Papaver somniferum* (Opium Poppy) both Turner. Better known as Cheesebowls, and as applied to *P rhoeas*. This is OE chespolle, chesboke, from AS ceosel, a pebble (the word still survives in Chesil Bank). So Cheesebowls, or its derivatives, would mean a ball of pebbly seeds.

CHESBOW 1. *Papaver rhoeas* (Red Poppy) Scot. B & H. 2. *Papaver somniferum* (Opium Poppy) B & H. See Chesboul.

CHESS *Lolium temulentum* (Darnel) W.Miller.

CHESS, Soft *Bromus mollis* (Soft Brome) USA Allan.

CHESS-APPLE *Sorbus aria* (Whitebeam) Ches, Lancs. Grigson. Cheese-apple?

CHESSE *Papaver somniferum* (Opium Poppy) B & H. See Chesboul.

CHESSES *Paeonia mascula* (Peony) Coles. B & H have Cheeses for this, and it is probably transferred from the poppy in error. See Cheesebowls.

CHESTNUT, American *Castanea dentata*. Chestnut is from OE cisten beam, eventually from the Latin castanea, through W Germ. kastinja. The Greek kastanea refers to a place name, either Castanea in Pontus, or Castana in Thessaly - though this is ancient folk etymology for a loan-word, something like the Armenian kaskeni, chestnut-tree.

CHESTNUT, Cape *Calodendrum capense*.

CHESTNUT, Chinese *Castanea mollissima*.

CHESTNUT, Earth 1. *Bunium bulbocastanum* (Great Earth-nut) Gerard. 2. *Conopodium majus* (Earth-nut) Prior. 3. *Lathyrus tuberosus* (Fyfield Pea) Douglas. They resemble chestnuts, and can be roasted like them.

CHESTNUT, Earth, Australian *Eleocharis sphacelata*.

CHESTNUT, European; CHESTNUT, French *Castanea sativa* (Sweet Chestnut) Brimble.1948. See also Italian and Spanish Chestnut, the latter being most often used, because, presumably, the best nuts were always said to be those exported from Spain.

CHESTNUT, Golden *Chrysolepis chrysophylla*. It is the underside of the evergreen leaves which is golden.

CHESTNUT, Horse *Aesculus hippocastanum*. 'Horse' usually denotes largeness and coarseness, and it is probably so here, but the old Latin name was Castanea equina, and Parkinson says "the horse chestnits are given in the East Country, and so through all Turkie, unto horses to cure them of the cough, shortnesse of winde, and such other diseases". Gerard also accounts for the name in this way. Skinner suggests the derivation from the likeness to a horse's hoof in the leaf cicatrix, and it may have been from the doctrine of signatures that the nuts, crushed as meal, were given to horses for various diseases.

CHESTNUT, Horse, American *Aesculus cal-*

ifornica (Californian Buckeye) Hatfield.
CHESTNUT, Horse, Indian *Aesculus indica.*
CHESTNUT, Horse, Japanese *Aesculus turbinata.*
CHESTNUT, Horse, Red *Aesculus x carnea.*
CHESTNUT, Italian *Castanea sativa* (Sweet Chestnut) Wit. Better known as Spanish Chestnut, but these names indicate its Mediterranean origin.
CHESTNUT, Japanese *Castanea crenata.*
CHESTNUT, Malayan, Greater *Chrysolepis megacarpa.*
CHESTNUT, Moreton Bay *Castanospermum australe.*
CHESTNUT, Rhodesian *Baikiaea plurijuga* (Rhodesian Teak) Palgrave. The upstanding flowers look quite like horse chestnut at a distance.
CHESTNUT, Spanish *Castanea sativa* (Sweet Chestnut) Rackham. Used about as much as Sweet Chestnut as the common name, but actually refers only to the importation of the nuts.
CHESTNUT, Sweet *Castanea sativa.*
CHESTNUT, Water 1. *Caltha palustris* (Marsh Marigold) Culpepper. 2. *Trapa natans.* Culpepper called Marsh Marigold Water Nut, as well. *Trapa natans* gets the name from the edible horny seeds.
CHESTNUT, Water, Chinese *Eleocharis tuberosus.*
CHESTNUT-LEAVED OAK *Quercus castaneifolia.*
CHESTNUT OAK 1. *Quercus montana* USA Zenkert. 2. *Quercus muhlenbergii.* 3. *Quercus prinus.*
CHESTNUT OAK, Japanese *Quercus acutissima.*
CHESTNUT ROSE 1. *Rosa microphylla* Bunyard. 2. *Rosa roxburghii* (Burr Rose) Thomas. In these cases, the hips are so covered with bristles as to resemble a sweet chestnut.
CHEVORELL *Anthriscus cerefolium* (Chervil) B & H.
CHEW-ROOT *Taraxacum kok-saghyz* (Russian Dandelion) Schery. This has been used as a substitute rubber plant; the use was discovered in Russia in 1931, and it was introduced into USA in 1942. The latex is taken from the roots.
CHEWBARK *Ulmus glabra* (Wych Elm) Berw. Grigson. Children chew the inner bark.
CHIA SAGE *Salvia columbariae.* Chia was a staple food with the Pacific coast Indians. A flour was made from the oily seed.
CHIBBLE; CHIBBAL; CHIBOUL(E) 1. *Allium ascalonium* (Shallot) Dev. Friend (Chibble); Northcote (Chibbal, Chiboule). 2. *Allium fistulosum* (Welsh Onion) T Hill (Chiboul). 3. *Allium porrum* (Leek) Dev. Choape. Cf. French ciboule, Italian cipolla. According to Choape, the name is given to young onions and young leeks indiscriminately.
CHIBE *Allium schoernoprasum* (Chives) N.Eng. Halliwell.
CHICHELINGS 1. *Lathyrus sylvestris* (Everlasting Pea) Gerard. 2. *Vicia sativa* (Common Vetch) N.Eng. B & H. Sometimes spelt Cichlings. These must be from Cicer, the genus known as Chick Peas.
CHICHESTER ELM *Ulmus x hollandica 'Vegeta'* (Huntingdon Elm) Mitchell.1974.
CHICHLING VETCH *Vicia lathyroides* (Spring Vetch) Camden/ Gibson.
CHICK PEA 1. *Cicer arietinum.* 2. *Vicia americana* (American Vetch) USA Cunningham.
CHICK WITTLES *Stellaria media* (Chickweed) Suff. Mabey. i.e. victuals for chickens.
CHICKASAW PLUM *Prunus angustifolia.*
CHICKEN CORN *Sorghum vulgare* (Kaffir Corn) USA Cunningham.
CHICKEN GRAPE *Vitis cordifolia.*
CHICKEN'S EVERGREEN; HEN'S EVERGREEN *Stellaria media* (Chickweed) Kent Tynan & Maitland.
CHICKEN'S MEAT *Stellaria media* (Chickweed) E.Ang. Forby.
CHICKENS *Saxifraga x urbinum* (London Pride) Dev. Friend. From its habit of send-

ing out runners from the parent plant - Hens-and-chickens is a more usual name.

CHICKENS, Hen-and- 1. *Anthyllis vulneraria* (Kidney Vetch) Som. Grigson. 2. *Aquilegia vulgaris* (Columbine) Norf. Grigson. 3. *Bellis perennis* the double garden variety (Daisy) Som. Elworthy. 4. *Butomus umbellatus* (Flowering Rush) Hunts. Marshall. 5. *Capsella bursa-pastoris* (Shepherd's Purse) Som. Macmillan. 6. *Cymbalaria muralis* (Ivy-leaved Toadflax) Kent Grigson. 7. *Geranium robertianum* (Herb Robert) Som. Macmillan. 8. *Glechoma hederacea* (Ground Ivy) Bucks. Grigson. 9. *Lotus corniculatus* (Bird's-foot Trefoil) Som. Macmillan; Glos, Oxf. Grigson. 10. *Narcissus pseudo-narcissus* (Daffodil) Dev. Friend. 11. *Saxifraga sarmentosa* (Mother-of-Thousands) Som. Macmillan; Wilts. Dartnell & Goddard. 12. *Saxifraga x urbinum* (London Pride) Dev. Friend; Wilts. Dartnell & Goddard. 13. *Sedum acre* (Biting Stonecrop) Som. Macmillan. 14. *Sempervivum tectorum* (Houseleek) Som. Macmillan. 15. *Tanacetum parthenium* (Feverfew) Tynan & Maitland. The name describes the picture of young plants growing around the parent, and is particularly apt for the Sedums and Saxifrages. What about Shepherd's Purse, though? Or daffodil?

CHICKENWEED 1. *Cerastium triviale* (Narrow-leaved Mouse-ear) Cumb. B & H. 2. *Portulaca quadrifida* Barbados Gooding. 3. *Senecio vulgaris* (Groundsel) Yks. Grigson. 4. *Stellaria media* (Chickweed) Norf, Lincs, Yks, Cumb, Scot. Grigson; Ches. Holland.

CHICKENWORT *Stellaria media* (Chickweed) Scot. Jamieson.

CHICKNYWEED *Stellaria media* (Chickweed) Dev. Grigson.

CHICKS, Goose *Salix capraea* catkins (Sallow catkins) Dev. Grigson.

CHICKS-AND-HENS *Sempervivum tectorum* (Houseleek) USA Cunningham. See Hen-and-chickens, rather.

CHICKWEED 1. *Lamium amplexicaule* (Henbit Deadnettle) B & H. 2. *Senecio vulgaris* (Groundsel) Som. Macmillan. 3. *Stellaria media*. Henbit Deadnettle was called Alsine by some of the early botanists, hence the name. In fact, the name chickweed was once applied to many small plants of a similar habit, such as various *Veronica* species, *Arenaria trinerva*, etc. Chickweed, as a name, is applied simply because the plant was given to fowls to eat, and to cage birds - *Stellaria media* is an iron-rich tonic, and groundsel as well was at one time regularly sold on the streets as food for cage-birds. Cf. too the names Birdseed, Canary's Food, or Canary's Seed for groundsel, as well as the varieties of chickweed.

CHICKWEED, Alpine *Cerastium alpinum* (Alpine Mouse-ear) Condry.

CHICKWEED, Barren *Cerastium velutinum*.

CHICKWEED, Blinking *Montia fontana* (Blinks) Prior. This is sometimes known too as Water Chickweed.

CHICKWEED, Daisy-leaved *Mollugo nudicaulis*.

CHICKWEED, Dwarf *Moenchia erecta*. Upright Chickweed, or Upright Pearlwort.

CHICKWEED, Field *Cerastium arvense* (Field Mouse-ear Chickweed) USA House.

CHICKWEED, Forked *Paronychia canadensis*.

CHICKWEED, Germander 1. *Veronica agrestis* (Green Field Speedwell) Prior. 2. *Veronica chamaedrys* (Germander Speedwell) B & H. Germander seems to be eventually from chamaedrys, a compound which means ground oak. Why did Prior insist on *V agrestis* for this name? Surely B & H are quite right in giving it to *V chamaedrys*?

CHICKWEED, Greater *Stellaria neglecta*.

CHICKWEED, Ground-ivy-leaved, Great *Lamium amplexicaule* (Henbit Deadnettle) Hulme.

CHICKWEED, Guernsey *Polycarpon tetraphyllum* (Four-leaved Allseed) McClintock. A name coined only in 1969.

CHICKWEED, Ivy *Veronica hederifolia* (Ivy-leaved Speedwell) Gerard.
CHICKWEED, Jagged *Holosteum umbellatum*.
CHICKWEED, Lesser *Stellaria pallida*.
CHICKWEED, Little *Sagina procumbens* (Common Pearlwort) Som. Macmillan.
CHICKWEED, Meadow *Cerastium arvense* (Field Mouse-ear Chickweed) USA House.
CHICKWEED, Mouse-ear *Cerastium sp* (Mouse-ear). i.e. Broad-leaved M E Chickweed *Cerastium glomeratum* Curtis. Clustered M E Chickweed Salisbury. Common M E Chickweed *holostoides* Curtis. Field M E Chickweed *arvense*. Least M E Chickweed *semidecandrum* Curtis. Marsh M E Chickweed *Myosoton aquaticum* Curtis. Narrow-leaved M E Chickweed *Cerastium triviale*. Wayside M E Chickweed *Cerastium triviale* J A MacCulloch.1905.
CHICKWEED, Nodding *Cerastium nutans*.
CHICKWEED, Plantain-leaved *Moehringia trinerva* (Three-leaved Sandwort) Curtis.
CHICKWEED, Red *Anagallis arvensis* (Scarlet Pimpernel) USA Grigson.
CHICKWEED, Sea *Honkenya peploides* (Sea Sandwort) B & H.
CHICKWEED, Star *Stellaria media* (Chickweed) Hatfield. Presumably to give it a distinguishing name from the Mouse-ear Chickweeds. After all, the genus is entitled *Stellaria*.
CHICKWEED, Thyme-leaved *Arenaria serpyllifolia* (Thyme-leaved Sandwort) Curtis.
CHICKWEED, Umbellate *Holosteum umbellatum* (Jagged Chickweed) Blamey.
CHICKWEED, Upright *Moenchia erecta*.
CHICKWEED, Water 1. *Montia fontana* (Blinks) Darling & Boyd. 2. *Myosoton aquaticum*.
CHICKWEED, Wood *Stellaria nemora*.
CHICKWEED WILLOWHERB *Epilobium alsinifolium*.
CHICKWEED WINTERGREEN 1. *Trientalis borealis* USA House. 2. *Trientalis europaeus*.
CHICORY *Cichorium intybus*. Mabey says Chicory is from the Arabic Chicouryeh - the root is still an Arab food, boiled and eaten. But this is not what ODEE says - the word is late ME cicoree, through Latin, and eventually from Greek. Succory is essentially the same word.
CHICORY, White *Cichorium endivia* (Endive) Candolle.
CHIDDIES *Solanum tuberosum* (Potato) Som. Macmillan. They are Tiddies in Dorset.
CHIER, Wild *Cheiranthus cheiri* (Wallflower) B & H. There are a number of variations on Cheiranthus. Besides this, there are Cheiry, or Keiry, Chare, even Churl in Shropshire.
CHIGGER-FLOWER *Asclepias tuberosa* (Orange Milkweed) Miss. Bergen.
CHIGGER-WEED *Anthemis cotula* (Maydweed) Indiana Bergen.
CHILBLAIN-BERRY *Tamus communis* (Black Bryony) Wilts. Wiltshire. A Wiltshire chilblain remedy was to steep the berries and roots in gin and then to apply them.
CHILDING CUDWEED *Filago germanica* (Cudweed) A S Palmer. An indication that cudweed must propagate itself by the 'hen-and-chickens' method.
CHILDING PINK *Kohlrauschia prolifer*. 'Childing' because this plant throws out younger and smaller flowers like a family of little children around it. Gerard called it Childing Sweet William.
CHILDING SWEET WILLIAM *Kohlrauschia prolifer* (Childing Pink) Gerard.
CHILDREN-OF-ISRAEL 1. *Malcomia sp* (Virginian Stock) Dev, Wilts. Wright; Dor Dacombe. 2. *Pulmonaria officinalis* (Lungwort) Dor. Grigson. 3. *Saxifraga x urbinum* Wilts. Macmillan. The Virginian Stock gets the name because of the large number of small flowers it has. As London Pride is also called Hen-and-chickens, this name probably is a reference to its method of reproduction by runners, the young plants clustered round the mother. Lungwort, though, probably earned the name from the

various colours on one flower (Cf. Abraham, Isaac and Joseph, etc).

CHILDREN'S CLOCK *Taraxacum officinalis* (Dandelion) Som. Macmillan.

CHILDREN'S DAISY *Bellis perennis* (Daisy) Yks. This must surely be more properly Childing Daisy, as for Childing Pink.

CHILDREN'S TOMATO *Solanum anomalum*.

CHILE *Capsicum annuum*. It is chilli in Nahuatl, the language of the Aztecs.

CHILE NUT TREE *Araucaria araucana* (Chile Pine) H & P.

CHILE PEPPER *Capsicum annuum* (Chile) Redfield & Villa.

CHILE STRAWBERRY *Fragaria chiloense*. This is a plant of the western maritime edge of the Americas, from Alaska to Patagonia, and gets its name because it was found as a garden plant on the island of Chiloé, off Chile.

CHILEAN BEET *Beta vulgaris var.cicla* (Swiss Chard) Bianchini.

CHILEAN BELL FLOWER *Lapageria rosea*.

CHILEAN CEDAR *Austrocedrus chilensis*.

CHILEAN FIG *Mesembryanthemum chilense*.

CHILEAN FIREBUSH *Embothrium coccineum*.

CHILEAN GLORY VINE *Eccremocarpus scaber* (Glory Flower) Ted Phillips.

CHILEAN GUM-BOX *Escallonia macrantha* W.Miller.

CHILEAN INCENSE CEDAR *Austrocedrus chilensis* (Chilean Cedar) Leathart.

CHILEAN JASMINE *Mandevilla laxa*.

CHILEAN PINE *Araucaria araucana*.

CHILEAN POTATO TREE *Solanum crispum*.

CHILEAN WINEBERRY *Aristotelia macqui*.

CHILLI *Capsicum annuum* (Chile) Gilliland.

CHILLI, Natal *Capsicum annuum* (Chile) S.Africa Watt.

CHILTERN GENTIAN *Gentianella germanica*. A very rare British plant, only found on the Chilterns.

CHIMNEY BELLFLOWER, CHIMNEY PLANT *Campanula pyramidalis*. Chimney Plant is in W.Miller. Because it used to be grown in pots as a summer ornament for the fireless grate.

CHIMNEY SMOCK *Anemone nemerosa* (Wood Anemone) Som. Macmillan.

CHIMNEY SWEEP 1. *Centaurea nigra* (Knapweed) Som. Macmillan. 2. *Luzula campestris* (Good Friday Grass) Ches, Lancs. B & H. 3. *Plantago lanceolata* (Ribwort Plantain) Som. Macmillan. War. B & H; N'thants. A.E.Baker; Lancs. Miss Formby. 4. *Plantago media* (Lamb's Tongue Plantain) Som. Macmillan. For knapweed, it is probably the sweep's brush shape of the flower which gives rise to this name. Cf. Brushes, Paintbrushes, both West Country names. With Good Friday Grass, it is certainly the colour, and with ribwort, it is probably the colour, which gave the name. It is known as Black Plantain, and there are many similar colour names, such as Blackie Tops, from Somerset, and the Wiltshire Black Boys. Given this, it is most likely that Chimney Sweeps for Lamb's Tongue Plantain is an error.

CHIMNEY SWEEPER 1. *Luzula campestris* (Good Friday Grass) Wilts. Dartnell. & Goddard; Ches, Lancs. B & H. 2. *Plantago lanceolata* (Ribwort Plantain) Wilts, War. Grigson; N'thants. A.E.Baker; Ches. Dyer. 3. *Typha latifolia* (False Bulrush) Som. Macmillan. See Chimney Sweep.

CHIMNEY SWEEPER'S BRUSH *Typha latifolia* (False Bulrush) Som. Macmillan.

CHINA, Golden Rose of *Rosa hugonis* (Father Hugo's Rose) Mattock.

CHINA, Pride of *Melia azedarach* (Chinaberry Tree) Mitton. There is Pride of India, too, for this.

CHINA, Rose of *Hibiscus rosa-sinensis*.

CHINA ASTER *Callistephus chinensis*. It was known at first as *Aster chinense*.

CHINA COCKLE *Saponaria officinalis*

(Soapwort) S.Africa Watt.
CHINA FIR *Cunninghamia lanceolata.*
CHINA GRASS *Roehmeria nivea* Wit.
CHINA JUTE *Abutilon avicennae.* The fibre was used in China for cloth and cordage.
CHINA ORANGE *Citrus aurantium var. amara sinensis.*
CHINA PRIVET *Lagerstroemeria indica.*
CHINA ROSE *Rosa indica.*
CHINA SQUASH *Cucurbita moschata* (Crook-neck Squash) Havard. But it is an American plant!
CHINA TEA, Wild *Sapindus drummondii* (Drummond's Soapberry) Vestal & Schultes.
CHINA-TREE 1. *Koelreuteria paniculata* (Golden Rain Tree) Barber & Phillips. 2. *Melia azedarach* (Chinaberry Tree) Chopra.
CHINA-WOOD *Aleurites fordii* (Tung-oil Tree) Schery.
CHINABERRY TREE *Melia azedarach.*
CHINAMAN'S BREECHES *Dicentra spectabilis* (Bleeding Heart) Coats. From the shape of the flowers. Dutchman's Breeches is the name usually given to close relative, *D cucullaria.*
CHINCHERINCHEE *Ornithogalum thyrsoides.* A South African plant; presumably this is of African derivation.
CHINCHONE *Senecio vulgaris* (Groundsel) B & H. It must be a version of Senecio, as the later forms Sension, etc., are.
CHINESE AMARANTH *Amaranthus tricolor* (Joseph's Coat) A.W.Smith.
CHINESE ANGELICA TREE *Aralia sinensis.*
CHINESE ANISE *Illicium verum* (True Anise) Grieve.1931.
CHINESE APPLE *Pyrus prunifolia* (Siberian Crab) Douglas.
CHINESE ARBOR-VITAE *Thuja orientalis.*
CHINESE ARROWHEAD *Sagittaria chinensis.*
CHINESE ARTICHOKE *Stachys affinis.* Edible tubers, which are eaten like potatoes.
CHINESE BANANA *Musa nana* (Canary Banana) Dalziel.
CHINESE BEECH *Fagus englerana.*

CHINESE BELLFLOWER *Platycodon grandiflorum* Coats.
CHINESE BUCKWHEAT *Polygonum cymosum.*
CHINESE CABBAGE 1. *Brassica sinensis* (Chinese Mustard) Usher. 2. *Brassica pekinensis.* *B pekinensis* is known as Pe-tsai, and *B sinensis* as Pak-choi, which is the Cantonese form of pe-tsai.
CHINESE CEDAR *Cedrela sinensis* (Toon) Leathart.
CHINESE CHESTNUT *Castanea mollissima.*
CHINESE CHIVES *Allium tuberosum* (Garlic Chives) G.B.Foster.
CHINESE CINNAMON *Cinnamonium cassia* (Cassia) Schery. Not confined to China, for it grows in Japan and S.E.Asia as well, but it is cultivated a lot in south China, especially in Kwangsi. Cf. Canton Cassia.
CHINESE CORK OAK *Quercus variabilis.*
CHINESE COW'S TAIL PINE *Cephalotaxus fortunei* (Chinese Plum-yew) Leathart.
CHINESE CRACKER FLOWER *Pyrostegia venusta.*
CHINESE DATE *Zizyphus jubajuba* (Jujube) Brouk.
CHINESE DATE-PLUM *Diospyros kaki.*
CHINESE ELM *Ulmus parvifolia.*
CHINESE FIG *Ficus carica* (Common Fig) F.P.Smith.
CHINESE FIG, Dwarf *Ficus pumila.*
CHINESE FLOWERING ASH *Fraxinus mariesii.*
CHINESE FRAGRANT FERN *Artemisia annua* (Sweet Wormwood) G.B.Foster. Feathery foliage, and very fragrant.
CHINESE GOLD LARCH *Pseudolarix amabilis.*
CHINESE GOOOSEBERRY *Actinidia chinensis.* The fruit bears a superficial resemblance to gooseberries. They are better known as Kiwis, for the vine has been cultivated in New Zealand for a long time.
CHINESE HAWTHORN 1. *Crataegus pentagyna.* 2. *Photinia serrulata.*

CHINESE HAZEL *Corylus chinensis*.
CHINESE HEMLOCK *Tsuga chinensis*.
CHINESE JASMINE *Trachelospermum jasminoides* (Confederate Jasmine) Goold-Adams.
CHINESE JUDAS TREE *Cercis chinensis*.
CHINESE JUJUBE *Zizyphus jubajuba* (Jujube) Douglas.
CHINESE JUNIPER *Juniperus chinensis*.
CHINESE KIDNEY BEAN *Wistaria sinensis* (Chinese Wistaria) Hay & Synge.
CHINESE LACQUER TREE *Rhus verniciflua* (Lacquer Tree) H & P.
CHINESE LANTERNS 1. *Dichrostachys glomerata*. 2. *Geranium robertianum* (Herb Robert) Som. Macmillan. 3. *Physalis alkekengi* (Winter Cherry) Macmillan. Physalis fruit is a berry which is completely enclosed in a ribbed covering, and this, from its appearance, accounts for the name.
CHINESE LARCH *Larix potaninii*.
CHINESE LIME *Tilia oliveri*.
CHINESE LONG BEAN *Vigna unguiculata* (Cowpea) Dalziel.
CHINESE MAHOGANY *Cedrula sinensis*.
CHINESE MEDLAR *Eriobotrya japonica* (Loquat) Bianchini. A Chinese fruit, though better known as Japanese Medlar, for it is in Japan that most cultivation occurs.
CHINESE MUGWORT *Artemisia verlotorum*.
CHINESE MUSTARD 1. *Brassica sinensis*. 2. *Brassica juncea* (Indian Mustard) Clapham.
CHINESE NARCISSUS *Narcissus canaliculatus var. orientalis* (Polyanthus Narcissus) Jenyns. Actually not Chinese - it is a Mediterranean plant, introduced into China some time before the 16th century, probably by way of Persia. By now it has a particular association with the Chinese New Year, when especially dwarfed forms are sold in thousands. It is grown in bowls filled with pebbles, and forced for new year, when it is supposed to confer good fortune for the next twelve months.
CHINESE NECKLACE POPLAR *Populus lasiocarpa*.
CHINESE ONION 1. *Allium chinense*. 2. *Allium fistulosum* (Welsh Onion) F.P.Smith. The actual homeland of *A fistulosum* is in Siberia.
CHINESE PARSLEY *Coriandrum sativum* (Coriander) Sanecki. Not a native Chinese plant, though they certainly knew it.
CHINESE PEAR; CHINESE PEAR-APPLE *Pyrus sinensis* (Sandy Pear) Candolle (Pear); Douglas (Pear-apple).
CHINESE PEONY *Paeonia albiflora* (White-flowered Peony) Douglas.
CHINESE PERSIMMON *Diospyros kaki* (Chinese Date-Plum) Barber & Phillips. or Japanese Persimmon. It was originally from China, but now grows in both countries.
CHINESE PINE *Pinus tabuliformis*.
CHINESE PLUM *Prunus triloba*.
CHINESE PLUM-YEW *Cephalotaxus fortunei*.
CHINESE POPPY, Yellow *Meconopsis integrifolia*.
CHINESE PRIMROSE *Primula sinensis*.
CHINESE PRIVET *Ligustrum lucidum* (Wax Tree) Wilkinson.1981.
CHINESE QUINCE *Chaenomeles speciosa* (Japanese Quince) Geng Junying.
CHINESE RED-BARKED BIRCH *Betula albo-sinensis*.
CHINESE REDBUD *Cercis chinensis* (Chinese Judas Tree) H & P.
CHINESE RHUBARB *Rheum officinale* (Turkey Rhubarb) Lloyd. Together with the name Russian Rhubarb, both the names above are given in accordance with the country through which it reached the market place from its native land, which seems to be Tibet, though it is said that it grows in parts of western and northwest China, too.
CHINESE ROSE *Hibiscus rosa-sinensis* (Rose of China) Perry.1979.
CHINESE SACRED LILY *Narcissus canaliculatus var. orientalis* (Polyanthus Narcissus) Jenyns. See Chinese Narcissus.
CHINESE ST JOHN'S WORT *Hypericum chinense*.

CHINESE SNOWDROP TREE *Chimonanthus fragrans* (Wintersweet) McDonald.
CHINESE SOUR CHERRY *Prunus cantabrigiensis.*
CHINESE SPRUCE *Picea asperata.*
CHINESE STRAWBERRY *Myrica rubra* F.P.Smith.
CHINESE SUGAR MAPLE *Sorghum saccharatum* (Sorgo) Grieve.1933. Sweet Sorghum, too. It is cultivated extensively for syrup production.
CHINESE SUMACH *Ailanthus glandulosa* (Tree of Heaven) Chopra.
CHINESE SWAMP CYPRESS *Glyptostrobus lineatus.*
CHINESE TREE PEONY *Paeonia suffruticosa.*
CHINESE TRUMPET FLOWER *Campsis grandiflora.*
CHINESE TULIP TREE *Liriodendron chinense.*
CHINESE WALNUT *Juglans cathayensis.*
CHINESE WATER CHESTNUT *Eleocharis tuberosus.*
CHINESE WHITE PINE *Pinus armandii.*
CHINESE WHITE POPLAR *Populus tomentosa.*
CHINESE WILLOW PATTERN TREE *Kohlreuteria paniculata.*
CHINESE WINGNUT *Pterocarya stenoptera.*
CHINESE WOLFBERRY *Lycium chinense* (Tea Tree) Douglas.
CHINESE YAM 1. *Dioscorea batatas*. 2. *Dioscorea esculenta* (Lesser Yam) Alexander & Coursey. 3. *Dioscorea japonica*. 4. *Dioscorea opposita*.
CHINESE YELLOW-WOOD *Cladrastris chinensis.*
CHINESE YEW *Taxus celebica.*
CHINKS *Ornithogalum thyrsoides* (Chincherinchee). Mentioned in The Times of 24 May, 1990.
CHINOIS *Citrus aurantium var.amara sinensis* (China Orange) Bianchini.
CHINQUAPIN *Castanea pumila*. An American Indian word, spelt in a number of ways, as Chinkapin, or Chincapin.
CHINQUAPIN, Giant *Chrysolepis chrysophylla* (Golden Chestnut) Leathart.
CHINQUAPIN, Water *Nelumbo lutea* (Yellow Lotus) USA Gilmore.
CHINQUAPIN OAK *Quercus muhlenbergii* (Chestnut Oak) USA Tolstead.
CHINY OYSTER *Callistephus chinensis* (China Aster) Wilts. Dartnell & Goddard. Is this a genuine name, or just an attempt at transliterating the local accent? We have Chany Oyster, too, as a Gloucestershire offering.
CHIPPLE *Allium fistulosum* (Welsh Onion) Dor. Dacombe. Jibble in Monmouthshire, Sybie or Sybow in Scotland. This must be the French ciboule, Italian cipolle, usually reserved for shallots.
CHIPS, Carpenter's *Nasturtium officinale* (Watercress) Glos. Grigson.
CHIR PINE, Himalayan; CHIR PINE, Indian *Pinus roxburghii* Leathart (Indian).
CHIRETTA *Swertia chirata.*
CHIRMS *Caltha palustris* (Marsh Marigold) N'thants. Grigson. Is this Churns, rather? That name is certainly given to Yellow Waterlily, and there could have been confusion.
CHIT-CHAT *Sorbus aucuparia* (Rowan) Wilts. Dartnell & Goddard. Further north, there is Sip-sap. Halliwell says the first sprouts of any plant are called chits.
CHITAH *Lewisia rediviva* (Bitter-root) USA Havard. An Indian name for the plant.
CHIVE GARLIC *Allium schoernoprasum* (Chives) Corn. Jago.
CHIVES *Allium schoernoprasum*. Chive is ultimately from cepa, onion, and existed in various forms - medieval siethes, sieves, sithes, and more recently cives, sives, syves, civet, etc. Why is this always in the plural?
CHIVES, Chinese *Allium tuberosum* (Garlic Chives) G.B.Foster.
CHIVES, Garlic *Allium tuberosum.*
CHIVI TREE *Guibourtia coleosperma.*
CHIVVEL *Allium schoernoprasum* (Chives)

Corn. Jago.

CHOCK-CHEESE *Malva sylvestris* (Common Mallow) Dev. Friend. The ripe fruit is similar in shape to a cheese. This name varies to Chakky-cheese in Cornwall, to Chucky-cheese in Somerset, all probably from Truckles-of-cheese.

CHOCOLATE, Indian *Geum rivale* (Water Avens) USA Leighton. When the roots are boiled up with milk and sweetened, they give a drink not unlike chocolate, apparently a great favourite with the American Indians.

CHOCOLATE BEAN; CHOCOLATE NUT; CHOCOLATE TREE *Theobroma cacao* (Cacao) Bennett & Bird; Pomet (Nut); Leathart (Tree). It is said that Chocolatl was the Aztec name for the beverage which they made by pounding the seed with maize grains, and then boiling the powder with water, adding capsicum pepper, but ODEE points out that chocolatl was an article of food made from the seeds, and the Europeans confused this with cacaua-atl, which was the actual name of the drink.

CHOCOLATE BERRY *Vitex payos* (Coffee Bean Tree) Palgrave. Edible beans, or berries rather, and children like them.

CHOCOLATE FLOWER *Geranium maculatum* (Spotted Cranesbill) Grieve.1931. An American species.

CHOCOLATE PLANT *Cosmos atrosanguineus*. From the distinctive fragrance?

CHOCOLATE ROOT 1. *Geum canadense* (White Avens) Lindley. 2. *Geum rivale* (Water Avens) USA Sanford. The roots, when boiled with milk and sweetened, give a beverage not unlike chocolate, apparently a favourite with the Indians, hence the name Indian Chocolate.

CHOCTAW ROOT *Apocynum cannabinum* (Indian Hemp) Grieve.1931.

CHOISY *Hepatica triloba* (American Liverwort) Le Strange.

CHOKE CHERRY *Prunus virginiana*. Presumably, 'choke' is an indication that these are inedible.

CHOKE CHERRY, Western *Prunus demissa*.

CHOKE PEAR *Sorbus torminalis* (Wild Service Tree) A S Palmer.

CHOKEBERRY, Black *Aronia melanocarpa*.

CHOKEBERRY, Red *Aronia arbutifolia*.

CHOKEWEED *Orobanche minor* (Lesser Broomrape) Turner. The broomrapes are parasites. Cf. Strangleweed (Somerset) for this.

CHOKKY-CHEESE *Malva sylvestris* (Common Mallow) Corn. Jago. Better known in the West Country as Chucky-cheese.

CHONGRAS *Phytolacca decandra* (Pokeroot) USA Watt.

CHONNOCKS *Brassica campestris var.rapa* (Turnip tops) Staffs. Wakelin.

CHOOKY-PIGS 1. *Antirrhinum majus* (Snapdragon) Som. Macmillan. 2. *Dactylorchis maculata* (Heath Spotted Orchid) Som. Macmillan.

CHOOP-ROSE; CHOOP-TREE *Rosa canina* (Dog Rose) both Cumb. Grigson (Rose); Earle (Tree).

CHOOPS; CHOUPS; CHOWPS *Rosa canina* fruit (Hips) N.Eng, Scot. B & H (Choops); Lancs. Nodal (Choups); Yks. B & H (Chowps).

CHOOPS, Cat; CHOOPS, Dog *Rosa canina* fruit (Hips) Cumb. Grigson (Cat); Yks. Morris (Dog).

CHOP NUT *Physostigma venenosum* (Calabar Bean) Le Strange. 'Chop' in the lethal sense - Cf. Ordeal Bean, etc for this.

CHOPPED EGGS *Linaria vulgaris* (Toadflax) Som. Macmillan Cumb. Grigson. Colour-descriptive.

CHOPS, Monkey 1. *Antirrhinum majus* (Snapdragon) Som. Macmillan. 2. *Glechoma hederacea* (Ground Ivy) Som. Macmillan.

CHOPS, Mutton *Chenopodium album* (Fat Hen) Dor. Grigson. Cf. Mutton Tops for the same plant in the same area.

CHOPS, Pig's 1. *Antirrhinum majus* (Snapdragon) Som. Macmillan. 2. *Linaria vulgaris* (Toadflax) Som. Macmillan. This is one of a series of 'mouth' names, from the shape of the flowers. Cf. Pig's Mouth, Lion's Mouth, etc.

CHOPS, Rabbit's *Linaria vulgaris* (Toadflax) Som. Macmillan.

CHRIST, Eye of *Veronica chamaedrys* (Germander Speedwell) Wales Wright.

CHRIST, Ladder of; CHRIST, School of *Hypericum perforatum* (St John's Wort) both Wales both Trevelyan.

CHRIST-AND-THE-APOSTLES *Passiflora caerulea* (Blue Passion Flower) Dev. Macmillan.

CHRIST-THORN *Paliuris spina-christi*. Moldenke said this could possibly be the genuine crown of thorns of the New Testament. The young stems are pliant, and could be woven into a crown.

CHRIST-THORN, Syrian *Zizyphus spina-christi* (Lotus Tree) Moldenke.

CHRIST'S EYE 1. *Caltha palustris* (Marsh Marigold) Fernie. 2. *Salvia horminoides* (Wild Sage) Culpepper. Wild Sage, "most blasphemously called Christ's Eye, because it cures Diseases of the Eye" (Culpepper). But this is medieval Oculus Christi. Clary becomes Clear-eye.

CHRIST'S HEEL 1. *Plantago lanceolata* (Ribwort Plantain) Wales Denham/. Hardy. 2. *Plantago major* (Great Plantain) Wales Denham/Hardy. Presumably this is an extension of the derivation of plantain from Latin planta, sole of the foot.

CHRIST'S HERB 1. *Helleborus niger* (Christmas Rose) Gerard. 2. *Polygala vulgaris* (Milkwort) Wales Grigson. The Welsh Llysian Crist probably stems from the same tradition that associates the milkwort with Rogation-tide. Christmas Rose easily becomes Christ's Herb.

CHRIST'S LADDER 1. *Centaureum erythraea* (Centaury) Vesey-Fitzgerald. 2. *Hypericum perforatum* (St John's Wort) Wales Trevelyan.

CHRIST'S SPEAR *Ophioglossum vulgatum* (Adder's Tongue) Leyel.1937.

CHRIST'S THORN 1. *Ilex aquifolium* (Holly) Yks. Grigson. 2. *Zizyphus spina-christi* (Lotus Tree) J.Smith. The thorny leaves and the association with Christmas is enough to give grounds for this name for holly.

CHRIST'S WORT *Helleborus niger* (Christmas Rose) B & H.

CHRISTIAN; CHRISTLINGS *Prunus domestica var. institia* fruit (Bullace) both Som. Macmillan (Christian); Elworthy (Christlings). Ultimately from something like the Cumbrian name for the tree - Crex; the fruits are known as Crickseys in E.Anglia, and there is also a variant, Crystals, known in the West Country, as well as other variants like Kerslins.

CHRISTIAN, Palm of *Ricinus communis* (Castor Oil Plant) S USA Puckett. i.e. Palma Christi, Christ's Hand.

CHRISTMAS; CHRISTMAS, Prickly *Ilex aquifolium* (Holly) Corn, Dev, Hants, Suss, Worcs, Suff, Lincs, Ches. Grigson; Som. Raymond; Wilts. Dartnell & Goddard; IOW Long; Corn. Grigson (Prickly Christmas). It seems to have been used as a name for holly as a Christmas decoration in Wiltshire - "Why, you haven't got a bit o' Christmas about the house yet" (Dartnell & Goddard). In Somerset, though,it seems that holly with red berries was always given the name.

CHRISTMAS, Holy Thorn of *Salix discolor* USA Bergen.1899. It is said to blossom out at old Christmas, and then the flowers "go in again".

CHRISTMAS ANTHEM *Chrysanthemum sinensis x indicus* (Chrysanthemum) Som. Macmillan.

CHRISTMAS BERRY 1. *Chironia baccifera*. 2. *Ilex aquifolium* (Holly) Dev, Som. Macmillan. 3. *Photinia salicifolia*. The last one is usually spelt as one word.

CHRISTMAS BUSH 1. *Alchornea cordifolia* Harley. 2. *Eupatorium odoratum* Barbados Gooding. 3. *Euphorbia pulcherrima* (Poinsettia) USA Nielsen. Poinsettia has of recent years become an accepted Christmas emblem in America. Evidently, it was regarded in that light before Dr Poinsett discovered it in 1828, for its Mexican name translates Flower of the Holy Night, and the Totonac at least use it as an altar decoration on Christmas Eve(Kelly & Palerm). The Eupatorium has a flowering period of a few weeks around Christmas.

CHRISTMAS CACTUS *Schlumbergera x buckleyi*.

CHRISTMAS CANDLES *Aesculus hippocastanum* flowers (Horse Chestnut) Som. Macmillan. Just as the sight of the trees in full bloom seems to be a reminder of the Christmas Tree, so the individual flower spikes are the Christmas candles.

CHRISTMAS CHERRY *Solanum capsicastrum* (Winter Cherry) Rochford & Gorer.

CHRISTMAS DAISY *Aster grandiflorus* Coats.

CHRISTMAS FERN *Polystichum acrostichoides*.

CHRISTMAS FLOWER 1. *Eranthis hyemalis* (Winter Aconite) B & H. 2. *Euphorbia pulcherrima* (Poinsettia) USA Watt. 3. *Helleborus niger* (Christmas Rose) Scot. B & H. Cf. New Year's Gift, an East coast name for Eranthis, and for Poinsettia its name of Christmas Bush. It is now part of the accepted Christmas decoration in America, and has travelled over the Atlantic, too.

CHRISTMAS-FLOWERING IRIS *Iris alata*.

CHRISTMAS HERB *Helleborus niger* (Christmas Rose) Folkard.

CHRISTMAS PEPPER *Capsicum annuum* (Chile) Rochford & Gorer.

CHRISTMAS PRIDE *Ruellia paniculata*.

CHRISTMAS ROSE 1. *Eranthis hyemalis* (Winter Aconite) Som. Macmillan. 2. *Helleborus niger*. They bloom December - March.

CHRISTMAS ROSE, Wild *Helleborus viridis* (Green Hellebore) Som. Macmillan.

CHRISTMAS TREE 1. *Aesculus hippocastanum* (Horse Chestnut) Som. Macmillan. 2. *Araucaria araucana* (Chile Pine) Som. Macmillan. 3. *Ilex aquifolium* (Holly) Suff. Grigson. 4. *Picea excelsa*. The Spruce Fir, the correct name for *Picea excelsa*, is the accepted Christmas Tree. Holly has the name, as well as Christmas, and Prickly Christmas, because of its use as a Christmas decoration. The sight of Horse Chestnut in bloom is a reminder to Somerset people in spring of a decorated Christmas tree - the flowers are called Christmas Candles. But surely the Chile Pine has never functioned as a Christmas tree?

CHRISTMAS TREE, New Zealand *Metrosideros excelsa* (Pohutukawa) Leathart.

CHRISTOPHER, Herb 1. *Actaea spicata*. 2. *Pulicaria dysenterica* (Yellow Fleabane) Prior. Grigson.1974 suggests *Actaea spicata* was so named either because it was used against the plague, for which St Christopher was invoked, or because the flowers are borne in a spike above the leaves like the infant Christ on St Christopher's shoulders.

CHRISTOPHER, Stinking 1. *Scrophularia aquatica* (Water Betony) Cumb. Grigson. 2. *Scrophularia nodosa* (Figwort) Cumb. Grigson. Stinking Roger, too, for both of these.

CHRYSANTHEMUM *Chrysanthemum sinensis x indicum*. This is the original cross, cultivated in the east for 2000 years, to which all varieties owe their origin, and to which the name Chrysanthemum is applied - the florists' name, in other words. The name itself is Greek, meaning gold flower.

CHRYSANTHEMUM, Annual; CHRYSANTHEMUM, Summer *Chrysanthemum coronarium* T H Kearney (Summer).

CHRYSANTHEMUM, Corn *Chrysanthemum segetum* (Corn Marigold) Curtis.

CHRYSANTHEMUM, Garland *Chrysanthemum coronarium* (Annual Chrysanthemum) Coats.

CHRYSANTHEMUM, Wild *Senecio jacobaea* (Ragwort) Dev. Macmillan.

CHUCKENWORT *Stellaria media* (Chickweed) Aber. B & H. This is Chickenwort, which goes through various changes in Scotland, e.g. Cickenwort, Cuckenwort, Cluckenwort, etc.

CHUCKY-CHEESE 1. *Crataegus monogyna* the young leaves (Hawthorn) Dev. B & H. 2. *Malva sylvestris* (Common Mallow) Som. Macmillan. The ripe fruit of mallow is similar in shape to a cheese, and there are a host of 'cheese' names to show it. Equally, there are a lot of 'cheese' names for hawthorn, of the 'bread-and-cheese' type - children love to eat the young leaves in spring.

CHUFA *Cyperus esculenta* (Tigernut) USA Havard.

CHUPAROSA *Beloperone californica*.

CHURCH BELLS 1. *Campanula medium* (Canterbury Bells) Som. Macmillan. 2. *Cymbalaria muralis* (Ivy-leaved Toadflax) Dor. Macmillan. 3. *Symphytum officinale* (Comfrey) Som. Macmillan.

CHURCH BROOM *Dipsacus fullonum* (Teasel) Som. Macmillan. From the resemblance of the flower heads in shape to the long "turk's-head" brooms that were once used for sweeping high places.

CHURCH STEEPLES 1. *Agrimonia eupatoria* (Agrimony) Som, Suss. Brownlow. 2. *Eupatorium cannabinum* (Hemp Agrimony) Friend.

CHURCHMAN'S GREETING *Viscum album* (Mistletoe) Som. Macmillan.

CHURCHWORT *Mentha pulegium* (Pennyroyal) Cockayne. Greenoak suggests the name is appropriate, for its sweet scent and flea-repellant properties would make it welcome when it was strewn in church. But was it ever used this way? Cf. Bishopwort.

CHURCHYARD ELDER *Capsella bursa-pastoris* (Shepherd's Purse) Som. Macmillan.

CHURL *Cheiranthus cheiri* (Wallflower) Shrop. B & H. Presumably this is a variation of Cheiranthus, which became Chier, and Chare, in places.

CHURL HEAD *Centaurea scabiosa* (Greater Knapweed) Tynan & Maitland.

CHURL HEMP *Cannabis sativa* (Hemp) B & H. Only the female plant, in spite of the name. There seems to have been some confusion about sex differentiation, for the Female Hemp of the old herbalists is in fact the male plant. Cf. Carl Hemp.

CHURL'S CRESS *Lepidium campestre* (Pepperwort) B & H.

CHURL'S HEAD *Centaurea nigra* (Knapweed) Prior.

CHURL'S TREACLE *Allium sativum* (Garlic) Prior. This means exactly the same as the other names, Clown's Treacle, Countryman's Treacle, Poor Man's Treacle. Treacle is from triacle, from therione, a name given to the viper, i.e. the superstition that viper's flesh will cure viper's bite. Philips, World of Words, defines treacle as a "physical compound made of vipers and other ingredients", and this was a favourite against all poisons. The word then became applied to any confection of sweet syrup, and finally and solely to the syrup of molasses.

CHURLICK *Sinapis arvensis* (Charlock) Hants. Cope. One of the many variations on charlock.

CHURMEL, Greater *Blackstonia perfoliata* (Yellow-wort) Cockayne. Lesser Churmel is the Centaury. Yellow-wort is also known as Greater Centaury.

CHURMEL, Lesser *Centaureum erythraea* (Centaury) Cockayne. The Greater Churmel is *Blackstonia perfoliata*. Curmel is perhaps more usual.

CHURN 1. *Narcissus pseudo-narcissus* (Daffodil) Lancs. Nodal. 2. *Nuphar lutea* (Yellow Waterlily) Oxf. Grigson. Descriptive perhaps, from the general outline shape of the flowers. But there is another reason, at least as far as daffodil is concerned. Lancashire children used to play a game by separating the corolla from the stem bearing the pistil, and working it up and down with a churning motion while repeating the rhyme: Churn, churn, chop, Butter cum ta t'top.

CHURN, Butter 1. *Nuphar lutea* (Yellow Waterlily) Mor. Grigson. 2. *Ranunculus acris* (Meadow Buttercup) War. Grigson. 3. *Ranunculus bulbosus* (Bulbous Buttercup)

War. Grigson. 4. *Ranunculus repens* (Creeping Buttercup) War. Grigson.

CHURN, Devil's *Euphorbia peplus* (Petty Spurge) Ire. Gomme. Almost certainly referring to the juice, for Devil's Milk is another name. Allocated to the devil because this juice is poisonous. But is it the growth that caused the name? In that case, churnstaff would make more sense.

CHURNSTAFF 1. *Euphorbia helioscopia* (Sun Spurge) Ches. Holland. 2. *Linaria vulgaris* (Toadflax) Ches. Holland.

CHURNSTAFF, Devil's *Euphorbia helioscopia* (Sun Spurge) Shrop. Burne; Ire. Grigson. Churnstaff becomes Kirnstaff in northern England and Scotland. It is the juice, or "milk", which causes its appropriation to the devil.

CHUSAN PALM *Trachycarpus excelsus*.

CIBBOLS *Allium ascalonium* (Shallot) B & H. This is French ciboule, Italian cipolla, and the word has variants like Chibbal, Chibble, etc.

CIBOULE 1. *Allium ascalonium* (Shallot) Northcote. 2. *Allium fistulosum* (Welsh Onion) F.P.Smith. This is a French word applied to both of these, actually from the Provençal cebula, the name for onion which originated in the Latin diminutive for cepa, onion, cepulla. The name seems to have been imported along with its variants, Cibbols, Chibble, etc.

CICELY; CICELY, Wild *Anthriscus sylvestris* (Cow Parsley) Yks. B & H. Better known as a name as Sweet Cicely, and not applied to this.

CICELY, Fool's *Aethusa cynapium* (Fool's Parsley) B & H. A poisonous plant,hence the 'Fool's'.

CICELY, Rough *Torilis japonica* (Hedge Parsley) B & H. Rough Chervil is another name.

CICELY, Silken *Asclepias syriaca* (Common Milkweed) Parkinson.1640. No connection with Cicely, except in Parkinson's mind. An American plant, and there is an American name, Silkweed.

CICELY, Smooth *Myrrhis odorata* (Sweet Cicely) Grieve.1931.

CICELY, Sweet *Myrrhis odorata*. Cicely is a confusion with a girl's name, but it comes by Latin from Greek seselis. The botanical name Sesili is given to a closely related plant, the Moon Carrot.

CICENDIA, Slender *Cicendia filiformis*.

CICH *Cicer arietinum* (Chick Pea) Turner.

CICH, Little *Lathyrus sativus* (Indian Pea) Turner.

CICHLINGS *Vicia sativa* (Common Vetch) N. Eng. B & H. This is the same as Chichelings, and is from the Latin cicer - chick pea.

CICKENWORT *Stellaria media* (Chickweed) Border B & H. Like Cichlings, for *Vicia sativa*, the 'Ci-' in the North Country and Border will produce the 'Chi-' sound more commonly found.

CIDER, Bread-and- *Crataegus monogyna* (Hawthorn) Som. Macmillan. Bread-and-cheese is the usual name for the young shoots that children eat.

CIDER, Bread-and-cheese-and- 1. *Anemone nemerosa* (Wood Anemone) Dor. Grigson. 2. *Malva sylvestris* (Common Mallow) Som. Grigson.

CIDER GUM; TASMANIAN CIDER TREE *Eucalyptus gunnii* W Miller (Cider Tree). The name looks a little odd, but the reason lies in its use in Tasmania as a substitute for honey, called cider there.

CIDERAGE; CYDERACH *Polygonum hydropiper* (Water Pepper) Prior (Ciderage); Langham (Cyderach). Prior says from a French word, cidrage, if that word actually exists.

CIGAR-BOX CEDAR *Cedrela odorata* (West Indian Cedar) N.Davey. It is used for making cigar boxes.

CIGAR FLOWER, Mexican *Cuphea platycentra*. A descriptive name - the flowers have ash-grey margins, and a black ring behind, like a cigar band.

CIGAR TREE 1. *Catalpa bignonioides* (Locust Bean) Alab. Harper. 2. *Catalpa spe-*

ciosa (Western Catalpa) USA Cunningham. The Indians are supposed to have smoked the capsules.

CILICIAN FIR *Abies cilicica*.

CINERARIA *Senecio cineraria*. The Latin word ciner means ashes. Cf. other names for this, such as Dusty Miller, Silver-frosted Plant, etc.

CINERARIA, Purple-leaved *Senecio cruentus*.

CINERARIA, Wild *Senecio elegans* (Jacobaea) S. Africa Watt.

CINKEFIELD *Potentilla reptans* (Cinquefoil) Parkinson.1640. A variation of Cinquefoil. Gerard has Sinkfield, and Turner Cynkfoly.

CINNAMON *Cinnamonium zeylandicum*. Cinnamon, the spice, is the inner bark of this tree. The name is eventuallly from Greek, but before that of Arabic origin. It was anomon, fragrant. With the prefix kin, it means 'fragrant plant of China' (Cf. Hebrew qinnamon). The name was originally given to another species, China Cinnamon. *C. zeylanicum* was not known in Europe until the 16th century.

CINNAMON, Australian *Cinnamonium oliveri* (Oliver Bark) Mitton.

CINNAMON, Bastard *Cinnamonium cassia* (Cassia) Moldenke. Because this is usually looked on as an inferior kind of cinnamon.

CINNAMON, Cassia *Cinnamonium cassia* (Cassia) Watt.

CINNAMON, Chinese *Cinnamonium cassia* (Cassia) Schery. Cassia is cultivated extensively in south China. Cf. Canton Cassia.

CINNAMON, Java *Cinnamonium burmani*.

CINNAMON, Mountain *Cinnamodendron corticosum* (False Winter's Bark) Le Strange.

CINNAMON, Saigon *Cinnamonium loureirii*.

CINNAMON, White 1. *Canella alba*. 2. *Drimys winteri* (Winter's Bark) Pomet. It was the Spanish who gave the cinnamon name to *Canella alba*, under the false impression that this was a genuine member of the genus.

CINNAMON, Wild; CINNAMON, Wild, West Indian *Canella alba* (White Cinnamon) Le Strange.

CINNAMON, Winter's *Drimys winteri* (Winter's Bark) Le Strange. Named in honour of Captain Winter of the *Elizabeth*, of Drake's overall command, who first used the bark as a spice, and as a medicine to combat scurvy in his crew.

CINNAMON FERN *Osmunda cinnamonea*.

CINNAMON ROOT *Inula conyza* (Ploughman's Spikenard) Gerard. The root has a spicy scent, and used to be hung up in cottages to scent the musty air - this practice of either hanging up the plant or burning the roots for the aromatic perfume accounts for the name Spikenard.

CINNAMON ROSE *Rosa cinnamonea*. Probably refers to the colour, rather than to the scent.

CINNAMON SEDGE; CINNAMON IRIS *Acorus calamus* (Sweet Flag) B & H (Sedge); Genders (Iris). Because it is an aromatic plant, once used for strewing on the floors of churches and in the houses of the rich. When the stems and leaves are crushed they release a scent like that of cinnamon, described also as violet-like.

CINNAMON WOOD *Sassafras variifolium* (Sassafras Tree) Pomet.

CINQUEFOIL *Potentilla reptans*. Five-leaf, for the leaves each have five leaflets. The name has gone through a number of variations, for instance Cynkfoly (Turner), Sinkfield(Gerard), and Cinkefield(Parkinson).

CINQUEFOIL, Alpine *Potentilla crantzii*.

CINQUEFOIL, Golden *Potentilla aurea*.

CINQUEFOIL, Grey *Potentilla cinerea*.

CINQUEFOIL, Himalayan *Potentilla atrosanguinea*.

CINQUEFOIL, Hoary *Potentilla argentea*.

CINQUEFOIL, Least *Sibbaldia procumbens*.

CINQUEFOIL, Marsh *Potentilla palustris*.

CINQUEFOIL, Norwegian *Potentilla norvegica*.

CINQUEFOIL, Pink *Potentilla nitida*.

CINQUEFOIL, Rock *Potentilla rupestris*.

CINQUEFOIL, Rough *Potentilla mon-*

speliensis.
CINQUEFOIL, Shrubby *Potentilla fruticosa*.
CINQUEFOIL, Silvery *Potentilla anserina* (Silverweed) Fernie.
CINQUEFOIL, Silvery-leaved *Potentilla argentea* (Hoary Cinquefoil) Hutchinson.
CINQUEFOIL, Snow *Potentilla nivea*.
CINQUEFOIL, Spring *Potentilla tabernae-montani*.
CINQUEFOIL, Sulphur *Potentilla recta*.
CINQUEFOIL, White 1. *Potentilla alba*. 2. *Potentilla rupestris* (Rock Cinquefoil).
CINQUEFOIL, White, Shrubby *Potentilla caulescens*.
CIPER NUT *Bunium bulbocastanum* (Great Earth-nut) Culpepper. This is the same as Gerard's Kipper-nut.
CIS, Sweet *Myrrhis odorata* (Sweet Cicely) Yks. Grigson. Cis is a common contraction of Cicely, itself a confusion with the girl's name, though it comes through Latin from Greek seselis.
CISS; CISWEED *Anthriscus sylvestris* (Cow Parsley) Lancs. Grigson (Ciss); Yks. B & H (Cisweed). Cf. Cicely, or Wild Cicely, for Cow Parsley.
CISTSAGE *Cistus salvifolius* (Sage-leaved Cistus) Turner. Turner merely abbreviated the clumsier common name.
CISTUS, Grey-leaved *Cistus albidus*. Cistus is a mythological reference. Greek Kistos was the name of a youth.
CISTUS, Gum *Cistus ladaniferus*. The gum is ladanum, an aromatic, much used for unguents, perfumes and medicines.
CISTUS, Hoary *Cistus incanus*.
CISTUS, Laurel-leaved *Cistus laurifolius*.
CISTUS, Marsh 1. *Andromeda polifolia* (Bog Rosemary) C P Johnson. 2. *Ledum palustre* (Marsh Rosemary) Schauenberg.
CISTUS, Rose *Cistus incanus* (Hoary Cistus) Whittle & Cook.
CISTUS, Sage-leaved *Cistus salvifolius*.
CISTUS, Sweet *Cistus ladaniferus* (Gum Cistus) Parkinson.1640.
CITHERWOOD; CITHERNWOOD *Artemisia abrotanum* (Southernwood) Ire. Tynan & Maitland. Two of many variants on Southernwood.
CITRON *Citrus medica var. bajana*. Or is it simply *Citrus medica* that is so-called?
CITRON, Buddha's Head *Citrus medica*.
CITY GOOSEFOOT *Chenopodium urbicum* (Upright Goosefoot) USA Zenkert.
CIVES; CIVET *Allium schoernoprasum* (Chives) Gerard (Cives); Turner (Civet). Chives was medieval siethes, sieves, or sithes - hence these forms, and Cf. Sives, or Syves. Civet must have been pronounced with a long 'i'.
CIVET BEAN *Phaseolus lunatus* (Lima Bean) USA Yanovsky. Better known as Sieva Bean, of which civet must be a corruption.
CLADEN; CLADER; CLEDEN *Galium aparine* (Goose-grass) Dor. Dacombe (Claden); Grigson (Clader); Dor. Barnes (Cleden). Cf. West Country Clider, Cliden. Probably all eventually from Cleavers, a common name. Halliwell spells these versions Cleden.
CLAGGERS; CLOGGIRS *Galium aparine* (Goose-grass) Yks. Grigson (Claggers); Ork. Montgomerie (Cloggirs). More in the Cleavers mould.
CLAITON *Galium aparine* (Goose-grass) Dor. B & H. Dorset names close to this include Clader and Claden.
CLAMMY LOCUST *Robinia viscosa*.
CLAMMY PLANTAIN *Plantago psyllium* (Fleawort) Fernie.
CLAMMY WEED *Cleome serrulata* (Rocky Mountain Bee Plant) USA Elmore.
CLAMOUN *Kalmia latifolia* (Calico Bush) USA House.
CLANWILLIAM CEDAR *Widdringtonia juniperoides*. It grows in the Cedarberg Mountains of Clanwilliam district, South Africa.
CLAPPEDEPOUCH 1. *Capsella bursa-pastoris* (Shepherd's Purse) Prior. 2. *Galium aparine* (Goose-grass) Som. Grigson. The 'pouch' part of the name is the same as the

'purse' of Shepherd's Purse. The whole thing is an allusion to the licensed begging of lepers, who stood at crossroads with a bell and clapper. Another name with the same allusion is Rattlepouch.

CLAPWEED *Silene cucubalis* (Bladder Campion) Hants. Grigson.

CLARIMOND TULIP *Tulipa praecox* Geldart.

CLARY 1. *Salvia horminoides* (Wild Sage) Curtis. 2. *Salvia sclarea*. 3. *Salvia turkestanica* Rohde. Clary is derived through OE slarie from med. Latin sclarea. Prior says it is from latin clarus, clear; then into clear-eye, which Culpepper has. Seebright is another name. Eye salves were made from it. The seeds swell up when put into water, and become mucilaginous. These were then put like drops into the eye to cleanse it.

CLARY, Meadow *Salvia pratensis*.

CLARY, Silver *Salvia argentea* (Silver Sage) Perry.

CLARY, White *Heliotropum angiospermum* Barbados Gooding.

CLARY, Whorled *Salvia verticillata*.

CLARY, Wild 1. *Heliotropum angiospermum* Barbados Gooding. 2. *Heliotropum indicum* (Indian Heliotrope) Barbados Gooding. 3. *Salvia horminoides* (Wild Sage) B & H. 4. *Salvia pratensis* (Meadow Clary) Physicians of Myddfai.

CLARY, Wild, Small *Heliotropum angiospermum* Barbados Gooding.

CLARY, Woolly *Salvia aethiopis*.

CLARY SAGE *Salvia sclarea* (Clary) Grieve.1931.

CLATTER MALLOCH *Trifolium pratense*.

CLATTERCLOGS *Tussilago farfara* (Coltsfoot) Cumb. Grigson. Presumably from the same source as Cleat - a dialect word meaning hoof.

CLAUT *Caltha palustris* (Marsh Marigold) Wilts. Dartnell & Goddard. Halliwell has this, too, without explanation, but it could be Clote, a name given to Yellow Waterlily. That word means 'hoof', and Horses' Hooves is a name for Marsh Marigold, the leaf shape being the reference.

CLAVER 1. *Galium aparine* (Goose-grass) Som. Macmillan; N. Eng. Brockett. 2. *Lotus corniculatus* (Bird's-foot Trefoil) N. Ire. Grigson. 3. *Trifolium pratense* (Red Clover) Dev, Dor, Glos, N.Eng, Ayr. Grigson. 4. *Trifolium repens* (White Clover) W & N Eng. Grigson. Claver is a widespread variation of clover, probably nearer to its original than the accepted modern word, for it is OE clafre and later claefre. ODEE says Claver was common 15th to 17th century. But Claver for Goose-grass is quite a different word - it is one of a large number of variations on Cleavers.

CLAVER, Hart *Melilotus officinalis* (Melilot) N.Eng. F.Grose. Hart's Clover also, later becoming Heartwort. But *Medicago arabica* was Heart Clover, too.

CLAVER, Heart *Medicago arabica* (Spotted Medick) Curtis. Also Heart Clover, Heart Medick, Heart Trefoil. And Heart Clover became corrupted to Heart Liver. Coles says it is so called "not only because the leaf is triangular like the heart of a man, but also because each leafe doth contain the perfection (or image) of an heart, and that in its proper colour, viz. a flesh colour. It defendeth the heart against the noisome vapour of the spleen".

CLAVER, Marsh *Menyanthes trifoliata* (Bogbean) Gerard.

CLAVER-GRASS; CLAVVER-GRASS *Galium aparine* (Goose-grass) both Cumb. B & H. These are not from clover, but cleavers, a very common name for the plant, from OE clife, claw.

CLAVER-SORREL *Oxalis acetosella* (Wood Sorrel) Langham.

CLAWS, Bird's *Lotus corniculatus* (Bird's-foot Trefoil) Dev. Macmillan.

CLAWS, Cat's 1. *Acacia greggii*. 2. *Anthyllis vulneraria* (Kidney Vetch) Som. Macmillan; Mor. Grigson. 3. *Lotus corniculatus* (Bird's-foot Trefoil) Som. Macmillan; Bucks. Grigson. 4. *Ranunculus repens* (Creeping Buttercup) Lancs. Grigson. 5.

Robinia neomexicana (New Mexican Locust) USA Elmore. 6. *Rubus fruticosus* (Bramble) Som. Macmillan.

CLAWS, Crab's 1. *Polygonum persicaria* (Persicaria) Som, Dor. Macmillan. 2. *Stratiotes aloides* (Water Soldier) Grieve.1931.

CLAWS, Crow(′s) 1. *Ranunculus arvensis* (Corn Buttercup) Hants, Suss, Ess. Grigson. 2. *Ranunculus repens* (Creeping Buttercup) Hants. Cope; Suss. B & H. Crowfoot is the normal name for the buttercup, and both that and the 'claws' name describe the spiny seeds.

CLAWS, Devil's 1. *Acacia greggii* (Cat's Claws) USA Jaeger. 2. *Lotus corniculatus* (Bird's-foot Trefoil) Dev. Friend. 3. *Phyteuma comosus*. 4. *Proboscidea louisiana* (Unicorn Plant) USA Bean. 5. *Ranunculus arvensis* (Corn Buttercup) Wilts. D.Grose; Hants, IOW Grigson. The Acacia is a notoriously thorny shrub.

CLAWS, Lark's *Delphinium ajacis* (Larkspur) Gerard.

CLAWS, Lion's *Crotalaria agatiflora*.

CLAWS, Lobster 1. *Clianthus puniceus* (Parrot's Bill) Simmons. 2. *Erythrina crista-galli* (Cock's Comb) Nicolaisen. 3. *Heliconia metallica*.

CLAWS, Owl's *Helenium hoopesii* (Orange Sneezeweed) USA Elmore.

CLAWS, Ram's 1. *Ranunculus acris* (Meadow Buttercup) Som. Elworthy. 2. *Ranunculus repens* (Creeping Buttercup) Som. Grigson Dor. Dacombe; Wilts. Dartnell & Goddard.1899. 3. *Stellaria media* (Chickweed) Som. Macmillan. 4. *Tussilago farfara* (Coltsfoot) Som. Macmillan. Surely at least the buttercup must have got this name originally from Raven's Claws. But the idea of claws on a ram may not be as crazy as it seems if the original was German Klaue, hoof. The name, if that were the case, would seem quite reasonable for coltsfoot, at least.

CLAWS, Raven's *Ranunculus repens* (Creeping Buttercup) Dor, Som, Wilts. Grigson.

CLAWS, Tiger's *Erythrina indica* (Indian Coral Tree) Malaya Gilliland.

CLAY *Galium aparine* (Goose-grass) Som. Macmillan. Cly is more likely.

CLAYT *Tussilago farfara* (Coltsfoot) Yks. B & H. Coltsfoot wine was called Clayt wine in Yorkshire. This is obviously the same as Cleat, a name also given in the area, which means hoof.

CLAYWEED; CLAYWORT *Tussilago farfara* (Coltsfoot) B & H (weed); Salisbury.1964 (wort). Either from its partiality to clay soils, or is this the same word as Clayt, i.e. cleat, in this case meaning hoof?

CLEANSING GRASS *Euphorbia lathyris* (Caper Spurge) Cockayne. Presumably from an obscure medicinal use, or perhaps simply for its purging quality.

CLEAR-EYE 1. *Salvia horminoides* (Wild Sage) B & H. 2. *Salvia sclarea* (Clary) Culpepper. i.e. Clary, which is from Latin sclarea, ultimately, says Prior, from clarus, clear. So eye salves were made from it. The seeds swell up when put into water, and become mucilaginous. These were then put like drops into the eye to cleanse it.

CLEAR-EYE, Wild *Salvia horminoides* (Wild Sage) B & H.

CLEAT 1. *Petasites hybridus* (Butterbur) Cumb. B & H; Yks. Carr. 2. *Tussilago farfara* (Coltsfoot) Lancs, Yks, Cumb. B & H. Cleat-leaves also for Butterbur, and there are similar names, such as Clayt amd Clote, for Coltsfoot. In both cases the name refers to the shape of the leaves, and cleat probably means hoof or something similar. Halliwell describes cleat as a piece of iron worn on shoes by country people; cleat-boards are broad pieces of wood fastened to the shoes to enable a person to walk on the mud without sinking into it.

CLEAT-LEAVES *Petasites hybridus* (Butterbur) Cumb. B & H. See Cleat.

CLEAVED GRASS *Galium aparine* (Goose-grass) Dawson.

CLEAVERS 1. *Arctium lappa* (Burdock) Som. Mabey. 2. *Dipsacus fullonum* (Teasel)

Som. Grigson. 3. *Galium aparine* (Goose-grass) common; USA Tolstead. Cleavers is probably OE clife, which may mean claw, but more to the point is the word which gave the verb to cleave, meaning to stick fast, adhere. Turner said it got the name because "it cleueth upon mennes clothes". Teasel's circumstances are similar.

CLEAVERS, Cross *Galium circaezans* USA Tolstead.

CLEAVERS, False *Galium spurium.*

CLEAVERS, Goose *Galium aparine* (Goose-grass) Lanark. Grigson.

CLEAVERS, Marsh *Menyanthes trifoliata* (Bogbean) B & H. In this case, cleaver is clover.

CLEAVERS, Spring *Galium aparine* (Goose-grass) USA Yarnell.

CLEEVE PINK *Dianthus gratianapolitanus* (Cheddar Pink) Som. Jennings. Cleeve is an old form of cliff. The plant only occurs at Cheddar Gorge, and it is called Cliff Pink in the area, too.

CLEMATIS, Bush *Clematis paniculata.*

CLEMATIS, Fragrant *Clematis flammula* (Upright Virgin's Bower) Polunin.1976.

CLEMATIS, Red *Ampelopsis hederacea* (Virginia Creeper) Dev. Friend.1882.

CLEMATIS, Wild 1. *Clematis virginiana* USA House. 2. *Clematis vitalba* (Old Man's Beard) Jones & Dillon.

CLEMATIS IRIS *Iris kaempferi* (Japanese Iris).

CLENCH *Ranunculus arvensis* (Corn Buttercup) N'thants. Grigson. This is possibly the same word as 'cling', and so would have reference to the spiny seeds.

CLERK, Parson-and- *Arum maculatum* (Cuckoo-pint) Dev, Som. Friend.1882. Spadix in spathe, so a double name in a series separate to the overtly sexual ones. Cf. Parson-in-the-pulpit, etc.

CLETHEREN; CLITHEREN *Galium aparine* (Goose-grass) Som. Macmillan (Cletheren); Prior (Clitheren). Prior notes the frequent change of v to th, i.e. clivers to clitheren, and the Cornish Clythers, etc.

CLEVER *Galium aparine* (Goose-grass) Gerard. i.e. Cleavers.

CLIDE; CLIDEN; CLIDER *Galium aparine* (Goose-grass) Fernie (Clide); Dev. Friend.1882 (Cliden); Corn. Courtney.1880 Dev, Som. B & H Wilts, Hants. Jones & Dillon; IOW Long (Clider). All part of the large group of names which starts with Cleavers.

CLIFF ELM *Ulmus racemosa* (Cork Elm) Wilkinson.1978.

CLIFF PINK *Dianthus gratianapolitanus* (Cheddar Pink) Som. Jennings. Cheddar Pink grows only on the side of the gorge. Cf. also Cleeve Pink for this.

CLIFF ROSE *Armeria maritima* (Thrift) Dev. Grigson. Habitat descriptive.

CLIMBERS *Clematis vitalba* (Old Man's Beard) Kent Grigson.

CLIME *Galium aparine* (Goose-grass) Som. Macmillan.

CLING-FINGERS *Orchis mascula* (Early Purple Orchis) Oxf. B & H. We also have Ring-finger from the same county, but, possibly more significantly, King-finger, from a wider but similar area. The name probably comes from the dark red blotches on the leaves, looking as if someone with bloody fingers had touched them, hence Bloody Fingers, etc.

CLING-RASCAL *Galium aparine* (Goose-grass) Dev. Friend.

CLINGING SWEETHEARTS *Galium aparine* (Goose-grass) Wilts. Macmillan. The burrs stick to your clothes like sweethearts. Yorkshire children used to say, when a burr is sticking to your dress, that you have got a sweetheart. In Alton, Hants., the schoolchildren say that the number of burrs sticking to a girl's clothes indicates the number of her sweethearts.

CLINKER BEECH *Nothofagus truncata* (Hard Beech) Brimble.1948.

CLIP-ME-DICK 1. *Euphorbia cyparissa* (Cypress Spurge) Ches. Holland; Lancs. Grigson. 2. *Polygonum convolvulus* (Black Bindweed) Ches. Holland. Probably a corruption of cyparissa, as is the Cheshire Kiss-me-dick. But in that case, why should Black Bindweed get the name? The answer probably lies in the OE word clyppan, to embrace, which is what all the bindweeds do.

CLITCH-BUTTONS 1. *Arctium lappa* (Burdock) Dev. B & H. 2. *Galium aparine* (Goose-grass) Dev. B & H. For clitch, see clite. The buttons refer to the burrs.

CLITE 1. *Arctium lappa* (Burdock) B & H. 2. *Galium aparine* (Goose-grass) Wilts. Jones & Dillon; Glos. J.D.Robertson; Oxf. Halliwell. This is OE clife, eventually from the word meaning claws. It exists in various forms for both these plants - Goose-grass has Clide, Clyte, Clits, etc., while Burdock has Clithe, Clitch Buttons, etc.

CLITHE; CLITHEREN; CLITHERS *Galium aparine* (Goose-grass) B & H (Clithe); Prior (Clitheren); Fernie (Clithers). All OE clite, eventually from the word meaning claw. Clitheren shows the plural in -en.

CLITS *Galium aparine* (Goose-grass) Dor. B & H. OE clite. Hair that is tangled is called clitty in parts of Wiltshire. Clits for tangles is also recorded.

CLIVERS; CLIVERS, Hedge; CLIVEN *Galium aparine* (Goose-grass) Gerard; Storms (Hedge Clivers); Dev. Friend.1882 (Cliven). i.e. Cleavers.

CLIVERS, Evergreen *Rubia peregrina* (Wild Madder) IOW Grigson. Madder is close to Goose-grass, hence the name.

CLOAK, Lady's *Cardamine pratensis* (Lady's Smock) Som. Macmillan. Cf. Lady's Mantle.

CLOBWEED *Centaurea nigra* (Knapweed) Glos. Grigson. Clob may be club, in which case this is another descriptive name, from the profile of the flower head.

CLOCK 1. *Plantago lanceolata* (Ribwort Plantain) Bucks. Grigson. 2. *Rhinanthus crista-galli* (Yellow Rattle) Scot. Grigson. 3. *Taraxacum officinale* (Dandelion) Som. Grigson; Wilts. Dartnell & Goddard; N'thants. A.E.Baker; Yks. Addy. Dandelion clocks are well-known, but Clock for the Ribwort is most probably a mistake for Cock, a reference to the game called Fighting Cocks, in which the heads are knocked off in a variation of conkers.

CLOCK, Children's *Taraxacum officinale* (Dandelion) Som. Macmillan.

CLOCK, Doonhead *Taraxacum officinale* (Dandelion) Scot. Grigson; USA Henkel.

CLOCK, Fairy *Taraxacum officinale* (Dandelion) Mabey.

CLOCK, Fairy's *Adoxa moschatellina* (Moschatel) Dev. Macmillan.

CLOCK, Farmer's *Taraxacum officinale* (Dandelion) Putnam.

CLOCK, Flora's *Tragopogon pratensis* (Goat's Beard) Tynan & Maitland. Cf. Jack-goes-to-bed-at-noon and many others of like kind.

CLOCK, Old Man's *Taraxacum officinale* (Dandelion) Putnam.

CLOCK, Peasant's *Taraxacum officinale* (Dandelion) Dyer.

CLOCK, Puff *Taraxacum officinale* (Dandelion) Som. Macmillan. Applied to the seed heads. Another example of children's time-telling names.

CLOCK, Shepherd's 1. *Anagallis arvensis* (Scarlet Pimpernel) Glos. Friend. 2. *Taraxacum officinale* (Dandelion) Mabey. 3. *Tragopogon pratensis* (Goat's Beard) Som, Suss. Grigson. From the folding of the petals of the first two, or the time-telling game with the seeds of dandelion.

CLOCK, Schoolboy's *Taraxacum officinale* (Dandelion) Mabey.

CLOCK, Town; CLOCK, Town-hall; CLOCK, Infirmary *Adoxa moschatellina* (Moschatel) Glos. Grigson (Town); Cumb. Grigson (Town-hall); Tynan & Maitland (Infirmary). Said to be descriptive - there are four faces to the flower, with a fifth pointing upwards. Fairy's Clock is recorded in Devonshire.

CLOCK, Weather *Taraxacum officinale* (Dandelion) Som. Macmillan. A mingling of two children's beliefs to produce a hybrid name. The seed tufts are used as barometers,

and of course, the globes are time-tellers.
CLOCK-FLOWER *Taraxacum officinale* (Dandelion) Som. Macmillan.
CLOCK-NEEDLE *Scandix pecten-veneris* (Shepherd's Needle) Bucks. Grigson.
CLOCK-POSY *Taraxacum officinale* (Dandelion) Lancs. Nodal.
CLOCKS, Watches-and- *Taraxacum officinale* (Dandelion) Som. Macmillan. Clocks-and-watches is probably more familiar than this.
CLOCKS-AND-WATCHES *Taraxacum officinale* (Dandelion) Mabey.
CLOCKVINE, Bengal *Thunbergia grandiflora* (Sky Vine) Simmons.
CLOCKVINE, Black-eyed *Thunbergia alata* Perry.
CLOD-BURR *Arctium lappa* (Burdock) Cumb, Yks. B & H. Clot-burr also, from the same area. This is OE clote.
CLODWEED 1. *Filago germanica* (Cudweed) W.Miller. 2. *Knautia arvensis* (Field Scabious) Bucks. B & H. There is also Clogweed for the scabious.
CLOFFING *Ranunculus sceleratus* (Celery-leaved Buttercup) Cockayne. B & H say this is probably a mistake, and that cloffing should apply to one of the hellebores (Halliwell too assigns the name to a hellebore). But the buttercup is glossed Cloftnunge on a C12 ms of Apuleius Barbarus, and surely it is OE clufu, bulb, where clufwyrt is buttercup.
CLOG PLANT *Hypocyrta glabra*. Is this descriptive of a shoe-shaped flower?
CLOGS, Pattens-and- 1. *Linaria vulgaris* (Toadflax) Som. Macmillan; Suss. B & H. 2. *Lotus corniculatus* (Bird's-foot Trefoil) Som. Macmillan; Glos. J.D.Robertson; Suss. Parish.
CLOGWEED 1. *Arctium lappa* (Burdock) Wilts. Mabey. 2. *Heracleum sphondyllium* (Hogweed) Som. Macmillan; Wilts. Dartnell & Goddard; Glos. J.D.Robertson. 3. *Knautia arvensis* (Field Scabious) Bucks. B & H. Burdock has a number of names akin to this, and the burrs themselves are known as Clots, Clouts, or Cluts in the north of England. Clogweed may simply be a mispronunciation of Hogweed, but Macmillan says it is the same as Keck-lock, the type of name which most umbellifers have. Field Scabious has Clodweed, too.
CLOGWEED, Cow *Heracleum sphondyllium* (Hogweed) Glos. B & H.
CLOSE SCIENCES; COSES SCIENCES *Hesperis matronalis* (Sweet Rocket) Gerard (Close); B & H (Coses). These are corruptions of Sciney, itself a corruption of damascene, the old specific name for the plant. Siney occurs in Devonshire as a name for this.
CLOT 1. *Arctium lappa* burrs (Burdock) Ches, N.Eng. B & H. 2. *Nuphar lutea* (Yellow Waterlily) B & H. 3. *Petasites hybridus* (Butterbur) Yks. Carr. 4. *Verbascum thapsus* (Great Mullein) Aubrey. Burrs are called Clouts, or Cluts in the north country, too. It is OE clote. Clote itself still occurs for the yellow waterlily in Devonshire, Dorset and Somerset, and both forms are recorded in Wiltshire for the mullein. Chaucer has clote lefe for either burdock or waterlily, and this probably refers to the shape of the leaf. Clote can mean wedge.
CLOT-BURR 1. *Agrimonia eupatoria* (Agrimony) Som. Macmillan. 2. *Arctium lappa* (Burdock) Cumb, Yks. B & H. 3. *Xanthium canadense* USA Elmore. 4. *Xanthium orientale* USA Zenkert. 5. *Xanthium pennsylvanicum* USA Gates. 6. *Xanthium spinosum* (Spiny Cocklebur) S.Africa Watt. 7. *Xanthium strumarium* (Cocklebur) Turner.
CLOT-BURR, Spiny *Xanthium spinosum* (Spiny Cocklebur) Watt.
CLOTE 1. *Galium aparine* (Goose-grass) Cockayne. 2. *Nuphar lutea* (Yellow Waterlily) Dev, Som. Macmillan; Dor. Dacombe. 3. *Petasites hybridus* (Butterbur) Grigson.1974. 4. *Tussilago farfara* (Coltsfoot) Ess, Norf. Grigson. 5. *Verbascum thapsus* (Great Mullein) Wilts. Dartnell & Goddard. For coltsfoot, clote is

probably the same as cleat, a dialect word meaning hoof, while the goose-grass version is one of a great variety of words seemingly derived from OE clife, claws. Not, though, according to Grigson.1974 - it was OE clate, originally meant for burdock, with the basic meaning of something which sticks (burrs), and then extended to other broad-leaved plants.

CLOTE, Fox's *Arctium lappa* (Burdock) Fernie.

CLOTE-BURR *Arctium lappa* (Burdock) Gerard.

CLOTH-LEAVED SALLOW *Salix pannosa*.

CLOTH OF GOLD *Baileya multiradiata* (Desert Baileya) USA Kingsbury. Bright yellow flowers.

CLOTH OF GOLD CROCUS *Crocus angustifolius*.

CLOTHES BRUSH; CLOTHIER'S BRUSH *Dipsacus fullonum* (Teasel) Som. Grigson; Wilts. Dartnell & Goddard (Clothes); Cumb. Grigson (Clothier's). Brushes for teasels is quite common. Cf. Barber's Brushes, Lady's Brushes, Combs-and-brushes, etc.

CLOTHES PEGS 1. *Digitalis purpurea* (Foxglove) Dor. Macmillan. 2. *Orchis mascula* (Early Purple Orchis) Som. Macmillan.

CLOUD PLANT *Gypsophila sp.*

CLOUDBERRY *Rubus chamaemorus*. Cloud is OE clud, rock or hill. Cf. Knotberry, Knoutberry, and Nub-berry, where the first elements in each case means a hill.

CLOUDED IRIS *Iris xiphium* (Spanish Iris) W.Miller. Miller also has Thunderbolt Iris for this.

CLOUT *Arctium lappa* burrs (Burdock) Ches, N.Eng. B & H. Clots, or clouts, in the same area. OE clote.

CLOVE, Madagascar *Ravensara aromatica*.

CLOVE BUSH *Buglossoides arvensis* (Corn Gromwell) S.Africa Watt.

CLOVE GILLIFLOWER; CLOVE JULY FLOWER *Dianthus caryophyllus* (Carnation) Gerard (Gilli-); Hill (July-). Gilliflower is a corruption of caryophyllus (clove).

CLOVE PINK *Dianthus caryophyllus* (Carnation) McClintock.

CLOVE-ROOT *Geum urbanum* (Avens) Vesey-Fitzgerald. Culpepper has Clove-wort. The roots have a sweet, spicy smell, which gave it the old shop name of Caryophyllata (under which name Gerard described the plant). Similar names occur in German, Spanish and Italian.

CLOVE-SCENTED BROOMRAPE *Orobanche caryophyllacea*.

CLOVE-TONGUE *Helleborus niger* (Christmas Rose) B & H. 'Cloven' rather than 'clove' makes some sense, given the shape of the leaves.

CLOVE TREE *Syzygium aromaticum*. Clove is from the French clou, nail. Cloves are the dried, unopened flower buds of this evergreen tree.

CLOVER 1. *Melilotus alba* (White Melilot) S.Africa Watt. 2. *Trifolium sp*. Clover is OE clafre, claefre, still surviving in place names like Clavering (Essex), Claverdon (War), Claverton (Ches). The form Claver was in common use from the 15th to the 17th century.

CLOVER, African, Red *Trifolium africanum* (Cape Clover) Watt.

CLOVER, Alsike *Trifolium hybridum*.

CLOVER, Bastard *Trifolium hybridum* (Alsike Clover) S.Africa Watt.

CLOVER, Beach *Trifolium fimbriatum*.

CLOVER, Beerseem *Trifolium alexandrinum* (Egyptian Clover) Schery.

CLOVER, Bird's-foot 1. *Lotus corniculatus* (Bird's-foot Trefoil). 2. *Trigonella ornithopodioides* (Fenugreek) Prior.

CLOVER, Bokhara *Melilotus alba* (White Melilot) B & H.

CLOVER, Broad *Trifolium pratense* (Red Clover) IOW Grigson. Macmillan has Broad-grass for this for Dorset.

CLOVER, Buffalo *Trifolium reflexum*.

CLOVER, Bur 1. *Medicago arabica* (Spotted Medick) Schery. 2. *Medicago polymorpha* (Hairy Medick) USA Kingsbury.

CLOVER, Burrowing *Trifolium subterraneum*.

CLOVER, Calvary *Medicago arabica* (Spotted Medick) McClintock. The spots on the leaves gave the name. Like many other plants with spotted leaves, and with names like this (Gethsemane, for instance) the legend that it grew on Calvary and that the spots are Christ's blood preserved for ever on the plant, is responsible for the name. A German superstition, taken to America, said that Calvary Clover would only germinate if sown on Good Friday.

CLOVER, Cape *Trifolium africanum*.

CLOVER, Cat's; CLOVER, Catten *Lotus corniculatus* (Bird's-foot Trefoil) N'thum. Grigson (Cat's); Scot. Grigson (Catten).

CLOVER, Clustered *Trifolium glomeratum*.

CLOVER, Cow 1. *Trifolium involucratum* (Multiflowered Clover) USA Elmore. 2. *Trifolium medium* (Zigzag Clover) Yks. B & H. 3. *Trifolium pratense* (Red Clover) B & H. Cow Grass is quite widespread for Red Clover, and there is a Scottish form Cowcloos.

CLOVER, Crimson *Trifolium incarnatum*.

CLOVER, Cuckoo's; CLOVER, Gowk's *Oxalis acetosella* (Wood Sorrel) N.Ire. Grigson (Cuckoo); N'thum. Grigson (Gowk).

CLOVER, Cup *Trifolium cyathiferum*.

CLOVER, Devil's *Medicago arabica* (Spotted Medick) Corn. Grigson.

CLOVER, Dog *Medicago lupulina* (Black Medick) Som. Grigson.

CLOVER, Dutch 1. *Trifolium pratense* (Red Clover) Hulme. 2. *Trifolium repens* (White Clover) Curtis. Apparently, the seed was largely imported from Holland. It was first introduced as a fodder plant by Sir Richard Western, ambassador to Holland, in 1645. The name is simplified to Dutch, for White Clover, in Somerset.

CLOVER, Egyptian *Trifolium alexandrinum*.

CLOVER, Field, Old *Trifolium arvense* (Hare's-foot Trefoil) USA House.

CLOVER, Framlington *Prunella vulgaris* (Self-heal) N'thum. Denham/Hardy. Long Framlington, in Northumberland.

CLOVER, Garden *Melilotus caerulea* (Swiss Melilot) Turner.

CLOVER, Hare's-foot *Trifolium arvense* (Hare's-foot Trefoil) Prior.

CLOVER, Hart's 1. *Melilotus altissima* (Tall Melilot) Yks, N.Eng. Gerard. 2. *Melilotus officinalis* (Melilot) Yks. B & H.

CLOVER, Heart *Medicago arabica* (Spotted Medick) B & H. There are also Heart Claver, Heart Medick and Heart Trefoil for this. Heart Liver is a corruption of Heart Clover. Coles says that it is so called "not only because the leaf is triangular like the heart of a man, but also because each leafe doth contain the perfection (or image) of an heart, and that therein its proper colour, viz a flesh colour. It defendeth the heart against the noisome vapour of the spleen".

CLOVER, Holy *Onobrychis sativa* (Sainfoin) Wit. Possibly from the same source as that which gave Lucerne, i.e. Latin lux, lucis, from the shining grains, but more likely from the misunderstanding that produced "Saintfoin" out of Sainfoin, even to the point of inventing an entirely spurious saint, Saint Foyne.

CLOVER, Hop 1. *Medicago lupulina* (Black Medick) Dev, Wilts. Grigson. 2. *Trifolium aureum* (Large Hop Trefoil) A.W.Smith. 3. *Trifolium campestre* (Hop Trefoil) Prior.

CLOVER, Hop, Least *Trifolium dubium* (Lesser Yellow Trefoil) USA Allan.

CLOVER, Horned 1. *Medicago lupulina* (Black Medick) Turner. 2. *Tetragonolobus maritimus* (Dragon's Teeth) Gerard.

CLOVER, Italian *Trifolium incarnatum* (Crimson Clover) Candolle.

CLOVER, Jamie Hedley's *Prunella vulgaris* (Self-heal) N'thum. Denham/Hardy. Hedley's father wanted to take a farm then vacant, but being blind sent his son Jamie to report on it. Jamie noticed one field particularly green, and, returning to his father, said it was grand land, full of clover. The "clover" was self-heal, indicative of sterility.

CLOVER, Kentish *Trifolium repens* (White

Clover) McClintock.

CLOVER, King's 1. *Melilotus altissima* (Tall Melilot) Scot. Grigson. 2. *Ranunculus acris* (Meadow Buttercup) Suff, Norf, Cumb. Grigson. 3. *Ranunculus bulbosus* (Bulbous Buttercup) Suff, Norf, Cumb. Grigson. 4. *Ranunculus repens* (Creeping Buttercup) Suff, Norf, Cumb. Grigson.

CLOVER, Knotted; CLOVER, Knotted, Soft *Trifolium striatum*. It is Hutchinson who calls it Soft Knotted Clover.

CLOVER, Ladino *Trifolium repens* (White Clover) S.Africa Watt.

CLOVER, Lady's *Oxalis acetosella* (Wood Sorrel) Perth. Grigson.

CLOVER, Maltese *Hedysarum coronarium* (French Honeysuckle) Pratt.

CLOVER, Marsh *Menyanthes trifoliata* (Buckbean) B & H.

CLOVER, Mary's *Lysimachia nemora* (Yellow Pimpernel) Ire. Grigson.

CLOVER, Meadow 1. *Trifolium medium* (Zigzag Clover) Hutchinson. 2. *Trifolium pratense* (Red Clover) Prior.

CLOVER, Multiflowered *Trifolium involucratum*.

CLOVER, Musk *Erodium moschatum* (Musk Storksbill) USA Grigson.

CLOVER, Persian *Trifolium resupinatum* (Reversed Clover) McClintock. From the country of origin.

CLOVER, Pin *Erodium cicutarium* (Storksbill) USA Elmore. Pin Grass also, in the States. Filaree is also an American name for this - it comes from Spanish alfiler, meaning a pin.

CLOVER, Pin-pointed *Trifolium gracilentum*.

CLOVER, Pink, Wild *Trifolium africanum* (Cape Clover) S.Africa Watt.

CLOVER, Plaister *Melilotus officinalis* (Melilot) Gerard. The plant has been used since Galen's time in poultices for dispersing tumours. "The fresh plant is excellent to mix in pultices to be applied to swellings" (Hill).

CLOVER, Puff *Trifolium fucatum*.

CLOVER, Purple *Trifolium pratense* (Red Clover) Prior.

CLOVER, Rabbit-foot *Trifolium arvense* (Hare's-foot Trefoil) USA House.

CLOVER, Red *Trifolium pratense*.

CLOVER, Reversed *Trifolium resupinatum*. It gets its name from the flower stalk being twisted so that the standard petal of the flower is below, and the keel above.

CLOVER, Rough *Trifolium scabrum*.

CLOVER, Russian *Coronilla varia* (Crown Vetch) USA House.

CLOVER, St Mawe's *Medicago arabica* (Spotted Medick) Corn. B & H.

CLOVER, Sea *Trifolium squamosum*.

CLOVER, Sleeping *Oxalis acetosella* (Wood Sorrel) Dor. Macmillan.

CLOVER, Small *Medicago lupulina* (Black Medick) Som. Macmillan.

CLOVER, Snail *Medicago sativa var.sativa* (Lucerne) Parkinson.

CLOVER, Soft *Trifolium striatum* (Knotted Clover) McClintock.

CLOVER, Soukie *Trifolium pratense* (Red Clover) Scot. B & H. Clovers have a number of 'honeysuckle' names, including honeysuckle itself. The 'soukie' of this name is the same as the 'suckle' of honeysuckle, and can appear by itself, as Gerard's Suckles, or Scottish Souks, or Sowkies, etc.

CLOVER, Sour *Oxalis acetosella* (Wood Sorrel) Berw. Grigson. 'Sour' in this context means the same as sorrel.

CLOVER, Spanish *Lotus americanus*.

CLOVER, Spotted *Medicago arabica* (Spotted Medick) Corn. Grigson.

CLOVER, Starry *Trifolium stellatum*.

CLOVER, Stinking 1. *Cleome sonorae* (Sonora Beeweed) USA Elmore. 2. *Melilotus indica* (Small-flowered Melilot) S.Africa Watt.

CLOVER, Stone *Trifolium arvense* (Hare's-foot Trefoil) USA House.

CLOVER, Strawberry *Trifolium frageriferum*.

CLOVER, Suckling *Trifolium dubium* (Lesser Yellow Trefoil) Polunin.

CLOVER, Suffocated *Trifolium suffocatum*.

CLOVER, Sulphur *Trifolium ochroleucum*.
CLOVER, Swedish *Trifolium hybridum* (Alsike Clover) S.Africa Watt.
CLOVER, Sweet 1. *Melilotus alba* (White Melilot) S.Africa Watt; New. Zealand Connor. 2. *Melilotus indica* (Small-flowered Melilot) S.Africa Watt.
CLOVER, Sweet, White *Melilotus alba* (White Melilot) USA Gates.
CLOVER, Sweet, Yellow *Melilotus officinalis* (Melilot) Kansas Gates.
CLOVER, Thousand-leaved *Achillea millefolium* (Yarrow) Border B & H; USA Henkel. The finely-divided leaves account for *millefolium* - the Thousand-leaved of the name. Cf. other names such as Milfoil, Mother-of-Thousands, etc.
CLOVER, Thunder *Ajuga reptans* (Bugle) Cockayne. There is also a Gloucestershire name, Thunder-and-lightning. Bugle growing on a building protected it from lightning, so it was said in Wiltshire.
CLOVER, Tomcat *Trifolium tridentatum*. From the smell?
CLOVER, Tree *Trifolium ciliatum*.
CLOVER, Twin-flowered *Trifolium bocconi*.
CLOVER, Upright *Trifolium strictum*.
CLOVER, White *Trifolium repens*.
CLOVER, Wild *Oxalis acetosella* (Wood Sorrel) Lincs. Grigson.
CLOVER, Winter *Mitchella repens* (Partridge-berry) Le Strange.
CLOVER, Woolly-headed *Trifolium eriocephalum*.
CLOVER, Yellow 1. *Lotus corniculatus* (Bird's-foot Trefoil) Yks. Grigson. 2. *Medicago lupulina* (Black Medick) B & H. 3. *Trifolium campestre* (Hop Trefoil) USA. 4. *Trifolium ochroleucum* (Sulphur Clover) Curtis.
CLOVER, Zigzag *Trifolium medium*.
CLOVER DEVIL; CLOVER DODDER *Cuscuta epithymum* (Dodder) Som. Macmillan (Devil); Ess. Grigson; USA House. (Dodder). In Haute Bretagne, it was believed that the devil spun the dodder at night to destroy the clover; clover was created by God, dodder was the devil's counter-plant. The plant was usually asssociated with the devil in this country, too.
CLOVER-KNOB *Centaurea nigra* (Knapweed) Notts. Grigson. Knob is the same as knap-, or knop. A knop is any small round protuberance.
CLOVER-RAPE *Orobanche minor* (Lesser Broomrape) Salisbury.1964.
CLOVER ROSE *Trifolium pratense* (Red Clover) Dev. Grigson.
CLOVEWORT 1. *Geum urbanum* (Avens) Culpepper. 2. *Ranunculus acris* (Meadow Buttercup) N'thants. Grigson. For Avens, see Clove-root.
CLOW *Dianthus caryophyllus* (Carnation) Norf, Suff, Scot. B & H. This must be the original of clove, which came from French clou, nail (English clout).
CLOWN'S ALLHEAL 1. *Stachys palustris* (Marsh Woundwort) Curtis. 2. *Stachys sylvatica* (Hedge Woundwort) Fernie. Clown in all these cases simply means a countryman, rustic. Any of the betony or woundwort group would be an allheal.
CLOWN'S LUNGWORT 1. *Lathraea squamosa* (Toothwort) B & H. 2. *Verbascum thapsus* (Great Mullein) B & H; USA Henkel. It is the flannel-like leaves of mullein that gives the name Lungwort. There are also Bullock's Lungwort, and Cow's Lungwort for this. In addition, Lungwort has become Longwort, or even Logwort, in places.
CLOWN'S MUSTARD 1. *Armoracia rusticana* (Horseradish) Parkinson.1640. 2. *Iberia amara* (Candytuft) B & H.
CLOWN'S TREACLE *Allium sativum* (Taper-leaved Garlic) B & H. Treacle is from triacle, an antidote to the bite of venomous animals. It is from the Greek theriake, from therione, a name given to the viper, i.e. the superstition that viper's flesh will cure viper's bite. Philips, *World of Words*, defines treacle as a "physical compound made of vipers and other ingredients", and this was a favourite against all poisons. The word then became applied to any confection of sweet syrup, and

finally and solely, to the syrup of molasses. Treacle appears in many names for garlic - as well as Clown's, there are Churl's, Countryman's, and Poor Man's.
CLOWN'S WOUNDWORT *Stachys palustris* (Marsh Woundwort) Gerard. Cf. Clown's Allheal for the same.
CLOWNS *Pinguicula vulgaris* (Butterwort) Roxb. Grigson.
CLUB, Hercules' *Aralia spinosa*. This is a descriptive name - Cf. Devil's Walking-stick.
CLUB, Shepherd's *Verbascum thapsus* (Great Mullein) IOW Long; Scot. Prior; USA Henkel. It is the height of the plant that causes the name - Cf. Jupiter's, and Shepherd's, Staff, Aaron's Rod, etc.
CLUB BUNCHES *Viburnum opulus* (Guelder Rose) Berks. Berkshire FWI. Particularly at Hagbourne, where they were used to decorate the president's chair at the club dinner held on Feast Day (2nd Tuesday after Whitsun).
CLUB-RUSH 1. *Scirpus lacustris* (Bulrush) H C Long.1910. 2. *Typha latifolia* (False Bulrush) H C Long.1910.
CLUB-RUSH, Sea *Scirpus maritimus*.
CLUB-SHAPED GOURD *Lagenaria siceraria* (Calabash) Bianchini.
CLUB WHEAT *Triticum compactum*.
CLUBER *Trifolium repens* (White Clover) Dawson.
CLUBMOSS *Lycopodium inundatum*.
CLUBMOSS, Alpine *Lycopodium alpinum*.
CLUBMOSS, Fir; CLUBMOSS, Mountain *Urostachys selago* USA Yarnell (Fir).
CLUBMOSS, Lesser *Selaginella selaginoides*.
CLUBMOSS, Marsh *Lycopodium inundatum* (Clubmoss) McClintock.
CLUBMOSS, Stag's-horn *Lycopodium clavatum*.
CLUBWEED *Centaurea nigra* (Knapweed) Halliwell. Cf. the Gloucestershire Clobweed.
CLUCKENWEED; CLUCKENWORT; CLUCKWEED *Stellaria media* all N'thum. all Grigson. Chickweed seems to be very variable. Besides these, there are Chuckenwort, Cuckenwort (both in Scotland), Cickenwort in the Border country and Schickenwir in Shetland, etc.
CLUSTER BEAN *Cyamopsis tetragonolobus*.
CLUSTER-CHERRIES *Prunus padus* fruit (Bird Cherry) B & H. Wild Cluster-berry is an old name for the tree.
CLUSTER PINE *Pinus pinaster* (Maritime Pine) Barton & Castle.
CLUSTER YAM *Disocorea dumetorum*.
CLUSTERBERRY *Vaccinium vitis-idaea* (Cowberry) Derb. Grigson.
CLUSTERBERRY, Wild *Prunus padus* (Bird Cherry) Camden/ Gibson.
CLUTCH *Polygonum aviculare* (Knotgrass) Som. Elworthy.
CLUTCH-BUTTON *Galium aparine* (Goose-grass) Dev. Friend. This is probably Clitch-button (I may have mis-recorded this). But there are such names as Grip-grass and Catchweed for this plant.
CLUTS *Arctium lappa* fruit (Burdock - burrs) N.Eng. B & H. They are also Clots, or Clouts, in N. England. Hence the plant itself is a Clot-burr, or Clote-burr. This is OE clote.
CLY; CLYDER; CLYTE; CLYTHERS *Galium aparine* (Goose-grass) Wessex Rogers (Cly); Corn. Jago; Dev. Grigson; Wilts. Dartnell & Goddard; Hants. Cope (Clyder); Wilts. Akerman (Clyte); Corn. Jago (Clythers). All these belong to the group starting with the common Cleavers - OE clife, claw.
CNOUTBERRY *Rubus chamaemorus* (Cloudberry) W.Miller. Another spelling of knoutberry, which sometimes becomes knotberry. It is from ME knot, hill. And 'cloud-' is from OE clud, a rock or hill, in spite of the Lancashire tradition that King Cnut was saved by eating this fruit.
COACH-AND-HORSES *Aconitum napellus* (Monkshood) Dor. Macmillan.
COACH-HORSES *Viola tricolor* (Pansy) Som. Elworthy.
COACH-WHEEL, Devil's *Ranunculus*

arvensis (Corn Buttercup) Hants. Cope. Cf. Devil's Curry-comb, Devil-on-both-sides, Hellweed, etc., all because of the prickly fruits, which make (or made) it a pest, for it is tall enough to be bound in with the corn.

COACHMAN'S BUTTONS *Knautia arvensis* (Field Scabious) Som. Macmillan.

COACHWOOD *Eucryphia moorei* Everett.

COAKUM, COCUM *Phytolacca decandra* (Poke-root) USA Henkel (Coakum); Watt (Cocum).

COAL OIL BRUSH *Tetradymia glabrata* (Littleleaf Horsebrush) USA Kingsbury.

COALIER, The *Hieraceum aurantiacum* (Orange Hawkweed) Border B & H. Cf. Grim the Collier for the same plant. "The stalks and cups of the flours are all set thicke with a blackishe downe or hairinesse as it were the dust of coles; whence the women who keepe it in gardens for novelties sake, have named it Grimm the Collier" (Gerard).

COAST *Tanacetum balsamita* (Balsamint) B & H. This and Cost are contractions of Alecost, and Alecoast, so given because according to Parkinson, it was used for flavouring ales. Cost, better known as Costmary, is from costus, an aromatic plant used in making perfumes in the East. The plant is dedicated to the Virgin Mary in most west European countries.

COAST LILY *Lilium maritimum*.

COAST VIOLET *Viola brittoniana*.

COAT, Jacob's *Acalypha wilkesiana* (Copperleaf) RHS. Meant for 'Joseph's Coat', obviously, for the leaves are multi-coloured. Cf. Match-me-if-you-can, for no two of them are ever alike.

COAT, Joseph's 1. *Alternanthera bettzickiana* (Joy Weed) USA Cunningham. 2. *Amaranthus tricolor*. 3. *Pulmonaria officinalis* (Lungwort) Brownlow. All these are "tricolor" - hence Joseph's Coat.

COAT, Prickly *Cirsium vulgare* (Spear Plume Thistle) Dor. Macmillan.

COB, King; COB, King's 1. *Caltha palustris* (Marsh Marigold) Berks, Hants. B & H. 2. *Ranunculus acris* (Meadow Buttercup) E.Ang, Cumb. Grigson. 3. *Ranunculus arvensis* (Corn Buttercup) E.Ang. Gerard. 4. *Ranunculus bulbosus* (Bulbous Buttercup) Suff, Norf, Cumb. Grigson. 5. *Ranunculus repens* (Creeping Buttercup) Suff, Norf, Cumb. Grigson. 'Cob' is presumably cup, but it may be halfway between cup and knob.

COB NUT *Corylus avellana* nut (Hazel) Wilkinson. The game (something like conkers) was called Cob-nuts. According to Hunter (*Hallamshire glossary*, 1829), the name was reserved for the winning nut. An older version of the game involved pitching with one nut against a heap of others arranged in a set pattern. All the nuts knocked down became the property of the pitcher. In this version, the nut used for pitching was called the Cob.

COBAN TREE; COBIN TREE *Viburnum lantana* (Wayfaring Tree) N'thum. Rowling (Coban); Dor. Grigson ((Cobin). Better known as Coven-tree, where coven is probably a meeting-place.

COBBEDY-CUT; COBBLY-CUT *Corylus avellana* (Hazel) both Corn. Grigson.

COBBLER'S PEGS 1. *Bidens bipinnata* (Spanish Needles) Australia Watt. 2. *Bidens pilosa* Australia Watt. 3. *Erigeron canadensis* (Canadian Fleabane) Australia Watt.

COBRA PLANT; COBRA LILY *Darlingtonia californica* Cunningham (Lily). Because the tubular leaves look like a snake.

COBWEB HOUSELEEK *Sempervivum arachnoideum*.

COCA *Erythroxilon coca*. This is the Spanish version of the Quechua cuca.

COCAINE PLANT *Erythroxilon coca* (Coca) USA Cunningham.

COCK, Cuckoo 1. *Arum maculatum* (Cuckoo-pint) Ess. Cockayne. 2. *Orchis mascula* (Early Purple Orchis) Ess. Grigson.

COCK-BATTLER; COCK-FIGHTER *Plantago lanceolata* (Ribwort Plantain) Corn. Courtney (Battler); N'thum. Denham/Hardy (Fighter). The flower heads are used as "soldiers" or "fighting cocks" by children. One child holds out a flower, and another tries to decapitate it with a second flower stalk..

COCK-BRAMBLE 1. *Rosa canina* (Dog Rose) Suff. Grigson. 2. *Rubus fruticosus* (Bramble) Norf, Suff. Grigson.

COCK-BRUMBLE *Rubus fruticosus* (Bramble) Norf, Suff. Forby.

COCK-DRUNKS *Sorbus aucuparia* fruit (Rowan berries) Cumb. B & H. Hen-drunks, too. The fruit has the reputation being able to make fowls drunk.

COCK-FLOWER *Orchis mascula* (Early Purple Orchis) Hants. B & H.

COCK-GRASS 1. *Lolium temulentum* (Darnel) Som. Macmillan; Cambs. Halliwell. 2. *Plantago lanceolata* (Ribwort Plantain) Dev, Som. Macmillan. 3. *Rhinanthus crista-galli* (Yellow Rattle) B & H.

COCK ROBIN 1. *Lychnis flos-cuculi* (Ragged Robin) Som. Macmillan. 2. *Melandrium rubrum* (Red Campion) Dev. Friend.1882.

COCK ROBIN, White *Silene cucubalis* (Bladder Campion) Som. Macmillan.

COCK-ROSE *Papaver rhoeas* (Red Poppy) Yks. Morris; Scot. Grigson. This is either the Celtic coch, red, or it is a mistake for Cop-rose. Cop is a button, like the shape of the capsule.

COCK-SORREL *Rumex acetosa* (Wild Sorrel) Yks. B & H; Hunts. Marshall.

COCK THISTLE *Onopordum acanthium* (Scotch Thistle) Dor. Macmillan.

COCK-UPON-PERCH 1. *Linaria vulgaris* (Yellow Toadflax) Som. Macmillan. 2. *Lotus corniculatus* (Bird's-foot Trefoil) Som. Macmillan.

COCK'S COMB 1. *Amaranthus hypondriachus* (Prince's Feathers) O.P.Brown. 2. *Amaranthus paniculatus* (Purple Amaranth) Watt. 3. *Celosia cristata*. 4. *Dactylorchis incarnata* (Early Marsh Orchid) Border B & H. 5. *Dipsacus fullonum* (Teasel) Dor. Macmillan. 6. *Erythrina crista-galli*. 7. *Heliotropum indicum* (Indian Heliotrope) Dalziel. 8. *Lychnis flos-cuculi* (Ragged Robin) Lanark. Grigson. 9. *Odontites verna* (Red Bartsia) Yks. B & H. 10. *Onobrychis sativa* (Sainfoin) Prior. 11. *Orchis mascula* (Early Purple Orchis) Border B & H. 12. *Papaver rhoeas* (Red Poppy) Berw. Grigson. 13. *Pedicularis palustris* (Red Rattle) Scot. Grigson. 14. *Pedicularis sylvatica* (Lousewort) Som, Scot. Grigson. 15. *Rhinanthus crista-galli* (Yellow Rattle) Som, Glos, Shrop. Grigson. All these are descriptive in some form or another - either colour or shape.

COCK'S COMB, Red *Amaranthus hypondriachus* (Purple Amaranth) O.P.Brown.

COCK'S FOOT *Aquilegia vulgaris* (Columbine) B & H; USA Hutchinson. There must be something to remind one of feet about this plant, for as well as Cock's, there are Dove's and Hawk's. And there are many "shoe" names, too.

COCK'S HEAD 1. *Antirrhinum majus* (Snapdragon) Tynan & Maitland. 2. *Onobrychis sativa* (Sainfoin) Som. Macmillan. 3. *Papaver argemone* (Rough Long-headed Poppy) Scot. B & H. 4. *Papaver dubium* (Smooth Long-headed Poppy) Scot. B & H. 5. *Papaver rhoeas* (Red Poppy) Scot. Grigson. 6. *Plantago lanceolata* (Ribwort Plantain) Som, Suss, Norf, Suff. Grigson. 7. *Trifolium pratense* (Red Clover) Gerard. In spite of appearances, 'cock' in these cases is probably the Celtic coch, red.

COCK'S HEAD, Purple *Astragalus danicus* (Purple Milk Vetch). B & H.

COCKENS *Papaver rhoeas* (Red Poppy) N'thum, Berw. Grigson. Possibly from Celtic coch, red.

COCKEREL *Agrostemma githago* (Corn Cockle) Suff. Grigson. Cockerel, like Cockleford in Gloucestershire and Cokeweed in Scotland, is a variation on Cockle, itself from Latin coccum, berry, from the large round seeds.

COCKFOOT *Chelidonium majus* (Greater Celandine) B & H. Presumably descriptive -

the shape of the leaves.
COCKHEAD 1. *Centaurea nigra* (Knapweed) N.Eng. Grigson. 2. *Stachys palustris* (Marsh Woundwort) Lanark. Grigson.
COCKILOORIE *Bellis perennis* (Daisy) Shet. B & H. Jamieson's derivation was from Germanic koka, sward, and lura, to lie hid, i.e. "what lies hidden during winter in the sward".
COCKLE 1. *Agrostemma githago* (Corn Cockle) Som. Macmillan. 2. *Arctium lappa* burrs (Burdock) Wilts. Dartnell & Goddard; Dor. Barnes; Hants. Cope. 3. *Lolium temulentum* (Darnel) B & H. 4. *Lotus corniculatus* (Bird's-foot Trefoil) Som. Macmillan. 5. *Melandrium album* (White Campion) Norf, Rut. Grigson. 6. *Saponaria officinalis* (Soapwort) S.Africa Watt. 7. *Silene cucubalis* (Bladder Campion) War. Grigson. 8. *Ulex europaeus* seeds (Furze) Som. Macmillan. 9. *Vinca major* (Greater Periwinkle) Dev. Friend.1882. 10. *Vinca minor* (Lesser Periwinkle) Glos. Bayley. As far as Corn Cockle, Soapwort and Furze seeds are concerned, this is probably Latin coccum, berry (the seeds in this case). The others may very well be cuckoo, rather. Burrs, for instance, have names variable from Cuckles to Cuckold's Buttons. Bird's-foot Trefoil is often known as Cuckoo's Stockings. The campions are certainly cuckoo flowers, while periwinkle remains doubtful. As far as darnel is concerned, the name is probably an acknowledgement of the damage it can do in the corn, being called after the equally pestilential Corn Cockle. But before Shakespeare's time, the name was applied to any noxious weed in the cornfields.
COCKLE, China *Saponaria officinalis* (Soapwort) S.Africa Watt. Or simply Cockle.
COCKLE, Corn *Agrostemma githago.* See Cockle.
COCKLE, Corn, Colorado *Vaccaria pyramidalis* (Cow-herb) Watt.
COCKLE, Corn, Gromwell *Buglossoides arvensis* (Corn Gromwell) S.Africa Watt.
COCKLE, Cow 1. *Saponaria officinalis* (Soapwort) S.Africa Watt. 2. *Vaccaria pyramidalis* (Cow-herb) Kansas Gates.
COCKLE, Purple *Agrostemma githago* (Corn Cockle) S.Africa Watt.
COCKLE, White 1. *Melandrium album* (White Campion) USA Allan. 2. *Silene cucubalis* (Bladder Campion) Berw. Grigson.
COCKLE-BELL; COCKLE-BUTTONS *Arctium lappa* burrs (Burdock) Corn. Jago (Bell); Corn. Courtney.1880; Dev. Friend.1882; USA Henkel (Buttons).
COCKLE DOCK *Dipsacus fullonum* (Teasel) Wilts. Goddard. Burdock seems a likelier candidate for this name, for Cuckold's Dock, and other similar ones, belong to that plant.
COCKLE-SHELL *Vinca minor* (Lesser Periwinkle) Som. Macmillan. They are Cockles next door in Gloucestershire.
COCKLEBUR; COCKLE-BUR 1. *Agrimonia eupatoria* (Agrimony) Fernie. 2. *Arctium lappa* (Burdock) Sanecki. 3. *Xanthium strumarium* and other species. With agrimony, the name occurs because the seed vessels cling by the hooked ends of their stiff hairs to anyone coming into contact with them. Cf. Sweethearts, Stickwort, etc. Presumably the American plant gets its name for the same reason. Cocklebur for burdock fits in the general run of cockle, cuckold and cuckoo names.
COCKLEBUR, Broad *Xanthium strumarium* (Cocklebur) Watt.
COCKLEBUR, Dagger *Xanthium spinosum* (Spiny Cocklebur) S.Africa Watt.
COCKLEBUR, Spiny *Xanthium spinosum.*
COCKLEFORD *Agrostemma githago* (Corn Cockle) Glos. Grigson. A variation of Cockle, as are Cockerel in Suffolk and Cokeweed in Scotland.
COCKLY-BUR *Arctium lappa* (Burdock) Cumb. B & H. Crockelty-bur occurs in the same area, but they are both probably from the series that includes Cuckold, Cuckold

Dock, etc.

COCKROSE *Papaver dubium* (Smooth Long-headed Poppy) Scot. Jamieson. Possibly Celtic coch, rose, but Coprose is recorded in Scotland as well, and this is taken to be from cop, a button (shape of the capsules).

COCKS 1. *Plantago lanceolata* (Ribwort Plantain) Suff, Armagh Denham/Hardy. 2. *Plantago major* (Great Plantain) Suff. Moor. Cocks for Great Plantain is probably an error, although other Ribwort names like Hardhead and Kemp are recorded. The name as applied to Ribwort refers to the "soldiers" or "fighting cocks" game children play. One child holds out a stem, while the other tries to decapitate it with his own flower stalk. The game was known by the same name in northern England.

COCKS, Bull's; COCKS, Dog *Arum maculatum* (Cuckoo-pint) Wilts. Dartnell & Goddard. Cf. Billycocks. The word has the same connotation as pintle, or the 'pint' of Cuckoo-pint.

COCKS, Fightee; COCKS, Fighting *Plantago lanceolata* (Ribwort Plantain) Suff, E.Ang, N'thants. Grigson (Fightee) Dev, Som, Shrop, Ches, Yks. Grigson; Wilts. Dartnell & Goddard; N'thants. Denham/Hardy; E Ang. Forby (Fighting). See Cocks.

COCKS, Johnny *Orchis mascula* (Early Purple Orchis) Dor. B & H. Cf. Cock-flower in Hampshire, and Cock's Combs, or Hen's Combs, in the Border country.

COCKS, Lent; COCKS, Leny; LENY-COCKS *Narcissus pseudo-narcissus* (Daffodil) Friend (Lent); Som. Tongue (Leny). These names seem to be an allusion to the practice of cock-throwing. When the custom died out, it was the flowers themselves that were decapitated.

COCKS-AND-HENS 1. *Geum rivale* (Water Avens) N'thum. Grigson. 2. *Lotus corniculatus* (Bird's-foot Trefoil) IOM Moore.1924. 3. *Plantago lanceolata* (Ribwort Plantain) Dev. Friend.1882; Som, N'thants. Grigson; Ire. Denham/Hardy. There are a lot of 'cock' names for Ribwort, some of them to do with the children's game of fighting cocks.

COCKSFOOT *Dactylis glomerata.*

COCKSPUR; COCKSPUR, Maltese *Centaurea melitensis* (Napa Thistle) S.Africa Watt; Watt (Maltese).

COCKSPUR CORAL TREE *Erythrina crista-galli* (Cock's Comb) RHS.

COCKSPUR THORN *Crataegus crus-galli.*

COCKWEED 1. *Agrostemma githago* (Corn Cockle) Halliwell. 2. *Lepidium campestre* (Pepperwort) Som. Macmillan.

COCKY-BABY *Arum maculatum* (Cuckoo-pint) IOW Long. Presumably cuckoo-baby, but Cuckoo-cocks, with the same meaning as Cuckoo-pint, is also recorded, and there may have been confusion between the two forms.

COCO DE MER *Lodoicea maldivica.*

COCO-YAM, Old *Colocasia esculenta* (Dasheen) Schery.

COCOA, Brazilian *Paullinia cupana* (Guarana) Mitton.

COCOA-BUTTONS *Arctium lappa* (Burdock) Som. Macmillan. Cocoa is presumably a corruption of cuckold.

COCOA-ROOT *Caladium bicolor.* The rhizomes are boiled and eaten as food in tropical countries under this name.

COCOA TREE *Theobroma cacao* (Cacao) Cunningham. Cocoa is from the Aztec name for the beverage, chocolatl.

COCOANUT *Cocos nucifera* (Coconut) Ackermann. A popular mistake for coconut, and nothing to do with cocoa, which is *Theobroma cacao.*

COCONUT *Cocos nucifera.* Spanish and Portuguese coco means a grinning face, grimace, an allusion to the monkey-like appearance of the base of the shell of the nut.

COCONUT, Double; COCONUT, Seychelles *Lodoicea maldivica* (Coco de mer) Turrill (Double); A.Huxley (Seychelles). The fruit is one of the largest known.

COCOWORT *Capsella bursa-pastoris* (Shepherd's Purse) B & H.

COCOYAM *Colocasia antiquorum* (Taro) Brouk.

COCOZELLE; CUCUZZI *Lagenaria siceraria* (Calabash) Bianchini (Cocozelle); A.W.Smith (Cucuzzi).

COD'S HEAD *Trifolium pratense* (Red Clover) Tynan & Maitland. Cock's Head, obviously. Is this genuine?

CODDLED APPLES *Epilobium hirsutum* (Great Willowherb) N'thants. A.E.Baker; Notts, Lincs. Grigson. To coddle means to stew. Cf. for the same plant Codlins-and-cream, Codlins, Sugar-codlins, as well as the various names like Apple-pie. Gerard, however, called it Codded Willow-herbe - "the flower groweth at the top of the stalke, comming out of the end of a small longe codde".

CODLINS 1. *Epilobium hirsutum* (Great Willowherb) Dev, Cumb, N'thum. Grigson; Glos. J.D.Robertson. 2. *Malus sylvestris* (Crabapple) Prior. Codlin, usually meant for a particular kind of apple, is the same as coddle, to stew. It used to be claimed that Great Willowherb got the name because it smells like apples when crushed, but there seems to be no factual basis for this. The probable origin is Gerard's named Codded Willowherb.

CODLINS, Sugar *Epilobium hirsutum* (Great Willowherb) Som, Wilts. Grigson.

CODLINS-AND-CREAM *Epilobium hirsutum* (Great Willowherb) Prior.

CODS, Penny *Umbilicus rupestris* (Wall Pennywort) Corn. Grigson.

CODS, Sweet *Spiranthes spiralis* (Lady's Tresses) *Notes and Queries*. vol 7. A reference to the testicle-like double tubers, more usually reserved for *Orchis mascula*.

CODWEED *Centaurea nigra* (Knapweed) Cockayne. Descriptive, from the shape of the head, cod meaning bag.

COFFEE *Rumex acetosa* (Wild Sorrel) Edin. Ritchie.

COFFEE, Abeokuta *Coffea liberica* (Liberian Coffee) Dalziel.

COFFEE, Abyssinian *Coffee arabica* (Arabian Coffee) Dalziel.

COFFEE, Arabian *Coffea arabica*. Coffee is supposed to be ultimately from Kaffa, the name of the part of Ethiopia where the plant is native, but the nearer derivation is from Turkish Kahveh, the local pronunciation of Arabic qahwah. In 1601 Anthony Sherley introduced Kahveh into England, and it was William Parry who anglicized this into 'coffee'.

COFFEE, Bush *Coffea stenophylla* (Narrow-leaved Coffee) Dalziel.

COFFEE, Liberian *Coffea liberica*.

COFFEE, Mogdad *Cassia occidentalis* (Coffee Senna) Lewin.

COFFEE, Monrovia *Coffee liberica* (Liberian Coffee) Dalziel.

COFFEE, Nandi *Coffea eugenioides*.

COFFEE, Narrow-leaved *Coffea stenophylla*.

COFFEE, Negro; COFFEE, Wild *Cassia occidentalis* (Coffee Senna) both Trinidad both Watt. The roasted seeds are a coffee substitute, though there is no caffeine content.

COFFEE, Rio Nuñez 1. *Coffea liberica* (Liberian Coffee) Dalziel. 2. *Coffea stenophylla* (Narrow-leaved Coffee) Dalziel. Presumably because it was exported from a port of that name in Guinea.

COFFEE, Robusta *Coffea canephora*.

COFFEE, Upland *Coffea stenophylla* (Narrow-leaved Coffee) Dalziel.

COFFEE, Wild, Prickly *Clerodendron aculeatum* Barbados Gooding.

COFFEE, Zambesi *Bauhinia petersiana* (Orchid Bauhinia) Palgrave. It was known to all early hunters and explorers by this name. The seeds were ground to powder and used as a coffee substitute.

COFFEE BEAN TREE *Vitex payos*. The fruits look like coffee beans.

COFFEE BERRY *Rhamnus californica* (Red Berry) USA Schenk & Gifford.

COFFEE FENCE *Clerodendron aculeatum* Barbados Gooding.

COFFEE FLOWER *Symphytum officinale* (Comfrey) Som. Macmillan. Are the roots used to make a coffee-like drink? You certainly make comfrey wine from the roots.

COFFEE ROOT *Cichorium intybus* (Chicory) USA Watt. Chicory is cultivated on the Continent for the root, which is used as a substitute for coffee, and is sometimes mixed with real coffee as an adulterant. There is no caffeine or tannin in the root.

COFFEE SENNA *Cassia occidentalis*. The roasted seeds are a coffee substitute, but there is no caffeine content.

COFFEE TREE *Boscia albitrunca*.

COFFEE TREE, Kentucky *Gymnocladus dioica*. Early settlers ground the seeds, roasted them, and used them as a coffee substitute. The Indians already ate the seeds like nuts.

COFFEEBEAN 1. *Sesbania drummondii*. 2. *Sesbania vesicaria* (Bagpod) USA Kingsbury.

COFFEEWEED *Cassia occidentalis* (Coffee Senna) Vestal & Schultes. The roasted seeds are a coffee substitute.

COFFIN-NAIL *Anacardium occidentale* (Cashew Nut) Willis.

COGWEED 1. *Medicago arabica* (Spotted Medick) Som. Grigson. 2. *Ranunculus arvensis* (Corn Buttercup) Yks. Grigson. Grigson explains Cogweed for Spotted Medick as descriptive of "the spiralling, spiny pods".

COGWOOD *Zizyphus chloroxylon*. This is a hard, tough wood, so presumably ideal for making cogs.

COHOBA *Anadenanthera peregrina*.

COHOSH, Black 1. *Actaea rubra* (Red Baneberry) USA House. 2. *Cimicifuga racemosa* (Black Snakeroot) USA House. Cohosh must be an Indian name. The Baneberry above is used as a substitute in medical practice for the real thing - Black Snakeroot.

COHOSH, Blue *Caulophyllum thalictroides*.

COHOSH, White *Actaea alba* (White Baneberry) Grieve.1931.

COIGUE *Nothofagus dombeyi*.

COIR PALM *Trachycarpus excelsus* (Chusan Palm) Hyatt.

COKAR-NUT *Cocos nucifera* (Coconut) Pomet.

COKE PINTEL *Arum maculatum* (Cuckoo-pint) Cockayne.

COKEWEED *Agrostemma githago* (Corn Cockle) Scot. Grigson. A variation of Cockle.

COL; COLANDER; COLIANDER *Coriandrum sativum* (Coriander) Ess. Fernie (Col); Turner (Colander); Fluckiger (Coliander). The ME version of coriander was coliandre, O. French the same, though coriander is clearly closer to the original Greek (koris).

COLE 1. *Brassica napus* (Rape) Clapham. 2. *Brassica oleracea var.capitata* (Cabbage) Turner. 3. *Crambe maritima* (Sea Kale) W D Cooper. Cole is the same word as Kale - OE caul, Latin caulis. The modern German is Kohl. The basic meaning, according to ODEE, of caulis, is "hollow stem".

COLE, Dog's *Mercurialis perennis* (Dog's Mercury) B & H. The plant is a poison, hence the 'dog's'.

COLE, Pig's *Heracleum sphondyllium* (Hogweed) Dev. Friend.1882.

COLE, Red 1. *Armoracia rusticana* (Horse Radish) N.Eng. Gerard. 2. *Brassica oleracea var.capitata* (Cabbage) Henslow.

COLE, Sea 1. *Brassica oleracea var. capitata* (Cabbage) Turner. 2. *Calystegia soldanella* (Sea Bindweed) B & H.

COLE, Wild *Sinapis arvensis* (Charlock) Turner.

COLESAT; COLESEED *Brassica napus* (Rape) W.Miller (Colesat) B & H (Coleseed). These two are synonymous. See German Kohlsaat, cabbage seed. Coltza has the same meaning.

COLEWORT 1. *Brassica napus* (Rape) Lincs. B & H. 2. *Brassica oleracea var.capitata* (Cabbage) B & H. 3. *Geum urbanum* (Avens) Som. Vesey-Fitzgerald. See Cole.

COLEWORT, Hare's *Sonchus oleraceus* (Sow Thistle) B & H. This is a virtual translation of one of the old Latin names - *Brassica leporina*, or *Lactuca leporina*. An older legend connected the plant with hares rather than with sows - it gave strength to hares when they were overcome with heat. This is from the Anglo-Saxon Herbal - "of this wort, it is said that the hare, when in summer by the vehement heat he is tired, doctors himself with this wort". And see the Grete Herball - "if the hare come under it, he is sure that no beast can touch hym". Cf. Hare's Lettuce, Hare's Thistle, Hare's Palace, for this.

COLEWORT, Sea 1. *Brassica oleracea var.capitata* (Cabbage) B & H. 2. *Calystegia soldanella* (Sea Bindweed) Pomet. 3. *Crambe maritima* (Sea Kale) Gerard.

COLIC ROOT 1. *Aletris farinosa* (Unicorn-root) Le Strange. 2. *Asclepias tuberosa* (Orange Milkweed) Grieve.1931. 3. *Liatris spicata* (Button Snakeroot) Grieve.1931. A medicinal use in America for these. The milkweed is also called Wind Root, and Pleurisy-root in the States, while Unicorn-root bears the name Ague-root, too.

COLICKWORT *Aphanes arvensis* (Parsley Piert) Heref. Grigson. Another name for this is Bowel-hive, or Bowel-hive Grass. Bowel-hive is an inflammation of the bowel, occurring in children, so virtually the same as colic.

COLING *Malus sylvestris* fruit (Crabs) Shrop. B & H. i.e. Codlin, so perhaps it was pronounced like this.

COLLARBIND *Aquilegia vulgaris* (Columbine) Wilts. Hudson. There are a number of attempts at Columbine among the local names - see Colourbine, Cullavine, and Currambine.

COLLARDS 1. *Brassica napus* (Rape) Prior. 2. *Brassica oleracea var. acephala* (Kale) USA Brouk. Collards in America are a particular type of kale in which the cabbage rosettes never develop into a compact head, but although the record says it is a name for Rape, surely collards are cabbage stalks in this country?

COLLETS; COLLUTS *Brassica oleracea var. capitata* (Cabbage) Prior (Collet); Berks. Lowsley (Collut). Probably colewort. The Berkshire name is only applied to young cabbages.

COLLIER, Grim the *Hieraceum aurantiacum* (Orange Hawkweed) B & H. Grim the Collier is the name of an Elizabethan comedy, and given to the plant because of its "black,smutty involucre". "The stalks and cupps of the flours are all set thicke with a blackishe downe or hairinesse as it were the dust of coles; whence the women who keepe it in gardens for novelties sake, have named it Grimm the Collier" (Gerard). This becomes simply The Coalier in the Border country.

COLLINHOOD *Papaver rhoeas* (Red Poppy) Lothian, Roxb. Grigson.

COLLOPS, Eggs-and- *Linaria vulgaris* (Toadflax) Som. Grigson; Lancs. Nodal; Yks. Carr. Double names like this derive from the two colours in the flower. Collops are slices of meat, or more usually, bacon. So this is another version of the better known Eggs-and-bacon.

COLMENIER *Dianthus barbatus* (Sweet William) Prior. This must be the same as Gerard's Tolmeiner, or its variant Tolmeneer. Parkinson divides it into Toll-me-neer. Prior thinks it most likely to be a corruption of something like d'Allemagne.

COLOCYNTH *Colocynthus vulgaris* (Bitter Gourd) Wit. Colocynth means "bitter gourd" - it *is* exceedingly bitter.

COLOMBIAN BARK *Cinchona lancifolia*.

COLOMBIAN MAHOGANY *Swietenia macrophylla* (Honduras Mahogany) Ackermann.

COLOQUINTIDA *Colocynthus vulgaris* (Bitter Gourd) Thornton. The older form.

COLORADO BARBERRY *Berberis fendleri*.

COLORADO BEE PLANT, Blue *Cleome serrulata* (Rocky Mountain Bee Plant) USA Elmore.

COLORADO BLUE SPRUCE *Picea pungens* (Colorado Spruce) Edlin.

COLORADO CORN COCKLE *Vaccaria pyramidata* (Cow-herb) Watt.

COLORADO DOUGLAS FIR *Pseudotsuga menziesii var. glauca*.

COLORADO FIR *Abies concolor* (White Fir) Everett.

COLORADO JUNIPER; COLORADO RED CEDAR *Juniperus scopulorum* (Western Red Cedar) USA Yanovsky (Juniper); Cunningham (Cedar).

COLORADO MILLET *Panicum texanum* (Texas Millet) Douglas.

COLORADO RIVER HEMP *Sesbania exaltata*.

COLORADO ROOT *Ligusticum filicinum*.

COLORADO RUBBERWEED *Hymenoxis richardsonii* (Pingue) USA Kingsbury. The Indians used to chew the pounded root skin, like chewing gum.

COLORADO SAGE *Artemisia frigida* (Fringed Wormwood) USA Painter.

COLORADO SPRUCE *Picea pungens*.

COLOSSEUM IVY *Cymbalaria muralis* (Ivy-leaved Toadflax) W.Miller. Presumably because of its liking for old walls.

COLOURBINE *Aquilegia vulgaris* (Columbine) Lincs. B & H; N'thants. A.E.Baker. One of the local attempts at Columbine - there are also Cullavine, Currambine, and Collarbind in various parts of the country.

COLT, Kicking *Impatiens capensis* (Orange Balsam) USA Grigson. Descriptive of the means of ejecting the seeds.

COLT-HERB *Tussilago farfara* (Coltsfoot) B & H.

COLT'S TAIL 1. *Equisetum arvense* (Common Horsetail) Dev. Macmillan. 2. *Erigeron canadensis* (Canadian Fleabane) USA Henkel. 3. *Hippuris vulgaris* (Mare's Tail) Dev. Macmillan. The fleabane is also Mare's Tail and Cow's Tail in America.

COLTSFOOT 1. *Asarum canadense* (Canadian Snakeroot) Grieve.1931. 2. *Petasites fragrans* (Winter Heliotrope) Chann.Is, Corn, Dev) McClintock.1975. 3. *Tussilago farfara*. It is the shape of the leaves which suggested the name Coltsfoot. Petasites was once included in Tussilago.

COLTSFOOT, Alpine *Homogyne alpina* (Purple Coltsfoot) RHS.

COLTSFOOT, Butterbur *Petasites hybridus* (Butterbur) Prior.

COLTSFOOT, Guernsey *Petasites fragrans* (Winter Heliotrope) Chann.Is. McClintock.1975.

COLTSFOOT, Japanese *Petasites japonicus* (Creamy Butterbur) Douglas.

COLTSFOOT, Purple *Homogyne alpina*.

COLTSFOOT, Sweet 1. *Petasites fragrans* (Winter Heliotrope) Geldart. 2. *Petasites palmatus*. 'Sweet', because of the fragrance, which, in *P fragrans*, is heliotrope-like.

COLTSFOOT, Water *Nuphar lutea* (Yellow Waterlily) B & H.

COL(T)ZA 1. *Brassica campestris* (Field Mustard) J Smith. 2. *Brassica napus* (Rape) B & H. This is German Kohlsaat, cabbage seed. Colesat in English is closest to the German, and Coleseed is also recorded.

COL(T)ZA, Indian *Brassica campestris var. sarson*.

COLUMBA'S HERB *Hypericum perforatum* (St John's Wort) Bonser.

COLUMBIA LILY *Lilium columbianum*.

COLUMBINE 1. *Aquilegia vulgaris*. 2. *Verbena officinalis* (Vervain) B & H. Columbine is from Latin columba, pigeon, because, it is said, the nectaries are like pigeon's heads in a ring round a dish, a favourite theme of old artists. It was used in heraldry, and occurs in the crest of the Barons Grey of Vitten. Vervain gets this name, and Pigeon's Grass, from thc Grcck name - peris tereon - doves, because they were fond of hovering round the plant; and Coles explains it as "because Pidgeons eat thereof as is supposed to clear their Eye sight" (an echo of swallows and celandine).

COLUMBINE, Alpine *Aquilegia alpina*.

COLUMBINE, Canadian *Aquilegia canadensis*.

COLUMBINE, Eastern *Aquilegia canadensis* Southwick.

COLUMBINE, Feathered; COLUMBINE, Tufted *Thalictrum aquilegifolium* B & H (Feathered); Parkinson (Tufted).

COLUMBINE, Pyrenean *Aquilegia pyre-*

naica.

COLUMBINE, Rocky Mountain *Aquilegia caerulea*.

COLUMBINE, Siberian *Aquilegia glandulosa*.

COLUMBINE, Wild *Aquilegia canadensis* USA House.

COLUMBUS GRASS *Stipa almum*.

COLUMN OF PEARLS *Haworthia chalwini* (Aristocrat Plant) USA Cunningham.

COMB *Scandix pecten-veneris* (Shepherd's Needle) Som. Grigson. *Pecten-veneris* is the comb of Venus, and Gerard's names include Lady's Comb and Venus' Comb. Shepherd's Comb is a Yorkshire name.

COMB, Cock's 1. *Amaranthus hypondriachus* (Prince's Feathers). 2. *Amaranthus paniculatus* (Purple Amaranth) Watt. 3. *Celosia cristata*. 4. *Dactylorchis incarnata* (Early Marsh Orchid) Border B & H. 5. *Dipsacus fullonum* (Teasel) Dor. Macmillan. 6. *Erythrina crista-galli*. 7. *Heliotropum indicum* (Indian Heliotrope) Dalziel. 8. *Lychnis flos-cuculi* (Ragged Robin) Lanark. Grigson. 9. *Odontites verna* (Red Bartsia) Yks. Grigson. 10. *Onobrychis sativa* (Sainfoin) Prior. 11. *Orchis mascula* (Early Purple Orchis) Border B & H. 12. *Papaver rhoeas* (Red Poppy) Berw. Grigson. 13. *Pedicularis palustris* (Red Rattle) Scot. Grigson. 14. *Pedicularis sylvatica* (Lousewort) Som, Scot. Grigson. 15. *Rhinanthus crista-galli* (Yellow Rattle) Som, Glos, Shrop. Grigson. All attributable to colour, except teasel (and that, in fact, may be doubtful), and Yellow Rattle.

COMB, Cock's, Red *Amaranthus hypondriachus* (Prince's Feathers) O.P.Brown.

COMB, Gypsy's 1. *Arctium lappa* (Burdock) Berks, Notts. Mabey. 2. *Dipsacus fullonum* (Teasel) Yks. Grigsom.

COMB, Hen's 1. *Dactylorchis maculata* (Heath Spotted Orchid) Berw. Grigson. 2. *Rhinanthus crista-galli* (Yellow Rattle) B & H.

COMB, Lady's *Scandix pecten-veneris* (Shepherd's Needle) Gerard. Venus' comb, of course, but here transferred, as was very common, to the Virgin.

COMB, St Bride's *Stachys officinalis* (Wood Betony) Wales Trevelyan. This must be a translation from the Welsh.

COMB, Shepherd's *Scandix pecten-veneris* (Shepherd's Needle) Yks. Grigson.

COMB, Venus's *Scandix pecten-veneris* (Shepherd's needle) Gerard. i.e. *pecten-veneris*.

COMB, Wolf's *Dipsacus fullonum* (Teasel) Blunt. Cockayne has Wolf's Teasel for this.

COMB-AND-BRUSH *Dipsacus fullonum* (Teasel) Wilts. Dartnell & Goddard.

COMB-AND-HAIRPINS *Taraxacum officinale* (Dandelion) Som. Macmillan.

COMBS, Brushes-and-; COMBS, Brushes-and-, Lady's *Dipsacus fullonum* (Teasel) Som. Macmillan; Dor. Dacombe; Som. Macmillan (Lady's).

COME-AND-KISS-ME *Viola tricolor* (Pansy) Heref. Leather. One of a number of similarly erotic names given to the pansy.

COMFORT, Traveller's *Galium aparine* (Goose-grass) Wilts. Dartnell & Goddard.1899.

COMFREY 1. *Bellis perennis* (Daisy) B & H. 2. *Symphytum officinale*. ODEE gives this as Anglo-Norman cumfirie, O. Fr. confire, or confiere, and eventually from Latin confervere, to heal, grow together. The name, now confined to this particular plant, once had a more general significance, in the same way as its other Med. Latin synonym conserva and consolida (English consound). Similarly, Symphytum is from Greek sumphuo, to grow together. In herbals, it was used for knitting bones, and gypsies still use it for this purpose. In medieval times, soldiers carried it about with them. It was known to the Crusaders as a wound herb.Daisy gets this as a book name, as well as Consound (Turner calls it consolida minor), because it, too, was believed to have bone-knitting qualities. "The Northern men call this herbe a Banwort because it helpeth bones to knyt agayne" (Turner).

COMFREY, Blue 1. *Symphytum caucasicum*. 2. *Symphytum peregrinum* (Russian Comfrey) Clapham.
COMFREY, Caucasian *Symphytum caucasicum* (Blue Comfrey) W.Miller.
COMFREY, Himalayan *Onosma pyramidale*.
COMFREY, Middle *Ajuga reptans* (Bugle) Clair. This is Turner's consolida media, just as he called daisy consolida minor. The name is given because bugle is a wound herb - "The Decoction of Bugle drunke dissolveth clotted or congealed bloud within the body, healeth and maketh sound all wounds of the body both inward and outward..." (Gerard).
COMFREY, Prickly; COMFREY, Rough *Symphytum asperum*. Prickly Comfrey is the usual name; Rough Comfrey is given by Clapham.
COMFREY, Russian *Symphytum peregrinum*.
COMFREY, Saracen's *Senecio fluviatilis* (Saracen's Woundwort) Gerard. Gerard called it Saracen's Consound, too, presumably because it was thought this was a Crusaders' wound herb, like the real comfrey.
COMFREY, Soft *Symphytum orientale*.
COMFREY, Spotted *Pulmonaria officinalis* (Lungwort) Gerard. Gerard calls it Wild Comfrey, too.
COMFREY, Tuberous *Symphytum tuberosum*.
COMFREY, Turkish *Symphytum orientale* (Soft Comfrey) Salisbury.1972. It grows in Turkey.
COMFREY, Wild 1. *Cynoglossum boreale* USA Upham. 2. *Pulmonaria officinalis* (Lungwort) Gerard.
COMFREY, Wild, Northern *Cynoglossum boreale* USA Zenkert.
COMFREY CONSOUND *Symphytum officinale* (Comfrey) Gerard. Consound is from Latin consolida, and the meaning is the same as that of comfrey. It sounds as if Gerard wanted to make quite sure that we got the right plant, especially as he called it Great Consound as well.

COMPACT RUSH *Juncus conglomeratus*.
COMPANION, Swamp's *Cardamina pratensis* (Lady's Smock) Som. Macmillan. An odd name, though of course it does grow in moist places.
COMPASS PLANT 1. *Lactuca serriola* (Prickly Lettuce) Salisbury.1972. 2. *Rosmarinus officinalis* (Rosemary) Grieve.1931. 3. *Silphium laciniatum*. All to do with the way the leaves are aligned. Prickly lettuce has its upper stem leaves standing vertical in a more or less north/south plane. Silphium, when growing in an exposed position, turns the edges of its leaves north to south. This probably is to avoid mid-day radiation. Rosemary must get the name for a similar reason, for as well as Compass-weed and Compass-plant, it has the name Polar Plant.
COMPASS-WEED *Rosmarinus officinalis* (Rosemary) Grieve.1931. See Compass-plant.
CONDOR VINE *Marsdenia condurango* (Eagle Vine) Le Strange.
CONE-DEAL APPLE *Picea abies* (Spruce Fir) E.Ang. B & H.
CONEFLOWER, Great *Rudbeckia maxima*. Presumably the Rudbeckias are called Coneflower because of the large centre boss.
CONEFLOWER, Green-headed *Rudbeckia laciniata* (Tall Coneflower) USA House.
CONEFLOWER, Hedgehog *Echinacea purpurea*.
CONEFLOWER, Purple, Narrow-leaved *Echinacea angustifolia*.
CONEFLOWER, Tall *Rudbeckia laciniata*.
CONEFLOWER, Thin-leaved *Rudbeckia triloba* (Three-lobed Coneflower).
CONEFLOWER, Three-lobed *Rudbeckia triloba*.
CONEFLOWER, Western *Rudbeckia occidentalis*.
CONEY PARSLEY *Anthriscus sylvestris* (Cow Parsley) Surr. B & H.
CONFEDERATE JASMINE *Trachelospermum jasminoides*. Why? It grows in China and Malaysia. Perhaps the reference is to the Federated Malay States.

CONFETTI *Chenopodium album* (Fat Hen) Som. Mabey.

CONFETTI TREE *Maytenus senegalensis*. It is the white flowers, which stay on for a long time, and appear quite fresh when they do fall, that give the name.

CONGER; CONGO; CUNGER *Cucumis sativa* (Cucumber) N'thants. A.E.Baker; Worcs. J Salisbury (Conger); N'thants. A E Baker (Congo); War. Halliwell (Cunger). They may look odd, but the modern French for cucumber is concombre, and that is the origin of these names.

CONGO ACACIA *Pentaclethra macrophylla* (Oil Bean) Dalziel.

CONGO BEAN, CONGO PEA *Cajanus indicus* (Pigeon Pea) Schery (Bean); Dalziel (Pea).

CONGO SNAKE *Sansevieria trifasciata* (Mother-in-law's Tongue) USA Cunningham.

CONKER TREE *Aesculus hippocastanum* (Horse Chestnut) Ches. Holland. It is the nuts which are conkers; so by extension Conker-tree for the tree itself is natural.

CONKERBERRY *Rosa canina* hips (Hips) Som, Dor. Macmillan. Conkers, too, in this area. The word is actually Canker in this case.

CONKERS 1. *Aesculus hippocastanum* nut (Horse Chestnut). 2. *Cucumis sativa* (Cucumber) N'thants. Sternberg. 3. *Plantago lanceolata* (Ribwort Plantain) Som. Macmillan. 4. *Rosa canina* hips (Hips) Som, Dor. Macmillan. This is Conquerors, still sometimes heard instead, except in the case of Rose hips, when the original was canker, and for cucumber, which is French concombre (Cf. Conger, etc).

CONKS *Aesculus hippocastanum* (Horse Chestnut) Wilts. Dartnell & Goddard. i.e. Conkers, itself from Conquerors.

CONNAUGHT MARSH ORCHID *Dactylorchis cruenta*.

CONQUER-MORE *Taraxacum officinale* (Dandelion) Dor. Macmillan. Presumably Canker-root, although Macmillan spells it - moor.

CONQUEROR-FLOWER *Plantago lanceolata* (Ribwort Plantain) Som. Grigson. See Conkers. The conqueror is the flower stem which succeeds in decapitating its opponent in the children's game.

CONQUERORS *Aesculus hippocastanum* nut Ches. B & H; Wilts. Dartnell & Goddard. More widely known simply as Conkers.

CONSOLATION FLOWER *Clematis vitalba* (Old Man's Beard) Tynan & Maitland.

CONSOUND 1. *Ajuga reptans* (Bugle) Prior. 2. *Bellis perennis* (Daisy) Prior. 3. *Symphytum officinale* (Comfrey) Prior. This is Latin consolida - ODEE has consolidation - "the uniting of fractured or wounded parts". Comfrey is the consound par excellence, but both Bugle and Daisy are wound plants, too.

CONSOUND, Comfrey; CONSOUND, Great *Symphytum officinale* (Comfrey) both Gerard.

CONSOUND, King's *Delphinium ajacis* (Larkspur) Lyte. I have no record of its being used as a consound, though it bears this name (*Consolida orientalis*). Possibly mistaken for something else in early medieval times.

CONSOUND, Middle *Ajuga reptans* (Bugle) Clair. Both Gerard and Turner have consolida media for this; daisy would be consolida minor, and, of course, comfrey is consolida major.

CONSOUND, Saracen's *Senecio fluviatilis* (Saracen's Woundwort) Gerard.

CONSTANTINOPLE, Campion of; CONSTANTINOPLE, Flower of *Lychnis chalcedonica* (Jerusalem Cross) B & H (Campion); Gerard (Flower).

CONSTANTINOPLE HAZEL *Corylus colurna* (Turkish Hazel) Ablett.

CONSUMPTION MOSS *Cetraria islandica* (Iceland Moss) USA Cunningham. It is a demulcent and expectorant, used for bronchial ailments like tuberculosis.

CONTORTED WILLOW *Salix matsudana var. tortuosa* (Corkscrew Willow) Mitchell.1974.

CONVAL LILY *Convallaria maialis* (Lily of the valley) Parkinson.

CONVICT GRASS *Centranthus ruber* (Spur Valerian) Dor. Macmillan. Name confined to Portland. Is there, or was there, a penal settlement there?

CONVOLVULUS, Creeping *Convolvulus arvensis* (Field Bindweed). Hatfield.

COOK'S SCURVY GRASS *Lepidium oleraceum*. A once-common plant in New Zealand, known as a good antidote for scurvy. Captain Cook had boatloads of it collected and taken aboard to be cooked.

COOKING BANANA *Musa paradisiaca*. This banana is not edible until it has been cooked.

COOL TANKARD *Borago officinalis* (Borage) N'thants. A.E.Baker. Recorded for N'thants., but not confined to that area. Borage forms one of the ingredients, with water, wine, lemon and sugar, in a drink called Cool Tankard, from which the name was transferred to the plant. Thornton says Cool Tankard was borage put into a mixture of cider, water, and lemon-juice, with some wine added.

COOLIBAH *Eucalyptus microtheca*. From an Australian aborigine word.

COOLWORT *Tiarella cordifolia* (Foam Flower) USA House.

COOMBS *Scandix pecten-veneris* (Shepherd's Needle) Suff. B & H. i.e. combs.

COON ROOT 1. *Hepatica triloba* (American Liverwort) S USA Puckett. 2. *Sanguinaria canadensis* (Bloodroot) Grieve.1931.

COONTIES *Zamia integrifolia* (Florida Arrowroot) USA Kingsbury. From the Indian name for the plant.

COOSIL *Tussilago farfara* (Coltsfoot) N.Eng. Potter & Sargent.

COOSLIP *Primula veris* (Cowslip) Border Grigson. Listed as if it were something more than just the local pronunciation.

COOTAMUNDRA WATTLE *Acacia baileyana*. Cootamundra is a town in Australia, roughly 60 miles northwest of Canberra.

COP-ROSE 1. *Papaver dubium* (Smooth Long-headed Poppy) Scot. Jamieson. 2. *Papaver rhoeas* (Red Poppy) Som, E.Ang, Yks, N'thum, Ire. Grigson; Scot. Jamieson. Cop is a button, like the shape of the capsule. From cop-rose, the name degenerated into Cup-rose, or Cock-rose.

COPAL, Gum *Copaifera copallifera*.

COPER, Mount *Dactylorchis incarnata* (Early Marsh Orchid) B & H.

COPEY *Clusia rosea* RHS.

COPIHUE *Lapageria rosea* (Chilean Bell Flower) Simmons.

COPPER BEECH *Fagus sylvatica var. cuprea*.

COPPER BRIAR, Austrian *Rosa foetida var.bicolor*.

COPPER IRIS *Iris fulva*.

COPPER ROSE; COPPEROZE *Papaver rhoeas* (Red Poppy) Halliwell (Copper Rose); Suff. Moor (Copperoze). Nothing to do with copper - this is cop-rose.

COPPERWEED *Oxytenia acerosa*.

COPSE LAUREL *Daphne laureola* (Spurge Laurel) Hants, IOW Cope. A woodland plant, also known as Wood Laurel.

CORAL *Corallorhiza odontorhiza* (Autumn Coral-root) USA Emboden.1974. This, as well as the common name, and the generic derive from the appearance of the roots.

CORAL-BARK WILLOW *Salix alba var. vitellina "britzensis"* (Scarlet Willow) Mitchell.1974.

CORAL BEAN 1. *Erythrina indica* (Indian Coral Tree) Geng Junying. 2. *Sophora secundiflora* (Mescal Bean) USA Norbeck. The beans are red in colour.

CORAL BEAN, Evergreen *Sophora secundiflora* (Mescal Bean) USA LaBarre.

CORAL BELLS *Heuchera sanguinea*. Colour descriptive.

CORAL BERRY 1. *Aechmea fulgens*. 2. *Ardisia crispa*. 3. *Symphoricarpos orbiculatus* (Red Wolfberry) Alab. Harper. Red berries.

CORAL DROPS *Bessera elegans*. Bright red

blooms.
CORAL HONEYSUCKLE *Lonicera sempervirens* (Trumpet Honeysuckle) USA House. Red flowers.
CORAL LILY *Lilium pumilum*.
CORAL MOSS *Nertera depressa* (Bead Plant) USA Cunningham.
CORAL PEA, Blue *Hardenbergia comptoniana/violacea*.
CORAL PEONY *Paeonia mascula* (Peony) Hutchinson. Another specific name for this is *corallina*.
CORAL PLANT 1. *Jatropha multifida*. 2. *Ribes sanguineum* (Flowering Currant) Ches. Holland.
CORAL ROOT 1. *Corallorhiza trifida*. 2. *Dentaria bulbifera*. From the appearance of the white, fleshy rhizomes.
CORAL-ROOT, Autumn *Corallorhiza odontorhiza*.
CORAL-ROOT, Spurred *Epipogium aphyllum* (Ghost Orchid) Turner Ettlinger.
CORAL-ROOT ORCHID *Corallorhiza trifida* (Coral-root) Turner Ettlinger.
CORAL SPURGE *Euphorbia corallioides*.
CORAL SUMACH *Metopium toxiferum* (Poison-wood) Usher.
CORAL TREE *Erythrina corallodendron*. It is named from the hard red seeds, which are made into necklaces so as to look like coral.
CORAL TREE, Cockspur *Erythrina cristagalli* (Cock's Comb) RHS.
CORAL TREE, Flame *Erythrina corallodendron* (Coral Tree) RHS.
CORAL TREE, Indian *Erythrina indica*.
CORAL TREE, Naked *Erythrina corallodendron* (Coral Tree) RHS.
CORAL TREE, Prickly *Pyracantha sp* (Fire Thorn) Parkinson.
CORAL VINE *Antigonon leptopus*.
CORAL-WOOD, African *Pterocarpus soyauxii* (Barwood) Dalziel.
CORAL-WORT *Dentaria bulbifera* (Coralroot) Gerard.
CORASEENA *Vaccinium myrtillus* (Whortleberry) Ire. Grigson.
CORCIR; KORKIR; CORK *Ochrolechia tartarea* (Cudbear Lichen) Scot. Bolton; A S Palmer (Cork).
CORD-GRASS *Spartina townsendii* (Ricegrass) McClintock.
CORD-GRASS, Lesser *Spartina maritima*.
CORIANDER *Coriandrum sativum*. This is Greek koris, bug, for when green, the seed has a very disagreeable taste and smell; it is only when ripe that the aromatic flavour comes out. The longer they are kept, the more fragrant they become.
CORIANDER, Roman *Nigella sativa* (Fennel-flower) Grieve.1931. The seeds are an eastern spicy flavouring, hence this name and others like Nutmeg-flower and Black Cumin.
CORIARIA, Mediterranean *Coriaria myrtifolia*.
CORIS, Yellow *Hypericum coris*.
CORITANIAN ELM *Ulmus coritana*. The reference is to the Coritanae, a British tribe who occupied that part of of what is now the Midlands near Leicester, where the tree was first identified in 1947.
CORK CORKER *Parmelia omphalodes* Highlands Bolton. A lichen.
CORK ELM *Ulmus racemosa*.
CORK LADY'S TRESSES *Spiranthes romanzoffiana* (Hooded Lady's Tresses) Turner Ettlinger. Cf. Irish Lady's Tresses. N. and W.Ireland are the most likely places in the British Isles to find this.
CORK OAK *Quercus suber*. This is the source of commercial cork.
CORK OAK, Chinese *Quercus variabilis*.
CORK TREE *Parinari mobola* (Sand Apple) Palgrave.
CORK TREE, Amur *Phellodendron amurense*.
CORKER, Cork *Parmelia omphalodes* Highlands Bolton. A lichen.
CORKLIT, KORKELIT *Ochrolechia tartarea* (Cudbear Lichen) Mabey.1977 (Corklit); Shet. Nicolson (Korkelit). Cf. Cork, Corcir, etc. The 'lit' syllable must mean 'dye'. Indigo was known in Shetland simply as lit.

CORKSCREW FLOWER *Strophanthus grandiflorus*. Presumably from its manner of growth - it is a climber.
CORKSCREW HAZEL *Corylus avellana* var. *contorta*.
CORKSCREW WILLOW *Salix matsudana* var. *tortuosa*.
CORKWOOD 1. *Musanga cecropioides* (Umbrella Tree) Kyerematen. 2. *Ochroma lagopus* (Balsa) Willis.
CORKWOOD, Marsh *Annona palustris* (Alligator Apple) Dalziel. The root wood is a bit heavier than cork, but still light enough for African fishermen to use as floats.
CORME *Sorbus torminalis* (Wild Service Tree) Halliwell.
CORMEILLE; CORMELE; COR-RAMEILLE *Lathyrus montanus* (Bittervetch) Scot. Pennant (Cormeille); B & H (Cormele); Grigson (Corrameille). These, and a number of other variants, are all from Gaelic caermeal, the name of the plant, which itself comes from corr, a crane, and meilg, a pod. In Welsh it is Pys y garanod, crane's peas.
CORN *Zea mays* (Maize) USA A.W.Smith.
CORN, Asiatic; CORN, Turkish; CORN, Welsh *Zea mays* (Maize) Lehner. Asiatic, and Turkish, because the 16th century herbalists believed the plant had been brought by the Turks from Asia. The Turks invaded Europe about this time, and brought many new items into the West. Anything unusual was labelled 'Turkish', or perhaps it just meant foreign. Certainly Welsh here means 'foreign'. Another theory, from Bianchini this time, suggests that there was confusion between maize and buckwheat, which was at one time labelled *turcicum* for some reason.
CORN, Broom 1. *Sorghum saccharatum* (Sorgo) USA J.Smith. 2. *Sorghum vulgare* (Kaffir Corn) Schery. *S saccharatum* was actually cultivated in parts of the USA, especially Ohio, for use as brooms. It is the inflorescences of both varieties that are used.
CORN, Chicken *Sorghum vulgare* (Kaffir Corn) USA Cunningham.

CORN, Crow *Aletris farinosa* (Unicorn-root) Le Strange.
CORN, Dent *Zea mays* var. *americana*. There is an indentation or depression on the top of the grain, caused by shrinkage of the endosperm, hence 'dent'.
CORN, Devil's *Stellaria holostea* (Greater Stitchwort) Shrop. Friend, Burne; Wales Fernie. The devil's plant in a number of places. It is Devil's Flower, or Devil's Nightcap in Somerset, Devil's Eyes in Dorset and in Wales, and Old Lad's Corn in Shropshire.
CORN, Drink *Hordeum sativum* (Barley) Tusser.
CORN, Fairy's *Lathyrus montanus* (Bittervetch) Donegal Grigson.
CORN, Flint; CORN, Field *Zea mays var. praecox* (Flint Maize). 'Flint', because of the hard endosperm.
CORN, Grass *Phalaris canariensis* (Canary Grass) Turner. "because it is partly lyke grasse and partly lyke corne".
CORN, Guinea *Sorghum vulgare* (Kaffir Corn) J.Smith.
CORN, Indian; CORN, Indy *Zea mays* (Maize). Indian Corn is common; Indy Corn occurs in Cumb. Rowling.
CORN, Kaffir *Sorghum vulgare*.
CORN, Pod *Zea mays* var. *tunicata*. Has a pod-like covering to the grain.
CORN, Saracen *Fagopyrum esculentum* (Buckwheat) Grieve.1931. In French it is sarrasino, because it was introduced into France by the Saracens. Note that Gerard calls it French Wheat.
CORN, Squirrel; CORN, Squirrel's *Dicentra canadensis* (Turkey Corn) USA Hewitt (Squirrel); USA Tolstead (Squirrel's).
CORN, Squirrel, American *Dicentra canadensis* (Turkey Corn) USA McDonald.
CORN, Sweet *Zea mays* var. *saccharata*.
CORN, Turkey 1. *Dicentra canadensis*. 2. *Dicentra formosa* (Western Bleeding Heart) USA Berdoe.
CORN, Welsh *Zea mays* (Maize) Lehner. Presumably 'Welsh' here just means 'for-

eign'. See Asiatic Corn.

CORN BEDSTRAW *Galium tricorne* (Small Goose-grass) Polunin.

CORN BELLFLOWER *Legousia hybrida* (Venus's Looking Glass).

CORN-BIN *Convolvulus arvensis* (Field Bindweed) Lincs. Peacock. Because bind, bindweed, was pronounced there with a short 'i', naturally producing 'bin'.

CORN-BIND 1. *Convolvulus arvensis* (Field Bindweed) N'thants. A.E.Baker. 2. *Polygonum convolvulus* (Black Bindweed) Turner; New. Zealand Connor. Usually as one word. Cf. Cornbine. The meaning is obvious - the plant that stifles the corn by winding round it.

CORN BINDWEED *Polygonum convolvulus* (Black Bindweed) Yks. B & H.

CORN-BINE *Convolvulus arvensis* (Field Bindweed) War, Oxf, Bucks, N'thants, Notts, Lincs, Yks Grigson. Usually as one word. Cf. Cornbind.

CORN BLINKS *Centaurea cyanus* (Cornflower) Dev. Grigson.

CORN-BOTTLE *Centaurea cyanus* (Cornflower) Dev. Macmillan; N'thants. A.E.Baker. There are also Bluebottle, Corn Bluebottle, and, from York-shire, Bottle-of-all-sorts. A bottle in Yorkshire dialect meant a bundle, of hay or straw, etc.

CORN BUTTERCUP *Ranunculus arvensis*.

CORN CAMOMILE *Anthemis arvensis*.

CORN CAMPION; CORN CHAMPION *Agrostemma githago* (Corn Cockle) B & H (Cockle); S.Africa Watt (Champion).

CORN CARRAWAY *Petroselinum segetum* (Corn Parsley) McClintock.

CORN CENTAURY *Centaurea cyanus* (Cornflower) B & H.

CORN CHRYSANTHEMUM *Chrysanthemum segetum* (Corn Marigold) Curtis.

CORN COCKLE *Agrostemma githago*. Cockle is OE coccul, Latin coccum, berry, referring in this case to the large round seeds. The word used to be applied to any noxious weed growing in corn, particularly darnel.

CORN COCKLE, Colorado *Vaccaria pyramidata* (Cow-herb) Watt. Cow-cockle is another name. Vaccaria is from Latin vacca, cow, because it was thought to be good cattle fodder.

CORN CROWFOOT *Ranunculus arvensis* (Corn Buttercup) Clapham.

CORN FEVERFEW 1. *Matricaria recutita* (Scented Mayweed) Curtis. 2. *Tripleurospermum maritimum* (Scentless Maydweed) Grieve.1931.

CORN FLAG 1. *Gladiolus sp* Gerard. 2. *Iris pseudo-acarus* (Yellow Iris) Som. Macmillan.

CORN GILLIFLOWER *Legousia hybrida* (Venus's Looking Glass) Gerard.

CORN GLADIN *Gladiolus sp* Gerard.

CORN HONEWORT *Petroselinum segetum* (Corn Parsley) Prior.

CORN HORSETAIL *Equisetum arvense* (Common Horsetail) Curtis.

CORN-LEAF *Umbilicus rupestris* (Wall Pennywort) Worcs. Grigson. The leaves of this are reckoned very efficacious as plasters for corns.

CORN LETTUCE *Valerianella locusta* (Cornsalad) Brouk. Used as a lettuce substitute when true lettuce is not available, in winter and early spring.

CORN LILY 1. *Calystegia sepium* (Great Bindweed) Yks. B & H. 2. *Convolvulus arvensis* (Field Bindweed) Yks. Grigson. 3. *Veratrum californicum* (False Hellebore) USA Schenk & Gifford.

CORN LILY, African *Ixia sp*.

CORN MARIGOLD *Chrysanthemum segetum*.

CORN MARIGOLD OF CANDY *Chrysanthemum coronarium* (Annual Chrysanthemum) Parkinson. It was introduced into this country from southern Europe some time before 1609.

CORN MIGNONETTE *Reseda phyteuma*.

CORN MINT 1. *Acinos arvensis* (Basil Thyme) Turner. 2. *Mentha arvensis*.

CORN PANSY *Viola tricolor* (Pansy) H C Long.1910. Presumably to distinguish it from

the Field Pansy, V arvensis.
CORN PARSLEY *Petroselinum segetum.*
CORN PHEASANT'S EYE *Adonis annua* (Pheasant's Eye) Flower.1859.
CORN PIMPERNEL *Anagallis arvensis* (Scarlet Pimpernel) Flower.1859.
CORN PINK 1. *Agrostemma githago* (Corn Cockle) N'thants. A.E.Baker. 2. *Legousia hybrida* (Venus's Looking-glass) Gerard.
CORN PLANT 1. *Dracaena deremensis.* 2. *Dracaena fragrans.*
CORN POP *Silene cucubalis* (Bladder Campion) Wilts. Dartnell & Goddard.
CORN POPPY *Papaver rhoeas* (Red Poppy) Corn. Grigson.
CORN ROCKET *Bunias erucago* (Crested Bunias) Bianchini.
CORN ROSE 1. *Agrostenna githago* (Corn Cockle) S.Africa Watt. 2. *Papaver rhoeas* (Red Poppy) Gerard. 3. *Rosa arvensis* (Field Rose) B & H.
CORN ROSE, Red *Papaver rhoeas* (Red Poppy) Turner.
CORN SEDGE *Gladiolus sp* Gerard.
CORN SILK *Zea mays* (Maize) Kourennoff. Obviously a name meant for the husk, but apparently applied to the plant as a whole, too.
CORN SNAPDRAGON *Misopates orontium* (Lesser Snapdragon) Salisbury.1972.
CORN SOW THISTLE *Sonchus arvensis.*
CORN SPEEDWELL 1. *Veronica arvensis* (Wall Speedwell) B & H. 2. *Veronica hederifolia* (Ivy-leaved Speedwell) B & H.
CORN THISTLE *Cirsium arvense* (Creeping Thistle) Cumb, N.Ire. B & H.
CORN VIOLET *Legousia hybrida* (Venus's Looking-glass) Prior.
CORN WOUNDWORT *Stachys arvensis* (Field Woundwort) Murdoch McNeill.
CORNALEE *Cornus sanguinea* (Dogwood) Ches. Holland. From Cornel, of course.
CORNBERRY *Vaccinium oxycoccus* (Cranberry) B & H. A corruption of cranberry, probably by way of the north-country Crone.
CORNEL *Cornus sanguinea* (Dogwood).

Cornel is as common as Dogwood for this. It comes from Latin cornix, crow - but the Latin cornus was the original name of the tree.
CORNEL, Dwarf 1. *Cornus canadensis* (Bunchberry) USA House. 2. *Cornus suecica.*
CORNEL, Female *Cornus sanguinea* (Dogwood) Gerard. *Cornus mas* is the Male Cornel.
CORNEL, Great-flowered *Cornus florida* (Flowering Dogwood) Whittle & Cook.
CORNEL, Low *Cornus canadensis* (Bunchberry) USA House.
CORNEL, Male *Cornus mas* (Cornel Cherry) Gerard. The Female Cornel is *Cornus sanguinea*, Dogwood.
CORNEL, Silky *Cornus amomum.*
CORNEL, Tame *Cornus mas* (Cornel Cherry) Gerard. Wild Cornel is the dogwood.
CORNEL, Wild *Cornus sanguinea* (Dogwood) B & H.
CORNEL CHERRY *Cornus mas.*
CORNELIA TREE; CORNELIAN CHERRY *Cornus mas* (Cornel Cherry) Gerard (Cornelia Tree); A.W.Smith (Cornelian Cherry).
CORNELIAN *Cornus sanguinea* (Dogwood) Bacon.
CORNEMELLAGH *Lathyrus montanus* (Bittervetch) Donegal Grigson. This is the same word as the Scottish Corrameille, and there are a number of other variants from the Gaelic caermeal, the name of the plant,which itself comes from corr, a crane, and meilg, a pod. In Welsh it is Pys y garanod,crane's peas.
CORNETS *Ononis repens* (Rest Harrow) Dor. Macmillan.
CORNFLOWER 1. *Agrostemma githago* (Corn Cockle) Ches. Grigson. 2. *Centaurea cyanus*. 3. *Centaurea scabiosa* (Greater Knapweed) Som. Macmillan. 4. *Knautia arvensis* (Field Scabious) Som. Macmillan. 5. *Papaver rhoeas* (Red Poppy) Dev. Macmillan. Because they grow in corn, logically. But does Knapweed really qualify, or is it included by relationship with *Centaurea*

cyanus?

CORNFLOWER, Golden; CORNFLOWER, Yellow *Chrysanthemum segetum* (Corn Marigold) both Gerard.

CORNFLOWER, Mountain *Centaurea montana* (Mountain Knapweed) Polunin.

CORNFLOWER, Perennial *Centaurea montana* (Mountain Knapweed) Blamey.

CORNFLOWER, Red 1. *Agrostemma githago* (Corn Cockle) W.Miller. 2. *Papaver rhoeas* (Red Poppy) W.Miller. Not members of genus *Centaurea*, but both cornfield plants.

CORNFLOWER-ASTER *Stokesia cyanus*.

CORNISH ELM *Ulmus stricta*. It is a west country species.

CORNISH FUZZ *Ulex gallii* (Dwarf Furze) Corn. Grigson. They also call it Tan-fuzz in Cornwall.

CORNISH HEATH *Erica vagans*. Confined to heaths in west Cornwall.

CORNISH MALLOW *Lavatera cretica*. A rare plant in this country, though not confined to Cornwall, for it grows in Pembrokeshire, Scilly and the Channel Islands.

CORNISH WHITEBEAM *Sorbus x latifolia* (French Hales) Brimble.1948.

CORNSALAD *Valerianella sp*.

CORNWOOD *Cornus sanguinea* (Dogwood) Som. Macmillan. Obviously a hybrid between Cornel and Dogwood.

CORONA REGIS *Melilotus officinalis* (Melilot) Fernie. i.e. King's Crown, which B & H prefer.

CORONATION; CORNATION *Dianthus caryophyllus* (Carnation) Lyte (Coronation); Grigson.1974 (Cornation). Closer to the original than is the modern name, which is in fact 'coronation' (Lat. corona, a chaplet).

COROON; KERROON *Prunus avium* (Wild Cherry) Herts. Jones-Baker.1974.

CORPSE FLOWER *Lathraea squamosa* (Toothwort) Yks. Grigson. It is because there is no chlorophyll in the plant that it gets this name, just as in America *Monotropa uniflora* is sometimes called Corpse Plant.

CORPSE PLANT *Monotropa uniflora* USA House. As with Corpse Flower for *Lathraea squamosa*, this name is given because of the lack of chlorophyll in the plant. Cf. Ghostflower for this.

CORR *Lathyrus montanus* (Bittervetch) Scot. Pennant. It is more usual to find both elements of the Gaelic caermeal for this - Caramaile, Cormele, Corrameille, etc. Caermeal comes from corr, a crane, and meilg, a pod.

CORSICAN HEATH *Erica stricta*.

CORSICAN HELLEBORE *Helleborus corsicus*.

CORSICAN MINT *Mentha requeni*.

CORSICAN PINE *Pinus nigra var. maritima*.

CORSICAN THYME *Thymus herba-baroni* (Carraway Thyme) Rohde.

CORTS *Daucus carota* (Carrot) Som. Halliwell. Carrots were called Corts, which must only be a local variant of the common name, in Somerset.

CORYDALIS, Climbing *Corydalis claviculata*. Greek koridalos - crested lark. The flowers are helmet-shaped, with a rounded spur, and the seeds are crested.

CORYDALIS, Golden *Corydalis aurea*.

CORYDALIS, Pink *Corydalis sempervirens*.

CORYDALIS, Purple *Corydalis bulbosa* (Bulbous Fumitory) Schauenberg.

CORYDALIS, Siberian; CORYDALIS, Kalmuk *Corydalis bulbosa* (Bulbous Fumitory) Douglas.

CORYDALIS, Yellow *Corydalis lutea*.

CORYMBOSE SQUILL *Scilla peruviana* (Peruvian Lily) Whittle & Cook. 'Corymbose' means having the flowers in a flattish-topped raceme.

COS LETTUCE *Lactuca sativa var. longifolia*. Called after the Greek island.

COSSACK ASPARAGUS *Typha latifolia* (False Bulrush) Lindley. It has been sometimes used as a food plant in this country, under this name.

COST; COSTMARY *Tanacetum balsamita* (Balsamint) both B & H. Cost is a contraction of Alecost, given because, according to Parkinson, it was used for flavouring ales. But Cost itself is from "costus", an aromatic plant used in making perfumes in the east. The plant is dedicated to the Virgin Mary in most west European countries. In France it is Herbe Sainte Marie, and Dawson has St Mary as a name.

COST, English; COSTMARY, English *Tanacetum vulgare* (Tansy) Cockayne (Cost); Storms (Costmary). The OE from which Storms got this is aenglisc cost, presumably to distinguish from the real Costmary, which is Balsamint.

COTINE *Asclepias fruticosa* (Shrubby Milkweed) S.Africa Watt. Cotton, perhaps? They call it Wild Cotton in S.Africa, too.

COTONEASTER, Entire-leaved *Cotoneaster integerrimus* (Great Orme Berry) Hutchinson. Cotoneaster is Latin cotonium, quince + Aster.

COTONEASTER, Herringbone; COTONEASTER, Fishbone *Cotoneaster horizontalis* Perry (Herringbone); McClintock.1974 (Fishbone). Descriptive.

COTONEASTER, Orange *Cotoneaster franchetti*.

COTONEASTER, Tree *Cotoneaster frigida*.

COTTAGE PINK *Dianthus caryophyllus* (Carnation) A.W.Smith.

COTTAGERS *Digitalis purpurea* (Foxglove) Ire. Grigson; USA Henkel.

COTTON, Adder's *Cuscuta epithymum* (Common Dodder) Corn. Grigson. The thread-like stems have given rise to a number of distinctive names, such as Devil's Thread, Hairweed, Red Tangle, etc.

COTTON, Bog *Eriophorum angustifolium* (Cotton-grass) McClintock.

COTTON, Dwarf; COTTON, Petty *Gnaphalium sylvaticum* (Heath Cudweed) both Culpepper. Gerard has Cottonweed, also.

COTTON, Hopi *Gossypium hopi*. An aboriginal cultivated cotton peculiar to the southwest of the USA.

COTTON, Kidney *Gossypium brasiliense*.

COTTON, Lavender *Santolina chamaecyparissa* (Ground Cypress) Webster. Cf. also Lavender Corn, and Silver Lavender, for this.

COTTON, Egyptian *Gossypium barbadense* (Sea-Island Cotton) H.G.Baker.

COTTON, Sea-Island *Gossypium barbadense*. Originally from tropical South America, but disseminated to the West Indies, and from there it was introduced into South Carolina and Georgia. So the name mirrors the view from the American plantations towards the West Indies.

COTTON, Silk, White *Cochlospermum religiosum*.

COTTON, Stinking *Senecio viscosus* (Sticky Groundsel) Le Strange.

COTTON, Upland *Gossypium punctatum*.

COTTON, Wild 1. *Apocynum cannabinum* (Indian Hemp) Grieve.1931. 2. *Asclepias fruticosa* (Shrubby Milkweed) S.Africa Watt. 3. *Ipomaea albivenia* S.Africa Watt.

COTTON BATTING PLANT *Gnaphalium chilense* USA Schenk & Gifford.

COTTON BUSH, Narrow-leaved *Asclepias fruticosa* (Shrubby Milkweed) S.Africa Watt.

COTTON FLOWER *Plantago media* (Lamb's Tongue Plantain) Som. Macmillan. A descriptive name - Cf. Hoary Plantain, and the Wiltshire Ashy Poker.

COTTON GRASS *Eriophorum angustifolium*. Or Bog Cotton. The heads were a crop on the Isle of Skye once, gathered by children, dried, and used to stuff pillows and quilts.

COTTON GRASS, Broad-leaved *Eriophorum latifolium*.

COTTON GRASS, Slender *Eriophorum gracile*.

COTTON GUM 1. *Nyssa aquatica* (Water Tupelo) USA Schery. 2. *Nyssa sylvatica* (Tupelo) Willis.

COTTON LAVENDER *Santolina chamaecyparissa*. Cotton Lavender is as common as Ground Cypress for this. Webster has Lavender Cotton.

COTTON SEDGE *Eriophorum angustifolium* (Cotton Grass) J.G.McKay.

COTTON THISTLE 1. *Cirsium eriophorum* (Woolly Thistle) Prior. 2. *Onopordum acanthium* (Scotch Thistle) Curtis. It is the thistle down which gives this name, or perhaps the generally hoary appearance of Scotch Thistle; H C Long,1910 described it as a "cottony" plant.

COTTON THORN *Tetradymia spinosa var. longispina.*

COTTON TREE 1. *Populus nigra* (Black Poplar) Suff. Grigson. 2. *Viburnum lantana* (Wayfaring Tree) Ess. Gerard; Lancs. Grigson. It is the female catkins which give the name for the poplar, and the white underside of the leaves as far as the Wayfaring Tree is concerned.

COTTON TREE, Small *Gossypium barbadense* (Sea-Island Cotton) Barbados Gooding.

COTTONER *Viburnum lantana* (Wayfaring Tree) Kent Grigson. As for Cotton-tree, the white underside of the leaves is the reason for the name. Rambosson records viorbe cotonneuse as one of the French names.

COTTONWEED 1. *Gnaphalium sylvaticum* (Heath Cudweed) Gerard. 2. *Gnaphalium uliginosum* (Marsh Cudweed) Grieve.1931. 3. *Otanthus maritimus.*

COTTONWEED, Mountain *Leontopodium alpinum* (Edelweiss) Parkinson.1640.

COTTONWOOD 1. *Populus balsamifera* (American Balsam Poplar) USA Tolstead. 2. *Populus deltoides* (Carolina Poplar) USA Harper. 3. *Populus tremuloides* (American Aspen) Canada Teit. 4. *Populus trichocarpa* (Western Balsam Poplar) USA E.Gunther.

COTTONWOOD, Black 1. *Populus angustifolia* (Narrowleaf Cottonwood) USA Elmore. 2. *Populus trichocarpa* (Western Balsam Poplar) USA Schenk & Gifford.

COTTONWOOD, Common *Populus angulata.*

COTTONWOOD, Eastern *Populus deltoides* (Carolina Poplar) USA Zenkert.

COTTONWOOD, Fremont *Populus fremontii.*

COTTONWOOD, Mountain *Populus angustifolia* (Narrowleaf Cottonwood) Robbins.

COTTONWOOD, Narrowleaf *Populus angustifolia.*

COTTONWOOD, Northern *Populus monolifera.*

COTTONWOOD, Plains *Populus sargentii* (Sargent Cottonwood) Cunningham.

COTTONWOOD, Sargent *Populus sargentii.*

COTTONWOOD, Swamp *Populus heterophylla.*

COTTONWOOD, Valley *Populus wislizeni.*

COUCH(GRASS) *Agropyron repens.* Couch is a variant of quitch, another name for the grass. They are from OE cwice, alive (Cf. Quick-grass).

COUCH(GRASS), Bearded *Agropyron caninum.*

COUCH, Black 1. *Agrostis tenuis* (Fine Bent) H C Long.1910. 2. *Agrostis stolonifera* (White Bent) Wilts. Dartnell & Goddard. 3. *Alopecurus myosuroides* (Slender Foxtail) Leyel.1948. For the creeping roots are nearly as troublesome as true couch in arable land, and they are black.

COUCH, Onion *Arrenatherium elatius* (False Oat) H C Long.1910. It gets the 'couch' name because it has the same ability to reproduce itself from bits of its rootstock as the real couch has.

COUCH(GRASS), Sand *Agropyron junceiforme.*

COUCH(GRASS), Sea *Agropyron repens.*

COUCH, White *Agropyron repens* (Couchgrass) Wilts. Dartnell & Goddard. *Agrostis stolonifera* is the usual Black Couch.

COUCHY-BENT *Agrostis stolonifera* (White Bent) Wilts. Dartnell & Goddard.

COUGHWORT *Tussilago farfara* (Coltsfoot) Culpepper. Coltsfoot is still smoked in all herbal tobaccos, as it is in Chinese medicine, for coughs. Bechion, the plant in Dioscorides taken to be coltsfoot, was smoked against the dry cough. It was smoked by the Cornish miners as a precaution against lung diseases, and gypsies smoke the dried leaves as beneficial for asthma and bronchitis, and they also drink a tea made from it for coughs.

COUL *Brassica oleracea* (Wild Cabbage)

Som. Halliwell. A local variant of Cole.

COULTER PINE *Pinus coulteri*.

COUNSELLORS *Cirsium vulgare* (Spear Plume Thistle) Ches. Holland. It is a reference to the burry seeds, which stick to one, like lawyers.

COUNTER-WOOD *Chlorophora excelsa* (African Oak) F N Howes. A favourite timber for shop counter-tops apparently.

COUNTRY LAWYERS *Rubus fruticosus* (Bramble) Leics. Grigson. Once you're caught up in it, it doesn't let you go very easily. The name Thief for the same plant has a similar nuance, though what the thief is taking is probably wool from the sheep.

COUNTRY PELLITORY *Erigeron affinis* Martinez. This is the False Pellitory, or Country Pellitory (falso peritre, or peritre del pais) of pharmacy(Martinez). It must be used in a similar way as true pellitory.

COUNTRY PEPPER *Sedum acre* (Biting Stonecrop) Gerard. Cf. Pepper Crop, Wall Pepper, or Poor man's Pepper, for the same plant.

COUNTRYMAN'S TREACLE 1. *Allium sativum* (Taper-leaved Garlic) B & H. 2. *Ruta graveolens* (Rue) B & H. 3. *Valeriana officinalis* (Valerian) Gerard. Treacle is from triacle, an antidote to the bite of venomous animals. It is from Greek theriake, from therione, a name given to the viper, i.e. the superstition that viper's flesh will cure viper's bite. Philips, World of Words, defines treacle as a "physical compound made of vipers and other ingredients", and this was a favourite against all poisons. The word then became applied to any confection of sweet syrup, and finally and solely to the syrup of molasses. Garlic, of course, is an antiseptic; as late as World War 1 it was put on sterilized swabs and applied to wounds to prevent them turning septic. Cf. Churl's Treacle, Clown's Treacle, or Poor Man's Treacle for garlic. Rue is another wound herb, and valerian in the Middle Ages was valued as a plague medicine - "the dry root is put into counterpoysons and medicines preservative against the pestilence, as are treacles, mithridates, and such like..." (Gerard). Names like Countryman's Treacle mean that this was a medicine known to, and available to, country people.

COUNTRYMAN'S WEATHERGLASS *Anagallis arvensis* (Scarlet Pimpernel) Swainson. Cf. also Farmer's, Ploughman's, Weatherglass, and a lot more, all based on pimpernel's reputation as a weather forecaster.

COURTSHIP AND MATRIMONY *Filipendula ulmaria* (Meadowsweet) Cumb. Wright. It is said that the name refers to the scent - very different before and after bruising. The Yorkshire name Bittersweet probably has a similar meaning.

COVE-KEYS *Primula veris* (Cowslip) Kent Grigson. The 'keys' part of the name arose from the fancied resemblance to a bunch of keys, the badge of St Peter, hence Herb Peter, and other 'key' names. 'Cove' is most probably culver (the name Culverkeys does occur both in Kent and Somerset) - culver ia a dove.

COVEN-TREE *Viburnum lantana* (Wayfaring Tree) Wilts. Dartnell & Goddard; Bucks. Grigson. Coban, and Cobin, are also recorded. The name apppears in the children's game, quoted by Northall:

Keppy-ball, keppy-ball,
Cobin-tree,
Come down and tell me
How many years old
Our (Jenny) shall be.

The number of keps or catches before the ball falls is the age. It seems that the game was played under this tree. But these names were given to any tree which stood before a Scottish mansion house; it was under this tree that the laird always went out to greet his visitors. So the word is covyne, a meeting or trysting place. The name was corrupted to Capon-tree in Cumbria.

COVENTRY BELLS 1. *Campanula medium* (Canterbury Bells). 2. *Campanula trachelium*

(Nettle-leaved Bellflower) Prior. 3. *Digitalis purpurea* (Foxglove) Som. Macmillan. 4. *Pulsatilla vulgaris* (Pasque-flower) Cambs. Gerard.

COVENTRY MARIAN; COVENTRY RAPES *Campanula medium* (Canterbury Bells) both Gerard. Marian, because there was an association of the plant with the Virgin. Gerard described the plant under the name Viola Marina, and there is a French name, Mariette. Rapes, presumably because of the large root, which was used in salads, or boiled and eaten with oil, vinegar and pepper in Gerard's time.

COVER-KEYS; COVEY-KEYS *Primula veris x vulgaris* (Oxlip) Kent B & H (Cover); Kent Grigson (Covey). Why only oxlip? This is the same as Cove-keys, one of a series applied to cowslips. But the word is really Culverkeys.

COVER-SHAME *Juniperus sabina* (Savin) B & H. The berries are abortifacients, hence this name. "...provocat etiam menstrua et fetum mortuum educit; decocta cum oleo idem melius operatur, si sit superposita..." (Circa Instans/Rufinus) (Thorndike). The leaves, according to Gerard, are just as efficacious. The use was apparently quite notorious. See Middleton, A game of chess, act 1, sc 2:

> To gather fruit, find nothing but the savin-tree,
> Too frequent in nuns' orchards, and there planted
> By all conjecture, to destroy fruit rather.

COVEY; COVEY, Sweet *Erodium moschatum* (Musk Storksbill) both B & H. From Muscovy - i.e. musk. The plant smells of it.

COW *Umbilicus rupestris* (Wall Pennywort) Corn. Grigson. Do the children "milk" them?

COW, Cushy *Rumex obtusifolius* (Broad-leaved Dock) Border B & H. The name is only given when the plant is in seed. The children "milk" them by drawing the stalks through their fingers. Cush, Cush was the call for the cows, and 'Cushy cow, bonny' is the start of a milking charm rhyme quoted in the Oxford dictionary of nursery rhymes.

COW BASIL 1. *Saponaria officinalis* (Soapwort) S.Africa Watt. 2. *Vaccaria pyramidata* (Cow-herb) B & H.

COW BASIL, Red *Centranthus ruber* (Spur Valerian) B & H.

COW BELL *Silene cucubalis* (Bladder Campion) Som. Macmillan.

COW-BELLY *Heracleum sphondyllium* (Hogweed) Som. Macmillan. Presumably Cow Billers.

COW-BUMBLE *Heracleum sphondyllium* (Hogweed) Som. Macmillan. Cow-mumble, rather.

COW CAKES; COW FLOPS 1. *Heracleum sphondyllium* (Hogweed) Scot. Grigson (cakes) Corn. Grigson (Flops). 2. *Pastinaca sativa* (Wild Parsnip) Roxb. Grigson (Cakes) Corn. Jago (Flops). Presumably descriptive, and referring to the flat flower heads, but cakes must surely be the same as Kex, and is a reference to the dried stems. Much closer are Cow-Keeks, or Cow-Keeps.

COW CHERVIL 1. *Anthriscus sylvestris* (Cow Parsley) Som. Mabey. 2. *Myrrhis odorata* (Sweet Cicely) Grieve.1931. The application of 'cow' here, and in a number of other cases, is to indicate a plant which looks something like chervil, but isn't. Cf. the name Mock Chervil for Sweet Cicely.

COW CLOGWEED *Heracleum sphondyllium* (Hogweed) Glos. B & H. Clog is probably the same as Kecklock, applied to any umbellifer, and particularly to the dried stems.

COW CLOOS *Trifolium pratense* (Red Clover) Scot. Jamieson. i.e. Cow Clover.

COW CLOVER 1. *Trifolium involucratum* (Multiflowered Clover) USA Elmore. 2. *Trifolium medium* (Zigzag Clover) Yks. B & H. 3. *Trifolium pratense* (Red Clover) B & H.

COW COCKLE 1. *Saponaria officinalis* (Soapwort) S.Africa Watt. 2. *Vaccaria pyramidata* (Cow-herb) Kansas Gates.

COW CRACKER *Silene cucubalis* (Bladder Campion) Dumf. Grigson.

COW CRANES *Caltha palustris* (Marsh Marigold) Oxf. Oxf.Annl.Rec.1951; N'thants. Grigson. There is Crow-cranes in Oxfordshire, too, and this is more likely (buttercups are Crowfeet).

COW CRESS 1. *Apium nodiflorum* (Fool's Watercress) Hants. Cope. 2. *Lepidium campestre* (Pepperwort) Prior. 3. *Veronica beccabunga* (Brooklime) Ess. Grigson.

COW EYES *Leucanthemum vulgare* (Ox-eye Daisy) Corn. Grigson.

COW-FAT *Vaccaria pyramidata* (Cow-herb) W.Miller. Vaccaria, which itself comes from Latin vacca, cow, was once much in demand as a cattle fodder.

COW FOOT *Saponaria officinalis* (Soapwort) S.Africa Watt.

COW GARLIC *Allium vineale* (Crow Garlic) B & H. 'Cow' is probably a misprint in the original, for 'crow'.

COW GRASS 1. *Plantago major* (Great Plantain) N.Ire. Grigson. 2. *Polygonum aviculare* (Knotgrass) B & H. 3. *Ranunculus flammula* (Lesser Spearwort) N.Ire. Grigson. 4. *Trifolium medium* (Zigzag Clover) N'thants. A.E.Baker; Ches, Lincs, Border, Scot. B & H. 5. *Trifolium pratense* (Red Clover) Som, IOW, Cumb, N'thum, Roxb. Grigson. Probably all because of the belief in their value as fodder plants.

COW HEAVE *Tussilago farfara* (Coltsfoot) Selk. B & H. Probably Cow-hoof. Cf. Horsehoof, Donnhoove, Tunhoof for the same plant, as well as all the 'foot' names. It is the shape of the leaves that gives the names.

COW-HERB *Vaccaria pyramidata*. Vaccaria is from Latin vacca, cow, apparently because this was thought to be excellent cattle fodder. The name Cow-fat emphasises this, and there are also Cow Basil and Cow Cockle. Cowherb seems to express this idea of being the cattle-fodder par excellence.

COW ITCH 1. *Polygonum hydropiper* (Water Pepper) Cockayne. 2. *Rhus radicans* (Poison Ivy) USA Hora. 3. *Rosa canina* seeds (Dog Rose seeds) Ches. Holland. 4. *Urera baccifera*. Cockayne had Cow-itch as a corruption of Culrage (culi rabies). Culrage is from the French, meaning the same as the common English name Arsmart. It is the irritant effect of the leaves which gives rise to these names. Dog Rose seeds are the original itching powder. *Urera baccifera* has stinging hairs on the leaves, and the plant is used as a cattle hedge in tropical America. There is Cowitch (one word) - it is *Mucuna pruriens*.

COW KEEKS; COW KEEPS *Heracleum sphondyllium* (Hogweed) both Border B & H. Keeks or Keeps must be the same word as Kex, which itself has a host of variants, and is a common name for almost any umbellifer, though it should be reserved for the dried stems.

COW LILY 1. *Caltha palustris* (Marsh Marigold) USA Grigson. 2. *Nymphaea advena* (Large Yellow Pond Lily) USA Tolstead.

COW MACK; COWMACK 1. *Melandrium album* (White Campion) N.Scot. B & H. 2. *Silene cucubalis* (Bladder Campion) Scot. Grigson. "Some husbands (to make the cow take the bull the sooner) do give her of the herb called Cow-mack, which groweth like a white gilly-flower among the corn" (quoted by B & H).

COW MUMBLE 1. *Anthriscus sylvestris* (Cow Parsley) Ess, Norf, Cambs. B & H. 2. *Chaerophyllum temulentum* (Rough Chervil) Suff. Grigson. 3. *Heracleum sphondyllium* (Hogweed) E.Ang. Nall.

COW PAIGLE; COW PEGGLE *Primula veris* (Cowslip) Herts. Grigson (Paigle); Herts. Jones-Baker.1974 (Peggle). Paigle is an old name for cowslip. Its derivation is unknown, perhaps from Fr. epingle, pin (describing the stigma) or Fr. paillole, spangle.

COW PAPS *Silene cucubalis* (Bladder Campion) Border B & H. Descriptive.

COW PARSLEY 1. *Aethusa cynapium* (Fool's Parsley) Som. Macmillan. 2. *Anthriscus sylvestris*. Cf. Cow Mumble, and Cow-weed, for Cow Parsley, as well as a number of variations on the parsley theme.

COW PARSNIP 1. *Heracleum lanatum* USA Schenk & Gifford. 2. *Heracleum maximum*. 3. *Heracleum sphondyllium* (Hogweed) Turner.
COW PARSNIP, Indian *Heracleum lanatum* Douglas.
COW POPS *Physalis angulata* Barbados Gooding.
COW QUAKES; COW QUAKERS *Briza media* (Quaking-grass) Som, Dor, Ches, N'thum Grigson.1959.
COW RATTLE 1. *Melandrium album* (White Campion) Bucks. Grigson. 2. *Silene cucubalis* (Bladder Campion) Bucks. Grigson. More likely to be Bladder Campion than the White, in fact. Cf. Bull Rattle, Rattle-bags, Rattleweed, etc., for *S cucubalis*.
COW SERL *Rumex acetosella* (Sheep's Sorrel) USA Bergen. An Americanization of the pronunciation of sorrel.
COW SINKIN 1. *Primula veris x vulgaris* (Oxlip) Cumb. Grigson. 2. *Primula vulgaris var.elatior* (Polyanthus) Cumb. B & H.
COW SOAPWORT *Saponaria officinalis* (Soapwort) S.Africa Watt.
COW STRIPLING; COW STRUPPLE *Primula veris* (Cowslip) Yks. Harland (Stripling); Yks, Cumb. Grigson (Strupple).
COW STROPPLE 1. *Primula veris* (Cowslip) Yks, Cumb. Grigson. 2. *Primula veris x vulgaris* (False Oxlip) Dur. Brockett.
COW VETCH *Vicia cracca* (Tufted Vetch) Glos. J.D.Robertson.
COW-WEED 1. *Anthriscus sylvestris* (Cow Parsley) B & H. 2. *Centaurea scabiosa* (Greater Knapweed) Tynan & Maitland. 3. *Myrrhis odorata* (Sweet Cicely) Yks. Grigson. 4. *Ranunculus aquatilis* (Water Crowfoot) Hants. Grigson.
COW-WHEAT 1. *Melampyrum pratense*. 2. *Rhinanthus crista-galli* (Yellow Rattle) Som. Macmillan; Cumb. Grigson. Said to be derived from the notion that the small seeds could be converted into wheat, probably because the plant seems to appear suddenly in cornfields.
COW-WHEAT, Crested *Melampyrum cristatum*.
COW-WHEAT, Eyebright *Odontites verna* (Red Bartsia) Prior. There are Eyebright, and Red Eyebright for this also.
COW-WHEAT, Field *Melampyrum arvense*.
COW-WHEAT, Narrow-leaved *Melampyrum lineare*.
COW-WHEAT, Purple *Melampyrum arvense* (Field Cow-wheat) Grigson.1959.
COW-WHEAT, Small *Melampyrum sylvaticum*.
COW-WHEAT, Wood *Melampyrum sylvaticum* (Small Cow-wheat) Clapham.
COW-WORT *Geum urbanum* (Avens) Leyel.1937.
COW'S EAR *Verbascum thapsus* (Mullein) Wales Grigson. It is Bull's Ear as well, in Wales, and elsewhere the ears belong to donkeys, or cuddies, and bunnies.
COW'S EYES *Leucanthemum vulgare* (Ox-eye Daisy) Som. Macmillan.
COW'S LICK *Bryonia dioica* (White Bryony) Norf. B & H. From its use as a horse and cow medicine.
COW'S LUNGWORT 1. *Helleborus niger* (Christmas Rose) Cockayne. 2. *Verbascum thapsus* (Mullein) B & H; USA Henkel. Shouldn't the name be applied to *Helleborus foetidus*, rather than *H niger*? The former was used often in cattle medicine. It is the leaves which give the name to mullein - it is better known as Bullock's Lungwort.
COW'S MOUTH *Primula veris* (Cowslip) Loth. Grigson.
COW'S PARSLEY *Pulmonaria officinalis* (Lungwort) Som. Macmillan. This must have started life as cowslip; lungwort has a number of names which include cowslip.
COW'S PARSNIP *Arum maculatum* (Cuckoo-pint) Som. Macmillan.
COW'S TAIL *Erigeron canadensis* (Canadian Fleabane) USA Henkel. It is known as Mare's Tail and Colt's Tail as well.
COW'S TAIL PINE *Cephalotaxus harringtonia var.drupacea* (Japanese Plum-Yew) Barber & Phillips.

COW'S TAIL PINE, Chinese *Cephalotaxus fortunei* (Chinese Plum-yew) Leathart.

COW'S THISTLE *Cirsium arvense* (Creeping Thistle) Som. Macmillan.

COW'S WEATHER-WIND; COW'S WITHYWIND *Stachys sylvatica* (Hedge Woundwort) both B & H.

COW'S WORT *Pedicularis palustris* (Red Rattle) Notts. Grigson.

COWAGE *Mucuna pruriens*. A name which has become Cowitch in places, with some justification. But Cowage is Hindi kavacs.

COWBANE 1. *Cicuta virosa*. 2. *Oenanthe crocata* (Hemlock Water Dropwort) Yks. Grigson. They are both "banes", being very poisonous. True Cowbane especially is very toxic to cattle, causing nausea, delirium and convulsions, followed by death from asphyxia.

COWBANE, American *Cicuta maculata*.

COWBANE, Spotted 1. *Cicuta maculata* (American Cowbane) USA Kingsbury.1967. 2. *Conium maculatum* (Hemlock) USA Henkel.

COWBEAN *Cicuta maculata* (American Cowbane) Le Strange. Cowbane, of course.

COWBERRY 1. *Potentilla palustris* (Marsh Cinquefoil) Scot. Grigson. 2. *Vaccinium myrtillus* (Whortleberry) Som. Macmillan; Yks. B & H. 3. *Vaccinium vitis-idaea*. Suspiciously, Crowberry is also given as a name for the last. It is probably more likely than Cowberry (but the generic name *Vaccinium* means pertaining to cows, or at any rate such a word as vaccinus does, and there may have been confusion).

COWBIND *Bryonia dioica* (White Bryony) B & H. Shelley used the name:

And in the warm hedge grew eglantine,
Green cowbind and moonlight-coloured May

(The Question). B & H give no other reference for the name - did Shelley coin it?

COWCUMBER; COW-CUMMER *Cucumis sativa* (Cucumber) Parkinson. (Cowcumber); Worcs. J Salisbury (Cowcummer). Cowcumber is a very common variant on cucumber, but it is surprising to find it as early as Parkinson, suggesting that it was standard pronunciation in his day.

COWER-SLOP *Primula veris* (Cowslip) Shrop. Grigson. Very local - Cowslop is quite widespread.

COWFAT *Centranthus ruber* (Spur Valerian) Halliwell.

COWFLOP 1. *Avena sativa* (Oats) Dev. Friend.1882. 2. *Digitalis purpurea* (Foxglove) Dev. Friend.1882. 3. *Heracleum sphondyllium* (Hogweed) Corn. Courtney.1880. 4. *Pastinaca sativa* (Wild Parsnip) Corn. Jago. 5. *Primula veris* (Cowslip) Dev, Som. Macmillan. Cow pat, of course. Clear enough for Cowslip, which is the same thing, but why the others? Wild Parsnip has Cow-cakes from Scotland, but the real reason for the name, as for foxglove, lies in the size of the leaves. Friend said that it was given to oats to distinguish them from Tartarian oats!.

COWFOOT *Senecio jacobaea* (Ragwort) Shrop. B & H. Crowfoot seems to be meant in this case. Not that crowfoot is a common name for ragwort, but the leaves are finely-cut, so that may be the reason.

COWHAGE *Mucuna pruriens* (Cowage) Leyel.1948.

COWHERB *Saponaria officinalis* (Soapwort) S.Africa Watt. There are a number of these 'cow' names for Soapwort from S.Africa. It can't be that they were given because it is a fodder plant, for it is toxic to animals. The taste is unpleasant, though, so they leave it alone.

COWITCH *Mucuna pruriens* (Cowage) Willis. The specific name is *pruriens*, so the 'itch' part of the name seems reasonable. In fact, the hooked hairs do cause itching, and were used in honey against intestinal worms. Cowage is from the Hindi kavacs.

COWL, Friar's *Arum maculatum* (Cuckoo Pint) B & H. Priest's Hood is another old book name, and they must both be viewed in the light of Parson-in-the-pulpit and its variants.

COWL, Monk's *Aconitum napellus* (Monkshood) Turner. It is said that the upper sepal resembles a monk's cowl. Cf. Friar's Cap, from Devonshire, and the French Capuchon de moine, as well as the common name, Monkshood.

COWLL *Corylus avellana* (Hazel) IOM Moore.1924. Manx coull, but compare similar names from Scotland.

COWPEA *Vigna unguiculata*.

COWPEA, Asparagus *Vigna sesquipedalus*.

COWQUAKE *Spergula arvensis* (Corn Spurrey) E.Ang. Halliwell.

COWRIE PINE *Agathis australis* (Kauri Pine) Willis. Another spelling of Kauri.

COWS *Umbilicus rupestris* (Wall Pennywort) Corn. Grigson. Presumably short for Milk-the-cows, also from Cornwall.

COWS, Bulls-and- *Arum maculatum* (Cuckoo Pint) Som. Macmillan; War. Morley; N'thants. A.E.Baker; Lincs. B & H; Yks. Nicholson; Lancs. Grigson.

COWS-AND-BULLS *Arum maculatum* (Cuckoo Pint) Wilts. Goddard. Another variation on the "double" names, which start off with Lords-and-ladies.

COWS-AND-CALVES; COWS-AND-KIES *Arum maculatum* (Cuckoo Pint) Wilts. Goddard; Glos. J.D.Robertson; N'thants. A.E.Baker; Dev, Dor, Som, Bucks, Shrop, Notts, Lincs, Yks, Cumb. Grigson; War. Morley (Calves); Yks. Grigson (Kies). Typically double names for cuckoo pint, but not male + female, as, for example, with the previous entry. Surely Cows-and-kies is pleonastic, for presumably kies means cows (kine, in fact).

COWSLAP; COWSLUP *Primula veris* (Cowslip) Turner (Cowslap); War, Worcs. Grigson (Cowslup). Only two of the cowslip variants - there are also Cowslop (applied to more than one plant), Cower-slop, Cowflop, Cowslek, etc.

COWSLEK *Primula veris* (Cowslip) B & H.

COWSLIP 1. *Anemone nemerosa* (Wood Anemone) N.Scot. B & H. 2. *Caltha palustris* (Marsh Marigold) USA House. 3. *Digitalis purpurea* (Foxglove) Dev. Friend. 4. *Fritillaria meleagris* (Snake's Head Lily) Hants. Grigson. 5. *Mertensia virginica* (Virginia Cowslip) USA A.W.Smith. 6. *Narcissus pseudo-narcissus* (Daffodil) Dev. B & H. 7. *Orchis mascula* (Early Purple Orchis) Rut. B & H. 8. *Primula auricula* (Auricula) Dev. Friend.1882. 9. *Primula veris*. 10. *Primula veris x vulgaris* (False Oxlip) Dor, Kent, E Angl, Midl, Herts. Grigson. 11. *Ranunculus acris* (Meadow Buttercup) Dev. B & H; Ire. O Suileabhain. 12. *Ranunculus bulbosus* (Bulbous Buttercup) Dev. Grigson. 13. *Ranunculus repens* (Creeping Buttercup) Dev. Grigson. Cowslip is OE cuslyppe, cow dung, or perhaps cow slobber (Potter & Sargent). Earle suggests that the 'lip' of cowslip may come from OE lyb, a purging drug, as in Cockayne's libcorn. A Gaelic saying may perhaps confirm this - cattle have an aversion to the cowslip, and they will refuse to eat it. It is thought to give them the cramp or colic, and they will become "elfshot". But it certainly is not cow's lip. Cf. Foxflop for Foxglove. The inference is that the plant has grown from the droppings.

COWSLIP, American *Dodecatheon meadia* (Shooting Star) McDonald.

COWSLIP, Bedlam 1. *Primula veris x vulgaris* (Oxlip) N'thants. A.E.Baker. 2. *Pulmonaria officinalis* (Lungwort) Som. Grigson. Lungwort bears its various cowslip names because of its primula-like flowers. Cf. Bethlehem Cowslip, Jerusalem Cowslip, Bethlehem, and Jerusalem Sage, etc. Oxlip, possibly, because it really is a mad cowslip.

COWSLIP, Bethlehem *Pulmonaria officinalis* (Lungwort) Coles.

COWSLIP, Blue *Pulmonaria longifolia* (Lungwort) Hants. Cope.

COWSLIP, Bugloss *Pulmonaria officinalis* (Lungwort) Prior. Bugloss leaves, primula flowers.

COWSLIP, Cape *Lachenalia aloides*.

COWSLIP, French *Primula auricula* (Auricula) Parkinson. Presumably he meant to indicate that it wasn't a native plant.
COWSLIP, Great *Primula vulgaris var. elatior* (Polyanthus) B & H.
COWSLIP, Himalayan *Primula florindae*.
COWSLIP, Jerusalem *Pulmonaria officinalis* (Lungwort) Gerard. Cf. Jerusalem Sage, Jerusalem Seeds, as well as Bethlehem attributes.
COWSLIP, Jerusalem, Narrow-leaved *Pulmonaria longifolia* (Lungwort) Gerard.
COWSLIP, Lady's *Gagea lutea* (Yellow Star of Bethlehem) B & H.
COWSLIP, Mountain *Primula auricula* (Auricula) Gerard.
COWSLIP, Virgin Mary's *Pulmonaria officinalis* (Lungwort) Glos, Worcs. Grigson; Shrop. Burne.
COWSLIP, Virginian 1. *Mertensia virginica*. 2. *Pulmonaria officinalis* (Lungwort). Mertensia is the American plant; lungwort is more easily identifiable as Virgin Mary's Cowslip.
COWSLOP 1. *Caltha palustris* (Marsh Marigold) USA Grigson. 2. *Digitalis purpurea* (Foxglove) Dev. B & H. 3. *Primula veris* (Cowslip) Wessex Rogers; N'thants. A.E.Baker; Lincs. Grigson; Ches. Holland. Cowslop is quite a common variant of cowslip. There are others, like Cowslap, Cowflop, etc. Cowslop becomes Cower-slop in Shropshire, and is reduced to Slop, or Slap, in Northants. Foxglove is more recognizable as Flops, or Floppy Dock.
COWTHWORT *Leonurus cardiacus* (Motherwort) Halliwell.
COXCOMB 1. *Amaranthus cruentus*. 2. *Rhinanthus crista-galli* (Yellow Rattle) Gerard. Cock's Comb, which is quite common for Yellow Rattle. The amaranth quoted here is an American plant.
COYNES *Cydonia vulgaris* (Quince) Ellacombe. Simply the old spelling of quince, that is itself a corruption of Cydonia. The word can be spelt with a 'y' or an 'i'. One of the transformations of kydonion on its way to quince was the Old French coin (the modern French is coing) - hence Coynes.
COYOTE TOBACCO *Nicotiana attenuata*.
CRAA-TAES; CRAA'S FOOT *Lotus corniculatus* (Bird's-foot Trefoil) both N'thum. both Grigson.
CRAB *Malus sylvestris*. Probably of Scandinavian origin. The older form, still known in the north, is scrab.
CRAB, Minshull *Mespilus germanica* (Medlar) B & H. It is given as a Cheshire name, but probably in error.
CRAB, Purple *Pyrus atropurpurea*.
CRAB, Siberian 1. *Malus baccata*. 2. *Pyrus prunifolia*.
CRAB APPLE; CRABAPPLE *Malus sylvestris*.
CRAB APPLE, Garland *Malus coronaria*.
CRAB APPLE, Oregon *Malus fusca*.
CRAB APPLE, Prairie *Malus ioensis*.
CRAB CHERRY *Prunus avium* (Wild Cherry) Bucks. Grigson. Presumably to indicate the wild kind, on the analogy of Crabapple.
CRAB GRASS 1. *Polygonum aviculare* (Knotgrass) E.Ang. Grigson. 2. *Salicornia europaea* (Glasswort) Gerard. There is Crabweed also for Knotgrass in Essex. Possibly Glasswort gets the name because it is a seashore plant.
CRAB-GRASS, Hairy; CRAB-GRASS, Large *Digitaria sanguinalis* (Hairy Finger-grass) USA (Large); Canada (Hairy) both Allan.
CRAB-GRASS, Silver *Eleusine indica* (Goosegrass) USA Allan.
CRAB-GRASS, Smooth *Digitaria ischaemum* (Smooth Finger-grass) USA Allan.
CRAB STOCK *Malus sylvestris* fruit (Crabapples) Dor, Suss. B & H.
CRAB'S CLAWS 1. *Polygonum persicaria* (Persicaria) Som, Dor. Macmillan. 2. *Stratiotes aloides* (Water Soldier) Grieve.1931.
CRAB'S EYE *Abrus precatorius* (Precatory Bean) USA Kingsbury; Barbados Gooding.

CRAB'S EYE LICHEN *Ochrolechia parella* (Crawfish Lichen) Bolton.
CRAB'S STONE *Abrus precatorius* (Precatory Bean) Chopra.
CRABWEED *Polygonum aviculare* (Knotgrass) E.Ang. Grigson. Cf. Crab Grass for this.
CRACHES *Stellaria media* (Chickweed) B & H.
CRACK-NUT *Corylus avellana* (Hazel) Dev. B & H. Cf. the Cornish Victor Nut. Cob-nuts is the name of a game played with them, much like conkers.
CRACK WILLOW *Salix fragilis*. Cf. Snap, or Brittle, Willow for this. A slight pressure at the base of a twig will separate it from the branch with a cracking sound.
CRADLE, Babe-and- *Fumaria officinalis* (Fumitory) Som. Macmillan. Babe-in-the-cradle, rather.
CRADLE, Babe-in-the- 1. *Arum maculatum* (Cuckoo Pint) Som. Macmillan. 2. *Fumaria officinalis* (Fumitory) Som. Macmillan. 3. *Scrophularia aquatica* (Water Betony) Som. Macmillan; Wilts. Goddard.
CRADLE, Baby's *Onobrychis sativa* (Sainfoin) Dor. Macmillan.
CRADLE, Fairy's *Tulipa sp* (Tulip) Dyer.
CRADLE, Moses-in-the- *Rhoeo spathacea* (Boat Lily) Goold-Adams. Given because of the way that the small flowers seem to be cradled in the bracts.
CRADLE-DOCK *Senecio jacobaea* (Ragwort) Ches. Grigson. Cheadle-dock is recorded in the same area. Both are probably from the commoner Keddle-dock, or Keedle-dock, themselves coming from Kedlock, a name associated more with umbelliferous plants.
CRADLE ORCHID *Anguloa clowesii*.
CRAE-NELS *Anthyllis vulneraria* (Kidney Vetch) N'thum. Grigson. i.e. crow beaks.
CRAID *Trifolium campestre* (Hop Trefoil) Scot. Grigson.
CRAIN *Ranunculus ficaria* (Lesser Celandine) N'thants Grigson.
CRAIN, Yellow; CRANE, Yellow 1.
Ranunculus ficaria (Lesser Celandine) N'thants A.E.Baker. 2. *Ranunculus flammula* (Lesser Spearwort) N'thants. Grigson.
CRAISEY *Potentilla sterilis* (Barren Strawberry) Worcs. Grigson. Cf. the 'Crazy' names given to various of the *Ranunculus sp*.
CRAKE NEEDLE *Scandix pecten-veneris* (Shepherd's Needle) N.Eng. Grigson. Crake is a North-country word for crow, and crow-needle occurs over a wide area of the south of the country.
CRAKEBERRY *Empetrum nigrum* (Crowberry) Yks. F K Robinson. Crake = crow.
CRAKEFEET 1. *Endymion nonscriptus* (Bluebell) Yks. B & H. 2. *Orchis mascula* (Early Purple Orchis) Yks. B & H. 3. *Orchis morio* (Green-winged Orchis) Yks. Friend.
CRAMBERRY *Vaccinium oxycoccus* (Cranberry) Ches. B & H. There are a lot of variations on Cranberry; besides this one, there are Crawnberry, Craneberry, Crannaberry, Crawberry, Croneberry, etc.
CRAMBLING ROCKET *Reseda lutea* (Wild Mignonette) B & H. Crambling probably means scrambling, or wandering.
CRAMMICK *Ononis repens* (Rest Harrow) Som. Macmillan. Cammick, or Cammock, is more usual.
CRAMP-BARK *Viburnum opulus var. americanus* (High-bush Cranberry) USA H.Smith.1945. Some of the American Indians drank the tea to cure stomach cramps.
CRAMP THISTLE *Cirsium arvense* (Creeping Thistle) War. Bloom. Some insect makes galls in the form of swellings on this thistle. These used to be carried round to prevent cramp, just like the wasp galls on Dog Rose, though that was for rheumatism.
CRAMP-WEED *Potentilla anserina* (Silverweed) Fernie. The tea has been used to cure stomach cramps.
CRAN-COMMER *Salix repens* (Creeping Willow) Donegal Grigson.
CRANBERRY 1. *Arctostaphylos uva-ursi* (Bearberry) N.Eng. Grigson. 2. *Lonicera conjugialis* (Bush Honeysuckle) Calif.

Spier. 3. *Vaccinium oxycoccus*. 4. *Vaccinium vitis-idaea* (Cowberry) N.Eng. Gilmour & Walters. According to ODEE, first used in England for the imported American *V macrocarpon* (American Cranberry), and then transferred to the native kind. It is an American word, from German kranbeere or Kranebere - crane-berry. This form still exists in the Scottish Craneberry.

CRANBERRY, American *Vaccinium macrocarpon*.

CRANBERRY, High; CRANBERRY, High-bush *Viburnum opulus var. americanus* or *Viburnum trilobum*. High-bush Cranberry is the usual name; High Cranberry occurs in USA Sanford; Cunningham. The berries are treated as cranberries are, and served as a sauce.

CRANBERRY, Large *Vaccinium macrocarpum* (American Cranberry) USA Yarnell.

CRANBERRY, Mountain *Vaccinium vitis-idaea* (Cowberry) Hay & Synge. The specific name suggests that the mountain is Mount Ida.

CRANBERRY, Upland *Arctostaphylos uva-ursi* (Bearberry) O.P.Brown.

CRANBERRY BUSH *Viburnum trifolium* USA Tolstead.

CRANBERRY BUSH, American *Viburnum trilobum*.

CRANBERRY BUSH, Rayless *Viburnum pauciflorum*.

CRANBERRY TREE 1. *Viburnum opulus* (Guelder Rose) Perry. 2. *Viburnum opulus var.americanus* (High-bush Cranberry) USA. Lloyd.

CRANBERRY-WIRE 1. *Vaccinium oxyoccus* (Cranberry) Cumb. Grigson. 2. *Vaccinium vitis-idaea* (Cowberry) Aber, Banff, Kinc, Mor. Grigson.

CRANE 1. *Empetrum nigrum* (Crowberry) Yks. Grigson. 2. *Vaccinium oxycoccus* (Cranberry) N'thum. Grigson; Cumb. B & H. Cranberry was originally craneberry. There is another use of Crane as a plant name for various species of *Ranunculus*; that version is often spelt Crain.

CRANE, Cow; CRANE, Crow; CRANE, Johnny *Caltha palustris* (Marsh Marigold) N'thants. Grigson (Cow); Oxf. Grigson (Crow); N'thants. Grigson (Johnny).

CRANEBERRY *Vaccinium oxycoccus* (Cranberry) Suth. B & H. This is a retention of the original form.

CRANEBERRY WIRE *Arctostaphylos uva-ursi* (Bearberry) Aber. Grigson.

CRANESBILL, American *Geranium maculatum* (Spotted Cranesbill) Grieve.1931. Geranium derives through Latin from Greek geranion, crane. The profile view of the seed vessels shows the reason for the name very well.

CRANESBILL, Blood-red; CRANESBILL, Bloody *Geranium sanguineum* Bloody Cranesbill is the usual name; Blood-red Cranesbill occurs in Darling & Boyd.

CRANESBILL, Blue *Geranium pratense* (Meadow Cranesbill) Gerard.

CRANESBILL, Bohemian *Geranium bohemicum*.

CRANESBILL, Crowfoot *Geranium pratense* (Meadow Cranesbill) Gerard.

CRANESBILL, Cut-leaved *Geranium dissectum*.

CRANESBILL, Dove's-foot *Geranium molle*. It is the shape of the leaf which gives Dove's-foot. It was known as Culverfoot in Anglo-Saxon times. Turner has Dove-foot, and Gerard Pigeon's Foot.

CRANESBILL, Dove's-foot, Perennial *Geranium pyrenaicum* (Mountain Cranesbill) Curtis.

CRANESBILL, Dusky *Geranium phaeum*. Purple to black flowers, so 'dusky'.

CRANESBILL, Foetid; CRANESBILL, Stinking; CRANESBILL, Strong-scented *Geranium robertianum* (Herb Robert) Barton & Castle (Foetid); Prior (Stinking); Curtis (Strong-scented).

CRANESBILL, French *Geranium endressi*.

CRANESBILL, Hedgerow *Geranium pyrenaicum* (Mountain Cranesbill) Blamey.

CRANESBILL, Hemlock-leaved *Erodium*

cicutarium (Storksbill) Curtis.
CRANESBILL, Italian *Geranium macrorrhizum*.
CRANESBILL, Jagged *Geranium dissectum* (Cut-leaved Cranesbill) Curtis.
CRANESBILL, Knotted *Geranium nodosum*.
CRANESBILL, Little *Geranium robertianum* (Herb Robert) Som. Macmillan. Probably a cross between a cranesbill and one of the names like Little Jack, or Little Jan.
CRANESBILL, Long-stalked *Geranium columbianum*.
CRANESBILL, Meadow *Geranium pratense*.
CRANESBILL, Mountain 1. *Geranium pyrenaicum*. 2. *Geranium sylvaticum* (Wood Cranesbill) Gilmour & Walters.
CRANESBILL, Musked *Erodium moschatum* (Musk Storksbill) Camden/Gibson.
CRANESBILL, Pencilled *Geranium versicolor* (Painted Lady) McClintock.
CRANESBILL, Purple *Geranium purpureum*.
CRANESBILL, Pyrenean *Geranium pyrenaicum* (Mountain Cranesbill) McClintock.
CRANESBILL, Round-leaved *Geranium rotundifolium*.
CRANESBILL, Shining *Geranium lucidum*.
CRANESBILL, Small *Geranium pusillum*.
CRANESBILL, Small-flowered *Geranium parviflorum*.
CRANESBILL, Spotted *Geranium maculatum*.
CRANESBILL, Streaked *Geranium versicolor* (Painted Lady) D.Grose.
CRANESBILL, Wild *Geranium maculatum* (Spotted Cranesbill) Grieve.1931.
CRANESBILL, Wood *Geranium sylvaticum*.
CRANNABERRY *Vaccinium oxycoccus* (Cranberry) Shrop. B & H. One of the many variations on cranberry.
CRANNOCK *Ulex europaeus* (Furze) Som. Macmillan.
CRANOPS *Raphanus raphanistrum* (Wild Radish) Berw. B & H.
CRAP 1. *Fagopyrum esculentum* (Buckwheat) Suss. W D Cooper; Worcs. Grigson. 2. *Lolium perenne* (Rye-grass) Suff, Suss. B & H. 3. *Lolium temulentum* (Darnel) Halliwell. 4. *Sinapis arvensis* seed (Charlock seed) Roxb. B & H. Probably the same word as crop; indeed, B & H have Crop as a name for buckwheat, and Culpepper calls darnel Crop.
CRAP, Gold 1. *Ranunculus bulbosus* (Bulbous Buttercup) Halliwell. 2. *Ranunculus repens* (Creeping Buttercup) Som. Grigson. Halliwell quotes Hollyband's Dictionarie, 1593, as his reference. He also has Gold-Knap for this. Could this actually be Craw?
CRAPEMYRTLE, Andaman *Lagerstroemeria hypoleuca*.
CRAPEMYRTLE, Battinan *Lagerstroemeria piriformus*.
CRAPEMYRTLE, Queen *Lagerstroemeria flos-reginae*.
CRASH; CRASH, Water *Nasturtium officinale* (Watercress) both Yks. Morris (Crash); B & H (Water Crash).
CRAW; CRAW, Yellow *Ranunculus bulbosus* (Bulbous Buttercup) both B & H. i.e. crow (the buttercups are called crowfoot, and it even results in Craw Crowfoot sometimes.
CRAW-CRAW PLANT 1. *Ageratum conyzoides* (Goatweed) Sierra Leone Dalziel. 2. *Cassia alata* (Candlestick Senna) Dalziel. They are used as a remedy for the disease of that name.
CRAW CROWFOOT *Ranunculus bulbosus* (Bulbous Buttercup) B & H.
CRAW PEA *Lathyrus pratensis* (Yellow Pea) Border B & H.
CRAW-SHAW *Menyanthes trifoliata* (Buckbean) Ork. Leask.
CRAW-TAES 1. *Endymion nonscriptus* (Bluebell) Border Turner. 2. *Orchis mascula* (Early Purple Orchis) Cumb. B & H. 3. *Ranunculus acris* (Meadow Buttercup) Scot. B & H. 4. *Ranunculus repens* (Creeping Buttercup) Scot, Donegal Grigson.
CRAWBERRY 1. *Empetrum nigrum* (Crowberry) Yks, N'thum. Grigson. 2.

Vaccinium oxycoccus (Cranberry) Border B & H.

CRAWCROOKS; CRAWCROUPS *Empetrum nigrum* (Crowberry) N'thum. Grigson (Crawcrooks); Scot. Jamieson (Crawcroups).

CRAWFISH LICHEN *Ochrolechia parella*.

CRAWFOOT 1. *Endymion nonscriptus* (Bluebell) Border B & H. 2. *Orchis mascula* (Early Purple Orchis) Yks. B & H. 3. *Orchis morio* (Green-winged Orchis) Yks. B & H. 4. *Ranunculus repens* (Creeping Buttercup) Yks, Scot. Grigson.

CRAWFOOT, Wood, Musk *Adoxa moschatellina* (Moschatel). B & H.

CRAWLERS *Cochlearia officinalis* (Scurvy-grass) Som. Macmillan.

CRAWLEY *Corallorhiza odontorhiza* (Autumn Coral-root) USA Emboden.1974. Said to derive in some way from the appearance of the root.

CRAWNBERRY *Vaccinium oxycoccus* (Cranberry) Cumb. Grigson.

CRAYFERY *Pulmonaria officinalis* (Lungwort) B & H. This name appears in the Grete Herball.

CRAYFISH; CRAYFISH LEOPARD'S BANE *Doronicum pardalianches* (Great Leopard's Bane) Gerard (Crayfish); W.Miller (Crayfish Leopard's Bane). Gerard invented the name Crayfish, from the shape of the roots.

CRAZY 1. *Caltha palustris* (Marsh Marigold) Som, Glos, Wilts. Macmillan. 2. *Nuphar lutea* (Yellow Waterlily) Som. Macmillan. 3. *Ranunculus acris* (Meadow Buttercup) Dev, Som, Wilts, Hants, Glos, Berks, Bucks, War, Worcs, Lancs) Grigson. 4. *Ranunculus bulbosus* (Bulbous Buttercup) Wilts. Prior. Dev, Som, Hants, Glos, Berks, Bucks, War, Worcs, Lancs. Grigson. 5. *Ranunculus ficaria* (Lesser Celandine) Wilts, Glos. Grigson. 6. *Ranunculus repens* (Creeping Buttercup) Dev, Som, Wilts, Hants, Glos, Berks, Bucks, War, Worcs, Lancs. Grigson. Prior thought Crazy to be a corruption of Christ's Eye, and A S Palmer of crow's eye, but country people believed that the smell of buttercup flowers caused madness. Marsh Marigolds have the name Drunkards, too; "probably from the way in which they suck up water" (Dartnell & Goddard), but the children's reason is that if you look long at them you will be sure to take to drink.

CRAZY, Creeping *Ranunculus repens* (Creeping Buttercup) Glos. J.D.Robertson.

CRAZY, Yellow *Caltha palustris* (Marsh Marigold) Wilts. Grigson.

CRAZY BESS; CRAZY BETSY; CRAZY BETTY *Caltha palustris* (Marsh Marigold) Dor, Wilts. Grigson (Bess); Dor. Grigson; Wilts. Macmillan (Betsy); Dor. Dacombe; Wilts. Dartnell & Goddard (Betty).

CRAZY BETS 1. *Caltha palustris* (Marsh Marigold) Dor. Grigson; Wilts. Dartnell & Goddard. 2. *Leucanthemum vulgare* (Ox-eye Daisy) Wilts. Dartnell. & Goddard. 3. *Nuphar lutea* (Yellow Waterlily) Som. Macmillan. 4. *Ranunculus acris* (Meadow Buttercup) Wilts. Grigson. 5. *Ranunculus bulbosus* (Bulbous Buttercup) Wilts. Grigson. 6. *Ranunculus ficaria* (Lesser Celandine) Wilts. Macmillan. 7. *Ranunculus repens* (Creeping Buttercup) Wilts. Grigson.

CRAZY CUP *Ranunculus ficaria* (Lesser Celandine) Som. Macmillan.

CRAZY LILY *Caltha palustris* (Marsh Marigold) Dor. Macmillan.

CRAZY-MORE; CRAZY-MOIR; CRAZY-MAR *Ranunculus repens* (Creeping Buttercup) all Wilts. Dartnell & Goddard (Crazy-more); Jones & Dillon (Crazy-moir, Crazy-mar). More usually means root. See Crazy for the meaning.

CRAZY-PATES *Caltha palustris* (Marsh Marigold) Hants. Boase. This must have been Crazy Bets originally.

CRAZY WEED 1. *Ranunculus acris* (Meadow Buttercup) Bucks. Grigson. 2. *Ranunculus bulbosus* (Bulbous Buttercup) Bucks. Grigson. 3. *Ranunculus repens* (Creeping Buttercup) Berks. Lowsley. 4.

Oxytropis lambertii (White Loco) USA Cunningham. See Crazy for the meaning - it certainly isn't what Lowsley says - "so called because it spreads about so wildly", referring to Creeping Buttercup.

CREACHEY *Ranunculus bulbosus* (Bulbous Buttercup) Halliwell. A variation of Crazy, which is much commoner.

CREAKING SALLOW *Salix strepida*.

CREAM, Curds-and- *Saxifraga x urbinum* (London Pride) Som. Macmillan.

CREAM-AND-BUTTER *Ranunculus ficaria* (Lesser Celandine) Dev. Macmillan. Colour descriptive. Cf. other 'butter' names.

CREAM CUPS *Platystemon californicus*.

CREAM NUT *Bertolletia excelsa* (Brazil Nut) Schery.

CREAM-OF-TARTAR TREE 1. *Adansonia digitata* (Baobab) Palgrave. 2. *Adansonia gregori* (Australian Baobab) J.Smith. The mealy part of the seed vessel contains tartaric acid, and this fruit pulp has long provided Africans with a refreshing drink. It is not only tartaric acid the fruit contains, but also citric acid, and potassium acid tartrate.

CREAMS, Yellow 1. *Ranunculus acris* (Meadow Buttercup) Som. Grigson. 2. *Ranunculus bulbosus* (Bulbous Buttercup) Som. Grigson. 3. *Ranunculus repens* (Creeping Buttercup) Som. Grigson.

CREASE, Water *Nasturtium officinale* (Watercress) London A S Palmer.

CREASHAK *Arctostaphylos uva-ursi* (Bearberry) Ross Grigson. A corruption of the Gaelic name.

CREED *Lemna minor* (Duckweed) Wilts. Dartnell & Goddard. Earle listed an old name - Greeds - for this. Greens, and Grains, are recorded, the latter by Gerard.

CREEK DOGWOOD *Cornus californica*.

CREEK SENECIO *Senecio douglasii* (Douglas Groundsel) USA Elmore.

CREEP-IVY; CREEPER *Calystegia sepium* (Great Bindweed) Tynan & Maitland (Creep-Ivy); Notts. B & H (Creeper).

CREEPER, American *Tropaeolum peregrinum* (Canary Creeper) Friend.

CREEPER, Canary *Tropaeolum peregrinum*.

CREEPER, Honolulu *Antigonon leptopus* (Coral Vine) Simmons.

CREES, Yellow 1. *Ranunculus bulbosus* (Bulbous Buttercup) Bucks, Herts. Grigson. 2. *Ranunculus repens* (Creeping Buttercup) Bucks, Herts. Grigson. Cf. Buttercreese, from the same area.

CREESE *Nasturtium officinale* (Watercress) Som. Macmillan. Cress, of course.

CREEVEREEGH *Crithmum maritimum* (Samphire) Donegal Grigson. Obviously Gaelic in origin, but what does it mean?

CREGGANS *Ulex europaeus* (Furze) IOM Gill.1963. Creggan is a rocky piece of land, so this name must be descriptive of habitat.

CREIVEL *Primula veris* (Cowslip) Dor. Grigson. This is a word which is more widespread in Wessex as Crewel.

CREME-DE-MENTHE MINT *Mentha requeni* (Corsican Mint) USA Cunningham.

CREOSOTE BUSH *Larrea mexicana*. There is a very peculiar smell to this plant, hence the name.

CRESS *Veronica anagallis-aquatica* (Water Speedwell) Bianchini. Probably a confusion with *Nasturtium officinale*, watercress.

CRESS, American *Barbarea verna* (American Land Cress) USA B & H.

CRESS, Bank 1. *Barbarea vulgaris* (Winter Cress) IOW B & H. 2. *Sisymbrium officinale* (Hedge Mustard) Prior. Because they grow on hedge banks?

CRESS, Bastard *Thlaspi arvense* (Field Pennycress) USA Cunningham.

CRESS, Belle Isle *Barbarea verna* (American Land Cress) B & H. Belle Isle is between Labrador and Newfoundland.

CRESS, Brook *Cardamine pennsylvanica*.

CRESS, Brown *Nasturtium officinale* (Watercress) B & H. Not brown - this is German Brunnen, a well, spring, or any running water.

CRESS, Butter *Ranunculus acris* (Meadow Buttercup) Bucks. B & H.

CRESS, Chamois *Hornungia petraea* (Hutchinsia) Polunin.

CRESS, Churl's *Lepidium campestre* (Pepperwort) B & H.
CRESS, Cow 1. *Apium nodiflorum* (Fool's Watercress) Hants. Cope. 2. *Lepidium campestre* (Pepperwort) Prior. 3. *Veronica beccabunga* (Brooklime) Ess. Grigson.
CRESS, Dock *Lapsana communis* (Nipplewort) Dor. Macmillan. Not a cress, of course, but there is another cress name assigned to this - Swine's Cress.
CRESS, Field *Lepidium campestre* (Pepperwort) Hutchinson.
CRESS, Fool's *Sium nodiflorum* (Procumbent Water Parsnip) Hulme.
CRESS, French *Barbarea vulgaris* (Winter Cress) B & H.
CRESS, Garden; CRESS, Garth *Lepidium sativum*. Garth = garden; the name is in B & H.
CRESS, Hairy *Cardaria draba* (Hoary Cress) Chopra. Possibly a misreading of Hoary Cress.
CRESS, Hen *Capsella bursa-pastoris* (Shepherd's Purse) Dawson.
CRESS, Hoary *Cardaria draba*.
CRESS, Hollow *Gentianella campestris* (Field Gentian) Cockayne.
CRESS, Horse *Veronica beccabunga* (Brooklime) Yks. Grigson.
CRESS, Indian; CRESS, Indian, Greater *Tropaeolum majus* (Nasturtium) Fernie (Greater). Cress because the leaves have a similar biting flavour to cress. Indian Cress was the name given by Gerard, because it came from the West Indies. This has the specific name majus, but what is the Lesser?
CRESS, Jesuits' *Tropaeolum majus* (Nasturtium) Bianchini. Because it was the Jesuits who first brought it to Europe, or so it is said, as a result of their world-wide travels in search of souls to be saved.
CRESS, Lamb's *Cardamine hirsuta* (Hairy Bittercress) Dev. Friend. This name also occurs in the AS Herbal.
CRESS, Land 1. *Barbarea vulgaris* (Winter Cress) Dev, Som. Macmillan. IOW Grigson. 2. *Cardamine amara* (Large Bittercress) Ches. Holland. 3. *Cardamine hirsuta* (Hairy Bittercress) Hants. Cope; War, Ches. B & H. Suspiciously, Lamb's Cress is a very old name for Hairy Bittercress. There may have been confusion.
CRESS, Land, American *Barbarea verna*.
CRESS, Land, Bitter *Barbarea vulgaris* (Winter Cress) Tynan & Maitland.
CRESS, Land, Small-flowered *Barbarea stricta*.
CRESS, Leek *Alliaria petiolata* (Jack-by-the-hedge) Cockayne.
CRESS, Meadow *Cardamine pratensis* (Lady's Smock) Ellacombe.
CRESS, Mediterranean *Morisia sp.*
CRESS, Mitre *Myagrum perfoliatum*.
CRESS, Mouse-ear *Arabidopsis thaliana* (Thale Cress) USA Zenkert. Curtis calls this Podded Mouse-ear.
CRESS, Narrow-leaved *Lepidium ruderale* (Narrow-leaved Pepperwort) Murdoch McNeill.
CRESS, Pepper 1. *Cardaria draba* (Hoary Cress) Salisbury.1964. 2. *Lepidium latifolium* (Dittander) C P Johnson. 3. *Lepidium ruderale* (Narrow-leaved Pepperwort) Salisbury.1964. 4. *Lepidium sativum* (Garden Cress) Dev. Friend.1882.
CRESS, Pie *Apium nodiflorum* (Fool's Watercress) Dev. Macmillan. West Country people used to collect it at one time, and cook it with meat in pies and pasties.
CRESS, Pig's 1. *Anthemis cotula* (Maydweed) Som, Dor. Macmillan. 2. *Lapsana communis* (Nipplewort) Som. Macmillan. Maydweed is Pig's Daisy, and Pig's Flower, in the same area, too, probably for the same reason that it is known as Dog Daisy, or Dog Camomile, and Horse Daisy, elsewhere. Another name, Stinking Camomile, tells why. Nipplewort is also called Swine's Cress in Somerset. Of course, neither of these plants are cresses in fact.
CRESS, Rock *Arabis sp.*
CRESS, Rock, Alpine *Arabis alpina*.
CRESS, Rock, Bristol *Arabis stricta*. Very rare, but it grows in the Avon Gorge, at

Bristol.
CRESS, Rock, Fringed *Arabis hirsuta* (Hairy Rock Cress) Clapham.
CRESS, Rock, Hairy *Arabis hirsuta*.
CRESS, Rock, Lyre-leaved *Arabis lyrata*.
CRESS, Rock, Mountain *Cardaminopsis petraea* (Northern Rock-cress) J A MacCulloch.1905.
CRESS, Rock, Prince's *Arabis pulchra*.
CRESS, Rock, Tower *Arabis turritis*.
CRESS, Shepherd's *Teesdalia nudicaulis*.
CRESS, Shepherd's, Lesser *Teesdalia coronopifolia*.
CRESS, Smith's *Lepidium smithii* (Smith's Pepperwort) McClintock.
CRESS, Spring *Barbarea verna* (American Land Cress) Salisbury.1964.
CRESS, Swine *Coronopus squamatus* (Wart-cress) Perring. i.e. a cress fit only for pigs.
CRESS, Swine's 1. *Apium nodiflorum* (Fool's Watercress) Yks, Ork. Grigson. 2. *Coronopus squamatus* (Wart-cress) Tradescant. 3. *Lapsana communis* (Nipplewort) Som. Macmillan. 4. *Polygonum aviculare* (Knotgrass) Dawson. 5. *Senecio jacobaea* (Ragwort) B & H. In the general sense of plants unfit for humans, but there were special reasons for giving knotgrass the name. It is glossed Swine's Grease on a 12th century ms of Apuleius Barbarus (Gunther), and it is widespread in various forms. It is, for instance, herbe à cochons in French. Gerard said "it is given unto Swine with good successe, when they are sicke and will not eat their meat; whereupon country people do call it Swinesgrasse, or Swine's Skir".
CRESS, Swine's, Lesser 1. *Coronopus didymus* (Slender Wart-cress) McClintock. 2. *Coronopus squamatus* (Wart-cress) Salisbury.1964.
CRESS, Taliesin's *Sedum telephium* (Orpine) Wales Grigson. Welsh Cerwr Taliesin.
CRESS, Thale *Arabidopsis thaliana*. The Thale here is Johannes Thal (1542-1583), a German botanist who described the plant.
CRESS, Thanet *Cardaria draba* (Hoary Cress) Salisbury.1964. It is called Thanet-weed as well. It is reputed to have been introduced after an ill-fated continental expedition. The fever-stricken soldiers were brought back to Ramsgate on mattresses stuffed with hay, and this was disposed of to a Thanet farmer, who ploughed it in for manure. Then the cress appeared, presumably from seeds contained in the hay.
CRESS, Tooth *Dentaria bulbifera* (Coral-root) Prior. Other members of the genus are called Toothwort.
CRESS, Tooth, Large *Dentaria maxima* (Large Toothwort) USA House.
CRESS, Tower 1. *Arabis turritis* (Tower Rock-cress) Clapham. 2. *Turritis glabra* (Tower Mustard) USA Zenkert. Tower is an indication of the tall straight stem.
CRESS, Town *Lepidium sativum* (Garden Cress) Dawson.
CRESS, Upland 1. *Barbarea verna* (American Land Cress) A.W.Smith. 2. *Barbarea vulgaris* (Winter Cress) USA E.Hayes.
CRESS, Wall 1. *Aubrietia purpurea*. 2. *Diplotaxis muralis* (Stinkweed) Salisbury.1964.
CRESS, Wall, Common *Arabidopsis thaliana* (Thale Cress) Clapham.
CRESS, Wart *Coronopus squamatus*. So named from the seed vessels.
CRESS, Wart, Slender *Coronopus didymus*.
CRESS, Water 1. *Cardamine amara* (Large Bittercress) Cumb. B & H. 2. *Nasturtium officinale*. Watercress is the more usual spelling.
CRESS, Water, Rocket *Nasturtium officinale* (Watercress) Turner.
CRESS, Well, Horse *Veronica beccabunga* (Brooklime) Scot. Grigson. Another Scottish name is Horse Well-grass, while there is also the Yorkshire Horse-cress. Horse usually means a large or coarse variety.
CRESS, Wild 1. *Lepidium virginicum* (Virginia Peppergrass) Gooding. 2. *Thlaspi arvense* (Field Penny-cress) B & H.
CRESS, Winter *Barbarea vulgaris*.

CRESS, Winter, Early *Barbarea intermedia*.
CRESS, Yellow 1. *Barbarea vulgaris* (Winter Cress) Yks. Grigson. 2. *Rorippa sp*.
CRESS, Yellow, Austrian *Rorippa austriaca*.
CRESS, Yellow, Creeping *Rorippa sylvestris*.
CRESS, Yellow, Greater *Rorippa amphibia*.
CRESS, Yellow, Marsh *Rorippa palustris*.
CRESS-ROCKET *Rorippa sylvestris* (Creeping Yellow Cress) Camden/Gibson.
CRESSET; CRESSIL; CRESSEL *Scrophularia aquatica* (Water Betony) Wilts. Dartnell & Goddard (Cresset, Cressil); Wilts. Jefferies (Cressel). A cresset is an open lamp fixed on a pole, carried in night processions. The name may be applied descriptively.
CRESTMARINE *Crithmum maritimum* (Samphire) Lyte. This was its London street cry.
CRETAN BRAKE *Pteris cretica* (Table Fern) RHS.
CRETAN MAPLE *Acer sempervirens*.
CRETAN PALM *Phoenix theophrasti*.
CRETAN ROSE *Cistus ladaniferus* (Gum Cistus) Genders.1972.
CRETE, Dittany of *Origanum dictamnus* (Dittany) Gerard. Dittany of Candy is another name of the period. The plant is called Dittany from Mount Dicte, in Crete.
CREWEL *Primula veris* (Cowslip) Dev, Som, Wilts. Macmillan Dor. Dacombe. Creivel is another Dorset name, and the word becomes simply Cruel in Somerset. Halliwell, besides the meaning of a textile, simply mentions it as a Somerset name for cowslip.
CREX *Prunus domestica var.institia* (Bullace) Cumb. Grigson. Possibly related to the East Anglian name for the fruit - Crickseys.
CRIBBLES *Allium cepa* (Onion) Wilts. Macmillan. Cf. Chibboles, a name given to young onions. But both of these are more likely to refer to shallots.
CRICKET BAT WILLOW *Salix alba x caerulea* (x caerulea). Cricket bats have been made commercially since about 1820 from this wood, but they have always been cut from willows, and the game was invented in the early 14th century. It is the female tree only which is used.
CRICKS; CRICKSEY *Prunus domestica var. institia* fruit (Bullace) Ess. Gepp; Ess, Norf, Cambs. Grigson (Cricksey). Cf. the Cumbrian Crex.
CRILLY-GREENS *Brassica oleracea var. acephala* (Kale) Dev. Choape. 'Crilly' means 'curly', referring to the leaves, of course.
CRIMEAN IRIS, Dwarf *Iris chamaeiris*.
CRIMEAN LIME *Tilia x euchlora*.
CRIMEAN PINE *Pinus nigra var. caramanica*.
CRIMSONBERRY *Phytolacca decandra* (Poke-root) W.Miller.
CRIMSONS *Matthiola incana var. annua* (Ten-week Stock) Wilts. Macmillan.
CRINKLEROOT *Dentaria diphylla* (Two-leaved Toothwort) USA House.
CRINOLINES *Fuchsia sp* Som. Macmillan.
CRISLINGS *Prunus domestica var. institia* fruit (Bullace) Som. Elworthy. Cf. Cristens, etc.
CRISPED MINT *Mentha crispa* (Curly Mint).
CRISTALDRE *Centaureum erythraea* (Centaury) Halliwell. This is Christ's Ladder, and is probably a mis-spelling.
CRISTENS *Prunus domestica var. institia* fruit (Bullace) Dor. Barnes. Cf. the Somerset Christians, obviously the same word. Both probably have Crickseys as their original.
CROCKELTY-BUR *Arctium lappa* (Burdock) Cumb. B & H.
CROCKS, Kettles-and-; CROCKS-AND-KETTLES *Buxus sempervirens* seed Som. Macmillan. Box seed has got a number of names like this. Pots-and-kettles, and Chairs-and-tables, are examples.
CROCODILE 1. *Clematis vitalba* (Old Man's Beard) Kent B & H. 2. *Ilex aquifolium* (Holly) Dev. Friend; Som. Grigson.
CROCUS, Autumn 1. *Colchicum autumnale* (Meadow Saffron) Grigson.1959. 2. *Crocus*

nudiflorus.
CROCUS, Celandine *Crocus korolkowi.*
CROCUS, Chilean *Tecophilaea cyanocrocus.*
CROCUS, Cloth of Gold *Crocus angustifolius.*
CROCUS, Dutch *Crocus purpureus* (Spring Crocus) RHS.
CROCUS, Fog *Colchicum autumnale* (Meadow Saffron) Yks. Grigson. Because this is the Autumn Crocus.
CROCUS, Golden *Crocus flavus.*
CROCUS, Meadow *Colchicum autumnale* (Meadow Saffron) Yks. Grigson.
CROCUS, Michaelmas *Colchicum autumnale* (Meadow Saffron) Som. Macmillan; Wilts. Dartnell & Goddard. Because it is autumn-flowering, Michaelmas being celebrated on 29 September.
CROCUS, Purple *Colchicum autumnale* (Meadow Saffron) Yks. Grigson.
CROCUS, Sand *Romulea columnae.*
CROCUS, Spring *Crocus purpureus.*
CROCUS, Yellow *Zephyranthes citrina* Barbados Gooding.
CROCUS JAPONICA *Kerria japonica* (Jew's Mallow) Som. Macmillan. The botanical name used to be *Corchorus japonica*. This is a corruption of that old name.
CROCUS TULIP *Tulipa pulchella.*
CROKERS *Solanum tuberosum* (Potato) Ire. Salaman. As early as 1640 potato roots were referred to under this name; it is said that it was because they had been first planted in Croker's field at Youghal.
CROMWELL *Lithospermum officinale* (Gromwell) C P Johnson.
CRON-REISH *Solanum dulcamara* (Woody Nightshade) IOM Moore.1924. Manx croanreisht.
CRONE *Vaccinium oxycoccus* (Cranberry) Lancs, West. Grigson. One of a number of variants for cranberry - Cf. croneberry, craneberry, crane, etc.
CRONEBERRY 1. *Vaccinium myrtillus* (Whortleberry) B & H. 2. *Vaccinium oxycoccus* (Cranberry) Lancs, West. Grigson. Cf. Crone, etc.

CRONESHANKS *Polygonum persicaria* (Persicaria) Cockayne. i.e. Crane's Shanks. Various 'knee' and 'shanks' names are applied to other species of *Polygonum*. It is the jointed look of the stalks that suggest them.
CROOK-NECK SQUASH; CROOK-NECK SQUASH, Canada *Cucurbita moschata*; Bianchini (Canada). From the shape of the gourds.
CROP 1. *Fagopyrum esculentum* (Buckwheat) B & H. 2. *Lolium perenne* (Rye-grass) Suss. B & H. 3. *Lolium temulentum* (Darnel) Suss. Culpepper. This appears as Crap occasionally. OE crop meant the top of anything, in particular the top of a plant, the part in this case that yields the crop in the modern sense.
CROP, Pepper; CROP, Rock *Sedum acre* (Biting Stonecrop) Gerard (Pepper); Corn. Jago (Rock).
CROPWEED *Centaurea nigra* (Knapweed) B & H.
CROSS, Herb of the *Verbena officinalis* (Vervain) Sanecki. A name in keeping with the idea that vervain is the "Holy Herb".
CROSS, Jerusalem *Lychnis chalcedonica*. Another name, Maltese Cross, suggests that the shape of the flower head gives rise to the name.
CROSS, King's *Cheiranthus cheiri* (Wallflower) Som. Macmillan. Cf. Cross-flower, Soldier's Cross, etc.
CROSS, Knight; CROSS, Maltese *Lychnis chalcedonica* (Jerusalem Cross) both B & H. The distribution of the petals gives these names. Maltese shows that the Knight of the name must be one of St John.
CROSS, Malta *Tribulus terrestris* (Puncture Vine) Moldenke.
CROSS, Scarlet *Lychnis chalcedonica* (Jerusalem Cross) B & H.
CROSS, Soldier's *Cheiranthus cheiri* (Wallflower) Som. Macmillan. Cf. King's Cross, and Cross-flower, from the same county.
CROSS CLEAVERS *Galium circaezans*

USA Tolstead. Presumably its habit is like the native Cross-wort, for this is an American plant.

CROSS FLOWER 1. *Cheiranthus cheiri* (Wallflower) Som. Macmillan. 2. *Endymion nonscriptus* (Bluebell) Dev. B & H. 3. *Orchis mascula* (Early Purple Orchis) Dev. B & H. 4. *Polygala vulgaris* (Milkwort) Gerard. Wallflower is a member of the Cruciferae, but the others are more difficult to account for. Perhaps bluebell has a connection with Easter which could explain it. Anyway, Early Purple Orchis and Bluebell share a number of local names, most of them highly imaginative. Cross-flower for Milkwort must be connected with the group of names which include Rogation-flower, and Procession-flower - but of course, the ascription may be wrong, and it may be *Polygala cruciata* that should have been identified.

CROSS GENTIAN; CROSS-LEAVED GENTIAN *Gentiana cruciata*. Cross Gentian is the usual name; Cross-leaved Gentian appears in Grieve.1931.

CROSS-LEAVED BEDSTRAW *Galium boreale* (Northern Bedstraw) J A MacCulloch.1905.

CROSS-LEAVED MILKWORT *Polygala cruciata*.

CROSS VINE *Bignonia capreolata*.

CROSSWORT 1. *Eupatorium perfoliatum* (Thoroughwort) USA Henkel. 2. *Galium cruciatum*. 3. *Lysimachia quadrifolia* (Whorled Loosestrife) USA House. *Galium cruciatum* - "the leaves are little, short, and smal, alwaies forme growing together and standing crosse-wise one right against another, making a Burgondion crosse" (Gerard).

CROSSWORT, Golden *Galium cruciatum* (Crosswort) Parkinson.

CROSSWORT GENTIAN *Gentiana cruciata* (Cross Gentian) Parkinson.

CROSTIL *Parmelia omphalodes* (Black Crottle) Highlands Bolton.

CROTAL *Ochrolechia tartarea* (Cudbear Lichen) Scot. Bolton. The word is spelt Crottle, too, but then seems to be applied to several lichens.

CROTON OIL PLANT *Ricinus communis* (Castor Oil Plant) Singer.1927. But Croton oil, according to Chambers Dictionary, is a powerful purgative got from the seeds of *Croton tiglium*.

CROTTLE 1. *Lobaria pulmonaria* (Lungwort Crottle) Scot. Bolton. 2. *Ochrolechia tartarea* (Cudbear Lichen) Scot. Bolton. 3. *Parmelia saxatilis* (Grey Stone Crottle) Scot. Bolton.

CROTTLE, Black *Parmelia omphalodes*.

CROTTLE, Dark *Hypogymnia physodes*.

CROTTLE, Hazel *Lobaria pulmonaria* (Lungwort Crottle) S.M. Robertson.

CROTTLE, Light *Ochrolechia parella* (Crawfish Lichen) Bolton.

CROTTLE, Lungwort *Lobaria pulmonaria*.

CROTTLE, Stone *Parmelia caperata*.

CROTTLE, Stone, Gray *Parmelia saxatilis*.

CROUPANS *Empetrum nigrum* (Crowberry) Scot. B & H.

CROW BELLS 1. *Endymion nonscriptus* (Bluebell) Prior. 2. *Narcissus pseudo-narcissus* (Daffodil) B & H. 3. *Ranunculus bulbosus* (Bulbous Buttercup) War. Grigson.

CROW BELLS, Yellow *Narcissus pseudo-narcissus* (Daffodil) B & H. Yellow, to distinguish from the ordinary Crow Bells, by which bluebell must be meant.

CROW('S) CLAWS 1. *Ranunculus arvensis* (Corn Buttercup) Hants, Suss, Ess. Grigson. 2. *Ranunculus repens* (Creeping Buttercup) Hants. Cope; Suss. B & H. The various 'claw' names refer to the prickly fruits of the buttercups.

CROW CORN *Aletris farinosa* (Unicorn-root) Le Strange.

CROW CRANE *Caltha palustris* (Marsh Marigold) Oxf. Grigson.

CROW FLOWER 1. *Caltha palustris* (Marsh Marigold) Som. Grigson. 2. *Endymion nonscriptus* (Bluebell) Dev. Friend.1882; Wilts. Dartnell & Goddard. 3. *Geranium sylvaticum* (Wood Cranesbill) Stir. B & H. 4. *Lychnis flos-cuculi* (Ragged Robin) Gerard. 5. *Orchis mascula* (Early

Purple Orchis) Dev. Friend.1882. 6. *Ranunculus acris* (Meadow Buttercup) Staffs, Rut, Midl. Grigson.

CROW FLOWER, Musk *Adoxa moschatellina* (Moschatel) Tynan & Maitland.

CROW GARLIC *Allium vineale*. There is Crow Onion too. It probably just means wild garlic.

CROW KILLER *Anamirta cocculus* India Chopra. That is exactly what it is used for in India. Cf. Fish Killer, for it is a fish poison, too.

CROW LEEK 1. *Endymion nonscriptus* (Bluebell) Prior. 2. *Ruscus aculeatus* (Butcher's Broom) R.T.Gunther. There are a lot of 'crow' names for bluebells - Crowbells, Crowfoot, Crow's Legs, etc. The name applied to Butcher's Broom occurs only as a gloss on a 12th century ms of Apuleius.

CROW LEEK, Great *Allium ursinum* (Ramsons) Cockayne.

CROW LING 1. *Calluna vulgaris* (Heather) Yks. B & H. 2. *Empetrum nigrum* (Crowberry) Yks. Grigson. 3. *Erica cinerea* (Bell Heather) Yks. Atkinson. 4. *Erica tetralix* (Cross-leaved Heath) Yks. Grigson.

CROW NEEDLE *Scandix pecten-veneris* (Shepherd's needle) Som. Grigson; Dor. Macmillan; IOW. Long; N'thants. A.E.Baker.

CROW ONION *Allium vineale* (Ramsons) Worcs. Grigson; War. Bloom. Cf. Crow Garlic for this.

CROW PACKLE; CROW PIKEL *Ranunculus bulbosus* (Bulbous Buttercup) N'thants. A.E.Baker. Packle and Pikel are variants of pightle, an old name for a meadow, or any small enclosed piece of land. Crow Pightle is shared by this and Lesser Celandine as a name from the same area.

CROW PARSNIP *Taraxacum officinale* (Dandelion) B & H.

CROW PEAS 1. *Empetrum nigrum* (Crowberry) Scot. B & H. 2. *Vicia sepium* (Bush Vetch) Cumb. Grigson.

CROW PECK 1. *Capsella bursa-pastoris* (Shepherd's Purse) Wilts. Grigson. 2. *Orchis mascula* (Early Purple Orchid) Hants. Tynan & Maitland. 3. *Ranunculus arvensis* (Corn Buttercup) Wilts. Dartnell & Goddard. 4. *Scandix pecten-veneris* (Shepherd's Needle) Som. Grigson; Wilts. Dartnell & Goddard; Hants. Cope.

CROW PIGHTLE 1. *Ranunculus bulbosus* (Bulbous Buttercup) N'thants. A.E.Baker. 2. *Ranunculus ficaria* (Lesser Celandine) Beds. Grigson. Pightle is an old word for a meadow, or any small piece of enclosed land. Cf. Crow Packle and Crow Pikel for the buttercup.

CROW POISON *Amaianthemum muscaetoxicum* (Staggergrass) USA Kingsbury.

CROW POTATO *Lycopus asper* (Bugleweed) USA Yarnell.

CROW ROCKET *Eupatorium cannabinum* (Hemp Agrimony) Donegal Grigson.

CROW SOAP; CROWTHER SOAP *Saponaria officinalis* (Soapwort) Cockayne; Le Strange (Crowther).

CROW TOES 1. *Endymion nonscriptus* (Bluebell) Turner. 2. *Lotus corniculatus* (Bird's-foot Trefoil) Som, Scot. Prior. 3. *Orchis mascula* (Early Purple Orchis) Cumb. B & H. 4. *Ranunculus repens* (Creeping Buttercup) Greenoak.1979.

CROW'S FLOWER *Dactylorchis maculata* (Heath Spotted Orchid) Som. Macmillan.

CROW'S FOOT 1. *Ranunculus bulbosus* (Bulbous Buttercup) Suss. B & H. 2. *Stellaria holostea* (Greater Stitchwort) Som. Macmillan. 3. *Tussilago farfara* (Coltsfoot) Som. Macmillan.

CROW'S FOOT, Globe *Trollius europaeus* (Globe-flower) Gerard.

CROW'S FOOT, Yellow *Anthyllis vulneraria* (Kidney Vetch) Bucks. Grigson. This comes from the same set of descriptive names that includes Lady's Fingers, Fingers-and-thumbs, etc., and such as Cat's Claws.

CROW'S FOOT PLANTAIN *Plantago coronopus* (Buck's Horn Plantain) Flower.1859. Cf. Crowfoot and Crowfoot Waybread, two early names.

CROW'S LEGS *Endymion nonscriptus*

(Bluebell) Wilts. Dartnell & Goddard. We already have Crow-bells, Crow-leek, Crowfoot, and other crow names, as well as a connection with other birds, including rooks, cuckoos and pigeons. Presumably it is because it is spring-flowering.

CROW'S NEST *Daucus carota* (Wild Carrot) Beds. Grigson. Bird's Nest, or Bee's Nest, are better known.

CROWBERRY 1. *Empetrum nigrum*. 2. *Phytolacca decandra* (Poke-root) Watt. 3. *Vaccinium myrtillus* (Whortleberry) Mor. Grigson; Ire. O Suileabhain. 4. *Vaccinium vitis-idaea* (Cowberry) Ellacombe. In these cases, it must be the black berries which attract the crow epithet.

CROWCUP *Fritillaria meleagris* (Snake's Head Lily) Bucks. Grigson. From the same area, there is Frockup. Grigson suggests that this may be frog-cup, but there is a rather odd-looking name, Frorechap, in Coats.

CROWDY-KIT 1. *Scrophularia aquatica* (Water Betony) Dev, Som. Dyer. 2. *Scrophularia nodosa* (Figwort) Dev. Grigson. A crowdy is a fiddle, possibly from the Welsh crwth. Kit means a violin,too - a small one, apparently. The word is descended from OE cythere, cittern. Children strip the stems of their leaves, and scrape them across each other, when they produce a squeaking sound.

CROWDY-KIT-O'-THE-WALL *Sedum acre* (Biting Stonecrop) Dev. Friend.1882. He claims that the leaves squeak when rubbed together, hence the crowdy reference.

CROWFEED *Capsella bursa-pastoris* (Shepherd's Purse) Wilts. D.Grose.

CROWFOOT 1. *Dactylorchis maculata* (Heath Spotted Orchis) Yks. B & H. 2. *Dentaria laciniata* (Cut-leaved Toothwort) USA Yarnell. 3. *Endymion nonscriptus* (Bluebell) Rad, Lancs, Cumb. B & H. 4. *Geranium maculatum* (Spotted Cranesbill) O.P.Brown. 5. *Lotus corniculatus* (Bird's-foot Trefoil) Som, Glos, Suff. Grigson. 6. *Luzula campestris* (Good Friday Grass) Yks. B & H. 7. *Mimulus moschatus* (Musk) Som. Macmillan. 8. *Orchis mascula* (Early Purple Orchis) Lincs, Yks. B & H. 9. *Orchis morio* (Green-winged Orchid) Lincs,Yks,Cumb. Grigson. 10. *Plantago coronopus* (Buck's Horn Plantain) B & H. 11. *Ranunculus flammula* (Lesser Spearwort) N.Ire. Grigson. 12. *Senecio jacobaea* (Ragwort) Shrop. Grigson.

CROWFOOT, Bristly *Ranunculus pennsylvanicus*.

CROWFOOT, Bulbous *Ranunculus bulbosus* (Bulbous Buttercup).

CROWFOOT, Bur *Ranunculus repens* (Creeping Buttercup) Yks. Grigson. The seed vessels are prickly, hence Bur.

CROWFOOT, Buttercup *Ranunculus acris* (Meadow Buttercup) Phillips. A pleonasm - all buttercups are Crowfoots, and vice versa.

CROWFOOT, Corn *Ranunculus arvensis* (Corn Buttercup) Clapham.

CROWFOOT, Craw *Ranunculus bulbosus* (Bulbous Buttercup) B & H.

CROWFOOT, Creeping *Ranunculus repens* (Creeping Buttercup) Curtis.

CROWFOOT, Cursed *Ranunculus sceleratus* (Celery-leaved Buttercup) USA Tolstead. Cursed, probably because this is reckoned the most poisonous of the buttercups, due to larger amounts of the toxic elements being present.

CROWFOOT, Flame-leaved *Ranunculus flammula* (Lesser Spearwort) Flower.1859. 'Flammula', in other words.

CROWFOOT, Glacier *Ranunculus glacialis*.

CROWFOOT, Globe *Trollius europaeus* (Globe-flower) Gerard.

CROWFOOT, Golden-haired *Ranunculus auriconus* (Wood Goldilocks) Gerard.

CROWFOOT, Hair, Dark *Ranunculus trichophyllus* (Thread-leaved Crowfoot) Wit.

CROWFOOT, Hooked *Ranunculus recurvatus*.

CROWFOOT, Hurtful *Ranunculus sceleratus* (Celery-leaved Buttercup) Flower.1859. Cf. the American Cursed Crowfoot for this.

CROWFOOT, Ivy; CROWFOOT, Ivy-leaved *Ranunculus hederaceus* (Ivy-leaved

Water Buttercup) Murdoch McNeill (Ivy); McClintock (Ivy-leaved).

CROWFOOT, Knobbed *Ranunculus bulbosus* (Bulbous Buttercup) Parkinson.1649.

CROWFOOT, Marsh *Ranunculus sceleratus* (Celery-leaved Buttercup) Grieve.1931.

CROWFOOT, Meadow, Upright *Ranunculus acris* (Meadow Buttercup) Barton & Castle.

CROWFOOT, Mud *Ranunculus tripartitus*.

CROWFOOT, Onion-rooted; CROWFOOT, Round-rooted *Ranunculus bulbosus* (Bulbous Buttercup) Gerard (Onion); Curtis (Round).

CROWFOOT, Pale-leaved *Ranunculus sardous* (Hairy Buttercup) Curtis.

CROWFOOT, Rape *Ranunculus bulbosus* (Bulbous Buttercup) Lyte.

CROWFOOT, Rigid-leaved *Ranunculus circinatus*.

CROWFOOT, River *Ranunculus fluitans*.

CROWFOOT, Snake-tongue *Ranunculus ophioglossifolius* (Serpent's-tongue Spearwort) Clapham.

CROWFOOT, Spear 1. *Ranunculus flammula* (Lesser Spearwort) B & H. 2. *Ranunculus lingua* (Greater Spearwort) B & H.

CROWFOOT, Sweet *Ranunculus auricomus* (Wood Goldilocks) Flower.1859.

CROWFOOT, Tall *Ranunculus acris* (Meadow Buttercup). H C Long.1924.

CROWFOOT, Thread-leaved *Ranunculus trichophyllus*.

CROWFOOT, Tongue-leaved *Ranunculus lingua* (Greater Spearwort) Flower.1859.

CROWFOOT, Urchin *Ranunculus arvensis* (Corn Buttercup) B & H. The achenes are covered with spines, and this has given a number of names, like Devil's Claws and Prickleback (as well as Hedgehogs).

CROWFOOT, Various-leaved *Ranunculus heterophyllus*.

CROWFOOT, Water 1. *Ranunculus aquatilis*. 2. *Ranunculus fluitans* (River Crowfoot) Clapham.

CROWFOOT, Water, Brackish *Ranunculus baudotii*.

CROWFOOT, Water, Le Normand's *Ranunculus lenormandi*.

CROWFOOT, Water, Long-leaved *Ranunculus fluitans* (River Crowfoot) Clapham.

CROWFOOT, Water, Seaside *Ranunculus baudotii* (Brackish Water Crowfoot) McClintock.1975.

CROWFOOT, Water, Short-leaved *Ranunculus trichophyllus* (Thread-leaved Crowfoot) Clapham.

CROWFOOT, Water, Stiff-leaved *Ranunculus circinatus* (Rigid-leaved Crowfoot) Clapham.

CROWFOOT, Water, Three-lobed *Ranunculus tripartitus* (Mud Crowfoot) Clapham.

CROWFOOT, White *Ranunculus aquatilis* (Water Crowfoot) Shrop. Grigson.

CROWFOOT, White, Double *Ranunculus aconitifolius fl pl* Gerard.

CROWFOOT, Wood 1. *Anemone nemorosa* (Wood Anemone) Fernie. 2. *Ranunculus auricomus* (Wood Goldilocks) Curtis.

CROWFOOT CRANESBILL *Geranium pratense* (Meadow Cranesbill) Gerard.

CROWFOOT PLANTAIN; CROWFOOT WAYBREAD *Plantago coronopus* (Buck's Horn Plantain) USA Bianchini (Plantain); Turner (Waybread). Crowfoot on its own is also recorded for this.

CROWFOOT VIOLET *Viola pedata* (Bird's-foot Violet) Whittle & Cook. Because of the way the leaves are palmately divided, looking very like the marks of birds' feet in mud or snow. Cf. Cut-leaved Violet.

CROWN, Friar's *Cirsium eriophorum* (Woolly Thistle) B & H. The reference is to the bald patch left after the seeds have flown. Dandelion shows a similar effect, hence Priest's Crown and Monkshead.

CROWN, King's 1. *Justicia carnea*. 2. *Melilotus officinalis* (Melilot) B & H. 3. *Trifolium pratense* (Red Clover) Som. Macmillan. 4. *Viburnum opulus* (Guelder Rose) Glos. B & H. For Guelder Rose, King's Crown because, it is said, the King of the May was crowned with it; it has indeed got a number of May names. Of the other two, melilot seems to have better claim to the name. It was Corona Regis a long time ago, and it also bears the name King's Clover.

CROWN, Owl's *Filago germanica* (Cudweed) W.Miller.

CROWN, Priest's *Taraxacum officinale* (Dandelion) Turner. Both this name and Monkshead refer to the appearance of the disc after the seeds have flown - like a shaven head. Cf. French couronne de prêtre.

CROWN ANEMONE *Anemone coronaria* (Poppy Anemone) Polunin. The specific name is *coronaria*, hence "crown".

CROWN BARK *Cinchona officinalis* Chopra.

CROWN DAISY *Chrysanthemum coronarium* (Annual Chrysanthemum) Coats. Note the specific name.

CROWN IMPERIAL *Fritillaria imperialis*. Descriptive: "the floures grow at the top of the stalke, incompassing it round, in forme of an Imperiall Crowne" (Gerard).

CROWN OF THE FIELD *Agrostemma githago* (Corn Cockle) Som. Macmillan. Cf. French couronne des blés; this sounds like a direct translation.

CROWN OF THORNS 1. *Euphorbia milli* USA Kingsbury. 2. *Nigella damascena* (Love-in-a-mist) Dor. Macmillan. 3. *Passiflora coerulea* (Blue Passion Flower) Dor. Macmillan. Passion Flower, ecclesiastical symbol for Holy Rood, has attracted names connected with the symbolism. Good Friday, Christ-and-the-Apostles, and the Ten Commandments are examples.

CROWN VETCH *Coronilla varia*.

CROWS *Primula veris* (Cowslip) Som. Macmillan. Sounds as if this was originally Crewel, a common enough name in the West Country for cowslips.

CROWTOES 1. *Ranunculus acris* (Meadow Buttercup) Dev. Grigson. 2. *Ranunculus bulbosus* (Bulbous Buttercup) Dev. Grigson. 3. *Ranunculus repens* (Creeping Buttercup) Dev. Grigson.

CROYD *Medicago lupulina* (Black Medick) Ayr. B & H. B & H said the name was probably Black Medick; Jamieson merely called it a yellow clover.

CRUCIFIX-FLOWER *Cheiranthus cheiri* (Wallflower) Dev. Macmillan. Any member of the Cruciferae is qualified to have this name, but it is natural that a favourite cottage garden flower should be so called. Cf. Crossflower.

CRUEL *Primula veris* (Cowslip) Dev. Halliwell; Som. B & H. Better recognized as Crewel.

CRUEL PLANT *Araujia sericifera*.

CRUMMOCK *Sium sisarum* (Skirret) Scot. Jamieson. From the Gaelic name for the plant - crumag.

CRUMPLE LILY 1. *Lilium martagon* (Turk's Cap Lily) Dev. Friend. 2. *Lilium tigrinum* (Tiger Lily) Dev. Friend. Presumably from the appearance of the petals.

CRUSADER'S SPEAR *Scilla maritima* (Squill) RHS.

CRY-BABY 1. *Anagallis arvensis* (Scarlet Pimpernel) Som. Macmillan. 2. *Geranium robertianum* (Herb Robert) Dev, Som. Macmillan. 3. *Chamaenerion angustifolium* (Rosebay Willowherb) Som. Macmillan.

CRY-BABY CRAB 1. *Anagallis arvensis* (Scarlet Pimpernel) Som. Grigson. 2. *Geranium robertianum* (Herb Robert) Som. Macmillan.

CRYSTAL *Prunus domestica var.institia* fruit (Bullace) Corn, Dev. Grigson. Probably from the same source that produced Crex and Crickseys.

CUBAN LILY *Scilla peruviana* (Peruvian Lily) Whittle & Cook.

CUBAN MAHOGANY *Swietenia mahogoni*.

CUBEB *Piper cubeba*.

CUBEB, Guinea *Piper guineense* (West African Black Pepper) Dalziel.

CUBS, Fox-and- *Hieraceum aurantiacum* (Orange Hawkweed) Coats.

CUCKENWORT *Stellaria media* (Chickweed) Scot. B & H. One of a whole range of words which are the same as chickweed, which progresses via Chickenweed, through Chickenwort, to Chuckenwort and Cuckenwort, with various stops on the way.

CUCKLE; CUCKLE BUTTONS; CUCKLE DOCK *Arctium lappa* (Burdock) Dor, Som. B & H (Cuckle); Dev. Friend.1882 (Button); Corn. Jago (Dock). Members of a group which include names around cockle, cuckoo and cuckold.

CUCKLEMOORS *Arctium lappa* (Burdock) Dor. B & H. Cf. Cuckle.

CUCKOLD; CUCKOLD DOCK; CUCKOLD('S) BUTTONS *Arctium lappa* (Burdock) Corn, Dor, Glos, Yks. B & H; Som. Jennings (Cuckold); Corn. Jago; Dev. Choape; Som. Macmillan; USA Henkel (Dock); Dev. Choape (Buttons). Cuckold here is cuckle, or rather, cockle, OE coccel.

CUCKOLD'S CAP *Aconitum napellus* (Monkshood) Cambs, Ess, Norf. B & H. Cap, from the helmet-shaped flowers. But why cuckold? The caps or bonnets, etc., are ascribed to grandmothers, old women, policemen, and friars among others.

CUCKOO 1. *Ajuga reptans* (Bugle) Dor. Macmillan. 2. *Anemone nemorosa* (Wood Anemone) Wilts. Dartnell & Goddard. 3. *Campanula rotundifolia* (Harebell) Dev. Hardy. 4. *Cardamine pratensis* (Lady's Smock) Dev, Som. Glos. Grigson. 5. *Endymion nonscriptus* (Bluebell) Corn. Courtney.1880. Dev. Grigson. 6. *Lychnis flos-cuculi* (Ragged Robin) Dev. Grigson. 7. *Melandrium rubrum* (Red Campion) Dev, Som, Notts. Grigson. 8. *Orchis mascula* (Early Purple Orchis) Suss, Bucks, Ess, Cambs, Herts, Norf. B & H; N'thants. Clare. 9. *Orchis morio* (Green-winged Orchis) Ess. B & H. 10. *Pinus sylvestris* cones (Pine cones) Mon. Waters; Ess, Yks. B & H. 11. *Primula veris* (Cowslip) Corn. Grigson. For the most part, these are spring flowers, and the significance of the name merely lies in that fact.

CUCKOO, Dry; CUCKOO, Dryland *Saxifraga granulata* (Meadow Saxifrage) both Wilts. both Dartnell & Goddard. To distinguish from the Wet Cuckoo, which is Lady's Smock.

CUCKOO, Water; CUCKOO, Wet *Cardamine pratensis* (Lady's Smock) both Wilts. both Dartnell & Goddard. To set against Dry Cuckoo, which is Meadow Saxifrage.

CUCKOO BABIES *Arum maculatum* (Cuckoo-pint) IOW Grigson. In the same area, it is called Cocky-baby, too.

CUCKOO BOOTS *Endymion nonscriptus* (Bluebell) Dor. Macmillan.

CUCKOO('S)-BREAD 1. *Cardamine pratensis* (Lady's Smock) Dev, Som. Grigson. 2. *Oxalis acetosella* (Wood Sorrel) Som. Macmillan; Dor. Barnes. 3. *Plantago major* (Great Plantain) Le Strange.

CUCKOO BREAD-AND-CHEESE *Oxalis acetosella* (Wood Sorrel) Cumb. Grigson.

CUCKOO-BUDS 1. *Cardamine pratensis* (Lady's Smock) W.Miller. 2. *Orchis mascula* (Early Purple Orchis) N'thants, Clare. 3. *Ranunculus acris* (Meadow Buttercup) Som, N'thants, Worcs. Grigson. 4. *Ranunculus bulbosus* (Bulbous Buttercup) Suss, N'thants. B & H; Som, N'thants, Worcs Grigson. 5. *Ranunculus repens* (Creeping Buttercup) Som, N'thants, Worcs. Grigson. A simple spring-time reference, perhaps, or is 'cuckoo' cuckold, rather? In the Aube district of France, buttercups used to be put over the door of a cuckold.

CUCKOO-BUTTONS 1. *Arctium lappa* burrs (Burdock) Som. Elworthy. 2. *Arctium minus* (Lesser Burdock) Dev. B & H. 3. *Cirsium vulgare* (Spear Plume Thistle) Som. Elworthy.

CUCKOO-CHEESE *Oxalis acetosella* (Wood Sorrel) Dev. Grigson.

CUCKOO-COCK 1. *Arum maculatum* (Cuckoo-pint) Ess. Cockayne. 2. *Orchis mascula* (Early Purple Orchis) Ess. Grigson. With the same meaning as Cuckoo-pint.

CUCKOO-FLOWER 1. *Anemone nemorosa* (Wood Anemone) Som. Friend; Wilts.

Dartnell & Goddard; Ches. Holland. 2. *Arum maculatum* (Cuckoo-pint) N'thants. Grigson. 3. *Caltha palustris* (Marsh Marigold) Staffs. Tynan & Maitland. 4. *Cardamine pratensis*. 5. *Dactylorchis incarnata* (Early Marsh Orchid) IOW. Grigson. 6. *Endymion nonscriptus* (Bluebell) Corn, Dev. Grigson. Som. Macmillan. 7. *Geranium dissectum* (Cut-leaved Cranesbill) War. Bloom. 8. *Geranium molle* (Dove's-foot Cranesbill) War. Bloom. 9. *Hottonia palustris* (Water Violet) Som. Macmillan. 10. *Lychnis flos-cuculi* (Ragged Robin) Dev. Friend.1882; Som. Grigson; Berks. Lowsley; Suff. Moor; USA House. 11. *Melandrium album* (White Campion) Som. Macmillan. 12. *Melandrium rubrum* (Red Campion) Dev. Friend.1882; Som, Leics. Grigson; N'thants. A.E.Baker. 13. *Orchis mascula* (Early Purple Orchis) Dev. Choape; Hants. Cope; E.Ang. Forby. 14. *Orchis morio* (Green-winged Orchid) Dev, Ess. B & H. E Ang. Forby. 15. *Oxalis acetosella* (Wood Sorrel) Wilts. Dartnell & Goddard.1899; Bucks, Kent, Notts, Yks. Grigson. 16. *Saxifraga granulata* (Meadow Saxifrage) Yks. Grigson. 17. *Stellaria holostea* (Greater Stitchwort) IOW Grigson. Because they are in bloom about the time the cuckoo arrives?

CUCKOO GILLIFLOWER *Lychnis flos-cuculi* (Ragged Robin) Lyte.

CUCKOO-GRASS *Luzula campestris* (Good Friday Grass) Scot. B & H.

CUCKOO-HOOD *Centaurea cyanus* (Cornflower) Scot. Grigson.

CUCKOO LILY *Arum maculatum* (Cuckoo-pint) Cambs. PLNN 6. This seems to be a fusion of two traditions. Cuckoo names for this plant are legion, but it is also known as some kind of lily over a wide area.

CUCKOO-MEAT 1. *Oxalis acetosella* (Wood Sorrel) Field. 2. *Rumex acetosa* (Wild Sorrel) Ches. B & H.

CUCKOO ORCHIS *Orchis mascula* (Early Purple Orchis) B & H. Cf. Cuckoo, Cuckooflower, Cuckoo-pint, Cuckoo's-Shoes-and-stockings and Gowk's Meat for this.

CUCKOO-PINT 1. *Arum maculatum*. 2. *Caltha palustris* (Marsh Marigold) Oxf. Oxf.Annl.Rec.1951. 3. *Cardamine pratensis* (Lady's Smock) Wilts, Suss, Leics. Grigson. 4. *Melandrium rubrum* (Red Campion) N'thants. Grigson. 5. *Orchis mascula* (Early Purple Orchis) Bucks. B & H. Cuckoo-pint is a shortened form of Cuckoo Pintle; OE pintel - penis), from the shape of the spadix. It appears in other forms of the name, e.g. Priest's Pintle, Pintelwort, etc.

CUCKOO-PINT, Large *Arum neglectum*.

CUCKOO-PINTLE 1. *Arum maculatum* (Cuckoo-pint) Turner. 2. *Cardamine pratensis* (Lady's Smock) Wilts, Suss, Leics. Grigson. The more accurate form of Cuckoo-pint.

CUCKOO-POINT *Arum maculatum* (Cuckoo-pint) Yks A S Palmer.

CUCKOO POTATOES *Conopodium majus* (Earthnut) Ire. Grigson. The tubers are edible, and best roasted like chestnuts. Another Irish name is Fairy Potatoes.

CUCKOO ROSE *Narcissus pseudo-narcissus* (Daffodil) Som. Elworthy. Like other 'cuckoo' names, this is to stress that it blooms in early spring (when the cuckoo arrives).

CUCKOO-SORROW *Rumex acetosa* (Wild Sorrel) Fernie. Sorrow = sorrel. It also has the name Cuckoo-meat.

CUCKOO-SPICE 1. *Cardamine pratensis* (Lady's Smock) Yks. Grigson. 2. *Oxalis acetosella* (Wood Sorrel) Swainson.1886.

CUCKOO-SPIT 1. *Anemone nemerosa* (Wood Anemone) Hants. Hants FWI; Worcs. Grigson. 2. *Arum maculatum* (Cuckoo-pint) B & H. 3. *Cardamine pratensis* (Lady's Smock) Dyer. 4. *Melandrium rubrum* (Red Campion) N'thants. A.E.Baker. Cuckoo-spit is the froth enveloping a pale green insect, and the name is sometimes transferred to the plants themselves.

CUCKOO'S BEADS *Crataegus monogyna* (Hawthorn) Shrop. Grigson.

CUCKOO'S BOOTS *Endymion nonscriptus* (Bluebell) Som. Macmillan.

CUCKOO'S BREAD; CUCKOO'S BREAD-AND-CHEESE *Oxalis acetosella* (Wood Sorrel) Prior (Bread); Som, Glos, Worcs, Shrop, Lancs, Cumb. Grigson; IOM Moore.1924 (Bread-and-cheese).

CUCKOO'S BREAD-AND-CHEESE TREE *Crataegus monogyna* (Hawthorn) Wilts. Dartnell & Goddard; Suss. Parish; Leics. Grigson. The young leaves and shoots are regularly eaten by children in the spring, and called Bread-and-cheese.

CUCKOO'S CAP 1. *Aconitum anglicum* Shrop. B & H. 2. *Aconitum napellus* (Monkshood) Ches. Holland. Just one of the many 'hood' and 'cap' names applied to monkshood.

CUCKOO'S CLOVER *Oxalis acetosella* (Wood Sorrel) N.Ire. Grigson. Wood Sorrel is well connected with the cuckoo, both here and on the Continent.

CUCKOO'S EYE *Geranium robertianum* (Herb Robert) Kent, Bucks. Grigson. The 'eye' part of the name appears many times - Bird's Eye, Robin's Eye, Cat's Eye, etc., all descriptive.

CUCKOO'S MEAT 1. *Geranium robertianum* (Herb Robert) Bucks. Grigson. 2. *Oxalis acetosella* (Wood Sorrel) Turner. 3. *Stellaria holostea* (Greater Stitchwort) Bucks. Grigson.

CUCKOO'S SHOES *Viola riviniana* (Dog Violet) Shrop. Grigson.

CUCKOO'S SHOES-AND-STOCKINGS 1. *Cardamine pratensis* (Lady's Smock) S.Wales Grigson. 2. *Orchis mascula* (Early Purple Orchis) Som. Macmillan.

CUCKOO'S SORREL; CUCKOO'S SOUROCKS *Oxalis acetosella* (Wood Sorrel) Friend (Sorrel); Border Hardy (Sourocks).

CUCKOO'S STOCKINGS 1. *Campanula rotundifolia* (Harebell) Hardy. 2. *Endymion nonscriptus* (Bluebell) Derb, Notts, Staffs. Grigson. 3. *Lotus corniculatus* (Bird's-foot Trefoil) Som, Suss, Shrop Grigson. 4. *Viola riviniana* (Dog Violet) Caith. Grigson.

CUCKOO'S VICTUALS 1. *Geranium robertianum* (Herb Robert) Bucks. Grigson. 2. *Oxalis acetosella* (Wood Sorrel) Dor, Yks. Grigson; Glos. J.D.Robertson. 3. *Stellaria holostea* (Greater Stitchwort) Bucks. Grigson.

CUCKOOSOUR *Oxalis acetosella* (Wood Sorrel) Cockayne. Sorrel comes from a word which means sour.

CUCKOW PINT *Arum maculatum* (Cuckoo-pint) USA Weiner. Is cuckow the American spelling?

CUCUMBER 1. *Cucumis sativus*. 2. *Iris pseudo-acarus* seed vessels (Yellow Iris) Dev. B & H. Cucumber seems to have been originally cucumer, superseded in the 15th century by cucumber. It must originally have come from Latin cucuma, which means a kettle - gourds have always been used for heating water.

CUCUMBER, African *Momordica charantia* (Wild Balsam-apple) Dalziel.

CUCUMBER, Bitter *Colocynthis vulgaris* (Bitter Gourd) Polunin.

CUCUMBER, Calabash *Lagenaria siceraria* (Calabash) Willis.

CUCUMBER, Globe *Cucumis prophetarium*.

CUCUMBER, Hairy *Cucumis chate*.

CUCUMBER, Horned *Cucumis metuliferus*.

CUCUMBER, Jerusalem *Cucumis anguria* (Gherkin) USA Candolle. Why is a West Indian plant given the epithet Jerusalem? Probably because it is simply a corruption of gherkin.

CUCUMBER, Lemon-scented *Cucumis melo var. dudaim* (Dudaim Melon) Howes.

CUCUMBER, Prickly *Cucumis metuliferus* (Horned Cucumber) Howes.

CUCUMBER, Serpent *Trichomanthes anguina* (Snake Gourd) Whittle & Cook. 'Anguina' in the specific name means snake-like.

CUCUMBER, Sikkim *Cucumis sativus var. sikkimensis*.

CUCUMBER, Snake *Cucumis melo var. flexuosus* (Serpent Melon) Grieve.1931.

CUCUMBER, Spiked *Citrullus caffer*.

CUCUMBER, Squirting *Ecballium elateri-*

um. Squirting, because when the fruit is fully ripe, it bursts open on being touched, so that the seeds, and a milky juice, are squirted out with some considerable force.

CUCUMBER, Trigona *Cucumis trigonus*.

CUCUMBER, Wild 1. *Bryonia dioica* (White Bryony) War. Grigson; Lincs. Rudkin. 2. *Ecballium elaterium* (Squirting Cucumber) Thornton. 3. *Cucumis anguria* (Gherkin) Barbados Gooding. 4. *Cucumis melo* (Melon) Watt. 5. *Momordica charantia* (Wild Balsam-apple) Sofowora. The inclusion of bryony in this list may seem strange, but it is a member of the *Cucurbitaceae*, as are all the rest.

CUCUMBER MAGNOLIA *Magnolia acuminata* (Cucumber Tree) Brimble.1948.

CUCUMBER TREE 1. *Averrhoa bilimbi* (Bilimbi) Douglas. 2. *Kigelia africana* (German Sausage Tree) Palgrave. 3. *Magnolia acuminata*. 4. *Magnolia tripetala* USA Harper. It is the shape and colour of the fruits which give the name.

CUCUMBER TREE, Japanese *Magnolia obovata*.

CUCUMER, Leaping *Ecballium elatorium* (Squirting Cucumber) Turner.

CUCUMMER *Cucumis sativus* (Cucumber) Turner. Closer to the original than the modern cucumber. The intrusive 'b' appeared in the 15th century.

CUCUZZI *Lagenaria siceraria* (Calabash) A.W.Smith.

CUDBEAR LICHEN *Ochrolechia tartarea*. Cudbear is Cuthbert. It seems that it was named by Cuthbert Gordon, the patentee of the process of preparing the dye from various lichens.

CUDDIE'S LUNGS; CUDDY'S LUGS *Verbascum thapsus* (Mullein) Fernie (Cuddie's Lungs); N'thum. Grigson (Cuddy's Lugs). Cuddy is a donkey. But which of these is right? Are the leaves its ears or its lungs? The texture of the leaves is the reference for the lungs (we have Bullock's Lungwort, Clown's Lungwort, etc). And the shape as well contributes to the idea of ears (Cf. Donkey's Ears, Cow's Ears, etc).

CUDDLE-ME; CUDDLE-ME-TO-YOU *Viola tricolor* (Pansy) Som. Macmillan; Grigson (longer version). Just two of a large number of pansy names, starting with Kiss-me-at-the-garden-gate.

CUDWEED 1. *Artemisia purshiana* Webster. 2. *Filago germanica*. 3. *Ochrolechia tartarea* (Cudbear Lichen) Scot. B & H. Leonard Mascall, *Government of Oxen*, 1587, says cudwort was given to cattle which had lost their cud. The Artemisia is an American species, but here too it seems the name was given from the idea that cattle browsed on it. As far as the lichen is concerned, Cudweed is evidently a mistaken extension of Cudbear.

CUDWEED, Alpine *Leontopodium alpinum* (Edelweiss) Folkard.

CUDWEED, Broad-leaved *Filago spathulata*.

CUDWEED, Cape *Filago undulata* or *Gnaphalium undulatum*.

CUDWEED, Childing *Filago germanica* (Cudweed) A S Palmer. An indication that this plant must propagate itself by the "hen-and-chickens" method.

CUDWEED, Dwarf *Gnaphalium supinum*.

CUDWEED, Heath *Gnaphalium sylvaticum*.

CUDWEED, Highland 1. *Gnaphalium norvegicum*. 2. *Gnaphalium sylvaticum* (Heath Cudweed) Flower.1859.

CUDWEED, Jersey *Gnaphalium luteo-album*.

CUDWEED, Least *Filago minima* (Small Cudweed) Flower.1859.

CUDWEED, Lion's *Leontopodium alpinum* (Edelweiss) Gerard.

CUDWEED, Lobed *Artemisia ludoviciana*.

CUDWEED, Marsh *Gnaphalium uliginosum*.

CUDWEED, Narrow *Filago gallica*.

CUDWEED, Red-tipped *Filago apiculata*.

CUDWEED, Slender *Filago minima* (Small Cudweed) Clapham.

CUDWEED, Small *Filago minima*.

CUDWEED, Spring *Antennaria plantaginifolia* (Plantain-leaved Everlasting) Grieve.1931.

CUDWEED, Wayside *Gnaphalium uliginosum* (Marsh Cudweed) McClintock.
CUDWEED, Wood *Gnaphalium sylvaticum* (Heath Cudweed) Clapham.
CUDWORT *Filago germanica* (Cudweed) Turner.
CUES, Lady *Primula veris* (Cowslip) Kent T Lloyd. 'Cues' is evidently a variation of 'keys', a common description of cowslips.
CUIRN *Sorbus aucuparia* (Rowan) IOM Moore. There is a range of Celtic names which seem to stem from Old Cornish caer, meaning a berry. In the Isle of Man, Cuirn is sometimes written Keirn.
CULL-ME-TO-YOU *Viola tricolor* (Pansy) B & H. Better known as Cuddle-me-to-you; even Call-me-to-you is known from Somerset. Are these anything to do with a mistaken belief in the aphrodisiac properties of pansies?
CULLAVINE; CULLENBEAM *Aquilegia vulgaris* (Columbine) Ches. B & H (Cullavine); Som, Wilts. Macmillan (Cullenbeam). Simple corruptions of columbine - there are a number of them, like Colourbine and Currambine.
CULLINS, Sweet *Spiranthes spiralis* (Lady's Tresses) *Notes. and Queries*. vol 7. That is how it is spelt in the reference, but 'cullions' is obviously intended.
CULLIONS Orchis sp B & H and *Dactylorchis incarnata* (Early Marsh Orchid) Donegal Grigson. A general name in England for the genus. Halliwell said a cullion meant a stupid fellow, but the word means testicles.
CULLIONS, Goat's *Himantoglossum hircinum* (Lizard Orchis) B & H. Cf. Goat Stones for this.
CULLIONS, Soldier's *Orchis militaris* (Soldier Orchis) Gerard.
CULRAGE; CULRACHE; CULERAGE; CURAGE *Polygonum hydropiper* (Water Pepper) B & H (Culrage, Culerage, Curage); Halliwell (Culrache). Culrage is from the French, and means the same as the common name, Arsmart (from the irritant effect of the leaves).
CULVER'S ROOT *Veronica virginica*. Culver means a pigeon.
CULVERFOOT *Geranium columbinum* (Long-stalked Cranesbill) Prior. Surely Dove's-foot Cranesbill is better suited for this name, in spite of the *columbinum* of the generic name.
CULVERKEYS 1. *Aquilegia vulgaris* (Columbine) Halliwell. 2. *Endymion nonscriptus* (Bluebell) Som. B & H. 3. *Fraxinus excelsior* seed (Ash keys) Kent B & H. 4. *Orchis mascula* (Early Purple Orchis) B & H. 5. *Primula veris* (Cowslip) Som, Kent, N'thants. Grigson. 6. *Vicia sylvatica* (Wood Vetch) Som, Kent, N'thants. Grigson. Culver is OE culfre - pigeon. Columbine comes from the Latin columba, meaning pigeon, and it has been suggested that it gets the name because the nectaries are like pigeon's heads in a ring round a dish, representing, in Christian symbolism, the seven gifts of the Holy Spirit. It doesn't seem to matter that columbine has five little doves, not seven. Bluebell has a lot of bird names, crows, rooks, and cuckoos mainly. Why? The keys part of the name is apt for Cowslip, for this is St Peter's Keys, and of course the seed vessels of the ash are always known as keys.
CULVERS *Endymion nonscriptus* (Bluebell) Ess, Norf. B & H. See Culverkeys.
CULVERWORT *Aquilegia vulgaris* (Columbine) McDonald. See Culverkeys.
CUMBERFIELD; CUMBERLAND *Polygonum aviculare* (Knotgrass). B & H (Cumberfield); Tynan & Maitland (Cumberland). This is from Bullein's Book of Simples, and must mean encumber-field, i.e. providing an encumbrance to good husbandry.
CUMBERLAND HAWTHORN *Sorbus aria* (Whitebeam) Gerard.
CUMFIRIE *Bellis perennis* (Daisy) B & H. i.e. comfrey, which also occurs, as does Consound, because it was used for much the same purpose as comfrey. "The Northern men call this herbe a Banwort because it helpeth

bones to knyt againe" (Turner).

CUMFURT *Symphytum officinale* (Comfrey) IOM Moore.1924.

CUMIN *Cuminum cyminum*. OE cymen, ultimately from Greek kuminon, which itself is probably from an Arabic source. ODEE points out that OE cymen was superseded by the cumin form. Cymen would have yielded kimmen.

CUMIN, Black *Nigella sativa* (Fennelflower) Wit. The very aromatic seeds caused various names used for other aromatics, like coriander and caraway, to be applied to Fennel-flower.

CUMIN ROYAL; CUMIN, Ethiopian *Ammi visnaga* (Bishop's Weed) Coles (Cumin Royal); Pomet (Ethiopian Cumin).

CUMQUAT-TREE *Fortunella japonica* or *margarita*. This is Cantonese kam kwat, which means gold orange.

CUNJEVOI *Alocasia brisbanensis*. An Australian species, with a name taken from an Aboriginal tongue.

CUNTBLOWS *Chamaemelon nobilis* (Camomile) E.Ang. Halliwell.

CUP 1. *Calystegia sepium* (Great Bindweed) Som. Macmillan. 2. *Ranunculus ficaria* (Lesser Celandine) Glos. Grigson.

CUP, Blister *Ranunculus acris* (Meadow Buttercup) Lincs. Grigson. A reference to the very acrid juice, which can indeed raise blisters.

CUP, Blood *Asperula odorata* (Woodruff) Dor. Macmillan.

CUP, Bull *Caltha palustris* (Marsh Marigold) Som. Macmillan. Bull-cup simply matches Kingcup. There is also, from Essex, Bull Buttercup, which just means "big buttercup".

CUP, Carlock *Caltha palustris* (Marsh Marigold) Som. B & H. Cf. Carlicup.

CUP, Crazy *Ranunculus ficaria* (Lesser Celandine) Som. Macmillan. Crazy, Crazy Bet and Crazy Cup were all West Country names for the lesser celandine, and Crazy appears in other buttercup names. A corruption of Christ's Eye, Prior thought, but country people have long thought that close association with buttercups could drive you mad.

CUP, Cream *Platystemon californicus*. Creamy-yellow flowers.

CUP, Custard *Epilobium hirsutum* (Great Willowherb) Shrop, Ches. Grigson. A name in line with the very common Apple-pie, and its many equivalents.

CUP, Dale *Caltha palustris* (Marsh Marigold) Som. Macmillan. Cf. the Dorset Dillcup for this plant.

CUP, Delty *Ranunculus repens* (Creeping Buttercup) Wilts. Dartnell & Goddard. Probably the same as Gilty-cup, from the same area (though not, it seems, for this particular buttercup).

CUP, Drinking, Golden *Ranunculus ficaria* (Lesser Celandine) Som. Macmillan.

CUP, Drinking, Soldier's *Sarracenia purpurea* (Pitcher Plant) Perry. There are a lot of other 'cup' names for this.

CUP, Eve's; CUP, Forefather's *Sarracenia purpurea* (Pitcher Plant) Grieve.1931 (Eve); Perry (Forefather).

CUP, Fairy's 1. *Campanula rotundifolia* (Harebell) Dor, Som. Grigson. 2. *Primula veris* (Cowslip) Som. Macmillan; Kent Friend.

CUP, Ganymede's *Narcissus triandrus* (Angel's Tears) Ire. Coats. Ganymede was the cup-bearer of the gods, and the connection with the tears of Angel's Tears lies in the fact that in primitive Greek religion, Ganymede seems to have been the deity responsible for sprinkling the earth with heaven's rain.

CUP, Gil *Ranunculus acris* (Meadow Buttercup) Dev. Friend. Probably better as one word. This is the same as Gilty-cup = gold cup.

CUP, Gilt *Ranunculus ficaria* (Lesser Celandine) Suss. Parish. Gilcup is a commoner form of this.

CUP, Gilted 1. *Ranunculus acris* (Meadow Buttercup) Som. Grigson. 2. *Ranunculus bulbosus* (Bulbous Buttercup) Som. Grigson. 3. *Ranunculus repens* (Creeping Buttercup) Som. Grigson.

CUP, Gilty 1. *Caltha palustris* (Marsh Marigold) Wilts. Dartnell & Goddard. 2. *Ranunculus acris* (Meadow Buttercup) Dev. Friend.1882; Dor, Som. Grigson. 3. *Ranunculus bulbosus* (Bulbous Buttercup) Dev, Dor, Som. Grigson. 4. *Ranunculus ficaria* (Lesser Celandine) Som. Elworthy. Dor. Dacombe. i.e. golden cup.

CUP, Go 1. *Caltha palustris* (Marsh Marigold) Som. Macmillan. 2. *Ranunculus acris* (Meadow Buttercup) Dev. Friend.1882; Som. Macmillan. 3. *Ranunculus ficaria* (Lesser Celandine) Som. Macmillan. The West country version of Gold-cup.

CUP, Gold 1. *Ranunculus acris* (Meadow Buttercup) Gerard. 2. *Ranunculus bulbosus* (Bulbous Buttercup) Grieve.1931. 3. *Ranunculus repens* (Creeping Buttercup) Corn, Dev, Som, Wilts, Hants, Kent (Grigson); Suss. Parish. 4. *Solandra maxima* (Chalice Vine) Perry.

CUP, Golden 1. *Caltha palustris* (Marsh Marigold) Som. Elworthy. 2. *Ranunculus acris* (Meadow Buttercup) Dev. B & H. 3. *Ranunculus ficaria* (Lesser Celandine) Dev, Som. Grigson.

CUP, Golding *Ranunculus ficaria* (Lesser Celandine) Dor. B & H.

CUP, Gulty; CUP, Guilty *Ranunculus acris* (Meadow Buttercup) both Dev. B & H (Gulty); Halliwell (Guilty). Gulty = gilty = golden.

CUP, Huntsman's *Sarracenia purpurea* (Pitcher Plant) Grieve.1931. Possibly a reference to the common folk tale of the fairy cup provided by them to refresh huntsmen.

CUP, Indian *Sarracenia purpurea* (Pitcher Plant) Perry.

CUP, Lent; CUP, Lenty *Narcissus pseudonarcissus* (Daffodil) Dev, Som. Macmillan. Cf. Lent Pitcher for this, from the same area. These names are in line with the better known versions like Lent Lily, or Lide Lily.

CUP, Monkey; CUP, Trumpet *Mimulus guttatus* (Monkey Flower) Dor. Grigson (Monkey); Som. Macmillan (Trumpet).

CUP, Painted 1. *Castilleia linariaeflora* (Long-leaved Paintbrush) USA. Whiting. 2. *Pedicularis sylvatica* (Lousewort) Tynan & Maitland.

CUP, Poison *Ranunculus sceleratus* (Celery-leaved Buttercup) Tynan & Maitland.

CUP, Snow-, Water *Ranunculus aquatilis* (Water Crowfoot) W.Miller.

CUP, Water 1. *Hydrocotyle vulgaris* (Pennywort) B & H. 2. *Nuphar lutea* (Yellow Waterlily) Som. Macmillan. 3. *Sarracenia purpurea* (Pitcher Plant) Grieve.1931.

CUP, White *Galanthus nivalis* (Snowdrop) Som. Macmillan.

CUP, Yellow 1. *Ranunculus acris* (Meadow Buttercup) Wilts, Hants, IOW, Bucks Grigson. 2. *Ranunculus arvensis* (Corn Buttercup) Hants. Cope; IOW Long. 3. *Ranunculus bulbosus* (Bulbous Buttercup) Wilts, Hants, IOW, Bucks. Grigson. 4. *Ranunculus repens* (Creeping Buttercup) Wilts, Hants, IOW, Bucks. Grigson.

CUP-AND-LADLE *Quercus robur* - acorn (Acorn) Roxb. B & H.

CUP-AND-SAUCER 1. *Caltha palustris* (Marsh Marigold) Som. Grigson. 2. *Quercus robur* - acorn (Acorn) Som. Elworthy Lincs. Peacock Yks. B & H. 3. *Umbilicus rupestris* (Wall Pennywort) Dev. Friend.1882. Som, Dor, Wilts. Grigson. 4. *Veronica chamaedrys* (Germander Speedwell) Tynan & Maitland. Perhaps Canterbury Bells should be included here, too, but it sounds very unlikely for Germander Speedwell.

CUP-AND-SAUCER, Devil's *Euphorbia amygdaloides* (Wood Spurge) Som. Macmillan.

CUP-AND-SAUCER CANTERBURY BELL *Campanula calycanthema*.

CUP-AND-SAUCER VINE; CUP-AND-SAUCER FLOWER *Cobaea scandens* A.W.Smith (Flower).

CUP CLOVER *Trifolium cyathiferum*.

CUP FLOWER *Nierambergia sp.*

CUP-MOSS *Ochrolechia tartarea* (Cudbear

Lichen) Banff. B & H.

CUP OF GOLD *Solandra maxima* (Chalice Vine) Perry.

CUP-OF-WINE *Taxus baccata* (Yew) Som. Macmillan.

CUP PLANT *Silphium perfoliatum.*

CUP-ROSE *Papaver rhoeas* (Red Poppy) N Eng. B & H. Probably a mistake for Coprose, where cop means a button.

CUP VINE, Silver *Solandra grandiflora.*

CUPID'S DARTS *Catananche caerulea.*

CUPID'S FLOWER *Viola tricolor* (Pansy) Shakespeare. The pansy was one of many flowers mistakenly thought at one time to have aphrodisiac qualities.

CUPIDONE *Catananche caerulea* (Cupid's Darts) Polunin.

CURACAO APPLE *Eugenia javanica*e.

CURCAS BEAN *Jatropha curcas* (Barbados Nut) USA Kingsbury.

CURDLY-GREENS *Brassica fimbricata* (Curled Kale) Som. Elworthy. 'Curly' is pronounced that way in Somerset.

CURDS-AND-CREAM *Saxifraga x urbinum* (London Pride) Som. Macmillan.

CURDWORT *Galium verum* (Lady's Bedstraw) Hants. B & H. This is one of the names referring to the rennet substitute use of this plant. Cf. Cheese-rennet, and similar names.

CURE-ALL 1. *Geum rivale* (Water Avens) B & H. 2. *Melissa officinalis* (Bee Balm) Hatfield. Where did B & H get this name for Water Avens? It certainly wasn't a cure-all, for the recorded medicinal uses are very limited. As for Bee Balm, given balm or balsam for a name, then Cure-all follows almost naturally.

CURL-LEAF MOUNTAIN MAHOGANY *Cercocarpus ledifolius.*

CURLANS; CURLY NUTS *Conopodium majus* (Earth-nut) IOM Moore.1924. Cf. the Scottish Curlun, Gourlin, Gowlin.

CURLDODDY 1. *Bellis perennis* (Daisy) S.W.Scot. Grigson. 2. *Dactylorchis maculata* (Heath Spotted Orchid) Shet. Nicolson. 3. *Knautia arvensis* (Field Scabious) Border Denham/Hardy. 4. *Plantago lanceolata* (Ribwort Plantain) Scot. Grigson. 5. *Plantago major* (Great Plantain) Forf. B & H. 6. *Succisa pratensis* (Devil's-bit) N.Eng, Scot. Dyer. 7. *Trifolium repens* (White Clover) Ork. Grigson. Apparently, in every case, a descriptive name - it is said that they get the name from the resemblance of the flower to a boy's curly head.

CURLDODDY, White *Trifolium repens* (White Clover) Scot. Jamieson.

CURLED KALE *Brassica fimbricata.*

CURLED MINT 1. *Mentha crispa* (Curly Mint). 2. *Mentha piperita* (Peppermint) Hatfield.

CURLED PARSLEY *Petroselinum crispum* (Parsley) Hemphill.

CURLEY-DODDIES *Senecio jacobaea* (Ragwort) North.

CURLIE-DODDIES 1. *Dactylorchis incarnata* (Early Marsh Orchid) Shet. Grigson. 2. *Dactylorchis maculata* (Heath Spotted Orchid) Shet. Grigson. 3. *Orchis mascula* (Early Purple Orchis) Shet. Grigson. 4. *Senecio jacobaea* (Ragwort) North.

CURLICK *Sinapis arvensis* (Charlock) Bucks, Oxf. B & H. One of a number of variants on charlock (they go from Cadlock to Warlock). In this case, it seems nearer to the OE (cerlic), than charlock itself.

CURLOCK *Raphanus raphanistrum* (Wild Radish) B & H. A charlock name attached to wild radish. There are in fact quite a number of them, ranging from charlock itself, through Cadlock to Ketlock.

CURLS, Blue 1. *Iris dichotoma* (Afternoon Iris) W.Miller. 2. *Prunella vulgaris* (Self-heal) Dev, Dor. Macmillan; USA Grigson.

CURLUNS *Conopodium majus* (Earth-nut) Kirk, Wigt. Grigson. From the same source as other Scottish names like Gourlins, or Gowlins.

CURLY-FLOWER *Brassica oleracea* var.botrytis (Cauliflower) Lincs. Peacock. A variation of cauliflower.

CURLY-LEAF MINT *Mentha spicata* (Spearmint) E.Hayes.

CURLY MINT *Mentha crispa*.
CURMEL, Lesser *Centaureum erythraea* (Centaury) Cockayne. Or Lesser Churmel. The name is given in Apuleius Barbarus under *C erythraea*, but is said to be *Blackstonia perfoliata*; but it looks like feverfew in the illustration, as the editor (Gunther) points out. The plants were probably mixed up very early.
CURN-FLOWER *Agrostemma githago* (Corn Cockle) Ches. Holland. Cornflower, obviously. Is it genuine?
CURNBERRIES *Ribes nigrum* (Blackcurrant) N.Eng. Halliwell. i.e. currantberries, the intermediate form being curranberries.
CURRANBINE *Aquilegia vulgaris* (Columbine) Ches. B & H. One of a number of corruptions of columbine recorded; there are also colourbine, cullavine, collarbind and cullenbeam.
CURRAN-PETRIS *Daucus carota* (Wild Carrot) Scot. Grigson.
CURRANBERRY, Black *Ribes nigrum* (Blackcurrant) Cumb. B & H.
CURRANT, Alaska *Ribes bracteosum*.
CURRANT, Alpine *Ribes alpinum*.
CURRANT, American *Ribes americanum*. Currant is from Corinth - they are in fact raisins de Corinthe, their original place of export. But these are grapes, and nothing to do with the genus *Ribes*.
CURRANT, Buffalo *Ribes aureum*.
CURRANT, Flowering 1. *Ribes aureum* (Buffalo Currant) USA Elmore. 2. *Ribes sanguineum*.
CURRANT, Fuchsia-flowered *Ribes speciosum* (Fuchsia Gooseberry) RHS.
CURRANT, Golden; CURRANT, Yellow-flowered *Ribes aureum* (Buffalo Currant) USA Elmore (Golden); Calif. Spier (Yellow-flowered).
CURRANT, Indian 1. *Callicarpa americana* Alab. Harper. 2. *Symphoricarpos orbiculatus* (Red Wolfberry) Alab. Harper.
CURRANT, Missouri *Ribes aureum* (Buffalo Currant) USA Elmore.

CURRANT, Mountain *Ribes alpinum*.
CURRANT, Prickly *Ribes lacustre* (Swamp Currant) Douglas.
CURRANT, Rhodesian *Rhus dentata* Watt.
CURRANT, Sierra, Wild *Ribes nevadense*.
CURRANT, Skunk 1. *Ribes glandulosum*. 2. *Ribes prostratum*.
CURRANT, Squaw *Ribes cereum* (Wax Currant) USA Cunningham.
CURRANT, Swamp *Ribes lacustre*.
CURRANT, Trailing *Ribes laxiflorum*.
CURRANT, Wax *Ribes cereum*.
CURRANT, Wild *Ribes inebrians*.
CURRANT-BERRY *Ribes rubrum* (Redcurrant) Yks. Morris.
CURRANT-DUMPLING *Epilobium hirsutum* (Great Willowherb) N'thum. Grigson. A name which is in line with other 'pie' and 'pudding' names which this flower has attracted.
CURRANT TREE, Hawthorn *Ribes oxyacanthoides* (Smooth Gooseberry) Notes and Queries, 1857.
CURRAWONG ACACIA *Acacia doratoxylon* (Brigalow) Usher.
CURRY *Curcuma longa* (Turmeric) Schauenberg. Actually, curry powder is a mixture of pepper, coriander, cinnamon, ginger, cloves, cardamom, pimento, cumin and mace, as well as turmeric, but it is the last that provides the colour.
CURRY-COMB, Devil's *Ranunculus arvensis* (Corn Buttercup) Shrop, Hants. Burne. A descriptive name (a curry-comb is an implement used to rub down horses). The achenes are covered with spines, hence the name, and such as Hedgehog, Prickleback, etc. It also accounts for the association with the devil, seen in such names as Devil-on-both-sides, Devil's Coachwheel, Hellweed, etc.
CURRY-PLANT *Helichrysum angustifolium*. From the scent. It is not the commercial curry, but sprigs can be used to flavour meat dishes.
CURSE, Hancock's *Polygonum cuspidatum* (Japanese Knotweed). So described, accord-

ing to a Times report, during the answering of a House of Lords question in July 1989. It was said that this name was current in the 1930's.

CURSE, Hell's *Amaranthus paniculatus* (Purple Amaranth) Watt.

CURSE, Paterson's *Echium plantagineum* (Purple Viper's Bugloss) Australia Salisbury.1964. It was introduced into Australia by the early settlers, and has now become a pestilential weed, except where it is a prized fodder plant in arid areas.

CURSE, Thompson's *Cardaria draba* (Hoary Cress) PLNN 10. It was claimed that Hoary Cress was first spread on to arable land by a farmer named Thompson. Apparently, it was not known in this country until 1809, when it was accidentally introduced after the ill-fated Walcheren expedition of that year. The fever-stricken soldiers were brought back to Ramsgate on mattresses stuffed with hay, and this was disposed of to a Thanet farmer who ploughed it in for manure. Then the cress appeared, presumably from seeds contained in the hay. The Thanet farmer must be the one named Thompson.

CURSE OF BARBADOS *Lantana camara* Nielsen.

CURSED CROWFOOT *Ranunculus sceleratus* (Celery-leaved Buttercup) USA Tolstead. This is the most poisonous of the buttercups.

CURSED THISTLE 1. *Carduus benedictus* (Holy Thistle) USA Henkel. 2. *Cirsium arvense* (Creeping Thistle) Curtis. Understandable for Creeping Thistle, but why should Holy Thistle have the name, when all the others given to it extol its virtues?

CURSHINS; CUSHINGS *Armeria maritima* (Thrift) Som. Elworthy (Curshins); Dev. Wright (Cushings). These must be cushions, a descriptive name common in a number of forms; it is the shape of the flower tufts that gives the name.

CURTAIN PLANT *Kalanchoe pinnata*.

CURTAINS, Gypsy 1. *Anthriscus sylvestris* (Cow Parsley) Som. Grigson. 2. *Conium maculatum* (Hemlock) Som. Macmillan.

CURTIS'S MOUSE-EAR *Cerastium pumilum*.

CUSH-CUSH *Dioscorea trifida*.

CUSHAG *Senecio jacobaea* (Ragwort) IOM Moore.1924. Manx cuishag vooar, great stalk. But cuise (dim. cuiseag) is the general name in Ireland for grasses, and the Scottish cuiseag means rye-grass.

CUSHAW SQUASH *Cucurbita moschata* (Crook-neck Squash) Bianchini. Cushaw, or Cashaw, was the Algonquian word for the squash.

CUSHIA *Heracleum sphondyllium* (Hogweed) Yks. B & H.

CUSHION 1. *Armeria maritima* (Thrift) Dev. Friend.1882; Som. Macmillan. 2. *Knautia arvensis* (Field Scabious) Wilts. Dartnell & Goddard. Descriptive of the flower tufts. Both these plants have attracted a number of 'cushion' names.

CUSHION, Eve's *Saxifraga hypnoides* (Dovedale Moss) Yks. Grigson. Not the flower heads this time, but the whole plant is like a cushion. Cf. Lady's, and Queen's, Cushion for this.

CUSHION, Lady *Armeria maritima* (Thrift) Ches. Leigh.

CUSHION, Lady's 1. *Anthyllis vulneraria* (Kidney Vetch) Wilts. Dartnell & Goddard. 2. *Arabis caucasica* (Garden Arabis) W.Miller. 3. *Armeria maritima* (Thrift) Gerard. 4. *Centaurea nigra* (Knapweed) Kent Grigson. 5. *Chrysosplenium oppositifolium* (Opposite-leaved Golden Saxifrage) Freethy. 6. *Knautia arvensis* (Field Scabious) Kent B & H. 7. *Lotus corniculatus* (Bird's-foot Trefoil) Dev, Wilts. Macmillan. 8. *Saxifraga hypnoides* (Dovedale Moss) Yks, Cumb. Grigson.

CUSHION, Our Lady's *Armeria maritima* (Thrift) Parkinson.

CUSHION, Queen's 1. *Saxifraga hypnoides* (Dovedale Moss) Dur. Grigson. 2. *Sedum acre* (Biting Stonecrop) Scot. Tynan & Maitland. 3. *Viburnum opulus* (Guelder Rose) Heref. Leather.

CUSHION, Red *Trifolium pratense* (Red Clover) Som. Macmillan.
CUSHION, Sea *Armeria maritima* (Thrift) Ches. Hole.1966.
CUSHION, Silver *Anthyllis vulneraria* (Kidney Vetch) Tynan & Maitland.
CUSHION CALAMINT *Clinopodium vulgare* (Wild Basil) Mabey. From the dome-shaped whorl of flowers.
CUSHION PINK 1. *Armeria maritima* (Thrift) Som. Grigson; Wilts. Dartnell & Goddard. 2. *Silene acaulis* (Moss Campion) Prior.
CUSHION SAXIFRAGE, One-flowered *Saxifraga burserana*.
CUSHION SPURGE *Euphorbia polychroma*.
CUSK *Papaver rhoeas* (Red Poppy) War. Halliwell. Probably from the shape of the flower. Cuskin is a provincial name for a drinking cup.
CUSTARD-APPLE 1. *Annona reticulata*. 2. *Annona squamosa* (Sweetsop) Candolle. Custard, because of the consistency of the fruit, something like ice cream. It is sometimes chilled, and eaten with a spoon.
CUSTARD-APPLE, Netted *Annona reticulata* (Custard-apple) Chopra.
CUSTARD-APPLE, Scaly *Annona squamosa* (Sugar-apple) India Dalziel.
CUSTARD-APPLE, Wild *Annona chrysophylla*.
CUSTARD-CHEESE *Malva sylvestris* (Common Mallow) Lincs. Grigson. Cheese from the shape of the ripe fruits. Custard, perhaps from the taste, but this is said to be more like that of peanuts. But of course, there is a great variety of these cheese names.
CUSTARD-CUPS *Epilobium hirsutum* (Great Willowherb) Shrop, Ches. Grigson. One of the series of names like Apple-pie, or Sugar Codlins.
CUSTARD SQUASH *Cucurbita pepo var.melopepo* (Scallop Gourd) Bianchini.
CUSTIN *Prunus domestica var.institia* (Bullace) Som. Halliwell. Macmillan has Kestin for this, from the same area.
CUT, Devil's 1. *Clematis vitalba* (Old Man's Beard) B & H. 2. *Clematis integrifolia* (Hungarian Climber). This is a tobacco cut, not a physical one. Boys used to smoke these porous stalks. Cf. Smoking Cane, Smokewood, Boy's Bacca, Gypsy's Tobacco, etc.
CUT-AND-COME-AGAIN *Dioscorea cayennensis* (Yellow Yam) W Indies Dalziel.
CUT-EYE BEAN 1. *Canavalia ensiformis* (Jack Bean) W Indies Dalziel. 2. *Dolichos ensiformis* (Horse-eye Bean) Beckwith. In the case of Jack Bean, Cf. One-eye Bean and Overlook Bean.
CUT-FINGER 1. *Scrophularia nodosa* (Figwort) Surr. Grigson. 2. *Umbilicus rupestris* (Wall Pennywort) Worcs. Grigson. 3. *Valeriana officinalis* (Valerian) Corn, Oxf. Grigson. 4. *Vinca major* (Greater Periwinkle) Nash. 5. *Vinca minor* (Lesser Periwinkle) Dor, Oxf. Grigson. Valerian is a heal-all - the Latin valere means to get well, or be in good health. B & H have Cut-heal for this, too. Periwinkle was used as a vulnerary at one time.
CUT-FINGER LEAF *Valeriana officinalis* (Valerian) Wilts. Dartnell & Goddard. See Cut-finger.
CUT-GRASS *Plantago major* (Great Plantain) Ire. Egan. Cf. Bleeding Grass for the same plant. Irish people used to put the leaves on cuts, to stop the bleeding, and to take the pain away. This is an ancient usage, occurring in the Anglo-Saxon version of Apuleius.
CUT-HEAL *Valeriana officinalis* (Valerian) B & H). Cf. Cut-finger for this. Prior saw OE cowth, womb, as the origin of 'cut' in this case, because it was used, he said, in uterine affections.
CUT-LEAF BEECH *Fagus sylvatica var. heterophylla*.
CUT-LEAF BLACKBERRY *Rubus laciniatus*.
CUT-LEAF NIGHTSHADE *Solanum triflorum* (Three-flowered Nightshade) USA Kingsbury.
CUTBERDILL; CUTBERDOLE *Acanthus*

mollis (Bear's Breech) both B & H.
CUTCH, Black; CATECHU, Black *Acacia catechu* (Cutch-tree) Hora (Cutch); Le Strange (Catechu). The earlier specific name was *nigrum*.
CUTCH, Gambier; CUTCH, White *Uncaria gambir* Hutchinson & Melville (White Cutch).
CUTCH-TREE *Acacia catechu*. Most of the catechu exported comes from the Gulf of Cutch, hence the name. It is an important dye plant, and the dyestuff itself is known as cutch, or sometimes as Gambir (see previous entry).
CYCLAMEN POPPY *Eonecon chionantha*.
CYNKFOLY *Potentilla reptans* (Cinquefoil) Turner. Merely an early variant spelling of cinquefoil.
CYPERUS, Black *Cyperus fuscus*.
CYPHEL 1. *Cherleria sedoides*. 2. *Sempervivum tectorum* (Houseleek) N.Eng. F.Grose. From Greek kuphella, hollows of the ear, perhaps reasonable for houseleek. But why Cherleria? Did kuphella mean something else, too?
CYPHEL, Mossy *Cherleria tectorum* (Cyphel) Clapham.
CYPRESS 1. *Cupressus sempervirens*. 2. *Tamarix anglica* (Tamarisk) Corn. Grigson. ME cipres, from French, and late Latin cupressus, and Greek kuparissos.
CYPRESS, Abrams *Cupressus abramsiana*.
CYPRESS, Alaska *Chamaecyparis nootkatensis* (Alaska Yellow Cedar) Edlin.
CYPRESS, Arizona, Rough-barked *Cupressus arizonica*.
CYPRESS, Arizona, Smooth *Cupressus glabra*.
CYPRESS, Baker *Cupressus bakeri*.
CYPRESS, Bald 1. *Taxodium distichum* (Swamp Cypress) A.W.Smith. 2. *Taxodium distichum var. mucronatum* (Mexican Bald Cypress) Mitchell.1972.
CYPRESS, Bald, Mexican *Taxodium distichum var. mucronatum*.
CYPRESS, Bhutan *Cupressus torulosa* (Himalayan Cypress) Mitchell.1972.

CYPRESS, Broom *Kochia scoparia* (Burning Bush) Geng Junying.
CYPRESS, Californian *Cupressus goveniana*.
CYPRESS, Dawn *Metasequoia glyptostroboides* (Dawn Redwood) Wilks.
CYPRESS, Dwarf; CYPRESS, Heath *Lycopodium alpinum* (Alpine Clubmoss) both Turner.
CYPRESS, Field *Ajuga chamaepitys* (Yellow Bugle) B & H. Cf. Ground Pine for this.
CYPRESS, Fire *Chamaecyparis obtusa*.
CYPRESS, Formosan *Chamaecyparis formosensis*.
CYPRESS, Funeral 1. *Cupressus funebris* (White Fir) J.Smith. 2. *Cupressus sempervirens* (Common Cypress) Polunin. Cypress has always been an emblem of death and mourning, the classical myth of Cyparissos having started it.
CYPRESS, Garden 1. *Artemisia maritima* (Sea Wormwood) Gerard. 2. *Santolina chamaecyparissus* (Ground Cypress) Lyte.
CYPRESS, Gowen *Cupressus goveniana* (Californian Cypress) Edlin.
CYPRESS, Ground *Santolina chamaecyparissus*.
CYPRESS, Guadalupe *Cupressus guadalupensis*.
CYPRESS, Himalayan *Cupressus torulosa*.
CYPRESS, Hinoki *Chamaecyparis obtusa* (Fire Cypress) Wit. The timber is known in Japan as hinoki-wood.
CYPRESS, Italian *Cupressus sempervirens* (Common Cypress) Mitchell.1974. Cf. Mediterranean Cypress.
CYPRESS, Kashmir *Cupressus cashmeriana*.
CYPRESS, Lawson *Chamaecyparis lawsoniana*. The name commemorates Peter Lawson, an Edinburgh nurseryman who often promoted botanical expeditions.
CYPRESS, Leyland *Cupressocyparis leylandii*. The hybrid, *Chamaecyparis nootkatensis* x *Cupressus macrocarpa*, was developed in 1888 at Leighton Hall, near Welshpool, by C J Leyland.

CYPRESS, Louisiana *Taxodium distichum* (Swamp Cypress) USA Brimble.1948.

CYPRESS, Magician's *Juniperus sabina* (Savin) Fernie. This, and another name, Devil's Tree, were probably applied because of its bad reputation as an abortifacient and contraceptive.

CYPRESS, Mediterranean *Cupressus sempervirens* (Common Cypress) Mitchell.1972. Cf. Italian Cypress.

CYPRESS, Mexican *Cupressus lusitanica*. Lusitanica, because it became better known in Europe through trees cultivated in Portugal; in fact, it was actuallly called Portuguese Cypress.

CYPRESS, Modoc *Cupressus bakeri* (Baker Cypress) Barber & Phillips.

CYPRESS, Monterey *Cupressus macrocarpa*. This is the Monterey Peninsula, S.California, where it grows.

CYPRESS, Montezuma *Taxodium distichum* (Swamp Cypress) Duncalf.

CYPRESS, Mourning; CYPRESS, Weeping *Cupressus funebris* (White Fir) Everett (Mourning); Wit (Weeping). Cf. Funeral Cypress, too.

CYPRESS, Nootka *Chamaecyparis nootkatensis* (Alaska Yellow Cedar) Mitchell.1972.

CYPRESS, Patagonian *Fitzroya cupressoides* (Alerce) Leathart.

CYPRESS, Pond *Taxodium ascendens*.

CYPRESS, Portuguese *Cupressus lusitanicus* (Mexican Cypress) J.Smith. Both the Portuguese and the lusitanica parts of the name result from the fact that the tree became best known through cultivation in Portugal.

CYPRESS, Prickly *Juniperus formosana* (Prickly Juniper) Mitchell.1972.

CYPRESS, Sawara *Chamaecyparis pisifera*.

CYPRESS, Sitka *Chamaecyparis nootkatensis* (Alaska Yellow Cedar) Willis.

CYPRESS, Smooth *Cupressus glabra* (South Arizone Cypress) RHS.

CYPRESS, Southern *Taxodium distichum* (Swamp Cypress) Duncalf.

CYPRESS, Standing *Gilia rubra* (Scarlet Gilia) Perry.

CYPRESS, Summer *Kochia scoparia* (Burning Bush) Thrimble.

CYPRESS, Swamp *Taxodium distichum*.

CYPRESS, Swamp, Chinese *Glyptostrobus lineatus*.

CYPRESS, Sweet; CYPRESS-ROOT *Cyperus longus* (Galingale) A S Palmer. i.e. Cyperus.

CYPRESS, Tecate *Cupressus forbesii*.

CYPRESS, White *Chamaecyparis thyoides* (White Cedar) Chaplin.

CYPRESS, Yellow *Chamaecyparis nootkatensis* (Alaska Yellow Cedar) Schery.

CYPRESS, Yunnan *Cupressus duclouxiana*.

CYPRESS-KNEES *Taxodium distichum* the aerial parts of the roots (Swamp Cypress) Polunin.1976.

CYPRESS OAK *Quercus robur var. fastigiata*.

CYPRESS OF INDIA; CYPRESS OF MALABAR; CYPRESS OF BABYLON *Curcuma longa* (Turmeric) Pomet.

CYPRESS PINE 1. *Callitris robusta*. 2. *Widdringtonia cupressoides*.

CYPRESS PINE, Northern *Callitris intratropica*.

CYPRESS PINE, White *Callitris glauca*.

CYPRESS SHRUB *Lawsonia inermis* (Henna) Dalziel.

CYPRESS SPURGE *Euphorbia cyparissa*. The specific name was given by Pliny, in the belief that Cyprus was its country of origin. Cypress, then, is Cyprus.

CYPRESS VINE *Ipomaea quamoclit*.

CYPRESS WORMWOOD *Artemisia pontica* (Roman Wormwood) Gerard.

CYPRUS CEDAR *Cedrus brevifolia*.

CYPRUS OAK *Quercus lusitanica*.

CYPRUS TURPENTINE *Pistacia terebinthus* (Terebinth) RHS.

CYRILLO *Stellaria media* (Chickweed) Canada Allan.

D

DAA-NETTLE; DEE-NETTLE *Lamium purpureum* (Red Deadnettle) both Shet. both Grigson. Both mean deaf nettle, as does the English Dea-nettle; with dumb nettle, they correspond to dead nettle.

DABBERY *Ribes uva-crispa* (Gooseberry) Kent Grigson. Dayberry is also recorded both in Kent and in Devon & Cornwall.

DADDY'S BEARD; DADDY MAN'S BEARD *Clematis vitalba* (Old Man's Beard) both Som. both Macmillan.

DADDY'S WHISKERS *Clematis vitalba* (Old Man's Beard) Som. Macmillan; Wilts. Dartnell & Goddard).

DADDY'S WHITE SHIRT *Calystegia sepium* (Great Bindweed) Som. Macmillan. Part of the imagery that gives bindweed the names of shirt, smock, chemise, etc.

DAFF *Narcissus pseudo-narcissus* (Daffodil). An extremely common contraction.

DAFF, Dilly *Narcissus pseudo-narcissus* (Daffodil) Som. Macmillan.

DAFF LILY; DAFT LILY *Narcissus pseudo-narcissus* (Daffodil) Tynan & Maitland.

DAFFADILLY; DAFFODILLY; DAFFADOWNDILLY; DAFFIDOWNDILLY; DAFFODOWNDILLY; DAFFADOONDILLY *Narcissus pseudo-narcissus* (Daffodil) all B & H.

DAFFANY *Daphne mezereon* (Lady Laurel) Dev. Friend.1882. However it is spelt, the inclusion of an extra vowel in Daphne is quite common.

DAFFODIL 1. *Daphne mezereon* (Lady Laurel) Yks. B & H. 2. *Fritillaria meleagris* (Snake's Head Lily) Hants. Cope. 3. *Narcissus pseudo-narcissus*. Affodill was the original of daffodil, from medieval Latin affodilus, from Greek asphodelus. The Oxford Dictionary says the d- is unexplained, but it is possible it came directly from the Dutch de affodil. The Yorkshire use of the name for Lady Laurel (it has all the usual variants as well) is explained by the faint resemblance of the name Daphne. Snake's Head Lily gets the name as an abbreviation, for it has long been known as Chequered Daffodil, or even Bastard Narcissus.

DAFFODIL, Bastard *Erythronium americanum* (Yellow Adder's Tongue) USA Leighton.

DAFFODIL, Chequered *Fritillaria meleagris* (Snake's Head Lily) Gerard. Parkinson called it the Chequered Lily, and there is also Chequered Tulip for this. 'Chequered', from the markings on the petals; Latin fritillus meant a dice-box, and by extension the chequered board on which the dice were thrown.

DAFFODIL, Dutch *Asphodelus ramosus* (White Asphodel) Turner.

DAFFODIL, Peruvian *Ismene calathium* USA Cunningham.

DAFFODIL, Queen Anne's *Narcissus eystettensis*.

DAFFODIL, Rush *Narcissus odorus* (Jonquil) McDonald.

DAFFODIL, Sea 1. *Ismene calathinum* McDonald. 2. *Pancratium maritimum*.

DAFFODIL, Spanish *Narcissus hispanicus*.

DAFFODIL, Tenby *Narcissus obvallaris*. Very rare in this country, growing in just a few fields near Tenby.

DAFFODIL, Virginian *Zephyranthes atamasco* (Atamasco Lily) Parkinson.

DAFFODIL, White *Narcissus maialis var. patellaris* (Pheasant's Eye) Turner.

DAFFODIL GARLIC *Allium neapolitanum*.

DAFFY; DAFFYDOWNDILLY *Narcissus pseudo-narcissus* (Daffodil) both Northall.

DAFT BERRIES *Atropa belladonna* (Deadly Nightshade) Som. Macmillan. The berries cause dizziness; Gerard said the plant "causeth sleep, troubleth the mind, bringeth madnesse if a few of the berries be inwardly taken...".

DAFT LILY *Narcissus pseudo-narcissus* (Daffodil) Tynan & Maitland. A shift from Daff Lily.

DAGGER 1. *Colchicum autumnale* (Meadow Saffron) Som. Grigson. 2. *Iris foetidissima*

(Foetid Iris) Dev. Friend.1882. 3. *Iris pseudo-acarus* (Yellow Flag) Dev. Friend.1882. It is the shape of the leaves that give the name, rather more sword-like than dagger-like, as witness the greater number of sword names.

DAGGER, Spanish 1. *Yucca baccata* (Datil) USA Elmore. 2. *Yucca schidigera* (Mohave Yucca) Jaeger.

DAGGER COCKLEBUR; DAGGERWEED *Xanthium spinosum* (Spiny Cocklebur) both S.Africa both Watt.

DAGGER FLOWER 1. *Iris foetidissima* (Foetid Iris) Som. Macmillan. 2. *Iris pseudo-acarus* (Yellow Flag) Som. Macmillan. 3. *Iris versicolor* (Purple Water Iris) Grieve.1931.

DAGGER PLANT 1. *Lobelia dortmanniana* (Water Lobelia) N.Wales. Grigson. 2. *Yucca aloifolia* (Aloe Yucca) USA Cunningham. That, at any rate, is what the Welsh Bidawglys translates, referring to Water Lobelia. The leaf shape accounts for the name for the yucca.

DAGGER WEED 1. *Xanthium spinosum* (Spiny Cocklebur) S.Africa Watt. 2. *Yucca glauca* (Soapweed) USA Elmore.

DA-HO *Anthriscus sylvestris* (Cow Parsley) N.Ire. B & H. This, with Ha-ho, and Hi-how, are pronunciation guides to the Gaelic names.

DAHOON *Ilex vomitoria* (Cassine) A.W.Smith. Unless there is a separate species Ilex cassine, in which case Dahoon should be reserved for this.

DAHURIAN LADYBELL *Adenophora verticillata*.

DAHURIAN LILY *Lilium dauricum*.

DAHURIAN LARCH *Larix gmelini*.

DAIKON *Raphanus sativus* (Garden Radish) A.W.Smith.

DAILY BREAD *Rosa canina* the young shoots (Dog Rose) Yks. B & H. Yorkshire children eat the young shoots,and call them Daily Bread,just as in Wiltshire they call them Sugar Candy,or in Somerset Bacon.

DAIMYO OAK *Quercus dentata*.

DAISIES, Bunch-o'- *Achillea millefolium*

(Yarrow) Dor. Macmillan. Descriptive, in the sense that any composite can be given this name.

DAISY 1. *Bellis perennis*. 2. *Senecio jacobaea* (Ragwort) Som. Macmillan. Daisy seems genuinely to be Day's Eye. The OE is daeges eage, and it got the name from its habit of covering the yellow disc in the evening, and showing it in the morning.

DAISY, African 1. *Arctotis stoechadifolia*. 2. *Dimorphoteca aurantiaca* (Star of the Veldt) Hay & Synge.

DAISY, Barbaton *Gerbera sp.*

DAISY, Big; DAISY, Great; DAISY, Greater *Leucanthemum. vulgare* (Ox-eye Daisy) Yks. B & H (Big); Cumb. B & H. (Great); Curtis (Greater).

DAISY, Blue 1. *Aster tripolium* (Sea Aster) Kent Gerard. 2. *Globularia vulgaris* (Globeflower) Schauenberg. 3. *Jasione montana* (Sheep's Bit Scabious) Ches. Grigson.

DAISY, Bog *Caltha palustris* (Marsh Marigold) Yks. Grigson.

DAISY, Bull *Leucanthemum vulgare* (Ox-eye Daisy) Ches. Holland; E.Ang, Yks, Cumb. Grigson. 'Bull' here denotes big, or coarse, as does 'horse' in Horse Daisy, and Horse Gowan. Cf. too Dog Daisy.

DAISY, Bush *Argyranthemum frutescens* (Paris Daisy) S.Africa Watt.

DAISY, Butter 1. *Leucanthemum vulgare* (Ox-eye Daisy) Dev, Dor. Macmillan. 2. *Ranunculus acris* (Meadow Buttercup) Bucks. B & H. 3. *Ranunculus bulbosus* (Bulbous Buttercup) Bucks. Grigson. 4. *Ranunculus repens* (Creeping Buttercup) Bucks. Grigson.

DAISY, Cabbage *Trollius europaeus* (Globe Flower) Scot. Grigson.

DAISY, Children's *Bellis perennis* (Daisy). More properly this is Childing Daisy, as *Dianthus prolifer* is Childing Pink. It means that the plant throws out younger and smaller flowers like a family of litle children around the parent.

DAISY, Christmas *Aster grandifolius* Coats. A late-flowering species.

DAISY, Crown *Chrysanthemum coronarium* (Annual Chrysanthemum) Coats. The specific name is *C coronarium*, hence "crown".

DAISY, Day *Leucanthemum vulgare* (Ox-eye Daisy). Reference lost. Possibly Dog Daisy, for the same reason as the giving of names like 'horse', or 'bull', or 'big' Daisy. On the other hand, the name Midsummer Daisy appears for this, and the Day may conceivably be this one.

DAISY, Devil 1. *Anthemis cotula* (Maydweed) Wilts. Macmillan. 3. *Leucanthemum vulgare* (Ox-eye Daisy) Middx. B & H. 2. *Tanacetum parthenium* (Feverfew) Som, Wilts. Macmillan. The smell of Maydweed probably gave rise to the name. But the others may possibly be explained by reference to other names. For ox-eye, see for instance Poverty-weed and Poor-land Daisy.

DAISY, Dicky; DAISY, Ducky *Bellis perennis* (Daisy) Ches. Holland (Dicky); Tynan & Maitland (Ducky).

DAISY, Dicky, Large *Leucanthemum vulgare* (Ox-eye Daisy) Ches. Holland.

DAISY, Dog 1. *Achillea millefolium* (Yarrow) N.Ire. Grigson. 2. *Anthemis cotula* (Maydweed) Dev, Shrop, Kent. Grigson. Yks. Robinson. 3. *Bellis perennis* (Daisy) Surr. Tynan & Maitland; Lincs, Cumb. Grigson; Lancs. Nodal; Yks. Addy. 4. *Leucanthemum vulgare* (Ox-eye Daisy) Grigson. 5. *Tripleurospermum maritimum* (Scentless Maydweed) Donegal. B & H. 'Dog' here usually means big or coarse. So why should the daisy be given this name over such a large area?

DAISY, Drummer *Leucanthemum vulgare* (Ox-eye Daisy) Som. Macmillan. Cf. Thunder Daisy, etc.

DAISY, Dun; DAISY, Dunder; DAISY, Dundle *Leucanthemum vulgare* (Ox-eye Daisy) Dev. Friend.1882 (Dun); Som. Macmillan (Dunder,Dundle). These are all 'Thunder' Daisy, which occurs in the same area. Is the reason possibly because this is also Midsummer Daisy?

DAISY, English *Bellis perennis* (Daisy) USA Hay & Synge.

DAISY, Ewe *Potentilla erecta* (Tormentil) Border Grigson.

DAISY, Fern-leaved *Tripleurospermum maritimum* (Scentless Maydweed) Som. Macmillan.

DAISY, Field 1. *Leucanthemum vulgare* (Ox-eye Daisy) Dev. Friend.1882. 2. *Tanacetum parthenium* (Feverfew) Som. Macmillan.

DAISY, Golden 1. *Chrysanthemum segetum* (Corn Marigold) Som. Macmillan. 2. *Ranunculus ficaria* (Lesser Celandine) N'thants. Clare.

DAISY, Gracy 1. *Bellis perennis* (Daisy) Dev. Grigson. 2. *Narcissus pseudo-narcissus* (Daffodil) Dev, Som. Macmillan. Gracy Daisy for the daffodil should probably be Gracy Day, and this in turn is probably a reference to Easter, which was once called Great Day.

DAISY, Ground *Townsendia scapigera*.

DAISY, Gypsies' *Leucanthemum vulgare* (Ox-eye Daisy) Fernie.

DAISY, Gypsy 1. *Leucanthemum vulgare* (Ox-eye Daisy) Dor, Som. Macmillan; Norf. Grigson. 2. *Knautia arvensis* Som. Macmillan. 3. *Tripleurospermum maritimum* (Scentless Maydweed) Som. Macmillan.

DAISY, Harvest *Leucanthemum vulgare* (Ox-eye Daisy Dor. Macmillan.

DAISY, Horse 1. *Anthemis cotula* (Maydweed) Bucks. Grigson. 2. *Leucanthemum vulgare* (Ox-eye Daisy) Dev. Friend.1882. Wilts. Dartnell & Goddard. 3. *Tripleurospermum maritimum* (Scentless Maydweed) Bucks. B & H.

DAISY, Horse, Yellow *Chrysanthemum segetum* (Corn Marigold) Corn. Grigson.

DAISY, Irish *Taraxacum officinale* (Dandelion) Yks. B & H; USA Henkel.

DAISY, Kingfisher *Felicia bergeriana*. Because of the blue flowers, I suppose.

DAISY, Livingstone *Mesembryanthemum crinifolium*.

DAISY, London *Leucanthemum vulgare* (Ox-eye Daisy) Som, Dor. Macmillan.

DAISY, March *Bellis perennis* (Daisy) N'thants. A.E.Baker.
DAISY, Marsh *Armeria maritima* (Thrift) Cumb. Grigson.
DAISY, Mat *Raoulia sp* Howes.
DAISY, Maudlin *Leucanthemum vulgare* (Ox-eye Daisy). Le Strange.
DAISY, Michaelmas 1. *Aster paniculatus*. 2. *Aster tripolium* (Sea Aster) Dev. Friend.1882; Som. Grigson; Ches. Holland. 3. *Tanacetum parthenium* (Feverfew) Dev. Friend.1882. Michaelmas, from the season of blooming.
DAISY, Midsummer 1. *Leucanthemum vulgare* (Ox-eye Daisy) Fernie. 2. *Tanacetum parthenium* (Feverfew) Dev, Som. Macmillan.
DAISY, Moon *Leucanthemum vulgare* (Ox-eye Daisy) Som. Tongue; Wilts. Dartnell & Goddard; Glos. J.D.Robertson; Worcs. J Salisbury. A descriptive name; there are also Moons, Moon-flower, Moon-penny and Moon's Eye.
DAISY, Mother *Leucanthemum vulgare* (Ox-eye Daisy) Som. Macmillan. It is also called Mothers in Herefordshire. Presumably these are the result of the name Maithen and its variants being applied to this instead of to Maydweed(*Anthemis cotula*).
DAISY, Mount Atlas *Anacyclus depressus*.
DAISY, Mowing *Leucanthemum vulgare* (Ox-eye Daisy) Som. Macmillan. Moon Daisy, perhaps?
DAISY, Namaqua(land) 1. *Dimorphoteca aurantiaca* (Star of the Veldt) Perry. 2. *Venidium fastuosum*.
DAISY, Orange *Erigeron aurantiacus*.
DAISY, Ox-eye *Leucanthemum vulgare*
DAISY, Painted *Pyrethrum roseum* (Persian Pellitory) USA Cunningham.
DAISY, Paper, Golden *Helichrysum bracteatum* (Everlasting Flower) Australia Howes.
DAISY, Paris *Argyranthemum frutescens*.
DAISY, Penny *Leucanthemum vulgare* (Ox-eye Daisy) Scot. PLNN25.
DAISY, Pig 1. *Anthemis cotula* (Maydweed) Dor. Macmillan. 2. *Pulicaria dysenterica* (Yellow Fleabane) Dor. Macmillan.
DAISY, Poison *Anthemis cotula* (Maydweed) Som. Macmillan Suss. Salisbury.1964. Poison, probably because of the smell.
DAISY, Poor-land *Leucanthemum vulgare* (Ox-eye Daisy) N'thants. A.E.Baker. Cf. Poverty-weed for this plant.
DAISY, Railway *Bidens pilosa*.
DAISY, Rain *Dimorphoteca pluvialis*.
DAISY, Rocky Mountain *Townsendia formosana*.
DAISY, St Peter Port *Erigeron mucronatus* (Mexican Fleabane) McClintock.1975. A Mexican native, but first recorded in the British Isles from St Peter Port, Guernsey.
DAISY, Scented *Tanacetum vulgare* (Tansy) Som. Macmillan.
DAISY, Sea *Armeria maritima* (Thrift) Corn, Suss. B & H Dev. Friend.1882.
DAISY, Shasta *Lueucanthemum maximum*.
DAISY, Shepherd's *Bellis perennis* (Daisy) N'thants. Grigson.
DAISY, Small *Bellis perennis* (Daisy) Dev. Friend.
DAISY, Stink *Tanacetum parthenium* (Feverfew) Som. Macmillan.
DAISY, Summer *Leucanthemum vulgare* (Ox-eye Daisy) N'thants. Clare. Cf. Midsummer Daisy for this.
DAISY, Sun 1. *Helianthemum chamaecistus* (Rock Rose) Lincs. B & H. 2. *Leucanthemum vulgare* (Ox-eye Daisy) Dor. Dacombe. But Cf. Moon Daisy for Ox-eye.
DAISY, Swan River *Brachycome iberidifolia*. An Australian composite.
DAISY, Tahotta *Machaeranthera tanacitifolia*.
DAISY, Thunder *Leucanthemum vulgare* (Ox-eye Daisy) Som. Macmillan. Why thunder? It is, of course, the Midsummer Daisy as well, so that it may get the name from an association of ideas. It is also called Dun Daisy in Somerset, dun being an abbreviation of thunder, and there is another Somerset name, Drummer Daisy, which surely is connected.
DAISY, Transvaal *Gerbera jamesonii*

(Barbaton Daisy) A.W.Smith.

DAISY, Turfing *Matricaria tchihatchensis.* Is this because camomile used to be grown quite a lot as a lawn? The maydweeds are often known as camomiles of one kind or another.

DAISY, White *Leucanthemum vulgare* (Ox-eye Daisy) Le Strange.

DAISY, Wild, Western *Aster integrifolius.*

DAISY, Yard *Tanacetum parthenium* (Feverfew) Som. Macmillan.

DAISY, Yellow 1. *Rudbeckia hirta* (Black-eyed Susan) USA House. 2. *Senecio jacobaea* (Ragwort) Dor. Grigson.

DAISY BUSH *Olearia haastii.*

DAISY FLEABANE 1. *Erigeron philadelphicus* (Philadelphia Fleabane) USA House. 2. *Erigeron ramosus* USA Tolstead.

DAISY GOLDINS *Leucanthemum vulgare* (Ox-eye Daisy) N.Ire. B & H. Part of the series containing the daisy names Gowan or Gowlan.

DAISY-LEAVED BITTERCRESS *Cardamine bellidifolia.*

DAISY-LEAVED CHICKWEED *Mollugo nudicaulis.*

DAISY-LEAVED TOADFLAX *Anarrhinum bellidifolium.*

DAKOTA POTATO *Apios americana* (Groundnut) Bianchini.

DALE-CUP 1. *Caltha palustris* (Marsh Marigold) Som. Macmillan. 2. *Ranunculus acris* (Meadow Buttercup) Som. Grigson. 3. *Ranunculus bulbosus* (Bulbous Buttercup) Som. Grigson. 4. *Ranunculus repens* (Creeping Buttercup) Som. Grigson. Probably better known as the Dorset Dillcup, which must be the same as Somerset Gilcup, i.e. gold cup.

DALMATIAN CAP *Tulipa sp* (Tulip) Gerard. In the right direction, at least,for Turk's Cap is closer to the original idea. Tulip comes from a Turkish word (tilband) meaning turban.

DALMATIAN IRIS *Iris pallida* (Pale Iris) RHS.

DALMATIAN PINE *Pinus nigra var. dalmatica.*

DAMASK, Autumn *Rosa damascena var. bifera.*

DAMASK, Painted *Rosa damascena 'Leda'.*

DAMASK FENNEL *Nigella damascena* (Love-in-a-mist) Schauenberg. *N sativa* is Fennel-flower.

DAMASK ROSE 1. *Rosa damascena.* 2. *Rosa gallica* (French Rose) Surr. Fluckiger. Damask because of the tradition that they were brought to Europe by the Crusaders from Damascus; but the word damask may once have referred to the colour - "damasked cheeks", for example. *Rosa gallica* was called Damask Rose at Mitcham in error - not such a bad mistake, for *Rosa damascena* is actually a hybrid - *R.gallica x R phoenicia.*

DAMIANA *Turnera diffusa.*

DAMMAR PINE *Agathis australis* (Kouri Pine) H & P.

DAMOCEEN *Chrysophyllum argenteum* (Antilles Pear) W Indies Howes.

DAMSEL; DAMSIL 1. *Prunus domestica var. institia* (*damascena*) (Damson) Som. Elworthy; Dor. Northover; Ches. Holland; Dur. Brockett (Damsel); Yks. B & H (Damsil). 2. *Prunus spinosa* fruit (Sloe) Dev. Friend.1882. Friend actually spelt the name with a 'z'. It must be a misapplication to the sloe.

DAMSEL OF THE WOOD *Betula verrucosa* (Birch) Scot. Cheviot. Cf. Lady of the Woods, and its quite common name, Lady Birch.

DAMSON, Wild *Prunus domestica var. institia* (Bullace) Yks. Grigson. Damson seems to be damascenum, i.e. of Damascus.

DAMSON-LEAVED SALLOW *Salix damascena.*

DAN'S CABBAGE *Senecio latifolius* S.Africa Watt.

DANCING LADY *Fuchsia* Som. Macmillan.

DANDELION 1. *Caltha palustris* (Marsh Marigold) Rad. B & H. 2. *Taraxacum officinale.* Dandelion is French dent de lion,itself a rendering of medieval Latin dens leonis, lion's tooth; an earlier name, Leontodon, means the same thing. All this is presumably a reference to the shape of the leaves.

DANDELION, Lesser; DANDELION, Small *Taraxacum laevigatum*. Usually known as the Lesser Dandelion; Salisbury.1964 has Small.

DANDELION, Marsh, Broad-leaved *Taraxacum spectabilis*.

DANDELION, Marsh, Lesser *Taraxacum paludosum* (Narrow-leaved Marsh Dandelion) Salisbury.1964.

DANDELION, Marsh, Narrow-leaved *Taraxacum paludosum*.

DANDELION, Pink *Crepis incana*.

DANDELION, Red-fruited *Taraxacum laevigatum* (Lesser Dandelion) Hepburn.

DANDELION, Russian *Taraxacum koksaghyz*.

DANDY, Rapper *Arctostaphylos uva-ursi* (Bearberry) N.Eng, Berw, Scot. Grigson.

DANDY GOSHEN *Orchis morio* (Green-winged Orchis) Wilts. B & H. More recognizable under another Wiltshire name, Dandy Goslings, shared with other Orchis species. Dandy here is Gander.

DANDY GOSLINGS 1. *Dactylorchis maculata* (Heath Spotted Orchid) Wilts. Grigson. 2. *Orchis mascula* (Early Purple Orchis) Wilts. Dartnell. & Goddard. 3. *Orchis morio* (Green-winged Orchis) Wilts. Dartnell & Goddard. Another Wiltshire name, Gandigoslings, comes closer to the original. Dandy, of course, is gander.

DANDY-GUSSET *Dactylorchis majalis* (Irish Marsh Orchid) Som. Macmillan.

DANE'S BLOOD 1. *Campanula glomerata* (Clustered Bellflower) Hants. Boase; Cambs. Dyer; Herts. Jones-Baker.1974; Ess, Norf. Grigson. 2. *Pulsatilla vulgaris* (Pasque Flower) Herts, Cambs, Ess, Norf. Grigson. 3. *Sambucus ebulus* (Dwarf Elder) Camden. All these plants are said in East Anglia to grow where Danish blood was shed in battle, or at least, on earthworks which are associated in some way with the Danes.

DANE'S ELDER; DANE'S WOOD *Sambucus ebulus* (Dwarf Elder) E.Ang, Lincs. W.Johnson (Dane's Elder); Folkard (Dane's Wood). See Daneball. Perhaps Dane's Wood is a misreading of Dane's Blood.

DANE'S FLOWER; DANE'S WEED *Pulsatilla vulgaris* (Pasque Flower) Cambs. Grigson; Herts. Coats (Dane's Flower); W.Johnson (Dane's Weed). See Dane's Blood.

DANEBALL; DANEWORT *Sambucus ebulus* (Dwarf Elder) Som. Macmillan (Daneball); Turner (Danewort). Daneball sometimes appears as two words. It is said to grow wherever blood has been shed, and particularly the blood of the Danes - wherever their blood was shed in battle, this plant afterwards sprang up. The fact that the stems turn red in September may possibly have given rise to the tradition. But Parkinson noted that as a purgative, it produced the danes, or diarrhoea, and it is suggested that the word Danewort means stink plant.

DANEWEED 1. *Eryngium campestre* (Field Eryngo) N'thants. A.E.Baker. 2. *Sambucus ebulus* (Dwarf Elder) Som, Suff. Dyer. The Eryngo gets the name from a similar reason as the Dwarf Elder - see Daneball.

DANGLE BELLS *Convallaria maialis* (Lily-of-the-valley) Som. Macmillan.

DANGLEBERRY *Gaylussacia frondosa*.

DANUBE GRASS *Phragmites australis* Howes.

DANZIG FIR; DANZIG PINE *Pinus sylvestris* (Scots Pine) Hutchinson & Melville. Cf. Baltic Yellow Deal - almost a trade name for timber imported from that area.

DAPHNE, February *Daphne mezereum* (Lady Laurel) H & P.

DAPHNE, Flax-leaved *Daphne cneorum*.

DARBOTTLE *Centaurea nigra* (Knapweed) Coles. It must be Tarbottle, which is recorded from Oxfordshire.

DARELE *Rumex hydrolapathum* (Great Water Dock) E.Gunther. Only as a gloss on a 12th century ms of Apuleius Barbarus.

DARLAHAD BARBERRY *Berberis aristata* (Nepal Barberry) Le Strange.

DARLING, Little *Reseda lutea* (Wild Mignonette) Som. Macmillan. Little darling would be an exact translation of mignonette.

DARLING OF APRIL *Primula vulgaris* (Primrose) Som. Macmillan.

DARLING PEA, Smooth *Swainsona galegifolia*. The reference is to the River Darling - this is an Australian plant.

DARN-GRASS *Anemone nemerosa* (Wood Anemone) Scot. Grigson. For giving to cattle for a disease known as the darn, according to Grigson.

DARNEL; DARNEL, Bearded *Lolium temulentum*. Bearded Darnel is in Barton & Castle. ODEE says the word is probably of northeast French origin, e.g. Walloon darnelle, the word being connected often with others denoting giddiness. So the plant is named from its stupefying properties.

DARNEL, Red; DARNEL, Great *Lolium perenne* (Rye-grass) Gerard (Great); B & H (Red).

DARNING-NEEDLE, Devil's 1. *Clematis virginiana* Howes. 2. *Scandix pecten-veneris* (Shepherd's Needle) Som. Macmillan.

DARNING-NEEDLE, Old Wife's *Scandix pecten-veneris* (Shepherd's Needle) Hants. Grigson.

DART, Poison *Aglaonema commutatum*.

DARWIN STRINGY-BARK *Eucalyptus tetradonta* (Stringy-bark) Hora. Presumably not Charles Darwin in this case, but the town of that name in Australia.

DARWIN'S BARBERRY *Berberis darwinii*. This was first found by Darwin on his 'Beagle' voyage.

DASH BAGGER *Chenopodium album* (Fat Hen) Dor. Northover.

DASHEEN *Colocasia esculenta*.

DASHEL 1. *Cirsium arvense* (Creeping Thistle) Dev. B & H. 2. *Cirsium vulgare* (Spear Plume Thistle) Dev. B & H. 3. *Taraxacum officinale* (Dandelion) Corn, Dev. Friend. Dashel is merely a local variation of thistle.

DASHEL, Horse *Cirsium vulgare* (Spear Plume Thistle) Dev. Macmillan.

DASHEL, Milky; DASSEL, Milky *Sonchus oleraceus* (Sow Thistle) both Corn, Dev. both B & H. Milky, because of the white sap.

DASHEL, Rough; DASHEL, Row *Onopordum acanthium* (Scotch Thistle) Dev. Grigson (Rough); Dev. Choape (Row). 'Row' is the same as 'rough'.

DASHEL FLOWER *Taraxacum officinale* (Dandelion) Corn, Dev. Friend.

DASO *Coleus rotundifolius* (Fra-Fra Potato) Schery.

DASSEL *Cirsium arvense* (Creeping Thistle) Dev. Friend.1882. Cf. Dashel.

DATCH *Vicia sativa* (Common Vetch) Som. Elworthy. An uncommon form - fetch for vetch is more usual.

DATE, Chinese *Ziziphus jubajuba* (Jujube) Brouk.

DATE, Desert *Balanites aegyptiaca* the ripe fruit (Jericho Balsam) Dale & Greenway.

DATE, Wild *Yucca baccata* (Datil) USA Elmore.

DATE PALM *Phoenix dactylifera*. Date is eventually from Greek daktulos, finger.

DATE PALM, Wild *Phoenix reclinata*.

DATE-PLUM *Diospyros lotus*.

DATE-PLUM, American *Diospyros virginiana* (Persimmon) J.Smith.

DATE-PLUM, Chinese *Diospyros kaki*. This is originally from China, but now frequent in both China and Japan.

DATE-PLUM, European *Diospyros lotus* (Date Plum) J.Smith. It grows in southern Europe.

DATIL *Yucca baccata*. Datil is the Spanish for date.

DATURA, Black *Datura metel* (Hairy Thorn-apple) Dalziel.

DATURA, Sacred *Datura meteloides* (Downy Thorn-Apple) USA Elmore. The plant was certainly sacred to the Californian Indian tribes.

DATURA, Small *Datura discolor* (Purple-stained Toloache) Jaeger.

DAUGHTER-BEFORE-MOTHER *Colchicum autumnale* (Meadow Saffron) Grigson.1959. The flowers appear before the leaves. There are a number of names to illustrate this. Son-before-the-father is better known than Daughter-before-mother, but Upstart, and Pop-ups, as well as the variants on Naked Ladies, all refer to the same phenomenon.

DAUKE *Daucus carota* (Wild Carrot) B & H.

DAVID LILY *Lilium davidii*. This is the French missionary & naturalist, Père Armand David (1827-1900), who collected the lily in 1869.

DAVID PINE *Pinus armandii* (Chinese White Pine) RHS.

DAVID'S HARP *Polygonatum multiflorum* (Solomon's Seal) B & H. Descriptive - the arching of the flowering stalks give some resemblance to medieval drawings of harps.

DAVID'S ROOT *Celastrus scandens* (American Bittersweet) W.Miller.

DAVID'S SEAL *Polygonatum multiflorum* (Solomon's Seal) Le Strange.

DAVIE(S), Stink 1. *Senecio jacobaea* (Ragwort) Fife Grigson. 2. *Taraxacum officinale* (Dandelion) Clack. Grigson. There are a number of 'stink' names for ragwort, not surprisingly - Stinking Alexander, Billy, Nancy or Willie, as well as simple Stinking Weed. Perhaps the Stink Davie given to dandelion should really refer to this, for what objection could there be to the smell of dandelions?

DAWDLE-GRASS *Briza media* (Quaking Grass) Suss. Tynan & Maitland. Dawdle is doddle, which in turn is dodder, meaning to tremble.

DAWN CYPRESS *Metasequoia glyptostroboides* (Dawn Redwood) Wilks.

DAWN FLOWER, Blue 1. *Ipomaea acuminata*. 2. *Ipomaea learii*. The Morning Glories open their flowers very early.

DAWN REDWOOD *Metasequoia glyptostroboides*. Presumably someone fancifully gave it this name as a reference to the dawn of time. Before the discovery of living specimens in Szechuan in 1945, it had only been known to science as a fossil.

DAWYCK BEECH *Fagus sylvatica var. fastigiata*. Dawyck, in Peebles, where the variety appeared round about 1860.

DAY, Eye of; DAY'S EYE *Bellis perennis* (Daisy) both Som. both Macmillan. This is exactly what daisy means.

DAY, Flower of a *Tradescantia virginica* (Spiderwort) Nash. Self-explanatory. Cf. Day Spiderwort.

DAY DAISY *Leucanthemum vulgare* (Ox-eye Daisy). References lost. Possibly this is Midsummer Day Daisy. It certainly can't have any connection with the length of flowering, and in any case, given the etymology of daisy, it sounds nonsense.

DAY FLOWER *Cistus ladaniferus* (Gum Cistus) Lincs. B & H. Given in B & H without comment.

DAY LILY 1. *Hemerocallis flava*. 2. *Taraxacum officinale* (Dandelion) Som. Macmillan. The dandelion name must refer to daytime flowering rather than duration. Dandelion flowers do open with the sun.

DAY LILY, Narrow-leaved *Hemerocallis minor*.

DAY LILY, Tawny *Hemerocallis fulva*.

DAY NETTLE 1. *Galeopsis tetrahit* (Hemp Nettle) Yks, Border, Aber, Mor. B & H. 2. *Lamium album* (White Deadnettle) Yks. Carr; Scot. Jamieson. 3. *Lamium purpureum* (Red Deadnettle) Yks. Carr. Day means something different again in this case. It is Dea-nettle, i.e. deaf nettle, an equivalent of dead nettle. A S Palmer's explanation looks doubtful, to say the least. He suggested that it meant the nettle injurious to labourers, OE deyes! It was believed to affect them with whitlows.

DAY SPIDERWORT *Tradescantia virginica* (Spiderwort) Parkinson. Cf. Flower of a Day for this.

DAYBERRY *Ribes uva-crispa* (Gooseberry) Corn. Jago; Dev, Kent Grigson. The sequence runs Dabbery, Dayberry, Deberry, and Dewberry, the last from Culpepper. A S Palmer suggests, reasonably, that this is thape, or theap, both names for gooseberry, plus berry.

DAYLIGHT, Shepherd's *Anagallis arvensis* (Scarlet Pimpernel) Som. Elworthy. Elworthy wasn't sure whether the word was 'delight' or 'daylight'. Either is possible, but the pronunciation would have been as 'day-light'.

DAZEG *Bellis perennis* (Daisy) Cumb. B & H. This is getting very close to the OE, where daisy was daeges eage.

DAZZLE 1. *Cirsium arvense* (Creeping Thistle) Dev. Friend.1882. 2. *Taraxacum officinale* (Dandelion) Corn, Dev. Friend. An appropriate enough name for the dandelion, but it actually means thistles, or at any rate a daisy-like flower of that family. Cf. Dashel, or Dashel-flower, from the same area.

DEA NETTLE 1. *Galeopsis tetrahit* (Hemp Nettle) N'thants. Sternberg; Scot. Jamieson. 2. *Lamium album* (White Deadnettle) Shrop, Yks, Cumb. Grigson. 3. *Lamium purpureum* (Red Deadnettle) Worcs, Yks, Cumb. Grigson. Deaf Nettle, which can appear as this, or Daa, Dee, etc.

DEAD CREEPERS *Bryonia dioica* (White Bryony) Lancs. Grigson. There are also Death Warrant, and Murren, which are all presumably mandrake names, for this plant became known as the English Mandrake.

DEAD MAN *Orobanche minor* (Lesser Broomrape) Wilts. Macmillan. Because of the colourless, deathly appearance.

DEAD MAN'S BELLOWS 1. *Ajuga reptans* (Bugle) Berw. Grigson. 2. *Digitalis purpurea* (Foxglove) N.Eng. Grigson. 3. *Pedicularis palustris* (Red Rattle) Som. Macmillan; Berw. Grigson. Bellows is the same word as pillies, i.e. male genitals.

DEAD MAN'S BELLS 1. *Digitalis purpurea* (Foxglove) Dor. Macmillan; Lancs, N'thum. Grigson; Scot. Jamieson. 2. *Fritillaria meleagris* (Snake's Head Lily) Friend. 3. *Silene maritima* (Sea Campion) Mor. PLNN22.

DEAD MAN'S BONES 1. *Linaria vulgaris* (Toadflax) USA Coats. 2. *Stellaria holostea* (Greater Stitchwort) Berw. Grigson. The stitchwort probably gets its 'bones' names from the way the stalks snap off.

DEAD MAN'S CREESH *Oenanthe crocata* (Hemlock Water Dropwort) Dumf. Grigson. It is also called Dead Man's Tongue, or Dead Tongue, "from its paralysing effect on the organs of voice" (Prior).

DEAD MAN'S FINGERS 1. *Arum maculatum* (Cuckoo-pint) Worcs. B & H. 2. *Dactylorchis incarnata* (Early Marsh Orchid) Border B & H. 3. *Dactylorchis maculata* (Heath Spotted Orchid) Border. B & H. 4. *Digitalis purpurea* (Foxglove) Som. Grigson. 5. *Gentianella amarella* (Autumn Gentian) Shet. Grigson. 6. *Lotus corniculatus* (Bird's-foot Trefoil) Hants. Grigson. 7. *Oenanthe crocata* (Hemlock Water Dropwort) North. 8. *Orchis mascula* (Early Purple Orchis) Shakespeare. 9. *Orchis morio* (Green-winged Orchis) Suss. B & H. Grigson describes the half-open buds of Autumn Gentian as looking like livid finger-nails protruding from the turf. For the orchids, the Dactyl-part of the generic name refers to the fact that the tubers are divided into several finger-like lobes. Hence the various 'hand' or 'finger' names applied to these. Foxglove and Hemlock Water Dropwort are both poisonous, and this probably is enough to account for the name. Bird's-foot Trefoil gets it possibly as mere description.

DEAD MAN'S FLESH *Anthriscus sylvestris* (Cow Parsley) Suff. PLNN 17. Probably from the same belief that gave names like Scabby hands, or Scabs.

DEAD MAN'S FLOURISH *Conium maculatum* (Hemlock) S Scot. Tynan & Maitland.

DEAD MAN'S GRIEF *Silene maritima* (Sea Campion) N'thum. Grigson.

DEAD MAN'S HAND 1. *Dactylorchis incarnata* (Early Marsh Orchid) Border B & H. 2. *Dactylorchis maculata* (Heath Spotted Orchid) Berw. Grigson. 3. *Orchis mascula* (Early Purple Orchis) Glos, Suss, War. B & H; Hants. Cope. The way orchids appear from the ground is enough to account for the connection with the dead. Dactylorchis has its

tubers shaped like a hand and fingers.
DEAD MAN'S THIMBLES *Digitalis purpurea* (Foxglove) Som. Macmillan; Ire. Skinner. Foxgloves are Dead Man's Fingers as well as thimbles.
DEAD MAN'S THUMB *Orchis mascula* (Early Purple Orchis) N.Eng. Dyer.
DEAD MAN'S TONGUE *Oenanthe crocata* (Hemlock Water Dropwort) Salisbury.1964. "from its paralysing effect on the organs of voice", according to Prior.
DEAD SEA APPLE 1. *Quercus lusitanicus* (Cyprus Oak) Howes. 2. *Solanum sodomeum* (Apple of Sodom) W.Miller.
DEADLY DWALE *Atropa belladonna* (Deadly Nightshade) B & H. Dwale is a word meaning torpor, or trance. It comes probably from a Norse word, dool, to delay or sleep. The name has been accepted since Chaucer's time as synonymous with opiate.
DEADLY NIGHTSHADE 1. *Atropa belladonna*. 2. *Solanum dulcamara* (Woody Nightshade) Forsyth. 3. *Solanum nigrum* (Black Bindweed) USA House. Quite erroneous for Woody Nightshade, but yet it is still called this in some parts of the country.
DEADNETTLE, Balm-leaved *Lamium orvala*.
DEADNETTLE, Cut-leaved *Lamium hybridum*. Deadnettle means the nettle which is dead, i.e. which does not sting.
DEADNETTLE, Hemp-leaved *Galeopsis tetrahit* (Hemp Nettle) Jamieson.
DEADNETTLE, Henbit *Lamium amplexicaule*.
DEADNETTLE, Hungary *Lamium orvala* (Balm-leaved Deadnettle) Whittle & Cook.
DEADNETTLE, Intermediate *Lamium molucellifolium*.
DEADNETTLE, Red *Lamium purpureum*.
DEADNETTLE, Spotted *Lamium maculatum*.
DEADNETTLE, White *Lamium album*.
DEADWEED *Erigeron canadensis* (Canada Fleabane) Jamaica Laguerre.
DEADWORT; DEATHWORT *Sambucus ebulus* (Dwarf Elder) Culpepper

(Deadwort); Folkard (Deathwort). It is said to grow where the blood of man has been shed. Note the Welsh name llysan gwaed gwyr, plant of the blood of man, and Cf. names like Bloodwort.
DEAF-AND-DUMB *Galeobdalon luteum* (Yellow Archangel) Som. Macmillan. A combination of two concepts - the deadnettles are often Deaf Nettles, or Dumb Nettles.
DEAF NETTLE 1. *Galeopsis tetrahit* (Hemp Nettle) Corn. Courtney.1880. 2. *Lamium album* (White Deadnettle) Dev. Friend.1882; Wilts. Dartnell & Goddard; Lincs, Yks. Grigson. 3. *Lamium purpureum* (Red Deadnettle) Dev. Friend.1882; Som. Elworthy; Yks. Grigson. Deaf Nettle is the same as Dead Nettle, Cf. also Blind, or Dumb, Nettle.
DEAL, White *Picea excelsa* (Spruce Fir) Polunin.1976. The name given to the timber, known as whitewood in the trade.
DEAL, Red; DEAL, Yellow; DEAL-TREE; DEAL-TREE, Fir *Pinus sylvestris* (Scots Pine) Wilkinson.1981 (Red Deal); S. Africa Watt (Yellow Deal); N'thants, E.Ang. B & H (Deal-Tree); N'thants. A.E.Baker (Fir Deal-tree). Deal is really the sawn timber, and comes from MLG dele, plank.
DEAL, Yellow, Baltic *Pinus sylvestris* (Scots Pine) Hutchinson & Melville. A lot of timber from this species has been imported from the Continent under names like this. Cf. Danzig Pine, etc.
DEAL-APPLE 1. *Picea abies* cones (Spruce Fir cones) B & H. 2. *Pinus sylvestris* cones (Pine cones) N'thants. B & H; Suff. Grigson.
DEALIES; DEALSEYS *Pinus sylvestris* cones (Pine cones) Suff. B & H (Dealies); Corn. Jago (Dealseys).
DEATH, Flower of *Vinca minor* (Lesser Periwinkle) Wales Trevelyan. There is a Welsh superstition that if you uproot a plant from a grave, the dead person beneath will appear to you, and you will have terrible dreams for a year. In Italy, garlands of these flowers were put on the coffins of dead chil-

dren, and the name fiore di morte applies there too.

DEATH, Young Man's *Convolvulus arvensis* (Field Bindweed) Perth. Vickery.1985. If a girl picks bindweed, her boy friend will die. Possibly because of the rapid fading of the flower?

DEATH ALDER *Euonymus europaeus* (Spindle Tree) Bucks. Grigson. An unlucky plant to bring into the house, but would this be enough to account for such a name? The berries are mildly poisonous, but surely not enough to justify the name? Apparently, though, Pliny said that too many spindle flowers foretells a plague.

DEATH BELL; DEITH BELL *Fritillaria meleagris* (Snake's Head Lily) Cumb. B & H (Death Bell); Friend (Deith Bell). "from the dingy, sad colour of the bell-shaped flowers" (B & H). Cf. Dead Man's Bell, etc.

DEATH CAMAS *Zigadenus gramineus*. A poisonous lily from America.

DEATH-COME-QUICKLY *Geranium robertianum* (Herb Robert) Cumb. Friend. Friend suggests that this refers to the superstition that if children pick the flowers, it will mean death to one of the parents. This superstition applies also to the campion - Herb Robert and Red Campion share a number of local names.

DEATH-FLOWER *Achillea millefolium* (Yarrow) Wales J.C.Davies. Trevelyan also has this. Yarrow is an unlucky plant, one of the large number that involve death (usually to a parent) if they are brought into the house. Cf. Mother-die.

DEATH OF MAN *Cicuta maculata* (American Cowbane) USA Weiner. A very reasonable name - it is highly poisonous.

DEATH VALLEY JOINT-FIR *Ephedra funerea*.

DEATH VALLEY LOCO *Astragalus funereus*.

DEATH VALLEY PHACELIA *Phacelia vallis-mortae*.

DEATH VALLEY SAGE *Salvia funerea*.

DEATH WARRANT *Bryonia dioica* (White Bryony) Dor. Macmillan. Why? Certainly the berries are poisonous, but not that poisonous.

DEATH'S FLOWER *Galanthus nivalis* (Snowdrop) Som. Macmillan. More than any other flower, snowdrops are associated with death. It is very unlucky to bring cut snowdrops into the house, something which would bring death with them. It is said this association results from the flower's resemblance to a shroud. Another reason is the fact that they are often found growing in old graveyards. Only Lily-of-the-valley, and white lilac has a reputation approaching it.

DEATH'S HERB; DEATHWEED *Atropa belladonna* (Deadly Nightshade) Folkard (Death's Herb); Tynan & Maitland. (Deathweed).

DEATHIN 1. *Cicuta virosa* (Cowbane) Scot. B & H. 2. *Oenanthe aquatica* (Horsebane) Border Grigson. Poisonous, of course, but less so than *Oe crocata*. But Cowbane is virulent, too.

DEBERRY *Ribes uva-crispa* (Gooseberry) Dev. Macmillan. Part of the sequence Dabbery, Dayberry, Deberry, Dewberry.

DECEIVER *Glechoma hederacea* (Ground Ivy) Som, Ess. Grigson.

DEER BRUSH *Ceanothus integerrimus*. Apparently because the deer are fond of eating it. Miwok hunters looked for the deer particularly where this grew.

DEER FERN *Blechnum spicant* (Hard Fern) USA Turner & Bell.

DEER('S)-GRASS 1. *Empetrum nigrum* (Crowberry) Donegal Grigson. 2. *Lycopodium inundatum* (Clubmoss) Scot. Gregor.1888. 3. *Trichophorum caespitosum*.

DEER NUT *Simmondsia chinensis* (Jajoba) USA T H Kearney.

DEER OAK *Quercus sadleriana*.

DEER VINE *Linnaea borealis var. americana* USA House.

DEER-WEED, Broom *Lotus scoparius*.

DEER-WEED, Pygmy *Lotus haydonii*.

DEER-WEED, Pale-leaved *Lotus leucophyllus*.

DEER'S EARS *Swertia radiata.*
DEER'S-FOOT BINDWEED *Convolvulus arvensis* (Field Bindweed) Chopra.
DEER'S MILK *Euphorbia amygdaloides* (Wood Spurge) Hants. Cope. The spurges have a milky juice.
DEER'S TONGUE *Liatris odoratissima.*
DEERBERRY 1. *Gaultheria procumbens* (Wintergreen) Grieve.1931. 2. *Mitchella repens* (Partridge-berry) Le Strange. 3. *Vaccinium stamineum* (Square Huckleberry) USA Zenkert.
DEFE NETTLE *Lamium album* (White Deadnettle) Corn. Jago.
DEITH-BELL *Fritillaria meleagris* (Snake's-head Lily) Cumb. B & H. Cf. Dead Man's Bell, both from the drooping flowers, as well as the sombre colouring.
DELICATE BESS 1. *Centranthus albus* Dev. B & H. 2. *Valeriana celtica* Dev. Friend.1882. Are these actually the same plant?
DELIGHT, Girl's; DELIGHT, Maiden's *Artemisia abrotanum* (Southernwood) Som. Macmillan (Girl's); Corn. Jago; Som. Macmillan (Maiden's). For the same reason as the common Boy's Love, and probably because a few branches of southernwood were generally added to the nosegay which courting youths gave the girls. A Fenland youth would cut some sprigs and put them in his buttonhole and then walk through groups of girls sniffing ostentatiously at them. If the girls went by and took no notice, he knew he had no chance, but if they turned and walked back slowly towards him, then he knew they had noticed his Lad's Love. He would then take his buttonhole and give it to the girl of his choice. If she was willing, she would also smell the southernwood and the two would set off together on their first courting stroll. Maiden's Delight, perhaps, but another name for this was Maiden's Ruin!
DELIGHT, Ladies' *Viola tricolor* (Pansy) Howes.
DELIGHT, Miller's *Centaurea cyanus* (Cornflower) Dor. Macmillan. There must be some obscure reason for this, to do with finding it among the wheat sent to him.
DELIGHT, Queen's *Stillingia treculeana.*
DELIGHT, Shepherd's 1. *Anagallis arvensis* (Scarlet Pimpernel) Som. Elworthy. 2. *Clematis vitalba* (Old Man's Beard) Som. Macmillan. 3. *Viburnum lantana* (Wayfaring Tree) Dor. Macmillan.
DELIGHT, Single *Moneses uniflora* (St Olaf's Candlesticks) USA Turner & Bell.
DELL *Pinus sylvestris* timber (Scots Pine) Dev. Choape. A variation on 'deal'.
DELL, Fairy 1. *Euphorbia helioscopia* (Sun Spurge) Dor. Macmillan; 2. *Euphorbia peplus* (Petty Spurge) Dor. Macmillan.
DELLCUP 1. *Ranunculus acris* (Meadow Buttercup) Som. Grigson. 2. *Ranunculus bulbosus* (Bulbous Buttercup) Som. Grigson. 3. *Ranunculus repens* (Creeping Buttercup) Som. Grigson. This word is better known as Dalecup, and even better known as Gilcup, i.e. gold cup.
DELSEED *Pinus sylvestris* cones (Pine cones) Corn. Jago. The first syllable must be the same as deal.
DELT-ORACH *Atriplex patula* (Orache) Prior. From the triangular leaves, like the Greek letter delta.
DELTA MAIDENHAIR *Adiantum raddianum.*
DELTY-CUP *Ranunculus repens* (Creeping Buttercup) Wilts. Dartnell & Goddard.1899.
DENT MAIZE; DENT CORN *Zea mays var. americana.* There is an indentation or depression on the top of the grain, caused by shrinkage of the endosperm.
DENTILINN; DENTELION; DENTYLION *Taraxacum officinale* (Dandelion) all Border, Scot. all B & H.
DEODAR; DEODAR CEDAR *Cedrus deodara.* Deodar Cedar is in B Thomas. Sanskrit Devadara, divine tree, or tree of the gods.
DEPTFORD PINK *Dianthus armeria.* "groweth in our pastures neere about London, and in other places, but especially in the great field next to Detford" (Gerard) - but he probably wasn't describing this species at all.

DERRIS *Derris elliptica*.
DESERT BAILEYA *Baileya multiradiata*.
DESERT BOX *Eucalyptus microtheca* (Coolibah) Howes.
DESERT JUNIPER *Juniperus californica* (California Juniper) Elmore.
DESERT MARIGOLD *Baileya multiradiata* (Desert Baileya) USA T H Kearney.
DESERT NUT PINE *Pinus monophylla* (Pinon) Mason.
DESERT SALTBUSH *Atriplex polycarpa* (Cattle Spinach) USA T H Kearney.
DESERT TRUMPET FLOWER *Datura meteloides* (Downy Thorn-apple) Safford.
DESHALGO *Tussilago farfara* (Coltsfoot) Scot. Tynan & Maitland. One of a number of the older attempts at Tussilago. Cf. Dishilago, Dishalga, etc., all apparently from Scotland.
DESSERT BANANA *Musa x paradisiaca* (Banana) H & P.
DEUTSA *Dicentra spectabilis* (Bleeding Heart) Dev. Friend.1882. Presumably an attempt at either *Dicentra* or *Dielytra*, an older name for the genus.
DEVIL, Angels-and- *Arum maculatum* (Cuckoo-pint) Som. Macmillan. Or Devils-and-angels.
DEVIL, Blue 1. *Echium vulgare* (Viper's Bugloss) Som. Macmillan; USA. Grigson. 2. *Iris foetidissima* (Foetid Iris) Som. Macmillan.
DEVIL, Chase-the- *Nigella damascena* (Love-in-a-mist) B & H. Probably for the same reason that produced Love-in-a-puzzle, etc.
DEVIL, Clover *Cuscuta epithymum* (Common Dodder) Som. Macmillan. Note the legend in Haute Bretagne that the devil spun the dodder at night to destroy the clover; clover was created by God, dodder was the devil's counter-plant. Dodder was usually associated with the devil in this country, too. Cf. Clover Dodder.
DEVIL, King 1. *Hieraceum florentinum*. 2. *Hieraceum pratense* USA Zenkert.
DEVIL, Knutsford *Calystegia sepium* (Great Bindweed) Ches. Holland.
DEVIL, Mexican *Eupatorium adenophorum*.
DEVIL, Prick *Scandix pecten-veneris* (Shepherd's Needle) Tynan & Maitland.
DEVIL, Tether 1. *Polygonum convolvulus* (Black Bindweed) Ches. Grigson. 2. *Solanum dulcamara* (Woody Nightshade) Ches. Holland. Tether-devil becomes Terrydevil, and then Terry-diddle.
DEVIL, Yellow *Iris pseudo-acarus* (Yellow Flag) Som. Macmillan. Blue Devil is the Foetid Iris.
DEVIL-AMONG-THE-TAILORS *Nigella damascena* (Love-in-a-mist) Som. Macmillan.
DEVIL-AND-ANGELS 1. *Arum maculatum* (Cuckoo-pint) Som. Macmillan; Dor. Dacombe. 2. *Plantago lanceolata* (Ribwort) Som. Macmillan.
DEVIL DAISY 1. *Anthemis cotula* (Maydweed) Wilts. Macmillan. 2. *Chrysanthemum parthenium* (Feverfew) Som. Macmiullan; Wilts. Dartnell & Goddard.
DEVIL-IN-A-BUSH; DEVIL-IN-THE-BUSH 1. *Centaurea cyanus* (Cornflower) Breck. M E Hartland. 2. *Nigella damascena* (Love-in-a-mist) Som. Elworthy. Glos. Dyer. 3. *Paris quadrifolia* (Herb Paris) Perth. B & H. Like Love-in-a-mist, a typical name for a flower which is half hidden by its own sepals.
DEVIL-IN-A-FRIZZLE; DEVIL-IN-A-FOG; DEVIL-IN-A-MIST *Nigella damascena* (Love-in-a-mist) All Tynan & Maitland.
DEVIL-IN-A-HEDGE *Nigella damascena* (Love-in-a-mist) Wilts. Dartnell & Goddard.
DEVIL-IN-CHURCH *Borago officinalis* (Borage) Dor. Macmillan.
DEVIL-IN-THE-PULPIT *Tradescantia virginica* (Spiderwort) Perry.
DEVIL-MAY-CARE 1. *Allium ursinum* (Ramsons) Gordon.1985. 2. *Cornus sanguinea* (Dogwood) Dor. Macmillan. It does not mean 'happy-go-lucky', just the opposite. Any mention of the devil in a plant name will be a comment on some unpleasant aspect, the

foul smell in the case of Ramsons. Dogwood is not so obvious, but it is an unlucky tree - it should be avoided, especially when it is in bloom.

DEVIL-ON-ALL-SIDES; DEVIL-ON-BOTH-SIDES *Ranunculus arvensis* (Corn Buttercup) Yks Grigson (all sides); Bucks, War, Dur. Dyer (both sides). The prickly fruits make this plant a pest, because it is tall enough to be bound with the corn. Cf. such names as Devil's Curry-comb, Hellweed, etc.

DEVIL TREE *Erythrina corallodendron* (Coral Tree) Howes.

DEVIL-WEED *Convolvulus arvensis* (Field Bindweed) Suss. A typical name for a much-hated plant. Cf. Hellweed.

DEVIL'S APPLE 1. *Datura stramonium* (Thorn-apple) USA Henkel. 2. *Mandragora officinalis* (Mandrake) W.Miller. 3. *Solanum aculeastrum* S.Africa Watt. Because the fruits are in each case narcotic, and they look like small apples, anyway, and smell strongly of them, too.

DEVIL'S APPLE-RINGIE *Matricaria recutita* (Scented Nayweed) B & H. "Devil's", because it is not the real Apple-Ringie, which is southernwood.

DEVIL'S APPLE TREE *Euphorbia helioscopia* (Sun Spurge) Clack. B & H.

DEVIL'S APRON *Urtica dioica* (Nettle) Ire. C.J.S.Thompson. The sting has given a number of associations with the devil - it is the Devil's Leaf, Plaything, etc.

DEVIL'S BANE 1. *Hypericum hirsutum* (Hairy St John's Wort) Som. Macmillan. 2. *Verbena officinalis* (Vervain) Wales Clair. These are both protective plants, with the gift of putting evil to flight.

DEVIL'S BEARD 1. *Anemone alpina* Friend. 2. *Pulsatilla vulgaris* (Pasque Flower) Tynan & Maitland. 3. *Sempervivum tectorum* (Houseleek) Wales C.J.S.Thompson. The appearance of houseleek attracted a number of 'beard' names, starting with Latin Iovis barba, and proceeding through Jove, Jupiter and Thor until Christianity relegated all of them to the devil.

The name for Pasque Flower seems doubtful, but it is claimed that the hairy stalks, or the bushy head of seeds, might account for it.

DEVIL'S BEGGAR-TICKS *Bidens frondosa* (Beggar-ticks) Allan.

DEVIL'S BERRIES 1. *Atropa belladonna* (Deadly Nightshade) Dyer. 2. *Hypericum androsaemum* (Tutsan) Corn. Grigson. 3. *Tamus communis* (Black Bryony) L Gordon. Deadly Nightshade and Bryony are obvious candidates for association with the devil, but why should such a plant as tutsan, which is toute saine, have the dubious honour? It was probably a case of someone playing safe - any succulent berries should be first regarded with suspicion.

DEVIL'S BIT 1. *Centaurea nigra* (Knapweed) Som. Macmillan. 2. *Hieraceum pilosellum* (Mouse-ear Hawkweed) Dyer. 3. *Knautia arvensis* (Field Scabious) Corn. Deane & Shaw. 4. *Succisa pratensis*. It is not clear why the hawkweed got the name, but the knapweed and Field Scabious certainly only got it by association. Devil's Bit, which has a short, blackish root, got it (the root and the name) because it was bitten off by the devil, out of spite to mankind, because he knew that otherwise it would be good for many purposes. The story is quite widespread on the Continent, too. Another version tells the opposite. The plant was a bane to mankind, and God took the power away from the devil, who bit the root off in vexation.

DEVIL'S BITE 1. *Liatris spicata* (Button Snakeroot) Grieve.1931. 2. *Succisa pratensis*. Obviously the same as Devil's Bit. Why did the American plant get the name? - presumably 'Button' in the common name is the clue.

DEVIL'S BLANKET *Verbascum thapsus* (Mullein) Wilts. Macmillan. Mullein's felty leaves seem to have served as a blanket or a flannel for a lot of people, from Adam onwards, but assigning it to the devil is completely out of character.

DEVIL'S BLOSSOM *Conium maculatum* (Hemlock) Dev. Macmillan. Poisonous, hence the allusion to the devil. It is Devil's Flower, too, in Somerset, and Devil's Oats in the northern counties.

DEVIL'S BOOTS *Sarracenia purpurea* (Pitcher Plant) Howes.

DEVIL'S BREAD *Conopodium majus* (Earth-nut) Yks. Grigson. Cf. Devil's Oatmeal, from the same area. Why the ascription, though? It is perfectly edible and wholesome.

DEVIL'S BRUSH *Pteridium aquilina* (Bracken) Waring.

DEVIL'S BUTTERCUP *Ranunculus scleratus* (Celery-leaved Buttercup) Tynan & Maitland. Presumably because this is the most poisonous of all the buttercups.

DEVIL'S BUTTON 1. *Geum rivale* (Water Avens) Tynan & Maitland. 2. *Knautia arvensis* (Field Scabious) Corn. Deane & Shaw. 3. *Succisa pratensis* (Devil's Bit) Corn. Courtney.1880. The Field Scabious gets the name by association, and the Devil's Bit from its more usual name, not from any sinister circumstance, though it is said that if you pick it the devil will appear by your bedside that night.

DEVIL'S CANDLE 1. *Mandragora officinalis* (Mandrake) C.J.S.Thompson. 2. *Zantedeschia aethiopica* (Arum Lily) Corn. Deane & Shaw. The association with funerals is enough to give the Arum Lily a bad name. Mandrake is more complicated.The name is actually only a translation from the Arabic. 'Devil's' because this is a plant with poisonous fruits and a very evil reputation; 'candle', because of an odd belief that it gave out a light at night. Flavius Josephus said that in the dungeons of the castle of Machaeras at Baras, there grew a root that was flame coloured and shone like lightning on persons who attempted to approach. In the 2nd century AD Aelian described a root which he called Aglaophotis, because it shone like a star by night. There may be some basis in fact for the belief that mandrake shines at night. For some reason its leaves are attractive to glowworms. This is from a 13th century Arab herbalist called Ebn Beita.

DEVIL'S CANDLESTICKS 1. *Glechoma hederacea* (Ground Ivy) Som, War. Dyer; Yks. Addy. 2. *Onobrychis sativa* (Sainfoin) Som. Macmillan. Sainfoin gets the name as a description of the bright pink flowers. Anyway, Lucerne, another name for this, comes from Latin lux, lucis, from the shiny grains.

DEVIL'S CHERRIES 1. *Atropa belladonna* (Deadly Nightshade) Northants. Tynan. & Maitland. 2. *Bryonia dioica* (White Bryony) Dyer. 3. *Solanum dulcamara* (Woody Nightshade) Som. Macmillan. Any plant with poisonous berries would get this name, and if you aren't sure and want to keep children away from them you call them this anyway.

DEVIL'S CHURN *Euphorbia peplus* (Petty Spurge) Ire. Gomme. Devil's Churnstaff is more usual, and makes more sense if this is a descriptive name - see *Euphorbia helioscopia*.

DEVIL'S CHURNSTAFF; DEVIL'S KIRN-STAFF *Euphorbia helioscopia* (Sun Spurge) Shrop, N.Ire. Burne.1914 (Churnstaff); Ayr, Lanark. Grigson (Kirnstaff).

DEVIL'S CLAWS 1. *Acacia greggii* (Cat's Claws) USA Jaeger. 2. *Lotus corniculatus* (Bird's-foot Trefoil) Dev. Friend. 3. *Phyteuma comosus*. 4. *Proboscidea louisiana* (Unicorn Plant) USA Bean. 5. *Ranunculus arvensis* (Corn Buttercup) Wilts. D.Grose. Hants, IOW. Grigson.

DEVIL'S CLOVER *Medicago arabica* (Spotted Medick) Corn. Grigson.

DEVIL'S COACHWHEEL *Ranunculus arvensis* (Corn Buttercup) Hants. Cope. Probably because of the prickly fruits, which used at one time to make the plant a pest in the corn.

DEVIL'S CORN *Stellaria holostea* (Greater Stitchwort) Shrop. Burne; Wales Fernie. In various parts of the country, stitchwort is the Devil's flower, nightcap and his eyes.

DEVIL'S CUPS AND SAUCERS *Euphorbia*

amygdaloides (Wood Spurge) Som. Macmillan. This must be an allusion to the way the leaves grow from the stem, leaving a little basin as they do.

DEVIL'S CURRY-COMB *Ranunculus arvensis* (Corn Buttercup) Hants, Shrop. Burne. The prickly fruits give this name.

DEVIL'S CUT 1. *Clematis integrifolia* (Hungarian Climber). 2. *Clematis vitalba* (Old Man's Beard) B & H. The 'cut' here is a cut of tobacco. Boys used to smoke the porous stalks. At least, it has to be presumed this is the reason for the name - for Devil's Guts also appears, a typical bindweed name.

DEVIL'S DAISY 1. *Leucanthemum vulgare* (Ox-eye Daisy) Middx B & H. 2. *Tanacetum parthenium* (Feverfew) Som, Wilts. Grigson.

DEVIL'S DARNING NEEDLE 1. *Clematis virginiana* Howes. 2. *Scandix pecten-veneris* (Shepherd's Needle) Som. Macmillan.

DEVIL'S DEW *Nigella damascena* (Love-in-a-mist) Som. Macmillan. The dew here must be the same as mist. Devil, presumably, because of the horned capsule. Cf. Devil-in-a-bush, etc., for this.

DEVIL'S DUNG *Ferula foetida* (Asafoetida) Mitton.

DEVIL'S DYE *Indigofera anil* (Indigo) C.J.S.Thompson. As a protection for woad, indigo was branded as "the food of the devil" in this country, but it was the German woad merchants who first designated it devil's dye.

DEVIL'S EAR 1. *Arisaema triphyllum* (Jack-in-the-pulpit) Le Strange. 2. *Stellaria holostea* (Greater Stitchwort) Leyel.1937.

DEVIL'S ENTRAILS *Convolvulus arvensis* (Field Bindweed) Wales Trevelyan. Trevelyan must have smartened this up - it is much better known as Devil's Guts (so are a number of other plants, all climbers or twiners).

DEVIL'S EYE 1. *Hyoscyamus niger* (Henbane) Som. Macmillan. 2. *Stellaria holostea* (Greater Stitchwort) Dor. Macmillan. Wales C.J.S.Thompson. 3. *Veronica chamaedrys* (Germander Speedwell) War. Grigson.

DEVIL'S FIERY POKER *Kniphofia uvaria* (Red Hot Poker) Som. Macmillan.

DEVIL'S FIG *Argemone mexicana* (Mexican Poppy) Perry.

DEVIL'S FINGERS 1. *Lotus corniculatus* (Bird's-foot Trefoil) Dev. Friend. 2. *Populus nigra* - male catkins (Black Poplar catkins). Grigson. 3. *Ranunculus arvensis* (Corn Buttercup) Jacob. The poplar catkins, at least, are unlucky - they will bring ill fortune if picked up.

DEVIL'S FLIGHT *Hypericum perforatum* (St John's Wort) Dev Hewett. This is the medieval book name Fuga Daemonum, and the plant is still known as chasse-diable in French. This is the reputation of St John's Wort, being calculated to drive away all "fantastical spirits".

DEVIL'S FLOWER 1. *Conium maculatum* (Hemlock) Som. Macmillan. 2. *Iris pseudoacarus* (Yellow Iris) Newfoundland Briggs.1980. 3. *Melandrium dioicum* (Red Campion) Ches. Friend. 4. *Spergula arvensis* (Corn Spurrey) Corn. Grigson. 5. *Stellaria holostea* (Greater Stitchwort) Som. Macmillan. 6. *Veronica chamaedrys* (Germander Speedwell) Dumf. B & H. Any poisonous plant would be a good candidate for the name, so that accounts for hemlock. Corn Spurrey can be a nuisance, as the name Beggarweed would indicate; any pestilential plant like this is often consigned to the devil. Stitchwort and Germander Speedwell have both got an unlucky side to their natures, hence the reference. That leaves the campion. It may be the association with Robin Goodfellow that is the cause - it is Robin's Flower, as well as the devil's. As far as Yellow Iris is concerned, it is probably the colour that invites the name - yellow is the devil's colour.

DEVIL'S GARTER 1. *Calystegia sepium* (Great Bindweed) Pemb, Ire. Grigson. 2. *Convolvulus arvensis* (Field Bindweed) Som. Macmillan; Ire. Friend.

DEVIL'S GILLOFER *Cheiranthus cheiri*

(Wallflower) Som. Macmillan. The Gillofer part of the name appears in various forms, but is best known as gilliflower. Why 'Devil's' though? Perhaps a connection with the other name, Bloody Warrior.

DEVIL'S GUTS 1. *Calystegia sepium* (Great Bindweed) Norf. Grigson. 2. *Clematis vitalba* (Old Man's Beard) Som. Macmillan; Dor. Dacombe; Wilts. Goddard. 3. *Convolvulus arvensis* (Field Bindweed) Som, IOW, Kent, Shrop, Beds, Notts, Lancs, N'thum. Grigson. 4. *Cuscuta epithymum* (Common Dodder) Som, Glos, Worcs, Shrop, Hants, IOW, Beds, Suss, Cumb, Lanark. Grigson. 5. *Cuscuta europaea* (Greater Dodder) Glos, Suss, Worcs, Shrop, Cambs. B & H. 6. *Ranunculus repens* (Creeping Buttercup) N'thum. Grigson. 7. *Succisa pratensis* (Devil's Bit) Som. Tongue. A most expressive description for most of these. Devil's Bit is the odd one out, and the twist in the story there is that when the devil bit off the taproot, all the small roots twisted round him, and nearly strangled him. Cf. Devil's Entrails for Field Bindweed.

DEVIL'S HAND *Dactylorchis maculata* (Heath Spotted Orchis) Emboden.1974. Devil's possibly because of the dark coloured tubers (Cf. Black Mary's Hand). Hand, like fingers, is descriptive of the tubers again.

DEVIL'S HATE 1. *Sorbus aucuparia* (Rowan) Wales T.G.Jones. 2. *Verbena officinalis* (Vervain) Wales T.G.Jones. A natural name for both of these. Vervain is the Holy Herb, par excellence, and rowan is the prime protector from all evil. But do either of them exist in the English? They are translations from the Welsh.

DEVIL'S HATTIES *Silene maritima* (Sea Campion) Mor. PLNN22. An unlucky plant, hence the ascription to the devil.

DEVIL'S HEAD 1. *Linaria vulgaris* (Toadflax) Le Strange. 2. *Plantago lanceolata* (Ribwort) Dyer. Cf. Devil's Ribbon for toadflax.

DEVIL'S HERB 1. *Atropa belladonna* (Deadly Nightshade) Clair. 2. *Datura stramonium* (Thorn-apple) Haining. 3. *Plumbago scandens* (Wild Plumbago) Perry. An obvious name for the two narcotics. The wild plumbago gets the name, it is said, from its ability to cause blistering.

DEVIL'S HORN *Proboscidea althaefolia*.

DEVIL'S IVY *Scindapsus pictus 'Argyraeus'*.

DEVIL'S LADIES AND GENTLEMEN *Arum maculatum* (Cuckoo-pint). A composite name - Lords-and-ladies, or Ladies-and-gentlemen, is quite common, and is one of a string of double names, like Parson-in-the-pulpit, or Bulls-and-cows, etc. The ascription to the devil here is probably a reflection of the mistrust often felt (with reason) for the shiny berries.

DEVIL'S LEAF *Urtica dioica* (Nettle) Som. Macmillan. There is Devil's Plaything, too, from the same area. It is, of course, a reference to the vicious sting the leaf gives.

DEVIL'S LINGELS *Polygonum aviculare* (Knotgrass) N'thum. Grigson. Lingels is probably tongues. Halliwell has, from admittedly a long way from Northumberland (from Oxfordshire in fact) the verb to linge, one of whose meanings is to loll out the tongue. Knotgrass has a number of other 'tongue' names - Bird's, and Sparrow's, Tongue, for instance.

DEVIL'S MEAL; DEVIL'S MEAT *Anthriscus sylvestris* (Cow Parsley) Dumb. B & H (Meal); Yks. Grigson (Meat). Meal from the powdery appearance of the flowers. But why the connection with the devil? It is such a popular and inoffensive plant. The answer is probably a misapplication from hemlock; the two plants share a number of names.

DEVIL'S MEN AND WOMEN *Arum maculatum* (Cuckoo Pint) Shrop. Burne. A name given to the fruits, but the whole plant is known as Devil's Ladies and Gentlemen, and there are various names of this kind, such as Knights-and-ladies, Bulls-and-cows, etc.

DEVIL'S MILK 1. *Chelidonium majus* (Greater Celandine) Yks. Grigson. 2. *Euphorbia helioscopia* (Sun Spurge) Wales

Trevelyan. Worcs, Middx. Grigson. 3. *Euphorbia peplus* (Petty Spurge) C.J.S.Thompson. Any plant with an acrid juice is liable to get a name like this.

DEVIL'S MILKPAIL; DEVIL'S MILK-PLANT *Taraxacum officinale* (Dandelion) Som. Macmillan (Milkpail); Kirk. Grigson (Milkplant). The milky sap gives the name.

DEVIL'S NET *Cuscuta epithymum* (Common Dodder) Kent Grigson. The strangling effect of the stems accounts for the name. Cf. Devil's Thread.

DEVIL'S NETTLE *Achillea millefolium* (Yarrow) Ches. B & H. An odd name for this - it isn't a nettle, of course. But it wasn't given the names Sneezewort and Nosebleed for nothing. There are a number of ascriptions to the devil for yarrow, and they may be tokens of disapproval of the plant's use in divinations and spells.

DEVIL'S NIGHTCAP 1. *Aconitum napellus* (Monkshood) War. F Savage. 2. *Calystegia sepium* (Great Bindweed) Som. Macmillan. 3. *Convolvulus arvensis* (Field Bindweed) N.Eng. Wakelin. 4. *Stellaria holostea* (Greater Stitchwort) Som. Macmillan. 5. *Torilis japonica* (Hedge Parsley) War. Palmer.

DEVIL'S OATMEAL 1. *Conopodium majus* (Earth-nut) Yks. Grigson. 2. *Heracleum sphondyllium* (Hogweed) War. Grigson. 3. *Petroselinum crispum* (Parsley) Wilts. Wiltshire. "Only the wicked can grow it", Mrs Wiltshire noted of parsley. And, of course, there are all the beliefs that parsley goes nine times to the devil before it will germinate. But why give earthnut the name? It is perfectly edible.

DEVIL'S OATS *Conium maculatum* (Hemlock) Cumb, N'thum, Dur. R.L.Brown. Cf. Devil's Blossoms and Devil's Flower from southern England.

DEVIL'S PAINKILLER *Hiercaeum aurantiacum* (Orange Hawkweed) Howes.

DEVIL'S PAINTBRUSH *Hieraceum aurantiacum* (Orange Hawkweed) USA House.

DEVIL'S PARSLEY *Anthriscus sylvestris* (Cow Parsley) Ches. B & H. It does seem to have some connection with the devil, possibly because of a confusion with hemlock. Cf. Satan's Bread, Devil's Meat, etc for this.

DEVIL'S PINCH *Polygonum persicaria* (Persicaria) Dor. Macmillan. It is more usual to find it called Virgin's Pinch, or Pinchweed. The legend is that the Virgin Mary plucked a root, left her mark om the leaf, and threw it aside as useless.

DEVIL'S PLAGUE *Daucus carota* (Wild Carrot) USA Howes.

DEVIL'S PLAYTHING 1. *Achillea millefolium* (Yarrow) Fernie. 2. *Stachys officinalis* (Wood Betony) Shrop. Grigson. 3. *Urtica dioica* (Nettle) Som. Macmillan. Why should betony get a name like this? It is one of the most esteemed of plants. Yarrow and nettle are understandable, though - see Devil's Nettle.

DEVIL'S POKER 1. *Kniphofia uvaria* (Red Hot Poker) Dev. Friend.1882. 2. *Typha latifolia* (False Bulrush) Som. Macmillan.

DEVIL'S POSY 1. *Allium ursinum* (Ramsons) Shrop. B & H. 2. *Allium vineale* (Crow Garlic) Shrop. B & H. 3. *Iris pseudoacarus* (Yellow Iris) Wales Trevelyan. If it is the smell that is responsible for the name, then surely the wrong iris has been nominated - it must be Stinking Iris? It is certainly appropriate for Ramsons and Crow Garlic.

DEVIL'S POTATO *Solanum carolinense* (Carolina Nightshade) Allan. Cf. Devil's Tomato for this.

DEVIL'S RATTLE *Achillea millefolium* (Yarrow) M.Baker. Cf. Devil's Nettle, or Plaything, for this. But is this a mis-reading of Nettle?

DEVIL'S RHUBARB 1. *Atropa belladonna* (Deadly Nightshade) Som. Macmillan. 2. *Petasites hybridus* (Butterbur) War,Worcs. F Savage.

DEVIL'S RIBBON 1. *Antirrhinum majus* (Snapdragon) Tynan & Maitland. 2. *Linaria vulgaris* (Yellow Toadflax) Dyer. Yellow is the devil's colour, so this name for toadflax makes some sense, but not for snapdragon, and it may be an error.

DEVIL'S ROOT *Orobanche minor* (Lesser Broomrape) Kent Grigson. There is Hellroot, too, from the same area. Both of them are presumably because of the plant's thoroughly unnatural appearance.

DEVIL'S SHIRT BUTTONS *Stellaria holostea* (Greater Stitchwort) Som. Macmillan. Shirt Buttons is alright - that is surely descriptive. But in spite of its cheerful nature, it seems to have been an un-popular plant, possibly because it is under the protection of the fairies.

DEVIL'S SHOESTRINGS 1. *Coronilla varia* (Crown Vetch) USA Puckett. 2. *Tephrosia virginiana*. 3. *Viburnum alnifolium* (Hobblebush) USA Perry. The lower branches of this American Viburnum droop to the ground and root at the tips, hence this name and others like Tangle-legs, etc.

DEVIL'S SPIT *Centaurea nigra* (Knapweed) Som. Macmillan. There is Devil's Bit, too, for this - that sounds more likely.

DEVIL'S SPOONS *Alisma plantago-aquatica* (Water Plantain) Scot. Jamieson.

DEVIL'S TETHER *Polygonum convolvulus* (Black Bindweed) Ches,Yks Grigson. Tether-devil is also known. They are both in the realm of bindweed names.

DEVIL'S THORN *Tribulus terrestris* (Puncture Vine) South Africa Dalziel.

DEVIL'S THREAD 1. *Clematis vitalba* (Old Man's Beard) Dyer. 2. *Cuscuta epithymum* (Dodder) Kent Grigson.

DEVIL'S TOBACCO *Heracleum sphondyllium* (Hogweed) Staffs. PLNN 17.

DEVIL'S TOMATO *Solanum carolinense* (Carolina Nightshade) Allan. It may look like a tomato, but it definitely isn't. Cf. Devil's Potato for this.

DEVIL'S TONGUE *Papaver rhoeas* (Red Poppy) Corn. Grigson.

DEVIL'S TORCH *Kniphofia uvaria* (Red Hot Poker) Som. Macmillan.

DEVIL'S TREE 1. *Bombax ceiba* (Red Silkcotton Tree) W. Indies HJ Bell. 2. *Juniperus sabina* (Savin) Fernie. In keeping with the bad reputation savin had. It is, of course, a well-known abortifacient, but another name, Magician's Cypress, suggests the reputation went further than this. The Silkcotton Tree gets the name because it is the tree most likely to harbour spirits, or jumbies.

DEVIL'S TRUMPET *Datura stramonium* (Thorn-apple) USA Henkel. A narcotic, hence the ascription to the devil.

DEVIL'S TURNIP *Bryonia dioica* (White Bryony) North. Navet du diable occurs in French, too. They are both to do with the use of the root as a mandrake substitute.

DEVIL'S TWINE 1. *Clematis vitalba* (Old Man's Beard) Skinner. 2. *Convolvulus arvensis* (Field Bindweed) N.Eng. Wakelin.

DEVIL'S VINE *Calystegia sepium* (Great Bindweed) Salisbury.1964.

DEVIL'S WALKING STICK *Aralia spinosa* (Hercules' Club) USA Kingsbury.

DEVIL'S WAND *Aethusa cynapium* (Fool's Parsley) Dor. Macmillan. Poisonous, hence the ascription to the devil.

DEVIL'S WOOD *Sambucus nigra* (Elder) Wilts. Wiltshire Folk Life, vol 3, no 10, 1983; Derb. Grigson.

DEVIL'S WORT *Eranthis hyemalis* (Winter Aconite) Som. Macmillan. "This herb is counted to be very dangerous and deadly", said Gerard. Certainly it is toxic to animals, but that doesn't seem enough to brand the flower in this way.

DEVIL'S YARN *Clematis vitalba* (Old Man's Beard) Trevelyan. Any climbing or twining plant like this merits the name, whether it is Devil's yarn, twine, or thread.

DEVILDUMS *Senecio jacobaea* (Ragwort) Dor. Macmillan.

DEVILWEED, Mexican *Aster spinosus* (Spiny Aster) USA Elmore. Because of the spines, presumably.

DEVILWOOD *Osmanthus americanus*.

DEVON EAVER; DEVON EVVER *Lolium temulentum* (Darnel) both Som. Elworthy. Or just Eaver in both Somerset and Devon.

DEVON PRIDE *Centranthus ruber* (Spur

Valerian) Dev. Macmillan.
DEVONSHIRE MYRTLE *Myrica gale* (Bog Myrtle) Som. Grigson.
DEW, Devil's *Nigella damascena* (Love-in-a-mist) Som. Macmillan. 'Dew' in this case must be the same as mist. Cf. also Devil-in-a-bush for this plant.
DEW, Sea *Rosmarinus officinalis* (Rosemary) Clair. This is a literal translation of Rosmarinus, and hence of Rosemary.
DEW BEAN *Phaseolus aconitifolius* (Moth Bean) Douglas.
DEW CUP; DEWCUP 1. *Alchemilla vulgaris* (Lady's Mantle) Scot. Grigson. 2. *Ranunculus acris* (Meadow Buttercup) Dor. Grigson. 3. *Ranunculus bulbosus* (Bulbous Buttercup) Dor. Grigson. 4. *Ranunculus repens* (Creeping Buttercup) Dor. Grigson. The dew on the leaves of Lady's Mantle was said to be neither rain nor dew, but surplus moisture exuded by the plant when its roots are taking up more than it needs. This "dew", in the Middle Ages, was carefully collected for use in the preparation of the philosopher's stone - hence Alchemilla.
DEW MANTLE *Alchemilla alpina* (Alpine Lady's Mantle) Sanecki. An obvious hybrid of a name, between Dew Cup and Lady's Mantle. It is not clear why it is only recorded for this species.
DEW PLANT 1. *Drosera rotundifolia* (Sundew) Grieve.1931. 2. *Mesembryanthemum crystallinum* (Sea Fig) Som. Macmillan. The "dew" of Sundew is actually glandular secretions. The Mesembryanthemum probably gets the name because it grows close to the sea.
DEWBERRY 1. *Ribes uva-crispa* (Gooseberry) Suss. Culpepper. 2. *Rubus caesius*. 3. *Rubus chamaemorus* (Cloudberry) Halliwell. 4. *Rubus flagellaris* USA Zenkert. 5. *Rubus idaeus* (Raspberry) Dor. Northover. 6. *Rubus trivialis* USA Harper. It is the dew-like bloom on the fruits which gives the name.
DEWBERRY, Swamp *Rubus hispidus* USA Harper.

DEWDROP *Galanthus nivalis* (Snowdrop) Som. Macmillan.
DEWDROP, Golden *Duranta repens*.
DEWTRY; DEUTERY *Datura stramonium* (Thorn-apple) Prior (Dewtry); B & H (Deutery). Names that started life as Datura.
DEWY LEAF *Alchemilla vulgaris* (Lady's Mantle) Tynan & Maitland.
DEWY PINE *Drosophyllum lusitanicum*.
DEYE NETTLE 1. *Galeopsis tetrahit* (Hemp Nettle) Yorks. B & H. 2. *Stachys sylvatica* (Hedge Woundwort) Border B & H. Deye Nettle is one of the many variants of Dead Nettle, and these are both deadnettle-like plants.
DHAK TREE *Butea monosperma* (Bastard Teak) India Howes.
DHAL *Cajanus indicus* (Pigeon Pea) Douglas.
DIAL, Husbandman's *Calendula officinalis* (Marigold) Coats. Because it follows the sun.
DIAL, Shepherd's *Anagallis arvensis* (Scarlet Pimpernel) Middx. Grigson.
DIAMOND FLOWER *Ionopsidium acaule* (Violet Cross) A.W. Smith.
DIAMOND PLANT *Mesembryanthemum crystallinum* (Sea Fig) Coats. Presumably for the same reason that the plant is known as Ice Plant.
DIAMOND WILLOW *Salix mackenzieana*.
DIAPENSIA *Diapensia lapponica*.
DIBBLE, Dog's *Arum maculatum* (Cuckoo Pint) Dev. B & H. Cf. Dog Spear, and Dog's Tassel, and more to the point, Dog's Dogger, for this. B & H quoted a rhyme (from the Barnstaple area): Dog's Dibble, Thick in the middle.
DICEL, Milky; DICKLE, Milky *Sonchus oleraceus* (Sow Thistle) Corn. Jago (Dicel); Dev. B & H (Dickle). Who decided it was to be spelt like this? It is obviously meant to rhyme with thistle. But then there is the Devonshire Dickle actually written out as such.
DICK, Blue *Brodiaea capitata* USA Schenk & Gifford.
DICK, Dirty *Chenopodium album* (Fat Hen)

Wilts. Grigson Ches. Holland. From the fact that it grows best on dunghills. Cf. Muckhill-weed, Dirtweed, etc.

DICKIES, Doddering *Briza media* (Quaking-grass) Yks. Drury.1992. Doddering, as with dothering, means trembling.

DICK(IES), Dothering *Briza media* (Quaking-grass) Dur, N'thum, Cumb, Yks. both Grigson.1959. Dothering means trembling.

DICK, Sleepy *Ornithogalum umbellatum* (Star-of-Bethlehem) Lancs. B & H. Because it opens and shuts with the sun. Hence such names as Jack-go-to-bed-at-noon, and this one.

DICKY BIRDS 1. *Acer pseudo-platanus* seeds (Sycamore keys) Som. Macmillan. 2. *Fumaria officinalis* (Fumitory) Wilts. Macmillan. Dor. Greenoak.1979. Sycamore keys get the name from the way they fly down from the tree. The bird imagery for fumitory is quite marked. Cf. Birds-on-the-bush, or Birds-on-the-bough.

DICKY DAISY; DUCKY DAISY *Bellis perennis* (Daisy) Ches. Holland (Dicky); Tynan & Maitland (Ducky).

DICKY DAISY, Large *Laucenthemum vulgare* (Ox-eye Daisy) Ches. Holland.

DICKY DILVER 1. *Vinca major* (Greater Periwinkle) Suff. B & H. 2. *Vinca minor* (Lesser Periwinkle) Som. Macmillan; Suff. Moor. Halliwell spells it Dick-a-dilver. Forby suggested that the word "dilver" was actually "delver", from the plant's habit of rooting (delving) at every joint. But it is far more likely that Dicky Dilver is nothing more than a play on Periwinkle.

DIDDER GRASS *Briza media* (Quaking-grass) Lancs. Grigson.1959; Cambs. Cope. Didder is the same as dodder, which when used as a verb means to tremble.

DIDDERY DOCK *Briza media* (Quaking-grass) Dur. Grigson.1959. See Didder grass.

DIGGER PINE *Pinus sabiniana*. The Digger of the name refers to the Digger Indians, who used to collect the cones in immense quantities, to store for winter food use.

DIKE-ROSE *Rosa canina* (Dog Rose) Cumb. Grigson. i.e. hedge rose.

DILL 1. *Anethum graveolens*. 2. *Torilis japonica* (Hedge Parsley) Halliwell. 3. *Vicia hirsuta* (Hairy Tare) Glos. J.D.Robertson. 4. *Vicia sepium* (Bush Vetch) Leics. Grigson. OE dill, which is O. Norse dilla, according to Folkard. It meant to lull. Another OE version was dile. The names may be identified with the modern dill, and if correct it is one of the few Anglo-Saxon drugs that have been retained in the modern pharmacopeias for their physiological action. The words dill and till, undoubtedly meaning this drug, were in use in Germany and Switzerland as early as AD 1000. The word only occurs in the Teutonic languages, and could be used both as an adjective and a verb - dull, in the sense of to lull (a decoction of the seed has always been used as a soothing remedy for children).

DILL, Dutch *Meum athamanticum* (Baldmoney) Turner.

DILL-CUP (probably better as one word) 1. *Caltha palustris* (Marsh Marigold) Dor. Macmillan. 2. *Ranunculus acris* (Meadow Buttercup) Dor, Hants. Grigson. 3. *Ranunculus arvensis* (Corn Buttercup) Hants. Cope. 4. *Ranunculus bulbosus* (Bulbous Buttercup) Dor, Wilts. Macmillan. 5. *Ranunculus ficaria* (Lesser Celandine) Dor. Macmillan; Wilts. Dartnell & Goddard. 6. *Ranunculus repens* (Creeping Buttercup) Dor, Hants. Grigson. It probably means goldcup, and goes with the sequence which includes Gilcup, though there is a Somerset version, Dale-cup, probably incorrect.

DILLFLOWERS *Nuphar lutea* (Yellow Waterlily) Dor. Vickery.1985. Probably more related to the last syllable of daffodil than to anything else.

DILLIES, Doddering; DILLIES, Dothering *Briza media* (Quaking-grass) N'thum. (Doddering); Dur. (Dothering) both Grigson.1959. To dodder means to tremble.

DILLY; DILLY, Down *Narcissus pseudo-narcissus* (Daffodil) Derb. B & H (Dilly);

Bucks. B & H (Down Dilly).
DILLY, White *Narcissus maialis var. patellaris* (Pheasant's Eye) Lancs. B & H.
DILLY DAFF; DILLY DALLY *Narcissus pseudo-narcissus* (Daffodil) both Som. Macmillan.
DILVER, Dicky 1. *Vinca major* (Greater Periwinkle) Suff. B & H. 2. *Vinca minor* (Lesser Periwinkle) Som. Macmillan; Suff. Moor. Dicky Dilver is probably just a play on the word 'periwinkle'.
DIME-A-BOTTLE PLANT *Gillenia trifoliata* (Indian Physic). USA Mitton. Indian Physic was the principal ingredient of "cure-all" nostrums sold by travelling medicine salesmen in America. So much of it was sold that the plant got the name Dime-a-bottle Plant.
DIMPLEWORT *Umbilicus rupestris* (Wall Pennywort) Dev. Macmillan. Descriptive.
DINDLE 1. *Sonchus oleraceus* (Sow Thistle) Norf. Halliwell. 2. *Taraxacum officinale* (Dandelion) E.Ang. Grigson.
DING-DONG *Campanula rotundifolia* (Harebell) Dor. Macmillan. An obvious name for a bellflower.
DINGLE, Sow *Sonchus oleraceus* (Sow Thistle) Lincs. Peacock. Dingle here is the same as dindle; they probably both come from dandelion.
DINGLE BELLS; DINGLE-DANGLES *Galanthus nivalis* (Snowdrop) Som. Grtigson (Dingle Bells); Midl. Tynan & Maitland (Dingle-dangles).
DINGLEBERRY *Vaccinium erythrocarpum.*
DINNER-BELL, Monkey *Hura crepitans* (Sandbox Tree) A. Huxley.
DIPPER GOURD *Lagenaria siceraria* (Calabash) Gilmore. Presumably for the same reason that it is called bottle gourd, for that is exactly what it is used for.
DIRT-A-BED *Taraxacum officinale* (Dandelion) Som. Macmillan. There are variations on this, but more usual are the names which confirm dandelion's reputation as a diuretic.
DIRTWEED; DIRTY DICK; DIRTY JACK
Chenopodium album (Fat Hen) Som, Lincs Grigson (Dirtweed); Wilts. Grigson; Ches. Holland (Dirty Dick); Ches. Grigson (Dirty Jack). These, plus Dirty John and such names as Muckhill-weed, come about because the muckhill is exactly the habitat which it enjoys best.
DIRTY JOHN 1. *Chenopodium album* (Fat Hen) Ches. Grigson. 2. *Chenopodium vulvaria* (Stinking Orach) Ches. B & H.
DISC CAMOMILE *Matricaria matricarioides* (Rayless Mayweed). Disc, because that is all you can see - it is rayless.
DISCIPLES, Twelve *Bellis perennis* (Daisy) Som. Macmillan. Children's Daisy, more properly Childing Daisy, conveys the same idea as Twelve Disciples, that of a central plant with offspring growing around it.
DISGRACE, Tree of *Sambucus nigra* (Elder) Hants. Boase. Probably for the same reason that produced the name Judas Tree - it is the tree from which Judas Iscariot hanged himself. But elder is an unlucky tree for a number of reasons; there is even a tradition that it is the tree of which the Cross was made.
DISH-CLOTH GOURD; DISH-RAG GOURD *Luffa cylindrica* (Loofah) A.W.Smith (Cloth); Whitaker & Davis (Rag). Cf. Towel Gourd for this.
DISHALAGA; DISHILAGO; DISHLAGO; DISHYLAGIE *Tussilago farfara* (Coltsfoot) Border. Scot. B & H (Dishalaga, Dishlago, Dishylagie); Scot. Jamieson (Dishilago). All are attempts at Tussilago.
DISLE, Milky *Sonchus oleraceus* (Sow Thistle) Corn. Quiller-Couch; Dev. B & H. Disle is an odd rendering of thistle. Milky, because of the white juice.
DISTAFF, Fairy Wives' *Typha latifolia* (False Bulrush) Scot. A. Stewart. This is the translation of the Gaelic Cuigeal-nam ban sith.
DISTAFF, Jupiter's *Salvia glutinosa.* What on earth is Jupiter, or for that matter any male deity, doing with a distaff?
DISTAFF THISTLE *Carthamus tinctoria*

(Safflower) A.W.Smith. It must be because this is a dyeplant.

DITCH BUR 1. *Arctium lappa* (Burdock) Parkinson.1640. 2. *Xanthium canadense* USA Elmore. 3. *Xanthium strumarium* (Cocklebur) Turner.

DITCH GRASS *Ruppia maritima* (Tassel Pondweed) USA Cunningham.

DITCH MOSS *Elodea canadensis* (Canadian Pondweed) USA Cunningham.

DITHERING GRASS; DITHERY DOTHER *Briza media* (Quaking-grass) Yks. (Dithering Grass); N'thum. (Dithery Dother) both Grigson.1959. Dither = didder = dodder, to tremble.

DITTANDER *Lepidium latifolium*. Dittander as a name seems to be a variant of Dittany.

DITTANY 1. *Lepidium latifolium* (Dittander) Turner. 2. *Origanum dictamnus*. It is said that the Origanum is called Dittany from Mount Dicte in Crete, hence Gerard's Dittanie of Crete, and Dittany of Candy mentioned in Henslow. The Oxford Dictionary is cautious, though; while recognizing that it comes from the Greek diktamnon, it goes on "perhaps from Dicte". Dittany is misapplied to Dittander - "some cal Lepidium also Dittany" (Turner), but Dittander as a name seems to be a variant of Dittany.

DITTANY, False 1. *Ballota acetabulosa* Howes. 2. *Dictamnus fraxinella*.

DITTANY, Bastard; DITTANY, White *Dictamnus fraxinella*. Gerard has Bastard Dittany, and Hill White Dittany.

DITTANY, Right *Origanum dictamnus* (Dittany) Turner. "for some cal Lepidium also Dittany".

DITTANY OF CANDY; DITTANY OF CRETE *Origanum dictamnus* (Dittany) Henslow (Candy); Gerard (Crete). Dittany apparently comes from Mount Dicte, in Crete.

DITTEN *Lepidium latifolium* (Dittander) N.Eng. Brockett.

DIVI-DIVI *Caesalpina coriaria*.

DOBBIN-IN-THE-ARK *Aconitum napellus* (Monkshood) Som. Macmillan. This name probably started as Doves-in-the-ark, much better known. There are a number of similar names, deriving from the dove-like shape of the nectaries.

DOCK 1. *Armoracia rusticana* (Horse Radish) Ess. PLNN 17. 2. *Malva sylvestris* (Common Mallow) Som. Jennings; Wilts. Akerman. 3. *Rumex sp.* OE docce. Mallow has often been called dock in the Somerset/ Wiltshire area. According to both Jennings and Akerman, it is mallow, and not Rumex, which is the anti-nettle sting plant here. Horse Radish leaves look very like those of dock. Is it true that it is the size of the leaf which determines whether a plant is to be called dock?

DOCK, Argentine *Rumex cuneifolius*. A South American species.

DOCK, Batter 1. *Petasites hybridus* (Butterbur) Ches. B & H. 2. *Rumex obtusifolius* (Broad-leaved Dock) Shrop. Grigson. Possibly butter dock: "...the countrey huswives were wont to wrap their butter in the large leaves thereof" (Coles). But there is also Blatterdock, so the reference to the size of the leaves may be more direct.

DOCK, Bitter *Rumex obtusifolius* (Broad-leaved Dock) USA Henkel. Presumably this started as Butter Dock, which is also recorded.

DOCK, Blood(y)-veined; DOCK, Bloody *Rumex sanguineus*. 'Blood-veined' is usual; 'Bloody-veined' is in J A MacCulloch.1905; 'Bloody' is recorded in Somerset (Elworthy). "...of the bloudy colour wherwith the whole plant is possest" (Gerard).

DOCK, Blunt-leaved *Rumex obtusifolius* (Broad-leaved Dock) Brockie USA Henkel.

DOCK, Broad-leaved *Rumex obtusifolius*.

DOCK, Butter 1. *Arctium lappa* (Burdock) Corn. Jago. 2. *Petasites hybridus* (Butterbur) Flück). 3. *Rumex alpinus* (Monk's Rhubarb) Grigson. 4. *Rumex longifolius*. 5. *Rumex obtusifolius* (Broad-leaved Dock) Shrop. B & H; Ches. Holland; USA Henkel. Because the large leaves of all these were used to wrap butter.

DOCK, Can 1. *Nuphar lutea* (Yellow Waterlily) Prior. 2. *Nymphaea alba* (White Waterlily) Som, War, Notts. Grigson.

DOCK, Cheadle *Senecio jacobaea* (Ragwort) Ches. Grigson. Cheadle is the name of a place in Cheshire, but the plant is called Cradle Dock there, too. Anyway, this name sounds very like those of the series which includes Kedlock, Kettle-dock, etc.

DOCK, Clustered *Rumex conglomeratus*.

DOCK, Cockle *Dipsacus fullonum* (Teasel) Wilts. Goddard.

DOCK, Cradle *Senecio jacobaea* (Ragwort) Ches. Grigson. Cf. Cheadle Dock, another Cheshire name, for this. But this name must be one of the series that includes Kadle Dock.

DOCK, Crisped *Rumex crispus* (Curled Dock) Salisbury.1964.

DOCK, Cuckle; DOCK, Cuckold *Arctium lappa* (Burdock) Corn. Jago (Cuckle); Corn. Jago; Dev. Choape; Som. Jennings; USA Henkel (Cuckold). Probably cockle, rather than cuckle - Latin coccum, berry, referring to the seeds, or burrs in this case. So cuckle dock means simply the plant with big leaves that has burrs.

DOCK, Curled; DOCK, Curled, Yellow; DOCK, Curly *Rumex crispus*. Usual name is Curled Dock; Yellow... is in Fernie; Curly Dock is an American name Gates.

DOCK, Diddery *Briza media* (Quakinggrass) Dur. Grigson.1959. 'Diddery' has a verb, 'to didder', the same as 'dodder', which in this kind of context means to tremble.

DOCK, Dothering *Briza media* (Quakinggrass) Yks, Cumb. Grigson.1959. Dother means to tremble.

DOCK, Dove *Tussilago farfara* (Coltsfoot) Caith. Grigson.

DOCK, Elf *Inula helenium* (Elecampane) Prior. There is a corruption of this - Elsedock, and it appears as Elfwort as well.

DOCK, Fiddle *Rumex pulcher*. Fiddle Dock because the basal leaves have an outline resembling a violin.

DOCK, Flabby; DOCK, Flap; DOCK, Flap-a-; DOCK, Flappy; DOCK, Flobby;

DOCK, Flop; DOCK, Flop-a-; DOCK, Floppy *Digitalis purpurea* (Foxglove) Fernie (Flabby); Dev. B & H; USA Henkel (Flap); Dev. Friend.1882 (Flap-a-); Dev. Friend.1882; Som. Macmillan (Flappy); Corn, Dev.; Dev. Friend.1882 (Flobby); USA Henkel (Flop); Dev. Friend.1882 (Flop-a-); Coats (Floppy). And there are many other names involving the word flop or flap.

DOCK, Flapper 1. *Digitalis purpurea* (Foxglove) Dev. Friend.1882. 2. *Petasites hybridus* (Butterbur) Grieve.1931. Cf. the Scottish Flatterbaw for Butterbur. There are many other similar names for foxglove.

DOCK, Flea *Petasites hybridus* (Butterbur) B & H.

DOCK, Floating 1. *Nuphar lutea* (Yellow Waterlily) Ches. Grigson. 2. *Nymphaea alba* (White Waterlily) Ches. Grigson.

DOCK, Foam *Saponaria officinalis* (Soapwort) Cockayne. 'Foam' because this is Soapwort - it produces a lather.

DOCK, Garden *Rumex patientia* (Patience Dock) Hill. Because this was introduced as a potherb.

DOCK, Gentle *Polygonum bistorta* (Bistort) Notts. Grigson; Yks. Julia Smith.

DOCK, Golden *Rumex maritimus*. It has golden-yellow fruits.

DOCK, Green *Rumex conglomeratus* (Clustered Dock) USA Schenk & Gifford.

DOCK, Kadle 1. *Anthriscus sylvestris* (Cow Parsley) Ches. B & H. 2. *Petasites hybridus* (Butterbur) Ches. Holland. 3. *Senecio jacobaea* (Ragwort) Ches. Grigson. This is Kedlock rather than any reference to dock. Kedlock in some form or other is applied to these plants, but more especially of course to charlock.

DOCK, Keddle; DOCK, Keedle *Senecio jacobaea* (Ragwort) Lancs. Nodal (Keddle); Ches, Lancs. Grigson (Keedle). See Kadle Dock.

DOCK, Kettle 1. *Anthriscus sylvestris* (Cow Parsley) Ches. B & H. 2. *Petasites hybridus* (Butterbur) Ches. B & H. 3. *Rumex obtusifolius* (Broad-leaved Dock) Lancs. B & H.

4. *Senecio jacobaea* (Ragwort) Ches. B & H. There seem to be two separate traditions here. Genuine dock names, in the sense of large-leaved plants, apply to the dock and to the butterbur. But the other two are Kedlock rather than any reference to dock.

DOCK, Marsh *Rumex palustris*.

DOCK, Mullein *Verbascum thapsus* (Mullein) Norf. B & H; USA Henkell. Dock to emphasise the large leaves.

DOCK, Mutton *Chenopodium bonus-henricus* (Good King Henry) Dor. Macmillan.

DOCK, Narrow; DOCK, Narrow-leaved 1. *Rumex crispus* (Curled Dock) USA Henkel (Narrow); Watt (Narrow-leaved). 2. *Rumex mexicanus* (Pale Dock) Johnston (Narrow-leaved).

DOCK, Northern *Rumex longifolius* (Butter Dock) Fitter.

DOCK, Pale *Rumex mexicanus*.

DOCK, Passion 1. *Polygonum bistorta* (Bistort) Derb, Yks, N'thum. Grigson. 2. *Rumex patientia* (Patience Dock) Denham/Hardy. As far as *Rumex patientia* is concerned, Passion = Patience = *lapathum*, which seems to be the Latin for sorrel. But bistort has a connection with Eastertide. Herb Pudding was made from its leaves on Easter Day, or more properly at Passion-tide, or even during the last two weeks of Lent. The plant was known as Easter Ledges, or Sedges, among other seasonal names.

DOCK, Patience 1. *Polygonum bistorta* (Bistort) Midl, N.Eng. Denham/Hardy. 2. *Rumex patientia*. For bistort, this should really be Passion Dock; but for *Rumex patientia*, the reference is to Latin lapathum, i.e. sorrel.

DOCK, Patient *Polygonum bistorta* (Bistort) Ches. Holland. Better as Patience Dock.

DOCK, Payshun *Rumex patientia* (Patience Dock) Lancs. Nodal.

DOCK, Pig *Aethusa cynapium* (Fool's Parsley) Som. Macmillan.

DOCK, Pop; DOCK, Pop-a-; DOCK, Poppy *Digitalis purpurea* (Foxglove) Corn. Jago; Som. Elworthy (Pop Dock); Som. Macmillan (Pop-a-Dock); Som. Elworthy (Poppy Dock). Poppy is a common name for foxglove - the unopened flowers can be popped. Pop-dock, which sometimes appears as Pop-a-dock, itself the same as Poppy-dock, is only one of a large number of names on this theme.

DOCK, Red 1. *Rumex conglomeratus* (Clustered Dock) Henslow. 2. *Rumex hydrolapathum* (Great Water Dock) R.A.Gunther.

DOCK, Red-veined *Rumex sanguineus* (Blood-veined Dock) McClintock.

DOCK, Rhubarb, Spanish *Rumex abyssinicus* Howes.

DOCK, Round *Malva sylvestris* (Common Mallow) Som. Prior. Given that it is mallow, and not dock, that receives the name in Somerset and Wiltshire, this seems to be an attempt to differentiate - this is the dock which has round leaves, in other words.

DOCK, Scabbit *Digitalis purpurea* (Foxglove) Corn. Grigson. The name is an indication that the leaves must have been used for skin complaints. In fact, an ointment made from them has long been used for just such a purpose.

DOCK, Scattle *Senecio jacobaea* (Ragwort) Lancs. Grigson. Of the same group that produces Kadle-dock and Kettle-dock, that is Kedlock.

DOCK, Scottish *Rumex aquaticus* (Trossachs Dock) Fitter.

DOCK, Sharp 1. *Rumex acetosa* (Wild Sorrel) B & H. 2. *Rumex conglomeratus* (Clustered Dock) Clapham. Sharp in taste, not in physical characteristics.

DOCK, Shield *Rumex scutatus* (Sorrel) Polunin. Latin scutum means a shield, so *scutatus* should mean armed with a shield, but it must signify shield-shaped here.

DOCK, Shore *Rumex rupestris*.

DOCK, Silver *Polygonum bistorta* (Bistort) Wilts. Macmillan.

DOCK, Smair *Rumex obtusifolius* (Broad-leaved Dock) Scot. B & H. The same as Butter Dock.

DOCK, Sorrel; DOCK, Sorrel, Sheep *Rumex acetosa* (Wild Sorrel) both USA

Upham (Sorrel); Gates (Sheep Sorrel).

DOCK, Sour 1. *Oxalis acetosella* (Wood Sorrel) Dev. Friend.1882. 2. *Rumex acetosa* (Wild Sorrel) Turner. 3. *Rumex acetosella* (Sheep's Sorrel) Watt. 4. *Rumex crispus* (Curled Dock) USA Henkel. 5. *Rumex obtusifolius* (Broad-leaved Dock) Donegal Grigson. Sour, because that is what sorrel means.

DOCK, Spatter *Nymphaea advena* (Large Yellow Pond Lily) USA Tolstead.

DOCK, Swamp *Rumex verticillatus*.

DOCK, Sweet *Polygonum bistorta* (Bistort) N.Eng. Schofield. Presumably 'sweet' can only be applied because people made Dock Pudding from it. But it was always said that the Easter Herb Pudding, also made from bistort, came about as a result of the bitter herbs of the Jewish Passover.

DOCK, Tanner's *Rumex hymenosepalus*. The roots are 25/35% tannin, and it is still cultivated in America for this.

DOCK, Trossachs *Rumex aquaticus*. A very rare plant in this country, known only on the eastern shore of Loch Lomond.

DOCK, Veined *Rumex venosus*.

DOCK, Velvet 1. *Inula helenium* (Elecampane) IOW Grigson. 2. *Verbascum thapsus* (Great Mullein) Dev. Choape; Som. Prior. Mullein, of course, has large velvety leaves. But has elecampane got velvety leaves? It sounds as if it might perhaps have come from Elsedock, that is, Elf-dock.

DOCK, Water *Arctium lappa* (Burdock) F Savage. 'Dock', as in most cases, is a reference to the large leaves.

DOCK, Water, Great 1. *Rumex britannicus* (Yellow-rooted Water Dock) USA Zenkert. 2. *Rumex hydrolapathum*.

DOCK, Water, Small *Rumex maritimus* (Golden Dock) Curtis.

DOCK, Water, Yellow-rooted *Rumex britannicus*.

DOCK, Willow-leaved *Rumex mexicanus* (Pale Dock) USA Zenkert.

DOCK, Wood 1. *Rumex acetosa* (Wild Sorrel) Cockayne. 2. *Rumex sanguineus* (Blood-veined Dock) McClintock.

DOCK, Yellow *Rumex crispus* (Curled Dock) USA Henkel.

DOCK CRESS *Lapsana communis* (Nipplewort) Dor. Macmillan. Neither a dock nor a cress, but the name persists, for there are also Swine's Cress and Pig's Cress. Dockerene for this is quite common.

DOCK FLOWER 1. *Chenopodium album* (Fat Hen) Som. Macmillan. 2. *Polygonum amphibium* (Amphibious Persicaria) Som. Macmillan.

DOCK-LEAVED SMARTWEED *Polygonum lapathifolium* (Pale Persicaria) USA Zenkert.

DOCK SORREL *Rumex acetosa* (Wild Sorrel) Watt. or, sometimes, Sorrel Dock.

DOCKEN, Butter *Rumex alpinus* (Monk's Rhubarb) Cumb. Grigson. It is called Butter Leaves too in Cumbria, in other words, leaves used to wrap butter in.

DOCKEN, Cap *Petasites hybridus* (Butterbur) Yks. B & H. "the leaves are very great, like to a round cap or hat, called in Latine Petasus..." (Gerard).

DOCKEN, Eldin 1. *Petasites hybridus* (Butterbur) Border B & H. 2. *Rumex aquaticus* (Trossachs Dock) Roxb. B & H. Eldin, or elden, means fuel. Poor people used to cut and dry the large leaves of both of these to use as fuel.

DOCKEN, Ell *Petasites hybridus* (Butterbur) Border B & H. Ell is eldin, which means fuel.

DOCKEN, Flop *Digitalis purpurea* (Foxglove) Yks. Grigson.

DOCKEN, Floss; DOCKEN, Flous; DOCKEN, Flowster *Digitalis purpurea* (Foxglove) all Yks. F K Robinson (Floss) Grigson (Flowster); Atkinson (Flous). i.e. showy dock.

DOCKEN, Flowery; DOCKEN, Mercury; DOCKEN, Smear *Chenopodium bonus-henricus* (Good King Henry) B & H (Flowery, Mercury); Scot. Grigson (Smear). Smear Docken means fat or grease dock, presumably because of its use for an ointment.

DOCKEN, Fox *Digitalis purpurea* (Foxglove) Yks. Grigson.

DOCKEN, Water *Petasites hybridus* (Butterbur) Cumb. B & H.

DOCKERENE *Lapsana communis* (Nipplewort).

DOCKIN, Soor *Rumex acetosa* (Wild Sorrel) Yks. Gutch.1911.

DOCKMACKIE *Viburnum acerifolium* A.W.Smith.

DOCKO *Artemisia vulgaris* (Mugwort) Berks. Grigson. This must be a reference to the general shape and size of the leaves, or to the size of the whole plant.

DOCKSEED *Rumex acetosa* (Wild Sorrel) Som. Macmillan.

DOCTOR, Black *Scrophularia aquatica* (Water Betony) Heref. Leather.

DOCTOR DOODLES *Caesalpina pulcherrima* (Barbados Pride) Rodway.

DOCTOR'S GUM *Metopium toxiferum* (Poison-wood) Usher. It is the resin which the stem yields which is known as Doctor's Gum, or Hog Gum, used as a violent purgative.

DOCTOR'S LOVE *Galium aparine* (Goose-grass) Som. Macmillan. An ambivalent name - do doctors love it because it cures everything, or because it sends patients to them? Probably the former, for goose-grass is a fairly innocuous plant.

DOCTOR'S MEDICINE 1. *Rubus fruticosus* fruit (Blackberry) Som. Macmillan. 2. *Rumex obtusifolius* leaves (Broad-leaved Dock) Som. Macmillan. That is why blackberries are called Brummelty Kites in Cumbria - too many of them lie uneasily on children's kites, and start to act like the best doctor's medicine. The dock must have got the name because of its traditional association with nettle stings.

DOD *Typha latifolia* (False Bulrush) Dev. Friend.1882. N'thants. Sternberg.

DODDER 1. *Cuscuta epithymum*. 2. *Convolvulus arvensis* (Field Bindweed) S.Africa Watt. 3. *Polygonum convolvulus* (Black Bindweed) Ches. Grigson. 4. *Spergula arvensis* (Corn Spurrey) Ches, Cumb. Grigson. 5. *Typha latifolia* (False Bulrush) N'thants. Sternberg. Fernie suggested that dodder is the plural of dodd, meaning a bunch of threads. It was ME doder, German dotter (which means the yolk of an egg).

DODDER, California *Cuscuta californica*.

DODDER, Clover *Cuscuta epithymum* (Dodder) Ess. Grigson USA House. Cf. the Somerset Clover Devil for this. There used to be a Breton belief that the devil spun the dodder at night to destroy the clover; clover was created by God, dodder was the devil's counter plant.

DODDER, Flax 1. *Cuscuta epithymum* (Dodder) S.Africa Watt. 2. *Cuscuta epilinum*. *C epilinum* is always parasitic on flax; Common Dodder usually only on gorse, thyme and ling.

DODDER, Greater *Cuscuta europaea*.

DODDER, Gronovius' *Cuscuta gronovii*.

DODDER, Lesser *Cuscuta epithymum* (Common Dodder) Grieve.1931. *Cuscuta europaea* is the Greater Dodder.

DODDER, Toothed *Cuscuta denticulata*.

DODDER-CAKE PLANT *Camelina sativa* (Gold of Pleasure). Probably something to do with its cultivation as an oil crop. See the name Oilseed.

DODDER GRASS; DODDLE GRASS *Briza media* (Quaking-grass) Norf. (Dodder Grass); Dev, Wilts, Suss. (Doddle Grass) both Grigson.1959. Dodder when used as a verb means to tremble.

DODDERING DILLIES; DODDERING DICKIES *Briza media* (Quaking-grass) N'thum. Grigson.1959 (Dillies); Yks. F K Robinson (Dickies). See Dodder Grass.

DODGER *Cirsium arvense* (Creeping Thistle) Shrop. B & H. Presumably a tribute to its persistence.

DODGEWEED *Gutierrezia microcephala* (Broomweed) USA Elmore. There is Slinkweed in America for this, too. What is the reference?

DODGILL-REEPAN *Dactylorchis incarnata*

(Early Marsh Orchid) Kirk. Grigson.
DODOL *Garcinia mangostana* (Mangosteen) Douglas.
DOES-MY-MOTHER-WANT-ME *Lolium perenne* (Rye-grass) Som. Grigson.1959. This is a name taken from the divination game of the Love-me, love-me-not kind, played by pulling off the alternating spikelets. The answer comes as yes or no, as is obvious in such names as the Somerset Yes-or-no, or the Gloucestershire Aye-no Bent.
DOG, Heal- *Lobularia maritima* (Sweet Alison) Friend. There is a tradition that this was once used for mad dogs' bites. Heal-bite is another name, and Madwort is also recorded. They probably all stem from the derivation of Alyssum, which is what Lobularia used to be classified under. It is Greek a, not, and lussa, madness.
DOG, Mad *Scutellaria laterifolia* (Virginian Skullcap) USA House. Cf. Mad-dog Herb, or Mad-dog Skullcap. It was used for hydrophobia, after a Dr Van Der Veer experimented with it in 1772. Interesting, because the British skullcap, *S.galericulata*, which is sedative and anti-spasmodic, has been used to treat anything from insomnia to madness and epilepsy.
DOG, Sour *Rumex acetosa* (Wild Sorrel) Som. Macmillan. 'Dog' in this case is dock.
DOG, Spotted 1. *Orchis mascula* (Early Purple Orchis) Som. Macmillan. 2. *Pulmonaria officinalis* (Lungwort) Coats. It is the spots on the leaves which give the name.
DOG-AND-BOBBIN; DOG BOBBINS *Arum maculatum* (Cuckoo Pint) N'thants. A.E.Baker. They say these are Northamptonshire lace-making references, the spadices being in shape like the lace-bobbins used there. There are quite a number of names involving the word. Why dog, though?
DOG BANNER; DOG BINDER *Anthemis cotula* (Maydweed) both Yks. Grigson. There are a lot of 'dog' names for this, all because of the bad smell, it is said.
DOG BERRIES 1. *Atropa belladonna* (Deadly Nightshade) Dur. Grigson. 2. *Rosa canina* fruit (Hips) Hants. Cope; Yks. Grigson. 3. *Sorbus aucuparia* fruit (Rowan berries) Ches, Cumb. Grigson. 'Dog' describing a plant often means poisonous, so the name is understandable as far as Deadly Nightshade is concerned. But why give it to the other two? *Rosa canina* is, after all, Dog Rose, and Dog-hips and its derivatives is common in the north of England and in Scotland, but the name isn't at all frequent for rowan berries.
DOG BREER *Rosa canina* (Dog Rose) Yks. Grigson.
DOG CAPER *Capparis canescens*.
DOG CHERRY *Cornus sanguinea* (Dogwood) Prior.
DOG CHOOPS *Rosa canina* fruit (Hips) Yks. Morris. Choops is a common word for the hips both in northern England and in Scotland. Adding 'dog' comes naturally in view of Dog Rose, Dog Hips and Dogberries. But Cat-choops is recorded from Cumbria as well.
DOG CLOVER *Medicago lupulina* (Black Medick) Som. Grigson. The addition of 'dog' to clover here seems to mean the clover which isn't really a clover.
DOG COCKS *Arum maculatum* (Cuckoo Pint) Wilts. Dartnell & Goddard. Dog Cocks probably comes under the same heading as Priest's Pintle, or indeed Cuckoo Pint itself.
DOG DAISY 1. *Achillea millefolium* (Yarrow) N.Ire. Grigson. 2. *Anthemis cotula* (Maydweed) Dev, Kent, Shrop. Grigson. Yks. Robinson. 3. *Bellis perennis* (Daisy) Surr. Tynan & Maitland; Lincs, Cumb. Grigson; Lancs. Nodal; Yks. Addy. 4. *Leucanthemum vulgare* common. 5. *Tripleurospermum maritimum* (Scentless Maydweed) Donegal. B & H. Ox-eye Daisy gets the name as a contrast to Daisy itself ('dog' would have the same significance as 'horse', i.e. big, or coarse). So why does daisy get the name so commonly? Yarrow is alright - the daisy-looking flower which isn't a daisy, and the maydweeds would qualify

anyway.

DOG DRAKE *Ligustrum vulgare* (Privet) Dor. Macmillan.

DOG ELLER 1. *Aegopodium podagraria* (Goutweed) Ches. Grigson. 2. *Viburnum opulus* (Guelder Rose) Ches. B & H. Ground Elder is a very common name for Goutweed, so Eller with the cautionary description is reasonable. The reference for Guelder Rose is probably Dogwood, rather than a flower which looks like elder.

DOG FENNEL 1. *Peucedanum officinale* (Hog's Fennel) R.T.Gunther. 2. *Peucedanum palustre* (Milk Parsley) B & H. Too much like Hog's Fennel to be coincidence.

DOG FINKLE *Anthemis cotula* (Maydweed) Yks. Grigson. Finkle is fennel. All the 'dog' names for maydweed seem to be because of the bad smell.

DOG FLOWER 1. *Leucanthemum vulgare* (Ox-eye Daisy) Cumb. Grigson. 2. *Mercurialis perennis* (Dog's Mercury) Som. Macmillan. Dog's Mercury, because it is a poison, i.e. a useless plant when compared with French Mercury. Ox-eye Daisy gets the name for a different reason - here dog means coarse, Dog Daisy in fact, when compared with *Bellis perennis*.

DOG GOWAN *Tripleurospermuum maritimum* (Scentless Maydweed) N.Scot. Jamieson. Cf. Dog Daisy, or Dog Camomile, for this.

DOG('S) GRASS *Agropyron repens* (Couch-grass) Yks. Carr USA Cunningham. Both this and Hound's Tooth were names given in deference to Pliny's *canaria*, which herbalists identified with couch. But never mind the classical references - this is the grass dogs most often eat when they feel they need a tonic.

DOG HEATHER *Calluna vulgaris* (Heather) Aber. Grigson. Possibly the same analogy as He-heather (for this) and She-heather, which is *Erica tetralix*, for the latter is also Cat Heather.

DOG HIPS; DOG HIPPANS; DOG('S) JOBS; DOG JUMPS *Rosa canina* fruit (Hips) Scot. Grigson (Hips); Aber. Grigson (Hippen); Yks. Grigson (Jobs); Yks. Atkinson (Jumps); Scot. Chapman (Jobs). Chapman says that 'jobs' is an old Scots word for thorns, hence the name is applicable not just to the hips (with which the word jobs may have been confused), but to the bush itself.

DOG LAUREL *Aucuba japonica* (Japanese Laurel) Polunin.1976. This shrub is often mistaken for the cherry-laurel, so this name puts it firmly back in its place.

DOG LEEK *Ornithogalum umbellatum* (Star-of-Bethlehem) Turner. He called it Dog's Onion, too.

DOG LICHEN *Peltigera canina* (Ash-coloured Ground Liverwort) Brightman.

DOG LIME *Sisymbrium canescens* (Pepper Grass) Douglas.

DOG LIVER *Kalanchoe crenata* Howes.

DOG MINT *Teucrium scorodonia* (Wood Sage) Tynan & Maitland.

DOG NETTLE 1. *Galeopsis tetrahit* (Hemp Nettle) Berw. B & H. 2. *Lamium purpureum* (Red Deadnettle) Ches. B & H. 3. *Urtica urens* (Small Nettle) USA Allan. Is this really Dog Nettle in the sense of an inferior nettle (the first two are both deadnettles), or is it just the result of a mishearing of one of the sequence, which goes fairly near, with dead, deaf, dumb, dunch, etc? Presumably Small Nettle gets the name simply because it is small.

DOG OAK *Acer campestre* (Field Maple) Som, Notts, Yks. Grigson. One can find Cat Oak in Yorkshire for this as well; quite often it is simply Oak in the West country.

DOG PARSLEY *Anthriscus sylvestris* (Cow Parsley) B & H.

DOG POISON *Aethusa cynapium* (Fool's Parsley) Som. Macmillan; USA Sanecki.

DOG POSY *Taraxacum officinale* (Dandelion) Lancs, Yks. Grigson. 'Posy' is probably daisy.

DOG RISE *Euonymus europaeus* (Spindle Tree) B & H. This seems to be a mixture - first of all, Spindle Tree is often referred to as dogwood (spindle and true dogwood share a number of names). Loudon said that a decoction of the leaves was used to wash dogs free from vermin (small boys' nits were treated in the same way). 'Rise', the other element in this name, can be traced back to Gatteridge, common as a name for spindle through the south of England. The word actually means goat-tree (OE gat), and Gadrise was one of its derivatives. From this it is an easy step to Dog Rise.

DOG ROSE 1. *Rosa canina*. 2. *Viburnum opulus* (Guelder Rose) Som. Macmillan. As far as the rose is concerned, this is a translation of both the Greek and Latin names for the plant. It has been claimed that it was so called because it was supposed to cure the bite of a mad dog. It is tempting to consider 'dog' merely as a contemptuous epithet until one remembers those classical names. It comes, too, as a surprise to find some authorities claiming that the 'dog' in Dog Rose is really 'dag', a reference to the dagger-like thorns! Guelder Rose has a number of 'dog' names - Dogberry, Dog-tree, Dogwood, etc.

DOG ROWAN *Viburnum opulus* (Guelder Rose) Scot. B & H. Cf. Dogberry, Dogwood, Dog-tree, etc.

DOG SENNA *Cassia obovata* (Italian Senna) Howes.

DOG SNOUT *Antirrhinum majus* (Snapdragon) Norf. B & H. 'Mouth' is a better-known descriptive name for snapdragons than snout. There actually is a Dog's Mouth, from Somerset.

DOG SPEAR *Arum maculatum* (Cuckoo Pint) Som. B & H. 'Spear' brings the name in line explicitly with the 'pint' of the common name.

DOG STALK; DOG STANDARD; DOG STANDERS *Senecio jacobaea* (Ragwort) Yks. Grigson (Stalk); Yks. Carr (Standard); Worcs, Yks, N.Eng. Grigson (Standers).

DOG STINKERS; DOG STINKS *Anthemis cotula* (Maydweed) Yks. Robinson (Stinkers); Cumb. Grigson (Stinks).

DOG TANSY *Potentilla anserina* (Silverweed) Scot. Jamieson. Goose Tansy (appropriate in view of the specific name) is quite widespread. So is Tansy by itself. Turner calls it Wild Tansy.

DOG THISTLE *Cirsium arvense* (Creeping Thistle) B & H.

DOG TIMBER 1. *Euonymus europaeus* (Spindle Tree) Dev, Som. Macmillan. 2. *Viburnum lantana* (Wayfaring Tree) Dev. Friend.1882. Dog Timber, or Dog's Timber, for Spindle is an extension of the common misnomer, Dogwood. Wayfaring Tree is often called Dogwood, too.

DOG TOOTH BERRY *Euonymus europaeus* fruit Surr. Grigson.

DOG TOOTH(ED) VIOLET 1. *Dentaria bulbifera* (Coral-root) Gerard. 2. *Erythronium dens-canis*. Descriptive, in both cases. Coral-root is also known as Toothed, or Tooth, Violet.

DOG TOOTH VIOLET, American *Erythronium americanum* (Yellow Adder's Tongue) Grieve.1931.

DOG TREE 1. *Alnus glutinosa* (Alder) Lancs, Yks. Halliwell. 2. *Cornus florida* (Flowering Dogwood) Grieve.1931. 3. *Cornus sanguinea* (Dogwood) Turner. 4. *Euonymus europaeus* (Spindle Tree) War. Grigson. 5. *Sambucus nigra* (Elder) Yks. Grigson. 6. *Viburnum opulus* (Guelder Rose) War. B & H; Yks. Addy.

DOG VIOLET *Viola riviniana*. 'Dog' here in the sense of inferior - this violet has no smell.

DOG VIOLET, American *Viola conspersa*.

DOG VIOLET, Heath *Viola canina*.

DOG VIOLET, Wood *Viola reichenbachiana*.

DOG'S AIRACH; DOG'S ORACH *Chenopodium vulvaria* (Stinking Orach) Culpepper (Airach); B & H (Orach). Because of the foul smell.

DOG'S BRIAR *Rosa canina* (Dog Rose) Hants. Grigson. A variation on Dog Rose.

DOG'S CAMOMILE 1. *Anthemis cotula* (Maydweed) B & H. 2. *Matricaria recutita* (Scented Mayweed) Prior). 3. *Tripleurospermum maritimum* (Scentless Maydweed) B & H.

DOG'S CAMOVYNE *Tripleurospermum maritimum* (Scentless Maydweed) Ross (Jamieson). i.e. Dog's Camomile, which is also recorded.
DOG'S CARVI *Anthriscus sylvestris* (Cow Parsley) Shet. Grigson. Carvi must mean carraway; Wild Carraway is a Gloucestershire name for this.
DOG'S CHERRIES *Bryonia dioica* (White Bryony) Som. Macmillan. Cf. Devil's Cherries.
DOG'S COLE *Mercurialis perennis* (Dog's Mercury) B & H. The plant is a poison, hence the 'dog'.
DOG'S DIBBLE *Arum maculatum* (Cuckoo Pint) Dev. B & H.
DOG'S DOGGER *Orchis mascula* (Early Purple Orchis) Clack. B & H. i.e. dog's dung.
DOG'S EARS *Artemisia vulgaris* (Mugwort) Pemb. Grigson.
DOG'S FENNEL *Anthemis cotula* (Maydweed) Prior. Fennel, presumably, from the finely cut leaves, but 'dog's' just to show it is useless.
DOG'S FINGER *Digitalis purpurea* (Foxglove) Som, Wales Grigson; USA Henkel.
DOG'S LEEK *Gagea lutea* (Yellow Star-of-Bethlehem) Turner.
DOG'S LUGS *Digitalis purpurea* (Foxglove) Fife. Grigson.
DOG'S MEDICINE *Mercurialis perennis* Som. Macmillan. Presumably 'medicine' is a misreading of 'mercury'. Dog's because the plant is a poison.
DOG'S MOUTH 1. *Antirrhinum majus* (Snapdragon) Som. Macmillan. 2. *Linaria vulgaris* (Toadflax) Som, Wilts. Rogers. Cf. Puppy Dog's Mouth.
DOG'S NOSE *Antirrhinum majus* (Snapdragon) Som. Macmillan.
DOG'S ONION *Ornithogalum umbellatum* (Star-of-Bethlehem) Turner.
DOG'S ORACHE *Chenopodium vulvaria* (Stinking Orach) B & H.
DOG'S PAISE *Anthyllis vulneraria* (Kidney Vetch) Banff. Grigson. i.e. dog's peas.
DOG'S PARSLEY 1. *Aethusa cynapium* (Fool's Parsley) B & H. 2. *Anthriscus sylvestris* (Cow Parsley) B & H.
DOG'S PENNIES; DOG'S SILLER *Rhinanthus crista-galli* (Yellow Rattle) Shet. Grigson (Pennies); Scot. Grigson (Siller). The rattling of the seeds in the pods are likened to the sound of money being shaken in a purse, but it is false money, hence 'dog's', as well as as 'gowk's' now and then.
DOG'S RIB *Plantago lanceolata* (Ribwort Plantain) Gerard.
DOG'S TASSEL; DOG'S TAUSLE *Arum maculatum* (Cuckoo Pint) Som. Elworthy (Tassel); Som. B & H (Tausle).
DOG'S THISTLE 1. *Arum maculatum* (Cuckoo Pint) Som. Grigson. 2. *Sonchus oleraceus* (Sow Thistle) Som. Macmillan; Surr. Grigson.
DOG'S THORN *Rosa canina* (Dog Rose) Gerard.
DOG'S TIMBER 1. *Cornus sanguinea* (Dogwood) Som. Macmillan. 2. *Euonymus europaeus* (Spindle Tree) Dev. Macmillan. See Dog Timber.
DOG'S TOE *Geranium robertianum* (Herb Robert) Donegal Grigson.
DOG'S TONGUE *Cynoglossum officinale* (Hound's Tongue) Turner.
DOG'S TOOTH *Erythronium dens-canis* (Dog Tooth Violet) Gerard.
DOG'S TOOTH LICHEN *Peltigera canina* (Ash-coloured Ground Liverwort) S.M.Robertson.
DOGBANE 1. *Apocynum cannabinum* (Indian Hemp) USA Kingsbury. 2. *Nerium oleander* (Rose Bay) Watt. With a name like this, there must be some belief that they are poisonous to dogs.
DOGBANE, Hemp *Apocynum cannabinum* (Indian Hemp) Thomson.
DOGBANE, Spreading *Apocynum androsaemifolium*.
DOGBERRY 1. *Arctostaphylos uva-ursi* (Bearberry) Aber. Grigson. 2. *Cornus sanguinea* (Dogwood) Gerard. 3. *Ribes cynos-*

bati (Prickly Gooseberry) USA Yarnell. 4. *Viburnum lantana* (Wayfaring Tree) Lincs. Grigson. 5. *Viburnum opulus* (Guelder Rose) Cumb. B & H.

DOGGER, Dog's *Orchis mascula* (Early Purple Orchis) Clack. B & H. i.e. dog's dung.

DOGGIES *Linaria vulgaris* (Toadflax) Aber. Grigson. The flowers have always been likened to an animal's face or snout, just as snapdragon's has. Cf. Dog's Mouth, or Puppy Dog's Mouth, in the south of England.

DOGMINT *Nepeta cataria* (Catmint) Pemb. Grigson.

DOGMOUTH *Antirrhinum majus* (Snapdragon) Lincs. B & H. The shape of the flowers has been likened to the mouths or snouts of rabbits, horses, tigers and lions, toads and frogs - even dragons.

DOGRISE *Euonymus europaeus* (Spindle Tree) Turner. Almost certainly, this was originally Gadrise, and means goat tree (OE gat). Gatteridge is the name which preserves this meaning.

DOGS, Cats-and- *Salix sp.* catkins (Willow catkins) Corn. Jago; Wilts. Macmillan.

DOGS-AND-CATS *Trifolium arvense* (Hare's-foot Trefoil) Mor. Grigson.

DOGSBANE *Apocynum androsaemifolium* (Spreading Dogbane) Grieve.1931.

DOGSTAIL, Crested *Cynosurus cristatus*.

DOGSTAIL, Rough *Cynosurus echinatus*.

DOGSTONES *Orchis mascula* (Early Purple Orchis) Gerard. The twin tubers have been likened to testicles from a very early date, and that is what 'stones' mean. Cf. Foxstones, Harestones or Goatstones for this.

DOGTOOTH LILY *Erythronium oregonum* (Giant Adder's Tongue) USA Kingsbury.

DOGWOOD 1. *Cornus sanguinea*. 2. *Euonymus europaeus* (Spindle Tree) B & H. 3. *Frangula alnus* (Alder Buckthorn) Corn, Hants, Suss, Kent. Grigson. 4. *Pyrus americana* USA Bergen. 5. *Solanum dulcamara* (Woody Nightshade) Lancs. Grigson. 6. *Viburnum lantana* (Wayfaring Tree) Som. Macmillan. 7. *Viburnum opulus* (Guelder Rose) Lancs. B & H. In spite of the etymology offered by Mrs Grieve, that it was called dogwood because a decoction was once used for washing mangy dogs, the true meaning is probably that given in Hart. The name derives from its traditional use for skewers - a 'dog' (OE dagge) means a sharp spike. Parkinson, though, said "because the berries are not fit to be eaten, or to be given to a dogge". All this refers to *Cornus sanguinea*; the others are confused, and probably get the name simply because they are about the shape and size of a true dogwood (Spindle gets the name almost indiscriminately). The odd one out is Woody Nightshade - the Scottish name Mad Dog's Berries might have some bearing on it.

DOGWOOD, Alternate-leaved *Cornus alternifolia*.

DOGWOOD, American *Cornus florida* Grieve.1931.

DOGWOOD, Black 1. *Frangula alnus* (Alder Buckthorn) North. 2. *Prunus padus* (Bird Cherry) Suss. Grigson.

DOGWOOD, Creek *Cornus californica*.

DOGWOOD, Creeping *Cornus canadensis* (Bunchberry) RHS.

DOGWOOD, Dwarf *Cornus canadensis* (Bunchberry) USA Turner & Bell.

DOGWOOD, Female *Cornus amomum* (Silky Cornel) Grieve.1931.

DOGWOOD, Flowering *Cornus florida*.

DOGWOOD, Flowering, Pacific *Cornus nuttallii* (Mountain Dogwood) E.Gunther.

DOGWOOD, Japanese *Cornus kousa*.

DOGWOOD, Male *Cornus mas* (Cornel Cherry) Gerard. *Cornus amomum* is the Female Dogwood. Why the differentiation?

DOGWOOD, Mountain *Cornus nuttallii*.

DOGWOOD, Nuttall's *Cornus nuttallii* (Mountain Dogwood) E. Gunther.

DOGWOOD, Pagoda *Cornus alternifolia* (Pagoda Tree) Howes.

DOGWOOD, Poison *Rhus vernix* (Swamp Sumach) USA Zenkert. Indeed, it is very poisonous.

DOGWOOD, Pond *Cephalanthus occidental-*

is (Buttonbrush) O.P.Brown.

DOGWOOD, Red *Cornus amomum* (Silky Cornel) Gilmore.

DOGWOOD, Red-barked *Cornus alba*.

DOGWOOD, Red Osier *Cornus stolonifera*.

DOGWOOD, Rough(-leaved) *Cornus asperifolia*.

DOGWOOD, Silky *Cornus amomum* (Silky Cornel) Yanovsky.

DOGWOOD, Striped *Acer pennsylvanicum* (Snake-bark Maple) W.Miller. Striped Maple, too. 'Striped' is a reference to the way the bark grows.

DOGWOOD, Swamp 1. *Cornus amomum* (Silky Cornel) USA Upham. 2. *Ptelea trifoliata* (Hop Tree) Leyel.1948.

DOGWOOD, Table *Cornus contraversa* (Wedding-cake Tree) Hora. 'Table', because of the same "tiered arrangement" that gave Wedding-cake Tree.

DOGWOOD, Tatarian *Cornus alba* (Red-barked Dogwood) Hora.

DOGWOOD, Virginian *Cornus florida* (Flowering Dogwood) Grieve.1931.

DOGWOOD, White *Viburnum opulus* (Guelder Rose) Lancs. B & H.

DOGWOOD NAVEL-SEED *Omphalodes cappadocica*.

DOLEFUL BELLS *Atropa belladonna* (Deadly Nightshade) Dor. Macmillan.

DOLEFUL BELLS OF SORROW *Fritillaria meleagris* (Snake's Head Lily) Oxf. Grigson. 'Sorrow' seems quite reasonable here, but the original was actually 'Sodom' - we have Mournful Bells of Sodom, Drooping Bells of Sodom, even Solemn Bells of Sodom, elsewhere. The reference is to the shape, and colour, of the flowers - "the dingy, sad colour of the bell-shaped flowers" (B & H).

DOLL CHEESES *Malva neglecta* (Dwarf Mallow) Yks. B & H. There are dozens of 'cheese' names for the mallows; it is the shape of the seed vessels that suggests the name.

DOLL'S EYES 1. *Actaea alba* (White Baneberry) H & P. 2. *Actaea rubra* (Red Baneberry) USA Kingsbury. 3. *Actaea spicata* (Herb Christopher) Tampion.

DOLLAR-LEAF *Pyrola rotundifolia* (Round-leaved Wintergreen) S USA Puckett.

DOLLS, Wax *Fumaria officinalis* (Fumitory) Som. Macmillan Hants, War, Yks, N'thum. Grigson; Kent Leyel.1937. Cf. such names as Babes-in-the-cradle, and Birds-on-the-bush, for fumitory.

DOLLY, Red *Papaver rhoeas* (Red Poppy) Som. Macmillan.

DOLLY SOLDIERS *Geranium molle* (Dove's-foot Cranesbill) Som. Macmillan.

DOLLY VARDEN *Tephrosia virginiana* (Devil's Shoestring) Howes.

DOLLY WINTER *Melandrium dioicum* (Red Campion) Corn. Grigson.

DOLLY'S APRON; DOLLY'S PINAFORE *Geranium robertianum* (Herb Robert) Dev, Som. Macmillan (Apron); Dev. Macmillan (Pinafore).

DOLLY'S BONNETS *Aquilegia vulgaris* (Columbine) Som. Macmillan. Better known as Granny's, rather than Dolly's, Bonnets.

DOLLY'S NIGHTCAP *Geranium robertianum* (Herb Robert) Dev. Macmillan.

DOLLY'S SHOES 1. *Aquilegia vulgaris* (Columbine) Som. Macmillan. 2. *Geranium robertianum* (Herb Robert) Dev. Macmillan.

DOLPHIN-FLOWER *Delphinium ajacis* (Larkspur) B & H. Delphinium is Greek delphinion, dolphin-like, and is a reference to the flower - the spur represents the long "beaked" head of the dolphin. Gerard's description is "the flowers, and especially before they be perfected, have a certaine shew and likenesse of those Dolphins, which old pictures and armes of certain antient families have expressed with a crooked and bending figure in shape, by which signe also the heavenly Dolphine is set forth".

DON'S TWITCH *Agropyron donianum*.

DONG BELLS *Narcissus pseudo-narcissus* (Daffodil) Som. Macmillan.

DONKEY 1. *Arctium lappa* (Burdock) Som.

Mabey. 2. *Galium aparine* (Goose-grass) Som. Macmillan.

DONKEY, Stick *Galium aparine* Som. Macmillan. If 'donkeys are the burrs, then 'stick' is a good adjective to use. It occurs a number of times, in such as Stickleback, or Stick-buttons, etc.

DONKEY EYE BEAN *Mucuna urens* (Florida Bean) Howes. Cf. Horse Eye Bean for this.

DONKEY'S EARS 1. *Kalanchoe tomentosa*. 2. *Plantago lanceolata* (Ribwort Plantain) Som. Macmillan. 3. *Stachys lanata* (Lamb's Ears) Dev. Friend.1882. 4. *Verbascum thapsus* (Great Mullein) Dor. Grigson. Most of these have a felted look to the leaves. But why has ribwort got the name?

DONKEY'S OATS 1. *Rumex acetosa* (Wild Sorrel) Dev. B & H. 2. *Rumex obtusifolius* (Broad-leaved Dock) Dev. Grigson.

DONKEY'S RHUBARB *Polygonum cuspidatum* (Japanese Knotweed) Corn. McClintock.1975.

DONKEY'S TAIL *Sedum morganianum* (Burro's Tail) Goold-Adams.

DONKEY'S THISTLE *Dipsacus fullonum* (Teasel) Som. Macmillan.

DONNHOVE *Tussilago farfara* (Coltsfoot) Grieve.1931. Apparently from donn, an old name for a horse - donkey is a little horse. 'Hove' is hoof.

DONNINETHELL *Galeopsis tetrahit* (Hemp Nettle) Gerard. This is Gerard's eccentric spelling of Dunny Nettle, a common enough variant of deadnettle.

DOODLES, Doctor *Caesalpina pulcherrima* (Barbados Pride) Rodway.

DOODYKE *Rumex spp* (Dock) N'thum. Grigson.

DOOM BARK *Erythrophlaeum guineense* (Ordeal Tree) Le Strange. Doom Bark is explained by the more common name. The bark provides the poison which used to be widely used by African witchdoctors as an ordeal poison. This is an emetic, prepared and administered to each suspect. The innocent will vomit, not necessarily immediately; the guilty will not, and will, therefore, so it is believed, die of the poison.

DOON-HEAD CLOCK *Taraxacum officinale* (Dandelion) Scot. Grigson; USA Henkel. One of the many 'clock' names given to dandelion seed globes.

DOORWEED; DOORYARD KNOTWEED *Polygonum aviculare* (Knotgrass) B & H, and USA Zenkert (Doorweed); S.Africa Watt (Dooryard Knotweed).

DOORYARD PLANTAIN *Plantago major* (Great Plantain) Kansas Gates.

DORNEL *Lolium temulentum* (Darnel) Scot. B & H.

D'ORSAY'S BUTTONHOLE ROSE *Rosa virginiana pl* (St Mark's Rose) Sitwell.

DORSET HEATH *Erica ciliaris*. Confined to the west country in Great Britain.

DOTHER 1. *Cuscuta epithymum* (Common Dodder) B & H. 2. *Polygonum convolvulus* (Black Bindweed) Ches. Holland. 3. *Spergula arvensis* (Corn Spurrey) Ches. Holland; Berw. Grigson. 4. *Vicia hirsuta* (Hairy Tare) Ches. Holland. Dother is an alternative spelling for dodder. In most cases, both forms are used. Cf. too Pother.

DOTHER, Dithery *Briza media* (Quaking-grass) N'thum. Grigson.1959. Both elements of the name mean trembling.

DOTHER GRASS; DOTHERING DICK; DOTHERING DICKIES; DOTHERING DILLIES; DOTHERING DOCK; DOTHERING GRASS; DOTHERING JOCKIES; DOTHERING NANCY; DOTHERY *Briza media* (Quaking-grass) Surr, Kent, Beds, Cumb. (Dother Grass); Dur, N'thum, Cumb, Yks. (Dothering Dick(ies)); Dur. (Dothering Dillies); Dur, Yks. (Dothering Dock); Lancs, Yks, Cumb. (Dothering Grass); Yks. (Dothering Jockies); Cumb. (Dothering Nancy, Dothery); all from Grigson.1959.

DOTHERUM; BOTHERUM 1. *Veronica chamaedrys* (Germander Speedwell) Ches. Grigson. 2. *Veronica hederifolia* (Ivy-leaved Speedwell) Ches. W. Miller. Both versons of the word mean trembling. It can be spelt in a number of ways; Morris, for Yorkshire,has the word dodderums in his glossary, and he defines it as a shaking or trembling.

DOTROA *Datura stramonium* (Thorn-apple) Aubrey.1686/7. Aubrey's stab at Datura.

DOTS-AND-DASHES *Saxifraga x urbinum* (London Pride) Som. Macmillan.

DOTTED THORN *Crataegus punctata.*

DOUBLE-LEAF *Listera ovata* (Twayblade) W.Miller. Double-leaf is exactly what Twayblade means.

DOUDLAR *Menyanthes trifoliata* - root (Buckbean) Roxb. Grigson.

DOUGH FIG *Ficus carica* (Common Fig) Dev. Friend.1882 Som. Macmillan. Probably to distinguish figs ("soft as dough") from ordinary raisins, which are called figs in Somerset.

DOUGLAS FIR *Pseudotsuga menziesii*. The Douglas of the common name was David Douglas, who sent seeds of the tree to England in 1827, after its discovery by the Scottish botanist Archibald Menzies, in 1791.

DOUGLAS FIR, Blue *Pseudotsuga menziesii glauca* (Colorado Douglas Fir) Hart.

DOUGLAS FIR, Colorado *Pseudotsuga menziesii glauca.*

DOUGLAS FIR, Green *Pseudotsuga menziesii* (Douglas Fir) Mitchell.

DOUGLAS FIR, Japanese *Pseudotsuga japonica.*

DOUGLAS FIR, Large-coned *Pseudotsuga macrocarpa.*

DOUGLAS GROUNDSEL *Senecio douglasii*. This is also known as Douglas Ragwort, or Douglas Squaw-weed.

DOUGLAS HACKBERRY *Celtis douglasii.*

DOUGLAS RAGWORT; DOUGLAS SQUAW-WEED *Senecio douglasii* Douglas Groundsel both USA Elmore.

DOUGLAS RABBIT BRUSH *Chrysothamnus confinis.*

DOUGLAS SPRUCE *Pseudotsuga menziesii* (Douglas Fir) USA Elmore.

DOVASTON YEW *Taxus baccata var. dovastoniana* (Westfelton Yew) Dallimore. The variety was found by John Dovaston, of Westfelton, in Shropshire.

DOVE CARRIAGE, Lady Lavinia's *Aconitum napellus* (Monkshood) Som. Grigson. There are other 'dove' names for this, some of them equally as odd as this one. What about Lady Dove and her coach and pair? All of them, it is claimed, come from the shape of the nectaries, described as "dove-like".

DOVE-DOCK *Tussilago farfara* (Coltsfoot) Caith. Grigson.

DOVE FLOWER *Aconitum napellus* (Monkshood) Suff. Jobson.

DOVE-FOOT; DOVE'S FOOT CRANESBILL *Geranium molle*. Dove's Foot Cranesbill is the usual name for this plant, deriving from the shape of the leaf. Dove-foot appears in Turner.

DOVE-PLANT *Aquilegia vulgaris* (Columbine) Ellacombe. Latin columba, which has given Columbine, means a pigeon. The nectaries are like pigeons' heads in a ring round a dish. One can always call a pigeon a dove.

DOVE TREE *Davidia involucrata* (Handkerchief Tree) Hay & Synge. Descriptive - what are handkerchiefs to one eye are doves to another.

DOVE-WEED 1. *Croton texensis*. 2. *Eremocarpus setigerus* (Turkey Mullein) USA Jaeger. 3. *Euphorbia prostrata* Barbados Gooding. Doves feed on the seeds of the Croton.

DOVE'S DUNG 1. *Ornithogalum nutans* (Drooping Star of Bethlehem) Whittle. & Cook. 2. *Ornithogalum umbellatum* (Star-of-Bethlehem) Grieve.1931. Descriptive, for both of these have white, dotted flowers that look just like bird droppings.

DOVE'S FOOT 1. *Aquilegia vulgaris* (Columbine) B & H. 2. *Geranium maculatum* (Spotted Cranesbill) O.P.Brown. There seems to be a mixing of imagery here.The cranesbill (it ought to be *Geranium molle*, really), gets the name from the shape of the leaves, while columbine gets the dove reference for quite a different reason.

DOVE'S FOOT CRANESBILL *Geranium molle*. Dove's foot, or Pigeon's foot, translates the medieval name, pes columbae, given from the shape of the leaf.

DOVE'S FOOT CRANESBILL, Perennial *Geranium pyrenaicum* (Mountain Cranesbill) Curtis.

DOVES, Turtle 1. *Aconitum napellus* (Monkshood) Som. Macmillan. 2. *Digitalis purpurea* (Foxglove) Som. Macmillan.

DOVES-AT-THE-FOUNTAIN, DOVES-ROUND-A-DISH *Aquilegia vulgaris* (Columbine) Som. Macmillan.

DOVES-IN-THE-ARK 1. *Aconitum napellus* (Monkshood) Dor. Dacombe. 2. *Aquilegia vulgaris* (Columbine) Som. Macmillan.

DOVEDALE MOSS *Saxifraga hypnoides*.

DOWN, Bloomy *Dianthus barbatus* (Sweet William) B & H.

DOWN, Vegetable *Asclepias fruticosa* (Shrubby Milkweed) Watt. Cf. the names Wild Cotton, or Narrow-leaved Cotton Bush, given in South Africa.

DOWN DILLY *Narcissus pseudo-narcissus* (Daffodil) Bucks. B & H. A contraction of the longer form, Daffy-down-dilly.

DOWN THISTLE 1. *Cirsium eriophorum* (Woolly Thistle) Gerard. 2. *Onopordum acanthium* (Scotch Thistle) W.Miller. Gerard said that the down of Scotch Thistle was collected for stuffing pillows and the like.

DOWN-YOU-GO *Viburnum alnifolium* (Hobble-bush) USA Perry. Because the lower branches droop to the ground and root at the tips.

DOWNIVINE *Clematis vitalba* (Old Man's Beard) Turner.

DOWNSCWOB *Caltha palustris* (Marsh Marigold) Som. Macmillan.

DOWNWEED *Filago germanica* (Cudweed) W.Miller.

DOWNY BIRCH *Betula pubescens*.

DOWNY GENTIAN *Gentiana puberula*.

DOWNY GOLDEN ROD *Solidago puberula*.

DOWNY GRAPE *Vitis cinerea*.

DOWNY HAWTHORN *Crataegus mollis* (Woolly Thorn) USA Everett.

DOWNY OAK *Quercus pubescens*.

DOWNY ROSE *Rosa tomentosa*.

DOWNY THORN-APPLE *Datura meteloides*.

DOWNY WILLOW *Salix lapponicum* (Lapland Willow) McClintock.

DOWNY YELLOW VIOLET *Viola pubescens*.

DRACAENA, Red *Cordyline terminalis* (Ti Plant) Perry.1979. The leaves are purplish-red.

DRAGANCE *Dracunculus vulgaris* (Dragon Arum) Dawson. A C14 version of Dragons.

DRAGGE; BRAGGE *Lolium temulentum* (Darnel) B & H. Probably connected with the name Drunk, which also appears as Drank, or Drake, and is a comment on the toxic nature of the seeds, causing giddiness or "drunkenness".

DRAGON *Artemisia dracunculus* (Tarragon) Storms. Dracunculus, the specific name, is the diminutive of draco, dragon, and tarragon is a corruption of dracunculus, through French estragon. The Arabic name also means little dragon, because it was supposed to heal the bites of venomous snakes, or, more likely, because of the resemblance of the roots to coiled snakes.

DRAGON, Great; DRAGON, Small *Arum maculatum* (Cuckoo Pint) both Suss. B & H. Both apparently for the same plant, in spite of the contradiction.

DRAGON ARUM *Dracunculus vulgaris*. Fancifully descriptive. See Gerard: "...spots of divers colours, like those of the adder or snake..." Hence, by doctrine of signatures, its use against snakebite, and hence also the attribute claimed for it of keeping snakes away.

DRAGON-BUSHES *Linaria vulgaris* (Toadflax) Bucks. Grigson.

DRAGON CLAW *Corallorhiza odontorhiza* (Coral-root) Howes.

DRAGON-FLOWER 1. *Iris foetidissima* (Foetid Iris) Dev. Friend.1882. 2. *Iris pseudo-acarus* (Yellow Iris) Dev. Friend.1882.

3. *Iris versicolor* (Purple Water Iris) Grieve.1931. 4. *Lamium orvala* (Balm-leaved Deadnettle) Whittle & Cook. Probably dagger-flower in the case of the Irises, from the shape of the leaves.

DRAGON-ROOT 1. *Arisaema triphyllum* (Jack-in-the-pulpit) Le Strange. 2. *Circaea lutetiana* (Enchanter's Nightshade) Ire. Grigson.

DRAGON SPRUCE *Picea asperata* (Chinese Spruce) Mitchell.1972. Presumably just because it is a Chinese tree.

DRAGON TREE *Dracaena draco*. When an elephant and a dragon engaged in mortal combat, the blood of the dragon would eventually soak into the earth, and a tree would grow from that spot; the sap of the tree would contain the dragon's blood. The dried sap is still known as dragon's blood, and it continued to be used long after the Middle Ages as a means of stopping bleeding, as an astringent mouthwash, and as an ingredient in the varnish applied to violins.

DRAGON TREE, Madagascar *Dracaena marginata*.

DRAGON'S BLOOD *Geranium robertianum* (Herb Robert) Som. Lawrence; Shrop. B & H. Cf. Bloody May, and Bloodwort, for this. It is the colour that has caused the name, but it was picked up by Culpepper for a wound herb, a use which continued right into the 19th century. As far as the ascription to dragons is concerned, Cf. Adder's Tongue, and Snake-flower as local names for Herb Robert.

DRAGON'S-CLAW WILLOW *Salix matsudana var. tortuosa* (Corkscrew Willow) RHS. More fanciful than the other names describing the contorted branches.

DRAGON'S FEMALE *Dracunculus vulgaris* (Dragon Arum) Som. Macmillan. An obscure name - what was the male? Is it really Female Dragons, certainly recorded for Water Arum?

DRAGON'S HEAD *Antirrhinum majus* (Snapdragon) Som. Macmillan.

DRAGON'S MOUTH 1. *Antirrhinum majus* (Snapdragon) Som. Macmillan; Lincs. B & H. 2. *Arethusa bulbosa*. 3. *Digitalis purpurea* (Foxglove) Suss. B & H.

DRAGON'S MUGWORT *Artemisia dracunculus* (Tarragon) Sanecki.

DRAGON'S PLANT, Common *Dracunculus vulgaris* (Dragon Arum) J.Smith.

DRAGON'S TEETH *Tetragonolobus maritimus*.

DRAGONFLIES *Acer pseudo-platanus* seeds (Som) Macmillan. Butterflies is another Somerset name for them - it is the illusion of flight that provides the imagery, hence the name Wings, and there are many more.

DRAGONHEAD *Physostegia virginiana*. The alternative generic name is *Dracocephalum*, and this is where the name comes from.

DRAGONHEAD, False 1. *Physostegia parviflora*. 2. *Physostegia virginiana* (Dragonhead) USA Zenkert.

DRAGONHEAD, Small-flowered *Physostegia parviflora* (False Dragonhead) Elmore.

DRAGONMOUTH *Horminum pyrenaicum*.

DRAGONS 1. *Artemisia dracunculus* (Tarragon) Storms. 2. *Arum maculatum* (Cuckoo-pint) B & H. 3. *Dracunculus vulgaris* (Dragon Arum) Cockayne. 4. *Polygonum bistorta* (Bistort) B & H. Bistort's writhed roots are responsible for the name. Cf. Snakeroot, Adderwort, etc., for this. By signature, bistort was used for snakebite and poison. The rest of the flowers listed above have ancient connections with dragons, as the specific names show.

DRAGONS, Female; DRAGONS, Water *Calla palustris* (Water Arum) both B & H.

DRAGONWORT 1. *Artemisia dracunculus* (Tarragon) Storms. 2. *Dracunculus vulgaris* (Dragon Arum) Gerard. 3. *Polygonum bistorta* (Bistort) B & H. See Dragons.

DRAKE; DRANK; DRUNK; DROKE; DRAWK *Lolium temulentum* (Darnel) E.Ang. (Drake) N.Eng. (Drank, Drunk) all Halliwell Yks. B & H (Droke); E Ang, Forby (Drawk). Cf. Drunken Plant. They are all connected with the toxic quality of darnel.

The French word ivraie, drunk, has given many other names for it, like Eaver.

DRAKE, Dog *Ligustrum vulgare* (Privet) Dor. Macmillan.

DRAKE'S FOOT *Orchis mascula* (Early Purple Orchis) Lincs. Gutch.1908.

DRAKES, Ducks-and- *Orchis mascula* (Early Purple Orchis) Dor. Macmillan.

DRALYER *Convolvulus arvensis* (Field Bindweed) Corn. Grigson. A Cornish word, draylyer, meaning a trailer.

DRANK; DRUNK *Lolium temulentum* (Darnel) N.Eng. Halliwell.

DRANT *Eruca sativa* (Rocket) Halliwell.

DRAPER'S TEASEL *Dipsacus fullonum* (Teasel) B & H. Draper usually conveys cloth dealer rather than cloth maker, but the meaning is clear.

DRAVICK *Lolium temulentum* (Darnel) B & H. Part of a sequence that includes Drake, Drank and Drunk.

DRAWK *Agrostemma githago* (Corn Cockle) B & H. A variation of Drake, given to Darnel as well. Perhaps it just means weeds in general, and not connected with 'drunk'.

DRESSES, Fairy's *Digitalis purpurea* (Foxglove) Som. Macmillan. Cf. Fairy's Petticoats.

DRINK CORN *Hordeum sativum* (Barley) Tusser.

DRISAG *Rubus fruticosus* (Blackberry) Donegal Grigson.

DROMEDARY 1. *Centaurea nigra* (Knapweed) Wilts. Dartnell & Goddard. 2. *Centaurea scabiosa* (Greater Knapweed) Wilts. Dartnell. & Goddard. Perhaps this is a confusion with Drumsticks, or better, Drummer Boys, descriptive names, both.

DROOPING BELLS; DROOPING HEADS; DROOPING LILY *Galanthus nivalis* (Snowdrop) All Som. Macmillan.

DROOPING BELLS OF SODOM *Fritillaria meleagris* (Snake's-head Lily) Dor. Grigson. Cf. Mournful, or Solemn, Bells of Sodom, all of them a comment on the shape and colouring of the flowers.

DROOPING SAXIFRAGE *Saxifraga cernua*.

DROOPING WILLOW *Salix alba var. vitellina "pendula"* (Weeping Willow) Dev. Macmillan.

DROP-TONGUE, Painted *Aglaonema crispum*.

DROPS *Glycyrrhiza glabra* (Liquorice) S.Africa Watt.

DROPS, Golden 1. *Laburnum anagyroides* (Laburnum) Lincs. B & H. 2. *Primula veris* (Cowslip) Som. Macmillan.

DROPS OF BLOOD *Anagallis arvensis* (Scarlet Pimpernel) Wilts. Grigson.

DROPS OF SNOW *Anemone nemerosa* (Wood Anemone) Suss. Grigson.

DROPWORT *Filipendula vulgaris*. According to Coles, it is called Dropwort because of its use in cases of strangury, and Culpepper says it is so called "because it helps such as piss by drops". The real etymology comes from the root, with its small drop-like tubers - so the medical use came from the doctrine of signatures. But Hill was still saying in the 19th century, "The root... is good in fits of the gravel".

DROPWORT, Water; DROPWORT, Water, Fine-leaved *Oenanthe aquatica* (Horsebane) Grieve.1931 (Water Dropwort); Storer (Fine-leaved).

DROPWORT, Water, Callous-fruited *Oenanthe pimpinelloides*.

DROPWORT, Water, Hemlock *Oenanthe crocata*.

DROPWORT, Water, Parsley 1. *Oenanthe lachenalii*. 2. *Oenanthe pimpinelloides* (Callous-fruited Water Dropwort). Storer.

DROPWORT, Water, River *Oenanthe fluviatilis*.

DROPWORT, Water, Tubular *Oenanthe fistulosa*.

DROPWORT, Water, Yellow *Oenanthe crocata* (Hemlock Water Dropwort) Grieve.1931. In this case, it is the juice from both stem and root which is markedly yellow, enough to stain the hands that colour.

DROTT, Yellow *Linaria vulgaris* (Yellow Toadflax) Dor. Northover. 'Drott' is 'throat'

in Dorset.
DRUCE'S PEARLWORT *Sagina normaniana*.
DRUM, Wattery *Senecio vulgaris* (Groundsel) Shet. Grigson.
DRUMMER BOY; DRUMMER HEAD *Centaurea nigra* (Knapweed) Dor. Macmillan (Boy); Som. Macmillan (Head). Descriptive of the profile view of the flowering heads.
DRUMMER DAISY *Leucanthemum vulgare* (Ox-eye Daisy) Som. Macmillan. Cf. Thunder Daisy, which surely has a connection with this name.
DRUMMOND PENNYROYAL *Hedeoma drummondii*.
DRUMMOND'S IRONWEED *Vernonia missurica*.
DRUMMOND'S SOAPBERRY *Sapindus drummondii*.
DRUMSTICK TREE 1. *Cassia fistula* (Indian Laburnum) Chopra. 2. *Moringa oleifera* (Horse Radish Tree) Chopra.
DRUMSTICKS 1. *Centaurea nigra* (Knapweed) Som. Macmillan; N'thants. A.E.Baker. 2. *Centaurea scabiosa* (Greater Knapweed) N'thants. A.E. Baker. 3. *Poterium sanguisorba* (Salad Burnet) Wilts. Macmillan; War. F Savage. 4. *Sanguisorba officinalis* (Great Burnet) Som, Glos. Grigson. All descriptive in their way.
DRUNK; DRANK *Lolium temulentum* (Darnel) N Eng. Halliwell. See Drunken Plant below.
DRUNKARD'S NOSE *Centranthus ruber* (Spur Valerian) Som. Macmillan. Colour descriptive.
DRUNKARDS 1. *Althaea officinalis* (Marsh Mallow) Dev. Fernie. 2. *Caltha palustris* (Marsh Marigold) Dev, Wilts. Wright. 3. *Centranthus ruber* (Spur Valerian) Dev. Sharman; Som. Macmillan. 4. *Gaultheria procumbens* the young leaves (Wintergreen) USA Sanford. 5. *Geranium robertianum* (Herb Robert) Som. Macmillan. 6. *Lychnis flos-cuculi* (Ragged Robin) Som. Macmillan. 7. *Melandrium rubrum* (Red Campion) Som. Macmillan. Drunkards is quite common in the west country, especially for Marsh Marigold, "probably from the way in which they suck up water when placed in a vase" (Dartnell & Goddard), but the reason according to the children is that if you look long at them you will be sure to take to drink. The group of names headed by Crazy is clearly related.
DRUNKEN ELM *Ulmus glabra* (Wych Elm) Lincs. Grigson. Perhaps from its looser habit, Grigson says.
DRUNKEN PLANT *Lolium temulentum* (Darnel) Dev. B & H. Interesting, because many of the names for darnel stem from French ivraie, drunk, for the alkaloid that makes it a dangerous pest causes giddiness. As Gerard said, it causes drunkenness in those who eat it in bread hot from the oven.
DRUNKEN SAILOR 1. *Centranthus ruber* (Spur Valerian) Dev. Friend.1882. 2. *Quisqualis indica* (Rangoon Creeper) Perry. Names like this are usually a commentary on the random or uncertain way the plant spreads.
DRUNKEN SLOTS *Valeriana officinalis* (Valerian) Som. Grigson. More likely to be Spur Valerian, rather than this, although. this listing was in Britten & Holland, too.
DRUNKEN WILLY *Centranthus ruber* (Spur Valerian) Dev, Som. Elworthy.
DRUNKITS *Centranthus ruber* (Spur Valerian) Som. Macmillan. Drunkards, of course.
DRY CUCKOO; DRYLAND CUCKOO *Saxifraga granulata* (Meadow Saxifrage) both Wilts. Dartnell & Goddard. In contrast to Wet Cuckoo, which is *Cardamina pratensis*; both of them are called Cuckoo-flower.
DRYLAND SCOUT *Heracleum sphondyllium* (Hogweed) Tyrone. B & H.
DRYMARY *Drymaria pachophylla*.
DUCHARTRE'S LILY *Lilium duchartrei*.
DUCK, Sour *Rumex acetosa* (Wild Sorrel) Som. Macmillan. For sour duck, read sorrel dock.
DUCK ACORN *Nelumbo lutea* (Yellow

Lotus) Perry.

DUCK POTATO *Sagittaria latifolia* (Broad-leaved Arrowhead) Douglas.

DUCK WILLOW *Salix alba* (White Willow) B & H.

DUCK'S BILL 1. *Dicentra spectabilis* (Bleeding Heart) Som. Elworthy. 2, *Iris pseudo-acarus* (Yellow Iris) Som. Macmillan. 3. *Syringa vulgaris* (Lilac) Dev. Friend.1882.

DUCK'S EYES *Cotula coronopifolia* (Buttonweed) Perry.

DUCK'S FOOT 1. *Alchemilla vulgaris* (Lady's Mantle) Dur, Berw. Grigson. 2. *Podophyllum peltatum* (May-apple) J.Smith. *Podophyllum* is from podos, foot, and phyllon, leaf - hence Duck's Foot.

DUCK'S MEAT *Lemna minor* (Duckweed) Gerard.

DUCK'S MOUTH *Digitalis purpurea* (Foxglove) Som. Macmillan.

DUCKGRASS *Eriocaulon septangulare* (Pipewort) USA Cunningham.

DUCKNUT *Juglans regia* nut (Walnut) War. F Savage. "Presumably on account of their large size", according to Savage. But what has that to do with it?

DUCKS-AND-DRAKES *Orchis mascula* (Early Purple Orchis) Dor. Macmillan.

DUCKWEED *Lemna minor*. Presumably just because it is an aquatic, but note Gerard's Duck's Meat.

DUCKWEED, Fat *Lemna gibba*.

DUCKWEED, Gibbous *Lemna gibba* (Fat Duckweed) Storer.

DUCKWEED, Great *Lemna polyrrhiza*.

DUCKWEED, Ivy; DUCKWEED, Ivy-leaved *Lemna trisulca* Gilmour & Walters (Ivy-leaved).

DUCKWEED, Least *Wolffia arrhiza*.

DUCKWEED, Lesser *Lemna minor* (Duckweed) Vickery.1983.

DUDAIM MELON *Cucumis dudaim*.

DUDGEON *Buxus sempervirens* (Box) Gerard. "Turners and cutlers, if I mistake not the matter, do call this wood dudgeon, wherewith they make dudgeon-hafted daggers."
Halliwell says it was the root of box which was called dudgeon, and the word came to mean homely, for wooden-handled daggers were certainly not used by the upper classes.

DUDLEY'S RUSH *Juncus dudleyi*.

DUFFIN BEAN *Phaseolus lunatus* (Lima Bean) Willis.

DUFFLE *Verbascum thapsus* (Mullein) Suff. Grigson. Duffle, whether it is spelt this way or as duffel, is a coarse woollen cloth with a thick nap, and so is a reference to the flannelly leaves. Duffle is named from a town in Brabant, Duffel.

DUIKER TREE *Pseudolachnostylis maprouneifolia*. Because antelopes are very fond of the ripe fruit.

DUKE OF ARGYLL'S TEA PLANT *Lycium halimifolium*. It is said that the 3rd Duke of Argyll received a tea plant, and one of these, with their labels exchanged, and so their identities mixed. He just referred to the *Lycium* as his tea-plant. It is still quite often cultivated under the name Tea Plant, though, as Young pointed out, we should be ill-advised to try making tea of it.

DUMB CAMMOCK *Ononis repens* (Rest Harrow) Som. Macmillan. Cammock is a common enough name for this plant, but why 'dumb'?

DUMB CANE *Dieffenbachia seguine*. Because the sap is highly irritating to the tongue; biting any part of it will prevent speech for some days. It is wishful thinking which produced in this connection the name Mother-in-law Plant.

DUMB NETTLE 1. *Lamium album* (White Deadnettle) Som, Wilts, Worcs, Herts, Ess. Grigson. 2. *Lamium purpureum* (Red Deadnettle) Som. Macmillan. Whether it is referred to as deaf, dumb or dead, it means the same thing - the nettle which has no sting.

DUMB NETTLE, Double *Ballota nigra* (Black Horehound) Wilts. Dartnell & Goddard.

DUMBLE DOR *Ophrys apifera* (Bee Orchid) Surr. B & H. A variation of Humble, or Bumble, bee.

DUMMIES *Petasites hybridus* (Butterbur) Hants. Cope.
DUMMY NETTLE *Lamium album* (White Deadnettle) Som. Macmillan; Berks. Lowsley. Given that this is a deadnettle, the use of 'dummy' is perfectly plain.
DUMPLING, Apple *Billardiera scandens* (Apple Berry) Perry.
DUMPLING, Currant *Epilobium hirsutum* (Great Willowherb) N'thum. Grigson. One of a series of names, the best known of which is Apple Pie, but the pie or dumpling will include a lot of other fruit as well. And all because, so it is claimed, of the fruity smell of the plant.
DUMPY CENTAURY *Centaureum capitatum*.
DUN DAISY *Leucanthemum vulgare* (Ox-eye Daisy) Dev. Friend.1882. This is an abbreviation of Thunder Daisy. It is also known as Thunder Flower in Somerset, and these are possibly a reflection of another name - Midsummer Daisy.
DUN NETTLE *Lamium album* (White Deadnettle) Shrop. Grigson. 'Dun' here is the same as 'dunch', or even 'dunny', and means dumb. Deadnettles are often dumb, or deaf, as well as dead.
DUN PEA *Pisum sativum var. arvense* (Field Pea) Howes.
DUNCH *Lamium album* (White Deadnettle) Wilts. Dartnell & Goddard. i.e. dumb, according to some, but it is more likely to mean deaf.
DUNCH NETTLE 1. *Lamium album* (White Deadnettle) Dor, Som. Macmillan; Wilts. Jones & Dillon. 2. *Lamium purpureum* (Red Deadnettle) Som. Grigson; Wilts. Macmillan.
DUNDER DAISY; DUNDLE DAISY *Leucanthemum vulgare* (Ox-eye Daisy) both Som. Macmillan. Cf. Dun Daisy. They all mean Thunder Daisy.
DUNG, Devil's *Ferula foetida* (Asafoetida) Mitton.
DUNG, Dove's 1. *Ornithogalum nutans* (Drooping Star-of-Bethlehem) Whittle. & Cook. 2. *Ornithogalum umbellatum* (Star-of-Bethlehem) Grieve.1931. Descriptive, for both of these have white, dotted flowers that look just like bird droppings.
DUNGWEED *Chenopodium album* (Fat Hen) Wilts. D Grose; Glos. Grigson. The reference here is to its favourite habitat, a dungheap. Cf. such names as Muckhill-weed, Dirtweed, etc.
DUNKELD LARCH *Larix x eurolepis*.
DUNKS *Zizyphus jubajuba* (Jujube) W Indies Howes.
DUNLUCE, Flower of *Geranium pratense* (Meadow Cranesbill) Antrim Grigson.
DUNNY-LEAF; DUNNY-WEED *Tussilago farfara* (Coltsfoot) both Herts. Grigson. Dunny must be donkey, so these names are conveying the same message as coltsfoot itself.
DUNNY-NETTLE 1. *Ballota nigra* (Black Horehound) Bucks. Grigson. 2. *Galeobdalon luteum* (Yellow Archangel) Bucks. B & H. 3. *Lamium album* (White Deadnettle) Oxf. B & H; Berks. Lowsley. 4. *Tussilago farfara* (Coltsfoot) Berks. Lowsley; Oxf. B & H. 'Dunny' for the horehound and archangel is in the same series as 'dunch' and all the rest of the words used to convey the sense of dumb, or deaf - in other words, dunny-nettle is deadnettle. But 'dunny' when used for a coltsfoot means a donkey.
DUNSE NETTLE 1. *Lamium album* (White Deadnettle) Wilts. Dartnell & Goddard. 2. *Lamium purpureum* (Red Deadnettle) Wilts. Dartnell & Goddard. As Dunch Nettle.
DUPPY BASIL *Ocymum micranthum* Barbados Gooding. Duppy, it is said, is a mosquito. The plant is actually called Mosquito-bush, too, from the fact that the essential oil, methyl cinnamate, seems to be repulsive to mosquitoes. But doesn't duppy usually mean a ghost?
DUPPY NEEDLES *Bidens pilosa* (Bur Marigold) Barbados Gooding.
DURANGO PINE *Pinus durangensis*. This is a Mexican species.

DURANGO ROOT *Datisca glomerata*.
DURHAM MUSTARD *Sinapis arvensis* (Charlock) Leyel.1937. For Durham mustard always used to be, and probably still is, made from charlock, and the seed was sold under this name.
DURIAN *Durio zibethinus*. This comes from a Malayan word, duri, meaning a thorn.
DURKENS *Pinus sylvestris* cones (Pine Cones) Nairn M H King.
DURMAST OAK *Quercus petraea*. Perhaps an error originally for dun-mast.
DUROBBY *Syzygium moorei* (Coolamon) Australia Wilson.
DURRA *Sorghum vulgare* (Kaffir Corn). Arabic dhurah.
DUST, Gold 1. *Aurinia saxatile*. 2. *Sedum acre* (Biting Stonecrop) Som. Macmillan.
DUST, Golden 1. *Aurinia saxatile* (Gold Dust) Som. Macmillan. 2. *Galium verum* (Lady's Bedstraw) Som. Macmillan. 3. *Sedum acre* (Biting Stonecrop) Corn, Suff. Grigson. 4. *Solidago virgaurea* (Golden Rod) Som. Macmillan.
DUSTER, Fairy *Calliandra eriophylla* (Hairy-leaved Calliandra) RHS. 'Duster' from the hairy leaves.
DUSTY BOB *Senecio cineraria* (Silver Ragwort) Som. Macmillan.
DUSTY HUSBAND 1. *Arabis alpina* (Alpine Rock-cress) Ches. B & H. 2. *Cerastium tomentosum* (Snow-in-summer) Ches. B & H.
DUSTY MILLER 1. *Artemisia stelleriana* (Beach Wormwood) Clapham. 2. *Centaurea candidissima* USA Cunningham. 3. *Centaurea gymnocarpa* USA Cunningham. 4. *Cerastium tomentosum* (Snow-in-summer) Lincs. B & H. 5. *Primula auricula* (Auricula) Aber, Loth. B & H. 6. *Senecio cineraria* (Silver Ragwort) USA Hay & Synge. In all cases, there is a mealy or hoary appearance to the foliage, hence this name.
DUTCH *Trifolium repens* (White Clover) Som. Elworthy Dor. Macmillan. See Dutch Clover.
DUTCH AGRIMONY 1. *Bidens cernua* (Nodding Bur-marigold) Gerard. 2. *Eupatorium cannabinum* (Hemp Agrimony) B & H.
DUTCH ARBEL *Populus alba* (White Poplar) Som. Grigson. Arbel, as well as Arbale, is a Somerset version of the more usual abele, a Dutch word, originally abeel. It comes from a diminutive of the Latin albus, white. Dutch, because these trees were imported from Holland in the 17th century.
DUTCH BARLEY *Hordeum vulgare* (Barley) Turner.
DUTCH BEECH *Populus alba* (White Poplar) B & H.
DUTCH BOX *Buxus sempervirens var. suffruticosa*. It is also called French Box. Have either of them significance as countries of origin, or are they used simply as 'foreign'?
DUTCH CHEESE *Malva neglecta* fruit (Dwarf Mallow) Ches. B & H. There are many 'cheese' names for the mallows, all descriptive.
DUTCH CLOVER 1. *Trifolium pratense* (Red Clover) Hulme. 2. *Trifolium repens* (White Clover) Curtis. Apparently, clover seed was largely imported from Holland. It was first introduced as a fodder plant in 1645 by Sir Richard Western, who was ambassador to Holland.
DUTCH CROCUS *Crocus purpureus* (Spring Crocus) RHS.
DUTCH DAFFODILL *Asphodelus ramosus* (White Asphodel) Turner.
DUTCH DILL *Meum athamanticum* (Baldmoney) Turner.
DUTCH ELDER *Aegopodium podagraria* (Goutweed) Wilts. Dartnell & Goddard. There is a superficial resemblance of the leaves to those of elder, so a very common name is Ground Elder, and there are a number of other 'elder' names. But why 'Dutch'?
DUTCH ELM *Ulmus x hollandica*. Traditionally linked with William of Orange, and supposedly brought from Holland at his accession in 1689. Actually it is said to be rare in Holland, but does occur in Picardy. Anyway, it could easily have originated in

England.
DUTCH GRASS *Eleusine indica* (Goosegrass) W Indies Howes.
DUTCH HEATH, Low *Erica tetralix* (Cross-leaved Heath) Parkinson.
DUTCH HONEYSUCKLE *Lonicera periclymenum* (Honeysuckle) McDonald.
DUTCH IRIS *Iris xiphium hybrids* (Spanish Iris). Confusing. because some of the hybrids of Spanish Iris are known as Dutch Iris.
DUTCH LAVENDER *Lavandula vera*. It is actually a Mediterranean plant.
DUTCH MEZERION *Daphne mezereon* (Lady Laurel) Lyte, Gerard.
DUTCH MICE *Lathyrus tuberosus* (Fyfield Pea) W.Miller.
DUTCH MORGAN *Leucanthemum vulgare* (Ox-eye Daisy) IOW Grigson. According to Halliwell, 'morgan' is defined as "tares in corn". Was ox-eye ever a cornfield pest?
DUTCH MYRTLE *Myrica gale* (Bog Myrtle) Som. Grigson.
DUTCH OSIER *Salix alba var. vitellina* (Golden Willow) Clapham.
DUTCH PINK PLANT *Reseda luteola* (Dyer's Rocket) W.Miller.
DUTCH RUSH *Equisetum hyemale*. Dutch, because it was imported in bundles from Holland as a domestic polisher.
DUTCHMAN, Flying *Acer pseudo-platanus* seeds (Sycamore keys) Som. Macmillan.
DUTCHMAN'S BREECHES *Dicentra cucullaria*. It is the profile view of the flowers that give the name (Cf. Chinaman's Breeches for *D spectabilis*).
DUTCHMAN'S PIPE 1. *Aristolochia californica* USA Barrett & Gifford. 2. *Aristolochia sipho*. From the shape of the tube of the flower, like the stem of a meerschaum pipe.
DWALE 1. *Atropa belladonna* (Deadly Nightshade) all B & H. 2. *Helleborus foetidus* (Stinking Hellebore) Voigts & Hudson. 3. *Solanum nigrum* (Black Bindweed) Voigts & Hudson. Dwale is an old word meaning torpor, or trance, probably from a Norse word dool, to delay or sleep.

The name has been accepted since Chaucerian times as synonymous with opiate. See Chaucer's Reeve's Tale - when the Miller and his wife went to bed, they had drunk so much ale, "hem needede no dwale". The word was originally used as a general term for sleeping draught, and only later was it restricted to a name for Deadly Nightshade, which was of course used for sleeping draughts. So were the other two bearing the name.
DWALE, Deadly; DWAY-BERRY *Atropa belladonna* (Deadly Nightshade) B & H.
DWINKLE *Vinca minor* (Lesser Periwinkle) Som. Macmillan. One of the many variants of periwinkle.
DWOSTLE, Dwarf *Mentha pulegium* (Pennyroyal) Cockayne.
DYE, Devil's *Indigofera anil* (Indigo) C.J.S.Thompson.1947. As a protection for woad, indigo as branded as "the food for the devil" in this country, but it was the German woad merchants who designated indigo devil's dye.
DYE BEDSTRAW *Asperula tinctoria* (Dyer's Woodruff) Usher. An earlier botanical name for this plant was Galium tinctorium, hence dye bedstraw. The roots yield a red dye.
DYEING CARTHAMUS *Carthamus tinctorius* (Safflower) Buhler.
DYER'S ALKANET *Alkanna tinctoria*. The roots give a red dye, used by pharmacists and perfumiers as a colouring agent, and also used to stain wood and marble.
DYER'S BROOM *Genista tinctoria* (Dyer's Greenweed) Prior.
DYER'S BUGLOSS *Anchusa tinctoria*. This is a famous dyeplant, giving a red dye used not only by dyers, but by cabinet-makers for staining wood, and by vintners for staining the corks of port bottles, even for colouring various liquids spuriously sold as port.
DYER'S CAMOMILE *Anthemis tinctoria* (Yellow Camomile) Coats. Quite popular at one time in Finland and Russia for a yellow dye.

DYER'S GRAPES *Phytolacca decandra* (Poke-root) W.Miller. Cf. such names as Red-ink Plant, Crimsonberry-plant, or Kermes-bush, for this.

DYER'S GREENING WEED *Genista tinctoria* (Dyer's Greenweed) B & H.

DYER'S GREENWEED 1. *Genista tinctoria*. 2. *Isatis tinctoria* (Woad) B & H. 3. *Reseda luteola* (Dyer's Rocket) Som. Grigson. All well-used dyeplants.

DYER'S HERB; DYER'S WOAD *Isatis tinctoria* (Woad) Pomet (Herb); Buhler.1948 (Woad).

DYER'S KNOTGRASS *Polygonum tinctorium*. One of the indigo-bearing dyeplants.

DYER'S MADDER *Rubia tinctoria* (Madder) Thornton.

DYER'S MULBERRY *Chlorophora tinctoria* (Fustic) Barton & Castle. Mulberry - *Chlorophora* is a part of the Moraceae, and an alternative name is *Morus tinctoria*. Fustic is the principal source of natural yellow pigments, and is obtained from this tree.

DYER'S OAK 1. *Quercus infectoria* (Aleppo Oak) Ablett. 2. *Quercus velutina* (Black Oak) J.Smith. The inner bark of Black Oak, stripped and prepared into an extract, called quercitron, is used as a dyestuff. It can give black, or yellow to orange, according to the recipe. Presumably Aleppo Oak is similarly used.

DYER'S ROCKET *Reseda luteola*. The plant has been cultivated since neolithic times for the dye obtained from it. The colour is yellow, but Lincoln green could be got by dyeing blue with woad, and then yellow from this.

DYER'S ROOT *Rubia tinctoria* (Madder) Leggett. The red pigment is obtained from the rootstock.

DYER'S SAFFRON *Carthamus tinctoria* (Safflower) Zoharry.

DYER'S SAVORY; DYER'S SAW-WORT *Serratula tinctoria* (Saw-wort) W.Miller (Savory); Flower.1859 (Saw-wort). Once used for dyeing woollen fabrics. Alum and the leaves were put in the dye-bath, and the leaves then gave a good green-yellow.

DYER'S WEED 1. *Asperula tinctoria* (Dyer's Woodruff) Webster. 2. *Genista tinctoria* (Dyer's Greenweed) Gerard. 3. *Isatis tinctoria* (Woad) B & H. 4. *Reseda luteola* (Dyer's Rocket) Gerard.

DYER'S WELD *Reseda luteola* (Dyer's Rocket) Howes.

DYER'S WOODRUFF *Asperula tinctoria*.

DYER'S YELLOW-WEED *Genista tinctoria* (Dyer's Greenweed) Grigson. In other words it provides the yellow to go with woad to make the green of Greenweed.

DYEWEED *Genista tinctoria* (Dyer's Greenweed) B & H. Cf. Dyer's Weed.

E

EACOR *Quercus robur* - acorn (Acorn) Dor. Barnes.

EAGLE, Bird; EEGLE, Bird's *Crataegus monogyna* fruit (Haws) Ches. Holland (Bird Eagle); Ches. Grigson (Bird's Eegle). Nothing to do with eagles - this is eggle, the diminutive of hague, the common name for the haw in Cheshire. The sequence is Bird's Eggle - Bird Eagle - even Bird's Eggs.

EAGLE FERN *Pteridiuum aquilina* (Bracken) Howes.

EAGLE VINE *Marsdenia condurango*.

EAGLEWOOD *Aquilaria agallocha*.

EAK *Quercus robur* (Oak) N.Eng. Halliwell. The attempts at representing the various local pronunciations of oak are idiosyncratic. They vary from Aik to Wuk, or even Yik.

EALIVER; EILEBER *Alliaria petiolata* (Jack-by-the-hedge) Cockayne (Ealiver); B & H (Eileber).

EAR, Ass *Symphytum officinale* (Comfrey) B & H. From the shape of the leaves. It is oreille d'âne in French, too.

EAR, Bear's 1. *Primula auricula* (Auricula) B & H. 2. *Saxifraga sarmentosa* (Mother-of-thousands) J.Smith. The 'ear' name for the saxifrage seems to be an aberration. The primula is different - auricula means little ear. Like most of these 'ear' names, it is the shape of the leaf that gives rise to it. Bear's Ears is corrupted to Boar's Ears, or Bores' Ears in Scotland, and in England there is further corruption to Baziers or Basiers in Lancashire, and to Bezors in Gloucestershire.

EAR, Boar's; EAR, Bore's *Primula auricula* (Auricula) both N.Scot. B & H). On the face of it, a variation from Bear's Ears. But Jamieson says a bear is called a boar in northern Scotland.

EAR, Bull's *Verbascum thapsus* (Mullein) Wales Grigson. It is rather the texture of the leaves that suggests an ear - they are called Donkey's, Bunny's, and Cow's, as well as Bull's.

EAR, Bunny's 1. *Stachys lanata* (Lamb's Ear) RHS. 2. *Verbascum thapsus* (Mullein) Dor. Macmillan.

EAR, Bunny Rabbit's *Lotus corniculatus* (Bird's-foot Trefoil) Som. Macmillan.

EAR, Cat's 1. *Agrostemma githago* (Corn Cockle) Dor. Macmillan. 2. *Antennaria dioica* (Cat's Foot) B & H. 3. *Hypochoeris radicata* (Long-rooted Cat's Ear) McClintock.

EAR, Cat's, Long-rooted *Hypochoeris radicata*.

EAR, Cat's, Smooth *Hypochoeris glabra*.

EAR, Cat's, Spotted *Hypochoeris maculata*.

EAR, Cow's *Verbascum thapsus* (Mullein) Wales Grigson. Cf. Bull's Ear for this.

EAR, Deer's *Swertia radiata*.

EAR, Devil's 1. *Arisaema triphyllum* (Jack-in-the-pulpit) Le Strange. 2. *Stellaria holostea* (Greater Stitchwort) Leyel.1937.

EAR, Dog's *Artemisia vulgaris* (Mugwort) Pemb. Grigson.

EAR, Donkey's 1. *Kalanchoe tomentosa*. 2. *Plantago lanceolata* (Ribwort Plantain) Som. Macmillan. 3. *Stachys lanata* (Lamb's Ears) Dev. Friend.1882. 4. *Verbascum thapsus* (Mullein) Dor. Grigson. The felty texture of the leaves of most of them are enough to account for the name, but the shape of Ribwort must have suggested it, though it sounds rather far-fetched.

EAR, Elephant's 1. *Bergenia cordifolia*. 2. *Colocasia antiquorum* (Taro). 3. *Enterolobium cyclocarpum* (Earpod) Prance. 4. *Philodendron domesticum*. In all cases, it is probably the size of the leaves (or in the case of *Enterolobium cyclocarpum* the pods) rather than any other quality which gives the names.

EAR, Hare's 1. *Bupleurum rotundifolium*. 2. *Conringia orientalis* (Hare's Ear Cabbage) Prior.

EAR, Hare's, Narrow *Bupleurum opacum*.

EAR, Hare's, Shrubby *Bupleurum fruticosum*.

EAR, Hare's, Sickle *Bupleurum falcatum.*
EAR, Hare's, Slender *Bupleurum tenuissimum.*
EAR, Hare's, Smallest *Bupleurum tenuissimum* (Slender Hare's Ear) Clapham.
EAR, Jew's 1. *Dicentra spectabilis* (Bleeding Heart) Som. Macmillan. 2. *Lycopersicon esculentum* (Tomato) Hants. Cope.
EAR, Lamb's 1. *Lamium purpureum* (Red Deadnettle) Som. Macmillan. 2. *Plantago media* (Lamb's Tongue Plantain) Cumb. Grigson. 3. *Potentilla anserina* (Silverweed) Som. Macmillan. 4. *Stachys lanata.* 5. *Succisa pratensis* (Devil's-bit) War. B & H. It is the texture of *Stachys lanata*, and the slightly woolly look of the rest that gives the name.
EAR, Lion's 1. *Leonotis leonurus* (Lion's Tail) RHS. 2. *Leonurus cardiaca* (Motherwort) Le Strange.
EAR, Mouse 1. *Cerastium sp.* 2. *Hieraceum pilosellum* (Mouse-ear Hawkweed) Turner. 3. *Stachys lanata* (Lamb's Ears) Dev. Friend.1882.
EAR, Mouse, Golden *Hieraceum aurantiacum* (Orange Hawkweed) Gerard.
EAR, Mouse, Little *Antennaria dioica* (Cat's Foot) Turner.
EAR, Mouse, Yellow *Hieraceum pilosellum* (Mouse-ear Hawkweed) Turner.
EAR, Pig's 1. *Cotyledon orbicularis* (Round-leaved Navelwort) Whittle & Cook. 2. *Sedum acre* (Biting Stonecrop) Friend.1882.
EAR, Pussy *Kalanchoe tomentosa* (Donkey's Ears) RHS.
EAR, Rabbit's *Stachys lanata* (Lamb's Ears) Som. Macmillan.
EAR, Sheep's 1. *Heliotropum appendiculatum* Howes. 2. *Stachys lanata* (Lamb's Ears) Som. Macmillan.
EAR-BOBS *Fuchsia sp* Som. Macmillan.
EAR-DROPS 1. *Fuchsia sp* Dev,Som,Suss. Wright. 2. *Dicentra spectabilis* (Bleeding Heart) Som. Macmillan.
EAR-DROPS, Lady's 1. *Dicentra spectabilis* (Bleeding Heart) Som. Macmillan. 2. *Fuchsia sp* Dev. Friend.1882.

EAR-DROPS, My Lady's *Fuchsia sp* Som. Macmillan.
EAR-RING FLOWER *Fuchsia sp* Lincs. B & H.
EAR-RINGS 1. *Fuchsia sp* Som. Macmillan. 2. *Laburnum anagyroides* (Laburnum) Ches. B & H.
EARACHE *Papaver rhoeas* (Red Poppy) Midl. Grigson. Headache is more usual for this, and it is more widespread. In Wiltshire, they said that if you picked poppies from the corn, you would either have a bad headache or there would be thunder and lightning. Earache was similarly rationalized, for it was said that if gathered and put to the ear, earache would follow. In Somerset, though, a poultice of poppy leaves laid against the ear was used to cure earache.
EARED SALLOW *Salix aurita.*
EARLE LOCO *Astragalus earlei.*
EARNING GRASS *Pinguicula vulgaris* (Butterwort) N'thum, Lanark. Grigson. Earning is a north country word for rennet; to earn is to curdle milk. "If the fresh-gathered leaves of this plant...are put into a strainer, through which warm milk from the cow is poured, it acquires a consistence and tenacity; the whey does not separate, nor does the cream. In which state it is an extremely grateful food..." (Taylor).
EARPOD *Enterolobium cyclocarpum.*
EARTH, Ball-of-the-; EARTH, Gall-of-the- *Centaureum eyrthraea.* Vesey-Fitzgerald (Ball); Fernie (Gall). Is Ball-of-the-earth a misprint? Gall-of-the-earth may possibly be genuine, though Cockayne's Earthgall is probably erroneous, and meant for a gentian. However, the name Bitter Herb for centaury seems established, and it was used at one time like ground ivy, to make beer keep.
EARTH, Man-of-the- *Ipomaea pandurata* (Wild Potato Vine) Kansas Gates. Cf. Man-in-the-ground for this.
EARTH, Star-of-the- *Geum urbanum* (Avens) Tynan & Maitland.
EARTH ALMOND *Cyperus esculentus* (Tigernut) Schery.

EARTH APPLE 1. *Cyclamen europaeum* (Sowbread) Henslow. 2. *Helianthus tuberosus* (Jerusalem Artichoke) Sanecki. The corm in the first place, and the tuber in the second, earn these plants the name.

EARTH BARK 1. *Potentilla anserica* (Silverweed) Shet. B & H. 2. *Potentilla erecta* (Tormentil) Ork. Martin; Shet. Marwick. The name is a reflection of the high tannin content of the roots, particularly of tormentil. They were actually used as a substitute for oak bark (hence the second part of the name) in tanning. The use is recorded in Ireland, the Western Isles, where fishermen tanned their nets with the roots, on Canna, where leather for shoes was tanned this way, and in the Northern Isles. Hibbert mentions it particularly in connection with the sheepskins used by Shetland fishermen over their ordinary dress.

EARTH CHESTNUT 1. *Bunium bulbocastanum* (Great Earthnut) Gerard. 2. *Conopodium majus* (Earthnut) Prior. 3. *Lathyrus tuberosus* (Fyfield Pea) Douglas. The tubers of the Earthnuts are known as nuts, and are edible; they can be roasted like chestnuts, and look rather like them.

EARTH CHESTNUT, Australian *Eleocharis sphacelata*.

EARTH IVY *Glechoma hederacea* (Ground Ivy) Dawson.

EARTH PEA *Arachis hypogaea* (Ground Nut) J.Smith.

FARTH QUAKES *Briza media* (Quaking-grass) N'thants. Sternberg.

EARTH STAR *Cryptanthus sp*. At least, that is what Rochford calls them.

EARTHBEET *Beta vulgaris* (Beet) Dawson.

EARTHGALL 1. *Blackstonia perfoliata* (Yellow-wort) Cockayne. 2. *Centaureum erythraea* (Centaury) Cockayne. There is some doubt that Cockayne meant centaury for this name, though Halliwell says the same. Gall - Cf. Bitter Herb; it was once used rather like ground ivy to make beer keep.

EARTHNUT 1. *Arachis hypogaea* (Ground Nut) USA Elmore. 2. *Conopodium majus*. 3. *Oenanthe pimpinelloides* (Callous-fruited Water Dropwort). Hants. Cope. *Conopodium majus* has a tuberous rootstock which is edible, especially when roasted like a chestnut, which it resembles. It never seems to have been cultivated, but has been known as a wild food since ancient times. Is the identification correct as far as the Dropwort is concerned?

EARTHNUT, Great *Bunium bulbocastanum*.

EARTHNUT, Pease *Lathyrus montanus* (Bittervetch) Gerard. The tubers are edible, and often roasted like chestnuts.

EARTHNUT PEA *Lathyrus tuberosus* (Fyfield Pea) Clapham.

EARTHSMOKE *Fumaria officinalis* (Fumitory) Friend. Fumitory is from O. French fumeterre, med. Latin fumus terrae, i.e. smoke of the earth. One reason is the belief that it did not spring up from seeds, but from the vapours of the earth. The root when fresh pulled up gives a strong gaseous smell like nitric acid, and this is probably the origin of the belief in its gaseous origin. Another suggestion is that the Greeks and Romans used the juice to clear the sight, and noted that, while doing so, it would make the eyes water, as smoke would. Probably it was just the wispy appearance that gave the name - from a distance it does look like smoke.

EARWIG *Convolvulus arvensis* (Field Bindweed) Som. Macmillan.

EASE, Traveller's *Achillea millefolium* (Yarrow) Wilts. Dartnell & Goddard. Why? Unless, of course, there is some now-forgotten usage involving putting a leaf inside the sock in a long journey, as with some other plants?

EAST AFRICAN CAMPHORWOOD *Ocotea usambarensis*.

EAST AFRICAN FIG *Ficus glumosa*.

EAST AFRICAN JUNIPER *Juniperus procera*.

EAST AFRICAN OLIVE *Olea hochstetteri*.

EAST AFRICAN SANDALWOOD *Osyris compressa*.

EAST AFRICAN YELLOW-WOOD *Podocarpus gracilior* (Podo) Dale & Greenway.
EAST ANGLIAN ELM *Ulmus x diversifolia* (Lock Elm) Wit.
EAST HIMALAYAN FIR *Abies spectabilis*.
EAST INDIAN BASIL *Ocymum gratissimum*.
EAST INDIAN MAHOGANY *Pterocarpus dalbergioides*.
EAST INDIAN ROSEWOOD *Dalbergia latifolia* (Blackwood) Willis.
EAST INDIAN WALNUT *Albizia lebbeck* (Siris Tree) Hora.
EAST INDIES ARROWROOT *Curcuma angustifolia* Howes.
EAST INDIES HEMP *Crotolaria juncea* (Sunn Hemp) Howes.
EAST LONDON BOX *Buxus macowanii* (Cape Box) Brimble.1948.
EASTER BELLS *Stellaria holostea* (Greater Stitchwort) Dev. Friend.1882. Cf. Easter-flower, a Sussex name for this, and also Whit-Sunday, or White Sunday, from Devon.
EASTER FLOWER 1. *Anemone nemerosa* (Wood Anemone) Dev, Dor. Macmillan. 2. *Euphorbia pulcherrima* (Poinsettia) USA Watt. 3. *Narcissus odorus* (Jonquil) Maryland Whitney & Bullock. 4. *Pulsatilla vulgaris* (Pasque Flower) Genders.1976. 5. *Stellaria holostea* (Greater Stitchwort) Suss. Grigson. The time of flowering accounts for the Wood Anemone and the stitchwort, and of course *Pulsatilla vulgaris* is the Pasque Flower, but Poinsettia is better known as a Christmas plant - in fact, it is known as Christmas-flower, or Christmas-bush in the States.
EASTER GIANTS; EASTERN GIANTS; EASTER MANGIANTS; EASTER MANTGIONS *Polygonum bistorta* (Bistort) Grigson (Giants); Yks, Cumb, N'thum. Grigson (Mangiants, Mantgions); N.Eng. Denham/Hardy (Eastern Giants). All, it seems, from French manger, to eat. For this was an Easter food, called Herb Pudding,or if not Easter, then the last two weeks of Lent, or Passion-tide. The leaves of bistort were boiled in broth with barley, chives, etc., and served to accompany veal and bacon. Sometimes, the name became Eastern Giants, and served to show that the user had forgotten the association with Easter.
EASTER LEDGERS; EASTER LEDGES; EASTER HEDGES; EASTER SEDGE *Polygonum bistorta* (Bistort) Cumb, Yks. Grigson (Ledgers, Ledges); Greenoak (Hedges); Cumb. Rowling (Sedge).
EASTER LILY 1. *Erythronium oregonum* (Giant Adder's Tongue) USA Turner. & Bell. 2. *Lilium longiflorum*. 3. *Narcissus pseudo-narcissus* (Daffodil) Dev. Friend; Som. Tongue. The true lily gets the name because of the huge numbers that are forced for Easter sale, though it is a summer-flowering plant. Daffodil is more obvious. Its time of flowering has given it such names as Lent, or Lide, Lily, as well as various permutations on these names.
EASTER ROSE 1. *Narcissus odorus* (Jonquil) Som. Tongue. 2. *Narcissus pseudo-narcissus* (Daffodil) Som. Elworthy. 3. *Primula vulgaris* (Primrose) Som. Macmillan. All spring-flowering, of course. The second syllable of primrose keeps manifesting itself erroneously as a separate entity - Cf. Butter Rose, First Rose, etc.
EASTERN BALSAM POPLAR *Populus balsamifera* (American Balsam Poplar) Mabey.1977.
EASTERN BORAGE *Trachystemon orientale* (Abraham, Isaac and Jacob) Polunin.
EASTERN COTTONWOOD *Populus deltoides* (Carolina Poplar) USA Zenkert. 'Eastern' in this case is the eastern side of North America.
EASTERN HEMLOCK *Tsuga canadensis*.
EASTERN LARKSPUR *Delphinium orientale*.
EASTERN PASQUE FLOWER *Pulsatilla patens*.
EASTERN RED CEDAR *Juniperus virginiana* (Virginian Juniper) Wit.
EASTERN REDBUD *Cercis canadensis*.

EASTERN ROCKET *Sisymbrium orientale*.
EASTERN SAVIN *Juniperus excelsa* (Grecian Juniper) Moldenke.
EASTERN STAR *Passiflora caerulea* (Blue Passion-flower) Som. Macmillan. It is called Star-of-Bethlehem in Somerset, too.
EASTERN WHITE CEDAR *Thuja occidentalis* (Arbor-vitae) RHS.
EASTERN WHITE PINE *Pinus strobus*. Eastern side of North America.
EASTWARD MARJORAM *Origanum vulgare* (Marjoram) Culpepper).
EASTWOOD LILY *Lilium nevadense* (Californian Panther Lily) Woodcock. Alice Eastwood, botanist.
EATIN BERRIES *Juniperus communis* (Juniper) Scot. Jamieson. This is the Scottish word Aiten, or Aitnagh, used for juniper.
EAU-DE-COLOGNE MINT *Mentha citrata* (Bergamot Mint) McClintock.
EAVER 1. *Lolium perenne* (Rye-grass) Corn. Quiller-Couch; Dev. Choape. 2. *Lolium temulentum* (Darnel) Dev. Halliwell; Som. Elworthy. It seems to be spelt Eever in Devonshire, but however it is spelt it can only be a variant of Haver, usually applied to Wild Oat.
EAVER, Devon *Lolium temulentum* (Darnel) Som. Elworthy.
EBBLE 1. *Populus alba* (White Poplar) E Ang. Nall. 2. *Populus tremula* (Aspen) E.Ang. Prior. This is abele, so the name is valid for White Poplar, but intrusive for aspen. Halliwell too quotes the name for aspen, though.
EBLE *Sambucus ebulus* (Dwarf Elder) R.T.Gunther. It is glossed as such on a 12th century ms of Apuleius Barbarus, and must be the French hièble, the usual name for Dwarf Elder in that country - or are they both just *ebulus*?
EBONY, African 1. *Dalbergia melanoxylon* (African Blackwood) Dale & Greenway. 2. *Diospyros mespiliformis*.
EBONY, Bombay *Diospyros montana*.
EBONY, Calabar *Diospyros mespiliformis* (African Ebony) Hora.
EBONY, Coromandel *Diospyros melanoxylon*.
EBONY, False *Laburnum anagyroides* (Laburnum) W.Miller.
EBONY, Lagos *Diospyros mespiliformis* (African Ebony) Hora.
EBONY, Mozambique *Dalbergia melanoxylon* (African Blackwood) Howes.
EBONY, North American *Diospyros virginiana* (Persimmon) Willis.
EBONY, Red *Rhamnus zeyheri* S.Africa Watt.
EBONY, Senegal; EBONY, Sudan *Dalbergia melanoxyla* (African Blackwood) Dalziel.
EBONY, Swamp *Diospyros mespiliformis* (African Ebony) Dalziel.
EBONY, Walking-stick *Diospyros monbuttensis* (Yoruba Ebony) Dalziel.
EBONY, Yoruba *Diospyros monbuttensis*.
EBONY, Zanzibar *Diospyros mespiliformis* the timber (African Ebony) Dalziel.
EBONY TREE 1. *Diospyros ebenaster* Bianchini. 2. *Diospyros ebenum*. 3. *Diospyros melanoxlyon* Bianchini.
ECCLE GRASS; EKKEL-GIRSE *Pinguicula vulgaris* (Butterwort) Ork. Jamieson (Eccle); Grigson (Ekkel).
ECKBERRY *Prunus padus* (Bird Cherry) Cumb. B & H. One of a series of names, all stemming from OE haga, hedge. Hackberry, or Heckberry, are the most widespread, but these become Hagberry and Hegberry, and so on. Even Eckberry has its variant, for in the same area Eggberry is recorded.
EDDER *Sambucus nigra* (Elder) Tusser.
EDDERWORT *Dracunculus vulgaris* (Dragon Arum) B & H. Edder is an OE word for adder, which is probably still pronounced this way in the northern counties. Cf. the old name Serpentine, and of course, there is the 'dragon' connection. Gerard says it has "spots of divers colours, like those of the adder or snake...".
EDDICK *Arctium lappa* (Burdock) Ches. B & H. Errick is also recorded from the same area.

EDDO *Colocasia antiquorum* (Taro) W Indies Brouk. A West African word originally.

EDELWEISS *Leontopodium alpinum*. Edel means noble, and weiss, white. The name was coined in all probability as a 19th century Alpine tourist attraction.

EDGEWEED *Oenanthe aquatica* (Horsebane) B & H. Apparently invented by Hill, 1769.

EDGING 1. *Armeria maritima* (Thrift) Dev. Friend.1882. 2. *Lobularia maritima* (Sweet Alison) Dev. Friend.1882. 3. *Saxifraga x urbinum* (London Pride) Dev. Friend.1882. Sweet Alison was known as Bordering, too. Garden usage, of course, but apparently the plants themselves were called by this name.

EEL-BEDS *Ranunculus fluitans* (River Crowfoot) W.Miller.

EEL-GRASS *Zostera marina*.

EEL-GRASS, Dwarf *Zostera nana*.

EEL-GRASS, Narrow-leaved *Zostera hornemanniana*.

EELWARE; EELWEED *Ranunculus aquatilis* (Water Crowfoot) N'thum. Grigson (Eelware); Donegal Grigson (Eelweed). A reasonable enough name for an aquatic plant.

EEVER 1. *Lolium perenne* (Rye-grass) Dev. B & H. 2. *Lolium temulentum* (Darnel) Dev. Halliwell. See Eaver.

EEVY *Hedera helix* (Ivy) B & H.

EGG, Bird's 1. *Crataegus monogyna* fruit (Haws) Ches. Holland. 2. *Silene cucubalis* (Bladder Campion) Shrop. Grigson. The first example is not 'egg' at all, but eggle or eagle, the diminutive of hague, the common name in Cheshire for haws. Presumably it is the shape of the bladder which gives *Silene cucubalis* the name.

EGG, Fried *Leucanthemum vulgare* (Ox-eye Daisy) Som, Wilts. Grigson. Descriptive.

EGG, Garden *Solanum melongena* (Egg Plant) Nielsen. Another example where the shape of the fruit provides the name.

EGG-AND-CHEESE *Oxalis acetosella* (Wood Sorrel) Suss. Vickery.1995.

EGG-IN-THE-PAN *Aurinia saxatile* (Gold Dust) Som. Macmillan.

EGG ORCHID *Cephalanthera damasonium* (Broad Helleborine) Young.

EGG-PEG BUSH *Prunus spinosa* (Blackthorn) Glos. J.D. Robertson. The 'egg' part of the name means hedge. Another name, Heg-peg Bush, comes a little closer to the original meaning. 'Peg' probably started off as 'pick'. Thus, picks is a Wiltshire name for sloes; so is Hedge-picks, which must be the original of Egg-peg.

EGG PLANT 1. *Solanum melongena*. 2. *Symphoricarpos rivularis* (Snowberry) Ches. B & H. The fruits are the reason for the name, the shape in one case, the colour in the other.

EGG TREE *Garcinia xanthocymus*.

EGGBERRY *Prunus padus* (Bird Cherry) Yks, Cumb. B & H. Eckberry is the immediate forbear of this name. The first syllable in each case means 'hedge'.

EGGCUPS *Tulipa sp* (Tulip) Som. Macmillan.

EGGERS *Crataegus monogyna* fruit (Haws) Suss. Parish. Cf. the Wessex Eglon, or the West Country Eglet, for haws.

EGGLE(BERRY); EGGLE, Bird's *Crataegus monogyna* fruit (Haws) Dev. Choape; Wessex Rogers (Eggle); Ches. Grigson (Bird's Eggle).

EGGREMUNNY *Agrimonia eupatoria* (Agrimony) Cumb. B & H.

EGGS, Bacon-and- 1. *Linaria vulgaris* (Toadflax) Wilts. Dartnell & Goddard. 2. *Narcissus odorus* (Jonquil) Som. Macmillan. 3. *Ranunculus aquatilis* (Water Crowfoot) Som. Macmillan.

EGGS, Butter-and- 1. *Leucanthemum vulgare* (Ox-eye Daisy) Tynan & Maitland. 2. *Linaria vulgaris* (Toadflax) Dev. Friend; Dor. Barnes; Wilts. Dartnell & Goddard; Glos. J.D.Robertson. Kansas Sackett & Koch. 3. *Lotus corniculatus* (Bird's-foot Trefoil) Som, Glos, War. Grigson; Suss. Parish. 4. *Narcissus incomparabilis* Dev, Surr, Lancs. B & H; Wilts. Dartnell &

Goddard. 5. *Narcissus maialis patellaris* (Pheasant's Eye) Corn. Quiller-Couch; Dev. Friend. 6. *Narcissus pseudo-narcissus* (Daffodil) Som. Elworthy; Dev, N'thum. Grigson. 7. *Oxalis acetosella* (Wood Sorrel) Som. Macmillan.

EGGS, Buttered *Lotus corniculatus* (Bird's-foot Trefoil) Cumb. Grigson.

EGGS, Buttery *Narcissus odorus* (Jonquil) Wilts. Macmillan.

EGGS, Chopped *Linaria vulgaris* (Toadflax) Som. Macmillan; Cumb. Grigson.

EGGS, Turkey *Fritillaria meleagris* (Snake's Head Lily) Berks. B & H. Guinea-hen, rather - the petals are spotted like this bird. Are turkey eggs spotted?

EGGS-AND-BACON 1. *Linaria vulgaris* (Toadflax) Dev. Friend.1882; Som. Tongue; Wilts. Dartnell & Goddard; Glos. J.D.Robertson. 2. *Lotus corniculatus* (Bird's-foot Trefoil) Som, Glos, Suss, Rut. Grigson; N'thants. A.E.Baker. 3. *Ranunculus arvensis* (Corn Buttercup) Ches. Grigson. 4. *Ranunculus bulbosus* (Bulbous Buttercup) Ches. Grigson. 5. *Ranunculus heterophyllus* (Various-leaved Buttercup) Som. Macmillan. 6. *Rubus caesius* the stalks (Dewberry) Som. Macmillan. 7. *Solanum tuberosum* flowers (Potato) Som. Macmillan.

EGGS-AND-BUTTER 1. *Linaria vulgaris* (Toadflax) Dev, Som, Wilts. Macmillan. 2. *Narcissus maialis var. patellaris* (Pheasant's Eye) Som. Tongue.1968. 3. *Ranunculus acris* (Meadow Buttercup) Ches. B & H. 4. *Ranunculus bulbosus* (Bulbous Buttercup) Ches. B & H.

EGGS-AND-COLLOPS 1. *Linaria vulgaris* (Toadflax) Som. Grigson; Lancs. Nodal; Yks. Carr. 2. *Lotus corniculatus* (Bird's-foot Trefoil) Yks. Grigson.

EGGS-EGGS *Crataegus monogyna* fruit (Haws) Wilts. Dartnell & Goddard. Cf. the Sussex Eggers, and the name Eglon, common throughout Wessex.

EGLANTINE 1. *Lonicera periclymenum* (Honeysuckle) Yks. Grigson. 2. *Rosa rubiginosa*. OE egla, or egle, meant a prickle or thorn, and was probably from Latin originally - aculentus means full of prickles (acus is a needle, and lentus, full). There are no thorns on honeysuckle, so it probably got the name by mistake, from the similarity of environment, and habit of growth.

EGLET; EGLON *Crataegus monogyna* fruit (Haws) Corn. Quiller-Couch; Dev, Som Grigson (Eglet); Wessex Rogers (Eglon).

EGREMOINE; EGREMOUNDE; EGRIMONY *Agrimonia eupatoria* (Agrimony) B & H (Agremoine, Egremounde); Gerard (Egrimony).

EGYPT, Bene of *Nelumbo nucifera* (Hindu Lotus) Turner.

EGYPTIAN *Geum rivale* (Water Avens) Wilts. Dartnell & Goddard.1899. It is the dusky colours of the flowers which earn this flower its 'gypsy' names. Cf. Egyptian Granny's Cap.

EGYPTIAN BEAN; EGYPTIAN KIDNEY BEAN *Dolichos lablab* (Lablab) Brouk (BEan); Dalziel (Kidney Bean). Also known as Indian Bean.

EGYPTIAN CLOVER *Trifolium alexandrinum*. It grows throughout the Near East.

EGYPTIAN COTTON *Gossypium barbadense* (Sea-Island Cotton) H.G.Baker.

EGYPTIAN GRANNY'S CAP *Geum rivale* (Water Avens) Wilts. Dartnell & Goddard.1899. This seems to be a fusion between Egyptian, i.e. gypsy, given because of the typical dusky colours, and Granny's Cap, a common name suggested by the shape of the flowers. Granny's Cap, Granny's Nightcap, or Old Woman's Bonnet, are all Wiltshire names.

EGYPTIAN HENBANE *Hyoscyamus muticus*. Grown there for its narcotic properties.

EGYPTIAN LETTUCE *Lactuca serriola* (Prickly Lettuce) Howes.

EGYPTIAN LOTUS *Nymphaea lotus*.

EGYPTIAN MALLOW *Malva parviflora* (Least Mallow) Howes.

EGYPTIAN MARJORAM *Origanum maru var. aegyptiacum*.

EGYPTIAN MELON, Round-leaved
Cucumis chate (Hairy Cucumber)
Moldenke. This is cultivated extensively in
Egypt.
EGYPTIAN MILLET 1. *Eleusine aegyptiaca*. 2. *Sorghum vulgare* (Kaffir Corn)
Leyel.1948.
EGYPTIAN MIMOSA 1. *Acacia arabica*
(Babul Tree) Dalziel. 2. *Acacia nilotica*
(Thorny Acacia) Moldenke.
EGYPTIAN MINT *Mentha rotundifolia*
(Apple Mint) Grieve.1931.
EGYPTIAN MULTIPLIER ONION *Allium cepa var. viviparum*. 'Multiplier', because it
produces onion-flavoured bulbils at the top of
the stalks.
EGYPTIAN MYROBALAN *Balanites aegyptiaca* the unripe fruit (Jericho Balsam)
Dale & Greenway.
EGYPTIAN ONION *Allium cepa var. aggregatum*.
EGYPTIAN PARADISE SEED *Amomum grana* (Grains of Paradise Plant) USA
Hurston.
EGYPTIAN PRIVET *Lawsonia inermis*
(Henna) Westermarck.
EGYPTIAN REED *Cyperus papyrus* (Paper
Sedge) Bonar.
EGYPTIAN ROSE 1. *Knautia arvensis*
(Field Scabious) IOW B & H. 2. *Scabiosa atropurpurea* (Sweet Scabious) IOW B &
H. It is the colour of the Sweet Scabious
which produced the name. Gypsy Rose is
another one. Field Scabious too has its
'gypsy' names, but probably only by association.
EGYPTIAN SESBAN *Sesbania sesban*.
EGYPTIAN THORN 1. *Acacia arabica*
(Babul Tree) Pomet. 2. *Acacia nilotica*
(Thorny Acacia) Hora. 3. *Acacia vera*. 4.
Pyracantha angustifolia (Orange Fire Thorn)
Ches. B & H.
EGYPTIAN'S HERB *Lycopus europaeus*
(Gypsywort) Grieve.1931. One can get a
black dye from this, used once for wool or
silk. But, more to the point, gypsies used to
stain their skins with it. "Some also thinke
good to call it Herba Aegyptia, because they
that feine themselves Aegyptians (such as
many times wander like vagabonds from citie
to citie in Germanie and other places) do use
with this herbe to give themselves a swart
colour, such as the Aegyptians and the people
of Affricke are of..." (Gerard).
EIGHT-DAY HEALING BUSH *Lobostemon fruticosus*.
EIKE-TREE; EKE-TREE *Quercus robur*
(Oak) Yks. B & H (Eike); Turner (Eke).
EILEBER; EALIVER *Alliaria petiolata*
(Jack-by-the-hedge). B & H (Eileber);
Cockayne (Ealiver).
EINCORN *Triticum monococcum*.
EISCH-KEYS *Fraxinus excelsior* fruit (Ash
keys) Lancs. B & H.
EITHIN *Ulex europaeus* (Furze) Wales
Sikes. But this is the Welsh for gorse, and not
an English name at all.
EKER *Nasturtium officinale* (Watercress) B
& H.
EKKEL-GIRSE; ECCLE-GRASS
Pinguicula vulgaris (Butterwort) Ork.
Grigson (Ekkel); Ork. Jamieson (Eccle).
EKKERN *Quercus robur* - acorn (Acorn)
Berks. Lowsley.
EKSIS-GIRSE *Taraxacum officinale*
(Dandelion) Shet. Grigson. Cf. Bitter Aks,
another Shetland name.
ELDEN; ELDIN *Petasites hybridus*
(Butterbur) N.Eng. Turner. Elden is fuel -
the plant was used as such among the poorer
people.
ELDEN, Dwarf *Sambucus ebulus* (Dwarf
Elder) Cumb. Grigson. A different meaning
for elden now - it is just a variant of elder.
ELDER 1. *Alnus glutinosa* (Alder) B & H.
2. *Sambucus nigra*. Elder is OE ellaern, from
a Germanic source. The older form is still
retained in such names as Eldern (which may
be the adjectival form), or Eller, Ellen etc.
Elder for alder is not nearly so common as
one would suspect, given the confusion
between the two names.
ELDER, Alpine *Sambucus racemosa* (Red
Elder) Polunin.

ELDER, American *Sambucus canadensis*.
ELDER, Autumn-flowering *Sambucus canadensis* (American Elder) W.Miller.
ELDER, Bishop's *Aegopodium podagraria* (Goutweed) IOW Grigson. Cf. Bishop's Weed, from the Border area. 'Bishop's', because it was so often found near old ecclesiastical ruins; it was said to have been introduced by the monks in the Middle Ages. Or was it because bishops were prone to gout?
ELDER, Black 1. *Eupatorium cannabinum* (Hemp Agrimony) Corn. Grigson. 2. *Valeriana officinalis* (Valerian) Corn. Grigson. 3. *Sambucus nigra* (Elder) Genders.
ELDER, Blackbead *Sambucus melanocarpa*.
ELDER, Blue 1. *Sambucus caerulea* (Blueberry Elder) Hora. 2. *Sambucus glauca*.
ELDER, Blueberry *Sambucus caerulea*.
ELDER, Box *Acer negundo*. Leaf shape, for the same reason as it is sometimes called Ash-leaved Maple.
ELDER, Bunchberry *Sambucus microbotrys*.
ELDER, California *Sambucus glauca* (Blue Elder) W.Miller.
ELDER, Churchyard *Capsella bursa-pastoris* (Shepherd's Purse) Som. Macmillan.
ELDER, Dane's *Sambucus ebulus* (Dwarf Elder) E.Ang, Lincs. Johnson. It is associated with the Danes - wherever their blood was shed in battle, this plant afterwards sprang up. The alternative tradition is that it was the Danes who brought it over, and who planted it on the battlefields and graves of their countrymen.
ELDER, Dutch *Aegopodium podagraria* (Goutweed) Wilts. Dartnell & Goddard. There are a number of elder names for this, all stemming from a superficial resemblance of the leaves.
ELDER, Dwarf 1. *Aegopodium podagraria* (Goutweed) Hants, IOW Grigson. 2. *Sambucus ebulus*.
ELDER, Golden *Sambucus nigra var. aurea*.
ELDER, Ground 1. *Aegopodium podagraria* (Goutweed) Salisbury.1964. 2. *Angelica sylvestris* (Wild Angelica) Ches. Holland. 3. *Sambucus ebulus* (Dwarf Elder) S.Eng. Grigson. Goutweed is probably better known under this name than its own, at least by frustrated gardeners. Wild Angelica gets the name because of similarity of appearance.
ELDER, Horse *Inula helenium* (Elecampane) Palaiseul. Elecampane only got this name by accident - the original OE was hors-elene, and there are several variations on this, including horseheal, or horseheel. It was given to horses as an appetite improver.
ELDER, Marsh *Viburnum opulus* (Guelder Rose) Gerard. He also called it Water Elder - it will only grow in damp hedgerows. There are other 'elder' names for this - White, Red, or Rose Elder are all recorded.
ELDER, Mexican *Sambucus mexicana*.
ELDER, Mountain *Sambucus pubens* (Scarlet Elder) Sanecki.
ELDER, Poison *Rhus vernix* (Swamp Sumach) USA Kingsbury.
ELDER, Prickly *Aralia spinosa* (Hercules' Club) Grieve.1931. It is also known as Prickly Ash.
ELDER, Red 1. *Sambucus ebulus* (Dwarf Elder) Som. Lawrence. 2. *Sambucus racemosa*. 3. *Viburnum opulus* (Guelder Rose) Scot. B & H. Dwarf Elder has red stems; the other two have red berries.
ELDER, Red-berried *Sambucus pubens* (Scarlet Elder) USA Yarnell.
ELDER, Rose *Viburnum opulus* (Guelder Rose) B & H.
ELDER, Round-leaved *Sambucus nigra var. rotundifolia*.
ELDER, Scarlet *Sambucus pubens*.
ELDER, Scarlet-berried *Sambucus racemosa* (Red Elder) Ablett.
ELDER, Stinking 1. *Sambucus nigra* (Elder) Dyer.1887. 2. *Sambucus pubens* (Scarlet Elder) USA Yarnell.
ELDER, Sweet *Sambucus canadensis* (American Elder) Sanecki.
ELDER, Water 1. *Sambucus ebulus* (Dwarf Elder) N'thants. A.E.Baker. 2. *Viburnum opulus* (Guelder Rose) Gerard.

ELDER, West Indian *Sambucus simpsonii*.
ELDER, White *Viburnum opulus* (Guelder Rose) B & H.
ELDER, Wild 1. *Aegopodium podagraria* (Goutweed) Bucks, Lincs. Grigson. 2. *Sambucus ebulus* (Dwarf Elder) R.T.Gunther.
ELDER, Witch *Fothergilla sp.*
ELDER, Wood *Sanicula europaea* (Sanicle) Som. Macmillan.
ELDER, Yellow *Tecoma stans.*
ELDER-FLOWERED ORCHID *Dactylorchis sambucina.*
ELDER ROSE; ELDER, Rose *Viburnum opulus* (Guelder Rose) Parkinson (Elder Rose); B & H (Rose Elder).
ELDER-TROT *Heracleum sphondyllium* (Hogweed) Wilts. Dartnell & Goddard.1899. This is a variation of Eltrot, which appears in other guises, too, such as Hilltrot.
ELDERBERRY, Hockle *Anamirta paniculata* (Levant Nut) Le Strange.
ELDERBERRY, Red *Sambucus callicarpa.*
ELDERBUSH *Piper dilatatum* Barbados Gooding.
ELDERN *Sambucus nigra* (Elder) Wilts. Akerman; Glos. J.D.Robertson; Hants. Cope; N'thants. Sternberg. Eldern for elder is very common, probably occurring in a much wider area than the references show. Groome said that the English gypsy name was Eldon.
ELDIN; ELDEN *Petasites hybridus* (Butterbur) N.Eng. Turner. Elden is fuel, and the plant was used as such among the poorer folk.
ELDIN-DOCKEN 1. *Petasites hybridus* (Butterbur) Border B & H. 2. *Rumex aquaticus* (Trossachs Dock) Roxb. B & H. Fuel, again - see Eldin. Ell-docken is another name for Butterbur.
ELDON *Sambucus nigra* (Elder) Groome. This is the English gypsy name quoted by Groome. It is probably eldern rather than eldon.
ELDROT 1. *Anthriscus sylvestris* (Cow Parsley) Dor. Grigson; 2. *Conium maculatum* (Hemlock) Dor. Northover. See Eltrot, possibly related to other words meaning fuel, and so referring to the dried stalks.
ELECAMPANE *Inula helenium*. This is a corruption of the medieval Latin enula campana - enula, inula, the plant name, and campana, of the fields.
ELECAMPANE, Helen's *Inula helenium* (Elecampane) Palaiseul. This is a mixture of the generic and specific names of the plant.
ELEM *Ulmus procera* (English Elm) Corn, Dev, Dor. Grigson; Som. Elworthy; Lincs. B & H. A simple and common variation of elm, which appears in many guises. Sometimes this version is offered with two l's, as Ellem in Sussex, or Ellum in Somerset.
ELEMI FRANKINCENSE *Boswellia frereana*. The stem yields a pale yellow, lemon-scented balsam known as African Elemi.
ELEPHANT BUSH *Portulacaria afra.*
ELEPHANT GARLIC *Allium ampeloprasum* (Wild Leek) Painter.
ELEPHANT GRASS *Pennisetum purpureum.*
ELEPHANT HEAD *Pedicularia groenlandica.*
ELEPHANT ORANGE *Strychnos spinosa* (Kaffir Orange) Palgrave.
ELEPHANT PUMPKIN *Cucurbita maxima* (Giant Pumpkin) W.Miller.
ELEPHANT TREE 1. *Boswellia papyrifera*. 2. *Bursera microphylla*. They say that elephants feed on the leaves of the Boswellia. But the reason given for the name as far as the Bursera is concerned is that the base of the trunk becomes swollen with water, and looks like a dropsical ankle, hence the elephant simile.
ELEPHANT WOOD *Bolusanthus speciosus* (Rhodesian Wisaria) H & P.
ELEPHANT YAM *Amorphophallus campanulatus.*
ELEPHANT'S EAR 1. *Bergenia cordifolia*. 2. *Colocasia antiquorum*. 3. *Enterolobium cyclocarpum* (Earpod) Prance. 4. *Philodendron domesticum*. It is the large leaves of all these that have earned them this

name.

ELEPHANT'S-EAR BEGONIA *Begonia scharfii*. From the size of the leaves.

ELEPHANT'S EAR TREE *Enterolobium cyclocarpum* (Earpod) Prance.

ELEPHANT'S FOOT 1. *Beaucarnea recurvata* (Pony Tail Plant) USA Whiting. 2. *Testudinaria elephantipes*. The latter plant has a large, woody base, hence the name, while Pony Tail Plant has a swollen base to its stem. Cf. Bottle Palm.

ELEVEN O'CLOCK, Lady; ELEVEN O'CLOCK LADY *Ornithogalum umbellatum* (Star-of-Bethlehem) Dev. Friend (Lady Eleven o'clock); Coats (Eleven O'clock Lady). The flowers open early and close early, though the time is a matter for speculation - eleven o'clock in this case, but some names quote noon, four o'clock, even six o'clock.

ELEVEN O'CLOCK FLOWER 1. *Ornithogalum umbellatum* (Star of Bethlehem) Dev. Friend. 2. *Portulaca grandiflora* (Sun Plant) H & P.

ELF, Rose *Claytonia virginica* (Narrow-leaved Spring Beauty) Usher.

ELF-DOCK; ELF-WORT *Inula helenium* (Elecampane) both Prior. Elf-dock became later Elsedock.

ELF-SHOT *Alchemilla vulgaris* (Lady's Mantle) Kirk, Wigt. Grigson. In Donegal, the "fairy doctor" used the juice of Lady's Mantle to cure cases of "elf-shot" in cattle. The juice was put in a pail of water scooped up against the flow of a stream where three townlands meet. The necessary coins and one flint were added to the mixture, and the cow had to swallow three cups of this. Some of it was poured down her spine, and massaged in with some of the dirt from her forefoot. The last drops were thrown into her ears. In view of the Scottish origin of the name, something similar must have gone on there.

ELGINS *Rumex aquaticus* (Trossachs Dock) Roxb. B & H. Probably eldin, which means fuel - the leaves were cut and dried for that purpose.

ELGON OLIVE *Olea welwitschii* (Loliondo) Dale & Greenway. The reference is to Mt Elgon, in Kenya.

ELICOMPANE *Inula helenium* (Elecampane) Corn, Ches, Lancs, Yks. Grigson.

ELK-BARK *Magnolia virginiana* (Sweet Bay) W.Miller.

ELK'S LIP *Caltha leptosepala* New Mexico T H Kearney.

ELKWEED *Swertia perennis* (Marsh Felwort) USA T H Kearney.

ELL-DOCKEN *Petasites hybridus* (Butterbur) Border B & H. Or Eldin-docken: docken for the big leaves, and eldin for fuel, for the leaves were at one time collected, dried and used as such.

ELL-SHINDERS; ELL-SHINDERS, Yellow *Senecio jacobaea* (Ragwort) Berw. Grigson (Ell-shinders); Border B & H (Yellow Ell-shinders). The word appears also as Elshinders.

ELLAR 1. *Alnus glutinosa* (Alder) Grigson. 2. *Sambucus nigra* (Elder) Suss, Kent, Lincs. Grigson.

ELLARNE; ELLERN *Sambucus nigra* (Elder) Shrop. B & H. (Ellarne); Dor. B & H; Glos. Brill (Ellern).

ELLEM; ELEM *Ulmus procera* (English Elm) Suss. B & H (Ellem); Corn, Dev, Dor, Som. Grigson; Lincs. B & H (Elem).

ELLEN, ELLEN-TREE 1. *Alnus glutinosa* (Alder). 2. *Sambucus nigra* (Elder) Yks. Earle.

ELLER 1. *Alnus glutinosa* (Alder) Yks. Morris; Dur. Dinsdale. 2. *Sambucus nigra* (Elder) Yks. Addy. Eller for alder is closer to the original, which was OE alor, and Germanic aller or alre, which then became elre, and also else.

ELLER, Dog 1. *Aegopodium podagraria* (Goutweed) Ches. Grigson. 2. *Viburnum opulus* (Guelder Rose) Ches. B & H.

ELLER, Ground *Aegopodium podagraria* (Goutweed) Hants, IOW, War, Worcs. Grigson.

ELLERN AUL *Sambucus nigra* (Elder)

Heref. G.C.Lewis. Aul is alder, while ellern is quite common for elder, so this name compounds the confusion between the two names.

ELLET *Sambucus nigra* (Elder) Suss. Halliwell.

ELLUM 1. *Sambucus nigra* (Elder) Heref. Leather. 2. *Ulmus procera* (English Elm) Som. Tongue.1965; Hants. Cope; Suff. GE Evans.1956.

ELM, American *Ulmus americana*. Elm is an OE borrowing from the Latin ulmus.

ELM, Bastard; ELM, False 1. *Celtis occidentalis* (Hackberry) USA both Vestal & Schultes. 2. *Celtis reticulata* (Western Hackberry) both USA both Elmore.

ELM, Bough *Ulmus glabra* (Wych Elm) Yks. Grigson.

ELM, Broad-leaved *Ulmus glabra* (Wych Elm) B & H.

ELM, Burl, Carpathian *Ulmus procera* timber (English Elm) USA Wilkinson.1978.

ELM, Campderdown *Ulmus glabra var. camperdownii*. The reference is to the house of that name near Dundee.

ELM, Caucasian *Zelkova carpinifolia*.

ELM, Cedar *Ulmus crassifolia*.

ELM, Chair *Ulmus glabra* (Wych Elm) Wilkinson.1978.

ELM, Chichester *Ulmus x hollandica 'vegeta'* (Huntingdon Elm) Mitchell.1974.

ELM, Chinese *Ulmus parvifolia*.

ELM, Cliff *Ulmus racemosa* (Cork Elm) Wilkinson.1978.

ELM, Coritanian *Ulmus coritana*. The reference is to the Coritanae, a British tribe that occupied an area in the Midlands near Leicester, where the tree was first identified in 1947.

ELM, Cork *Ulmus racemosa*.

ELM, Cornish *Ulmus stricta*.

ELM, Drunken *Ulmus glabra* (Wych Elm) Lincs. Grigson. Perhaps from its looser habit, Grigson says.

ELM, Dutch *Ulmus x hollandica*. Traditionally associated with William of Orange, and supposedly brought from Holland at his accession in 1689. Actually, it is said to be rare in Holland, but does occur in Picardy. Anyway, it is traditionally common on the Suffolk/Essex border, and could very well have originated in this country.

ELM, East Anglian *Ulmus x diversifolia* (Lock Elm) Wit. It does seem to be confined to East Anglia.

ELM, English *Ulmus procera*. Probably not indigenous to Britain, in spite of the name.

ELM, European *Ulmus procera* timber (English Elm) Wilkinson.1978.

ELM, Exeter *Ulmus glabra var. exoniensis*.

ELM, False, African *Celtis integrifolia* (African Nettle Tree) Dalziel.

ELM, Field *Ulmus minor* (Smooth-leaved Elm) Hora.

ELM, Fluttering *Ulmus laevis*. Because the flowers are borne on long stalks, which flutter in the wind.

ELM, Ford's *Ulmus glabra var. exoniensis* (Exeter Elm) Wilkinson.1978.

ELM, Grey *Ulmus americana* timber (American Elm) Wilkinson.1978.

ELM, Guernsey *Ulmus stricta var. sarniensis*.

ELM, Hairy *Ulmus canescens*.

ELM, Hickory *Ulmus racemosa* (Cork Elm) Everett.

ELM, Himalayan, West *Ulmus villosa*.

ELM, Huntingdon *Ulmus x hollandica var. vegeta*. Because it came from a nursery near there.

ELM, Indian *Ulmus fulva* (Slippery Elm) Hatfield. North American Indian, of course.

ELM, Japanese *Ulmus japonica*.

ELM, Jersey *Ulmus stricta var. sarniensis* (Guernsey Elm) Barber & Phillips.

ELM, Lock *Ulmus x diversifolia*. 'Lock', apparently, in the sense of locking a saw, for the timber is that rough (all elm wood is rough).

ELM, Midland *Ulmus x elegantissima*.

ELM, Moose *Ulmus fulva* (Slippery Elm) W.Miller.

ELM, Mountain *Ulmus glabra* (Wych Elm) Prior.

ELM, Nave *Ulmus procera* timber (English

Elm) Wilkinson.1978.

ELM, Plot's *Ulmus plotii.*

ELM, Red 1. *Ulmus alata* (Winged Elm) USA Harper. 2. *Ulmus fulva* (Slippery Elm) USA Tolstead. 3. *Ulmus procera* timber (English Elm) Wilkinson.1978.

ELM, Rock 1. *Chlorophora excelsa* (African Oak) Dalziel. 2. *Ulmus fulva* (Slippery Elm) USA Vestal & Schultes. 3. *Ulmus racemosa* (Cork Elm) USA Tolstead.

ELM, Scotch *Ulmus glabra* (Wych Elm) Border B & H. It is certainly happier growing in the west and north, and is the only common elm in Scotland.

ELM, Siberian 1. *Ulmus pumila*. 2. *Zelkova carpinifolia* (Caucasian Elm) Hora.

ELM, Slippery *Ulmus fulva*. The mucilaginous inner bark, or cambium, is the medicinal slippery elm.

ELM, Slippery, Californian *Fremontia californica* (Flannel Bush) Le Strange.

ELM, Small-leaved *Ulmus x diversifolia* (Lock Elm) Clapham.

ELM, Smooth-leaved *Ulmus minor.*

ELM, Soft *Ulmus americana* timber (American Elm) Wilkinson.1978.

ELM, Spanish *Cordia gerascanthus* W.Indies Roys.

ELM, Swamp *Ulmus americana* timber (American Elm) Wilkinson.1978.

ELM, Sweet *Ulmus fulva* (Slippery Elm) USA Vestal & Schultes.

ELM, Switch *Ulmus glabra* (Wych Elm) Yks. Grigson. Often produced as one word, Switchelm. Wych means switchy, or pliant.

ELM, Water *Ulmus americana* timber (American Elm) Wilkinson.1978.

ELM, Wheatley *Ulmus stricta var. sarniensis* (Guernsey Elm) Barber & Phillips.

ELM, White 1. *Ulmus americana* (American Elm) USA Harper. 2. *Ulmus laevis* (Fluttering Elm) Wilkinson.1978. So called from its grey trunk. Or does it refer to the white wood?

ELM, Winged *Ulmus alata.*

ELM, Witan *Ulmus glabra* (Wych Elm) Shrop. Grigson. 'Witan' must be a derivative of OE wice, supple.

ELM, Witch 1. *Sorbus aucuparia* (Rowan) Kelly. 2. *Ulmus glabra* (Wych Elm) Som. Elworthy; Yks. Nicholson. 'Witch' for 'wych' is common, in spite of the fact that there is only one reference down. It must have arisen when the word 'wych' was no longer understood, and had the effect of creating a belief that witches dread it, so Yorkshire carters used to put a sprig on their horses, and carried a piece of wood in their pockets. Or were they really carrying rowan, which is a genuine anti-witch tree? The interchange of the name between the two trees is quite understandable, though 'witch' for rowan is the result of misunderstanding its true origin, which is OE cuic, alive, and so from quite a different source.

ELM, Wych *Ulmus glabra*. The dictionaries say that witch and wych are the same words, but surely wych means pliable, switchy (see Switch Elm for this same tree).

ELM, Wych, Japanese *Ulmus laciniata.*

ELM, Yoke *Carpinus betula* (Hornbeam) Gerard. Yokes were once made of it, but this tree is more often called a beech than an elm.

ELM-LEAVED SUMACH *Rhus coriaria* (European Sumach) Hora.

ELM-WYCH *Ulmus glabra* (Wych Elm) N'thum. Grigson.

ELMEN; ELMIN *Ulmus procera* (English Elm) W.Eng. B & H (Elmen); Wilts. Akerman; Hants. Cope (Elmin). Elmen must be the old adjectival form (as oaken, ashen, etc) eventually used as a noun.

ELMLEAF BRAMBLE *Rubus ulmifolius.*

ELNORNE *Sambucus nigra* (Elder) B & H. There is an obtrusive 'n' here, but otherwise the name is familiar as Ellarne, from Shropshire, or as the West-country Ellern.

ELPHAMY *Bryonia dioica* (White Bryony) N.Eng. Halliwell. If this actually means the abode of fairies, them it is strange, for fairies do not usually associate themselves with a poisonous plant, which is the devil's own in most cases.

ELREN *Sambucus nigra* (Elder) N.Eng.

Halliwell.

ELSEDOCK *Inula helenium* (Elecampane) B & H. This is a corruption of Elf-dock. Elfwort too is a name for this plant.

ELSHANDER, Stinking *Tanacetum vulgare* (Tansy) Perth. Grigson. Alexander? But Cf. Ell-shinder for Ragwort. Cf. too Stinking Willie for tansy, another Scottish name. On the other hand, there is the Devonshire Scented Fern, a much more restrained comment.

ELSHINDER, Stinking *Senecio jacobaea* (Ragwort) B & H.

ELSHINS; ELSHINS, De'il's *Scandix pecten-veneris* (Shepherd's needle) both Berw. both B & H. Elshins apparently means awls, so the names correspond to the Somerset Devil's Darning Needle.

ELTROT 1. *Aegopodium podagraria* (Goutweed) Grieve.1931. 2. *Anthriscus sylvestris* (Cow Parsley) Dor. B & H. 3. *Daucus carota* (Wild Carrot) Hants. Grigson. 4. *Heracleum sphondyllium* (Hogweed) Som, Dor. B & H. Wilts Dartnell & Goddard. 5. *Oenanthe crocata* (Hemlock Water Dropwort) Som, Wilts. Grigson. These are all umbelliferous plants, and the name seems to apply itself to the dried stalks, much as Kex and its variants do. Perhaps the name has some relationship to other words for fuel.

ELVEN *Ulmus procera* (English Elm) Suss. Parish; Kent, War, Worcs. Grigson.

ELWAND, King's *Digitalis purpurea* (Foxglove) N'thum. Grigson.

EMBROIDERY, Lady's *Stellaria holostea* (Greater Stitchwort) Som. Macmillan; Wilts. Tynan & Maitland.

EMERALD RIPPLE *Peperomia caperata* (Rat Tail Plant) USA Cunningham.

EMETIC ROOT *Euphorbia corollata* (Flowering Spurge) Le Strange. Cf. Purging Root. Actually, spurge means a purging plant (Latin expurgare).

EMETIC WEED *Lobelia inflata* (Indian Tobacco) W.Miller. The dried leaves and tops are a poison which act upon the nervous system and bowels, causing vomiting. Cf. Vomitwort and Pukeweed as American names for this.

EMLET, Blood-drop *Mimulus luteus* Scot. Grigson.

EMMAL, EMMEL *Ulmus procera* (English Elm) both Cumb. Grigson. Consonantal transpositions like this are very common. Emmal is recorded also for *Ulmus glabra* (Wych Elm) Cumb. Grigson.

EMMER *Triticum dicoccum.*

EMMET'S STALK *Lythrum salicaria* (Purple Loosestrife) Som. Macmillan.

EMONY, Blue *Pulsatilla vulgaris* (Pasque Flower) Rut. Grigson.

EMPRESS TREE *Paulownia tomentosa* (Foxglove Tree) Hora.

EMU BUSH *Eremophila longifolia.*

ENANCIA; ANANCIA *Geum urbanum* (Avens) Fernie.

ENCHANTER'S NIGHTSHADE *Circaea lutetiana.* In France, its names include herbe à la magicienne, enchantresse, herbe aux sorciers. It is Hexenkraut in German, too, and was certainly linked with witchcraft there, but there is no evidence of any connection in this country before the publication of the herbals. But Storms used the name Witchwort, and his book was entitled *Anglo-Saxon magic.*

ENDIVE 1. *Cichorium endivia.* 2. *Cichorium intybus* (Chicory) B & H. This word comes directly from Latin intybum. So the specific names of both these plants are in fact the same word.

ENDIVE, Belgian; ENDIVE, French *Cichorium intybus* (Chicory) Hemphill (Belgian); Schery (French).

ENDIVE, Blue *Cichorium intybus* (Chicory) Som. Macmillan.

ENDIVE, Green 1. *Lactuca serriola* (Prickly Lettuce) Turner. 2. *Lactuca virosa* (Great Prickly Lettuce) B & H.

ENDIVE, White *Cichorium endivia* (Endive) Turner.

ENELL, Ground *Scandix pecten-veneris* (Shepherd's Needle). B & H. This mistakenly appears as Ground Evil in Halliwell.

ENEMY *Anemone nemorosa* (Wood Anemone) Lincs. B & H; Som. Elworthy.

ENGELMANN SPRUCE *Picea engelmannii.*

ENGLISH, German-and- *Ranunculus scleratus* (Celery-leaved Buttercup) Som. Macmillan.

ENGLISH ALDER *Alnus glutinosa* (Alder) Le Strange.

ENGLISH ARROWROOT *Solanum tuberosum* (Potato) Howes.

ENGLISH BRYONY *Bryony dioica* (White Bryony) Parkinson.1640. Perhaps because he had in mind English Mandrake, already in use in his time.

ENGLISH CATCHFLY *Silene anglica* (Small-flowered Catchfly) Hutchinson.

ENGLISH COST; ENGLISH COSTMARY *Tanacetum vulgare* (Tansy) Cockayne (Cost); Storms (Costmary). OE aenglisc cost. Presumably to distinguish from real Costmary, which is Balsamint.

ENGLISH DAISY *Bellis perennis* (Daisy) USA Hay & Synge.

ENGLISH ELM *Ulmus procera*. Probably not indigenous to this country, in spite of the name.

ENGLISH FLYTRAP *Drosera rotundifolia* (Sundew) Som. Macmillan. 'Flytrap' is exactly what Sundew is.

ENGLISH GREENWEED *Genista anglica* (Needle Whin) Flower.1859.

ENGLISH HARE'S BELL, ENGLISH HAREBELL *Endymion nonscriptus* (Bluebell) Parkinson (Hare's Bell); Gerard (Harebell).

ENGLISH HYACINTH *Endymion nonscriptus* (Bluebell) Gerard.

ENGLISH IRIS *Iris xiphioides*. For hundreds of years it has flourished in the Bristol neighbourhood, and the Dutch bulb merchants thought it was native to Britain - hence the name.

ENGLISH IVY *Hedera helix* (Ivy) USA Allan.

ENGLISH LAUREL *Prunus laurocerasus* (Cherry Laurel) USA Cunningham. Presumably because it was so beloved of Victorian owners of shrubberies.

ENGLISH LAVENDER *Lavandula spica.*

ENGLISH MANDRAKE *Bryonia dioica* (White Bryony) Brownlow. It was this plant that was taken for the mandrake in England. The reference was to the anthropomorphic nature of the roots, and steps were taken to achieve the shape by artificial means, either by carving, or by the use of a mould.

ENGLISH MAPLE *Acer campestre* (Field Maple) Brimble.1948.

ENGLISH MARJORAM *Origanum vulgare* (Marjoram) B & H.

ENGLISH MARQUERY; ENGLISH MERCURY *Chenopodium bonus-henricus* (Good King Henry) Fernie (Marquery); Gerard (Mercury).

ENGLISH MASTERWORT *Aegopodium podagraria* (Goutweed) Grieve.1931. Fernie has Wild Masterwort for this, too.

ENGLISH MEADOWSWEET *Filipendula ulmaria* (Meadowsweet) USA Sanecki. A distingushing name, for our Meadowsweet was taken to America by the early settlers, and is now naturalised in the eastern states.

ENGLISH MULBERRY *Morus nigra* (Mulberry) Hora.

ENGLISH ORRIS *Valeriana officinalis* (Valerian) Webster. This is the name given to the powder made from the roots.

ENGLISH PALM *Salix capraea* (Sallow) Staffs. Poole. N'thants. A E Baker. This is the tree which gives "palm" for Palm Sunday in this country.

ENGLISH PASSION FLOWER *Arum maculatum* (Cuckoo-pint) Fernie. The blotched leaves account for the name - the familiar legend of a plant growing at the foot of the Cross and receiving some drops of Christ's blood is applied to cuckoo-pint as well as virtually any other plant which has spotted leaves.

ENGLISH PEONY *Paeonia mascula* (Peony) W.Miller.

ENGLISH PLANTAIN 1. *Plantago lanceolata* (Ribwort Plantain) USA Upham. 2. *Plantago major* (Great Plantain) USA

Gates; Barbados. Gooding. There is a well-known legend of its persistently following the tracks of man. More specifically, one superstition says that it follows Englishmen, and springs up in whatever part of the world he makes his home. In this case, "White Man's Foot", which is what the American Indians call the plant (see Longfellow's Hiawatha), becomes Englishman's Foot.

ENGLISH POPLAR, Old *Populus nigra* (Black Poplar) B & H.

ENGLISH RHUBARB 1. *Rheum rhaponticum* (Rhubarb) Parkinson.1640. 2. *Thalictrum flavum* (Meadow Rue) Gerard. False Rhubarb, and Bastard Rhubarb, too, for Meadow Rue, "which names are taken of the colour, and taste of the roots" (Gerard). But they have a medicinal effect to emulate rhubarb, and this is probably the real reason that the name occurs - it was, in fact, a Poor Man's Rhubarb.

ENGLISH ROSE *Rosa canina* (Dog Rose) Genders.

ENGLISH SARSAPARILLA *Potentilla erecta* (Tormentil) Grieve.1931. Because it was so renowned for its tonic properties.

ENGLISH SEA GRAPE *Salicornia europaea* (Glasswort) B & H. Or, Sea Grape on its own.

ENGLISH SERPENTARY *Polygonum bistorta* (Bistort) Culpepper. It is the writhed roots which are responsible for the serpentine image.

ENGLISH SORREL *Rumex acetosa* (Wild Sorrel) Hemphill.

ENGLISH STONECROP *Sedum anglicum.*

ENGLISH THYME *Thymus serpyllum* (Breckland Wild Thyme) Sanecki.

ENGLISH TREACLE 1. *Alliaria petiolata* (Jack-by-the-hedge) B & H. 2. *Teucrium chamaedrys* (Germander) Turner. 3. *Teucrium scordium* (Water Germander) Coles. 4. *Teucrium scorodonia* (Wood Sage) Turner. Treacle in its original sense, of course - that of a particular medicinal preparation.

ENGLISH VIOLET *Viola odorata* (Sweet Violet) Scot. B & H.

ENGLISH WALNUT *Juglans regia* (Walnut) Schery. Though it actually comes from Persia, and the very name walnut means that it is a foreign tree.

ENGLISH YEW *Taxus baccata.*

ENGLISHMAN'S FOOT *Plantago major* (Great Plantain) Leyel. There is a well-known legend of the way it persistently follows the tracks of man. One story is more specific, and says that it follows Englishmen, and springs up wherever in the world he makes his home. In this case, "White Man's Foot" (which is what the American Indians called the plant), becomes Englishman's Foot.

ENTRAILS, Devil's *Convolvulus arvensis* (Field Bindweed) Wales Trevelyan. It sounds as if Marie Trevelyan gentrified the name a little, for it is much more familiar as Devil's Guts.

EPAULETTE TREE *Pterostyrax hispida.*

EPIPHANY *Cuscuta epithymum* (Common Dodder) Grigson. This is quite common, and is simply a corruption of *epithymum*.

EPS *Populus tremula* (Aspen) Kent Grigson. Eps, or its much more widespread counterpart, Aps, is probably nearer to the original form. The transposition of consonants occurred later to give asp, then aspen and its variants.

EREWORT *Sempervivum tectorum* (Houseleek) Grigson. A name from the Middle Ages, when the plant was used for ear-drops, and to cure deafness. Cf. Cyphel, which is from Greek kuphella, meaning the hollow of the ear.

ERINGO *Eryngium maritimum* (Sea Holly) Culpepper. Better spelt directly from the Latin, as eryngo. The name was also used for the candied roots of this plant.

ERNUT 1. *Bunium bulbocastanum* (Great Earthnut) Prior. 2. *Conopodium majus* (Earthnut) Prior. i.e. Earthnut. Arnut is the same word.

ERRICK *Arctium lappa* (Burdock) Ches. B & H. Eddick is noted from the same county.

ERRIE; EERIE *Achillea millefolium* (Yarrow) Aber. Rorie.1994 (Errie); B & H

(Eerie). Yarrow appears in a number of forms - these are but extreme versions.

ERRIF *Galium aparine* (Goose-grass) Ches. B & H. There is quite a long series of names which starts with Hayriff, the first element meaning 'hedge'. It then varies into such as Airup, Haireve, etc. The nearest to the name quoted is probably the early Heyriffe, quoted in Storms.

ERSMART *Polygonum hydropiper* (Water Pepper) B & H. See Arsmart.

ERYNGO, Alpine *Eryngium alpinum.*

ERYNGO, Blue *Eryngium amethystinum* (Oliver's Sea Holly) Polunin.

ERYNGO, Field *Eryngium campestre.*

ERYSIPELAS PLANT *Heliotropum indicum* (Indian Heliotrope) Howes. The plant has been used in native medicine to cure the complaint.

ESCAROLE *Cichorium endivia* (Endive) Bianchini. This is the same as Gerard's Scariole, and is usually reserved for the broad-leaved kind.

ESCH; ESH *Fraxinus excelsior* (Ash) Scot. Jamieson (Esch) Lincs, Cumb, Border, Scot. B & H; Yks. Morris; Dur. Dinsdale; E Ang. Forby (Esh).

ESCHALLOT *Allium ascalonicum* (Shallot) E Hayes. This seems to be the French name, eschalotte. But both come straight from Latin ascalonia.

ESH, Wild *Aegopodium podagraria* (Goutweed) Cumb. Grigson. In this case, the 'ash' reference is to ache rather than to the tree. Ache means parsley, and it appears in various forms like aise and aish.

ESP; ESP, Quakin *Populus tremula* (Aspen) Cumb. Grigson Yks. Carr (Esp); N.Ire. Grigson (Quakin Esp).

ESPARSETTE *Onobrychis sativa* (Sainfoin) USA Zenkert.

ESPARTO GRASS *Stipa tenacissima.*

ESPIBAWN *Leucanthemum vulgare* (Ox-eye Daisy) N Ire. B & H. An englishing of Gaelic easbog bán, white bishop. Cf. another Irish name, Bishop's Posy.

ESPIN *Populus tremula* (Aspen) Yks. Grigson. Cf. Esp.

ETERNAL FLOWER *Helichrysum stoechas* (Golden Cassidony) Grieve.1931. More recognizable as Everlasting Flower.

ETHIOPIAN BANANA *Ensete ventricosum* (Abyssinian Banana) H & P.

ETHIOPIAN CUMIN *Ammi visnaga* (Bishop's Weed) Pomet.

ETHIOPIAN OLIVE *Elaeagnus angustifolia* (Narrow-leaved Oleaster) R T Gunther.1934.

ETHIOPIAN PEPPER *Xylopia aethiopica.*

ETHIOPIAN SOUR BREAD *Adansonia digitata* (Baobab) Douglas.

ETNAGH-BERRIES *Juniperus communis* (Juniper) Angus B & H. Etnagh is rendered Aiten, or Aitnagh, even Eatin.

ETRUSCAN HONEYSUCKLE *Lonicera etrusca.*

ETTLE; ETTLEY *Urtica dioica* (Nettle) Wilts. Akerman; Glos. J.D.Robertson; Worcs. J Salisbury; N'thants. Sternberg (Ettle); Heref. G C Lewis (Ettley).

EUGH *Taxus baccata* (Yew) Aubrey. Natural Hist. Wilts. Presumably only Aubrey's version of the sound, not appearing elsewhere.

EUPATOR'S AGRIMONY *Agrimonia eupatoria* (Agrimony) Le Strange.

EUPHORBIA, Finger *Euphorbia tirucalli* (Pencil Tree) Dale & Greenway. Pencil Tree and Finger Euphorbia both describe the same growth phenomenon.

EUPHORBIA, Gum *Euphorbia resinifera.*

EUPHORBIA, Tree *Euphorbia grandidens.*

EUPHRASY; EWFRACE; EWFRAS *Euphrasia officinalis* (Eyebright) T.Wright (Euphrasy, Ewfrace); B & H (Ewfras). Euphrasia means cheerfulness.

EUPHRATES ASPEN; EUPHRATES BALSAM-TREE *Populus euphratica* (Euphrates Poplar) Moldenke.

EUPHRATES POPLAR *Populus euphratica.*

EUREKA LILY *Lilium occidentale.*

EUROPEAN AGRIMONY *Agrimonia eupatoria* (Agrimony) Le Strange.

EUROPEAN ALDER *Alnus glutinosa* (Alder) USA Zenkert.

EUROPEAN BELLFLOWER *Campanula*

rapunculoides (Creeping Bellflower) USA House.
EUROPEAN BIRCH *Betula verrucosa* (Birch) Flück.
EUROPEAN BOX *Buxus sempervirens* timber (Boxwood) Brimble.1948.
EUROPEAN BOXTHORN *Lycium europaeum*.
EUROPEAN CHESTNUT *Castanea sativa* (Sweet Chestnut) Brimble.1948.
EUROPEAN DATE PLUM *Diospyros lotus* (Date Plum) J.Smith.
EUROPEAN ELM *Ulmus procera* timber (English Elm) Wilkinson.1978.
EUROPEAN FAN-PALM *Chamaerops humilis* (Dwarf Fan-Palm) RHS.
EUROPEAN HOP-HORNBEAM *Ostrya carpinifolia*.
EUROPEAN LARCH *Larix decidua* (Larch) Clapham.
EUROPEAN POTATO *Solanum tuberosum* (Potato) USA Brouk. This and Irish Potato were names given when potatoes from Great Britain were introduced to the English colonists on the American mainland; they are still in use today.
EUROPEAN SNAKEROOT *Aristolochia clematitis* (Birthwort) Thomson.1978.
EUROPEAN SUMACH *Rhus coriaria*.
EUROPEAN TEA *Veronica officinalis* (Common Speedwell) Mitton.
EUROPEAN WILLOW *Salix alba* (White Willow) Hatfield.
EVE, Adam-and- 1. *Aconitum napellus* (Monkshood) Som, Wilts. Macmillan; Norf. B & H. 2. *Arum maculatum* (Cuckoo-pint) Som, Leics, Lincs, Yks. B & H. 3. *Dactylorchis incarnata* (Early Marsh Orchid) B & H. 4. *Dactylorchis maculata* (Heath Spotted Orchid) N'thants. B & H; IOM Moore.1924; Yks. Carr. 5. *Lamium album* (White Deadnettle) Boland.1977. 6. *Orchis mascula* (Early Purple Orchis) Corn. B & H; N'thants. A E Baker. 7. *Pulmonaria longifolia* (Lungwort) Grigson. 8. *Pulmonaria officinalis* (Lungwort) Grigson. B & H said the name referred to the colour of the spadices of cuckoo-pint - the dark ones represented Adam, and the light ones Eve. But this is actually the archetypal male + female name suggested by the form of the plant, the spadix in the spathe, the emblem of copulation, and accounting for a host of other, similar, names, like Lords-and-ladies and Bulls-and-cows. The various orchids have the name applied to the tubers, which are in pairs, so these double names are given. The pairs of tubers also give rise to rich dialect names meaning testicles. As far as monkshood in concerned, the explanation is offered that "when the hood (of the flower) is lifted up, there is an appearance of two little figures". The hood of the name is of course the upper petals. There is a similar explanation for White Deadnettle - "because in the flower, if you hold it upside down, the stamens can be seen lying for all the world like two people side by side in a curtained bed" (Boland.1977).
EVE, Herb 1. *Ajuga chamaepitys* (Yellow Bugle) B & H. 2. *Coronopus squamatus* (Wart-cress) Prior. 3. *Plantago coronopus* (Buck's-horn Plantain) Gerard. Ivy, probably. In each case, Herb Ivy, or Herb Ive, are also recorded.
EVE'S APPLE TREE *Ervatamia dichotoma*. The fruits look tempting enough, but they are very poisonous. Moslems call it "the forbidden fruit of Eden".
EVE'S CUPS *Sarracenia purpurea* (Pitcher Plant) Grieve.1931. 'Cups' are ascribed to a huntsman, a soldier, and a forefather, as well as to Eve.
EVE'S CUSHION *Saxifraga hypnoides* (Dovedale Moss) Yks. Grigson. Descriptive - there are also Lady's, and Queen's, Cushion.
EVE'S TEARS *Galanthus nivalis* (Snowdrop) Som. Macmillan. Only in Somerset are the drooping flowers likened to teardrops.
EVENING, Pride-of-the-; EVENING PRIDE *Lonicera periclymenum* (Honeysuckle) both Dev, Dor. Macmillan.

Because the scent of the honeysuckle is much more obvious in the evening.

EVENING, Shades-of-; EVENING CLOSE *Melandrium album* (White Campion) both Dor. Macmillan.

EVENING CAMPION *Melandrium album* (White Campion) H C Long.1910. Cf. Evening Close, and Shades of Evening.

EVENING PRIMROSE 1. *Euphorbia hirta* Kourennoff. 2. *Oenothera sp.*

EVENING PRIMROSE, Fragrant *Oenothera stricta*.

EVENING PRIMROSE, Golden *Oenothera brevipes*.

EVENING PRIMROSE, Large *Oenothera erythrosepala*.

EVENING PRIMROSE, Least *Oenothera parviflora*.

EVENING PRIMROSE, Lesser *Oenothera biennis*.

EVENING PRIMROSE, Northern *Oenothera muricata*.

EVENING PRIMROSE, Small-flowering *Oenothera cruciata*.

EVENING PRIMROSE, White 1. *Oenothera breviflora*. 2. *Oenothera speciosa*.

EVENING SNOW *Gilia dichotoma*.

EVENING STAR 1. *Oenothera biennis* (Lesser Evening Primrose) USA Bergen. 2. *Oenothera erythrosepala* (Large Evening Primrose) Dev. Friend.

EVENING TRUMPET FLOWER *Gelsenium sempervirens* (Carolina Jessamine) USA Kingsbury.

EVENING TWILIGHT 1. *Anemone nemerosa* (Wood Anemone) Dor. Macmillan. 2. *Oxalis acetosella* (Wood Sorrel) Dor. Macmillan.

EVER 1. *Conium maculatum* (Hemlock) Heref. Havergal. 2. *Lolium perenne* (Rye-grass) Dev. Halliwell. Hemlock is called Hever in Herefordshire, and there is a similar variation for Rye-grass in Cornwall - Heaver Hayver.

EVER-BLOOMING IRIS *Iris ruthenica* (Russian Iris) W.Miller.

EVER-WHITE *Anaphalis margaretacea* (Pearly Immortelle) N'thants. A.E.Baker.

EVERBEARING STRAWBERRY *Fragaria vesca* (Wild Strawberry) USA Schery.

EVERFERN *Polypodium vulgare* (Common Polypody) B & H.

EVERLASTING, American *Anaphalis margaretacea* (Pearly Immortelle) Grieve.1931.

EVERLASTING, Blunt-leaved *Gnaphalium polycephalum* (Mouse-ear Everlasting) Grieve.1931.

EVERLASTING, Californian *Gnaphalium decurrens var. californicum*.

EVERLASTING, Common *Gnaphalium polycephalum* (Mouse-ear Everlasting) USA Youngken.

EVERLASTING, Life 1. *Anaphalis margaretacea* (Pearly Immortelle) B & H. 2. *Antennaria dioica* (Cat's Foot) Grieve.1931. 3. *Antennaria plantaginifolia* (Plantain-leaved Everlasting). Grieve.1931.

EVERLASTING, Life, Pearl-flowered *Anaphalis margaretacea* (Pearly Immortelle) Grieve.1931.

EVERLASTING, Life, Sweet-scented *Gnaphalium polycephalum* (Mouse-ear Everlasting) O.P.Brown.

EVERLASTING, Marsh *Gnaphalium uliginosum* (Marsh Cudweed) Grieve.1931.

EVERLASTING, Moor; EVERLASTING, Mountain *Antennaria dioica* (Cat's Foot) N.Eng. Grigson (Moor); McClintock (Mountain).

EVERLASTING, Mouse-ear *Gnaphalium polycephalum*.

EVERLASTING, Pearl *Anaphalis margaretacea* (Pearly Immortelle) McClintock.

EVERLASTING, Pease *Lathyrus latifolius* (Garden Everlasting Pea) Parkinson. Everlasting, in this case, because the plant is perennial.

EVERLASTING, Plantain-leaved *Antennaria plantaginifolia*.

EVERLASTING, Shining *Anaphalis margaretacea* (Pearly Immortelle) Salisbury.1964.

EVERLASTING, Shrubby *Helichrysum stoechas* (Golden Cassidony) Grieve.1931.

EVERLASTING, Sweet *Gnaphalium polycephalum* (Mouse-ear Everlasting) USA Upham.

EVERLASTING, White *Gnaphalium microcephalum.*

EVERLASTING, Winged *Ammobium alatum* (Everlasting Sand Flower) A.W.Smith.

EVERLASTING FLOWER *Helichrysum bracteatum.*

EVERLASTING FRIENDSHIP *Galium aparine* (Goose-grass) Hutchinson & Melville. Is this a genuine name? It smacks of Gerard's remark, that the plant is called "of some, Philanthropos, as though he should say, a man's friend, because it taketh hold of a man's garments." Cf. such names as Sweethearts, and Huggy-me-close.

EVERLASTING GRASS *Onobrychis sativa* (Sainfoin) Oxf. Grigson.

EVERLASTING PEA *Lathyrus sylvestris.*

EVERLASTING PEA, Garden *Lathyrus latifolius.*

EVERLASTING PEA, Narrow-leaved *Lathyrus sylvestris* (Everlasting Pea) Clapham. These Everlasting Pea names are given because the plants are perennial.

EVEROCKS; EVRON *Rubus chamaemorus* (Cloudberry) Scot. B & H (Everocks); Banff, Mor. Grigson (Evron). The type word for a series of Scottish names seems to be Averen. This varies into Averin, or Evron, as well as Aivern, or Aiverin. Everocks must be an extension of these.

EVERY-GRASS *Lolium perenne* (Rye-grass) Dor. Barnes. Cf. Ever, and the Cornish Heaver Hayver, both for this grass.

EVEWEED *Hesperis matronalis* (Sweet Rocket) Hill. Apparently invented by Hill, Family Herbal, 1769, using *Hesperis* as his example. In much the same way, Vesperflower was produced, too. Another name, Night-smelling Rocket, explains the reason why.

EVVER, Devon *Lolium temulentum* (Darnel) Som. Macmillan. Eaver, or eever, is more usual.

EWE *Taxus baccata* (Yew) Gerard.

EWE-BRAMBLE *Rubus fruticosa* (Blackberry) Som. B & H. This is almost certainly He-bramble. He-brimmel is a Somerset name, as is Yoe-brimmel. Halliwell defines He-brimmle as a bramble of more than one year's growth.

EWE-DAISY *Potentilla erecta* (Tormentil) N'thum, Berw. Grigson.

EWE-GA; EWE-GOLLAN; EWE-GOWAN *Bellis perennis* (Daisy) Yks. Grigson (Ewe-ga); N.Eng, Scot. Grigson (Ewe-gollan); N'thum. Brockett; Scot. Jamieson (Ewe-gowan).

EWGH *Taxus baccata* (Yew) Aubrey Nat.Hist. Wilts.

EXETER ELM *Ulmus glabra var. exoniensis.*

EXILE TREE *Thevetia peruviana* (Yellow Oleander) Dalziel.

EYE, Blainy *Euphorbia lateriflora* Dalziel.

EYE OF CHRIST *Veronica chamaedrys* (Germander Speedwell) Wales E. Wright. Cf. too God's Eye, from Devonshire and Lincolnshire.

EYEBANE *Euphorbia maculata* (Spotted Spurge) USA Kingsbury. Presumably this has an irritant juice, dangerous if put accidentally in the eye, though it has apparently been used to remove corneal opacities. A name like this would not inspire confidence.

EYEBERRY *Rubus pubescens* (Dwarf Raspberry) USA Yarnell.

EYEBRIGHT 1. *Anagallis arvensis* (Scarlet Pimpernel) Corn. Grigson. 2. *Anchusa officinalis* (Alkanet) Som. Elworthy. 3. *Chamaenerion angustifolium* (Rosebay) Dev. Friend.1882. 4. *Epilobium parviflorum* (Small-flowered Willowherb) Som. Friend. 5. *Euphrasia officinalis*. 6. *Houstonia caerulea* (Bluets) USA House. 7. *Lobelia inflata* (Indian Tobacco) USA House. 8. *Papaver rhoeas* (Red Poppy) Som. Macmillan. 9. *Stellaria holostea* (Greater Stitchwort) Som. Elworthy. 10. *Stellaria longifolia* USA Grigson. 11. *Veronica chamaedrys* (Germander Speedwell) Som. Elworthy; Shrop, Yks, N'thum, N.Ire) Grigson. 12. *Viola tricolor* (Pansy) Som.

Macmillan. Most of these are descriptive, like virtually all the 'eye' names, and owe the ascription to an "eye" in the centre of the flower, either of a different colour to the rest, or to the tube-like neck which gives the illusion of being a different colour. *Euphrasia officinalis*, of course, is associated with eye leechdoms, but the use is almost certainly doctrine of signatures, on the strength of the black pupil-like spot on its corolla. Milton has the Archangel using eyebright to clear the vision of the first man and woman:

"Then purged with euphrasy and rue
His visual orbs, for he had much to see".

Curiously, the one plant one would have expected to carry the name, i.e. Deadly Nightshade, doesn't have it.

EYEBRIGHT, Red *Odontites verna* (Red Bartsia) Curtis.

EYEBRIGHT COW-WHEAT *Odontites verna* (Red Bartsia) Prior.

EYEGLASSES *Pinus sylvestris* (Scots Pine) Som. Macmillan.

EYELASH BEGONIA *Begonia bowerae*. Long, curling hairs grow from the leaf edges, hence the 'eyelash' of the name.

EYES, Adder's *Anagallis arvensis* (Scarlet Pimpernel) Herts. Grigson.

EYES, Angel's *Veronica chamaedrys* (Germander Speedwell) Som. Macmillan; Dev. Grigson; Oxf. Thompson. Speedwell is a typical flower with 'eye' names. They seem to have to be small, with a pale spot in the middle to qualify.

EYES, Beaty; EYES, Beedy's *Viola tricolor* (Pansy) Som. Macmillan (Beaty); Som. Jennings (Beedy's). This is Bridget (i.e. St Bride). The sequence runs Biddy, Beedy, Beaty. But see Eyes, Biddy's.

EYES, Biddy's 1. *Viola arvensis* (Field Pansy) Som. Macmillan. 2. *Viola tricolor* (Pansy) Som. Tongue. Biddy means a chick.

EYES, Bird *Callicarpa americana* (French Mulberry) Alab. Harper.

EYES, Bird's 1. *Anagallis arvensis* (Scarlet Pimpernel) Wilts, Oxf, Bucks. Grigson. 2. *Cardamine pratensis* (Lady's Smock) Shrop, Yks, Cumb. Grigson. 3. *Euphrasia officinalis* (Eyebright) Som. Macmillan. 4. *Geranium robertianum* (Herb Robert) Corn, Bucks, N'thum. Grigson; Dev. Macmillan. 5. *Gilia tricolor*. 6. *Glechoma hederacea* (Ground Ivy) Oxf, Bucks, N'thants. Grigson. 7. *Lotus corniculatus* (Bird's-foot Trefoil) Som. Macmillan. 8. *Melandrium dioicum* (Red Campion) Dev, Som. Macmillan. 9. *Myosotis arvensis* (Field Forget-me-not) Som. Macmillan. Hants, N'thants. Grigson. 10. *Myosotis palustris* (Water Forget-me-not) Som, Hants. Grigson; N'thants. A.E.Baker. 11. *Pentaglottis sempervirens* (Evergreen Alkanet) Som. Elworthy. 12. *Primula farinosa* (Bird's-eye Primrose) Prior. 13. *Sagina procumbens* (Common Pearlwort) Suss. Grigson; USA Allan. 14. *Saxifraga x urbinum* (London Pride) Dev. Friend.1882. 15. *Stellaria holostea* (Greater Stitchwort) Worcs, Derb, Yks, Dur. Grigson. 16. *Stellaria media* (Chickweed) Som. Macmillan. 17. *Veronica beccabunga* (Brooklime) Som, Dor. Macmillan. 18. *Veronica chamaedrys* (Germander Speedwell) common. Grigson (inc. Wilts. Dartnell & Goddard). 19. *Veronica hederifolia* (Ivy-leaved Speedwell) Ess. B & H. 20. *Veronica officinalis* (Common Speedwell) Wilts. Dartnell & Goddard. 21. *Veronica persica* (Common Field Speedwell) Wilts. Dartnell & Goddard; Bucks. B & H. 22. *Viola tricolor* (Pansy) Som. Macmillan.

EYES, Bird's, Blue *Veronica chamaedrys* (Germander Speedwell) Suss, Oxf. Grigson; Bucks. B & H.

EYES, Bird's, Bonny *Cardamine pratensis* (Lady's Smock) Cumb. Grigson.

EYES, Bird's, Pink *Geranium robertianum* (Herb Robert) Som. Macmillan; Bucks. Grigson.

EYES, Bird's, Red 1. *Anagallis arvensis* (Scarlet Pimpernel) Som. Macmillan. 2. *Geranium robertianum* (Herb Robert) Prior.

3. *Melandrium dioicum* (Red Campion) Rad. Grigson.

EYES, Bird's, Water *Veronica beccabunga* (Brooklime) Dev. Macmillan.

EYES, Bird's, White 1. *Stellaria holostea* (Greater Stitchwort) Rad. Grigson. 2. *Stellaria media* (Chickweed) Bucks. Grigson.

EYES, Blind *Papaver rhoeas* (Red Poppy) N'thants. A.E.Baker; Bucks. Harman. There was quite a widespread superstition that if you put poppies to the eyes, they would cause blindness. Cf. the Wiltshire Blind Man, and the Yorkshire Blindy-buff. The superstition is probably related to the one which says the plants cause headaches.

EYES, Blue *Veronica chamaedrys* (Germander Speedwell) W.Eng. Grigson.

EYES, Bobby's *Veronica chamaedrys* (Germander Speedwell) Som. Grigson; Hants. Cope.

EYES, Bobby's, Red *Geranium robertianum* (Herb Robert) Wilts. Dartnell & Goddard; Hants. Grigson.

EYES, Bobby's, White *Stellaria holostea* (Greater Stitchwort) Hants. Grigson.

EYES, Bright; EYES, Bright, Billy *Veronica chamaedrys* (Germander Speedwell) Som. Grigson; Ire. Grigson (Billy).

EYES, Bull's 1. *Aesculus hippocastanum* nuts (Horse Chestnut) Illinois. Dorson.1964. 2. *Althaea officinalis* (Marsh Mallow) Fernie. 3. *Caltha palustris* (Marsh Marigold) Dev, Som. Macmillan. Dor. Dacombe; Scot. Friend. 4. *Hypericum perforatum* (St John's Wort) Som. Macmillan. 5. *Leucanthemum vulgare* (Ox-eye Daisy) Corn, Ches. Grigson; Som. Macmillan. 6. *Malva sylvestris* (Common Mallow) Dev. Macmillan. 7. *Melandrium dioicum* (Red Campion) Dev. Friend.1882. 8. *Nuphar lutea* (Yellow Waterlily) Som, Dor. Macmillan. 9. *Papaver rhoeas* (Red Poppy) Som. Macmillan.

EYES, Bullock's 1. *Sedum acre* (Biting Stonecrop) N.Eng. Friend. 2. *Sempervivum tectorum* (Houseleek) N.Eng. Best & Brightman in Albertus Magnus.

EYES, Cat's 1. *Chamaenerion angustifolium* (Rosebay) Shrop. B & H. 2. *Geranium robertianum* (Herb Robert) Dev. Tynan & Maitland; Dor, Hants. Grigson. 3. *Hottonia palustre* (Water Violet) Som. Macmillan. 4. *Myosotis sylvatica* (Wood Forget-me-not) Herts. B & H. 5. *Veronica chamaedrys* (Germander Speedwell) Corn, Glos, Cumb. B & H; Dev. Friend.1882; Hants. Cope. 6. *Veronica persica* (Common Field Speedwell) Ess. B & H.

EYES, Cats-and- *Fraxinus excelsior* fruit (Ash keys) Yks. Morris. For 'eyes', read 'keys' for the original of this.

EYES, Christ's 1. *Caltha palustris* (Marsh Marigold) Fernie. 2. *Salvia horminoides* (Wild Sage) Culpepper. Note the medieval Oculus Christi for Wild Sage ("most blasphemously called Christ's Eye, because it cures Diseases of the Eye")(Culpepper).

EYES, Clear- 1. *Salvia horminoides* (Wild Sage) B & H. 2. *Salvia sclarea* (Clary) (Culpepper).

EYES, Clear-, Wild *Salvia horminoides* (Wild Sage) B & H.

EYES, Cow; EYES, Cow's *Leucanthemum vulgare* (Ox-eye Daisy) Corn. Grigson (Cow); Som. Macmillan (Cow's).

EYES, Crab's *Abrus precatorius* (Precatory Bean) USA Kingsbury; Barbados Gooding.

EYES, Cuckoo's *Geranium robertianum* (Herb Robert) Bucks, Kent Grigson.

EYES, Day's; EYE OF DAY *Bellis perennis* (Daisy) both Som. Macmillan. This is actually what daisy means.

EYES, Devil's 1. *Hyoscyamus niger* (Henbane) Som. Macmillan. 2. *Stellaria holostea* (Greater Stitchwort) Dor. Macmillan. Wales C.J.S.Thompson. 3. *Veronica chamaedrys* (Germander Speedwell) War. Grigson. The ascription is obvious for such a poisonous plant as henbane, but both stitchwort and speedwell have their darker sides, and bear other names with 'devil' associations.

EYES, Doll's 1. *Actaea alba* (White

Baneberry) H & P. 2. *Actaea rubra* (Red Baneberry) USA Kingsbury. 3. *Actaea spicata* (Herb Christopher) Tampion.

EYES, Duck's *Cotula coronopifolia* (Buttonweed) Perry.

EYES, Gardener's *Lychnis coronaria* (Rose Campion) B & H. Gerard called it Gardener's Delight, too.

EYES, God's 1. *Salvia sclarea* (Clary) B & H. 2. *Veronica chamaedrys* (Germander Speedwell) Dev. Friend.1882. It is odd to find both Devil's Eyes, and God's Eyes, for the same flower. Clary, though, gets the name from the esteem with which it used to be regarded for eye diseases - clary means 'clear-eye'.

EYES, Goody's *Salvia sclarea* (Clary) B & H. This may just possibly be God's Eyes, for both this and wild sage had Oculus Christi as a name in medieval times.

EYES, Green *Scleranthus annuus* (Knawel) Dor. Macmillan.

EYES, Hare's *Melandrium dioicum* (Red Campion) B & H.

EYES, Heaven's *Myosotis palustris* (Water Forget-me-not) Dev. Tynan & Maitland.

EYES, Indian *Dianthus plumarius* (Pink) B & H. The Oxford dictionary suggests that the word 'pink' may possibly be a shortening of 'pink-eyed', which means having narrow, or half-shut, eyes. However tenuous the connection is, we still have Indian Eye to retain the imagery.

EYES, Jupiter's *Sempervivum tectorum* (Houseleek) Gerard.

EYES, Lark's 1. *Veronica chamaedrys* (Germander Speedwell) Som. Macmillan. 2. *Viola tricolor* (Pansy) Som. Macmillan.

EYES, Milkmaid's *Veronica chamaedrys* (Germander Speedwell) N'thum. Grigson.

EYES, Moon's *Leucanthemum vulgare* (Ox-eye Daisy) Som. Macmillan. This name represents a mixture of two sets of imagery.

EYES, Mouse- *Bernardia myricaefolia*.

EYES, Old Woman's *Vinca minor* (Lesser Periwinkle) Dor. Macmillan.

EYES, Open *Bellis perennis* (Daisy) Som. Macmillan.

EYES, Owl's *Anagallis arvensis* (Scarlet Pimpernel) Som. Macmillan. Cf. Bird's Eye and Pheasant's Eye for this; there is even Adder's Eye.

EYES, Ox 1. *Adonis vernalis* Grieve.1931. 2. *Anagallis arvensis* (Scarlet Pimpernel) R.T.Gunther. 3. *Leucanthemum vulgare*. 4. *Chrysanthemum segetum* (Corn Marigold) Som, Yks. Grigson. 5. *Heliopsis helianthoides* (False Sunflower) USA House. Scarlet Pimpernel only claims the name by virtue of a gloss on a 12th century ms of Apuleius Barbarus.

EYES, Ox, Yellow *Chrysanthemum segetum* (Corn Marigold) Fernie.

EYES, Pheasant's 1. *Adonis annua*. 2. *Anagallis arvensis* (Scarlet Pimpernel) Som. Macmillan. 3. *Dianthus caryophyllus* (Carnation) Som. Macmillan. 4. *Narcissus maialis var. patellaris*. 5. *Pentaglottis sempervirens* (Evergreen Alkanet) Som. Elworthy.

EYES, Pheasant's, Corn *Adonis annua* (Pheasant's Eye) Flower.1859.

EYES, Pheasant's, Large *Adonis flammea*.

EYES, Pheasant's, Summer *Adonis aestivalis*.

EYES, Pheasant's, Yellow *Adonis vernalis*.

EYES, Pig's *Cardamine pratensis* (Lady's Smock) Ess. Grigson.

EYES, Pigeon's *Cardamine pratensis* (Lady's Smock) Yks. Grigson.

EYES, Robin's 1. *Geranium robertianum* (Herb Robert) Dev. Friend.1882; Wilts. Dartnell & Goddard; Suff. B & H. 2. *Melandrium dioicum* (Red Campion) Dev. Friend. 3. *Myosotis arvensis* (Field Forget-me-not) Hants. Grigson. 4. *Polygala vulgaris* (Milkwort) Hants. Cope.

EYES, Robin's, Small *Geranium robertianum* (Herb Robert) Glos. Grigson.

EYES, Robin's, White *Stellaria holostea* (Greater Stitchwort) Wilts. Goddard.

EYES, St Candida's *Vinca minor* (Lesser Periwinkle) Dor. Dacombe. St Candida's Well is at Morcombe Lake, in Dorset, and the water is said to be a certain cure for sore eyes; it is close to here that wild periwinkles are called St Candida's Eyes. Who was the saint, though? Possibly St Blanche, of Brittany, or the Welsh St Gwen? Or the English St Wita? They all mean white.

EYES, Sore *Veronica chamaedrys* (Germander Speedwell) Norf. V G Hatfield.1994. The flowers were used in Norfolk to make an eyebath to relieve sore eyes.

EYES, Sun's *Helianthus annuus* (Sunflower) Som. Macmillan.

EYES, Starry *Ornithogalum umbellatum* (Star-of-Bethlehem) Som. Macmillan.

EYES, Wolf's *Lycopsis arvensis* (Small Bugloss) Som. Macmillan.

EYESEED *Salvia horminoides* (Wild Sage) Lincs. Gutch. Clary (*S.horminum*) seeds swell up when put in water, and become mucilaginous. These used to be put into the eye like drops to cleanse it. Were Wild Sage seeds used in the same way, or is this mistaken identification?

EYLES; AILES; ILES *Hordeum sativum* (Barley) Wilts. Dartnell & Goddard.

F

FABE; FABERRY; FAEBERRY *Ribes uva-crispa* (Gooseberry) E.Ang. Prior (Fabe); Yks. Addy (Faberry); E.Ang. Prior (Faeberry). These are three in quite an extensive series of names, of which Feaberry is probably the best known. The series includes such forms as Fape, or Thape, which may mean the fruit from a prickly bush - OE thefe.

FABIRAMA *Coleus rotundifolius* (Fra-fra Potato) Schery.

FACE, Bull *Deschampsia cespitosa* (Tufted Hair-grass) H C Long.1910. The tufts of this grass are called Bull faces, or Bull Pates, in some districts.

FACE, Cat's *Viola tricolor* (Pansy) Som, Suss. Grigson.

FACE, Freckle *Hypoestes phyllostachya* (Polka Dot Plant) Bonar. Because of the pink dots that cover the leaves.

FACE, Freckled *Primula veris* (Cowslip) Som. Macmillan. Descriptive, obviously, but is it the recognition of the signature for the well-known ointment for the complexion made from cowslips?

FACE, Funny 1. *Tropaeolum majus* (Nasturtium) Som. Macmillan. 2. *Viola tricolor* (Pansy) Som. Macmillan.

FACE, Granny's *Viola tricolor* (Pansy) Som. Macmillan.

FACE, Men's *Viola tricolor* (Pansy) Som. Macmillan.

FACE, Monkey('s) 1. *Antirrhinum majus* (Snapdragon) Wilts. Macmillan. 2. *Linaria vulgaris* (Toadflax) Som. Macmillan. 3. *Viola tricolor* (Pansy) Suss. Grigson.

FACE, Old Man's 1. *Antirrhinum majus* (Snapdragon) Dev. Macmillan. 2. *Viola tricolor* (Pansy) Som. Macmillan.

FACE, Pig *Mesembryanthemum equilaterale* (Fig Marigold) Tasmania H.L.Roth.

FACE, Pussy *Viola tricolor* (Pansy) Som. Macmillan.

FACE, Shame *Viola tricolor* (Pansy) Som. Macmillan. More in line with the "modesty" names for the violet than for pansy.

FACE-AND-HOOD *Viola tricolor* (Pansy) Ess, Cambs, Norf. B & H. This was originally Three-faces-in-a-hood, which was one of the names Gerard used.

FACE-IN-HOOD *Aconitum napellus* (Monkshood) Norf. B & H.

FADDY-TREE *Acer pseudo-platanus* (Sycamore) Corn. Deane & Shaw. The Helston Furry was at one time known as the Faddy (possibly OE fade, to go). The sycamore got the name because boys made whistles from its branches to use at the festival.

FAIR DAYS *Potentilla anserina* (Silverweed) N'thum, Berw. Grigson. Surely not because "it expands its bright flowers only in clear weather and in sunshine", because that is simply not true. Fair Grass is another name, and they are possibly applied for the same reason as Silverweed is - the whitish underside of the leaf.

FAIR GRASS 1. *Potentilla anserina* (Silverweed) Roxb. Grigson. 2. *Ranunculus bulbosus* (Bulbous Buttercup) Roxb. Grigson. See Fair Days. Jamieson quoted the name for the buttercup, but B & H did not believe the ascription.

FAIR LADY *Atropa belladonna* (Deadly Nightshade). i.e. *belladonna*.

FAIR MAIDS 1. *Galanthus nivalis* (Snowdrop) Norf. B & H. 2. *Saxifraga granulata* (Meadow Saxifrage) Bucks. Grigson. Snowdrops are better known as Fair Maids of February, while the saxifrage has the name Pretty Maids, perhaps the pretty maids all in a row, from Mary, Mary, quite contrary.

FAIR MAIDS, February; FAIR MAIDS OF FEBRUARY *Galanthus nivalis* (Snowdrop) Som. Macmillan (February Fair Maids). Fair Maids of February is in Notes & Queries;1871. Candlemas Bells is another name, and this helps to explain the Fair Maid epithet. Snowdrops were dedicated to the

Purification of the Virgin, Candlemas, that is, on February 2nd, and it was the custom for young women dressed in white to walk in procession at the feast.

FAIR MAIDS OF FRANCE 1. *Achillea ptarmica* (Sneezewort) Middx. B & H. 2. *Leucanthemum vulgare* (Ox-eye Daisy) Som, Wilts. Grigson. 3. *Lychnis flos-cuculi* (Ragged Robin) Som. Grigson. 4. *Ranunculus aconitifolius* (White Buttercup) B & H. 5. *Saxifraga granulata* (Meadow Saxifrage) Bucks. Grigson. It would seem that they are so called because they are white, except Ragged Robin.

FAIR MAIDS OF KENT *Ranunculus aconitifolius* (White Buttercup) B & H.

FAIR WEATHER THISTLE *Carlina acaulis* (Stemless Carline Thistle) D Parish. The bracts are outspread in fine weather, whereas in damp conditions they fold over to protect the flowers.

FAIRIES *Arum maculatum* (Cuckoo-pint) Som. Macmillan.

FAIRY BASINS *Primula veris* (Cowslip) Som. Macmillan.

FAIRY BELLS 1. *Campanula rotundifolia* (Harebell) Dor, Som. Grigson; IOM Moore.1924. 2. *Convallaria maialis* (Lily-of-the-valley) Som. Macmillan. 3. *Digitalis purpurea* (Foxglove) Som. Grigson; N'thum, Ire. Denham/Hardy; USA Henkel. 4. *Disporum sp.* 5. *Narcissus pseudo-narcissus* (Daffodil) Dor. Grigson. 6. *Oxalis acetosella* (Wood Sorrel) Wales Northcote. 7. *Primula veris* (Cowslip) Som. Grigson. Wood Sorrel may not be so obvious, but the rest are, with distinctive bell-shaped flowers.

FAIRY BOATS *Nuphar lutea* (Yellow Waterlily) Som. Macmillan.

FAIRY CAPS; FAIRY HATS *Digitalis purpurea* (Foxglove) Som. Grigson; Lancs. Harland & Wilkinson.1873; Ire. Starkie; USA Henkel(Caps); Dor. Macmillan(Hats). Cf. Granny's Bonnets. The flowers of foxglove can be seen as gloves, thimbles, or hats, even dresses.

FAIRY CHEESE 1. *Malva neglecta* (Dwarf Mallow) Yks. B & H. 2. *Malva sylvestris* fruit (Common Mallow) Som, Yks. Friend. Cheese is a name almost universally given to the seed vessels of the mallow. It is the shape that suggests it, but they are edible, or children at any rate think so.

FAIRY CHEESECAKE *Medicago lupulina* (Black Medick) Wilts. Macmillan.

FAIRY CLOCK *Taraxacum officinale* (Dandelion) Mabey. In line with the very many 'clock' names for this. Children tell the time by blowing the seeds off the head, the number of blows needed before all of them are away being the time on the clock.

FAIRY CUPS *Primula veris* (Cowslip) Som. Macmillan; Kent Friend; Lincs. Dyer.

FAIRY DELL 1. *Euphorbia helioscopia* (Sun Spurge) Dor. Macmillan; 2. *Euphorbia peplus* (Petty Spurge) Dor. Macmillan.

FAIRY DRESSES *Digitalis purpurea* (Foxglove). Cf. Fairy Petticoats.

FAIRY DUSTER *Calliandra eriophylla* (Hairy-leaved Calliandra) RHS. 'Duster' from the hairy leaves, of course.

FAIRY FERN *Azolla filiculoides.*

FAIRY FINGERS *Digitalis purpurea* (Foxglove) Som. Macmillan; Hants. Boase; Ire. O Suileabhain.

FAIRY FLAX 1. *Euphrasia officinalis* (Eyebright) Donegal Grigson. 2. *Linum catharticum.* The fairies used it for their clothes; presumably they say the same thing of eyebright in Donegal.

FAIRY FLOWER 1. *Heuchera sanguinea* (Coral Bells). 2. *Melandrium rubrum* (Red Campion) IOM Moore.1924. 3. *Primula veris* (Cowslip) Som. Grigson.

FAIRY GRASS *Briza media* (Quaking-grass) Ire. Grigson.1959.

FAIRY HAIR *Cuscuta epithymum* (Common Dodder) Jersey Grigson. Cf. Hairweed, etc.

FAIRY HORSE *Senecio jacobaea* (Ragwort) Ire. Grigson. Both the ragwort and the St John's Wort are believed to be fairy horses in disguise. If you tread them down after sunset a horse will arise from the root of each injured plant and will gallop away with you.

The witches used ragwort as a horse, too. There is a saying in the Isle of Man: "as arrant a witch as ever rode on ragwort".

FAIRY KEYS *Primula veris x vulgaris* (False Oxlip) Dev. Macmillan. Cf. the cowslip's series of names allocating the keys to St Peter rather than to the fairies.

FAIRY LAMPS *Arum maculatum* (Cuckoo-pint) Ire. Porter. The pollen of the flowers does give a faint light at dusk, and it was the Irish labourers who came to work in the Fens during the famines in the 19th century who named the plants Fairy Lamps. Similarly, they are called Shiners in the same area.

FAIRY LANTERN 1. *Calochortus albus*. 2. *Disporum smithii* USA Schenk & Gifford.

FAIRY LANTERN, Golden *Calochortus amabilis*.

FAIRY LARCH *Equisetum sylvaticum* (Wood Horsetail) N.Eng. Salisbury.1936.

FAIRY LINT *Linum catharticum* (Fairy Flax) N'thum. Denham/Hardy. See Fairy Flax.

FAIRY('S) PETTICOATS *Digitalis purpurea* (Foxglove) Som. Macmillan; Ches. Holland. Cf. Fairy Dresses.

FAIRY POPS *Trifolium pratense* (Red Clover) Dor. Macmillan. Pops means the same as sweets, so this name is one of the Honeysuckle series.

FAIRY POTATOES *Conopodium majus* (Earth-nut) Ire Grigson. The tubers are edible, and often eaten. They are more often likened to chestnuts than to potatoes, though. This is something that belongs to the fairies in Ireland, and especially to the leprechaun.

FAIRY PRIMROSE *Primula malacoides* Nicolaisen.

FAIRY QUEEN *Viola tricolor* (Pansy) Som. Macmillan. There must have been more than a casual nod towards Shakespeare in the bestowing of this name.

FAIRY RINGERS *Campanula rotundifolia* (Harebell) Dor, Som. Macmillan.

FAIRY SOAP *Polygala vulgaris* (Milkwort) Donegal Grigson. It was believed that the fairies make a lather from the roots and leaves.

FAIRY TABLES *Hydrocotyle vulgaris* (Pennywort) Ches. Holland. It is the flat round leaves that suggest the name.

FAIRY TRUMPETS 1. *Calystegia sepium* (Great Bindweed) Som. Macmillan. 2. *Lonicera periclymenum* (Honeysuckle) Som. Macmillan.

FAIRY WEED *Digitalis purpurea* (Foxglove) Hants. Boase. Ire. Grigson. It is the property of the fairies (as well as the fox) in many different guises, their gloves, fingers, thimbles, caps and hats, even bells.

FAIRY WIVES' DISTAFF *Typha latifolia* (False Bulrush) Scot. A.Stewart.

FAIRY WOMAN'S FLAX *Linum catharticum* (Fairy Flax) Spence.1949b. This is a translation from the Gaelic. The fairies made their clothes from this flax.

FAIRY'S BASIN 1. *Ranunculus acris* (Meadow Buttercup) Dev. Grigson. 2. *Ranunculus bulbosus* (Bulbous Buttercup) Dev. Grigson. 3. *Ranunculus repens* (Creeping Buttercup) Dev. Grigson.

FAIRY'S BATH *Geum rivale* (Water Avens) Dor. Macmillan.

FAIRY'S BEDS *Scrophularia nodosa* (Figwort) Dor. Macmillan.

FAIRY'S BELLS *Endymion nonscriptus* (Bluebell) Som. Macmillan.

FAIRY'S BROOM *Dipsacus fullonum* (Teasel) Som. Macmillan.

FAIRY'S CAP *Campanula rotundifolia* (Harebell) Dor, Som. Grigson; Wilts. Macmillan.

FAIRY'S CLOCK *Adoxa moschatellina* (Moschatel) Dev. Macmillan. 'Clock' names for moschatel are suggested by the way the five flowers it bears face in different directions, like miniature clock faces.

FAIRY'S CORN *Lathyrus montanus* (Bittervetch) Donegal Grigson. The tubers are edible, and are eaten like potatoes, roasted.

FAIRY'S CRADLE *Tulipa sp* (Tulip) Dyer.1889.

FAIRY'S CUP *Campanula rotundifolia*

(Harebell) Dor, Som. Grigson.
FAIRY'S FIRE 1. *Chaenomeles speciosa* (Japanese Quince) Dyer.1889. 2. *Dipsacus fullonum* (Teasel) Som. Macmillan.
FAIRY'S GLOVES *Digitalis purpurea* (Foxglove) Som. Macmillan; Shrop. Burne. Cf. Lady's, Granny's, even Witches', Gloves.
FAIRY'S LANTERNS *Linaria vulgaris* (Toadflax) Som. Macmillan.
FAIRY'S PAINTBRUSHES *Vinca minor* (Lesser Periwinkle) Som. Macmillan.
FAIRY'S THIMBLES 1. *Campanula pusilla*. 2. *Campanula rotundifolia* (Harebell) Dor, Som. Macmillan. IOM Moore.1924. 3. *Digitalis purpurea* (Foxglove) Som. Macmillan; USA. Henkel. The flowers of foxglove can be thimbles as well as gloves.
FAIRY'S UMBRELLAS *Convolvulus arvensis* (Field Bindweed) Som. Macmillan.
FAIRY'S WAND 1. *Agrimonia eupatoria* (Agrimony) Dor. Macmillan. 2. *Verbascum thapsus* (Mullein) Dor. Macmillan.
FAIRY'S WINDFLOWER *Anemone nemerosa* (Wood Anemone) Dor. Macmillan.
FAIRY'S WINECUPS *Convolvulus arvensis* (Field Bindweed) Som. Macmillan.
FAITH, HOPE AND CHARITY *Pulmonaria officinalis* (Lungwort) Dor. Udal. A multiple name, suggested by the multiple colours in the flowers. The best known of these names is probably Abraham, Isaac and Jacob.
FALCON'S CLAW ACACIA *Acacia polyacantha*. Descriptive of the hooked thorns.
FALFALARIES *Fritillaria meleagris* (Snake's Head Lily) Yks. Grigson. Presumably originally from fritillaries.
FALL MEADOW RUE *Thalictrum polyganum*.
FALL PANICUM *Panicum dichotomiflorum*.
FALLING STARS 1. *Ceratophyllum demersum* (Hornwort) Tynan & Maitland. 2. *Crocosmia aurea*.
FALSE ANEMONE *Anemonopsis macrophylla*.
FALSE CAMOMILE, Sweet *Matricaria recutita* (Scented Mayweed). H & P.

FALSE DITTANY *Ballota acetabulosa* Howes.
FALSE ELM 1. *Celtis occidentalis* (Hackberry) USA Vestal & Schultes. 2. *Celtis reticulata* (Western Hackberry) USA Elmore.
FALSE ELM, African *Celtis integrifolia* (African Nettle Tree) Dalziel.
FALSE GARLIC *Allium vineale* (Crow Garlic) Howes.
FALSE HEMP *Crotalaria juncea* (Sunn Hemp) Chopra.
FALSE INDIGO *Amorpha californica* (Indigo-bush) USA T H Kearney.
FALSE IROKO 1. *Antiaris africana* (Barkcloth Tree) Dalziel. 2. *Antiaris toxicaria* (Upas Tree) Dale & Greenway.
FALSE JALAP *Mirabilis jalapa* Howes. False, because it is used as an adulterant.
FALSE JERUSALEM CHERRY *Solanum diflorum*.
FALSE KOLA *Garcinia kola* (Bitter Kola) Dalziel.
FALSE MALLOW, Red *Malvastrum coccineum*.
FALSE NUTMEG *Monodora myristica* (Calabash Nutmeg) Dalziel.
FALSE OXLIP *Primula veris x vulgaris*. For the true Oxlip is *Primula elatior*.
FALSE PARSLEY *Cicuta maculata* (American Cowbane) USA Kingsbury.1967.
FALSE PELLITORY *Achillea ptarmica* (Sneezewort) Culpepper. Cf. Gerard's Wild Pellitory, and Tradescant's Bastard Pellitory.
FALSE PELLITORY-OF-SPAIN *Peucedanum ostruthium* (Masterwort) Leyel.1948.
FALSE ROSEWOOD *Thespesia populnea* (Portia Tree) Dalziel.
FALSE RUBBER TREE *Funtumia africana*.
FALSE SOLOMON'S SEAL *Smilacina racemosa*.
FALSE SPIKENARD *Smilacina racemosa* (False Solomon's Seal USA H H Smith.
FALSE TARRAGON *Artemisia dracunculoides* (Russian Tarragon) USA T H Kearney.

FALSE VALERIAN *Senecio aureus* (Swamp Squaw-weed) O P Brown.

FAMINTERRY *Fumaria officinalis* (Fumitory) IOM Grigson. All sorts of variations on fumitory are recorded, mostly as wide of the mark as this is.

FAN-LEAVED BUTTERCUP *Ranunculus flabellatus* (Jersey Buttercup) Clapham.

FAN PALM 1. *Rhapis excelsa*. 2. *Trachycarpus excelsus*. 3. *Washingtonia filifera*. Fan Palm is a name given to more than one genus of the Palmaceae, and refers to the spread of the leaves.

FAN PALM, Dwarf; FAN-PALM, European *Chamaerops humilis* RHS (European).

FANCY *Viola tricolor* (Pansy) B & H. The East Anglian Tittle-my-fancy would seem to be the original of this.

FANCY-LEAVED CALADIUM *Caladium bicolor 'Splendens'*.

FANSEL *Areca catechu* (Areca Palm) Pomet.

FANWEED *Thlaspi arvense* (Field Pennycress) USA Kingsbury.

FAPE *Ribes uva-crispa* (Gooseberry) Suff. Moor. One of the group which includes Fabe and Thape, and starts off with a word something like Fea, possibly a fruit from a prickly bush, OE thefe. But according to Forby the name was only applied to the immature fruits - "nobody ever talks of a ripe fape".

FARENUT *Conopodium majus* (Earthnut) Corn. Jago. This appears as Varenut, too. It looks to be related to Harenut, a widespread form of Earthnut. But on the other hand, fare is a young pig, OE fearh.

FAREWELL, Summer's 1. *Aster tripolium* (Sea Aster) Dev, Som. Wright. 2. *Filipendula ulmaria* (Meadowsweet) Dev, Dor. Macmillan. 3. *Senecio jacobaea* (Ragwort) Dev, Glos. Grigson. All late summer flowering plants. Phlox ought to be included in the list, as their flowers are always a sign of summer coming to an end.

FAREWELL-(TO)-SUMMER 1. *Aster paniculatus* (Michaelmas Daisy) Wilts. Dartnell. & Goddard. 2. *Parnassia palustris* (Grass of Parnassus) Scot. Tynan. & Maitland. 3. *Saponaria officinalis* (Soapwort) Mor. Grigson. 4. *Solidago virgaurea* (Golden Rod) Som. Grigson.

FAREWELL-TO-SPRING 1. *Godetia biloba* USA Barrett & Gifford. 2. *Godetia viminea* USA Barrett & Gifford.

FARGES FIR *Abies fargesii*.

FARGES LILY *Lilium fargesii*. Paul Guillaume Farges (1844-1912), who collected many specimens of Chinese plants for the Paris Museum.

FARKLEBERRY *Vaccinium arboreum*.

FARMER'S CLOCK *Taraxacum officinale* (Dandelion) Putnam.

FARMER'S PLAGUE *Aegopodium podagraria* (Goutweed) N.Ire. Grigson. Under the name of Ground Elder, it is a plague in gardens, too.

FARMER'S RUIN *Spergula arvensis* (Corn Spurrey) Leyel.1948. Cf. Poverty, for it grows in the poorest soil, and its presence on farm land is a bad sign indeed.

FARMER'S WEATHERGLASS *Anagallis arvensis* (Scarlet Pimpernel) Som. Macmillan.

FARNAMBUCK *Caesalpina bonduc* (Brazilian Redwood) Pomet. A version of Pernambuco, from where the wood was shipped. Cf. Pernambuco Redwood, or Pernambuco Wood.

FARRAIN *Heracleum sphondyllium* (Hogweed) IOM Moore.1924.

FART, Mare *Senecio jacobaea* (Ragwort) Ches. Grigson.

FARTHING ROT *Hydrocotyle vulgaris* (Pennywort) Norf. B & H. The cost rises - it is Shilling Rot in Scotland. There was a firm belief that the plant caused foot rot in sheep.

FASELLES *Phaseolus vulgaris* (Kidney Bean) Turner. Phaseolus, presumably.

FAT, Cow- *Vaccaria pyramidata* (Cow-herb) W.Miller. The generic name, *Vaccaria*, is from Latin vacca, cow, and was given because it was thought to be very good cattle fodder.

FAT, Mule *Baccharis viminea*.

FAT, Sheep- *Atriplex confertifolia* (Shadscale) USA Elmore. The Navajo use it to provide salt for the sheep in winter.

FAT, Winter *Eurotia lanata* (Winter Sage) Jaeger.

FAT BELLIES *Silene cucubalis* (Bladder Campion) Som. Macmillan. Descriptive.

FAT DUCKWEED *Lemna gibba*.

FAT GOOSE *Chenopodium bonus-henricus* (Good King Henry) Cambs. Porter.1974. Fat Hen is *C.album*; they are both the equivalents of the medieval Latin tota bona, one of the names by which this plant was known to Tudor botanists.

FAT GRASS *Veronica anagallis-aquatica* (Water Speedwell) Bianchini.

FAT HEN 1. *Artemisia vulgaris* (Mugwort) Bucks. Grigson. 2. *Atriplex patula* (Orache) Dev, E.Ang, Yks, N'thum. Grigson. 3. *Capsella bursa-pastoris* (Shepherd's Purse) Glos. J.D. Robertson. 4. *Chenopodium album*. 5. *Chenopodium bonus-henricus* (Good King Henry) Suss, Kent, Berks, Surr, Ches. Grigson. 6. *Chenopodium vulvaria* (Stinking Orach) Norf. B & H. 7. *Chrysanthemum segetum* (Corn Marigold) Hants. Grigson. 8. *Galium aparine* (Goose-grass) Yks. Nicholson. 9. *Glechoma hederacea* (Ground Ivy) Bucks. Grigson. 10. *Fagopyrum esculentum* (Buckwheat) Bucks. Grigson. 11. *Polygonum persicaria* (Persicaria) Som. Macmillan. 12. *Sedum telephium* (Orpine) Tynan & Maitland. *Chenopodium album* is the plant which claims the right of using Fat Hen as its common name. But is the name really Fat Henry? If it is, then it should be reserved for *C.bonus-henricus*. 'Fat' probably refers to its predilection to middens, and would account for such other names as Bacon-weed and Pigweed. Perhaps the only one in the list which does not fit into the midden pattern is *Chrysanthemum segetum*; however did that get the name? Buckwheat seems unlikely, too. It does not need rich soil.

FAT PORK TREE *Clusia rosea* RHS.

FAT SOLOMON *Smilacina amplexicaulis* Schenk & Gifford.

FATA *Solanum tuberosum* (Potato) Ire. Salaman. Probably from patata, which would account for Pratie too.

FATCH *Vicia sativa* (Common Vetch) Glos, War. Grigson. Fatch, usually in the plural, as fatches, is quite a common rendering of vetch.

FATCH, Meadow *Onobrychis sativa* (Sainfoin) B & H.

FATHER BIGFACE *Carduus nutans* (Musk Thistle) Som. Macmillan.

FATHER HUGO'S ROSE *Rosa hugonis*.

FATHER OF HEATH *Erica tetralix* (Cross-leaved Heath) Yks. Grigson.

FATHER TIME *Clematis vitalba* (Old Man's Beard) Som. Macmillan. Another name on the 'greybeard' theme.

FATTAHS *Crataegus monogyna* fruit (Haws) Glos. J.D. Robertson.

FAVA BEAN *Vicia faba* (Broad Bean) Kingsbury.

FAVEROLE *Dracunculus vulgaris* (Dragon Arum) W.Miller.

FAYBERRY 1. *Ribes uva-crispa* (Gooseberry) Lancs. Nodal. 2. *Vaccinium myrtillus* (Whortleberry) Shrop. B & H. One of a number of names for gooseberry apparently meaning fruit of a prickly bush.

FEA; FEABE; FEAP; FEAPBERRY *Ribes uva-crispa* (Gooseberry) Grieve.1931 (Fea); E.Ang. Prior (Feabe); Fernie (Feap); E.Ang. Prior (Feapberry). See Fayberry.

FEABERRY 1. *Ribes uva-crispa* (Gooseberry) Grigson. 2. *Vaccinium oxycoccus* (Cranberry) Shrop. Grigson. See Fayberry.

FEASIL *Phaseolus vulgaris* (Kidney Bean) W.Eng. Halliwell. Evidently an attempt at *Phaseolus*.

FEATHER FERN *Spiraea japonica* Dev. Friend.1882.

FEATHER GERANIUM *Chenopodium botrys* (Jerusalem Oak) USA Tolstead. They call it Oak-leaf Geranium in America, too.

FEATHER GRASS *Stipa pennata*.

FEATHERBEDS, Lady's *Saxifraga granulata* (Meadow Saxifrage) Suss. Grigson.

FEATHERBOW; FEATHERFALL; FEATHERFEW; FEATHERFOE; FEATHERFOLD; FEATHERFOWL; FEATHERFOY; FEATHERFULL; FEATHERWHEELIE *Tanacetum parthenium* (Feverfew) Corn. Jago (Featherbow); Friend (Featherfall, Featherfew); Som. Macmillan (Featherfoe); Som. Macmillan; Heref. G.C.Lewis (Featherfold); Yks. Carr (Featherfowl); Glos. J.D.Robertson (Featherfoy); Fernie (Featherfull); Scot. Jamieson (Featherwheelie). These are all variations on Feverfew, which is OE feberfugen, Latin febris, fever, and fugare, to drive away.

FEATHERED PINK *Dianthus plumarius* (Pink) Gerard. Descriptive of the indented petals; the specific name is plumarius, which means the same thing.

FEATHERFOIL 1. *Hottonia inflata* USA Upham. 2. *Hottonia palustre* (Water Violet) Grigson. 3. *Tanacetum parthenium* (Feverfew) Dawson. There are two traditions here - the first is as a variation of Feverfew; the second means literally feathery leaf (foil - folium).

FEATHERFOIL, Bog; FEATHERFOIL, Marsh; FEATHERFOIL, Water *Hottonia palustre* (Water Violet) B & H (Bog); Flower.1859 (Marsh); Prior (Water).

FEATHERS 1. *Amaranthus caudatus* (Love-lies-bleeding) Som. Macmillan. 2. *Gynerium argenteum* (Pampas Grass) Som. Macmillan.

FEATHERS, Capon's 1. *Aquilegia vulgaris* (Columbine) B & H. 2. *Valeriana officinalis* (Valerian) B & H.

FEATHERS, Gay 1. *Liatris scariosa* Le Strange. 2. *Liatris spicata* (Button Snakeroot) Grieve.1931.

FEATHERS, Hedge *Clematis vitalba* (Old Man's Beard) Yks. Grigson.

FEATHERS, Kansas *Liatris pychnostachnya*.

FEATHERS, King's *Saxifraga umbrosa* (Pyrenean Saxifrage) W. Miller. Given for this plant, but without a doubt meant for *S spathularis x umbrosa* (London Pride).

FEATHERS, Pheasant's *Saxifraga x urbinum* (London Pride) Suss. B & H.

FEATHERS, Plume *Gynerium argenteum* (Pampas Grass) Som. Macmillan.

FEATHERS, Prince's 1. *Amaranthus hypondriachus*. 2. *Amaranthus retroflexus* (Pigweed) Australia Watt. 3. *Gynerium argenteum* (Pampas Grass) Dev. Friend. 4. *Leucanthemum vulgare* (Ox-eye Daisy) Tynan & Maitland. 5. *Onobrychis sativa* (Sainfoin) Wilts. Tynan & Maitland. 6. *Polygonum orientale* USA Zenkert. 7. *Potentilla anserina* (Silverweed) Som. Macmillan. 8. *Prunella vulgaris* (Self-heal) N'thum. Grigson. 9. *Saxifraga x urbinum* (London Pride) Dev. Friend.1882. 10. *Syringa vulgaris* (Lilac) Dev, Rut. B & H. 11. *Umbilicus rupestris* (Wall Pennywort) Yks. Grigson.

FEATHERS, Prince of Wales's 1. *Centranthus ruber* (Spur Valerian) Dor. Macmillan. 2. *Syringa vulgaris* (Lilac) Dev. Macmillan.

FEATHERS, Queen's 1. *Filipendula ulmaria* (Meadowsweet) Som. Macmillan. 2. *Saxifraga x urbinum* (London Pride) Ches. B & H. 3. *Syringa vulgaris* (Lilac) Dor. Dacombe.

FEATHERS, Robin Hood's *Clematis integrifolia* (Hungarian Climber) Cumb. Wright.

FEATHERS, Silver *Potentilla anserina* (Silverweed) Oxf. Grigson.

FEATHERS, Soldier's 1. *Amaranthus caudatus* (Love-lies-bleeding) Som. Macmillan. 2. *Syringa vulgaris* (Lilac) Som. Tynan & Maitland.

FEATHERS, Water *Hottonia palustris* (Water Violet) Perry.

FEATHERY PLUM *Gynerium argenteum* (Pampas Grass) Som. Macmillan.

FEATHYFEW *Tanacetum parthenium* (Feverfew) Friend. This version is just straying a little from the series which includes Featherfew. It is of course a variation on Feverfew.

FEBERRY *Ribes uva-crispa* (Gooseberry) Yks. Addy. Feaberry is seen more often than this, but it exists in other forms. such as Faberry, or Feapberry, etc. The derivation seems to be OE thefe, probably a prickly bush.

FEBRUARY, Fair Maids of; FEBRUARY FAIR-MAID *Galanthus nivalis* (Snowdrop) Notes & Queries, 1871 (Fair Maids of February); Som. Macmillan (February Fair-Maid). The reference is to the time of flowering. and the consequent dedication of the plant to the Purification of the Virgin, 2 February - Candlemas, in other words.

FECHTER *Plantago lanceolata* (Ribwort Plantain) Scot. Grigson. Soldiers, Fighting Cocks, Conkers, etc are all references to the game Scottish children play with the flower heads - one child holds out his fechter, and the other tries to decapitate it with another. The winning ribwort stacks up its victories by being called Bully of two, etc., just like conkers.

FEDDERFEW *Tanacetum parthenium* (Feverfew) Gerard. Fedderfew is Featherfew, itself a corruption of Feverfew.

FEET, Fox *Urostachys selago* (Mountain Clubmoss) B & H.

FELLBLOOM 1. *Lotus corniculatus* (Bird's-foot Trefoil) Scot. Grigson. 2. *Medicago lupulina* (Black Medick) Scot. Jamieson.

FELLON-BERRY *Bryonia dioica* (White Bryony) B & H. A felon, or fellon, is a whitlow, or any skin eruption like a sore. Whitlows were called in Latin furunculi, or little thieves - felons, in other words, according to Prior. It follows that any plant which has fellon as part of its name would either be regarded as causing the sore place, or, more likely, would be used to cure it. But bryony does not seem to have been so used, though very early on it was reckoned to be an antidote to leprosy.

FELLON-GRASS 1. *Geranium robertianum* (Herb Robert) Yks. Grigson. 2. *Helleborus foetidus* (Stinking Hellebore) Yks. Hartley. & Ingilby. 3. *Helleborus niger* (Christmas Rose) B & H. 4. *Peucedanum ostruthium* (Masterwort) Yks, Cumb. Grigson. Masterwort has Fellon-wood and Fellonwort also; the latter is also one of the names for Herb Robert, which was certainly used for ulcers, erysipelas, and the like.

FELLON-HERB 1. *Artemisia vulgaris* (Mugwort) Som. B & H; Wilts. Wiltshire. 2. *Cerastium holostioides* (Mouse-ear) Corn. Courtney.1880. 3. *Helleborus viridis* (Green Hellebore) Cumb. Grigson. 4. *Hieraceum pilosellum* (Mouse-ear Hawkweed) Corn. Jago.

FELLON-WEED *Senecio jacobaea* (Ragwort) Salisbury.1964. Martin records the use for a boil in the western Isles.

FELLON-WOOD 1. *Peucedanum ostruthium* (Masterwort) Cumb. Grigson. 2. *Solanum dulcamara* (Woody Nightshade) Corn. Grigson. Cumb, Yks. B & H; Lancs. Nodal. Masterwort has Fellon-grass, and Fellon-wort; Woody Nightshade has the latter, too. "Country people commonly use to take the Berries of (Woody Nightshade), and having bruised them, they apply them to Felons, and thereby soon rid their Fingers of such troublesome Guests" (Culpepper). Warwickshire people used to rub the berries on chilblains, too.

FELLON-WORT 1. *Artemisia vulgaris* (Mugwort) Fernie. 2. *Chelidonium majus* (Greater Celandine) B & H. 3. *Geranium robertianum* (Herb Robert) Yks. Grigson. 4. *Peucedanum ostruthium* (Masterwort) Cumb. Grigson. 5. *Solanum dulcamara* (Woody Nightshade) Culpepper. 6. *Solanum nigrum* (Black Nightshade) Jordan. Greater Celandine is a wart-remover, its yellow juice being outwardly applied not only to warts, but to corns as well. It has a whole series of names to commemorate the use - besides Fellon-wort, there are Tetter-wort, Wart-flower, Wartwort, etc.

FELLWORT 1. *Centaureum erythraea* (Centaury) Vesey-Fitzgerald. 2. *Gentiana lutea* (Yellow Gentian) Perry. 3. *Gentianella amarella* (Autumn Gentian) Browning. The word is spelt Felwort for the gentians, and it seems that it is a name used for all of them, but more especially for Autumn Gentian. 'Fel' is OE feld, field (Cf. the book name Fieldwort for *G.amarella*). Centaury of course belongs to the Gentian family, and this too has Fieldwort applied as a book name.

FELLWORT, Marsh *Swertia perennis*. This, too, is in the *Gentianaceae*, and should properly have the name spelt Felwort.

FELLWORT, Spring *Gentiana verna* (Spring Gentian) Tradescant.

FELT BUSH *Kalanchoe beharensis*. The leaves have a characteristically felted appearance.

FELT-THORN, Bald-leaved *Tetradymia glabrata* (Littleleaf Horsebrush) USA Jaeger.

FELT-THORN, Grey *Tetradymia canescens* (Spineless Horsebrush) USA Jaeger.

FELT-THORN, Narrow-scaled *Tetradymia stenolepis*.

FELT-THORN, White *Tetradymia comosa*.

FELTWORT *Verbascum thapsus* (Mullein) Cockayne; USA Henkel. The texture of the leaves is responsible for the name; they have been likened to flannel and blankets as well.

FEMALE, Dragon's *Dracunculus vulgaris* (Dragon Arum) Som. Macmillan. It looks rather obscure like this. Could it be Female Dragons, a name certainly recorded for Water Arum?

FEMALE CORNEL *Cornus sanguinea* (Dogwood) Gerard. *C mas* is the Male Cornel.

FEMALE DOGWOOD *Cornus amomum* (Silky Cornel) Grieve.1931.

FEMALE HANDED ORCHID *Dactylorchis maculata* (Heath Spotted Orchid) Jamieson. The plant is "handed", because of the finger-like lobes of the tubers, and 'female' because it lacks the obvious male characteristic of twin tubers.

FEMALE NETTLE *Urtica dioica* (Nettle) Gerard. *U pilulifera* is the Male Nettle in Gerard's view, but it is not clear what he had in mind.

FEMALE PIMPERNEL *Anagallis foemina* (Blue Pimpernel) Gerard. Scarlet Pimpernel is the male, though it is not clear what the reason is.

FEMALE REGULATOR *Senecio aureus* (Swamp Squaw-weed) O P Brown. The American Indians use it - it "exerts a very powerful and peculiar influence upon the reproductive organs of females" (O P Brown). In Alabama, it was given during confinements, and it was used to bring on menstrual flow in young girls when it had stopped because of colds.

FEMALE SATYRION *Dactylorchis maculata* (Heath Spotted Orchid) Gerard. The "dactyl" part of the generic name, and the various 'hand' and 'finger' names refer to the fact that the tubers are divided into several finger-like lobes. Presumably this is why Gerard called it Female Satyrion, for Male Satyrion would be for the *Orchis* species with two testicle-like tubers.

FEMBLE; FIMBLE *Cannabis sativa* (Hemp) Ess, Suss. (Fimble) both B & H. Applied only to the male plant.

FEN RAGWORT, Great *Senecio paludosus*. This was once a native of East Anglia, but now apparently extinct there.

FENBERRY *Vaccinium oxycoccus* (Cranberry) Prior. This is a habitat name - Cf. others such as Bogberry, and Marshwort.

FENCE, Coffee *Clerodendron aculeatum* Barbados Gooding.

FENCE, Flower *Caesalpina pulcherrima* (Barbados Pride) Barbados Gooding.

FENCKELL; FENKEL *Foeniculum vulgare* (Fennel) Gerard (Fenckell); Prior (Fenkel). Both of these are closer to the original than the modern fennel. The OE was fingel, and forms like Finkle are still widespread.

FENNEL 1. *Anthemis cotula* (Maydweed) USA H.Smith.1923. 2. *Foeniculum vulgare*. The Latin is foeniculum, the diminutive of faenum, hay, but it comes to modern times through its OE version, fingel. A number of names survive that show the midway point between ancient and modern - Fenkel, for instance, or Finkel. There is even an odd hybrid - Fennel-Finkle.

FENNEL, Black *Foeniculum vulgare var. nigra.*

FENNEL, Damask *Nigella damascena* (Love-in-a-mist) Schauenberg. *N sativa* is Fennel-flower.

FENNEL, Dog 1. *Peucedanum officinale* (Hog's Fennel) R.T.Gunther. 2. *Peucedanum palustre* (Milk Parsley) B & H.

FENNEL, Dog's *Anthemis cotula* (Maydweed) Prior.

FENNEL, Dwarf *Foeniculum azoricum.*

FENNEL, False *Anethum graveolens* (Dill) Polunin.

FENNEL, Florence *Foeniculum dulce* (Finocchio) Pomet.

FENNEL, French *Foeniculum dulce* Grieve.1933. Both this and Florence Fennel, and yet another name - Roman Fennel, show that it was known that this plant came from S. France and Italy. It was introduced into this country in Stuart times.

FENNEL, Giant *Ferula communis.*

FENNEL, Hog's 1. *Peucedanum officinale.* 2. *Peucedanum palustre* (Milk Parsley) Clapham.

FENNEL, Hog's, Marsh 1. *Peucedanum officinale* (Hog's Fennel) Grieve.1931. 2. *Peucedanum palustre* (Milk Parsley) C P Johnson.

FENNEL, Hog's, Sea *Peucedanum officinale* (Hog's Fennel) Barton & Castle.

FENNEL, Hound's *Anthemis cotula* (Maydweed) Henslow. Cf. Dog's Fennel, as well as Dog's Camomile, Dog Stinks, etc., all because of the bad smell.

FENNEL, Italian *Foeniculum vulgare var. piperitum* (Sicilian Fennel) E.Hayes.

FENNEL, Roman *Foeniculum dulce* (Finocchio) Sanecki. Cf. Florence Fennel for this.

FENNEL, Sea *Crithmum maritimum* (Samphire) Coles. The leaves do vaguely taste like fennel.

FENNEL, Sicilian *Foeniculum vulgare var. piperitum.*

FENNEL, Small *Nigella sativa* (Fennel-flower) Chopra.

FENNEL, Sow *Peucedanum officinale* (Hog's Fennel) Culpepper.

FENNEL, Sweet 1. *Anethum graveolens* (Dill) Thornton. 2. *Foeniculum dulce* (Finocchio) E Hayes.

FENNEL, Water 1. *Callitriche verna* (Water Starwort) W.Miller. 2. *Oenanthe aquatica* (Horsebane) B & H.

FENNEL-FINKLE *Foeniculum vulgare* (Fennel) Dyer.1887. An artificial name which brings together both main elements of the word.

FENNEL FLOWER *Nigella sativa.* It is probably the finely-divided leaves which are responsible for the name, but the seeds are used as a spice (and there are various spicy names for the plant) in the Mediterranean area, and for seasoning curries in the East.

FENNEL FLOWER, Garden *Nigella damascena* (Love-in-a-mist) Friend. *Nigella sativa* is the Fennel Flower.

FENNEL-LEAVED PEONY *Paeonia tenuifolia* (Fringed Peony) W.Miller.

FENUGREEK 1. *Chelidonium majus* (Greater Celandine) Fernie. 2. *Trigonella ornithopodioides.* Fenugreek is foenum-graecum, Grecian hay, because it was apparently used to scent inferior hay. But why should Greater Celandine ever get the name? Fernie gives both fenugreek and Grecian Hay for it.

FENUGREEK, Bird's-foot *Trigonella ornithopodioides* (Fenugreek) Clapham. Bird's-foot, from *ornithopodioides*, the specific name.

FERN, Adder's *Polypodium vulgare* (Common Polypody) Hants. Cope. Adders are suggested by the rows of orange-coloured spores.

FERN, Asparagus *Asparagus officinale.*

FERN, Bird's-nest *Asplenium nidus.* For all the leaves emerge from a basal rosette, likened to a bird's nest.

FERN, Brake *Pteridium aquilenum.*

FERN, Bristle *Polystichum aculeatum* (Hard Shield Fern) USA Cunningham.

FERN, Buckler *Dryopteris dilatata.*
FERN, Buckler, Crested *Dryopteris cristata.*
FERN, Buckler, Hay-scented *Dryopteris aemula.*
FERN, Buckler, Narrow 1. *Dryopteris carthusiana* (Prickly Buckler Fern) RHS. 2. *Dryopteris spinulosa.*
FERN, Buckler, Prickly *Dryopteris carthusiana.*
FERN, Buckler, Rigid *Dryopteris villarsii.*
FERN, California *Conium maculatum* (Hemlock) USA Kingsbury. Cf. Nebraska Fern. Fern-like foliage?
FERN, Christmas *Polystichum acrostichoides.*
FERN, Cinnamon *Osmunda cinnamonea.*
FERN, Deer *Blechnum spicant* (Hard Fern) USA Turner & Bell.
FERN, Eagle *Pteridium aquilinum* (Bracken) Howes. In keeping with the specific name.
FERN, Fairy *Azolla filiculoides.*
FERN, Feather *Spiraea japonica* Dev. Friend.1882.
FERN, Finger *Ceterach officinarum* (Rustyback) Turner.
FERN, Fragrant, Chinese *Artemisia annua* (Sweet Wormwood). G B Foster.
FERN, Hard *Blechnum spicant.*
FERN, Hare's-foot *Davallia canariensis.*
FERN, Hen-and-chicken *Asplenium bulbiferum.* 'Hen-and-chicken', because of its habit of producing little ferns from bulbils on the mature fronds.
FERN, Holly *Polystichum lonchitis.* 'Holly', because the leaflets have spiny teeth.
FERN, Horseshoe *Marattia fraxinea.*
FERN, Lady *Athyrium filix-femina.* It is described as more graceful and delicate than the Male Fern, which is Dryopteris filix-mas.
FERN, Licorice *Polypodium vulgare* (Comon Polypody) USA Turner & Bell. For the rhizomes have a strong liquorice taste, and are sweet enough to have been used as a sugar subsitute at one time.
FERN, Maidenhair *Adiantum capillus-veneris.*
FERN, Male *Dryopteris filix-mas.*

FERN, Male, Golden-scaled *Dryopteris borreri.*
FERN, Male, Small *Dryopteris abbreviata.*
FERN, Moss *Polypodium vulgare* (Common Polypody) B & H.
FERN, Mother *Asplenium bulbiferum* (Hen-and-chickens Fern) USA Cunningham. For the same reason as 'hen-and-chickens' - see above.
FERN, Nebraska *Conium maculatum* (Hemlock) USA Kingsbury. Cf. California Fern. Fern-like foliage?
FERN, Oak 1. *Polypodium vulgare* (Common Polypody) B & H. 2. *Pteridium aquilina* (Bracken) Som. Elworthy. The reason, Elworthy says, is that if "the stalk is cut across near the root there are dark markings on the section which strongly resemble a very symmetrical oak tree".
FERN, Parsley *Tanacetum vulgare* (Tansy) Dev. Friend.1882; USA Henkel.
FERN, Rabbit's-foot *Davallia fijiensis.*
FERN, Rattlesnake *Botrychium virginianum.*
FERN, Ribbon *Pteris cretica* (Table Fern) Simmons. 'Ribbon', because of the arrangement of its spores, which are borne in continuous bands along the lower leaf margins.
FERN, Royal *Osmunda regalis.*
FERN, Rue *Asplenium ruta-muraria* (Wall Rue) Dev. Friend.1882.
FERN, Scale *Ceterach officinarum* (Rustyback) Turner.
FERN, Scented *Tanacetum vulgare* (Tansy) Dev. Friend.1882 Som. B & H; USA Henkel. The scent in this case is usually likened to that of ginger.
FERN, Shield *Dryopteris cristata* (Crested Buckler Fern) USA Yarnell.
FERN, Shield, Hard; FERN, Shield, Prickly, Hard *Polystichum aculeatum* The first is the usual name; the second is in. J A MacCulloch.1905.
FERN, Shield, Japanese *Dryopteris erythrosora.*
FERN, Shield, Soft; FERN, Shield, Prickly, Soft *Polystichum lonchitis* The first is the usual name; the second is in. J A

MacCulloch.1905.
FERN, Silver *Potentilla anserina* (Silverweed) Som. Macmillan; Wilts. Dartnell & Goddard.
FERN, Snake 1. *Blechnum spicant* (Hard Fern) Hants. Cope. 2. *Osmunda regalis* (Royal Fern) Hants. Cope. 3. *Pteridium aquilina* (Bracken) Wilts. Dartnell & Goddard. Descriptive of the way the fronds rear up to unfold.
FERN, Squirrel's-foot *Davallia bullata*.
FERN, Sweet 1. *Myrica asplenifolia*. 2. *Myrrhis odorata* (Sweet Cicely) Coles. It is the fern-like leaves plus the aromatic quality that gives the name in both cases.
FERN, Sword *Polystichum munitum*.
FERN, Table *Pteris cretica*.
FERN, Tree, Black *Cyathea medullaris*.
FERN, Tree, Silver *Cyathea dealbata*.
FERN, Tree, West Indian *Cyathea arborea*.
FERN, Wall *Polypodium vulgare* (Common Polypody) B & H.
FERN, Water *Azolla filiculoides*.
FERN, Wood *Polypodium vulgare* (Common Polypody) Norf. B & H.
FERN, Wood, Crested *Dryopteris cristata* (Crested Buckler Fern) USA Yarnell.
FERN, Wood, Giant *Dryopteris goldiana*.
FERN BUTTERCUP *Potentilla anserina* (Silverweed) Wilts. Dartnell & Goddard.
FERN-LEAF BEECH *Fagus sylvatica var. heterophylla* (Cut-leaf Beech) Brimble.1948.
FERN-LEAF TREE *Filicium decipiens*.
FERN-LEAVED DAISY *Tripleurospermum maritimum* (Scentless Maydweed) Som. Macmillan.
FERN-LEAVED PARSLEY *Petroselinum crispum var. filicinum*.
FERN MINT *Monarda fistulosa* (Purple Bergamot) Howes.
FERN NUT *Conopodium majus* (Earth-nut) Corn. Grigson. In this case, fern is the same word as earth. Earth-nut goes through various modifications, the nearest to the present name being Gernut, or Yernut.
FERN PINE *Podocarpus gracilior* (Podo)

USA Cunningham.
FERN TREE *Jacaranda filicifolia*.
FERNLEAF YARROW *Achillea filipendula* (Yellow Yarrow) Painter.
FESCUE, Arizona *Festuca arizonica*.
FESCUE, Blue *Festuca glauca*.
FESCUE, Giant *Festuca gigantea*. From Latin festuca, which means a straw.
FESCUE, Meadow *Festuca pratensis*.
FESCUE, Red *Festuca rubra*.
FESCUE, Rush-leaved *Festuca juncifolia*.
FESCUE, Sheep's *Festuca ovina*.
FESCUE, Tall *Festuca arundinacea*.
FESCUE, Viviparous *Festuca vivipara*.
FESCUE, Wood *Festuca altissima*.
FESTIKE NUT *Pistacia vera* (Pistachio) Turner.
FETCH *Vicia sativa* (Common Vetch) Shrop. Grigson. Fetch is a common variation of vetch, often appearing in the plural, as Fetches.
FETCH, Medick *Onobrychis sativa* (Sainfoin) Culpepper.
FETCH, Wild *Vicia cracca* (Tufted Vetch) Scot. Grigson.
FETCHLING, Red *Onobrychis sativa* (Sainfoin) Gerard.
FETTER, Robin Hood's 1. *Clematis integrifolia* (Hungarian Climber) Cumb. Wright. 2. *Clematis vitalba* (Old Man's Beard) Cumb. B & H. Fetter is feather here.
FETTERBUSH *Pieris floribunda*.
FETTERFOE *Tanacetum parthenium* (Feverfew) Halliwell. An extreme example in the series which starts with feverfew, and proceeds through Fedderfew and Featherfoe.
FEVER-BARK *Alstonia constricta*. This is the Australian tree whose bark is used in the same way as quinine - this is often called the Australian Quinine.
FEVER BUSH 1. *Garrya fremontii* (Bear Brush) Grieve.1931. 2. *Ilex verticillata* (Winterberry) Le Strange. Winterberry bark is a quinine substitute, and the Garryas are similarly used.
FEVER BUSH, Californian *Garrya fremontii* (Bear Brush) Grieve.1931. Another mem-

ber of the genus, *G. flavescens*, is known as Quinine-bush.

FEVER LEAF *Ocymum viride* (Fever Plant) Dalziel.

FEVER NUT *Caesalpina bonduc* (Brazilian Redwood) Watt. Both in Africa and in India, the seed is used as a febrifuge, particularly in malaria.

FEVER PLANT 1. *Achillea millefolium* (Yarrow) Yks. PLNN 17. 2. *Ocymum viride*. *Ocymum viride* is a West African plant, whose leaf decoction is drunk as a tea for fever over all the area. But yarrow is thought of as causing fever rather than curing it.

FEVER ROOT *Cicuta maculata* (American Cowbane) USA Kingsbury.1967. A dangerous name for the misguided to use. The plant is extremely poisonous.

FEVER TREE 1. *Acacia xanthophloea*. 2. *Cinchona sp* (Duran-Reynals. 3. *Eucalyptus globosus* (Blue Gum) Palaiseul. 4. *Lippia javanica* Watt. Blue Gum is a tree which absorbs great quantities of water. It has been used in Europe to dry up and purify regions that were a breeding ground for fevers (for instance, in an area about 3 miles from Rome, where the deserted monastic buildings were malaria-infested). Hence the name Fever Tree. The leaf decoction is used by herbalists for fevers, too. *Lippia javanica* is a native remedy for blackwater fever and malaria in Zimbabwe, and the Acacia gets the name because it grows in the previously malaria-infested areas of southern Africa. Of course, Cinchona is the Fever Tree par excellence.

FEVER-TWIG *Celastrus scandens* (American Bittersweet) W.Miller. Presumably this is used as a fever remedy, but I have no record.

FEVER WEED *Verbena stricta* (Hoary Vervain) Ill. Bergen.1899. It is used as an Illinois folk remedy for fever and ague.

FEVERBERRY *Ribes uva-crispa* (Gooseberry) Grieve.1931. This may sound reasonable, and one may be tempted to accept this exactly for what it says, but the real derivation is from Fea, and Feaberry, and a number of other variants. The word would seem to mean fruit from a prickly bush.

FEVERFEW 1. *Centaureum erythraea* (Centaury) Ches. B & H. 2. *Tanacetum parthenium*. Feverfew is OE feberfugen, Latin febris, fever, and fugare, to drive away. This is a famous medicinal plant - according to one folk belief from Derbyshire, all you had to do to cure a fever was to put a piece in the bed. Centaury, too, is still used by herbalists for fever, as well as loss of appetite, etc.

FEVERFEW, Camomile *Matricaria recutita* (Scented Mayweed) C P Johnson.

FEVERFEW, Corn 1. *Matricaria recutita* (Scented Mayweed) Curtis. 2. *Tripleurospermum maritimum* (Scentless Mayweed) Flower.1859.

FEVERFEW, Scentless *Tripleurospermum maritimum* (Scentless Mayweed) Flower.1859.

FEVERFOULLIE *Tanacetum parthenium* (Feverfew) Scot. Jamieson. One of many corruptions of feverfew.

FEVERFUGE *Centaureum erythraea* (Centaury) Cockayne. Feverfew is a recorded name for centaury; this is an earlier version of the name, closer to the Latin original.

FEVERTORY *Fumaria officinalis* (Fumitory) Wilts. Rogers. There is no 'fever' connection here - this is one of a number of versions of fumitory, including Faminterry and Hemitory.

FEVERWORT 1. *Centaureum erythraea* (Centaury) Vesey-Fitzgerald. 2. *Eupatorium perfoliatum* (Thoroughwort) USA Henkel. Centaury is a well-used fever plant. So in America is Thoroughwort.

FEWS *Sempervivum tectorum* (Houseleek) Scot. Jamieson. Fouets was the French name, and Fews is a version of this. It only occurs in northern England and, more particularly, in Scotland. Versions of the word include Foose, or Fooz, Fuets, and Fullen.

FIBRE-TREE *Securidaca longepedunculata* (Wild Violet Tree). A good fibre can be got from the bark, often used in East Africa to make fishing nets.

FIDDLE 1. *Caltha palustris* (Marsh Marigold) Banff Grigson. 2. *Daucus carota* (Wild Carrot) Lincs. Grigson. 3. *Scrophularia aquatica* (Water Betony) Dev. Fernie; Som. Macmillan. 4. *Scrophularia nodosa* (Figwort) Som. Macmillan; Yks. Grigson. Figwort and Water Betony get the name because children manage to get a squeaking noise out of the leaves, by stripping them and scraping one across another, like a fiddle.

FIDDLE, Monkey *Euphorbia tirucalli* (Pencil Tree) USA Kingsbury.

FIDDLE, Snake's *Iris foetidissima* (Foetid Iris) IOW Grigson.

FIDDLE-BACK FIG; FIDDLE-LEAF FIG *Ficus lyrata*. Cf. Banjo Fig, another name for this. The reference is to the fiddle-shaped leaves.

FIDDLE-CASES *Rhinanthus crista-galli* (Yellow Rattle) IOW Grigson.

FIDDLE DOCK *Rumex pulcher*. Because the basal leaves have an outline resembling that of a violin.

FIDDLE-GRASS *Epilobium hirsutum* (Great Willowherb) Yks. Grigson.

FIDDLE-LEAF FIG *Ficus lyrata* (Fiddle-back Fig) RHS.

FIDDLEHEADS *Osmunda cinnamonea* the immature fronds (Cinnamon Fern) USA Yarnell. Descriptive, of course.

FIDDLENECK *Amsinckia intermedia* (Tarweed) USA Kingsbury.

FIDDLENECK, Checker *Amsinckia tessellata*.

FIDDLENECK, Shiny-seeded *Amsinckia vernicosa*.

FIDDLER'S TRUMPET *Sarracenia drummondii*.

FIDDLESTICKS 1. *Scrophularia aquatica* (Water Betony) Dev. Grigson; Wilts. Goddard. 2. *Scrophularia nodosa* (Figwort) Dev. Friend. See Fiddle.

FIDDLESTRINGS 1. *Plantago major* (Great Plantain) Wilts. Dartnell & Goddard. 2. *Scrophularia aquatica* (Water Betony) Som. Macmillan. The ribs of plantain leaves, when pulled out, are the fiddle-strings. For Water Betony, see Fiddle.

FIDDLEWOOD *Scrophularia aquatica* (Water Betony) Dev, Yks. Grigson; Som. Fernie.

FIELDHOVE *Tussilago farfara* (Coltsfoot) Grieve.1931.

FIELDWORT 1. *Centaureum erythraea* (Centaury) Cockayne. 2. *Centaureum pulchellum* (Slender Centaury) Cockayne. 3. *Gentianella amarella* (Autumn Gentian) B & H. Better known as Felwort, a generic name, almost, for gentians. Fel is OE feld, field. Note that the centauries are gentian family.

FIG, Adam's *Musa paradisiaca* (Cooking Banana) Bianchini.

FIG, Banjo *Ficus lyrata* (Fiddle-back Fig) Rochford.

FIG, Bearded *Ficus citrifolia*.

FIG, Bengal; FIG, Indian *Ficus benghalensis* (Banyan) Rochford.

FIG, Broad *Ficus carica* (Common Fig) Dev. Friend.1882.

FIG, Cape *Ficus capensis*.

FIG, Chilean *Mesembryanthemum chilense*.

FIG, Chinese *Ficus carica* (Common Fig) F.P.Smith.

FIG, Chinese, Dwarf *Ficus pumila*.

FIG, Cluster *Ficus racemosa*.

FIG, Common *Ficus carica*. Fig is from French figue, and thence from Latin ficus.

FIG, Creeping *Ficus pumila* (Dwarf Chinese Fig) Nicolaisen.

FIG, Devil's *Argemone mexicana* (Mexican Poppy) Perry. Presumably, 'devil's', because the seeds are narcotic.

FIG, Dough *Ficus carica* (Common Fig) Dev. Friend.1882; Som. Macmillan. Probably to distinguish it ("soft as dough") from ordinary raisins, which are always called figs in Somerset.

FIG, East African *Ficus glumosa*.

FIG, Fiddle-back; FIG, Fiddle-leaf *Ficus lyrata*. Cf. Banjo Fig for this.

FIG, Gooseberry *Mesembryanthemum acinaciforme*.

FIG, Hottentot *Mesembryanthemum edule*. The seed vessels are known as figs; they are edible, and quite tasty. Hottentots did actually use them - at least, they fried the leaves in butter, if they did not actually use the figs.

FIG, Indian; FIG, Bengal *Ficus benghalensis*.

FIG, Java *Ficus benjamina* (Weeping Fig) Howes.

FIG, Kaffir *Mesembryanthemum edule* (Hottentot Fig) McClintock.

FIG, Mistletoe 1. *Ficus deltoides*. 2. *Ficus diversifolia*.

FIG, Moreton Bay *Ficus macrophylla*. Moreton Bay, by Brisbane - this is an Australian fig.

FIG, Natal *Ficus natalensis*.

FIG, Pharaoh's *Ficus sycomorus* (Sycamore Fig) Hora.

FIG, Sea 1. *Mesembryanthemum chilense* (Chilean Fig) USA Cunningham. 2. *Mesembryanthemum crystallinum*.

FIG, Silk *Musa x paradisiaca* (Banana) H & P.

FIG, Strangling *Ficus aurea*.

FIG, Sycamore *Ficus sycomorus*. Sycamore means fig-mulberry, hence the other name for this - Mulberry Fig. The leaves and wood closely resemble those of mulberry.

FIG, Tufted *Ficus benjamina var. comosa*.

FIG, Turkey *Ficus carica* (Common Fig) Dev. Friend.1882.

FIG, Weeping *Ficus benjamina*.

FIG, Wild 1. *Ficus capensis* (Cape Fig) Palgrave. 2. *Ficus platypoda*.

FIG BEAN *Lupinus sp* (Lupin) Coles.

FIG-LEAF GOURD *Cucurbita ficifolia* (Malabar Gourd) Whitaker & Davis.

FIG-LEAVED GOOSEFOOT *Chenopodium ficifolium*.

FIG MARIGOLD 1. *Mesembryanthemum edule* (Hottentot Fig) A.W.Smith. 2. *Mesembryanthemum equilaterale*. i.e. Marigold-like flowers with fig-like seed vessels.

FIG MARIGOLD, Green *Aridaria viridiflora*.

FIG MARIGOLD, Jagged-leaved *Aethephyllum pinnatifidum*.

FIG-MULBERRY; FIG, Mulberry *Ficus sycomorus*. Fig-Mulberry is in Moldenke, and Mulberry Fig in Turner. The reason for the name lies in the '-morus' part of the specific name. Anyway, sycamore means fig-mulberry.

FIG NUT *Jatropha curcas* (Barbados Nut) W Africa Dalziel.

FIG-NUT FLOWER, Red *Jatropha gossypifolia* (Bellyache Bush) Dalziel.

FIG-TREE, Wild *Acer pseudo-platanus* (Sycamore) Evelyn.

FIGHTEE COCKS; FIGHTING COCKS; FIGHTERS, Cock *Plantago lanceolata* (Ribwort Plantain) Suss, E.Ang, N'thants. Denham/Hardy (Fightee Cocks); Dev, Som, N'thants, Shrop, Ches, Yks. Denham/Hardy; Wilts. Dartnell & Goddard; E Ang. Forby (Fighting Cocks); N'thum. Denham/Hardy (Cock Fighters). These are references to a children's game - the flower heads are used as "soldiers" or "fighting cocks". One child holds a flower head out, and the other tries to decapitate it with his own. Scores are much like those of conkers.

FIGWORT 1. *Ranunculus ficaria* (Lesser Celandine) Turner. 2. *Scrophularia aquatica* (Water Betony) B & H. 3. *Scrophularia nodosa*. 'Fig' in this context means piles, and these would have been used to treat the complaint, relying heavily on the doctrine of signatures for any effect. *Nodosa* in the specific name of Figwort explains the reason - the tubers are knobbly. In a similar way, celandine roots have little excrescences growing on them, like piles.

FIGWORT, Balm-leaved *Scrophularia scorodonia*.

FIGWORT, Cape 1. *Lachenalia aloides* (Cape Cowslip) Howes. 2. *Phygelius capensis*.

FIGWORT, French *Scrophularia canina*.

FIGWORT, Great *Scrophularia nodosa* (Figwort) Gerard.

FIGWORT, Hare *Scrophularia lanceolata*.

FIGWORT, Knotted *Scrophularia nodosa* (Figwort) Grieve.1931. The specific name is *nodosa*, knotted.
FIGWORT, Nettle-leaved *Scrophularia peregrina*.
FIGWORT, Water; FIGWORT, Marsh *Scrophularia aquatica* (Water Betony) B & H (Water); H C Long.1910 (Marsh).
FIGWORT, Western *Scrophularia umbrosa*. It grows in damp, shady places in the western side of this country.
FIGWORT, Yellow *Scrophularia vernalis*.
FIJI ARROWROOT *Tacca pinnatifida* Douglas.
FILAERA 1. *Eupatorium cannabinum* (Hemp Agrimony) Berw Grigson. 2. *Valeriana officinalis* (Valerian) N Eng, N Ire. Grigson. A corruption of Valerian - it becomes more obvious when others are considered, like Villera, and Valara.
FILAGO, Common; FILAGO, German *Filago germanica* (Cudweed) both Flower.1859.
FILAREE, Red-stemmed *Erodium cicutarium* (Storksbill) USA Schenk & Gifford. Filaree comes from Spanish alfiler, pin. Note the American Pin Grass, or Pin Clover for this. There are 'needle' names, as well, for storksbill, all deriving from the sharp-ended seed vessels. How is Filaree pronounced? Filairy would sound more reasonable than as written.
FILAREE, White-stemmed *Erodium moschatum* (Musk Storksbill) USA Grigson.
FILBEARD; FILBERD; FILBORD *Corylus avellana* (Hazel) Shrop, War, Worcs, Glos, Heref, IOW, Ches, Leics, N'thants, Oxf. Grigson (Filbeard); Gerard (Filberd); Evelyn (Filbord). Filbeard, as the sequence shows, is a form of Filbert, i.e. St Philibert, whose feast day is 22 August, when, it is claimed, the nuts are ripe. Actually, the nuts would be soft and taste-less at this time in August, and certainly not ripe. Anyway, it is a Norman-French word, appearing as philbert in the 13th century, and still in use in Normandy patois at the beginning of this century. Another explanation is that filbert means "full beard", and is a reference to the fringed husk. The name appears as Beard-tree, also.
FILBERT 1. *Corylus avellana* (Hazel). 2. *Corylus maxima*.
FILBERT, Indian *Areca catechu* (Areca Palm) Pomet.
FILEWORT *Filago minima* (Small Cudweed) Halliwell.
FILLYFINDILLAN *Filipendula vulgaris* (Dropwort) Ire. B & H.
FIMBLE; FEMBLE *Cannabis sativa* (Hemp) Ess, Suss. (Fimble) both B & H. Applied only to the male plant.
FIN; FINWEED *Ononis repens* (Rest Harrow) both N'thants. A.E.Baker (Fin); Sternberg (Finweed).
FINCKLE *Foeniculum vulgare* (Fennel) Prior. Fennel is OE fingel, hence forms like this, or, more wide-spread, without the 'c'.
FINELEAF *Viola odorata* (Sweet Violet) Lincs. Halliwell.
FINGER-CAP; FINGER-HUT *Digitalis purpurea* (Foxglove) Som. Macmillan. "Finger-hut" looks strange, but "hut" is probably the same as "hud", a local word apparently meaning a finger-stall.
FINGER EUPHORBIA *Euphorbia tirucalli* (Pencil Tree) Dale & Greenway. Given for the same reason as Pencil Tree as a descriptive title.
FINGER FERN *Ceterach officinarum* (Rusty-back) Turner.
FINGER-FLOWER; FINGER-GLOVES *Digitalis purpurea* (Foxglove) Parkinson.1629; USA Henkel (Finger-flower); Tynan & Maitland (Finger-gloves).
FINGER-GRASS, Hairy *Digitaria sanguinalis*.
FINGER-GRASS, Smooth *Digitaria ischaemum*.
FINGER-LEAVED GOURD *Cucurbita digitata*.
FINGER MILLET *Eleusine coracana*.
FINGER ORCHIS *Dactylorchis maculata* (Heath Spotted Orchid) Gerard. The 'dactyl'

part of the generic name, and the various 'hand' and 'finger' names given to the plant refer to the fact that the tubers are divided into several finger-like lobes.

FINGER POPPY MALLOW *Callirhoe digitata.*

FINGER-ROOT; FINGER-TIPS *Digitalis purpurea* (Foxglove) Suss, War. Grigson (Finger-root); Wilts. Macmillan. (Finger-tips).

FINGERBERRY *Rubus villosus* (American Blackberry) Grieve.1931.

FINGERED SAXIFRAGE *Saxifraga tridactylites* (Rue-leaved Saxifrage) McClintock. Cf. the Somerset name Three-fingered Jack for this.

FINGERED SPEEDWELL *Veronica triphyllos.*

FINGERS *Digitalis purpurea* (Foxglove) Som. Elworthy.

FINGERS, Bloody 1. *Arum maculatum* (Cuckoo-pint) Fernie. 2. *Digitalis purpurea* (Foxglove) Som. Elworthy; Heref. Havergal; Yks, Cumb. Grigson; Scot. Aitken. 3. *Orchis mascula* (Early Purple Orchis) Glos. J D Robertson. For Cuckoo-pint, it is the blotch on the leaves that suggests fingermarks, and the same applies to the orchid.

FINGERS, Bloody Man's 1. *Arum maculatum* (Cuckoo-pint) Friend. 2. *Digitalis purpurea* (Foxglove) Heref. B & H. 3. *Orchis mascula* (Early Purple Orchis) Glos. J.D. Robertson. 4. *Orchis morio* (Green-winged Orchid) Ches. Holland.

FINGERS, Blue *Vinca major* (Greatyer Periwinkle) Le Strange.

FINGERS, Buttered *Anthyllis vulneraria* (Kidney Vetch) Som. Macmillan. Descriptive.

FINGERS, Cling *Orchis mascula* (Early Purple Orchis) Oxf. B & H.

FINGERS, Dead Man's 1. *Arum maculatum* (Cuckoo-pint) Worcs. B & H. 2. *Dactylorchis incarnata* (Early Marsh Orchid) Border B & H. 3. *Dactylorchis maculata* (Heath Spotted Orchid) Border. B & H. 4. *Digitalis purpurea* (Foxglove) Som. Grigson. 5. *Gentianella amarella* (Autumn Gentian) Shet. Grigson. 6. *Lotus corniculatus* (Bird's-foot Trefoil) Hants. Grigson. 7. *Oenanthe crocata* (Hemlock Water Dropwort) North. 8. *Orchis mascula* (Early Purple Orchis) Shakespeare. 9. *Orchis morio* (Green-winged Orchis) Suss. B & H. For most of these, it is the method of growth that suggests the imagery. Foxglove, from the shape of its flowers, is completely associated with fingers and gloves. Not only that, but there was once a belief in Scotland that where it grows someone has been buried, the colour of the flowers probably instrumental in fostering the belief. The odd one is Hemlock Water Dropwort, and this is probably an error for Dead Man's Tongue, "from its paralysing effect on the organs of voice" (Prior). For Cuckoo-pint and the orchids, the fingermark blotches on the leaves suggest the name. Cf. Bloody Fingers, and Bloody Man's Fingers.

FINGERS, Devil's 1. *Lotus corniculatus* (Bird's-foot Trefoil) Dev. Friend. 2. *Populus nigra* male catkins (Black Poplar) Grigson. 3. *Ranunculus arvensis* (Corn Buttercup) Jacob.

FINGERS, Dog's *Digitalis purpurea* (Foxglove) Som, Wales Grigson; USA Henkel.

FINGERS, Fairy *Digitalis purpurea* (Foxglove) Som. Macmillan; Hants. Boase.

FINGERS, Five 1. *Lotus corniculatus* (Bird's-foot Trefoil) Cambs, Ess, Norf Grigson. 2. *Panax quinquefolium* (Ginseng) Leyel.1948. 3. *Potentilla erecta* (Tormentil) Suff. Grigson. 4. *Potentilla reptans* (Cinquefoil) Som. Macmillan; Suss, Ess. Grigson. 5. *Primula veris x vulgaris* (False Oxlip) E.Ang. Halliwell. 6. *Primula vulgaris var. elatior* (Oxlip) Suff. B & H. The division of the leaf of the cinquefoils and ginseng accounts for the name. Do the oxlips have five blossoms per head?

FINGERS, Fox *Digitalis purpurea* (Foxglove) Yks. Grigson.

FINGERS, Gentleman's *Arum maculatum*

(Cuckoo-pint) Wilts. Dartnell & Goddard. Cf. Lady's Fingers, Gentlemen's and Ladies' Fingers, etc.

FINGERS, Gentlemen's and Ladies' *Arum maculatum* (Cuckoo-pint) Wilts. Dartnell & Goddard. Cf. Lords' and Ladies' Fingers, etc.

FINGERS, King 1. *Orchis mascula* (Early Purple Orchis) Bucks, Leics, War. B & H. 2. *Orchis morio* (Green-winged Orchis) Bucks, War. B & H; N'thants. A.E.Baker. 3. *Oxalis acetosella* (Wood Sorrel) Bucks. Grigson.

FINGERS, King's 1. *Dactylorchis incarnata* (Early Marsh Orchid) Tynan & Maitland. 2. *Dactylorchis maculata* (Heath Spotted Orchid) Tynan & Maitland. 3. *Lotus corniculatus* (Bird's-foot Trefoil) Bucks. Grigson.

FINGERS, Lady's 1. *Abelmoschus esculentus* fruit (Okra) Argenti & Rose. 2. *Anthyllis vulneraria* (Kidney Vetch) Dor, Midl. Dyer. 3. *Arum maculatum* (Cuckoo-pint) Wilts. Dartnell & Goddard. Glos. B & H; Kent Grigson. 4. *Digitalis purpurea* (Foxglove) Dev, Worcs, Shrop, N'thum, Ire. Grigson; Som. Elworthy. 5. *Hippocrepis comosa* (Horseshoe Vetch) Wilts. Dartnell. & Goddard. 6. *Laburnum anagyroides* (Laburnum) Yks. B & H. 7. *Lathyrus pratensis* (Yellow Pea) Wilts. Dartnell & Goddard; Yks. Grigson. 8. *Lonicera periclymenum* (Honeysuckle) Yks, Dur, N'thum, Roxb. Grigson. 9. *Lotus corniculatus* (Bird's-foot Trefoil) Wilts. Dartnell; Hants. Cope; Suss, Herts, Hunts, Lincs, Lancs, Yks. Grigson; N'thants. A.E.Baker. 10. *Orchis mascula* (Early Purple Orchis) Berks. B & H. Som. Elworthy. 11. *Potentilla reptans* (Cinquefoil) Som. Macmillan. 12. *Primula veris* (Cowslip) Scot. Grigson. 13. *Primula veris x vulgaris* (False Oxlip) Norf, Suff, Yks. Grigson. 14. *Ranunculus arvensis* (Corn Buttercup) Yks. Grigson. 15. *Senecio vulgaris* (Groundsel) Wilts. Macmillan. Most of these are descriptive in some way or another - the long pods of those members of the pea family, or the long flowers of honeysuckle or the drooping flowers of cowslip, can all call to mind fingers. But groundsel?

FINGERS, Lords' and Ladies' *Arum maculatum* (Cuckoo-pint) War. B & H. Cf. Gentlemen's and Ladies' Fingers, etc.

FINGERS, Poison *Arum maculatum* (Cuckoo-pint) Dor. Macmillan. The berries really are poisonous.

FINGERS, Purple *Digitalis purpurea* (Foxglove) B & H.

FINGERS, Queen's 1. *Dactylorchis maculata* (Heath Spotted Orchid) War. Grigson. 2. *Orchis morio* (Green-winged Orchid) War. Grigson. All members of this genus have tubers with finger-like lobes, hence this name and many others, like King's Fingers and Ring-fingers.

FINGERS, Red *Trifolium incarnatum* (Crimson Clover) Som. Macmillan.

FINGERS, Ring 1. *Dactylorchis maculata* (Heath Spotted Orchid) Bucks. Grigson. 2. *Orchis mascula* (Early Purple Orchis) Bucks. B & H. Cf. Queen's Fingers.

FINGERS, Tom Thumb's Thousand *Rumex acetosa* (Wild Sorrel) Kent B & H.

FINGERS, Thumbs-and- *Ulex europaeus* (Furze) Som. Macmillan.

FINGERS, Virgin's *Digitalis purpurea* (Foxglove) Som. Macmillan.

FINGERS-AND-THUMBS 1. *Anthyllis vulneraria* (Kidney Vetch) Som, Dor. Macmillan. 2. *Corydalis lutea* (Yellow Fumitory) Dor. Macmillan. 3. *Cypripedium calceolus* (Lady's Slipper) Dev. Friend.1882. 4. *Digitalis purpurea* (Foxglove) Som. Macmillan. 5. *Hippocrepis comosa* (Horseshoe Vetch) Som. Macmillan. 6. *Lathyrus pratensis* (Yellow Pea) Som. Macmillan. 7. *Linaria vulgaris* (Toadflax) Som. Macmillan. 8. *Lotus corniculatus* (Bird's-foot Trefoil) Dev, Som. Macmillan; Hants. Cope. 9. *Medicago lupulina* (Black Medick) Dor. Dacombe. 10. *Melandrium rubrum* (Red Campion) Som. Macmillan. 11. *Ulex europaeus* (Furze) Wilts. Dartnell & Goddard. 12. *Vicia cracca* (Tufted Vetch) Som. Macmillan. 13. *Vicia sepium* (Bush Vetch) Som. Macmillan.

FINGERS-AND-THUMBS, Double *Anthyllis vulneraria* (Kidney Vetch) Som. Grigson.

FINGERS-AND-THUMBS, God's 1. *Fumaria officinalis* (Fumitory) Dor. Grigson. 2. *Lotus corniculatus* (Bird's-foot Trefoil) Som. Macmillan.

FINGERS-AND-THUMBS, God Almighty's *Anthyllis vulneraria* (Kidney Vetch) Dor. Grigson.

FINGERS-AND-THUMBS, Lady's *Lotus corniculatus* (Bird's-foot Trefoil) Som. Macmillan; Wilts. Dartnell & Goddard.

FINGERS-AND-THUMBS, Lady's Double 1. *Anthyllis vulneraria* (Kidney Vetch) Wilts. Dartnell. & Goddard. 2. *Lotus corniculatus* (Bird's-foot Trefoil) Wilts. Dartnell & Goddard.

FINGERS-AND-THUMBS, Tom Thumb's *Lotus corniculatus* (Bird's-foot Trefoil) Som. Macmillan.

FINGERS-AND-THUMBS, Yellow *Anthyllis vulneraria* Dor, Som. Grigson.

FINGERS-AND-TOES 1. *Anthyllis vulneraria* (Kidney Vetch) Lincs. Tynan & Maitland. 2. *Lotus corniculatus* (Bird's-foot Trefoil) Dev. Macmillan; Ess, Cambs, Norf. Grigson.

FINITERRY *Fumaria officinalis* (Fumitory) Pollock. An unusual variation on Fumitory.

FINKLE *Foeniculum vulgare* (Fennel) Kent, Lincs, N.Eng. Grigson; Yks. Addy. Closer to the OE original, fingel, than modern fennel is.

FINKLE, Dog *Anthemis cotula* (Maydweed) Yks. Grigson. Dog's Fennel, or Hound's Fennel, are older names. All the 'dog' names for this plant come about because of the bad smell.

FINKLE, Fennel *Foeniculum vulgare* (Fennel) Dyer.1887. Pleonastic, of course - both elements are the same word.

FINNIGIG *Trigonella ornithopodioides* (Fenugreek) Suff. G E Evans.1960. Fenugreek seed has always been used by vets as a medicine for horses, and Evans suggests that this is a deliberate corruption of the common name on the part of Suffolk horsemen, so that the true identity of the ingredient they were buying would be unrecognisable to third parties.

FINNISH WHITEBEAM *Sorbus x hybrida*.

FINOCCHIO *Foeniculum dulce*. Finocchio is Italian for fennel, but it seems to be applied as the usual name for this species here.

FINTOCK *Rubus chamaemorus* (Cloudberry) Perth. Grigson.

FINWEED; FIN *Ononis repens* (Rest Harrow) N'thants. Sternberg.

FINZACH *Polygonum aviculare* (Knotgrass) Banff. Grigson.

FIORIN *Agrostis stolonifera* (White Bent) H C Long.1910. This is Irish fiorthán.

FIR, Algerian *Abies numidica*. Fir is OE fyrh. Cf. German Föhre.

FIR, Alpine *Abies lasiocarpa*.

FIR, Archangel *Pinus sylvestris* (Scots Pine) Wilkinson.1981. One of the names under which the timber was imported.

FIR, Balm of Gilead *Abies balsamea* (Balsam Fir) Grieve.1931.

FIR, Balsam *Abies balsamea*. Blisters on the bark are the source of the resin known as Canada Balsam, used as a stimulant and diuretic, but better known because it has the same refractive index as glass, making it useful for cementing glass in optical instruments.

FIR, Balsam, Fraser's *Abies fraseri*.

FIR, Balsam, Southern *Abies fraseri* (Fraser's Balsam Fir) Brimblc.1948.

FIR, Beautiful *Abies amabilis* (Red Silver Fir) Leathart.

FIR, Cascade *Abies amabilis* (Red Silver Fir) Everett.

FIR, Caucasian *Abies nordmanniana*. It grows in the Caucasus and Asia Minor.

FIR, China *Cunninghamia lanceolata*.

FIR, Cilician *Abies cilicica*. A native of Asia Minor and northern Syria.

FIR, Colorado *Abies concolor* (White Fir) Everett. Called Colorado Fir, but it can be found throughout the western part of North America, from Mexico to Oregon.

FIR, Corkbark *Abies lasiocarpa var. arizoni-*

ca.

FIR, Danzig *Pinus sylvestris* (Scots Pine) Hutchinson & Melville. Cf. Baltic Yellow Deal, etc - almost a trade name for timber imported from that area.

FIR, Douglas *Pseudotsuga menziesii*. The specific name celebrates the Scottish botanist Archibald Menzies, who first discovered the tree in 1791. But it was David Douglas who sent seeds to England, in 1827.

FIR, Douglas, Blue *Pseudotsuga menziesii var. glauca* Hart. Better known as **Colorado Douglas Fir**.

FIR, Douglas, Green *Pseudotsuga menziesii* (Douglas Fir) Mitchell.1972.

FIR, Douglas, Japanese *Pseudotsuga japonica*.

FIR, Douglas, Large-coned *Pseudotsuga macrocarpa*.

FIR, Farges *Abies fargesii*.

FIR, Flaky *Abies squamata*.

FIR, Giant; FIR, Grand *Abies grandis* (Lowland Fir) Clapham (Giant); Wilks (Grand). 'Giant' is reasonable - on Vancouver Island, these can grow as tall as 300 ft, and elsewhere up to 170 ft is normal.

FIR, Greek *Abies cephalonica*.

FIR, Hedgehog *Abies pinsapo* (Spanish Fir) Mitchell.1972.

FIR, Hemlock *Tsuga heterophylla* (Western Hemlock) Brimble.1948.

FIR, Himalayan, East *Abies spectabilis*.

FIR, Himalayan, West *Abies pindrow*.

FIR, Joint, California *Ephedra californica*.

FIR, Joint, Death Valley *Ephedra funerea*.

FIR, Joint, Long-leaved *Ephedra trifurca*.

FIR, Joint, Mountain *Ephedra viridis*.

FIR, Joint, Nevada *Ephedra nevadensis*.

FIR, King Boris's *Abies borisii-regis*. A Balkans species, or perhaps hybrid, hence this name.

FIR, Korean *Abies koreana*.

FIR, Lowland *Abies grandis*.

FIR, Manchurian *Abies holophylla*.

FIR, Marie's *Abies mariesii*.

FIR, Min *Abies recurvata*. The reference is to the Min River, China, where it grows.

FIR, Momi *Abies firma*.

FIR, Nikko *Abies homolepis*.

FIR, Noble *Abies procera*.

FIR, Norway *Pinus sylvestris* (Scots Pine) Wilkinson.1981. The timber has been imported from Scandinavia and the Baltic under a number of different names, including Danzig, Riga, and Archangel Fir.

FIR, Oregon 1. *Abies grandis* (Lowland Fir) USA Brimble.1948. 2. *Pseudotsuga menziesii* (Douglas Fir) Wilkinson.1981.

FIR, Prince Albert's *Tsuga heterophylla* (Western Hemlock) Brimble.1948.

FIR, Red 1. *Abies alba* (Silver Fir) Turner. 2. *Abies procera* (Noble Fir) Hay & Synge. 3. *Pseudotsuga menziesii* (Douglas Fir) USA Elmore.

FIR, Red, California *Abies magnifica*.

FIR, Riga *Pinus sylvestris* (Scots Pine) Wilkinson.1981. Cf. Danzig Fir, and a number of other names under which the timber was imported.

FIR, Rocky Mountain *Abies lasiocarpa* (Alpine Fir) Usher.

FIR, Sacred *Abies religiosa*. Religiosa, or 'Sacred', because the branches are used for church decoration at religious festivals.

FIR, Sakhalin *Abies sacchalinensis*.

FIR, Santa Lucia *Abies venusta*. Santa Lucia Mountain, in California, where they grow.

FIR, Scotch; FIR, Scots *Pinus sylvestris* (Scots Pine). Hutchinson has Scots Fir, but it is in America that it is known as Scotch Fir.

FIR, Siberian *Abies sibirica*.

FIR, Silver *Abies alba*.

FIR, Silver, American *Abies balsamea* (Balsam Fir) Grieve.1931.

FIR, Silver, Himalayan *Abies pindrow* (West Himalayan Fir) Usher.

FIR, Silver, Indian *Abies spectabilis* (East Himalayan Fir) Brimble.1948.

FIR, Silver, Japanese *Abies firma* (Momi Fir) Usher.

FIR, Silver, Nikko *Abies homolepis* (Nikko Fir) Usher.

FIR, Silver, Pacific *Abies amabilis* (Red Silver Fir) Schery.

FIR, Silver, Red *Abies amabilis.*

FIR, Spanish *Abies pinsapo.*

FIR, Spruce *Picea abies.* Spruce comes from Pruce, once the English name for Prussia, from where presumably the first examples of Norway Spruce were brought to Britain.

FIR, Spruce, Norway *Picea abies* Thornton.

FIR, White 1. *Abies concolor.* 2. *Cupressus funebris.*

FIR, White, Lowland *Abies grandis* (Lowland Fir) Schery.

FIR, Yellow *Pseudotsuga menziesii* (Douglas Fir) USA Elmore.

FIR-APPLE 1. *Picea abies* Surr, Herts, Lincs, Cumb. (Spruce Fir) B & H. 2. *Pinus sylvestris* (Scots Pine) Hants. Cope; N'thants. A.E.Baker.

FIR-BALL; FIR-BOB *Pinus sylvestris* cones (Scots Pine cones) Shrop. (Fir-ball); Shrop, Ches, Leics. (Fir-bob) B & H.

FIR CLUBMOSS *Urostachys selago* (Mountain Clubmoss) USA Yarnell.

FIR-DEAL TREE; FIR-DALE TREE *Pinus sylvestris* (Scots Pine) N'thants. A.E.Baker (Fir-Deal); N'thants. Clare (Fir-Dale). Deal-tree is probably more common. Fir-Dale Tree is recorded, but could it have been a misprint?

FIR-RAPE *Monotropa hypopitys* (Yellow Bird's Nest). The name was given in the belief that the plant is a parasite upon the roots of Scots Pine and beech. Actually, it has its own roots.

FIR-TOP *Pinus sylvestris* cones (Pine cones) Scot. B & H.

FIRE, Burning *Taraxacum officinale* (Dandelion) Som. Macmillan. Possibly the sun is meant here. Golden Sun is another name for the dandelion, and there is Sonnewirbel in German. Legend has it that the dandelion was born of the dust raised by the chariot of the sun, which is why it opens at dawn and closes at dusk. In Gaelic it is Little Flame of God.

FIRE, Fairy's 1. *Chaenomeles speciosa* (Japanese Quince) Dyer. 2. *Dipsacus fullonum* (Teasel) Som. Macmillan.

FIRE, Indian *Salvia coccinea* (Red Sage) USA Howes.

FIRE, Light *Pteridium aquilina* (Bracken) (Lincs) Tynan & Maitland. Dried bracken was often used as kindling.

FIRE, Maori *Pennantia corymbosa.* The Maoris used it for friction fire making.

FIRE, Strike- *Veronica chamaedrys* (Germander Speedwell) N'thants. Grigson. It is just possible that the name refers to the superstition that picking the flowers will cause a thunderstorm.

FIRE, Wild *Achillea ptarmica* (Sneezewort) Donegal Grigson.

FIRE-BALL LILY *Haemanthus cinnabarinus.*

FIRE-BUSH, Chilean *Embothreum coccineum.* The brilliant crimson-scarlet flowers suggest the name.

FIRE-BUSH, Mexican *Kochia scoparia* (Burning Bush) A.W.Smith.

FIRE-CHERRY *Prunus pennsylvanica* (Pin Cherry) USA Kingsbury. Because this is a red cherry, perhaps?

FIRE CYPRESS *Chamaecyparis obtusa.*

FIRE-GRASS 1. *Aphanes arvensis* (Parsley Piert) B & H. 2. *Plantago lanceolata* (Ribwort Plantain) Som. Grigson. Cf. Fire-leaf for Ribwort.

FIRE-LEAF 1. *Plantago lanceolata* (Ribwort Plantain) Som, Glos. Grigson. 2. *Plantago media* (Lamb's Tongue Plantain) Glos. J.D. Robertson.

FIRE LIGHTS *Viola odorata* (Violet) Worcs. J Salisbury. A rather extremne form of mispronunciation of the common name.

FIRE LILY *Lilium tigrinum* (Tiger Lily) Dev. Macmillan.

FIRE-O'-GOLD *Caltha palustris* (Marsh Marigold) Bucks. Grigson. A reference to the golden-yellow flowers, of course, as is the Scottish name Wildfire.

FIRE-PINK *Silene virginica.*

FIRE-PLANT *Plumbago indica.*

FIRE-PLANT, Mexican *Euphorbia pulcherrima* (Poinsettia) Watt. Cf. Mexican Flame Tree, and Mexican Flame Leap.

FIRE-SCREEN *Tropaeolum speciosum* (Scotch Flame Flower) Dor. Macmillan. Colour descriptive.

FIRE-THORN *Pyracantha sp.* This is a straight translation of *Pyracantha*.

FIRE-THORN, Orange *Pyracantha angustifolia*.

FIRE-THORN, Red *Pyracantha coccinea*.

FIRE-WHEEL TREE *Stenocarpus sinuatus*. It bears scarlet and orange flowers.

FIRE WILLOW *Salix scouleriana* (Nuttall Willow) USA T H Kearney. Because of the rapidity with which it colonises areas burnt out by forest fires.

FIRECRACKER, Utah *Penstemon utahensis*.

FIRECRACKER PLANT *Brodiaea idamaea* USA Schenk & Gifford.

FIREFLOUT *Papaver rhoeas* (Red Poppy) Som, N'thum. Grigson.

FIRESTICKS *Asclepias fruticosa* (Shrubby Milkweed) S.Africa Watt.

FIREWEED 1. *Chamaenerion angustifolium* (Rosebay) Dor. Grigson; USA House. 2. *Datura stramonium* (Thorn-apple) USA Henkel. 3. *Epilobium adenocaulon* (American Willowherb) USA Elmore. 4. *Erigeron canadensis* (Canadian Fleabane) USA Henkel. 5. *Plantago lanceolata* (Ribwort Plantain) Tynan & Maitland. 6. *Plantago media* (Lamb's Tongue Plantain) B & H. Rosebay (and presumably the willowherb too) get the name because its seeds are almost the first to colonize an area cleared of vegetation by fire (like the plant itself, the name is of American origin). Thorn-apple, too, will be quick to germinate after a fire. Fleabane probably got it because of its ability to spread quickly (like wildfire).

FIREWEED, Mexican *Kochia scoparia* (Burning Bush) USA Gates.

FIREWEED, Narrow-leaf *Chamaenerion angustifolium* (Rosebay) USA Elmore.

FIRST OF MAY *Saxifraga granulata* (Meadow Saxifrage) Ches. Holland.

FIRST ROSE *Primula vulgaris* (Primrose) Som. Macmillan. They persist in regarding the second syllable of primrose as 'rose' in its own right. Cf. Easter Rose, Butter Rose, etc., for this.

FISH BELLY 1. *Cirsium eriophorum* (Woolly Thistle) Cumb. B & H. 2. *Cirsium heterophyllum* (Melancholy Thistle) Cumb. Grigson. Possibly from the white underside of the leaves, with Melancholy Thistle a likelier candidate for the name.

FISH BONES 1. *Aesculus hippocastanum* (Horse Chestnut) Som. Macmillan. 2. *Potentilla anserina* (Silverweed) Som. Macmillan. Fish Bones is said to be a name children give to horse chestnut leaves when the green parts have been pulled off. It must be a reference to the leaves of silverweed, too.

FISH KILLER *Anamirta cocculus* India Chopra. The berries are used there as a fish poison, and they are used too to kill crows, hence Crow Killer.

FISH MINT 1. *Mentha aquatica* (Water Mint) Gerard. 2. *Mentha spicata* (Spearmint) Grieve.1933. Gerard had Mackerel-mint for the latter as well.

FISH POISON BEAN *Tephrosia vogelii*.

FISHBERRY 1. *Anamirta cocculus* Chopra. 2. *Anamirta paniculata* (Levant Nut) Le Strange. The berries were thrown into ponds or river to stupefy the fish.

FISHBONE COTONEASTER *Cotoneaster horizontalis* McClintock.1975. Or Herringbone Cotoneaster. Descriptive.

FISHING RODS, Angel's *Sparaxis pulcherrima* (Wand Flower) Chaplin.

FIT PLANT *Monotropa uniflora* O.P.Brown. There must be some medicinal use which would explain this name.

FITCH *Vicia sativa* (Common Vetch) Suss, War, Shrop, N'thants, Yks, Cumb, Ire. Grigson; Ches. Holland. Vetch can appear as fitch, fatch or fetch; fitch is the most widespread of them.

FITCH, Medick *Onobrychis sativa* (Sainfoin) B & H. Cf. Medick Fetch, Medick Fitchling, and Medick Vetchling for this.

FITCH, Tar, Blue *Vicia cracca* (Tufted

Vetch) Ches. Holland. Tar-fitch = tare vetch.

FITCH, Tare 1. *Vicia cracca* (Tufted Vetch) W.Eng, Shrop, Ches. Grigson. 2. *Vicia hirsuta* (Hairy Tare) W.Eng, Shrop, Ches. Grigson.

FITCH, Wild 1. *Vicia hirsuta* (Hairy Tare) Cumb. Grigson. 2. *Vicia sativa* (Common Vetch) Cumb. Grigson.

FITCHACKS 1. *Vicia cracca* (Tufted Vetch) Aber. Grigson. 2. *Vicia sativa* (Common Vetch) Aber, Mor. Grigson.

FITCHLING, Medick; FITCHLING, Red *Onobrychis sativa* (Sainfoin) Gerard (Medick Fitchling); Culpepper (Red Fitchling).

FITSROOT *Astragalus glycophyllus* (Milk Vetch) Howes. As many of the genus are called Loco, which is Spanish for mad, this must refer to a treatment for epilepsy.

FITTONIA, Silver-nerve *Fittonia verschaffeltii*.

FITWEED *Eryngium foetidum* Howes.

FIVE-FACED BISHOP *Adoxa moschatellina* (Moschatel) Polunin. Four of the five flowers it bears face in different directions, with the fifth pointing straight upwards.

FIVE-FINGERS 1. *Lotus corniculatus* (Bird's-foot Trefoil) E Ang. Grigson. 2. *Panax quinquefolium* (Ginseng) Leyel.1948. 3. *Potentilla erecta* (Tormentil) Suff. Grigson. 4. *Potentilla reptans* (Cinquefoil) Som. Macmillan; Suss, Ess. Grigson. 5. *Primula vulgaris var. elatior* (Polyanthus) Suff. B & H. 6. *Primula veris x vulgaris* (False Oxlip) E.Ang. Grigson. The fingers in the case of the Potentillas and Ginseng are the leaves, cinquefoil meaning five leaves, while for the Primulas, the reference must be to the flower heads.

FIVE-FINGER, Marsh *Potentilla palustris* (Marsh Cinquefoil) USA H.Smith.1945.

FIVE-FINGER, Rock *Potentilla saxosa*.

FIVE-FINGER, Slender *Potentilla gracilis* (Slender Goose-grass) USA Elmore.

FIVE-FINGER BLOSSOM 1. *Potentilla erecta* (Tormentil) Suff. Grigson. 2. *Potentilla reptans* (Cinquefoil) Suff. Grigson.

FIVE-FINGER GRASS 1. *Potentilla erecta* (Tormentil) Glos, IOW Grigson. 2. *Potentilla reptans* (Cinquefoil) Turner.

FIVE-FINGERED ROOT *Oenanthe crocata* (Hemlock Water Dropwort) Barton & Castle.

FIVE FINGERS OF MARY *Potentilla reptans* (Cinquefoil) Ire. Grigson. Irish cuig mhear Mhaire. Lady's Fingers is a Somerset name for cinquefoil.

FIVE-LEAVED GRASS 1. *Erodium cicutarium* (Storksbill) Hants. Grigson. 2. *Fritillaria meleagris* (Snake's Head Lily) Oxf. Grigson. 3. *Potentilla erecta* (Tormentil) Worcs, War, Bucks, Lincs, Notts. Grigson. 4. *Potentilla reptans* (Cinquefoil) Gerard. The number is only accurate as far as Cinquefoil (and possibly Tormentil) are concerned. It is only approximate in the case of Snake's Head Lily, though 'grass' is apt enough. Five leaves for the Storksbill is quite inaccurate, and is probably a mistake.

FIVE-LEAVES *Ampelopsis hederacea* (Virginian Creeper) O.P. Brown.

FIVE SISTERS *Euphorbia helioscopia* (Sun Spurge) Limerick Westropp. If the name refers to the number of branches of the plant, there is some dispute with co Clare, where it is usually known as Seven Sisters.

FIVE-SPOT BABY *Nemophila maculata*.

FIZZ-GIGG *Senecio jacobaea* (Ragwort) Berw. Grigson.

FLABBY DOCK *Digitalis purpurea* (Foxglove) Fernie. See Floppy Dock.

FLAG, Bastard *Acorus calamus* (Sweet Flag) Culpepper. From the iris-like leaves.

FLAG, Blue 1. *Iris germanica*. 2. *Iris versicolor* (Purple Water Iris) USA H.Smith.1928. The etymology of 'flag' used in this sense is not known, but it must be related to the Dutch word spelt in the same way. Such forms as Flaggon, Flagger or even Fligger occur for the Yellow Iris.

FLAG, Blue, Narrow *Iris prismatica*.

FLAG, Blue, Western *Iris missouriensis* (Missouri Iris) USA Elmore.

FLAG, Cat-tail 1. *Iris versicolor* (Purple

Water Iris) Howes. 2. *Typha latifolia* (False Bulrush) USA Elmore.

FLAG, Corn 1. *Gladiolus sp* Gerard. 2. *Iris pseudo-acarus* (Yellow Iris) Som. Macmillan.

FLAG, Garden *Iris germanica* (Blue Flag) Le Strange.

FLAG, Mexican *Dichorisandra albo-marginata*. Probably the colours are the same as that of the Mexican national flag.

FLAG, Myrtle 1. *Acorus calamus* (Sweet Flag) B & H. 2. *Iris pseudo-acarus* (Yellow Iris) Culpepper. Sweet Flag is so called because of its aromatic qualities, usually likened to that of violets, but obviously, given this name, some have thought the smell to be like that of myrtle, or even of cinnamon, for it is also called Cinnamon Sedge, or Cinnamon Iris.

FLAG, Poison *Iris versicolor* (Purple Water Iris) Grieve.1931. Cf. Snake Lily and Dragon-flower for this.

FLAG, Spanish *Mina lobata*.

FLAG, Sweet *Acorus calamus*. It combines iris-like leaves (calamus = reed) with violet-like scent. It was used for strewing on the floors of churches once, for the fragrance is brought out as they are trodden on.

FLAG, Sword 1. *Gladiolus sp* Gerard. 2. *Iris pseudo-acarus* (Yellow Iris) Grieve.1931. It is the shape of the leaves which provides the 'sword' imagery; in fact, gladiolus is the diminutive of Latin gladius, which means a sword.

FLAG, Water 1. *Iris pseudo-acarus* (Yellow Iris) Border Gerard. 2. *Iris versicolor* (Purple Water Iris) Grieve.1931.

FLAG, Water, Purple *Iris versicolor*.

FLAG, Water, Yellow; FLAG, Yellow *Iris pseudo-acarus* (Yellow Iris) Thornton (Water); Ire. O Suileabhain (Yellow Flag).

FLAG IRIS *Iris germanica* (Blue Flag) Le Strange.

FLAG LILY 1. *Iris pseudo-acarus* (Yellow Iris) Som. Macmillan. 2. *Iris versicolor* (Purple Water Iris) Grieve.1931.

FLAG SEDGE *Iris pseudo-acarus* (Yellow Iris) Pratt.

FLAGGERS; FLAGONS *Iris pseudo-acarus* (Yellow Iris) IOM Killip; Ire. O Suileabhain (Flaggers); N.Ire B & H (Flagons).

FLAGROOT *Acorus calamus* (Sweet Flag) USA Sanford.

FLAGROOT, Poison *Iris prismatica* (Narrow Blue Flag) USA House.

FLAGS 1. *Calystegia sepium* (Great Bindweed) Cambs. PLNN 11. 2. *Typha latifolia* leaves (False Bulrush) Hants. Cope.

FLAKE FLOWER *Centaurea cyanus* (Cornflower) Skinner.

FLAKY FIR *Abies squamata*. 'Flaky', to agree with the specific name, *squamata*.

FLAKY JUNIPER *Juniperus squamata* (Himalayan Juniper) RHS.

FLAME, Namdi *Spathodea campanulata* (African Tulip Tree) Kenya Perry. Scarlet flowers.

FLAME CLIMBER; FLAME NASTURTIUM *Tropaeolum speciosum* (Scotch Flame Flower) Dor. Macmillan. Scarlet flowers.

FLAME CORAL TREE *Erythrina corallodendron* (Coral Tree) RHS.

FLAME FLOWER 1. *Kniphofia uvaria* (Red Hot Poker) Som. Macmillan. 2. *Pyrostegia venusta* (Chinese Cracker Flower) RHS. 3. *Viola tricolor* (Pansy) Fernie.

FLAME FLOWER, Scotch *Tropaeolum speciosum*. The scarlet flowers give this name.

FLAME GOLD-TIPS *Leucospermum discolor* S.Africa Perry.

FLAME HEATH *Erica flammea*.

FLAME LEAP, Mexican; FLAME TREE, Mexican *Euphorbia pulcherrima* (Poinsettia) USA Watt. Scarlet bracts.

FLAME LILY *Lilium philadelphicum* (Philadelphia Lily) USA Woodcock.

FLAME-LEAVED CROWFOOT *Ranunculus flammula* (Lesser Spearwort) Flower.1859. The reference in this case is to the shape of the leaves.

FLAME NASTURTIUM *Tropaeolum specio-*

sum (Scotch Flame Flower) Macmillan.
FLAME NETTLE *Coleus sp* Blackwood.
FLAME OF GOD, Little *Taraxacum officinale*. This is a translation of one of the Gaelic names. Burning Fire is a Somerset name, and Mabey lists Golden Sun as well.
FLAME OF THE FOREST 1. *Butea monosperma* (Bastard Teak) H & P. 2. *Delonix regia*. 3. *Spathodea campanulata* (African Tulip Tree) Perry. All of these have red flowers, crimson in the Delonix, and scarlet in the others.
FLAME PLANT *Anthurium scherzerianum* (Flamingo Plant) H & P.
FLAME TREE 1. *Rhododendron arboreum* J.Smith. 2. *Spathodea campanulata* (African Tulip Tree) Dalziel.
FLAME VINE *Pyrostegia venusta* (Chinese Cracker Flower) Simmons.
FLAME VINE, Mexican *Senecio confusus*.
FLAME VIOLET *Episcia cupreata*.
FLAMES, Flower *Tropaeolum speciosum* (Scotch Flame Flower) Som. Macmillan.
FLAMING KATY *Kalanchoe blossfeldiana* (Brilliant Star) RHS.
FLAMING SWORD *Kniphofia uvaria* (Red Hot Poker) Dor. Macmillan.
FLAMINGO PLANT *Anthurium scherzerianum*.
FLAMY *Viola tricolor* (Pansy) Genders.1976. He suggests that the "flame" names for pansy are from the colours, which resemble those of burning wood.
FLANDERS POPPY *Papaver rhoeas* (Red Poppy) Howes.
FLANNEL *Verbascum thapsus* (Mullein) Som, Suss. Grigson. All the 'flannel' names for mullein refer to the texture of the leaves.
FLANNEL, Aaron's *Verbascum thapsus* (Mullein) Dor. Macmillan.
FLANNEL, Adam's 1. *Verbascum lychnites* (White Mullein) Lincs. Peacock. 2. *Verbascum thapsus* (Mullein) Som. Friend; N'thants. A.E.Baker; Ches. Holland; Yks. Carr; USA Henkel.
FLANNEL, Lady's; FLANNEL, Our Lady's *Verbascum thapsus* (Mullein) both Som. Grigson (Lady's); Macmillan (Our Lady's).
FLANNEL, Old Man's *Verbascum thapsus* (Mullein) Som. B & H.
FLANNEL, Our Lord's 1. *Echium vulgare* (Viper's Bugloss) Kent Grigson. 2. *Verbascum thapsus* (Mullein) Kent Grigson; USA Henkel.
FLANNEL, Our Saviour's 1. *Echium vulgare* (Viper's Bugloss) Kent Grigson. 2. *Verbascum thapsus* (Mullein) Kent Grigson.
FLANNEL, Poor Man's *Verbascum thapsus* (Mullein) Som, Glos, Bucks Grigson.
FLANNEL-BUSH *Fremontia californica*.
FLANNEL FLOWER; FLANNEL LEAF; FLANNEL MULLEIN; FLANNEL PLANT *Verbascum thapsus* (Mullein) Som. B & H; Kent Pratt (Flower); Som. Macmillan; USA Henkel (Leaf); Corn. Grigson; Hants. Cope; IOW Long (Plant): N'thants. Clare.
FLANNEL JACKET *Verbascum thapsus* (Mullein) Norf. Grigson.
FLANNEL PETTICOAT *Verbascum thapsus* (Mullein) Som. Macmillan.
FLAP-A-DOCK *Digitalis purpurea* (Foxglove) Dev. Friend.1882.
FLAPDICK *Digitalis purpurea* (Foxglove) Som. Elworthy. This is Flap-dock, better known as Flop-dock, or Floppy Dock. Flops, without any adjective,is quite common in the West country. The meaning becomes clear when another name, Cow-flops, is considered - Cf. cowslip.
FLAPDOCK *Digitalis purpurea* (Foxglove) Dev. B & H; USA Henkel.
FLAPPER BAGS *Arctium lappa* (Burdock) Scot. B & H. Probably a reference to the large leaves.
FLAPPER DOCK 1. *Digitalis purpurea* (Foxglove) Dev. Friend.1882. 2. *Petasites hybridus* (Butterbur) Grieve.1931.
FLAPPY DOCK *Digitalis purpurea* (Foxglove) Dev. Friend.1882; Som. Macmillan. See Flapdick.
FLAT-CROWN TREE *Albizia gummifera* Dalziel.

FLAT-TOPS *Vernonia noveboracensis* (Ironweed) W.Miller.

FLATTERBAW *Petasites hybridus* (Butterbur) Scot. Aitken. The sequence seems to run Butter-dock, Batter-dock, Blatterdock, to Flatterbaw, the last syllable here possibly the 'bur' of Butterbur.

FLATTERDOCK 1. *Nuphar lutea* (Yellow Waterlily) Ches. Holland. 2. *Nymphaea alba* (White Waterlily) Ches. Holland. 3. *Polygonum amphibium* (Amphibious Persicaria) Ches. B & H. Floating Dock, apparently, a name which is recorded as well as Flatterdock.

FLATWEED *Hypochoeris radicata* (Long-rooted Cat's Ear) Australia Berney. It is only an assumption that this is the cat's ear mentioned in the article.

FLAW-FLOWER 1. *Anemone nemerosa* (Wood Anemone) Fernie. 2. *Pulsatilla vulgaris* (Pasque Flower) Prior. A flaw is a gust of wind. Anemones are known generally as windflowers, a name which is explained by saying that some of them flourish in open exposed places, or that they would not open till the March winds began to blow. The belief is from Pliny: Flos numquam se aperit nisi vento spirante, unde et nomen eius. Greek anemos is the wind, and the name anemone means literally daughter of the wind.

FLAX, Alpine *Linum alpinum.*

FLAX, Bitter *Linum catharticum* (Fairy Flax) H C Long.1910.

FLAX, Blue *Linum lewisii* (Rocky Mountain Flax) Jaeger.

FLAX, Cathartic *Linum catharticum* (Fairy Flax) J A MacCulloch.1905. Or Purging Flax. It is an effective, though dangerous, purge.

FLAX, Common *Linum usitatissimum.* 'Flax' is derived either from a word in Old Teutonic meaning to plait, or to fray, or, as seems more likely, from medieval Latin filassium, yarn, this coming from the verb filare, to spin.

FLAX, Dwarf; FLAX, Ground *Linum catharticum* (Fairy Flax). B & H (Dwarf); Fernie (Ground).

FLAX, Fairy 1. *Euphrasia officinalis* (Eyebright) Donegal Grigson. 2. *Linum catharticum*. The fairies used it for their clothes, and its small bells make music which cannot be heard by human ears.

FLAX, Fairy Woman's *Linum catharticum* (Fairy Flax) Spence.1949b.

FLAX, False *Camelina sativa* (Gold of Pleasure) Dimbleby. Gold of Pleasure is often associated with flax, probably occurring as a weed in flax fields.

FLAX, Golden *Linum flavum.*

FLAX, Holy *Santolina rosmarinifolia.*

FLAX, Lewis *Linum lewisii* (Rocky Mountain Flax) USA Elmore.

FLAX, Maori *Phormium tenax* (New Zealand Flax) Andersen.

FLAX, Mountain 1. *Centaureum eyrthraea* (Centaury) Cumb. B & H. 2. *Linum catharticum* (Fairy Flax) Shrop, Derb, Yks, Cumb. Grigson. 3. *Phormium colensoi*. 4. *Polygala senega* (Senega Snakeroot) USA House. 5. *Spergula arvensis* (Corn Spurrey) Dor, Shrop, Yks. Grigson.

FLAX, Narrow-leaved *Linum bienne* (Pale Flax) D.Grose. *Angustifolium* is an alternative specific name.

FLAX, New Zealand *Phormium tenax.* This was used by the Maori as a textile fibre just as flax was.

FLAX, Pale *Linum bienne.*

FLAX, Perennial *Linum perenne.*

FLAX, Pine, Yellow *Linum neomexicanum.*

FLAX, Prairie *Linum lewisii* (Rocky Mountain Flax) USA Elmore.

FLAX, Purging *Linum catharticum* (Fairy Flax) Fernie. It is still used in herbal medicine as a purge.

FLAX, Rocky Mountain *Linum lewisii.*

FLAX, Spurge 1. *Daphne cneorum* (Flax-leaved Daphne) Gerard. 2. *Daphne mezereum* (Lady Laurel) Gerard.

FLAX, Wild 1. *Cuscuta epilinum* (Flax Dodder) B & H. 2. *Linaria vulgaris* (Toadflax) Lyte. 3. *Linum africanum* S.Africa Watt. 4. *Linum thunbergii* S.Africa

Watt.

FLAX, Yellow 1. *Linum campanulatum*. 2. *Linum puberulum*. 3. *Linum rigidum*.

FLAX, Yellow, Wild *Linum virginianum*.

FLAX DODDER 1. *Cuscuta epilinum*. 2. *Cuscuta epithymum* (Common Dodder) S.Africa Watt. Cf. Wild Flax for *C epilinum*. It seems out of place for common dodder, which is *epithymum* rather than *epilinum*.

FLAX-LEAVED DAPHNE *Daphne cneorum*.

FLAX-LEAVED ST JOHN'S WORT *Hypericum linarifolium*.

FLAX LILY 1. *Dianella caerulea*. 2. *Phormium tenax* (New Zealand Flax) W.Miller.

FLAX SPURGE *Euphorbia paralias* (Sea Spurge) B & H.

FLAXFIELD CATCHFLY *Silene linicola*.

FLAXSEED; FLAXSEED,Thyme-leaved *Radiola linoides* (Allseed) McClintock (Flaxseed); Macmillan (Thyme-leaved Flaxseed).

FLAXTAIL *Typha latifolia* (False Bulrush) W.Miller.

FLAXWEED *Linaria vulgaris* (Toadflax) Gerard. Toadflax seems to mean a wild, useless flax, i.e. a flax fit only for toads. It was, in fact, a destructive weed in the flax fields.

FLEA DOCK *Petasites hybridus* (Butterbur) B & H.

FLEA MINT *Mentha pulegium* (Pennyroyal) Sanecki. The *pulegium* of the specific name comes from Latin pulices, fleas, because it is good for destroying them. The common name, pennyroyal, is a corruption of Puliol Royal, so the first element is a reference to fleas again.

FLEA NIT; FLEA NUT *Senecio jacobaea* (Ragwort) both Ches. both Grigson. The name occurs as Flee-dod, or Fly-dod, in Cheshire too. Nearby, in Lancashire, Fleawort was recorded. Presumably the plant was used as an aid to getting rid of fleas.

FLEABANE 1. *Artemisia vulgaris* (Mugwort) Chopra. 2. *Inula conyza* (Ploughman's Spikenard) Hill. 3. *Plantago psyllium* (Fleawort) Lyte. Anything with a name like fleabane must be of benefit in ridding a house of fleas, even if flea reads fly. Parkinson derives *conyza* from Greek knops, gnat. In Devonshire, they say that a handful of this burnt each day during the summer, will keep house-flies out of the house. Mugwort has a similar derivation.

FLEABANE, Alaska *Erigeron salsuginosus*.

FLEABANE, Alpine *Erigeron borealis* (Highland Fleabane) Polunin.1972.

FLEABANE, Blue *Erigeron acris*.

FLEABANE, Canadian *Erigeron canadensis*. As you would expect from the name, American Indians used it to destroy fleas and gnats, by burning the herb, and so smoking the pests.

FLEABANE, Common *Pulicaria dysenterica* (Yellow Fleabane) Curtis. It has an insect-repellent smell, and was often burnt as well, so that the smoke in the room had a similar effect.

FLEABANE, Daisy 1. *Erigeron philadelphicus* (Philadelphia Fleabane) USA House. 2. *Erigeron ramosus* USA Tolstead.

FLEABANE, Great *Inula conyza* (Ploughman's Spikenard) Le Strange. It is not clear whether "great" describes size or efficacy.

FLEABANE, Highland *Erigeron borealis*.

FLEABANE, Irish *Inula salicina*.

FLEABANE, Mexican *Erigeron mucronatus*.

FLEABANE, Middle *Pulicaria dysenterica* (Yellow Fleabane) Le Strange. For it is not so big as Ploughman's Spikenard, which is the Great Fleabane.

FLEABANE, Nevada *Erigeron nevadensis*.

FLEABANE, Philadelphia *Erigeron philadelphicus*.

FLEABANE, Purple 1. *Centranthemum anthelminticum*. 2. *Erigeron acris* (Blue Fleabane) Curtis. *Centranthemum* is an Indian plant. There, to get rid of fleas, they either roast the plant in a room, or powder it to throw on the floor.

FLEABANE, Pygmy *Erigeron uncialis*.

FLEABANE, Small *Pulicaria vulgaris*.

FLEABANE, Tidy *Erigeron concinnus.*
FLEABANE, Yellow *Pulicaria dysenterica.*
FLEABANE-MULLET *Pulicaria dysenterica* (Yellow Fleabane) Prior. Prior says 'mullet' is the French mollet, soft, a reference to the texture of the leaves.
FLEABITE *Melandrium dioicum* (Red Campion) Corn. Grigson. Cf. Lousy Beds, a Cumbrian name for this.
FLEAS-AND-LICE *Cymbalaria muralis* (Ivy-leaved Toadflax) Som. Macmillan.
FLEASEED *Plantago psyllium* (Fleawort) Le Strange.
FLEAWEED *Galium verum* (Lady's Bedstraw) Suff. Grigson. Lady's Bedstraw is another of the plants used to keep down the flea population, hence, of course, its use as a bedstraw.
FLEAWOOD *Myrica gale* (Bog Myrtle) N'thum. Grigson. Bog Myrtle is a well-known flea repellent in the north. Highlanders used at one time to sleep on beds of it. It is a good moth repellent, too.
FLEAWORT 1. *Erigeron canadensis* (Blue Fleabane) Grieve.1931. 2. *Inula conyza* (Ploughman's Spikenard) Prior. 3. *Plantago psyllium.* 4. *Senecio jacobaea* (Ragwort) Lancs. Fernie.
FLEAWORT, Field *Senecio integrifolius.*
FLEAWORT, Jacoby *Senecio jacobaea* (Ragwort) Leyel.1937. Jacoby, and the specific name, *jacobaea*, are references to St James. Perhaps it is because the plant would be in full flower on St James's Day, which is 25 July, but more likely because St James is the patron saint of horses, and ragwort has much to do with horses, either in causing or curing sickness.
FLEAWORT, Marsh *Senecio congestus var. palustris.*
FLEAWORT, Spathulate *Senecio spathulifolius.*
FLEE-DOD *Senecio jacobaea* (Ragwort) Ches. Holland. Cf. other Cheshire names like Fly-dod, Fleanut,etc.
FLEECE FLOWER *Polygonum amplexicaule.*

FLEECE-FLOWER, Japanese *Polygonum cuspidatum* (Japanese Knotweed) Howes.
FLESH-AND-BLOOD *Potentilla erecta* (Tormentil) Berw. Grigson. From a similar area, it is known as Blood-root. The reference must be the use of the roots to give a red dye.
FLESHY-FRUITED YUCCA *Yucca baccata* (Datil) Jaeger.
FLEX *Linum usitatissimum* (Flax) Henslow. Closer to the OE flaex than the modern word, Cf. Flix.
FLIBBERTY-GIBBET *Malva sylvestris* (Common Mallow) Som. Macmillan.
FLIG *Stellaria media* (Chickweed) IOM Moore.1924.
FLIGGERS *Iris pseudo-acarus* (Yellow Flag) E.Ang. Halliwell. The starting point for Fliggers was probably Flag, the common name for any Iris. There is a version from Ireland, Flaggers, which suggests this.
FLIGHT, Devil's *Hypericum perforatum* (St John's Wort) Dev. Hewett. This may have been recorded in Devonshire, but it is simply a translation of the old book name Fuga daemonum. The French have chasse-diable, too. Bassardus Viscontus said that the plant gathered on a Friday, about the full moon in July, and worn round the neck, is a cure for melancholy, and calculated to drive away all "fantastical spirits".
FLINT MAIZE; FLINT CORN *Zea mays var. praecox.* So called because of the hard endosperm.
FLINTWOOD *Eucalyptus pilularis* New South Wales J.Smith.
FLINWORT, Mace *Leucanthemum vulgare* (Ox-eye Daisy) Fernie. The 'mace' part of this name is probably akin to the Lincolnshire Maise, or Maze, themselves originating from something like the West country Maithen, or Mauthern.
FLIRTWORT *Tanacetum parthenium* (Feverfew) Dev. Friend.1882; Som. Fernie.
FLIX *Linum usitatissimum* (Flax) Dor. B & H. Cf. Flex.
FLIXWEED; FLIXWORT *Descurainia sophia.* Flixweed is the usual name; Flixwort

is in Lyte. The names point to a former medicinal use for dysentery, which was once known as flix, or flux.

FLOBBY DOCK; FLABBY DOCK *Digitalis purpurea* (Foxglove) Dev. Friend.1882 (Flobby); Fernie (Flabby). Floppy Dock is more usual, but Cf. Flap-dock.

FLOCK, Lady's *Cardamine pratensis* (Lady's Smock) Notts. Grigson. 'Flock' for 'smock' is quite common, but that is not its meaning. The word means a tuft of wool, and is very descriptive. The plant, when seen at a distance, can have this appearance.

FLOCK, Shepherd's *Arabis caucasica* (Garden Arabis) Som. Macmillan.

FLOOKWORT; FLOWKWORT *Hydrocotyle vulgaris* (Pennywort) Prior (Flookwort); Norf. Gerard (Flowkwort). It is believed to cause fluke worms of liver rot in sheep.

FLOP *Digitalis purpurea* (Foxglove) Dev. Choape; Som. Macmillan. Cf. fox, cow or goose, flop.

FLOP, Cow 1. *Digitalis purpurea* (Foxglove) Dev. Friend.1882. 2. *Pastinaca sativa* (Wild Parsnip) Corn. Jago.

FLOP, Goose 1. *Digitalis purpurea* (Foxglove) Dev. Friend. 2. *Narcissus pseudo-narcissus* (Daffodil) Dev, Som. Elworthy.

FLOP, Pig's *Heracleum sphondyllium* (Hogweed) Dev. Macmillan. 'Flop' is a reference to the large leaves (Cf. similar names for foxglove). So the name means 'large leaves that are fed to pigs'. But 'flop' can also mean droppings, just as the 'slip' of cowslip does.

FLOP-A-DOCK; FLOP-DOCK; FLOP-DOCKEN; FLOPPY DOCK; FLOS-DOCKEN *Digitalis purpurea* (Foxglove) Dev. Friend.1882; Wilts. Dartnell & Goddard) (Flop-a-dock); Dev. Friend.1882 (Flop-dock); USA Henkel (Flop-dock); Coats (Floppy Dock); Yks. Grigson (Flos-docken). 'Dock' is nearly always a comment on the size of the leaves. Cf. Flobby, Flabby or Flap Dock.

FLOP-POPPY *Digitalis purpurea* (Foxglove) Dev. Friend.1882.

FLOP-TOP *Digitalis purpurea* (Foxglove) Som. Macmillan.

FLORA'S CLOCK *Tragopogon pratensis* (Goat's Beard) Tynan & Maitland.

FLORAMOR; FLORIMER 1. *Amaranthus caudatus* (Love-lies-bleeding) Prior. 2. *Amaranthus tricolor* (Joseph's Coat) Gerard. From French fleur d'amour, a mistranslation of amaranthus. Florimer is in Prior.

FLORAMOUR *Amaranthus tricolor* (Joseph's Coat) Parkinson.

FLORENCE COURT YEW *Taxus baccata var. fastigiata* (Irish Yew) Brimble.1948. The variety was found growing on the mountains of co Fermanagh, near Florence Court.

FLORENCE FENNEL *Foeniculum dulce* (Finocchio) Pomet. Cf. Roman Fennel.

FLORENCE FLEUR-DE-LUCE; FLORENTINE IRIS *Iris germanica var. florentina* Gerard (Fleur-de-luce); Thornton (Iris).

FLORIDA ARROWROOT *Zamia integrifolia*.

FLORIDA BEAN *Mucuna urens*.

FLORIDA SUGAR MAPLE *Acer floridanum*.

FLORIDA VELVET BEAN 1. *Mucuna pruriens* (Cowage) Willis. 2. *Stizolobium deeringianum* (Velvet Bean) Schery.

FLOSS *Juncus conglomeratus* (Compact Rush) Shet. Nicolson.1920. If this is a correct ascription - the authority refers to it simply as 'common rush'.

FLOSS-DOCKEN; FLOUS-DOCKEN; FLOWSTER-DOCKEN *Digitalis purpurea* (Foxglove) all Yks. F K Robinson (Floss); Grigson (Flowster); Atkinson (Flous). i.e showy dock.

FLOSS FLOWER *Ageratum houstonianum*. Descriptive.

FLOSS-SEAVES *Eriophorum angustifolium* (Cotton-grass) Yks. F K Robinson.

FLOSS SILK TREE, Brazilian *Chorisia speciosa*.

FLOSSY, Silky *Salpiglossis sinuata* (Velvet

Trumpet Flower) Som. Macmillan. An apt name, suggested by the word Salpiglossis itself.

FLOTE GRASS *Glyceria fluitans*. An obsolete form of 'float', retained in the name.

FLOTE GRASS, Small *Glyceria declinata*.

FLOUR-DE-LUCE, Yellow *Iris pseudoacarus* (Yellow Iris) Turner.

FLOUR MAIZE *Zea mays var. amylacea*.

FLOURISH, Dead Man's *Conium maculatum* (Hemlock) Tynan & Maitland.

FLOWER AMOR *Amaranthus tricolor* (Joseph's Coat) Langham. See rather Floramor.

FLOWER ARMOUR *Amaranthus caudatus* (Love-lies-bleeding) A S Palmer. This was a corruption of Floramor that appeared in Tusser, Five hundred pointes of good husbandrie, 1577.

FLOWER-DE-LUCE 1. *Iris germanica var. florentina* (Orris). 2. *Lilium candidum* (Madonna Lily) (Rohde). French fleur de lys.

FLOWER-DE-LUCE, Bastard *Iris pseudoacarus* (Yellow Iris) Gerard.

FLOWER-DE-LUCE, Florence *Iris germanica var. florentina* (Orris) Gerard.

FLOWER-DE-LUCE, Velvet *Hermodactylus tuberosus* (Snakeshead Iris) Gerard.

FLOWER-DE-LUCE, Water; FLOWER-DE-LUCE, Water, Yellow *Iris pseudoacarus* (Yellow Iris) Gerard; Wesley (Yellow).

FLOWER GENTLE *Amaranthus tricolor* (Joseph's Coat) Prior. Prior's explanation was that the plant looks like the plumes once worn by people of rank.

FLOWER-OF-A-DAY *Tradescantia virginica* (Spiderwort) Nash. Cf. Parkinson's name for this plant - Day Spiderwort.

FLOWER-OF-AN-HOUR 1. *Hibiscus trionum* (Bladder Katmia) Wit. 2. *Malva sylvestris* (Common Mallow) Som. Macmillan. Cf. Goodnight-at-noon for Bladder Katmia.

FLOWER VELURE *Amaranthus caudatus* (Love-lies-bleeding). Passevelure, the French name, was in use in England (B & H). So were Velvet-flower, or Purple Velvet-flower (Turner).

FLOWER VELURE, Golden *Amaranthus luteus*.

FLOWERING CURRANT *Ribes sanguineum*.

FLOWSTER-DOCKEN *Digitalis purpurea* (Foxglove) Yks. Grigson. i.e. showy dock.

FLOX *Digitalis purpurea* (Foxglove) Dev. Friend.1882. It seems to be a mix of 'fox' and 'flop'.

FLUE BRUSHES *Typha latifolia* (False Bulrush) Som. Macmillan. A good enough descriptive name, considering the sooty appearance. Cf. Chimney Sweeper, Chimney Sweeper's Brush, etc.

FLUELLA, Female *Kickxia spuria* (Round-leaved Fuella) Prior. Prior said that the soft, velvety leaves were responsible for the 'female' part of this name.

FLUELLA, Round-leaved; FLUELLIN, Round-leaved *Kickxia spuria*. It is Curtis who has Fluellin.

FLUELLA, Sharp-leaved; FLUELLIN, Sharp-leaved *Kickxia elatior*.

FLUELLEN *Veronica officinalis* (Common Speedwell) Browne. "In Welch it is called Fluellen, and the Welch people attribute great virtues to the same" (Gerard). The Welsh is actually Llysiau Llywelyn - the herb of (St) Llywelyn, by which is meant a herb which flowers around 7 April, the feast day, a date which seems a little early for this.

FLUELLEN, Upright; FLUELLEN, Tree *Veronica spicata* (Spiked Speedwell) both Gerard.

FLUFF, White *Menyanthes trifoliata* (Buckbean) Norf. Grigson.

FLUFFWEED *Verbascum thapsus* (Mullein) Som. Grigson. The texture of the leaves is responsible for this name, as well as for all the 'flannel' and 'blanket' names.

FLUFFY BUTTONS *Salix capraea* (Goat Willow) Som. Macmillan. It must be the catkins that are referred to here.

FLUFFY-PUFFY *Taraxacum officinale* (Dandelion) Som. Macmillan. The seed heads, which are 'puffed' by children to tell the time.

FLUTTERING ELM *Ulmus laevis*. So called because of the long-stalked flowers, which flutter in the wind.

FLUXWEED *Descurainia sophia* (Flixweed) Prior. Dysentery was once known as flix, or flux. "The seed...drunke with wine or smiths water, stoppeth the bloudy flix, the laske, and all other issues of bloud..." (Gerard).

FLY-ANGELS; FLYING-ANGELS *Acer pseudo-platanus* seeds (Sycamore keys) Som. Macmillan. The peculiar method adopted by both the maples and the ash for seed disposal has given rise to some brilliantly descriptive names. In addition to these, there are similar ones, like Wings, or Butterflies, even Aeroplanes, or Propellers.

FLY-DOD *Senecio jacobaea* (Ragwort) Ches. Grigson. Ragwort is an insecticide, or at least is a flea deterrent. Cf. the names Fleawort, Fleanit, etc., all from the Cheshire area.

FLY FLOWER *Prunella vulgaris* (Self-heal) Glos. Grigson.

FLY HONEYSUCKLE *Lonicera xylosteum*. According to Prior, these honeysuckles got the name in error, by confusion with another, fly-catching, plant.

FLY HONEYSUCKLE, American *Lonicera canadensis* (Early Fly Honeysuckle) Youngken.

FLY HONEYSUCKLE, Blue *Lonicera caerulea*.

FLY HONEYSUCKLE, Early *Lonicera canadensis*.

FLY HONEYSUCKLE, Mountain 1. *Lonicera caerulea* (Blue Fly Honeysuckle) USA House. 2. *Lonicera villosa*.

FLY HONEYSUCKLE, Swamp *Lonicera oblongifolia*.

FLY ORCHIS *Ophrys insectifera*.

FLY POISON *Amaianthemum muscaetoxicum* (Staggergrass) USA Kingsbury. Fly poison is a translation of the specific name, and is self-explanatory.

FLYAWAYS *Acer pseudo-platanus* seeds (Sycamore keys) Som. Macmillan. See Fly-angels.

FLYCATCHER 1. *Arum maculatum* (Cuckoo-pint) Wilts. Macmillan. 2. *Drosera rotundifolia* (Sundew) Dev. Macmillan. 3. *Pinguicula vulgaris* (Butterwort) Dor. Macmillan. 4. *Sarracenia purpurea* (Pitcher Plant) Grieve.1931. Sundew, Butterwort and Pitcher Plant are all insectivorous, while Cuckoo-pint relies on trapping midges inside the spathe so that they fertilise the female flowers.

FLYING DUTCHMEN *Acer pseudo-platanus* seeds (Sycamore keys) Macmillan. See fly-angels.

FLYTRAP 1. *Apocynum androsaemifolium* (Spreading Dogbane) Grieve.1931. 2. *Drosera rotundifolia* (Sundew) Som. Macmillan. 3. *Sarracenia purpurea* (Pitcher Plant) Grieve.1931.

FLYTRAP, English *Drosera rotundifolia* (Sundew) Som. Macmillan.

FLYTRAP, Venus *Dionaea muscipula*. The leaves have marginal spines, and trap insects by the two halves of the blade closing like a hook as soon as one of three central bristles is stimulated. A protein-dissolving ferment is secreted once the insect is caught. After digestion, the leaves open up again.

FOAL'S FOOT 1. *Glechoma hederacea* (Ground Ivy) B & H. 2. *Tussilago farfara* (Coltsfoot) Kent, Suss. Grigson; N'thants. A.E.Baker; Leics. A.B.Evans; Yks. Nicholson. It is the shape of the leaves that suggests names like this.

FOALFOOT 1. *Asarum europaeum* (Asabaracca) Turner. 2. *Ranunculus ficaria* (Lesser Celandine) Ayr. Grigson. 3. *Tussilago farfara* (Coltsfoot) Gerard.

FOALFOOT, Sea *Calystegia soldanella* (Sea Bindweed) Turner. Where did Turner get this? There are no other names which suggest the imagery.

FOALSWORT *Tussilago farfara* (Coltsfoot) Grieve.1931. This name probably started as Foal's Foot.

FOAM, Meadow *Limnanthes douglassi*.

FOAM DOCK *Saponaria officinalis* (Soapwort) Cockayne. i.e. a large-leaved plant which can be used to produce a lather.

FOAM FLOWER *Tiarella cordifolia*.

FOAM OF MAY *Spiraea arguta*.

FODDER, Medick *Medicago sativa var. sativa* (Lucerne) Parkinson.

FODDER BURNET *Poterium polygamum*. This was the burnet which Purton, in *Flora of the Midland counties*, said formed almost the whole of the vegetation on the sheep walks of Wiltshire in sight of Salisbury Cathedral spire.

FODDER VETCH *Vicia villosa* (Lesser Tufted Vetch) Blamey.

FOETID CAMOMILE *Anthemis cotula* (Maydweed) Australia Watt.

FOETID CASSIA *Cassia tora* (Sicklepod) Dalziel.

FOETID IRIS *Iris foetidissima*.

FOETID WILD GOURD *Cucurbita foetidissima* (Buffalo Gourd) Youngken.

FOG, Devil-in-a- *Nigella damascena* (Love-in-a-mist) Tynan & Maitland.

FOG, Indian 1. *Sedum glaucum* Donegal B & H. 2. *Sedum reflexum* (Yellow Stonecrop) Ire. Grigson.

FOG, Yorkshire *Holcus lanatus*. 'Fog' is a descriptive term, firstly because this very common grass has downy leaves and sheaths, and secondly because of the variable colour of the panicles - the colours blend well, and when there is a lot of this grass in a pasture, then the name 'fog' becomes well understandable. Such a pasture is said to be fogged, or foggy.

FOG CROCUS *Colchicum autumnale* (Meadow Saffron) Yks. B & H. Presumably because it flowers in autumn, the season of fogs. It isn't a crocus, of course.

FOGFRUIT 1. *Lippia cuneifolia* Kansas Gates. 2. *Lippia lanceolata* Kansas Gates.

FOGWEED *Atriplex expansa*.

FOGWORT *Ranunculus ficaria* (Lesser Celandine) Dor. Macmillan. Undoubtedly a mistake for Figwort, either on the part of the user or of the collector.

FOILE-FOOT *Tussilago farfara* (Coltsfoot) Yks. B & H. Foal's-foot, rather.

FOLESFOTH *Glechoma hederacea* (Ground Ivy) B & H. Again, foal's foot.

FOLLAN-FING *Galium saxatile* (Heath Bedstraw) IOM Moore.1924.

FOLLOW-MY-LAD *Galium aparine* (Goose-grass) Scot. Tynan & Maitland.

FOLLOWERS 1. *Galium aparine* (Goose-grass) Scot. Tynan & Maitland. 2. *Rubus fruticosa* (Bramble) Suff. Parish. This name for bramble conveys the same idea as does Lawyers, or Country Lawyers - once you have been caught, they never let you go. For goose-grass, though, the name means sweethearts, a reference to the clinging habit of the burrs.

FOLLY'S FLOWER *Aquilegia vulgaris* (Columbine) Dor, Som. Macmillan. In the language of flowers, columbine was always associated with folly, possibly because it was also an emblem of cuckoldom.

FONTAINEBLEAU, Service Tree of *Sorbus x latifolia* (French Hales) Polunin.1976.

FOO *Sempervivum tectorum* (Houseleek) PLNN 7. One of the many variations on a word which is usually given as Fews. Kay Sanecki gave one of them as Fow, which, she said, means to cleanse. But Calvin Podd, in PLNN 7, says that these names all mean drunk in northern dialect. It is not very clear why houseleek should have such an association, but there is, of course, the most famous plant name of them all - Welcome-home-husband-though-never-so-drunk.

FOOD, Adder's *Arum maculatum* (Cuckoo-pint) Friend; Wilts. Goddard. Adder is OE attor, poison, and the berries are indeed poisonous. Cf. Adder's Meat, and Snake's Meat.

FOOD, Canary 1. *Plantago major* (Great Plantain) Som. Grigson. 2. *Senecio vulgaris* (Groundsel) Som. Grigson. Literally apt - both were collected for bird seed, groundsel particularly.

FOOD, Pig's *Heracleum sphondyllium* (Hogweed) Dor. Macmillan. Because it was actually used as a pig food.

FOOD, Rabbit's *Oxalis acetosella* (Wood Sorrel) Lancs. Grigson. Cf. Rabbit's Meat, Hare's Meat and Fox's Meat for this.

FOOD, Snake's 1. *Allium ursinum* (Ramsons) Som. Macmillan. 2. *Arum maculatum* (Cuckoo-pint) Dev. B & H. 3. *Iris foetidissima* fruit (Foetid Iris) Dev. Macmillan. 4. *Linaria vulgaris* (Toadflax) Dor. Macmillan. 5. *Mercurialis perennis* (Dog's Mercury) Dor. Macmillan. 6. *Petasites hybridus* (Butterbur) Som. Macmillan. 7. *Solanum dulcamara* (Woody Nightshade) Som. Macmillan. 8. *Synphytum officinale* (Comfrey) Som. Macmillan. 9. *Tamus communis* (Black Bryony) Som. Macmillan. They are either poisonous (Cuckoo-pint, Dog's Mercury, Bryony), or at least look suspicious (Foetid Iris, Woody Nightshade). Foetid Iris has a double reason for the name - the very bad smell would warrant a name like this, and that is why Ramsons carries it. Butterbur possibly gets the name by association with the plague; it may have been used as a remedy, but in some places they say it sprang from the graves of the victims. Again, the large leaves, like those of comfrey, may have been suspicious as a possible hiding place for snakes. Toadflax seems to be the odd one out, but a possible explanation is that the plant was originally called bubonium, because it cured buboes (the plague connection, again). Bubonium became corrupted into Bufonium - Latin bufo, toad, and so into the meaningless Toadflax.

FOOD, Turkey's *Galium aparine* (Goose-grass) Som. Macmillan. The name Goose-grass, and indeed all the 'goose-' names, are given because the plant was used as a food for goslings - so why not for turkeys as well?

FOOL GOOSEBERRY *Pulmonaria officinalis* (Lungwort) B & H.

FOOL'S BALLOCKS; FOOL'S STONES *Orchis morio* (Green-winged Orchis) Lyte (Ballocks); Gerard (Stones). The testicle shape of the twin tubers has given a lot of names like this.

FOOL'S CAP *Solanum dulcamara* (Woody Nightshade) Som. Macmillan. From the shape of the flowers.

FOOL'S CHERRY *Prunus padus* fruit (Bird Cherry) B & H.

FOOL'S CRESS *Sium nodiflorum* (Procumbent Water Parsnip) Hulme.

FOOL'S HUCKLEBERRY *Menziesia ferruginea* (Mock Azalea) USA E Gunther.

FOOL'S ONION *Triteleia hyacinthina*.

FOOL'S PARSLEY *Aethusa cynapium*. Only a fool would mistake it, for it is extremely poisonous.

FOOL'S STONES *Orchis mascula* (Early Purple Orchis) B & H.

FOOL'S WATERCRESS *Apium nodiflorum*. This often grows with watercress, but is quite different, and is in fact not a cress at all, but an umbelifer.

FOOLWORT *Tussilago farfara* (Coltsfoot) Le Strange. A misreading, obviously - the first syllable must have been 'foal'.

FOOSE *Sempervivum tectorum* (Houseleek) N.Eng, Scot. Grigson. See above, under Foo.

FOOT, Alexander's *Anacyclus pyrethrum* (Pellitory-of-Spain) B & H. This is French pied d'Alexandre.

FOOT, Ass's *Tussilago farfara* (Coltsfoot) Prior. The shape of the leaves gives coltsfoot, as well as this and foal's foot.

FOOT, Bear's 1. *Aconitum napellus* (Monkshood) Notts. B & H. 2. *Alchemilla vulgaris* (Lady's Mantle) Hants, N'thum. Grigson; Derb. Camden/Gibson. 3. *Helleborus foetidus* (Stinking Hellebore) Gerard. 4. *Helleborus niger* (Christmas Rose) Hill. 5. *Helleborus viridis* (Green Hellebore) Dor. Dacombe; Glos. Grigson. Monkshood and the hellebores have leaves which could be described as having a claw outline, but Lady's Mantle is not so obvious.

FOOT, Bird's *Lotus corniculatus* (Bird's foot Trefoil) Som. Grigson.

FOOT, Boar's *Helleborus viridis* (Green Hellebore) Bucks. Grigson. An error for Bear's Foot, or at least given when the meaning of Bear's Foot had been forgotten.

FOOT, Bull's *Tussilago farfara* (Coltsfoot) Turner.
FOOT, Calf's 1. *Allium vineale* (Crow Garlic) B & H. 2. *Arum maculatum* (Cuckoo-pint) Gerard. 3. *Sinapis arvensis* (Charlock) Glos. J.D.Robertson; War. Grigson. 4. *Tussilago farfara* (Coltsfoot) Som. Macmillan. The shape of the leaf is the usual criterion for 'foot' names, but that can hardly be the case with Crow Garlic.
FOOT, Cat's 1. *Antennaria dioica*. 2. *Antennaria neglecta* USA Zenkert. 3. *Glechoma hederacea* (Ground Ivy) N.Eng. Prior. 4. *Gnaphalium polycephalum* (Mouse-ear Everlasting) Grieve.1931. Sometimes written as one word, i.e. Catsfoot.
FOOT, Cat's, Lesser *Antennaria neodioica*.
FOOT, Cock *Chelidonium majus* (Greater Celandine) B & H.
FOOT, Cock's *Aquilegia vulgaris* (Columbine) B & H; USA Hutchinson.
FOOT, Cow *Saponaria officinalis* (Soapwort) S.Africa Watt.
FOOT, Craa's *Lotus corniculatus* (Bird's-foot Trefoil) N'thum. Grigson.
FOOT, Crow 1. *Endymion nonscriptus* (Bluebell) Rad, Lancs, Cumb. B & H. 2. *Lotus corniculatus* (Bird's-foot Trefoil) Som, Glos, Suff. Grigson. 3. *Ranunculus sp.* All the buttercups are known as some kind of Crowfoot - it is usual to spell it as one word rather than two.
FOOT, Crow, Crane's-bill *Geranium pratense* (Meadow Cranesbill) Prior.
FOOT, Crow, Yellow *Anthyllis vulneraria* (Kidney Vetch) Bucks. Grigson.
FOOT, Crow's *Ranunculus bulbosus* (Bulbous Buttercup) Suss. B & H.
FOOT, Dove *Geranium molle* (Dove's-foot Cranesbill) Turner.
FOOT, Dove's 1. *Aquilegia vulgaris* (Columbine) B & H. 2. *Geranium maculatum* (Spotted Cranesbill) O P Brown.
FOOT, Drake's *Orchis mascula* (Early Purple Orchis) Lincs. Gutch.
FOOT, Duck's 1. *Alchemilla vulgaris* (Lady's Mantle) Dur, Berw. Grigson. 2. *Podophyllum peltatum* (May-apple) J Smith. The shape of the leaf accounts for the name as far as Lady's Mantle is concerned. The generic name for May-apple, *Podophyllum*, comes from podos, foot, and phyllon, leaf.
FOOT, Elephant's 1. *Beaucarnea recurvata* (Pony Tail Plant) USA Whiting. 2. *Testudinaria elephantipes*. The latter plant has a large, woody base, hence the name, and Pony Tail Plant has a similar swollen base.
FOOT, Englishman's *Plantago major* (Great Plantain) Leyel. Cf. the name White Man's Footprints. There is a well-known legend of its persistently following the tracks of man. More specifically, one superstition says that it follows Englishmen, and springs up in whatever part of the world he makes his home.
FOOT, Foal's 1. *Glechoma hederacea* (Ground Ivy) B & H. 2. *Tussilago farfara* (Coltsfoot) Kent, Suff. Grigson; N'thants. A.E.Baker; Leics. A.B.Evans; Yks. Nicholson. In both cases, it is the leaf shape which suggests the name.
FOOT, Foile *Tussilago farfara* (Coltsfoot) Yks. B & H. Foal-foot, presumably.
FOOT, Frog's 1. *Ranunculus bulbosus* (Bulbous Buttercup) Grieve.1931. 2. *Ranunculus ficaria* (Lesser Celandine) Som. Grigson.
FOOT, Goat's 1. *Aegopodium podagraria* (Goutweed) Dev. Macmillan. 2. *Helianthemum chamaecistus* (Rock Rose) Som. Macmillan. 3. *Tragopogon pratensis* (Goat's Beard) Som. Macmillan. All probably in error for something else - 'Gout' can easily become 'goat', and a goat's 'beard' can become its 'foot' if the reason for neither is apparent.
FOOT, Hawk's *Aquilegia vulgaris* (Columbine) B & H.
FOOT, Hen's 1. *Caucalis daucoides* (Bur Parsley) Gerard. 2. *Corydalis claviculata* (Climbing Corydalis) B & H. Hen's Foot for the Corydalis is a translation of Pliny's name for it, pes gallinaceus.
FOOT, Lamb's 1. *Alchemilla vulgaris* (Lady's Mantle) Lancs. Grigson. 2. *Anthyllis*

vulneraria (Kidney Vetch) Som. Macmillan. 3. *Lotus corniculatus* (Bird's-foot Trefoil) Lancs. Grigson. 4. *Plantago major* (Great Plantain) Cumb. Grigson.

FOOT, Lion's 1. *Alchemilla vulgaris* (Lady's Mantle) Cockayne. 2. *Helleborus niger* (Christmas Rose) B & H. 3. *Leontopodium alpinum* (Edelweiss) Parkinson.1640. 4. *Prenanthes alba* (White Lettuce) USA Cunningham.

FOOT, Padda *Teucrium africanum* S.Africa Watt.

FOOT, Pig's *Lotus corniculatus* (Bird's-foot Trefoil) Suff. Grigson.

FOOT, Pigeon's 1. *Geranium molle* (Dove's-foot Cranesbill) Gerard. 2. *Salicornia europaea* (Glasswort) Howes.

FOOT, Pussy *Trifolium repens* (White Clover) Som. Macmillan.

FOOT, Ram's 1. *Geum urbanum* (Avens) Dev. Friend.1882. 2. *Ranunculus aquatilis* (Water Crowfoot) B & H.

FOOT, Rat's *Glechoma hederacea* (Ground Ivy) Dev. Macmillan.

FOOT, Sheep *Lotus corniculatus* (Bird's-foot Trefoil) Cumb. Grigson.

FOOT, Snake's *Polygonum bistorta* (Bistort) Som. Macmillan. This must surely be Snake's Food.

FOOT, Sow *Tussilago farfara* (Coltsfoot) Yks. Grigson.

FOOT, Traveller's *Plantago major* (Great Plantain) War. Grigson. See Foot, Englishman's.

FOOTPRINTS, White Man's *Plantago major* (Great Plantain) USA Watt. See Foot, Englishman's.

FOR-BETE; FORBITTEN MORE; FOREBIT *Succisa pratensis* (Devil's Bit) Halliwell (For-bete); Prior (Forbitten more); Gerard (Forebit). The reference is to the legend that the short, blackish root was originally bitten off by the devil, because he knew that otherwise it would be good for many purposes. Another version says that the plant was a bane to mankind, and that God took the power away from the devil, who bit the root off in vexation. 'More', in Forbitten-more, has the sense of 'root', strictly mor. Ofbit, or Ofbitten, are medieval names for this plant.

FORD'S ELM *Ulmus glabra var. exoniensis* (Exeter Elm) Wilkinson.1978.

FOREFATHER'S CUP *Sarracenia purpurea* (Pitcher Plant) Perry.

FOREST YAM *Dioscorea praehensilis* (Bush Yam) Dalziel.

FORGET-ME-NOT 1. *Ajuga chamaepitys* (Yellow Bugle) Henslow. 2. *Myosotis sp.* 3. *Veronica chamaedrys* (Germander Speedwell) Dev, Ches, Yks, Cumb, Border B & H. 4. *Viola tricolor* (Pansy). There are many stories of the origin of the Forget-me-not (Myosotis). One of the best-known of them concerns a knight and his lady, strolling on the banks of the Danube. They saw a spray of blue flowers floating on the water. She wanted them, and was sorry to see them float past, so he dived in, but the current was too strong. He got the flowers, and flung them to the bank, and was then swept away, calling "Forget-me-not".

FORGET-ME-NOT, Alpine *Myosotis alpestris.*

FORGET-ME-NOT, Bur *Lappula myosotis.*

FORGET-ME-NOT, Cape *Anchusa capensis* (Cape Alkanet) W.Miller. Forget-me-not - like flowers.

FORGET-ME-NOT, Changing *Myosotis discolor.* The flowers start yellow, and then change to blue.

FORGET-ME-NOT, Chatham Isle *Myosotidium hortensia.* A New Zealand native.

FORGET-ME-NOT, Chinese *Cynoglossum amabile* Howes.

FORGET-ME-NOT, Creeping 1. *Myosotis repens.* 2. *Omphalodes verna* (Winter Forget-me-not) W.Miller.

FORGET-ME-NOT, Early *Myosotis hispida.*

FORGET-ME-NOT, Field *Myosotis arvensis.*

FORGET-ME-NOT, Jersey *Myosotis sicula.* Confined to Jersey.

FORGET-ME-NOT, Little *Myosotis arvensis* (Field Forget-me-not) Som. Macmillan.

FORGET-ME-NOT, Mountain *Myosotis sylvatica* (Wood Forget-me-not) Manker.
FORGET-ME-NOT, Tufted *Myosotis caespitosa*.
FORGET-ME-NOT, Water 1. *Myosotis palustris*. 2. *Myosotis scorpioides*.
FORGET-ME-NOT, Water, Creeping *Myosotis secunda*.
FORGET-ME-NOT, Water, Northern *Myosotis brevifolia*.
FORGET-ME-NOT, Winter *Omphalodes verna*.
FORGET-ME-NOT, Wood *Myosotis sylvatica*.
FORGET-ME-NOT, Yellow *Amsinckia intermedia* (Tarweed) USA Salisbury.1964.
FORK, Knife-and- *Geranium robertianum* (Herb Robert) Som. Macmillan; Bucks. Grigson.
FORKS, Knives-and- *Acer pseudo-platanus* (Sycamore) Kent Grigson.
FORMOSA LILY *Lilium formosanum*.
FORMOSAN CYPRESS *Chamaecyparis formosensis*.
FORMOSAN JUNIPER *Juniperus formosana*.
FORSYTHIA, Weeping *Forsythia suspensa*.
FORSYTHIA, White *Abeliophyllum distichum*.
FORTUNE-TELLER *Taraxacum officinale* (Dandelion) B & H; USA Henkel. Children use the seed globes to find out more than the time - how many years before they are married, how many children they will have, or how many years they have to live. American children will blow a dandelion globe three times. If all the seeds blow away it is a sign that their mother doesn't want them. Another American divination is to award each puff to blow off the seeds a letter of the alphabet; the letter which ends the blowing is the initial of the name of the person you will marry.
FOSSIL TREE *Metasequioa glyptostroboides* (Dawn Redwood) Hart. Before the discovery of living specimens in 1945, it had only been known to science as a fossil - hence this name.

FOUETS *Sempervivum tectorum* (Houseleek) Scot. Jamieson. An extremely variable name from Scotland and the northern counties of England. Kay Sanecki said that Fow was a word which meant to cleanse; but Calvin Podd, in PLNN7, says that these names all mean drunk in northern dialect. Interesting when the famous name Welcome-home-husband-though-never-so-drunk is considered.
FOUL-GRASS *Agropyron repens* (Couchgrass) Norf. V G Hatfield.1994. An understandable reaction by anyone trying to eradicate it.
FOULRUSH *Euonymus europaeus* (Spindle Tree) Bucks. Grigson. Whatever the meaning of the first syllable, the second means a shrub. The OE was risce, which accounts for the 'rise' of Gadrise, too, for Spindle.
FOUNTAIN DRACAENA *Dracaena indivisa*.
FOUNTAIN FLOWER *Ceropegia sandersonii*.
FOUNTAIN GRASS, African *Pennisetum setaceum*.
FOUNTAIN TREE 1. *Casuarina stricta* (She Oak) A Huxley. 2. *Spathodea campanulata* (African Tulip Tree) Howes. This name,and that of Bleeding Tree, are given to the She Oak because of the tree's ability to yield water from the trunk and branches.
FOUR-LEAVED CLOVER *Oxalis deppei* Hohn. Cf. Good Luck Plant, another American name for this plant.
FOUR-LEAVED GRASS; FOUR-LEAVED TRUELOVE *Paris quadrifolia* (Herb Paris) B & H (Grass); Midl. Tynan & Maitland (Truelove). '*Quadrifolia*' is four-leaves.
FOUR-O'CLOCK 1. *Mirabilis dichotoma* (Marvel of Peru) Friend. 2. *Mirabilis jalapa* A W Smith. 3. *Oenothera biennis* (Lesser Evening Primrose) Dev. B & H. 4. *Ornithogalum umbellatum* (Star of Bethlehem) New Engl. Leighton. 5. *Taraxacum officinale* (Dandelion) Dev, Som. Macmillan. Mirabilis flowers sometimes open very late in the day, so of course do the evening primroses. Star of Bethlehem

gets the name for the opposite reason - it opens early, and closes early. Four-o'clock for the dandelion has nothing to do with either of them - this is a name given from the children's use of the seed globes to find out what the time is.

FOUR-O'CLOCK, Giant *Mirabilis froebellii*.

FOUR-O'CLOCK, Pale-stemmed *Mirabilis aspera*.

FOUR-O'CLOCK, Wild 1. *Mirabilis linearis* USA Gates. 2. *Mirabilis nyctaginea* USA Gates.

FOUR-O'CLOCK MILKWEED *Asclepias nyctaginifolia*.

FOUR-SEEDED VETCH *Vicia tetrasperma* (Smooth Tare) Hutchinson.

FOUR SISTERS *Polygala vulgaris* (Milkwort) Ire Grigson. From the different colours in the flower.

FOUR-WING SALTBUSH *Atriplex canescens* (Chamizo) Kluckhohn.

FOUR-WING SOPHORA *Sophora tetraptera* (Yellow Kowhai) Hora.

FOUSE; FOUZE; FOW *Sempervivum tectorum* (Houseleek) PLNN7 (Fouse, Fouze); Sanecki (Fow). Kay Sanecki said that Fow meant to cleanse, but Calvin Podd, in PLNN7, pointed out that the variations all meant 'drunk' in some way in northern dialects. Cf. the famous name, Welcome-home-husband-though-never-so-drunk.

FOWLER'S SERVICE *Sorbus aucuparia* (Rowan) B & H. Explained as a reference to the berries being used as bait to catch blackbirds, etc. This sounds reasonable, for *aucuparia* is from Latin auceps, a fowler. 'Service' is the name given to other members of the genus, and is from serves, plural of serve, from OE syrfe, itself from Latin sorbus, the generic name.

FOWLFOOT *Eleusine indica* (Goosegrass) Trinidad Laguerre.

FOX-AND-CUBS *Hieraceum aurantiacum* (Orange Hawkweed) Coats.

FOX-AND-HOUNDS *Linaria vulgaris* (Toadflax) Lincs. Grigson.

FOX-AND-LEAVES; FOX-LEAVES; FOX-TER-LEAVES *Digitalis purpurea* (Foxglove) Donegal B & H; Roxb. B & H (Foxter).

FOX BERRY *Callicarpa americana* (French Mulberry) Alabama Harper.

FOX DOCKEN *Digitalis purpurea* (Foxglove) Yks. Grigson.

FOX FEET *Urostachys selago* (Mountain Clubmoss) B & H.

FOX FINGER *Digitalis purpurea* (Foxglove) Yks. Grigson. A variation on the foxglove theme.

FOX GERANIUM; FOX GRASS *Geranium robertianum* (Herb Robert) Border B & H. 'Fox' in this case probably has something to do with the smell. The leaves, said Gerard, are "...of a most loathsome stinking smell".

FOX GRAPE *Vitis labrusca*.

FOX POISON *Daphne laureola* (Spurge Laurel) Lincs. Grigson. The fruits, of course, are poisonous.

FOX ROSE *Rosa pimpinellifolia* (Burnet Rose) War. B & H.

FOX'S BRUSH *Centranthus ruber* (Spur Valerian) Som. Macmillan; N.Ire. Grigson. Descriptive.

FOX'S CLOTE *Arctium lappa* (Burdock) Fernie.

FOX'S GLOVE *Digitalis purpurea* (Foxglove) Oxf. Grigson.

FOX'S MEAT *Oxalis acetosella* (Wood Sorrel) Corn. Grigson. Wood Sorrel, in folklore, seems to provide food for all sorts of creatures - fox, hare, rabbit, but more widespread, the cuckoo.

FOX'S MOUTH *Aconitum napellus* (Monkshood) Som. Macmillan.

FOXBANE *Aconitum vulparia* (Wolfsbane) Howes.

FOXBERRY *Vaccinium vitis-idaea* (Cowberry) Brouk.

FOXES, Smell 1. *Anemone nemerosa* (Wood Anemone) Som, Bucks. Grigson; Hants. Fernie. 2. *Geranium robertianum* (Herb Robert) Tynan & Maitland.

FOXFLOP *Digitalis purpurea* (Foxglove)

Som. Macmillan. The inference of this name is that the plant has grown from fox droppings. Cf. Cowslip and Cowflop, where 'slip' and 'slop' have exactly the same meaning.

FOXGLOVE 1. *Campanula latifolia* (Giant Bellflower) Yks. Grigson. 2. *Digitalis purpurea*. 3. *Gladiolus sp* Dev. Friend.1882. 4. *Verbascum thapsus* (Mullein) B & H. Of all plant names, foxglove seems to have exercised the minds of etymogolists the most. The OE is foxesglafa, which has been translated both Fox's and Folk's (i.e. fairy's) glove. Friend suggested that the OE may have been foxes gleow (fox music), gleow meaning glee, and referring to a tintinnabulum. Certainly the plant exactly resembles this; and Cf. the Norwegian Reveleika, which also means fox music. The other two plants quoted above get the name from similarity of growth.

FOXGLOVE, Bastard *Maurandya scandens*. It was Curtis who gave it this name; the flowers do bear some resemblance to that of a foxglove.

FOXGLOVE, Blue *Campanula trachelium* (Nettle-leaved Bellflower) Shrop. Grigson.

FOXGLOVE, Fairy *Erinus alpinus* McClintock.

FOXGLOVE, Lady's *Verbascum thapsus* (Mullein) B & H; USA Henkel.

FOXGLOVE, Mountain *Ourisia macrocarpa*.

FOXGLOVE, Rusty *Digitalis ferruginea*. 'Rusty', presumably, because of the brown veining of the flowers.

FOXGLOVE, White *Campanula latifolia* (Giant Bellflower) Lancs. Grigson.

FOXGLOVE, Wild *Penstemon grandiflorus* (Large-flowered Beardtongue) USA Gilmore.

FOXGLOVE, Woolly *Digitalis lanata*.

FOXGLOVE, Yellow, Large *Digitalis ambigua*.

FOXGLOVE, Yellow, Small *Digitalis lutea*.

FOXGLOVE TREE *Paulownia tomentosa*. "A name suggested both by the shape of the flowers and by the way they grow".

FOXSTONES *Orchis mascula* (Early Purple Orchis) B & H. One of a number of names whose origin lie in the twin tubers, reminding one inevitably of testicles.

FOXTAIL 1. *Corylus avellana* catkins (Hazel catkins) Som. Macmillan. 2. *Equisetum arvense* (Common Horsetail) USA Kingsbury. 3. *Equisetum palustre* (Marsh Horsetail) USA Kingsbury. 4. *Stellaria media* (Chickweed) Som. Macmillan.

FOXTAIL, Alpine *Alopecurus alpinus*.

FOXTAIL, Bristly 1. *Setaria barbata*. 2. *Setaria viridis* (Bottle Grass) USA Cunningham.

FOXTAIL, Bulbous *Alopecurus bulbosus*. Bulbous, in the sense that there is a marked swelling at the base of the stem.

FOXTAIL, Field *Alopecurus myosuroides* (Slender Foxtail) H C Long.1910.

FOXTAIL, Floating *Alopecurus geniculatus* (Marsh Foxtail) H C Long.1910.

FOXTAIL, Green *Setaria viridis* (Bottle-grass) USA Allan.

FOXTAIL, Marsh *Alopecurus geniculatus*.

FOXTAIL, Meadow *Alopecurus pratensis*.

FOXTAIL, Orange *Alopecurus aequalis*.

FOXTAIL, Slender *Alopecurus myosuroides*.

FOXTAIL BROME *Bromus rubens*.

FOXTAIL LILY *Eremurus sp.*

FOXTAIL MILLET *Setaria italica*.

FOXTAIL PINE 1. *Pinus aristata* (Bristlecone Pine) T H Kearney. 2. *Pinus balfouriana*.

FOXTER; FOXTER-LEAVES; FOXTREE *Digitalis purpurea* (Foxglove) Scot. Grigson (Foxter); Roxb. B & H (Foxter-leaves); Dalyell (Foxtree).

FOXWORT *Ranunculus ficaria* (Lesser Celandine) Som. Macmillan.

FOXY; FOXY-LEAVES *Digitalis purpurea* (Foxglove) Ire. Grigson (Foxy); Tynan & Maitland (Foxy-leaves).

FRA-FRA POTATO; FURA-FURA POTATO *Coleus rotundifolius* Schery (Fura-Fura). This is extensively cultivated in West Africa for the tubers, hence the potato epithet.

FRAGHAN; FRUGHAN; FRAUGHAN; FRAWN *Vaccinium myrtillus* W.Miller (Fraghan, Frughan); O Suileabhain (Fraughan); Fernie (Frawn). All spelling variations on the Gaelic fraochain.

FRAGRANT BEDSTRAW *Galium triflorum.*

FRAGRANT BITTERWEED *Hymenoxis odorata* (Bitterweed) USA T H Kearney.

FRAGRANT EVENING PRIMROSE *Oenothera stricta.*

FRAGRANT GOLDEN ROD *Solidago graminifolia.*

FRAMBOISE BUSH; FRAMBOYS *Rubus idaeus* (Raspberry) Gerard (Framboise); B & H (Framboys). Framboise is the French name for raspberry, a word "of disputed origin", according to ODEE.

FRAMLINGTON CLOVER *Prunella vulgaris* (Self-heal) N'thum. Denham/Hardy. Longframlington, in Northumberland, is a village not far from Rothbury.

FRANCE, Fair Maid of 1. *Achillea ptarmica* (Sneezewort) Middx. B & H. 2. *Leucanthemum vulgare* (Ox-eye Daisy) Som, Wilts. Grigson. 3. *Lychnis flos-cuculi* (Ragged Robin) B & H. 4. *Ranunculus aconitifolius* (White Buttercup) B & H. 5. *Saxifraga granulata* (Meadow Saxifrage) Bucks. Grigson. The link would seem to be the white flowers, but then Ragged Robin is included here.

FRANCKE SPURRY; FRANCKING SPURWORT; FRANKE *Spergula arvensis* (Corn Spurrey) Parkinson.1641 (Francke Spurry, Francking Spurwort); Prior (Franke). Franke seems to mean a stall in which cattle were shut up to be fattened. Corn Spurrey was certainly grown to fatten cattle, and still is on the Continent.

FRANGIPANI *Plumeria alba.* Legend has it that a 12th century Italian marquis named Frangipani created an exquisite perfume for scenting gloves by combining certain volatile oils. European settlers in the Caribbean, four hundred years later, discovered a plant whose flower had a similar perfume, so it was naturally called Frangipani.

FRANKFURT ROSE *Rosa francofurtiana.*

FRANKINCENSE *Boswellia thurifera.* From the Old French franc encens; frank seems to mean 'of superior quality'.

FRANKINCENSE, Bible *Boswellia carteri.*

FRANKINCENSE, Elemi *Boswellia frereana.* The stem yields a pale yellow, lemon-scented balsam known as African Elemi.

FRANKINCENSE, Indian *Boswellia serrata.*

FRANKINCENSE, Male *Boswellia thurifera* (Frankincense) Pomet.

FRANKINCENSE PINE *Pinus taeda* (Shortleaf Pine) Barton & Castle. The frankincense used in Europe in modern times is a mixture of the gums of this + *Picea abies* + *Thuja occidentalis.*

FRAP *Digitalis purpurea* (Foxglove) IOM Moore.1924. Onomatopeic, and akin to the many 'pop' names for foxglove, and given for the same reason.

FRASER'S BALSAM FIR *Abies fraseri.*

FRATA *Solanum tuberosum* (Potato) Ire. Salaman. Pratie, presumably the same word, is more common. They are both derived via "prata" from patata.

FRAUGHANS; FRAWNS *Vaccinium myrtillus* (Whortleberry) Ire. O Suileabhain (Fraughans); Fernie (Frawns). From Gaelic fraochain.

FRAW-CUP; FROCKUP *Fritillaria meleagris* (Snake's-head Lily) PLNN 17 (Fraw-cup); Bucks. Grigson (Frockup). Grigson suggested frog-cup as the original, but Alice Coats had Frorechap, which sounds the same. Nevertheless, the existence of Crowcup in the same area suggests that Fraw-cup could be a mistake for Crawcup.

FRAXINELL *Polygonatum multiflorum* (Solomon's Seal) W.Miller.

FRAXINELLA *Dictamnus fraxinella* (False Dittany) (Moldenke). "because of the resemblance of them unto young Ashes, in their winged leaves" (Parkinson).

FRAY *Ulex europaeus* (Furze) Leyel.1937.

FRECKLE FACE *Hypoestis phyllostachya*

(Polka Dot Plant) Bonar. Because of the pink dots that cover the leaves.

FRECKLED FACE *Primula veris* (Cowslip) Som. Macmillan. Descriptive, presumably, but is this the recognition of the signature for the well-known ointment for the complexion made from cowslips?

FREE HOLLY *Ilex aquifolium* (Holly) Shrop. B & H. Applied to the smooth-leaved variety, so it is either saying that this is the kind which is 'free' of prickles, or else it is a corruption of She-holly, which is a common enough name for this variety.

FREISER *Fragaria vesca* (Wild Strawberry) Halliwell.

FREMONT COTTONWOOD *Populus fremontii*.

FREMONT HOLLY-GRAPE *Berberis fremontii* (Desert Barberry) USA Jaeger.

FREMONT THORNBUSH *Lycium fremontii*.

FRENCH *Onobrychis sativa* (Sainfoin) Wilts. Dartnell & Goddard.1899. Cf. French Grass, or French Hay, from the same area. The word sainfoin is French, and means healthy hay, i.e. a dried crop good for cattle.

FRENCH ALPINE JUNIPER *Juniperus thurifera* (Spanish Juniper) Grey-Wilson.

FRENCH-AND-ENGLISH GRASS; FRENCH-AND-ENGLISH WEED; FRENCH-AND-ENGLISH SOLDIERS *Plantago lanceolata* (Ribwort Plantain) All Tynan & Maitland. 'French and English' must be part of the children's game played with the flowering stalks, on the principle of conkers.

FRENCH ARTICHOKE *Cynara cardunculus var. scolymus* (Globe Artichoke) Schery.

FRENCH ASH *Laburnum anagyroides* (Laburnum) Derb. B & H. Is there some resemblance to ash leaves?

FRENCH ASPARAGUS *Ornithogalum pyrenaicum* (Spiked Star-of-Bethlehem) Wilts. D. Grose. French Sparrow-grass, and French Grass, occur too, but this is better known as Bath Asparagus, for it grows around Bath, and is still picked to be sold in markets in Bath and Bristol, to be eaten like asparagus.

FRENCH BAY *Chamaenerion angustifolium* (Rosebay) Sanecki. A name apparently given by William Robinson, *The English flower garden*.

FRENCH BEAN *Phaseolus vulgaris* (Kidney Bean) Wit. Because they are commonly grown in France, but, more to the point, they were introduced from there; French Bean as a name was already being used by 1572.

FRENCH-BERRIES *Rhamnus cathartica* the unripe berries (Buckthorn) B & H.

FRENCH BOX *Buxus sempervirens var. suffruticosa* (Dutch Box) Genders.1971.

FRENCH BRUM *Laburnum anagyroides* (Laburnum) Shrop. B & H. Cf. French Ash for this. Presumably in this case 'brum' is not broom, but a contraction of, or approximation to, laburnum.

FRENCH CHESTNUT *Castanea sativa* (Sweet Chestnut) Brimble.1948. One finds them labelled Italian and Spanish, as well as French. They simply mean that this is a European Chestnut, and that the countries mentioned were the ones who exported the nuts. It was said, though, that the best nuts came from Spain.

FRENCH COWSLIP *Primula auricula* (Auricula) Parkinson. 'French', to show it is a foreign plant.

FRENCH CRANESBILL *Geranium endressii*. A plant of the western French Pyrenees.

FRENCH CRESS *Barbarea vulgaris* (Winter Cress) B & H.

FRENCH ENDIVE *Cichorium intybus* (Chicory) Schery. Belgian Endive, sometimes, too.

FRENCH FENNEL *Foeniculum dulce* (Finocchio) Grieve.1933.

FRENCH FIGWORT *Scrophularia canina*.

FRENCH FUZZ *Ulex europaeus* (Gorse) Corn, Dev. Ire. Grigson.

FRENCH GARLIC *Allium scordoprasum* (Sand Leek) G.B.Foster.

FRENCH GRASS 1. *Onobrychis sativa* (Sainfoin) Som. Macmillan; Wilts. Dartnell & Goddard; Hants. Grigson. 2.

Ornithogalum pyrenaicum (Spiked Star-of-Bethlehem) Som. B & H; Wilts. D Grose. Sainfoin gets the name (and also French, or French Hay) because it is recognised that this is a foreign introduction. 'Grass' because this is a fodder crop; but 'grass' for the Star-of-Bethlehem is a corruption of 'asparagus'. French Asparagus, or French Sparrow-grass, was the name under which the sprouts were sold in the Bath markets, to be eaten as if they were indeed asparagus.

FRENCH GUAVA *Cassia alata* (Candlestick Senna) Chopra.

FRENCH GUAVA, Wild *Cassia occidentalis* (Coffee Senna) Barbados Gooding.

FRENCH HALES *Sorbus latifolia*. 'French', because this is a tree of western and central Europe, not confined to France though. Note the other name, Service Tree of Fontainebleau. 'Hales' is apparently halse, i.e. hazel.

FRENCH HARDHEAD *Centaurea jacea*. 'French' probably just because (from the English point of view), this is a foreign plant. But it is very widespread, occurring all over Europe, North Africa, North and West Asia. 'Hardhead' is a common name for any of the knapweeds, and comes from the boll below the flowers.

FRENCH HAY *Onobrychis sativa* (Sainfoin) Wilts. D.Grose. Cf. French, French Grass, for this.

FRENCH JASMINE *Jasminum officinalis* (Jasmine) Turner.

FRENCH JUJUBE *Zizyphus jubajuba* fruit (Willis). Jujube is a version of the specific name, *jubajuba*. The berries have been famous since ancient times for cold cures and bronchitis. They were made up into lozenges and widely exported - such lozenges are still called jujubes, whether there is any content or not of these berries.

FRENCH LAVENDER 1. *Lavandula dentata* (Toothed Lavender) RHS. 2. *Lavandula stoechas* (Spanish Lavender) Gerard.

FRENCH LEEK *Allium porrum* (Leek) Lyte.

FRENCH LILAC *Galega officinalis* (Goat's Rue) Clapham.

FRENCH LUNGWORT *Hieraceum murorum* (Wall Hawkweed) Gerard.

FRENCH MALLOW 1. *Lavatera olbia*. 2. *Malva sylvestris* (Common Mallow) Corn. Grigson.

FRENCH MARIGOLD *Tagetes patula*. Not French, of course, any more than the African Marigold, *Tagetes erecta*, is African. These are Mexican species.

FRENCH MARJORAM *Origanum onites*.

FRENCH MERCURY *Mercurialis annua*.

FRENCH MIGNONETTE 1. *Dianthus chinensis* (Indian Pink) W.Miller. 2. *Saxifraga umbrosa* (Pyrenean Saxifrage) W.Miller. The latter should probably be London Pride rather than the species quoted by Miller.

FRENCH MOSS *Sedum acre* (Biting Stonecrop). Moss is fairly obvious to describe this, and occurs several times in different combinations.

FRENCH MULBERRY *Callicarpa americana*. Not French, and not a mulberry either. It comes from the southern states of North America, and doesn't even belong to the same family as the mulberry.

FRENCH MUSTARD *Erysimum cheiranthoides* (Treacle Mustard) Folkard.

FRENCH NETTLE *Lamium purpureum* (Red Deadnettle) Shrop. Grigson.

FRENCH NUT 1. *Castanea sativa* (Sweet Chestnut) B & H. 2. *Juglans regia* (Walnut) Dev. Earle; Som. Jennings. Just as the 'wal' part of walnut means foreign, so does 'French' in both cases. The qualification is necessary, for a nut in English could only mean hazel.

FRENCH ONION *Scilla maritima* (Sea Squill) Turner.

FRENCH PAIGLE *Pulmonaria officinalis* (Lungwort) Ess. Gepp. Cf. the many 'cowslip' names given to the plant.

FRENCH PHYSIC NUT *Jatropha curcas* (Barbados Nut) Perry.

FRENCH PINK 1. *Armeria maritima* (Thrift) Dev. Friend.1882. 2. *Dianthus chi-*

nensis (Indian Pink) Dev. Friend.1882; Som. Elworthy. 'French', again, in the sense of not being native - a foreign plant, no matter whence it actually comes.

FRENCH PLANTAIN *Musa x paradisiaca* (Banana) H & P.

FRENCH POPPY *Verbascum thapsus* (Mullein) Dev. Macmillan.

FRENCH PRIMROSE *Primula veris x vulgaris* (False Oxlip) Som. Macmillan.

FRENCH ROSE *Rosa gallica*.

FRENCH RYE GRASS *Arrhenatherium elatius* (False Oat) Howes.

FRENCH SALLY *Salix pentandra* (Bay-leaved Willow) Donegal Grigson.

FRENCH SAUGH *Chamaenerion angustifolium* (Rosebay) Lanark Grigson. French Willow, too, this time from Somerset. There are lots of 'willow' names for this, including of course Willowherb; it is the resemblance of the leaves to those of the willow which inspires the names.

FRENCH SAXIFRAGE *Saxifraga clusii*.

FRENCH SNOWDROP *Ornithogalum umbellatum* (Star-of-Bethlehem) A.E.Baker. Cf. the American name Snowdrop, and the Devonshire Snowflake, or Summer Snowflake.

FRENCH SORREL 1. *Oxalis acetosella* (Wood Sorrel) B & H. 2. *Rumex scutatus* (Sorrel) Brownlow. Sorrel is very popular in France for flavouring omelettes and sauces, hence this name.

FRENCH SPARROW-GRASS *Ornithogalum pyrenaicum* (Spiked Star-of-Bethlehem) W.Miller. Sparrow-grass = asparagus. This grows around Bath, and is often known and used as Bath asparagus.

FRENCH SPIKENARD *Valeriana celtica* Turner.

FRENCH SPINACH 1. *Atriplex hortensis* (Garden Orache) G.B.Foster. 2. *Rumex scutatus* (Sorrel) Wit. Sorrel grows in clumps, like spinach; French, because it is so popular there.

FRENCH TAMARISK *Tamarix gallica* Brimble.1948.

FRENCH TARRAGON *Artemisia dracunculus* (Tarragon) Rohde.1936. This is more often used in French cooking than in English, hence the differentiation.

FRENCH THYME *Thymus zygis* Brouk.

FRENCH TOADFLAX *Linaria arenaria* (Sand Toadflax) McClintock. The plant's main range is on coastal dunes in France.

FRENCH WHEAT *Fagopyrum esculentum* (Buckwheat) Gerard. 'French' to indicate a foreign plant. It is Asian, as another name, Saracen's Corn, shows.

FRENCH WILLOW 1. *Chamaenerion angustifolium* (Rosebay) Som. Grigson. 2. *Salix triandra*. 3. *Thevetia peruviana* (Yellow Oleander) Howes.

FRENCH WORMWOOD *Artemisia maritima* (Sea Wormwood) Turner.

FRENCHWEED *Thlaspi arvense* (Field Pennycress) USA Gates.

FREYN *Fraxinus excelsior* (Ash) B & H.

FREZ *Ulex europaeus* (Furze) N'thum. B & H. There are a lot of different ways of spelling furze, according to local pronunciation.

FRIAR'S CAP *Aconitum napellus* (Monkshood) Dev. Friend.1882.

FRIAR'S COWL *Arum maculatum* (Cuckoo-pint) B & H. Descriptive. Cf. Priest's Hood.

FRIAR'S CROWN *Cirsium eriophorum* (Woolly Thistle) B & H. The reference is to the bald patch left after the seeds have flown. Dandelion has similar names.

FRIED CANDLESTICKS *Orchis mascula* (Early Purple Orchis) Som. Macmillan. Candlesticks is probably Gander-gosses, which occurs in various forms, like Gandigoslings, Giddy Gander, etc.

FRIED EGGS *Leucanthemum vulgare* (Ox-eye Daisy) Som, Wilts. Macmillan. Descriptive.

FRIEND, Old Man's *Anagallis arvensis* (Scarlet Pimpernel) Som. Macmillan.

FRIEND, Poor Man's *Clematis vitalba* (Old Man's Beard) Som. Macmillan. Cf. Shepherd's Delight, from the same area. Is it because the stalks were a kind of tobacco substitute, when times were hard?

FRIEND, Shepherd's *Sorbus aucuparia* (Rowan) Dor. Macmillan. There might be any number of reasons for this. After all, rowan was the universal protector.

FRIEND, Vagabond's *Polygonatum multiflorum* (Solomon's Seal) Cumb. Wright. Possibly because an ointment used to be made from the leaves to apply to a bruise or a black eye. Gypsies certainly used it thus, and so did Fenland people.

FRIENDSHIP, Everlasting *Galium aparine* (Goose-grass) Hutchinson & Melville. Is this a genuine name? It smacks of Gerard's remark, that the plant is called "of some, Philanthropos, as though he should say, a man's friend, because it taketh hold of a man's garments." Cf. such names as Sweethearts, and Huggy-me-close.

FRIENDSHIP PLANT 1. *Billbergia nutans* (Queen's Tears) USA Cunningham. 2. *Pilea cadierei* (Aluminium Plant) Oplt. 3. *Pilea involucrata*.

FRIJOLKO *Sophora secundiflora* (Mescal Bean) Texas Kingsbury.

FRILLS, White *Bellis perennis* (Daisy) Putnam.

FRINGE, Mountain 1. *Adlumia fungosa* (Allegheney Vine) H & P. 2. *Artemisia frigida* (Fringed Wormwood) USA Painter.

FRINGE, Purple *Cotinus coggygria* (Venetian Sumach) W.Miller.

FRINGE, Water *Nymphoides peltatum* (Fringed Waterlily) RHS.

FRINGE-FLOWER *Schizanthus sp* W.Miller.

FRINGE-TREE *Chionanthus virginicus*.

FRINGED BROME *Bromus ciliatus*.

FRINGED GENTIAN *Gentiana crinita*.

FRINGED GENTIAN, Smaller *Gentiana procera*.

FRINGED ONION *Allium fimbriatum*.

FRINGED PINK *Dianthus monspessulamus*.

FRINGED RUE *Ruta chalepensis*.

FRIT *Fritillaria meleagris* (Snake's Head Lily) Berks. Grigson. A shortened version of Fritillary.

FRITH *Crataegus monogyna* (Hawthorn) Wilts. Dartnell & Goddard.

FRITILLARY; FRITILLARY, Common *Fritillaria meleagris* (Snake's Head Lily) McClintock; Curtis (Common). Latin fritillus meant a dice-box; the allusion here is to the chequer-board on which the dice were thrown - the blossoms are chequered. "Fritillaria, of the table or board upon which men play at Chesse, which square checkers the floure doth very much resemble..." (Gerard).

FRIZ *Ulex europaeus* (Furze) F Savage. An example of misplacement of consonants, very common in dialect.

FRIZZLE, Devil-in-a- *Nigella damascena* (Love-in-a-mist) Tynan & Maitland.

FROCK, Smock *Stellaria holostea* (Greater Stitchwort) Dev, Bucks. Grigson. Like bindweed, the flowers are compared to shirts, chemises, and petticoats as well as to smocks.

FROCKEN *Vaccinium myrtillus* (Whortleberry) W.Miller. Better known as Fraughan, or something like it. It is Gaelic fraochain.

FROCKUP *Fritillaria meleagris* (Snake's Head Lily) Bucks. Grigson. Grigson suggested frog-cup for this name, but Alice Coats had Frorechap, which sounds the same word.

FROG BITES *Hydrocharis morsus-ranae* (Frogbit) Som. Macmillan.

FROG CHEESE *Malva sylvestris* (Common Mallow) Oxf. B & H. 'Cheese' is suggested by the shape of the ripe seed vessels, and is a very common name in various forms for this.

FROG GRASS *Salicornia herbacea* (Glasswort) Gerard.

FROG ORCHID *Habenaria viridis*.

FROG ORCHID, White *Leucorchis albida* (Small White Orchid) Turner Ettlinger.

FROG'S FOOT 1. *Ranunculus bulbosus* (Bulbous Buttercup) Grieve.1931. 2. *Ranunculus ficaria* (Lesser Celandine) Som. Grigson.

FROG'S MEAT *Arum maculatum* (Cuckoo-pint) Dor. Macmillan.

FROG'S MOUTH 1. *Antirrhinum majus* (Snapdragon) Dor. Macmillan. 2. *Mimulus luteus* Scot. Grigson. 3. *Orchis mascula* (Early Purple Orchis) Som. Macmillan.

FROGBIT *Hydrocharis morsus-ranae.* Frogbit is of course the same as the specific name, *morsus-ranae.*

FROGFOOT *Verbena officinalis* (Vervain) Prior.

FROGWORT 1. *Orchis mascula* (Early Purple Orchis) B & H. 2. *Orchis morio* (Green-winged Orchis) B & H.

FRORECHAP *Fritillaria meleagris* (Snake's Head Lily) Coats.

FROST-BLITE *Chenopodium album* (Fat Hen) B & H. 'Blite', a name given in particular to Good King Henry, derives from bliton, meaning insipid. Calling this Frost-blite presumably means simply "white form of Good King Henry", and this is exactly what *Chenopodium album* means - White Goosefoot.

FROSTED ORACHE *Atriplex laciniata.* The whole plant is covered with white scales, giving it a silvery, so 'frosted' appearance.

FROSTWEED 1. *Helianthemum canadense.* 2. *Tuberaria guttata* (Annual Rock Rose) USA Cunningham. Frostweed, because crystals of ice shoot from its cracked bark during cold weather. The annual rock rose presumably gets the name by association, if indeed it is correct identification.

FROTHY POPPY *Silene cucubalis.* From the froth, called cuckoo-spit, often found round it. The Cumbrian name, Spatling Poppy, means the same thing.

FROZ *Ulex europaeus* (Gorse) N'thants. Sternberg. A familiar displacement of consonants in dialect - the original is furze.

FRUGHAN; FRUOG *Vaccinium myrtillus* (Whortleberry) W.Miller (Frughan); Ire. E E Evans.1942 (Fruog). From Gaelic fraochain.

FRUIT SALAD PLANT *Feijoa sellowiana* (Pineapple Guava) Bonar.

FRUTILLA *Lycium brevipes.*

FUCHSIA, Australian *Correa ssp.*

FUCHSIA, Californian *Zauschneria californica.*

FUCHSIA GOOSEBERRY; FUCHSIA-FLOWERED CURRANT *Ribes speciosum.* The latter is in RHS.

FUETS *Sempervivum tectorum* (Houseleek) Scot. J.Smith. See Fouets.

FUIGHANS *Vaccinium myrtillus* (Whortleberry) B & H. Presumably this is the same as Fraughans or its many variants.

FULL; FULLEN *Sempervivum tectorum* (Houseleek) N.Eng. Brockett (Full); N.Eng. Grigson (Fullen). This must belong to the same series as Fouets and the like. Fullen is probably no more than the old plural of Full.

FULL MOON MAPLE *Acer japonicum* (Japanese Maple) USA Hay & Synge.

FULLER'S GRASS; FULLER'S HERB *Saponaria officinalis* (Soapwort) B & H. Saponaria is from Latin sapo, soap. A lather can be got by rubbing the leaves in water, and so would be the fuller's standby for washing cloth without causing any damage to the fibres. It is still used for cleaning and restoring old tapestries.

FULLER'S TEASEL; FULLER'S THISTLE *Dipsacus fullonum var. fullonum* Fuller's teasel is the usual name, while Fuller's Thistle was used by Loudon in 1844. The name commemorates the still continuing use of the teasel in the cloth-making industry. The dried flower heads are used to raise the nap on woollen cloth, to give a softer feel to the fabric. Teasel is from OE taesel, taesan, our verb to tease, in the sense of pluck or pull.

FUMITERRE; FUMITORY *Fumaria officinalis* Fumitory is the common name; Fumiterre is Gerard's name, closer to the original, which is Old French fumeterre, medieval Latin fumus terrae, i.e. smoke of the earth. One reason is the belief that it did not spring up from seeds, but from the vapours of the earth. The root when freshly pulled up gives a strong gaseous smell like nitric acid, and this is probably the origin of the belief in its gaseous origin. Another suggestion is that the Greeks and Romans used the juice to clear the sight, and noted that, while doing so, it would make the eyes water, as smoke would.

FUMITORY, American *Fumaria indica*.
FUMITORY, Bastard's *Fumaria bastardii*.
FUMITORY, Boreau's *Fumaria muralis boraei*.
FUMITORY, Bulbous *Corydalis bulbosa*.
FUMITORY, Climbing 1. *Adlumia fungosa* (Allegheny Vine) RTHS. 2. *Corydalis claviculata* (Climbing Corydalis) Clapham. 3. *Fumaria occidentalis*.
FUMITORY, Indian *Fumaria indica* (American Fumitory) Le Strange.
FUMITORY, Martin's *Fumaria martinii*.
FUMITORY, Rampant; FUMITORY, Ramping *Fumaria capreolata* Clapham (Ramping).
FUMITORY, Wall *Fumaria muralis var. muralis*.
FUMITORY, Yellow *Corydalis luteus*.
FUNERAL CYPRESS 1. *Cupressus funebris* (White Fir) J.Smith. 2. *Cupressus sempervirens* (Common Cypress) Polunin. Cypress has always been an emblem of death and mourning. The classical myth tells of Cyparissos, who, stricken with sorrow at having killed his favourite stag, begged the gods to let him mourn for ever, so he was transformed by Apollo into a cypress tree, and in this form he still mourns. Both the Greeks and Romans called it the "mournful tree", because it was sacred to the rulers of the underworld. As such, it became the custom to plant it by the grave, and in the event of a death to put it by the house to warn those about to perform a sacred rite against going into a place polluted by a dead body. And the wood was used in classical times for making coffins - it is insect-proof, and incorruptible.
FUNNY-FACE 1. *Tropaeolum majus* (Nasturtium) Som. Macmillan. 2. *Viola tricolor* (Pansy) Som. Macmillan.
FUR; FURRA *Ulex europaeus* (Gorse) Lincs. Grigson (Fur). Norf. Grigson (Furra). i.e. furze.
FURA-FURA POTATO *Coleus rotundifolius* (Fra-Fra Potato) Schery. Edible tubers, and extensively cultivated in West Africa for them.
FURNITURE *Buxus sempervirens* (Box) Som. Macmillan. A tribute to the quality of the timber.
FURROW-WEED *Fumaria officinalis* (Fumitory) G;los. PLNN35. A habitat description.
FURROWED SAXIFRAGE *Saxifraga exarata*.
FURZE 1. *Ononis repens* (Rest-harrow) Gerard. 2. *Ulex europaeus*. Furze is OE fyrs, possibly related to OE fyrh, furh, Fir-tree.
FURZE, Autumnal, Dwarf *Ulex minor* (Small Furze) Flower.1851.
FURZE, Bed 1. *Ulex gallii* (Dwarf Furze) Hants. Grigson. 2. *Ulex minor* (Small Furze) Hants. B & H. Surprising as it might sound, furze has been used for bedding, especially in Ireland.
FURZE, Dwarf 1. *Ulex gallii*. 2. *Ulex minor* (Small Furze) Hutchinson.
FURZE, Great *Ulex europaeus* (Furze) Lyte.
FURZE, Ground *Ononis repens* (Rest-harrow) Gerard.
FURZE, Lesser *Ulex minor* (Small Furze) McClintock.
FURZE, Needle *Genista anglica* (Needle Whin) Gerard.
FURZE, Small *Ulex minor*.
FURZE, Spanish *Genista hispanica* (Spanish Gorse) W.Miller.
FURZE, Tam *Ulex minor* (Small Furze) Corn. Quiller-Couch. 'Tam' here means dwarf.
FURZEN *Ulex europaeus* (Furze) Gerard.
FUSTIC 1. *Maclura tinctoria* J.Smith. 2. *Chlorophora tinctoria*. Are these botanical names synonymous? Fustic is the principal source of natural yellow pigments, and C.tinctoria wood is the one from which it is obtained. The word is from French fustoc, and is ultimately Greek pistake, pistachio.
FUSTIC, Hungarian; FUSTIC, Young *Cotinus coggygria* (Venetian Sumach) both

Watt. It is the yellow dye which is obtainable from the wood that is known as young fustic. J Smith suggested that the name was given to distinguish it from Maclura tinctoria. Is this a synonym for Chlorophora tinctoria? The pigment from that is certainly known as Old Fustic.

FUSTIC, Old *Chlorophora tinctoria* (Fustic). To distinguish from Young Fustic, which is the pigment from Cotinus coggyria.

FUSTIC, Zante *Cotinus coggygria* (Venetian Sumach) Hutchinson & Melville. Zante is one of the Ionian islands.

FUZZ *Ulex europaeus* (Furze) Dev. Friend.1882; Wilts. Akerman; Dor, Herts, Worcs. B & H; Glos. Brill; N'thants. A.E.Baker.

FUZZ, Cornish *Ulex gallii* (Dwarf Furze) Corn. Grigson.

FUZZ, French *Ulex europaeus* (Furze) Corn, Dev, Ire. Grigson.

FUZZ, Tam *Ulex gallii* (Dwarf Furze) Corn. Quiller-Couch. Tam here means dwarf, sometimes written as 'tame'.

FUZZEN *Ulex europaeus* (Furze) Dor. B & H; N'thants. Sternberg. Probably an old plural in '-en'.

FUZZY-WEED *Artemisia dracunculoides* (Russian Tarragon) USA Gilmore.

FYFIELD PEA *Lathyrus tuberosus*. It only occurs in this country near Fyfield, in Essex.

FYNEL *Foeniculum vulgare* (Fennel) Henslow.

FYRRYS *Ulex europaeus* (Furze) B & H. One of the more extreme of the many different ways of spelling furze.

G

GA, Ewe *Bellis perennis* (Daisy) Yks. Grigson. Ewe-gowan, or Ewe-gollan, is the original.

GABON YELLOW-WOOD *Pentaclethra macrophylla* (Oil Bean) Dalziel.

GABRIEL, Angel *Lilium tigrinum* (Tiger Lily) Som. Macmillan. Presumably the reference is to trumpets.

GABRIEL'S TRUMPET *Solandra longiflora*.

GAD ROUGE *Euonymus europaeus* (Spindle Tree) Fernie. The nearest variation is Gadrise, which is itself a variant of Gatteridge, a common name for spindle. It means goat tree.

GADGEVRAW; GADJERWRAW *Leucanthemum vulgare* (Ox-eye Daisy) both Corn. Bottrell, B & H. From Cornish, meaning 'great daisy'.

GADRISE 1. *Cornus sanguinea* (Dogwood) Turner. 2. *Euonymus europaeus* (Spindle-tree) E Ang. Nall. 3. *Viburnum opulus* (Guelder Rose) E Ang. Nall. The original is Gatteridge, a common name for spindle in particular, but names for dogwood and spindle are generally interchangeable. It means goat-tree.

GAGROOT *Lobelia inflata* (Indian Tobacco) USA Henkel. Other names like Vomitwort and Pukeweed explain this.

GAITBERRY; GAITERBERRY *Rubus fruticosus* fruit (Blackberries) Scot. B & H. Two examples of a Scottish series which starts with Gatter-berries, and finishes up with Garter-berries.

GAITER-TREE *Cornus sanguinea* (Dogwood) N.Eng. B & H. This is OE gat, goat. Cf. Gatteridge, Gadrise, etc.

GAITRE *Cornus sanguinea* (Dogwood). This is Chaucer's rendering of the previous entry.

GAL; GALE; GALES *Myrica gale* (Bog Myrtle) Turner (Gal). Aber. B & H (Gales). This is the proper English name, from OE gagel, which entered a good many place names, e.g. Gailes in Staffordshire and Galsworthy, in Devon.

GALE, Scotch *Myrica gale* (Bog Myrtle) Scot. Jamieson.

GALE, Sweet 1. *Myrica asplenifolia* (Sweet Fern) USA Zenkert. 2. *Myrica gale* (Bog Myrtle) Lincs, Yks, Renf. Grigson; E Ang. Nall. Both are very fragrant, hence 'sweet'. Sometimes in the north of England, Bog Myrtle is known simply as Sweet.

GALINGALE; GALANGALE *Cyperus longus*. The name is given to the rootstock of this sedge (and to the whole plant, of course), but it really refers to the rootstock of some East Indian plants of the ginger family, for it was used much like ginger is. The name seems to have come into English via O Fr. galingal, then from Arabic khalanjan, and eventually from Chinese ko-liang-kiang. The first element of the Chinese name is Ko, a district near Canton, and the rest means mild ginger.

GALINGALE, Round-rooted *Kaempferia rotunda* Curtis.

GALL *Myrica gale* (Bog Myrtle) Scot. B & H. Gale, of course.

GALL-OF-THE-EARTH *Centaureum erythraea* (Centaury) Fernie. Cockayne had Earthgall for this. Both of them are probably erroneous. Was it a gentian? But centaury is certainly bitter.

GALL-SICK BUSH, Albany *Conyza ivaefolia* S Africa Watt.

GALL, Ram('s) *Menyanthes trifoliata* (Buckbean) Storms. True, this is a bitter plant, bitter enough to be used in brewing instead of hops.

GALL, Snotter *Taxus baccata* fruit (Yew berries) Wilts. Dartnell & Goddard; Berks. Grigson.

GALLA; GALLUC *Symphytum officinale* (Comfrey) Henslow (Galla); Cockayne (Galluc).

GALLBERRY *Ilex glabra*.

GALLIGASKINS *Primula veris* (Cowslip)

Parkinson.

GALLION 1. *Galium aparine* (Goose-grass) B & H. 2. *Galium verum* (Lady's Bedstraw) Culpepper. Gallion = Galium, of course.

GALLIPOLI ROSE *Cistus salvifolius* (Sage-leaved Cistus) New Zealand Perry. Because New Zealand troops took seeds back with them after World War I.

GALLON *Petasites hybridus* (Butterbur) Ire. B & H.

GALLOW-GRASS *Cannabis sativa* (Indian Hemp) Som. Elworthy. Neckweed is another name for this, and both of them came about because the gallows rope was made from the fibre.

GALLOWAY WHIN *Genista anglica* (Needle Whin) Scot. Grigson.

GALLOWS-FRUIT *Fuchsia sp* Som. Macmillan.

GALLWOOD *Artemisia vulgaris* (Mugwort) Scot. Grigson. All the Artemisias are bitter - consider wormwood, for example.

GALLWORT *Linaria vulgaris* (Toadflax) Suss. Culpepper. "In Sussex we call it Gallwort, and lay it in our Chickens Water to cure them of the Gall, I think; I am sure it relieves them when they are drooping" (Culpepper).

GAMBEL'S OAK *Quercus gambelii*.

GAMBIAN BUSH TEA *Lippia multiflora*.

GAMBIAN KINO *Pterocarpus erinaceus*. Kino is actually the coloured resin that is formed from the tree.

GAMBIAN MAHOGANY *Khaya senegalensis* (Dry-zone Mahogany) Dalziel.

GAMBIER CUTCH; GAMBIER-PLANT; GAMBIR PLANT *Uncaria gambir*. Gambier Cutch is the usual name. The others are in Le Strange.

GAMBLE-WEED *Sanicula menziesii* USA Schenk & Gifford.

GAMBO HEMP *Hibiscus cannabinus* (Ambari Hemp) Hutchinson & Melville.

GAMBOGE, Bengal *Garcinia venulosa*. Gamboge is Cambodia, from where the plant was brought around 1600.

GAMBOGE, Black *Garcinia nigro-lineata*.

GAMBOGE TREE *Garcinia morella*. True gamboge is the gum resin obtained from this species.

GAMBOL, Christmas *Rivea corymbosa* (Morning Glory) West Indies Howes.

GANDER, Giddy 1. *Orchis mascula* (Early Purple Orchis) Dor. B & H. 2. *Orchis morio* (Green-winged Orchid) Dor. B & H.

GANDER, Goose-and- 1. *Melandrium rubrum* (Red Campion) Som. Macmillan. 2. *Vicia cracca* (Tufted Vetch) Som. Macmillan.

GANDER, Goosey 1. *Endymion nonscriptus* (Bluebell) Dev, Dor. Macmillan. 2. *Orchis mascula* (Early Purple Orchis) Dor, Glos. B & H. 3. *Orchis morio* (Green-winged Orchid) Dor,Som. Grigson.

GANDER, Goosey-goosey- *Orchis mascula* (Early Purple Orchis) Som. Macmillan.

GANDER-GAUZE; GANDERGLASS *Orchis mascula* (Early Purple Orchis) Wilts. Macmillan (Gander-gauze); A S Palmer (Ganderglass). Gander-gosse is more usual, as the next entry.

GANDER-GOSSE(S) 1. *Dactylorchis maculata* (Heath Spotted Orchid) Leland. 2. *Orchis mascula* (Early Purple Orchis) B & H.

GANDER-GRASS *Potentilla anserina* (Silverweed) Macmillan. There is a 'goose' connection as far as this plant is concerned, if only by reference to the specific name, *anserina*. But compare also Goose-grass, Goosewort, Goosefoot, etc, all given to it over wide areas.

GANDERGOOSE 1. *Orchis morio* (Green-winged Orchid) RHS. 2. *Senecio jacobaea* (Ragwort) Halliwell. See the other 'gander' names for the orchid.

GANDIGOSLINGS *Orchis mascula* (Early Purple Orchis) Wilts. Dartnell & Goddard. Cf. Giddy Gander, Gander-gosses, etc.

GANG-FLOWER *Polygala vulgaris* (Milkwort) Gerard. The reference here is to Gang Days - Rogation-tide. "...of which floures the maidens which use in the countries to walke the Procession do make themselves garlands and Nosegaies" (Gerard). Gang-flower, Cross-flower, Procession-flower, Rogation-flower, all names for milk-wort, refer to the use of the flower in the Rogation-tide processions, a tradition which probably started on the Continent.

GANYMEDE'S CUP *Narcissus triandrus* (Angel's Tears) Ire. Coats. Accounted for by Alice Coats by suggesting that Ganymede, the cup-bearer of the gods, was in primitive Greek religion the deity responsible for sprinkling the earth with heaven's rain. Hence, too, the connection with the name Angel's Tears.

GAP-MOUTH 1. *Antirrihinum majus* (Snapdragon) Som. Macmillan. 2. *Digitalis purpurea* (Foxglove) Som. Macmillan. 3. *Linaria vulgaris* (Toadflax) Som. Macmillan. 4. *Mimulus guttatus* (Monkey-flower) Som. Macmillan.

GAPING JACK *Linaria vulgaris* (Toadflax) Som. Macmillan. Cf. Gap-mouth.

GARBANZO *Cicer arietinum* (Chick Pea) USA Anderson. Garbanzo is the Spanish word for chick-pea.

GARCLIVE *Agrimonia eupatoria* (Agrimony) B & H.

GARDEN GATES 1. *Geranium robertianum* (Herb Robert) Bucks. Grigson. 2. *Saxifraga x urbinum* (London Pride) Dev. Friend.1882; Glos. J D Robertson. 3. *Viola tricolor* (Pansy) Bucks, Ess, Suff, Norf, Cambs. B & H; Worcs. J Salisbury. An abbreviation of the picturesque but long-winded Kiss-me-love-at-the-garden-gate.

GARDEN'S PLAGUE *Aegopodium podagraria* (Goutweed) N.Ire. Grigson. Goutweed is better known among gardeners as Ground Elder - and this Ulster name is perfect, for once it gets into a garden it is well-nigh irradicable.

GARDENER'S DELIGHT; GARDENER'S EYE *Lychnis coronaria* (Rose Campion) both Gerard.

GARDENER'S GARTERS *Phalaris arundinacea* (Reed-grass) Tynan & Maitland.

GARDENIA, Wild *Gardenia spathulifolia*.

The genus is named in honour of Dr Alexander Garden, an 18th century American botanist.

GARDENIA, Wild, Small *Gardenia asperula*.

GARGET; GARGET-PLANT *Phytolacca decandra* (Poke-root) USA Sanford (Garget); Lloyd (Garget-plant). The early American settlers applied the plant as a poultice to the inflammatory condition of a cow's udder which prevents the free flow of milk, in the disease known as garget.

GARGUT-ROOT *Helleborus foetidus* (Stinking Hellebore) Norf. B & H. Gargut is a cattle disease. Surely it is the same word as garget?

GARLAND CHRYSANTHEMUM *Chrysanthemum coronarium* (Annual Chrysanthemum) Coats. 'Garland' from the specific name, *coronarium*. Cf. Crown Daisy.

GARLAND CRABAPPLE *Malus coronaria*.

GARLAND FLOWER *Daphne cneorum* (Flax-leaved Daphne) Hay & Synge.

GARLAND THORN *Paliuris spina-christi* (Christ-thorn) Howes.

GARLETE *Allium sativum* (Garlic) B & H.

GARLIC, Bear's *Allium ursinum* (Ramsons) B & H. Garlic is OE garleac, where gar means spear, and leac is the familiar word leek. The ascription to the bear in this name arises from the specific name, *ursinum*, though it is not clear why it was given. Tabernaemontanus apparently said that "ursi eo delectantur". Cf. Ursine Garlic.

GARLIC, Broad-leaved *Allium ursinum* (Ramsons) Folkard.

GARLIC, Canada *Allium canadense*.

GARLIC, Chive *Allium schoernoprasum* (Chives) Corn. Jago.

GARLIC, Crow; GARLIC, Cow *Allium vineale* B & H (Cow). Presumably 'cow' is a misprint for 'crow' in the original.

GARLIC, Daffodil *Allium neapolitanum*.

GARLIC, Elephant *Allium ampeloprasum* (Wild Leek) Painter.

GARLIC, False *Allium vineale* (Crow Garlic) Howes.

GARLIC, Field 1. *Allium oleraceum*. 2. *Allium vineale* (Crow Garlic) USA Zenkert.
GARLIC, French *Allium scordoprasum* (Sand Leek) G.B.Foster.
GARLIC, German *Allium senescens*.
GARLIC, Giant 1. *Allium ampeloprasum* (Wild Leek) Painter. 2. *Allium scordoprasum* (Sand Leek) E.Hayes.
GARLIC, Golden *Allium moly*.
GARLIC, Hog's *Allium ursinum* (Ramsons) B & H.
GARLIC, Hedge *Alliaria petiolata* (Jack-by-the-hedge) Cumb. Grigson. The leaves have a garlic-like taste.
GARLIC, Keeled *Allium carinatum*. 'Keeled' is a reference to the shape of the underside of the leaves.
GARLIC, Levant *Allium ampeloprasum* (Wild Leek) Douglas.
GARLIC, Meadow *Allium canadense* USA Tolstead.
GARLIC, Naples *Allium neapolitanum* (Daffodil Garlic) Howes.
GARLIC, Rosy *Allium roseum*.
GARLIC, Round-headed 1. *Allium ampeloprasum* (Wild Leek) Barton & Castle. 2. *Allium sphaerocephalum* (Round-headed Leek) Camden/Gibson.
GARLIC, Rush *Allium schoernoprasum* (Chives) Lyte. Rush-like leaves.
GARLIC, Serpentine *Allium ursinum* (Ramsons) T Hill. Cf. Snake-flower, Snake's Food, etc.
GARLIC, Sorcerer's *Allium moly* (Golden Garlic) Gerard. The reference here is classical myth, and to the magic herb moly, given to Odysseus by Hermes to enable him to restore his crew to their human shape after they had been turned into pigs by Circe. But is this the herb moly? It may bear the specific name, but Golden Garlic is almost certainly not moly. Ordinary garlic is a much likelier candidate.
GARLIC, Spanish *Allium scordoprasum* (Sand Leek) H & P.
GARLIC, Taper-leaved *Allium sativum*.
GARLIC, Triquetrous *Allium triquetrum* (Three-cornered Leek) Grigson. 'Triquetrous' means triangular,or three-cornered,as the common name has it.
GARLIC, Turkey, Great *Allium scordoprasum* (Sand Leek) Tradescant.
GARLIC, Ursine *Allium ursinum* (Ramsons) Geldart. See Bear's Garlic.
GARLIC, Wild 1. *Allium canadense* USA H.Smith.1928. 2. *Allium ursinum* (Ramsons) Dev. Friend; Glos. J.D. Robertson; Ches. Holland. 3. *Allium vineale* (Crow Garlic) Turner.
GARLIC, Wood *Allium ursinum* (Ramsons) Howes.
GARLIC, Yellow *Allium moly* (Golden Garlic) Turner.
GARLIC CHIVES *Allium tuberosum*.
GARLIC GERMANDER 1. *Teucrium scordium* (Water Germander) B & H. 2. *Teucrium scorodonia* (Wood Sage) B & H. Water Germander seems better suited to this name than Wood Sage. Scordium, the specific name, comes from a Greek word meaning garlic, and one is assured that there is a resemblance in the smell.
GARLIC MUSTARD; GARLIC TREACLE-MUSTARD *Alliaria petiolata* (Jack-by-the-hedge) B & H (Mustard); Barton & Castle (Treacle-mustard). "...if it be bruised or stamped it smelleth altogether like Garlick..." (Gerard). Treacle is from Greek theriake, from therion, a name given to the viper. The point was that, according to ancient superstition, viper's flesh will cure viper's bite, and treacle was certainly originally a compound made of vipers among other things. It was a favourite against all poisons. The word then became applied to any confection of sweet syrup, and finally and solely to the syrup of molasses.
GARLIC SAGE *Teucrium scorodonia* (Wood Sage) Gerard. Cf. Garlic Germander.
GARLICKWORT *Alliaria petiolata* (Jack-by-the-hedge) B & H.
GARLOCK *Sinapis arvensis* (Charlock) B & H.
GARNESIE VIOLET *Matthiola incana*

(Stock) Lyte.

GARNET-BERRY 1. *Ribes nigrum* (Blackcurrant) Leyel.1937. 2. *Ribes rubrum* (Redcurrant) Fernie. For blackcurrant, this is probably an error - garnet must surely refer to the colour.

GARRY OAK *Quercus garryana* (Oregon Oak) USA Turner & Bell.

GARTEN-BERRY; GARTEN-BERRY, Lady's *Rubus fruticosus* fruit (Blackberry) Scot. Jamieson. This name seems to start at Gatter-berry (the bush is called a gatter-tree). From that point, one gets Gaitberry, or Gaiterberry, and then Garten-berry, before it finally reaches Lady's Garter-berry.

GARTER, Star-and- *Ornithogalum umbellatum* (Star-of-Bethlehem) Wilts. D Grose.

GARTER-BERRY; GARTER-BERRY, Lady's *Rubus fruticosus* fruit (Blackberry) Roxb. Grigson. See Garten-berry.

GARTERS, Devil's 1. *Calystegia sepium* (Great Bindweed) Pemb, Ire. Grigson. 2. *Convolvulus arvensis* (Field Bindweed) Som. Macmillan; Ire. Friend. A most appropriate name, as most gardeners would agree.

GARTERS, Gardener's *Phalaris arundinacea* (Reed-grass) Tynan & Maitland.

GARTERS, Lady's 1. *Phalaris arundinacea* (Reed-grass) Som. Elworthy. 2. *Rubus fruticosus* (Bramble) Roxb. Grigson.

GARTH CRESS *Lepidium sativum* (Garden Cress) B & H. i.e. garden cress.

GAS PLANT *Dictamnus fraxinella* (False Dittany) W.Miller. Such names as Gas Plant, Burning Bush and Candle Plant for the False Dittany arise because of the volatile oil which it exudes, and which will readily take fire if anybody approaches with a light.

GASKINS 1. *Primula veris* (Cowslip) B & H. 2. *Prunus avium* (Wild Cherry) Som. Macmillan; Suss, Kent. Grigson. 3. *Ribes uva-crispa* (Gooseberry) Scot. Jamieson. The name means Gascon. Jamieson said it was "a name commonly given to a rough green gooseberry, originally brought from Gascony." There was a Gascoigne Cherry, too, very popular in Elizabethan times. Parish said wild cherry was called geen or gaskin, "having been brought from France by Joan of Kent when her husband, the Black Prince, was commanding in Guienne and Gascony." So, geen seems to be Guienne, and gaskin, Gascony. But what is cowslip doing in this company?

GATTEN-TREE 1. *Cornus sanguinea* (Dogwood) Lyte. 2. *Euonymus europaeus* (Spindle-tree) Prior. 3. *Viburnum opulus* (Guelder Rose) Prior. The first syllable is OE gat, goat. Not according to Jordan, though, for he says the first syllable is gad, or goad. This seems not unreasonable, considering the use of the wood for skewers.

GATTER-BERRY *Rubus fruticosus* fruit (Blackberry) Roxb. Grigson. Cf. Gartenberry, garter-berry, etc.

GATTER-BUSH 1. *Cornus sanguinea* (Dogwood) E Ang. Nall. 2. *Euonymus europaeus* (Spindle-tree) Kent Grigson; E Ang. Nall. 3. *Viburnum opulus* (Guelder Rose) E.Ang. Nall. Cf. Gatten-bush, and other forms like Gatteridge.

GATTER-TREE 1. *Cornus sanguinea* (Dogwood) Lincs. Grigson. 2. *Rubus fruticosus* (Bramble) Roxb. Grigson. Not from the same derivation in each case. See Gartenberry and the like for the bramble entry.

GATTERIDGE-TREE 1. *Cornus sanguinea* (Dogwood) E.Ang. B & H. 2. *Euonymus europaeus* (Spindle-tree) Kent Grigson. 3. *Viburnum opulus* (Guelder Rose) E.Ang. B & H. As Gatter-bush, etc.

GATTRIDGE-BERRY *Euonymus europaeus* fruit (Spindle-tree) S.Eng. Grose.

GATTRIDGE TREE 1. *Euonymus europaeus* (Spindle Tree) S.Eng. Grose. 2. *Viburnum opulus* (Guelder Rose) E Ang. Forby.

GAUL; GAULE *Myrica gale* (Bog Myrtle) Cumb, Scot. B & H. Camden/Gibson (Gaule). This just reflects the local pronunciation of Gale, which appears in various forms as the general name for the plant.

GAUL-NUT *Juglans regia* nut (Walnut) Ablett. Cf. French Nut for this. Ablett says that it was thought to have been introduced into England from France, and that before 1562 Gaul-nut was the name always given to it. But isn't this just a variation of walnut, where the first syllable actually means 'foreign'? On the other hand, there is a French verb, gauler, which has the specialized meaning of 'to thrash a walnut tree'.

GAUN; GAUND *Petasites hybridus* (Butterbur) Lanark, Dumf. B & H.

GAUZE FLOWER *Gypsophila sp.* Descriptive name much used by florists.

GAY FEATHER 1. *Liatris scariosa* Le Strange. 2. *Liatris spicata* (Button Snakeroot) Grieve.1931.

GAY FEATHER, Kansas *Liatris callilepsis.*

GAZEL; GAZLE 1. *Crataegus monogyna* fruit (Haws) Tynan & Maitland. 2. *Ribes nigrum* (Blackcurrant) Suss, Kent W D Cooper. 3. *Ribes rubrum* (Redcurrant) Kent Grigson. Gazel is a corruption of French groseille, usually applied to the redcurrant. In turn, groseille is from the late Latin acricella, a reference to the sharp taste. Gazle, in this spelling, is recorded by Fernie, for blackcurrant only, in Sussex and Kent. Gazels for haws is different, though. Here it is just an abbreviation of Hog Hazel.

GEAGLES *Heracleum sphondyllium* (Hogweed) Banff. Grigson.

GEAL-GOWAN; GEAL-SEED *Chrysanthemum segetum* (Corn Marigold) both Donegal B & H. 'Geal', which can appear as 'gill' as well in Ireland, is the same as the Scots 'gule', so such a combination as geal-gowan would mean, if anything, goldgold.

GEAN; GEEN *Prunus avium* (Wild Cherry) Grigson; Scot. Jamieson (Geen). From French guigne, sweet cherry.

GEANUCANACH'S PIPES *Quercus robur* acorn (Acorn) Ire. O Suileabhain. They are simple 'Pipes' in various places in England, but this must be a reference to fairy pipes.

GECKDOR *Galium aparine* (Goose-grass) B & H. This name is in Halliwell, too - with no comment in either case.

GEESE-AND-GULLIES *Salix caprea* catkins (Sallow) Shrop, Ches. Grigson. Presumably descriptive in a way. 'Goslings' is better known, and where this name occurs, there is an association with the real thing. It is unlucky to bring the catkins into the house, for they would be fatal to the real goslings.

GEIGER TREE *Cordia sebestana* Howes.

GELDERS ROSE *Viburnum opulus* (Guelder Rose) Gerard.

GELL ALFRED *Cheiranthus cheiri* (Wallflower) Som. Macmillan. A corruption of Gilofer, which is Gilliflower, itself originally from caryophyllus, clove - see carnation.

GELSEMINE *Jasminum officinale* (Jasmine) Gerard. One of a number of variations of jasmine, through the form jessamine. The group even finished with Gethsemane.

GEMSBUCK BEAN *Bauhinia esculenta* (Tsin Bean) Howes.

GENET *Cytisus scoparius* (Broom) Schery. This is the French genêt, i.e. Genista, the generic name of some of the brooms.

GENEVA PLANT *Juniperus communis* (Juniper) Ches. Holland. Geneva is O. Dutch genever, O. French genevre, and from Latin juniperus. It is a confusion to associate the Swiss city name with this.

GENTIAN 1. *Centaureum erythraea* (Centaury) Suss, Scot. B & H. 2. *Gentiana sp.* The name comes from Gentius, a king of Illyria in the second century BC. He was the man credited with the discovery of the medicinal properties of the plant, at any rate according to Pliny. Centaury belongs to the same family as the gentians, hence the adoption of the name.

GENTIAN, Alpine *Gentiana nivalis.* Cf. Snow Gentian.

GENTIAN, American *Gentiana catesbei.*

GENTIAN, Autumn *Gentianella amarella.*

GENTIAN, Bitter *Gentiana lutea* (Yellow Gentian) Howes.

GENTIAN, Bladder *Gentiana utriculosa.*

GENTIAN, Brown *Gentiana pannonica.*

GENTIAN, Centaury *Centaureum erythraea* (Centaury) Grieve.1931.

GENTIAN, Chiltern *Gentianella germanica*. A very rare plant in this country, confined to the Chilterns.

GENTIAN, Closed 1. *Gentiana affinis*. 2. *Gentiana andrewsii*.

GENTIAN, Crested *Gentiana septemfida*.

GENTIAN, Cross; GENTIAN, Cross-leaved *Gentiana cruciata*. Cross Gentian is the usual name; Cross-leaved Gentian is in Grieve.1931.

GENTIAN, Cross-wort *Gentiana cruciata* (Cross Gentian) Parkinson.

GENTIAN, Downy *Gentiana puberula*.

GENTIAN, Early *Gentianella anglica*.

GENTIAN, Field *Gentianella campestris*.

GENTIAN, Field, Large *Gentianella germanica* (Chiltern Gentian) Polunin.1969.

GENTIAN, Fringed *Gentiana crinita*.

GENTIAN, Fringed, Smaller *Gentiana procera*.

GENTIAN, Green *Swertia perennis* (Marsh Felwort) USA T H Kearney.

GENTIAN, Hungarian *Gentiana pannonica* (Brown Gentian) Polunin.1969. This grows in the Austrian mountains.

GENTIAN, Indian *Swertia chirata* (Chiretta) Mitton. Same family as the gentians, and like them, it furnishes a bitter tonic, much used to restore a poor appetite.

GENTIAN, Japanese *Gentiana scabra*.

GENTIAN, Marsh 1. *Gentiana ochroleuca* (Sampson Snakeweed) Le Strange. 2. *Gentiana pneumonanthe*.

GENTIAN, Northern *Gentianella aurea*.

GENTIAN, Pleated *Gentiana affinis* (Closed Gentian) T H Kearney.

GENTIAN, Prairie *Gentiana affinis* (Closed Gentian) Johnston.

GENTIAN, Purple *Gentiana purpurea*.

GENTIAN, Pyrenean *Gentiana pyrenaica*.

GENTIAN, Red *Centaureum erythraea* (Centaury) Shaw.

GENTIAN, Rough *Gentiana scabra* (Japanese Gentian) Hyatt. 'Scabra' is the specific name, hence 'rough'.

GENTIAN, Scottish *Gentianella septentrionalis*.

GENTIAN, Small *Gentiana nivalis* (Alpine Gentian) Clapham.

GENTIAN, Snow *Gentiana nivalis* (Alpine Gentian) Polunin.1969.

GENTIAN, Soapwort *Saponaria officinalis* (Soapwort) Lyte.

GENTIAN, Southern *Gentiana alpina*.

GENTIAN, Spring *Gentiana verna*.

GENTIAN, Star *Gentiana verna* (Spring Gentian) Grigson.

GENTIAN, Stiff *Gentiana quinquefolia*.

GENTIAN, Swallow-wort *Gentiana asclepiadea* (Willow Gentian) Parkinson. The specific name comes from a Greek name meaning swallow-wort, and it was from that lead that Parkinson used it in English.

GENTIAN, Trumpet *Gentiana acaulis*.

GENTIAN, Trumpet, Stemless *Gentiana clusii*. Why single out this particular gentian for the name, when the previous entry has the specific name *acaulis*.

GENTIAN, Welsh *Gentianella uliginosa*. It only occurs, and that very rarely, on dunes in south Wales, as far as Britain is occurred.

GENTIAN, Wild *Chironia baccifera* (Christmas Berry) Perry.

GENTIAN, Willow *Gentiana asclepiadea*.

GENTIAN, Yellow 1. *Blackstonia perfoliata* (Yellow-wort) Leyel.1948. 2. *Gentiana lutea*.

GENTIAN SAGE *Salvia patens*. Presumably because the flowers are bright blue.

GENTIANELLA *Microcala filiformis*.

GENTLE, Flower *Amaranthus tricolor* (Joseph's Coat) Prior. From the resemblance of the flower to the plumes once worn by. people of rank, i.e. "gentle" folk, or so said Prior.

GENTLE, Lavender *Lavandula stoechas* (Spanish Lavender). Turner.

GENTLE DOCK *Polygonum bistorta* (Bistort). Notts. Grigson. Yks. Julia Smith.

GENTLE THISTLE *Cirsium dissectum* (Meadow Thistle) Hill. Hill actually coined the name, in 1769.

GENTLEMAN, Turn-again 1. *Lilium chalcedonicum* (Scarlet Turk's-cap Lily) Coats. 2. *Lilium martagon* (Turk's-cap Lily) Glos,

Heref, Worcs, Bucks, N'thants. B & H. A reference to the reverted petals. Cf. the name Turncap.
GENTLEMAN JOHN; GENTLEMAN TAILOR *Viola tricolor* (Pansy) Fernie (Gentleman John); Dor. Macmillan (Gentleman Tailor).
GENTLEMAN'S CAP *Tulipa sp* (Tulip) Som. Macmillan.
GENTLEMAN'S FINGER *Arum maculatum* (Cuckoo-pint) Wilts. Dartnell & Goddard. Descriptive, and politely so, in view of the other names.
GENTLEMAN'S PINCUSHION *Knautia arvensis* (Field Scabious) Som. Macmillan.
GENTLEMAN'S PURSE *Capsella bursa-pastoris* (Shepherd's Purse) Som. Macmillan.
GENTLEMAN'S TORMENTORS *Galium aparine* (Goose-grass) Suff. Grigson. The reference must be to the burrs.
GENTLEMEN, Ladies-and- 1. *Arum maculatum* (Cuckoo-pint) Wilts. Dartnell & Goddard. Shrop. Burne; N'thants. A.E.Baker. 2. *Viola tricolor* (Pansy) Som. Macmillan. Lords-and-ladies is a better-known name for the cuckoo-pint, but both of them, and a host of others, are typically sexually oriented, and refer to the method of growth, the spadix in the spathe. Pansy, too, has aphrodisiac and erotic connections reflected in a number of the names.
GENTLEMEN, Melancholy 1. *Astrantia major* (Pink Masterwort) Keble Martin. 2. *Hesperis tristis* Coats.
GENTLEMEN, Roundabout *Fritillaria imperialis* (Crown Imperial) Dor. Dacombe.
GENTLEMEN-AND-LADIES *Arum maculatum* (Cuckoo-pint) Som. Macmillan; Oxf. Grigson.
GENTLEMEN'S-AND-LADIES' FINGERS *Arum maculatum* (Cuckoo-pint) Wilts. Dartnell & Goddard.
GENTLEMEN'S BUTTONS *Succisa pratensis* (Devil's-bit) Shrop. B & H.
GENTLEMEN'S CAP-AND-FRILLS *Ranunculus ficaria* (Lesser Celandine Som. Macmillan.
GEORDIE, Water; GEORGIE, Water *Caltha palustris* (Marsh Marigold) both Som. Grigson (Geordie); Macmillan (Georgie).
GEORGE, John *Caltha palustris* (Marsh Marigold) Bucks. M Baker.1974.
GEORGE LILY *Vallota speciosa* (Scarborough Lily) S.Africa Watt.
GERAFLOUR *Cheiranthus cheiri* (Wallflower) Scot. B & H. A version of gillyflower.
GERANIUM, Bassinet *Geranium pratense* Coats. Bassinet exists by itself, too, but the names are not explained. Geranium comes through Latin from Greek geranion, crane - hence the usual name, cranesbill.
GERANIUM, Feather *Chenopodium botrys* (Jerusalem Oak) USA Tolstead.
GERANIUM, Fox *Geranium robertianum* (Herb Robert) Border B & H. There is Fox Grass for this, too, and they both seem to be related to the unpleasant smell. The leaves, Gerard said, are "of a most loathsome stinking smell".
GERANIUM, Hanging *Saxifraga sarmentosa* (Mother-of-Thousands) Wilts. Dartnell & Goddard. Because that's the way they always used to be seen in cottage windows. Cf. the Devonshire name, Poor Man's Geranium.
GERANIUM, High *Hydrangea x macrophylla* (Hydrangea) Dev. Choape. A local variaition on Hydrangea.
GERANIUM, Ivy-leaved *Pelargonium peltatum.*
GERANIUM, Low *Geranium pusillum* (Small Cranesbill) USA E.Gunther.
GERANIUM, Mint *Tanacetum balsamita* (Balsamint) Mabey.1977.
GERANIUM, Nutmeg-scented *Pelargonium x fragrans.*
GERANIUM, Oak-leaf 1. *Chenopodium botrys* (Jerusalem Oak) USA G B Foster. 2. *Pelargonium quercifolium.* The Pelargonium is called Scented Oak, as well. A further 'geranium' name for Jerusalem Oak in America is Feather Geranium.

GERANIUM, Peppermint *Pelargonium tomentosum*.
GERANIUM, Polecat *Lantana sellowiana*.
GERANIUM, Poor Man's *Saxifraga sarmentosa* (Mother-of-Thousands) Dev. Friend. Cf. the Wiltshire Hanging Geranium.
GERANIUM, Rose *Pelargonium graveolens*. The leaves have got a rose-like perfume to them.
GERANIUM, Scotch *Geranium robertianum* (Herb Robert) Forf. B & H.
GERANIUM, Smelly *Geranium robertianum* (Herb Robert) Tynan & Maitland. Cf. the 'foxy' names for Herb Robert.
GERANIUM, Spear-leaved *Pelargonium glaucum*.
GERANIUM, Strawberry *Saxifraga granulata* (Meadow Saxifrage) W Miller. Is it because the flowers are vaguely similar to those of strawberry, or does it have the same method of propagation, by runners?
GERANIUM, Strawberry-leaved *Saxifraga sarmentosa* (Mother-of-thousands) Wilts. Dartnell & Goddard.
GERANIUM, Wild 1. *Althaea officinalis* (Marsh Mallow) (Vesey-Fitzgerald). 2. *Geranium maculatum* (Spotted Cranesbill) Grieve.1931. 3. *Geranium robertianum* (Herb Robert) Dev. Friend.1882. 4. *Melandrium dioicum* (Red Campion) Ches. B & H.
GERARD, Herb *Aegopodium podagraria* (Goutweed) Fernie. Not a dedication to John Gerard, but to Saint Gerard, the patron saint of gout sufferers. "Herb Gerard with his roots stamped, and laid upon members that are troubled or vexed with the gout, swageth the paine..." "The very bearing of it about one easeth the Pains of the Gout, and defends him that bears it from the Disease". That is from Culpepper, while the first quote is from Gerard.
GERARD'S PINE *Pinus gerardiana*.
GEREKEN *Cucumis anguria* (Gherkin) Evelyn.
GERMAN-AND-ENGLISH *Ranunculus sceleratus* (Celery-leaved Buttercup) Som.

Macmillan.
GERMAN ASPHODEL *Tofieldia calyculatum*.
GERMAN CAMOMILE *Matricaria recutita* (Mayweed) Brownlow. Cf. Hungarian Camomile for this.
GERMAN CELERY *Apium graveolens var. rapaceum* (Celeriac) Bianchini. Presumably because as a vegetable this is much more popular on the Continent than it ever has been in this country.
GERMAN FILAGO *Filago germanica* (Cudweed) Flower.1851.
GERMAN GARLIC *Allium senescens*.
GERMAN IRIS *Iris germanica* (Blue Flag) W Miller.
GERMAN IVY *Senecio mikanioides*. German, because it was first grown as a house plant in that country.
GERMAN LILAC *Centranthus ruber* (Spur Valerian) Lincs. Friend.1882.
GERMAN MADWORT *Asperugo procumbens* (Madwort) Prior.
GERMAN MILLET *Setaria italica* (Foxtail Millet) Brouk. Italian, German, Hungarian or Siberian Millet, according to the country of origin.
GERMAN SAUSAGE TREE *Kigelia africana*. The reference is to the large fruits, often as much as a metre long, and hanging just like sausages from the branches.
GERMAN SPURGE OLIVE *Daphne mezereon* (Lady Laurel) Gerard.
GERMAN TAMARISK *Myricaria germanica* (False Tamarisk) Polunin.1969.
GERMAN THYME *Thymus serpyllum var. citriodorus* (Lemon Thyme) Sanecki.
GERMAN VELVET-GRASS *Holcus mollis* (Creeping Soft-grass) USA Allan. It is a fairly recent introduction in America from Europe.
GERMANDER *Teucrium chamaedrys*. Jaeger suggested that germander may be a corruption of Scamander, a Trojan river - but it must surely be a corruption of the specific name, chamaedrys, where Greek chamai means on the ground, and drys, oak, a reference to the

shape of the leaves.
GERMANDER, American *Teucrium canadense*.
GERMANDER, Cut-leaved *Teucrium botrys*.
GERMANDER, Garlick 1. *Teucrium scordium* (Water Germander) Turner. 2. *Teucrium scorodonia* (Wood Sage) B & H. *T scordium* has the better right to the name. Scordium comes from a Greek word meaning garlic, and there is indeed a resemblance in the smell.
GERMANDER, Hairy *Teucrium occidentale*.
GERMANDER, Large-leaved *Teucrium scorodonia* (Wood Sage) Grieve.1931.
GERMANDER, Low *Teucrium depressum*.
GERMANDER, Marsh *Teucrium scordium* (Water Germander) Parkinson.1640.
GERMANDER, Mediterranean *Teucrium poly* (Poly) Howes.
GERMANDER, Mountain *Teucrium montanum*.
GERMANDER, Sage-leaved *Teucrium scorodonia* (Wood Sage) Brownlow.
GERMANDER, Shrubby *Teucrium fruticans* (Tree Germander) Polunin.1976.
GERMANDER, Tall; GERMANDER, Tree *Teucrium fruticans* Painter (Tall). 'Tree' is a bit ambitious for this. It is a shrub from the Mediterranean area.
GERMANDER, Wall *Teucrium chamaedrys* (Germander) C P Johnson.
GERMANDER, Water *Teucrium scordium*.
GERMANDER, Wild *Veronica chamaedrys* (Germander Speedwell) Gerard.
GERMANDER, Wood *Teucrium scorodonia* (Wood Sage) B & H.
GERMANDER CHICKWEED 1. *Veronica agrestis* (Green Field Speedwell) Prior. 2. *Veronica chamaedrys* (Germander Speedwell) B & H.
GERMANDER SPEEDWELL *Veronica chamaedrys*. It is the specific name chamaedrys that gave rise to the word germander.
GERNETER *Punica granatum* (Pomegranate) Halliwell. Halliwell quotes this from a plant list in ms Sloane 5, f3.

GERNUT *Conopodium majus* (Earthnut) N'thum, Yks. Grigson. The brief series is Earthnut - Ernut - Gernut - Yernut, with variations all the way.
GESMINE; GESSE; GESSEMINE *Jasminum officinale* (Jasmine) Parkinson (Gesmine); Gerard (Gesse, Gessemine). Jasmine, and of course, these are the result of the Latinization of the Persian yasmin into Jasminum.
GESSLINGS, Goose-and- *Salix caprea* catkins (Goat Willow catkins) N'thants. Clare.
GETHSEMANE 1. *Arum maculatum* (Cuckoo-pint) Ches. B & H. 2. *Jasminum officinale* (Jasmine) Turner. 3. *Orchis mascula* (Early Purple Orchis) Ches. B & H. Cuckoo-pint and the Orchis get the name from the blotched leaves - the legend is that blood dripped from the Cross on to them. Jasmine used to be called Gethesemane from the similarity of the names, particularly in the older versions, such as Gessemine.
GEUKY FLOWER 1. *Melandrium dioicum* (Red Campion) Dev. Friend.1882. 2. *Orchis mascula* (Early Purple Orchis) Dev. Friend.1882. Geuky is the same as gowky, from gowk, cuckoo. That, Cuckoo-flower, and other similar names are recorded for these, so presumably this name means the same thing, in spite of the fact that gowk is a north country or Scottish word, and has little to do with Devonshire.
GEW-GOG *Ribes uva-crispa* (Gooseberry) Suff. Moor. The local equivalent of Goosegog.
GEZLINS *Salix caprea* catkins (Sallow catkins) Ches. Holland. This is local orthography for goslings, a common name for willow catkins of all kinds.
GHERKIN *Cucumis anguria*. Unlikely as it may sound, the specific name is the original for the English name. Gherkin apparently comes from an earlier form of the Dutch augurk, and this in turn comes from a Slavonic original (there is still the Polish ogórek, for example). But it seems that a

Greek word, aggoúrion, was the pattern. From this Greek word comes Spanish angurria and the Italian anguria - hence the specific name used currently. All these words, incidentally, just mean gherkin.

GHERKIN, Bur *Cucumis anguria* (Gherkin) Whitaker & Davis.

GHERKIN, West Indian *Cucumis anguria* (Gherkin) Barbados Gooding.

GHOST, Holy 1. *Angelica archangelica* (Archangel) Palaiseul. 2. *Angelica sylvestris* (Wild Angelica) (Dyer). "in high Dutch...des heileghen Geyst wurzel, that is, Spiritus sancti radix, the root of the Holy Ghost, as Fuchsius witnesseth" (Gerard). Rhodes suggested that Holy Ghost pie, used in the Black Mass, was an angelica-flavoured cake, therefore a host. But doctors used to refer to it by this name as a means to puff up its qualities, its "great and divine properties".

GHOST, Prickly *Ulex europaeus* (Gorse) Dor. Macmillan. Ghost here is a misuse of gorse.

GHOST, Single *Orchis mascula* (Early Purple Orchis) Som, Wilts Macmillan. There is another Somerset name, Single-guss. They are without doubt both related to such names as Sammy Gussets, or Standing Gussets, and ultimately to the much more widespread Gander-gosses. 'Ghost', in other words, is actually 'goose'.

GHOST GRASS *Gynerium argenteum* (Pampas Grass) Som. Macmillan. In this case, 'ghost' is a form of description, rising from the white appearance of the tall grasses.

GHOST GUM *Eucalyptus papuana*. The tree has a smooth white bark - hence the name.

GHOST-KEX *Angelica sylvestris* (Wild Angelica) Yks. Grigson. This is nothing to do with the appearance, but the remains of the early name, Holy Ghost, coupled with the typical 'kex' for a hollow-stalked umbellifer.

GHOST-TREE *Davidia involucrata* (Hamdkerchief Tree) Hay & Synge. Presumably from the large white bracts which also account for the more common Handkerchief Tree.

GHOSTFLOWER *Monotropa uniflora* USA A.W.Smith. Cf. Corpse Plant for this.

GIANT'S RATTLE *Entada phaseoloides* (Nicker Bean) Chopra.

GIANTS, Easter; GIANTS, Eastern *Polygonum bistorta* (Bistort) Grigson (Easter); Denham/Hardy (Eastern). 'Easter' because a herb pudding used to be made from bistort leaves on Easter Day, or more properly, at Passion-tide. 'Eastern', of course, is simply a mis-reading of the correct name. 'Giants' is originally Mangiants, from the French manger, eat.

GIBBLES; GIBLETS *Allium fistulosum* (Welsh Onion) Wilts. Jones & Dillon. This occurs elsewhere in various forms, like Chipples or Jibbles. They are all originally cepa, the Latin for onion, though these two names should be reserved for spring onions. Nevertheless, in the west country gibbles was used for young onions generally.

GIBBLES, Gypsy's *Allium ursinum* (Ramsons) Som. Macmillan.

GIBRALTAR MINT *Mentha pulegium* (Pennyroyal) Howes.

GIBS *Salix capraea* catkins (Sallow catkins) Lincs. Peacock.

GICKS *Heracleum sphondyllium* (Hogweed) Wilts. Jefferies. A local variation of Kex.

GIDDY-GANDER 1. *Orchis mascula* (Early Purple Orchis) Dor. B & H. 2. *Orchis morio* (Green-winged Orchis) Dor. B & H.

GIFT, New Year's *Eranthis hyemalis* (Winter Aconite) Ess. Wright; Lincs. (Coats). It will be in flower in early January.

GIGGARY *Narcissus pseudo-narcissus* (Daffodil) Dev. Friend. Probably Gregory, which also is recorded in the same county.

GILAWFER *Matthiola incana* (Stock) Som. Elworthy. A local variant of gilliflower which can also be spelt as Jilloffer.

GILCUP 1. *Caltha palustris* (Marsh Marigold) Som. Macmillan. 2. *Ranunculus acris* (Meadow Buttercup) Dev. Friend.1882; Dor, Som, Wilts, Hants. Grigson. 3. *Ranunculus bulbosus* (Bulbous Buttercup) Dor. Macmillan. Dev, Som,

Wilts, Hants. Grigson. 4. *Ranunculus ficaria* (Lesser Celandine) Som, Wilts. Macmillan. 5. *Ranunculus repens* (Creeping Buttercup) Dev, Dor, Som, Wilts, Hants. Grigson. i.e. gold cup. Cf. Golden Cup, Goldicup, etc.,as well as the next entry.

GILDCUP 1. *Caltha palustris* (Marsh Marigold) Hants. Cope. 2. *Ranunculus acris* (Meadow Buttercup) Som. Grigson. 3. *Ranunculus bulbosus* (Bulbous Buttercup) Som. Grigson. 4. *Ranunculus repens* (Creeping Buttercup) Som. Grigson.

GILEAD, Balm of 1. *Cedronella triphylla* Moldenke. 2. *Commiphora opobalsamum*. 3. *Melittis melisophyllum* (Bastard Balm) Wilts. Dartnell & Goddard. 4. *Populus x candicans* (Balsam Poplar) USA Brownlow. 5. *Populus trichocarpa* (Western Balsam Poplar) USA Mason. *Commiphora opobalsamum* is the true Balm of Gilead, from Arabia. The others get the name by comparison. 'Balm' is the same word as 'balsam', probably of Hebrew origin, and perhaps meaning something like sweet-smelling. "Balm" is usually a gum obtained by making incisions in the stem and branches. The exuding sap soon hardens into small nodules (the American Balsam Poplars behave in the same way, hence the name).

GILIA, Broad-leaved *Gilia latifolia*.
GILIA, Golden *Gilia aurea*.
GILIA, Humble *Gilia demissa*.
GILIA, Long-tubed *Gilia brevicula*.
GILIA, Scarlet 1. *Gilia aggregata*. 2. *Gilia attenuata*. 3. *Gilia rubra*.
GILIA, Spotted *Gilia punctata*.
GILIA, Tooth-leaved *Gilia leptomeria*.
GILIA, White *Gilia longiflora*.
GILL *Glechoma hederacea* (Ground Ivy) Glos. J D Robertson; Worcs. B & H. Soft 'g' usually.
GILL, Snoder *Taxus baccata* fruit (Yew berries) Hants. B & H. See Snotty Gogs, rather.
GILL-ALE *Glechoma hederacea* (Ground Ivy) Dev. Friend. Gill, it is claimed, is from French guiller, to ferment beer - this does not tie in with the fact that the English word has a soft 'g'. However, ground ivy certainly used to be a feature in brewing, owing to its tonic bitterness. Alehoof, another old name for the plant, has precisely the same meaning, for it signifies that which will cause ale to heave, or work. The place where such medicated beer was sold was known as a gill-house.

GILL-CREEP-BY-THE-GROUND; GILL-GO-BY-GROUND; GILL-GO-BY-THE-GROUND; GILL-GO-BY-THE-HEDGE; GILL-GO-ON-THE-GROUND; GILL-OVER-THE-GROUND; GILL-RUN-ALONG-THE-GROUND; GILL-RUN-BY-THE-GROUND *Glechoma hederacea* (Ground Ivy) Culpepper (Gill-creep-by-the-ground); Hants. Cope (Gill-go-by-ground); Cumb. (Prior) (Gill-go-by-the-ground); Grieve.1931 (Gill-go-by-the-hedge); Som. Macmillan (Gill-go-on-the-ground); USA Upham (Gill-over-the-ground; Som. Macmillan (Gill-run-along-the-ground); Som, Bucks. Grigson; Lincs. Peacock (Gill-run-by-the-ground).

GILL GOWAN *Chrysanthemum segetum* (Corn Marigold) N.Ire. B & H. In this case, 'gill' is 'gull', or better, 'gule', a word which means gold. So indeed does gowan, so the name actually means gold gold.

GILL HEN *Glechoma hederacea* (Ground Ivy) Sanecki.

GILL-RUN-BY-THE-STREET *Saponaria officinalis* (Soapwort) Suss, Kent Parkinson.

GILLIES, Whitsun *Hesperis matronalis* (Sweet Rocket) War. Bloom. Whitsun Gilliflower, of course, a name usually reserved for the double variety.

GILLIFLOWER 1. *Armeria maritima* (Thrift) Usher. 2. *Cheiranthus cheiri* (Wallflower). 3. *Polemonium caeruleum* (Jacob's Ladder) Dev. Friend. Gilliflower is from Old French girofle, and thence from Greek karyophyllon, clove-tree. So one would expect the name to be reserved for the carnation, which is *Dianthus caryophyllus*.

But it was used for any flower which smelled like cloves. The name exists in a large number of variations - wallflower alone has 20 of them. Curiously, the pleonasm Clove Gilliflower was always reserved for the carnation, never Gilliflower itself. Thrift, which is also known as the Sea Gilliflower, once had the name *Caryophyllum marinus*, under which Gerard described the plant.

GILLIFLOWER, Castle *Mathiola incana* (Stock) Lyte.

GILLIFLOWER, Clove *Dianthus caryophyllus* Chaucer. If 'gilliflower' is indeed from caryophyllus, then the expression Clove Gilliflower must be a pleonasm. Yet as early as Chaucer's time, this was the common name for a carnation.

GILLIFLOWER, Corn *Legousia hybrida* (Venus's Looking Glass) Gerard.

GILLIFLOWER, Cuckoo *Lychnis flos-cuculi* (Ragged Robin) Lyte. Cf. Cuckoo, or Cuckoo-flower, still provincial local names for this.

GILLIFLOWER, Dame's *Hesperis matronalis* (Sweet Rocket). B & H. In spite of the specific name *matronalis*, 'Dame's' is a corruption of Damask. This plant was Viola Damascena, which in French became Violette de Damas, misspelt to Violette des Dames, and so into English Dame's Violet. The name proliferated into Dame's Gilliflower, and, in America, to Dame's Rocket.

GILLIFLOWER, Marsh *Lychnis flos-cuculi* (Ragged Robin) Lyte.

GILLIFLOWER, Mock *Saponaria officinalis* (Soapwort) Lyte.

GILLIFLOWER, Queen's *Hesperis matronalis* (Sweet Rocket) Gerard. Presumably as a compliment to Queen Elizabeth, for the plant was very popular in the 16th century.

GILLIFLOWER, Rogue's *Hesperis matronalis* (Sweet Rocket) Gerard. 'Rogue' is possibly 'rouge' here.

GILLIFLOWER, Sea *Armeria maritima* (Thrift) Gerard. A reasonable name from Gerard's point of view, for he described the plant under the title Caryophyllum marinus.

GILLIFLOWER, Stock *Mathiola incana* (Stock) Prior.

GILLIFLOWER, Stock, Blue *Matthiola incana* (Stock) Turner.

GILLIFLOWER, Stock, Yellow *Cheiranthus cheiri* Gerard. Cf. Yellow Gilliflower.

GILLIFLOWER, Turkey *Tagetes erecta* (African Marigold) Gerard.

GILLIFLOWER, Wall *Cheiranthus cheiri* (Wallflower) Turner.

GILLIFLOWER, Water *Hottonia palustre* (Water Violet) B & H.

GILLIFLOWER, Whitsun *Hesperis matronalis* (Sweet Rocket) Som. Elworthy; Dor. Dacombe. Similarly, in Warwickshire, it was called Whitsun Gillies.

GILLIFLOWER, Whitsuntide *Cardamine pratensis* (Lady's Smock) Glos. B & H.

GILLIFLOWER, Wild *Dianthus caryophyllus* (Carnation) Turner.

GILLIFLOWER, Winter 1. *Cheiranthus cheiri* (Wallflower) Gerard. 2. *Galanthus nivalis* (Snowdrop) Tynan & Maitland. 3. *Hesperis matronalis* (Sweet Rocket) Gerard. Why should Gerard call Sweet Rocket by a name like this? It is a plant of high summer, and the name is probably a corruption of Whitsun Gilliflower.

GILLIFLOWER, Yellow *Cheiranthus cheiri* (Wallflower) W.Miller. Cf. Yellow Stock-Gilliflower.

GILLIVER; GILVER *Cheiranthus cheiri* (Wallflower) Derb. Porteous; Lancs. Nodal (Gilliver); IOM Moore.1924 (Gilver). Cf. Jilliver etc.

GILLOFER, Devil's *Cheiranthus cheiri* (Wallflower) Som. Macmillan. There is no indication from anywhere else of an association with the devil. Wallflower is otherwise a genial plant.

GILLOFLOWER *Cardamina pratensis* (Cuckoo-flower) Dur. Hardy.

GILLOVER, Stock *Matthiola incana* (Stock) Parkinson.

GILLY *Cheiranthus cheiri* (Wallflower) Som. Macmillan. A shortened form of

'gilliflower'.
GILLYFER; GILLYVER *Cheiranthus cheiri* (Wallflower) both N'thants. A.E.Baker.
GILRUMBITHGROUND *Glechoma hederacea* (Ground Ivy) Langham. An eccentric form of the more usual Gill-run-by-the-ground or its variants.
GILT CUP *Ranunculus ficaria* (Lesser Celandine) Suss. Parish.
GILTED CUP 1. *Ranunculus acris* (Meadow Buttercup) Som. Grigson. 2. *Ranunculus bulbosus* (Bulbous Buttercup) Som. Grigson.
GILTY-CUP 1. *Caltha palustris* (Marsh Marigold) Wilts. Dartnell & Goddard. 2. *Ranunculus acris* (Meadow Buttercup) Dev. Friend.1882; Dor, Som. Grigson. 3. *Ranunculus bulbosus* (Bulbous Buttercup) Dev, Som, Dor. Grigson. 4. *Ranunculus ficaria* (Lesser Celandine) Som. Elworthy; Dor. Dacombe. 5. *Ranunculus repens* (Creeping Buttercup) Dev, Dor, Som. Grigson. Cf. Gilcup or Gildcup, and of course Golden Cup and its variants.
GINGELLY *Sesamum indicum* (Sesame) India Dalziel.
GINGER 1. *Sedum acre* (Biting Stonecrop) Som. Macmillan; Suff, Yks. Grigson. 2. *Sedum reflexum* (Yellow Stonecrop) Kent Grigson. 3. *Tanacetum vulgare* (Tansy) Bucks, Herts, Kent, Leics. Grigson. 4. *Zingiber officinale*. As far as Tansy is concerned,the whole plant is strongly aromatic, with a smell of ginger. Presumably stonecrop gets the name from its 'biting' taste. The word 'ginger' comes from the Latin, zingiber, and that came originally from a Sanskrit word which meant horn-like body, a reference to its antler-shaped. root.
GINGER, Bob *Polygonum hydropiper* (Water Pepper) Cumb. Grigson. Acrid, hence the 'pepper' of the common name, and 'ginger'. But there are a lot of others, like Bity Tongue, and Arsmart.
GINGER, Green 1. *Artemisia absinthium* (Wormwood) Grieve.1931. 2. *Artemisia vulgaris* (Mugwort) Lincs. Grigson.

GINGER, Indian *Asarum canadense* (Canadian Snakeroot) Grieve.1931. They say it has the flavour of true ginger.
GINGER, Wall *Sedum acre* (Biting Stonecrop) Grieve.1931.
GINGER, Wild *Asarum canadense* (Canadian Snakeroot) Grieve.1931. See Indian Ginger.
GINGER-BREAD PLUM *Parinari macrophylla*.
GINGER GRASS *Cymbopogon maritima*.
GINGER-PLANT *Tanacetum vulgare* (Tansy) Bucks, Herts, Kent Grigson; USA Henkel.
GINGER MINT *Mentha x gentilis* (Bushy Mint) Brownlow.
GINGLERS, Silver *Briza media* (Quaking-grass) Roxb. Grigson.1959.
GINNY-HEN FLOWER *Fritillaria meleagris* (Snake's-head Lily) Genders. Guinea-hen, of course, and so corresponding to the specific name.
GINSENG 1. *Panax quinquefolium*. 2. *Panax schinseng*. *P.schinseng* is the true Chinese ginseng. *P quinquefolium* is the North American member of the genus, used as a substitute when the real thing became rare. The word is the Chinese Jin-chen, which means man-like - the forked root was treated like the human form (as with mandrake). The more closely the root resembled the human form, the more valuable the Chinese considered it, and well-formed roots are worth their weight in gold. Because of the form, the doctrine of signatures worked to say that the plant healed all parts of the body.
GINSENG, Dwarf *Panax trifolium*.
GIRASOLE *Helianthus tuberosus* (Jerusalem Artichoke) A.W.Smith. A sunflower name, for this is also known as the Sunflower Artichoke.
GIRL'S CURLY LOVE *Artemisia abrotanum* (Southernwood) Tynan & Maitland. Cf. Lad's Love, etc.
GIRL'S DELIGHT *Artemisia abrotanum* (Southernwood) Som. Macmillan. Better known as Lad's Love, but "Lads' love is lass-

es' delight, And if the lads don't love, Lasses will flite". A few branches of southernwood generally used to be added to the nosegay that courting youths used to give the girls. Or in some parts, notably in the Fenlands, he would put them in his buttonhole, hoping the girls would notice his Lad's love.

GIRL'S LOVE *Rosmarinus officinalis* (Rosemary) Som. Macmillan.

GIRL'S MERCURY *Mercurialis annua* (French Mercury) Lyte. Lyte was not very consistent about this. He gave the name Boy's Mercury, which he explained by referring to an old belief that when "dronken (it) causeth to engender male children". But then he also gave Girl's Mercury and Maiden Mercury for the same plant, with no explanation.

GIRLS, Boys-and- 1. *Mercurialis annua* (French Mercury) USA Kingsbury. 2. *Primula veris* (Cowslip) Dor. Macmillan. 3. *Primula vulgaris* (Primrose) Wilts. Dartnell & Goddard. This name for French Mercury seems to compound the confusion mentioned under Girl's Mercury. Wiltshire people say that the long-pistilled or "pin-eyed" primroses are "boys", while the short-pistilled or "thrum-eyed" ones are "girls".

GIRSE, Blue *Vicia cracca* (Tufted Vetch) Shet. Grigson.

GIRSE, Well *Nasturtium officinale* (Watercress) Scot. Grigson. 'Girse' here must be cress.

GIRSE, Yule *Filipendula ulmaria* (Meadowsweet) Shet. Nicolson.

GIT, Herb *Nigella damascena* (Love-in-a-mist) Turner. The word appears as Gith elsewhere (as in the specific name of Corn Cockle, *githago*), said to be from a Hebrew word, gesah.

GITH 1. *Agrostemma githago* (Corn Cockle) Prior. 2. *Lolium temulentum* (Darnel) J Harvey. 3. *Nigella damascena* (Love-in-a-mist) Coles. 4. *Nigella sativa* (Fennelflower) J Harvey. The fact that this OE name, originally given to *Nigella*, is awarded to both Darnel and Cockle may mean that it is used as a name for any noxious weed among the corn.

GITHCORN *Daphne laureola* fruit (Spurge Laurel - fruit) Cockayne.

GIX, GIXY *Conium maculatum* the dried stalks Wilts. Akerman (Gix); Som, Wilts. Macmillan (Gixy). A local variation on the more usual Kex.

GLACIER CROWFOOT *Ranunculus glacialis*.

GLACIER LILY *Erythronium grandiflorum*.

GLADDEN; GLADDON 1. *Acorus calamus* (Sweet Flag) Norf. Leyel.1937. 2. *Iris foetidissima* (Foetid Iris) Dawson. Latin gladius means a sword, a reference in both the Irises and the Gladiolus to the shape of the leaves.

GLADDON, Stinking *Iris foetidissima* (Foetid Iris) Gerard.

GLADDY *Iris foetidissima* (Foetid Iris) Som. Macmillan.

GLADE LILY *Lilium philadelphicum* (Philadelphia Lily) USA Woodcock.

GLADEN *Iris pseudo-acarus* (Yellow Flag) Turner.

GLADENE *Scilla maritima* (Squill) R T Gunther. This only occurs as a gloss, and presumably an incorrect one, on a manuscript of Apuleius Barbarus.

GLADER *Iris sp*.

GLADIN, Corn *Gladiolus sp* Gerard.

GLADINE; GLADING ROOT; GLADYNE *Iris foetidissima* (Foetid Iris) Halliwell (Gladine); Ire. B & H (Glading Root). Dawson (Gladyne).

GLADIOLE, Water 1. *Butomus umbellatus* (Flowering Rush) Curtis. 2. *Lobelia dortmanniana* (Water Lobelia) Cumb. Grigson. Cf. the Welsh name which translates dagger plant for the latter plant.

GLADIOLUS, Magpie *Acidanthera bicolor* McDonald.

GLADWIN *Iris pseudo-acarus* (Yellow Flag) Wesley.

GLADWIN, Stinking *Iris foetidissima* (Foetid Iris) Culpepper.

GLADWYN *Iris foetidissima* (Foetid Iris)

Hill.

GLASS, Ram's *Ranunculus acris* (Meadow Buttercup) Som. Macmillan. Possibly a corruption of Ram's Claws, which is actually recorded for Creeping Buttercup (*Ranunculus repens*).

GLASS EYE, Old Man's *Anagallis arvensis* (Scarlet Pimpernel) Som. Grigson. A name which started off as Old Man's Weatherglass, for pimpernel is a weather forecaster - fine or wet according to whether the flowers are open or closed.

GLASS SALTWORT *Salicornia europaea* (Glasswort) Gerard. Gerard compromised here, for this plant is sometimes called glasswort, and sometimes saltwort (the former being more usual).

GLASSWORT *Salicornia europaea*. The ashes of the plant were used in glassmaking, providing an impure carbonate of soda for mixing with the sand.

GLASSWORT, Bushy *Salicornia dolicostachys*.

GLASSWORT, Creeping *Salicornia perennis*.

GLASSWORT, Fragile *Salicornia pusilla*.

GLASSWORT, Jointed *Salicornia europaea* (Glasswort) Gerard USA House.

GLASSWORT, Prickly *Salsola kali* (Saltwort) Prior. Like Salicornia, saltwort was burnt for its fixed salt, used in glassmaking.

GLASSWORT, Seablite *Salicornia prostrata*.

GLASSWORT, Slender *Salicornia europaea* (Glasswort) USA House.

GLASSWORT, Starry *Cerastium arvense* (Field Mouse-ear) USA Elmore.

GLASSWORT, Twiggy *Salicornia ramosissima*.

GLASSWORT, Western *Salicornia rubra*.

GLASTONBURY THORN *Crataegus monogyna var. praecox*. The legend is that Joseph of Arimathea, on his way to Glastonbury, arrived at Wearyall Hill, south of the town, and rested there after having thrust his staff into the ground. The stick took root, and blossomed each year on the anniversary of the birth of Christ (old style). In later times, the legend was used to show that the true Christmas was the old style, 6 January, and not 25 December. The legend seems to have been complete in the early 18th century - it was a local innkeeper who first launched the Wearyall Hill story. Plants grown from the haws of the Glastonury Thorn do not retain the characteristics of the parent. The only way of achieving this is by grafting or budding on to other roots. It seems that the tree was common in Palestine, where it bloomed at the same time. It flowers twice, once in the winter, and again in the spring. The winter flowers produce no fruit, though the spring ones do.

GLEAMING STAR *Saxifraga spathularis x umbrosa* (London Pride) Som. Macmillan.

GLENNIES 1. *Ranunculus acris* (Meadow Buttercup) Wilts. Grigson. 2. *Ranunculus bulbosus* (Bulbous Buttercup) Wilts. Grigson. 3. *Ranunculus repens* (Creeping Buttercup) Wilts. Grigson.

GLENS *Narcissus pseudo-narcissus* (Daffodil) Ayr. B & H.

GLOBE AMARANTH 1. *Amaranthus hypondriachus* (Prince's Feathers). 2. *Gomphrenia globosa*.

GLOBE ARTICHOKE *Cynara cardunculus var. scolymus*.

GLOBE CROWFOOT *Trollius europaeus* (Globe Flower) Gerard.

GLOBE CUCUMBER *Cucumis prophetarium*.

GLOBE FLOWER 1. *Cephalanthus occidentalis* (Buttonbrush) O P Brown. 2. *Globularia vulgaris*. 3. *Trollius europaeus*.

GLOBE FLOWER, American *Trollius laxus*.

GLOBE FLOWER, Asiatic *Trollius asiaticus* (Orange Globe Flower) Grieve.1931.

GLOBE FLOWER, Orange *Trollius asiaticus*.

GLOBE LILY *Calochortus albus* (Fairy Lantern) RHS.

GLOBE RANUNCULUS *Trollius europaeus*.

GLOBE THISTLE *Echinops ritro*.

GLOBE TULIP, Golden *Calochortus amabilis* (Golden Fairy Lantern) RHS.
GLOBES *Trollius europaeus* (Globe Flower) Som. Elworthy.
GLODEN *Helianthus annuus* (Sunflower) Lincs. B & H. A misprint, perhaps? But it appears in the correct place as it stands in B & H, so perhaps it is genuine.
GLORILESS *Adoxa moschatellina* (Moschatel) Som. Macmillan. This is the English equivalent of the Greek adoxa.
GLORY, Morning 1. *Calystegia sepium* (Large Bindweed) Som. Macmillan. 2. *Convolvulus arvensis* (Field Bindweed) Som. Macmillan. 3. *Convolvulus major*. 4. *Rivea corymbosa*.
GLORY, Morning, Bush *Ipomaea leptophylla*.
GLORY, Morning, Field *Convolvulus arvensis* (Field Bindweed) USA Elmore.
GLORY BUSH *Tibouchina urvilleana*.
GLORY FLOWER; GLORY VINE, Chilean *Eccremocarpus scaber* Ted Phillips (Chilean Glory Vine).
GLORY LILY *Gloriosa superba*.
GLORY OF THE SUN *Leucocoryne ixioides*.
GLORY PEA *Clianthus formosus* (Sturt's Desert Pea) Perry.
GLORYBOWER 1. *Clerodendron splendens*. 2. *Clerodendron thomsoniae* (Bleeding Heart Vine) Bonar.
GLORYBOWER, Blue *Clerodendron ugadense*.
GLOSSY PRIVET *Ligustrum lucidum* (Wax Tree) Tampion.
GLOVES, Boxing *Lotus corniculatus* (Bird's-foot Trefoil) Som. Macmillan.
GLOVES, Fairy's *Digitalis purpurea* (Foxglove) Som. Macmillan; Shrop. Burne.
GLOVES, Finger *Digitalis purpurea* (Foxglove) Tynan & Maitland.
GLOVES, Goblin's *Digitalis purpurea* (Foxglove) Wales.
GLOVES, Granny's *Digitalis purpurea* (Foxglove) Som. Macmillan.
GLOVES, Lady's 1. *Campanula medium* (Canterbury Bells) Som. Macmillan; Herts. Jones-Baker.1974. 2. *Cardamine pratensis* (Lady's Smock) N'thants. Grigson. 3. *Digitalis purpurea* (Foxglove) Som, Shrop. Burne; USA. Henkel. 4. *Fumaria officinalis* (Fumitory) Wilts. Tynan & Maitland. 5. *Inula conyza* (Ploughman's Spikenard) W Miller. 6. *Lotus corniculatus* (Bird's-foot Trefoil) Dor, Wilts. Macmillan; N'thants. A E Baker. 7. *Lotus uliginosus* (Narrow-leaved Bird's-foot Trefoil) Wilts. D Grose.
GLOVES, Pop *Digitalis purpurea* (Foxglove) Corn. Courtney.1880. The 'gloves' are the flowers of the Foxglove,which are 'popped' by children.
GLOVES, Witches' *Digitalis purpurea* (Foxglove) Skinner.
GLOVEWORT *Convallaria maialis* (Lily-of-the-valley) Cockayne. The sap was recommended for sore hands.
GLOW, Golden 1. *Rudbeckia laciniata* (Tall Coneflower) Kansas Gates. 2. *Solidago virgaurea* (Golden Rod) Som. Macmillan.
GLOXINIA *Sinningia speciosa*. All modern Gloxinias are hybrid forms of this.
GLUTTON, Ground *Senecio vulgaris* (Groundsel) Scot. Grieve.1931. In the sense of groundsel's original meaning (according to some) of ground swallower. OE grundswelge.
GLUTTONY, Plant of *Cornus suecica* (Dwarf Cornel) W Miller. This is a translation from the Gaelic, and the name is given to celebrate the virtues of the berries, known to be a fine tonic for the appetite.
GLYCINE, Purple *Hardenbergia violacea* (Blue Coral Pea) Whittle & Cook. This plant was formerly put with the Glycines, as G virens, or G bimaculata.
GNAPPERTS *Lathyrus montanus* (Bittervetch) Mor. B & H. It also appears as Knapperts, and then varies into Napperty, Napple, Nipper, even Kipper.
GNAT FLOWER *Centaurea nigra* (Knapweed) Som. Macmillan.
GNAT ORCHID *Gymnadenia conopsea* (Scented Orchid) Turner Ettlinger.
GO-CUP 1. *Caltha palustris* (Marsh Marigold) Som. Macmillan. 2. *Ranunculus*

acris (Meadow Buttercup) Dev. Friend.1882; Som. Macmillan. 3. *Ranunculus ficaria* (Lesser Celandine) Som. Macmillan. i.e. gold cup.

GO-TO-BED-AT-NOON *Tragopogon pratensis* (Goat's Beard) Gerard. One of a series of names, the best known of which is Jack-go-to-bed-at-noon, to illustrate the plant's habit of opening its flowers at sunrise, and closing them again at noon.

GO-TO-SLEEP-AT-NOON *Colchicum autumnale* (Meadow Saffron) Som. Macmillan.

GOA, Cedar of *Cupressus lusitanica* (Mexican Cypress) Mitchell.1972. Presumably because of the Portuguese connection. This is an American tree, which only became known in Europe through examples cultivated in Portugal.

GOA BEAN *Psophocarpus tetragonolobus*.

GOAT APPLE *Datura stramonium* (Thornapple) S Africa Watt. It is probably the awful smell that accounts for the name. Cf. Stinkweed, etc.

GOAT LEAF HONEYSUCKLE *Lonicera caprifolium* (Perfoliate Honeysuckle) McDonald. Caprifolium means goat leaf alright; but it is a mistake. It was originally capparifolium, caper leaf.

GOAT NUT *Simmondsia chinensis* (Jajoba) USA T H Kearney.

GOAT PEPPER *Capsicum annuum* (Chile) Lindley.

GOAT-SCENTED TUTSAN *Hypericum hircinum* (Stinking Tutsan) McClintock.1975.

GOAT STONES 1. *Himantoglossum hircinum* (Lizard Orchid) Gerard. 2. *Orchis mascula* (Early Purple Orchis) Fluckiger. 'Stones' from the testicle-looking twin tubers; Lizard Orchid gets the 'goat' appellation because of the smell while in bloom.

GOAT TREE *Lonicera periclymenum* (Honeysuckle) Dev, Som. Grigson. A name which must have arisen out of a memory of the old Latin name caprifolium, even though the latter was an error.

GOAT WILLOW *Salix capraea*. Apparently because goats are fond of it, and readily browse on the leaves.

GOAT'S AIRACH *Chenopodium vulvaria* (Stinking Orache) Culpepper. The plant stinks, so the attribution to goats is natural.

GOAT'S BEAN *Menyanthes trifoliata* (Buckbean) Fernie. Buckbean is a translation of the Flemish books boonen, goats' beans.

GOAT'S BEARD 1. *Astilbe sp*. 2. *Cichorium endivia* (Endive) Som. Macmillan. 3. *Filipendula ulmaria* (Meadowsweet) Dev, Dor. Macmillan. 4. *Hypericum perforatum* (St John's Wort) USA Watt. 5. *Spiraea aruncus* Broadwood. 6. *Tragopogon porrifolius* (Salsify) Australia Watt. 7. *Tragopogon pratensis*.

GOAT'S BEARD, Purple *Tragopogon porrifolius* (Salsify) Grieve.1931.

GOAT'S CULLIONS *Himantoglossum hircinum* (Lizard Orchid). B & H. Cullions, in an early sense, meant testicles. Cf. Goat Stones for this.

GOAT'S FOOT 1. *Aegopodium podagraria* (Goutweed) Dev. Macmillan. 2. *Helianthemum chamaecistus* (Rock Rose) Som. Macmillan. 3. *Tragopogon pratensis* (Goat's Beard) Som. Macmillan.

GOAT'S FOOT IPOMAEA *Ipomaea pes-caprae*.

GOAT'S HERB *Aegopodium podagraria* (Goutweed) Le Strange. *Aegopodium* means goat's foot.

GOAT'S LEAF *Lonicera periclymenum* (Honeysuckle) Dev, Som. Macmillan. From the old Latin name caprifolium.

GOAT'S MARJORAM *Origanum vulgare* (Marjoram) Tynan & Maitland.

GOAT'S RUE *Galega officinalis*. Probably so called because of its extremely disagreeable taste - it is very bitter indeed.

GOAT'S STONES, Great *Himantoglossum hircinum* (Lizard Orchid) Camden/Gibson. Cf. Goat Stones.

GOAT'S THORN *Astragalus tragacantha* Thornton.

GOATWEED 1. *Aegopodium podagraria*

(Goutweed) B & H. 2. *Ageratum conyzoides*. 3. *Hypericum perforatum* (St John's Wort) USA Watt. 4. *Polygonum convolvulus* (Black Bindweed) Wilts. Macmillan. The first one is a simple mistake - goatweed for goutweed.

GOB, Hay *Polygonum convolvulus* (Black Bindweed) War. Halliwell.

GOBBLETY-GUTS *Rumex acetosa* (Wild Sorrel) Yks. Addy.

GOBBO *Abelmoschus esculentus* (Okra) Howes. Gombo was a West African name for the plant, and Gobbo is just one of several versions of it in English.

GOBLIN'S GLOVES; GOBLIN'S THIMBLES *Digitalis purpurea* (Foxglove) Wales (Gloves); Hants. Boase (Thimbles). Cf. Fairy's Gloves and Fairy's Thimbles.

GOD, Hand of *Lapsana communis* (Nipplewort) Wilts. Powell.

GOD ALMIGHTY'S BREAD-AND-CHEESE Oxalis acetosella (Wood Sorrel) Som. Elworthy.

GOD ALMIGHTY'S FINGERS-AND-THUMBS *Anthyllis vulneraria* (Kidney Vetch) Dor. Grigson.

GOD ALMIGHTY'S FLOWER *Lotus corniculatus* (Bird's-foot Trefoil) Dev. Macmillan.

GOD ALMIGHTY'S THUMB-AND-FINGER *Lotus corniculatus* (Bird's-foot Trefoil) Hants. Cope.

GOD'S EYE 1. *Salvia sclarea* (Clary) B & H. 2. *Veronica chamaedrys* (Germander Speedwell) Dev. Friend.1882; Lincs, Glos. Grigson. Clary is 'clear-eye', hence such names as this. Speedwell has its 'eye' names as a description.

GOD'S FINGERS-AND-THUMBS 1. *Fumaria officinalis* (Fumitory) Dor. Grigson. 2. *Lotus corniculatus* (Bird's-foot Trefoil) Som. Macmillan.

GOD'S FLOWER *Helichrysum stoechas* (Golden Cassidony). W Miller. God, in this case, is Gold.

GOD'S MEAT *Crataegus monogyna* (Hawthorn) Yks. B & H.

GOD'S STINKING TREE *Sambucus nigra* (Elder) Dor. Dacombe.

GOD'S TREE 1. *Artemisia abrotanum* (Southernwood) Le Strange. 2. *Ilex aquifolium* (Holly) English gypsy Groome.

GODFATHERS AND GODMOTHERS *Viola tricolor* (Pansy) Dur. B & H.

GODS, Tree of the *Ailanthus glandulosa* (Tree of Heaven) Brimble.1948.

GOG, Horse- *Prunus spinosa* fruit (Sloe) Yks. Nicholson.

GOG, Snotty *Taxus baccata* fruit (Yew berries) Wilts. Macmillan; Suss. Parish. All the local names for yew berries are variations on this.

GOG, Sour *Rumex acetosa* (Wild Sorrel) Bucks. Harman.

GOGGLES *Ribes uva-crispa* (Gooseberry) Lincs. Grigson. Cf. Goosegog for the fruit.

GOGGLES, Blue *Endymion nonscriptus* (Bluebell) Wilts. Dartnell & Goddard. In this case, 'goggles' sounds as if it were a variation on another local name, Greggles, or Griggles.

GOGGLES, Water *Caltha palustris* (Marsh Marigold) Oxf. Grigson; Bucks. Harman. Possibly a mixture of 'gold' and 'bubbles', with a reference to the blistering qualities of the juice.

GOINERS *Helianthus tuberosus* - tubers (Jerusalem Artichoke) H Smith.1928.

GOLD 1. *Calendula officinalis* (Marigold) Parkinson. 2. *Caltha palustris* (Marsh Marigold) Prior. 3. *Chrysanthemum segetum* (Corn Marigold) Som. Macmillan; Midl. Grigson. 4. *Helianthus annuus* (Sunflower) B & H. 5. *Laburnum anagyroides* (Laburnum) Tynan & Maitland. 6. *Myrica gale* (Bog Myrtle) Som. Jennings.

GOLD, Carr *Chrysanthemum segetum* (Corn Marigold) Lancs. Grigson. Carr means boggy ground, but this is not the kind of environment that would suit this plant.

GOLD, Cloth of *Baileya multiradiata* (Desert Baileya) USA Kingsbury.

GOLD, Cup of; GOLD CUP *Solandra maxima* (Chalice Vine) Perry.

GOLD, Garden *Calendula officinalis* (Marigold) Tynan & Maitland. A sensible differentiation. Gold (see above) applies to half a dozen plants, but this name could only describe the one.

GOLD, Horse *Ranunculus arvensis* (Corn Buttercup) Herts. Grigson; N'thants. A E Baker.

GOLD, Mary *Calendula officinalis* (Marigold) Dev, Som. Macmillan. That is how it would be pronounced in the west country.

GOLD, Mary's *Caltha palustris* (Marsh Marigold) Som. Macmillan.

GOLD, Rods *Calendula officinalis* (Marigold) B & H. Ruddes was a common name for marigolds at one time. The word means red, so Rods-gold is either a way of pinpointing the flower, for 'gold' may mean Corn Marigold among others, or it was a more accurate way of identifying the colour.

GOLD, Vegetable *Coptis trifolia* (Goldthread) Howes.

GOLD, White *Leucanthemum vulgare* (Ox-eye Daisy) Lancs, Cumb. Grigson. For proper Gold (or Yellow Gold) is the other Chrysanthemum, Corn Marigold.

GOLD, Yellow *Chrysanthemum segetum* (Corn Marigold) Cumb. Grigson. See White Gold.

GOLD BASKET *Alyssum saxatile* (Gold Dust) W.Miller.

GOLD BELLS *Narcissus pseudo-narcissus* (Daffodil) Wilts. Macmillan.

GOLD BLOOM *Calendula officinalis* (Marigold) B & H.

GOLD CHAIN *Sedum acre* (Biting Stonecrop) B & H.

GOLD CHAIN, Wild *Melilotus altissima* (Tall Melilot) Som. Grigson.

GOLD-CRAP 1. *Ranunculus acris* (Meadow Buttercup) Som. Grigson. 2. *Ranunculus bulbosus* (Bulbous Buttercup) Halliwell. 'Crap' is a word used to describe a plant which is unwanted, like darnel or buckwheat. Or the name could have been Gold-cup originally.

GOLD CUP; CUP OF GOLD *Solandra maxima* (Chalice Vine) Perry.

GOLD DUST 1. *Alyssum saxatile*. 2. *Sedum acre* (Biting Stonecrop) Som. Macmillan.

GOLD DUST TREE *Aucuba japonica* (Japanese Laurel) Perry.1979.

GOLD FLOWER 1. *Gnaphalium orientale* Gerard. 2. *Helichrysum stoechas* (Golden Cassidony) W.Miller.

GOLD-IN-GREEN *Chrysogonum virginianum* (Golden Knee) Howes.

GOLD KNOP *Ranunculus bulbosus* (Bulbous Buttercup) Glos. Halliwell. A reference to the flower buds. 'Knob' is more usual than 'knop'. The word even becomes 'knot', as in the next entry.

GOLD KNOTS *Ranunculus acris* (Meadow Buttercup) North.

GOLD OF PLEASURE *Camelina sativa*. "said to bear ironical reference to the disappointment of its first cultivators here, who found their investment in it about as profitable as gold squandered on 'pleasure' usually proves" (C P Johnson).

GOLD POPPY, Desert *Escholtzia glyptosperma*.

GOLD POPPY, Little *Escholtzia minutiflora*.

GOLD STAR *Geum urbanum* (Avens) Som. Macmillan.

GOLD-TIPS, Flame *Leucospermum discolor* S.Africa Perry.

GOLD WATCH *Hypericum calycinum* (Rose of Sharon) Som. Macmillan.

GOLD WIRE *Hypericum concinnum*.

GOLD(EN) WITHY *Myrica gale* (Bog Myrtle) Hants, IOW Cope.

GOLDBAND LILY *Lilium auratum* (Japanese Lily) Woodcock.

GOLDCROP *Ranunculus acris* (Meadow Buttercup) Som. Grigson.

GOLDCUP *Ranunculus bulbosus* (Bulbous Buttercup) Hants. Cope; Corn,Dev,Som,Wilts,Suss,Kent Grigson. Probably, the other common buttercups should be included, too.

GOLDEN APPLE MINT *Mentha x gentilis* (Bushy Mint) Webster.

GOLDEN BALLS 1. *Trollius europaeus* (Globe Flower) Som. Macmillan; Ches, Lancs. Grigson. 2. *Viburnum opulus* (Guelder Rose) Som. Elworthy.

GOLDEN BARBERRY *Berberis stenophylla*.

GOLDEN BELLS *Primula veris* (Cowslip) Som. Macmillan.

GOLDEN BLOSSOM *Potentilla reptans* (Cinquefoil) Dev. Friend,1882.

GOLDEN BUTTERCUP *Caltha palustris* (Marsh Marigold) Som. Macmillan.

GOLDEN BUTTONS *Tanacetum vulgare* (Tansy) Dev. Grigson.

GOLDEN CAMOMILE *Anthemis tinctoria* (Yellow Camomile) Howes.

GOLDEN CARPET *Sedum acre* (Biting Stonecrop) Som. Grigson.

GOLDEN CASSIA *Cassia fasciculata*.

GOLDEN CASSIDONY *Helichrysum stoechas*.

GOLDEN CHAIN *Laburnum anagyroides* (Laburnum) Wright. (inc. Worcs. J Salisbury).

GOLDEN CHAIR *Sarothamnus scoparius* (Broom) Dor, Som. Grigson.

GOLDEN CHESTNUT *Chrysolepis chrysophylla*. It is the underside of the evergreen leaves that is golden.

GOLDEN CINQUEFOIL *Potentilla aurea*.

GOLDEN CORNFLOWER *Chrysanthemum segetum* (Corn Marigold) Gerard.

GOLDEN CUP *Caltha palustris* (Marsh Marigold) Som. Elworthy. Cf. Golden Kingcup. *Caltha* is from Greek calathos, a cup or goblet.

GOLDEN CURRANT *Ribes aureum* (Buffalo Currant) USA Elmore.

GOLDEN DAISY 1. *Chrysanthemum segetum* (Corn Marigold) Som. Macmillan. 2. *Ranunculus ficaria* (Lesser Celandine) N'thants. Clare.

GOLDEN DOCK *Rumex maritimus*.

GOLDEN DROPS 1. *Laburnum anagyroides* (Laburnum) Lincs. B & H. 2. *Onosma echioides*. 3. *Primula veris* (Cowslip) Som. Macmillan.

GOLDEN DUST 1. *Aurinia saxatile* (Gold Dust) Som. Macmillan. 2. *Galium verum* (Lady's Bedstraw) Som. Macmillan. 3. *Sedum acre* (Biting Stonecrop) Corn, Suff. Grigson. 4. *Solidago virgaurea* (Golden Rod) Som. Macmillan.

GOLDEN ELDER *Sambucus nigra var. aurea*.

GOLDEN FLAX *Linum flavum*.

GOLDEN FLOWER 1. *Chrysanthemum segetum* (Corn Marigold) Som. Macmillan. 2. *Potentilla anserina* (Silverweed) Som. Macmillan.

GOLDEN GARLIC *Allium moly*.

GOLDEN GILIA *Gilia aurea*.

GOLDEN GLOW *Rudbeckia laciniata* (Tall Coneflower) Kansas Gates.

GOLDEN GORSE *Ulex europaeus* (Gorse) Grieve.1931.

GOLDEN GRAIN *Verbascum thapsus* (Mullein) Dev. Friend.1882.

GOLDEN GROUNDSEL *Senecio aureus* (Swamp Squaw-weed) Grieve.1931.

GOLDEN GUINEA TREE *Dillenia alata* (Red Beech) Robson. Bright yellow flowers that last for just one day.

GOLDEN HERB *Atriplex patula* (Orache) N.Eng. Halliwell.

GOLDEN KINGCUP *Caltha palustris* (Marsh Marigold) Som. Macmillan.

GOLDEN KNOB *Caltha palustris* (Marsh Marigold) Som, Berks. Grigson.

GOLDEN LOCKS; GOLDEN MAIDENHAIR *Polypodium vulgare* (Common Polypody) Heref. (Locks); Kent, Heref. (Maidenhair) both B & H.

GOLDEN MIMOSA *Acacia baileyana* (Cootamundra Wattle) Hora.

GOLDEN MOSS *Sedum acre* (Biting Stonecrop) Som, Oxf, War, Yks. Grigson; USA Upham.

GOLDEN NIGGER *Helianthus annuus* (Sunflower) Som. Macmillan.

GOLDEN OSIER *Myrica gale* (Bog Myrtle) IOW Grigson.

GOLDEN PAPER DAISY *Helichrysum bracteatum* (Everlasting Flower) Australia

Howes.
GOLDEN PEA *Thermopsis rhombifolia* (False Lupin) Johnston.
GOLDEN PINE 1. *Grevillea robusta* (Silky Oak) Hora. 2. *Pseudolarix amabilis* (Chinese Gold Larch) Howes.
GOLDEN POLYPODY *Polypodium vulgare* (Common Polypody) Kent B & H.
GOLDEN PRIVET *Ligustrum ovalifolium* (Japanese Privet) Howes.
GOLDEN PURSLANE *Portulaca sativa* (Purslane) Clair.
GOLDEN RAGWORT *Senecio aureus* (Swamp Squaw-weed) USA House.
GOLDEN RAIN *Laburnum anagyroides* (Laburnum) North.
GOLDEN ROD 1. *Agrimonia eupatoria* (Agrimony) Dor. Grigson. 2. *Hypericum maculatum* (Imperforate St John's Wort) Dev, Som. Grigson. 3. *Hypericum tetrapterum* (St Peter's Wort) Dev, Som. Grigson. 4. *Sarothamnus scoparius* (Broom) Som. Macmillan. 5. *Solidago virgaurea*. 6. *Verbascum thapsus* (Mullein) Dev, Som. Friend.1882.
GOLDEN ROD, Blue-stemmed *Solidago caesia*.
GOLDEN ROD, Broad-leaved *Solidago flexicaulis* (Zigzag Golden Rod) USA House.
GOLDEN ROD, Canadian *Solidago altissima*.
GOLDEN ROD, Downy *Solidago puberula*.
GOLDEN ROD, Early *Solidago gigantea*.
GOLDEN ROD, Fragrant *Solidago graminifolia*.
GOLDEN ROD, Hardleaf *Solidago rigida*.
GOLDEN ROD, Pale *Solidago bicolor* (White Golden Rod) USA House.
GOLDEN ROD, Rock *Solidago altissima* (Canadian Golden Rod) USA House.
GOLDEN ROD, Seaside *Solidago sempervirens*.
GOLDEN ROD, Tall *Solidago altissima* (Canadian Golden Rod) Youngken. 'Tall', for the specific name is *altissima*.
GOLDEN ROD, Three-ribbed, Smooth *Solidago gigantea* (Early Golden Rod) Grieve.1931.
GOLDEN ROD, White *Solidago bicolor*.
GOLDEN ROD, Wreath *Solidago caesia* (Blue-stemmed Golden Rod) USA House.
GOLDEN ROD, Zigzag *Solidago flexicaulis*.
GOLDEN ROSE *Primula vulgaris* (Primrose) Som. Grigson.
GOLDEN ROSE OF CHINA *Rosa hugonis* (Father Hugo's Rose) Mattock.
GOLDEN SAMPHIRE *Inula crithmoides*.
GOLDEN SEAL *Hydrastia canadensis*.
GOLDEN SHOWER 1. *Cassia fistula* (Indian Laburnum) Everett. 2. *Laburnum anagyroides* (Laburnum) Shrop. B & H. 3. *Pyrostegia venusta* (Chinese Cracker Flower) RHS.
GOLDEN SLIPPER 1. *Cypripedium calceolus* (Lady's Slipper) USA Cunningham. 2. *Lotus corniculatus* (Bird's-foot Trefoil) Hants. Boase.
GOLDEN SOVEREIGNS *Potentilla anserina* (Silverweed) Som. Macmillan.
GOLDEN SPANIARD *Aciphylla aurea*.
GOLDEN SPLEENWORT *Chrysosplenium oppositifolium* (Opposite-leaved Golden Saxifrage). *Chrysosplenium* is from Greek chrusos, gold, and splen, spleen.
GOLDEN STAR 1. *Chrysogonum virginianum* (Golden Knee) A W Smith. 2. *Primula vulgaris* (Primrose) Som. Macmillan. 3. *Ranunculus ficaria* (Lesser Celandine) Som. Macmillan.
GOLDEN STONEBREAK *Chrysosplenium oppositifolium* (Opposite-leaved Golden Saxifrage) Freethy.
GOLDEN SUN *Taraxacum officinale* (Dandelion) Som. Grigson. An appropriate name, both descriptively and mythologically. For there is a legend that the dandelion was born of the dust raised by the chariot of the sun, which is why it opens at dawn and closes at dusk. In Gaelic it is Little Flame of God, and there used to be a German name, Sonnewirbel.
GOLDEN SWAN *Crocosmia masonorum*.
GOLDEN THISTLE 1. *Scolymus hispanicus* (Spanish Salsify) Schery. 2. *Scolymus macu-*

latus.

GOLDEN TIMOTHY GRASS *Setaria sphacelata* S Africa Dalziel.

GOLDEN TRUMPET *Narcissus pseudo-narcissus* (Daffodil) Som. Macmillan.

GOLDEN TRUMPET FLOWER *Allemanda cathartica* (Yellow Allemande) Simmons.

GOLDEN WATTLE *Acacia pycnantha.*

GOLDEN WATTLE, Sydney *Acacia longifolia.*

GOLDEN WILLOW *Salix alba var. vitellina.*

GOLDEN WIRE *Sedum acre* (Biting Stonecrop) N'thants. Clare.

GOLDEN WITHY *Myrica gale* (Bog Myrtle) Hants, IOW Grigson.

GOLDEN WONDER *Cassia didymobotrya.*

GOLDENBUSH, Slender *Haplopappus gracilis.*

GOLDENEYE, Annual *Viguiera annua.*

GOLDFISH PLANT *Columnea gloriosa.*

GOLDICUP *Caltha palustris* (Marsh Marigold) Corn. Grigson. Cf. Golden Cup and other forms such as Go-cup or Gilcup.

GOLDILOCKS 1. *Amaranthus luteus* Folkard. 2. *Caltha palustris* (Marsh Marigold) Som. Macmillan. 3. *Helichrysum stoechas* (Golden Cassidony) Grieve.1931. 4. *Linosyris vulgaris.* 5. *Ranunculus auricomus* (Wood Goldilocks) Glos, Scot. B & H. 6. *Trollius europaeus* (Globe Flower) West. Grigson. 7. *Verbascum thapsus* (Mullein) Clare.

GOLDILOCKS, Wood *Ranunculus auricomus.*

GOLDING CUP *Ranunculus ficaria* (Lesser Celandine) Dor. B & H.

GOLDINGS 1. *Calendula officinalis* (Marigold) Lincs. Tynan & Maitland. 2. *Chrysanthemum segetum* (Corn Marigold) N'thants. Sternberg; Ches. Halliwell. 3. *Leucanthemum vulgare* (Ox-eye Daisy) Fernie. Obviously, the name belongs to the marigolds; Ox-eye Daisy only gets it by association.

GOLDINS, Camomile *Tripleurospermum maritimum* (Scentless Maydweed) B & H.

GOLDINS, Daisy *Leucanthemum vulgare* (Ox-eye Daisy) N Ire. Grigson.

GOLDINS, Marigold *Chrysanthemum segetum* (Corn Marigold) N Ire. Grigson.

GOLDLOCK *Sinapis arvensis* (Charlock) Wilts. Macmillan. A variation on charlock itself, which can become chadlock, which Gerard used, or kedlock, thus letting in such a hybrid as goldlock.

GOLDTHREAD *Coptis trifolia.*

GOLDWEED 1. *Ranunculus acris* (Meadow Buttercup) Som. Grigson. 2. *Ranunculus arvensis* (Corn Buttercup) Hants. Cope. 3. *Ranunculus bulbosus* (Bulbous Buttercup) Som. Grigson. 4. *Ranunculus repens* (Creeping Buttercup) Som. Grigson.

GOLDWORT *Calendula officinalis* (Marigold) Henslow.

GOLDY 1. *Geum urbanum* (Avens) B & H. 2. *Ranunculus acris* (Meadow Buttercup) Som. Macmillan. 3. *Ranunculus bulbosus* (Bulbous Buttercup) Som. Grigson. 4. *Ranunculus repens* (Creeping Buttercup) Som. Grigson.

GOLDY KNOB 1. *Ranunculus acris* (Meadow Buttercup) Oxf. B & H. 2. *Ranunculus ficaria* (Lesser Celandine) Oxf. Grigson.

GOLE 1. *Chrysanthemum segetum* (Corn Marigold) Scot. B & H. 2. *Myrica gale* (Bog Myrtle) Surr. Grigson. In the first case, this is better known as Gules, though there are many variants ranging from Gold to Gull. For Bog Myrtle, the name is simply a misreading of gale. This, too, has a number of variants.

GOLFOB *Ribes uva-crispa* (Gooseberry) Derb. Grigson.

GOLLAN, Ewe *Bellis perennis* (Daisy) N Eng, Scot. Grigson. Gollan is the same as gowan, a much-used name in the northern counties and Scotland for any kind of daisy-like flower. In fact, Ewe-gowan is commoner than Ewe-gollan.

GOLLAN, Yellow 1. *Ranunculus acris* (Meadow Buttercup) N'thum, Scot. Grigson. 2. *Ranunculus bulbosus* (Bulbous

Buttercup) N'thum, Scot. Grigson. 3. *Ranunculus repens* (Creeping Buttercup) N'thum, Scot. Grigson.

GOLLAND 1. *Caltha palustris* (Marsh Marigold) Lancs, N'thum, Caith. Grigson. 2. *Chrysanthemum segetum* (Corn Marigold) Yks. Grigson. 3. *Leucanthemum vulgare* (Ox-eye Daisy) N'thum. Grigson. 4. *Ranunculus acris* (Meadow Buttercup) Yks, N'thum. Grigson. 5. *Ranunculus bulbosus* (Bulbous Buttercup) Turner. 6. *Ranunculus repens* (Creeping Buttercup) Yks, N'thum, Berw. Grigson. 7. *Trollius europaeus* (Globe Flower) Cumb. Grigson.

GOLLAND, Lucken *Caltha palustris* (Marsh Marigold) Border Turner.

GOLLAND, Water 1. *Caltha palustris* (Marsh Marigold) N.Eng. Grigson. 2. *Nuphar lutea* (Yellow Waterlily) Yks. Grigson.

GOLLIN; GOLLEN, May *Caltha palustris* (Marsh Marigold) Lancs. Nodal (Gollin); Lancs. Lancs FWI (May Gollen).

GOLLYWOG *Papaver rhoeas* (Red Poppy) Som. Macmillan.

GOOBER; GOOBER PEA *Arachis hypogaea* (Ground Nut) S USA Puckett (Goober); Dalziel (Goober Pea). The original of this must be the name, given as gooba or guba, by which it is known throughout Africa.

GOOCOO-FLOWER; GOOKOO BUTTONS *Cardamine pratensis* (Lady's Smock) Dor. Barnes (flower); Som. B & H (buttons). This was Barnes's rendering of Cuckoo-flower.

GOOD FRIDAY *Adoxa moschatellina* (Moschatel) Wilts. Grigson.

GOOD FRIDAY FLOWER 1. *Adoxa moschatellina* (Moschatel) Dor, Som. Macmillan. 2. *Passiflora caerulea* (Blue Passion Flower) Som. Macmillan. The three styles of Passion-flower represent the three nails, the ovary is a sponge soaked in vinegar. The stamens are the wounds of Christ, and the crown, located above the petals, stands for the crown of thorns. The petals and sepals indicate the Apostles. The story is well-known - when the Spaniards first saw the flower, they took it as an omen that the Indians would be converted to Christianity. So, besides this name and the commonly accepted one, there are such as Story-of-the-Cross, Easter Star, Christ-and-the-Apostles, etc.

GOOD FRIDAY GRASS *Luzula campestris*.

GOOD FRIDAY PLANT *Pulmonaria officinalis* (Lungwort) Som. Macmillan. From the time of year it comes into flower, presumably. But there are other names like Twelve Apostles, besides names which stem from the legend that some of the Virgin's milk fell on the leaves, so causing the white blotches on them.

GOOD HENRY *Chenopodium bonus-henricus* (Good King Henry) Gerard. A straight translation of *bonus-henricus*, without the insertion of 'king'.

GOOD KING HARRY *Chenopodium bonus-henricus* (Good King Henry) Cambs. Gerard.

GOOD KING HENRY 1. *Chenopodium bonus-henricus*. 2. *Rumex obtusifolius* (Broad-leaved Dock) Som. Macmillan. Good King Henry, it is said, is the 16th century German Guter Heinrich (Heinrich is the name of an elf with a knowledge of healing plants).

GOOD LUCK *Oxalis acetosella* (Wood Sorrel) Som. Macmillan.

GOOD LUCK FLOWER *Schizopetalon walkeri*.

GOOD LUCK PLANT 1. *Cordyline terminalis* (Ti Plant) RHS. 2. *Oxalis deppei* USA Cunningham. Because the Oxalis is 'tetraphylla',or four-leaved,the luck being in the fact that this genus is often called a clover. Ti Plant is a protective shrub, planted in Malaysian graveyards to drive away ghosts and demons.

GOOD LUCK TREE *Thevetia peruviana* (Yellow Oleander) Howes.

GOOD NEIGHBOURHOOD 1. *Centranthus ruber* (Spur Valerian) Wilts. B & H; Glos.

J D Robertson. 2. *Chenopodium bonus-henricus* (Good King Henry) Wilts. Dartnell & Goddard.

GOOD NEIGHBOURS *Centranthus ruber* (Spur Valerian) Som. Elworthy; Wilts, Glos, Oxf. Grigson.

GOOD-NIGHT-AT-NOON 1. *Hibiscus trionum* (Bladder Katmia) Coats. 2. *Malva sylvestris* (Mallow) Som. Macmillan. But applied to any plant whose flowers close early. Cf. Flower-of-an-hour for Bladder Katmia.

GOODBYE, Summer's; GOODBYE, Autumn's *Parnassia palustris* (Grass of Parnassus) Tynan & Maitland. 'Autumn's Goodbye' seems out of correct dating,for the plant blooms from July to September,reasonable enough for 'Summer's Goodbye'.

GOODE *Leucanthemum vulgare* (Ox-eye Daisy) Lancs. B & H.

GOODY'S EYE *Salvia sclarea* (Clary) Som. B & H. A reference to the fact that eye salves used to be made from it. The seeds swell up when put into water, and become mucilaginous. These were then put like drops into the eye to cleanse it.

GOOLS 1. *Calendula officinalis* (Marigold) Prior. 2. *Caltha palustris* (Marsh Marigold) Prior. 3. *Chrysanthemum segetum* (Corn Marigold) Scot. Guthrie. 4. *Sinapis arvensis* (Charlock) Potter & Sargent. Better known as Gules, but the name exists in a lot of disguises. It means gold.

GOOSE, Fat *Chenopodium bonus-henricus* (Goood King Henry) Cambs. Porter.1974. As an answer to Fat Hen (*Chenopodium album*) presumably.

GOOSE, Grey *Potentilla anserina* (Silverweed) Tynan & Maitland.

GOOSE-AND-GANDER 1. *Melandrium rubrum* (Red Campion) Som. Macmillan. 2. *Vicia cracca* (Tufted Vetch) Som. Macmillan. The vetch probably gets the name merely in deference to the fact that the flowers are in two parts (standard and wings). The reference to geese is purely fortuitous - fingers-and-thumbs would do just as well, and is in fact another Somerset name for the same plant. But then campion has these two names as well in the same area.

GOOSE-AND-GESSLINGS *Salix caprea* catkins (Goat Willow catkins) N'thants. Clare.

GOOSE-AND-GOSLINGS 1. *Orchis mascula* (Early Purple Orchis) Som. Grigson. 2. *Orchis morio* (Green-winged Orchis) Prior. 3. *Salix caprea* catkins (Goat Willow catkins) Midl. Grigson.

GOOSE-AND-GUBBLIES *Salix caprea* catkins (Goat Willow catkins) Shrop. Grigson.

GOOSE BILL 1. *Galium aparine* (Goose-grass) Halliwell. 2. *Geranium robertianum* (Herb Robert) Som. Macmillan. The Geraniums are rather better named as Cranesbills than as Goose bills.

GOOSE CHICKS *Salix capraea* catkins (Goat Willow catkins) Dev. Grigson. 'Goslings' is more frequent than this.

GOOSE CLEAVERS *Galium aparine* (Goose-grass) Lanark Grigson. A confused mixture of the two main names for the plant - Goose-grass and Cleavers.

GOOSE FLOP 1. *Digitalis purpurea* (Foxglove) Dev. Friend. 2. *Narcissus pseudo-narcissus* (Daffodil) Som. Elworthy. Foxglove has a great many 'flop' names, like Cow-flops, Flop-poppy, Flop-dock, etc. But this is the only example of one for daffodil, and the emphasis is probably on the 'goose' part of the name rather than 'flop', for there seems to be a connection between daffodils and geese, or rather their goslings - see under Goose Leek.

GOOSE GRASS 1. *Eleusine indica*. 2. *Galium aparine*. 3. *Galium verum* (Lady's Bedstraw) B & H. 4. *Polygonum aviculare* (Knotgrass) F P Smith. 5. *Polygonum bistorta* (Bistort) Som. Macmillan. 6. *Potentilla anserina* (Silverweed) Som, Glos, Hants, Herts, Lincs, Yks, Border Grigson; Ire O Suilleabhain. 7. *Tanacetum vulgare* (Tansy) Wilts. Macmillan. Goose grass is "excellent food for goslings, who are very fond of it", as

Akerman said. Apparently, so must have been Silver-weed, with such a specific name as that (Latin anser means goose); there are quite a lot of 'goose' or 'gander' names for it. Was tansy fed to geese in Wiltshire?

GOOSE GRASS, Purple *Sherardia arvensis* (Field Madder) Turner.

GOOSE GRASS, Slender *Potentilla gracilis*.

GOOSE GRASS, Small *Galium tricorne*.

GOOSE-GRASS BEDSTRAW *Galium aparine* (Goose-grass) USA T H Kearney.

GOOSE GRAY *Potentilla anserina* (Silverweed) Fernie. A hybrid, mixing the anserine element with silverweed.

GOOSE GREASE *Galium aparine* (Goose-grass) Gerard. One assumes he meant goose-grass; but goose grease was always in demand as a base for home-made ointments.

GOOSE HEIRIFF *Galium aparine* (Goose-grass) Coles. Fernie has this as one word, Gooseheriff, and Turner mentioned Goosehareth as well. Heiriff, in one form or another, ranging from Airess to Haireve, was common and widespread for goose-grass, so the form Goose Heiriff is unnecessarily explicit.

GOOSE LEEK *Narcissus pseudo-narcissus* (Daffodil) IOM. Hare. This is Manx Las-ny-guiy. As with primroses, there used to be a superstition about bringing two or three daffodils into the house in early spring, before the goslings were hatched. A Cornish belief was that if a goose saw a daffodil before hatching its goslings, it would destroy them when hatched. A Dorset compromise was that you must always take care that the first daffodils brought into the house each season are of good quality, for otherwise something is sure to go wrong with your poultry. Cf. a similar belief attached to primroses.

GOOSE NEST *Monotropa hypopitys* (Yellow Bird's Nest) Gerard. Gerard described it as having "many tangling roots platted or crossed one over another very intricately, which resembleth a Crows nest made of sticks".

GOOSE PLUM *Prunus americana* (Canada Plum) USA Elmore.

GOOSE-SHARE *Galium aparine* (Goose-grass) A S Palmer. A variation on Goosehareth, or Goose Heiriff.

GOOSE TANSY *Potentilla anserina* (Silverweed) Northants. M Baker; Norf, Lincs, Cumb. Grigson.

GOOSE-TONGUE 1. *Achillea millefolium* (Yarrow) Som. Macmillan. 2. *Achillea ptarmica* (Sneezewort) War, Shrop. B & H; N'thants. A.E.Baker; Yks. Carr. 3. *Galium aparine* (Goose-grass) Som. Macmillan; Ches. Grigson. 4. *Ranunculus flammula* (Lesser Spearwort) Scot. Grigson. 5. *Tanacetum balsamita* (Balsamint) Conway. All descriptive in their way, mostly in reference to the rasp-like texture of the leaves, though in the case of Balsamint it. is their shape.

GOOSE WITHY *Salix caprea* (Goat Willow) Lyte.

GOOSEBERRY *Ribes uva-crispa*.

GOOSEBERRY, American *Ribes cynosbati* (Prickly Gooseberry) Douglas.

GOOSEBERRY, Australian *Muehlenbeckia adpressa*.

GOOSEBERRY, Black *Ribes nigrum* (Blackcurrant) B & H.

GOOSEBERRY, Cape 1. *Physalis bunyardii*. 2. *Physalis pubescens* (Husk Tomato) Bianchini.

GOOSEBERRY, Chinese *Actinidia chinensis*.

GOOSEBERRY, Fool *Pulmonaria officinalis* (Lungwort) B & H.

GOOSEBERRY, Fuchsia *Ribes speciosum*.

GOOSEBERRY, Indian *Phyllanthus acidus*.

GOOSEBERRY, Missouri *Ribes missouriense*.

GOOSEBERRY, Northern *Ribes oxyacanthoides* (Smooth Gooseberry) USA Yarnell.

GOOSEBERRY, Oak-belt *Ribes quercetorum*.

GOOSEBERRY, Orange *Ribes pinetorum*.

GOOSEBERRY, Otaheite *Phyllanthus distichus*.

GOOSEBERRY, Prickly *Ribes cynosbati*.

GOOSEBERRY, Red *Ribes rubrum* (Redcurrant) B & H.
GOOSEBERRY, Smooth *Ribes oxyacanthoides.*
GOOSEBERRY, Straggly *Ribes divaricatum.*
GOOSEBERRY, White-stem *Ribes inerma.*
GOOSEBERRY, Wild 1. *Physalis bunyardii* (Cape Gooseberry) S.Africa Watt. 2. *Physalis minima* S.Africa Watt. 3. *Ribes inebrians* (Wild Currant) USA Elmore.
GOOSEBERRY CURRANT *Ribes montigenum.*
GOOSEBERRY FIG *Mesembryanthemum acinaciforme.*
GOOSEBERRY PIE 1. *Epilobium hirsutum* (Great Willowherb) Dev. Friend.1882. 2. *Melandrium album* (White Campion) Som. Macmillan. 3. *Ononis repens* (Rest-harrow) Som. Macmillan. 4. *Symphytum officinale* (Comfrey) Dev, Dor, Wilts, Suff. Grigson. 5. *Valeriana dioica* (Marsh Valerian) Dev, Dor, Suff. Grigson; Wilts. Dartnell & Goddard. More common for the willowherb than for any of the others. The name has something to do with the smell, so it is usually claimed.
GOOSEBERRY PUDDING *Epilobium hirsutum* (Great Willowherb) Suss. Grigson.
GOOSEBERRY TOMATO *Physalis bunyardii* (Cape Gooseberry) Willis.
GOOSECHITE *Agrimonia eupatoria* (Agrimony) B & H.
GOOSEFOOT 1. *Chenopodium sp.* 2. *Potentilla anserina* (Silverweed) Palaiseul. All the Chenopodiums are known as Goosefoot - it is the shape of the leaf that provides the name.
GOOSEFOOT, City *Chenopodium urbicum* (Upright Goosefoot) USA Zenkert.
GOOSEFOOT, Fig-leaved *Chenopodium ficifolium.*
GOOSEFOOT, Fine-leaf *Chenopodium leptophyllum.*
GOOSEFOOT, Glaucous *Chenopodium glaucum* (Oak-leaved Goosefoot) Clapham.
GOOSEFOOT, Green *Chenopodium suecicum.*
GOOSEFOOT, Grey *Chenopodium opulifolium.*
GOOSEFOOT, Many-seeded *Chenopodium polyspermum.*
GOOSEFOOT, Maple-leaved *Chenopodium hybridum* (Sowbane) USA Zenkert.
GOOSEFOOT, Mercury *Chenopodium bonus-henricus* (Good King Henry) Grieve.1931. Other 'mercury' names are False Mercury, Mercury Docken, English Mercury, or just Mercury.
GOOSEFOOT, Nettle-leaved *Chenopodium murale.*
GOOSEFOOT, Oak-leaved *Chenopodium glaucum.*
GOOSEFOOT, Perennial *Chenopodium bonus-henricus* (Good King Henry) Jordan.
GOOSEFOOT, Red *Chenopodium rubrum.*
GOOSEFOOT, Red, Small *Chenopodium botryodes.*
GOOSEFOOT, Round-leaved *Chenopodium polyspermum* (Many-seeded Goosefoot) Flower.1859.
GOOSEFOOT, Small-seeded Chenopodium rubrum (Red Goosefoot) Curtis.
GOOSEFOOT, Sticky *Chenopodium botrys* (Jerusalem Oak) Polunin.
GOOSEFOOT, Stinking *Chenopodium vulvaria* (Stinking Orach) McClintock.
GOOSEFOOT, Strawberry *Chenopodium foliosum.*
GOOSEFOOT, Thorn-apple-leaved *Chenopodium hybridum* (Sowbane) Curtis.
GOOSEFOOT, Upright *Chenopodium urbicum.*
GOOSEFOOT, White *Chenopodium album* (Fat Hen) Curtis.
GOOSEFOOT PLANT *Syngonium podophyllum.*
GOOSEGOB; GOOSEGOG *Ribes uva-crispa* (Gooseberry) Dev. Friend.1882; Herts. Jones-Baker.1977; Derb. Grigson; Lancs. Nodal (Goosegob); Som, Hants, Middx, N'thants, E Ang, Ches, Yks. Grigson; IOW Long (Goosegog). There is still another form of this - Gew-gog, recorded in Suffolk.
GOOSEHARETH; GOOSEHERIFF;

GOOSEHARE *Galium aparine* (Goose-grass) Turner (Goosehareth); Fernie (Gooseheriff); Culpepper (Goosehare). The second element in these names usually appears on its own, the plant being called by such names as Airess, Airif, or Hairiff, and their many variants.

GOOSEWEED 1. *Galium aparine* (Goose-grass) Som. Macmillan. 2. *Potentilla anserina* (Silverweed) C P Johnson.

GOOSEWORT *Potentilla anserina* (Silverweed) Grieve.1931.

GOOSEY-GANDER 1. *Endymion nonscriptus* (Bluebell) Dev, Dor. Macmillan. 2. *Orchis mascula* (Early Purple Orchis) Dor. Dacombe; Wilts. Dartnell & Goddard; Glos. B & H. 3. *Orchis morio* (Green-winged Orchid) Dor, Som. Grigson.

GOOSEY-GOOSEY-GANDER 1. *Endymion nonscriptus* (Bluebell) Som. Macmillan. 2. *Orchis mascula* (Early Purple Orchis) Som. Macmillan.

GOPHERWOOD 1. *Cladestris lutea* (Yellow-wood) USA Moldenke. 2. *Cupressus sempervirens* (Common Cypress) Grigson.1974. It is the name given to the timber of the cypress (var. horizontalis, according to Moldenke). This is Hebrew gopher, from which Greek kuparissos, and so English cypress, are possibly derived. It was from Gopherwood that Noah built the Ark, as tradition has it.

GORDOLABA *Achillea millefolium* (Yarrow) USA Henkel.

GORE-THETCH *Vicia sativa* (Common Vetch) B & H.

GORGON'S HEAD *Euphorbia gorgonis*.

GORSE *Ulex europaeus*. This is the standard English name, related to Latin hordeum, which means spiked barley, and eventually to Latin horrere, to bristle.

GORSE, Dwarf; GORSE, Lesser *Ulex minor* (Small Furze) Hutchinson (Dwarf); McClintock (Lesser).

GORSE, Golden *Ulex europaeus* (Gorse) Grieve.1931.

GORSE, Hen 1. *Odontites verna* (Red Bartsia) Ches. B & H. 2. *Ononis repens* (Rest Harrow) Midl, Ches, N Eng. Grigson.

GORSE, Manx *Ulex minor* (Small Furze) IOM Moore.1924.

GORSE, Needle *Genista anglica* (Needle Whin) Howes.

GORSE, Spanish *Genista hispanica*.

GORSE, Western *Ulex gallii* (Dwarf Furze) McClintock. In the British Isles, it is confined to western England and Ireland.

GORST 1. *Juniperus communis* (Juniper) B & H. 2. *Ulex europaeus* (Gorse) Shrop, Ches. B & H.

GORSTBERRY *Ribes uva-crispa* (Gooseberry) Heref. G C Lewis.

GOSHEN, Dandy *Orchis morio* (Green-winged Orchid) Wilts. B & H. Evidently a variation of Gander-gosling, some form of which is common both for this orchid and for *Orchis mascula*.

GOSLER *Ribes uva-crispa* (Gooseberry) Lancs. Lancs FWI.

GOSLING GRASS; GOSLING-WEED *Galium aparine* (Goose-grass) Glos, Heref. Grigson; N'thants. A E Baker (Grass); Halliwell (Weed). "This weed is considered excellent food for goslings, who are very fond of it" (Akerman).

GOSLING SCRATCH *Galium aparine* (Goose-grass) Ess, Norf, Cambs. Grigson.

GOSLING TREE *Salix caprea* (Goat Willow) Wilts. Dartnell & Goddard. The catkins would be the "goslings".

GOSLINGS 1. *Orchis mascula* (Early Purple Orchid) Wilts. Jones & Dillon. 2. *Salix caprea* catkins (Goat Willow catkins) Wilts. Macmillan; Heref. Leather; Suff. Grigson. 3. *Salix pentandra* catkins (Bay-leaved Willow catkins). Yks. Grigson.

GOSLINGS, Dandy 1. *Dactylorchis maculata* (Heath Spotted Orchid) Wilts. Grigson. 2. *Orchis mascula* (Early Purple Orchid) Wilts. Dartnell & Goddard. 3. *Orchis morio* (Green-winged Orchid) Wilts. Dartnell & Goddard.

GOSLINGS, Goose-and- 1. *Orchis mascula* (Early Purple Orchid) Som. Grigson. 2.

Orchis morio (Green-winged Orchid) Prior. 3. *Salix caprea* catkins (Goat Willow catkins) Midl. Grigson.

GOSLINGS, Granfer 1. *Dactylorchis maculata* (Heath Spotted Orchid) Wilts. Dartnell & Goddard. 2. *Orchis mascula* (Early Purple Orchid) Wilts. Jean. Philpot.

GOSLINGS, May *Salix caprea* catkins (Goat Willow catkins) Yks. Grigson.

GOSS 1. *Ononis repens* (Rest Harrow) Wilts. Dartnell & Goddard. 2. *Phragmites communis* (Reed) Corn. Quiller-Couch. 3. *Typha latifolia* (False Bulrush) Corn. Courtney.1880. 4. *Ulex europaeus* (Gorse) Shrop, Leics, Lincs, Kent B & H. N'thants. A E Baker; War. Palmer. Obviously a variation of 'gorse' for Rest Harrow and Gorse itself. But for bulrush and reed there must be a different (possibly Cornish?) derivation.

GOSSIPS; GOSSIPS, Little *Orchis mascula* (Early Purple Orchid) Heref. B & H (Gossips); Som. Macmillan (Little Gossips).

GOST *Ulex europaeus* (Gorse) Heref. B & H.

GOTTRIDGE *Viburnum opulus* (Guelder Rose) Suff. B & H. A version of Gatteridge, which is a dogwood name. A lot of Guelder Rose names are borrowed from those of dogwood or spindle. This one derives from OE gat, goat.

GOUD *Isatis tinctoria* (Woad) B & H. Probably the same word as woad.

GOULAN 1. *Calendula officinalis* (Marigold) Prior. 2. *Caltha palustris* (Marsh Marigold) Prior. 3. *Chrysanthemum segetum* (Corn Marigold) Prior. Better known as Gowan, or Golland, possibly connected with the word 'gold'.

GOULD; GOULD-WEED *Chrysanthemum segetum* (Corn Marigold) Cumb. B & H.

GOULE *Myrica gale* (Bog Myrtle) Jamieson. In this case, goule is gale, the proper English name for Bog Myrtle. It appears in various forms, like Gall, Goyle, Gaul or Gow.

GOULON, Locker *Trollius europaeus* (Globe Flower) Camden. Gowan is a yellow flower, and Locken means closed in, or locked in - this accounts for the more usual form Locken Gowan. Locker Goulon is a variant of this, and there are others, such as Lockren Gowlan, which is Gerard's version, or Lockyer-goldens, a Yorkshire form.

GOULS *Chrysanthemum segetum* (Corn Marigold) Midl. Grigson. One of a long series of names, all meaning 'gold' - gold, golding, gowlan, gules, gull, etc.

GOURD, Bitter 1. *Colocynthus vulgaris*. 2. *Momordica charantia* (Wild Balsam-apple) USA Kingsbury. 'Gourd' is eventually from Latin cucurbita, now used as a generic name for a lot of the gourds. Colocynthus is aptly named - in fact it actually means 'bitter gourd'. It *is* exceedingly bitter; the gall of the Bible often refers to this.

GOURD, Bottle *Lagenaria siceraria* (Calabash) Wit. Because the ripe fruit is hollowed out, cleared of pulp, and used as a bottle (and even, at one time, as a gunpowder flask). But this really applies to all gourds - one wonders, in this case, whether it is rather the peculiar (bottle) shape of the fruit which gave the name.

GOURD, Buffalo *Cucurbita foetidissima*.

GOURD, Club-shaped *Lagenaria siceraria* (Calabash) Bianchini.

GOURD, Dipper *Lagenaria siceraria* (Calabash) Elmore.

GOURD, Dishcloth; GOURD, Dishrag; GOURD, Washrag *Luffa cylindrica* (Loofah) A W Smith (Dishcloth); Whitaker & Davis (Dishrag); Chopra (Washrag).

GOURD, Fig-leaf *Cucurbita ficifolia* (Malabar Gourd) Whitaker & Davis.

GOURD, Finger-leaved *Cucurbita digitata*.

GOURD, Leprosy *Momordica charantia* (Wild Balsam-apple) Howes.

GOURD, Malabar *Cucurbita ficifolia*. This is actually a Central American plant, but it is grown throughout the tropics, and presumably that is why the name 'Malabar' is given.

GOURD, Missouri *Cucurbita foetidissima* (Buffalo Gourd) USA Vestal & Schultes.

GOURD, Pilgrim's *Lagenaria siceraria*

(Calabash) Chopra. Cf. Bottle Gourd - that is why it earns the name 'Pilgrim's'.
GOURD, Scallop *Cucurbita pepo var. melopepo.*
GOURD, Snake *Trichosanthes anguina.*
GOURD, Sour *Adansonia digitata* (Baobab) J Smith.
GOURD, Spanish *Cucurbita maxima* (Giant Pumpkin) Lindley.
GOURD, Sponge *Luffa cylindrica* (Loofah) Chopra.
GOURD, Towel *Luffa cylindrica* (Loofah) Candolle. Cf. Dishcloth Gourd, etc.
GOURD, Trumpet *Lagenaria siceraria* (Calabash) Lindley. Descriptive of the shape.
GOURD, Viper *Trichosanthes anguina* (Snake Gourd) Whittle & Cook.
GOURD, Wax, Chinese *Benincasa hispida.*
GOURD, White *Benincasa cerifera.*
GOURD, White-flowered *Lagenaria siceraria* (Calabash) Whitaker & Davis.
GOURD, Wild, Foetid *Cucurbita foetidissima* (Buffalo Gourd) Youngken.
GOURD TREE *Adansonia gregorii* (Australian Baobob) Howes. Cf. Bottle Tree. Perhaps from the general appearance of the trunk, but also because they have actually been used for water storage.
GOURDE *Cucurbita pepo* (Vegetable Marrow) Turner.
GOURLINS; GOWLINS *Conopodium majus* (Earthnut) both Scot. Grigson. Cf. Curluns.
GOUROU NUT *Cola nitida* (Kola) Notes & Queries.1902. Hausa goro, but gourou in other West African dialects, where it also appears as garru, which is presumably the origin of another name for Kola - Karoo Nut.
GOUT IVY *Ajuga chamaepitys* (Yellow Bugle) Prior. This was once used for gout, and formed an ingredient of the once famous Portland powder. Langham recommended the distilled water for gout.
GOUTWEED; GOUTWORT *Aegopodium podagraria.* Goutweed is the normal name for this; Goutwort is in Culpepper. Long associated with the painful malady, it is even called Herb Gerard, which is not a dedication to the herbalist, but to St Gerard, the patron saint of gout sufferers.
GOUTY TREE *Adansonia gregori* (Australian Baobab) J Smith. So called, it seems, by Allan Cunningham in 1818, this is a perfectly good descriptive name. Baobabs have this swollen apprearance, as if full of fluid.
GOVERNOR PLUM *Flacourtia indica* (Kaffir Plum) Howes.
GOW *Myrica gale* (Bog Myrtle) B & H. A rendering of Gale, which is the proper English name.
GOWAN 1. *Bellis perennis* (Daisy) Scot. Jamieson. 2. *Calendula officinalis* (Marigold) Prior. 3. *Caltha palustris* (Marsh Marigold) Cumb, N'thum. Grigson. 4. *Chrysanthemum segetum* (Corn Marigold) Prior. 5. *Ranunculus acris* (Meadow Buttercup) Wigt. Grigson. 6. *Taraxacum officinale* (Dandelion) Glos. PLNN35. Apparently a form of 'gollan', perhaps connected with the word 'gold', in which case gowan for the daisy seems misplaced. But the word seems to have been used for any yellow daisy, and the common daisy got the name by association. The Gloucestershire record for the dandelion seems to be rather too far south to be entirely credible.
GOWAN, Dog *Tripleurospermum maritimum* (Scentless Maydweed) N.Scot. Kamieson. Cf. Dog Daisy, Dog's Camomile for this - presumably because it is scentless.
GOWAN, Ewe *Bellis perennis* (Daisy) N'thum. Brockett; Scot. Jamieson.
GOWAN, Geal; GOWAN, Gill; GOWAN, Gule *Chrysanthemum segetum* (Corn Marigold) All N Ire. all B & H. Pleonastic, presumably - any of these would mean 'gold gold'.
GOWAN, Hawkweed *Taraxacum officinale* (Dandelion) Scot. Pratt.
GOWAN, Horse 1. *Leucanthemum vulgare* (Ox-eye Daisy) Edin. Ritchie. 2. *Matricaria recutita* (Scanted Mayweed) Border, Banff, Berw, Dumb. B & H; USA Watt. 3. *Taraxacum officinale* (Dandelion) Scot. B

& H; USA. Henkel.

GOWAN, Lapper; GOWAN, Lockan; GOWAN, Locken; GOWAN, Lockin- ma-; GOWAN, Lopper; GOWAN, Luckan; GOWAN, Lukin *Trollius europaeus* (Globe Flower) N Eng, Scot. Grigson (Lapper, Lockan, Locken); Cumb. Grigson (Lockin-ma-); Clyde B & H (Lopper); Scot. Jamieson (Luckan, Lukin). 'Locken' means closed in, or locked in, and gowan means a yellow flower, resulting in good descriptive names for the Globe Flower.

GOWAN, Luckie (Lucky) *Trollius europaeus* (Globe Flower) Scot. Tynan & Maitland.

GOWAN, Mary 1. *Bellis perennis* (Daisy) N'thum, Berw. Grigson. 2. *Calendula officinalis* (Marigold) N'thum. B & H. Mary Gowan means the same as marigold.

GOWAN, May *Bellis perennis* (Daisy) W Miller.

GOWAN, Meadow *Caltha palustris* (Marsh Marigold) Ayr B & H.

GOWAN, Milk *Taraxacum officinale* (Dandelion) Forf. B & H. In other words, the yellow flower with the milky juice.

GOWAN, Open *Caltha palustris* (Marsh Marigold) Cumb. B & H. To distinguish it from the closed one, or Locken Gowan, the Globe Flower.

GOWAN, Sheep's *Trifolium repens* (White Clover) Scot. Grigson. This is out of the usual run of 'gowan' names - it is neither yellow nor is it a daisy.

GOWAN, Tushy-lucky *Tussilago farfara* (Coltsfoot) Dumf. Grigson. Tushy-lucky, of course, is a version of *Tussilago*.

GOWAN, Water *Caltha palustris* (Marsh Marigold) Cumb. B & H.

GOWAN, White *Leucanthemum vulgare* (Ox-eye Daisy) Scot. Gregor. An anomaly, this, trying to say a white-flowered yellow daisy. But Ox-eye's near relation, Corn Marigold, is Gowan par excellence. Cf. White Gowlan and White Gold.

GOWAN, Witch 1. *Taraxacum officinale* (Dandelion) Scot. B & H. 2. *Trollius europaeus* (Globe Flower) Grigson. It is not clear why either of these should have the witch association.

GOWAN, Yellow 1. *Chrysanthemum segetum* (Corn Marigold) Caith. Grigson. 2. *Ranunculus acris* (Meadow Buttercup) Berw. B & H. 3. *Taraxacum officinale* (Dandelion) Scot. B & H; USA. Henkel. Pleonastic - gowan means yellow anyway.

GOWEN CYPRESS *Cupressus goveniana* (Californian Cypress) Edlin.

GOWK'S CLOVER *Oxalis acetosella* (Wood Sorrel) N'thum. Grigson. Gowk is the usual north country and Scottish word for cuckoo. Wood Sorrel is Cuckoo's bread, or cheese; even in Latin it is panis cuculi.

GOWK'S HOSE 1. *Campanula latifolia* (Giant Bellflower) Scot. Grigson. 2. *Campanula medium* (Canterbury Bells) S Scot. Hardy. 3. *Campanula rotundifolia* (Harebell) Dumb. Hardy. 4. *Endymion non-scriptus* (Bluebell) Dumb. B & H.

GOWK('S) MEAT 1. *Orchis mascula* (Early Purple Orchis) B & H. 2. *Orchis morio* (Green-winged Orchis) B & H. 3. *Oxalis acetosella* (Wood Sorrel) Scot. Swainson. 4. *Rumex acetosa* (Wild Sorrel) Leyel.1937. Cf. Cuckoo Bread for the Wood Sorrel, and the Latin panis cuculi.

GOWK'S SHILLINGS; GOWK'S SILLER; GOWK'S SIXPENCES *Rhinanthus crista-galli* (Yellow Rattle) Lanark (Shillings) N'thum, Berw. Roxb. (Siller); N'thum, Berw. (Sixpences) all Grigson. Presumably these names owe their existence to the same set of circumstances as produced 'Rattle' - the rattling of the seeds in the capsules, like coins in a purse.

GOWK'S THUMB *Campanula rotundifolia* (Harebell) N.Scot. Gregor.

GOWKS, Scab *Dactylorchis maculata* (Heath Spotted Orchid) Dur. Grigson.

GOWLAN 1. *Bellis perennis* (Daisy) Derb. B & H. 2. *Calendula officinalis* (Marigold) N'thum. B & H. 3. *Chrysanthemum segetum* (Corn Marigold) N.Eng. B & H. 4. *Leucanthemum vulgare* (Ox-eye Daisy) N'thum. Grigson. Gowlan seems to be a mixture of 'gowan' and 'golland', and of course means the same thing.

GOWLAN, Locken; GOWLAN, Lockren *Trollius europaeus* (Globe Flower) both Gerard. They mean a closed-in yellow flower.

GOWLAN, Mary 1. *Bellis perennis* (Daisy) Berw, Ayr. Grigson. 2. *Calendula officinalis* (Marigold) N'thum. B & H. 3. *Chrysanthemum segetum* (Corn Marigold) N'thum. Grigson. i.e. marigold.

GOWLAN, White *Leucanthemum vulgare* (Ox-eye Daisy) N'thum. Grigson. An oddity, as is White Gowan - it would mean 'white gold'.

GOWLAN, Yellow *Caltha palustris* (Marsh Marigold) Cumb, N'thum. Grigson. It seems tautological, given that gowan, or in this case gowlan, means 'gold'.

GOWLAND *Chrysanthemum segetum* (Corn Marigold) Yks. Morris.

GOWLAND, Ling *Chrysanthemum segetum* (Corn Marigold) Yks. F K Robinson.

GOWLAND, Water 1. *Caltha palustris* (Marsh Marigold) N.Eng. Grigson. 2. *Nuphar lutea* (Yellow Waterlily) Yks. Morris.

GOWLES 1. *Calendula officinalis* (Marigold) Prior. 2. *Caltha palustris* (Marsh Marigold) Prior. 3. *Chrysanthemum segetum* (Corn Marigold) Prior.

GOWLINS; GOURLINS *Conopodium majus* (Earthnut) both Scot. Grigson.

GOZILL 1. *Ribes rubrum* (Redcurrant) Kent B & H. 2. *Ribes uva-crispa* (Gooseberry) Kent B & H.

GOZZLE-GRASS *Galium aparine* (Goosegrass) Glos. Brill. i.e. gosling-grass.

GOYLE *Myrica gale* (Bog Myrtle) Corn. B & H. Gale is the proper English name for the plant - this is a local variant.

GRAB; GRAB-APPLE *Malus sylvestris* fruit (Crabapple) Corn, Som. B & H; Dev. Friend.1882; Glos. J D Robertson. (Grab); Dev. Choape (Grab-apple). Crab, of course. The tree itself would be "Grabstock". Som. B & H; Dor. Barnes.

GRABS, Sour 1. *Malus sylvestris* fruit (Crabapple) Som. Mabey. 2. *Oxalis acetosella* (Wood Sorrel) Dev. Friend. 3. *Rumex acetosa* (Wild Sorrel) Fernie. In the third case, this is more likely to be Sour Grass.

GRACE, Ave; GRACE, Herb *Ruta graveolens* (Rue) Loudon (Ave Grace); Turner (Herb Grace). Holy water was sprinkled from brushes made of rue. And, as Ave Grace, it was used in the rites of exorcism of the Catholic Church.

GRACE, Herb of 1. *Ruta graveolens* (Rue) Gerard. 2. *Verbena officinalis* (Vervain) McDonald. Vervain, too, is the Holy Herb.

GRACE, Odin's *Geranium pratense* (Meadow Cranesbill) Perry.

GRACE OF GOD 1. *Geranium pratense* (Meadow Cranesbill) Gerard. 2. *Hypericum perforatum* (St John's Wort) B & H. 3. *Hyssopus officinalis var. ruber* (Hedge Hyssop) Folkard. 4. *Plantago coronopus* (Buck's-horn Plantain) Som. Macmillan.

GRACY DAISY 1. *Bellis perennis* (Daisy) Dev. Grigson. 2. *Narcissus pseudo-narcissus* (Daffodil) Dev, Som. Macmillan. Probably the reference is to Easter, which was once called Great Day. Both these flowers have Easter names - Cf. the French Pâquerette for daisy. Daffodil, besides the well-known Lent Lily and similar names, is actually called Easter Lily in the area in which Gracy Daisy occurs.

GRACY DAY *Narcissus pseudo-narcissus* (Daffodil) Dev. Friend. As for Gracy Daisy.

GRAIN, Golden *Verbascum thapsus* (Mullein) Dev. Friend.

GRAIN, Red *Cajanus indicus* (Pigeon Pea) Douglas.

GRAIN TREE *Quercus ilex* (Holm Oak) Hora. 'Grain' is the Latin grana tinctorum, of the same import as 'kerm'.

GRAINS *Lemna minor* (Duckweed) Gerard. Is this 'greens', which is also recorded, or is it the other way round?

GRAINS, Guinea *Aframomum melegueta* (Melegueta Pepper) Dalziel.

GRAINS OF PARADISE 1. *Aframomium*

melegueta seeds (Melegueta Pepper). 2. *Amomum grana*. The seeds of Melegueta Pepper are the grains of paradise, from which the Grain Coast (Liberia approximately) got its name. The very pungent seeds are like pepper, or better as some claim, but were displaced from favour when the true peppers were introduced from America. *Amomum grana* is listed separately, but presumably these are the same plant.

GRAM, Bengal *Cicer arietinum* (Chick Pea) Douglas.

GRAM, Black *Phaseolus mungo*. Gram is Portuguese graõ, Latin granum, seed, or grain.

GRAM, Golden *Phaseolus aureus* (Green Gram) Brouk.

GRAM, Green *Phaseolus aureus*.

GRAMMER GREYGLE *Endymion non-scriptus* (Bluebell) Dor. Macmillan. 'Greygle' seems to mean greyish-blue, and appears in many forms, sometimes in its own, but usually attributed in one way or another to Granfer rather than Grammer.

GRAMOPHONE 1. *Lonicera periclymenum* (Honeysuckle) Som. Macmillan. 2. *Tropaeolum majus* (Nasturtium) Som. Macmillan. Descriptive, and more obvious in the next entry.

GRAMOPHONE HORNS *Lonicera periclymenum* (Honeysuckle) Som. Macmillan.

GRANADILLA 1. *Passiflora caerulea* (Blue Passion Flower) S.Africa. Watt. 2. *Passiflora edulis* (Passion Fruit) S.Africa Watt. 3. *Passiflora quadrangularia*. Granadilla is "little pomegranate".

GRANADILLA, Apple-fruited *Passiflora maliformis* (Sweet Calabash) Grieve.1931.

GRANDAVY *Glechoma hederacea* (Ground Ivy) Scot. R.M.Robertson.

GRANDFATHER GRIGGLES *Orchis mascula* (Early Purple Orchis) Som. Macmillan. Early Purple Orchis shares a number of names with the bluebell. If greygle really does mean greyish-blue, how can it be applied in this case?

GRANDFATHER'S BEARD 1. *Clematis vitalba* (Old Man's Beard) Som. Macmillan. 2. *Equisetum arvense* (Common Horsetail) Dev. Macmillan. 3. *Geum triflorum* (Prairie Smoke) USA T H Kearney. The Geum has silvery feathered tails to its fruits, hence the descriptive name. The other two are too well-known for comment.

GRANDFATHER'S BUTTONS *Caltha palustris* (Marsh Marigold) Som. Macmillan. Applied to the flower buds, of course.

GRANDFATHER'S WEATHERGLASS *Anagallis arvensis* (Scarlet Pimpernel) Dev. Macmillan. Cf. Ploughman's, Old Man's, Countryman's Weatherglass, etc.

GRANDFATHER'S WHISKERS *Clematis vitalba* (Old Man's Beard) Corn, Som. Grigson.

GRANDFY'S BEARD *Clematis vitalba* (Old Man's Beard) Som. Macmillan.

GRANDMOTHER *Leucanthemum vulgare* (Ox-eye Daisy) Notts. Grigson.

GRANDMOTHER'S BONNET *Aconitum napellus* (Monkshood) Yks. Grigson.

GRANDMOTHER'S DARNING NEEDLE *Scandix pecten-veneris* (Shepherd's Needle) Tynan & Maitland. The 'needle' reference is to the long beaked fruits.

GRANDMOTHER'S HAIR *Festuca heterophylla*.

GRANDMOTHER'S NEEDLE *Valeriana officinalis* (Valerian) Bucks. Harman.

GRANDMOTHER'S NIGHTCAP 1. *Aconitum napellus* (Monkshood) Yks. Grigson. 2. *Calystegia sepium* (Great Bindweed) Som. Grigson. 3. *Silene maritima* (Sea Campion) Tynan & Maitland.

GRANDMOTHER'S PINCUSHION *Knautia arvensis* (Field Scabious) Suss. Parish.

GRANDMOTHER'S SPECTACLES *Lunaria annua* (Honesty) Som. Macmillan.

GRANDMOTHER'S TOENAILS *Lotus corniculatus* (Bird's-foot Trefoil) Dev, Som. Macmillan.

GRANDSIR-GREYBEARD *Chionanthus virginicus* (Fringe Tree) Howes. Cf. Old

Man's Beard for this.
GRANFER-GOSLINGS 1. *Dactylorchis maculata* (Heath Spotted Orchid) Wilts. Dartnell & Goddard. 2. *Orchis mascula* (Early Purple Orchis) Wilts. Jean. Philpot.
GRANFER-GREGOR *Endymion nonscriptus* (Bluebell) Dev. B & H.
GRANFER-GREGOR, Red *Orchis mascula* (Early Purple Orchis) Dor. B & H.
GRANFER-GREYGLES 1. *Melandrium dioicum* (Red Campion) Dor. Macmillan. 2. *Orchis mascula* (Early Purple Orchis) Dor. B & H.
GRANFER-GREYGLES, Blue *Endymion nonscriptus* (Bluebell) Dor. B & H. It sounds unnecessary if, as it is said, 'greygle' means bluish-grey. But, to judge from the previous entry, Dorset people had forgotten this.
GRANFER-GREYGLES, Red *Melandrium dioicum* (Red Campion) Dor. B & H.
GRANFER-GRIDDLE-GOOSEY-GANDER *Orchis mascula* (Early Purple Orchis) Wilts. Dartnell & Goddard. This extraordinary name results from the fusion of two distinct traditions, the Granfer-Greygle series and the Goosey-gander series. Granfer-goslings, listed above, is a shorter version.
GRANFER GRIGG 1. *Endymion nonscriptus* (Bluebell) Som. Macmillan. 2. *Orchis mascula* (Early Purple Orchis) Wilts. Macmillan. But Granfer-grigg means a woodlouse in Wiltshire.
GRANFER GRIGGLE-STICKS 1. *Endymion nonscriptus* (Bluebell) Som. Macmillan. 2. *Orchis mascula* (Early Purple Orchis) Som. Macmillan. 3. *Taraxacum officinale* (Dandelion) Som. Macmillan. Dandelion is the odd man out in this list.
GRANFER-GRIGGLES 1. *Endymion nonscriptus* (Bluebell) Dor. Dacombe. 2. *Melandrium dioicum* (Red Campion) Dor. Macmillan. 3. *Orchis mascula* (Early Purple Orchis) Dor. Grieve.1931.
GRANFER-GRIGGLES, Scotch *Prunella vulgaris* (Self-heal) Dev. Macmillan.
GRANFER-GRIZZLE *Anthyllis vulneraria* (Kidney Vetch) Som. Macmillan.
GRANFER JAN *Melandrium dioicum* (Red Campion) Dor, Wilts. Macmillan.
GRANNY-BONNETS *Aquilegia vulgaris* (Columbine) Suss. Doris Cooper; Wilts. in common use.
GRANNY-GRIGGLES *Endymion nonscriptus* (Bluebell) Dor. Grieve.1931. Granfer-griggles is commoner than this.
GRANNY-JUMP-OUT-OF-BED 1. *Aconitum napellus* (Monkshood) Wilts. Powell. 2. *Aquilegia vulgaris* (Columbine) Wilts. Grigson. Powell considered this, and other similar names, to be. children's folk stories condensed into a plant name.
GRANNY-THREAD-THE-NEEDLE 1. *Anemone nemerosa* (Wood Anemone) Som. Macmillan. 2. *Geranium robertianum* (Herb Robert) Som. Macmillan.
GRANNY-THREADS *Ranunculus repens* (Creeping Buttercup) B & H.
GRANNY'S BONNET. 1 *Aconitum napellus* (Monkshood) Som. Macmillan. 2. *Antirrhinum majus* (Snapdragon) Som. Macmillan. 3. *Aquilegia vulgaris* (Columbine) Corn. Grigson; Dor. Dacombe; Hants. Boase. 4. *Calystegia sepium* (Great Bindweed) Wilts. Dartnell & Goddard. 5. *Delphinium ajacis* (Larkspur) Som. Macmillan. 6. *Digitalis purpurea* (Foxglove) Som. Macmillan. 7. *Geranium pratense* (Meadow Cranesbill) Som. Macmillan. 8. *Geum rivale* (Water Avens) Dor. Macmillan. 9. *Physalis alkekengi* (Winter Cherry) Som. Macmillan.
GRANNY'S CAP; GRANNY'S CAP, Egyptian *Geum rivale* (Water Avens) Wilts. Dartnell & Goddard.1899.
GRANNY'S FACE *Viola tricolor* (Pansy) Som. Macmillan.
GRANNY'S NIGHT BONNET *Calystegia sepium* (Great Bindweed) Som. Macmillan.
GRANNY'S NIGHTCAP 1. *Aconitum napellus* (Monkshood) Som. Macmillan; Wilts. Dartnell & Goddard; Glos. J D Robertson. 2. *Anemone nemerosa* (Wood Anemone) Som, War. Grigson; Wilts. Dartnell & Goddard. 3.

Antirrhinum majus (Snapdragon) Som. Macmillan. 4. *Aquilegia vulgaris* (Columbine) Dev. Friend.1882; Som, Glos. Grigson; Dor. Dacombe; Wilts. Dartnell & Goddard. 5. *Borago officinalis* (Borage) Som. Macmillan. 6. *Calystegia sepium* (Great Bindweed) Som. Macmillan. Wilts. Dartnell & Goddard. 7. *Convolvulus arvensis* (Field Bindweed) Wilts. Dartnell. & Goddard. 8. *Delphinium ajacis* (Larkspur) Som. Macmillan. 9. *Geranium robertianum* (Herb Robert) Dev. Macmillan. 10. *Geum rivale* (Water Avens) Wilts. Dartnell & Goddard. 11. *Hemerocallis flava* (Day Lily) Som. Macmillan. 12. *Melandrium album* (White Campion) Som. Friend. 13. *Solanum dulcamara* (Woody Nightshade) Dev. Macmillan. 14. *Stellaria holostea* (Greater Stitchwort) Dor. Grigson. 15. *Stellaria graminea* (Lesser Stitchwort) Som. Macmillan.

GRANNY'S SHOES *Aconitum napellus* (Monkshood) Dor. Macmillan.

GRANNY'S SLIPPER-SLOPPERS *Lathyrus pratensis* (Yellow Pea) Dor. Grigson. Cf. Lady's Slippers for this, from Wessex again.

GRANNY'S SLIPPERS 1. *Aconitum napellus* (Monkshood) Dor. Macmillan. 2. *Lotus corniculatus* (Bird's-foot Trefoil) Hants. Hants FWI.

GRANNY'S TEARS *Campanula rotundifolia* (Harebell) Som. Macmillan.

GRANNY'S THIMBLE *Aquilegia vulgaris* (Columbine) Som. Macmillan.

GRANYAGH *Spergula arvensis* (Corn Spurrey) Ire. Grigson.

GRAPE, Bear's 1. *Arctostaphylos uva-ursi* (Bearberry) H & P. 2. *Phytolacca decandra* (Poke-root) S M Robertson.

GRAPE, Canyon *Vitis arizonica* (Arizona Wild Grape) USA Elmore.

GRAPE, Chicken *Vitis cordifolia*.

GRAPE, Downy *Vitis cinerea*.

GRAPE, Dyer's *Phytolacca decandra* (Poke-root) W Miller. The Navajo used the root bark, with plum root, for a purple-brown dye. The berries, with alum, give red to tan colours.

GRAPE, False *Ampelopsis hederacea* (Virginia Creeper) O P Brown.

GRAPE, Fox *Vitis labrusca*.

GRAPE, Hedge *Bryonia dioica* (White Bryony) Worcs. B & H. Cf. Grapewort.

GRAPE, Holly, Cluster *Berberis pinnata*.

GRAPE, Holly, Creeping *Berberis repens*.

GRAPE, Holly, Fremont *Berberis fremontii* (Desert Barberry) USA Jaeger.

GRAPE, Holly, Longleaf *Berberis nervosa*.

GRAPE, Holly, Oregon *Mahonia aquifolium* (Oregon Grape) Yanovsky.

GRAPE, Holly, Red *Berberis haematocarpa*.

GRAPE, Mountain; GRAPE, Rocky Mountain *Mahonia aquifolium* (Oregon Grape) Schenk & Gifford (Mountain); Hutchinson (Rocky Mountain).

GRAPE, Oregon *Mahonia aquifolium*. The Karok name means 'Oregon Indian's grape'.

GRAPE, Sea 1. *Ephedra sinica* Thomson.1978. 2. *Salicornia europaea* (Glasswort) (B & H). 3. *Salsoli kali* (Saltwort) Bianchini.

GRAPE, Sea, English *Salicornia europaea* (Glasswort) B & H.

GRAPE, Wild, Arizona *Vitis arizonica*.

GRAPE, Winter, Sweet *Vitis cinerea* (Downy Grape) Vestal & Schultes.

GRAPE FLOWER; GRAPE FLOWER, Musk *Muscari botryoides* (Grape Hyacinth) Gerard (Grape Flower); Parkinson (Musk Grape Flower).

GRAPE HERB, Mexican *Chenopodium ambrosioides* (American Wormseed) Watt.

GRAPE HYACINTH *Muscari botryoides*.

GRAPE IVY, Miniature *Cissus striata* (Ivy of Uruguay) RHS.

GRAPE VINE *Vitis vinifera*.

GRAPE VINE IVY *Cissus rhombifolia*.

GRAPEFRUIT *Citrus paradisii*. The fruit grows in clusters - like bunches of grapes?

GRAPES *Sedum acre* (Biting Stonecrop) Som. Macmillan.

GRAPEWORT 1. *Actaea spicata* (Herb Christopher) Turner. 2. *Bryonia dioica*

(White Bryony) Lyte. Cf. Hedge Grape for vryony.

GRASS *Asparagus officinalis* (Asparagus) Lincs. Peacock. Asparagus is often rendered in various forms of Sparra-grass, and this gets contracted to Grass.

GRASS, Black *Alopecurus myosuroides* (Slender Foxtail) Howes.

GRASS, Sea *Armeria maritima* (Thrift) Gerard.

GRASS, Water, Great *Glyceria maxima*.

GRASS, Wood *Luzula sylvatica* (Great Woodrush) Scot. B & H.

GRAVEL-ROOT; GRAVELWEED *Eupatorium purpureum* (Purple Boneset) USA Bergen; Grieve.1931 (Gravelweed). Gravelweed tea for gallstones, a domestic remedy from Alabama, must be a fair indication of the origin of this.

GRAVELWIND *Convolvulus arvensis* (Field Bindweed) Wessex Rogers.

GRAVEYARD WEED *Euphorbia cyparissa* (Cypress Spurge) USA Kingsbury.

GRAWMPY GRIGGLE *Endymion nonscriptus* (Bluebell) Som. Macmillan. Granfer Griggle is more usual.

GRAY ROOT *Aster hesperus*.

GRAY'S LILY *Lilium grayi*. To commemorate the American botanist Asa Gray, who first collected this lily.

GRAYLICK *Sinapis arvensis* (Charlock) Suff. Jobson.

GRAZY *Caltha palustris* (Marsh Marigold) Som. Macmillan. A variation on 'crazy', of course.

GREASE, Goose- *Galium aparine* (Goosegrass) Gerard. How many more times does 'grease' appear for 'grass'? Gerard was fond of recommending goose-grease as a base for ointments, and it must have slipped in as a name.

GREASE, Pig's *Veronica beccabunga* (Brooklime) Dor. Grigson. Does 'grease' mean 'cress'? There are plenty of other 'cress' names for Brooklime. On the other hand, the word may simply mean plant (there is an entry in Halliwell for 'gres', which suggests this).

GREASE, Swine's *Polygonum aviculare* (Knotgrass) R T Gunther. Both Swine's Grass and Swine's Cress appear for this plant. The name appears as a gloss on a 12th century manuscript of Apuleius Barbarus.

GREASEWOOD; GREASEWOOD, Big; GREASEWOOD, Black; GREASEWOOD, Caterpillar *Sarcobatus vermicularis*. Greasewood is the usual name; the others are all in Jaeger.

GREAT ORME BERRY *Cotoneaster integerrimus*. It grows, albeit very rarely, on Great Orme Head.

GRECIAN HAY 1. *Chelidonium majus* (Greater Celandine) Fernie. 2. *Trigonella ornithopodioides* (Fenugreek) Grieve.1931. Grecian Hay is what Fenugreek, i.e. foenumgraecum, means. It was apparently used to scent inferior hay. Fernie called Greater Celandine Fenugreek, as well as Grecian Hay, though it is not clear why.

GRECIAN JUNIPER *Juniperus excelsa*. Certainly not confined to Greece, for it appears all over the Balkans and eastward to the Caucasus.

GRECIAN LAUREL *Laurus nobilis* (Bay) USA Cunningham.

GRECIAN SCABIOUS *Scabiosa pterocephala*.

GREEDS *Lemna minor* (Duckweed) Earle. Probably the same as the name Creed, recorded in Wiltshire, but it seems to appear as Greens, or Groves, even Grozen, as well.

GREEK FIR *Abies cephalonica*.

GREEK HAYSEED *Trigonella ornithopodioides* (Fenugreek) Clair. See Grecian Hay.

GREEK MALLOW *Kerria japonica* (Jew's Mallow) Salisbury.1936. Why Greek? Or Jew's, for that matter? It is a Chinese plant.

GREEK MAPLE *Acer heldreichii*.

GREEK NETTLE *Urtica pilulifera* (Roman Nettle) Lyte. Southern European, rather than specifically Greek, or Roman. It may very well be extinct, anyway.

GREEK SAND SPURREY *Spergularia bocconii* (Red Sand Spurrey) McClintock.1975.

GREEK STRAWBERRY TREE *Arbutus andrachne*.

GREEK VALERIAN 1. *Polemonium caeruleum* (Jacob's Ladder) Gerard. 2. *Polemonium reptans* (Abscess Root) USA House. 'Valerian' for Jacob's Ladder - it is said that cats are nearly as fond of this as they are of the real valerian.

GREEK VALERIAN, American Polemonium reptans (Abscess Root) Hatfield. American Greek Valerian as a name sounds the ultimate in confused nomenclature!

GREEK YARROW *Achillea ageratifolia*.

GREEN GROWER *Euphorbia amygdaloides* (Wood Spurge) Som. Macmillan.

GREENBRIAR *Smilax aspera*.

GREENBRIER, Broadleaf; GREENBRIER, Round-leaved *Smilax rotundifolia* (Cat Brier) Yanovsky (Broadleaf); USA Zenkert (Round-leaved).

GREENBRIER, Hispid *Smilax hispida*.

GREENBRIER, Laurel *Smilax laurifolia*.

GREENGAGE *Prunus domestica var. italica*. Named, it is said, after Sir William Gage, of Hengrave Hall, near Bury St Edmunds, who brought back plum trees of this group in 1724.

GREENHEART *Nectandra rodiaei*.

GREENING-WEED; GREENING-WEED, Dyer's *Genista tinctoria* (Dyer's Greenweed) Gerard; B & H (Dyer's).

GREENS 1. *Crataegus monogyna var. praecox* (Glastonbury Thorn) Ches. Parkinson. 2. *Lemna minor* (Duckweed) W.Miller. It sounds obvious, but Gerard knew Duckweed as Grains, and earlier than that Greeds is recorded. Parkinson explained the name for the thorn by saying that it grew "neare unto Nantwiche in Cheshire by a place called White Greene, which tooke the name as it was thought from the white bushes of thornes which there they call Greenes".

GREENS, Slick *Brassica oleracea var. capitata* the young plants (Cabbage) Glos. J D Robertson.

GREENS, Winter *Brassica fimbricata* (Curled Kale) Som. Macmillan.

GREENTEETH, Jenny *Lemna minor* (Duckweed) War. B & H. Roy Vickery argued convincingly that the nursery bogey called Jenny Greenteeth, who is supposed to drag children down into quiet pools, is Duckweed, which completely covers the surface of stagnant water. In Cheshire, it was understood that it was the weed which held children under, but they were told to beware of Jenny Greenteeth. Perhaps the small leaves are rather like tiny teeth.

GREENWEED 1. *Genista tinctoria* (Dyer's Greenweed) Gerard. 2. *Reseda luteola* (Dyer's Rocket) Suss, Kent Grigson. Dyer's Greenweed, when mixed with woad, gave the colour known as Kendal Green.

GREENWEED, Dyer's 1. *Genista tinctoria*. 2. *Isatis tinctoria* (Woad) B & H. It is probably the association of the two dyes in producing a green colour that resulted in Woad getting Dyer's Greenweed's name.

GREENWEED, English *Genista anglica* (Needle Whin) Flower.1859.

GREENWEED, Hairy *Genista pilosa*.

GREENWEED, Needle *Genista anglica* (Needle Whin) Som. Macmillan.

GREENWOOD 1. *Cytisus scoparius* (Broom) Som. Macmillan. 2. *Genista tinctoria* (Dyer's Greenweed) B & H.

GREET-WORT *Colchicum autumnale* (Meadow Saffron) Cockayne.

GREGGLE *Endymion nonscriptus* (Bluebell) Wilts. Dartnell & Goddard. This appears as Griggle, even Greygole, but is better known as some form of Granfer-Griggles.

GREGORY *Narcissus pseudo-narcissus* (Daffodil) Dev. B & H. It seems to grow wild in the neighbourhood near Torrington, where there was a monastery which belonged to the Canons of St Gregory. Nevertheless, Gregory is more likely to be some sort of corruption of Great Day - Easter, that is.

GREINS; GREINS OF PARIS *Aframomum melegueta* seeds (Melegueta Pepper) Halliwell. Greins is grains, and Paris should read Paradise - hence the better known name

for the seeds - Grains of Paradise.
GRENADILLE; GRANADILLA 1. *Passiflora caerulea* (Blue Passion Flower) S Africa. Watt. 2. *Passiflora edulis* (Passion fruit) S Africa Watt. Granadilla is "little pomegranate".
GRENADILLO *Pterocarpus draco.*
GRESS, Water *Nasturtium officinale* (Watercress) N.Ire. Grigson. Local pronunciation only.
GREVILLEA, Holly *Grevillea wickhamii.*
GREY-BEARD *Clematis vitalba* (Old Man's Beard) Wilts. D Grose; Hants. Grigson.
GREYGLE 1. *Endymion nonscriptus* (Bluebell) Dor. B & H; Wilts. Dartnell & Goddard. 2. *Orchis mascula* (Early Purple Orchis) Dor. B & H. It is suggested that Greygle and its many variants mean greyish-blue.
GREYGLE, Grammer *Endymion nonscriptus* (Bluebell) Dor. Macmillan.
GREYGLE, Granfer *Orchis mascula* (Early Purple Orchis) Dor. B & H.
GREYGLE, Granfer, Blue *Endymion nonscriptus* (Bluebell). B & H.
GREYGOLE *Endymion nonscriptus* (Bluebell) Dor. Halliwell.
GREYMILL *Lithospermum officinale* (Gromwell) Barton & Castle. Cf. Grey Millet, and Grey Myle for this.
GRIBBLE 1. *Malus sylvestris* fruit (Crabapple) Dor, Som. Mabey. 2. *Prunus spinosa* (Blackthorn) Dor. Barnes. Barnes said this was the name given to the tree as well as to crabapples. The walking sticks made from branches of blackthorn were called gribbles in Dorset, too.
GRIG 1. *Calluna vulgaris* (Heather) Corn, Heref, Wales, Norf. Grigson; Shrop. Halliwell; Ches. Holland. 2. *Erica tetralix* (Cross-leaved Heath) Ches. B & H. This is Welsh grug; the Cornish word for heather, gruglan, accounts for Griglans and Griglum.
GRIGG, Granfer 1. *Endymion nonscriptus* (Bluebell) Som. Macmillan. 2. *Orchis mascula* (Early Purple Orchis) Wilts. Macmillan.

GRIGGLE 1. *Endymion nonscriptus* (Bluebell) Som, Dor. Macmillan. 2. *Orchis mascula* (Early Purple Orchis) Som, Dor. Macmillan.
GRIGGLE, Grandfather *Orchis mascula* (Early Purple Orchis) Som. Macmillan.
GRIGGLE, Granfer 1. *Endymion nonscriptus* (Bluebell) Dor. Dacombe. 2. *Melandrium dioicum* (Red Campion) Som. Grigson. 3. *Orchis mascula* (Early Purple Orchis) Dor. Grieve.1931.
GRIGGLE, Granny *Endymion nonscriptus* (Bluebell) Dor. Grieve.1931.
GRIGGLE, Grawmpy *Endymion nonscriptus* (Bluebell) Som. Macmillan.
GRIGGLE-STICKS, Granfer 1. *Endymion nonscriptus* (Bluebell) Som. Macmillan. 2. *Orchis mascula* (Early Purple Orchis) Som. Macmillan. 3. *Taraxacum officinale* (Dandelion) Som. Macmillan.
GRIGLANS; GRIGLINGS; GRIGLUM *Calluna vulgaris* (Heather) all Corn. Bottrell (Griglans); Courtney.1880 (Griglings); Grigson (Griglum). Gruglan is the Cornish for heather. Cf. Grig.
GRIM THE COLLIER *Hieraceum aurantiacum* (Orange Hawkweed) B & H. Grim the Collier is the name of an Elizabethan comedy, and the hawkweed gets the name from its "black, smutty involucre". "The stalks and cups of the flours are all set thicke with a blackishe downe or hairinesse as it were the dust of coles; whence the women who keepe it in gardens for novelties sake, have named it Grimm the Collier" (Gerard).
GRINDELIA, Scaly *Grindelia squarrosa* (Gumweed) USA Henkel.
GRINDSTONE APPLE *Malus sylvestris* fruit (Crabapple) Wilts. Mabey. Not, apparently, because they are so hard, for Richard Jefferies said that country lads used to sharpen their knives by "drawing the blade slowly to and fro through a crab-apple; the acid of the fruit eats the steel like aqua fortis".
GRINGEL *Echium vulgare* (Viper's Bugloss) Hants. Cope.
GRINNEL, Neddy *Rosa canina* (Dog Rose)

Worcs. J Salisbury.

GRINNING SWALLOW *Senecio vulgaris* (Groundsel) Scot. Grigson. The Scottish sequence seems to run Groundsel, Groundie Swallow, Grundy Swallow, Grinning Swallow, and depends on the derivation of groundsel as ground swallower.

GRINSEL *Senecio vulgaris* (Groundsel) Wilts. Macmillan; Ches. Holland.

GRIP-GRASS *Galium aparine* (Goose-grass) N'thum. Grigson. A good descriptive name - Cf. such as Catchweed. All the variants of Cleavers are from the same mould.

GRISONS SALLOW *Salix grisonensis*.

GRISTLINGS *Prunus domestica var. institia* (Bullace) Dev. Hewett.1892. Cf. the Somerset Crislings, and others of the sequence.

GRIZZLE *Ribes uva-crispa* (Gooseberry) Dumf. B & H. This is French groseille, Old French grozelle, the usual name for a currant.

GROATS *Avena sativa* (Oats) Thomson.1978.

GROMALY; GROMELL; GROMVEL *Lithospermum officinale* (Gromwell) Halliwell (Gromaly); Gerard (Gromell); Hill (Gromvel).

GROMWELL *Lithospermum officinale*. Gromwell is from Old French gromil (in modern French it is grémil), meaning obscure, but the second syllable must be Latin milium, millet. In medieval Latin, gromwell was named milium solis, millet of the sun. Note the Somerset name Grey Millet, and some other variants.

GROMWELL, American *Lithospermum latifolium*.

GROMWELL, Blue *Buglossoides purpurocaeruleum* (Purple Gromwell) Clapham.

GROMWELL, Corn *Buglossoides arvensis*.

GROMWELL, Narrow-leaf *Lithospermum angustifolium* (Puccoon) USA Elmore.

GROMWELL, Purple *Buglossoides purpurocaeruleum*.

GROMWELL, Smooth, Seaside *Mertensia maritima* (Smooth Lungwort) Young.

GROMWELL, Stoneseed *Lithospermum canescens* (Hoary Puccoon) USA Elmore. All the gromwells have got these very hard, stony seeds. In fact, they are the signature for medicinal use against the stone.

GROMWELL CORN COCKLE *Buglossoides arvensis* (Corn Gromwell) S Africa Watt.

GRONOVIUS'S DODDER *Cuscuta gronovii*.

GROSEL; GROSER; GROSERT; GROSART; GROSIER; GROSSET *Ribes uva-crispa* (Gooseberry) C P Johnson (Grosel); N'thum, Scot Turner (Groser); Scot. Jamieson (Grosert); Dodson (Grosart) N.Eng. Grose (Grosier); Scot. Grigson (Grosset). All from French groseille, the usual name for a currant.

GROSSBERRY *Ribes uva-crispa* (Gooseberry) Yks. B & H. French groseille.

GROUND ASH *Angelica sylvestris* (Wild Angelica) N Eng, Berw. Grigson. Cf. Ground Elder for this.

GROUND CEDAR *Juniperus communis* (Juniper) USA Brimble.1948.

GROUND CHERRY 1. *Physalis bunyardii* (Cape Gooseberry) USA Watt. 2. *Physalis fendleri*. 3. *Physalis heterophylla*.

GROUND CHERRY, Clammy *Physalis heterophylla* (Ground Cherry) USA House.

GROUND CHERRY, Hairy *Physalis pubescens* (Husk Tomato) Jaeger.

GROUND CHERRY, Ivy-leaved *Physalis hederaefolia*.

GROUND CHERRY, Prairie *Physalis lanceolata*.

GROUND CHERRY, Smooth *Physalis subglabrata*.

GROUND CHERRY, Virginia *Physalis virginiana*.

GROUND ELDER *Angelica sylvestris* (Wild Angelica) Ches. Grigson. Cf. Ground Ash for this.

GROUND GLUTTON *Senecio vulgaris* (Groundsel) Scot. Grieve.1931. A result of the derivation of groundsel from OE grundswelge, which would give it the meaning of ground swallower. Grundswelge became Groundie Swallow and Grundy Swallow in Scotland, later developing into Grinning Swallow, and Swallow Grundy.

GROUND-HALE 1. *Lithospermum officinale* (Gromwell) Halliwell. 2. *Veronica officinalis* (Common Speedwell) Lyte. 'Hale' is probably 'Heale'.

GROUND-HEELE *Veronica officinalis* (Common Speedwell) Prior. One of Prior's more unlikely-sounding explanations is that both Ground-Heale and Ground-Hale for the common Speedwell are references back to a leprosy-curing legend associated with the plant. He says that leprosy in German is Grind (which actually means scab, or mange), and that the name came into English via a German Grundheil.

GROUND HOLLY *Gaultheria procumbens* (Wintergreen) Genders.1972.

GROUND IVY 1. *Convolvulus arvensis* (Field Bindweed) Corn, Dev. Wakelin. 2. *Glechoma hederacea*.

GROUND IVY-LEAVED CHICKWEED, Great *Lamium amplexicaule* (Henbit Deadnettle) Hulme.

GROUND LILY 1. *Convolvulus arvensis* (Field Bindweed) Wessex Rogers. 2. *Trillium erectum* (Bethroot) Grieve.1931.

GROUND PEA *Arachis hypogaea* (Groundnut) USA Elsmore.

GROUND PLUM 1. *Astragalus caryocarpus* USA Tolstead. 2. *Astragalus crassicarpus*.

GROUND-SILL *Glechoma hederacea* (Ground Ivy) Grose.

GROUNDIE-SWALLOW *Senecio vulgaris* (Groundsel) Scot. Jamieson. A result of the derivation from OE grundswelge, ground swallower.

GROUNDNUT 1. *Apios americana*. 2. *Arachis hypogaea*.

GROUNDNUT, American *Apios americana* (Groundnut) Bianchini.

GROUNDNUT PEAVINE *Lathyrus tuberosus*.

GROUNDSEL *Senecio vulgaris*. There are two schools of thought as to the derivation. The more obvious is that it is from OE grundswelge (it appears as groundswilly in the Dawson text), which would mean ground swallower. Groundsel may be an unwanted weed, but it hardly warrants that kind of abuse. The other derivation is from an entry dated about AD 700, where the name appears as gundae swelgiae, from gund, pus, and swelgan, swallow, absorb. The notion receives some support from the old practice of using the leaves in poultices for sores and abscesses. Grundeswelge only appears about AD 1000.

GROUNDSEL, Broad-leaved *Senecio fluviatilis* (Saracen's Woundwort) Murdoch McNeill.

GROUNDSEL, Broom *Senecio spartioides*.

GROUNDSEL, Bush *Baccharis halimifolia*.

GROUNDSEL, Douglas *Senecio douglasii*.

GROUNDSEL, Golden *Senecio aureus* (Swamp Squaw-weed) Grieve.1931.

GROUNDSEL, Heath *Senecio sylvaticus*.

GROUNDSEL, Hoary *Senecio erucifolius* (Hoary Ragwort) Grieve.1931.

GROUNDSEL, Mohave *Senecio mohavensis*.

GROUNDSEL, Mountain *Senecio sylvaticus* (Heath Groundsel) Murdoch McNeill.

GROUNDSEL, Purple *Senecio elegans* (Jacobaea) S Africa Watt.

GROUNDSEL, Riddell's *Senecio riddelli*.

GROUNDSEL, Sticky *Senecio viscosus*.

GROUNDSEL, Stinking *Senecio viscosus* (Sticky Groundsel) Polunin.

GROUNDSEL, Thread-leaf *Senecio longilobus* (Woolly Groundsel) USA Kingsbury.

GROUNDSEL, Tree 1. *Baccharis halimifolia* (Bush Groundsel) H & P. 2. *Senecio stanleyi*. 'Tree' is right - these can grow to about 30 feet tall on the slopes of Mount Ruwenzori.

GROUNDSEL, Wood *Senecio sylvaticus* (Heath Groundsel) Clapham.

GROUNDSEL, Woolly *Senecio longilobus*.

GROUNDSWELL *Senecio vulgaris* (Groundsel) Langham. A result of the derivation from OE grundswelge.

GROUNDWILL *Senecio vulgaris* (Groundsel) Dev. Grigson.

GROUSE WHORTLEBERRY *Vaccinium scoparium* (Low Huckleberry) Yanovsky.

GROVE MARJORAM *Origanum vulgare* (Marjoram) B & H.

GROVE-NUT *Conopodium majus* (Earth-nut) Corn. Jago. Is this simply ground-nut?

GROVES *Lemna minor* (Duckweed) Som. Macmillan. Presumably from the series that produced Grains, Greeds and Greens.

GROZEN *Lemna minor* (Duckweed) Som. Macmillan. This must be a plural in 'en'.

GROZER; GROZET; GROZZLE *Ribes uva-crispa* (Gooseberry) Dur. F M T Palgrave (Grozer); Scot. Jamieson (Grozet); B & H (Grozzle). From French groseille, usually retaining the 's', but these are three examples of the hard 's' being rendered with a 'z'.

GRUMBELL; GRUMMEL *Lithospermum officinale* (Gromwell) Tradescant (Grumbell); Turner (Grummel).

GRUMSEL *Taraxacum officinale* (Dandelion) Dev. B & H. Groundsel, of course, but this is in Halliwell, too, and he quite clearly calls it dandelion.

GRUNDAVY *Glechoma hederacea* (Ground Ivy) N'thum, Scot. Grigson.

GRUNDSEL *Senecio vulgaris* (Groundsel) Bucks, Derb, Lincs, Yks, Cumb. B & H.

GRUNDSWAITH; GRUNDSWATHE *Senecio jacobaea* (Groundsel) Cumb. B & H (Grundswaith); Cumb. Grigson (Grundswathe).

GRUNDY, Swallow; GRUNDY-SWALLOW *Senecio vulgaris* (Groundsel) N.Eng, Scot. Grigson. See the derivation of groundsel, as ground-swallower.

GRUNNISHULE; GRUNNISTULE *Senecio vulgaris* (Groundsel) Scot. B & H.

GRUNNUT *Conopodium majus* (Earth-nut) Prior. Groundnut, of course.

GRUNSEL; GRUNSIL 1. *Senecio jacobaea* (Ragwort) Cumb. Grigson. 2. *Senecio vulgaris* (Groundsel) Bucks, Derb, Lincs, Yks, Cumb. B & H; Dur. Dinsdale.

GRUNTY ASH *Aegopodium podagraria* (Goutweed) Bucks. Harman. This must be a corruption of Ground Ash, but Harman said that grunty meant tough.

GRUZEL; GRUZZLE *Ribes uva-crispa* (Gooseberry) Scot. Grigson (Gruzel); Dumf, Roxb. B & H (Gruzzle). Like a lot of other examples, this is from French groseille, the word for a currant.

GUADALUPE CYPRESS *Cupressus guadalupensis*. It grows not only on Guadalupe Island, but also in California and Mexico.

GUAIACUM *Guaiacum officinale* (Lignum Vitae) Hutchinson & Melville.

GUAJILLO *Acacia berlandieri*. Guaje is the Spanish-American name; guajillo would be diminutive of this.

GUANABANA *Annona muricata*.

GUANO-WEED *Spergula arvensis* (Corn Spurrey) Corn. Grigson.

GUARANA *Paullinia cupana*. This is a Tupi word that signifies a creeping or climbing plant.

GUARD-ROBE *Rosmarinus officinalis* (Rosemary) Friend. It was put in with clothes against moths. See Gerard - the people "do put it into chests and presses among clothes, to preserve them from moths and other vermin".

GUASHACKS; GUASHICKS *Arctostaphylos uva-ursi* (Bearberry). B & H (Guashacks); Banff, Mor. Grigson (Guashicks).

GUATEMALA RHUBARB *Jatropha podagrica*.

GUAVA *Psidium guajava*. Spanish guayaba, a word of South American origin.

GUAVA, French *Cassia alata* (Candlestick Senna) Chopra.

GUAVA, French, Wild *Cassia occidentalis* (Coffee Senna) Barbados Gooding.

GUAVA, Pineapple *Feijoa sellowiana*.

GUAVA, Wild *Careya arborea*.

GUAVA BERRY *Eugenia lineata*.

GUAYULE *Parthenium argenteum*. A Spanish word, of Nahuatl origin.

GUBBLIES, Goose-and- *Salix caprea* catkins (Sallow catkins) Shrop. B & H. Cf.

Geese-and-gullies, from the same general area. There are a lot of 'goose' or 'gosling' names for the catkins, presumably descriptive in a way, but there is an association with the real thing: it is unlucky to bring them into the house, for they would prove fatal to the real goslings.

GUBGUB *Vigna unguiculata* (Cowpea) W Indies Howes.

GUCKOO-FLOWER; GUCKOOS *Endymion nonscriptus* (Bluebell) Corn. Jago (Guckoo-flower); Corn. Courtney.1880 (Guckoos).

GUELDER ROSE *Viburnum opulus*. Introduced from Gueldres, it is said, hence the name.

GUELDER ROSE, Mealy *Viburnum lantana* (Wayfaring Tree) Clapham.

GUERNSEY CHICKWEED *Polycarpon tetraphyllum* (Four-leaved Allseed) McClintock.1975.

GUERNSEY ELM *Ulmus stricta var. sarniensis*.

GUERNSEY LILY *Nerine sarniensis*. Not a native, but said to have grown in Guernsey from bulbs washed ashore from a wreck of a ship from Japan about 1659. But there is another local legend concerning a Guernsey girl named Michelle de Garis, who married a fairy man and went off with him. She asked him for something to leave her family to let them have a token with which to remember her, and he gave her a bulb to plant in the sand above Vazon Bay. This was the Guernsey Lily, which, it was said, would grow in none of the other Channel Islands.

GUERNSEY STAR-OF-BETHLEHEM *Allium neapolitanum* (Daffodil Garlic) McClintock.1975.

GUILD 1. *Berberis vulgaris* (Barberry) Selk. Grigson. 2. *Calendula officinalis* (Marigold) Prior. 3. *Caltha palustris* (Marsh Marigold) Prior. 4. *Chrysanthemum segetum* (Corn Marigold) N Scot. B & H. The same as 'gold'. It appears in all sorts of guises, Gules probably being the most common.

GUILD TREE *Berberis vulgaris* (Barberry) Selk. Grigson.

GUILD WEED *Chrysanthemum segetum* (Corn Marigold) Salisbury.1964.

GUILE; GUILLS *Chrysanthemum segetum* (Corn Marigold) Mor. Gregor (Guile); Som. Macmillan (Guills).

GUILTY-CUP; GULTY-CUP *Ranunculus acris* (Meadow Buttercup) Dev. Halliwell (Guilty); Dev. B & H (Gulty). More often spelt Gilty-cup, so making the meaning, Gold-cup, more obvious.

GUIMAUVE *Althaea officinalis* (Marsh Mallow) W Miller. The 'mauve' part of the name would be the same as mallow, the Malva of the other genus.

GUIND *Prunus avium* (Wild Cherry) Scot. B & H. This is French guigne, sweet cherry, better known as Gean.

GUINEA, Golden *Ranunculus ficaria* (Lesser Celandine) N'thants. A E Baker. Golden flowers, of course.

GUINEA CORN *Sorghum vulgare* (Kaffir Corn) J Smith.

GUINEA CUBEB *Piper guineense* (West African Black Pepper) Dalziel.

GUINEA FLOWER 1. *Fritlllaria meleagris* (Snake's Head Lily) Hudson. 2. *Hibbertia scandens* (Snake Vine) Simmons. 3. *Kerria japonica* (Jew's Mallow) Lincs. B & H. For the Fritillaria, this name refers to the Guinea-hen, and is descriptive. A different tradition accounts for the name for Kerria. Guinea in this case is a coin - Cf. Sovereign-flower and Guinea-plant, all from Lincolnshire.

GUINEA FLOWER, Twining *Hibbertia dentata*.

GUINEA GOLD VINE *Hibbertia scandens* (Snake Vine) Simmons.

GUINEA GRAINS *Aframomum melegueta* (Melegueta Pepper) Dalziel.

GUINEA GRASS 1. *Panicum jumentorum*. 2. *Panicum maximum*. They are of African origin, hence this name.

GUINEA HEN *Fritillaria meleagris* (Snake's Head Lily) B & H. The petals are spotted like the feathers of a guinea-hen. In fact, the specific name, *meleagris*, is from Greek meaning guinea-hen. Cf. Ginny-hen Flower.

GUINEA HEN WEED *Petiveria alliacea*.

GUINEA PEA *Abrus precatorius* (Precatory Bean) Chopra.

GUINEA PEPPER 1. *Aframomum melegueta* (Melegueta Pepper) Dalziel. 2. *Capsicum annuum* (Chile) Watt.

GUINEA PLANT *Kerria japonica* (Jew's Mallow) Lincs. B & H. See Guinea Flower.

GUINEA PLUM *Parinari excelsa* (Rough-skinned Plum) Dalziel.

GUINEA-PODS *Capsicum annuum* (Chile) Dalziel.

GUINEA SORREL *Hibiscus sabdariffa* (Rozelle) Dalziel.

GUINEA YAM, White *Dioscorea rotundata* (White Yam) Alexander & Coursey.

GUINEA YAM, Yellow *Dioscorea cayenensis* (Yellow Yam) Alexander & Coursey.

GULES 1. *Calendula officinalis* (Marigold) Prior. 2. *Caltha palustris* (Marsh Marigold) Prior. 3. *Chrysanthemum segetum* (Corn Marigold) Prior. The word means Gold.

GULES-GOWAN *Chrysanthemum segetum* (Corn Marigold) B & H. Both elements of the name mean gold.

GULL 1. *Chrysanthemum segetum* (Corn Marigold) Cumb. Grigson. 2. *Salix sp* catkins (Willow catkins) Suss. Parish. As far as Corn Marigold is concerned, the name is but one of the many variants of Gules. But Gull for the willow catkins probably means 'gosling' (Cf. Goose-and-gullies), vaguely descriptive, but, more to the point, there is an association with the real goslings. You should never bring willow catkins into the house, for such an act would spell death to the goslings.

GULL, Pea *Primula veris* (Cowslip) J Smith. This must be the ultimate variant in the group which starts with Paigle.

GULL, White *Leucanthemum vulgare* (Ox-eye Daisy) Cumb. B & H. To distinguish from the Yellow Gull, *Chrysanthemum segetum*.

GULL, Yellow *Chrysanthemum segetum* (Corn Marigold) Cumb. B & H. If Gull, i.e. Gules, means gold, then the only justification for such a name as this lies in the fact that they call Ox-eye White Gull.

GULL-GRASS *Galium aparine* (Goose-grass) Heref, Glos. Grigson. Gull is a word which was once used in the southern half of the country for a gosling.

GULLIES, Geese-and-; GULLS *Salix caprea* catkins (Goat Willow catkins) Shrop, Ches. Grigson (Geese-and-gullies); Heref. Leather (Gulls). Gullies and Gulls mean goslings. The names are descriptive in a way, but the connection is with the belief that bringing the catkins into the house would prove fatal to the real goslings. Similar beliefs are recorded with other spring flowers, snowdrops and daffodils, for instance. Care had to be taken about picking them, or the young poultry would be harmed.

GULLY-ROOT *Petiveria alliacea* (Guinea-hen Weed) Guyana Laguerre.

GULSA GIRSE *Menyanthes trifoliata* (Buckbean) Shet. Grigson. i.e. jaundice-grass, obviously the result of some medicinal usage.

GULSECK-GIRSE *Teucrium scorodonia* (Wood Sage) Ork. Leask. Another example of 'jaundice-grass'. Wood Sage was certainly used in Orkney for that ailment.

GUM, Barbary *Acacia gummifera* (Mogador Acacia) Hora.

GUM, Black *Nyssa sylvatica* (Tupelo) USA Harper.

GUM, Blue *Eucalyptus globosus*.

GUM, Blue, Sydney *Eucalyptus saligna*.

GUM, Blue, Tasmanian *Eucalyptus globosus* (Blue Gum) RHS.

GUM, Budge *Bursera simaruba* W Indies Hora.

GUM, Cabbage *Eucalyptus pauciflora*.

GUM, Candle-bark *Eucalyptus rubida*.

GUM, Cape *Acacia horrida*.

GUM, Cedar *Eucalyptus gunnii* (Cider Gum) Barber & Phillips.

GUM, Cider *Eucalyptus gunnii*. 'Cider', it seems, is the same as honey in Tasmania, and the gum was used as a substitute for honey there.

GUM, Cotton 1. *Nyssa aquatica* (Water Tupelo) USA Schery. 2. *Nyssa sylvatica* (Tupelo) Willis.

GUM, Doctor's *Metopium toxiferum* (Poison-wood) Usher. The stem yields a resin, known as Doctor's Gum, or Hog Gum, which is used as a violent purgative.

GUM, Flowering, Red *Eucalyptus ficifolia* (Red Gum) Duncalf.

GUM, Ghost *Eucalyptus papuana*. The smooth white bark accounts for the name.

GUM, Hog *Metopium toxiferum* (Poison-wood) Usher. See Doctor's Gum.

GUM, Karaya *Stercularia urens*.

GUM, Manna *Eucalyptus viminalis*.

GUM, Morocco *Acacia gummifera*.

GUM, Peppermint *Eucalyptus coccifera* (Mount Wellington Gum) Wilkinson.1981.

GUM, Red 1. *Copaifera copallifera* (Gum Copal) Dalziel. 2. *Eucalyptus camaldulensis* (River Red Gum) Polunin.1976. 3. *Eucalyptus ficifolia*. 4. *Liquidambar styraciflua* (Sweet Gum) USA Harper.

GUM, Red, Murray *Eucalyptus camaldulensis* (River Red Gum) RHS. The river of the common name is specified as the Murray in this.

GUM, Red, River *Eucalyptus camaldulensis*.

GUM, Ribbon *Eucalyptus viminalis* (Manna Gum) Polunin.1976.

GUM, Sap *Liquidambar styraciflua* (Sweet Gum) Brimble.1948.

GUM, Snow *Eucalyptus pauciflora* (Cabbage Gum) Hora.

GUM, Snow, Australian *Eucalyptus niphophila*.

GUM, Snow, Tasmanian *Eucalyptus coccifera* (Mount Wellington Peppermint) Wilkinson.1981.

GUM, Sour *Nyssa sylvatica* (Tupelo) USA Schery. Presumably a reference to the fruits, which are very acid.

GUM, Spinning *Eucalyptus perriniana*.

GUM, Sweet *Liquidambar styracifolia*.

GUM, Tahl *Acacia seyal*.

GUM, Tingiringi *Eucalyptus glaucescens*.

GUM, Urn *Eucalyptus urnigera*. Urn, from the shape of the fruits.

GUM, Yellow 1. *Copaifera copallifera* (Gum Copal) Dalziel. 2. *Eucalyptus leucoxylon*. 3. *Liquidambar styraciflua* (Sweet Gum) Brimble.1948.

GUM ACACIA; GUM ARABIC *Acacia senegal*.

GUM COPAL; GUM COPAL, Sierra Leone *Copaifera copallifera* Dalziel.

GUM EUPHORBIA *Euphorbia resinifera*.

GUM-LAC TREE; LAC TREE *Schleichera oleosa* (Ceylon Oak) Chopra.

GUM PLANT 1. *Grindelia robusta*. 2. *Symphytum officinale* (Comfrey) Grieve.1931. For comfrey, this is a reference to the glutinous matter of the root, used as a kind of splint when set. Cf. the name Slippery Root.

GUM THISTLE 1. *Euphorbia resinifera* (Gum Euphorbia) Le Strange. 2. *Onopordum acanthium* (Scotch Thistle) Turner.

GUMBO *Abelmoschus esculentus* fruit (Okra) Grigson.1974. Gombo seems to be more accurate - that was a West African name, and these pod-like fruits are called gomboes, or Lady's Fingers.

GUMBOLIMBO *Bursera simaruba* Florida Roys.

GUMWEED *Grindelia squarrosa*. Cf. Tarweed, and Sticky-head, for this plant.

GUNBRIGHT *Equisetum hyemale* (Dutch Rush) Maine Bergen. It is said to have been used by the Indians in polishing their guns. Not so far-fetched as it sounds, for the horsetails have a high silica content (the ashes of common Horsetail are 80% silica). Dutch Rush has been used for polishing wood, ivory and brass, and Common Horsetail was a well-known pot-scourer.

GUNPOWDER PLANT *Pilea microphylla* Rochford. There are Pistol Plant and Artillery Plant as well, all because if the plant is shaken when in flower, it will produce clouds of pollen.

GUNS *Capsella bursa-pastoris* (Shepherd's Purse) Som. Macmillan. Because of the explosive seed distribution?

GUSSES, Single *Orchis mascula* (Early Purple Orchis) Som. Jennings; Wilts. Grigson. Gusses started out as geese (there are a number of Goosey-gander type names for this). Guss in turn became 'ghost' - there is Single Ghost recorded, and Single Castle as well.

GUSSES, Standing *Arum maculatum* (Cuckoo-pint) Som. Grigson. Is it a corruption of Gethsemane? The plant was said to have been growing at the foot of the Cross and to have received some drops of Christ's blood on its leaf. The other plant to have "Gusses" names is Early Purple Orchis, which also has blotched leaves. Nevertheless, the likely origin of "guss" is "geese".

GUSSETS *Orchis mascula* (Early Purple Orchis) Dor. B & H.

GUSSETS, Dandy *Dactylorchis majalis* (Irish Marsh Orchid) Som. Macmillan. From the Gandigoslings sequence. The attribution in this case must be very doubtful.

GUSSETS, Standing *Orchis mascula* (Early Purple Orchis) Dev. Macmillan.

GUSSIES, Single *Endymion nonscriptus* (Bluebell) Som. Macmillan. Bluebell is another plant which has Goosey-gander names.

GUSSIPS *Orchis mascula* (Early Purple Orchis) Som. Macmillan. Gossips is also recorded, but that must have been as an extension of Gussips rather than the other way round, for this must surely be part of the 'gusses' series.

GUT, Cat *Tephrosia virginiana* (Devil's Shoestring) Howes.

GUTS, Devil's 1. *Calystegia sepium* (Great Bindweed) Norf. Grigson. 2. *Clematis vitalba* (Old Man's Beard) Som. Macmillan; Dor. Dacombe; Wilts. Goddard. 3. *Convolvulus arvensis* (Field Bindweed) Som, IOW, Kent, Beds, Shrop, Leics, Notts, N'thum. Grigson. 4. *Cuscuta epithymum* (Dodder) Som, Glos, Hants, IOW, Suss, Beds, Worcs, Shrop, Beds, Cumb, Lanark. Grigson. 5. *Cuscuta europaea* (Greater Dodder) Glos, Suss, Worcs, Shrop, Cambs.

B & H. 6. *Ranunculus repens* (Creeping Buttercup) N'thum. Grigson. 7. *Succisa pratensis* (Devil's-bit) Som. Tongue. Either because they strangle, as with the bindweeds, old man's beard, and dodder, or because of the creeping rootstock, as in Creeping Buttercup. Devil's Bit seems to be the odd one out, but the story is that when the devil bit off the taproot, all the small roots twisted round him and nearly strangled him.

GUTTA-PERCHA TREE *Excoecaria parviflora.*

GUTWEED *Sonchus arvensis* (Corn Sow Thistle) Ess. B & H. The long, creeping roots suggest the name.

GUYANA ARROWROOT *Dioscorea alata* (Greater Yam) Howes.

GUZZLEBERRY *Ribes uva-crispa* (Gooseberry) Wilts. Dartnell & Goddard. It may not look like it, but this is still presumably from French groseille, like a lot of the names for gooseberries.

GYE 1. *Agrostemma githago* (Corn Cockle) Suff, Lincs. Grigson. 2. *Galium aparine* (Goose-grass) Norf, Suff. Grigson. 3. *Papaver rhoeas* (Red Poppy) Suff. Grigson. 4. *Ranunculus arvensis* (Corn Buttercup) Ess, Suff, Norf, Lincs. Grigson. 5. *Ranunculus repens* (Creeping Buttercup) Ess. Gepp. This is pronounced with a soft 'g', if we take into account the Essex variant, Joy. Whatever the word's original meaning, it clearly refers to any weed growing in the corn.

GYPSY 1. *Geranium pratense* (Meadow Cranesbill) Wilts. Grigson. 2. *Geranium robertianum* (Herb Robert) Som. Macmillan. 3. *Luzula campestris* (Good Friday Grass) Wilts. Dartnell & Goddard.1899. 4. *Plantago lanceolata* (Ribwort Plantain) Som. Grigson. 5. *Senecio jacobaea* (Ragwort) Som. Macmillan.

GYPSY, Black *Plantago lanceolata* (Ribwort Plantain) Som. Grigson.

GYPSY COMB *Arctium lappa* (Burdock) Berks, Notts. Mabey.

GYPSY CURTAINS 1. *Anthriscus sylvestris*

(Cow Parsley) Som. Grigson. 2. *Conium maculatum* (Hemlock) Som. Grigson. It is the lacy effect of the plant that suggests curtains.
GYPSY DAISY 1. *Leucanthemum vulgare* (Ox-eye Daisy) Som, Dor. Macmillan; Norf. Grigson. 2. *Knautia arvensis* (Field Scabious) Som. Macmillan. 3. *Tripleurospermum maritimum* (Scentless Maydweed) Som. Macmillan.
GYPSY FLOWER 1. *Aethusa cynapium* (Fool's Parsley) Dor. Northover. 2. *Anthriscus sylvestris* (Cow Parsley) Som. Grigson. 3. *Cirsium arvense* (Creeping Thistle) Som. Macmillan. 4. *Conium maculatum* (Hemlock) Som. Macmillan. 5. *Cynoglossum officinale* (Hound's Tongue) Glos. J D. Robertson; USA Tolstead. 6. *Geranium pratense* (Meadow Cranesbill) Wilts. Dartnell. & Goddard.1899. 7. *Geranium robertianum* (Herb Robert) Som. Macmillan. 8. *Knautia arvensis* (Field Scabious) IOW, Ess, Norf, Cumb, Yks. Grigson. 9. *Lychnis flos-cuculi* (Ragged Robin) Som. Macmillan. 10. *Melandrium dioicum* (Red Campion) Som. Macmillan. 11. *Tripleurospermum maritimum* (Scentless Maydweed) Som. Macmillan.
GYPSY LACE *Galium palustre* (Water Bedstraw) Som. Macmillan.
GYPSY LACES *Anthriscus sylvestris* (Cow Parsley) Dor, Som. Macmillan.
GYPSY MAIDS *Centranthus ruber* (Spur Valerian) Som. Macmillan.
GYPSY NUTS 1. *Crataegus monogyna* haws (Haws) Wilts. Goddard. 2. *Rosa canina* hips (Hips) Wilts. Goddard.
GYPSY ONION *Allium ursinum* (Ramsons) S Eng. B & H.
GYPSY PEAS *Vicia sativa* (Common Vetch) Som. Macmillan.
GYPSY PINK *Dianthus caryophyllus* (Carnation) Som. Macmillan.
GYPSY ROSE 1. *Knautia arvensis* (Field Scabious) IOW, Ess, Norf, Yks, Cumb. B & H. 2. *Rosa arvensis* (Field Rose) Halliwell. 3. *Scabiosa atropurpurea* (Sweet Scabious) Dev. Friend; Som. Coats; Wilts. Dartnell & Goddard; IOW B & H. 4. *Succisa pratensis* (Devil's-bit) Som. Friend. There is some doubt about including Field Rose here - Halliwell calls it Corn Rose, and there lies the doubt.
GYPSY VIOLET *Viola riviniana* (Dog Violet) Som. Macmillan.
GYPSY'S BACCA 1. *Clematis vitalba* (Old Man's Beard) Som. Grigson. 2. *Rumex acetosa* (Wild Sorrel) Som. Mabey. 3. *Teucrium scorodonia* (Wood Sage) Dor. Macmillan. Presumably they were all smoked in one way or another. Old Man's Beard certainly was - boys used to smoke its porous stalks. Cf. names like Smoking Cane and Devil's Cut.
GYPSY'S COMB *Dipsacus fullonum* (Teasel) Yks. Grigson.
GYPSY'S CURTAINS *Conium maculatum* (Hemlock) Som. Macmillan. Cf. the various 'lace' and 'needlework' names for hemlock.
GYPSY'S DAISY *Leucanthemum vulgare* (Ox-eye Daisy) Fernie.
GYPSY'S GIBBLES *Allium ursinum* (Ramsons) Som. Macmillan. Gibbles, or chipples, is the Somerset name for young onions.
GYPSY'S HAT *Convolvulus arvensis* (Field Bindweed) Som. Macmillan.
GYPSY'S LACE(S) *Heracleum sphondyllium* (Hogweed) Som. Macmillan. Most of the white umbellifers have names like this.
GYPSY'S MONEY *Caltha palustris* (Marsh Marigold) Som. Macmillan. It is the gold colour that accounts for the name.
GYPSY'S PARSLEY 1. *Anthriscus sylvestris* (Cow Parsley) Som. Grigson. 2. *Geranium robertianum* (Herb Robert) Som. Macmillan.
GYPSY'S RHUBARB 1. *Arctium lappa* (Burdock) Som. Macmillan. 2. *Petasites hybridus* (Butterbur) Som. Macmillan. The large leaves are likened to those of rhubarb.
GYPSY'S SAGE *Teucrium scorodonia* (Wood Sage) Dor. Macmillan.
GYPSY'S TOBACCO 1. *Clematis vitalba*

(Old Man's Beard) Dor. Macmillan. 2. *Rumex acetosa* (Wild Sorrel) Som. Macmillan. See Gypsy's Bacca.

GYPSY'S TREACLE *Sambucus nigra* (Elder) Ire. Jean Philpot. Treacle, in the sense of medicine.

GYPSY'S UMBRELLA *Anthriscus sylvestris* (Cow Parsley) Som. Grigson.

GYPSYWEED 1. *Lycopus europaeus* (Gypsywort) Grieve.1931. 2. *Lycopus virginicus* (Virginia Bugleweed) Grieve.1931. 3. *Veronica oficinalis* (Common Speedwell) USA House.

GYPSYWORT 1. *Ballota nigra* (Black Horehound) Vesey-Fitzgerald. 2. *Lycopus europaeus*. *L europaeus* yields a black dye once used for wool and silk, but also used by gypsies to stain their skins, hence the name. "Some also thinke good to call it Herba Aegyptia, because they that feine themselves Aegyptians (such as many times wander like vagabonds from citie to citie in Germanie and other places) do use with this herbe to give themselves a swart colour..." (Gerard).

H

HA-HO 1. *Anthriscus sylvestris* (Cow Parsley) Ire. Grigson. 2. *Heracleum sphondyllium* (Hogweed) Ire. B & H. Obviously a pronunciation guide to a Gaelic name. The form exists also as Hi-how and as Da-ho.

HAB-NABS *Crataegus monogyna* fruit (Haws) Som. Macmillan.

HACK; HACKER *Prunus padus* fruit (Bird Cherry) N Eng. Grigson. This must be OE haga, hedge, also the basis of haw and its many variants.

HACKBERRY 1. *Celtis occidentalis*. 2. *Prunus padus* (Bird Cherry) Cumb, Berw, Dumf, Perth, Roxb. Grigson.

HACKBERRY, American *Celtis occidentalis* (Hackberry). W Miller.

HACKBERRY, Desert *Celtis pallida*.

HACKBERRY, Douglas *Celtis douglasii*.

HACKBERRY, Net-leaf *Celtis reticulata* (Western Hackberry) USA T H Kearney.

HACKBERRY, Sugar *Celtis laevigata* (Sugarberry) Barber & Phillips.

HACKBERRY, Western *Celtis reticulata*.

HACKMATACK 1. *Larix laricina* (American Larch) USA Everett. 2. *Populus balsamifera* (American Balsam Poplar) Howes. This name is used for the timber. It comes from the American Indian name for the larch, as does the commoner Tamarack. Cf. also Tacamahac.

HACKWOOD *Prunus padus* (Bird Cherry) Cumb. B & H. Hack, or Hacker, is used for the fruit - so Hackwood is the tree.

HACKYMORE; HICKYMORE *Centaurea nigra* (Knapweed) Som. Macmillan. Hardhack is a Wiltshire name for this plant. Cf. such as Hardheads, or Hardiron.

HADDER; HADDYR; HEDDER *Calluna vulgaris* (Heather) N Eng, Scot. Grose (Hadder); Scot. Jamieson (Haddyr); E Ang, Yks, Cumb. Grigson (Hedder). Hadder is the older Scots name for heather. It later appeared as hather (heather did not appear till the 18th century in England, with Hedder probably the prototype).

HAG; HAG TREE *Crataegus monogyna* (Hawthorn) Dev, Som Grigson; Hants. Cope (for haws). OE haga, hedge, which was also responsible for 'haw' and its many variants.

HAG BUSH; HAHS-BUSH *Crataegus monogyna* (Hawthorn) Yks. Grigson (Hag Bush); Bucks. Harman (Hahs-bush).

HAG-HAW *Crataegus monogyna* fruit (Haw) Bucks. Harman. An oddity, for it would appear that both the elements mean the same thing, coming as they do from OE haga, hedge.

HAG-LEAF; HAG-TAPER; HAG'S TAPER *Verbascum thapsus* (Mullein) Som, Bucks. Grigson (Hag-leaf); Bucks, Herts, N.Ire. B & H (Hag-taper); C P Johnson (Hag's Taper). Presumably 'hag' means 'hedge' in this case, too. 'Taper' is a reference to the old country custom of dipping the stems and leaves of mullein in tallow or suet, and burning it to give light at outdoor gatherings, or even in the home.

HAG-ROPE *Clematis vitalba* (Old Man's Beard) Som. Macmillan. i.e. hedge-rope.

HAGAG *Crataegus monogyna* fruit (Haw) Wessex Rogers.

HAGBERRY 1. *Crataegus monogyna* fruit (Haws) Hants. Cope. 2. *Prunus padus* (Bird Cherry) N.Eng, Scot. Grigson; Lancs Nodal; Yks. Morris; Roxb. Denham/Hardy. 3. *Sorbus torminalis* (Wild Service Tree) Yks. Grigson.

HAGGAS; HAGGIL *Crataegus monogyna* fruit (Haw) Berks. Lowsley (Haggas); Hants. Cope (Haggil).

HAGISSES *Rosa canina* fruit (Hips) Hants. Cope.

HAGTHORN *Crataegus monogyna* (Hawthorn) Dev. Friend.1882; Som. Macmillan; Yks. Grigson.

HAGUE; HAIG; HAIGH *Crataegus monogyna* fruit (Haw) Ches. B & H; Lancs. Nodal; Yks. Carr (Hague) or Addy (Haigh); N Eng. Ellacombe (Haig).

HAGWORM FLOWER; AGWORM FLOWER *Stellaria holostea* (Greater Stitchwort) Yks. Grigson. i.e. adder flower. Cf. Adder's Meat, Adder's Spit, Snakeflower, etc.

HAILS; HALES *Crataegus monogyna* fruit (Haws) Som. Macmillan.

HAILWEED; HALE-WEED 1. *Cuscuta epithymum* (Common Dodder) B & H. 2. *Cuscuta europaea* (Greater Dodder) B & H. Cf. Hellweed, widespread in England and Scotland.

HAINBERRY *Rubus idaeus* (Raspberry) Roxb. B & H. Probably Hindberry. There is an intermediate stage, Hineberry, from northern England.

HAIR, Cat's 1. *Euphorbia hirta* (Hairy Spurge) Sofowora. 2. *Spergula arvensis* (Corn Spurrey) Corn. Grigson.

HAIR, Fairy *Cuscuta epithymum* (Common Dodder) Jersey Grigson. Cf. Hairweed, etc.

HAIR, Lady's 1. *Adiantum capillus-veneris* (Maidenharir Fern) Leyel.1945. 2. *Briza media* Som, E Ang, Lincs, N'thum. Grigson.1959.

HAIR, Maid's *Galium verum* (Lady's Bedstraw) Yks, N Eng. Turner.

HAIR, Maiden *Galium aparine* (Goosegrass) Yks. B & H.

HAIR, Maiden's 1. *Briza media* (Quakinggrass) IOW, Norf. Grigson.1959. 2. *Clematis vitalba* (Old Man's Beard) Bucks. Grigson. 3. *Cuscuta epithymum* (Common Dodder) IOW Grigson. 4. *Galium cruciatum* (Crosswort) N'thum. Grigson. 5. *Galium verum* (Lady's Bedstraw) Glos. PLNN35. 6. *Narthecium ossifragum* (Bog Asphodel) Lancs. Putnam.

HAIR, Venus *Adiantum capillus-veneris* (Maidenhair Fern) Turner. This is a straight translation of the specific name.

HAIR, Virgin's *Briza media* (Quaking Grass) Tynan & Maitland. Cf. Maiden's Hair.

HAIR-BEARD *Luzula campestris* (Good Friday Grass) N'thants. B & H.

HAIR-GRASS, Alpine *Deschampsia alpina*.

HAIR-GRASS, Bog *Deschampsia setacea*.

HAIR-GRASS, Grey *Corynephorus canescens*.

HAIR-GRASS, Purple *Muhlenbergia pungens*.

HAIR-GRASS, Silver *Aira caryphyllea*.

HAIR-GRASS, Small *Aira praecox*.

HAIR-GRASS, Tufted *Deschampsia cespitosa*.

HAIR-GRASS, Wavy *Deschampsia flexuosa*.

HAIRBELL; AIRBELL *Campanula rotundifolia* (Harebell) Potter & Sargent (Hairbell); A S Palmer (Airbell). It doesn't matter which way you spell harebell, for 'hare' is. O E hara, hair.

HAIRBRUSH *Dipsacus fullonum* (Teasel) Som. Macmillan. There are many other "brush" names for teasel, all descriptive. Cf. Brushes, Lady's Brushes, etc.

HAIREVE; HAIRIFF; HAIRITCH *Galium aparine* (Goose-grass) Glos. Halliwell (Haireve); Yks. Nicholson (Hairiff); Prior (Hairitch). These names are extremely variable, including a dozen or so forms from Airup through to Heyriffe. They are all apparently from an OE form hege-reafa, which would mean hedge-robber (robber presumably from the action of the burrs in attaching themselves to anything they touch). Cf. Gripgrass, etc.

HAIRHOOF, Scented; HAIRHOOF, Sweet *Asperula odorata* (Woodruff) both Yks. Grigson.

HAIRHOUND *Ballota nigra* (Black Horehound) Berw. Grigson. Hairhound is horehound.

HAIRLESS WILLOW *Salix glabra*.

HAIRPINS, Combs-and- *Taraxacum officinale* (Dandelion) Som. Macmillan.

HAIROUGH; HAY-ROUGH; HAIRUP; AIRUP *Galium aparine* (Goose-grass) Yks. Nicholson (Hairup); A S Palmer (Hairough, Hay-rough). See Haireve.

HAIRWEED 1. *Cuscuta epithymum* (Common Dodder) Hants, Beds, Norf. Grigson. 2. *Cuscuta europaea* (Greater Dodder) Beds. B & H. 3. *Galium aparine* (Goose-grass) B & H. Descriptive. Cf. Fairy

Hair, Maiden's Hair, etc., for dodder.

HAIRWOOD *Acer pseudo-platanus* (Sycamore) MacQuoid. Or Harewood, when the timber is stained grey with iron salts.

HAIRY BIND *Cuscuta epithymum* (Common Dodder) Hants. Grigson.

HAIRY BINDWEED *Calystegia pulchra.*

HAIRY BIRCH *Betula pubescens* (Downy Birch) Hart.

HAIRY BITTERCRESS *Cardamine hirsuta.*

HAIRY CRESS *Cardaria draba* (Hoary Cress) Chopra. Possibly a misreading of Hoary Cress.

HAIRY CUCUMBER *Cucumis chate.*

HAIRY ELM *Ulmus canescens.*

HAIRY GREENWEED *Genista pilosa.*

HAIRY HAWKBIT *Leontodon taraxacoides.*

HAIRY LADY'S SMOCK *Cardamine hirsuta* (Hairy Bittercress) Curtis.

HAIRY MEDICK *Medicago polymorpha.*

HAIRY MOUNTAIN LAUREL *Kalmia hirsuta.*

HAIRY ST JOHN'S WORT *Hypericum hirsutum.*

HAIRY SPURGE 1. *Euphorbia hirta.* 2. *Euphorbia pilosa.*

HAIRY TARE *Vicia hirsuta.*

HAIRY THORN-APPLE *Datura metel.*

HAIRY VETCH 1. *Vicia hirsuta* (Hairy Tare) Hutchinson. 2. *Vicia villosa* (Lesser Tufted Vetch) USA Kingsbury.

HAIRY WILD SUNFLOWER *Helianthus mollis.*

HAISH *Fraxinus excelsior* (Ash) B & H. The word 'ash' assumes various forms - Ache, Esh, Esch, etc. They are all from OE aesc.

HAKERNE *Quercus robur* acorn (Acorn) Halliwell.

HALBERD-LEAVED ORACHE *Atriplex hastata.*

HALBERD-LEAVED TEAR-THUMB *Polygonum arifolium.*

HALCUP *Caltha palustris* (Marsh Marigold) Hants. Cope.

HALEHOUSE *Glechoma hederacea* (Ground Ivy) Fernie. A variant of Alehoof, which means literally that which will cause ale to heave, or work.

HALENUT *Corylus avellana* (Hazel) Corn. Grigson.

HALES 1. *Corylus avellana* (Hazel) Corn, Som. Grigson. 2. *Crataegus monogyna* haws (Haws) Som, Dor. Macmillan. Sometimes spelt Hails, but this is more usual. It varies into ales or even isles, and can be identified as the second syllable in such names Pigall, Pigale, Peggy-ailes - even Pigshell. But Hazel itself provides the original for Hales, for metathesis of the common name gave Halse, itself further misspelt into this.

HALES, French 1. *Pyrus scandica* Dev. Friend. 2. *Sorbus latifolia.* French, because it was first found growing at Fontainebleau.

HALES, Pig's *Crataegus monogyna* haws (Haws) Som. Jennings. 'Pig's' is possibly a result of confusion of 'hog' and 'hag'; if the latter, then it is OE haga, hedge.

HALF-AND-HALF *Crataegus monogyna* haws (Haws) Som. Macmillan. Elsewhere haws are called Halves, or Harves, even Haves. This must be a corruption when the meaning was lost.

HALF-MEN *Pachypodium namaquanum.*

HALFMOON LOCO *Astragalus argillophyllus.*

HALFPENNIES-AND-PENNIES; HAPPENNIES-AND-PENNIES *Umbilicus rupestris* (Wall Pennywort) Dev. Macmillan (Halfpennies); Dev. Grigson (Happennies). It is the round leaves which provide the analogy, which is repeated many times in such names as Penny-cakes, Penny-caps, etc.

HALFSMART *Galium verum* (Lady's Bedstraw) Bucks. Grigson. Possibly from its use as a fleaweed (that name actually occurs in Suffolk). But see Polygonum hydropiper (arsesmart, which seems more likely to be the original of halfsmart).

HALFWOOD 1. *Clematis vitalba* (Old Man's Beard) Glos. J D Robertson. 2. *Lycium chinense* (Tea Tree) War. Bloom. 3. *Solanum dulcamara* (Woody Nightshade)

War. Palmer; Worcs. Grigson. The suggestion is that these are, though woody, only shrubs - half trees.

HALIWORT *Corydalis lutea* (Bulbous Fumitory) A S Palmer. i.e. Holy Wort, which itself is actually Holewort, or Hollow-wort.

HALL-NUT *Corylus avellana* nut (Hazel nut) Corn. Jago. There is a 'hazel' form which gives 'haul', too.

HALLELUJAH *Oxalis acetosella* (Wood Sorrel) Dev. Friend; Som, Wales B & H. There is a legend that the colour of the flowers were from drops of blood shed by Christ, and the plant was often placed in the foreground of Italian paintings of the Crucifixion, especially by Fra Angelico. It is called Alleluia in Italy, too. But it is more probable that the time of blooming (about Easter) accounts for the various renderings of Alleluia.

HALSE 1. *Corylus avellana* (Hazel) Dev, Som, Ire. Grigson. 2. *Ulmus glabra* (Wych Elm) Som. Grigson. This is a true hazel name, derived from a familiar consonantal misplacement of hazel. The Wych Elm only gets it by confusion with Witch Hazel.

HALSE, Nut *Corylus avellana* (Hazel) Dev. Choape.

HALSE, Witch 1. *Corylus avellana* (Hazel) Corn. Grigson. 2. *Ulmus glabra* (Wych Elm) Som. Elworthy.

HALSE, Wych *Ulmus glabra* (Wych Elm) Corn, Som. Grigson.

HALSE-BUSH *Alnus glutinosa* (Alder) Dev. Grigson. Cf. Alls-bush, also from Devonshire. This has a different derivation from that halse used for hazel, through Allen,a common name for alder.

HALVE *Crataegus monogyna* haws (Haws) Dev. Friend.1882. This version of haw or hag also occurs as Harve, Haves, Howes, or even Half-and-half in Somerset.

HAMBURG PARSLEY *Petroselinum fusiformis*.

HAMMERWORT *Parietaria diffusa* (Pellitory-of-the-wall). B & H.

HANCOCK'S CURSE *Polygonum cuspidatum* (Japanese Knotweed). A name quoted in a House of Lords answer to a question in 1989. A "curse" because it is a "pestilential plant", virtually impossible to eradicate.

HAND, Black Mary's *Dactylorchis maculata* (Heath Spotted Orchid) Emboden.1974. See Dead Man's Hand. The root is black.

HAND, Blind Man's *Ajuga reptans* (Bugle) Hants. Grigson.

HAND, Bloody Man's *Orchis mascula* (Early Purple Orchis) Kent B & H.

HAND, Dead Man's 1. *Dactylorchis incarnata* (Early Marsh Orchid) Border. B & H. 2. *Dactylorchis maculata* (Heath Spotted Orchid) Berw. Grigson. 3. *Orchis mascula* (Early Purple Orchis) Glos, War, Suss. B & H; Hants. Cope. "Hand" for Dactylorchis seems appropriate. Both names come from the fact that, whereas Orchis has twin tubers, Dactylorchis has finger-like roots. "Dead Man", or "Bloody Man", because of the purple spots on the leaves.

HAND, Devil's *Dactylorchis maculata* (Heath Spotted Orchid) Emboden.1974. Cf. Black Mary's Hand for this. It is because the tubers are so dark coloured that they are given these names.

HAND, Lady's *Briza media* (Quaking Grass) N'thum. Grigson.1959.

HAND, Mary's, Black *Dactylorchis maculata* (Heath Spotted Orchid) Emboden.1974. Black, because of the very dark tubers.

HAND OF GOD *Lapsana communis* (Nipplewort) Wilts. Powell.

HAND ORCHIS; HAND SATYRION *Dactylorchis maculata* (Heath Spotted Orchis) Grieve.1932 (Hand Orchis); Turner (Hand Satyrion). The "dactyl" part of the generic name, and the various "hand" or "finger" names, refer to the fact that the tubers are divided into several finger-like lobes.

HANDED ORCHID, Female *Dactylorchis maculata* (Heath Spotted Orchid) Jamieson. See Hand Orchis. 'Female', because it does not have the obvious male characteristic of twin tubers.

HANDFLOWER *Cheiranthus cheiri*

(Wallflower) Fernie. Cheiranthus means handflower. In the Middle Ages, wallflowers were carried in the hand at festivals; in other words, nosegays were habitually made of them - as Parkinson said, "the sweetness of the flowers causes them to be used in nosegays".

HANDKERCHIEF, Lace, Queen Anne's 1. *Anthriscus sylvestris* (Cow Parsley) Dor. Grigson. 2. *Torilis anthriscus* (Hedge Parsley) Dor. Dacombe. Descriptive - all the white umbellifers have "lace" or "needlework" names.

HANDKERCHIEF TREE; HANDKERCHIEF TREE, Pocket *Davidia involucrata*. Hay & Synge call it Pocket Handkerchief Tree. The name comes from the appearance of the conspicuous white bracts.

HANDS, Scabby 1. *Anthriscus sylvestris* (Cow Parsley) Som. Grigson. 2. *Conium maculatum* (Hemlock) Som. Macmillan. 3. *Conopodium majus* (Earthnut) Cumb. Grigson. 4. *Heracleum sphondyllium* (Hogweed) Som. Grigson. Headington children dislike Cow Parsley, and believe that if they touch it, they will get sore hands. Something similar must account for the Somerset name. Earthnut has Lousy Arnut as one of its names - if you eat too many of the Lousy Arnuts, lice will crowd into your hair. There must be something similar here, too, to account for Scabby Hands.

HANDS-IN-POCKETS *Ampelopsis quinquefolia* (Virginia Creeper) Som. Macmillan. It is the petioles of the leaves which are so called, because children use them to whip the knuckles of others, when they say "Hands in pockets".

HANGDOWNS *Malus sylvestris* fruit (Crabapples) Som. Macmillan.

HANGING GERANIUM *Saxifraga sarmentosa* (Mother-of-thousands) Wilts. Dartnell & Goddard. Because that's the way they always used to be seen in cottage windows.

HANNAH, Aunt *Arabis alpina* (Alpine Rock-cress) E.Ang. Wright.

HANSON LILY *Lilium hansoni*. Hanson was an amateur lily grower in Brooklyn.

HAPS *Rosa canina* fruit (Hips) N.Scot. B & H.

HARBER; HARBUR *Carpinus betula* (Hornbeam) Norf, Suff. B & H. Hard Beam, presumably, which is what Gerard called it.

HARBS; HERBS *Crataegus monogyna* fruit (Haws) Ess. Gepp. Cf. Harves, or Harvies, also from Essex.

HARCHER *Cheiranthus cheiri* (Wallflower) Dor. Macmillan.

HARD BEECH *Nothofagus truncata*.

HARD-GRASS, Curved *Parapholis incurva*.

HARD-GRASS, Sea *Parapholis strigosus*.

HARDBACK *Collinsonia canadensis* (Stone Root) Le Strange.

HARDBEAM *Carpinus betula* (Hornbeam) Gerard. Hornbeam means the tree whose wood is tough like horn; Hardbeam conveys the same impression.

HARDEWES *Cichorium intybus* (Chicory) Turner.

HARDHACK 1. *Centaurea nigra* (Knapweed) Wilts. Macmillan. 2. *Spiraea douglasii* USA E Gunther. 3. *Spiraea tomentosa*.

HARDHAY 1. *Hypericum perforatum* (St John's Wort) Clair. 2. *Hypericum tetrapterum* (St Peter's Wort) Gerard.

HARDHEAD 1. *Achillea ptarmica* (Sneezewort) Ayr. Grigson. 2. *Agrostemma githago* (Corn Cockle) N'thum. Grigson. 3. *Centaurea nigra* (Knapweed) Dev. Friend.1882. 4. *Centaurea picris* (Russian Knapweed) S Africa Watt. 5. *Centaurea scabiosa* (Greater Knapweed) Glos. J D. Robertson. 6. *Heracleum sphondyllium* (Hogweed) Glos. B & H. 7. *Lolium temulentum* (Darnel) Som. Macmillan. 8. *Plantago lanceolata* (Ribwort Plantain) Dev. Macmillan; Lancs. Denham/Hardy. 9. *Plantago major* (Great Plantain) Dev, Dor, Wilts, Worcs, Lancs, Yks. Grigson. 10. *Sanguisorba officinalis* (Great Burnet) Heref. Havergal. 11. *Succisa pratensis* (Devil's-bit) Lancs. Nodal. Descriptive, more obvious in the case of the knapweeds and devil's-bit than some of the others. But almost certainly darnel and the plantains would be used in some kind of divination game that involved knocking off the heads.

HARDHEAD, French *Centaurea jacea*.
HARDHEAD, Horse *Centaurea nigra* (Knapweed) B & H.
HARDHEAD, Slender *Centaurea nemoralis*.
HARDHEAD, Woolly *Succisa pratensis* (Devil's-bit) B & H.
HARDHEAD KNAPWEED *Centaurea nigra* (Knapweed) Phillips.
HARDHOW *Calendula officinalis* (Marigold) B & H.
HARDINE 1. *Centaurea nigra* (Knapweed) Ches, Lancs, Notts, Staffs. Grigson. 2. *Ranunculus arvensis* (Corn Buttercup) Midl. Grigson.
HARDING GRASS *Phalaris aquatica*.
HARDIRON 1. *Atriplex patula* (Orache) Leics. A B Evans. 2. *Centaurea nigra* (Knapweed) Ches, Lancs, Notts, Staffs. Grigson. 3. *Ranunculus arvensis* (Corn Buttercup) Midl. Grigson. Different attributions - in the case of orache, it is the root which is hard as iron (Cf. Iron-root). But in the case of the others, it is the buds.
HARDLEAF GOLDEN ROD *Solidago rigida*.
HARDOCK; HARDOKE *Arctium lappa* (Burdock) Prior; USA Henkel (Hardock); Grigson.1974.
HARDWAY *Hypericum tetrapterum* (St Peter's Wort) Tynan & Maitland. Cf. Hardhay. Is this genuine, or just a misprint?
HARE FIGWORT *Scrophularia lanceolata*.
HARE'S BALLOCKS *Himantoglossum hircinum* (Lizard Orchid) Lyte. The twin tubers inevitably call to mind a pair of testicles. They are assigned to goats as well as to hares.
HARE'S BEARD *Verbascum thapsus* (Mullein) Gerard; USA Henkel.
HARE'S BELLS, English *Endymion non-scriptus* (Bluebell) Parkinson. 'English', presumably, because the Scottish bluebell is *Campanula rotundifolia*.

HARE'S COLEWORT *Sonchus oleraceus* (Sow Thistle) B & H. One of the old Latin names was *Brassica leporina*. This is a direct translation.
HARE'S EAR 1. *Bupleurum rotundifolium*. 2. *Conringia orientalis* (Hare's Ear Cabbage) Prior. 'Ear' names usually have reference to the shape of the leaves.
HARE'S EAR, Narrow *Bupleurum opacum*.
HARE'S EAR, Shrubby *Bupleurum fruticosum*.
HARE'S EAR, Sickle *Bupleurum falcatum*.
HARE'S EAR, Slender *Bupleurum tenuissimum*.
HARE'S EAR, Smallest *Bupleurum tenuissimum* (Slender Hare's Ear) Clapham.
HARE'S EAR CABBAGE; HARE'S EAR MUSTARD *Conringia orientalis* USA Allan (Mustard).
HARE'S EYE *Melandrium dioicum* (Red Campion) B & H.
HARE'S FOOT CLOVER *Trifolium arvense* (Hare's Foot Trefoil) Prior.
HARE'S FOOT FERN *Davallia canariensis*.
HARE'S FOOT PLANTAIN *Plantago lagopus*.
HARE'S FOOT TREFOIL *Trifolium arvense*. It is leporis pes in Apuleius (Earle).
HARE'S LETTUCE 1. *Mycelis muralis* (Wall Lettuce) Cockayne. 2. *Sonchus oleraceus* (Sow Thistle) Henslow. The inclusion of Wall Lettuce here is probably the result of an incorrect reading. Sow Thistle has an alternative traditional association - that with the hare. An old legend tells how this plant gave strength to hares when they were overcome with the heat. From the AS Herbal, "of this wort, it is said that the hare, when in summer by the vehement heat he is tired, doctors himself with this wort." The Grete Herball has a garbled version of this - "if the hare come under it, he is sure that no beast can touch hym".
HARE'S MEAT *Oxalis acetosella* (Wood Sorrel) Corn. Jago Som. Macmillan. Better known as cuckoo's meat, but foxes and rabbits eat it, apparently, as well as hares.

HARE'S PALACE; HARE'S THISTLE *Sonchus oleraceus* (Sow Thistle) Dyer. See Hare's Lettuce.

HARE'S PARSLEY 1. *Anthriscus sylvestris* (Cow Parsley) Som, Wilts. Grigson. 2. *Conium maculatum* (Hemlock) Som. Macmillan.

HARE'S TAIL 1. *Eriophorum angustifolium* (Cotton-grass) Som. Macmillan. 2. *Eriophorum vaginatum*.

HAREBELL 1. *Campanula rotundifolia*. 2. *Digitalis purpurea* (Foxglove) Ire. Grigson. 3. *Endymion nonscriptus* (Bluebell) Dev. Friend.1882. 'Hair', really, for this is OE hara, hairy, a reference to the thin stalks. Bluebell and harebell seem to be interchangeable in the west country.

HAREBELL, Australian *Wahlenbergia gracilis*.

HAREBELL, Californian *Campanula prenanthoides*.

HAREBELL, English *Endymion nonscriptus* (Bluebell) Gerard. Scottish bluebell is the harebell, so why not call the bluebell English harebell?

HAREBELL, One-flowered *Campanula uniflora* (One-flowered Bellflower) USA Elmore.

HAREBELL POPPY *Meconopsis quintuplinerva*. The colour probably acounts for the 'harebell'.

HAREBOTTLE *Centaurea nigra* (Knapweed) B & H. The word occurs also as Tarbottle, or Darbottle. 'Bottle' itself means a bundle (of hay, etc), a reference to the tightly packed composite arrangement of the flowers. Bluebottle is the best known name incorporating this. 'Hare' obviously has nothing to do with the animal. It is probably the same word that helped produce Hardheads and its derivatives.

HAREBURR *Arctium lappa* (Burdock) B & H. This name is recorded as Hurr Burr in some places, but the connection is with Hardock.

HAREFOOT 1. *Geum urbanum* (Avens) B & H. 2. *Trifolium arvense* (Hare's Foot Trefoil) Turner.

HARENUT *Conopodium majus* (Earthnut) Dor, Lancs, Ire. Grigson; Yks. Halliwell. This is Earthnut, via forms such as Arnut. William Barnes had this down as Heare-nut, which is the way it must be pronounced.

HARESTAIL *Eriophorum vaginatum*. See Hare's Tail.

HARESTONES *Orchis mascula* (Early Purple Orchis) Fluckiger. They are Dogstones, Foxstones, Goatstones, as well as Bull's Bags.

HARESTRONG; HARSTRONG *Peucedanum officinale* (Hog's Fennel) Barton & Castle (Harestrong); B & H (Harstrong). Apparently from the German Haarstrang, which would mean a hank or rope of hair. The name appears as Horestrong in Gerard, who also has Horestrange, or, as in Culpepper, Hoarstrong.

HAREWOOD *Acer pseudo-platanus* (Sycamore) Nickerson. See Hairwood.

HAREWORT *Lepidium latifolium* (Dittander) Storms.

HARICOT BEAN *Phaseolus vulgaris* (Kidney Bean) Schery. Possibly from the Nahuatl ayacotl? At least, the poet Heredia said it was. It was certainly in use as a French word by 1640.

HARIF; HARRIFF; HARITCH; HAROFE *Galium aparine* (Goose-grass) Prior (Harif, Haritch); Fernie (Harriff); Halliwell (Harofe).

HARLEQUIN, Rock *Corydalis sempervirens* (Pink Corydalis) USA Cunningham.

HARLEQUIN FLOWER *Sparaxis tricolor* S Africa Perry.

HARLOCK 1. *Arctium lappa* (Burdock) W Miller. 2. *Sinapis arvensis* (Charlock) Ess. Grigson. Hardock, or something like it, is more usual for the burdock. For charlock, harlock is simply one of a large number of variants on the usual name. They go through such as Callock, or Kellock, Kedlock, Careluck, to Warlock, with a dozen or so stops on the way.

HARMAL; HARMEL *Peganum harmala* (African Rue) Chopra. These are from the

Arabic, as is the specific name.

HARP, David's *Polygonatum multiflorum* (Solomon's Seal) B & H. Descriptive, in some way, presumably, but there is a tendency to keep the ascription to close biblical terms.

HARPING JOHNNY *Sedum telephium* (Orpine) E.Ang. Grigson. Not immediately obvious, but the first element is in fact a corruption of orpine. An intermediate form, Orphan John, makes it clear.

HARROW, Rest; HARROW, Wrest; HARROW-REST *Ononis repens*. Rest Harrow is the common form. Murdoch McNeill has Wrest Harrow, which makes the meaning clearer. Harrow-rest is a Lincolnshire variant, quoted by Grigson. 'Rest' is actually 'arrest' - the name was brought in from remore aratri - plough hindrance. There was also arest bovis, "because it maketh the Oxen whilst they be in plowing to rest or stand still" (Gerard). The proper English name is Cammock, but this one is picturesque enough, and is a reference to the plant's toughness. It was difficult to get out of arable land with primitive tools.

HARRUL *Alnus glutinosa* (Alder) Glos. J D Robertson. This must be a local version of Arl, which has a wide distribution.

HARRY, Old Uncle *Artemisia vulgaris* (Mugwort) Som. Grigson. Possibly connected with Old Man, a southernwood name?

HARRY DOBS *Dianthus caryophyllus* (Carnation) Som. Macmillan.

HARRY NETTLE *Stachys officinalis* (Wood Betony) Dor. Macmillan. Hairy nettle, perhaps?

HARSHWEED *Centaurea scabiosa* (Greater Knapweed) Hill.

HARSHWEED, Knapweed *Centaurea jacea* (French Hardhead) Grieve.1931.

HARSY *Crataegus monogyna* haws (Haws) Ess. Grigson.

HART CLAVER *Melilotus officinalis* (Melilot) N Eng. F Grose.

HART MINT *Mentha spicata* (Spearmint) Parkinson.

HART'S CLOVER 1. *Melilotus altissima* (Tall Melilot) N'thum, Yks. Gerard. 2. *Melilotus officinalis* (Melilot) Yks. B & H.

HART'S TREE *Melilotus officinalis* (Melilot) Leyel.1937.

HARTBERRY *Vaccinium myrtillus* (Whortleberry) Som, Dor. Macmillan. 'Hart' here is a variant of 'whort'. It appears in a number of forms, of which 'hurts' is the best known.

HARTS *Vaccinium myrtillus* (Whortleberry) Wilts. Balch. See Hartberry above.

HARTSEASE *Viola tricolor* (Pansy) Langham. See Heartease, rather.

HARTSHORN 1. *Coronopus squamatus* (Wart-cress) T Hill. 2. *Plantago coronopus* (Buck's Horn Plantain) Turner. 3. *Rhamnus cathartica* (Buckthorn) Lloyd. Buck's Horn Plantain gets the name on account of the deeply cut leaves. 'Buck' and 'hart' are more or less interchangeable. Hartshorn for buckthorn shows confusion between 'horn' and 'thorn'. It appears more accurately as Hartsthorn, as well.

HARTSHORN PLANTAIN *Plantago coronopus* (Buck's Horn Plantain) B & H.

HARTSTHORN *Rhamnus cathartica* (Buckthorn) Fluckiger.

HARTSTONGUE *Phyllitis scolopendrium*.

HARTWORT *Tordylium maximum*.

HARVES; HARVIES *Crataegus monogyna* haws (Haws) Ess. Gepp. See also Haves, Halves, and Harsies.

HARVEST, Snow-in- 1. *Cerastium tomentosum* (Snow-in-summer) Som. Elworthy. Wilts. Dartnell & Goddard; Leics. B & H. 2. *Clematis vitalba* (Old Man's Beard) Som. Macmillan; N'thants. Grigson. 3. *Lobularia maritima* (Sweet Alison) Bucks. Harman; N'thants. B & H.

HARVEST BELLS 1. *Campanula rotundifolia* (Harebell) N'thants. Clare. 2. *Gentiana pneumonanthe* (Marsh Gentian) Gerard. Only because the gentian is an autumn-flowering species, sometimes known as Autumn Bells or Bellflower.

HARVEST DAISY *Leucanthemum vulgare* (Ox-eye Daisy) Dor. Macmillan.

HARVEST FLOWER 1. *Chrysanthemum segetum* (Corn Marigold) Som. Macmillan. 2. *Pulicaria dysenterica* (Yellow Fleabane) Corn. Grigson.
HARVEST LICE 1. *Agrimonia eupatoria* seed vessels (Agrimony) Hants. Cope. 2. *Galium aparine* seed vessels (Goose-Grass) Hants. Cope.
HARVEST LILY *Calystegia sepium* (Great Bindweed) Surr. Grigson.
HARYHOUND *Marrubium vulgare* (White Horehound) Norf. V G Hatfield.1994.
HASILL TREE *Corylus avellana* (Hazel) Scot. B & H.
HASKETT 1. *Acer campestre* (Field Maple) Dor. Macmillan. 2. *Corylus avellana* (Hazel) Dor. Macmillan.
HASPEN *Populus tremula* (Aspen) Glos. B & H.
HASKWORT 1. *Campanula latifolia* (Giant Bellflower) Prior. 2. *Campanula trachelium* (Nettle-leaved Bellflower) Gerard. Cf. Throatwort for both of these plants. They were used for the hask (inflamed trachea in modern terms), a word retained still in huskiness. Nettle-leaved Bellflower in particular secretes a yellow latex, its signature of its value against sore throat and tonsilitis.
HASSOCK *Juncus conglomeratus* (Compact Rush) Lancs. A Burton. The name is applied to reeds as well, or to any coarse grass. Only secondarily is it applied to things made with rushes, etc., i.e. the mats or cushions on which people kneel in church.
HASSOCK GRASS *Deschampsia cespitosa* (Tufted Hair-grass). The tufts give the name - they are hassock-like.
HASTATE KNOTGRASS *Polygonum arifolium* (Halberd-leaved Tear-. thumb) Grieve.1931.
HASTATE ORACHE *Atriplex hastata* (Halberd-leaved Orache) Darling & Boyd.
HASTY PEACH *Prunus armeniaca* (Apricot) Turner. Because apricot blooms and fruits much earlier than the peach. Cf. Precocious-tree, which derives from the Latin praecox, early. Indeed, the last syllable of apricot has the same derivation.
HASTY ROGER *Scrophularia nodosa* (Figwort) West. Grigson.
HAT, Bishop's *Epimedium alpinum* (Barrenwort) W Miller.
HAT, Fairy *Digitalis purpurea* (Foxglove) Dor. Macmillan.
HAT, Gypsy's *Convolvulus arvensis* (Field Bindweed) Som. Macmillan.
HAT, Monkey *Tropaeolum majus* (Nasturtium) Ches. Holland.
HAT, Penny *Umbilicus rupestris* (Wall Pennywort) Dev. Friend.1882.
HAT, Summer *Viola tricolor* (Pansy) Som. Macmillan.
HATBAND, Robin Hood's *Lycopodium inundatum* (Clubmoss) Halliwell.
HATCH-HORN *Quercus robur* acorn (Acorn) Lancs. B & H. Acorn is from Latin acer, field. The soft 'c' of the original is retained in this name, and similar forms, like Atchern or Achorn.
HATCHETS, Hooks-and-; HATCHETS-AND-BILLHOOKS *Acer campestre* seeds (Field Maple) Som. Macmillan (Hooks-and-hatchets); Wilts. Williams.1981 (Hatchets-and-billhooks).
HATHER *Calluna vulgaris* (Heather) Gerard.
HATPINS, Lady's *Knautia arvensis* (Field Scabious) Som. Macmillan.
HATTIES, Devil's *Silene maritima* (Sea Campion) Mor. PLNN22. An unlucky plant, hence the ascription to the devil.
HAUL *Corylus avellana* (Hazel) Som. B & H. Sometimes written as Hall. The word is hazel, which with consonantal misplacement becomes halse, and then hale, or hall, etc.
HAUSA POTATO *Coleus rotundifolius* (Frafra Potato) Douglas.
HAUTBOIS STRAWBERRY; HAUTBOY *Fragaria moschata*. The usual name is Hautboy. Clapham has Hautbois Strawberry. Hautbois means 'high wood', for *Fragaria moschata* lifts its fruit above the leaves, in contrast to the common strawberry.
HAUTH; HAWTH; HOTH *Ulex europaeus*

(Gorse) all Suss. Grigson (Hauth, Hoth); Fernie (Hawth).

HAV *Avena sativa* (Oats) Dev, Dor. Friend.1882. Cf. Havers for Wild Oat.

HAVELS; AVELS *Hordeum sativum* awns (Barley) E Angl. G E Evans.1956 (Havels); Forby (Avels).

HAVERDRIL *Narcissus pseudo-narcissus* (Daffodil) Ches. B & H.

HAVERS *Avena fatua* (Wild Oat) H C Long.1910. This is a north country word, and is the German Hafer. Both come originally from Old Norse hafrar.

HAVES 1. *Crataegus monogyna* haws (Haws) Dor. Macmillan. 2. *Rosa canina* hips (Hips) Dor. Barnes. This appears in various forms, usually variants of Halves, or Harves.

HAW 1. *Avena sativa* (Oat) Kent Friend.1882. 2. *Crataegus monogyna* (Hawthorn) Dumf, Selk. Grigson. 3. *Crataegus monogyna* fruit. 4. *Rosa canina* fruit (Hips) Dor. Grigson. Only in Scotland is the word used on its own for the tree. Haw is OE haga, hedge. Hips and haws have always been fairly interchangeable in the popular tongue - in fact, haws are sometimes called Hip-haws - a way of hedging one's bets.

HAW, Black 1. *Viburnum lentago* USA Gilmore. 2. *Viburnum prunifolium*. 3. *Viburnum rufidulum* USA Harper.

HAW, Bull *Crataegus monogyna* fruit (Haw) Yks. B & H.

HAW, Butter *Crataegus monogyna* fruit (Haw) Oxf, Norf. Grigson.

HAW, Cat *Crataegus monogyna* fruit (Haw) N'thants. A E Baker; Yks. Morris; Lincs, Cumb. B & H; Dur. F M T Palgrave.

HAW, Hag *Crataegus monogyna* fruit (Haw) Bucks. Harman. Both elements of this name mean 'hedge'.

HAW, Hip; HAW, Hipperty *Crataegus monogyna* fruit (Haw) Oxf, West. (Hip); Oxf, Shrop. (Hipperty) both Grigson.

HAW, Hog *Crataegus monogyna* fruit (Haw) Hants. Grigson. There may be 'pig' names for haws (in fact, Pig-haw is recorded), but it is probable that 'pig' is the result of folk etymology, and that 'hog' is OE haga, hedge.

HAW, Pig('s) *Crataegus monogyna* fruit (Haw) Som, Wilts, Hants. Grigson; Som. Macmillan (Pig's). See Hog haw.

HAW, Possum *Viburnum nudum* USA Harper.

HAW BUSH; HAW TREE *Crataegus monogyna* (Hawthorn) Dumf. Grigson (Bush); Glos, N'thum, Scot. Grigson (Tree).

HAW GAW *Crataegus monogyna* fruit (Haw) Surr. Grigson.

HAWAIIAN ARROWROOT *Tacca hawaiiensis*.

HAWAIIAN FALSE NETTLE *Boehmeria grandis*. Same family as nettle, and most of the genus are fibre plants, though this one is not of the quality of Ramie.

HAWBERRY 1. *Crataegus monogyna* fruit (Haw) Wessex Rogers; Ches. B & H. 2. *Crataegus rivularis* Br.Col. Teit.

HAWK-NUT *Conopodium majus* (Earthnut) Prior. There is a Scottish name for Earthnut, Hornecks, which sounds vaguely like this, but both of them, as well as Harenut, probably owe their origin to some form of Earthnut, or if not that, Hog-nut.

HAWK-YOUR-MOTHER'S-EYES-OUT *Veronica chamaedrys* (Germander Speedwell) Dor. Dacombe. There are other names like this, all from the west country. But the belief made explicit in the names is found all over England e.g. Yorkshire, where it is said that if a child gathers germander speedwell its mother will die during the year, or the Lincolnshire superstition that if anyone picks the flower, his eyes will be eaten. Cf. Blind-flower, from co Durham, where they used to say that if you look steadily at the flower for an hour, you will become blind. Of course, a descriptive 'eye' forms part of a very large number of names for the speedwell.

HAWK'S BEARD *Crepis sp.*

HAWK'S BILL BRAMBLE *Rubus fruticosus* (Bramble) E Ang. Grigson. It must be

the shape of the thorns that acocunt for the name.
HAWK'S FOOT *Aquilegia vulgaris* (Columbine) B & H. Cf. Cock's Foot and Dove's Foot. The spur is responsible for the imagery.
HAWKBERRY 1. *Prunus avium* (Wild Cherry) Stir. Grigson. 2. *Prunus padus* (Bird Cherry) Stir. Grigson.
HAWKBIT, Autumn *Leontodon autumnalis*.
HAWKBIT, Greater *Leontodon hispidus*.
HAWKBIT, Hairy *Leontodon taraxacoides*.
HAWKBIT, Rough *Leontodon hispidus* (Greater Hawkbit) Salisbury.1962.
HAWKWEED, Alpine *Hieraceum alpinum*.
HAWKWEED, Bushy *Hieraceum umbellatum* (Umbellate Hawkweed) Curtis. Hieraceum is Latin, from Greek hierakion, from hierax, hawk. Hawkweed, then, is a translation of Hieraceum. The name arose from the belief that hawks used the juice of the plant to clear their sight. This is from Pliny, although the name was originally given to some other, now unidentifiable, plant. Hare said that eagles ate it, too, varied with wild lettuce, to keep keen eyesight. Certainly, it was given to feed hawks in the old art of falconry. So hawkweed became the emblem of quick-sightedness, and so it was used in early medical practice as a remedy for dimness of sight.
HAWKWEED, Few-leaved *Hieraceum murorum* (Wall Hawkweed) Blamey.
HAWKWEED, Great *Picris hieracioides* (Hawkweed Ox-tongue) Turner.
HAWKWEED, Leafy *Hieraceum umbellatum* (Umbellate Hawkweed) Blamey.
HAWKWEED, Long-rooted *Hypochoeris radicata* (Long-rooted Cat's Ear) Curtis.
HAWKWEED, Mouse-ear *Hieraceum pilosellum*.
HAWKWEED, Northern *Hieraceum boreale* (Shrubby Hawkweed).
HAWKWEED, Orange *Hieraceum aurantiacum*.
HAWKWEED, Purple *Saussurea alpina*.
HAWKWEED, Rough *Hieraceum scabrum*.

HAWKWEED, Scarlet *Hieraceum aurantiacum* (Orange Hawkweed) Salisbury.1936.
HAWKWEED, Shaggy *Hieraceum villosum*.
HAWKWEED, Shrubby *Hieraceum boreale*.
HAWKWEED, Small-flowered *Hypochoeris glabra* (Smooth Cat's Ear) Curtis.
HAWKWEED, Umbellate *Hieraceum umbellatum*.
HAWKWEED, Wall *Hieraceum murorum*.
HAWKWEED GOWAN *Taraxacum officinale* (Dandelion) Scot. Pratt.
HAWKWEED OX-TONGUE *Picris hieracioides*.
HAWKWEED SAXIFRAGE *Saxifraga hieracifolia*.
HAWP *Rosa canina* fruit (Hip) N Scot. Grigson.
HAWTHORN *Crataegus monogyna*. Literally, hedge thorn, 'haw' being the OE haga, hedge.
HAWTHORN, American *Crataegus mollis* (Woolly Thorn) Douglas.
HAWTHORN, Black *Crataegus douglasii*.
HAWTHORN, Chinese 1. *Crataegus pentagyna*. 2. *Photinia serrulata*.
HAWTHORN, Cumberland *Sorbus aria* (Whitebeam) Gerard.
HAWTHORN, Downy *Crataegus mollis* (Woolly Thorn) USA Everett.
HAWTHORN, Evergreen *Pyracantha sp* Parkinson.
HAWTHORN, Japanese *Raphiolepis sp*.
HAWTHORN, May *Crataegus rufula* Alab. Harper.
HAWTHORN, Midland *Crataegus laevigata* (Woodland Hawthorn) Clapham.
HAWTHORN, Parsley *Crataegus apiifolia*.
HAWTHORN, Quebec *Crataegus submollis*.
HAWTHORN, River *Crataegus rivularis*.
HAWTHORN, Sugar *Crataegus spathulata*.
HAWTHORN, Thicket *Crataegus coccinea*.
HAWTHORN, Washington *Crataegus phaeonpyrum*.
HAWTHORN, Water *Aponogeton distachyum* (Cape Asparagus) RHS.
HAWTHORN, Woodland *Crataegus laevigata*.

HAWTHORN CURRANT TREE *Ribes oxyacanthoides* (Smooth Gooseberry) Notes and Queries, 1857.

HAWTHORN MAPLE *Acer crataegifolium*.

HAY, Black *Medicago lupulina* (Black Medick) Clapham. 'Hay' will usually indicate a fodder plant.

HAY, Burgundy *Medicago sativa var. sativa* (Lucerne) B & H.

HAY, French *Onobrychis sativa* (Sainfoin) Wilts. D Grose. Cf. French, or French Grass, for this.

HAY, Grecian 1. *Chelidonium majus* (Greater Celandine) Fernie. 2. *Trigonella ornithopodioides* (Fenugreek) Grieve.1931. Fenugreek is foenum graecum, Grecian hay, given apparently because it was used to scent inferior hay. Fernie gave both Fenugreek and Grecian Hay as names for Greater Celandine, but on what authority?

HAY, Hard *Hypericum quadrangulare* (St Peter's Wort) Gerard. Gerard wrote it as one word.

HAY, Holy *Medicago sativa var. sativa* (Lucerne) Prior. This is the result of the legend that dried hay was used in Christ's manger. It blossomed and grew to encircle the baby's head.

HAY, Indian *Cannabis sativa* (Hemp) USA Watt.

HAY, Ladies-in-the- *Asperula odorata* (Woodruff) Wilts. Grigson.

HAY, New-mown 1. *Asperula odorata* (Woodruff) Som, Notts. Grigson. 2. *Filipendula ulmaria* (Meadowsweet) Som. Grigson. Woodruff develops a smell of hay as it dries - so does Meadowsweet.

HAY, Newmade *Asperula odorata* (Woodruff) Wilts. Goddard.

HAY, Sweet *Filipendula ulmaria* (Meadowsweet) Suss. Grigson; Dor. Macmillan.

HAY BEECH 1. *Carpinus betula* (Hornbeam) Bucks. Harman. 2. *Fagus sylvatica* (Beech) B & H. There is a difference in the meaning of 'hay' now. When it means dried grass, it comes from OE hieg, but this version comes from OE haga, and means a hedge.

HAY GOB *Polygonum convolvulus* (Black Bindweed) War. Halliwell.

HAY HOA *Glechoma hederacea* (Ground Ivy) Hants. Cope. 'Hay' is hedge; 'hoa' is a version of 'hove' or hoof'. Alehoof is a well-known old name for Ground Ivy, and 'hoof' is the same word as 'heave'. So the whole means that which causes ale to ferment.

HAY PLANT *Asperula odorata* (Woodruff) N.Ire. Grigson. See New-mown Hay.

HAY RATTLE 1. *Rhinanthus crista-galli* (Yellow Rattle) Rackham. 2. *Silene cucubalis* (Bladder Campion) Dev. PLNN 17.

HAY ROPE *Clematis vitalba* (Old Man's Beard) Som. Grigson. Cf. Hag Rope. Both mean hedge rope.

HAY-SCENTED BUCKLER FERN *Dryopteris aemula*. Apparently, the hay scent from the dried leaves is not all that obvious.

HAY SHACKLE *Rhinanthus crista-galli* (Yellow Rattle) Som. Macmillan.

HAY SHAKERS *Briza media* (Quaking-grass) Ches. Grigson.1959.

HAYCOCKS, Buttered *Linaria vulgaris* (Toadflax) Yks. Grigson.

HAYHOUSE, HAYHOVE *Glechoma hederacea* (Ground Ivy) Fernie (Hayhouse); Dawson (Hayhove). The second part of these names is better known as 'hoof', as in Alehoof, and means that which causes beer to "heave", or ferment.

HAYMAIDS; HAYMAIDENS *Glechoma hederacea* (Ground Ivy) Glos. PLNN35; Prior (Haymaids); Dev, Dor. B & H; Som. Jennings (Haymaidens).

HAYRIFF 1. *Filipendula ulmaria* (Meadowsweet) Dev. Friend.1882; Shrop. Grigson. 2. *Galium aparine* (Goose-grass) Som. Macmillan; Worcs. J Salisbury. 3. *Polygonum convolvulus* (Black Bindweed) Dor. Macmillan.

HAYSEED, Greek *Trigonella ornithopodioides* (Fenugreek) Clair. Almost exactly what fenugreek means - foenum graecum - Greek hay.

HAYTHORN *Crataegus monogyna* (Hawthorn) T Hill. i.e. hedge thorn.

HAYVER, Heaver *Lolium perenne* (Ryegrass) Corn. Jago. Both parts of the name mean 'oats'. Cf. Havers for wild Oat - German Hafer. Or were these two different names?

HAZEL 1. *Corylus avellana*. 2. *Crataegus monogyna* fruit (Haws) Dev. Friend.1882. 3. *Lobaria pulmonaria* (Lungwort Crottle) Bolton. Hazel is OE haesel, which comes by a roundabout route from Latin corylus. Hazel is also applied to the typical light-brown colour, hence the ascription to the lichen, which is a dyestuff.

HAZEL, Beaked *Corylus rostrata*.

HAZEL, California *Corylus rostrata var. californica*.

HAZEL, Chinese *Corylus chinensis*.

HAZEL, Constantinople *Corylus colurna* (Turkish Hazel) Ablett.

HAZEL, Corkscrew *Corylus avellana var. contorta*.

HAZEL, Hog- *Crataegus monogyna* fruit (Haw) Surr, Suss, Kent Wakelin. The meaning becomes clearer when another form of the name - Hogazel - is considered. This is hog-gazel, gazel meaning a berry. So Hog-hazel actually means exactly the same as Pig-haw, which is the west country version of the name, except that 'hog' is not a pig at all, but a version of 'hedge', from OE haga. Pig-haw was coined when that fact was forgotten.

HAZEL, Turkish *Corylus colurna*.

HAZEL, Wild *Simmondsia chinensis* (Jajoba) USA T H Kearney.

HAZEL, Witch 1. *Carpinus betula* (Hornbeam) Gerard. 2. *Hamamelis virginica*. 3. *Sorbus aucuparia* (Rowan) Yks. Denham/Hardy. 4. *Ulmus glabra* (Wych Elm) B & H. 'Witch' here is a popular misreading - the word is 'wych', which means pliant, or bending. Twigs of Hamamelis virginica were used as a divining rod, just as those of hazel were in England. As divining was looked on in America as the result of occult power, the name Witch Hazel was given. As far as rowan is concerned, the word derives from OE cuic, alive, and only by corruption did it apply, through one of the versions of Quickbeam, probably Wicken. The misreading appears again in Witch-elm.

HAZEL, Witch, Japanese *Hamamelis japonica*.

HAZEL, Winter *Corylopsis sp*.

HAZEL, Wych 1. *Carpinus betula* (Hornbeam) Ess. Grigson. 2. *Ulmus glabra* (Wych Elm) Dev, Som, Wilts, Worcs, Ches. Grigson. The word 'wych' is applied to trees and shrubs which have pliant branches - OE wice.

HAZEL ALDER *Alnus rugosa* (Smooth Alder) Youngken.

HAZEL CROTTLE; HAZEL-RAW *Lobaria pulmonaria* (Lungwort Crottle) S M Robertson (Hazel Crottle); N Ire. Bolton (Hazel-raw). The colour is the reference here.

HAZEL NUT, Australian *Macadamia ternifolia*. They look like hazels.

HAZEL PINE *Liquidambar styraciflua* - sapwood timber (Sweet Gum) Wilkinson.1981. The sapwood used to be sold in Europe under this name.

HAZELNUT, Snapping *Hamamelis virginica* (Witch Hazel) Grieve.1931. Because of the explosive way it ejects its seeds from the nuts.

HAZELWORT *Asarum europaeum* (Asarabacca) Gerard.

HAZLE *Crataegus monogyna* fruit (Haw) Dev. Grigson. Cf. Hog-hazel, Hogazel.

HAZZLE *Corylus avellana* (Hazel) Ches. Holland.

HE-BARFOOT *Helleborus foetidus* (Stinking Hellebore) Grigson. She-Barfoot is the Green Hellebore, H viridis. Barfoot is bear's-foot, a reference to the digitale form of the leaves.

HE-BRIMMEL *Rubus fruticosus* (Bramble) Som. Grigson. Halliwell defined He-brimmel as a bramble of more than one year's growth.

HE-BROOM *Laburnum anagyroides* (Laburnum) Fife B & H. In all probability,

High Broom is meant, to distinguish it from Low Broom, which is *Genista tinctoria*.

HE-BULKISHAWN *Senecio jacobaea* (Ragwort) Ire. Crooke. 'Bulkishawn' is another version of the names stemming from Gaelic buaghallan. She-Bulkishawn is tansy.

HE-HEATHER *Calluna vulgaris* (Heather) Border Grigson. She-heather would be the Cross-leaved Heath, *Erica tetralix*, or *Erica cinerea*, Bell Heather. Why the differentiation?

HE-HOLLY *Ilex aquifolium* (Holly). Reserved for the prickly variety; the smooth-leaved kind is always She-holly.

HE-YEW *Juniperus sabina* (Savin) E Ang. G E Evans.

HEAD, Blue *Succisa pratensis* (Devil's Bit) Shrop. B & H.

HEAD, Bull; HEAD, Bully *Centaurea nigra* (Knapweed) Som. Grigson. Not 'bull', but 'boll', an OE word which meant any circular body. Cf. Ballweed, Belweed, Bollwood, Bow-weed, etc., all from the same source.

HEAD, Bull's *Trapa bicornuta*. It gets the name because the nut has only two horns as against the four of Water Chestnut, *Trapa natans*.

HEAD, Cat's *Salix sp* catkins (Willow catkins) Ess, Cambs, Norf. B & H.

HEAD, Churl *Centaurea scabiosa* (Greater Knapweed) Tynan & Maitland.

HEAD, Churl's *Centaurea nigra* (Knapweed) Prior.

HEAD, Cock's 1. *Antirrhinum majus* (Snapdragon) Tynan & Maitland. 2. *Onobrychis sativa* (Sainfoin) Som. Macmillan. 3. *Papaver argemone* (Rough Long-headed Poppy) Scot. B & H. 4. *Papaver dubium* (Smooth Long-headed Poppy) Scot. B & H. 5. *Papaver rhoeas* (Red Poppy) Scot. Grigson. 6. *Plantago lanceolata* (Ribwort) Som, Suss, E Ang. Grigson. 7. *Trifolium pratense* (Red Clover) Gerard. Ribwort is the odd one out, for the rest have a red top to them to justify the name. 'Cock' in these cases is probably the Celtic coch, red.

HEAD, Cock's, Purple *Astragalus danicus* (Purple Milk Vetch) B & H.

HEAD, Cod's *Trifolium pratense* (Red Clover) Tynan & Maitland. Cock's Head, obviously. Is this genuine?

HEAD, Devil's 1. *Linaria vulgaris* (Toadflax) Le Strange. 2. *Plantago lanceolata* (Ribwort) Dyer.

HEAD, Dragon's *Antirrhinum majus* (Snapdragon) Som. Macmillan.

HEAD, Drooping *Galanthus nivalis* (Snowdrop) Som. Macmillan.

HEAD, Drummer *Centaurea nigra* (Knapweed) Som. Macmillan. Drumsticks, from the same area, makes better sense.

HEAD, Elephant *Pedicularis groenlandica*.

HEAD, Gorgon's *Euphorbia gorgonis*.

HEAD, Hairy *Centaurea nigra* (Knapweed) Som. Macmillan.

HEAD, Hound's *Misopates orontium* (Lesser Snapdragon) Cockayne.

HEAD, Iron *Centaurea nigra* (Knapweed) Prior. A variation of the more usual Hardheads, but this form existed in OE as isonheerde, and there are also Iron-knobs and Ironweed, as well as the hybrid Iron-hard.

HEAD, Lamb's, Yellow *Tropaeolum majus* (Nasturtium) Fernie.

HEAD, Leopard's *Fritillaria meleagris* (Snake's-head Lily) Leyel.1948. 'Leopard' here is 'leper'. Cf. Leopard Lily.

HEAD, Maiden's *Sanguisorba officinalis* (Great Burnet) Yks. Grigson.

HEAD, Nigger's 1. *Plantago lanceolata* (Ribwort) Dev. Grigson. 2. *Plantago media* (Lamb's Tongue Plantain) Dev. Macmillan. Cf. Black Plantain and many other names like Blackie Tops for Ribwort. The name is quite out of place for Lamb's Tongue Plantain, and is probably an error.

HEAD, Parrot *Sarracenia psittacina*.

HEAD, Pheasant's *Fritillaria meleagris* (Snake's-head Lily) Leyel.1948.

HEAD, Red *Sanguisorba officinalis* (Great Burnet) Yks. Grigson.

HEAD, Scabby 1. *Anthriscus sylvestris* (Cow Parsley) Oxf. Oxf.Annl.Rec.1951. 2. *Torilis* *japonica* (Hedge Parsley) Ches. Holland. Scabs, Scab-flower and Scabby Hands occur too for Cow Parsley. As far as the Oxfordshire record is concerned, it is said that Headington children disliked Cow Parsley, and believed that if they touched it they would get sore hands.

HEAD, Snake's 1. *Potentilla erecta* (Tormentil) Wilts. Dartnell & Goddard. 2. *Verbascum thapsus* (Mullein) Dor. Grigson. Mullein is Snake's Flower in Wiltshire, too. In this area children were told that a snake might be hiding under the leaves. That is quite irrelevant as far as Tormentil is concerned, and there is no obvious reason for the name.

HEAD, Sticky *Grindelia squarrosa* (Gumweed) Gilmore. Gumweed, Tarweed, and Sticky-head - all for the same reason.

HEAD, Toad's *Fritillaria meleagris* (Snake's Head Lily) Wilts. Dartnell & Goddard.

HEAD, Turk's 1. *Fritillaria meleagris* (Snake's Head Lily) War. Grigson. 2. *Lilium tigrinum* (Tiger Lily) Som. Macmillan.

HEAD, Woolly *Anemone nemorosa* (Wood Anemone) Som. Macmillan.

HEAD, Yellow *Senecio vulgaris* (Groundsel) Som. Macmillan.

HEAD BETONY *Pedicularis canadensis* USA House. 'Head' is probably a mishearing of 'wood' in this case.

HEAD LETTUCE *Lactuca sativa var. capitata* (Cabbage Lettuce) Brouk. Because this is grown to come to a head, like a cabbage, hence *var. capitata*.

HEADACHE 1. *Cardamina pratensis* (Lady's Smock) Glos, Cumb. Grigson. 2. *Geranium robertianum* (Herb Robert) Som. Macmillan; Surr. Tynan & Maitland. 3. *Papaver dubium* Ches. Holland. 4. *Papaver rhoeas* (Red Poppy) Som, Ches, Derb, Leics, Rut, Notts, Cumb, Ire. Grigson; Suss. Parish; N'thants. Sternberg; Lincs. Peacock; Yks. Nicholson; E Ang. Forby. 5. *Stellaria holostea* (Greater Stitchwort) Cumb. Grigson. It is really red poppy that the name concerns, not the others. They were used to cure headache, but the underlying folklore behind names like this is that they cause it. John Clare:

Corn poppys that in crimson dwell
Calld 'head achs' from their sickly smell.

In Wiltshire, they said that if you picked poppies from the corn, you would either have a bad headache or there would be thunder and lightning (hence the 'thunder' names. Cf. too Ear-ache).

HEADACHE TREE *Umbellularia californica* (Californian Laurel) Hora. The strong scent given off by the tree is reputed to cause headaches in those who sit under it too long.

HEADACHER *Papaver rhoeas* (Red Poppy) N'thants. A E Baker. See Headache.

HEADMAN *Plantago lanceolata* (Ribwort) Perth B & H. Headman, in two senses, firstly in the game of "soldiers", when one flower stalk is used to knock off the head of an opponent. Secondly, as the winner of the game.

HEADRIDGE *Sinapis arvensis* (Charlock) Pemb. Grigson.

HEADWARKE; HEADWORK *Papaver rhoeas* (Red Poppy) Fernie (Headwarke); Grose (Headwork). i.e. headache.

HEAL-ALL 1. *Arum maculatum* (Cuckoo-pint) Cambs. Porter. 2. *Collinsonia canadensis* (Stone Root) Le Strange. 3. *Prunella vulgaris* (Self-heal) USA House. 4. *Sedum telephium* (Orpine) N Ire. Grigson. 5. *Valeriana officinalis* (Valerian) Corn, Oxf. Grigson. Self-heal and valerian are certainly candidates for a name like heal-all, for both were very popular in folk medicine. Orpine looks unlikely until one considers a passage in the Anglo-Saxon version of Dioskorides in which it is said that it "healeth manifold infirmities of the body". A healing ointment made from orpine is still available commercially. But cuckoo-pint seems extremely unlikely to warrant this name.

HEAL-BITE; HEAL-DOG *Lobularia maritima* (Sweet Alison) Friend. Alyssum, and so

Alison, is Greek a, not and lussa, madness. Cf. Madwort for this plant, which does seem to have been recommended as a remedy against dogs' bites both in England and in Ireland.

HEALING BLADE 1. *Plantago major* (Great Plantain) Scot. Denham/Hardy. 2. *Sempervivum tectorum* (Houseleek) Clack. Grigson. Plantain in particular was a favourite wound plant, and has been used since ancient times to stop bleeding.

HEALING BUSH, Eight-day *Lobostemon fruticosus.*

HEALING LEAF 1. *Plantago major* (Great Plantain) Scot. Denham/Hardy. 2. *Sedum telephium* (Orpine) Scot. Grigson. 3. *Sempervivum tectorum* (Houseleek) Clack. Grigson. See Heal-all, Healing Blade.

HEARE-NUT *Conopodium majus* (Earthnut) Dor. Barnes. Harenut is more usual, but this is presumably the way it is pronounced. Either form probably owes its existence to earth-nut originally.

HEART, Bleeding 1. *Cheiranthus cheiri* (Wallflower) Dev. Friend; Wilts. Akerman; Glos. J D Robertson; Berks. Lowsley. 2. *Dicentra spectabilis.* 3. *Viola tricolor* (Pansy) Hants. B & H. Fancifully descriptive, as far as *Dicentra spectabilis* is concerned, but the other two are heart medicines, pansy mistakenly, for the name and virtue probably belong to some other plant. Perhaps it is wallflower, which also has been given the name Heartsease.

HEART, Bleeding, Fringed *Dicentra eximea.*

HEART, Bleeding, Western *Dicentra formosa.*

HEART, Broken *Clerodendron thomsoniae* (Bleeding Heart Vine) Howes.

HEART, Bullock's *Annona reticulata* (Custard Apple) A W Smith.

HEART, Lion's *Physostegia virginica* (Dragonhead) USA House.

HEART, Mother's *Capsella bursa-pastoris* (Shepherd's Purse) Border Dyer. Pick-your-mother's-heart-out is the full version. Probably from the heart-shaped capsules. Children used to play a game with the seed pod. Dyer described it as holding the capsule out to their companions, inviting them to take "take a hand o'that". It immediately cracks, and then follows a triumphant shout "You've broken your mother's heart". But there is another Scottish name, Mother-die, given too to Cow Parsley. Both plants are unlucky, and both have the similar superstition that if you give them to your mother, she will die.

HEART, Our Lady's *Dicentra spectabilis* (Bleeding Heart) Som. Macmillan.

HEART, Rosy *Dicentra spectabilis* (Bleeding Heart) Som. Macmillan.

HEART, Sea *Entada phaseoloides* (Nicker Bean) Hebr. Duncan.

HEART CHERRY, Wild *Prunus avium* (Wild Cherry) Camden/Gibson.

HEART CLAVER; HEART CLOVER; HEART TREFOIL *Medicago arabica* (Spotted Medick) Curtis (Claver); B & H (Clover, Trefoil)). Cf. Heart Medick. Coles says it was so called "not only because the leaf is triangular like the heart of a man, but also because each leafe doth contain the perfection (or image) of an heart, and that in its proper colour, viz a flesh colour. It defendeth the heart against the noisome vapour of the spleen".

HEART FEVER GRASS *Taraxacum officinale* (Dandelion) Donegal Black. The leaves used to be eaten there as a cure for "heart-fever", whatever that may be. Three leaves had to be eaten on successive mornings, pointing to a magical rather than to a practical cure.

HEART-LEAF 1. *Hepatica acutiloba* (Heart Liverleaf) Carolina Bergen. 2. *Hepatica triloba* (American Liverwort) Carolina Bergen. 3. *Philodendron cordatum.* For in the Carolina mountains, the liverwort is used as a love-philtre. In the case of the philodendron, it is the shape of the leaf that gives the name.

HEART-LEAVED ASTER *Aster cordifolius.*

HEART-LEAVED TWAYBLADE *Listera cordata.*

HEART-LEAVED WILLOW *Salix cordata.*

HEART-LIVER *Medicago arabica* (Spotted Medick) B & H. This is a corruption of Heart Clover.
HEART LIVERLEAF *Hepatica acutiloba*.
HEART MEDICK *Medicago arabica* (Spotted Medick) Curtis. See Heart Clover.
HEART NUT *Juglans ailantifolia var. cordiformis*.
HEART-O'-THE-EARTH *Prunella vulgaris* (Self-heal) E Ang. Grigson; Border, Scot. B & H; USA Freethy.
HEART PANSY *Viola tricolor* (Pansy) B & H.
HEART PEA *Cardiospermum halicacabum* (Blister Creeper) Chopra.
HEART SNAKEROOT *Asarum virginicum*.
HEART VINE *Ceropegia woodii* (Rosary Vine) RHS.
HEARTLEAF *Asarum virginicum* (Heart Snakeroot) USA Cunningham.
HEARTLEAF LILY *Cardiocrinum cordatum*.
HEARTMINT *Mentha spicata* (Spearmint) Leyel.1937. See rather Hart Mint.
HEARTS 1. *Oxalis acetosella* (Wood Sorrel) N'thum. Grigson. 2. *Vaccinium myrtillus* (Whortleberry) Hants. Cope. Leaf shape accounts for wood sorrel, but for whortleberry, 'hearts' is hurts, or whorts, but gathering the fruit is called hearting, or harting.
HEARTS, Lady's *Dicentra spectabilis* (Bleeding Heart) Som. Macmillan.
HEARTS, String-of- *Ceropegia woodii* (Rosary Vine) Goold-Adams. The hanging stems look just like strings, with many heart-shaped leaves on them.
HEARTS-AND-HONEY VINE *Ipomaea x multifida* (Cardinal Climber) RHS.
HEARTS AT EASE *Viola tricolor* (Pansy) Friend. A variant of Heartsease. Could it have been of Friend's making?
HEARTS-ENTANGLED *Ceropegia woodii* (Rosary Vine) Bonar. The reference is to the small heart-shaped leaves growing down the mass of stalks.
HEARTS-ON-STRINGS *Dicentra spectabilis* (Bleeding Heart) Som. Macmillan.
HEARTSEASE 1. *Cheiranthus cheiri* (Wallflower) Corn. B & H. 2. *Polygonum persicaria* (Persicaria) USA Bergen. 3. *Prunella vulgaris* (Self-heal) Donegal Grigson; USA. Bergen. 4. *Symphytum officinale* (Comfrey) Lancs. PLNN 17. 5. *Viola tricolor*. Self-heal at least was thought to cure heart disease, while persicaria has heart-shaped markings on the leaves, reckoned to be its signature for heart trouble. But if pansy was ever used medicinally for heart disease, it was by mistake, and hearts-ease is more likely to be a mirror of its supposed aphrodisiac qualities.
HEARTSEED 1. *Cardiospermum halicacabum* (Blister Creeper) Chopra. 2. *Polygonum pennsylvanicum*. 3. *Viola tricolor* (Pansy) Friend. A corruption of Heartsease as far as pansy is concerned, but descriptive for *Cardiospermum*, as well as being a straight translation of the generic name.
HEARTWOOD *Cassia abbreviata* (Longtail Cassia) Palgrave.
HEARTWORT 1. *Melilotus officinalis* (Melilot) B & H. 2. *Polygonum persicaria* (Persicaria) Dawson. Possibly Hartwort for the melilot, for there is also Hart's Clover. Persicaria has heart-shaped markings on the leaves.
HEARTWORT, Long *Aristolochia longa* (Long Birthwort) Turner.
HEARTWORT, Small *Aristolochia clematitis* (Birthwort) Turner.
HEARTYCHOKE *Cynara cardunculus var. scolymus* (Globe Artichoke) Potter & Sargent.
HEATH 1. *Calluna vulgaris*. 2. *Empetrum nigrum* (Crowberry) Derb. Grigson. 3. *Tamarix gallica* (Tamarisk) B & H.
HEATH, Alpine *Erica carnea*.
HEATH, Besom 1. *Erica scoparia* (Green heather) RHS. 2. *Erica tetralix* (Cross-leaved Heath) Hulme.
HEATH, Bell *Erica tetralix* (Cross-leaved Heath) Som. Grigson; Hants. Cope.
HEATH, Black *Erica cinerea* (Bell Heather)

Hants. Cope.

HEATH, Black-berried *Empetrum nigrum* (Crowberry) Camden/Gibson.

HEATH, Bloody *Erica cruenta*.

HEATH, Broom *Erica tetralix* (Cross-leaved Heath) Hulme.

HEATH, Carlin *Erica cinerea* (Bell Heather) Yks, Scot. Grigson. Carlin Heath, or Carlin Heather, but also known as She-heather, as is *Erica tetralix*, as distinct from He-heather, which is *Calluna vulgaris*.

HEATH, Channelled *Erica canaliculata*.

HEATH, Ciliate *Erica ciliaris* (Dorset Heath) Wit. Or Fringed Heath, which means the same thing.

HEATH, Corsican *Erica stricta*. Not wholly confined to Corsica, for it is found in Sardinia, southern Italy, southwest Spain and northwest Morocco as well.

HEATH, Cross-leaved; HEATH, Crossed *Erica tetralix*. Gerard has Crossed Heath - "small leaves set at certain spaces two upon one side, and two on the other, opposite, one answering another, even as do the leaves of Crossewort".

HEATH, Dorset *Erica ciliaris*. A west country species.

HEATH, Dutch, Low *Erica tetralix* (Cross-leaved Heath) Parkinson.

HEATH, False *Fabiana imbricata*.

HEATH, Father-of- *Erica tetralix* (Cross-leaved Heath) Yks. Grigson. An odd name for this species, for *E tetralix* is She-heather.

HEATH, Fringed *Erica ciliaris*.

HEATH, Fine-leaved *Erica cinerea* (Bell Heather) Curtis.

HEATH, Flame *Erica flammea*.

HEATH, Giant *Erica arborea* (Tree Heath) Dale & Greenway.

HEATH, Grey *Erica cinerea* (Bell Heather) Hay & Synge. The specific name is *cinerea*, ashy.

HEATH, Irish 1. *Daboecia cantabrica* (Irish Heather) W.Miller. 2. *Erica erigena*. Both these are only found in these islands in the west of Ireland, and then only very locally.

HEATH, Long *Calluna vulgaris* (Heather)

Lyte. Or was 'ling' meant?

HEATH, Lusitanian *Erica lusitanica* (Western Heath) Polunin.1976.

HEATH, Mackay's *Erica mackaiana*.

HEATH, Moor *Erica vagans* (Cornish Heath) B & H.

HEATH, Mountain, Blue *Phyllodoce caerulea* (Menziesia) Polunin.

HEATH, Porcelain *Erica ventricosa*. Curtis's name - for the flowers look like little glazed jars, just as if they were made of porcelain.

HEATH, Portuguese *Erica lusitanica* (Western Heath) Hay & Synge.

HEATH, Prickly *Pernettya mucronata*.

HEATH, Purple *Erica purpurea*.

HEATH, Red *Calluna vulgaris* (Heather) Hants. Cope.

HEATH, St Dabeoc's *Daboecia cantabrica*. The generic name is spelt *Daboecia* in spite of the fact that it is St Dabeoc's heath. Who was this saint?

HEATH, Sea *Frankenia laevis*.

HEATH, Small *Calluna vulgaris* (Heather) Hants. Cope. Presumably it is the size of the flowers that is described, for another name for heather is Long Heath.

HEATH, Small-leaved *Erica cinerea* (Bell Heather) Gerard.

HEATH, Spanish *Erica australis*.

HEATH, Spike *Bruckenthalia spiculifolia*.

HEATH, Tree *Erica arborea*. Usually up to 12 feet in height, but it can grow to as much as 60 feet tall.

HEATH, Upright *Erica stricta* (Corsican Heath) W Miller.

HEATH, Western *Erica lusitanica*.

HEATH, White *Aster ericoides* (Many-flowered Aster) Kansas) Gates.

HEATH, Winter 1. *Erica carnea* (Alpine Heath) RHS. 2. *Erica erigena* (Irish Heath) Hay & Synge.

HEATH ASTER, White *Aster ericoides* (Many-flowered Aster) USA Zenkert.

HEATH BELLFLOWER *Campanula rotundifolia* (Harebell) Curtis.

HEATH BROOM *Erica tetralix* (Cross-leaved Heath) B & H. Or Broom Heath. Cf.

Besom Heath.
HEATH CUDWEED *Gnaphalium sylvaticum.*
HEATH GROUNDSEL *Senecio sylvaticus.*
HEATH URTS *Empetrum nigrum* (Crowberry) Som. Macmillan.
HEATHBELL *Campanula rotundifolia* (Harebell) Dor. Grigson; N'thants. Clare.
HEATHBERRY *Vaccinium myrtillus* (Whortleberry) Storms.
HEATHER *Calluna vulgaris.* The older Scots word is hadder, but the origin of heather is unknown, though probably it is a mixture of hadder and heath.
HEATHER, Arctic *Andromeda floribunda* (Free-flowering Andromeda) McDonald.
HEATHER, Beach *Hudsonia tomentosa.* It is also called Poverty Grass, because it grows on poor sandy soils and beaches.
HEATHER, Bell *Erica cinerea.*
HEATHER, Bog *Erica tetralix* (Cross-leaved Heath) A W Smith.
HEATHER, Bull *Erica tetralix* (Cross-leaved Heath) Scot. MacColl & Seeger.
HEATHER, Carlin *Erica cinerea* (Bell Heather) Scot. Jamieson. 'Carlin' means an old woman. Cf. She-heather for this.
HEATHER, Cat 1. *Erica cinerea* (Bell Heather) Scot. Grigson. 2. *Erica tetralix* (Cross-leaved Heath) Aber. B & H.
HEATHER, Cornish *Erica vagans.* Virtually only on heaths in west Cornwall as far as Britain is concerned, though it can be found also in western France and northern Spain.
HEATHER, Dog *Calluna vulgaris* (Heather) Aber. Grigson. Bell Heather is She- and Cat-heather, while the true heather is He- or Dog-heather.
HEATHER, Green *Erica scoparia.*
HEATHER, He *Calluna vulgaris* (Heather) N'thum, Berw. Grigson. Bell Heather, or Cross-leaved Heath (both of these are in a different genus, *Erica*) is She-heather.
HEATHER, Irish *Daboecia cantabrica.*
HEATHER, Lapp *Phyllodoce caerulea* (Menziesia) Manker.
HEATHER, Ling *Calluna vulgaris* (Heather) A W Smith. Ling is a very common alternative name for heather; Ling-heather mixes the two.
HEATHER, Monnaghs; HEATHER, Monox; HEATHER, Monnocs *Empetrum nigrum* (Crowberry) W Miller (Monnaghs, Monox); N.Ire. Grigson (Monnocs). Another Ulster name, Moonogs, shows that the original was the Gaelic name for the plant.
HEATHER, Purple *Erica cinerea* (Bell Heather) Wit.
HEATHER, Ringe *Erica tetralix* (Cross-leaved Heath) Scot. Jamieson. A ringe is a whisk. Cf. Besom, or Broom, Heather for this.
HEATHER, Scotch *Erica cinerea* (Bell Heather) W.Miller.
HEATHER, She 1. *Erica cinerea* (Bell Heather) N'thum, Berw. Grigson. 2. *Erica tetralix* (Cross-leaved Heath) N'thum, Berw. Grigson. Bell Heather is the real owner of the name - it opposes the He-heather which is given to *Calluna vulgaris*. Cf. Cat-heather for *Erica cinerea*, too.
HEATHER BELL 1. *Campanula rotundifolia* (Harebell) Dor. Macmillan. 2. *Erica tetralix* (Cross-leaved Heath) Scot. B & H.
HEATHER-FUE *Eriophorum angustifolium* (Cotton-grass) Som. Macmillan.
HEATHER WHIN 1. *Genista anglica* (Needle Whin) Berw. Grigson. 2. *Genista tinctoria* (Dyer's Greenweed) Border B & H.
HEATHERBERRY *Vaccinium myrtillus* (Whortleberry) Donegal Maire MacNeill.
HEAVE, Cow- *Tussilago farfara* (Coltsfoot) Selk. B & H. Probably cow-hoof, from the shape of the leaves, as is coltsfoot.
HEAVEN, Breath of *Diosma ericoides.*
HEAVEN, Flower of *Ceratophyllum demersum* (Hornwort) Tynan & Maitland.
HEAVEN, Keys of *Primula veris* (Cowslip) Grieve.1931. There is a fancied resemblance to a bunch of keys, hence a reference to St Peter, and by extension names like this. The legend is that St Peter once dropped the keys of heaven, and the first cowslips grew up

where they fell.

HEAVEN, Ladder-to- 1. *Convallaria maialis* (Lily of the valley) Barton & Castle. 2. *Polemonium caeruleum* (Jacob's Ladder) Lanark. Grigson. Jacob's Ladder is also quoted for lily of the valley. The real one, *Polemonium caeruleum*, owes the 'ladder' imagery to the fact that the leaflets are arranged in successive pairs - in ladder-like formation, in fact.

HEAVEN, Road-to- *Polemonium caeruleum* (Jacob's Ladder) Dor. Macmillan.

HEAVEN, Tree of *Ailanthus glandulosa*. *Ailanthus* is a version of the Chinese meaning "strong enough to reach heaven".

HEAVEN'S EYE *Myosotis palustris* (Water Forget-me-not) Dev. Tynan & Maitland.

HEAVER HAYVER *Lolium perenne* (Rye Grass) Corn. Jago. A version of the commoner Haver (usually in the plural, and applied to Wild Oats). The German for oat is Hafer. Are these in reality two different names?

HEAVY-WOODED PINE *Pinus ponderosa* (Yellow Pine) Ablett.

HEBON *Taxus baccata* (Yew) Dyer.1887. Dyer used it, quoting Spenser. The German is Eiben.

HECH-HOW *Conium maculatum* (Hemlock) Scot. Grigson; USA Henkel. Presumably from a Gaelic original.

HECKBERRY *Prunus padus* (Bird Cherry) Yks, Cumb. Halliwell; Dur. Dinsdale. The word means hedgeberry, which is also recorded for this tree.

HEDDER *Calluna vulgaris* (Heather) Grigson. It looks to be half-way between the original, Hadder, and the modern Heather, though in fact the modern name is probably a hybrid of Heath and Hadder.

HEDGE; HEDGE-BALL *Maclura pomifera* (Osage Orange) Kansas Sackett & Koch (Hedge); Illinois H M Hyatt (Hedge-ball). Called simply Hedge because it is so often used for one.

HEDGEBERRY *Prunus padus* (Bird Cherry) Lancs, Cumb. B & H.

HEDGEHERIFF *Galium aparine* (Goose-grass) Fernie.

HEDGEHOG BROOM *Erinacea anthyllis*. It was Clusius who named it Erinacea, after the Latin for hedgehog.

HEDGEHOG CONEFLOWER *Echinacia purpurea*.

HEDGEHOG FIR *Abies pinsapo* (Spanish Fir) Mitchell. Descriptive - the stiff leaves radiate perpendicularly around the stem.

HEDGEHOG PARSLEY *Caucalis daucoides* (Bur Parsley) McClintock.

HEDGEHOG PARSLEY, Broad-leaved *Caucalis latifolia* (Great Bur Parsley) McClintock.1975.

HEDGEHOGS 1. *Arctium lappa* seeds (Burs) Som. Macmillan. 2. *Galium aparine* (Goose-grass) Som. Macmillan. 3. *Ranunculus arvensis* (Corn Buttercup) Wilts. Dartnell. & Goddard; Suss, Surr, Kent B & H. 4. *Scandix pecten-veneris* (Shepherd's needle) Som. Dor. Macmillan; Suss. Parish. It is the spiny seed vessels in their different ways that account for this name.

HEDGEMAIDS *Glechoma hederacea* (Ground Ivy) Corn, Dev, Som, Dor, Glos, E Ang, Cumb. Friend.

HEDGEPEGS *Prunus spinosa* fruit (Sloe) Wilts. Dartnell & Goddard.

HEDGEPICKS *Prunus spinosa* fruit (Sloe) Hants. Cope; Wilts. Macmillan.

HEDGEROW CRANESBILL *Geranium pyrenaicum* (Mountain Cranesbill) Blamey.

HEDGESPEAKS 1. *Prunus spinosa* fruit (Sloe) Wilts. Grigson. 2. *Rosa canina* fruit (Hips) Glos. Grigson.

HEDGESPECKS 1. *Prunus spinosa* fruit (Sloe) Wilts. Macmillan. 2. *Crataegus monogyna* fruit (Haws) Wilts. Macmillan.

HEDGEWEED *Sisymbrium officinale* (Hedge Mustard) Hill.

HEDGY-PEDGY *Rosa canina* fruit (Hips) Wilts. Grigson.

HEEL, Christ's 1. *Plantago lanceolata* (Ribwort Plantain) Wales Denham/Hardy. 2. *Plantago major* (Great Plantain) Wales

Denham/Hardy.

HEEL, Lark's *Delphinium ajacis* (Larkspur) Gerard.

HEEL, Lark's, Yellow *Tropaeolum majus* (Nasturtium) Parkinson.

HEEL, Ox *Helleborus foetidus* (Stinking Hellebore) Prior. More properly Ox-heal. Cf. Bear's foot for this. But the use by cattle doctors explains the name.

HEEL, Pig's *Crataegus monogyna* fruit (Haw) Som. Macmillan. An aberrant form of Pigall via one of its Somerset variants, Pig's Hales.

HEEL, Shoemaker's *Chenopodium bonus-henricus* (Good King Henry) Shrop, Rad. Grigson.

HEEL TROT *Pastinaca sativa* (Wild Parsnip) Wilts. D.Grose. Eltrot was the original of this.

HEETHEN-BERRY *Crataegus monogyna* fruit (Haws) Ches. Grigson. Heethen here is an extreme form of hawthorn.

HEG-BEG *Urtica dioica* (Nettle) Scot. Grigson. Apparently not Heg-Peg, but they are Hidgy-pidgies in Devonshire, and that is a name for sloes and haws.

HEG-PEG 1. *Crataegus monogyna* fruit (Haws) Glos. Grigson. 2. *Prunus spinosa* fruit (Sloes) Glos. Grigson. The general meaning of Heg-peg is hedgeberry.

HEG-PEG BUSH 1. *Crataegus monogyna* (Hawthorn) Glos. Grigson. 2. *Prunus spinosa* (Blackthorn) Glos. B & H.

HEGBERRY *Prunus padus* (Bird Cherry) Cumb. Gerard.

HEIHOW *Glechoma hederacea* (Ground Ivy) B & H. The first syllable is more usual as 'hay', and means 'hedge'. Cf. this with Hay-hoa, Hayhove, etc.

HEIRIFF; HEIRIFFE *Galium aparine* (Goose-grass) Prior; N'thants. A E Baker. Members of a series of names which include variants from Airess to Haireve.

HEIRIFF, Goose *Galium aparine* (Goose-grass) Coles.

HELEN FLOWER *Helenium sp* W Miller. To account for the name of the genus, one tradition says that these are the flowers that sprang from the tears of Helen of Troy.

HELEN'S ELECAMPANE *Inula helenium* (Elecampane) Palaiseul. Elecampane seems to be a corruption of medieval Latin enula campana, enula showing a possible connection with helenium.

HELIOTROPE, Garden *Valeriana officinalis* (Valerian) Schery. The smell is similar to that of heliotrope, hence, too, the Wiltshire name Cherry Pie.

HELIOTROPE, Indian *Heliotropum indicum*.

HELIOTROPE, Seaside *Heliotropum curassavicum*.

HELIOTROPE, Winter *Petasites fragrans*. Heliotrope-like scent.

HELL'S CURSE *Amaranthus paniculatus* (Purple Amaranth) Watt. This may be a South African name, and might be more valid in that country than in its native central American habitat, for it is an introduced plant there. The assumption is that it may have got out of hand.

HELLBIND *Cuscuta epithymum* (Common Dodder) Herts. Grigson. Hellweed is more widespread as a name for dodder, but other names for it, like Devil's Guts, or Devil's Net, make the imagery quite clear.

HELLEBORE, American *Veratrum viride* (American White Hellebore) USA Lloyd. Hellebore is from Gk elein, to injure, and bora, food - a reference to the fact that hellebores are poisonous.

HELLEBORE, Bastard *Helleborus viridis* (Green Hellebore) Lyte.

HELLEBORE, Black 1. *Astrantia major* (Pink Masterwort) B & H. 2. *Epipactis purpurata* (Violet Helleborine) Wilts. D.Grose. 3. *Helleborus niger* (Christmas Rose) Gerard. The reference to 'black' for the Christmas Rose is to the colour of the root.

HELLEBORE, Black, Wild *Helleborus viridis* (Green Hellebore) Gerard.

HELLEBORE, Corsican *Helleborus corsicus*.

HELLEBORE, False 1. *Adonis annua*

(Pheasant's Eye). 2. *Adonis vernalis* Mitton. 3. *Veratrum californicum*. 4. *Veratrum viride* (American White Hellebore) USA Zenkert.

HELLEBORE, Green 1. *Helleborus viridis*. 2. *Veratrum viride* (American White Hellebore) Fluckiger.

HELLEBORE, Indian *Veratrum viride* (American White Hellebore) USA Turner & Bell.

HELLEBORE, Stinking *Helleborus foetidus*. The flowers hang their heads in shame because of the unpleasant smell they emanate, so the legend has it. Actually the stink is caused by trimethylamine, which attracts midges and bluebottles for the essential process of the plant's pollination.

HELLEBORE, Swamp *Veratrum viride* (American White Hellebore) USA O P Brown.

HELLEBORE, White *Veratrum album*.

HELLEBORE, White, American *Veratrum viride*.

HELLEBORE, Winter 1. *Eranthis hyemalis* (Winter Aconite) Prior. 2. *Helleborus niger* (Christmas Rose) Jones-Baker.1974. They are both winter-flowering.

HELLEBORINE *Epipactis helleborine* (Broad-leaved Helleborine) Turner Ettlinger. Does helleborine mean 'like a hellebore'?

HELLEBORINE, Broad 1. *Cephalanthera damasonium*. 2. *Epipactis helleborine* (Broad-leaved Helleborine) Clapham.

HELLEBORINE, Clustered *Epipactis purpurata* (Violet Helleborine) Turner Ettlinger.

HELLEBORINE, Dune *Epipactis dunensis*.

HELLEBORINE, False, White *Veratrum album* White Hellebore Polunin.

HELLEBORINE, Giant *Epipactis gigantea* (Stream Orchis) Usher.

HELLEBORINE, Green; HELLEBORINE, Green-flowered *Epipactis phyllanthes* Turner Ettlinger (Green).

HELLEBORINE, Green-leaved *Epipactis leptochila* (Narrow-lipped Helleborine) Turner Ettlinger.

HELLEBORINE, Long-leaved *Cephalanthera longifolia* (Narrow Helleborine) Clapham.

HELLEBORINE, Marsh *Epipactis palustris*.

HELLEBORINE, Narrow *Cephalanthera longifolia*.

HELLEBORINE, Narrow-lipped *Epipactis leptochila*.

HELLEBORINE, Pendulous-flowered *Epipactis phyllanthes* (Green-flowered Helleborine) Turner Ettlinger.

HELLEBORINE, Purple *Epipactis atrorubens* (Dark Red Helleborine) Turner Ettlinger.

HELLEBORINE, Purple-washed *Epipactis purpurata* (Violet Helleborine) Turner Ettlinger.

HELLEBORINE, Red *Cephalanthera rubra*.

HELLEBORINE, Red, Dark *Epipactis atrorubens*.

HELLEBORINE, Slender-lipped *Epipactis leptochila* (Narrow-lipped Helleborine) Turner Ettlinger.

HELLEBORINE, Small-flowered *Epipactis atrorubens* (Dark Red Helleborine) Turner Ettlinger.

HELLEBORINE, Sword-leaved *Cephalanthera longifolia* (Narrow Helleborine) Turner Ettlinger.

HELLEBORINE, Violet *Epipactis purpurata*.

HELLEBORINE, White *Cephalanthera damasonium* (Broad Helleborine) Clapham. or, Large White Helleborine.

HELLROOT 1. *Orobanche minor* (Lesser Broomrape) Kent Grigson. 2. *Smyrnium olusatrum* (Alexanders) Dor. Grigson. When applied to Alexanders, this name sometimes appears as Helrut, which probably indicates that it was originally heal-root. That, at least, was what Barnes thought. As far as the broomrape is concerned, the word probably means what it says. A parasitic plant looking so different from everything else would arouse suspicions.

HELLS, Pig's *Crataegus monogyna* fruit (Haws) Som. Macmillan. This is Pig's Hales, itself a variation on the common name Pigaul.

HELLTROT *Heracleum lanatum* USA Bergen. Probably because the original name given to the native British hogweed, Eltrot, was misunderstood in America.

HELLWEED 1. *Calystegia sepium* (Great Bindweed) N'thants. Grigson. 2. *Convolvulus arvensis* (Field Bindweed) N'thants. Sternberg. 3. *Cuscuta epithymum* (Common Dodder) Suss, Kent, Beds, Berks, Herts, Cambs, N'thants, Lanark. Grigson. 4. *Ranunculus arvensis* (Corn Buttercup) Yks. Grigson. Bindweed and dodder are strangling plants, and a thorough nuisance, hence the name. Corn Buttercup is equally unwanted.

HELLYCOMPANE *Inula helenium* (Elecampane) Corn. Jago. Cf. Helen's Elecampane.

HELM 1. *Ulmus glabra* (Wych Elm) Glos. B & H. 2. *Ulmus procera* (English Elm) Glos. B & H.

HELMET, Policeman's 1. *Aconitum napellus* (Monkshood) Som. Macmillan. 2. *Impatiens glandulifera* (Himalayan Balsam) Clapham. Helmet-flowers, in both cases.

HELMET-FLOWER 1. *Aconitum napellus* (Monkshood) Gerard. 2. *Lamium album* (White Deadnettle) Som. Grigson. 3. *Scutellaria galericulata* (Skullcap) Grieve.1931.

HELRUT *Smyrnium olusatrum* (Alexanders) Dor. Macmillan. Cf. Hellroot - but both versions are probably heal-root.

HELTROT *Heracleum sphondyllium* (Hogweed) Hants. Cope. Eltrot is more usual, but it exists as Altrot, Elder-trot, Old-rot and even Hill Trot.

HELVER *Ilex aquifolium* (Holly) Suff. B & H. Hulver is the better known East Anglian form.

HEMITORY *Fumaria officinalis* (Fumitory) Kent Grigson. A local variant on Fumitory.

HEMLOCK 1. *Anthriscus sylvestris* (Cow Parsley) Norf. Vickery.1985. 2. *Conium maculatum*. 3. *Heracleum sphondyllium* (Hogweed) Banff. Grigson. 4. *Tsuga sp.* OE hymlice, which means that the north country Humlick is closer to the original than the modern standard name. The vowel change in the second syllable must have occurred later, for no known reason (but charlock went through the same process, from OE cerlic). Cow Parsley gets the name because of the similarity in appearance - the two plants do in fact share a number of names. But why should the large trees of *Tsuga* species have the name? They are not related in any way to *Conium maculatum*, nor are they poisonous. Geoffrey Grigson suggested that the name was awarded from some fancied resemblance of the branches to giant hemlock leaves, and Chambers Dictionary says the same. The name once given to these trees, Hemlock Spruce certainly gives the impression of being descriptive. One more offering, this time from Mitchell - he says it is the smell that gives the name. If you crush the leaves they will remind you of hemlock leaves (which traditionally smell of mice).

HEMLOCK, Canada *Tsuga canadensis* (Eastern Hemlock) Yanovsky.

HEMLOCK, Carolina *Tsuga caroliniana*.

HEMLOCK, Chinese *Tsuga chinensis*.

HEMLOCK, Eastern *Tsuga canadensis*.

HEMLOCK, Ground *Taxus canadensis* (Canadian Yew) USA H Smith. It must be a dwarf species.

HEMLOCK, Himalayan *Tsuga dumosa*.

HEMLOCK, Japanese, Northern *Tsuga diversifolia*.

HEMLOCK, Japanese, Southern *Tsuga sieboldii*.

HEMLOCK, Lesser; HEMLOCK, Smaller *Aethusa cynapium* (Fool's Parsley) Barton & Castle (Lesser); Grieve.1931 (Smaller).

HEMLOCK, Mountain 1. *Ligusticum scoticum* (Lovage) W.Miller. 2. *Tsuga mertensiana*.

HEMLOCK, Northern *Tsuga canadensis* (Eastern Hemlock). SM Robertson.

HEMLOCK, Poison 1. *Cicuta maculata* (American Cowbane) USA Kingsbury.1967. 2. *Conium maculatum* (Hemlock) USA E Gunther. Stressing 'poison' here would seem an unnecessary complication.

HEMLOCK, Southern *Tsuga caroliniana* (Carolina Hemlock). S M Robertson.

HEMLOCK, Spotted 1. *Cicuta maculata* (American Cowbane) USA Kingsbury.1967. 2. *Conium maculatum* (Hemlock) USA Kingsbury.

HEMLOCK, Water 1. *Cicuta maculata* (American Cowbane) USA Weiner. 2. *Cicuta virosa* (Cowbane) Prior. 3. *Heracleum sphondyllium* (Hogweed) Banff. Grigson. 4. *Oenanthe aquatica* (Horsebane) B & H. 5. *Oenanthe crocata* (Hemlock Water Dropwort) Suss, Cumb. Grigson. 6. *Sium latifolium* (Great Water Parsnip) Grieve.1931. Hemlock is appropriate, for they are all poisonous, with the possible exception of the Water Parsnip, which is only said to be toxic, and the certain exception of Hogweed.

HEMLOCK, Water, Western *Cicuta douglasi.*

HEMLOCK, Western *Tsuga heterophylla.*

HEMLOCK, Yunnan *Tsuga yunnanensis.*

HEMLOCK CHERVIL *Torilis japonica* (Hedge Parsley) B & H.

HEMLOCK-FIR *Tsuga heterophylla* (Western Hemlock) Brimble.1948.

HEMLOCK-LEAVED CRANESBILL *Erodium cicutarium* (Storksbill) Curtis.

HEMLOCK SPRUCE *Tsuga sp* Barber & Phillips.

HEMLOCK STORKSBILL *Erodium cicutarium* (Storksbill).

HEMLOCK WATER DROPWORT *Oenanthe crocata.*

HEMMING-AND-SEWING *Achillea millefolium* (Yarrow) Hants Grigson.

HEMP 1. *Ajuga chamaepitys* (Yellow Bugle) Cockayne. 2. *Cannabis sativa.* Hemp is OE haenep, Latin cannabis, Greek Kannabis. Yellow Bugle is glossed Hemp on a 12th century ms of Apuleius Barbarus. Any plant or tree which bears the name would yield textile fibres.

HEMP, African 1. *Sansevieria guineensis.* 2. *Sparmannia africana.*

HEMP, Ambari *Hibiscus cannabinus.*

HEMP, American; HEMP, Canadian *Apocynum cannabinum* (Indian Hemp) both Grieve.1931.

HEMP, Bahama *Agave sisalina* (Sisal) Roys. A Mexican plant, introduced into the West Indies for commercial cultivation.

HEMP, Barren *Cannabis sativa* (Hemp) Lyte. Applied only to the male plant.

HEMP, Bastard 1. *Eupatorium cannabinum* (Hemp Agrimony) B & H. 2. *Galeopsis tetrahit* (Hemp Nettle) Gerard.

HEMP, Bombay; HEMP, Madras *Crotolaria juncea* (Sunn Hemp) Howes.

HEMP, Bowstring *Sansevieria metallica.*

HEMP, Carl; HEMP, Churl *Cannabis sativa* (Hemp) W Miller (Carl); B & H (Churl). i.e. male hemp. But it in fact applies only to the female plant. There must have been a lot of confusion, for the Female Hemp of the old herbalists is the male. Carl Hemp, though, was defined by Halliwell as "late grown hemp".

HEMP, Chinese *Abutilon avicennae* (China Jute) H & P.

HEMP, Colorado River *Sesbania exaltata.*

HEMP, East Indies *Crotalaria juncea* (Sunn Hemp) Howes.

HEMP, False *Crotalaria juncea* (Sunn Hemp) Chopra.

HEMP, Gambo *Hibiscus cannabinus* (Ambari Hemp) Hutchinson & Melville.

HEMP, Indian 1. *Apocynum cannabinum.* 2. *Cannabis sativa* (Hemp) Watt. 3. *Crotalaria juncea* (Sunn Hemp) Hutchinson & Melville. A mixture of Indians here. The first one belongs to the American Indians, while the third is Asian. Probably, the Indians referred to in the second hemp name are American, too, for the plant is known as Indian Hay in the States. Perhaps not, for it may very well be a description of the variety 'indica' which usually is grown here.

HEMP, Indian, Black *Apocynum cannabinum* (Indian Hemp) Grieve.1931.

HEMP, Manila *Musa textilis.* Indigenous to the Philippines, and in use there by the early 16th century. Not utilized outside the

Philippines until the 19th century, but now grown commercially in central America, as well as in Indonesia and Burma.

HEMP, Mauritius *Furcraea gigantea*. Grown commercially on the island of Mauritius, though it is actually a native of Brazil.

HEMP, Nettle *Galeopsis tetrahit* (Hemp Nettle) Gerard.

HEMP, New Zealand *Phormium tenax* (New Zealand Flax) Schery.

HEMP, San; HEMP, Sunn *Crotalaria juncea*. Sunn Hemp is the usual name. Schery has San.

HEMP, Swedish *Urtica dioica* (Nettle) South Africa Watt.

HEMP, Thistle *Cannabis sativa* (Hemp) B & H.

HEMP, Water 1. *Bidens tripartita* (Bur Marigold) Prior. 2. *Eupatorium cannabinum* (Hemp Agrimony) Turner.

HEMP, Wild *Galeopsis tetrahit* (Hemp Nettle) Gerard.

HEMP AGRIMONY *Eupatorium cannabinum*. Hemp, from the resemblance of the leaves of those of hemp.

HEMP AGRIMONY, Trifid *Bidens tripartita* (Bur Marigold) Curtis.

HEMP AGRIMONY, Water, Nodding *Bidens cernua* (Nodding Bur Marigold) Curtis.

HEMP BROOMRAPE *Orobanche ramosus* (Branching Broomrape) Rackham.1986.

HEMP DOGBANE *Apocynum cannabinum* (Indian Hemp) Thomson.

HEMP-LEAVED DEADNETTLE *Galeopsis tetrahit* (Hemp Nettle) Jamieson.

HEMP-LEAVED HIBISCUS *Hibiscus cannabinus* (Ambari Hemp) Dalziel.

HEMP NETTLE *Galeopsis tetrahit*.

HEMP NETTLE, Downy *Galeopsis dubia*.

HEMP NETTLE, Large *Galeopsis speciosa*.

HEMP NETTLE, Large-flowered *Galeopsis versicolor*.

HEMP NETTLE, Narrow-leaved *Galeopsis angustifolia* (Red Hemp Nettle) Salisbury.

HEMP NETTLE, Pyrenean *Galeopsis pyrenaica*.

HEMP NETTLE, Red *Galeopsis angustifolia*.

HEMP TREE *Vitex agnus-castus* (Chaste Tree) Turner.

HEMPER *Rhinanthus crista-galli* (Yellow Rattle) Som, Yks, Cumb, N'thum. Grigson.

HEMPSEED *Eupatorium cannabinum* (Hemp Agrimony) B & H.

HEN, Gill *Glechoma hederacea* (Ground Ivy) Sanecki.

HEN-AND-CHICKEN FERN *Asplenium bulbiferum*. Because this fern produces little ones from bulbils on the mature fronds. Cf. Mother Fern, or Mother Spleenwort.

HEN-AND-CHICKENS 1. *Anthyllis vulneraria* (Kidney Vetch) Som. Grigson. 2. *Aquilegia vulgaris* (Columbine) Norf. Grigson. 3. *Bellis perennis* (Daisy) the double garden variety Som. Elworthy. 4. *Butomus umbellatus* (Flowering Rush) Hunts. Marshall. 5. *Capsella bursa-pastoris* (Shepherd's Purse) Som. Macmillan. 6. *Cymbalaria muralis* (Ivy-leaved Toadflax) Kent Grigson. 7. *Geranium robertianum* (Herb Robert) Som. Macmillan. 8. *Glechoma hederacea* (Ground Ivy) Bucks. Grigson. 9. *Lotus corniculatus* (Bird's-foot Trefoil) Som. Macmillan; Glos, Oxf. Grigson. 10. *Narcissus pseudo-narcissus* (Daffodil) Friend. 11. *Saxifraga sarmentosa* (Mother-of-Thousands) Som. Macmillan; Wilts. Dartnell & Goddard. 12. *Saxifraga x urbinum* (London Pride) Dev. Friend; Wilts. Dartnell & Goddard. 13. *Sedum acre* (Biting Stonecrop) Som. Macmillan. 14. *Sempervivum tectorum* (Houseleek) Som. Grigson. 15. *Tanacetum parthenium* (Feverfew) Tynan & Maitland. Usually applied to plants which reproduce themselves by runners, giving the effect of a mother plant with children around her. Umbel-like plants, like Flowering Rush, could give the same impression. Others, like Herb Robert or Columbine, have bird connections of their own. Daffodil seems to be the odd one out.

HEN APPLE *Sorbus aria* (Whitebeam) Mor. Grigson.

HEN BELL *Hyoscyamus niger* (Henbane) B & H. Henne-belle in some early ME texts. OE haennebelle.

HEN DRUNKS *Sorbus aucuparia* fruit (Rowan berries) Cumb. B & H. Cf. Cock-drunks, too. It is claimed that the berries will make fowls drunk if they eat them.

HEN GORSE 1. *Odontites verna* (Red Bartsia) Ches. B & H. 2. *Ononis repens* (Rest Harrow) Midl, Ches, N Eng. Grigson.

HEN PEAS *Zea mays* (Maize) Derb. Porteous.

HEN PLANT *Plantago lanceolata* (Ribwort Plantain) Le Strange.

HEN'S BILL *Onobrychis sativa* (Sainfoin) B & H.

HEN'S COMB 1. *Dactylorchis maculata* (Heath Spotted Orchid) Berw. Grigson. 2. *Rhinanthus crista-galli* (Yellow Rattle) B & H. 'Comb' is 'crest', as in *crista-galli*. It is more familiar as cock's comb.

HEN'S EVERGREEN; CHICKEN'S EVERGREEN *Stellaria media* (Chickweed) Kent Tynan & Maitland.

HEN'S FOOT 1. *Caucalis daucoides* (Bur Parsley) Gerard. 2. *Corydalis claviculata* (Climbing Corydalis) B & H.

HEN'S KAMES *Dactylorchis maculata* (Heath Spotted Orchid) Berw. Grigson. Kames = combs.

HENBANE *Hyoscyamus niger*. There are other 'hen' names for this - Henpen, Henpenny, Henkam and Henbell. The OE was haennebelle, or haennepol, and it appears as henne-belle in some early Middle English texts. There is another name, Belene, which is probably bellen, i.e. furnished with bells. Nevertheless, henbane probably means what it says. Cf. the French mort-aux-poules. And in Gaelic it is coach-nan-cearc, that which blinds the hen. Henbane is certainly deadly to poultry.

HENBANE, Black *Hyoscyamus niger* (Henbane) Duncalf.

HENBANE, Egyptian *Hyoscyamus muticus*. It is grown there for its narcotic properties.

HENBANE, Indian *Nicotiana tabacum* (Tobacco) Parkinson.

HENBANE, Russian *Hyoscyamus albus* (White Henbane) Grieve.1931.

HENBANE, White *Hyoscyamus albus*. 'White', because it has paler flowers than H niger.

HENBANE, Yellow *Hyoscyamus luteus*.

HENBIT 1. *Ballota nigra* (Black Horehound) Som. Macmillan. 2. *Lamium amplexicaule* (Henbit Deadnettle) Friend.

HENBIT, Greater *Lamium amplexicaule* (Henbit Deadnettle) Prior.

HENBIT, Lesser *Veronica hederifolia* (Ivy-leaved Speedwell) Gerard.

HENBIT DEADNETTLE *Lamium amplexicaule*.

HENCRESS *Capsella bursa-pastoris* (Shepherd's Purse) Dawson. Cf. the Wiltshire Crow-feed for this plant.

HENDERSON SHOOTING STAR *Dodecatheon hendersonii*.

HENEP *Ajuga chamaepitys* (Yellow Bugle) Gunther. That is a gloss on a 12th century ms of Apuleius Barbarus.

HENEQUEN *Agave fourcroydes*.

HENKAM *Hyoscyamus niger* (Hanbane) Halliwell.

HENNA *Lawsonia inermis*. Arabic hinna.

HENPEN 1. *Hyoscyanus niger* (Henbane) Cumb. Grigson. 2. *Rhinanthus crista-galli* (Yellow Rattle) Halliwell.

HENPENNY 1. *Hyoscyanus niger* (Henbane) Cumb. Grigson. 2. *Melampyrum pratense* (Common Cow-wheat) Lancs. Grigson. 3. *Rhinanthus crista-galli* (Yellow Rattle) Yks. F K Robinson; N'thum. Grigson.

HENRY, Good *Chenopodium bonus-henricus* (Good King Henry) Gerard.

HENRY LILY *Lilium henryi*. The name commemorates Professor Augustine Henry (1857-1930), who was the original discoverer of this lily.

HENS *Dactylorchis maculata* (Heath Spotted Orchid) Border. B & H.

HENS, Chicks-and- *Sempervivum tectorum* (Houseleek) USA Cunningham. See Hen-and-chickens.

HENS, Cocks-and- 1. *Geum rivale* (Water Avens) N'thum. Grigson. 2. *Lotus corniculatus* (Bird's-foot Trefoil) IOM Moore.1924. 3. *Plantago lanceolata* (Ribwort Plantain) Dev. Friend.1882; Som, N'thum. Grigson; Ire. Denham/Hardy. Ribwort has a number of similar names, all referring to the game known as "soldiers" or "fighting cocks", rather like conkers, but the aim is to decapitate one flowering stem by using another.

HEP *Rosa canina* fruit (Hip) Dor. Macmillan; Lancs. Nodal; Yks. Addy.

HEP, Cat *Rosa pimpinellifolia* (Burnet Rose) Berw, N Eng. Grigson. The shrub itself is often known as Cat-rose in the north, presumably as a distinction from the Dog-rose.

HEP BRIAR; HEP TREE *Rosa canina* (Dog Rose) Ches. B & H (Briar); Gerard (Tree).

HEP-THORN *Crataegus monogyna* (Hawthorn) Tynan & Maitland.

HERB-FLOWER *Ajuga reptans* (Bugle) Dor. Grigson.

HERB-LILY *Alstroemaria aurantiaca* (Peruvian Lily) Salisbury.

HERB-OF-THE-SEVEN-CURES *Achillea millefolium* (Yarrow) Ire. Wood-Martin.

HERB OF VINE *Asperula cynanchica* (Squinancywort) Gerard.

HERBGRASS; HERB-A-GRASS; HERBY-GRASS *Ruta graveolens* (Rue) Lupton (Herbgrass); Yks. Addy (Herb-a-grass); Clair (Herbygrass). All originating as Herb of Grace. Holy water was sprinkled from brushes made of rue.

HERCULES' CLUB 1. *Aralia spinosa*. 2. *Xanthoxylum clava-herculis* (Southern Prickly Ash) Howes.

HERCULES' WOUNDWORT *Prunella vulgaris* (Self-heal) O P Brown.

HERD'S GRASS *Phleum pratense* (Timothy) USA Cunningham.

HERIF(F) *Galium aparine* (Goose-grass) Hants. Cope; War. Bloom. One of a long sequence, including Airif, Hairiff, Hairup, etc.

HERNIARY *Herniaria ciliata* (Rupture-wort) Le Strange.

HERON'S BILL 1. *Erodium cicutarium* (Storksbill) S Africa Watt; USA. Elmore. 2. *Erodium moschatum* (Musk Storksbill) S Africa Watt. 3. *Erodium reichardii*.

HERON'S BILL, Desert *Erodium texanum*.

HERRINGBONE COTONEASTER *Cotoneaster horizontalis* Perry.

HERTECLOWRE *Teucrium chamaedrys* (Germander) Halliwell.

HERTFORDSHIRE WEED *Sambucus nigra* (Elder) Herts. Jones-Baker. In the same vein as Wiltshire Weed for elm, or Sussex Weed for oak.

HERTS *Vaccinium myrtillus* (Whortleberry) W Eng. Halliwell. It is more usual to find this as Hurts; both are variants of whorts.

HERTWORT *Fraxinus excelsior* (Ash) B & H.

HESSEL *Corylus avellana* (Hazel) Yks. Addy. To emphasise local pronunciation, i.e. soft 's' instead of hard 'z', which is the usual sound.

HETTLE *Urtica dioica* (Nettle) Glos. J D Robertson; War. F Savage. Apparently genuine, and not just a misreading of nettle. After all, Ettle is quite widespread, so why not Hettle?

HEVER; EVER *Conium maculatum* (Hemlock) Dor. Grigson; Heref. Havergal.

HEXHAM SCOUT *Melilotus indica* (Small-flowered Melilot) S.Africa Watt.

HEYHOVE; HEYHOWN *Glechoma hederacea* (Ground Ivy) I B Jones (Heyhove); Dawson (Heyhown). Hey = hay = hedge (OE haga). The second syllable is the same as that in Alehoof, and is the same as the modern verb, to heave, meaning in this case to ferment.

HEYRIFFE *Galium aparine* (Goose-grass) Storms.

HEZZLE *Corylus avellana* (Hazel) Yks, Cumb. Nicholson.

HI-HOW *Anthriscus sylvestris* (Cow Parsley) Ire. Grigson. Cf. Ha-ho, Da-ho, both Irish names, and obviously pronunciation guides to a Gaelic name.

HIBA *Thujopsis dolobrata.* A Japanese tree, and a Japanese name.

HIBBERT BEAN *Phaseolus lunatus* (Lima Bean) Chopra.

HIBBIN *Hedera helix* (Ivy) IOM Moore. Cf. Hyvin, or Hyven.

HIBISCUS, Hemp-leaved *Hibiscus cannabinus* (Ambari Hemp) Dalziel.

HIBISCUS, Norfolk Island *Lagunaria patersonii.* Not confined to Norfolk Island, but that was where it was first discovered, in 1792.

HIBISCUS, Tree, Rhodesian *Thespesia garckeana.*

HICBERRY *Prunus padus* (Bird Cherry) West B & H. A variant of Hackberry, which appears in many guises, e g Heckberry, Hagberry, Hawkberry, Eggberry, etc. The first syllable means hedge (OE haga).

HICKEN *Sorbus aucuparia* (Rowan) N Eng. Grigson. i.e. Whicken, which is the same as Quicken (OE cuic - alive).

HICKORY, Big-bud *Carya tomentosa* (White Hickory) S M Robertson. Hickory is a clipped version of the Virginian powcahicora or powicherry, which seems to have been the liquor expressed from the nuts. The Aztec was apparently xicali, modified by the Spaniards into jicara. The things that are made from the hard shells of hickory fruits are also known as jicara, which became chichara in Italian, for a teacup (see Potter & Sargent).

HICKORY, Bitternut *Carya cordiformis* (Swamp Hickory) USA Cunningham. Butternut is more likely; it is probably a misprint - there are plenty in Cunningham & Côté.

HICKORY, Pignut *Carya cordiformis* (Swamp Hickory) Schery.

HICKORY, Red *Carya ovalis.*

HICKORY, Shagbark *Carya ovata.* Shagbark, because the bark curls away into strips.

HICKORY, Shellbark *Carya ovata* (Shagbark Hickory) Everett.

HICKORY, Shellbark, Big *Carya laciniosa.*

HICKORY, Swamp *Carya cordiformis.*
HICKORY, Water *Carya aquatica.*
HICKORY, White *Carya tomentosa.*
HICKORY ELM *Ulmus racemosa* (Cork Elm) Everett.

HICKORY PINE *Pinus aristata* (Bristlecone Pine) Everett.

HICKYMORE *Centaurea nigra* (Knapweed) Som. Macmillan. Better known as Hackymore, the first element of which appears again in the Wiltshire Hardhack, which in turn links the name with Hardhead and its variants.

HIDGY-PIDGY *Urtica dioica* (Nettle) Dev. Macmillan. Cf. the Scottish version, heg-peg. In both cases, the name is better kept for the hedge plant par excellence, hawthorn, and for its fruits in particular.

HIG-TAPER; HIGH-TAPER; HIGGIS-TAPER *Verbascum thapsus* (Mullein) Gerard (Hig); Parkinson.1640 (High); Turner (Higgis). Cf. Hag-taper, Hedge Taper, which is what all the variants must mean. Taper, because the stems and leaves were dipped in tallow or suet, and burnt to give light at outdoor country gatherings, or in the home. They were used at funerals, too.

HIGAN CHERRY *Prunus cerasus var. subhirtella.*

HIGH GERANIUM *Hydrangea x macrophylla* (Hydrangea) Dev. Choape. A local variation on Hydrangea.

HIGH MALLOW *Malva sylvestris* (Common Mallow) O P Brown.

HIGHBUSH BLACKBERRY *Rubus allegheniensis.*

HIGHBUSH BLUEBERRY *Vaccinium corymbosum* (Highbush Huckleberry) USA Yarnell.

HIGHBUSH CRANBERRY *Viburnum trilobum* (American Cranberry-bush) USA Cunningham.

HIGHBUSH HUCKLEBERRY *Vaccinium corymbosum.*

HIGHCLERE HOLLY *Ilex x altaclarensis.* Highclere, in Berkshire, where this particular hybrid holly apparently first arose.

HIGHLAND BREADFRUIT *Ficus dammaropsis.*
HIGHLAND CUDWEED *Gnaphalium norvegicum.*
HIGHLAND FLEABANE *Erigeron borealis.*
HIGHLAND SAXIFRAGE *Saxifraga rivularis.*
HIGHWAY THORN *Rhamnus cathartica* (Buckthorn) Fernie. Cf. the Shropshire name, Waythorn.
HILDER *Sambucus nigra* (Elder) Norf. Halliwell. A name with Scandinavian influence - Cf. Danish hylde, and then such mythological figures as Frau Helde.
HILDER, Blood *Sambucus ebulus* (Dwarf Elder) Norf. Fluckiger. The elder with the blood-red stems.
HILES *Hordeum sativum* (Barley) Corn. Courtney.1880. A variant of Ailes or Iles.
HILLERNE *Sambucus nigra* (Elder) Halliwell.
HILTROT 1. *Anthriscus sylvestris* (Cow Parsley) Wilts. Dartnell & Goddard. 2. *Daucus carota* (Wild Carrot) Hants. Cope. 3. *Heracleum sphondyllium* (Hogweed) Wilts. Dartnell & Goddard. 4. *Oenanthe crocata* (Hemlock Water Dropwort) Wilts. Dartnell & Goddard. Eltrot is more frequent than this version, which is spelt with one 'l' or two.
HILLWORT 1. *Mentha pulegium* (Pennyroyal) B & H. 2. *Thymus drucei* (Wild Thyme) Dawson. Presumably in both cases connected with pulegium montanum, or is it serphyllum montanum?
HILPS *Prunus spinosa* fruit (Sloes) Wilts. Dartnell & Goddard.
HIMALAYAN BALSAM *Impatiens glandulifera.*
HIMALAYAN BIRCH *Betula utilis.*
HIMALAYAN BIRCH, White-barked *Betula jacquemontii.*
HIMALAYAN BLUE PINE *Pinus wallichiana* (Bhutan Pine) Polunin.1976.
HIMALAYAN BLUE POPPY *Meconopsis betonicifolia.*
HIMALAYAN CEDAR *Cedrus deodara.*

HIMALAYAN CHIR PINE *Pinus roxburghii.*
HIMALAYAN CINQUEFOIL *Potentilla atrosanguinea.*
HIMALAYAN COMFREY *Onosma pyramidale.*
HIMALAYAN COWSLIP *Primula florindae.*
HIMALAYAN CYPRESS *Cupressus torulosa.*
HIMALAYAN ELM, West *Ulmus villosa.*
HIMALAYAN HEMLOCK *Tsuga dumosa.*
HIMALAYAN HOLLY *Ilex dipyrena.*
HIMALAYAN HONEYSUCKLE *Leycesteria formosa.*
HIMALAYAN HOUND'S TONGUE *Cynoglossum nervosum.*
HIMALAYAN JUNIPER *Juniperus squamosa.*
HIMALAYAN KNOTWEED 1. *Polygonum campanulatum.* 2. *Polygonum polystachyum.*
HIMALAYAN LABURNUM *Sophora mollis.*
HIMALAYAN LARCH *Larix griffithii.*
HIMALAYAN LILAC *Syringa emodi.*
HIMALAYAN MANNA ASH *Fraxinus floribunda.*
HIMALAYAN PEONY *Paeonia emodi.*
HIMALAYAN PINE *Pinus wallichiana* (Bhutan Pine) Barber & Phillips.
HIMALAYAN PRIMROSE *Primula rosea.*
HIMALAYAN SILVER FIR *Abies pindrow* (West Himalayan Fir) Usher.
HIMALAYAN SPRUCE *Picea smithiana.*
HIMALAYAN WEEPING JUNIPER *Juniperus recurva* (Drooping Juniper) RHS.
HIMALAYAN WHITEBEAM *Sorbus cuspidata.*
HIMALAYAN YEW *Taxus wallichiana.*
HIN *Chenopodium album* (Fat Hen) Ess. Gepp. Local pronunciation of hen.
HIND-HEAL 1. *Eupatorium cannabinum* (Hemp Agrimony) Cockayne. 2. *Teucrium scorodonia* (Wood Sage) Prior. Maud Grieve did suggest that there was a belief that the hind used Wood Sage when sick or wounded, but surely 'hind' here is the same as 'peasant', so that the whole would be the

same as the much commoner Clown's All-heal?

HINDBERRY *Rubus idaeus* (Raspberry) Turner. Turner used the name, and it is recorded regionally from Staffs, Yks, Lancs, Cumb,N'thum. and Scot. (Grigson). Hineberry, and Hainberry, are used as well.

HINDHEEL *Tanacetum vulgare* (Tansy) Halliwell.

HINDU LOTUS *Nelumbo nucifera*.

HINEBERRY *Rubus idaeus* (Raspberry) Yks. Addy; Cumb. B & H. See Hindberry.

HINNYSICKLE *Lonicera periclymenum* (Honeysuckle) Border B & H. It should not really be included as a separate name at all - this is just the result of local pronunciation.

HINOKI CYPRESS *Chamaecyparis obtusa* (Fire Cypress) Wilks. The timber is known in Japan as hinoki wood.

HIP *Rosa canina* fruit. OE hēope.

.HIP, Cat *Rosa canina* (Dog Rose) B & H.

HIP, Dog *Rosa canina* fruit (Hip) Scot. Grigson.

HIP BRAMBLE *Rosa canina* (Dog Rose) Young.

HIP BRIAR; HIP ROSE; HIP TREE *Rosa canina* (Dog Rose) Glos, Shrop. Grigson (Briar, Rose); Worcs. J Salisbury (Briar); Glos, N'thum. Grigson USA Fogel (Tree).

HIP HAW; HIPPERTY HAW *Crataegus monogyna* fruit (Haw) Oxf, West. both Grigson. So, in Shropshire, the hawthorn is "Hipperty-Haw Tree".

HIPPAN; HIPPAN, Dog *Rosa canina* fruit (Hip) Mor. Grigson (Hippan); Aber. Grigson (Dog Hippan).

HIPPO ROOT *Euphorbia corollata* (Flowering Spurge) Le Strange.

HIPSON *Rosa canina* fruit (Hip) Oxf. Grigson. This sounds very like a double plural, one in 's' and the other in 'en'.

HIPTYPIPS *Rosa canina* fruit (Hip) Herts. Jones-Baker.1974.

HIPWORT *Umbilicus rupestris* (Wall Pennywort) Coles. Either from the resemblance of the leaf to the hip socket, which is Prior's contention, or "for that it easeth the paines of the hippes", which is what Coles said. Actually, doctrine of signatures would have produced Coles's prescription, so both are right.

HIRSE *Panicum miliaceum* (Millet) Turner.

HISSING TREE *Parinari mobola* (Sand Apple) Palgrave. It is believed that if the tree is chopped, it hisses.

HOAR WITHY *Sorbus aria* (Whitebeam) Hants. Boase. "Grey willow", not at all a bad description of whitebeam.

HOARHOUND *Marrubium vulgare* (White Horehound) S Africa Watt; USA O P Brown. 'Hoar' is as good as 'hore', for the name derives from OE har, hoar, or grey. 'Hound' is a plant name, hune.

HOARHOUND, Water *Lycopus americanus*.

HOARSTRANGE; HOARSTRONG *Peucedanum officinale* (Hog's Fennel) both Culpepper. The name exists in several other forms - Horestrong, Horestrange, Harstrong, or Harestrong, all apparently from the German Haarstrang, literally hair rope.

HOARY ALDER *Alnus incana*.

HOARY BASIL 1. *Ocymum americanum* (American Basil) Dalziel. 2. *Ocymum canum*. Is 'americanum' a misreading for 'canum'?

HOARY CISTUS *Cistus incanus*.

HOARY GROUNDSEL *Senecio erucifolius* (Hoary Ragwort) Grieve.1931.

HOARY MULLEIN *Verbascum pulverulentum*.

HOARY RAGWORT *Senecio erucifolius*.

HOARY VERVAIN *Verbena stricta*.

HOARY WILLOW *Salix candida* (Sage Willow) Youngken.

HOBBLE, Witch; HOBBLE-BUSH *Viburnum alnifolium* Hobble-bush is the usual name; Witch-hobble is in A W Smith. Other names like Tangle-legs, Down-you-go, etc., occur because the lower branches droop to the ground and root at the tips, so the 'hobble' names must derive from the same phenomenon.

HOBBLE-GOBBLES 1. *Arum maculatum* (Cuckoo-pint) Kent Grigson. 2. *Caltha* *palustris* (Marsh Marigold) Tynan & Maitland.

HOBBLY-FLOWER *Aesculus hippocastanum* the flowers (Horse Chestnut) Som. Macmillan. From Hobbly-Honkers, a Somerset name for the nuts, and a variation of Conkers.

HOBLIONKERS; HOBBLY-HONKERS *Aesculus hippocastanum* nut (Horse Chestnut) Worcs. J Salisbury (Hoblionkers); Som. Macmillan (Hobbly-honkers). Better known perhaps as Oblionkers - conkers in other words, itself from conquerors, from the game played with them.

HOBURN SAUGH *Laburnum anagyroides* (Laburnum) Scot. Jamieson. 'Hoburn' is auburn, the colour of the wood.

HOCK 1. *Althaea rosea* (Hollyhock) B & H. 2. *Malva sylvestris* (Common Mallow) Dawson. OE hoc meant mallow.

HOCK, Mallow *Malva sylvestris* (Common Mallow) Som. Macmillan. A pleonasm, as hock means mallow.

HOCK-HOLLER *Althaea rosea* (Hollyhock) Som. Macmillan. The 'holly' of hollyhock means 'holy'. It seems that this is a reference to the fact that it was brought to Europe from the Holy Land.

HOCK-LEAF; HOCKHERB *Malva sylvestris* (Common Mallow) Cockayne (Hock-leaf); Clair (Hockherb).

HOCKERIE-TOPNER *Sempervivum tectorum* (Houseleek) Dumf. Grigson. The second element of the name may be the same as the Irish Tourpin, or Turpeen.

HOCKLE ELDERBERRY *Anamirta paniculata* (Levant Nut) Le Strange.

HOCKWEED *Pastinaca sativa* (Wild Parsnip) Wilts. D Grose.

HOD-THE-RAKE *Ranunculus repens* (Creeping Buttercup) Cumb. B & H. Obviously a gardener's name!

HODROD *Primula veris* (Cowslip) Dor. Grigson. This appears in the same county as Oddrod, or Holrod.

HOG-A-BACK *Succisa pratensis* (Devil's-bit) Cumb. B & H.

HOG APPLE *Podophyllum peltatum* (Mayapple) Hatfield.

HOG-BERRY 1. *Crataegus monogyna* fruit (Haw) Hants. Cope. 2. *Prunus padus* fruit (Bird Cherry) Hants. Grigson. In a lot of cases, 'hog' is OE hag, hedge.

HOG BRAKE 1. *Ambrosia artemisiaefolia* (American Wormwood) Howes. 2. *Pteridium aquilinum* (Bracken) USA Cunningham.

HOG CHERRY *Prunus padus* (Bird Cherry) W Miller.

HOG CRANBERRY *Arctostaphylos uva-ursi* (Bearberry) H & P.

HOG DOCTOR TREE *Bursera simaruba*.

HOG-GOSSE *Crataegus monogyna* fruit (Haw) Suss. Grigson. There is another form of this, Hogasses, which makes it easier to see, for this is more properly hog-hawses, a double plural.

HOG-GRASS *Coronopus squamatus* (Wart-cress) War. Grigson. Cf. Sow-grass, Swine's-cress, etc. The message in the names is that it is a cress fit only for pigs.

HOG GUM 1. *Clusia flava*. 2. *Metopium toxiferum* (Poison-wood) Usher. The stem of Poison-wood yields a resin, and this is what is known as Hog Gum, or Doctor's Gum, for it is a violent purgative. As far as Clusia flava is concerned, Frances Perry said it was so called because injured swine rub against the plant, and the resin heals them.

HOG-HAGHES; HOG-HAW *Crataegus monogyna* fruit (Haw) both Hants. Cope (Hog-haghes); Grigson (Hog-haws). Cf. Hog-gosse, etc.

HOG-HAZEL *Crataegus monogyna* fruit (Haw) Surr, Suss, Kent Wakelin. As in the previous entry.

HOG MILLET *Panicum miliaceum* (Millet) Brouk.

HOG PEANUT *Amphicarpa bracteata*.

HOG PLUM 1. *Prunus americana* (Canada Plum) Elmore. 2. *Prunus umbellata*. 3. *Spondias monbin* (Monbin) Dalziel. 4. *Spondias purpurea*. Pigs like the fruit of *Spondias purpurea*, and fatten upon them. But why should Canada Plum get the derogatory name, for the fruits are very popular in that country, cooked in a variety of ways?

HOG'S BEANS 1. *Aster tripolium* (Sea Aster) Ess. Gerard. 2. *Hyoscyamus niger* (Henbane) Som. Macmillan. "About Harwich", Gerard said, talking about Sea Aster, "it is called Hogs Beans, for that the swine do greatly desire to feed thereon, as also for that the knobs about the roots do somewhat resemble the garden bean". For henbane, the name is sometimes spelt as one word, Hogbean. Cf. the French fève à cochons. Actually they are both translations of Hyoscyamus. Dioscorides gave it this name, because although to man and many animals it is poisonous, pigs could apparently eat it without harm.

HOG'S FENNEL 1. *Peucedanum officinale*. 2. *Peucedanum palustre* (Milk Parsley) Clapham.

HOG'S FENNEL, Marsh 1. *Peucedanum officinale* (Hog's Fennel) Grieve.1931. 2. *Peucedanum palustre* (Milk Parsley) C P Johnson.

HOG'S FENNEL, Sea Peucedanum officinale (Hog's fennel) Barton & Castle.

HOG'S GARLIC *Allium ursinum* (Ramsons) B & H.

HOGAIL; HOGARVE; HOGASSES; HOGAZEL *Crataegus monogyna* fruit (Haw) IOW Long (Hogail); Wilts. Tynan & Maitland; Surr. Grigson; Suss. Parish (Hogarve); Wakelin (Hogasses); Suss. Parish (Hogazel). All contractions of their counterparts - Hog Hazel, etc.

HOGBEAN *Hyoscyamus niger* (Henbane) Cumb. Grigson. See Hog's Beans rather.

HOGGAN; HOGGIN *Crataegus monogyna* fruit (Haw) both Corn. Bottrell (Hoggan); Briggs.1967 (Hoggin).

HOGNUT *Conopodium majus* (Earth-nut) B & H. Pignut is better known than this.

HOGWEED 1. *Erigeron canadensis* (Canadian Fleabane) S Africa Watt. 2. *Heracleum sphondyllium*. 3. *Papaver rhoeas* (Corn Poppy) E Ang. Grigson. 4. *Polygonum aviculare* (Knotgrass) Beds. Grigson; Norf. Halliwell. 5. *Sonchus arvensis* (Corn Sow Thistle) Northants. A E Baker. 6. *Torilis japonica* (Hedge Parsley) Glos. Grigson. 7. *Tussilago farfara* (Coltsfoot) Yks. Grigson. Presumably, all these have the name because they have been used as pig food - Hogweed proper certainly was, and so was Sow Thistle. Knotgrass has a number of 'pig' names in addition to this one - Pigweed, Pig-rush, Pig-grass, Swine-grass, etc. (it is herbe à cochons in French, too).

HOGWEED, Giant *Heracleum montegazzianum*. 'Giant' is right - this can be 10 feet tall or more.

HOGWORT *Croton texensis*.

HOKY-POKY *Urtica dioica* (Nettle) Dev. Macmillan.

HOLDFAST *Ononis repens* (Rest Harrow) Tynan & Maitland. A name with the same import as the common name, "because it maketh the Oxen whilst they be in plowing to rest or stand still" (Gerard).

HOLE *Corylus avellana* (Hazel) Leyel.1948. Usually rendered as Haul, which is a variant of hazel, through Halse and Hale.

HOLENE; HOLYN *Ilex aquifolium* (Holly) both Scot. Jamieson.

HOLES, Hundred *Hypericum perforatum* (St John's Wort) Fernie. They aren't holes, of course, in spite of the *perforatum* of the specific name. The glandular dots which can be seen when held up to the light on the under side of the leaves look like perforations, but these are the oil sacs which give the plant its aromatic odour when bruised.

HOLES, Thousand *Hypericum hirsutum* (Hairy St John's Wort) Yks. Grigson. A tenfold improvement on the previous entry.

HOLEWORT 1. *Adoxa moschatellina* (Moschatel) W Miller. 2. *Corydalis tuberosa* Turner. They are both "*tuberosa*", hence the name, and also Hollow-wort, or Hollow-root.

HOLIGOLD *Calendula officinalis* (Marigold) Cockayne.

HOLLARD *Alnus glutinosa* (Alder) Som. B & H. It must be one of the series which includes Aller, Ouler, Howler, etc. for alder.

HOLLAND SMOCK *Calystegia sepium* (Great Bindweed) Som. Macmillan. The 'smock' part of the name is because they remind one of clothes spread out on the bushes to dry - they have been seen as smocks, shirts, or shimmies. Is 'Holland' here a kind of material, or is it rather the same as 'Our Lady'?

HOLLEN; HOLLIN *Ilex aquifolium* (Holly) N'thum. Grose (Hollen); Shrop, Ches, IOM, Lancs, Notts, Lincs, Derb, N'thum, Scot. Grigson; Yks. Morris; Dur. Dinsdale (Hollin). Hollin seems to be the usual North Country name for holly. Cf. Hullin, the version recorded in the Isle of Man.

HOLLICK; OLLICK *Sempervivum tectorum* (Houseleek) Corn. Jago.

HOLLIN, Prick *Ilex aquifolium* (Holly) Lincs, Yks. Nicholson.

HOLLIN-TRAIE *Eryngium maritimum* (Sea Holly) IOM Moore.1924. Manx hollyn hraie, 'strand holly', with the same meaning as sea holly.

HOLLIOCK; HOLLOAK *Althaea rosea* (Hollyhock) Lupton (Holliock); Hill (Holloak).

HOLLOND *Ilex aquifolium* (Holly) Lincs. B & H.

HOLLOW-CRESS *Gentianella campestris* (Field Gentian) Cockayne.

HOLLOW-ROOT; HOLLOW-WORT 1. *Adoxa moschatellina* (Moschatel) W Miller. 2. *Corydalis tuberosa* Lyte, Gerard. Cf. Holewort.

HOLLY *Ilex aquifolium*. OE holegn, from which names like Holyn, Hollin, etc., are recognisable.

HOLLY, American *Ilex opaca*.

HOLLY, Berry *Ilex aquifolium* (Holly) Som, Wilts. Grigson.

HOLLY, Box *Ruscus aculeatus* (Butcher's Broom) Bianchini.

HOLLY, Box-leaved *Ilex crenata*.

HOLLY, California *Photinia arbutifolia*.

HOLLY, Canary *Ilex perado*. Or Madeira Holly.

HOLLY, Cape *Ilex mitis*.

HOLLY, Free *Ilex aquifolium* (Holly) Shrop. B & H. Reserved for the smooth-leaved variety, often known as She-holly. Free Holly is either a corruption of She-holly, or given. to indicate that this kind is free of prickles.

HOLLY, Ground 1. *Chimophila umbellata* (Umbellate Wintergreen) Le Strange. 2. *Gaultheria procumbens* (Wintergreen) Genders.1972.

HOLLY, He *Ilex aquifolium* (Holly) N'thum. Denham/Hardy. He-holly if with prickles, and She-holly if without.

HOLLY, Highclere *Ilex x altaclarensis*. A hybrid between common Holly and Canary Holly, apparently arising at Highclere in Berkshire.

HOLLY, Himalayan *Ilex dipyrena*.

HOLLY, Horned *Ilex cornuta*. "Horned", because the leaf spines at the apex of the leaf are arranged so that they look like horns.

HOLLY, Japanese *Ilex crenata* (Box-leaved Holly) RHS.

HOLLY, Jungle *Taxotrophis ilicifolia* Malaya Gilliland. Because of its use at Christmas time, Gilliland says, but the specific name is *ilicifolia*, anyway.

HOLLY, Knee *Ruscus aculeatus* (Butcher's Broom) Halliwell. Cf. Knee Holme, Knee Hulver, etc.

HOLLY, Madeira *Ilex perado* (Canary Holly) Hora.

HOLLY, Mountain *Nemopanthus mucronata*.

HOLLY, New Zealand *Olearia macrodonta*.

HOLLY, Perny's *Ilex pernyi*.

HOLLY, Prick *Ilex aquifolium* (Holly) Lincs, Yks. Nicholson.

HOLLY, Rhodesian *Psorospermum febrifugum var. ferrugineum*. It has got small red berries, like a holly.

HOLLY, Sea *Eryngium maritimum*.

HOLLY, Sea, Levant *Eryngium campestre* (Field Eryngo) Gerard.

HOLLY, Sea, Oliver's *Eryngium amethystianum*.

HOLLY, She *Ilex aquifolium* (Holly) N'thum. Denham/Hardy. Reserved for the kind which has no prickles, just as He-holly refers only to the prickly sort.

HOLLY FERN *Polystichum lonchitis*. 'Holly', because the leaflets have spiny teeth.

HOLLY GRAPE, Cluster *Berberis pinnata*.

HOLLY GRAPE, Creeping *Berberis repens*.

HOLLY GRAPE, Fremont *Berberis fremontii* (Desert Barberry) USA Jaeger.

HOLLY GRAPE, Longleaf *Berberis nervosa*.

HOLLY GRAPE, Oregon *Mahonia aquifolium* (Oregon Grape) Yanovsky.

HOLLY GRAPE, Red *Berberis haematocarpa*.

HOLLY GREVILLEA *Grevillea wickhamii*.

HOLLY-LEAF BUR-SAGE *Franseria icifolia*.

HOLLY-LEAF CHERRY *Prunus ilicifolia*.

HOLLY-LEAF SPURGE *Tetracoccus ilicifolius*.

HOLLY-LEAVED BARBERRY *Mahonia aquifolium* (Oregon Grape) Grieve.1931.

HOLLY OAK 1. *Althaea rosea* (Hollyhock) A S Palmer. 2. *Quercus ilex* (Holm Oak) Gerard. 3. *Quercus pungens* (Scrub Oak) USA Elmore. Holly Oak for Hollyhock arose via the Holioke spelling of the word.

HOLLY ROSE *Cistus sp* Gerard.

HOLLY ROSE, Sweet *Cistus ladaniferus* (Gum Cistus) Parkinson.

HOLLYANDERS *Althaea rosea* (Hollyhock) Som. Macmillan.

HOLLYHOCK 1. *Althaea rosea*. 2. *Argemone mexicana* (Mexican Poppy) Barbados Gooding. 3. *Digitalis purpurea* (Foxglove) Dev. Macmillan. 4. *Lavatera arborea* (Tree Mallow) J Harvey. Hock means a mallow plant, so Hollyhock means Holy Mallow, which is actually recorded as a name. It is thought to be derived from the plant having been brought back to Europe by the Crusaders. The wild plant does still grow profusely in Palestine. Potter & Sargent say there is no evidence that the plant was originally brought from the Holy Land, but go on to say there *is* evidence of a connection with St Cuthbert, and suggest there was a medieval Latin name malva benedicta, which would translate easily into Holy Mallow.

HOLLYHOCK, Yellow *Argemone mexicana* (Mexican Poppy) Barbados Gooding.

HOLLYSEDGE *Eryngium maritimum* (Sea Holly) Cockayne.

HOLM 1. *Ilex aquifolium* (Holly) Dev. Friend.1882; Som. Raymond; Dor. Dacombe; Shrop, Ches, Notts, Derb, Lincs, Lancs, Cumb, Yks. Grigson. 2. *Ulmus procera* (Elm) Cumb, Yks. Grigson. Holm replaces holly in a number of areas - they are both from the same OE holegn. Holm for elm is quite different, and is just a variation of the pronunciation.

HOLM, Berry *Ilex aquifolium* (Holly) Som. Grigson.

HOLM OAK; HOLM OAK, Scarlet *Quercus ilex*. Scarlet Holm Oak was Gerard's name for it. The kerm, an insect which this tree harbours, is not unlike a holly berry to look at, hence the name.

HOLME *Ulmus glabra* (Wych Elm) Cumb, Yks. Grigson. Holme is a variation of elm.

HOLME, Knee *Ruscus aculeatus* (Butcher's broom) Gerard.

HOLME, Sea *Eryngium maritimum* (Sea Holly) Gerard.

HOLN *Ilex aquifolium* (Holly) Dev. B & H. Cf. Holm, a widespread version of holly.

HOLROD *Primula veris* (Cowslip) Dor. Macmillan. Cf. Hodrod, or Oddrod, from the same county.

HOLTROT *Heracleum sphondyllium* (Hogweed) Wilts. Goddard.

HOLY BASIL *Ocymum sanctum*. This is the sacred basil of India, a protector from all misfortunes and disease. From time out of mind, it has been cultivated as tulsi, a plant sacred to Vishnu.

HOLY CLOVER *Onobrychis sativa* (Sainfoin) Wit.

HOLY FLAX *Santolina rosmarinifolia*.

HOLY GHOST 1. *Angelica archangelica* (Archangel) Grieve. 2. *Angelica sylvestris* (Wild Angelica) Dyer. This was the medieval radix Sancti Spiritus, given because of its "great and divine properties". Sometimes one comes across mention of Holy Ghost pie, used apparently in the Black Mass.

HOLY GRASS *Hierochloe odorata* (Sweet Grass) Vestal & Schultes.

HOLY HAY *Medicago sativa var. sativa* (Lucerne) Prior. This is the result of the legend that dried hay was used in Christ's manger. It began to blossom, and grew to encircle the baby's head.

HOLY HERB 1. *Hyssopus officinalis* (Hyssop) Hatfield. 2. *Verbena officinalis* (Vervain) Som. Grigson. Hyssop is OE ysope, from Latin hyssopus, Greek hussopus, from Hebrew esob, hence holy herb. But vervain is the holy herb par excellence, and it has always been known as such.

HOLY INNOCENTS *Crataegus monogyna* (Hawthorn) Wilts. Macmillan.

HOLY MALLOW *Althaea rosea* (Hollyhock) Blunt.1957. The 'hock' of hollyhock means mallow.

HOLY PLANT *Angelica sylvestris* (Wild Angelica) Jacob. See Holy Ghost.

HOLY POKER *Typha latifolia* (False Bulrush) Dev. Macmillan.

HOLY ROPE 1. *Eupatorium cannabinum* (Hemp Agrimony) Prior. 2. *Galeopsis tetrahit* (Hemp Nettle) B & H. 'Hemp' is the link here. Both bear some resemblance in their leaves to those of hemp, and the name Holy Rope is a remembrance of the rope with which Jesus was bound.

HOLY ROSE *Rosa sancta* (Abyssinian Rose) Grigson.1976. Grigson explains that this rose, a form of Summer Damask, comes from Ethiopia, where it was grown in Christian premises in the province of Tigre, around the holy city of Axum.

HOLY ROSE, Marsh *Andromeda polifolia* (Bog Rosemary) Prior.

HOLY THISTLE *Carduus benedictus*. When a plant is called 'holy', or 'blessed', it means it has the power of counteracting poison, or so it was supposed. Cf. Blessed Thistle.

HOLY THORN 1. *Berberis vulgaris* (Barberry) Vesey-Fitzgerald. 2. *Crataegus monogyna var. praecox* (Glastonbury Thorn) Hole.1957b. Glastonbury Thorn earns the name by its time of flowering, traditionally on Christmas Eve, old style.

HOLY THORN OF CHRISTMAS *Salix discolor* (American Black Willow) USA Bergen. It is said to blossom out at old Christmas, and then the flowers "go in again".

HOLY TREE *Ilex aquifolium* (Holly) Pratt. From 'holly', of course,but given weight because churches were decorated with it at Christmas.

HOLY VERVAIN *Verbena officinalis* (Vervain) Northall. It is the Holy Herb par excellence.

HOLY WATER SPRINKLE *Equisetum arvense* (Common Horsetail) Greenoak. The name likens the plant to a holy water brush.

HOLYN; HOLENE *Ilex aquifolium* (Holly) both Scot. Jamieson.

HOLYOKE *Althaea rosea* (Hollyhock) Langham.

HOME *Ilex aquifolium* (Holly) Dev. Friend.1882. It should really be holm - home is a dubious spelling locally.

HOMEWORT *Sempervivum tectorum* (Houseleek) W Miller.

HOMLOCK *Conium maculatum* (Hemlock) Turner.

HONDO SPRUCE; HONSHU SPRUCE *Picea jezoensis* (Yeddo Spruce) Leathart (Hondo); Wilkinson.1981 (Honshu).

HONDURAS MAHOGANY *Swietenia macrophylla*.

HONESTY 1. *Clematis vitalba* (Old Man's Beard) Wilts. Dartnell & Goddard; Glos. J D Robertson; Berks. Lowsley; Heref. Havergal; Oxf, War, Worcs, E Ang, Lancs, Yks. Grigson. 2. *Lunaria biennis*. It is said that wherever Purple Honesty flourishes in a garden, the gardener is exceptionally honest.

HONESTY, Maiden's 1. *Clematis vitalba* (Old Man's Beard) Aubrey. 2. *Lunaria annua* (Honesty) Halliwell. Cf. Virgin's Bower for Old Man's Beard, apparently coined as a tribute to Queen Elizabeth I.

HONESTY, Small *Dianthus plumarius*

(Pink) Gerard.

HONEWORT 1. *Aphanes arvensis* (Parsley Piert) Vesey-Fitzgerald. 2. *Sison ammonium* (Stone Parsley) Fernie. 3. *Trinia glauca*. "Painful swellings are in some parts of the kingdom called hones" (Hill). Fernie particularises the swellings as boils on the cheek. So these plants were used in some way to treat them.

HONEWORT, Corn *Petroselinum segetum* (Corn Parsley) Prior.

HONEY-BALM *Melittis melissophyllum* (Bastard Balm) McDonald.

HONEY-BEE *Lamium album* (White Deadnettle) Dev. Macmillan. Bee Nettle, or White Bee Nettle, has a much wider spread.

HONEY-BEE FLOWER *Ophrys apifera* (Bee Orchis) Hulme.

HONEY-BLOB *Ribes uva-crispa* (Gooseberry) Scot. B & H. The name exists also simply as Blob, which usually means a blister, and must be a reference to the shape of the fruit. Perhaps 'blob' just means round.

HONEY-CAP *Pedicularis palustris* (Red Rattle) Grigson.

HONEY-FLOWER 1. *Filipendula ulmaria* (Meadowsweet) Som. Macmillan. 2. *Lamium album* (White Deadnettle) Som. Macmillan. 3. *Ophrys apifera* (Bee Orchis) Kent B & H. 4. *Trifolium pratense* (Red Clover) Tynan & Maitland.

HONEY-FLOWER, Cape *Melianthus major*.

HONEY-FLOWER, Sham *Anacamptis pyramidalis* (Pyramidal Orchis) Som. Macmillan. So the 'real' Honey-flower is the Bee Orchis.

HONEY-LOCUST *Gleditschia triacanthos*. 'Honey' from the succulent pulp in which the seeds are embedded.

HONEY MESQUITE *Prosopis juliflora* (Mesquite) USA T H Kearney.

HONEY MYRTLE, Bracelet *Melalauca armillaris*.

HONEY-PLANT *Melissa officinalis* (Bee Balm) Clair.

HONEY-PLANTAIN *Plantago media* (Lamb's-tongue Plantain) Som. Macmillan.

HONEY TREE *Schleichera oleosa* (Ceylon Oak) Chopra.

HONEYBALL *Buddleia globosa* (Orange Ball Tree) Som. Elworthy.

HONEYBELLS *Campanula medium* (Canterbury Bells) Som. Macmillan.

HONEYBIND *Lonicera periclymenum* (Honeysuckle) Oxf. Grigson.

HONEYBOTTLE 1. *Calluna vulgaris* (Heather) Wilts. Macmillan. 2. *Erica tetralix* (Cross-leaved Heath) Som, Wilts. Macmillan. 3. *Ulex europaeus* (Gorse) Som. Macmillan; Wilts. Macmillan.

HONEYBUSH *Melianthus major* (Cape Honey Flower) RHS.

HONEYSOOKIES *Pedicularis sylvatica* (Lousewort) Shet. Grigson.

HONEYSTALK 1. *Trifolium pratense* (Red Clover) Shakespeare. 2. *Trifolium repens* (White Clover) War. Palmer.

HONEYSTICK *Clematis vitalba* (Old Man's Beard) Glos, War. B & H.

HONEYSUCK 1. *Lonicera periclymenum* (Honeysuckle) Dor, Som. Grigson. Hants. Cope. 2. *Trifolium pratense* (Red Clover) Som. Elworthy; Hants, Glos, War. Grigson; N'thants. A E Baker.

HONEYSUCKLE 1. *Ajuga reptans* (Bugle) Dor. Macmillan. 2. *Aquilegia canadensis* (Canadian Columbine) H & P. 3. *Azalea nudiflorum* (Pink Azalea) Alab. Harper. 4. *Beloperone californica* (Chuparosa) USA T H Kearney. 5. *Calystegia sepium* (Great Bindweed) Dev. Friend.1882. 6. *Cornus suecica* (Dwarf Cornel) Yks. B & H. 7. *Lamium album* (White Deadnettle) Wilts. Dartnell & Goddard. 8. *Lonicera periclymenum*. 9. *Lotus corniculatus* (Bird's-foot Trefoil) Ches. Holland. 10. *Pedicularis palustris* (Red Rattle) Donegal Grigson. 11. *Pedicularis sylvatica* (Lousewort) Hants. Grigson. 12. *Rhinanthus crista-galli* (Yellow Rattle) B & H. 13. *Trifolium pratense* (Red Clover) Gerard. 14. *Trifolium repens* (White Clover) Som, Midl. Grigson; Wilts. Dartnell & Goddard. As is said in Chambers' dictionary, the name would be applied because honey is readily sucked from the flower (by long-tongued insects only). *Lonicera* is the real honeysuckle, but the clovers in particular have large numbers of similar names.

HONEYSUCKLE, Alpine *Lonicera alpigena*.

HONEYSUCKLE, Bearberry *Lonicera involucrata*.

HONEYSUCKLE, Black-berried *Lonicera nigra*.

HONEYSUCKLE, Burmese, Great *Lonicera hildebrandiana*.

HONEYSUCKLE, Bush 1. *Diervilla lonicera*. 2. *Lonicera conjugialis*.

HONEYSUCKLE, Californian *Lonicera hispidula var. californica*.

HONEYSUCKLE, Cape *Tecomaria capensis*.

HONEYSUCKLE, Chaparral *Lonicera interrupta*.

HONEYSUCKLE, Coral *Lonicera sempervirens* (Trumpet Honeysuckele) USA House.

HONEYSUCKLE, Dutch *Lonicera periclymenum* (Honeysuckle) McDonald.

HONEYSUCKLE, Dwarf *Cornus suecica* (Dwarf Cornel) B & H.

HONEYSUCKLE, Etruscan *Lonicera etrusca*.

HONEYSUCKLE, Fly *Lonicera xylostemon*. All the species called Fly Honeysuckle get the name in error, for it was so dubbed in confusion with another fly-catching plant.

HONEYSUCKLE, Fly, American *Lonicera canadensis* (Early Fly Honeysuckle) Youngken.

HONEYSUCKLE, Fly, Blue *Lonicera caerulea*.

HONEYSUCKLE, Fly, Early *Lonicera canadensis*.

HONEYSUCKLE, Fly, Mountain 1. *Lonicera caerulea* (Blue Fly Honeysuckle) USA House. 2. *Lonicera villosa*.

HONEYSUCKLE, Fly, Swamp *Lonicera oblongifolia*.

HONEYSUCKLE, French *Hedysarum coronarium*.

HONEYSUCKLE, Goat-leaf *Lonicera caprifolium* (Perfoliate Honeysuckle) McDonald. The specific name, *caprifolium*, means goat-leaf, but it is a mistake - it was originally capparifolium, caper-leaf.

HONEYSUCKLE, Ground *Lotus corniculatus* (Bird's-foot Trefoil) Ches. Holland.

HONEYSUCKLE, Hairy *Lonicera hirsuta*.

HONEYSUCKLE, Himalayan *Leycesteria formosa*.

HONEYSUCKLE, Italian *Lonicera caprifolium* (Perfoliate Honeysuckle) Grieve.1931. Country of origin?

HONEYSUCKLE, Jamaica *Passiflora laurifolia*.

HONEYSUCKLE, Japanese *Lonicera nitida*.

HONEYSUCKLE, Little *Trifolium pratense* (Red Clover) Som. Macmillan.

HONEYSUCKLE, Meadow *Trifolium pratense* (Red Clover) Hulme.

HONEYSUCKLE, Orange *Lonicera ciliosa*.

HONEYSUCKLE, Perfoliate; HONEYSUCKLE, Perfoliate, Pale *Lonicera caprifolium* Barton & Castle (Pale).

HONEYSUCKLE, Pyrenean *Lonicera pyrenaica*.

HONEYSUCKLE, Red 1. *Banksia serrata*. 2. *Hedysarum coronarium* (French Honeysuckle) B & H. 3. *Trifolium pratense* (Red Clover) Culpepper.

HONEYSUCKLE, Scarlet *Lonicera sempervirens* (Trumpet Honeysuckle) Watt.

HONEYSUCKLE, Silvery *Grevillea striata* (Turraie) Everett.

HONEYSUCKLE, Swamp 1. *Azalea viscosa*. 2. *Lonicera involucrata* (Bearberry Honeysuckle) USA E Gunther.

HONEYSUCKLE, Tartarian *Lonicera tatarica*.

HONEYSUCKLE, Tom Thumb's *Lotus corniculatus* (Bird's-foot Trefoil) Wilts. Dartnell & Goddard.

HONEYSUCKLE, Tree *Banksia integrifolia* (White Honeysuckle) Edlin.

HONEYSUCKLE, Trumpet 1. *Campsis radicans* (Trumpet Creeper) RHS. 2. *Lonicera* *sempervirens*.

HONEYSUCKLE, Virgin Mary's *Pulmonaria officinalis* (Lungwort) Shrop, Ches. Burne. The association with the Virgin Mary lies with the spotted leaves rather than with honeysuckle connections. Cf. such names as Virgin Mary's Milkdrops, Lady's Milkwort, or just Virgin Mary. The reference is to a legend that during the flight into Egypt, some of the Virgin's milk fell on the leaves while she was nursing the infant Jesus, causing the white blotches on them.

HONEYSUCKLE, White 1. *Banksia integrifolia*. 2. *Trifolium repens* (White Clover) Culpepper.

HONEYSUCKLE, Wild *Pedicularis palustris* (Red Rattle) Donegal Grigson.

HONEYSUCKLE HONEY Fungus on *Rhododendron nudiflorum* USA Bergen.

HONEYSUCKLE TREFOIL *Trifolium pratense* (Red Clover) Flower.1859. In line with a number of names for this clover ranging from Honeysuckle itself to such as Sowkie Soo or Sugar-plums.

HONEYSWEET *Filipendula ulmaria* (Meadowsweet) Som. Storer. Cf. Honeyflower.

HONEYWORT 1. *Cerinthe major*. 2. *Galium cruciatum* (Crosswort) B & H.

HONG KONG LILY *Lilium brownii var. australe*. Not just Hong Kong, of course, but it is confined to that part of southern China.

HONITON LACE 1. *Anthriscus sylvestris* (Cow Parsley) Dev. Grigson. 2. *Conium maculatum* (Hemlock) Dev. Macmillan. 3. *Torilis japonica* (Hedge Parsley) Dev. Macmillan. Descriptive in each case. The general appearance of the flowers has given a number of picturesque names, like Queen Anne's Lace, Lady's Needlework, etc. The Devonshire name would naturally tend towards the local lace product, for Honiton used to be a great lace-making centre there.

HONOLULU CREEPER *Antigonon leptopus* (Coral Vine) Simmons.

HOOD, Cuckoo *Centaurea cyanus* (Cornflower) Scot. Grigson.

HOOD, Face-and- *Viola tricolor* (Pansy) Cambs, Ess, Norf. B & H. A contraction of Two-faces-in-a-hood, a name which Turner uses. Some see an extension of the erotic theme in pansy names in this one. Anyway, it takes on another aspect of the names by becoming Three-faces-under-, or -in-, -a hood, for pansy is also the Herb Trinity, from its three petals.

HOOD, Face-in- *Aconitum napellus* (Monkshood) Norf. B & H.

HOOD, Granny *Aquilegia vulgaris* (Columbine) Yks. Grigson. Better known further south as Granny Bonnets.

HOOD, Jenny; HOOD, John *Geranium robertianum* (Herb Robert) Dev, Som (Jenny); Som (John) both Macmillan. Whatever the ultimate origin, a name like Herb Robert inevitably leads to association with Robin Hood, hence presumably these, even by confusion. Jenny Hood, for instance, seems to be a mixture of Jenny-flower, or Jenny-wren, and Robin Hood. John, of course, has its own association with Robin Hood.

HOOD, King's *Geranium sylvaticum* (Wood Cranesbill) Border B & H.

HOOD, Monkey's 1. *Aconitum anglicum* (Wild Monkshood) Dev. B & H. 2. *Aconitum napellus* (Monkshood) Friend. A corruption of Monkshood.

HOOD, Old Wife *Aconitum napellus* (Monkshood) Cumb. B & H.

HOOD, Priest's *Arum maculatum* (Cuckoopint) B & H. Cf. Friar's Cowl for this - both are descriptive.

HOOD, Riding, Red *Melandrium dioicum* (Red Campion) Dev, Som, Dor. Grigson.

HOOD, Riding, White; HOOD, White *Silene cucubalis* (Bladder Campion) Som. Grigson (Riding); Som. Macmillan.

HOOD, Robin 1. *Agrostemma githago* (Corn Cockle) Dor. Grigson. 2. *Anemone coronaria* (Poppy Anemone) B & H. 3. *Geranium robertianum* (Herb Robert) Dev. Grigson; Som. Tongue. 4. *Lychnis flos-cuculi* (Ragged Robin) Dev, Dor, Som, Dur.

Grigson. Herb Robert begs the name by virtue of its common name. But what does 'Robert' mean? It is dedicated to St Robert on 29 April, but Robert could be an 11th century Abbot of Molesne of that name, or it could be Robert, Duke of Normandy. There is a disease which used to be known in Germany as Ruprechtsplage, said to take its name from that Duke, and this plant was used to cure it. Or Robert could be simply Latin ruber, red, so that Herb Robert could just mean a red-flowered plant. The likeliest explanation would be to equate Robin Hood with Robin Goodfellow - the seed vessels, with their sharp needles, are known as Pook Needles.

HOOD, Three-faces-in-a-; HOOD, Three-faces-under-a-; HOOD, Two-faces-in-a- *Viola tricolor* (Pansy) Gerard (Three-in-) N'thants. A E Baker (Three-under-); Turner (Two-in-). See Face-in-hood.

HOODED PITCHER-PLANT *Sarracenia minor*.

HOODWORT 1. *Scutellaria galericulata* (Skullcap) Grieve.1931. 2. *Scutellaria laterifolia* (Virginian Skullcap) O P Brown. Descriptive - Hoodwort conveys the same idea exactly as skullcap does.

HOOF, Cat's *Glechoma hederacea* (Ground Ivy) Tynan & Maitland.

HOOF, Horse 1. *Caltha palustris* (Marsh Marigold) Shet. Grigson. 2. *Tussilago farfara* (Coltsfoot) Turner. From the shape of the leaf in both cases, but more insistent in Coltsfoot.

HOOFS *Tussilago farfara* (Coltsfoot) Fernie.

HOOK-HEAL; HOOKWEED *Prunella vulgaris* (Self-heal) Gerard (Hook-heal); Som. Grigson (Hookweed). Self-heal had a reputation for healing wounds - the corolla is shaped like a billhook, and, from the doctrine of signatures, it was supposed to heal wounds from edged tools. Hence Hook-heal, and a number of similar names, like Sicklewort, Carpenter's Herb, etc.

HOOK THORN TREE *Acacia campylacantha*.

HOOKER, Joey *Galinsoga parviflora* (Kew Weed) A Huxley. After the Kew director of the time when it was introduced. He presumably answered queries about it. Or was Joey the director's son?

HOOKS-AND-HATCHETS *Acer campestre* seed vessels (Field Maple keys) Som. Macmillan.

HOOP ASH *Fraxinus nigra* (Black Ash) Grieve.1931.

HOOP PETTICOAT 1. *Narcissus bulbocodium*. 2. *Narcissus pseudo-narcissus* (Daffodil) Dor. Grigson. The general shape of *N bulbocodium* flowers certainly suggests the name.

HOOP PINE *Araucaria cunninghamii*. Because of the horizontal cracks in the encircling bands of the bark, so that it often sheds rings, or "hoops".

HOOVES *Tussilago farfara* (Coltsfoot) Glos. Grigson. Leaf shape gives this and other 'foot' names.

HOP 1. *Bryonia dioica* (White Bryony) Glos. J D Robertson. 2. *Humulus sp*. The Dutch name is Hop, while the OE word was hymele, which is responsible for the botanical name Humulus.

HOP, American *Humulus americanus*.

HOP, Bog *Menyanthes trifoliata* (Buckbean) N Eng. Grigson. From its use in beer-making.

HOP, Field *Achillea millefolium* (Yarrow) Skinner. This too is a reference to its use in beer-making.

HOP, Japanese *Humulus japonicus*.

HOP, Oregon *Leycesteria formosa* (Himalayan Honeysuckle) Ted Phillips.

HOP, Wild 1. *Bryonia dioica* (White Bryony) IOW, Yks. Grigson. 2. *Humulus lupulus*. 3. *Polygonum convolvulus* (Black Bindweed) Ches. Holland. 4. *Stachys officinalis* (Wood Betony) Worcs. Grigson. 5. *Tamus communis* (Black Bryony) Fernie. Betony was certainly one of the herbs used before true hops came in to make beer keep. The rest seem to get the name by their climbing habit.

HOP BUSH *Dodonea viscosa* (Switch

Sorrel) Australia Dalziel. Bitter, hence both 'hop' and 'sorrel'. It has actually been used as a substitute for hops.

HOP CLOVER 1. *Medicago lupulina* (Black Medick) Dev, Wilts. Grigson. 2. *Trifolium aureum* (Large Hop Trefoil) A W Smith. 3. *Trifolium campestre* (Hop Trefoil) Prior.

HOP CLOVER, Least *Trifolium dubium* (Lesser Yellow Trefoil) USA Allan.

HOP HORNBEAM, American *Ostrya virginiana*. The catkins look like those of hops, while the leaves resemble those of hornbeam.

HOP HORNBEAM, European *Ostrya carpinifolia*.

HOP HORNBEAM, Japanese *Ostrya japonica*.

HOP MARJORAM 1. *Mentha pulegium* (Pennyroyal) Fernie. 2. *Origanum dictamnus* (Dittany) USA Cunningham.

HOP MEDICK *Medicago lupulina* (Black Medick) N'thum. Grigson.

HOP PLANT *Origanum dictamnus* (Dittany) Brownlow. Because the bracts and flowers look like miniature pink hops.

HOP TREE *Ptelea trifoliata*. The bitter fruits are used in America to make home-brewed beer.

HOP TREFOIL *Trifolium campestre*.

HOP TREFOIL, Large *Trifolium aureum*.

HOP TREFOIL, Lesser *Trifolium dubium* (Lesser Yellow Trefoil) Salisbury.1964.

HOP-O'-MY-THUMB 1. *Geranium robertianum* (Herb Robert) Som. Macmillan. 2. *Lotus corniculatus* (Bird's-foot Trefoil) Som. Macmillan.

HOPES *Matthiola incana* (Stock) Norf. B & H.

HOPI COTTON *Gossypium hopi*. An aboriginal cultivated cotton peculiar to the southwest of the USA.

HOPWEED *Salvia occidentalis* Barbados Gooding.

HORE CALAMINT *Calamintha ascendens* (Calamint) Turner.

HOREHOUND, Black *Ballota nigra*. OE hare hune. Hune was the plant name, and har is the modern hoar.

HOREHOUND, Foetid; HOREHOUND, Stinking *Ballota nigra* (Black Horehound) Barton & Castle (Foetid); Lyte, Turner (Stinking). The strong smell is responsible for the generic name. Ballota is from the Greek meaning 'rejected', because animals refuse to eat it, probably owing to that strong smell, which seems to have some effect on the nervous system of cats, rather as catmint does.

HOREHOUND, Marsh *Lycopus europaeus* (Gypsywort) B & H.

HOREHOUND, Silver *Marrubium candidissima*.

HOREHOUND, Virginia; HOREHOUND, Water, Virginia *Lycopus virginicus* USA Zenkert (Virginia); Grieve.1931 (Water Virginia).

HOREHOUND, Water 1. *Lycopus americanus*. 2. *Lycopus europaeus* (Gypsywort) Gerard.

HOREHOUND, White *Marrubium vulgare*.

HOREHOUND, Wild *Eupatorium verbenaefolium* (Rough Thoroughwort) Grieve.1931.

HORESTRANGE; HORESTRONG *Peucedanum officinale* (Hog's Fennel) both Gerard. Apparently, the German Haarstrang, hair rope. But not according to A S Palmer, who offers Harnstrang, strangury, with the statement that the plant was used for the complaint.

HOREWORT *Filago germanica* (Cudweed) Halliwell.

HORN, Bog *Petasites hybridus* (Butterbur) Lincs. Dyer. Children used the hollow stalks as horns or trumpets, so Dyer said.

HORN, Buck's 1. *Plantago coronopus* (Buck's Horn Plantain) Parkinson. 2. *Plantago maritima* (Sea Plantain) B & H. Or Buck's Horn Plantain.

HORN, Devil's *Proboscidea althaefolia*.

HORN, Gramophone *Lonicera periclymenum* (Honeysuckle) Som. Grigson. Descriptive.

HORN, Huntsman's *Sarracenia flava* (Trumpet-leaf) L Gordon.

HORN, Powder *Cerastium arvense* (Field Mouse-ear) USA Elmore.

HORN, Ram's 1. *Allium ursinum* (Ramsons) Glos. J D Robertson. 2. *Arum maculatum* (Cuckoo-pint) Suss. B & H. 3. *Orchis mascula* (Early Purple Orchis) Fernie. 4. *Orchis morio* (Green-winged Orchid) Suss. B & H. The substitution of Ram's Horns for Ramsons is understandable enough, but the rest have sexual origins.

HORN BEECH *Carpinus betula* (Hornbeam) B & H.

HORN BIRCH *Ulmus glabra* (Wych Elm) Som. Grigson.

HORN FLOWER *Zantedeschia aethiopica* (Arum Lily) Folkard. From the shape of the calyx. Cf. Trumpet Lily for this.

HORN-NUT *Trapa natans* (Water Chestnut) Pratt.

HORN OF PLENTY *Datura metel* USA Cunningham.

HORN POPPY *Glaucium flavum* (Horned Poppy) Dev. Grigson.

HORNBEAM 1. *Carpinus betula*. 2. *Ulmus glabra* (Wych Elm) Som. Elworthy. Hornbeam means the tree whose wood is tough like horn (Gerard called it Hard Beam). It is tough enough for many of the cogs and wheels of old windmills to have been made of it, and is stronger than oak, they say. The derivation is not, as C P Johnson thought, "formerly in great demand for beams or yokes for oxen", quite true, "whence the English name of the tree".

HORNBEAM, American *Carpinus caroliniana* (Blue Beech) USA Zenkert.

HORNBEAM, Hop, American *Ostrya virginiana*.

HORNBEAM, Hop, European *Ostrya carpinifolia*.

HORNBEAM, Hop, Japanese *Ostrya japonica*.

HORNBEAM, Japanese *Carpinus japonica*.

HORNBEAM, Oriental *Carpinus orientalis*.

HORNBEAM-LEAVED SALLOW *Salix carpinifolia*.

HORNECKS, HORNICKS *Conopodium majus* (Earthnut) Scot. Grigson (Hornecks); Scot. Gibson (Hornicks).

HORNED BEECH *Carpinus betula* (Hornbeam) Bucks. Harman.

HORNED CLOVER *Medicago lupulina* (Black Medick) Turner.

HORNED CUCUMBER *Cucumis metuliferus*.

HORNED MILKWORT *Polygala cornuta*.

HORNED PONDWEED *Zannichellia palustris*.

HORNED POPPY *Glaucium flavum*.

HORNED POPPY, Red *Glaucium corniculatum*.

HORNED POPPY, Sea *Glaucium flavum* (Horned Poppy) Gerard.

HORNED TULIP *Tulipa acuminata*.

HORNED VIOLET *Viola cornuta* (Butterfly Viola) RHS.

HORNS *Hordeum sativum* (Barley) E Angl. Forby. Awns, of course.

HORNS, Antelope *Asclepiodora decumbens* (Spider Milkweed) USA Elmore.

HORNWORT *Ceratophyllum demersum*.

HORNWORT, Rigid *Ceratophyllum demersum* (Hornwort) McClintock.1975.

HORNWORT, Spineless *Ceratophyllum submersum*.

HORONE *Marrubium vulgare* (White Horehound) Halliwell. Presumably an old version of hoarhound.

HORSE(S), Coach *Viola tricolor* (Pansy) Som. Elworthy.

HORSE, Fairy *Senecio jacobaea* (Ragwort) Ire. Grigson. They used to say that both the ragwort and St John's Wort were fairy horses in disguise. If you trod them down after sunset a horse would arise from the root of each injured plant, and would gallop away with you. Witches used it, too, to ride on. "As arrant a witch as ever rode on ragwort" is claimed as a saying from the Isle of Man.

HORSE-AND-HOUNDS *Ajuga reptans* (Bugle) Dor. Macmillan.

HORSE BEAN *Canavalia ensiformis* (Jack Bean) Douglas.

HORSE BEAN, Little-leaved *Cercidium microphyllum*.

HORSE BEECH *Carpinus betula*

(Hornbeam) Kent Evelyn; Hants. Cope; Suss. Parish. 'Horse' here is probably 'hurst', which means a small wood. The name Hurst Beech is quite common in the same area.

HORSE BLOBS 1. *Caltha palustris* (Marsh Marigold) Wilts. D Grose; Som, War, Leics, Notts. Grigson; N'thants. A E Baker. 2. *Leucanthemum vulgare* (Ox-eye Daisy) N'thants. Grigson. Horse Blobs for the Ox-eye does not sound right, though 'horse' does, for usually this epithet means large, or even coarse. Horse Daisy for this plant sums it up. But Horse Blobs for Marsh Marigold simply conveys large buttercup. Blobs means blisters, and the acrid juice of any of the buttercups will produce them. Cf. Mare-blobs for Marsh Marigold.

HORSE BRAMBLE *Rosa canina* (Dog Rose) E Ang. Halliwell.

HORSE BREATH *Ononis repens* (Rest-harrow) Som, Worcs. Grigson.

HORSE BRIER *Smilax rotundifolia* (Cat Brier) USA Cunningham.

HORSE BUCKLE *Primula veris* (Cowslip) Wilts. Macmillan; Kent Grigson. 'Buckle' is probably paigle, an old name for cowslips.

HORSE BUTTERCUP *Caltha palustris* (Marsh Marigold) Dev, Som. Elworthy. 'Horse' in the sense of large.

HORSE BUTTONS *Malva sylvestris* (Common Mallow) Donegal Grigson.

HORSE CASSIA *Cassia grandis*.

HORSE CHESTNUT *Aesculus hippocastanum* 'Horse' probably denotes largeness and coarseness here as usual, but there is an old Latin name Castanea equina, and Parkinson says "the horse chestnits are given in the East Country, and so through all Turkie, unto horses to cure them of the cough, shortnesse of winde, and such other diseases". Gerard also accounts for the name in this way. Skinner suggests the derivation from the likeness to a horse's hoof in the leaf cicatrix, and it may have been from the doctrine of signatures that the nuts, crushed as meal, were given to horses for various diseases. Anyway, we are told that horses don't seem to like them, though deer and cattle do.

HORSE CHESTNUT, American *Aesculus californica* (Californian Buckeye) Hatfield.

HORSE CHESTNUT, Indian *Aesculus indica*.

HORSE CHESTNUT, Japanese *Aesculus turbinata*.

HORSE CHESTNUT, Red *Aesculus x carnea*.

HORSE CRESS *Veronica beccabunga* (Brooklime) Yks. Grigson.

HORSE DAISY 1. *Anthemis cotula* (Maydweed) Bucks. Grigson. 2. *Leucanthemum vulgare* (Ox-eye Daisy) Dev. Friend.1882. Wilts. Dartnell & Goddard. 3. *Tripleurospermum maritimum* (Scentless Mayweed) Bucks. B & H. 'Horse' in each case to highlight the contrast between these and the true daisy, *Bellis perennis*.

HORSE DAISY, Yellow *Chrysanthemum segetum* (Corn Marigold) Corn. Grigson.

HORSE DASHEL *Cirsium vulgare* (Spear Plume Thistle) Dev. Macmillan.

HORSE ELDER *Inula helenium* (Elecampane) Palaiseul. The result of a misreading somewhere. The original was probably Horselene or one of its variants, which would have the same meaning as Horseheal. Elecampane was apparently given to horses as an appetite improver.

HORSE-EYE BEAN 1. *Dolichos ensiformis*. 2. *Mucuna urens* (Florida Bean) Le Strange.

HORSE FLOWER *Melampyrum sylvaticum* (Small Cow-wheat) Prior.

HORSE-FLY WEED *Baptisia tinctoria* (Wild Indigo) Grieve.1931. This must be in recognition of its use as an antiseptic.

HORSE GOG *Prunus spinosa* fruit (Sloe) Yks. Nicholson.

HORSE GOLD 1. *Ranunculus arvensis* (Corn Buttercup) Herts. Grigson; N'thants. A E Baker. 2. *Ranunculus bulbosus* (Bulbous Buttercup) N'thants. A E Baker.

HORSE GOWAN 1. *Leucanthemum vulgare* (Ox-eye Daisy) Edin. Ritchie. 2. *Matricaria recutita* (Scented Mayweed) Border, Berw,

Banff, Dumb. B & H; USA Watt. 3. *Taraxacum officinale* (Dandelion) Scot. B & H; USA. Henkel.

HORSE GRAIN *Dolichos uniflorus* (Horsegram) Dalziel.

HORSE HARDHEAD *Centaurea nigra* (Knapweed) Dev. B & H. Hardhead is a common enough name for knapweed. Cf. Horse-knop, etc.

HORSE HOOF 1. *Caltha palustris* (Marsh Marigold) Shet. Grigson. 2. *Tussilago farfara* (Coltsfoot) Turner. Leaf shape, in each case.

HORSE KNOB; HORSE NOP *Centaurea nigra* (Knapweed) Dor. Macmillan; N Eng. B & H (Knob); Lancs. Nodal (Nop).

HORSE KNOB, Great *Centaurea scabiosa* (Greater Knapweed) Yks. B & H.

HORSE KNOP 1. *Centaurea cyanus* (Cornflower) N Eng. Brockett. 2. *Centaurea nigra* (Knapweed) Dor. Macmillan; Yks. Nicholson. 3. *Centaurea scabiosa* (Greater Knapweed) Cumb, Yks. B & H. Cf. Knopweed or Knobweed for knapweed. Cf. too Bullweed, where 'bull' is OE boll, a circular body.

HORSE KNOT 1. *Centaurea nigra* (Knapweed) Prior. 2. *Centaurea scabiosa* (Greater Knapweed) Tynan & Maitland.

HORSE MAY *Ulmus procera* (English Elm) Corn. B & H. 'Horse' to convey coarse, or inferior? Given when the real stuff was not available?

HORSE MINT 1. *Mentha aquatica* (Water Mint) Som. Macmillan; Glos, War, IOW, Lincs, N'thum. Grigson. 2. *Mentha longifolia*. 3. *Mentha rotundifolia* (Apple Mint) Glos. J D Robertson. 4. *Monarda fistulosa* (Purple Bergamot) USA Elmore. 5. *Monarda pectinata*. 6. *Monarda punctata* (American Horse Mint) USA Elmore.

HORSE MINT, American *Monarda punctata*.

HORSE MINT, Blue *Monarda mollis*.

HORSE MINT, Spotted *Monarda punctata* (American Horse Mint) USA Tolstead.

HORSE NETTLE *Solanum carolinense* (Carolina Nightshade) USA Tolstead.

HORSE NETTLE, White; HORSE NETTLE, Silver *Solanum eleagnifolium* (Silverleaf Nightshade) USA Kingsbury. (White); USA T H Kearney (Silver).

HORSE NICKER *Caesalpina bonduc* Barbados Gooding. Nicker Nut, or Nicker Bean, are usual for this.

HORSE-NUT TREE *Aesculus hippocastanum* (Horse Chestnut) Som. Macmillan.

HORSE PARSLEY 1. *Heracleum sphondyllium* (Hogweed) Som. Macmillan. 2. *Smyrnium olusatrum* (Alexanders) Prior.

HORSE PEASE *Vicia orobus* (Bitter-vetch) Cumb. Grigson.

HORSE PEN *Rhinanthus crista-galli* (Yellow Rattle) Cumb. Grigson. 'Pen' - pennies - a lot of names commemorate the idea of seeds rattling like money in a purse.

HORSE PENNIES 1. *Leucanthemum vulgare* (Ox-eye Daisy) Derb. Grigson. 2. *Rhinanthus crista-galli* (Yellow Rattle) Derb, Lancs. Grigson; Yks. Carr.

HORSE PEPPER *Angelica sylvestris* (Wild Angelica) E Angl. V G Hatfield.1994.

HORSE PEPPERMINT *Ajuga reptans* (Bugle) Wilts. Dartnell & Goddard.1899.

HORSE PURSLANE *Trianthema pentandra*.

HORSE RADISH *Armoracia rusticana*.

HORSE RADISH TREE *Moringa oleifera*. All the parts are pungent, and it has actually been used as a horse radish substitute.

HORSE SAVING *Juniperus communis* (Juniper) Cumb. Grigson. Saving is savin.

HORSE-SEED BUSH *Dodonea viscosa* (Switch Sorrel) Howes.

HORSE SNAP *Centaurea nigra* (Knapweed) Dev. B & H. Presumably part of the Horse Knop series.

HORSE SORREL *Rumex hydrolapathum* (Great Water Dock) Lyte.

HORSE-STRONG *Peucedanum officinale* (Hog's Fennel) A S. Palmer. A corruption - this should be Horestrong.

HORSE SUGAR *Symplocos tinctoria* (Sweetleaf) Wit.

HORSE THISTLE 1. *Cirsium vulgare*

(Spear Plume Thistle) Dev, Som. Grigson. 2. *Lactuca virosa* (Great Prickly Lettuce) B & H. The prickles account for 'thistle', while 'horse' is the equivalent of the 'great' in Great Prickly Lettuce.

HORSE THYME 1. *Clinopodium vulgare* (Wild Basil) Turner. 2. *Thymus vulgaris* (Wild Thyme) N'thants. Sternberg.

HORSE VIOLET 1. *Viola riviniana* (Dog Violet) Dev. Friend.1882. 2. *Viola tricolor* (Pansy) Dev. Friend.1882. For the same reason that it is called Dog Violet - because it is inferior in that it has no perfume. Giving the name to pansy seems to be an aberration.

HORSE WELL-CRESS; HORSE WELL-GRASS *Veronica beccabunga* (Brooklime) Scot. Grigson (Cress); Jamieson (Grass).

HORSE WICKY *Pieris nitida* USA Harper.

HORSE'S HOOVES *Caltha palustris* (Marsh Marigold) Shet. Grigson. Leaf shape.

HORSE'S MOUTH *Antirrhinum majus* (Snapdragon) Som. Macmillan. The shape of the flowers has prompted the likening to mouths of rabbits, dogs, lions, tigers, toads, frogs and dragons, as well as horses.

HORSE'S TOOTH *Melilotus officinalis* (Melilot) Henslow.

HORSEBALM *Collinsonia canadensis* (Stone Root) Le Strange.

HORSEBANE 1. *Oenanthe aquatica*. 2. *Oenanthe crocata* (Hemlock Water Dropwort) Fernie. 'Bane' because they are very poisonous.

HORSEBEAN *Vicia faba* (Broad Bean) Kelly & Palerm. This may be an American name, but curiously, there is a Sussex name for the bean - Housebean. Is it a coincidence?

HORSEBRUSH, Little-leaf *Tetradymia glabrata*.

HORSEBRUSH, Stemless *Tetradymia canescens*.

HORSECHIRE *Teucrium chamaedrys* (Germander) B & H.

HORSEFOOT *Tussilago farfara* (Coltsfoot) Pratt.

HORSEGOWL, Marsh *Caltha palustris* (Marsh Marigold) Fernie. This seems to be an exact transcription of marsc meargealle, the original of marsh marigold.

HORSEGRAM *Dolichos uniflorus*.

HORSEHEAL; HORSEHEEL; HORSEHELNE; HORSELENE *Inula helenium* (Elecampane) Dawson (Horseheal); USA Tolstead (Horseheel) Fernie (Horsehele); Dawson (Horselene). Veterinary usage accounts for these. The American Horseheel can only be a misunderstanding of "... heal".

HORSES, Chariot-and- 1. *Aconitum anglicum* (Wild Monkshood) Herts. B & H. 2. *Aconitum napellus* (Monkshood) Le Strange.

HORSES, Coach-and- *Aconitum napellus* (Monkshood) Dor. Macmillan.

HORSES-AND-CHARIOTS *Briza media* (Quaking Grass) Som. Grigson.1959.

HORSESHOE FERN *Marattia fraxinea*.

HORSESHOE VETCH *Hippocrepis comosus*. So called from the shape of the pods. It had the reputation of being able to unshoe any horse that trod on it, a superstition that has been recorded in France, too. Probably, the reason is that it grows in the sort of stony ground which leads to accidents.

HORSESHOES *Acer pseudo-platanus* fruit (Sycamore keys) Wilts. Dartnell & Goddard.

HORSETAIL, Barren, Marsh *Hippuris vulgaris* (Mare's Tail) Parkinson.

HORSETAIL, Boston *Equisetum ramosissimum*. Boston, Lincolnshire, that is. A riverbank there is the only station in this country.

HORSETAIL, Common *Equisetum arvense*.

HORSETAIL, Corn; HORSETAIL, Field *Equisetum arvense* (Common Horsetail) Curtis (Corn); Palaiseul (Field).

HORSETAIL, Female *Hippuris vulgaris* (Mare's Tail) Gerard.

HORSETAIL, Giant *Equisetum telmateja*.

HORSETAIL, Great *Equisetum telmateja* (Giant Horsetail) Perring.

HORSETAIL, Marsh *Equisetum palustre*.

HORSETAIL, Rough *Equisetum hyemale* (Dutch Rush) Barton & Castle. All the horsetails are rough, but this one more than most.

Not for nothing is it also called Scouring Rush. It is even said that the North American Indians used it for polishing their guns, hence the name there of Gunbright.

HORSETAIL, Shady *Equisetum pratense.*

HORSETAIL, Variegated *Equisetum variegatum.*

HORSETAIL, Water *Equisetum limosum.*

HORSETAIL, Wood *Equisetum sylvaticum.*

HORSETAIL TREE *Casuarina equisetifolia* (Beefwood) Everett.

HORSEWEED 1. *Collinsonia canadensis* (Stone Root) Le Strange. 2. *Erigeron canadensis* (Canadian Fleabane) USA Harper. Cf. Colt's Tail, or Mare's Tail, for the latter, and Horsebalm for the former.

HORSHELNE *Inula helenium* (Elecampane) Dawson. Also as Horselene, Horseheel, etc.

HORSHONE *Glechoma hederacea* (Ground Ivy) Fernie.

HORST BEECH *Carpinus betula* (Hornbeam) W Miller. 'Horst' is 'hurst', a small wood.

HORTICHOKE *Cynara cardunculus var. scolymus* (Globe Artichoke) Potter & Sargent. Choking the garden? See Artichoke for the etymology.

HORTLEBERRY; HORTS *Vaccinium myrtillus* (Whortleberry) Grose (Hortleberry); Dev. Grigson (Horts). Hurts is commoner than Horts, in line with the pronunciation.

HOSE, Gowk's 1. *Campanula latifolia* (Giant Bellflower) Scot. Grigson. 2. *Campanula medium* (Canterbury Bells) S Scot. Hardy. 3. *Campanula rotundifolia* (Harebell) Dumb. Hardy. 4. *Endymion nonscriptus* (Bluebell) Dumb. B & H. Further south, the name would be Cuckoo's Stockings.

HOSE-DOUP; HOW-DOUP *Mespilus germanica* (Medlar) Scot. Grigson.

HOT CROSS BUN *Euonymus europaeus* fruit (Spindle berries) Dor. Macmillan. Quite descriptive of the peculiar shape of these berries.

HOT WATER PLANT *Achimenes longiflora.* So called because the way to start them is to put them in the warm, and up against the hot water pipes in a greenhouse is the ideal.

HOTH; HAWTH; HAUTH *Ulex europaeus* (Gorse) Suss. Grigson (Hoth, Hauth); Fernie (Hawth).

HOTTENTOT BREAD *Dioscorea elephantipes.*

HOTTENTOT CHERRY *Maurocenia capensis.*

HOTTENTOT FIG; KAFFIR FIG *Mesembryanthemum edule* McClintock (Kaffir).

HOTTENTOT'S BEDDING *Helichrysum crispum* S Africa Watt.

HOTTONIA, Water *Hottonia palustre* (Water Violet) Curtis.

HOUND'S FENNEL *Anthemis cotula* (Maydweed) Henslow. All the 'dog' names given to this are the result of the bad smell.

HOUND'S HEAD *Misopates orontium* (Lesser Snapdragon) Cockayne.

HOUND'S MIE; HOUND'S PISS *Cynoglossum officinale* (Hound's Tongue) Cockayne, Gerard. "...for in the world there is not anything that smelleth so like Dogs pisse as the leaves of this plant doe" (Gerard).

HOUND'S TONGUE 1. *Cynoglossum officinale.* 2. *Lappula virginiana* USA Youngken. 3. *Stachys palustris* (Marsh Woundwort) Mor. Grigson. Hound's Tongue is a translation of the generic name, *Cynoglossum*, and has been derived variously from the texture of the leaf, from its "doggy" smell, and by association, from the fact that it was used to cure dog bites.

HOUND'S TONGUE, Green *Cynoglossum germanicum.*

HOUND'S TONGUE, Himalayan *Cynoglossum nervosum.*

HOUND'S TOOTH *Agropyron repens* (Couchgrass) Grigson.1959. Cf. Dog-grass. Given perhaps because it is a dog's favourite tonic grass, or perhaps it simply infers whiteness (one of the Wiltshire names is White Couch).

HOUND'S TREE; HOUNDBERRY TREE *Cornus sanguinea* (Dogwood) both Gerard.

HOUNDS, Fox-and- *Linaria vulgaris* (Toadflax) Lincs. Grigson. Double names like this are common for toadflax, e.g. Fingers-and-thumbs, or Pattens-and-clogs, usually referring to the two colours. But this one may have something to do with the smell.

HOUNDSBANE *Marrubium vulgare* (White Horehound) USA Henkel. Recorded in America, apparently, but Halliwell has Houndbene for this.

HOUNDSBERRY 1. *Atropa belladonna* (Deadly Nightshade) Dawson. 2. *Solanum nigrum* (Black Nightshade) Cockayne. Perhaps a way of saying that these berries are toxic. Deadly Nightshade berries are known as Dogberries in co Durham.

HOUR, Flower-of-an- 1. *Hibiscus trionum* (Bladder Katmia) Wit. 2. *Malva sylvestris* (Common Mallow) Som. Macmillan.

HOUSE BEAN; HOUSEHOLD BEAN *Vicia faba* (Broad Bean) both Suss. Parish.

HOUSEGREEN *Sempervivum tectorum* (Houseleek) War, N'thants. Grigson; Ches. Holland. A mixture of two names, houseleek and some version of singreen.

HOUSELEEK *Sempervivum tectorum*.

HOUSELEEK, Cobweb *Sempervivum arachnoideum*.

HOUSELEEK, Dwarf *Sedum reflexum* (Yellow Stonecrop) B & H.

HOUSELEEK, Least; HOUSELEEK, Little *Sedum acre* (Biting Stonecrop) Hill (Least); N'thum, Cumb. Grigson (Little).

HOUSELEEK, Mountain *Sempervivum montanum*.

HOUSELEEK, Roof *Sempervivum tectorum* (Houseleek) RHS. 'Roof', in acknowledgement of its favoured habitat, which is why it bears the specific name *tectorum*.

HOUSELEEK, Small *Sedum album* (White Stonecrop) Leyel.1937.

HOUSELEEK, Water *Stratiotes aloides* (Water Soldier) Grieve.1931. Cf. Water Sengreen, which means the same thing.

HOUSTOUNIA, Fringed *Houstounia ciliolata*.

HOUSTOUNIA, Long-leaved *Houstounia longifolia*.

HOVE; HOVE, Red *Glechoma hederacea* (Ground Ivy) Cockayne, Storms respectively. From the evidence of such names as Alehoof, or Allhoove, it would seem that Hove means to ferment (the modern verb to heave is relevant here). Or is it from OE hufe, a crown, alluding to the chaplet which crowned the ale-stake, as Friend thought? Earle has another explanation - he says ael(el)-hofe is "another sort of hofe", which he explains is the true name for the violet.

HOW-DOUP *Mespilus germanica* (Medlar) Scot. Grigson. 'Doup' is Scottish for buttocks; 'how' must be hollow, so the whole name bears exactly the same message as the English Open-arse, which was current in Turner's time, and an obvious comment on the effects of the fruit.

HOWES *Crataegus monogyna* fruit (Haws) Suff. Moor.

HOWKA *Meum athamanticum* (Baldmoney) N'thum. Grigson.

HOWLER *Alnus glutinosa* (Alder) Leics, Ches, Lancs, Yks. B & H. OE alor, which accounts for alder, is also responsible for Aller, common in the southern part of the country, or, in Scotland, Allar. This develops easily into such forms as Ouler or Owler, and then into Howler.

HOYLES, OILS *Avena sativa* (Oats) Dev. Friend.1882. Cf. Ailes.

HUANUCO LEAF *Erythroxylon coca* (Coca) Thomson.

HUBBARD SQUASH *Cucurbita maxima* (Giant Pumpkin) Watt.

HUCKLEBERRY 1. *Gaylussacia sp.* 2. *Vaccinium myrtillus* (Whortleberry) Clapham. Gaylussacia is the true huckleberry, not Vaccinium. The word is probably a variant of hurtleberry, though, through French hurte.

HUCKLEBERRY, Bear *Gaylussacia ursina*.

HUCKLEBERRY, Black *Gaylussacia baccata*.

HUCKLEBERRY, Blue 1. *Vaccinium ovali-*

folium (Blue Whortleberry) E Gunther. 2. *Vaccinium pennsylvanicum*.
HUCKLEBERRY, Box *Gaylussacia brachycera*.
HUCKLEBERRY, California *Vaccinium ovatum*.
HUCKLEBERRY, Dwarf *Gaylussacia dumosa*.
HUCKLEBERRY, Evergreen *Vaccinium ovatum* (California Huckleberry) USA E Gunther.
HUCKLEBERRY, Fool's *Menziesia ferruginea* (Mock Azalea) USA E Gunther.
HUCKLEBERRY, Garden *Solanum intrusum*. It may be called huckleberry, but the berries are still as poisonous as those of its parent, *S nigrum*.
HUCKLEBERRY, High-bush *Vaccinium corymbosum*.
HUCKLEBERRY, Low *Vaccinium scoparium*.
HUCKLEBERRY, Red 1. *Vaccinium parvifolium* (Red Bilberry) E Gunther. 2. *Vaccinium vitis-idaea* (Cowberry) Wit.
HUCKLEBERRY, Squaw *Vaccinium stamineum*.
HUCKLEBERRY LILY *Lilium philadelphicum* (Philadelphia Lily) USA Woodcock.
HUDS *Rumex hydrolapathum* (Great Water Dock) Fernie.
HUFF-CAP *Agropyron repens* (Couch-grass) Shrop. Grigson.1959; Heref. Halliwell. A huff-cap is a bully, "a swaggering fellow", as Halliwell defined it. The name must have been given by some exasperated gardener.
HUGGABACK *Vicia cracca* (Tufted Vetch) Cumb. Grigson.
HUGGAN *Rosa canina* fruit (Hips) Yks. Grigson.
HUGGY-ME-CLOSE *Galium aparine* (Goose-grass) Dor. Macmillan. Cf. Sweethearts for this. They both measure the clinging properties of the burrs.
HUISACHE *Acacia farnesiana* (Cassie) Polunin.
HULL, HULLIN *Ilex aquifolium* (Holly) Halliwell (Hull); IOM Moore (Hullin). Cf.

Hollin, or Hollen.
HULM *Ilex aquifolium* (Holly) Som. Macmillan. Holm is more usual, and common.
HULST *Ilex aquifolium* (Holly) Prior.
HULVER *Ilex aquifolium* (Holly) E Ang. Grose. Cf. Helver, a Suffolk version.
HULVER, Knee; HULYER, Knee *Ruscus aculeatus* (Butcher's Broom) Gerard (Hulver); Fernie (Hulyer). There is Knee Holly, or Knee Holme for this. Is Hulyer genuine? This is the only occasion it has been used, and might be a misreading.
HULVER, Sea *Eryngium maritimum* (Sea Holly) Turner.
HULVER OAK *Quercus ilex* (Holm Oak) Gerard.
HULWORT *Teucrium polium* B & H.
HUMACK *Rosa canina* (Dog Rose) Som. Grigson.
HUMBLE *Trifolium campestre* (Hop Trefoil) Cockayne.
HUMBLE-BEE *Ophrys apifera* (Bee Orchid) B & H.
HUMBLE GILIA *Gilia demissa*. The specific name means humble, presumably in the sense of low-growing, or at least downward hanging.
HUMBLE-PLANT *Mimosa pudica* (Sensitive Plant) J Smith.
HUMBOLDT LILY *Lilium humboldtii*. Not found by Humboldt, but named after him apparently because it was found on his birthday.
HUMILITY *Saxifraga sarmentosa* (Mother-of-thousands) B & H.
HUMLICK *Conium maculatum* (Hemlock) N Eng. Brockett. Closer to the original than the modern name hemlock, for this OE hymlice.
HUMLICK, Sweet *Myrrhis odorata* (Sweet Cicely) N.Eng. Grigson.
HUMLOCK 1. *Anthriscus sylvestris* (Cow Parsley) Yks. B & H. 2. *Conium maculatum* (Hemlock) Norf, Lincs, Cumb, N'thum, Scot. Grigson; Ess. Gepp; Yks. Addy; Dur. Dinsdale.

HUMLY *Conium maculatum* (Hemlock) Roxb. Grigson. A shortened version of the N English name Humlick.

HUMMINGBIRD TREE *Chelone glabra* (Turtlehead) Le Strange.

HUMPBACKS *Viola odorata* (Sweet Violet) Som. Macmillan. Descriptive of the way the flowers are held; the same habit accounts for such names as Shame-faced Maiden, or Miss Modesty.

HUMPY-SCRUMPLE *Heracleum sphondyllium* (Hogweed) Dev. Macmillan. Other similar West Country names are Lumper-scrump, or Limper-scrimp, and Rumpet-scrumps. They are all a reference to the hollow stalks, used for drinking cider through.

HUNDRED HOLES *Hypericum perforatum* (St John's Wort) Fernie. The specific name is *perforatum*, and the mistake lies in thinking that the glandular dots, which are actually oil sacs, on the underside of the leaves are holes.

HUNDRED-LEAVED GRASS *Achillea millefolium* (Yarrow) Berw. Grigson; Aber. McNeill. More conservative than the specific name.

HUNDREDS-AND-THOUSANDS 1. *Cymbalaria muralis* (Ivy-leaved Toadflax) Som. Macmillan. 2. *Heliotropum europaeum* (Cherry-pie) Som. Macmillan. 3. *Malcolmia maritima* (Virginian Stock) Som. Macmillan. 4. *Pulmonaria officinalis* (Lungwort) Brownlow. 5. *Rumex acetosa* seeds (Wild Sorrel) Som. Macmillan. 6. *Saxifraga x urbinum* (London Pride) Som. Macmillan. 7. *Sedum acre* (Biting Stonecrop) Som. Grigson. 8. *Sempervivum tectorum* (Houseleek) Som. Macmillan. A name like this usually implies prolific reproduction.

HUNGARIAN CAMOMILE *Matricaria recutita* (Scented Mayweed) Watt. Apparently because of the popularity of the tea in Hungary, but it is also known as German Camomile.

HUNGARIAN CLIMBER *Clematis integrifolia*.

HUNGARIAN FUSTIC *Cotinus coggyria* (Venetian Sumach) Watt. Cf. Young Fustic, both being names for the pigment obtainable from the tree.

HUNGARIAN GENTIAN *Gentiana pannonica* (Brown Gentian) Polunin.

HUNGARIAN LILAC *Syringa josikaea*.

HUNGARIAN MILLET *Setaria italica* (Foxtail Millet) Brouk. It is ascribed in name to Italy, Germany and Siberia, as well as to Hungary.

HUNGARIAN OAK *Quercus frainetto*.

HUNGARIAN SPURGE *Euphorbia esula*.

HUNGARIAN THORN *Crataegus nigra*.

HUNGARY, Star of *Ornithogalum umbellatum* (Star of Bethlehem) Grieve.1931.

HUNGARY DEADNETTLE *Lamium orvala* (Balm-leaved Deadnettle) Whittle & Cook.

HUNGER-WEED 1. *Alopecurus myosuroides* (Slender Foxtail) H C Long.1910. 2. *Ranunculus acris* (Meadow Buttercup) Fernic. 3. *Ranunculus arvensis* (Corn Buttercup) Glos. J D Robertson; Norf. Grigson. Hunger-weed surely implies a weed which will grow on very poor soil, but that does not seem to apply to the buttercups. Foxtail gets the name in all probability because it used to be such a troublesome pest in the growing crops.

HUNGIN *Allium cepa* (Onion) Hunts. Marshall.

HUNGRY GRASS *Alopecurus myosuroides* (Slender Foxtail) Leyel.1948. Cf. Hungerweed.

HUNGRY RICE *Digitaria exilis*. Presumably because the grain is so very small.

HUNTER, Jubilee *Rubus caesius* (Dewberry) Wilts. Dartnell & Goddard.

HUNTERS *Rumex acetosa* seeds (Wild Sorrel seeds) Som. Macmillan.

HUNTINGDON ELM *Ulmus x hollandica* '*vegeta*'. Because it came from a nursery near there.

HUNTINGDON WILLOW *Salix alba* (White Willow) Brimble.1948.

HUNTSMAN, Red *Papaver rhoeas* (Corn Poppy) Som. Macmillan.

HUNTSMAN'S CAP *Scrophularia aquatica*

(Water Betony) Corn. B & H. Descriptive of the shape of the flowers.

HUNTSMAN'S CUP *Sarracenia purpurea* (Pitcher Plant) Grieve.1931. The "pitcher" can easily be thought of as an outsize cup, and it appears in a lot of the names for this plant.

HUNTSMAN'S HORN; HUNTSMAN'S TRUMPET *Sarracenia flava* (Trumpet-leaf) L Gordon (horn); Howes (trumpet).

HUON PINE *Dacrydium franklinii*. Because it was first found on the Huon River in Tasmania.

HUPEH ROWAN *Sorbus hupehensis*.

HUPEH VINE *Actinidia chinensis* (Chinese Gooseberry) Douglas.

HURDREVE *Centaureum erythraea* (Centaury) Halliwell.

HURDS; HURS, Black *Vaccinium myrtillus* (Whortleberry) co Sligo (Hurds); co Cork (Black Hurs) both Maire Mac Neill.

HURR-BURR *Arctium lappa* (Burdock) Shrop. B & H; Leics. A B Evans; USA Henkel. This is also recorded as Hareburr.

HURRAH-BUSH *Pieris nitida* USA Harper.

HURST BEECH *Carpinus betula* (Hornbeam) Hants, Suss, Kent, Norf. Grigson. 'Hurst' means a small wood. The name appears also as Horst Beech, and then corrupted into Horse Beech, and finally Husbeech.

HURT-SICKLE 1. *Centaurea cyanus* (Cornflower) Gerard. 2. *Centaurea nigra* (Knapweed) Worcs. B & H. Presumably because coming up against one of these turns the edge of reapers' sickles.

HURTFUL CROWFOOT *Ranunculus scleratus* (Celery-leaved Buttercup) Flower.1851. It always used to be said that this was the most poisonous of all the buttercups.

HURTLEBERRY *Vaccinium myrtillus* (Whortleberry) Dev, Som. Macmillan.

HURTS *Vaccinium myrtillus* (Whortleberry) Corn. Bottrell Dev. Crossing; Som. Potter & Sargent; E Ang. Grigson. Newfoundland Narváez.

HURTS, Sweet *Vaccinium angustifolium* (Low Sweet Blueberry) USA Yarnell.

HUSBAND, Dusty 1. *Arabis alpina* (Alpine Rock-cress) Ches. B & H. 2. *Cerastium tomentosum* (Snow-in-summer) Ches. B & H. Arising from the typically hoary appearance of both these.

HUSBANDMAN'S DIAL *Calendula officinalis* (Marigold) Coats. For marigold is a solsequium, a sun follower, and has actually been called sunflower before now. Cf. the Guernsey name Soucique.

HUSBANDMAN'S TREE 1. *Fraxinus excelsior* (Ash) Leyel.1937. 2. *Sorbus aucuparia* (Rowan) Tynan & Maitland. Rowan is probably given the name in error, because of the other name, Mountain Ash. The real ash gets it in commemoration of the many uses to which the timber has been put.

HUSBANDMAN'S WEATHERGLASS; HUSBANDMAN'S WEATHER-WARNER *Anagallis arvensis* (Scarlet Pimpernel) Tynan & Maitland.

HUSBANDMAN'S WOUNDWORT 1. *Stachys palustris* (Marsh Woundwort) Gerard. 2. *Stachys sylvatica* (Betony) Fernie.

HUSBEECH *Carpinus betula* (Hornbeam) Hants. Cope; Suss. W D Cooper; Norf. B & H. i.e. Hurst Beech.

HUSK TOMATO 1. *Physalis aequata* USA Driver. 2. *Physalis pubescens*. Husk, because the berries are covered with a papery calyx, or husk.

HUSLOCK *Sempervivum tectorum* (Houseleek) Sanecki. A version of houseleek.

HUT, Finger *Digitalis purpurea* (Foxglove) Som. Macmillam. 'Hut' is presumably the same as Somerset 'hud', a finger-stall.

HUTCHINSIA *Hornungia petraea*.

HYACINTH 1. *Hyacinthus sp.* 2. *Iris pseudoacarus* (Stinking Iris) Som. Macmillan. The name comes from Greek mythology, for the flower was supposed to have sprung from the blood of Hyacinthus, a youth accidentally killed by Apollo. But the Greek hyakinthos is reckoned to be a species of Scilla.

HYACINTH, Alpine *Hyacinthus amethysti-*

nus.

HYACINTH, Bog *Orchis mascula* (Early Purple Orchis) Scot. Grigson.

HYACINTH, Californian *Brodiaea laxa*.

HYACINTH, Cape *Galtonia candicans* Coats.

HYACINTH, English *Endymion nonscriptus* (Bluebell) Gerard.

HYACINTH, Feathered *Muscari botryoides* (Grape Hyacinth) McDonald.

HYACINTH, Grape *Muscari botryoides*.

HYACINTH, Musk *Muscari botryoides* (Grape Hyacinth) McDonald.

HYACINTH, Purple *Orchis mascula* (Early Purple Orchis) Som. Macmillan.

HYACINTH, Spanish *Endymion nonscriptus* (Bluebell) Parkinson.

HYACINTH, Star *Scilla verna* (Spring Squill) Prior.

HYACINTH, Star, Autumnal *Scilla autumnalis* (Autumnal Squill) Camden/Gibson.

HYACINTH, Starch *Muscari racemosum* Grieve.1931. The juice from the stems was apparently used in Elizabethan times to stiffen ruffs.

HYACINTH, Tassel *Muscari comosum*.

HYACINTH, Wild 1. *Brodiaea capitata* Calif. Perry. 2. *Brodiaea coronaria* (Harvest Brodiaea) USA Turner & Bell. 3. *Endymion nonscriptus* (Bluebell) Gerard, and still used in Scot. Gilmour & Walters. 4. *Orchis mascula* (Early Purple Orchis) Som. Macmillan.

HYACINTH, Winter *Scilla autumnalis* (Autumnal Squill). W Miller.

HYACINTH, Wood *Endymion nonscriptus* (Bluebell) Tynan & Maitland.

HYDRANGEA *Hydrangea macrophylla*. Hydrangea is a word latinized from the Greek hudor, water, and aggeion, a jar or vase, from the shape of the capsule.

HYDRANGEA, Climbing *Hydrangea anomala var. petiolaris*.

HYDRANGEA, Oak-leaved *Hydrangea quercifolia*.

HYDRANGEA, Rough *Hydrangea aspera*.

HYDRANGEA, Tree *Hydrangea arborescens*.

HYDRANGEA, Wild 1. *Rumex venosus* (Veined Dock) Kansas Gates. 2. *Viburnum lantana* (Wayfaring Tree) Pratt. They know the dock as Wild Begonia in America, too. Wayfaring tree obviously got the name because of a superficial resemblance in the flowers.

HYPOCRITES *Viola riviniana* (Dog Violet) Som. Macmillan. Presumably because it masquerades as something it isn't - a scentless violet would seem an anomaly, hence the 'dog' of Dog Violet.

HYSSOP *Hyssopus officinalis*. Hyssop is OE ysope, from Latin hyssopus, Greek hussopos, which is possibly from Hebrew esob. But it has been pointed out that hyssop doesn't grow in Palestine or Egypt. Perhaps there was no linguistic connection at all between hyssopus and esob, just a similarity of sound.

HYSSOP, Anise- *Agastache anethiodora*.

HYSSOP, Giant 1. *Agastache nepetoides*. 2. *Agastache urticaefolius* (Nettle-leaf Hyssop) USA Elmore.

HYSSOP, Giant, Fragrant *Agastache anethioides* (Anise-Hyssop) USA Youngken.

HYSSOP, Giant, Mexican *Agastache mexicana*.

HYSSOP, Hedge 1. *Hyssopus officinalis var. ruber*. 2. *Polygala vulgaris* (Milkwort) Gerard. 3. *Scutellaria minor* (Lesser Skullcap) Prior.

HYSSOP, Nettle-leaf *Agastache urticaefolius*.

HYSSOP, Pink *Hyssopus officinalis var. ruber* (Hedge Hyssop) Webster.

HYSSOP, Stitch *Genista anglica* (Needle Whin) Hants. Cope. 'Stitch' on the analogy of the 'needle' of the common name?

HYSSOP, Syrian *Origanum maru* (Syrian Marjoram) Zohary. This is probably the hyssop of the Old Testament, in spite of the fact that it is not a hyssop at all.

HYSSOP, Water *Hyssopus officinalis var. ruber* (Hedge Hyssop) Fernie.

HYSSOP, White *Hyssopus officinalis var. albus*.

HYSSOP, Wild *Verbena hastata* (Blue Vervain) O P Brown.
HYSSOP HEDGE NETTLE *Stachys hyssopifolia* Sanecki.
HYVEN; HYVIN *Hedera helix* (Ivy) both Cumb, Yks. B & H. Hyven is also in Harland for Yks. Cf. the Manx Hibbin.

I

IACINTH *Hyacinthus sp* (Hyacinth) Parkinson.

IBOGA *Tabernanthe iboga*. Iboga is the Fang name for this West African plant. Iboga is really only an approximation - eboka appears more often.

ICE-LEAF, Wild *Verbascum thapsus* (Mullein) Bucks. B & H; USA Henkel.

ICE-PLANT 1. *Mesembryanthemum crystallinum* (Sea Fig) Thrimble. 2. *Monotropa uniflora* O P Brown. 3. *Saxifraga sarmentosa* (Mother-of-thousands) Dev. Friend.1882. 4. *Sedum spectabile*. 5. *Umbilicus rupestris* (Wall Pennywort) Dev. Macmillan. Probably something to do with the way the dew collects on the plant, though Brown gave the explanation for *Monotropa uniflora* that when it is handled it seems to melt away like ice.

ICELAND LICHEN; ICELAND MOSS *Cetraria islandica*. Iceland Moss is the usual name. Iceland Lichen is used by Flück.

ICELAND POPPY *Papaver nudicaule*.

ICELAND WATERCRESS *Rorippa palustris* (Marsh Yellow Cress) Hutchinson.

ICHANG LILY *Lilium leucanthum*.

IDOL, Love; IDOLS, Living; IDOLS, Loving *Viola tricolor* (Pansy) Grieve.1931 (Love); Tynan & Maitland (Living); Wessex Rogers; also Berks. Notes & Queries.8.1871 (Loving). Some of the many variations on Love-in-idleness.

IFE; YFE *Taxus baccata* (Yew) E Ang. both Nall.

IGNATIUS BEAN *Strychnos ignatii*.

ILES 1. *Drosera rotundifolia* (Sundew) Corn. Carew. 2. *Hordeum sativum* (Barley) Wilts. Dartnell & Goddard. i.e. for Sundew, liver-rot. Cf. the French herbe de l'igler, Danish iglegras. Liver-rot in cattle is supposed to be caused by their eating sundew, an effect shared with Pinguicula and Hydrocotyle. The name for the awns of barley is a variation of Ailes, or Eyles (it becomes Piles in Devon).

ILEX *Quercus ilex* (Holm Oak) Clapham.

ILL-SCENTED SUMACH *Rhus trilobata* (Three-leaf Sumach) Le Strange.

IMBREKE *Sempervivum tectorum* (Houseleek) W Miller.

IMMORTELLE *Helichrysum bracteatum* (Everlasting Flower) RHS. Immortelle - everlasting.

IMMORTELLE, Pearly *Anaphalis margaretacea*.

IMMORTELLE, Swamp *Erythrina glauca*.

IMPATIENT LADY'S SMOCK *Cardamine impatiens* (Narrow-leaved Bittercress) Curtis. Impatient, because of the explosive method of seed destribution.

IMPERIAL, Crown; IMPERIAL LILY *Fritillaria imperialis* Gerard has Imperial Lily.

IMPUDENT LAWYER *Linaria vulgaris* (Toadflax) USA Coats.

INCARNATION *Dianthus caryophyllus* (Carnation) B & H. The earlier, and more accurate, form of carnation. Grigson pointed out that it may have been a religious name deriving from the association of the word and the flower - "God made flesh", etc.

INCARNATION ROSE *Rosa alba 'Maiden's Blush'* Rohde.

INCA WHEAT *Amaranthus caudatus* (Love-lies-bleeding) Perry. Is this a correct recording? This is not a native American plant, although it is grown now in the Andean region of Bolivia, Peru and northern Argentina. Frances Perry said this was the Achita, Jataco or Quihuicha of the Aztecs, and eighteen of the Aztec empire's granaries were kept filled with the seeds. Surely it is more likely to be some other Amaranth, *A paniculatus*, for instance?

INCAS, Lily of the *Alstroemaria peregrina*.

INCENSE-BEARING JUNIPER *Juniperus thurifera* (Spanish Juniper) Ablett.

INCENSE CEDAR *Libocedrus decurrens*.

INCENSE CEDAR, Chilean *Austrocedrus chilensis* (Chilean Cedar) Leathart.

INCENSE ROSE *Rosa primula.* So called because of the scent of the young foliage.
INCENSIER *Rosmarinus officinalis* (Rosemary) Hatfield. It was burned as incense.
INCH PLANT *Tradescantia fluminensis* Gannon. Its other name, Wandering Jew, supplies the reason for Inch Plant as a name.
INCH PLANT, Silver *Zebrina pendula.*
INDIA, Bean of *Nelumbo nucifera* (Hindu Lotus) Whittle & Cook. Cf. Bean of Egypt, Turner's name for it.
INDIA, Cypress of *Curcuma longa* (Turmeric) Pomet.
INDIA, Pride of 1. *Koelreuteria paniculata* (Golden Rain Tree) Barber & Phillips. 2. *Lagerstroemeria flos-reginae* (Queen Crapemyrtle) RHS. 3. *Melia azederach* (Chinaberry Tree) Hogarth.
INDIA RUBBER PLANT; INDIA RUBBER TREE, Bengal *Ficus elastica* Lindley (the longer name).
INDIA WHEAT *Fagopyrum tataricum* (Green Buckwheat) Douglas.
INDIAN ACONITE 1. *Aconitum ferox* W Miller. 2. *Aconitum laciniatum.*
INDIAN AGUE-WEED *Eupatorium perfoliatum* (Thorough-wort) W Miller. Cf. Ague-weed and Feverwort for this. A tea made from the dried leaves and flowers is a well-known American domestic medicine for feverish colds.
INDIAN ALMOND 1. *Arisaema triphyllum* (Jack-in-the-pulpit) Douglas. 2. *Terminalia catappa* (Myrobalan) Dale & Greenway. North American Indian for Jack-in-the-pulpit, but presumably Asiatic Indian for Myrobalan.
INDIAN APPLE *Datura meteloides* (Downy Thorn-apple) USA Elmore. North American Indian, of course, in this case.
INDIAN ARROW-WOOD *Euonymus atropurpureus* (Wahoo) Howes.
INDIAN ARROWROOT; BOMBAY ARROWROOT *Curcuma angustifolia* Howes.
INDIAN AZALEA *Rhododendron simsii.*
INDIAN BALM *Trillium erectum* (Bethroot) Grieve.1931.
INDIAN BALMONY *Swertia chirata* (Chiretta) Leyel.1948.
INDIAN BALSAM *Impatiens glandulifera* (Himalayan Balsam) Howes.
INDIAN BARBERRY 1. *Berberis aristata* (Nepal Barberry) Chopra. 2. *Berberis asiatica.*
INDIAN BEAN 1. *Catalpa bignonoides* (Locust Bean) USA Harper. 2. *Dolichos lablab* (Lablab) Brouk. 3. *Erythrina flabelliformis* (Western Coral Bean) USA T H Kearney. Lablab is said by some to be a native of India, but others say it must have originated in tropical West Africa. The others, of course, are American trees, so the Indian referred to is a North American one.
INDIAN BEECH *Pongamia pinnata.*
INDIAN BERRY 1. *Anamirta cocculus* Chopra. 2. *Anamirta paniculata* (Levant Nut) Le Strange.
INDIAN BINDWEED *Convolvulus major* (Morning Glory) Grieve.1931.
INDIAN BORAGE *Coleus amboinicus.*
INDIAN BREADROOT *Psoralea esculenta* (Breadroot) Johnston.
INDIAN BUTTER BEAN *Dolichos lablab* (Lablab) Dalziel.
INDIAN CABBAGE *Yucca glauca* (Soapweed) USA Elmore. The central part of the spike used to be eaten as "cabbage" by the Indians.
INDIAN CANE *Canna indica* (Indian Shot) Tradescant. Cane - Canna.
INDIAN CEDAR *Cedrus deodara* (Deodar) Dyer.
INDIAN CHERRY 1. *Rhamnus caroliniana* USA Harper. 2. *Rhamnus crocea.*
INDIAN CHICKWEED *Mollugo verticillata* (Carpet-weed) USA T H Kearney.
INDIAN CHIR PINE *Pinus roxburghii* (Himalayan Chir Pine) Leathart.
INDIAN CHOCOLATE *Geum rivale* (Water Avens) USA Leighton. Cf. the name Chocolate-root. The roots, when boiled with milk and sweetened, give a beverage not unlike chocolate, a great favourite with

American Indians, apparently.

INDIAN COLZA *Brassica campestris var. sarson.*

INDIAN CORAL TREE *Erythrina indica.*

INDIAN CORN *Zea mays* (Maize).

INDIAN COW PARSNIP *Heracleum lanatum* Douglas.

INDIAN CRESS; INDIAN CRESS, Greater *Tropaeolum majus* (Nasturtium) Gerard, and Fernie respectively. Gerard gave it this name because it came from the West Indies. Cress, because the leaves have the same biting flavour as the true cress.

INDIAN CUP *Sarracenia purpurea* (Pitcher Plant) Perry.

INDIAN CURRANT *Symphoricarpos orbiculatus* (Red Wolfberry) USA Harper.

INDIAN ELM *Ulmus fulva* (Slippery Elm) Hatfield.

INDIAN EYE *Dianthus plumarius* (Pink) B & H.

INDIAN FIG *Ficus benghalensis* (Banyan) Rochford. Cf. Bengal Fig for this.

INDIAN FILBERT *Areca catechu* (Areca Palm) Pomet.

INDIAN FIRE *Salvia coccinea* (Red Sage) USA Howes.

INDIAN FLOWERING REED *Canna indica* (Indian Shot) Parkinson.

INDIAN FOG 1. *Sedum glaucum* Donegal B & H. 2. *Sedum reflexum* (Yellow Stonecrop) Ire. Grigson.

INDIAN FRANKINCENSE *Boswellia serrata.* From northwest India.

INDIAN FUMITORY *Fumaria indica* (American Fumitory) Le Strange.

INDIAN GENTIAN *Swertia chirata* (Chiretta) Mitton.

INDIAN GINGER *Asarum canadense* (Canadian Snakeroot) Grieve.1931. No relation, but this has the flavour of true ginger.

INDIAN GOOSEBERRY *Phyllanthus acidus.*

INDIAN GRASS 1. *Corchorus capsularis* (or *C olitorius*) (Jute) Gibson. 2. *Gynerium argenteum* (Pampas Grass) Suss. Friend. This is what East India Company representatives called jute when they first sent a parcel to Europe in the 18th century.

INDIAN HAY *Cannabis sativa* (Hemp) USA Watt.

INDIAN HELIOTROPE *Heliotropum indicum.*

INDIAN HELLEBORE *Veratrum viride* (American White Hellebore) USA Turner & Bell.

INDIAN HEMP 1. *Apocynum cannabinum.* 2. *Cannabis sativa* (Hemp) Watt. 3. *Crotalaria juncea* (Sunn Hemp) Hutchinson & Melville.

INDIAN HEMP, Black *Apocynum cannabinum* (Indian Hemp) Grieve.1931.

INDIAN HENBANE *Nicotiana tabacum* (Tobacco) Parkinson.

INDIAN HORSE CHESTNUT *Aesculus indica.*

INDIAN JALAP *Ipomaea hederacea* (Ivy-leaf Morning Glory) Chopra.

INDIAN KIDNEY TEA *Orthosiphon stamineus* (Java Tea) Thomson.1978.

INDIAN LABURNUM *Cassia fistula.*

INDIAN LILAC 1. *Lagerstroemeria indica* (China Privet) Chopra. 2. *Melia azedarach* (Chinaberry Tree) Le Strange.

INDIAN LIQUORICE *Abrus precatorius* (Precatory Bean) Grieve.1931. Wild Liquorice is another name for this.

INDIAN MADDER *Rubia cordifolia.*

INDIAN MALLOW *Abutilon avicennae* (China Jute) Schery.

INDIAN MAY-APPLE *Podophyllum hexandrum.*

INDIAN MOSS 1. *Saxifraga hypnoides* (Dovedale Moss) Grigson. 2. *Sedum reflexum* (Yellow Stonecrop) Donegal Grigson.

INDIAN MULBERRY *Moringa citrifolia.*

INDIAN MUSTARD *Brassica juncea.*

INDIAN NIGHT JASMINE *Nyctanthes arbortristis.*

INDIAN OAK *Barringtonia acutangula.*

INDIAN OLEANDER *Nerium indicum.*

INDIAN OLIBANUM *Boswellia serrata* (Indian Frankincense) Usher.

INDIAN OLIVE *Olea ferruginea.*

INDIAN ORRIS *Saussurea lappa.*
INDIAN PAINT 1. *Chenopodium capitatum* (Strawberry Blite) USA Yarnell. 2. *Lithospermum angustifolium* (Puccoon) USA Elmore. 3. *Lithospermum canescens* (Hoary Puccoon) Miss. Bergen. 4. *Sanguinaria canadensis* (Bloodroot) Grieve.1931. The reference in the last three cases is to the red pigment obtained from the roots.
INDIAN PAINTBRUSH 1. *Asclepias tuberosa* (Orange Milkweed) USA Cunningham. 2. *Castilleia integra.* 3. *Castilleia angustifolia* (Desert Paintbrush) USA E Gunther.
INDIAN PAPER BIRCH *Betula bhojpattra.*
INDIAN PEA *Lathyrus sativus.*
INDIAN PENNYWORT *Hydrocotyle asiatica.*
INDIAN PHYSIC 1. *Apocynum cannabinum* (Indian Hemp) Grieve.1931. 2. *Gillenia trifoliata.* In other words, plants which the American Indians used as medicine.
INDIAN PINK 1. *Dianthus chinensis.* 2. *Lychnis flos-cuculi* (Ragged Robin) Glos. Grigson. 3. *Silene californica* USA Schenk & Gifford. 4. *Spigelia marilandica* (Pinkroot) Willis.
INDIAN PIPEWEED *Eriogonum inflatum* (Desert Trumpet) USA T H Kearney.
INDIAN PLANTAIN *Cacalia atriplicifolia.*
INDIAN POKE *Veratrum viride* (American White Hellebore) Fluckiger.
INDIAN POPPY 1. *Glaucium flavum* (Horned Poppy) Som. Macmillan. 2. *Meconopsis cambrica* (Welsh Poppy) Som. Macmillan.
INDIAN POSY *Gnaphalium polycephalum* (Mouse-ear Everlasting) O P Brown.
INDIAN POTATO 1. *Apios americana* (Ground Nut) H & P. 2. *Solanum tuberosum* (Potato) USA H Smith.
INDIAN RAPE *Brassica campestris var. toria.*
INDIAN RHUBARB *Saxifraga peltata.*
INDIAN RICE *Zizania aquatica* (Wild Rice) USA Howes.

INDIAN ROOT 1. *Aristolochia watsoni* USA T H Kearney. 2. *Asclepias curassivica* (Blood Flower) Barbados Gooding.
INDIAN ROSEWOOD *Dalbergia latifolia* (Blackwood) Howes.
INDIAN SAFFRON *Curcuma longa* (Turmeric) Clair. The reference is to the yellow dye. But in adition, *Curcuma*, the generic name, is from Persian kurkum, which means saffron.
INDIAN SAGE *Eupatorium perfoliatum* (Thoroughwort) USA Henkel.
INDIAN SCABIOUS *Scabiosa atropurpurea* (Sweet Scabious) Folkard. The dusky colour prompts the name, and was also responsible for such others as Egyptian Rose, or Gypsy Rose.
INDIAN SHAMROCK *Trillium erectum* (Bethroot) Grieve.1931.
INDIAN SHOT *Canna indica.* From the round, bullet-like seeds, which were also used to make rosaries in India.
INDIAN SILVER FIR *Abies spectabilis* (East Himalayan Fir) Brimble.1948.
INDIAN SPINACH *Basella alba.*
INDIAN SUNFLOWER *Helianthus annuus* (Sunflower) Gerard.
INDIAN SWEDE *Terminalia catappa* (Myrobalan) Douglas.
INDIAN TOBACCO 1. *Antennaria plantaginifolia* (Plantain-leaved Everlasting). Grieve.1931. 2. *Lobelia inflata.* 3. *Nicotiana bigelovii var. exaltata* USA Schenk & Gifford. The non-tobaccos were used as substitutes, the everlasting presumably being a mild smoke, for the name Ladies' Tobacco is also recorded.
INDIAN TRAVELLER'S JOY *Clematis gouriana.*
INDIAN TURNIP 1. *Arisaema triphyllum* (Jack-in-the-pulpit) USA Kingsbury. 2. *Psoralea tenuiflora* USA Kingsbury. 3. *Terminalia catappa* (Myrobalan) Douglas. The corms of Jack-in-the-pulpit at least used to be eaten.
INDIAN TURNSOLE *Heliotropum indicum* (Indian Heliotrope) Dalziel.

INDIAN VALERIAN *Nardostachys jatamansi* (Spikenard) Leyel.1937.
INDIAN WALNUT TREE *Aleurites moluccana* (Candlenut Tree) Leyel.1948.
INDIAN WATER NAVELWORT *Hydrocotyle asiatica* (Indian Pennywort) Schauenberg.
INDIAN WHEAT 1. *Fagopyrum esculentum* (Buckwheat) N America Grieve.1931. 2. *Zea mays* (Maize) Cobbett.
INDIAN WHEAT, Woolly *Plantago purshii* (Pursh Plantain) USA Elmore.
INDIAN-WOOD *Haematoxylon campechianum* (Logwood) Pomet. West Indian, in this case.
INDIAN WORMSEED *Chenopodium ambrosioides* (American Wormseed) Dalziel.
INDIAN'S PIPE *Monotropa uniflora* USA Tolstead. It is said that the single bell-shaped flower dropping from the tip of a practically leafless stem resembles a white miniature Indian pipe of peace.
INDIGO *Indigofera anil*. Indigo is Latin indicum, and India was the oldest centre of indigo dyeing.
INDIGO, Bastard 1. *Amorpha fruticosa* (Indigo-bush) H & P. 2. *Tephrosia purpurea*.
INDIGO, Blue *Baptisia australis* (False Indigo) Perry.
INDIGO, Creeping *Indigofera endocaphylla*.
INDIGO, False 1. *Amorpha fruticosa* (Indigo-bush) USA T H Kearney. 2. *Baptisia australis*.
INDIGO, Java *Indigofera arrecta* (Natal Indigo) Dalziel. It was introduced into Java from Natal.
INDIGO, Natal *Indigofera arrecta*.
INDIGO, Wild 1. *Baptisia tinctoria*. 2. *Tephrosia purpurea* (Bastard Indigo) Chopra. *Baptisia* is taken from Greek bapto, to dye.
INDIGO, Wild, Yoruba *Lonchocarpus cyanescens* (Indigo Vine) Picton & Mack. A favourite blue dye plant for Yoruba textiles.
INDIGO-BROOM; INDIGO-WEED *Baptisia tinctoria*.

INDIGO-BUSH *Amorpha fruticosa*.
INDIGO VINE *Lonchocarpus cyanescens*.
INDISH PEPPER *Capsicum annuum* (Chile) Turner.
INDY 1. *Lychnis flos-cuculi* (Ragged Robin) Glos. Grigson. 2. *Melandrium dioicum* (Red Campion) Glos. Grigson. Cf. Gypsy-flower for these.
INDY CORN *Zea mays* (Maize) Cumb. Rowling.
INDY PINK *Dianthus caryophyllus* (Carnation) Glos. J D Robertson.
INE *Allium cepa* (Onion) Ellacombe. The older English name for onion. It is OE Yne, or Ynne.
INEYUN *Allium cepa* (Onion) IOW Long. A rather extreme form of pronunciation presentation.
INFANT ONION *Allium schoernoprasum* (Chives) Mabey. Because of its mildness and almost complete absence of bulb.
INFIRMARY CLOCK *Adoxa moschatellina* (Moschatel) Tynan & Maitland. Cf. Town Clock, or Town Hall Clock; all because of the way the flowers grow like the faces on a clock.
INGAN; ININ; INING; INION *Allium cepa* (Onion) Suff. Moor; Scot. Ritchie (Ingan); Som. Jennings (Inin); Som. Macmillan (Ining); N'thants. A E Baker (Inion).
INK, Wall *Veronica beccabunga* (Brooklime) B & H. This name also occurs as Wellink, which makes more sense. Cf. the Scottish Horse Well-cress or Well-grass for this plant.
INK ROOT 1. *Limonium carolinianum* (American Sea Lavender) Grieve.1931. 2. *Limonium vulgare* (Sea Lavender) Fernie. There must be a very good reason for this name, but I haven't yet found a record of it.
INK VINE *Passiflora suberosa* Barbados Gooding.
INKBERRY 1. *Ilex glabra* (Gallberry) northern USA Harper. 2. *Lonicera involucrata* (Bearberry Honeysuckle) USA. T H Kearney. 3. *Phytolacca decandra* (Pokeroot) USA Sanford. The coloured juice of Pokeroot has been used as ink, and to give

the red stain that the Indians used to colour their ornaments. It is also known as Red-ink Plant, or Red-ink Berry.

INKBERRY, Red *Phytolacca decandra* (Pokeroot) USA Henkel.

INKBUSH *Suaeda fruticosa* (Shrubby Seablite) S Africa Watt.

INKWEED 1. *Phytolacca octandra*. 2. *Suaeda torreyana var. ramosissima* (Torrey Seablite) USA Jaeger. A black ink can be made from the leaves of the Seablite.

INKWOOD *Drymaria pachyphylla* (Drymary) USA Kingsbury. A purplish juice can be squeezed from the seed capsules before they are ripe.

INLAND JERSEY TEA *Ceanothus ovatus*.

INNION; INYUN *Allium cepa* (Onion) Som. Macmillan (Innion); Hants. Cope; IOW Long (Inyun).

INNOCENCE; INNOCENCE, Blue *Houstounia caerulea* (Bluets) USA House (Innocence); W Miller (Blue Innocence).

INNOCENT 1. *Bellis perennis* (Daisy) Som. Macmillan. 2. *Convallaria maialis* (Lily of the valley) Dor. Macmillan.

INNOCENTS, Holy *Crataegus monogyna* (Hawthorn) Wilts. Macmillan.

INON; INUN *Allium cepa* (Onion) Dor. Barnes; Wilts. Akerman Heref. Havergal; Hants. Cope; Berks. Lowsley; N'thants. Sternberg (Inon); Worcs. J Salisbury (Inun). The older English name for onion was Ine. These are variations, or even hybrids with onion.

INSANE-ROOT *Hyoscyamus niger* (Henbane) Safford.1920.

INSECT POWDER PLANT *Pyrethrum cinearifolium* (Pyrethrum Flower) Brownlow. Pyrethrum is famous as the Insect Powder Plant, for the dried and powdered flowers are used as insecticide, and against moths. The main growing country these days is Kenya, which supplies 70% of the world demand.

INUL *Inula helenium* (Elecampane) B & H.

INULE, Rigid *Inula conyza* (Ploughman's Spikenard) Potter & Sargent.

IODINE WEED *Suaeda torreyana var. ramosissima* (Torrey Seablite) USA Jaeger. This is very strong in tannin, so presumably that is the reason for the name.

IPECAC, American *Euphorbia ipecacuanhae*.

IPECACUANHA; IPECAC *Cephaelis ipecacuanha*. Ipecacuanha apparently means "wayside-plant-emetic", or "sick-making plant" in a South American Indian tongue, the original being given as ipe-kaa-guene. That was taken into Portuguese to give the name still in use.

IPECACUANHA, American *Gillenia stipulacea*.

IPECACUANHA, Bastard *Asclepias curassivica* (Blood Flower) Le Strange.

IPECACUANHA, West Indian *Asclepias curassavica* (Blood Flower) Chopra.

IPECACUANHA, Wild 1. *Asclepias curassavica* (Blood Flower) Barbados Gooding. 2. *Euphorbia ipecacuanha* (American Ipecac) Le Strange.

IPO *Carum gardneri* (Yampa) Le Strange.

IPOMAEA, Goat's-foot *Ipomaea pes-caprae*.

IPOMAEA, Star *Ipomaea coccinea* (Red Morning Glory) RHS.

IRANIAN BOX; PERSIAN BOX *Buxus sempervirens* timber (Boxwood) Brimble.1948.

IRBY-DALE GRASS *Euphorbia helioscopia* (Sun Spurge) Lincs. B & H. Irby-dale is a place in Lincolnshire, near Laceby, where apparently a lot of sun spurges used to grow.

IRELAND, Bells of *Molucella laevis* (Molucca Balm) Perry.

IRIS, Afternoon *Iris dichotoma*.

IRIS, Algerian *Iris stylosa*.

IRIS, Bulbous, Great *Iris xiphioides* (English Iris) W Miller.

IRIS, Butterfly *Iris spuria*.

IRIS, Christmas-flowering *Iris alata*.

IRIS, Cinnamon *Acorus calamus* (Sweet Flag) Genders. When the stems and leaves are crushed, they release a scent rather like that of cinnamon. Cinnamon Sedge is another version of the name.

IRIS, Clematis *Iris kaempferi* (Japanese Iris).

IRIS, Clouded *Iris xiphium* (Spanish Iris) W Miller.
IRIS, Copper *Iris fulva.*
IRIS, Crimean, Dwarf *Iris chamaeiris.*
IRIS, Dalmatian *Iris pallida* (Pale Iris) RHS.
IRIS, Dutch *Iris xiphium hybrids.*
IRIS, Dwarf *Iris pumila.*
IRIS, English *Iris xiphioides.* For hundreds of years, it has flourished in the Bristol neighbourhood, and the Dutch bulb merchants thought it was a native of Britain - hence the name.
IRIS, Ever-blooming *Iris ruthenica* (Russian Iris) W Miller.
IRIS, Flag *Iris germanica* (Blue Flag) Le Strange.
IRIS, Florentine *Iris germanica var. florentina* (Orris) Thornton.
IRIS, Foetid; IRIS, Stinking *Iris foetidissima* RHS (Stinking). Only foetid when the leaves are rubbed, then it becomes, in Gerard's words, "of a stinking smell very lothsome".
IRIS, German *Iris germanica* (Blue Flag) W Miller.
IRIS, Golden *Iris monnieri.*
IRIS, Ground *Iris macrosiphon.*
IRIS, Japanese *Iris kaempferi.*
IRIS, Juno *Iris scorpiris.*
IRIS, Lake, Dwarf *Iris lacustris.*
IRIS, Missouri *Iris missouriensis.*
IRIS, Mourning *Iris susiana.* Grey flowers, with purple-black veins, hence the name.
IRIS, Onion *Hermodactylis tuberosus* (Snakeshead Iris) W Miller.
IRIS, Pale *Iris pallida.*
IRIS, Peacock *Moraea pavonia* (Peacock Flower) Simmons.
IRIS, Rocky Mountain *Iris missourensis* (Missouri Iris) USA T H Kearney.
IRIS, Roof, Japanese *Iris tectorum* (Wall Iris) RHS.
IRIS, Russian *Iris ruthenica.*
IRIS, Sad-flowered *Iris susiana* (Mourning Iris) W Miller.
IRIS, Scarlet-seeded *Iris foetidissima* (Foetid Iris) A W Smith.
IRIS, Scorpion *Iris alata* (Christmas-flowering Iris) W Miller.
IRIS, Siberian *Iris siberica.*
IRIS, Smooth *Iris laevigata.*
IRIS, Snakeshead *Hermodactylis tuberosus.*
IRIS, Spanish *Iris xiphium.*
IRIS, Spotted, Great *Iris susiana* (Mourning Iris) W Miller.
IRIS, Thunderbolt *Iris xiphium* (Spanish Iris) W Miller.
IRIS, Vesper *Iris dichotoma* (Afternoon Iris). The corolla never expands before afternoon, hence the name.
IRIS, Violet *Iris verna* (Dwarf American Iris) USA Cunningham.
IRIS, Water, Purple *Iris versicolor.*
IRIS, Widow *Hermodactylus tuberosus* (Snakeshead Iris). A W Smith. Both the common names derive from the colour of the flowers, which is greenish-black.
IRIS, Wild, Western *Iris missouriensis* (Missouri Iris).
IRIS, Winter 1. *Iris stylosa* (Algerian Iris) RHS. 2. *Marica northiana* (Apostle Plant) Gannon.
IRIS, Yellow *Iris pseudo-acarus.*
IRIS, Yellowband *Iris ochroleuca.*
IRISH BROOM *Cytisus scoparius* (Broom) USA (?) Mitton.
IRISH BUTTERWORT *Pinguicula grandiflora* (Great Butterwort) W Miller. A plant from northern Spain and the Pyrenees, very locally established in S W Ireland.
IRISH DAISY *Taraxacum officinale* (Dandelion) Yks. B & H USA Henkel.
IRISH FLEABANE *Inula salicina.* Only on the shores of Lough Derg in these islands.
IRISH HEATH 1. *Daboecia cantabrica* (St Dabeoc's Heath) W Miller. 2. *Erica erigena.* Both confined to wet moors in the west of Ireland.
IRISH JUNIPER *Juniperus communis var. hibernica.*
IRISH LADY'S TRESSES *Spiranthes romanzoffiana* (Hooded Lady's Tresses) Genders.
IRISH MAHOGANY *Alnus glutinosa*

(Alder) Ire. Grigson. Or perhaps Scotch Mahogany. It is the red timber that is the reference.

IRISH MARSH ORCHID *Dactylorchis majalis*.

IRISH ORCHID *Neotinea intacta* (Dense-flowered Orchid) Turner Ettlinger. Not quite confined to Mayo and Galway, for it occurs also in the Isle of Man.

IRISH POTATO *Solanum tuberosum* (Potato) USA H Smith. The name was given when potatoes from Great Britain were introduced to the English colonists on the American mainland, and it is still in use today.

IRISH ST JOHN'S WORT *Hypericum canadense* (Canadian St John's Wort) Blamey. It occurs very locally in the west of Ireland.

IRISH SPURGE *Euphorbia hiberna*.

IRISH TEA *Prunus spinosa* (Blackthorn) Ire. O Suilleabhain. Blackthorn leaves have been used as a tea adulterant for a long time.

IRISH VINE *Lonicera periclymenum* (Honeysuckle) Ire. Grigson.

IRISH WORTS *Daboecia cantabrica* (St Dabeoc's Heath) B & H.

IRISH YEW *Taxus baccata var. fastigiata*.

IRISHMAN, Wild *Discaria toumatou*. This New Zealand shrub is described as stiff and thorny. Is that the reason for the name?

IROKO; IROKO FUSTIC WOOD *Chlorophora excelsa* (African Oak) Wit (Iroko); Usher (Iroko Fustic Wood). Iroko is the Yoruba name for the tree.

IROKO, False 1. *Antiaris africana* (Barkcloth Tree) Dalziel. 2. *Antiaris toxicaria* (Upas Tree) Dale & Greenway.

IRON CROSS BEGONIA *Begonia masoniana*. From the markings on the leaves.

IRON-FLOWER 1. *Allium ursinum* (Ramsons) Som. Macmillan. 2. *Jasione montana* (Sheep's-bit Scabious) Ches. Holland.

IRON-GRASS *Polygonum aviculare* (Knotweed) Herts. Grigson. Cf. Wire-grass, Wireweed, etc. And Armstrong, generally taken to be from the fact that it is so difficult to pull up, conveys the same idea.

IRON-HARD 1. *Centaurea nigra* (Knapweed) W Miller. 2. *Lythrum salicaria* (Purple Loosestrife) Hunts. Marshall. 3. *Verbena officinalis* (Vervain) Storms. For knapweed, Cf. Hardhead, and all its derivatives. It is the roots of Purple Loosestrife that are so hard, hard enough to bend a shovel if the digger hits one.

IRON-HEAD; IRON-KNOB *Centaurea nigra* (Knapweed) Prior (Iron-head); Ches. Holland (Iron-knob).

IRON OAK *Quercus stellata* (Post Oak) USA Vestal & Schultes.

IRON-PEAR *Sorbus aria* (Whitebeam) Wilts. Dartnell & Goddard. Whitebeam is sometimes included in the same genus as pear, *Pyrus*. 'Iron' must be a comment on the hardness of the fruits.

IRON-ROOT *Atriplex patula* (Orache) Clapham.

IRON-WEED 1. *Buglossoides arvensis* (Corn Gromwell) S Africa Watt. 2. *Centaurea nigra* (Knapweed) Som. Macmillan; N'thants. A E Baker. 3. *Echium vulgare* (Viper's Bugloss) Beds. Grigson. 4. *Vernonia noveboracensis*. Presumably they all have tough stems.

IRON-WEED, Drummond's *Vernonia missurica*.

IRONBARK *Eucalyptus sideroxylon*.

IRONBARK, White *Eucalyptus leucoxylon* (Yellow Gum) Perry.

IRONHEART TREE *Swartzia madagascariensis* (Snake Bean Tree) Palgrave.

IRONWOOD 1. *Acacia estrophiolata* Australia Meggitt. 2. *Carpinus caroliniana* (Blue Beech) USA Harper. 3. *Olea paniculata*. 4. *Ostrya virginiana* (American Hophornbeam) Everett. 5. *Prosopis africana* Gambia Dalziel. All hardwoods.

IRONWOOD, Black *Olea laurifolia*.

IRONWOOD, Ceylon *Mesua ferrea*.

IRONWOOD, Jamaica *Erythroxylon areolatum*.

IRONWOOD, North American *Ostrya vir-

giniana (American Hop-Hornbeam) J Smith.

IRONWOOD, Rhodesian *Copaifera mopane* (Mopane) Palgrave.

IRONWOOD, South Sea *Casuarina equisetifolia* (Beefwood) Hora.

IROQUOIS BREADFRUIT *Arisaema triphyllum* (Jack-in-the-pulpit) USA Yanovsky. Because the corms were eaten. Cf. Indian Turnip for this.

ISAAC, Wild *Eupatorium perfoliatum* (Thoroughwort) USA Henkel.

ISABEL, Nodding *Briza media* (Quaking Grass) Lancs. Grigson.1959.

ISLAY *Prunus ilicifolia* (Hollyleaf Cherry) Leathart.

ISLE OF WIGHT VINE 1. *Bryonia dioica* (White Bryony) IOW B & H. 2. *Tamus communis* (Black Bryony) W Miller. Given on the same basis as Wiltshire Weed for elm in that county.

ISPAGHUL PLANTAIN *Plantago ovata*.

ISRAEL, Children of 1. *Malcomia maritima* (Virginian Stock) Dev, Wilts. Wright Dor. Dacombe. 2. *Pulmonaria officinalis* (Lungwort) Dor. Grigson. 3. *Saxifraga spathularia x umbrosa* (London Pride) Wilts. Macmillan. Lungwort may get the name because of the different colours in the flower, but the other two are known as Children of Israel because of the very large amount of little flowers.

ISRAELITES *Malcomia maritima* (Virginian Stock) Wilts. Macmillan.

ISTLE, Jaumarve *Agave funkiana*. Istle is the fibre, and the word comes through Mexican Spanish ixtle from Nahuatl ichtli.

ISTLE, Palma *Yucca carnerosana*.

ITALIAN ALDER *Alnus cordata*.

ITALIAN ALKANET *Anchusa italica*.

ITALIAN ASTER *Aster amellus*.

ITALIAN BELLFLOWER *Campanula isophylla*.

ITALIAN BUCKTHORN *Rhamnus alaternus* (Mediterranean Buckthorn) RHS.

ITALIAN CATCHFLY *Silene italica*.

ITALIAN CELERY, Upright *Apium graveolens* (Celery) G M Taylor.

ITALIAN CHESTNUT *Castanea sativa* (Sweet Chestnut) Wit. Better known as Spanish Chestnut, but these names betray the tree's Mediterranean origin.

ITALIAN CLOVER *Trifolium incarnatum* (Crimson Clover) Candolle.

ITALIAN CRANESBILL *Geranium macrorrhizum*.

ITALIAN CYPRESS *Cupressus sempervirens* (Common Cypress) Mitchell.1974. A tree of the eastern Mediterranean, but it was introduced into Italy in antiquity. Cf. Roman Cypress.

ITALIAN FENNEL *Foeniculum vulgare var. piperitum*.

ITALIAN HONEYSUCKLE *Lonicera caprifolium* (Perfoliate Honeysuckle) Grieve.1931.

ITALIAN JASMINE 1. *Jasminum grandiflorum* (Spanish Jasmine) Howes. 2. *Jasminum humile*.

ITALIAN MAPLE *Acer opalus*.

ITALIAN MILLET *Setaria italica* (Foxtail Millet) Brouk. But it is also known as German, Hungarian, or Siberian Millet.

ITALIAN NARCISSUS *Narcissus canaliculatum* Lindley.

ITALIAN PARSLEY *Petroselinum crispum var. neapolitanum*.

ITALIAN POPLAR, Black *Populus x canadensis 'serotina'*.

ITALIAN ROCKET *Reseda lutea* (Wild Mignonette) B & H.

ITALIAN RYE-GRASS; ITALIAN RAY-GRASS *Lolium multiflorum* Ray-grass (B & H).

ITALIAN SAINFOIN *Hedysarum coronarium* (French Honeysuckle) Polunin.

ITALIAN SENNA 1. *Cassia obovata*. 2. *Cassia obtusifolia* Barbados Gooding. *C obovata* used to be cultivated in Italy.

ITALIAN TOADFLAX *Linaria angustissima*. Not confined to Italy, in spite of the name. It is a European alpine.

ITALIAN TURNIP BROCCOLI *Brassica ruvo*.

ITALIAN VETCH *Galega officinalis* (Goat's Rue) Parkinson.

ITCH; ITYO *Gingko triloba* (Maidenhair Tree) S Africa Watt. Icho is the Japanese name.

ITCH, Cow 1. *Polygonum hydropiper* (Water Pepper) Cockayne. 2. *Rhus radicans* (Poison Ivy) Hora. 3. *Rosa canina* seeds (Dog Rose seeds) Ches. Holland. 4. *Urera baccifera*. The first three are itching plants in their different ways, but the *Urera* is probably Cowage, which means something quite different.

ITCHING BERRY; ITCHY-BACKS *Rosa canina* fruit (Hips) Lancs. Grigson. Dog Rose seeds are prime itch-producers, a fact well known to small boys, who delight in putting them down each others' necks.

ITCHWEED *Veratrum viride* (American White Hellebore) USA Lloyd.

ITHURIEL'S SPEAR *Brodiaea laxa* (Californian Hyacinth) Simmons.

IVE, Herb 1. *Ajuga chamaepitys* (Yellow Bugle) Henslow. 2. *Coronopus squamatus* (Wart-cress) Prior. 3. *Plantago coronopus* (Buck's Horn Plantain) Cambs. Turner.

IVER *Lolium perenne* (Rye-grass) Corn. B & H. Eventually from the French ivraie (drunk), and more applicable to darnel, which causes giddiness. Cf. such forms as Eaver or Heaver.

IVERY; IVORY *Hedera helix* (Ivy) Herts, Ess, E Ang, Lincs, Notts, Rut, Scot. (Grigson (Ivery); Som. Raymomd; Lincs. Peacock (Ivory).

IVIN *Hedera helix* (Ivy) Ches, Derb, Cumb, N'thum. Grigson; Lancs. Nodal; Lincs. Peacock; Yks. Atkinson. Cf. Hyvin, or Hyven.

IVORY, Red *Rhamnus zeyheri* S Africa Watt.

IVORY, Sea *Ramalina siliquosa*.

IVORY PALM, Vegetable *Hyphaene benguellensis*. 'Ivory', because the nut is so extremely hard.

IVRAY *Lolium temulentum* (Darnel) Lyte. French ivraie, drunk, which is the original of a number of names for darnel, including eventually, Rye-grass.

IVVIN, Ground *Glechoma hederacea* (Ground Ivy) Ches, Lancs. Grigson.

IVVY *Hedera helix* (Ivy) Heref. Havergal. Written thus to indicate pronunciation.

IVY 1. *Hedera helix*. 2. *Kalmia latifolia* (Calico Bush) S.USA Harper. A word with obscure etymology. The OE was ifig, through a Germanic word, probably with the sense of climber.

IVY, American *Ampelopsis hederacea* (Virginia Creeper) Grieve.1931.

IVY, Black *Hedera helix* (Ivy) Lyte.

IVY, Boston *Ampelopsis tricuspidata*.

IVY, Broad-leaved *Kalmia latifolia* (Calico Bush) House.

IVY, Cape *Senecio mikanioides* (German Ivy) Australia Watt.

IVY, Colosseum *Cymbalaria muralis* (Ivy-leaved Toadflax) W Miller.

IVY, Creep *Calystegia sepium* (Great Bindweed) Tynan & Maitland.

IVY, Devil's *Scindapsus pictus 'Argyraeus'*.

IVY, Earth *Glechoma hederacea* (Ground Ivy) Dawson.

IVY, English *Hedera helix* (Ivy) USA Allan.

IVY, Five-leaved *Ampelopsis hederacea* (Virginia Creeper). B & H.

IVY, German *Senecio mikanioides*. A South African plant, known as German Ivy only because it was first grown as a house plant there.

IVY, Gout *Ajuga chamaepitys* (Yellow Bugle) Prior. It was at one time used for gout, and was one of the ingredients of the once famous Portland Powder.

IVY, Grape, Miniature *Cissus striata* (Ivy of Uruguay) RHS.

IVY, Grape-vine *Cissus rhombifolia*.

IVY, Ground 1. *Ajuga chamaepitys* (Yellow Bugle) B & H. 2. *Calystegia sepium* (Great Bindweed) Dev. Friend. 3. *Convolvulus arvensis* (Field Bindweed) Corn, Dev. Wakelin. 4. *Glechoma hederacea*. 5. *Vinca minor* (Lesser Periwinkle) B & H.

IVY, Hedge *Tamus communis* (Black Bryony) Lincs. Rudkin.
IVY, Herb 1. *Ajuga chamaepitys* (Yellow Bugle) Gerard. 2. *Coronopus squamatus* (Wart-cress) Prior. 3. *Plantago coronopus* (Buck's-horn Plantain).
IVY, Japanese *Ampelopsis tricuspidata* (Boston Ivy) RHS.
IVY, Kenilworth *Cymbalaria muralis* (Ivy-leaved Toadflax) USA Zenkert.
IVY, Persian *Hedera colchica*.
IVY, Poison *Rhus radicans*.
IVY, Red *Ampelopsis hederacea* (Virginia Creeper) Som. Macmillan.
IVY, Underground 1. *Cymbalaria muralis* (Ivy-leaved Toadflax) Som. Macmillan. 2. *Glechoma hederacea* (Ground Ivy) Som. Macmillan.
IVY BELLS *Caltha palustris* (Marsh Marigold) Som. Macmillan.
IVY BINDWEED *Polygonum convolvulus* (Black Bindweed) B & H.
IVY BROOMRAPE *Orobanche hederae*.
IVY CAMPANULA *Wahlenbergia hederifolia* (Ivy-leaved Bellflower) Clapham.
IVY CHICKWEED *Veronica hederifolia* (Ivy-leaved Speedwell) Gerard.
IVY CROWFOOT; IVY-LEAVED CROWFOOT *Ranunculus hederaceus* (Ivy-leaved Water Buttercup) Murdoch McNeill (Ivy); McClintock (Ivy-leaved).
IVY DUCKWEED; IVY-LEAVED DUCKWEED *Lemna trisulca* Gilmour & Walters (Ivy-leaved).
IVY FLOWER 1. *Euonymus europaeus* (Spindle Tree) Som. Macmillan. 2. *Hepatica caerulea* (Hepatica) Glos. B & H. Spindle flowers are quite like those of ivy.
IVY-LEAF MORNING GLORY *Ipomaea hederacea*.
IVY-LEAVED GERANIUM *Pelargonium peltatum*.
IVY-LEAVED GROUND CHERRY *Physalis hederaefolia*.
IVY-LEAVED SCURVY-GRASS *Cochlearia danica* (Danish Scurvy-grass) Le Strange.
IVY-LEAVED SNAPDRAGON *Cymbalaria muralis* (Ivy-leaved Toadflax) Carr.
IVY-LEAVED SPEEDWELL *Veronica hederifolia*.
IVY-LEAVED TOADFLAX *Cymbalaria muralis*.
IVY-LEAVED VIOLET *Viola hederacea* (Australian Violet) RHS.
IVY-LEAVED WATER BUTTERCUP *Ranunculus hederaceus*.
IVY OF URUGUAY *Cissus striata*.
IVYBUSH *Kalmia latifolia* (Calico Bush) USA Kingsbury. Cf. Ivy, Broad-leaved Ivy, for this.
IVYWORT *Cymbalaria muralis* (Ivy-leaved Toadflax) Parkinson.

J

JACA *Artocarpus heterophyllus* (Jackfruit) F P Smith.

JACARANDA, Rhodesian *Stereospermum kuntheanum*.

JACK 1. *Cheiranthus cheiri* (Wallflower) B & H. 2. *Dianthus caryophyllus* (Carnation) B & H. Jack was the old slang word for the single carnation, according to B & H. Presumably wallflower gets it as an extension of the confusion that used to reign about just what was a gilliflower, for both plants bore the name indiscriminately. But it would seem from the various names that follow that Jack was a typical West Country name for a flower (rather than a plant, that is). In other cases, Jack is clearly a euphemism for the devil.

JACK, Black 1. *Bidens bipinnata* (Spanish Needles) S Africa Watt. 2. *Bidens pilosa* S Africa Watt. 3. *Plantago lanceolata* (Ribwort Plantain) Shrop. Grigson.

JACK, Blue 1. *Centaurea cyanus* (Cornflower) Som. Grigson. 2. *Centaurea nigra* (Knapweed) Som. Grigson. 3. *Vinca major* (Greater Periwinkle) Hatfield. 4. *Vinca minor* (Lesser Periwinkle) Som. Macmillan.

JACK, Break 1. *Stellaria graminea* (Lesser Stitchwort) Dor. Macmillan. 2. *Stellaria holostea* (Greater Stitchwort) Dor. Grigson. The reference is to the way they snap off at the joints, giving rise to a host of names like Breakbones, Snap-jacks, etc. Break-Jack seems like a hybrid between the last two mentioned.

JACK, Canterbury *Bryonia dioica* (White Bryony) Kent Grigson.

JACK, Creeping *Sedum acre* (Biting Stonecrop) Som, Wilts. Macmillan; Ches. Holland.

JACK, Dirty *Chenopodium album* (Fat Hen) Ches. Grigson. 'Dirty', because its favourite growing place is a dungheap. Cf. names like Muckweed, Dungweed, etc.

JACK, Gaping *Linaria vulgaris* (Toadflax) Som. Grigson. The various 'mouth' names convey the same idea.

JACK, Jumping *Impatiens noli-me-tangere* (Touch-me-not) Suss. Parish. Reflecting the same habit as Touch-me-not does, for it has an explosive method of seed distribution, triggered as soon as the capsule is touched.

JACK, Little *Geranium robertianum* (Herb Robert) Dev. Macmillan.

JACK, Mischievous *Stellaria media* (Chickweed) Som. Macmillan.

JACK, Monkey *Mimulus guttatus* (Monkey-flower) Dor. Macmillan.

JACK, Naked *Colchicum autumnale* (Meadow Saffron) Som. Macmillan. The flowers appear before the leaves, hence this, and such as Son-before-the-father, Upstarts, etc.

JACK, Pop *Stellaria holostea* (Greater Stitchwort) Som. Macmillan. The reference is to the way they are liable to snap off at the joints - Cf. Snap-jacks, Breakbones, etc.

JACK, Rag 1. *Chenopodium album* (Fat Hen) Ches. Holland. 2. *Primula veris x vulgaris* (False Oxlip) Lincs. Grigson.

JACK, Ragged 1. *Lychnis flos-cuculi* (Ragged Robin) Som. Elworthy; IOW Long; Suss. Parish; Kent, Ess. Grigson. 2. *Senecio jacobaea* (Ragwort) Yks. Grigson. Descriptive in each case.

JACK, Rattle *Rhinanthus crista-galli* (Yellow Rattle) Lincs. Peacock.

JACK, Red *Melandrium dioicum* (Red Campion) Ches. Holland.

JACK, Runaway *Glechoma hederacea* (Ground Ivy) Som. Grigson; Glos. J D Robertson.

JACK, Saucy *Centaurea melitensis* (Napa Thistle) S Africa Watt.

JACK, Shaggy *Lychnis flos-cuculi* (Ragged Robin) Dev, Som. Macmillan.

JACK, Snap *Stellaria holostea* (Greater Stitchwort) Dev. Grigson; Dor. Dacombe; Wilts. Dartnell & Goddard. Cf. Pop-jack etc.

JACK, Sticky *Arctium lappa* (Burdock) Som. Mabey.

JACK, Three-fingered *Saxifraga tridactylites* (Rue-leaved Saxifrage) Som. Macmillan.

JACK, Wandering *Cymbalaria muralis* (Ivy-leaved Toadflax) Som. Macmillan.

JACK, Wiry *Sisymbrium officinale* (Hedge Mustard) Palaiseul.

JACK, Yellow *Narcissus odorus* (Jonquil) Surr. B & H.

JACK-ABED-AT-NOON *Tragopogon pratensis* (Goat's Beard) Som. Macmillan.

JACK-AN-APES-ON-HORSEBACK 1. *Bellis perennis* (Dairy) Parkinson. 2. *Calendula officinalis* (Marigold) Gerard. "called of the vulgar sort of woman, Jack-an-apes-on-horseback..." (Gerard). He was describing a marigold that produced a lot of small flowers from the base of the main bloom.

JACK-AND-THE-BEANSTALK *Polygala vulgaris* (Milkwort) Som. Macmillan.

JACK-AT-THE-HEDGE *Galium aparine* (Goose-grass) Cumb. Grigson.

JACK BEAN *Canavalla ensiformis*.

JACK-BEHIND-THE-GARDEN-GATE *Viola tricolor* (Pansy) Suff. Grigson.

JACK-BY-THE-GROUND *Glechoma hederacea* (Ground Ivy) Tynan & Maitland.

JACK-BY-THE-HEDGE 1. *Alliaria petiolata*. 2. *Barbarea vulgaris* (Winter Cress). 3. *Chaenorhinum minum* (Small Toadflax) Berw. B & H. 4. *Geranium robertianum* (Herb Robert) Som. Macmillan. 5. *Glechoma hederacea* (Ground Ivy) Page.1978. 6. *Melandrium dioicum* (Red Campion) Som, Suss. Grigson. 7. *Tragopogon pratensis* (Goat's Beard) Suss. B & H.

JACK-BY-THE-HEDGESIDE *Alliaria petiolata* (Jack-by-the-hedge) Culpepper.

JACK-DURNALS *Conopodium majus* (Earthnut) Cumb. Grigson. The name appears too as Jacky Jurnals, or Jocky Jurnals. A common form of 'earthnut' is Yernut, which sometimes changes into Jurnut. Hence the series which includes Jack Durnals.

JACK-FLOWER *Geranium robertianum* (Herb Robert) Dor. Macmillan.

JACK-GO-TO-BED-AT-NOON 1. *Anagallis arvensis* (Scarlet Pimpernel) Dor. Dacombe. 2. *Ornithogalum umbellatum* (Star-of-Bethlehem) Ches. B & H. 3. *Tragopogon pratensis* (Goat's Beard) Clapham.

JACK-HORNER *Geranium robertianum* (Herb Robert) Dor. Macmillan.

JACK-IN-A-LANTERN *Physalis bunyardii* (Cape Gooseberry) Som. Macmillan. Typical lantern shape of the dried pods - this is probably Japanese Lantern, anyway.

JACK-IN-PRISON *Nigella damascena* (Love-in-a-mist) Suss. Parish; Lincs. B & H. Cf. Devil-in-a-bush, Spider-in-his-web, and a lot of other combinations for this.

JACK-IN-THE-BOX 1. *Arum maculatum* (Cuckoo-pint) Som. Elworthy; Bucks. B & H; N Ire. Grigson. 2. *Scrophularia nodosa* (Figwort) Dor. Macmillan. 3. *Stellaria holostea* (Greater Stitchwort) Dor. Macmillan.

JACK-IN-THE-BUSH 1. *Alliaria petiolata* (Jack-by-the-hedge) Glos. J D. Robertson; Heref. Grigson. 2. *Lapsana communis* (Nipplewort) Glos. Grigson. 3. *Umbilicus rupestris* (Pennywort) Scot. Jamieson.

JACK-IN-THE-BUTTERY *Sedum acre* (Biting Stonecrop) Som. Macmillan.

JACK-IN-THE-GREEN 1. *Adonis annua* (Pheasant's Eye) Wilts. Dartnell & Goddard. 2. *Arum maculatum* (Cuckoo-pint) Som. Macmillan. 3. *Primula vulgaris var. elatior* (Polyanthus) Dor. Barnes.

JACK-IN-THE-HEDGE 1. *Alliaria petiolata* (Jack-by-the-hedge) Som. Tongue; Lincs. Peacock. 2. *Bryonia dioica* (White Bryony) Hants. Cope. 3. *Glechoma hederacea* (Ground Ivy) Page.1978. 4. *Melandrium dioicum* (Red Campion) Som. Macmillan; Suss. Parish. 5. *Stellaria holostea* (Greater Stitchwort) Som. Macmillan.

JACK-IN-THE-LANTERN 1. *Melandrium dioicum* (Red Campion) Dor. Macmillan. 2. *Stellaria holostea* (Greater Stitchwort) Som. Macmillan.

JACK-IN-THE-PULPIT 1. *Arisaema triphyllum*. 2. *Arum maculatum* (Cuckoo-pint) Corn, Som, Lincs. Grigson. The 'jack' is the spadix, and the 'pulpit' the spathe.

JACK-JENNETS *Conopodium majus* (Earthnut) Yks. Grigson. 'Jennets' is a corruption of Earthnut, through some such form as Jurnut.

JACK-JUMP-ABOUT 1. *Aegopodium podagraria* (Goutweed) Herts, Oxf. Grigson; N'thants. A E Baker. 2. *Angelica archangelica* (Archangel) Fernie. 3. *Angelica sylvestris* (Wild Angelica) N'thants. A E Baker. 4. *Lotus corniculatus* (Bird's-foot Trefoil) N'thants. A E Baker. For angelica, and perhaps for the other umbellifers, the name comes from a game children play in which a piece of grass is threaded through the stalk and made to "emerge and disappear somewhat after the manner of a jack-in-the-box".

JACK-JUMP-UP-AND-KISS-ME *Viola tricolor* (Pansy) Corn. Hazlitt.

JACK-O'-BOTH-SIDES; JACK-O'-TWO-SIDES *Ranunculus arvensis* (Corn Buttercup) Leics, Notts. Grigson (both sides); Shrop. Grigson (two sides).

JACK-O'-LANTERN 1. *Anemone nemorosa* (Wood Anemone) Dor. Macmillan. 2. *Eupatorium cannabinum* (Hemp Agrimony) Dor. Macmillan.

JACK-OF-THE-BUTTERY *Sedum acre* (Biting Stonecrop) Gerard.

JACK-OF-THE-HEDGE *Alliaria petiolata* (Jack-by-the-hedge) Turner.

JACK PINE *Pinus banksiana* (Gray Pine) USA Zenkert.

JACK-RUN-ALONG-BY-THE-HEDGE *Alliaria petiolata* (Jack-by-the-hedge) Wilts. Dartnell & Goddard.

JACK-RUN-IN-THE-HEDGE *Calystegia sepium* (Great Bindweed) Som. Macmillan.

JACK-RUN-IN-THE-COUNTRY 1. *Calystegia sepium* (Great Bindweed) Yks. B & H. 2. *Convolvulus arvensis* (Field Bindweed) Yks. Grigson. Aptly descriptive for the bindweeds. Cf. Robin-run-in-the-field, etc.

JACK-RUN-UP-DYKE *Galium aparine* (Goose-grass) N'thum. Grigson.

JACK-SNAPS; JACK SPRAT *Stellaria holostea* (Greater Stitchwort) both Som. both Macmillan.

JACK-STRAWS *Plantago lanceolata* (Ribwort Plantain) Yks. Grigson.

JACK'S CHEESES *Malva sylvestris* fruit (Common Mallow) Som. Macmillan. Cf. Jacky Cheeses, or Jacky's Cheeses, from the same county, all from Chucky Cheese or something similar, just as they are from. Truckles-of-cheese.

JACK'S LADDER *Phaseolus coccineus* (Scarlet Runner) Som. Macmillan.

JACKET, Flannel *Verbascum thapsus* (Great Mullein) Norf. Grigson.

JACKET, Thousand *Gaya lyalli* (Mountain Ribbonwood) Perry.

JACKETS, Blue *Polemonium caeruleum* (Jacob's Ladder) N Ire. Grigson.

JACKETS, Soldiers' *Orchis mascula* (Early Purple Orchis) Dor. B & H.

JACKFRUIT *Artocarpus heterophyllus*.

JACKWEED *Ranunculus arvensis* (Corn Buttercup) Oxf. Grigson.

JACKWOOD *Prunus padus* (Bird Cherry) Cumb, N'thum. Grigson.

JACKY('S) CHEESES *Malva sylvestris* (Common Mallow) Som. Macmillan. Cf. Jack's Cheeses.

JACKY-JURNALS *Conopodium majus* (Earthnut) Cumb. Grigson. The 'jurnals' part of this name was originally earthnut, and arrived at this point via some form like Yernut. Cf. Job-Jurnals and Jocky-Jurnals.

JACOB, Running *Tropaeolum majus* (Nasturtium) Dor. Macmillan.

JACOB'S CHARIOT *Aconitum napellus* (Monkshood) Ess. B & H. It is usually likened to Venus's chariot, rather than to Jacob's.

JACOB'S COAT 1. *Acalypha wilkesiana* (Copperleaf) RHS. 2. *Alternanthera bettzickiana* (Joy Weed) Howes. Copperleaf has multi-coloured leaves, hence the name. Cf. Match-me-if-you-can, for no two of them are ever alike.

JACOB'S LADDER 1. *Aconitum napellus* (Monkshood) Tynan & Maitland. 2. *Althaea rosea* (Hollyhock) Som. Macmillan. 3. *Antirrhinum majus* (Snapdragon) Som. Macmillan. 4. *Atropa belladonna* (Deadly Nightshade) Ayr. Grigson. 5. *Chelidonium majus* (Greater Celandine) Shrop. Grigson. 6. *Convallaria maialis* (Lily-of-the-valley) Grieve.1931. 7. *Delphinium ajacis* (Larkspur) Dev. B & H. 8. *Impatiens noli-me-tangere* (Touch-me-not) Som. Macmillan. 9. *Polemonium caeruleum*. 10. *Polygonatum multiflorum* (Solomon's Seal) Som, Berks. Grigson; Wilts. Dartnell & Goddard. 11. *Sedum telephium* (Orpine) Kent Grigson. 12. *Smilax herbacea* (Carrion-flower) USA Gilmore. In the best examples, the name relies on an opposite positioning of the leaves round the stalk. Hollyhock gets the name, of course, simply because of its height.

JACOB'S LADDER, False *Polemonium reptans* (Abscess-root) Hatfield.

JACOB'S STAFF 1. *Asphodeline lutea* (Yellow Asphodel) Grieve.1931. 2. *Verbascum thapsus* (Mullein) Cumb. B & H; USA Henkel.

JACOB'S STEE *Atropa belladonna* (Deadly Nightshade) Lincs. Grigson. A 'stee' is a ladder - see Jacob's Ladder.

JACOB'S SWORD *Iris pseudo-acarus* (Yellow Flag) Aber. B & H. The shape of the leaves ensures that it gets a lot of 'sword' or 'dagger' names.

JACOB'S TEARS *Convallaria maialis* (Lily-of-the-valley). Le Strange.

JACOB'S WALKING STICK *Polemonium caeruleum* (Jacob's Ladder) Hants. Wright.

JACOBAEA *Senecio elegans*.

JACOBEAN AMARYLLIS *Sprekelia formosissima* (Jacobean Lily) Whittle & Cook. For *Amaryllis formosissima* is an alternative botanical name.

JACOBEAN LILY *Sprekelia formosissima*.

JACOBITE ROSE *Rosa alba*.

JACOBY; JACOBY FLEAWORT *Senecio jacobaea* (Ragwort) Fernie (Jacoby); Leyel.1937 (Jacoby Fleawort). Jacoby, and the specific name *jacobaea*, are references to St James, and the plant is also known as St James' Wort, or St James' Ragwort. James is the patron saint of horses, and it is either this, and the use of the plant in veterinary practice, or the fact that it is in full bloom on St James' Day, 25 July, that accounts for the name.

JADE PLANT 1. *Crassula argentea*. 2. *Crassula portulacea* (Money-plant) Nicolaisen.

JADE VINE *Strongylodon macrobotrys*.

JAGGED CRANESBILL *Geranium dissectum* (Cut-leaved Cranesbill) Curtis.

JAGGED MALLOW *Althaea rosea* (Hollyhock) Turner.

JAGGED SEA ORACHE *Atriplex laciniata* (Frosted Orache) Clapham.

JAGGY NETTLE *Urtica dioica* (Nettle) Edin. Ritchie.

JAK-FRUIT *Artocarpus heterophyllus* (Jackfruit) F P Smith.

JALAP; JALAP, False *Mirabilis jalapa* Barbados Gooding; Howes (False). Jalap is Jalapa, a city in Mexico, and should refer to the Jalapa Purge, *Ipomaea purga*. In any event, it is used as an adulterant, which would be enough to earn the adjective 'false'.

JALAP, Cancer *Phytolacca decandra* (Poke-root) USA Henkel. It is also known as Cancer-root in America.

JALAP, Indian *Ipomaea hederacea* (Ivy-leaf Morning Glory) Chopra.

JALAP, Wild *Ipomaea pandurata* (Wild Potato Vine) O P Brown.

JALAP BINDWEED *Ipomaea purga* (Jalapa Purge) Thornton.

JALAPA PURGE *Ipomaea purga*. It is certainly a purge, and has been well-known as such since ancient times. Jalapa is a city in Mexico, where the plant grows.

JAM TARTS 1. *Fumaria officinalis* (Fumitory) Som. Macmillan. 2. *Geranium molle* (Dove's-foot Cranesbill) Som. Macmillan. 3. *Geranium robertianum* (Herb

Robert) Som. Macmillan.

JAMAICA ALLSPICE *Eugenia pimenta* W Miller. Allspice is the dried unripened berry of this tree.

JAMAICA HONEYSUCKLE *Passiflora laurifolia*.

JAMAICA IRONWOOD *Erythroxylon areolatum*.

JAMAICA LIQUORICE *Asclepias syriaca* (Common Milkweed) McDonald.

JAMAICA MIGNONETTE *Lawsonia inermis* (Henna) Dalziel.

JAMAICA MOUNTAIN SAGE *Lantana camara*.

JAMAICA PEPPER; JAMAICA PIMENTO *Pimento dioica* (Allspice) both Schery.

JAMAICA PLUM *Spondias mombin* (Mombin) Howes.

JAMAICA SENNA *Cassia obovata* (Italian Senna) Dalziel.

JAMAICA VERVAIN *Verbena jamaicensis*.

JAMAICA-WOOD *Haematoxylon campechianum* (Logwood) Pomet.

JAMAICAN RED CEDAR *Cedrela odorata* (West Indian Cedar) Brimble.1948.

JAMAICAN ROSE *Blakea trinerva*.

JAMAICAN SORREL *Hibiscus sabdariffa* (Rozelle) Perry. An Indian shrub, but introduced into central America and the West Indies - very popular in Jamaica, hence the name.

JAMBA *Eruca sativa* (Rocket) India Howes.

JAMBERBERRY *Physalis ixiocarpa* (Coats).

JAMBOLAN *Syzygium cumini*.

JAMBUL *Eugenia jambolana*.

JAMES'S WEED *Senecio jacobaea* (Ragwort) Shrop. Grigson. St James, of course, as is implicit in the specific name. He was the patron saint of horses, and the plant used to be widely employed in veterinary practice, particularly to cure the staggers in horses, a case of homeopathic magic, for ragwort actually causes the disease. Anyway, perhaps the dedication is simply because it is in full bloom by St James's Day (25 July).

JAMESTOWN LILY; JAMESTOWN WEED *Datura stramonium* (Thorn-apple) USA both Henkel. Both these, and their corruption Jimsonweed, come from the story of the soldiers sent to Jamestown to quell the uprising known as Bacon's Rebellion, in 1676. They gathered young Thorn-apple plants and cooked them as a potherb - "the effect of which was a very pleasant Comedy; for they turn'd natural Fools upon it for several days...".

JAMIE HEDLEY'S CLOVER *Prunella vulgaris* (Self-heal) N'thum. Denham/Hardy. Jamie Hedley's father wanted to take a farm then vacant, but being blind, sent his son to report on it. Jamie noticed one field particularly green, and, returning to his father, said it was grand land, full of clover. The "clover" was self-heal, a surer sign of sterile ground than of fertile.

JAN, Granfer *Melandrium dioicum* (Red Campion) Som, Wilts. Macmillan.

JAN, Little *Geranium robertianum* (Herb Robert) Dev, Som. Macmillan.

JAN THE CROWDER *Lychnis flos-cuculi* (Ragged Robin) Tynan & Maitland.

JANE, Creeping *Lysimachia nummularia* (Creeping Jenny) Wilts. Dartnell & Goddard.

JANE, Lady; JANE, Mary *Geranium robertianum* (Herb Robert) both Dev. Macmillan.

JANE, Little *Anagallis arvensis* (Scarlet Pimpernel) Som. Macmillan.

JANE, Poor 1. *Geranium robertianum* (Herb Robert) Som. Macmillan. 2. *Melandrium album* (White Campion) Som. Macmillan. 3. *Melandrium dioicum* (Red Campion) Som. Macmillan.

JANE, Red *Melandrium dioicum* (Red Campion) Som. Macmillan.

JANE, Salvation *Echium plantagineum* (Purple Viper's Bugloss) Australia Salisbury. From the resemblance of the flowers to Salvation Army bonnets.

JANET-FLOWER *Caltha palustris* (Marsh Marigold) A S Palmer. There is Jonettes, too. They are both from French jaunette, so are

simply a comment on the colour.
JAPAN PEPPER *Xanthoxylum piperita*.
JAPAN VARNISH TREE *Ailanthus glandulosa* (Tree of Heaven) Chopra.
JAPANESE ALLSPICE *Chimonanthus fragrans* Salisbury.1936.
JAPANESE ANDROMEDA *Andromeda japonica*.
JAPANESE ANGELICA TREE *Aralia elata*.
JAPANESE APRICOT *Prunus mume*.
JAPANESE ARTICHOKE *Stachys sieboldii*.
JAPANESE BAMBOO *Polygonum cuspidatum* (Japanese Knotweed) Gorer. It is referred to as Mexican Bamboo, as well.
JAPANESE BANANA *Musa basjoo*.
JAPANESE BARBERRY *Berberis thunbergii*.
JAPANESE BARNYARD MILLET *Echinochloa crus-galli frumentacea*.
JAPANESE BASIL *Ocymum crispum*.
JAPANESE BLACK PINE *Pinus thunbergii*.
JAPANESE BUNCHING ONION *Allium fistulosum* (Welsh Onion). A W Smith.
JAPANESE CAMPHOR *Cinnamomium camphora* (Camphor) S Africa Watt.
JAPANESE CHERRY *Prunus serrulata*.
JAPANESE CHESTNUT *Castanea crenata*.
JAPANESE CHESTNUT OAK *Quercus acutissima*.
JAPANESE COLTSFOOT *Petasites japonicus* (Creamy Butterbur) Douglas.
JAPANESE CYPRESS *Chamaecyparis obtusa* (Fire Cypress) Howes.
JAPANESE DOGWOOD *Cornus kousa*.
JAPANESE DOUGLAS FIR *Pseudotsuga japonica*.
JAPANESE ELM *Ulmus japonica*.
JAPANESE EVERGREEN OAK *Quercus acuta*.
JAPANESE FLEECE FLOWER *Polygonum cuspidatum* (Japanese Knotweed) Howes.
JAPANESE GENTIAN *Gentiana scabra*.
JAPANESE HEMLOCK, Northern *Tsuga diversifolia*.
JAPANESE HEMLOCK, Southern *Tsuga sieboldii*.

JAPANESE HOLLY *Ilex crenata* (Box-leaved Holly) RHS.
JAPANESE HONEYSUCKLE *Lonicera nitida*.
JAPANESE HOP *Humulus japonicus*.
JAPANESE HOP HORNBEAM *Ostrya japonica*.
JAPANESE HORNBEAM *Carpinus japonica*.
JAPANESE HORSE CHESTNUT *Aesculus turbinata*.
JAPANESE IRIS *Iris kaempferi*.
JAPANESE IVY *Ampelopsis tricuspidata* (Boston Ivy) RHS.
JAPANESE KNOTWEED *Polygonum cuspidatum*.
JAPANESE LACQUER *Rhus verniciflua* (Lacquer Tree) Howes.
JAPANESE LANTERNS 1. *Campanula medium* (Canterbury Bells) Som. Macmillan. 2. *Physalis alkekengi* (Winter Cherry) Tampion.
JAPANESE LARCH *Larix kaempferi*.
JAPANESE LARGE-LEAVED PODOCARP *Podocarpus macrophylla*.
JAPANESE LAUREL *Aucuba japonica*.
JAPANESE LILY 1. *Lilium auratum*. 2. *Lilium japonicum*.
JAPANESE MAIZE *Zea japonica*.
JAPANESE MAPLE 1. *Acer japonicum* (Downy Japanese Maple) Hay & Synge. 2. *Acer palmatum*.
JAPANESE MEDLAR 1. *Diospyros kaki* (Chinese Date Plum) Bianchini. 2. *Eriobotrya japonica* (Loquat) Perry.
JAPANESE MILLET *Echinochloa utilis*.
JAPANESE MINT *Mentha arvensis var. piperascens*.
JAPANESE MOUNTAIN ASH *Sorbus matsumarama*.
JAPANESE NUTMEG *Torreya nucifera* (Kaya) Hora.
JAPANESE ONION *Allium senescens* (German Garlic) Douglas. An odd name for a central European species!
JAPANESE ORANGE *Citrus reticulata* (Tangerine) Howes.

JAPANESE PERSIMMON *Diospyros kaki* (Chinese Date Plum).
JAPANESE PINK *Dianthus chinensis var. heddewiggii.*
JAPANESE PLUM 1. *Eriobotrya japonica* (Loquat) Perry. 2. *Prunus salicina.*
JAPANESE PLUM-YEW *Cephalotaxus drupacea.*
JAPANESE POPLAR *Populus maximowiczii.*
JAPANESE PRIVET *Ligustrum ovalifolium.*
JAPANESE QUINCE *Chaenomeles speciosa.*
JAPANESE RAY LILY *Lilium auratum* (Ray Lily) Grant White.
JAPANESE RED CEDAR *Cryptomeria japonica* (Japanese Cedar) Polunin.1976.
JAPANESE RED PINE *Pinus densiflora* (Red Pine) Mitchell.
JAPANESE ROOF IRIS *Iris tectorum* (Wall Iris) RHS.
JAPANESE ROSE *Rosa rugosa.*
JAPANESE ROWAN *Sorbus commixta.*
JAPANESE RUBBER PLANT *Crassula portulacea* (Money Plant) USA Cunningham.
JAPANESE SHIELD FERN *Dryopteris erythrosora.*
JAPANESE SILVER FIR *Abies firma* (Momi Fir) Usher.
JAPANESE SPINDLE *Euonymus japonicus* (Evergreen Spindle) Polunin.1976.
JAPANESE TALLOW *Rhus succedanea* (Japanese Wax Tree) H & P.
JAPANESE TEA PARTY *Anemone hupehensis* Wilts. Macmillan.
JAPANESE WALNUT *Juglans ailantifolia.*
JAPANESE WAX TREE *Rhus succedanea.*
JAPANESE WHITE PINE *Pinus parviflora.*
JAPANESE WINEBERRY *Rubus phoenocolasius.*
JAPANESE WINGNUT *Pterocarya rhoifolia.*
JAPANESE WITCH HAZEL *Hamamelis japonica.*
JAPANESE WYCH ELM *Ulmus laciniata.*
JAPANESE YELLOW-WOOD *Cladrastris platycarpa.*
JAPANESE YEW *Taxus cuspidata.*
JAPONICA *Chaenomeles speciosa* (Japanese Quince) Wilkinson.1981.
JAPONICA, Crocus *Kerria japonica* (Jew's Mallow) Som. Macmillan. The botanical name used to be *Corchorus japonica*, and this is a corruption of the old name.
JARRAH *Eucalyptus marginata.* An Australian aborigine word.
JASMINE *Jasminum officinale.* Jasmine is the result of the Latinization of the Persian yasmin into Jasminum.
JASMINE, Arabian *Jasminum sambac.*
JASMINE, Cape *Gardenia jasminoides.* It may be called 'Cape' jasmine, but this comes from China and Japan.
JASMINE, Catalonian *Jasminum grandiflorum* (Spanish Jasmine) Grieve.1931. Not Spanish or Catalan, but from the Himalayas. The names probably arise from the fact that this is the jasmine of the perfumery trade - was it centred in Barcelona, perhaps?
JASMINE, Cayenne *Catharanthus rosea* (Madagascar Periwinkle).
JASMINE, Chilean *Mandevilla laxa.*
JASMINE, Chinese *Trachelospermum jasminoides* (Confederate Jasmine) Goold-Adams.
JASMINE, Confederate *Trachelospermum jasminoides.* Is 'Confederate' a reference to the Federated Malay States?
JASMINE, False *Gelsenium sempervirens.*
JASMINE, French *Jasminum officinale* (Jasmine) Turner.
JASMINE, Ground *Rasserina stelleri.*
JASMINE, Italian 1. *Jasminum grandiflorum* (Spanish Jasmine) Howes. 2. *Jasminum humile.* Neither are Italian plants at all, but native to Indonesia in the case of J humile, and the Himalayas in the case of Spanish Jasmine. So it is not Spanish either!
JASMINE, Madagascar *Stephanotis floribunda.* Not a jasmine, of course, but it has white flowers which are very fragrant.
JASMINE, Night, Indian *Nyctanthes arbortristis.*

JASMINE, Red *Plumeria rubra.*
JASMINE, Rock *Androsace sp.*
JASMINE, Roman *Philadelphus coronarius* (Mock Orange) Dor. Macmillan.
JASMINE, Royal *Jasminum grandiflorum* (Spanish Jasmine) RHS.
JASMINE, Spanish 1. *Jasminum grandiflorum.* 2. *Plumeria acuminata* (Pagoda Tree) Chopra. *J grandiflorum* sometimes appears as Catalonian Jasmine, too.
JASMINE, Star *Trachelospermum jasminoides* (Confederate Jasmine) Simmons.
JASMINE, Star, Southern *Jasminum gracillimum.*
JASMINE, West Indian *Plumeria rubra* (Red Jasmine) Bonar.
JASMINE, Wild 1. *Anemone nemerosa* (Wood Anemone) B & H. 2. *Jasminum fruticans.*
JASMINE, Winter *Jasminum nudiflorum.*
JASMINE, Yellow *Jasminum humile* (Italian Jasmine) RHS.
JASMINE, Yellow, True *Jasminum odoratissimum.*
JASMINE TREE 1. *Holarrhena febrifuga.* 2. *Plumeria acuminata* (Pagoda Tree) Chopra. Heavily scented white flowers.
JAUMARVE ISTLE *Agave funkiana.*
JAUNDERS-BERRY; JAUNDERS-TREE; JAUNDICE-TREE *Berberis vulgaris* (Barberry) Som. (Jaunders-berry); Corn, Som. Elworthy (Jaunders-tree, Jaundice-tree). A result of the doctrine of signatures, for the yellow bark became a specific for the yellow disease, jaundice. It was taken as a decoction in ale or white wine, or as a tea. An Irish remedy was to brew it to a strong drink, and take it every morning, fasting, for nine successive mornings. There are plenty of similar receipts.
JAVA BEAN *Phaseolus lunatus* (Lima Bean) Kingsbury.
JAVA CINNAMON *Cinnamonium burmani.*
JAVA FIG *Ficus benjamina* (Weeping Fig) Howes.
JAVA INDIGO *Indigofera arrecta* (Natal Indigo) Dalziel. It was introduced into Java from Natal.
JAVA PLUM *Eugenia jambolana* (Jambul) Grieve.1931.
JAVA TEA *Orthosiphon stamineus.*
JAWS, Monkey *Cymbalaria muralis* (Ivy-leaved Toadflax) Som. Macmillan.
JAWS, Open *Antirrhinum majus* (Snapdragon) Som. Macmillan.
JAWS, Squeeze *Linaria vulgaris* (Yellow Toadflax) Som. Macmillan.
JAYWEED *Anthemis cotula* (Maydweed) Suss. Grigson.
JEALOUSY *Sedum roseum* (Rose-root) Shrop. B & H.
JEELICO *Angelica sylvestris* (Wild Angelica) Border B & H. A corruption of angelica.
JEETRYM-JEES *Hippuris vulgaris* (Mare's Tail) IOM Moore.1924. Manx jeetdrym-jeeas.
JEFFREY PINE *Pinus ponderosa var. jeffreyi.*
JELLY-FLOWER 1. *Cheiranthus cheiri* (Wallflower) Dev. Friend.1882. 2. *Matthiola incana* (Stock) Corn. Jago; Dor. Macmillan. Gilliflower, of course.
JELLY-STOCK *Cheiranthus cheiri* (Wallflower) Som. Macmillan. Definitely claimed for wallflower, but surely in error. Is it not a simple reversal of Stock Gilliflower, and so a name for *Matthiola annua*, rather?
JEN, Little *Geranium robertianum* (Herb Robert) Som. Macmillan. Cf. Little Jan and Little Jack.
JENEPER *Juniperus communis* (Juniper) Turner.
JENNY, Creeping 1. *Ampelopsis quinquefolia* (Virginia Creeper) Som. Macmillan. 2. *Calystegia sepium* (Great Bindweed) Som. Grigson. 3. *Chrysosplenium oppositifolium* (Opposite-leaved Golden. Saxifrage) Freethy. 4. *Convolvulus arvensis* (Field Bindweed) USA Tolstead. 5. *Cymbalaria muralis* (Ivy-leaved Toadflax) Dev, Som, Dor. Macmillan; Wilts. Dartnell & Goddard. 6. *Galium verum* (Lady's Bedstraw) Som. Macmillan. 7. *Glechoma hederacea* (Ground

Ivy) Som. Macmillan; Lincs. B & H. 8. *Lysimachia nummularia*. 9. *Lysimachia vulgaris* (Yellow Loosestrife) Som. Macmillan. 10. *Potentilla reptans* (Cinquefoil) Som. Macmillan. 11. *Sedum acre* (Biting Stonecrop) Dev. Friend.1882. 12. *Sedum reflexum* (Yellow Stonecrop) Heref. Grigson.

JENNY, Roving 1. *Cymbalaria muralis* (Ivy-leaved Toadflax) Dev. Grigson. IOW Long. 2. *Saxifraga sarmentosa* (Mother of thousands) Dev. B & H.

JENNY, Running; JENNY, Wandering *Lysimachia nummularia* (Creeping Jenny) Grieve.1931 (Running); Som. Macmillan (Wandering).

JENNY, Spinning *Acer campestre* (Field Maple) Som. Macmillan. From the way the 'keys' spin through the air.

JENNY, Stinking 1. *Allium ursinum* (Ramsons) Som. Macmillan. 2. *Geranium robertianum* (Herb Robert) Som. Macmillan. Herb Robert has a characteristic smell when bruised, hence this name and many others like it - Cf. Stinking Bob, Stink Flower, etc. Ramsons, of course, is inherently evil-smelling.

JENNY CREEPER *Lysimachia nummularia* (Creeping Jenny) Som. Macmillan.

JENNY FLOWER; JENNY HOOD *Geranium robertianum* (Herb Robert) Som. (Jenny Flower); Dev, Som. (Jenny Hood) both Macmillan. Jenny Hood seems to be a hybrid between the 'Jenny' names and the 'Robin Hood' names.

JENNY GREENTEETH *Lemna minor* (Duckweed) War. B & H. Jenny Greenteeth is the nursery bogey who is supposed to drag children down into quiet pools. The plant covers the surface of stagnant water. In Cheshire, it was understood that it was the weed that held children under, but they were told to beware of Jenny Greenteeth. Perhaps the small leaves are rather like tiny teeth. The theory was that Jenny enticed little children into the ponds by making them look like grass, and so safe to walk on. As soon as the child stepped on to the green, it parted, and of course the child fell through into Jenny's clutches and was drowned. The green weed then closed over, hiding all traces of the child's ever being there. The teeth analogy was carried further in Lancashire - children were told that if they didn't clean their teeth, they would one day be dragged into one of the pools by Jenny Greenteeth (see Vickery).

JENNY NETTLE *Urtica dioica* (Nettle) IOM S Morrison.

JENNY-RUN-BY-THE-GROUND *Glechoma hederacea* (Ground Ivy) Lincs. Grigson. One of a host of affectionately similar names from different parts of the country. Cf. Gill-go-by-the-ground, etc.

JENNY WREN *Geranium robertianum* (Herb Robert) Dev. B & H; Som. Elworthy. This in turn produced Wren, or Wren-flower. The name probably originated from the hybrid Jenny Hood.

JEQUIRITY BEAN *Abrus precatorius* (Precatory Bean) USA Kingsbury. Presumably a corruption of 'precatory'.

JERICHO, Rose of *Anastatica hierochuntica*. Presumably, it is suggested, a reference to the legend that all plants opened up and bloomed when Christ was born. It is that legend that accounts for such names as Mary's Flower, or Rose of the Virgin, but it is not clear what the connection is with Rose of Jericho.

JERICHO BALSAM *Balanites aegyptiaca*. A plant that grows in desert regions of Palestine, as well as in North Africa and Egypt. The gum, so it is claimed, is the original balm of Gilead.

JERICHO POTATO *Solanum incanum* (Hoary Nightshade) Moldenke. Cf. Palestine Nightshade, and Apple of Sodom, for this.

JERKIN, Miller's *Verbascum thapsus* (Mullein) Wales Grigson. A reference to the floury appearance of the leaves.

JERLIN *Lolium temulentum* (Darnel) IOM Moore.1924.

JEROFFLERIS *Cheiranthus cheiri* (Wallflower) Scot. Jamieson. A version of Giraflour, part of the Gilliflower sequence, but surely a plural form?

JERRYMANDER *Veronica chamaedrys* (Germander Speedwell) Ches. Grigson. *Chamaedrys* means dwarf oak. It comes through medieval Latin from the Greek chamaedrua, which is a corruption of chamai, on the ground, and drus, oak. Germander means the same thing, for it is just an Englishing of some stage of the development which produced chamaedrys. Jerrymander, of course, is germander. But why call the speedwell a dwarf oak anyway? Shape of the leaves, perhaps?

JERSEY BUTTERCUP *Ranunculus flabellatus*.

JERSEY CUDWEED *Gnaphalium luteo-album*.

JERSEY ELM *Ulmus stricta var. sarniensis* (Guernsey Elm) Barber & Phillips.

JERSEY FORGET-ME-NOT *Myosotis sicula*.

JERSEY LILY 1. *Amaryllis belladonna* (Belladonna Lily) H & P. 2. *Vallota speciosa* (Scarborough Lily) Som. Elworthy.

JERSEY LIVE-LONG *Gnaphalium luteo-album* (Jersey Cudweed) Prior.

JERSEY ORCHID *Orchis laxiflora* (Loose-flowered Orchid) Clapham.

JERSEY PINK *Dianthus gallicus* (Western Pink) McClintock.1975.

JERSEY TEA *Gaultheria procumbens* (Wintergreen) USA Sanecki. The leaves have been used as a tea substitute, hence this, and a number of other 'tea' names, like Mountain, Canadian or Ground, Tea. Jersey in this case must be New Jersey.

JERSEY TEA, Inland *Ceanothus ovatus*.

JERSEY THISTLE *Centaurea calcitrapa* (Star Thistle) Pratt.

JERSEY THRIFT *Armeria arenaria* (Broad-leaved Thrift) Polunin. Only from sandy turf in southwest Jersey.

JERSEY TOADFLAX *Linaria pelisseriana*.

JERUSALEM, Apple of *Momordica charantia* (Wild Balsam-apple) W Miller.

JERUSALEM, Ash of *Isatis tinctoria* (Woad) Turner.

JERUSALEM, Oak of *Teucrium botrys* (Cut-leaved Germander) Prior.

JERUSALEM, Star of *Tragopogon pratensis* (Goat's Beard) Gerard.

JERUSALEM ARTICHOKE *Helianthus tuberosus*. Jerusalem in this case is a corruption of the Italian girasole, turn-sun. The plant is a sunflower, and is sometimes known as Sunflower Artichoke.

JERUSALEM ARTICHOKE, White *Alstroemaria edulis* Coats.

JERUSALEM ASH 1. *Isatis tinctoria* (Woad) W Miller. 2. *Reseda luteola* (Dyer's Rocket) W Miller.

JERUSALEM CHERRY 1. *Solanum elaeagnifolium* (Silverleaf Nightshade) USA White. 2. *Solanum pseudocapsicum*.

JERUSALEM CHERRY, False *Solanum diflorum*.

JERUSALEM CROSS *Lychnis chalcedonica* Prior. The reference is to the shape of the flowers, resembling the badge of the Crusaders, hence too Knight's Cross, or Maltese Cross.

JERUSALEM COWSLIP *Pulmonaria officinalis* (Lungwort) Gerard. The names are full of similar references - Cf. Bethlehem, or Bedlam, Cowslip, or Jerusalem Sage, etc.

JERUSALEM COWSLIP, Narrow-leaved *Pulmonaria longifolia* (Narrow-leaved Lungwort) Gerard.

JERUSALEM CUCUMBER *Cucumis anguria* (Gherkin) USA Candolle.

JERUSALEM NETTLE *Lamium maculatum* (Spotted Deadnettle) Lincs Gutch.

JERUSALEM OAK *Chenopodium botrys*. The young leaves look like miniature oak leaves.

JERUSALEM PINE *Pinus halepensis* (Aleppo Pine) Everett.

JERUSALEM PRIMROSE *Pulmonaria officinalis* (Lungwort) Tynan & Maitland. Cf. Jerusalem Cowslip.

JERUSALEM SAGE 1. *Phlomis fruticosa*. 2. *Pulmonaria officinalis* (Lungwort).

JERUSALEM SEEDS *Pulmonaria officinalis* (Lungwort) Dev, Som. Elworthy.

Presumably 'seeds' is 'sage'.
JERUSALEM STAR 1. *Cerastium tomentosum* (Snow-in-summer) War. B & H. 2. *Hypericum calycinum* (Rose of Sharon) Shrop. B & H. 3. *Senecio cineraria* (Silver Ragwort) Wilts. Macmillan.
JERUSALEM TEA *Chenopodium ambrosioides* (American Wormseed) USA Henkel. Cf. Jesuit Tea, which was probably its original.
JERUSALEM THORN *Paliuris spina-christi* (Christ-thorn) Polunin.1976.
JERUSALEM WORMSEED *Chenopodium botrys* (Jerusalem Oak) Watt. An anthelmintic, like a number of the Chenopodiums.
JESHAL *Eupatorium cannabinum* (Hemp Agrimony) IOM Moore.1924.
JESSAMINE; JESSAMY 1. *Jasminum officinale* (Jasmine) McDonald (Jessamine); Dev. *Dev. Asscn. Trans.* vol 65; 1933 (Jessamy). 2. *Orchis mascula* (Early Purple Orchis) War. Grigson. Jessamine is a recognised alternative to jasmine, but there is. no conceivable reason why the orchid should get the name, unless it be a corruption of Gethsemane,q v.
JESSAMINE, Carolina *Gelsemium sempervirens*.
JESSAMINE, Day-blooming *Cestrum diurnum*.
JESSAMINE, Green *Cestrum parqui*.
JESSAMINE, Night-blooming *Cestrum nocturnum*.
JESSAMINE, Poet's *Jasminum officinale* (Jasmine) Whittle & Cook.
JESSAMINE, Roman *Philadelphus coronarius* (Mock Orange) Dev. Choape.
JESSAMINE, Wild 1. *Anemone nemerosa* (Wood Anemone) B & H. 2. *Gelsemium sempervirens* (Carolina Jessamine) OP Brown.
JESSAMINE, Willow-leaved *Cestrum parqui* (Green Cestrum) USA Kingsbury.
JESSAMINE, Yellow *Gelsemium sempervirens* (Carolina Nightshade) USA Kingsbury).

JESSE; JESSE'S FLOWER *Jasminum officinale* (Jasmine) B & H (Jesse); A S Palmer (Jesse's Flower).
JESUIT TEA 1. *Chenopodium ambrosioides* (American Wormseed) USA Henkel. 2. *Ilex paraguayensis* (Maté) Schery.
JESUIT'S BARK *Cinchona sp* (Peruvian Bark) Duran-Reynard. Given as a memorial to the Jesuits' part in bringing quinine to Europe.
JESUIT'S BARK, Bastard *Croton eleutheria* (Sweet-wood) Leyel.1937.
JESUIT'S BRAZIL TEA *Ilex paraguayensis* (Maté) Le Strange.
JESUIT'S CRESS *Tropaeolum majus* (Nasturtium) Bianchini. Nasturtium was brought to Europe by the Jesuits from the West Indies.
JESUIT'S NUT *Trapa natans* (Water Chestnut) Brouk.
JESUIT'S POWDER *Cinchona sp* (Peruvian Bark) Duran-Reynard. Cf. Jesuit's Bark for this.
JESUIT'S TEA *Psoralea glandulosa*.
JET BEAD BUSH *Rhodotypos sp*.
JEW, Wandering 1. *Cymbalaria muralis* (Ivy-leaved Toadflax) Suss. B & H. 2. *Glechoma hederacea* (Ground Ivy) Donegal Grigson. 3. *Saxifraga sarmentosa* (Mother-of-thousands) W Miller. 4. *Tradescantia crassifolia* USA Bergen. 5. *Tradescantia fluminensis* Gannon. 6. *Tradescantia fluviatilis* Rochford. 7. *Tradescantia virginica* (Spiderwort) Harvey. 8. *Zebrina pendula* Newall.1973. Usually because of a trailing habit.
JEW'S APPLE *Solanum melongena* (Egg Plant) W Miller.
JEW'S BEARD *Sempervivum tectorum* (Houseleek) A S Palmer. A Latin name for houseleek was Iovis barba, Jove's Beard, and this is the original of Jew's Beard for houseleek.
JEW'S EAR 1. *Dicentra spectabilis* (Bleeding Heart) Som. Macmillan. 2. *Lycopersicon esculentum* (Tomato) Hants. Cope.

JEW'S MALLOW 1. *Corchorus olitorius* Moldenke. 1. *Kerria japonica*.
JEW'S MYRTLE *Ruscus aculeatus* (Butcher's Broom) Grieve.1931. Because it was sold to Jews for use during the feast of the Tabernacle, according to Folkard. Other 'myrtle' names for this are Wild Myrtle, and Shepherd's Myrtle.
JEW'S PLUM *Spondias dulcis* (Ambarella) Howes.
JEWEL ORCHID *Anoectochilus discolor*.
JEWEL ORCHID, Royal *Anoectochilus regalis*.
JEWEL-WEED 1. *Impatiens capensis* (Orange Balsam) USA Grigson. 2. *Impatiens pallida* (Pale Touch-me-not) USA House.
JEWEL-WEED, Pale (Pale Touch-me-not) USA Yarnell.
JEWELS, Speckled *Impatiens capensis* (Orange Balsam) Grieve.1931.
JIBA *Erythroxylon havanense*.
JIBBLES *Allium fistulosum* (Welsh Onion) Mon. Waters. A version of chibbles and its variants, usually reserved for shallots. The word should be compared with French ciboule, or Italian cipolla, onion.
JICANA *Pachyrrhizus palmatilobus*. This is the Mexican name, accented on the first syllable, thus jícana.
JILL *Glechoma hederacea* (Ground Ivy) Worcs. B & H. A spelling of Gill, which is pronounced with a soft "g".
JILAFFER; JILLIVER *Cheiranthus cheiri* (Wallflower) Dev. Macmillan (Jilaffer); Yks. B & H (Jilliver). i.e. gilliflower.
JILLOFFER 1. *Cheiranthus cheiri* (Wallflower) Som, Dor. Macmillan. 2. *Matthiola incana* (Stock) Som. Macmillan.
JILLOFFER STOCK *Matthiola incana* (Stock) Som. Macmillan. Better known as Stock Gillyflower.
JILLY OFFERS *Cheiranthus cheiri* (Wallflower) Som. Macmillan.
JIMMY, Shivering *Briza media* (Quaking-grass) Suss. Grigson.1959. Cf. Shiver-shakes, or Shivery-shakers, etc.

JIMMY WEED *Haplopappus heterophyllus* (Rayless Golden Rod) USA Kingsbury.
JIMSON, Western *Datura meteloides* (Downy Thorn-apple) Jaeger. See next entry.
JIMSONWEED 1. *Oxalis corniculata* (Sleeping Beauty) Watt. 2. *Datura stramonium* (Thorn-appele) USA Henkel. This is the more usual rendering of Jamestown Weed, or Jamestown Lily. The story is that soldiers sent to Jamestown to quell the uprising known as Bacon's Rebellion, in 1676, gathered young plants of this species and cooked them as a potherb - "the effect of which was a very pleasant Comedy; for they turn'd natural Fools upon it for several days...".
JIMSONWEED, Purple *Datura stramonium* (Thorn-apple) Watt.
JIMSY-WEED *Datura stramonium* (Thorn-apple) Maryland Whitney & Bullock. See Jimson, or Jamestown, Weed.
JINIFER *Juniperus communis* (Juniper) Scot. MacColl & Seeger.
JINNY JOES *Taraxacum officinale* seed heads (Dandelion clocks) Dublin PLNN16.
JOAN, Bobbin; JOAN, Bobbin-and-; JOAN, Bobby-and- *Arum maculatum* (Cuckoo-pint) all Northants Sternberg (Bobbin Joan); A E Baker (the rest). Plus a Corn. reference Grigson, for Bobbin Joan. The Northamptonshire lace industry is the source of these names - the spadices look like the lace bobbins. Bobbin-Joan is also the name of an old country dance. Bobbin Joan is also recorded as a name for *Lychnis flos-cuculi* (Ragged Robin) Dev. Tynan & Maitland.
JOAN'S RIBBON *Phalaris arundinacea* (Reed Grass) Tynan & Maitland.
JOAN'S SILVER PIN *Papaver rhoeas* (Corn Poppy) Coles. Joan's silver pin is a tawdry ornament worn ostentatiously by a sloven. The flower is "fair without and foul within", a reference to the yellow juice.
JOB, Dog 1. *Rosa canina* (Dog Rose) Scot. Chapman. 2. *Rosa canina* hips (Hips) Yks. Grigson. 'Jobs' sounds very like the north country 'choops', hence Dog-choops in Yorkshire for hips.

JOB'S TEARS 1. *Polygonatum multiflorum* (Solomon's Seal) Corn. Grigson. 2. *Pulicaria dysenterica* (Yellow Fleabane) Som. Macmillan. The first is descriptive, and the second is from a tradition that Job applied the plant to his boils, and got relief thereby.

JOBARBE *Sempervivum tectorum* (Houseleek) Sanecki. From Latin jovis barba, through French joubarbe - Jupiter's Beard.

JOBO *Spondias mombin* (Mombin) W Indies Howes.

JOCKIES, Dothering; JOCKIES, Trembling; JOCKS, Trembling *Briza media* (Quaking-grass) all Yks. Grigson.1959 (Dothering); F K Robinson (Trembling) (which would be pronounced as "trimmling"); Wright (Jocks).

JOCKY JURNALS *Conopodium majus* (Earthnut) Grigson. The name appears as Jacky Jurnals, too. 'Jurnals' is one of the many forms of 'earthnut', through such a version as Yernut.

JOE, Poor *Dioidia teres*. Because it thrives in poor soils.

JOE PYE WEED 1. *Eupatorium falcatum* USA Zenkert. 2. *Eupatorium maculatum* USA Zenkert. 3. *Eupatorium purpureum* (Purple Boneset) USA Tolstead. Joe Pye was apparently an Indian doctor who was supposed to have cured typhus with this.

JOE STANLEY *Geranium robertianum* (Herb Robert) Dor. Macmillan.

JOEY HOOKER *Galinsoga parviflora* (Kew-weed) A Huxley. Sir Joseph Hooker was the director of Kew at the time when this was introduced there in 1796. Presumably he was the one who answered queries about it. Or was Joey the director's son?

JOHAN *Hypericum perforatum* (St John's Wort) Halliwell.

JOHN, Blue *Taxus stricta* W Miller.

JOHN, Dirty 1. *Chenopodium album* (Fat Hen) Ches. Grigson. 2. *Chenopodium vulvaria* (Stinking Orach) Ches. Because its favourite habitat is on a dungheap.

JOHN, Gentleman *Viola tricolor* (Pansy) Fernie.

JOHN, Herb; JOHN, Penny *Hypericum perforatum* (St John's Wort) Gerard (Herb); Norf. Grigson (Penny).

JOHN, Little *Stellaria holostea* (Greater Stitchwort) Som. Macmillan.

JOHN, Orphan *Sedum telephium* (Orpine) E Ang. Grigson. Orphan here is a corruption of orpine.

JOHN, Pink-eyed; JOHN, Pink-of-my-; JOHN, Pinkeney *Viola tricolor* (Pansy) Wright (Pink-eyed); Beds. Grigson, Leics. A B Evans (Pink-of-my-); N'thants. A E Baker (Pinkeney). Presumably Pink-of-my-John is the original. Cf. Gentleman John. They must all be part of the long series of erotic names for pansy.

JOHN, Sweet *Dianthus barbatus* (Sweet William) Gerard.

JOHN GEORGES *Caltha palustris* (Marsh Marigold) Bucks. M Baker.1974. Cf. Johnny Cranes for this, nearby in Northamptonshire. Perhaps the Somerset 'Geordies', in the form of Water Geordies, or Georgies, is relevant.

JOHN-GO-TO-BED-AT-NOON 1. *Anagallis arvensis* (Scarlet Pimpernel) Midl. Wright. 2. *Ornithogalum umbellatum* (Star-of-Bethlehem) Shrop. B & H. 3. *Tragopogon pratensis* (Goat's Beard) Dor. Macmillan. All have flowers that close up, if not at noon, then very early in comparison with others.

JOHN HOOD *Geranium robertianum* (Herb Robert) Som. Macmillan. A result of connecting Robert, and thence Robin, with Hood, almost inevitably. There is Jenny Hood, as well as John Hood, for this.

JOHN-THAT-GOES-TO-BED-AT-NOON 1. *Anagallis arvensis* (Scarlet Pimpernel) Northants. A E Baker. 2. *Tragopogon pratensis* (Goat's Beard) Northants. A E Baker.

JOHN'S CABBAGE *Hydrophyllum virginianum* (Virginian Waterleaf) Perry.

JOHN'S FEAST-DAY WORT *Artemisia vulgaris* (Mugwort) IOM Moore. Manx Bellar feaill-Eoin. It is John's herb in Welsh, too - Llysiau Ifan.

JOHN'S FLOWER *Geranium lucidum*

(Shining Cranesbill) Som. Macmillan.

JOHN'S PLANT *Silene cucubalis* (Bladder Campion) Vesey-Fitzgerald.

JOHN'S WORT *Hypericum perforatum* (St John's Wort) Som. Elworthy.

JOHNNY, Harping *Sedum telephium* (Orpine) E Ang. Grigson. 'Harping' is a corruption of orpine. Cf. Orphan John.

JOHNNY COCKS *Orchis mascula* (Early Purple Orchis) Dor. B & H.

JOHNNY CRANES *Caltha palustris* (Marsh Marigold) N'thants. Grigson.

JOHNNY-GO-TO-BED *Tragopogon pratensis* (Goat's Beard) Tynan & Maitland.

JOHNNY-JUMP-UP 1. *Viola palmata* (Early Blue Violet) Le Strange. 2. *Viola pedunculata* (Yellow Pansy Violet) Perry. 3. *Viola tricolor* (Pansy) Wilts. Wiltshire.

JOHNNY-JUMP-WELL 1. *Viola rafinesquii* USA Gates. 2. *Viola tricolor* (Pansy) USA Upham.

JOHNNY Mac GOREY *Crataegus monogyna* fruit (Haws) Ire. B & H.

JOHNNY O'NEELE 1. *Chenopodium album* (Fat Hen) Shrop. B & H. 2. *Chenopodium bonum-henricum* (Good King Henry) Shrop. Grigson.

JOHNNY PRICK-FINGER *Dipsacus fullonum* (Teasel) Dor. Macmillan. Descriptive, obviously. Cf. Pricky-back, or Prickly Beehive, the latter descriptive by both character and shape.

JOHNNY-RUN-THE-STREET *Viola tricolor* (Pansy) Som. Macmillan.

JOHNNY WOODS *Melandrium dioicum* (Red Campion) Dor. Macmillan.

JOHNSMAS FLOWERS; JOHNSMAS PAIRS 1. *Plantago lanceolata* (Ribwort Plantain) Shet. John Nicolson. 2. *Plantago major* (Great Plantain) Shet. Grigson. On St John's Day in Shetland it used to be the custom to pick two stems of plantain, one for the boy, the other for the girl, to foretell if they would love and marry. The procedure was to pick the florets and then lay the heads under a flat stone. If the florets reappeared before the heads withered, they would be sure to marry.

It is usually reckoned that the Ribwort is the plant to be used, rather than Great Plantain.

JOHNSON GRASS *Sorghum halepense*.

JOINT-FIR, Long-leaved *Ephedra trifurca*.

JOINT-GRASS 1. *Galium verum* (Lady's Bedstraw) Midl, E Ang. Grigson; N Eng. Halliwell. 2. *Hippuris vulgaris* (Mare's Tail) Dev, Herts. Grigson.

JOINT-PINE 1. *Ephedra distachya*. 2. *Ephedra fragilis*. 3. *Ephedra trifurca* (Long-leaved Joint-Fir) USA Elmore.

JOINTS; JOINTWEED *Equisetum arvense* (Common Horsetail) Gerard (Joints); Som. Elworthy (Jointweed).

JOINTS, Blue *Agropyron repens* (Couch-grass) USA Chamberlin.

JOINTS, Nine *Polygonum aviculare* (Knotweed) Coles.

JOINTS, Red *Polygonum persicaria* (Persicaria) Dor. Macmillan.

JOJOBA *Simmondsia chinensis*.

JOLLY SOLDIERS *Orchis mascula* (Early Purple Orchis) Dev. Macmillan.

JONETTE *Caltha palustris* (Marsh Marigold) Scot. B & H. Sounding more Scottish, there is also Janet-flower. They are both from French jaunette, so are just a comment on the colour.

JONQUIL *Narcissus odorus*. French jonquille, from Latin juncus, rush, descriptive of the leaves.

JONQUIL, Queen Anne's Irish *Narcissus minimus*.

JONQUIL, Wild *Narcissus pseudo-narcissus* (Daffodil) Yks. Grigson.

JOPI WEED *Eupatorium purpureum* (Purple Boneset) Grieve.1931. Jopi is Joe Pye, an Indian doctor who was supposed to have cured typhus with this.

JORDAN ALMOND *Prunus amygdalus var.dulcis* Hatfield. 'Jordan' here is jardin, garden - in other words, this is the kind that is grown in gardens.

JOSEPH'S COAT 1. *Alternanthera bettzickiana* (Joy Weed) USA Cunningham. 2. *Amaranthus tricolor*. 3. *Pulmonaria officinalis* (Lungwort) Brownlow. Joseph's coat was, it will be remembered, "of many colours", so this is a perfectly good descriptive name for these flowers.

JOSEPH'S FLOWER *Tragopogon pratensis* (Goat's Beard) Gerard.

JOSEPH'S WALKING STICK *Polemonium caeruleum* (Jacob's Ladder) Hants. Cope. Cf. Jacob's Walking Stick.

JOSEPH-AND-MARY 1. *Pulmonaria longifolia* (Long-leaved Lungwort). 2. *Pulmonaria officinalis* (Lungwort) Som. Tongue.1968; Dor. Grigson; Wilts. Dartnell & Goddard; Hants. Cope.

JOSEPHINE'S LILY *Brunsvigia josephinae*.

JOSHUA TREE *Yucca brevifolia*.

JOVE'S BEARD *Sempervivum tectorum* (Houseleek) Conway. From a Latin name, jovis barba, which also accounts for the variants of Joubarbe.

JOVE'S NUT *Quercus robur* - acorn (Acorn) Som. Halliwell. Is this really a Somerset name? It smacks too much of the classical tradition of the oak being Jove's own tree.

JOY 1. *Euphrasia officinalis* (Eyebright) Som. Macmillan. 2. *Ranunculus arvensis* (Corn Buttercup) Ess. B & H. As far as the buttercup is concerned, this must be Gye, a name given to goose-grass as well. So it is probably a reference to the burry seeds.

JOY, Porter's *Ipomaea learii* (Blue Dawn Flower) India Perry. It is also called Railway Creeper in India.

JOY, Simpler's *Verbena officinalis* (Vervain) Northall. Because it is a heal-all.

JOY-OF-THE-GROUND *Vinca minor* (Lesser Periwinkle). This is a medieval book name.

JOY-OF-THE-MOUNTAIN *Origanum vulgare* (Marjoram) Som. Macmillan. That is the meaning of Origanum.

JOY WEED *Alternanthera bettzickiana*.

JOYNSON'S REMEDY CHEESE *Tanacetum vulgare* (Tansy) Bucks. B & H. A mysterious name - Britten & Holland offer no explanation whatsoever.

JUBARBE; JUBARD *Sempervivum tectorum* (Houseleek) Sanecki (Jubarbe); Halliwell (Jubard). These started off as jovis barba, Jove's Beard.

JUBILEE HUNTER *Rubus caesius* (Dewberry) Wilts. Dartnell & Goddard.1899.

JUDAEA TREE *Cercis siliquastrum* (Judas Tree) Barber & Phillips. It is suggested that Judas Tree should actually be Judea Tree, which was what it was called in France, and it was from France that the tree came to England. It had become Judas Tree by Shakespeare's time.

JUDAS' BAG *Adansonia digitata* (Baobab) Howes.

JUDAS PENCE *Lunaria annua* (Honesty) Coats. Many people have equated the silver seed vessels with coins, so we have Money-flower, Penny-flower, Shillings, and a host of other small change.

JUDAS TREE 1. *Cercis siliquastrum*. 2. *Sambucus nigra* (Elder) Kent Grigson. According to one account, Judas Tree should be Judaea Tree. It seems that the name first appeared in German. Elder has a Judas connection too, for legend has it that this was the tree he hanged himself on, Any tree with a Judas association would have a corresponding witch connection, as indeed both of these have. There is a superstition that it would be death to fall into a Judas Tree, an obvious re-enactment of the hanging of Judas.

JUDAS TREE, American *Cercis canadensis* (Redbud) Brimble.1948.

JUDAS TREE, Chinese *Cercis chinensis*.

JUDEAN SAGE *Salvia judaica*.

JUG, Cat *Rosa canina* - hips (Hips) Yks, Dur. B & H. 'Jug' seems to be a form of the common north country word for hips - choops. It appears as 'jobs' and jumps', too.

JUJUBE *Zizyphus jubajuba*. The berries are known as **French Jujubes**. They have been famous since ancient times for cold cures and bronchitis. They were made up into lozenges and widely exported, and such lozenges are still called jujubes.

JUJUBE, Chinese *Zizyphus jubajuba*

(Jujube) Douglas).

JULIAN *Narcissus pseudo-narcissus* (Daffodil) Herts. B & H.

JULY-FLOWER 1. *Cheiranthus cheiri* (Wallflower) B & H. 2. *Dianthus caryophyllus* (Carnation) Gilbert. Gilliflower, of course.

JULY-FLOWER, Clove *Dianthus caryophyllus* (Carnation) Hill. See Clove Gilliflower.

JULY-FLOWER, Wall *Cheiranthus cheiri* (Wallflower) Wesley.

JUM *Lolium temulentum* (Darnel) Culpepper.

JUMBIE BALSAM *Ocymum micranthum* Trinidad Laguerre.

JUMBIE TREE *Bombax ceiba* (Red Silkcotton Tree) W Indies H J Bell. Jumbie is a ghost, or at least a spirit of some kind, and this is the tree most likely to harbour one.

JUMBY-BUBBY *Solanum mammosum* (Amoise) Rodway. The seeds are narcotic, and are smoked for the effect. Presumably, Jumby-bubby is a reference to this.

JUMP, Dog *Rosa canina* - hips (Hips) Yks. Atkinson. Cf. Dog-jobs, Cat-jugs, etc.

JUMP-ABOUT *Aegopodium podagraria* (Goutweed) Oxf, War. Grigson. Ground Elder's creeping rootstock makes it do just what the name says.

JUMP-UP-AND-KISS-ME 1. *Portulaca pilosa* (Red Wild Purslane) W Indies Howes. 2. *Viola tricolor* (Pansy) Hants. Cope Suss. Parish. There are a lot of names for the pansy very similar to this, all connected with the erotic aspect of pansy belief.

JUMPING BETTY; JUMPING JACK *Impatiens noli-me-tangere* (Touch-me-not) both Suss. Parish. A comment on the plant's method of seed distribution.

JUMPLING, Bull *Trollius europaeus* (Globe-flower) Kinross Grigson.

JUNCTION VINE *Aristolochia odoratissima* (Sweet-scented Birthwort) Barbados Gooding.

JUNE-FLOWER *Anthriscus sylvestris* (Cow Parsley) Som. Macmillan.

JUNEBERRY *Amelanchier intermedia*. The fruit of all the Amelanchiers is called Juneberry, not just this one.

JUNEBERRY, Prune-leaved *Amelanchier prunifolia* (Prune-leaved Serviceberry) USA Elmore.

JUNEBERRY, Round-leaved *Amelanchier sanguinea*.

JUNEBERRY, Smooth *Amelanchier laevis*.

JUNEBUSH *Amelanchier alnifolia* (Western Service-berry) USA Turner & Bell.

JUNGLE RICE *Echinochloa crus-galli* (Japanese Barnyard Millet) Howes.

JUNIPER 1. *Chamaecyparis thyoides* (White Cedar) USA Harper. 2. *Juniperus communis*. 3. *Larix americana* USA Bergen. 4. *Vaccinium myrtillus* (Whortleberry) Ire O Suilleabhain. Ablett gave as his version of the derivation of juniper - junior, younger and pario, to produce. The berries remain on the plant for two years, i.e. while some of the fruit is ripe, a younger crop is in course of being produced.

JUNIPER, Alligator; JUNIPER, Alligator-bark *Juniperus pachyphlaea* USA Elmore (A-bark). It has a thick bark, divided into scaly squares, somewhat resembling an alligator skin.

JUNIPER, Alpine, French *Juniperus thurifera* (Spanish Juniper) Grey-Wilson.

JUNIPER, Black *Juniperus wallichiana*.

JUNIPER, Bonin Isles *Juniperus procumbens*. The Bonin Isles lie to the south of Japan.

JUNIPER, Brown-fruited *Juniperus oxycedrus* (Prickly Juniper) Grieve.1931.

JUNIPER, California *Juniperus californica*.

JUNIPER, Canary Island *Juniperus cedrus*.

JUNIPER, Cherrystone *Juniperus monosperma* (One-seeded Juniper) USA Yanovsky. Presumably because it is "one-seeded".

JUNIPER, Chinese *Juniperus chinensis*.

JUNIPER, Colorado *Juniperus scopulorum* (Western Red Cedar) USA Yanovsky.

JUNIPER, Creeping 1. *Juniperus communis* (Juniper) USA Tolstead. 2. *Juniperus horizontalis*.

JUNIPER, Desert *Juniperus californica* (California Juniper) USA Elmore.
JUNIPER, Drooping *Juniperus recurva.*
JUNIPER, Dwarf *Juniperus communis var. nana.*
JUNIPER, East African *Juniperus procera.*
JUNIPER, Flaky *Juniperus squamata* (Himalayan Juniper) RHS.
JUNIPER, Formosan *Juniperus formosana.*
JUNIPER, Great *Tetraclinus articulata* (Sandarac) Pomet.
JUNIPER, Grecian *Juniperus excelsa.*
JUNIPER, Himalayan *Juniperus squamata.*
JUNIPER, Incense-bearing *Juniperus thurifera* (Spanish Juniper) Ablett.
JUNIPER, Irish *Juniperus communis var. hibernica.*
JUNIPER, Large *Juniperus oxycedrus* (Prickly Juniper) Grieve.1931.
JUNIPER, Low *Juniperus communis var. depressa.*
JUNIPER, Mexican *Juniperus flaccida.*
JUNIPER, Mountain *Juniperus sibirica.*
JUNIPER, Mountain, Red *Juniperus scopulorum* (Western Red Cedar) USA Elmore.
JUNIPER, Needle *Juniperus rigida* (Temple Juniper) Hora.
JUNIPER, One-seeded *Juniperus monosperma.*
JUNIPER, Pencil *Juniperus virginiana* (Virginian Juniper) Howes. Better known as Pencil Cedar.
JUNIPER, Phoenician *Juniperus phoenicia.*
JUNIPER, Prickly *Juniperus oxycedrus.*
JUNIPER, Red 1. *Juniperus scopulorum* (Western Red Cedar) USA Elmore. 2. *Juniperus virginiana* (Virginian Juniper) Taylor.
JUNIPER, Rocky Mountain *Juniperus scopulorum* (Western Red Cedar) USA Weiner.
JUNIPER, Sierra 1. *Juniperus occidentalis.* 2. *Pinus occidentalis.*
JUNIPER, Spanish 1. *Juniperus thurifera.* 2. *Tetraclinus articulata* (Sandarac) Pomet.
JUNIPER, Stinking *Juniperus foetidissima.*
JUNIPER, Swedish *Juniperus communis var. suecica.*
JUNIPER, Szechuan *Juniperus detans.*
JUNIPER, Syrian *Juniperus drupacea.*
JUNIPER, Temple *Juniperus rigida.*
JUNIPER, Utah 1. *Juniperus californica* (California Juniper) USA Elmore. 2. *Juniperus osteosperma.*
JUNIPER, Virginian *Juniperus virginiana.*
JUNIPER, Weeping, Himalayan *Juniperus recurva* (Drooping Juniper) RHA.
JUNIPER, Western *Juniperus occidentalis* (Sierra Juniper) USA Elmore.
JUNIPER MISTLETOE *Phoradendron juniperum.*
JUNIPER SAVIN *Juniperus sabina* (Savin) Fernie.
JUNO IRIS *Iris scorpiris.*
JUNO'S ROSE *Lilium candidum* (Madonna Lily) Parkinson. There is a legend that Jupiter, to make his infant son Hercules immortal, put him to the breast of Juno. The drops of milk that fell to the ground became white lilies, and those that went into the sky became the Milky Way. Of course, the transition from Juno to the Madonna is natural with a change of religion.
JUNO'S TEARS *Verbena officinalis* (Vervain) Halliwell.
JUPITER'S BEAN *Hyoscyamus niger* (Henbane) Jordan.
JUPITER'S BEARD 1. *Anthyllis vulneraria* (Kidney Vetch) Friend. 2. *Centranthus ruber* (Spur Valerian) A W Smith. 3. *Pulsatilla vulgaris* (Pasque Flower) Tynan & Maitland. 4. *Sempervivum tectorum* (Houseleek) Gerard. The Romans called houseleek Diapetes, iovis caulis, or iovis barba, which translated immediately to Jove's Beard. This is the French joubarbe, which in turn appeared in English as Jobarbe or Jubarbe, and eventually was responsible for Jew's Beard.
JUPITER'S DISTAFF *Salvia glutinosa.* This is more likely to be Jupiter's Staff. One cannot really imagine Jupiter with a distaff!
JUPITER'S EYE *Sempervivum tectorum* (Houseleek) Gerard.
JUPITER'S STAFF *Verbascum thapsus* (Mullein) Parkinson (and in USA Henkel).

The upright growth gives the name - it appears also as Shepherd's, Peter's, Jacob's Staff, Aaron's Rod, etc.

JUR-NUT *Conopodium majus* (Earthnut) Northants. A E Baker. A variation of earthnut. Cf. such forms as Gernut, Yernut, etc.

JURNALS, Jacky; JURNALS, Jocky; JURNALS, Job *Conopodium majus* (Earthnut) all Cumb. Grigson. Jurnals comes by a devious route from earthnut, presumably via Jurnut.

JUTE *Corchorus capsularis* or *C olitorius*. The word is probably an adaptation of the Bengali word jhuto, eventually from a Sanskrit word meaning matted hair.

JUTE, China *Abutilon avicennae*.

JUTE, Long-fruited *Corchorus olitorius*.

JUTE, Manchurian *Abutilon avicennae* (China Jute) Howes.

K

KADAMBA *Nauclea cadamba.*
KADLE DOCK 1. *Anthriscus sylvestris* (Cow Parsley) Ches. B & H. 2. *Petasites hybridus* (Butterbur) Ches. Holland. 3. *Senecio jacobaea* (Ragwort) Ches. B & H. 'Kadle' sometimes appears in the same county as 'Kettle'. Butterbur has the necessary large leaves to warrant being called a dock, but the others have not, and the name is probably the same as Kedlock, which is one of the variants of charlock.
KADLOCK *Sinapis arvensis* (Charlock) N'thants. A E Baker.
KAFFIR CORN *Sorghum vulgare.*
KAFFIR FIG *Mesembryanthemum edule* (Hottentot Fig) McClintock.
KAFFIR LILY *Schizostylis coccinea.*
KAFFIR NUT *Harpephyllum caffrum.*
KAFFIR ORANGE *Strychnos spinosa.* The yellow fruits are about the size of an orange.
KAFFIR ORANGE, Small *Strychnos cocculoides.*
KAFFIR PLUM *Flacourtia indica.*
KAFFIR POTATO 1. *Coleus barbatus* S Africa Watt. 2. *Coleus esculentus* S Africa Watt. Both are cultivated for their tubers in South Africa.
KAFFIR THORN *Lycium afrum.*
KAFFIR TREE *Erythrina caffra.*
KAGER; KAIYER 1. *Daucus carota* (Wild Carrot) Corn. Jago. 2. *Pastinaca sativa* (Parsnip) Corn. Courtney.1880. The Cornish word is kegys, applied to any umbelliferous plant. It sounds very like 'kex' and all its variants, a ubiquitous word for the dried stalks of umbelifers.
KAHIKATEA *Podocarpus dacrydioides.* That is the Maori name for the tree.
KAIL, Bow *Brassica oleracea var. capitata* (Cabbage) Scot. Jamieson.
KAKA; KAKEZIE *Conium maculatum* (Hemlock) Ork. Grigson. (Kaka); Dev, Som. Grigson (Kakezie). Presumably a 'kex' or 'kecksy' name.
KAKEE *Diospyros kaki* (Chinese Date-Plum) Barber & Phillips. A version of the specific name, which is Japanese in origin.
KALE *Brassica oleracea var. acephala.* The northern forn of 'cole'.
KALE, Corn; KALE, Field *Sinapis arvensis* (Charlock) Cumb. B & H (Corn); Cumb. Grigson (Field).
KALE, Curled *Brassica fimbricata.*
KALE, Muggert *Artemisia vulgaris* (Mugwort) Scot. Grigson.
KALE, Pencuir *Polygonum bistorta* (Bistort) Ayr. Grigson.
KALE, Roman *Beta vulgaris var. cicla* (Swiss Chard) Bianchini.
KALE, Ruvo *Brassica ruvo* (Italian Turnip Broccoli) Usher. An Italian vegetable, Ruvo being a town near Bari, in the south of the country.
KALE, Sea *Crambe maritima.*
KALE, Wild 1. *Brassica oleracea* (Wild Cabbage) N'thum. B & H. 2. *Sinapis arvensis* (Charlock) Scot, N Ire. Grigson.
KALE, Will *Sinapis arvensis* (Charlock) Lanark. Grigson. Presumably, 'will' is 'wild'.
KALMUK CORYDALIS; SIBERIAN CORYDALIS *Corydalis bulbosa* (Bulbous Fumitory) Douglas.
KAMES, Hen's *Dactylorchis maculata* (Heath Spotted Orchid) Berw. Grigson. Kames is combs, so this is a descriptive name.
KAMICS; KAMMICK *Ononis repens* (Rest-harrow) both Som. Macmillan. Better known as Cammock, which is OE cammoc, commuc, and still quite common for Rest-harrow. The form in 'k' appears again in Somerset as Kemmick, and Krammics.
KANGA BUTTER *Pentadesma butyracea* Howes. The bark and fruit exude a yellow, oily latex. Such names as this are a little misplaced, for this is in no sense a greasy latex.
KANGAROO APPLE 1. *Solanum aviculare.*

2. *Solanum laciniatum*. Australian species, both.

KANGAROO PAW *Anigozanthos flavida*.

KANGAROO PAW, Black *Macropidia fuliginosa*.

KANGAROO PAW, Golden *Anigozanthos pulcherrima*.

KANGAROO PAW, Red-and-green *Anigozanthos manglesii*.

KANGAROO THORN *Acacia armata*. Surely this is an African tree?

KANGAROO VINE *Cissus antartica*.

KANGRA BUCKWHEAT *Fagopyrum tataricum* (Green Buckwheat) Howes.

KANSAS FEATHER *Liatris pycnostachyna*.

KANSAS GAYFEATHER *Liatris callilepsis*.

KANSAS THISTLE *Solanum rostratum* (Prickly Nightshade) USA Kingsbury.

KAPOK TREE *Ceiba pentandra*. A Malay word, kapoq. The commodity itself is the floss of the seed pods.

KAPUR *Dryobalanops camphora* (Camphor of Borneo) N Davey.

KARAKA *Corynocarpus laevigata*. The Maori name (this is a New Zealand plant).

KARAMYLE *Lathyrus montanus* (Bittervetch) Scot. B & H. More usual with an initial 'c' rather than with a 'k'. Whatever the spelling, the word is Caermeal in Gaelic, the name of the plant, which itself comes from corr, crane, and meilg, pod. It is Pys y garanod, crane's peas, in Welsh.

KARAYA GUM *Stercularia urens*.

KARELA *Momordica charantia* (Wild Balsam-apple) Douglas.

KARNIP *Arum maculatum* (Cuckoo-pint) Skinner.

KARO *Pittosporum crassifolium*.

KAROO NUT *Cola nitida* (Kola) Notes & Queries.1902. One of the dialectal forms of West African gourou (see Gourou Nut) is garru, hence presumably this form in English.

KARRI *Eucalyptus diversicolor*.

KARRO *Dioscorea sativa* Berdoe.

KARSE, Well *Nasturtium officinale* (Watercress) N'thum, Scot. Grigson. Karse is cress.

KASHMIR CYPRESS *Cupressus cashmeriana*.

KASHMIR ROWAN *Sorbus cashmiriana*.

KASSOD TREE *Cassia siamea*.

KATHERINE'S FLOWER *Nigella damascena* (Love-in-a-mist). The reference is to St Katherine of Siena, who suffered martyrdom on a wheel, and the comparison is to the wheel-shaped flower. They say the seed should be sown on her day (30 April), and then it will be in bloom on 14 September, the feast day of her namesake, St Katherine of Genoa.

KATMIA, Bladder *Hibiscus trionum*.

KATMIA, Syrian *Hibiscus syriacus*.

KATSURA *Cercidiphyllum japonica*.

KATTY KEYS *Fraxinus excelsior* seed vessels (Ash keys) Dur. Dinsdale. The winged seed vessels are known as 'keys' over a wide area. This variation is more usual as Kitty Keys, but it also appears as Ketty, Kite or Kit Keys.

KATY, Flaming *Kalanchoe blossfeldiana* (Brilliant Star) RHS.

KAURI PINE *Agathis australis*. The Maori name, anglicised into Cowrie sometimes.

KAURI PINE, Queensland *Agathis robusta*.

KAVA PEPPER *Piper methysticum*.

KAWAKA *Libocedrus plumosa*. The Maori name.

KAYA *Torreya nucifera*. A Japanese tree.

KAYS 1. *Acer pseudo-platanus* seed vessels (Sycamore keys) Ches. Holland. 2. *Fraxinus excelsior* seed vessels (Ash keys) Ches. Holland. The local pronunciation of 'keys'.

KEAKI *Zelkova serrata*.

KEATLEGS 1. *Orchis mascula* (Early Purple Orchis) Kent B & H. 2. *Orchis morio* (Green-winged Orchis) Kent Friend. This appears also as Skeat-legs, which makes the derivation from OE sceat, swathe, more plausible. But another form from Kent is given as Neatlegs.

KEBLOCK 1. *Brassica campestre var. rapa* (Turnip) Suss. Parish. 2. *Sinapis arvensis* (Charlock) N Eng. Halliwell. This sounds too much like one of the names for charlock to be taken too seriously as a name for turnip. Halliwell, though, assigns the name to the wild turnip, which surely must be taken to mean charlock.

KECK 1. *Anthriscus sylvestris* (Cow Parsley) Som, Dor, War, Bucks, Suss, Northants, Yks. B & H. 2. *Heracleum sphondyllium* (Hogweed) Suss, Surr, Ches, Yks. B & H. It may be applied to the plants, but Keck or something like it is a country-wide word for the dried stalks of any of the umbellifers.

KECK, Broad-leaved *Heracleum sphondyllium* (Hogweed) B & H.

KECK, Trumpet *Angelica sylvestris* (Wild Angelica) Cumb. B & H. Boys make trumpets of the hollow stems.

KECKERS *Heracleum sphondyllium* (Hogweed) Som. Elworthy.

KECKLOCK *Sinapis arvensis* (Charlock) Leics. B & H; Yks. Addy. This is not a 'keck' word, but a variation on charlock. There is a whole series of these variants, ranging from Cadlock to Warlock.

KECKSIES, Winter *Prunus spinosa* fruit (Sloe) IOW Long. Cf. the Hampshire Kex for sloes.

KECKSY *Conium maculatum* the dried stalk (Hemlock) Wilts. Akerman. See Kex.

KECKSY, Kelk- 1. *Angelica sylvestris* (Wild Angelica) Yks. B & H. 2. *Conium maculatum* (Hemlock) Yks. Harland. Listed as Wild Angelica, but this could very well refer actually to Hogweed, *Heracleum sphondyllium*.

KEDDLE-DOCK; KEEDLE-DOCK *Senecio jacobaea* (Ragwort) Lancs. Nodal (Keddle); Ches, Lancs. Grigson (Keedle). A number of these charlock names have attached themselves to ragwort, including Kedlock, Kadle-dock and Kettle-dock.

KEDLACK; KEDLET *Sinapis arvensis* (Charlock) both Leics. A.B.Evans (Kedlack); PLNN 18 (Kedlet).

KEDLOCK 1. *Angelica sylvestris* (Wild Angelica) Ches. Grigson. 2. *Anthriscus sylvestris* (Cow Parsley) Derb. B & H. 3. *Heracleum sphondyllium* (Hogweed) Ches.

B & H. 4. *Raphanus raphanistrum* (Wild Radish) Glos, Staffs. B & H. 5. *Senecio jacobaea* (Ragwort) Lancs. Grigson. 6. *Sinapis arvensis* (Charlock) Glos, Shrop, Derb, Leics. B & H; N'thants. Sternberg; Yks. Addy.

KEEK-LEGS *Orchis mascula* (Early Purple Orchis) Kent Grigson.

KEEKS *Conium maculatum* the dried stalks (Hemlock) Wilts. Akerman. See Kex.

KEEKS, Cow; KEEPS, Cow *Heracleum sphondyllium* (Hogweed) Border B & H.

KEELE *Brassica oleracea var. capitata* (Cabbage) Turner. A form of kale, i.e. cole.

KEELE, Sea *Crambe maritima* (Sea Kale) Evelyn.

KEER *Sorbus aucuparia* (Rowan) Corn, Dev. Grigson. A Celtic word, which appears in Cornish as caer, resulting in Care or Keer, or Cayer in Pembrokeshire. It originally meant berry.

KEESHION *Anthriscus sylvestris* (Cow Parsley) Ire. B & H.

KEESLIP *Galium verum* (Lady's Bedstraw) Scot. Grigson. Not cowslip, which at first glance seems likely, but cheeselip, or rennet, for Lady's Bedstraw used to be a well-known substitute for rennet in cheese-making.

KEGGAS 1. *Heracleum sphondyllium* (Hogweed) Corn. Grigson. 2. *Pastinaca sativa* (Wild Parsnip) Corn. Jago. Cornish kegys, a word for any umbelliferous plant.

KEGGERS *Daucus carota* (Wild Carrot) Corn. Jago.

KEGLUS 1. *Angelica sylvestris* (Cow Parsley) Ches. Grigson. 2. *Heracleum sphondyllium* (Hogweed) Ches. Grigson.

KEICE, KEICER *Conium maculatum* (Hemlock) both Yks. Robinson. Kex is probably the original word, but these seem to be local variants of Kesh, or Keish.

KEIRN; CUIRN *Sorbus aucuparia* (Rowan) IOM Paton (Keirn) IOM Moore (Cuirn). See Keer.

KEIRY *Cheiranthus cheiri* (Wallflower) Coats. A version of the specific name.

KEISH *Conium maculatum* (Hemlock) Lancs. Nodal. Cf. Kesh.

KELIAGE *Polygonum hydropiper* (Water Pepper) Halliwell. A form of Culrage (culi rabies), a name resulting from the irritant effect of the acrid leaves. The name has several variants, as Killridge, Culerage, Curage or Culrache.

KELK 1. *Aethusa cynapium* (Fool's Parsley) Wilts, Suss, Surr, Kent, Yks, N'thum, Dur. Grigson. 2. *Anthriscus sylvestris* (Cow Parsley) Yks. Morris; Cumb. B & H. 3. *Conium maculatum* (Hemlock) Wilts, Suss, Surr, Kent, Yks, N'thum, Dur. Grigson. 4. *Sinapis arvensis* (Charlock) Wilts, Surr, Kent Grigson; Suss Parish. Presumably only a variation of 'keck', or 'kex', but it is suprising that it has such a wide spread. Perhaps 'kellock' is the imtermediate stage.

KELK, Broad *Heracleum sphondyllium* (Hogweed) Yks. B & H.

KELK-KECKSY 1. *Angelica sylvestris* (Wild Angelica) Yks. B & H. 2. *Conium maculatum* (Hemlock) Yks. Harland. Listed as Wild Angelica, but this could very well refer actually to *Heracleum sphondyllium*.

KELLAS; KELLY *Conopodium majus* (Earthnut) Corn. Grigson. Cornish keleren. Cf. Killimore.

KELLOCK 1. *Anthriscus sylvestris* (Cow Parsley) Lincs. B & H. 2. *Raphanus raphanistrum* (Wild Radish) Glos. B & H. 3. *Sinapis arvensis* (Charlock) Glos, Yks. B & H.

KELLOGG LILY *Lilium kelloggii*. Dr Albert Kellogg, one of the founders of the California Academy of Sciences.

KELP *Salsola kali* (Saltwort) Pomet.

KEMANG *Mangifera caesia* (Binjai) Usher.

KEMMICK *Ononis repens* (Rest-harrow) Som. Elworthy. 'Cammock' is the usual rendering.

KEMPS 1. *Plantago lanceolata* (Ribwort Plantain) N Eng, Scot. Denham/Hardy. 2. *Plantago major* (Great Plantain) N Eng. Denham/Hardy. OE caempa, a warrior. The reference is to a children's game, using the flower heads as "soldiers". One child holds out a stalk, and the other tries to decapitate it with one of his own. Hence Hardheads, Conkers, Fechters, Cock-fighters, etc.

KEMPS, Sea *Plantago maritima* (Sea Plantain) Border B & H.

KEMPSEED *Plantago lanceolata* (Ribwort Plantain) N Eng, Scot. Grigson.

KENANGA *Canangium odoratum* (Ylang-Ylang) Leyel.1937.

KENILWORTH IVY *Cymbalaria muralis* (Ivy-leaved Toadflax) USA Zenkert.

KENNIN HERB; KENNING HEAL *Ranunculus arvensis* (Corn Buttercup) both Corn. Jago (Kennin herb); Courtney (Kenning heal). The plant was used in charms for the cure of kennings, or kernels, or in modern terms, ulcers in the eye. Ker is a Cornish word meaning the peel, or skin, of anything.

KENNING-WORT *Chelidonium majus* (Greater Celandine) USA Leighton. It is suggested that this is the Scottish verb to ken, in the sense of to see. Celandine has always had the reputation of improving the eyesight, from the legend of swallows (the Greek for which gave the name celandine) restoring their sight by resorting to celandine juice. But see the note on kennings, above.

KENRY *Sinapis arvensis* (Charlock) Som. Grigson.

KENT, Fair Maid of *Ranunculus aconitifolius* B & H. Probably both this aand Fair Maids of France result simply from the white flowers.

KENTIA PALM *Howeia forsteriana*. Kentia is the capital of Lord Howe Island, where the palm grows, and from which the genus takes its name.

KENTISH BALSAM *Mercurialis perennis* (Dog's Mercury) Kent Grigson. On the same lines as Wiltshire Weed for elm?

KENTISH CLOVER *Trifolium repens* (White Clover) McClintock.

KENTISH MILKWORT *Polygala austriaca*.

KENTUCKY BLUE-GRASS *Poa pratensis* (Smooth Meadow-grass) USA Allan.

KENTUCKY COFFEE TREE *Gymnocladus dioica*. Early settlers ground the seeds and used them as a coffee substitute.
KERK *Angelica sylvestris* (Wild Angelica) Leics. A B Evans.
KERLACK; KERLECK; KERLICK; KERLOCK *Sinapis arvensis* (Charlock) Northants. Sternberg (Kerlack); Northants. B & H (Kerleck); Bucks, Oxf. B & H (Kerlick); Glos. J D Robertson (Kerlock).
KERM OAK; KERMES OAK *Quercus ilex* (Holm Oak) Graves (Kerm); Grigson.1973 (Kermes). i.e. the oak on which the kermes insect is found, a scarlet insect not unlike a holly berry in appearance (hence Holm Oak).
KERMES BUSH *Phytolacca decandra* (Poke-root) Watt. Cf. Red-ink Plant, Redweed, etc. The coloured juice has been used as an ink, and the root-bark to give genuine red dyes.
KERN *Sorbus aucuparia* (Rowan) IOM Moore.1924.
KERNELWORT *Scrophularia nodosa* (Figwort) Gerard. The 'fig' of Figwort means piles; kernels may not be exactly the same thing as piles, but the general meaning is clear - this plant was used as a cure for them.
KERROON *Prunus avium* (Wild Cherry) Herts. Jones-Baker.1974. Sometimes written as Coroon.
KERSE 1. *Lepidium sativum* (Garden Cress) Turner. 2. *Nasturtium officinale* (Watercress) Som. Grigson. Cress, of course - the OE existed in two different forms, cresse and cerse, the latter presumably the earlier.
KERSE, Well *Nasturtium officinale* (Watercress) N'thum, Scot Grigson.
KERSLINS *Prunus domestica var. institia* fruit (Bullace) Som. Elworthy. The Cumbrian name for the tree is Crex, and this has produced Crickseys for the fruit (though this is an East Anglian word). Crystals, from Devon and Cornwall, seems to be the next step, and this then becomes Christians, Christlings or Crislings. Kerslins is the same word with a misplaced 'r'.

KERSOUNS *Nasturtium officinale* (Watercress) N Eng. Halliwell.
KESH 1. *Aegopodium podagraria* (Goutweed) Cumb. Grigson. 2. *Conium maculatum* (Hemlock) Lancs, Yks, Cumb. Grigson. 3. *Heracleum sphondyllium* (Hogweed) Yks, Cumb. B & H. A local variant of Kex. Cf. Keish.
KESH, Dry *Heracleum sphondyllium* (Hogweed) Cumb. B & H.
KESH, Smooth *Angelica sylvestris* (Wild Angelica) Cumb. Mabey.
KESH, Water *Angelica sylvestris* (Wild Angelica) Cumb. B & H. It certainly prefers damp places, but it is not a water plant.
KESK 1. *Angelica sylvestris* (Wild Angelica) Cumb. Grigson. 2. *Anthriscus sylvestris* (Cow Parsley) Cumb. B & H. 3. *Heracleum sphondyllium* (Hogweed) Cumb. B & H.
KESLINGS; KESTINS *Prunus domestica var. institia* fruit (Bullace) Dev. Friend (Keslings); Som. Macmillan (Kestins). See Kerslins.
KESSELRING LILY *Lilium kesselringianum*.
KETLACK *Sinapis arvensis* (Charlock) Lancs. B & H.
KETLOCK 1. *Raphanus raphanistrum* (Wild Radish) Glos, Yks. B & H. 2. *Sinapis arvensis* (Charlock) Glos, Lancs, Lincs. B & H. Yks. Nicholson.
KETT *Agropyron repens* (Couch-grass) Scot. Grigson.1959. Ket(t) is an old name meaning filth, so this sums up the gardener's detestation of couchgrass very well.
KETTLE CAP *Orchis mascula* (Early Purple Orchis) IOW Long.
KETTLE-CASE 1. *Dactylorchis maculata* (Heath Spotted Orchid) Som. Macmillan. 2. *Orchis mascula* (Early Purple Orchis) S Eng. Halliwell. 3. *Petasites hybridus* (Butterbur) B & H. Presumably an extension of the 'Kedlock' series of names, but where do the orchids fit?
KETTLE-DOCK 1. *Anthriscus sylvestris* (Cow Parsley) Ches. B & H. 2. *Rumex obtusifolius* (Broad-leaved Dock) Lancs. B

& H. 3. *Senecio jacobaea* (Ragwort) Ches. B & H. 4. *Sinapis arvensis* (Charlock) Lancs. A Burton. Certainly an extension of the 'Kedlock' series.

KETTLE-PAD *Orchis mascula* (Early Purple Orchis) Hants. Cope.

KETTLE-SMOCKS 1. *Convolvulus arvensis* (Field Bindweed) Wilts. Macmillan. 2. *Lonicera periclymenum* (Honeysuckle) Som. Macmillan. 3. *Melandrium dioicum* (Red Campion) Som. Macmillan. 'Kettle-smock' is an old name for a farm labourer's smock, as Halliwell assures us. Nevertheless, 'kettle' may very well be part of the sequence that includes 'Kadle'.

KETTLES, Crocks-and-; KETTLES-AND-CROCKS *Buxus sempervirens* seed (Box) both Som. Macmillan. Cf. Pots-and-kettles, Chairs-and-tables, etc.

KETTLES, Pots-and- *Buxus sempervirens* seed (Box) Wilts. Dartnell & Goddard.

KETTY-KEYS 1. *Acer campestre* seed vessels (Field Maple keys) Yks. Grigson. 2. *Fraxinus excelsior* seed vessels (Ash keys) Yks. Nicholson. 'Keys', for the kind of seed vessel produced by maples, ash, sycamore, etc., are universally known as such. The variation comes in the first element; they can be Cat's, Cattle, Katty, Kitty, even Kite, etc.

KEW WEED *Galinsoga parviflora*. A South American plant, introduced into Kew Gardens in 1796, and then established between Kew and Richmond by 1863. Joey Hooker was another name given to it - he was the director of Kew at the time it was introduced, so presumably he was the one who answered queries about it.

KEWSIES 1. *Angelica sylvestris* (Wild Angelica) Lincs. Grigson. 2. *Anthriscus sylvestris* (Cow Parsley) Lincs. B & H. 3. *Conium maculatum* (Hemlock) Lancs, Lincs, Yks. Grigson. 4. *Heracleum sphondyllium* (Hogweed) Lincs. B & H. Nothing to do with Kew - this is just a local variant of Kecksies. Cf. Kous, etc.

KEX 1. *Anthriscus sylvestris* (Cow Parsley) B & H. 2. *Conium maculatum* (Hemlock)

Gerard. 3. *Daucus carota* (Wild Carrot) Som. Macmillan. 4. *Heracleum sphondyllium* (Hogweed) Suss, Surr, Ches, Yks. B & H. 5. *Prunus spinosa* fruit (Sloes) Hants. Cope. Strictly, Kex refers to the dried stalks of any umbelliferous plant, but it and its many variants are applied to the plant itself as well. Kex seems quite out of place as a name for sloes, but it is probably related to the Cumbrian name for bullace, Crex. That name goes through a number of changes, and finally reaches Keslings in Devonshire.

KEX, Ghost *Angelica sylvestris* (Wild Angelica) Yks. Grigson. 'Holy Ghost' originally - it is called Angelica from its angel-like properties, hence "Spiritus sancti radix, the root of the Holy Ghost".

KEX, Red *Torilis japonica* (Hedge Parsley) Yks. Grigson.

KEX, Rough *Heracleum sphondyllium* (Hogweed) Corn. Grigson.

KEXIES *Conium maculatum* (Hemlock) Som. Macmillan.

KEY-BALLS *Pinus sylvestris* cones (Pine cones) Dev. Friend.1882.

KEY-FLOWER *Primula veris* (Cowslip) Brownlow. All the 'key' names that are applied to cowslip result from a supposed resemblance to the badge of St Peter - a bunch of keys. The legend is that Peter once dropped the keys of heaven, and the first cowslips grew up where they fell. So we have Herb Peter, or St Peter's Herb, as well as the 'key' names.

KEYN 1. *Fraxinus excelsior* fruit (Ash keys) Wilts. Dartnell & Goddard. 2. *Acer pseudo-platanus* fruit (Sycamore keys) Wilts. Dartnell & Goddard. The old plural in '-n' is retained in this name.

KEYS 1. *Acer campestre* fruit (Field Maple keys) Dor. B & H. 2. *Acer pseudo-platanus* fruit (Sycamore keys) Dor. Grigson; Wilts. Dartnell & Goddard. 3. *Fraxinus excelsior* fruit (Ash keys) Wilts, Shrop, N'thants, Suff, Lincs, Scot. B & H.

KEYS, Bunch of 1. *Fraxinus excelsior* fruit (Ash keys) Som. Macmillan. 2. *Primula*

veris (Cowslip) Som. Macmillan. 3. *Ulex europaeus* (Gorse) Som. Macmillan.
KEYS, Bunch of, Lady's *Primula veris* (Cowslip) Som. Grigson.
KEYS, Cats-and- 1. *Acer campestre* fruit (Maple keys) Dev. Friend.1882. 2. *Acer pseudo-platanus* fruit (Sycamore keys) Dev. B & H. 3. *Fraxinus excelsior* fruit (Ash keys) Dev. Friend.1882.
KEYS, Cat's; KEYS, Cattle *Fraxinus excelsior* fruit (Ash keys) Norf. B & H (Cat's); Derb. Addy (Cattle).
KEYS, Cove *Primula veris* (Cowslip) Kent Grigson.
KEYS, Cover; KEYS, Covey *Primula veris x vulgaris* (False Oxlip) both Kent B & H (Cover); Grigson (Covey).
KEYS, Eisch *Fraxinus excelsior* fruit (Ash keys) Lancs. B & H.
KEYS, Fairies' *Primula veris x vulgaris* (False Oxlip) Dev. Macmillan.
KEYS, Katty *Fraxinus excelsior* fruit (Ash Keys) Dur. Dinsdale.
KEYS, Ketty 1. *Acer campestre* fruit (Field Maple Keys) Yks. Grigson. 2. *Fraxinus excelsior* fruit (Ash Keys) Yks. Nicholson.
KEYS, Kit 1. *Acer campestre* fruit (Field Maple Keys) Yks. Grigson. 2. *Fraxinus excelsior* fruit (Ash Keys) Berks. Lowsley.
KEYS, Kite *Fraxinus excelsior* fruit (Ash Keys) Gerard.
KEYS, Kitty 1. *Acer campestre* fruit (Field Maple Keys) Derb. Addy; Yks. Grigson. 2. *Fraxinus excelsior* fruit (Ash Keys) Derb, Yks. Addy. 3. *Sorbus aucuparia* fruit (Rowan berries) Som. Macmillan. Out of place for rowan berries. A misreading, perhaps?
KEYS, Lady's 1. *Arum maculatum* (Cuckoo-pint) Kent Grigson. 2. *Endymion nonscriptus* (Bluebell) Som. Macmillan. 3. *Acer pseudo-platanus* fruit (Sycamore Keys) Wilts. Macmillan.
KEYS, Locks-and- 1. *Acer campestre* fruit (Field Maple keys) Dev. Friend.1882; Som. Macmillan. 2. *Acer pseudo-platanus* fruit (Sycamore keys) Som. Elworthy; Cambs, Ess, Norf. Grigson; Lincs. Peacock. 3. *Dicentra spectabilis* (Bleeding Heart) Som. Friend; Wilts. Dartnell & Goddard. 4. *Endymion nonscriptus* (Bluebell) Som. Macmillan. 5. *Fraxinus excelsior* fruit (Ash Keys) Dev. Friend.1882. Notts, Cambs, Ess, Norf, Lincs, Cumb. B & H. 6. *Orchis mascula* (Early Purple Orchis) Som. Macmillan. 7. *Rosa canina* fruit (Hips) Som. Macmillan. 8. *Tilia europaea* fruit (Lime) Som. Macmillan.
KEYS, Peter *Fraxinus excelsior* fruit (Ash Keys) Coles.
KEYS, St Peter's *Primula veris* (Cowslip) Som. Macmillan. See Key-flower.
KEYS-OF-HEAVEN *Primula veris* (Cowslip) Glos. PLNN35.
KEYWORT *Primula veris* (Cowslip) Skinner.
KHAKI WEED; KHAKI, Bur *Alternanthera repens* Dalziel. South African names, both.
KHASIA BERRY *Cotoneaster simonsii*.
KHASYA PINE *Pinus insularis*.
KHAT *Catha edulis*. Khat, or quat, chat, tschat, tchai are all names for a "tea", which in Arabia predated coffee by over a hundred years, made from the buds, twigs and fresh leaves.
KHELLA *Ammi visnaga* (Bishop's Weed) Thomson.1978.
KIAWE BEAN *Prosopis juliflora* (Mesquite) USA Kingsbury.
KICKING COLT *Impatiens capensis* (Orange Balsam) USA Grigson. Presumably a reference to the explosive seed-release action common to the whole genus.
KICKS *Aethusa cynapium* (Fool's Parsley) Fernie. 'Kex', of course, but it is unusual to find it for this plant. But at least in theory, kex applies to the dried stalks of any of the umbellifers.
KICKSY *Conium maculatum* (Hemlock) Ess. Gepp. One of the many 'kex' variants.
KIDNEY, Mary's *Mucuma urens* (Florida Bean) Hebrides Shaw. The translation of Gaelic Airne Mhoire. The bean is sometimes found washed up on the Atlantic shore of the

islands, where women used to hold it in their hand during childbirth.

KIDNEY BEAN *Phaseolus vulgaris*. Descriptive of the shape.

KIDNEY BEAN, Chinese *Wistaria sinensis* (Wistaria) Hay & Synge.

KIDNEY BEAN, Egyptian *Dolichos lablab* (Lablab) Dalziel.

KIDNEY BEAN, Scimitar-podded *Physalis lunatus* (Lima Bean) Candolle.

KIDNEY COTTON *Gossypium brasiliense*.

KIDNEY-LEAVED BUCKWHEAT *Eriogonum reniforme*.

KIDNEY LIVERLEAF *Hepatica americana*.

KIDNEY SAXIFRAGE *Saxifraga hirsuta*.

KIDNEY SORREL *Oxyria digyna* (Mountain Sorrel) Hutchinson.

KIDNEY TEA, Indian *Orthosiphon stamineus* (Java Tea) Thomson.1978.

KIDNEY VETCH *Anthyllis vulneraria*. Because it was used for kidney troubles?

KIDNEYWEED *Umbilicus rupestris* (Wall Pennywort) Som. Elworthy.

KIDNEYWORT 1. *Hepatica triloba* (American Liverwort) USA Le Strange. 2. *Saxifraga stellaria* (Star Saxifrage) B & H. 3. *Umbilicus rupestris* (Wall Pennywort) Gerard.

KIERAN *Sorbus aucuparia* (Rowan) IOM Gill.1932. The original is a Celtic word, something like the Cornish caer, which means a berry. This form appears in Anglo-Manx also as Keirn, or Cuirn, and Kern.

KIES, Cows-and- *Arum maculatum* (Cuckoo-pint) Yks. Grigson. Kies are heifers. This is one of the very numerous double names belonging to cuckoo-pint. This one is unusual in that it is one of the very few that is not male plus female, like Lords-and-ladies, Bulls-and-cows, etc.

KIKUYU GRASS *Pennisetum clandestinum*.

KILK; KELK *Sinapis arvensis* (Charlock) Som. Macmillan; Suss. Parish; Surr, Kent B & H (Kilk); Wilts, Surr, Kent Grigson; Suss. Parish (Kelk). See Kex.

KILL-BASTARD *Juniperus sabina* (Savin) Yks. B & H. See Bastard-killer.

KILL-YOUR-MOTHER-QUICK *Anthriscus sylvestris* (Cow Parsley) Ess. Vickery.1985. There is another version of this - Mother-die. It is an unlucky flower to bring indoors, and more immediately, note the Sussex superstition: "Never give cow parsley to your mother, or she will die". A strange belief to associate with such a harmless plant, but it may apply to any white-flowered umbellifer, simply because it resembles hemlock.

KILL-WART *Chelidonium majus* (Greater Celandine) Dev. Hay & Synge. Cf. Wart-flower, Wart Plant, Wartweed, Wart-curer, etc. The juice of Greater Celandine has always been applied to warts, with varying degrees of success.

KILLARNEY STRAWBERRY-TREE *Arbutus unedo* (Strawberry Tree) Hay & Synge. A Mediterranean tree, but it grows wild in some parts of Ireland, Killarney obviously being one of them.

KILLER, Bastard *Juniperus communis* (Juniper) Som. Grigson. Probably in error for Savin, *J sabina*, which is the real abortifacient.

KILLIMORE *Conopodium majus* (Earthnut) Corn. Jago. Is Cornish keleren the ancestor of this? Cf. Kellas, and Kelly.

KILLIN *Ilex aquifolium* (Holly) Corn. Grigson. Is this connected with the Manx Hullin, and such older forms as Hollin?

KILLRIDGE *Polygonum hydropiper* (Water Pepper) Halliwell. This is Culrage (see also Culerage, Curage, Keliage). It is from the French, and means the same as that other well-known name for Water Pepper - Arsmart.

KILMARNOCK WILLOW *Salix capraea* var. *pendula*.

KING, Moor *Pedicularis sceptrum-carolinum*.

KING ANTHURIUM *Anthurium veitchii*.

KING-COB 1. *Caltha palustris* (Marsh Marigold) Berks, Hants. B & H. 2. *Ranunculus arvensis* (Corn Buttercup) E Ang. Gerard. It must be a variation of Kingcup.

KING-DEVIL 1. *Hieraceum florentinum*. 2. *Hieraceum pratense* USA Zenkert. Presumably because they are such weeds in America.

KING-FINGER 1. *Orchis mascula* (Early Purple Orchis) War, Leics, Bucks. B & H. 2. *Orchis morio* (Green-winged Orchid) War, Bucks. B & H; Northants. A E Baker. 3. *Oxalis acetosella* (Wood Sorrel) Bucks. Grigson.

KING HARRY, Good *Chenopodium bonus-henricus* (Good King Henry) Cambs. Gerard.

KING HENRY, Good *Chenopodium bonus-henricus*. Note that the specific name is good Henry, nothing about a King. In 16th century German, it was guter Heinrich, the name of an elf with a knowledge of healing plants.

KING ISLAND MELILOT *Melilotus indica* (Small-flowered Melilot) Connor. Presumably this is the King Island between Victoria and Tasmania.

KING KONG *Caltha palustris* (Marsh Marigold) Som. Macmillan. Kingcup appears in some places as King-cob, and then in Somerset King Kong.

KING OF THE ALPS *Eritrichium nanum*.

KING OF THE MEADOW *Thalictrum polygonum* (Fall Meadow Rue) Howes.

KING OF THE WOODS *Aralia racemosa* (American Spikenard) S USA Puckett.

KING ORANGE *Citrus reticulata* (Tangerine) Howes.

KING WILLIAM PINE *Athrotaxus selaginoides*.

KING'S BUSH *Ulex europaeus* (Gorse) Tynan & Maitland.

KING'S CLOVER 1. *Melilotus altissima* (Tall Melilot) Scot. Grigson. 2. *Ranunculus acris* (Meadow Buttercup) Suff, Norf, Cumb. Grigson. 3. *Ranunculus bulbosus* (Bulbous Buttercup) Suff, Norf, Cumb. Grigson. 4. *Ranunculus repens* (Creeping Buttercup) Suff, Norf, Cumb. Grigson.

KING'S COB 1. *Ranunculus acris* (Meadow Buttercup) Suff, Norf, Cumb. Grigson. 2. *Ranunculus bulbosus* (Bulbous Buttercup) Suff, Norf, Cumb. Grigson. 3. *Ranunculus repens* (Creeping Buttercup) Suff, Norf, Cumb. Grigson.

KING'S CONSOUND *Delphinium ajacis* (Larkspur) Lyte. *Consolida* was an old specific name for larkspur, hence Consound in this name.

KING'S CROSS *Cheiranthus cheiri* (Wallflower) Som. Macmillan. Wallflower is cruciferous, hence this and other names like it, e.g. Cross-flower, Soldier's Cross, Crucifix-flower, etc.

KING'S CROWN 1. *Justicia carnea*. 2. *Melilotus officinalis* (Melilot) B & H. 3. *Trifolium pratense* (Red Clover) Som. Macmillan. 4. *Viburnum opulus* (Guelder Rose) Glos. B & H.

KING'S EVIL *Ranunculus ficaria* (Lesser Celandine) Som. Macmillan. The root nodules that gave 'ficaria' as a specific name - fig, that is, an old word for piles, must also have called to mind the scrophulous glands that were once known as the King's Evil.

KING'S FEATHER *Saxifraga umbrosa* (Pyrenean Saxifrage) W Miller. Given by Miller to this species, but more likely to belong to London Pride.

KING'S FINGERS 1. *Dactylorchis incarnata* (Early Marsh Orchid) Tynan & Maitland. 2. *Dactylorchis maculata* (Heath Spotted Orchid) Tynan & Maitland. 3. *Lotus corniculatus* (Bird's-foot Trefoil) Bucks. Grigson. There are all sorts of 'fingers' and 'thumbs' names for Bird's-foot Trefoil, and of course the tubers of these orchids are digitate.

KING'S HOOD *Geranium sylvaticum* (Wood Cranesbill) Border B & H.

KING'S KNOBS 1. *Ranunculus acris* (Meadow Buttercup) Glos. B & H. 2. *Ranunculus bulbosus* (Bulbous Buttercup) Som. Macmillan.

KING'S MANTLE *Thunbergia erecta* Howes.

KING'S NOBS *Ranunculus bulbosus* (Bulbous Buttercup) Som. Macmillan.

KING'S ROD *Narthecium ossifragum* (Bog Asphodel) Tynan & Maitland.

KING'S SPEAR 1. *Asphodeline lutea* (Yellow Asphodel) Gerard. 2. *Asphodelus ramosus* (White Asphodel) Grieve.1931. 3. *Eremurus sp* (Foxtail Lily) RHS. 4. *Narcissus pseudo-narcissus* (Daffodil) Som. Macmillan. Cf. Royal Staff and Silver Rod for Asphodel. Presumably daffodil gets this name because it is the same word as asphodel.

KING'S TAPER *Verbascum thapsus* (Mullein) Som. Macmillan.

KING'S WINEGLASS *Tulipa sp* (Tulip) Som. Macmillan.

KING'S WORT *Sambucus nigra* (Elder) R T Gunther.

KINGCUP 1. *Caltha palustris* (Marsh Marigold). 2. *Ranunculus acris* (Meadow Buttercup) Corn, Dev, Som, Suss, Bucks, Ess, Cambs, Norf, Cumb. Grigson. 3. *Ranunculus bulbosus* (Bulbous Buttercup). 4. *Ranunculus ficaria* (Lesser Celandine) Dev. Grigson. 5. *Ranunculus repens* (Creeping Buttercup) Corn, Dev, Som, Suss, Bucks, Ess, Cambs, Norf, Cumb. Grigson. 6. *Trollius europaeus* (Globe Flower) Som. B & H.

KINGCUP, Golden *Caltha palustris* (Marsh Marigold) Som. Macmillan.

KINGCUP, Persian *Ranunculus asiaticus* Coats.

KINGFISHER *Orchis morio* (Green-winged Orchid) War. B & H; Leics. Ruddock.

KINGFISHER DAISY *Felicia bergeriana*. A way of describing the peculiar shade of steel-blue of this daisy.

KINGS-AND-QUEENS *Arum maculatum* (Cuckoo-pint) Som, Dev. Macmillan; Dur. B & H. Male plus female name, on the lines of Lords-and-ladies.

KINKLE *Sinapis arvensis* (Charlock) Kent B & H. Presumably a variation of Kilk, which occurs in the same area.

KINNIKINNIK 1. *Arctostaphylos uva-ursi* (Bearberry) USA E Gunther. 2. *Cornus amomum* (Silky Cornel) Grieve.1931. 3. *Cornus stolonifera* (Red-osier Dogwood) Gilmore. Are the two species of *Cornus* one and the same? Kinnikinnik comes from an Algonquin word that means "that which is mixed". Tobacco mixtures in the eastern USA and Canada were kinnikinnik, and the inference must be that the plants that bear the name were recognised tobacco adulterants.

KINO *Pterocarpus marsupium* (Bastard Teak) O P Brown. Kino is really the tree's juice or resin, used as an astringent. The word is apparently of West African origin.

KINO, African; KINO, Gambian *Pterocarpus erinaceus* Dalziel.

KINO, Australian *Eucalyptus amygdalina* (Tasmanian Peppermint) Howes.

KINO, Malabar *Pterocarpus marsupium* (Bastard Teak) Howes.

KIPPERNUT 1. *Bunium bulbocastanum* (Great Earthnut) Gerard. 2. *Lathyrus montanus* (Bittervetch) B & H. For Bittervetch, this is the same word as Gnapperts, which went through various transformations - Knapperts, Knapperty, Knipper-nut, and thence to Kippernut. Culpepper called it Ciper Nut.

KIRNSTAFF; KIRNSTAFF, Devil's *Euphorbia helioscopia* (Sun Spurge) Ches, Lancs, Yks, Cumb, Kirk, Wigt. Grigson (Kirnstaff) ; Lanark, Ayr. Grigson (Devil's Kirnstaff).

KISK *Conium maculatum* (Hemlock) Suff, Ess. Moor. A local form of Kex.

KISKIES *Heracleum sphondyllium* (Hogweed). Corn. Grigson.

KISS, Blue *Succisa pratensis* (Devil's-bit) Suss. B & H.

KISS-AND-GO *Viscum album* (Mistletoe) Dor. Macmillan.

KISS-AND-LOOK-UP *Viola tricolor* (Pansy) Som. Grigson. The first in a long series denoting the aphrodisiac and erotic beliefs in pansy.

KISS-AT-THE-GARDEN-GATE *Viola tricolor* (Pansy) Suff. Grigson.

KISS-BEHIND-THE-GARDEN-GATE 1. *Saxifraga x urbinum* (London Pride) Wilts. Dartnell & Goddard; Glos. J D Robertson;

Oxf. Oxf.Ann.Rec.1951. 2. *Viola tricolor* (Pansy) War. Grigson.

KISS-BEHIND-THE-PANTRY-DOOR *Centranthus ruber* (Spur Valerian) Som. Macmillan.

KISS-I'-MY-CORNER *Artemisia abrotanum* (Southernwood) Bucks. Harman. Presumably for the same reason as that for Lad's Love - this was a courting herb.

KISS-ME 1. *Centranthus ruber* (Spur Valerian) Dev. Grigson. 2. *Geranium robertianum* (Herb Robert) Dev. Macmillan. 3. *Saxifraga x urbinum* (London Pride) Dev. Friend.1882. 4. *Viola tricolor* (Pansy) Suss. Parish; Lincs. Peacock.

KISS-ME-AND-GO *Artemisia abrotanum* (Southernwood) Som. Macmillan. Cf. Lad's Love.

KISS-ME-AT-THE-GARDEN-GATE *Viola tricolor* (Pansy) Dev. Grigson; Herts. Jones-Baker.1974; Northants. A E Baker.

KISS-ME-BEHIND-THE-GARDEN-GATE *Viola tricolor* (Pansy) Dev. Macmillan; War. Grigson.

KISS-ME-DICK *Euphorbia cyparissa* (Cypress Spurge) Ches. Holland. This name, and Clip-me-dick, are corruptions of the specific name, cyparissa.

KISS-ME-ERE-I-RISE *Viola tricolor* (Pansy) Halliwell.

KISS-ME-JOHN-AT-THE-GARDEN-GATE *Viola tricolor* (Pansy) Dor. Dacombe.

KISS-ME-LOVE 1. *Centranthus ruber* (Spur Valerian) Dev. Grigson. 2. *Geranium robertianum* (Herb Robert) Dev. Grigson. 3. *Saxifraga x urbinum* (London Pride) Dev. Friend.1882. 4. *Viola tricolor* (Pansy) Norf. Grigson.

KISS-ME-LOVE-AT-THE-GARDEN-GATE 1. *Geranium robertianum* (Herb Robert) Bucks. Grigson. 2. *Saxifraga x urbinum* (London Pride) Dev. Macmillan. 3. *Viola tricolor* (Pansy) Dev. Grigson.

KISS-ME-LOVE-BEHIND-THE-GARDEN-GATE *Saxifraga x urbinum* (London Pride) Dev. B & H.

KISS-ME-NOT *Saxifraga x urbinum* (London Pride) Som. Macmillan.

KISS-ME-OVER-THE-GARDEN-GATE *Viola tricolor* (Pansy) War. Bloom; Norf. Grigson.

KISS-ME-QUICK 1. *Amaranthus caudatus* (Love-lies-bleeding) Som. Macmillan. 2. *Arctium minus* (Lesser Burdock) Som. Macmillan. 3. *Artemisia abrotanum* (Southernwood) Som. Macmillan. 4. *Asperula odorata* (Woodruff) Som, Dor. Macmillan. 5. *Centranthus ruber* (Spur Valerian) Dev, Som. Macmillan Wilts. Dartnell & Goddard; Dor. Dacombe. 6. *Filipendula ulmaria* (Meadowsweet) Som. Macmillan. 7. *Galium aparine* (Goose-grass) Som, Wilts, Dor. Macmillan. 8. *Geranium robertianum* (Herb Robert) Dev, Wilts. Friend. 9. *Saxifraga x urbinum* (London Pride) Dev. Friend.1882. 10. *Silene cucubalis* (Bladder Campion) Som. Macmillan. 11. *Viola tricolor* (Pansy) Dev, Som. Elworthy.

KISS-ME-QUICK-AND-GO *Artemisia abrotanum* (Southernwood) Dev. Wright. Cf. Lad's Love.

KISS-ME-TWICE-BEFORE-I-RISE *Nigella damascane* (Love-in-a-mist) Coles.

KISS-THE-GARDEN-DOOR *Centranthus ruber* (Spur Valerian) Dor. Macmillan.

KISSES 1. *Arctium lappa* (Burdock) Som. Mabey; Dor. Northover. 2. *Galium aparine* (Goose-grass) Som. Macmillan. 3. *Viola tricolor* (Pansy) E Ang. B & H. The first two refer to the clinging properties of the burs, and as far as pansy is concerned, this is only one of an immense number of 'kissing' names connected with a supposed aphrodisiac quality.

KISSES, Lovers' *Galium aparine* (Goose-grass) Som. Macmillan.

KISSING KIND *Centranthus ruber* (Spur Valerian) Dor. Macmillan.

KIT KEYS 1. *Acer campestre* fruit (Field Maple keys) Yks. Grigson. 2. *Fraxinus excelsior* fruit (Ash keys) Berks. Lowsley.

KIT-RUN-THE-FIELDS *Viola tricolor* (Pansy) Som. Macmillan.

KIT WILLOW *Salix triandra* (French Willow) Northants. Sternberg.

KITE, Black *Rubus fruticosus* fruit (Blackberry) Cumb. Grigson.

KITE, Brannel; KITE, Brumble; KITE, Brummel *Rubus fruticosus* fruit (Blackberry) Dur. Grigson (Brannel); Hants, Yks, N'thum. Grigson; Cumb. Grose; Dur. Denham/Hardy (Brumble); Cumb. Grigson (Brummel). 'Kite' in Yorkshire dialect meant stomach, and, according to Morris, 'bumble' means to hum (there is Bumblekite, too), on the face of it saying that the fruit does not lie easily on the stomach. But surely, the first element in all these names is simply a local variant of bramble. Kite, then, must be a word for berry.

KITE KEYS *Fraxinus excelsior* fruit (Ash keys) Gerard. See Kit Keys, rather.

KITE'S LEGS 1. *Colchicum autumnale* (Meadow Saffron) Kent Grigson. 2. *Orchis mascula* (Early Purple Orchis) Kent Grigson.

KITE'S PAN *Dactylorchis maculata* (Heath Spotted Orchis) Wilts. Dartnell & Goddard.

KITLINGS, Cats-and- *Salix sp* catkins (Willow catkins) Northants A E Baker.

KITTENS' TAILS *Corylus avellana* catkins (Hazel catkins) Dor. Macmillan.

KITTENS, Cats-and- 1. *Corylus avellana* catkins (Hazel catkins) Worcs. B & H. 2. *Salix sp* catkins (Willow catkins) Northants. A E Baker.

KITTY-COME-DOWN-THE-LANE-JUMP-AND-KISS-ME *Arum maculatum* (Cuckoo-pint) Kent Grigson. More in keeping with Pansy names rather than Cuckoo-pint, but both have sexual connotations, anyway.

KITTY KEYS 1. *Acer campestre* fruit (Field Maple keys) Yks. Grigson. 2. *Fraxinus excelsior* fruit (Ash keys) Derb, Yks. Addy. 3. *Sorbus aucuparia* fruit (Rowan berries) Som. Elworthy. Out of place for rowan berries, and possibly a misreading.

KITTY-RUN-THE-STREET *Viola tricolor* (Pansy) Som, Kent Grigson; Wilts. Macmillan.

KITTY-TWO-SHOES *Lotus corniculatus* (Bird's-foot Trefoil) Dor. Macmillan. There are a lot of 'shoe' names for this, to go with the 'foot' names.

KIWI *Actinidia chinensis* fruit (Chinese Gooseberry) USA H G Baker. Because they are imported into the States from New Zealand, where they have been cultivated for some time. The name has stuck, for that is what they are called here, now that they are available in the shops.

KIX 1. *Conium maculatum* (Hemlock) Lincs. Peacock. 2. *Prunus domestica var. institia* fruit (Bullace) S Eng. Halliwell. Kix for hemlock is a variation of Kex, but for bullace it has a different source, and must be included in the sequence that includes Krickseys.

KLAMATH WEED *Hypericum perforatum* (St John's Wort) USA Schenk & Gifford. The Klamath Indians of California. The plant was introduced into the USA, and has become a pest, just as it has in Australia and New Zealand.

KLONGER; KLUNGER *Rosa canina* (Dog Rose) Shet. Grigson.

KNAP *Trifolium pratense* (Red Clover) Dor. Grigson. Knap means a hill, or a protuberance, OE cnaepp. It is easy enough to see where the "protuberance" comes in when considering the other plants with the name, including Knapweed, but with clover, the reference must be to the flowerheads themselves.

KNAP, Gold *Ranunculus bulbosus* (Bulbous Buttercup) Halliwell.

KNAP-BOTTLE *Silene cucubalis* (Bladder Campion) Prior. Prior said 'knap' was Dutch knappe, to snap, or crack, but the OE cnaepp is its origin, and refers to the "bladder".

KNAPPERS *Quercus robur* acorn (Acorn) N Scot. B & H.

KNAPPERTS; KNAPPERTY *Lathyrus montanus* (Bittervetch) Scot. Grigson (Knapperts); N Ire. Grigson (Knapperty).

There are various forms of these names, ranging from Gnapperts to Napple, Nipper and even Kipper.

KNAPPERTY, Sheep's *Potentilla erecta* (Tormentil) N Ire. Grigson.

KNAPWEED *Centaurea nigra*.

KNAPWEED, Black *Centaurea nigra* (Knapweed) Fernie.

KNAPWEED, Brown; KNAPWEED, Brown-headed; KNAPWEED, Brown-rayed *Centaurea jacea* (French Hardhead) USA Zenkert (Brown); Salisbury (Brown-headed); Clapham (Brown-rayed).

KNAPWEED, Greater *Centaurea scabiosa*. Lesser Knapweed is *Centaurea nigra*.

KNAPWEED, Hardhead *Centaurea nigra* (Knapweed) Phillips.

KNAPWEED, Lesser *Centaurea nigra* (Knapweed) Darling & Boyd. Greater Knapweed is *Centaurea scabiosa*.

KNAPWEED, Mountain *Centaurea montana*.

KNAPWEED, Panicled *Centaurea paniculata*.

KNAPWEED, Russian *Centaurea picris*.

KNAPWEED, Yellow *Centaurea salonitana*.

KNAPWEED BROOMRAPE *Orobanche elatior* (Tall Broomrape).

KNAPWORT HARSHWEED *Centaurea jacea* (French Hardhead) Grieve.1931.

KNAUPERTS *Empetrum nigrum* (Crowberry) Banff. Grigson.

KNAWEL *Scleranthus annuus*. Knawel apparently is German Knäuel, a ball of thread, or tangle.

KNEE, Golden *Chrysogonum virginianum*.

KNEE, Pigeon's *Cardiospermum halicacabum* (Blister Creeper) Chopra.

KNEE HOLLY; KNEE HULVER; KNEE HULYER; KNEEHOLY *Ruscus aculeatus* (Butcher's Broom) Halliwell (Holly); Gerard (Hulver, Hulyer); Culpepper (Kneeholy).

KNEED HOLLY *Ruscus aculeatus* (Butcher's Broom) Ess. Gepp.

KNEES, Cypress *Taxodium distichum* (Swamp Cypress) Polunin.1976. Applied only to the aerial parts of the roots, and very descriptive it is.

KNEES, Many *Polygonatum multiflorum* (Solomon's Seal) Fernie. Polygonatum, "of many Knees, for so the Greeke word doth import" (Gerard).

KNEES, Red 1. *Polygonum hydropiper* (Water Pepper) Prior. 2. *Polygonum persicaria* (Persicaria) Ches. Holland; Lancs. (Grigson). Cf. Red Joints for Persicaria, which makes Red Knees more understandable.

KNICKER TREE *Gymnocladus dioica* (Kentucky Coffee Tree) Hora.

KNIFE ACACIA; KNIFE-LEAF WATTLE *Acacia cultriformis*.

KNIFE-AND-FORK *Geranium robertianum* (Herb Robert) Som. Grigson; Bucks. Grigson.

KNIGHT CROSS *Lychnis chalcedonica* B & H. Cf. Maltese Cross, Jerusalem Cross, etc for this.

KNIGHT'S MILFOIL *Achillea millefolium* (Yarrow) Grieve.1931. A crusaders' name, linking the plant to a number of "woundwort" ascriptions. Yarrow was often used as a wound herb, being quite styptic in action. Staunch-grass and Sanguinary are names that illustrate this.

KNIGHT'S PONDWEED *Stratiotes aloides* (Water Soldier) Grieve.1931.

KNIGHT'S SPUR *Delphinium ajacis* (Larkspur) Lyte.

KNIGHT'S STAR LILY *Amaryllis belladonna* (Belladonna Lily) McDonald.

KNIGHTS-AND-LADIES *Arum maculatum* (Cuckoo-pint) Som. Macmillan. One of a number of male plus female names, on the lines of Lords-and-ladies.

KNIPPER NUT *Lathyrus montanus* (Bittervetch) B & H. Cf. Kippernut, or Nipper-nut, coming from words like Knapperty, or Knapperts.

KNITBACK; KNITBONE *Symphytum officinale* (Comfrey) Gerard (Knitback); Brownlow (Knitbone). For Comfrey was used as a consound. Symphytum comes from

Greek sumphuo, to grow together. Whether this is really the plant which the Greeks named and described as useful for knitting bones is doubtful, but it was certainly taken to be such in medieval times, and has been used for the purpose ever since. It was the glutinous matter of the root that used to be grated for a plaster which set hard over a fracture.

KNIVES-AND-FORKS 1. *Acer pseudo-platanus* fruit (Sycamore keys) Kent. Grigson. 2. *Geranium robertianum* (Herb Robert) Som. Elworthy.

KNIVES-AND-FORKS, Lady's *Lycopodium inundatum* (Clubmoss) Som. Elworthy.

KNOB 1. *Caltha palustris* (Marsh Marigold) Berks. Wright. 2. *Lavandula vera* (Dutch Lavender) Ches. B & H. The name is a reasonably good description for Marsh Marigold, or for that matter, any buttercup. It is the flowerheads of lavender that are given the name Knobs, and that is obviously the same as Neps, used in the same area. And that sounds very much as if it were catmint (Nepeta), whose flower heads and those of lavender are very similar. Cf. Golden Knob for Marsh Marigold.

KNOB, Clover *Centaurea nigra* (Knapweed) Notts. Grigson. 'Knob' and 'Knap' are the same word. But why 'clover'? The possibility that it is the same as the first element of Clobweed, a West Country name for knapweed, should be considered.

KNOB, Gold 1. *Ranunculus acris* (Meadow Buttercup) Gerard. 2. *Ranunculus repens* (Creeping Buttercup) Som. Macmillan.

KNOB, Golden 1. *Caltha palustris* (Marsh Marigold) Som, Berks. Grigson. 2. *Ranunculus calandrinoides* B & H. A name like this for a white-flowered plant like *R calandrinoides* is very odd.

KNOB, Goldy 1. *Ranunculus acris* (Meadow Buttercup) Oxf. B & H. 2. *Ranunculus ficaria* (Lesser Celandine) Oxf. Grigson.

KNOB, Horse *Centaurea nigra* (Knapweed) Dor. Macmillan; N Eng. B & H. Cf. Horse-knot, Horse-knop, Horse-nop, etc.

KNOB, Horse, Great *Centaurea scabiosa* (Greater Knapweed) B & H.

KNOB, Iron *Centaurea nigra* (Knapweed) Ches. Holland. In the same vein as the common form, Hardheads. Cf. Ironweed, Ironhard, etc.

KNOB, King's 1. *Ranunculus acris* (Meadow Buttercup) Glos. B & H. 2. *Ranunculus bulbosus* (Bulbous Buttercup) Som. Macmillan. Is this a fanciful reference to the royal orb?

KNOB, Red 1. *Poterium sanguisorba* (Salad Burnet) Notts. Grigson. 2. *Sanguisorba officinalis* (Great Burnet) Notts. Grigson.

KNOB-ROOT *Collinsonia canadensis* (Stone Root) Le Strange. Descriptive. The roots are still used in tincture for the treatment of piles and bladder complaints.

KNOB-CONE PINE *Pinus tuberculata*. An odd-looking name, but strictly descriptive. The tree's bark actually grows over the cones, which cannot be released until the tree has died, fallen and decayed.

KNOB-THORN *Acacia nigrescens*.

KNOBBED CROWFOOT *Ranunculus bulbosus* (Bulbous Buttercup) Parkinson.

KNOBWEED 1. *Centaurea cyanus* (Cornflower) N'thants. A E Baker. 2. *Centaurea nigra* (Knapweed) N'thants. A E Baker. 3. *Centaurea scabiosa* (Greater Knapweed) A E Baker. 4. *Collinsonia canadensis* (Stone Root) Le Strange.

KNOCK-A-NIDLES, Little *Viola tricolor* (Pansy) Som. Macmillan. Quite a long way from its original, Love-in-idleness.

KNOCKHEADS *Plantago lanceolata* (Ribwort Plantain) B & H. A reference to the game children used to play, rather like Conkers, which involved holding a stalk of ribwort, and the opponent trying to decapitate it with his stalk.

KNOLL *Brassica campestris var. rapa* (Turnip) Kent Halliwell. A knoll is a small rounded hill, so the analogy is obvious.

KNOLLES; KNOLLS 1. *Brassica campestris var. rapa* (Turnip) Kent Halliwell. 2. *Brassica napus* (Rape) Kent

Grose. A knoll is a small rounded hill, and the analogy seems reasonable, if the derivation is correct.

KNOOP *Rubus chamaemorus* (Cloudberry) N Eng. Brockett. Or Noops, which a little farther north becomes Nub, i.e. knub or knob, in the sense of a hill.

KNOP, Gold 1. *Ranunculus acris* (Meadow Buttercup) Glos. B & H. 2. *Ranunculus bulbosus* (Bulbous Buttercup) Glos. J D. Robertson. 3. *Ranunculus repens* (Creeping Buttercup) Glos. J D. Robertson.

KNOP, Horse 1. *Centaurea cyanus* (Cornflower) N Eng. Brockett. 2. *Centaurea nigra* (Knapweed) Dor. Macmillan; Yks. Nicholson. 3. *Centaurea scabiosa* (Greater Knapweed) Yks, Cumb. B & H.

KNOPWEED 1. *Centaurea cyanus* (Cornflower) Notts. Grigson. 2. *Centaurea nigra* (Knapweed) Cockayne. 3. *Centaurea scabiosa* (Greater Knapweed) Tynan & Maitland. Cf. Knobweed, etc.

KNOT, Horse 1. *Centaurea nigra* (Knapweed) Prior. 2. *Centaurea scabiosa* (Greater Knapweed) Tynan & Maitland.

KNOT, Top *Centaurea nigra* (Knapweed) Berw. Grigson. 'Knot' should be 'knop' in this case.

KNOT, Liquory *Lathryrus montanus* (Bittervetch) Berw. Dyer. It is said that the tubers, when dried, taste not unlike liquorice. But the general opinion is that they are more like chestnuts.

KNOT, Ninety *Polygonum aviculare* (Knotgrass) Shrop. Grigson. Ten more joints, according to names like Centinode. But Coles had Nine-joints.

KNOT, Sailor's *Geranium robertianum* (Herb Robert) Bucks. Grigson.

KNOT, Shepherd's *Potentilla erecta* (Tormentil) N'thum, Scot. Grigson.

KNOT, Tristram's *Cannabis sativa* (Hemp) B & H. This is from Bullein's *Book of Simples*. It sounds as if it had something to do with hemp's use for a gallows rope. Cf. Gallow-grass and Neckweed.

KNOT, True Lover's *Paris quadrifolia* (Herb Paris) Midl. Tynan & Maitland. Cf. the various "truelove" names for this.

KNOT OAT-GRASS *Arrhenatherium elatius* (False Oat) H C Long.1910. The 'knots' are little tubers, produced at the nodes, sometimes called 'pearls', too - hence Pearl-grass.

KNOTBERRY *Rubus chamaemorus* (Cloudberry) N Eng. Gerard. Or Knoutberry (sometimes Cnoutberry). Knout is ME knot, meaning a hill, and is a comment on habitat.

KNOTBONE *Symphytum officinale* (Comfrey) Vesey-Fitzgerald. A variation on Knitbone.

KNOTGRASS 1. *Centaurea nigra* (Knapweed) Hants. Grigson. 2. *Polygonum aviculare*. There have been some wild speculations as to why *P aviculare* should be so called. Dyer suggested it was from some unrecorded character by the doctrine of signatures, that it stops growth in children (presumably if they eat it). Cf. Beaumont & Fletcher, Burning Pestle: "...and say they would put him into a strait pair of gaskins, 'twere worse than knotgrass: he would never grow after it". Shakespeare used it too (MND): "Get gone, you dwarf, You minimus, of hindering knot-grass made, You bead, you acorn...". Whatever the superstition was, surely the name is given from the "knots", or nodes, of the intricately jointed stem.

KNOTGRASS, Bird('s) *Polygonum aviculare* (Knotgrass) Som. Macmillan (Bird); B & H (Bird's). The specific name is *aviculare*. Birds seem to have a particular liking for the stems.

KNOTGRASS, Dyer's *Polygonum tinctorium*. An indigo-bearing species.

KNOTGRASS, Erect *Polygonum erectum* (Weedy Knotweed) Grieve.1931.

KNOTGRASS, Hastate *Polygonum arifolium* (Halberd-leaved Tear-thumb) Grieve.1931.

KNOTGRASS, Ray's *Polygonum raii* (Slender Sea Knotgrass) Clapham.

KNOTGRASS, Russian *Polygonum erectum* (Weedy Knotweed) Grieve.1931.

KNOTGRASS, Sea *Polygonum maritimum*.

KNOTGRASS, Sea, Slender *Polygonum raii*.

KNOTS, Gold *Ranunculus acris* (Meadow Buttercup) North. 'Knots' is 'knops' here, and is a description of the tight golden buds.

KNOTS, Lovers' *Galium aparine* (Goosegrass) Wilts. Macmillan. A reference to the clinging habit of both the plant as a whole, and particularly the burrs. 'Knot' may be 'knop' (the round burrs), or the name may just be a cliché.

KNOTTED CLOVER *Trifolium striatum*.

KNOTTED CLOVER, Soft *Trifolium striatum* (Knotted Clover) Hutchinson.

KNOTTED CRANESBILL *Geranium nodosum*.

KNOTTED FIGWORT *Scrophularia nodosa* (Figwort) Grieve.1931.

KNOTTED MARJORAM *Origanum majorana*. 'Knotted', because the flowers are gathered into roundish close heads like knots.

KNOTTED PEARLWORT *Sagina nodosa*.

KNOTTED SPURRY *Sagina nodosa* (Knotted Pearlwort) Murdoch McNeill.

KNOTTY MEAL *Conopodium majus* (Earthnut) Inv. Grigson. Nutty?

KNOTWEED 1. *Centaurea cyanus* (Cornflower) N'thants. Denham/Hardy. 2. *Centaurea nigra* (Knapweed) N'thants. A E Baker. 3. *Centaurea scabiosa* (Greater Knapweed) N'thants. Denham/Hardy. 4. *Polygonum aviculare* (Knotgrass) Palaiseul. Only with the Polygonum is knot really knot (the joints, or nodes); the others are knops.

KNOTWEED, Alpine 1. *Polygonum alpinum*. 2. *Polygonum viviparum* (Alpine Bistort) Grieve.1931.

KNOTWEED, Climbing *Polygonum scandens*.

KNOTWEED, Dooryard *Polygonum aviculare* (Knotgrass) S Africa Watt.

KNOTWEED, Giant *Polygonum sachalinense*. Giant is right - this can grow ten feet high.

KNOTWEED, Himalayan *Polygonum campanulatum*.

KNOTWEED, Japanese *Polygonum cuspidatum*.

KNOTWEED, Spotted *Polygonum persicaria* (Persicaria) Murdoch McNeill.

KNOTWEED, Weedy *Polygonum erectum*.

KNOTWEED, Whorled *Illecebrum verticillatum* (Coral Necklace) Hutchinson.

KNOTWORT *Polygonum aviculare* (Knotgrass) B & H.

KNOUTBERRY *Rubus chamaemorus* (Cloudberry) N Eng. Gerard. See Knotberry.

KNUCKLE-BLEEDERS *Aesculus hippocastanum* petioles (Horse Chestnut) B & H. For a boys' game, not conkers, apparently, so presumably something like the 'soldiers' game played with stalks of Ribwort Plantain. The name seems graphically descriptive.

KNUTSFORD DEVIL *Calystegia sepium* (Great Bindweed) Ches. Holland. Are they any worse in Cheshire than elsewhere in the country?

KOBUS MAGNOLIA *Magnolia kobus*.

KOCHANG LILY *Lilium distichum*.

KOHLRABI *Brassica oleracea caulo-rapa*. Straight from German, but this is Latin caulis, cabbage and rapa, turnip. It is sometimes called Cabbage Turnip.

KOHUHU *Pittosporum tenuifolium*.

KOKKELURI, Muckle *Leucanthemum vulgare* (Ox-eye Daisy) Shet. Grigson.

KOLA *Cola nitida*.

KOLA, Bitter *Garcinia kola*.

KOLA, False; KOLA, Male; KOLA, Orogbo *Garcinia kola* (Bitter Kola) Dalziel.

KOLA, Monkey *Cola caricifolia*. Actually, any false kola may be called by this name.

KOLING *Malus sylvestris* fruit (Crabapple) Shrop. B & H. Also spelt as Colins, which makes one suspect that this is the same as Codlins.

KOMBE *Strophanthus kombe*.

KOOCHLA TREE *Strychnos nux vomica* (Snakewood) Duncalf.

KOOYAH *Valeriana edulis* (Edible Valerian) USA Havard. Evidently the Indian name.

KORAKAN *Eleusine corakana* (Finger Millet) Brouk.

KOREAN BOX *Buxus microphylla*.

KOREAN FIR *Abies koreana*.
KOREAN LILY *Lilium amabile*.
KOREAN PINE *Pinus koraiensis*.
KORKELIT; KORKALETT; CORKLIT
Ochrolechia tartarea (Cudbear Lichen)
Shet. Nicolson (Korkelit); B & H
(Korkalett); Mabey.1977 (Corklit). The 'lit'
syllable must mean dye. Indigo was known in
Shetland simply as lit.
KOUS; KOUSHE; KOUSHLE *Conium
maculatum* (Hemlock) Lancs. B & H
(Kous); Lincs, Lancs Grigson (Koushe,
Koushle). They look rather exotic, but they
are simple variants of one of the forms of
Kex. Cf. Kewsies.
KOUSSO *Brayera anthelmintica*.
KOWHAI *Sophora microphylla*.
KOWHAI, Red *Clianthus puniceus*.
KOWHAI, Yellow *Sophora tetraptera*.
KRAMICS *Ononis repens* (Rest Harrow)
Som. Macmillan. A version of Cammock -
the Somerset spellings all seem to begin with
'k'.
KRICKSIES *Conium maculatum* (Hemlock)
Sanecki. One of the more elaborate forms in
the series that starts with Kex, but the 'r' of
Kricksies seems intrusive. One would expect
this word to be Kicksies, from the Kicks variant of Kex.
KRISLINGS *Prunus domestica var. institia*
fruit (Bullace) Som. Elworthy. The
sequence seems to be Crex, Crickseys,
Crystals, Christians, Christlings or Crislings,
Keslins, Keslings and Krislings, though this
does not complete the list.
KUDZU *Pueraria thunbergiana*.
KUMQUOT *Fortunella japonica/margarita*
(Cumquat) Cunningham.
KURWINI MANGO *Mangifera odorata*.
KUSA *Poa cynosuroides*.
KUSAM *Schleichera oleosa* (Ceylon Oak)
Chopra. The Arabic name.
KWIGGA *Agropyron repens* (Couch-grass)
Grigson.1959. OE cwice, cuic, - alive. The
names go through a lot of variations, and are
strikingly similar to those for rowan from the
same source.

KYERLIC *Sinapis arvensis* (Charlock) Oxf.
B & H. A form of charlock.

L

LABRADOR TEA 1. *Ledum groenlandicum.* 2. *Ledum palustre* (Marsh Rosemary) Gilmour & Walters.

LABRADOR TEA, Pacific *Ledum columbianum.* A name that offers a rare geographical mix!

LABRADOR TEA, Western *Ledum glandulosum.*

LABLAB *Dolichos lablab.*

LABURNUM *Laburnum anagyroides.* Prior said Laburnum is derived from Latin labor, work, and made of it a symbol of the hours of work. The leaflets close at night, and open by day. Chambers dictionary is notably silent on the matter of its derivation.

LABURNUM, African 1. *Cassia afrofistula.* 2. *Cassia sieberiana.*

LABURNUM, Alpine *Laburnum alpinum* (Scotch Laburnum) Brimble.1948.

LABURNUM, Himalayan *Sophora mollis.*

LABURNUM, Indian *Cassia fistula.*

LABURNUM, New Zealand *Sophora tetraptera* (Yellow Kowhai) Barber & Phillips. This is the New Zealand national flower.

LABURNUM, Scotch *Laburnum alpinum.* Actually, it is a native of the southern Alps, Apennines and central Europe.

LABURNUM, Wild 1. *Melilotus altissima* (Tall Melilot) Surr. Grigson. 2. *Melilotus officinalis* (Melilot) Surr. B & H.

LAC TREE; GUM-LAC TREE *Schleichera oleosa* (Ceylon Oak) Chopra.

LACE, Gypsy *Galium palustre* (Water Bedstraw) Som. Macmillan.

LACE(S), Gypsy's *Heracleum sphondyllium* (Hogweed) Som. Macmillan. 'Lace', in one form or another, is a common and much-loved descriptive name for most of the umbellifers.

LACE, Honiton 1. *Anthriscus sylvestris* (Cow Parsley) Dev. Grigson. 2. *Conium maculatum* (Hemlock) Dev. Macmillan. 3. *Torilis japonica* (Hedge Parsley) Dev. Macmillan. Honiton used to be a famous lace-making centre in Devonshire.

LACE('S), Lady's 1. *Achillea millefolium* (Yarrow) Glos. PLNN 18. 2. *Aethusa cynapium* (Fool's Parsley) Som. Macmillan. 3. *Anthriscus sylvestris* (Cow Parsley) Som. Grigson; Oxf. Oxf Ann Rec.1951. 4. *Conium maculatum* (Hemlock) Som. Macmillan. 5. *Cuscuta epithymum* (Common Dodder) Som. Friend. 6. *Torilis japonica* (Hedge Parsley) Som. Macmillan.

LACE, My Lady's *Anthriscus sylvestris* (Cow Parsley) Dor. Macmillan.

LACE, Old Lady's *Anthriscus sylvestris* (Cow Parsley) Som. Macmillan; Wilts. D Grose.

LACE, Queen Anne's 1. *Anthriscus sylvestris* (Cow Parsley) Hemphill. 2. *Daucus carota* (Wild Carrot) USA Gates. 3. *Daucus pusillus* USA Elmore. 4. *Didiscus caerulea* (Blue Lace Flower) Rochford & Gorer.

LACE, St Audre's *Cannabis sativa* (Hemp) B & H. This is from Bullein's *Book of Simples.*

LACE-BARK PINE *Pinus bungeana.*

LACE-BARK TREE 1. *Hoheria populnea* (Ribbonwood) RHS. 2. *Lagetta lintearia* (Lace Tree) Usher.

LACE CURTAINS *Aethusa cynapium* (Fool's Parsley) Som. Grigson.

LACE-FLOWER 1. *Conium maculatum* (Hemlock) Som. Macmillan. 2. *Torilis japonica* (Hedge Parsley) Som. Macmillan.

LACE-FLOWER, Blue *Didiscus caerulea.*

LACE HANDKERCHIEF, Queen Anne's *Torilis japonica* (Hedge Parsley) Dor. Dacombe.

LACE TREE *Lagetta lintearia.* The bast fibres are removed from the stem by some process like maceration, to form a network for making dresses, etc.

LACEBARK *Gaya lyalli* (Mountain Ribbonwood) Everett. The reference is to the inner bark, which has typically a lacy appearance.

LACES, Gypsy *Anthriscus sylvestris* (Cow Parsley) Dor, Som. Mabey.
LACES, Lady's 1. *Cuscuta epithymum* (Dodder) Som. Friend. 2. *Phalaris arundinacea* (Reed-grass) Dev. Friend.1882.
LACEWOOD 1. *Platanus x acerifolia* timber (London Plane) Hart. 2. *Platanus occidentalis* timber (Western Plane) Everett.
LACEY, White 1. *Arabis alpina* (Alpine Rock-cress) Som. Macmillan 1. *Sedum album* (White Stonecrop) Som. Macmillan.
LACQUER TREE *Rhus verniciflua*. Or Varnish Tree. This is a Chinese tree, and they have been tapped for the resin for a very long time.
LACQUER TREE, Chinese; LACQUER TREE, Japanese *Rhus verniciflua* (Lacquer Tree) H & P (Chinese); Howes (Japanese).
LACY PHACELIA *Phacelia tanacetifolia*.
LAD-LOVE-LASS *Artemisia abrotanum* (Southernwood) Lincs. Peacock. See Lad's Love.

Lad's love is lasses' delight
And if the lads don't love
Lasses will flite.

'Flite' is an old North country word meaning to scold.
LAD SAVOUR *Artemisia abrotanum* (Southernwood) Lancs. B & H.
LAD'S LOVE *Artemisia abrotanum* (Southernwood) Dev, Dor, Som. B & H; E Ang. Forby. A few branches of southernwood were generally added to the nosegays that courting youths used to give the girls. Or they would put some in their buttonholes, and sniff ostentatiously at it when walking towards a group of girls. If the girls kept on walking he knew he had no chance, but if they stopped and walked slowly back, he knew they had noticed his Lad's Love.
LADDER, Christ's *Centaureum erythraea* (Centaury) Vesey-Fitzgerald. A name given in the 14th century (Christi scala) in mistake for Christ's Cup (Christi schale), an allusion to the bitter draught given to Christ on the cross. This version also appears as Cristaldre.
LADDER, Jack's *Phaseolus coccineus* (Scarlet Runner) Som. Macmillan.
LADDER, Jacob's 1. *Aconitum napellus* (Monkshood) Tynan & Maitland. 2. *Althaea rosea* (Hollyhock) Som. Macmillan. 3. *Antirrhinum majus* (Snapdragon) Som. Macmillan. 4. *Atropa belladonna* (Deadly Nightshade) Ayr. Grigson. 5. *Chelidonium majus* (Greater Celandine) Shrop. Grigson. 6. *Convallaria maialis* (Lily-of-the-valley) Grieve.1931. 7. *Delphinium ajacis* (Larkspur) Dev. B & H. 8. *Impatiens noli-me-tangere* (Touch-me-not) Som. Macmillan. 9. *Polemonium caeruleum*. 10. *Polygonatum multiflorum* (Solomon's Seal) Som, Berks. Grigson; Wilts. Dartnell & Goddard. 11. *Sedum telephium* (Orpine) Kent Grigson. 12. *Smilax herbacea* (Carrion-flower) USA Gilmore. The name is given for various reasons - for the real Jacob's Ladder, it is because the leaves are arranged in successive pairs. In other cases, it is the arrangement of the flowers that suggests a ladder, as with Solomon's Seal, or Lily-of-the-valley. Or perhaps, as with hollyhock, it is just the fact that it is tall.
LADDER, Jacob's, American *Polemonium van-bruntiae*.
LADDER, Jacob's, False *Polemonium reptans*.
LADDER LOVE *Centaurea cyanus* (Cornflower) Som. Macmillan. Lad's Love, perhaps? And wrongly applied?
LADDER OF CHRIST *Hypericum perforatum* (St John's Wort) Wales Trevelyan.
LADDER TO HEAVEN 1. *Convallaria maialis* (Lily-of-the-valley) Barton & Castle. 2. *Polemonium caeruleum* (Jacob's Ladder) Lanark. Grigson. 3. *Polygonatum multiflorum* (Solomon's Seal) B & H.
LADIES, Gentlemen-and- *Arum maculatum* (Cuckoo-pint) Som. Macmillan; Oxf. Grigson. Cf. Lords-and-ladies.
LADIES, Naked 1. *Cardamine pratensis* (Lady's Smock) Som. Macmillan. 2.

Colchicum autumnale (Meadow Saffron) Prior. For Meadow Saffron, the name arises from the way the flowers rise naked from the earth, before the leaves appear.

LADIES-AND-GENTLEMEN 1. *Arum maculatum* (Cuckoo-pint) Wilts. Dartnell & Goddard. Shrop. Burne; Northants. A E Baker. 2. *Viola tricolor* (Pansy) Som. Macmillan. For Cuckoo-pint, this is one of a whole series of male plus female names, with consequent erotic overtones, all arising from the way the spadix (male) grows inside the spathe (female). Pansy, of course, has equally erotic symbolism, for less obvious reasons.

LADIES-AND-GENTLEMEN, Devil's *Arum maculatum* (Cuckoo-pint) Friend. It sounds like a puritanical comment on the name.

LADIES-IN-A-SHIP *Aconitum napellus* (Monkshood) Som. Macmillan. The 'ship' referred to was usually Noah's Ark. Cf. Doves-in-the-ark, or Dobbin-in-the-ark.

LADIES-IN-THE-SHADE *Nigella damascena* (Love-in-a-mist) Som. Macmillan.

LADIES-IN-WHITE *Saxifraga x urbinum* (London Pride) Som. Macmillan.

LADIES' FINGERS, Gentlemen-and- *Arum maculatum* (Cuckoo-pint) Wilts. Dartnell & Goddard.

LADIES' FLOWER *Viola tricolor* (Pansy) Skinner.

LADIES' LORDS *Arum maculatum* (Cuckoo-pint) Kent Grigson. Better known as Lords-and-ladies.

LADIES' THISTLE *Carduus benedictus* (Holy Thistle) Halliwell. A plural form that makes nonsense of the attribution to the Virgin.

LADON-SHRUB *Cistus ladaniferus* (Gum Cistus) Turner. Ladanum, or Labdanum, is an aromatic gum exuded by the shrub. The similarity of the name with laudanum (Turner calls it Laudan as well as Ladon), has given rise to confusion.

LADINO CLOVER *Trifolium repens* (White Clover) S Africa Watt.

LADSAVVUR *Artemisia abrotanum* (Southernwood) Lancs. Nodal. Lad Savour, in other words, and connected with Lad's Love.

LADY, Biggoty *Impatiens noli-me-tangere* (Touch-me-not) Som. Macmillan. Biggoty means saucy, apparently connected in some way with the explosive seed distribution.

LADY, Butterfly *Papaver rhoeas* (Corn Poppy) Dor. Macmillan.

LADY, Crimson *Dianthus caryophyllus* (Carnation) Som. Macmillan.

LADY, Dancing *Fuchsia sp* Som. Macmillan.

LADY, Eleven o'clock; LADY ELEVEN O'CLOCK *Ornithogalum umbellatum* (Star-of-Bethlehem) Coats; Dev. Friend. A lot of the names for Star-of-Bethlehem reflect its habit of opening early, and shutting equally early, whether it is at ten, eleven o'clock, noon, or one, for all these times figure in the local names.

LADY, Fair *Atropa belladonna* (Deadly Nightshade) Friend. i.e. belladonna, and all to do with an ancient belief that this shrub is the form of a witch called Atropai. At certain times, the plant takes the form of a very beautiful enchantress, dangerous to look upon. But more immediately, it used to be the custom on the Continent for women to use it as a cosmetic to make the eyes sparkle.

LADY, Lonesome; LADY, Naked *Cardamine pratensis* (Lady's Smock) Dev. Macmillan (Lonesome); Som. Grigson (Naked).

LADY, Painted 1. *Castilleia sp* (Paintbrush) Willis. 2. *Dianthus barbatus* (Sweet William Dyer. 3. *Geranium versicolor*. 4. *Lathyrus odoratus* (Sweet Pea) Wilts. Dartnell & Goddard; IOM Moore.1924. 5. *Saxifraga x urbinum* (London Pride) Som. Macmillan.

LADY, Pretty *Saxifraga x urbinum* (London Pride) Wilts. Macmillan.

LADY, Quaker *Spiraea latifolia* USA House.

LADY, Ragged *Nigella damascena* (Love-in-a-mist) Wit.

LADY, Sullen *Fritillaria meleagris* (Snake's-head Lily) Coats. 'Sullen' is probably 'solemn', from the rather dingy colouring of the flowers.

LADY, White 1. *Galanthus nivalis* (Snowdrop) B & H. 2. *Kalanchoe thyrsiflora* S Africa Watt. The ascription of this name to snowdrop is perhaps doubtful.

LADY BETTY *Centranthus ruber* (Spur Valerian) Dor. Dacombe. Cf. Bouncing Bess, Sweet Betsy, Saucy Bet, etc.

LADY BIRCH *Betula verrucosa* (Birch) B & H. Coleridge called the birch the Lady of the Woods, and there is a Scottish name, Damsel of the Wood.

LADY CAKES *Oxalis acetosella* (Wood Sorrel) Dumf. Grigson.

LADY CUES *Primula veris* (Cowslip) Kent T Lloyd. Evidently a variation on Lady's Keys.

LADY CUSHION *Armeria maritima* (Thrift) Ches. Leigh.

LADY DOVE AND HER COACH AND PAIR *Aconitum napellus* (Monkshood) Dor. Dacombe. It sounds garbled - it is the doves that draw the coach. Hence Lady Lavinia's Dove Carriage, also from Dorset, and Venus's chariot drawn by doves, with a wider spread. All these references to doves are from a fancied resemblance of the nectaries to the birds.

LADY FERN *Athyrium filix-femina*. It is described as more graceful and delicate than the Male Fern, which is Dryopteris filix-mas.

LADY GARTEN BERRIES *Rubus fruticosus* fruit (Blackberry) Scot. Jamieson. Probably Gatter-berries originally. That proliferated into Gaitberries, Gaiterberries, Garten-berries, Lady Garten-berries, even Lady's Garter-berries.

LADY-IN-THE-BOAT; OUR LADY-IN-A-BOAT *Dicentra spectabilis* (Bleeding Heart) Som. Macmillan (Lady); Coats (Our Lady). There is also **LADY-IN-THE-BATH** for this Perry.

LADY-IN-THE-BOWER *Nigella damascena* (Love-in-a-mist) W Miller.

LADY JANE *Geranium robertianum* (Herb Robert) Dev. Macmillan.

LADY LAUREL *Daphne mezereon*.

LADY LAVINIA'S DOVE CARRIAGE *Aconitum napellus* (Monkshood) Dor. Macmillan. Supposedly from the "dove-like" nectaries.

LADY LILY *Lilium candidum* (Madonna Lily) Bucks. Friend. As in a good number of other cases, the 'Lady' of the name is the Virgin Mary. This is her emblem.

LADY MARY'S TEARS *Pulmonaria officinalis* (Lungwort) Dor. Macmillan. The white spots on the leaves are either taken to be the Virgin's tears, or drops of her milk.

LADY-MY-LORD *Arum maculatum* (Cuckoo-pint) Ess. Gepp. See Lords-and-ladies.

LADY NEVER-FADE *Anaphalis margaretacea* (Pearly Immortelle) Glos. J D Robertson.

LADY NUT *Castanea sativa* (Chestnut) Som. Macmillan.

LADY OF SPRING *Taraxacum officinale* (Dandelion) Dor. Macmillan.

LADY OF THE LAKE *Nymphaea alba* (White Waterlily) Som. Macmillan.

LADY OF THE MEADOW *Filipendula ulmaria* (Meadowsweet) S Scot. B & H.

LADY OF THE NIGHT 1. *Brunsfelsia tastevini*. 2. *Mirabilis dichotoma* (Marvel of Peru) Friend. The flowers of Marvel of Peru open very late in the afternoon, hence the name. But with Brunsfelsia it is the scent that matters - it is most noticeable at night.

LADY OF THE WOODS 1. *Betula verrucosa* (Birch) Coleridge. 2. *Fagus sylvatica* (Beech) Hart. Not necessarily just an acknowledgement of the gracefulness of a birch tree, for, at least in Somerset, birch is traditionally the tree of death, "the one with the white hand", a spirit that haunted the moorland near Taunton.

LADY ORCHID *Orchis purpurea*.

LADY POPLAR 1. *Populus alba* (White Poplar) Ches. Grigson. 2. *Populus nigra var. italica* (Lombardy Poplar) Ches. Holland.

LADY SMOCK *Cardamine pratensis* (Lady's Smock) Ches. Gerard.

LADY TRACES *Spiranthes autumnalis* (Lady's Tresses) Turner.

LADY TULIP *Tulipa clusiana*.

LADY WHIN *Ononis repens* (Rest-harrow) Som. Macmillan; Scot. Grigson.

LADY'S BALL *Centaurea nigra* (Knapweed) Som. Grigson; Wilts. Dartnell & Goddard. OE boll, a circular body.

LADY'S BED *Galium verum* (Lady's Bedstraw) Dev. Macmillan.

LADY'S BEDSTRAW 1. *Galium cruciatum* (Cross-wort) Yks. Grigson. 2. *Galium verum*. The medieval legend was that the Virgin lay on a bed of bracken and *Galium verum* during the Nativity. Bracken refused to acknowledge the child, and lost its flowers. *G verum* welcomed the child, and, blossoming at that moment, found its flowers had changed from white to gold. Another version says that it was the only plant in the stable that the donkey did not eat. Sudetenland women used to put it in their beds to make child-birth easier and safer, and since women who have just had a child are susceptible to attack from demons, they will not go out unless they have some Lady's Bedstraw with them in their shoes. Its astringency probably also brought it in the bed against fleas.

LADY'S BELT, My *Filipendula ulmaria* (Meadowsweet) Scot. B & H.

LADY'S BONNET *Aquilegia vulgaris* (Columbine) Som. Macmillan. Cf. Granny Bonnets, etc.

LADY'S BOOTS 1. *Cypripedium calceolus* (Lady's Slipper) Dev. Friend.1882. 2. *Lotus corniculatus* (Bird's-foot Trefoil) Dev. Friend. There are any number of 'boots', 'shoes', 'stockings', and 'slippers' names for these.

LADY'S BOUQUET *Galium triflorum* (Fragrant Bedstraw) Gilmore.

LADY'S BOWER *Clematis vitalba* (Old Man's Beard) Som. Macmillan. Cf. Virgin's Bower, which, so at least one writer has claimed, Gerard coined as a tribute to Queen Elizabeth. But, at any rate in German legend, Mary and the Child sheltered under the Virgin's Bower on the flight into Egypt.

LADY'S BRUSH-AND-COMB; LADY'S BRUSHES *Dipsacus fullonum* (Teasel) Som. Macmillan (Brush-and-comb); Putnam (Brush).

LADY'S BUNCH-OF-KEYS *Primula veris* (Cowslip) Som. Macmillan. Descriptive, but they are usually ascribed to St Peter, as the keeper of the keys of heaven.

LADY'S BUTTONS *Stellaria holostea* (Greater Stitchwort) Som. Macmillan.

LADY'S CANDLE *Verbascum thapsus* (Mullein) Som. Macmillan. Cf. the Irish Virgin Mary's Candle, but there are a lot of other 'taper' or 'torch' names. The stems and leaves were dipped in tallow or suet and burnt to give light at outdoor country gatherings, or even in the home.

LADY'S CANDLESTICKS *Primula vulgaris var. elatior* (Polyanthus) N. Eng. B & H.

LADY'S CHAIN *Laburnum anagryoides* (Laburnum) Dev, Dor. Macmillan.

LADY'S CHEESE *Malva sylvestris* (Common Mallow) Dor. Grigson. It is usually the seed vessels that are given the 'cheese' names.

LADY'S CHEMISE 1. *Anemone nemerosa* (Wood Anemone) Som. Macmillan. 2. *Calystegia sepium* (Great Bindweed) Som. Macmillan. 3. *Stellaria holostea* (Greater Stitchwort) Som. Macmillan.

LADY'S CLOAK *Cardamine pratensis* (Lady's Smock) Som. Macmillan. Cf. Lady's Mantle.

LADY'S CLOVER *Oxalis acetosella* (Wood Sorrel) Perth. Grigson.

LADY'S COMB *Scandix pecten-veneris* (Shepherd's Needle) Gerard.

LADY'S COWSLIP *Gagea lutea* (Yellow Star-of-Bethlehem) B & H.

LADY'S CUSHION 1. *Anthyllis vulneraria* (Kidney Vetch) Wilts. Dartnell & Goddard. 2. *Arabis caucasica* (Garden Arabis) W Miller. 3. *Armeria maritima* (Thrift) Gerard.

4. *Centaurea nigra* (Knapweed) Kent Grigson. 5. *Chrysosplenium oppositifolium* (Opposite-leaved Golden Saxifrage) Freethy. 6. *Knautia arvensis* (Field Scabious) Kent B & H. 7. *Lotus corniculatus* (Bird's-foot Trefoil) Dev, Wilts. Macmillan. 8. *Saxifraga hypnoides* (Dovedale Moss) Yks, Cumb. Grigson.

LADY'S EAR-DROPS 1. *Dicentra spectabilis* (Bleeding Heart) Som. Macmillan. 2. *Fuchsia sp* (Fuchsia) Dev. Friend.1882.

LADY'S FEATHERBEDS *Saxifraga granulata* (Meadow Saxifrage) Suss. Grigson.

LADY'S FINGERS 1. *Abelmoschus esculentus* fruit (Okra) Argenti & Rose. 2. *Anthyllis vulneraria* (Kidney Vetch) Dor, Midl. Dyer. 3. *Arum maculatum* (Cuckoopint) Wilts. Dartnell & Goddard; Glos. B & H; Kent Grigson. 4. *Digitalis purpurea* (Foxglove) Dev, Worcs, Shrop,. N'thum, Ire. Grigson; Som. Elworthy. 5. *Hippocrepis comosa* (Horseshoe Vetch) Wilts. Dartnell. & Goddard. 6. *Laburnum anagryoides* (Laburnum) Yks. B & H. 7. *Lathyrus pratensis* (Yellow Pea) Wilts. Dartnell & Goddard; Yks. Grigson. 8. *Lonicera periclymenum* (Honeysuckle) Yks, Dur, N'thum, Roxb Grigson. 9. *Lotus corniculatus* (Bird's-foot Trefoil) Wilts. Dartnell & Goddard; Hants. Cope; Surr, Herts, Hunts, Lancs, Lincs, Yks. Grigson; N'thants. A E Baker. 10. *Orchis mascula* (Early Purple Orchis) Berks. B & H; Som. Elworthy. 11. *Potentilla reptans* (Cinquefoil) Som. Macmillan. 12. *Primula veris* (Cowslip) Scot. Grigson. 13. *Primula veris x vulgaris* (False Oxlip) Norf, Suff, Yks. Grigson. 14. *Ranunculus arvensis* (Corn Buttercup) Yks. Nicholson. 15. *Senecio vulgaris* (Groundsel) Wilts. Macmillan.

LADY'S FINGERS-AND-THUMBS *Lotus corniculatus* (Bird's-foot Trefoil) Som. Macmillan; Wilts. Dartnell & Goddard.

LADY'S FINGERS-AND-THUMBS, Double 1. *Anthyllis vulneraria* (Kidney Vetch) Wilts. Dartnell & Goddard. 2. *Lotus corniculatus* (Bird's-foot Trefoil) Wilts. Dartnell & Goddard.

LADY'S FLANNEL *Verbascum thapsus* (Mullein) Som. Grigson. The texture of the leaves - there are any number of 'flannel' or 'blanket' names.

LADY'S FLOCK *Cardamine pratensis* (Lady's Smock) Notts. Grigson. 'Flock' means a tuft of wool, and is very descriptive. Lady's Smock, when seen at a distance, can have this appearance.

LADY'S FOXGLOVE *Verbascum thapsus* (Mullein) B & H (also USA Henkel.

LADY'S GARTER-BERRIES *Rubus fruticosa* fruit (Blackberry) Scot. Jamieson.

LADY'S GARTERS 1. *Phalaris arundinacea* (Reed-grass) Som. Elworthy. 2. *Rubus fruticosus* fruit (Blackberry) Roxb. Grigson. As far as blackberry is concerned, this name started as Gatter-berry.

LADY'S GLOVE 1. *Campanula medium* (Canterbury Bell) Som. Macmillan; Herts. Jones-Baker.1974. 2. *Cardamine pratensis* (Lady's Smock) N'thants. Grigson. 3. *Digitalis purpurea* (Foxglove) Shrop, Som. Burne; USA. Henkel. 4. *Fumaria officinalis* (Fumitory) Wilts. Tynan & Maitland. 5. *Inula conyza* (Ploughman's Spikenard) W Miller. 6. *Lotus corniculatus* (Bird's-foot Trefoil) Dor, Wilts. Macmillan; N'thants. A E Baker. 7. *Lotus uliginosus* (Narrow-leaved Bird's-foot Trefoil). Wilts. D Grose.

LADY'S GOLDEN BEDSTRAW *Galium verum* (Lady's Bedstraw) Yks. Grigson.

LADY'S GRASS *Phalaris arundinacea* (Reed-grass) Dev. Friend.1882.

LADY'S HAIR 1. *Adiantum capillus-veneris* (Maidenhair Fern) Leyel.1948. 2. *Briza media* (Quaking Grass) Som, E Angl, Lincs, N'thum. Grigson.1959. Cf. Lady's Tresses.

LADY'S HANDS *Briza media* (Quaking Grass) N'thum. Grigson.1959.

LADY'S HATPINS *Knautia arvensis* (Field Scabious) Som. Macmillan.

LADY'S HEARTS *Dicentra spectabilis* (Bleeding Heart) Som. Macmillan.

LADY'S KEYS 1. *Acer pseudo-platanus*

fruit (Sycamore keys) Wilts. Macmillan. 2. *Arum maculatum* (Cuckoo-pint) Kent Grigson. 3. *Primula veris* (Cowslip) Som. Macmillan; Wilts, Kent. Grigson.

LADY'S KNIVES-AND-FORKS *Lycopodium inundatum* (Clubmoss) Som. Elworthy.

LADY'S LACE 1. *Achillea millefolium* (Yarrow) Glos. PLNN 18. 2. *Aethusa cynapium* (Fool's Parsley) Som. Macmillan. 3. *Anthriscus sylvestris* (Cow Parsley) Som. Grigson; Oxf. Oxf.Ann.Rec.1951. 4. *Conium maculatum* (Hemlock) Som. Macmillan. 5. *Cuscuta epithymum* (Common Dodder) Som. Friend. 6. *Torilis japonica* (Hedge Parsley) Som. Macmillan. Descriptive, especially for the umbellifers, and for yarrow at a pinch. Dodder, though, gets the name for different descriptive reasons.

LADY'S LACES 1. *Cuscuta epithymum* (Dodder) Som. Friend. 2. *Phalaris arundinacea* (Reed-grass) Dev. Friend.1882.

LADY'S LEEK *Allium cernuum* (Nodding Wild Onion) Douglas.

LADY'S LINT *Stellaria holostea* (Greater Stitchwort) Dev. Friend.1882. Lint - flax. There is a Manx name for this meaning fairy flax.

LADY'S LOCKET 1. *Acer campestre* fruit (Field Maple keys) Som. Macmillan. 2. *Dicentra spectabilis* (Bleeding Heart) Coats. 3. *Fumaria officinalis* (Fumitory) Som. Macmillan. 4. *Lunaria annua* (Honesty) Som. Macmillan. 5. *Polygonatum multiflorum* (Solomon's Seal) Som. Macmillan.

LADY'S LOOKING-GLASS *Legousia hybrida* (Venus' Looking-glass) Prior.

LADY'S MAID *Artemisia chamaemelifolia* Brownlow.

LADY'S MANTLE 1. *Adoxa moschatellina* (Moschatel) Som. Macmillan. 2. *Alchemilla vulgaris*. 3. *Cardamine pratensis* (Lady's Smock) Som. Macmillan. 4. *Plantago lanceolata* (Ribwort Plantain) Som. Macmillan. Alchemilla got the name because of the way the leaves are folded, reminiscent of the mantle of Tudor times. Cf. Lady's Cloak for Lady's Smock.

LADY'S MANTLE, Alpine *Alchemilla alpina*.

LADY'S MANTLE, Field *Aphanes arvensis*.

LADY'S MANTLE, Five-leaved *Alchemilla alpina* (Alpine Lady's Mantle) Sanecki.

LADY'S MEAT *Crataegus monogyna* (Hawthorn) Fernie. Presumably for the same reason that children call it Bread-and-cheese Tree - they regularly eat the young leaves and shoots.

LADY'S MILK *Silybum marianum* (Milk Thistle) B & H. The white veins in the leaves are the Virgin's milk.

LADY'S MILKCANS *Anemone nemerosa* (Wood Anemone) Som. Macmillan.

LADY'S MILKING-STOOLS *Stellaria graminea* (Lesser Stitchwort) Som. Macmillan.

LADY'S MILKSILE 1. *Cardamine pratensis* (Lady's Smock) Yks. B & H. 2. *Pulmonaria officinalis* (Lungwort) Ches. Grigson. 'Sile' is the same word as 'soil', i.e. stain. During the flight into Egypt, some of the Virgin's milk fell on the lungwort's leaves while she was nursing the infant Jesus, causing the white blotches on them. But, in connection with Lady's Smock, B & H point out that in Yorkshire 'sile' meant a strainer (to sile milk was to strain it). But what about the flower could remind anyone of a strainer?

LADY'S MILKWORT *Pulmonaria officinalis* (Lungwort) Lyte.

LADY'S MINT *Mentha spicata* (Spearmint) Gerard. A name that is more understandable on the Continent than here, for it is Menthe de Notre-Dame in France, Erba Santa Maria in Italy, and Unser Frauen Muntz in Germany.

LADY'S NAVEL *Umbilicus rupestris* (Wall Pennywort).

LADY'S NEEDLE *Scandix pecten-veneris* (Shepherd's Needle) Tynan & Maitland. The 'needle' reference is to the long, beaked fruits.

LADY'S NEEDLEWORK 1. *Alliaria petio-*

lata (Jack-by-the-hedge) Som. Macmillan. 2. *Anthriscus sylvestris* (Cow Parsley) Glos. Mabey. 3. *Asperula odorata* (Woodruff) Som. Macmillan. 4. *Centranthus ruber* (Spur Valerian) Corn, Worcs. Grigson. Som. Macmillan. 5. *Conium maculatum* (Hemlock) Som. Macmillan. 6. *Lobularia maritima* (Sweet Alison) Som. Macmillan. 7. *Scabiosa atropurpurea* (Sweet Scabious) Som. Macmillan. 8. *Stellaria holostea* (Greater Stitchwort) Som. Macmillan. Wilts. Tynan & Maitland. 9. *Torilis japonica* (Hedge Parsley) Ches. Holland. Cf. Lady's Lace.

LADY'S NIGHTCAP 1. *Anemone nemerosa* (Wood Anemone) Glos. J D Robertson; Heref. B & H. 2. *Calystegia sepium* (Great Bindweed) Som. Grigson; Wilts. Dartnell & Goddard; Hants. Cope. 3. *Campanula medium* (Canterbury Bells) Dev. Friend. 4. *Convolvulus arvensis* (Field Bindweed) Wilts. Akerman. 5. *Convolvulus major* (Morning Glory) Wilts. Friend. In contrast, note Devil's, or Old Man's, Nightcap.

LADY'S PARASOL *Daucus carota* (Wild Carrot) Tynan & Maitland.

LADY'S PETTICOATS 1. *Anemone nemerosa* (Wood Anemone) Wilts. Dartnell & Goddard. 2. *Aquilegia vulgaris* (Columbine) Som. Macmillan.

LADY'S PETTICOATS, White *Stellaria holostea* (Greater Stitchwort) Corn, Heref. Grigson.

LADY'S PINCUSHION 1. *Anthyllis vulneraria* (Kidney Vetch) Wilts. Macmillan. 2. *Armeria maritima* (Thrift) Hants. Cope. 3. *Corydalis lutea* (Yellow Corydalis) Dev. Friend.1882. 4. *Knautia arvensis* (Field Scabious) Som. Macmillan; Hants. Boase. 5. *Lotus corniculatus* (Bird's-foot Trefoil) Dor. Grigson. 6. *Pulmonaria officinalis* (Lungwort) Ches. Grigson; Yks. Addy. 7. *Scabiosa atropurpurea* (Sweet Scabious) Som. Macmillan. Suff. B & H.

LADY'S POSY *Trifolium pratense* (Red Clover) Som. Macmillan.

LADY'S PRIDE *Cardamine pratensis* (Lady's Smock) Som. Macmillan.

LADY'S PURSE 1. *Anemone nemerosa* (Wood Anemone) Dev. Grigson. 2. *Aquilegia vulgaris* (Columbine) Dor. Macmillan. 3. *Calceolaria sp* (Slipperwort) Ches. Holland. 4. *Capsella bursa-pastoris* (Shepherd's Purse) E Ang, Berw. Grigson. 5. *Dicentra spectabilis* (Bleeding Heart) Som. Macmillan.

LADY'S RIBANDS *Phalaris arundinacea* (Reed-grass) Dev. Friend.1882.

LADY'S RUFFLES 1. *Filipendula vulgaris* (Dropwort) Grigson.1959. 2. *Narcissus pseudo-narcissus* (Daffodil) Wilts. Grigson.

LADY'S SEAL 1. *Bryonia dioica* (White Bryony) Friend. 2. *Polygonatum multiflorum* (Solomon's Seal) Wright. 3. *Tamus communis* (Black Bryony) Gerard. Bryony has a bad reputation. Why should there be a dedication to the Virgin Mary, not only in this name, but also in fact, at the feast of her nativity?

LADY'S SHAKES *Briza media* (Quakinggrass) Yks. Grigson.1959.

LADY'S SHIMMEY 1. *Anemone nemerosa* (Wood Anemone) Glos. Grigson. 2. *Calystegia sepium* (Great Bindweed) Som. Grigson. Shimmey is chemise.

LADY'S SHOES 1. *Aquilegia vulgaris* (Columbine) Som. Macmillan; E Ang. Grigson. 2. *Cardamine pratensis* (Lady's Smock) Som. Macmillan. 3. *Fumaria officinalis* (Fumitory) Som. Macmillan; Wilts. Dartnell & Goddard. 4. *Lotus corniculatus* (Bird's-foot Trefoil) Hants. Grigson.

LADY'S SHOES-AND-STOCKINGS *Lotus corniculatus* (Bird's-foot Trefoil) Som. Macmillan; Kent Grigson.

LADY'S SIGNET 1. *Polygonatum multiflorum* (Solomon's Seal) Prior. 2. *Tamus communis* (Black Bryony) Prior.

LADY'S SLIPPER 1. *Aconitum napellus* (Monkshood) Som. Macmillan. 2. *Anthyllis vulneraria* (Kidney Vetch) War. Grigson. 3. *Antirrhinum majus* (Snapdragon) Dor. Macmillan. 4. *Aquilegia vulgaris*

(Columbine) Som, Wilts. Macmillan. 5. *Arum maculatum* (Cuckoo-pint) Wilts. Grigson. 6. *Cypripedium calceolus*. 7. *Cytisus scoparius* (Broom) Som. Macmillan. 8. *Digitalis purpurea* (Foxglove) Som. Macmillan. 9. *Hippocrepis comosa* (Horseshoe Vetch) Som. Macmillan. 10. *Lathyrus pratensis* (Yellow Pea) Som, Wilts. Macmillan. 11. *Linaria vulgaris* (Toadflax) Som. Macmillan. 12. *Lotus corniculatus* (Bird's-foot Trefoil) Som, Wilts. Macmillan; Hants, Yks. Grigson. 13. *Ranunculus acris* (Meadow Buttercup) Som. Macmillan.

LADY'S SLIPPER, Downy *Cypripedium parviflorum var. pubescens* (Larger Yellow Lady's Slipper) USA House.

LADY'S SLIPPER, Pink *Cypripedium acaule* (Stemless Lady's Slipper) USA Bingham.

LADY'S SLIPPER, Queen *Cypripedium reginae* (Showy Lady's Slipper) USA Bingham.

LADY'S SLIPPER, Ram's-head *Cypripedium arietinum*.

LADY'S SLIPPER, Showy *Cypripedium reginae*.

LADY'S SLIPPER, Stemless *Cypripedium acaule*.

LADY'S SLIPPER, White, Small *Cypripedium candidum*.

LADY'S SLIPPER, Wild *Impatiens capensis* (Orange Balsam) Grieve.1931. Slipperweed is also recorded for this.

LADY'S SLIPPER, Yellow, Larger *Cypripedium parviflorum var. pubescens*.

LADY'S SLIPPER, Yellow, Small *Cypripedium parviflorum*.

LADY'S SMOCK 1. *Anemone nemerosa* (Wood Anemone) N'thants. Clare. 2. *Artemisia lactiflora* (White Wormwood) Dyer. 3. *Arum maculatum* (Cuckoo-pint) Dor, Som. Macmillan; Hants. Cope. 4. *Calystegia sepium* (Great Bindweed) Corn, Dev, War, Notts, Yks. Grigson; Som. Jennings; Suss. Parish; N'thants. A E Baker. 5. *Campanula medium* (Canterbury Bell) Halliwell. 6. *Cardamine pratensis*. 7. *Convolvulus arvensis* (Field Bindweed) Som. Macmillan. 8. *Stellaria holostea* (Greater Stitchwort) Corn, Som. Grigson; Dor. Macmillan. It is the white flowers in each case that engendered the name, except in the case of cuckoo-pint, where the name had probably a more erotic origin, and also in the case of Canterbury Bell.

LADY'S SMOCK, Bitter *Cardamine amara* (Large Bittercress) Curtis.

LADY'S SMOCK, Hairy *Cardamine hirsuta* (Hairy Bittercress) Curtis.

LADY'S SMOCK, Impatient *Cardamine impatiens* (Narrow-leaved Bittercress) Camden/Gibson. Impatient, because of the explosive method of seeds destruction.

LADY'S SORREL *Oxalis acetosella* (Wood Sorrel) USA. H Smith.1923. It is Lady's Clover in Scotland.

LADY'S SUNSHADE *Convolvulus arvensis* (Field Bindweed) Som. Macmillan. Cf. Lady's Umbrella. Both are suggested by the shape of the flowers.

LADY'S TAPER *Verbascum thapsus* (Mullein) Som. Macmillan.

LADY'S TEARS *Convallaria maialis* (Lily-of-the-valley) Som. Dyer.

LADY'S THIMBLE 1. *Campanula medium* (Canterbury Bell) Som. Macmillan. 2. *Campanula rotundifolia* (Harebell) Som. Elworthy; E Ang, N Scot, Ire. Grigson. 3. *Digitalis purpurea* (Foxglove) Som. Macmillan; Norf,. N'thum. Grigson. 4. *Endymion nonscriptus* (Bluebell) Ire. Tynan & Maitland. 5. *Stellaria holostea* (Greater Stitchwort) Dor. Macmillan. 6. *Veronica chamaedrys* (Germander Speedwell) Lancs. Grigson. Canterbury Bell, Harebell (Bluebell is probably the same as this, though listed separately) and Foxglove can all be visualised as something that could fit over a finger, but stitchwort and speedwell cannot.

LADY'S THISTLE *Silybum marianum* (Milk Thistle) Som. Macmillan; Lincs. Grigson.

LADY'S THUMB *Polygonum persicaria* (Persicaria) USA Tolstead. Cf. Virgin's

Pinch. The legend is that that the Virgin plucked a root, left her mark on the leaf, and threw it aside, saying, "This is useless", and it has been useless ever since.

LADY'S THUMB, Bigroot *Polygonum muhlenbergii* (Swamp Persicaria) USA Yanovsky.

LADY'S THUMBS-AND-FINGERS *Lotus corniculatus* (Bird's-foot Trefoil) Som. Macmillan.

LADY'S TOBACCO 1. *Antennaria dioica* (Cat's Foot) USA Leighton. 2. *Antennaria plantaginifolia* (Plantain-leaved Everlasting). New Engl. Sanford. 3. *Gnaphalium polycephalum* (Mouse-ear Everlasting) Emboden. Presumably all of them are tobacco substitutes, and found to be very mild in character, hence "Lady's".

LADY'S TRESSES 1. *Arum maculatum* (Cuckoo-pint) Som. Macmillan. 2. *Briza media* (Quaking-grass) L Gordon. 3. *Listera ovata* (Twayblade) Wilts. Tynan & Maitland. 4. *Galium verum* (Lady's Bedstraw) Som. Macmillan. 5. *Spiranthes spiralis*.

LADY'S TRESSES, American *Spiranthes romanzoffiana*.

LADY'S TRESSES, Autumn *Spiranthes spiralis* (Lady's Tresses) Clapham.

LADY'S TRESSES, Broad-leaved *Spiranthes lucida* (Wide-leaved Lady's Tresses) USA Zenkert.

LADY'S TRESSES, Irish; LADY'S TRESSES, Cork *Spiranthes romanzoffiana* (American Lady's Tresses) Genders (Irish); Turner Ettlinger (Cork). An American plant, right enough, but it very rarely occurs in Ireland, and in Western Scotland.

LADY'S TRESSES, Nodding *Spiranthes cernua*.

LADY'S TRESSES, Shining *Spiranthes lucida* (Wide-leaved Lady's Tresses) USA Zenkert.

LADY'S TRESSES, Slender *Spiranthes gracilis*.

LADY'S TRESSES, Summer *Spiranthes aestivalis*.

LADY'S TRESSES, Wide-leaved *Spiranthes lucida*.

LADY'S TWO-SHOES *Lotus corniculatus* (Bird's-foot Trefoil) Ess. Gepp. A hybrid - probably Lady's Shoes x Kitty-two-shoes.

LADY'S UMBRELLA 1. *Calystegia sepium* (Great Bindweed) Som. Macmillan. 2. *Convolvulus arvensis* (Field Bindweed) Dor. Macmillan. 3. *Solanum dulcamara* (Woody Nightshade) Som. Macmillan. Cf. Lady's Sunshade. Both are suggested by the shape of the flowers.

LADYBELL, Bush *Adenophora potaninii*.

LADYBELL, Dahurian *Adenophora verticillata*.

LADYBIRD *Anagallis arvensis* (Scarlet Pimpernel) Som. Macmillan.

LAGOS SILK-RUBBER TREE *Funtumia elastica*.

LAGWORT *Petasites hybridus* (Butterbur) B & H.

LAISTER *Iris pseudo-acarus* (Yellow Flag) Corn. Jago.

LAKENS, Lamb's 1. *Arum maculatum* (Cuckoo-pint) N'thants. A E Baker; N'thum. Grigson. 2. *Cardamine pratensis* (Lady's Smock) Cumb. Grigson. Lakens, or Lakins, seems to mean a toy, OE laecon being the word's original.

LAKEPYPIE *Watsonia meriana* S Africa Whittle & Cook.

LAKEWEED 1. *Polygonum amphibium* (Amphibious Persicaria) Shrop. B & H. 2. *Polygonum hydropiper* (Water Pepper) Ches. Prior. 3. *Polygonum persicaria* (Persicaria) Ches. Holland.

LAMB-IN-A-PULPIT *Arum maculatum* (Cuckoo-pint) Dev. Friend; Wilts. Grigson.

LAMB-LAKINS *Arum maculatum* (Cuckoo-pint) N'thants. B & H.

LAMB-MINT 1. *Mentha x piperita* (Peppermint) Dev. B & H. 2. *Mentha spicata* (Spearmint) Dev. B & H. Mint sauce has always been the proper accompaniment to roast lamb. The name appears in Somerset as Lammint.

LAMB-TONGUE 1. *Chenopodium urbicum*

(Upright Goosefoot) Som. Elworthy. 2. *Phyllitis scolopendrium* (Hartstongue) Som. Elworthy.

LAMB'S CRESS *Cardamine hirsuta* (Hairy Bittercress) Dev. Friend. Cf. Land Cress.

LAMB'S EAR 1. *Lamium purpureum* (Red Deadnettle) Som. Macmillan. 2. *Plantago media* (Lamb's-tongue Plantain) Cumb. Grigson. 3. *Potentilla anserina* (Silverweed) Som. Macmillan. 4. *Stachys lanata*. 5. *Succisa pratensis* (Devil's-bit Scabious) War. B & H.

LAMB'S FOOT 1. *Alchemilla vulgaris* (Lady's Mantle) Lancs. Grigson. 2. *Anthyllis vulneraria* (Kidney Vetch) Som. Macmillan. 3. *Lotus corniculatus* (Bird's-foot Trefoil) Lancs. Grigson. 4. *Plantago major* (Great Plantain) Cumb. Grigson.

LAMB'S HEAD, Yellow *Tropaeolum majus* (Nasturtium) Fernie.

LAMB'S LAKENS 1. *Arum maculatum* (Cuckoo-pint) N'thants. A E Baker; N'thum. Grigson. 2. *Cardamine pratensis* (Lady's Smock) Cumb. Grigson. Lakens, or Lakins, seems to mean a toy, OE laecon being the word's original.

LAMB'S LETTUCE 1. *Lapsana communis* (Nipplewort) Som. Macmillan. 2. *Poterium sanguisorba* (Salad Burnet) Som. Macmillan. 3. *Valerianella locusta* (Cornsalad) McClintock. McClintock accounted for the name by saying that Cornsalad is at its best during the lambing season.

LAMB'S PUMMY *Alliaria petiolata* (Jack-by-the-hedge) Som. Grigson.

LAMB'S QUARTERS 1. *Atriplex patula* (Orache) Som. Macmillan. 2. *Chenopodium album* (Fat Hen) Som. Macmillan; IOW, N Eng Grigson; USA Tolstead. 3. *Chenopodium murale* (Nettle-leaved Goosefoot) S Africa Watt. 4. *Trillium erectum* (Bethroot) Grieve.1931. A S Palmer suggested that this might be Lammas Quarter, but surely 'quarters' is food - the stuff the quartermaster is responsible for.

LAMB'S QUARTERS, Desert *Chenopodium fremontii*.

LAMB'S QUARTERS, Narrow-leaved *Chenopodium leptophyllum* (Fine-leaf Goosefoot) Stevenson.

LAMB'S SUCKLINGS 1. *Lotus corniculatus* (Bird's-foot Trefoil) Yks. B & H. 2. *Trifolium pratense* (Red Clover) Cumb. B & H. 3. *Trifolium repens* (White Clover) Yks, Cumb. Grigson. Sucklings is a name often given in different forms to clover. It conveys the same meaning as the 'suckle' of honeysuckle.

LAMB'S TAILS 1. *Alnus glutinosa* - catkins (Alder catkins) Som. Macmillan. 2. *Anthyllis vulneraria* (Kidney Vetch) Som. Macmillan. 3. *Corylus avellana* - catkins (Hazel catkins) Dev. Friend.1882; Wilts. Dartnell & Goddard. 4. *Cotyledon oppositifolia*. 5. *Plantago lanceolata* (Ribwort Plantain) Som. Macmillan. 6. *Plantago major* (Great Plantain) Norf, Cumb, Scot. Grigson. 7. *Salix capraea* - catkins (Sallow catkins) Grigson. 8. *Salix fragilis* - catkins (Crack Willow catkins) Corn. Jago. All catkins qualify for the name, for obvious descriptive reasons.

LAMB'S TOE 1. *Anthyllis vulneraria* (Kidney Vetch) Som. Macmillan; N'thants. Sternberg; Rut. Grigson. 2. *Lotus corniculatus* (Bird's-foot Trefoil) Midl. Grigson. 3. *Medicago lupulina* (Black Medick) Staffs. Grigson.

LAMB'S TONGUE 1. *Chenopodium album* (Fat Hen) Dev, Ches. B & H. 2. *Mentha arvensis* (Corn Mint) Scot. Grigson. 3. *Plantago lanceolata* (Ribwort Plantain) Som. Macmillan; Hants, Suss, Shrop, N'thum. Grigson. 4. *Plantago media* (Lamb's-tongue Plantain) Som. Macmillan; Suss. Grigson. 5. *Polygonum persicaria* (Persicaria) Dev. Grigson. 6. *Rhinanthus crista-galli* (Yellow Rattle) Som. Macmillan. 7. *Stachys lanata* (Lamb's Ears) Dev. Friend.

LAMB'S TONGUE PLANTAIN *Plantago media* Som. Macmillan.

LAMBERT'S PINE *Pinus lambertiana* (Sugar Pine) Ablett.

LAMBKILL *Kalmia angustifolia* (Sheep Laurel) USA House. Also Sheepkill and Calfkill. " It kills sheep and lesser animals, when they eat of it plentifully..." (Taylor).

LAMBKINS *Corylus avellana* - catkins (Hazel catkins) Wilts. Dartnell & Goddard.

LAMBS *Aesculus hippocastanum* (horse Chestnut) Dev. B & H. Given to the flowers only.

LAMBS, Baa; LAMBS, Baccy *Corylus avellana* - catkins (Hazel catkins) both Som. Macmillan.

LAMBS, Bay *Pinus sylvestris* (Scots Pine) Yks. B & H. Applied to the male flowers.

LAMMIE SOUROCK *Rumex acetosa* (Wild Sorrel) Border B & H. Sourock is sorrel. Cf. the name with Sheep's Sorrel, usually reserved for *Rumex acetosella*.

LAMMIES *Larix decidua* - cones (Larch cones) Aber. B & H.

LAMMINT *Mentha spicata* (Spearmint) Som. Macmillan. Lamb Mint.

LAMP, Fairy's *Arum maculatum* (Cuckoopint) Ire. Porter. Apparently, the pollen of the flowers gives a faint light at dusk. Cf. the Fenland name Shiners.

LAMPFLOWER *Lychnis chalcedonica* (Jerusalem Cross) Tynan & Maitland. Cf. Great Candlesticks for this plant.

LAMPS OF SCENT *Lonicera periclymenum* (Honeysuckle) Som. Macmillan.

LAMPSHADE POPPY *Meconopsis integrifolia* (Yellow Chinese Poppy) RHS.

LAMPWICK GRASS *Juncus conglomeratus* (Compact Rush) USA Hyatt. This is the kind from which rush lights are made, hence this name.

LANCASHIRE ASPHODEL *Narthecium ossifragum* (Bog Asphodel) Prior. It was at one time given the same Asphodelus lancastriae.

LANCE-LEAF SAGE *Salvia lanceolata*.

LANCE-LEAF SANDALWOOD *Santalum lanceolatum* (Australian Sandalwood) Usher.

LANCE-LEAVED TICKSEED *Coreopsis lanceolata*.

LANCE TREE *Lonchocarpus capassa* (Rain Tree) Palgrave.

LAND CRESS 1. *Cardamine amara* (Large Bittercress) Ches. Holland. 2. *Cardamine hirsuta* (Hairy Bittercress) Hants. Cope; War, Ches. B & H. Cf. Lamb Cress.

LAND CRESS, Bitter *Barbarea vulgaris* (Winter cress) Tynan & Maitland.

LAND LUNG *Peltigera canina* (Ashcoloured Ground Liverwort) Halliwell.

LAND WHIN *Ononis repens* (Rest Harrow) E. Ang. Halliwell.

LANG DE BEEF; LANGLEY BEEF *Picris echioides* (Bristly Ox-tongue) Prior, A.S. Palmer.

LANGWORT *Veratrum album* (White Hellebore) Halliwell. Gerard called this "Lingwort", surely the same name?

LANKONG LILY *Lilium langkongense*. Lankong is in NW China, about 20 miles NW of Tali lake.

LANTERN-LEAVES *Ranunculus repens* (Creeping Buttercup) Som. Macmillan.

LANTERNS *Carpinus betula* seed capsules (Hornbeam seeds) Freethy.

LANTERNS, Chinese 1. *Dichrostachys glomerata*. 2. *Geranium robertianum* (Herb Robert) Som. Macmillan. 3. *Physalis alkekengi* (Winter Cherry) Macmillan.

LANTERNS, Fairy 1. *Calochortus albus*. 2. *Disporum smithii* USA Schenk & Gifford.

LANTERNS, Fairy, Golden *Calochortus amabilis*.

LANTERNS, Fairy's *Linaria vulgaris* (Toadflax) Som. Macmillan.

LANTERNS, Jack-o'- *Anemone nemerosa* (Wood Anemone) Dor. Grigson.

LANTERNS, Japanese 1. *Campanula medium* (Canterbury Bells) Som. Macmillan. 2. *Physalis alkekengi* (Winter Cherry) Tampion. Descriptive, of course, and not an acknowledgement of place of origin.

LANTERNS, Robber's *Aesculus hippocastanum* flowers (Horse Chestnut flowers) Dor. Macmillan.

LANTERNWEED *Physalis heterophylla* (Ground Cherry) USA Gates.
LANTHORN LILY *Narcissus pseudo-narcissus* (Daffodil) War. A S Palmer.
LAPLAND ROSE BAY *Rhododendron lapponicum*.
LAPLAND WILLOW *Salix lapponicum*.
LAPLOVE 1. *Convolvulus arvensis* (Field Bindweed) Midl, N Eng, Scot. Grigson. 2. *Polygonum convolvulus* (Black Bindweed) Midl, Roxb. Grigson. A S Palmer derived Laplove from lap-leaves, that which laps or enfolds the leaves, but it doesn't sound very convincing.
LAPP HEATHER *Phyllodoce caerulea* (Menziesia) Manker.
LAPPER-GOWAN; LOPPER-GOWAN *Trollius europaeus* (Globe-flower) Cumb. Grigson; Roxb. B & H (Lapper); Clyde. B & H (Lopper).
LAPWEED *Polygonum convolvulus* (Black Bindweed) Ches. Holland. Cf. Laplove for this.
LARCH *Larix decidua*. Immediately, larch derives through German from Latin larix, but it is said that it comes originally from a Celtic word, lar, meaning 'fat', a reference to the resin it exudes. Dioscorides says Larix is the Gallic name for resin.
LARCH, American *Larix laricina*.
LARCH, Chinese *Larix potaninii*.
LARCH, Dahurian *Larix gmelini*.
LARCH, Dunkeld *Larix x eurolepis*.
LARCH, European *Larix decidua* (Larch) Clapham.
LARCH, Fairy *Equisetum sylvaticum* (Wood Horsetail) N Eng. Salisbury.1936.
LARCH, Gold, Chinese *Pseudolarix amabilis*.
LARCH, Himalayan *Larix griffithii*.
LARCH, Japanese *Larix kaempferi*.
LARCH, Polish *Larix polonica*.
LARCH, Siberian *Larix sibirica*.
LARCH, Sikkim *Larix griffithii* (Himalayan Larch) Mitchell.
LARCH, Western *Larix occidentalis*.

LARD, Bladder of *Silene cucubalis* (Bladder Campion) Som. Macmillan.
LAREABELL *Helianthus annuus* (Sunflower) Lincs. B & H.
LARICK, LARIX TREE *Larix decidua* (Larch) Inv, Renf. B & H (Larick); Gerard (Larix Tree).
LARK'S CLAWS; LARK'S HEEL; LARK'S TOE *Delphinium ajacis* (Larkspur) all Gerard. See Larkspur.
LARK'S EYE 1. *Veronica chamaedrys* (Germander Speedwell) Som. Macmillan. 2. *Viola tricolor* (Pansy) Som. Macmillan. Not so specific, usually - Bird's Eye will do.
LARK'S HEEL, Yellow *Tropaeolum majus* (Nasturtium) Parkinson.
LARKSEED *Plantago major* (Great Plantain) Wilts. Dartnell & Goddard. Self-explanatory, and the practice of feeding cage birds with the seeds was quite widespread. Cf. Bird's Meat from the north of Scotland to such names as Canary-flower from Somerset.
LARKSPUR 1. *Delphinium ajacis*. 2. *Delphinium decorum* USA Schenk & Gifford. 3. *Delphinium scaposum* USA Mason. 4. *Linaria vulgaris* (Toadflax) Bucks. Grigson. The lark has a long hind claw, and larkspur has a long calyx-spur.
LARKSPUR, Branching; LARKSPUR, Forking *Delphinium ajacis* (Larkspur) Lindley (Branching).
LARKSPUR, Eastern *Delphinium orientale*.
LARKSPUR, Field *Delphinium ajacis* (Larkspur) Wit.
LARKSPUR, Palmated *Delphinium staphisagria* (Stavesacre) Thornton.
LARKSPUR, Prairie *Delphinium virescens*.
LARKSPUR LETTUCE *Lactuca pulchella*.
LATHERWORT *Saponaria officinalis* (Soapwort).
LAUDAN *Cistus ladaniferum* (Gum Cistus) Turner. Laudan instead of Ladan gave rise to some confusion between ladanum and laudanum. Ladanam, or Labdanum, is the gum yielded by this.
LAUGHING PARSLEY *Pulsatilla vulgaris* (Pasque-flower) Coats.

LAUGHTER-BRINGER *Anagallis arvensis* (Scarlet Pimpernel) Som. Macmillan. Presumably because the plant was used as a simple against melancholy.

LAUNCELEY *Plantago lanceolata* (Ribwort Plantain) Halliwell.

LAUREL 1. *Daphne laureola* (Spurge Laurel) B & H. 2. *Laurus nobilis* (Bay) Turner. 3. *Umbellularia californica* (Californian Laurel) Calif. C.Grant.

LAUREL, Alexandria(n) 1. *Calophyllum inophyllum*. 2. *Ruscus racemosus*.

LAUREL, Bay; LAUREL-BAY *Laurus nobilis* (Bay) W Miller (Bay Laurel); Dawson (Laurel-Bay).

LAUREL, Black 1. *Gordonia lasianthus* (Loblolly-Bay) Whittle & Cook. 2. *Leucothoe davisiae* (Sierra Laurel) USA Kingsbury.

LAUREL, Bog *Kalmia polifolia* (Swamp Laurel) USA Kingsbury.

LAUREL, Broad-leaved *Kalmia latifolia* (Calico Bush) Grieve.1931.

LAUREL, California *Umbellularia californica*.

LAUREL, Camphor *Cinnamonium camphora* (Camphor) Fluckiger.

LAUREL, Canary Island *Laurus canariensis*.

LAUREL, Cherry *Prunus laurocerasus*.

LAUREL, Cherry, American *Prunus caroliniana*.

LAUREL, Copse *Daphne laureola* (Spurge Laurel) Hants, IOW Cope. Cf. Wood Laurel.

LAUREL, Dog *Aucuba japonica* (Japanese Laurel) Polunin.1976.

LAUREL, Dwarf 1. *Daphne mezereon* (Lady Laurel) Le Strange. 2. *Kalmia angustifolia* (Sheep Laurel) Taylor.

LAUREL, English *Prunus laurocerasus* (Cherry Laurel) USA Cunningham. 'English', presumably, because it was so often planted in Victorian shrubberies.

LAUREL, Great *Rhododendron maximum* (Rose Bay) USA House.

LAUREL, Grecian *Laurus nobilis* (Bay) USA Cunningham.

LAUREL, Japanese *Aucuba japonica*.

LAUREL, Lady *Daphne mezereon*.

LAUREL, Madrona *Arbutus menziesii* (Madrona) Brimble.1948.

LAUREL, Mountain 1. *Kalmia latifolia* (Calico Bush) USA Harper. 2. *Rhus ovata* (Sugar Sumach) Arizona T H Kearney. 3. *Sophora secundiflora* (Mescal Bean) USA Howard.

LAUREL, Mountain, Hairy *Kalmia hirsuta*.

LAUREL, Narrow-leaved *Kalmia angustifolia* (Sheep Laurel) Grieve.1931.

LAUREL, Noble *Laurus nobilis* (Bay) Grieve.1931.

LAUREL, Pale *Kalmia polifolia* (Swamp Laurel) USA House.

LAUREL, Portuguese *Prunus lusitanica*.

LAUREL, Red-berry *Daphne mezereon* (Lady Laurel) Yks. Grigson.

LAUREL, Roman *Laurus nobilis* (Bay) Prior.

LAUREL, Rose *Nerium oleander* (Rose Bay) Turner.

LAUREL, Sheep *Kalmia angustifolia*. For this is highly poisonous in the wild to sheep and other stock. Cf. Sheepkill, Lambkill, etc.

LAUREL, Sierra *Leucothoe davisiae*.

LAUREL, Sparked; LAUREL, Spotted *Aucuba japonica* (Japanese Laurel) Som. Elworthy (Sparked); Hibberd (Hibberd). 'Sparked' means variegated.

LAUREL, Spurge 1. *Daphne laureola*. 2. *Daphne mezereum* (Lady Laurel) Gerard.

LAUREL, Swamp 1. *Kalmia polifolia*. 2. *Magnolia virginiana* (Sweet Bay) Genders.1972.

LAUREL, True *Laurus nobilis* (Bay) Grieve.1931.

LAUREL, Wood; LAUREL-WOOD *Daphne laureola* (Spurge Laurel) Som. Macmillan; Glos, IOW, Bucks. Grigson; Hants. Cope (Wood Laurel); Glos. J D Robertson; (Laurel-wood). Cf. too Copse Laurel.

LAUREL CHERRY *Prunus caroliniana* (American Cherry Laurel) USA Harper.

LAUREL GREENBRIER *Smilax laurifolia*.

LAUREL-LEAVED CISTUS *Cistus laurifolius.*
LAUREL MAGNOLIA *Magnolia grandiflora* (Bull Bay) USA Wit.
LAUREL SABINO *Magnolia splendens.*
LAUREOLE *Daphne laureola* (Spurge Laurel) Halliwell.
LAURERE *Laurus nobilis* (Bay) Scot. Jamieson.
LAURIEL; LAURY *Daphne laureola* (Spurge Laurel) Turner (Lauriel); Prior (Laury).
LAURIER ROSE *Nerium oleander* (Rose Bay) Watt.
LAURUSTINE *Viburnum tinus*. i.e. *Laurus tinus*. Cf. Gerard's Wild Bay for this.
LAUS TIBI *Narcissus maialis var. patellaris* (Pheasant's Eye) Turner.
LAUTER *Laurus nobilis* (Bay) Halliwell.
LAVATERA, Annual *Lavatera trimestris.*
LAVENDER 1. *Lavandula sp.* 2. *Polygonum persicaria* (Persicaria) Som. Macmillan. It has been suggested that Lavandula comes from lavare, to wash, and that the Romans used lavender for scenting baths. But an earlier Latin form of the name was livendula, perhaps, as Chambers's dictionary says, connected with lividus, livid.
LAVENDER, Cotton; LAVENDER COTTON *Santolina chamaecyparissus* (Ground Cypress) Webster (Lavender Cotton).
LAVENDER, Cut-leaved *Lavandula multifida.*
LAVENDER, Desert *Hyptis emoryi.*
LAVENDER, Dutch *Lavandula vera.*
LAVENDER, English *Lavandula spica.*
LAVENDER, French 1. *Lavandula dentata* (Toothed Lavender) RHS. 2. *Lavandula stoechas* (Spanish Lavender) Gerard. 3. *Santolina chamaecyparissus* (Ground Cypress) Webster.
LAVENDER, Marsh *Limonium vulgare* (Sea Lavender) Gerard.
LAVENDER, Scotch *Asperula odorata* (Woodruff) Scot. Swire.1963.
LAVENDER, Sea *Limonium vulgare.*
LAVENDER, Sea, Alderney *Limonium lychnidifolium.*
LAVENDER, Sea, American *Limonium carolinianum.*
LAVENDER, Sea, Lax *Limonium humile.*
LAVENDER, Sea, Matted *Limonium bellidifolium.*
LAVENDER, Sea, Remote-flowered *Limonium humile* (Lax Sea Lavender) Hepburn.
LAVENDER, Sea, Rock *Limonium binervosum.*
LAVENDER, Seaside *Limonium carolinianum* (American Sea Lavender) USA House.
LAVENDER, Seaside, Small *Heliotropum curassavicum* (Barbados) Gooding.
LAVENDER, Silver *Santolina chamaecyparissus* Shrop. B & H.
LAVENDER, Spanish *Lavandula stoechas.*
LAVENDER, Toothed *Lavandula dentata.*
LAVENDER, Wild *Heliotropum curassavicum* Barbados Gooding.
LAVENDER CORN; LAVENDER COTTON *Santolina chamaecyparissus* (Ground Cypress) B & H (Corn); Webster (Cotton).
LAVENDER GENTLE *Lavandula stoechas* (Spanish Lavender) Turner.
LAVENDER SPIKE *Lavandula spica* (English Lavender) Tusser.
LAVENDER SNIPS *Cymbalaria muralis* (Ivy-leaved Toadflax) Hants. Grigson.
LAVENDER THRIFT *Limonium vulgare* (Sea Lavender) B & H.
LAVER *Iris pseudo-acarus* (Yellow Flag) Lyte. OE leafer, preserved as Lever or Levver in the south of England.
LAVEROCK; LAVEROCK, Three-leaved *Oxalis acetosella* (Wood Sorrel) both Yks. Grigson.
LAVEROCK'S LINT *Linum catharticum* (Fairy Flax) Lanark. Grigson. Laverock is an old Scottish form of 'lark'.
LAWRELL *Daphne laureola* (Spurge Laurel) Gerard.
LAWSON CYPRESS *Chamaecyparis lawsoniana.* The name commemorates Peter

Lawson, an Edinburgh nurseryman, who often promoted botanical expeditions, but it was actually introduced by Andrew Murray in 1854.

LAWYER, Bush 1. *Rubus australis* New Zealand Goldie. 2. *Rubus cissoides*. Presumably for the same reason that brambles are called Country Lawyers in this country.

LAWYER, Impudent *Linaria vulgaris* (Toadflax) USA Coats.

LAWYER-WEED 1. *Ranunculus acris* (Meadow Buttercup) Som. Grigson. 2. *Ranunculus bulbosus* (Bulbous Buttercup) Som. Grigson. 3. *Ranunculus repens* (Creeping Buttercup) Som. Grigson. The prickly seeds, especially those of Corn Buttercup, which doesn't appear on this list, are difficult to get rid of, just like lawyers (cleavers and brambles earn the name, too).

LAWYERS 1. *Rosa canina* (Dog Rose) Surr, War. Grigson. 2. *Rubus fruticosus* (Bramble) Suss, Worcs. Grigson; Berks Lowsley.

LAWYERS, Country *Rubus fruticosus* (Bramble) Lincs. Grigson.

LAXATIVE RAM *Rhamnus cathartica* (Buckthorn) Gerard. Very reasonable - the berries are a powerful purgative. " A rough purge, but a very good one", Hill said.

LAY-A-BED *Taraxacum officinale* (Dandelion) Mabey. It sounds rather a hybrid, or perhaps just a politeness - Piss-a-bed it ought to be.

LAYLOCK 1. *Cardamine pratensis* (Lady's Smock) Dev. Grigson; Wilts. Dartnell & Goddard. 2. *Syringa vulgaris* (Lilac) Som. Elworthy; Dor. Dacombe; Wilts. Dartnell & Goddard; Yks. Nicholson.

LAZARUS-BELL *Fritillaria meleagris* (Snake's-head Lily) Dev. Friend. The reference is to the bells that lepers once had to carry about with them. Leopard's Lily is another reference, for 'leopard' here is a corruption of leper.

LAZY-BONES *Potentilla sterilis* (Barren Strawberry) Dor. Macmillan.

LEAD, Red 1. *Dactylorchis incarnata* (Early Marsh Orchid) B & H. 2. *Dactylorchis maculata* (Heath Spotted Orchid) B & H. 3. *Orchis mascula* (Early Purple Orchis) B & H. 4. *Orchis morio* (Green-winged Orchid) B & H.

LEAD PLANT *Amorpha canescens*.

LEADWORT 1. *Ceratostigma plumbaginoides*. 2. *Minuartia verna* (Spring Sandwort) Yks. Condry. Leadwort is explained by Ceratostigma's synonym, Plumbago, while the sandwort gets the name because it is often to be found growing around old lead workings.

LEADWORT, Cape *Plumbago capensis*.

LEADWORT, Ceylon *Plumbago zeylanica*.

LEADWORT, Rosy-flowered *Plumbago indica* (Fire Plant) Chopra.

LEADWORT, White-flowered *Plumbago zeylanica* (Ceylon Leadwort) Watt.

LEAP-UP-AND-KISS-ME *Viola tricolor* (Pansy) Som, Hants, Suss. Grigson. One of a multitude of names of the Kiss-me-at-the-garden-gate type.

LEAPING CUCUMER *Ecballium elaterium* (Squirting Cucumber) Turner.

LEATHER *Iris pseudo-acarus* (Yellow Flag) Som. Macmillan. OE leafer, better preserved as Levver, or Laver.

LEATHER FLOWER *Clematis virona* USA H Smith.1928.

LEATHER-LEAVED ASH *Fraxinus velutina var. coriacea*.

LEATHERLEAF *Chamaedaphne calyculata*. i.e. with leathery leaves.

LEATHERWOOD *Eucryphia billardieri*.

LEBANON CEDAR *Cedrus libani*.

LEBANON OAK *Quercus libani*.

LEDGER, Easter; LEDGES, Easter; LEDGES, Waster *Polygonum bistorta* (Bistort) All Cumb. Grigson (Easter); B & H (Waster). "Herb Pudding" was made from bistort leaves on Easter Day, or more properly at Passion-tide, boiled in broth with barley, chives, etc, and served to accompany veal and bacon. The usual name was Ledger Pudding. But what does ledger mean? It is just possible that it may be sedge, for that is an alternative name for the pudding.

LEDGER BARK *Cinchona ledgeriana*.

LEECHWORT *Sanicula europaea* (Sanicle) Storms. 'Leech' in the sense of a medical recipe, for sanicle comes from Latin *sanare*, to heal.

LEEK *Allium porrum*. OE leac, and very important in Anglo-Saxon times, as can be judged from the fact that the leac-tun (leek garden) was the common name for a kitchen garden, and the leac-ward (leek-keeper) meant a gardener.

LEEK, Brook *Arum dracunculus* (Dragon Arum) W Miller.

LEEK, Crow *Endymion nonscriptus* (Bluebell) Prior.

LEEK, Crow, Great *Allium ursinum* (Ramsons) Cockayne.

LEEK, Dog *Ornithogalum umbellatum* (Star-of-Bethlehem) Turner.

LEEK, Dog's *Gagea lutea* (Yellow Star-of-Bethlehem) Turner.

LEEK, Few-flowered *Allium paradoxum*.

LEEK, French *Allium porrum* (Leek) Lyte.

LEEK, Goose *Narcissus pseudo-narcissus* (Daffodil) IOM Hare. Manx Las-ny guiy. There is an old superstition there that if two or three flowers are brought into the house in early spring, before the goslings are hatched, it would be bad luck indeed for the poultry-keeper. Cf. cowslip and primrose. In Cornwall, too, they used to say that if a goose saw a daffodil before hatching its goslings, it would kill them when they did hatch. A Dorset compromise says that you must take care always to bring indoors only daffodils of good quality, otherwise something is sure to go wrong with your poultry.

LEEK, Lady's *Allium cernuum* (Nodding Wild Onion) Douglas.

LEEK, Lily- *Allium moly* (Golden Garlic) W Miller.

LEEK, Meadow *Allium canadense* (Canada Garlic) Douglas.

LEEK, Round-headed *Allium sphaerocephalum*.

LEEK, Rose *Allium canadense* (Canada Garlic) Douglas.

LEEK, Rush *Allium vineale* (Crow Garlic) USA Grigson.

LEEK, Sand *Allium scorodoprasum*.

LEEK, Sea *Scilla maritima* (Squill) Pomet. Cf. Sea Onion for this.

LEEK, Sour *Rumex acetosa* (Wild Sorrel) Roxb, Ire. B & H.

LEEK, Stone *Allium fistulosum* (Welsh Onion) W Miller.

LEEK, Three-cornered *Allium triquetrum*.

LEEK, Water *Allium ursinum* (Ramsons) Som. Macmillan.

LEEK, Wild 1. *Allium ampeloprasum*. 2. *Allium schoernoprasum* (Chives) S Africa Watt. 3. *Allium tricoccum* (Wood Leek) USA Tolstead. 4. *Allium ursinum* (Ramsons) Border B & H. 5. *Allium vineale* (Crow Garlic) Henslow.

LEEK, Wood *Allium tricoccum*.

LEEK-CRESS *Alliaria petiolata* (Jack-by-the-hedge) B & H. Another name, Garlic Mustard, explains this.

LEEK-LEAVED SALSIFY *Tragopogon porrifolius* (Salsify) G M Taylor.

LEEMERS *Corylus avellana* nuts (Hazel nuts) Lancs. Nodal. Only given when the nuts are ripe.

LEGS, Crow's *Endymion nonscriptus* (Bluebell) Wilts. Dartnell & Goddard. Bluebell has a lot of associations with the crow (and the cuckoo). It is Crow-bells, or Crow-leek, Crowfoot and Crow-toes, as well as simply Crowflower.

LEGS, Keek *Orchis mascula* (Early Purple Orchis) Kent Grigson.

LEGS, Kite's 1. *Colchicum autumnale* (Meadow Saffron) Kent Grigson. 2. *Orchis mascula* (Early Purple Orchis) Kent Grigson. Presumably a reference to the "naked" look of the flower stems, which appear before the leaves.

LEGS, Long *Primula veris* (Cowslip) Som. Macmillan.

LEGS, Red 1. *Polygonum aviculare* (Knotgrass) W Eng, Norf. Friend. 2. *Polygonum bistorta* (Bistort) Prior. 3.

Polygonum persicaria (Persicaria) Som. Macmillan; Ches. Grigson.

LEGS, Skeat 1. *Dactylorchis maculata* (Heath Spotted Orchid) Kent. Grigson. 2. *Orchis mascula* (Early Purple Orchis) Kent B & H. Apparently from OE sceat, meaning "any description of wrapping or swathing". The leg, or stem, of these plants are partially wrapped in a sheathing leaf.

LEGS, Tangle- *Viburnum alnifolium* (Hobble-bush) USA Perry. There are a number of names for this, like Down-you-go, or Devil's Shoestrings, and indeed Hobble-bush itself, that result from the observation of how the lower branches droop to the ground and root at the tips.

LEGS, Twiny *Bartsia odontites* (Red Bartsia) Dev. Friend.1882.

LEGWORT *Ranunculus ficaria* (Lesser Celandine) Som. Macmillan.

LEICHTLIN LILY *Lilium leichtlinii*. Max Leichtlin (1831-1910), the foremost nineteenth century lily grower.

LEMMAROSA *Citrus aurantium var. bergamia* (Bergamot Orange) Lindley.

LEMON 1. *Citrus limonium*. 2. *Lippia citriodora* (Lemon Verbena) Dev. Friend.1882. There is some confusion about the word. One is not sure, for instance, whether the French limon means a lemon or a lime.

LEMON, Ground *Podophyllum peltatum* (May-apple) Douglas.

LEMON, Sweet *Citrus limetta* (Sweet Lime) Douglas).

LEMON, Water *Passiflora laurifolia* (Jamaica Honeysuckle) Thrimble.

LEMON, Water, Wild *Passiflora foetida* J Smith.

LEMON, Wild *Podophyllum peltatum* (May-apple) E Hayes.

LEMON BALM *Melissa officinalis* (Bee Balm) New Engl. Sanford. The leaves have a lemon fragrance.

LEMON BASIL *Ocymum citriodorum*.

LEMON BOTTLEBRUSH *Callistemon citrinus*. Lemon-scented leaves.

LEMON FLOWER; LEMONADE *Agrimonia eupatoria* (Agrimony) Som. Macmillan. Because of the colour, perhaps? Surely there is no lemon fragrance?

LEMON GRASS *Cymbopogon citratus*. The lemon flavour is in the bulb, and it is grown commercially for lemon oil for use in soap powders, flavourings and cosmetics.

LEMON LILY 1. *Hemerocallis flava* (Day Lily) Coats. 2. *Lilium parryi*.

LEMON MINT 1. *Monarda citriodora* (Prairie Bergamot) A W Smith. 2. *Monarda pectinata* (Horse Mint) USA Elmore. 3. *Monarda punctata* (American Horse Mint) A W Smith.

LEMON MONARDA *Monarda pectinata* (Horse Mint) Elmore.

LEMON PAEONY *Paeonia mlokosewitschii*. From the colour of the flowers.

LEMON PLANT; LEMON TREE *Lippia citriodora* (Lemon Verbena) Som. Elworthy (Plant); B & H (Tree).

LEMON-SCENTED CUCUMBER *Cucumis melo var. dudaim* (Dudaim Melon) Howes.

LEMON SUMACH *Rhus aromatica* (Sweet Sumach) Howes.

LEMON THYME *Thymus serphyllus var. citriodorus*.

LEMON VERBENA 1. *Hedeoma drummondii* (Drummond Pennyroyal) USA Elmore. 2. *Lippia citriodora*.

LEMONADE; LEMON FLOWER *Agrimonia eupatoria* (Agrimony) Som. Macmillan.

LEMONADE BERRY; LEMONADE SUMACH *Rhus trilobata* (Three-leaf Sumach) USA Elmore (Berry); Yanovsky (Sumach). The berries are crushed to make a cooling drink. Lemonade-berry is also used for *Rhus integrifolia* (Sourberry) H & P.

LEMONADE TREE 1. *Rhus integrifolia* (Sourberry) Le Strange. 2. *Rhus typhina* (Stag's-horn Sumach) Le Strange. Those sour berries are made by the Indians into cooling drinks.

LEMONWOOD, New Zealand *Pittosporum eugenioides*. It is the leaves that have the lemon scent.

LENGA *Nothofagus pumilo*.

LENORMAND'S WATER CROWFOOT *Ranunculus lenormandii*.

LENSCALE *Atriplex lentiformis* (Quailbrush) Jaeger.

LENT *Narcissus pseudo-narcissus* (Daffodil) Friend. A commemoration of the time of flowering.

LENT-COCKS; LENYCOCKS *Narcissus pseudo-narcissus* (Daffodil) Friend (Lent-cocks); Som. Tongue (Lenycocks). An apparent reference to cock-throwing. When that barbarous sport fell into desuetude, it was the daffodils that were decapitated instead.

LENT CUP; LENTY CUPS; LENT PITCHER *Narcissus pseudo-narcissus* (Daffodil) Greenoak (Cup); Som. Macmillan (Lenty Cups) Dev, Som. Elworthy (Pitcher).

LENT LILY; LENTY LILY *Narcissus pseudo-narcissus* (Daffodil) Corn. Courtney; Dev. Friend; Dor. Dacombe; Som. Elworthy; Herts. Jones-Baker; Ches. Leigh (all Lent Lily) Greenoak (Lenty Lily).

LENT ROSE 1. *Narcissus x biflorus* (Primrose Peerless) Dev. Friend.1882. 2. *Narcissus pseudo-narcissus* (Daffodil) Friend. 3. *Primula vulgaris* (Primrose) Dev. Macmillan.

LENTEN *Tilia europaea* (Common Lime) Halliwell. In this case,a version of Linden.

LENTEN ROSE *Helleborus orientalis*. Blooms February to April, but probably given to distinguish it from the Christmas Rose, of which this is probably a natural hybrid.

LENTIGO *Lemna minor* (Duckweed) Henslow. Cf. Gerard's Water Lentils for this.

LENTIL 1. *Lens esculenta*. 2. *Narcissus pseudo-narcissus* (Daffodil) Dev. B & H. Latin lens, lentis. But not for the daffodil, for this is just a misapplication of the various 'lent' names.

LENTIL, Water *Lemna minor* (Duckweed) Gerard.

LENTIL, Wild *Astragalus cicer*.

LENTISK *Pistacia lentiscus*. Probably better known as Mastick Tree.

LEOPARD LILY 1. *Dieffenbachia segnine* (Dumb-cane) RHS. 2. *Fritillaria meleagris* (Snake's-head Lily) Dev. Friend. 3. *Lilium catesbaei* (Catesby's Lily) USA Woodcock. 4. *Lilium pardalinum* (Panther Lily) Woodcock. 5. *Sansevieria trifasciata* (Mother-in-law's Tongue) USA. Cunningham. Not, as may be thought, because of the spots on Snake's-head Lily - this is a corruption of Leper's Lily. Cf. Lazarus Bell - because of the resemblance to the bells that lepers were once forced to carry. Presumably, the Sansevieria actually has got spots.

LEOPARD PLANT *Ligularia tussilaginea*.

LEOPARD'S BANE 1. *Arnica montana* (Arnica) Lloyd. 2. *Doronicum plantagineum*. 3. *Paris quadrifolia* (Herb Paris) Friend. Turner has Libardbayne for Herb Paris.

LEOPARD'S BANE, Crayfish *Doronicum pardalianches* (Great Leopard's Bane) W Miller. Crayfish, from the shape of the roots, and probably invented by Gerard.

LEOPARD'S BANE, Great *Doronicum pardalianches*. "It is reported and affirmed, that it killeth Panthers, Swine, Wolves, and all kindes of wilde beasts, being given them with flesh..." (Gerard).

LEOPARD'S HEAD *Fritillaria meleagris* (Snake's-head Lily) Leyel.1948. 'Leopard' here is 'leper'. Cf. Leopard Lily.

LEST-WE-FORGET *Reseda lutea* (Wild Mignonette) Som. Macmillan.

LETTUCE *Lactuca sativa*. From Lactuca, eventually from lac, milk - a reference to the milky juice.

LETTUCE, Acrid; LETTUCE, Bitter *Lactuca virosa* (Great Prickly Lettuce) Grieve.1931 (Acrid); Usher (Bitter).

LETTUCE, Blue 1. *Lactuca perennis*. 2. *Lactuca pulchella* (Larkspur Lettuce) USA Allan.

LETTUCE, Blue, Tall *Lactuca spicata*.

LETTUCE, Canker *Pyrola rotundifolia* (Round-leaved Wintergreen) O P Brown.

LETTUCE, Corn *Valerianella locusta* (Cornsalad) Brouk.

LETTUCE, Cos *Lactuca sativa var. longifo-*

lia. Called after the Greek island of that name.

LETTUCE, Hare's 1. *Mycelis muralis* (Wall Lettuce) Cockayne. 2. *Sonchus oleraceus* (Sow Thistle) Henslow. Cockayne tells the same story of Wall Lettuce as most others do of Sow Thistle - "of this wort, it is said that the hare, when in summer by the vehement heat he is tired, doctors himself with this wort" (Anglo-Saxon Herbal).

LETTUCE, Lamb's 1. *Lapsana communis* (Nipplewort) Som. Macmillan. 2. *Poterium sanguisorba* (Salad Burnet) Som. Macmillan. 3. *Valerianella locusta* (Cornsalad) Prior. The name is given to Cornsalad, so it is said, because it is at its best during the lambing season.

LETTUCE, Larkspur *Lactuca pulchella*.

LETTUCE, Least *Lactuca saligna*.

LETTUCE, Miner's 1. *Claytonia alsinoides* (Pink Purslane) Usher. 2. *Claytonia perfoliata* (Spring Beauty) USA Yanovsky. So called because it saved many Californian gold miners from scurvy. The Indians had used it as greens, and apparently used to lay quantities of the leaves near red ants' nests. The ants circulate all through it, and when after a while the leaves are shaken clear of them, they ate the leaves - the ants gave a sour taste to the leaves, as if vinegar had been put on them. Why go to all the bother? Why not just add a little vinegar, anyway?

LETTUCE, Miner's, Narrow-leaved *Montia spathulata var. tenuifolia*.

LETTUCE, Opium; LETTUCE, Poisonous *Lactuca virosa* (Great Prickly Lettuce) Brownlow (Opium); Salisbury (Poisonous). All lettuces are slightly narcotic and soporific. But none more so than this one, which has juice that hardens when exposed to the air, and which produces a gum called lactucaria, or lettuce opium.

LETTUCE, Prickly *Lactuca serriola*.

LETTUCE, Prickly, Great *Lactuca virosa*.

LETTUCE, Purple *Prenanthes purpurea*.

LETTUCE, Romaine *Lactuca sativa var. longifolia* (Cos Lettuce) Brouk.

LETTUCE, Slender *Lactuca saligna* (Least Lettuce) Salisbury.

LETTUCE, Spanish *Claytonia perfoliata* (Spring Beauty) Salisbury.

LETTUCE, Strong-scented *Lactuca virosa* (Great Prickly Lettuce) Grieve.1931.

LETTUCE, Wall *Mycelis muralis*.

LETTUCE, Water *Pistia stratiotes*.

LETTUCE, White *Prenanthes alba*.

LETTUCE, Wild 1. *Claytonia perfoliata* (Spring Beauty) USA Powers. 2. *Lactuca virosa* (Great Prickly Lettuce) Lyte.

LETTUCE, Willow *Lactuca saligna* (Least Lettuce) W Miller.

LETTUCE, Wood *Lactuca serriola* (Prickly Lettuce) Cockayne.

LETTUCE-LEAF BASIL *Ocymum crispum* (Japanese Basil) Painter.

LEVANT BERRY 1. *Anamirta cocculus* Chopra. 2. *Anamirta paniculata* (Levant Nut) Le Strange.

LEVANT GARLIC *Allium ampeloprasum* (Wild Leek) Douglas.

LEVANT NUT *Anamirta paniculata*. But this comes from India, Ceylon and Indonesia.

LEVANT ROSE *Rosa foetida* double form (Capucine Rose) Whittle & Cook.

LEVANT SEA HOLLY *Eryngium campestre* (Field Eryngo) Gerard.

LEVANT WORMSEED 1. *Artemisia cina*. 2. *Artemisia maritima* (Sea Wormwood) Chopra. The flowers of Artemisia cina (small enough to be known as "seeds") are used for intestinal worms, and often made up into tablets - though large doses can be dangerous.

LEVER *Iris pseudo-acarus* (Yellow Flag) Fernie. OE leafer.

LEVER-WOOD *Ostrya virginiana* (American Hop-Hornbeam) Willis.

LEVVER 1. *Iris pseudo-acarus* (Yellow Flag) Som. Friend.1882; Dor. Macmillan; IOW Long. 2. *Typha latifolia* (False Bulrush) Som. Macmillan. See Lever.

LEWIS FLAX *Linum lewisii* (Rocky Mountain Flax) USA Elmore.

LEWTE *Ononis repens* (Rest-harrow) Som. Halliwell.

LEY *Dianthus caryophyllus* (Carnation) Lancs. Nodal.

LEYLAND CYPRESS *Cupressocyparis leylandii*. The hybrid was raised in 1888 at Leighton Hall, near Welshpool, by C J Leyland.

LEZZORY; LIZZORY *Sorbus torminalis* (Wild Service Tree) Glos. J D Robertson.

LIBBARDINE; LIBBARD'S BANE *Aconitum napellus* (Monkshood) Halliwell. i.e leopard's bane.

LIBERIAN COFFEE *Coffea liberica*.

LIBLONG *Sedum telephium* (Orpine) Gerard. Presumably Livelong, which Gerard uses as well. There is an alternative name to this - Lovelong. But they all refer to a divination practice - pieces were hung up and named for a girl's boyfriends, and the piece that lived the longest determined the successful one.

LICE, Beggar 1. *Galium aparine* (Goose-grass) Northants. A E Baker. 2. *Lappula virginiana* USA Youngken.

LICE, Beggar's 1. *Daucus pusillus* USA Elmore. 2. *Galium aparine* (Goose-grass) Glos, Wilts, Hants, Bucks,. Northants Dyer. Lice for the burrs seems particularly appropriate, both for their appearance and for the difficulty in getting rid of them.

LICE, Buckie *Rosa canina* seeds (Dog Rose seeds) S Scot, Ire. Grigson. Children put the seeds down each other's backs - these are the Itchy-backs of the Midlands.

LICE, Fleas-and- *Cymbalaria muralis* (Ivy-leaved Toadflax) Som. Macmillan.

LICE, Harvest 1. *Agrimonia eupatoria* fruit (Agrimony) Hants. Grigson. 2. *Galium aparine* fruit (Goose-grass) Hants. Grigson. Descriptive, of course.

LICE, Lops-and- *Rosa canina* hips (Dog Rose hips) Dur. F M T Palgrave.

LICEBANE *Delphinium staphisagria* (Stavesacre) Prior. As long ago as Pliny's time, the powdered seeds were used for destroying vermin on the head and body. Cf. Lousebane and Lousewort.

LICHEN, Beard *Usnea barbata*.

LICHEN, Crab's-eye *Ochrolechia parella* (Crawfish Lichen) Bolton.

LICHEN, Crawfish *Ochrolechia parella*.

LICHEN, Cudbear *Ochrolechia tartarea*. 'Cudbear' is a reference to the purple dyestuff obtained from this lichen, commercialized by Dr Cuthbert (hence Cudbear) Gordon.

LICHEN, Dog *Peltigera canina* (Ash-coloured Ground Liverwort) Brightman.

LICHEN, Hoary, Ragged *Evernia prunaseris* (Stag's-horn Lichen) Bolton.

LICHEN, Iceland *Cetraria islandica* (Iceland Moss) Flück. Not confined to Iceland, of course, for it has been eaten by the Lapps since ancient times.

LICHEN, Stag's-horn *Evernia prunaseris*.

LICHEN, Wall, Yellow, Common *Xanthelia parientina*.

LICHWALE, LYTHEWALE *Lithospermum officinale* (Gromwell) Gerard (Lichwale); Dyer (Lythewale).

LICHWORT *Parietaria diffusa* (Pellitory-of-the-wall) Halliwell.

LICK *Allium porrum* (Leek) Dev. Friend.1882. Local pronunciation. "As green as a lick" used to be a common Devonshire simile.

LICORICE *Astragalus glycyphyllus* (Milk Vetch) Perring. A variant on the more usual Liquorice. This plant, of course, is not Liqurice, but it is known as Liquorice Vetch, or Wild Liquorice.

LIDE LILY *Narcissus pseudo-narcissus* (Daffodil) Friend. Lide is the OE word for the month of March, Hlyd-monath. Cf. such names as Lent Lily.

LIE-ABED, Little *Senecio vulgaris* (Groundsel) Som. Macmillan. Presumably to distinguish it fom the larger flowers of dandelion and hawkbit, Macmillan suggests.

LIFE, Thread-of- 1. *Cymbalaria muralis* (Ivy-leaved Toadflax) Bucks. Harman. 2. *Saxifraga sarmentosa* (Mother-of-thousands) Northants. B & H. Descriptive of the runners and method of reproduction.

LIFE EVERLASTING 1. *Anaphalis margaretacea* (Pearly Immortelle) B & H. 2.

Antennaria dioica (Cat's Foot) Grieve.1931. 3. *Antennaria plantaginifolia* (Plantain-leaved Everlasting). Grieve.1931.
LIFE EVERLASTING, Pearl-flowered
Anaphalis margaretacea (Pearly Immortelle) Grieve.1931.
LIFE EVERLASTING, Sweet-scented
Gnaphalium polycephalum (Mouse-ear Everlasting) O P Brown.
LIFE-LONG *Sedum telephium* (Orpine) Hants. Read. Livelong is the more usual form.
LIFE-OF-MAN 1. *Hemerocallis flava* (Day Lily) Pemb. Awbery. 2. *Tradescantia virginica* (Spiderwort) Dor. Macmillan. An ironic comment on the very short-lived flowers.
LIFE-ROOT *Senecio aureus* (Swamp Squaw-weed) O P Brown.
LIFT-UP-YOUR-HEAD-AND-I'LL-KISS-YOU *Dicentra spectabilis* (Bleeding Heart) Worcs. Wright. A name that is reminiscent of the erotic series of names given to the pansy.
LIGHT-FIRE *Pteridium aquilina* (Bracken) Lincs. Tynan & Maitland. The dried fronds were often used as kindling.
LIGHTNING; LIGHTNING-FLOWER
Papaver rhoeas (Corn Poppy) N'thum. Grigson. Colour descriptive.
LIGHTNING, Scarlet 1. *Centranthus ruber* (Spur Valerian) Dev, Hants. Grigson. 2. *Lychnis chalcedonica* (Rose Campion) Dev. Friend.1882; Dor. Dacombe; Glos. J D Robertson. Presumably the name must have started as Scarlet Lychnis.
LIGHTNING, Thunder-and- 1. *Ajuga reptans* (Bugle) Glos. Grigson. 2. *Pulmonaria officinalis* (Lungwort) Banff. Grigson. Bugle growing on a building protected it from lightning, so it was said in Wiltshire. Much more widespread is the same belief connected with houseleek. Cockayne has Thunder Clover for this. There seems to be no similar belief linking lungwort with lightning, and in this case, it is probably one of a lot of double names suggesting the two-coloured flowers.

LIGHTWORT *Mertensia maritima* (Smooth Lungwort) Hill. Apparently, Hill coined the name as an equivalent for *Pneumaria*, a name he proposed for the plant.
LIGNUM VITAE *Guaiacum officinale*.
LIKIANG LILY *Lilium papilliferum*.
LILAC 1. *Cardamine pratensis* (Lady's Smock) N'thants Clare. 2. *Syringa vulgaris*. From the Persian word nilak, where nil means blue. The word was introduced with the shrub. For Lady's Smock, this can only be a comment on the colour of the flowers.
LILAC, American *Centranthus ruber* (Spur Valerian) Dev. Friend.1882. Cf. German Lilac and Ground Lilac for this.
LILAC, California *Ceanothus integerrimus* (Deer-brush) Calif. Mason.
LILAC, French *Galega officinalis* (Goat's Rue) Clapham. It comes from southern Europe - Cf. Italian Vetch.
LILAC, German *Centranthus ruber* (Spur Valerian) Lincs. Friend.1882.
LILAC, Ground *Centranthus ruber* (Spur Valerian) Lincs. Grigson.
LILAC, Himalayan *Syringa emodi*.
LILAC, Hungarian *Syringa josikae*. It grows in the mountains of Roumania and the Ukraine.
LILAC, Indian 1. *Lagerstoemeria indica* (China Privet) Chopra. 2. *Melia azedarach* (Chinaberry Tree) Le Strange.
LILAC, Persian 1. *Melia azedarach* (Chinaberry Tree) Polunin.1976. 2. *Syringa persica*. Chinaberry grows in northern India and Afghanistan - hence the other name, Indian Lilac.
LILAC, Rouen *Syringa x chinensis*.
LILAC, Senegal *Lonchocarpus sericeus*.
LILAC, Summer 1. *Buddleia davidii* Grigson. 2. *Hesperis matronalis* (Sweet Rocket) Som. B & H.
LILAC, Vine *Hardenbergia violacea* (Blue Coral Pea) RHS.
LILAC, Wall *Centranthus ruber* (Spur Valerian) Dev. Macmillan.
LILAC-FLOWER *Mentha aquatica* (Water

Mint) Som. Macmillan. The colour is the reference in this case.

LILLY-CONVALLY *Convallaria maialis* (Lily-of-the-valley) Aubrey(1857). One of a number of variations on Lily plus *Convallaria*.

LILLY-PILLY, Australian *Eugenia smithii*.

LILY 1. *Arum maculatum* (Cuckoo-pint) Wilts. Dartnell & Goddard. 2. *Calystegia sepium* (Great Bindweed) Wilts. Dartnell & Goddard. 3. *Convolvulus arvensis* (Field Bindweed) Hants, Suss. Grigson. 4, *Iris foetidissima* (Stinking Iris) Som. Macmillan. 5. *Lilium sp.* 6. *Narcissus maialis* (Pheasant's Eye) War. B & H. 7. *Narcissus pseudo-narcissus* (Daffodil) Scot. Grigson. 8. *Polygonum convolvulus* (Black Bindweed) Hants, IOW Cope. 9. *Ranunculus aquatilis* (Water Crowfoot) Wilts. Goddard. Perhaps 'lily' for the daffodil is just a sequel to 'Dilly'. In the same way, Pheasant's Eye is sometimes known as White Dilly, so there should be a form White Lily somewhere. But 'lily' for the bindweeds is an appreciation of the appearance of the flowers.

LILY, African *Agapanthus africanus*.

LILY, Amazon *Eucharis grandiflora*.

LILY, Annunciation *Lilium candidum* (Madonna Lily) Woodcock.

LILY, Arum 1. *Arum maculatum* (Cuckoo-pint) Cambs. Porter. 2. *Zantedeschia aethiopica*.

LILY, Atamasco *Zephyranthes atamasco*.

LILY, Aztec *Sprekelia formosissima* (Jacobean Lily) Simmons.

LILY, Baker *Lilium bakerianum*.

LILY, Barbados *Hippeastrum equestre*.

LILY, Belladonna *Amaryllis belladonna*.

LILY, Bermuda *Lilium longiflorum var. eximium*. A Japanese plant, but much grown since last century in the Bermuda lily industry.

LILY, Bine 1. *Calystegia sepium* (Great Bindweed) Dor. Macmillan. 2. *Convolvulus arvensis* (Field Bindweed) Dor. Macmillan. Cf. Lily-bind. The 'lily' designation is joined to the 'bind' element in these names.

LILY, Black *Fritillaria camschatcensis* Woodcock.

LILY, Black Sea *Lilium ponticum*. 'Ponticum' - the ancient kingdom of Pontus was in northeast Asia Minor, the Black Sea area.

LILY, Blood *Scadoxus multiflorus var. katharinae*.

LILY, Blue *Agapanthus africanus* (African Lily) S Africa Whittle & Cook.

LILY, Boat *Rhoeo spathacea*.

LILY, Bourbon *Lilium candidum* (Madonna Lily) Woodcock.

LILY, Bush *Astelia nervosa*.

LILY, Calla *Zantedeschia aethiopica* (Arum Lily) Simmons.

LILY, Canada *Lilium canadense*.

LILY, Cape *Crinum sp*.

LILY, Carolina *Lilium michauxii*.

LILY, Catesby's *Lilium catesbaei*. This is Mark Catesby (c1680-1749), who first described the lily in his *Natural History of Carolina*.

LILY, Cathay *Cardiocrinum cathayanum*.

LILY, Caucasian *Lilium monadelphum*.

LILY, Chalice *Narcissus pseudo-narcissus* (Daffodil) Tynan & Maitland. Or Chalice Flower.

LILY, Chaparral *Lilium rubescens*.

LILY, Chatham Isle *Myosotidium hortensia* (Chatham Isle Forget-me-not) New Zealand Perry.

LILY, Chequered *Fritillaria meleagris* (Snake's-head Lily) Parkinson. Cf. Chequered Tulip and Chequered Daffodil for this. It is the peculiar colouring of the flowers that suggests the name.

LILY, Climbing *Gloriosa superba* (Glory Lily) Tampion.

LILY, Coast *Lilium maritimum*.

LILY, Cobra *Darlingtonia californica* (Cobra Plant) USA Cunningham.

LILY, Columbia *Lilium columbianum*.

LILY, Conval *Convallaria maialis* (Lily-of-the-valley) Parkinson.

LILY, Coral *Lilium pumilum*.

LILY, Corn 1. *Calystegia sepium* (Great Bindweed) Yks. B & H. 2. *Convolvulus* *arvensis* (Field Bindweed) Yks. Grigson. 3. *Veratrum californicum* (False Hellebore) USA Schenk & Gifford.

LILY, Corn, African *Ixia sp.*

LILY, Cow 1. *Caltha palustris* (Marsh Marigold) USA Grigson. 2. *Nymphaea advena* (Large Yellow Pond Lily) USA Tolstead.

LILY, Crazy *Caltha palustris* (Marsh Marigold) Dor. Macmillan. A whole group of names for Marsh Marigold begins with Crazy, and continues with Crazy Bess, Crazy Bet, etc. They are probably related to Drunkards, another name common in the west country. If you look long at them, you will be sure to take to drink. There must surely be a similar belief to account for Crazy.

LILY, Crumple 1. *Lilium martagon* (Turk's Cap Lily) Dev. Friend. 2. *Lilium tigrinum* (Tiger Lily) Dev. Friend.

LILY, Cuban *Scilla peruviana* (Peruvian Lily) Whittle & Cook.

LILY, Cuckoo *Arum maculatum* (Cuckoo-pint) Cambs. PLNN6. A hybrid name, mixing the 'cuckoo' nomenclature with Lily.

LILY, Dahurian *Lilium dauricum.*

LILY, David *Lilium davidii.* This is the French missionary and naturalist Père Armand David (1826-1900), who collected the lily in 1869.

LILY, Day 1. *Hemerocallis flava.* 2. *Taraxacum officinale* (Dandelion) Som. Macmillan.

LILY, Day, Narrow-leaved *Hemerocallis minor.*

LILY, Day, Tawny *Hemerocallis fulva.*

LILY, Desert *Hesperocallis undulata.*

LILY, Dog-tooth *Erythronium oregonum* (Giant Adder's Tongue) USA Kingsbury.

LILY, Drooping *Galanthus nivalis* (Snowdrop) Som. Macmillan.

LILY, Duchartre's *Lilium duchartrei.*

LILY, Easter 1. *Lilium longiflorum.* 2. *Narcissus pseudo-narcissus* (Daffodil) Dev. Friend; Som. Tongue. The true lily gets the name because it is forced in great quantity for Easter sale.

LILY, Eastwood *Lilium nevadense* (Californian Panther Lily) Woodcock. Alice Eastwood, botanist.

LILY, Eureka *Lilium occidentale.*

LILY, Farges *Lilium fargesii.* Paul Farges collected many specimens of Chinese plants for the Paris Museum.

LILY, Fawn 1. *Erythronium americanum* (Yellow Adder's Tongue) Le Strange. 2. *Erythronium oregonum* (Giant Adder's Tongue) USA Kingsbury.

LILY, Field *Iris foetidissima* (Stinking Iris) Dor. Grigson.

LILY, Fire *Lilium tigrinum* (Tiger Lily) Dev. Macmillan.

LILY, Fire-ball *Haemanthus cinnabarinus.*

LILY, Flag 1. *Iris pseudo-acarus* (Yellow Flag) Som. Macmillan. 2. *Iris versicolor* (Purple Water Iris) Grieve.1931.

LILY, Flame *Lilium philadelphicum* (Philadelphia Lily) USA Woodcock.

LILY, Flax 1. *Dianella caerulea.* 2. *Phormium tenax* (New Zealand Flax) W Miller.

LILY, Formosa *Lilium formosanum.*

LILY, Foxtail *Eremurus sp.*

LILY, George *Vallota speciosa* (Scarborough Lily) S Africa Watt.

LILY, Giant *Cardiocrinum giganteum.* 'Giant' is right - these can be anything up to 12 feet tall.

LILY, Glacier *Erythronium grandiflorum.*

LILY, Glade *Lilium philadelphicum* (Philadelphia Lily) USA Woodcock.

LILY, Glory *Gloriosa superba.*

LILY, Goldband *Lilium auratum* (Japanese Lily) Woodcock.

LILY, Gray's *Lilium grayi.* The name commemorates the American botanist Asa Gray, who first collected this lily.

LILY, Green *Helleborus viridis* (Green Hellebore) Wilts. Dartnell & Goddard.1899.

LILY, Ground 1. *Convolvulus arvensis* (Field Bindweed) Wessex Rogers. 2. *Trillium erectum* (Bethroot) Grieve.1931.

LILY, Guernsey *Nerine sarniensis.* Said to have grown in Guernsey from bulbs washed ashore from a wreck of a ship from Japan about 1659, but the local legend is that a Guernsey girl, Michelle de Garis, married a fairy man and went off with him. She asked him for something to leave her family to let them have a token with which to remember her, and he gave her a bulb to plant in the sand above Vazon Bay. This was the Guernsey Lily, which, it was said, would grow in none of the other Channel Islands.

LILY, Hanson *Lilium hansoni*. Hanson was an amateur lily grower in Brooklyn.

LILY, Harvest *Calystegia sepium* (Great Bindweed) Surr. Grigson.

LILY, Heartleaf *Cardiocrinum cordatum*.

LILY, Hedge *Calystegia sepium* (Great Bindweed) Som. Macmillan; Hants, IOW Cope.

LILY, Henry *Lilium henryi*. Professor Augustine Henry was the original discoverer of this lily.

LILY, Herb *Alstroemaria aurantiaca* (Peruvian Lily) Salisbury.

LILY, Hong Kong *Lilium brownii var. australe*. Not confined to Hong Kong, though it is only found in that part of southern China.

LILY, Huckleberry *Lilium philadelphicum* (Philadelphia Lily) USA Woodcock.

LILY, Humboldt *Lilium humboldtii*. Not collected by Humboldt, but named after him apparently because it was found on his birthday.

LILY, Ichang *Lilium leucanthum*.

LILY, Imperial *Fritillaria imperialis* (Crown Imperial) Gerard.

LILY, Jacobean *Sprekelia formosissima*.

LILY, Jamestown *Datura stramonium* (Thorn-apple) USA Henkel. From the well-known story of the soldiers sent to Jamestown to quell the uprising known as Bacon's Rebellion (1676), who cooked some of the young plants as potherbs - "the effect of which was a very pleasant Comedy; for they turn'd natural Fools upon it for several days...".

LILY, Japanese 1. *Lilium auratum*. 2. *Lilium japonicum*.

LILY, Jersey *Vallota speciosa* (Scarborough Lily) Som. Macmillan. By association with the Guernsey Lily, presumably, but as the common name suggests, nothing to do with the Channel Islands. The link is the fact that both plants owe their names to bulbs being washed ashore from a wrecked ship.

LILY, Kaffir *Schizostylis coccinea*.

LILY, Kellogg *Lilium kelloggii*. Dr Albert Kellogg was one of the founders of the California Academy of Sciences.

LILY, Kerry *Simethis planifolia*. Only to be found in co Kerry in these islands.

LILY, Kesselring *Lilium kesselringianum*.

LILY, Knysna *Vallota speciosa* (Scarborough Lily) S Africa Watt.

LILY, Kochang *Lilium distichum*.

LILY, Korean *Lilium amabile*.

LILY, Lady *Lilium candidum* (Madonna Lily) Bucks. Friend.

LILY, Lankong *Lilium langkongense*. Lankong is in NW China, about 20 miles NW of Tali lake.

LILY, Lanthorn *Narcissus pseudo-narcissus* (Daffodil) War. A S Palmer. It is tempting to see this as a descriptive name, but it is actually a corruption of Lenten Lily.

LILY, Leichtlin *Lilium leichtlinii*. Max Leichtlin (1831-1910), the foremost 19th century lily grower.

LILY, Lemon 1. *Hemerocallis flava* (Day Lily) Coats. 2. *Lilium parryi*.

LILY, Lent; LILY, Lenty *Narcissus pseudo-narcissus* (Daffodil) Corn. Courtney; Dev. Friend; Som. Elworthy Dor. Dacombe; Herts. Jones-Baker; Ches. Leigh (Lent) Greenoak (Lenty).

LILY, Leopard 1. *Dieffenbachia segnine* (Dumb-cane) RHS. 2. *Fritillaria meleagris* (Snake's-head Lily) Dev. Friend. 3. *Lilium catesbaei* (Catesby's Lily) Woodcock. 4. *Lilium pardalinum* (Panther Lily) Woodcock. 5. *Sansevieria trifasciata* (Mother-in-law's Tongue) USA. Cunningham. Not as it seems for Snake's Head Lily, for this is a corruption of Leper's Lily. The shape of the flowers is a reminder of the bells that lepers were once made to carry about. Presumably, mother-in-law's tongue actually has got spots, otherwise the name is meaningless.

LILY, Lide *Narcissus pseudo-narcissus* (Daffodil) Friend. Lide is OE hlyd - Hlydmonath was the month of March, so in essence it means the same as Lent Lily.

LILY, Likiang *Lilium papilliferum*.

LILY, Liver *Iris versicolor* (Purple Water Iris) Grieve.1931. It is not clear whether this is a comment on the colour, or a reference to the fact that the root contains iridin, which has a powerful effect on the liver.

LILY, Loddon *Leucojum aestivum*. Loddon is the river of that name in Berkshire.

LILY, Lough *Nymphaea alba* (White Waterlily) Donegal Grigson.

LILY, Madonna *Lilium candidum*. The lily of sacred art is always the Madonna Lily, the name showing that it was dedicated to the Virgin, and it has the same dedication in Buckinghamshire under the name Lady Lily. It is the symbol of purity, and also, according to Bede, the emblem of the Resurrection of the Virgin.

LILY, Male *Convallaria maialis* (Lily-of-the-valley) Culpepper. i.e. May Lily.

LILY, March *Amaryllis belladonna* (Belladonna Lily) S Africa Watt.

LILY, Mariposa *Callicore venustus*.

LILY, Marsh *Caltha palustris* (Marsh Marigold) Dor. Macmillan.

LILY, May 1. *Convallaria maialis* (Lily-of-the-valley) Gerard. 2. *Maianthemum bifolium*. 3. *Syringa vulgaris* (Lilac) Dev. Tynan & Maitland.

LILY, Mayflower *Convallaria maialis* (Lily-of-the-valley) Tynan & Maitland.

LILY, Meadow 1. *Lilium canadense* (Canada Lily) USA Upham. 2. *Lilium candidum* (Madonna Lily) O P Brown. *Lilium canadense* certainly grows in moist meadows, but 'meadow' sounds very like 'Madonna' in the second record.

LILY, Methonica *Gloriosa superba* (Glory Lily) Dalziel.

LILY, Michigan *Lilium michiganense*.

LILY, Midsummer *Lilium candidum* (Madonna Lily) Som. Macmillan.

LILY, Missouri *Tritelia uniflora*.

LILY, Morning Star *Lilium concolor*.

LILY, Mountain *Lilium martagon* (Turk's-cap Lily) Gerard.

LILY, Nankeen *Lilium x testaceum*.

LILY, Nepal *Lilium nepalense*.

LILY, Nilgiri *Lilium neilgherrense*. Nilgiri, in southern India.

LILY, Noble *Lilium nobilissimum*.

LILY, Nodding 1. *Lilium canadense* (Canada Lily) USA House. 2. *Lilium cernuum*. 3. *Lilium superbum* (Turk's Cap Lily) USA Woodcock.

LILY, Oaksey *Fritillaria meleagris* (Snake's-head Lily) Wilts. Grigson. Meadows near Oaksey are some of the few places where this plant grows wild.

LILY, Ochre *Lilium primulinum*.

LILY, Orange 1. *Anagallis arvensis* (Scarlet Pimpernel) Dumf. B & H. 2. *Lilium bulbiferum var. croceum*.

LILY, Orange, Wild *Lilium philadelphicum* (Philadelphia Lily) USA Woodcock.

LILY, Orange-cup *Lilium philadelphicum* (Philadelphia Lily) USA Yanovsky.

LILY, Orange-cup, Western *Lilium umbellatum*.

LILY, Orange River *Crinum bulbispermum* S Africa Watt.

LILY, Oregon *Lilium columbianum* (Columbia Lily) Woodcock.

LILY, Palm *Dracaena indivisa* (Fountain Dracaena) Andersen.

LILY, Panther *Lilium pardalinum*.

LILY, Panther, Californian *Lilium nevadense*.

LILY, Paradise *Papaver rhoeas* (Red Poppy) Som. Macmillan.

LILY, Park, Great *Convallaria maialis* (Lily-of-the-valley) B & H.

LILY, Peace *Spathiphyllum patinii* (White Sails) Bonar.

LILY, Persian 1. *Fritillaria imperialis* (Crown Imperial) Parkinson. 2. *Fritillaria* *persica* Gerard. Parkinson too describes Persian Lily under *F persica*. The name is accurate enough for Crown Imperial - it is a native of Persia.

LILY, Peruvian 1. *Alstroemaria aurantiaca*. 2. *Scilla peruviana*.

LILY, Pheasant *Fritillaria meleagris* (Snake's-head Lily) Cumb. B & H. Better as a Guinea-hen than a pheasant. The petals are spotted like the former bird, hence Guinea-hen, Guinea-flower, Turkey-hen Flower, etc.

LILY, Philadelphia *Lilium philadelphicum*.

LILY, Philippine *Lilium philippense*.

LILY, Pig *Arum maculatum* (Cuckoo-pint) Som. Macmillan.

LILY, Pine *Lilium catesbaei* (Catesby's Lily) Woodcock.

LILY, Pixy *Stellaria holostea* (Greater Stitchwort) Tynan & Maitland.

LILY, Plantain *Funkia sp*.

LILY, Pompon *Lilium pomponium*. Obscure - perhaps there is some connection with the Roman family Pomponius, or perhaps it is from the old French word pompon, meaning a melon, or something similar.

LILY, Pond 1. *Begonia feasti* Gannon. 2. *Iris pseudo-acarus* (Yellow Flag) Dev. Macmillan.

LILY, Pond, White *Nymphaea odorata*.

LILY, Pond, Yellow, Large *Nymphaea advena*.

LILY, Pot of Gold *Iris iridollae*. The mythical pot of gold to be found at the end of the rainbow. The specific name is iris, rainbow, plus olla, pot, and the colour of the flowers is responsible for the name.

LILY, Pyrenean *Lilium pyrenaicum*.

LILY, Rain *Zephyranthes atamasco* (Atamasco Lily) USA Kingsbury.

LILY, Ray, Japanese *Lilium auratum* (Japanese Lily) Grant White.

LILY, Red *Lilium philadelphicum* (Philadelphia Lily) USA House.

LILY, Red, Southern *Lilium catesbaei* (Catesby's Lily) USA Woodcock.

LILY, Regal *Lilium regale*.

LILY, Rock *Arthropodium cirrhatum*.

LILY, Sacred, Chinese *Narcissus canaliculatus var. orientalis* (Polyanthus Narcissus). Introduced into China from the Mediterranean area some time before the 16th century. This is the flower specially connected with the Chinese new year.

LILY, St Anthony's *Lilium candidum* (Madonna Lily) Whittle & Cook.

LILY, St Bernard's *Anthericum liliago*.

LILY, St Bruno's *Anthericum liliastrum*.

LILY, St Catherine's *Lilium candidum* (Madonna Lily) Haig. This is St Catherine of Siena, who is usually represened with a lily. Catherine is from Greek katharos, in the sense of purity, i.e. it has the same meaning as the symbolism of the lily, anyway.

LILY, Sargent *Lilium sargentiae*.

LILY, Scarborough *Vallota speciosa*. Like the Guernsey Lily, it got its name through bulbs being washed ashore from a wrecked ship.

LILY, Sea *Pancratium maritimum* (Sea Daffodil) RHS.

LILY, Sierra *Lilium parvum*.

LILY, Slimstem *Lilium callosum*.

LILY, Snake 1. *Brodiaea volubilis*. 2. *Erythronium mesochoreum* (Spring Lily) USA Gilmore. 3. *Iris versicolor* (Purple Water Iris) Grieve.1931.

LILY, Snake's-head *Fritillaria meleagris*.

LILY, Snowdon *Lloydia serotina*.

LILY, Spider *Tradescantia virginica* (Spiderwort) USA House.

LILY, Spider, Golden *Lycoris aurea*.

LILY, Spire *Galtonia candicans*.

LILY, Spring *Erythronium mesochoreum*.

LILY, Star, Knight's *Amaryllis belladonna* (Belladonna Lily) McDonald.

LILY, Stink *Fritillaria imperialis* (Crown Imperial) Coats. Because, so it is said, when the root is rubbed it smells like a fox.

LILY, Stinking *Allium ursinum* (Ramsons) Som. Grigson; Yks. L Gordon.

LILY, Sulphur *Lilium sulphureum*.

LILY, Swamp 1. *Lilium superbum* (Turk's Cap Lily) USA Upham. 2. *Zephyranthes sp*.

LILY, Swamp, Southern *Lilium michauxii*

(Carolina Lily) Woodcock. But apparently it grows on dry slopes, and never in a swamp.

LILY, Sword 1. *Gladiolus triphyllus*. 2. *Iris pseudo-acarus* (Yellow Flag) Som. Macmillan. In each case, it is the shape of the leaves that provides the imagery. Gladiolus is the diminutive of Latin gladius, sword.

LILY, Tali *Lilium taliense*. The Tali range in Ne Yunnan, China, where it was discovered in 1883.

LILY, Thimble *Lilium bolanderi*.

LILY, Thorn *Catesbaea spinosa*.

LILY, Tiger 1. *Lilium catesbaei* (Catesby's Lily) USA Woodcock. 2. *Lilium columbianum* (Columbia Lily) USA E Gunther. 3. *Lilium pardalinum* (Panther Lily) USA Schenk & Gifford. 4. *Lilium tigrinum*.

LILY, Tiger, Wild *Lilium superbum* (Turk's Cap Lily) USA Woodcock.

LILY, Toad 1. *Tricyrtis hirta*. 2. *Trillium sessile* (Toadshade) Perry.

LILY, Torch *Kniphofia uvaria* (Red Hot Poker) Dev. Friend.

LILY, Trout *Erythronium americanum* (Yellow Adder's Tongue) Le Strange.

LILY, Trout, White *Erythronium albidum* (White Adder's Tongue) USA Yanovsky.

LILY, Trumpet *Zantedeschia aethiopica* (Arum Lily) J Smith.

LILY, Tsingtau *Lilium tsingtauense*.

LILY, Turk's-cap 1. *Lilium martagon*. 2. *Lilium superbum*. Descriptive - Martagon is apparently derived from a Turkish word meaning turban.

LILY, Turk's-cap, American *Lilium superbum* (Turk's Cap Lily) Yanovsky.

LILY, Turk's-cap, Scarlet *Lilium chalcedonicum*.

LILY, Turk's-cap, Western *Lilium michiganense*.

LILY, Turk's-cap, Yellow *Lilium pyrenaicum* (Pyrenean Lily) Le Strange.

LILY, Valley *Convallaria maialis* (Lily-of-the-valley) Gerard.

LILY, Voodoo *Sauromatum guttatum*.

LILY, Wallich *Lilium wallichianum*.

LILY, Ward *Lilium wardii*.

LILY, Washington *Lilium washingtonianum*.

LILY, Water 1. *Caltha palustris* (Marsh Marigold) Wilts. Dartnell & Goddard. 2. *Cardamine pratensis* (Lady's Smock) Norf. Grigson. 3. *Iris pseudo-acarus* (Yellow Flag) Dev. Friend.1882. 4. *Nuphar sp.* 5. *Nymphaea sp.* 6. *Nymphoides sp.* 7. *Ranunculus aquatilis* (Water Crowfoot) Som,Wilts. Macmillan; Donegal Grigson. 8. *Zantedeschia aethiopica* (Arum Lily) Ches. Holland. In most cases, better as one word, Waterlily.

LILY, Water, American *Nymphaea odorata* (White Pond Lily) Gorer.

LILY, Water, Fringed *Nymphoides peltatum*.

LILY, Water, Royal *Victoria amazonica*.

LILY, Water, White *Nymphaea alba*.

LILY, Water, White, Lesser *Nymphaea alba var. minor*.

LILY, Water, Yellow *Nuphar lutea*.

LILY, Water, Yellow, Hybrid *Nuphar x intermedia*.

LILY, Water, Yellow, Least *Nuphar pumila*.

LILY, Wheel *Lilium medeoloides*. 'Wheel', because of the whorled arrangement of the leaves.

LILY, White 1. *Calystegia sepium* (Great Bindweed) Cumb. Grigson. 2. *Lilium candidum* (Madonna Lily) Woodcock. 3. *Narcissus maialis var. patellaris* (Pheasant's Eye) B & H.

LILY, Wild 1. *Arum maculatum* (Cuckoo-pint) Dev. Friend. 2. *Calystegia sepium* (Great Bindweed) Wilts. D Grose.

LILY, Wood 1. *Cephalanthera damasonium* (Broad Helleborine) Wilts. Tynan & Maitland. 2. *Convallaria maialis* (Lily-of-the-valley) Fernie. 3. *Lilium philadelphicum* (Philadelphia Lily) USA House. 4. *Trillium grandiflorum*.

LILY, Wood, Painted *Trillium undulatum* (Painted Trillium) RHS.

LILY, Yellow *Narcissus pseudo-narcissus* (Daffodil) B & H.

LILY, Yellow, Wild *Lilium canadense* (Canada Lily) USA House.

LILY-AMONG-THORNS *Lonicera pericly-*

menum (Honeysuckle) Prior. A quote from Canticles, ii. 2, understood, so Prior assures us, to be this honeysuckle (Lilium inter spinas).

LILY-BIND 1. *Calystegia sepium* (Great Bindweed) Yks. Grigson. 2. *Convolvulus arvensis* (Field Bindweed) Yks. B & H. See Bine-Lily.

LILY-CAN *Nuphar lutea* (Yellow Waterlily) Fife, Perth. Grigson. Cf. Can-dock, and Water-can, for this.

LILY-CONFANCY; LILY-CONSTANCY *Convallaria maialis* (Lily-of-the-valley) Som. Macmillan (Confancy); Hatfield (Constancy). Just two of a whole series of fanciful variations that include Liricumfancy and Lilly-convally, all of which are a play on lily and the specific name, *convallaria*.

LILY-FLOWER *Calystegia sepium* (Great Bindweed) Hants. Cope.

LILY-GRASS *Arum maculatum* (Cuckoo-pint) Suss. B & H. Cf. Lily, Wild Lily, etc.

LILY-LEEK *Allium moly* (Golden Garlic) W Miller.

LILY-OAK *Syringa vulgaris* (Lilac) Scot. B & H. A rather extreme variation of lilac.

LILY-OF-THE-INCAS *Alstroemaria peregrina*.

LILY-OF-THE-MOUNTAIN *Polygonatum multiflorum* (Solomon's Seal) War. Grigson. Presumably, as this was once known botanically as *Convallaria multiflora*, to distinguish it from *Convallaria maialis*, the Lily-of-the-valley.

LILY-OF-THE-NILE 1. *Agapanthus africanus* (African Lily) Bonar. 2. *Zantedeschia aethiopica* (Arum Lily) Grigson.1959.

LILY-OF-THE-VALLEY *Convallaria maialis*. The specific name is from convallis, valley.

LILY-OF-THE-VALLEY, False *Maianthemum canadense*.

LILY-OF-THE-VALLEY, Wild 1. *Maianthemum canadense* (False Lily-of-the-valley) USA. Yarnell. 2. *Maianthemum dilatatum*. 3. *Pyrola rotundifolia* (Round-leaved Wintergreen) RHS.

LILY-OF-THE-VALLEY TREE 1. *Clethra arborea* McDonald. 2. *Pieris japonica*.

LILY-PILLY, Australian *Eugenia smithii*.

LILY POTATO *Asphodeline lutea* (Yellow Asphodel) Douglas.

LILY-ROYAL 1. *Lilium superbum* (Turk's Cap Lily) USA Woodcock. 2. *Mentha pulegium* (Pennyroyal) S Eng. Halliwell. For pennyroyal, a corruption of Puliall-royal. The French is Menthe pouliot, and there is a Dutch form, polei that occurs in English as Poley.

LILY-TREE *Magnolia denudata* (Yulan) Hay & Synge. Owing to the superficial resemblance of the flower to a lily. Any of the Magnolias can bear the name.

LIMA BEAN *Phaseolus lunatus*. They have been named Java Beans, Burma, Sieva, Carolina, Cape or Madagascar, but Lima seems the most accurate. The plant originated in Central America, and they have been found in Peruvian excavations dated to 6000-5000BC.

LIMBER PINE *Pinus flexilis*. Presumably 'linber' in this sense means the same as the specific name, *flexilis*.

LIME 1. *Citrus aurantiifolia*. 2. *Tilia eruopaea*. Two entirely different derivations. Lime for the citrus came via Spanish lima from Arabic limah, while the familiar European tree gets the name as an (unexplained) alteration of line, which is OE lind, giving the alternative name linden.

LIME, Broadleaf *Tilia platyphylla* (Large-leaved Lime) Barber & Phillips.

LIME, Caucasian *Tilia x euchlora* (Crimean Lime) Duncalf.

LIME, Chinese *Tilia oliveri*.

LIME, Crimean *Tilia x euchlora*.

LIME, Dog; LIME, Hedge *Sisymbrium canescens* (Pepper Grass) Douglas.

LIME, Large-leaved *Tilia platyphylla*.

LIME, Musk *Citrus microcarpa*.

LIME, Ogeechee *Nyssa sylvatica* (Tupelo) Willis.

LIME, Red 1. *Tilia cordata* (Small-leaved Lime) McClintock. 2. *Tilia platyphylla* (Large-leaved Lime) B & H. 3. *Tilia rubra*.
LIME, Red-twigged *Tilia platyphylla* (Large-leaved Lime) Brimble.1948.
LIME, Rough-skinned *Citrus hystrix*.
LIME, Silver *Tilia tomentosa*.
LIME, Silver, Weeping *Tilia x petiolaris*.
LIME, Small-leaved *Tilia cordata*.
LIME, Sweet *Citrus limetta*.
LIME, White *Tilia argentea*. White, because of the silvery undersurface to the leaves.
LIME, Wild 1. *Xanthoxylum fagara* Bahamas Roys. 2. *Ximenia americana* (Wild Oliver) Sofowora.
LIMESTONE BUGLE *Ajuga pyramidalis*. Habitat descriptive.
LIMESTONE WOUNDWORT *Stachys alpina*.
LIMETTE *Citrus limetta* (Sweet Lime) Rimmel.
LIMEWORT 1. *Silene armeria* (Catchfly) Gerard. 2. *Veronica beccabunga* (Brooklime) B & H. Gerard explained the name for Catchfly - "floures...covered over with a most thicke and clammy matter like unto Birdlime...". But 'lime' in both Limewort and Brooklime means something different. The OE was broc hleomoc, the second element being the plant name (and so the same as "bung" in the specific name). It eventually became 'lime', but Gerard was still calling the flower Brooklem.
LIMODORE *Limodorum abortivum*.
LIMPER-SCRIMP; LIMPERNSCRIMP *Heracleum sphondyllium* (Hogweed) both Som. B & H (Limper-scrimp); Elworthy (Limpernscrimp). Or Lumper-scrump, Humpy-scrumple, etc.
LIMPET-SCRIMP *Lemna minor* (Duckweed) Som. Macmillan.
LIMPWORT *Veronica beccabunga* (Brooklime) Heref. Grigson. Cf. Limewort. Broclempe and Brokelemp are early forms that explain Limpwort. 'Limp' is the plant name, and this is added to 'wort'.
LIN; LINN; LINE 1. *Linum usitatissimum* (Flax) Yks. Carr; Scot. B & H. (Lin); Turner (Line). 2. *Tilia europaea* (Lime) Yks. F K Robinson (Lin); Halliwell (Linn); Gerard (Line). Two different origins here - it is OE lin for flax, and OE linde for lime.
LINARY *Linaria vulgaris* (Toadflax) Turner.
LINCOLNSHIRE ASPARAGUS; LINCOLNSHIRE SPINACH *Chenopodium bonus-henricus* (Good King Henry) both Sanecki. This was once under cultivation in Lincolnshire.
LIND; LINDEN *Tilia europaea* (Lime) Scot. Jamieson (Lind). Linden is a common alternative to lime - both come from the OE linde.
LINDEN, African *Mitragyna stipulosa* Dalziel.
LINDEN, American *Tilia americana* (Basswood) Everett.
LINDEN, Indoor *Sparrmania africana* (African Hemp) Bonar.
LINEN-BUTTONS *Convallaria maialis* (Lily-of-the-valley) Som. Macmillan.
LING 1. *Calluna vulgaris*. 2. *Tamarix gallica* (Tamarisk) Turner. 3. *Ulex europaeus* (Gorse) Derb, N Eng. Grigson. O Norse lyng.
LING, Bell *Erica cinerea* (Bell Heather) Yks. Grigson.
LING, Black *Calluna vulgaris* (Heather) Yks. Atkinson.
LING, Crow 1. *Calluna vulgaris* (Heather) Yks. B & H. 2. *Empetrum nigrum* (Crowberry) Yks. Grigson. 3. *Erica cinerea* (Bell Heather) Yks. Atkinson. 4. *Erica tetralix* (Cross-leaved Heath) Yks. Grigson.
LING, Red *Calluna vulgaris* (Heather) Hants. Grigson.
LING, Wire 1. *Empetrum nigrum* (Crowberry) Yks. Atkinson. 2. *Erica tetralix* (Cross-leaved Heath) Yks. Grigson.
LING GOWLAND *Chrysanthemum segetum* (Corn Marigold) Yks. F K Robinson.
LING HEATHER *Calluna vulgaris* (Heather) A W Smith.
LINGBERRY 1. *Empetrum nigrum* (Crowberry) Cumb, Yks. Grigson. 2.

Vaccinium myrtillus (Whortleberry) IOM Maire Mac Neill. 3. *Vaccinium vitis-idaea* (Cowberry) Cumb, Yks. Grigson.

LINGELS, Devil's *Polygonum aviculare* (Knotweed) N'thum. Grigson. 'Lingels' are thongs, or according to Halliwell, shoemakers' threads.

LINGWORT 1. *Angelica archangelica* (Archangel) Fernie/Grieve. 2. *Veratrum album* (White Hellebore) Gerard.

LINK *Cytisus scoparius* (Broom) Fernie.

LINK MOSS; LINKS, Love *Sedum reflexum* (Yellow Stonecrop) Shrop B & H (Link Moss); Scot. Grigson (Love Links).

LINKS, Lovers' *Umbilicus rupestris* (Wall Pennywort) Scot. Jamieson.

LINSEED *Linum usitatissimum* (Flax). OE lin, flax.

LINT 1. *Linum usitatissimum* (Flax) Turner. 2. *Vicia sativa* (Common Vetch) Derb, Lincs, Yks. Grigson.

LINT, Fairy *Linum catharticum* (Fairy Flax) N'thum. Denham/Hardy.

LINT, Lady's *Stellaria holostea* (Greater Stitchwort) Dev.Friend.1882.

LINT, Laverock's *Linum catharticum* (Fairy Flax) Lanark. Grigson. Laverock is a Scottish word meaning lark.

LINT-BOW *Linum usitatissimum* (Flax) Scot. B & H.

LINT-SPURGE *Euphorbia esula* (Hungarian Spurge) Turner.

LINTELS *Vicia hirsuta* (Hairy Tare) N'thants. Grigson. Lentils, apparently. Clare uses the word:

The bearded rye was in the row
The lintel in the pod.

LINTIN *Vicia sativa* (Common Vetch) Yks. Nicholson. Cf. Lint. This is the old plural in '-n'.

LION'S CLAW *Crotalaria agatiflora*.

LION'S CUDWEED *Leontopodium alpinum* (Edelweiss) Gerard.

LION'S EAR 1. *Leonotis leonurus* (Lion's Tail) RHS. 2. *Leonurus cardiaca* (Motherwort) Le Strange.

LION'S FOOT 1. *Alchemilla vulgaris* (Lady's Mantle). 2. *Helleborus niger* (Christmas Rose) B & H. 3. *Leontopodium alpinum* (Edelweiss) Parkinson.1640. 4. *Prenanthes alba* (White Lettuce) USA Cunningham. Leontopodium = lion's foot. They have all of them leaves vaguely resembling an animal's paw.

LION'S HEART *Physostegia virginiana* (Dragonhead) USA House.

LION'S HERB *Aquilegia vulgaris* (Columbine) Skinner. Columbine was once known as Herba leonis, from a belief that it was the lion's favourite plant, and there was another belief that if you rub columbine on your hands, you will have the courage of a lion.

LION'S MOUTH 1. *Antirrhinum majus* (Snapdragon) Suss. B & H. 2. *Digitalis purpurea* (Foxglove) Suss. Grigson; USA. Henkel. 3. *Glechoma hederacea* (Ground Ivy) Suss. Parish. 4. *Linaria vulgaris* (Toadflax) Fernie. 5. *Misotapes orontium* (Lesser Snapdragon). All descriptive, in their way.

LION'S PAW *Alchemilla vulgaris* (Lady's Mantle) Gerard. The leaf shape suggests the name.

LION'S SNAP 1. *Antirrhinum majus* (Snapdragon) Folkard. 2. *Galeobdalon luteum* (Yellow Archangel) Som. Macmillan. 3. *Lamium amplexicaule* (Henbit Deadnettle) W Miller.

LION'S TAIL 1. *Agastache mexicanum* Brownlow. 2. *Leonurus cardiacus* (Motherwort) Brownlow. 3. *Leonotis leonurus*. Leonurus = lion's tail.

LION'S TEETH 1. *Mycelis muralis* (Wall Lettuce) Som. Macmillan. 2. *Taraxacum officinale* (Dandelion) B & H. In both cases, it is the shape of the leaves that provides the imagery.

LION'S TONGUE *Linaria vulgaris* (Toadflax) Dev. Macmillan. Cf. Lion's Mouth.

LIP, Elk's *Caltha leptosepala* T H Kearney.
LIPSTICK PLANT *Aeschynanthus pulcher*. Bright red flowers.
LIQUID STORAX *Liquidambar styraciflua* (Sweet Gum) Leyel.1937.
LIQUORICE *Glycyrrhiza glabra*. Liquorice is a corruption of Glycyrrhiza (there is a transitional medieval form, Gliquiricia. So it just means 'sweet root' - glukos, sweet and riza, root.
LIQUORICE, American *Glycyrrhiza lepidota*.
LIQUORICE, Black *Cardiospermum halicacabum* (Blister Creeper) Chopra.
LIQUORICE, Chenook *Lupinus littoralis*.
LIQUORICE, Indian *Abrus precatorius* (Precatory Bean) Grieve.1931. Cf. Wild Liquorice for this.
LIQUORICE, Jamaica *Asclepias syriaca* (Common Milkweed) McDonald.
LIQUORICE, Mock *Galega officinalis* (Goat's Rue) Turner.
LIQUORICE, Russian *Glycyrrhiza glabra* var. *glandulifera*.
LIQUORICE, Spanish *Glycyrrhiza glabra* var. *typica*. Preseumably because it was first cultivated in Spain, though now it is grown not only there, but also in France, Germany and in America. It is native of a wide band across southern Europe, to the Caucasus and northern Iran.
LIQUORICE, Wild 1. *Abrus precatorius* (Precatory Bean) Grieve.1931. 2. *Aralia nudicaulis* (American Sarsaparilla) Grieve.1931. 3. *Astragalus glycyphyllus* (Milk Vetch) McClintock. 4. *Cephalanthus occidentalis* (Buttonbrush) Grieve.1931. 5. *Galium circaezans* USA Zenkert. 6. *Galium lanceolatum* USA Zenkert. 7. *Ononis repens* (Rest Harrow) Cumb, Yks, Dumf, Inv, Mor. Grigson.
LIQUORICE PLANT *Ononis repens* (Rest Harrow) Som. Macmillan. Cf. Wild Liquorice in the north for this, and similar names like Spanish-root and Liquory-stick. This is the only example in the south of the country. North-country children used to dig up the root and eat it, or else they used to put it in a bottle of water, shake it and drink the result - Spanish water.
LIQUORICE VETCH *Astragalus glycyphyllus* (Milk Vetch). B & H. Cf. Wild Liquorice, or just Licorice, for this. The reason lies in another name - Sweet Milk Vetch.
LIQUORY-KNOT *Lathyrus montanus* (Bittervetch) Berw. Dyer. The tubers are edible, and when dried, the taste is not unlike liquorice (they are more like chestnuts when roasted).
LIQUORY-STICK *Ononis repens* (Rest Harrow) Roxb. B & H. See Liquorice Plant.
LIRICON-FANCY; LIRICUM-FANCY; LIRY-CONFANCY *Convallaria maialis* (Lily-of-the-valley) all B & H. All variations on something like Aubrey's Lily-convally.
LISAMOO *Heracleum sphondyllium* (Hogweed) Corn. Grigson. Cornish les-anmogh, i.e. pigweed.
LITHEWORT 1. *Myosotis arvensis* (Field Forget-me-not) Halliwell. 2. *Sambucus ebulus* (Dwarf Elder) Cockayne. 3. *Viburnum lantana* (Wayfaring Tree) Storms. Lithe must mean pliant, at least as far as the two shrubs or trees are concerned.
LITHWORT *Sambucus ebulus* (Dwarf Elder) Bonser. As Lithewort.
LITHY-TREE *Viburnum lantana* (Wayfaring Tree) Prior. As Lithewort. As Gerard said, "the branches are long, tough, and easie to be bowed, and hard to be broken". There are 'whip' names for this, too.
LITTLE-AND-PRETTY 1. *Agrostemma githago* (Corn Cockle) Som. Macmillan. 2. *Malcomia maritima* (Virginian Stock) Dor. B & H; Dev. Friend.1882; Som, Wilts. Macmillan. 3. *Saxifraga x urbinum* (London Pride) Dev. Friend; Dor. B & H. 4. *Veronica chamaedrys* (Germander Speedwell) Som. Macmillan. Surely Corn Cockle would be the last plant to be given an affectionate name like this!
LITTLE FAIR ONE *Cytisus scoparius* (Broom) Dev. Macmillan.
LITTLE-GOODIE; LITTLE-GUID; LIT-

TLEGOOD *Euphorbia helioscopia* (Sun Spurge) N'thum. Grigson (Little-goodie); Scot. Gregor (Little-guid); Scot. Jamieson (Littlegood). A euphemism for the devil, according to Grigson.

LITTLE JANE *Anagallis arvensis* (Scarlet Pimpernel) Som. Macmillan.

LITTLE JOHN *Stellaria holostea* (Greater Stitchwort) Som. Macmillan.

LITTLE JOHN ROBIN HOOD *Geranium robertianum* (Herb Robert) Dor. Grigson.

LITTLE LIE-ABED *Senecio vulgaris* (Groundsel) Som. Macmillan.

LITTLEWALE *Lithospermum officinale* (Gromwell) Halliwell. Cf. Lichwale, Lythewale.

LIVE-FOREVER 1. *Antennaria dioica* (Cat's Foot) USA Leighton. 2. *Gnaphalium uliginosum* (Marsh Cudweed) Som. Macmillan. 3. *Sedum purpureum* USA Upham. 4. *Sedum telephium* (Orpine) USA Bergen. For these will make an immortelle.

LIVE-IN-IDLENESS *Viola tricolor* (Pansy) Glos, Oxf, N'thants. Grigson. For Love-in-idleness.

LIVE-LONG 1. *Anaphalis margaretacea* (Pearly Immortelle) B & H. 2. *Filago germanica* (Cudweed) *Notes and Queries*. vol 7. 3. *Sedum telephium* (Orpine) Gerard. 4. *Vangueriopsis lanciflora* (Wild Medlar) Palgrave. 5. *Lannea discolor*. Either everlasting, or nearly so, as far as Immortelle and Orpine are concerned. But the two African trees get the name (as well as Never-die), because they strike so easily. Lannea poles are used for fences, which very quickly become growing trees, and Vangueriopsis truncheons stuck in the ground invariably grow, as well.

LIVE-LONG, Jersey *Gnaphalium luteo-album* (Jersey Cudweed) Prior. Apparently not entirely confined to the Channel Islands, for it can occur in East Anglia too.

LIVE-LONG-AND-LOVE-LONG; LIVE-LONG-LOVE-LONG *Sedum telephium* (Orpine) Putnam (with the 'and'); Suss. J Simpson (without the 'and'). The 'live-long' part of the names refers to the undoubted longevity of the flowering stems. This recommended itself to those planning love divinations, hence the 'love-long' part. A girl could hang up a piece for each of her boy-friends. The piece that lives the longest dertermines the successful one. Other divinations are by the bending of the leaves to the right or left, so telling whether a lover were true or false; or, as in America, to tell from what quarter the lover will come. If gathered by two people on Midsummer Eve, and the slips planted, they will know their fortune by the growing or otherwise of the slips. If they leaned towards each other, the couple would marry; if one withered, the person represented would die, etc, etc.

LIVE-LONG SAXIFRAGE *Saxifraga aizoon*.

LIVER 1. *Caltha palustris* (Marsh Marigold) Dor. Macmillan. 2. *Iris pseudo-acarus* (Yellow Flag) Dor. Macmillan. The Iris gets the name from OE leafer, which appears also as Lever, or Levver, Laver, even Leather. Probably, liver for marsh marigold comes from the same source.

LIVER, Heart *Medicago arabica* (Spotted Medick) B & H. A corruption from Heart Clover. Coles said it is so-called "not only because the leaf is triangular like the heart of a man, but also because each leafe doth contain the perfection (or image) of an heart, and that in its proper colour, viz. a flesh colour. It defendeth the heart against the noisome vapour of the spleen".

LIVER-LEAF 1. *Hepatica acutiloba* (Heart Liver-leaf) USA Tolstead. 2. *Hepatica americana* (Kidney Liver-leaf) USA Zenkert. 3. *Hepatica hepatica*. 4. *Hepatica triloba* (American Liverwort) USA Le Strange. The liver-shaped leaves of these plants ensured that, by the doctrine of signatures, they would be used for ailments of the liver.

LIVER-LEAF, Heart *Hepatica acutiloba*.

LIVER-LEAF, Kidney *Hepatica americana*.

LIVER LILY *Iris versicolor* (Purple Water Iris) Grieve.1931. The root contains iridin, which acts powerfully on the liver. But at the same time, the OE word for iris, leafer, should be borne in mind.

LIVERWORT 1. *Agrimonia eupatoria* (Agrimony) Lyte. 2. *Eupatorium cannabinum* (Hemp Agrimony) Cockayne. 3. *Pulmonaria officinalis* (Lungwort) Dor. Udal. Agrimony and Hemp Agrimony both have 'eupatoria' in their names, and besides being a memorandum of the first use by Mithridates Eupator, king of Pontus, is also said to denote its use in liver complaints. They have always been an ingredient in herbal liver medicines. But Lungwort hasn't, and the very local usage possibly stems from a misreading somewhere.

LIVERWORT, American *Hepatica triloba*.

LIVERWORT, Ground, Ash-coloured *Peltigera canina*. A lichen.

LIVERWORT, Noble *Hepatica caerulea* (Hepatica) Gerard. It is also known as Noble Agrimony.

LIVERWORT, Three-leaved *Hepatica caerulea* (Hepatica) B & H.

LIVERWORT, Water *Ranunculus aquatilis* (Water Crowfoot). B & H.

LIVERWORT, White *Parnassia palustris* (Grass of Parnassus) Gerard. For Gerard reckoned this to be some form of Hepatica,for it was used as a liver medicine. But whether the name or the usage came first is difficult to decide. It was still being used in the Highlands for liver complaints in comparatively recent times.

LIVING IDOLS *Viola tricolor* (Pansy) Tynan & Maitland. From Love-in-idleness, via Loving Idols.

LIVINGSTONE DAISY *Mesembryanthemum crinifolium*.

LIZARD FLOWER; LIZARD ORCHID *Himantoglossum hircinum*. Lizard Orchid is the usual name; Lizard Flower appears in Camden/Gibson.

LIZZIE, Busy *Impatiens walleriana*.

LIZZIE-BY-THE-HEDGE *Glechoma hederacea* (Ground Ivy) Sanecki.

LIZZIE-IN-THE-HEDGE *Galium aparine* (Goose-grass) N'thum. Grigson. Robin, in or round, the hedge is more common.

LIZZIE-RUN-THE-HEDGE; LIZZIE-RUN-UP-THE-HEDGE *Glechoma hederacea* (Ground Ivy) B & H (run-); Fernie (run-up-).

LIZZORY; LEZZORY *Sorbus torminalis* (Wild Service Tree) Glos. J D Robertson.

LOAVES-OF-BREAD 1. *Hyoscyanum niger* (Henbane) N'thants. B & H. 2. *Malva sylvestris* (Common Mallow) Som, Dor. Macmillan. Applied to the seed vessel in each case. Those of mallow are better known as 'cheeses', with all its variants. But. nevertheless, the Arabic name for the plant is hubeize, which comes from hubez, bread.

LOB-GRASS; LOP-GRASS *Bromus mollis* (Soft Brome) Som. Elworthy (Lob-grass); Hants. Cope (Lop-grass).

LOBEL'S MAPLE *Acer lobelii*.

LOBELIA, Acrid *Lobelia urens*. The genus is named after Mathias de l'Obel, of Lille, botanist and physician to James I.

LOBELIA, Brook *Lobelia kalmii*.

LOBELIA, Garden *Lobelia erinus*.

LOBELIA, Great *Lobelia syphilitica*.

LOBELIA, Heath *Lobelia urens* (Acrid Lobelia) McClintock.1956.

LOBELIA, Water *Lobelia dortmanniana*.

LOBELIA, Wild *Polygala vulgaris* (Milkwort) Dor. Macmillan.

LOBLOLLY 1. *Amaranthus caudatus* (Love-lies-bleeding) Som. Macmillan. 2. *Magnolia grandiflora* (Bull Bay) USA Harper. 3. *Valerianella locusta* (Cornsalad) Gerard. In America, loblolly is a muddy swamp, or perhaps just a moist depression in the ground. It comes from an English seaman's usage for a thick gruel served in sick bay. So a Loblolly boy is a surgeon's mate.

LOBLOLLY BAY *Gordonia lasianthus*.

LOBLOLLY PINE *Pinus taeda* (Shortleaf Pine) Fluckiger.

LOBSTER-CLAW 1. *Clianthus puniceus* (Parrot's Bill) Simmons. 2. *Erythrina cristagalli* (Cock's Comb) Nicolaisen. 3. *Heliconia metallica*.

LOBSTER FLOWER *Euphorbia pulcherrima* (Poinsettia) USA Watt.

LOCK ELM *Ulmus x diversifolia*. 'Lock', apparently, in the sense of locking a saw - the wood is that rough (all elm timber is).

LOCKAN GOWAN; LOCKEN GOWAN; LOCKEN GOWLAN; LOCKERGOULON; LOCKIN-MA-GOWAN; LOCKREN GOWLAN *Trollius europaeus* (Globe Flower) N Eng, Scot. Grigson (Lockan Gowan, Locken Gowan); Gerard (Locken Gowlan); Camden/Gibson (Locker-Goulon); Cumb. Grigson (Lockin-ma-Gowan); Gerard (Lockren-Gowlan). Gowan is a yellow flower, and locken means closed in, or locked in.

LOCKET, Lady's 1. *Acer campestre* seeds (Field Maple keys) Som. Macmillan. 2. *Dicentra spectabilis* (Bleeding Heart) Coats. 3. *Fumaria officinalis* (Fumitory) Som. Macmillan. 4. *Lunaria annua* (Honesty) Som. Macmillan. 5. *Polygonatum multiflorum* (Solomon's Seal) Som. Macmillan.

LOCKET, Lucy *Cardamine pratensis* (Lady's Smock) Rut. Grigson; Derb. Northall.

LOCKETS-AND-CHAINS *Dicentra spectabilis* (Bleeding Heart) Som. Macmillan. Cf. Locks-and-keys.

LOCKETS-AND-CHAINS, Gold *Laburnum anagryoides* (Laburnum) Som. Macmillan.

LOCKS, Golden 1. *Laburnum anagryoides* (Laburnum) Dev. Macmillan. 2. *Polypodium vulgare* (Common Polypody) Heref. B & H.

LOCKS-AND-KEYS 1. *Acer campestre* fruit (Field Maple keys) Dev. Friend.1882; Som. Macmillan. 2. *Acer pseudo-platanus* fruit (Sycamore keys) Som. Elworthy; Ess, Norf, Cambs. Grigson; Lincs. Peacock. 3. *Dicentra spectabilis* (Bleeding Heart) Som. Friend; Wilts. Dartnell & Goddard. 4. *Endymion nonscriptus* (Bluebell) Som. Macmillan. 5. *Fraxinus excelsior* fruit (Ash keys) Dev. Friend.1882. Ess, Norf, Lincs, Cambs, Notts. B & H. 6. *Orchis mascula* (Early Purple Orchis) Som. Macmillan. 7. *Rosa canina* fruit (Dog Rose hips) Som. Macmillan. 8. *Tilia europaea* seeds (Common Lime seeds) Som. Macmillan.

LOCKYER-GOLDEN *Trollius europaeus* (Globe Flower) Yks. Grigson. Cf. the more usual Lockan-gowan, etc.

LOCO, Blue *Astragalus diphysus*. Presumably Loco is the Spanish word meaning 'mad'. But why? The name is applied only to those American species of the southwest of the country.

LOCO, Death Valley *Astragalus funereus*.

LOCO, Earle *Astragalus earlei*.

LOCO, Gray *Astragalus calycosus*.

LOCO, Half-moon *Astragalus argillophilus*.

LOCO, Mohave *Astragalus mohavensis*.

LOCO, Purple *Astragalus mollissimus* (Woolly Loco) Kingsbury.

LOCO, Sheep *Astragalus nothoxys*.

LOCO, Speckled *Astragalus lentiginosus*.

LOCO, Stemless *Oxytropis lamberti* (White Loco) USA Kingsbury.

LOCO, Thurber *Astragalus thurberi*.

LOCO, White *Oxytropis lamberti*.

LOCO, Whitepoint *Oxytropis lamberti* (White Loco) USA Kingsbury.

LOCO, Woolly *Astragalus mollissimus*.

LOCO, Wooton *Astragalus wootoni*.

LOCO-WEED *Oxytropis lamberti* (White Loco) Grigson.1974.

LOCUST, Black *Robinia pseudacacia* (False Acacia) USA Tolstead. Latin locusta meant a lobster, thence by inference a locust. The reference must be to the shape of the pods of these leguminous trees.

LOCUST, Bristly *Robinia hispida* (Rose Acacia) USA Zenkert.

LOCUST, Caspian *Gleditsia caspica*.

LOCUST, Clammy *Robinia viscosa*.

LOCUST, Honey *Gleditsia triacanthos*. Honey, fron the succulent pulp in which the seeds are embedded.

LOCUST, New Mexican *Robinia neomexicana*.

LOCUST, Shipmast *Robinia pseudacacia*

var. rectissima. 'Rectissima' in the name of the variety explains the common name.

LOCUST, Sweet *Gleditsia triacanthos* (Honey Locust) USA Elmore. See Honey Locust.

LOCUST, Water *Gleditsia aquatica*.

LOCUST, West Indian *Hymenaea courbaril*.

LOCUST, Yellow *Robinia pseudacacia* (False Acacia) Grigson.1959.

LOCUST BEAN 1. *Catalpa bignonoides*. 2. *Ceratonia siliqua* pods (Carob) Wit.

LOCUST TREE *Ceratonia siliqua* (Carob) Polunin.

LOCUSTS *Rhinanthus crista-galli* (Yellow Rattle) Bucks. Grigson. There was a local belief in Buckinghamshire that Yellow Rattle was the locusts eaten by John the Baptist in the wilderness.

LODDON LILY *Leucojum aestivum*. Loddon is the river of that name in Berkshire.

LODEWORT *Ranunculus aquatilis* (Water Crowfoot) Halliwell.

LODGEPOLE PINE *Pinus contorta* (Beach Pine) N America Dimbleby. Because it was the favourite wood for making tipi poles.

LOG, Cheese *Malva sylvestris* (Common Mallow) Bucks. B & H.

LOGANBERRY *Rubus loganobaccus*. Whether it is a natural or deliberate hybrid, it originated in the garden of Judge Logan, of Santa Cruz, California.

LOGGERHEADS 1. *Centaurea montana* (Mountain Knapweed) Glos. B & H. 2. *Centaurea nigra* (Knapweed) Som, Glos, Oxf. Grigson. 3. *Centaurea scabiosa* (Greater Knapweed) Tynan & Maitland. Cf. such names as Ironheads, and Hardheads for knapweed. A loggerhead is actually a round mass of iron with a long handle attached to it, and as such the name is a good descriptive one.

LOGGERUM *Centaurea nigra* (Knapweed) Wilts. Dartnell & Goddard. See Loggerheads.

LOGWOOD *Haemotoxylon campechianum*. Apparently because it was exported in logs.

LOGWORT *Verbascum thapsus* (Mullein) Lupton.

LOKKI'S-OO *Eriophorum angustifolium* (Cotton-grass) Shet. Marwick. i.e. Loki's wool, a commemoration of the disruptive god of Scandinavian mythology.

LOLIONDO *Olea welwitschii*.

LOLLILOP PLANT *Pachystachys lutea*.

LOMBARDY POPLAR *Populus nigra var. italica*.

LONACHIES *Agropyron repens* (Couch-grass) Grigson.1959. It sounds Scottish, but Grigson did not identify it.

LONDON, Pride-o'- *Saxifraga x urbinum* (London Pride) IOM Moore.1924.

LONDON BASKET 1. *Geum rivale* (Water Avens) Yks. Grigson. 2. *Geum urbanum* (Avens) Yks. B & H.

LONDON BELLS *Calystegia sepium* (Great Bindweed) Dev. Macmillan.

LONDON BOBS 1. *Dianthus barbatus* (Sweet William) Lancs. Nodal. 2. *Geum rivale* (Water Avens) Tynan & Maitland. Perhaps an acknowledgement to nurserymen in distant parts as far as Sweet William is concerned.

LONDON BOTTLES *Prunella vulgaris* (Self-heal) Ayr. Grigson.

LONDON DAISY *Leucanthemum vulgare* (Ox-eye Daisy) Som, Dor. Macmillan. Possibly Thunder Daisy, via a form like Dunder, or Dundle, Daisy, both of which are recorded in Somerset.

LONDON PINK *Geranium robertianum* (Herb Robert) Glos. J D Robertson.

LONDON PLANE *Platanus x acerifolia*. "London" notwithstanding, it very likely originated in Oxford about 1673. But, of course, it thrives in London, because it sheds its bark, so it used to be said, the argument being that in doing so it exposed a fresh surface unclogged by soot, so the tree could "breathe". That is quite fallacious, of course - trees don't "breathe" through their bark.

LONDON PRETTY *Saxifraga x urbinum* (London Pride) Dor. Macmillan. A hybrid name - London Pride x None-so-pretty.

LONDON PRIDE 1. *Artemisia abrotanum* (Southernwood) Som. Macmillan. 2. *Dianthus barbatus* (Sweet William) N'thants, Yks B & H. 3. *Drosera rotundifolia* (Sundew) Som. Macmillan. 4. *Lychnis chalcedonica* (Rose Campion) Glos. B & H. 5. *Saxifraga x urbinum*. 6. *Sedum acre* (Biting Stonecrop) Som. Elworthy. At least for the flower usually known by this name, the origin is an 18th century Royal Gardener, Mr London, who introduced it (so it ought to be known rather as London's Pride).

LONDON PRIDE, Wild 1. *Circaea lutetiana* (Enchanter's Nightshade) Som. Macmillan. 2. *Sanicula europaea* (Sanicle) Som. Macmillan. It may be stretching a point, but the flowers could possibly be interpreted as wild versions of the real London Pride.

LONDON ROCKET *Sisymbrium irio*. It is said that it first appeared in London in the spring following the Great Fire.

LONDON ROCKET, False *Sisymbrium loeslii*.

LONDON TUFT 1. *Dianthus barbatus* (Sweet William) E Ang. B & H. 2. *Saxifraga x urbinum* (London Pride) N'thants. A E Baker. Cf. London Bobs for Sweet William, and of course London Pride.

LONESOME LADY *Cardamine pratensis* (Lady's Smock) Dev. Macmillan.

LONG CHERRY *Cornus mas* (Cornel Cherry) Turner.

LONG-LEGS *Primula veris* (Cowslip) Som. Grigson.

LONG PEASEN *Phaseolus vulgaris* (Kidney Bean) Turner.

LONG PLANTAIN *Plantago lanceolata* (Ribwort Plantain) Le Strange.

LONG PURPLES 1. *Arum maculatum* (Cuckoo-pint) War. Grigson. 2. *Digitalis purpurea* (Foxglove) Som. Macmillan. 3. *Lythrum salicaria* (Purple Loosestrife) Dev, Som, N'thants Grigson; Wilts. Dartnell & Goddard. 4. *Orchis mascula* (Early Purple Orchis) Shakespeare.

LONGER-THE-DEARER, The *Viola tricolor* (Pansy) Skinner.

LONGLEAF *Falcaria vulgaris*.

LONGWORT 1. *Anacyclus pyrethrum* (Pellitory-of-Spain) Halliwell. 2. *Pulmonaria officinalis* (Lungwort) Tusser. 3. *Verbascum thapsus* (Mullein) Turner. Lungwort, perhaps, for mullein? That is certainly recorded for it.

LOOFAH *Luffa cylindrica*. Arabic lufah.

LOOK-UP-AND-KISS-ME 1. *Saxifraga x urbinum* (London Pride) Dev. B & H. 2. *Viola tricolor* (Pansy) Som. Macmillan. One of a number of amorous names given, particularly to pansy, and to a lesser extent to London Pride.

LOOKING-GLASS *Stellaria holostea* (Greater Stitchwort) Dor. Macmillan.

LOOKING-GLASS, Lady's *Legousia hybrida* (Venus's Looking-glass) Prior.

LOOKING-GLASS, Old Man's *Anagallis arvensis* (Scarlet Pimpernel) Som. Macmillan. Looking-glass in this case is a corruption, probably of Weatherglass, one of the many "weather forecasting" names for pimpernel.

LOOKING-GLASS, Venus's *Legousia hybrida*. From the shining seeds, like, in Richard Mabey's words, "brilliantly polished brass mirrors".

LOOSESTRIFE, Bulb-bearing *Lysimachia terrestris*. The name Loosestrife arises from a mistranslation of late Latin lysimachis, from the Greek personal name Lusimakhos (the king who was said to have been the first to discover the medicinal properties of the plant). That name was taken from the Greek adjective lusimakhos, which meant loosing, that is ending, strife, and was meant to describe the political prowess of the king. The plant name was translated literally, instead of as a personal name.

LOOSESTRIFE, Creeping *Lysimachia nummularia* (Creeping Jenny) USA House.

LOOSESTRIFE, Dotted *Lysimachia punctata*.

LOOSESTRIFE, Garden *Lysimachia punctata* (Dotted Loosestrife) RHS.

LOOSESTRIFE, Golden *Lysimachia vulgaris* (Yellow Loosestrife) Lyte.
LOOSESTRIFE, Hairy *Lysimachia ciliata*.
LOOSESTRIFE, Purple; LOOSESTRIFE, Purple-spiked *Lythrum salicaria*. Purple-spiked Loosestrife is Curtis's name.
LOOSESTRIFE, Red *Lythrum salicaria* (Purple Loosestrife) Turner.
LOOSESTRIFE, Slender *Lythrum virgatum*.
LOOSESTRIFE, Spiked *Lythrum salicaria* (Purple Loosestrife) Thomson.1978.
LOOSESTRIFE, Tufted *Naumbergia thyrsiflora*.
LOOSESTRIFE, Whorled *Lysimachia quadrifolia*.
LOOSESTRIFE, Wood *Lysimachia nemora* (Yellow Pimpernel) Curtis.
LOOSESTRIFE, Yellow *Lysimachia vulgaris*.
LOOSESTRIFE, Yellow, Large *Lysimachia punctata* (Dotted Loosestrife) Polunin.
LOP-GRASS; LOB-GRASS *Bromus mollis* (Soft Brome) Som. Elworthy (Lob); Hants. Cope (Lop).
LOPPER-GOWAN; LAPPER-GOWAN *Trollius europaeus* (Globe-flower) Clyde B & H (Lopper); Cumb. Grigson; Roxb. B & H (Lapper).
LOPPY-MAJOR *Arctium lappa* (Burdock) Culpepper. Loppy = lappa; Major, to distinguish it from *Arctium minus*, the Lesser Burdock. Hence, too, Gerard's Great Bur.
LOPS-AND-LICE *Rosa canina* hips (Dog Rose hips) Dur. F M T Palgrave. Strictly Hips and Haws.
LOQUAT *Eriobotrya japonica*. A Chinese tree, and the name comes from the Chinese luh kwat.
LOQUAT, Wild *Uapaca kirkiana*. This is an African tree.
LORD ANSON'S PEA *Lathyrus magellanicus*. The cook on Lord Anson's ship Centurion collected the seeds in 1744.
LORDS, Ladies' *Arum maculatum* (Cuckoo-pint) Kent Grigson. A garbled version of Lords-and-ladies.
LORDS-AND-LADIES 1. *Arum maculatum*. 2. *Orchis mascula* (Early Purple Orchis) Dor. Macmillan. 3. *Plantago lanceolata* (Ribwort Plantain) Norf. Grigson. 4. *Plantago media* (Lamb's-tongue Plantain) Norf. Grigson. A typical male + female name, especially for Cuckoo-pint, with erotic overtones. Cf. Ladies-and-gentlemen, and many others.
LORDS-AND-LADIES' FINGERS *Arum maculatum* (Cuckoo-pint) War. B & H.
LORDWOOD *Liquidambar orientalis*.
LOREL *Daphne laureola* (Spurge Laurel) Turner.
LORER; LORRY *Laurus nobilis* (Bay) Chaucer (Lorer); Halliwell (Lorry).
LOT-TREE *Sorbus aria* (Whitebeam) W Miller. Can the reference be to Lot's wife, turned into a pillar of salt, and likened to the general white appearance of the tree?
LOTE, Wild *Melilotus officinalis* (Melilot) Turner.
LOTE TREE *Celtis australis* (Nettle Tree) Turner. For the Romans referred to the tree as the lotus.
LOTUS, American *Nelumbo lutea* (Yellow Lotus) USA Yarnell.
LOTUS, Blue *Nymphaea caerulea*.
LOTUS, Egyptian *Nymphaea lotus*.
LOTUS, Hairy *Lotus tomentellus*.
LOTUS, Hill *Lotus humistratus*.
LOTUS, Hindu *Nelumbo nucifera*.
LOTUS, Sacred 1. *Nelumbium nelumbo*. 2. *Nelumbo nucifera* (Hindu Lotus) Brouk.
LOTUS, Stiff-haired *Lotus strigosus*.
LOTUS, Winged *Lotus tetragonolobus* (Asparagus Pea) Whittle & Cook.
LOTUS, Wright *Lotus wrightii*.
LOTUS, Yellow *Nelumbo lutea*.
LOTUS BUSH *Zizyphus lotus*.
LOTUS FLOWER *Butomus umbellatus* (Flowering Rush) Som. Notes and Queries, vol 168, 1935.
LOTUS TREE 1. *Diospyros kaki* (Chinese Date Plum) J Smith. 2. *Zizyphus spina christi*.
LOUISIANA CYPRESS *Taxodium distichum* (Swamp Cypress) USA Brimble.1948.

LOUISIANA WORMWOOD *Artemisia ludoviciana* (Lobed Cudweed) USA Yanovsky.

LOURY *Daphne laureola* (Spurge Laurel) Turner.

LOUSE BUR *Arctium minus* (Small Burdock) Parkinson.1640. Is this a descriptive name, or just a variation of Less(er) Burdock?

LOUSEBANE; LOUSE-HERB *Delphinium staphisagria* (Stavesacre) Cockayne (Lousebane); Pomet (Louse-herb). Pliny mentioned the use of the powdered seeds for destroying vermin on the head and body, and it is still so used. The fine powder is put into the hair each night, and combed out the following morning.

LOUSEBERRY *Anamirta cocculus* Chopra. Descriptive of function. The berries are poisonous, used in India for destroying lice, as well as killing crows and fish.

LOUSEBERRY TREE *Euonymus europaeus* (Spindle Tree) Fernie. Loudon said that a decoction of the leaves was used to wash dogs free from vermin. That has always been the accepted reason for the name Dogwood (for this and for *Cornus sanguinea*). Small boys' nits were treated, as well as dogs' coats. As Evelyn said, "...the powder made of the berry, being bak'd, kills nits, and cures scurfy heads".

LOUSEWORT 1. *Delphinium staphisagria* (Stavesacre) Clair. 2. *Pedicularis canadensis* USA House. 3. *Pedicularis palustris* (Red Rattle) Gerard. 4. *Pedicularis sylvatica*. Lousewort is the exact equivalent of the generic name *Pedicularis*, so called because "it filleth sheep and other cattell that feed in medowes where this groweth, full of lice", according to Gerard, but it is a belief that can still be found.

LOUSEWORT, Leafy *Pedicularis foliosa*.

LOUSEWORT, Marsh *Pedicularis palustris* (Red Rattle) Thrimble.

LOUSEWORT, Swamp *Pedicularis lanceolata*.

LOUSY ARNOT *Conopodium majus* (Earth-nut) Scot. Jamieson.

LOUSY ARNUT 1. *Bunium bulbocastanum* (Great Earth-nut) Scot. Jamieson. 2. *Conopodium majus* (Earth-nut) Aber, Perth. B & H. 'Lousy', because it was once thought that eating them breeds lice. Arnot, or Arnut, are two of the many variations of Earth-nut. Cf. Lucy Arnut.

LOUSY BEDS *Melandrium dioicum* (Red Campion) Cumb. Grigson. Cf. the Cornish Fleabites.

LOUSY GRASS *Spergula arvensis* (Corn Spurrey) Leyel.1948.

LOVACHE *Ligusticum scoticum* (Lovage) W Miller.

LOVAGE 1. *Ligusticum apiodorum* USA Schenk & Gifford. 2. *Ligusticum scoticum*. 3. *Smyrnium olusatrum* (Alexanders) R Gunther. Lovage is a corruption of the generic name, Ligusticum, via French livèche. Alexanders gets the name because of the similarity of the leaves, but Turner differentiated it by calling it Black Lovage.

LOVAGE, Black *Smyrnium olusatrum* (Alexanders) Turner. "because it hath leaves lyke Lovage, and yet blacke seede".

LOVAGE, Scotch *Ligusticum scoticum* (Lovage) Grieve.1931.

LOVAGE, Sea *Ligusticum scoticum* (Lovage) Le Strange.

LOVAGE, Water 1. *Oenanthe crocata* (Hemlock Water Dropwort) B & H. 2. *Oenanthe fistulosa* (Tubular Water Dropwort) Grieve.1931.

LOVE 1. *Clematis vitalba* (Old Man's Beard) Parkinson.1640. 2. *Galium aparine* (Goose-grass) Som. Macmillan. Presumably because, in their different ways, they cling.

LOVE, Boys' *Artemisia abrotanum* (Southernwood) Corn. Jago; Dor. Dacombe; Som. Jennings; Wilts. Akerman. Hants. Cope. Cf. Lad's Love.

LOVE, Cats' *Valeriana officinalis* (Valerian) Wilts. Macmillan; Yks. Grigson. Cats are said to be very fond of valerian. "The root of the herb Valerian, is very like to the eye of a cat and wheresoever it groweth, if cats come thereunto, they instantly dig it up for the love thereof, as I myself have seen in mine own garden, for it smelleth moreover like a cat" (Topsell, Four-footed beasts, 1607).

LOVE, Curly, Girl's *Artemisia abrotanum* (Southernwood) Tynan & Maitland. Of the Lad's Love genre.

LOVE, Doctor's *Galium aparine* (Goosegrass) Som. Macmillan.

LOVE, Girls' *Rosmarinus officinalis* (Rosemary) Som. Macmillan.

LOVE, Lad's *Artemisia abrotanum* (Southernwood) Dev, Som, Dor. Prior; N'thants. A E Baker; E Ang. Forby. St Francis de Sales refers to sprigs of southernwood being included in bouquets given to each other by lovers, the message being fidelity even in bitter circumstances. A few branches of southernwood were always included in the nosegays that courting youths used to give the girls. And Fenland youths used to wear it in their buttonholes. He would smell it ostentatiously as he walked up to a group of girls, and give it to the girl of his choice. If she was willing, she would also smell the herb, and the two would set off together on their first country walk.

LOVE, Ladder *Centaurea cyanus* (Cornflower) Som. Macmillan.

LOVE, Maids' *Artemisia abrotanum* (Southernwood) N'thants. A E Baker. Cf. Boys' Love, and Lads' Love.

LOVE, Measure of *Bellis perennis* (Daisy) Coats. From the "he loves me, he loves me not" divination.

LOVE, Old Man's *Artemisia abrotanum* (Southernwood) Berks. Lowsley; N'thum. B & H. Out of place, given Boy's Love, and Lad's Love, etc., but it probably came from the widespread Old Man as a hybrid.

LOVE, Seven Years' 1. *Achillea millefolium* (Yarrow) Glos. Fernie. 2. *Achillea ptarmica* (Sneezewort) W Eng. Friend. The origin probably lies in the divination games played with yarrow, with rhymes like

Green arrow, green arrow, you bears a white blow,
If my love love me my nose will bleed now;
If my love don't love me, it 'ont bleed a drop: If my love do love me, 'twill bleed every drop.

"The leaves being put in the nose do cause it to bleed, and easeth the pain of the megrim" (Gerard). It was this propensity that was used to test a lover's fidelity.

LOVE, Shepherd's *Polygala calcarea* (Chalk Milkwort) Wilts. D Grose. Presumably because its presence in the turf of chalk downland indicated good pasture for sheep. Cf. Shepherd's Thyme, and Shepherd's Blue Thyme, both from Wiltshire, too.

LOVE, True *Lunaria annua* (Honesty) Briggs. There was a divination game played with Honesty. Two girls must sit together in a room by themselves, from midnight to 1 o'clock, in silence. During this hour each must take as many hairs from her head as she is years old, and put them in a linen cloth with some of the herb true-love. As soon as the clock strikes 1 o'clock, she must burn every hair separately, saying

I offer this my sacrifice
To him most precious in my eyes,
I charge thee now come forth to me,
That I this minute may thee see.

Upon which, the future husband will appear, walk round the room, and then vanish. The same thing happens to both girls, but neither see the other's lover (Halliwell).

LOVE-A-LI-DELL; LOVE-A-LI-DO *Viola tricolor* (Pansy) both Wilts. Wiltshire (Love-a-li-dell); Clarice Merrett (Love-a-li-do. As with the versions that follow, these started off as Love-in-idleness.

LOVE-AND-IDLE; LOVE-AND-IDLE, Wild *Viola tricolor* (Pansy) Som, Wilts, Dor, Glos, Oxf, Berks. Grigson; Glos. B & H (Wild).

LOVE-AND-IDLENESS *Viola tricolor* (Pansy) War. Grigson.

LOVE-AND-IDOLS; LOVE-IDOL *Viola*

tricolor (Pansy) Wilts. Dartnell & Goddard (Love-and-idols); Grieve.1931 (Love-idol).

LOVE-AND-TANGLE *Trifolium pratense* (Red Clover) Dor. Grigson.

LOVE-APPLE 1. *Epilobium hirsutum* (Great Willowherb) War. Grigson. 2. *Lycopersicon esculentum* (Tomato) Gerard. 3. *Mandragora officinalis* (Mandrake) Hartland.1909. The original Italian name for the tomato was pomo dei mori (apple of the Moors) and this later became pomo d'ore (golden apple). It was introduced to France as an aphrodisiac, and the French misspelled its name as pomme d'amour. So the tomato eventually reached England under the name pome amoris - love-apple, which name went back to America with the colonists. Mandrake, too, was reckoned to be an aphrodisiac. Dudaim, the Hebrew name for it, means love-apple. Great Willowherb apparently got the name simply by association with the various 'apple' names already given to it - such as Apple-pie, Coddled Apples, etc.

LOVE-BIND *Clematis vitalba* (Old Man's Beard) B & H. Cf. Love-entangled for this.

LOVE-ENTANGLE *Nigella damascena* (Love-in-a-mist) Corn. Jago.

LOVE-ENTANGLED 1. *Clematis vitalba* (Old Man's Beard) Dev. B & H. 2. *Lotus corniculatus* (Bird's-foot Trefoil) Corn. B & H. 3. *Nigella damascena* (Love-in-a-mist) Dev. Friend.1882. 4. *Sedum acre* (Biting Stonecrop) Som. Macmillan.

LOVE-FLOWER, Fiery *Eranthemum igneum*.

LOVE-IN-A-CHAIN *Sedum reflexum* (Yellow Stonecrop) Cumb. Grigson. Cf. the Scottish Love-links.

LOVE-IN-A-HEDGE *Nigella damascena* (Love-in-a-mist) Tynan & Maitland.

LOVE-IN-A-MIST 1. *Cerastium tomentosum* (Snow-in-Summer) N'thants. B & H. 2. *Foeniculum vulgare* (Fennel) Potter & Sargent. 3. *Nigella damascena*. 4. *Passiflora foetida* Barbados Gooding.

LOVE-IN-A-PUZZLE *Nigella damascena* (Love-in-a-mist) Dev. Friend.1882.

LOVE-IN-A-TANGLE *Sedum acre* (Biting Stonecrop) Cumb. Grigson.

LOVE-IN-IDLE *Viola tricolor* (Pansy) Prior.

LOVE-IN-IDLENESS *Viola tricolor* (Pansy) Shakespeare. Pansies were once thought to be aphrodisiac, an idea mistaken enough to engender this name, which surely can mean no more than love-in-vain.

LOVE-IN-VAIN *Viola tricolor* (Pansy) Som. Grigson.

LOVE-IN-WINTER *Chimaphila umbellata* (Umbellate Wintergreen) Le Strange. *Chimaphila* suggests the name.

LOVE-LEAVES *Arctium lappa* (Burdock) Grieve.1931. Presumably because of the clinging nature of burrs. Cf. the Somerset Kisses.

LOVE-LIES-BLEEDING 1. *Adonis annua* (Pheasant's Eye) Glos. J D Robertson. 2. *Amaranthus caudatus*. 3. *Amaranthus retroflexus* (Pigweed) Australia Watt. 4. *Dicentra spectabilis* (Bleeding Heart) Dev. Friend. 5. *Lunaria annua* (Honesty) Dor. Macmillan. 6. *Viola tricolor* (Pansy) Grieve.1931.

LOVE-LINKS *Sedum relexum* (Yellow Stonecrop) Scot. Grigson. Cf. Love-in-a-chain.

LOVE-LONG; LOVE-LONG, Live-long-and- *Sedum telephium* (Orpine) Radford; the longer version is in Putnam. All to do with the love divinations carried out with the plant. Pieces of it are hung up and named for a girl's boy friends. The piece that lives the longest represents the successful suitor.

LOVE-MAN; LOVEMAN 1. *Clematis vitalba* (Old Man's Beard) Tynan & Maitland. 2. *Galium aparine* (Goose-grass) Vesey-Fitzgerald. Because of the way the burrs cling to every passing person, and. the way that Old Man's Beard embraces whatever it is growing over.

LOVE-ME 1. *Clematis vitalba* (Old Man's Beard) Tynan & Maitland. 2. *Myosotis arvensis* (Field Forget-me-not) Yks. Grigson.

LOVE-ME, LOVE-ME-NOT *Lolium perenne* (Rye-grass) Som. Grigson.1959. Because of the children's divination game of that name, played by pulling off the leaflets one by one.

LOVE-ME-NOT *Veronica chamaedrys* (Germander Speedwell) Bucks. Grigson. Not in keeping with Speedwell itself, and certainly not with Forget-me-not, another name for it, but perhaps it should be linked with such names as Mother-breaks-her-heart. If a child gathers germander speedwell its mother will die.

LOVE-ROOT *Iris germanica var. florentina* (Orris) Valiente. It was a love-charm in some way or other.

LOVE-SEED *Cogswellia daucifolia*. Plains Indian tribes once used the seeds (along with those of Wild Columbine and the dried roots of Ginseng and Cardinal Flower), as a love charm.

LOVE-STONE *Hedera helix* (Ivy) Leics. Grigson. Merely a comment on its habit.

LOVE-TREE *Cercis siliquastrum* (Judas Tree) W Miller. Why? With the Judas connection, this tree could be nothing but ill-omened.

LOVE-TROTH *Paris quadrifolia* (Herb Paris) Midl. Tynan & Maitland. Cf. the various 'truelove' names for this.

LOVE-TRUE *Viola tricolor* (Pansy) N'thants. Grigson.

LOVE-VINE 1. *Cuscuta americana* Barbados Gooding. 2. *Cuscuta compacta* USA Bergen. 3. *Cuscuta gronovii* (Gronovius's Dodder) USA Tolstead. 4. *Cuscuta indecora* Barbados Gooding. 5. *Cuscuta paradoxa* USA Gilmore. All used for love-charms. *C compacta* is used in Tennessee by breaking off a piece of the vine, and trailing it round the head three times, and then dropping it on a bush behind one. If it grows, the lover is true; if not, false.

LOVE'S TEST *Antennaria plantaginifolia* (Plantain-leaved Everlasting) Indiana Bergen. The love divination in this case is played by taking a leaf by its ends, while the diviner thinks of someone of the opposite sex. The leaf is then pulled apart. If the down on the under-side is drawn out long, much love is indicated. Or, both ends may be named, and the one whose end has the longer down is the more ardent lover.

LOVEACHE *Ranunculus scleratus* (Celery-leaved Buttercup) Henslow. Lovage is the type, but 'ache' is the word for any parsley-leaved plant.

LOVEIDOLDS *Viola tricolor* (Pansy) Wilts. Jones & Dillon. Yet another version of Love-in-idleness.

LOVER, Hedge *Geranium robertianum* (Herb Robert) Som. Vesey-Fitzgerald.

LOVER'S KISSES; LOVER'S KNOTS *Galium aparine* (Goose-grass) Som. Macmillan (Kisses); Wilts. Macmillan (Knots). A reference to the clinging burrs.

LOVER'S LINKS *Umbilicus rupestris* (Wall Pennywort) Scot. Jamieson.

LOVER'S PRIDE *Polygonum persicaria* (Persicaria) Suss. Grigson.

LOVER'S STEPS *Lolium temulentum* (Darnel) Suss. B & H. The florets grow like steps up the stalk, and they are removed one by one in a Love-me, love-me-not game.

LOVER'S THOUGHTS *Viola tricolor* (Pansy) Som. Macmillan.

LOVER'S WANTON 1. *Dactylorchis incarnata* (Early Marsh Orchid) Aber. B & H. 2. *Dactylorchis maculata* (Heath Spotted Orchid) Aber. B & H. 3. *Orchis mascula* (Early Purple Orchis) Aber. B & H. There are a lot of erotic names for Early Purple Orchis, all resulting from the double, testicle-like tubers.

LOVERS, Silly *Arum maculatum* (Cuckoo-pint) Som. Grigson. Cf. Sweethearts, also from Somerset. They are two of the many names with sexual implications given to Cuckoo-pint. 'Silly', in its archaic sense, meant happy.

LOVING ANDREWS *Geranium pratense* (Meadow Cranesbill) Wilts. Dartnell & Goddard.

LOVING IDOLS; LOVING LYDLES *Viola*

tricolor (Pansy) Wessex Rogers; also Berks. Notes & Queries. 8;1871 (Loving Idols); Wilts. Macmillan (Loving Lydles). i.e. Love-in-idleness.

LOW BELIA *Lobelia inflata* (Indian Tobacco) USA Henkel.

LOWLAND FIR; LOWLAND WHITE FIR *Abies grandis* Schery (White).

LOWLAND SPRUCE *Picea sitchensis* (Sitka Spruce).

LOWRIES, Sturdy *Daphne laureola* (Spurge Laurel) Dur. Grigson. A variation of Spurge Laurel (Lowries was often used for laurel).

LOWRY *Daphne laureola* (Spurge Laurel) Prior.

LOXA BARK *Cinchona officinalis* Chopra.

LUBBER-LUB *Menyanthes trifoliata* (Buckbean) IOM Gill.1929.

LUCE, Flower de 1. *Iris germanica var. florentina*. 2. *Lilium candidum* (Madonna Lily) Chaucer. Not only in Chaucer, but later, in the 16th century, St Francis de Sales. Flower de Luce is only an Englishing of fleur de lys, where 'lys' is the French lis, lily.

LUCE, Flower de, Bastard *Iris pseudo-acarus* (Yellow Flag) Gerard.

LUCE, Flower de, Florence *Iris germanica var. florentina* (Orris) Gerard.

LUCE, Flower de, Water *Iris pseudo-acarus* (Yellow Flag) Gerard.

LUCE, Flower de, Water, Yellow *Iris pseudo-acarus* (Yellow Flag) Wesley.

LUCE, Flower de, Yellow *Iris pseudo-acarus* (Yellow Flag) Lyte.

LUCERNE 1. *Medicago sativa var. sativa*. 2. *Onobrychis sativa* (Sainfoin) B & H. French luzerne, but the word is from Latin lux, lucis = light, a reference as far as sainfoin is concerned, to the shiny grains.

LUCERNE, Sweet *Melilotus officinalis* (Melilot) Leyel.1937.

LUCIFER *Medicago sativa var. sativa* (Lucerne) Corn. Grigson. Lucifer here is a corruption of Lucerne.

LUCIFER MATCHES *Sisymbrium officinale* (Hedge Mustard) Worcs. B & H. Presumably descriptive.

LUCK *Anthyllis vulneraria* (Kidney Vetch) Norf. Nall.

LUCK, Good *Oxalis acetosella* (Wood Sorrel) Som. Macmillan.

LUCK-HERB *Hypericum pulchrum* (Upright St John's Wort) IOM Moore.1924. Used, in some unspecified way, to bring good luck.

LUCKEN GOLLAND *Caltha palustris* (Marsh Marigold) Border Turner.

LUCKEN GOWAN *Trollius europaeus* (Globe-flower) Scot. Jamieson. 'Locken' is more usual. It means closed in, or locked in, and Gowan is a yellow flower.

LUCKIE GOWAN; LUCKY GOWAN *Trollius europaeus* (Globe-flower) Scot. Tynan & Maitland.

LUCKIE'S MUTCH *Aconitum napellus* (Monkshood) Lanark. B & H. Descriptive - 'mutch' is a cap, so the whole name means the same as Granny's Bonnet, or Granny's Nightcap, which are also recorded.

LUCKY BEAN TREE *Erythrina abyssinica* Palgrave.

LUCKY MOON *Umbilicus rupestris* (Wall Pennywort) Dor. Dacombe.

LUCOMBE OAK *Quercus x hispanica lucombeana*. This variety was raised by an Exeter nurseryman called Lucombe in the 1760's.

LUCY, Patient *Impatiens walleriana* (Busy Lizzie) Harvey. Cf. Patience Plant.

LUCY ARNUT *Conopodium majus* (Earthnut) Fife Gibson. Lousy Arnut is its immediate forebear. Arnut is earthnut. and the 'lousy' epithet comes from the belief that if you eat too many of them, lice will crowd into your hair.

LUCY LOCKET *Cardamine pratensis* (Lady's Smock) Rut. Grigson; Derb. Northall.

Lucy Locket, lost her pocket,
In a shower of rain.
Milner fun' it, milner grun' it,
In a peck of grain

according to one version of the nursery rhyme. But it is not clear what it has to do with Lady's Smock.

LUG *Iris pseudo-acarus* (Yellow Flag) B & H.

LUGS, Cat's *Primula auricula* (Auricula) Roxb. B & H. Bear's Ears is more common than this. That kind of name is suggested by auricula itself.

LUGS, Cuddy's *Verbascum thapsus* (Mullein) N'thum. Grigson. i.e. donkey's ears. Those flannelly leaves have also suggested Bunny's Ears, Bull's Ears and Cow's Ears.

LUGS, Dog's *Digitalis purpurea* (Foxglove) Fife Grigson.

LUJULA *Oxalis acetosella* (Wood Sorrel) B & H. This is a variant of Hallelujah, which can appear in various forms. It is called Alleluia in Italy, too, and the plant was often put in the foreground of Italian paintings of the Crucifixion.

LUKES *Brassica campestris var. rapa* the leaves (Turnip tops) S Eng. Halliwell.

LUKIN GOWAN *Trollius europaeus* (Globeflower) Scot. Jamieson. See Lucken-golland and the like.

LUMPER-SCRUMP; LIMPER-SCRIMP *Heracleum sphondyllium* (Hogweed) Som. B & H.

LUNARY; LUNARY, Great *Lunaria annua* (Honesty) Drayton (Lunary); Turner (Great Lunary). Lunaria is from the half-moon shape of the segments of the fronds. Cf. Moonwort, and Moon-flower.

LUNARY, Little *Botrychium lunaria* (Moonwort) Turner. Great Lunary is Honesty. These moon names occur by reason of the crescent shape of this fern's leaflets.

LUNDI, Star of the *Pachypodium saundersii*.

LUNG, Land *Peltigera canina* (Ash-coloured Ground Liverwort) Halliwell.

LUNG, Oak *Lobaria pulmonaria* (Lungwort Crottle) Cullum. At one time it was made into a jelly and given to those suffering from lung trouble (hence *pulmonaria*). Or it was given as a tisane for bronchial catarrh.

LUNG-FLOWER *Gentiana pneumonanthe* (Marsh Gentian) Culpepper. Of some medicinal significance, given the specific name *pneumonanthe*.

LUNG MOSS *Lobaria pulmonaria* (Lungwort Crottle) Flück. See Oak Lung.

LUNGS, Cuddy's *Verbascum thapsus* (Mullein) Fernie. Lugs or lungs are equally relevant as descriptions of the leaves, but Cuddy's Lugs is already recorded, and one suspects that this is a misreading.

LUNGWORT 1. *Lathraea squamaria* (Toothwort) Gerard. 2. *Lobaria pulmonaria* (Lungwort Crottle) Bolton. 3. *Pulmonaria officinalis*. 4. *Verbascum thapsus* (Mullein) Som. Macmillan. Lungwort gets its name from the spotted leaves, which by doctrine of signatures were used as a remedy for diseased lungs. In fact it is of some value, and is still used in infusion for lung infections and respiratory disorders.

LUNGWORT, Bullock's *Verbascum thapsus* (Mullein) B & H; USA Henkel.

LUNGWORT, Clown's 1. *Lathraea squamaria* (Toothwort) B & H. 2. *Verbascum thapsus* (Mullein) B & H; USA Henkel.

LUNGWORT, Cow's 1. *Helleborus niger* (Christmas Rose) Cockayne. 2. *Verbascum thapsus* (Mullein) B & H; USA Henkel. Listed as Christmas Rose, but should it not be Stinking Hellebore, *Helleborus foetidus*? Other names for this include Setter, Setterwort, etc. When cattle coughed, an issue was made through the dewlap with a setter, or thread, and a length of hellebore root inserted to irritate the flesh and keep it running.

LUNGWORT, French; LUNGWORT, Golden *Hieraceum murorum* (Wall Hawkweed) both Gerard.

LUNGWORT, Narrow-leaved *Pulmonaria longifolia*.

LUNGWORT, Sea 1. *Mertensia maritima* (Smooth Lungwort) Gilmour & Walters. 2.

Pulmonaria maritima (Sea Bugloss) Curtis. Are these the same plant?

LUNGWORT, Smooth *Mertensia maritima*.

LUNGWORT CROTTLE *Lobaria pulmonaria*. 'Pulmonaria' (and Lungwort) because at one time it was made into a jelly and given to those suffering from lung trouble. Or sometimes it was given as a tisane for bronchial catarrh.

LUPIN, Big-bend *Lupinus leucopsis*. Lupin means literally wolfish (Latin lupus), and there are references to wolves in some of the old names. The Dutch name means wolf's bean, for example.

LUPIN, Blue *Lupinus angustifolius*.

LUPIN, Broad-leaved *Lupinus latifolius*.

LUPIN, False *Thermopsis rhombifolia*.

LUPIN, Garden *Lupinus polyphyllus*.

LUPIN, Low *Lupinus pusillus*.

LUPIN, Plumas *Lupinus onustas*.

LUPIN, Nootka *Lupinus nootkatensis* (Scottish Lupin) Hutchinson. 'Nootka' sounds more reasonable than 'Scottish'. It got the latter name because it happens to grow in river shingle in the central Highlands.

LUPIN, Rose *Lupinus densiflorus*.

LUPIN, Scottish *Lupinus nootkatensis*. See Nootka Lupin.

LUPIN, Silky *Lupinus sericeus*.

LUPIN, Silvery *Lupinus argenteus*.

LUPIN, Spurred, Douglas *Lupinus laxiflorus*.

LUPIN, Spurred, Kellogg's *Lupinus caudatus*.

LUPIN, Sweet *Lupinus luteus* (Yellow Lupin) Le Strange.

LUPIN, Tree *Lupinus arboreus*.

LUPIN, White *Lupinus albus*.

LUPIN, Wild *Lupinus perennis*.

LUPIN, Woolly-leaved *Lupinus leucophyllus*.

LUPIN, Yellow *Lupinus luteus*.

LURGADISH; LURGEYDISH; LURKEY-DISH; LURK-IN-DITCH *Mentha pulegium* (Pennyroyal) Ches, IOM Garrad (Lurgadish); IOM Moore.1924 (Lurgeydish); Ches. Holland (Lurkey-dish). Fernie (Lurk-in-ditch).

LUS NA LAOCH *Sedum roseum* (Roseroot) Ire. Grigson.

LUS-Y-VOLLEY *Galium verum* (Lady's Bedstraw) IOM Grigson. i.e. herb of the sweet smell.

LUSITANIAN HEATH *Erica lusitanica* (Western Heath) Polunin.1976. It grows on the western seaboard of Europe.

LUSMORE *Digitalis purpurea* (Foxglove) Ire. B & H. i.e. the great herb.

LUSS-NY-OLLEE *Pinguicula vulgaris* (Butterwort) IOM Moore. 'plant of the cattle', literally.

LUSTWORT *Drosera rotundifolia* (Sundew) Prior. Cf. Youthwort for this. Both are medieval names, presumably from the aphrodisiac and strengthening power of the distillation. Cows are said to have their copulative instincts excited by eating even a small quantity of the plant.

LYCHEE *Litchi chinensis*. All varieties of the name are from Pekin Mandarin li-chih. It is, of course, a Chinese tree, but grown chiefly in Bengal.

LYCHNIS, Meadow *Lychnis flos-cuculi* (Ragged Robin) Curtis.

LYME-GRASS *Elymus arenarius*.

LYND *Tilia europaea* (Lime) Scot. Jamieson. Lind(en), of course.

LYNE *Linum usitatissimum* (Flax) Fernie.

LYRE-FLOWER *Dicentra spectabilis* (Bleeding Heart) Coats. Descriptive, encompassing the whole flowering stem rather then the individual flower.

LYRE-LEAVED ROCK-CRESS *Arabis lyrata*.

LYRE-TREE *Liriodendron tulipifera* (Tulip-tree) W Miller.

LYTHEWALE; LITHEWALE *Lithospermum officinale* (Gromwell) Dyer (Lythewale); Halliwell (Lithewale). The earlier version of Gerard's Lichwale.

LYVER *Typha latifolia* (False Bulrush) Som. Macmillan. Levver is also recorded in Somerset. These, rightly belonging to Yellow Flag, are from OE leafer.

M

MACARTNEY ROSE *Rosa bracteata.* Introduced from northwest China in 1765 by Lord Macartney, and named after him.

MACASSAR OIL TREE *Canangium odoratum* (Ylang-Ylang) Leyel.1937.

MACAW-BUSH *Solanum mammosum* (Amoise) Howes.

MACE 1. *Fagus sylvatica* fruit (Beech mast) Som. Elworthy. 2. *Myristica fragrans* (Nutmeg) Grieve.1931. 3. *Quercus robur* acorn (Acorn) Som. Jennings; Dev. Dev. Asscn. Trans. vol 65, 1933. 4. *Tanacetum balsamita* (Balsamint) Brownlow. Mace is the thin shell that surrounds the seed of Nutmeg, and serves as a spice in its own right. Balsamint is an aromatic that also has the name Allspice. When applied to beechmast and acorns, it is just a misreading, or local rendering, of mast.

MACE FLINWORT *Leucanthemum vulgare* (Ox-eye Daisy) Fernie.

MACEDONIAN OAK *Quercus trojana.*

MACEDONIAN PINE *Pinus peuce.*

MACEY; MACEY, Oak *Quercus robur* acorn (Acorn) both Som. Macmillan. Macey = mace = mast.

MACEY-TREE *Quercus robur* (Oak) Som. Macmillan.

MACKAY BEAN *Entada phaseoloides* (Nicker Bean) Australia Chopra.

MACKAY'S HEATH *Erica mackaiana.*

MACKEREL MINT *Mentha spicata* (Spearmint) Gerard. It is also known as Fish Mint. But it is better known, and used as, Lamb Mint.

MAD-APPLE 1. *Datura stramonium* (Thornapple) USA Henkel. 2. *Solanum insanum* (?) Gerard. There is the well-known story of the soldiers sent to Jamestown to quell the uprising known as Bacon's Rebellion (1676), who gathered young plants of Datura and cooked them as a potherb - "the effect of which was a very pleasant Comedy; for they turn'd natural Fools upon it for several days...".

MAD DOG WEED *Alisma plantago-aquatica* (Water Plantain) Grieve.1931. At one time it was considered in Russia a proper cure for hydrophobia, and if there is an English name like this, that belief could not have been confined to Russia.

MAD DOG'S BERRIES *Solanum dulcamara* (Woody Nightshade) Mor. Grigson. Probably because they are poisonous - they are labelled as such quite consistently by names like Poison-berries, Snakeberries, etc. And Devil's Cherries carries the same warning.

MAD NIP *Bryonia dioica* (White Bryony) Pomet. "The root of this plant is so violent, that the Peasants call it the Mad Nip; which, if they happen to eat thro' Inadventure, it makes them frantick, and sometimes they run the risque of Death itself" (Pomet). Besides being given to catmint as an abbreviation of Nepeta, the word nip, or nep, was often used for a large root - in fact, it used to be a quite general name for a turnip in the north of England.

MAD WOMAN'S MILK *Euphorbia helioscopia* (Sun Spurge) Bucks. B & H. There are lots of 'milk' names for this, references to the milky juice. 'Mad woman' perhaps because this is also the devil's plant.

MADAGASCAR BEAN *Phaseolus lunatus* (Lima Bean) Bianchini. There doesn't seem to be any consensus as to what part of the world in which to fix this bean. Besides Madagascar and Lima, it is ascribed to Java, Burma, Sieva, Carolina and the Cape.

MADAGASCAR CLOVE *Ravensara aromatica.*

MADAGASCAR DRAGON TREE *Dracaena marginata.*

MADAGASCAR NUTMEG *Ravensara aromatica* (Madagascar Clove) Howes.

MADAGASCAR PERIWINKLE *Vinca rosea.*

MADAGASCAR PLUM *Flacourtia indica* (Kaffir Plum) Howes.

MADAM UGLY *Fritillaria meleagris* (Snake's-head Lily) Coats. From the drab colouring of the flowers - but that is a matter of taste.

MADARIN *Tripleurospermum maritimum* (Scentless Mayweed) Dor. Macmillan.

MADBERRY *Veratrum album* (White Hellebore) Cockayne. Presumably because this plant is highly poisonous.

MADDER 1. *Anthemis cotula* (Maydweed) Hants. Cope. 2. *Asperula odorata* (Woodruff) Wilts. Dartnell & Goddard. 3. *Rubia tinctoria*. OE maeddre, maedere, apparently from an O Norse word, mathra. But Madder for Maydweed is quite different - it is mather, OE maegtha.

MADDER, Dyer's *Rubia tinctoria* (Madder) Thornton.

MADDER, Field *Sherardia arvensis*.

MADDER, Indian *Rubia cordifolia*.

MADDER, Levant *Rubia peregrina* (Wild Madder) Howes.

MADDER, Red *Rubia tinctoria* (Madder) Gerard. It is a red dye.

MADDER, Wild 1. *Galium mollugo* (White Bedstraw) B & H; USA Upham. 2. *Rubia peregrina*.

MADDERN; MADDERS *Anthemis cotula* (Maydweed) Wilts. D Grose (Maddern); Hants. Grigson; Wilts. Dartnell & Goddard; Dor. Barnes (Madders). The same word, for Maddern is the old plural in 'n'. They are the same as Maithen, Mather, Mauthern, etc.

MADDERWORT 1. *Asperugo procumbens* (Madwort) W Miller. 2. *Lobularia maritima* (Sweet Alison) W Miller. Although Sweet Alison was said to be a remedy against mad dogs' bites (Madwort, Heal-dog, Heal-bite are also recorded for it), and Alyssum, the old generic name, is Greek a, not, and lussa, madness, the reference is probably to madder, anyway.

MADEIRA BROOM *Genista virgata*.

MADEIRA HOLLY *Ilex perado* (Canary Holly) Hora.

MADEIRA VINE *Anredera cordifolia*.

MADERN *Leucanthemum vulgare* (Ox-eye Daisy) Wilts. Goddard. A lot of the Maydweed names appear again in Wiltshire for the white daisy. Cf., for instance, Mathern, Maithen, Mauthern, which are all forgotten old plurals in "-n".

MADERWORT *Artemisia vulgaris* (Mugwort) Halliwell. Cf. Maidenwort.

MADNEP *Heracleum sphondyllium* (Hogweed) Gerard. Probably mead-nep, according to A S Palmer.

MADONNA LILY *Lilium candidum*. The lily of sacred art is always the Madonna Lily, and is, of course, dedicated to the Virgin Mary; it is the symbol of purity and chastity.

MADRAS BEAN *Dolichos uniflorus* (Horsegram) Douglas.

MADRAS HEMP *Crotolaria juncea* (Sunn Hemp) Howes.

MADRON *Tanacetum parthenium* (Feverfew) Dor. Macmillan.

MADRONA; MADRONA LAUREL *Arbutus menziesii* Madrona is the usual name; Madrona Laurel is in Brimble.1948.

MADWEED 1. *Ballota nigra* (Black Horehound) Vesey-Fitzgerald. 2. *Marrubium vulgare* (White Horehound) Vesey-Fitzgerald. Horehound is supposed to act as an antidote to the bite of a mad dog.

MADWORT 1. *Asperugo procumbens*. 2. *Ballota nigra* (Black Horehound) Clair. 3. *Lobularia maritima* (Sweet Alison). See Madderwort, and also Madweed.

MADWORT, German *Asperugo procumbens* (Madwort) Prior.

MAGELLAN RAGWORT *Senecio smithii*.

MAGEROUM *Origanum vulgare* (Marjoram) T Wright. A garbled version of marjoram.

MAGGIE, Sleeping *Trifolium pratense* (Red Clover) N'thum. Grigson.

MAGHET *Tanacetum parthenium* (Feverfew) R T Gunther.1905. Cf. Maithes, Maids, and Mayweed.

MAGIC HERB *Datura stramonium* (Thorn-apple) Haining.

MAGICIAN'S CYPRESS *Juniperus sabina*

(Savin) Fernie. It was also known as the Devil's Tree.

MAGNOLIA, Cucumber *Magnolia acuminata* (Cucumber Tree) Brimble.1948.

MAGNOLIA, Kobus *Magnolia kobus*.

MAGNOLIA, Laurel *Magnolia grandiflora* (Bull Bay) USA Wit. 'Laurel' because of the evergreen leaves.

MAGNOLIA, Saucer *Magnolia x soulangeana*.

MAGNOLIA, Southern *Magnolia grandiflora* (Bull Bay) Leathart. It is a native of the southern states of America.

MAGNOLIA, Sweet *Magnolia virginiana* (Sweet Bay) Taylor.

MAGNOLIA LAUREL *Magnolia virginiana* (Sweet Bay) Howes.

MAGPIE GLADIOLUS *Acidanthera bicolor* McDonald. Bicolor, hence 'magpie'.

MAGUEY 1. *Agave americana* (Mescal) Emboden). 2. *Agave utahensis* USA Elmore. Maguey is the Spanish name.

MAGWEED *Leucanthemum vulgare* (Oxeye Daisy) A S Palmer. It may be a misprint for Mayweed, but Palmer said it was a corruption of marguerite.

MAHAW *Crataegus monogyna* (Hawthorn) Ire. Grigson.

MAHOE *Thespesia populnea* (Portia Tree) RHS.

MAHOGANY, African *Khaya ivorensis*.

MAHOGANY, African, Smooth-barked *Khaya anthoteca* (White Mahogany) Dalziel.

MAHOGANY, Benin *Khaya grandifoliola*. Cf. Benin Wood.

MAHOGANY, Borneo *Calophyllum inophyllum* (Alexandrian Laurel) Howes.

MAHOGANY, Broad-leaved *Khaya grandifoliola* (Benin Mahogany) Dalziel.

MAHOGANY, Cape *Trichilia emetica*.

MAHOGANY, Ceylon *Melia dubium* Australia Hora.

MAHOGANY, Chinese *Cedrula sinensis*.

MAHOGANY, Colombian *Swietenia macrophylla* (Honduras Mahogany) Ackermann.

MAHOGANY, Cuban *Swietenia mahogoni*. This is the "true" mahogany.

MAHOGANY, Dry-zone *Khaya senegalensis*.

MAHOGANY, East Indian *Pterocarpus dalbergioides*.

MAHOGANY, Gambian *Khaya senegalensis* (Dry-zone Mahogany) Dalziel.

MAHOGANY, Honduras *Swietenia macrophylla*.

MAHOGANY, Irish *Alnus glutinosa* (Alder) Ire. Grigson. It is the reddish timber which is the reference.

MAHOGANY, Mountain 1. *Cercocarpus montanus*. 2. *Betula lenta* (Cherry Birch) Grieve.1931. Cf. Mahogany Birch for the latter.

MAHOGANY, Mountain, Alder-leaf *Cercocarpus montanus* (Mountain Mahogany) USA T H Kearney.

MAHOGANY, Mountain, Birch-leaved *Cercocarpus betuloides*.

MAHOGANY, Mountain, Curl-leaf *Cercocarpus ledifolius*.

MAHOGANY, Mountain, Little-leaved *Cercocarpus intricatus*.

MAHOGANY, Natal *Kiggelaria dregeana*.

MAHOGANY, Pod *Afzelia cuanzenis*.

MAHOGANY, Red *Khaya nyasica*.

MAHOGANY, Rhodesian *Afzelia cuanzensis* (Pod Mahogany) Dalziel.

MAHOGANY, St Domingo *Swietenia mahagoni* (Cuban Mahogany) Ackermann.

MAHOGANY, Scotch *Alnus glutinosa* (Alder) Ablett. Cf. Irish Mahogany for alder.

MAHOGANY, Singapore *Melanorrhoea curtisii* Gimlette.

MAHOGANY, Small *Turraea nilotica*.

MAHOGANY, Spanish *Swietenia mahagoni* (Cuban Mahogany) Ackermann.

MAHOGANY, Swamp 1. *Chlorophora excelsa* (African Oak) Dalziel. 2. *Eucalyptus robustus*.

MAHOGANY, Tabasco *Swietenia macrophylla* (Honduras Mahogany) Ackermann.

MAHOGANY, White 1. *Khaya anthoteca*. 2.

Trichilia emetica (Cape Mahogany) Palgrave.
MAHOGANY BEAN *Afzelia cuanzensis* (Pod Mahogany) Dale & Greenway.
MAHOGANY BIRCH *Betula lenta* (Cherry Birch) Grieve.1931. Cf. Mountain Mahogany for this.
MAHUA *Bassia latifolia*.
MAID, Fair, February *Galanthus nivalis* (Snowdrop) Som. Macmillan. Fair Maids of February is the better-known form.
MAID, Lady's *Artemisia chamaemelifolia* Brownlow.
MAID, May *Glechoma hederacea* (Ground Ivy) Rohde. Probably originally Hay-maid, where 'hay' means 'hedge'.
MAID, Meadow *Filipendula ulmaria* (Meadowsweet) Pemb. Grigson.
MAID, Old *Catharanthus rosea* (Madagascar Periwinkle) Howes.
MAID-IN-THE-MEADOW; MAIDEN-THE-MEADOW *Ranunculus bulbosus* (Bulbous Buttercup) both Som. Macmillan (Maid); Grigson (Maiden).
MAID-IN-THE-MIST *Umbilicus rupestris* (Wall Pennywort) Scot. Jamieson.
MAID-OF-THE-MEAD; MAID-OF-THE-MEADOW *Filipendula ulmaria* (Meadowsweet) Ches. (Maid); Som. (Meadow) both B & H.
MAID-OF-THE-MIST *Thalictrum fendleri* USA Elmore.
MAID'S HAIR; MAIDEN'S HAIR *Galium verum* (Lady's Bedstraw) Turner (Maid's); Glos. PLNN35 (Maiden's).
MAID'S LOVE *Artemisia abrotanum* (Southernwood) N'thants. A E Baker. See rather Lad's Love.
MAIDEN *Anthemis cotula* (Maydweed) Dor. Macmillan. There are many variations of a modern form of OE maegde, including Mather or Mauther. Mayweed is another version, and the first syllable there is not 'may'.
MAIDEN, Modest *Viola odorata* (Sweet Violet) Som. Macmillan. A descriptive name - there is always a tendency for violets to hang their heads.

MAIDEN, Shame-faced 1. *Anemone nemerosa* (Wood Anemone) Wilts. Dartnell & Goddard. 2. *Ornithogalum umbellatum* (Star of Bethlehem) Wilts. Dartnell & Goddard.
MAIDEN, Yellow *Narcissus pseudo-narcissus* (Daffodil) Som. Macmillan.
MAIDEN MERCURY *Mercurialis annua* (French Mercury) Lyte. Or Girl's Mercury. Why? For there is also Boy's Mercury quoted by Lyte. He explains it by referring to an old belief that when "dronken (it) causeth to engender male children". But that doesn't help to interpret Maiden Mercury.
MAIDEN OAK *Quercus petraea* (Durmast Oak) Hants. Grigson.
MAIDEN PINK *Dianthus deltoides*. "The floures are of a blush colour, whereof it tooke its name" (Gerard), but Maiden in this case is more likely to be "meadow". It is said that the name may be a mistake for mead-pink.
MAIDEN'S BLUSH *Momordica charantia* (Wild Balsam-apple) Howes.
MAIDEN'S BREATH *Gypsophila paniculata* Clapham.
MAIDEN'S DELIGHT *Artemisia abrotanum* (Southernwood) Corn. Jago; Som. Macmillan.

Lads' love is lasses' delight,
And if the lads don't love,
Lasses will flite.

A North Country rhyme (flite means scold).
MAIDEN'S HAIR 1. *Briza media* (Quaking Grass) IOW, Norf. Grigson.1959. 2. *Clematis vitalba* (Old Man's Beard) Bucks. Grigson. 3. *Cuscuta epithymum* (Dodder) IOW Grigson. 4. *Galium cruciatum* (Crosswort) N'thants. Grigson. 5. *Galium verum* (Lady's Bedstraw) Glos. PLNN35. 6. *Narthecium ossifragum* (Bog Asphodel) Lancs. Putnam. Descriptive, in their various ways, except for Bog Asphodel. The reference here is to an old use as a hair dye.
MAIDEN'S HEADS *Sanguisorba officinalis* (Great Burnet) Yks. Grigson.

MAIDEN'S HONESTY 1. *Clematis vitalba* (Old Man's Beard) Aubrey. 2. *Lunaria annua* (Honesty) Halliwell.

MAIDEN'S RUIN *Artemisia abrotanum* (Southernwood) Friend. 'Lad's Love' may very well result in 'Maiden's Ruin'. Friend thought that the French name Armoise au Rone may have given rise to it, but that seems doubtful.

MAIDEN'S TEARS *Silene cucubalis* (Bladder Campion) Howes.

MAIDENHAIR 1. *Galium aparine* (Goose-grass) Yks. B & H. 2. *Glechoma hederacea* (Ground Ivy) Coles. 3. *Thalictrum minus var. adiantifolium*.

MAIDENHAIR, Black *Asplenium adiantum-nigrum* (Black Spleenwort) Clare.

MAIDENHAIR, Delta *Adiantum raddianum*.

MAIDENHAIR, Golden *Polypodium vulgare* (Common Polypody) Kent,Heref. B & H.

MAIDENHAIR, Rose *Adiantum hispidulum*. 'Rose', because the young fronds are a distinct pink colour.

MAIDENHAIR FERN *Adiantum capillus-veneris*.

MAIDENHAIR SPLEENWORT *Asplenium trichomanes* (Common Spleenwort) Brightman.

MAIDENHAIR TREE *Gingko biloba*. Because each leaf resembles a magnified leaflet of the Maidenhair Fern.

MAIDENS, Milk *Primula veris* (Cowslip) Lincs. Grigson. Colour descriptive, presumably.

MAIDENS, Milky *Cardamine pratensis* (Lady's Smock) Dev. Grigson. The colour of the flowers is said to resemble milkmaids' complexions. Cf. Milkmaids, Milking Maids, etc.

MAIDENS, Naked 1. *Colchicum autumnale* (Meadow Saffron) Dor. Dacombe. 2. *Galanthus nivalis* (Snowdrop) Som. Macmillan. The flowers of Meadow Saffron rise naked from the earth. Cf. Naked Ladies, Naked Virgins, Naked Nannies and Naked Boys.

MAIDENWORT *Artemisia vulgaris* (Mugwort) Fernie. Cf. Motherwort. There was an early form, Moderwort, which may possibly be from its use in uterine diseases; in fact old herbalists called mugwort Mater herbarum.

MAIDS *Tanacetum parthenium* (Feverfew).

MAIDS, Fair; MAIDS, Pretty *Saxifraga granulata* (Meadow Saxifrage) Bucks. Grigson (Fair); Berks. Dyer (Pretty). It is probably Fair Maids of France, which is also recorded in the same county. Pretty Maids may very well be the Pretty maids all in a row, from Mary, Mary, quite contrary.

MAIDS, Milking *Cardamine pratensis* (Lady's Smock) Som. Macmillan.

MAIDS, Red *Calandrina ciliata*.

MAIDSWEET *Filipendula ulmaria* (Meadowsweet) Limerick Fitzgerald. Not 'maid'in this case, but 'mead'. But Cf. Maid of the Meadow.

MAIDWEED *Anthemis cotula* (Maydweed) Lupton.

MAIDWORT *Artemisia abrotanum* (Southernwood) Tynan & Maitland. A name of the Lad's Love variety.

MAIKEN; MAIKIN *Iris pseudo-acarus* (Yellow Flag) Lancs. B & H (Maiken); NodaL (Maikin). This appears as Mekkin in Cumbria.

MAILKES, Red *Papaver rhoeas* (Red Poppy) Halliwell, who records it without comment.

MAILS *Chenopodium album* (Fat Hen) Ayr. B & H. Cf. Meldweed, Myles, or Melgs, all Scottish names for Fat Hen.

MAINGAY'S OAK *Quercus maingayi*.

MAISE 1. *Anthemis cotula* (Maydweed) Dor, Shrop, Lincs. Grigson. 2. *Leucanthemum vulgare* (Ox-eye Daisy) Lincs. Grigson. Obviously connected with the Maithes and Mayweed sequence of names. The name varies into Maze in the same area.

MAIT-BANES *Vicia faba* (Broad-bean) Corn. Jago.

MAITEN *Maytenus boaria*.

MAITHEN 1. *Anthemis cotula* (Maydweed)

Wilts, Dor, Hants, IOW Grigson. 2. *Leucanthemum vulgare* (Ox-eye Daisy) Glos, Wilts. Grigson.

MAITHER *Anthemis cotula* (Maydweed) N Eng. Prior. Prior said that Mather is a word used in the North country to denote a working-class girl. But the name in some form or other is not confined to the north - Mather, for instance, is recorded from Dorset. The plant, to quote Hill, "is good in all hysteric complaints, and it promotes the menses".

MAITHES *Tanacetum parthenium* (Feverfew) Prior.

MAITHES, Red *Adonis annua* (Pheasant's Eye) Prior.

MAITHEWEED *Anthemis cotula* (Maydweed) B & H.

MAIZE *Zea mays*. Both the botanical name, mays, and the common name, maize, come from the Arawak-Carib name, mahiz.

MAIZE, Dent *Zea mays var. americana*. There is an indentation or depression on the top of the grain, caused by shrinkage of the endosperm.

MAIZE, Flint *Zea mays var. praecox*. It gets the name from the hard endosperm.

MAIZE, Flour *Zea mays var. amylacea*.

MAIZE, Japanese *Zea japonica*.

MAIZE, Soft *Zea mays var. amylacea* (Flour Maize).

MAKEBATE *Polemonium caeruleum* (Jacob's Ladder) Prior. Because, so it was said, "if it is put into the bed of a married couple, it sets them quarrelling"; but presumably it is because of the name polemonium, from polemis, war. This is of course the earlier meaning of the word bate, as strife (still to be found as 'debate').

MAKEBEGGARS *Sagina prucumbens* (Common Pearlwort) Halliwell. Cf. the Norfolk name Poverty - the reference is to the kind of ground in which it thrives.

MAKEPEACE *Betula verrucosa* (Birch) Coles. "hath an admirable influence upon (children) to quiet them when they are out of order, and therefore some call it makepeace". He was talking about whipping them with a birch, of course.

MAKINBOY *Euphorbia hiberna* (Irish Spurge) B & H. An anglicized version of the Irish, meaning yellow parsnip.

MALABAR, Cypress of *Curcuma longa* (Turmeric) Pomet.

MALABAR GOURD; MALABAR SQUASH *Cucurbita ficifolia* Whitaker & Davis (Squash).

MALABAR KINO *Pterocarpus marsupium* (Bastard Teak) Howes.

MALABAR NIGHTSHADE *Basella alba* (Indian Spinach) Dalziel.

MALABAR NUT *Adhatoda vasica*.

MALABAR ROSEWOOD *Dalbergia latifolia* (Blackwood) Howes.

MALABAR TREE *Euphorbia tirucalli* (Pencil Tree) USA Kingsbury.

MALAGETTA PEPPER; MELEGUETA PEPPER *Amomum grana* (Grains of Paradise) Lindley (Malagetta); J Smith (Melegueta).

MALAY APPLE *Eugenia malaccensis*.

MALAY NETTLE-TREE *Laportea stimulans*. Apparently, they sting so much that walking with bare bodies through them in wet weather can be fatal.

MALAYAN CHESTNUT, Greater *Chrysolepis megacarpa*.

MALAYSIA, Camphor of *Dryobalanops camphora* (Camphor of Borneo) Tosco.

MALE *Taraxacum officinale* (Dandelion) Dor. Halliwell.

MALE CORNEL; MALE DOGWOOD *Cornus mas* (Cornel Cherry) Gerard (Cornel); W Miller (Dogwood).

MALE FERN *Dryopteris filix-mas*.

MALE FRANKINCENSE *Boswellia thurifera* (Frankincense) Pomet.

MALE KOLA *Garcinia kola* (Bitter Kola) Dalziel.

MALE LILY *Convallaria maialis* (Lily-of-the-valley) Culpepper. He must mean May-lily.

MALE NETTLE *Urtica pilulifera* (Roman Nettle) Gerard. Gerard called the common nettle, *U dioica*, Female Nettle, for no very clear reason.

MALE ORCHIS *Orchis mascula* (Early Purple Orchis) Thornton.

MALE PIMPERNEL; TOM PIMPERNEL *Anagallis arvensis* (Scarlet Pimpernel) Gerard (Male); Yks. Grigson (Tom). The Blue Pimpernel, *A foemina*, is the Female Pimpernel. In either case, the sex is awarded for no clear reason, but Gerard was sure of it, for he described pimpernel under *Anagallis mas*.

MALE SPEEDWELL *Veronica officinalis* (Common Speedwell) Curtis.

MALLACE *Malva sylvestris* (Common Mallow) Dev, Som, IOW, Bucks. Grigson; Hants. Cope. Also spelt Mallus in the Isle of Wight (Long).

MALLARD *Althaea officinalis* (Marsh Mallow) Hatfield.

MALLEE *Eucalyptus perriniana* (Spinning Gum) Wilkinson.1981. Many of the smaller members of the genus have the name, which is from an aborigine original.

MALLEE, Blue *Eucalyptus gamophylla*.

MALLICE, Marsh 1. *Althaea officinalis* (Marsh Mallow) Som. Elworthy; IOW. Grigson. 2. *Malva sylvestris* (Common Mallow) Dev, Som, Shrop, Cumb, N'thum. Grigson; IOW Long; Glos. J D Robertson.

MALLOCH, Clatter *Trifolium pratense* (Red Clover) Scot. B & H.

MALLOW *Malva sylvestris*. Mallow is from OE mealwe, Latin malva, which is from the Greek malake, to soften, perhaps because of soft leaves, or, more likely, because it has emollient qualities. Gerard quoted the opinion of his time, that it came from the Hebrew malluach, "of the saltnesse, because the Mallow groweth in saltish and old ruinous places, as in dung-hills and such like".

MALLOW, Blue *Malva sylvestris* (Common Mallow) Kourennoff.

MALLOW, Chinese *Malva verticillata* (Whorled Mallow) Fitter.

MALLOW, Common 1. *Malva neglecta* (Dwarf Mallow) USA Allan. 2. *Malva sylvestris*.

MALLOW, Cornish *Lavatera cretica*.

MALLOW, Dwarf *Malva neglecta*.

MALLOW, Egyptian *Malva parviflora* (Least Mallow) Howes.

MALLOW, False, Red *Malvastrum coccineum*.

MALLOW, Field *Malva sylvestris* (Common Mallow) Gerard.

MALLOW, French 1. *Lavatera olbia*. 2. *Malva sylvestris* (Common Mallow) Corn. Grigson.

MALLOW, Garden *Althaea rosea* (Hollyhock) Gerard.

MALLOW, Great *Althaea rosea* (Hollyhock) Albertus Magnus.

MALLOW, Greek 1. *Kerria japonica* (Jew's Mallow) Salisbury.1936. 2. *Sidalcea sp*. Far from being Greek, *Sidalcea* is an American species. If it comes to that, *Kerria japonica* comes from China.

MALLOW, High *Malva sylvestris* (Common Mallow) USA O P Brown.

MALLOW, Holy *Althaea rosea* (Hollyhock) Blunt.1957. The 'holy' of this name is the same as the 'holly' of hollyhock. It was thought that the origin lay in the plant having been brought back to Europe by the Crusaders, but there is no evidence of that, although there is apparently some evidence of a connection with Saint Cuthbert. There was probably a medieval name malva benedicta.

MALLOW, Indian *Abutilon avicennae* (China Jute) Schery.

MALLOW, Jagged *Althaea rosea* (Hollyhock) Turner.

MALLOW, Jew's 1. *Corchorus olitorius* Moldenke. 2. *Kerria japonica*.

MALLOW, Least *Malva parviflora*.

MALLOW, Low *Malva neglecta* (Dwarf Mallow) Le Strange.

MALLOW, Marsh 1. *Althaea officinalis*. 2. *Caltha palustris* (Marsh Marigold) Oxf. Oxf.Annl.Rept.1951; Yks. B & H. 3. *Lavatera arborea* (Tree Mallow) Som. Macmillan. 4. *Malva neglecta* (Dwarf

Mallow) Ches. Holland. 5. *Malva sylvestris* (Common Mallow) Ches. Holland; Lincs. Rudkin. Marsh Marigold is the odd one in this company. It is probably a simple misnomer.

MALLOW, Marsh, Hispid *Althaea hirsuta*.

MALLOW, Moorish *Althaea officinalis* (Marsh Mallow) Gerard. Presumably 'Moorish' is 'marsh', unless this is connected with Hollyhock, *Althaea rosea*, which traditionally did come from the Near East via the Crusaders.

MALLOW, Musk 1. *Hibiscus abelmoschus* (Muskseed) Dalziel. 2. *Malva moschata*.

MALLOW, Poppy *Callirhoe papaver*.

MALLOW, Poppy, Finger *Callirhoe digitata*.

MALLOW, Poppy, Purple *Callirhoe involucrata*.

MALLOW, Prairie *Sidalcea malvaeflora*.

MALLOW, Prairie, White *Sidalcea candida*.

MALLOW, Purple *Callirhoe involucrata*.

MALLOW, Rose *Hibiscus rosa-sinensis* (Rose of China) Howes.

MALLOW, Rose, Swamp *Hibiscus moschatus*.

MALLOW, Rough *Althaea hirsuta* (Hispid Marsh Mallow) McClintock.

MALLOW, Round-leaved *Malva neglecta* (Dwarf Mallow) Allan.

MALLOW, Tall *Malva sylvestris* (Common Mallow) H & P.

MALLOW, Thorny *Hibiscus sabdariffa* (Rozelle) Schery.

MALLOW, Tree *Lavatera arborea*.

MALLOW, Tree, Lesser *Lavatera cretica* (Cornish Mallow) Wit.

MALLOW, Tree, Sea *Lavatera arborea* (Tree Mallow) Camden/Gibson.

MALLOW, Venice *Hibiscus trionum* (Bladder Katmia) Coats. It grows in central Africa, so why 'Venice'?

MALLOW, Water *Althaea officinalis* (Marsh Mallow) Turner.

MALLOW, White *Althaea officinalis* (Marsh Mallow) Gerard.

MALLOW, Whorled *Malva verticillata*.

MALLOW-HOCK *Malva sylvestris* (Common Mallow) Som. Macmillan.

MALLOW-LEAVED BINDWEED *Convolvulus althaeoides*.

MALLOW-ROCK *Sempervivum tectorum* (Houseleek) Som. Macmillan.

MALLUS; MALLY *Malva sylvestris* (Common Mallow) IOW Long. (Mallus); Tynan & Maitland.

MALLY-GOWL *Calendula officinalis* (Marigold) Yks. B & H. A variation of Mary Gowlan.

MALTA CROSS *Tribulus terrestris* (Puncture Vine) Moldenke.

MALTA LILY *Sprekelia formosissima* (Jacobean Lily) Howes.

MALTA ORANGE *Citrus aurantium var. sinensis*. Also known as the Portuguese Orange.

MALTA THISTLE; MALTESE THISTLE; MALTESE STAR THISTLE *Centaurea melitensis* (Napa Thistle) Watt (Malta, Maltese) (the latter being the Australian name); Clapham (Maltese Star). A Mediterranean plant.

MALTESE CLOVER *Hedysarum coronarium* (French Honeysuckle) Pratt.

MALTESE CROSS *Lychnis chalcedonica* A W Smith. A descriptive name - Cf. Jerusalem Cross, Knight Cross, etc.

MALTESE COCKSPUR *Centaurea melitensis* (Napa Thistle) Watt.

MAMEY *Mammea americana*. Or Mammee Apple, Mammey Apple.

MAMMA'S MILK *Euphorbia helioscopia* (Sun Spurge) Bucks. B & H. From the milky juice.

MAMMEE APPLE; MAMMEY APPLE *Mammea americana* (Mamey) Candolle (Mammee); Brouk (Mammey).

MAMMOTH TREE *Sequoia gigantea* (Wellingtonia) Wit.

MAMRE, Oak of *Quercus pseudo-coccifera* (Abraham's Oak) Hutchinson & Melville. Said to be the oak under which Abraham entertained the three angels.

MAN, Black 1. *Plantago lanceolata*

(Ribwort Plantain) Som, Dor. Grigson. 2. *Plantago media* (Lamb's-tongue Plantain) Som. Macmillan. 3. *Prunella vulgaris* (Self-heal) Yks. Fernie.

MAN, Blind *Papaver rhoeas* (Corn Poppy) Wilts. Grigson. There was a belief at one time that if poppies were put to the eyes, they would cause blindness, hence Blind Eyes, or Blindy-buff, as well as this. Cf. this with the various Headache names.

MAN, Old 1. *Artemisia abrotanum* (Southernwood) Wilts. Dartnell & Goddard; Middx, Cambs, Norf, N'thants, Shrop. Grigson; Oxf. Thompson; Ess. Gepp. 2. *Rosmarinus officinalis* (Rosemary) Som. Macmillan; Suss. B & H. 3. *Senecio douglasii* (Douglas Groundsel) USA Elmore.

MAN-AND-WOMAN *Arum maculatum* (Cuckoo-pint) Som. Grigson. One of a series of names (Cf. Cows-and-bulls) which are male plus female. The spadix in the spathe provides the imagery.

MAN-IN-THE-GROUND; MAN-OF-THE-EARTH *Ipomaea pandurata* (Wild Potato Vine) O P Brown (Man-in-the-ground); USA Gates (Man-of-the-earth).

MAN-IN-THE-PULPIT *Arum maculatum* (Cuckoo-pint) Som. Macmillan. Better known as Parson-in-the-pulpit. Once again, the upright spadix in the spathe provides the imagery.

MAN-OF-WAR *Plantago lanceolata* (Ribwort Plantain) Dev. Grigson. There are a few names based on the children's game like conkers played with the flower heads of ribwort, which are 'Soldiers', or 'Fighters'.

MAN ORCHID 1. *Aceras anthropophorum*. 2. *Dactylorchis incarnata* (Early Marsh Orchid) B & H. 3. *Dactylorchis maculata* (Heath Spotted Orchid) B & H. 4. *Listera ovata* (Twayblade) Som. Macmillan. 5. *Orchis morio* (Green-winged Orchid) B & H.

MAN ORCHID, Green *Aceras anthropophorum* (Man Orchid) Turner Ettlinger.

MAN-TIE *Polygonum aviculare* (Knotgrass) Dev. Friend.1882.

MAN'S MOTHERWORT *Ricinus communis* (Castor Oil Plant) Halliwell.

MANACA *Brunfelsia hopeana*.

MANCHESTER POPLAR *Populus nigra var. betulifolia*.

MANCHINEEL TREE *Hippomane mancinella*.

MANCHURIAN CHERRY *Prunus maackii*.

MANCHURIAN FIR *Abies holophylla*.

MANCHURIAN JUTE *Abutilon avicennae* (China Jute) Howes.

MANCHURIAN WALNUT *Juglans mandschurica*.

MANDARIN; MANDARIN ORANGE *Citrus reticulata*.

MANDRAGON *Mandragora officinalis* (Mandrake) Gerard.

MANDRAKE 1. *Arum maculatum* (Cuckoo-pint) Yks. B & H. 2. *Bryonia dioica* (White Bryony) Dor. Dacombe; IOW Long; Cambs. Porter; War. Palmer; Norf. Grieve.1931; Ches. Holland; Dev, Wilts, Herts, Heref, Shrop, Leics, Lincs, Yks. Grigson. 3. *Circaea lutetiana* (Enchanter's Nightshade) Dev. Grigson. 4. *Mandragora officinalis*. 5. *Tamus communis* (Black Nightshade) Lincs. Rudkin. The 'drake' part of the name means 'dragon', and the association of 'man' and 'drake' is probably an allusion to the man-like form the roots often take.The Bryonies get the name because they were (especially White Bryony) taken as such. The vaguely anthropomorphic roots were the sign that this was as good as the real thing. "...for the roots [of mandrake] which are carried about by impostors to deceive unfruitful women, are made of the roots of canes, briony, and other plants; for in these, yet fresh and virent, they carve out the figures of men and women, first sticking therein grains of barley or millet where they intend the hair should grow; then bury them in sand until the grains shoot forth their roots which, at the longest, will happen in twenty days; they afterwards clip and trim those tender strings in the fashion of beards and other hairy teguments..." (Browne). If the 'drake' part of the name means 'dragon', then cuckoo-pint gets the name on that account. There are Dragons, or Great Dragons, among its names (and of course there is *Arum dracunculus*).

MANDRAKE, American; MANDRAKE, Wild *Podophyllum peltatum* (May-apple) E Hayes (American); O P Brown (Wild). The root gives the name - this is, in fact, the modern mandrake root of the pharmacists.

MANDRAKE, English *Bryonia dioica* (White Bryony) Brownlow. See Mandrake.

MANELET *Chrysanthemum segetum* (Corn Marigold) Scot. Jamieson.

MANGELWURZEL *Beta vulgaris* (Beet) Grigson.1974. This is German Mangold, beet plus würzel, root. The error of calling it Mangelwürzel rather than Mangoldwürzel gave rise to the name Scarcity Root, for German Mangel means scarcity.

MANGETTI TREE *Ricinodendron rautanenii*.

MANGIANTS, Easter; MANTGIONS, Easter *Polygonum bistorta* (Bistort) Yks, Cumb, N'thum. Grigson. From French manger, to eat, the connection with Easter being the Herb Pudding made from the leaves of bistort on Easter Day, or more properly at Passion-tide, served to accompany veal and bacon. It was more usually known as Ledger Pudding, and bistort had other names of relevance, like Easter Ledger. Mangiants produced Giants, so we also have Easter Giants, even Eastern Giants.

MANGO *Mangifera indica*. According to Chambers's Dictionary, this is eventually from Tamil man-kay, the name for the fruit, and came into European language via Portuguese manga.

MANGO, Bachang *Mangifera foetida*.

MANGO, Black *Pentadesma butyracea* Dalziel.

MANGO, Kurwini *Mangifera odorata*.

MANGO, Monjet *Mangifera laurino*.

MANGO, Pahutan *Mangifera altissima*.

MANGOSTEEN *Garcinia mangostana*. Malayan manggistan.

MANGROVE, Red *Rhizophora mangle*.

MANGWE *Terminalia sericea*.

MANICON *Atropa belladonna* (Woody Nightshade) B & H.

MANILA BEAN *Psophocarpus tetragonolobus* (Goa Bean) Douglas.

MANILA HEMP *Musa textilis*. Grown elsewhere now, but this is actually indigenous to the Philippines.

MANIO *Podocarpus nubigena*.

MANIOC *Manihot esculenta*. The name can also be Mandioc, which is nearer to the original Tupí - mandioca.

MANIOC BEAN *Pachyrrhizus tuberosus* (Yam Bean) Schery.

MANITOBA MAPLE *Acer negundo* (Box Elder) Johnston.

MANKIL ROSE-APPLE *Eugenia javanica* (Curacao Apple) Douglas.

MANNA *Lecanora esculenta var. mannifera*. At least, this is what is often accepted as being the biblical Manna.

MANNA ASH *Fraxinus ornus*. The manna in this case is the saccharine exudation of this tree. This is the manna of European medicine, used as a gentle laxative suitable for children and infants.

MANNA ASH, Himalayan *Fraxinus floribunda*.

MANNA-GRASS, Sea *Puccinellia maritima* (Sea Meadow-grass) Hepburn.

MANNA GUM *Eucalyptus viminalis*. The manna in this case is a sweet gummy substance secreted by the bark, perfectly edible.

MANNA OAK *Quercus cerris* (Turkey Oak) Howes.

MANNA TAMARISK *Tamarix gallica var. mannifera*. The manna in this case is caused by the insect Coccus manniparus, which punctures the branches, so that little honey-like drops are exuded, and solidify. Perhaps this, after all, was the biblical manna, for the little drops eventually fall to the ground, so that bushes and the ground round the tree look from a distance to be covered with hoar frost.

MANTGIONS, Easter; MANGIANTS,

Easter *Polygonum bistorta* (Bistort) both Yks, Cumb, N'thum. Grigson. See Mangiants.

MANTLE, Dew *Alchemilla alpina* (Alpine Lady's Mantle) Sanecki.

MANTLE, King's *Thunbergia erecta* Howes.

MANTLE, Lady's 1. *Adoxa moschatellina* (Moschatel) Som. Macmillan. 2. *Alchemilla vulgaris*. 3. *Cardamine pratensis* (Lady's Smock) Som. Macmillan. 4. *Plantago lanceolata* (Ribwort Plantain) Som. Macmillan. True Lady's Mantle gets the name from the way the leaves are folded, like a mantle of Tudor times.

MANTLE, Lady's, Five-leaved *Alchemilla alpina* (Alpine Lady's Mantle) Sanecki.

MANTLE, Thor's 1. *Arctium lappa* (Burdock) Friend. 2. *Digitalis purpurea* (Foxglove). 3. *Potentilla erecta* (Tormentil) Wright. Whatever the origin of the others, *Potentilla erecta*'s name is a simple play on the common one, Tormentil.

MANUKA *Leptospermum scoparium*.

MANX GORSE *Ulex minor* (Small Furze) IOM Moore.1924.

MANY-FEET; MEG MANY-FEET *Ranunculus repens* (Creeping Buttercup) Yks. Carr (Many-feet); Cumb. Grigson (Meg Many-feet). These are witty comments on the plant's creeping tendencies, and the way it roots so readily.

MANY-KNEES *Polygonatum multiflorum* (Solomon's Seal) Fernie. Polygonatum, "of many knees, for so the Greeke word doth import" (Gerard).

MANZANITA *Arctostaphylos manzanita*. It is Spanish for little apple.

MANZANITA, Green *Arctostaphylos patula*.

MAORI BULRUSH *Typha angustifolia* (Lesser Bulrush) Andersen.

MAORI CELERY *Apium australe*.

MAORI FIRE *Pennantia corymbosa*. The Maoris used it for friction fire making.

MAORI FLAX *Phormium tenax* (New Zealand Flax) Andersen.

MAORI MINT *Mentha cunninghami*.

MAP TREE *Euphorbia grandidens*.

MAPLE 1. *Acer campestre*. 2. *Acer pseudo-platanus* (Sycamore) Cumb. Grigson. OE mapul.

MAPLE, Ash-leaved *Acer negundo* (Box Elder) Zenkert.

MAPLE, Balkan *Acer hyrcanum*.

MAPLE, Black *Acer saccharum var. nigrum*.

MAPLE, Broad-leaf *Acer macrophyllum* (Large-leaved Maple) USA E Gunther.

MAPLE, Canadian, Scarlet *Acer rubrum* (Red Maple) Hay & Synge.

MAPLE, Cappadocian *Acer cappadocicum*.

MAPLE, Cretan *Acer sempervirens*.

MAPLE, Curled *Acer rubrum* (Red Maple) Taylor. "Some of these trees are called the Curled Maple on account of the wood being marbled, as it were, within...".

MAPLE, English *Acer campestre* (Field Maple) Brimble.1948.

MAPLE, Field *Acer campestre*.

MAPLE, Full Moon *Acer japonicum* (Downy Japanese Maple) USA Hay & Synge.

MAPLE, Great *Acer pseudo-platanus* (Sycamore) Scot. Ellacombe.

MAPLE, Greek *Acer heidreichii*.

MAPLE, Hard *Acer saccharum* (Sugar Maple) USA H Smith.

MAPLE, Hawthorn *Acer crataegifolium*.

MAPLE, Hedge *Acer campestre* (Field Maple) A W Smith.

MAPLE, Horned *Acer diabolicum*. There are little horn-like projections on the seed vessels, hence *diabolicum*.

MAPLE, Italian *Acer opalus*. Not confined to Italy, though, for it grows in France and Spain as well.

MAPLE, Japanese 1. *Acer palmatum*. 2. *Acer japonicum* (Downy Japanese Maple) Hay & Synge.

MAPLE, Japanese, Downy *Acer japonicum*.

MAPLE, Large-leaved 1. *Acer macrophyllum*. 2. *Acer pseudo-platanus* (Sycamore) C P Johnson.

MAPLE, Manitoba *Acer negundo* (Box Elder) Johnston.

MAPLE, Montpelier *Acer monspessulanum*.

MAPLE, Mountain *Acer spicatum*.
MAPLE, Nikko *Acer nikoense*.
MAPLE, Norway *Acer platanoides*.
MAPLE, Oregon *Acer macrophyllum* (Large-leaved Maple) Mason.
MAPLE, Pacific *Acer macrophyllum* (Large-leaved Maple) Brimble.1948.
MAPLE, Paper-bark *Acer griseum*.
MAPLE, Red *Acer rubrum*.
MAPLE, Rock *Acer saccharum* (Sugar Maple) Brimble.1948. Cf. Hard Maple for this.
MAPLE, Rocky Mountain *Acer glabra*.
MAPLE, Scottish *Acer pseudo-platanus* (Sycamore) Howes.
MAPLE, Silver *Acer saccharinum*. 'Silver' because of the down on the reverse of the leaf-blades.
MAPLE, Small-leaved *Acer campestre* (Field Maple) Brimble.1948. The 'large-leaved maple' is *A macrophyllum*.
MAPLE, Snake-bark *Acer pennsylvanicum*.
MAPLE, Soft 1. *Acer rubrum* (Red Maple) Brimble.1948. 2. *Acer saccharinum* (Silver Maple) USA Harper.
MAPLE, Striped *Acer pennsylvanicum* (Snake-bark Maple) Zenkert.
MAPLE, Sugar *Acer saccharum*.
MAPLE, Sugar, Black *Acer nigrum*.
MAPLE, Sugar, Chinese *Sorghum saccharatum* (Sorgo) Grieve.1931. A courtesy title only, for it is not a maple, of course.
MAPLE, Sugar, Florida *Acer floridanum*.
MAPLE, Swamp *Acer rubrum* (Red Maple) USA Kingsbury.
MAPLE, Sycamore *Acer pseudo-platanus* (Sycamore) A W Smith.
MAPLE, Tatarian *Acer tataricum*.
MAPLE, Trident *Acer buergerianum*.
MAPLE, Vine 1. *Acer circinatum*. 2. *Menispermum canadense* (Moonseed) Le Strange. The fruit of the latter looks like small purple grapes, but they are poisonous.
MAPLE, White *Acer saccharinum* (Silver Maple) USA Harper.
MAPLE-LEAVED GOOSEFOOT *Chenopodium hybridum* (Sowbane) USA Zenkert.
MAPLE PEA *Pisum sativum var. arvense* (Field Pea) Howes.
MAPLE SERVICE; MAPLE TREE *Sorbus torminalis* (Wild Service) B & H (Maple Service); Brimble.1948 (Maple Tree). For the leaves are lobed, rather like those of a maple.
MAPLELEAF BEGONIA *Begonia dregei*.
MAPLIN *Acer campestre* (Field Maple) Glos. Grigson.
MAPNEY *Dioscorea trifida* (Cush-cush) Alexander & Coursey.
MARACOCK *Passiflora caerulea* (Blue Passion Flower) Halliwell.
MARANTA, Ribboned *Calathea vittata*.
MARBLE, Andaman *Diospyros kurzii*.
MARBLES VINE *Dioclea reflexa*. All over West Africa, children use the seeds in a game played like marbles.
MARCARAM *Chenopodium bonus-henricus* (Good King Henry) Yks. Grigson. This is Mercury, often used in one form or another for Good King Henry.
MARCH 1. *Apium graveolens* (Wild Celery) Lyte. 2. *Petroselinum crispum* (Parsley) Fernie.
MARCH AND MAY *Arabis alpina* (Alpine Rock-cress) N'thants. B & H.
MARCH DAISY *Bellis perennis* (Daisy) N'thants. A E Baker.
MARCH LILY *Amaryllis belladonna* (Belladonna Lily) S Africa Watt.
MARCH VIOLET *Viola odorata* (Sweet Violet) B & H.
MARCHE, Wood *Sanicula europaea* (Sanicle) Payne.
MARE BLOB *Caltha palustris* (Marsh Marigold) Som. Macmillan; Glos, War, Derb. Grigson; N'thants. A E Baker. Blob is a dialect word meaning a blister, and is a reference to the acrid juice and its blister-raising capabilities. Horse Blob is just as widespread, but 'mare' may just possibly be 'May', as in May Blob, etc.
MARE FART *Senecio jacobaea* (Ragwort) Ches. Holland.

MARE'S FAT *Pulicaria dysenterica* (Yellow Fleabane) E Ang. Forby. Cf. the previous entry.

MARE'S MILK *Euphorbia helioscopia* (Sun Spurge) Scot. Dyer. There are a number of 'milk' names for this, given because of the whitish sap.

MARE'S TAIL 1. *Euphorbia amygdaloides* (Wood Spurge) Ire. B & H. 2. *Hippuris vulgaris*. 3. *Equisetum arvense* (Common Horsetail) Som. Elworthy. 4. *Erigeron canadense* (Canadian Fleabane) S Africa Watt. All descriptive.

MAREGALL, Marsh *Gentiana pneumonanthe* (Marsh Gentian) Cockayne. Cockayne identified the plant as Marsh Gentian, but a name like Marsh Maregall seems to call to mind Marsh Marigold rather than gentian. After all, Marsh Marigold came from OE marsc meargealle, where mearh is horse, and gealla a blister. The local names offer many derivatives from this base.

MARES, Stallions-and- *Arum maculatum* (Cuckoo-pint) Yks. B & H. One of the many male plus female names for this plant. Cf. Lords-and-ladies, Bulls-and-cows, and a lot more.

MARG; MURG *Anthemis cotula* (Maydweed) Hants. Grigson (Marg); B & H (Murg). Cf. Morgan. They must all belong to the series based on OE maegde, giving the common name itself, and also such as Maither, Maiden, etc.

MARGAN *Anthemis cotula* (Maydweed) Halliwell. Cf. Morgan, and other local names probably meaning 'maiden'.

MARGARET; MARGUERITE *Leucanthemum vulgare* (Ox-eye Daisy) Yks. Grigson (Margaret); Som. Grigson (Marguerite). It is said that it got the name because it was taken as the emblem of Margaret of Anjou, the queen of Henry VI, but surely it is the colour that accounts for the name. Greek margarites means a pearl.

MARGARET, Herb *Bellis perennis* (Daisy) Dyer. Dyer puts forward St Margaret of Antioch and St Margaret of Cortona as patrons, but on the whole prefers the latter. Actually, St Margaret of Cortona's Day, 22 February, used to be reckoned as the first day of spring, and this is the probable reason for the name.

MARGERY 1. *Chenopodium bonus-henricus* (Good King Henry) Lincs. Grigson. 2. *Origanum vulgare* (Marjoram) Corn. Bottrell. It is easy enough to see Margery as the equivalent of marjoram, but when applied to Good King Henry, it is not marjoram that is the starting point, but mercury - this was known as English Mercury.

MARGOSA TREE 1. *Melia azedarach* (Chinaberry Tree) Lincoln. 2. *Melia indica* (Neem Tree) Chopra.

MARGRET *Leucanthemum vulgare* (Ox-eye Daisy) Som. Macmillan. See Margaret.

MARGUERITE See Margaret.

MARGUERITE, Blue *Felicia amelloides*.

MARGUERITE, Golden *Anthemis tinctoria* (Yellow Chamomile) Webster.

MARIAN 1. *Campanula medium* (Canterbury Bells). 2. *Cirsium vulgare* (Spear Plume Thistle) Forf. B & H. 3. *Silybum marianum* (Milk Thistle) Scot. Dyer. Milk Thistle bears the dedication to the Virgin Mary, the point of reference being the white veins of the leaves, believed to be due to the milk spilled from her breast. Children blow the thistle-down away, while singing:

"Marian, Marian, what's the time of day?
One o'clock, two o'clock, it's time we were away".

Canterbury Bells are also associated with the Virgin, not so obviously this time. Cf. the French Mariette, and Viola Marina, under which name Gerard described the plant.

MARIAN, Coventry *Campanula medium* (Canterbury Bells). Cf. Coventry Bells, which at one time was almost as common a name as the modern Canterbury Bells.

MARIE'S FIR *Abies mariesii*.

MARIET *Campanula medium* (Canterbury Bells). See Marian.

MARIGOLD 1. *Calendula officinalis*. 2.

Caltha palustris (Marsh Marigold) Yks. Grigson. 3. *Chrysanthemum segetum* (Corn Marigold) Cumb. B & H. The gold dedicated to the Virgin Mary, gold, like gowan, being a general name for a yellow daisy.

MARIGOLD, African 1. *Dimorphoteca pluvialis* (Rain Daisy) Coats. 2. *Tagetes erecta*. The latter is not African at all, of course, but comes from Mexico.

MARIGOLD, Aztec *Tagetes erecta* (African Marigold) RHS. A more reasonable name than African Marigold - the genus is Mexican.

MARIGOLD, Bur 1. *Bidens bipinnata* (Spanish Needles) Dalziel. 2. *Bidens tripartita*.

MARIGOLD, Bur, Nodding *Bidens cernua*.

MARIGOLD, Bur, Small *Bidens cernua* (Nodding Bur Marigold) USA House.

MARIGOLD, Bur, Trifid; MARIGOLD, Bur, Tripartite *Bidens tripartita* all Le Strange.

MARIGOLD, Cape *Dimorphoteca aurantiaca* (Star of the Veldt) Perry.

MARIGOLD, Corn *Chrysanthemum segetum*.

MARIGOLD, Corn, of Candy *Chrysanthemum coronarium* (Annual Chrysanthemum) Parkinson.

MARIGOLD, Desert *Baileya multiradiata* (Desert Baileya) USA T H Kearney.

MARIGOLD, Field *Chrysanthemum segetum* (Corn Marigold) Shrop. Grigson.

MARIGOLD, Fig 1. *Mesembryanthemum edule* (Hottentot Fig) A W Smith. 2. *Mesembryanthemum equilaterale*.

MARIGOLD, Fig, Green *Aridaria viridiflora*.

MARIGOLD, Fig, Jagged-leaved *Aethephyllum pinnatifidum*.

MARIGOLD, French *Tagetes patula*. Not French, any more than the African Marigold comes from Africa.

MARIGOLD, Marsh *Caltha palustris*. Probably from some confusion, for the name is OE marsc meargealle. Mearh is horse, probably because this is bigger and coarser than the average buttercup, and gealla, blister. The local names offer many examples of derivations from this base. Like most of the members of the Ranunculaceae, its juice is acrid, hence the use of the term for blister.

MARIGOLD, Mexican *Tagetes lucida*. A *Tagetes* name showing the correct country of origin!

MARIGOLD, Mountain *Senecio lyalli*.

MARIGOLD, Pot *Calendula officinalis* (Marigold) E Hayes. The flowers used to be a common potherb.

MARIGOLD, Water *Bidens cernua* (Nodding Bur Marigold) USA T H Kearney.

MARIGOLD, Wild 1. *Chrysanthemum segetum* (Corn Marigold) N Ire. Grigson. 2. *Pulicaria dysenterica* (Yellow Fleabane) Prior.

MARIGOLD-GOLDINS *Chrysanthemum segetum* (Corn Marigold) N Ire. Grigson. Note the repetition of 'gold'.

MARIGOLD OF PERU *Helianthus annuus* (Sunflower) Gerard. Well, Gerard knew it came from somewhere out there. The probable area of origin is Mexico.

MARIPOSA, Desert *Calochortus kennedyi*. Mariposa = butterfly.

MARIPOSA, Straggling *Calochortus flexuosus*.

MARIPOSA LILY *Calochortus venustus*.

MARIPOSA LILY, Yellow *Calochortus luteus*.

MARITIME PINE *Pinus pinaster*.

MARJOLAINE, La *Origanum vulgare* (Marjoram) Guernsey MacCulloch.

MARJORAM *Origanum vulgare*. Origin doubtful, Chambers's dictionary admits, but the Old French was majorane.

MARJORAM, Bastard; MARJORAM, Eastward *Origanum vulgare* (Marjoram) Parkinson (Bastard); Culpepper (Eastward). The same, perhaps?

MARJORAM, Egyptian *Origanum maru aegyptiacum*.

MARJORAM, English *Origanum vulgare* (Marjoram) B & H.

MARJORAM, French *Majorana onites*.

Apparently, Sicily is its homeland, not France.

MARJORAM, Garden *Majorana hortensis* (Knotted Marjoram) Sanecki.

MARJORAM, Goat's *Origanum vulgare* (Marjoram) Tynan & Maitland.

MARJORAM, Golden *Origanum aureum*.

MARJORAM, Grove *Origanum vulgare* (Marjoram) B & H.

MARJORAM, Hop 1. *Mentha pulegium* (Pennyroyal) Fernie. 2. *Origanum dictamnus* (Dittany) USA Cunningham.

MARJORAM, Knotted *Majorana hortensis*. 'Knotted', because the flowers are gathered into roundish close heads like knots.

MARJORAM, Perennial *Majorana onites* (French Marjoram) Grieve.

MARJORAM, Pot 1. *Majorana onites* (French Marjoram) Rohde. 2. *Origanum vulgare* (Marjoram) Lincs. B & H.

MARJORAM, Spanish *Urtica pilulifera var. dodartii*. A joke name, for it was grown in 18th century, and given this name to trap the unwary. It has no appearance of a nettle, but when the victims were invited to smell it, they were severely stung.

MARJORAM, Sweet 1. *Majorana hortensis* (Knotted Marjoram) Webster. 2. *Origanum heracleoticum* (Winter Marjoram) G N Taylor.

MARJORAM, Syrian *Origanum maru*.

MARJORAM, Unsavoury *Prunella vulgaris* (Self-heal) Turner.

MARJORAM, Wild *Origanum vulgare* (Marjoram) B & H.

MARJORAM, Winter 1. *Majorana onites* (French Marjoram) Sanecki. 2. *Origanum heracleoticum*.

MARKE, Green *Oenanthe crocata* (Hemlock Water Dropwort) Turner.

MARKER, Pie *Abutilon avicennae* (China Jute) USA Bergen. Cf. Butter-print, and Pie-print. The pods and leaves were used to stamp butter or pie-crust.

MARKERY; MARQUERY, English; MARQUERRY *Chenopodium bonus-henricus* (Good King Henry) Cambs. Porter.1974 (Markery); Fernie (English Marquery; Lincs. Peacock (Marquerry). Variations on 'mercury'. Gerard described this under the name English Mercury, and there are a number of other versions.

MARKING-NUT *Anacardium occidentale* (Cashew Nut) Codrington. The juice makes an indelible marking ink.

MARKWEED *Rhus radicans* (Poison Ivy) Kingsbury.

MARL-GRASS 1. *Trifolium medium* (Zigzag Clover) B & H. 2. *Trifolium pratense* (Red Clover) Som, Wilts. Grigson.

MARMALADE, Bread-and- *Sinapis arvensis* (Charlock) Som. Macmillan.

MARMARITAN *Paeonia mascula* (Peony) B & H.

MARRAM *Ammophila arenaria*. From O Norse marr, sea, and halmr, haulm.

MARROM *Ammophila arenaria* (Marram) Ire. E E Evans.1942.

MARROW, Vegetable *Cucurbita pepo*. Presumably marrow is used in the same sense as bone-marrow, i.e. soft tissue inside bones - the analogy is fairly obvious.

MARRUBE *Marrubium vulgare* (White Horehound) USA Henkel.

MARSHLOCKS *Potentilla palustris* (Marsh Cinquefoil) USA Densmore.

MARSHWEED *Equisetum palustre* (Marsh Horsetail) Dev. Friend.1882; Som. Macmillan.

MARSHWORT 1. *Apium inundatum*. 2. *Vaccinium oxycoccus* (Cranberry) Prior.

MARSHWORT, Least *Apium inundatum* (Marshwort) Murdoch McNeill.

MARSHWORT, Procumbent *Apium nodiflorum* (Fool's Watercress) Murdoch McNeill.

MARTAGON *Listera ovata* (Twayblade) Turner.

MARTAGON, Scarlet *Lilium chalcedonicum* (Scarlet Turk's Cap Lily) Perry.

MARTAGON IMPERIAL *Lilium martagon* (Turk's Cap Lily) Parkinson. OED says that martagon is derived from a Turkish word for a kind of turban. But Grigson demurred, and said it is possibly herba Martis, herb of Mars.

MARTHA; MARTHUS *Anthemis arvensis* (Corn Camomile) both Som. Macmillan.

MARULA *Sclerocarya caffra*.

MARVEL 1. *Ballota nigra* (Black Horehound) Suss. Parish. 2. *Marrubium vulgare* (White Horehound) Suss. B & H; USA. Henkel. Eventually from *Marrubium*; the sequence seems to run Marrube, Mauroll, Mawroll, Marvel, all of which have been used at one time or another.

MARVEL OF PERU *Mirabilis dichotoma*.

MARY, Bloody *Geranium robertianum* (Herb Robert) Yks. Grigson. From the red stems, and the generally red tinge of the leaves. Cf. Bloodwort.

MARY, Blue-eyed 1. *Collinsia verna* USA Zenkert. 2. *Omphalodes verna* (Winter Forget-me-not) McClintock.

MARY, Herb *Tanacetum balsamita* (Balsamint) Mabey.1977. From another name for the plant, Costmary, where 'cost' is from 'costus', an aromatic plant used in making perfumes in the East. The dedication is to the Virgin in most European countries.

MARY, Herb of; MARY, Spotted *Pulmonaria officinalis* (Lungwort) Wales Trevelyan (Herb of Mary); Rad. Grigson (Spotted Mary). There are many names for lungwort involving the Virgin, all referring to the legend that during the flight into Egypt, some of her milk fell on the leaves while she was nursing the infant Jesus, causing the white blotches on them.

MARY, Joseph-and- *Pulmonaria officinalis* (Lungwort) Hants. Cope. This type of name refers to the bi-, or even tri-, coloured flowers. The names exist in all sorts of forms - Adam and Eve, Soldiers-and-sailors, etc.

MARY, Virgin *Eupatorium cannabinum* (Hemp Agrimony) Corn. Grigson.

MARY, William-and- 1. *Malcomia maritima* (Virginian Stock) Som. Macmillan. 2. *Pulmonaria officinalis* (Lungwort) Brownlow. See Joseph-and-Mary among others.

MARY ALONE *Lapsana communis* (Nipplewort) Glos. Grigson.

MARY-AT-THE-COTTAGE-GATE *Stellaria holostea* (Greater Stitchwort) Som. Macmillan.

MARY GOLD *Calendula officinalis* (Marigold) Dev, Som. Macmillan.

MARY GOOLES *Calendula officinalis* (Marigold) Leyel.1937. Cf. Mary Gowlan, and better still, Mary Gold.

MARY GOWAN *Bellis perennis* (Daisy) Ayr, Berw. Grigson. Cf. Mary Gowlan, or just Gowan on its own, which usually refers to a yellow daisy.

MARY GOWLAN 1. *Bellis perennis* (Daisy) N'thum, Berw. Grigson. 2. *Calendula officinalis* (Marigold). 3. *Chrysanthemum segetum* (Corn Marigold) N'thum. Grigson.

MARY JANE 1. *Geranium robertianum* (Herb Robert) Dev. Macmillan. 2. *Melandrium rubrum* (Red Campion) Som. Macmillan.

MARY SPINK *Primula vulgaris* (Primrose) Scot. Grigson. On the face of it, this is just a version of Mayspink. In the south of England, primroses are the Darlings of April, to quote a Somerset name. But further north it is May before they are out, hence Mayflowers, and Mayspinks.

MARY'S CLOVER *Lysimachia nemora* (Yellow Pimpernel) Ire. Grigson.

MARY'S FLOWER *Anastatica hierochuntica* (Rose of Jericho) Moldenke. Because of a legend that tells that all the plants of this species expanded, became green and blossomed again at the birth of Jesus, and still do so in commemoration. Cf. Rose of the Virgin.

MARY'S GOLD *Caltha palustris* (Marsh Marigold) Som. Macmillan.

MARY'S KIDNEY *Mucuma urens* (Florida Bean) Hebrides Shaw. The translation of Gaelic Airne Mhoire. The bean is sometimes found washed up on the Atlantic shore of the islands, where women used to hold it in their hand during childbirth.

MARY'S NUT *Entada phaseoloides* (Nicker Bean) Hebrides Shaw. The translation of Gaelic Cnò Mhoire. The bean is sometimes found washed up on the Atlantic shore of the islands, where it brings good luck to the finder.

MARY'S REST *Veronica chamaedrys* (Germander Speedwell) Tynan & Maitland.

MARY'S SEED *Sonchus oleraceus* (Sow Thistle) B & H. It must be accountable to the milky juice.

MARY'S TAPER *Galanthus nivalis* (Snowdrop) Coats.

MARY'S TEARS *Pulmonaria officinalis* (Lungwort) Dor. J J Foster. This must be the result of another version of the story given to account for the white blotches on the leaves - it is usually her milk, rather than her tears, that are the cause.

MARY'S THISTLE *Silybum marianum* (Milk Thistle) S Africa Watt.

MARYBOUT *Caltha palustris* (Marsh Marigold) Lancs. Grigson. The 'bout' part of the name occurs on its own, or as Boots. It comes from the French bouton, and is a comment on the knob- or button-like unopened flowers.

MARYBUD 1. *Calendula officinalis* (Marigold) Dyer.1883. 2. *Caltha palustris* (Marsh Marigold) Dor. Dacombe; War. Grigson. 3. *Ranunculus acris* (Meadow Buttercup) Som. Grigson. 4. *Ranunculus bulbosus* (Bulbous Buttercup) Shakespeare. 5. *Ranunculus repens* (Creeping Buttercup) Som. Grigson. Cf. Marybout, which must have been the original of this name.

MARYGOLD, Mountain *Doronicum pardalianches* (Great Leopard's Bane) Tradescant.

MASER TREE; MAZER *Acer campestre* (Field Maple) Scot. B & H. This must be of Norse derivation - the modern Icelandic for maple is mösurr.

MASH-CORNS *Potentilla anserina* (Silverweed) Ire. Grigson. Cf. the Scottish Moss-corns, where moss is the same as moor.

MASHUA *Tropaeolum tuberosum*.

MASK *Quercus robur* acorns (Acorn) Dev. Friend.1882. Mast, of course.

MASK-FLOWER 1. *Alonsoa sp*. 2. *Mimulus luteus*.

MASKERT; MASKERT, Swine's *Stachys palustris* (Marsh Woundwort) both Scot. Jamieson. Somebody suggested 'mask-wort' for these, but there is another form - Swine's Mosscorts.

MASS *Quercus robur* acorn (Acorn) Som. Grose. Mast, of course.

MASSLIN *Viscum album* (Mistletoe) Suff. Grigson. This is just a variation on the 'mistle' part of the common name. Cf. Mislin-bush, also from East Anglia.

MAST BEECH *Fagus sylvatica* (Beech) Culpepper. Mast is really the name for the fruit of any forest tree, but these days it seems to be applied to beechnuts only. Cf. the Somerset Mace.

MASTERWORT 1. *Angelica archangelica* (Archangel) Fernie. 2. *Angelica atropurpurea* (Purple-stemmed Angelica) O P Brown. 3. *Heracleum maximum* (Cow Parsnip) USA Leighton. 4. *Peucedanum ostruthium*.

MASTERWORT, Black *Astrantia major* (Pink Masterwort) Gerard.

MASTERWORT, English; MASTERWORT, Wild *Aegopodium podagraria* (Goutweed) Grieve.1931 (English); Fernie (Wild).

MASTERWORT, Great *Astrantia major* (Pink Masterwort) Polunin.

MASTERWORT, Pink *Astrantia major*.

MASTIC, Peruvian *Schinus molle*. The name should be reserved for the gum collected from the shrub, but the plant itself usually bears it.

MASTIC TREE *Pistacia lentiscus* (Lentisk) Polunin.1976. Mastic is actually the gum resin exuded by this and other trees, used for varnish, etc., as well as for chewing gum and also to flavour a Greek wine, known by the same name.

MASTICK THYME *Thymus mastichenus*.

MAT, Grey *Halimione portulacoides* (Sea Purslane) Loewenfeld.

MAT BEAN *Phaseolus aconitifolius* (Moth Bean) Schery.

MAT DAISY *Raoulia sp* Howes.

MAT REED *Typha latifolia* (False Bulrush) W Miller. Presumably quite literal.

MAT SPURGE *Euphorbia glyptosperma*.

MATAI *Podocarpus spicata*.

MATCH-ME-IF-YOU-CAN *Acalypha wilkesiana* (Copperleaf) Perry. The reference is to the great variations in leaf colouring, no two leaves being exactly alike.

MATCHWEED *Gutierrezia microcephala* (Broomweed) USA Elmore.

MATCHES, Lucifer *Sisymbrium officinale* (Hedge Mustard) Worcs. B & H. Descriptive.

MATÉ *Ilex paraguayensis*. Or yerba maté, Paraguay Tea.

MATFELLON 1. *Centaurea nigra* (Knapweed) Gerard. 2. *Centaurea scabiosa* (Greater Knapweed) Turner.

MATFELLON, Black *Centaurea nigra* (Knapweed) Halliwell.

MATGRASS 1. *Ammophila arenaria* (Marram) Hepburn. 2. *Nardus stricta*. 3. *Polygonum aviculare* (Knotgrass) S Africa Watt. For centuries, the people of Newborough, in Anglesey, have made a scanty living by plaiting the marram into mats for haystacks, barn roofs, etc. The other two probably get the name as a description of growth.

MATHER; MAITHER 1. *Anthemis cotula* (Maydweed) Dor. Macmillan (Mather); N.Eng. Prior (Maither). 2. *Leucanthemum vulgare* (Ox-eye Daisy) Halliwell. OE maegde, and all the derivatives mean girl (Cf. modern maiden) in one way or another. Mather was a word used in the north country when speaking of a working class girl. And the plant appeared in the early herbals as "commended against the infirmities of the mother..." (Gerard). Mather has been used too for Blue Whortleberry (*Vaccinium ovalifolium*), according to Howes, and certainly for Ox-eye (*Leucanthemum vulgare*).

MATHERN 1. *Anthemis cotula* (Maydweed) Wilts. Jefferies); Dor. Macmillan. 2. *Leucanthemum vulgare* (Ox-eye Daisy) Wilts. Dartnell & Goddard; Heref. G C Lewis. No more than the plural in "-n" of Mather.

MATHES, Stinking *Anthemis cotula* (Maydweed) Gerard.

MATHET, Red *Adonis annua* (Pheasant's Eye) Lyte. Turner had Red Maydweed for this, and Cf. Red Maithes, or Red Maythe.

MATILIJA POPPY *Romneya coulteri* (Californian Bush Poppy) Wit.

MATRICARY, Scentless *Tripleurospermum maritimum* (Scentless Maydweed) Clair. An alternative botanical name is *Matricaria inodora*.

MATRIMONY, Courtship-and- *Filipendula ulmaria* (Meadowsweet) Cumb. Wright. Apparently taken from the scent of the flowers before and after bruising.

MATRIMONY VINE 1. *Lycium halimifolium* (Duke of Argyll's Tea Plant) USA. Kingsbury. 2. *Lycium pallidum* (Rabbit Thorn) USA Elmore.

MATRIMONY VINE, Chinese *Lycium chinense* (Tea Tree) Howes.

MATURA TEA *Cassia auriculata* (Tanner's Cassia) Howes.

MATWEED 1. *Ammophila arundinacea* (Marram) C P Johnson. 2. *Nardus stricta* (Matgrass) H C Long.1910. At least as far as marram is concerned, the name is self-explaining - a weed to make mats from. Cf. Matgrass.

MAUDELINE *Tanacetum balsamita* (Costmary) Painter. A version of Gerard's Maudlinwort, perhaps referring to Mary Magdalene, but more likely to be one of the 'mather' series. In any case, Herb Mary, or St Mary, and indeed Costmary itself, would point to an association with the Virgin.

MAUDLIN; MAUDLIN DAISY *Leucanthemum vulgare* (Ox-eye Daisy) Wilts. Dartnell & Goddard (Maudlin); Le Strange. (Maudlin Daisy). One of the 'mather' series, if it is not a reference to Mary Magdalene.

MAUDLIN, Sweet *Achillea millefolium* (Yarrow) Tynan & Maitland.

MAUDLINWORT 1. *Leucanthemum vulgare*

(Ox-eye Daisy) Notes & Queries.1873. 2. *Tanacetum balsamita* (Balsamint) Langham.

MAUL, MAULE *Malva sylvestris* (Common Mallow) W Miller (Maul); Prior (Maule). From Latin malva - Cf. Italian and Spanish maula.

MAUL OAK *Quercus chrysopelis*.

MAUPLE *Acer campestre* (Field Maple) Heref. G C Lewis.

MAURANDYA, Blue *Maurandya antirrhiniflora*.

MAURANDYA, Rock *Maurandya petrophila*.

MAURITIUS ALOE; MAURITIUS HEMP *Furcraea gigantea*. Mauritius Hemp is the usual name; Mauritius Aloe is in Schery. The plant is a native of Brazil, but is grown commercially on the island of Mauritius.

MAURITIUS THORN *Caesalpina decepetala* (Myrose Thorn) Dale & Greenway.

MAUROLE; MAWROLL *Marrubium vulgare* (White Horehound) B & H. (Maurole); W Miller (Mawroll).

MAUTHER *Anthemis cotula* (Maydweed) W Miller. One of the 'mather' series.

MAUTHERN 1. *Anthemis cotula* (Maydweed) Wilts. Dartnell & Goddard. 2. *Leucanthemum vulgare* (Ox-eye Daisy) Wilts. Akerman. An old plural in "-en" of mauther? See Mather.

MAUVE *Malva sylvestris* (Common Mallow) Clair. Malva is the origin of the name (and the colour).

MAVIN *Anthemis cotula* (Maydweed) Suss. Grigson.

MAWS *Malva sylvestris* (Common Mallow) Notts, N'thum, Scot. Grigson. Obviously a version of Maul, itself from Malva.

MAWS, Wild *Papaver rhoeas* (Corn Poppy) Derb. Grigson.

MAWSEED *Papaver somniferum* (Opium Poppy) Grieve.1931. Apparently the German Mahsaat, Mah being a German name for poppy. This is the name under which the seeds used to be sold for caged bird feed.

MAWTH; MAWTHEM *Anthemis cotula* (Maydweed) Halliwell (Mawth); IOW Grigson (Mawthem). Cf. Mauther, etc.

MAY 1. *Acer pseudo-platanus* (Sycamore) Corn. Grigson. 2. *Anthemis cotula* (Maydweed) Hants, Suss. Grigson. 3. *Arabis alpina* (Alpine Rock-cress) Dev. Friend.1882. 4. *Caltha palustris* (Marsh Marigold) N Ire. PLNN 19. 5. *Crataegus monogyna*. 6. *Syringa vulgaris* (Lilac) Dev. B & H. 7. *Ulmus procera* (English Elm) Dev. B & H. 8. *Viburnum tinus* (Laurustine) Dev. Macmillan. Hawthorn is the symbolisation of the May, the bough brought in on the first of May, but in Cornwall, on the evidence of the name, sycamore served the purpose - and so it did in co Cork, where it was given the name of Summer tree. Presumably lilac and laurustine had similar purposes in Devonshire, as elm certainly did. But for Maydweed, the name is simply a contraction of the common name, and perhaps the garden Arabis got it for similarity, or again, perhaps it was the white flowers that suggested it.

MAY, Blackthorn *Prunus spinosa* (Blackthorn) Middx B & H. A name that seems to compound the confusion between Blackthorn and Whitethorn.

MAY, First of *Saxifraga granulata* (Meadow Saxifrage) Ches. Holland.

MAY, Foam-of- *Spiraea arguta*.

MAY, Garden *Viburnum tinus* (Laurustine) Dor. Macmillan. Or May, simply, in Devonshire.

MAY, Horse *Ulmus procera* (English Elm) Corn. B & H. 'Horse' in a plant name usually means large, or coarse. So this probably means a second-best May, used when the more usual hawthorn bough was not available.

MAY, March-and- *Arabis alpina* (Alpine Rock-cress) N'thants. B & H.

MAY, White *Arabis alpina* (Alpine Rock-cress) Ches. Holland.

MAY-APPLE 1. *Passiflora incarnata* (Maypops) Schauenberg. 2. *Podophyllum peltatum*.

MAY-BALL *Viburnum opulus* (Guelder

Rose) Dor. Dacombe. Cf. May Rose, May Tassels, etc for Guelder Rose.

MAY-BLOB 1. *Caltha palustris* (Marsh Marigold) Wilts. Dartnell & Goddard; Glos. Brill; Leics. Ruddock. 2. *Cardamine pratensis* (Lady's Smock) N'thants. A E Baker. 3. *Ranunculus sceleratus* (Celery-leaved Buttercup) N'thants. A E Baker. 4. *Trollius europaeus* (Globe Flower) Leics. Grigson. A 'blob' is a blister, the kind that can be raised by the juice of any of the *Ranunculaceae*. So why does Lady's Smock get the name?

MAY-BLOSSOM *Convallaria maialis* (Lily-of-the-valley) Barton & Castle.

MAY-BLUB *Caltha palustris* (Marsh Marigold) Wilts. Dartnell & Goddard. See May-blob.

MAY BREAD-AND-CHEESE BUSH; MAY BREAD-AND-CHEESE TREE *Crataegus monogyna* (Hawthorn) Som, Hants, E Angl.(Bush); Norf, Lincs (Tree) Grigson.

MAY-BUBBLES *Caltha palustris* (Marsh Marigold) Som. Macmillan; Wilts. Dartnell & Goddard. The bubble in this case would be a blister. Cf. May-blob, etc.

MAY-BUSH *Crataegus monogyna* (Hawthorn) .

MAY-FLOWER 1. *Caltha palustris* (Marsh Marigold) W Eng, Ire. Grigson. 2. *Cardamine pratensis* (Lady's Smock) Som, Hants. Grigson; Ches. Holland; Lancs. Nodal; Yks. Addy. 3. *Chelidonium majus* (Greater Celandine) Ulster Foster. 4. *Crataegus monogyna* (Hawthorn). 5. *Primula veris* (Cowslip) B & H. 6. *Primula vulgaris* (Primrose) Shet. Grigson; Ire. O Suilleabhain. 7. *Stellaria holostea* (Greater Stitchwort) Cumb. Grigson. 8. *Syringa vulgaris* (Lilac) Corn. Dyer; Dev, Som. Macmillan. It is odd to find Lady's Smock with this name, for that was the one flower that was almost universally excluded from the May garland. The others are connected with the May festival in one way or another - Greater Celandine was hung up with rowan on May Eve to protect the people and cattle from harm, and so was Marsh Marigold. But the May par excellence is still the hawthorn. Stitchwort was always reckoned to be under fairy protection, and that may be the reason for the name given, for the fairies were most active on May Eve.

MAY-FLOWER LILY *Convallaria maialis* (Lily-of-the-valley) Tynan & Maitland.

MAY-FRUIT *Crataegus monogyna* fruit (Haws) Yks. Grigson.

MAY GOLLEN *Caltha palustris* (Marsh Marigold) Lancs. Lancs FWI. 'Gollen' is one of a long series of names that starts off with 'Gowan', a word evidently meaning yellow. All these names are therefore colour descriptive. This is the only one of the series prefaced by 'May', but May-flower for this plant is common in the west country, and also in Ireland.

MAY-GOSLINGS *Salix pentandra* catkins (Bay-leaved Willow) Yks. Grigson.

MAY GOWAN *Bellis perennis* (Daisy) W Miller.

MAY-GRASS *Stellaria holostea* (Greater Stitchwort) Shrop. Grigson. See May.

MAY-HAWTHORN *Crataegus rufula* USA Harper.

MAY-LILY 1. *Convallaria maialis* (Lily-of-the-valley) Gerard. 2. *Syringa vulgaris* (Lilac) Dev. Tynan & Maitland. Cf. May-blossom for Lily-of-the-valley. The flowers are a customary May Day gift in Paris. Both the flower and the scent, muguet in French, are widely advertised as May Day draws near, and huge quantities of them are sold for gifts.

MAY MAID *Glechoma hederacea* (Ground Ivy) Rohde. Probably originally Haymaid, where 'hay' means 'hedge' - OE haga.

MAY-OF-THE-MEADOW *Filipendula ulmaria* (Meadowsweet) War. Grigson. It sounds awkward, and was probably originally Maids-of-the-meadow, which does occur elsewhere.

MAY-PINK *Dianthus caryophyllus* (Carnation) Dev. Friend.1882.

MAY-ROSE 1. *Rosa cinnamonea* (Cinnamon

Rose) McDonald. 2. *Viburnum opulus* (Guelder Rose) B & H.

MAY-SPINK *Primula vulgaris* (Primrose) Scot. B & H.

MAY-TASSELS; MAY TOSSELS; MAY-TOSTY *Viburnum opulus* (Guelder Rose) Dev. Macmillan (Tassels, Tossels); Som. Macmillan (Tosty). It is usually cowslips that form the Tisty-tosties, but Guelder Rose flowers must have been used in the same way.

MAY-THORN *Crataegus monogyna* (Hawthorn) Tynan & Maitland.

MAY-TREE 1. *Acer pseudo-platanus* (Sycamore) Corn. Grigson. 2. *Crataegus monogyna* (Hawthorn). Hawthorn is the "may" par excellence, but sycamore was often brought in with it in Cornwall.

MAYBERRY 1. *Crataegus monogyna* fruit (Haws) Wessex Rogers. 2. *Rubus idaeus var. strigosus* (Red Raspberry) Howes.

MAYBUDS 1. *Ranunculus acris* (Meadow Buttercup) Som. Grigson. 2. *Ranunculus bulbosus* (Bulbous Buttercup) Som. Grigson. 3. *Ranunculus repens* (Creeping Buttercup) Som. Grigson.

MAYDWEED *Anthemis cotula*. OE maegde + weed.

MAYDWEED, Rayless *Matricaria matricoides*.

MAYDWEED, Red; MAYTHE, Red *Adonis annua* (Pheasant's Eye) Turner (Red Maydweed); Gerard (Red Maythe).

MAYDWEED, Scentless *Tripleurospermum maritimum*.

MAYPOLE *Viburnum opulus* (Guelder Rose) Dev. Macmillan. Cf. May Rose, May Tassels, etc., for this.

MAYPOLE-ING TREE *Acer campestre* (Field Maple) Glos, Wilts. A Williams. 'Maplin' is the starting point for this. Williams noted it as a folk-singer's variant in the well-known wassail song that includes the line: "Our bowl it is made of a maplin tree".

MAYPOPS *Passiflora incarnata* fruit Bianchini.

MAYTHEM; MAYTHIG *Anthemis cotula* (Maydweed) N Eng. Prior (Maythem); Shrop. B & H (Maythig).

MAYWEED 1. *Anthemis cotula* (Maydweed) Chopra. 2. *Anthriscus sylvestris* (Cow Parsley) Worcs. Mabey. 3. *Leucanthemum vulgare* (Ox-eye Daisy) Suff. Grigson. 4. *Tanacetum parthenium* (Feverfew). Mayweed for Cow Parsley must be one of the few occasions when the month of May is being celebrated. The rest seem to be nearer to maiden than to may (OE maegde).

MAYWEED, Corn *Tripleurospermum maritimum* (Scentless Maydweed).

MAYWEED, Rayless *Matricaria matricarioides*.

MAYWEED, Scented *Matricaria recutita*.

MAYWEED, Scentless *Tripleurospermum maritimum* (Scentless Maydweed) Darling & Boyd.

MAYWEED, Sea *Tripleurospermum maritimum* (Scentless Maydweed) McClintock.1975.

MAYWEED, Stink; MAYWEED, Stinking *Anthemis cotula* (Maydweed) Murdoch McNeill (Stink); S Africa Watt (Stinking).

MAYWORT 1. *Artemisia vulgaris* (Mugwort) McDonald. 2. *Galium cruciatum* (Cross-wort) Som. Macmillan. Probably a misreading of Mugwort in each case, for Cross-wort is also known as Mugwort. On the other hand, in the case of mugwort, the original may have been Maydwort, in which case another name, Motherwort, would be relevant.

MAZALIUM *Daphne mezereon* (Lady Laurel) Bucks. B & H. There are all sorts of variations on Mezereon for this. The word comes from the Persian Madzaryoun, destroyer of life. The red berries are poisonous, and can be fatal.

MAZAR-TREE *Prunus padus* (Bird Cherry) Dev. Grigson. Usually applied to Wild Cherry, *P avium*, when there are quite a lot of different forms of the name.

MAZEERIE *Daphne mezereon* (Lady Laurel) Lincs. Rudkin. A corruption of 'mezereon'.

MAZELL; MEZELL *Daphne mezereon* (Lady Laurel) Hants. Grigson (Mazell); Gerard (Mezell).

MAZER; MASER *Acer campestre* (Field Maple) both Scot. Jamieson (Mazer); B & H (Maser). See Maser.

MAZES 1. *Anthemis cotula* (Maydweed) Lincs. Grigson. 2. *Leucanthemum vulgare* (Ox-eye Daisy) Lincs. Peacock. This must be local rendering of something like Maithes.

MAZZARD 1. *Prunus avium* (Wild Cherry) Corn. Barton; Dev. Friend.1882; Glos. Grose; Herts. Jones-Baker.1974. 2. *Prunus padus* (Bird Cherry) Lincs. Grigson. Prior said that this and its variants are from a late Latin form, manzar, which would mean bastard or spurious, ultimately of Hebrew origin. But it is more likely to be a variant of mazer, which would be a maple. Does the wood have markings such as maple wood does?

MAZZARD, Brandy *Prunus avium* (Wild Cherry) Dev. Grigson.

MAZZUD *Prunus avium* (Wild Cherry) Dev. Friend.1882.

MEABERRY *Vaccinium oxycoccus* (Cranberry) Yks. Grigson. Moorberry, perhaps?

MEAD, Maid of the *Filipendula ulmaria* (Meadowsweet) Ches. Grigson.

MEADEN *Anthemis cotula* (Maydweed) Dor. Macmillan.

MEADOW; MEADOW-FLOWER *Cardamine pratensis* (Lady's Smock) Yks. B & H (Meadow); Cumb. Grigson (Meadow-flower).

MEADOW, Bright *Caltha palustris* (Marsh Marigold) W Miller.

MEADOW, Maid of the; MEADOW, May of the; MEADOW, Queen of the *Filipendula ulmaria* (Meadowsweet) Som. Grigson (Maid); War. Grigson (May); W Eng. Dacombe, Friend, Storer; Scot. Jamieson (Queen).

MEADOW BITTERCRESS; MEADOW CRESS *Cardamine pratensis*.

MEADOW-BOUT *Caltha palustris* (Marsh Marigold) Shrop, Ches, Lancs. Grigson. Bout is French bouton, and it appears as Boots as well. The derivation must include the "button" names.

MEADOW BUTTERCUP *Ranunculus acris*.

MEADOW-GRASS, Alpine *Poa alpina*.

MEADOW-GRASS, Annual *Poa annua*.

MEADOW-GRASS, Procumbent *Puccinellia rupestris*.

MEADOW-GRASS, Reed *Glyceria maxima* (Great Water Grass) Pratt.

MEADOW-GRASS, Reflexed *Puccinellia distans*.

MEADOW-GRASS, Sea *Puccinellia maritima*.

MEADOW-GRASS, Smooth *Poa pratensis*.

MEADOW-GRASS, Wavy *Poa flexuosa*.

MEADOW LILY 1. *Lilium canadense* (Canada Lily) USA Upham. 2. *Lilium candidum* (Madonna Lily) O P Brown.

MEADOW MAID *Filipendula ulmaria* (Meadowsweet) Pemb. Grigson. Cf. Maid of the Meadow.

MEADOW PARSNIP *Heracleum sphondyllium* (Hogweed) Gerard.

MEADOW PINK *Cardamine pratensis* (Lady's Smock) Dev. Macmillan.

MEADOW QUEEN *Filipendula ulmaria* (Meadowsweet) Perth. Grigson. Cf. Queen of the Meadow.

MEADOWBRIGHT *Caltha palustris* (Marsh Marigold) N'thants. Sternberg.

MEADOWFLOWER *Cardamine pratensis* (Lady's Smock) Cumb. Grigson.

MEADOWSOOT *Filipendula ulmaria* (Meadowsweet) Wilts. Dartnell & Goddard. Apparently just a mispronunciation of 'sweet', but this is not so, 'soot' is the old word sote, sweet, as in Chaucer's "Whan that Aprille with his shoures soote...".

MEADOWSWEET 1. *Filipendula ulmaria*. 2. *Filipendula vulgaris* (Dropwort) N'thants. B & H. 3. *Spiraea latifolia* USA House. 4. *Spiraea salicifolia* (Bridewort) USA H Smith.1928. Meadowsweet was originally Meadwort, from the drink mead (medo'), later associated with meadow, though Fitzgerald said it seems to be meadowsweat (there is a Welsh name, Chwys Arthur, Arthur's sweat.

MEADOWSWEET, English *Filipendula ulmaria* (Meadowsweet) USA Sanecki. The plant was taken over by the early settlers, and is now naturalised in the eastern states. So this name is given to distinguish between it and the native species.

MEADSWEET *Filipendula ulmaria* (Meadowsweet) Corn, Ches, Border B & H.

MEADUART; MEADWORT *Filipendula ulmaria* (Meadowsweet) Scot. Grigson (Meaduart); Som. Macmillan (Meadwort).

MEAL, Devil's *Anthriscus sylvestris* (Cow Parsley) Dumb. B & H. 'Meal', because of the powdery appearance of the flowers. The ascription to the devil is probably a confusion with hemlock, but Cow Parsley does have a lot of them.

MEAL, Knotty *Conopodium majus* (Earthnut) Inv. Grigson. Is it "nutty (oat) meal"?

MEAL TREE; MEALY-TREE *Viburnum lantana* (Wayfaring Tree) C P Johnson (Meal); Prior (Mealy). It is the typical dusty appearance that accounts for this name, as well as for the common name, Wayfaring Tree.

MEALBERRY *Arctostaphylos uva-ursi* (Bearberry) B & H. Cf. the American Mealy-plum for this. But should it not be Meaberry, as recorded for cranberry?

MEALIE *Zea mays* (Maize) S Africa Watt. 'Mealie' is Cape Dutch milje.

MEALY-CUP SAGE *Salvia farinacea* (Mealy Sage) H & P.

MEALY-PLUM *Arctostaphylos uva-ursi* (Bearberry) New Engl. Sanford. Cf. Mealberry.

MEALY PRIMROSE *Primula farinosa* (Birdseye Primrose) Hutchinson. The specific name *farinosa* explains it, and it is a good descriptive name.

MEALY SAGE *Salvia farinacea*.

MEASLE-FLOWER *Calendula officinalis* (Marigold) Wilts. Morrison. Mrs Morrison said that children were warned not to pick marigolds from the garden, for they would give them measles. The reason is more likely to be the opposite, for nearby, in Dorset, it was used to cure the disease.

MEASURE OF LOVE *Bellis perennis* (Daisy) Coats. Rather high-flown, but this is a reference to the "he loves me, he loves me not" divination game that children play.

MEAT, Adder's 1. *Anthriscus sylvestris* (Cow Parsley) Som. Mabey. 2. *Arum maculatum* (Cuckoo-pint) Corn. B & H; Dev, Som. Grigson; Glos. PLNN35. 3. *Mercurialis perennis* (Dog's Mercury) Herts. Grigson. 4. *Stellaria holostea* (Greater Stitchwort) Som. Grigson; Suss. Parish. 5. *Tamus communis* (Black Bryony) Dev. Friend.1882. Dog's Mercury and Black Bryony are poisonous plants, and that could explain the reference to adders. But the origin is OE attor, which meant poison. Cuckoo-pint berries are likewise poisonous. Cow Parsley shares a lot of names with Hemlock, so that may be the explanation for the 'snake' name, but what about Stitchwort, which has an equally large number of 'snake' names? Cornish children say they will be bitten by an adder if they pick it, but this is a perfectly harmless plant.

MEAT, Bird's 1. *Crataegus monogyna* fruit (Haws) Dev. Choape; Som. Grigson. 2. *Plantago major* (Great Plantain) Aber. Grigson. Cf. Canary-seed, Larkseed, etc for the plantain.

MEAT, Chicken's *Stellaria media* (Chickweed) E Ang. Grigson.

MEAT, Cuckoo's 1. *Geranium robertianum* (Herb Robert) Bucks. Grigson. 2. *Oxalis acetosella* (Wood Sorrel) Turner. 3. *Rumex acetosa* (Wild Sorrel) Ches. B & H. 4. *Rumex acetosella* (Sheep's Sorrel) Ches. B & H. 5. *Stellaria holostea* (Greater Stitchwort) Bucks. Grigson. Anything to do with a cuckoo in a plant name usually is an indicator of the time of flowering - when the cuckoos arrive.

MEAT, Devil's *Anthriscus sylvestris* (Cow

Parsley) Yks. Grigson. 'Meal', rather than 'meat'.

MEAT, Duck's *Lemna minor* (Duckweed) Gerard.

MEAT, Fox's *Oxalis acetosella* (Wood Sorrel) Corn. Grigson. Insubstantial food for a fox - it is better known as food for cuckoos.

MEAT, Frog's *Arum maculatum* (Cuckoo-pint) Dor. Macmillan.

MEAT, God's *Crataegus monogyna* (Hawthorn) Yks. B & H. Cf. Ladies' Meat, for the young leaves and shoots, regularly eaten by children in spring, and called Bread-and-cheese.

MEAT, Gowk's 1. *Orchis mascula* (Early Purple Orchis) B & H. 2. *Orchis morio* (Green-winged Orchis) B & H. 3. *Oxalis acetosella* (Wood Sorrel) Scot. Swainson. 4. *Rumex acetosa* (Wild Sorrel) Leyel.1937. Cf. Cuckoo's Meat.

MEAT, Hare's *Oxalis acetosella* (Wood Sorrel) Corn. Jago Som. Macmillan .

MEAT, Hog, Poison *Aristolochia gigas* (Pelican Flower) Emboden.1974. The awful smell is usually enough to keep animals away, but when wild pigs, out of necessity, do eat it, they are killed.

MEAT, Lady's 1. *Crataegus monogyna* (Hawthorn) Fernie. 2. *Oxalis acetosella* (Wood Sorrel) Clack. B & H. See God's Meat.

MEAT, Pigeon's *Verbena officinalis* (Vervain) Sanecki. Cf. Pigeon's Grass and Columbine for this. The Greek name was peristereon, doves, because they were supposed to be fond of hovering round the plant. Coles explains it as "because Pidgeons eat thereof as is supposed to clear their Eye sight".

MEAT, Rabbit 1. *Alternanthera ficoides* (Parrot-leaf) West Indies Howes. 2. *Heracleum sphondyllium* (Hogweed) Lincs. Gutch.

MEAT, Rabbit's 1. *Heracleum sphondyllium* (Hogweed) Som. Macmillan. 2. *Lamium purpureum* (Red Deadnettle) Shrop. Grigson. 3. *Oxalis acetosella* (Wood Sorrel)

Corn, Dev, Som. Grigson. 4. *Sonchus oleraceus* (Sow Thistle) Som. Macmillan. 5. *Taraxacum officinale* (Dandelion) Som. Macmillan.

MEAT, Snake's 1. *Arum maculatum* fruit (Cuckoo-pint) B & H. 2. *Heracleum sphondyllium* (Hogweed) Dev. Macmillan. 3. *Iris foetidissima* (Foetid Iris) Dev. Friend.1882. 4. *Mercurialis perennis* (Dog's Mercury) Som. Grigson. 5. *Prunella vulgaris* (Self-heal) Dev. Macmillan. 6. *Solanum dulcamara* (Woody Nightshade) Som. Grigson. 7. *Tamus communis* (Black Bryony) Dev. Friend.1882. Some of these are poisonous, accounting very well for the name, and the seed pods of Foetid Iris look as if they ought to be, but Hogweed is perfectly alright, and so is Self-heal.

MEAT, Toad's *Arum maculatum* (Cuckoo-pint) Corn. Grigson.

MEAT, White *Anthriscus sylvestris* (Cow Parsley) Yks. Grigson. Once again, 'meat' here is probably 'meal'.

MEAT NUT *Castanea sativa* (Sweet Chestnut) Dev. Friend.1882. Presumably, this name just means an edible nut.

MECCA ASTER *Aster cognatus*. Why? This is an American plant.

MECCA BALSAM *Commiphora opobalsamum* (Balm of Gilead) Howes.

MEDIAN APPLE *Citrus medica bajoura* Brouk. *Medica*, in the specific name, means the same as Median, i.e. coming from Media, the area of northwest Persia.

MEDICINE, Bitter *Sambucus nigra* (Elder) Som. Macmillan. Cf. Bitter-flower, also from Somerset.

MEDICINE, Doctor's 1. *Rubus fruticosus* fruit (Blackberry) Som. Macmillan. 2. *Rumex obtusifolius* leaves (Broad-leaved Dock) Som. Macmillan. Too many blackberries has the same effect as physic, and dock is a country remedy for all sorts of ailments.

MEDICINE, Dog's *Mercurialis perennis* (Dog's Mercury) Som. Macmillan.

MEDICK, Black *Medicago lupulina*. Medick

comes from herba medica, the Median, or Persian, herb. Nevertheless, a lot of the species are native to this country.

MEDICK, Bur 1. *Medicago minima*. 2. *Medicago polymorpha* (Hairy Medick) New Zealand Connor.

MEDICK, Hairy *Medicago polymorpha*.

MEDICK, Heart *Medicago arabica* (Spotted Medick) Curtis. Cf. Heart Clover, or Claver, Heart Trefoil. Coles says it is so called "not only because the leaf is triangular like the heart of a man, but also because each leafe doth contain the perfection (or image) of an heart, and that in its proper colour, viz. a flesh colour. It defendeth the heart against the noisome vapour of the spleen".

MEDICK, Hop *Medicago lupulina* (Black Medick) N'thum. Grigson.

MEDICK, Purple *Medicago sativa var. sativa* (Lucerne) Hatfield.

MEDICK, Sickle *Medicago sativa var. falcata*.

MEDICK, Small *Medicago minima* (Bur Medick) Clapham.

MEDICK, Spotted *Medicago arabica*.

MEDICK, Toothed *Medicago polymorpha* (Hairy Medick) McClintock.1974.

MEDICK, Tree *Medicago arborea*. This Mediterranean species can grow as a shrub to about 12 feet high.

MEDICK-FETCH; MEDICK-FITCH; MEDICK-FITCHLING *Onobrychis sativa* (Sainfoin) Culpepper (Fetch); B & H (Fitch); Gerard (Fitchling).

MEDICK-FODDER *Medicago sativa var. sativa* (Lucerne) Parkinson.

MEDICK-VETCHLING *Onobrychis sativa* (Sainfoin) Camden/Gibson.

MEDITERRANEAN ALOE *Aloe vera*.

MEDITERRANEAN BUCKTHORN *Rhamnus alaternus*.

MEDITERRANEAN CORIARIA *Coriaria myrtifolia*.

MEDITERRANEAN CYPRESS *Cupressus sempervirens* (Cypress) Mitchell. Cf. Italian, or Roman, Cypress.

MEDITERRANEAN GERMANDER *Teucrium polium* (Poly) Howes.

MEDITERRANEAN MEDLAR *Crataegus azarolus* (Azarole) Polunin.

MEDITERRANEAN MEZEREON *Daphne gnidium*.

MEDLAR *Mespilus germanica*. Medlar is from O French medler, mesler, which is Latin mespilum, and which in turn is from Greek mespilon.

MEDLAR, Chinese *Eriobotrya japonica* (Loquat) Bianchini.

MEDLAR, Dutch *Mespilus germanica* (Medlar) B & H.

MEDLAR, False *Sorbus chamaemespilus*.

MEDLAR, Japanese 1. *Diospyros kaki* (Chinese Date Plum) Bianchini. 2. *Eriobotrya japonica* (Loquat) Perry.

MEDLAR, Mediterranean *Crataegus azarolus* (Azarole) Polunin.

MEDLAR, Wild *Vangueriopsis lanciflora*.

MEDLAR-BUSH *Amelanchier ovalis* (Snowy Mespilus) W Miller.

MEDNIP *Bryonia dioica* (White Bryony) Heref. Havergal. Usually reserved for the root. It must be Mad Nip, which has been used.

MEDUART, MEDWART *Filipendula ulmaria* (Meadowsweet) Scot. Jamieson. i.e. Meadwort.

MEEDLES *Atriplex patula* (Orache) Halliwell.

MEEKS *Polygonum bistorta* (Bistort) Notts. Grigson; Yks. Julia Smith.

MEET-ME-LOVE 1. *Saxifraga x urbinum* (London Pride) Dev. Macmillan. 2. *Viola tricolor* (Pansy) Dev. Wright.

MEET-ME-LOVE-BEHIND-THE-GARDEN-DOOR *Viola tricolor* (Pansy) Dev. Wright.

MEET-HER-IN-THE-ENTRY-KISS-HER-IN-THE-BUTTERY Viola tricolor (Pansy) Lincs. Wright. There are a lot of similar names with erotic intent that signal the view once held that pansy was an aphrodisiac.

MEETING HOUSE SEEDS *Anethum grave-*

olens seeds (Dill seeds) New Engl. M Baker.1980.

MEETING HOUSES *Aquilegia canadensis* (Canadian Columbine) H & P.

MEETING SEEDS *Foeniculum vulgare* seeds (Fennel seeds) Wilts. Wiltshire. Seed heads of both fennel and dill were taken to church at one time, so people could nibble them during the over-long sermons. The name evidently travelled to America, and was used by the Puritan settlers in New England.

MEG-MANY-FEET *Ranunculus repens* (Creeping Buttercup) Cumb. Grigson. An appropriate name, considering its pernicious creeping habit.

MEGWEED *Smyrnium olusatrum* (Alexanders) Suss. Grigson.

MEKILWORT *Atropa belladonna* (Deadly Nightshade) Scot. Jamieson.

MEKKIN *Iris pseudo-acarus* (Yellow Flag) Cumb. B & H. A version of the Lancashire Maikin, or Maiken.

MEL GRASS *Ammophila arenaria* (Marram) Howes.

MEL-SYLVESTRE *Lonicera periclymenum* (Honeysuckle) Halliwell. It must mean something like woodland honey.

MELAMPODE *Helleborus niger* (Christmas Rose) Grieve.1931. There is a legend, concerning either this plant or H orientalis, that the daughters of King Praetus of Argos were cured of madness by the soothsayer and physician Melampus. But doesn't this just mean 'black root'?

MELANCHOLY *Achillea millefolium* (Yarrow) Shet. Grigson.

MELANCHOLY GENTLEMEN 1. *Astrantia major* (Pink Masterwort) Keble Martin. 2. *Hesperis tristis* (Coats).

MELANCHOLY THISTLE *Cirsium heterophyllum*. The reference is to the hanging flower heads, and this became the "signature" of the plant. Hence the use for melancholy, as in Gerard: "The decoction of the Thistle in Wine being drunk, expels superfluous Melancholy out of the Body, and makes a man as merry as a Cricket...".

MELAXO, Rose of *Rosa damascena* (Damask Rose) Gerard. "a city in Asia, from whence some have thought it was first brought...".

MELDE *Atriplex patula*. OE melde.

MELDWEED *Chenopodium album* (Fat Hen) Scot. Grigson.

MELEGUETA PEPPER 1. *Aframomum melegueta*. 2. *Amomum grana* (Grains of Paradise Plant) J Smith. Are these one and the same?

MELGS *Chenopodium album* (Fat Hen) Mor. Grigson. Cf. other Scottish names for this plant, like Myles and Mails.

MELILOT *Melilotus officinalis*. Melilot is from the Greek, meaning literally a clover rich in honey.

MELILOT, Golden *Melilotus altissima* (Tall Melilot) McClintock.

MELILOT, King Island *Melilotus indica* (Small-flowered Melilot) Connor. Presumably this is the King Island between Victoria and Tasmania.

MELILOT, Ribbed *Melilotus officinalis* (Melilot) Blamey.

MELILOT, Sicilian *Melilotus messinansis*. The specific name confirms that this is a plant that belongs to Messina, in Sicily.

MELILOT, Small-flowered *Melilotus indica*.

MELILOT, Swiss *Melilotus caerulea*.

MELILOT, Tall *Melilotus altissima*.

MELILOT, White *Melilotus alba*.

MELILOT-TREFOIL *Melilotus officinalis* (Melilot) Barton & Castle.

MELLICE, Mesh *Althaea officinalis* (Marsh Mallow) Dev. Fernie.

MELMONT BERRIES; MELMOT BERRIES *Juniperus communis* (Juniper) Mor. B & H.

MELON *Cucumis melo*. The Greek original from which melon comes means an apple, but they used the word for different kinds of tree-fruit, without a distinguishing epithet.

MELON, Bitter 1. *Citrullus naudinianus*. 2. *Momordica charantia* (Wild Balsam-apple) Sofowora.

MELON, Cantaloup *Cucumis melo* var. can-

talupensis. First grown in Europe at the castle of Cantalupo, in Italy, from which it gets its name.

MELON, Coyote *Cucurbita palmata*.

MELON, Dudaim *Cucumis dudaim*.

MELON, Egyptian, Round-leaved *Cucumis chate* (Hairy Cucumber) Moldenke.

MELON, Great *Cucurbita pepo* (Marrow) Parkinson.

MELON, Musk *Cucumis melo* (Melon) Parkinson.

MELON, Neapolitan *Cucumis melo var. inodorus* (Winter Melon) Bianchini.

MELON, Netted *Cucumis melo var. scandens*. Its other specific name, reticulata, goes to show that this is not the result of putting the fruit into nets to ripen.

MELON, Pocket, Queen Anne's *Cucumis dudaim* (Dudaim Melon) Grieve.1931.

MELON, Rock *Cucumis melo var. cantalupensis* (Cantaloup Melon) Bianchini.

MELON, Serpent *Cucumis melo var. flexuosus*.

MELON, Spanish *Cucumis melo var. inodorus* (Winter Melon) Bianchini. Cf. Neapolitan Melon.

MELON, Sweet *Cucumis melo* (Melon) Watt.

MELON, Tree *Carica papaya* (Paw-paw) Chopra.

MELON, Tsama *Citrullus vulgaris* (Water Melon) Lee. The Kalahari variety gets this name.

MELON, Water *Citrullus citrullus*.

MELON, Winter *Cucumis melo var. inodorus*.

MELON PUMPKIN *Cucurbita maxima* (Giant Pumpkin) Watt.

MELON TREE *Carica papaya* (Paw-paw) Grieve.1931. The fruits are as big as a melon.

MEMORY, Herb of *Rosmarinus officinalis* (Rosemary) J Smith. Rosemary was used in various charms to strengthen the memory, and it is a symbol of remembrance. Ophelia says, in Hamlet, "there's rosemary - that's for remembrance". Sir Thomas More said "As for Rosmarine, I lett it runne all over my garden walls, not onlie because my bees love it, but because 'tis the herb sacred to remembrance,..." Its use by mourners at funerals made it a token to wear in remembrance of the dead, and in the thirties, there was a demand for it for Armistice Day ceremonies, along with symbolic poppies.

MEN, Blue *Knautia arvensis* (Field Scabious) Bucks. B & H.

MEN, Naked *Colchicum autumnale* (Meadow Saffron) Dor. Macmillan. The flowers appear before the leaves, and this accounts for the imagery. But the sex appears indeterminate - Cf. Naked Ladies in particular.

MEN-AND-WOMEN; MEN-AND-WOMEN, Devil's *Arum maculatum* (Cuckoo-pint) Som. Macmillan (Men-and-women); Shrop. Burne (Devil's Men-and-women). Members of the series of names with sexual connotations that starts with the well-known Lords-and-ladies. They are all male + female.

MEN'S FACES *Viola tricolor* (Pansy) Som. Macmillan.

MENZIES SPRUCE *Picea sitchensis* (Sitka Spruce) Brimble.1948.

MENZIESIA *Phyllodoce caerulea*. An alternative botanical name for this is *Menziesia caerulea*.

MEON *Meum athamanticum* (Baldmoney) Langham.

MERCH *Apium graveolens* (Wild Celery) B & H. The OE word meant parsley, and other versions of this name are shared with the latter plant.

MERCURY 1. *Atriplex patula* (Orache) Halliwell. 2. *Chenopodium bonus-henricus* (Good King Henry) Lincs, Yks, Cumb, N'thum. Grigson. 3. *Mercurialis sp*. Presumably from the Roman messenger of the gods, the same as the Greek Hermes.

MERCURY, Annual *Mercurialis annua* (French Mercury) Clapham.

MERCURY, Blood of *Verbena officinalis* (Vervain) Clair. Called elsewhere Mercury's Moist Blood.

MERCURY, Boy's *Mercurialis annua* (French Mercury) Lyte. Explained by him by referring to an old belief that when "dronken (it) causeth to engender male children". But in that case what about Girl's Mercury, or Maiden Mercury, for the same plant?

MERCURY, Dog's *Mercurialis perennis*. A poisonous plant, hence "dog's", i.e. a useless plant when compared with *Mercurialis annua*.

MERCURY, English; MERCURY, False *Chenopodium bonus-henricus* (Good King Henry) Gerard (English); B & H (False). There are a lot of Mercury names for this, even branching out into variations like Markery and Margery.

MERCURY, French *Mercurialis annua*.

MERCURY, Garden *Mercurialis annua* (French Mercury) Grieve.1931. Presumably meaning a plant that one would allow in the garden, as against the poisonous Dog's Mercury, which one would not.

MERCURY, Girl's; MERCURY, Maiden *Mercurialis annua* (French Mercury) both Lyte. Cf. Boy's Mercury, and note the name Boys-and-girls.

MERCURY, Herb *Mercurialis perennis* (Dog's Mercury) North.

MERCURY, Scotch *Digitalis purpurea* (Foxglove) Berw. Grigson; USA Henkel. Cf. Wild Mercury.

MERCURY, Vegetable *Brunfelsia hopeana* Le Strange.

MERCURY, Wild *Digitalis purpurea* (Foxglove) Berw. Grigson. Cf. Scotch Mercury.

MERCURY-DOCKEN; MERCURY GOOSEFOOT *Chenopodium bonus-henricus* (Good King Henry) B & H (Docken); Grieve.1931 (Goosefoot).

MERCURY'S MOIST BLOOD *Verbena officinalis* (Vervain) Fernie. Cf. Blood of Mercury.

MERCURY'S VIOLET 1. *Campanula medium* (Canterbury Bells) Gerard). 2. *Campanula trachelium* (Nettle-leaved Bellflower) Gerard.

MERECROP *Anagallis arvensis* (Scarlet Pimpernel) Halliwell.

MERICH *Petroselinum crispum* (Parsley) Fernie. March is probably the original for this.

MERMAIDS *Nymphaea alba* (White Waterlily) PLNN 19.

MERRICK *Medicago sativa var sativa* (Lucerne) Dev. Friend.1882. A variation of medick, obviously.

MERRY; MERRY-TREE *Prunus avium* (Wild Cherry) Hants. B & H. The word is from French merise, wild cherry, with merry becoming a spurious singular, just as cherry arises from cerise.

MERRY, Black 1. *Prunus avium* (Wild Cherry) Hants. B & H. 2. *Prunus padus* (Bird Cherry) Hants. Grigson.

MERRY-GO-ROUNDS *Calendula officinalis* (Marigold) Dor. Macmillan. Is it descriptive? Or did it just originate with 'marigold'? The latter seems likelier, especially from some form like Mary Gowlan.

MERU OAK *Vitex keniensis*. Meru is a district of Kenya.

MESCAL 1. *Agave americana*. 2. *Agave deserti* Bean. 3. *Agave utahensis* USA Elmore. *A americana* is the true mescal. Probably, the intoxicating qualities of pulque, which is made from this, account for the fact that the name mescal is also applied to other plants that have a similar effect, not through alcohol but through a hallucinogen - *Sophora secundiflora* is known as Mescal Bean, and peyote, the cactus *Lophophora williamsii*, has also been called Mescal Button. This is a Spanish word that comes from the Nahuatl mexcalli.

MESCAL BEAN *Sophora secundiflora*.

MESCAT ACACIA *Acacia constricta*.

MESPILUS, Snowy 1. *Amelanchier arborea* (Tree Amelanchier) J Smith. 2. *Amelanchier ovalis*.

MESQUITE *Prosopis juliflora*.

MESQUITE, False *Calliandra eriophylla* (Hairy-leaved Calliandra) T H Kearney.

MESQUITE, Honey *Prosopis juliflora* (Mesquite) USA T H Kearney.
MESQUITE, Screw *Prosopis pubescens* (Screwbean) USA Youngken.
MESQUITE GRASS, Vine *Panicum obtusum*.
MESS *Quercus robur* acorn (Acorn) Dev. Friend.1882. Mast, of course.
MESS-A-BED *Taraxacum officinale* (Dandelion) Som. Macmillan. Cf. the very much more common Pissabed, but there are lots of other names all arising from the well-known diuretic effects of the plant.
MESSENGER, Spring *Ranunculus ficaria* (Lesser Celandine) Som. Macmillan.
MESSMATE *Eucalyptus obliqua* (Stringybark) Hora.
METAL-LEAF BEGONIA *Begonia metallica*.
METHONICA LILY *Gloriosa superba* (Glory Lily) Dalziel.
MEU; MEW *Meum athamanticum* (Baldmoney) B & H (Meu); Turner (Mew).
MEXICAN ASTER, Purple *Cosmos bipinnatus*.
MEXICAN BALD CYPRESS *Taxodium mucronatum*.
MEXICAN BAMBOO *Polygonum cuspidatum* (Japanese Knotweed). A W Smith. It is sometimes known as Japanese Bamboo, too.
MEXICAN BLUE OAK *Quercus oblongifolia*.
MEXICAN BUSH SAGE *Salvia leucantha*.
MEXICAN CAMPION *Silene laciniata* (Mexican Catchfly) USA Elmore.
MEXICAN CATCHFLY *Silene laciniata*.
MEXICAN CIGAR FLOWER *Cuphea platycentra*. 'Cigar', because the scarlet flowers have ash-grey margins, and a black ring behind, like a cigar band.
MEXICAN CYPRESS *Cupressus lusitanica*. 'Mexican' is quite right, in spite of the 'lusitanica' of the specific name, which it got because the trees became known in Europe through those cultivated in Portugal, and they have been so long established there that it was thought they were indigenous.

MEXICAN DEVIL *Eupatorium adenophorum*.
MEXICAN DEVILWEED *Aster spinosus* (Spiny Aster) USA Elmore.
MEXICAN ELDER *Sambucus mexicana*.
MEXICAN FIRE BUSH *Kochia scoparia* (Burning Bush) A W Smith.
MEXICAN FIRE PLANT *Euphorbia pulcherrima* (Poinsettia) Watt.
MEXICAN FLAG *Dichorisandra albo-marginata*. The reference is probably to the colours of the national flag.
MEXICAN FLAME LEAF; MEXICAN FLAME TREE *Euphorbia pulcherrima* (Poinsettia) both Watt.
MEXICAN FLAME VINE *Senecio confusus*.
MEXICAN FLEABANE *Erigeron mucronatus*.
MEXICAN GIANT HYSSOP *Agastache mexicana*.
MEXICAN GRAPE HERB *Chenopodium ambrosioides* (American Wormseed) USA Henkel.
MEXICAN GROUND CHERRY *Physalis ixiocarpa* Howes.
MEXICAN HAT PLANT *Kalanchoe daigremontianum*.
MEXICAN JUNIPER *Juniperus flaccida*.
MEXICAN MARIGOLD *Tagetes lucida*.
MEXICAN MUGWORT *Artemisia mexicana*.
MEXICAN MULBERRY *Morus microphylla*.
MEXICAN NUT PINE *Pinus cembroides*.
MEXICAN ORANGE FLOWER *Choisya ternata*.
MEXICAN OXALIS *Oxalis latifolia*.
MEXICAN PINE *Pinus patula* (Spreadinglife Pine) Leathart.
MEXICAN POPPY 1. *Argemone mexicana*. 2. *Kallstroemeria grandiflora* (Arizona Poppy) USA. T H Kearney.
MEXICAN SUNFLOWER *Tithonia rotundifolia*.
MEXICAN TEA 1. *Ephedra trifurca* (Long-leaved Joint-fir) USA Elmore. 2. *Chenopodium ambrosioides* (American

Wormseed) USA Henkel. There are a lot of 'tea' names for both of these, indicating medicinal use in the case of the Wormseed, and in the case of the *Ephedra*, a beverage drink, made by infusing the branches.

MEXICAN THISTLE *Argemone mexicana* (Mexican Poppy) Barbados Gooding.

MEXICAN WHITE PINE *Pinus ayacahuite*.

MEXICAN WHORLED MILKWEED *Asclepias mexicana*.

MEZARD *Prunus avium* (Wild Cherry) Som. Macmillan. See Mazzard, rather.

MEZELL; MAZELL *Daphne mezereon* (Lady Laurel) Gerard (Mezell); Hants. Grigson (Mazell).

MEZEREON, Dutch *Daphne mezereon* (Lady Laurel) Gerard.

MEZEREON, Mediterranean *Daphne gnidium*.

MEZEREON SPURGE; MEZEREON TREE *Daphne mezereon* (Lady Laurel) Le Strange (Spurge); Bacon (Tree). It is derived from the Persian Madzaryoun, destroyer of life. The red berries are poisonous, severely so.

MICE, Blue *Viola riviniana* (Dog Violet) Som. Macmillan.

MICE, Dutch *Lathyrus tuberosus* (Fyfield Pea) W Miller.

MICE, Rats-and- *Cynoglossum officinale* (Hound's Tongue) Wilts. Macmillan. It is the characteristic smell of the plant that reminds one of mice.

MICHAELMAS CROCUS *Colchicum autumnale* (Meadow Saffron) Som. Macmillan; Wilts. Dartnell & Goddard.

MICHAELMAS DAISY 1. *Aster paniculatus*. 2. *Aster tripolium* (Sea Aster) Dev. Friend.1882; Som. Grigson; Ches. Holland. 3. *Tanacetum parthenium* (Feverfew) Dev. Friend.1882. The name is applied to several American asters, but it really belongs to *A paniculatus*.

MICHAUX'S BLUE-EYED GRASS *Sisyrinchium mucronatum*.

MICHEN; MICKEN *Meum athamanticum* (Baldmoney) Scot. Jamieson (Michen); Grigson (Micken). That name varies into Moiken sometimes.

MICHIGAN LILY *Lilium michiganese*.

MIDDAY MARVEL *Oroxylum indicum* Malaya Gilliland.

MIDDAY STARS *Hypoxis stellata* var. *gawleri*. Because of its habit of opening only in the sun. This is the South African name.

MIDDEN MYLIES 1. *Chenopodium album* (Fat Hen) N Scot. Grigson. 2. *Chenopodium bonus-henricus* (Good King Henry) Selk. Grigson. A comment on its favourite habitat. Cf. such names as Muckweed, Dungweed, etc.

MIDGE PLANT; MIDGEWORT *Artemisia vulgaris* (Mugwort) Cullum (Midge Plant); Cockayne (Midgewort). Exactly what Mugwort means. The first part of the name means a fly or gnat - midge is the same word. It actually does keep midges away.

MIDGET, Rock *Mimulus rupicola*.

MIDLAND ELM *Ulmus x elegantissima*.

MIDLAND HAWTHORN *Crataegus laevigata* (Woodland Hawthorn) Clapham.

MIDNIGHT, Golden *Lotus corniculatus* (Bird's-foot Trefoil) Som. Macmillan.

MIDNIGHT HORROR *Oroxylum indicum* Malaya Gilliland. So called because of its foul-smelling, bat-pollinated flowers. The source should be checked, for it is not clear whether Midday Marvel and Midnight Horror are two separate names, or whether it is one name, viz Midnight Horror, Midday Marvel.

MIDSHIPMAN'S BUTTER; SUBALTERN'S BUTTER *Persea gratissima* (Avocado Pear) both Grigson.1974.

MIDSUMMER DAISY 1. *Leucanthemum vulgare* (Ox-eye Daisy) Fernie. 2. *Tanacetum parthenium* (Feverfew) Dev, Som. Macmillan.

MIDSUMMER FAIRMAID *Armeria maritima* (Thrift) Som. Macmillan.

MIDSUMMER LILY *Lilium candidum* (Madonna Lily) Som. Macmillan.

MIDSUMMER MEN 1. *Centranthus ruber* (Spur Valerian) Som. Macmillan. 2. *Orobanche minor* (Lesser Broomrape)

Wilts. Macmillan. 3. *Sedum roseum* (Roseroot) Clapham. 4. *Sedum telephium* (Orpine) Brand. Connected with Orpine more than with the others, and mainly because of the divination practiced on Midsummer Eve: girls would hang up pieces of it and name it for their boy friends. The piece that lived the longest determined the successful one. Another divination was for two people to gather it on Midsummer Eve, and plant the slips. They would know their fortune by checking whether the slips grew towards one another - if they did, the couple would marry. If one withered, the person represented would die.

MIDSUMMER SILVER *Potentilla anserina* (Silverweed) Surr. Grigson.

MIDSUMMER THISTLE *Centaurea solstitialis* (St Barnaby's Thistle) Parkinson. St Barnabas's Day is June 11, near enough to the solstice to count.

MIE, Hound's *Cynoglossum officinale* (Hound's Tongue) Cockayne. Or, as Gerard had it, Hound's Piss, "for in the world there is not any thing that smelleth so like Dogs pisse as the leaves of this plant doe".

MIGNONETTE, Corn *Reseda phyteuma*. Mignonette is the diminutive of French mignon, darling. The name of endearment was given to Wild Mignonette by Lord Bateman in 1742.

MIGNONETTE, French 1. *Dianthus chinensis* (Indian Pink) W Miller. 2. *Saxifraga umbrosa* (Pyrenean Saxifrage) W Miller.

MIGNONETTE, Jamaica; MIGNONETTE, West Indian; MIGNONETTE, Tree *Lawsonia inermis* (Henna) Dalziel; Howes (West Indian).

MIGNONETTE, Ploughman's *Euphorbia cyparissias* (Cypress Spurge) Coats.

MIGNONETTE, Sweet-scented *Reseda odorata*.

MIGNONETTE, Upright *Reseda alba* (White Mignonette) Clapham.

MIGNONETTE, White *Reseda alba*.

MIGNONETTE, Wild *Reseda lutea*.

MIGNONETTE VINE *Anredera cordifolia* (Madeira Vine) RHS.

MIGWORT *Artemisia vulgaris* (Mugwort) Dor. Macmillan. i.e. midge wort, which is what Mugwort means.

MILDER *Chenopodium bonus-henricus* (Good King Henry) Wilts. Wiltshire. Cf. the Scottish Midden Mylies.

MILE *Apium graveolens* (Wild Celery) Roxb. B & H.

MILE-A-MINUTE 1. *Ipomaea caisica* Australia Watt. 2. *Polygonum baldschaunicum* (Russian Vine) RHS. *Ipomaea caisica* is an African plant, so the Australian name must refer to an introduced plant, and must also be a comment on its rapid growth. It certainly is in the case of Russian Vine, as anyone foolhardy enough to plant it will know.

MILE TREE *Casuarina equisetifolia* (Beefwood) Howes.

MILFOIL *Achillea millefolium*. Milfoil is *millefolium*, thousand-leaf, which does exist as a book name. It is descriptive, of course, of the deeply cut leaves. Cf. Thousand-leaf, Thousand-weed,etc., as well as the more restrained Hundred-leaved Grass.

MILFOIL, Hooded *Utricularia vulgaris* (Greater Bladderwort) Prior.

MILFOIL, Knight's *Achillea millefolium* (Yarrow) Grieve.1931. A crusaders' name, for yarrow was a wound herb - it has styptic qualities. There is a tradition that it was always carried by the Greek and Roman armies.

MILFOIL, Water 1. *Hottonia palustris* (Water Violet) Grieve.1931. 2. *Myriophyllum sp.*

MILFOIL, Yellow *Achillea tomentosa*.

MILGIN *Cucurbita maxima* (Giant Pumpkin) Norf. Halliwell. Milgin pies are pumpkin pies.

MILION *Cucumis melo* (Melon) Parkinson.

MILITARY ORCHID *Orchis militaris* (Soldier Orchid) Polunin.

MILK, Bread-and- 1. *Cardamine pratensis* (Lady's Smock) Fernie. 2. *Oxalis acetosella* (Wood Sorrel) Som. Macmillan. White, or at any rate, whitish, flowers account for the epithet 'milk'.

MILK, Cat's 1. *Euphorbia amygdaloides* (Wood Spurge) Glos. PLNN35. 2. *Euphorbia helioscopia* (Sun Spurge) Worcs. Prior. In the case of the spurges, it is the milky sap that is the reference point. Cf. Deer's Milk, etc.

MILK, Deer's *Euphorbia amygdaloides* (Wood Spurge) Hants. Cope. Cf. Cat's Milk.

MILK, Devil's 1. *Chelidonium majus* (Greater Celandine) Yks. Grigson. 2. *Euphorbia helioscopia* (Sun Spurge) Worcs, Middx. Grigson; Wales Trevelyan. 3. *Euphorbia peplus* (Petty Spurge) C J S Thompson. 'Devil's' in the recognition that the milky sap is dangerous.

MILK, Lady's *Silybum marianum* (Milk Thistle) B & H.

MILK, Mad Woman's *Euphorbia helioscopia* (Sun Spurge) Bucks. B & H.

MILK, Mamma's *Euphorbia helioscopia* (Sun Spurge) Bucks. B & H.

MILK, Mare's *Euphorbia helioscopia* (Sun Spurge) Scot. Dyer.

MILK, Mother Mary's *Polygala vulgaris* (Milkwort) Som. Macmillan.

MILK, Mouse *Euphorbia helioscopia* (Sun Spurge) Yks. B & H.

MILK, Virgin's 1. *Silybum marianum* (Milk Thistle) Page.1978. 2. *Sonchus oleraceus* (Sow Thistle) Som. Macmillan.

MILK, Witch's *Hippuris vulgaris* (Mare's Tail) Lancs, Scot. Grigson.

MILK, Wolf's 1. *Euphorbia cyparissias* (Cypress Spurge) Kourennoff. 2. *Euphorbia helioscopia* (Sun Spurge) Prior. 3. *Euphorbia paralias* (Sea Spurge) Gerard. Cf. Devil's Milk - 'wolf' conveys the same warning.

MILK BUSH *Euphorbia tirucalli* (Pencil Tree) USA Kingsbury.

MILK GOWAN *Taraxacum officinale* (Dandelion) Forf. B & H. Literally, the yellow daisy with the milky juice.

MILK FLOWER *Melandrium album* (White Campion) Wilts. Dartnell & Goddard.

MILK MAIDENS 1. *Primula veris* (Cowslip) Lincs. Grigson. 2. *Stellaria holostea* (Greater Stitchwort) Dev, Som, IOW, Yks. Grigson.

MILK MOUNTAIN *Linum catharticum* (Fairy Flax) Grigson. Mill-mountain is the original, possibly from *Chamaelinum montanum*, mountain ground-flax, with the centre syllable of *Chamaelinum* providing the 'mill'.

MILK PARSLEY *Peucedanum palustre*. Cf. Milkweed for this.

MILK PARSLEY, False *Selinum carvifolia* (Cambridge Parsley) Genders.

MILK PLANT, Devil's; MILKPAIL, Devil's *Taraxacum officinale* (Dandelion) Kirk. Grigson (Milk Plant); Som. Macmillan (Milkpail).

MILK PURSLANE 1. *Euphorbia hypericifolia*. 2. *Euphorbia supina*.

MILK SPURGE *Euphorbia maculata* (Spotted Spurge) USA Gates.

MILK THE COWS *Umbilicus rupestris* (Wall Pennywort) Corn. Grigson. Presumably children pull the flower or seed heads off by running them through the fingers, in a "milking" action. Certainly docks were treated this way.

MILK THISTLE 1. *Silybum marianum*. 2. *Sonchus oleraceus* (Sow Thistle) Som, War, Dor, Lincs. Grigson. Milk Thistle proper gets the name from the white veins of the leaves, due, according to legend, to the Virgin Mary spilling milk from her breast on to them. That is why it was once reckoned to be the proper diet for wet nurses. Sow Thistle gets the name for a quite different reason, the 'milk' being the white juice.

MILK THISTLE, Spiny *Sonchus asper* (Prickly Sow Thistle) Clapham.

MILK TREE *Alstonia scholaris*. A latex-exuding tree, hence the name.

MILK VETCH *Astragalus glycyphyllus*.

MILK VETCH, Alpine *Astragalus alpinus*.

MILK VETCH, Alpine, Yellow *Astragalus frigidus*.

MILK VETCH, Canadian *Astragalus carolinianus*.

MILK VETCH, Mountain, Purple *Oxytropis halleri*.
MILK VETCH, Purple *Astragalus danicus*.
MILK VETCH, Sweet *Astragalus glycyphyllus* (Milk Vetch). D Grose.
MILK VETCH, Yellow *Oxytropis campestris*.
MILKCANS *Stellaria holostea* (Greater Stitchwort) Wilts. Macmillan; Ches. Holland.
MILKCANS, Lady's *Anemone nemorosa* (Wood Anemone) Som. Macmillan.
MILKDROPS, Virgin Mary's *Pulmonaria officinalis* (Lungwort) Wilts. Macmillan; Mon. Grigson. The white blotches on the leaves are accounted for in legend by saying that they are the result of the Virgin's milk dropping on them during the flight into Egypt.
MILKGIRL; MILKIES *Cardamine pratensis* (Lady's Smock) Dev. Friend.1882 (Milkgirl); Dev. Macmillan (Milkies). Cf. Milkmaids, etc.
MILKGRASS *Valerianella locusta* (Cornsalad) Sanecki.
MILKING MAIDS *Cardamine pratensis* (Lady's Smock) Som. Macmillan.
MILKING STOOLS, Lady's *Stellaria graminea* (Lesser Stitchwort) Som. Macmillan.
MILKMAID, Trefoil *Cardamine trifolia* Salisbury.1964.
MILKMAIDS 1. *Alliaria petiolata* (Jack-by-the-hedge) Kent Tynan & Maitland. 2. *Anemone nemorosa* (Wood Anemone) Som. Macmillan. 3. *Calystegia sepium* (Great Bindweed) Surr. Grigson; Suss. Parish. 4. *Cardamine pratensis* (Lady's Smock) Dev, Som, Middx, Ess. Grigson; Wilts. Dartnell & Goddard; Suss. Parish. 5. *Lotus corniculatus* (Bird's-foot Trefoil) Suss. Parish. 6. *Melandrium album* (White Campion) Som. Macmillan. 7. *Polygala vulgaris* (Milkwort) Som. Macmillan. 8. *Primula veris x vulgaris* (False Oxlip) Yks. Friend.1882. 9. *Saxifraga granulata* (Meadow Saxifrage) Dor. Macmillan. Hants. Hants FWI. 10. *Stellaria holostea* (Greater Stitchwort) Dev. Friend.1882; Wilts. Dartnell & Goddard.1899; Yks. Fernie. Suggested by the colour, in most cases. See also Milky Maidens.
MILKMAID'S EYE *Veronica chamaedrys* (Germander Speedwell) N'thum. Grigson. One of many 'eye' names for this, all descriptive.
MILKPAIL, Devil's *Taraxacum officinale* (Dandelion) Som. Macmillan.
MILKPANS *Stellaria holostea* (Greater Stitchwort) Ches. Grigson.
MILKSILE *Cardamine pratensis* (Lady's Smock) Yks. B & H.
MILKSILE, Lady's 1. *Cardamine pratensis* (Lady's Smock) Yks. B & H. 2. *Pulmonaria officinalis* (Lungwort) Ches. Grigson. 'Sile' means soil, or stain, the idea being to account for the white patches on lungwort's leaves, caused by the Virgin's milk dropping on them. But 'sile' can also mean a strainer, it seems.
MILKSTOOLS *Buxus sempervirens* flowers (Box) Dor. Macmillan.
MILKWEED 1. *Apocynum androsaemifolium* (Spreading Dogbane) O P Brown. 2. *Apocynum cannabinum* (Indian Hemp) Grieve.1931. 3. *Asclepias syriaca*. 4. *Euphorbia corollata* (Flowering Spurge) O P Brown. 5. *Euphorbia glomifera* Barbados Gooding. 6. *Euphorbia helioscopia* (Sun Spurge) Hants, Ess, Suff, Norf. B & H. 7. *Euphorbia peplus* (Petty Spurge) North. 8. *Peucedanum palustre* (Milk Parsley) W Miller. 9. *Sonchus oleraceus* (Sow Thistle) Som. Elworthy; Leics. R Palmer.1985. The milky juice accounts for most of these.
MILKWEED, Black *Euphorbia prostrata* Barbados Gooding.
MILKWEED, Blunt-leaved *Asclepias amplexicaulis*.
MILKWEED, Broadleaf *Asclepias latifolia*.
MILKWEED, Butterfly *Asclepias tuberosa* (Orange Milkweed) Bergen.
MILKWEED, Desert *Asclepias erosa*.

MILKWEED, Four-leaved *Asclepias quadrifolia*.
MILKWEED, Four-o'clock *Asclepias nyctaginifolia*.
MILKWEED, Hairy *Asclepias incarnata var. pulchra*.
MILKWEED, Labriform *Asclepias labriformis*.
MILKWEED, Orange *Asclepias tuberosa* .
MILKWEED, Poke 1. *Asclepias exaltata*. 2. *Asclepias phytolaccoides* USA Kingsbury. 'Poke' in this name is a reference to *Phytolacca decandra*, known as Poke-root.
MILKWEED, Polk *Asclepias exaltata* (Poke Milkweed) USA House.
MILKWEED, Purple *Asclepias purpurascens*.
MILKWEED, Red *Euphorbia heterophylla* Barbados Gooding.
MILKWEED, Seaside *Glaux maritima* (Sea Milkwort) W Miller.
MILKWEED, Showy *Asclepias speciosa*.
MILKWEED, Shrubby *Asclepias fruticosa*.
MILKWEED, Spider *Asclepiodora decumbens*.
MILKWEED, Swamp *Asclepias incarnata*.
MILKWEED, Swamp, Hairy *Asclepias incarnata var. pulchra*.
MILKWEED, Tall *Asclepias exaltata* (Poke Milkweed) USA House.
MILKWEED, White *Asclepias variegata*.
MILKWEED, White-stemmed *Asclepias albicans*.
MILKWEED, Whorled *Asclepias verticillata*.
MILKWEED, Whorled, Low; MILKWEED, Whorled, Plains *Asclepias pumila* USA Kingsbury (Plains).
MILKWEED, Whorled, Mexican *Asclepias mexicana*.
MILKWEED, Woolly-pod *Asclepias eriocarpa*.
MILKWORT 1. *Campanula rotundifolia* (Harebell) N Scot. B & H. 2. *Euphorbia helioscopia* (Sun Spurge) Som, Ess. Grigson. 3. *Euphorbia peplus* (Petty Spurge) Wilts. Dartnell & Goddard. 4. *Polygala vulgaris*. 5. *Polygala cornuta* (Horned Milkwort) USA Schenk & Gifford. 6. *Sonchus oleraceus* (Sow Thistle) Dor. Macmillan; Leics. R Palmer.1985. *Polygala* means much milk. The roots secrete a milky fluid, so the herb became recognized as a specific for increasing the milk of nursing mothers. It is the roots of harebell, too, that are known by this name. The spurges and the sow thistle have a milky sap, hence the name.
MILKWORT, Black *Astragalus glycyphyllus* (Milk Vetch) Gerard.
MILKWORT, Box-leaved *Polygala chamaebuxus* (Bastard Box) Polunin.
MILKWORT, Chalk *Polygala calcarea*.
MILKWORT, Common *Polygala serpyllifolia* (Heath Milkwort) Clapham.
MILKWORT, Cross-leaved *Polygala cruciata*.
MILKWORT, Field *Polygala viridescens*.
MILKWORT, Fringed *Polygala paucifolia*.
MILKWORT, Heath *Polygala serpyllifolia*.
MILKWORT, Horned *Polygala cornuta*.
MILKWORT, Kentish *Polygala austriaca*. Only to be found in this country, and then but rarely, in open chalk grassland in Kent.
MILKWORT, Lady's *Pulmonaria officinalis* (Lungwort) Lyte. Any plant given the possessive 'Lady's' is dedicated in some way to the Virgin Mary. In this case the reference is to the legend that the white blotches on the leaves were caused by some of the Virgin's milk dropping on them.
MILKWORT, Marsh *Polygala cruciata* (Cross-leaved Milkwort) USA House.
MILKWORT, Orange *Polygala lutea*.
MILKWORT, Purple *Polygala viridescens* (Field Milkwort) USA House.
MILKWORT, Sea *Glaux maritima*.
MILKWORT, Shrubby *Polygala chamaebuxus* (Bastard Box) Polunin.
MILKWORT, Spiny *Polygala subspinosa*.
MILKWORT, White *Polygala alba*.
MILKWORT, Yorkshire *Polygala amara*. Confined in this country to a few areas in Upper Teesdale and Craven Pennines.

MILKY BELLFLOWER *Campanula lactiflora.*

MILKY DASHEL; MILKY DASSEL; MILKY DICEL; MILKY DICKLE; MILKY DISLE. *Sonchus oleraceus* (Sow Thistle) Corn, Dev. Grigson (Dashel, Dassel); Corn. Jago (Dicel); Corn, Dev. Grigson (Dickle); Corn. Quiller-Couch; Dev. B & H (Disle).

MILKY MAIDEN *Cardamine pratensis* (Lady's Smock) Dev. Grigson. In some parts, it is said that the name Milkmaids was given because the colour of the flowers resembled the milkmaids' complexion. Milky Maidens seems to embody this idea.

MILKY TASSEL *Sonchus oleraceus* (Sow Thistle) Corn. B & H.

MILKY THISTLE, Field *Sonchus arvensis* (Corn Sow Thistle) Clapham.

MILKY THISTLE, Striped *Silybum marianum* (Milk Thistle) Coles.

MILL MOUNTAIN 1. *Calamintha ascendens* (Calamint) Grieve.1931. 2. *Linum catharticum* (Fairy Flax) Gerard. Cf. Milk Mountain. Latin chamaelinum montanum is Prior's explanation of the derivation. The middle syllable of chamaelinum is the 'mill' of this name.

MILLE FLOWER *Achillea millefolium* (Yarrow) Oxf. Thompson. *Millefolium* - thousand leaves, rather than a thousand flowers.

MILLER, Dusty 1. *Artemisia stelleriana* (Beach Wormwood) Clapham. 2. *Centaurea candidissima* USA Cunningham. 3. *Centaurea gymnocarpa* USA Cunningham. 4. *Cerastium tomentosum* (Snow-in-summer) Lincs. B & H. 5. *Primula auricula* (Auricula) E Scot. B & H. 6. *Senecio cineraria* (Silver Ragwort) USA Hay & Synge. There is a characteristic mealy appearance in all these to account for the name.

MILLER'S DELIGHT *Centaurea cyanus* (Cornflower) Dor. Macmillan.

MILLER'S JERKIN *Verbascum thapsus* (Mullein) Wales Grigson. The leaves have a mealy, flannelly look about them, like a miller's apron.

MILLER'S STAR *Stellaria holostea* (Greater Stitchwort). Suss B & H.

MILLET 1. *Panicum miliaceum.* 2. *Sorghum vulgare* (Kaffir Corn) Brouk. Although it often gets the name, millet for Sorghum is erroneous. Latin milium is the origin of the word.

MILLET, Abyssinian *Eleusine tocussa.*

MILLET, African *Eleusine coracana* (Finger Millet) Brouk.

MILLET, Barnyard, Japanese *Echinochloa crus-galli var. frumentacea.*

MILLET, Bird's foot *Eleusine coracana* (Finger Millet) Brouk.

MILLET, Broom *Panicum miliaceum* (Millet) Brouk.

MILLET, Bulrush *Pennisetum glaucum* (Pearl Millet) Brouk.

MILLET, Colorado *Panicum texanum* (Texas Millet) Douglas.

MILLET, Egyptian *Eleusine aegyptiaca.*

MILLET, Finger *Eleusine coracana.*

MILLET, Foxtail *Setaria italica.*

MILLET, German *Setaria italica* (Foxtail Millet) Brouk. But also known as Italian, Hungarian or Siberian Millet, according to country of origin.

MILLET, Grey *Lithospermum officinale* (Gromwell) Som. Grigson. Cf. Stone Millet, and also Grey Mill, or Greymyle.

MILLET, Hog *Panicum miliaceum* (Millet) Brouk.

MILLET, Hungarian; MILLET, Italian *Setaria italica* (Foxtail Millet) Brouk. See German Millet.

MILLET, Japanese *Echinochloa utilis.*

MILLET, Pearl *Pennisetum glaucum.*

MILLET, Proso *Panicum miliaceum* (Millet) Brouk. 'Proso' is the Russian word for millet.

MILLET, Red *Digitaria ischaemum* (Smooth Finger-grass) Allan.

MILLET, Sanwa *Echinochloa crus-galli var. frumentacea* (Japanese Barnyard Millet) Brouk.

MILLET, Siberian *Setaria italica* (Foxtail Millet) Brouk.
MILLET, Stone *Lithospermum officinale* (Gromwell) Watt.
MILLET, Texas *Panicum texanum*.
MILLET, Turkish *Zea mays* (Maize) Turner.
MILLION *Cucurbita maxima* (Giant Pumpkin) E Ang. Forby. The word is the same as 'melon'; Cf. Melon Pumpkin for this.
MILLION, Musk *Cucumis melo* (Melon) Tusser.
MILLIONS, Mother-of- 1. *Cymbalaria muralis* (Ivy-leaved Toadflax) Dev. Friend.1882; Som. Macmillan. 2. *Geranium molle* (Dove's-foot Cranesbill) Yks. Grigson. This kind of name usually means that the plant is a prolific seed producer.
MILNER FLOWER *Epilobium hirsutum* (Great Willowherb) Grigson.
MIMOSA *Acacia dealbata* (Silver Wattle) Polunin. Mimosa is the florists' name, an extension from *Mimosa pudica*, the Sensitive Plant. The word comes from Greek mimos, a mimic.
MIMOSA, Bailey's *Acacia baileyana* (Cootamundra Wattle) Hora.
MIMOSA, Egyptian 1. *Acacia arabica* (Babul Tree) Dalziel. 2. *Acacia nilotica* (Thorny Acacia) Moldenke.
MIMOSA, Golden *Acacia baileyana* (Cootamundra Wattle) Hora.
MIMOSA, Myrtle-leaved *Acacia myrtifolia*.
MIMOSA THORN *Acacia karroo*.
MIN FIR *Abies recurvata*. A Chinese species, growing in forests on the mountains above the Min River.
MINARET-FLOWER *Leonotis leonurus* (Lion's Tail) Howes.
MINASTER *Geum urbanum* (Avens) Ire. B & H.
MIND-YOUR-OWN-BUSINESS *Helxine soleirolii*.
MINER'S LETTUCE 1. *Claytonia alsinoides* (Pink Purslane) Usher. 2. *Claytonia perfoliata* (Spring Beauty) USA Yanovsky. The leaves are eaten as a vegetable; so, too, are the roots. Cf. Spanish Lettuce. It got the name because it saved many Californian gold miners from scurvy.
MINER'S LETTUCE, Narrow-leaved *Montia spathulata var. tenuifolia*.
MINER'S LETTUCE, Siberian *Montia sibirica*.
MINER'S TEA *Ephedra nevadensis* (Nevada Joint-Fir) USA Bean. Cf. Mormon Tea and Teamster's Tea. Presumably it was made as a beverage, and certainly it was used for a medicinal tea.
MINGWORT *Artemisia absinthium* (Wormwood) N Eng. F Grose. It sounds as if it were a misreading of Mugwort, in which case *Artemisia vulgaris* would be more appropriate.
MINORCA BOX *Buxus balearica*. The species is not confined to Minorca, being native to, on one side of the Balearics, southern Spain, and on the other, Sardinia.
MINSHULL CRAB *Mespilus germanica* (Medlar) B & H.
MINT, Apple 1. *Mentha arvensis* (Corn Mint) Ire. E E Evans.1942. 2. *Mentha rotundifolia*.
MINT, Apple, American; MINT, Apple, Golden *Mentha x gentilis* (Bushy Mint) G B Foster (American); Webster (Golden).
MINT, Apple, Large *Mentha alopacuroides*.
MINT, Apple, Silver *Mentha rotundifolia variegata* Webster.
MINT, Balm 1. *Mentha longifolia* (Horse Mint) Turner. 2. *Mentha x piperita* (Peppermint) Hatfield.
MINT, Basil-smelling *Mentha citrata* (Bergamot Mint) Tradescant.
MINT, Bergamot *Mentha citrata*.
MINT, Brandy 1. *Mentha x piperita* (Peppermint) West B & H. 2. *Mentha spicata* (Spearmint) War. F Savage. Mint Julep requires a sprig of mint in the whisky. Is there a recipe for its use in brandy, too?
MINT, Brown *Mentha spicata* (Spearmint) Gerard.
MINT, Bushy *Mentha x gentilis*.

MINT, Canadian *Mentha arvensis var. canadensis.*
MINT, Capitate *Mentha aquatica* (Water Mint) H C Long.1910.
MINT, Corn 1. *Acinos arvensis* (Basil Thyme) Turner. 2. *Mentha arvensis.*
MINT, Corsican *Mentha requeni.*
MINT, Creeping *Mentha requeni* (Corsican Mint) Salisbury.1964.
MINT, Crisped *Mentha crispa* (Curly Mint) Minag Bull.76.
MINT, Curled 1. *Mentha crispa* (Curly Mint) Minag Bull.76. 2. *Mentha x piperita* (Peppermint) Hatfield.
MINT, Curly *Mentha crispa.*
MINT, Curly-leaf *Mentha spicata* (Spearmint) E Hayes.
MINT, Dog *Teucrium scorodonia* (Wood Sage) Tynan & Maitland.
MINT, Eau-de-Cologne *Mentha citrata* (Bergamot Mint) McClintock.
MINT, Egyptian *Mentha rotundifolia* (Apple Mint) Grieve.1931.
MINT, Fern *Monarda fistulosa* (Purple Bergamot) Howes.
MINT, Field 1. *Mentha arvensis* (Corn Mint) H C Long.1910. 2. *Nepeta cataria* (Catmint) USA Henkel.
MINT, Fish 1. *Mentha aquatica* (Water Mint) Gerard. 2. *Mentha spicata* (Spearmint) Grieve.1931.
MINT, Flea *Mentha pulegium* (Pennyroyal) Sanecki. Good for destroying fleas, as the specific name indicates - pulegium is Latin pulices, fleas.
MINT, Garden *Mentha spicata* (Spearmint) Sanecki.
MINT, Gibraltar *Mentha pulegium* (Pennyroyal) Howes.
MINT, Ginger *Mentha x gentilis* (Bushy Mint) Brownlow.
MINT, Green *Mentha spicata* (Spearmint) C P Johnson.
MINT, Hairy *Mentha aquatica* (Water Mint) Wilts. Dartnell & Goddard.
MINT, Hart *Mentha spicata* (Spearmint) Parkinson.

MINT, Horse 1. *Mentha aquatica* (Water Mint) Som. Macmillan; Glos, IOW, War, Lincs, Yks, N'thum. Grigson. 2. *Mentha longifolia.* 3. *Mentha rotundifolia* (Apple Mint) Glos. J D Robertson. 4. *Monarda fistulosa* (Purple Bergamot) USA Elmore. 5. *Monarda pectinata* (American Horse Mint) USA Elmore.
MINT, Horse, American *Monarda pectinata.*
MINT, Horse, Blue *Monarda mollis.*
MINT, Horse, Spotted *Monarda punctata.*
MINT, Japanese *Mentha arvensis var. piperascens.*
MINT, Lady's *Mentha spicata* (Spearmint) Gerard. A rare association in this country, but it is dedicated to the Virgin in France, under the name of Menthe de Notre-Dame, and in Italy, where it is called Erba Santa Maria. It is Unser Frauen Muntz in Germany.
MINT, Lamb 1. *Mentha piperita* (Peppermint) Dev. B & H. 2. *Mentha spicata* (Spearmint) Dev. B & H. Of course, mint is the necessary accompaniment to lamb in English cooking.
MINT, Lemon 1. *Melissa oficinalis* (Bee Balm) Howes. 2. *Mentha citrata* (Bergamot Mint) Bardswell. 3. *Monarda citriodora* (Prairie Bergamot) A W Smith. 4. *Monarda pectinata* (Horse Mint) USA Elmore. 5. *Monarda punctata* (American Horse Mint) A W Smith.
MINT, Mackerel *Mentha spicata* (Spearmint) Gerard. Cf. Fish-mint, also. It shows that at one time mint was a "must" with fish.
MINT, Maori *Mentha cunninghami.*
MINT, Marsh *Mentha aquatica* (Water Mint) Grieve.1931.
MINT, Monkey *Melampyrum pratense* (Common Cow-wheat) Wilts. Goddard.
MINT, Mountain 1. *Calaminta ascendens* (Calamint) Culpepper. 2. *Koellia virginica.* 3. *Monarda didyma* (Red Bergamot) W Miller.
MINT, Orange *Mentha citrata* (Bergamot Mint) Webster.

MINT, Our Lady's *Tanacetum balsamita* (Balsamint) Tynan & Maitland.

MINT, Pale *Mentha pulegium* (Pennyroyal) Potter & Sargent.

MINT, Pea *Mentha spicata* (Spearmint) Brownlow. Mint is the traditional accompaniment to peas.

MINT, Pennyroyal *Mentha pulegium* (Pennyroyal) Thornton.

MINT, Pineapple *Mentha rotundifolia var. variegata* G B Foster.

MINT, Red 1. *Mentha aquatica* (Water Mint) Turner. 2. *Monarda didyma* (Red Bergamot) G B Foster. What had Turner in mind? The flowers are lilac to purple.

MINT, Rock *Teucrium scorodonia* (Wood Sage) Som. Tynan & Maitland.

MINT, Round-leaved *Mentha rotundifolia* (Apple Mint) Minag Bull 76.

MINT, Smith's *Mentha smithiana* (Tall Mint).

MINT, Spanish *Mentha requeni* (Corsican Mint) Ire. Coats. There is a tradition that it was first introduced into Ireland at the time of the Spanish Armada.

MINT, Spear; MINT, Spire *Mentha spicata* (Spearmint) Prior (Spire). Spire Mint is really the correct name, rather than Spearmint. Anyway, both words have the same meaning in this context.

MINT, Tall *Mentha smithiana*.

MINT, Tule *Mentha arvensis var. canadensis* (Canadian Mint) USA Elmore.

MINT, Water 1. *Mentha aquatica*. 2. *Mentha longifolia* (Horse Mint) Turner.

MINT, Whirl *Mentha pulegium* (Pennyroyal) Hants. Grigson.

MINT, Whorled *Mentha x verticillata*.

MINT, Wild 1. *Ajuga reptans* (Bugle) Berks. Grigson. 2. *Mentha longifolia* (Horse Mint) S Africa Watt. 3. *Mentha rotundifolia* (Apple Mint) Turner.

MINT, Wild, American *Mentha arvensis var. canadensis* (Canadian Mint) USA House.

MINT, Winter *Mentha cordifolia*. This species does not die down in winter as most of the mints do.

MINT, Woodland *Mentha longifolia* (Horse Mint) Genders.1971.

MINT, Woolly *Mentha rotundifolia* (Apple Mint) Webster.

MINT, Woolly, White *Mentha longifolia* (Horse Mint) Bardswell.

MINT GERANIUM *Tanacetum balsamita* (Balsamint) Mabey.1977. It is described as having a scent which is a cross between mint and lemon.

MINT OF LIFE *Mentha aquatica* (Water Mint) N'thants. Clare.

MINTDROP; MINTDROP, Red *Melandrium rubrum* (Red Campion) both N'thum. Grigson.

MINTDROP, White *Silene cucubalis* (Bladder Campion) N'thum. Grigson.

MINTLEAF BEE BALM *Monarda fistulosa* (Purple Bergamot) USA Elmore.

MINTWEED *Salvia reflexa* (Annual Sage) USA Kingsbury.

MIP *Brassica campestris var. rapa* (Turnip) Heref. G C Lewis. An abbreviated form - Mit and Nip are also recorded.

MIRACLE TREE *Weltwischia mirabilis*. Because of its strange habit of burying its woody stem deep in the ground. It grows under extremely dry conditions, and all that appears above ground level is the top of the trunk, possibly a yard wide, with only two huge leaves, which may be up to 12 feet long.

MIRBECK OAK *Quercus canariensis* (Algerian Oak) Duncalf.

MIRE BLOB *Caltha palustris* (Marsh Marigold) Northants. B & H. 'Blob' means blister, a reference to the acridity of the juice. So the name means the blister-producing plant that grows in the mud by water.

MIRO *Podocarpus ferruginea*.

MIRROR-OF-VENUS *Ophrys speculum* (Mirror Orchid) Polunin.

MIRROR ORCHID *Ophrys speculum*.

MIRROT *Daucus carota* (Wild Carrot) Ross, Suth. Jamieson.

MISCELDIN; MISCELTO *Viscum album* (Mistletoe) both Turner.

MISCHIEVOUS JACK *Stellaria media* (Chickweed) Som. Macmillan.
MISLE; MISLIN-BUSH *Viscum album* (Mistletoe) both E Ang. Grigson (Misle); Halliwell (Mislin-bush).
MISS MODESTY 1. *Bellis perennis* (Daisy) Som. Macmillan. 2. *Viola odorata* (Sweet Violet) Dor. Macmillan.
MISS SCENTY *Viola odorata* (Sweet Violet) Som. Macmillan.
MISSANDA *Erythrophlaeum guineense* (Ordeal Tree) Dale & Greenway.
MISSEL; MISSELTO; MISSELTOE; MISSLETOE; MIZZELTOE *Viscum album* (Mistletoe) Gerard (Missel); Wesley (Misselto); Gerard (Misseltoe); Roberts (Missletoe); Clare (Mizzeltoe).
MISSELTOW *Viscum album* (Mistletoe) Drayton.
MISSION GRASS *Pennisetum polystachyon*.
MISSOURI CURRANT *Ribes aureum* (Buffalo Currant) USA Elmore.
MISSOURI GOOSEBERRY *Ribes missouriense*.
MISSOURI GOURD *Cucurbita foetidissima* (Buffalo Gourd) USA Vestal & Schultes.
MISSOURI IRIS *Iris missouriensis*.
MISSOURI LILY *Tritelia uniflora*.
MISSY-MOOSEY *Pyrus americana* Howes.
MIST, Devil-in-a- *Nigella damascena* (Love-in-a-mist) Tynan & Maitland.
MIST, Mountain *Calluna vulgaris* (Heather) Som. Macmillan.
MIST FLOWER *Eupatorium riparium*.
MISTEL; MISTIL *Clinopodium vulgare* (Wild Basil) Gunther. So it is glossed on a 12th century manuscript of Apuleius.
MISTLETOE 1. *Phoradendron juniperinum* USA Youngken. 2. *Phoradendron villosum* USA Kingsbury. 3. *Viscum album*. Mistletoe is OE mistiltan, from mistil, the name of the plant, and tan, twig. The German mist means bird droppings, and one tradition says that its seed was deposited on certain trees by birds, the messengers of the gods. A similar story is that a certain bird, known as the "missel-bird", fed upon a particular kind of seed that it could not digest. It evacuated it whole, and the seed, falling upon the boughs of trees, germinated and produced the mistletoe. One superstition says the mistletoe seed can only germinate after it has passed through a bird. Popularly, the bird can only be the mistle-thrush.
MISTLETOE, Birdlime *Viscum album* (Mistletoe) Flück. "...white translucent berries...whereof the best Bird-lime is made, far exceeding that which is made of the Holm or Holly bark..." (Gerard). It is made by drying, pounding, soaking in water for 12 days, and pulverizing again. It was used up to medieval times for taking small birds on the branches of trees, and also for catching hawks, which were decoyed by a bird tethered between an arched stick coated with birdlime - hence, by the way, the name Viscum.
MISTLETOE, Cape *Viscum capense*.
MISTLETOE, Juniper *Phoradendron juniperum*.
MISTLETOE FIG 1. *Ficus deltoides*. 2. *Ficus diversifolia*.
MIT *Brassica campestris var. rapa* (Turnip) Heref. G C Lewis. An abbreviated form of Turmit - see also Mip and Nip.
MITCHELL'S CAPER *Capparis mitchelli*.
MITHRIDATE MUSTARD 1. *Lepidium campestre* (Pepperwort) W Miller. 2. *Thlaspi arvense* (Field Pennycress) B & H. Mithridate came to mean an antidote to poison, after Mithridates, King of Pontus, who according to tradition made himself proof against poisons.
MITHRIDATE PEPPERWORT *Lepidium campestre* (Pepperwort) Prior.
MITHRIDATUM *Thlaspi arvense* (Field Pennycress) Dawson.
MITRE CRESS *Myagrum perfoliata*.
MITREWORT, False *Tiarella cordifolia* (Foam Flower) USA House.
MLANJE CEDAR *Widdringtonia whytei*. Mount Mlanje, in Malawi, is the best environment for these trees.
MOCASSIN, Pink; MOCASSIN FLOWER

Cypripedium acaule (Stemless Lady's Slipper) USA Zenkert (Pink); USA H Smith 2 (Flower).

MOCK CHERVIL *Myrrhis odorata* (Sweet Cicely) Grieve.1931.

MOCK-EEL ROOT; MUSKRAT WEED; MUSQUASH ROOT *Cicuta maculata* (American Cowbane) USA Kingsbury.1967.

MOCK GILLIFLOWER *Saponaria officinalis* (Soapwort) Lyte.

MOCK LOCUST *Amorpha californica* USA T H Kearney.

MOCK ORANGE 1. *Philadelphus coronaria*. 2. *Pittosporum tobiana* RHS.

MOCK PLANE *Acer pseudo-platanus* (Sycamore) Prior.

MOCK SAFFRON *Carthamus tinctorius* (Safflower) Turner.

MOCKERNUT *Carya tomentosa* (White Hickory) Wit. Barber & Phillips suggested that the name was given because the nut is deceptive (?).

MODEST MAIDEN *Viola odorata* (Sweet Violet) Som. Macmillan.

MODESTY, Miss 1. *Bellis perennis* (Daisy) Som. Macmillan. 2. *Viola odorata* (Sweet Violet) Dor. Macmillan.

MODOC CYPRESS *Cupressus bakeri* (Baker Cypress) Barber & Phillips.

MOGADOR ACACIA *Acacia gummifera*.

MOGDAD COFFEE *Cassia occidentalis* (Coffee Senna) Lewin.

MOGUE TOBIN *Chrysanthemum segetum* (Corn Marigold) Ire (co Carlow) Vickery.1995. Mogue Tobin, so we are told, was a farmer who was driven out of his holding when all he could grow was corn marigold.

MOGVURD *Artemisia vulgaris* (Mugwort) Som. Macmillan. Merely an extreme presentation of mugwort.

MOHAVE ASTER *Aster abatus*.

MOHAVE BUCKBRUSH *Ceanothus vestitus*.

MOHAVE GROUNDSEL *Senecio mohavensis*.

MOHAVE LOCO *Astragalus mohavensis*.

MOHAVE PENNYROYAL *Monardella exilis*.

MOHAVE ROSE *Rosa mohavensis*.

MOHAVE RUBBERBRUSH *Chrysothamnus nauseosus var. mohavensis*.

MOHAVE SAGE *Salvia mohavensis*.

MOHAVE SALTBUSH *Atriplex spinifera*.

MOHAVE YUCCA *Yucca schidigera*.

MOIKEN *Meum athamanticum* (Baldmoney) Perth B & H. Cf. Micken, or Michen.

MOITHERN *Anthemis cotula* (Maydweed) Shrop. B & H. This is one of the many variants of Mather, meaning a girl. The nearest to this one is probably the Wiltshire Mauthern.

MOLAYNE *Verbascum thapsus* (Mullein) Fernie.

MOLE-PLANT; MOLE-TREE *Euphorbia lathyris* (Caper Spurge) USA G B Foster (Plant); Bergen (Tree). These may be American names, but the practice of planting Caper Spurges in the garden to discourage moles is certainly as well known in this country as in America.

MOLEERY-TEA 1. *Achillea millefolium* (Yarrow) Caith. Grigson. 2. *Achillea ptarmica* (Sneezewort) Caith. Grigson.

MOLL-BLOB 1. *Caltha palustris* (Marsh Marigold) Worcs. Grigson; Northants. A E Baker. 2. *Cardamine pratensis* (Lady's Smock) Northants. Grigson. Literally, cow-blister, more appropriate to Marsh Marigold, given the blister-raising propensities of most of the *Ranunculaceae*, than to Lady's Smock. Cf. Molly-blob.

MOLL-O'-THE-WOODS 1. *Anemone nemorosa* (Wood Anemone) Dor. Macmillan: War. Grigson. 2. *Arum maculatum* (Cuckoo-pint) War. Grigson.

MOLLY, Red *Kochia americana*. But this is also referred to as **GREEN MOLLY** in America see T H Kearney.

MOLLY-BLOB *Caltha palustris* (Marsh Marigold) Northants. Sternberg. It has been pointed out that one of the German names for the plant is Mollesblume, where Molle means a cow. Blob, of course, means a blister, and is a reference to the very acrid juice.

MOLUCCA BALM *Molucella laevis.*
MOLUCCAN BRAMBLE *Rubus molucannus.*
MOLY *Allium ursinum* (Ramsons) Dev, Som. Macmillan. It was given to this plant, but there is an *Allium moly*, the Golden Garlic. But even this is only a guess to the identification of the ancient Greek moly.
MOMI FIR *Abies firma.*
MONARCH BIRCH *Betula maximowicziana.*
MONARCH-OF-THE-EAST *Sauromatum guttatum* (Voodoo Lily) RHS.
MONARDA, Lemon *Monarda pectinata* (Horse Mint) Elmore.
MONARDELLA, Narrow-leaved *Monardella linoides.*
MONARCH OF THE VELDT *Venidium fastuosum* (Namaqualand Daisy) Hay & Synge.
MONBIN *Spondias monbin.*
MONEY 1. *Lunaria annua* (Honesty) Dev, Som. Macmillan. 2. *Rhinanthus crista-galli* (Yellow Rattle) Som. Macmillan. Bucks. Grigson; Northants. A E Baker. 3. *Sedum acre* (Biting Stonecrop) Leyel.1937. With Honesty, it is the silvery capsules that remind one of money, but with Yellow Rattle it is the money-box effect of the seeds themselves inside the capsule.
MONEY, Blue *Pulsatilla vulgaris* (Pasque Flower) W Miller. 'Money' in this case is simply a version of anemone.
MONEY, Gypsy's *Caltha palustris* (Marsh Marigold) Som. Macmillan. Presumably a reference to the golden colour of the flowers.
MONEY, Red *Centranthus ruber* (Spur Valerian) Som. Macmillan.
MONEY, White *Lobularia maritima* (Sweet Alison) Bucks. Harman.
MONEY BAGS 1. *Capsella bursa-pastoris* (Shepherd's Purse) Som. Macmillan. 2. *Impatiens noli-me-tangere* (Touch-me-not) Tynan & Maitland.
MONEY BOX *Scrophularia nodosa* (Figwort) Som. Macmillan.
MONEY BUSH *Cassia bicapsularis* Barbados Gooding.

MONEY FLOWER *Lunaria annua* (Honesty) Gerard.
MONEY GRASS *Rhinanthus crista-galli* (Yellow Rattle) Leics. Grigson. Cf. Money, Penny-weed and a lot of other 'money' names for this, all dependent on the "money-box" effect of the rattling seeds.
MONEY-IN-BOTH-POCKETS 1. *Acer campestre* seeds (Field Maple keys) Som. Macmillan. 2. *Agrimonia eupatoria* (Agrimony) Dor. Macmillan. 3. *Lunaria annua* (Honesty) Wright; inc. Wilts. Tynan & Maitland.
MONEY-IN-EVERY-POCKET; MONEY-POCKETS *Lunaria annua* (Honesty) both Som. Macmillan.
MONEY-IN-YOUR-POCKET *Lunaria annua* (Honesty) Wilts. Tynan & Maitland.
MONEY PENNY *Umbilicis rupestris* (Wall Pennywort) Dev. Grigson; Glos. PLNN35. From the round leaves, presumably.
MONEY PLANT 1. *Crassula portulacea.* 2. *Lunaria annua* (Honesty) Dev. B & H. 3. *Thlaspi arvense* (Field Penny-cress) Som. Macmillan.
MONEY TREE *Thlaspi arvense* (Field Penny-cress) Som. Macmillan.
MONEYWORT 1. *Anagallis tenella* (Bog Pimpernel) B & H. 2. *Lunaria annua* (Honesty) Prior. 3. *Lysimachia nummularia* (Creeping Jenny) Gerard. Creeping Jenny's leaves provide the 'money' imagery, while those of Bog Pimpernel resemble them enough to condone the book use of the name.
MONEYWORT, Cornish *Sibthorpia europaea.* Cf. Penny-pies for this.
MONEYWORT, Wood *Lysimachia nemora* (Yellow Pimpernel) Curtis.
MONGONGO *Ricidendron rautanenii* (Mangetti Tree) Lee. Mongongo is the Tswana name for the tree, while Mangetti is the Herero version.
MONK'S BASIL *Ocymum sanctum* (Holy Basil) W Miller.
MONK'S BEARD *Cichorium intybus* (Chicory) Sanecki. Probably from the French name, barbe de capucin.

MONK'S COWL *Aconitum napellus* (Monkshood) Turner. The imagery here is from the helmet-, or cowl-, shape of the flowers.

MONK'S HERB *Mentha rotundifolia* (Apple Mint) Rohde. Presumably because it was grown in monks' gardens.

MONK'S PEPPER *Vitex agnus-castus* (Chaste Tree) Leyel.1937. Cf. Small, or Wild, Pepper.

MONK'S RHUBARB 1. *Rumex alpinus*. 2. *Rumex obtusifolius* (Broad-leaved Dock) Flück. 3. *Rumex patientia* (Patience Dock) Gerard. "Because it seems some Monke or other hath used the root hereof in stead of Rubarb" was Gerard's comment for Patience Dock. *Rumex alpinus* is the one that the name should be reserved for, but Broad-leaved Dock is sometimes used medicinally in the same way, and then it is called Monk's Rhubarb.

MONKEY, Puzzle- *Araucaria araucana* (Chile Pine) Som. Elworthy. Monkey-puzzle is the usual one, but occasionally Puzzle-monkey is met in Somerset.

MONKEY APPLE 1. *Annona glabra* Barbados Gooding. 2. *Annona palustris* (Alligator Apple) W Indies Dalziel. 3. *Strychnos spinosa* (Kaffir Orange) Palgrave.

MONKEY BALLS *Strychnos spinosa* fruit (Kaffir Orange) S Africa Dalziel.

MONKEY BELLS *Caltha palustris* (Marsh Marigold) Som. Macmillan.

MONKEY BLOSSOM *Mimulus guttatus* (Monkey Flower) Corn. Grigson.

MONKEY BREAD TREE 1. *Adansonia digitata* (Baobab) Everett. 2. *Adansonia madagascariensis*.

MONKEY CHOPS 1. *Antirrhinum majus* (Snapdragon) Som. Macmillan. 2. *Glechoma hederacea* (Ground Ivy) Som. Macmillan. In line with the large number of 'mouth' names given to both these plants.

MONKEY CUP; MONKEY JACK *Mimulus guttatus* (Monkey Flower) both Dor. Grigson(Cup); Macmillan (Jack).

MONKEY DINNER-BELL *Hura crepitans* (Sandbox Tree) A Huxley.

MONKEY('S) FACES 1. *Antirrhinum majus* (Snapdragon) Wilts. Macmillan. 2. *Linaria vulgaris* (Toadflax) Som. Macmillan; Dor. Grigson. 3. *Viola tricolor* (Pansy) Som. Macmillan.

MONKEY FIDDLE *Euphorbia tirucalli* (Pencil Tree) USA Kingsbury.

MONKEY FLOWER 1. *Antirrhinum majus* (Snapdragon) Som. Macmillan. 2. *Glechoma hederacea* (Ground Ivy) Som. Macmillan. 3. *Linaria vulgaris* (Toadflax) Yks. Grigson. 4. *Mimulus guttatus*.

MONKEY FLOWER, Square-stemmed *Mimulus ringens*.

MONKEY GUAVA *Diospyros mespiliformis* (African Ebony) Dalziel.

MONKEY HAT *Tropaeolum majus* (Nasturtium) Ches. Holland.

MONKEY JAW *Cymbalaria muralis* (Ivy-leaved Toadflax) Som. Macmillan.

MONKEY KOLA *Cola caricifolia*. Actually, any false kola may be called this.

MONKEY MINT *Melampyrum pratense* (Cow-wheat) Wilts. Goddard.

MONKEY MOUTH 1. *Antirrhinum majus* (Snapdragon) Som. Macmillan. 2. *Cymbalaria muralis* (Ivy-leaved Toadflax) Som. Macmillan.

MONKEY MUSK 1. *Antirrhinum majus* (Snapdragon) Dev, Som. Macmillan. 2. *Mimulus guttatus* (Monkey Flower) Dev. Friend.1882; Wilts. Dartnell & Goddard.

MONKEY NOSES *Antirrhinum majus* (Snapdragon) Dor. Macmillan.

MONKEY NUT 1. *Arachis hypogaea* (Ground Nut) Grigson.1974. 2. *Poa annua* (Annual Meadow-grass) Wilts. Dartnell & Goddard. The only reason that Grigson could think of for giving the name to Ground Nuts was that they were bought for feeding the monkeys at the Zoo. The Wiltshire name for the grass is mysterious.

MONKEY ORCHID *Orchis simea*.

MONKEY PLANT 1. *Linaria vulgaris* (Toadflax) Yks. Grigson. 2. *Mimulus gutta-*

tus (Monkey Flower) Dev. Grigson; Wilts. Macmillan. 3. *Rhinanthus crista-galli* (Yellow Rattle) Dor. Macmillan.

MONKEY-PUZZLE *Araucaria araucana* (Chile Pine). The name is said to have arisen from a remark by Charles Austin, during the ceremonial planting of one of these trees in the gardens at Pencarrow, in Cornwall, in 1834 - "it would be a puzzle for a monkey to climb that tree". Cf. Puzzle-Monkey, and Monkey Tree, which is just a shortened version of Monkey-Puzzle.

MONKEY STAR-APPLE *Chrysophyllum perpulchrum*. The fruit is reckoned inferior to the other Star-apples, hence 'Monkey'.

MONKEY STICKS *Antirrhinum majus* (Snapdragon) Som. Macmillan.

MONKEY TREE *Araucaria araucana* (Chile Pine) Som. Elworthy. A shortened version of Monkey-Puzzle.

MONKEY VINE *Ipomaea nil*.

MONKEY'S HOOD 1. *Aconitum anglicum* (Wild Monkshood) Dev. B & H. 2. *Aconitum napellus* (Monkshood) Dev. Friend.1882. A corruption of monkshood, of course.

MONKSHEAD *Taraxacum officinale* (Dandelion) B & H. The disc looks like a shaven head after the seeds have flown. Turner called it Priest's Crown.

MONKSHOOD 1. *Aconitum napellus*. 2. *Delphinium ajacis* (Larkspur) Parkinson. Descriptive of the helmet-shaped flowers.

MONKSHOOD, Western *Aconitum columbianum*.

MONKSHOOD, Wild *Aconitum anglicum*.

MONKSHOOD, Yellow 1. *Aconitum anthora*. 2. *Aconitum vulparia* (Wolfsbane) Schauenberg & Paris.

MONKSWOOD *Aconitum napellus* (Monkshood) Ches. Holland.

MONNAGHS HEATHER; MONOX HEATHER; MONNOCS HEATHER *Empetrum nigrum* (Crowberry) W Miller; N.Ire. Grigson (Monnocs).

MONNIES *Leucanthemum vulgare* (Ox-eye Daisy) Som. Grigson. It sounds like a cross between some form of Maithen and Moons, both common West-country names for Ox-eye.

MONROVIA COFFEE *Coffea liberica* (Liberian Coffee) Dalziel.

MONSTER PLANT *Monstera deliciosa*.

MONTBRETIA *Crocosmia x crocosmiiflora*.

MONTEREY CYPRESS *Cupressus macrocarpa*. The reference is to the Monterey Peninsula, southern California.

MONTEREY PINE *Pinus radiata*.

MONTEZUMA CYPRESS 1. *Taxodium distichum* (Swamp Cypress) Duncalf. 2. *Taxodium mucronatum* (Bald Cypress) Hora.

MONTEZUMA PINE *Pinus montezumae*. A Mexican species, as one would guess.

MONTHLY ROSE *Rosa damascena var. semperflorens*.

MONTPELIER BROOM *Cytisus monspessulana*.

MONTPELIER MAPLE *Acer monspessulanum*.

MOOCH; MOOCHERS *Rubus fruticosus* fruit (Blackberry) Glos. Grigson (Mooch); Wilts. Dartnell & Goddard; Glos. J D Robertson (Moochers). To mooch means to play truant, and the names must be comments on children's activities while the blackberries are in season.

MOOGARD *Artemisia vulgaris* (Mugwort) Caith. Grigson. Possibly from Gaelic.

MOON 1. *Chrysanthemum segetum* (Corn Marigold) Dor, Worcs, War. Grigson; Northants. A E Baker. 2. *Leucanthemum vulgare* (Ox-eye Daisy) Wilts. Dartnell & Goddard; Glos. J D Robertson; Berks. Grigson; Worcs. J Salisbury. Cf. Moon Daisy, Moon-flower, Moon Penny, etc., for Ox-eye. But Corn Marigold has no other 'moon' names.

MOON, Lucky *Umbilicus rupestris* (Wall Pennywort) Dor. Dacombe.

MOON, Yellow *Chrysanthemum segetum* (Corn Marigold) Dor, War, Worcs. Grigson.

MOON CARROT *Seseli libanoticum*.

MOON CREEPER *Ipomaea bona-nox*

(Moon-flower Vine) W Miller. Presumably because the white flowers only open late in the evening.

MOON DAISY *Leucanthemum vulgare* (Ox-eye Daisy) Som. Tongue; Wilts. Dartnell & Goddard; Glos. J D Robertson. Worcs. J Salisbury. Cf. Moon, etc.

MOON FLOWER 1. *Anemone nemerosa* (Wood Anemone) Worcs. Grigson. 2. *Leucanthemum vulgare* (Ox-eye Daisy) Som, Worcs. Grigson. 3. *Crataegus monogyna* (Hawthorn) Som. Grigson. 4. *Lunaria annua* (Honesty) Dor. Macmillan. 5. *Stellaria holostea* (Greater Stitchwort) Worcs. Grigson. White flowers, with the exception of Honesty, and in that case it is the seed capsules that earn the name.

MOON-FLOWER VINE *Ipomaea bona-nox*. Cf. Moon-creeper.

MOON PENNY *Leucanthemum vulgare* (Ox-eye Daisy) Ches. Holland; Yks. Addy. Cf. Moon Daisy and Penny Daisy.

MOON PLANT 1. *Asclepias acida* Berdoe. 2. *Sarcostemma acidum*.

MOON-TOBY *Leucanthemum vulgare* (Ox-eye Daisy) Midl. Tynan & Maitland.

MOON TREFOIL *Medicago arborea* (Tree Medick) RHS.

MOON'S EYE *Leucanthemum vulgare* (Ox-eye Daisy) Som. Macmillan.

MOONLIGHT *Anthriscus sylvestris* (Cow Parsley) Wilts. Mabey.

MOONLIGHT BROOM *Cytisus scoparius var. sulphureus*.

MOONS; MOONS, Silver *Lunaria annua* (Honesty) Tynan & Maitland.

MOONSEED *Menispermum canadense*. A translation of the generic name.

MOONSEED, Canadian; MOONSEED, Texas *Menispermum canadense* (Moonseed) both Le Strange.

MOONSEED SARSAPARILLA *Menispermum canadense* (Moonseed) Le Strange. Sarsaparilla presumably because the root is used for a tonic that restores appetite and energy. It is also called Yellow Sarsaparilla and American Sarsaparilla.

MOONSHINE *Anaphalis margaretacea* (Pearly Immortelle) USA House.

MOONOGS 1. *Empetrum nigrum* (Crowberry) N Ire. Grigson. 2. *Vaccinium vitis-idaea* (Cowberry) Ire. Grigson. Cf. Monnocs-heather, Monox Heather or Monnaghs Heather for the crowberry.

MOONVINE *Ipomaea bona-nox* (Moon-flower Vine) Perry.

MOONWORT 1. *Andromeda polifolia* (Bog Rosemary) Prior. 2. *Botrychium lunaria*. 3. *Leucanthemum vulgare* (Ox-eye Daisy) Glos. Tynan & Maitland. 4. *Lunaria annua* (Honesty) Friend. 5. *Stellaria holostea* (Greater Stitchwort) Yks. Grigson. Honesty and Stitchwort are obvious 'moon' plants, from the colour of the flowers in one case, or the seed capsules in the other. The fern has moon-shaped leaflets, but Moonwort for Bog Rosemary is not at all obvious, and is likely to be a misreading of Moorwort.

MOONWORT, Blue *Soldanella alpina* (Alpine Snowbell) Parkinson.

MOOR 1. *Brassica campestris var. rapa* (Turnip) Halliwell. 2. *Calluna vulgaris* (Heather) Yks. B & H. 3. *Potentilla anserina* - roots (Silverweed) Yks. Fernie. Naming a plant after its habitat is not uncommon, but 'moor' for silverweed roots is something different, in spite of another name for the plant - Moor-grass. Surely calling the roots 'moor' is a direct reversion to the OE moru, more, used for root crops like carrots or parsnips, as is the case in this example for turnips.

MOOR-BERRY *Vaccinium oxycoccus* (Cranberry) B & H.

MOOR-EVERLASTING *Antennaria dioica* (Cat's Foot) N Eng. Grigson. It makes an immortelle - Cf. Life Everlasting and the American Live-for-ever.

MOOR-GLOOM *Drosera rotundifolia* (Sundew) Yks. Grigson.

MOOR-GRASS 1. *Drosera rotundifolia* (Sundew) Cumb, Yks. Grigson. 2. *Potentilla anserina* (Silverweed) Scot. Grigson.

MOOR-GRASS, Purple *Molinia caerulea*.

MOOR MYRTLE *Myrica gale* (Bog Myrtle) Yks. Grigson.

MOOR-WHIN; MOSS-WHIN *Genista anglica* (Needle-whin) both Berw. Grigson.

MOOR-WORT *Andromeda polifolia* (Bog Rosemary) B & H.

MOORHENS *Potentilla anserina* (Silverweed) Tynan & Maitland.

MOOSE ELM *Ulmus fulva* (Slippery Elm) W Miller.

MOOSEBERRY *Viburnum pauciflorum* (Rayless Cranberry-bush) Douglas.

MOOSEFLOWER *Trillium grandiflorum* (Wood Lily) Perry. Because it is found in typical moose country.

MOOSEWOOD 1. *Acer pennsylvanicum* (Snake-bark Maple) USA Zenkert. 2. *Viburnum alnifolium* (Hobble-bush) USA Perry.

MOP *Centaurea scabiosa* (Greater Knapweed) Dev. Macmillan. Descriptive.

MOP-HEADED ACACIA *Robinia pseudacacia var. inermis*.

MOPANE *Copaifera mopane*.

MORANE *Heracleum sphondyllium* (Hogweed) IOM Moore.1924.

MORBEAM *Morus nigra* (Mulberry) Bonser.

MORCROP *Anagallis arvensis* (Scarlet Pimpernel) Halliwell. Cf. Merecrop.

MORE *Daucus carota* (Wild Carrot) Earle. The name actually comprises all taproots, including parsnips, radish, etc. Parsnip is Wealmore in OE, and carrot Feldmore. It is also Mohre in German, and in Russian morkovi. The Welsh moron (plural) means taproots or root vegetables.

MOREL; MOREL, Petty *Solanum nigrum* (Black Nightshade) Gerard. From Italian morello, diminutive of moro, Moor, so called from the black berries. Petty Morel, to distinguish it from the Great Morel, which is Deadly Nightshade.

MOREL, Great *Atropa belladonna* (Deadly Nightshade) Prior.

MORELLO CHERRY *Prunus cerasus*. Derivation as for Morel.

MORETON BAY CHESTNUT *Castanospermum australe*. Moreton Bay is in Queensland, by Brisbane.

MORETON BAY FIG *Ficus macrophylla*.

MORETON BAY PINE *Araucaria cunninghamii* (Hoop Pine) J Smith. Moreton Bay, just north of Brisbane, for this is an Australian species.

MORETTE *Atropa belladonna* (Deadly Nightshade) Putnam.

MORGAN *Anthemis cotula* (Maydweed) Hants. Cope; IOW Long; Suss. Parish. Part of a sequence of names, all derived from OE maegde, and including Maiden. The medieval herbals said the plant was good for hysterical girls.

MORGAN, Dutch *Leucanthemum vulgare* (Ox-eye Daisy) IOW Grigson. Halliwell defined 'morgan' as "tares in corn". But see the previous entry, too.

MORGELINE *Veronica hederifolia* (Ivy-leaved Speedwell) Prior. It is Latin morsus gallinae, i.e. henbit. That sounds a little odd for a speedwell, but Gerard called it Lesser Henbit.

MORINDA SPRUCE *Picea smithiana* (Himalayan Spruce) Mitchell.

MORMON TEA 1. *Ephedra nevadensis* (Nevada Joint-Fir) USA Bean. 2. *Ephedra trifurca* (Long-leaved Joint-Fir) USA Elmore. Cf. Brigham Tea, Brigham Young Tea, Desert Tea, Mexican Tea, Teamster's Tea, etc. All arise from the popularity locally of a tea made by infusing the branches, taken either as a beverage, or as a medicinal tea.

MORMON TEA, Green *Ephedra viridis* (Mountain Joint-Fir) USA Elmore.

MORMON TEA, Torrey *Ephedra torreyana*.

MORNING GLORY 1. *Calystegia sepium* (Great Bindweed) Som. Macmillan. 2. *Convolvulus arvensis* (Corn Bindweed) Som. Macmillan. 3. *Convolvulus major*. 4. *Rivea corymbosa*. A name like this usually refers to the length of time the flowers are open.

MORNING GLORY, Bush *Ipomaea leptophylla*.

MORNING GLORY, False *Heliotropum convolvulaceum*.

MORNING GLORY, Field *Convolvulus arvensis* (Corn Bindweed) USA Elmore.

MORNING GLORY, Ivy-leaf *Ipomaea hederacea*.

MORNING GLORY, Red *Ipomaea coccinea*.

MORNING GLORY, Wild *Calystegia sepium* (Great Bindweed) USA Upham.

MORNING STAR *Stellaria holostea* (Greater Stitchwort) Dev. Macmillan.

MORNING STAR LILY *Lilium concolor*.

MOROCCO, Red *Adonis annua* (Pheasant's Eye) Curtis. "...it is one of those plants which are annually cried about our streets, under the name of Red Morocco".

MOROCCO GUM *Acacia gummifera* (Mogador Acacia) Hora.

MORREL, Petty *Aralia racemosa* (American Spikenard) Howes.

MORTAL *Solanum dulcamara* (Woody Nightshade) Culpepper. Labelled as such by Culpepper, but of course, it is not so lethal as Deadly Nightshade.

MORTHEN *Knautia arvensis* (Field Scabious) Heref. Havergal. A doubtful ascription - he says " a plant, probably the field scabious" - but it is much likelier to be maydweed.

MORTIFICATION ROOT *Althaea officinalis* (Marsh Mallow) Hatfield. The powdered roots made a poultice which would remove obstinate inflammation and prevent "mortification" in external or internal injuries.

MORUB *Polygonum persicaria* (Persicaria) Halliwell. Quoted in Halliwell without comment.

MOSAIC PLANT *Fittonia verschaffeltii* (Silver-nerved Fittonia) USA Cunningham. 'Mosaic', from the pattern on the leaves.

MOSCHATEL *Adoxa moschatellina*. The reference in this name is to the musky smell of the flowers. Moschatel is French moscatelle, Italian moscatella, from moscato, musk.

Curtis referred to this as "tuberous moschatel".

MOSES, Prickly *Acacia verticillata*. 'Mimosa', presumably.

MOSES-AND-THE-BULRUSHES *Tradescantia virginica* (Spiderwort) Harvey.

MOSES-IN-THE-BULRUSHES 1. *Sagittaria sagittifolia* (Water Archer) Som. Macmillan. 2. *Tradescantia virginica* (Spiderwort) Dor. Macmillan.

MOSES-IN-THE-CRADLE *Rhoeo spathacea* (Boat Lily) Goold-Adams. Because of the way the small flowers seem to be cradled in their bracts.

MOSES'S BLANKET *Verbascum thapsus* (Mullein) Som. Macmillan. The texture of the leaves ensures its names of blanket, flannel, etc.

MOSQUITO BUSH; MOSQUITO PLANT *Ocymum micranthum* Barbados Gooding (Bush); Dalziel (Plant). Cf. Duppy Basil, where duppy apparently in this case means mosquito. The essential oil is methyl cinnamate, pleasant-smelling, but repulsive to mosquitoes.

MOSS; MOSS, French; MOSS, Golden *Sedum acre* (Biting Stonecrop) War. Grigson (Moss); Grigson (French Moss); Som, Oxf, War, Yks. Grigson; USA Upham (Golden Moss).

MOSS, Ball *Tillandsia recurvata*.

MOSS, Consumption *Cetraria islandica* (Iceland Moss) USA Cunningham. It is a demulcent and expectorant, used for chest complaints like tuberculosis.

MOSS, Coral *Nertera depressa* (Bead Plant) USA Cunningham.

MOSS, Cup *Ochrolechia tartarea* (Cudbear Lichen) Banff. B & H.

MOSS, Cypress *Lycopodium alpinum* (Alpine Clubmoss) Parkinson. Cf. Dwarf Cypress, or Heath Cypress.

MOSS, Ditch *Elodea canadensis* (Canadian Pondweed) USA Cunningham.

MOSS, Iceland *Cetaria islandica*.

MOSS, Indian *Sedum reflexum* (Yellow

Stonecrop) Donegal Grigson. Cf. Indian Fog, also from Ireland.

MOSS, Link *Sedum reflexum* (Yellow Stonecrop) Shrop. B & H. It begins to make sense when compared first with the Scottish 'Love-links' and then with the Cumbrian 'Love-in-a-chain'. There must have been some marriage divination connected with it.

MOSS, Lung *Lobaria pulmonaria* (Lungwort Crottle) Flück. At one time, it was made into a jelly and given to those suffering from lung trouble. Or it was given as a tisane, with a handful of lichen to a pint of water, boiled, and left to stand. It was taken for bronchial catarrh.

MOSS, Oak *Evernia prunastri* (Stag's Horn Lichen) S M Robertson.

MOSS, Pustulous *Umbilicaria pustulata*.

MOSS, Rose *Portulaca grandiflora* Hay & Synge.

MOSS, Silver *Cerastium tomentosum* (Snow-in-summer) War. B & H.

MOSS, Spanish *Tillandsia usneoides*. It grows on the Gulf coast of North America, so 'Spanish' is not to be taken literally.

MOSS, Stag's-horn *Lycopodium inundatum* (Clubmoss) Som Elworthy.

MOSS, Tree 1. *Lycopodium alpinum* (Alpine Clubmoss) B & H. 2. *Urostachys selago* (Mountain Clubmoss) B & H.

MOSS, Wall *Sedum acre* (Biting Stonecrop) Yks. B & H.

MOSS-BERRY 1. *Vaccinium myrtillus* (Whortleberry) Donegal Grigson. 2. *Vaccinium oxycoccus* (Cranberry) Yks. Grigson. 'Moss' in this case is the same as 'moor'.

MOSS CAMPION *Silene acaulis*.

MOSS-CORNS *Potentilla anserina* (Silverweed) Scot. Jamieson. Cf. the Irish Mash-corns. But more to the point is the Scottish Moss-crop and Moor-grass.

MOSS-CROP 1. *Eriophorum angustifolium* (Cotton-grass) Lancs. Nodal; Yks. Carr. 2. *Pedicularis palustris* (Red Rattle) Ches. Grigson; Yks. Addy. 3. *Potentilla anserina* (Silverweed) Scot. Jamieson.

MOSS FERN *Polypodium vulgare* (Common Polypody) B & H.

MOSS-FLOWER *Pedicularis palustris* (Red Rattle) Ches. Holland.

MOSS LOCUST *Robinia hispida* (Rosa Acacia) Howes.

MOSS-MILLIONS; MOSS-MINGIN *Vaccinium oxycoccus* (Cranberry) Scot. B & H (Millions); Grigson (Mingin).

MOSS PINK *Phlox subulata* (Ground Phlox) USA House.

MOSS ROSE *Rosa centifolia var. muscosa*. The angel who takes care of flowers, and sprinkles upon them the dew in the still night, slumbered on a spring day in the shade of a rose bush, and she woke, said, "Most beautiful of my children, I thank thee for thy refreshing odour and cooling shade: could you now ask me a favour, how willingly would I grant it". "Adorn me then with a new charm", said the spirit of the rose-bush, and the angel adorned it with the simple moss. Thus Dyer. Another legend says that it was the blood of Christ, falling on the moss at the foot of the cross, that gave birth to the moss-rose.

MOSS THISTLE *Cirsium palustre* (Marsh Thistle) Scot. B & H.

MOSS-WHIN *Genista anglica* (Needle Whin) Berw. Grigson.

MOSS WYTHAM *Myrica gale* (Bog Myrtle) Cumb. Grigson.

MOSSCORTS, Swine's *Stachys palustris* (Marsh Woundwort) Scot. Jamieson. Cf. Maskert, or Swine's Maskert (mask-wort?).

MOTH, Ware- *Artemisia absinthium* (Wormwood) Coats. Warmot was the original, a version of wormwood, but when the origin had become confused, it very easily turned into Ware-moth, especially as wormwood is very useful for keeping clothes-moths at bay.

MOTH BEAN *Phaseolus aconitifolius*.

MOTH MULLEIN *Verbascum blattaria*. "Moths and Butter-flies, and all other small flies and bats, do resort to the place where these herbs are layd or strewed" (Gerard).

MOTH ORCHID *Phalaenopsis schilleriana*.
MOTH PLANT *Verbascum thapsus* (Mullein) Som. Macmillan.
MOTHAN *Sagina procumbens* (Pearlwort) Scot. Campbell.1900. Others believe that the magical plant, mothan, is butterwort, *Pinguicula vulgaris*. It protects from fairy changing, from witchcraft or from fire.
MOTHER; MOTHER DAISY *Leucanthemum vulgare* (Ox-eye Daisy) Heref. Grigson (Mother); Som. Macmillan (Mother Daisy). Mather, rather than Mother.
MOTHER-BREAKS-HER-HEART *Veronica chamaedrys* (Germander Speedwell) Corn. Grigson. This and similar names are from the west country, but they mirror a belief that used to be current all over England. For example, from Yorkshire - if a child gathers Germander Speedwell its mother will die during the year. Picking the flowers always brings dire results - see the Lincolnshire superstition that if anyone picks the flower, his eyes will be eaten.
MOTHER-DEE 1. *Melandrium rubrum* (Red Campion) Cumb. Grigson. 2. *Torilis japonica* (Hedge Parsley) Ches. Grigson. i.e. Mother-die. See next entry.
MOTHER-DIE 1. *Achillea millefolium* (Yarrow) PLNN21. 2. *Aethusa cynapium* (Fool's Parsley) Ches. Vickery.1985. 3. *Anthriscus sylvestris* (Cow Parsley) Vickery.1984. 4. *Capsella bursa-pastoris* (Shepherd's Purse) Perth. Vickery. 5. *Crataegus monogyna* (Hawthorn) Lancs. Vickery.1985. 6. *Melandrium album* (White Campion) Cumb. Radford. 7. *Melandrium rubrum* (Red Campion) Cumb. Grigson. Cow Parsley is an unlucky plant to bring indoors. More immediately, there is a Sussex superstition: never give cow parsley to your mother, or she will die. Possibly, any white-flowered umbellifer is unlucky, simply because it resembles hemlock. Fool's Parsley probably gets the name in error for cow parsley. There is the same belief regarding Shepherd's Purse, with the addition of a children's game played with the seed pod. Children hold it out to their companions, inviting them to "take a hand o' that". It immediately cracks, and the first child shouts "You've broken your mother's heart" (Cf. the names Mother's Heart and Pick-your-mother's-heart-out). Hawthorn and Yarrow are two more of the white flowers unlucky to bring indoors. Exactly the same belief applies to the campions - if you pick them, the result will be the death of one of your parents - father if it is Red Campion, mother if white. All these probably arose by parental proscription, to keep children away from something harmful - again, hemlock is the probable original.
MOTHER FERN *Asplenium bulbiferum* (Hen-and-chickens Fern) USA Cunningham. For the same reason as the commoner name - see Hen-and-chickens.
MOTHER-IN-LAW PLANT *Dieffenbachia sp* Rochford. Dieffenbachias are very poisonous; biting any part of it will prevent speech for some days. Hence this wishful name.
MOTHER-IN-LAW'S TONGUE *Sanseviera trifasciata*. Possibly because of the length of the tongue-like leaves.
MOTHER MARY'S MILK *Polygala vulgaris* (Milkwort) Som. Macmillan. Polygala means 'much milk'. The roots secrete a milky fluid, so the herb became recognized as a specific for increasing the milk of nursing mothers.
MOTHER-OF-MILLIONS 1. *Cymbalaria muralis* (Ivy-leaved Toadflax) Dev. Friend.1882; Som. Macmillan. 2. *Geranium molle* (Dove's-foot Cranesbill) Yks. Grigson. The name is a reflection on their reproductive capabilities.
MOTHER-OF-THE-EVENING *Hesperis matronalis* (Sweet Rocket) Howes.
MOTHER-OF-THOUSANDS 1. *Achillea millefolium* (Yarrow) Dor. Macmillan. 2. *Corydalis lutea* (Yellow Corydalis) Dev. Macmillan. 3. *Cymbalaria muralis* (Ivy-leaved Toadflax) Hulme (inc. Wilts. Dartnell & Goddard). 4. *Helxine soleirolii* (Mind-your-own-business) McClintock. 5.

Malcomia maritima (Virginian Stock) Wilts, Dor. Macmillan. 6. *Saxifraga sarmentosa*. 7. *Saxifraga x urbinum* (London Pride) Wilts. Macmillan. 8. *Tolmiea menziesii* Gannon. A more modest version of the previous entry. But for yarrow the meaning is different. It is an echo of *millefolium*, thousand-leaf.

MOTHER-OF-THYME; MOTHER-THYME 1. *Acinos arvense* (Basil Thyme) Som. B & H. 2. *Thymus drucei* (Wild Thyme) Gerard; Som. Grigson. (Motherthyme). It means 'womb-thyme', from the supposed effect of the plant on the womb.

MOTHER-OF-WHEAT *Veronica hederifolia* (Ivy-leaved Speedwell) Border B & H. Perhaps because it seems to precede wheat in disturbed ground.

MOTHER SHIMBLE'S SNICK-NEEDLES *Stellaria holostea* (Greater Stitchwort) Wilts. Dartnell & Goddard. 'Shimble' is probably 'thimble'. Mother's Thimble is another name for the same plant.

MOTHER SPLEENWORT *Asplenium bulbiferum* (Hen-and-chicken. Fern) Bonar. Both names arise from the fact that as the fronds mature they develop bulbils in large quantities.

MOTHER-THREAD-MY-NEEDLE *Geranium robertianum* (Herb Robert) Som. Macmillan. There are a number of 'needle' names for this, all from the shape of the seed pods.

MOTHER-WILL-DIE *Crataegus monogyna* (Hawthorn) Hants PLNN21. See Mother-die.

MOTHER'S HEART *Capsella bursa-pastoris* (Shepherd's Purse) Glos. J D Robertson; Border Dyer. See Mother-die.

MOTHER'S NIGHTCAP *Calystegia sepium* (Great Bindweed) Dev. Macmillan.

MOTHER'S THIMBLE *Stellaria holostea* (Greater Stitchwort) Wilts. Grigson.

MOTHERWOOD *Artemisia abrotanum* (Southernwood) Lincs. Fernie. Probably Southernwood.

MOTHERWORT 1. *Artemisia vulgaris* (Mugwort) Prior. 2. *Atriplex patula* (Orache) Som. Macmillan. 3. *Hesperis matronalis* (Sweet Rocket) Tynan & Maitland. 4. *Leonurus cardiacus*. 5. *Lysimachia nummularia* (Creeping Jenny) Dev. Friend.1882. 6. *Thymus vulgaris* (Wild Thyme) Tynan & Maitland. For Creeping Jenny, this is probably a corruption of Moneywort. *Leonurus cardiacus* was always regarded as a womb plant, "used for them that are in hard travell with childe" (Gerard). It is still in use for all female complaints, and Mugwort was also so used. The specific name *matronalis* would suggest the name for Sweet Rocket.

MOTHERWORT, Man's *Ricinus communis* (Castor Oil Plant) Halliwell.

MOTHERWORT, Stinking *Chenopodium vulvaria* (Stinking Orach) Gerard.

MOTHWEED *Gnaphalium orientale* Gerard. "The branches and leaves laid amongst clothes keepeth them from moths...".

MOTHWORT 1. *Gnaphalium orientale* Gerard. 2. *Helichrysum stoechas* (Golden Cassidony) B & H.

MOTHWORT, Golden *Gnaphalium orientale*.

MOUCHERS; MOUCHERS, Blackberry; MOUCHERS, Penny *Rubus fruticosus* fruit (Blackberry) all Wilts. Dartnell & Goddard (Mouchers); Jones & Dillon (the others). See Moochers.

MOULDS *Glechoma hederacea* (Ground Ivy). Rut. Grigson.

MOUNT ATLAS DAISY *Anacyclus depressus*.

MOUNT COPER *Dactylorchis incarnata* (Early Marsh Orchid) B & H.

MOUNT ETNA BROOM *Genista aetnensis*.

MOUNT MORGAN WATTLE *Acacia podalyrifolia*.

MOUNT MORRISON SPRUCE *Picea morrisonicola*. Mount Morrison, in what used to be called Formosa.

MOUNT WELLINGTON PEPPERMINT *Eucalyptus coccifera*.

MOUNTAIN ASH 1. *Populus tremula* (Aspen) Inv. B & H. 2. *Sorbus aucuparia*.

MOUNTAIN ASTER *Aster acuminatus*.

MOUNTAIN AVENS *Dryas octopetala*.
MOUNTAIN AVENS, Purple *Geum rivale* (Water Avens) Camden/Gibson.
MOUNTAIN BALM *Calamintha ascendens* (Calamint) Webster. Cf. Mountain Mint.
MOUNTAIN BAMBOO *Arundinaria alpina*.
MOUNTAIN BRAMBLE *Rubus chamaemorus* (Cloudberry) B & H.
MOUNTAIN CALAMINT *Calamintha ascendens* (Calamint) Gerard.
MOUNTAIN CAMPION *Silene acaulis* (Moss Campion) Camden/Gibson.
MOUNTAIN CEDAR *Libocedrus bidwillii*.
MOUNTAIN COTTONWOOD *Populus angustifolia* (Narrowleaf Cottonwood) Robbins.
MOUNTAIN CURRANT *Ribes alpinum*.
MOUNTAIN DOGWOOD *Cornus nuttalli*.
MOUNTAIN ELM *Ulmus glabra* (Wych Elm) Prior. Presumably because the older specific name was 'montana'.
MOUNTAIN FLAX 1. *Centaureum erythraea* (Centaury) Cumb. B & H. 2. *Linum catharticum* (Fairy Flax) Shrop, Ches, Yks, Cumb. Grigson. 3. *Phormium colensoi*. 4. *Polygala senega* (Seneca Snakeroot) USA House. 5. *Spergula arvensis* (Corn Spurrey) Dor, Shrop, Yks. Grigson.
MOUNTAIN FLOWER *Geranium sylvaticum* (Wood Cranesbill).
MOUNTAIN GROUNDSEL *Senecio sylvaticum* (Heath Groundsel) Murdoch McNeill.
MOUNTAIN LILY *Lilium martagon* (Turk's Cap Lily) Gerard.
MOUNTAIN MAHOGANY 1. *Betula lenta* (Cherry Birch) Grieve.1931. 2. *Cercocarpus montanus*. Cherry Birch gets the name in deference to its much sought-after wood. Cf. Mahogany Birch.
MOUNTAIN MINT *Calamintha ascendens* (Calamint) Culpepper.
MOUNTAIN MIST *Calluna vulgaris* (Heather) Som. Grigson.
MOUNTAIN PINE *Pinus mugo*.
MOUNTAIN PINK 1. *Centaureum beyrichii* (Rock Centaury) USA Kingsbury. 2. *Centaureum calycosum* (Buckley Centaury) USA Kingsbury.
MOUNTAIN SNOW *Euphorbia marginata*.
MOUNTAIN THYME *Thymus serpyllum* (Breckland Wild Thyme) Webster.
MOUNTAIN WHITE PINE *Pinus monticola*.
MOUNTAIN YEW *Juniperus communis* (Juniper) Campbell. Gaelic Iubha-beinne.
MOUNTAIN, Joy-of-the- *Origanum vulgare* (Marjoram) Som. Macmillan. Given as a local Somerset name, but it can't be, for Joy-of-the-mountain is what Origanum means.
MOUNTAIN, Milk *Linum catharticum* (Fairy Flax). Mill-mountain is what is meant.
MOUNTAIN, Mill- 1. *Calamintha ascendens* (Calamint) Grieve.1931. 2. *Linum catharticum* (Fairy Flax) Coles. Prior said it was Latin chamaelinum montanum, ground flax.
MOUNTAIN, Pell-a-; MOUNTAIN, Penny; MOUNTAIN, Puliall *Thymus vulgaris* (Wild Thyme) Gerard (Pell and Puliall); B & H (Penny). Either corruptions of serphyllum montanum, or pulegium montanum, both old book names for thyme, though the latter sounds likelier to refer to pennyroyal. On the other hand, Gerard did have Puliall-mountain.
MOURNFUL BELLS OF SODOM *Fritillaria meleagris* (Snake's-head Lily) Dor. Dacombe. The drooping flowers and their sombre colour are the reason for this and a number of other similar names, such as Solemn Bells of Sodom and Doleful Bells of Sorrow.
MOURNFUL WIDOW 1. *Geranium phaeum* (Dusky Cranesbill) Yks, Lancs. Grigson. 2. *Scabiosa atropurpurea* (Sweet Scabious) Wright. Because of the mourning colours of the flowers.
MOURNING BRIDE 1. *Knautia arvensis* (Field Scabious) Wilts. Macmillan. 2. *Scabiosa atropurpurea* (Sweet Scabious) Som. Tongue. Field Scabious gets the name by association only.
MOURNING CYPRESS *Cupressus funebris* (White Fir) Everett. Cf. Funeral Cypress, and Weeping Cypress.

MOURNING IRIS *Iris susiana*. The flowers are grey with purple-violet and black veins.
MOURNING WIDOW 1. *Geranium phaeum* (Dusky Cranesbill) Coats. 2. *Knautia arvensis* (Field Scabious) Dev. Friend.1882. 3. *Scabiosa atropurpurea* (Sweet Scabious) Dor. Macmillan. See Mournful Widow and Mourning Bride.
MOUSE BARLEY *Hordeum murinum* (Wall Barley) A S Palmer.
MOUSE EAR 1. *Cerastium holostoides*. 2. *Hieraceum pilosella* (Mouse-ear Hawkweed) Turner. 3. *Stachys lanata* (Lamb's Ears) Friend.1882.
MOUSE EAR, Alpine *Cerastium alpinum*.
MOUSE EAR, Arctic *Cerastium edmonstonii*.
MOUSE EAR, Field *Cerastium arvense*.
MOUSE EAR, Golden *Hieraceum aurantiacum* (Orange Hawkweed) Gerard.
MOUSE EAR, Grey, Dark *Cerastium atrovirens*.
MOUSE EAR, Little 1. *Antennaria dioica* (Cat's Foot) Turner. 2. *Cerastium semidecandrum*.
MOUSE EAR, Narrow-leaved *Cerastium triviale*.
MOUSE EAR, Podded *Arabidopsis thaliana* (Thale Cress) Curtis.
MOUSE EAR, Starwort *Cerastium cerastoides*.
MOUSE EAR, Sticky *Cerastium glomeratum*.
MOUSE EAR, Yellow *Hieraceum pilosellum* (Mouse-ear Hawkweed) Turner.
MOUSE EAR CHICKWEED, Broad-leaved *Cerastium glomeratum* (Sticky Mouse Ear) Curtis.
MOUSE EAR CHICKWEED, Clustered *Cerastium glomeratum* (Sticky Mouse Ear) Salisbury).
MOUSE EAR CHICKWEED, Common *Cerastium holostoides* (Mouse Ear) Curtis).
MOUSE EAR CHICKWEED, Least *Cerastium semidecandrum* (Little Mouse Ear) Curtis).
MOUSE EAR CHICKWEED, Marsh *Myosoton aquaticum* (Water Chickweed) Curtis.
MOUSE EAR CHICKWEED, Wayside *Cerastium triviale* (Narrow-leaved Mouse-ear) J A MacCulloch.1905.
MOUSE EAR CRESS *Arabidopsis thaliana* (Thale Cress) USA Zenkert.
MOUSE EAR EVERLASTING *Gnaphalium polycephalum*.
MOUSE EAR HAWKWEED *Hieraceum pilosellum*.
MOUSE EAR SCORPION GRASS *Myosotis arvensis* (Field Forget-me-not) Browning. Mouse-ear is a translation of Myosotis, Greek muos otis, from the small oval leaves. Scorpion-grass was apparently suggested by the curve of the raceme.
MOUSE-EYE *Bernardia myricaefolia*.
MOUSE MILK *Euphorbia helioscopia* (Sun Spurge) Yks. B & H. The milky juice accounts for all the 'milk' names - Cf. Cat's Milk, Mare's Milk, etc.
MOUSE PEA 1. *Lathyrus montanus* (Bittervetch) Berw. B & H. 2. *Lathyrus pratensis* (Yellow Pea) Donegal Grigson. 3. *Vicia hirsuta* (Hairy Tare) Donegal Grigson.
MOUSE PLANT *Arisarum proboscideum*.
MOUSE'S MOUTH; MICE'S MOUTH *Linaria vulgaris* (Toadflax) Wilts. Grigson (Mouse); D & G (Mice).
MOUSE'S PEASE *Vicia cracca* (Tufted Vetch) N'thum, Mor. Grigson. Another Scottish name for this is Cat's Peas, and there is a Gaelic name, peasair radan, rat's peas.
MOUSETAIL 1. *Myosurus minimus*. 2. *Sedum acre* (Biting Stonecrop) Turner.
MOUTH, Adder's 1. *Iris foetidissima* (Foetid Iris) Som. Macmillan; Suss. Grigson. 2. *Orchis mascula* (Early Purple Orchis) Som. Macmillan.
MOUTH, Bunny 1. *Antirrhinum majus* (Snapdragon) Suss. B & H. 2. *Misotapes orontium* (Lesser Snapdragon).
MOUTH, Bunny Rabbit's 1. *Antirrhinum majus* (Snapdragon) Som. Macmillan. 2. *Cymbalaria muralis* (Ivy-leaved Toadflax)

Som. Macmillan. 3. *Digitalis purpurea* (Foxglove) Som. Macmillan.

MOUTH, Cow's *Primula veris* (Cowslip) Loth. Grigson.

MOUTH, Dog *Antirrhinum majus* (Snapdragon) Lincs. B & H.

MOUTH, Dog's 1. *Antirrhinum majus* (Snapdragon) Som. Macmillan. 2. *Linaria vulgaris* (Toadflax) Som, Wilts. Rogers.

MOUTH, Dragon's 1. *Antirrhinum majus* (Snapdragon) Som. Macmillan; Lincs. B & H. 2. *Arethusa bulbosa*. 3. *Digitalis purpurea* (Foxglove) Suss. B & H. See also Lion's and Tiger's Mouth for Foxglove.

MOUTH, Duck's *Digitalis purpurea* (Foxglove) Som. Macmillan. Descriptive, likening the flower to a mouth. There are also Lion's and Tiger's, even Dragon's mouth.

MOUTH, Fox's *Aconitum napellus* (Monkshood) Som. Macmillan.

MOUTH, Frog's 1. *Antirrhinum majus* (Snapdragon) Dor. Macmillan. 2. *Mimulus luteus* Scot. Grigson. 3. *Orchis mascula* (Early Purple Orchis) Som. Macmillan.

MOUTH, Gap- 1. *Antirrhinum majus* (Snapdragon) Som. Macmillan. 2. *Digitalis purpurea* (Foxglove) Som. Macmillan. 3. *Linaria vulgaris* (Toadflax) Som. Macmillan. 4. *Mimulus guttatus* (Monkey Flower) Som. Macmillan.

MOUTH, Horse's *Antirrhinum majus* (Snapdragon) Som. Macmillan.

MOUTH, Lion's 1. *Antirrhinum majus* (Snapdragon) Suss. B & H. 2. *Digitalis purpurea* (Foxglove) Suss. Grigson; USA. Henkel. 3. *Glechoma hederacea* (Ground Ivy) Suss. Parish. 4. *Linaria vulgaris* (Toadflax) Fernie. 5. *Misopates orontium* (Lesser Snapdragon). See also Duck's Mouth for Foxglove.

MOUTH, Monkey 1. *Antirrhinum majus* (Snapdragon) Som. Macmillan. 2. *Cymbalaria muralis* (Ivy-leaved Toadflax) Som. Macmillan.

MOUTH, Mouse's; MOUTH, Mice's *Linaria vulgaris* (Toadflax) Wilts. Grigson (Mouse); D & G (Mice).

MOUTH, Nanny Goat's *Cymbalaria muralis* (Ivy-leaved Toadflax) Dev. Macmillan.

MOUTH, Open *Antirrhinum majus* (Snapdragon) Som. Macmillan.

MOUTH, Pig's 1. *Antirrhinum majus* (Snapdragon) Som. Macmillan 1. *Linaria vulgaris* (Toadflax) Som. Macmillan.

MOUTH, Puppy Dog's *Linaria vulgaris* (Toadflax) Wilts. Wright.

MOUTH, Rabbit's 1. *Antirrhinum majus* (Snapdragon) B & H. 2. *Cymbalaria muralis* (Ivy-leaved Toadflax) Dev, Som. Macmillan. 3. *Glechoma hederacea* (Ground Ivy) Dev. Macmillan. 4. *Linaria vulgaris* (Toadflax) Som. Macmillan.

MOUTH, Rat's 1. *Glechoma hederacea* (Ground Ivy) Dev. Macmillan. 2. *Lamium album* (White Deadnettle) Dev. Macmillan. 3. *Lamium purpureum* (Red Deadnettle) Dev. Macmillan.

MOUTH, Tiger's 1. *Antirrhinum majus* (Snapdragon) Suss, Suff. B & H. 2. *Digitalis purpurea* (Foxglove) Suss. Grigson.

MOUTH, Toad's 1. *Antirrhinum majus* (Snapdragon) B & H. 2. *Fritillaria meleagris* (Snake's Head Lily) Wilts. Dartnell & Goddard.1899. 3. *Misopates orontium* (Lesser Snapdragon).

MOUTH, Yap *Antirrhinum majus* (Snapdragon) Som. Macmillan.

MOUTHROOT *Coptis trifolia* (Goldthread) USA Cunningham. Most 'mouth' names derive from the shape of the flowers, but this is different. Both the Indians and the early settlers used the root as a remedy for sore and ulcerated mouths, and for mouth cankers. Cf. the name Canker Root for this plant.

MOWING DAISY *Leucanthemum vulgare* (Ox-eye Daisy) Som. Macmillan. It must be a corruption of Moon Daisy.

MOXA WEED *Artemisia moxa*. Moxa is well-known for its use in Chinese medicine, though the name itself is from a Japanese word. The effect of acupuncture is increased when a cylinder, or little pyramid made of the dried leaves, is put on the appropriate spot on the body, and set on fire.

MOXIE PLUM *Gaultheria hispidula*.

MOZAMBIQUE EBONY *Dalbergia melanoxylon* (African Blackwood) Howes.

MU OIL TREE *Aleurites montana*.

MUCK, Blue *Endymion nonscriptus* (Bluebell) IOM Moore.1924.

MUCKHILL WEED *Chenopodium album* (Fat Hen) War. Grigson. Because this is its favourite habitat, though perhaps such a name merely denotes rich, fat land. Cf. Muckweed, Dungweed, Dirtweed, and a series of names like Dirty Dick.

MUCKIES *Rosa canina* fruit (Hips) Inv. PLNN21. Probably 'buckies'.

MUCKWEED *Chenopodium album* (Fat Hen) Glos, Yks. Grigson; Suff. Moor; Norf. Halliwell; N'thum. Brockett. See Muckhillweed.

MUD CROWFOOT *Ranunculus tripartitus*.

MUDWEED *Apium inundatum* (Marshwort) B & H.

MUDWORT *Limosella sp*. Accurate enough - they grow in mud.

MUGGER *Artemisia vulgaris* (Mugwort) Scot. Grigson. One of the Scottish variations of mugwort. See also Muggert and Muggert Kale.

MUGGERT 1. *Artemisia vulgaris* (Mugwort) Scot. Grigson. 2. *Senecio jacobaea* (Ragwort) Cumb. Grigson.

MUGGERT KALE *Artemisia vulgaris* (Mugwort) Scot. Grigson.

MUGGET 1. *Convallaria maialis* (Lily-of-the-valley) Prior. 2. *Viburnum opulus* (Guelder Rose) Dev, Som. Macmillan. Not at all the same word as Muggert, which comes from OE meaning a gnat. Mugget is a French word, muguet, which is applied there to both the scent and the flower, and seems to mean a dandy.

MUGGET, Golden *Galium cruciatum* (Cross-wort) Gerard.

MUGGET, Petty *Galium verum* (Lady's Bedstraw) Gerard. French petit muguet.

MUGGET ROSE *Viburnum opulus* (Guelder Rose) Dev, Som. Macmillan.

MUGGLES *Cannabis sativa* (Hemp) USA Watt.

MUGGONS; MUGGINS; MUGGURTH; MUGGWITH *Artemisia vulgaris* (Mugwort) Scot. Grigson (Muggons); Scot. Beith (Muggins); Ire. Davidson (Muggurth); Ire. Grigson (Muggwith).

MUGO PINE *Pinus mugo*.

MUGS-WITHOUT-HANDLES *Campanula medium* (Canterbury Bells) Som. Macmillan. Cf. the various 'Cup-and-saucer' names for this.

MUGUET, Golden *Galium cruciatum* (Cross-wort) Lyte.

MUGWEED 1. *Artemisia vulgaris* (Mugwort) Ches. Grigson. 2. *Asperula odoratum* (Woodruff) A S Palmer. 3. *Galium cruciatum* (Cross-wort) Camden/Gibson. A variation on Mugwort for the first entry, but the other two owe the name to the French muguet.

MUGWET *Asperula odorata* (Woodruff) B & H. French muguet.

MUGWOOD *Artemisia vulgaris* (Mugwort) Shrop, Yks. Grigson; Dur. Dinsdale.

MUGWORT 1. *Artemisia absinthium* (Wormwood) Grigson. 2. *Artemisia dracunculoides* (Russian Tarragon) Canada. Jenness.1935. 3. *Artemisia vulgaris*. 4. *Galium cruciatum* (Cross-wort) Clapham. OE muogwyrt, from a Germanic base meaning a fly or gnat. Midge is the same word. Of course, mugwort keeps the flies away. You can either wear a sprig, or keep an infusion to sponge over the face and arms.

MUGWORT, Breckland *Artemisia campestris*.

MUGWORT, Californian *Artemisia vulgaris var. heterophylla*.

MUGWORT, Chinese *Artemisia verlotorum*.

MUGWORT, Dragon's *Artemisia dracunculus* (Tarragon) Sanecki.

MUGWORT, Fine-leaved *Artemisia abrotanum* (Southernwood) Camden/Gibson.

MUGWORT, Mexican *Artemisia mexicana*.
MUGWORT, Sea *Artemisia maritima* (Sea Wormwood) Turner.
MUGWORT, Verlot's *Artemisia verlotorum* (Chinese Mugwort) Genders.
MUGWORT, Western *Artemisia ludoviciana* (Lobed Cudweed) USA Elmore.
MUGWORT, White 1. *Artemisia gnaphalodes*. 2. *Artemisia lactiflora* (White Wormwood) Webster. 3. *Artemisia multellina*.
MUHLY, Mountain *Muhlenbergia montana*.
MUHLY, New Mexican *Muhlenbergia pauciflora*.
MUHLY, Ring *Muhlenbergia torreyi*.
MUHLY, Sand-hill *Muhlenbergia pungens* (Purple Hair-grass) USA T H Kearney.
MUHLY, Spike *Muhlenbergia wrightii*.
MUKUMARI *Cordia abyssinica*.
MULBERRY 1. *Cuscuta epithymum* (Common Dodder) Som. Grigson. 2. *Morus nigra*. 3. *Rubus caesius* (Dewberry) Wilts. D Grose. 4. *Rubus fruticosus* fruit (Blackberry) Suff, Norf. G E Evans.1969. 5. *Sorbus aria* (Whitebeam) Aber. Grigson. 6. *Vaccinium myrtillus* (Whortleberry) Donegal Maire Mac Neill. Those that are not true mulberries presumably get the name by reference to their colour, though it is difficult to fit dodder into that pattern, although it is not at all a bad description. Actually, all fruits like the blackberry (raspberry,strawberry etc) were known in East Anglia as mulberries.
MULBERRY, American *Morus rubra* (Red Mulberry) Hora.
MULBERRY, Black *Morus nigra* (Mulberry) Hora.
MULBERRY, Dyer's *Chlorophora tinctoria* (Fustic) Barton & Castle. Sometimes known by the botanical name *Morus tinctoria*, hence Dyer's Mulberry.
MULBERRY, English *Morus nigra* (Mulberry) Hora.
MULBERRY, Fig *Ficus sycamorus* (Sycamore Fig) Moldenke. Cf. Mulberry Fig. It is the fig with leaves like those of mulberry.
MULBERRY, French *Callicarpa americana*. Not French, and not a mulberry either. This is an American plant of the Verbena family.
MULBERRY, Ground *Rubus chamaemorus* (Cloudberry) Halliwell. 'Chamaemorus' means dwarf mulberry.
MULBERRY, Indian *Moringa citrifolia*.
MULBERRY, Mexican *Morus microphylla*.
MULBERRY, Paper *Broussonetia papyrifera*. The bark can be stripped off in sheets, and tapa, or bark cloth can be made from this.
MULBERRY, Persian *Morus nigra* (Mulberry) Hora.
MULBERRY, Red *Morus rubra*.
MULBERRY, Russian *Morus alba var. tatarica*.
MULBERRY, Texas *Morus microphylla*.
MULBERRY, West African *Chlorophora excelsa* (African Oak) Dalziel.
MULBERRY, White *Morus alba*.
MULBERRY, Yellow, Giant *Myrianthus holstii*. 'Giant', presumably, because of the size of the mulberry-like fruits - they are about 36 mm diameter.
MULBERRY FIG *Ficus sycamorus* (Sycamore Fig) Turner.
MULBERRY WILLOW *Salix cramacile*.
MULE-FAT *Baccharis viminea*.
MULGA *Acacia aneura*.
MULL *Trifolium repens* (White Clover) Corn. Grigson.
MULLEIN, Black *Verbascum nigricum*. Mullein is perhaps from French mol, soft, a reference to the woolly leaves. The Anglo-French version was moleine.
MULLEIN, Blanket *Verbascum thapsus* (Great Mullein) Ches. Grigson. The leaves again, on both counts. Cf. the various 'flannel' names with at least half a dozen 'blanket' ones.
MULLEIN, Broussa *Verbascum broussa*. Named from Bursa, or Brusa, in Asia Minor, whence it was introduced in 1930.
MULLEIN, Flannel *Verbascum thapsus* (Mullein) N'thants. Clare.
MULLEIN, Great; MULLEIN, Broad-

leaved, Great *Verbascum thapsus*. Great Mullein is the usual common name; Great Broad-leaved Mullein is in Thornton.
MULLEIN, Hoary *Verbascum pulverulenta*.
MULLEIN, Large-flowered *Verbascum virgatum*.
MULLEIN, Little *Eremocarpus setigerus* (Turkey Mullein) Powers.
MULLEIN, Moth *Verbascum blattaria*. Gerard said, "Moths and Butter-flies, and all other small flies and bats,do resort to the place where these herbs are layd or strewed".
MULLEIN, Orange *Verbascum phlomoides*.
MULLEIN, Petty *Primula veris* (Cowslip) Gerard. Gerard reckoned all the primulas "among the kindes of Mulleines".
MULLEIN, Purple *Verbascum phoenicium*.
MULLEIN, Turkey *Eremocarpus setigerus*.
MULLEIN, Twiggy *Verbascum virgatum* (Large-flowered Mullein) Clapham.
MULLEIN, Velvet *Verbascum thapsus* (Great Mullein) USA House.
MULLEIN, White 1. *Verbascum lychnitis*. 2. *Verbascum thapsus* (Great Mullein) B & H.
MULLEIN DOCK *Verbascum thapsus* (Great Mullein) Norf. B & H; USA Henkel. Any large-leaved plant is liable to be given the name 'dock'.
MULLEIN-LEAVED SAGE *Salvia verbascifolia*.
MULLEIN PINK 1. *Agrostemma githago* (Corn Cockle) S Africa Watt. 2. *Lychnis coronaria* (Rose Campion) USA Zenkert.
MULLEN *Verbascum thapsus* (Great Mullein) Lupton.
MULTIFLORA BEAN *Phaseolus coccineus* (Scarlet Runner) USA Brouk. An alternative botanical name is *Phaseolus multiflorus*.
MUMBLE, Cow 1. *Anthriscus sylvestris* (Cow Parsley) Cambs, Ess, Norf. B & H. 2. *Chaerophyllum temulentum* (Rough Chervil) Suff. Grigson. 3. *Heracleum sphondyllium* (Hogweed) E Ang. Nall. See also Cow Bumble.
MUNG BEAN *Phaseolus aureus* (Green Gram) Anderson. Mung is the Hindi name.
MUNSHOOK *Vaccinium vitis-idaea* (Cowberry) Scot. B & H. From a Gaelic word meaning either mountain-, or moss-, berry.
MURFEYS *Solanum tuberosum* (Potato) Worcs. J Salisbury An odd rendering of Murphy.
MURG *Anthemis cotula* (Maydweed) B & H. This must be related to Morgan, recorded in the south of England, and Marg, from Hampshire.
MURRAIN-GRASS *Scrophularia nodosa* (Figwort) B & H.
MURRAY PINE; MURRAY RIVER PINE *Callitris sp*. *C columellaris* is the species usually called Murray River Pine.
MURRAY RED GUM *Eucalyptus camaldulensis* (River Red Gum) RHS.
MURREN 1. *Bryonia dioica* (White Bryony) Hants, Norf, Yks. Grigson. 2. *Stellaria media* (Chickweed) Yks. Grigson. Murren for chickweed is from French mouron. Chickweed is mouron des oiseaux. But for bryony the word is quite different. Usually spelt 'murrain', it means pestilence, especially a disease of cattle, and is from O French morine, with the same meaning.
MURREN-BERRIES; MMURRAIN-BERRIES 1. *Bryonia dioica* (White Bryony) IOW B & H. 2. *Tamus communis* (Black Bryony) IOW Long.
MURRILL, Swine's 1. *Scilla verna* (Spring Squill) Shet. Grigson. 2. *Stachys palustris* (Marsh Woundwort) Shet. Grigson. Murrill means roots, so the name suggests that pigs like to search for the bulbs in the case of the squill, and the rhizome in the case of the woundwort.
MURUN *Stellaria media* (Chickweed) B & H.
MUSCLE TREE *Carpinus caroliniana* (Blue Beech) Wit.
MUSCOVY *Erodium moschatum* (Musk Storksbill) Dev. Grigson. Cf. Covey, and Sweet Covey, derivatives from Muscovy.
MUSH *Rubus fruticosus* fruit (Blackberry) Dev. Grigson.
MUSHQUASH ROOT 1. *Cicuta maculata*

(American Cowbane) USA Kingsbury.1967.
2. *Conium maculatum* (Hemlock) USA
Skinner. Cf. Beaver Poison, also from
America.
MUSHROOM, Early *Petasites hybridus*
(Butterbur) Dor. Macmillan.
MUSINE *Croton megalocarpus*.
MUSK 1. *Erodium moschatum* (Musk
Storksbill) Prior. 2. *Mimulus moschatus*.
MUSK, Monkey 1. *Antirrhinum majus*
(Snapdragon) Dev, Som. Macmillan. 2.
Mimulus guttatus (Monkey Flower) Dev.
Friend.1882; Wilts. Dartnell & Goddard. The
name is explained for *Mimulus guttatus* by
saying that, though scentless, they are like
musk, so 'monkey' means 'mock'.
MUSK, Water *Mimulus guttatus* (Monkey
Flower) Gilmour & Walters.
MUSK, Wild *Erodium cicutarium*
(Storksbill) Beds. B & H.
MUSK, Yellow *Mimulus luteus* (Mask-
flower) RHS.
MUSK BUTTON *Aster tripolium* (Sea Aster)
Glos. Grigson.
MUSK CLOVER *Erodium moschatum*
(Musk Storksbill) USA Grigson. It sounds
suspiciously like 'Muscovy', already record-
ed for this.
MUSK CROWFLOWER *Adoxa moschatel-
lina* (Moschatel) Tynan & Maitland.
**MUSK GRAPE FLOWER; MUSK
HYACINTH** *Muscari botryoides* (Grape
Hyacinth) Parkinson (Grape Flower);
McDonald (Hyacinth).
MUSK LIME 1. *Citrus microcarpa*. 2. *Citrus
mitis* (Calamondin) Hora.
MUSK MALLOW 1. *Hibiscus abelmoschus*
(Muskseed) Dalziel. 2. *Malva moschata*.
MUSK MILLION *Cucumis melo* (Melon)
Tusser.
MUSK PLANT *Malva moschata* USA
House.
MUSK MELON *Cucumis melo* (Melon)
Parkinson.
MUSK ORCHIS *Herminium monorchis*.
MUSK-ROOT *Ferula sumbul* Mitton.

MUSK ROSE 1. *Rosa arvensis* (Field Rose)
Genders.1976. 2. *Rosa moschata*.
MUSK STORKSBILL *Erodium moschatum*.
MUSK THISTLE *Carduus nutans*.
MUSK WILLOW *Salix aegyptiaca* (Calaf of
Persia's Willow) RHS.
MUSK WOOD CROWFOOT *Adoxa
moschatellina* (Moschatel) B & H.
MUSKED CRANESBILL *Erodium moscha-
tum* (Musk Storksbill) Camden/Gibson.
**MUSKRAT WEED; MUSQUASH ROOT;
MOCK-EEL ROOT** *Cicuta maculata*
(American Cowbane) USA Kingsbury.1967.
MUSKROOT *Adoxa moschatellina*
(Moschatel) B & H; USA Tolstead.
MUSKSEED *Hibiscus abelmoschus*. A per-
fume called ambrette is made from these
seeds.
MUSKWEED *Adoxa moschatellina*
(Moschatel) Yks. Grigson.
MUSKY SAXIFRAGE, White *Saxifraga
exarata* (Furrowed Saxifrage) Grey-Wilson.
MUSSEL-SHELL PEA *Clitoria ternatea*
(Butterfly Pea) Simmons.
MUSSOORIE-BERRY *Coriaria nepalensis*.
MUSTARD, Abyssinian *Brassica carinata*.
Apparently it is grown there occasionally as a
potherb. Mustard came originally from Latin
mustus, must, because the condiment was
prepared with must.
MUSTARD, Ball *Neslia paniculata*.
MUSTARD, Bastard *Cleome gynandra*.
MUSTARD, Black *Brassica nigra*.
MUSTARD, Boor's *Thlaspi arvense* (Field
Pennycress) Turner. Cf. Churl's Mustard, etc.
MUSTARD, Bowyer's 1. *Lepidium ruderale*
(Narrow-leaved Pepperwort) Prior. 2. *Thlaspi
arvense* (Field Pennycress) B & H.
'Bowyer's' here is a corruption of 'boor's' -
see previous entry.
MUSTARD, Brown 1. *Brassica juncea*
(Indian Mustard) S Africa Watt. 2. *Brassica
nigra* (Black Mustard) USA Henkel.
MUSTARD, Buckler *Biscutella laevigata*.
MUSTARD, Candy *Iberis amara*
(Candytuft) Prior. Candy is a reference to
Crete, from where the plant was brought.

MUSTARD, Chinese 1. *Brassica chinensis*. 2. *Brassica juncea* (Indian Mustard) Clapham.
MUSTARD, Clown's 1. *Armoracia rusticana* (Horse Radish) Parkinson. 2. *Iberis amara* (Candytuft) B & H.
MUSTARD, Common *Sinapis arvensis* (Charlock) Canada Allan.
MUSTARD, Corn *Sinapis arvensis* (Charlock) B & H.
MUSTARD, Dish *Thlaspi arvense* (Field Pennycress) Turner.
MUSTARD, Durham *Sinapis arvensis* (Charlock) Leyel.1937. For Durham mustard always used to be, and probably still is, made from charlock, and the seed was sold under this name.
MUSTARD, Field 1. *Brassica campestris*. 2. *Sinapis arvensis* (Charlock) USA Tolstead.
MUSTARD, French *Erysimum cheiranthoides* (Treacle Mustard) Folkard.
MUSTARD, Garlic *Alliaria petiolata* (Jack-by-the-hedge). Cf. Hedge Garlic and Garlickwort. "Sauce alone [another name for it] hath affinity with Garlicke in name, not because it is like it in forme, but in smell: for if it be bruised or stamped it smelleth altogether like Garlicke..." (Gerard).
MUSTARD, Green *Lepidium latifolium* (Dittander) B & H.
MUSTARD, Hare's-ear *Conringia orientalis* (Hare's-ear Cabbage) USA Allan.
MUSTARD, Hedge 1. *Sisymbrium altissimum* (Tumbling Mustard) USA Gates. 2. *Sisymbrium canescens* (Peppergrass) USA Chamberlin. 3. *Sisymbrium incisum* (Western Tansy Mustard) W Miller. 4. *Sisymbrium officinale*.
MUSTARD, Hedge, Broad-leaved *Sisymbrium irio* (London Rocket) Flower.1859.
MUSTARD, Hedge, Fine-leaved *Descurainia sophia* (Flixweed) Flower.1859.
MUSTARD, Hedge, Stinking *Alliaria petiolata* (Jack-by-the-hedge) Thornton.
MUSTARD, Hedge, Winter *Barbarea vulgaris* (Winter Cress) Flower.1859.

MUSTARD, Hoary *Hirschfeldia incana*.
MUSTARD, Indian *Brassica juncea*. Presumably American Indian, for this is a common plant all over North America.
MUSTARD, Leaf *Brassica juncea* (Indian Mustard) Schery.
MUSTARD, Mithridate 1. *Lepidium campestre* (Pepperwort) W Miller. 2. *Thlaspi arvense* (Field Pennycress) B & H. The 'Mithridate' of the name must mean they are used medicinally, and as both of these are also known as Treacle Mustard, the medicinal use would have been as an antidote to poison (see Treacle).
MUSTARD, Old Man's *Achillea millefolium* (Yarrow) Lincs. Grigson. Cf. Old Man's Pepper. Yarrow was used as a cheap snuff once.
MUSTARD, Poor Man's *Alliaria petiolata* (Jack-by-the-hedge) Lincs. Grigson.
MUSTARD, Red *Brassica nigra* (Black Mustard) USA Henkel.
MUSTARD, Sand *Diplotaxis muralis* (Stinkweed) Hutchinson.
MUSTARD, Sarepta *Brassica juncea* (Indian Mustard) S Africa Watt.
MUSTARD, Spanish *Brassica perviridis*. But the plant comes from eastern Asia.
MUSTARD, Tansy *Descurainia pinnata*.
MUSTARD, Tansy, Northern *Sisymbrium brachycarpum*.
MUSTARD, Tansy, Western *Sisymbrium incisum*. Northern and Western in these two names refer to North America.
MUSTARD, Tower *Turritis glabra*. 'Tower' is descriptive of the tall, straight stem.
MUSTARD, Tower, Smooth *Turritis glabra* (Tower Mustard) Curtis.
MUSTARD, Towers *Turritis glabra* (Tower Mustard) Gerard.
MUSTARD, Treacle 1. *Erysimum cheiranthoides*. 2. *Erysimum repandum* USA Gates. 3. *Lepidium campestre* (Pepperwort) B & H. 4. *Thlaspi arvense* (Field Pennycress) Turner. 'Treacle' is from Greek theriake, from therion, a name given to the viper, and is a reference to the superstition that a viper's

flesh would cure a viper's bite. Philips, *World of words*, defines treacle as a "physical compound made of vipers, and other ingredients", and this was a favourite against all poisons. The word then became applied to any confection or sweet syrup, and finally and solely to the syrup of molasses.

MUSTARD, Treacle, Garlic *Alliaria petiolata* (Jack-by-the-hedge) Barton & Castle.

MUSTARD, Treacle, Swiss *Erysimum helveticum*.

MUSTARD, Tumble 1. *Sisymbrium altissimum* (Tumbling Mustard) W Miller. 2. *Sisymbrium officinale* (Hedge Mustard) USA Upham.

MUSTARD, Tumbling *Sisymbrium altissimum*.

MUSTARD, Wall *Diplotaxis muralis* (Stinkweed) Clapham.

MUSTARD, Wallflower *Erysimum cheiranthoides* (Treacle Mustard) USA Zenkert.

MUSTARD, White *Sinapis alba*.

MUSTARD, Wild 1. *Cleome icosandra* Chopra. 2. *Cleome viscosa* W Africa Dalziel. 3. *Raphanus raphanistrum* (Wild Radish) Border B & H. 4. *Sinapis arvensis* (Charlock) Cumb, Scot. Grigson.

MUSTARD, Wormseed *Erysimum cheiranthoides* (Treacle Mustard) USA Zenkert. It is a vermifuge - Cf. Wormseed and Treacle Wormseed for this. Actually the seeds can be quite dangerous, to animals in particular.

MUSTARD, Yellow 1. *Brassica campestre* (Field Mustard) USA Elmore. 2. *Sinapis alba* (White Mustard) USA Henkel.

MUSTARD-TIPS 1. *Medicago lupulina* (Black Medick) Som. Macmillan. 2. *Trifolium campestre* (Hop Trefoil) Som. Macmillan. In these cases, it is the colour of the flowers that gives the name.

MUSTARD TREE *Salvadora persica* (Toothbrush Tree) Dalziel. Because the berries are slightly aromatic, and pungent, rather like cress.

MUSTARD WEED *Sinapis arvensis* (Charlock) J Smith.

MUTCH, Luckie's; MUTCH, Old Wives' *Aconitum napellus* (Monkshood) Lanark. B & H (Luckie's); Perth. B & H (Old Wives'). Mutch is a Scots word meaning a close-fitting cap. Cf. Granny's Bonnet and similar names from more southern parts.

MUTTER PEA 1. *Lathyrus sativus* (Indian Pea) Forsyth. 2. *Pisum sativum var. arvense* (Field Pea) Howes.

MUTTON CHOPS 1. *Chenopodium album* (Fat Hen) Dor. Grigson. 2. *Galium aparine* (Goose-grass) Som, Dor. Grigson. Sheep apparently love browsing on Goose-grass, so that might account for the name. But not for Fat Hen, which is also called Mutton Tops in the same area - something to do with grease, perhaps? Cf. the Scottish Smear Docken, which certainly suggests grease for an ointment base.

MUTTON DOCK *Chenopodium bonus-henricus* (Good King Henry) Dor. Macmillan.

MUTTON GRASS *Poa fendleriana*.

MUTTON ROSE *Trifolium repens* (White Clover) Corn. PLNN21.

MUTTON TOPS *Chenopodium album* (Fat Hen) Som, Dor. Macmillan. See Mutton Chops.

MY LADY'S EARDROPS *Fuchsia magellanica* (Fuchsia) Som. Macmillan.

MY LADY'S LACE *Anthriscus sylvestris* (Cow Parsley) Dor. Macmillan. Cf. Queen Anne's Lace, Honiton Lace, etc.

MY LADY'S SMOCK *Cardamine pratensis* (Lady's Smock) Dor. Macmillan.

MY LADY'S WASHING BOWL *Saponaria officinalis* (Soapwort) Mabey.1977. A number of names celebrate its lathering qualities - see, for instance, Latherwort, Foam Dock, Ground Soap, Crowsoap, and indeed Soapwort itself.

MYALL-WOOD *Acacia pendula*. From an Australian aborigine word.

MYLE, Grey *Lithospermum officinale* (Gromwell) Turner. Cf. Grey Millet and Stone Millet. There was a Latin name milium solis, but the names are directly from the Old French.

MYLES *Chenopodium album* (Fat Hen)

Berw. Grigson. OE melde - Cf. Meldweed, Melgs and Mails for Fat Hen.

MYLIES, Midden 1. *Chenopodium album* (Fat Hen) N Scot. Grigson. 2. *Chenopodium bonus-henricus* (Good King Henry) Selk. Grigson.

MYPE 1. *Brassica napus* (Rape) Prior. 2. *Pastinaca sativa* (Wild Parsnip) Gerard.

MYROBALAN 1. *Prunus cerasifera* (Cherry-plum) Barber & Phillips. 2. *Terminalia catappa*. Apparently from French myrobolan, from Latin myrobalanum, Greek myrobalanos, "unguent fruit".

MYROBALAN, Bastard *Terminalia bellirica*.

MYROBALAN, Egyptian *Balanites aegyptiaca* the unripe fruit (Jericho Balsam) Dale & Greenway.

MYRRH *Myrrhis odorata* (Sweet Cicely) IOM Paton; Cumb, Aber. Grigson. Eventually from an Arabic word, murr, which means bitter. It should be reserved for the gum exuded from the bark of *Commiphora myrrha*.

MYRRH, African *Commiphora africana*.

MYRRH, British *Myrrhis odorata* (Sweet Cicely) Grieve.1931.

MYRRH TREE *Commiphora myrrha*. In most cases, the myrrh of the Bible is the gum exuded from this. The Hebrew word for bitter is very similar - mar.

MYRT TREE *Myrtus communis* (Myrtle) Turner.

MYRTLE 1. *Myrica gale* (Bog Myrtle) Scot. Grigson. 2. *Myrtus communis*. 3. *Vinca minor* (Lesser Periwinkle) USA Zenkert. Myrtle came from Old French myrtil, which is the diminutive of myrt, from Latin myrtus, and eventually from Greek myrtos.

MYRTLE, Blue *Vinca minor* (Lesser Periwinkle) USA Upham.

MYRTLE, Bog *Myrica gale*.

MYRTLE, Burren *Arctostaphylos uva-ursi* (Bearberry) Ire. Grigson. 'Burren' here is a reference to the Burren limestone, where this grows.

MYRTLE, Devonshire; MYRTLE, Dutch *Myrica gale* (Bog Myrtle) both Som. Grigson.

MYRTLE, Grass *Acorus calamus* (Sweet Flag) B & H.

MYRTLE, Honey, Bracelet *Melaleuca armillaris*.

MYRTLE, Jew's *Ruscus aculeatus* (Butcher's Broom) Folkard. Because, he says, it was sold to Jews for use during the feast of the Tabernacle.

MYRTLE, Moor *Myrica gale* (Bog Myrtle) Yks. Grigson.

MYRTLE, Oregon *Umbellularia californica* (Californian Laurel) Hora.

MYRTLE, Running *Vinca minor* (Lesser Periwinkle) USA Perry.

MYRTLE, Sand *Leiophyllum buxifolium*.

MYRTLE, Shepherd's *Ruscus aculeatus* (Butcher's Broom) Bianchini.

MYRTLE, Sweet *Acorus calamus* (Sweet Flag) Grieve.1931.

MYRTLE, Tasmanian *Nothofagus cunninghamii* (Myrtle Tree) N Davey.

MYRTLE, Tree, Weeping *Eugenia ventenattii*.

MYRTLE, Wax *Myrica pennsylvanica* (Bayberry) USA Zenkert. The berries are boiled, and the scum rising to the surface is skimmed off and moulded into fragrant candles, still made in Cape Cod and Nantucket. The wax is produced in greater quantity than in Bog Myrtle.

MYRTLE, Wild 1. *Myrica gale* (Bog Myrtle) Jennings. 2. *Ruscus aculeatus* (Butcher's Broom) W Miller.

MYRTLE, Yellow *Lysimachia nummularia* (Creeping Jenny) USA Upham.

MYRTLE BEECH *Nothofagus cunninghamii* (Myrtle Tree) Howes.

MYRTLE FLAG; MYRTLE GRASS 1. *Acorus calamus* (Sweet Flag) B & H. 2. *Iris pseudo-acarus* (Yellow Flag) Culpepper.

MYRTLE-LEAVED GRASS PEA *Lathyrus myrtifolius*.

MYRTLE-LEAVED MIMOSA *Acacia myrtifolia*.

MYRTLE SEDGE *Acorus calamus* (Sweet

Flag) B & H. Cf. Sweet Myrtle, for the smell of the plant has been likened to that of myrtle, though it is more usually compared to violets, but even cinnamon has been called to mind.

MYRTLE SPURGE *Euphorbia lathyris* (Caper Spurge) Turner.

MYRTLE TREE *Nothofagus cunninghamii*.

MYSORE THORN *Caesalpina decepetala*.

MYSTERIOUS PLANT *Daphne mezereon* (Lady Laurel) Derb. B & H. 'Mysterious' is a corruption of mezereon.

N

NA-HOW *Conium maculatum* (Hemlock) IOM Moore.1924.
NAGI *Podocarpus nagi*.
NAILS *Bellis perennis* (Daisy) Wilts. Dartnell & Goddard.
NAILWORT 1. *Erophila verna* (Whitlow Grass) Gerard. 2. *Paronychia jamesii*. 3. *Saxifraga tridactylites* (Rue-leaved Saxifrage) Prior. Of Whitlow Grass, Gerard said "as touching the qualitie hereof, we have nothing to set downe; only it hath beene taken to heale the disease of the nailes called a Whitlow, whereof it tooke his name". Note that this Saxifrage is also known sometimes as Whitlow Grass, and Paronychia is Whitlow-wort.
NAIVANSHA THORN *Acacia xanthophloea* (Fever Tree) Dale & Greenway.
NAKED BOYS 1. *Colchicum autumnale* (Meadow Saffron) Som, Wilts. Aubrey; E Ang. Nall. 2. *Crocus nudiflorus* (Autumn Crocus) Ches. Holland. 3. *Ranunculus fluitans* (River Crowfoot) Dor. Macmillan. Names like this usually imply that the flowers appear before the leaves, as is explicit in the specific name of Autumn Crocus. Both this and Meadow Saffron actually do look quite naked as they bloom, but plant names are not sure what their sex is - see Naked Ladies, Maidens, Men, Nannies and Virgins. The flowers of River Crowfoot stand alone, too, to justify the name.
NAKED CORAL TREE *Erythrina corallodendron* (Coral Tree) RHS.
NAKED INDIAN TREE *Bursera simaruba* (Hog Doctor Tree) Prance.
NAKED JACK *Colchicum autumnale* (Meadow Saffron) Som. Macmillan.
NAKED LADIES 1. *Cardamine pratensis* (Lady's Smock) Som. Macmillan. 2. *Colchicum autumnale* (Meadow Saffron) Prior. 3. *Crocus nudiflorus* (Autumn Crocus) McClintock.
NAKED MAIDENS 1. *Colchicum autumnale* (Meadow Saffron) Dor. Dacombe. 2. *Galanthus nivalis* (Snowdrop) Som. Macmillan.
NAKED MEN *Colchicum autumnale* (Meadow Saffron) Dor. Macmillan.
NAKED NANNIES 1. *Colchicum autumnale* (Meadow Saffron) Wilts. Dartnell & Goddard. 2. *Orchis mascula* (Early Purple Orchis) Som. Macmillan.
NAKED VIRGINS *Colchicum autumnale* (Meadow Saffron) Ches. B & H.
NAMAQUA DAISY *Dimorphoteca aurantiaca* (Star of the Veldt) Perry.
NAMAQUALAND DAISY *Venidium fastuosum*.
NAMARA POTATO *Helianthus tuberosus* (Jerusalem Artichoke) S Africa Watt.
NAN WADE *Verbena officinalis* (Vervain) IOM Gill in *Notes and Queries*, 1941. Nan Wade was the Isle of Man's most famous white witch, and vervain was her "trump card".
NANAN WOOD *Lagerstroemia lanceolata*.
NANCE, Sweet *Stellaria holostea* (Greater Stitchwort) Som. Grigson.
NANCY 1. *Anemone nemorosa* (Wood Anemone) Dor. Macmillan. 2. *Narcissus pseudo-narcissus* (Daffodil) Tynan & Maitland. 3. *Stellaria holostea* (Greater Stitchwort) Som. Macmillan.
NANCY, Dothering *Briza media* (Quaking Grass) Cumb. Grigson.1959.
NANCY, Pretty 1. *Saxifraga x urbinum* (London Pride) Som. Macmillan. 2. *Stellaria holostea* (Greater Stitchwort) Som. Macmillan.
NANCY, Stinking 1. *Matricaria recutita* (Scented Mayweed) Leics. PLNN29. 2. *Succisa pratensis* (Devil's-bit) Ches. B & H.
NANCY, Sweet 1. N*arcissus maialis patellaris* (Pheasant's Eye) War. Bloom; Ches. Leigh; Norf. B & H. 2. *Narcissus pseudo-narcissus* (Daffodil) Tynan & Maitland. 3. *Stellaria holostea* (Greater Stitchwort) Som. Macmillan.

NANCY, White; NANCY, Wild *Narcissus maialis patellaris* (Pheasant's Eye) Ches, Staffs. B & H (White); Ches. Holland (Wild).

NANCY-PRETTY 1. *Agrostemma githago* (Corn Cockle) Som. Macmillan. 2. *Malcomia maritima* (Virginian Stock) Som. Elworthy. 3. *Saxifraga x urbinum* (London Pride) Dev, Dor. Macmillan. Ches, Staffs. B & H; Yks. Nicholson; Scot. A S Palmer. 4. *Stellaria holostea* (Greater Stitchwort) Som. Grigson. 5. *Veronica chamaedrys* (Germander Speedwell) Som. Macmillan. Is Nancy-pretty none-so-pretty?

NANCY-NONE-SO-PRETTY *Saxifraga x urbinum* (London Pride) Lincs. B & H. A hybrid, fusing the original Nancy-pretty with its derivative, None-so-pretty.

NANDI COFFEE *Coffea eugenioides*.

NANDI FLAME *Spathodea campanulata* (African Tulip Tree) Kenya Perry.

NANKEEN LILY *Lilium x testaceum*.

NANNIES, Naked 1. *Colchicum autumnale* (Meadow Saffron) Wilts. Dartnell & Goddard. 2. *Orchis mascula* (Early Purple Orchis) Som. Macmillan.

NANNY, Stinking 1. *Anthemis cotula* (Maydweed) Leics. PLNN29. 2. *Matricaria chamomilla* (Scented Mayweed) Leics. PLNN29. 3. *Senecio jacobaea* (Ragwort) Notts. Grigson; Lincs. PLNN 31. 4. *Tripleurospermum maritimum* (Scentless Maydweed) Leics. PLNN29.

NANNY GOAT'S MOUTH *Cymbalaria muralis* (Ivy-leaved Toadflax) Dev. Macmillan.

NANNYBERRY 1. *Orchis mascula* (Early Purple Orchis) Som. Macmillan. 2. *Viburnum lentago* USA Tolstead.

NANPIE *Paeonia mascula* (Peony) Yks. B & H. A reversal of some form like Pie-nanny, also from Yorkshire.

NAP, Red *Papaver rhoeas* (Corn Poppy) Som. Macmillan. Red Cap, perhaps, which is also a Somerset name?

NAP-AT-NOON 1. *Ornithogalum umbellatum* (Star-of-Bethlehem) Howes. 2. *Tragopogon porrifolius* (Salsify) Lancs. Nodal; Yks. Carr. 3. *Tragopogon pratensis* (Goat's Beard) Shrop, Cumb. B & H. Goat's Beard is just as well known under the name of Jack-go-to-bed-at-noon, hence this name and many other similar ones. The flowers, of course, open at sunrise, and close at noon.

NAPA THISTLE *Centaurea melitensis*.

NAPE *Brassica napus* (Rape) Corn. B & H.

NAPIER'S FODDER *Pennisetum purpureum* (Elephant Grass) Dalziel.

NAPLES GARLIC *Allium neapolitanum* (Daffodil Garlic) Howes.

NAPOLEON *Trifolium incarnatum* (Crimson Clover) IOW, Suff. Grigson. Napoleon is a corruption of Trifolium.

NAPPERTY, NAPPLE, NAPPLE-ROOT *Lathyrus montanus* (Bittervetch) N Ire. B & H (Napperty); Scot. B & H (Napple); Scot. Aitken (Napple-root). Cf. Knapperts, Knapperty, etc., even Gnapperts, all from Scotland and Ulster.

NARCISSUS, Bastard *Fritillaria meleagris* (Snake's-head Lily) Hudson. For they have been called Daffodils in some parts, as well as Chequered Daffodils.

NARCISSUS, Bunch-flowered *Narcissus canaliculatus var. orientalis* (Polyanthus Narcissus) McDonald.

NARCISSUS, Chinese *Narcissus canaliculatus var. orientalis* (Polyanthus Narcissus) Jenyns. Not a Chinese plant at all, for it is a native of the Mediterranean area and the Near East. But it was introduced into China some time before the 16th century, probably by way of Persia, and it is the flower particularly connected with the Chinese New Year. Specially dwarfed forms are sold in their thousands, to be grown in bowls filled with pebbles, and forced for the New Year. They are said to confer good fortune for the coming year.

NARCISSUS, Italian *Narcissus canaliculatus* Lindley.

NARCISSUS, Poet's *Narcissus maialis var. patellaris* (Pheasant's Eye) Genders.

NARCISSUS, Polyanthus *Narcissus canaliculatus var. orientalis.*

NARCISSUS, Rush-leaved *Narcissus juncifolius.*

NARCISSUS, Two-flowered *Narcissus x biflorus* (Primrose Peerless) Hulme.

NARCISSUS-FLOWERED ANEMONE *Anemone narcissiflora.*

NARD *Nardostachys jatamansi* (Spikenard) Moldenke. Nard is not just a contraction of Spikenard, which is spica nardi, Nardus being the name of the plant.

NARD, Celtic *Valeriana celtica* RT Gunther.1934. Oil from *Valeriana officinalis* is sometimes substituted for the true spikenard, *Nardostachys jatamansi.*

NARD, Wild *Asarum europaeum* (Asabaracca) Barton & Castle. "Some call it *Nardus sylvestris*" (R T Gunther.1934).

NARD GRASS *Nardus stricta* (Mat Grass) Howes.

NASEBERRY *Manilkara achras* (Sapotilla) Brouk.

NASHAG *Arctostaphylos uva-ursi* (Bearberry) Caith. Grigson.

NASTURTIUM *Tropaeolum majus.* Properly, Nasturtium should be reserved for the genus of that name, the watercress, but in popular use it is applied to *Tropaeolum*, which would otherwise probably be known as Indian Cress. The name is a tribute to the pungency of both plants, for it is from Latin nasus, nose and torquere, to twist.

NASTURTIUM, Flame *Tropaeolum speciosum* (Scotch Flame Flower) Macmillan.

NASTURTIUM, Wild *Rorippa sylvestris* (Creeping Yellow-cress) Hutchinson.

NATAL CHILI *Capsicum annuum* (Chile) S Africa Watt.

NATAL FIG *Ficus natalensis.*

NATAL INDIGO *Indigofera arrecta.* It is a native of southern and eastern Africa.

NATAL MAHOGANY *Kiggelaria dregeana.*

NATAL VINE *Cissus rhombifolia* (Grape Vine Ivy) Nicolaisen.

NATURAL GRASS *Medicago lupulina* (Black Medick) Skye Grigson.

NAUGHTY MAN *Artemisia vulgaris* (Mugwort) L Gordon. 'Naughty Man' in plant names is always a euphemism for the devil. Cf. Old Uncle Harry for Mugwort, a Somerset name.

NAUGHTY MAN'S CHERRIES *Atropa belladonna* (Deadly Nightshade) Som, Bucks. Grigson. If 'Naughty Man' is the devil, then anything at all noxious would be assigned to him.

NAUGHTY MAN'S OATMEAL *Anthriscus sylvestris* (Cow Parsley) War. B & H. There is nothing noxious about Cow Parsley, but this name, and the one following, probably referred originally to hemlock.

NAUGHTY MAN'S PARSLEY *Anthriscus sylvestris* (Cow Parsley) War. Palmer.

NAUGHTY MAN'S PLAYTHING 1. *Capsella bursa-pastoris* (Shepherd's Purse) War. Palmer. 2. *Urtica dioica* (Nettle) Som. Grigson; Suss. Parish.

NAVAJO TEA *Heuchera bracteata* USA Elmore.

NAVE ELM *Ulmus procera* timber (English Elm) Wilkinson.1978.

NAVEL, Lady's; NAVEL OF THE EARTH *Umbilicus rupestris* (Wall Pennywort) Parkinson (Navel of the Earth). Descriptive.

NAVELSEED, Dogwood *Omphalodes cappadocica.*

NAVELWORT 1. *Cynoglossum officinale* (Hound's Tongue) Wilts. Macmillan. 2. *Umbilicus rupestris* (Wall Pennywort) Som, Worcs, Leics, Yks. Grigson.

NAVELWORT, Bastard *Cymbalaria muralis* (Ivy-leaved Toadflax) Tradescant.

NAVELWORT, Round-leaved *Cotyledon orbiculata. Cotyledon* will explain Navelwort. It is from Gk cotyle, a cavity or dish, a description of the hollowed leaf common to the genus.

NAVELWORT, Venus's 1. *Omphalodes linifolia.* 2. *Umbilicus rupestris* (Wall Pennywort) B & H. The older botanical name of Wall Pennywort was Cotyledon umbilicus-veneris.

NAVELWORT, Water, Indian *Hydrocotyle*

asiatica (Indian Pennywort) Schauenberg & Paris.

NAVET; NAVET-GENTLE *Brassica napus* (Rape) B & H (Navet); Turner (Navet-gentle). Navet, and navew, come from Latin napus, through French naveau. Navet is still the modern French word for turnip. Cf. Navew.

NAVEW 1. *Brassica campestris var. rapa* (Turnip) Clapham. 2. *Brassica napus* (Rape) B & H. See Navet. Navew is closer to the old French word naveau.

NAVEW, Wild *Bryonia dioica* (White Bryony) Pomet. A tribute to the turnip-size roots that bryony produces. Not for nothing is it called Bigroot in East Anglia. See Navet.

NAVY BEAN *Phaseolus vulgaris* (Kidney Bean) Howes.

NEAPOLITAN MELON *Cucumis melo var. inodorus* (Winter Melon) Bianchini.

NEAPOLITAN STAR OF BETHLEHEM *Ornithogalum nutans* (Drooping Star of Bethlehem) Whittle & Cook. Clusius named it *Ornithogalum neapolitanum* simply because a consignment of bulbs reached him from Naples.

NEATLEGS 1. *Orchis mascula* (Early Purple Orchis) Kent Fernie. 2. *Orchis morio* (Green-winged Orchis) Kent Friend. Cf. Keatlegs, or Skeatlegs, again from Kent.

NEBRASKA FERN *Conium maculatum* (Hemlock) USA Kinsgbury. Cf. California Fern.

NECKLACE, Coral *Illecebrum verticillatum*.

NECKLACE-BEARING POPLAR *Populus monilifera* (Northern Cottonwood) Ablett.

NECKLACE POPLAR *Populus deltoides* (Carolina Poplar) USA Zenkert.

NECKLACE POPLAR, Chinese *Populus lasiocarpa*.

NECKLACE TREE *Ormosia dasycarpa*. Because the hard seeds are used as such.

NECKLACE WEED *Veronica peregrina* (American Speedwell) Howes.

NECKWEED 1. *Cannabis sativa* (Hemp) Som. Elworthy; E Ang. Forby. 2. *Veronica peregrina* (American Speedwell) W Miller. The speedwell was apparently used as a charm round the neck, but hemp got the name for a very different reason. Another name, Gallows-grass, shows why - rope for the gallows was made from hemp.

NECKWORT 1. *Campanula trachelium* (Nettle-leaved Bellflower) Storms. 2. *Narcissus pseudo-narcissus* (Daffodil) Storms. The bellflower perhaps got the name by the virtue that gave Throatwort, Haskwort and Uvula-wort - "of the vertue it hath against the paine and swelling thereof". There is a yellow latex which is got from the plant, regarded as the signature of its value against sore throat and tonsilitis. So, too, with daffodils - their long stems (or necks, as it were) allowed the doctrine of signatures to ensure their use against diseases of the neck. According to the Lacnunga, they were specifically used against erysipelas in that area.

NECTARINE *Prunus persica var. nectarina*.

NEDCUSHION *Anemone nemerosa* (Wood Anemone) Donegal Grigson.

NEDDY GRINNEL *Rosa canina* (Dog Rose) Worcs. J Salisbury.

NEEDLE, Adam's 1. *Scandix pecten-veneris* (Shepherd's Needle) Som. Macmillan; N'thum, Berw. Grigson. 2. *Yucca sp*, and particularly *Yucca filamentosa*. 'Needles' are the leaves in the yuccas, and the seed vessels in the rest of the examples to follow.

NEEDLE, Bagger's *Erodium moschatum* (Musk Storksbill) Worcs. Allies.

NEEDLE, Beggar's 1. *Erodium moschatum* (Musk Storksbill) Worcs. Allies. 2. *Scandix pecten-veneris* (Shepherd's Needle) Som, Midl. Grigson.

NEEDLE, Clock *Scandix pecten-veneris* (Shepherd's Needle) Bucks. Grigson.

NEEDLE, Crake; NEEDLE, Crow *Scandix pecten-veneris* (Shepherd's Needle) N Eng. Grigson (Crake); Som. Grigson Dor. Macmillan; IOW Long; N'thants. A E Baker (Crow).

NEEDLE, Darning, Devil's 1. *Scandix pecten-veneris* (Shepherd's Needle) Som. Macmillan. 2. *Clematis virginiana* Howes.

NEEDLE, Darning, Old Wife's *Scandix pecten-veneris* (Shepherd's Needle) Yks. Grigson.

NEEDLE, Grandmother's *Valeriana officinalis* (Valerian) Bucks. Harman.

NEEDLE, Granny's *Geranium robertianum* (Herb Robert) Som. Grigson.

NEEDLE, Granny-thread-the- 1. *Anemone nemerosa* (Wood Anemone) Som. Grigson. 2. *Geranium robertianum* (Herb Robert) Som. Grigson.

NEEDLE, Ground *Erodium moschatum* (Musk Storksbill) B & H.

NEEDLE, Lady's *Scandix pecten-veneris* (Shepherd's Needle) Tynan & Maitland.

NEEDLE, Mother-thread-my- *Geranium robertianum* (Herb Robert) Som. Grigson.

NEEDLE, Old Woman's *Scandix pecten-veneris* (Shepherd's Needle) Hants. Cope.

NEEDLE, Pick *Erodium moschatum* (Musk Storksbill) Allies.

NEEDLE, Pink 1. *Erodium cicutarium* (Storksbill) Turner. 2. *Erodium moschatum* (Musk Storksbill) B & H. These various 'pick', 'pink', 'poke', 'pook' etc needles all probably mean Puck.

NEEDLE, Poke *Scandix pecten-veneris* (Shepherd's Needle) Som. B & H.

NEEDLE, Pook 1. *Agrostemma githago* (Corn Cockle) Suss. W D Cooper. 2. *Erodium cicutarium* (Storksbill) Bell. 3. *Scandix pecten-veneris* (Shepherd's Needle) Hants. Grigson; Suss. Latham. The needles are the seed vessels in the case of the last two, but probably the long teeth of Cockle's calyx.

NEEDLE, Pound *Scandix pecten-veneris* (Shepherd's Needle). B & H.

NEEDLE, Powk *Erodium cicutarium* (Storksbill) B & H.

NEEDLE, Prick *Agrostemma githago* (Corn Cockle) Suss. Grigson, who reckons that it must be a reference to the long teeth of the calyx.

NEEDLE, Puck 1. *Agrostemma githago* (Corn Cockle) Suss. Grigson. 2. *Scandix pecten-veneris* (Shepherd's Needle) Hants. Cope. See Needle, Pook. In either case, they must be fairy needles, with this name.

NEEDLE, Pucker *Scandix pecten-veneris* (Shepherd's Needle) Tynan & Maitland. Puck Needle, rather.

NEEDLE, Shepherd's 1. *Myrrhis odorata* (Sweet Cicely) Grieve.1931. 2. *Scandix pecten-veneris*.

NEEDLE, Tailor's 1. *Scandix pecten-veneris* (Shepherd's Needle) Corn. Jago. Dev. Grigson. 2. *Spergula arvensis* (Corn Spurrey) Corn. Grigson.

NEEDLE, Venus's *Scandix pecten-veneris* (Shepherd's Needle) B & H.

NEEDLE, Witches' *Scandix pecten-veneris* (Shepherd's Needle) Berw. Grigson.

NEEDLE-AND-THREAD *Stipa comata*.

NEEDLE-AND-THREAD, Adam's *Yucca filamentosa* Som. Macmillan. Because the leaves have thread-like fibres on their margins. The sharp points of the leaves account for the 'needle'.

NEEDLE-CASES *Symphytum officinale* (Comfrey) Dor. Macmillan.

NEEDLE CHERVIL *Scandix pecten-veneris* (Shepherd's Needle) W Miller.

NEEDLE FURZE, NEEDLE GREENWEED, NEEDLE GORSE *Genista anglica* (Needle Whin) Gerard (Furze); Som. Macmillan (Greenweed); Howes (Gorse).

NEEDLE JUNIPER *Juniperus rigida* (Temple Juniper) Hora.

NEEDLE POINTS *Scandix pecten-veneris* (Shepherd's Needle) Ess. Grigson.

NEEDLE WHIN *Genista anglica*. 'Needle' is a reference to the spines.

NEEDLES *Scandix pecten-veneris* (Shepherd's Needle) E Ang. Nall.

NEEDLES, Darning, Grandmother's *Scandix pecten-veneris* (Shepherd's Needle) Tynan & Maitland.

NEEDLES, Duppy *Bidens pilosa* Barbados Gooding. Duppy here seems to mean mosquito, although its usual meaning is a corpse, or ghost.

NEEDLES, Pins-and- 1. *Knautia arvensis* (Field Scabious) Som. Macmillan. 2.

Saxifraga x urbinum (London Pride) Som. Macmillan. 3. *Scandix pecten-veneris* (Shepherd's Needle) Som. Macmillan. 4. *Stellaria holostea* (Greater Stitchwort) Som. Macmillan. 5. *Ulex europaeus* (Furze) Som. Grigson.

NEEDLES, Spanish 1. *Bidens bipinnata*. 2. *Bidens pilosa* Watt.

NEEDLES-AND-PINS 1. *Ulex europaeus* (Furze) Som. Macmillan. 2. *Viola tricolor* (Pansy) Dor. Macmillan.

NEEDLEWEED *Scandix pecten-veneris* (Shepherd's Needle) Halliwell.

NEEDLEWORK, Lady's 1. *Alliaria petiolata* (Jack-by-the-hedge) Som. Macmillan. 2. *Anthriscus sylvestris* (Cow Parsley) Glos. Mabey. 3. *Asperula odorata* (Woodruff) Som. Macmillan. 4. *Centranthus ruber* (Spur Valerian) Corn, Worcs. Grigson. Som. Macmillan. 5. *Conium maculatum* (Hemlock) Som. Macmillan. 6. *Lobularia maritima* (Sweet Alison) Som. Macmillan. 7. *Scabiosa atropurpurea* (Sweet Scabious) Som. Macmillan. 8. *Stellaria holostea* (Greater Stitchwort) Som. Macmillan; Wilts. Tynan & Maitland. 9. *Torilis japonica* (Hedge Parsley) Ches. Holland. The flower heads of the umbellifers always deserve this name, and its other common description - lace. The others on this list get the name for similar reasons - valerian and sweet alison, for instance, remind one of umbellifers, and hence needlework.

NEEDLEWORK, Old Woman's *Centranthus ruber* (Spur Valerian) Som. Macmillan.

NEEDLEWORK, Queen Anne's 1. *Centranthus ruber* (Spur Valerian) Som. Macmillan. 2. *Geranium versicolor* (Painted Lady) N'thants. A E Baker.

NEEDLEWORK, St Anne's *Saxifraga x urbinum* (London Pride) Friend.

NEEM TREE *Melia indica*.

NEEPS, NEAPS *Brassica campestris var. rapa* (Turnip) G M Taylor (Neeps); Corn. Halliwell (Neaps). Usually confined to Scotland, (except for the Cornish spelling), but this is OE naep, from Latin napus.

NEESEWORT 1. *Achillea ptarmica* (Sneezewort) Gerard. 2. *Veratrum album* (White Hellebore) Gerard. 'Neese' means the same as 'sneeze', and probably is an earlier form of it. Sneezewort has the virtue, if that is what it is, of making one sneeze, and the powdered leaves were used at one time for just that purpose, either medicinally or as a cheap snuff substitute. The specific name ptarmica conveys the same message. White Hellebore was used for the same purpose, but in this case it was the roots rather than the leaves that were administered, powdered.

NEESEWORT, White *Veratrum album* (White Hellebore) Turner.

NEESING ROOT *Veratrum album* (White Hellebore) Gerard.

NEGRO COFFEE *Cassia occidentalis* (Coffee Senna) Trinidad Watt. The roasted seeds are a coffee substitute, though there is no caffeine content.

NEGRO YAM *Dioscorea cayennensis* (Yellow Yam) Willis. The roots are black.

NEGRO'S SLIPPERS *Euphorbia myrtifolia* W Miller.

NEGUNDO, Ash-leaved *Acer negundo* (Box Elder) Hibberd. Cf. Ash-leaved Maple for this.

NEIDPATH YEW *Taxus baccata var. erecta* Lowe.

NEIGHBOURHOOD, Good 1. *Centranthus ruber* (Spur Valerian) Som. Elworthy; Wilts. Macmillan; Glos. J D Robertson. 2. *Chenopodium bonus-henricus* (Good King Henry) Wilts. Dartnell & Goddard. There are a lot of 'neighbours' names for the valerian, but this is the only one recorded for Good King Henry.

NEIGHBOURS *Centranthus ruber* (Spur Valerian) Oxf. Oxf.Annl.Rec. 1951.

NEIGHBOURS, Good; NEIGHBOURS, Quiet *Centranthus ruber* (Spur Valerian) Som, Wilts, Glos, Oxf. Grigson (Good); Wilts. Dartnell & Goddard (Quiet).

NELE *Agrostemma githago* (Corn Cockle) B

& H. French nielle, and the same word as Nigella.

NELSON'S BUGLE *Ajuga reptans* (Bugle) Som. Macmillan.

NEMINY *Anemone nemorosa* (Wood Anemone) Lancs. Nodal.

NEMONY, Water *Ranunculus aquatilis* (Water Crowfoot) Som, Wilts. Grigson.

NENUFAR 1. *Nuphar lutea* (Yellow Waterlily) Turner. 2. *Nymphaea alba* (White Waterlily) Turner. Nuphar is Persian nufar, a reduced form of ninufar, which eventually comes from Sanskrit meaning blue lotus.

NENUFAR, Petty *Caltha palustris* (Marsh Marigold) Turner.

NEP 1. *Brassica campestris var. rapa* (Turnip) N'thum. Grose. 2. *Glechoma hederacea* (Ground Ivy) Suss. Grigson. 3. *Lavandula vera* (Dutch Lavender) Ches, Lancs. B & H. 4. *Nepeta cataria* (Catmint) Ess, Suff, Yks, N'thum. Grigson. This can also appear as Nip; presumably both are from Nepeta, the Latin for catmint, so the flowers of lavender and Ground Ivy can easily fit in here. The odd one is turnip, where there is a different derivation - in this case it is from 'napus', the specific name for rape.

NEP, Wild 1. *Bryonia dioica* (White Bryony) Gerard. 2. *Tamus communis* (Black Bryony) Gunther. These must be in the 'napus' sense, and a reference to their very large roots, the reason for their being likened to the mandrake.

NEP-IN-A-HEDGE *Nepeta cataria* (Catmint) Ess, Suff, Yks, N'thum. Grigson.

NEPAL ACONITE 1. *Aconitum ferox* W Miller. 2. *Aconitum laciniatum* (Indian Aconite) Grieve.1931.

NEPAL BARBERRY *Berberis aristata*.

NEPAL LILY *Lilium nepalense*.

NEPE, NEPT *Nepeta cataria* (Catmint) Turner (Nepe); Dawson (Nept). See Nep.

NEPKIN *Prunus persica var. nectarina* (Nectarine) Som. Halliwell.

NEPT, Wild *Bryonia dioica* (White Bryony) Dawson. See Wild Nep.

NERVE PLANT *Fittonia verschaffeltii* (Silver-nerved Fittonia) USA Cunningham.

NERVE-ROOT 1. *Cypripedium acaule* (Stemless Lady's Slipper) Mass. Bergen. 2. *Cypripedium parviflorum var. pubescens* (Larger Yellow. Lady's Slipper) O P Brown. An infusion of the powdered root of these, collected after flowering, is an official drug used as a nervine. For the same reason, they are also known as Valerian, or American Valerian.

NESPITE *Calamintha ascendens* (Calamint) B & H.

NESPOLI *Eriobotrya japonica* (Loquat) Rochford.

NEST, Bee's 1. *Daucus carota* (Wild Carrot) Gerard. 2. *Heracleum sphondyllium* (Hogweed) Som. Macmillan. These 'nest' names are highly descriptive - the umbels of wild carrot close in towards the centre after flowering.

NEST, Bird's 1. *Daucus carota* (Wild Carrot) Gerard. 2. *Listera ovata* (Twayblade) Som. Macmillan.

NEST, Bird's, Yellow *Monotropa hypopitys*. "Birds nest hath many tangling roots platted or crossed one over another very intricately, which resembleth a Crows nest made of sticks" (Gerard).

NEST, Crow's *Daucus carota* (Wild Carrot) Beds. Grigson.

NEST, Goose *Monotropa hypopitys* (Yellow Bird's Nest) Gerard.

NET, Devil's *Cuscuta epithymum* (Common Dodder) Kent Grigson. Descriptive, even to the ascription to the devil.

NET-LEAF, Painted *Fittonia verschaffeltii* (Silver-nerve Fittonia) RHS.

NET-LEAF HACKBERRY *Celtis reticulata* (Western Hackberry) USA T H Kearney.

NET-LEAF OAK *Quercus reticulata*.

NET-LEAVED WILLOW, NETTED WILLOW, NET-VEINED WILLOW *Salix reticulata* Warren-Wren (Netted); RHS (Net-veined).

NETTED CUSTARD APPLE *Annona reticu-*

lata (Custard Apple) Chopra. The specific name, *reticulata*, means netted.

NETTED MELON *Cucumis melo var. scandens.*

NETTLE *Urtica dioica.* OE netele.

NETTLE, Annual *Urtica urens* (Small Nettle) Salisbury.

NETTLE, Bee 1. *Galeobdalon luteum* (Yellow Archangel) Ches, Notts. Grigson. 2. *Galeopsis tetrahit* (Hemp Nettle) B & H. 3. *Galeopsis versicolor* (Large-flowered Hemp Nettle) B & H. 4. *Lamium album* (White Deadnettle) Som, Leics, Notts, Lincs. Grigson; Bucks. Harman. 5. *Lamium purpureum* (Red Deadnettle) Notts, Lincs. Grigson. Is this really 'bee'? Or did it start off as Dea Nettle, i.e. Dead Nettle?

NETTLE, Bee, Red *Lamium purpureum* (Red Deadnettle) War. Grigson.

NETTLE, Bee, White *Lamium album* (White Deadnettle) War. Grigson.

NETTLE, Big *Urtica dioica* (Nettle) USA Yanovsky.

NETTLE, Bigstring *Urtica dioica* (Nettle) Douglas. Presumably a hark back to nettle's textile uses.

NETTLE, Blind 1. *Galeopsis tetrahit* (Hemp Nettle) Corn. Jago; Dev. B & H. 2. *Lamium album* (White Deadnettle) Dev, Som. Macmillan. 3. *Stachys sylvatica* (Hedge Woundwort) B & H. 'Blind', in the same sense as 'dead' - it does not sting.

NETTLE, Bull 1. *Jatropha stimuloca* (Spurge Nettle) Tampion. 2. *Solanum carolinense* (Carolina Nightshade) USA Gates. 3. *Solanum elragnigolium* (Silverleaf Nightshade) USA Stevenson.

NETTLE, Burning 1. *Urtica pilulifera* (Roman Nettle) B & H. 2. *Urtica urens* (Small Nettle) Gerard.

NETTLE, Carolina *Solanum carolinense* (Carolina Nightshade) USA Allan.

NETTLE, Coast *Urtica californica.*

NETTLE, Daa *Lamium purpureum* (Red Deadnettle) Shet. Grigson. Cf. Dee Nettle, from Shetland also; they mean 'deaf'.

NETTLE, Day 1. *Galeopsis tetrahit* (Hemp Nettle) Yks, Border, Aber, Mor. B & H. 2. *Lamium album* (White Deadnettle) Scot. Jamieson. Another in the series meaning either 'dead' or 'deaf' nettle.

NETTLE, Dea 1. *Galeopsis tetrahit* (Hemp Nettle) N'thants. Sternberg; Scot. Jamieson. 2. *Lamium album* (White Deadnettle) Shrop, Yks, Cumb, N'thum. Grigson. 3. *Lamium purpureum* (Red Deadnettle) Worcs, Yks, Cumb. Grigson. 4. *Stachys palustris* (Marsh Woundwort) Cumb. Grigson.

NETTLE, Dead; DEADNETTLE *Lamium sp.* The symbolism is obvious - these 'dead' nettles are the ones that do not sting. "Deadnettle" is preferred to "Dead Nettle".

NETTLE, Dead, Cut-leaved *Lamium hybridum.*

NETTLE, Dead, Hedge *Stachys sylvatica* (Hedge Woundwort) Fernie.

NETTLE, Dead, Hemp-leaved *Galeopsis tetrahit* (Hemp Nettle) Jamieson.

NETTLE, Dead, Henbit *Lamium amplexicaule.*

NETTLE, Dead, Red *Lamium purpureum.*

NETTLE, Dead, Spotted *Lamium maculatum.*

NETTLE, Dead, White *Lamium album.*

NETTLE, Dead, Yellow *Galeobdalon luteum* (Yellow Archangel) Prior.

NETTLE, Deaf 1. *Lamium album* (White Deadnettle) Dev. Friend.1882; Wilts. Dartnell & Goddard; Lincs, Yks. Grigson. 2. *Lamium purpureum* (Red Deadnettle) Dev. Friend.1882; Som, Yks. Grigson.

NETTLE, Dee *Lamium purpureum* (Red Deadnettle) Shet. Grigson. Cf. Daa Nettle.

NETTLE, Defe *Lamium album* (White Deadnettle) Corn. Jago.

NETTLE, Devil's *Achillea millefolium* (Yarrow) Ches. B & H. Cf. Devil's Plaything, Devil's Rattle, etc. They probably arise from disapproval of the plant's use in divinations and spells. But children used to draw the yarrow across their faces just for the pleasure of the tingling sensation it would produce.

NETTLE, Deye 1. *Galeopsis tetrahit* (Hemp

Nettle) Yks. B & H. 2. *Stachys sylvatica* (Hedge Woundwort) Border B & H.

NETTLE, Dog 1. *Galeopsis tetrahit* (Hemp Nettle) Berw. B & H. 2. *Lamium purpureum* (Red Deadnettle) Ches. B & H. 3. *Urtica urens* (Small Nettle) USA Allan.

NETTLE, Dumb 1. *Galeobdalon luteum* (Yellow Archangel) Oxf. Grigson. 2. *Lamium album* (White Deadnettle) Som, Worcs, Ess, Herts. Grigson; Wilts. Dartnell & Goddard; Glos. J D Robertson. 3. *Lamium purpureum* (Red Deadnettle) Som. Macmillan.

NETTLE, Dumb, Double *Ballota nigra* (Black Horehound) Wilts. Dartnell & Goddard.

NETTLE, Dummy *Lamium album* (White Deadnettle) Som. Macmillan; Berks. Lowsley.

NETTLE, Dun *Lamium album* (White Deadnettle) Shrop. Grigson.

NETTLE, Dunch 1. *Lamium album* (White Deadnettle) Som, Dor, Hants. Grigson; Wilts. Jones & Dillon. 2. *Lamium purpureum* (Red Deadnettle) Dor. Grigson; Wilts. Jones & Dillon.

NETTLE, Dunny 1. *Ballota nigra* (Black Horehound) Bucks. Grigson. 2. *Galeobdalon luteum* (Yellow Archangel) Bucks. Grigson. 3. *Lamium album* (White Deadnettle) Oxf. B & H; Berks. Lowsley. 4. *Tussilago farfara* (Coltfoot) Berks. Lowsley; Oxf. B & H. The first three of these are in the "dead" nettle series, but 'dunny' for the coltsfoot means a donkey.

NETTLE, Dunse 1. *Lamium album* (White Deadnettle) Wilts. Dartnell & Goddard. 2. *Lamium purpureum* (Red Deadnettle) Wilts. Dartnell & Goddard.

NETTLE, False 1. *Boehmeria cylindrica*. 2. *Urticastrum divaricatum*.

NETTLE, False, Hawaiian *Boehmeria grandis*.

NETTLE, Female *Urtica dioica* (Nettle) Gerard. To distinguish it from the Male Nettle, *Urtica pilulifera*, in Gerard's vocabulary.

NETTLE, Field *Stachys arvensis* (Field Woundwort) USA Kingsbury.

NETTLE, Flame *Coleus sp* Blackwood.

NETTLE, French *Lamium purpureum* (Red Deadnettle) Shrop. Grigson.

NETTLE, Grass; NETTLE, Grass, Wild *Stachys sylvatica* (Grass Nettle) W Miller; N'thants. A E Baker (Wild).

NETTLE, Great *Urtica dioica* (Nettle) Gerard.

NETTLE, Greek *Urtica pilulifera* (Roman Nettle) Lyte. Given the common name, this is excusable for the classically-minded, but it does not come from Rome at all, but from Romney, in Kent, where it used to grow. One should add that it is a weed from southern Europe.

NETTLE, Harry *Stachys officinalis* (Betony) Dor. Macmillan. "Hairy", perhaps?

NETTLE, Hedge 1. *Galeopsis tetrahit* (Hemp Nettle) Tradescant. 2. *Stachys bullata* USA Schenk & Gifford. 3. *Stachys sylvatica* (Hedge Woundwort) Curtis. 4. *Stachys hyssopifolia* USA Upham. 5. *Stachys tennuifolia* USA Zenkert. 6. *Stachys palustris* (Marsh Woundwort) USA Densmore.

NETTLE, Hedge, Hyssop *Stachys hyssopifolia* Sanecki.

NETTLE, Hedge, Purple *Lamium purpureum* (Red Deadnettle) Barton & Castle.

NETTLE, Hemp *Galeopsis tetrahit*.

NETTLE, Hemp, Downy *Galeopsis dubia*.

NETTLE, Hemp, Large *Galeopsis speciosa*.

NETTLE, Hemp, Large-flowered *Galeopsis versicolor*.

NETTLE, Hemp, Narrow-leaved *Galeopsis angustifolia* (Red Hemp Nettle) Salisbury.

NETTLE, Hemp, Pyrenean *Galeopsis pyrenaica*.

NETTLE, Hemp, Red *Galeopsis angustifolia*.

NETTLE, Horse *Solanum carolinense* (Carolina Nightshade) USA Tolstead.

NETTLE, Horse, White; NETTLE, Horse, Silver *Solanum eleagnifolium* (Silverleaf Nightshade) USA Kingsbury. (White); USA T H Kearney (Silver).

NETTLE, Jaggy; NETTLE, Jenny *Urtica dioica* (Nettle) Edin. Ritchie (Jaggy); IOM S Morrison (Jenny).

NETTLE, Jerusalem *Lamium maculatum* (Spotted Deadnettle) Lincs. Gutch. There is a legend that it became spotted from a drop of the Virgin's milk falling on it.

NETTLE, Male *Urtica pilulifera* (Roman Nettle) Gerard. *Urtica dioica* is the Female Nettle, in Gerard's parlance.

NETTLE, Pill *Urtica pilulifera* (Roman Nettle) Hatfield.

NETTLE, Roman *Urtica pilulifera*. A weed of southern Europe, which used to grow at Romney, Kent. The name, in fact, is derived from Romney, not Roman, in spite of the tradition that it owed its introduction to the Romans, who used it like spinach.

NETTLE, Slender *Urtica gracilis*.

NETTLE, Small *Urtica urens*.

NETTLE, Spurge *Jatropha stimuloca*.

NETTLE, Sting; NETTLE, Stingy *Urtica dioica* (Nettle) Som. Elworthy (Sting); Dev, N'thants. B & H (Stingy).

NETTLE, Sting, White *Lamium album* (White Deadnettle) Dev. Macmillan. In spite of the fact that it does not sting!

NETTLE, Stinging 1. *Galeopsis tetrahit* (Hemp Nettle) Ire. B & H. 2. *Urtica dioica* (Nettle). Of course, Hemp Nettle does not sting.

NETTLE, Stingless *Boehmeria cylindrica* (False Nettle) USA Yarnell.

NETTLE, Tall *Urtica dioica* (Nettle) Watt.

NETTLE, Tanging *Urtica dioica* (Nettle) Yks. Grigson. Tang is an old North Country word for a sting.

NETTLE, Tree *Urtica ferox*. Not quite a 'tree', but at least a shrub about 6 feet tall. The specific name is accurate, for the stings it can give are ferocious enough to be lethal to both man and animal.

NETTLE, Variegated *Lamium maculatum* (Spotted Deadnettle) Ches. Holland.

NETTLE, White *Lamium album* (White Deadnettle) Glos. J D Robertson.

NETTLE, Wild *Urtica pilulifera* (Roman Nettle) Lyte.

NETTLE, Wood *Laportea canadensis*. Designated 'nettle' because of its stinging hairs.

NETTLE FIBRE *Urtica dioica* (Nettle) S Africa Watt. Nettle fibres used to be an important textile at one time.

NETTLE HEMP *Galeopsis tetrahit* (Hemp Nettle) Gerard.

NETTLE-LEAF HYSSOP *Agastache urticifolius*.

NETTLE-LEAVED BELLFLOWER *Campanula trachelium*.

NETTLE-LEAVED FIGWORT *Scrophularia peregrina*.

NETTLE-LEAVED GOOSEFOOT *Chenopodium murale*.

NETTLE TREE 1. *Celtis australis*. 2. *Celtis reticulata* (Western Hackberry) USA Elmore.

NETTLE TREE, African *Celtis integrifolia*.

NETTLE TREE, Caucasian *Celtis caucasica*.

NETTLE TREE, Malay *Laportea stimulans*. Their sting is so bad that walking with bare bodies through these nettle-trees in wet weather can prove fatal.

NETTLEFOOT *Stachys sylvatica* (Hedge Woundwort) Ches. B & H.

NEVADA FLEABANE *Erigeron nevadensis*.

NEVADA JOINT-FIR *Ephedra nevadensis*.

NEVADA NUT PINE *Pinus monophylla* (Piñon) Mason.

NEVADA WILD ALMOND *Prunus andersonii*.

NEVER-DIE 1. *Kalamchoe crenata*. 2. *Kalanchoe pinnata* Dalziel. 3. *Lannea discolor* (Livelong) Palgrave. 4. *Moringa oleifera* (Horse-radish Tree) W Africa Dalziel. 5. *Vangueriopsis lanciflora* (Wild Medlar) Palgrave. Lannea and Vangueriopsis apparently get this name, as well as Livelong, because truncheons stuck into the ground, for fencing, etc., invariably strike and grow.

NEVER-NEVER PLANT *Ctenanthe oppenheimiana*.

NEW CALEDONIAN PINE *Araucaria columnaris*.
NEW CHAPEL FLOWER *Orobanche minor* (Lesser Broomrape) N'thum. Turner. "this herbe is called about Morpeth, in Northumberland, new-chappell floure, because it grew in a chappell there, in a place called Bottel Bankes, where as the unlearned people dyd worshippe the Image of Saynt Mary, and reckoned that the herbe grewe in that place by virtue of that Image".
NEW ENGLAND ASTER *Aster novae-angliae*.
NEW ENGLAND BOXWOOD *Cornus florida* (Flowering Dogwood) Grieve.1931.
NEW GUINEA ASPARAGUS *Setaria palmifolia* (Palm Grass) Rappaport.
NEW GUINEA ROSEWOOD *Pterocarpus indicus* (Amboyna Wood) Howes.
NEW JERSEY TEA *Ceanothus americanus*. During both the War of Independence and the Civil War, the leaves were used as a tea substitute, although they contain no caffeine. This was for a beverage, but of course, the tea had been used medicinally for a long time.
NEW MEXICAN LOCUST *Robinia neomexicana*.
NEW MEXICAN MUHLY *Muhlenbergia pauciflora*.
NEW MEXICO THISTLE *Cirsium neomexicanum*.
NEW MOWN HAY 1. *Asperula odorata* (Woodruff) Som,Notts. Grigson. 2. *Filipendula ulmaria* (Meadowsweet) Som. Macmillan. 3. *Rhinanthus crista-galli* (Yellow Rattle) Som. Macmillan. When dried, Meadowsweet and Woodruff smell strongly of new-mown hay. So, presumably, does Yellow Rattle.
NEW YEAR'S GIFT *Eranthis hyemalis* (Winter Aconite) Ess. Wright; Lincs. Coats. From the time of flowering, of course. Cf. Christmas Flower, or Christmas Rose.
NEW YEAR'S ROSE *Helleborus niger* (Christmas Rose) B & H.
NEW YORK ASTER *Aster novae-belgae*.

NEW ZEALAND BEECH *Nothofagus menziesii*.
NEW ZEALAND BLUEBELL *Wahlenbergia albomarginata*.
NEW ZEALAND CEDAR *Libocedrus bidwilli* (Mountain Cedar) Grieve.1931.
NEW ZEALAND CHRISTMAS TREE *Metrosideros excelsa* (Pohutukawa) Leathart.
NEW ZEALAND FLAX *Phormium tenax*.
NEW ZEALAND HEMP *Phormium tenax* (New Zealand Flax) Schery.
NEW ZEALAND HOLLY *Olearia macrodonta*.
NEW ZEALAND LABURNUM *Sophora tetraptera* (Yellow Kowhai) Barber & Phillips. It is the New Zealand national flower.
NEW ZEALAND LEMONWOOD *Pittosporum eugenioides*.
NEW ZEALAND PINCUSHION *Raoulia sp* Howes.
NEW ZEALAND RED PINE *Dacrydium cupressinum* (Rimu) Duncalf.
NEW ZEALAND SATIN FLOWER *Libertia grandiflora*.
NEW ZEALAND SPINACH *Tetragonia tetragonioides*.
NEW ZEALAND WILLOWHERB 1. *Epilobium nerterioides* (Prostrate Willowherb) McClintock.1975. 2. *Epilobium pedunculare*.
NEWMADE HAY *Asperula odorata* (Woodruff) Wilts. Goddard. New-mown Hay is more widespread as a name. The hay smell that it develops as it dries is due to the coumarin content, useful for anti-coagulant drugs used for heart disease.
NGAIO *Myoporum laetum*.
NGAIO, Tasmanian *Myoporum insulare*.
NICKER, Grey; NICKER, Horse *Caesalpina bonduc* (Brazilian Redwood) both Barbados Gooding.
NICKER BEAN *Entada phaseoloides*.
NICKER BEAN, Yellow *Caesalpina bonduc* (Brazilian Redwood) A W Smith.

NICKER NUT *Caesalpina bonduc* (Brazilian Redwood) Watt.

NIDDLE *Urtica dioica* (Nettle) Dor. Northover.

NIG, NIGGER *Anemone coronaria* (Poppy Anemone) Coats.

NIGELLA, Bastard; NIGELLA, Field; NIGELLA, Wild *Agrostemma githago* (Corn Cockle) Watt (Bastard); Gerard (Field, Wild).

NIGER PLUM *Flacourtia flavescens*.

NIGGER 1. *Anemone coronaria* (Poppy Anemone) Coats. 2. *Plantago lanceolata* (Ribwort Plantain) Som. Macmillan. Cf. Nig for Poppy Anemone. But the name for Ribwort is different - Cf. all the 'black' names for it.

NIGGER, Golden *Helianthus annuus* (Sunflower) Som. Macmillan.

NIGGER-TOE *Bertholletia excelsa* (Brazil Nut) Schery. Probably for the same reason, descriptive, as the Devonshire Shoe Nut.

NIGGER'S HEAD 1. *Plantago lanceolata* (Ribwort Plantain) Dev. Grigson. 2. *Plantago media* (Lamb's-tongue Plantain) Dev. Macmillan.

NIGHT, Lady-of-the- *Mirabilis dichotoma* (Marvel of Peru) Friend. Sometimes the flowers open very late in the day.

NIGHT, Shady *Solanum dulcamara* (Woody Nightshade) Lancs. Grigson. A simple reversal of nightshade.

NIGHT-BONNETS *Bryonia dioica* (White Bryony) Dor. Macmillan.

NIGHT-BONNETS, Granny's *Calystegia sepium* (Great Bindweed) Som. Macmillan. Descriptive.

NIGHT VIOLET *Platanthera chlorantha* (Butterfly Orchid) Wilts. Dartnell & Goddard.

NIGHTCAP 1. *Aquilegia vulgaris* (Columbine) Som. Macmillan; Wilts. Dartnell & Goddard. 2. *Calystegia sepium* (Great Bindweed) Som, Lancs. Grigson. Wilts. Dartnell & Goddard. 3. *Dactylorchis maculata* (Heath Spotted Orchid) Derb. B & H. 4. *Orchis mascula* (Early Purple Orchis) Derb. B & H.

NIGHTCAP, Devil's 1. *Aconitum napellus* (Monkshood) War. F Savage. 2. *Calystegia sepium* (Great Bindweed) Som. Macmillan. 3. *Convolvulus arvensis* (Field Bindweed) N Eng. Wakelin. 4. *Stellaria holostea* (Greater Stitchwort) Som. Macmillan. 5. *Torilis japonica* (Hedge Parsley) War. Palmer. Descriptive, in each case. 'Devil's' is understandable in the case of the bindweeds, and presumably Hedge Parsley gets the name in mistake for hemlock. Greater Stitchwort is generally speaking an unlucky plant, associated with snakes for some reason. Cf. Old Man's Nightcap.

NIGHTCAP, Dolly's *Geranium robertianum* (Herb Robert) Dev. Macmillan.

NIGHTCAP, Grandmother's 1. *Aconitum napellus* (Monkshood) Yks. Grigson. 2. *Calystegia sepium* (Great Bindweed) Dev, Suss. Grigson. 3. *Silene maritima* (Sea Campion) Tynan & Maitland.

NIGHTCAP, Granny's 1. *Aconitum napellus* (Monkshood) Som. Macmillan; Glos. J D Robertson. 2. *Anemone nemorosa* (Wood Anemone) Som, War. Grigson; Wilts. Dartnell & Goddard. 3. *Antirrhinum majus* (Snapdragon) Som. Macmillan. 4. *Aquilegia vulgaris* (Columbine) Dev, Som, Glos. Grigson. Dor. Dacombe; Wilts. Dartnell & Goddard. 5. *Borago officinalis* (Borage) Som. Macmillan. 6. *Calystegia sepium* (Great Bindweed) Som. Macmillan; Wilts. Dartnell & Goddard. 7. *Convolvulus arvensis* (Field Bindweed) Wilts. Dartnell. & Goddard. 8. *Delphinium ajacis* (Larkspur) Som. Macmillan. 9. *Geranium robertianum* (Herb Robert) Dev. Macmillan. 10.*Geum rivale* (Water Avens) Wilts. Dartnell & Goddard. 11.*Hemerocallis flava* (Day Lily) Som. Macmillan. 12.*Melandrium album* (White Campion) Som. Friend. 13.*Solanum dulcamara* (Woody Nightshade) Dev. Macmillan. 14.*Stellaria graminea* (Lesser Stitchwort) Som. Macmillan. 15.*Stellaria* *holostea* (Greater Stitchwort) Dor. Grigson. All descriptive.

NIGHTCAP, Lady's 1. *Anemone nemorosa* (Wood Anemone) Glos. J D Robertson; Heref. B & H. 2. *Calystegia sepium* (Great Bindweed) Som. Grigson. Wilts. Dartnell & Goddard; Hants. Cope. 3. *Campanula medium* (Canterbury Bells) Dev. Friend. 4. *Convolvulus arvensis* (Field Bindweed) Wilts. Friend. 5. *Convolvulus major* (Morning Glory) Wilts. Friend.

NIGHTCAP, Mother's 1. *Aconitum napellus* (Monkshood) Breck. M E Hartland. 2. *Calystegia sepium* (Great Bindweed) Dev. Macmillan.

NIGHTCAP, Old Maid's *Geranium maculatum* (Spotted Cranesbill) Grieve.1931.

NIGHTCAP, Old Man's 1. *Calystegia sepium* (Great Bindweed) Som. Macmillan; Suss. Parish; Surr. Grigson. 2. *Convolvulus arvensis* (Field Bindweed) Som, Wilts. Macmillan. 3. *Convolvulus major* (Morning Glory) Suss. Friend. 'Old Man' is usually the devil. Sure enough, there is Devil's Nightcap too, though not in the same area.

NIGHTCAP, Old Woman's 1. *Aconitum napellus* (Monkshood) Som. Macmillan; Bucks. B & H. 2. *Calystegia sepium* (Great Bindweed) Som. Macmillan. 3. *Campanula medium* (Canterbury Bells) Som. Macmillan.

NIGHTCAP, Our Lady's *Calystegia sepium* (Great Bindweed) Som. Macmillan.

NIGHTCAP, Woman's *Oxalis acetosella* (Wood Sorrel) Som. Macmillan.

NIGHTINGALE 1. *Arum maculatum* (Cuckoo-pint) Ess. B & H. 2. *Geranium robertianum* (Herb Robert) Bucks. Grigson. 3. *Stellaria holostea* (Greater Stitchwort) Wilts. Dartnell & Goddard.

NIGHTINGALE FLOWER *Cardamine pratensis* (Lady's Smock) Wilts. Goddard; Hants. Grigson.

NIGHTSHADE *Datura stramonium* (Thornapple) Barbados Gooding.

NIGHTSHADE, American *Phytolacca decandra* (Poke-root) USA Henkel; S Africa Watt.

NIGHTSHADE, Bindweed *Circaea lutetiana* (Enchanter's Nightshade) Gerard.

NIGHTSHADE, Bitter *Solanum dulcamara* (Woody Nightshade) USA House. 'Bitter', from the 'amara' part of the specific name. Cf. Bitterwseet.

NIGHTSHADE, Black 1. *Solanum nigrum*. 2. *Solanum nodiflorum* S Africa Watt.

NIGHTSHADE, Carolina *Solanum carolinense*.

NIGHTSHADE, Climbing *Solanum dulcamara* (Woody Nightshade) USA House. Although it scrambles rather than climbs.

NIGHTSHADE, Common *Solanum nigrum* (Black Nightshade) USA Kingsbury.

NIGHTSHADE, Cutleaf *Solanum triflorum* (Three-flowered Nightshade) USA Kingsbury.

NIGHTSHADE, Deadly 1. *Atropa belladonna*. 2. *Solanum dulcamara* (Woody Nightshade) Forsyth. 3. *Solanum nigrum* (Black Nightshade) USA House.

NIGHTSHADE, Edible *Solanum tuberosum* (Potato) Young.

NIGHTSHADE, Enchanter's *Circaea lutetiana*. Both the Greeks and Romans assigned certain plants to Circe, and the name was given to this one in the 16th century. It was certainly linked with witchcraft in France and Germany, but there is no evidence of any connection in this country before the publication of the herbals.

NIGHTSHADE, Garden *Solanum nigrum* (Black Nightshade) Gerard.

NIGHTSHADE, Graceful *Solanum gracile*.

NIGHTSHADE, Gray *Solanum incanum* (Hoary Nightshade) Zohary.

NIGHTSHADE, Green *Solanum sarrachoides*.

NIGHTSHADE, Hairy *Solanum villosum*.

NIGHTSHADE, Hoary *Solanum incanum*.

NIGHTSHADE, Malabar *Basella alba* (Indian Spinach) Dalziel.

NIGHTSHADE, Palestine *Solanum incanum* (Hoary Nightshade) Moldenke.

NIGHTSHADE, Prickly *Solanum rostratum.*
NIGHTSHADE, Purple *Solanum dulcamara* (Woody Nightshade) Gilmour & Walters.
NIGHTSHADE, Red *Physalis alkekengi* (Winter Cherry) Langham.
NIGHTSHADE, Red-fruited *Solanum miniatum.*
NIGHTSHADE, Silverleaf *Solanum eleagnifolium.*
NIGHTSHADE, Sleeping *Atropa belladonna* (Deadly Nightshade) Gerard. The plant was once used for sleeping draughts. Cf. Dwale, etc.
NIGHTSHADE, Three-flowered *Solanum triflorum.*
NIGHTSHADE, Three-leaved *Trillium erectum* (Bethroot) Le Strange. Cf. Trinity Flower.
NIGHTSHADE, Tree *Solanum capsicastrum* (Winter Cherry) Parkinson.
NIGHTSHADE, Tuberous *Solanum tuberosum* (Potato) Palaiseul.
NIGHTSHADE, Wood *Solanum dulcamara* (Woody Nightshade) Parkinson.
NIGHTSHADE, Woody 1. *Solanum dulcamara.* 2. *Solanum nigrum* (Black Nightshade) S Africa Watt.
NIGHTSHADE, Woolly *Solanum mauritianum.*
NIGHTSHADE BINDWEED *Circaea lutetiana* (Enchanter's Nightshade) B & H.
NIGHTSHIRT *Calystegia sepium* (Great Bindweed) Som. Macmillan.
NIKKO FIR, NIKKO SILVER FIR *Abies homolepis* Usher (Silver).
NIKKO MAPLE *Acer nikkoense.*
NILE, Lily-of-the- 1. *Agapanthus africanus* (African Lily) Bonar. 2. *Zantedeschia aethiopica* (Arum Lily) Grigson.1973.
NILE TAMARISK *Tamarix nilotica.*
NILGIRI LILY *Lilium neilgherrense.* Nilgiri, southern India.
NIMBLE WILL *Muhlenbergia schreberi.*
NINE HOOKS *Alchemilla vulgaris* (Lady's Mantle) Grigson. The leaves are cut into 7, or more usually 9, toothed lobes.
NINE JOINTS *Polygonum aviculare* (Knotgrass) Coles. The 'joints', 'knots' or 'nodes' are a feature of this plant, and of all the genus. The number claimed is modest, for in Shropshire it was known as Ninety-knot, and Centinode, or Centynody are old book names.
NINE-LEAVED BITTERCRESS *Cardamine enneaphyllos.*
NINEBARK *Physocarpus opulifolius.*
NINETY-KNOT *Polygonum aviculare* (Knotgrass) Shrop. Grigson. See Nine Joints.
NINNYVER *Nymphaea alba* (White Waterlily) Halliwell. Cf. Nenufar.
NIP 1. *Brassica campestris var. rapa* (Turnip) Heref. Lewis. Suff. Halliwell. 2. *Nepeta cataria* (Catmint) Ess, Yks, N'thum. Grigson. E Ang. Forby. 3. *Rosa canina* fruit (Hips) Ches. Holland. The second syllable of turnip, of course, often used by itself in a number of disguises. Cf. Mip, Mit, Neeps, etc. For Catmint it is a variation of the more usual Nep, and presumably Nip for Hip is merely an approximation of the usual form.
NIP, Mad *Bryonia dioica* (White Bryony) Pomet. "the root of this plant is so violent that the Peasants call it the Mad Nip; which, if they happen to eat thro' Inadvertence, it makes them frantick, and sometimes they run the risque of Death itself." But it is also known as Mednip, which is not necessarily the same thing. 'Nip' is 'Nep ', of course, and White Bryony is the Wild Nep.
NIPBONE *Symphytum officinale* (Comfrey) Vesey-Fitzgerald. Knitbone is more usual, for comfrey is a consound, used for helping to knit broken bones.
NIPP *Nepeta cataria* (Catmint) USA Leighton.
NIPPERNAIL *Rosa canina* fruit (Hips) Ches. Holland. Cf. Nips, from the same county.
NIPPERNUT *Lathyrus montanus* (Bittervetch) Som. Macmillan. One of a series, mostly Scottish, that includes Gnapperts, or Knapperts, Napperty, Napple, even Kippernut.

NIPPLE, Virgin Mary's 1. *Euphorbia amygdaloides* (Wood Spurge) Som. Grigson. 2. *Euphorbia helioscopia* (Sun Spurge) Dev. Friend. Because of the milky juice, typical of all the spurges.

NIPPLEWORT 1. *Lapsana communis*. 2. *Umbilicus rupestris* (Wall Pennywort) Suss. Grigson. Descriptive, but according to Ray, it got the name from its efficacy in curing sore nipples.

NISEWORT, Black *Helleborus foetidus* (Stinking Hellebore) Turner.

NISSOL'S VETCHLING *Lathyrus nissolia* (Grass Pea) Flower.1859.

NIT CLICKERS *Calystegia sepium* (Great Bindweed) Som. Macmillan.

NITS *Stellaria holostea* (Greater Stitchwort) Som. Macmillan. Probably in the same class, and for the same reason, as names like Poppers, or Snaps. The seed vessels burst with a pop.

NO-EYE PEA *Cajanus indicus* (Pigeon Pea) W Indies Dalziel.

NOAH'S ARK 1. *Aconitum anglicum* (Wild Monkshood) Glos. B & H. 2. *Aconitum napellus* (Monkshood) Glos. Grigson. 3. *Aquilegia vulgaris* (Columbine) Som. Macmillan. 4. *Cypripedium parviflorum var. pubescens* (Larger Yellow. Lady's Slipper) O P Brown. 5. *Delphinium ajacis* (Larkspur) Dor. Macmillan.

NOBLE FIR *Abies procera*. *A nobilis* is the earlier botanical name.

NOBLE LILY *Lilium nobilissimum*.

NOBODY'S-FLOWER *Calendula officinalis* (Marigold) Wilts. Macmillan. Perhaps because it is so common that nobody bothers to claim ownership.

NOBS, King's *Ranunculus bulbosus* (Bulbous Buttercup) Som. Grieve.1931. The reference is to the unopened flower buds, i.e. knobs.

NODDING BUCKWHEAT *Eriogonum cernuum*.

NODDING CATCHFLY *Silene pendula*.

NODDING LILY 1. *Lilium canadense* (Canada Lily) USA House. 2. *Lilium cernuum*. 3. *Lilium superbum* (Turk's Cap Lily) USA Woodcock.

NODDING PINCUSHION *Leucospermum cordifolium*.

NODDING SAGE *Salvia nutans*.

NODDING SPURGE *Euphorbia maculata* (Spotted Spurge) USA Allan.

NODDING THISTLE *Carduus nutans* (Musk Thistle) Salisbury.

NODDING TRILLIUM *Trillium cernuum*.

NODDING WILD ONION *Allium cernuum*.

NOGS *Cannabis sativa* (Hemp) Shrop. B & H.

NOISETTE ROSE *Rosa moschata x chinensis*.

NONE-SO-PRETTY 1. *Gnaphalium polycephalum* (Mouse-ear Everlasting) Grieve.1931. 2. *Malcomia maritima* (Virginian Stock) Dev. B & H. 3. *Primula vulgaris elatior* (Polyanthus) Som. Tongue. 4. *Saxifraga spathularia x umbrosa* (London Pride) Wright. 5. *Silene armeria* (Catchfly) USA Upham.

NONESUCH; NONESUCH, Black *Medicago lupulina* (Black Medick) Hants. Cope; Suss. Grigson (Nonesuch); Norf. Grigson (Black Nonesuch). Nonesuch means unique.

NONESUCH, White *Lolium perenne* (Ryegrass) Norf. B & H.

NONGO *Albizia zygia*.

NONSUCH *Lychnis chalcedonica* Gerard.

NOON-DAY FLOWER *Tragopogon pratensis* (Goat's Beard) Folkard. Goat's Beard is one of the flowers that open at sunrise, and close at noon.

NOON-FLOWER 1. *Bellis perennis* (Daisy) Scot. Tynan & Maitland. 2. *Tragopogon pratensis* (Goat's Beard) Prior.

NOON PEEPERS *Ornithogalum umbellatum* (Star-of-Bethlehem) Wilts. D Grose. A flower with similar habits to Goat's Beard. Cf. Twelve o'clock, and Jack-go-to-bed-at-noon, etc.

NOONTIDE *Tragopogon pratensis* (Goat's Beard) Gerard. See Noon-day Flower.

NOOPS *Rubus chamaemorus* (Cloudberry)

Berw. Grigson. Sometimes appearing as Knoops, this must be the same as Nub, and it means a hill. So does the 'cloud' of Cloudberry.

NOOTKA CYPRESS *Chamaecyparis nootkatensis* (Alaska Yellow Cedar) Mitchell. Nootka Island is off Vancouver Island, British Columbia.

NOOTKA SOUND ARBOR-VITAE *Thuja plicata* (Giant Cedar) W Miller.

NOP, Horse *Centaurea nigra* (Knapweed) Lancs. Nodal. Cf. Horse-knot, Horse Knob, Horse-knop, etc.

NORFOLK ISLAND HIBISCUS *Lagunaria patersonii*. Not confined to Norfolk Island, but that is where it was first discovered, in 1792.

NORFOLK ISLAND PINE *Araucaria heterophylla*. Norfolk Island is in the Pacific, between Australia and New Zealand.

NORFOLK REED *Phragmites communis* (Reed) J G Jenkins.1976. This is the name by which it is known to thatchers.

NORMANDY CRESS *Barbarea verna* (American Land Cress) H & P.

NORTH, Star-of-the- *Geum urbanum* (Avens) *Notes and Queries* vol 7. 'North' here is a misreading of 'earth'. Star-of-the-earth is a genuine book name for avens.

NORTH AMERICAN EBONY *Diospyros virginiana* (Persimmon) Willis.

NORTH AMERICAN IRONWOOD *Ostrya virginiana* (American Hop-Hornbeam) J Smith. A hardwood.

NORTH AMERICAN PERSIMMON *Diospyros virginiana* (Persimmon) Willis.

NORTHERN BAYBERRY *Myrica pennsylvanica* (Bayberry) Zenkert.

NORTHERN BILBERRY *Vaccinium uliginosum*.

NORTHERN BUGLEWEED *Lycopus virginicus* (Virginia Bugleweed).

NORTHERN CATCHFLY *Silene wahlbergella*.

NORTHERN COTTONWOOD *Populus monolifera*.

NORTHERN DOCK *Rumex longifolius* (Butter Dock) Fitter.

NORTHERN EVENING PRIMROSE *Oenothera muricata*.

NORTHERN GENTIAN *Gentianella aurea*.

NORTHERN GOOSEBERRY *Ribes oxyacanthoides* (Smooth Gooseberry) USA Yarnell.

NORTHERN HAWKSBEARD *Crepis mollis* (Blunt-leaved Hawksbeard) Blamey.

NORTHERN HEMLOCK *Tsuga canadensis* (Eastern Hemlock) S M Robertson.

NORTHERN MARSH ORCHID *Dactylorchis purpurella*.

NORTHERN PITCH PINE *Pinus rigida* (Pitch Pine) Mitchell.

NORTHERN PITCHER PLANT *Sarracenia purpurea* (Pitcher Plant) Youngken.

NORTHERN PRICKLY ASH *Xanthoxylum americanum* (Prickly Ash) W Miller.

NORTHERN RED OAK *Quercus maxima* (Red Oak) Schery.

NORTHERN STITCHWORT *Stellaria borealis*.

NORTHERN TANSY MUSTARD *Sisymbrium brachycarpum*.

NORTHERN VIOLET *Viola selkirkii*.

NORTHERN WATER FORGET-ME-NOT *Myosotis brevifolia*.

NORTHERN WHITE CEDAR *Thuja occidentalis* (Arbor-vitae) A W Smith.

NORTHERN WILLOWHERB *Epilobium adenocaulon* (American Willowherb) USA House.

NORWAY FIR *Pinus sylvestris* (Scots Pine) Wilkinson.1981.

NORWAY MAPLE *Acer platanoides*.

NORWAY PINE 1. *Pinus resinosa* (Red Pine) USA Zenkert. 2. *Pinus sylvestris* (Scots Pine) Murdoch McNeill. 'Norway' in the case of the Red Pine is a reference to the Maine village of that name, but for Scots Pine it is probably an acknowledgement of the country of origin of imports of the timber. Cf. Riga Pine.

NORWAY SPRUCE *Picea abies* (Spruce Fir).
NORWAY SPRUCE-FIR *Picea abies* (Spruce Fir) Thornton.
NORWAY WILLOW, Golden-flowered *Salix chrysanthis*.
NORWEGIAN CINQUEFOIL *Potentilla norvegica*.
NORWEGIAN GREY ALDER *Alnus incana* (Hoary Alder) Brimble.1948.
NORWEGIAN WINTERGREEN *Pyrola norvegica*.
NOSE, Dog's *Antirrhinum majus* (Snapdragon) Som. Macmillan. 'Nose' names for snapdragon are of course descriptive. Tne botanical name means 'like a nose'.
NOSE, Drunkard's *Centranthus ruber* (Spur Valerian) Som. Macmillan. The colour, presumably.
NOSE, Monkey *Antirrhinum majus* (Snapdragon) Dor. Macmillan.
NOSE, Parson's *Orchis morio* (Green-winged Orchid) Dev. Friend.1882.
NOSE, Pig's *Rosa canina* fruit (Hips) Dev. Grigson.
NOSE, Weasel's *Galeobdalon luteum* (Yellow Archangel) Dor. Macmillan. Galeobdala means weasel's snout.
NOSE-SMART, NOSE-TICKLER, NOSE-TWITCHER *Tropaeolum majus* (Nasturtium) Som. (Nose-smart, Nose-tickler); Dor. (Nose-twitcher) both Macmillan. In reference to nasturtium's cress-like qualities - after all, nasturtium is nasus torsus, twisted nose.
NOSEBLEED 1. *Achillea millefolium* (Yarrow) Som, Suss, Suff, N'thum. Grigson; Lincs. Rudkin; USA Henkel. 2. *Conium maculatum* (Hemlock) Som. Macmillan. 3. *Tanacetum parthenium* (Feverfew) Kent B & H. 4. *Trillium erectum* (Bethroot) Le Strange. "The leaves [of yarrow] put in the nose do cause it to bleed, and easeth the pain of the megrim" (Gerard), hence Nosebleed and such names as Sneezewort or Sneezings. The French, too, have saigne-nez for it. Prior claimed that it got this application by mistranslation, the plant actually referred to being the horsetail. Nevertheless, it is firmly fixed in yarrow's folklore. The propensity was used to test a lover's fidelity. In East Anglia, for instance, a girl would tickle the inside of a nostril with a yarrow leaf, saying at the same time:

Yarroway, yarroway, bear a white blow;
If my love love me, my nose will bleed now.

A similar rhyme is used in America. Feverfew gets the name in mistake for yarrow, presumably.
NOSEWORT *Achillea ptarmica* (Sneezewort) T Hill. Another form of Neesewort, itself a variant of Sneezewort.
NOTCHWEED *Chenopodium vulvaria* (Stinking Orach) Coles. 'Notch' is an old word for the female genitalia. Cf. Stinking Motherwort, etc.
NOTTINGHAM CATCHFLY *Silene nutans*. It was common growing on the rocks upon which Nottingham Castle stands, or so it is claimed.
NUB, NUB-BERRY *Rubus chamaemorus* (Cloudberry) Dumf. Grigson. Nub is knub, or knob, in the sense of a hill, in which case nub-berry is the same as cloudberry, for cloud is from OE clud, a rock or hill.
NUBIAN SENNA *Cassia acutifolia* (Alexandrian Senna) Dalziel.
NUFFIN IDOLS *Viola tricolor* (Pansy) Wessex Rogers. An extreme example of a corruption from the series that includes Love-in-idleness.
NUMMAN-IDLES *Viola tricolor* (Pansy) Glos. J D Robertson.
NUMPINOLE *Anagallis arvensis* (Scarlet Pimpernel) Som. Macmillan; Wilts. Dartnell & Goddard. A corruption of Pimpernel.
NUN *Orchis morio* (Green-winged Orchid) Notts. B & H.
NUN, Bleeding *Cyclamen europaeum* (Sowbread) Skinner. In Palestine it is dedicated to the Virgin, because the sword of sor-

row that pierced her heart is symbolised in the blood drop at the heart of the flower. The name Bleeding Nun occurs for the same reason.

NUN OF THE FIELD *Campanula rotundifolia* (Harebell) W Eng. C Chapman.

NUN'S SCOURGE *Amaranthus caudatus* (Love-lies-bleeding) Skinner. Descriptive, since it suggests the flagellation endured by penitents. The same name is given in France.

NURSE-GARDEN *Malus sylvestris* (Crabapple Tree) Halliwell. It is obviously 'nursery-garden', but the connection is not at all clear.

NUT, Acacia *Entada phaseoloides* (Nicker Bean) Hebr. Duncan.

NUT, Australian *Macadamia ternifolia* (Queensland Nut) Schery.

NUT, Bahera *Terminalia bellirica* (Bastard Myrobalan) Chopra.

NUT, Barbados *Jatropha curcas*.

NUT, Bedda *Terminalia bellirica* (Bastard Myrobalan) Chopra.

NUT, Bissy *Cola nitida* (Kola) Jamaica Howes.

NUT, Bladder *Staphylea pinnata*. From the inflated capsules, and hence, by the doctrine of signatures, used for bladder complaints.

NUT, Bog *Menyanthes trifoliata* (Buckbean) Scot. Grigson.

NUT, Brazil *Bertholettia excelsa*.

NUT, Bur *Tribulus terrestris* (Puncture Vine) Moldenke.

NUT, Bush *Macadamia ternifolia* (Queensland Nut) Wilson.

NUT, Cashew *Anacardium occidentale*. Cashew is from Portuguese cajú, itself from the Tupi Indian name, acaju.

NUT, Chaste *Castanea sativa* (Sweet Chestnut) Ire. A S Palmer. 'Chaste' is 'chestnut'. Cf. the early name Chastey.

NUT, Chocolate *Theobroma cacao* (Cocoa) Pomet.

NUT, Chop *Physostigma venenosum* (Calabar Bean) Le Strange.

NUT, Ciper *Bunium bulbocastanum* (Great Earth-nut) Culpepper. Gerard called it, probably more accurately, Kipper-nut.

NUT, Cob *Corylus avellana* nuts (Hazel nut) common.

NUT, Coker *Cocos nucifera* (Coconut) Pomet.

NUT, Crack *Corylus avellana* nuts (Hazel nut) Dev. B & H.

NUT, Cream *Bertholletia excelsa* (Brazil Nut) Schery.

NUT, Deer *Simmondsia chinensis* (Jajoba) USA T H Kearney.

NUT, Earth 1. *Arachis hypogaea* (Ground Nut) USA Elsmore. 2. *Conopodium majus*.

NUT, Earth, Great *Bunium bulbocastanum*.

NUT, Fern *Conopodium majus* (Earth Nut) Corn. Grigson.

NUT, Festike *Pistacia vera* (Pistachio) Turner.

NUT, Fever *Caesalpina bonduc* (Brazilian Redwood) Watt. The powdered seed is given for all kinds of fevers, both in South America and the West Indies, but also in areas like Mozambique where it has been introduced. Even in India, it is still given for malaria.

NUT, Fig *Jatropha curcas* (Barbados Nut) W Africa Dalziel.

NUT, Flea *Senecio jacobaea* (Ragwort) Ches. Grigson. Usually written as one word, when its relationship to its origin becomes more obvious, for this is Fleanit. Cf. also Fleawort and Flea-dod, from the same general area.

NUT, French 1. *Castanea sativa* (Sweet Chestnut) Dor. B & H. 2. *Juglans regia* nut (Walnut) Corn. Courtney.1880. Dev. Earle; Som. Jennings. 'French' probably signifies nothing more then 'foreign'. Neither of these are native British trees, and in English 'nut' could only mean hazel.

NUT, Gaul *Juglans regia* nut (Walnut) Ablett. An odd hybrid. 'Gaul' presumably does not mean French here, but is just a corruption of Walnut. On the other hand, there is a French verb 'gauler' with the specialized meaning of 'to thrash a walnut tree'.

NUT, Goat *Simmondsia chinensis* (Jajoba) USA T H Kearney.
NUT, Gourou *Cola nitida* (Kola) Notes & Queries.1902. Hausa goro, but other West African dialects have gourou.
NUT, Grass *Brodiaea laxa* (Californian Hyacinth) USA Schenk & Gifford.
NUT, Grass, Climbing *Brodiaea volubilis* (Snake Lily) Powers.
NUT, Grass, Long-leaved *Brodiaea congesta*.
NUT, Ground 1. *Apios americana*. 2. *Arachis hypogaea*. 3. *Bunium bulbocastanum* (Great Earth-nut) Culpepper. 4. *Conopodium majus* (Earth-nut) Som, Herts. Grigson.
NUT, Grove *Conopodium majus* (Earth-nut) Corn. Jago.
NUT, Gypsy 1. *Crataegus monogyna* fruit (Haws) Wilts. Goddard. 2. *Rosa canina* fruit (Hips) Wilts. Goddard.
NUT, Hare *Conopodium majus* (Earth-nut) Dor, Lancs, Yks, Ire. Grigson. Probably from some sort of variation of Earth-nut, which can appear in all sorts of guises; Yernut is one of them that sounds not very far off Hare-nut.
NUT, Hawk *Conopodium majus* (Earth-nut) Prior. Possibly from Hog-nut?
NUT, Hedge *Corylus avellana* (Hazel) Gerard.
NUT, Hog *Conopodium majus* (Earth-nut) Dev, Som. Grigson.
NUT, Horn *Trapa natans* (Water Chestnut) Pratt.
NUT, Husked *Castanea sativa* (Sweet Chestnut) Mabey.
NUT, Jesuit's *Trapa natans* (Water Chestnut) Brouk.
NUT, Jove's *Quercus robur* acorn (Acorn) Som. B & H. Quite logical - oak is Zeus's own tree.
NUT, Kaffir *Herpephyllum caffrum*.
NUT, Karoo *Cola nitida* (Kola) Notes & Queries.1902. One of the West African dialectal forms of gourou (see Gourou Nut) is garru, hence presumably this name in English.
NUT, Knipper *Lathyrus montanus* (Bittervetch) B & H. Cf. Nipper-nut and Kippernut. They belong to a series that starts with Gnapperts, or Knapperts, and goes on to Napple.
NUT, Lady 1. *Castanea sativa* (Sweet Chestnut) Som. Macmillan. 2. *Entada phaseoloides* (Nicker Bean) Chopra.
NUT, Levant *Anamirta paniculata*. But the plant comes from India, Sri Lanka and Indonesia.
NUT, Malabar *Adhatoda vasica*.
NUT, Marking *Anacardium occidentale* (Cashew Nut) Codrington. The juice makes an indelible marking ink.
NUT, Mary's *Entada phaseoloides* (Nicker Bean) Hebrides Shaw. The translation of Gaelic Cnò Mhoire. The bean is sometimes found washed up on the Atlantic shore of the islands, where it brings good luck to the finder.
NUT, Meat 1. *Castanea sativa* (Sweet Chestnut) Dev. Friend.1882. 2. *Conopodium majus* (Earth-nut) Som. Grigson.
NUT, Monkey 1. *Arachis hypogaea* (Ground Nut) Grigson.1973. 2. *Poa annua* (Annual Meadow-grass) Wilts. Dartnell & Goddard.
NUT, Nicker *Caesalpina bonduc* (Brazilian Redwood) Watt.
NUT, Oak *Quercus robur* acorn (Acorn) Som. Macmillan.
NUT, Oil, Portia *Thespesia populnea* (Portia Tree) RHS.
NUT, Pará *Bertholletia excelsa* (Brazil Nut) Brouk.
NUT, Paradise *Lecythis usitata*.
NUT, Physic 1. *Jatropha curcas* (Barbados Nut) USA Kingsbury. 2. *Jatropha multifida* (Coral Plant) USA Kingsbury. Aptly named, for they are laxative in the extreme.
NUT, Pig *Jatropha curcas* (Barbados Nut) W Africa Dalziel.
NUT, Poison *Strychnos nux-vomica* (Snakewood) Le Strange.
NUT, Polly *Castanea sativa* (Sweet Chestnut) Som. Macmillan.
NUT, Purge *Jatropha curcas* (Barbados Nut) USA Kingsbury. Cf. Physic Nut.
NUT, Queensland *Macadamia ternifolia*.

NUT, Rough *Castanea sativa* (Sweet Chestnut) Ches. Holland.

NUT, Rush *Cyperus esculentus* (Tigernut) Dalziel.

NUT, St Anthony's 1. *Conopodium majus* (Earth-nut) Som. Macmillan. 2. *Staphylea pinnata* (Bladder Nut) Gerard.

NUT, Sardian *Castanea sativa* (Sweet Chestnut) Mabey.

NUT, Shoe *Bertholletia excelsa* (Brazil Nut) Dev. Friend.1882. The general shape of the nut suggests the name.

NUT, Singhara *Trapa bispinosa*.

NUT, Spanish *Sisyrinchium angustifolium* (Blue-eyed Grass) Parkinson.

NUT, Stock *Corylus avellana* (Hazel) Prior.

NUT, Stover *Castanea sativa* (Sweet Chestnut) Dev. Friend.1882. It must be a very local name, for 'Stover' is Stover Park, the estate of the Duke of Somerset near Newton Abbot. Chestnut is so called because a lot of them were planted there.

NUT, Sweet *Achillea millefolium* (Yarrow) Dor. Macmillan.

NUT, Underground *Conopodium majus* (Earth-nut) Corn. Grigson; Dev. Choape; Som. Macmillan.

NUT, Victor *Corylus avellana* nut (Hazel) Corn. Grigson. Cob-nuts used to be really the name for a game played with hazel nuts, much like conkers. In Yorkshire,the name cob-nuts was reserved for the winner. Hence the Cornish usage with Victor nuts.

NUT, Vomit *Jatropha curcas* (Barbados Nut) Howes.

NUT, Walsh; NUT, Welsh *Juglans regia* nut (Walnut) Gerard, Evelyn respectively. As with walnut itself, these names simply mean foreign, deriving from OE wealh, and they are a recognition that the tree is not a native of this country.

NUT, Water *Caltha palustris* (Marsh Marigold) Culpepper.

NUT, Wood *Corylus avellana* (Hazel) Yks. B & H.

NUT BUSH *Corylus avellana* (Hazel) Border B & H.

NUT GRASS *Cyperus rotundus*.

NUT GRASS, Yellow; NUT SEDGE, Yellow *Cyperus esculentus* (Tigernut) both USA Allan.

NUT HALL, NUT HALSE *Corylus avellana* (Hazel) Corn. Jago Dev. Friend.1882 (Nut Hall); Dev. Choape (Nut Halse). i.e. hall, or haul, hazel that is, nut reversed.

NUT PALM *Corylus avellana* catkins (Hazel catkins) Berw. B & H. Palm is quite common for the catkins, from their use as a Palm Sunday substitute.

NUT PINE *Pinus sabiniana* (Digger Pine) Watt.

NUT PINE, Arizona *Pinus edulis*.

NUT PINE, Desert; NUT PINE, Nevada *Pinus monophylla* (Piñon) Jaeger (Desert); Mason (Nevada).

NUT PINE, Mexican *Pinus cembroides*.

NUT RAGS *Corylus avellana* catkins (Hazel catkins) Ches. B & H. Cf. Rags, from Yorkshire. Both are presumably descriptive.

NUT STOWELL *Corylus avellana* (Hazel) Wilts. Hayward.

NUT TREE *Corylus avellana* (Hazel) Gerard.

NUTGALL OAK *Quercus infectoria* (Aleppo Oak) Lindley.

NUTMEG *Myristica fragrans*. The 'meg' part of the name apparently comes eventually from Latin muscus, musk.

NUTMEG, African *Monodora myristica* (Calabash Nutmeg) Dalziel.

NUTMEG, Calabash *Monodora myristica*.

NUTMEG, California *Torreya californica*. The fruit of this bears some superficial resemblance to a nutmeg.

NUTMEG, Chinese *Torreya grandis*.

NUTMEG, False 1. *Monodora myristica* (Calabash Nutmeg) Dalziel. 2. *Pycnanthus kombo*.

NUTMEG, Japanese *Torreya nucifera* (Kaya) Hora.

NUTMEG, Madagascar *Ravensara aromatica* (Madagascar Clove) Howes.

NUTMEG, West African *Monodora myristica* (Calabash Nutmeg) Howes.

NUTMEG, Wild *Pycnanthus kombo* (False Nutmeg) Dalziel.
NUTMEG, Woolly, Great *Knema hookeriana*.
NUTMEG, Yellow-flowered *Monodora brevipes*.
NUTMEG FLOWER *Nigella sativa* (Fennel-flower) Coats. The seeds smell like nutmeg, but according to the names recorded, they remind people of cumin, coriander and caraway as well.
NUTMEG-SCENTED GERANIUM *Pelargonium x fragrans*.
NUTMUG *Myristica fragrans* (Nutmeg) N Eng. Brockett.
NUTTALL *Corylus avellana* (Hazel) Corn. Jago; Dev. Friend.1882. Or, more accurately, Nut Hall, where haul is hazel.
NUTTALL WILLOW *Salix scouleriana*.
NUTTALL'S DOGWOOD *Cornus nuttallii* (Mountain Dogwood) E Gunther.
NYMPH,Wood *Nones uniflora* USA Yanovsky.

O

OAK 1. *Acer campestre* (Field Maple) Dev. Friend.1882; Som. Macmillan. 2. *Quercus robur*.

OAK, Abraham's *Quercus pseudo-coccifera*. Or the Oak of Mamre, said to be the one under which Abraham entertained the three angels.

OAK, African 1. *Chlorophora excelsa*. 2. *Oldfieldia africana*.

OAK, Aleppo *Quercus infectoria*. A native of western Asia and Cyprus, hence 'Aleppo'.

OAK, Algerian *Quercus canariensis*.

OAK, Armenian *Quercus pontica* (Pontine Oak) RHS. It is from the Caucasus.

OAK, Australian *Eucalyptus regnans* (Australian Ash) Howes.

OAK, Bamboo-leaved *Quercus myrsinaefolia*.

OAK, Bartram's *Quercus x heterophylla*.

OAK, Basket 1. *Quercus michauxii*. 2. *Quercus prinus* (Chestnut Oak) Mitchell.1974. They provided basketry materials for the Indians.

OAK, Bay *Quercus petraea* (Durmast Oak) Berks. B & H.

OAK, Belote *Quercus gramuntia*.

OAK, Black 1. *Quercus californica*. 2. *Quercus robur* (Oak) Hants. B & H. 3. *Quercus velutina*.

OAK, Black, California *Quercus kelloggii*.

OAK, Blackjack 1. *Quercus marilandica*. 2. *Quercus nigra* (Water Oak) USA Vestal & Schultes.

OAK, Blue *Quercus gambelii* (Gambel's Oak) USA Elmore.

OAK, Blue, Mexican *Quercus oblongifolia*.

OAK, Bull *Casuarina rotundifolia* (Beefwood) Howes.

OAK, Bur 1. *Quercus lobata* (Valley Oak) Powers. 2. *Quercus macrocarpa*.

OAK, Bush *Chlorophora excelsa* (African Oak) Dalziel. Not an oak that grows as a bush, but 'bush' in the sense of growing in the country usually known as bush.

OAK, Canyon *Quercus chrysolepis* (Maul Oak) USA Schenk & Gifford.

OAK, Cat *Acer campestre* (Field Maple) Yks. Grigson. In the same area it is called Dog Oak, and in Devon and Somerset these maples are known simply as oaks.

OAK, Caucasian *Quercus macranthera*.

OAK, Ceylon *Schleichera oleosa*.

OAK, Champion *Quercus maxima* (Red Oak) J Smith.

OAK, Chestnut 1. *Quercus montana* USA Zenkert. 2. *Quercus muhlenbergii*. 3. *Quercus prinus*.

OAK, Chestnut, Japanese *Quercus acutissima*.

OAK, Chestnut-leaved *Quercus castaneifolia*.

OAK, Chinquapin *Quercus muhlenbergii* (Chestnut Oak) USA Tolstead.

OAK, Cork *Quercus suber*.

OAK, Cork, Chinese *Quercus variabilis*.

OAK, Cypress *Quercus robur var. fastigiata*.

OAK, Cyprus *Quercus lusitanica*.

OAK, Deer *Quercus sadleriana*.

OAK, Desert 1. *Casuarina decaisneana*. 2. *Quercus wislienzi* (Large Scrub Oak) Barrows.

OAK, Dog *Acer campestre* (Field Maple) Som, Notts, Yks. Grigson. Cf. Cat Oak, also from Yorkshire.

OAK, Downy *Quercus pubescens*.

OAK, Durmast *Quercus petraea*.

OAK, Dyer's 1. *Quercus infectoria* (Aleppo Oak) Ablett. 2. *Quercus velutina* (Black Oak) J Smith. In the case of the Aleppo Oak, it is the galls that are used, but with Black Oak the inner bark, called Quercitron, is the dyestuff, used to give black, or yellow to orange, according to the recipe.

OAK, Evergreen 1. *Quercus ilex* (Holm Oak) B & H. 2. *Quercus undulata* (Wavyleaf Oak) USA Elmore.

OAK, Evergreen, Japanese *Quercus acuta*.

OAK, Forest *Casuarina equisetifolia* (Beefwood) Wit.

OAK, Gambel's *Quercus gambelii*.
OAK, Garry *Quercus garryana* (Oregon Oak) USA Turner & Bell.
OAK, Ground *Teucrium chamaedrys* (Germander) Barton & Castle. A literal translation of the specific name. The leaves are indented like oak leaves - hence too the French name Petit Chêne.
OAK, Holly 1. *Althaea rosea* (Hollyhock) A S Palmer. 2. *Quercus ilex* (Holm Oak) Gerard. 3. *Quercus pungens* (Scrub Oak) USA Elmore. The first example is a simple corruption of Hollyhock. But Holm, i.e. Holly, Oak gets the name because of the scarlet kerm insects that infest it, looking not unlike holly berries.
OAK, Holm *Quercus ilex*. See Holly Oak.
OAK, Holm, Scarlet *Quercus ilex* (Holm Oak) Gerard.
OAK, Hulver *Quercus ilex* (Holm Oak) Gerard.
OAK, Hungarian *Quercus frainetto*. It actually grows in the Balkans and Italy.
OAK, Indian *Barringtonia acutangula*.
OAK, Iron *Quercus stellata* (Post Oak) USA Vestal & Schultes.
OAK, Jerusalem *Chenopodium botrys*. The young leaves look rather like miniature oak leaves.
OAK, Kerm; OAK, Kermes *Quercus ilex* (Holm Oak) Graves (Kerm); Grigson.1973 (Kermes). i.e. the oak on which the kermes insect is found.
OAK, Lebanon *Quercus libani*.
OAK, Lily *Syringa vulgaris* (Lilac) Scot. B & H. Lily Oak is just a corruption of Lilac.
OAK, Live 1. *Quercus chrysolepis* (Maul Oak) Kroeber. 2. *Quercus virginiana*.
OAK, Live, California *Quercus agrifolia*.
OAK, Live, Coast *Quercus agrifolia* (California Live Oak) USA Bean.
OAK, Lucombe *Quercus x hispanica lucombeana*. Raised by an Exeter nurseryman called Lucombe in the 1760's.
OAK, Macedonian *Quercus trojana*.
OAK, Maiden *Quercus petraea* (Durmast Oak) Hants. Grigson.

OAK, Maingay's *Quercus maingayi*.
OAK, Manna *Quercus cerris* (Turkey Oak) Howes.
OAK, Maul *Quercus chrysolepis*.
OAK, Meru *Vitex keniensis*. Meru is a district in Kenya.
OAK, Mirbeck *Quercus canariensis* (Algerian Oak) Duncalf. An alternative specific name is *mirbeckii*.
OAK, Mossy-cup 1. *Quercus cerris* (Turkey Oak) J Smith. 2. *Quercus macrocarpa* (Bur Oak) USA Zenkert. Descriptive.
OAK, Net-leaf *Quercus reticulata*.
OAK, Nutgall *Quercus infectoria* (Aleppo Oak) Lindley.
OAK, Oregon *Quercus garryana*.
OAK, Overcup *Quercus lyrata* (Swamp Post Oak) USA Harper.
OAK, Page *Rhus radicans* (Poison Ivy) H & P.
OAK, Pedunculate *Quercus robur* (Oak) Hart. Pedunculate, i.e. with acorns on a stalk. Durmast Oak is sessile, with no stalks.
OAK, Pin 1. *Quercus durandii*. 2. *Quercus palustris*.
OAK, Poison 1. *Rhus diversiloba*. 2. *Rhus radicans* (Poison Ivy) USA Zenkert.
OAK, Poison, Californian; OAK, Poison, Western *Rhus diversiloba* (Poison Oak) both Le Strange.
OAK, Pontine *Quercus pontica*.
OAK, Portuguese *Quercus faginea*.
OAK, Possum *Quercus nigra* (Water Oak) Howes.
OAK, Post *Quercus stellata*.
OAK, Post, Swamp *Quercus lyrata*.
OAK, Pyrenean *Quercus pyrenaica*.
OAK, Quebec *Quercus alba* (White Oak) Grieve.1931.
OAK, Quercitron *Quercus velutina* (Black Oak) Mitchell.1974. Quercitron is an extract of the inner bark, used as a dyestuff.
OAK, Red *Quercus maxima*.
OAK, Red, Northern *Quercus maxima* (Red Oak) Schery.
OAK, River, Australian *Casuarina cunninghamia* (River She-Oak) Leathart.

OAK, Rocky Mountain *Quercus undulata* (Wavyleaf Oak) Chamberlin.

OAK, Round-leaved *Quercus rotundifolia*.

OAK, Royal, Red *Viburnum lantana* (Wayfaring Tree) Dor. Grigson.

OAK, Sawtooth *Quercus acutissima* (Japanese Chestnut Oak) RHS. Descriptive of the leaves, of course.

OAK, Scarlet *Quercus ilex* (Holm Oak) Gerard. Scarlet being the kermes insect that feeds on the tree.

OAK, Scented *Pelargonium quercifolium* (Oak-leaved Geranium) Sanecki.

OAK, Scrub 1. *Quercus dumosa*. 2. *Quercus pungens*. 3. *Quercus undulata* (Wavyleaf Oak) USA Elmore.

OAK, Scrub, Large *Quercus wislienzi*.

OAK, Sessile *Quercus petraea* (Durmast Oak) McClintock. Sessile, meaning stalkless, a reference to the way the acorns grow. The common oak is pedunculate, which means the acorns have stalks.

OAK, She *Casuarina stricta*.

OAK, She, Fire *Casuarina cunninghamia* (River She Oak) Everett.

OAK, She, Forest *Casuarina torulosa*.

OAK, She, River *Casuarina cunninghamia*.

OAK, She, Swamp *Casuarina equisetifolia* (Beefwood) Everett.

OAK, Shin *Quercus gambeli* (Gambel's Oak) Howes.

OAK, Shingle 1. *Casuarina equisetifolia* (Beefwood) Duncalf. 2. *Quercus imbricaria*. This oak was so called because the early settlers used the timber for roof shingles.

OAK, Short-stalked *Quercus petraea* (Durmast Oak) Lindley. Cf. Sessile Oak.

OAK, Silky *Grevillea robusta*.

OAK, Silky, Red *Stenocarpus salignus*.

OAK, Silky, White *Grevillea hilleana*.

OAK, Singapore *Quercus conocarpa*. Common on Singapore Island, less so on the mainland.

OAK, Spanish 1. *Quercus falcata*. 2. *Quercus ilex* (Holm Oak) USA Harper. In spite of the name, *Q falcata* is a North American species of the southern states.

OAK, Swamp *Casuarina glauca*.

OAK, Tan *Quercus densiflora*.

OAK, Tasmanian *Eucalyptus obliqua*.

OAK, Turkey *Quercus cerris*. A native of southwest Asia and the adjacent parts of Europe.

OAK, Turkish *Quercus aegilops*.

OAK, Valley *Quercus lobata*.

OAK, Valonia; OAK, Velanidi *Quercus macrolepis* Argenti & Rose (Velanidi). Valonia was the trade term in use from the early eighteenth century for the large acorn cups and acorns of this tree - they were exported from Valonia, an Albanian port, to be used for their tannin in the preparation of leather.

OAK, Wainscot *Quercus cerris* (Turkey Oak) J Smith.

OAK, Wallich's *Quercus wallichiana*.

OAK, Water *Quercus nigra*.

OAK, Wavyleaf *Quercus undulata*.

OAK, White 1. *Quercus alba*. 2. *Quercus garryana* (Oregon Oak) Kroeber. 3. *Quercus lobata* (Valley Oak) USA Barrett & Gifford. 4. *Quercus petraea* (Durmast Oak) Hants,IOW Grigson.

OAK, White, Arizona *Quercus arizonica*.

OAK, White, Swamp *Quercus bicolor*.

OAK, Willow *Quercus phellos*.

OAK, Winter *Quercus ilex* (Holm Oak) Rodd. Because it is evergreen.

OAK, Yellow *Quercus muhlenbergii* (Chestnut Oak) USA Harper.

OAK ATCHERN *Quercus robur* fruit (Acorn) Ches. B & H.

OAK-BELT GOOSEBERRY *Ribes quercetorum*.

OAK-CORN *Quercus robur* acorn (Acorn) Halliwell. An aberrant form,quoted in the form of Oke-corne in Halliwell. In fact, acorn is not "corn of the oak" at all, for OE aecorn came from Latin acer, field, probably a general term for mast.

OAK-FERN 1. *Polypodium vulgare* (Common Polypody) B & H. 2. *Pteridium aquilina* (Bracken) Som. Elworthy.

OAK-LEAF GERANIUM 1. *Chenopodium*

botrys (Jerusalem Oak) USA G B Foster. 2. *Pelargonium quercifolium*.

OAK-LEAVED GOOSEFOOT *Chenopodium glaucum*.

OAK-LEAVED HYDRANGEA *Hydrangea quercifolia*.

OAK LUNG *Lobaria pulmonaria* (Lungwort Crottle) Cullum. The 'lung' reference is to its use at one time for lung disease, bronchial catarrh, etc.

OAK-MACEY *Quercus robur* fruit (Acorn) Som. Macmillan. 'Macey' is 'mast', technically the fruit of all forest trees, perhaps now reserved for that of beech.

OAK MOSS *Evernia prunastri* (Stag's Horn Lichen) S M Robertson.

OAK NUT (or written as one word) *Quercus robur* fruit (Acorn) Som. Macmillan; Border B & H.

OAK OF CAPPADOCIA *Ambrosia elatior*.

OAK OF JERUSALEM *Teucrium botrys* (Cut-leaved Germander) Prior.

OAK OF MAMRE *Quercus pseudo-coccifera* (Abraham's Oak) Hutchinson & Melville.

OAK OF PARADISE *Chenopodium botrys* (Jerusalem Oak) E Hayes.

OAK-RAG *Lobaria pulmonaria* (Lungwort Crottle) Bolton.

OAKBERRY *Quercus robur* fruit (Acorn) Yks. B & H.

OAKSEY LILY *Fritillaria meleagris* (Snake's Head Lily) Wilts. Grigson. Meadows near Oaksey, in the north of Wiltshire, are one of the few places where Snake's Head Lily grows wild.

OAT *Avena sativa*. Oftener in the plural. OE ate.

OAT, Black *Avena strigosa*.

OAT, Bristle-pointed *Avena strigosa* (Black Oat) H C Long.1910).

OAT, False *Arrhenatherium elatius*.

OAT, Poor *Avena fatua* (Wild Oat) Som. Elworthy.

OAT, Wild *Avena fatua*.

OAT-GRASS, Knot *Arrhenatherium elatius* (False Oat) H C Long.1910.

OAT THISTLE *Onopordum acanthium* (Scotch Thistle) Turner.

OATMEAL, Bad Man's 1. *Anthriscus sylvestris* (Cow Parsley) Yks. Mabey. 2. *Capsella bursa-pastoris* (Shepherd's Purse) Dur. Grigson. 3. *Conium maculatum* (Hemlock) Yks, Dur, N'thum. Grigson; USA Henkel. Cow Parsley and Hemlock often share names. One would have thought it would have been well to know the difference between the two, but the 'oatmeal' in the name is a reference to the characteristic mealy appearance as the flowers spread their pollen.

OATMEAL, Beggarman's *Alliaria petiolata* (Jack-by-the-hedge) Leics. Grigson.

OATMEAL, Devil's 1. *Conopodium majus* (Earth-nut) Yks. Grigson. 2. *Heracleum sphondyllium* (Hogweed) War. Grigson. 3. *Petroselinum crispum* (Parsley) Wilts. Wiltshire. "Only the wicked can grow it", Mrs Wiltshire said of parsley. There are a lot of sayings about the seed going nine times to the devil. But why should something as wholesome as Earth-nut have the ascription?

OATMEAL, Naughty Man's *Anthriscus sylvestris* (Cow Parsley) War. B & H. Cf. Bad Man's Oatmeal. Both bad man and naughty man are euphemisms for the devil.

OATS, Devil's *Conium maculatum* (Hemlock) Cumb, Dur, N'thum. R L Brown. Cf. Bad Man's Oatmeal.

OATS, Donkey's 1. *Rumex acetosa* (Wild Sorrel) Dev. B & H. 2. *Rumex obtusifolius* (Broad-leaved Dock) Dev. Grigson.

OBEDIENCE PLANT *Maranta arundinacea* (Arrowroot) USA Cunningham.

OBEDIENT FLOWER *Physostegia virginiana* (Dragonhead) USA House.

OBLIONKER; HOBLIONKER; HOBBLY-HONKER *Aesculus hippocastanum* fruit (Horse Chestnut) Worcs. B & H (Oblionker); Worcs. J Salisbury (Hoblionker); Som. Macmillan (Hobbly-Honker). From conker, originally. The tree is the "Oblionker Tree".

OCA *Oxalis tuberosa*.

OCHRE LILY *Lilium primulinum*.
OCTOBER PLANT *Sedum sieboldii*.
ODDROD *Primula veris* (Cowslip) Dor. Grigson. Cf. Hodrod, or Holrod, all from Dorset.
ODE *Isatis tinctoria* (Woad) Turner.
ODE, Pope's *Aconitum anglicum* (Wild Monkshood) N'thants. B & H. 'Ode' here is 'hood'.
ODHRAN *Heracleum sphondyllium* (Hogweed) N Ire. Grigson.
ODIN'S GRACE *Geranium pratense* (Meadow Cranesbill) Perry. Gerard called it Grace of God.
OFBIT; OFBITEN *Succisa pratensis* (Devil's-bit Scabious) both B & H. These are medieval names, and are references to the short, blackish root, said to have been bitten off by the devil out of spite to mankind, because he knew that otherwise it would be good for many purposes.
OFFICINAL SQUILL *Scilla maritima* (Squill) Thornton. 'Official' means that it was in the Pharmacopeia. Thornton prescribed it as an expectorant and diuretic.
OGEECHEE LIME *Nyssa sylvatica* (Tupelo) Willis.
OHIA *Eugenia malaccensis* (Malay Apple) Schery.
OHIO BUCKEYE *Aesculus glabra*.
OIL BEAN *Pentaclethra macrophylla*.
OIL-NUT, Wood, African *Ricinodendron africanum*.
OIL PALM *Elaeis guineensis*. The species that is the provider of palm oil, used widely in the manufacture of soap, margarine and candles, and also as a lubricant.
OIL PLANT 1. *Drosera rotundifolia* (Sundew) Donegal Grigson. 2. *Sesamum indicum* (Sesame) Laguerre.
OIL POPPY *Papaver somniferum* (Opium Poppy) Hutchinson & Melville. The name acknowledges the value of the oil got from the seeds.
OIL-SEED RAPE *Brassica campestris var. rapa 'Oleifera'*.

OILNUT *Juglans cinerea* (White Walnut) Grieve.1931. Cf. Butternut for this.
OILS 1. *Avena sativa* (Oat) Dev. Friend.1882. 2. *Hordeum sativum* (Barley - the beards) Hants. Cope. Cf. Ailes, or Hoyles, also from Devonshire. Barley also has similar names, including Iles.
OILSEED 1. *Camelina sativa* (Gold of Pleasure) Prior. 2. *Sesamum indicum* (Sesame) Turner.
OILY-GRAIN PLANT *Sesanum indicum* (Sesame) W Miller. i.e. oil-seed.
OISTERLOIT *Polygonum bistorta* (Bistort) Gerard. Evidently related to another book name, Osterick, and Turner's Astrologia. Osterick is med. Lat. ostriacum, apparently a corruption from Latin aristolochia, the name now given to birthwort, and for which Turner misread Astrologia.
OKKERDI *Lamium purpureum* (Red Deadnettle) Shet. Grigson.
OKRA *Abelmoschus esculentus*. From a West African name for the plant.
OKRA, Bush *Corchorus olitorius* (Long-fruited Jute) Dalziel.
OKRA, Wild *Hibiscus vitifolius* Barbados Gooding.
OLD FOLKS' HERB *Drosera rotundifolia* (Sundew) USA Conway. Because of its use in the treatment of coughs and chest complaints in the elderly.
OLD LAD'S CORN *Stellaria holostea* (Greater Stitchwort) Shrop. Grigson. 'Old lad' is the devil. It is odd that a plant like this should have so many ascriptions to the devil - there are Devil's Flower, Devil's Nightcap, Devil's Eyes, and Devil's Corn, as well as this entry.
OLD LADY *Artemisia borealis* Brownlow. Presumably given as a counter to Southernwood's 'Old Man'.
OLD LADY'S LACE *Anthriscus sylvestris* (Cow Parsley) Som. Macmillan; Wilts. D Grose.
OLD LADY'S SMOCK *Calystegia sepium* (Great Bindweed) Som. Macmillan.
OLD MAID 1. *Catharanthus rosea*

(Madagascar Periwinkle) Howes. 2. *Cheiranthus cheiri* (Wallflower) Som. Macmillan.

OLD MAID'S FLOWER; OLD MAID'S LAST FRIEND *Viola tricolor* (Pansy) both Som. Macmillan. The aphrodisiac superstition once attached to pansy would account for these.

OLD MAID'S NIGHTCAP *Geranium maculatum* (Spotted Cranesbill) Grieve.1931.

OLD MAID'S PINK *Agrostemma githago* (Corn Cockle) S Africa Watt.

OLD MAID'S SCENT *Tanacetum parthenium* (Feverfew) Som. Macmillan. Cf. Stink Daisy.

OLD MAN 1. *Anagallis arvensis* (Scarlet Pimpernel) Wilts. Dartnell. & Goddard. 2. *Artemisia abrotanum* (Southernwood) Wilts. Dartnell & Goddard; Oxf. Thompson; Worcs. J Salisbury; Shrop, Middx, Cambs, Norf, N'thants, Ches. B & H; Ess. Gepp. 3. *Clematis vitalba* (Old Man's Beard) Som. Macmillan; Suss. Grigson. 4. *Rosmarinus officinalis* (Rosemary) Som. Macmillan; Suss. B & H. 5. *Senecio douglasii* (Douglas Groundsel) USA Elmore.

OLD MAN'S BEARD 1. *Artemisia abrotanum* (Southernwood) Dor, Wilts. Macmillan. 2. *Chionanthus virginicus* (Fringe Tree) USA Thomson.1978. 3. *Clematis brachiata* S Africa Watt. 4. *Clematis vitalba*. 5. *Equisetum arvense* (Horsetail) Som. Elworthy. 6. *Filipendula ulmaria* (Meadowsweet) Som. Macmillan. 7. *Hippuris vulgaris* (Mare's Tail) Dor. Grigson. 8. *Hypericum calycinum* (Rose of Sharon) Dev. Friend.1882. 9. *Nigella damascena* (Love-in-a-mist) Coles. 10. *Parmelia saxatilis* (Grey Stone Crottle) Nicolson. 11. *Rosa canina* the galls that grow on them Dev. Friend.1882. 12. *Saxifraga sarmentosa* (Mother-of-thousands) Dev. Friend.1882. All descriptive in their way, whether of colour or method of growth (e g the bunches of stamens on Rose of Sharon), or the general appearance of the seed heads of Clematis vitalba, etc. The name as applied to Southernwood recalls one of the explanations of that very common name Lad's Love; it is said that the herb was used in an ointment that young men used to promote the growth of a beard. Gerard, among others, talked of southernwood's anti-baldness properties.

OLD MAN'S BELL 1. *Campanula rotundifolia* (Harebell) Scot. Tynan & Maitland. 2. *Endymion nonscriptus* (Bluebell) Tynan & Maitland. Probably the same plant is meant. i.e. harebell, in spite of the fact that it is firmly listed under *Endymion*.

OLD MAN'S BREAD-AND-CHEESE *Malva sylvestris* (Common Mallow) Som. Macmillan.

OLD MAN'S BUTTONS 1. *Arctium lappa* burrs (Burdock) Dev. Macmillan. 2. *Caltha palustris* (Marsh Marigold) Som, Dor. Macmillan. 3. *Ranunculus acris* (Meadow Buttercup) Som. Grigson. 4. *Ranunculus bulbosus* (Bulbous Buttercup) Som. Grigson. 5. *Ranunculus repens* (Creeping Buttercup) Som. Grigson. The tight buds of the buttercups provide the imagery, while the burrs are an obvious point of reference.

OLD MAN'S CLOCK *Taraxacum officinale* (Dandelion) Putnam.

OLD MAN'S FACE 1. *Antirrhinum majus* (Snapdragon) Dev. Macmillan. 2. *Viola tricolor* (Pansy) Som. Macmillan.

OLD MAN'S FLANNEL *Verbascum thapsus* (Mullein) Som. B & H; USA Henkel. The reference is to the texture of the leaves.

OLD MAN'S FRIEND *Anagallis arvensis* (Scarlet Pimpernel) Som. Macmillan.

OLD MAN'S LOOKING GLASS *Anagallis arvensis* (Scarlet Pimpernel) Som. Macmillan. A rather garbled version of one of the "weather forecasting" names for pimpernel, probably Old Man's Weatherglass, an even more garbled version of which is Old Man's Glass Eye.

OLD MAN'S LOVE *Artemisia abrotanum* (Southernwood) Berks. Lowsley; N'thum. B & H. A hybrid - Old Man x Lad's Love.

OLD MAN'S MUSTARD *Achillea millefolium* (Yarrow) Lincs. Grigson.

OLD MAN'S NIGHTCAP 1. *Calystegia sepium* (Great Bindweed) Som. Macmillan; Surr. Grigson; Suss. Parish. 2. *Convolvulus arvensis* (Field Bindweed) Som, Wilts. Macmillan. 3. *Convolvulus major* (Morning Glory) Suss. Friend. Cf. Devil's Nightcap.

OLD MAN'S PEPPER 1. *Achillea millefolium* (Yarrow) Som. Macmillan. 2. *Poterium sanguisorba* (Salad Burnet) Som. Macmillan.

OLD MAN'S PEPPER-BOX *Achillea ptarmica* (Sneezewort) Som. Macmillan. The common name, Sneezewort, is sufficient explanation of this name.

OLD MAN'S PLAYTHING *Pimpinella saxifraga* (Burnet Saxifrage) Shrop. Grigson.

OLD MAN'S PULPIT *Arum maculatum* (Cuckoo-pint) Som. Macmillan. Cf. Parson-in-the-pulpit, etc.

OLD MAN'S SHIRT 1. *Calystegia sepium* (Great Bindweed) Som. Macmillan. 2. *Stellaria holostea* (Greater Stitchwort) Corn. Grigson. Dor. Dacombe.

OLD MAN'S WEATHERGLASS; OLD MAN'S GLASS EYE *Anagallis arvensis* (Scarlet Pimpernel) Som. Macmillan (Weatherglass) Som. Grigson (Glass Eye).

OLD MAN'S WHISKERS *Geum triflorum* (Priarie Smoke) USA T H Kearney. Because of the silvery feathered tails to the fruit. Cf. Grandfather's Beard.

OLD MAN'S WOOZARD *Clematis vitalba* (Old Man's Beard) Bucks. Grigson.

OLD SOW *Anaphalis margaritacea* (Pearly Immortelle) Norf. Grigson.

OLD UNCLE HARRY *Artemisia vulgaris* (Mugwort) Som. Macmillan. Is this an extended family version of southernwood's Old Man, or wormwood's Old Woman? Actually, it is more likely that Old Uncle Harry is the devil. Cf. Naughty Man.

OLD WARRIOR *Artemisia pontica* (Roman Wormwood) Brownlow. Old Woman, perhaps?

OLD WIFE'S THREADS *Ranunculus repens* (Creeping Buttercup) Yks. Grigson.

OLD WITCH GRASS *Panicum capillare*.

OLD WIVES' MUTCHES *Aconitum napellus* (Monkshood) Perth. B & H. Mutches are close-fitting caps, so the name is very descriptive.

OLD WIVES' TONGUES *Populus tremula* (Aspen) Roxb. Grigson. Never still, like the leaves.

OLD WOMAN 1. *Artemisia absinthium* (Wormwood) Oxf. Oxf.Annl.Rec.1951. 2. *Artemisia maritima* (Sea Wormwood) Grieve.1931. 3. *Artemisia pontica* (Roman Wormwood) G B Foster. 4. *Artemisia stelleriana* (Beach Wormwood) Webster. 5. *Lavandula stoechas* (Spanish Lavender) Heref. Leather. For the Artemisias, the name must be regarded as a companion to Old Man, which is Southernwood, presumably.

OLD WOMAN THREADING THE NEEDLE *Geranium robertianum* (Herb Robert) Som. Macmillan. Cf. Granny-thread-the-needle, Mother-thread-my-needle, etc.

OLD WOMAN'S BONNETS 1. *Campanula medium* (Canterbury Bells) Som. Macmillan. 2. *Capsella bursa-pastoris* (Shepherd's Purse) Som. Macmillan. 3. *Geum rivale* (Water Avens) Wilts. Dartnell & Goddard.

OLD WOMAN'S EYE *Vinca minor* (Lesser Periwinkle) Dor. Macmillan.

OLD WOMAN'S NEEDLE *Scandix pecten-veneris* (Shepherd's Needle) Hants. Cope.

OLD WOMAN'S NEEDLEWORK *Centranthus ruber* (Spur Valerian) Som. Macmillan.

OLD WOMAN'S NIGHTCAP 1. *Aconitum napellus* (Monkshood) Bucks. B & H. 2. *Calystegia sepium* (Great Bindweed) Som. Macmillan. 3. *Campanula medium* (Canterbury Bells) Som. Macmillan.

OLD WOMAN'S ORCHID *Orchis purpurea* (Lady Orchid) B & H.

OLD WOMAN'S PENNY *Lunaria annua* (Honesty) Som. Macmillan. A member of the series that includes Money-flower, Penny-flower, etc. The imagery is provided by the silver capsules.

OLD WOMAN'S PETTICOATS *Papaver rhoeas* (Red Poppy) Som. Grigson.

OLD WOMAN'S PINCUSHION *Dactylorchis maculata* (Heath Spotted Orchid) Wilts. Dartnell & Goddard. The tightly packed flowers are probably the source of a 'pincushion' epithet.

OLD WOMAN'S PURSE *Impatiens noli-me-tangere* (Touch-me-not) Cumb. Grigson. From the shape of the hanging flowers.

OLDROT 1. *Anthriscus sylvestris* (Cow Parsley) Som. Grigson. 2. *Heracleum sphondyllium* (Hogweed) Som. B & H. This name varies between Eltrot and Hill-trot, and probably, like Kexies, refers to the dried stalks.

OLEANDER *Nerium oleander* (Rose Bay) Watt.

OLEANDER, Bastard *Thevetia peruviana* (Yellow Oleander) Chopra.

OLEANDER, Indian *Nerium indicum*.

OLEANDER, Wild *Adina microcephala*.

OLEANDER, Yellow *Thevetia peruviana*.

OLEASTER, Narrow-leaved *Elaeagnus angustifolia*. The name is connected with Olea, olive - this has been known as the Wild Olive, and the botanical name itself confirms this, for Greek elaia is the origin of Latin olea.

OLER; OLERN *Alnus glutinosa* (Alder) Grigson.

OLIBANUM *Boswellia thurifera* (Frankincense) Lindley. The name of the resin rather than that of the plant.

OLIBANUM, Indian *Boswellia serrata* (Indian Frankincense) Usher.

OLIVE *Olea europaea*.

OLIVE, Bastard *Buddleia saligna* S Africa Watt.

OLIVE, Brown *Olea africana*.

OLIVE, Californian *Umbellularia californica* (Californian Laurel) Howes.

OLIVE, East African *Olea hochstetteri*.

OLIVE, Elgon *Olea welwitschii* (Loliondo) Dale & Greenway. The reference is to Mt Elgon, in Kenya.

OLIVE, Ethiopian *Elaeagnus angustifolia* (Narrow-leaved Oleaster) R T Gunther.1934.

OLIVE, Fragrant *Osmanthus fragrans* McDonald.

OLIVE, Indian *Olea ferruginea*.

OLIVE, Russian *Elaeagnus angustifolia* (Narrow-leaved Oleaster) USA Cunningham.

OLIVE, Spurge 1. *Daphne laureola* (Spurge Laurel) Jordan. 2. *Daphne mezereon* (Lady Laurel) B & H.

OLIVE, Spurge, German *Daphne mezereon* (Lady Laurel) Gerard.

OLIVE, Tea *Osmanthus fragrans* (Fragrant Olive) Howes.

OLIVE, Wild 1. *Elaeagnus angustifolia* (Narrow-leaved Oleaster) Moldenke. 2. *Olea africana* (Brown Olive) Hora.

OLIVE, Wild, American *Osmanthus americanus* (Devilwood) Howes.

OLIVE-LEAVED SALLOW *Salix oleifolia*.

OLIVE PLUM *Elaeodendron glaucum* (Ceylon Tea Tree) Chopra. An oily fruit, much the same shape as an olive, a fact noted in the generic name, which is from Gk elaia, olive, and dendron, tree.

OLIVER, Wild *Ximenia americana*.

OLIVER BARK *Cinnamonium oliveri*.

OLIVER'S SEA HOLLY *Eryngium amethystinum*.

OLIVER CROMWELL'S CREEPING COMPANION *Helxine soleirolii* (Mind-your-own-business) Som. Macmillan.

OLLAR *Alnus glutinosa* (Alder) Grigson. Cf. Oler, Olern, Owlorn etc.

OLLICK 1. *Allium porrum* (Leek) Corn. B & H. 2. *Sempervivum tectorum* (Houseleek) Corn. Jago. Cf. the Devonshire Lick for leek, the result of local pronunciation.

OLYMPIC BELLFLOWER *Campanula grandis*. Far from Olympia, this actually comes from Siberia.

OME-TREE 1. *Ulmus glabra* (Wych Elm) Cumb. B & H. 2. *Ulmus procera* (English Elm) Cumb. B & H.

ONE-BERRY 1. *Celtis reticulata* (Western Hackberry) USA Elmore. 2. *Mitchella* *repens* (Partidge-berry) Le Strange. 3. *Paris quadrifolia* (Herb Paris) Turner.

ONE-BLADE *Maianthemum bifolium* (May Lily) Prior.

ONE-DAY FLOWER *Tradescantia virginica* (Spiderwort) Howes. Cf. Flower-of-a-day.

ONE-EYE BEAN *Canavalia ensiformis* (Jack Bean) W Indies Dalziel. Cf. Cut-eye Bean and Overlook Bean for this.

ONE-FLOWERED CUSHION SAXIFRAGE *Saxifraga burseriana*.

ONE-I-EAT *Arbutus unedo* (Strawberry Tree) Som. Macmillan. A translation of 'unedo'.

ONE-LEAF ORCHID *Orchis rotundifolia* (Small Round-leaved Orchid) USA Bingham.

ONE-LEAF PINE *Pinus monophylla* (Piñon) Mason.

ONE-O'CLOCK 1. *Ornithogalum umbellatum* (Star-of-Bethlehem) Som. Macmillan. 2. *Stellaria holostea* (Greater Stitchwort) Dev. Macmillan. 3. *Taraxacum officinale* (Dandelion) Som. Mamillan; Lancs. B & H; USA Henkel. 4. *Tragopogon pratensis* (Goat's Beard) Som. Macmillan. 5. *Tussilago farfara* (Coltsfoot) Som. Macmillan. Some, like Star-of-Bethlehem and Goat's Beard, have the reference because of the early opening and closing of the flowers. But dandelion gets the name from the children's game played with the clock - the seed heads. Probably coltsfoot gets the name for the same reason, although there is no record of children actually playing with them.

ONE, TWO, THREE; ONE-TWO-THREE, PEE-A-BED *Taraxacum officinale* (Dandelion) Freethy (One, two, three); Som. Grigson (One-two-three, pee-a-bed). One, two, three is a reference to children's games blowing the seed heads, and the other one mixes the traditions by adding a reference to dandelion's well-known diuretic qualities.

ONEBERRY *Celtis occidentalis* (Hackberry) USA Vestal & Schultes.

ONION *Allium cepa*. French oignon, from Latin unio. But the OE was Yne, or Ynne, and the older Engish name was Ine, hence hybrids like Inon, once common over southern England.

ONION, Ainu *Allium nipponicum*.

ONION, Assam *Allium rubellum*.

ONION, Ballhead *Allium sphaerocephalum* (Round-head Leek) Douglas. 'Ballhead' is a translation of *sphaerocephalum*.

ONION, Bunching, Japanese *Allium fistulosum* (Welsh Onion) A W Smith.

ONION, Chinese 1. *Allium chinense*. 2. *Allium fistulosum* (Welsh Onion) F P Smith.

ONION, Crested *Allium cristatum*.

ONION, Crow *Allium vineale* (Crow Garlic) War. Bloom; Worcs. Grigson.

ONION, Desert *Allium deserticola*.

ONION, Dog's *Ornithogalum umbellatum* (Star-of-Bethlehem) Turner.

ONION, Egyptian *Allium cepa var. aggregatum*.

ONION, Field, Wild *Ornithogalum umbellatum* (Star-of-Bethlehem) Lyte.

ONION, Fool's *Triteleia hyacinthina*.

ONION, French *Scilla maritima* (Squill) Turner.

ONION, Fringed *Allium fimbriatum*.

ONION, Gypsy *Allium ursinum* (Ramsons) S Eng. B & H.

ONION, Infant *Allium schoernoprasum* (Chives) Mabey. Because of its mildness and almost total absence of bulb.

ONION, Japanese *Allium senescens* (German Garlic) Douglas. Why Japanese? This is a central European species.

ONION, Multiplier, Egyptian *Allium cepa var. viviparum*. It produces onion-flavoured bulbils at the top of the stalks.

ONION, Pearl, Bunching *Allium ampeloprasum* (Wild Leek) Brouk.

ONION, Pink *Allium roseum* (Rosy Garlic) Salisbury.

ONION, Pitted *Allium lacunosum*.

ONION, Potato 1. *Allium cepa* (Onion) M Baker.1977. 2. *Allium cepa var. viviparum* (Egyptian Multiplier Onion). G B Foster.

ONION, Red *Allium carinatum* (Keeled Garlic) Salisbury.

ONION, St Thomas's 1. *Allium cepa* (Onion) M Baker. 2. *Allium cepa var. aggregatum* (Egyptian Onion) M Baker. 3. *Allium fistulosum* (Welsh Onion) M Baker. The onion is said to be sacred to St Thomas. On the vigil of the saint's day, divinations were practiced, using an onion. Derbyshire girls would peel one, and stick nine pins in it, one in the centre, and the rest radially. As they were put in, they said the following rhyme:

Good St Thomas, do me right,
Send me my true love tonight,
In his clothes and his array,
Which he weareth every day,
That I may see him in the face,
And in my arms may him embrace.

The Pennsylvania Germans used the divination, too, but this time the girl would take 4 onions, give each a name, and put them under the bed or the stove. The one that has sprouted next morning bears the name of the future husband.

ONION, Sea 1. *Endymion nonscriptus* (Bluebell) Gerard. 2. *Scilla maritima* (Squill) Turner. 3. *Scilla verna* (Spring Squill) IOW Grigson. All bulbous plants were apt to be given the name onion in one form or another.

ONION, Spanish *Allium fistulosum* (Welsh Onion) USA Cunningham.

ONION, Spring *Allium fistulosum* (Welsh Name) Candolle. In spite of the name, this is not the species that is usually grown as Spring Onions.

ONION, Squill *Scilla maritima* (Squill) T Hill.

ONION, Stinking 1. *Allium triquetrum* (Three-cornered Leek) McClintock. 2. *Allium ursinum* (Ramsons) Freethy. Cf. Onion Stinker for Ramsons.

ONION, Top *Allium cepa var. viviparum* (Egyptian Multiplier Onion) Howes. "Top", because of its habit of producing onion-flavoured bulbils at the top of its stalk.

ONION, Tree *Allium cepa var. viviparum* (Egyptian Multiplier Onion) Painter.

ONION, Welsh *Allium fistulosum*. 'Welsh' here means foreign, the German walch. The plant was introduced through Germany from Siberia.

ONION, Wild 1. *Allium acuminatum* USA Schenk & Gifford. 2. *Allium bolanderi* USA Schenk & Gifford. 3. *Allium canadense* (Canada Garlic) USA Kingsbury. 4. *Allium ursinum* (Ramsons) Som. Macmillan. 5. *Allium vineale* (Crow Garlic) War. Grigson.

ONION, Wild, Nodding *Allium cernuum*.

ONION, Wild, Prairie *Allium stellatum*.

ONION, Winter *Allium cepa* (Onion) Turner.

ONION COUCH *Arrhenatherium elatius* (False Oat) H C Long.1910. It has a bulbous-looking rootstock in the sense that it is tuberous at the nodes, producing several "bulbs", "pearls", or "knots". Hence, too, Knot Oatgrass, or Pearl Grass.

ONION-FLOWER *Allium ursinum* (Ramsons) Som. Macmillan.

ONION IRIS *Hermodactylus tuberosus* (Snakeshead Iris). W Miller.

ONION-LEAVED ASPHODEL *Asphodelus fistulosus*.

ONION-ROOTED CROWFOOT *Ranunculus bulbosus* (Bulbous Buttercup) Gerard.

ONION STINKER *Allium ursinum* (Ramsons) Som. Macmillan. Cf. Stinking Onion.

ONION WEED *Allium triquetrum* (Three-cornered Leek) New Zealand Painter.

ONTARIO POPLAR *Populus x candicans* (Balsam Poplar) Hay & Synge.

OOKOW *Brodiaea pulchella* Calif. Schenk & Gifford.

OOSTER-MUNATH-JONNUMS *Polygonum bistorta* (Bistort) Cumb. Rowling. 'Ooster-munath' must be Easter month, and the whole must be a reference to the herb puddings once eaten at Easter. Cf. Easter Ledges, etc.

OOZE APPLE *Datura meteloides* (Downy Thorn-apple) USA Elmore.

OPEN-AND-SHUT *Ornithogalum umbellatum* (Star-of-Bethlehem) Wilts. Macmillan. A reference to the short time that the flowers are open. They are better known as Ten, Eleven, Twelve or One-o'clock, depending on the original observer.

OPEN-ARCE; OPEN-ARSE; OPEN-ASS; OPENER; OPYMAN *Mespilus germanica* (Medlar) B & H (Arce); Turner (Arse); Corn. Quiller-Couch; Som. Elworthy; Worcs. Grigson (Ass); B & H (Opener & Opyman). All comments on the effect eating the fruit has.

OPEN-GOWAN *Caltha palustris* (Marsh Marigold) Cumb. B & H.

OPEN-JAWS; OPEN-MOUTH *Antirrhinum majus* (Snapdragon) Som. Macmillan.

OPHTHALMIC BARBERRY *Berberis aristata* (Nepal Barberry) Le Strange.

OPIER *Viburnum opulus* (Guelder Rose) Turner. "it is called in frenche...opier, and so may it be also in englishe tyl we fynde a better name".

OPIUM LETTUCE *Lactuca virosa* (Great Prickly Lettuce) Brownlow. It has the milky juice common to all the lettuces, but this species is actually cultivated for it. The juice hardens when exposed to the air, and produces a gum known as lactucaria, or lettuce opium, which has distinct narcotic and soporific qualities. All lettuces are slightly narcotic, but none more so than this one. Cf. Sleepwort.

OPIUM POPPY *Papaver somniferum*.

OPLE-TREE; OPPLE-TREE *Viburnum opulus* (Guelder Rose) W Miller (Ople); Gerard (Opple).

OPOPANAX *Acacia farnesiana* (Cassie) Polunin. Greek opos, juice, and panax, a panacea.

OPOSSUM-WOOD *Halesia caroliniana* (Snowdrop Tree) Howes.

ORACHE 1. *Atriplex argentea* (Saltbush) USA Gates. 2. *Atriplex hortensis* (Garden Orache). 3. *Atriplex patula*. Orache comes through French arroche from Latin atriplex. It has also been said that the name is a corruption of aurum, gold, because the seeds, mixed with wine, were supposed to cure the yellow jaundice.

ORACHE, Asiatic *Atriplex hortensis* (Garden Orache) Schery.

ORACHE, Babington's *Atriplex glabriuscula*.

ORACHE, Delt- *Atriplex patula* (Orache) Prior. From the triangular leaves, like the Greek letter delta.

ORACHE, Dog's *Chenopodium vulvaria* (Stinking Orache) B & H.

ORACHE, Frosted *Atriplex laciniata*. The plant is covered with white scales, giving it a silvery, or 'frosted' appearance.

ORACHE, Garden *Atriplex hortensis*.

ORACHE, Grass-leaved *Atriplex littoralis*.

ORACHE, Halberd-leaved; ORACHE, Hastate *Atriplex hastata*. Halberd-leaved Orache is the usual name; Hastate (which means spear-shaped) Orache is in Darling & Boyd.

ORACHE, Red 1. *Atriplex hortensis* (Garden Orache) Salisbury/1936. 2. *Atriplex rosea* (Redscale) Jaeger.

ORACHE, Sea, Jagged *Atriplex laciniata* (Frosted Orache) Camden/Gibson.

ORACHE, Shore *Atriplex littoralis* (Grass-leaved Orache) Clapham.

ORACHE, Shrubby *Atriplex halimus* (Tree Purslane) Moldenke.

ORACHE, Silver *Atriplex argentea* (Saltbush) USA Elmore.

ORACHE, Spear-leaved *Atriplex hastata* (Halberd-leaveed Orache) Curtis.

ORACHE, Spreading *Atriplex patula* (Orache) Grieve.1931.

ORACHE, Stinking *Chenopodium vulvaria*.

ORACHE, Wild *Chenopodium album* (Fat Hen) Gerard.

ORAGE *Atriplex hortensis* (Garden Orache) T Hill.

ORANGE *Citrus aurantium*. A French word, ultimately from the Arabic naranj.

ORANGE, Bergamot *Citrus aurantium var.*

bergamia. The name 'Bergamot' is from the town of Bergama, in Turkey. Oil of bergamot is made from the rinds of this fruit (not from bergamot, though the scents are almost indistinguishable).

ORANGE, Blood *Citrus aurantium var. melitensis*.

ORANGE, China *Citrus aurantium amara var. sinensis*.

ORANGE, Elephant *Strychnos spinosa* (Kaffir Orange) Palgrave.

ORANGE, Japanese *Citrus reticulata* (Tangerine) Howes.

ORANGE, Kaffir *Strychnos spinosa*. It has yellow fruits, about the size of an orange.

ORANGE, Kaffir, Small *Strychnos cocculoides*.

ORANGE, King *Citrus reticulata* (Tangerine) Howes.

ORANGE, Malta *Citrus sinensis*.

ORANGE, Mandarin *Citrus reticulata* (Tangerine) Hora.

ORANGE, Mock 1. *Maclura pomifera* (Osage Orange) USA Puckett. 2. *Philadelphus coronarius*. 3. *Pittosporum tobiana* RHS. 4. *Prunus caroliniana* (American Cherry Laurel) USA Harper.

ORANGE, Panama *Citrus mitis* (Calamondin) Grigson.1974. Why, when it is a Filipino tree?

ORANGE, Portuguese *Citrus sinensis* (Malta Orange) Willis.

ORANGE, Seville; ORANGE, Sour *Citrus aurantium var. bigaradia* A W Smith (Sour Orange).

ORANGE, Sweet *Citrus aurantium var. sinensis* (Malta Orange) Willis.

ORANGE, Wild 1. *Capparis mitchelli* (Mitchell's Caper). 2. *Strychnos cocculoides* (Small Kaffir Orange) Lee.

ORANGE BLOSSOM *Philadelphus coronarius* flowers (Mock Orange) Dev. Friend.1882.

ORANGE-FLOWER, American; ORANGE-FLOWER, Mexican *Choisya ternata* McDonald (American).

ORANGE FLOWER TREE *Philadelphus coronarius* (Mock Orange) Lincs. B & H.

ORANGE MINT *Mentha citrata* (Bergamot Mint) Webster.

ORANGE RIVER LILY *Crinum bulbispermum* S Africa Watt.

ORANGE THYME *Thymus fragrantissimum*.

ORANGE WILLOW *Lippia citriodora* (Lemon Verbena) Dev. Friend.1882.

ORCANETTE *Anchusa officinalis* (Alkanet) Pomet. Orcanette is the French from which Alkanet is derived.

ORCHANET 1. *Anchusa officinalis* (Alkanet) Gerard. 2. *Lycopsis arvensis* (Small Bugloss) Turner.

ORCHARD GRASS *Dactylis glomerata* (Cocksfoot) USA T H Kearney.

ORCHARD WEED *Anthriscus sylvestris* (Cow Parsley) B & H.

ORCHID, Angel, White *Platanthera chlorantha* (Butterfly Orchid) Som. Macmillan.

ORCHID, Bee *Ophrys apifera*.

ORCHID, Bee, Brown *Ophrys fusca*.

ORCHID, Bird's Nest *Neottia nidus-avis*. Because of the tangled fibrous roots.

ORCHID, Bird's Nest, Violet *Limorodum abortivum* (Limodore) D Parish.

ORCHID, Blunt-leaf *Habenaria obtusa* (Small Northern Bog Orchis) USA Bingham.

ORCHID, Bog *Hammanbya paludosa*.

ORCHID, Bog, Northern, Small *Habenaria obtusata*.

ORCHID, Bog, White, Tall *Habenaria dilatata*.

ORCHID, Broad-leaved *Dactylorchis majalis* (Irish Marsh Orchid) Polunin.

ORCHID, Brown-winged *Orchis purpurea* (Lady Orchid) Turner Ettlinger.

ORCHID, Bug *Orchis coriophora*. The flowers smell of bed bugs.

ORCHID, Bumble-bee *Ophrys bombyliflora*.

ORCHID, Burnt; ORCHID, Burnt-stick *Orchis ustulata* (Burnt-tip Orchid) D Grose (Burnt); Clapham (Burnt-stick).

ORCHID, Burnt-tip *Orchis ustulata*.

ORCHID, Butterfly *Platanthera chlorantha*.

ORCHID, Butterfly, Lesser *Platanthera biflora*.

ORCHID, Butterfly, Pink *Orchis papilionacea*.

ORCHID, Close-flowered *Neotinia intacta* (Dense-flowered Orchid) Turner Ettlinger.

ORCHID, Club-spear, Little *Habenaria clavellata* (Small Green Wood Orchid) USA Bingham. 'Club-spear', from the shape of the spur.

ORCHID, Cradle *Anguloa clowesii*.

ORCHID, Cuckoo *Orchis mascula* (Early Purple Orchis) B & H.

ORCHID, Dark-winged *Orchis ustulata* (Burnt-tip Orchid) Clapham.

ORCHID, Dense-flowered *Neotinia intacta*.

ORCHID, Dwarf *Orchis ustulata* (Burnt-tip Orchid) Genders.

ORCHID, Egg *Cephalanthera damasonium* (Broad Helleborine) Young. The shape of the white flowers.

ORCHID, Elder-flowered *Dactylorchis sambucina*.

ORCHID, Fen *Liparis loeslii*.

ORCHID, Finger *Dactylorchis maculata* (Heath Spotted Orchid) Gerard. The 'dactyl' part of the specific name and the various "hand" or "finger" names refer to the fact that the tubers are divided into several finger-like lobes.

ORCHID, Fly *Ophrys insectifera*.

ORCHID, Fragrant *Gymnadenia conopsea*.

ORCHID, Fragrant, Short-spurred *Gymnadenia odoratissima*.

ORCHID, Fragrant, White *Leucorchis albida* (Small White Orchid) Turner Ettlinger.

ORCHID, Fringed, Purple, Large *Habenaria fimbriata*.

ORCHID, Fringed, Ragged *Habenaria lacora*.

ORCHID, Fringed, White *Habenaria blephariglottis*.

ORCHID, Fringed, White, Prairie *Habenaria psycodes*.

ORCHID, Fringed, Yellow *Habenaria ciliaris*.

ORCHID, Frog *Coeloglossum viride*.

ORCHID, Frog, White *Leucorchis albida* (Small White Orchid) Turner Ettlinger.

ORCHID, Ghost *Epipogium aphyllum*.

ORCHID, Giant *Epipactis gigantea* (Stream Orchid) Usher.

ORCHID, Gnat *Gymnadenia conopsea* (Scented Orchid) Turner Ettlinger.

ORCHID, Great *Orchis militaris* (Soldier Orchid) Curtis.

ORCHID, Green, Bracted *Coeloglossum viride* (Frog Orchid) USA Yarnell.

ORCHID, Green, Northern, Tall *Habenaria hyperborea*.

ORCHID, Green-veined *Orchis morio* (Green-winged Orchid) Turner Ettlinger.

ORCHID, Green-winged *Orchis morio*.

ORCHID, Hand; ORCHID, Handed, Female *Dactylorchis maculata* (Heath Spotted Orchid) Grieve.1931 (Hand); Jamieson (Female Handed). See Finger Orchid. 'Female' presumably because it lacks the "male" signs of twin tubers of the other orchids.

ORCHID, Heath *Dactylorchis fuchsii* (Common Spotted Orchid) Storer.

ORCHID, Hooker's *Habenaria hookeri*.

ORCHID, Irish *Neotinea intacta* (Dense-flowered Orchid) Turner Ettlinger. A very rare native to these islands, and confined to cos Mayo and Galway (hence this name) and the Isle of Man.

ORCHID, Jersey *Orchis laxiflora* (Loose-flowered Orchid) Clapham. Very rare in Jersey and Guernsey.

ORCHID, Jewel *Anoectochilus discolor*.

ORCHID, Jewel, Royal *Anoectocholus regalis*.

ORCHID, Lady *Orchis purpurea*.

ORCHID, Leafless *Epipogium aphyllum*.

ORCHID, Lizard *Himantoglossum hircinum*. The long central lobe looks like a lizard.

ORCHID, Long-bracted *Habenaria bracteata*.

ORCHID, Loose-flowered *Orchis laxiflora*.

ORCHID, Male *Orchis purpurea* (Early Purple Orchis) Thornton.

ORCHID, Man 1. *Aceras anthropophorum*.

2. *Dactylorchis incarnata* (Early Marsh Orchid) B & H. 3. *Dactylorchis maculata* (Heath Spotted Orchid) B & H. 4. *Listera ovata* (Twayblade) Som. Macmillan. 5. *Orchis morio* (Green-winged Orchid) B & H.
ORCHID, Man, Green *Aceras anthropophorum* (Man Orchid) Turner Ettlinger.
ORCHID, Marsh, Broad-leaved *Dactylorchis majalis* (Irish Marsh Orchid).
ORCHID, Marsh, Connaught *Dactylorchis cruenta.*
ORCHID, Marsh, Dwarf *Dactylorchis purpurella* (Northern Marsh Orchid) Turner Ettlinger.
ORCHID, Marsh, Early *Dactylorchis incarnata.*
ORCHID, Marsh, Irish *Dactylorchis majalis.*
ORCHID, Marsh, Northern *Dactylorchis purpurella.*
ORCHID, Marsh, Pugley's *Dactylorchis traunsteineri.*
ORCHID, Marsh, Purple *Dactylorchis praetermissa* (Southern Marsh Orchid) Turner Ettlinger.
ORCHID, Marsh, Southern *Dactylorchis praetermissa.*
ORCHID, Marsh, Wicklow *Dactylorchis traunsteineri* (Pugley's Marsh Orchid) Turner Ettlinger.
ORCHID, Meadow 1. *Orchis mascula* (Early Purple Orchid) Geldart. 2. *Orchis morio* (Green-winged Orchid) Curtis.
ORCHID, Meadow, Green-winged *Orchis morio* (Green-winged Orchid) Pratt.
ORCHID, Military *Orchis militaris* (Soldier Orchid) Polunin.
ORCHID, Mirror *Ophrys speculum.*
ORCHID, Monkey *Orchis simea.*
ORCHID, Moth *Phalaenopsis schilleriana.*
ORCHID, Mountain, White *Leucorchis albida* (Small White Orchid) Turner Ettlinger.
ORCHID, Musk *Herminium monorchis.*
ORCHID, Old Woman's *Orchis purpurea* (Lady Orchid) B & H.
ORCHID, One-leaf *Orchis rotundifolia* (Small Round-leaved Orchid) USA Bingham.
ORCHID, Pale-flowered *Orchis pallens.*
ORCHID, Pigeon *Dendrobium crumenatum.*
ORCHID, Ploughshare *Serapias vomeracea.* The strange, wedge-shaped flowers are likened to a ploughshare, and that is echoed in the specific name, for Latin vomer also means a ploughshare.
ORCHID, Poor Man's 1. *Iris xiphioides* (English Iris) Staffs. Marrow. 2. *Schizanthus sp* (Butterfly Flower) Rochford.
ORCHID, Purple, Dwarf *Dactylorchis purpurella* (Northern Marsh Orchid) Turner Ettlinger.
ORCHID, Purple, Early *Orchis mascula.*
ORCHID, Pyramidal *Anacamptis pyramidalis.*
ORCHID, Rein 1. *Coeloglossum viride* (Frog Orchid) USA Yarnell. 2. *Habenaria elegans* USA Schenk & Gifford.
ORCHID, Rose-purple, Andrews's *Habenaria andrewsii.*
ORCHID, Round-leaved, Large *Habenaria orbiculata.*
ORCHID, Round-leaved, Small *Orchis rotundifolia.*
ORCHID, Scented *Gymnadenia conopaea.*
ORCHID, Scented, Larger *Gymnadenia densiflora.*
ORCHID, Showy *Orchis spectabilis.*
ORCHID, Slipper, Lady's *Cypripedium calceolus.*
ORCHID, Soldier *Orchis militaris.*
ORCHID, Spider *Ophrys sphegodes.*
ORCHID, Spider, Late *Ophrys fuciflora.*
ORCHID, Spotted, Common *Dactylorchis fuchsii.*
ORCHID, Spotted, Heath *Dactylorchis maculata.*
ORCHID, Spotted, Heath, Rhum *Dactylorchis maculata var. rhoumensis.*
ORCHID, Spotted, Moorland *Dactylorchis maculata* (Heath Spotted Orchid) Turner Ettlinger.
ORCHID, Spotted, Woodland *Dactylorchis*

fuchsii (Common Spotted Orchid) Turner Ettlinger.
ORCHID, Star of Bethlehem *Angraecum sesquipedale*.
ORCHID, Stream *Epipactis gigantea*.
ORCHID, Sweet *Herminium monorchis* (Musk Orchis) Camden/Gibson.
ORCHID, Toothed *Orchis tridentata*.
ORCHID, Tubercled *Habenaria flava*.
ORCHID, Vanilla, Black *Nigritella nigra*.
ORCHID, Wasp *Ophrys apifera var. trollii*.
ORCHID, White, Small *Leucorchis albida*.
ORCHID, Winged *Orchis purpurea* (Lady Orchid) Genders.
ORCHID, Wood, Green, Small *Habenaria clavellata*.
ORCHID, Woodcock *Ophrys scolopax*.
ORCHID, Woodland *Habenaria clavellata* (Small Green Wood Orchid) USA Cunningham.
ORCHID BAUHINIA *Bauhinia petersiana*.
ORCHID TREE *Bauhinia purpurea*.
ORDEAL BEAN *Physostigma venenosum* (Calabar Bean) Le Strange. Very poisonous, and used by some West African peoples as an ordeal - the prisoner had to eat a certain amount of the seed, and if he vomited within a specified period, thus ridding himself of the poison, he was acquitted as innocent, and, of course, guilty if the seeds were retained. Certain types of crime were also punishable by forcing the guilty to drink an infusion of the crushed seed, which usually resulted in death within the hour.
ORDEAL TREE 1. *Cerbera tanghin*. 2. *Erythrophloeum guineense*. Again, used as an ordeal to ascertain guilt. The innocent will vomit, the guilty will not, and will, therefore, so it is believed, die of poison. Anyone accused of being a witch and pleading not guilty could demand the right of trial by *Erythrophloeum* poison. This is West African, while Cerbera comes from Madagascar.
ORECH; OREGE 1. *Atriplex patula* (Orach) Turner.
OREGANO 1. *Coleus amboinicus* (Indian Borage) S America G B Foster. 2. *Lippia graveolens*. 3. *Origanum vulgare* (Marjoram) Hemphill. 4. *Oreganum vulgare var. viride*. Oregano refers to a flavour rather than to a plant. All these plants have this particular flavour, and are used in similar ways.
OREGON ALDER *Alnus rubra* (Red Alder) Mitchell.1974. The earlier specific name was *oregona*, before *rubra* was preferred.
OREGON ASH *Fraxinus oregana*.
OREGON BALSAM POPLAR *Populus trichocarpa* (Western Balsam Poplar) Brimble.1948.
OREGON CEDAR *Chamaecyparis lawsoniana* (Lawson Cypress) Wit.
OREGON CRABAPPLE *Malus fusca*.
OREGON FIR 1. *Abies grandis* (Lowland Fir) USA Brimble.1948. 2. *Pseudotsuga menziesii* (Douglas Fir) Wilkinson.1981.
OREGON GRAPE *Mahonia aquifolium*.
OREGON HOLLY GRAPE *Mahonia aquifolium* (Oregon Grape) Yanovsky.
OREGON HOP *Leycesteria formosa* (Himalayan Honeysuckle) Phillips.
OREGON LILY *Lilium columbianum* (Columbia Lily) Woodcock.
OREGON MAPLE *Acer macrophyllum* (Large-leaved Maple) Mason.
OREGON MYRTLE *Umbellularia californica* (Californian Laurel) Hora.
OREGON OAK *Quercus garryana*.
OREGON PINE *Pseudotsuga menziesii* (Douglas Fir) USA Elmore.
ORGAL *Mentha pulegium* (Pennyroyal) Corn. Jago. Organ, Organy and Orgal are corruptions of oreganum, which is marjoram.
ORGAMENT *Origanum vulgare* (Marjoram) B & H.
ORGAN 1. *Mentha pulegium* (Pennyroyal) Dev. Choape. 2. *Origanum vulgare* (Marjoram) Turner.
ORGAN-HERB *Mentha pulegium* (Pennyroyal) W Eng. Grigson.
ORGANY 1. *Mentha pulegium* (Pennyroyal) Gerard. 2. *Origanum vulgare* (Marjoram) Dev, Worcs. Grigson; Wilts Dartnell & Goddard.
ORGLON *Crataegus monogyna* fruit

(Haws) Corn. Courtney.1880. This is actually Awglen, or Awglon, the first syllable being a rendering of OE haga, hedge.

ORHAMWOOD *Ulmus americana* timber (American Elm) Canada Wilkinson.1978.

ORIENTAL BEECH *Fagus orientalis*. Not all that far east, though, for this is a native of eastern Europe.

ORIENTAL BERRY *Anamirta cocculus* Chopra. Cf. Levant Berry for this.

ORIENTAL BITTERSWEET *Celastrus orbiculatus*.

ORIENTAL PLANE *Platanus orientalis*.

ORIENTAL POPPY *Papaver orientale*.

ORIENTAL THORN *Crataegus laciniata*. It doesn't seem to be all that oriental, for this is a native of southeast Europe.

ORIGANE *Oreganum vulgare* (Marjoram) Culpepper.

ORL *Alnus glutinosa* (Alder) Som. Macmillan; Glos. J D Robertson; Heref. Lewis; Worcs, Shrop. B & H. A variant of Aul, or Arl.

ORLAYA, Large-flowered *Orlaya grandiflora*.

OROBANCH *Orobanche minor* (Lesser Broomrape) Gerard.

OROGBO KOLA *Garcinia kola* (Bitter Kola) Dalziel.

ORPHAN JOHN *Sedum telephium* (Orpine) E Ang. Grigson. Cf. Harping Johnny. Both 'orphan' and 'harping' are corruptions of Orpine. The 'John' part of the name is a reference to a Midsummer Eve divination (St John's Eve, that is). A girl would hang up pieces of the plant, each named after her boy friends. The piece that lives the longest determines the successful one. There were other divination games played with it, according to which way the leaves bent, or determining whether the lover were true or false, etc.

ORPHYNE; ORPIES *Sedum telephium* (Orpine) Jones-Baker.1974 (Orphyne); Dur, N'thum, Berw, Roxb. Grigson (Orpies).

ORPINE *Sedum telephium*. French orpin, apparently from orpiment, which is Latin auripigmentum - gold pigment. Was it originally applied to another Sedum, yellow stonecrop, for instance?

ORPINE, St Giles's *Sedum telephium* (Orpine) Geldart. Because it was appropriated to St Giles's Day, 1 September.

ORPINE, Stone *Sedum reflexum* (Yellow Stonecrop) Prior.

ORPY-LEAF *Sedum telephium* (Orpine) Scot. Grigson.

ORRICE; ORRICE-ROOT *Iris germanica var. florentina* (Orris) both B & H.

ORRIS *Iris germanica var. florentina*. Perhaps orris is iris.

ORRIS, English *Valeriana officinalis* (Valerian) Webster. That is what the powder made from the roots was called.

ORRIS, Indian *Saussurea lappa*.

ORRIS-ROOT, Purple *Iris germanica* (Blue Flag) Lindley.

ORRIS-ROOT, White *Iris germanica var. florentina* (Orris) Lindley.

ORYELLE *Alnus glutinosa* (Alder) B & H. It sounds as though the West Midlands form Orl may have been responsible for this.

OSAGE ORANGE *Maclura pomifera*. The fruit looks like an orange, but it is not edible. 'Osage' in the name refers to the Osage Indians, in whose territory it was found.

OSHA *Ligusticum filicinum* (Colorado Root) USA H Smith.

OSIER *Salix viminalis*.

OSIER, American, Red *Cornus amomum* (Silky Cornel) Grieve.1931.

OSIER, Dutch *Salix alba var. vitellina* (Golden Willow) Chaplin.

OSIER, Golden *Myrica gale* (Bog Myrtle) IOW Grigson.

OSIER, Green 1. *Cornus alternifolia* (Pagoda Tree) USA Yarnell. 2. *Cornus florida* (Flowering Dogwood) W Miller.

OSIER, Purple *Salix purpurea*.

OSIER, Stone *Salix purpurea* (Purple Osier) Wilts. Dartnell & Goddard.

OSIER, Wimmer's *Salix calodendron*.

OSTERICK *Polygonum bistorta* (Bistort) Prior. Med. Latin ostriacum, apparently a corruption from Latin aristolochia, the name now given to birthwort.

OSWEGO TEA *Monarda didyma* (Red Bergamot) Brownlow. The leaves, fresh or dried, by themselves, as tea, or as an addition to the usual brew. Oswego is the name of a place on the shores of Lake Ontario, and the plant is common in that area.

OTAHEITE APPLE *Eugenia malaccensis* (Malay Apple) Howes.

OTAHEITE GOOSEBERRY *Phyllanthus distichus*.

OTAHEITE POTATO *Dioscorea bulbifera* (Potato Yam) Howes.

OULER *Alnus glutinosa* (Alder) Ches. Leigh. The OE was alor, aler (the 'd' being intrusive in the modern name). Cf. other forms such as Owler, Howler, etc.

OUR-LADY-IN-A-BOAT *Dicentra spectabilis* (Bleeding Heart) Coats.

OUR LADY'S BASIN *Dipsacus fullonum* (Teasel) Som. Macmillan. The 'basin' of the name is the cup formed by the fusing together of the plant's opposite leaves, "so fastened that they hold dew and raine water in manner of a little bason" (Gerard). That water was much prized for cosmetic use.

OUR LADY'S CANDLE *Verbascum thapsus* (Mullein) Som. Macmillan.

OUR LADY'S CUSHION *Armeria maritima* (Thrift) Parkinson.

OUR LADY'S FLANNEL *Verbascum thapsus* (Mullein) Som. Macmillan. A reference to the texture of the leaves.

OUR LADY'S HEART *Dicentra spectabilis* (Bleeding Heart) Som. Macmillan.

OUR LADY'S MINT *Tanacetum balsamita* (Balsamint) Tynan & Maitland.

OUR LADY'S NIGHTCAP *Calystegia sepium* (Great Bindwed) Som. Macmillan.

OUR LADY'S PINCUSHION *Capsella bursa-pastoris* (Shepherd's Purse) Thorpe.

OUR LADY'S SEAL *Bryonia dioica* (White Bryony) Geldart. See Lady's seal.

OUR LADY'S SMOCK *Calystegia sepium* (Great Bindwed) Som. Macmillan.

OUR LADY'S TEARS *Lithospermum officinale* (Gromwell) Presumably the reference is to the pearly seeds.

OUR LADY'S THIMBLE *Campanula rotundifolia* (Harebell) Som. Macmillan.

OUR LADY'S THISTLE *Carduus benedictus* (Holy Thistle) USA Henkel. Cf. Virgin Mary's Thistle, from East Anglia.

OUR LADY'S VINE *Tamus communis* (Black Bryony) Tynan & Maitland.

OUR LORD'S CANDLE *Yucca whipplei* USA Elmore.

OUR LORD'S FLANNEL; OUR SAVIOUR'S FLANNEL *Verbascum thapsus* (Mullein) both Kent Grigson.

OUTENIQUA YELLOW-WOOD *Podocarpus falcata*.

OUW *Hydrocotyle vulgaris* (Pennywort) IOM Moore.1924.

OVA-OVA *Monotropa uniflora* W Miller.

OVEN BUSH *Conyza ivaefolia* S Africa Watt.

OVERCUP OAK *Quercus lyrata* (Swamp Post Oak) USA Harper.

OVERENYIE *Artemisia abrotanum* (Southernwood) A S Palmer. i.e. Averoyne, from French armoise au rone = abrotanum.

OVERLOOK BEAN *Canavalia ensiformis* (Jack Bean) W Indies Dalziel. Cf. Cut-eye Bean and One-eye Bean, all descriptive, and from the West Indies.

OVEST *Quercus robur* fruit (Acorn) Hants. B & H.

OWD LAD'S PEASCODS *Laburnum anagyroides* fruit (Laburnum) Yks. Very poisonous, hence the attribution to the Owd Lad - Satan.

OWL'S CLAWS *Helenium hoopesii* (Orange Sneezeweed) USA Elmore.

OWL'S CROWN *Filago germanica* (Cudweed) W Miller.

OWL'S EYE *Anagallis arvensis* (Scarlet Pimpernel) Som. Macmillan. Cf. Bird's Eye, Pheasant's Eye, Eyebright, etc.

OWL'S SOCKS *Sarracenia purpurea* (Pitcher Plant) Canada Jenness.1935.

OWLDER *Alnus glutinosa* (Alder) Yks. Addy. These Owler or Ouler forms are usually without a 'd', but the 'd' of Alder returns in this variation.

OWLER 1. *Alnus glutinosa* (Alder) Derb. Grose; Yks. Hunter; Lancs. Nodal. 2. *Populus tremula* (Aspen) Wilts. Macmillan. Cf. Ouler. The Aspen record must have been a mis-reading.

OWLORN *Alnus glutinosa* (Alder) Grigson. Cf. Ollar, Oler and Olern.

OWM *Ulmus procera* (English Elm) Yks. Nicholson.

OX-BALM *Collinsonia canadensis* (Stone Root) Le Strange.

OX-EYE 1. *Adonis vernalis* (Yellow Pheasant's Eye) Grieve.1931. 2. *Anagallis arvensis* (Scarlet Pimpernel) Gunther. 3. *Chrysanthemum segetum* (Corn Marigold) Suff, Yks. Grigson. 4. *Heliopsis helianthoides* (False Sunflower) USA House. 5. *Leucanthemum vulgare*.

OX-EYE, American *Heliopsis laevis*.

OX-EYE, Girt *Leucanthemum vulgare* (Ox-eye Daisy) Cumb. B & H. Large (girt) in comparison with the common daisy, Bellis perennis, that is.

OX-EYE, Rough *Heliopsis scaber*.

OX-EYE, Yellow *Chrysanthemum segetum* (Corn Marigold) Fernie.

OX-EYE CAMOMILE *Anthemis tinctoria* (Yellow Camomile) Coats.

OX-HEAL; OX-HEEL *Helleborus foetidus* (Stinking Hellebore) Prior. 'Heal', rather than 'heel', for this was well known for its veterinary use.

OX-TONGUE 1. *Anchusa officinalis* (Alkanet) Lyte. 2. *Borago officinalis* (Borage) Lyte. 3. *Lycopsis arvensis* (Small Bugloss). In each case, it is the shape and texture of the leaves that suggest the imagery.

OX-TONGUE, Bristly *Picris echioides*.

OX-TONGUE, Hawkweed *Picris hieracioides*.

OXALIS, Blue *Parochetus communis* (Shamrock Pea) A W Smith.

OXALIS, Mexican *Oxalis latifolia*.

OXALIS, Pale *Oxalis incarnata*.

OXALIS, Pink *Oxalis floribunda*.

OXALIS, Tree *Oxalis ortgiesi*.

OXBERRY 1. *Arum maculatum* (Cuckoo-pint) Worcs. Grigson. 2. *Tamus communis* (Black Bryony) Fernie.

OXFORD AND CAMBRIDGE BUSH *Psoralea foliosa*. The flowers are bicoloured, light blue and dark blue.

OXFORD RAGWORT *Senecio squalidus*. It established itself first round the Oxford colleges in the 18th century, for it was grown in Oxford Botanic Garden from about 1690.

OXFORD WEED *Cymbalaria muralis* (Ivy-leaved Toadflax) Berks, Oxf. Grigson. Is it any commoner there than anywhere else?

OXLIP *Primula elatior*.

OXLIP, Bardfield *Primula elatior* (Oxlip) Genders.1976. This name is a measure of the shrinkage of oxlip's natural habitat as the old woodlands disappear. It suggests that it is only to be found near the Essex village of that name.

OXLIP, Common *Primula veris x vulgaris* (False Oxlip) Grigson.

OXLIP, False *Primula veris x vulgaris*.

OXSLIP *Primula veris x vulgaris* (False Oxlip) Suff, Norf, Yks. Grigson.

OYSTER 1. *Pinus sylvestris* cones (Pine cones) Dev. Friend.1882. 2. *Syringa vulgaris* (Lilac) Dev. Friend.1882.

OYSTER, Chiny *Callistephus chinensis* (China Aster) Wilts. Dartnell & Goddard. i.e. Chinese Aster.

OYSTER, Vegetable *Tragopogon porrifolius* (Salsify) A W Smith. The root has the flavour of oysters, they say. Cf. the American name Oyster Plant.

OYSTER BAY PINE *Callitris rhomboidea*.

OYSTER PLANT 1. *Mertensia maritima* (Smooth Lungwort) Border B & H. 2. *Rhoeo spathacea* (Boat Lily) Tampion. 3. *Tragopogon porrifolius* (Salsify) USA Zenkert. 4. *Tragopogon pratensis* (Goat's Beard) USA Gates. They say salsify roots taste like oysters, and they also say the same about the leaves of Smooth Lungwort.

OYSTERNUT *Telfairia pedata*.

OZARK SUNDROPS *Oenothera missouriensis*.

P

PACIFIC FLOWERING DOGWOOD
Cornus nuttallii (Mountain Dogwood) E Gunther.
PACIFIC LABRADOR TEA *Ledum columbianum*. A built-in contradiction in the name, but this is not the only species that grows on the west coast of America.
PACIFIC MAPLE *Acer macrophyllum* (Large-leaved Maple) Brimble.1948.
PACIFIC RED CEDAR *Thuja plicata* (Giant Cedar) Wit.
PACIFIC SILVER FIR *Abies amabilis* (Red Silver Fir) Schery.
PACIFIC WILLOW *Salix lasiandra* (Yellow Willow) USA T H Kearney. 'Pacific', because it is a native of the western side of North America.
PACIFIC YEW *Taxus brevifolia* (Western Yew) Yamovsky.
PACKLE, Crow *Ranunculus bulbosus* (Bulbous Buttercup) N'thants. A E Baker. Cf. from the same county, Crow Pightle and Crow Pikel. Pightle is an old word for a meadow.
PADDA FOOT *Teucrium africanum* S Africa Watt.
PADDIE PIPES *Equisetum palustre* (Marsh Horsetail) Scot. Gibbings.1940. Paddie is the diminutive of paddock, usually a toad, but probably meant to include frogs as well. The croaking of frogs is said to be the playing, by the frogs, of these hollow pipes, like musical instruments.
PADDOCK CHEESE *Asparagus officinalis* (Asparagus) Halliwell.
PADDOCK FLOWER *Caltha palustris* (Marsh Marigold) Potter & Sargent. 'Paddock' is a toad. But what is the connection with this plant? It is just possible that the large leaves are relevant, as a hiding place for toads, perhaps.
PADDOCK PIPE 1. *Equisetum arvense* (Common Horsetail) H C Long.1910. 2. *Equisetum palustre* (Marsh Horsetail) Scot. Jamieson. See Paddie's Pipes.
PADDOCK'S PIPES 1. *Equisetum arvense* (Common Horsetail) Allan. 2. *Hippuris vulgaris* (Mare's Tail) Cumb. Grigson.
PADDOCK'S SPINDLE 1. *Orchis mascula* (Early Purple Orchid) Perth. B & H. 2. *Orchis morio* (Green-winged Orchid) Perth. B & H.
PADELION *Alchemilla vulgaris* (Lady's Mantle) Gerard. French pas de lion, from the resemblance of the leaf to the impress of a lion's foot. But it is also Bear's, Lamb's, and even Duck's Foot.
PAGE OAK *Rhus radicans* (Poison Ivy) H & P.
PAGGLE *Primula veris* (Cowslip) Tusser. See Paigle.
PAGODA DOGWOOD *Cornus alternifolia* (Pagoda Tree) Howes.
PAGODA FLOWER *Clerodendron paniculatum*.
PAGODA TREE 1. *Cornus alternifolia*. 2. *Plumeria acuminata*. 3. *Sophora japonica*. It is the general shape of the Cornus that occasions the name, but the Sophora is a Far Eastern tree anyway.
PAHUTAN MANGO *Mangifera altissima*.
PAHUTEA *Libocedrus bidwilli* (Mountain Cedar) Leathart.
PAIGLE 1. *Cardamine pratensis* (Lady's Smock) Suff. Grigson. 2. *Primula elatior* (Oxlip) Clapham. 3. *Primula veris*. 4. *Primula veris x vulgaris* (False Oxlip) Gerard. 5. *Primula vulgaris* (Primrose) Ches. Leigh. 6. *Ranunculus acris* (Meadow Buttercup) Suff. Grigson. 7. *Ranunculus bulbosus* (Bulbous Buttercup) Suff. B & H. 8. *Ranunculus repens* (Creeping Buttercup) Suff. Grigson. 9. *Stellaria holostea* (Greater Stitchwort) B & H. Some botanists prefer the name Paigle for Oxlip, but it is most commonly recorded for Cowslip. The derivation is unknown, but a verb 'to paggle' is sometimes quoted. It apparently meant to bulge, or swell. But Grigson said 'to paggle', when applied to a cow's neck, meant to hang and

shake - hence an analogy with the loosely hanging flowers. Yet another attempt at the derivation saw the original as the French paillette, a spangle.

PAIGLE, Cow *Primula veris* (Cowslip) Hants. Grigson.

PAIGLE, French *Pulmonaria officinalis* (Lungwort) Ess. Gepp. Cf. the various 'cowslip' names for lungwort.

PAIGYA BEAN *Phaseolus lunatus* (Lima Bean) H C Long.1924.

PAIN-KILLER, Devil's *Hieraceum aurantiacum* (Orange Hawkweed) Howes.

PAINE, Tom *Quercus robur* (Oak) Yks. Grigson. Because of oak's sterling, all-round qualities, perhaps?

PAINT, Indian 1. *Chenopodium capitatum* (Strawberry Blite) USA Yarnell. 2. *Lithospermum angustifolium* (Puccoon) USA Elmore. 3. *Lithospermum canescens* (Hoary Puccoon) Miss. Bergen. 4. *Sanguinaria canadensis* (Bloodroot) Grieve.1931. The puccoons and the bloodroot yield a red dye from their rootstocks, used for face paint too, and strawberry blite was also used by the Indians for a dream dance paint.

PAINTBRUSH 1. *Centaurea nigra* (Knapweed) Dev. Macmillan. 2. *Tragopogon pratensis* seeds (Goat's Beard) Som. Macmillan.

PAINTBRUSH, Desert *Castilleia angustifolia*.

PAINTBRUSH, Devil's *Hieraceum aurantiacum* (Orange Hawkweed) USA House.

PAINTBRUSH, Fairy's *Vinca minor* (Lesser Periwinkle) Som. Macmillan.

PAINTBRUSH, Indian 1. *Asclepias tuberosa* (Orange Milkweed) USA Cunningham. 2. *Castilleia angustifolia* (Desert Paintbrush) USA. E Gunther. 3. *Castilleia integra*.

PAINTBRUSH, Long-leaved *Castilleia linariaefolia*.

PAINTBRUSH, Sessile *Castilleia sessiliflora*.

PAINTBRUSH, White *Haemanthus albiflos*.

PAINTBRUSH, Woolly *Castilleia foliolosa*.

PAINTBRUSH, Yellow *Castilleia plagiotoma*.

PAINTED CALADIUM *Caladium picturatum*.

PAINTED CUP 1. *Castilleia linariaeflora* (Long-leaved Paintbrush) USA. Whiting. 2. *Pedicularis sylvatica* (Lousewort) Tynan & Maitland.

PAINTED DAISY *Pyrethrum roseum* (Persian Pellitory) USA Cunningham.

PAINTED DAMASK *Rosa damascena* 'Leda'.

PAINTED DROPTONGUE *Aglaonema crispum*.

PAINTED GRASS *Phalaris arundinacea* (Reed-grass) Howes. Because of the two-coloured leaves.

PAINTED LADY 1. *Castilleia sp* Willis. 2. *Dianthus barbatus* (Sweet William) Dyer. 3. *Geranium versicolor*. 4. *Lathyrus odoratus* (Sweet Pea) Wilts. Dartnell & Goddard; IOM Moore.1924. 5. *Saxifraga x urbinum* (London Pride) Som. Macmillan.

PAINTED LEAF *Euphorbia heterophylla* USA Cunningham.

PAINTED NET-LEAF *Fittonia verschaffeltii* (Silver-nerve Fittonia) RHS.

PAINTED PINK *Dianthus furcatus*.

PAINTED SPURGE *Euphorbia heterophylla* USA T H Kearney.

PAINTED TONGUE *Salpiglossis sinuata* (Velvet Trumpet Flower) Perry.

PAINTED TRILLIUM *Trillium undulatum*.

PAINTED WOOD LILY *Trillium undulatum* (Painted Trillium) RHS.

PAINTER'S PALETTE 1. *Anthurium andreanum*. 2. *Anthurium scherzerianum* (Flamingo Plant) Rochford.

PAIUTE CABBAGE *Stanleya pinnata* (Prince's Plume) USA Jaeger. The Indians used to eat the young leaves as greens.

PAK-CHOI *Brassica sinensis* (Chinese Mustard) Usher. Pak-choi is the Cantonese form of pai-tsai, white vegetable.

PALACE, Hare's *Sonchus oleraceus* B & H. An old legend tells how this plant gave strength to hares when they were overcome with the heat. Surely it is far more likely that hares (and rabbits) actually eat the plant.

Cowper said that his pet hares were very fond of sow thistle.

PALAY RUBBER *Cryptostegia grandiflora* (Rubber Vine) Schery.

PALESTINE ANEMONE *Anemone coronaria* (Poppy Anemone) Moldenke.

PALESTINE BRAMBLE *Rubus sanctus.*

PALESTINE BUCKTHORN *Rhamnus palaestina.*

PALESTINE NIGHTSHADE *Solanum incanum* (Hoary Nightshade) Moldenke.

PALESTINE TEREBINTH *Pistacia terebinthus vae.palaestina.*

PALESTINE WILLOW *Salix safsaf.*

PALESTINIAN TUMBLEWEED *Anastatica hierochuntica* (Rose of Jericho) Moldenke. Because of its "tumbleweed" habit of forming itself into a ball when the seeds have ripened, and allowing itself to be blown over the desert until it finds moisture.

PALETTE, Painter's 1. *Anthurium andreanum.* 2. *Anthurium scherzerianum* (Flamingo Plant) Rochford.

PALM 1. *Buxus sempervirens* (Box) Dllimore. 2. *Corylus avellana* catkins (Hazel catkins) B & H. 3. *Salix capraea* (Goat Willow) Wilts, Oxf, Norf. Grigson. Heref. Leather. 4. *Taxus baccata* (Yew) Notes & Queries, 1858. All used as Palm Sunday substitutes for the real thing.

PALM, Betel; PALM, Beetle-nut *Areca catechu* (Areca Palm) H & P (Betel); USA Cunningham. Beetle = betel.

PALM, Bottle *Beaucarnea recurvata* (Pony Tail Plant) USA Goold-Adams. The swollen base of the stem accounts for the name. Cf. the American name Elephant's Foot.

PALM, Canary *Phoenix canariensis.*

PALM, Carnauba Wax *Copernicia cerifera.*

PALM, Chusan *Trachycarpus excelsus.*

PALM, Coir *Trachycarpus excelsus* (Chusan Palm) Hyatt.

PALM, Cretan *Phoenix theophrasii.*

PALM, Date *Phoenix dactylifera.*

PALM, Date, Wild *Phoenix recunata.*

PALM, English *Salix capraea* (Goat Willow) N'thants. A E Baker; Staffs. Poole. This is the tree that gives "palm" for Palm Sunday. Perhaps any catkin-bearing tree could serve, for hazel was used, too. But those of Goat Willow have always been the English embodiment of "palm".

PALM, Fan 1. *Rhapis excelsa.* 2. *Trachycarpus excelsus.* 3. *Washingtonia filifera.* The name refers to the spread of the leaves.

PALM, Fan, Dwarf; PALM, FAN, European *Chamaerops humilis* RHS (European).

PALM, Ivory, Vegetable *Hyphaene benguellensis.*

PALM, Kentia *Howeia forsteriana.* Kentia is the capital of Lord Howe Island, where the palm grows, and after which the genus is named.

PALM, Nut *Corylus avellana* catkins (Hazel catkins) B & H. A "palm" for Palm Sunday decorations.

PALM, Oil *Elaeis guineensis.* Palm oil is extracted from the fleshy part of the fruit, and used in soap, margarine and candle manufacture, as well as for some lubricants.

PALM, Panama Hat *Carludovica palmata.* Strips of the leaves are woven into Panama hats, an industry which is centred in Ecuador.

PALM, Paradise *Howeia forsteriana* (Kentia Palm) RHS.

PALM, Parlour *Chamaedorea elegans.*

PALM, Raffia *Raphia ruffia.* This is the palm from whose dried leaves bast, or raffia, is made.

PALM, Sentry *Howeia forsteriana* (Kentia Palm) RHS.

PALM, Thatch-leaf *Howeia forsteriana* (Kentia Palm) RHS.

PALM, Tie *Dracaena indivisa* (Fountain Tree) Howes.

PALM, Windmill *Trachycarpus excelsus.*

PALM GRASS *Setaria palmifolia.*

PALM LILY *Dracaena indivisa* (Fountain Dracaena) Andersen.

PALM OF CHRISTIAN *Ricinus communis* (Castor Oil Plant) southern USA Puckett. i.e. Palma Christi.

PALM WILLOW *Salix capraea* (Goat Willow) Leics. Grigson. See English Palm.

PALMA CHRISTI 1. *Dactylorhis maculata* (Heath Spotted Orchid) Henslow. 2. *Ricinus communis* (Castor Oil Plant) USA Kingsbury. 'Palma' is the palm of the hand, so the name for the orchid comes from the shape of the tubers, which are divided into several finger-like lobes. Hence the many "hand" or "finger" names for it. Castor Oil Plant gets the name from the leaves rather than the roots - these are palmately divided. Actually this name for Ricinus, though from an American source,comes from Turner.

PALMA ISTLE *Yucca carnerosana*. 'Istle' should really be reserved for the fibre. The word is Mexican Spanish ixtle, which in turn came from Nahuatl ichtli.

PALMER *Salix capraea* (Goat Willow) Dor. Grigson. i.e. the plant that provides the "palm" for Palm Sunday in England.

PALMEROSA OIL GRASS *Cymbopogon maritima* (Ginger Grass) Zohary.

PALMILLA *Yucca elata*.

PALO VERDE, Border *Cercidium floridum*. 'Palo verde' means green stick.

PALSY-CURER *Cardiospermum halicacabum* (Blister Creeper) Chopra.

PALSY PLANT *Primula vulgaris* (Primrose) Fernie. In mistake for cowslip, presumably - see Palsywort.

PALSYWORT *Primula veris* (Cowslip) Gerard. Cowslips formed a medieval cure for palsy. Gerard - "cowslips are commended against...the slacknesse of the sinues, which is the palsy". It must have been the trembling or nodding of the flowers that suggested it.

PAMPHREY *Brassica oleracea var. capitata* (Cabbage) N Ire. PLNN25.

PAMPAS GRASS *Gynerium argenteum*.

PAN, Kite's *Dactylorchis maculata* (Heath Spotted Orchid) Wilts. Dartnell & Goddard.

PANAMA HAT PALM *Carludovica palmata*. Strips of leaf are woven into Panama hats, an industry that is centred in Ecuador.

PANAMA ORANGE *Citrus mitis* (Calamondin) Grigson.1974. Why? When this is a Filipino tree?

PANAMA RUBBER *Castilla elastica*.

PANCAKE *Umbilicus rupestris* (Wall Pennywort) Dev. Friend.1882; Som. Macmillan. Descriptive - from the flat, round leaves.

PANCAKE PLANT; PANS-AND-CAKES *Malva sylvestris* (Common Mallow) Som. Macmillan; Lincs. Rudkin (Pancake Plant); Som. Macmillan (Pans-and-cakes). "Owing to the "pie-shaped leaves", according to Rudkin.

PANCE; PAUNCE *Viola tricolor* (Pansy) Fernie (Pance); B & H (Paunce).

PANES *Pastinaca sativa* (Parsnip) Corn. Jago; Dor. Macmillan.

PANICLED KNAPWEED *Centaurea paniculata*.

PANICK, Petty *Phalaris canariensis* (Canary Grass) Turner. By this name, he is likening it to a small kind of millet (*Panicum*).

PANICUM, Fall *Panicum dichotomiflorum*. June to October is its flowering time.

PANIKE *Panicum italica* Turner.

PANSY 1. *Prunella vulgaris* (Self-heal) W Miller. 2. *Viola tricolor*. Pansy is the French word pensée, thought. Shakespeare has Ophelia say "...and there is pansies, that's for thoughts". Pansy for self-heal is probably just an extension of Heartsease, given because it was thought to cure heart disease.

PANSY, Corn *Viola tricolor* (Pansy) H C Long.1910.

PANSY, Dwarf *Viola kitaibeliana*.

PANSY, Field 1. *Viola arvensis*. 2. *Viola rafinesquii* USA T H Kearney.

PANSY, Heart *Viola tricolor* (Pansy) Dev. B & H. Cf. Heartease, both suggesting that pansies can make a cordial. It must have been given in mistake, for the reputation as a medicine for heart trouble is not borne out by domestic practice, and must properly belong to another plant, perhaps the wallflower.

PANSY, Mountain *Viola lutea*.

PANSY, Sand *Viola tricolor var. curtisii* (Seaside Pansy) Hepburn.
PANSY, Seaside *Viola tricolor var. curtisii*.
PANSY, Shepherd's *Viola lutea* (Mountain Pansy) N'thum. Grigson.
PANSY, Wild 1. *Viola arvensis* (Field Pansy) USA Upham. 2. *Viola rafinesquii* USA Gates.
PANSY VIOLET, Yellow *Viola pedunculata*.
PANTHER LILY *Lilium pardalinum*.
PANTHER LILY, Californian *Lilium nevadense*.
PAPAW *Asimina triloba*. Probably a variant of papaya, even though this is an entirely different genus.
PAPAYA *Carica papaya* (Pawpaw) Wit. A Spanish word, but probably Carib originally.
PAPER, Ag-; PAPER, Rag *Verbascum thapsus* (Mullein) both Bucks. B & H. Probably a mishearing of Ag-taper, where 'ag' (and 'rag') mean hedge, and taper is a reference to the old country practice of dipping the stems of mullein in tallow or suet and burning them to provide light at outdoor country gatherings, or even in the home. Cf. Hag-taper and its variants.
PAPER-BAG BUSH *Salazaria mexicana*. The reference is to the papery pods.
PAPER-BARK, Scented *Melaleuca squarrosa*.
PAPER-BARK (TREE) 1. *Melaleuca lasiandra* Australia Meggitt. 2. *Melaleuca leucadendron* (Cajuput) Chopra. 3. *Melaleuca quinquenervia*.
PAPER-BARK MAPLE *Acer griseum*.
PAPER-BARK THORN *Acacia woodii*.
PAPER BEECH *Betula verrucosa* (Birch) Wilts. Macmillan.
PAPER BIRCH 1. *Betula papyrifera*. 2. *Betula verrucosa* (Birch) Wilts. Dartnell & Goddard.
PAPER BIRCH, Indian *Betula bhojpattra*.
PAPER DAISY, Golden *Helichrysum bracteatum* (Everlasting Flower) Australia Howes.
PAPER FLOWER *Lunaria annua* (Honesty) Tynan & Maitland.

PAPER MULBERRY *Broussonettia papyrifera*. This is the tree whose bark provides tapa, or bark cloth.
PAPER SEDGE; PAPER REED *Cyperus papyrus*. The leaf stalks contain the material that is made into papyrus, the original writing material of the ancient Egyptians.
PAPER TREE *Commiphora marlothii*. The 'paper' peels from the trunk. It is rather brittle, but can be written on with a very soft pencil.
PAPER TREE, Red *Albizia rhodesica*.
PAPPLE *Agrostemma githago* (Corn Cockle) Scot. Grigson. Cf. Pawple and Popille, also from Scotland, and the widespread Popple for Cockle.
PAPOOSE-ROOT *Caulophyllum thalictroides* (Blue Cohosh) USA Grieve.1931. Both this and Squawroot seem to occur because of its old use as an aid in childbirth. An infusion of the root in warm water was drunk as a tea for a week or two prior to the expected date of delivery, in order to promote an easy birth.
PAPRIKA *Capsicum annuum var. longum*. Paprika is a Hungarian word, from the modern Greek piperi, pepper.
PAPWORT *Mercurialis perennis* (Dog's Mercury) Halliwell.
PAPYRUS PLANT *Cyperus papyrus* (Paper Sedge) Bonar.
PARA NUT *Bertholletia excelsa* (Brazil Nut) Brouk.
PARA RHATANY *Krameria argentea*.
PARACHUTE PLANT *Ceropegia sandersonii* (Fountain Flower) RHS. The petals are flared widely at the tips to form "parachutes".
PARACHUTES *Campanula medium* (Canterbury Bells) Som. Macmillan.
PARADISE, Bird of 1. *Aconitum napellus* (Monkshood) Som. Macmillan. 2. *Caesalpina pulcherrima* (Barbados Pride) S Africa Watt. 3. *Poinciana gilliesii* (Poinciana) USA Kingsbury.
PARADISE, Grains of 1. *Aframomum melegueta* seeds. 2. *Elettaria cardamomum* (Cardamom) Langham.

PARADISE, Oak of *Chenopodium botrys* (Jerusalem Oak) E Hayes. The young leaves look like miniature oak leaves.

PARADISE APPLE 1. *Lycopersicon esculentum* (Tomato) S Africa Watt. 2. *Malus sylvestris var. paradisiaca*.

PARADISE FLOWER 1. *Caesalpina pulcherrima* (Barbados Pride) Howes. 2. *Solanum wendlandii* (Giant Potato Creeper) Gooding.

PARADISE LILY *Papaver rhoeas* (Red Poppy) Som. Macmillan.

PARADISE NUT *Lecythis usitata*.

PARADISE PALM *Howeia forsteriana* (Kentia Palm) RHS.

PARADISE PLANT *Daphne mezereon* (Lady Laurel) Som, Glos. Grigson. Alice Coats suggested that this is a reference to the old concept of a "paradise garden", an enclosure or park planted with trees and shrubs, where exotic animals were kept.

PARADISE SEED, Egyptian *Amomum grana* (Grains of Paradise Plant) USA Hurston.

PARAGUAY TEA *Ilex paraguayensis* (Maté) Schery. Cf. Brazil Tree.

PARALYSY, Herb *Primula veris* (Cowslip) Gerard. See Palsywort.

PARANA PINE *Araucaria angustifolia* (Brazilian Pine) Wit. This is the name under which the timber is exported.

PARANG BEAN *Canavalia gladiata*.

PARASOL 1. *Convolvulus arvensis* (Field Bindweed) Som. Macmillan. 2. *Poterium sanguisorba* (Salad Burnet) Wilts. Tynan & Maitland. 3. *Sanguisorba officinalis* (Great Burnet) Wilts. Dartnell. & Goddard. Obviously descriptive for bindweed, but the imagery for the burnets is not at all clear.

PARASOL, Lady's *Daucus carota* (Wild Carrot) Tynan & Maitland.

PARASOL PINE *Sciadopitys verticillata* (Umbrella Pine) Willis.

PARASOL TREE *Schefflera arboricola*.

PARCHMENT BARK *Pittosporum crassifolium* (Karo) Howes.

PAREIRA *Chondodendron tomentosum*.

PARILLA, Yellow *Menispermum canadense* (Moonseed) Mitton. Parilla is short for sarsaparilla, which appears in a number of names for this.

PARIS, Herb *Paris quadrifolia*. 'Paris' here is the Latin meaning pair, hence herba paris, "pair herb", i.e. herb of equality, "from the numerical harmony of its parts" (Grigson).

PARIS DAISY *Argyranthemum frutescens*.

PARK LEAVES *Hypericum androsaemum* (Tutsan) Som. Macmillan. Probably a corruption of Hypericum.

PARLOUR PALM *Chamaedorea elegans*.

PARMACETTY, Poor Man's *Capsella bursa-pastoris* (Shepherd's Purse) Gerard. According to Dyer, a joke on the latin bursa, purse, which to a poor man is always the best remedy for his bruises.

PARNASSIA, Rocky Mountain *Parnassia fimbriata*.

PARNASSUS, Grass of *Parnassia palustris*. The name was apparently coined by Lyte in 1578, and is a translation of gramen Parnasium, itself from Greek agrestis en Parnasso, the agrestis growing on Parnassus, which was taken to be this.

PARNELL *Viburnum opulus* (Guelder Rose) Shrop. B & H.

PARNELL, Prattling *Saxifraga x urbinum* (London Pride) Wright.

PARRITORY *Parietaria diffusa* (Pellitory-of-the-wall) Grigson. The older name for this plant, through Old French and ultimately from the Latin paries, wall.

PARROT FEATHER 1. *Myriophyllum aquaticum*. 2. *Myriophyllum brasiliense*.

PARROT HEAD *Sarracenia psittacina* (Parrot Pitcher Plant) Howes.

PARROT LEAF *Alternanthera ficoidea*. An acknowledgment of the brightly coloured foliage.

PARROT PITCHER-PLANT *Sarracenia psittacina*.

PARROT-SEED *Carthamus tinctorius* (Safflower) Leyel.1948.

PARROT-WEED 1. *Bocconia frutescens*

(Tree Celandine) W Miller. 2. *Eryngium foetidum* Jamaica Beckwith.
PARROT'S BILL *Clianthus puniceus*. Scarlet flowers, shaped like a parrot's bill.
PARROT'S FLOWER *Heliconia psittacorum*.
PARROT'S PLANTAIN *Heliconia psittacorum* (Parrot's Flower RHS.
PARRY PINE *Pinus quadrifolia*.
PARSIL; PARCEL *Petroselinum crispum* (Parsley) Yks. Carr. (Parsil); Halliwell (Parcel). The earlier forms were Persele or Percely.
PARSLEY *Petroselinum crispum*. The OE was petersilie, modified by French persil, and both were from the Latin petroselinum, itself from Greek petroselinon, where petros means rock, and selinon is the name of the herb.
PARSLEY, Alexandrian *Smyrnium olusatrum* (Alexanders) Genders.
PARSLEY, Ass *Aethusa cynapium* (Fool's Parsley) Prior. Not just another rendering of Fool's Parsley - 'ass' is probably ache, in which case the name just means parsley-parsley.
PARSLEY, Bastard *Caucalis daucoides* (Bur Parsley) Gerard.
PARSLEY, Beaked 1. *Anthriscus caucalis* (Bur Chervil). 2. *Anthriscus cerefolium* (Chervil) Barton & Castle. 3. *Anthriscus sylvestris* (Cow Parsley). 4. *Myrrhis odorata* (Sweet Cicely) Leyel.1937.
PARSLEY, Bur 1. *Caucalis daucoides*. 2. *Torilis japonica* (Hedge Parsley) Salisbury.
PARSLEY, Bur, Corn; PARSLEY, Bur, Field *Torilis arvensis* (Spreading Bur Parsley) both Salisbury.
PARSLEY, Bur, Greater *Turgenia latifolia*.
PARSLEY, Bur, Knotted *Torilis nodosa*.
PARSLEY, Bur, Spreading *Torilis arvensis*.
PARSLEY, Cambridge *Selinum carvifolia*. Very rare, occurring only in one or two fens in Cambridgeshire.
PARSLEY, Chinese *Coriandrum sativum* (Coriander) Sanecki.
PARSLEY, Coney *Anthriscus sylvestris* (Cow Parsley) Surr. B & H.

PARSLEY, Corn 1. *Petroselinum segetum*. 2. *Trinia glauca* (Honewort) Gerard.
PARSLEY, Cow 1. *Aethusa cynapium* (Fool's Parsley) Som. Macmillan. 2. *Anthriscus sylvestris*.
PARSLEY, Curled *Petroselinum crispum* (Parsley) Hemphill.
PARSLEY, Devil's *Anthriscus sylvestris* (Cow Parsley) Ches. B & H. Why this attribution to a relatively harmless plant? The answer probably lies in the confusion between this and hemlock. Quite a lot of names are shared between the two.
PARSLEY, Dog's 1. *Aethusa cynapium* (Fool's Parsley) B & H. 2. *Anthriscus sylvestris* (Cow Parsley) B & H.
PARSLEY, False 1. *Aethusa cynapium* (Fool's Parsley) Shrop. Grigson. 2. *Cicuta maculata* (American Cowbane) USA Kingsbury.1967.
PARSLEY, Fern-leaved *Petroselinum crispum var. filicinum*.
PARSLEY, Fool's *Aethusa cynapium*. The young leaves of this look vaguely like parsley, but the effects of mistaking them can be dire - it is poisonous.
PARSLEY, Gypsy's 1. *Anthriscus sylvestris* (Cow Parsley) Som. Grigson. 2. *Geranium robertianum* (Herb Robert) Som. Macmillan.
PARSLEY, Hamburg *Petroselinum fusiformis*.
PARSLEY, Hare's 1. *Anthriscus sylvestris* (Cow Parsley) Som, Wilts. Grigson. 2. *Conium maculatum* (Hemlock) Som. Macmillan.
PARSLEY, Hedge *Torilis anthriscus*.
PARSLEY, Hedge, Knotted *Torilis nodosa*.
PARSLEY, Hedge, Spreading *Torilis arvensis*.
PARSLEY, Hedge, Upright *Torilis japonica*.
PARSLEY, Hedgehog *Caucalis daucoides* (Bur Parsley) McClintock. 'Hedgehog' for 'bur' is a common simile.
PARSLEY, Hedgehog, Broad-leaved *Caucalis latifolia* (Great Bur Parsley) McClintock.1976.

PARSLEY, Horse 1. *Heracleum sphondyllium* (Hogweed) Som. Macmillan. 2. *Smyrnium olusatrum* (Alexanders) Prior. 'Horse' in this kind of context always means big, or coarse.

PARSLEY, Italian *Petroselinum crispum var. neapolitanum.*

PARSLEY, Laughing *Pulsatilla vulgaris* (Pasque Flower) Coats. A name apparently given by William Salmon in 1710 (English herbal).

PARSLEY, Macedonian *Smyrnium olusatrum* (Alexanders) Turner. This is possibly the reason for the name Alexanders; Macedonia was Alexander's country. Or perhaps it is just a corruption of olusatrum.

PARSLEY, Marsh *Apium graveolens* (Wild Celery) Gerard.

PARSLEY, Milk *Peucedanum palustre.*

PARSLEY, Milk, False *Selinum carvifolia* (Cambridge Parsley) Genders.

PARSLEY, Naughty Man's *Anthriscus sylvestris* (Cow Parsley) War. Palmer. Cf. Devil's Parsley.

PARSLEY, Pig's 1. *Daucus carota* (Wild Carrot) Corn. Grigson; Dor. Macmillan. 2. *Torilis japonica* (Upright Hedge Parsley) Som. Elworthy.

PARSLEY, Poison *Conium maculatum* (Hemlock) USA Henkel.

PARSLEY, Rose *Anemone hortensis* Turner.

PARSLEY, Scotch *Ligusticum scoticum* (Lovage) Pennant.

PARSLEY, Sea *Ligusticum scoticum* (Lovage) Scot. Grigson.

PARSLEY, Sheep's 1. *Anthriscus sylvestris* (Cow Parsley) Kent, Norf. B & H. 2. *Chaerophyllum temulentum* (Rough Chervil) Suff. Grigson.

PARSLEY, Spotted 1. *Cicuta maculata* (American Cowbane) USA Kingsbury.1967. 2. *Conium maculatum* (Hemlock) USA Henkel.

PARSLEY, Square *Bunium bulbocastanum* (Great Earth-nut) Turner.

PARSLEY, Stone *Sison amomum.*

PARSLEY, Turnip-rooted *Petroselinum fusiformis* (Hamburg Parsley) Clair.

PARSLEY, Water *Apium graveolens* (Wild Celery) Gerard.

PARSLEY, Welsh *Cannabis sativa* (Hemp) Coles.

PARSLEY, White-flowered *Cogswellia orientalis.*

PARSLEY, Wild 1. *Anthriscus sylvestris* (Cow Parsley) Lincs. Mabey; USA Allan. 2. *Scandix pecten-veneris* (Shepherd's Needle) Bucks. Grigson. 3. *Sison amomum* (Stone Parsley) Turner. 4. *Smyrnium olusatrum* (Alexanders) Henslow. They are all umbellifers, so would qualify for the name.

PARSLEY BREAKSTONE 1. *Alchemilla vulgaris* (Lady's Mantle) Fernie. 2. *Aphanes arvensis* (Parsley Piert) Suff, Scot. B & H; Staffs. Grigson. Neither of them are umbellifers - in fact the name Parsley Piert comes from French perce-pierre, meaning breakstone. By sympathy, it was much used against stones in the bladder. Parsley in this name refers only to the form of the leaves.

PARSLEY FERN *Tanacetum vulgare* (Tansy) Dev. Friend.1882 USA Henkel.

PARSLEY HAWTHORN *Crataegus apiifolia.*

PARSLEY PEART; PARSLEY PEAT; PARSLEY PERK; PARSLEY PIERCESTONE *Aphanes arvensis* (Parsley Piert) Yks. Addy (Peat); Ches. Holland (Perk); Grieve.1931 (Piercestone); Notes and Queries, 1858 (Peart).

PARSLEY PIERT 1. *Aphanes arvensis*. 2. *Scleranthus annuus* (Knawel) Gerard. It comes from French perce-pierre, meaning breakstone, because it grows on gravelly soil. That fact, using the doctrine of signatures, ensured that it was much used against stones in the bladder. Gypsies use an infusion of the dried herb for gravel and other bladder troubles, and it was in great demand in the 1939-1945 war for that purpose.

PARSLEY VLIX *Aphanes arvensis* (Parsley Piert) Dor. Grigson.

PARSLEY WATER DROPWORT 1. *Oenanthe lachenalii.* 2. *Oenanthe pimpinelloides* (Callous-fruited Water Dropwort) Storer.
PARSNEP, Wild *Pastinaca sativa* (Wild Parsnip) Gerard.
PARSNIP, Cow 1. *Heracleum lanatum* USA Schenk & Gifford. 2. *Heracleum maximum.* 3. *Heracleum sphondyllium* (Hogweed) Turner.
PARSNIP, Cow, Indian *Heracleum lanatum* USA Douglas.
PARSNIP, Cow's *Arum maculatum* (Cuckoo-pint) Som. Macmillan.
PARSNIP, Crow *Taraxacum officinale* (Dandelion) B & H.
PARSNIP, Meadow *Heracleum sphondyllium* (Hogweed) Gerard.
PARSNIP, Pig's *Heracleum sphondyllium* (Hogweed) Som. Elworthy; Shrop. B & H.
PARSNIP, Rough *Heracleum sphondyllium* (Hogweed) Turner.
PARSNIP, Tree *Heracleum villosum* Jacob.
PARSNIP, Water 1. *Berula erecta* (Lesser Water Parsnip) G N Taylor. 2. *Sium nodiflorum* (Procumbent Water Parsnip) Hulme. 3. *Sium suave.* 4. *Stratiotes aloides* (Water Soldier) Grieve.1931.
PARSNIP, Water, Broad-leaved *Sium latifolium* (Greater Water Parsnip) Watt.
PARSNIP, Water, Creeping *Sium nodiflorum* (Procumbent Water Parsnip) Thornton.
PARSNIP, Water, Greater *Sium latifolium.*
PARSNIP, Water, Lesser *Berula erecta.*
PARSNIP, Water, Narrow-leaved *Berula erecta* (Lesser Water Parsnip) Clapham.
PARSNIP, Water, Procumbent *Sium nodiflorum.*
PARSNIP, Wild 1. *Cicuta maculata* (American Cowbane) USA Kingsbury.1967. 2. *Heracleum lanatum* USA Spier. 3. *Pastinaca sativa.* Latin pastinaca meant a carrot, and the verb pastinare to dig. Hence pastinum was a garden fork. The Old French was pasnie, which became in Middle English passenep.
PARSON-AND-CLERK; PARSON-IN-HIS-SMOCK *Arum maculatum* (Cuckoo-pint) Dev, Som. Friend.1882 (Parson-and-clerk); Lincs. Grigson (Parson-in-his-smock). The spadix in the spathe, besides giving rise to many names of sexual import, also accounted for another set of descriptive names like these.
PARSON-IN-THE-PULPIT 1. *Aconitum anglicum* (Wild Monkshood) Dev. B & H. 2. *Aconitum napellus* (Monkshood) Dev. Friend.1882. 3. *Arisaema triphyllum* (Jack-in-the-pulpit) Whittle & Cook. 4. *Arum maculatum* (Cuckoo-pint) Dev. Friend.1882. For cuckoo-pint, Cf. Parson-and-clerk etc.
PARSON'S BILLYCOCK; PARSON'S PILLYCODS *Arum maculatum* (Cuckoo-pint) Som. Grigson (Billycock); Yks. B & H (Pillycods). Of exactly the same import as the name Cuckoo-pint.
PARSON'S NOSE *Orchis morio* (Green-winged Orchid) Dev. Friend.1882.
PARTRIDGE-BERRY 1. *Gaultheria procumbens* (Wintergreen) USA Lloyd. 2. *Mitchella repens.*
PARTRIDGE-BREASTED ALOE *Aloe variegata.*
PARTRIDGE-PEA 1. *Cassia fasciculata* USA Zenkert. 2. *Cassia marilandica* (American Senna) USA H.Smith.1928. 3. *Pisum sativum var. arvense* (Field Pea) Howes.
PASCHAL FLOWER; PASQUE FLOWER *Pulsatilla vulgaris* Gerard (Paschal). It was apparently originally Passe-flower, with or without the 'e', perhaps with the meaning surpassing flower, and it was Gerard who altered it to Pasque, or sometimes Paschal ("they floure for the most part about Easter, which hath moved mee to name it Pasque-floure").
PASH-LEAF *Plantago lanceolata* (Ribwort Plantain) Pemb. Grigson.
PASMENT; PASMET *Pastinaca sativa* (Wild Parsnip) Som. Macmillan (Pasment); Wilts. Akerman; Berks. Lowsley Hants. Cope (Pasmet). In a similar way, the 'nip' of turnip becomes 'met', to give turmet.

PASQUE-FLOWER, Eastern *Pulsatilla patens*.
PASQUE-FLOWER, Pale *Pulsatilla vernalis* (Spring Anemone) Blamey.
PASQUE-FLOWER, Small *Pulsatilla pratensis*.
PASSE-FLOWER *Pulsatilla vulgaris* (Pasque-flower) Gerard. The original name, possibly meaning the 'surpassing flower', before Gerard changed it to Pasque-flower.
PASSERAGE *Lepidium sativum* (Garden Cress) Prior. Because it was believed to drive away madness ('rage').
PASSEVELOURS *Amaranthus caudatus* (Love-lies-bleeding). B & H. The French name, but used in England with its near translation, Velvet-flower.
PASSION 1. *Polygonum bistorta* (Bistort) Ches. Gerard. 2. *Rumex patientia* (Patience Dock) B & H. 'Patience' for the Dock is here Latin lapathium, with the general meaning of large-leaved. 'Passion' is a late corruption, but for Bistort the association with Easter is quite clear. Herb Pudding, better known as Ledger-pudding, is made from its leaves on Easter Day, or more properly, at Passion-tide.
PASSION DOCK 1. *Polygonum bistorta* (Bistort) Derb, Yks, N'thum. Grigson. 2. *Rumex patientia* (Patience Dock) Denham/Hardy.
PASSION FLOWER 1. *Arum maculatum* (Cuckoo-pint) Skinner. 2. *Passiflora sp*. It is said that when the Spaniards first saw the Passion Flower, they took it as an omen that the Indians would be converted to Christianity. The three styles represent the three nails, the ovary is a sponge soaked in vinegar. The stamens are the wounds of Christ, and the crown (located above the petals) stands for the crown of thorns. The petals and sepals indicate the Apostles. For Cuckoo-pint, the tradition is different. It was said to have been growing at the foot of the Cross, and to have received some drops of Christ's blood on its leaf. Cf. the Cheshire name Gethsemane.

PASSION FLOWER, Blue *Passiflora caerulea*.
PASSION FLOWER, English *Arum maculatum* (Cuckoo-pint) Fernie. See Passion Flower.
PASSION FLOWER, Stinking *Passiflora foetida*.
PASSION FRUIT *Passiflora edulis*.
PASSION FRUIT, Banana- 1. *Passiflora antioqiensis*. 2. *Passiflora mollissima*.
PASSMENT *Pastinaca sativa* (Parsnip) Som. Macmillan. Cf. Pasmet, Pasment.
PASSPER *Crithmum maritimum* (Samphire) Scot. Grigson. This is the French perce-pierre. Cf. Pierce-stone. Samphire (herbe de St Pierre) carries the same import.
PASTEL *Isatis tinctoria* (Woad) Lyte. This is still the French name for woad.
PASTURE BRAKE *Pteridium aquilinum* (Bracken) USA Cunningham.
PASTURE ROSE *Rosa humilis*.
PASTURE THISTLE *Cirsium pumilum*.
PATAGONIAN BEAN *Canavalia ensiformis* (Jack Bean) Chopra.
PATAGONIAN CYPRESS *Fitzroya cupressoides* (Alerce) Leathart.
PATCHOULI *Pogostemon patchouli*. Hindi pacholi, from Tamil paccilai, green leaf. Patchouli is a well-known Eastern perfume. The scent of the plant is very powerful, and in its unadulterated form smells very unpleasantly of goats. But when the otto is diluted with attar of roses and dissolved in spirit, the unpleasant quality goes completely.
PATERNOSTER BEAN *Abrus precatorius* (Precatory Bean) Howes.
PATERSON'S CURSE *Echium plantagineum* (Purple Viper's Bugloss) Australia Salisbury. Some of the early settlers in Australia, of whom Paterson was presumably one, were misguided enough to introduce it there. Now it is a pestilential weed.
PATES, Bull *Deschampsia cespitosa* (Tufted Hair-grass) H C Long.1910. Cf. Bull Faces, Bullpolls, or Bullpulls.
PATES, Crazy- *Caltha palustris* (Marsh Marigold) Hants. Boase. Probably Crazy

Bets originally. Cf. Crazy or Grazy, Crazy Lilies, etc., as well as the series Crazy Bets, Betsies, Betties, or Bess. The name must be connected with the west country name Drunkards, which see.

PATIENCE 1. *Polygonum bistorta* (Bistort) B & H. 2. *Rumex patientia* (Patience Dock) Turner. Patience is "Passion" for Bistort - a reference to the Easter practice of eating herb pudding, of which bistort is an ingredient. But for *Rumex patientia*, the derivation is from Latin lapathium, referring to the size of the leaves.

PATIENCE, Garden 1. *Rumex alpinus* (Monk's Rhubarb) Flück. 2. *Rumex patientia* (Patience Dock) Parkinson.

PATIENCE, Herb *Rumex patientia* (Patience Dock) Rohde.

PATIENCE DOCK 1. *Polygonum bistorta* (Bistort) Midl, N Eng. Denham/Hardy. 2. *Rumex patientia*. See Patience.

PATIENCE PLANT; PATIENT LUCY *Impatiens walleriana* (Busy Lizzie) both Harvey. Odd, when it is really 'Impatiens', given because of the explosive way it scatters its seeds.

PATIENT DOCK *Rumex bistorta* (Bistort) Ches. Holland. See Patience.

PATS, Butter *Viola canina* fruit (Heath Dog Violet) Lancs. B & H.

PATTENS-AND-CLOGS 1. *Linaria vulgaris* (Toadflax) Som. Macmillan; Suss. B & H. 2. *Lotus corniculatus* (Bird's-foot Trefoil) Som. Macmillan; Glos. J D Robertson; Suss. Parish. Cf. Boots-and-shoes, Shoes-and-stockings for Bird's-foot Trefoil, and such names as Lady's Slippers for both of them.

PATTERN WOOD *Alstonia congensis*.

PATTIKEYS *Fraxinus excelsior* fruit (Ash keys) N'thants. Sternberg. Presumably from such a name as Katty Keys, which has a number of variants. Coles called them Peter Keys.

PATTY-CAREY *Hepatica caerulea* (Hepatica) Wilts. Macmillan. Patty-carey is a corruption of Hepatica caerulea.

PATTY-PAN SQUASH *Cucurbita pepo var. melopepo* (Scallop Gourd) Bianchini. A good description.

PATTY-PANS *Nuphar lutea* (Yellow Waterlily) Worcs. Grigson.

PAUL'S BETONY 1. *Veronica officinalis* (Common Speedwell) Browne. 2. *Veronica serpyllifolia* (Thyme-leaved Speedwell) Curtis. Of Common Speedwell "the common people have some conceit in reference to St Paul; wheras indeed, that name is derived from Paulus Aegineta, an ancient physician of Aegina..." (Browne).

PAULOWNIA, Royal *Paulownia tomentosa* (Foxglove Tree) Edlin.

PAUNCE; PAUNSY *Viola tricolor* (Pansy) B & H (Paunce); Gerard (Paunsy).

PAW, Cat's 1. *Antennaria dioica* (Cat's Foot) Cumb. Grigson. 2. *Glechoma hederacea* (Ground Ivy) Folkard. 3. *Lotus corniculatus* (Bird's-foot Trefoil) Wilts. D Grose.

PAW, Kangaroo *Anigozanthus flavida*.

PAW, Kangaroo, Black *Macropidia fuliginosa*.

PAW, Kangaroo, Golden *Anigozanthus pulcherrima*.

PAW, Kangaroo, Red-and-green *Anigozanthus manglesii*.

PAW, Lion's *Alchemilla vulgaris* (Lady's Mantle) Gerard. The shape of the leaf determines the name. Cf. Lion's Foot and Padelion.

PAW-PAW *Carica papaya*.

PAWPLE *Agrostemma githago* (Corn Cockle) Scot. Grigson. Popple is probably the original, but it appears as Papple and Popille, as well as Poppy.

PAYSHUN-DOCK *Rumex patientia* (Patience Dock) Lancs. Nodal.

PE-TSAI *Brassica pekinensis* (Chinese Cabbage) Schery. The Chinese name, often used in this country, and more particularly, in America.

PEA, Anchovy *Grias cauliflora*.

PEA, Angola *Cajanus indicus* (Pigeon Pea) Dalziel.

PEA, Asparagus *Lotus tetragonolobus*.

PEA, Australian *Dolichos lablab* (Lablab) RHS.
PEA, Austrian *Pisum sativum var. arvense* (Field Pea) Kingsbury.
PEA, Beach *Lathyrus maritimus* (Sea Pea) USA Zenkert.
PEA, Black *Lathyrus niger*.
PEA, Black-eyed *Vigna unguiculata* (Cowpea) Wit.
PEA, Blue *Clitoria ternatea* (Butterfly Pea) Howes.
PEA, Butterfly *Clitoria ternata*.
PEA, Caley *Lathyrus hirsutus* (Hairy Pea) USA Kingsbury.
PEA, Canada *Vicia cracca* (Tufted Vetch) Howes.
PEA, Cape *Phaseolus vulgaris* (Kidney Bean) S Africa Watt.
PEA, Cat 1. *Lotus corniculatus* (Bird's-foot Trefoil) Corn. Grigson. 2. *Vicia cracca* (Tufted Vetch) Scot. Grigson.
PEA, Cat's *Cytisus scoparius* (Broom) Corn. Grigson.
PEA, Chick 1. *Cicer arietinum*. 2. *Vicia americana* (American Vetch) USA Cunningham.
PEA, Congo *Cajanus indicus* (Pigeon Pea) Dalziel.
PEA, Coral *Abrus precatorius* (Precatory Bean) Howes.
PEA, Coral, Blue *Hardenbergia comptoniana/violacea*.
PEA, Craw *Lathyrus pratensis* (Yellow Pea) Border B & H.
PEA, Crow 1. *Empetrum nigrum* (Crowberry) Scot. B & H. 2. *Vicia sepium* (Bush Vetch) Cumb. Grigson.
PEA, Darling, Smooth *Swaisona galegifolia*. An Australian species, so the reference is to the River Darling.
PEA, Desert, Sturt's *Clianthus formosus*.
PEA, Dun; PEA, Grey *Pisum sativum var. arvense* (Field Pea) Howes.
PEA, Earth-nut *Lathyrus tuberosus* (Fyfield Pea) Clapham. Cf. Earth Chestnut, and Groundnut Peavine, and indeed the specific name itself.

PEA, Everlasting *Lathyrus sylvestris*. 'Everlasting' in the sense that it is perennial.
PEA, Everlasting, Garden *Lathyrus latifolius*.
PEA, Everlasting, Narrow-leaved *Lathyrus sylvestris* (Everlasting Pea) Clapham.
PEA, Field *Pisum sativum var. arvense*.
PEA, Flat *Lathyrus sylvestris* (Everlasting Pea) USA Kingsbury.
PEA, Fyfield *Lathyrus tuberosus*. Fyfield is an Essex village near which this grows - the only location in this country.
PEA, Garden *Pisum sativum*. Pea and Pease are from Pisum, thought to derive from pin-sum, a mortar, in which peas were pounded into small pieces. The conversion from pease into pea may be from the French pois being pronounced "pay", or from the mistaken belief that pease was a plural form. The original plural would have been peasen. A similar case is that of cherry, from cerise.
PEA, Glory *Clianthus formosus* (Sturt's Desert Pea) Perry.
PEA, Golden *Thermopsis rhombifolia* (False Lupin) Johnston.
PEA, Goober *Arachis hypogaea* (Ground Nut) Dalziel. Or simply Goober, which must be from an African source - it is known as gooba or guba in West Africa.
PEA, Grass *Lathyrus nissolia*.
PEA, Grass-leaved *Lathyrus nissolia* (Grass Pea) Hutchinson.
PEA, Ground *Arachis hypogaea* (Ground Nut) USA Elsmore.
PEA, Guinea *Abrus precatorius* (Precatory Bean) Chopra.
PEA, Gypsy *Vicia sativa* (Common Vetch) Som. Macmillan.
PEA, Hairy *Lathyrus hirsutus*.
PEA, Heart *Cardiospermum halicacabum* (Blister Creeper) Chopra.
PEA, Heath *Lathyrus montanus* (Bittervetch) Prior.
PEA, Hen *Zea mays* (Maize) Derb. Porteous.
PEA, Indian *Lathyrus sativus*.
PEA, Lord Anson's *Lathyrus magellanicus*.

The cook on Lord Anson's ship Centurion collected the seeds in 1744.
PEA, Maple *Pisum sativum var. arvense* (Field Pea) Howes.
PEA, Marsh *Lathyrus palustris*.
PEA, Marsh, Myrtle-leaved *Lathyrus myrtifolius*.
PEA, Meadow *Lathyrus pratensis* (Yellow Pea) McClintock.1977.
PEA, Mouse('s) 1. *Lathyrus montanus* (Bittervetch) Berw. B & H. 2. *Lathyrus pratensis* (Yellow Pea) Donegal Grigson. 3. *Vicia cracca* (Tufted Vetch) Scot, Border Grigson. 4. *Vicia hirsuta* (Hairy Tare) Donegal Grigson.
PEA, Mussel-shell *Clitoria ternatea* (Butterfly Pea) Simmons.
PEA, Mutter 1. *Lathyrus sativus* (Indian Pea) Forsyth. 2. *Pisum sativum var. arvense* (Field Pea) Howes.
PEA, No-eye *Cajanus indicus* (Pigeon Pea) W Indies Dalziel.
PEA, Partridge 1. *Cassia fasciculata* USA Zenkert. 2. *Cassia marilandica* (American Senna) USA H Smith.1928. 3. *Pisum sativum var. arvense* (Field Pea) Howes.
PEA, Pharaoh's *Lathyrus sylvestris* (Everlasting Pea) N'thants. PLNN3. There is a legend that someone from the village of Weebly in Northants went to Egypt and brought home some seeds from a royal tomb. Of course, this plant does not grow in Egypt.
PEA, Pigeon 1. *Cajanus indicus*. 2. *Laburnum anagyroides* (Laburnum) Tynan & Maitland.
PEA, Rabbit *Tephrosia virginiana* (Devil's Shoestring) Howes.
PEA, Rock, Desert *Lotus rigidus*.
PEA, Rosary *Abrus precatorius* (Precatory Bean) USA Kinsgbury. Cf. Prayer Beads.
PEA, Scurf *Psoralea argophylla*. i.e. Scurvy Pea.
PEA, Scurf, Beaverbread *Psoralea castorea* (Beaver-dam Breadroot) Douglas.
PEA, Scurf, Yellow *Psoralea lanceolata*.
PEA, Scurvy *Psoralea argophylla* (Scurf Pea) USA Yarnell.
PEA, Sea *Lathyrus maritimus*.

PEA, Seaside *Lathyrus maritimus* (Sea Pea) USA House.
PEA, Shamrock *Parochetus communis*.
PEA, Show *Lathyrus odoratus* (Sweet Pea) Norf. B & H.
PEA, Singletary 1. *Lathyrus hirsutus* (Hairy Pea) USA Kingsbury. 2. *Lathyrus pusillus*.
PEA, Spanish *Cicer arietinum* (Chick Pea) Howes.
PEA, Spring *Lathyrus vernus*.
PEA, Stinking *Cassia occidentalis* (Coffee Senna) Dalziel.
PEA, Sweet *Lathyrus odoratus*.
PEA, Sweet, Prairie *Lathyrus venosus*.
PEA, Sweet, Swamp *Lathyrus palustris* (Marsh Pea) USA Tolstead.
PEA, Sweet, White *Lathyrus ochroleucus*.
PEA, Sweet, Wild 1. *Hedysarum mackenzii*. 2. *Ononis repens* (Rest-harrow) Som. Macmillan.
PEA, Tangier *Lathyrus tangitanus*.
PEA, Tuberous *Lathyrus tuberosus* (Fyfield Pea) Hutchinson.
PEA, Turkey 1. *Dicentra canadensis* (Turkey Corn) Grieve.1931. 2. *Dicentra formosa* (Western Bleeding Heart) USA Berdoe. 3. *Sanicula tuberosa*. 4. *Tephrosia virginiana* (Devil's Shoestrings) USA Youngken.
PEA, Tutty *Lathyrus odoratus* (Sweet Pea) Som. Macmillan. A 'tutty' is a posy.
PEA, Veiny *Lathyrus venosus* (Prairie Sweet Pea) USA H Smith.
PEA, Winged *Tetragonolobus maritimus* (Dragon's Teeth) A W Taylor.
PEA, Winter, Wild *Lathyrus hirsutus* (Hairy Pea) USA Kingsbury.
PEA, Wood 1. *Lathyrus montanus* (Bittervetch) Curtis. 2. *Lathyrus sylvestris* (Everlasting Pea) Hutchinson.
PEA, Yellow *Lathyrus pratensis*.
PEA GULL *Primula veris* (Cowslip) J Smith. An extreme form of Paigle, which is a common name for cowslip.
PEA MINT *Mentha spicata* (Spearmint) Brownlow. A recognition that spearmint is the right and proper herb to cook with peas.
PEA THATCH *Lotus corniculatus* (Bird's-

foot Trefoil) Som. Macmillan. 'Thatch' in this case means 'vetch'.

PEA TREE *Laburnum anagyroides* (Laburnum) Ayr, Loth. Grigson.

PEACE-AND-PLENTY *Saxifraga x urbinum* (London Pride) Som. Macmillan; Wilts. Wright. Probably an equivalent of Hens-and-chickens, i.e. a comment on the way it reproduces itself by runners.

PEACE LILY *Spathiphyllum patinii* (White Sails) Bonar.

PEACE PLANT *Arisaema triphyllum* (Jack-in-the-pulpit) southern USA Puckett.

PEACH *Prunus persica*. Originally meaning Persicum malum, Persian apple. That, in fact, was Pliny's mistake - it comes from China, not Persia.

PEACH, Hasty *Prunus armeniaca* (Apricot) Turner. 'Hasty', because it blooms so early, earlier than peach.

PEACH, Wild 1. *Eucarva persicarius*. 2. *Prunus andersonii* (Nevada Wild Almond) USA Jaeger. 3. *Prunus caroliniana* (American Cherry Laurel) USA Harper.

PEACH BELLS *Campanula persicifolia* (Peach-leaved Bellflower) Dev. Friend.1882.

PEACH-LEAVED BELLFLOWER *Campanula persicifolia*.

PEACH-THORN *Lycium cooperi*.

PEACHWOOD *Caesalpina sappan* (Brazil-wood) Willis.

PEACHWORT *Polygonum persicaria* (Persicaria) Culpepper.

PEACOCK ANEMONE *Anemone pavonia*.

PEACOCK FLOWER 1. *Caesalpina pulcherrima* (Barbados Pride) S Africa Watt. 2. *Moraea pavonia*. 3. *Tigridia pavonia* (Shell-flower) Simmons.

PEACOCK PLANT 1. *Calathea mackoyana*. 2. *Episcia cupreata* (Flame Violet) USA Cunningham.

PEAGLE 1. *Primula veris* (Cowslip) Hazlitt. 2. *Primula vulgaris* (Primrose) Fernie. 3. *Ranunculus arvensis* (Corn Buttercup) Suff. B & H. Paigle is the original, for cowslip. It varies to Pigle, or Peggle,

as well, even to Pea Gull. The other two plants get the name by analogy only.

PEANUT *Arachis hypogaea* (Groundnut). As common as Groundnut for this.

PEANUT, Hog *Amphicarpa bracteata*.

PEAPUL *Ficus religiosa*. Cf. Pipal, and Pippala, the latter probably the berries. It is the Hindi word pipul.

PEAR, Alligator *Persea gratissima* (Avocado) Everett. The Aztecs called it ahuacatl; the Conquistadors changed this to aguacate, thence corrupted to avigato, and finally to alligator.

PEAR, Almond-leaved *Pyrus amygdaliformis*.

PEAR, Antilles *Chrysophyllum argenteum*. A Caribbean tree.

PEAR, Bachelor's *Solanum mammosum* (Amoise) Barbados Gooding. Perhaps the name has some connection with the "love-magic" for which the narcotic seeds are used, if not in Barbados, then certainly in Haiti.

PEAR, Balsam *Momordica charantia* (Wild Balsam-apple) USA Kingsbury.

PEAR, Bird's 1. *Crataegus monogyna* fruit (Haws) Som. Macmillan; Dor. Barnes. 2. *Rosa canina* fruit (Hips) Som. Macmillan.

PEAR, Chinese *Pyrus sinensis* (Sandy Pear) Candolle.

PEAR, Choke 1. *Pyrus communis*. 2. *Sorbus torminalis* A S Palmer. 'Choke' implies inedible, or at least too hard to eat raw. Cf. Iron Pear for the whitebeam.

PEAR, Iron *Sorbus aria* (Whitebeam) Wilts. Dartnell & Goddard.

PEAR, Pig's; PEAR, Pigsy *Crataegus monogyna* fruit (Haws) both Som. Macmillan. Not 'pigs', but 'pixies'.

PEAR, Pixy 1. *Crataegus monogyna* fruit (Haws) Som, Dor. Grigson. 2. *Rosa canina* fruit (Hips) W Eng. Friend.

PEAR, Plymouth *Pyrus cordata*. Now only growing in a couple of sites in Plymouth, plus a few more in Cornwall.

PEAR, Sandy *Pyrus sinensis*.

PEAR, Snow *Pyrus nivalis*.

PEAR, Swallow *Sorbus torminalis* (Wild

Service Tree) Prior. It seems to be the exact opposite of 'Choke Pear', also recorded for this.

PEAR, Vegetable *Sechium edule* (Chayote) Howes.

PEAR, Water *Ilex mitis* (Cape Holly) S Africa Watt. Cf. Water Tree, and Waterwood.

PEAR, Whitbin; PEAR, Widbin *Sorbus aria* (Whitebeam) Chaplin (Whitbin); Bucks. Harman (Widbin).

PEAR, Whitty 1. *Sorbus domestica* (True Service Tree) Rackham.1986. 2. *Sorbus torminalis* (Wild Service Tree) Worcs. Grigson.

PEAR, Wild 1. *Pyrus communis* (Choke Pear) Gerard. 2. *Pyrus pyraster*. 3. *Sorbus aria* (Whitebeam) Derb. Grigson.

PEAR, Willow-leaved *Pyrus salicifolia*.

PEAR, Witten; PEAR, Withy *Sorbus domestica* (True Service Tree) Fernie (Witten); Worcs. Gent Mag: PS (Withy).

PEAR-APPLE, Chinese *Pyrus sinensis* (Sandy Pear) Douglas.

PEAR CASHEW *Anacardium occidentale* (Cashew Nut). 'Pear', from the shape of the fruit.

PEAR-LEAF WINTERGREEN *Pyrola rotundifolia* (Round-leaved Wintergreen) O P Brown.

PEAR THORN *Crataegus tomentosa*.

PEARL BUSH *Exochorda macrantha*.

PEARL CUDWEED *Anaphalis margaretacea* (Pearly Immortelle) Howes.

PEARL EVERLASTING; PEARL-FLOWERED EVERLASTING *Anaphalis margaretacea* (Pearly Immortelle) McClintock (Pearl); Grieve.1931 (Pearl-flowered).

PEARL GRASS *Arrhenatherium elatius* (False Oat) H C Long.1910. The rootstock is often tuberous at the nodes, producing several "bulbs", "pearls", or "knots", as they are variously called.

PEARL MILLET *Pennisetum glaucum*.

PEARL OF SPAIN *Muscari botryoides var. album* B & H.

PEARL ONION, Bunching *Allium ampeloprasum* (Wild Leek) Brouk.

PEARL PLANT 1. *Haworthia margaretifera*. 2. *Lithospermum officinale* (Gromwell) Gerard. Gromwell's 'pearls' are the hard, stony seeds.

PEARLS, Primrose *Narcissus maialis var. patellaris* (Pheasant's Eye) Som. Macmillan. It sounds as if there were some confusion with Primrose Peerless, which is the name usually given to Narcissus x biflorus.

PEARLS, Column of *Haworthia chalwini* (Aristocrat Plant) USA Cunningham.

PEARLS, String of *Senecio rowleyianus* (String-of-beads Plant) USA Cunningham.

PEARLWORT *Sagina procumbens*.

PEARLWORT, Alpine *Sagina saginoides*.

PEARLWORT, Alpine, Lesser *Sagina intermedia*.

PEARLWORT, Annual *Sagina apetala*.

PEARLWORT, Awl-leaved *Sagina subulata* (Heath Pearlwort) Clapham.

PEARLWORT, Balearic *Arenaria balearica* (Balearic Sandwort) Clapham.

PEARLWORT, Boyd's *Sagina boydii*.

PEARLWORT, Druce's *Sagina normaniana*.

PEARLWORT, Fringed *Sagina ciliata*.

PEARLWORT, Heath *Sagina subulata*.

PEARLWORT, Knotted *Sagina nodosa*.

PEARLWORT, Mossy *Sagina procumbens* (Pearlwort).

PEARLWORT, Sea *Sagina maritima*.

PEARLWORT, Trailing *Sagina procumbens* (Pearlwort) Campbell.1900.

PEARLWORT, Upright *Moenchia erecta* (Dwarf Chickweed) Curtis.

PEARLY IMMORTELLE *Anaphalis margaretacea*.

PEART, Parsley *Aphanes arvensis* (Parsley Piert) Notes & Queries, 1858.

PEASANT'S CLOCK *Taraxacum officinale* (Dandelion) Dyer. Children tell the time by puffing at the seed-head.

PEASCODS, Owd Lad's *Laburnum anagyroides* (Laburnum) B & H. The reference is to the seed pods, and the ascription to the Owd Lad, the devil, that is, is an acknowledgement that the seeds are very poisonous.

PEASE *Pisum sativum* (Pea) Putnam. The original version of "pea", from the Latin pisum. When this fact was forgotten, it was thought that this was a plural, so the singular was amended to pea. The original plural would have been peasen.

PEASE, Bull's *Rhinanthus crista-galli* (Yellow Rattle) Donegal Grigson.

PEASE, Horse *Vicia orobus* (Bitter Vetch) Cumb. Grigson.

PEASE, Mouse's *Vicia cracca* (Tufted Vetch) N'thum, Mor. Grigson.

PEASE EARTH-NUT *Lathyrus montanus* (Bittervetch) Gerard.

PEASE EVERLASTING *Lathyrus latifolius* (Garden Everlasting Pea) Parkinson.

PEASEN, Long *Phaseolus vulgaris* (Kidney Bean) Turner.

PEAT, Parsley *Aphanes arvensis* (Parsley Piert) Yks. Addy. Parsley Piert is from French perce-pierre, meaning breakstone. There are a few variations, of which this is one.

PEAVINE, Groundnut *Lathyrus tuberosus* (Fyfield Pea) Douglas.

PEBBLE-VETCH *Vicia sativa* (Common Vetch) B & H.

PECAN *Carya pecan*. From an American Indian name, paccan.

PECULIAR *Petunia sp* Oxf. A S Palmer.

PEDELION *Helleborus niger* (Christmas Rose) Halliwell. Lion's foot, in other words, and given by comparing the shape of the leaves with animals' feet. Bear's foot is more usual for the hellebores.

PEDLAR, Shepherd's *Capsella bursa-pastoris* (Shepherd's Purse) Wilts. Dartnell & Goddard.

PEDLAR'S BASKET 1. *Cymbalaria muralis* (Ivy-leaved Toadflax) Som. Elworthy. Derb, Lancs. Grigson; Ches. Holland; Yks. Carr. 2. *Saxifraga sarmentosa* (Mother-of-Thousands) Shrop, Ches, Lancs. B & H.

PEDUNCULATE OAK *Quercus robur* (Oak) Hart. i.e. the oak that carries its flower, and so its acorn, on a peduncle, or stalk. Sessile Oak for Quercus petraea emphasises the difference, for that has no stalk.

PEE-A-BED; ONE, TWO, THREE, PEE-A-BED *Taraxacum officinale* (Dandelion) Dev. B & H; Som. Grigson. Dandelion is a famous diuretic. But the name shows that the effect is superstitious as well as medical. Children who even gather the flowers, so it is said, will experience the symptoms. Pissabed is better known than these, but Cf. French pissenlit, and German Pissblume.

PEE-BED; PEE-IN-BED *Taraxacum officinale* (Dandelion) Cumb. (Pee-bed); Suss, Norf. (Pee-in-bed) both PLNN 25.

PEEP-BO, Little; PEEPERS, Little *Anagallis arvensis* (Scarlet Pimpernel) Som. Macmillan (Peep-bo); Som. Grigson (Peepers). Connected with the item of weather lore that says that the pimpernel will only open its flowers when the weather is going to stay fine. In fact, it is the countryman's weather forecaster, as many of the local names show.

PEEP-O'-DAY; PEEPERS, Noon *Ornithogalum umbellatum* (Star of Bethlehem) Shrop. B & H (Peep-o'-day); Wilts. D Grose (Noon Peepers). Cf. such names as Sunflower - "these flowers open themselves at the rising of the Sunne, and shut againe at the Sun-setting..." (Gerard).

PEEPS; PIPS *Primula veris* flowers (Cowslip) Heref. Havergal.

PEEPUL; PEAPUL *Ficus religiosa*.

PEESEWEEP GRASS *Luzula campestris* (Good Friday Grass) Scot. B & H. Peeseweep is the same as peewit further south. It is probably a comment on the time of flowering. Cf. Cuckoo-grass.

PEEWEETS *Euphrasia officinalis* (Eyebright) Dev. Grigson.

PEG-NUT *Conopodium majus* (Earth-nut) Notes & Queries, 1872. Pig-nut, of course. This is just local pronunciation of 'pig', perpetuated in print.

PEGGALL-BUSH *Crataegus monogyna* (Hawthorn) Wilts. Grigson.

PEGGLE 1. *Crataegus monogyna* fruit

(Haws) Wilts. Jefferies. 2. *Primula veris* (Cowslip) Grieve.1931. A variation of paigle as far as cowslip is concerned, but peggle for haws is just one of a large number of variations on Pig Haw.

PEGGLE, Cow *Primula veris* (Cowslip) Herts. Jones-Baker.1974.

PEGGY-AILES; PEGGYILES *Crataegus monogyna* fruit (Haws) Wilts. Tanner (Peggy-ailes); Hayward (Peggyiles). Cf. Pig-ales, also from Wiltshire.

PEGROOTS *Helleborus viridis* (Green Hellebore) Prior. From the use of the roots by cattle doctors in the operation of pegging, or settering, according to Prior.

PEGS, Clothes 1. *Digitalis purpurea* (Foxglove) Dor. Macmillan. 2. *Orchis mascula* (Early Purple Orchid) Som. Macmillan.

PEGS, Cobbler's 1. *Bidens bipinnata* (Spanish Needles) Australia Watt. 2. *Erigeron canadensis* (Canadian Fleabane) Australia Watt.

PEGWOOD 1. *Euonymus europaea* (Spindle Tree) Dev. Macmillan. 2. *Cornus sanguinea* (Dogwood) W Miller. Gypsies always used to choose these two shrubs for wood to make pegs.

PEGYLL *Primula vulgaris* (Primrose) Fernie. Meant for paigle, which is a name for cowslip rather than primrose.

PEIPINO *Solanum muricatum*.

PEKIN WILLOW *Salix matsudana*.

PELETIR *Parietaria diffusa* (Pellitory-of-the-wall) B & H.

PELICAN FLOWER *Aristolochia gigas*. Because the young, unopened flowers look like the head of a pelican.

PELLAS *Malva neglecta* (Dwarf Mallow) Corn. B & H. Applied to the fruit only.

PELL-A-MOUNTAIN *Thymus vulgaris* (Wild Thyme) Gerard. Either a corruption of serphyllum montanum, or from Pulegium montanum, both old book names for thyme.

PELLITORY *Parietaria pennsylvanica* USA Upham.

PELLITORY, Bastard *Achillea ptarmica* (Sneezewort) Tradescant. Cf. False, or Wild, Pellitory.

PELLITORY, Country *Erigeron affinis* Martinez.

PELLITORY, False 1. *Achillea ptarmica* (Sneezewort) Culpepper. 2. *Erigeron affinis* Martinez.

PELLITORY, Persian *Pyrethrum roseum*.

PELLITORY, Wall *Parietaria diffusa* (Pellitory-of-the-wall) Hutchinson.

PELLITORY, Wild *Achillea ptarmica* (Sneezewort) Gerard. "The whole plant is sharp, biting the tongue and mouth like Pellitorie-of-Spaine, for which cause some have called it wild pellitorie...".

PELLITORY-OF-SPAIN *Anacyclus pyrethrum*.

PELLITORY-OF-SPAIN, Bastard; PELLITORY-OF-SPAIN, False *Peucedanum ostruthium* (Masterwort) Gerard (Bastard); Leyel.1948 (False).

PELLITORY-OF-THE-WALL *Parietaria diffusa*. The older name was Parritory, through Old French, and from Latin paries, wall. Hence, too, the generic name, Parietaria. This is the plant that grows on old walls.

PEMBINA *Viburnum opulus var. americanus* (High-bush Cranberry) USA Gilmore.

PENCE *Rhinanthus crista-galli* (Yellow Rattle) Northants. A E Baker. The seeds rattle about like pennies in a purse, hence this and a host of other names involving coins of various denominations.

PENCE, Judas; PENCE, Peter's *Lunaria annua* (Honesty) Coats (Judas); Som. Macmillan (Peter's). Another plant with a lot of 'money' names; it is the flat white seedcases that suggest it.

PENCIL CEDAR; PENCIL CEDAR, Virginian *Juniperus virginiana* (Virginian Juniper) Watt; H Davey (Virginian). It is said that no other wood has been found that has just the right physical properties for the casing of lead pencils.

PENCIL CEDAR, African *Juniperus procera* (East African Juniper) N Davey.

PENCIL JUNIPER *Juniperus virginiana* (Virginian Juniper) Howes. Better known as Pencil Cedar.

PENCILLED CRANESBILL *Geranium versicolor* (Painted Lady) McClintock.

PENCUIR KALE *Polygonum bistorta* (Bistort) Ayr. Grigson. Cf. the Lancashire Poor Man's Cabbage. The leaves are edible.

PENITERRY *Parietaria diffusa* (Pellitory-of-the-wall) Ire. Wood-Martin.

PENNIES *Lunaria annua* (Honesty) Tynan & Maitland.

PENNIES, Dog's *Rhinanthus crista-galli* (Yellow Rattle) Shet. Grigson.

PENNIES, Halfpennies-and-; PENNIES, Happenies-and- *Umbilicus rupestris* (Wall Pennywort) Dev. Macmillan; Grigson (Happenies).

PENNIES, Horse 1. *Leucanthemum vulgare* (Ox-eye Daisy) Derb. Grigson. 2. *Rhinanthus crista-galli* (Yellow Rattle) Derb, Lancs. Grigson; Yks. Carr.

PENNIES, Silver *Bellis perennis* (Daisy) N'thants. A E Baker.

PENNIES, Two-, in-a-purse *Lunaria annua* (Honesty) Coats.

PENNIES-AND-HAPPENIES 1. *Lysimachia nummularia* (Creeping Jenny) Som. Macmillan. 2. *Rhinanthus crista-galli* (Yellow Rattle) Som. Macmillan.

PENNIWINKLE *Vinca major* (Greater Periwinkle) Le Strange.

PENNSYLVANIA PERSICARIA *Polygonum pennsylvanicum* (Heartseed) USA Youngken.

PENNSYLVANIA SUMACH *Rhus glabra* (Smooth Sumach) Grieve.1931.

PENNSYLVANIA TEA *Monarda didyma* (Red Bergamot) Lewin.

PENNY, Bawd's *Meum athamanticum* (Baldmoney) Jason Hill. Cf. Bawd-money, which is the link between Baldmoney and Bawd's Penny. See Baldmoney for the derivation.

PENNY, Marsh *Hydrocotyle vulgaris* (Pennywort) Le Strange.

PENNY, Money *Umbilicus rupestris* (Wall Pennywort) Dev. Grigson; Glos. PLNN35.

PENNY, Moon *Leucanthemum vulgare* (Ox-eye Daisy) Ches. Holland; Yks. Addy. Cf. Moon Daisy, Moon-flower, etc.

PENNY, Old Woman's *Lunaria annua* (Honesty) Som. Macmillan.

PENNY, Rattle *Rhinanthus crista-galli* (Yellow Rattle) Dur, N'thum. Grigson.

PENNY, Royal *Umbilicus rupestris* (Wall Pennywort) Som. Macmillan. Presumably, Pennyroyal was in someone's mind. It is in fact sometimes called Wall Pennyroyal.

PENNY, Silver *Lunaria annua* (Honesty) Som. Macmillan.

PENNY BOX *Rhinanthus crista-galli* (Yellow Rattle) Tynan & Maitland.

PENNY CAKE; PENNY CAP; PENNY COD *Umbilicus rupestris* (Wall Pennywort) Corn. Jago; Dev, Pemb. Grigson (Penny Cake); Dev, Som. Grigson (Penny Cap); Corn. Grigson. (Penny Cod).

PENNY CRESS, Alpine *Thlaspi alpestris*.

PENNY CRESS, Cotswold *Thlaspi perfoliatum*.

PENNY CRESS, Field *Thlaspi arvense*. Cf. Money Plant, and Money-tree, both Somerset names for this.

PENNY DAISY *Leucanthemum vulgare* (Ox-eye Daisy) Scot. PLNN25.

PENNY FLOWER 1. *Lunaria annua* (Honesty) Gerard. 2. *Umbilicus rupestris* (Wall Pennywort) Som. Macmillan.

PENNY GIRSE; PENNY GRASS *Rhinanthus crista-galli* (Yellow Rattle) Shet. Grigson (Penny Girse); Gerard (Penny Grass).

PENNY HATS; PENNY LEAVES *Umbilicus rupestris* (Wall Pennywort) Dev. Friend.1882 (Hats); Dev, Som, Ire. Grigson (Leaves).

PENNY HEDGE; PENNY IN THE HEDGE *Alliaria petiolata* (Jack-by-the-hedge) Worcs, Norf. Grigson.

PENNY JOHN *Hypericum perforatum* (St John's Wort) Norf. Grigson.

PENNY MOUNTAIN *Thymus vulgaris*

(Wild Thyme) Prior. Cf. Pell-a-mountain and Puliall-mountain. Either corruptions from serphyllum montanum or Pulegium montanum, both old book names.

PENNY MOUCHERS *Rubus fruticosus* fruit (Blackberries) Wilts. Jones & Dillon. Moochers were truants, originally applied to children who went mooching (or mouching) from school in the blackberry season, then applied to the blackberries themselves.

PENNY PIE 1. *Sibthorpia europaea* (Cornish Moneywort) Corn. Grigson. 2. *Umbilicus rupestris* (Wall Pennywort) Corn. Grigson.

PENNY PIE, Small *Sibthorpia europaea* (Cornish Moneywort) Corn. Grigson.

PENNY PLATE *Umbilicus rupestris* (Wall Pennywort) Dev. Grigson.

PENNY RATTLE *Rhinanthus crista-galli* (Yellow Rattle) Som. Macmillan; Suss. Parish.

PENNY ROT *Hydrocotyle vulgaris* (Pennywort) Shrop. Grigson. Cf. Rot-grass, Sheep-rot, White Rot, etc. "...husbandmen know well, that it is noisome unto Sheepe, and other Cattell that feed thereon, and for the most part bringeth death unto them..." (Gerard). In fact, it is liable to cause inflammation of the digestive tract and haematuria.

PENNYGRASS 1. *Hydrocotyle vulgaris* (Pennywort) Lyte. 2. *Umbilicus rupestris* (Wall Pennywort) Cockayne.

PENNYRINKLE; PENNYWINKLE *Vinca minor* (Lesser Periwinkle) Dev. Friend.1882 (Pennyrinkle); Som. Macmillan; Hants. Fernie; Berks. Lowsley (Pennywinkle).

PENNYROYAL 1. *Mentha longifolia* (Horse Mint) S Africa Watt. 2. *Mentha pulegium*. Pennyroyal is a corruption of Puliol Royal, from Latin pulices, fleas, because it is good for destroying the pests. The 'royal' part of the name, it is said, came from the fact that they were used in royal palaces for just this purpose.

PENNYROYAL, American *Hedeoma pulegioides*.

PENNYROYAL, Drummond *Hedeoma drummondii*.

PENNYROYAL, Mock *Hedeoma nana*.

PENNYROYAL, Mohave *Monardella exilis*.

PENNYROYAL, Rock *Monardella robisonii*.

PENNYROYAL, Thyme *Hedeoma thymoides*.

PENNYROYAL, Wall *Umbilicus rupestris* (Wall Pennywort) Grieve.1931. Cf. Royal Penny, a Somerset name.

PENNYROYAL, Wood *Veronica officinalis* (Common Speedwell) Turner.

PENNYROYAL MINT *Mentha pulegium* (Pennyroyal) Thornton.

PENNYROYAL THYME *Thymus pulegioides* (American Pennyroyal) Painter.

PENNYWALL *Umbilicus rupestris* (Wall Pennywort) IOM Moore.1924.

PENNYWEED *Rhinanthus crista-galli* (Yellow Rattle) Midl. Grigson.

PENNYWORT 1. *Cymbalaria muralis* (Ivy-leaved Toadflax) B & H. 2. *Hydrocotyle vulgaris*. 3. *Lunaria annua* (Honesty) Tynan & Maitland. 4. *Lysimachia nummularia* (Creeping Jenny) Genders.1976.

PENNYWORT, Asiatic *Hydrocotyle asiatica* (Indian Pennywort) Chopra.

PENNYWORT, Floating *Hydrocotyle ranunculoides*.

PENNYWORT, Indian *Hydrocotyle asiatica*.

PENNYWORT, Marsh; PENNYWORT, Water *Hydrocotyle vulgaris* (Pennywort) Curtis (Marsh); Gerard (Water).

PENNYWORT, Rose *Saxifraga virginiensis*.

PENNYWORT, Wall *Umbilicus rupestris*.

PEONY *Paeonia mascula*. Paeonia (and of course Peony) as the name was first established by Threophrastus, who chose it in honour of Paeon, the one who first used the plant medicinally, and who was said to have cured with it the wounds which the gods received during the Trojan war. Paeon first received the flower from the mother of Apollo on Mount Olympus, and with it cured Pluto of a wound he had received in a fight with Hercules. Paeon's success in healing aroused so much jealousy in Asklepios that he secret-

ly compassed his death, but Pluto transferred his body into the flower, which he called Paeony.
PEONY, Chinese *Paeonia albiflora* (White-flowered Peony) Douglas.
PEONY, Coral *Paeonia mascula* (Peony) Hutchinson.
PEONY, English *Paeonia mascula* (Peony) W Miller.
PEONY, Fennel-leaved *Paeonia tenuifolia* (Fringed Peony). W Miller.
PEONY, Fine-leaved *Paeonia tenuifolia* (Fringed Peony) Whittle & Cook.
PEONY, Fringed *Paeonia tenuifolia*.
PEONY, He; PEONY, Male *Paeonia mascula* (Peony) Lupton (He) Gerard (Male). The specific name is 'mascula'. Why?
PEONY, Himalayan *Paeonia emodi*.
PEONY, Lemon *Paeonia mlokosewitschii*. 'Lemon' refers to the colour of the flowers in this case.
PEONY, Slender-leaved *Paeonia tenuifolia* (Fringed Peony). W Miller.
PEONY, Tree *Paeonia arborea*.
PEONY, Tree, Chinese *Paeonia suffruticosa*.
PEONY, White-flowered *Paeonia albiflora*.
PEONY ROSE *Paeonia emodi* (Himalayan Peony) Chopra.
PEPEROMIA, Silver *Peperomia argyraea*.
PEPILLARY; PEPLAR *Populus nigra* (Black Poplar) Ches. B & H (Pepillary); Gerard (Peplar). Cf. Popilary, etc. They must be influenced by Old French poplier.
PEPILLES *Peplis portula* (Water Purslane) Halliwell.
PEPLAR, White *Populus alba* (White Poplar) Lyte.
PEPPER, Alligator *Aframomum melegueta* (Melegueta Pepper) Dalziel. 'Alligator' here is a corruption of 'Melegueta'.
PEPPER, Ashanti *Piper guineense* (West African Black Pepper) Dalziel.
PEPPER, Bell 1. *Capsicum annuum* (Chile) USA Cunningham. 2. *Capsicum grossum*.
PEPPER, Benin *Piper guineense* (West African Black Pepper) Dalziel.
PEPPER, Bird 1. *Capsicum annuum var. longum* (Paprika) Hatfield. 2. *Capsicum baccatum*. 'Bird', because the seeds are used by canary fanciers to improve the birds' plumage.
PEPPER, Black *Piper nigrum*.
PEPPER, Black, West African *Piper guineense*.
PEPPER, Brazilian *Schinus terebinthifolius*.
PEPPER, Bush *Piper guineense* (West Afreican Black Pepper) Dalziel.
PEPPER, Cherry *Capsicum cerasiforme*.
PEPPER, Chile *Capsicum annuum* (Chile) Redfield & Villa.
PEPPER, Christmas *Capsicum annuum* (Chile) Rochford.
PEPPER, Country *Sedum acre* (Biting Stonecrop) Gerard. Cf. Pepper Crop, Poor Man's Pepper, etc.
PEPPER, Ethiopian *Xylopia aethiopica*.
PEPPER, False *Xanthoxylum alatum* Bianchini. The fruits can be used as a pepper substitute.
PEPPER, Goat *Capsicum annuum* (Chile) Lindley.
PEPPER, Green *Capsicum annuum* (Chile) Thomson.1978.
PEPPER, Guinea 1. *Aframomum melegueta* (Melegueta Pepper) Dalziel. 2. *Capsicum annuum* (Chile) Watt.
PEPPER, Horse *Angelica sylvestris* (Wild Angelica) E Angl. V G Hatfield.1994.
PEPPER, Indish *Capsicum annuum* (Chile) Turner.
PEPPER, Jamaica *Pimento dioica* (Allspice) Schery.
PEPPER, Kava *Piper methysticum*.
PEPPER, Malagetta *Amomum grana* (Grains of Paradise Plant) Lindley.
PEPPER, Marsh *Polygonum hydropiper* (Water Pepper) Allan.
PEPPER, Meleguetta *Aframomum melegueta*.
PEPPER, Monk's *Vitex agnus-castus* (Chaste Tree) Leyel.1937. Cf. Small, or Wild, Pepper.
PEPPER, Mountain 1. *Daphne mezereon* (Lady Laurel) Pratt. 2. *Drimys lanceolata*.

PEPPER, Old Man's 1. *Achillea millefolium* (Yarrow) Som. Macmillan. 2. *Achillea ptarmica* (Sneezewort) Som. Macmillan. 3. *Poterium sanguisorba* (Salad Burnet) Som. Macmillan. Yarrow and Sneezewort have nosebleeding propensities, as well as sneezing. Yarrow was even used as a cheap snuff once.

PEPPER, Pod *Capsicum annuum* (Chile) J Smith.

PEPPER, Poor Man's 1. *Lepidium campestre* (Pepperwort) War. B & H; USA Upham. 2. *Lepidium latifolium* (Dittander) Camden/Gibson. 3. *Lepidium virginicum* (Virginia Peppergrass) USA Allan. 4. *Poterium sanguisorba* (Salad Burnet) Dor. Macmillan. 5. *Sedum acre* (Biting Stonecrop) Suss, Notts. Grigson. 6. *Valeriana dioica* (Marsh Valerian) Wilts. Macmillan. Valerian looks a bit strange in this company - it is poisonous.

PEPPER, Red; PEPPER, Scarlet *Capsicum annuum* (Chile) Schauenberg.

PEPPER, Salt-and- *Agrimonia eupatoria* (Agrimony) Grigson.

PEPPER, Saturday's; PEPPER, Saturday Night's *Euphorbia helioscopia* (Sun Spurge) both Wilts. Dartnell & Goddard.

PEPPER, Small *Vitex agnus-castus* fruit (Chaste Tree) Pomet. Also Wild Pepper - "as well because their round Figure renders them like that sort of Pepper, as because their Taste is a little biting and aromatic".

PEPPER, Spanish *Capsicum annuum* (Chile) Schauenberg.

PEPPER, Spur *Capsicum annuum* (Chile) Barbados Gooding.

PEPPER, Sweet *Clethra alnifolia*.

PEPPER, Tabasco *Capsicum annuum* (Chile) Watt.

PEPPER, Tailed *Piper cubeba* (Cubeb) Mitton.

PEPPER, Wall *Sedum acre* (Biting Stonecrop) Gerard.

PEPPER, Water 1. *Elatine hexandra* (Waterwort) Murdoch McNeill. 2. *Elatine hydropiper* (Eight-stamened Waterwort). 3. *Polygonum hydropiper*.

PEPPER, Water, Least *Polygonum minus*.

PEPPER, Water, Tasteless *Polygonum mite*.

PEPPER, Wild 1. *Achillea millefolium* (Yarrow) Berw. Grigson. 2. *Daphne mezereon* (Lady Laurel) Le Strange. 3. *Vitex agnus-castus* fruits (Chaste Tree) Pomet. Cf. Sneezewort and Sneezings for yarrow, as well as its nosebleed propensities.

PEPPER-AND-SALT *Capsella bursa-pastoris* (Shepherd's Purse) Middx. Grigson. It is known as Pepper Plant, and Peppergrass, elsewhere.

PEPPER-BOX 1. *Papaver rhoeas* (Red Poppy) Som. Macmillan. 2. *Rhinanthus crista-galli* (Yellow Rattle) Som. Macmillan. In both cases, the name comes from the way the seeds rattle in their cases.

PEPPER-BOX, Old Man's *Achillea ptarmica* (Sneezewort) Som. Macmillan. In this case the 'pepper-box' analogy is different. It is a reference to the virtues that also gave the name Sneezewort.

PEPPER CRESS 1. *Cardaria draba* (Hoary Cress) Salisbury. 2. *Lepidium latifolium* (Dittander) C P Johnson. 3. *Lepidium ruderale* (Narrow-leaved Pepperwort) Salisbury. 4. *Lepidium sativum* (Garden Cress) Dev. Friend.1882.

PEPPER CROP *Sedum acre* (Biting Stonecrop) Gerard.

PEPPER-GIRSE *Achillea ptarmica* (Sneezewort) Shet. Grigson. Cf. Old Man's Pepper Box, from Somerset, and other similar names.

PEPPER-GRASS 1. *Capsella bursa-pastoris* (Shepherd's Purse) New Engl. Skinner. 2. *Sisymbrium canescens*.

PEPPER-GRASS, Perennial *Cardaria draba* (Hoary Cress) USA Gates.

PEPPER-GRASS, Tall *Lepidium virginicum* (Virginia Pepper-grass) USA Schenk & Gifford.

PEPPER-GRASS, Virginia *Lepidium virginicum*.

PEPPER-GRASS, Wild 1. *Lepidium densi-*

florum. 2. *Lepidium virginicum* (Virginia Pepper-grass) USA Upham.

PEPPER-GRASS, Yellow *Lepidium flavum*.

PEPPER-PLANT 1. *Capsella bursa-pastoris* (Shepherd's Purse) S Africa. Watt. 2. *Polygonum hydropiper* (Water Pepper) Yks. Grigson.

PEPPER-ROOT 1. *Dentaria diphylla* (Two-leaved Toothwort) Perry. 2. *Dentaria laciniata* (Cut-leaved Toothwort) USA House.

PEPPER SAXIFRAGE *Silaum silaus*.

PEPPER TREE; PEPPER TREE, California; PEPPER TREE, Peruvian *Schinus molle* (Peruvian Mastic) Perry (Pepper Tree); Kearney (California); RHS (Peruvian). Pepper Tree, possibly because it was thought at one time that the fruits would produce pepper, but the foliage gives off a peppery, pungent smell, anyway. It is Peruvian, but much planted in California as an ornamental street tree.

PEPPER TURNIP *Arisaema triphyllum* (Jack-in-the-pulpit) Le Strange. From the acrid corms, presumably - they are edible after boiling.

PEPPERIDGE 1. *Berberis vulgaris* (Barberry) Suff. Moor. 2. *Nyssa sylvatica* (Tupelo) USA Harper. Pepperidge for barberry comes from French pepin, pip and rouge, red - an allusion to the berries. Cf. Turner's Piprige, and also Piprage and Pipperidge. Tupelo's berries are black, though, so why is this name given?

PEPPERMINT 1. *Mentha piperita*. 2. *Mentha spicata* (Spearmint) Dev. Friend.1882. 3. *Pedicularis sylvatica* (Lousewort) Dev. Tynan & Maitland.

PEPPERMINT, Black *Mentha piperita var. officinalis*. The stems and leaves are a dark purplish-brown, hence the 'black' of the name.

PEPPERMINT, Horse *Ajuga reptans* (Bugle) Wilts. Dartnell & Goddard.

PEPPERMINT, Mount Wellington *Eucalyptus coccifera*. Cf. Peppermint Gum.

PEPPERMINT, Tasmanian *Eucalyptus amygdalina*. Cf. Peppermint Tree.

PEPPERMINT, White *Mentha piperita var. vulgaris*. A much paler plant than the one known as Black Peppermint, hence the 'white' of this name.

PEPPERMINT, Wild *Mentha aquatica* (Water Mint) Bardswell.

PEPPERMINT GERANIUM *Pelargonium tomentosum*.

PEPPERMINT GUM *Eucalyptus coccifera* (Mount Wellington Peppermint) Wilkinson.1981.

PEPPERMINT TREE 1. *Eucalyptus amygdalina* (Tasmanian Peppermint) J Smith. 2. *Eucalyptus odorata*.

PEPPERWEED, Desert *Lepidium fremontii* (Desert Alyssum) Douglas.

PEPPERWEED, Virginia *Lepidium virginicum* (Virginia Peppergrass) USA Allan.

PEPPERWORT 1. *Lepidium campestre*. 2. *Lepidium latifolium* (Dittander) Gerard. 3. *Silaum silaus* (Pepper Saxifrage) Salisbury.

PEPPERWORT, Broad-leaved *Lepidium latifolium* (Dittander) Clapham.

PEPPERWORT, Hoary *Cardaria draba* (Hoary Cress) McClintock. At one time, the seeds were ground as a pepper substitute. Cf. Whitlow Pepperwort.

PEPPERWORT, Mithridate *Lepidium campestre* (Pepperwort) Prior. Cf. Mithridate Mustard, and also Treacle Mustard for this, both 'treacle' and 'Mithridate' indicating its medicinal use.

PEPPERWORT, Narrow-leaved *Lepidium ruderale*.

PEPPERWORT, Smith's *Lepidium smithii*.

PEPPERWORT, Whitlow *Cardaria draba* (Hoary Cress) PLNN10. Cf. Hoary Pepperwort.

PEPPILARY *Populus nigra* (Black Poplar) Ches. Holland. Cf. Popillary, Poppilery etc., all presumably from French poplier.

PERCELY *Petroselinum crispum* (Parsley) Fernie.

PERCEPIER *Aphanes arvensis* (Parsley Piert) Prior. i.e. Perce-pierre, breakstone. Parsley Piert means the same.

PERILLA *Perilla frutescens*.

PERIWINKLE 1. *Endymion nonscriptus* (Bluebell) Dor. Macmillan. 2. *Ruscus aculeatus* (Butcher's Broom) Gunther. 3. *Vinca sp.* Butcher's Broom is glossed Crowleek and Periwinkle on a C12 ms of Apuleius. It sounds as if Bluebell were meant, for that does occasionally bear the name, as in the Dorset example quoted. The name is from OE peruince, from Latin pervinca.

PERIWINKLE, Greater *Vinca major*.

PERIWINKLE, Lesser *Vinca minor*.

PERIWINKLE, Madagascar *Vinca rosea*.

PERIWINKLE, Small *Vinca minor* (Lesser Periwinkle) Curtis.

PERIWINKLE, South African *Vinca rosea* (Madagascar Periwinkle) Mitton.

PERMACITY, Whoreman's *Capsella bursa-pastoris* (Shepherd's Purse) Culpepper. A garbled version of Gerard's Poor Man's Parmacetty. According to Dyer, it is a joke on bursa of the specific name, for to a poor man his purse is always the best remedy for his bruises.

PERMANENT WAVE TREE *Salix matsudana var. tortuosa* (Contorted Willow) Everett.

PERNAMBUCO REDWOOD; PERNAMBUCO WOOD *Caesalpina bonduc* (Brazilian Redwood) both Watt.

PERNEL *Anagallis arvensis* (Scarlet Pimpernel) Halliwell. i.e. pimpernel, with the loss of the first syllable.

PERNY'S HOLLY *Ilex pernyi*.

PERRYMEDOLL *Campanula pyramidalis* (Chimney Bellflower) Suff. Moor. i.e. Pyramidal.

PERSELE; PERSIL *Petroselinum crispum* (Parsley) Gerard (Persele); Scot. Jamieson (Persil).

PERSIA, Pride of *Melia azedarach* (Chinaberry Tree) Howes.

PERSIA, Star of *Allium christophii*.

PERSIAN ACACIA *Albizia julibrissin*.

PERSIAN APPLE *Citrus medica var. bajoura* Brouk. The 'medica' of the specific name means coming from Media.

PERSIAN BOX; IRANIAN BOX *Buxus sempervirens* timber (Boxwood) Brimble.1948.

PERSIAN BRIAR *Rosa foetida* (Capucine Rose) Plowden.

PERSIAN CANDYTUFT *Aethionema sp.*

PERSIAN CLOVER *Trifolium resupinatum* (Reversed Clover) McClintock.1974. Probably a native of Asia somewhere, 'Persian' being a stab at the country of origin.

PERSIAN IVY *Hedera colchica*.

PERSIAN KINGCUP *Ranunculus asiaticus* Coats.

PERSIAN LILAC 1. *Melia azedarach* (Chinaberry Tree) Polunin.1976. 2. *Syringa persica*.

PERSIAN LILY 1. *Fritillaria imperialis* (Crown Imperial). 2. *Fritillaria persica* Gerard.

PERSIAN MULBERRY *Morus nigra* (Mulberry) Hora.

PERSIAN PELLITORY *Pyrethrum roseum*.

PERSIAN SAXIFRAGE *Saxifraga cymbalaria*. 'Persian' is not quite accurate, for it grows in an area from the Carpathians, through Asia Minor to the Caucasus, and also in Algeria.

PERSIAN VIOLET *Exacum affine*. Cf. Arabian Violet.

PERSIAN WALNUT *Juglans regia* (Walnut) Schery. Accurate enough - it is actually a native of Persia.

PERSIAN WILLOW *Chamaenerion angustifolium* (Rosebay Willowherb) Prior. In spite of this name, this is actually an American plant. The 'willow' ascription is because there is some resemblance in the leaves to those of the willow - Cf. Willow, Bay Willow, Blooming Willow, and a number of others.

PERSICARIA *Polygonum persicaria*. Persicaria means a peach-tree, from the resemblance in the leaves. Cf. Culpepper's Peachwort.

PERSICARIA, Amphibious *Polygonum amphibium*.

PERSICARIA, Biting *Polygonum hydropiper* (Water Pepper) Barton & Castle.

PERSICARIA, Knotted *Polygonum nodosum*.

PERSICARIA, Pale *Polygonum lapathifolium*.

PERSICARIA, Pennsylvania *Polygonum pennsylvanicum* (Heartseed) USA Youngken.

PERSICARIA, Spotted 1. *Polygonum nodosum* (Knotted Persicaria) Allan. 2. *Polygonum persicaria* (Persicaria) Curtis.

PERSICARIA, Swamp *Polygonum muhlenbergii*.

PERSICARIA, Willow *Polygonum lapathifolium* (Pale Persicaria) USA Upham.

PERSIMMON 1. *Diospyros kaki* (Chinese Date Plum) Willis. 2. *Diospyros virginiana*. 'Persimmon' is apparently from an American Indian word.

PERSIMMON, Chinese; PERSIMMON, Japanese *Diospyros kaki* (Chinese Date Plum) both Barber & Phillips.

PERSIMMON, North American *Diospyros virginiana* (Persimmon) Willis.

PERU, Apple of 1. *Datura stramonium* (Thorn-apple) Gerard; USA Henkel. 2. *Nicandra physalodes* (Shoo-fly Plant) USA Zenkert.

PERU, Marigold of *Helianthus annuus* (Sunflower) Gerard. The sunflower was actually a native of Mexico.

PERU, Marvel of *Mirabilis dichotoma*.

PERUVIAN BARK *Cinchona sp* (Quinine) Duran-Reynald.

PERUVIAN CHERRY *Physalis bunyardii* (Cape Gooseberry) Howes.

PERUVIAN DAFFODIL *Ismene calathium* USA Cunningham.

PERUVIAN LEAF *Erythroxylon truxillense* Thomson.

PERUVIAN LILY 1. *Alstroemeria aurantiaca*. 2. *Scilla peruviana*.

PERUVIAN MASTIC *Schinus molle*. The name really should be applied to the gum collected from this shrub, also known as American Mastic, but the plant itself carries the name.

PERUVIAN PEPPER TREE *Schinus molle* (Peruvian Mastic) RHS.

PERUVIAN RHATANY *Krameria triandra*. 'Rhatany' is from a Quechua original, through the Spanish ratania. The plant grows on the western slopes of the Andes in Peru.

PERUVIAN SWAMP LILY *Zephyranthes candida* (Zephyr Lily) Howes.

PERUVIAN TURNSOLE *Heliotropum europaeum* (Cherry-pie) J Smith. In spite of the specific name "europaeum", this is indeed a Peruvian plant, introduced here some time before 1757. 'Turnsole' means 'turning towards the sun'.

PERVENKLE; PERVINCA; PERVINKLE; PERWINKLE *Vinca minor* (Lesser Periwinkle) Fernie (Pervenkle, Pervinca); Gerard (Pervinkle, Perwinkle).

PESTILENCE-WORT; PESTILENT-WORT *Petasites hybridus* (Butterbur) B & H (Pestilence); Fernie (Pestilent). "... in high Dutch Pestilenzwurts" (Gerard), because of its use as a plague remedy, though they say in Cornwall that it is because the plant sprang from the graves of plague victims in Veryan churchyard.

PESTLE, Marsh *Typha latifolia* (False Bulrush) W Miller.

PET, Baby's *Bellis perennis* (Daisy) Som. Macmillan.

PETER, Blue *Aconitum napellus* (Monkshood) Yks. Barbour.

PETER, Herb *Primula veris* (Cowslip) Gerard. From a supposed resemblance to the badge of St Peter. The legend is that St Peter once dropped the keys of heaven, and the first cowslips grew up where they fell.

PETER KEYS *Fraxinus excelsior* fruits (Ash keys) Coles. Given that the seed vessels are called Keys, it was perhaps inevitable that they would be linked with St Peter.

PETER'S GRASS *Rhinanthus crista-galli* (Yellow Rattle) Tynan & Maitland.

PETER'S PENCE *Lunaria annua* (Honesty) Dor. Macmillan. The reference is to the tax of a silver penny paid to the Pope. The pennies are the seed vessels.

PETER'S STAFF *Verbascum thapsus* (Mullein) B & H; USA Henkel. Mullein is a tall, upright plant that in folklore has provided a staff for Jupiter, Jacob and Aaron, as well as for Peter.

PETERKIN *Primula veris* (Cowslip) Tynan & Maitland. i.e. Peter's Keys.

PETHWINE *Clematis vitalba* (Old Man's Beard) Suss. Parish. More widespread as Bethwine, or Bethwind. The more usual 'b' has become a 'p' in Sussex - Cf. also Pithywind for this plant.

PETIGREE; PETTIGREE, Prickly; PETTIGRUE; PETTIGRUE, Prickly *Ruscus aculeatus* (Butcher's Broom) Gerard (Petigree); Barton & Castle (Prickly Pettigree); Turner (Pettigrue); Fernie (Prickly Pettigrue). French, meaning little holly, from the prickly leaves.

PETREA, Twining *Petrea volubilis* (Purple Wreath Vine) Whittle & Cook.

PE-TSAI *Brassica pekinensis* (Chinese Cabbage) Schery.

PETTICOATS, Fairy's *Digitalis purpurea* (Foxglove) Som. Macmillan; Ches. Holland. Cf. Fairy's Dresses.

PETTICOATS, Flannel *Verbascum thapsus* (Mullein) Som. Macmillan.

PETTICOATS, Hoop 1. *Narcissus bulbocodium*. 2. *Narcissus pseudo-narcissus* (Daffodil) Dor. Grigson.

PETTICOATS, Lady's 1. *Anemone nemerosa* (Wood Anemone) Wilts. Dartnell & Goddard. 2. *Aquilegia vulgaris* (Columbine) Som. Macmillan.

PETTICOATS, Old Woman's; PETTICOATS, Red *Papaver rhoeas* (Red Poppy) Som. Grigson (Old Woman's); Kent Grigson (Red).

PETTICOATS, White, Lady's *Stellaria holostea* (Greater Stitchwort) Corn, Heref. Grigson.

PETTIGREW *Hyssopus officinalis* (Hyssop) Palaiseul. Is this correct? In one form or another, this is a name usually associated with Butcher's Broom.

PETTITOES, Pig's *Lotus corniculatus* (Bird's-foot Trefoil) Suss. Parish.

PETTY MORREL *Aralia racemosa* (American Spikenard) Howes.

PETTY PANICK *Phalaris canariensis* (Canary Grass) Turner. He is likening it, by this name, to a small kind of millet (Panicum).

PETTY SPURGE *Euphorbia peplus*. Petty means small.

PETTY WHIN 1. *Genista anglica* (Needle Whin) Gerard. 2. *Ulex minor* (Small Furze) Jones & Dillon.

PEVIT *Ligustrum vulgare* (Privet) Corn. Grigson.

PEWEEP-TREE *Acer pseudo-platanus* (Sycamore) Corn. Grigson. Did children make whistles from it, as with elder twigs?

PEWTER-WORT *Equisetum arvense* (Common Horsetail) Gerard. He also knew it as Shave-grass - "not without cause named Asprelle, of his ruggedness, which is not unknown to women, who scoure their pewter and wooden things of the kitchen therwith". It was used as a pot-scourer long after Gerard's time.

PEY *Pisum sativum* (Garden Pea) Yks. Addy. Pronounced 'pay'.

PEYOTE *Lophophora williamsii*. Aztec peyotl.

PHACELIA, Death Valley *Phacelia vallismortae*.

PHACELIA, Lacy; PHACELIA, Tansy *Phacelia tanacetifolia*. Lacy Phacelia seems to be the usual name. Tansy Phacelia, which seems appropriate in view of the specific name, is in Usher.

PHACELIA, Weasel *Phacelia mustelina*. 'Weasel', because of the smell.

PHARAOH'S FIG *Ficus sycomorus* (Sycamore Fig) Hora.

PHARAOH'S PEAS *Lathyrus sylvestris* (Everlasting Pea) N'thants. PLNN3. Quoted as *L sylvestris*, but more likely to be *L latifolius*. There is a legend that someone from the village of Weebly went to Egypt and brought home some seeds from a royal tomb. Of course, this plant does not grow in Egypt.

PHARBITIS SEED *Ipomaea hederacea* (Ivy-leaf Morning Glory) Chopra. The seeds contain a rosin from which a glucooside named pharbitin has been isolated. These seeds are toxic, and a drastic purgative.

PHEASANT LILY *Fritillaria meleagris* (Snake's-Head Lily) Cumb. B & H. The colouring of the flower accounts for this name. But the specific name, *meleagris*, means a guinea-hen, and there are several Guinea-hen names recorded.

PHEASANT'S EYE 1. *Adonis annua*. 2. *Anagallis arvensis* (Scarlet Pimpernel) Som. Macmillan. 3. *Dianthus caryophyllus* (Carnation) Som. Macmillan. 4. *Narcissus maialis var. patellaris*. 5. *Pentaglottis sempervirens* (Evergreen Alkanet) Som. Elworthy.

PHEASANT'S EYE, Corn *Adonis annua* Flower.1859.

PHEASANT'S EYE, Large *Adonis flammea*.

PHEASANT'S EYE, Pyrenean *Adonis pyrenaica*.

PHEASANT'S EYE, Yellow *Adonis vernalis*.

PHEASANT'S FEATHERS *Saxifraga x urbinum* (London Pride) Suff. B & H. Cf. Prince's Feather and Queen's Feather for this.

PHEASANT'S HEAD *Fritillaria meleagris* (Snake's-head Lily) Leyel.1948.

PHILADELPHIA FLEABANE *Erigeron philadelphicus*.

PHILADELPHIA LILY *Lilium philadelphicum*.

PHILANTHROPOS *Agrimonia eupatoria* (Agrimony) Hatfield. The seed vessels cling by the hooked ends of their stiff hairs to anyone coming into contact with the plant. Hence this name, and also Sweethearts.

PHILIPPINE LILY *Lilium philippense*.

PHILIPPINE VIOLET *Barleria cristata*.

PHILTREWORT *Circaea lutetiana* (Enchanter's Nightshade) Yks. Grigson.

PHLOX, Blue, Wild *Phlox divaricata*.

PHLOX, Downy *Phlox pilosa*.

PHLOX, Ground *Phlox subulata*.

PHLOX, Pink *Phlox caespitosa*.

PHLOX, Prairie *Phlox pilosa* (Downy Phlox) USA House.

PHLOX, Prickly *Gilia californica*.

PHLOX, Wild *Epilobium hirsutum* (Great Willowherb) Dev. Grigson. There is perhaps a resemblance in the flowers of the two species.

PHLOX, Yellow *Erysimum asperum* (Desert Wallflower) USA Elmore.

PHOENICIAN JUNIPER *Juniperus phoenicia*.

PHOENICIAN ROSE *Rosa phoenicia*.

PHORAMS, PHORANS *Rumex sp* (Dock) Ire. Grigson.

PHU(E) *Valeriana officinalis* (Valerian) Langham. The word is Greek,and means foul-smelling.

PHYSIC, Indian 1. *Apocynum cannabinum* (Indian Hemp) Grieve.1931. 2. *Gillenia trifoliata*.

PHYSIC NUT 1. *Jatropha curcas* (Barbados Nut) USA Kingsbury. 2. *Jatropha multifida* (Coral Plant) USA Kingsbury. Cf. Purge Nut and Castor-oil Bean for Jatropha curcas. The plant is well-known as a drastic, even dangerous, purge.

PHYSIC NUT, French *Jatropha curcas* (Barbados Nut) Perry. Why French? This is a tropical American plant. Perhaps the connection lies in Haiti.

PHYSIC ROOT *Veronica virginica* (Culver's Root) USA Lloyd. The drug obtained from it is a violent purgative.

PIANET *Paeonia mascula* (Paeony) Shrop, Ches, Lincs, Cumb. B & H. This is but one of a number of variations on the name 'paeony'. Cf. Pie-nanny, Nanpie, Piny, etc.

PIANO-ROSES *Paeonia mascula* (Paeony) Ire. B & H.

PICHI *Fabiana imbricata* (False Heath) Mitton.

PICHON *Dodonea viscosa* (Switch Sorrel) West Indies Howes.

PICK CHEESE; PICK-A-CHEESE *Malva sylvestris* (Common Mallow) Herts, Norf.

B & H (Pick); Norf. V G Hatfield.1994 (Pick-a-).

PICKS; PICKS, Hedge *Prunus spinosa* fruit (Sloes) Wilts. Dartnell & Goddard (Picks); Wilts. Macmillan; Hants. Grigson (Hedge Picks).

PICKS, Winter (or as one word, Winterpicks) *Prunus spinosa* -fruit (Sloes) Hulme.

PICK-A-BACK PLANT; PIGGY-BACK PLANT *Tolmiea menziesii* Gannon (Pick-a-back); USA Cunningham (Piggy-back).

PICK-FOLLY *Cardamine pratensis* (Lady's Smock) N'thants. Grigson. It must mean that it is folly to pick them. And so it is, for this is a fairy plant, and it would be bad luck to pick one. But not necessarily, for there is a divination game of the "Rich man, poor man" type, involving picking the leaves one by one.

PICK-NEEDLE *Erodium moschatum* (Musk Storksbill) B & H. Or Pink-needle. In each case, so Allies says, 'Pick' or 'Pink' means Puck.

PICK-YOUR-MOTHER'S-EYES-OUT *Veronica chamaedrys* (Germander Speedwell) Dor. Grigson. Cf. Hawk-your-mother's-eyes-out, also from Dorset, and the Devonshire Tear-your-mother's-eyes-out. The names are from the West country, but the belief is found all over England, e.g. Yorkshire, where it is said that if a child gathers germander speedwell its mother will die during the year, or the Lincolnshire superstition that if anyone picks the flower, his eyes will be eaten. The insistence on 'eyes' is because the flower itself has a prominent "eye".

PICK-YOUR-MOTHER'S-HEART-OUT *Capsella bursa-pastoris* (Shepherd's Purse) War. Grigson. Cf. Mother's Heart, and Mother-die. 'Heart', possibly because of the heart-shaped capsules, but children play a sort of game with the pod. They hold it out to their companions, inviting them to "take a hand o' that". It immediately cracks, and then follows a triumphant shout, "You've broken your mother's heart" (Dyer).

PICKEREL-WEED *Ranunculus aquatilis* (Water Crowfoot) E Ang. Grigson. A pickerel is a young pike.

PICKLE-PLANT 1. *Cucumis anguria* (Gherkin) USA H M Hyatt. 2. *Salicornia europaea* (Glasswort) Cumb. B & H. Steeped in malt vinegar, the young shoots of Glasswort make a good pickle, and were often used as a substitute for samphire, though they say they are inferior to the proper stuff.

PICKPOCKET 1. *Alliaria petiolata* (Jack-by-the-hedge) Dev. Macmillan. 2. *Capsella bursa-pastoris* (Shepherd's Purse) Prior; Lincs. Tynan & Maitland. 3. *Cardamine pratensis* (Lady's Smock) War. F Savage. 4. *Conium maculatum* (Hemlock) Som. Macmillan. 5. *Cymbalaria muralis* (Ivy-leaved Toadflax) Som. Macmillan. 6. *Lolium perenne* (Rye-grass) Dev. B & H. 7. *Prunella vulgaris* (Self-heal) Ess. Grigson. 8. *Ranunculus repens* (Creeping Buttercup) Ess. Gepp. 9. *Sedum acre* (Biting Stonecrop) Dor. Macmillan. 10. *Spergula arvensis* (Corn Spurrey) Ches. Grigson. 11. *Stellaria holostea* (Greater Stitchwort) Dev. Friend.1882; Som, Kent Grigson. 12. *Veronica persica* (Common Field Speedwell) Som. Macmillan. Pickpocket is usually given either because the plant so named flourishes in unproductive (in the profit sense) land, or because it is such a rampant weed as to damage the farmer's chance of profit. But in the case of Greater Stitchwort, the "pick" of the name is probably "pixy". Cf. Pixies, Piskies etc., for this plant.

PICKPOCKET-TO-LONDON *Capsella bursa-pastoris* (Shepherd's Purse) Yks. Grigson.

PICKPURSE 1. *Capsella bursa-pastoris* (Shepherd's Purse) Gerard. 2. *Spergula arvensis* (Corn Spurrey) E Ang, Lincs. Hutchinson.

PICNIC THISTLE *Cirsium acaulon*. The Stemless Thistle, the kind you inevitably sit on when having a picnic.

PICRIS BROOMRAPE *Orobanche picridis*.

PICTOU DISEASE *Senecio latifolius* Nova

Scotia Watt. Pictou disease is cirrhosis of the liver in cattle, caused by this plant, and by ragwort in this country.

PIDDLY-BED *Taraxacum officinale* (Dandelion) Devlin. One of the many variants of Piss-a-bed, all given because of the well-known diuretic qualities of the plant.

PIE, Apple 1. *Artemisia vulgaris* (Mugwort) Ches. B & H. 2. *Cardamine pratensis* (Lady's Smock) Yks. Grigson. 3. *Chamaenerion angustifolium* (Rosebay Willowherb) Dev. Friend. 4. *Epilobium hirsutum* (Great Willowherb) Som, Herts, Suff, Yks, N'thum. Grigson; Dev. Friend.1882; Glos. J D Robertson; Hants. Cope; Suss. Parish; Ches. Holland; Ess. Gepp. It is said that Great Willowherb smells like apples when it is crushed, but it is probably Gerard's Codded Willowherb, with its resemblance to the name Codlins, which is the origin.

PIE, Cherry 1. *Chamaenerion angustifolium* (Rosebay Willowherb) Dor. B & H. 2. *Epilobium hirsutum* (Great Willowherb) Dor, Som, Worcs, Notts. Grigson. 3. *Heliotropum europaeum.* 4. *Valeriana dioica* (Marsh Valerian) Wilts. Macmillan. 5. *Valeriana officinalis* (Valerian) Wilts. Dartnell & Goddard. Because of the smell, in all cases.

PIE, Gooseberry 1. *Epilobium hirsutum* (Great Willowherb) Dev. Friend.1882. 2. *Ononis repens* (Rest-harrow) Som. Macmillan. 3. *Melandrium album* (White Campion) Som. Macmillan. 4. *Symphytum officinale* (Comfrey) Dev, Dor, Wilts, Suff. Grigson. 5. *Valeriana dioica* (Marsh Valerian) Dev, Dor, Suff. Grigson; Wilts. Dartnell & Goddard. Again, something to do with the smell.

PIE, Penny *Umbilicus rupestris* (Wall Pennywort) Dev, Som. Grigson. The round leaves, likened to both pies and pennies.

PIE-ANNA *Paeonia mascula* (Paeony) IOM Moore.1924.

PIE-APPLE *Pinus sylvestris* cones (Pine cones) N'thants. A E Baker. Cf. Pin-apples and Pur-apples.

PIE-CHERRY *Prunus cerasus* (Morello Cherry) USA Leighton. This is the sour cherry, used mainly for cooking, as a filling for tarts, etc.

PIE-CRESS *Apium nodiforum* (Fool's Watercress) Dev. Macmillan. West country people used to collect it at one time, and cook it with meat in pies and pasties, more particularly, it seems, in the neighbourhood of Polperro, in Cornwall.

PIE-MARKER; PIE-PRINT *Abutilon avicennae* (China Jute) USA Bergen. Both the pods and leaves were used to stamp butter or pie crust.

PIE-NANNY *Paeonia mascula* (Paeony) Yks. B & H. Cf. Pie-Anna, and several other variations on Paeony.

PIE-PLANT *Rheum rhaponticum* (Rhubarb) Kansas Sackett & Koch. It is also known as Tart Rhubarb in America.

PIECESTONE *Alchemilla vulgaris* (Lady's Mantle) Fernie. 'Piercestone' was the original, the plant being confused with Parsley Piert.

PIERCE-SNOW *Galanthus nivalis* (Snowdrop) Som. Macmillan. Cf. Snow-piercer, also from Somerset.

PIERCE-STONE *Crithmum maritimum* (Samphire) B & H. A play on the name Samphire, which is Saint Pierre. 'Pierre' also means stone, so if we disregard the attribution to St Peter, we can see that samphire means simply a plant that grows on a rock, and so by extension a herb that "pierces" the rock.

PIERCE-STONE, Parsley *Aphanes arvensis* (Parsley Piert) Grieve.1931. Parsley Piert itself means breakstone, from French perce-pierre. By sympathy, it was much used against stones in the bladder, even right into World War II times.

PIG-ALES *Crataegus monogyna* fruit (Haws) Som. Macmillan. There are a number of 'pig' names for haws. In all probability, they come from 'hog', as a folk etymology. In fact, the 'hog' names come from OE haga, hedge.

PIG-BERRY *Crataegus monogyna* fruit (Haws) Som. Macmillan; Wilts. Dartnell & Goddard.
PIG DAISY *Pulicaria dysenterica* (Yellow Fleabane) Dor. Macmillan.
PIG DOCK *Aethusa cynapium* (Fool's Parsley) Som. Macmillan.
PIG-FACES *Mesembryanthemum equilaterale* (Fig Marigold) Tasmania H L Roth.
PIG GRASS 1. *Polygonum aviculare* (Knotgrass) Shrop, Notts, Lincs, Yks. Grigson. 2. *Polygonum persicaria* (Persicaria) Lincs. Grigson.
PIG HAW *Crataegus monogyna* fruit (Haws) Som, Hants. Grigson; Wilts. Akerman.
PIG-IN-THE-HEDGE *Prunus spinosa* (Blackthorn) Hants. Grigson. 'Pig' here is probably 'pick', a common West country name for sloes.
PIG LAUREL *Kalmia angustifolia* (Sheep Laurel) Howes.
PIG LEAVES 1. *Cirsium dissectum* (Meadow Thistle) Yks. B & H. 2. *Onopordum acanthium* (Scotch Thistle) Yks, N Eng. Halliwell.
PIG LILY *Arum maculatum* (Cuckoo-pint) Som. Macmillan.
PIG NUT *Jatropha curcas* (Barbados Nut) W Africa Dalziel.
PIG-O'-THE-WALL *Antirrhinum majus* (Snapdragon) Som. Macmillan.
PIG ROSE 1. *Papaver rhoeas* (Red Poppy) Corn. Grigson. 2. *Rosa canina* (Dog Rose) Corn. Grigson.
PIG RUSH *Polygonum aviculare* (Knotgrass) Shrop. Grigson. Cf. Pigweed, Hogweed, Swine-grass, etc.
PIG VIOLET *Viola riviniana* (Dog Violet) Corn. Grigson.
PIG'S AILES *Crataegus monogyna* fruit (Haws) Som. Macmillan.
PIG'S BUBBLES; PIG'S COLE *Heracleum sphondyllium* (Hogweed) Som. Elworthy (Bubbles); Dev. Friend.1882 (Cole). There are a number of 'pig' names for Hogweed, presumably because it was actually used as a pig food.
PIG'S CHOPS 1. *Antirrhinum majus* (Snapdragon) Som. Macmillan. 2. *Linaria vulgaris* (Toadflax) Som. Macmillan. Cf. Pig's Mouth.
PIG'S CRESS 1. *Anthemis cotula* (Maydweed) Som. Macmillan. 2. *Lapsana communis* (Nipplewort) Som. Macmillan.
PIG'S DAISY 1. *Anthemis cotula* (Maydweed) Dor. Macmillan. 2. *Pulicaria dysenterica* (Yellow Fleabane) Dor. Grigson.
PIG'S EAR 1. *Cotyledon orbiculata* (Round-leaved Navelwort) Whittle & Cook. 2. *Sedum acre* (Biting Stonecrop) Friend.1882.
PIG'S EYE *Cardamine pratensis* (Lady's Smock) Ess. Grigson. Cf. Pigeon's Eye.
PIG'S FLOP *Heracleum sphondyllium* (Hogweed) Dev. Macmillan. 'Flop' refers to the large leaves - Cf. the various similar names for foxglove.
PIG'S FLOWER *Anthemis cotula* (Maydweed) Dor. Macmillan.
PIG'S FOOD *Heracleum sphondyllium* (Hogweed) Dor. Macmillan.
PIG'S FOOT *Lotus corniculatus* (Bird's-foot Trefoil) Suff. Grigson.
PIG'S GREASE *Veronica beccabunga* (Brooklime) Dor. Grigson. 'Grease' might just possibly have something to do with ointments (the plant was certainly used to make some), or it could very well just mean plant (there is an entry in Halliwell for gres, which suggests this).
PIG'S HALES; PIG'S HAWS *Crataegus monogyna* fruit (Haws) Som. Jennings (Hales); Som. Macmillan (Haws).
PIG'S HEELS; PIG'S HELLS; PIG'S ISLES *Crataegus monogyna* fruit (Haws) all Som. Macmillan. An extremely variable name, probably starting with Pigall. See Pig-Ales.
PIG'S MOUTH 1. *Antirrhinum majus* (Snapdragon) Som. Macmillan. 2. *Linaria vulgaris* (Toadflax) Som. Macmillan. Cf. Pig's Chops.

PIG'S NOSE *Rosa canina* fruit (Hips) Dev. Grigson.

PIG'S PARSLEY 1. *Daucus carota* (Wild Carrot) Corn. Grigson; Dor. Macmillan. 2. *Torilis japonica* (Hedge Parsley) Som. Elworthy.

PIG'S PARSNIP *Heracleum sphondyllium* (Hogweed) Som. Elworthy; Shrop. B & H.

PIG'S PEARS *Crataegus monogyna* fruit (Haws) Som. Macmillan.

PIG'S PETTITOES *Lotus corniculatus* (Bird's-foot Trefoil) Suss. Parish.

PIG'S RHUBARB *Arctium lappa* (Burdock) Dor. Mabey.

PIG'S ROSE *Rosa canina* (Dog Rose) Dev. Grigson.

PIG'S SNOUT *Antirrhinum majus* (Snapdragon) Som. Macmillan.

PIGALES *Crataegus monogyna* fruit (Haws) Wilts. Jones & Dillon.

PIGALL *Crataegus monogyna* fruit (Haws) Som. Macmillan Wilts. Akerman; Hants. Grigson.

PIGAUL; PIGHAU *Crataegus monogyna* fruit (Haws) both Hants. Cope.

PIGEON-BERRY 1. *Duranta repens* (Golden Dewdrop) Perry. 2. *Lonicera involucrata* (Bearberry Honeysuckle) USA. T H Kearney. 3. *Phytolacca decandra* (Poke-root) Brownlow. 4. *Rhamnus californica* (Redberry) USA T H Kearney. 5. *Rivina humilis*.

PIGEON ORCHID *Dendrobium crumentatum*.

PIGEON PEA 1. *Cajanus indicus*. 2. *Laburnum anagyroides* (Laburnum) Tynan & Maitland.

PIGEON'S EYE *Cardamine pratensis* (Lady's Smock) Yks. Grigson. Cf. Pig's Eye.

PIGEON'S FOOT 1. *Geranium molle* (Dove's-foot Cranesbill) Gerard. 2. *Salicornia europaea* (Glasswort) Howes.

PIGEON'S GRASS; PIGEON'S MEAT *Verbena officinalis* (Vervain) Prior (Grass); Sanecki (Meat). There is also Columbine for this. The Greek name was peristereon, doves, because they were fond of hovering round the plant. But Coles explains it as "because Pidgeons eat thereof as is supposed to clear their Eye sight".

PIGEON'S KNEE *Cardiospermum halicacabum* (Blister Creeper) Chopra.

PIGEONS *Acer pseudo-platanus* fruit (Sycamore keys) Son. Macmillan.

PIGEONWOOD *Hedycarya arborea*.

PIGGLE *Primula veris* (Cowslip) Som. Macmillan. The old name Paigle is the original of this.

PIGGY-WIGGY *Antirrhinum majus* (Snapdragon) Som. Macmillan.

PIGHTLE, Crow 1. *Ranunculus bulbosus* (Bulbous Buttercup) N'thants. A E Baker. 2. *Ranunculus ficaria* (Lesser Celandine) Beds. Grigson. Pightle is an old name for a small meadow. The name also occurs in the same county as Crow Pikel, or Crow Packle.

PIGLE *Stellaria holostea* (Greater Stitchwort) Dawson. Paigle, the usual name for a cowslip, also occurs for this.

PIGNUT 1. *Bunium bulbocastanum* (Great Earth-nut) J Smith. 2. *Carya glabra*. 3. *Conopodium majus* (Earth-nut) common McClintock (including Wiltshire Dartnell & Goddard). Cf. Pegnut, which is simply local pronunciation.

PIGNUT, Sweet *Carya ovalis* (Red Hickory) Wilkinson.1981.

PIGNUT HICKORY *Carya cordiformis* (Swamp Hickory) Schery.

PIGS, Hedge *Prunus spinosa* fruit (Sloes) Glos. J D Robertson. 'Picks' is the more recognisable form.

PIGSHELL *Crataegus monogyna* fruit (Haws) Wessex Rogers. One of the series that includes Peggall.

PIGSY-PEARS *Crataegus monogyna* fruit (Haws) Som. Macmillan. Cf. Pixy-Pears.

PIGTAIL *Galium aparine* (Goose-grass) N'thants. A E Baker; Notts. Grigson.

PIGTAIL PLANT *Anthurium scherzerianum* (Flamingo Plant) USA Cunningham.

PIGTOES *Lotus corniculatus* (Bird's-foot

Trefoil) Suss. PLNN 25. Cf. Pig's Foot and Pig's Pettitoes.

PIGWEED 1. *Aegopodium podagraria* (Goutweed) Grieve.1931. 2. *Amaranthus retroflexus*. 3. *Amaranthus viridis* (Green Amaranth) Barbados Gooding. 4. *Chenopodium album* (Fat Hen) Som. Grigson; Hants. Cope; Canada Grieve.1931. 5. *Chenopodium glaucum* (Oak-leaved Goosefoot) S Africa. Watt. 6. *Chenopodium nuttalliae* USA Driver. 7. *Chenopodium rubrum* (Red Goosefoot) Prior. 8. *Heracleum sphondyllium* (Hogweed) Oxf. B & H. 9. *Polygonum aviculare* (Knotgrass) Hants. Cope; Worcs. Grigson. 10. *Portulaca oleracea* (Green Purslane) Australia Dalziel. 11. *Portulaca sativa* (Purslane) Grieve.1931. 12. *Symphytum officinale* (Comfrey) Wilts. Dartnell & Goddard.

PIGWEED, Amaranth *Amaranthus retroflexus* (Pigweed) USA Zenkert.

PIGWEED, Bushy *Amaranthus albus* (White Pigweed) USA Upham.

PIGWEED, Green 1. *Amaranthus chlorostachys*. 2. *Amaranthus retroflexus* (Pigweed) USA Watt.

PIGWEED, Prostrate *Amaranthus blitoides* (Tumbleweed) USA Allan.

PIGWEED, Redroot *Amaranthus retroflexus* (Pigweed) Watt.

PIGWEED, Rough *Amaranthus retroflexus* (Pigweed) S Africa Watt.

PIGWEED, Slender *Amaranthus hybridus*.

PIGWEED, Spreading *Amaranthus blitoides* (Tumbleweed) USA Elmore.

PIGWEED, Sweet *Chenopodium ambrosioides* (American Wormseed) Dalziel.

PIGWEED, Thorny *Amaranthus spinosus*.

PIGWEED, White 1. *Amaranthus albus*. 2. *Chenopodium album* (Fat Hen) USA Kluckhohn.

PIKEL, Crow *Ranunculus bulbosus* (Bulbous Buttercup) N'thants. A E Baker. See Pightle.

PILES *Hordeum sativum* awns (Barley) Dev. Friend.1882. Worcs. J Salisbury. Nothing to do with the ailment of the same name. This is obviously the same word as those used in Wiltshire for the awns - they vary between Ailes, Eyles or Iles.

PILEWORT *Ranunculus ficaria* (Lesser Celandine) Gerard; Cambs. Porter. There are small tubers on the roots, and these serve for the signature of piles, hence the widespread use as a country remedy for the ailment. There does seem to be some curative effect.

PILEWORT, Great *Scrophularia nodosa* (Figwort) Coles. In this context, figwort and pilewort are synonymous. Apparently, a tincture of the fresh plant is still recommended for piles, as it has been since medieval times.

PILEWORT, Marsh *Ranunculus ficaria* (Lesser Celandine) N'thants. A E Baker.

PILGRIM'S GOURD *Lagenaria siceraria* (Calabash) Chopra. Cf. Bottle Gourd - the ripe fruit is hollowed out, cleared of pulp and used as a bottle.

PILGRIM'S TREE *Ravenala madagascariensis* (Traveller's Tree) Tosco. The leaf bases fill with water, and they are often picked to provide drinking water, hence the names.

PILIOL *Thymus vulgaris* (Wild Thyme) B & H. Presumably the result of an old name, serphyllum montanum. Gerard had Puliall-mountain.

PILL NETTLE *Urtica pilulifera* (Roman Nettle) Hatfield.

PILLAR APPLE *Malus tschonoskii*. An erect growing species.

PILLERDS *Hordeum sativum* (Barley) Corn. Halliwell.

PILLOW, Robin's *Rosa canina* the galls that grow on it Page.1978.

PILLUS *Avena fatua* (Wild Oat) Corn. Quiller-Couch.

PILLWORT *Pilularia globulifera*.

PILLY, Priest's; PILLYCODS, Parson's *Arum maculatum* (Cuckoo-pint) West. Grigson (Pilly); Yks. B & H (Pillycods). Pilly in this context is the same as the 'pint' of Cuckoo-pint, i.e. pintle, or penis.

PILOT-WEED *Silphium laciniatum* (Compass Plant) USA Youngken.

Presumably for the same reason as the more usual name, Compass Plant. Its leaves turn their edges north to south, and so avoid midday radiation.

PIMENTARY *Melissa officinalis* (Balm) B & H.

PIMENTO 1. *Capsicum annuum* (Chile) USA Grigson.1974. 2. *Eugenia pimenta* fruit Rimmel. 3. *Pimento dioica* (Allspice) Schery. Portuguese pimenta, Spanish pimiento, pepper. In the case of allspice, because the dried fruits look like peppercorns.

PIMENTO, Jamaica *Pimento dioica* (Allspice) Schery. Cf. Jamaica Pepper for this.

PIMENTO ROYAL *Myrica gale* (Bog Myrtle) Leyel.1937.

PIMMEROSE *Primula vulgaris* (Primrose) Heref. Havergal.

PIMPERNEL *Poterium sanguisorba* (Salad Burnet) B & H. Not the usual plant to be called pimpernel, which always calls to mind members of the genus Anagallis. But Italian pimpinella means burnet, and the word derives ultimately from Latin piper, pepper (the fruits of burnet bear some resemblance to a peppercorn.

PIMPERNEL, Bastard *Centunculus minimus* (Chaffweed) Curtis.

PIMPERNEL, Blue 1. *Anagallis foemina*. 2. *Anagallis linifolia* (Shrubby Pimpernel) H & P.

PIMPERNEL, Bog *Anagallis tenella*.

PIMPERNEL, Common *Anagallis arvensis* (Scarlet Pimpernel) Forsyth.

PIMPERNEL, Corn *Anagallis arvensis* (Scarlet Pimpernel) Flower.1859.

PIMPERNEL, Female *Anagallis foemina* (Blue Pimpernel) Gerard. Scarlet Pimpernel is the male.

PIMPERNEL, Male *Anagallis arvensis* (Scarlet Pimpernel) Lyte.

PIMPERNEL, Red *Anagallis arvensis* (Scarlet Pimpernel) Dawson.

PIMPERNEL, Round *Samolus valerandi* (Brookweed) B & H. i.e. round-leaved.

PIMPERNEL, Scarlet *Anagallis arvensis*. Pimpernel really has significance to burnet, and in the Romance languages the word still means burnet. Quite how it transferred in English to this plant is not clear.

PIMPERNEL, Seaside *Honkenya peploides* (Sea Sandwort) W Miller.

PIMPERNEL, Shrubby *Anagallis linifolia*.

PIMPERNEL, Tom *Anagallis arvensis* (Scarlet Pimpernel) Grigson. Cf. Male Pimpernel for this. The blue-flowered *A foemina* is the female.

PIMPERNEL, Water 1. *Samolus floribundus* USA Zenkert. 2. *Samolus valerandi* (Brookweed) B & H. 3. *Veronica anagallis-aquatica* (Water Speedwell) B & H. 4. *Veronica beccabunga* (Brooklime) Prior.

PIMPERNEL, Water, Round-leaved *Samolus valerandi* (Brookweed) Curtis. Cf. Round Pimpernel.

PIMPERNEL, Wood 1. *Lysimachia nemora* (Yellow Pimpernel) Hutchinson. 2. *Lysimachia vulgaris* (Yellow Loosestrife) Le Strange.

PIMPERNEL, Yellow *Lysimachia nemora*.

PIMPERNEL ROSE *Rosa pimpinellifolia* (Burnet Rose) B & H.

PIMROSE *Primula vulgaris* (Primrose) common, including Dev. Choape; Wilts. Grigson.

PIN, Silver, Joan's *Papaver rhoeas* (Red Poppy) Coles. Joan's Silver Pin is a tawdry ornament worn ostentatiously by a sloven. The reference is to the flower being "fair without but foul within", the yellow juice being the "foul" element.

PIN-BURR *Galium aparine* (Goose-grass) Beds. Grigson.

PIN CHERRY *Prunus pennsylvanica*.

PIN-CLOVER; PIN GRASS *Erodium cicutarium* (Storksbill) USA Elmore. There are other 'needle' names for this plant, and Filaree, another name for it, comes from Spanish alfiler, pin. The references are from the sharpened end of the long seed vessels.

PIN OAK 1. *Quercus durandii*. 2. *Quercus palustris*.

PIN-POINTED CLOVER *Trifolium gracilentum.*

PINAFORE, Baby's; PINAFORE, Dolly's; PINAFORE, Pink; PINAFORE, Print *Geranium robertianum* (Herb Robert) Dev. (Baby's, Dolly's); Dor. (Pink, Print) all Macmillan.

PINCH, Devil's; PINCH, Virgin's *Polygonum persicaria* (Persicaria) Dor. Macmillan (Devil's); Berks. Grigson (Virgin's). Cf. also Pinchweed. The legend is that the Virgin Mary plucked a root, left her mark on the leaf, and threw it aside, saying, "This is useless", and it has been useless ever since. In fact, it is sometimes called Useless in Scotland. Presumably, the devil did likewise, and with malice aforethought.

PINCH-ME-TIGHT *Orchis mascula* (Early Purple Orchis) Som. Macmillan.

PINCHWEED *Polygonum persicaria* (Persicaria) Oxf, N'thum. Grigson. see above.

PINCUSHION 1. *Anthyllis vulneraria* (Kidney Vetch) Wilts. Dartnell & Goddard. 2. *Armeria maritima* (Thrift) Dev. Friend.1882; Som. Macmillan. 3. *Centaurea cyanus* (Cornflower) Suff. Grigson. 4. *Dipsacus fullonum* (Teasel) Som. Macmillan. 5. *Euonymus europaea* fruit (Spindle berries) Glos. J D. Robertson; War. Grigson. 6. *Corydalis lutea* (Yellow Corydalis) Dev, Suff. B & H. 7. *Knautia arvensis* (Field Scabious) Som. Macmillan; Wilts. Dartnell & Goddard; Ess. Gepp; Norf, Notts. B & H. 8. *Lotus corniculatus* (Bird's-foot Trefoil) Dor. Macmillan. 9. *Polygonum persicaria* (Persicaria) Som. Macmillan. 10. *Scabiosa atropurpurea* (Sweet Scabious) Dev. Friend.1882; Som. Macmillan; Bucks. B & H. 11. *Scabiosa columbaria* (Small Scabious) S Africa Watt. 12. *Succisa pratensis* (Devil's-bit) Som. Macmillan. 13. *Viburnum opulus* (Guelder Rose) Heref. B & H. All descriptive, and fairly obviously so.

PINCUSHION, Double *Anthyllis vulneraria* (Kidney Vetch) Wilts. Dartnell & Goddard.

PINCUSHION, Gentleman's *Knautia arvensis* (Field Scabious) Som. Macmillan.

PINCUSHION, Grandmother's *Knautia arvensis* (Field Scabious) Suss. Parish.

PINCUSHION, Lady's 1. *Anthyllis vulneraria* (Kidney Vetch) Som. Macmillan. 2. *Armeria maritima* (Thrift) Hants. Cope. 3. *Corydalis lutea* (Yellow Corydalis) Dev. Friend. 4. *Knautia arvensis* (Field Scabious) Som. Macmillan; Hants. Boase. 5. *Lotus corniculatus* (Bird's-foot Trefoil) Dor. Grigson. 6. *Pulmonaria officinalis* (Lungwort) Ches. Grigson; Yks. Addy. 7. *Scabiosa atropurpurea* (Sweet Scabious) Som. Macmillan. Suff. B & H.

PINCUSHION, New Zealand *Raoulia sp* Howes.

PINCUSHION, Nodding *Leucospermuum cordifolium.*

PINCUSHION, Old Woman's *Dactylorchis maculata* (Heath Spotted Orchid) Wilts. Dartnell & Goddard.

PINCUSHION, Our Lady's *Capsella bursa-pastoris* (Shepherd's Purse) Thorpe.

PINCUSHION, Queen's *Viburnum opulus* (Guelder Rose) N'thants. A E Baker.

PINCUSHION, Robin's 1. *Knautia arvensis* (Field Scabious) Dor. Macmillan. 2. *Rosa canina* the galls that grow on it Page.1978.

PINCUSHION PLANT *Cotula barbata.*

PINCUSHION SHRUB *Euonymus europaeus* (Spindle Tree) Bucks. Grigson.

PINCUSHION TREE 1. *Buddleia globosa* (Orange Ball Tree) Salisbury.1936. 2. *Viburnum opulus* (Guelder Rose) War, Oxf, Bucks. B & H.

PINDER *Arachis hypogaea* (Ground Nut) W Indies Howes.

PINE, Air *Aechmea fasciata* (Urn Plant) Oplt.

PINE, Aleppo *Pinus halepensis.*

PINE, Amber *Pinus succinifera.* This was one of the ancient sources of amber. The resin, over millions of years, loses its liquids and gases, and is left as a solid mixture of succinic acid and other resins and oils. This mixture is succinic amber.

PINE, Armand *Pinus armandii* (Chinese White Pine) RHS.
PINE, Arolla *Pinus cembra* (Swiss Stone Pine) Polunin.
PINE, Australian 1. *Araucaria heterophylla* (Norfolk Island Pine) USA. Cunningham. 2. *Casuarina stricta* (She Oak) A Huxley. Not wholly accurate for the *Araucaria*, for this is generalised in the South Pacific, as the more usual name, Norfolk Island Pine, shows.
PINE, Austrian *Pinus nigra*.
PINE, Balkan *Pinus heldreichii*.
PINE, Beach *Pinus contorta*.
PINE, Bhutan *Pinus griffithii*.
PINE, Big-cone *Pinus coulteri* (Coulter Pine) Mitchell. This has the biggest cones of all - they can be 35cm long, and 2kg in weight.
PINE, Bishop *Pinus muricata*. Coulter found this in 1832 at San Luis Obispo, hence the name 'Bishop'.
PINE, Black 1. *Callitris calcarata*. 2. *Pinus nigra* (Austrian Pine) Wit. 3. *Pinus serotina*. 4. *Podocarpus amara*. 5. *Podocarpus spicata* (Matai) New Zealand Everett.
PINE, Black, Japanese *Pinus thunbergii*.
PINE, Blue, Himalayan *Pinus wallichiana* (Bhutan Pine) Polunin.1976.
PINE, Bosnian *Pinus leucodermis*.
PINE, Bournemouth *Pinus pinaster* (Maritime Pine) Barber & Phillips. A Mediterranean pine, introduced here some time before 1596, and naturalized in the Bournemouth area.
PINE, Brazilian *Araucaria angustifolia*.
PINE, Bristlecone *Pinus aristata*.
PINE, British Columbian *Pseudotsuga menziesii* (Douglas Fir) N Davey.
PINE, Brown *Podocarpus alata*.
PINE, Buddhist *Podocarpus sp* Goold-Adams.
PINE, Bull 1. *Pinus ponderosa* (Yellow Pine) Wilkinson.1981. 2. *Pinus sabiniana* (Digger Pine) Watt.
PINE, Bunya-bunya *Araucaria bidwillii*. The Bunya-bunya fruit is its seed, about 2" long, and tasting like roasted chestnuts.
PINE, Calabrian *Pinus brutia*.

PINE, Canary *Pinus canariensis*.
PINE, Cedar *Pinus glabra* (Spruce Pine) Everett.
PINE, Celery; PINE, Celery-top *Phyllocladus rhomboidalis*, or *P trichomanoides* (Tanekaha) Duncalf.
PINE, Cembrian *Pinus cembra* (Swiss Stone Pine) Ablett.
PINE, Chilean *Araucaria araucana* (Monkey Puzzle Tree).
PINE, Chinese *Pinus tabuliformis*.
PINE, Chir, Himalayan *Pinus roxburghii*.
PINE, Chir, Indian *Pinus roxburghii* (Himalayan Chir Pine) Leathart.
PINE, Cluster *Pinus pinaster* (Maritime Pine) Barton & Castle.
PINE, Corsican *Pinus nigra var. maritima*. Not confined to Corsica, for it also occurs in southern Italy and in Sicily.
PINE, Coulter *Pinus coulteri*.
PINE, Cow's-tail *Cephalotaxus harringtonia var. drupacea* (Japanese Plum-yew) Barber & Phillips.
PINE, Cow's-tail, Chinese *Cephalotaxus fortunei* (Chinese Plum-yew) Leathart.
PINE, Cowrie *Agathis australis* (Kauri Pine) Willis.
PINE, Crimean *Pinus nigra var. caramanica*.
PINE, Cypress 1. *Callitris robusta*. 2. *Widdringtonia cupressoides*.
PINE, Cypress, Black; PINE, Cypress, Red *Callitris endlicheri* Hora (Red).
PINE, Cypress, Northern *Callitris intratropica*. The 'northern' of the name refers to northern Australia.
PINE, Cypress, Tasmanian *Callitris oblonga*.
PINE, Cypress, White 1. *Callitris columellaris* (Murray River Pine) Hora. 2. *Callitris glauca*.
PINE, Dalmatian *Pinus nigra var. dalmatica*.
PINE, Dammar *Agathis australis* (Kauri Pine) H & P.
PINE, Danzig *Pinus sylvestris* (Scots Pine) Hutchinson & Melville. A lot of pine timber has been imported from the Continent under this name, or Danzig Fir, Baltic Yellow Deal or Baltic Redwood.

PINE, David *Pinus armandii* (Chinese White Pine) RHS.
PINE, Dewy *Drosophyllum lusitanicum.*
PINE, Digger *Pinus sabiniana.* The 'Digger' of the name are the Digger Indians.
PINE, Durango *Pinus durangensis.* A Mexican species.
PINE, Fern *Podocarpus gracilior* (Podo) USA Cunningham.
PINE, Foxtail 1. *Pinus aristata* (Bristlecone Pine) T H Kearney. 2. *Pinus balfouriana.*
PINE, Frankincense *Pinus taeda* (Shortleaf Pine) Barton & Castle.
PINE, Gerard's *Pinus gerardiana.*
PINE, Gigantic *Pinus lambertiana* (Sugar Pine) Ablett. It is the largest of all the pines, anything up to 245 ft tall, and 18 ft in diameter.
PINE, Golden 1. *Grevillea robusta* (Silky Oak) Hora. 2. *Pseudolarix amabilis* (Chinese Gold Larch) Howes.
PINE, Grey 1. *Pinus banksiana.* 2. *Pinus sabiniana* (Digger Pine) Watt.
PINE, Greyleaf *Pinus sabiniana* (Digger Pine) Watt.
PINE, Ground *Ajuga chamaepitys* (Yellow Bugle) Turner.
PINE, Hazel *Liquidambar styraciflua* sapwood timber (Sweet Gum) Wilkinson.1981. The sapwood used to be sold in Europe under this name.
PINE, Heavy-wooded *Pinus ponderosa* (Yellow Pine) Ablett.
PINE, Hickory *Pinus aristata* (Bristlecone Pine) Everett.
PINE, Himalayan *Pinus wallichiana* (Bhutan Pine) Barber & Phillips.
PINE, Hoop *Araucaria cunninghamii.* 'Hoop Pine', because of the horizontal cracks in the encircling bands of the bark, so that it often sheds rings, or "hoops".
PINE, Huon *Dacrydium franklinii.* For it was first found on the Huon River in Tasmania.
PINE, Jack *Pinus banksiana* (Gray Pine) USA Zenkert.
PINE, Jeffrey *Pinus ponderosa var. jeffreyi.*

PINE, Jerusalem *Pinus halepensis* (Aleppo Pine) Everett.
PINE, Joint- 1. *Ephedra distachya.* 2. *Ephedra fragilis.* 3. *Ephedra trifurca* (Long-leaved Joint-Fir) USA Elmore.
PINE, Kauri *Agathis australis.* Cowrie is another spelling for Kauri.
PINE, Kauri, Queensland *Agathis robusta.*
PINE, Khasya *Pinus insularis.*
PINE, King William *Athrotaxus selaginoides.*
PINE, Knob-cone *Pinus tuberculata.* 'Knob-cone', for the tree's bark actually grows over the cones, which cannot be released until the tree has died, fallen and decayed.
PINE, Korean *Pinus koranensis.*
PINE, Lace-bark; PINE, Lace *Pinus bungeana.* The scales of the bark peel off to leave a variegated effect, white to greyish-purple.
PINE, Lambert's *Pinus lambertiana* (Sugar Pine) Ablett.
PINE, Limber *Pinus flexilis.*
PINE, Loblolly *Pinus taeda* (Shortleaf Pine) Fluckiger. A 'loblolly' is a local expression for a moist depression in the ground, and these pines like such situations.
PINE, Lodgepole *Pinus murrayana.* 'Lodgepole', because it was the favourite timber for tipi poles.
PINE, Lofty *Pinus wallichiana* (Bhutan Pine) Ablett.
PINE, Longleaf *Pinus palustris.*
PINE, Macedonian *Pinus peuce.*
PINE, Maritime *Pinus pinaster.*
PINE, Marsh *Pinus serotina* (Black Pine) USA Cunningham.
PINE, Mexican *Pinus patula* (Spreading-leaf Pine) Leathart.
PINE, Monterey *Pinus radiata.* Confined to the Monterey Peninsula, California.
PINE, Montezuma *Pinus montezumae.*
PINE, Moreton Bay *Araucaria cunninghamii* (Hoop Pine) J Smith. Moreton Bay, just north of Brisbane, for this is an Australian species.
PINE, Mountain 1. *Halocarpus bidwilli*

(Tarwood) H & P. 2. *Pinus mugo*. 3. *Pinus uncinata*.

PINE, Mountain, Dwarf *Pinus mugo* (Mountain Pine) Mitchell. Below the tree line, it has quite a normal appearance. But above the tree line, it becomes dwarf, spreading into mat-like growth.

PINE, Mountain, Swiss *Pinus mugo* (Mountain Pine) RHS.

PINE, Mugo *Pinus mugo* (Mountain Pine) Wit.

PINE, Murray (River) *Callitris columellaris*.

PINE, New Caledonian *Araucaria columnaris*.

PINE, Norfolk Island *Araucaria heterophylla*. Norfolk Island, in the South Pacific, where it was discovered by Captain Cook.

PINE, Norway 1. *Pinus resinosa* (Red Pine) USA Zenkert. 2. *Pinus sylvestris* (Scots Pine) Murdoch McNeill. In the case of *Pinus resinosa*, 'Norway' refers to the Maine village of that name.

PINE, Nut *Pinus sabiniana* (Digger Pine) Watt.

PINE, Nut, Arizona *Pinus edulis*.

PINE, Nut, Desert; PINE, Nut, Nevada *Pinus monophylla* (Piñon) Jaeger (Desert); Mason (Nevada).

PINE, Nut, Mexican *Pinus cembroides*.

PINE, Oneleaf *Pinus monophylla* (Piñon) Merrill.

PINE, Oregon *Pseudotsuga menziesii* (Douglas Fir) USA Elmore.

PINE, Oyster Bay *Callitris rhomboidea*.

PINE, Parana *Araucaria angustifolia* (Brazilian Pine) Wit. This is the name under which the timber is exported.

PINE, Parasol *Sciadopitys verticillata* (Umbrella Pine) Willis.

PINE, Parry *Pinus quadrifolia*.

PINE, Pitch 1. *Pinus palustris* (Longleaf Pine) Watt. 2. *Pinus ponderosa* (Yellow Pine) Willis. 3. *Pinus rigida*. These are all very resinous, and an old source of pitch.

PINE, Pitch, Northern *Pinus rigida* (Pitch Pine) Mitchell.

PINE, Pond *Pinus serotina* (Black Pine) USA Clemson.

PINE, Port Jackson *Callitris rhomboidea* (Oyster Bay Pine) Hora.

PINE, Prickle *Pinus pungens*.

PINE, Prince's *Chimaphila umbellata* (Umbellate Wintergreen) USA Corlett.

PINE, Pyrenean *Pinus nigra var. cedennensis*.

PINE, Red 1. *Dacrydium cupressinum* (Rimu) Everett. 2. *Pinus densiflora*. 3. *Pinus resinosa*. 4. *Pseudotsuga menziesii* (Douglas Fir) Wilkinson.1981.

PINE, Red, Japanese *Pinus densiflora* (Red Pine) Mitchell.

PINE, Red, New Zealand *Dacrydium cupressinum* (Rimu) Duncalf.

PINE, Richmond River *Araucaria cunninghamii* (Hoop Pine) Everett.

PINE, Riga *Pinus sylvestris* (Scots Pine) Murdoch McNeill. Cf. Danzig Pine, Baltic Yellow Deal, etc. - a lot of timber was imported from the Continent, and known by these names.

PINE, Rock *Pinus brachyptera* (Rocky Mountain Yellow Pine) USA Youngken.

PINE, Rottnest Island *Callitris robusta* (Cypress Pine) Hora.

PINE, Sabine's *Pinus sabiniana* (Digger Pine) Watt.

PINE, Scots *Pinus sylvestris*.

PINE, Screw 1. *Pandanus sp*. 2. *Pinus murrayana* (Beach Pine) Leathart. Pandanus gets the name because of the corkscrew-like trunks. 'Pine' because the leaves are like those of a pineapple. Beach Pine is so called because it gets very gnarled and "contorta".

PINE, Scrub *Pinus virginiana* (Virginia Pine) Everett.

PINE, She *Podocarpus elatus*.

PINE, Shore *Pinus contorta* (Beach Pine) Everett.

PINE, Short-leaf 1. *Pinus echinata*. 2. *Pinus taeda*.

PINE, Shortneedle *Pinus echinata* (Shortleaf Pine) USA Cunningham.

PINE, Siberian *Pinus cembra* (Swiss Stone

Pine) Ablett. The seeds, which are edible, are popular in Siberia, from where they are exported, too, via Archangel. They are known as Russian nuts in Norwegian ports.

PINE, Slash *Pinus elliottii*.

PINE, Spreading-leaf *Pinus patula*.

PINE, Spruce *Pinus glabra*.

PINE, Star *Pinus pinaster* (Maritime Pine) Notes and Queries. vol 4, 1851. 'Pinaster' was taken to mean 'star pine', and that name was actually used, but in error, for the suffix '-aster' means a poor imitation (poetaster, for example).

PINE, Stone *Pinus pinea*. 'Stone', because the seeds are enclosed in a very hard shell; in fact, they are usually referred to as nuts.

PINE, Stone, Swiss *Pinus cembra*.

PINE, Sugar *Pinus lambertiana*.

PINE, Swamp *Pinus palustris* (Longleaf Pine) Fluckiger.

PINE, Tenasserim *Pinus merkusii*. An Indonesian member of the genus.

PINE, Torrey *Pinus torreyana*.

PINE, Umbrella 1. *Pinus pinea* (Stone Pine) Watt. 2. *Sciadopitys verticillata*. 'Umbrella', because that is what they both look like. Cf. the French name for Pinus pinea - pin parasol.

PINE, Virginia *Pinus virginiana*.

PINE, Water *Stratiotes aloides* (Water Soldier) Ches. Holland.

PINE, Water, Chinese *Glyptostrobus lineatus* (Chinese Swamp Cypress) Hora.

PINE, Weymouth *Pinus strobus* (Eastern White Pine) Geldart. This species was introduced into England by Lord Weymouth, who planted them at Longleat in 1710 (though the Duchess of Beaufort had cultivated them at Badminton in 1705).

PINE, Whistling *Casuarina equisetifolia* (Beefwood) Kenya Dale & Greenway.

PINE, White 1. *Pinus flexilis* (Limber Pine) Elmore. 2. *Podocarpus dacrydioides* (Kahikatea) Leathart.

PINE, White, Chinese *Pinus armandii*.

PINE, White, Eastern *Pinus strobus*. The east of North America, not the Far East.

PINE, White, Japanese *Pinus parviflora*.

PINE, White, Mexican *Pinus ayacahuite*.

PINE, White, Mountain *Pinus monticola*.

PINE, White, Rocky Mountain *Pinus flexilis* (Limber Pine) Howes.

PINE, White, Western *Pinus monticola* (Mountain White Pine) USA Schery. The western side of North America.

PINE, Whitebark *Pinus albicaulis*.

PINE, Yellow *Pinus ponderosa*.

PINE, Yellow, Canadian *Pinus strobus* (Eastern White Pine) H Davey.

PINE, Yellow, Rocky Mountain *Pinus brachyptera*.

PINE, Yellow, Western 1. *Pinus pinea* (Stone Pine) Watt. 2. *Pinus ponderosa* (Yellow Pine) USA Kluckhohn.

PINE-APPLE 1. *Picea abies* (Spruce Fir) B & H. 2. *Pinus sylvestris* cones (Pine cones) N'thants. B & H.

PINE FLAX, Yellow *Linum neomexicanuum*.

PINE LILY *Lilium catesbaei* (Catesby's Lily) USA Woodcock.

PINE-SAP *Monotropa hypopitys* (Yellow Bird's Nest) Prior. This name, and also Firrape, were given because it was believed that it was parasitic on the roots of Scots Pine and beech. Actually, though, the plant has its own roots invested by a fungus.

PINE SPURGE *Euphorbia esula* (Hungarian Spurge) Parkinson.

PINE-STRAWBERRY 1. *Fragaria collina* (Hill Strawberry) Bianchini. 2. *Fragaria grandifolia*.

PINEAPPLE *Ananas comosus*. Because of the pine-cone like form of the fruit.

PINEAPPLE, Red; PINEAPPLE, Wild *Ananas bracteatus*.

PINEAPPLE FLOWER *Eucomis comosa*. Because it grows a tuft of narrow leaves on top of the flower spike, in much the same way as the tuft is produced on top of pineapple fruit.

PINEAPPLE GUAVA *Feijoa sellowiana*.

PINEAPPLE MINT *Mentha rotundifolia* 'variegata' G B Foster.

PINEAPPLE SAGE *Salvia rutilans*.

PINEAPPLE WEED *Matricaria matricarioides* (Rayless Weed) McClintock; USA Zenkert. Given because of the smell of the crushed leaves. Hence its value at one time as a strewing herb.

PINEASTER *Pinus pinaster* (Maritime Pine) Ablett.

PINEMAT *Arctostaphylos nevadensis*.

PINGWING *Hymenoxis richardsonii* (Pingui) USA. T H Kearney.

PINHA *Annona squamosa* (Sugar Apple) Schery.

PINHEADS 1. *Centranthus ruber* (Spur Valerian) Som. Macmillan. 2. *Matricaria recutita* (Scented Mayweed) Palaiseul.

PINJANE *Zostera marina* IOM Moore.1924. Pinjane (or binjean) is also applied in Anglo-Manx dialect to the food junket - hence the assumption that Eel Grass is edible.

PINK 1. *Agrostemma githago* (Corn Cockle) N Ire. Grigson. 2. *Armeria maritima* (Thrift) Dev. Friend.1882. 3. *Cardamine pratensis* (Lady's Smock) N'thum. Hardy. 4. *Dianthus plumarius*. 5. *Saxifraga x urbinum* (London Pride) Dev. Friend.1882. Pink (at least for *Dianthus*) is a puzzle. It may be the same word as the verb 'to pink', or pierce, which originally seems to have meant cutting cloth to show the colour underneath, as in the ornamental openings cut in Elizabethan dress, and only later came to mean giving the cloth a scalloped edge. The reference here is the indented petals that pinks undoubtedly have. Or pink may be from the Dutch Pinkster, neaning Whitsun-tide. Or yet again, it may be from Dutch pinken, to wink. Perhaps the word is a shortening of "pink-eyed", which means having narrow, or half-shut, eyes. The connection seems tenuous, but the "eye" imagery is retained in another name, Indian Eye, and all plants that have a centre in a different colour from the rest of the flower, as most wild pinks have, will have a whole variety of "eye" names.

PINK, Alpine *Dianthus alpinus*.

PINK, Bearded *Dianthus barbatus* (Sweet William) Macmillan. A translation of *Dianthus barbatus*.

PINK, Burst-belly *Dianthus plumarius* (Pink) Som. Macmillan. Cf. Burst-bellies, or Busters - all from observation of the flower buds getting larger and larger.

PINK, Carthusian *Dianthus carthusianorum*.

PINK, Cheddar *Dianthus gratianopolitanus*. Only to be found at Cheddar Gorge in this country (the nearest known occurrence to the British station is 600 miles away, in the Belgian Ardennes).

PINK, Cliff; PINK, Cleeve *Dianthus gratianopolitanus* (Cheddar Pink) both Som. Jennings. 'Cleeve' is a local word for 'cliff'.

PINK, Childing *Kohlrauschia prolifera*. 'Childing', because it throws out younger and smaller plants like a family of little children around it.

PINK, Clove *Dianthus caryophyllus* (Carnation) McClintock.

PINK, Corn 1. *Agrostemma githago* (Corn Cockle) N'thants. A E Baker. 2. *Legousia hybrida* (Venus's Looking-glass) Gerard.

PINK, Cottage *Dianthus caryophyllus* (Carnation) A W Smith.

PINK, Cushion 1. *Armeria maritima* (Thrift) Som. Grigson; Wilts. Dartnell & Goddard. 2. *Silene acaulis* (Moss Campion) Prior.

PINK, Deptford *Dianthus armeria*. Gerard claimed he knew it "...in the great field next to Detford". But the description he gave makes it fairly certain that he wasn't talking about this species at all.

PINK, Feathered *Dianthus plumarius* (Pink) Gerard.

PINK, Fire *Silene virginica*.

PINK, French 1. *Armeria maritima* (Thrift) Dev. Friend.1882. 2. *Dianthus chinensis* (Indian Pink) Dev. Friend.1882; Som. Elworthy. 'French' simply means foreign, without any implication of actual origin.

PINK, Fringed *Dianthus monspessulamus*.

PINK, Grass *Limodorum tuberosum*.

PINK, Ground *Phlox subulata* (Ground Phlox) USA House.

PINK, Gypsy *Dianthus caryophyllus*

(Carnation) Som. Macmillan. A name that was kept for the striped varieties.

PINK, Hedge *Saponaria officinalis* (Soapwort) Hants. Grigson.

PINK, Indian 1. *Dianthus chinensis*. 2. *Lychnis flos-cuculi* (Ragged Robin) Glos. Grigson. 3. *Melandrium rubrum* (Red Campion) Glos. Grigson. 4. *Silene californica* USA Schenk & Gifford. 5. *Spigelia marilandica* (Pinkroot) Willis.

PINK, Indy *Dianthus caryophyllus* (Carnation) Glos. J D Robertson.

PINK, Japanese *Dianthus chinensis heddewiggii*.

PINK, Jersey *Dianthus gallicus* (Western Pink) McClintock.1975. 'Western', referring to the Atlantic seaboard of Europe.

PINK, Little *Saxifraga x urbinum* (London Pride) Som. Macmillan.

PINK, London *Geranium robertianum* (Herb Robert) Glos. J D Robertson.

PINK, Maiden *Dianthus deltoides*. "The floures are of a blush colour, whereof it tooke its name" (Gerard). But it is more likely that Maiden in this case means "meadow", and is said to be a mistake for mead-pink.

PINK, May *Dianthus caryophyllus* (Carnation) Dev. Friend.1882.

PINK, Meadow 1. *Cardamine pratensis* (Lady's Smock) Dev. Macmillan. 2. *Dianthus deltoides* (Maiden Pink) B & H. 3. *Lychnis flos-cuculi* (Ragged Robin) Dev, Som. Macmillan.

PINK, Moss *Phlox subulata* (Ground Phlox) USA House.

PINK, Mountain 1. *Centaureum beyrichii* (Rock Centaury) USA Kingsbury. 2. *Centaureum calycosum* (Buckley Centaury) USA Kingsbury.

PINK, Mullein 1. *Agrostemma githago* (Corn Cockle) S Africa Watt. 2. *Lychnis coronaria* (Rose Campion) USA Zenkert.

PINK, Old Maid's *Agrostemma githago* (Corn Cockle) S Africa Watt.

PINK, Painted *Dianthus furcatus*.

PINK, Poverty *Prunella vulgaris* (Self-Heal) Berw. Denham/Hardy. Because it grows in poor soil. An Essex name, Pickpocket, carries the same message. So does Heart-o'-the-earth, recorded from East Anglia up into Scotland - the farmers believed that it ate away the substance of the soil.

PINK, Proliferous *Kohlrauschia prolifera* (Childing Pink) Clapham.

PINK, Scawfell *Armeria maritima* (Thrift) Cumb. Grigson.

PINK, Sea *Armeria maritima* (Thrift) Dev, Staffs, Border, Ireland (B & H) Ches, Yks, Cumb. Grigson.

PINK, Swamp *Arethusa bulbosa* (Dragon's Mouth) Emboden.1974.

PINK, Western *Dianthus gallicus*. Because it grows on the Atlantic seaboard of Europe.

PINK, Wild 1. *Dianthus crenatus* S Africa Watt. 2. *Dianthus deltoides* (Maiden Pink) Border B & H. 3. *Dianthus scaber* S Africa Watt. 4. *Geranium robertianum* (Herb Robert) Glos. J D Robertson. 5. *Stellaria holostea* (Greater Stitchwort) Bucks. Grigson.

PINK, Wood *Dianthus sylvestris*.

PINK-EYED JOHN; PINK-O'-MY-JOHN; PINKENEY JOHN *Viola tricolor* (Pansy) Midl. Wright (Pink-eyed John); Beds. Grigson; Leics. A B Evans (Pink-o'-my-John); Northants. A E Baker (Pinkeney John). 'Pink-eyed' means small-eyed, according to Halliwell, or possibly with half-shut eyes. But the name is very variable.

PINK-NEEDLE 1. *Erodium cicutarium* (Storksbill) B & H. 2. *Erodium moschatum* (Musk Storksbill) B & H. 3. *Scandix pecten-veneris* (Shepherd's Needle) B & H. Probably Pook-needle, or Puck-needle, in other words fairy's needle. The 'needle' is the long, pointed seed vessel.

PINK PLANT, Dutch *Reseda luteola* (Dyer's Rocket) W Miller.

PINKIES *Trifolium pratense* (Red Clover) Dor. Macmillan.

PINKROOT *Spigelia marilandica*.

PINKSTER-FLOWER; PINXTER-FLOWER *Azalea nudiflora* (Pink Azalea)

USA House (Pinkster); USA Zenkert (Pinxter).

PINKWEED *Polygonum aviculare* (Knotgrass) Dor. Macmillan Yks. Coles.

PINKWOOD 1. *Eucryphia billardieri* (Leatherwood) Everett. 2. *Eucryphia moorei* Everett.

PINNER, Pug-in-a- *Primula vulgaris var. elatior* (Polyanthus) Coats.

PIÑON *Pinus monophylla*.

PIÑON, Two-leaf *Pinus edulis* (Arizona Nut Pine) USA Elmore.

PIÑON STRANGLEROOT *Orobanche fasciculata*.

PINPATCH *Vinca minor* (Lesser Periwinkle) Suss. Grigson.

PINS, Needles-and- 1. *Ulex europaeus* (Furze) Som. Macmillan. 2. *Viola tricolor* (Pansy) Dor. Macmillan. Furze is an obvious candidate for a name like this, but pansy is not clear at all.

PINS-AND-NEEDLES 1. *Knautia arvensis* (Field Scabious) Som. Macmillan. 2. *Saxifraga x urbinum* (London Pride) Som. Macmillan. 3. *Scandix pecten-veneris* (Shepherd's needle) Som. Macmillan. 4. *Stellaria holostea* (Greater Stitchwort) Som. Macmillan. 5. *Ulex europaeus* (Furze) Som. Macmillan. Shepherd's Needle and Furze carry the name with obvious descriptive reason. Scabious may have been given it by inference, for this is a "pincushion" plant (Cf. Pincushion, Lady's Pincushion, etc). Stitchwort has other 'needle' names, like Mother Shimble's Snick Needles, and London Pride may have been given the name simply from cottage garden affection; many of these double names were so given.

PINT, Cuckoo 1. *Arum maculatum*. 2. *Cardamine pratensis* (Lady's Smock) Wilts, Suss, Leics. Grigson. 3. *Orchis mascula* (Early Purple Orchis) Bucks. B & H. 'Pint' is the OE pintel, penis, and the name is one of a number of highly descriptive and bawdy epithets for the Arum. It has to be assumed that Lady's Smock got the name accidentally, and because it is often known as Cuckooflower.

PINT, Cuckow *Arum maculatum* (Cuckoopint) USA Weiner.

PINTEL, Coke *Arum maculatum* (Cuckoopint) Cockayne.

PINTEL, Cuckoo 1. *Arum maculatum* (Cuckoo-pint) Turner. 2. *Cardamine pratensis* (Lady's Smock) Wilts, Suss, Leics. Grigson.

PINTEL, Priest's 1. *Arum maculatum* (Cuckoo-pint) Gerard. 2. *Orchis mascula* (Early Purple Orchis) B & H. 3. *Sedum roseum* (Rose-root) Banff. Grigson.

PINTEL, Wake *Arum maculatum* (Cuckoopint) Prior.

PINTELWORT 1. *Arum maculatum* (Cuckoo-pint) Cockayne. 2. *Euphorbia lathyris* (Caper Spurge) R T Gunther. The name (for Caper Spurge) appears as a gloss on a 12th century copy of Apuleius. It sounds unlikely, but the illustration is certainly not of *Arum maculatum*.

PINWEED *Helianthemum villosum* USA Tolstead.

PINY; PIONY *Paeonia mascula* (Paeony) Corn. Hawke; Dev. Friend.1882; Glos. B & H (Piny); Ches. B & H (Piony).

PIPAL *Ficus religiosa*. Pippala originally, probably the berry of the tree. Also rendered as Peapul, this is the Hindi word pipul.

PIPE *Quercus robur* acorn (Acorn) Som. Macmillan; Yks. B & H. Descriptive, and obviously so.

PIPE, Blow-,Tree; PIPE, Blue *Syringa vulgaris* (Lilac) Skinner (Blow); Gerard (Blue). In view of the other name, presumably 'blow' is 'blue' in this case.

PIPE, Bum *Taraxacum officinale* (Dandelion) Banff, Lanark Grigson.

PIPE, Dutchman's 1. *Aristolochia californica* USA Barrett & Gifford. 2. *Aristolochia sipho*. From the shape of the tube of the flower, like the stem of a meerschaum pipe.

PIPE, Geanucanach's *Quercus robur* acorn (Acorn) Ire. O Suilleabhain. A fairy, presumably, but which one?

PIPE, Indian's *Monotropa uniflora* USA Tolstead. It is said that the single bell-shaped flower dropping from a practically leafless stem resembles a white miniature Indian pipe of peace.

PIPE, Paddie *Equisetum palustre* (Marsh Horsetail) Scot. Gibbings. 'Paddie' - see Paddock.

PIPE, Paddock 1. *Equisetum arvense* (Common Horsetail) H C Long.1910. 2. *Equisetum palustre* (Marsh Horsetail) Scot. Jamieson. Paddock, or puddock, is a toad or frog, OE pade, padde, and O Norse padda, archaic in England but still in use in Scotland, where they say that the croaking of frogs is the playing (by the frogs) of these hollow pipes, like musical instruments.

PIPE, Paddock's *Hippuris vulgaris* (Mare's Tail) Cumb. Grigson.

PIPE, Paddy's 1. *Equisetum arvense* (Common Horsetail) Allan. 2. *Hippuris vulgaris* (Mare's Tail) Cumb. Grigson.

PIPE, Pudding-, Tree *Cassia fistula* (Indian Laburnum) Barker. "Because the cod is like a Pudding", according to Coles's explanation.

PIPE, Snake *Equisetum arvense* (Common Horsetail) Som. Macmillan.

PIPE, Toad 1. *Equisetum arvense* (Common Horsetail) H C Long.1910. 2. *Equisetum limosum* (Water Horsetail) Lincs. Peacock; Yks. Carr.

PIPE-STEM *Clematis lasiantha* USA Schenk & Gifford.

PIPE-PRIVIT *Syringa vulgaris* (Lilac) Tynan & Maitland.

PIPE-TREE 1. *Cassia fistula* (Indian Laburnum) Watt. 2. *Syringa vulgaris* (Lilac) Parkinson.

PIPE-TREE, Pudding *Cassia fistula* (Indian Laburnum) Barker.

PIPE-VINE *Aristolochia sipho* (Dutchman's Pipe) W Miller. From the shape of the tube of the flower.

PIPER *Achillea millefolium* (Yarrow) Canada Allan. Presumably this started as 'pepper', so it would come into the same category as various English names like Wild Pepper, or Old Man's Mustard.

PIPEWEED, Indian *Eriogonum inflatum* (Desert Trumpet) USA T H Kearney.

PIPEWORT *Eriocaulon septangulare*.

PIPLIN *Populus alba* (White Poplar) Som. Halliwell.

PIPPALA TREE *Ficus religiosa* (Peapul) Philpot. Pippala is probably the berry of the tree.

PIPPERIDGE BUSH *Berberis vulgaris* (Barberry) Ess. Grose. Cf. Piprige, etc. The word is from French pepin, pip, and rouge, red - an allusion to the berries.

PIPPLE 1. *Populus alba* (White Poplar) Som. Halliwell. 2. *Populus tremula* (Aspen) Dev, Som, IOW Grigson. A 'poplar' name.

PIPRAGE; PIPRIGE *Berberis vulgaris* (Barberry) Prior (Piprage); Turner (Piprige). See Pipperidge.

PIPS *Primula veris* flowers (Cowslip) Heref. Havergal. Cf. Peeps, both usually reserved for the dried flowers from which cowslip wine ought to be made.

PIPSISSEWA *Chimaphila umbellata* (Umbellate Wintergreen) Le Strange. This is the Algonquin name for this plant, which grows both in North America and in Europe.

PIPSISSEWA, Striped *Chimaphila maculata* (Spotted Wintergreen) Le Strange.

PIRRI-PIRRI BUR *Acaena anserinifolia*. Piripiri is the Maori word for this New Zealand plant.

PISGIE-FLOWER *Stellaria holostea* (Greater Stitchwort) Corn. Grigson. Fairies protect the stitchwort, and it must not be gathered, or the offender will be "fairy-led" into swamps and thickets at night. The Cornish 'pisgie' becomes 'pixy' in Devon, where Pixies is recorded as a name for this flower.

PISHAMOOLAG *Taraxacum officinale* (Dandelion) Donegal Mabey.

PISKIES 1. *Heracleum sphondyllium* - dried stems (Hogweed) Corn. B & H. 2. *Stellaria holostea* (Greater Stitchwort) Dev. Grigson. Piskies for hogweed stems looks like a mishearing of one of the variants of Kex, most likely Kecksies, or nearer still, Kiskies. But the fairy name seems genuine for stitchwort - see Pisgie-flower.

PISMIRE; PISSIMIRE *Taraxacum officinale* (Dandelion) both Yks. B & H (Pismire); Nicholson (Pissimire). Examples of the Pissabed series.

PISS, Hound's *Cynoglossum officinale* (Hound's Tongue) Gerard. Hound's Mie in Cockayne. To quote Gerard, "for in the world there is not any thing that smelleth so like Dogs pisse as the leaves of this plant doe".

PISSABED 1. *Ranunculus calandrinoides* B & H. 2. *Taraxacum officinale* (Dandelion) common. Dandelion, of course, is a known diuretic. "Children that eat it in the evening experience its diuretic effects, which is the reason that other European nations, as well as ourselves, vulgarly call it Pissabed". But it is said that if children even gather the flowers, they will experience the symptoms. There are many variants of the name. The name does not seem very appropriate for the buttercup, though.

PISSIBED *Taraxacum officinale* (Dandelion) Yks. PLNN 25.

PISSPOT *Calystegia sepium* (Great Bindweed) Suss. Grigson. Presumably descriptive of the shape, but far-fetched.

PISTACHIO *Pistacia vera*.

PISTACIA, Wild *Staphylea pinnata* (Bladder Nut) Gerard.

PISTEL *Dipsacus fullonum* (Teasel) R Gunther. Possibly a thistle name. It appears as a gloss on a 12th century manuscript of Apuleius.

PISTOL PLANT *Pilea microphylla* Rochford. Cf. Artillery Plant, Gunpowder Plant. All because when the plant is shaken when in flower, it will produce clouds of pollen.

PIT-PIT *Saccharum edule*.

PITANGA; PITANJA CHERRY *Eugenia uniflora* (Surinam Cherry) Schery; Dalziel (Pitanja Cherry).

PITCH APPLE *Clusia rosea* RHS.

PITCH PINE 1. *Pinus palustris* (Longleaf Pine) Watt. 2. *Pinus ponderosa* (Yellow Pine) Willis. 3. *Pinus rigida*.

PITCH PINE, Northern *Pinus rigida* (Pitch Pine) Mitchell.

PITCHER, Lent *Narcissus pseudo-narcissus* (Daffodil) Dev, Som. Elworthy. 'Lent', from the time of flowering - Cf. Lent Lily, Lent Rose, etc., and then the series of 'Easter' names.

PITCHER, Marsh *Heliamphora nutans*.

PITCHER PLANT *Sarracenia purpurea*. Named from the extraordinary shape of the flowers, built to attract insects, which are then caught by the secretions that gloss the inside of the cup, or pitcher.

PITCHER PLANT, Australian *Cephalotus follicularis*.

PITCHER PLANT, California *Darlingtonia californica* (Cobra Plant) USA Cunningham.

PITCHER PLANT, Hooded *Sarracenia minor*.

PITCHER PLANT, Northern *Sarracenia purpurea* (Pitcher Plant) Youngken.

PITCHER PLANT, Parrot *Sarracenia psittacina*.

PITCHERY; PETGERY; BEDGERY *Duboisia hopwoodii* (Pituri) Lewin. All anglicizations of the native name, Pituri.

PITCHFORK 1. *Bidens bipinnata* (Spanish Needles) Canada Watt. 2. *Bidens pilosa* Canada Watt.

PITHYWIND *Clematis vitalba* (Old Man's Beard) Suss. Cooper. An aberrant form, from Bithywind. But there are many variations on this, and it must have started from Withywind, or something like it.

PITTED ONION *Allium lacunosum*.

PITTLE-BED *Taraxacum oficinale* (Dandelion) Suff. Grigson. See Pissabed.

PITURI *Duboisia hopwoodii*. The Australian aboriginal name for the plant, englished into Pitchery, Petgery or Bedgery.

PIVOT; PEVIT *Ligustrum vulgare* (Privet) Corn. Grigson. Both local variations of Privet.

PIXIES *Stellaria holostea* (Greater Stitchwort) Dev. Folkard; Som. Tongue. Cf. Piskies or Pisgie-flower from the West country, too.

PIXY-GRASS *Eriophorum angustifolium* (Cotton-grass) Dev. St Leger-Gordon.

PIXY LILY *Stellaria holostea* (Greater Stitchwort) Tynan & Maitland.

PIXY-PEAR 1. *Crataegus monogyna* fruit (Haws) Dor, Som. Grigson. 2. *Rosa canina* fruit (Hips) W Eng. Friend.

PLACKETT *Avena sativa* (Oats) IOM Gill.1963.

PLAGGIS *Primula veris* (Cowslip) Halliwell.

PLAGUE, Devil's *Daucus carota* (Wild Carrot) USA Howes.

PLAGUE, Farmer's; PLAGUE, Garden's *Aegopodium podagraria* (Goutweed) both N Ire. Grigson. Any gardener would agree. Under the name Ground Elder, this is an unadulterated pest once it gets in a garden.

PLAGUE FLOWER; PLAGUE WORT *Petasites hybridus* (Butterbur) Fernie (Plague Flower); Corn. Deane & Shaw (Plague Wort). Presumably because of its use as a plague remedy. But they say in Cornwall that Plague Wort was given because the plant sprang from the graves of plague victims in Veryan churchyard.

PLAISTER CLOVER *Melilotus officinalis* (Melilot) Gerard. It has been used since Galen's time in poultices for dispersing tumours - and more down to earth, for boils and swellings.

PLANE 1. *Acer pseudo-platanus* (Sycamore) Yks, Cumb. Grigson; Scot. Jamieson. 2. *Platanus sp.* It is from the French plane, itself from Latin platanus.

PLANE, American *Platanus occidentalis* (Western Plane) Everett.

PLANE, Californian *Platanus racemosa*.

PLANE, Dwarf *Viburnum opulus* (Guelder Rose) Gerard. There is a certain resemblance in the leaf shape.

PLANE, London *Platanus x acerifolia*. It used to be said that it thrived in London because it shed its bark, the argument being that in doing so it exposed a fresh surface unclogged by soot, so the tree could "breathe" - a fallacy, of course, trees do not "breathe" through their bark. Nevertheless, the tree is good-tempered enough to thrive in cities.

PLANE, Mock *Acer pseudo-platanus* (Sycamore) Prior.

PLANE, Oriental *Platanus orientalis*.

PLANE, Scots *Acer pseudo-platanus* (Sycamore) Murdoch McNeill.

PLANE, Western *Platanus occidentalis*.

PLANT *Plantago major* (Great Plantain) Dev. B & H.

PLANTAIN *Musa x paradisiaca*. Of doubtful origin, according to the dictionary, but certainly not the same as plantain for members of the genus Plantago. That comes from Latin planta, the sole of the foot, a comment on the shape of the leaves.

PLANTAIN, Alpine *Plantago alpina*.

PLANTAIN, Black *Plantago lanceolata* (Ribwort Plantain) Salisbury.

PLANTAIN, Black-seed *Plantago rugelii*.

PLANTAIN, Branched *Plantago indica*.

PLANTAIN, Broad-leaved *Plantago major* (Great Plantain) Hill.

PLANTAIN, Buck's-horn 1. *Plantago coronopus*. 2. *Plantago maritima* (Sea Plantain) B & H. The name is given from the deeply-cut leaves.

PLANTAIN, Clammy *Plantago psyllium* (Fleawort) Fernie.

PLANTAIN, Common *Plantago major* (Great Plantain) Curtis.

PLANTAIN, Crow's-foot; PLANTAIN, Crowfoot *Plantago coronopus*. (Bucks'-horn Plantain) Flower.1859 (Crow's-foot); USA Bianchini (Crowfoot). Presumably the finely cut leaves that gave Buck's-horn Plantain are also responsible for these.

PLANTAIN, Dooryard *Plantago major* (Great Plantain) USA Gates.

PLANTAIN, English 1. *Plantago lanceolata* (Ribwort Plantain) USA Upham. 2. *Plantago major* (Great Plantain) USA

Gates; Barbados. Gooding. Great Plantain, in particular, bears the well-known legend of its persistently following the tracks of man. More specifically, one superstition says that it follows Englishmen, and springs up in whatever part of the world he makes his home. Hence this name, and also Englishman's Foot, or White Man's Footprint (plantain is from planta, the sole of the foot).

PLANTAIN, French *Musa x paradisiaca* (Banana) H & P.

PLANTAIN, Great *Plantago major*.

PLANTAIN, Hare's-foot *Plantago lagopus*.

PLANTAIN, Hartshorn *Plantago coronopus* (Buck's-horn Plantain) B & H.

PLANTAIN, Hoary *Plantago media* (Lamb's-tongue Plantain) McClintock.

PLANTAIN, Honey *Plantago media* (Lamb's-tongue Plantain) Som. Macmillan. Presumably a misreading of 'hoary'.

PLANTAIN, Indian 1. *Cacalia atriplicifolia*. 2. *Cacalia suaveolens* USA Zenkert.

PLANTAIN, Ispaghul *Plantago ovata*.

PLANTAIN, Lamb's-tongue *Plantago media*. The Greek name for plantain was arnoglossos, hence lamb's-tongue.

PLANTAIN, Long; PLANTAIN, Narrow-leaved *Plantago lanceolata* (Ribwort Plantain) Le Strange (Long); Curtis (Narrow-leaved). 'Long' in the sense that this has lanceolate leaves, unlike Great Plantain, whose leaves are ovate.

PLANTAIN, Parrot's *Heliconia psittacorum* (Parrot's Flower) RHS.

PLANTAIN, Pond; PLANTAIN, Swimming *Potamogeton natans* (Broad-leaved Pondweed) both Turner.

PLANTAIN, Poor Robin's *Hieraceum venosum* (Rattlesnake Weed) USA House.

PLANTAIN, Pursh *Plantago purshii*.

PLANTAIN, Rat-tail *Plantago major* (Great Plantain) Clapham. Cf. Rat's Tails, from Norfolk, Cumbria, Isle of Man and Scotland, and Ratten Tails from Northumberland.

PLANTAIN, Rattlesnake 1. *Epipactis sp*. 2. *Goodyera repens* (Creeping Lady's Tresses) USA Weiner. Goodyera repens has white-veined leaves, which looked like snake skin to the early herbalists, who took them as the sign for virtues against snake bite.

PLANTAIN, Rib; PLANTAIN, Ribwort *Plantago lanceolata*. Ribwort Plantain is the common name; Gerard used Rib Plantain. Cf. the very widespread Ribgrass.

PLANTAIN, Robin's *Erigeron pulchellus* USA Tolstead.

PLANTAIN, Rugel's *Plantago rugelii* (Black-seed Plantain) Canada Allan.

PLANTAIN, Sea 1. *Plantago maritima*. 2. *Spergularia salina* (Sea Spurrey) Darling & Boyd.

PLANTAIN, Seaside *Plantago maritima* (Sea Plantain) Murdoch McNeill.

PLANTAIN, Shrubby *Plantago sempervirens*.

PLANTAIN, Snake *Plantago lanceolata* (Ribwort Plantain) Flück.

PLANTAIN, Stag's-horn *Plantago coronopus* (Buck's-horn Plantain) Hepburn.

PLANTAIN, Water *Alisma plantago-aquatica*.

PLANTAIN, Water, Lesser *Baldella ranunculoides*.

PLANTAIN, White *Antennaria plantaginifolia* (Plantain-leaved Everlasting) USA Grieve.1931.

PLANTAIN-LEAVED EVERLASTING *Antennaria plantaginifolia*.

PLANTAIN-LEAVED PONDWEED *Potamogeton coloratus*.

PLANTAIN-LEAVED PUSSY TOES *Antennaria plantaginifolia* (Plantain-leaved Everlasting) USA Youngken.

PLANTAIN-LEAVED THRIFT *Armeria arenaria* (Broad-leaved Thrift) Polunin.

PLANTAIN LILY *Hosta sp*. 'Plantain', from the leaf shape.

PLANTAIN SHOREWEED *Littorella uniflora* (Shoreweed) B & H.

PLANTINE *Plantago major* (Great Plantain) Cambs. Porter.

PLASTER, Adam's *Polygonum persicaria* (Persicaria) Newfoundland Grigson. Presumably because they used it for a poultice there. In Gerard's time, it was certainly used for "inflammations and hot swellings".

PLATES, Penny *Umbilicus rupestris* (Wall Pennywort) Dev. Macmillan.

PLAYTHING, Bad Man's *Achillea millefolium* (Yarrow) Fernie.

PLAYTHING, Devil's 1. *Achillea millefolium* (Yarrow) Fernie. 2. *Stachys officinalis* (Betony) Shrop. Grigson. 3. *Urtica dioica* (Nettle) Som. Macmillan. An understandable name for nettle, but not at all obvious for betony, which was always a most desired herb for all medicinal purposes. The leaf shape probably led to confusion. There is a note of reproval in the adoption of the name for yarrow, probably because of the divinations practiced with it.

PLAYTHING, Naughty Man's 1. *Capsella bursa-pastoris* (Shepherd's Purse) War. Palmer. 2. *Urtica dioica* (Nettle) Som. Grigson; Suss. Parish. 'Naughty Man' (and 'Bad Man') are euphemisms for the devil. It may have been thought unlucky to say the real name.

PLAYTHING, Old Man's *Pimpinella saxifraga* (Burnet Saxifrage) Shrop. Grigson.

PLEASANT-IN-SIGHT 1. *Cardamine pratensis* (Lady's Smock) Tynan & Maitland. 2. *Lychnis flos-cuculi* the double-flowered form (Ragged Robin) B & H. Given without comment in each of the authorities.

PLEATED GENTIAN *Gentiana affinis* (Closed Gentian) T H Kearney.

PLENTY *Sedum acre* (Biting Stonecrop) Som. Macmillan.

PLENTY, Peace-and- *Saxifraga x urbinum* (London Pride) Wilts. Grigson.

PLEURISY-ROOT *Asclepias tuberosa* (Orange Milkweed) USA House. It used to be boiled up and used as a poultice for pleurisy. The Indians made a root tea, or just ate the root, raw or dried, for a pneumonia remedy.

PLOT'S ELM *Ulmus plotii*.

PLOUGHMAN'S MIGNONETTE *Euphorbia cyparissias* (Cypress Spurge) Coats. It is difficult to see why anyone, let alone a ploughman, should regard this plant as a little darling, for that is what mignonette means.

PLOUGHMAN'S SPIKENARD 1. *Inula conyza*. 2. *Petasites fragrans* (Winter Heliotrope) Dor. Dacombe. The root of *Inula conyza* has a spicy scent, and used to be hung up in cottages to scent the musty air. Sometimes, too, the roots were burnt to scent the room. Hence, of course, the name spikenard, which is Latin spica nardi,the true owner of which is the plant *Nardostachys jatamansi*. The name really means 'poor man's spikenard. Winter Heliotrope gets the name by association, for this is 'fragrans', also.

PLOUGHMAN'S WEATHERGLASS *Anagallis arvensis* (Scarlet Pimpernel) Wilts. Dartnell & Goddard. One of a very long series of names, all relating to the plant's habit of closing its flowers when wet weather is likely.

PLOUGHSHARE ORCHID *Serapias vomeracea*. The flowers are wedge-shaped, likened in the name to a ploughshare, an image that is echoed in the specific name, for Latin vomer means ploughshare.

PLUM *Prunus domestica*. OE plume, from Latin prunum.

PLUM, American *Prunus americana* (Canada Plum) Elmore.

PLUM, Annam *Flacourtia sepiaria*.

PLUM, Assyrian *Cordia myxa* (Sapistan) Dalziel.

PLUM, August *Prunus americana* (Canada Plum) USA Elmore.

PLUM, Balkan *Prunus cerasifera* (Cherry-plum) Anderson. This is the fruit for slibowitz, the plum brandy made by Balkan peasants.

PLUM, Batoka *Flacourtia indica* (Kaffir Plum) Palgrave.

PLUM, Beach *Prunus maritima*.

PLUM, Black *Vitex doniana* E Africa Dale & Greenway.

PLUM, Californian *Prunus salicina* (Japanese Plum) Howes.
PLUM, Canada *Prunus americana*.
PLUM, Cherry *Prunus cerasifera*.
PLUM, Chickasaw *Prunus angustifolia*.
PLUM, Chinese 1. *Prunus salicina* (Japanese Plum) Howes. 2. *Prunus triloba*.
PLUM, Date- *Diospyros Lotus*.
PLUM, Date-, American *Diospyros virginiana* (Persimmon) J Smith.
PLUM, Date-, Chinese *Diospyros kaki*.
PLUM, Date-, European *Diospyros lotus* (Date-plum) J Smith.
PLUM, Ginger-bread *Parinari macrophylla*.
PLUM, Goose *Prunus americana* (Canada Plum) USA Elmore.
PLUM, Governor *Flacourtia indica* (Kaffir Plum) Howes.
PLUM, Grey *Parinari excelsa* (Rough-skinned Plum) Dalziel.
PLUM, Ground 1. *Astragalus caryocarpus* (Buffalo Bean) USA Tolstead. 2. *Astragalus crassicarpus*.
PLUM, Guinea *Parinari excelsa* (Rough-skinned Plum) Dalziel.
PLUM, Hog 1. *Prunus americana* (Canada Plum) USA Elmore. 2. *Prunus umbellata*. 3. *Spondias monbin* (Monbin) Dalziel. 4. *Spondias purpurea*. Pigs like the fruit of Spondias purpurea, and fatten upon them.
PLUM, Jamaica *Spondias mombin* (Mombin) Howes.
PLUM, Japanese 1. *Eriobotrya japonica* (Loquat) Perry. 2. *Prunus salicina*. Japan is the largest producer of loquats, and the fruit is very popular there.
PLUM, Java *Eugenia jambolana* (Jambul) Grieve.1931.
PLUM, Jew's *Spondias dulcis* (Ambarella) Howes.
PLUM, Kaffir *Flacourtia indica*. An African tree, in spite of the specific name.
PLUM, Madagascar *Flacourtia indica* (Kaffir Plum) Howes.
PLUM, Mealy *Arctostaphylos uva-ursi* (Bearberry) New Engl. Sanford.
PLUM, Moxie *Gaultheria hispidula*.

PLUM, Natal *Carissa grandiflora*.
PLUM, Niger *Flacourtia flavescens*.
PLUM, Olive *Elaeodendron glaucum* (Ceylon Tea Tree) Chopra. An oily fruit, much like an olive in shape, a fact noted in the generic name, which is from two Gk words meaning olive tree.
PLUM, Prairie *Prunus americana* (Canada Plum) USA Harper.
PLUM, Purple-leaved *Prunus pissardii*.
PLUM, Red *Prunus americana* (Canada Plum) USA Elmore.
PLUM, River *Prunus americana* (Canada Plum) USA Elmore.
PLUM, Rough-skinned *Parinari excelsa*.
PLUM, Seaside *Ximenia americana* (Wild Oliver) Soforowa.
PLUM, Sebestan *Cordia myxa* (Sapistan) Dalziel.
PLUM, Sierra *Prunus subcordata* USA Schenk & Gifford.
PLUM, Sour *Ximenia caffra*.
PLUM, Spanish *Spondias purpurea* (Hog Plum) Douglas.
PLUM, Spanish, Yellow *Spondias monbin* (Monbin) Jamaica Dalziel.
PLUM, Spiny *Prunus spinosa* (Blackthorn) Flower.1859.
PLUM, Star *Chrysophyllum cainito* (Star-apple) Howes.
PLUM, Sugar *Uapaca guineensis* fruit Liberia Dalziel.
PLUM, Thorn *Prunus americana* (Canada Plum) USA Elmore.
PLUM, Wild *Prunus domestica var. institia* (Bullace) Yks. Grigson.
PLUM, Wild, American *Prunus americana* (Canada Plum) Usher.
PLUM, Yellow *Prunus americana* (Canada Plum) USA Elmore.
PLUM, Yucatan *Spondias lutea*.
PLUM-BUSH *Canthium latifolium*.
PLUM-FRUITED YEW *Podocarpus andinus*. The fruits look like damsons.
PLUM-LEAF CRAB *Malus prunifolia* (Chinese Crabapple) Hora.

PLUM-LEAVED WILLOW *Salix arbuscula* (Small Tree-Willow) McClintock.
PLUM PUDDING 1. *Epilobium hirsutum* (Great Willowherb) Ches. Holland. 2. *Melandrium album* (White Campion) Suff. Grigson. 3. *Melandrium rubrum* (Red Campion) Som. Macmillan; Ess. Gepp. Willowherb in particular has a number of similar names, like Apple-pie, Cherry-pie, etc. It is said that the plant smells like apples when it is crushed. Probably, though, Gerard's Codded Willowherb, and the similarity of the word to Codlins, is the origin of the belief.
PLUM-YEW, Chinese *Cephalotaxus fortunei*.
PLUM-YEW, Japanese *Cephalotaxus drupacea*.
PLUMBAGO 1. *Plumbago sp.* 2. *Polygonum persicaria* (Persicaria) Culpepper. Plumbago is Pliny's translation of the Greek name molybdaina, lead. The reference is to the blue flowers. It is difficult to tell what Culpepper had in mind when he gave the name to Persicaria.
PLUMBOY *Rubus pubescens* (Dwarf Raspberry) USA Yarnell.
PLUME, Apache *Fallugia paradoxa*. Because of the fancied resemblance of the feathery seed clusters to the plumed war bonnets of the Apache Indians.
PLUME, Desert; PLUME, Prince's *Stanleya pinnata*. Prince's Plume is the common name; Desert Plume is in Jaeger.
PLUME, Feathery *Gynerium argenteum* (Pampas Grass) Som. Macmillan.
PLUME, Queen Anne's *Gynerium argenteum* (Pampas Grass) Som. Macmillan.
PLUME, Scarlet *Euphorbia pulcherrima* (Poinsettia).
PLUME ALBIZIA *Albizia lophantha* (Brush Wattle) H & P.
PLUME FEATHERS *Gynerium argenteum* (Pampas Grass) Som. Macmillan.
PLUME POPPY *Bocconia cordata*.
PLUME THISTLE, Marsh *Cirsium dissectum* (Meadow Thistle) Clapham.

PLUME THISTLE, Spear *Cirsium vulgare*.
PLUME THISTLE, Tuberous *Cirsium tuberosum* (Tuberous Thistle) Flower.1857.
PLUMROCKS *Primula vulgaris* (Primrose) Scot. B & H. Presumably a corruption of primrose.
PLUMWOOD 1. *Eucryphia moorei*. 2. *Santalum lanceolatum* (Australian Sandalwood) Howes.
PLUSH ANEMONE *Anemone coronaria* (Poppy Anemone) Sir Thomas Hanmer, Garden Book, 1659.
PLYMOUTH PEAR *Pyrus cordata*. Now only growing in a couple of sites in Plymouth, plus a few more in Cornwall.
PLYVENS *Trifolium pratense* (Red Clover) Scot. Grigson.
POACHED-EGG FLOWER *Limnanthes douglassi* (Meadow Foam) Phillips. The flowers are white, with a yellow centre.
POACHED-EGG PLANT *Cephalanthera damasonium* (Broad Helleborine) Turner Ettlinger.
POCAN *Phytolacca decandra* (Poke-root) USA Henkel. Evidently the same as Poke, which is either from puccoon, in one of the American Indian languages, or else from pocon, apparently given to any plant yielding a red or yellow dye.
POCKET, Shepherd's *Capsella bursa-pastoris* (Shepherd's Purse) Bucks. Grigson.
POCKET-HANDKERCHIEF TREE *Davidia involucrata* (Handkerchief Tree) Hay & Synge. The reference is to the white bracts.
POCKWOOD TREE *Guaiacum officinale* (Lignum vitae) Trinidad Laguerre.
POD CORN *Zea mays var. tunicata*. It has a pod-like covering to the grain, and is of no use commercially.
POD MAHOGANY *Afzelia cuanzensis*.
POD THISTLE *Cirsium acaulon* (Picnic Thistle) N'thants. A E Baker. Probably a misreading of "Pad" Thistle, unless "pod" is a local variant.
PODDER; POTHER 1. *Cuscuta epithymum* (Dodder) B & H. 2. *Cuscuta europaea*

(Greater Dodder) B & H. Podder is perhaps a misprint in B & H's source, as they acknowledge, or it may be a hybrid between Dodder, Pother and Dother.

PODO *Podocarpus gracilior*. The botanical name, of which this is an abbreviation, is from the Greek podos, foot and karpos, fruit, a reference to the fleshy, foot-like stalk on the fruit.

PODOCARP, Large-leaved, Japanese *Podocarpus macrophylla*.

PODOCARP, Rusty *Podocarpus ferruginea* (Miro) Leathart.

PODOCARP, Tasmanian *Podocarpus alpinus*.

PODOCARP, Willow *Podocarpus saligna*.

POET'S JESSAMINE *Jasminum officinale* (Jasmine) Whittle & Cook.

POET'S LAUREL *Laurus nobilis* (Bay) Bardswell. The bay was sacred to Apollo, and since he was the god of poetry, the crown of bay leaves became the customary award in the universities to graduates in rhetoric and poetry.

POET'S NARCISSUS; POET'S DAFFODIL; POETICUS *Narcissus maialis var. patellaris* (Pheasant's Eye) Genders (Poet's Narcissus); RHS (Poet's Daffodil).

POHA *Physalis bunyardii* (Cape Gooseberry) USA Kingsbury.

POHUTUKAWA *Metrosideros excelsa*.

POI *Basella alba* (Indian Spinach) West Indies Howes.

POINCIANA *Poinciana gilliesii*.

POINSETTIA *Euphorbia pulcherrima*. Dr J R Poinsett was apparently the man who discovered it, in Mexico, 1828.

POINSETTIA, Desert *Euphorbia eriantha*.

POINT, Cuckoo *Arum maculatum* (Cuckoo-pint) Culpepper.

POISON, Adder's *Tamus communis* (Black Bryony) Dev. Friend.1882. It is not supposed to be confined to the adder. Attor was OE for poison, so the name looks pleonastic.

POISON, Beaver 1. *Cicuta maculata* (American Cowbane) USA Kingsbury.1967. 2. *Conium maculatum* (Hemlock) USA Skinner. Is there any suggestion that they were actually used to poison beavers?

POISON, Blackbeetle *Lamium album* (White Deadnettle) Som. Macmillan.

POISON, Blue *Ligustrum vulgare* (Privet) Som. Macmillan. A reference to the berries, which are certainly poisonous.

POISON, Bushman's *Acocanthera oblongifolia* (Wintersweet) Howes.

POISON, Crow *Amaianthemum muscaetoxicum* (Staggergrass) USA Kingsbury. Cf. Fly Poison for this.

POISON, Dog *Aethusa cynapium* (Fool's Parsley) Som. Macmillan; USA Sanecki. Certainly a poison, but why 'dog'? Dog Parsley is another of the names, in the sense of a false parsley, and that is probably the origin.

POISON, Fly *Amaianthemum muscaetoxicum* (Staggergrass) USA Kingsbury. i.e. the 'muscaetoxicum' of the specific name.

POISON, Fox *Daphne laureola* (Spurge Laurel) Lincs. Grigson.

POISON, Snake's *Iris foetidissima* (Foetid Iris) Dev. B & H. People have always viewed the scarlet fruits and smelt the awful smell with suspicion. Hence the attribution to poison generally and snakes in particular.

POISON APPLE *Solanum panduraeforme* S Africa Watt.

POISON ASH 1. *Chionanthus virginicus* (Fringe Tree) Leyel.1948. 2. *Rhus vernix* (Swamp Sumach) Le Strange.

POISON BAY *Illicium floridanum*.

POISON BERRY 1. *Anamirta cocculus* Chopra. 2. *Arum maculatum* (Cuckoo-pint) Dev. Friend.1882; Yks. B & H. 3. *Ilex aquifolium* fruit (Holly berry) Yks. B & H. 4. *Iris foetidissima* (Foetid Iris) Dev, Som. Macmillan. 5. *Solanum dulcamara* (Woody Nightshade) Barton & Castle. 6. *Sorbus aucuparia* fruit (Rowan berries) Som. Macmillan; Yks, N'thum. Grigson. 7. *Tamus communis* (Black Bryony) Dev. Friend.1882; Wilts. Dartnell & Goddard; Suss. Parish. The result of accurate observation in some cases, and of prejudice, or at least caution, in others.

POISON BUSH 1. *Acocanthera oblongifolia* (Wintersweet) S Africa Howes. 2. *Taxus baccata* (Yew) Som. Notes and Queries. vol 168, 1935.

POISON CUP *Ranunculus sceleratus* (Celery-leaved Buttercup) Tynan & Maitland.

POISON DAISY *Anthemis cotula* (Maydweed) Som. Macmillan Suss. Salisbury. Probably because of the tainting effect it has on milk, in Salisbury's opinion.

POISON DART *Aglaonema commutatum*.

POISON DOGWOOD; POISON ELDER *Rhus vernix* (Swamp Sumach) USA Zenkert (Dogwood); Kingsbury (Elder).

POISON FINGERS *Arum maculatum* (Cuckoo-pint) Dor. Macmillan. Cf. Poison Root.

POISON FLAG *Iris versicolor* (Purple Water Iris) Grieve.1931. Cf. Snake Lily and Dragon Flower for this.

POISON FLOWER *Solanum dulcamara* (Woody Nightshade) Herts. Grigson.

POISON HEMLOCK 1. *Cicuta maculata* (American Cowbane) USA Kingsbury.1967. 2. *Conium maculatum* (Hemlock) USA E Gunther.

POISON HOG MEAT *Aristolochia gigas* (Pelican Flower) Emboden.1974. The awful smell keeps animals away, but when wild pigs do eat it out of necessity, they are killed.

POISON IVY *Rhus radicans*.

POISON NUT *Strychnos nux-vomica* (Snakewood) Le Strange.

POISON OAK 1. *Rhus diversiloba*. 2. *Rhus radicans* (Poison Ivy) Zenkert.

POISON OAK, Californian; POISON OAK, Western *Rhus diversiloba* (Poison Oak) both Le Strange.

POISON PARSLEY *Conium maculatum* (Hemlock) USA Henkel.

POISON POPPY *Papaver rhoeas* (Red Poppy) Bucks. Grigson. All the poppies are poisonous to some extent, and though this one has been used medicinally, the name sug-

gests doubt as to its virtues. Cf. the Cornish Devil's Tongue, etc.

POISON RHUBARB *Petasites hybridus* (Butterbur) Yks. B & H. 'Rhubarb' from the size and shape of the leaves. 'Poison' to deter you from using them as such. Cf. Snake's Rhubarb, Snake's Food, etc.

POISON ROOT *Arum maculatum* (Cuckoo-pint) Wilts. Macmillan. Cf. Poison Fingers.

POISON SEGO *Zygadenus nuttalli*.

POISON SUMACH *Rhus vernix* (Swamp Sumach) Zenkert.

POISON VINE *Rhus radicans* (Poison Ivy) USA Kingsbury.

POISON WOOD *Metopium toxiferum*. Of the same family as the genus Rhus, this yields a resin that is a violent purgative, and the sap itself is a severe skin poison.

POISONING BERRIES 1. *Bryonia dioica* (White Bryony) Yks. Grigson. 2. *Solanum dulcamara* (Woody Nightshade) Yks. Grigson. Accurate, in both cases.

POISONOUS LETTUCE *Lactuca virosa* (Great Prickly Lettuce) Salisbury. All lettuces have some narcotic effect, largely bred out of the cultivated kinds. *L virosa*, though, has juice that is distinctly narcotic and soporific.

POISONOUS TEA PLANT *Solanum dulcamara* (Woody Nightshade) Oxf. B & H. It sounds as if the name was awarded to distinguish this plant from *Lycium halimifolium*,which is often cultivated under the name Tea Plant. The latter is a hedge plant with scarlet berries, and so may be mistaken for the nightshade.

POKE 1. *Phytolacca decandra* (Poke-root) USA Sanford. 2. *Veratrum viride* (American White Hellebore) New Engl. Sanford. Poke is either from puccoon, in one of the American Indian languages, or pocon, apparently given to any plant yielding a red or yellow dye.

POKE, Indian *Veratrum viride* (American White Hellebore) Fluckiger.

POKE, Virginian *Phytolacca decandra* (Poke-root) Brownlow.

POKE MILKWEED 1. *Asclepias exaltata*. 2. *Asclepias phytolaccoides* USA Kingsbury.

POKE-NEEDLE *Scandix pecten-veneris* (Shepherd's Needle) Suss. B & H. 'Poke' here is different. It must be Pook, or Puck, and the long, beaked fruits are the fairies' needle.

POKE-ROOT *Phytolacca decandra*.

POKEBERRY *Phytolacca decandra* (Poke-root) USA Gates.

POKER 1. *Arum maculatum* (Cuckoo-pint) Som. Macmillan. 2. *Kniphofia uvaria* (Red-hot Poker) Som. Macmillan. 3. *Typha latifolia* (False Bulrush) Som. Macmillan. All descriptive, with either red-hot ends, or, in the last case, black.

POKER, Ashy *Plantago media* (Lamb's Tongue Plantain) Wilts. Macmillan. 'Ashy' because this is also the Hoary Plantain.

POKER, Baillie Nicoll Jarvie's *Kniphofia uvaria* (Red Hot Poker) W Scot. Coats. After the incident in *Rob Roy*.

POKER, Black *Typha latifolia* (False Bulrush) Som. Notes and Queries, vol 168, 1935.

POKER, Devil's 1. *Kniphofia uvaria* (Red Hot Poker) Dev. Friend.1882. 2. *Typha latifolia* (False Bulrush) Som. Macmillan.

POKER, Fiery, Devil's *Kniphofia uvaria* (Red Hot Poker) Som. Macmillan.

POKER, Holy *Typha latifolia* (False Bulrush) Dev. Macmillan.

POKER, Pull *Typha latifolia* (False Bulrush) Ess. Gepp.

POKER, Red Hot 1. *Arum maculatum* (Cuckoo-pint) Som. Macmillan; Suss. Doris Cooper. 2. *Kniphofia uvaria*. 3. *Plantago lanceolata* (Ribwort Plantain) Som. Macmillan.

POKER-PLANT *Typha latifolia* (False Bulrush) Suss. Parish.

POKEWEED 1. *Phytolacca decandra* (Poke-root) USA Gates. 2. *Stellaria media* (Chickweed) Som. Macmillan.

POLANT *Primula vulgaris var. elatior* (Polyanthus) N'thants. A E Baker. Just a contraction of Polyanthus, albeit strange-looking.

POLAR PLANT *Rosmarinus officinalis* (Rosemary) Grieve.1931. Cf. Compass-weed, or Compass-plant, all presumably to do with the leaf orientation.

POLE BEAN *Phaseolus lunatus* (Lima Bean) Howes.

POLE-REED *Phragmites communis* Som. Elworthy. Possibly a corruption of pool-reed, but Elworthy describes pole-reed as "a long stout reed used for ceilings instead of laths".

POLECAT BUSH *Rhus triloba* (Three-leaf Sumach) USA Elmore. Because of the unpleasant smell. Cf. Skunk Bush, and Ill-scented Sumach.

POLECAT GERANIUM *Lantana sellowiana*.

POLECAT TREE *Illicium floridanum* (Poison Bay) Howes.

POLECAT-WEED *Symplocarpus foetidus* (Skunk Cabbage) Le Strange. Again, because of the foul smell.

POLEY *Mentha pulegium* (Pennyroyal) Watt. The French is menthe pouliot, and there is a Dutch name, polei, that seems to be responsible for the English poley. Cf. too Gerard's Puliall Royal. The Latin word they come from is pulices, fleas, for this herb is fine for getting rid of them.

POLEYE; POLION *Teucrium polium* R Gunther. It is glossed both poleye and polion on a C12 ms of Apuleius Barbarus, but the illustration does not bear much resemblance to any kind of labiate.

POLICEMAN'S BUTTONS *Caltha palustris* (Marsh Marigold) Som. Macmillan. It is the unopened buds that are likened to buttons of many kinds.

POLICEMAN'S HELMET 1. *Aconitum napellus* (Monkshood) Som. Macmillan. 2. *Impatiens glandulifera* (Himalayan Balsam) Clapham.

POLISH LARCH *Larix polonica*.

POLISHED WILLOW *Salix laevigata* (Red Willow) Everett.

POLK MILKWEED *Asclepias exaltata*

(Poke Milkweed) USA House. 'Polk' for 'poke' is unusual.

POLKA DOT PLANT *Hypoestes phyllostachya*. The spotted leaves are the reference.

POLLARD-FLOWERS *Tilia europaea* (Lime) Dor. Macmillan.

POLLY, Auntie; POLLY ANN; POLLY ANDREWS; POLLYANDICE *Primula vulgaris var. elatior* (Polyanthus) Som. Macmillan (Auntie Polly); Heref. Havergal (Polly Ann); Wilts. Dartnell & Goddard; Glos. J D Robertson (Polly Andrews); Som. Jennings (Pollyandice).

POLLY BAKER *Lychnis flos-cuculi* (Ragged Robin) Som. Macmillan.

POLLY NUT *Castanea sativa* (Sweet Chestnut) Som. Macmillan.

POLLY PODS *Lunaria annua* (Honesty) Dor. Macmillan.

POLSTEAD CHERRY *Prunus padus* (Bird Cherry) Suff. B & H. There is a Suffolk village of that name, not far from Hadleigh.

POLTATE *Solanum tuberosum* (Potato) Corn. Halliwell.

POLTERS *Digitalis purpurea* (Foxglove) IOM Moore.1924. 'Polter' has the same meaning as 'pop' in many of the English names like Pop-dock.

POLY *Teucrium polium* Turner.

POLY, Grass *Lythrum hyssopifolia*.

POLY-MOUNTAIN 1. *Acinos arvense* (Basil Thyme) B & H. 2. *Bartsia alpina* (Alpine Bartsia). 3. *Clinopodium vulgare* (Wild Basil) Camden/Gibson. Cf. Poley for Pennyroyal, where the name seems to come from Latin pulices, fleas.

POLYANTHA ROSE *Rosa multiflora*.

POLYANTHUS *Primula vulgaris var. elatior*. It means 'many flowers'.

POLYANTHUS NARCISSUS *Narcissus canaliculatum var. orientalis*. Many flowers, again.

POLYPODY, Common *Polypodium vulgare*. It comes from the Greek meaning 'many feet', so named from the appearance of the rhizomes.

POLYPODY, Golden *Polypodium vulgare* (Common Polypody) Kent B & H. 'Golden', presumably because of the bright orange spores.

POLYPODIUM *Campanula latifolia* (Giant Bellflower) Yks. B & H.

POMEGARNET; POMEGRANATE *Punica granatum* Parkinson (Pomegarnet). Pomegranate is Latin pomum granatum, apple + seed, i.e. an apple with so many seeds in it.

POMELO *Citrus x paradisi* (Grapefruit) Grigson.1974. Perhaps a shortening of the Dutch Pompelnous, which is Dutch pomp, melon plus limoes, a Malayan borrowing from Portuguese limão, lemon.

POMELO, Wild *Citrus maxima*.

POMERACK *Eugenia malaccensis* Everett.

POMPELMOUS *Citrus maxima* (Wild Pomelo) Grigson.1974. This is a Dutch word, from pomp, also Dutch, and meaning melon, and limoes, a Malayan borrowing from Portuguese limão, lemon.

POMPION *Cucurbita pepo* (Vegetable Marrow) Gerard.

POMPON LILY *Lilium pomponium*. An obscure name. Can it be connected with the old French word pompon, melon or something similar (as in the previous entry)? Or is there some connection with the Roman family Pomponius, Woodcock asks?

POND CYPRESS *Taxodium ascendens*.

POND DOGWOOD *Cephalanthus occidentalis* (Buttonbrush) O P Brown.

POND GRASS *Commelina diffusa*.

POND PINE *Pinus serotina* (Black Pine) S.Car. Clemson.

POND PLANTAIN *Potamogeton natans* (Broad-leaved Pondweed) Turner.

PONDWEED, Bog *Potamogeton polygonifolius*.

PONDWEED, Broad-leaved *Potamogeton natans*.

PONDWEED, Canadian *Elodea canadensis*. Canadian it may be, but this is the weed that used to be such a serious pest on most of the canal systems of this country.

PONDWEED, Cape *Aponogeton distachyum* (Cape Asparagus) RHS.
PONDWEED, Coiled *Ruppia spiralis*.
PONDWEED, Fen *Potamogeton coloratus* (Plantain-leaved Pondweed) Godwin.
PONDWEED, Horned *Zannichellia palustris*.
PONDWEED, Knight's *Stratiotes aloides* (Water Soldier) Grieve.1931.
PONDWEED, Plantain-leaved *Potamogeton coloratus*.
PONDWEED, Reddish *Potamogeton alpinus*.
PONDWEED, Tassel *Ruppia maritima*.
PONTEFRACT ROOT *Glycyrrhiza glabra* (Liquorice) Leighton. Liquorice has been cultivated in the Pontefract district of Yorkshire since the latter part of the 16th century, when the Black Friars started growing it there. The place was famous for Pontefract cakes, the precursors of liquorice allsorts.
PONTIC RHODODENDRON *Rhododendron ponticus* (Common Rhododendron) Forsyth.
PONTIC WORMWOOD *Artemisia pontica* (Roman Wormwood) Turner.
PONTINE OAK *Quercus pontica*.
PONY BEE BALM *Monarda pectinata* (Horse Mint) USA Elmore.
PONY TAIL PLANT *Beaucarnea recurvata*.
PONY'S TAIL *Plantago major* (Great Plantain) Dev. Macmillan.
POOK-NEEDLE 1. *Agrostemma githago* (Corn Cockle) Suss. W D Cooper. 2. *Erodium cicutarium* (Storksbill) Bell. 3. *Scandix pecten-veneris* (Shepherd's Needle) Hants. Grigson; Suss. Latham. i.e. fairy's needle, from the sharpened end of the seed-vessels of the last two, but probably a reference to the long teeth of Cockle's calyx.
POOLROOT *Sanicula europaea* (Sanicle) Grieve.1931.
POOR JAN'S LEAF *Sempervivum tectorum* (Houseleek) Dev. Macmillan. 'Poor Jan' is the archetypal countryman, who used houseleek as a cure for many ills.

POOR JANE *Geranium robertianum* (Herb Robert) Som. Macmillan.
POOR JOE *Dioidia teres*. An American weed, called Poor Joe because it flourishes in poor soil.
POOR-LAND DAISY *Leucanthemum vulgare* (Ox-eye Daisy) N'thants. A E Baker.
POOR MAN'S BACCY *Tussilago farfara* (Coltsfoot) Som. Macmillan. Coltsfoot, of course, has been smoked since ancient times, and is still smoked in herbal mixtures, for coughs, particularly.
POOR MAN'S BEER *Humulus lupulus* (Hop) Dor. Macmillan.
POOR MAN'S BLANKET; POOR MAN'S FLANNEL *Verbascum thapsus* (Mullein) Donegal Grigson(Blanket); Som, Glos, Bucks. Grigson (Flannel). A reference to the felty leaves.
POOR MAN'S BLOOD *Orchis mascula* (Early Purple Orchis) Kent Grigson.
POOR MAN'S BRUSH *Dipsacus fullonum* (Teasel) Som. Macmillan.
POOR MAN'S CABBAGE 1. *Polygonum bistorta* (Bistort) Lancs. Grigson. 2. *Rumex patientia* (Patience Dock) Lancs. Nodal. Both have been used as a potherb.
POOR MAN'S FRIEND *Clematis vitalba* (Old Man's Beard) Som. Macmillan. Cf. Shepherd's Delight for this.
POOR MAN'S GERANIUM *Saxifraga sarmentosa* (Mother-of-thousands) Dev. Friend.1882. Cf. Hanging Geranium - for that was the way they always used to be seen in cottage windows.
POOR MAN'S HERB *Rumex acetosa* (Wild Sorrel) Ire. Logan.
POOR MAN'S MUSTARD *Alliaria petiolata* (Jack-by-the-hedge) Lincs. Grigson.
POOR MAN'S ORCHID 1. *Iris xiphioides* (English Iris) Staffs. Marrow. 2. *Schizanthus sp* (Butterfly Flower) Rochford.
POOR MAN'S PARMACETTY; POOR MAN'S PURSE *Capsella bursa-pastoris* (Shepherd's Purse) Gerard (Parmacetty); Som. Macmillan (Purse). According to Dyer, Poor Man's Parmacetty is a joke on Latin

bursa, purse, to a poor man always the best remedy for his bruises.

POOR MAN'S PEPPER 1. *Lepidium campestre* (Pepperwort) War. B & H; USA. Henkel. 2. *Lepidium latifolium* (Dittander) Camden/Gibson. 3. *Lepidium virginicum* (Virginia Peppergrass) USA Allan. 4. *Poterium sanguisorba* (Salad Burnet) Dor. Macmillan. 5. *Sedum acre* (Biting Stonecrop) Suss, Notts. Grigson. 6. *Valeriana dioica* (Marsh Valerian) Wilts. Macmillan.

POOR MAN'S REMEDY *Valeriana officinalis* (Valerian) B & H. Cf. Countryman's Treacle. Valerian is a "heal-all".

POOR MAN'S RHUBARB *Thalictrum flavum* (Meadow Rue) W Miller.

POOR MAN'S SALVE *Scrophularia aquatica* (Water Betony) Dev. B & H. Applied to cuts, sores and ulcers.

POOR MAN'S TEA *Veronica chamaedrys* (Germander Speedwell) Cumb. Grigson. Curtis said at the end of the eighteenth century that "the leaves have been recommended by some writers as a substitute for Tea".

POOR MAN'S TOBACCO *Tussilago farfara* (Coltsfoot) Som. Macmillan.

POOR MAN'S TREACLE 1. *Alliaria petiolata* (Jack-by-the-hedge) B & H. 2. *Allium sativum* (Garlic) B & H. 3. *Allium vineale* (Crow Garlic). 4. *Teucrium scorodonia* (Wood Sage) Tynan & Maitland. 'Treacle' here is from triacle, an antidote to the bite of venomous animals. It is from the Greek theriake, from therione, a name given to the viper. So the superstition was that viper's flesh would cure viper's bite. Treacle was certainly defined as a "physical compound made of vipers and other ingredients", and was a favourite against all poisons.

POOR MAN'S WEATHERGLASS *Anagallis arvensis* (Scarlet Pimpernel) Corn. Deane & Shaw; Som. Macmillan; Hants. Cope; Ches. Holland; USA Whitney & Bullock. The flowers open when it is going to be sunny, and close when it is going to rain.

POOR OATS *Avena fatua* (Wild Oats) Som. Elworthy.

POOR ROBERT *Geranium robertianum* (Herb Robert) Dev, Som. Macmillan.

POOR ROBIN 1. *Geranium robertianum* (Herb Robert) Dev, Som. Macmillan. 2. *Melandrium rubrum* (Red Campion) Dev. Friend.1882.

POOR ROBIN'S PLANTAIN *Hieraceum venosum* (Rattlesnake Weed) USA House.

POP, Green *Digitalis purpurea* (Foxglove) Corn. Grigson. The 'pop' names for Foxglove, and there are a lot of them, refer to the children's habit of inflating the flowers and then popping them.

POP ASH *Fraxinus caroliniana* USA Harper.

POP-BELLS; POP-BLADDER; POP-GLOVE *Digitalis purpurea* (Foxglove) Som. Macmillan (Bells); Dor. Macmillan (Bladder); Corn. Courtney.1880 (Glove). See Poppy.

POP-DOCK; POP-A-DOCK *Digitalis purpurea* (Foxglove) Corn. Jago; Som. Macmillan; USA Henkel. or Poppy Dock. See Poppy.

POP-GUN 1. *Digitalis purpurea* (Foxglove) Wright. 2. *Silene cucubalis* (Bladder Campion) Dor. Macmillan. 3. *Stellaria holostea* (Greater Stitchwort) Som. Macmillan. For foxglove, see Poppy.

POP-JACK *Stellaria holostea* (Greater Stitchwort) Som. Macmillan.

POP-SHELLS *Hedera helix* - berries (Ivy) Som. Macmillan.

POP-UP *Colchicum autumnale* (Meadow Saffron) Som. Macmillan. A comment on the way the flowers appear before any leaves. Cf. Upstart, and all the Naked Ladies type of names.

POP-VINE; POPS *Physalis pubescens* (Husk Tomato) Barbados Gooding.

POPCORN *Zea mays var. rostrata*. Called popcorn because of the way the grains swell and burst when heated, turning themselves inside out.

POPCORNS *Euonymus europaea* (Spindle

berries) Som. Grigson. Descriptive. Cf. Hot Cross Buns.

POPE *Papaver rhoeas* (Red Poppy) Northants. A E Baker. Presumably only a shortening of 'poppy'.

POPE'S ODE *Aconitum anglicum* (Wild Monkshood) Northants. B & H. 'Ode' is 'hood', as in monkshood.

POPILARY; POPILLARY *Populus nigra* (Black Poplar) both Ches. B & H.

POPILLE *Agrostemma githago* (Corn Cockle) Scot. Grigson. 'Popple' is a widespread name for cockle. It appears in Scotland as Papple and Pawple, as well as Popille.

POPINAC *Acacia farnesiana* (Cassie) Polunin. Cf. Opopanax.

POPLAIN *Populus alba* (White Poplar) W Eng. Halliwell.

POPLAR, Balsam *Populus x candicans*. 'Poplar' comes through O French poplier from Latin populus. Balsam is a reference to the medical uses to which the resinous buds are put.

POPLAR, Balsam, American *Populus balsamifera*.

POPLAR, Balsam, Eastern *Populus balsamifera* (American Balsam Poplar) Mabey.

POPLAR, Balsam, Western; POPLAR, Balsam, Oregon *Populus trichocarpa* Brimble.1948 (Oregon).

POPLAR, Berlin *Populus x berolinensis*. 'Berlin', because that was where the hybrid was produced, about 1800.

POPLAR, Berry-bearing *Populus monilifera* (Northern Cottonwood.

POPLAR, Black 1. *Populus monilifera* (Northern Cottonwood) B & H. 2. *Populus nigra*. 3. *Populus tremula* (Aspen) Som. B & H.

POPLAR, Black, American *Populus deltoides* (Carolina Poplar) Polunin.1976.

POPLAR, Black, Carolina *Populus angulata* (Common Cottonwood) Brimble.1948.

POPLAR, Canadian 1. *Populus x canadensis*. 2. *Populus monilifera* (Northern Cottonwood) Ablett.

POPLAR, Carolina *Populus deltoides*.

POPLAR, Catfoot *Populus nigra* (Black Poplar) Lancs. Grigson. A reference, so it is said, to the dark knots in the timber.

POPLAR, Downy *Populus alba* (White Poplar) Grigor in N D G James.

POPLAR, English, Old *Populus nigra* (Black Poplar) B & H.

POPLAR, Euphrates *Populus euphratica*.

POPLAR, Grey *Populus canescens*.

POPLAR, Italian, Black *Populus x canadensis var. serotina*.

POPLAR, Japanese *Populus maximowiczii*.

POPLAR, Lady 1. *Populus alba* (White Poplar) Ches. Grigson. 2. *Populus nigra* (Black Poplar) Ches. Holland.

POPLAR, Lombardy *Populus nigra var. italica*.

POPLAR, Manchester *Populus nigra var. betulifolia*.

POPLAR, Necklace *Populus deltoides* (Carolina Poplar) USA Zenkert.

POPLAR, Necklace, Chinese *Populus lasiocarpa*.

POPLAR, Necklace-bearing *Populus monilifera* (Northern Cottonwood) Ablett. Cf. Berry-bearing Poplar for this.

POPLAR, Ontario *Populus x candicans* Hay & Synge.

POPLAR, Railway *Populus x canadensis* 'Regenerata' Mitchell.1974. Mitchell felt this was a very appropriate name for it, because it is so common by railway lines into cities, and in coal-yards and sidings.

POPLAR, Silver; POPLAR, Silver-leaf *Populus alba* (White Poplar) Som. Grigson (Silver); USA Tolstead (Silver-leaf).

POPLAR, Tana River *Populus ilicifolia*. The Tana River runs through eastern Kenya.

POPLAR, Trembling 1. *Populus tremula* (Aspen) Prior. 2. *Populus tremuloides* (American Aspen) USA Zenkert.

POPLAR, Tulip *Liriodendron tulipifera* (Tulip Tree) USA Harper.

POPLAR, Water *Populus nigra* (Black Poplar) Som. Elworthy.

POPLAR, White 1. *Liriodendron tulipifera*

(Tulip Tree) USA Harper. 2. *Populus alba.* 3. *Populus tremuloides* (American Aspen) Grieve.1931.
POPLAR, White, Chinese *Populus tomentosa.*
POPLAR, Willow *Populus nigra* (Black Poplar) Cambs. Grigson.
POPLAR, Yellow *Liriodendron tulipifera* (Tulip Tree) USA Harper. This is what the wood is often known as in the timber trade.
POPLIN *Populus alba* (White Poplar) IOM Moore.1924.
POPOLO *Solanum sodomeum* (Apple of Sodom) USA Kingsbury.
POPPER 1. *Digitalis purpurea* (Foxglove) Hants. Cope. 2. *Silene cucubalis* (Bladder Campion) Som, Wilts. Grigson. 3. *Stellaria holostea* (Greater Stitchwort) Wilts. Grigson. Foxglove flowers are blown up by children, and popped; the inflated calyx of bladder campion is likewise satisfyingly popped, and greater stitchwort seed vessels burst with a pop of their own accord.
POPPET *Papaver rhoeas* (Red Poppy) War. Grigson.
POPPILERY *Populus nigra* (Black Poplar) Ches. B & H. Cf. Popillary, etc.
POPPLE 1. *Agrostemma githago* (Corn Cockle) E Ang, Lincs, Cumb, N'thum, N Ire. Grigson; Yks. Atkinson; Scot. Jamieson. 2. *Papaver rhoeas* (Red Poppy) Glos, Oxf, N'thants, Notts, Kent Grigson; Yks. F K Robinson. 3. *Populus nigra* (Black Poplar) Suff. Jobson. 4. *Populus tremuloides* (American Aspen) USA Schery. 5. *Salix alba* (White Willow) Grigson. 6. *Sinapis arvensis* (Charlock) Cumb. Grigson. A mixture of 'poppy' and 'poplar' names here, with Charlock being the odd one out.
POPPY 1. *Agrostemma githago* (Corn Cockle) Ches. Holland. 2. *Digitalis purpurea* (Foxglove) Corn. Courtney.1880. Dev. Friend.1882; Wilts. Dartnell & Goddard. 3. *Papaver sp.* 4. *Silene cucubalis* (Bladder Campion) Wilts. Dartnell & Goddard. 5. *Stellaria holostea* (Greater Stitchwort) Wilts. Dartnell & Goddard. OE popig, and thence Latin papaver. But there is confusion with 'pop' and 'poppy' names in this list, particularly for foxglove, for which 'poppy' means 'pop'. Children have always amused themselves by bursting the flower buds so that they make the characteristic "pop" sound.
POPPY, Alpine *Papaver alpinum.*
POPPY, Alpine, Yellow *Papaver rhaeticum.*
POPPY, Arctic *Papaver radicatum* (Yellow Poppy) Polunin.
POPPY, Arizona *Kallstroemeria grandiflora.*
POPPY, Babington's *Papaver lecoqii.*
POPPY, Bastard, Rough-headed *Papaver argemone* (Rough Long-headed Poppy) Camden/Gibson.
POPPY, Bastard, Smooth-headed *Papaver dubium* (Smooth Long-headed Poppy) Camden/Gibson.
POPPY, Bastard, Wild, Yellow *Meconopsis cambrica* (Welsh Poppy) Camden/Gibson.
POPPY, Blue *Centaurea cyanus* (Cornflower) Som. Macmillan.
POPPY, Blue, Himalayan *Meconopsis betonicifolia.*
POPPY, Bristly *Papaver hybridum.*
POPPY, Bush, Californian *Romneya sp.*
POPPY, Californian 1. *Escholtzia californica.* 2. *Romneya sp* (Californian Bush Poppy) Salisbury.1936.
POPPY, Carnation *Papaver somniferum* (Opium Poppy) Chopra.
POPPY, Celandine *Chelidonium majus* (Greater Celandine) USA Kingsbury. The plant is a member of the poppy family.
POPPY, Chinese, Yellow *Meconopsis integrifolia.*
POPPY, Corn *Papaver rhoeas.*
POPPY, Cyclamen *Eonecon chioantha.*
POPPY, Field *Papaver rhoeas* (Red Poppy) North.
POPPY, Flanders *Papaver rhoeas* (Red Poppy) Howes.
POPPY, Flop *Digitalis purpurea* (Foxglove) Dev. Friend.1882. In this case, it is 'popper' rather than 'poppy' - the unopened flowers

can be popped. Cf. Green Pops, which can be become Green Poppies in the same area.

POPPY, French *Verbascum thapsus* (Mullein) Dev. Macmillan.

POPPY, Hedge; POPPY, Hill 1. *Digitalis purpurea* (Foxglove) both Som. Macmillan. 2. *Verbascum thapsus* (Mullein) both Som. Macmillan.

POPPY, Frothy *Silene cucubalis* (Bladder Campion) Gerard. The reference is to the cuckoo-spit often found around it. Cf. Spatling Poppy, which means the same.

POPPY, Gold, Desert *Escholtzia glytosperma*.

POPPY, Gold, Little *Escholtzia minutiflora*.

POPPY, Green; POPS, Green *Digitalis purpurea* (Foxglove) both Corn. Grigson. See Flop Poppy above.

POPPY, Harebell *Meconopsis quintuplinerva*.

POPPY, Horn; POPPY, Horned *Glaucium flavum*. Horned Poppy is the usual name; Horn Poppy occurs in Dev. Grigson.

POPPY, Horned, Purple *Roemeria hybrida* (Violet Horned Poppy) Salisbury.

POPPY, Horned, Red *Glaucium corniculatum*.

POPPY, Horned, Violet *Roemeria hybrida*.

POPPY, Iceland *Papaver nudicaule*.

POPPY, Indian 1. *Glaucium flavum* (Horned Poppy) Som. Macmillan. 2. *Meconopsis cambrica* (Welsh Poppy) Som. Macmillan.

POPPY, Lampshade *Meconopsis integrifolia* (Yellow Chinese Poppy) RHS.

POPPY, Long-headed; POPPY, Long-headed, Smooth *Papaver dubium*. Clapham prefers the name without 'smooth'.

POPPY, Long-headed, Rough *Papaver argemone*. This species has bristles on its seed pods (so 'Rough'), while those of *P dubium* are hairless (so 'Smooth').

POPPY, Matilija *Romneya coulteri* (Californian Bush Poppy) Wit.

POPPY, Mexican 1. *Argemone mexicana*. 2. *Kallstroemeria grandiflora* (Arizona Poppy) USA T H Kearney.

POPPY, Oil *Papaver somniferum* (Opium Poppy) Hutchinson & Melville. The name acknowledges the value of the oil got from its seeds.

POPPY, Opium *Papaver somniferum*.

POPPY, Oriental *Papaver orientale*.

POPPY, Pale *Papaver argemone* (Rough Long-headed Poppy) Hutchinson.

POPPY, Plume *Bocconia cordata*.

POPPY, Poison *Papaver rhoeas* (Red Poppy) Bucks. Grigson.

POPPY, Prickly 1. *Argemone mexicana* (Mexican Poppy) USA Kingsbury. 2. *Papaver argemone* (Rough Long-headed Poppy) Blamey.

POPPY, Prickly-headed, Long *Papaver argemone* (Rough Long-headed Poppy) Clapham.

POPPY, Prickly-headed, Round *Papaver hybridum* (Bristly Poppy) Clapham. It is the seed pods that are 'prickly'.

POPPY, Prickly-headed, Smooth *Papaver dubium* (Smooth Long-headed Poppy) Clapham.

POPPY, Pyrenean *Papaver suaveolens*.

POPPY, Red *Papaver rhoeas*.

POPPY, Rhaetian *Papaver rhaeticum* (Yellow Alpine Poppy) Grey-Wilson.

POPPY, Rough *Papaver hybridum* (Bristly Poppy) Blamey.

POPPY, Round-headed *Papaver hybridum* (Bristly Poppy) D Grose. 'Round-headed', as distinct from the 'long-headed' poppies.

POPPY, Round-headed, Smooth *Papaver rhoeas* (Red Poppy) Curtis.

POPPY, Satin *Meconopsis nepalense*.

POPPY, Sea; POPPY, Sea, Horned *Glaucium flavum* (Horned Poppy) both Gerard.

POPPY, Smooth-headed, Long *Papaver dubium* (Smooth Long-headed Poppy) Curtis.

POPPY, Snow *Eonecon chionantha* (Cyclamen Poppy) A W Smith.

POPPY, Spatling *Silene cucubalis* (Bladder Campion) Cumb. B & H. A reference to cuckoo-spit. Cf. Frothy Poppy.

POPPY, Summer 1. *Kallstroemeria grandiflora* (Arizona Poppy) USA T H Kearney.

2. *Papaver rhoeas* (Red Poppy) Som. Macmillan.

POPPY, Tulip *Papaver glaucum*.

POPPY, Velvet *Verbascum thapsus* (Mullein) Corn. Grigson. 'Velvet' is a description of the leaves.

POPPY, Water *Melandrium rubrum* (Red Campion) Lincs. Grigson.

POPPY, Welsh *Meconopsis cambrica*. Not entirely confined to Wales, for it can be found in similar habitats in the west of England.

POPPY, White *Papaver somniferum* (Opium Poppy) Barton & Castle.

POPPY, Wild, Bastard *Papaver argemone* (Rough Long-headed Poppy) Gerard.

POPPY, Yellow 1. *Argemone mexicana* (Mexican Poppy) Chopra. 2. *Glaucium flavum* (Horned Poppy) Gerard. 3. *Papaver radicatum*.

POPPY ANEMONE *Anemone coronaria*.

POPPY DOCK *Digitalis purpurea* (Foxglove) Turner. Cf. Pop-dock and a number of other similar names, all from the children's practice of 'popping' the flowers.

POPPY MALLOW *Callirhoe papaver*.

POPPY MALLOW, Finger *Callirhoe digitata*.

POPPY MALLOW, Purple *Callirhoe involucrata*.

POPPYWORT *Meconopsis betonicifolia* (Himalayan Blue Poppy) Grant White. e.

POPS 1. *Digitalis purpurea* (Foxglove) Som. Elworthy; Hants. Cope. 2. *Physalis pubescens* (Husk Tomato) Barbados Gooding. 3. *Physalis turbinata* Barbados Gooding. 4. *Stellaria holostea* (Greater Stitchwort) Wilts. Jones. & Dillon. Cf. Pop Vine for the Physalis, and the various Popper and Poppy names for the others.

POPS, Cow *Solanum angulata* Barbados Gooding.

POPS, Fairy *Trifolium pratense* (Red Clover) Dor. Macmillan. 'Pops' in this case apparently means sweets.

POPS, Green; POPPY, Green *Digitalis purpurea* (Foxglove) both Corn. Grigson.

Children used to have a favourite pastime of "popping" the unopened buds of foxglove.

PORCELAIN FLOWER *Hoya carnosa* (Wax Flower) Harvey. Descriptive of the texture of the flowers.

PORCELAIN HEATH *Erica ventricosa*. Curtis's name - for the flowers look like little glazed jars, just as if they were made of porcelain.

PORCELANE *Portulaca sativa* (Purslane) Gerard.

PORO-PORO 1. *Solanum aviculare* (Kangaroo Apple) New Zealand Perry. 2. *Solanum laciniatum* (Kangaroo Apple) New Zealand Perry. Maori poporo.

PORCUPINE GRASS *Stipa spartea* (Speargrass) USA Johnston.

PORCUPINE PLANT *Solanum aviculare* (Kangaroo Apple) New Zealand Howes.

PORT JACKSON PINE *Callitris rhomboidea* (Oyster Bay Pine) Hora.

PORT ORFORD CEDAR *Chamaecyparis lawsoniana* (Lawson Cypress) USA Schenk & Gifford. Port Orford, on the Oregon coast, where it thrives particularly well.

PORT ROYAL SENNA *Cassia obovata* (Italian Senna) Dalziel.

PORTER'S JOY *Ipomaea learii* (Blue Dawn Flower) India Perry. Perhaps another name, Railway Creeper, may explain this one.

PORTIA TREE; PORTIA OIL NUT *Thespesia populnea* RHS (Nut).

PORTLAND ARROWROOT; PORTLAND SAGO; PORTLAND STARCH *Arum maculatum* (Cuckoo-pint) W Miller (Arrowroot, Starch); Fernie (Sago). Once cultivated on the Isle of Portland for the tubers, which when cooked yielded a starch known by these names. It was used as a substitute for arrowroot, but the starch was "most hurtfull to the hands of the Laundresse that hath the handling of it, for it chappeth, blistereth, and maketh the hands rough and rugged, and withall smarting" (Gerard).

PORTLAND SPURGE *Euphorbia portlandica*. Not confined to Portland, for it can be found locally on dunes and other coastal sites from Chichester Harbour to Galloway, and in Ireland, too.

PORTUGAL BROOM, White *Cytisus albus*.

PORTUGUESE CYPRESS *Cupressus lusitanicus* (Mexican Cypress) J Smith. An American tree, but 'lusitanicus' was given to it because it became known in Europe through trees cultivated in Portugal, and so long established there that they were thought indigenous.

PORTUGUESE HEATH *Erica lusitanica*.

PORTUGUESE LAUREL *Prunus lusitanicus*.

PORTUGUESE OAK *Quercus faginea*.

PORTUGUESE ORANGE *Citrus aurantium var. sinensis* (Malta Orange) Willis.

PORTUGUESE QUINCE *Cydonia oblonga* (Quince) Howes.

PORTUGUESE SUNDEW *Drosophyllum lusitanicum* (Dewy Pine) Simmons.

POSSET *Filipendula ulmaria* (Meadowsweet) Lancs. Nodal.

POSSUM HAW *Viburnum nudum* USA Harper.

POSSUM OAK *Quercus nigra* (Water Oak) Howes.

POST OAK *Quercus stellata*.

POST OAK, Swamp *Quercus lyrata*.

POSY *Paeonia mascula* (Peony) Wilts. B & H.

POSY, Bad Man's; POSY, Black Man's *Lamium purpureum* (Red Deadnettle) Yks, N'thum. (Bad Man); Cumb. (Black Man) both Grigson. These names are the only real sign that red deadnettle was ever viewed with distrust - both of them mean the devil.

POSY, Bishop's *Leucanthemum vulgare* (Ox-eye Daisy) Donegal Grigson.

POSY, Boggart *Mercurialis perennis* (Dog's Mercury) Yks. Grigson. Cf. Boggart-flower, also from Yorkshire, boggart being a North country form of bogle, a goblin.

POSY, Cat *Bellis perennis* (Daisy) Som. Macmillan; Cumb. Grigson.

POSY, Clock *Taraxacum officinale* (Dandelion) Lancs. Nodal.

POSY, Devil's 1. *Allium ursinum* (Ramsons) Shrop. B & H. 2. *Allium vineale* (Crow Garlic) Shrop. Burne. 3. *Iris pseudo-acarus* (Yellow Flag) Wales Trevelyan. The bad smell of the two garlics accounts for its association with the devil. The name for Yellow Flag is probably an error for *Iris foetidissima*, which has an equally bad smell.

POSY, Dog *Taraxacum officinale* (Dandelion) Lancs, Yks. Grigson.

POSY, Indian *Gnaphalium polycephalum* (Mouse-ear Everlasting) O P Brown. This is an American plant, so the 'Indian' of the name refers to the American Indian.

POSY, Lady's *Trifolium pratense* (Red Clover) Som. Macmillan.

POT-ASH *Aegopodium podagraria* (Goutweed) Dev. Macmillan. Better known as Ground Ash, and even better as Ground Elder.

POT-HERB, Desert *Calandrinia ambigua*. Edible, and used for greens, hence the name.

POT-OF-GOLD LILY *Lilium iridollae*. The reference is to the pot of gold to be found at the end of the rainbow. The specific name 'iridollae' is made up of 'iris', rainbow and 'olla', pot.

POTATO *Solanum tuberosum*. 'Potato' is from Spanish patata, and that came from Haitian batata.

POTATO, Air *Dioscorea bulbifera* (Potato Yam) Schery. Cf. Aerial Yam, because aerial tubers are produced in the leaf axils.

POTATO, Canada *Helianthus tuberosus* (Jerusalem Artichoke) Parkinson. Parkinson would have no truck with the idea that 'Jerusalem' was the place of origin. He was quite right in ascribing it to Canada.

POTATO, Carib; POTATO, Otaheite *Dioscorea bulbifera* (Potato Yam) Howes.

POTATO, Chinese *Dioscorea batatas* (Chinese Yam) Howes.

POTATO, Crow *Lycopus asper* (Bugleweed) USA Yarnell. It produces edible tubers, which were dried and boiled by the American Indians.

POTATO, Cuckoo *Conopodium majus* (Earthnut) Donegal Grigson. Potato, because the tubers are edible, and best roasted like chestnuts. Cf. Fairy Potato.

POTATO, Dakota *Apios americana* (Groundnut) Bianchini.

POTATO, Devil's *Solanum carolinense* (Carolina Nightshade) Allan. Cf. Devil's Tomato. Both names show that the roots and fruit are not what they seem to be, and better left alone.

POTATO, Duck *Sagittaria latifolia* (Broad-leaved Arrowhead) Douglas. Some of the American Indians called this wild potato; they were gathered in autumn and dried, and later boiled for use.

POTATO, European *Solanum tuberosum* (Potato) USA Brouk. An anomaly, like the American use of Irish Potato. They came about when potatoes were shipped from Britain to the English colonists on the American mainland.

POTATO, Fairy *Conopodium majus* (Earthnut) Ire. Grigson. Cf. Cuckoo Potato.

POTATO, Fra-Fra; POTATO, Fura-Fura *Coleus rotundifolius* Schery (Fura). Extensively cultivated in West Africa for the tubers.

POTATO, Hausa *Coleus rotundifolius* (Fra-Fra Potato) Douglas.

POTATO, Indian 1. *Apios americana* (Ground Nut) H & P. 2. *Solanum tuberosum* (Potato) USA H Smith.

POTATO, Irish *Solanum tuberosum* (Potato) USA H Smith. See European Potato.

POTATO, Jericho *Solanum incanum* (Hoary Nightshade) Moldenke. Cf. Palestine Nightshade.

POTATO, Kaffir 1. *Coleus barbatus* S Africa Watt. 2. *Coleus esculentus* S Africa Watt. Both of these have edible tubers, and are grown in central and West Africa.

POTATO, Lily *Asphodeline lutea* (Yellow Asphodel) Douglas. Apparently widely cultivated in Roman times.

POTATO, Namara *Helianthus tuberosus* (Jerusalem Artichoke) S Africa Watt.

POTATO, Native *Coleus esculentus* S Africa Watt.

POTATO, Prairie *Psoralea esculenta* (Breadroot) USA Havard.

POTATO, Spanish *Ipomaea batatas* (Sweet Potato) Dalziel. It was originally thought that the introduction of sweet potato into the Pacific area had been made by Spanish explorers, but in fact it was being cultivated there a long time before any European contact.

POTATO, Sweet *Ipomaea batatas.*

POTATO, Sweet, Wild *Phytolacca heptandra* (Umbra Tree) S Africa Watt.

POTATO, Tule *Sagittaria latifolia* (Broad-leaved Arrowhead) USA Bean. See Duck Potato.

POTATO, White *Solanum tuberosum* (Potato) USA Brouk.

POTATO, Wild 1. *Chlorogalum pomeridianum* (Soap Plant) USA Havard. 2. *Ipomaea pandurata* (Wild Potato Vine) O P Brown. 3. *Ipomaea tiliacea* Barbados Gooding. 4. *Sagittaria arifolia* (Arum-leaved Arrowhead) Calif. Spier. 5. *Sagittaria latifolia* (Broad-leaved Arrowhead) USA. Turner & Bell. 6. *Solanum fendleri.* Soap Plant has an edible bulb, hence the name. Arrowhead's rhizome is also edible, and known to the Indians as "white potatoes". The others on the list have genuine claims to the name.

POTATO BEAN *Apios americana* (Groundnut) Douglas.

POTATO CREEPER *Solanum seaforthianum.*

POTATO CREEPER, Giant *Solanum wendlandii.*

POTATO FLOWER, Wild *Solanum dulcamara* (Woody Nightshade) Som. Macmillan.

POTATO ONION 1. *Allium cepa* (Onion) M Baker.1977. 2. *Allium cepa var. viviparum* (Egyptian Multiplier Onion). G B Foster. The latter produces onion-flavoured bulbils at the tops of the stalks.

POTATO TREE *Solanum wrightii.*

POTATO TREE, Chilean *Solanum crispum.*

POTATO VINE *Solanum jasminoides.*

POTATO VINE, Wild *Ipomaea pandurata.*

POTATO YAM *Dioscorea bulbifera*. Cf. Air Potato and Aerial Yam. All these names come about because of the plant's habit of forming aerial tubers in the leaf axils.

POTATOES-IN-THE-DISH 1. *Euphorbia amygdaloides* (Wood Spurge) Dor. Macmillan. 2. *Euphorbia helioscopia* (Sun Spurge) Dor. Macmillan. Descriptive.

POTENTILLA, Sibbald's *Sibbaldia procumbens* (Least Cinquefoil) Hutchinson.

POTENTILLA, Strawberry-leaved *Potentilla sterilis* (Barren Strawberry) Hutchinson.

POTHER *Cuscuta epithymum* (Dodder) B & H. Dother is another form of this.

POTHERB, Black *Smyrnium olusatrum* (Alexanders) Culpepper.

POTHERB, White *Valerianella locusta* (Cornsalad) Gerard.

POTS-AND-KETTLES *Buxus sempervirens* fruit (Box) Wilts. Dartnell & Goddard.

POTTAGE HERB *Brassica napus* (Rape) N Eng. B & H.

POUCH, Shepherd's 1. *Capsella bursa-pastoris* (Shepherd's Purse) Gerard. 2. *Orobanche minor* (Lesser Broomrape) IOW W H Long.

POUCH, Witches' *Capsella bursa-pastoris* (Shepherd's Purse) Scot. B & H.

POUMILLERE, La *Helleborus viridis* (Green Hellebore) Guernsey MacCulloch.

POUNCE, Shepherd's *Capsella bursa-pastoris* (Shepherd's Purse) Culpepper. A pounce-bag, or pounce box, was a perforated bag or box for sprinkling pounce, a powder used either to make a pattern or to prepare a writing surface.

POUND-GARNET *Punica granatum* (Pomegranate) Haining.

POUND NEEDLE *Scandix pecten-veneris* (Shepherd's Needle). B & H.

POVERTY 1. *Ononis repens* (Rest-Harrow) Som. Macmillan. 2. *Polemonium caeruleum* (Jacob's Ladder) Cumb. Grigson. 3. *Rhinanthus crista-galli* (Yellow Rattle) Som. Macmillan. 4. *Sagina procumbens* (Pearlwort) Norf. Grigson. 5. *Spergula arvensis* (Corn Spurrey) Suss. Parish. Such a name usually means that the plants grow in the poorest soil, and hence are a sign to the farmer that they may well keep him in poverty.

POVERTY-GRASS 1. *Hudsonia tomentosa* (Beach Heather) Perry. 2. *Plantago major* (Great Plantain) Som. Grigson. Beach Heather is an American plant that grows on poor, sandy soils, or on the beach.

POVERTY-PINK *Prunella vulgaris* (Selfheal) Berw. Denham/Hardy. Again, a plant of poor soil.

POVERTY-PURSE *Capsella bursa-pastoris* (Shepherd's Purse) Lincs. Grigson.

POVERTY-WEED 1. *Leucanthemum vulgare* (Ox-eye Daisy) Glos. PLNN35. Ches. Holland; Canada H C Long.1910. 2. *Melampyrum arvense* (Field Cow-wheat) Som. Grigson; IOW W H Long. 3. *Rhinanthus crista-galli* (Yellow Rattle) Som. Macmillan. Cow-wheat and Yellow Rattle are undesirables in the corn. They contain glycosides that contaminate the wheat, and so are poverty makers. For ox-eye, Cf. another name - Poor-land Daisy.

POWDER-HORN *Cerastium arvense* (Field Mouse-ear) USA Elmore.

POWDERED BEAN *Primula farinosa* (Bird's-eye Primrose) N Scot. B & H. Cf. Mealy Primrose. The specific name accounts for the epithets.

POWER-WORT *Ranunculus ficaria* (Lesser Celandine) Som. Macmillan. Probably Pilewort originally.

POWK-NEEDLE *Erodium cicutarium* (Storksbill) B & H. i.e. fairy needle, powk being pook, or puck. The seed vessels have long sharp points.

POYI *Dalbergia melanoxylon* (African Blackwood) Dale & Greenway. This is actually the trade name for the timber.

PRAIRIE, Queen of the *Filipendula rubra*.

PRAIRIE ACACIA *Acacia angustissima*.

PRAIRIE APPLE *Psoralea esculenta* (Breadroot) Howes.

PRAIRIE BERGAMOT *Monarda citriodora*.
PRAIRIE BULRUSH *Scirpus paludosus*.
PRAIRIE CLOVER *Lotus americanus* (Spanish Clover) Howes.
PRAIRIE CRABAPPLE *Malus ioensis*.
PRAIRIE CROCUS *Anemone patens* (Spreading Anemone) Johnston.
PRAIRIE FLAX *Linum lewisii* (Rocky Mountain Flax) USA Elmore.
PRAIRIE GENTIAN *Gentiana affinis* (Closed Gentian) Johnston.
PRAIRIE GROUND CHERRY *Physalis lanceolata*.
PRAIRIE LARKSPUR *Delphinium virescens*.
PRAIRIE MALLOW *Sidalcea malvaeflora*.
PRAIRIE MALLOW, White *Sidalcea candida*.
PRAIRIE PLUM *Prunus americana* (Canada Plum) USA Harper.
PRAIRIE POTATO, PRAIRIE TURNIP *Psoralea esculenta* (Breadroot) USA Havard.
PRAIRIE ROSE 1. *Rosa arkansana* USA Tolstead. 2. *Rosa humilis*. 3. *Rosa pratincola*. 4. *Rosa setigera*.
PRAIRIE ROCKET *Erysimum asperum* (Desert Wallflower) USA Elmore.
PRAIRIE SAGE *Artemisia gnaphalodes* (White Mugwort) USA Vestal & Schultes.
PRAIRIE SAGEWORT *Artemisia frigida* (Fringed Wormwood) USA Yarnell.
PRAIRIE SMOKE *Geum triflorum*.
PRAIRIE SWEET PEA *Lathyrus venosus*.
PRAIRIE WILLOW *Salix humilis*.
PRASE; PRATIE *Solanum tuberosum* (Potato) IOM Moore.1924 (Prase); Ire. Salaman (Pratie). Pratie is claimed to be Gaelic, but is almost certainly derived via "prata" from patata, the Spanish word that itself came from Haitian batata. The same applies to the Manx prase, and also pridda.
PRATTLING PARNELL *Saxifraga x urbinum* (London Pride) Wright.
PRAYER BEADS *Abrus precatorius* (Precatory Bean) Grieve.1931. Cf. Rosary Pea, an indication of their use.
PRAYER PLANT *Maranta kerchovena*. So called from the way the young leaves fold together at night.
PREACHER-IN-THE-PULPIT *Arum maculatum* (Cuckoo-pint) Som. Macmillan. One of a number of names that reflect the manner of growth - spadix in the spathe.
PRECATORY BEAN *Abrus precatorius*. 'Precatory' is from the Latin for to pray - Cf. Rosary Vine, and Prayer Beads for this.
PRECOCIOUS TREE *Prunus armeniaca* (Apricot) Dyer. Because it flowers and fruits earlier than the peach, hence Turner's Hasty Peach for the apricot.
PRETTY, Little-and- 1. *Agrotemma githago* (Corn Cockle) Som. Macmillan. 2. *Saxifraga x urbinum* (London Pride) Dev. Friend; Dor. B & H. An inappropriate name for corn cockle. Little-and-pretty seems to imply affection, in a cottage-garden sort of way, and affection would be the last emotion shown by country people for this plant.
PRETTY, London *Saxifraga x urbinum* (London Pride) Dor. Macmillan.
PRETTY, Nancy 1. *Agrostemma githago* (Corn Cockle) Som. Macmillan. 2. *Malcomia maritima* (Virginian Stock) Som. Elworthy. 3. *Saxifraga x urbinum* (London Pride) Glos, Staffs. B & H; Yks. Nicholson. 4. *Stellaria holostea* (Greater Stitchwort) Som. Grigson. 5. *Veronica chamaedrys* (Germander Speedwell) Som. Macmillan.
PRETTY, None-so- *Saxifraga x urbinum* (London Pride) Wright.
PRETTY, Small-and- *Malcomia maritima* (Virginian Stock) Som. Macmillan.
PRETTY-AND-LITTLE *Malcomia maritima* (Virginian Stock) Dev. Friend.
PRETTY BETSY 1. *Centranthus ruber* (Spur Valerian) Hudson. 2. *Saxifraga x urbinum* (London Pride) Suff. B & H.
PRETTY BETTY *Saxifraga x urbinum* (London Pride) Dor. Macmillan; Worcs. J Salisbury.

PRETTY LADY *Saxifraga x urbinum* (London Pride) Wilts. Macmillan.

PRETTY MAIDS *Saxifraga granulata* (Meadow Saxifrage) Berks. Dyer. Is this the Pretty Maids all in a row, from Mary, Mary quite contrary?

PRETTY NANCY 1. *Saxifraga spathularis x umbrosa* (London Pride) Som Macmillan. 2. *Stellaria holostea* (Greater Stitchwort) Som. Macmillan.

PRETTY WILLIE *Dianthus barbatus* (Sweet William) Som. Macmillan.

PRICK-BUSH; PRICK-HOLLIN; PRICK-HOLLY *Ilex aquifolium* (Holly) Lincs. Grigson (Prick-Bush); Lincs, Yks. Nicholson (Prick-Hollin, Prick-Holly).

PRICK-DEVIL *Scandix pecten-veneris* (Shepherd's Needle) Tynan & Maitland.

PRICK-MADAM 1. *Sedum acre* (Biting Stonecrop) Prior. 2. *Sedum album* (White Stonecrop) Prior. 3. *Sedum anglicum* (English Stonecrop) Coats. 4. *Sedum reflexum* (Yellow Stonecrop) Cumb. Grigson. 5. *Sempervivum tectorum* (Houseleek) Som. Macmillan. Apparently from French triquemadame, for traicque à madame, i.e. Lady's Treacle.

PRICK-MY-NOSE *Nigella damascena* (Love-in-a-mist) Friend.

PRICK-NEEDLE *Agrostemma githago* (Corn Cockle) Suss. Grigson, who reckoned that it must be a reference to the long teeth of the calyx.

PRICK-TIMBER; PRICK-WOOD 1. *Cornus sanguinea* (Dogwood) Gerard (Timber); B & H. (Wood). 2. *Euonymus europaeus* (Spindle Tree) Gerard (Timber); Som, Suss, Cumb. Grigson (Wood). The reference in both cases is to the use of the wood for skewers, hence a commoner form, Skewerwood.

PRICK-TREE *Cornus sanguinea* (Dogwood) B & H.

PRICK-TREE, Butcher's 1. *Euonymus europaeus* (Spindle Tree) B & H. 2. *Frangula alnus* (Alder Buckthorn) B & H. Again, from the use of the wood of both trees for skewers.

PRICK WILLOW *Hippophae rhamnoides* (Sea Buckthorn) Turner.

PRICKET *Sedum acre* (Biting Stonecrop) Gerard. See Prick-Madam.

PRICKLE PINE *Pinus pungens*.

PRICKLEBACK *Ranunculus arvensis* (Corn Buttercup) Yks. B & H. A reference to the spiny seed vessels.

PRICKLY AMARANTH *Amaranthus spinosus* (Thorny Piugweed) Dalziel.

PRICKLY ASH 1. *Aralia spinosa* (Hercules' Club) Everett. 2. *Xanthoxylum americanum*.

PRICKLY ASH, Northern *Xanthoxylum americanum* (Prickly Ash) W Miller.

PRICKLY ASH, Southern *Xanthoxylum clava-herculis*.

PRICKLY BEEHIVE *Dipsacus fullonum* (Teasel) Dev. Macmillan. 'Beehive', from the general shape of the flower heads.

PRICKLY BROOM 1. *Genista anglica* (Needle Whin) Gerard. 2. *Ulex europaeus* (Gorse) B & H.

PRICKLY CATERPILLAR *Amaranthus spinosus* (Thorny Pigweed) Barbados Gooding.

PRICKLY CEDAR *Juniperus oxycedrus* (Prickly Juniper) Grieve.1931.

PRICKLY CHRISTMAS *Ilex aquifolium* (Holly) Corn. Grigson.

PRICKLY COATS *Cirsium vulgaris* (Spear Plume Thistle) Dor. Macmillan.

PRICKLY COMFREY *Symphytum asperum*.

PRICKLY CUCUMBER *Cucumis metuliferus* (Horned Cucumber) Howes.

PRICKLY CURRANT *Ribes lacustre* (Swamp Currant) Douglas.

PRICKLY CYPRESS *Juniperus formosana* (Formosan Juniper) Mitchell.

PRICKLY ELDER *Aralia spinosa* (Hercules' Club) Grieve.1931.

PRICKLY GHOST *Ulex europaeus* (Furze) Dor. Macmillan. 'Ghost' here is 'Gorst', a common variant of gorse.

PRICKLY GLASSWORT *Salsola kali* (Saltwort) Prior. Like Salicornia, the plant

was burnt for its fixed salt, used in glass-making.
PRICKLY GOOSEBERRY *Ribes cynosbati*.
PRICKLY JUNIPER *Juniperus oxycedrus*.
PRICKLY MOSES *Acacia verticillata*. 'Moses' is presumably 'Mimosa'.
PRICKLY NIGHTSHADE *Solanum rostratum*.
PRICKLY PETTIGREE; PRICKLY PETTIGRUE *Ruscus aculeatus* (Butcher's Broom) Barton & Castle (Pettigree); Fernie (Pettigree).
PRICKLY PHLOX *Gilia californica*.
PRICKLY POPPY 1. *Argemone mexicana* (Mexican Poppy) USA Kingsbury. 2. *Papaver argemone* (Rough Long-headed Poppy) Blamey.
PRICKLY SALTWORT *Salsola kali* (Saltwort) Hutchinson.
PRICKLY SHIELD FERN, Hard *Polystichum aculeatum* (Hard Shield Fern) J A MacCulloch.1905.
PRICKLY SHIELD FERN, Soft *Polystichum setiferum* (Soft Shield Fern) J A MacCulloch.1905.
PRICKLY SOW THISTLE *Sonchus asper*.
PRICKLY THISTLE *Cirsium arvense* (Creeping Thistle) Yks. B & H.
PRICKSONG-WORT *Lunaria annua* (Honesty) Gerard.
PRICKWIND *Smilax aspera* (Greenbriar) Turner.
PRICKWOOD, Butcher's *Frangula alnus* (Alder Buckthorn) Prior. Because the wood was used for making skewers. Cf. Butcher's Prick-tree.
PRICKYBACK *Dipsacus fullonum* (Teasel) Lincs. Grigson. Cf. Johnny-prick-finger.
PRIDDA *Solanum tuberosum* (Potato) IOM Moore.1924. Evidently derived via 'prata' from patata, the original Haitian name.
PRIDE *Helianthus annuus* (Sunflower) Som. Macmillan. Because they stand so tall.
PRIDE, Aaron's *Saxifraga spathularis x umbrosa* (London Pride) Som. Macmillan.
PRIDE, Barbados *Caesalpina pulcherrima*.
PRIDE, Christmas *Ruellia paniculata*.
PRIDE, Devon *Centranthus ruber* (Spur Valerian) Dev. Macmillan.
PRIDE, Evening; PRIDE OF THE EVENING *Lonicera periclymenum* (Honeysuckle) both Dev, Dor. Macmillan.
PRIDE, Lady's *Cardamine pratensis* (Lady's Smock) Som. Macmillan.
PRIDE, London 1. *Artemisia abrotanum* (Southernwood) Wilts. Macmillan. 2. *Dianthus barbatus* (Sweet William) N'thants, Yks. B & H. 3. *Drosera rotundifolia* (Sundew) Som. Macmillan. 4. *Lychnis chalcedonica* (Jerusalem Cross) Glos. B & H. 5. *Saxifraga x urbinum*. 6. *Sedum acre* (Biting Stonecrop) Som. Elworthy. 'London', at least in the case of the plant we usually know as London Pride, is Mr London, an 18th century royal gardener, who introduced it. So perhaps it ought to be called London's Pride. Sweet William also has London Tufts, and London Bobs.
PRIDE, Lovers' *Polygonum persicaria* (Persicaria) Suss. Grigson.
PRIDE, Soldier's *Centranthus ruber* (Spur Valerian) Salisbury. Presumably the red colour dictates this kind of name.
PRIDE, Texan *Phlox drummondii* W Miller.
PRIDE, Venus's, Blue *Houstonia caerulea* (Bluets) W Miller.
PRIDE-O'-LONDON *Saxifraga x urbinum* (London Pride) IOM Moore.1924.
PRIDE OF CHINA *Melia azederach* (Chinaberry Tree) Mitton.
PRIDE OF INDIA 1. *Koelreuteria caerulea* (Golden Rain Tree) Barber & Phillips. 2. *Lagerstroemeria flos-reginae* (Queen Crapemyrtle) RHS. 3. *Melia azederach* (Chinaberry Tree) Hogarth.
PRIDE OF PERSIA *Melia azedarach* (Chinaberry Tree) Howes. Cf. Pride of India and Pride of China, but the specific name 'azedarach' is of Persian origin, suggesting that this version should have pride of place.
PRIDE OF SUSSEX *Phyteuma orbiculare* (Round-headed Rampion) Suss. Parish. Because it is confined to the chalk downs in the south.

PRIDE OF THE MEADOW *Filipendula ulmaria* (Meadowsweet) Som. Macmillan.

PRIDE OF THE THAMES *Butomus umbellatus* (Flowering Rush) Som, Dor. Macmillan.

PRIDE OF THE WOODS *Endymion nonscriptus* (Bluebell) Som. Macmillan.

PRIDEWEED *Erigeron canadensis* (Canadian Fleabane) USA Henkel.

PRIE *Ligustrum vulgare* (Privet) Tusser.

PRIEST *Dactylorchis maculata* (Heath Spotted Orchid) Yks. Grigson. Probably a truncated version of something longer. Cf. Priest-in-the-pulpit, etc., for *Arum maculatum*.

PRIEST-AND-PULPIT; PRIEST-IN-THE-PULPIT *Arum maculatum* (Cuckoo-pint) both Som. Macmillan. Two of a number of double names suggested by the spadix in the spathe of Cuckoo-pint.

PRIEST'S CROWN *Taraxacum oifficinale* (Dandelion) Turner. Probably suggested by the bald patch left when the seeds have all blown away, like a tonsure. Cf. French couronne de prêtre.

PRIEST'S HOOD; PRIEST'S PILLY *Arum maculatum* (Cuckoo-pint) Lancs. B & H (Priest's Hood); West Grigson (Priest's Pilly). Cf. Parson's Pillycods with the second name. They both mean penis, and are but two of a number of erotic names suggested by the shape of the spadix. Priest's Hood, though, seems to owe its imagery to asexual matters. Cf. Parson-in-the-pulpit.

PRIEST'S PINTLE 1. *Arum maculatum* (Cuckoo-pint) Gerard. 2. *Orchis mascula* (Early Purple Orchid) B & H. 3. *Sedum roseum* (Rose-root) Banff. Grigson. 'Pintle' is the same as the 'pint' of Cuckoo-pint. Cf. Parson's Pillycods.

PRIESTIES *Arum maculatum* (Cuckoo-pint) Lancs. B & H. 'Pintle' or 'Pillycods' is understood, but there are other "priestly" names that do not have this connotation.

PRIM; PRIMET *Ligustrum vulgaris* (Privet) Parkinson (Prim) Prior (Primet).

PRIMAROSE; PRIMMY ROSE *Primula vulgaris* (Primrose) Som. Reeves (Primarose); Glos. J D Robertson (Primmy Rose).

PRIMMILY *Primula auricula* (Auricula) Suff. Moor. Primula, presumably.

PRIMP; PRIMPRINT *Ligustrum vulgare* (Privet) Lincs. Peacock (Primp); Turner (Primprint).

PRIMROSE 1. *Bellis perennis* (Daisy) Lupton. 2. *Ligustrum vulgare* (Privet) B & H. 3. *Oenothera pumila* (Small Sundrops) USA Corlett. 4. *Primula vulgaris*. Primrose is Latin prima, feminine of primus, first. The diminutive of this is primula. Medieval French turned this into primerole, and then into primrose. That still sounds substantially as imported, in the West country Primarose and Primmy Rose. Privet arrives at the name through its series Prim, Primprint and Primwort.

PRIMROSE, Bird's-eye *Primula farinosa*.

PRIMROSE, Cape *Streptocarpus sp*.

PRIMROSE, Canadian, Dwarf *Primula mistassini*.

PRIMROSE, Chinese *Primula sinensis*.

PRIMROSE, Evening 1. *Euphorbia hirta* Kourennoff. 2. *Oenothera sp*.

PRIMROSE, Evening, Fragrant *Oenothera stricta*.

PRIMROSE, Evening, Golden *Oenothera brevipes*.

PRIMROSE, Evening, Large *Oenothera erythrosepala*.

PRIMROSE, Evening, Least *Oenothera parviflora*.

PRIMROSE, Evening, Lesser *Oenothera biennis*.

PRIMROSE, Evening, Northern *Oenothera muricata*.

PRIMROSE, Evening, Small-flowering *Oenothera cruciata*.

PRIMROSE, Evening, White 1. *Oenothera breviflora*. 2. *Oenothera speciosa*.

PRIMROSE, Fairy *Primula malacoides* Nicolaisen.

PRIMROSE, French *Primula veris x vulgaris* (False Oxlip) Som. Macmillan.

PRIMROSE, Jerusalem *Pulmonaria officinalis* (Lungwort) Tynan & Maitland. Cf. Jerusalem Cowslip, as well as Jerusalem Sage.
PRIMROSE, Least *Primula minima*.
PRIMROSE, Mealy *Primula farinosa* (Bird's-eye Primrose) Hutchinson.
PRIMROSE, Natal *Thunbergia atriplicifolia* S Africa Watt.
PRIMROSE, Poison *Primula obconica* USA Cunningham. It can give some people dermatitis.
PRIMROSE, Scotch *Primula scotica*.
PRIMROSE, Tree *Oenothera biennis* (Lesser Evening Primrose) Grieve.1931.
PRIMROSE PEARLS *Narcissus maialis var patellaris* (Pheasant's Eye) Som, Wilts. Macmillan.
PRIMROSE PEERLESS *Narcissus x biflorus*.
PRIMROSE SOLDIERS *Aquilegia vulgaris* (Columbine) Wilts. Dartnell & Goddard.
PRIMWORT *Ligustrum vulgare* (Privet) B & H. Cf. Prim, Primprint,and Primrose.
PRINCE ALBERT YEW *Saxegothea conspicua*.
PRINCE ALBERT'S FIR *Tsuga heterophylla* (Western Hemlock) Brimble.1948.
PRINCE OF WALES'S FEATHERS 1. *Centranthus ruber* (Spur Valerian) Dor. Macmillan. 2. *Syringa vulgaris* (Lilac) Dev. Macmillan.
PRINCE OF WALES'S FEATHERS TREE *Brachystegia boehmii*. In this case, the name is given from the shape of the freshly burst leaf buds.
PRINCE'S FEATHERS 1. *Amaranthus hypondriachus*. 2. *Amaranthus retroflexus* (Pigweed) Australia Watt. 3. *Gynerium argenteum* (Pampas Grass) Dev. Friend. 4. *Leucanthemum vulgare* (Ox-eye Daisy) Tynan & Maitland. 5. *Onobrychis sativa* (Sainfoin) Wilts. Tynan & Maitland. 6. *Polygonum orientale* USA Zenkert. 7. *Potentilla anserina* (Silverweed) Som. Macmillan. 8. *Prunella vulgaris* (Self-heal) N'thum. Grigson. 9. *Saxifraga x urbinum* (London Pride) Dev. Friend.1882. 10. *Syringa vulgaris* (Lilac) Dev, Rut. B & H. 11. *Umbilicus rupestris* (Wall Pennywort) Yks. Grigson. All descriptive in some way or another, some more obvious than others.
PRINCE'S FLOWER *Calystegia soldanella* (Sea Bindweed) Eriskay Freethy. The local legend is that the Young Pretender brought the seeds with him from France in 1745.
PRINCE'S PINE *Chimaphila umbellata* (Umbellate Wintergreen) USA Corlett.
PRINCE'S PLUME *Stanleya pinnata*.
PRINCE'S ROCK-CRESS *Arabis pulchra*.
PRINCESS BEAN *Psophocarpus tetragonolobus* (Goa Bean) Douglas.
PRINCESS LEAF *Mirabilis dichotoma* (Marvel of Peru) Malaya Friend.
PRINCESS TREE *Paulownia tomentosa* (Foxglove Tree) RHS. 'Princess', for the generic name is for the daughter of Czar Paul I, Anna Paulownia.
PRINCESS'S ROBE *Narcissus pseudo-narcissus* (Daffodil) Som. Macmillan.
PRINCEWOOD *Cordia gerascanthus* the timber Willis.
PRINT *Ligustrum vulgare* (Privet) Gerard.
PRINT, Butter; PRINT, Pie *Abutilon avicennae* (China Jute) USA Bergen. The pods were used to stamp butter or pie-crust; so, according to Mabey, were the leaves, but that does not sound very feasible.
PRINT PINAFORE *Geranium robertianum* (Herb Robert) Dor. Macmillan. Also in the West country, there are Baby's, or Dolly's Pinafore, and Dolly's Apron.
PRIPET *Ligustrum vulgare* (Privet) Som. Macmillan.
PRIVET *Ligustrum vulgare*. Of unknown origin, according to the dictionary, but certainly related to primprint, an equally early name, and its abbreviations Prim, Primp and Print.
PRIVET, China *Lagerstroemeria indica*.
PRIVET, Chinese *Ligustrum lucidum* (Wax Tree) Wilkinson.1981.
PRIVET, Egyptian *Lawsonia inermis* (Henna) Westermarck.

PRIVET, Glossy *Ligustrum lucidum* (Wax Tree) Tampion.
PRIVET, Golden *Ligustrum ovalifolium* (Japanese Privet) Howes.
PRIVET, Japanese *Ligustrum ovalifolium*.
PRIVIT, Pipe *Syringa vulgaris* (Lilac) Tynan & Maitland. Pipe-tree, Blow-pipe tree, and Blue Pipe are all early names for lilac.
PRIVY; PRIVY SAUGH *Ligustrum vulgare* (Privet) Shrop, War, Suff, Yks, Scot. B & H Ches. Holland (Privy); Yks. Grigson Scot. Jamieson (Privy Saugh).
PROCESSION FLOWER *Polygala vulgaris* (Milkwort) Gerard. The procession of the name is the Rogation-tide perambulation, in which the flower featured strongly. Cf. the names Cross-flower, Gang-flower and Rogation-flower. Gerard comments that milkwort is the plant "of which floures the maidens which use in the countries to walke the Procession do make themselves garlands and Nosegaies".
PROFOLIUM *Armeria maritima* (Thrift) Hants. Grigson.
PROMOTION-NUT *Anacardium occidentale* nut (Cashew Nut) Willis.
PROPELLERS 1. *Acer campestre* seeds (Field Maple) Som. Macmillan. 2. *Acer pseudo-platanus* seeds (Sycamore) Som. Macmillan. The description of the winged seed vessels is obvious in the name.
PROPHET FLOWER *Arnebia pulchra*.
PROSO MILLET *Panicum miliaceum* (Millet) Brouk. Proso is the Russian word for millet.
PROUD CARPENTER *Prunella vulgaris* (Self-heal) Ches. Holland. Cf. Carpenter's Grass, Carpenter's Herb, etc. The corolla is shaped rather like a billhook, and from the doctrine of signatures the plant was supposed to heal wounds from edged tools.
PROVENCAL ROSE *Rosa centifolia* (Cabbage Rose) F Savage.
PROVENCE ROSE, Yellow *Rosa hemisphaerica* (Sulphur Rose) Thomas. But it bears no resemblance to the true Provence Rose.

PROVINS, Rose of *Rosa gallica* (French Rose). A rose of the warmer parts of Europe, Asia Minor and the Caucasus, but it was brought by Thibaud le Chansonnier from Palestine to Provence in the 13th century. Or is this the town of Provins, some 60 miles southeast of Paris?
PRUNE-LEAVED JUNEBERRY; PRUNE-LEAVED SERVICE-BERRY; PRUNE-LEAVED SHADBERRY *Amelanchier prunifolia*. Prune-leaved Service-berry is the usual name; the other two are both USA Elmore.
PRUNE TREE *Prunus domestica* (Plum) Lindley.
PRUNELL *Prunella vulgaris* (Self-heal) Gerard.
PRUSHIA *Sinapis arvensis* (Charlock) Ire. Drury.
PRUSSIAN ASPARAGUS *Ornithogalum pyrenaicum* (Spiked Star of Bethlehem) B & H. Cf. French Asparagus. Better known as Bath Asparagus, for it is quite common around Bath, where it is sold to be eaten like asparagus.
PRY *Tilia cordata* (Small-leaved Lime) Ess. Rackham.1986. He uses this name as if it were the common one.
PUBLICANS *Caltha palustris* (Marsh Marigold) Oxf, Yks. Grigson. Publicans and sinners are *Caltha palustris* and *Ranunculus* species growing alongside each other. It is too facile to associate this name with the commoner Drunkards.
PUCCOON 1. *Lithospermum angustifolium*. 2. *Sanguinaria canadensis* (Bloodroot) USA Speck. Apparently the Virginian Indian name.
PUCCOON, Hoary *Lithospermum canescens*.
PUCCOON, Red *Sanguinaria canadensis* (Bloodroot) USA Kingsbury.
PUCCOON, Spring *Lithospermum canescens* (Hoary Puccoon) Perry.
PUCCOON, Yellow *Lithospermum canescens* (Hoary Puccoon) Howes.
PUCELLAGE *Vinca minor* (Lesser Periwinkle) Fernie. i.e. virgin flower. But why?

PUCK NEEDLE 1. *Agrostemma githago* (Corn Cockle) Suss. Grigson. 2. *Scandix pecten-veneris* (Shepherd's Needle) Som. Grigson; Hants. Cope. The fairies' needle, of course, and a reference to the long, pointed seed vessels. 'Needle' names for cockle probably refer, Grigson says, to the long teeth of the calyx.

PUCKER NEEDLE *Scandix pecten-veneris* (Shepherd's needle) Tynan & Maitland. A variation on Puck Needle; so are Pook-, Poke-, and Pink-needle.

PUDDING *Melandrium rubrum* (Red Campion) Som. Grigson. They are also known as Plum Puddings in the same area.

PUDDING, Black *Typha latifolia* (False Bulrush) Som. Macmillan; IOW Long. Descriptive, of course.

PUDDING, Gooseberry *Epilobium hirsutum* (Great Willowherb) Suss. Grigson. There are all sorts of pies and puddings for willowherb, all probably stemming from the claim that the plant smells like apples when it is crushed. Perhaps it does, but it is probably Gerard's Codded Willowherb that is the origin, someone mistaking 'codded' for 'codlin'.

PUDDING, Plum 1. *Epilobium hirsutum* (Great Willowherb) Ches. Holland. 2. *Melandrium album*. 3. *Melandrium rubrum* (Red Campion) Som. Macmillan.

PUDDING-BERRY *Cornus canadensis* (Bunchberry) W Miller. Does this mean what it says?

PUDDING-GRASS; PUDDING-HERB *Mentha pulegium* (Pennyroyal) Turner (Pudding-grass); Yks. F K Robinson (Pudding-herb). Pennyroyal is used to make stuffings for meat, and stuffings used to be called puddings.

PUDDING-PIPE TREE *Cassia fistula* (Indian Laburnum) Coles. "Because the Cod is like a Pudding", in Coles's words.

PUFF CLOCKS *Taraxacum officinale* the seed heads (Dandelion) Som. Macmillan.

PUFF CLOVER *Trifolium fucatum*.

PUG-IN-A PINNER *Primula vulgaris 'elatior'* (Polyanthus) Coats.

PUGLEY'S MARSH ORCHID *Dactylorchis traunsteineri*.

PUKAPUKA *Brachyglottis repanda* (Rangiora) RHS.

PUKEWEED *Lobelia inflata* (Indian Tobacco) USA Henkel. Cf. Gagroot, Vomitwort, and Emetic-weed. The dried leaves and tops are used as an expectorant, and it acts upon the nervous system and bowels, causing vomiting.

PULIALL-MOUNTAIN *Thymus drucei* (Wild Thyme) Gerard. A corruption of serphyllum montanum, according to Prior. But perhaps it was pulegium montanum, an old name. Cf. also Pell-a-mountain, and Penny Mountain.

PUKOTEA *Laurelia novae-zelandiae*.

PULIALL-ROYAL *Mentha pulegium* (Pennyroyal) Gerard. Pennyroyal is a corruption of this name, which is from Latin pulices, fleas, because it is good for destroying them. The 'royal' part of the name, it is said, came from the fact that they were used in royal palaces for just this purpose.

PULL-DAILIES *Dactylorchis incarnata* (Early Marsh Orchid) Scot. B & H. Balderry is the type name in Scotland for this kind of orchid. Pull-dailies is a variation of a corruption of that name, Bull-daisy.

PULL-POKER *Typha latifolia* (False Bulrush) Ess. Gepp.

PULPIT, Devil-in-the- *Tradescantia virginica* (Spiderwort) Perry. Cf. Moses-in-the-bulrushes. Presumably they are both reflections on the way the flower hides itself in the foliage.

PULPIT, Jack-in-the- 1. *Arisaema triphyllum*. 2. *Arum maculatum* (Cuckoo-pint) Corn, Som, Lincs. Grigson. In both cases, the spadix is the jack, and the spathe the pulpit.

PULPIT, Lamb-in-a-; PULPIT, Man-in-the-; PULPIT, Old Man's *Arum maculatum* (Cuckoo-pint) Dev. Friend.1882; Wilts. Grigson (Lamb-in-a-pulpit); Som. Macmillan (both Man-in-the-pulpit and Old Man's

Pulpit). The spathe is the pulpit, and the spadix is the incumbent.

PULPIT, Parson-in-the- 1. *Aconitum anglicum* (Wild Monkshood) B & H. 2. *Aconitum napellus* (Monkshood) Dev. Friend.1882. 3. *Arisaema triphylla* (Jack-in-the-pulpit) Whittle & Cook. 4. *Arum maculatum* (Cuckoo-pint) Dev. Friend.1882. For cuckoo-pint Cf. Parson-and-clerk, etc.

PULPIT, Preacher-in-the-; PULPIT, Priest-and-; PULPIT, Priest-in-the- *Arum maculatum* (Cuckoo-pint) all Som. Macmillan.

PUMMELO *Citrus maxima* (Wild Pomelo) Brouk. Both names are probably a shortening of the Dutch Pompelmous, Pompelnous, from Dutch pomp, melon plus limoes, a Malayan borrowing from Portuguese Limão, lemon.

PUMMY, Lamb's *Alliaria petiolata* (Jack-by-the-hedge) Som. Grigson.

PUMPERNEL *Anagallis arvensis* (Scarlet Pimpernel) Som. Macmillan.

PUMPION *Cucumis sativus* (Cucumber) Ellacombe.

PUMPKIN 1. *Cucurbita moschata* (Crook-neck Squash) Whitaker & Davis. 2. *Cucurbita pepo* (Vegetable Marrow) USA Corlett. Pumpkin is Old French pompon, hence forms like Pompion and Pumpion. All of them derive from Latin pepo, Greek pepon.

PUMPKIN, American *Cucurbita foetidissima* (Buffalo Gourd) Bianchini.

PUMPKIN, Chinese *Benincasa hispida* (Chinese Wax Gourd) Howes.

PUMPKIN, Elephant *Cucurbita maxima* (Giant Pumpkin) W Miller.

PUMPKIN, Field, Yellow *Cucurbita maxima* (Giant Pumpkin) Watt.

PUMPKIN, Giant *Cucurbita maxima*.

PUMPKIN, Melon *Cucurbita maxima* (Giant Pumpkin) Watt.

PUMPKIN, Seminole *Cucurbita moschata* (Crook-neck Squash) USA Howes. The Seminole Indians lived in the far south of Florida.

PUMPKIN, White *Lagenaria siceraria* (Calabash) Dalziel.

PUMPKIN, Wild *Cucurbita foetidissima* (Buffalo Gourd) USA Howes.

PUMPY, Water *Veronica beccabunga* (Brooklime) Grieve.1931. Water Pimpernel, evidently. It went through Water Purpie and Water Purple before reaching this name.

PUNCTURE VINE *Tribulus terrestris*. Because of the spiny fruits, sharp enough to puncture a bicycle tyre if wheeled over land where the weed is prevalent.

PUPPY DOG'S MOUTH *Linaria vulgaris* (Toadflax) Wilts. Wright. Cf. Dog's Mouth.

PUR-APPLE *Pinus sylvestris* cones (Pine cones) N'thants. A E Baker. It seems to be a cross between Pin-apple and Fir-apple, both from the same area, but there is also Pie-apple.

PURCELLAINE *Portulaca sativa* (Purslane) Turner. Closer to the original than the modern name, for it came through Old French porcelaine into English.

PURGE, Jalapa *Ipomaea purga*. Jalapa is the name of a Mexican city in the general area in which this plant grows.

PURGE NUT *Jatropha curcas* (Barbados Nut) USA Kingsbury. Cf. Physic Nut. It is the oil from the nut that is used as a purgative, though too much is toxic. The leaf too is drastic.

PURGING BUCKTHORN *Rhamnus cathartica* (Buckthorn) North. The berries make a powerful purgative, dangerous to children. "It is a rough purge", said Hill, "but a very good one".

PURGING CASSIA *Cassia fistula* (Indian Laburnum) Willis. Cassia pod is the purgative, and a root decoction is also made that is rigorous.

PURGING FLAX *Linum catharticum* (Fairy Flax) Fernie. Cf. Cathartic Flax. It is certainly an effective purge, though, like many another herb, thoroughly dangerous to use.

PURGING ROOT *Euphorbia corollata* (Flowering Spurge) Le Strange. An American plant, used by some of the Indians as a laxative. Anyway, spurge actually means a purging plant (Latin expurgare).

PURGING THORN *Rhamnus cathartica* (Buckthorn) Parkinson.
PURIFICATION FLOWER *Galanthus nivalis* (Snowdrop) Prior. The snowdrop was dedicated to the Purification of the Virgin, otherwise Candlemas, 2 February, when it was the custom for young women dressed in white to walk in procession at the feast. Hence, too, Candlemas Bells, Fair Maids of February, shortened to Fair Maids.
PURPIE, Water; PURPLE, Water *Veronica beccabunga* (Brooklime) both Scot. Jamieson (Purpie); Grigson (Purple).
PURPLE-BERRY *Sambucus nigra* (Elder) Som. Macmillan.
PURPLE-BUSH *Halliophytum hallii*.
PURPLE-GRASS 1. *Lythrum salicaria* (Purple Loosestrife) B & H. 2. *Medicago arabica* (Spotted Medick) B & H.
PURPLE-TOP *Verbena bonariensis* (South American Vervain) H & P.
PURPLES; PURPLES, Long *Orchis mascula* (Early Purple Orchis) B & H. Shakespeare used Long Purples for this orchid.
PURRET *Allium porrum* (Leek) Prior. From the Latin porrum, possibly through French porrette.
PURSE *Rhinanthus crista-galli* (Yellow Rattle) Som. Macmillan.
PURSE, Gentleman's *Capsella bursa-pastoris* (Shepherd's Purse) Som. Macmillan.
PURSE, Lady's 1. *Anemone nemerosa* (Wood Anemone) Dor. Grigson. 2. *Aquilegia vulgaris* (Columbine) Dor. Macmillan. 3. *Calceolaria sp* (Slipperwort) Ches. Holland. 4. *Capsella bursa-pastoris* (Shepherd's Purse) Glos. PLNN35; E Angl, Berw. Grigson. 5. *Dicentra spectabilis* (Bleeding Heart) Som. Macmillan.
PURSE, Old Woman's *Impatiens noli-me-tangere* (Touch-me-not) Cumb. Grigson.
PURSE, Pick *Capsella bursa-pastoris* (Shepherd's Purse) Gerard. Prior explains this by saying that it robs the farmer by stealing the goodness of his land. Cf. Pickpocket.
PURSE, Poor Man's; PURSE, Poverty *Capsella bursa-pastoris* (Shepherd's Purse) Som. Macmillan (Poor Man); Lincs. Grigson (Poverty).
PURSE, Shepherd's 1. *Capsella bursa-pastoris*. 2. *Lotus corniculatus* (Bird's-foot Trefoil) Som. Macmillan. 3. *Lunaria annua* (Honesty) Som. Macmillan. 4. *Polygala vulgaris* (Milkwort) Wilts. D Grose. 5. *Rhinanthus crista-galli* (Yellow Rattle) Som. Macmillan. Cumb. Grigson. In most cases, the name depends on the shape of the seed pods.
PURSE, Two-pennies-in-a- *Lunaria annua* (Honesty) Coats.
PURSE, Wild; PURSE FLOWER *Capsella bursa-pastoris* (Shepherd's Purse) both Som. Macmillan.
PURSH PLANTAIN *Plantago purshii*.
PURSH RIBGRASS *Plantago purshii* (Pursh Plantain) USA Elmore.
PURSLAIN; PURSLAND *Portulaca sativa* (Purslane) both S Africa Watt.
PURSLANE *Portulaca sativa*. From Old French porcelaine, and the Latin portulaca, literally "carrying milk", a reference to the milky sap.
PURSLANE, Garden; PURSLANE, Golden *Portulaca sativa* (Purslane) Grieve.1931 (Garden); Clair (Golden).
PURSLANE, Green *Portulaca oleracea*.
PURSLANE, Hampshire *Ludwigia palustris*. Very rare in this country, and only to be found on wet ground in the New Forest.
PURSLANE, Horse *Trianthema pentandra*.
PURSLANE, Iceland *Koenigia islandica*.
PURSLANE, Milk 1. *Euphorbia hypericifolia*. 2. *Euphorbia supina*.
PURSLANE, Pink *Claytonia alsinoides*.
PURSLANE, Pot *Portulaca oleracea* (Green Purslane) McClintock.1978.
PURSLANE, Rock *Calandrina umbellata*.
PURSLANE, Sea 1. *Atriplex halimus* (Tree Purslane) Moldenke. 2. *Halimione portulacoides*. 3. *Honkenya peploides* (Sea Sandwort) Murdoch McNeill.
PURSLANE, Sea, Spanish *Atriplex halimus* (Tree Purslane) Moldenke.

PURSLANE, Tree 1. *Atriplex halimus*. 2. *Portulacaria afra*.
PURSLANE, Water *Peplis portula*.
PURSLANE, White *Euphorbia corollata* (Flowering Spurge) Le Strange.
PURSLANE, Wild *Euphorbia peplis* (Purple Spurge) B & H.
PURSLANE, Wild, Red *Portulaca pilosa*.
PURSLANE, Winter *Claytonia perfoliata* (Spring Beauty) USA Grigson.
PURSLANE SPEEDWELL *Veronica peregrina* (American Speedwell) USA Upham.
PURSLEY 1. *Portulaca oleracea* (Green Purslane) USA H M Hyatt. 2. *Portulaca sativa* (Purslane) USA Gates.
PUSLEY 1. *Portulaca quadrifida* S Africa Watt. 2. *Portulaca sativa* (Purslane) USA Upham.
PUSLEY, Chinese *Heliotropum curassavicum* (Seaside Heliotrope) USA T H Kearney.
PUSSIES *Typha latifolia* (False Bulrush) Som. Macmillan.
PUSSIES, Golden *Salix capraea* catkins (Sallow catkins) Som. Macmillan.
PUSSLEY 1. *Portulaca oleracea* (Green Purslane) USA Elmore. 2. *Portulaca sativa* (Purslane) USA Gates.
PUSSY CAT'S TAIL *Corylus avellana* catkins (Hazel catkins) Dev. B & H.
PUSSY CATS 1. *Corylus avellana* catkins (Hazel catkins) Wilts. Dartnell & Goddard; Berks. Lowsley. 2. *Salix caprea* catkins (Goat Willow catkins) War. F Savage.
PUSSY EARS *Kalanchoe tomentosa* (Donkey's Ears) RHS.
PUSSY FACE *Viola tricolor* (Pansy) Som. Macmillan.
PUSSY FOOT *Trifolium repens* (White Clover) Som. Macmillan.
PUSSY TOES *Antennaria neglecta* USA Zenkert. Cf. Cat's Foot for this.
PUSSY TOES, Plantain-leaved *Antennaria plantaginifolia* (Plantain-leaved Everlasting) USA Youngken.
PUSSY WILLOW 1. *Salix caprea* (Sallow) Friend. 2. *Salix discolor* (American Black Willow) USA Upham.

PUSSY'S TAIL *Typha latifolia* (False Bulrush) Som. Macmillan.
PUSSY'S TOES *Antennaria dioica* (Cat's Foot) USA Leighton.
PUSTULOUS MOSS *Umbilicaria pustulata*.
PUZZLE, Love-in-a- *Nigella damascena*.
PUZZLE, Monkey *Araucaria araucana*. Said to have arisen from a remark by Charles Austin, during the ceremonial planting of one of these trees in the gardens at Pencarrow, in Cornwall, in 1834 - "it would be a puzzle for a monkey to climb that tree".
PUZZLE, Solomon's *Sedum telephium* (Orpine) London Grigson. It was the London flower sellers' name for the plant.
PUZZLE-MONKEY *Araucaria araucana* (Monkey-puzzle) Som. Elworthy. See Monkey-Puzzle.
P W D TREE *Ceiba pentandra* (Kapok Tree) Malaya Gilliland. Because of its peculiar telegraph-pole shape; P W D stands for Public Works Department.
PYANOT *Paeonia mascula* (Peony) Lancs. Nodal. Pronounced, at least round Sheffield, with a long 'a'. There are many variations of Peony, and this is not in any way extreme. What about Pie-nanny, for example?
PYGMY-WEED *Tillaea erecta*.
PYRAMIDAL ORCHID *Anacamptis pyramidalis*.
PYRAMIDAL SAXIFRAGE *Saxifraga cotyledon*.
PYRENEAN ASPHODEL *Asphodelus albus* var. *pyrenaeus*.
PYRENEAN AVENS *Geum pyrenaicum*.
PYRENEAN BELLFLOWER *Campanula speciosa*.
PYRENEAN BROOM *Cytisus purgans*.
PYRENEAN BUTTERCUP 1. *Ranunculus amplexicaulis*. 2. *Ranunculus pyrenaeus*.
PYRENEAN COLUMBINE *Aquilegia pyrenaica*.
PYRENEAN CRANESBILL *Geranium pyrenaicum* (Mountain Cranesbill) McClintock.
PYRENEAN GENTIAN *Gentiana pyrenaica*.

PYRENEAN HEMP NETTLE *Galeopsis pyrenaica*.
PYRENEAN HONEYSUCKLE *Lonicera pyrenaica*.
PYRENEAN LILY *Lilium pyrenaicum*.
PYRENEAN OAK *Quercus pyrenaica*.
PYRENEAN PHEASANT'S EYE *Adonis pyrenaica*.
PYRENEAN PINE *Pinus nigra var. cedennensis*.
PYRENEAN POPPY *Papaver suaveolens*.
PYRENEAN SAXIFRAGE 1. *Saxifraga longifolia*. 2. *Saxifraga umbrosa*.
PYRENEAN SNAKESHEAD *Fritillaria pyrenaica*.
PYRENEAN SQUILL *Scilla liliohyacinthus*.
PYRENEAN THISTLE *Carduus carlinoides*.
PYRENEAN TOADFLAX *Linaria pyrenaica*.
PYRENEAN VALERIAN *Valeriana pyrenaica* (Giant Valerian) Clapham.
PYRETHRUM FLOWER *Pyrethrum cinerarifolium*.
PYRRIE *Pyrus communis* (Choke Pear) Leyel.1937. A variation of the generic name.
PYRUL, Woolly *Phaseolus mungo* (Black Gram) Brouk.
PYTHAGOREAN BEAN *Nelumbo nucifera* (Hindu Lotus) Leyel.1948.

Q

QUACK *Agropyron repens* (Couch-grass) H C Long.1910. In this case, 'quack' is a variant of 'quick', OE cwice, which gave quitch, or squitch, a common name for couch.

QUACKSALVER'S SPURGE; QUACKSALVER'S TURBITH *Euphorbia esula* (Hungarian Spurge) both W Miller. Quacksalver means the same as the modern quack (one who quacks about his salves). Turbith seems to convey the general idea of a medicinal root. That name was originally applied to one of the Ipomaeas of the East, and the whole thing seems to imply a rough-and-ready purge given by someone who did not really know what he was doing.

QUAIL GRASS *Celosia argentea* Howes.

QUAIL PLANT *Heliotropum curassavicum*.

QUAKE GRASS; QUAKER GRASS *Briza media* (Quaking-grass) Suss, N'thants. Grigson.1959 (Quake-grass); Worcs. Halliwell; Yks. F K Robinson (Quaker-grass).

QUAKER BUTTONS *Strychnos nux-vomica* (Snakewood) Le Strange.

QUAKER LADY 1. *Houstonia caerulea* (Bluets) Howes. 2. *Spiraea latifolia* USA House.

QUAKERS 1. *Arum maculatum* (Cuckoo-pint) Lancs. B & H. 2. *Briza media* (Quaking-grass) Wilts. Dartnell & Goddard; Glos, Lincs. Grigson.1959; Heref. Havergal. An odd name for Cuckoo-pint, given, so the authority says, when the spadices are dull-coloured. But for the grass, Cf. Shakers.

QUAKERS, Cow; QUAKES, Cow *Briza media* (Quaking-grass) Som, Dor, Ches, N'thum. Grigson.1959.

QUAKES, Earth *Briza media* (Quaking-grass) N'thants. Grigson.1959.

QUAKIN ESP *Populus tremula* (Aspen) N Ire. Grigson.

QUAKING ASH *Populus tremula* (Aspen) Scot. Jamieson.

QUAKING ASPEN *Populus tremuloides* (American Aspen) Wit.

QUAKING-GRASS *Briza media*. Wiltshire children used to be told that if ever the spikelets stop trembling they will change into silver sixpences or shillings.

QUAMASH *Camassia esculenta*.

QUARTERS *Thespesia garckeana* (Rhodesia Tree Hibiscus) Palgrave.

QUARTERVINE *Bignonia capreolata* (Cross-vine) H & P.

QUASH *Cucumis melo* (Melon) Halliwell.

QUAT-VESSEL *Cirsium vulgare* (Spear-plume Thistle) Hants. Cope. 'Quat' is an old word for a pimple. Presumably, there must have been a lotion, now forgotten, made from thistles, for skin eruptions.

QUATER *Vinca major* (Greater Periwinkle) RHS.

QUEBEC HAWTHORN *Crataegus submollis*.

QUEBEC OAK *Quercus alba* (White Oak) Grieve.1931.

QUEBRACHO *Schinopsis lorentzii*. Literally, axe-breaker, from Spanish quebrar, to break, and hacha, axe. It is a reference to the exceedingly hard wood, one of the hardest and heaviest known.

QUEEN, Fairy *Viola tricolor* (Pansy) Som. Macmillan.

QUEEN, Meadow *Filipendula ulmaria* (Meadowsweet) Perth. B & H. Or Queen of the Meadow.

QUEEN, White *Galanthus nivalis* (Snowdrop) Som. Macmillan.

QUEEN CRAPEMYRTLE *Lagerstroemia flos-reginae*.

QUEEN OF THE MIST *Saxifraga x urbinum* (London Pride) Dor. Macmillan.

QUEEN OF THE WOOD *Ilex aquifolium* the smooth-leaved kind (Holly) Ire. Kinahan. Evidence of a dedication to the Virgin Mary. Legend has it that St Maelrubha introduced "the sacred smooth-leaf holly to out-rival the Druidical oaks", around Loch Maree, and dedicated them to the Virgin.

QUEEN STOCK *Matthiola incana* (Stock) Turrill.

QUEEN'S CUSHION 1. *Saxifraga hypnoides* (Dovedale Moss) Dur. Grigson. 2. *Sedum acre* (Biting Stonecrop) Scot. Tynan & Maitland. 3. *Viburnum opulus* (Guelder Rose) Heref. Leather. Descriptive either of the method of growth, as with the Saxifrages and Sedums, or the flower balls, as with Guelder Rose.

QUEEN'S DELIGHT *Stillingia treculeana*.

QUEEN'S FEATHER 1. *Filipendula ulmaria* (Meadowsweet) Som. Macmillan. 2. *Saxifraga x urbinum* (London Pride) Ches. B & H. 3. *Syringa vulgaris* (Lilac) Dor. Dacombe.

QUEEN'S FINGERS 1. *Dactylorchis maculata* (Heath Spotted Orchid) War. Grigson. 2. *Orchis morio* (Green-winged Orchid) War. Grigson. The reference here is to the roots - not the twin tubers of members of the genus Orchis (in spite of the inclusion of *O morio*), but the fingers of the tubers of Dactylorchis. Cf. King's Fingers, and Ringfingers.

QUEEN'S FLOWER *Lagerstroemeria flos-reginae* (Queen Crapenyrtle) Howes.

QUEEN'S GILLIFLOWER *Hesperis matronalis* (Sweet Rocket) Gerard. Presumably as a compliment to Queen Elizabeth, as it was very popular in the 16th century.

QUEEN'S PINCUSHION *Viburnum opulus* (Guelder Rose) N'thants. A E Baker. Cf. Queen's Cushion.

QUEEN'S ROGUES *Hesperis matronalis* (Sweet Rocket) Tynan & Maitland. Cf. Rogue's Gilliflower, perhaps 'rouge' originally, i.e. Red Gilliflower.

QUEEN'S ROOT *Stillingia treculeana* (Queen's Delight) Grieve.1931.

QUEEN'S TEARS *Billbergia nutans*.

QUEEN'S WREATH *Petraea volubilis* (Purple Wreath Vine) West Indies Perry.

QUEEN ANNE'S DAFFODIL *Narcissus eystettensis*.

QUEEN ANNE'S FLOWER *Narcissus pseudo-narcissus* (Daffodil) Norf. Wright.

QUEEN ANNE'S IRISH JONQUIL *Narcissus minimus*.

QUEEN ANNE'S LACE 1. *Anthriscus sylvestris* (Cow Parsley) Hemphill. 2. *Daucus carota* (Wild Carrot) USA Gates. 3. *Daucus pusillus* USA Elmore. 4. *Didiscus caerulea* (Blue Lace Flower) Rochford. Descriptive, of course, and best appreciated when Cow Parsley is massed along the roadsides in full flower.

QUEEN ANNE'S LACE HANDKERCHIEF 1. *Anthriscus sylvestris* (Cow Parsley) Dor. Grigson. 2. *Torilis japonica* (Hedge Parsley) Dor. Dacombe.

QUEEN ANNE'S NEEDLEWORK 1. *Centranthus ruber* (Spur Valerian) Som. Macmillan. 2. *Geranium versicolor* (Painted Lady) N'thants. A E Baker.

QUEEN ANNE'S PLUMES *Gynerium argenteum* (Pampas Grass) Som. Macmillan.

QUEEN ANNE'S POCKET MELON *Cucumis dudaim* (Dudaim Melon) Grieve.1931.

QUEEN ANNE'S THIMBLE FLOWER *Gilia capitata*.

QUEEN ANNE'S THISTLE *Carduus nutans* (Musk Thistle) Berw. B & H.

QUEEN ELIZABETH IN HER BATH *Dicentra spectabilis* (Bleeding Heart) Som. Macmillan. Cf. Lady-in-the-bath and Lady-in-the-boat.

QUEEN FLOWER *Syringa vulgaris* (Lilac) Dor. Macmillan. Possibly a result of the names Queen's Feather, or Prince of Wales's Feather.

QUEEN LADY'S SLIPPER *Cypripedium reginae* (Showy Lady's Slipper) USA Bingham.

QUEEN MARY'S THISTLE *Onopordum acanthium* (Scotch Thistle) N'thants. Grigson. Mary, Queen of Scots is meant here. Was she not imprisoned in the county?

QUEEN OF HEARTS *Hypericum calycinum* (Rose of Sharon) Dor. Macmillan.

QUEEN OF THE ALPS *Eryngium alpinum* (Alpine Eryngo) Polunin.

QUEEN OF THE FIELDS *Filipendula ulmaria* (Meadowsweet) Parkinson.

QUEEN OF THE MARSHES *Iris pseudoacarus* (Yellow Flag) Som. Macmillan.

QUEEN OF THE MEADOW 1. *Eupatorium purpureum* (Purple Boneset) O P Brown. 2. *Filipendula ulmaria* (Meadowsweet) Dev. Friend.1882; Dor. Dacombe; Som. Storer; Scot. Jamieson.

QUEEN OF THE MIST *Saxifraga spathularia x umbrosa* (London Pride) Dor. Macmillan.

QUEEN OF THE PRAIRIE *Filipendula rubra*.

QUEEN OF THE RIVER *Nuphar lutea* (Yellow Waterlily) Som. Macmillan.

QUEENS, Kings-and- *Arum maculatum* (Cuckoo-pint) Dev, Som. Macmillan; Dur. B & H. A bisexual name, better known as Lords-and-ladies, occasioned by the sexual imagery of a spadix in a spathe.

QUEENSLAND ASTHMA WEED *Euphorbia hirta* Australia Watt. It is known as Asthma-plant in South Africa. Gimlette, in 1915, noted its use for asthma and bronchitis.

QUEENSLAND KAURI PINE *Agathis robusta*.

QUEENSLAND NUT *Macadamia ternifolia*.

QUEENSLAND PYRAMIDAL TREE *Lagunaria patersonii* (Norfolk Island Hibiscus) RHS.

QUEENSLAND SILVER WATTLE *Acacia podalyrifolia* (Mount Morgan WQattle) H & P.

QUEENSLAND UMBRELLA TREE *Brassaia actinophylla*.

QUERCITRON OAK *Quercus velutina* (Black Oak) Mitchell.1974. Quercitron is the extract of the inner bark, stripped and prepared for use as a dyestuff (Hence Dyer's Oak, also). Black, yellow and orange can be obtained according to the recipe.

QUICK 1. *Agropyron repens* (Couchgrass) Suff. Forby. 2. *Crataegus monogyna* (Hawthorn) Som, Shrop, Ches, Lincs, Ire. Grigson. O E cwice, alive, because it contrasted with the dead material used in early fences and hedges. Couch has the same derivation - its aliveness is proverbial.

QUICK-IN-THE-HAND *Impatiens noli-me-tangere* (Touch-me-not) Prior. i.e. alive-in-the-hand, a reference to the explosive seed distribution this plant employs.

QUICK-TREE *Sorbus aucuparia* (Rowan) Turner.

QUICKBANE *Sorbus aucuparia* (Rowan) N Eng. Denham/Hardy. A misreading of Quickbeam, presumably, otherwise the name would make very little sense.

QUICKBEAM 1. *Populus tremula* (Aspen) Ellacombe. 2. *Sorbus aria* (Whitebeam) Herts. Grigson. 3. *Sorbus aucuparia* (Rowan) Gerard. Quick means alive, and beam, tree - hence Turner's Quick-tree. Aspen seems to be a better candidate for the name than the more usual rowan.

QUICKEN *Crataegus monogyna* (Hawthorn) Glos. P W F Brown.

QUICKEN-BERRY *Sorbus aucuparia* fruit (Rowan berries) Ire. Grigson.

QUICKEN-GRASS; QUICKGRASS *Agropyron repens* (Couch-grass) Suff. G E Evans.1960; N Eng. Halliwell (Quicken-grass); Flück (Quickgrass). Cf. Quitch-grass,or Squitch, and even Twitch-grass. They are all from OE cwice, alive, as any gardener will appreciate. 'Couch' is also of the same derivation.

QUICKEN-TREE 1. *Sorbus aucuparia* (Rowan) Lincs, Yks. Denham/Hardy. 2. *Ulmus glabra* (Wych Elm) War. Grigson.

QUICKENWOOD *Sorbus aucuparia* (Rowan) Lincs, Yks. Grigson.

QUICKSET; QUICKTHORN; QUICK-WOOD *Crataegus monogyna* (Hawthorn) Prior (Quickset); Lancs, Yks. Grigson (Quickthorn); Lincs, Yks. B & H (Quickwood).

QUICKSILVER WEED *Thalictrum dioicum* (Early Meadow Rue) Howes.

QUIET NEIGHBOURS *Centranthus ruber* (Spur Valerian) Wilts. Dartnell & Goddard.

Cf. Neighbours, Good Neighbours, and Good Neighbourhood, all from the same general area.

QUILLET *Trifolium repens* (White Clover) Corn. Jago.

QUINCE *Cydonia vulgaris*. An example of a forgotten plural. The original was quine, so that the plural has now become a singular. Eventually, the word came from Greek kydonion, and the original of that was a placename, Kydonia, in Crete.

QUINCE, Bengal *Aegle marmelos* (Bael) Brouk.

QUINCE, Chinese *Chaenomeles speciosa* (Japanese Quince) Geng Junying.

QUINCE, Japanese *Chaenomeles speciosa*.

QUINCE, Portuguese *Cydonia oblonga* (Quince) Howes.

QUININE *Cinchona sp*. This is a Quechua word, kina, or kinkina - bark. The generic name, though, is given in honour of the Countess of Chinchon, who is credited with ensuring that the common people of Lima could be treated for malaria with the bark.

QUININE, Australian *Alstonia constricta* (Fever Bark) Mitton. The bark is commonly used in Australia, in much the same way as true quinine, to combat malaria bouts.

QUININE-BUSH *Garrya flavescens*.

QUININE PLANT *Simmondsia chinensis* (Jajoba) USA T H Kearney.

QUINOA *Chenopodium quinoa*.

QUINSEY-BERRY *Ribes nigrum* (Blackcurrant) B & H. Blackcurrants were used in folk medicine against sore throats (a quinsey, or squinancy), long before cultivation, and are still so used. A wine or jelly used to be made in Yorkshire from the fruit, and set aside for sore throats.

QUINSEY-WORT *Asperula cynanchica* (Squinancywort) Grieve.1931. The plant provides an astringent gargle that was used to treat quinsey, which is a form of tonsilitis. Squinancywort is a corruption of quinseywort.

QUITCH; QUITCH-GRASS *Agropyron repens* (Couch-grass) both Som. Elworthy. Couch is actually a variant of quitch, which is OE cwice, alive. Quitch is rendered as Squitch quite often, while Twitch and Quick have the same derivation.

QUITCH, Black *Alopecurus myosuroides* (Slender Foxtail) Leyel.1948.

QUIVER-LEAF *Populus tremuloides* (American Aspen) Howes.

QUIVER-TREE *Aloe dichotoma*. Bushmen made their quivers by hollowing out a section of the branch of this plant.

R

RABBIE-RINNIE-HEDGE *Galium aparine* (Goose-grass) Ayr. Grigson. The English equivalents are Robin-run-i'-the-hedge and its many variants.

RABBIT 1. *Antirrhinum majus* (Snapdragon) Dev. Friend.1882; Som. Macmillan; Kent Tynan & Maitland. 2. *Cymbalaria muralis* (Ivy-leaved Toadflax) Dev. Grigson. 3. *Linaria vulgaris* (Toadflax) Dev. Friend.1882; Som. Grigson. 4. *Misopates orontium* (Lesser Snapdragon) Dev. Friend. All descriptive.

RABBIT, Arb *Geranium robertianum* (Herb Robert) Dev. B & H. Local pronunciation of Herb Robert.

RABBIT FLOWER 1. *Cymbalaria muralis* (Ivy-leaved Toadflax) Dev. Macmillan. 2. *Dicentra spectabilis* (Bleeding Heart) Wilts. Dartnell. & Goddard. 3. *Digitalis purpurea* (Foxglove) Dev. Grigson; USA. Henkel. 4. *Linaria vulgaris* (Toadflax) Dev. Macmillan; USA. Grigson. 5. *Misopates orontium* (Lesser Snapdragon) Dev. Friend.

RABBIT-FOOT CLOVER *Trifolium arvense* (Hare's-foot Trefoil) USA House.

RABBIT-MEAT 1. *Alternanthera ficoides* (Parrot-leaf) West Indies Howes. 2. *Anthriscus sylvestris* (Cow Parsley) Suss, Yks. B & H. 3. *Heracleum sphondyllium* (Hogweed) Lincs. Gutch. 4. *Lamium purpureum* (Red Deadnettle) Shrop. Grigson.

RABBIT PEA *Tephrosia virginiana* (Devil's Shoestring) Howes.

RABBIT-ROOT *Aralia nudicaulis* (American Sarsaparilla) Grieve.1931.

RABBIT-THORN *Lycium pallidum*.

RABBIT TRACKS *Maranta leuconeura* USA Cunningham.

RABBIT-WEED *Bigelovia graveolens*.

RABBIT'S CHOPS *Linaria vulgaris* (Toadflax) Som. Macmillan. Cf. Rabbit's Mouth.

RABBIT'S EARS *Stachys lanata* (Lamb's Ears) Som. Macmillan.

RABBIT'S FOOD *Oxalis acetosella* (Wood Sorrel) Lancs. Grigson.

RABBIT'S-FOOT FERN *Davallia fijiensis*.

RABBIT'S MEAT 1. *Heracleum sphondyllium* (Hogweed) Som. Macmillan. 2. *Lamium purpureum* (Red Deadnettle) Shrop. Grigson. 3. *Oxalis acetosella* (Wood Sorrel) Corn, Dev, Som. Grigson. 4. *Sonchus oleraceus* (Sow Thistle) Som. Macmillan. 5. *Taraxacum officinale* (Dandelion) Som. Macmillan.

RABBIT'S MOUTH 1. *Antirrhinum majus* (Snapdragon) B & H. 2. *Cymbalaria muralis* (Ivy-leaved Toadflax) Dev, Som. Macmillan. 3. *Glechoma hederacea* (Ground Ivy) Dev. Macmillan. 4. *Linaria vulgaris* (Toadflax) Som. Macmillan.

RABBIT'S VICTUALS *Sonchus oleraceus* (Sow Thistle) Som. Grigson.

RABBITBERRY *Shepherdia argentata* (Buffaloberry) Howes.

RABBITBRUSH 1. *Bigelovia graveolens* (Rabbit Weed) Los Angeles. 2. *Chrysothamnus nauseosus var. bigelovii*.

RABBITBRUSH, Blackbanded *Chrysothamnus paniculatus*.

RABBITBRUSH, Douglas *Chrysothamnus confinis*.

RABBITBRUSH, Rubber *Chrysothamnus nauseosus*.

RABBITBRUSH, Spring *Tetradymia glabrata* (Littleleaf Horsebrush) USA Kingsbury.

RABONE *Raphanus sativus* (Garden Radish) Gerard. Cf. Rawbone. They are both from Spanish rábano, itself from Latin raphanus.

RACADAL *Armoracia rusticana* (Horseradish) Scot. A R Forbes. Cf. the Scottish name Rotcoll, and the North country Red-cole.

RACALUS *Primula auricula* (Auricula) Glos. J D Robertson. A variation on 'auricula'.

RACCONALS 1. *Primula veris* (Cowslip) Ches. Holland. 2. *Primula veris x vulgaris* (False Oxlip) Ches. Holland.

RACCOON-BERRY *Podophyllum peltatum* (May-apple) O P Brown.
RACERS *Vicia sativa* (Common Vetch) B & H.
RACK *Agropyron repens* (Couchgrass) Grigson.1959. Rack is an old Suffolk word meaning weeds, or rubbish.
RACKELER; RACKLISS *Primula auricula* (Auricula) N'thants. A E Baker (Rackeler); Som. Elworthy (Rackliss). Variations on 'auricula'. So is Reckless, much commoner than these two.
RADICE, Alman; RADICE, Rape *Raphanus sativus* (Garden Radish) both Turner.
RADISH, Garden *Raphanus sativus*. Radish is eventually Latin radix, root.
RADISH, Green *Armoracia rusticana* (Horseradish) N Eng. Parkinson.
RADISH, Horse *Armoracia rusticana*. 'Horse' presumably to signify a large, or coarse, radish.
RADISH, Mountain *Armoracia rusticana* (Horseradish) Gerard.
RADISH, Rat-tailed *Raphanus caudatum*.
RADISH, Sea *Raphanus maritimus*.
RADISH, Summer *Raphanus sativus* (Garden Radish) Watt.
RADISH, Wild *Raphanus raphanistrum*.
RADISH-LEAVED BITTERCRESS *Cardamine raphanifolia*.
RADISH TREE, Horse *Moringa oleifera*. Named so because all its parts are pungent, and it is actually used as a horseradish substitute in India and Malaysia, where it grows.
RAFF-A-ROBBER; REAF-A-ROBBER *Arctium lappa* (Burdock) Tynan & Maitland. 'Reaf' is an old Devonshire word meaning to unravel, unwind.
RAFFIA PALM *Raphia ruffia*. This is the palm from whose dried leaves bast, or raffia, is made.
RAFORT *Raphanus sativus* (Garden Radish) Halliwell. Scots equivalents are Reefort, or Ryfart.
RAG-A-TAG *Lychnis flos-cuculi* (Ragged Robin) Shet. Grigson.
RAG-JACK 1. *Chenopodium album* (Fat Hen) Ches. Holland. 2. *Primula veris x vulgaris* (False Oxlip) Lincs. Grigson.
RAG-PAPER *Verbascum thapsus* (Mullein) Bucks. B & H. Presumably a corruption of Hag-taper, used in the same area. Mullein stems and leaves used to be dipped in tallow or suet, and burnt to give light at outdoor country gatherings, or indeed in the home.
RAG-ROSE *Primula veris x vulgaris* (False Oxlip) Grigson.
RAGGED JACK 1. *Lychnis flos-cuculi* (Ragged Robin) Som. Elworthy; IOW. Long; Suss. Parish; Kent, Ess. Grigson. 2. *Senecio jacobaea* (Ragwort) Yks. Grigson.
RAGGED LADY *Nigella damascena* (Love-in-a-mist) Wit.
RAGGED ROBIN 1. *Lychnis flos-cuculi*. 2. *Senecio jacobaea* (Ragwort) Yks. Grigson. 3. *Silene cucubalis* (Bladder Campion) Som. Macmillan.
RAGGED SHIRT *Convolvulus arvensis* (Field Bindweed) Som. Macmillan.
RAGGED URCHIN; RAGGED WILLIE *Lychnis flos-cuculi* (Ragged Robin) Dev. Macmillan (Urchin); Shet. Grigson (Willie).
RAGI *Eleusine corocana* (Finger Millet) Hutchinson & Melville.
RAGONET *Areca catechu* (Areca Palm) Som. Elworthy.
RAGS 1. *Corylus avellana* catkins (Hazel catkins) Yks. B & H. 2. *Lobaria pulmonaria* (Lungwort Crottle) Bolton.
RAGS, Nut *Corylus avellana* catkins (Hazel catkins) Ches. B & H.
RAGS, Oak *Lobaria pulmonaria* (Lungwort Crottle) Bolton. Also called Rags.
RAGS, Red *Papaver rhoeas* (Red Poppy) Dor. Macmillan.
RAGS, Tassel *Salix caprea* catkins (Sallow catkins) Ches. B & H.
RAGS-AND-TATTERS 1. *Aquilegia vulgaris* (Columbine) Som. Macmillan. 2. *Malva sylvestris* (Common Mallow) Dor, Som. Macmillan.
RAGWEED 1. *Ambrosia artemisifolia* (American Wormwood) USA Clapham. 2. *Ambrosia elatior* (Oak of Cappadocia) USA

Youngken. 2. *Senecio jacobaea* (Ragwort) Gerard.
RAGWEED, Giant; RAGWEED, Great *Ambrosia trifida* USA Allan (Giant).
RAGWEED, Roman *Ambrosia artemisifolia* (American Wormwood) Polunin.
RAGWEED, Western *Ambrosia psilostachya* (Black Sage) USA Vestal & Schultes.
RAGWEED, White *Franseria discolor*.
RAGWORT 1. *Orchis mascula* (Early Purple Orchis) B & H. 2. *Senecio jacobaea*. 'Rag' for *S jacobaea* is a result of the cut leaves, but that reason cannot apply to the orchid.
RAGWORT, Alpine 1. *Senecio fuchsii*. 2. *Senecio nemorensis*.
RAGWORT, Broad-leaved *Senecio fluviatilis* (Saracen's Woundwort) Clapham.
RAGWORT, Douglas *Senecio douglasii* (Douglas Groundsel) USA Elmore.
RAGWORT, Fen, Great *Senecio paludosus*.
RAGWORT, Golden *Senecio aureus* (Swamp Squaw-weed) USA House.
RAGWORT, Hoary *Senecio erucifolius*.
RAGWORT, Magellan *Senecio smithii*.
RAGWORT, Marsh *Senecio aquaticus*.
RAGWORT, Oxford *Senecio squalidus*. Actually a native of Sicily, but it was named for Oxford because it was grown in the botanic garden there from about 1690, and had become established around the Oxford colleges during the 18th century.
RAGWORT, St James's *Senecio jacobaea* (Ragwort) Flower.1859. Note the specific name. St James is the patron saint of horses, and there may be a possible connection here. But more likely is the fact that ragwort is usually in full flower on St James's Day (25 July).
RAGWORT, Sea *Senecio cineraria* (Silver Ragwort) Hibberd.
RAGWORT, Silver *Senecio cineraria*.
RAGWORT, Tansy *Senecio jacobaea* (Ragwort) USA Allan. Tansy and Ragwort are superficially quite similar.
RAGWORT, Welsh *Senecio cambrensis*.
RAGWORT, Yellow *Senecio jaobaea* (Ragwort) Shaw.

RAI *Brassica juncea* (Indian Mustard) Grieve.1933. Rai is really the oil obtained from the seeds.
RAIFORT, Great *Armoracia rusticana* (Horseradish) Gerard. That is actually French, and means 'strong root'.
RAILWAY CREEPER *Ipomaea learii* (Blue Dawn Flower) India Perry. Cf. Porter's Joy. They must be cultivated around railway stations.
RAILWAY DAISY *Bidens pilosa* Howes.
RAILWAY POPLAR *Populus x canadensis* 'Regenerata' Mitchell. Given by Mitchell because it is so common by main railway lines into cities, and in coalyards and sidings.
RAIN, Golden *Laburnum anagyroides* (Laburnum) North.
RAIN DAISY *Dimorphoteca pluvialis*. Appropriately enough, considering both the specific and common names, this is dedicated to St Swithin, and there is a saying that "if it opens not its flowers before seven, it will rain that day, or thunder".
RAIN LILY *Zephyranthes atamasco* (Atamasco Lily) USA Kingsbury.
RAIN TREE 1. *Enterolobum saman*. 2. *Lonchocarpus capassa*. The African tree, *Lonchocarpus capassa* does actually "rain" for a week or more during the hot, dry days just before the summer rains. The ground below the tree can be saturated, even enough for cattle to leave footprints in the mud. It is an insect that is responsible for this "raining", a kind of frog-hopper, the nymph of which covers itself with foam, like cuckoo-spit. The insects get nourishment by piercing the wood of the tree with their stylets, and sucking up the sap at a great speed, ejecting the almost pure water just as fast - this causes the "raining" of the tree.
RAIN TREE, Golden *Koelreuteria paniculata*.
RAINBERRY-THORN *Rhamnus cathartica* (Buckthorn) B & H. A variation of Rhineberry, apparently a translation of the German name, but probably from Rhamnus anyway.

RAINBOW STAR *Cryptanthus bromelioides*.
RAISIN, Desert *Solanum ellipticum* Australia Meggitt.
RAISIN TREE *Ribes rubrum* (Redcurrant) Turner.
RAIT *Ranunculus aquatilis* (Water Crowfoot) W Eng and Midl. Grigson. The name must be from the habitat, for to rait something is to soak it. The verb is probably better recognised as to ret. Flax, or hemp, is softened by soaking it in water, a process known as retting.
RAM, Laxative *Rhamnus cathartica* (Buckthorn) Gerard. A fair comment. The berries are a powerful purgative. " A rough purge", said Hill, "but a good one".
RAM('S)-GALL *Menyanthes trifoliata* (Buckbean) Storms. 'Gall', for it is indeed excessively bitter, and 'ram' possibly is used instead of 'buck'.
RAM'S CLAWS 1. *Ranunculus acris* (Meadow Buttercup) Som. Elworthy. 2. *Ranunculus repens* (Creeping Buttercup) Som. Grigson; Wilts. Dartnell & Goddard.1899; Dor. Dacombe. 3. *Stellaria media* (Chickweed) Som. Macmillan. 4. *Tussilago farfara* (Coltsfoot) Som. Macmillan. Rams don't have claws; it is suggested that the name should actually be raven's claws. Elworthy said of Meadow Buttercup that they were only called Ram's Claws when overgrown, with long, coarse stalks without leaves, but still with flowers on top.
RAM'S FOOT 1. *Geum urbanum* (Avens) Dev. Friend.1882. 2. *Ranunculus aquatilis* (Water Crowfoot) B & H. Presumably Ram's Foot for Avens is a version of Hare's Foot, which was the Saxon name. The root of the plant is said to resemble one. For Water Crowfoot, can it have started as Raven's Foot (on the analogy of Crowfoot being the usual name for all the buttercups)?
RAM'S GLASS *Ranunculus acris* (Meadow Buttercup) Som. Macmillan. Probably a corruption of Ram's Claws.

RAM'S HEAD LADY'S SLIPPER *Cypripedium arietinum*.
RAM'S HORNS 1. *Allium ursinum* (Ramsons) Glos. J D Robertson. 2. *Arum maculatum* (Cuckoo-pint) Suss. B & H. 3. *Orchis mascula* (Early Purple Orchis) Fernie. 4. *Orchis morio* (Green-winged Orchid) Suss. B & H. In the first instance, the name is simply a corruption of the common name, Ramsons. Probably in all cases, the offensive smell has something to do with the name.
RAMA *Hibiscus sabdariffa* (Rozelle) Schery.
RAMANAS ROSE *Rosa rugosa* (Japanese Rose) Mattock. Probably from the Japanese meaning 'blooming profusely'.
RAMIE *Boehmeria nivea*. There are various spellings, including rami and ramee. It is a Malay word, rami.
RAMP *Allium tricoccum* (Wood Leek) USA Yarnell. Obviously taken from an English original for *Allium ursinum*, Ramsons.
RAMPE *Arum maculatum* (Cuckoo-pint) Turner. The word means wanton, and is a clear reference to its supposed aphrodisiac powers.
RAMPION; RAMPION BELLFLOWER *Campanula rapunculus*. The reference is to the root, which is edible. Rapunculus is from the Latin meaning small turnip.
RAMPION, Dark *Phyteuma ovatum*.
RAMPION, Horned *Phyteuma scheuchzeri*.
RAMPION, Large *Oenothera biennis* (Lesser Evening Primrose) W Miller. Presumably because the tap root is edible, and has been used as a vegetable.
RAMPION, Round-headed *Phyteuma orbiculare*.
RAMPION, Spiked *Phyteuma spicata*.
RAMPION, Spiked, Blue *Phyteuma betonicifolium*.
RAMPS 1. *Allium ursinum* (Ramsons) Som, Dor. Macmillan; Lancs. Nodal; Yks. Hartley & Ingilby; Border, Ire. B & H. 2. *Campanula rapunculus* (Rampion Bellflower) Yks. B & H.

RAMS 1. *Allium ursinum* (Ramsons) Turner. 2. *Colchicum autumnale* (Meadow Saffron) Yks. B & H.

RAMSEY 1. *Allium ursinum* (Ramsons) Dev. Friend.1882; Norf. B & H. 2. *Ononis repens* (Rest-harrow) Dev. Grigson.

RAMSIES; RAMSINS; RAMSDEN *Allium ursinum* (Ramsons) Grigson (Ramsies); Gerard (Ramsins); IOW Grigson (Ramsden). Cf. Ramsons.

RAMSONS 1. *Allium ursinum*. 2. *Arum maculatum* (Cuckoo-pint) Cumb. Grigson. 3. *Ononis repens* (Rest-harrow) Dev. Grigson. A double plural. OE hramason was the plural of hrama, and was later taken as a singular. So a new plural was formed with 's'. The word occurs in several place names, such as Ramsey (ramson island), and Ramsbottom (ramson valley). For cuckoo pint, the name probably comes from the same source as Turner's Rampe, which means wanton, from the plant's supposed aphrodisiac powers.

RAMSTEAD *Linaria vulgaris* (Toadflax) USA House.

RAMSTHORN *Rhamnus cathartica* (Buckthorn) Jordan. Presumably suggested by the generic name Rhamnus, and nothing to do with the zoological ram.

RAMTIL *Guizotia abissinica*.

RAN-TREE *Sorbus aucuparia* (Rowan) N Eng. Denham/Hardy. A version from rowan. Cf. Rantry.

RANGIORA *Brachyglottis repanda*.

RANGOON BEAN *Phaseolus lunatus* (Lima Bean) H C Long.1924. It originated in Central America, but it has attracted names like this from all over the world, e.g. Cape Bean, Madagascar Bean, Burma Bean, etc.

RANGOON CREEPER *Quisqualis indica*.

RANSOMS *Allium ursinum* (Ramsons) Som. Macmillan. A mispronunciation of the usual name, and probably much more widespread than the solitary Somerset record.

RANTING WIDOW *Chamaenerion angustifolium* (Rosebay Willowherb) Ches. Wright. 'Widow' here is 'widdy', or willow. The leaves resemble somewhat those of the willow, hence a number of names like French Willow, Tame Withy, Blooming Willow, etc.

RANTIPOLE *Daucus carota* (Wild Carrot) Wilts. Dartnell & Goddard; Hants. Cope. In Yorkshire, a rantypole is a seesaw, but according to Halliwell the west country rantipole is a "rude romping child".

RANTLE-TREE; RANTRY *Sorbus aucuparia* (Rowan) Scot. Davidson (Rantle-tree); Yks. Robinson; Scot. Gentleman's Magazine.1784 (Rantry). Cf. Ran-tree.

RANTY-BERRY *Sorbus aucuparia* fruit (Rowan berries) Ire. B & H.

RANTYTANTY *Rumex obtusifolius* (Broad-leaved Dock) Ayr. Grigson. It sounds as if it should be the name of a children's game.

RAOUL; RAULI *Nothofagus procera*.

RAPE 1. *Brassica campestris* (Field Mustard) Gerard. 2. *Brassica napus*. 3. *Sinapis arvensis* (Charlock) Yks. B & H. Rape is Latin rapa, rapum, turnip, which is a close enough relative to *Brassica napus*, whose specific name also means turnip. The other meaning of Rape, as in Clover Rape and Fir Rape below, has a different derivation. This time it comes from the Latin verb rapere, snatch, and is used in the sense of parasitism.

RAPE, Bird *Brassica campestris* (Field Mustard) Usher.

RAPE, Clover *Orobanche minor* (Lesser Broomrape) Salisbury. Because it is parasitic on clover.

RAPE, Indian *Brassica campestris var. toria*.

RAPE, Fir *Monotropa hypopitys* (Yellow Bird's-nest). Given in the belief that the plant was parasitic upon the roots of Scots Pine and beech. Actually it has its own roots invested by a fungus.

RAPE, Long *Brassica napus* (Rape) Langham. From what is the name differentiated by calling it 'long'?

RAPE, Oil-seed *Brassica campestris var. rapa 'Oleifera'*.

RAPE, St Anthony's *Ranunculus bulbosus* (Bulbous Buttercup) Gerard. 'Rape' as usual here means turnip, a reference to the bulb. Cf.

Rape-Crowfoot and St Anthony's Turnip. 'Rape' became corrupted to 'rope', so we find St Anthony's Rope for this, too.

RAPE, Wild *Sinapis arvensis* (Charlock) Gerard.

RAPE CROWFOOT *Ranunculus bulbosus* (Bulbous Buttercup) Lyte.

RAPE RADICE *Raphanus sativus* (Garden Radish) Turner.

RAPE VIOLET *Cyclamen europaeus* (Sowbread) Turner. "because it hath a roote lyke a Rape and floures lyke a Violet".

RAPESEED 1. *Brassica napus* (Rape) Gerard. 2. *Brassica rapa var. oleifera*.

RAPPER-DANDY *Arctostaphylos uva-ursi* (Bearberry) N Eng, Berw, Scot. Grigson.

RAPPERS *Digitalis purpurea* flowers (Foxglove) Wilts. Dartnell & Goddard.1899. The unopened flowers can be popped, and rap is the Wiltshire word for pop. Cf. Poppers, and a lot of other similar names.

RASCAL, Cling- *Galium aparine* (Goosegrass) Dev. B & H.

RASP 1. *Rubus idaeus* (Raspberry) N'thants, Lincs, Scot, Ire. Grigson; Yks. Addy. 2. *Rubus saxatilis* (Stone Bramble) Camden/Gibson. The earlier form is Raspis.

RASP, Wood *Rubus idaeus* (Raspberry) Selk. Grigson.

RASPBERRIES-AND-CREAM *Eupatorium cannabinum* (Hemp Agrimony) IOW Grigson. Colour descriptive.

RASPBERRY 1. *Rubus idaeus*. 2. *Ribes rubrum* (Redcurrant) B & H.

RASPBERRY, American *Rubus idaeus var. strigosus* (Red Raspberry) USA Elsmore.

RASPBERRY, Arctic *Rubus arcticus*.

RASPBERRY, Black 1. *Rubus leucodermis* (Western Raspberry) USA Turner & Bell. 2. *Rubus occidentalis* USA Tolstead.

RASPBERRY, Cane, Purple *Rubus parviflorus* (Thimbleberry) Howes.

RASPBERRY, Dwarf *Rubus pubescens*.

RASPBERRY, Purple *Rubus neglectus*.

RASPBERRY, Purple-flowering *Rubus odoratus*.

RASPBERRY, Red *Rubus idaeus var. strigosus*.

RASPBERRY, Red, Arizona *Rubus arizonicus*.

RASPBERRY, Virginian *Rubus occidentalis* (Thimble-berry) Grieve.1931.

RASPBERRY, Western *Rubus leucodermis*.

RASPBERRY ACACIA; RASPBERRY JAM TREE *Acacia acuminata*. The wood is scented, like raspberry jam.

RASPICE; RASPIS *Rubus idaeus* (Raspberry) A S Palmer (Raspice); Gerard (Raspis). This is the earlier form of Raspberry.

RASSELS *Ononis repens* (Rest-harrow) Suff. Nall. Probably connected with the "arrest" of Rest-Harrow.

RASTEWORT *Peucedanum officinale* (Hog's Fennel) Gunther. This name appears as a gloss on a C12 ms of Apuleius Barbarus.

RAT-TAIL PLANTAIN *Plantago major* (Great Plantain) Clapham. Cf. Rat's Tails, and Ratten-tails, mostly from the North country.

RAT-TAIL PLANT *Peperomia caperata*.

RAT-TAILED RADISH *Raphanus caudatum*.

RAT'S BANE *Anthriscus sylvestris* (Cow Parsley) Som. Potter & Sargent. Not very obvious for something like Cow Parsley. Should it really refer to hemlock?

RAT'S FOOT *Glechoma hederacea* (Ground Ivy) Dev. Macmillan. The shape of the leaves suggests all sort of 'foot' names, usually cats', which is probably the original of this name. Rat's Foot seems unlikely, but there is a genuine resemblance to a cat's foot.

RAT'S MOUTH 1. *Glechoma hederacea* (Ground Ivy) Dev. Macmillan. 2. *Lamium album* (White Deadnettle) Dev. Macmillan. 3. *Lamium purpureum* (Red Deadnettle) Dev. Macmillan. They are all labiates, hence the 'mouth' name.

RAT'S TAIL 1. *Agrimonia eupatoria* (Agrimony) Wilts. Macmillan. 2. *Plantago lanceolata* (Ribwort Plantain) Cumb. Grigson. 3. *Plantago major* (Great Plantain)

Norf, Cumb, Scot Grigson; IOM Moore.1924.

RATS-AND-MICE *Cynoglossum officinale* (Hound's Tongue) Wilts. Macmillan. The acetamide in the leaves gives this plant a typical mouse smell.

RATTEN-TAIL *Plantago major* (Great Plantain) N'thum. Grigson.

RATTLE, Baby's 1. *Ajuga reptans* (Bugle) Som. Macmillan. 2. *Melampyrum pratense* (Cow-wheat) Som. Macmillan. 3. *Rhinanthus crista-galli* (Yellow Rattle) Som. Macmillan. Most 'rattle' names mean just what they say - the sound that seeds make in an inflated calyx. Sometimes, as presumably with bugle, the name is purely descriptive.

RATTLE, Bull 1. *Melandrium album* (White Campion) Bucks. Harman. 2. *Melandrium rubrum* (Red Campion) Bucks. Grigson. 3. *Silene cucubalis* (Bladder Campion) IOW, Bucks B & H.

RATTLE, Cow 1. *Melandrium album* (White Campion) Bucks. Grigson. 2. *Silene cucubalis* (Bladder Campion) Bucks. Grigson.

RATTLE, Devil's *Achillea millefolium* (Yarrow) M Baker. Cf. Devil's Nettle, which may be the origin of this name, Devil's Plaything, etc. The names seem to convey hints of disapproval of the plant's use in divinations and spells. But this name may very well be a misreading of 'nettle'.

RATTLE, Giant's *Entada phaseoloides* (Nicker Bean) Chopra.

RATTLE, Hay 1. *Rhinanthus crista-galli* (Yellow Rattle) Rackham; Dev. PLNN17. 2. *Silene cucubalis* (Bladder Campion) Dev. PLNN17.

RATTLE, Meadow *Rhinanthus crista-galli* (Yellow Rattle). B & H.

RATTLE, Penny *Rhinanthus crista-galli* (Yellow Rattle) Som. Macmillan; Suss. Parish. Cf. Penny-weed, Rattle-penny, etc. In these cases, the seeds in the calyx are likened to coins in a purse.

RATTLE, Red 1. *Odontites verna* (Red Bartsia) Darling & Boyd. 2. *Pedicularis palustris*. 3. *Pedicularis sylvatica* (Lousewort) Dev, Som. Macmillan.

RATTLE, Red, Dwarf *Pedicularis sylvatica* (Lousewort) J A MacCulloch.1905.

RATTLE, Red, Marsh *Pedicularis palustris* (Red Rattle) J A MacCulloch.1905.

RATTLE, White *Rhinanthus crista-galli* (Yellow Rattle) Gerard. Why 'white'?

RATTLE, Yellow *Rhinanthus crista-galli*.

RATTLE-BAGS 1. *Rhinanthus crista-galli* (Yellow Rattle) Dor. Macmillan. 2. *Silene cucubalis* (Bladder Campion) Corn, Dev. Grigson.

RATTLE-BASKET 1. *Briza media* (Quaking-grass) Som. Grigson.1959. 2. *Pedicularis sylvatica* (Lousewort) Som. Macmillan. 3. *Rhinanthus crista-galli* (Yellow Rattle) Som. Macmillan; Wilts. Dartnell & Goddard.

RATTLE-BOX 1. *Crotalaria sagittalis*. 2. *Rhinanthus crista-galli* (Yellow Rattle) Shrop, Ire. Grigson. 3. *Sesbania drummondii* (Coffeebean) USA Kingsbury.

RATTLE-BOX, Purple *Sesbania punicea* (Purple Sesbane) USA Kingsbury.

RATTLE-CAPS *Rhinanthus crista-galli* (Yellow Rattle) Som. Grigson.

RATTLE-GRASS 1. *Briza media* (Quaking-grass) IOW Grigson.1959. 2. *Pedicularis palustris* (Red Rattle) Gerard. 3. *Rhinanthus crista-galli* (Yellow Rattle) Som. Macmillan.

RATTLE-GRASS, Red *Pedicularis palustris* (Red Rattle) Gerard.

RATTLE-JACK; RATTLE-PENNY *Rhinanthus crista-galli* (Yellow Rattle) Lincs. Peacock (Jack); N Eng. Grigson (Penny).

RATTLE-POUCH *Capsella bursa-pastoris* (Shepherd's Purse) Grieve.1931.

RATTLE-ROOT; RATTLE-SNAKEROOT *Cimicifuga racemosa* (Black Snakeroot) USA both Lloyd. As with the other plants bearing the name 'rattle', the dried spikes carrying the seed rattle in the wind.

RATTLE-TRAPS *Rhinanthus crista-galli* (Yellow Rattle) Dor. Macmillan.
RATTLEBRUSH *Sesbania drummondii* (Coffeebean) USA Kingsbury.
RATTLEBUSH *Baptisia tinctoria* (Wild Indigo) Grieve.1931.
RATTLEPOD *Pedicularis sylvatica* (Lousewort) Som. Macmillan.
RATTLEPOD, Black *Baptisia bracteata*.
RATTLEPOD, Little *Astragalus carolinianus* (Canadian Milk Vetch) Gilmore.
RATTLEPOD, Many-coloured *Astragalus allochrous* (Many-coloured Rattleweed) USA Elsmore.
RATTLESNAKE FERN *Botrychium virginianum*.
RATTLESNAKE GRASS *Glyceria canadensis*.
RATTLESNAKE('S)-MASTER 1. *Eryngium aquaticum* (Button Snakeroot) Le Strange. 2. *Eryngium yuccifolium* (Button Snakeroot) USA H Smith. 3. *Liatris scariosa* (Blazing Star) Howes. 'Snakeroot' gives the answer - the coiled roots provide the signature for its use against snake bite, and particularly rattlesnake bite.
RATTLESNAKE PLANT *Calathea insignis*. In this case, it is the patterning of the leaves that suggests the name.
RATTLESNAKE PLANTAIN *Goodyera repens* (Creeping Lady's Tresses) USA Weiner. Again, it is the white-veined leaves that looked like rattlesnakes to the early settlers, who took it as the sign for virtues against snake bite.
RATTLESNAKE PLANTAIN, Downy *Epipactis pubescens*.
RATTLESNAKE PLANTAIN, Lesser *Epipactis ophioides*.
RATTLESNAKE PLANTAIN, Menzies' *Epipactis decipiens*.
RATTLESNAKE-ROOT 1. *Cimicifiga racemosa* (Black Snakeroot) USA Sanford. 2. *Polygala senega* (Senega Snake-root) USA Barton & Castle. Senega Snake-root was a remedy for snakebite, presumably from the doctrine of signatures. One account says it grows wherever there are rattlesnakes, so that it was used as a remedy for their bite. Black Snake-root, though, probably gets the name by a simple addition of 'rattle' and 'snake' - Cf. Rattle-weed and Rattle-root.
RATTLESNAKE-WEED 1. *Daucus pusillus* USA Barrett & Gifford. 2. *Echinacea angustifolia* (Narrow-leaved Purple Coneflower). USA Vestal & Schultes. 3. *Euphorbia albomarginata* USA T H Kearney. 4. *Euphorbia ocellata* USA Barrett & Gifford. 5. *Hieraceum venosum*. 6. *Lycopus sinuatus* USA Bergen. Miwok Indians chewed *Daucus pusillus* and then put it on a snake-bite, just as they did with the mashed leaves of *Euphorbia ocellata*. Similarly, *Lycopus sinuatus* enjoyed a reputation as an antidote to snake-bite. Perhaps the latter was doctrine of signatures. After all, sinuatus might suggest a connection with snakes.
RATTLEWEED 1. *Astragalus diphysus* (Blue Loco) USA Kingsbury. 2. *Cimicifuga racemosa* (Black Snakeroot) USA Lloyd. 3. *Pedicularis palustris* (Red Rattle) Norf. Grigson. 4. *Silene cucubalis* (Bladder Campion) Wilts. Dartnell & Goddard.
RATTLEWEED, Many-coloured *Astragalus allochrous*.
RATTLEWORT *Rhinanthus crista-galli* (Yellow Rattle) Cockayne.
RATTLING ASP *Populus tremula* (Aspen) Lyte. Gerard also knew this name, for he says "it has received a name amongst the low-country men, from the noise and rattling of the leaves, viz: Rateeler".
RAULI; RAOUL *Nothofagus procera*.
RAUNTREE; RAWNTREE *Sorbus aucuparia* (Rowan) Dale-Green (Rauntree); Aber. Banks (Rawntree). From rowan-tree, via Rountree.
RAUPO *Typha angustifolia* (Lesser Bulrush) New Zealand Howes.
RAVEN-TREE *Sorbus aucuparia* (Rowan) Scot. A S Palmer. Presumably a corruption of one of the 'rowan' variations.
RAVEN'S CLAWS *Ranunculus repens* (Creeping Buttercup) Som, Dor, Wilts.

Grigson. All the 'claw' names for the buttercups must have been suggested for the same reasons as 'crowfoot'.

RAWBONE; RABONE *Raphanus sativus* (Garden Radish) A S Palmer (Rawbone); Gerard (Rabone). From Spanish rábano, Latin raphanus.

RAWHEADS *Ranunculus aquatilis* (Water Crowfoot) Shrop. Grigson.

RAWP *Brassica napus* (Rape) Yks. Atkinson.

RAXEN *Butomus umbellatus* (Flowering Rush) Som. Macmillan.

RAY *Agrostemma githago* (Corn Cockle) B & H. Connected with the same name for darnel, as is another of cockle's names, Drawk. Perhaps they both mean dangerous, noxious weeds.

RAY, Red *Lolium perenne* B & H. The old Latin name was Lolium rubrum.

RAY, Sun's *Helianthus annuus* (Sunflower) Som. Macmillan.

RAY-GRASS 1. *Lolium perenne* (Rye-grass) J Smith. 2. *Lolium temulentum* (Darnel) Barton & Castle. The earlier form, from French ivraie, drunk. The grain of darnel is toxic, producing a state of confused perception.

RAY-GRASS, Italian *Lolium multiflorum* B & H.

RAY'S KNOTGRASS *Polygonum raii*.

RAYLESS MAYWEED *Matricaria matricarioides*.

REATE *Ranunculus aquatilis* (Water Crowfoot). Origin obscure, according to Chambers's dictionary, but surely it must be the same word as Rait, which is a reference to the plant's habitat. To rait something is to soak it (better recognised as to ret).

RECKLESS *Primula auricula* (Auricula) Dor. Dacombe; Staffs. Marrow; Lincs, Cumb. B & H Yks. Addy. A corruption of 'auricula'.

RED BARK *Cinchona succirubra*.

RED BEAN *Sophora secundiflora* (Mescal Bean) USA Norbeck.

RED BERRY 1. *Cornus florida* (Flowering Dogwood) Grieve.1931. 2. *Rosa canina* fruit (Hips) Yks. Grigson. An unusually simple name for hips, considering the large number of more picturesque examples to hand.

RED HEAD *Asclepias curassivica* (Blood Flower) Barbados Gooding.

RED HOT POKER TREE *Erythrina abyssinica* Dale & Greenway. From the scarlet flowers.

RED INK BERRY; RED INK PLANT *Phytolacca decandra* (Poke-root) USA Henkel (Berry); Brownlow (Plant). Cf. Inkberry, etc. The coloured juice has been used as an ink, and to give a red stain that the Indians used to colour their ornaments.

RED LEAD 1. *Dactylorchis incarnata* (Early Marsh Orchid) B & H. 2. *Dactylorchis maculata* (Heath Spotted Orchid) B & H.

RED STEM *Erodium cicutarium* (Storksbill) USA Elmore.

REDBERRY 1. *Cornus florida* (Flowering Dogwood) Grieve.1931. 2. *Panax quinquefolium* (Ginseng) Leyel.1948. 3. *Rhamnus californica*. 4. *Rhamnus crocea* (Indian Cherry) USA Bean.

REDBREAST: REDBREAST, Robin *Geranium robertianum* (Herb Robert) Som, N'thants Grigson (Redbreast); N'thants. Grigson (Robin Redbreast). Given that Robin is an alternative to Robert, as in the names Poor Robin, Round Robin, etc., Redbreast is a simple association of ideas.

REDBUD *Acer rubrum* (Red Maple) USA Harper.

REDBUD, California *Cercis occidentalis* (Western Redbud) Brimble.1948.

REDBUD, Chinese *Cercis chinensis* (Chinese Judas Tree) H & P.

REDBUD, Eastern *Cercis canadensis*.

REDBUD, Texas *Cercis reniformis*.

REDBUD, Western *Cercis occidentalis*.

REDCAP; REDCUP *Papaver rhoeas* (Red Poppy) both Som. Macmillan.

REDCO *Armoracia rusticana* (Horseradish) Turner. Gerard has a north country name, Red Cole, and presumably this is the same.

REDCURRANT *Ribes rubrum.*
REDCURRANT, Rock *Ribes petraeum* (Rock Redcurrant) Polunin.
REDCURRANT, Upright *Ribes petraeum.*
REDOUL *Coriaria myrtifolia* (Mediterranean Coriaria) Polunin.1976.
REDROOT 1. *Amaranthus palmeri* (Careless-weed) USA T H Kearney. 2. *Amaranthus powellii* New Zealand Connor. 3. *Amaranthus retroflexus* (Pigweed) USA Gates. 4. *Cannabis sativa* (Hemp) USA Watt. 5. *Ceanothus americanus* (New Jersey Tea) USA Harper. 6. *Ceanothus ovatus* (Inland Jersey Tea) USA Tolstead. 7. *Eriogonum jamesii* (Antelope Sage) USA Vestal & Schultes. 8. *Lacnanthes tinctoria* (Bloodroot) USA Kingsbury. 9. *Sanguinaria canadensis* (Bloodroot) Grieve.1931.
REDROOT, Small *Ceanothus ovatus* (Inland Jersey Tea) USA Yarnell.
REDROOT PIGWEED *Amaranthus retroflexus* (Pigweed) Watt.
REDSCALE *Atriplex rosea.*
REDSHANKS 1. *Ceanothus americanus* (New Jersey Tea) USA Harper. 2. *Geranium robertianum* (Herb Robert) N Eng. Prior. 3. *Polygonum amphibium* (Amphibious Persicaria) N'thum. B & H. 4. *Polygonum hydropiper* (Water Pepper) Lincs, Cumb, Yks, N'thum. Grigson. 5. *Polygonum persicaria* (Persicaria) Lincs, Yks, Cumb, N'thum. Grigson. 6. *Rumex acetosa* (Wild Sorrel) Roxb. B & H. 7. *Rumex obtusifolia* (Broad-leaved Dock) N Scot. B & H.
REDSHANKS, Robin *Geranium robertianum* (Herb Robert) Yks. Grigson. Cf. Robin Redbreast.
REDWEED 1. *Anagallis arvensis* (Scarlet Pimpernel) Dor. Macmillan. 2. *Geranium robertianum* (Herb Robert) Ches. Holland. 3. *Papaver rhoeas* (Red Poppy) Wilts. Macmillan; Dor. Barnes; IOW, Norf. B & H; Berks. Lowsley; Hants. Cope. 4. *Phytolacca decandra* (Poke-root) USA Henkel. 5. *Polygonum aviculare* (Knotgrass) Dev, Som. Friend.1882. 6. *Polygonum hydropiper* (Water Pepper) Ches. B & H. 7. *Polygonum persicaria* (Persicaria) Ches. Grigson.
REDWOOD 1. *Adina microcephala* (Wild Oleander) Palgrave. 2. *Ceanothus americanus* (New Jersey Tea) J Smith. 3. *Pinus sylvestris* (Scots Pine) N Davey. 4. *Pterocarpus soyauxii* (Barwood) Dalziel. 5. *Sequoia sempervirens.*
REDWOOD, Baltic *Pinus sylvestris* (Scots Pine) Hutchinson & Melville. A lot of timber from this species has been imported from the Continent under such names as Baltic Redwood, Baltic Yellow Deal, Danzig Pine, etc.
REDWOOD, Brazilian *Caesalpina bonduc.*
REDWOOD, Californian *Sequoia sempervirens* (Redwood) Howes.
REDWOOD, Coast *Sequoia sempervirens* (Redwood) Wilks.
REDWOOD, Dawn *Metasequoia glyptostroboides.*
REDWOOD, Pernambuco *Caesalpina bonduc* (Brazilian Redwood) Watt.
REDWOOD, Sierra *Sequoia gigantea* (Wellingtonia) Leathart.
REDWOOD, Zambesi *Baikiaea plurijuga* (Rhodesian Teak) Palgrave.
REDWORT *Brassica oleracea var. capitata* (Cabbage) Henslow. Cf. Red Cole, also in Henslow. They must strictly be only applicable to red cabbage, presumably *B oleracea var. capitata f. purpurea.*
REED *Phragmites communis.*
REED, Egyptian *Cyperus papyrus* (Paper Sedge) Bonar.
REED, Norfolk *Phragmites communis* (Reed) J G Jenkins.1976. That is the name by which thatchers know it.
REED, Paper; SEDGE, Paper *Cyperus papyrus.* The leaf stalks contain the material that is made into papyrus, the original writing material of the ancient Egyptians.
REED, Pole *Phragmites communis* (Reed) Som. Elworthy. Possibly a corruption of pool-reed.
REED, Sea *Ammophila arenaria* (Marram) Wales J G Jenkins.1976.

REED-GRASS *Phalaris arundinacea*.
REED MEADOW GRASS *Glyceria maxima* (Great Water Grass) Pratt.
REEDBIND *Convolvulus arvensis* (Field Bindweed) Wessex Rogers.
REEDMACE, Great *Typha latifolia* (False Bulrush) McClintock. Apparently because the Italian painters, in Ecce Homo pictures, depicted Christ as holding this reed as a mace or sceptre.
REEDMACE, Lesser *Typha angustifolia* (Lesser Bulrush) McClintock.
REEFERS *Cannabis sativa* (Hemp) USA Watt. Properly reserved for the cigarettes made up of the drug, but occasionally used for the plant itself.
REEFORT; RYFART *Raphanus sativus* (Garden Radish) Scot. Jamieson. Presumably straight from Latin Raphanus.
REFLEXED STONECROP *Sedum reflexum* (Yellow Stonecrop) D. Grose.
REGAL LILY *Lilium regale*.
REGALS *Orchis mascula* (Early Purple Orchis) Dor. Grigson.
REGULATOR, Female *Senecio aureus* (Swamp Squaw-weed) O P Brown. "Exerts a very powerful and peculiar influence upon the reproductive organs of females" (O P Brown). In Alabama, it is used to bring on menstrual flow in young girls when it has stopped because of colds.
REIN ORCHIS *Coeloglossum viride* (Frog Orchid) USA Yarnell. Apparently, it is as it stands, and is not "Rain".
REMCOPE *Succisa pratensis* (Devil's-bit) B & H. A name taken from the Grete Herball.
REMEDY, Poor Man's *Valeriana officinalis* (Valerian) B & H. Cf. Countryman's Treacle. Valerian is a "heal-all".
REMEMBER-ME *Veronica chamaedrys* (Germander Speedwell) Yks. Grigson. Forget-me-not is also used, and is quite widespread. The suggestion is that these names illustrate the fragility of the flowers - Speedwell has the same effect.
RENGAS *Stagmaria verniciflua*.
RENNET; RENNET, Cheese *Galium verum* (Lady's Bedstraw) Kent, Herts, Cumb. Grigson (Rennet); Som, War, Cumb, Ire. (Cheese Rennet) Prior. This has been used as a rennet substitute since Dioscorides' time.
REPENTANCE, Herb of *Ruta graveolens* (Rue) A S Palmer. Rue has always been the symbol of regret and repentance. After all, to rue means to be sorry. The word is possibly derived from the same root as Ruth, meaning sorrow or remorse.
REPS *Ribes rubrum* (Redcurrant) Grieve.1931. Probably originally from Raisin-tree, of which there are Scottish variants like Rizzles. They have been reduced to Reps, Ribs or Risp.
RESCUE GRASS *Bromus catharticus*.
RESHES *Juncus inflexus* (Hard Rush) Yks. F K Robinson.
RESPIS *Rubus idaeus* (Raspberry) Fernie. Raspis is more usual. Rasp is the shortened version of this, or perhaps it is the other way round, with Raspis being the plural of Rasp.
REST, Bee's *Caltha palustris* (Marsh Marigold) Som. Macmillan.
REST, Mary's *Veronica chamaedrys* (Germander Speedwell) Tynan & Maitland.
REST, Traveller's *Tanacetum vulgare* (Tansy) Wilts. Dartnell & Goddard.
REST HARROW *Ononis repens*. 'Arrest harrow', of course, brought in from remore aratri, plough hindrance. There was also arest bovis, "because it maketh the Oxen whilst they be in plowing to rest or stand still" (Gerard).
REST HARROW, Erect *Ononis spinosa* (Spiny Rest Harrow) Genders.
REST HARROW, Round-leaved *Ononis rotundifolia*.
REST HARROW, Small *Ononis reclinata*.
REST HARROW, Spiny *Ononis spinosa*.
REST HARROW, Thorny *Ononis spinosa* (Spiny Rest Harrow) Flower.1859.
REST HARROW, Yellow, Large *Ononis natrix*.
REST HAVEN *Oenothera biennis* (Lesser Evening Primrose) Dor. Macmillan.
RESURRECTION PLANT; RESURREC-

TION FLOWER 1. *Anastatica hierocauntica* (Rose of Jericho) Perry. 2. *Kalanchoe pinnata*. As the seeds of Anastatica ripen during the dry season, the leaves fall off and the branches curve inwards to make a round, lattice-type ball, which is blown out of the soil to roll about the desert until it reaches a moist spot, or the rainy season begins. Another reason for the name is that it is supposed to have first blossomed at Christ's birth, closed at his Crucifixion, and blossomed again at Easter. The generic name is taken directly from the Greek anastasia, resurrection.

RETICULATE WILLOW *Salix reticulata* (Net-leaved Willow) Darling & Boyd.

RHAETIAN POPPY *Papaver rhaeticum* (Yellow Alpine Poppy) Grey-Wilson.

RHATANY *Krameria triandra*. The name is from a Quechua original, through the Spanish ratania.

RHATANY, Brazilian; RHATANY, Para *Krameria argentea*.

RHATANY, Peruvian; RHATANY, Red *Krameria triandra* (Rhatany) Le Strange.

RHATANY, Range *Krameria parviflora*.

RHATANY, Savanilla *Krameria ixene*.

RHATANY, Texan *Krameria lanceolata*.

RHATANY, White *Krameria grayi*.

RHENOSTER *Ranunculus multifidus* S Africa Watt.

RHEUM WEED, Salt *Chelone glabra* (Turtlehead) O P Brown.

RHEUMATISM-ROOT *Jeffersonia diphylla* (Twin-leaf) USA Berdoe. It is a popular American rheumatism remedy.

RHEUMATISM WEED 1. *Apocynum cannabinum* (Indian Hemp) Grieve.1931. 2. *Chimaphila umbellata* (Umbellate Wintergreen) Le Strange. The leaves of the wintergreen are used for rheumatic complaints among other ailments.

RHINEBERRY *Rhamnus cathartica* (Buckthorn) B & H. Apparently a translation from the German name. It also got corrupted to Rainberry.

RHODE ISLAND BENT *Agrostis canina* (Brown Bent) H & P.

RHODES'S VIOLET *Securidaca longipedunculata* (Wild Violet Tree) S Africa Dalziel. 'Violet', because the flowers smell of them.

RHODESIAN ASH *Burkea africana* (Wild Syringa) Palgrave.

RHODESIAN CHESTNUT *Baikiaea plurijuga* (Rhodesian Teak) Palgrave. The upstanding flowers look quite like horse chestnut at a distance.

RHODESIAN CURRANT *Rhus dentata* Watt.

RHODESIAN HOLLY *Psorospermum febrifugum ferrugineum*.

RHODESIAN IRONWOOD *Copaifera mopane* (Mopane) Palgrave.

RHODESIAN JACARANDA *Stereospermum kuntheanum*.

RHODESIAN MAHOGANY *Afzelia cuanzensis* (Pod Mahogany) Dalziel.

RHODESIAN RUBBER *Diplorhynchus condylocarpon mossambicensis*.

RHODESIAN TEAK *Baikiaea plurijuga*.

RHODESIAN TIMOTHY *Setaria sphacelata* S Africa Dalziel.

RHODESIAN TREE HIBISCUS *Thespesia garckeana*.

RHODESIAN WATTLE *Peltophorum africanum*.

RHODESIAN WISTARIA *Bolusanthus speciosus*.

RHODODAPHNE; RHODOPHANES *Nerium oleander* (Rose Bay) T Hill. Of the same import as Turner's Rose Laurel, or its inverted form, Laurier Rose.

RHODODENDRON, Arctic *Rhododendron lapponicum* (Lapland Rose Bay) Blamey.

RHODODENDRON, Common *Rhododendron ponticum*.

RHODODENDRON, Pontic *Rhododendron ponticum* (Common Rhododendron) Forsyth.

RHUBARB 1. *Rheum raponticum*. 2. *Rubus fruticosus* (Bramble) Som. Macmillan. 3. *Rumex alpinus* (Monk's Rhubarb) B & H. Rhubarb is said to come from the river Rha, now the Volga, on whose banks it grows, with the Greek barbaron, foreign. The dried rhizome of Monk's Rhubarb is used as a laxative, for it has similar effects as medicinal Rhubarb.

RHUBARB, Bastard 1. *Rumex patientia* (Patience Dock) Parkinson. 2. *Thalictrum flavum* (Meadow Rue) Gerard. Both have laxative effects, hence the name. Meadow Rue has other 'rhubarb' names, like False, or English, Rhubarb, "which names are taken of the colour, and taste of the roots", according to Gerard, but he nevertheless acknowledges that they are as laxative as true Rhubarb.

RHUBARB, Bastard, Spanish, Great *Thalictrum aquilegifolium* (Greater Meadow Rue) Gerard.

RHUBARB, Bog *Petasites hybridus* (Butterbur) Som. Macmillan; Lincs. B & H. In this case it is the size of the leaves that earns it the name. Cf. Wild, and Poison, Rhubarb.

RHUBARB, Chinese *Rheum officinale* (Turkey Rhubarb) Lloyd. It is also Russian Rhubarb. The names were given in accordance with the country through which it reached the market from its native land, which is Tibet and neighbouring parts of China.

RHUBARB, Devil's 1. *Atropa belladonna* (Deadly Nightshade) Som. Macmillan. 2. *Petasites hybridus* (Butterbur) War, Worcs. F Savage.

RHUBARB, Donkey's *Polygonum cuspidatum* (Japanese Knotweed) Corn. McClintock.1975.

RHUBARB, English 1. *Rheum rhaponticum* (Rhubarb) Parkinson. 2. *Thalictrum flavum* (Meadow Rue) Gerard.

RHUBARB, False *Thalictrum flavum* (Meadow Rue) B & H.

RHUBARB, Guatemala *Jatropha podagrica*. All the members of this genus are purgatives, hence the allusion to rhubarb.

RHUBARB, Gypsy's 1. *Arctium lappa* (Burdock) Som. Macmillan. 2. *Petasites hybridus* (Butterbur) Som. Macmillan. The reference here is to the large leaves.

RHUBARB, Indian *Saxifraga peltata*. They are the North American Indians, and presumably they used this in some way as a purge.

RHUBARB, Monk's 1. *Rumex alpinus*. 2. *Rumex obtusifolius* (Broad-leaved Dock) Flück). 3. *Rumex patientia* (Patience Dock) B & H.

RHUBARB, Pig's *Arctium lappa* (Burdock) Dor. Mabey. Cf. Wild Rhubarb, Gypsy's Rhubarb and Turkey Rhubarb for this. It must be the large leaves that are responsible for these names.

RHUBARB, Poison *Petasites hybridus* (Butterbur) Yks. B & H.

RHUBARB, Poor Man's *Thalictrum flavum* (Meadow Rue). W Miller. Cf. False Rhubarb, English Rhubarb, Bastard Rhubarb.

RHUBARB, Russian *Rheum officinale* (Turkey Rhubarb) Lloyd. See Chinese Rhubarb.

RHUBARB, Shensi *Rheum officinale* (Turkey Rhubarb) Camp. See Chinese Rhubarb.

RHUBARB, Snake's 1. *Arctium lappa* (Burdock) Dor. Macmillan. 2. *Petasites hybridus* (Butterbur) Dor. Macmillan.

RHUBARB, Sorrel *Rheum palmatum*.

RHUBARB, Tart *Rheum rhaponticum* (Rhubarb) USA Sanecki. Cf. Pie-plant, also from America.

RHUBARB, Turkey 1. *Petasites hybridus* (Butterbur) Som. Macmillan. 2. *Rheum officinale*.

RHUBARB, Wild 1. *Arctium lappa* (Burdock) Som. Mabey. 2. *Petasites hybridus* (Butterbur) War, Worcs. F Savage. 3. *Rumex hymenosepalus* (Tanner's Dock) USA Kluckhohn. 4. *Rumex obtusifolius* (Broad-leaved Dock) USA Turner & Bell. 5. *Tussilago farfara* (Coltsfoot) Som. Macmillan.

RHUBARB DOCK, Spanish *Rumex abyssinicus* Howes.

RHUBARB-WEED *Petasites fragrans* (Winter Heliotrope) Dor. Northover.

RHUM HEATH SPOTTED ORCHID *Dactylorchis maculata var. rhoumensis*. It

grows on the machair on Rhum and some of the other Inner Hebridean islands.
RIB *Nasturtium officinale* (Watercress) E Ang. Halliwell.
RIB, Dog's *Plantago lanceolata* (Ribwort Plantain) Gerard.
RIB GRASS *Plantago major* (Great Plantain) S Africa Watt.
RIB PLANTAIN *Plantago lanceolata* (Ribwort Plantain) Gerard.
RIB-SEEDED SAND-MAT *Euphorbia glyptosperma* (Mat Spurge) Jaeger.
RIBANDS, Lady's *Phalaris arundinacea* (Reed-grass) Dev. Friend.1882.
RIBBON, Devil's 1. *Antirrhinum majus* (Snapdragon) Tynan & Maitland. 2. *Linaria vulgaris* (Toadflax) Dyer.
RIBBON FERN *Pteris cretica* (Table Fern) Simmons. Because of the arrangement of its spores,which are borne in continuous bands along the lower leaf margins.
RIBBON GRASS *Phalaris arundinacea* (Reed-grass) USA Cunningham.
RIBBON GUM *Eucalyptus viminalis* (Manna Gum) Polunin.1976.
RIBBON-TREE *Betula verrucosa* (Birch) Lincs. Peacock. The bark of young trees can be pulled off in long strips like ribbons.
RIBBONED MARANTA *Calathea vittata*.
RIBBONWOOD *Hoheria populnea*. Because the inner bark can be peeled in narrow strips.
RIBBONWOOD, Mountain *Gaya lyalli*.
RIBGRASS *Plantago lanceolata* (Ribwort Plantain) Som, E Ang, Staffs, Ches, Yks, Cumb, N'thum, Kirk, Wigt Denham/Hardy; Hants. Cope.
RIBGRASS, Pursh *Plantago purshii* (Pursh Plantain) USA Elmore.
RIBS *Ribes rubrum* (Redcurrant) Grieve.1931. Cf. Reps.
RIBWORT 1. *Coronopus squamatus* (Wartcress) E Ang. Nall. 2. *Plantago lanceolata* (Ribwort Plantain) Watt. 3. *Plantago major* (Great Plantain) S Africa Watt.
RIBWORT, Soldier's *Plantago lanceolata* (Ribwort Plantain) Le Strange. Presumably a result of the children's game involving the decapitation of a stem of Ribwort. The plant is known as Soldiers, Fechters, or Kemps, all meaning the same thing.
RIBWORT PLANTAIN *Plantago lanceolata*.
RIBWORTH *Plantago lanceolata* (Ribwort Plantain) Denham/Hardy.
RICE 1. *Asperula odorata* (Woodruff) Dor. Macmillan. 2. *Galium cruciatum* (Crosswort) Dor. Macmillan. 3. *Galium palustre* (Marsh Bedstraw) Dor. Macmillan. 4. *Oryza sativa*. There are two distinct etymologies here. Oryza sativa gets the name through Old French ris from the Latin oryza, which in turn comes from the Greek and ultimately from somewhere eastern. The other derivation is from OE hris, brushwood, and that applies to all the other plants mentioned here.
RICE, Hungry *Digitaria exilis*. Presumably because the grain is so very small.
RICE, Indian *Zizania aquatica* (Wild Rice) USA Howes.
RICE, Jungle *Echinochloa crus-galli* (Japanese Barnyard Millet) Howes.
RICE, White *Sorbus aria* (Whitebeam) Hants, IOW Cope. Cf. White-leaf Tree. Presumably 'rice' here means brushwood.
RICE, Wild; RICE, Wild, Canada *Zizania aquatica*.
RICE BEAN *Phaseolus calcareus*.
RICE FLOWER 1. *Asperula odorata* (Woodruff) Som. Macmillan. 2. *Scabiosa columbaria* (Small Scabious) S Africa Watt.
RICE GRASS *Spartina townsendii*.
RICE PAPER PLANT 1. *Fatsia japonica* (Castor Oil Plant). 2. *Tetrapanax papyriferum*. Chinese "rice-paper" was made by beating out the pith of the latter shrub.
RICE-ROOT *Fritillaria lanceolata* (Chocolate Lily) USA Turner & Bell. Probably because the bulbs, which are edible, are glutinous, with a slightly sweet taste.
RICHLEAF *Collinsonia canadensis* (Stoneroot) Le Strange. Cf. Richweed.
RICHMOND RIVER PINE *Araucaria cunninghami* (Hoop Pine) Everett.
RICHWEED 1. *Cimicifuga racemosa* (Black

Snakeroot) USA Lloyd. 2. *Collinsonia canadensis* (Stone-root) Le Strange. 3. *Eupatorium urticaefolium* (White Snakeroot) USA Kingsbury. Richweed was given as a name to Black Snakeroot by Gronovius in 1752, because the plant frequents rich woodlands. Stone-root is also known as Richleaf.

RIDDELL'S GROUNDSEL *Senecio riddelli*.

RIDING HOOD, Red *Melandrium rubrum* (Red Campion) Dev, Som. Grigson; Dor. Dacombe.

RIDING HOOD, White 1. *Melandrium album* (White Campion) Dev. Macmillan. 2. *Silene cucubalis* (Bladder Campion) Som. Macmillan.

RIDWEED *Papaver rhoeas* (Red Poppy) IOW Long. 'Redweed', of course.

RIELY *Lolium temulentum* (Darnel) Ire. B & H. Perhaps from the same source as Ray-grass, i.e. French ivraie, drunk, describing the toxic effects.

RIENGA LILY *Arthropodium cirrhatum* (Rock Lily) RHS.

RIFLE-THE-LADIES'-PURSE *Capsella bursa-pastoris* (Shepherd's Purse) Banff. Grigson. It seems a hybrid between the 'purse' names and those inferring robbery, as with Pickpocket, or Pick-purse, explained by Prior as describing the way it robs the farmer by stealing the goodness of his land.

RIGA PINE; RIGA FIR *Pinus sylvestris* (Scots Pine) Murdoch McNeill (Pine); Wilkinson.1981 (Fir). A lot of timber has been imported from the Baltic region under such a name. Cf. Baltic Redwood, Danzig Pine, etc.

RIGGLERS *Primula auricula* (Auricula) Som. Macmillan. A corruption of auriculas.

RILTS *Berberis vulgaris* fruit (Barberry) S E Eng. B & H.

RIM ASH *Celtis occidentalis* (Hackberry) W Miller.

RIMU *Dacrydium cupressinum*. The Maori name for this New Zealand tree.

RIMU, Mountain *Dacrydium laxifolium*.

RING-FINGER 1. *Dactylorchis maculata* (Heath Spotted Orchid) Bucks. Grigson. 2. *Orchis mascula* (Early Purple Orchid) Bucks. B & H. Cf. Cling-finger, and King-finger.

RING MUHLY; RING GRASS *Muhlenbergia torreyi* USA T H Kearney.

RING-O'-BELLS *Endymion nonscriptus* (Bluebell) Lancs. Rohde.

RINGE HEATHER *Erica tetralix* (Cross-leaved Heath) Scot. Jamieson. 'Ringe' is a whisk of heather.

RINGERS, Fairy *Campanula rotundifolia* (Harebell) Dor, Som. Macmillan.

RINGIE, Apple *Artemisia abrotanum* (Southernwood) Scot. Simpson. Conflicting origins are given. One says that 'apple' is from the old 'aplen', church, with 'ringie' as Saint Rin's, or St Ninian's, wood. Another is that the name stems from Appelez Ringan, pray to Ringan. It went through Appleringan to Apple-ringie. Jamieson, though, said it was from French apile, strong, and auronne, southernwood, from abrotanum.

RINGWORM SHRUB *Cassia alata* (Candlestick Senna) Chopra. The leaves are used in India as a local application in skin diseases, especially ringworm.

RINGWORM ROOT *Rhinacanthus nastulus*. In the late 18th century, it was considered a sovereign cure for ringworm.

RIO NUÑEZ COFFEE 1. *Coffea liberica* (Liberian Coffee) Dalziel. 2. *Coffea stenophylla* (Narrow-leaved Coffee) Dalziel. Presumably because it was exported from the port of that name in Guinea.

RIO ROSEWOOD *Dalbergia nigra* (Rosewood) Tampion. Rosewood, because of the faint smell of roses when the wood is freshly cut. Rio, because this is a Brazilian tree.

RIPPLE, Emerald *Peperomia caperata* (Rat Tail Plant) USA Cunningham.

RIPPLE GRASS 1. *Plantago lanceolata* (Ribwort Plantain) Galloway. Denham/Hardy. 2. *Plantago major* (Great Plantain) Scot. Grigson. At first glance this would seem to be a variant of ribwort, but the existence of Rippling Grass in Scotland, and

Rupple Grass in Ireland, seems to indicate otherwise.

RIPPLING GRASS *Plantago lanceolata* (Ribwort Plantain) Lanark. Denham/Hardy.

RISHES *Juncus sp* (Rushes) Hants. Cope. The old pronunciation of rushes.

RISING SUN *Leucanthemum vulgare* (Ox-eye Daisy) Som. Macmillan. Cf. Sun Daisy.

RISP *Ribes rubrum* (Redcurrant) Grieve.1931. Probably originally Raisin Tree, which was Turner's name for the shrub. Scottish variants of this include Rizzles and Rizards. These were reduced to Ribs, Risp and Reps.

RIVERY *Lolium temulentum* (Darnel) Ire. B & H. As with Rye-grass, which was originally Ray-grass, Rivery comes from French ivraie, drunk, a comment on the toxic effects of the plant.

RIX *Phragmites communis* (Reed) Dev. Halliwell.

RIZARDS; RIZZER-BERRIES; RIZZLES *Ribes rubrum* (Redcurrant) all Scot. Jamieson (Rizards, Rizzer-berries); Grigson (Rizzles). All traceable back to 'raisin'.

ROAD-TO-HEAVEN *Polemonium caeruleum* (Jacob's Ladder) Dor. Macmillan. Cf. Ladder-to-heaven.

ROAN-TREE *Sorbus aucuparia* (Rowan) N Eng. Kelly.

ROANOKE BELLS *Mertensia virginica* (Virginia Cowslip) Howes.

ROAST BEEF *Iris foetidissima* (Foetid Iris) Prior. From the smell of the bruised leaf.

ROB ROY *Melandrium rubrum* (Red Campion) Som. Macmillan. A mistaken appellation. It comes from some form of Red Robin, or Poor Robin.

ROB-RUN-UP-DYKE 1. *Galium aparine* (Goose-grass) Cumb. Grigson. 2. *Glechoma hederacea* (Ground Ivy) Cumb. Grigson. Both plants have a host of similar names, as Robin-run-in-the grass, Rabbie-rinnie-hedge, etc.

ROBBER'S LANTERNS *Aesculus hippocastanum* flowers (Horse Chestnut flowers) Dor. Macmillan. Descriptive. Cf. Candles, Christmas Candles, etc.

ROBE, Princess's *Narcissus pseudo-narcissus* (Daffodil) Som. Macmillan. Descriptive. Cf. Hoop Petticoat, Lady's Ruffles, etc.

ROBE, Virgin's *Borago officinalis* (Borage) Cambs. Porter.

ROBERT; ROBERT, Bob *Geranium robertianum* (Herb Robert) Dev. Grigson (Robert); Dor. Macmillan (Bob Robert). Robert was ascribed by Halliwell to *Erodium cicutarium*, Storksbill. But did he really mean this, and not Herb Robert?

ROBERT, Herb 1. *Geranium robertianum*. 2. *Salvia coccinea* (Red Sage) Dev, Som. Friend.1882. Who is Robert? It is apparently dedicated to St Robert on 29 April, but that may be a late connection. Perhaps it is Robin Hood, or Robin Goodfellow (the seed vessels, with their sharp needles, are known as Pook Needles). In Germany, the plant was used to cure a disease known as Ruprechtsplage, from Robert, Duke of Normandy. Yet another possibility is Robert, an 11th century Abbot of Molesne. And then again, Robert may simply be ruber, red.

ROBERT, Poor; ROBERT, Stinking *Geranium robertianum* (Herb Robert) Dev, Som. Macmillan (Poor Robert); Donegal Grigson (Stinking Robert). 'Stinking' is a reference to the characteristic smell when bruised.

ROBIN 1. *Geranium lucidum* (Shining Cranesbill) Dev. Grigson. 2. *Geranium robertianum* (Herb Robert) Dev. Friend.1882; Wilts. Grigson. 3. *Melandrium rubrum* (Red Campion) Dev. Friend.1882.

ROBIN, Blue *Borago officinalis* (Borage) Som. Macmillan.

ROBIN, Bob *Melandrium rubrum* (Red Campion) Corn. Grigson; Som, Wilts. Macmillan.

ROBIN, Cant *Rosa pimpinellifolia* (Burnet Rose) Fife. Grigson.

ROBIN, Cock 1. *Lychnis flos-cuculi* (Ragged Robin) Som. Macmillan. 2. *Melandrium* *rubrum* (Red Campion) Corn. Grigson; Dev. Friend.1882; Som. Macmillan.

ROBIN, Cock, White *Silene cucubalis* (Bladder Campion) Som. Macmillan.

ROBIN, Herb *Geranium robertianum* (Herb Robert) Usher.

ROBIN, Little 1. *Geranium purpureum* (Purple Cranesbill) McClintock. 2. *Geranium robertianum* (Herb Robert) Friend.

ROBIN, Poor 1. *Geranium robertianum* (Herb Robert) Dev, Som. Macmillan. 2. *Lychnis flos-cuculi* (Ragged Robin) Som. Grigson. 3. *Melandrium rubrum* (Red Campion) Dev. Friend.1882. 4. *Odontites verna* (Red Bartsia) N'thum. Grigson.

ROBIN, Ragged 1. *Epilobium hirsutum* (Great Willowherb) Som. Macmillan. 2. *Geranium robertianum* (Herb Robert) Bucks. Grigson. 3. *Lychnis flos-cuculi*. 4. *Lythrum salicaria* (Purple Loosestrife) Wilts. Macmillan. 5. *Senecio jacobaea* (Ragwort) Yks. Grigson. 6. *Silene cucubalis* (Bladder Campion) Som. Macmillan.

ROBIN, Red 1. *Geranium robertianum* (Herb Robert) Som, Glos. Grigson USA House. 2. *Lychnis flos-cuculi* (Ragged Robin) Som. Grigson. 3. *Melandrium rubrum* (Red Campion) Dev. Friend.1882; Dor, Som. Grigson; Wilts. Dartnell & Goddard. 4. *Orchis mascula* (Early Purple Orchis) Herts, Ess, Norf, Cambs, Pemb. Grigson. 5. *Polygonum aviculare* (Knotgrass) Friend.

ROBIN, Red, Little *Geranium robertianum* (Herb Robert) Som. Macmillan.

ROBIN, Round 1. *Geranium robertianum* (Herb Robert) Dev, Kent Grigson. 2. *Melandrium rubrum* (Red Campion) Dev. Friend.1882.

ROBIN, Rough *Lychnis flos-cuculi* (Ragged Robin) Cumb. Grigson.

ROBIN, Stick *Galium aparine* (Goose-grass) Tynan & Maitland.

ROBIN, Wake 1. *Arum maculatum* (Cuckoo-pint) Turner. 2. *Melandrium rubrum* (Red Campion) Yks. Grigson. 3. *Trillium erectum* (Bethroot) Grieve.1931. 4. *Trillium grandiflorum* (Wood Lily) USA Densmore. 5. *Orchis mascula* (Early Purple Orchid) Ches. Holland. The meaning of this name is rather different to the rest of the 'robin' names. 'Wake' means alive, and 'robin', if we are to believe Prior, is the French robinet, which once meant penis (the modern meaning is a tap, so that connotation is very possible). So Wake Robin means 'erect penis'. Cuckoo-pint means the same, and there are a lot of other names for it with the same import. So there are for Early Purple Orchid.

ROBIN, Wake, American *Arisaema triphylla* (Jack-in-the-pulpit) Le Strange.

ROBIN, White *Melandrium album* (White Campion) Dor. Macmillan.

ROBIN-CATCH-THE-HEDGE *Galium aparine* (Goose-grass) Tynan & Maitland.

ROBIN FLOWER 1. *Geranium robertianum* (Herb Robert) Tynan & Maitland. 2. *Melandrium rubrum* (Red Campion) Som. Tongue.

ROBIN HOOD 1. *Agrostemma githago* (Corn Cockle) Dor. Grigson. 2. *Anemone coronaria* (Poppy Anemone) B & H. 3. *Geranium robertianum* (Herb Robert) Dev. Grigson; Som. Tongue. 4. *Lychnis flos-cuculi* (Ragged Robin) Dev, Som, Dor, Dur. Grigson. 5. *Melandrium rubrum* (Red Campion) W Eng. Macmillan.

ROBIN HOOD, Red *Melandrium rubrum* (Red Campion) Wilts. Dartnell & Goddard.

ROBIN HOOD, White 1. *Melandrium album* (White Campion) Som. Macmillan. 2. *Silene cucubalis* (Bladder Campion) Som. Grigson; Wilts. Dartnell & Goddard.

ROBIN HOOD; LITTLE JOHN *Geranium robertianum* (Herb Robert) Dor. Grigson.

ROBIN HOOD'S FEATHER *Clematis integrifolia* (Hungarian Climber) Cumb. Wright.

ROBIN HOOD'S FETTER 1. *Clematis integrifolia* (Hungarian Climber) Cumb. Wright. 2. *Clematis vitalba* (Old Man's Beard) Cumb. B & H.

ROBIN HOOD'S HATBAND *Lycopodium inundatum* (Clubmoss) Halliwell.

ROBIN-I'-THE-HEDGE 1. *Geranium robertianum* (Herb Robert) Yks. Carr. 2. *Melandrium rubrum* (Red Campion) Yks. Grigson.

ROBIN-IN-THE-HOUSE *Melandrium rubrum* (Red Campion) Tynan & Maitland.

ROBIN REDBREAST 1. *Geranium robertianum* (Herb Robert) N'thants. Grigson. 2. *Melandrium rubrum* (Red Campion) Corn, Dev. Grigson.

ROBIN REDBREAST'S CUSHION *Rosa canina* the galls that grow on it Suss. Latham.

ROBIN REDSHANKS *Geranium robertianum* (Herb Robert) Yks. Grigson.

ROBIN-ROUND-THE-HEDGE; ROBIN-RUN-I'-THE-HEDGE; ROBIN-RUN-BY-THE-GRASS; ROBIN-RUN-IN-THE-GRASS *Galium aparine* (Goose-grass) Vesey-Fitzgerald (run-by-the-grass); Lancs. Nodal (run i'-the-hedge).

ROBIN-RUN-IN-THE-FIELD; ROBIN-RUN-THE-DYKE *Convolvulus arvensis* (Field Bindweed) Som. Macmillan (Field); N Eng. Wakelin (Dyke). Cf. Jack-run-in-the-country. Dyke would mean a hedge, or perhaps a wall, in the northern counties.

ROBIN-RUN-THE-HEDGE 1. *Calystegia sepium* (Great Bindweed) Hants, Ches, Lancs, N. Ire Grigson. 2. *Galium aparine* (Goose-grass) Grigson. 3. *Galium verum* (Lady's Bedstraw) Yks. Addy. 4. *Glechoma hederacea* (Ground Ivy) Som, Suss, Worcs, Ches, Leics, Notts, Derb, Lancs, Ire. Grigson. 5. *Melandrium rubrum* (Red Campion) Dor. Macmillan. 6. *Solanum dulcamara* (Woody Nightshade) Lancs. Grigson.

ROBIN-UNDER-THE-HEDGE *Cerastium holostoides* (Mouse-ear) Vesey-Fitzgerald.

ROBIN'S EYE 1. *Geranium robertianum* (Herb Robert) Dev. Friend.1882; Wilts. Dartnell & Goddard; Suff. B & H. 2. *Melandrium rubrum* (Red Campion) Dev. Friend.1882. 3. *Myosotis arvensis* (Field Forget-me-not) Dev. Grigson. 4. *Polygala vulgaris* (Milkwort) Hants. Cope.

ROBIN'S EYE, Small *Geranium robertianum* (Herb Robert) Glos. Grigson.

ROBIN'S EYE, White *Stellaria holostea* (Greater Stitchwort) Wilts. Goddard.

ROBIN'S FLOWER 1. *Geranium robertianum* (Herb Robert) Dev, Som. Macmillan. 2. *Melandrium rubrum* (Red Campion) Corn, Dev. Grigson.

ROBIN'S PILLOW; ROBIN'S PINCUSHION galls on *Rosa canina* (Dog Rose) Page.

ROBIN'S PLANTAIN *Erigeron pulchellis* USA Tolstead.

ROBLE (BEECH) *Nothofagus obliqua*.

ROBLUM *Prunus cerasifera* (Cherry-Plum) Ess. Gepp. A simplification of Myrobalan, which is another name for this.

ROBUSTA COFFEE *Coffea canephora*. Robusta is the alternative specific name.

ROCAMBOLE; ROCCOMBOLE *Allium scorodoprasum* (Sand Leek) Prior (Rocambole); Raper (Roccombole). German Roggen, rye and Bolle, onion, i.e. an onion growing in fields of rye. Earlier, Candolle had said that the derivation meant an onion growing among rocks.

ROCHLIS *Rhinanthus crista-galli* (Yellow Rattle) Heref. Grigson. Rochlis apparently means death rattle in the dialect of Hereford and Pembroke. Grigson compares the Flemish rochel.

ROCK, White 1. *Arabis alpina* (Alpine Rock-cress) Ches. B & H. 2. *Cerastium tomentosum* (Snow-in-summer) Som. Elworthy.

ROCK-BUSH *Piper dilatatum* Barbados Gooding.

ROCK-CRESS *Arabis sp.*

ROCK-CRESS, Alpine *Arabis alpina*.

ROCK-CRESS, Annual *Arabis recta*.

ROCK-CRESS, Bristol *Arabis stricta*. The Avon Gorge is its only British station.

ROCK-CRESS, Fringed *Arabis hirsuta* (Hairy Rock-cress) Clapham.

ROCK-CRESS, Hairy *Arabis hirsuta*.

ROCK-CRESS, Lyre-leaved *Arabis lyrata*.

ROCK-CRESS, Mountain *Cardaminopsis*

petraea (Northern Rock-cress) J A MacCulloch.1905.

ROCK-CRESS, Naked-stalked *Teesdalia nudicaulis* (Shepherd's Cress) Curtis.

ROCK-CRESS, Prince's *Arabis pulchra*.

ROCK-CRESS, Tall *Cardaminopsis arenosa*.

ROCK-CRESS, Tower *Arabis turritis*.

ROCK-CROP *Sedum acre* (Biting Stonecrop) Corn. B & H.

ROCK ELM 1. *Chlorophora excelsa* (African Oak) Dalziel. 2. *Ulmus racemosa* (Cork Elm) USA Tolstead.

ROCK PLANT *Sedum acre* (Biting Stonecrop) Dev. Friend.1882.

ROCK STONECROP *Sedum reflexum* (Yellow Stonecrop) Polunin. There seems to be an unnecessary repetition of 'rock' and 'stone' with this name.

ROCKET 1. *Eruca sativa*. 2. *Erysimum asperum* (Desert Wallflower) USA Gates. 'Rocket' is Latin eruca -through Italian ruca, dim. ruchetta, French roquette.

ROCKET, Austrian *Sisymbrium austriacum*.

ROCKET, Base *Reseda luteola* (Dyer's Rocket) Prior. 'Base' presumably is from its use in dyeing, though that word is used to describe a mordant. On the other hand, it may just mean low-growing.

ROCKET, Bastard *Sinapis arvensis* (Charlock) B & H.

ROCKET, Blue 1. *Aconitum anglicum* (Wild Monkshood) Ire. B & H. 2. *Aconitum napellus* (Monkshood) Som. Macmillan; South. Africa Watt. 3. *Dactylorchis incarnata* (Early Marsh Orchid) Donegal. Grigson. 4. *Endymion nonscriptus* (Bluebell) Ire. B & H.

ROCKET, Corn *Bunias erucago* (Crested Bunias) Bianchini.

ROCKET, Crambling *Reseda lutea* (Wild Mignonette) B & H. Crambling probably means scrambling, or wandering.

ROCKET, Cress *Rorippa sylvestris* (Creeping Yellow Cress) Camden/Gibson.

ROCKET, Crow *Eupatorium cannabinum* (Hemp Agrimony) Donegal Grigson.

ROCKET, Dame's *Hesperis matronalis* (Sweet Rocket) USA Upham. 'Dame's' here is a corruption of Damask. The plant was Viola Damascena, which became in French Violette de Dames, misspelt to Violette des Dames, and so into English Dame's Violet.

ROCKET, Dyer's *Reseda luteola*. One of the oldest of vegetable dyes, certainly in use since neolithic times.

ROCKET, Eastern *Sisymbrium orientale*.

ROCKET, Garden 1. *Eruca sativa* (Rocket) Tradescant. 2. *Hesperis matronalis* (Sweet Rocket) Prior.

ROCKET, Hairy *Erucastrum gallicum*.

ROCKET, Italian *Reseda lutea* (Wild Mignonette) B & H.

ROCKET, London *Sisymbrium irio*. It is said that it first appeared in London in the spring following the Great Fire.

ROCKET, London, False *Sisymbrium loeslii*. Another rare London native.

ROCKET, Meadow *Dactylorchis incarnata* (Early Marsh Orchid) Dumf. Grigson. Cf. Blue Rocket.

ROCKET, Night-smelling *Hesperis matronalis* (Sweet Rocket).

ROCKET, Prairie *Erysimum asperum* (Desert Wallflower) USA Elmore.

ROCKET, Purple *Chamaenerion angustifolium* (Rosebay Willowherb) Leyel.1937.

ROCKET, Red *Hesperis matronalis* (Sweet Rocket) Ches. B & H. Given to the lilac-flowered variety.

ROCKET, Roman *Eruca sativa* (Rocket) Parkinson.

ROCKET, Salad *Eruca sativa* (Rocket) A W Smith.

ROCKET, Sand *Diplotaxis muralis* (Stinkweed) D Grose.

ROCKET, Scrambling *Sisymbrium officinale* (Hedge Mustard) B & H.

ROCKET, Sea *Cakile maritima*.

ROCKET, Sea, American *Cakile edulenta*.

ROCKET, Sweet *Hesperis matronalis*.

ROCKET, Tall *Sisymbrium altissimum* (Tumbling Mustard) Clapham.

ROCKET, Tansy-leaved *Sisymbrium tanacetifolium*.

ROCKET, Wall 1. *Diplotaxis muralis* (Stinkweed) Clapham. 2. *Diplotaxis tenuifolia*.
ROCKET, Wall, White *Diplotaxis erucoides*.
ROCKET, Water; ROCKET, Water, Creeping *Rorippa sylvestris* (Creeping Yelow Cress) Hutchinson (Water Rocket); Curtis (Creeping Water Rocket).
ROCKET, White *Hesperis matronalis* (Sweet Rocket) B & H.
ROCKET, Wild 1. *Cardamine pratensis* (Lady's Smock) Border Hardy. 2. *Diplotaxis muralis* (Stinkweed) Curtis. 3. *Reseda luteola* (Dyer's Rocket) S Africa Watt.
ROCKET, Winter *Barbarea vulgaris* (Winter Cress) B & H.
ROCKET, Wound *Barbarea vulgaris* (Winter Cress) Turner. "for it is good for a wound".
ROCKET, Yellow 1. *Barbarea vulgaris* (Winter Cress) Som. Macmillan. 2. *Lysimachia vulgaris* (Yellow Loosestrife) Bucks. B & H. 3. *Reseda luteola* (Dyer's Rocket) N Eng, Scot. B & H.
ROCKET, Yellow, Early-flowering *Barbarea verna* (American Land Cress) Clapham.
ROCKET, Yellow, Intermediate *Barbarea intermedia* (Early Winter Cress) Clapham.
ROCKET, Yellow, Small-flowered *Barbarea stricta* (Small-flowered Land Cress) Clapham.
ROCKET CRESS *Barbarea vulgaris* (Winter Cress) H & P.
ROCKET GENTLE *Eruca sativa* (Rocket) Parkinson.
ROCKET WATER CRESS *Nasturtium officinale* (Watercress) Turner.
ROCKFOIL *Saxifraga sp.*
ROCKSPRAY *Cotoneaster microphylla* (Rose Box Tree) McClintock. Cf. Wallspray for this.
ROCKWEED *Helianthus salicifolius* (Willow-leaved Wild Sunflower) USA Gates.
ROCKWOOD *Asperula odorata* (Woodruff) Dev. Friend.1882. Puzzling at first glance, but this is really a version of Woodruff with the syllables interchanged.
ROCKY MOUNTAIN ASPEN *Populus tremuloides* (American Aspen) USA Elmore.
ROCKY MOUNTAIN BEE PLANT; ROCKY MOUNTAIN BEEWEED *Cleome serrulata* Bee Plant is the usual name; Beeweed is in Elmore.
ROCKY MOUNTAIN BIRCH *Betula fontinalis* (Streamside Birch) USA Elmore.
ROCKY MOUNTAIN CEDAR; ROCKY MOUNTAIN JUNIPER *Juniperus scopulorum* (Western Red Cedar) USA Elmore (Cedar); USA Weiner (Juniper).
ROCKY MOUNTAIN FIR *Abies lasiocarpa* (Alpine Fir) Usher.
ROCKY MOUNTAIN FLAX *Linum lewisii*.
ROCKY MOUNTAIN GRAPE *Mahonia aquifolium* (Oregon Grape) Hutchinson.
ROCKY MOUNTAIN IRIS *Iris missouriensis* (Missouri Iris) USA T H Kearney.
ROCKY MOUNTAIN MAPLE *Acer glabra*.
ROCKY MOUNTAIN OAK *Quercus undulata* (Wavyleaf Oak) Chamberlin.
ROCKY MOUNTAIN PARNASSIA *Parnassia fimbriata*.
ROCKY MOUNTAIN SAGE 1. *Artemisia tridentata* (Sagebrush) Robbins. 2. *Salvia reflexa* (Annual Sage) USA T H Kearney.
ROCKY MOUNTAIN WHITE PINE *Pinus flexilis* (Limber Pine) Howes.
ROCKY MOUNTAIN WHORTLEBERRY *Vaccinium oreophibium*.
ROCKY MOUNTAIN YELLOW PINE *Pinus brachyptera*.
ROD, Aaron's 1. *Agrimonia eupatoria* (Agrimony) Som. Grigson. 2. *Arum maculatum* (Cuckoo-pint) Skinner. 3. *Kniphofia uvaria* (Red Hot Poker) Som. Macmillan. 4. *Solidago virgaurea* (Golden Rod) Corn, Som, Shrop, War. Vesey-Fitzgerald. 5. *Verbascum thapsus* (Mullein) Som, Glos, Lincs, Midl, Scot. Grigson; USA Henkel. 6. *Verbascum virgatum* (Large-flowered Mullein) S Africa. Watt. In most cases, it is the upright growth that provides the imagery.

The odd one is Cuckoo-pint, the sequence being Arum, Aron to Aaron. 'Rod' has no significance, unless the shape of the spadix is relevant.

ROD, Bloody *Cornus sanguinea* (Dogwood) B & H. Cf. Bloody Twig. Both names are from the dark crimson winter colouring of the twigs.

ROD, Golden 1. *Agrimonia eupatoria* (Agrimony) Dor. Grigson. 2. *Hypericum maculatum* (Imperforate St John's Wort) Dev,Som. Grigson. 3. *Hypericum tetrapterum* (St Peter's Wort) Dev, Som. Grigson. 4. *Sarothamnus scoparius* (Broom) Som. Macmillan. 5. *Solidago virgaurea*. 6. *Verbascum thapsus* (Mullein) Dev, Som. Friend.1882. The specific name of *Solidago virgaurea* is Latin virga, rod. All these are descriptive, and obviously so, with the possibly exception of broom.

ROD, Golden, Blue-stemmed *Solidago caesia*.

ROD, Golden, Broad-leaved *Solidago flexicaulis* (Zigzag Golden Rod) USA House.

ROD, Golden, Canadian *Solidago altissima*.

ROD, Golden, Downy *Solidago puberula*.

ROD, Golden, Early *Solidago gigantea*.

ROD, Golden, Fragrant *Solidago graminifolia*.

ROD, Golden, Hardleaf *Solidago rigida*.

ROD, Golden, Late *Solidago bicolor* (White Golden Rod) USA House.

ROD, Golden, Rayless *Haplopappus heterophyllus*.

ROD, Golden, Rock *Solidago altissima* (Canadian Golden Rod) USA House.

ROD, Golden, Seaside *Solidago sempervirens*.

ROD, Golden, Tall *Solidago altissima* (Canadian Golden Rod) Youngken.

ROD, Golden, Three-ribbed, Smooth *Solidago gigantea* (Early Golden Rod) Grieve.

ROD, Golden, White *Solidago bicolor*.

ROD, Golden, Wreath *Solidago caesia* (Blue-stemmed Golden Rod) USA House.

ROD, Golden, Zigzag *Solidago flexicaulis*.

ROD, King's *Narthecium ossifragum* (Bog Asphodel) Tynan & Maitland. Is the identification correct? It sounds as if *Asphodelus ramosus* is more likely.

ROD, Shepherd's 1. *Dipsacus fullonum* (Teasel) Som. Macmillan. 2. *Dipsacus pilosum* (Small Teasel) McClintock. The old name for teasel was virga pastoris, of which Shepherd's Rod is a direct translation.

ROD, Silver 1. *Asphodelus ramosus* (White Asphodel) J Smith. 2. *Solidago bicolor* (White Golden Rod) USA House.

ROD, Yellow *Linaria vulgaris* (Toadflax) Ches. B & H.

RODDEN; RODDIN; RODDON *Sorbus aucuparia* (Rowan) all Scot. McNeill (Rodden); Grigson (Roddin); B & H (Roddon). Either a version of Rowan, or a reference to the colour of the berries.

RODEN; RODIN; RODON *Sorbus aucuparia* (Rowan) all Scot. Denham/Hardy (Roden); B & H (Rodin, Rodon).

RODEN-QUICKEN; RODEN-QUICKEN-ROWAN (Rowan) both Leland.

RODEWORT; RODSGOLD *Calendula officinalis* (Marigold) Cockayne (Rodewort); B & H (Rodsgold). 'Ruddes' was a commonly used name for marigolds at one time; Gerard uses it, for example. Rodsgold may be a way of pinpointing the plant, for 'gold' can mean Corn Marigold as well, or it may be a way of describing the colour more accurately, for rud is red.

ROE-BRIAR *Rosa canina* (Dog Rose) Som. Grigson.

ROEBUCK *Rubus saxatilis* (Stone Bramble) Cumb. Grigson; Scot. Jamieson.

ROEBUCK-BERRY *Rubus chamaemorus* (Cloudberry) Hutchinson.

ROGATION FLOWER *Polygala vulgaris* (Milkwort) Gerard. "of which floures the maidens which use in the countries to walke the Procession do make themselves garlands and Nosegaies". Cf. Cross-flower, Procession-flower, Gang Flower. All these names refer to the use of the flower in the

Rogation-tide processions, probably a continental tradition, according to Grigson.

ROGER, Hasty 1. *Lapsana communis* (Nipplewort) Dev. Grigson. 2. *Scrophularia nodosa* (Figwort) W Eng. Grigson.

ROGER, Stinking 1. *Ballota nigra* (Black Horehound) Shrop. Grigson. 2. *Geranium robertianum* (Herb Robert) Lancs. Grigson. 3. *Hyoscyanum niger* (Henbane) Cumb. Grigson. 4. *Frangula alnus* (Alder Buckthorn) Ches. Grigson. 5. *Scrophularia aquatica* (Water Betony) Ches. Holland; Cumb, Ayr, N Ire. Grigson. 6. *Scrophularia nodosa* (Figwort) Ches, Cumb, Ayr, N Ire. Grigson.

ROGUERY *Centranthus ruber* (Spur Valerian) Som. Macmillan.

ROGUES, Queen's; ROGUE'S GILLIFLOWER *Hesperis matronalis* (Sweet Rocket) Tynan & Maitland (Queen's Rogues); Gerard (Rogue's Gilliflower). Perhaps 'rouge' originally, in which case Rogue's Gilliflower would mean simply Red Gilliflower.

ROLLINGWEED, White *Amaranthus albus* (White Pigweed) USA Elmore. It is one of those plants that break off at ground level when the fruits are ripe, and scatter their seeds as the wind blows them away. Cf. Tumbleweed, a better known name, not confined to this plant.

ROMAINE LETTUCE *Lactuca sativa var. longifolia* (Cos Lettuce) Brouk.

ROMAN CAMOMILE *Chamaemelon nobile* (Camomile) Clair. First so named by the 16th century German humanist Joachim Canerarius, who found it growing near Rome.

ROMAN CANDLES 1. *Aesculus hippocastanum* flowers (Horse Chestnut flowers). Som. Macmillan. 2. *Kniphofia uvaria* (Red Hot Poker) Som. Macmillan. Descriptive, in both cases.

ROMAN CORIANDER *Nigella sativa* (Fennel-flower) Grieve.1931.

ROMAN CYPRESS *Cupressus sempervirens* (Common Cypress) F Savage. Cf. Mediterranean, or Italian, Cypress.

ROMAN FENNEL *Foeniculum dulce* (Finocchio) Sanecki. Cf. Florence Fennel.

ROMAN JASMINE; ROMAN JESSAMINE *Philadelphus coronarius* (Mock Orange) Dor. Macmillan (Jasmine); Dev. Choape (Jessamine). A reasonable name, for this is a shrub from southern Europe.

ROMAN KALE *Beta vulgaris var. cicla* (Swiss Chard) Bianchini.

ROMAN LAUREL *Laurus nobilis* (Bay) Prior.

ROMAN NETTLE *Urtica pilulifera*. A weed of southern Europe, which used to grow at Romney, in Kent. The name, in fact, is derived from Romney, not Roman, in spite of a tradition that it owed its introduction to the Romans, who used it like spinach, and also to rub and chafe their limbs with, to fight against the cold.

ROMAN PLANT 1. *Chenopodium bonus-henricus* (Good King Henry) Lancs. Grigson. 2. *Myrrhis odorata* (Sweet Cicely) Lancs. Grigson.

ROMAN RAGWEED *Ambrosia artemisifolia* (American Wormwood) Polunin. It is an American plant, in spite of this name.

ROMAN ROCKET *Eruca sativa* (Rocket) Parkinson.

ROMAN WILLOW *Syringa vulgaris* (Lilac) Lincs. B & H.

ROMAN WORMWOOD 1. *Ambrosia artemisiaefolia* (American Wormwood) Allan. 2. *Artemisia pontica*. Cf. Roman Ragwort for Ambrosia. But it is an American plant.

ROMMY; ROMS *Allium ursinum* (Ramsons) both Yks. B & H. From Ramsons via Rams.

RONE *Sorbus aucuparia* (Rowan) Scot. Jamieson.

RONNOCHS *Agropyron repens* (Couchgrass) Grigson.1959.

ROOF HOUSELEEK *Sempervivum tectorum* (Houseleek) RHS. 'Roof' in acknowledgement of its favoured habitat,which is why it bears the specific name *tectorum*.

ROOK'S FLOWER *Endymion nonscriptus*

(Bluebell) Dev. Macmillan. Crows rather than rooks are the usual birds mentioned in association with bluebells.

ROOSTER FLOWER *Aristolochia brasiliensis*.

ROOT-CORN, Sweet *Calathea allovia*.

ROPE, Hag; ROPE, Hay *Clematis vitalba* (Old Man's Beard) both Som. Elworthy (Hag); Grigson (Hay). Both of these are from OE haga, hedge.

ROPE, Holy 1. *Eupatorium cannabinum* (Hemp Agrimony) Prior. 2. *Galeopsis tetrahit* (Hemp Nettle) B & H. Named after the rope with which Jesus was bound.

ROPE, St Anthony's *Ranunculus bulbosus* (Bulbous Buttercup) Gerard. 'Rope' here should be 'rape'.

ROPE, Skipping *Clematis vitalba* (Old Man's Beard) Wilts. Dartnell & Goddard.

ROPEWIND 1. *Calystegia sepium* (Great Bindweed) Salisbury. 2. *Convolvulus arvensis* (Field Bindweed) Som. Macmillan. In other words, a rope-like winding plant.

ROPPERS *Digitalis purpurea* (Foxglove) Wilts. Dartnell & Goddard. A version of Rapper, 'rap' being the Wiltshire word for 'pop'. Cf. the many names on that theme.

ROQUETTE *Hesperis matronalis* (Sweet Rocket) Hemphill. i.e. Rocket.

ROSAMUND *Allium ursinum* (Ramsons) Ches. Holland. This must surely be an attempt at Ransoms, probably through the local form Rosems.

ROSARY PEA *Abrus precatorius* (Precatory Bean) USA Kingsbury. Cf. Prayer Beads, etc.

ROSARY ROOT *Apios americana* (Groundnut) USA Yarnell. Descriptive of the way the nuts are formed along the roots.

ROSARY VINE *Ceropegia woodi*. It has hanging strings of tiny leaves. Cf. String-of-hearts.

ROSE, Abyssinian *Rosa sancta*.

ROSE, Alpine *Rosa pendula*.

ROSE, Apothecaries' *Rosa gallica* (French Rose) Rohde. It was commonly used for medicinal purposes, and from a very remote period. Later, an infusion of red rose petals, acidulated with sulphuric acid, and slightly sweetened, came to be used as a vehicle for other medicines.

ROSE, Apple; ROSE, Apple-fruited *Rosa pomifera*.

ROSE, Arkansas *Rosa arkansana*.

ROSE, Ayrshire *Rosa arvensis* (Field Rose) Bunyard.

ROSE, Barrow; ROSE, Burrow *Rosa pimpinellifolia* (Burnet Rose) Pemb. B & H (Barrow); Pemb. Grigson (Burrow). Sand-dunes are meant by barrow, or burrow.

ROSE, Bell *Narcissus pseudo-narcissus* (Daffodil) Som. B & H.

ROSE, Black *Helleborus niger* (Christmas Rose) J Smith. 'Black' to conform with the specific name, and so called, wrote Reginald Farrer, "because its heart, or root, is black, while its face shines with a blazing white innocence, unknown to the truly pure of heart". But the name is a hybrid - Black Hellebore x Christmas Rose.

ROSE, Bobby *Trifolium repens* (White Clover) Corn. Grigson.

ROSE, Bourbon *Rosa bourboniana*. The reference is to the island of Bourbon, now known as Réunion, east of Madagascar, where the original plant came from (1819). It is an accidental cross of China Rose and Damask Rose.

ROSE, Boursault *Rosa pendula x chinensis*.

ROSE, Briar *Rosa canina* (Dog Rose) N Eng. Grigson.

ROSE, Brid *Rosa pimpinellifolia* (Burnet Rose) Ches. Holland. i.e. bird-rose.

ROSE, Burnet *Rosa pimpinellifolia*. 'Burnet', because the leaves are like those of the Salad Burnet.

ROSE, Burr *Rosa roxburghii*.

ROSE, Butter 1. *Primula veris* (Cowslip) Dev. Friend. 2. *Primula vulgaris* (Primrose) Dev. Fernie. 3. *Ranunculus acris* (Meadow Buttercup)Dev. Friend. 4. *Ranunculus bulbosus* (Bulbous Buttercup) Dev. Grigson. 5. *Ranunculus repens* (Creeping Buttercup) Dev. Grigson.

ROSE, Buttonhole, D'Orsay's *Rosa virginiana pl* (St Mark's Rose) Sitwell.

ROSE, Cabbage 1. *Paeonia mascula* (Peony) Som. Macmillan. 2. *Rosa centifolia.*

ROSE, Canker 1. *Papaver rhoeas* (Red Poppy) Norf, Suff. B & H. 2. *Rosa canina* (Dog Rose) Dev. Friend.1882; Som, Kent. Grigson; Wilts. Dartnell & Goddard; Herts. Jones-Baker. 'Canker' is a name given to the galls on a wild rose, so it is not clear whether the name refers to that or the rose itself.

ROSE, Capucine *Rosa foetida.*

ROSE, Carnation *Rosa fimbriata.* An older specific name is *dianthiflora*. That accounts for 'carnation'.

ROSE, Cat 1. *Rosa arvensis* (Field Rose) Ches. B & H. 2. *Rosa canina* (Dog Rose) Ches. Grigson. 3. *Rosa pimpinellifolia* (Burnet Rose) Yks. Grigson.

ROSE, Ceylon *Nerium oleander* (Rose Bay) Watt.

ROSE, Chestnut 1. *Rosa microphylla* Bunyard. 2. *Rosa roxburghii* (Burr Rose) Thomas. The hips of *R microphylla* are so covered with bristles that the resemblance is made to a sweet chestnut. Similarly with the other one, hence the epithet 'Burr'.

ROSE, China *Rosa indica.*

ROSE, Chinese *Hibiscus rosa-sinensis* (Rose of China) Perry.1979.

ROSE, Choop *Rosa canina* (Dog Rose) Cumb. Grigson. Cf. Choop-tree. Choops are hips.

ROSE, Christmas 1. *Eranthis hyemalis* (Winter Aconite) Som. Macmillan. 2. *Helleborus niger.* Cf. Christmas-flower, New Year's Rose and Winter Rose for the hellebore.

ROSE, Christmas, Wild *Helleborus viridis* (Green Hellebore) Som. Macmillan.

ROSE, Cinnamon *Rosa cinnamonea.*

ROSE, Cliff *Armeria maritima* (Thrift) Dev. Friend.1882.

ROSE, Clover *Trifolium pratense* (Red Clover) Dev. Grigson.

ROSE, Cock; ROSE, Cop; ROSE, Copper; ROSE, Cup *Papaver rhoeas* (Red Poppy) Yks. Morris; Scot. Grigson (Cock); Som, E Ang, Yks, N'thum, Scot. Grigson (Cop); Halliwell (Copper); N Eng. B & H (Cup). In spite of the fact that there are other 'cock' names for the poppy, the original is probably 'cop'. It is a button, like the shape of the capsules. 'Cup' here is the same as 'cop'.

ROSE, Corn 1. *Agrostemma githago* (Corn Cockle) S Africa Watt. 2. *Papaver rhoeas* (Red Poppy) Gerard. 3. *Rosa arvensis* (Field Rose) B & H.

ROSE, Corn, Red *Papaver rhoeas* (Red Poppy) Turner.

ROSE, Cretan *Cistus ladaniferus* (Gum Cistus) Genders.1972.

ROSE, Cuckoo *Narcissus pseudo-narcissus* (Daffodil) Som. Elworthy.

ROSE, Damask 1. *Rosa damascena.* 2. *Rosa gallica* (French Rose) Surr. Fluckiger. *Rosa damascena* was so named because of the tradition that they were brought to Europe by the Crusaders from Damascus. But the word 'damask' may once have referred to the colour ("damasked cheeks", for example).

ROSE, Damask, Autumn *Rosa damascena var. bifera.*

ROSE, Damask, Painted *Rosa damascena* 'Leda'.

ROSE, Desert *Adenium coetanum.*

ROSE, Dike *Rosa canina* (Dog Rose) Cumb. Grigson. 'Dike' here means hedge.

ROSE, Dog 1. *Rosa canina.* 2. *Viburnum opulus* (Guelder Rose) Som. Macmillan. Dog Rose for *Rosa canina* is a translation of both the Greek and Latin names for the plant. Some say it is so called because it was supposed to cure the bite of a mad dog. Others, though, say that 'dog' may actually be 'dag', a reference to the dagger-shaped thorns. But 'dog' usually means an inferior species when put in front of the name.

ROSE, Downy *Rosa tomentosa.*

ROSE, Early *Primula vulgaris* (Primrose) Som. Macmillan.

ROSE, Easter 1. *Narcissus odorus* (Jonquil) Som. Tongue. 2. *Narcissus pseudo-narcissus*

(Daffodil) Som. Elworthy. 3. *Primula vulgaris* (Primrose) Som. Macmillan.

ROSE, Egyptian 1. *Knautia arvensis* (Field Scabious) IOW B & H. 2. *Scabiosa atropurpurea* (Sweet Scabious) IOW B & H. 'Egyptian' means 'Gypsy'. Cf. Gypsy Rose for Sweet Scabious. It is the dark colours that provide the reference both for these names, and for such as Mournful Widow.

ROSE, Elder *Viburnum opulus* (Guelder Rose) Parkinson. Cf. Water Elder, Marsh Elder, White Elder, etc.

ROSE, English *Rosa canina* (Dog Rose) Genders.

ROSE, Father Hugo's *Rosa hugonis*.

ROSE, Field *Rosa arvensis*.

ROSE, First *Primula vulgaris* (Primrose) Som. Macmillan. Cf. Early Rose.

ROSE, Fox *Rosa pimpinellifolia* (Burnet Rose) War. B & H.

ROSE, Frankfurt *Rosa francofurtana*.

ROSE, French *Rosa gallica*.

ROSE, Gallipoli *Cistus salvifolius* (Sage-leaved Cistus) New Zealand Perry. Because New Zealand troops took seeds back after World War I.

ROSE, Gelders *Viburnum opulus* (Guelder Rose) Gerard.

ROSE, Golden *Primula vulgaris* (Primrose) Som. Grigson.

ROSE, Golden, of China *Rosa hugonis* (Father Hugo's Rose) Mattock.

ROSE, Green *Rosa viridiflora*.

ROSE, Ground *Rosa spithamaea*.

ROSE, Guelder *Viburnum opulus*. Introduced from Gueldres, hence the name.

ROSE, Guelder, Mealy *Viburnum lantana* (Wayfaring Tree) Clapham.

ROSE, Gypsy 1. *Knautia arvensis* (Field Scabious) IOW, Ess, Norf, Yks, Cumb B & H. 2. *Rosa arvensis* (Field Rose) Halliwell. 3. *Scabiosa atropurpurea* (Sweet Scabious) Dev. Friend; Som. Coats; Wilts. Dartnell & Goddard; IOW B & H. 4. *Succisa pratensis* (Devil's Bit) Som. Friend. See Egyptian Rose.

ROSE, Hip *Rosa canina* (Dog Rose) Glos, Shrop. Grigson.

ROSE, Holly *Cistus sp* Gerard.

ROSE, Holly, Sweet *Cistus ladaniferus* (Gum Cistus) Parkinson.

ROSE, Holy *Rosa sancta* (Abyssinian Rose) Grigson.1976. This is a form of the Summer Damask from Ethiopia, where it was grown in Christian premises in the province of Tigra, around the holy city of Axum.

ROSE, Holy, Marsh *Andromeda polifolia* (Bog Rosemary) Prior.

ROSE, Incarnation *Rosa alba* 'Maiden's Blush' Rohde.

ROSE, Incense *Rosa primula*. Incense, because of the smell of the young foliage.

ROSE, Jacobite *Rosa alba*.

ROSE, Jamaican *Blakea trinerva*.

ROSE, Japanese *Rosa rugosa*.

ROSE, Juno's *Lilium candidum* (Madonna Lily) Parkinson.

ROSE, Laurier *Nerium oleander* (Rose Bay) Watt. Cf. Turner's Rose Laurel.

ROSE, Lent 1. *Narcissus pseudo-narcissus* (Daffodil) Friend. 2. *Primula vulgaris* (Primrose) Dev. Macmillan.

ROSE, Lenten *Helleborus orientalis*.

ROSE, Levant *Rosa foetida* double form (Capucine Rose) Whittle & Cook.

ROSE, Macartney *Rosa bracteata*. Named after Lord Macartney, who introduced it from China in 1765.

ROSE, May 1. *Rosa cinnamonea* (Cinnamon Rose) McDonald. 2. *Viburnum opulus* (Guelder Rose) B & H. Cf. May Ball, May Tassels, May Tosty, and even Maypole for Guelder Rose.

ROSE, Meadow *Rosa blanda* (Smooth Rose) USA Tolstead.

ROSE, Mohave *Rosa mohavensis*.

ROSE, Monthly *Rosa damascena* 'Semperflorens'. 'Monthly', because '*semperflorens*'.

ROSE, Moss *Rosa centifolia var. muscosa*. A legend accounts for the name by saying that the angel who takes care of flowers, and sprinkles on them the dew in the night, slept on a spring day in the shade of a rose bush. When she woke, she said "Most beautiful of my children, I thank thee for thy refreshing odour and cooling shade; could you now ask me a favour, how willingly would I grant it." "Adorn me with a new charm", said the spirit of the rose-bush, and the angel adorned it with the simple moss. Another legend says that it was the blood of Christ, falling on the moss at the foot of the cross, that gave birth to the moss-rose.

ROSE, Mugget *Viburnum opulus* (Guelder Rose) Dev, Som. Macmillan.

ROSE, Musk 1. *Rosa arvensis* (Field Rose) Genders.1976. 2. *Rosa moschata*.

ROSE, Mutton *Trifolium repens* (White Clover) Corn. PLNN 21.

ROSE, New Year's *Helleborus niger* (Christmas Rose) B & H.

ROSE, Noisette *Rosa moschata x chinensis*.

ROSE, Pasture *Rosa humilis*.

ROSE, Peony *Paeonia emodi* (Himalayan Peony) Chopra.

ROSE, Phoenician *Rosa phoenicia*.

ROSE, Piano *Paeonia mascula* (Peony) Ire. B & H. 'Piano' is a corruption of 'peony'. It goes through a lot of transformations - Cf. Pie-anna, Pie-nanny, Piny, etc.

ROSE, Pig 1. *Papaver rhoeas* (Red Poppy) Corn. Grigson. 2. *Rosa canina* (Dog Rose) Corn. Grigson.

ROSE, Pig's *Rosa canina* (Dog Rose) Dev. Grigson.

ROSE, Pimpernel *Rosa pimpinellifolia* (Burnet Rose) B & H. The reference, as with the name Burnet Rose, is to the fact that the leaves resemble those of the Salad Burnet, and thence by analogy to the Burnet Saxifrage, *Pimpinella saxifraga*.

ROSE, Polyantha *Rosa multiflora*. 'Polyantha' means many flowers, hence the latinized 'multiflora'.

ROSE, Prairie 1. *Rosa arkansana* USA Tolstead. 2. *Rosa humilis*. 3. *Rosa pratincola*. 4. *Rosa setigera*.

ROSE, Primmy *Primula vulgaris* (Primrose) Glos. J D Robertson.

ROSE, Provençal *Rosa centifolia* (Cabbage Rose) F Savage.

ROSE, Provence, Yellow *Rosa hemisphaerica* (Sulphur Rose) Thomas.

ROSE, Rag *Primula veris x vulgaris* (False Oxlip) Lincs. Grigson. Cf. Rag Jack. Presumably a comment on the appearance of the flowers.

ROSE, Ramanas *Rosa rugosa* (Japanese Rose) Mattock. Probably from the Japanese meaning 'blooming profusely'.

ROSE, Rock 1. *Armeria maritima* (Thrift) Dev. Grigson. 2. *Cistus sp*. 3. *Helianthemum chamaecistus*.

ROSE, Rock, Annual *Tuberaria guttata*.

ROSE, Rock, Hoary *Helianthemum canum*.

ROSE, Rock, Spotted *Tuberaria guttata* (Annual Rock Rose) Polunin.

ROSE, Rock, White *Helianthemum apenninum*.

ROSE, Rock, White-leaved *Cistus incanus* (Hoary Cistus) Whittle & Cook.

ROSE, Rosin 1. *Hypericum calycinum* (Rose of Sharon) Yks. B & H. 2. *Hypericum perforatum* (St John's Wort) Yks, USA Grigson. From the rosin-like smell of the oil contained in the sacs - the ones that look like dots (or even holes) in the leaves of St John's Wort.

ROSE, Sacramento *Rosa stellata var. mirifica*.

ROSE, Sage *Cistus salvifolius* (Sage-leaved Cistus) Parkinson.

ROSE, St David's *Rosa pimpinellifolia* (Burnet Rose) Pemb. Freethy. It was the emblem of the see, displayed in St David's Cathedral.

ROSE, St Mark's *Rosa virginiana pl*.

ROSE, Sacramento *Rosa stellata var. mirifica*.

ROSE, Salt *Armeria maritima* (Thrift) Yks. Tynan & Maitland.

ROSE, Scotch; ROSE, Scots *Rosa pimpinellifolia* (Burnet Rose) Hutchinson (Scotch); Bunyard (Scots).

ROSE, Sea *Armeria maritima* (Thrift) Yks. Tynan & Maitland.

ROSE, Selon's *Nerium oleander* (Rose Bay) Watt. i.e. Ceylon Rose, also recorded.

ROSE, Sheep-shearin' *Paeonia mascula* (Peony) Worcs. B & H.

ROSE, Shining *Rosa nitida*.

ROSE, Silver *Potentilla anserina* (Silverweed) Freethy.

ROSE, Smooth *Rosa blanda*.

ROSE, Snowball *Viburnum opulus* (Guelder Rose) Suss. Parish.

ROSE, Snowdon *Sedum roseum* (Rose-root) Wales Grigson.

ROSE, South Sea *Nerium oleander* (Rose Bay) Watt.

ROSE, Stock *Sparrmania africana* (African Hemp) Howes.

ROSE, Sulphur *Rosa hemisphaerica*.

ROSE, Summer *Kerria japonica* (Jew's Mallow) Dev. Friend.1882.

ROSE, Sun *Helianthemum chamaecistus*.

ROSE, Sun, Alpine *Helianthemum alpestre*.

ROSE, Sun, Showy *Helianthemum venustum*.

ROSE, Threepenny-bit *Rosa farreri var. persetosa*. The name refers to the tiny, single blooms.

ROSE, Virginia *Lupinus luteus* (Yellow Lupin) Coles.

ROSE, Water 1. *Nuphar lutea* (Yellow Waterlily) Turner. 2. *Nymphaea alba* (White Waterlily) Turner.

ROSE, Whitsun *Viburnum opulus* (Guelder Rose) Dev. Macmillan. Cf. Whitsun-boss, from Gloucestershire. Guelder Rose is the ecclesiastical symbol for Whitsuntide.

ROSE, Wild, Swamp *Rosa palustris*.

ROSE, Wind *Papaver argemone* (Rough Long-headed Poppy) Gerard.

ROSE, Winter *Helleborus niger* (Christmas Rose) Dev. Friend.1882.

ROSE, Yellow *Kerria japonica* (Jew's Mallow) Dev. B & H; Som. Macmillan.

ROSE, York and Lancaster *Rosa damascena 'versicolor'*.

ROSE-A-RUBY *Adonis annua* (Pheasant's Eye) Gerard. This is Latin rosa rubra.

ROSA ACACIA *Robinia hispida*.

ROSE-AMONG-THE-THORNS *Nigella damascena* (Love-in-a-mist) Som. Macmillan. Of the same class as Devil-in-a-bush, etc.

ROSE-APPLE 1. *Eugenia jambolana* (Jambul) Grieve.1931. 2. *Syzygium moorei* (Coolamon) Australia Wilson. There is a fragrance of apples when Jambul fruit is being eaten.

ROSE-APPLE, Mankil; ROSE-APPLE, Samarang *Eugenia javanica* (Curacao Apple) Douglas.

ROSE BAY 1. *Nerium oleander*. 2. *Rhododendron maximum*. Cf. Rose Laurel for Nerium oleander.

ROSE BAY, Californian *Rhododendron macrophyllum*.

ROSE BAY, Lapland *Rhododendron lapponicum*.

ROSE CAMPION 1. *Agrostemma githago* (Corn Cockle) S Africa Watt. 2. *Lychnis chalcedonica* (Jerusalem Cross) Gerard. 3. *Lychnis coronaria*.

ROSE ELDER; ELDER ROSE *Viburnum opulus* (Guelder Rose) B & H. Cf. Water Elder, Marsh Elder, White Elder, etc.

ROSE GERANIUM *Pelargonium graveolens*.

ROSE LAUREL *Nerium oleander* (Rose Bay) Turner.

ROSE MALLOW 1. *Althaea rosea* (Hollyhock) Som. Macmillan. 2. *Hibiscus rosa-sinensis* (Rose of China) Howes.

ROSE MALLOW, Swamp *Hibiscus moscheutos*.

ROSE MOSS *Portulaca grandiflora* Hay & Synge.

ROSE NOBLE 1. *Cynoglossum officinale* (Hound's Tongue) N Ire. Grigson. 2. *Scrophularia aquatica* (Water Betony) Ire. Wilde. 3. *Scrophularia nodosa* (Figwort) Dumf, Wigt, Kirk, N Ire. Grigson. Rose noble means the king's golden coin, and must have been used in lieu of the king's own touch in the treatment of the King's Evil.

ROSE OF A HUNDRED LEAVES *Rosa centifolia* (Cabbage Rose). A literal translation of the botanical name.

ROSE OF CHINA *Hibiscus rosa-sinensis*.
ROSE OF JERICHO *Anastatica hierochuntica*. Not a rose, of course, and it has little to do with Jericho. The reference is to the legend that this plant expanded, became green and blossomed again at the birth of Jesus. Cf. Rose of the Virgin, and Mary's Flower.
ROSE OF MELAXO *Rosa damascena* (Damask Rose) Gerard. "a city in Asia, from whence some have thought it was first brought...".
ROSE OF PROVINS *Rosa gallica* (French Rose). The rose was brought, so tradition says, by Thibault le Chansonnier from Palestine to Provence in the 13th century. But isn't this the town Provins, about 60 miles southeast of Paris?
ROSE OF SHARON 1. *Hibiscus syriacus* (Syrian Katmia) Wit. 2. *Hypericum androsaemum* (Tutsan) Ire. Fernie. 3. *Hypericum calycinum*. 4. *Narcissus canaliculatus* Hutchinson & Melville. The reference is to Song of Solomon 2:1 "I am the rose of Sharon, and the lily of the valleys."
ROSE OF THE VIRGIN *Anastatica hierochuntica* (Rose of Jericho) Ackermann. Cf. Mary's Flower, both because of a legend which tells that all the plants of this species expanded, became green and blossomed again at the birth of Jesus, and still do so in commemoration.
ROSE PARSLEY *Anemone hortensis* Turner.
ROSE PENNYWORT *Saxifraga virginiensis*.
ROSE ROOT (or as one word) *Sedum roseum*. When the root is cut, a rose-like smell is given out. A rose-scented water can be made from them, too.
ROSE SAGE *Salvia pachyphylla*.
ROSE TREE *Nerium oleander* (Rose Bay) Gerard.
ROSEBAY; ROSEBAY WILLOWHERB *Chamaenerion angustifolium*.
ROSEBUD CHERRY *Prunus cerasus var. subhirtella* (Higan Cherry) Mitchell.1974.
ROSECAMPI *Lychnis coronaria* (Rose Campion) Turner.
ROSEMARY 1. *Croton linearis* Jamaica Beckwith. 2. *Rosmarinus officinalis*. 3. *Sonchus arvensis* (Corn Sow Thistle) Lincs. B & H. 4. *Turnera diffusa* (Damiana) Bahamas Laguerre. Rosemary means "dew of the sea" (it often grows on sea shores).
ROSEMARY, Bog 1. *Andromeda glaucophylla* USA H Smith.1945. 2. *Andromeda polifolia*.
ROSEMARY, Creeping *Rosmarinus prostratus*.
ROSEMARY, Marsh 1. *Ledum palustre*. 2. *Limonium carolinianum* (American Sea Lavender) USA House. 3. *Limonium vulgare* (Sea Lavender) Fernie.
ROSEMARY, Wild 1. *Andromeda polifolia* (Bog Rosemary) Som. Macmillan. 2. *Galium verum* (Lady's Bedstraw) Culpepper. 3. *Ledum palustre* (Marsh Rosemary) Fernie.
ROSEMS *Allium ursinum* (Ramsons) Staffs, Yks. B & H. Presumably a variation of the name 'ramsons', involving a transposition of consonants through some form like Roms, another Yorkshire name.
ROSEWOOD *Dalbergia nigra*. Because of the faint smell of roses when the wood is freshly cut.
ROSEWOOD, African *Pterocarpus erinaceus* Dalziel.
ROSEWOOD, Bornean *Melanorrhoea curtisii* Gimlette.
ROSEWOOD, Brazilian; ROSEWOOD, Bahia *Dalbergia nigra* (Rosewood) Howes.
ROSEWOOD, Burmese *Pterocarpus indicus* (Amboyna Wood) Howes.
ROSEWOOD, East Indian *Dalbergia latifolia* (Blackwood) Willis.
ROSEWOOD, False *Thespesia populnea* (Portia Tree) Dalziel.
ROSEWOOD, Malabar; ROSEWOOD, Indian *Dalbergia latifolia* (Blackwood) Howes.
ROSEWOOD, New Guinea *Pterocarpus indicus* (Amboyna Wood) Howes.
ROSEWOOD, Rio *Dalbergia nigra* (Rosewood) Tampion.

ROSIN-BRUSH *Baccharis sarothroides* (Broom Baccharis) USA T H Kearney.

ROSIN ROSE 1. *Hypericum calycinum* (Rose of Sharon) Yks. B & H. 2. *Hypericum perforatum* (St John's Wort) Yks, USA Grigson. From the rosin-like smell of the oil contained in the sacs which look like dots in the leaves of St John's Wort.

ROSTEWORT *Peucedanum officinale* (Hog's Fennel) Gunther. It is glossed as such on a 12th century ms of Apuleius.

ROSY HEART *Dicentra spectabilis* (Bleeding Heart) Som. Macmillan.

ROSY MORN *Lotus corniculatus* (Bird's-foot Trefoil) Som. Macmillan.

ROSYDANDRUM *Rhododendron ponticum* (Common Rhododendron) Staffs. Marrow.

ROT, Farthing; ROT, Penny; ROT, Shilling *Hydrocotyle vulgaris* (Pennywort) Norf. B & H (Farthing); Shrop. Grigson (Penny); Ayr. B & H (Shilling). The coinage refers to the round leaves, and 'rot' - "husbandmen know well, that it is noisome unto Sheepe, and other Cattell that feed thereon, and for the most part bringeth death unto them..." (Gerard).

ROT, Red *Drosera rotundifolia* (Sundew) Som. Macmillan; N Eng, Scot. Grigson. Another local name, Iles, means liver rot. The disease is known variously as Red-water, Red-swelling, Red-disease, or Red murrain, and it is supposed to be caused by eating sundew.

ROT, Sheep 1. *Hydrocotyle vulgaris* (Pennywort) War, Cumb, N'thum, Caith. Grigson. 2. *Pinguicula vulgaris* (Butterwort) N Scot. Grigson. Butterwort was another plant supposed to cause liver rot in animals.

ROT, Sheet *Hydrocotyle vulgaris* (Pennywort) N Scot. Grigson. Is this genuine? It can only be a misreading of 'sheep'.

ROT, Water *Hydrocotyle vulgaris* (Pennywort) Ches. Holland.

ROT, White 1. *Hydrocotyle vulgaris* (Pennywort) N Eng. Gerard. 2. *Pinguicula vulgaris* (Butterwort) B & H.

ROT-GRASS 1. *Hydrocotyle vulgaris* (Pennywort) N'thum. Grigson. 2. *Pinguicula vulgaris* (Butterwort) Cumb, N'thum, Berw. Grigson.

ROTCOLL *Armoracia rusticana* (Horseradish) Scot. Jamieson. The English equivalent is Red Cole.

ROTTLE-PENNY *Rhinanthus crista-galli* (Yellow Rattle) Dor. Macmillan. A local variant of Rattle-penny.

ROTTNEST ISLAND PINE *Callitris robusta* (Cypress Pine) Hora.

ROUEN LILAC *Syringa x chinensis*.

ROUGE PLANT 1. *Carthamus tinctorius* (Safflower) Howes. 2. *Rivina humilis* (Pigeonberry) Wit. Cf. Bloodberry for Pigeonberry. Both the names come from the scarlet berries, which are the source of a red dye, while safflower gets the name more directly, for the red dye is often used in India to make a brush paint for cosmetic rouge.

ROUGH(-LEAVED) DOGWOOD *Cornus asperifolia*.

ROUGH PARSNIP *Heracleum sphondyllium* (Hogweed) Turner.

ROUGHWEED *Stachys palustris* (Marsh Woundwort) N Ire. Grigson.

ROUN-TREE *Sorbus aucuparia* (Rowan) N Eng. Denham/Hardy.

ROUND ROBIN *Geranium robertianum* (Herb Robert) Dev, Kent Grigson.

ROUND TOWERS *Lythrum salicaria* (Purple Loosestrife) Dev. Macmillan.

ROUNDABOUT GENTLEMEN *Fritillaria imperialis* (Crown Imperial) Dor. Dacombe. Ingeniously descriptive of the flowering head.

ROUNDBERRY *Sorbus aucuparia* (Rowan) Swainson. 'Round' here is a variation on 'rowan'.

ROUNDWOOD *Pyrus americana*.

ROVING JENNY *Saxifraga sarmentosa* (Mother-of-thousands) Dev. B & H.

ROW DASHLE *Onopordum acanthium* (Scotch Thistle) Dev. Choape. i.e. rough thistle.

ROWAN *Sorbus aucuparia*. Derived from Danish roun, Swedish runn, traceable to O Norse rune, a charm.

ROWAN, Dog *Viburnum opulus* (Guelder Rose) Scot. B & H.
ROWAN, Hupeh *Sorbus hupehensis*.
ROWAN, Japanese *Sorbus commixta*.
ROWAN, Kashmir *Sorbus cashmiriana*.
ROWAN, Roden-quicken- *Sorbus aucuparia* (Rowan) Leland. An attempt of putting all the names for rowan in one.
ROWAN ASH *Sorbus aucuparia* (Rowan) Leland.
ROWBERRY 1. *Bryonia dioica* (White Bryony) Som. Grigson. 2. *Tamus communis* (Black Bryony) Dev. Friend.1882. It can be pronounced, at least in Devonshire, either to rhyme with 'no' or with 'cow'. It is OE hreow, rue (Rueberry also occurs), perhaps because they are poisonous?
ROWNTREE *Sorbus aucuparia* (Rowan) Scot. Dalyell.
ROYAL FERN *Osmunda regalis*.
ROYAL JASMINE *Jasminum grandiflorum* (Spanish Jasmine) RHS.
ROYAL JEWEL ORCHID *Anoectochilus regalis*.
ROYAL OAK, Red *Viburnum lantana* (Wayfaring Tree) Dor. Grigson.
ROYAL PAULOWNIA *Paulownia tomentosa* (Foxglove Tree) Edlin. Paulownia is botanists' latinization of Paulovna, the tree having been named in honour of Anna Paulovna, daughter of Czar Paul I.
ROYAL PENNY *Umbilicus rupestris* (Wall Pennywort) Som. Macmillan. This is probably just an inversion of 'pennyroyal', for Wall Pennyroyal is another name for this plant.
ROYAL RED BUGLER *Aeschynanthus pulcher* (Lipstick Plant) RHS.
ROYAL STAFF *Asphodelus ramosus* (White Asphodel) Grieve.1931. Cf. King's Spear and Silver Rod for this.
ROYAL WALNUT *Juglans regia* (Walnut) Ablett. Simply a translation of Juglans regia.
ROYAL WATERLILY *Victoria amazonica*.
ROYAN; ROYNE-TREE *Sorbus aucuparia* (Rowan) both N Eng. Denham/Hardy.
ROZELLE; ROSELLE *Hibiscus sabdariffa*.
RUBBER *Havea brasiliensis*.

RUBBER, Palay *Cryptostegia grandiflora* (Rubber Vine) Schery.
RUBBER, Ceara *Manihot glaziovii*.
RUBBER, Panama *Castilla elastica*.
RUBBER, Rhodesian *Diplorhynchus condylocarpon var. mossambicensis*.
RUBBER PLANT *Ficus elastica* (India Rubber Plant) A W Smith.
RUBBER PLANT, Japanese *Crassula portulacea* (Money Plant) USA Cunningham.
RUBBER PLANT, South American *Helenium hoopesii* (Orange Sneezeweed) USA Elmore.
RUBBER RABBITBRUSH *Chrysothamnus nauseosus*.
RUBBER TREE, False *Funtumia africana*.
RUBBER TREE, India, Bengal *Ficus elastica* (India Rubber Plant) Lindley.
RUBBER TREE, Lagos *Funtumia elastica* (Lagos Silk-Rubber Tree) Howes.
RUBBER TREE, West African *Funtumia elastica* (Lagos Silk-Rubber Tree) Dalziel.
RUBBER VINE *Cryptostegia grandiflora*.
RUBBERBRUSH, Mohave *Chrysothamnus nauseosus var. mohavensis*.
RUBBERWEED, Bitter *Hymenoxis odorata* (Bitterweed) USA Kingsbury.
RUBBERWEED, Colorado *Hymenoxis richardsonii* (Pingue) USA Kingsbury.
RUBWORT *Geranium robertianum* (Herb Robert) Ches. Holland. Presumably 'rub' is 'rob', short for Robert.
RUBYWOOD *Pterocarpus santalinus* (Red Sandalwood) Grieve.1931. Colour descriptive.
RUD; RUDDES *Calendula officinalis* (Marigold) IOW B & H (Rud); Gerard (Ruddes). 'Rud' means red, not very accurate a description for marigold, hence one version, Rods-gold. But it was a very common name for the plant at one time.
RUE *Ruta graveolens*. Through French rue from Latin ruta, and eventually from Greek rhute, originally a Peloponnesian word, according to ODEE.
RUE, African *Peganum harmala*. Cf. Syrian Rue for this.

RUE, Arabian *Ruta tuberculata*.
RUE, Dwarf *Ruta patavina*.
RUE, Fen *Thalictrum flavum* (Meadow Rue) Gerard.
RUE, Fringed *Ruta chalepensis*.
RUE, Garden *Ruta graveolens* (Rue) Watt.
RUE, Goat's *Galega officinalis*. Probably so called because of its disagreeable taste - it is extremely bitter.
RUE, Meadow *Thalictrum flavum*.
RUE, Meadow, Alpine *Thalictrum alpinum*.
RUE, Meadow, Early *Thalictrum dioicum*.
RUE, Meadow, Fall *Thalictrum polygamum*.
RUE, Meadow, Flat-fruited *Thalictrum sparsiflorum*.
RUE, Meadow, Greater 1. *Thalictrum aquilegifolium*. 2. *Thalictrum major*.
RUE, Meadow, Lesser *Thalictrum minus* (Small Meadow Rue) Clapham.
RUE, Meadow, Purple *Thalictrum dasycarpum*.
RUE, Meadow, Small *Thalictrum minus*.
RUE, Meadow, Tall *Thalictrum dasycarpum* (Purple Meadow Rue). Johnston.
RUE, Meadow, Waxy *Thalictrum revolutum*.
RUE, Rock *Euphrasia officinalis* (Eyebright) Donegal Grigson.
RUE, Stone *Asplenium ruta-muraria* (Wall Rue) Turner.
RUE, Syrian *Peganum harmala* (African Rue) Emboden.
RUE, Wall *Asplenium ruta-muraria*. Cf. Rue Fern. One sometimes finds cases of using this Rue Fern instead of the real rue in ritual, as when a jilted girl throws a handful at the man when he was being married, and cursing him with "May you rue this day as long as you live".
RUE, Wild 1. *Oenanthe crocata* (Hemlock Water Dropwort) Dor. Grigson. 2. *Peganum harmala* (African Rue) Chopra.
RUE ANEMONE *Anemone thalictroides*. The reference here is to Meadow Rue.
RUE FERN *Asplenium ruta-muraria* (Wall Rue) Dev. Friend.1882.
RUE-LEAVED SAXIFRAGE *Saxifraga tridactylites*.

RUE-WEED *Thalictrum flavum* (Meadow Rue) Som. Macmillan.
RUEBERRY *Tamus communis* (Black Bryony) Dev. Friend.1882.
RUFFET *Ulex europaeus* (Furze) Dor. B & H.
RUFFLES, Lady's 1. *Filipendula vulgaris* (Dropwort) Grigson.1959. 2. *Narcissus pseudo-narcissus* (Daffodil) Wilts. Grigson.
RUGBY FOOTBALL PLANT *Peperomia argryreia* (Watermelon Peperomia) Nicolaisen. Presumably for the same reason as it is dubbed 'Watermelon'.
RUGEL'S PLANTAIN *Plantago rugelii* (Black-seed Plantain) Canada Allan.
RUM CHERRY *Prunus virginiana* (Choke Cherry) USA Zenkert. Country people in America infuse these cherries in brandy as a flavouring, and this name suggests that it is not only brandy that is treated so.
RUMP *Raphanus raphanistrum* (Wild Radish) Oxf. B & H. Is there a connection with the Scottish name Runch?
RUMPET-SCRUMPS *Heracleum sphondyllium* (Hogweed) Som. Macmillan. One of a number of names given, we are told, because of the habit of drinking cider through the hollow stalks of Hogweed.
RUN-BY-THE-GROUND *Mentha pulegium* (Pennyroyal) Fernie. Cf. Lurk-in-ditch and its derivatives.
RUNAGATES, Meadow *Lysimachia nummularia* (Creeping Jenny) N'thants. Grigson. A runagates is someone who can't mind his own business, but a runagate boy in Yorkshire is one who climbs trees.
RUNAWAY JACK *Glechoma hederacea* (Ground Ivy) Som. Grigson; Glos. J D Robertson.
RUNCH 1. *Raphanus raphanistrum* (Wild Radish) Cumb. B & H; Scot. Jamieson. 2. *Sinapis alba* (White Mustard) Yks. F LK Robinson. 3. *Sinapis arvensis* (Charlock) N Eng. F Grose; Scot. Jamieson.
RUNCH, White *Raphanus raphanistrum* (Wild Radish) Yks. Atkinson. 'White', to

distinguish it from Charlock, also called Runch in the area.

RUNCH-BALLS *Sinapis arvensis* (Charlock) N Eng. F Grose. Dried charlock, according to Halliwell.

RUNCHIE; RUNCHIK; RUNGY *Sinapis arvensis* (Charlock) Scot, Ork. Grigson (Runchie, Runchik); Shet. B & H (Rungy).

RUNNER, Blue *Glechoma hederacea* (Ground Ivy) Bucks. Grigson.

RUNNER, Scarlet *Phaseolus coccineus.*

RUNNIDYKE *Glechoma hederacea* (Ground Ivy) Cumb. Grigson. Cf. Rob-run-up-dyke, Robin-run-in-the-hedge, and a lot of other similar names describing the plant's habit.

RUNNING JACOB *Tropaeolum majus* (Nasturtium) Dor. Macmillan.

RUNNING THYME *Thymus vulgaris* (Wild Thyme) Turner.

RUPPLE-GRASS *Plantago lanceolata* (Ribwort Plantain) Donegal Denham/Hardy. A version of Ripple, or Rippling, Grass, recorded in Scotland.

RUPTURE-WORT *Herniaria ciliata.* "It is reported that being drunke it is singular good for Ruptures, and that very many that have been bursten were restored to health by the use of this herbe..." (Gerard). Herbalists still prescribe it for ruptures among other ailments.

RUPTURE-WORT, Hairy *Herniaria hirsuta.*

RUPTURE-WORT, Smooth *Herniaria glabra.*

RUPTURE-GRASS *Herniaria ciliata* (Rupture-wort) Gerard.

RUSH, Bog *Juncus stygius.* Rush is OE risce, German Risch.

RUSH, Bulbous *Juncus bulbosus.*

RUSH, Cat *Equisetum arvense* (Common Horsetail) Ches. Holland.

RUSH, Compact *Juncus conglomeratus.*

RUSH, Dudley's *Juncus effusus.*

RUSH, Dutch *Equisetum hyemale.* 'Dutch', because it was imported in bundles from Holland as a domestic polisher.

RUSH, Flowering *Butomus umbellatus.*

RUSH, Hard *Juncus inflexus.*

RUSH, Heath *Juncus squarrosus.*

RUSH, Jointed *Juncus articulatus.*

RUSH, Mud *Juncus gerardi* (Saltmarsh Rush) Hepburn.

RUSH, Saltmarsh 1. *Juncus gerardi.* 2. *Scirpus maritimus* (Sea Club-rush) Pratt.

RUSH, Scouring 1. *Equisetum arvense* (Common Horsetail) USA Youngken. 2. *Equisetum hyemale* (Dutch Rush) Leighton. Silica-rich plants, particularly Dutch Rush, hence their use as scourers and polishers.

RUSH, Sea *Juncus maritimus.*

RUSH, Sea, Greater *Juncus acutus* (Sharp Rush) Hepburn.

RUSH, Sharp *Juncus acutus.*

RUSH, Soft *Juncus effusus.*

RUSH, Sweet *Acorus calamus* (Sweet Flag) Grieve.1931.

RUSH, Toad *Juncus bufonius.*

RUSH, Wire 1. *Juncus balticus.* 2. *Juncus inflexus* (Hard Rush) F K Robinson.

RUSH, Wood *Equisetum sylvaticum* (Wood Horsetail) USA Youngken.

RUSH DAFFODIL *Narcissus odorus* (Jonquil) McDonald. Jonquil is from French jonquille, Spanish jonquillo, the diminutive of junco, rush - hence this name.

RUSH GARLIC *Allium schoernoprasum* (Chives) Lyte. Cf. Rushleek.

RUSH-LEAVED FESCUE *Festuca juncifolia.*

RUSH-LEAVED NARCISSUS *Narcissus juncifolius.*

RUSH NUT *Cyperus esculentus* (Tigernut) Dalziel.

RUSHLEEK 1. *Allium vineale* (Crow Garlic) USA Grigson. 2. *Allium schoernoprasum* (Chives) Clair. This is a literal translation of schoernoprasum. Cf. the Dutch biesloek.

RUSOT *Berberis asiatica* (Indian Barberry) Le Strange.

RUSSIAN ACONITE *Aconitum orientale.*

RUSSIAN COMFREY *Symphytum peregrinum.*

RUSSIAN DANDELION *Taraxacum koksaghyz.*

RUSSIAN HENBANE *Hyoscyamus albus.*

RUSSIAN IRIS *Iris ruthenica*.
RUSSIAN KNAPWEED *Centaurea picris*.
RUSSIAN KNOTGRASS *Polygonum erectum* (Weedy Knotweed) Grieve.1931.
RUSSIAN LIQUORICE *Glycyrrhiza glabra var. glandulifera*.
RUSSIAN MULBERRY *Morus alba var. tatarica*.
RUSSIAN OLIVE *Elaeagnus angustifolia* (Narrow-leaved Oleaster) USA Cunningham. It is actually a Near Eastern plant. 'Olive' (Cf. Wild Olive, and Ethiopian Olive) because some Biblical references to the olive-tree probably refer to this.
RUSSIAN RHUBARB *Rheum officinale* (Turkey Rhubarb) Lloyd. Names for this tend to be in accordance with the country through which it reached the market from its native Tibet, hence this name, or the common name, or Chinese Rhubarb.
RUSSIAN TARRAGON *Artemisia dracunculoides*.
RUSSIAN THISTLE 1. *Salsola kali* (Saltwort) S Africa Watt. 2. *Salsola pestifera*.
RUSSIAN TUMBLEWEED *Salsola kali* (Saltwort) S Africa Watt.
RUSSIAN TURNIP *Brassica napus var. napobrassica* (Swede) Bianchini. 'Russian' and 'Swedish' are offered, but neither is right - the plant comes from Bohemia.
RUSSIAN VETCH *Vicia villosa* (Lesser Tufted Vetch) Howes.
RUSSIAN VINE *Polygonum baldschuanicum*.
RUSSIAN WORMWOOD *Artemisia sacrorum*.
RUSSLES *Ribes rubrum* (Redcurrant) Kirk, Wigt. Grigson. A version of Rizzles, which has a number of variations, all apparently from 'raisin' (Turner called it Raisin-tree).
RUSTBURN *Ononis repens* (Rest-harrow) Yks. B & H. Perhaps rest-bourn.
RUSTY-BACK *Ceterach officinarum*. Perfectly descriptive.
RUSTY FOXGLOVE *Digitalis ferruginea*. 'Rusty', presumably from the brown veining of the flowers.
RUSTY-LEAVED BEECH *Fagus grandifolia* (American Beech) J Smith.
RUSTY PODOCARP *Podocarpus ferruginea* (Miro) Leathart.
RUSTYLEAF *Menziesia ferruginea* (Mock Azalea) USA Kingsbury.
RUTABAGA 1. *Brassica campestris* (Field Mustard) USA Gates. 2. *Brassica napus var. napobrassica* USA Brouk. Swedish rotbagga, ram's root.
RUTLAND BEAUTY *Calystegia sepium* (Great Bindweed) USA Grigson.
RUVO KALE *Brassica ruvo* (Italian Turnip Broccoli) Usher. Ruvo is a town near Bari, in southern Italy.
RYE, Wild *Geum urbanum* (Avens) Fernie.
RYE, Wild, Giant *Elymus cinereus*.
RYE BROME *Bromus secalinus*.
RYE-GRASS *Lolium perenne*. Ray-grass is the earlier name, from which this one is derived. It comes from French ivraie, drunk. The poisonous darnel may be more appropriate for the name, though Rye-grass does contain alkaloids dangerous to animals, who become totally unco-ordinated when affected.
RYE GRASS, French *Arrhenatherium elatius* (False Oat) Howes.
RYE-GRASS, Italian *Lolium multiflorum*.
RYE-GRASS, Wild *Lolium temulentum* (Darnel) Wasson.1978.
RYFART *Raphanus sativus* (Garden Radish) Scot. Jamieson. Sometimes written as Reefort, but whatever the spelling they must be versions of Raphanus.

S

SABI, Star of the *Adenium multiflorum*.
SABIN *Juniperus sabina* (Savin) B & H.
SABINE'S PINE *Pinus sabiniana* (Digger Pine) Watt.
SABINO, Laurel *Magnolia splendens*.
SABRE BEAN *Canavalia ensiformis* (Jack Bean) Willis. Cf. Sword Bean for this.
SABS, Sour 1. *Oxalis acetosella* (Wood Sorrel) Dev. Grigson. 2. *Rumex acetosa* (Wild Sorrel) Corn. Jago; Dev. B & H. 'Sour Sabs' sometimes becomes 'Sour Saps', even 'Sour Grabs'.
SACK TREE *Antiaris innoxia*.
SACRAMENTO ROSE *Rosa stellata var. mirifica*.
SACRED BEAN *Nelumbo nucifera* (Hindu Lotus) Leyel.1948.
SACRED DATURA *Datura meteloides* (Downy Thorn-apple) USA Elmore. The very nature of the Daturas, intoxicant and hypnotic, would put them firmly in the sacred class, particularly to the Californian Indian tribes.
SACRED FIR *Abies religiosa*. 'Sacred' (and 'religiosa') because the branches are used for church decoration at religious festivals in Mexico.
SACRED LOTUS 1. *Nelumbium nelumbo*. 2. *Nelumbo nucifera* (Hindu Lotus) Brouk.
SACYMORE; SECYMORE *Acer pseudoplatanus* (Sycamore) Yks. Addy.
SAD-FLOWERED IRIS *Iris susiana* (Mourning Iris) W Miller. From the colouring of the flower - grey veined purple-violet and black.
SAD STOCK *Mathiola fruticosa*. Some of them have dull brownish pink flowers, sombre enough to warrant the adjective 'sad', and the old specific name, tristis.
SAD TREE *Nyctanthes arbortristis* (Indian Night Jasmine) Gordon.1985. A translation of the specific name, and a sentimental comment on the fact that the scented flowers open at sunset and have fallen by dawn.
SADDLE-TREE *Liriodendron tulipifera* (Tulip Tree) W Miller.
SADLER'S WILLOW *Salix sadleri*.
SAFFEN *Juniperus sabina* (Savin) E Angl. G E Evans Presumably just a variation of savin. Cf. Saffern.
SAFFERN 1. *Crocus sativus* (Saffron) Corn. Courtney.1890. 2. *Juniperus sabina* (Savin) Fernie. Mispronunciations in both cases.
SAFFLOWER *Carthamus tinctorius*. Surely it must be related to saffron, but the connection is to O French saffleur.
SAFFLOWER, Wild *Carthamus oxyacanthus*.
SAFFORNE *Crocus sativus* (Saffron) Turner.
SAFFRON 1. *Crocus nudiflorus* (Autumn Crocus) Shrop. B & H. 2. *Crocus sativus*. 3. *Juniperus sabina* (Savin) Vickery.1995. Saffron is O French safran, from Arabic za'-faran. In the case of Savin, 'saffron' is 'savin', the sequence being Savin - Saffern - Saffron.
SAFFRON, Bastard; SAFFRON, False; SAFFRON, Mock *Carthamus tinctorius* (Safflower) Pomet (Bastard); A W Smith (False) Turner (Mock).
SAFFRON, Catalonia; SAFFRON, Spanish *Carthamus tinctorius* (Safflower) Parkinson.
SAFFRON, Dyer's *Carthamus tinctoria* (Safflower) Zohary.
SAFFRON, Indian *Curcuma longa* (Turmeric) Clair. Because Turmeric gives a yellow dye, as does saffron. But there is another reason. The generic name, *Curcuma*, is from Persian kurkum, which means saffron.
SAFFRON, Meadow *Colchicum autumnale*.
SAFFRON, Spring *Crocus purpureus* (Spring Crocus) Gerard.
SAFFRON, Wild 1. *Carthamus tinctorius* (Safflower) Parkinson. 2. *Colchicum autumnale* (Meadow Saffron) Turner.
SAFFRON SPIKE *Aphelandra squarrosa* (Zebra Plant) H & P.

SAFFRON THISTLE *Carthamus tinctorius* (Safflower) Leyel.1948.

SAFFRON TREE *Sassafras variifolium* (Sassafras Tree) Suff. G E Evans.1960. "Probably" identified as this tree by Evans.

SAFFRON-WOOD from *Cassine crocea* Willis.

SAGAPENE, Herb *Ferula communis* (Giant Fennel) Turner.

SAGE 1. *Artemisia vulgaris* (Mugwort) USA Gates. 2. *Atriplex polycarpa* (Cattle Spinach) USA T H Kearney. 3. *Lantana trifolia* Barbados Gooding. 4. *Salvia officinalis*. Sage derives through Old French sauge from Latin salvia. Salvus means safe.

SAGE, Annual *Salvia reflexa*.

SAGE, Antelope *Eriogonum jamesii*.

SAGE, Apple-bearing *Salvia pomifera*. The "apple" is a gall that results from an insect's action. It is edible, with a spicy taste.

SAGE, Aromatic *Salvia africana-caerulea* S Africa Watt.

SAGE, Autumn *Salvia greggii*.

SAGE, Bethlehem *Pulmonaria officinalis* (Lungwort) Gerard. Cf. Jerusalem Sage,etc.

SAGE, Black *Ambrosia psilostachya*.

SAGE, Bladder *Salazaria mexicana* (Paper-bag Bush) USA T H Kearney.

SAGE, Bog *Salvia uliginosa*.

SAGE, Bur, Holly-leaf *Franseria ilicifolia*.

SAGE, Bur, White *Franseria dumosa*.

SAGE, Bur, Woolly *Franseria eriocentra*.

SAGE, Bush *Cistus salvifolius* (Sage-leaved Cistus) Turner.

SAGE, Bush, Mexican *Salvia leucantha*.

SAGE, Chia *Salvia columbariae*. Chia is the name of the food prepared from this, a staple for the Pacific coast Indians.

SAGE, Clary *Salvia sclarea* (Clary) Grieve.1931.

SAGE, Colorado *Artemisia frigida* (Fringed Wormwood) USA Painter.

SAGE, Death Valley *Salvia funerea*.

SAGE, Garlick *Teucrium scorodonia* (Wood Sage) Gerard.

SAGE, Gentian *Salvia patens*. The flowers are gentian blue in colour.

SAGE, Gray *Artemisia ludoviciana* (Lobed Cudweed) USA D E Jones.

SAGE, Green *Artemisia forwoodii*.

SAGE, Gypsy's *Teucrium scorodonia* (Wood Sage) Dor. Macmillan.

SAGE, Indian *Eupatorium perfoliatum* (Thoroughwort) USA Henkel.

SAGE, Jerusalem 1. *Phlomis fruticosa*. 2. *Pulmonaria officinalis* (Lungwort) Gerard.

SAGE, Judean *Salvia judaica*.

SAGE, Lance-leaf *Salvia lanceolata*.

SAGE, Meadow *Salvia pratensis* (Meadow Clary) McClintock.

SAGE, Mealy(-cup) *Salvia farinacea*.

SAGE, Mohave *Salvia mohavensis*.

SAGE, Mountain 1. *Pulmonaria officinalis* (Lungwort) Som. Grigson. 2. *Teucrium scorodonia* (Wood Sage) Ches, Cumb. Grigson.

SAGE, Mountain, Jamaica *Lantana camara*.

SAGE, Mullein-leaved *Salvia verbascifolia*.

SAGE, Nodding *Salvia nutans*.

SAGE, Pineapple *Salvia rutilans*. The leaves have the scent and flavour of pineapples.

SAGE, Prairie *Artemisia gnaphalodes* (White Mugwort) USA Vestal & Schultes.

SAGE, Purple *Salvia africana-caerulea* S Africa Watt.

SAGE, Purple-leaved *Salvia officinalis* 'purpurea'.

SAGE, Red 1. *Kochia americana* (Red Molly) USA T H Kearney. 2. *Lantana camara* (Barbados) Gooding. 3. *Salvia coccinea*. 4. *Salvia officinalis* 'purpurea' (Purple-leaved Sage). 5. *Salvia splendens* Clair.

SAGE, Red-topped *Salvia horminum*.

SAGE, Rock *Lantana involucrata* (Barbados) Gooding.

SAGE, Rocky Mountain 1. *Artemisia tridentata* (Sagebrush) Robbins. 2. *Salvia reflexa* (Annual Sage) USA T H Kearney.

SAGE, Rose *Salvia pachyphylla*.

SAGE, Russian *Perowskia atriplicifolia* (Silver Sage) Brownlow.

SAGE, Santa Rosa *Salvia eremostachya*.

Only growing in the Santa Rosa Mountains, America.
SAGE, Scarlet *Salvia coccinea* (Red Sage) Roys.
SAGE, Seaside *Croton balsamifer* (Yellow Balsam) Barbados Gooding.
SAGE, Shining *Salvia leonuroides*.
SAGE, Silver 1. *Artemisia filifolia* (Sand Sagebrush) Robbins. 2. *Perowskia atriplicifolia*. 3. *Salvia argentea*.
SAGE, South American *Salvia coccinea* (Red Sage) Watt.
SAGE, Sweet *Poliomentha incana*.
SAGE, Texas *Salvia coccinea* (Red Sage) Watt.
SAGE, Thistle *Salvia carduacea*.
SAGE, Vervain *Salvia horminoides* (Wild Sage) Clair.
SAGE, Wall *Parietaria diffusa* (Pellitory-of-the-wall) War. Grigson.
SAGE, Wand *Salvia vaseyi*.
SAGE, White 1. *Artemisia ludoviciana* (Lobed Cudweed) USA H Smith.1943. 2. *Lantana camara* (Jamaica Mountain Sage) West Indies. Laguerre. 3. *Lantana involucrata* Barbados Gooding. 4. *Eurotia lanata* (Winter Sage) Jaeger.
SAGE, Whorled *Salvia verticillata*.
SAGE, Wild 1. *Artemisia gnaphalodes* (White Mugwort) USA Gilmore. 2. *Eupatorium perfoliatum* (Thoroughwort) USA House. 3. *Lantana camara* Nielsen. 4. *Salvia africana-caerulea* S Africa Watt. 5. *Salvia horminoides*. 6. *Salvia nemerosa*. 7. *Teucrium scorodonia* (Wood Sage) Gerard.
SAGE, Wild, Little *Artemisia frigida* (Fringed Wormwood) USA Gilmore.
SAGE, Winter *Eurotia lanata*.
SAGE, Wood 1. *Prunella vulgaris* (Self-heal) N Ire. Grigson. 2. *Teucrium canadense* (American Germander) Webster. 3. *Teucrium occidentale* (Hairy Germander) USA House. 4. *Teucrium scorodonia*.
SAGE-LEAVED CISTUS *Cistus salvifolius*.
SAGE-LEAVED GERMANDER *Teucrium scorodonia* (Wood Sage) Brownlow.

SAGE OF BETHLEHEM *Mentha spicata* (Spearmint) Lincs. Gordon.1895.
SAGE PLANT *Artemisia tridentata* (Sagebrush) J Smith.
SAGE ROSE *Cistus salvifolius* (Sage-leaved Cistus) Parkinson.
SAGE WILLOW *Salix candida*.
SAGEBRUSH 1. *Artemisia tridentata*. 2. *Atriplex canescens* (Chamizo) USA T H Kearney. 3. *Atriplex polycarpa* (Cattle Spinach) USA T H Kearney.
SAGEBRUSH, Arctic *Artemisia frigida* (Fringed Wormwood) USA Elmore.
SAGEBRUSH, Basin *Artemisia tridentata* (Sagebrush) USA Elmore.
SAGEBRUSH. Big *Artemisia tridentata* (Sagebrush) USA Elmore.
SAGEBRUSH, Black; SAGEBRUSH, Blue *Artemisia tridentata* (Sagebrush) USA Elmore.
SAGEBRUSH, Bud *Artemisia spinescens*.
SAGEBRUSH, Fringed; SAGEBRUSH, Mountain; SAGEBRUSH, Pasture *Artemisia frigida* (Fringed Wormwood) USA Elmore.
SAGEBRUSH, Prairie *Artemisia frigida* (Fringed Wormwood) USA Yarnell.
SAGEBRUSH, Sand *Artemisia filifolia*.
SAGEBRUSH, Silver *Artemisia filifolia* (Sand Sagebrush) USA Elmore.
SAGEBRUSH, Three-lobed *Artemisia trifida*.
SAGEBRUSH, Three-tip *Artemisia tripartita*.
SAGEBRUSH, Three-toothed *Artemisia tridentata* (Sagebrush) USA Elmore.
SAGGAN; SAGGIN *Iris pseudo-acarus* (Yellow Flag) Ire. B & H (Saggan); Evans (Saggin). Evidently the same word as Seggin in England. OE segg meant a small sword, hence the various 'sword' and 'dagger' names for Iris.
SAGGAN, Yellow *Iris pseudo-acarus* (Yellow Flag) Sanecki.
SAGO, Portland *Arum maculatum* (Cuckoo-pint) Fernie. It was once cultivated in the Isle of Portland for the tubers, which when

cooked yielded Portland Sago, used as a substitute for arrowroot. It was quite popular at one time because it was thought to be aphrodisiac,but Gerard reported it simply as a "pure and white starch, but most hurtfull to the hands of the Laundresse that hath the handling of it, for it chappeth, blistereth, and maketh the hands rough and rugged, and withall smarting".

SAGO, Wild 1. *Plantago lanceolata* (Ribwort Plantain) S Africa Watt. 2. *Plantago major* (Great Plantain) S Africa Watt.

SAIGON CINNAMON *Cinnamonium loureirii*.

SAILOR, Blue *Cichorium intybus* (Chicory) USA Zenkert.

SAILOR, Bovisand *Centranthus ruber* (Spur Valerian) Dev. Friend.1882.

SAILOR, Climbing *Cymbalaria muralis* (Ivy-leaved Toadflax) Dumf. Grigson.

SAILOR, Creeping 1. *Cymbalaria muralis* (Ivy-leaved Toadflax) Som, Suss. Grigson. 2. *Saxifraga sarmentosa* (Mother-of-thousands) Shrop. B & H. 3. *Sedum acre* (Biting Stonecrop) Shrop. Grigson.

SAILOR, Drunken 1. *Centranthus ruber* (Spur Valerian) Dev. Friend.1882. 2. *Quisqualis indica* (Rangoon Creeper) Perry.

SAILOR, Rambling 1. *Cymbalaria muralis* (Ivy-leaved Toadflax) Som. Macmillan; Lancs. Grigson; IOM Moore.1924. 2. *Lysimachia nummularia* (Creeping Jenny) Hants. Grigson. 3. *Saxifraga sarmentosa* (Mother-of-thousands) Dev. B & H.

SAILOR, Roving 1. *Cymbalaria muralis* (Ivy-leaved Toadflax) Corn, IOW. Grigson; Dev. Friend.1882; Som. Macmillan. 2. *Saxifraga sarmentosa* (Mother-of-thousands) Dev. B & H.

SAILOR, Wandering 1. *Cymbalaria muralis* (Ivy-leaved Toadflax) Dev. Friend.1882; Som. Elworthy. 2. *Lysimachia nummularia* (Creeping Jenny) Dev. Macmillan. 3. *Saxifraga sarmentosa* (Mother-of-thousands) Coats.

SAILOR BUTTONS 1. *Knautia arvensis* (Field Scabious) Som. Macmillan. 2. *Melandrium rubrum* (Red Campion) Som. Tynan & Maitland. 3. *Stellaria holostea* (Greater Stitchwort) Hants. Grigson. 4. *Succisa pratensis* (Devil's-bit Scabious) Dev. Macmillan.

SAILOR'S KNOT *Geranium robertianum* (Herb Robert) Bucks. Grigson.

SAILOR'S TOBACCO *Artemisia vulgaris* (Mugwort) Hants. Grigson. The leaves often used to be smoked, not necessarily by sailors.

SAILORS, Soldiers-and- 1. *Arum maculatum* (Cuckoo-pint) Som. Macmillan. 2. *Echium vulgare* (Viper's Bugloss) Leyel.1937. 3. *Iris pseudo-acarus* (Yellow Flag) Dor. Macmillan. 4. *Pulmonaria officinalis* (Lungwort) Som. Macmillan; Suff. Grigson. Cuckoo-pint has a lot of these double names, as Lord-and-ladies. Most of them are sexual in nature, but a few, like this one, have no such overtones, unless that is, we no longer understand the innuendo. Another use of the double name is to mirror the two or more colours of the flower, as with lungwort and viper's bugloss here. That may be the reason for the ascription to yellow flag, too, for another of these names, though from a different area, is Butter-and-eggs, and that is almost always colour descriptive.

SAILS, White *Spathiphyllum patinii*.

SAINFOIN 1. *Medicago lupulina* (Black Medick) Hants. Grigson. 2. *Medicago sativa var. sativa* (Lucerne) B & H. 3. *Onobrychis sativa*. French, meaning healthy hay. In other words, a dried crop good for cattle.

SAINFOIN, False *Vicia onobrychioides*.

SAINFOIN, Italian; SAINFOIN, Spanish *Hedysarum coronarium* (French Honeysuckle) Polunin (Italian); Candolle (Spanish).

ST ANNE'S NEEDLEWORK *Saxifraga x urbinum* (London Pride) Friend.

ST ANTHONY'S LILY *Lilium candidum* (Madonna Lily) Whittle & Cook.

ST ANTHONY'S NUT 1. *Conopodium majus* (Earth-nut) Som. Macmillan. 2. *Staphylea pinnata* (Bladder-nut) Gerard.

ST ANTHONY'S RAPE; ST ANTHONY'S ROPE; ST ANTHONY'S TURNIP *Ranunculus bulbosus* (Bulbous Crowfoot) all Gerard. Rape is turnip, so these are references to the so-called bulb of this buttercup.

ST AUDRE'S LACE *Cannabis sativa* (Hemp) B & H.

ST BARBARA'S HERB; ST BARBARA'S CRESS *Barbarea vulgaris* (Winter Cress) Som. Grigson (Herb); Dyer (Cress). Cf. Herb Barbara. The dedication is also to be found in France. The names were given probably because the leaves were gathered as early as St Barbara's feast day, 4 December.

ST BARNABY'S THISTLE *Centaurea solstitialis*. 11 June is St Barnabas's Day, but that seems a bit early for the flowering - even 'old style' would not fit, for the usual time is late July.

ST BENEDICT'S HERB *Geum urbanum* (Avens) Friend. Cf. also Herb Bennett and Way Bennett for this plant. They all come from latin herba benedicta, which means simply blessed herb. But legend links it with St Benedict, founder of the Benedictine order. He was a hard ruler, apparently, and his monks planned to murder him by poisoning his wine. St Benedict made the sign of the cross over the glass, and it flew into pieces. The legend does not say what avens has to do with this, but the plant became known as an antidote to poison.

ST BENEDICT'S THISTLE *Carduus benedictus* (Holy Thistle) USA Henkel. A wrong interpretation of 'benedictus'. It means blessed.

ST BENNET(T)'S HERB 1. *Conium maculatum* (Hemlock) B & H; USA Henkel. 2. *Valeriana officinalis* (Valerian) Prior. Cf. Gerard's Herb Bennett for hemlock. Very odd for such a poisonous plant, but herba benedicta appears in a 13th century vocabulary. Valerian is a "heal-all", so as such it is "benedicta".

ST BERNARD'S LILY *Anthericum liliago*.

ST BRIDE'S COMB *Stachys officinalis* (Wood Betony) Wales Trevelyan. A translation from the Welsh Cribau Sant Ffraid.

ST BRIDE'S FORERUNNER *Taraxacum officinale* (Dandelion) Bayley. Dandelion is "the plant of Bride" in the Highlands, and is one of her insignia as the goddess of spring.

ST BRIDGET'S WORT *Plantago coronopus* (Buck's-horn Plantain) IOM Grigson.

ST BRUNO'S LILY *Anthericum liliastrum*.

ST CANDIDA'S EYES *Vinca minor* (Lesser Periwinkle) Dor. Dacombe. St Candida's Well is at Morcombe Lake, Dorset, and the water is said to be a certain cure for sore eyes; it is on Stonebarrow Hill that the wild periwinkles are called St Candida's Eyes. Candida is white - on that basis the saint is the same as St Blanche, of Brittany, or the Welsh St Gwen, or even the English St Wita.

ST CATHERINE'S LILY *Lilium candidum* (Madonna Lily) Haig. This is Saint Catherine of Siena, who is usually represented with a lily. But it should be noted that Catherine is from Greek Katharos, in the sense of purity. So it has the same meaning as the symbolism of the lily anyway.

ST COLUMCILLE'S PLANT *Lysimachia nemora* (Yellow Pimpernel) Scot. Grigson. This is the translation from the Gaelic.

ST DABEOC'S HEATH *Daboecia cantabrica* (Irish Heather). B & H. Daboecia, i.e. 'oe' in spite of the 'eo' of the saint's name.

ST DAVID'S PLANT *Allium ampeloprasum* (Wild Leek) Corn. Davey.

ST DAVID'S ROSE *Rosa pimpinellifolia* (Burnet Rose) Pemb. Freethy. It is the emblem of the see of St David's, displayed in the cathedral.

ST DOMINGO APRICOT *Mammea americana* (Mamey) Wit. Cf. Wild Apricot for this.

ST DOMINGO MAHOGANY *Swietenia mahagoni* (Cuban Mahogany) Ackermann.

ST FOYNE *Onobrychis sativa* (Sainfoin) Herts Jones-Baker. A pseudo saint, of course - this is just Sainfoin, another version of which appears as Saintpoin.

ST FRANCIS'S THORN *Eryngium maritimum* (Sea Holly).

ST FRANCIS'S WOOD *Lonicera nigra* (Black-berried Honeysuckle) Polunin.

ST GEORGE'S BELLS; ST GEORGE'S FLOWER *Endymion nonscriptus* (Bluebell) Tynan & Maitland.

ST GEORGE'S HERB *Valeriana officinalis* (Valerian) Howes.

ST GILES'S ORPINE *Sedum telephium* (Orpine) Geldart. Orpine is appropriated to St Giles's Day - 1 September.

ST JAMES'S RAGWORT *Senecio jacobaea* (Ragwort) Flower.1851. Perhaps it is because the plant is in full flower on St James's Day, which is 25 July, but there is another possibility. St James is the patron saint of horses, and the plant was certainly used in veterinary practice. The specific name, jacobaea, also links it to the saint.

ST JAMES'S TEA *Ledum groenlandicum* (Labrador Tea) Grieve.1931.

ST JAMES'S WORT 1. *Capsella bursa-pastoris* (Shepherd's Purse) Glos, Middx, Lancs. Vesey-Fitzgerald. 2. *Senecio jacobaea* (Ragwort) Fernie. Presumably Shepherd's Purse earns the name by reference to the leather pouches carried by the poorer pilgrims to the saint's shrine at Compostella.

ST JOHN, Blood of *Hieraceum pilosellum* (Mouse-ear Hawkweed). Apparently it was gathered on St John's Eve, to bring good luck. This name is merely a translation of the German Johannisblut.

ST JOHN'S BLOOM *Chrysanthemum segetum* (Corn Marigold) Notes & Queries. vol 7.

ST JOHN'S BREAD; ST JOHN'S SWEETBREAD *Ceratonia siliqua* fruit (Carob) Wit(Bread); Kourenoff (Sweetbread). A reference to the tradition that the pods were the "locusts" eaten by St John the Baptist. They are certainly edible.

ST JOHN'S BUSH *Psychotria nervosa* Guyana Laguerre.

ST JOHN'S GRASS *Hypericum perforatum* (St John's Wort) Gerard.

ST JOHN'S GRASS, Great; ST JOHN'S GRASS, Square *Hypericum tetrapterum* (St Peter's Wort) both Turner.

ST JOHN'S HERB *Eupatorium cannabinum* (Hemp Agrimony) Friend.

ST JOHN'S WORT 1. *Artemisia vulgaris* (Mugwort) Genders. 2. *Chelidonium majus* (Greater Celandine) Dev. Grigson. 3. *Hypericum perforatum*. The red sap of *Hypericum perforatum* represents the blood of St John, or according to another legend, the glandular dots that look like holes and account for the specific name *perforatum* are drops of the saint's blood, appearing every year on St John's Day, 24 June. Another tradition says they appear on 29 August, the day of the beheading of the saint. But this is a typical St John's herb, used much in divinations on Midsummer Eve, i.e. St John's Eve. Presumably Greater Celandine gets the name in Devonshire simply because it looks rather like the real thing, but mugwort is a genuine herb of St John, worn by him as a girdle in the wilderness, according to one legend. It does keep the flies away if worn, and grows in the right area. It is used ritually at the Midsummer festival.

ST JOHN'S WORT, Bog *Hypericum elodes*.

ST JOHN'S WORT, Canadian *Hypericum canadense*.

ST JOHN'S WORT, Chinese *Hypericum chinense*.

ST JOHN'S WORT, Flax-leaved *Hypericum linarifolium*.

ST JOHN'S WORT, Giant 1. *Hypericum ascyrion* (Great St John's Wort) USA House. 2. *Hypericum lanceolatum*. 'Giant' is right in the latter case - it is a tree, often about 40 feet high.

ST JOHN'S WORT, Great 1. *Hypericum ascyrion*. 2. *Hypericum tetrapterum* (St Peter's Wort) Gerard.

ST JOHN'S WORT, Hairy *Hypericum hirsutum*.

ST JOHN'S WORT, Imperforate *Hypericum maculatum*.

ST JOHN'S WORT, Irish *Hypericum*

canadense (Canadian St John's Wort) Blamey.

ST JOHN'S WORT, Large-flowered *Hypericum calycinum* (Rose of Sharon) Dartnell & Goddard.

ST JOHN'S WORT, Marsh *Hypericum elodes* (Bog St John's Wort) Polunin.

ST JOHN'S WORT, Mountain *Hypericum montanum* (Pale St John's Wort) Clapham.

ST JOHN'S WORT, Pale 1. *Hypericum ellipticum* USA House. 2. *Hypericum montanum*.

ST JOHN'S WORT, Slender *Hypericum pulchrum* (Upright St John's Wort) Fitter,Fitter & Blamey.

ST JOHN'S WORT, Square *Hypericum tetrapterum* (St Peter's Wort).

ST JOHN'S WORT, Stinking *Hypericum hircinum* (Stinking Tutsan) Clapham.

ST JOHN'S WORT, Tall *Hypericum elatum*.

ST JOHN'S WORT, Trailing *Hypericum humifusum*.

ST JOHN'S WORT, Upright *Hypericum pulchrum*.

ST JOHN'S WORT, Wavy *Hypericum undulatum*.

ST JOSEPH'S FLOWER *Veronica chamaedrys* (Germander Speedwell) Tynan & Maitland.

ST JOSEPH'S LILY *Lilium candidum* (Madonna Lily) Howes.

ST JOSEPH'S STAFF 1. *Narcissus pseudonarcissus* (Daffodil) Tynan & Maitland. 2. *Solidago virgaurea* (Golden Rod) Scot. Tynan & Maitland. Cf. Aaron's Rod, a typical south of England name. They both owe much to the specific name, *virgaurea*.

ST JOSEPH'S WORT 1. *Ocymum basilicum* (Sweet Basil) Sanecki. 2. *Ocymum minimum* (Bush Basil) Sanecki.

ST LUCIE CHERRY *Prunus mahaleb*.

ST MARK'S ROSE *Rosa virginiana pl.*

ST MARY *Tanacetum balsamita* Balsamint Dawson. Probably deriving from the name Costmary - French coste amère, Latin costus amarus. So 'Mary' here is a simple mistake that did not stop the plant from being dedicated to the Virgin in most west European countries.

ST MARY'S SEAL *Polygonatum multiflorum* (Solomon's Seal) Le Strange. Cf. Lady's Seal, or Lady's Signet.

ST MAWE'S CLOVER *Medicago arabica* (Spotted Medick) Corn. B & H.

ST OLAF'S CANDLESTICKS *Moneses uniflora*.

ST PATRICK'S CABBAGE *Saxifraga spathularia*. Perhaps because it grows in the west of Ireland, where the saint lived. But it may be a mis-translation from the Gaelic. It is claimed in *Cybele Hibernica* (2nd ed 1898) that the genitive forms of the Irish for 'Patrick' and 'fox' are similar enough to be confused in this name (see Britten).

ST PATRICK'S SPIT; ST PATRICK'S STAFF *Pinguicula vulgaris* (Butterwort) Antrim PLNN1.

ST PETER PORT DAISY *Erigeron mucronatus* (Mexican Fleabane) McClintock.1975. It was first recorded in the British Isles from St Peter Port, Guernsey.

ST PETER'S BELL *Narcissus pseudo-narcissus* (Daffodil) Wales Grigson.

ST PETER'S CABBAGE *Saxifraga virginiensis* (Rose Pennywort) Howes.

ST PETER'S FLOWER *Rhinanthus crista-galli* (Yellow Rattle) Tynan & Maitland. Cf. Peter's Grass. It is suggested that the clue lies in its resemblance to a cock's comb.

ST PETER'S HERB; ST PETER'S KEYS *Primula veris* (Cowslip) Yks. Grigson (Herb); Som. Macmillan (Keys). "Keys", from the supposed resemblance to the badge of St Peter - a bunch of keys. The legend is that St Peter once dropped the keys of heaven, and the first cowslips grew up where they fell.

ST PETER'S WORT 1. *Hypericum ascyrion* (Great St John's Wort) Le Strange. 2. *Hypericum elodes* (Bog St John's Wort) B & H. 3. *Hypericum tetrapterum*. 4. *Symphoricarpos rivularis* (Snowberry) Gorer. 5. *Tanacetum parthenium* (Feverfew) Bucks. B & H. St Peter's Day is 29 June, just 5 days after that of St John. But why bring St Peter in anyway?

ST THOMAS'S ONION 1. *Allium cepa* (Onion) M Baker.1977. 2. *Allium cepa var. aggregatum* (Egyptian Onion) M Baker.1977. 3. *Allium fistulosum* (Welsh Onion) M Baker.1977. Onions are sacred to St Thomas, and used in a divination for St Thomas's Eve, when girls peeled a large red onion and stuck nine pins in it, one in the centre being named for the man she really wanted, and the rest put radially round. Was she supposed to dream of the future husband? Or they looked for onions sprouting next morning, when the one that had would reveal, presumably by having been named, who she would marry.

ST THOMAS'S TREE *Bauhinia tomentosa* Dalziel.

SAINTFOIN *Onobrychis sativa* (Sainfoin) A S Palmer. A common mis-spelling, that could even appear as St Foyne.

SAKHALIN FIR *Abies sacchalinensis*.

SAKHALIN SPRUCE *Picea glehnii*.

SAL *Shorea robusta* (Saul Tree) W Miller.

SALAD BURNET *Poterium sanguisorba*.

SALAD BURNET, Prickly *Poterium polygamum* (Fodder Burnet) Genders.

SALAD CHERVIL *Anthriscus cerefolium* (Chervil) A W Smith.

SALAD ROCKET *Eruca sativa* (Rocket) A W Smith.

SALADINE *Chelidonium majus* (Greater Celandine) Ches, Yks, Cumb. Grigson.

SALAL 1. *Gaultheria humifusa* (Teaberry) USA Elmore. 2. *Gaultheria shallon* (Shallon) Hutchinson.

SALARY 1. *Apium graveolens* (Wild Celery) Som. Elworthy; Yks. A S Palmer. 2. *Rumex acetosa* (Sorrel) Oxf. B & H. 'Salary' for 'celery' is understandable. The same word for 'sorrel' is a little more recondite.

SALE, Solomon's *Polygonatum multiflorum* (Solomon's Seal) Lancs. Nodal. Not a misprint, just a rendering of local pronunciation.

SALEP 1. *Dactylorchis incarnata* (Early Marsh Orchid Prior. 2. *Orchis mascula* (Early Purple Orchid) Prior. 3. *Orchis morio* (Green-winged Orchid) Prior. Salep is a starch-like substance got from the tubers of orchids. It was once imported from the East, and was used as an article of diet, and said to be very nutritious, so much so that is was once part of every ship's provisions, to prevent famine at sea. It was made into a soft drink, too, long before the introduction of coffee houses, and was mentioned in Victorian books as a common beverage for manual workers. From the Arabic Sahlab, but more immediately from Turkish (most of the tubers for its manufacture were imported from Turkey).

SALIGO *Caltha palustris* (Marsh Marigold) Culpepper.

SALLEE, White *Eucalyptus pauciflora* (Cabbage Gum) Wilkinson.1981.

SALLET *Rumex acetosa* (Sorrel) Bucks. B & H.

SALLIT *Lactuca sativa* (Lettuce) Yks. Robinson. Lettuce being the embodiment of salads.

SALLOW *Salix capraea*. OE sealh, willow, stll preserved in the Wiltshire place name, Zeals.

SALLOW, Anson's *Salix ansoniana*.

SALLOW, Cloth-leaved *Salix pannosa*.

SALLOW, Creaking *Salix strepida*.

SALLOW, Damson-leaved *Salix damascena*.

SALLOW, Eared *Salix aurita*.

SALLOW, Great *Salix capraea* (Sallow) Clapham.

SALLOW, Grey *Salix atrocinerea*.

SALLOW, Grey-leaved *Salix grisophylla*.

SALLOW, Grisons *Salix grisonensis*.

SALLOW, Hairy-branched *Salix hirta*.

SALLOW, Hornbeam-leaved *Salix carpinifolia*.

SALLOW, Lake *Salix lacustris*.

SALLOW, Long-leaved *Salix acuminata*.

SALLOW, Mountain, Glaucous *Salix forsteriana*.

SALLOW, Mountain, Green *Salix andersoniana*.

SALLOW, Olive-leaved *Salix oleifolia*.

SALLOW, River *Salix rivularis*.
SALLOW, Round-eared *Salix aurita* (Eared Sallow) Warren-Wren.
SALLOW, Rock *Salix petra*.
SALLOW, Rock, Silky *Salix rupestris*.
SALLOW, Southern *Salix australis*.
SALLOW, Swiss *Salix helvetica*.
SALLOW, Thick-leaved *Salix crassifolia*.
SALLOW, Trailing *Salix aurita* (Eared Sallow) Warren-Wren.
SALLOW, Vaudois *Salix vaudensis*.
SALLOW, Water *Salix aquatica*.
SALLOW, White *Acacia longifolia* (Sydney Golden Wattle) Polunin.1969.
SALLOW, White-leaved *Salix incanescens*.
SALLOW THORN *Hippophae rhamnoides* (Sea Buckthorn) Hutchinson. Cf. Willowthorn for this.
SALLOW WILLOW, Great *Salix capraea* (Sallow) Hutchinson.
SALLY; SALLY, Black *Salix capraea* (Sallow) Wilts. Dartnell & Goddard; Shrop. Grigson (both Black Sally).
SALLY, Blooming 1. *Chamaenerion angustifolium* (Rosebay) N Ire. B & H. 2. *Epilobium hirsutum* (Great Willowherb) Ire. Grigson. 3. *Lythrum salicaria* (Purple Loosestrife) Grieve.1931.
SALLY, Broad-leaved *Eucalyptus camphera*.
SALLY, Weeping *Eucalyptus mitchelliana*.
SALLY, Flowering *Lythrum salicaria* (Purple Loosestrife) Grieve.1931. There are many 'willow' names for loosestrife. The leaves are similar to those of the willows.
SALLY, French *Salix pentandra* (Bay-leaved Willow) Donegal Grigson.
SALLY, Red *Lythrum salicaria* (Purple Loosetsrife) Lancs. Grigson.
SALLY, Sour *Oxalis acetosella* (Wood Sorrel) Som. Macmillan; Mon. Waters. 'Sally' here is 'sorrel', probably a mixture of 'sorrel' and 'alleluia', one of the names for this plant.
SALLY, White *Eucryphia moorei* Everett.
SALLY-MY-HANDSOME *Carpobrotus edulis* (Sea Fig) Coats. This is a Cornish variation of Mesembryanthemum.

SALLY WITHY *Salix capraea* (Sallow) Wilts, Heref, Shrop. Grigson.
SALLY-WOOD *Salix capraea* (Sallow) Wales Hornell.
SALMON-BERRY 1. *Rubus nutkanus*. 2. *Rubus spectabilis*.
SALSIFY 1. *Tragopogon porrifolius*. 2. *Tragopogon pratensis* (Goat's Beard) G M Taylor. French salsifis, from Italian sassefrica, though there is always a possibility that salsify is a corruption of solsequium.
SALSIFY, Black *Scorzonera hispanica*.
SALSIFY, Leek-leaved *Tragopogon porrifolius* (Salsify) G M Taylor. i.e. 'porrifolius'.
SALSIFY, Purple *Tragopogon porrifolius* (Salsify) Watt.
SALSIFY, Spanish 1. *Scolymus hispanicus*. 2. *Scorzonera hispanica* (Black Salsify) Leyel.1937.
SALSIFY, Wild; SALSIFY, Yellow *Tragopogon pratensis* (Goat's Beard) A Huxley (Wild); USA Gates (Yellow).
SALT, Pepper-and- *Capsella bursa-pastoris* (Shepherd's Purse) Middx. Grigson. Cf. Pepper Plant and Peppergrass for this. It was at one time used as a potherb.
SALT-AND-PEPPER *Agrimonia eupatoria* (Agrimony) Corn. Grigson.
SALT BUSH *Salvadora persica* (Toothbrush Tree) Dalziel. In the Lake Chad area, a vegetable salt is prepared from the ashes of this tree.
SALT CEDAR 1. *Tamarix glauca* USA Elmore. 2. *Tamarix pentandra*.
SALT-CELLAR *Oxalis acetosella* (Wood Sorrel) Dor. Macmillan.
SALT-MARSH RUSH *Scirpus maritimus* (Sea Club-rush) Pratt.
SALT RHEUMWEED *Chelone glabra* (Turtlehead) O P Brown.
SALT ROSE *Armeria maritima* (Thrift) Yks. Tynan & Maitland. Because it is a "sea" plant.
SALTBUSH 1. *Atriplex argentea*. 2. *Atriplex canescens* (Chamizo) Stevenson.

SALTBUSH, Desert *Atriplex polycarpa* (Cattle Spinach) USA T H Kearney.
SALTBUSH, Four-wing *Atriplex canescens* (Chamizo) Kluckhohn.
SALTBUSH, Hoary *Atriplex canescens* (Chamizo) Jaeger.
SALTBUSH, Mohave *Atriplex spinifera*.
SALTBUSH, Spiny *Atriplex confertifolia* (Shadscale) Elmore.
SALTBUSH, Torrey *Atriplex torreyi*.
SALTWEED *Juncus bufonius* (Toad Rush) Halliwell.
SALTWORT 1. *Salicornea europaea* (Glasswort) Gerard. 2. *Salsola kali*. Both of these plants were burned for their fixed salt, used in glass-making.
SALTWORT, Black *Glaux maritima* (Sea Milkwort) Prior.
SALTWORT, Glass *Salicornia europaea* (Glasswort) Gerard.
SALTWORT, Prickly *Salsola kali* (Saltwort) Hutchinson.
SALTWORT, Shrubby *Suaeda fruticosa* (Shrubby Seablite) Watt.
SALVADOR SISAL *Agave letonae*.
SALVATION JANE *Echium plantagineum* (Purple Viper's Bugloss) Australia Salisbury.1964. Apparently from the resemblance of the flowers to Salvation Army bonnets.
SALVE, Poor Man's 1. *Scrophularia aquatica* (Water Betony) Dev. B & H. 2. *Scrophularia nodosa* (Figwort) Dev. Grigson. The leaves are anodyne, and ease pain wherever they are applied. They are put on cuts and bruises, and burns, and the tuber is put on a sore.
SALVES, Sour *Rumex acetosa* (Wild Sorrel) Dev. Grigson.
SAMARANG ROSE-APPLE *Eugenia javanica* (Curacao Apple) Douglas.
SAMFER; SANFER *Crithmum maritimum* (Samphire) N'thum. PLNN2 (Samfer); Lancs. Lancs FWI (Sanfer). Variations on Samphire.
SAMI *Prosopis spicigera*.
SAMMY GUSSETS *Orchis mascula* (Early Purple Orchid) Som. Macmillan. There are many versions of this, all of them probably stemming from Gandigoslings or something like it. 'Gussets' also appears as 'Gussips' or 'Gossips'.
SAMPER *Crithmum maritimum* (Samphire) IOW Long. Cf. Semper, Samfer, etc.
SAMPHIRE 1. *Crithmum maritimum*. 2. *Salicornia europaea* (Glasswort) Lincs. B & H; USA. Chamberlin. Samphire is from the French herbe de St Pierre. 'Pierre' also means stone, so, disregarding the attribution to St Peter, the word simply means a plant that grows on a rock, a fact that led to its medicinal use, following the doctrine of signatures, against the stone. Glasswort gets the name, as well as a number of variations on it, because the young shoots make a good pickle, steeped in malt vinegar, and were often used as a substitute for samphire, though they say it is inferior to the proper stuff.
SAMPHIRE, Golden *Inula crithmoides*. Once used in the same way as the true samphire, but it has not got the aromatic qualities of that plant.
SAMPHIRE, Marsh 1. *Salicornia europaea* (Glasswort) N'thum. B & H. 2. *Salsola kali* (Saltwort) Bianchini.
SAMPHIRE, Rock 1. *Crithmum maritimum* (Samphire) B & H. 2. *Salicornia europaea* (Glasswort) N'thum. B & H.
SAMPIER; SAMPIER, Rock *Crithmum maritimum* (Samphire) both Gerard.
SAMPION *Salicornia europaea* (Glasswort) Ches. B & H.
SAMPKINS *Crithmum maritimum* (Samphire) N Wales PLNN1. A version of Samphire.
SAMPSON *Echinacea angustifolia* (Narrow-leaved Purple Coneflower) Schauenberg.
SAMPSON SNAKEWEED *Gentiana ochroleuca*. 'Snakeweed', because it was used by the American Indians as an antidote to snakebite.
SAMPSON'S SNAKEROOT *Psoralea pedunculata* S USA Puckett.

SAN HEMP *Crotolaria juncea* (Sunn Hemp) Schery.
SANCTUARY 1. *Centaureum erythraea* (Centaury) Shrop, Leics, Ches. B & H; Yks. Jeffrey. 2. *Odontites verna* (Red Bartsia) B & H. 'Sanctuary' is a widespread variation on 'centaury'.
SANCTUARY, Yellow *Blackstonia perfoliata* (Yellow-wort) Ches. Holland.
SAND APPLE *Parinari mobola*.
SAND BRIER *Solanum carolinense* (Carolina Nightshade) USA Tolstead.
SAND BUR *Solanum rostratum* (Prickly Nightshade) USA House.
SAND CROCUS *Romulea columnae*.
SAND CHERRY 1. *Prunus besseyi*. 2. *Prunus pumila*.
SAND FLOWER *Armeria maritima* (Thrift) Dor. Macmillan.
SAND FLOWER, Everlasting *Ammobium alatum*.
SAND-MAT, Rib-seedeed *Euphorbia glyptosperma* (Mat Spurge) Jaeger.
SAND-MAT, Small-seeded *Euphorbia polycarpa*.
SAND ROCKET *Diplotaxis muralis* (Stinkweed) D Grose.
SAND SAGEBRUSH *Artemisia filifolia*.
SAND TOADFLAX *Linaria arenaria*.
SAND WILLOW *Salix arenaria*.
SANDALWOOD 1. *Santalum album*. 2. *Thespesia populnea* (Portia Tree) Englert. The name is eventually from Sanskrit candama, burning wood, in other words, wood for incense.
SANDALWOOD, African *Spirostachys africana* (Tambootie) Summers.
SANDALWOOD, Australian; SANDALWOOD, Lance-leaf *Santalum lanceolatum* Usher (Lance-leaf).
SANDALWOOD, East African *Osyris compressa*.
SANDALWOOD, Red *Pterocarpus santalinus*.
SANDARAC *Tetraclinus articulata*.
SANDBAR WILLOW 1. *Salix longifolia* Usher. 2. *Salix sessifolia*.

SANDBELL *Jasione montana* (Sheep's-bit Scabious) Perry.
SANDBERRY *Arctostaphylos uva-ursi* (Bearberry) USA T H Kearney.
SANDBOX TREE *Hura crepitans*. The fruits used to be wired together and used as sandboxes before blotting paper was invented.
SANDERS; SAUNDERS; SAUNDERS, Red *Pterocarpus santalinus* (Red Sandalwood) A S Palmer (Sanders, Saunders); Grieve.1931 (Red Saunders). Not Alexanders, as one would expect in this case, but that name probably influenced this abbreviation of Sandalwood.
SANDPAPER LEAF *Ficus exasperata* Harley. The leaf is actually used by carpenters as a substitute for sandpaper.
SANDPAPER TREE 1. *Cordia ovalis*. 2. *Ficus asperifolia*. These too have very rough leaves, the Ficus actually being used in Africa to smooth wood, etc.
SANDPAPER VINE *Petraea volubilis* (Purple Wreath Vine) USA Perry.
SANDWEED *Spergula arvensis* (Corn Spurrey) Norf. Halliwell.
SANDWORT, Alpine *Minuartia rubella* (Red Sandwort) Clapham.
SANDWORT, Balearic *Arenaria balearica*.
SANDWORT, Bog *Minuartia stricta* (Teesdale Sandwort) Clapham.
SANDWORT, Curved *Minuartia recurva*.
SANDWORT, Fine-leaved *Minuartia tenuifolia*.
SANDWORT, Fries' *Arenaria gothica* (Yorkshire Sandwort) Clapham.
SANDWORT, Hairy *Arenaria ciliata* (Irish Sandwort) Polunin.
SANDWORT, Irish *Arenaria ciliata*.
SANDWORT, Mossy 1. *Arenaria balearica* (Balearic Sandwort) McClintock.1978. 2. *Moehringia muscosa*.
SANDWORT, Norwegian *Arenaria norvegica* (Scottish Sandwort) Clapham.
SANDWORT, Ovate *Honkenya peploides* (Sea Sandwort) Grieve.1931.
SANDWORT, Red 1. *Minuartia rubella*. 2. *Spergularia rubra* (Sand Spurrey) Phillips.

SANDWORT, Rock *Minuartia stricta* (Teesdale Sandwort) USA Tolstead.
SANDWORT, Scottish *Arenaria norvegica*.
SANDWORT, Sea *Honkenya peploides*.
SANDWORT, Slender *Minuartia tenuifolia* (Fine-leaved Sandwort) Salisbury.
SANDWORT, Spring *Minuartia verna*.
SANDWORT, Teesdale *Minuartia stricta*.
SANDWORT, Three-veined *Moehringia trinervum*.
SANDWORT, Thyme-leaved *Arenaria serpyllifolia*.
SANDWORT, Vernal *Minuartia verna* (Spring Sandwort) Condry.
SANDWORT, Wood *Arenaria laterifolia*.
SANDWORT, Yorkshire *Arenaria gothica*.
SANGUINARY 1. *Achillea millefolium* (Yarrow) B & H. 2. *Capsella bursa-pastoris* (Shepherd's Purse) Dawson. Both of these plants have been used to stay bleeding, and Shepherd's Purse is still used, especially for internal haemorrhages.
SANICLE 1. *Ajuga reptans* (Bugle) Fernie. 2. *Sanicula europaea*. Sanicle means a healing herb, from the Latin sanare, to heal, although attempts have been made to see it as a corruption of St Nicholas!
SANICLE, American *Heuchera americana* (Alum-root) W Miller.
SANICLE, Great *Alchemilla vulgaris* (Lady's Mantle) Gerard.
SANICLE, Mountain *Astrantia major* (Pink Masterwort) Polunin.
SANICLE, Poison *Sanicula bipinnata*. Given the meaning of 'Sanicle', the two parts of this name seem mutually exclusive. Is it poisonous? Perhaps the reason for the name lies in American Indian uses against snakebite.
SANICLE, Purple *Sanicula bipinnatifida*.
SANICLE, Wood *Sanicula europaea* (Sanicle) Thrimble.
SANICLE, Yorkshire *Pinguicula vulgaris* (Butterwort) Gerard.
SANIKER *Sanicula europaea* (Sanicle) Gerard.
SANTA LUCIA FIR *Abies venusta*. This is Santa Lucia Mountain, California.

SANTA ROSA SAGE *Salvia eremostachya*. Santa Rosa Mountains, USA, which is the only place that they grow.
SANTONICA *Artemisia maritima* (Sea Wormwood) Chopra. A reference to the drug santonin, used extensively as a vermifuge, of which this plant is the source.
SANWA MILLET *Echinochloa crus-galli var. frumentacea* (Japanese Barnyard Millet) Brouk.
SAP GUM *Liquidambar styraciflua* the sapwood (Sweet Gum) Brimble.1948. One of the names under which the sapwood is sold.
SAP TREE *Sorbus aucuparia* (Rowan) Yks. Atkinson. Cf. Sip-sap, another Yorkshire name.
SAPELE *Entandrophragma cylindrica*.
SAPISTAN *Cordia myxa*.
SAPONILLA *Manilkara achras*.
SAPPAN *Caesalpina sappan* (Brazil-wood) Barker.
SAPPHIRE-BERRY *Symplocos paniculata*.
SAPWORT, Water *Oenanthe crocata* (Hemlock Water Dropwort) Grigson.
SARACEN BIRTHWORT *Aristolochia clematitis* (Birthwort) Pomet. 'Saracen' probably means 'foreign', without any claim of more exact origin. Birthwort is a plant of central and southern Europe, introduced here into medieval physic gardens, and now an escape.
SARACEN CORN *Fagopyrum esculentum* (Buckwheat) Grieve.1931. Probably brought in by the Crusaders. It is blés sarrasin, or sarrasino, in French.
SARACEN'S WOUNDWORT; SARACEN'S CONFREY; SARACEN'S CONSOUND *Senecio fluviatilis* Saracen's Woundwort is the usual name; the other two are in Gerard. Presumably the Crusaders brought it back to England. Gerard reckoned it "not inferiour to any of the wound-herbes whatsoever, being inwardly ministred, or outwardly applied in ointments or oyles". The fact that he called it also Comfrey and Consound implies that it could also be used for helping to repair broken bones.

SARAH JANE; MARY JANE *Melandrium rubrum* (Red Campion) Som. Macmillan.
SARANA, Black *Fritillaria camschatcensis* (Black Lily) RHS.
SARDIAN NUT *Castanea sativa* (Sweet Chestnut) Mabey.
SAREPTA MUSTARD *Brassica juncea* (Indian Mustard) S Africa Watt.
SARGENT COTTONWOOD *Populus sargentii*.
SARGENT LILY *Lilium sargentiae*.
SARGENT SPRUCE *Picea brachytyla*.
SAROCK *Rumex acetosa* (Sorrel) Scot. Jamieson. Another form of Sourock, or Sourack.
SARRAT *Genista tinctoria* (Dyer's Greenweed) W Eng. Grigson.
SARSAPARILLA *Smilax sarsaparilla*. Originally zarza parilla, zarza meaning bramble, and parilla, the diminutive of parra, vine. So the word means little prickly vine.
SARSAPARILLA, American 1. *Aralia nudicaulis*. 2. *Menispermum canadense* (Moonseed) Le Strange.
SARSAPARILLA, Australian *Hardenbergia violacea* (Blue Coral Pea) RHS.
SARSAPARILLA, English *Potentilla erecta* (Tormentil) Grieve.1931.
SARSAPARILLA, False 1. *Aralia nudicaulis* (American Sarsaparilla) Grieve.1931. 2. *Hardenbergia violacea* (Blue Coral Pea) Howes.
SARSAPARILLA, Lesser *Euphorbia lateriflora* Dalziel.
SARSAPARILLA, Moonseed; SARSAPARILLA, Yellow *Menispermum canadense* (Moonseed) Le Strange.
SARSAPARILLA, Wild 1. *Aralia nudicaulis* (American Sarsaparilla) Grieve.1931. 2. *Smilax kraussiana* S Africa Watt.
SASKATOON *Amelanchier alnifolia* (Western Service-tree) USA Gilmore. The name is given to any of the Amelanchiers or their fruit, and comes from a Cree word, misáskwatomin.
SASSAFRAS, Australian; SASSAFRAS, Tasmanian *Atherosperma moschatum* RHS.

SASSAFRAS, Californian *Umbellularia californica* (Californian Laurel) Le Strange.
SASSAFRAS, Nepal *Cinnamonium glanduliferum*.
SASSAFRAS, Swamp *Magnolia virginiana* (Sweet Bay) O P Brown.
SASSAFRAS TREE *Sassafras variifolium*. 'Sassafras' is the same word as 'saxifrage', that is a plant that can break up stone, for which Sassafras was given.
SASSIFRAX *Saxifraga granulata* (Meadow Saxifrage) Som. Grigson.
SASSWOOD; SASSYBARK *Erythrophloeum guineense* (Ordeal Tree) Harley (Sasswood); Verdcourt (Sassy Bark) (better as one word).
SATAN'S APPLE *Mandragora officinalis* fruit (Mandrake) Clair. Cf. Devil's Apple. 'Apple', because the fruit looks rather like a small apple, and smells strongly of them, too. 'Satan's', because they are poisonous.
SATAN'S BREAD *Anthriscus sylvestris* (Cow Parsley) Lincs. Gutch. Cf. Devil's Meal, Bad Man's Oatmeal, etc - 'meal' from the powdery appearance of the flowers. But these ascriptions to the devil apply rather to hemlock.
SATAN'S CHERRY *Atropa belladonna* (Deadly Nightshade) Yks. Grigson. Poisonous, of course, hence the attribution to the devil.
SATES *Crataegus monogyna* (Hawthorn) Shrop. B & H. Given to hedging plants - 'sets' is the usual pronunciation.
SATIN; SATIN, White *Lunaria annua* (Honesty) Gerard (Satin); Friend (White Satin).
SATIN-BALLS *Calluna vulgaris* (Heather) Som. Macmillan.
SATIN FLOWER 1. *Lunaria annua* (Honesty) Northants, Norf. B & H. 2. *Sisyrinchium sp.* 3. *Stellaria holostea* (Greater Stitchwort) Fernie. 4. *Stellaria media* (Chickweed) USA Watt.
SATIN FLOWER, New Zealand *Libertia grandiflora*.

SATIN FLOWER, Red *Hedysarum coronarium* (French Honeysuckle) Parkinson.
SATIN FLOWER, White *Lunaria annua* (Honesty) Gerard.
SATIN POPPY *Meconopsis nepalense*.
SATINS, Silks-and- *Lunaria annua* (Honesty) Dev. Friend.1882.
SATIN WALNUT *Liquidambar styraciflua* timber (Sweet Gum) Wilkinson.1981. The timber was sold in Europe under this name.
SATINWOOD 1. *Chloroxylon swietenia*. 2. *Cedrus atlantica* (Atlas Cedar) Grieve.1931. 3. *Terminalia ivorensis* (Yellow Terminalia) Dalziel. Because of the satin sheen of the wood.
SATURDAY NIGHT'S PEPPER; SATURDAY'S PEPPER *Euphorbia helioscopia* (Sun Spurge) both Wilts. Dartnell & Goddard.
SATYRION, Female; SATYRION, Hand; SATYRION ROYAL *Dactylorchis maculata* (Heath Spotted Orchid) Gerard (Female); Turner (Hand, Royal). The reference to a satyr stems from the aphrodisiac properties that orchids were once supposed to have. Presumably this is why Gerard called this one Female Satyrion, for Male Satyrion would be reserved for those species with the particularly male twin tubers. 'Hand', in the second of these names, is given because the tubers are divided into several finge-like lobes, hence the "dactyl" part of the generic name.
SAUCE, Green 1. *Oxalis acetosella* (Wood Sorrel) Corn. Courtney.1880; Dev. Friend. 2. *Rumex acetosa* (Sorrel) Corn. Jago; Dev. Macmillan; Glos, War, Leics, Notts, Lincs. B & H; Ches. Holland; Lancs. Nodal; Yks. Addy. 3. *Rumex acetosella* (Sheep's Sorrel) Ches. Holland.
SAUCE, Green, London *Rumex acetosa* (Wild Sorrel) Lancs. Grigson.
SAUCE, Sour *Rumex acetosa* (Sorrel) Corn. Jago.; Lincs. B & H. Sorrel comes from a word meaning 'sour'.
SAUCE-ALONE 1. *Alliaria petiolata* (Jack-by-the-hedge) Turner. 2. *Heracleum sphondyllium* (Hogweed) Som. Macmillan.

"Divers eat the stamped leaves [of Jack-by-the-hedge] ... with Salt-fish for a sauce, as they do those of Ramsons" (Gerard). The 'alone' part of the name means garlic, and comes either from the French ail, ailloignon, or Italian aglio, alione.
SAUCER MAGNOLIA *Magnolia soulangeana*.
SAUCERS, Summer *Melandrium album* (White Campion) Som. Macmillan.
SAUCHWEED *Polygonum persicaria* (Persicaria) Ayr. B & H. 'Sauch' means 'willow'. Cf. such names as Willow-weed amd the Wiltshire Blind Withy. Perhaps Amphibious Persicaria is meant, for the leaves of that plant much more resemble willow leaves than common Persicaria.
SAUCY BARK *Erythrophloeum guineense* Le Strange. Evidently a variation of Sassy Bark.
SAUCY BET *Centranthus ruber* (Spur Valerian) Corn. Grigson.
SAUCY JACK *Centaurea melitensis* (Napa Thistle) S Africa Watt.
SAUF *Salix caprea* (Sallow) Grose.
SAUGH, SAUCH 1. *Salix alba* (White Willow) Scot. Aitken. 2. *Salix caprea* (Sallow) N Eng, Scot, Ire. Grigson. Saugh is cognate with English Sallow, or Sally. All derive from OE sealh, willow. Sauch is given for the Scottish name in Gibson.
SAUGH, French *Chamaenerion angustifolium* (Rosebay) Lanark. Grigson. Cf. French Willow, and a number of other 'willow' names,all because of the resemblance of the leaves to those of willow.
SAUGH, Hoburn *Laburnum anagyroides* (Laburnum) Scot. Jamieson. Cf. Weeping Willow and Drooping Willow. 'Hoburn' is auburn, the colour of the wood.
SAUGH, Privy *Ligustrum vulgare* (Privet) Yks. Grigson; Scot. Jamieson.
SAUGH, Yellow *Lysimachia vulgaris* (Yellow Loosestrife) Yks. Grigson. Willow-like leaves, and so there are a number of 'willow' names.
SAUGH-TREE 1. *Salix alba* (White Willow)

Scot. Aitken. 2. *Salix caprea* (Sallow) Grigson.

SAUL-TREE *Shorea robusta*. Cf. Sal for this.

SAUNDERS; SANDERS; SAUNDERS, Red *Pterocarpus santalinus* (Red Sandalwood) A S Palmer; Grieve.1931 (Red Saunders). All corruptions of the usual name.

SAUSAGE TREE; SAUSAGE TREE, German *Kigelia africanus* (German Sausage Tree) Palgrave. Perfectly descriptive - the fruits are often a metre long.

SAUVE *Salix caprea* (Sallow) Dur. Dinsdale. It is also Sauf, in the same area. Both are variations of Saugh.

SAVAGER, Wild *Agrostemma githago* (Corn Cockle) B & H.

SAVANILLA RHATANY *Krameria ixene*.

SAVE 1. *Hypericum perforatum* (St John's Wort) Sanecki. 2. *Salvia officinalis* (Sage) Halliwell. A 12th century name for St John's Wort, probably meaning an all-heal. 'Save' for sage is easier - it is the Latin Salvia.

SAVE-WHALLOP *Rosa canina* galls (Dog Rose galls) Yks. F K Robinson. Yorkshire schoolboys used to reckon the galls to be a charm against a flogging.

SAVIN 1. *Artemisia maritima* (Sea Wormwood) Suss. B & H. 2. *Juniperius sabina*. Latin sabina, Sabine. Sea Wormwood was either given the name in error, or, equally possible, because it was used in the same way as Savin, i.e. for abortion. It is certainly anthelmintic, so that is quite possible.

SAVIN, Eastern *Juniperus excelsa* (Grecian Juniper) Moldenke.

SAVIN, Juniper *Juniperus sabina* (Savin) Fernie.

SAVIN, Virginia *Juniperus virginiana* (Virginian Juniper) USA Elmore.

SAVING, Horse *Juniperus communis* (Juniper) Cumb. Grigson. 'Saving' is 'savin', of course, but it is not clear why anyone should want to abort mares. Perhaps, though, this is just a case of "horse" in a plant name meaning large or coarse.

SAVING-TREE *Juniperus sabina* (Savin) Scot. A S Palmer.

SAVIOUR'S BLANKET *Stachys lanata* (Lamb's-ears) Suss. Friend. "Blanket", because of the woolly leaves. Cf. Blanket-leaf, from Devonshire.

SAVONETTE *Satureia hortensis* (Summer Savory) Howes.

SAVORY, Dyer's *Serratula tinctoria* (Saw-wort) W Miller. Cf. Dyer's Saw-wort. It was once used for dyeing woollen fabrics a green-yellow, with alum.

SAVORY, Mountain *Satureia montana* (Winter Savory) Hatfield. It is tempting, given the peppery taste of the plant, to equate 'savory' with 'savoury'. But apparently the word is a descendant of the Latin Satureia.

SAVORY, Summer *Satureia hortensis*.

SAVORY, Winter *Satureia montana*.

SAVOUR, Lad *Artemisia abrotanum* (Southernwood) Lancs. B & H.

SAVOY; SAVOY CABBAGE *Brassica oleracea var. bullata*. Presumably, Savoy is the place of origin.

SAW-GRASS *Cladium mariscus* (Sedge) USA Rackham.1986. A comment on the sharply saw-edged leaves. Thatching with sedge is very effective, but can be a very bloody business.

SAW-WORT; SAW-WORT, Dyer's *Serratula tinctoria* Flower.1849 (Dyer's). Saw-wort is a descriptive name, from the saw-toothed leaves.

SAWARA CYPRESS *Chamaecyparis pisifera*.

SAWG *Salix caprea* (Goat Willow) Yks. Hunter. Indicating the local pronunciation of 'Saugh'.

SAWTOOTH OAK *Quercus acutissima* (Japanese Chestnut Oak) RHS. Descriptive of the leaves, of course.

SAXIFRAGE, Arctic; SAXIFRAGE, Alpine *Saxifraga nivalis* Darling & Boyd (Alpine). Saxifrage is derived from two Latin words - saxum, rock and frangere, to break. The name is given because the plant often grows in clefts of rocks, leading to the mistaken impression that the roots had actually broken the rock.

SAXIFRAGE, Brook *Saxifraga rivularis* (Highland Saxifrage) Clapham.
SAXIFRAGE, Bulbous *Saxifraga granulata* (Meadow Saxifrage) Wit.
SAXIFRAGE, Burnet *Pimpinella saxifraga*.
SAXIFRAGE, Burnet, Greater *Pimpinella major*.
SAXIFRAGE, Cushion, One-flowered *Saxifraga burserana*.
SAXIFRAGE, Drooping *Saxifraga cernua*.
SAXIFRAGE, Fingered *Saxifraga tridactylites* (Rue-leaved Saxifrage) McClintock.
SAXIFRAGE, French *Saxifraga clusii*.
SAXIFRAGE, Furrowed *Saxifraga exarata*.
SAXIFRAGE, Golden *Chrysosplenium sp.*
SAXIFRAGE, Hawkweed *Saxifraga hieracifolia*.
SAXIFRAGE, Highland *Saxifraga rivularis*.
SAXIFRAGE, Kidney *Saxifraga hirsuta*.
SAXIFRAGE, Livelong *Saxifraga aizoon*.
SAXIFRAGE, Marsh, Yellow *Saxifraga hirculus*.
SAXIFRAGE, Meadow 1. *Saxifraga granulata*. 2. *Silaum silaus* (Pepper Saxifrage) B & H.
SAXIFRAGE, Mountain, Yellow *Saxifraga aizoides*.
SAXIFRAGE, Mossy *Saxifraga hypnoides* (Dovedale Moss) McClintock.
SAXIFRAGE, Musky *Saxifraga muscoides*.
SAXIFRAGE, Musky, White *Saxifraga exarata* (Furrowed Saxifrage) Grey-Wilson.
SAXIFRAGE, Pepper *Silaum silaus*.
SAXIFRAGE, Persian *Saxifraga cymbalaria*.
SAXIFRAGE, Purple *Saxifraga oppositifolia*.
SAXIFRAGE, Pyramidal *Saxifraga cotyledon*.
SAXIFRAGE, Pyrenean 1. *Saxifraga longifolia*. 2. *Saxifraga umbrosa*.
SAXIFRAGE, Rue-leaved *Saxifraga tridactylites*.
SAXIFRAGE, Scree *Saxifraga androsacea*.
SAXIFRAGE, Spoon-leaved *Saxifraga cuneifolia* (Wedge-leaved Saxifrage) Grey-Wilson.
SAXIFRAGE, Star(ry) *Saxifraga stellaria*.
SAXIFRAGE, Thick-leaved *Saxifraga callosa*.
SAXIFRAGE, Tufted *Saxifraga caespitosa*.
SAXIFRAGE, Water *Saxifraga aquatica*.
SAXIFRAGE, Wedge-leaved *Saxifraga cuneifolia*.
SAXIFRAGE, White *Saxifraga granulata* (Meadow Saxifrage) Gerard.
SAXIFRAGE, Wood *Saxifraga umbrosa* (Pyrenean Saxifrage) Polunin.
SAXIFRAGE, Yellow *Saxifraga aretioides*.
SCAB FLOWER *Anthriscus sylvestris* (Cow Parsley) Cumb. Mabey. Cf. Scabby Hands, or Heads. Apparently, children dislike Cow Parsley, and believe that if they touch it they will get sore hands (but that came from Oxfordshire).
SCABBIT-DOCK *Digitalis purpurea* (Foxglove) Corn. Grigson. A name that can only mean that the plant may be used to treat skin diseases. In fact, an ointment has long been made from the leaves for just such a purpose.
SCABBY HANDS 1. *Anthriscus sylvestris* (Cow Parsley) Som. Grigson. 2. *Conium maculatum* (Hemlock) Som. Macmillan. 3. *Conopodium majus* (Earthnut) Inv. Grigson. 4. *Heracleum sphondyllium* (Hogweed) Som. Grigson.
SCABBY HEADS 1. *Anthriscus sylvestris* (Cow Parsley) Oxf. Oxf.Annl.Rpt.1951. 2. *Torilis japonica* (Upright Hedge Parsley) Ches. Holland.
SCABGOWK 1. *Dactylorchis maculata* (Heath Spotted Orchid) Dur. Grigson. 2. *Orchis mascula* (Early Purple Orchid) Dur. B & H. There are a number of other 'cuckoo' names for these two orchids, probably given as a generalised springtime appellation, rather than as a reference to the bird itself.
SCABIOUS *Erigeron canadensis* (Canadian Fleabane) USA Henkel. Scabious is from Latin scabere, to scratch. Scabies either got its name from the plant, which was used as a remedy, or, more likely, the other way round, the plant being named from the scab.

Medieval Latin scabiosa herba is named as a specific against the itch.

SCABIOUS, Devil's-bit *Succisa pratensis*. The legend is that the short, blackish root was originally bitten off by the devil, out of spite to mankind, because he knew that otherwise it would be good for many purposes. Another version states the opposite, that it was a bane to mankind. So God took the power away from the devil, who bit off the root in vexation.

SCABIOUS, Field 1. *Centaurea scabiosa* (Greater Knapweed) H C Long.1910. 2. *Knautia arvensis*.

SCABIOUS, Giant *Cephalaria tartarica*. 'Giant' is reasonable - this grows about 5 feet in height.

SCABIOUS, Grass-leaved *Scabiosa graminifolia*.

SCABIOUS, Grecian *Scabiosa pterocephala*.

SCABIOUS, Grey *Scabiosa canescens*.

SCABIOUS, Indian *Scabiosa atropurpurea* (Sweet Scabious) Folkard.

SCABIOUS, Meadow *Succisa pratensis* (Devil's-bit Scabious) Curtis.

SCABIOUS, Premorse *Succisa pratensis* (Devil's-bit Scabious) Grieve.1931. A reference to the "bitten-off" appearance of the root.

SCABIOUS, Sheep's-bit; SCABIOUS, Sheep's, Hairy *Jasione montana*. Hairy Sheep's Scabious was Curtis's name.

SCABIOUS, Shining *Scabiosa lucida*.

SCABIOUS, Small *Scabiosa columbaria*.

SCABIOUS, Sweet 1. *Erigeron annuus* (White Top) Polunin. 2. *Scabiosa atropurpurea*.

SCABIOUS, Syrian *Cephalaria syriaca*.

SCABIOUS, Yellow 1. *Cephalaria tatarica* (Giant Scabious) RHS. 2. *Scabiosa ochroleuca*.

SCABRIDGE; SCABRIL; SCAYBRIL *Knautia arvensis* (Field Scabious) Ches. B & H; Halliwell (Scabridge).

SCABS 1. *Alliaria petiolata* (Jack-by-the-hedge) Wilts. Dartnell & Goddard. 2. *Anthriscus sylvestris* (Cow Parsley) Som. Grigson. See Scab-flower for Cow Parsley. The record of the name for Jack-by-the-hedge is unique, and may very well be in error for Cow Parsley.

SCABWORT *Inula helenium* (Elecampane) Gerard. It was certainly used to treat scab in sheep, but it was invaluable for human skin complaints as well.

SCAD; SCAD-TREE *Prunus domestica var. institia* (Bullace) Suss. Halliwell; Kent, Lincs. Grigson. Cf. the Buckinghamshire Skeg.

SCADDIE *Urtica dioica* (Nettle) Forf. B & H.

SCALD; SCALDWEED 1. *Cuscuta epithymum* (Dodder) Cambs. Grigson. 2. *Cuscuta europaea* (Greater Dodder) Cambs. B & H. i.e. scab, or scabies, weed. It is not clear whether it was thought dodder caused the itch, or whether it relieved it. Perhaps, in the principles of homeopathic magic, both could apply.

SCALD-HEAD 1. *Cynoglossum officinale* (Hound's Tongue) Suff. Grigson. 2. *Rubus fruticosus* (Bramble) Grieve.1931. Perhaps blackberries produce the complaint known as scaldhead in children who eat too much of the fruit, or because the fruit has a curative effect on this scalp disease, or even because the leaves are externally applied to scalds, and has nothing to do with the scalp.

SCALDBERRY *Rubus fruticosus* (Bramble) Prior.

SCALDRICK *Sinapis arvensis* (Charlock) Scot. Jamieson. Cf. Skellocks, Skellies and Scallock, all Scottish variants of the familiar English versions of the OE cerlic.

SCALE FERN *Ceterach officinarum* (Rustyback) Turner.

SCALIES *Sinapis arvensis* (Charlock) Forf, Stir. B & H. Cf. Scaldrick, etc.

SCALLION 1. *Allium ascalonium* (Shallot) Shrop. B & H; Suff. Moor. 2. *Allium porrum* (Leek) Yks. Harland; Scot. B & H. Shallot is from French eschalotte, Latin ascalonia, hence so is Scallion, but the word is usually reserved for shallots. Cf. Scullion.

SCALLOCK *Sinapis arvensis* (Charlock) Stir. B & H. Cf. Scaldrick, etc.

SCALLOP GOURD *Cucurbita pepo var. melopepo*. Cf. Pattypan Squash; they are both from the characteristic shape of the fruit.

SCALY CUSTARD APPLE *Annona squamosa* (Sugar Apple) India Dalziel.

SCAMINE *Convolvulus scammonia* (Scammony) Halliwell.

SCAMMONY; SCAMMONY BINDWEED *Convolvulus scammonia* Thornton (Scammony Bindweed).

SCAMMONY, Levant *Convolvulus scammonia* (Scammony) Howes.

SCARB TREE *Malus sylvestris* (Crabapple) B & H. Evidently a consonantal shift from Scrab, which is a common name for it in the northern counties and Scotland. Skrab is an Old Norse word meaning sour wild apple.

SCARBAGRACE *Cochlearia officinalis* (Scurvy-grass) Pollock. A version of scurvy-grass, which appeared in a number of different forms, such as Scruby, or Scrubby, grass.

SCARCITY ROOT *Beta vulgaris* (Beet) Woodforde. The entry in Woodforde's diary for June 29, 1787 reads "...Sir William Jernegan sent me by Mr Custance a Treatise on the Plant called Scarcity Root." The name came about by the confusion of German Mangel, want, with Mangold, beet.

SCARET-ROOT *Sium sisarum* (Skirret) Halliwell.

SCAREWEED *Gutierrezia microcephala* (Broomweed) USA Elmore.

SCARIBEUS; SQUARRIB *Scrophularia nodosa* (Figwort) Wilts. Dartnell & Goddard. i.e. square rib, perhaps from the shape of the leaves.

SCARIOLE *Cichorium endivia* (Endive) Gerard. Cf. Escarole, given to the broad-leaved kind.

SCATTLE-DOCK *Senecio jacobaea* (Ragwort) Lancs. Grigson. Part of a series of names that includes Kedlock and Kettle-dock.

SCAW; SCAW-TREE; SCAWEN *Sambucus nigra* (Elder) all Corn. Jago (Scaw); Grigson (Scaw-tree, Scawen). Scaw is Cornish for elder, and it is found also as Skaw or Scow. The flowers are Scawsy-buds.

SCAW-COO *Solanum dulcamara* (Woody Nightshade) Corn. Grigson. Cornish scaw cough, red elder.

SCAW-DOWER *Scrophularia aquatica* (Water Betony) Corn. Courtney.1880. It means water-elder.

SCAWFELL PINK *Armeria maritima* (Thrift) Cumb. Grigson.

SCAW(N)SY-BUDS *Sambucus nigra* flowers (Elder) Corn. Jago; Courtney.1887 (Scawnsy). See Scaw.

SCENT, Lamps-of- *Lonicera periclymenum* (Honeysuckle) Som. Macmillan.

SCENT, Old Maid's *Tanacetum parthenium* (Feverfew) Som. Macmillan. Stink Daisy also occurs in Somerset. Is the name unkind enough for that?

SCENT BOTTLE *Plantago media* (Lamb's-tongue Plantain) Dor, Som. Grigson.

SCENTY, Miss *Viola odorata* (Sweet Violet) Som. Macmillan.

SCHICKENWIR *Stellaria media* (Chickweed) Shet. Grigson. Chickenweed, presumably.

SCHOLAR-TREE *Sophora japonica* (Pagoda-tree) Barber & Phillips.

SCHOOL BELL *Campanula rotundifolia* (Harebell) Wilts. Dartnell & Goddard.

SCHOOL OF CHRIST *Hypericum perforatum* (St John's Wort) Wales Trevelyan. Cf. Ladder of Christ, also from the Welsh.

SCHOOLBOYS' CLOCK *Taraxacum officinale* (Dandelion) Mabey. Children, of course, tell the time by the number of puffs it takes to blow the seeds from the seedhead.

SCHOOLMASTER *Arum maculatum* (Cuckoo-pint) Suss. B & H. Probably owing something to the Parson-in-the-pulpit set of names.

SCIATICA-CRESS *Iberis amara* (Candytuft) Prior. "The leaves are recommended greatly in the sciatica; they are to be applied externally, and repeated as they grow dry" (Hill).

SCIENCES, Close; SCIENCES, Coses *Hesperis matronalis* (Sweet Rocket) Gerard (Close Sciences); B & H (Coses Sciences). Corruptions of Sciney, itself a corruption of damascena, the old specific name for the plant (hence Damask Violet). It is pointed out that Close Sciences would be the double variety, as opposed to Single Sciences.

SCILLY BUTTERCUP *Ranunculus muricatus*.

SCIMITAR-PODDED KIDNEY BEAN *Phaseolus lunatus* (Lima Bean) Candolle.

SCINEY; SINEY *Hesperis matronalis* (Sweet Rocket) B & H. See Sciences.

SCISSOR PLANT *Iris dichotoma* (Afternoon Iris) Friend.

SCOBE *Cytisus scoparius* (Broom) Donegal Grigson.

SCOKE; SKOKE *Phytolacca decandra* (Poke-root) USA Henkel (Scoke); Sanford (Skoke). Cf. Shoke, Coakum, etc.

SCORDION *Teucrium scordium* (Water Germander) Gerard.

SCORPION-GRASS, Changeable *Myosotis discolor* (Changing Forget-me-not) Hulme. The forget-me-nots got the old name Scorpion-grass from the curve of the one-sided raceme. The doctrine of signatures made this into the medicine to cure the sting of a scorpion.

SCORPION-GRASS, Field *Myosotis arvensis* (Field Forget-me-not) J A MacCulloch.1905.

SCORPION-GRASS, Field, Early *Myosotis hispida* (Early Forget-me-not) Hulme.

SCORPION-GRASS, Mouse-ear *Myosotis arvensis* (Field Forget-me-not) Browning. The small oval leaves are responsible for the Mouse-ear part of the name. Myosotis is derived from Greek muos otis - mouse ear.

SCORPION-GRASS, Parti-coloured *Myosotis discolor* (Changing Forget-me-not) J A MacCulloch.1905.

SCORPION-GRASS, Water, Creeping *Myosotis secunda* (Creeping Water Forget-me-not) J A MacCulloch.1905.

SCORPION-GRASS, Wood *Myosotis sylvatica* (Wood Forget-me-not) Hulme.

SCORPION IRIS *Iris alata* (Christmas-flowering Iris) W Miller.

SCORPION SENNA *Coronilla emerus*.

SCORPION-WEED *Phacelia glandulosa* (Waterleaf) USA Elmore.

SCORPION'S TAIL *Heliotropum europaeum* (Cherry-pie) Turner.

SCORZONER *Scorzonera humilis* (Viper's Grass) Gerard.

SCOTCH BEAN *Vicia faba* (Broad Bean) Schery.

SCOTCH BROOM 1. *Cytisus scoparius* (Broom) USA Zenkert. 2. *Genista canariensis*. Why Scotch in both cases? Common broom is not confined to Scotland, and anyway has Irish Broom as another of its names.

SCOTCH CAMOMILE *Chamaemelon nobile* (Camomile) B & H.

SCOTCH CROCUS *Crocus biflorus*.

SCOTCH ELM *Ulmus glabra* (Wych Elm) Border B & H. Not confined to Scotland, but this is the only common elm there.

SCOTCH FLAME FLOWER *Tropaeolum speciosum*. Why 'Scotch'? It is a native of Chile.

SCOTCH GALE *Myrica gale* (Bog Myrtle) Scot. Jamieson. 'Gale' is the proper English name for the plant, from OE gagel. Not confined to Scotland, of course, but at least it is a plant of northerly areas.

SCOTCH GERANIUM *Geranium robertianum* (Herb Robert) Forf. B & H.

SCOTCH GRANFER-GRIGGLES *Prunella vulgaris* (Self-heal) Dev. Macmillan. 'Granfer-griggles' is a name usually given to the common native orchids. It probably started as something like Gandigoslings. Why 'Scotch', and in Devonshire, too?

SCOTCH HEATHER *Erica cinerea* (Bell Heather) W Miller.

SCOTCH LABURNUM *Laburnum alpinum*. Actually, it is a native of the southern Alps, Apennines and central Europe.

SCOTCH LAVENDER *Asperula odorata* (Woodruff) Scot. Swire.1963.

SCOTCH LOVAGE *Ligusticum scoticum* (Lovage) Grieve.1931.

SCOTCH MAHOGANY *Alnus glutinosa* (Alder) Ablett. The reddish timber is the reference for this (and Irish Mahogany).

SCOTCH MERCURY *Digitalis purpurea* (Foxglove) Berw. Grigson. Cf. Wild Mercury, from the same area. They must bear witness to some now-forgotten medical usage.

SCOTCH PARSLEY *Ligusticum scoticum* (Lovage) Grieve.1931.

SCOTCH PRIMROSE *Primula scotica*.

SCOTCH ROSE; SCOTS ROSE *Rosa pimpinellifolia* (Burnet Rose) Hutchinson (Scotch); Bunyard (Scots).

SCOTCH THISTLE 1. *Cirsium acaulon* (Picnic Thistle) Som. Macmillan; N Ire. Vickery.1995. 2. *Cirsium vulgare* (Spear Plume Thistle) Worcs. Grigson. 3. *Onopordum acanthium*. The tradition is that when a Scottish garrison was in danger of being surprised by a Danish enemy, a barefooted Dane, creeping along in the darkness, trod on the sharp prickles of a thistle, and, yelling with the pain, aroused the drowsy sentry. The garrison, thus saved, adopted the emblem and motto which afterwards became that of the whole nation.

SCOTINO *Cotinus coggyria* (Venetian Sumach) W Miller. Surely a version of the generic name,Cotinus.

SCOTLAND, Blue Bells of *Campanula rotundifolia* (Harebell).

SCOTS FIR *Pinus sylvestris* (Scots Pine) Hutchinson.

SCOTS PINE *Pinus sylvestris*.

SCOTS PLANE *Acer pseudo-platanus* (Sycamore) Murdoch McNeill.

SCOTS WHITEBEAM *Sorbus intermedia* (Cut-leaved Whitebeam) Brimble.1948.

SCOTTISH ASPHODEL *Tofieldia pusilla*.

SCOTTISH BLUEBELL *Campanula rotundifolia* (Harebell) Polunin.

SCOTTISH DOCK *Rumex aquaticus* (Trossachs Dock) Fitter.

SCOTTISH GENTIAN *Gentianella septentrionalis*.

SCOTTISH MAPLE *Acer pseudo-platanus* (Sycamore) Howes.

SCOTTISH SCURVY-GRASS *Calystegia soldanella* (Sea Bindweed) Barton & Castle. It was used as an anti-scorbutic.

SCOTTISH WORMWOOD *Artemisia norvegica*.

SCOURGE, Nun's *Amaranthus caudatus* (Love-lies-bleeding) Skinner. Descriptive - it suggested the flagellations endured by penitents. The name also occurs in France, and this is probably no more than a translation of it.

SCOURING RUSH 1. *Equisetum arvense* (Common Horsetail) USA Youngken. 2. *Equisetum hyemale* (Dutch Rush) Leighton. The stems of both of these have been used like sandpaper, and as a pot scourer. The reason is that the plant is very rich in silica.

SCOURWORT *Saponaria officinalis* (Soapwort) Turner. In this case, "scour" does not mean quite the same thing. The reference here is the fact that a lather can be got by rubbing the leaves in water, and it has been used as a specialist fabric soap since ancient times.

SCOUT, Dryland *Heracleum sphondyllium* (Hogweed) Tyrone. B & H.

SCOUT, Hexham *Melilotus indica* (Small-flowered Melilot) S Africa Watt.

SCOUTCH *Agropyron repens* (Couch-grass) Heref. Havergal. Scutch or Scoutch are the local names, both the same word as squitch, which is itself derived from OE cwice, alive.

SCOW *Sambucus nigra* (Elder) Corn. Hawke. Scaw is the Cornish for elder, and scow is a derivative.

SCRAB TREE *Malus sylvestris* (Crabapple) N Eng, Scot. B & H. Skrab is an Old Norse word meaning sour wild apple.

SCRABS *Malus sylvestris* (Crabapples) N Eng. Brockett.

SCRAG *Crataegus monogyna* (Hawthorn) Dor. Opie & Tatem; Wilts. Notes & Queries.1941.

SCRAMBLING ROCKET *Sisymbrium officinale* (Hedge Mustard) B & H.
SCRAPE-CLEAN *Senecio jacobaea* (Ragwort) Lincs. Grigson.
SCRATCH, Gosling *Galium aparine* (Goose-grass) Ess, Norf, Cambs. Grigson. 'Scratch' is obvious when the burrs are considered. 'Gosling', because the plant was used for feeding them.
SCRATCH-BUR *Ranunculus arvensis* (Corn Buttercup) Beds. Grigson.
SCRATCH COCO *Colocasia esculenta* (Dasheen) Howes.
SCRATCHGRASS; SCRATCHWEED *Galium aparine* (Goose-grass) Herts. Grigson (grass); N'thants. Sternberg; N'thum. Prior (weed).
SCREE SAXIFRAGE *Saxifraga androsacea*.
SCREEN, Fire *Tropaeolum speciosum* (Scotch Flame Flower) Dor. Macmillan.
SCREW AUGER *Spiranthes cernua* (Nodding Lady's Tresses) USA Howes.
SCREW PINE 1. *Pandanus sp.* 2. *Pinus contorta* (Beach Pine) Leathart. Pandanus, the true Screw Pines, gets the name because of the corkscrew-like trunks, and they are called Pine because the leaves are like those of a pineapple. Beach Pine earns the name because they get very gnarled and "contorta".
SCREW MESQUITE *Prosopis pubescens* (Screwbean) USA Youngken.
SCREWBEAN *Prosopis pubescens*.
SCRIP, Shepherd's *Capsella bursa-pastoris* (Shepherd's Purse) Gerard. 'Scrip' is an old word for a small bag, often used for a pilgrim's pouch.
SCROFULA-PLANT *Scrophularia nodosa* (Figwort) Grieve.1931. "Fig-wort is good against... Scrophulas, that is, the Kings Evill..." (Gerard). It is a good example of the doctrine of signatures, for the knobbly tubers were looked on as the signature of scrophularious glands.
SCROG(G) 1. *Crataegus monogyna* (Hawthorn) Yks. F K Robinson; Ire. Grigson. 2. *Malus sylvestris* fruit (Crabapple) Dur, N'thum, Berw, Roxb.

Mabey. 3. *Prunus spinosa* (Blackthorn) Barton & Castle.
SCROG-BUSH *Crataegus monogyna* (Hawthorn) Ire. Grigson.
SCROITA *Xantheria parietina* (Common Yellow Wall Lichen) Shet. Nicolson. Presumably a variation of Crottal, a usual name for any lichen.
SCRUB OAK 1. *Quercus dumosa*. 2. *Quercus pungens*. 3. *Quercus undulata* (Wavyleaf Oak) USA Elmore.
SCRUB OAK, Large *Quercus wislienzi*.
SCRUB PINE *Pinus virginiana* (Virginia Pine) Everett.
SCRUBBY GRASS; SCRUBY GRASS; SCROOBY GRASS *Cochlearia officinalis* (Scurvy-grass) Yks. Carr (Scrubby); Yks. Hunter; Scot. B & H (Scruby); Grose (Scrooby).
SCRUMP *Malus domestica* (Apple) Dor. Dacombe. The origin must have been the same as that for crabapple's name Scrab.
SCROTTYIE *Parmelia saxatilis* (Grey Stone Crottle) Shet. Bolton. A variation on Crottle.
SCRYLE *Agropyron repens* (Couchgrass) W Eng. Halliwell. Probably the same as Cornish Stroil, or Stroyl.
SCULLION *Allium cepa* (Onion) Corn. Jago. More likely to be the shallot, whose botanical name is Allium ascalonium. It is named for the town of Ascalon, in Palestine, for the shallot is a native of that area. The Latin word was ascalonia.
SCURF PEA *Psoralea argophylla*. Cf. Scurvy Pea, which is more likely to be the right name.
SCURF PEA, Beaverbread *Psoralea castorea* (Beaver-dam Breadroot) Douglas.
SCURF PEA, Yellow *Psoralea lanceolata*.
SCURVY-BEAN *Menyanthes trifoliata* (Buckbean) Fernie. Buckbean used to be of prime importance in the treatment of scurvy.
SCURVY-BERRIES 1. *Maianthemum canadense* (False Lily-of-the-valley) USA Yarnell. 2. *Smilacina racemosa* (False Solomon's Seal) USA Weiner.
SCURVY-GRASS 1. *Cochlearia officinalis*.

2. *Galium aparine* (Goose-grass) Ches. Grigson; Yks. Robinson. 3. *Oxalis enneaphylla*. 4. *Stellaria holostea* (Greater Stitchwort) Worcs. Grigson. *Cochlearia officinalis* was used at one time to combat scurvy at sea. It was taken on board in the form of dried bundles or distilled extracts, disguised with spices to hide the taste. Goose-grass also had similar country uses, so, presumably, did stitchwort.

SCURVY-GRASS, Alpine *Cochlearia alpina*.

SCURVY-GRASS, Cook's *Lepidium oleraceum*. This is a New Zealand plant, and a good scurvy antidote.

SCURVY-GRASS, Danish *Cochlearia danica*.

SCURVY-GRASS, English *Cochlearia anglica*.

SCURVY-GRASS, Ivy-leaved *Cochlearia danica* (Danish Scurvy-grass) Le Strange.

SCURVY-GRASS, Mountain *Cochlearia alpina* (Alpine Scurvy-grass) Clapham.

SCURVY-GRASS, Northern *Cochlearia groenlandica*.

SCURVY-GRASS, Scottish 1. *Calystegia soldanella* (Sea Bindweed) Barton & Castle. 2. *Cochlearia groenlandica* (Northern Scurvy-grass) Clapham. 3. *Cochlearia micacea*. The bindweed was used in a similar way as the scurvy-grasses.

SCURVY-GRASS, Sea *Cochlearia anglica* (English Scurvy-grass) Le Strange.

SCURVY-GRASS, Stalked *Cochlearia danica* (Danish Scurvy-grass) Hepburn.

SCURVY PEA *Psoralea argophylla* (Scurf Pea) USA Yarnell. 'Scurf' in the common name is the same as 'scurvy'.

SCU(T)CH GRASS *Agropyron repens* (Couch-grass) Heref. G C Lewis; IOM Moore.1924; Lancs. A Burton. Cf. Scoutch. They both come from OE cwice, alive.

SEA GRASS 1. *Armeria maritima* (Thrift) Gerard. 2. *Salicornia europaea* (Glasswort) Gerard.

SEA GREEN *Stratiotes aloides* (Water Soldier) Grieve.1931.

SEA ISLAND COTTON *Gossypium barbadense*.

SEA TURF *Armeria maritima* (Thrift) Dev. Grigson.

SEABLITE *Suaeda maritima*.

SEABLITE, Hairy *Bassia hirsuta*.

SEABLITE, Shrubby *Suaeda fruticosa*.

SEABLITE, Torrey *Suaeda torreyana ramosissima*.

SEABLITE GLASSWORT *Salicornia prostrata*.

SEAKALE BEET *Beta vulgaris var. cicla* (Swiss Chard) Brouk.

SEAL, David's *Polygonatum multiflorum* (Solomon's Seal) Le Strange. Cf. David's Harp, a descriptive name.

SEAL, Golden *Hydrastia canadensis*.

SEAL, Lady's 1. *Bryonia dioica* (White Bryony) Friend. 2. *Polygonatum multiflorum* (Solomon's Seal) Wright. 3. *Tamus communis* (Black Bryony) Gerard.

SEAL, Our Lady's *Bryonia dioica* (White Bryony) Geldart. Dedicated to the Virgin, at the feast of her nativity. An ascription to the Virgin seems strange, considering the plant's bad reputation.

SEAL, St Mary's *Polygonatum multiflorum* (Solomon's Seal) Le Strange. Cf. Lady's Seal.

SEAL, Solomon's 1. *Hypericum calycinum* (Rose of Sharon) N'thants. B & H. 2. *Polygonatum commutatum* (Smooth Solomon's Seal) USA. Tolstead. 3. *Polygonatum multiflorum*. On cutting the root of *Polygonatum multiflorum* transversely, scars resembling the device known as Solomon's Seal, a 6-pointed star, can be seen. So, from the doctrine of signatures, it was used for sealing wounds.

SEAL, Solomon's, Angular *Polygonatum odoratum*.

SEAL, Solomon's, False *Smilacina sp*.

SEAL, Solomon's, Giant; SEAL, Solomon's, Great *Polygonatum commutatum* (Smooth Solomon's Seal) both USA Zenkert (Great); House (Giant).

SEAL, Solomon's, Hairy *Polygonatum biflorum*.

SEAL, Solomon's, Scented *Polygonatum odoratum* (Angular Solomon's Seal) Genders.

SEAL, Solomon's, Small *Polygonatum pubescens*.

SEAL, Solomon's, Smooth *Polygonatum commutatum*.

SEAL, Solomon's, Whorled *Polygonatum verticillatum*.

SEAL FLOWER *Dicentra spectabilis* (Bleeding Heart) W Miller.

SEALE TREE 1. *Salix alba var. vitellina* (Golden Willow) Cumb. B & H. 2. *Salix caprea* (Goat Willow) Yks. Halliwell. An archaism, clearly the OE sealh, from which the word sallow is derived.

SEALWORT *Polygonatum multiflorum* (Solomon's Seal) Wright.

SEARCHLIGHT *Linaria vulgaris* (Toadflax) Som. Macmillan. At the other end of the light scale from Fairies' Lanterns, the earlier Somerset name.

SEASIDE ALDER *Alnus maritima*.

SEASIDE ASTER *Aster spectabilis*.

SEASIDE BALSAM *Croton eleutheria* (Sweet-wood) Howes.

SEASIDE GOLDEN ROD *Solidago sempervirens*.

SEASIDE HELIOTROPE *Heliotropum curassavicum*.

SEASIDE LAVENDER *Limonium carolinianum* (American Sea Lavender) USA House.

SEASIDE LAVENDER, Small *Heliotropum curassavocum* Barbados Gooding.

SEASIDE MILKWEED *Glaux maritima* (Sea Milkwort) W Miller.

SEASIDE PANSY *Viola tricolor var. curtisii*.

SEASIDE PEA *Lathyrus maritimus* (Sea Pea) USA House.

SEASIDE PIMPERNEL *Honkenya peploides* (Sea Sandwort) W Miller.

SEASIDE PLANTAIN *Plantago maritima* (Sea Plantain) Murdoch McNeill.

SEASIDE PLUM *Ximenia americana* (Wild Oliver) Sofowora.

SEASIDE SAGE *Croton balsamifer* (Yellow Balsam) Barbados Gooding.

SEASIDE SMOOTH GROMWELL *Mertensia maritima* (Smooth Lungwort) Young.

SEASIDE SPURGE *Euphorbia mesembryanthemifolia* Barbados Gooding.

SEASIDE SWORD BEAN *Canavalia obtusifolia*.

SEASIDE THISTLE *Carduus tenuiflorus*.

SEASIDE WATER CROWFOOT *Ranunculus baudotii* (Brackish Water Crowfoot) McClintock.1975.

SEASIDE YAM *Ipomaea pes-caprae brasiliensis* Barbados Gooding.

SEAVES; SIEVES *Juncus conglomeratus* (Compact Rush) Yks. Hartley & Ingilby (Seaves); Cumb. Rollinson (Sieves).

SEAVES, Floss *Eriophorum angustifolium* (Cotton-grass) Yks. F K Robinson.

SEBESTAN PLUM *Cordia myxa* (Sapistan) Dalziel.

SECENTION *Senecio vulgaris* (Groundsel) E Angl. V G Hatfielld.1995. One of a number of variations of Senecio.

SECYMORE *Acer pseudo-platanus* (Sycamore) Yks. Addy.

SEDOCKE *Acanthus mollis* (Bear's Breech) B & H.

SEDGE 1. *Cladium mariscus*. 2. *Iris pseudoacarus* (Yellow Flag) Lancs. Miss Formby. Sedge is extended to iris, and there are many variations on it for this plant. See for instance Seggs, Saggan, Cagge. Segg was OE for a small sword, so this is an entirely descriptive name for the leaves.

SEDGE, Cinnamon; SEDGE, Myrtle *Acorus calamus* (Sweet Flag) B & H. 'Sweet' in Sweet Flag is a reference to the aromatic properties, reckoned to be like violets - or cinnamon, or myrtle, if we are to believe these names.

SEDGE, Corn *Gladiolus sp* Gerard. 'Gladiolus' too means a small sword - it is

the diminutive of gladius. Cf. Sword Flag for it.
SEDGE, Cotton *Eriophorum angustifolium* (Cotton Grass) J G McKay.
SEDGE, Easter *Polygonum bistorta* (Bistort) Cumb. Rowling. In this case, 'sedge' is a mistake for 'ledge' or 'ledger'. Herb pudding made from bistort among other ingredients was known as Ledger pudding.
SEDGE, Flag *Iris pseudo-acarus* (Yellow Iris) Pratt.
SEDGE, Nut, Yellow *Cyperus esculentus* (Tigernut) USA Allan.
SEDGE, Paper *Cyperus papyrus*. The leaf stalks contain the material that is made into papyrus, the original writing material of the ancient Egyptians.
SEDGE, Sweet *Acorus calamus* (Sweet Flag) B & H.
SEDGE, Yellow *Iris pseudo-acarus* (Yellow Flag) Border. B & H.
SEE-ME-CONTRACT *Eryngium foetidum* Jamaica Beckwith.
SEEBRIGHT *Salvia sclarea* (Clary) B & H. Clary seeds swell up when put into water, and become mucilaginous. They were then put like drops into the eye to cleanse it.
SEED-CAKE THYME *Thymus herba-baroni* (Carraway Thyme) Grant White. The proper seeds for seed-cake are carraway, and this thyme has a scent like carraway, hence the name.
SEEDER *Humulus lupulus* (male plant) Suss. B & H.
SEEDLING *Lobularia maritima* (Sweet Alison) Dev. B & H. These are Seedlings par excellence, just as in the same county, they are called Bordering, and Edging.
SEEDS 1. *Carum carvi* (seeds) (Carraway) Minag Bull 76. 2. *Lolium perenne* (Rye Grass) Oxf, Suss. B & H. Seeds for seed-cake can only be those of carraway, while Seeds, or Sids, for Rye-grass are related to other names like Crop, or Crap.
SEEP WILLOW *Baccharis glutinosa*.
SEEPWEED *Suaeda suffrutescens*.

SEERSUCKER PLANT *Dichorisandra mosaica* USA Rochford.
SEG; SEGG; SEGGINS *Iris pseudo-acarus* (Yellow Flag) Yks. Carr; Scot. Jamieson (Seg); N Eng. Gerard; Scot. Jamieson (Segg); Cumb, Scot. B & H (Seggins).
SEG, Sweet *Acorus calamus* (Sweet Flag) E Ang. B & H.
SEG, Water *Iris pseudo-acarus* (Yellow Flag) Yks. B & H.
SEG(G), Bull 1. *Orchis mascula* (Early Purple Orchis) Scot. B & H. 2. *Orchis morio* (Green-winged Orchid) Scot. B & H. 3. *Typha latifolia* (False Bulrush) Scot. Jamieson. For explanation, see under BULL SEG.
SEGG, Stinking *Iris foetidissima* (Foetid Iris) Langham.
SEGGINS, Blue; SEGGINS, Purple *Iris foetidissima* (Foetid Iris) Ayr. B & H (Blue); Scot. Tynan & Maitland (Purple).
SEGGRUM *Senecio jacobaea* (Ragwort) Yks. Atkinson. See Seggy.
SEGGY 1. *Acer pseudo-platanus* (Sycamore) Yks. Nicholson. 2. *Senecio jacobaea* (Ragwort) Yks. Grigson. Probably a derivative of sycamore in the one case, but quite different in the other. It is claimed that ragwort was applied as a vulnerary to newly castrated bulls, called seggs or staggs. Hence Seggy, or Seggrum.
SEGO, Poison *Zygadenus nuttalli*.
SEGO LILY *Calochortus nuttalli*. From the American Indian name for the plant, si-go.
SEGUMBER *Acer pseudo-platanus* (Sycamore) Dev. Grigson. Cf. the Yorkshire Seggy above.
SELADINE *Chelidonium majus* (Greater Celandine) Turner. Cf. Saladine, still widespread in northern England, and the Irish Sollandine.
SELF-HEAL 1. *Ajuga reptans* (Bugle) Prior. 2. *Pimpinella saxifraga* (Burnet Saxifrage) B & H. 3. *Prunella vulgaris*. 4. *Sanicula europaea* (Sanicle) Grieve.1931. *Prunella vulgaris* is the true Self-heal, a name given because of the many uses it has in folk medi-

cine. Bugle gets the name because it is very similar in appearance to Prunella. But sanicle is included in its own right, as it were. Sanicle means a healing herb, from Latin sanare, to heal.

SELF-HEAL, Cut-leaved *Prunella laciniata.*

SELF-HEAL, Large *Prunella grandiflora.*

SELGREEN *Sempervivum tectorum* (Houseleek) Dev. Friend.1882; Dor. Dacombe. See Sengreen, rather.

SELLY *Salix caprea* (Goat Willow) Yks. Atkinson. A local variant of Sally.

SELON'S ROSE *Nerium oleander* (Rose Bay) Watt. A corruption of Ceylon Rose.

SEMAPHORE PLANT *Desmodium gyrans* (Telegraph Plant) Howes.

SEMINOLE PUMPKIN *Cucurbita moschata* (Crook-neck Squash) Howes. The Seminole Indians lived in the far south of Florida.

SEMPER 1. *Crithmum maritimum* (Samphire) Yks. Nicholson. 2. *Salicornia europaea* (Glasswort) Grigson. A version of samphire, of course. Glasswort gets the name because the young shoots often used to be pickled as a substitute for samphire.

SEMPER, Rock *Crithmum maritimum* (Samphire) Yks, N'thum. Grigson. 'Rock Samphire' to distinguish it from 'Marsh Samphire', often given to Glasswort.

SENE *Colutea arborescens* (Bladder Senna) Turner.

SENECA GRASS *Hierochloe odorata* (Sweet Grass) Vestal & Schultes.

SENECIO, Creek *Senecio douglassi* (Douglas Groundsel) USA Elmore.

SENEGA SNAKEROOT *Polygala senega.* The Seneca Indians always used the root as a snakebite remedy. Actually, it is the doctrine of signatures at work here. One account says it grows wherever there are rattlesnakes, so that was reason enough for using them.

SENEGAL EBONY *Dalbergia melanoxyla* (African Blackwood) Dalziel.

SENEGAL LILAC *Lonchocarpus sericeus.*

SENEGAL SENNA *Cassia obovata* (Italian Senna) Dalziel.

SENEKA *Polygala senega* (Senega Snakeroot) Lindley.

SENGREEN 1. *Saxifraga granulata* (Meadow Saxifrage) B & H. 2. *Sedum acre* (Biting Stonecrop) Friend. 3. *Sempervivum tectorum* (Houseleek) Barton & Castle. 4. *Vinca minor* (Lesser Periwinkle) Som, IOW Grigson.

SENGREN *Sempervivum tectorum* (Houseleek) E Ang. G E Evans.

SENGREN, Water *Stratiotes aloides* (Water Soldier) Grieve.1931. Cf. Water Houseleek for this.

SENNA *Cassia angustifolia.* Senna is from an Arabic word, sana.

SENNA, Aden *Cassia acutifolia* (Alexandrian Senna) Dalziel.

SENNA, Alexandrian 1. *Cassia acutifolia.* 2. *Cassia angustifolia* (Senna) Thomson.1978.

SENNA, American *Cassia marilandica.*

SENNA, Bastard; SENNA, Bladder *Colutea arborescens.* Bladder Senna is the usual name. It was Parkinson who dubbed it Bastard Senna.

SENNA, Candlestick *Cassia alata.*

SENNA, Coffee *Cassia occidentalis.* The roasted seeds are used as a coffee substitute, though there is no caffeine content.

SENNA, Dog *Cassia obovata* (Italian Senna) Howes.

SENNA, Four-leaved *Cassia absus.*

SENNA, Italian 1. *Cassia obovata.* 2. *Cassia obtusifolia* Barbados Gooding. *Cassia obovata* used to be cultivated in Italy.

SENNA, Jamaica *Cassia obovata* (Italian Senna) Dalziel.

SENNA, Nubian *Cassia acutifolia* (Alexandrian Senna) Dalziel.

SENNA, Port Royal *Cassia obovata* (Italian Senna) Dalziel.

SENNA, Scorpion *Coronilla emerus.*

SENNA, Senegal *Cassia obovata* (Italian Senna) Dalziel.

SENNA, Sickle *Cassia tora* (Sicklepod) Hora.

SENNA, Spanish *Cassia obovata* (Italian Senna) Howes.

SENNA, Tea *Cassia mimosoides*.
SENNA, Tripoli *Cassia obovata*.
SENNA, Wild *Cassia marilandica* (American Senna) USA Zenkert.
SENNA-PODS *Centaureum erythraea* (Centaury) Som. Macmillan. Presumably just a children's name based on Centaury.
SENSHON; SENSION; SENTION; SENCION *Senecio vulgaris* (Groundsel) Fernie (Senshon); E Ang. Grigson (Sension); Hulme (Sention); Nall (Sencion). All variations on Senecio, even venturing into Ascension.
SENSITIVE PLANT *Mimosa pudica*.
SENTRY PALM *Howeia forsteriana* (Kentia Palm) RHS.
SENVIE; SENVY 1. *Brassica nigra* (Black Mustard) B & H. 2. *Sinapis alba* (White Mustard) B & H. 3. *Sinapis arvensis* (Charlock) B & H. This name varies into Zenvy or Zenry, and Sinvey in Somerset.
SEPTFOIL *Potentilla erecta* (Tormentil) W Miller. As against *Potentilla reptans*, which is Cinquefoil.
SEQUIER'S BUTTERCUP *Ranunculus seguieri*.
SERAPIA'S TURBITH *Aster tripolium* (Sea Aster) Gerard. Turbith seems to mean a root, or a medicine extracted from it. The word is from Persian or Arabic. Richard Mabey, in his glossary to Thomas Hill's *The gardener's labyrinth*, suggested that it was probably an extract of the roots of Giant Fennel, *Ferula communis*. Serapia is presumably Serapis, or Sarapis, a god of the Greeks of Egypt, identified with Apis and Osiris.
SERB 1. *Sorbus aria* fruit (Whitebeam) Suss. Parish. 2. *Sorbus aucuparia* fruit (Rowan berries) Suss. Parish. 3. *Sorbus torminalis* fruit (Wild Service berries). Serb is a variation of Service, OE syrfe, and ultimately from Latin sorbus.
SERBIAN SPRUCE *Picea omorika*.
SERGE *Typha latifolia* (False Bulrush) Ches. Holland. As printed, but is this just a misprint for Sedge?
SERGEANT, Hasty *Lapsana communis* (Nipplewort) Som. Dor. Grigson. Cf. Hasty Roger for this.
SERL, Cow *Rumex acetosella* (Sheep's Sorrel) USA Bergen. 'Serl' is 'sorrel'.
SERMOUNTAIN *Laserpitium latifolium*.
SERPENT APPLE *Annona palustris* (Alligator Apple) W Indies Dalziel. Judging by these two names, the fruit must have a particularly scaly skin.
SERPENT CUCUMBER *Trichomanthes anguina* (Snake Gourd) Whittle & Cook. 'Anguina' of the specific name means snaky.
SERPENT MELON *Cucumis melo var. flexuosum*. 'Flexuosum' explains the name.
SERPENT TONGUE 1. *Erythronium americanum* (Yellow Adder's Tongue) O P Brown. 2. *Ophioglossum vulgatum* (Adder's Tongue) Tynan & Maitland. Serpent Tongue is a translation of Ophioglossum.
SERPENT'S TONGUE SPEARWORT *Ranunculus ophioglossifolius*.
SERPENTARY *Aristolochia serpentaria* (Virginian Snakeroot) Fluckiger.
SERPENTARY, English *Polygonum bistorta* (Bistort) Culpepper. Bistort's writhed roots suggest a whole range of 'snake' names, e g Adderwort, Snakeweed, Dragons, etc.
SERPENTINE *Dracunculus vulgaris* (Dragon Arum) T Wright. The spadix is black, "...with spots of divers colours, like those of the adder or snake" (Gerard). Not surprisingly, with that kind of signature, the plant was widely used for snakebite.
SERPENTINE GARLIC *Allium ursinum* (Ramsons) T Hill. Cf. Snake-flower, Snake's Food, etc.
SERPOLET *Thymus serpyllum* (Breckland Wild Thyme) Rimmel.
SERRADELLA *Ornithopus sativus*.
SERVICE, Fowler's *Sorbus aucuparia* (Rowan) B & H. 'Service' comes through OE syrfe from the Latin sorbus. The full botanical name accounts for Fowler's Service. As Coles explains, the berries were used as bait to catch blackbirds and the like. It sounds reasonable, for 'aucuparia' is from auceps, a fowler.

SERVICE, Maple *Sorbus torminalis* (Wild Service) B & H.
SERVICE, Right *Sorbus domestica* (True Service) Hill. 'Right' in the sense of 'correct' - see next entry.
SERVICE, True *Sorbus domestica*.
SERVICE, Wild 1. *Sorbus aucuparia* (Rowan) Ablett. 2. *Sorbus torminalis*.
SERVICE-BERRY 1. *Amelanchier laevis* (Smooth June-berry) USA Zenkert. 2. *Sorbus aria* (Whitebeam) Mor. Grigson.
SERVICE-BERRY, Alder-leaved *Amelanchier alnifolia* (Western Service-berry) USA Elmore.
SERVICE-BERRY, Prune-leaved *Amelanchier prunifolia*.
SERVICE-BERRY, Western *Amelanchier alnifolia*.
SERVICE-TREE, Bastard *Sorbus x hybrida* (Finnish Whitebeam) Mitchell.1974.
SERVICE TREE OF FONTAINEBLEAU *Sorbus x latifolia* (French Hales) Polunin.1976. This hybrid was found at Fontainebleau some time before 1750.
SERVOILE *Lonicera periclymenum* (Honeysuckle) B & H. It seems to have been chêvrefeuille originally, and that was from Latin caprifolium. Cf. Caprifoy, Goat's Leaf, Goat Tree, etc.
SESAME *Sesamum indicum*. It is a Greek word, sesamon.
SESBAN, Egyptian *Sesbania sesban*.
SESBANE, Purple *Sesbania punicea*.
SETFOIL 1. *Aegopodium podagraria* (Goutweed) Lyte. 2. *Potentilla erecta* (Tormentil) Gerard. Cf. Septfoil. In other words, it is 'seven leaves', distinguishing it from cinquefoil, which has five.
SETON-GRASS *Helleborus foetidus* (Stinking Hellebore) Yks. F K Robinson. Seton is the more usual word, to be used instead of "setter", which see.
SETTER; SETTERGRASS; SETTER-WORT *Helleborus foetidus* (Stinking Hellebore) or *Helleborus viridis* (Green Hellebore) Grigson (Setter); Gerard (Settergrass,Setterwort). When cattle coughed, cowmen used to make a hole through the dewlap with a setter, or thread, and then a length of hellebore root would be inserted to irritate the flesh and keep it running. The modern word is seton, instead of setter. Cf. Seton-grass.
SETWALL 1. *Valeriana officinalis* (Valerian) Gerard. 2. *Valeriana pyrenaica* (Giant Valerian) B & H. The name is a lot earlier than Gerard, though. Chaucer knew it - "he himself was swete as any Setwall". Cf. Cetywall.
SEVEN-BARK 1. *Hydrangea arborescens* (Tree Hydrangea) USA Cunningham. 2. *Hydrangea quercifolia* USA Harper.
SEVEN-CURES, Herb-of-the- *Achillea millefolium* (Yarrow) Ire. Wilde. Probably a "woundwort" name.
SEVEN-LEAF *Potentilla erecta* (Tormentil) Cockayne. To distinguish it from Five-leaf, which is Potentilla reptans. Cf. Setfoil.
SEVEN SISTERS 1. *Euphorbia helioscopia* (Sun Spurge) Ire. Grigson. 2. *Euphorbia peplus* (Petty Spurge) Ire. B & H. Given, apparently, from the seven branches of the stem, though in Limerick they can only count five, if we are to believe the name there, Five Sisters.
SEVEN-YEARS'-LOVE 1. *Achillea millefolium* (Yarrow) Glos. Fernie. 2. *Achillea ptarmica* (Sneezewort) W Eng. Friend. A reference to the marriage divinations practised with yarrow. More specifically, it is a reference to the country custom of putting yarrow in the bridal wreath. It was said that this guaranteed seven years of married bliss.
SEVILLE ORANGE *Citrus aurantium var. bigaradia*.
SEYCHELLES COCONUT *Lodoicea maldivica* (Coco-de-mer). A Huxley.
SEYNY-TREE *Laburnum anagyroides* (Laburnum) Shrop. B & H. The leaves are thought to resemble senna leaves, hence this name.
SGEACH *Crataegus monogyna* (Haws) Ire. Grigson. Irish seach, a bush or bramble. The name varies from Skeeog, through Skayug, to

Shiggy. It is the isolated or "fairy" thorns that get these names.

SHA ROOT *Vigna denteri*. 'Sha' is the San name.

SHABUB *Lunaria annua* (Honesty) Turner.

SHACKLE, Silver *Briza media* (Quaking-grass) Scot. Grigson.1959. A long way from Scotland, but Wiltshire children used to be admonished to watch the grass until it was quite still, and then they would see them turn into silver coins. It was a way of saying that they would never be still. 'Shackle', in spite of spawning 'shekel' - Silver Shekels is another Scots name for this grass, has the same import as 'Rattle'.

SHACKLE-BACKLES *Silene cucubalis* (Bladder Campion) Som. Macmillan.

SHACKLE-BAG *Rhinanthus crista-galli* (Yellow Rattle) Som, Dor. Macmillan.

SHACKLE-BASKET 1. *Briza media* (Quaking-grass) Som. Grigson.1959. 2. *Rhinanthus crista-galli* (Yellow Rattle) Dor. Dacombe.

SHACKLE-BOX 1. *Briza media* (Quaking-grass) Som. Grigson.1959. 2. *Pedicularis sylvatica* (Lousewort) Dev. Macmillan. Lousewort is closely related to Red Rattle, hence this name.

SHACKLE-CAP *Rhinanthus crista-galli* (Yellow Rattle) Som. Macmillan.

SHACKLE-GRASS *Briza media* (Quaking-grass) Som. Grigson.1959.

SHACKLERS 1. *Acer campestre* fruit (Field Maple keys) Dev. Macmillan. 2. *Fraxinus excelsior* fruit (Ash keys) Dev. B & H. 3. *Rhinanthus crista-galli* (Yellow Rattle) Som. Macmillan.

SHACKLERS, Shickle *Briza media* (Quaking-grass) Som. Grigson.1959.

SHACKLES *Rhinanthus crista-galli* (Yellow Rattle) Dor. Grigson.

SHAD BLOW *Amelanchier alnifolia* (Western Service-berry) USA Elmore.

SHADBERRY, Prune-leaved *Amelanchier prunifolia* (Prune-leaved Service-berry) USA Elmore.

SHADBUSH; SHADBUSH, Smooth *Amelanchier laevis* (Smooth Juneberry) USA Zenkert.

SHADDOCK *Citrus maxima* (Wild Pomelo) Candolle. Apparently the name of a sea captain who first introduced the species into the West Indies (it is an Asian plant).

SHADES OF EVENING *Melandrium album* (White Campion) Som. Macmillan.

SHADOW, Shaking; SHADOW, Trembling *Briza media* (Quaking-grass) both Tynan & Maitland.

SHADOW GRASS *Luzula sylvatica* (Great Woodrush) B & H.

SHADSCALE 1. *Atriplex confertifolia*. 2. *Atriplex canescens* (Chamizo) USA Elmore.

SHADY NIGHT *Solanum dulcamara* (Woody Nightshade) Lancs. Grigson. A reversal of the usual name.

SHAGBARK HICKORY *Carya ovata*. Because the bark curls away in strips, shag being the noun whose adjective is shaggy.

SHAGGY JACK *Lychnis flos-cuculi* (Ragged Robin) Dev, Som. Macmillan.

SHAKERS; SHAKERS, Hay; SHAKERS, Silver *Briza media* (Quaking-grass) Wilts, Shrop, Ches, Lancs. Grigson.1959 (Shakers); Ches. Grigson.1959 (Hay Shakers); Scot. Grigson.1959 (Silver Shakers).

SHAKES, Lady's; SHAKES, Shiver- *Briza media* (Quaking-grass) Yks. (Lady's Shakes); Som. (Shiver-Shakes) both Grigson.1959.

SHAKING ASP *Populus tremula* (Aspen) Ches. Holland.

SHAKING GRASS; SHAKY GRASS *Briza media* (Quaking-grass) Dev, Som, Bucks, Shrop. Grigson.1959 (Shaking); Dev. Friend.1882; Bucks. Grigson.1959 (Shaky).

SHAKING SHADOW; TREMBLING SHADOW *Briza media* (Quaking-grass) both Tynan & Maitland.

SHALDER; SHALDON *Iris pseudo-acarus* (Yellow Flag) Dev. Choape; Som. Macmillan(Shalder); Sanecki (Shaldon).

SHALLON *Gaultheria shallon*.

SHALLOT *Allium ascalonium*. Shallot is

through French eschalotte, from Latin ascalonia.

SHAM HONEY-FLOWER *Anacanptis pyramidalis* (Pyramidal Orchid) Som. Macmillan. 'Sham', for the "real" Honey-flower is the Bee Orchid.

SHAME, Cover *Juniperus sabina* (Savin) B & H. Cf. Bastard Killer. The berries are a notorious abortifacient.

SHAME-BUSH *Mimosa pudica* (Sensitive Plant) Barbados Gooding.

SHAME-FACE 1. *Geranium maculatum* (Spotted Cranesbill) Grieve.1931. 2. *Viola tricolor* (Pansy) Som. Macmillan. More in line with the "modesty" names for the violet than with the brassier pansy.

SHAME-FACED MAIDEN 1. *Anemone nemerosa* (Wood Anemone) Wilts. Dartnell & Goddard. 2. *Ornithogalum umbellatum* (Star of Bethlehem) Wilts. Dartnell & Goddard.

SHAMEWEED *Mimosa pudica* (Sensitive Plant) S. USA Puckett.

SHAMROCK 1. *Medicago lupulina* (Black Medick) Dyer. 2. *Nasturtium officinale* (Watercress) Fernie. 3. *Oxalis acetosella* (Wood Sorrel) Ire. Grigson. 4. *Trifolium dubium* (Lesser Yellow Trefoil) Ackermann. 5. *Trifolium pratense* (Red Clover) Gerard. 6. *Trifolium repens* (White Clover) Dyer. 7. *Veronica chamaedrys* (Germander Speedwell). Shamrock is Irish seamrog, the diminutive of seamar, clover. Just which of these is the real shamrock has exercised the minds of antiquaries for a long time. Granted that the requirement is that it is a trefoil, it would seem that Watercress and Germander Speedwell are usurpers. According to the legend, it was St Patrick who was the first to use clover as an illustration of the Holy Trinity, and it is used still as the emblem of the Trinity.

SHAMROCK, Blue *Parochetus communis* (Shamrock Pea) W Miller.

SHAMROCK, Indian *Trillium erectum* (Bethroot) Grieve.1931. 'Indian' in this case is North American Indian.

SHAMROCK, Wild 1. *Adoxa moschatellina* (Moschatel) Som. Macmillan. 2. *Medicago lupulina* (Black Medick) Som. Macmillan. 3. *Oxalis acetosella* (Wood Sorrel) Som. Macmillan.

SHAMROCK PEA *Parochetus communis*.

SHAMSHER *Crithmum maritimum* (Samphire) Corn. Grigson. A local version of Samphire.

SHANDER-GRASS *Briza media* (Quaking-grass) Tynan & Maitland. Is this the same as Shackle-grass? Or is it used in the same sense as Shekel-basket? (shand is a very old slang word for a coin).

SHANTUNG CABBAGE *Brassica sinensis* (Chinese Mustard) Howes.

SHARE *Acer pseudo-platanus* (Sycamore) W Eng. Halliwell.

SHAREWORT *Aster tripolium* (Sea Aster) Prior. Given, Prior says, because it was supposed to cure diseases of the share, or groin, called buboes.

SHARON, Rose of 1. *Hibiscus syriacus* (Syrian Katmia) Wit. 2. *Hypericum androsaemum* (Tutsan) Ire. Fernie. 3. *Hypericum calycinum*. 4. *Narcissus canaliculatus* Hutchinson & Melville. The reference is to the Rose of Sharon of the Song of Solomon.

SHARON TULIP *Tulipa sharonensis*. This is, according to Moldenke, probably the Rose of Sharon of the Song of Solomon.

SHARPBIND *Smilax aspera* (Greenbriar) Turner. Turner also knew it as Prickwind.

SHASLAGH *Ammophila arenaria* (Marram) IOM Moore, Morrison & Goodwin.

SHASTA DAISY *Leucanthemum maximum*.

SHAVEGRASS 1. *Equisetum arvense* (Common Horsetail) Gerard. 2. *Equisetum hyemale* (Dutch Rush) Barton & Castle. The mature plants contain a large amount of silica, which makes them extremely hard when dried, hard enough to use as pot scourers, or as fine grade "sandpaper" for cabinet makers. Hence this name and others like Scouring Rush.

SHAVEGRASS, Short *Hippuris vulgaris* (Mare's Tail) Turner.
SHAVING BRUSH *Centaurea nigra* (Knapweed) Shrop. Grigson. Descriptive.
SHAWL, Spanish *Schizocentron elegans* Gannon.
SHE-BARFOOT *Helleborus viridis* (Green Hellebore) War. Grigson. He-Barfoot is the Stinking Hellebore, *Helleborus foetidus*, and barfoot is bear's-foot, the common name for the hellebores in earlier times.
SHE-BALSAM *Abies fraseri* (Fraser's Balsam Fir) Brimble.1948.
SHE-BROOM *Genista tinctoria* (Dyer's Greenweed) Ches, Yks. Grigson.
SHE-BULKISHAWN *Tanacetum vulgare* (Tansy) Ire. Crooke. Bulkishawn is one of the versions of Gaelic buaghallan, ragwort, which is He-Bulkishawn.
SHE-HEATHER 1. *Empetrum nigrum* (Crowberry) Donegal Grigson. 2. *Erica cinerea* (Bell Heather) N'thum, Berw. Grigson. 3. *Erica tetralix* (Cross-leaved Heath) N'thum, Berw. Grigson. As distinct from He-heather, which is Calluna vulgaris.
SHE-HOLLY *Ilex aquifolium* (Holly) N'thum. Denham/Hardy. The smooth-leaved variety; the one with prickles is He-holly. The antagonism between the two kinds is well brought by the Midlands superstition that if He-Holly is brought into the house first on Christmas Eve, the master of the house *is* master, and mistress is master if She-Holly wins.
SHE-OAK *Casuarina stricta*.
SHE-OAK, Fire *Casuarina cunninghamia* (River She-Oak) Everett.
SHE-OAK, Forest *Casuarina torulosa*.
SHE-OAK, River *Casuarina cunninghamia*.
SHE-OAK, Swamp 1. *Casuarina equisetifolia* (Beefwood) Everett. 2. *Casuarina glauca* Everett.
SHE-PINE *Podocarpus elatus*.
SHEA BUTTERNUT *Butyrospermum parkii*. Shea was Mungo Park's spelling of the Mandingo (West Africa) name, si.

SHEAR-GRASS *Hippuris vulgaris* (Mare's Tail) Dor. Notes & Queries.1871.
SHEEGIE THIMBLES; SHILLY THIMBLES *Digitalis purpurea* (Foxglove) Donegal Grigson. i.e. fairy thimbles. They are both from the Irish sidhe.
SHEEP *Pinus sylvestris* cones (Pine cones) Yks. B & H.
SHEEP, Vegetable *Raoulia sp* Salisbury.1936.
SHEEP BELLS *Campanula rotundifolia* (Harebell) Dor. Grigson.
SHEEP FAT *Atriplex confertifolia* (Shadscale) USA Elmore.
SHEEP FOOT *Lotus corniculatus* (Bird's-foot Trefoil) Cumb. Grigson. Bird's Foot is the usual imagery, but as well as sheep, they are likened to both lamb's and pig's feet, all inconsequential when the shape of birds' feet is considered.
SHEEP-KILLING PENNY-GRASS *Hydrocotyle vulgaris* (Pennywort) Lyte. "husbandmen know well, that it is noisome unto Sheepe, and other Cattell that feed thereon, and for the most part bringeth death unto them... (Gerard).
SHEEP LAUREL *Kalmia angustifolia*. See rather Sheepkill.
SHEEP LOCO *Astragalus nothoxys*.
SHEEP-ROOT *Pinguicula vulgaris* (Butterwort) Roxb. B & H. Probably an error for Sheep-rot, but Jamieson explained it away by saying that "when turned up by the plough, the sheep greedily feed on it".
SHEEP-ROT 1. *Hydrocotyle vulgaris* (Pennywort). 2. *Pinguicula vulgaris* (Butterwort) N Scot. Jamieson. Both of these plants are traditionally dangerous to sheep, and said to cause rot in them. See Sheep-killing Penny-grass, and Cf. such names for Butterwort as Rot-grass.
SHEEP-SHEARING ROSE *Paeonia mascula* (Peony) Worcs. B & H.
SHEEP-SOORAG *Oxalis acetosella* (Wood Sorrel) Caith. Grigson. 'Soorag' is sorrel, and Sheep's Sorrel is quite a widespread name for this.

SHEEP-SORREL DOCK *Rumex acetosella* (Sheep's Sorrel) USA Gates.
SHEEP-SOUCE *Rumex acetosa* (Sorrel) Dor, Border B & H.
SHEEP'S-BIT SCABIOUS *Jasione montana*.
SHEEP'S BRISKEN *Stachys palustris* (Marsh Woundwort) Ire. Grigson. Briosclan was the Gaelic name for some unspecified edible root.
SHEEP'S CHEESE *Agropyron repens* (Couch-grass) Grigson.1959.
SHEEP'S EARS 1. *Heliotropum appendiculatum*. 2. *Stachys lanata* (Lamb's Ears) Som. Macmillan. *Stachys lanata* is the plant with the furry leaves, suitably described by various animals' ears.
SHEEP'S FESCUE *Festuca ovina*.
SHEEP'S GOWAN *Trifolium repens* (White Clover) Scot. Grigson.
SHEEP'S HERB *Plantago maritima* (Sea Plantain) Grieve.1931.
SHEEP'S KNAPPERTY *Potentilla erecta* (Tormentil) N Ire. Grigson. 'Knapperty' is usually reserved for Bittervetch.
SHEEP'S PARSLEY 1. *Anthriscus sylvestris* (Cow Parsley) Kent, Norf. B & H. 2. *Chaerophyllum temulentum* (Rough Chervil) Suff. Grigson.
SHEEP'S SCABIOUS, Hairy *Jasione montana* (Sheep's-bit Scabious) Curtis.
SHEEP'S SORREL 1. *Oxalis acetosella* (Wood Sorrel) Dor, E Ang, Ire. Grigson. 2. *Rumex acetosella*.
SHEEP'S SORREL, Slender *Rumex tenuifolius*.
SHEEP'S SOURACK *Rumex acetosa* (Sorrel) Scot. B & H.
SHEEP'S SOUROCK *Rumex acetosella* (Sheep's Sorrel) Scot. Jamieson.
SHEEP'S TAILS *Corylus avellana* catkins (Hazel catkins) Som. Macmillan. Cf. Lamb's Tails.
SHEEP'S THISTLE *Cirsium arvense* (Spear Plume Thistle) Som. Macmillan.
SHEEP'S THYME *Thymus serpyllum* (Breckland Wild Thyme) Dor. Macmillan.

SHEEPBANE *Hydrocotyle vulgaris* (Pennywort) Leyel.1948.
SHEEPBERRY *Viburnum lentago* USA Upham.
SHEEPBINE *Convolvulus arvensis* (Field Bindweed) Herts. Grigson; Ess. Gepp. Very odd. The 'binding' must refer to something vegetable - withies, usually. But "sheep"?
SHEEPKILL *Kalmia angustifolia* (Sheep Laurel) USA Kingsbury. Literally true - Cf. Lambkill and Calfkill.
SHEEPWEED *Gutierrezia microcephala* (Broomweed) USA Elmore.
SHEKEL BASKET; SHEKEL BOX 1. *Briza media* (Quaking-grass) Som. Notes & Queries. vol. 168, 1925. 2. *Rhinanthus cristagalli* (Yellow Rattle) Dor. Macmillan. Cf. Money-grass, Penny-rattle, and a good many more for Yellow Rattle. The rattling of the seeds inside the capsule is likened to coins rattling in a money box.
SHEKELS, Silver *Briza media* (Quaking-grass) Som. Grigson.1959. Cf. Silver Shakers, and Silver Shackles. Presumably, then, shekels in this case is the same as shackles. But Wiltshire children used to be told to watch the grasses, to see them turn into silver coins when they were perfectly still, an event that could almost never happen.
SHELL FLOWER 1. *Chelone glabra* (Turtlehead) Grieve.1931. 2. *Molucella laevis* (Molucca Balm) Perry. 3. *Tigridia pavonia*.
SHELLBARK HICKORY *Carya ovata* (Shagbark Hickory) Everett.
SHELLBARK HICKORY, Big *Carya laciniosa*.
SHELLS, Cockle *Vinca minor* (Lesser Periwinkle) Som. Macmillan. The reference is to the nursery rhyme "Mary, Mary, quite contrary".
SHEMIS *Ligusticum scoticum* (Lovage) W Scot, Hebr. Grigson. It is more likely that the name is applied to the greens as a vegetable.
SHENSI RHUBARB *Rheum officinale* (Turkey Rhubarb) Camp. It was called Turkey, Russian, Chinese and Shensi

Rhubarb in accordance with the country through which it reached the market from its native land.

SHEPHERD, Underground *Orchis mascula* (Early Purple Orchid) Wilts. Jones & Dillon.

SHEPHERD'S BAG *Capsella bursa-pastoris* (Shepherd's Purse) Turner.

SHEPHERD'S BAROMETER *Anagallis arvensis* (Scarlet Pimpernel) Som. Macmillan. Pimnpernel is a weather forecaster, and it is the shepherd's dial, glass, sundial, warning and calendar as well as his barometer.

SHEPHERD'S BEDSTRAW *Asperula cynanchica* (Squinancywort) Glos. J D Robertson. The Asperulas used to be included in the Galiums, hence the name Bedstraw.

SHEPHERD'S BLUE THYME *Polygala calcarea* (Chalk Milkwort) Wilts. Dartnell & Goddard.1899.

SHEPHERD'S BODKIN *Scandix pecten-veneris* (Shepherd's Needle) Halliwell. All the "needle" names for this come from the shape of the long, beaked fruits.

SHEPHERD'S CALENDAR *Anagallis arvensis* (Scarlet Pimpernel).

SHEPHERD'S CLOCK 1. *Anagallis arvensis* (Scarlet Pimpernel) Glos. Friend. 2. *Taraxacum officinale* (Dandelion) Mabey. 3. *Tragopogon pratensis* (Goat's Beard) Som, Suss. Grigson. Dandelion and Goat's Beard from their "clock" seedheads, and Scarlet Pimpernel presumably from its weather forecasting capability.

SHEPHERD'S CLUB *Verbascum thapsus* (Great Mullein) IOW Long; Scot. Prior; USA Henkel.

SHEPHERD'S COMB *Scandix pecten-veneris* (Shepherd's Needle) Yks. Grigson.

SHEPHERD'S CRESS *Teesdalia nudicaulis*.

SHEPHERD'S CRESS, Lesser *Teesdalia coronopifolia*.

SHEPHERD'S DAISY *Bellis perennis* (Daisy) N'thants. Grigson.

SHEPHERD'S DAYLIGHT *Anagallis arvensis* (Scarlet Pimpernel) Som. Elworthy. Shepherd's Delight, obviously - it is pronounced that way in some parts of Somerset.

SHEPHERD'S DELIGHT 1. *Anagallis arvensis* (Scarlet Pimpernel) Som. Elworthy. Lincs. Grigson. 2. *Clematis vitalba* (Old Man's Beard) Som. Macmillan. 3. *Viburnum lantana* (Wayfaring Tree) Dor. Macmillan.

SHEPHERD'S DIAL *Anagallis arvensis* (Scarlet Pimpernel). Cf. Shepherd's Barometer - a reference to the pimpernel as a weather fcorecaster.

SHEPHERD'S FLOCK *Arabis caucasica* (Garden Arabis) Som. Macmillan. Because of the white flowers.

SHEPHERD'S FRIEND *Sorbus aucuparia* (Rowan) Dor. Macmillan.

SHEPHERD'S GLASS; SHEPHERD'S JOY *Anagallis arvensis* (Scarlet Pimpernel) Norf, Rut. Grigson (Glass); Som, Dor. Macmillan (Joy). 'Glass' - barometer.

SHEPHERD'S KNOT *Potentilla erecta* (Tormentil) N'thum, Scot. Grigson.

SHEPHERD'S LOVE *Polygala calcarea* (Chalk Milkwort) Wilts. D Grose.

SHEPHERD'S MYRTLE *Ruscus aculeatus* (Butcher's Broom) Bianchini. Cf. Wild Myrtle and Jew's Myrtle.

SHEPHERD'S NEEDLE 1. *Myrrhis odorata* (Sweet Cicely) Grieve.1931. 2. *Scandix pecten-veneris*. Scandix gets the name from its long pointed seed vessels, but the name seems unlikely for Sweet Cicely, though its seed vessels are certainly long.

SHEPHERD'S PANSY *Viola lutea* (Mountain Pansy) N'thum. Grigson.

SHEPHERD'S PEDLAR; SHEPHERD'S POCKET *Capsella bursa-pastoris*. (Shepherd's Purse) Wilts. Dartnell & Goddard (Pedlar); Bucks. Grigson (Pocket).

SHEPHERD'S POUCH 1. *Capsella bursa-pastoris* (Shepherd's Purse) Gerard. 2. *Orobanche minor* (Lesser Broomrape) IOW Long.

SHEPHERD'S POUNCE *Capsella bursa-pastoris* (Shepherd's Purse) Culpepper.

SHEPHERD'S PURSE 1. *Capsella bursa-*

pastoris. 2. *Lotus corniculatus* (Bird's-foot Trefoil) Som. Macmillan. 3. *Lunaria annua* (Honesty) Som. Macmillan. 4. *Polygala vulgaris* (Milkwort) Wilts. D Grose. 5. *Rhinanthus crista-galli* (Yellow Rattle) Som. Macmillan. Cumb. Grigson. The shape of Capsella's seed pods gives rise to the name, and it is easy to see why Honesty and Yellow Rattle get the name, for there is a lot of imagery of money in a purse for both of these. But Bird's-foot Trefoil and Milkwort are not so obvious, unless the shape of Milkwort's flowers could call to mind a purse.

SHEPHERD'S ROD 1. *Dipsacus fullonum* (Teasel) Som. Macmillan. 2. *Dipsacus pilosus* (Small Teasel) McClintock.

SHEPHERD'S STAff 1. *Dipsacus fullonum* (Teasel) Som. Macmillan. 2. *Verbascum thapsus* (Mullein) Cumb. B & H.

SHEPHERD'S SCRIP *Capsella bursa-pastoris* (Shepherd's Purse) Gerard. 'Scrip' is a pouch, or satchel.

SHEPHERD'S SUNDIAL *Anagallis arvensis* (Scarlet Pimpernel) Suff. Grigson.

SHEPHERD'S THYME 1. *Polygala calcarea* (Chalk Milkwort) Wilts. Dartnell & Goddard. 2. *Polygala vulgaris* (Milkwort) Wilts. Dartnell & Goddard. 3. *Thymus serpyllum.*

SHEPHERD'S WARNING; SHEPHERD'S WATCH *Anagallis arvensis* (Scarlet Pimpernel) Som. Macmillan; Lincs. Rudkin (Warning); E Angl. Grigson (Watch). In the first case, it is warning of the weather that is implied; in the second, the time of day, calculated by the closing of the flowers.

SHEPHERD'S WEATHERGLASS 1. *Anagallis arvensis* (Scarlet Pimpernel) Som. Macmillan. Wilts. Dartnell & Goddard; N'thants. A E Baker. 2. *Stellaria holostea* (Greater Stitchwort) Lancs. Grigson. An unusual role for stitchwort, but pimpernel is a well-known weather forecaster.

SHEPHERD'S YARD *Dipsacus fullonum* (Teasel) Dawson.

SHERVES *Sorbus torminalis* fruit (Wild Service) Suss. Parish. A version of 'service'.

SHICK-SHACK TREE *Quercus robur* (Oak) Oxf. Parker. The sprigs of oak, preferably with an oakapple attached, that were carried about on Royal Oak Day (29 May) used to be known throughout Wessex by colourful names like Shick-shack, shit-sack, sheet-shack, and many similar. Just outside the area, in at least one village in Oxfordshire, these names were actually transferred to the tree itself.

SHICKLE-SHACKLERS *Briza media* (Quaking-grass) Som. Grigson.1959.

SHIELD DOCK *Rumex scutatus* (Sorrel) Polunin. The specific name is from Latin scutum, shield, a reference to the shape of the leaves.

SHIELD FERN *Dryopteris cristata* (Crested Buckler Fern) USA Yarnell.

SHIELD FERN, Hard; SHIELD FERN, Prickly, Hard *Polystichum aculeatum* The first is the usual name; the second is in J A MacCulloch.1905.

SHIELD FERN, Japanese *Dryopteris erythrysora.*

SHIELD FERN, Soft; SHIELD FERN, Prickly, Soft *Polystichum setiferum* The first is the usual name; the second is in J A MacCulloch.1905.

SHIELD LICHEN *Hypogymnia physodes* (Dark Crottle) S M Robertson.

SHIGGY *Crataegus monogyna* (Hawthorn) Ire. Grigson. It must be a version, though a long way from it, of Irish sceach, a bush, or bramble, from which such Irish hawthorn names as Sgeach, or Skeeog, derive, unless this is a word that means 'fairy', Irish sidhe.

SHILLERS, Brown *Corylus avellana* nuts (Hazel) Yks. Hunter. Given only when the nuts are ripe.

SHILLING-GRASS; SHILLING-ROT *Hydrocotyle vulgaris* (Pennywort) both Ayr. B & H. Cf. Penny-rot and Farthing-rot, the coinage being likened to the shape of the leaves. 'Rot' because it is "noisome unto Sheepe and other Cattell that feed thereon...".

SHILLINGS *Lunaria annua* (Honesty) Dor. Macmillan; War. B & H. 'Shillings' rather than 'pennies', because of the silvery colour of the seed capsules - in fact, Honesty is called the Silver Shilling Flower sometimes.

SHILLINGS, Gowk's *Rhinanthus crista-galli* (Yellow Rattle) Lanark. Grigson.

SHILLY THIMBLES; SHEEGIE THIMBLES *Digitalis purpurea* (Foxglove) Donegal Grigson. i.e. fairy thimbles, for the first element in the two names is the Irish sidhe.

SHIMMIES 1. *Calystegia sepium* (Great Bindweed) Dev, Dor, Som. Macmillan; Wilts. Dartnell & Goddard. 2. *Stellaria holostea* (Greater Stitchwort) Som. Grigson. 'Shimmies' - chemises. The imagery is that of white clothes hung out to dry.

SHIMMIES, Shirts-and- *Convolvulus arvensis* (Field Bindweed) Som. Macmillan.

SHIMMIES-AND-SHIRTS 1. *Calystegia sepium* (Great Bindweed) Dev, Wilts. Macmillan; Dor. Dacombe. 2. *Stellaria holostea* (Greater Stitchwort) Som. Grigson.

SHIMMY, Lady's 1. *Anemone nemorosa* (Wood Anemone) Glos. Grigson. 2. *Calystegia sepium* (Great Bindweed) Som. Grigson.

SHIMMY-AND-BUTTONS *Calystegia sepium* (Great Bindweed) Dor. Grigson.

SHIMMY-SHIRTS *Calystegia sepium* (Great Bindweed) Dev, Dor, Wilts. Macmillan.

SHIN 1. *Prunus avium* (Wild Cherry) N Eng. Halliwell. 2. *Quercus havardii*. Shin for the cherry must surely be a variation of Gean.

SHIN-LEAF 1. *Pyrola elliptica* USA Tolstead. 2. *Pyrola rotundifolia* (Round-leaved Wintergreen) O P Brown. 3. *Ramischia secunda* (Yavering Bells) USA Zenkert.

SHIN OAK *Quercus gambeli* (Gambel's Oak) Howes.

SHINERS *Arum maculatum* (Cuckoo-pint) Cambs. Porter. The pollen of the flowers gives a faint light at dusk, so it is said. Cf. the Irish Fairy Lamps.

SHINGLE OAK 1. *Casuarina equisetifolia* (Beefwood) Duncalf. 2. *Quercus imbricaria*. 'Shingle', because that was what the early settlers used it for.

SHINGLE PLANT *Monstera deliciosa* (Swiss Cheese Plant) A W Smith.

SHINGLE-WOOD *Terminalia ivorensis* (Yellow Terminalia) Dalziel.

SHINING SAGE *Salvia leonuroides*.

SHINING SUMACH *Rhus copallina* (Black Sumach) Schery.

SHINING WILLOW *Salix lucida*.

SHINLOCK *Eruca sativa* (Rocket) B & H.

SHINNERY *Quercus havardii* (Shin) USA Kingsbury.

SHIPMAST LOCUST *Robinia pseudacacia var. rectissima*. 'Rectissima' in the name of the variety explains the common name.

SHIR *Sorbus torminalis* (Wild Service Tree) Som. Grigson.

SHIRT 1. *Brassica napus* seeds (Rape seeds) Scot. Jamieson. 2. *Convolvulus arvensis* (Field Bindweed) Som. Macmillan. 3. *Sinapis arvensis* seeds (Charlock) Roxb. B & H. Bindweed is often likened to clothes put on the bushes to dry. Cf. Shimmies, for example. But for the other two, shirt must mean something different.

SHIRT, Old Man's 1. *Calystegia sepium* (Great Bindweed) Som. Macmillan. 2. *Stellaria holostea* (Greater Stitchwort) Corn. Grigson.

SHIRT, Ragged *Convolvulus arvensis* (Field Bindweed) Som. Macmillan.

SHIRT, Shimmies-and- *Calystegia sepium* (Great Bindweed) Dev, Dor, Wilts. Macmillan.

SHIRT, Shimmy- *Calystegia sepium* (Great Bindweed) Dev, Wilts. Macmillan; Dor. Dacombe.

SHIRT, White; SHIRT, White, Daddy's *Calystegia sepium* (Great Bindweed) Som. Macmillan.

SHIRT BUTTONS 1. *Achillea ptarmica* (Sneezewort) Coats. 2. *Melandrium album*

(White Campion) Som. Macmillan. 3. *Stellaria holostea* (Greater Stitchwort) Dor. Dacombe. Descriptive.

SHIRT BUTTONS, Devil's *Stellaria holostea* (Greater Stitchwort) Som. Macmillan. Why the ascription to the devil in this case? Stitchwort seems to have been regarded with some suspicion, for not only is it the Devil's Flower, or his eyes, nightcap and corn, but it is also referred to as Snakeweed and other similar names.

SHIRTS-AND-SHIMMIES *Convolvulus arvensis* (Field Bindweed) Som. Macmillan. Cf. Shirts, etc.

SHIT-A-BED *Taraxacum officinale* (Dandelion) Corn. Grigson; Wilts. Dartnell & Goddard. It is its diuretic rather than its laxative properties that are usually celebrated in the local names.

SHITTIM BARK *Rhamnus purshiana* (California Buckthorn) Calif. Maddox. The reference is biblical. The early settlers in California said that this was the Shittim wood from which the Ark was made.

SHIVE *Allium schoernoprasum* (Chives) IOM Moore.1924.

SHIVER-GRASS; SHIVER-SHAKES *Briza media* (Quaking-grass) Hants. Cope; Bucks (Grigson) (Grass); Som. (Shakes) Grigson. Cf. Shivery-shakeries, etc.

SHIVER-TREE 1. *Populus nigra* (Black Poplar) Lincs. Gutch. 2. *Populus tremula* (Aspen) R L Brown.

SHIVERING JIMMY *Briza media* (Quaking-grass) Suss. Grigson.1959.

SHIVERY-SHAKERIES; SHIVERY-SHAKES *Briza media* (Quaking-grass) Wilts. (Shivery-shakeries); Som, Wilts. Shivery-shakes) both Grigson.1959. Cf. Shiver-grass, etc.

SHOE NUT *Bertholletia excelsa* (Brazil Nut) Dev. Friend.1882. The general shape of the nut suggests the name.

SHOEFLOWER *Hibiscus rosa-sinensis* (Rose of China) Wit. The red of the flowers becomes black when bruised, and is then used for blacking shoes (or colouring the eyebrows). Cf. Blacking Plant.

SHOEMAKER'S HEELS *Chenopodium bonus-henricus* (Good King Henry) Shrop,Rad. Grigson.

SHOES *Geranium robertianum* (Herb Robert) Dev. Grigson.

SHOES, Baby's 1. *Ajuga reptans* (Bugle) Wilts. Dartnell & Goddard. 2. *Aquilegia vulgaris* (Columbine) Som. Macmillan.

SHOES, Boots-and- 1. *Cypripedium calceolus* (Lady's Slipper) Dev. Friend.1882. 2. *Fraxinus excelsior* fruit (Ash keys) Som. Macmillan. 3. *Lotus corniculatus* (Bird's-foot Trefoil) Dev. Friend.1882; Som. Macmillan.

SHOES, Crimson *Lathyrus nissolia* (Grass Pea) Suss. Grigson.

SHOES, Cuckoo's *Viola riviniana* (Dog Violet) Shrop. Grigson.

SHOES, Dolly's 1. *Aquilegia vulgaris* (Columbine) Som. Grigson. 2. *Geranium robertianum* (Herb Robert) Som. Grigson.

SHOES, Granny's *Aconitum napellus* (Monkshood) Dor. Macmillan.

SHOES, Horse *Acer pseudo-platanus* fruit (Sycamore keys) Wilts. Dartnell & Goddard.

SHOES, Lady's 1. *Aquilegia vulgaris* (Columbine) Som. Macmillan; Ess, Cambs, Norf. Grigson; USA Hutchinson. 2. *Cardamina pratensis* (Lady's Smock) Som. Macmillan. 3. *Fumaria officinalis* (Fumitory) Som. Macmillan; Wilts. Dartnell & Goddard. 4. *Lotus corniculatus* (Bird's-foot Trefoil) Hants. Grigson.

SHOES, Stockings-and- 1. *Aquilegia vulgaris* (Columbine) Som. Macmillan. 2. *Arum maculatum* (Cuckoo-pint) Suss. PLNN31. 3. *Lotus corniculatus* (Bird's-foot Trefoil) Dev. Friend.

SHOES, Whip-poor-will's *Cypripedium reginae* (Showy Lady's Slipper) USA House.

SHOES-AND-SLIPPERS *Anemone nemerosa* (Wood Anemone) Som. Macmillan.

SHOES-AND-SOCKS 1. *Aquilegia vulgaris* (Columbine) Som. Macmillan. 2. *Lotus cor-*

niculatus (Bird's-foot Trefoil) Kent Grigson.

SHOES-AND-STOCKINGS 1. *Aquilegia vulgaris* (Columbine) Som. Macmillan. 2. *Cardamina pratensis* (Lady's Smock) Bucks. Grigson. 3. *Lamium album* (White Deadnettle) Som. Macmillan. 4. *Lathyrus nissolia* (Grass Pea) Suss. Grigson. 5. *Linaria vulgaris* (Toadflax) Corn. Grigson. 6. *Lotus corniculatus* (Bird's-foot Trefoil) Dev, Som, Glos, Bucks. Grigson; Hants. Cope; Suss. Parish. 7. *Plantago media* (Lamb's Tongue Plantain) Som. Macmillan. 8. *Viola riviniana* (Dog Violet) Pemb. Grigson. 9. *Viola tricolor* (Pansy) Som. Macmillan.

SHOES-AND-STOCKINGS, Cuckoo's 1. *Cardamina pratensis* (Lady's Smock) S Wales Grigson. 2. *Orchis mascula* (Early Purple Orchis) Som. Macmillan.

SHOES-AND-STOCKINGS, Lady's *Lotus corniculatus* (Bird's-foot Trefoil) Som. Macmillan; Kent Grigson.

SHOESTRINGS, Devil's 1. *Coronilla varia* (Crown Vetch) USA Puckett. 2. *Tephrosia virginiana*. 3. *Viburnum alnifolium* (Hobble-bush) USA Perry. Hobble-bush gets the name, as well as others such as Tangle-legs, because the lower branches droop to the ground and root at the tips.

SHOKE *Phytolacca decandra* (Poke-root) Watt. Cf. Skoke, or Scoke, Coakum or Cocum etc., all obviously related to Poke, which is either from puccoon, or else pocon, apparently a name given to any plant yielding a red or yellow dye.

SHOLGIRSE; SJOLGIRSE *Achillea ptarmica* (Sneezewort) Shet. Grigson.

SHOO-FLY PLANT *Nicandra physalodes*. Obviously, a fly repeller.

SHOOTING STAR *Dodecatheon meadia*.

SHOOTING STAR, Henderson *Dodecatheon hendersonii*.

SHORE ORACHE *Atriplex littoralis* (Grass-leaved Orache) Clapham.

SHORE PINE *Pinus contorta* (Beach Pine) Everett.

SHOREGRASS; SHOREWEED; SHOREWEED, Plantain *Littorella uniflora*. Shoreweed is the usual name; Shoregrass is in Prior and Plantain Shoreweed in B & H.

SHOREWORT, Northern *Mertensia maritima* (Sea Lungwort) Clapham.

SHORTLEAF PINE 1. *Pinus echinata*. 2. *Pinus taeda*.

SHORTNEEDLE PINE *Pinus echinata* (Shortleaf Pine) USA Cunningham.

SHOT, Indian *Canna indica*. 'Shot', from the round, bullet-like seeds.

SHOTBUSH *Aralia nudicaulis* (American Sarsaparilla) Grieve.1931.

SHOTER; SHOOTER YEW *Taxus baccata* (Yew) Halliwell (Shoter); Dallimore (Shooter Yew). Shoter is given without further comment than to say it is A/S, but the existence of Shooter Yew shows the reference to be with bow-making.

SHOUPS *Rosa canina* hips (Dog Rose Hips) Yks. Carr. Spelt Choups as well, but the usual spelling is Choops.

SHOW PEA *Lathyrus odorata* (Sweet Pea) Norf. B & H.

SHOWER, Golden 1. *Cassia fistula* (Indian Laburnum) Everett. 2. *Laburnum anagyroides* (Laburnum) Shrop. B & H. 3. *Pyrostegia venusta* (Chinese Cracker Flower) RHS.

SHRIMP PLANT *Beloperone guttata*.

SHUNAS *Ligusticum scoticum* (Lovage) Skye B & H. Also written as Siunas and Shemis, probably not applied to the plant, but only to the greens as a vegetable.

SHY WIDOWS *Fritillaria meleagris* (Snake's-head Lily) War. Grigson. The drooping flowers, coupled with their sombre colouring, account for the name, and also similar names like Weeping Widow.

SIAM WEED *Eupatorium odoratum* Malaya Gilliland.

SIBERIAN BUGLOSS *Brunnera macrophylla*.

SIBERIAN CEDAR *Pinus cembra* (Swiss Stone Pine) Hora.

SIBERIAN CORYDALIS *Corydalis bulbosa* (Bulbous Fumitory) Douglas.
SIBERIAN CRAB *Malus baccata.*
SIBERIAN ELM 1. *Ulmus pumila.* 2. *Zelkova carpinifolia* (Caucasian Elm) Hora.
SIBERIAN FIR *Abies siberica.*
SIBERIAN IRIS *Iris siberica.*
SIBERIAN LARCH *Larix siberica.*
SIBERIAN MILLET *Setaria italica* (Foxtail Millet) Brouk. It is known as Italian, German, Hungarian or Siberian Millet according to the country of origin.
SIBERIAN MINERS' LETTUCE *Montia siberica.*
SIBERIAN PINE *Pinus cembra* (Swiss Stone Pine) Ablett. The edible seeds are popular in Siberia, from where they are exported via Archangel (they are known as Russian Nuts in Norway).
SIBERIAN SPRUCE *Picea obovata.*
SIBERIAN THISTLE *Cirsium oleraceum* (Cabbage Thistle) Douglas.
SIBERIAN VETCH *Vicia villosa* (Lesser Tufted Vetch) Howes.
SIBERIAN WALLFLOWER *Sisymbrium allionii.*
SICILIAN BEET *Beta vulgaris var. cicla* (Swiss Chard) Bianchini. But it is also Spanish, or Chilean, Beet, and Roman Kale is also used.
SICILIAN FENNEL *Foeniculum vulgare var. piperitum.*
SICILIAN MELILOT *Melilotus messanensis.*
SICILIAN SUMACH *Rhus coriaria* (European Sumach) Grieve.1931.
SICKLE, Hurt 1. *Centaurea cyanus* (Cornflower) Gerard. 2. *Centaurea nigra* (Knapweed) Worcs. B & H. Cf. Blunt-sickle for cornflower, which had the reputation of turning the edge of reapers' sickle.
SICKLE GRASS *Polygonum arifolium* (Halberd-leaved Tear-thumb) Grieve.1931.
SICKLE MEDICK *Medicago sativa var. falcata.*
SICKLE-POD 1. *Arabis canadensis.* 2. *Cassia tora.*

SICKLE SENNA *Cassia tora* (Sicklepod) Hora.
SICKLEWORT 1. *Ajuga chamaepitys* (Yellow Bugle) B & H. 2. *Ajuga reptans* (Bugle) Gerard. 3. *Prunella vulgaris* (Selfheal) Gerard. In these cases, the reference is to the wounds caused by sickles rather than to the shape. These are all wound herbs, Selfheal particularly having a number of names from that derivation.
SIDESADDLE FLOWER *Sarracenia purpurea* (Pitcher Plant) USA Sanford. Descriptive. In Whittle & Cook's words, women used to hook their right leg over the pummel of a saddle, in much the same way as the petals hang over the expanded style of this plant.
SIDS *Lolium perenne* (Rye-grass) Oxf, Suss. B & H. 'Sids' is seeds.
SIERRA CURRANT, Wild *Ribes nevadense.*
SIERRA JUNIPER 1. *Juniperus occidentalis.* 2. *Pinus occidentalis.*
SIERRA LAUREL *Leucothoe davisiae.*
SIERRA LEONE GUM COPAL *Copaifera copallifera* (Gum Copal) Dalziel.
SIERRA LILY *Lilium parvum.*
SIERRA PLUM *Prunus subcordata* USA Schenk & Gifford.
SIERRA REDWOOD *Sequoia gigantea* (Wellingtonia) Leathart.
SIEVA BEAN *Phaseolus lunatus* (Lima Bean) Kingsbury.
SIEVES; SEAVES *Juncus conglomeratus* (Compact Rush) Cumb. Rollinson (Sieves); Yks. Hartley & Ingilby (Seaves).
SIGNET, Lady's 1. *Polygonatum multiflorum* (Solomon's Seal) Prior. 2. *Tamus communis* (Black Bryony) Prior. The seal in Solomon's Seal is contained in the roots. On cutting them transversely, scars resembling the device known as Solomon's Seal, a 6-pointed star, can be seen. In this case, both of the plants are here ascribed to the Virgin Mary.
SIGRIM *Senecio jacobaea* (Ragwort) B & H. More usually Seggrum, given, according to Prior, because ragwort was applied to newly castrated bulls, called seggs, or staggs.

SIKKIM CUCUMBER *Cucumis sativus var. sikkimensis.*

SIKKIM LARCH *Larix griffithii* (Himalayan Larch) Mitchell.

SIKKIM SPRUCE *Picea spinulosa.*

SILGREEN *Sempervivum tectorum* (Houseleek) Dev. Friend.1882; Wilts. Dartnell & Goddard; Dor. Barnes; Heref. Havergal. It means evergreen, and is one of a whole series of names of which Singreen is the most widespread.

SILK, Corn *Zea mays* (Maize) Kourennoff.

SILK, Virginian *Asclepias syriaca* (Common Milkweed) Parkinson. Cf. Silkweed and Silken Cicely, the latter another of Parkinson's names.

SILK FIG *Musa x paradisiaca* (Banana) H & P.

SILK GRASS *Yucca filamentosa* (Adam's Needle) Coats.1975.

SILK-RUBBER TREE, Lagos *Funtumia elastica.*

SILK TASSEL, Yellow-leaf *Garrya flavescens* (Quinine-bush) Jaeger.

SILK TASSEL BUSH *Garrya elliptica.*

SILK TREE 1. *Albizia julibrissin* (Persian Acacia) Everett. 2. *Eucalyptus naudiniana.*

SILK TREE, Floss, Brazilian *Chorisia speciosa.*

SILK VINE *Periploca graeca.*

SILKCOTTON, White *Cochlospermum religiosum.*

SILKCOTTON TREE *Ceiba pentandra* (Kapok Tree) Kyerematen. Because of the fine silky fibres enveloping the seeds.

SILKCOTTON TREE, Red *Bombax ceiba.* 'Red' applies to the flowers.

SILKEN CICELY *Asclepias syriaca* (Common Milkweed) Parkinson.

SILKS AND SATINS *Lunaria annua* (Honesty) Dev. Friend.1882. There are a number of 'satin' names for this, all referring to the seed capsules.

SILKWEED *Asclepias syriaca* (Common Milkweed) USA House. The silky hairs from the seed pods are still used in hat-making, and for stuffing beds and pillows.

SILKWEED, Swamp *Asclepias incarnata* (Swamp Milkweed) Grieve.1931.

SILKWEED, Whorled *Asclepias verticillata* (Whorled Milkweed) USA Elmore.

SILKY CORNEL; SILKY DOGWOOD *Cornus amomum.*

SILKY FLOSSY *Salpiglossis sinuata* (Velvet Trumpet Flower) Som. Macmillan. Presumably a play on 'Salpiglossis'.

SILKY OAK *Grevillea robusta.*

SILKY OAK, Red *Stenocarpus salignus.*

SILKY OAK, White *Grevillea hilliana.*

SILKY ROCK SALLOW *Salix rupestris.*

SILKY WILLOW *Salix alba var. sericea.*

SILKY WILLOW, Dwarf *Salix repens* (Creeping Willow) Brimble.1948.

SILKY WISTARIA *Wistaria venusta.*

SILL GREEN; SILLY GREEN *Sempervivum tectorum* (Houseleek) Glos. J D Robertson; Worcs. J Salisbury (Sill); Heref. Havergal (Silly). Cf. the Wiltshire Silgreen; they are both part of the sequence of which Singreen is the most widespread, and they all mean 'evergreen'.

SILLER, Dog's *Rhinanthus crista-galli* (Yellow Rattle) Scot. Jamieson. Cf. Gowk's Siller, and many others. The coins are the seeds rattling in their capsule.

SILLER TASSELS *Briza media* (Quaking-grass) Perth. Grigson.1959.

SILVER, Midsummer *Potentilla anserina* (Silverweed) Surr. Grigson.

SILVER BELLS 1. *Anemone nemorosa* (Wood Anemone) Som. Macmillan. 2. *Viburnum opulus* (Guelder Rose) Wilts. Dartnell & Goddard.

SILVER BIRCH *Betula verrucosa.*

SILVER BUSH 1. *Anthyllis vulneraria* (Kidney Vetch) Folkard. 2. *Clematis vitalba* (Old Man's Beard) Jersey B & H.

SILVER CASSIA *Cassia artemisioides.*

SILVER CHAIN *Robinia pseudacacia* (False Acacia) B & H.

SILVER CLARY *Salvia argentea* (Silver Sage) Perry.

SILVER CUSHION *Anthyllis vulneraria* (Kidney Vetch) Tynan & Maitland.

SILVER FEATHER; SILVER FERN *Potentilla anserina* (Silverweed) Oxf. Grigson (Feather); Som. Macmillan; Wilts. Dartnell & Goddard (Fern).
SILVER FIR *Abies alba.*
SILVER FIR, American *Abies balsamea* (Balsam Fir) Grieve.1931.
SILVER FIR, Himalayan *Abies pindrow* (West Himalayan Fir) Usher.
SILVER FIR, Indian *Abies spectabilis* (East Himalayan Fir) Brimble.1948.
SILVER FIR, Japanese *Abies firma* (Momi Fir) Usher.
SILVER FIR, Nikko *Abies homolepis* (Nikko Fir) Usher.
SILVER FIR, Pacific *Abies amabilis* (Red Silver Fir) Schery.
SILVER FIR, Red *Abies amabilis.*
SILVER-FROSTED PLANT *Senecio cineraria* (Cineraria) Hibberd.
SILVER GRASS *Potentilla anserina* (Silverweed) Wilts. Dartnell & Goddard.
SILVER HOREHOUND *Marrubium candidissimum.*
SILVER HORSE NETTLE *Solanum elaeagnifolium* (Silverleaf Nightshade) USA T H Kearney.
SILVER INCH PLANT *Zebrina pendula.*
SILVER-LEAF 1. *Gnaphalium polycephalum* (Mouse-ear Everlasting) Grieve.1931. 2. *Spiraea tomentosa* (Hardhack) Grieve.1931. 3. *Stillingia treculeana* (Queen's Delight) Grieve.1931.
SILVER-LEAF POPLAR *Populus alba* (White Poplar) USA Tolstead.
SILVER-LEAF TREE *Leucodendron argenteum* (Silver Tree) Edlin.
SILVER-LEAVED TREE 1. *Betula verrucosa* (Birch) Som. Macmillan. 2. *Populus alba* (White Poplar) Som. Macmillan.
SILVER LEAVES *Potentilla anserina* (Silverweed) Som. Grigson.
SILVER LIME *Tilia tomentosa.*
SILVER MAPLE *Acer saccharinum.*
SILVER MOONS *Lunaria annua* (Honesty) Tynan & Maitland.

SILVER MOSS *Cerastium tomentosum* (Snow-in-summer) War. B & H.
SILVER-NERVE FITTONIA *Fittonia verschaffentii.*
SILVER PENNIES *Bellis perennis* (Daisy) N'thants. A E Baker.
SILVER PLATE *Lunaria annua* (Honesty) Gerard.
SILVER POPLAR *Populus alba* (White Poplar) Som. Grigson.
SILVER RAGWORT *Senecio cineraria.*
SILVER ROD 1. *Asphodelus ramosus* (White Asphodel) J Smith. 2. *Solidago bicolor* (White Golden Rod) USA House.
SILVER ROSE *Potentilla anserina* (Silverweed) Freethy.
SILVER SAGE 1. *Artemisia filifolia* (Sand Sagebrush) USA Robbins. 2. *Salvia argentea.*
SILVER SAGEBRUSH *Artemisia filifolia* (Sand Sagebrush) USA Elmore.
SILVER SHAKERS; SILVER SHACKLES; SILVER SHEKELS *Briza media* (Quaking-grass) Scot. Tynan & Maitland. All reminiscent of the Wiltshire children's admonition to watch the grass until it is completely still, and then see them turn into silver coins.
SILVER SHILLING FLOWER *Lunaria annua* (Honesty) Tynan & Maitland.
SILVER SPRUCE *Picea sitchensis* (Sitka Spruce) Brimble.1948.
SILVER TREE *Leucodendron argenteum.*
SILVER VASE PLANT *Aechmea fasciata* (Urn Plant) RHS.
SILVER VINE 1. *Actinidia polygama.* 2. *Scindapsus pictus 'Argyraeus'* (Devil's Ivy) Perry.1979.
SILVER WATTLE *Acacia dealbata.* Silver, because of the down that coats the foliage.
SILVER WATTLE, Queensland *Acacia podalyrifolia* (Mount Morgan Wattle) H & P.
SILVER WILLOW *Salix alba var. sericea* (Silver Willow) Barber & Phillips.
SILVERBELL, Mountain *Halesia monticola.*

SILVERBELL TREE *Halesia carolina* (Snowdrop Tree) USA Chaplin.
SILVERBERRY 1. *Elaeagnus argentea*. 2. *Elaeagnus commutata*.
SILVERLEAF 1. *Lunaria annua* (Honesty) Dor. Macmillan. 2. *Potentilla anserina* (Silverweed) Som. Macmillan. 3. *Stachys lanata* (Lamb's Ears) Som. Macmillan.
SILVERLEAF NIGHTSHADE *Solanum eleagnifolium*.
SILVERWEED *Potentilla anserina*.
SILVERY BINDWEED *Convolvulus cneorum* (Shrubby Bindweed) W Miller.
SILVERY CINQUEFOIL *Potentilla anserina* (Silverweed) Fernie.
SILVERY WORMWOOD *Artemisia filifolia* (Sand Sagebrush) USA Elmore.
SIMMERIN; SIMMEREN *Primula vulgaris* (Primrose) Yks. Grigson.
SIMPLER'S JOY *Verbena officinalis* (Vervain) Corn, Som, War. Northall. Because, if we are to believe all the accounts, vervain was used to combat practically every ailment.
SIMPSON *Senecio vulgaris* (Groundsel) Ess. Grose. One of a number of names, that include Simson and Sinsion, that are variations on the generic name, Senecio.
SIMSON 1. *Erigeron canadensis* (Canadian Fleabane) Culpepper. 2. *Senecio vulgaris* (Groundsel) Suff. Moor.
SINCAMA *Pachyrrhizus tuberosus* (Yam Bean) Schery.
SINCLE, White *Pinguicula vulgaris* (Butterwort). Hants. B & H. 'Sincle' is 'sanicle', i.e. a healing herb.
SINEY; SCINEY *Hesperis matronalis* (Sweet Rocket) Dev. B & H. They are corruptions of damascena, the old specific name for the plant. It had already given Damask Violet, used by Gerard, and that had resulted in further corruptions of 'Damask' into 'Dame's', as in Dame's Violet, Dame's Rocket, etc. Siney is also given to Bladder Nut (*Staphylea pinnata*) according to Halliwell.
SINFULL *Sempervivum tectorum* (Houseleek) C J S Thompson. Much earlier than the other "evergreen" names of this kind, for this appears in Anglo-Saxon accounts.
SINGAPORE MAHOGANY *Melanorrhoea curtisii* Gimlette.
SINGAPORE OAK *Quercus conocarpa*.
SINGHARA NUT *Trapa bispinosa*. 'Singhara' apparently means ' horny'.
SINGLE CASTLE *Orchis morio* (Green-winged Orchid) Dor. Macmillan. A long way from the original, which must have been something like Gandergosses.
SINGLETARY PEA 1. *Lathyrus hirsutus* (Hairy Pea) USA Kingsbury. 2. *Lathyrus pusillus*.
SINGREEN *Sempervivum tectorum* (Houseleek) Som. Macmillan; Wilts, Hants, Suss, Kent, Bucks, Worcs, Shrop. Grigson; IOW Long. 'Singreen' means evergreen, and is one of a long series of similar names.
SINKFIELD *Potentilla reptans* (Cinquefoil) Gerard. A variation of Cinquefoil.
SINKIN, Cow 1. *Primula veris x vulgaris* (False Oxlip) Cumb. Grigson. 2. *Primula vulgaris var. elatior* (Polyanthus) Cumb. B & H.
SINNEGAR; ZINEGAR *Matthiola incana* (Stock) both Som. Macmillan.
SINSION *Senecio vulgaris* (Groundsel) E Ang. Grigson. One of a whole series of names from East Anglia that are corruptions of 'Senecio'.
SINVEY *Sinapis arvensis* (Charlock) Som. Notes & Queries. vol 168, 1935. Cf. Senvie, or Zenvy.
SION *Apium nodiflorum* (Fool's Watercress) B & H. i.e. Sium, another umbelliferous genus, containing the Water Parsnips.
SIP-SAP *Sorbus aucuparia* (Rowan) Lancs, Yks. Grigson. Cf. Sap-tree, from Lincolnshire & Yorkshire. But the name is connected with Whistlewood, for boys used to make whistles in spring when the sap rose, and while doing so they would beat a suitably shaped branch with the haft of their penknives in an attempt to loosen the wetted

bark, and at the same time chant rhymes, such as

Sip, sip, say,
Sip, sip, say

or

Sip sap, sip sap,
Willie, Willie Whitecap.

SIRIS, Pink *Albizia julibrissin* (Persian Acacia) Polunin.
SIRIS, White *Albizia procera*.
SIRIS TREE *Albizia lebbeck*.
SISAL *Agave sisalana*. Sisal is the name of a place in Mexican Yucatán from which the fibre was exported.
SISAL, Salvador *Agave letonae*.
SISTER-IN-LAW *Viola tricolor* (Pansy) Skinner. One of a group of "family" names that includes Stepmothers, or Stepdaughters. In north-east Scotland, it is said that the largest of the lower petals is the stepmother, while on each side of her stand her own two daughters, and the upper petals are the stepdaughters.
SISTER VIOLET *Viola sororia* (Woolly Blue Violet) USA House. See the specific name, *sororia*.
SISTERS, Five *Euphorbia helioscopia* (Sun Spurge) Limerick Westropp. Cf. Seven Sisters. The number refers to the branches of the stem, five in co Limerick, and seven in co Clare, apparently.
SISTERS, Four *Polygala vulgaris* (Milkwort) Ire. Grigson. From the different colours in the flower, according to Grigson.
SISTERS, Seven 1. *Euphorbia helioscopia* (Sun Spurge) co Clare Westropp. 2. *Euphorbia peplus* (Petty Spurge) Ire. B & H. Cf. Five Sisters.
SIT-SICKER; SIT-SICCAR 1. *Ranunculus acris* (Meadow Buttercup) S Scot. Grigson. 2. *Ranunculus bulbosus* (Bulbous Buttercup) S Scot. Grigson. 3. *Ranunculus repens* (Creeping Buttercup) Scot. B & H; Rorie.1994 (Sit-Siccar). Siccar, in either spelling, is an old Scots word meaning sure, or certain, a comment on the way Creeping Buttercup in particular holds itself to the ground, with consequent difficulty in removal. Cf. Hod-the-rake, and more immediately, Sitfast.
SITFAST 1. *Ononis repens* (Rest Harrow) Mor. Jamieson. 2. *Ranunculus repens* (Creeping Buttercup) Dumf, Ire. Grigson. A comment on the difficulty in pulling them up, especially Rest Harrow, whose name (= arrest harrow) conveys the same message.
SITHES *Allium schoernoprasum* (Chives) Som. Macmillan. A relic of the medieval version of the Latin cepa, onion.
SITKA CYPRESS *Chamaecyparis nootkatensis* (Alaska Yellow Cedar) Willis.
SITKA SPRUCE *Picea sitchensis*.
SIUNAS; SHUNAS *Ligusticum scoticum* (Lovage) Skye B & H. Probably only applied to the greens as a vegetable.
SIVES *Allium schoernoprasum* (Chives) Corn,Dor. B & H; Som. Elworthy.
SIVVEN *Rubus idaeus* (Raspberry) Scot. Jamieson. A version of Gaelic suibhean.
SIX O'CLOCK; SIX O'CLOCK FLOWER *Ornithogalum umbellatum* (Star of Bethlehem) N'thants. A E Baker (Six o'clock); Midl. B & H (Six o'clock Flower). References to the time of opening of the flower. The trouble is that we can find such variation - they are four, six, ten, eleven, twelve and one o'clock flowers.
SIX WEEKS' GRASS *Poa annua* (Annual Meadow-grass) USA Cunningham. Presumably this is a measure of the length of time it takes to get a crop.
SIXPENCES, Gowk's *Rhinanthus crista-galli* (Yellow Rattle) N'thum, Berw. Grigson. The sixpences are the seeds rattling about in the seed-case, and the gowk is a cuckoo.
SJOLGIRSE; SHOLGIRSE *Achillea ptarmica* (Sneezewort) Shet. Grigson.
SKALLY *Agropyron repens* (Couch-grass) Grigson.1959. Probably with a similar deriva-

tion to the West country Scryle, which is (eventually) OE cwice, alive.

SKAW; SCAW *Sambucus nigra* (Elder) Corn. B & H (Skaw); Jago (Scaw). Scaw seems the more likely spelling; it is the Cornish for elder.

SKAYUG *Crataegus monogyna* fruit (Haw) Ire. Grigson. Irish sceach, a bush or bramble. It appears as Sgeach, Skeeog, even Shiggy.

SKEAT-LEGS 1. *Dactylorchis maculata* (Heath Spotted Orchid) Kent Grigson. 2. *Orchis mascula* (Early Purple Orchid) Kent B & H. Cf. Keatlegs, Neatlegs, also from Kent. Britten & Holland explain the name as being from OE sceat, meaning any description of wrapping or swathing, the stem or 'leg' of the plant being partially enveloped in a sheathing leaf.

SKEEOG *Crataegus monogyna* fruit (Haws) Ire. Grigson. See Skayug above.

SKEDGE; SKEDGEWITH *Ligustrum vulgare* (Privet) Corn. Grigson (Skedge); Jago (Skedgewith). From a Cornish word meaning 'shade-tree'.

SKEDLOCK 1. *Senecio jacobaea* (Ragwort) Lancs. A Burton. 2. *Sinapis arvensis* (Charlock) Lancs. Nodal. A typical charlock name. The attribution to ragwort was probably an error.

SKEETS 1. *Heracleum sphondyllium* (Hogweed) Corn. Grigson. 2. *Smyrnium olusatrum* (Alexanders) Corn. Jago. Perhaps a variation of kex, but more likely to be allied to skeiters.

SKEG 1. *Crataegus monogyna* fruit (Haws) IOM Moore.1924. 2. *Iris pseudo-acarus* (Yellow Flag) Yks. B & H. 3. *Prunus domestica var. institia* (Bullace) Bucks. Harman. N'thants. Sternberg. 4. *Prunus spinosa* (Blackthorn) War, N'thants. B & H.

SKEG, Water *Iris pseudo-acarus* (Yellow Iris) Pratt.

SKEITERS, Ait *Angelica sylvestris* (Wild Angelica) Scot. Dyer. Children shoot oats through the hollow stems, like peas through a pea-shooter.

SKEITERS, Bear 1. *Angelica sylvestris* (Wild Angelica) Scot. Dyer. 2. *Heracleum sphondyllium* (Hogweed) Mor. B & H. 'Bear' here would be bere, barley, shot through the hollow stems.

SKELDICK; SKELDOCK *Raphanus raphanistrum* (Wild Radish) Scot. B & H. Cf. Skellock and Skillock.

SKELETON WEED *Eriogonum deflexum*.

SKELLIES *Sinapis arvensis* (Charlock) Scot. Jamieson. The name exists in a number of versions, including Scaldricks and Skedlock.

SKELLOCK 1. *Raphanus raphanistrum* (Wild Radish) Scot. B & H. 2. *Sinapis arvensis* (Charlock) Scot. Jamieson. Another of the versions mentioned under Skellies.

SKERRET *Sium sisarum* (Skirret) B & H.

SKERRISH *Ligustrum vulgare* (Privet) Corn. Jago. Evidently another version of Skedge (see above).

SKEVISH *Erigeron philadelphicus* (Philadelphia Fleabane) USA House.

SKEW *Sambucus nigra* (Elder) Corn. Courtney.1880. Scaw is the Cornish for elder, and there are a few variants on this, as Skaw, or Scow.

SKEWERWOOD 1. *Cornus sanguinea* (Dogwood) Heref. Havergal; Bucks. B & H. 2. *Euonymus europaeus* (Spindle Tree) Dev, Dor, IOW, Glos, Berks, Yks. Grigson; Wilts. Dartnell & Goddard. 3. *Sambucus nigra* (Elder) Heref. Havergal. It means exactly what it says. Butchers preferred to have skewers made from dogwood and spindle, because they did not taint the meat. The name is unusual for elder, and it may very well be a mistake for spindle, unless, of course, it is an echo of the Cornish name Scaw. Cf. Skiver, Skiverwood, etc.

SKIDGY *Ligustrum vulgare* (Privet) Corn. Grigson. Cf. Skedge and Skerrish, all Cornish names.

SKILLOCK 1. *Raphanus raphanistrum* (Wild Radish) Scot. B & H. 2. *Sinapis arvensis* (Charlock) Donegal Grigson.

SKILLOG *Sinapis arvensis* (Charlock) Donegal Grigson.

SKIPPING ROPE *Clematis vitalba* (Old Man's Beard) Wilts. Dartnell & Goddard. The long bines are referred to here.

SKIRRET *Sium sisarum*. According to Grigson, it comes by various ways from medieval Latin carvi, caraway.

SKIRRET, Water *Sium nodiflorum* (Procumbent Water Parsnip) Barton & Castle.

SKIRRET, Wild *Potentilla anserina* (Silverweed) Freethy.

SKIRT BUTTONS 1. *Stellaria holostea* (Greater Stitchwort) Som. Grigson. 2. *Stellaria media* (Chickweed) Dor. Grigson.

SKIRWITS; SKIRWORT *Sium sisarum* (Skirret) Langham (Skirwits); Turner (Skirwort).

SKIT; SKEET *Smyrnium olusatrum* (Alexanders) Corn. Jago.

SKIVER 1. *Cornus sanguinea* (Dogwood) Som. Macmillan. 2. *Euonymus europaeus* (Spindle Tree) Som, Wilts. Grigson. 'Skiver' is 'skewer', both trees being used to make them.

SKIVER-TIMBER *Euonymus europaeus* (Spindle Tree) Som. Elworthy; Wilts. Grigson.

SKIVER-TREE 1. *Cornus sanguinea* (Dogwood) Corn, Dev, Dor, Som, IOW, Glos, Bucks Grigson. 2. *Euonymus europaeus* (Spindle Tree) Dev. Grigson.

SKIVER-WOOD 1. *Cornus sanguinea* (Dogwood) IOW B & H. 2. *Euonymus europaeus* (Spindle Tree) Dor. Barnes; Som. Grigson; IOW Long; Glos. Grigson.

SKIVVER *Cornus sanguinea* (Dogwood) Wilts. Dartnell & Goddard.1899. Spelt this way to indicate the pronunciation.

SKIWET *Pastinaca sativa* (Parsnip) Graves. Probably related to 'skirret', another umbellifer.

SKOKE *Phytolacca decandra* (Poke-root) New Engl. Sanford. Also rendered as Scoke, or as Shoke. Coakum or Cocum are obviously related.

SKULLCAP 1. *Aquilegia vulgaris* (Columbine) Corn. Grigson. 2. *Scutellaria galericulata*. Descriptive, though as far as columbine is concerned, the likeness is usually rendered as bonnets rather than skullcaps.

SKULLCAP, Blue; SKULLCAP, Hooded *Scutellaria galericulata* (Skullcap) both USA House.

SKULLCAP, Lesser *Scutellaria minor*.

SKULLCAP, Mad-dog *Scutellaria lateriflora* (Virginian Skullcap) USA Zenkert. Cf. Mad-dog and Mad-dog Herb. It was used for hydrophobia after Van Derveer experimented with it in 1772.

SKULLCAP, Marsh *Scutellaria galericulata* (Skullcap) USA House.

SKULLCAP, Side-flowering *Scutellaria lateriflora* (Virginian Skullcap) O P Brown. A translation of the botanical name.

SKULLCAP, Virginian *Scutellaria lateriflora*.

SKULLCAP SPEEDWELL *Veronica scutellata* (Marsh Speedwell) USA House. The name is a translation of the specific name.

SKUNK BUSH 1. *Garrya fremontii* (Bear Brush) Grieve.1931. 2. *Rhus trilobata* (Three-leaf Sumach) USA Elmore. In the case of the sumach, Cf. the names Polecat Bush, and Ill-scented Sumach.

SKUNK CABBAGE 1. *Symplocarpus foetidus*. 2. *Veratrum californicum* (False Hellebore) USA Kingsbury. The name is given to Symplocarpus because of the foul smell when the plant is disturbed.

SKUNK CURRANT 1. *Ribes glandulosum*. 2. *Ribes prostratum*.

SKUNKWEED 1. *Cleome serrulata* (Rocky Mountain Bee Plant) USA. T H Kearney. 2. *Cleome sonorae* (Sonora Beeweed) USA Elmore. 3. *Symplocarpus foetidus* (Skunk Cabbage) Le Strange. They are all foul-smelling.

SKY FLOWER *Duranta repens* (Golden Dewdrop) Perry.

SKY PLANT *Tillandsia ionantha*.

SKY VINE *Thunbergia grandiflora*. Sky-blue flowers.

SKYROCKET 1. *Gilia aggregata* (Scarlet Gilia) USA Schenk & Gifford. 2. *Gilia attenuata* (Scarlet Gilia) USA Elmore.

SKYSCRAPER *Helianthus annuus* (Sunflower) Som. Macmillan. The sheer height of the plant earns the name.

SKYTES 1. *Angelica sylvestris* (Wild Angelica) Scot. Grigson. 2. *Heracleum sphondyllium* (Hogweed) Scot. B & H. Cf. Skeets. Probably the reference is to the hollow stems, through which boys used to shoot oats or barley, in the manner of a pea-shooter, i.e. Ait-skeiters.

SLAA-THORN; SLAA-TREE *Prunus spinosa* (Blackthorn) Yks, Cumb. Grigson; Scot. B & H (Thorn); Yks, Cumb. Grigson (Tree). 'Slaa' is the word in the North for 'sloe', but it comes through Old Scandinavian languages.

SLACEN-BUSH *Prunus spinosa* (Blackthorn) N'thants. A E Baker.

SLAE *Prunus spinosa* (Blackthorn) Border B & H.

SLAG *Prunus spinosa* fruit (Sloe) Oxf. Grigson.

SLAIGH *Prunus spinosa* (Blackthorn) Lancs. B & H.

SLAM *Prunus spinosa* fruit (Sloe) War. F Savage. It is doubtful if this is a genuine local name, and it may even be a simple misprint for 'slan'.

SLAN 1. *Prunus spinosa* (Blackthorn) Wilts. Akerman. 2. *Prunus spinosa* fruit (Sloe) Wilts, Shrop, Kent. A E Baker; Glos. J D Robertson; Oxf. Oxf.Annl.Rec.1951. A modern singular of OE plural slan, the singular of which was sla(h).

SLANLUS *Plantago major* (Great Plantain) Scot, N Ire. Denham/Hardy. It is Gaelic, and means healing herb.

SLAP *Primula veris* (Cowslip) N'thants. A E Baker. Sometimes the 'cow' of cowslip if left out, as in this instance, leaving Slap or Slop to define the flower.

SLASH PINE *Pinus elliottii*.

SLATH *Prunus domestica var. institia* (Bullace) Lancs. Grigson.

SLAUN-BUSH; SLAUN-TREE *Prunus spinosa* (Blackthorn) N'thants. A E Baker (Bush); Leics. A B Evans (Tree). A variation on Slon-bush, 'slon' being a plural form of sloe.

SLAVESACRE *Delphinium staphisagria* (Stavesacre) Lupton. Is this a simple misprint?

SLAW *Prunus spinosa* (Blackthorn) Dev, Glos, N'thants. B & H.

SLAWNES *Prunus spinosa* fruit (Sloe) Worcs. J Salisbury. A double plural, as can be found with many of the names for sloes.

SLEA *Prunus spinosa* fruit (Sloe) N'thants. A E Baker.

SLEEP-AT-NOON *Tragopogon pratensis* (Goat's Beard) B & H. The flowers open at sunrise and close at noon, giving rise to a host of picturesque local names.

SLEEPING BEAUTY 1. *Oxalis acetosella* (Wood Sorrel) Dor. Macmillan. 2. *Oxalis corniculata*. Presumably given because of the way the leaves fold back.

SLEEPING CLOVER *Oxalis acetosella* (Wood Sorrel) Dor. Macmillan. It is a trefoil, hence 'Clover'.

SLEEPING MAGGIE *Trifolium pratense* (Red Clover) N'thum. Grigson.

SLEEPING NIGHTSHADE *Atropa belladonna* (Deadly Nightshade) Gerard. 'Sleeping' is in a different context here - it is given because the plant causes sleep.

SLEEPWORT 1. *Lactuca sativa* (Lettuce) Ellacombe. 2. *Lactuca virosa* (Great Prickly Lettuce) Prior. 3. *Mycelis muralis* (Wall Lettuce) Gunther. All the lettuces are narcotic, but none more so than *Lactuca virosa*, which is sometimes called Opium Lettuce. The soporific qualities are commemorated in this name.

SLEEPY CATCHFLY *Silene antirrhina*.

SLEEPY DICK *Ornithogalum umbellatum* (Star of Bethlehem) Hants. B & H.

SLEEPY DOSE *Senecio jacobaea* (Ragwort) Banff. Grigson. There is no record of the use of ragwort as a sedative. But what else could this mean?

SLEEPY GRASS *Stipa robusta*. 'Sleepy', because it has narcotic effects on horses, though not on cattle, apparently.

SLEEPYHEAD 1. *Papaver rhoeas* (Red Poppy) Som. Macmillan. 2. *Tragopogon pratensis* (Goat's Beard) Som. Macmillan. Given for different reasons. In the case of the poppy, it is a recognition of its narcotic properties. As for Goat's Beard, it commemorates the fact that it goes to bed at noon.

SLENDER NETTLE *Urtica gracilis*.

SLENDER TARE *Vicia tenuissima*.

SLICK GREENS *Brassica oleracea var. capitata* the young leaves Gloos. J D Robertson.

SLIMSTEM LILY *Lilium callosum*.

SLINKWEED *Gutierrezia microcephala* (Broomweed) USA Kingsbury. Probably another of its names, Snakeweed, gives the clue to the meaning of this.

SLIPPER, Golden 1. *Cypripedium calceolus* (Lady's Slipper) USA Cunningham. 2. *Lotus corniculatus* (Bird's-foot Trefoil) Som. Macmillan; Hants. Boase.

SLIPPER, Grandmother's *Lotus corniculatus* (Bird's-foot Trefoil) Hants. Grigson.

SLIPPER, Granny's 1. *Aconitum napellus* (Monkshood) Dor. Macmillan. 2. *Lotus corniculatus* (Bird's-foot Trefoil) Hants. Hants. FWI.

SLIPPER, Lady's 1. *Aconitum napellus* (Monkshood) Som. Macmillan. 2. *Anthyllis vulneraria* (Kidney Vetch) War. Grigson. 3. *Antirrhinum majus* (Snapdragon) Dor. Macmillan. 4. *Aquilegia vulgaris* (Columbine) Som, Wilts. Macmillan. 5. *Arum maculatum* (Cuckoo-pint) Wilts. Grigson. 6. *Cypridium calceolus*. 7. *Cytisus scoparius* (Broom) Som. Macmillan. 8. *Digitalis purpurea* (Foxglove) Som. Macmillan. 9. *Hippocrepis comosa* (Horseshoe Vetch) Som. Macmillan. 10. *Lathyrus pratensis* (Yellow Pea) Som, Wilts. Macmillan. 11. *Linaria vulgaris* (Yellow Toadflax) Som. Macmillan. 12. *Lotus corniculatus* (Bird's-foot Trefoil) Som, Wilts. Macmillan; Hants, Yks. Grigson. 13. *Ranunculus acris* (Meadow Buttercup) Som. Macmillan.

SLIPPER, Lady's, Downy *Cypripedium parviflorum* (Small Yellow Lady's Slipper) USA House.

SLIPPER, Lady's, Pink *Cypripedium acaule* (Stemless Lady's Slipper) USA Bingham.

SLIPPER, Lady's, Queen *Cypripedium reginae* (Showy Lady's Slipper).

SLIPPER, Lady's, Ram's Head *Cypripedium arietinum*.

SLIPPER, Lady's, Showy *Cypripedium reginae*.

SLIPPER, Lady's, Stemless *Cypripedium acaule*.

SLIPPER, Lady's, White, Small *Cypripedium candidum*.

SLIPPER, Lady's, Wild *Impatiens capensis* (Orange Balsam) Grieve.1931.

SLIPPER, Lady's, Yellow, Larger *Cypripedium parviflorum var. pubescens*.

SLIPPER, Lady's, Yellow, Small *Cypripedium parviflorum*.

SLIPPER-SLOPPERS, Granny's *Lathyrus pratensis* (Yellow Pea) Dor. Grigson.

SLIPPERS, Negro's *Euphorbia myrtifolia* W Miller.

SLIPPERS, Silver *Nigella damascena* (Love-in-a-mist) Som. Macmillan.

SLIPPERS, Shoes-and- *Anemone nemorosa* (Wood Anemone) Som. Macmillan.

SLIPPERWEED *Impatiens capensis* (Orange Balsam) Grieve.1931.

SLIPPERWORT *Calceolaria sp.*

SLIPPERY ELM *Ulmus fulva*. Named from the mucilaginous inner bark, used in medicine for stomach ulcers among other complaints.

SLIPPERY ELM, Californian *Fremontia californica* (Flannel Bush) Le Strange. The bark is said to have the same qualities as the real Slippery Elm.

SLIPPERY ROOT *Symphytum officinale* (Comfrey) Grieve.1931. Cf. Gum Plant. Both names are references to the mucilaginous matter in the root, used as a kind of splint when set.

SLIPS, Wild *Ipomaea tiliacea* Barbados Gooding.

SLITE *Cyclamen europaeum* (Sowbread) Cockayne.

SLOANES *Prunus spinosa* fruit (Sloe) Corn. Hawke. Spelt thus, but this must be a case of a double plural. See Slon, rather.

SLOE 1. *Prunus americana* fruit (Canada Plum) USA Elmore. 2. *Prunus spinosa* fruit. 3. *Prunus umbellata* fruit (Hog Plum) USA Harper.

SLOE, American *Viburnum prunifolium* (Black Haw) Mitton.

SLOE-BUSH *Prunus spinosa* (Blackthorn) Som. Macmillan.

SLOETHORN *Prunus spinosa* (Blackthorn) Grattan & Singer.

SLON; SLONE *Prunus spinosa* fruit (Sloe) Wilts. Dartnell & Goddard; N'thants. A E Baker (Slon); Hazlitt (Slone). The old plural in '-n'.

SLON-BUSH; SLON-TREE *Prunus spinosa* (Blackthorn) Leics, N'thants Grigson (Bush); Corn, Som, Leics Grigson (Tree).

SLOON *Prunus spinosa* fruit (Sloe) Suff. Nall.

SLOP *Primula veris* (Cowslip) N'thants. A E Baker. Slap and Slop are used to define the whole flower name. The original was OE cuslyppe, cow dung.

SLOTS, Drunken *Valeriana officinalis* (Valerian) Som. Grigson.

SLOUGH-HEAL; SLOUGHWORT *Prunella vulgaris* (Self-heal) Prior (Slough-heal); Tynan & Maitland (Sloughwort). Slough-heal was taken for a corruption of Self-heal, but Prior did not think so, "the 'slough' being that which is thrown off from a foul sore, and not that which is healed, the wound itself".

SLOVENWOOD *Artemisia abrotanum* (Southernwood) E Angl. Forby. A corruption of 'southernwood', which produced some very odd variations, like the Orkney Cedar-wood.

SLOW *Prunus spinosa* fruit (Sloe) T Hill.

SLUE *Prunus spinosa* fruit (Sloe) Wilts. Dartnell & Goddard.

SLUMBER, Sweet *Sanguinaria canadensis* (Bloodroot) Grieve.1931. This name suggests that the plant must be a sedative, but there is no record of this kind of use.

SMAIR-DOCK *Rumex obtusifolius* (Broad-leaved Dock) Border B & H. Cf. Butter Dock - the leaves were used to wrap butter.

SMALACH; SMALLACHE; SMALL-ADGE; SMALLAGE; SMALLEDGE *Apium graveolens* (Wild Celery) B & H (Smalach, Smallache); Parkinson (Smalladge, Smalledge); Brownlow (Smallage). Ach is possibly from an Old French word, directly from Apium, and actually meant for parsley. Smallage is the accepted version.

SMALL-AND-PRETTY *Malcomia maritima* (Virginia Stock) Som. Macmillan.

SMALLWORT *Ranunculus ficaria* (Lesser Celandine) Grieve.1931.

SMARA; SMOORA *Trifolium repens* (White Clover) Shet. Grigson.

SMARTARSE *Polygonum hydropiper* (Water Pepper) Dev, Som. Elworthy. Arsmart is perhaps the more usual form. They both owe their existence to the irritant effect of the leaves. Cockayne said "it derives its name from its use in that practical education of simple Cimons, which village jokers enjoy to impart". Culrage is another version of the same thing.

SMARTWEED 1. *Polygonum hydropiper* (Water Pepper) Norf, USA Grigson. 2. *Polygonum lapathifolium* (Pale Persicaria) Stevenson. 3. *Polygonum persicaria* (Persicaria) Norf. Grigson; USA. Bergen. 4. *Polygonum punctatum* USA Tolstead.

SMARTWEED, Alpine *Polygonum bistortioides*.

SMARTWEED, Dock-leaved;
   SMARTWEED, Pale *Polygonum lapathifolium* (Pale Persicaria) both USA Zenkert.

SMARTWEED, Swamp *Polygonum muhlenbergii* (Swamp Persicaria) USA Gates.

SMARTWEED, Water 1. *Polygonum*

hydropiper (Water Pepper) USA Zenkert. 2. *Polygonum punctatum* USA Upham.

SMEAR-DOCKEN *Chenopodium bonus-henricus* (Good King Henry) Scot. Grigson. Cf. Smearwort. From its use as an ointment.

SMEARWORT 1. *Aristolochia clematitis* (Birthwort) Cockayne. 2. *Chenopodium bonus-henricus* (Good King Henry) Scot. Grigson. Again, from their use in ointments.

SMEARWORT, All-good *Chenopodium bonus-henricus* (Good King Henry) Leyel.1937. It may be a misprint, as All-good and Smearwort are separately recorded as names for this plant.

SMELL BADGER *Geranium robertianum* (Herb Robert) Tynan & Maitland. The plant has a characteristic smell when bruised, likened to foxes as well as badgers. Actually the "foxy" smell is more of a musky aroma.

SMELL-FOXES 1. *Anemone nemerosa* (Wood Anemone) Bucks, Scot. Grigson. 2. *Geranium robertianum* (Herb Robert) Som. Macmillan; Hants. Fernie. Does Wood Anemone have this musky smell as well?

SMELL-SMOCK; SMICK-SMOCK *Cardamine pratensis* (Lady's Smock) Glos. J D Robertson; War. Morley;Bucks. Harman; Kent Grigson (Smell); Glos. J D Robertson; Hants, Oxf. Grigson (Smick). See under 'Smock'.

SMELLING-WOOD *Artemisia abrotanum* (Southernwood) Oxf. B & H. Perhaps from the ostentatious sniffing that a hopeful youth made at his buttonhole of Lad's Love. Though the name may very well be a variation of southernwood.

SMELLY GERANIUM *Geranium robertianum* (Herb Robert) Tynan & Maitland. Cf. Smell-foxes, as well as the many "Stinking" names given to it.

SMERE *Trifolium pratense* (Red Clover) Fernie.

SMEREWORT 1. *Aristolochia clematitis* (Birthwort) B & H. 2. *Mercurialis perennis* (Dog's Mercury) Fernie. Cf. Smearwort.

SMIDDY-LEAVES *Chenopodium bonus-henricus* (Good King Henry) Berw. Grigson. It has been suggested that this name shows that one of the plant's favourite haunts was about a blacksmith's smithy. But surely 'smiddy' is the same word as 'midden'? And that is a much better guide to habitat.

SMITE, Baalam's *Stellaria holostea* (Greater Stitchwort) Suff. Grigson.

SMITH'S CRESS; SMITH'S PEPPER-CRESS *Lepidium smithii*.

SMOCK 1. *Calystegia sepium* (Great Bindweed) Dor, Som. Macmillan. 2. *Stellaria holostea* (Greater Stitchwort) Bucks. Grigson. Smock originally was an undergarment, white, and made of linen. It was later called a shift, and later still a chemise. The imagery is that of garments laid out on bushes to dry.

SMOCK, Blue *Vinca minor* (Lesser Periwinkle) Som. Macmillan.

SMOCK, Chimney *Anemone nemerosa* (Wood Anemone) Som. Macmillan. It sounds very odd, but this plant does have Small Smock and Lady's Shimmey as names.

SMOCK, Holland *Calystegia sepium* (Great Bindweed) Som. Macmillan. 'Holland' is a coarse linen fabric, used these days for upholstery, though originally it was not coarse at all, and is described as " a fine kind of linen first made in Holland".

SMOCK, Kettle 1. *Convolvulus arvensis* (Field Bindweed) Wilts. Macmillan. 2. *Lonicera periclymenum* (Honeysuckle) Som. Macmillan. 3. *Melandrium rubrum* (Red Campion) Som. Macmillan. 'Kettle smock' is an old name for a farm labourer's smock, but 'kettle' in this context may very well be a different word altogether, i.e. Kadle, or something like it. See, for instance, Kadle Dock.

SMOCK, Lady *Cardamine pratensis* (Lady's Smock) Ches. Gerard.

SMOCK, Lady's 1. *Anemone nemerosa* (Wood Anemone) N'thants. Clare. 2. *Artemisia lactiflora* (White Wormwood) Dyer. 3. *Arum maculatum* (Cuckoo Pint) Dor, Som. Macmillan; Hants. Cope. 4. *Calystegia sepium* (Great Bindweed) Corn,

War, Notts, Yks. Grigson; Dev. Friend.1882; Som. Jennings; Suss. Parish; N'thants. A E Baker. 5. *Campanula medium* (Canterbury Bells) Halliwell. 6. *Cardamine pratensis*. 7. *Convolvulus arvensis* (Field Bindweed) Som. Macmillan. 8. *Stellaria holostea* (Greater Stitchwort) Corn, Som. Grigson; Dor. Macmillan. See Smock.

SMOCK, Lady's, Bitter *Cardamine amara* (Large Bittercress) Curtis.

SMOCK, Lady's, Hairy *Cardamine hirsuta* (Hairy Bittercress) Curtis.

SMOCK, Lady's, Impatient *Cardamine impatiens* (Narrow-leaved Bittercress Camden/Gibson. 'Impatient', because of its explosive method of seed dispersal, as with Touch-me-not, which also occurs as a name for this plant.

SMOCK, My Lady's *Cardamine pratensis* (Lady's Smock) Dor. Macmillan. 'My Lady' is the same as 'Our Lady'.

SMOCK, Old Lady's *Calystegia sepium* (Great Bindweed) Som. Macmillan. 'Old Lady' must originally have been 'Our Lady'.

SMOCK, Our Lady's *Calystegia sepium* (Great Bindweed) Som. Macmillan.

SMOCK, Parson-in-his- *Arum maculatum* (Cuckoo-pint) Lincs. Grigson. The spadix in the spathe provides the imagery, and the 'smock' in this case is a surplice.

SMOCK, Smell; SMOCK, Smick *Cardamine pratensis* (Lady's Smock) Glos. J D Robertson; War. Morley; Bucks. Harman; Kent Grigson (Smell-smock); Glos. J D Robertson; Hants, Oxf. Grigson (Smick-smock). 'Smock' eventually came to have coarse associations, and was replaced in polite language by 'shift'. These two names belong to the coarse vocabulary. 'Smick' presumably comes from a verb 'to smicker', meaning to leer, and have amorous intentions.

SMOCK, Small *Anemone nemorosa* (Wood Anemone) Hants. B & H. Is it really 'small'? Or 'smell'?

SMOCK, White 1. *Calystegia sepium* (Great Bindweed) Dev. Macmillan. 2. *Convolvulus arvensis* (Field Bindweed) Dev. Grigson.

SMOCK FROCK *Stellaria holostea* (Greater Stitchwort) Dev, Bucks. Grigson.

SMOKE, Prairie *Geum triflorum*.

SMOKE PLANT; SMOKE TREE *Cotinus coggyria* (Venetian Sumach). Descriptive.

SMOKEWOOD; SMOKING CANE *Clematis vitalba* (Old Man's Beard) Friend (Smokewood); Dev. Friend.1882; Som, Hants. Grigson. Suss. Parish (Smoking Cane). Boys used to smoke the porous stalks. Cf. Boys' Bacca, Gypsy's Bacca and Devil's Cut.

SMOORA; SMARA *Trifolium repens* (White Clover) Shet. Grigson.

SMOTHERWEED *Bassia hyssopifolia*.

SMOTHERWOOD *Artemisia vulgaris* (Mugwort) Lincs. Grigson. Is this a corruption of southernwood? It sounds like it, yet there is also Motherwort recorded for mugwort, and the herb was certainly used in uterine diseases.

SMUTS *Luzula campestre* (Good Friday Grass) Bucks. B & H. The grass has blackish-brown flowers, hence this name, and also Sweeps and names like that.

SNAFFLES *Rhinanthus crista-galli* (Yellow Rattle) Kent Grigson.

SNAG-BUSH *Prunus spinosa* (Blackthorn) Som, Dor, Hants. Grigson; Wilts. Dartnell & Goddard. 'Snags' is a common Wessex name for sloes.

SNAGGS *Silene cucubalis* (Bladder Campion) Som. Macmillan.

SNAGS 1. *Prunus spinosa* (Blackthorn) Hants. Cope. 2. *Prunus spinosa* fruit (Sloe) Dor. Grigson; Som. Jennings; Wilts. Dartnell & Goddard; Hants. Cope. Sloes are often known by this name, but it is unusual to find the tree itself so called.

SNAIL CLOVER *Medicago sativa var. sativa* (Lucerne) Parkinson.

SNAIL PLANT *Vigna caracalla*.

SNAKE *Symphytum officinale* (Comfrey) Dor. Grigson. Many 'snake' names are given, as in this case, because of writhed roots.

SNAKE, Congo *Sanseviera trifasciata*

(Mother-in-law's Tongue) USA Cunningham.
SNAKE-BARK MAPLE *Acer pennsylvanicum*.
SNAKE BEAN TREE *Swartzia madagascariensis*.
SNAKE CUCUMBER *Cucumis melo var. flexuosum* (Serpent Melon) Grieve.1931.
SNAKE FERN 1. *Blechnum spicant* (Hard Fern) Hants. Cope. 2. *Osmunda regalis* (Royal Fern) Hants. Cope. 3. *Pteridium aquilina* (Bracken) Wilts. Dartnell & Goddard. Presumably because of the way the fronds unfurl.
SNAKE FLOWER; SNAKEFLOWER 1. *Allium ursinum* (Ramsons) Som. Macmillan. 2. *Capsella bursa-pastoris* (Shepherd's Purse) Som. Macmillan. 3. *Dactylorchis maculata* (Heath Spotted Orchid) Wilts. Goddard. 4. *Fritillaria meleagris* (Snake's-head Lily) B & H. 5. *Geranium robertianum* (Herb Robert) Glos. PLNN29. 6. *Orchis mascula* (Early Purple Orchis) Som. Grigson. 7. *Prunella vulgaris* (Self-heal) Hants. Hants FWI. 8. *Solanum dulcamara* (Woody Nightshade) Som. Grigson. 9. *Stellaria graminea* (Lesser Stitchwort) Dor. Macmillan. 10. *Stellaria holostea* (Greater Stitchwort) Som, Hants. Grigson; Wilts. Dartnell & Goddard.
SNAKE GOURD *Trichomanthes anguina*. 'Anguina' in the specific name means snake-like.
SNAKE-GRASS *Stellaria holostea* (Greater Stitchwort) Hants. Grigson.
SNAKE LILY 1. *Brodiaea volubilis*. 2. *Erythronium mesochoreum* (Spring Lily) USA Gilmore. 3. *Iris versicolor* (Purple Water Iris) Grieve.1931.
SNAKE PLANT 1. *Allium ursinum* (Ramsons) Som. Macmillan. 2. *Arum maculatum* (Cuckoo-pint) Suss. Cooper. Cuckoo-pint berries are poisonous. There are a number of 'poison' and 'snake' names for the plant, and Ramsons was called Serpentine Garlic early on.

SNAKE PLANTAIN *Plantago lanceolata* (Ribwort Plantain) Flück.
SNAKE-TONGUE CROWFOOT *Ranunculus ophioglossifolius* (Serpent Tongue Spearwort) Clapham. The specific name translates as 'leaves in the shape of a snake's tongue'.
SNAKE VINE *Hibbertia scandens*.
SNAKE VIOLET 1. *Viola reichenbachiana* (Wood Dog Violet) Dor. Macmillan. 2. *Viola riviniana* (Dog Violet) Dor. Grigson.
SNAKE'S BERRY *Iris foetidissima* fruit (Foetid Iris) Dor. Northover. What with the bad smell and the spectacular looking seed vessels, it is no wonder that country people looked on this plant with suspicion. There are many 'snake' and 'poison' names for it.
SNAKE'S BIT *Mercurialis perennis* (Dog's Mercury) Suss. Grigson. A poisonous plant, hence the 'snake' ascription.
SNAKE BUGLOSS *Echium vulgare* (Viper's Bugloss) Gerard.
SNAKE'S CHERRIES *Cornus sanguinea* fruit (Dogwood berries) Som. Macmillan.
SNAKE'S FIDDLE *Iris foetidissima* (Foetid Iris) IOW Grigson. A hybrid name, the 'snake' part of it referring to the scarlet seed-cases, always viewed with suspicion by country people. The 'fiddle' part probably refers to the squeaking noise that can be coaxed out of the leaves.
SNAKE'S FLOWER 1. *Allium ursinum* (Ramsons) Som, Dor. Macmillan. 2. *Anemone nemorosa* (Wood Anemone) Som, Dor. Macmillan. 3. *Calystegia sepium* (Great Bindweed) Som. Macmillan. 4. *Colchicum autumnale* (Meadow Saffron) Som. Macmillan. 5. *Echium vulgare* (Viper's Bugloss) IOW, USA Grigson. 6. *Endymion nonscriptus* (Bluebell) Som. Macmillan. 7. *Fritillaria meleagris* (Snake's-head Lily) B & H. 8. *Geranium robertianum* (Herb Robert) Som. Macmillan. 9. *Knautia arvensis* (Field Scabious) Som. Macmillan. 10. *Lamium album* (White Deadnettle) Cambs, Ess, Norf. Grigson. 11. *Lychnis flos-cuculi* (Ragged Robin) Som. Macmillan.

12. *Mercurialis perennis* (Dog's Mercury) Som, Dor. Macmillan. 13. *Melandrium album* (White Campion) Middx Grigson. 14. *Orchis mascula* (Early Purple Orchid) Som. Macmillan; Cumb. B & H. 15. *Pulmonaria longifolia* (Narrow-leaved Lungwort) Hants. Cope. 16. *Stellaria graminea* (Lesser Stitchwort) Dor. Macmillan. 17. *Stellaria holostea* (Greater Stitchwort) Wilts. D & G. 18. *Verbascum thapsus* (Mullein) Dor, Wilts. Macmillan.

SNAKE'S FOOD 1. *Allium ursinum* (Ramsons) Som. Macmillan. 2. *Arum maculatum* fruit (Cuckoo-pint) Dev. B & H. 3. *Geranium robertianum* (Herb Robert) Dor. PLNN29. 4. *Iris foetidissima* fruit (Foetid Iris) Dev. Macmillan. 5. *Linaria vulgaris* (Yellow Toadflax) Dor. Macmillan. 6. *Mercurialis perennis* (Dog's Mercury) Dor. Macmillan. 7. *Petasites hybridus* (Butterbur) Som. Macmillan. 8. *Solanum dulcamara* (Woody Nightshade) Som. Macmillan. 9. *Symphytum officinale* (Comfrey) Som. Macmillan. 10. *Tamus communis* (Black Bryony) Som. Macmillan. Usually it is the berries that give rise to this name - they may be poisonous, or merely brightly coloured, and so suspicious to country people's eyes.

SNAKE'S FOOT *Polygonum bistorta* (Bistort) Som. Macmillan. Writhed roots, in this case.

SNAKE'S GRASS 1. *Achillea millefolium* (Yarrow) Dor. Macmillan. 2. *Myosotis arvensis* (Field Forget-me-not) Yks. Grigson. 3. *Stellaria holostea* (Greater Stitchwort) Hants. Grigson.

SNAKE'S HEAD 1. *Potentilla erecta* (Tormentil) Wilts. Dartnell & Goddard. 2. *Verbascum thapsus* (Mullein) Dor. Grigson.

SNAKE'S HEAD, Pyrenean *Fritillaria pyrenaica*.

SNAKE'S HEAD IRIS *Hermodactylus tuberosus*. From the colour of the flowers, which are greenish-black. That also accounts for Widow Iris, another name.

SNAKE'S HEAD LILY *Fritillaria meleagris*. Both shape and colour contribute to this, though the main descriptive element is the flower bud.

SNAKE'S MEAT 1. *Arum maculatum* fruit (Cuckoo-pint berries) Glos. B & H. 2. *Heracleum sphondyllium* (Hogweed) Dev. Macmillan. 3. *Iris foetidissima* (Foetid Iris) Dev. Friend.1882. 4. *Mercurialis perennis* (Dog's Mercury) Som. Grigson. 5. *Prunella vulgaris* (Self-heal) Dev. Macmillan. 6. *Solanum dulcamara* (Woody Nightshade) Som. Grigson. 7. *Tamus communis* (Black Bryony) Dev. Friend.1882. Self-heal is the odd one out in this list. The rest qualify for the name by being poisonous, or at least suspicious enough to be regarded as such.

SNAKE'S PIPE *Equisetum arvense* (Common Horsetail) Som. Macmillan. Cf. Toad's Pipe, and Pipeweed.

SNAKE'S PLANT 1. *Allium ursinum* (Ramsons) Som. Macmillan. 2. *Sansevieria trifasciata* (Mother-in-law's Tongue) USA. Cunningham.

SNAKE'S POISON *Iris foetidissima* (Foetid Iris) Dev. B & H.

SNAKE'S POISON-FOOD *Solanum dulcamara* (Woody Nightshade) Som. Grigson.

SNAKE'S RHUBARB 1. *Arctium lappa* (Burdock) Dor. Macmillan. 2. *Petasites hybridus* (Butterbur) Dor. Macmillan. 'Rhubarb' in each case because of the large leaves, large enough for snakes to hide under.

SNAKE'S SKIN WILLOW *Salix triandra* (French Willow) Wilts. Dartnell & Goddard. This is a tree that sheds its bark, hence this name.

SNAKE'S TONGUE 1. *Erythronium americanum* (Yellow Adder's Tongue) Mitton. 2. *Ranunculus flammula* (Lesser Spearwort) Berw. Grigson.

SNAKE'S VICTUALS 1. *Arum maculatum* fruit (Cuckoo-pint berries) Wilts. Jefferies; Glos. B & H. 2. *Mercurialis perennis* (Dog's Mercury) Suss. Grigson.

SNAKE'S VIOLET *Viola reichenbachiana* (Wood Dog Violet) Dor. Macmillan.

SNAKEBERRY 1. *Actaea rubra* (Red Baneberry) H & P. 2. *Bryonia dioica* (White

Bryony) Som. Notes & Queries, vol 168. 3. *Maianthemum dilatatum* (Wild Lily-of-the-valley) USA. E Gunther. 4. *Solanum dulcamara* (Woody Nightshade) Suff. Grigson.

SNAKEBITE *Sanguinaria canadensis* (Bloodroot) Grieve.1931. None of the authorities have any record of its use to treat snake bite, but why else should it be given the name? A mistake, perhaps?

SNAKEHEAD *Chelone glabra* (Turtlehead) USA House. Descriptive.

SNAKEROOT 1. *Actaea alba* (White Baneberry) USA House. 2. *Cicuta maculata* (American Cowbane) USA Kingsbury.1967. 3. *Eupatorium urticaefolium* (White Snakeroot) USA Bergen. 4. *Polygonum bistorta* (Bistort) Clapham. 5. *Rauwolfia serpentina*. 6. *Sanicula europaea* (Sanicle) Thomson.1978.

SNAKEROOT, Black 1. *Cimicifuga racemosa*. 2. *Sanicula marilandica*.

SNAKEROOT, Button 1. *Eryngium aquaticum*. 2. *Eryngium yuccifolium*. 3. *Liatris spicata*.

SNAKEROOT, Button, Dotted *Liatris punctata*.

SNAKEROOT, Canadian *Asarum canadense*.

SNAKEROOT, Clustered *Sanicula gregaria*.

SNAKEROOT, European *Aristolochia clematitis* (Birthwort) Thomson.1978.

SNAKEROOT, Heart *Asarum virginicum*. 'Heart' in this case is a reference to the shape of the leaf.

SNAKEROOT, Sampson's *Psoralea pedunculata* S.USA Puckett.

SNAKEROOT, Senega *Polygala senega*.

SNAKEROOT, Texas *Aristolochia reticulata*.

SNAKEROOT, Virginian *Aristolochia serpentaria*.

SNAKEROOT, White 1. *Eupatorium occidentale purpureum* (Purple Thoroughwort). Elmore. 2. *Eupatorium urticaefolium*.

SNAKES-AND-ADDERS 1. *Anemone nemerosa* (Wood Anemone) Som. Macmillan. 2. *Ophrys apifera* (Bee Orchid) Som. Macmillan.

SNAKETONGUE CROWFOOT *Ranunculus ophioglossifolius* (Serpent-tongue Spearwort).

SNAKEWEED 1. *Cicuta maculata* (American Cowbane) USA Kingsbury.1967. 2. *Conium maculatum* (Hemlock) USA Cunningham. 3. *Euphorbia hirta* USA Watt. 4. *Fagopyrum esculentum* (Buckwheat) B & H. 5. *Galium aparine* (Goose-grass) Som. Macmillan. 6. *Gutierrezia filifolia*. 7. *Mercurialis perennis* (Dog's Mercury) Shrop. Grigson. 8. *Plantago major* (Great Plantain) USA Le Strange. 9. *Polygonum aviculare* (Knotgrass) Som. Macmillan. 10. *Polygonum bistorta* (Bistort) Gerard. 11. *Stellaria holostea* (Greater Stitchwort) Som. Macmillan. 12. *Tamus communis* (Black Bryony) Dev. Macmillan.

SNAKEWEED, Broom; SNAKEWEED, Perennial *Gutierrezia microcephala* (Broomweed) USA Weiner (Broom).

SNAKEWEED, Sampson *Gentiana ochroleuca*.

SNAKEWOOD *Strychnos nux-vomica*.

SNAKEWORT *Artemisia bracteata* Sofowora.

SNAP, Horse *Centaurea nigra* (Knapweed) Dev. B & H.

SNAP, Lion's 1. *Antirrhinum majus* (Snapdragon) Folkard. 2. *Galeobdalon luteum* (Yellow Archangel) Som. Macmillan. 3. *Lamium amplexicaule* (Henbit Deadnettle) W Miller. In each case, it is the mouth-like appearance of the flowers that is responsible for the name.

SNAP-LION *Antirrhinum majus* (Snapdragon) Tynan & Maitland. Cf. Lion's Snap.

SNAP WILLOW *Salix fragilis* (Crack Willow) Wilts. Dartnell & Goddard; Kent Grigson. A slight pressure at the base of a twig will separate it from the branch with quite a loud cracking sound.

SNAPCRACKER *Stellaria holostea* (Greater Stitchwort) Ess. Grigson. One of the many 'snap' or 'pop' names for this plant, either because they tend to break off at the joints, or because they burst their seed-vessels with a pop.

SNAPDRAGON 1. *Antirrhinum majus*. 2. *Aquilegia vulgaris* (Columbine) Dev. Friend.1882; USA. Hutchinson. 3. *Cymbalaria muralis* (Ivy-leaved Toadflax) Som. Macmillan. 4. *Digitalis purpurea* (Foxglove) Dev. Friend.1882. 5. *Fumaria officinalis* (Fumitory) N'thants. Grigson. 6. *Linaria vulgaris* (Yellow Toadflax) Dev, Ches, Yks. B & H. 7. *Misopates orontium* (Lesser Snapdragon) Dev. Friend.1882. Cf. Snapjack for foxglove.

SNAPDRAGON, Corn *Misopates orontium* (Lesser Snapdragon) Salisbury.

SNAPDRAGON, Creeping *Asarina procumbens*.

SNAPDRAGON, Lesser; SNAPDRAGON, Small *Misopates orontium* Lesser Snapdragon is the usual name; Small Snapdragon is used by Curtis.

SNAPDRAGON, Soft *Antirrhinum molle*.

SNAPDRAGON, Twining *Antirrhinum filipes*.

SNAPDRAGON, Wild 1. *Chamaenerion angustifolium* (Rosebay Willowherb) Glos. Grigson. 2. *Linaria vulgaris* (Yellow Toadflax) Wilts. D Grose; Som, Oxf, Lincs. Grigson.

SNAPJACK 1. *Anagallis arvensis* (Scarlet Pimpernel) Som. Macmillan. 2. *Antirrhinum majus* (Snapdragon) Som. Macmillan. 3. *Digitalis purpurea* (Foxglove) Som. Macmillan. 4. *Geranium robertianum* (Herb Robert) Som. Macmillan. 5. *Linaria vulgaris* (Yellow Toadflax) Som. Macmillan. 6. *Melandrium album* (White Campion) Som. Macmillan. 7. *Stellaria holostea* (Greater Stitchwort) Dev, Som. Elworthy; Dor. Dacombe; Wilts. D & G. Either a variation of 'mouth' names, as with Snapdragon and Toadflax, or a comment on the way some plants snap off at the joints, as with Stitchwort and Campion. The odd one appears to be foxglove, where 'snap' is the same as 'pop' - children squeeze the unopened buds, to hear them pop.

SNAPJACK, White *Silene maritima* (Sea Campion) Som. Macmillan.

SNAPPER-FLOWER *Stellaria holostea* (Greater Stitchwort) Som. Grigson.

SNAPPERS 1. *Silene cucubalis* (Bladder Campion) Suss, Kent Grigson. 2. *Stellaria graminea* (Lesser Stitchwort) Glos. PLNN35. 3. *Stellaria holostea* (Greater Stitchwort) Dev. Fernie; Heref. Leather.

SNAPS 1. *Digitalis purpurea* (Foxglove) Som. Elworthy. 2. *Linaria vulgaris* (Yellow Toadflax) Som. Grigson. 3. *Stellaria holostea* (Greater Stitchwort) Wilts. Dartnell & Goddard. See Snapjacks.

SNAPS, Brandy; SNAPS, Jack *Stellaria holostea* (Greater Stitchwort) Suss. Grigson (Brandy); Som. Macmillan (Jack).

SNAPSEN *Populus tremula* (Aspen) IOW Grigson. Consonantal misplacement accounts for the 'apsen' part of the name, which is a hybrid, possibly connected with 'rattling' epithets, such as Lyte's Rattling Asp.

SNAPSTALKS; SNAPWORT *Stellaria holostea* (Greater Stitchwort) Ches. Holland (Snapstalks); Kent Grigson (Snapwort). See Snapjacks.

SNAPWEED *Impatiens capensis* (Orange Balsam) USA Grigson.

SNARES *Galium aparine* (Goose-grass) Yks. Grigson.

SNATBERRY *Taxus baccata* fruit (Yew berries) N'thants. Sternberg. All the names for yew berries are on the same theme - Cf. Snots, Snotty Gogs, etc.

SNAUPER *Digitalis purpurea* (Foxglove) Glos. J D Robertson. Presumably allied to the 'snap' names, like Snapdragon, Snapjack, etc.

SNEEZEWEED *Helenium autumnale*. The American Indians used to dry the flower heads and keep them in a loose bunch, hung from the rafters of the house. When wanted for use, they are pulverized and used as snuff - it makes one sneeze violently several times, and so can loosen up a cold in the head.

SNEEZEWEED, Fine-leaved *Helenium tenuifolium*.

SNEEZEWEED, Orange *Helenium hoopesii.*
SNEEZEWORT 1. *Achillea millefolium* (Yarrow) Glos. J D Robertson. 2. *Achillea ptarmica.* 3. *Parietaria diffusa* (Pellitory-of-the-wall) Genders.1972. 4. *Veratrum album* (White Hellebore) Kourennoff. Ground ivy, camomile and pellitory were the principal ingredients of Elizabethan snuffs, and about the same time, Gerard said that White Hellebore got the name because "the powder of the roots is used to procure neesing". But the great sneezing plant is *Achillea ptarmica.* The powdered leaves of this, and to a lesser extent, yarrow, were once in common use to prucure sneezing, either medicinally, or as a cheap snuff substitute. The French equivalent is herbe à éternuer; the specific name tells the same story, for ptarmica is from Greek stamos, sneezing.
SNEEZINGS *Achillea millefolium* (Yarrow) Glos. Grigson.
SNEGS *Prunus spinosa* fruit (Sloes) Dor. Dacombe. A slight variation on the more widespread Snags.
SNICK-NEEDLES, Mother Shimble's *Stellaria holostea* (Greater Stitchwort) Wilts. Dartnell & Goddard. Cf. Mother's Thimble, and Lady's Thimble.
SNITCHBACK *Cyclamen europaeum* (Sowbread) Dor. Macmillan.
SNOB, Green 1. *Oxalis acetosella* (Wood Sorrel) War. Grigson. 2. *Rumex acetosa* (Wild Sorrel) War. Grigson.
SNODER GILL *Taxus baccata* fruit (Yew berries) Hants. Cope. See Snot Berry.
SNOD-GOGGLES *Taxus baccata* fruit (Yew berries) Ess. Gepp. See Snotgoggles.
SNOT, Red *Taxus baccata* fruit (Yew berries) Kent. PLNN 31. Red berries, and the 'snot' is what is inside them.
SNOT BERRY; SNOTS; SNOTTER GALL *Taxus baccata* fruit (Yew berries) Bucks. Harman; N'thants. Grigson (Snot Berry); Som. Macmillan (Snots); Wilts. Dartnell & Goddard (Snotter Gall). All the local names for yew berries are variations on the 'snot' theme, and comments on the mucilaginous covering of the seeds.
SNOTTERBERRIES 1. *Symphoricarpos rivularis* (Snowberry) Som. Macmillan. 2. *Taxus baccata* fruit (Yew berries) Som. Macmillan.
SNOTTERGOBS; SNOTGOBBLES; SNOTTLE BERRY; SNOTTY GOG *Taxus baccata* fruit (Yew berries) N'thants. Harman (Snottergobs); Beds. PLNN 31 (Snotgobbles); Yks. Grigson (Snottle Berry); Wilts. Macmillan; Suss. Parish (Snotty Gog).
SNOTTYGOBBLES *Taxus baccata* fruit (Yew berries) Som. Macmillan.
SNOUT, Calf's 1. *Antirrhinum majus* (Snapdragon) Tynan & Maitland. 2. *Misopates orontium* (Lesser Snapdragon) Turner.
SNOUT, Calf's, Broad *Antirrhinum majus* (Snapdragon) Turner.
SNOUT, Dog *Antirrhinum majus* (Snapdragon) Norf. B & H.
SNOUT, Pig's *Antirrhinum majus* (Snapdragon) Som. Macmillan.
SNOUT, Swine's *Taraxacum officinale* (Dandelion) B & H.
SNOUT, Turkey's *Amaranthus caudatus* (Love-lies-bleeding) Som. Macmillan. Very descriptive.
SNOUT, Weasel's 1. *Galeobdalon luteum* (Yellow Archangel) Som. Elworthy. Glos. J D Robertson; Yks. Grigson. 2. *Linaria vulgaris* (Yellow Toadflax) Kent Grigson. 3. *Misopates orontium* (Lesser Snapdragon).
SNOW *Stellaria holostea* (Greater Stitchwort) Suss. Grigson.
SNOW, Drops-of- *Anemone nemerosa* (Wood Anemone) Suss. B & H.
SNOW, Evening *Gilia dichotoma.*
SNOW, Glory-of-the- *Chionodoxa sp.*
SNOW, Mountain 1. *Arabis caucasica* (Garden Arabis) W Miller. 2. *Euphorbia marginata.*
SNOW BUTTERCUP *Ranunculus glacialis.*
SNOW CINQUEFOIL *Potentilla nivea.*

SNOW-CUPS, Water *Ranunculus aquatilis* (Water Crowfoot). W Miller.
SNOW-DRIFT *Lobularia maritima* (Sweet Alison) Dev. Friend.1882.
SNOW GENTIAN *Gentiana nivalis* (Alpine Gentian) Polunin.
SNOW GUM *Eucalyptus pauciflora* (Cabbage Gum) Hora.
SNOW GUM, Australian *Eucalyptus niphophila*.
SNOW GUM, Tasmanian *Eucalyptus coccifera* (Mount Wellington Peppermint) Wilkinson.1981.
SNOW-IN-HARVEST 1. *Arabis caucasica* (Garden Arabis) Som. Macmillan. 2. *Cerastium tomentosum* (Snow-in-summer) Som. Elworthy; Wilts. Dartnell & Goddard; Leics. B & H. 3. *Clematis vitalba* (Old Man's Beard) Som. Macmillan; N'thants. Grigson. 4. *Lobularia maritima* (Sweet Alison) Bucks. Harman; N'thants. B & H.
SNOW-IN-SUMMER 1. *Arabis caucasica* (Garden Arabis) Hay & Synge. 2. *Cerastium tomentosum*. 3. *Euphorbia marginata* (Mountain Snow) RHS.
SNOW-ON-THE-MOUNTAIN 1. *Arabis alpina* (Alpine Rock-cress) Glos, Heref, Suss. B & H. 2. *Cerastium tomentosum* (Snow-in-Summer) Som. B & H. 3. *Euphorbia marginata* (Mountain Snow) USA Tolstead. 4. *Lobularia maritima* (Sweet Alison) Dev. Friend. 5. *Saxifraga granulata* (Meadow Saxifrage) Wilts. Dartnell & Goddard. 6. *Stellaria holostea* (Greater Stitchwort) Wilts. Macmillan.
SNOW PEAR *Pyrus nivalis*.
SNOW PIERCER *Galanthus nivalis* (Snowdrop) Som. Macmillan. Cf. Pierce-snow, from the same area.
SNOW-PLANT *Cerastium tomentosum* (Snow-in-summer) W Miller.
SNOW POPPY *Eonecon chionantha* (Cyclamen Poppy) A W Smith.
SNOW TOSS *Viburnum opulus* (Guelder Rose) Som. Macmillan. 'Toss' here is the same as Tisty-tosty, or May Tosty, as with cowslips. Children used to make flower balls of them and throw them to and fro in a set game pattern.
SNOW TRILLIUM *Trillium nivale*.
SNOWBALL 1. *Taraxacum officinale* seed heads (Dandelion seed heads). Tynan & Maitland. 2. *Symphoricarpos rivularis* (Snowberry) Dev. Friend.1882. 3. *Viburnum opulus* (Guelder Rose) Halliwell.
SNOWBALL, Wild *Ceanothus americanus* (New Jersey Tea) Grieve.1931.
SNOWBALL ROSE *Viburnum opulus* (Guelder Rose) Suss. Parish.
SNOWBALL TREE *Viburnum opulus var. sterilis*.
SNOWBALL TREE, Japanese *Viburnum tomentosum*.
SNOWBAW *Viburnum opulus* (Guelder Rose) Ches. Holland.
SNOWBELL, Alpine *Soldanella alpina*.
SNOWBELL, Dwarf *Soldanella pusilla*.
SNOWBELLS 1. *Allium triquetrum* (Three-cornered Leek) Grigson. 2. *Galanthus nivalis* (Snowdrop) Som. Macmillan. 3. *Styrax officinalis* (Snowdrop Bush) USA Hay & Synge.
SNOWBERRY 1. *Gaultheria antipoda*. 2. *Symphorcarpos orbiculatus* (Red Wolfberry) USA Elmore. 3. *Symphoricarpos rivularis*.
SNOWBERRY, Creeping *Gaultheria hispidula* (Moxie Plum) USA Yarnell.
SNOWBERRY, Long-flowered *Symphorcarpos longiflorus*.
SNOWDON ROSE *Sedum roseum* (Rose-root) Wales Grigson.
SNOWDROP 1. *Fritillaria meleagris* (Snake's-head Lily) Hants. Grigson. 2. *Galanthus nivalis*. 3. *Ornithogalum umbellatum* (Star-of-Bethlehem) USA. Kingsbury. 'Drop' usually refers to an ear ornament, so this would be 'snowy drop'. There seems to have been doubt in Hampshire as to what Snake's-head Lily really was, for besides snowdrop it was also known as cowslip there.
SNOWDROP, Barbados *Zephyranthes tubispatha*.
SNOWDROP, French *Ornithogalum umbellatum* (Star of Bethlehem) Baker.

SNOWDROP, Mountain; SNOWDROP, Summer *Leucojum aestivum* (Summer Snowflake) War. Grigson.
SNOWDROP, Yellow *Erythronium americanum* (Yellow Adder's Tongue) Grieve.1931.
SNOWDROP-ANEMONE *Anemone sylvestris* (Snowdrop Windflower) A W Smith.
SNOWDROP-BERRY *Symphoricarpos rivularis* (Snowberry) Ches. B & H.
SNOWDROP-BUSH *Styrax officinalis*.
SNOWDROP-TREE 1. *Chionanthus virginicus* (Fringe Tree) Thomson.1978. 2. *Halesia carolina*.
SNOWDROP-TREE, Chinese *Chimonanthus fragrans* (Wintersweet) McDonald.
SNOWDROP TREE, Mountain *Halesia monticola* (Mountain Silverbell) Hora.
SNOWDROP-WINDFLOWER *Anemone sylvestris*.
SNOWDROPPER *Galanthus nivalis* (Snowdrop) Glos. J D Robertson.
SNOWFLAKE 1. *Crataegus monogyna* flowers (Hawthorn flowers) Som. Macmillan. 2. *Lamium album* (White Deadnettle) Allan. 3. *Ornithogalum umbellatum* (Star-of-Bethlehem) Dev. Friend.1882. 4. *Stellaria holostea* (Greater Stitchwort) Suss. Grigson. 5. *Viburnum opulus* (Guelder Rose) Wilts. Macmillan.
SNOWFLAKE, Autumn *Leucojum autumnale*.
SNOWFLAKE, Spring *Leucojum vernum*.
SNOWFLAKE, Summer *Leucojum aestivum*.
SNOWFLOWER 1. *Chionanthus virginicus* (Fringe Tree) Leyel.1948. 2. *Galanthus nivalis* (Snowdrop) B & H.
SNOWY MESPILUS 1. *Amelanchier arborea* (Tree Amelanchier) J Smith. 2. *Amelanchier ovalis*.
SNOXUMS *Digitalis purpurea* (Foxglove) Glos. J D Robertson. The West country word 'snock' means a knock or a smart blow, and is a reference to the children's pastime of inflating and popping foxglove flowers and buds.
SNUFF-CANDLE *Galeobdalon luteum* (Yellow Archangel) Wilts. Dartnell & Goddard.1899. It must be the profile view of the flowers that suggests this.
SNUFFBOX BEAN *Entada phaseoloides* (Nicker Bean) Chopra. The seeds are hollowed out, and made into small boxes.
SOAP, Black 1. *Centaurea nigra* (Knapweed) Dev, Glos. Friend.1882. 2. *Knautia arvensis* (Field Scabious) Dev. Friend.1882.
SOAP, Crow; SOAP, Crowther *Saponaria officinalis* (Soapwort) Cockayne (Crow); Le Strange (Crowther). See Soapwort.
SOAP, Fairy *Polygala vulgaris* (Milkwort) Donegal Grigson. The belief was that the fairies made a lather from the roots and leaves.
SOAP, Ground *Saponaria officinalis* (Soapwort) Cockayne. See Soapwort.
SOAP-BARK *Acacia instia*. The bark is used in India as soap.
SOAP-NUT TREE 1. *Sapindus mukorossi*. 2. *Sapindus saponaria* (Soapberry) Tampion. 3. *Sapindus trifoliatus*. The fruits of *Sapindus mukorossi* are still used for washing fabrics in India, and by women as a hair shampoo. It is the pulp of the fruits of all the species that can lather with water, and so can be used as a soap substitute.
SOAP-PLANT *Chlorogalum pomeridianum*. In this case, it is the root that gives a soapy lather.
SOAP POD *Acacia concinna*. The pods of this species are used in India for washing fabrics, and the hair - even sometimes for cleaning silver.
SOAPBARK TREE *Pithecellobium bigemimum*.
SOAPBERRY 1. *Sapindus saponaria*. 2. *Shepherdia canadensis*. The fruit pulp of *Sapindus saponaria* lathers with water, and can be used as a soap substitute, much used in Mexico for washing clothes.

SOAPBERRY, Drummond's *Sapindus drummondii.*

SOAPBERRY TREE *Balanites aegyptiaca* (Jericho Balsam) Dalziel. The fruit pulp of a related species (*B roxburghii*) is used in India for cleaning silk, and presumably this species is similarly used.

SOAPROOT 1. *Chlorogalum pomeridianum* (Soap-Plant) Schery. 2. *Saponaria officinalis* (Soapwort) Grieve.1931. 3. *Yucca glauca* (Soapweed) USA Weiner. Literally correct in each case.

SOAPROOT, Narrow-leaved *Yucca glauca* (Soapweed) USA Elmore.

SOAPROOT, White *Gypsophila arrostii.* The root contains large amounts of saponin.

SOAPWEED 1. *Salsola kali* (Saltwort) Pomet. 2. *Yucca baccata* (Datil) USA Elmore. 3. *Yucca glauca.* Saltwort probably earned the name by its soda content, but *Yucca glauca* is the real soapweed, used by the American Indians by stripping off the root bark, and pounding the rest in cold water to produce suds. Blankets were washed in this way; in fact, the Navajo, in washing wool, prefer yucca roots, because there is no grease or fatty substance in it. As with *Yucca baccata*, there is a special use as a hair shampoo.

SOAPWEED, Narrow-leaved *Yucca glauca* (Soapwed) USA Elmore.

SOAPWORT *Saponaria officinalis.* A lather can be got by rubbing Soapwort leaves in water, and it has been used thus since Greek and Roman times. It is valuable for cleaning and restoring old tapestry, without damaging the fabric.

SOAPWORT, Bladder *Saponaria officinalis* (Soapwort) S Africa Watt.

SOAPWORT, Cow 1. *Saponaria officinalis* (Soapwort) S Africa Watt. 2. *Vaccaria pyramidata* (Cow-herb) USA T H Kearney.

SOAPWORT, Rock *Saponaria ocymoides.*

SOAPWORT, Spoon-leaved *Saponaria bellidifolia.*

SOAPWORT, Yellow *Saponaria lutea.*

SOAPWORT GENTIAN *Saponaria officinalis* (Soapwort) Lyte.

SOBS, Sour *Oxalis pes-caprae* (Bermuda Buttercup) Australia Watt. The name is meant to convey 'sorrel'. Cf. Sour Sab or Sap for Wood Sorrel.

SOCKS, Owl's *Sarracenia purpurea* (Pitcher Plant) Canada Jenness.1935.

SOCKS, Shoes-and- *Aquilegia vulgaris* (Columbine) Som. Macmillan.

SOCKS, Water *Nymphaea alba* (White Waterlily) B & H.

SOD APPLE *Epilobium hirsutum* (Great Willowherb) Som, Wilts. Jefferies. Boiled apples, in other words. Coddled has the same meaning. It is said that the plant smells like apples when it is crushed, but the origin is more like to be "codded" rather than "coddled". As Gerard said, "the flower groweth at the top of the stalke, comming out of the end of a small longe codde".

SODA APPLE *Solanum aculeastrum* USA Kingsbury. 'Soda' is 'Sodom' - see below.

SODA PLANT *Salsola kali* (Saltwort) Prior. Cf. Sowdwort. It is a common source of soda.

SODGE, Sour; SOG, Sour *Rumex acetosa* (Wild Sorrel) both Bucks. Heather (Sodge); Harman (Sog). Cf. the West Country Sour Sabs, or Sour Suds. The 'sour' part of the name is the same as sorrel.

SODOM, Apple of 1. *Atropa belladonna* (Deadly Nightshade) Emboden. 2. *Solanum aculeastrum* S Africa Watt. 3. *Solanum incanum* (Hoary Nightshade) Moldenke. 4. *Solanum panduraeforme* S Africa Watt. 5. *Solanum sodomeum. S sodomeum* is a Mediterranean species, known also as the Dead Sea Apple. So 'Sodom' is merely a Palestinian epithet. But the name does imply a poisonous fruit.

SODOM, Bells of, Drooping; SODOM, Bells of, Mournful; SODOM, Bells of, Solemn *Fritillaria meleagris* (Snake's-head Lily) all Dor. Grigson (Drooping); Dacombe (Mournful); Macmillan (Solemn). Drooping flowers, with sombre colouring, are enough to account for the name.

SOE THISTLE *Sonchus oleraceus* (Sow Thistle) S Africa Watt.

SOFT ELM *Ulmus americanus* timber (American Elm) Wilkinson.1978.
SOFT-GRASS, Creeping *Holcus mollis*.
SOFT-GRASS, Meadow *Holcus lanatus* (Yorkshire Fog) Allan.
SOFT MAPLE *Acer rubrum* (Red Maple) Brimble.1948.
SOL-FLOWER *Helianthemum chamaecistus* (Rock Rose) Mor. B & H. 'Sol' - 'sun'. Cf. Sun-rose or Sunflower for this.
SOLANDINE *Chelidonium majus* (Greater Celandine) Ire. Grigson.
SOLDIER-AND-HIS-WIFE *Pulmonaria officinalis* (Lungwort) IOW Grigson. A double name like this is usually a comment on the bi-coloured flowers.
SOLDIER BOYS *Centranthus ruber* (Spur Valerian) Som. Macmillan. Because of the red flowers, presumably.
SOLDIER ORCHID *Orchis militaris*.
SOLDIER'S BUTTONS 1. *Anemone nemerosa* (Wood Anemone) Som. Macmillan. 2. *Aquilegia vulgaris* (Columbine) Som, Wilts. Macmillan. 3. *Arctium lappa* (Burdock) Som. Macmillan; Wilts. Dartnell & Goddard. 4. *Caltha palustris* (Marsh Marigold) Som. Macmillan. 5. *Galium aparine* (Goose-grass) Cumb. Grigson. 6. *Geranium robertianum* (Herb Robert) Bucks. Grigson. 7. *Geum rivale* (Water Avens) Wilts. Macmillan. 8. *Helianthemum chamaecistus* (Rock Rose) Som. Macmillan. 9. *Knautia arvensis* (Field Scabious) Som. Macmillan. 10. *Melandrium rubrum* (Red Campion) Yks. Grigson. 11. *Nymphaea alba* (White Waterlily) Som. Macmillan. 12. *Nuphar lutea* (Yellow Waterlily) Som. Grigson. 13. *Prunella vulgaris* (Self-heal) Som. Macmillan. 14. *Ranunculus acris* (Meadow Buttercup) Som. Macmillan. 15. *Ranunculus bulbosus* (Bulbous Buttercup) Som. Grigson. 16. *Ranunculus repens* (Creeping Buttercup) Som. Grigson. 17. *Rosa pimpinellifolia* (Burnet Rose) Kirk. Grigson. 18. *Stellaria holostea* (Greater Stitchwort) Bucks. Harman.

SOLDIER'S CAP 1. *Aconitum anglicum* (Wild Monkshood) N'thants. B & H. 2. *Aconitum napellus* (Monkshood) N'thants. A E Baker. 3. *Orchis mascula* (Early Purple Orchid) Som. Macmillan.
SOLDIER'S CROSS *Cheiranthus cheiri* (Wallflower) Som. Macmillan. Cf. Cross-flower, Crucifix-flower, etc. For wallflower is a member of the Cruciferae.
SOLDIER'S CULLIONS *Orchis militaris* (Soldier Orchid) Gerard. 'Cullions' means 'testicles' - the twin tubers provide the reference.
SOLDIER'S DRINKING CUP *Sarracenia purpurea* (Pitcher Plant) Perry.
SOLDIER'S FEATHERS 1. *Amaranthus caudatus* (Love-lies-bleeding) Som. Macmillan. 2. *Syringa vulgaris* (Lilac) Som. Tynan & Maitland.
SOLDIER'S JACKET *Orchis mascula* (Early Purple Orchid) Dor. B & H.
SOLDIER'S PRIDE *Centranthus ruber* (Spur Valerian) Salisbury.
SOLDIER'S RIBWORT; SOLDIER'S TAPPIE *Plantago lanceolata* (Ribwort Plantain) Le Strange (Ribwort); Ayr. Grigson (Tappie). In this case, the reference is to a game played by children, something like 'conkers', in which each stalk of Ribwort is called a "soldier". Cf. Soldiers, as well as a number of names like Hardheads, Fighting Cocks, etc.
SOLDIER'S TEARS *Verbascum thapsus* (Mullein) Dor. Dacombe.
SOLDIER'S WOUNDWORT *Achillea millefolium* (Yarrow) Friend; USA Henkel. Yarrow is a wound herb, carried, so it is traditionally claimed, by the Greek and Roman armies.
SOLDIER'S YARROW *Stratiotes aloides* (Water Soldier) W Miller.
SOLDIERS 1. *Anemone nemerosa* (Wood Anemone) Bucks. B & H. 2. *Arum maculatum* (Cuckoo-pint) Som. Macmillan; Wilts. Goddard. 3. *Dactylorchis maculata* (Heath Spotted Orchid) Dor. Macmillan. 4. *Digitalis purpurea* (Foxglove) Dor. Macmillan. 5. *Geranium robertianum* (Herb

Robert) Dev. Macmillan. 6. *Kniphofia uvaria* (Red-hot-poker) S Africa Watt. 7. *Lythrum salicaria* (Purple Loosestrife) Norf. Grigson. 8. *Melandrium rubrum* (Red Campion) Ches. B & H; Yks. Grigson. 9. *Ophrys apifera* (Bee Orchid) Som. Macmillan. 10. *Orchis mascula* (Early Purple Orchid) Dor. Dacombe. 11. *Papaver rhoeas* (Red Poppy) Wilts. Dartnell & Goddard; N'thants. A E Baker; Norf. Grigson. 12. *Plantago lanceolata* (Ribwort Plantain) Som. Elworthy; Notts, Ches, Cumb, Scot. Denham/Hardy. 13. *Rosa canina* fruit (Hips) Kent Grigson. 14. *Rumex acetosa* (Wild Sorrel) Som. Mabey. 15. *Trifolium incarnatum* (Crimson Clover) Som. Macmillan. 16. *Verbascum thapsus* (Mullein) USA Sanford. Mostly red or upstanding, or both. The exceptions are Ribwort Plantain, where the reference is a game (see Soldier's Ribwort), Bee Orchid, in which case the name has simply been transferred from the other native orchids that qualify for the name, and Wood Anemone, completely out of the reckoning. Cf. White Soldiers, from the same county.

SOLDIERS, Dolly *Geranium molle* (Dove's-foot Cranesbill) Som. Macmillan.

SOLDIERS, French-and-English *Plantago lanceolata* (Ribwort Plantain) Tynan & Maitland. Another example from the children's game. In this case, the adversaries are named.

SOLDIERS, Freshwater *Stratiotes aloides* (Water Soldier) Grieve.1931.

SOLDIERS, Gallant *Galinsoga parviflora* (Kew-weed) London Grigson. The whole name is a garbled version of the generic name, Galinsoga.

SOLDIERS, Jolly *Orchis mascula* (Early Purple Orchid) Dev. Macmillan.

SOLDIERS, Primrose *Aquilegia vulgaris* (Columbine) Wilts. Dartnell & Goddard.

SOLDIERS, Red 1. *Melandrium rubrum* (Red Campion) N Eng. Grigson. 2. *Papaver rhoeas* (Red Poppy) Som. Macmillan.

SOLDIERS, Shaggy *Galinsoga ciliata*.

SOLDIERS, Water *Stratiotes aloides*.

SOLDIERS, White *Anemone nemerosa* (Wood Anemone) Bucks. B & H.

SOLDIERS-AND-ANGELS *Arum maculatum* (Cuckoo-pint) Dev. Macmillan. A double name of the 'Lords-and-ladies' type.

SOLDIERS-AND-SAILORS 1. *Arum maculatum* (Cuckoo-pint) Som. Macmillan. 2. *Echium vulgare* (Viper's Bugloss) Leyel.1937. 3. *Iris pseudo-acarus* (Yellow Flag) Dor. Macmillan. 4. *Pulmonaria officinalis* (Lungwort) Som. Macmillan; Suff. Grigson. In the case of Viper's Bugloss and Lungwort, this is merely a double name, one of many, to celebrate the two-coloured flowers. Cuckoo-pint, too, has many double names, best known of all being Lords-and-ladies.

SOLDIERS-OF-THE-QUEEN *Galinsoga parviflora* (Kew-weed) London Grigson. Cf. Gallant Soldiers, which is a corruption of Galinsoga.

SOLDIERS-SAILORS-TINKERS-TAILORS *Lolium perenne* (Rye-grass) Wilts. Dartnell & Goddard. From the children's game, of the Love-me, Love-me-not variety, played by pulling off the alternating spikelets. Cf. Tinker-tailor Grass, or Yes-or-no, etc.

SOLEMN BELLS OF SODOM *Fritillaria meleagris* (Snake's-head Fritillary) Dor. Macmillan. Cf. Drooping Bells of Sodom, etc., all because of the bell-shaped flowers and their sombre colours.

SOLLANDINE *Chelidonium majus* (Greater Celandine) Ire. Grigson. A spelling to accomodate local pronunciation.

SOLLAR *Salix caprea* (Goat Willow) Northall. 'Sollar' ia a variation of 'sallow', OE sealh, willow.

SOLLERY *Apium graveolens* (Wild Celery) Worcs. J Salisbury. A variation of 'celery'.

SOLLOP *Rumex acetosa* (Wild Sorrel) Bucks. Harman. It seems like a shortening of some form like Sour Sop.

SOLOMON, Fat *Smalicina amplexicaulis* USA Schenk & Gifford.

SOLOMON'S PUZZLE *Sedum telephium*

(Orpine) London Grigson. That used to be a London flower-sellers' name.

SOLOMON'S SALE *Polygonatum multiflorum* (Solomon's Seal) Lancs. Nodal.

SOLOMON'S SEAL 1. *Hypericum calycinum* (Rose of Sharon) N'thants. B & H. 2. *Polygonatum multiflorum*. 3. *Polygonatum commutatum* (Smooth Solomon's Seal) USA Tolstead. If you cut the roots of true Solomon's Seal transversely, scars resembling the device that bears this name can be seen. It is like a 6-pointed star.

SOLOMON'S SEAL, Angular *Polygonatum odoratum*.

SOLOMON'S SEAL, False *Smilacina racemosa*.

SOLOMON'S SEAL, False, Three-leaved *Smilacina trifolia*.

SOLOMON'S SEAL, False, Two-leaved *Maianthemum canadense*. (False Lily-of-the-valley) USA Zenkert.

SOLOMON'S SEAL, Giant; SOLOMON'S SEAL, Great *Polygonatum commutatum* (Smooth Solomon's Seal) both USA House (Giant) Zenkert (Great).

SOLOMON'S SEAL, Hairy *Polygonatum biflorum*.

SOLOMON'S SEAL, Scented *Polygonatum odoratum* (Angular Solomon's Seal) Genders.

SOLOMON'S SEAL, Small *Polygonatum pubescens*.

SOLOMON'S SEAL, Smooth *Polygonatum commutatum*.

SOLOMON'S SEAL, Whorled *Polygonatum verticillatum*.

SOLWHERF *Achillea tomentosa* (Yellow Milfoil) Cockayne.

SOMALI TEA *Catha edulis* (Khat) Dale & Greenway. A "tea" is made from the buds, twigs and fresh leaves, usually known as khat. or quat, or other names like it. It is a native of northeast Africa, hence this name. Cf. also Arabian Tea, Abyssinian Tea.

SON, Stepmother's *Viola tricolor* (Pansy) Yks. Grigson. Cf. Stepmothers, and Stepdaughters, for this.

SON-BEFORE-THE-FATHER 1. *Colchicum autumnale* (Meadow Saffron) B & H. 2. *Epilobium hirsutum* (Great Willowherb) B & H. 3. *Petasites hybridus* (Butterbur) Scot. B & H. 4. *Tussilago farfara* (Coltsfoot) Wilts. Wiltshire; Cumb, Scot. Grigson. Any plant whose flowers appear before the leaves will qualify for this name. But butterbur and the willowherb do not. So what is the thinking behind their inclusion?

SONORA BEEWEED *Cleome sonorae*.

SONORA STINK FLOWER *Cleome sonorae* (Sonora Beeweed) USA Elmore.

SOOKIE *Pedicularis sylvatica* (Lousewort) Shet. Grigson. Such a name implies sweetness - 'honeysuckle', in fact, which is included in the names for lousewort.

SOOKIE, Bee *Pedicularis sylvatica* (Lousewort) Shet. Grigson.

SOOKIE, White *Trifolium repens* (White Clover) Berw. Grigson. Cf. Honeysuckle, Sucklings, etc.

SOOKIE-SOORACKS 1. *Oxalis acetosella* (Wood Sorrel) Caith. Grigson. 2. *Taraxacum officinale* (Dandelion) Inv. PLNN29. Because of the sour, milky juice from the stems of dandelion ('sooracks' will refer to sorrel).

SOOPALALIE *Shepherdia canadensis* (Soapberry) USA Turner & Bell. From the Chinook name for the plant.

SOOR DOCKIN *Rumex acetosa* (Wild Sorrel) Yks. Gutch.1911.

SOORACKS, Sookie *Taraxacum officinale* (Dandelion) Inv. PLNN29.

SOORAG, Sheep *Oxalis acetosella* (Wood Sorrel) Caith. Grigson. 'Soorack' or 'Soorag' is the equivalent of 'sorrel'.

SOORIK *Rumex acetosa* (Wild Sorrel) Shet. Grigson.

SOPHIA, Herb *Descurainia sophia* (Flixweed) USA Zenkert. 'Sophia' - wisdom. There is actually a name for this plant using the idea - Wisdom of Surgeons.

SOPHORA, Four-wing *Sophora tetraptera* (Yellow Kowhai) Hora.

SOPHORA, Silky *Sophora sericea*.

SOPOR *Datura stramonium* (Thorn-apple) Cambs. Porter. Evidently meaning unnaturally deep sleep. Even the scent of the "apples", so they used to say in the Fen country, could produce coma.

SOPS, Sour 1. *Annona muricata* (Guanabana) Barbados Gooding. 2. *Rumex acetosa* (Wild Sorrel) Corn. Jago. *Annona muricata* gets the sorrel name in contrast to *A squamosa*, which is Sweetsop.

SOPS-IN-WINE *Dianthus caryophyllus* (Carnation) Spenser. "Bring of coronations and sops-in-wine, Worn of paramours". It is generally taken to be carnations, but that quotation seems to suggest two different plants. Brand said that sops-in-wine were a species of flower "among the smaller kind of pinks". Anyway, it was used to give a spicy flavour to ale and wine.

SORB, Wild *Sorbus aucuparia* (Rowan).

SORB-APPLE 1. *Sorbus domestica* fruit (True Service Tree) Turner. 2. *Sorbus torminalis* fruit (Wild Service Tree) Gerard. 'Sorb' is straight from the Latin sorbus, the name of the genus, from which Service also derives.

SORB-TREE *Sorbus domestica* (True Service Tree) Gerard.

SORCERER'S GARLIC *Allium moly* (Golden Garlic) Gerard. If this really is the ancient 'moly', then the name is understandable. It was this herb, given to Odysseus by Hermes, that enabled him to restore his crew to their human shape, after they had been turned into pigs by Circe.

SORCERER'S VIOLET *Vinca minor* (Lesser Periwinkle) Friend. Because of its use in charms against the evil eye. Worn in the buttonhole, or carried dried in a sachet, it was a great protection against any witch not carrying it herself.

SORE EYES *Veronica chamaedrys* (Germander Speedwell) Norf. V G Hatfield.1994. The flowers were used in Norfolk to make an eyebath to relieve sore eyes, probably on the basis of the doctrine of signatures.

SORGHUM, Sweet *Sorghum saccharatum* (Sorgo) Schery. Cf. Chinese Sugar Maple. It is cultivated extensively for syrup production.

SORGO *Sorghum saccharatum*.

SORREL 1. *Hibiscus sabdariffa* (Rozelle) Schery. 2. *Oxalis acetosella* (Wood Sorrel) Ches. Grigson. 3. *Rumex acetosella* (Sheep's Sorrel) S Africa Watt. 4. *Rumex scutatus*. From O French surale, itself from a word meaning 'sour'. Rozelle has an acid nature, hence the name.

SORREL, Broad-leaved *Oxalis latifolia* (Mexican Oxalis) Allan.

SORREL, Buckler-shaped *Rumex scutatus* (French Sorrel) Leyel.1937. 'Scutatus' means shaped like a buckler.

SORREL, Claver *Oxalis acetosella* (Wood Sorrel) Langham. 'Claver' - 'clover', from the trefoil leaf.

SORREL, Cock *Rumex acetosa* (Wild Sorrel) Yks. B & H; Hunts. Marshall.

SORREL, Cuckoo('s) 1. *Oxalis acetosella* (Wood Sorrel) Friend. 2. *Rumex acetosa* (Wild Sorrel) Ire. Grigson. 3. *Rumex acetosella* (Sheep's Sorrel) Ire. B & H. Most plants ascribed to the cuckoo are spring-flowering, in other words, when the cuckoo arrives.

SORREL, Dock *Rumex acetosa* (Wild Sorrel) Watt.

SORREL, English *Rumex acetosa* (Wild Sorrel) Hemphill. Presumably to distinguish it from the true sorrel (*Rumex scutatus*), which is certainly not English.

SORREL, Field *Rumex acetosella* (Sheep's Sorrel) Watt.

SORREL, French 1. *Oxalis acetosella* (Wood Sorrel) B & H. 2. *Rumex scutatus* (Sorrel) Brownlow. Sorrel certainly grows in France, though not exclusively so. But its use in French cooking probably accounts for the name.

SORREL, Garden *Rumex acetosa* (Wild Sorrel) USA Allan.

SORREL, Green 1. *Oxalis acetosella* (Wood Sorrel) Bucks. Grigson. 2. *Rumex acetosa*

(Wild Sorrel) Bucks. Grigson; USA Sanford.

SORREL, Guinea; SORREL, Jamaican *Hibiscus sabdariffa* (Rozelle) Dalziel (Guinea); Perry (Jamaican). It is an Indian plant, but was introduced into both West Africa (hence Guinea Sorrel), and the West Indies. It is apparently particularly popular in Jamaica.

SORREL, Horse *Rumex hydrolapathum* (Great Water Dock) Lyte. 'Horse' in a plant name usually denotes large, or coarse. This plant would obviously qualify.

SORREL, Kidney *Oxyria digyna* (Mountain Sorrel) Hutchinson. An alternative specific name is reniformis - hence 'kidney'.

SORREL, Lady's *Oxalis corniculata* (Sleeping Beauty) USA H Smith.

SORREL, Meadow *Rumex acetosa* (Wild Sorrel) Harman.

SORREL, Mountain *Oxyria digyna*.

SORREL, Red 1. *Hibiscus sabdariffa* (Rozelle) Dalziel. 2. *Rumex acetosella* (Sheep's Sorrel) USA Gates.

SORREL, Sheep's 1. *Oxalis acetosella* (Wood Sorrel) Dor, E Ang, Ire. Grigson. 2. *Rumex acetosella*.

SORREL, Sheep's, Slender *Rumex tenuifolius*.

SORREL, Sow *Rumex acetosa* (Wild Sorrel) Hants. B & H.

SORREL, Switch *Dodonea viscosa*. Bitter leaves, hence 'sorrel'. 'Switch' presumably means it has long, pliant shoots.

SORREL, Water *Rumex hydrolapathum* (Great Water Dock) B & H.

SORREL, Wild 1. *Oxalis pes-caprae* (Bermuda Buttercup) S Africa Watt. 2. *Rumex acetosa*.

SORREL, Wood 1. *Oxalis acetosella*. 2. *Oxalis corniculata* (Sleeping Beauty) S Africa Watt. 3. *Oxalis pes-caprae* (Bermuda Buttercup) Australia Watt.

SORREL, Wood, True *Oxalis acetosella* (Wood Sorrel) USA House.

SORREL, Wood, White *Oxalis acetosella* (Wood Sorrel) USA House.

SORREL, Yellow, Procumbent *Oxalis corniculata* (Sleeping Beauty) Clapham.

SORREL, Yellow, Upright *Oxalis europaea*.

SORREL DOCK *Rumex acetosa* (Wild Sorrel) USA Upham.

SORREL DOCK, Sheep *Rumex acetosella* (Sheep's Sorrel) USA Gates.

SORREL RHUBARB *Rheum palmatum*.

SORREL TREE 1. *Andromeda arborea*. 2. *Oxydendron arboreum*. *Oxydendron* is a genuine tree, reaching 50 feet in the USA. The bitter leaves are chewed to allay thirst. Cf. Sourwood Tree.

SORROW *Rumex acetosa* (Wild Sorrel) Glos. J D Robertson; IOW, Suss. B & H. 'Sorrow' for 'sorrel' is quite widespread - probably more so than the records show.

SORROW, Bells of, Doleful *Fritillaria meleagris* (Snake's-head Lily) Oxf. Grigson. The drooping flowers with their sombre colouring are the reason for this and similar names. In this case, 'sorrow' is the same as 'Sodom'.

SORROW, Cuckoo *Rumex acetosa* (Wild Sorrel) Fernie.

SOT-WEED *Nicotiana tabacum* (Tobacco) Halliwell. Is this a comment on its narcotic properties?

SOUCE, Green; SOUCE, Sheep *Rumex acetosa* (Wild Sorrel) Bucks. B & H (Green); Dor, Border B & H (Sheep).

SOUCIQUE *Calendula officinalis* (Marigold) Guernsey MacCulloch. Latin solsequium - Marigold is a sun follower.

SOUKIE CLOVER; SOUKS *Trifolium pratense* (Red Clover) both Scot. B & H (Soukie Clover); Jamieson (Souks). Cf. Sowks, and Sowkie Soo from Scotland. Further south, the corresponding name is Sucklings or Suckles, or indeed Honeysuckle.

SOUR; SOWER *Allium cepa* (Onion) Derb. F Grose.

SOUR BREAD, Egyptian; SOUR GOURD *Adansonia digitata* (Baobab) Douglas (Bread); J Smith (Gourd). The white powder surrounding the trees can be made into a kind of porridge, and the fruit pulp has for long provided Africans with a refreshing drink, for it contains tartaric and other acids.

SOUR CHERRY *Eugenia crynantha*.

SOUR CHERRY, Chinese *Prunus cantabrigiensis*.

SOUR CLOVER 1. *Melilotus indica* (Small-flowered Melilot) USA T H Kearney. 2. *Oxalis acetosella* (Wood Sorrel) Berw. Grigson.

SOUR CREEPER *Sarcostemma acidum* (Moon Plant) Chopra.

SOUR DOCK 1. *Oxalis acetosella* (Wood Sorrel) Dev. Friend.1882. 2. *Rumex acetosa* (Wild Sorrel) Dev. Friend.1882; Som, Dor, Ches, Lancs, Yks, Cumb. B & H. 3. *Rumex acetosella* (Sheep's Sorrel) Watt. 4. *Rumex crispus* (Curled Dock) USA Henkel. 5. *Rumex obtusifolius* (Broad-leaved Dock) Donegal Grigson.

SOUR DOG; SOUR DUCK *Rumex acetosa* (Wild Sorrel) Som. Macmillan. Both 'dog' and 'duck' are substitutes for 'dock'.

SOUR GOGS; SOUR SOGS *Rumex acetosa* (Wild Sorrel) Bucks. Harman.

SOUR GRABS 1. *Malus sylvestris* fruit (Crabapple) Som. Mabey. 2. *Oxalis acetosella* (Wood Sorrel) Dev. Friend.1882. 3. *Rumex acetosa* (Wild Sorrel) Fernie. 'Grab' for 'crab' is quite widespread in southern England. Grabstock is used interchangeably for crabstock for the tree.

SOUR GRASS 1. *Oxalis acetosella* (Wood Sorrel) Dev, Yks. Grigson. 2. *Rumex acetosa* (Wild Sorrel) Norf, Yks. B & H. 3. *Rumex acetosella* (Sheep's Sorrel) Watt.

SOUR GREENS *Rumex venonus* (Veined Dock) USA Cunningham.

SOUR GUM *Nyssa sylvatica* (Tupelo) USA Schery.

SOUR LEAVES *Rumex acetosa* (Wild Sorrel) Som. Macmillan (Leaves); Roxb, Ire. B & H (Leek).

SOUR LEEK 1. *Rumex acetosa* (Wild Sorrel) B & H. 2. *Rumex acetosella* (Sheep's Sorrel) B & H.

SOUR LEEK, Red *Rumex acetosa* (Wild Sorrel) N Ire. Grigson.

SOUR PLUM *Ximenia caffra*.

SOUR ORANGE *Citrus aurantium var. bigaradia* (Seville Orange) A W Smith.

SOUR SAB 1. *Oxalis acetosella* (Wood Sorrel) Dev. Friend.1882. 2. *Rumex acetosa* (Wild Sorrel) Corn. Jago; Dev. B & H.

SOUR SALLY; SOUR SAP *Oxalis acetosella* (Wood Sorrel) Som. Macmillan; Mon. Waters (Sally); Dev. Macmillan (Sap).

SOUR SALVES *Rumex acetosa* (Wild Sorrel) Dev. Grigson.

SOUR SAUCE; SOUR SODGE; SOUR SOG *Rumex acetosa* (Wild Sorrel) Corn. Jago; Lincs. B & H (Sauce); Bucks. Heather (Sodge); Bucks. Harman (Sog).

SOUR SOBS *Oxalis pes-caprae* (Bermuda Buttercup) Australia Watt.

SOUR SOP; SOURSOP 1. *Annona muricata* (Guanabana) Barbados Gooding; Malaya. Gilliland. 2. *Rumex acetosa* (Wild Sorrel) Corn. Jago. Guanabana has acid fruit juice, but *Annona squamosa* is Sweetsop.

SOUR SOP, Wild *Annona chrysophylla* (Wild Custard Apple) Dale & Greenway.

SOUR-SOUR *Hibiscus sabdariffa* (Rozelle) Sierra Leone Dalziel. Acid, hence other 'sorrel' names for this.

SOUR SUDS 1. *Oxalis acetosella* (Wood Sorrel) Dev. Friend.1882. 2. *Rumex acetosa* (Wild Sorrel) Dev. B & H.

SOUR TOP *Vaccinium canadense* (Canada Blueberry).

SOUR TOP BLUEBERRY *Vaccinium myrtilloides*.

SOUR TREFOIL *Oxalis acetosella* (Wood Sorrel) Gerard.

SOURACH, Sookie *Oxalis acetosella* (Wood Sorrel) Inv. Grigson.

SOURACK 1. *Rumex acetosa* (Wild Sorrel) Scot. B & H. 2. *Rumex acetosella* (Sheep's Sorrel) Watt.

SOURACK, Sheep's *Rumex acetosa* (Wild Sorrel) Scot. B & H.

SOURBERRY *Rhus integrifolia*. Acid-tasting fruits, made by the Indians into cooling drinks, hence also Lemonade Tree.

SOUROCK 1. *Oxalis acetosella* (Wood

Sorrel) Donegal Grigson. 2. *Polygonum persicaria* (Persicaria) Donegal Grigson. 3. *Rumex acetosa* (Wild Sorrel) Scot. Jamieson.

SOUROCK, Cuckoo's *Oxalis acetosella* (Wood Sorrel) Border Hardy. Ascription to the cuckoo merely means that it is in flower when the cuckoo arrives. Most spring flowers are similarly assigned.

SOUROCK, Lammie 1. *Rumex acetosa* (Wild Sorrel) Border B & H. 2. *Rumex acetosella* (Sheep's Sorrel) Scot. B & H.

SOUROCK, Sheep's *Rumex acetosella* (Sheep's Sorrel) Scot. Jamieson.

SOURRICK, SOORIK *Rumex acetosa* (Wild Sorrel) Scot. Shet. (Soorik) Jamieson (Sourick); Gent. Mag. (Sorrick); Grigson (Soorik).

SOURWEED *Rumex acetosella* (Sheep's Sorrel) USA Zenkert.

SOURWOOD TREE *Oxydendron arboreum* (Sorrel Tree) USA. It is not the wood that is sour, but the leaves, which are chewed to allay thirst.

SOUTH AFRICAN ASPARAGUS *Asparagus laricinus*.

SOUTH AFRICAN BLACKBERRY *Rubus pinnatus*.

SOUTH AFRICAN PERIWINKLE *Vinca rosea* (Madagascar Periwinkle) Mitton.

SOUTH AMERICAN AGAVE *Agave americana* (Mescal) O P Brown.

SOUTH AMERICAN RUBBER PLANT *Helenium hoopesii* (Orange Sneezeweed) USA Elmore. Chewing gum is made from the roots.

SOUTH AMERICAN SAGE *Salvia coccinea* (Red Sage) Watt. Not confined to South America, though, for it grows in the southern part of the USA as well.

SOUTH AMERICAN VERVAIN *Verbena bonariensis*.

SOUTH SEA IRONWOOD *Casuarina equisetifolia* (Beefwood) Hora.

SOUTH SEA ROSE *Nerium oleander* (Rose Bay) Watt.

SOUTH SEA TEA *Ilex vomitoria* (Cassine) J Smith. But this is a New World species.

SOUTHERN BALSAM FIR *Abies fraseri* (Fraser's Balsam Fir) Brimble.1948.

SOUTHERN BEECH *Nothofagus procera* (Raoul) RHS.

SOUTHERN CYPRESS *Taxodium distichum* (Swamp Cypress) Duncalf.

SOUTHERN GENTIAN *Gentiana alpina*.

SOUTHERN HEMLOCK *Tsuga caroliniana* (Carolina Hemlock) S M Robertson.

SOUTHERN MARSH ORCHID *Dactylorchis praetermissa*.

SOUTHERN RED CEDAR *Juniperus silicicola*.

SOUTHERN RED LILY *Lilium catesbaei* (Catesby's Lily) Woodcock.

SOUTHERN SALLOW *Salix australis*.

SOUTHERN STAR JASMINE *Jasminum gracillimum*.

SOUTHERN SWAMP LILY *Lilium michauxii* (Carolina Lily) Woodcock.

SOUTHERN TWAYBLADE *Listera australis*.

SOUTHERN WHITE CEDAR *Chamaecyparis thyoides* (White Cedar) Howes.

SOUTHERN WILLOWHERB *Epilobium lamyi*.

SOUTHERNWOOD *Artemisia abrotanum*. i.e., one must assume, southern wormwood. It is sometimes called that.

SOUTHERNWOOD, Field *Artemisia campestris* (Breckland Mugwort) Clapham.

SOVEREIGN FLOWER *Kerria japonica* (Jew's Mallow) Lincs. B & H. Cf. Guinea-flower, or Guinea-plant.

SOVEREIGNS, Golden *Potentilla anserina* (Silverweed) Som. Macmillan. The golden-yellow flowers supply the imagery.

SOVEREIGNS, Strings-of- *Lysimachia nummularia* (Creeping Jenny) Ire. A descriptive name. Cf. Gerard's Moneywort, and Turner's Herb Twopence and Twopenny Grass.

SOW, Old 1. *Anaphalis margaretacea* (Pearly Immortelle) Norf. Grigson. 2. *Melilotus caerulea* (Swiss Melilot) B & H.

SOW DINGLE *Sonchus oleraceus* (Sow Thistle) Lincs. Peacock. 'Thistle' appears as dindle, or dingle, in East Anglia.

SOW FENNEL *Peucedanum officinale* (Hog's Fennel) Culpepper.

SOW FLOWER *Sonchus oleraceus* (Sow Thistle) Wilts. Dartnell & Goddard.

SOW FOOT *Tussilago farfara* (Coltsfoot) Yks. Grigson. The shape of the leaves has suggested a good many feet besides this - those of bulls, calves, foals, donkeys, etc.

SOW GRASS *Coronopus squamatus* (Wart-cress) Yks. Grigson. Cf. Hog-grass, Swine's Cress, etc., all implying a cress fit only for pigs.

SOW SORREL *Rumex acetosa* (Wild Sorrel) Hants. B & H. "Sow" must have been "sour" originally.

SOW-TEAT BLACKBERRY *Rubus allegheniensis* (Highbush Blackberry) USA Yarnell.

SOW-TEAT STRAWBERRY *Fragaria vesca var. americana*.

SOW THISTLE *Sonchus oleraceus*.

SOW THISTLE, Alpine *Cicerbita alpina*.

SOW THISTLE, Blue *Cicerbita macrophylla*.

SOW THISTLE, Corn *Sonchus arvensis*.

SOW THISTLE, Creeping *Sonchus arvensis* (Corn Sow Thistle) Salisbury.

SOW THISTLE, Marsh *Sonchus palustris*.

SOW THISTLE, Mountain *Cicerbita alpina* (Alpine Sow Thistle) RHS.

SOW THISTLE, Prickly; SOW THISTLE, Spiny *Sonchus asper* Clapham (Spiny).

SOW THISTLE, Smooth *Sonchus oleraceus* (Sow Thistle) McClintock.

SOW THISTLE, Tree 1. *Sonchus arvensis* (Corn Sow Thistle) B & H. 2. *Sonchus palustris* (Marsh Sow Thistle) Curtis. Surely, this name should be reserved for *Sonchus palustris*, which can grow up to 10 feet tall.

SOW THRISTLE *Sonchus oleraceus* (Sow Thistle) Berw. Grigson. Thristle is a local version of thistle.

SOW'S TITS *Polygonatum multiflorum* (Solomon's Seal) Dor. Grigson. Graphically descriptive.

SOWBANE 1. *Chenopodium hybridum*. 2. *Chenopodium rubrum* (Red Goosefoot) Grieve.1931. Curtis said of *C hybridum* that "Tragus mentions it as a plant fatal to swine".

SOWBREAD 1. *Cyclamen europaeum*. 2. *Sonchus oleraceus* (Sow Thistle) Kent Grigson. Cyclamen was used as a food for swine once. Sowbread is also a name for a truffle.

SOWD-WORT *Salsola kali* (Saltwort) Prior. 'Sowd' here is the same as 'soda', for this plant was once a common source of soda.

SOWER; SOUR *Allium cepa* (Onion) Derb. Grose.

SOWKIE; SOWKIE SOO *Trifolium pratense* (Red Clover) N'thum, Scot. Grigson. The equivalent elsewhere of Sucklings and various other versions of Honeysuckle, all given to commemorate the attraction of clover for bees.

SOWLERS *Avena fatua* (Wild Oat) Halliwell.

SOYBEAN; SOYA BEAN *Glycine max*. Apparently through the Japanese sho-yu, thence Chinese shi-yu, which means salt bean oil.

SPADE-LEAF *Philodendron domesticum* (Elephant's Ear) RHS.

SPADES *Polygonum convolvulus* (Black Bindweed) Border B & H. From the shape of the leaves, perhaps.

SPADIC BUSH *Erythroxylon coca* (Coca) W Miller.

SPAIN, Pearls of *Muscari botryoides var. album* B & H.

SPAIN, Pellitory-of- *Anacyclus pyrethrum*. True pellitory comes from the Latin parietaria, which in turn comes from paries, wall. But this pellitory apparently comes from pyrethrum.

SPANGLED BEAU *Mesembryanthemum crystallinum* (Sea Fig) Coats.

SPANIARD *Aciphylla squarrosa*.

SPANIARD, Giant *Aciphylla scott-thomsonii*. 'Giant' is right - this can be up to 14ft tall.

SPANIARD, Golden *Aciphylla aurea*.

SPANISH ASH *Syringa vulgaris* (Lilac) Glos. B & H. Lilac is often referred to as Ash in Gloucestershire, either as Ash by itself, or with an adjective. Blue Ash is also recorded there.

SPANISH BASTARD RHUBARB, Great *Thalictrum aquilegifolium* (Greater Meadow Rue) Gerard.

SPANISH BAYONET 1. *Yucca aloifolia* (Aloe Yucca). 2. *Yucca baccata* (Datil) USA Elmore. 3. *Yucca glauca* (Soapweed) USA Gilmore. 'Bayonet' is a typical description of sharp yucca leaves. It is as common as 'dagger'.

SPANISH BEET *Beta vulgaris var. cicla* (Swiss Chard) Bianchini. Spanish Beet and Leaf Beet are names given to a variety of Swiss Chard that has long green petioles that are used with the leaf like spinach.

SPANISH BLUEBELL *Endymion hispanicus*. 'Spanish', but it actually grows through southern Europe and down to West Africa.

SPANISH BROOM *Spartium junceum*.

SPANISH BROOM, White *Cytisus multiflorus*.

SPANISH BUGLOSS *Anchusa officinalis* (Alkanet) Culpepper.

SPANISH BUTTONS *Centaurea nigra* (Knapweed) USA Cunningham.

SPANISH CAMPION; SPANISH CATCHFLY *Silene otites* (Breckland Catchfly) W Miller (Campion); Clapham (Catchfly).

SPANISH CARNATION *Caesalpina pulcherrima* (Barbados Pride) Barbados Gooding.

SPANISH CEDAR *Cedrela odorata* (West Infdian Cedar) Hora.

SPANISH CHERVIL *Myrrhis odorata* (Sweet Cicely) Grieve.1931. A corruption of Anise Chervil, the name often given to the cultivated plant.

SPANISH CHESTNUT *Castanea sativa* (Sweet Chestnut). The nuts sold in this country are nearly always imported from Spain, for it was said that the best nuts came from there.

SPANISH CLOVER *Lotus purshiana*.

SPANISH DAFFODIL *Narcissus hispanicus*.

SPANISH DAGGER 1. *Yucca baccata* (Datil) USA Elmore. 2. *Yucca schidigera* (Mohave Yucca) Jaeger. Cf. Spanish Bayonet and other names that commemorate the sharp-pointed leaves.

SPANISH ELM *Cordia gerascanthus* W Indies Roys.

SPANISH FIR *Abies pinsapo*.

SPANISH FLAG *Mina lobata*. Presumably Spanish Flag is a reference to the national flag of Spain (this is a Mexican plant). It is certainly not a synonym for Iris, for this plant belongs to the Convolvulaceae.

SPANISH FURZE *Genista hispanica* (Spanish Gorse) W Miller.

SPANISH GARLIC *Allium scordoprasum* (Sand Leek) H & P.

SPANISH GORSE *Genista hispanica*.

SPANISH GOURD *Cucurbita maxima* (Giant Pumpkin) Lindley.

SPANISH HEATH *Erica australis*.

SPANISH HYACINTH *Endymion nonscriptus* (Bluebell) Parkinson. Why should he label it 'Spanish'? Gerard stuck to English Hyacinth for the bluebell.

SPANISH IRIS *Iris xiphium*. Confusingly, it is *Iris xiphioides*, called English Iris, that originated in the Pyrenees. They are both garden hybrids these days, and for Spanish, one might as well read Dutch.

SPANISH JASMINE 1. *Jasminum grandiflorum*. 2. *Plumeria acuminata* (Pagoda Tree) Chopra. Cf. Catalonian Jasmine for *Jasminum grandiflorum*. Actually, it is a Himalayan plant. And Pagoda Tree comes from tropical America.

SPANISH JUNIPER 1. *Juniperus thurifera*. 2. *Tetraclinis articulata* (Sandarac) Pomet.

SPANISH LETTUCE *Claytonia perfoliata* (Spring Beauty) Salisbury. Cf. Miner's Lettuce, which is an American name. 'Lettuce', because it has edible leaves.

SPANISH LIQUORICE *Glycyrrhiza glabra*

var. *typica*. Perhaps because it is cultivated in Spain, but certainly not exclusively.

SPANISH MARJORAM *Urtica pilulifera var. dodartii* Grigson. It was grown in 18th century gardens for practical jokes. Called Spanish Marjoram, and having leaves without teeth and no appearance of a nettle, the unwary were invited to smell it in the flower bed, and were severely stung.

SPANISH MELON *Cucumis melo var. inodorus*.

SPANISH MINT *Mentha requeni* (Corsican Mint) Ire. Coats. There is a tradition in Ireland that it was first introduced at the time of the Spanish Armada.

SPANISH MOSS *Tillandsia usneoides*.

SPANISH MUSTARD *Brassica perviridis*. Actually, it comes from eastern Asia.

SPANISH NEEDLES 1. *Bidens bipinnata*. 2. *Bidens pilosa* (Bur Marigold) Watt. Probably the whole genus carries this name.

SPANISH NUT *Sisyrinchium angustifolium* (Blue-eyed Grass) Parkinson.

SPANISH OAK 1. *Quercus falcata*. 2. *Quercus ilex* (Jolm Oak) USA Harper.

SPANISH ONION *Allium fistulosum* (Welsh Onion) USA Cunningham. Given that 'Welsh' in this context just means 'foreign', then probably 'Spanish' will have the same connotation.

SPANISH PEA *Cicer arietinum* (Chick Pea) Howes.

SPANISH PEPPER *Capsicum annuum* (Chile) Schauenberg.

SPANISH PLUM *Spondias purpurea* (Hog-Plum) Douglas.

SPANISH PLUM, Yellow *Spondias mombin* (Mombin) Jamaica Dalziel.

SPANISH POTATO *Ipomaea batatas* (Sweet Potato) Dalziel. It was originally thought that the introduction of sweet potato into the Pacific area had been made by Spanish explorers, but in fact it was being cultivated there a long time before any European contact.

SPANISH RHUBARB DOCK *Rumex abyssinica* Howes.

SPANISH ROOT *Ononis repens* (Rest-harrow) Cumb. Grigson. The reference here is to liquorice. Cf. Wild Liquorice and Liquory-stick. Children dig up the root to eat, or to put in a bottle of water, shake, and drink the result - Spanish Water.

SPANISH SAFFRON *Carthamus tinctorius* (Safflower) Parkinson. He also called it Catalonian Saffron. Perhaps the reason lies in the tendency to call anything 'Spanish' that came from the Spanish sphere of influence.

SPANISH SAINFOIN *Hedysarum coronarium* (French Honeysuckle) Candolle. It grows throughout southern Europe, hence the geographical spread of the names, which include Italian Sainfoin as well.

SPANISH SALSIFY 1. *Scolymus hispanicus*. 2. *Scorzonera hispanica* (Black Salsify) Leyel.1937.

SPANISH SEA PURSLANE *Atriplex halimus* (Tree Purslane) Moldenke. A Mediterranean plant, not confined to Spain.

SPANISH SENNA *Cassia obovata* (Italian Senna) Howes.

SPANISH SHAWL *Schizocentron elegans* Gannon.

SPANISH TEA *Chenopodium ambrosioides* (American Wormseed) USA Henkel. Cf. Mexican Tea, Jerusalem Tea, Jesuit Tea. The tea is medicinal.

SPANISH THISTLE *Centaurea iberica*.

SPANISH THYME *Thymus zygis* Brouk. A Mediterranean plant, also known as French Thyme.

SPANISH TUFTS *Thalictrum aquilegifolium* (Greater Meadow Rue) Parkinson.

SPARAGE; SPARAGUS *Asparagus officinalis* (Asparagus) Gunther (Sparage); Culpepper (Sparagus).

SPARKED GRASS *Phalaris arundinacea* (Reed-grass) Som. Elworthy; Wilts. Dartnell & Goddard. Sparked means variegated.

SPARKED LAUREL *Aucuba japonica* (Japanese Laurel) Som. Elworthy.

SPARRA GRASS; SPARRA GRACE *Asparagus officinalis* (Asparagus) Yks.

Nicholson (Sparra grass); Berks. Lowsley (Sparra grace).

SPARROW-BIRDS *Geranium robertianum* (Herb Robert) Dev,Som. Elworthy.

SPARROW-GRASS *Asparagus officinalis* (Asparagus) Som. Elworthy.

SPARROW-GRASS, French *Ornithogalum pyrenaicum* (Spiked Star-of-Bethlehem) W Miller. This was the name under which the sprouts of the plant were sold in the Bath markets, to be eaten as asparagus. Grass was once a common greengrocers' name for asparagus.

SPARROW-WEED *Ranunculus lingua* (Greater Spearwort) Ire. B & H. 'Sparrow' is probably 'spear' in this case.

SPARROW-WORT *Erica passerina*.

SPARROW-WORT, Shaggy *Passerina hirsuta*.

SPARROW'S TONGUE *Polygonum aviculare* (Knotgrass) Som, Norf. Cockayne. Cf. Bird's Tongue, Bird Knotgrass,etc - all from the shape of the leaf.

SPATLING POPPY *Silene cucubalis* (Bladder Campion) Cumb. Cf. Gerard's Frothy Poppy. Both the names are references to the froth, called cuckoo spit, often found round it.

SPATLUM *Lewisia rediviva* (Bitter-root) W Miller.

SPEAK 1. *Lavandula spica* (English Lavender) Som. Macmillan. 2. *Rosmarinus officinalis* (Rosemary) Som. Macmillan. 'Speak' here is 'spike', appearing in a number of guises, including Spick, Speck and Speke. Presumably rosemary is being confused with, or likened to, lavender when the names are given to it.

SPEAK, Hedge 1. *Prunus spinosa* fruit (Sloe) Wilts. Dartnell & Goddard. 2. *Rosa canina* hips (Hips) Glos. Grigson. In this case, 'speak' is 'pick', which is another Wiltshire name for sloes. Cf. Winter Picks, and Hedge-picks, which is also rendered as Hedgespecks. 'Picks' also becomes 'pegs', as in Hedge-pegs, or Heg-Pegs; it also easily becomes 'pigs', and Hedge-pigs is recorded, too. Another sloe name, Hedgy-pedgies, is given to hips, and one finds a number of 'haw' names mistakenly given to sloes, too.

SPEAR 1. *Gynerium argenteum* (Pampasgrass) Som. Macmillan. 2. *Phragmites communis* (Reed) Dor. Nash. The tall, tufted stems are often likened to spears, or feathers.

SPEAR-Cast-the- *Solidago virgaurea* (Golden Rod) Dor. Macmillan.

SPEAR, Christ's *Ophioglossum vulgatum* (Adder's Tongue) Leyel.1937.

SPEAR, Crusader's *Scilla maritima* (Squill) RHS.

SPEAR, Dog *Arum maculatum* (Cuckoopint) Som. Elworthy. 'Spear' in this case has the same meaning as the 'pint' of Cuckoopint.

SPEAR,Ithuriel's *Brodiaea laxa* (Californian Hyacinth) Simmons.

SPEAR, King's 1. *Asphodeline lutea* (Yellow Asphodel) Gerard. 2. *Asphodelus ramosus* (White Asphodel) Grieve.1931. 3. *Eremurus sp* (Foxtail Lily) RHS. 4. *Narcissus pseudonarcissus* (Daffodil) Som. Macmillan.

SPEAR, Sword-and- *Plantago lanceolata* (Ribwort Plantain) Dor. Macmillan. Evidently an extension of 'Soldiers' and the other names invented to describe a children's game with the flowering stalks, played like conkers.

SPEAR CROWFOOT 1. *Ranunculus flammula* (Lesser Spearwort) B & H. 2. *Ranunculus lingua* (Greater Spearwort) B & H.

SPEAR-GRASS 1. *Aciphylla squarrosa* (Spaniard) H & P. 2. *Agropyron repens* (Couch-grass) E Angl. G E Evans.1969. 3. *Stipa spartea*.

SPEAR-GRASS, Low (Annual Meadowgrass) *Poa annua* USA Allan.

SPEAR-LEAVED GERANIUM *Pelargonium glaucum*.

SPEAR-LEAVED ORACHE *Atriplex hastata* (Halberd-leaved Orache) Curtis. Hastate Orache is also used - hastate means spearshaped, describing the leaves in this case.

SPEAR-LEAVED WILLOW *Salix hastata*.

SPEAR-LEAVED WILLOWHERB *Epilobium lanceolatum*.

SPEAR PLUME THISTLE; SPEAR THISTLE *Cirsium vulgare* The former is the usual name; Spear Thistle is in Perring.

SPEAR-WOOD *Acacia doratoxylon* (Brigalow) Usher. In this case, the name means a wood out of which spears are made - by the Australian aborigines, who also make boomerangs from it.

SPEARGRASS *Ranunculus flammula* (Lesser Spearwort) Turner.

SPEARMINT *Mentha spicata*. 'Spire' mint is really the correct name, rather than 'spear' mint.

SPEARSCALE *Atriplex patula* (Orache) USA Tolstead.

SPEARWORT *Inula helenium* (Elecampane) Cockayne. Presumably because it is a tall, upstanding plant.

SPEARWORT, Adder's Tongue *Ranunculus ophioglossifolius* (Serpent's-tongue Spearwort) McClintock.

SPEARWORT, Creeping *Ranunculus reptans*.

SPEARWORT, Greater *Ranunculus lingua*.

SPEARWORT, Lesser; SPEARWORT, Small *Ranunculus flammula*. Cf. Speargrass, etc. Then 'spear' varies into 'spire', and even 'spur', as in the Devonshire Spurwood. Curtis called it Small Spearwort.

SPEARWORT, Serpent's Tongue *Ranunculus ophioglossifolius*.

SPEARWORT, Water *Ranunculus laxicaulis*.

SPECK 1. *Lavandula spica* (English Lavender) Dor. Dacombe. 2. *Lunaria annua* (Honesty) Scot. Tynan & Maitland. 'Speck' for lavender is 'spike', also recorded as 'Spick', or 'Speak'.

SPECKLED ALDER *Alnus incana* (Hoary Alder) USA Tolstead.

SPECTACLE-POD *Dithyraea wislizeni*. There is some doubt about this - could it be 'rod' instead of 'pod'?

SPECTACLES, Grandmother's *Lunaria annua* (Honesty) Som. Macmillan. Perfectly descriptive - the seed capsules are exactly the shape of old-fashioned spectacles.

SPEEDWELL, Alpine *Veronica alpina*. 'Speedwell' means goodbye. We are asked to believe that the plants got this widely used name because the blossoms when full fall off and fly away. Forget-me-not, now applied to quite a different plant, was apparently once given to the speedwells, for the same reason. But Andrew Young suggests that if 'speed' means the same as in "speed the plough", then perhaps the name means good luck.

SPEEDWELL, American *Veronica peregrina*.

SPEEDWELL, Bog *Veronica scutellata* (Marsh Speedwell) Curtis.

SPEEDWELL, Breck *Veronica praecox*. Breck - Breckland, where the plant is confined in this country.

SPEEDWELL, Buxbaum's *Veronica persica* (Common Field Speedwell) McClintock. Johann Christian Buxbaum (1683-1730) was a German botanist, and he it was who first described the plant, from fields near Constantinople.

SPEEDWELL, Common *Veronica officinalis*.

SPEEDWELL, Corn 1. *Veronica arvensis* (Wall Speedwell) B & H. 2. *Veronica hederifolia* (Ivy-leaved Speedwell) B & H.

SPEEDWELL, Creeping *Veronica filiformis* (Slender Speedwell) Polunin.

SPEEDWELL, Female *Kickxia spuria* (Round-leaved Fluella) Parkinson. This plant was also referred to as Female Fluella, from the round, velvety leaved, Prior said. Male Speedwell is *Veronica officinalis*.

SPEEDWELL, Field, Common; SPEEDWELL, Field, Large *Veronica persica*. Clapham prefers the second name.

SPEEDWELL, Field, Green *Veronica agrestis*.

SPEEDWELL, Field, Grey *Veronica polita*.

SPEEDWELL, Fingerd *Veronica triphyllos*.

SPEEDWELL, Garden; SPEEDWELL, Garden, Procumbent *Veronica agrestis*

(Green Field Speedwell) Yks. B & H (Garden); Curtis (Procumbent).
SPEEDWELL, Germander *Veronica chamaedrys*. 'Germander' means the same as the specific name, chamaedrys. It comes through medieval Latin from the Greek chamaedrus, which is a corruption of chamai, on the ground, and drus, oak.
SPEEDWELL, Heath *Veronica officinalis* (Common Speedwell) McClintock.
SPEEDWELL, Ivy-leaved *Veronica hederifolia*.
SPEEDWELL, Large *Veronica austriaca*.
SPEEDWELL, Long-leaved *Veronica longifolia*.
SPEEDWELL, Male *Veronica officinalis* (Common Speedwell) Curtis. It is difficult to know why Curtis gave it this name, though certainly there is a Female Speedwell, given not to a member of the Veronica genus, but to *Kickxia spuria*.
SPEEDWELL, Marsh *Veronica scutellata*.
SPEEDWELL, Mountain 1. *Veronica montana*. 2. *Veronica tenella*.
SPEEDWELL, Purslane *Veronica peregrina* (American Speedwell) USA Upham.
SPEEDWELL, Rock *Veronica fruticans*.
SPEEDWELL, Round-leaved *Veronica filiformis* (Slender Speedwell) Allan.
SPEEDWELL, Skullcap *Veronica scutellata* (Marsh Speedwell) USA House.
SPEEDWELL, Slender *Veronica filiformis*.
SPEEDWELL, Smooth, Little *Veronica serpyllifolia* (Thyme-leaved Speedwell) Curtis.
SPEEDWELL, Spiked *Veronica spicata*. 'Spiked' in the sense of bearing an upright spike of flowers.
SPEEDWELL, Spring *Veronica verna*.
SPEEDWELL, Thyme-leaved *Veronica serpyllifolia*.
SPEEDWELL, Trifid *Veronica triphyllos* (Fingered Speedwell) Curtis.
SPEEDWELL, Upright *Veronica spicata* (Spiked Speedwell) Camden/Gibson.
SPEEDWELL, Wall *Veronica arvensis*.
SPEEDWELL, Water 1. *Veronica anagallis-aquatica*. 2. *Veronica beccabunga* (Brooklime) Thornton.
SPEEDWELL, Water, Pink *Veronica catenata*.
SPEEDWELL, Wood *Veronica montana*.
SPEEK 1. *Lavandula spica* (English Lavender) Som. Macmillan. 2. *Rosmarinus officinalis* (Rosemary) Som. Macmillan. 'Speek' is 'spike' - note the specific name, and Cf. Spick, Speck and Speak. The name as given to rosemary was probably an error originally.
SPEKE; SPICK *Lavandula spica* (English Lavender) Dev. Friend.1882 (Speke); Dor. Dacombe (Spick).
SPEKNEL *Meum athamanticum* (Baldmoney) Turner. Spignel is the modern version, but it exists also as Spikenel or Spicknel.
SPERAGE 1. *Asparagus officinalis* (Asparagus) Gerard. 2. *Ornithogalum pyrenaicum* (Spiked Star-of-Bethlehem) B & H. The Star-of-Bethlehem gets the name because this is Bath Asparagus, sold in the markets there at one time as such.
SPICE, Black *Rubus fruticosus* fruit (Blackberry) Yks. Carr. Possibly from the series that starts with Blaggs, or Bleggs, and which includes such names as Blackbides or Black Boyds for the fruit.
SPICE, Cuckoo 1. *Cardamine pratensis* (Lady's Smock) Yks. Grigson. 2. *Oxalis acetosella* (Wood Sorrel) Swainson. Cuckoo-spit is close to this, and that is a name given to Lady's Smock; in fact, there are a number of 'cuckoo' names given to both plants. The only connection seems to be that they are in flower at the time in spring when the cuckoo arrives.
SPICE BUSH *Lindera benzoin*.
SPICKNEL *Meum athamanticum* (Baldmoney) Prior. Probably the same word as 'spike'.
SPICY WINTERGREEN *Gaultheria procumbens* (Wintergreen) USA House.
SPIDER FLOWER *Chleome spinosa*. Plants with 'spider' names usually have their flow-

ers nestling close to the leaves, giving a 'spider in its web' effect.

SPIDER-IN-HIS-WEB *Nigella damascena* (Love-in-a-mist) Som. Macmillan. One of a number of descriptive names, of which the common name is as good as any.

SPIDER LILY *Tradescantia virginica* (Spiderwort) USA House.

SPIDER LILY, Golden *Lycoris aurea*.

SPIDER MILKWEED *Asclepiodora decumbens*.

SPIDER ORCHID *Ophrys sphegodes*.

SPIDER ORCHID, Late *Ophrys fuciflora*.

SPIDER PLANT 1. *Anthericum variegatum*. 2. *Chlorophytum comosum var. variegatum*. 3. *Cleome serrulata* (Rocky Mountain Bee Plant) USA Elmore. 4. *Saxifraga sarmentosa* (Mother of thousands) Dev. Friend.1882.

SPIDER PLANT, Giant *Cleome spinosa* (Spider-flower) USA Elmore.

SPIDER'S WEB *Nigella damascena* (Love-in-a-mist) Dor. Macmillan.

SPIDERWORT 1. *Nigella damascena* (Love-in-a-mist) Som. Macmillan. 2. *Tradescantia virginica*.

SPIDERWORT, Day *Tradescantia virginica* (Spiderwort) Parkinson. 'Day', because each flower lasts only a day.

SPIDERWORT, Western *Tradescantia occidentalis*.

SPIDERWORT, Yellow *Cleome lutea*.

SPIGNEL 1. *Foeniculum vulgare* (Fennel) Som. Grigson. 2. *Meum athamanticum* (Baldmoney) B & H. Prior has the derivation Spanish espiga, spike, and eneldo, dill (Latin anethum). He refers to Baldmoney, and says that the plant was imported from Spain under that name, which could vary to Spicknel or Spikenel. Spignel for fennel sometimes appears as Spingel.

SPIGNEL-MEU *Meum athamanticum* (Baldmoney) McClintock. A hybrid of Spignel and Meum.

SPIKE; SPIKE, Lavender; SPIKE, Lesser *Lavandula spica* (English Lavender) Rimmel (Spike); Tusser (Lavender Spike); Parkinson (Lesser Spike).

SPIKE BENT *Agrostis exarata*.

SPIKE-HEATH *Bruckenthalia spiculifolia*. One of the Ericaceae, and the specific name *spiculifolia* explains the first part of the name.

SPIKE MUHLY *Muhlenbergia wrightii*.

SPIKE-RUSH *Scirpus palustris*. A close relative of the bulrush, which is enough to make 'spike' understandable.

SPIKED CUCUMBER *Citrullus caffer*.

SPIKED LOOSESTRIFE *Lythrum salicaria* (Purple Loosestrife) Thomson.1978. Curtis called it Purple-spiked Loosestrife.

SPIKENARD 1. *Anthoxanthum odoratum* (Sweet Vernal Grass) Wilts. Dartnell & Goddard. 2. *Centaureum erythraea* (Centaury) Som. Macmillan. 3. *Lavandula spica* (English Lavender) Som. Macmillan; Wilts. Dartnell & Goddard. 4. *Nardostachys jatamansi*. 5. *Sison amonum* (Stone Parsley) Hants. Cope. Only *Nardostachys* has any right to the name, which is Latin spica nardi, the flower spike of the nard. 'Spike', of course, is associated with the lavender - after all, the specific name says so. Presumably the scent of the others accounts for the name.

SPIKENARD, American *Aralia racemosa*.

SPIKENARD, Caucasian *Inula glandulosa*.

SPIKENARD, Celtic *Valeriana celtica* Gerard.

SPIKENARD, False *Smilacina racemosa* (False Solomon's Seal) USA H Smith.

SPIKENARD, French *Valeriana celtica* Turner. Or the Celtic Nard.

SPIKENARD, Ploughman's 1. *Inula conyza*. 2. *Petasites fragrans* (Winter Heliotrope) Dor. Dacombe. For 'ploughman', read 'poor man', who could use the roots of *Inula conyza* to hang in his cottage to scent the musty air.

SPIKENARD, Small *Aralia nudicaulis* (American Sarsaparilla) Grieve.1931.

SPIKENEL; SPICKNEL; SPIGNEL *Meum athamanticum* (Baldmoney) Prior (Spikenel, Spicknel); B & H (Spignel). See Spignel.

SPIKY-FLOWERS *Cardamine hirsuta* (Hairy Bittercress) Som. Macmillan.
SPINACH 1. *Amaranthus dubius* Barbados Gooding. 2. *Spinacea oleracea*. Apparently from an Arabic original, isfanadoch, esbanach, sepanach. Spinach is probably indigenous to southwest Asia, and was introduced to Europe by the Moors about 1000AD.
SPINACH, American *Phytolacca decandra* (Poke-root) Le Strange. Poke greens were a common potherb in the early spring in the eastern USA.
SPINACH, Cattle *Atriplex polycarpa*.
SPINACH, Chinese *Amaranthus tricolor* (Joseph's Coat) H & P.
SPINACH, French 1. *Atriplex hortensis* (Garden Orache) G B Foster. 2. *Rumex scutatus* (Sorrel) Wit. Garden Sorrel grows in clumps rather like spinach.
SPINACH, Indian *Basella alba*.
SPINACH, Lincolnshire *Chenopodium bonus-henricus* (Good King Henry) Sanecki. It used to be much cultivated in Lincolnshire - Cf. Lincolnshire Asparagus for this plant.
SPINACH, Mountain *Atriplex hortensis* (Garden Orache) Rohde.
SPINACH, New Zealand *Tetragonia tetragonioides*. The leaves are cooked like spinach, and taste rather like it, too.
SPINACH, Sea *Beta vulgaris var. maritima* (Beet) IOW Grigson.
SPINACH, Wild 1. *Beta vulgaris* (Beet) IOW Grigson. 2. *Campanula latifolia* (Giant Bellflower) Yks. Grigson. 3. *Chenopodium album* (Fat Hen) Midl. Grigson. 4. *Chenopodium bonus-henricus* (Good King Henry) Som, IOW Grigson; Hants. Cope. 5. *Mercurialis annua* (French Mercury) North. Mercury is the odd one out - the rest have edible leaves, broadly treated like spinach, but surely no-one treats French Mercury as a potherb?
SPINACH BEET *Beta vulgaris var. cicla* (Swiss Chard) Bianchini.

SPINDLE, Broad-leaved *Euonymus latifolius*.
SPINDLE, Evergreen; SPINDLE, Japanese *Euonymus japonicus* Polunin.1976 (Japanese).
SPINDLE, Paddock's 1. *Orchis mascula* (Early Purple Orchid) Perth. B & H. 2. *Orchis morio* (Green-winged Orchid) Perth. B & H. 'Spindle' has a different significance in this case. It is the same word as 'pintle', and the 'pint' of 'Cuckoo-pint', i.e. penis. So the name means toad's penis.
SPINDLE, Rough-stemmed *Euonymus verrucosus*.
SPINDLE-TREE *Euonymus europaeus*. It seems that spindles were actually made from this wood, for which purpose it is ideal, but not necessarily in this country, where there is no great tradition of spinning. The name is actually imported, for Turner mentioned the Dutch name Spilboome, without implying an English equivalent.
SPINDLE-TREE, Winged *Euonymus alata*.
SPINDLEWOOD *Euonymus europaeus* (Spindle-tree) Glos. Grigson.
SPINGEL *Foeniculum vulgare* (Fennel) Som. B & H. Spignel, of course, and usually reserved for *Meum athamanticum*. One has to assume its genuineness, and to assign it to displaced consonants.
SPINK 1. *Cardamine pratensis* (Lady's Smock) N'thum. Grigson. 2. *Dianthus deltoides* (Maiden Pink) Scot. Jamieson. 'Pink', rather, which is used for Lady's Smock as well as the genuine pink. But does Maiden Pink grow in Scotland?
SPINK, Mary *Primula vulgaris* (Primrose) Scot. Grigson. Cf. Mayspink, which is the original of this name. They are called Mayflowers, too, in Shetland, and also in Ireland.
SPINK, Meadow *Lychnis flos-cuculi* (Ragged Robin) Scot. Grigson.
SPINNAGE *Spinacea oleracea* (Spinach) Raper.
SPINNING GUM *Eucalyptus perriniana*.
SPINNING JENNY *Acer campestre* (Field

Maple) Som. Macmillan. From the way the seeds spin through the air.

SPINY AMARANTH *Amaranthus spinosus* (Thorny Pigweed) Dalziel.

SPINY THRIFT *Armeria fasciculata*.

SPIRAEA, Turf *Spiraea caespitosa*.

SPIRAEA, Willow; SPIRAEA, Willow-leaved *Spiraea salicifolia* (Bridewort) McClintock (Willow); Hutchinson (Willow-leaved).

SPIRE LILY *Galtonia candicans*. Descriptive.

SPIRE MINT *Mentha spicata* (Spearmint) Prior. 'Spire' in this case is 'spear', though it is actually the other way round, for 'Spire' is correct, and 'spear' derivative.

SPIRES *Phalaris arundinacea* (Reed-grass) Hants. Cope. So the fields where they grow are called spire-beds, or spear-beds.

SPIREWORT *Ranunculus flammula* (Lesser Spearwort) Hebr. Martin. 'Spire' for 'spear'.

SPIRIT PLANT *Lacnanthes tinctoria* (Bloodroot) Mitton. 'Spirit', because it is used in voodoo.

SPIRITWEED *Eryngium foetidum* Jamaica Beckwith. Probably because the smoke from burning this was reckoned in Jamaica to keep duppies away.

SPIT, Adder's 1. *Listera ovata* (Twayblade) Tynan & Maitland. 2. *Pteridium aquilina* (Bracken) Tynan & Maitland. 3. *Stellaria holostea* (Greater Stitchwort) Corn. Grigson. Glos. PLNN35. All these have other 'snake' connections. Twayblade is Adder's Tongue, bracken is Snake Fern in Wiltshire, and Greater Stitchwort is Adder's Meat among other snake names, for no very obvious reason. However, Cornish children say they will be bitten by an adder if they pick stitchwort.

SPIT, Cuckoo 1. *Anemone nemerosa* (Wood Anemone) Worcs. Grigson. 2. *Arum maculatum* (Cuckoo-pint) B & H. 3. *Cardamine pratensis* (Lady's Smock) Dyer. 4. *Melandrium rubrum* (Red Campion) N'thants. A E Baker. Cuckoo spit is the froth secreted by frog-hoppers on plants to enclose the larvae. But all these plants have been given other 'cuckoo' names, usually to emphasise that they are spring flowers.

SPIT, Devil's *Centaurea nigra* (Knapweed) Som. Macmillan. Cf. Devil's Bit, from the same area.

SPIT, St Patrick's *Pinguicula vulgaris* (Butterwort) Antrim PLNN1. It is also known as St Patrick's Staff there.

SPIT, Yellow *Chelidonium majus* (Greater Celandine) Hants. Grigson.

SPLEENWORT, Black *Asplenium adiantum nigrum*.

SPLEENWORT, Common *Asplenium trichomanes*. Given because it was used for spleen disorders.

SPLEENWORT, Forked *Asplenium septentrionalis*.

SPLEENWORT, Golden *Chrysosplenium oppositifolium* (Opposite-leaved Golden Saxifrage) Freethy.

SPLEENWORT, Green *Asplenium viride*.

SPLEENWORT, Lanceolate *Asplenium obovatum*.

SPLEENWORT, Maidenhair *Asplenium trichomanes* (Common Spleenwort) Brightman.

SPLEENWORT, Mother *Asplenium bulbiferum* Bonar. Cf. Hen-and-chicken Fern for this, which, taken with the specific name *bulbiferum*, is enough to explain 'mother' in the name.

SPOGEL SEED *Plantago ovata* (Ispaghul Plantain) Le Strange. 'Spogel' is a variation of 'Ispaghul'.

SPOIL *Agropyron repens* (Couchgrass) Grigson.1959.

SPOKEWOOD *Euonymus europaeus* (Spindle Tree) B & H. A useful wood, whether for spokes or spindles. But see also Pegwood, and Skewerwood too.

SPONGE, Bath *Luffa cylindrica* (Loofah) Howes.

SPONGE, Vegetable; SPONGE, Washrag; SPONGE GOURD *Luffa cylindrica* (Loofah) all Chopra. The comparison with sponges is obvious.

SPOON-LEAVED SAXIFRAGE *Saxifraga cuneifolia* (Wedge-leaved Saxifrage) Grey-Wilson.

SPOON-LEAVED SOAPWORT *Saponaria bellidifolia*.

SPOONS, Devil's 1. *Alisma plantago-aquatica* (Water Plantain) Scot. Jamieson. 2. *Potamogeton natans* (Broad-leaved Pondweed) Gordon. From the shape of the leaves.

SPOONWOOD *Kalmia latifolia* (Calico Bush) USA Harper.

SPOONWORT *Cochlearia officinalis* (Scurvy-grass) Gerard. A translation of Cochlearia, which comes from the word meaning spoon, so named from the shape of the leaves.

SPOOTS *Angelica sylvestris* (Wild Angelica) Shet. Mabey.

SPOTTED DOG *Orchis mascula* (Early Purple Orchid) Som. Macmillan. From the spotted leaves.

SPOTTED GRASS *Medicago arabica* (Spotted Medick) Corn. Grigson.

SPRAY, Sea *Rosmarinus officinalis* (Rosemary) Som. Macmillan. Not a bad stab at *Rosmarinus*. It actually means 'dew of the sea'.

SPRAY, Wall *Cotoneaster horizontalis* RHS.

SPREADING-LEAF PINE *Pinus patula*.

SPREESPRINKLE 1. *Dactylorchis maculata* (Heath Spotted Orchid) Ches. Holland. 2. *Orchis mascula* (Early Purple Orchid) Ches. B & H. A corruption of Priest's Pintle.

SPREUSIDANY *Peucedanum officinale* (Hog's Fennel) Prior. An approximation of *Peucedanum*.

SPRING, Flower of *Bellis perennis* (Daisy) Som. Macmillan.

SPRING, Lady of *Taraxacum officinale* (Dandelion) Dor. Macmillan. The 'Lady' must be St Bride, as the goddess of spring. Dandelion was known as St Bride's Forerunner, and in Gaelic it is "the plant of Bride".

SPRING ANEMONE *Pulsatilla vernalis*.

SPRING BEAUTY *Claytonia perfoliata*.

SPRING BEAUTY, Carolina *Claytonia caroliniana* (Wide-leaved Spring Beauty) USA House.

SPRING BEAUTY, Narrow-leaved *Claytonia virginica*.

SPRING BEAUTY, Virginia *Claytonia virginica* (Narrow-leaved Spring Beauty) Yanovsky.

SPRING BEAUTY, Western *Claytonia lanceolata*.

SPRING BEAUTY, Wide-leaved *Claytonia caroliniana*.

SPRING BELLS *Sisyrinchium douglassi* W Miller.

SPRING CARROT *Potentilla anserina* (Silverweed) Freethy. The roots of silverweed have been used in Scotland and Ireland as a carrot-like vegetable, eaten either roasted or boiled, or even raw. They were cultivated, too, and records of this go back to prehistoric times, and the Anglo-Saxons are known to have used silverweed as a root crop.

SPRING CINQUEFOIL *Potentilla tabernaemontanii*.

SPRING CLEAVERS *Galium aparine* (Goose-grass) USA Yarnell.

SPRING CRESS 1. *Barbarea verna* (American Land Cress) Salisbury. 2. *Dentaria bulbifera* (Coral-root) Howes.

SPRING CROCUS *Crocus purpureus*.

SPRING CUDWEED *Antennaria plantaginifolia* (Plantain-leaved Everlasting) Grieve.1931.

SPRING FELWORT *Gentiana verna* (Spring Gentian) Tradescant.

SPRING-FLOWER 1. *Geranium robertianum* (Herb Robert) Dor. Macmillan. 2. *Primula vulgaris var. elatior* (Polyanthus) Wilts. Dartnell & Goddard.

SPRING GENTIAN *Gentiana verna*.

SPRING GRASS *Anthoxanthum odoratum* (Sweet Vernal Grass) Leyel.1937.

SPRING LILY *Erythronium mesochoreum*.

SPRING MESSENGER *Ranunculus ficaria* (Lesser Celandine) Dor, Som. Macmillan.

SPRING ONION *Allium fistulosum* (Welsh Onion) Candolle. It may bear the name, but this is not the species that is grown as Spring Onions.

SPRING PEA *Lathyrus vernus*.

SPRING PUCCOON *Lithospermum canescens* (Hoary Puccoon) Perry.

SPRING SAFFRON *Crocus purpureus* (Spring Crocus) Gerard.

SPRING SANDWORT *Minuartia verna*.

SPRING SPEEDWELL *Veronica verna*.

SPRING SQUILL *Scilla verna*.

SPRING VETCH 1. *Vicia lathyroides*. 2. *Vicia sativa* (Common Vetch) USA E Gunther.

SPRING VIOLET *Gentiana verna* (Spring Gentian) Dur. Grigson.

SPRING WHITLOW GRASS *Erophila verna* (Whitlow Grass) Clapham.

SPRINGWORT *Euphorbia lathyris* (Caper Spurge) Cockayne. Springwort is a mystical (or mythical) plant, associated by both Cockayne and Grimm with Caper Spurge, which has the power of opening doors and locks.

SPRIT *Hordeum sativum* (Barley - the awns) N'thants. Sternberg.

SPRUCE, Alcock *Picea bicolor*. Spruce seems to be the same word as Prussian; Pruce used to be the English name for Prussia, from where presumably the first examples of Norway Spruce were brought to Britain.

SPRUCE, Bigcone *Pseudotsuga macrocarpa* Brimble.1948. 'Bigcone' is a translation of macrocarpa.

SPRUCE, Black *Picea mariana*.

SPRUCE, Blue *Picea pungens var. glauca*.

SPRUCE, Blue, Colorado *Picea pungens* (Colorado Spruce) Edlin.

SPRUCE, Canadian *Picea canadensis* (White Spruce) N Davey.

SPRUCE, Caucasian *Picea orientalis* (Priental Spruce) Mitchell.

SPRUCE, Chinese *Picea asperata*.

SPRUCE, Colorado *Picea pungens*.

SPRUCE, Douglas *Pseudotsuga menziesii* (Douglas Fir) USA Elmore.

SPRUCE, Dragon *Picea asperata* (Chinese Spruce) Mitchell.

SPRUCE, Engelmann *Picea engelmannii*.

SPRUCE, False *Pseudotsuga menziesii* (Douglas Fir) USA Elmore.

SPRUCE, Hemlock *Tsuga sp* (Hemlock) Barber & Phillips.

SPRUCE, Himalayan *Picea smithiana*.

SPRUCE, Hondo; SPRUCE, Honshu *Picea jezoensis* (Yeddo Spruce) Leathart (Hondo); Wilkinson.1981 (Honshu).

SPRUCE, Lowland *Picea sitchensis* (Sitka Spruce).

SPRUCE, Menzies *Picea sitchensis* (Sitka Spruce) Brimble.1948.

SPRUCE, Morinda *Picea smithiana* (Himalayan Spruce) Mitchell.

SPRUCE, Mount Morrison *Picea morrisonicola*. Named after what used to be known as Mount Morrison, in Taiwan.

SPRUCE, Mountain *Picea engelmannii* (Engelmann Spruce) RHS.

SPRUCE, Norway *Picea excelsa* (Spruce Fir).

SPRUCE, Oriental *Picea orientalis*.

SPRUCE, Red *Picea rubens*.

SPRUCE, Sakhalin *Picea glehnii*.

SPRUCE, Sargent *Picea brachytyla*.

SPRUCE, Serbian *Picea omorika*. Confined to the Drina valley, in what was Yugoslavia.

SPRUCE, Siberian *Picea obovata*.

SPRUCE, Sikkim *Picea spinulosa*.

SPRUCE, Silver 1. *Picea pungens* (Colorado Spruce) USA Elmore. 2. *Picea sitchensis* (Sitka Spruce) Brimble.1948.

SPRUCE, Sitka *Picea sitchensis*. It takes the name from Sitka, in Alaska, even though it grows also far to the south.

SPRUCE, Taiwan *Picea morrisonicola* (Mount Morrison Spruce) RHS.

SPRUCE, Tideland *Picea sitchensis* (Sitka Spruce) USA Brimble.1948.

SPRUCE, Tien Shan *Picea schrenkiana*.

SPRUCE, Tigertail *Picea polita*. A Japanese species.

SPRUCE, White *Picea canadensis*.

SPRUCE, White, Alberta *Picea canadensis var. albertiana*.

SPRUCE, Wilson *Picea wilsonii*.

SPRUCE, Yeddo *Picea jezoensis*.
SPRUCE FIR *Picea abies*.
SPRUCE FIR, Norway *Picea abies* (Spruce Fir) Thornton.
SPRUCE PINE *Pinus glabra*.
SPRUCEBERRY *Gaultheria procumbens* (Wintergreen) USA Sanecki.
SPUD *Solanum tuberosum* (Potato) Salaman. The word originally meant some kind of spade or digging fork, more particularly the three-pronged fork used to raise the potato crop. Spuddy is a slang name for a man who sells bad potatoes.
SPUR, Carlin- *Genista anglica* (Needle Whin) Scot. Jamieson. From the spines, of course.
SPUR, Knight's *Delphinium ajacis* (Larkspur) Lyte.
SPUR PEPPER *Capsicum annuum* (Chili) Barbados Gooding.
SPURGE, Balsam *Euphorbia balsamifera*. 'Spurge' is the same word as 'purge', from Latin expurgare.
SPURGE, Blooming *Euphorbia corollata* (Flowering Spurge) O P Brown.
SPURGE, Broad; SPURGE, Broad-leaved *Euphorbia platyphyllos*.
SPURGE, Caper *Euphorbia lathyris*. The fruits are quite often used green as a caper substitute - a dangerous practice, for they are as poisonous as all the other spurges.
SPURGE, Coral *Euphorbia corallioides*.
SPURGE, Cushion *Euphorbia polychroma*.
SPURGE, Cut-lobed *Euphorbia schizoloba*.
SPURGE, Cypress *Euphorbia cyparissa*. The specific name cyparissias was given by Pliny in the belief that Cyprus was its country of origin. So 'Cypress' in the English version is quite wrong.
SPURGE, Dwarf *Euphorbia exigua*.
SPURGE, Flax *Euphorbia paralias* (Sea Spurge) B & H. 'Flax', presumably from the shape of the leaves.
SPURGE, Flowering 1. *Daphne laureola* (Spurge Laurel) Jordan. 2. *Daphne mezereon* (Lady Laurel) Folkard. 3. *Euphorbia corollata*.

SPURGE, Garden, Small *Euphorbia peplus* (Petty Spurge) Curtis.
SPURGE, Glaucous, Broad-leaved *Euphorbia myrsinites*.
SPURGE, Great *Ricinus communis* (Castor Oil Plant) Parkinson.
SPURGE, Hairy 1. *Euphorbia hirta*. 2. *Euphorbia pilosa*.
SPURGE, Holly-leaf *Tetracoccus ilicifolius*.
SPURGE, Hungarian *Euphorbia esula*.
SPURGE, Irish *Euphorbia hiberna*. Locally common in the west of Ireland, but it is very occasionally found in southwest England, too.
SPURGE, Leafy *Euphorbia esula* (Hungarian Spurge) USA Gates.
SPURGE, Lint; SPURGE, Little *Euphorbia esula* (Hungarian Spurge) both Turner.
SPURGE, Mat *Euphorbia glyptosperma*.
SPURGE, Mezereon *Daphne mezereon* (Lady Laurel) Le Strange.
SPURGE, Milk *Euphorbia maculata* (Spotted Spurge) USA Gates. But all the spurges are noted for their milky sap.
SPURGE, Mountain *Euphorbia montana*.
SPURGE, Myrtle *Euphorbia lathyris* (Caper Spurge) Turner.
SPURGE, Nodding *Euphorbia maculata* (Spotted Spurge) USA Allan.
SPURGE, Painted *Euphorbia heterophylla* USA T H Kearney.
SPURGE, Petty *Euphorbia peplus*.
SPURGE, Pine *Euphorbia esula* (Hungarian Spurge) Parkinson.
SPURGE, Portland *Euphorbia portlandica*. Not confined to the Isle of Portland, for it can be found on dunes and other coastal sites from Chichester Harbour round to Galloway, and in Ireland, too.
SPURGE, Prostrate *Euphorbia supina* (Milk Purslane) Allan.
SPURGE, Purple; SPURGE, Sea, Purple *Euphorbia peplis*. Purple Sea Spurge is in Camden/Gibson.
SPURGE, Quacksalver's *Euphorbia esula* (Hungarian Spurge) W Miller. 'Quacksalver' means the same as the modern 'quack'.

SPURGE, Sea *Euphorbia paralias*.

SPURGE, Seaside *Euphorbia mesembryanthemifolia* Barbados Gooding.

SPURGE, Small *Euphorbia exigua* (Dwarf Spurge) Curtis.

SPURGE, Spotted 1. *Euphorbia maculata*. 2. *Euphorbia supina* (Milk Purslane) New Zealand Connor.

SPURGE, Sun; SPURGE, Sun-following *Euphorbia helioscopia*. The longer Sun-following Spurge was Turner's name. The names are given because of what Gerard described as its habit "of turning or keeping time with the Sunne".

SPURGE, Thyme-leaved *Euphorbia serpyllifolia*.

SPURGE, Tintern *Euphorbia stricta*. Very rare in this country, and confined to clearings in limestone woods in the Wye Valley, and one other site near Bath.

SPURGE, Upright *Euphorbia stricta* (Tintern Spurge) Clapham.

SPURGE, Wart *Euphorbia helioscopia* (Sun Spurge) Prior. Probably the best known of all wart remedies. Generations of children have gathered Sun Spurge to put the milky juice on their warts. Cf. also Wartweed, Warty-grass and a lot of variations on the theme.

SPURGE, Warted, Broad-leaved *Euphorbia platyphyllos* (Broad Spurge) Flower.1859.

SPURGE, White-margined *Euphorbia marginata* (Mountain Snow) Vestal & Schultes.

SPURGE, Wood *Euphorbia amygdaloides*.

SPURGE FLAX 1. *Daphne cneorum* (Flax-leaved Daphne) Gerard. 2. *Daphne mezereon* (Lady Laurel) Gerard (Flax).

SPURGE LAUREL 1. *Daphne laureola*. 2. *Daphne mezereum* (Lady Laurel) Gerard.

SPURGE NETTLE *Jatropha stimuloca*. All the Jatrophas are strongly purgative; this one earns the 'nettle' epithet from the fact that the spiny hairs can cause painful skin irritation. Cf. Bull-nettle.

SPURGE OLIVE 1. *Daphne laureola* (Spurge Laurel) Jordan. 2. *Daphne mezereon* (Lady Laurel) B & H.

SPURGE OLIVE, German *Daphne mezereon* (Lady Laurel) Gerard.

SPURGE THYME *Euphorbia peplus* (Petty Spurge) Turner.

SPURGE TREE *Euphorbia tirucalli* (Pencil Tree) USA Kingsbury.

SPURGEWORT *Iris foetidissima* (Foetid Iris) Dor. Gerard. In Gerard's words, "it is used by many Country People to purge corrupt Phlegm and Choler, which they do by drinking the Decoction of the Roots; and some, to make it more gentle do but infuse the sliced roots in Ale; and some take the Leaves, which serve well for the weaker Stomachs...".

SPURREY, Cliff *Spergularia rupicola* (Rock Sand Spurrey) McClintock. 'Spurrey' comes from a Dutch word, spurrie.

SPURREY, Corn *Spergula arvensis*.

SPURREY, Sand 1. *Spergularia rubra*. 2. *Suaeda maritima* (Seablite) Darling & Boyd.

SPURREY, Sand, Bocconi's; SPURREY, Sand, Greek; SPURREY, Sand, Red *Spergularia bocconi* Red Sand Spurrey is the usual name; Clapham prefers Bocconi's, and McClintock.1975 Greek.

SPURREY, Sand, Rock *Spergularia rupicola*.

SPURREY, Sand, Salt-marsh *Spergularia marginata*.

SPURREY, Sand, Shore *Spergularia salina*.

SPURREY, Sea *Spergularia salina*.

SPURREY, Sea, Greater *Spergularia marginata*.

SPURREY, Sea, Lesser *Spergularia salina* (Sea Spurrey) Darling & Boyd.

SPURRY *Spergula arvensis* (Corn Spurrey) S Africa Watt.

SPURRY, Franck *Spergula arvensis* (Corn Spurrey) Parkinson. Parkinson also called it Francking Spurwort, and Prior also has Franke. This plant used to be grown to fatten cattle; a franke is a stall in which they were shut up to be fattened.

SPURRY, Knotted *Sagina nodosa* (Knotted Pearlwort) Murdoch McNeill.

SPURS, Carlin *Genista anglica* (Needle Whin) Scot. Jamieson.

SPURWOOD *Ranunculus flammula* (Lesser Spearwort) Dev. B & H. 'Spear' varies locally; for instance it becomes 'spire' sometimes (Spirewort is recorded in the Hebrides), and it also becomes 'spur', as in this Devonshire name.

SPURWORT *Sherardia arvensis* (Field Madder) Prior.

SPURWORT, Francking *Spergula arvensis* (Corn Spurrey) Parkinson. Cf. Franck Spurry.

SQUARE, Carpenter's *Scrophularia nodosa* (Figwort) Grieve.1931.

SQUARE ST JOHN'S WORT; SQUARE ST JOHN'S GRASS *Hypericum tetrapterum* (St Peter's Wort) Gerard, Turner.

SQUARE TREE *Euonymus europaeus* (Spindle Tree) Turner. Probably meant to describe the berries.

SQUARRIB; SCARIBEUS *Scrophularia nodosa* (Figwort) Wilts. Dartnell & Goddard.1899. i.e. square rib, perhaps from the shape of the stem.

SQUASH *Cucumis melo* (Melon) Watt.

SQUASH, Autumn *Cucurbita maxima* (Giant Pumpkin) USA Elmore. Cf. Winter Squash.

SQUASH, Barbary *Cucurbita moschata* (Crook-neck Squash) Wit.

SQUASH, China *Cucurbita moschata* (Crook-neck Squash) Havard. But it is an American plant!

SQUASH, Crook-neck *Cucurbita moschata*.

SQUASH, Crook-neck,Canada *Cucurbita moschata* (Crook-neck Squash) Bianchini.

SQUASH, Cushaw *Cucurbita moschata* (Crook-neck Squash) Bianchini.

SQUASH, Custard *Cucurbita pepo var. melopepo* (Scallop Gourd). Cf. also Pattypan Squash, all descriptive.

SQUASH, Great *Cucurbita maxima* (Giant Pumpkin) Watt.

SQUASH, Hubbard *Cucurbita maxima* (Giant Pumpkin) Watt.

SQUASH, Malabar *Cucurbita ficifolia* (Malabar Gourd) Whitaker & Davis.

SQUASH, Pattypan *Cucurbita melo var. melopepo* (Scallop Gourd) Bianchini. Descriptive - Cf. Custard Squash.

SQUASH, Summer *Cucurbita pepo var. condensa*.

SQUASH, Turban *Cucurbita maxima* (Giant Pumpkin) Whitaker & Davis. Imaginatively descriptive.

SQUASH, Winter 1. *Cucurbita maxima* (Giant Pumpkin) USA Elmore. 2. *Cucurbita moschata* (Crook-neck Squash) Whitaker & Davis.

SQUAT; SQUATMORE *Glaucium flavum* (Horned Poppy) S Eng. Grigson. A squat is a bruise. At one time it was thought to be a remedy for them. 'More' in Squatmore means root.

SQUAW-BERRY; SQUAW-BUSH 1. *Rhus aromatica* (Sweet Sumach) Chamberlin (Squaw-berry). 2. *Rhus trilobata* (Three-leaf Sumach) both USA Elmore. Presumably because the berries were collected by American Indian women. They used to dry them, then grind them and mix with sugar and water, like jam. They crush the fresh fruit, too, to make into a cooling drink.

SQUAW-BUSH 1. *Cornus stolonifera* (Red-osier Dogwood) USA Bergen. 2. *Rhus trilobata* (Three-leaf Sumach) USA Elmore.

SQUAW CURRANT 1. *Ribes cereum* (Wax Currant) USA Cunningham. 2. *Ribes inebriens* (Wild Currant) USA T H Kearney.

SQUAW HUCKLEBERRY *Vaccinium stamineum*.

SQUAW LETTUCE, Western *Hydrophyllum occidentale*.

SQUAW TEA *Ephedra trifurca* (Long-leaved Joint-Fir) USA Elmore. An infusion of the branches is a popular local beverage in the areas in which it grows. That tea has its medicinal uses, too. Cf. Mormon Tea, Brigham Young Tea, and a number of others.

SQUAW-THORN *Lycium torreyi*.

SQUAW-VINE *Mitchella repens* (Partridge-berry) USA Mitton. Indian women used it for several weeks before childbirth, to ensure a safe and easy delivery. They also boiled down strong decoctions of the leaves to apply to sore breasts and cracked nipples. European immigrants learned from this, and used it similarly.

SQUAW-WEED, Douglas *Senecio douglasii* (Douglas Groundsel).

SQUAW-WEED, Round-leaved *Senecio oblovatus*.

SQUAW-WEED, Swamp *Senecio aureus*.

SQUAW'S CARPET *Ceanothus prostratus*.

SQUAWROOT 1. *Carum gardneri* (Yampa) USA Schenk & Gifford. 2. *Caulophyllum thalictroides* (Blue Cohosh) Grieve.1931. 3. *Cimicifuga racemosa* (Black Snakeroot) USA Lloyd. 4. *Trillium erectum* (Bethroot) RHS. Because they were all used by Indian women for female complaints. Blue Cohosh and Bethroot were used as aids in childbirth, the former by drinking a root infusion in warm water for a week or two prior to the expected date of delivery, in order to promote an easy birth, and the latter as a uterine stimulant.

SQUEAKER *Scrophularia aquatica* (Water Betony) Dev, Dor. Macmillan. Children used to strip the stems of their leaves, and scrape them across each other, when they produce a squeaking sound. Other names, like Fiddles and Crowdy-kits, which means the same thing, were given for that reason too.

SQUEEZE-JAWS *Linaria vulgaris* (Yellow Toadflax) Som. Macmillan. Descriptive, not only of this plant, but also of all the snapdragon type of flower.

SQUILL; SQULL ONION; SQUILL, Sea *Scilla maritima*. Latin squilla, from scilla, and then Greek skilla, the name of the plant. Squill Onion appears in T Hill and Sea Squill in RHS.

SQUILL, Alpine *Scilla bifolia*.

SQUILL, Autumnal *Scilla autumnalis*.

SQUILL, Corymbose *Scilla peruviana* (Peruvian Lily) Whittle & Cook.

SQUILL, Lebanon *Pushkinia libanotica* (Striped Squill) Howes.

SQUILL, Officinal *Scilla maritima* (Squill) Thornton. 'Officinal' (now altered to 'official') means that it was recognised in the pharmacopeia.

SQUILL, Pyrenean *Scilla liliohyacinthus*.

SQUILL, Red *Scilla maritima* (Squill).

SQUILL, Spring *Scilla verna*.

SQUILL, Striped *Pushkinia libanotica*.

SQUILL, Wild 1. *Scilla cooperi* S Africa Watt. 2. *Scilla natalensis* S Africa Watt.

SQUINANCY; SQUINANCY-BERRY *Ribes nigrum* Fernie (Squinancy); Ess, Lancs, Cumb. Grigson (Squinancy-berry). Blackcurrant was used in folk medicine against sore throat (a quinsy, or squinancy), long before cultivation, and it is still so used.

SQUINANCYWORT; SQUINANCY WOODRUFF *Asperula cynanchica* Squinancywort is the usual name; Squinancy Woodruff is in Barton & Castle. The word itself is a corruption of quinsey-wort, though both versions are from late Latin quinancia. The plant provides an astringent gargle which was used to treat the complaint. The version with 'woodruff' is merely an acknowledgement of the plant's close relationship with *Asperula odorata*, the woodruff.

SQUINANTYKE, Herb *Asperula cynanchica* (Squinancywort) Gerard.

SQUINTER-PIP *Geranium robertianum* (Herb Robert) Shrop. Grigson.

SQUIRREL-CORN; SQUIRREL-CORN, American; SQUIRREL'S CORN *Dicentra canadensis* (Turkey Corn) USA Hewitt (Squirrel-corn); McDonald (American); USA Tolstead (Squirrel's Corn).

SQUIRREL-TAIL GRASS *Hordeum marinum* (Wall Barley) Hepburn.

SQUIRREL'S FOOT FERN *Davallia bullata*.

SQUIRT, Water *Angelica sylvestris* (Wild Angelica) Som. Mabey. A reference to the hollow tubes, and to what children get up to with them. In Scotland, they are known as Ait-skeiters, or Bear-skeiters, 'bear' meaning oats, which is what children shoot through the stems, like a pea-shooter.

SQUIRTERS 1. *Acer rubrum* (Red Maple) USA H M Hyatt. 2. *Symphoricarpos rivularis* (Snowberry) Som. Macmillan. Children used to "squirt" the juice from the maple pods into each others' faces in the spring. The juice was thought to cause freckles.

SQUIRTING CUCUMBER *Ecballium elaterium*. 'Squirting', because when the fruit is fully ripe, it bursts open on being touched, so that the seeds, and a milky juice, are squirted out with some considerable force.

SQUITCH *Agropyron repens* (Couchgrass) Grigson.1959. Squitch, or Quitch, is interchangeable with Couch. They are all from OE cwice, alive.

STABWORT; STUBWORT; STOBWORT *Oxalis acetosella* (Wood Sorrel) Prior (Stabwort, Stubwort); B & H (Stobwort). According to Parkinson, so called "because it is singular good in all wounds and stabbes into the body". But Stubwort shows that it grows round the stubs of trees.

STAFF, Jacob's 1. *Asphodeline lutea* (Yellow Asphodel) Grieve.1931. 2. *Verbascum thapsus* (Mullein) Cumb. B & H; USA Henkel. A tall plant like mullein attracts 'staff' and 'rod' names - Cf. Jupiter's, Shepherd's and Peter's Staff, and Aaron's Rod, Golden Rod, etc. Even the asphodel has been likened to a spear as well as a staff.

STAFF, Jupiter's *Verbascum thapsus* (Mullein) Parkinson; USA Henkel.

STAFF, Peter's *Verbascum thapsus* (Mullein) B & H; USA Henkel.

STAFF, Royal *Asphodelus ramosus* (White Asphodel) Grieve.1931. Cf. King's Spear, which repeats the regal attribute.

STAFF, St Joseph's 1. *Narcissus pseudo-narcissus* (Daffodil) Tynan & Maitland. 2. *Solidago virgaurea* (Golden Rod) Tynan & Maitland. St Joseph's staff was chiefly memorable as the origin of the Glastonbury Thorn. What it is doing in this company is not known, particularly for daffodil. At least, Golden Rod has the 'rod' name to refer to 'staff'.

STAFF, St Patrick's *Pinguicula vulgaris* (Butterwort) Antrim PLNN1.

STAFF, Shepherd's 1. *Dipsacus fullonum* (Teasel) Som. Macmillan. 2. *Verbascum thapsus* (Mullein) Cumb. B & H.

STAFF TREE *Celastrus scandens* (American Bittersweet) J Smith.

STAFF VINE 1. *Celastrus orbiculatus* (Oriental Bittersweet) RHS. 2. *Celastrus scandens* (American Bittersweet) W Miller.

STAG'S HORN CLUBMOSS *Lycopodium clavatum*.

STAG'S HORN LICHEN *Evernia prunastris*.

STAG'S HORN MOSS *Lycopodium inundatum* (Clubmoss) Som. Elworthy.

STAG'S HORN PLANTAIN *Plantago coronopus* (Buck's Horn Plantain) Hepburn.

STAG'S HORN SUMACH *Rhus typhina*.

STAGBERRY *Symphoricarpos orbicularis* (Red Wolfberry) USA Elmore.

STAGBUSH *Viburnum prunifolium* (Black Haw) USA Thomson.1978.

STAGGERGRASS *Amaianthemum muscaetoxicum*.

STAGGERS WEED *Matricaria nigellaefolia* S Africa Watt.

STAGGERWEED 1. *Dicentra canadensis* (Turkey Corn) Grieve.1931. 2. *Stachys arvensis* (Field Woundwort) New Zealand Connor. Field Woundwort may cause "staggers" in cattle that eat it, but there are no records as to why Turkey Corn should bear the name.

STAGGERWORT; STAGGWORT *Senecio jacobaea* (Ragwort) both Prior. There is some confusion with these names. One reads that ragwort cures the staggers in horses, and one also reads that the disease is actually caused by it, which is more likely. Prior, though, said that ragwort was applied as a vulnerary to newly castrated bulls, called seggs or staggs.

STAINCH *Ononis repens* (Rest-harrow) N Eng. B & H.

STAINLESS BAY *Laurus nobilis* (Bay) Som. Macmillan. Bay is the same as laurel, so presumably this is to distinguish it from the other laurel, *Prunus laurocerasus*, which has spotted leaves.

STALEWORT *Artemisia abrotanum* (Southernwood) Radford. Why? No-one seems to complain of the smell. Smelling-wood, also recorded, sounds more congenial.

STALLIONS; STALLIONS-AND-MARES *Arum maculatum* (Cuckoo-pint) both Yks. B & H. An example of the male plus female double names that start with the best-known, Lords-and-ladies, and all brought about by the imagery of the spadix in the spathe. Sometimes, as with Stallions, only the first element of these double names is used. Cf. Bulls.

STAMMERWORT *Senecio jacobaea* (Ragwort) Culpepper. Le Strange said that ragwort was prescribed for speech impediments, but this looks too like staggerwort to be accepted without question.

STANLEY, Joe *Geranium robertianum* (Herb Robert) Dor. Macmillan.

STANCHE *Capsella bursa-pastoris* (Shepherd's Purse) Cockayne. Shepherd's Purse was celebrated as a vulnerary, and it is still used to stop bleeding.

STANDARD, Dog; STANDERS, Dog *Senecio jacobaea* (Ragwort) Yks. Carr (Standard); Worcs, Yks, N Eng Grigson (Standers).

STANDELWELKS *Orchis mascula* (Early Purple Orchis) Halliwell. Cf. Standlegrass, etc.

STANDER-GRASS 1. *Orchis mascula* (Early Purple Orchid) Lyte. 2. *Spiranthes spiralis* (Lady's Tresses) Notes & Queries. vol 7. Cf. Stannen-grass, Standerwort, Standlewort and Standlegrass for Early Purple Orchid.

STANDERWORT; STANDLE-GRASS; STANDLEWORT *Orchis mascula* (Early Purple Orchid) Prior (Standerwort); Parkinson (Standle-grass, Standlewort). To do with the aphrodisiac qualities ascribed to the plant, according to Prior.

STANDING GUSSES *Arum maculatum* (Cuckoo-pint) Som. Grigson. Perhaps a corruption of Gethsemane, which also occurs as a name for Cuckoo-pint, but more likely to be a sexual name allied to cuckoo-pint itself.

STANDING GUSSETS *Orchis mascula* (Early Purple Orchid) Dev. Macmillan. A variation of the better-known Gandergosses, which becomes Gandigoslings in Wiltshire.

STANDING CYPRESS *Gilia rubra* (Scarlet Gilia) Perry.

STANEY-RAW *Parmelia saxatilis* (Gray Stone Crottle) Scot. Bolton.

STANMARCH *Smyrnium olusatrum* (Alexanders) B & H. 'March' is an old name for parsley (and also for Wild Celery), so this seems to be Stone Parsley; but that is *Sison amomum*.

STANNEN-GRASS *Orchis mascula* (Early Purple Orchid) Lyte. Cf. Stander-grass.

STAR 1. *Bellis perennis* (Daisy) Som. Macmillan. 2. *Campanula glomerata* (Clustered Bellflower) Wilts. Dartnell & Goddard. 3. *Geranium robertianum* (Herb Robert) Som. Macmillan. 4. *Lysimachia nummularia* (Creeping Jenny) Wilts. Grigson. 5. *Potentilla erecta* (Tormentil) Wilts. Grigson. 6. *Stellaria holostea* (Greater Stitchwort) Wilts. Dartnell & Goddard. 7. *Syringa vulgaris* (Lilac) Som. Macmillan. Clustered Bellflower seems to be the odd one out, for the rest have flowers that seem to fit in with the general appellation.

STAR, Aniseed *Illicium verum* (True Anise) Grieve. It is sometimes known as the Star-Anise.

STAR, Bethlem *Ornithogalum umbellatum* (Star-of-Bethlehem) B & H. Descriptive of the flowers.

STAR, Blazing 1. *Aletris farinosa* (Unicorn-root) USA Le Strange. 2. *Liatris cylindracea* USA Zenkert. 3. *Liatris punctata* (Dotted Button-Snakeroot) USA Gates. 4. *Liatris scabiosa*. 5. *Liatris spicata* (Button-Snakeroot) Hay & Synge.

STAR, Blue 1. *Amsonia salicifolia*. 2. *Veronica chamaedrys* (Germander Speedwell) Stir. B & H.

STAR, Brilliant *Kalanchoe blossfeldiana.*
STAR, Earth *Cryptanthus sp.*
STAR, Earth, Green *Cryptanthus acaulis.*
STAR, Eastern *Passiflora caerulea* (Passion Flower) Som. Macmillan. Cf. Star-of-Bethlehem for this. Indeed nearly all of the names have some bearing on Christ's life and Passion.
STAR, Evening 1. *Oenothera biennis* (Lesser Evening Primrose) USA Bergen. 2. *Oenothera erythrosepala* (Large Evening Primrose) Friend.
STAR, Falling 1. *Ceratophyllum demersum* (Hornwort) Tynan & Maitland. 2. *Crocosmia aurea.*
STAR, Gleaming *Saxifraga x urbinum* (London Pride) Som. Macmillan.
STAR, Gold *Geum urbanum* (Avens) Som. Macmillan.
STAR, Golden 1. *Chrysogonum virginianum* (Golden Knee) A W Smith. 2. *Primula vulgaris* (Primrose) Som. Macmillan. 3. *Ranunculus ficaria* (Lesser Celandine) Som. Macmillan.
STAR, Jerusalem 1. *Cerastium tomentosum* (Snow-in-summer) War. B & H. 2. *Hypericum calycinum* (Rose-of-Sharon) Shrop. B & H. 3. *Senecio cinerea* (Cineraria) Wilts. Macmillan.
STAR, Little 1. *Bellis perennis* (Daisy) Som. Macmillan. 2. *Vinca minor* (Lesser Periwinkle) Som. Macmillan.
STAR, Midday *Hypoxis stellata.* A South African plant, given this name there because of its habit of opening only in the sun.
STAR, Miller's *Stellaria holostea* (Greater Stitchwort) Suss. B & H.
STAR, Morning *Stellaria holostea* (Greater Stitchwort) Dev. Macmillan.
STAR, Open *Leucanthemum vulgare* (Ox-eye Daisy) Som. Grigson.
STAR, Open, Little *Bellis perennis* (Daisy) Som. Macmillan.
STAR, Rainbow *Cryptanthus bromelioides.*
STAR, Shooting *Dodecatheon meadia.*
STAR, Shooting, Henderson *Dodecatheon hendersonii.*

STAR, Twinkle *Stellaria holostea* (Greater Stitchwort) Som. Macmillan.
STAR, Yellow 1. *Eranthis hyemalis* (Winter Aconite) Tynan & Maitland. 2. *Tussilago farfara* (Coltsfoot) Som. Macmillan.
STAR-ANISE; ANISEED STAR *Illicium verum* (True Anise) both Grieve.
STAR-AND-GARTER *Ornithogalum umbellatum* (Star-of-Bethlehem) Wilts. D Grose.
STAR-APPLE *Chrysophyllum cainito.*
STAR-APPLE, African *Chrysophyllum africanum.*
STAR-APPLE, Monkey *Chryosphyllum perpulchrum.* The fruit is reckoned to be inferior to that of the other Star-apples, hence 'monkey' in the common name.
STAR-APPLE, White *Chrysophyllum allbidum.*
STAR-BUR *Centaurea calcitrapa* (Star Thistle) S Africa Watt.
STAR CHICKWEED *Stellaria media* (Chickweed) Hatfield.
STAR-FLOWER 1. *Allium ursinum* (Ramsons) Fernie. 2. *Borago officinalis* (Borage) Dev. B & H. 3. *Lysimachia nemora* (Yellow Pimpernel) Wilts. Dartnell. & Goddard. 4. *Ornithogalum umbellatum* (Star of Bethlehem) Bucks. B & H. 5. *Potentilla erecta* (Tormentil) Wilts. Dartnell & Goddard. 6. *Ranunculus ficaria* (Lesser Celandine) Som. Macmillan. 7. *Sedum acre* (Biting Stonecrop) Som. Macmillan. 8. *Stellaria holostea* (Greater Stitchwort) Suss, Lancs. Grigson. 9. *Trientalis borealis* USA House. 10. *Trientalis europaea var. latifolia* USA Schenk & Gifford. All descriptive by shape of flower.
STAR-FLOWER, Arabian *Ornithogalum arabicum* Tradescant.
STAR-FLOWER VINE *Petrea volubilis* (Purple Wreath Vine) Simmons.
STAR-FRUIT *Damasonium alisma.*
STAR GENTIAN *Gentiana verna* (Spring Gentian) Grigson.
STAR-GRASS 1. *Aletris farinosa* (Unicorn-root) USA Le Strange. 2. *Callitriche verna* (Water Starwort) Prior. 3. *Asperula odorata*

(Woodruff) Cumb, N'thum. Grigson. 4. *Stellaria holostea* (Greater Stitchwort) Yks. Grigson.

STAR HYACINTH *Scilla verna* (Spring Squill) Prior.

STAR HYACINTH, Autumnal *Scilla autumnalis* (Autumnal Squill) Camden/Gibson.

STAR IPOMAEA *Ipomaea coccinea* (Red Morning Glory) RHS.

STAR JASMINE *Trachelospermum jasminoides* (Confederate Jasmine) Simmons.

STAR JASMINE, Southern *Jasminum gracillimum*.

STAR LILY *Eucharis grandiflora* (Amazon Lily) Howes.

STAR LOTUS *Nymphaea odorata* (White Pond Lily) Howes.

STAR NAKED BOYS; STAR NAKED LADIES *Colchicum autumnale* (Meadow Saffron) Norf. B & H (Boys); Ess. Gepp (Ladies). 'Stark naked', of course. There are many similar names, all referring to the way the flowers appear before the leaves.

STAR OF BETHLEHEM 1. *Anagallis arvensis* (Scarlet Pimpernel) Som. Macmillan. 2. *Anemone nemerosa* (Wood Anemone) Som. Macmillan. 3. *Campanula isophylla* (Italian Bellflower) Gannon. 4. *Hypericum calycinum* (Rose of Sharon) B & H. 5. *Ornithogalum thyrsoides* (Chincherinchee) S Africa Watt. 6. *Ornithogalum umbellatum*. 7. *Passiflora caerulea* (Passion Flower) Som. Macmillan. 8. *Stellaria holostea* (Greater Stitchwort) Dev, Som. Macmillan; Wilts. Dartnell & Goddard.1899; E Ang, N Ire. Grigson. 9. *Tragopogon pratensis* (Goat's Beard) Tynan & Maitland. All with more or less starry shaped flowers, except Goat's Beard, which seems the odd one out. Some are not white, as with pimpernel and Rose of Sharon, and Star of Bethlehem for Italian Bellflower seems a modern florists' name.

STAR OF BETHLEHEM, Drooping *Ornithogalum nutans*.

STAR OF BETHLEHEM, Guernsey *Allium neapolitanum* (Daffodil Garlic) McClintock.1975.

STAR OF BETHLEHEM, Neapolitan *Ornithogalum nutans* (Drooping Star of Bethlehem) Whittle & Cook. Clusius named it *O neapolitanum*, simply because a consignment of bulbs reached him from Naples.

STAR OF BETHLEHEM, Spiked *Ornithogalum pyrenaicum*.

STAR OF BETHLEHEM, Yellow *Gagea lutea*.

STAR OF BETHLEHEM ORCHID *Angraecum sesquipedale*.

STAR OF HUNGARY *Ornithogalum umbellatum* (Star of Bethlehem) Grieve.1931.

STAR OF JERUSALEM *Tragopogon pratensis* (Goat's Beard) Gerard. It also used to be known as Star of Bethlehem. Why?

STAR OF PERSIA *Allium christophii*.

STAR OF THE EARTH 1. *Coronopus squamatus* (Wart-cress) Suss. Grigson. 2. *Geum urbanum* (Avens) Tynan & Maitland. 3. *Plantago coronopus* (Buck's-horn Plantain) Aubrey.

STAR OF THE LUNDI *Pachypodium saundersii*. A Central African shrub. The Lundi is a river in Zimbabwe.

STAR OF THE NIGHT *Clusia rosea* Howes.

STAR OF THE NORTH *Geum urbanum* (Avens) Notes & Queries.7.

STAR OF THE SABI *Adenium multiflorum*.

STAR OF THE VELDT *Dimorphoteca aurantiaca*. A South African plant.

STAR OF THE WOOD *Stellaria holostea* (Greater Stitchwort) Dev, Som. Macmillan.

STAR PINE *Pinus pinaster* (Maritime Pine) Notes & Queries. vol 4. It has been claimed that 'pinaster' means star pine. It does not - the suffix '-aster' means a poor imitation. This name must surely have been given as a translation of the specific name when the wrong meaning was ascribed to it.

STAR PLUM *Chrysophyllum cainito* (Star-apple) Howes.

STAR(RY) SAXIFRAGE *Saxifraga stellata*.

STAR THISTLE *Centaurea calcitrapa*.
STAR THISTLE, Maltese *Centaurea melitensis* (Napa Thistle) Clapham.
STAR THISTLE, Red *Centaurea calcitrapa* (Star Thistle) Blamey.
STAR THISTLE, Rough *Centaurea aspera*.
STAR THISTLE, Yellow *Centaurea solstitialis* (St Barnaby's Thistle) Blamey.
STARCH, Portland *Arum maculatum* (Cuckoo-pint) W Miller. It was once cultivated in the Isle of Portland for the tubers, which when cooked yielded Portland Starch, better known as Portland Sago, used as a substitute for arrowroot. It was " a pure and white starch", Gerard said, "but most hurtfull to the hands of the Laundresse that hath the handling of it, for it chappeth, blistereth, and maketh the hands rough and rugged, and withall smarting".
STARCH HYACINTH *Muscari racemosum* Grieve.1931. In Elizabethan times, the juice from the stems was used to stiffen ruffs.
STARCHWORT *Arum maculatum* (Cuckoo-pint) Gerard. See Portland Starch.
STARFISH FLOWER *Stapelia variegata*. Star-shaped flowers.
STARLIGHT *Geranium molle* (Dove's-foot Cranesbill) Bucks. Grigson.
STARRY EYES *Ornithogalum umbellatum* (Star of Bethlehem) Som. Macmillan.
STARRY GLASSWORT *Cerastium arvense* (Field Mouse-ear) USA Elmore.
STARVE-ACRE *Ranunculus arvensis* (Meadow Buttercup) Prior. Cf. Hungerweed. Given either because it impoverishes the land, or because the presence of the plant indicates poor soil.
STARVED ASTER *Aster lateriflorus*.
STARWEED *Stellaria media* (Chickweed) USA Allan.
STARWORT 1. *Aster macrophyllus* (Large-leaved Aster) USA Yarnell. 2. *Aster novae-angliae* (Michaelmas Daisy) Tynan & Maitland. 3. *Aster tripolium* (Sea Aster) Geldart. 4. *Liatris punctata* (Dotted Button-Snakeroot) USA Vestal. & Schultes. 5. *Stellaria graminea* (Lesser Stitchwort) Glos. J D Robertson. 6. *Stellaria holostea* (Greater Stitchwort) Glos. J D Robertson; Dev, Som, Surr, E Ang, N Ire. Grigson. 7. *Stellaria media* (Chickweed) Cambs. Porter.
STARWORT, Autumnal *Callitriche autumnalis* (Narrow Water Starwort) Clapham.
STARWORT, Bog *Stellaria uliginosa* (Bog Stitchwort) Murdoch McNeill.
STARWORT, Common *Callitriche stagnalis*.
STARWORT, Great *Stellaria holostea* (Greater Stitchwort) Hutchinson.
STARWORT, Hairy *Aster canescens* (Hairy Aster) USA Elmore.
STARWORT, Sea *Aster tripolium* (Sea Aster) Gerard.
STARWORT, Water *Callitriche verna*.
STARWORT, Water, Narrow *Callitriche autumnalis*.
STARWORT MOUSE-EAR *Cerastium cerastoides*.
STATICE, Candlewick *Limonium suworowii*.
STAUNCH 1. *Anthyllis vulneraria* (Kidney Vetch) B & H. 2. *Capsella bursa-pastoris* (Shepherd's Purse) Cockayne. Shepherd's Purse "staieth bleeding in any part of the body... It healeth greene or bleeding wounds..." (Gerard). Herbalists still use it to stop bleeding, and they also still use kidney vetch to treat wounds. The specific name '*vulneraria*' should be enough to proclaim the usage.
STAUNCH-GIRS; STAUNCH-GRASS *Achillea millefolium* (Yarrow) Scot. B & H, Fernie. Yarrow is a wound herb, applied simply as a kind of poultice. Cf. Woundwort, used in Somerset, and the old book name Sanguinary.
STAVERWORT *Senecio jacobaea* (Ragwort) B & H. Stavers, Prior said, is another form of staggers, which is the more usual word, and the plant is better known in this context as Staggerwort. See the last-named for the explanation.
STAVESACRE *Delphinium staphisagria*. 'Staphisagria' means wild grape, the leaves being similar to those of the vine. Stavesacre is simply an englishing of that Latin name.

STAY-PLOUGH *Ononis repens* (Rest-harrow) Prior. Stay-plough is of exactly the same import as Rest-harrow. Much older than either of them is remore aratri, plough hindrance. There was also arest bovis, "because it maketh the Oxen whilst they be in plowing to rest or stand still" (Gerard).

STEDFAST *Ricinus communis* (Castor Oil Plant) B & H.

STEE, Jacob's *Atropa belladonna* (Deadly Nightshade) Lincs. Grigson. 'Stee' is a local name for a ladder. Jacob's Ladder is applied to this plant in Scotland.

STEEPGRASS; STEEPWEED; STEEPWORT *Pinguicula vulgaris* (Butterwort) Mackenzie (Steepgrass); N Ire. Grigson (Steepweed, Steepwort). 'Steep' is a word for rennet, for the leaves act like it. Cf. such names as Earning-grass (to earn is North country for to curdle milk) and Thickening Grass.

STEEPLE; STEEPLEWORT *Agrinomia eupatoria* (Agrimony) both Tynan & Maitland. Cf. Church Steeples, given to describe the long spikes of the flowers.

STEEPLE, Church 1. *Agrimonia eupatoria* (Agrimony) Som, Suss. Brownlow. 2. *Eupatorium cannabinum* (Hemp Agrimony) Friend.

STEEPLE BELLFLOWER; STEEPLE BELLS *Campanula pyramidalis* (Chimney Bellflower) Parkinson, B & H.

STEEPLEBRUSH *Spiraea tomentosa* (Hardhack) USA House.

STEPDAUGHTERS *Viola tricolor* (Pansy) Yks. B & H. See Stepmothers.

STEPMOTHER 1. *Stellaria holostea* (Greater Stitchwort) Som. Macmillan. 2. *Viola tricolor* (Pansy) Yks. Carr. In northeast Scotland, the largest of the three lower petals of pansies is the stepmother, while on each side of her stand her own two daughters, and the two upper petals are the step-daughters. Cf. also Sisters-in-law and Godfathers-and-godmothers.

STEPMOTHER'S BLESSING *Anthriscus sylvestris* (Cow Parsley) Shrop.

Vickery.1985. Cf. the various names centered on Mother-die. Never give cow parsley to your mother, or she will die. In the light of this, Stepmother's Blessing has some significance. Possibly, any white-flowered umbellifer is unlucky, simply because it resembles hemlock.

STEPMOTHER'S SON *Viola tricolor* (Pansy) Yks. Grigson.

STERTION *Nasturtium officinale* (Watercress) Fernie. An attempt at Nasturtium, of course.

STICADORE; STICADOVE *Lavandula stoechas* (Spanish Lavender) Brownlow (Sticadore); Parkinson (Stickadove). Taken from the specific name, stoechas.

STICKABACK; STICKLEBACK *Galium aparine* (Goose-grass) Ches, Lancs, Cumb. Grigson (Stickaback); Ches. Grigson (Stickleback). "Stick" names for the burrs provide obvious imagery.

STICK-BUTTON 1. *Arctium lappa* (Burdock) Som. Macmillan; USA Henkel. 2. *Galium aparine* (Goose-grass) Som. Macmillan.

STICK-DONKEY; STICK-ROBIN *Galium aparine* Som. Macmillan (Donkey); Tynan & Maitland (Robin).

STICKERS *Arctium lappa* burrs (Burrs) Som. Macmillan.

STICKLEWORT; STICKWORT *Agrimonia eupatoria* (Agrimony) Fernie (Sticklewort); Grieve.1931 (Stickwort). Because, like burrs, the seed vessels cling by the hooked ends of their stiff hairs to anyone coming into contact with them.

STICKS, Black *Typha latifolia* (False Bulrush) Som. Macmillan.

STICKTIGHT 1. *Bidens bipinnata* (Spanish Needles) USA Elmore. 2. *Bidens cernua* (Nodding Bur Marigold) USA Zenkert. 3. *Bidens frondosa* (Beggar-ticks) USA Gates.

STICKY-BACK 1. *Arctium lappa* burrs (Burrs) Dev. Macmillan. 2. *Drosera rotundifolia* (Sundew) Som. Macmillan. 3. *Galium aparine* (Goose-grass) Som. Macmillan; Cumb. Grigson. Sundew is viscous, hence 'sticky'. The other two plants use the word in a different sense, that of hooking themselves on to clothes, etc., to stick there.

STICKY-BALLS 1. *Arctium lappa* burrs (Burrs) War. F Savage. 2. *Galium aparine* (Goose-grass) Som. Macmillan.

STICKY-BUDS 1. *Aesculus hippocastanum* leaf buds. 2. *Cynoglossum officinale* (Hound's Tongue) Dor. Macmillan.

STICKY-BUTTONS 1. *Arctium lappa* (Burdock) Dev. B & H. 2. *Galium aparine* (Goose-grass) Wessex Rogers.

STICKY CATCHFLY *Viscaria vulgaris* (Red Catchfly) Blamey.

STICKY CATCHFLY, White *Silene viscosa*.

STICKY CURRANT *Ribes viscosissimum*.

STTCKY GOOSEFOOT *Chenopodium botrys* (Jerusalem Oak) Polunin.

STICKY GRASS 1. *Cladium jamaicense*. 2. *Galium mollugo* (White Bedstraw) Cumb. B & H.

STICKY GROUNDSEL *Senecio viscosus*.

STICKY-HEAD *Grindelia squarrosa* (Gumweed) Gilmore.

STICKY JACK *Arctium lappa* (Burdock) Som. Mabey.

STICKY STORKSBILL *Erodium cicutarium* (Storksbill) McClintock.1985.

STICKY-WEED *Galium aparine* (Goose-grass) Tynan & Maitland.

STICKY WILLIE *Galium aparine* (Goose-grass) Scot. Fairweather.

STICKY WILLOW *Arctium lappa* (Burdock) Argyll PLNN.29.

STIKE-PILE *Erodium cicutarium* (Storksbill) Halliwell.

STING NETTLE; STINGING NETTLE *Urtica dioica* (Nettle) Som. Elworthy (Sting).

STING NETTLE, White *Lamium album* (White Deadnettle) Dev. Grigson. Given even though it does not sting.

STING-NETTLE FLOWER *Ajuga reptans* (Bugle) Som. Macmillan.

STINGING NETTLE TREE *Obetia pinnatifida*. It belongs to the nettle family, and the veins on the upper side of the leaf bear many stinging hairs, hence the name.

STINGY NETTLE *Urtica dioica* (Nettle) Dev, N'thants. B & H.

STINGY-WINGIES *Galeobdalon luteum* (Yellow Archangel) Dor. Macmillan. Vaguely nettle shaped leaves, but they do not sting.

STINK CELTIS *Celtis cinnamonea*.

STINK DAISY *Tanacetum parthenium* (Feverfew) Som. Macmillan.

STINK DAVIE *Taraxacum officinale* (Dandelion) Clack. Grigson. Probably misapplied, for no-one else objects to the smell, Gerard even describing it as "sweet in smell". It probably should be Ragwort, for there can be genuine objections to the smell of that, and it is known as Stinking Davies, among other epithets.

STINK FLOWER 1. *Cleome serrulata* (Rocky Mountain Bee Plant) USA Elmore. 2. *Conium maculatum* (Hemlock) Som. Macmillan. 3. *Geranium robertianum* (Herb Robert) Som. Macmillan.

STINK FLOWER, Sonora *Cleome sonorae* (Sonora Beeweed) USA Elmore.

STINK LILY *Fritillaria imperialis* (Crown Imperial) Coats. It is always claimed that they smell like a fox when the root is rubbed.

STINK MAYWEED *Anthemis cotula* (Maydweed) Murdoch McNeill. Cf. Stinking Camomile, Stinking Mathes, etc.

STINK PLANT *Allium ursinum* (Ramsons) Lincs. B & H. Cf. Onion Stinker, Stinking Onion, Stinking Jenny, Stinking Lily, all most appropriate.

STINK TREE *Viburnum opulus* (Guelder Rose) IOW B & H.

STINKER, Dog *Anthemis cotula* (Maydweed) Yks. Robinson.

STINKER, Onion *Allium ursinum* (Ramsons) Som. Macmillan. Cf. Stink Plant, etc.

STINKER BOB; STINKING BOB *Geranium robertianum* (Herb Robert) Som. Macmillan (Stinker); Midl. Grigson (Stinking).

STINKING AIRACH; STINKING ARAG; STINKING ARRACH *Chenopodium vulvaria* (Stinking Orach) Culpepper (Airach); Scot. Graham (Arag); Hill (Arrach).

STINKING ALISANDERS; STINKING ELSHINDERS *Senecio jacobaea* (Ragwort) N'thum, Stir Grigson (Alisanders); B & H (Elshinders). Cf. Stinking Billy, Stinking Willie etc.

STINKING ASH *Ptelea trifoliata* (Hop Tree) Hora.

STINKING BENJAMIN *Trillium erectum* (Bethroot) Howes.

STINKING BILLY *Senecio jacobaea* (Ragwort) Lincs. Grigson.

STINKING BLITE *Chenopodium vulvaria* (Stinking Orach) Curtis.

STINKING BUSH *Cassia ocidentalis* (Coffee Senna) Barbados Gooding. Cf. Stinking Weed, Stinking Pea and Stinkweed.

STINKING CAMOMILE *Anthemis cotula* (Maydweed) Yks. B & H.

STINKING CHRISTOPHER 1. *Scrophularia aquatica* (Water Betony) Cumb. Grigson. 2. *Scrophularia nodosa* (Figwort) Cumb. Grigson.

STINKING CLOVER 1. *Cleome sonorae* (Sonora Beeweed) USA Elmore. 2. *Melilotus indica* (Small-flowering Melilot) S Africa. Watt.

STINKING COTTON *Senecio viscosus* (Sticky Groundsel) Le Strange.

STINKING CRANESBILL *Geranium robertianum* (Herb Robert) Friend.

STINKING DAVIES *Senecio jacobaea* (Ragwort) Friend.

STINKING ELDER 1. *Sambucus nigra* (Elder) Dyer. 2. *Senecio pubens* USA Yarnell.

STINKING ELSHANDER *Tanacetum vulgare* (Tansy) Perth. Grigson.

STINKING GLADDON; STINKING GLADWYN *Iris foetidissima* (Foetid Iris) Gerard (Gladdon); Culpepper (Gladwyn). 'Gladdon' is one of the old names for an iris. It means sword, and is a description of the plant's leaves.

STINKING GOOSEFOOT *Chenopodium vulvaria* (Stinking Orach) McClintock.

STINKING GROUNDSEL *Senecio viscosus* (Sticky Groundsel) Polunin.

STINKING HEDGE MUSTARD *Alliaria petiolata* (Jack-by-the-hedge) Thornton.

STINKING HELLEBORE *Helleborus foetidus*.

STINKING IRIS *Iris foetidissima* (Foetid Iris) RHS.

STINKING JENNY 1. *Allium ursinum* (Ramsons) Som. Macmillan. 2. *Geranium robertianum* (Herb Robert) Som. Macmillan. Cf. Stink Plant, etc., for Ramsons.

STINKING JUNIPER *Juniperus foetidissima*.

STINKING LILY *Allium ursinum* (Ramsons) Som. Grigson; Yks. L Gordon. Cf. Stink Plant.

STINKING MAYWEED *Anthemis cotula* (Maydweed) S Africa Watt.

STINKING MOTHERWORT *Chenopodium vulvaria* (Stinking Orach) Gerard.

STINKING NANCY 1. *Matricaria recutita* (Scented Mayweed) Leics. PLNN29. 2. *Senecio jacobaea* (Ragwort) Notts. Grigson. 3. *Succisa pratensis* (Devil's-bit) Ches. B & H. It is odd to find Devil's-bit in this company. There are no further comments on the smell.

STINKING NANNY 1. *Anthemis cotula* (Maydweed) Leics. PLNN29. 2. *Matricaria chamomilla* (Scented Mayweed) Leics. PLNN29. 3. *Tripleurospermum maritimum* (Scentless Mayweed) Leics. PLNN29. Obviously the plurality is the result of confusion in identification. The last one is scentless, anyway.

STINKING ONION 1. *Allium triquetrum* (Three-cornered Leek) McClintock.1975. 2. *Allium ursinum* (Ramsons) Freethy. Cf. Stink Plant, etc., for Ramsons.

STINKING ORACH *Chenopodium vulvaria*.

STINKING PASSION FLOWER *Passiflora foetida*.

STINKING PEA *Cassia occidentalis*

(Coffee Senna) Dalziel. Cf. Stinkweed, Stinking Weed and Stinking Bush for this.

STINKING ROBERT *Geranium robertianum* (Herb Robert) Donegal Grigson. Cf. Stinking Bob, Stinker Bob, etc. There is no doubt of the bad smell of Herb Robert, hence too such names as Fox Grass. The hairs of the plant release this foxy or musky smell when handled.

STINKING ROGER 1. *Ballota nigra* (Black Horehound) Shrop. Grigson. 2. *Frangula alnus* (Alder Buckthorn) Ches. Grigson. 3. *Geranium robertianum* (Herb Robert) Lancs. Grigson. 4. *Hyoscyamus niger* (Henbane) Cumb. Grigson. 5. *Scrophularia aquatica* (Water Betony) Ches. Holland; Cumb, Ayr, Ire. Grigson. 6. *Scrophularia nodosa* (Figwort) Ches, Cumb, Ayr, Ire. Grigson.

STINKING ST JOHN'S WORT *Hypericum hircinum* (Stinking Tutsan) Clapham.

STINKING SEGG *Iris foetidissima* (Foetid Iris) Langham.

STINKING TAM; STINKING TOMMY *Ononis repens* N'thum. Grigson (Tam); Fernie (Tommy).

STINKING TREE, God's *Sambucus nigra* (Elder) Dor. Grigson. A gypsy name, apparently. The legend is that the Lord hid in an elder, which betrayed him. The punishment was that "you shall always stink".

STINKING TUTSAN *Hypericum hircinum*. The specific name shows that it is goats that the plant stinks of. Cf. Goat-scented Tutsan, and Stinking St John's Wort.

STINKING WEED 1. *Cassia occidentalis* (Coffee Senna) USA Watt. 2. *Lamium purpureum* (Red Deadnettle) Cumb. Grigson. 3. *Senecio jacobaea* (Ragwort) Scot. Jamieson.

STINKING WILLIE 1. *Senecio jacobaea* (Ragwort) Hants. Hants FWI; Mor. Grigson; USA Kingsbury. 2. *Tanacetum vulgare* (Tansy) Scot. Grigson.

STINKING WILLOW *Amorpha californica* USA T H Kearney.

STINKROOT *Datura stramonium* (Thorn-apple) S Africa Watt. Cf. Stinkweed and Stinkwort.

STINKS, Dog *Anthemis cotula* (Maydweed) Cumb. Grigson.

STINKWEED 1. *Cassia occidentalis* (Coffee Senna) Dalziel. 2. *Cleome serrulata* (Rocky Mountain Bee Plant) USA Elmore. 3. *Datura stramonium* (Thorn-apple) USA Henkel. 4. *Diplotaxis muralis*. 5. *Thlaspi arvense* (Field Pennycress) USA Gates; Canada. Salisbury.

STINKWEED, Purple *Datura stramonium* (Thorn-apple) S Africa Watt.

STINKWEED, Vineyard *Diplotaxis viminea*.

STINKWEED, White 1. *Celtis africana*. 2. *Datura stramonium* (Thorn-apple) Watt.

STINKWOOD 1. *Celtis kraussiana*. 2. *Ocotea bullata*. The name comes from the powerful smell of the timber when newly cut.

STINKWOOD, Black *Ocotea bullata* (Stinkwood) Howes.

STINKWOOD, White *Celtis africana*. The timber, when freshly cut, has an unpleasant smell.

STINKWORT *Datura stramonium* (Thorn-apple) USA Henkel. Cf. Stinkroot and Stinkweed.

STITCH, Stone- *Lithospermum officinale* (Gromwell) Prior. Stony seeds (after all, that is what Lithospermum means), used by the doctrine of signatures to combat the stone.

STITCH HYSSOP *Genista anglica* (Needle Whin). Hants. Cope.

STITCHWORT, Bog *Stellaria uliginosa*.

STITCHWORT, Greater *Stellaria holostea*. Stitchwort is so named because it was once thought to be good for stitch in the side.

STITCHWORT, Heath *Stellaria graminea* (Lesser Stitchwort) Hutchinson.

STITCHWORT, Lesser *Stellaria graminea*.

STITCHWORT, Marsh *Stellaria palustris*.

STITCHWORT, Northern *Stellaria borealis*.

STITCHWORT, Water *Myosoton aquaticum*.

STITCHWORT, Wood *Stellaria nemora*.

STITSON *Hypericum androsaemum* (Tutsan)

Dev. B & H. Stitson is described as a corruption of Tutsan.

STOBWORT *Oxalis acetosella* (Wood Sorrel) B & H. Probably Stubwort, for it grows round the stubs of trees, or so Prior claims. But there is also Stabwort in Parkinson, "because it is singular good in all wounds and stabbes into the body".

STOCK 1. *Matthiola incana*. 2. *Trollius europaeus* (Globe Flower) Cumb. Grigson. It is difficult to see why Globe Flower should ever get the name, for Stock is OE stocc, a stick.

STOCK, Hoary *Matthiola incana* (Stock) Blamey.

STOCK, Jilloffer *Matthiola incana* (Stock) Som. Macmillan. 'Jilloffer' is Gillyflower. The standard name for stocks used to be Stock Gillyflower.

STOCK, Jelly *Cheiranthus cheiri* (Wallflower) Som. Macmillan. Gilliflower served for a number of different flowers, perhaps modified for convenient identification, as in Clove Gilliflower for carnation. Wallflowers were called Wall Gilliflowers, but a form like Yellow Stock-Gilliflower also occurred. It is this last name that got shortened into Jelly Stock, among others.

STOCK, Night-scented *Matthiola bicornis*.

STOCK, Queen *Matthiola incana* (Stock) Turrill.

STOCK, Sad *Matthiola fruticosa*. Some of them have dull brownish pink flowers, sombre enough to warrant the adjective 'sad', and the old specific name *tristis*.

STOCK, Sea *Matthiola sinuata*.

STOCK, Ten-week *Matthiola annua*.

STOCK, Virginian *Malcomia maritima*. Called 'Virginian' in error. The earlier name was Dwarf Annual Stock.

STOCK GILLIFLOWER; STOCK GILLOVER *Matthiola incana* (Stock) Prior (Gilliflower); Parkinson (Gillover).

STOCK GILLIFLOWER, Blue *Matthiola incana* (Stock) Turner.

STOCK GILLIFLOWER, Yellow *Cheiranthus cheiri* (Wallflower) Gerard.

STOCK NUT *Corylus avellana* (Hazel) Prior. "From its growing on a stick,..., and not on a tree like the walnut" (Prior). 'Stock' is OE stocc, a stick.

STOCK ROSE *Sparrmania africana* (African Hemp) Howes.

STOCKINGS, Boots-and- *Plantago media* (Lamb's-tongue Plantain) Som. Macmillan. Cf. Shoes-and-stockings, from the same area.

STOCKINGS, Cuckoo's 1. *Campanula rotundifolia* (Harebell) Hardy. 2. *Endymion nonscriptus* (Bluebell) Derb, Notts, Staffs. Grigson. 3. *Lotus corniculatus* (Bird's-foot Trefoil) Som, Shrop, Suss. Grigson. 4. *Viola riviniana* (Dog Violet) Caith. Grigson.

STOCKINGS, Shoes-and- 1. *Aquilegia vulgaris* (Columbine) Som. Macmillan. 2. *Cardamine pratensis* (Lady's Smock) Bucks. Grigson. 3. *Lamium album* (White Deadnettle) Som. Macmillan. 4. *Lathyrus nissolia* (Grass Pea) Suss. Grigson. 5. *Linaria vulgaris* (Yellow Toadflax) Corn. Grigson. 6. *Lotus corniculatus* (Bird's-foot Trefoil) Dev, Glos, Bucks. Grigson; Hants. Cope; Suss. Parish. 7. *Plantago media* (Lamb's-tongue Plantain) Som. Macmillan. 8. *Viola riviniana* (Dog Violet) Pemb. Grigson. 9. *Viola tricolor* (Pansy) Som. Macmillan.

STOCKINGS-AND-SHOES 1. *Aquilegia vulgaris* (Columbine) Som. Macmillan. 2. *Arum maculatum* (Cuckoo-pint) Suss. PLNN31. 3. *Lotus corniculatus* (Bird's-foot Trefoil) Dev. Friend.1882.

STOKES'S ASTER *Stokesia laevis*.

STONE, Love- *Hedera helix* (Ivy) Leics. Grigson. Descriptive of its habit.

STONE BASIL *Clinopodium vulgare* (Wild Basil) B & H.

STONE BRAMBLE *Rubus saxatilis*.

STONE-BREAK *Saxifraga granulata* (Meadow Saxifrage) B & H. Really a name for all the saxifrages, for that is exactly what the general name means.

STONE-BREAK, Golden *Chrysoplenium oppositifolium* (Opposite-leaved Golden

Saxifrage) Freethy. 'Stone-break' is just a translation of Saxifrage.

STONE-BREAK, White *Saxifraga granulata* (Meadow Saxifrage) Gerard.

STONE CLOVER *Trifolium arvense* (Hare's-foot Trefoil) USA House.

STONE CROTTLE *Parmelia caperata*.

STONE CROTTLE, Grey *Parmelia saxatilis*.

STONE-HORE; STONE-HOT *Sedum reflexum* (Yellow Stonecrop) both Prior. Stone-hore, or Stonor, he says, are corruptions of Stone Orpine; Stone-hot, or Stonnord, though,started as Stonewort.

STONE LEEK *Allium fistulosum* (Welsh Onion) W Miller.

STONE LICHEN, Green *Parmelia caperata*.

STONE MILLET *Lithospermum officinale* (Gromwell) Watt. 'Stone' in this case is a reference to the extremely hard seeds. 'Millet', which sometimes appears as 'mil' or 'myle', as in Turner's Grey Myle, is the Latin milium, millet. In medieval Latin this plant was named milium solis, millet of the sun.

STONE ORPINE *Sedum reflexum* (Yellow Stonecrop) Prior. Orpine is Fr orpin, apparently from orpiment, which is Latin auripigmentum, pigment of gold, and is usually reserved for *Sedum telephium*. But in view of the derivation, perhaps it was intended for this or some other yellow stonecrop.

STONE OSIER *Salix purpurea* (Purple Osier) Wilts. Dartnell & Goddard.

STONE PINE *Pinus pinea*. 'Stone', because the seeds are enclosed in a very hard shell; in fact, they are often referred to as nuts.

STONE PINE, Swiss *Pinus cembra*.

STONE-ROOT *Collinsonia canadensis*. So called because the roots are prescribed by herbalists in the treatment of bladder complaints. Cf. Knob-root and Knobweed.

STONE RUE *Asplenium ruta-muraria* (Wall Rue) Turner.

STONE-STITCH; STONE-SWITCH *Lithospermum officinale* (Gromwell) Prior (Stitch); Dyer (Switch). 'Stone' is a reference to the very hard seeds, which were used in medicine to cure the stone. "Indeed there is no plant which so instantaneously proclaims, at the mere sight of it, the medicinal purposes for which it was originally intended" (Pliny).

STONEBERRY *Rubus saxatilis* (Stone Bramble) (Donegal) Grigson. A habitat name - it grows on rocks and ledges.

STONECROP, Annual *Sedum annuum*.

STONECROP, Biting *Sedum acre*. 'Biting' in the sense of peppery. Cf. the names Pepper Crop, Country Pepper, etc.

STONECROP, English *Sedum anglicum*.

STONECROP, Caucasian *Sedum spurium*.

STONECROP, Creamish *Sedum ochroleucum*.

STONECROP, Golden *Sedum acre* (Biting Stonecrop) Berks. Grigson.

STONECROP, Great *Umbilicis rupestris* (Wall Pennywort) B & H.

STONECROP, Hairy *Sedum villosum* (Pink Stonecrop) Clapham.

STONECROP, Insipid *Sedum sexangulare*. Another name is Tasteless Yellow Stonecrop.

STONECROP, Large *Sedum telephium* (Orpine) Loewenfeld.

STONECROP, Little *Sedum acre* (Biting Stonecrop) Turner.

STONECROP, Mossy *Crassula tillaea*.

STONECROP, Pink *Sedum villosum*.

STONECROP, Reddish *Sedum anacampseros*.

STONECROP, Reflexed *Sedum reflexum* (Yellow Stonecrop) D Grose.

STONECROP, Rock 1. *Sedum forsterianum*. 2. *Sedum reflexum* (Yellow Stonecrop) Polunin.

STONECROP, Swamp, Australian *Crassula helmsii*.

STONECROP, Thick-leaved *Sedum dasyphyllum*.

STONECROP, Wall *Sedum acre* (Biting Stonecrop) Thornton.

STONECROP, Welsh, Small *Sedum forsterianum* (Rock Stonecrop) Phillips.

STONECROP, White *Sedum album*.

STONECROP, Yellow *Sedum reflexum*.

STONECROP, Yellow, Common *Sedum acre* (Biting Stonecrop) Curtis.

STONECROP, Yellow, Crooked;
STONECROP, Yellow, Large *Sedum reflexum* (Yellow Stonecrop) Grieve.1931 (Crooked); Phillips (Large).

STONECROP, Yellow, Tasteless *Sedum sexangulare* (Insipid Stonecrop) Grieve.1931.

STONEHORE *Sedum acre* (Biting Stonecrop) Gerard.

STONES, Fool's *Orchis mascula* (Early Purple Orchis) B & H.

STONESEED GRONWELL *Lithospermum canescens* (Hoary Puccoon) USA Elmore. 'Stoneseed' is a translation of Lithospermum.

STONEWEED 1. *Polygonum aviculare* (Knotgrass) Hants. Cope; Suff. Halliwell. 2. *Polygonum persicaria* (Persicaria) Som, Wilts. Macmillan.

STONEWORT *Sedum reflexum* (Yellow Stonecrop) Prior.

STONNORD, STONOR *Sedum reflexum* (Yellow Stonecrop) both Prior.

STONY-HARD *Lithospermum officinale* (Gromwell) N Eng. Grigson. It is the seeds that are 'stony-hard'.

STONY-IN-THE-WELL *Capsella bursa-pastoris* (Shepherd's Purse) Lincs. Grigson.

STOOL WOOD *Alstonia congensis* (Pattern Wood) Dalziel. The wood is commonly used for stools throughout Ghana, Nigeria and the Cameroons.

STORAX, Levant *Liquidambar orientalis* (Lordwood) Howes.

STORAX, Liquid *Liquidambar styraciflua* (Sweet Gum) Leyel.1937.

STORKBILL, STORKS *Geranium robertianum* (Herb Robert) N Eng. B & H (Storkbill); Dor. Macmillan (Storks). The Geraniums are usually Cranesbills, and the Erodiums Storksbills, but there is always interchanging of the names, for they are given for exactly the same reason - the seed cases are long and pointed, and therefore likened to a bird's bill.

STORKSBILL 1. *Erodium cicutarium*. 2. *Geranium maculatum* (Spotted Cranesbill) Grieve.1931. 3. *Geranium robertianum* (Herb Robert) Som. Macmillan.

STORKSBILL, Hemlock *Erodium cicutarium* (Storksbill). 'Hemlock' from a similarity between the leaves. Cf. Hemlock-leaved Cranesbill for this.

STORKSBILL, Musk *Erodium moschatum*.

STORKSBILL, Sea *Erodium maritimum*.

STORKSBILL, Sticky *Erodium cicutarium* (Storksbill) McClintock.1975.

STORKSBILL WINDFLOWER *Anemone apennina* (Blue Anemone) Gerard.

STORSHIN; STORSHINER *Tropaeolum majus* (Nasturtium) IOM Moore.1924. Both attempts at Nasturtium.

STORY-OF-THE-CROSS *Passiflora caerulea* (Passion Flower) Som. Macmillan. It is said that when the Spaniards first saw the Passion Flower, they took it as an omen that the Indians would be converted to Christianity, for each part of the flower was taken to represent the symbols of the Passion. The ten white petals show Christ's innocence, the outer circlet of filaments symbolized his disciples, the inner circlet the crown of thorns, the styles the three nails, and so on.

STORY-TELLERS *Potentilla sterilis* (Barren Strawberry) Som. Macmillan.

STOVER-NUT *Castanea sativa* (Sweet Chestnut) Dev. Friend.1882. A very local name, for 'Stover' is Stover Park, the estate of the Duke of Somerset, near Newton Abbot. It is so called because he has a lot of Sweet Chestnuts growing there.

STOWEL, Nut *Corylus avellana* (Hazel) Wilts. Hayward.

STRAGGLY GOOSEBERRY *Ribes divaricatum*.

STRAINER VINE *Luffa cylindrica* (Loofah) Dalziel.

STRAMONY *Datura stramonium* (Thornapple) Geldart.

STRAND CABBAGE *Crambe maritima* (Sea Kale) Donegal Grigson.

STRANGLE TARE 1. *Cuscuta epithymum*

(Common Dodder) Grieve.1931. 2. *Cuscuta europaea* (Greater Dodder) B & H. 3. *Vicia hirsuta* (Hairy Tare) Salisbury. Dodder is a parasite, so the name is apt, and Hairy Tare can destroy crops by strangling them. Cf. Bindweed for the latter, and also Dother, for that is the same as dodder.

STRANGLER, Burro-weed *Orobanche ludoviciana var. cooperi*.

STRANGLER VETCH *Vicia hirsuta* (Hairy Tare) Blamey.1980. See Strangle Tare.

STRANGLEROOT, Piñon *Orobanche fasciculata*.

STRANGLEWEED 1. *Calystegia sepium* (Great Bindweed) Som. Macmillan. 2. *Cuscuta epithymum* (Common Dodder) Som. Macmillan. 3. *Orobanche minor* (Lesser Broomrape) Som. Macmillan. Broomrape looks a little out of place. Certainly it is a parasite, and Turner called it Chokeweed, but the others really do earn the 'strangling' epithet.

STRANGLEWORT *Cynanchum acutum*. This is a climber.

STRANGLING FIG *Ficus aurea*.

STRAP-GRASS *Agropyron repens* (Couch-grass) Hants. Cope.

STRAPWORT *Corrigiola littoralis*.

STRATHMORE WEED *Pimelia prostrata*. Not the Scottish Strathmore, for this is a New Zealand plant.

STRAW, Jack *Plantago lanceolatum* (Ribwort Plantain) Yks. Grigson.

STRAW-FLOWER 1. *Helichrysum bracteatum* (Everlasting Flower) RHS. 2. *Helichrysum stoechas* (Golden Cassidony). This is what they called the latter in the television programme, Victorian kitchen garden, but Everlasting flowers have always been so called. The name is of course a comment on the texture of the flowers.

STRAWBED *Galium verum* (Lady's Bedstraw) Dev. Friend.1882.

STRAWBERRY, Alpine *Fragaria vesca var. semperflorens*. Barton & Castle maintained that the word strawberry was actually strayberry, from the straying habit of its runners, but this is not true, for the OE was clearly streawberige, though ODEE admits it does not know why. Anyway, there never was a word 'strayberry'. Can it be a gardener's name, as early as OE times? The familiar practice of putting straw round the plants to protect the fruit does not sound much like OE usage, but who knows?

STRAWBERRY, Barren 1. *Potentilla sterilis*. 2. *Waldsteinia fragarioides* (Dry Strawberry) USA House. Wild Strawberry is often confused with Barren Strawberry, which in fact blooms much earlier, and is not even a member of the same genus.

STRAWBERRY, Bog *Potentilla palustris* (Marsh Cinquefoil) IOM Grigson.

STRAWBERRY, Chile *Fragaria chiloense*. This is a plant of the western maritime edge of the Americas, from Alaska to Patagonia. It was found as a garden plant on the island of Chiloé, off the coast of Chile, and may very well have been an Inca garden plant.

STRAWBERRY, Chinese *Myrica rubra* F P Smith.

STRAWBERRY, Dry *Waldsteinia fragarioides*.

STRAWBERRY, Garden *Fragaria x ananassa*.

STRAWBERRY, Hautbois *Fragaria moschata* (Hautboy) Clapham. 'Hautbois' means 'high wood', for *F moschata* lifts its fruit above the leaves, in contrast with the common strawberry.

STRAWBERRY, Hill *Fragaria collina*.

STRAWBERRY, Pine 1. *Fragaria collina* (Hill Strawberry) Bianchini. 2. *Fragaria grandifolia*.

STRAWBERRY, Scarlet *Fragaria virginiana*.

STRAWBERRY, Sow-teat *Fragaria vesca var. americana*.

STRAWBERRY, Virginian *Fragaria virginiana*.

STRAWBERRY, Wild *Fragaria vesca*.

STRAWBERRY, Winter *Arbutus unedo* (Strawberry Tree) Dev. Choape; Som. Macmillan.

STRAWBERRY, Wood 1. *Fragaria californica* USA Schenk & Gifford. 2. *Fragaria vesca* (Wild Strawberry) Grigson.1974.

STRAWBERRY, Yellow *Geum urbanum* (Avens) Som. Macmillan. The flowers and leaves look not unlike those of the strawberry.

STRAWBERRY BEGONIA; STRAWBERRY-LEAVED GERANIUM; STRAWBERRY PLANT *Saxifraga sarmentosa* (Mother-of-thousands) Gannon (Begonia) Wilts. Dartnell & Goddard (-leaved); Dev. B & H (Plant). The reference is to the method of propagation by runners.

STRAWBERRY BLITE *Chenopodium capitatum*.

STRAWBERRY BUSH *Euonymus americanus*.

STRAWBERRY CLOVER *Trifolium frageriferum*.

STRAWBERRY GERANIUM *Saxifraga granulata* (Meadow Saxifrage) W Miller. This is the species quoted, but it is better applicable to *S. sarmentosa*, which reproduces itself by means of runners, as does strawberry.

STRAWBERRY GOOSEFOOT *Chenopodium foliosum*.

STRAWBERRY-LEAVED POTENTILLA *Potentilla sterilis* (Barren Strawberry) Hutchinson.

STRAWBERRY PLANT 1. *Potentilla sterilis* (Barren Strawberry) Dev. Macmillan. 2. *Saxifraga sarmentosa* (Mother-of-thousands) Dev. B & H.

STRAWBERRY SHRUB *Calycanthus floridus* (Allspice) Howes. There is a strong scent of strawberries to the wood when crushed.

STRAWBERRY TOMATO 1. *Physalis alkekengi* (Winter Cherry) Bianchini. 2. *Physalis bunyardii* (Cape Gooseberry) USA Watt. 3. *Physalis pubescens* (Husk Tomato) Bianchini.

STRAWBERRY TREE *Arbutus unedo*. It bears a spherical fruit, rather like a strawberry in appearance.

STRAWBERRY TREE, Killarney *Arbutus unedo* (Strawberry Tree) Hay & Synge. It is a Mediterranean species, but grows apparently wild in some stations in Ireland, Killarney in particular.

STRAWBERRY TREE, Trailing *Arctostaphylos uva-ursi* (Bearberry) Jamieson. Thornton referred to this species as Trailing Arbutus, hence the reference to Strawberry Tree here.

STRAWBERRY TREE, Greek *Vaccinium arctostaphylos*.

STRAWBONNETS *Aquilegia vulgaris* (Columbine) Bucks. Evans. There are a lot of 'bonnet' names for columbine, all descriptive, the best known of which is Granny Bonnets.

STREAMSIDE BIRCH *Betula fontinalis*.

STRIKE *Linaria vulgaris* (Tellow Toadflax) Dor. Macmillan.

STRIKE-FIRE *Veronica chamaedrys* (Germander Speedwell) N'thants. Grigson. Cf. Thunderbolt, a Cheshire name. Picking the flowers can cause a thunderstorm. In Germany, too, the belief was that picking speedwell would cause a storm. That belief may be implicit in the name Strike-fire.

STRING BEAN *Phaseolus vulgaris* (Kidney Bean) Gooding. The bean that has to be supported on strings?

STRING-OF-BEADS PLANT; STRING-OF PEARLS *Senecio rowleyanus* String of Pearls is the American name - see Cunningham. The small spherical leaves are strung out like a necklace.

STRING-OF-HEARTS *Ceropegia woodii*. It has hanging strings of tiny, heart-shaped leaves, hence another name, Hearts Entangled.

STRINGS, Bonnet *Agrostis stolonifera* (White Bent) Som. Elworthy. In this case, 'bonnet' is the same as 'bent'; they both come from OE beonet.

STRINGY-BARK 1. *Eucalyptus obliqua*. 2. *Eucalyptus tetradonta*. These are the species that the Aborigines use for bark painting. The bark is peeled off during the wet season when the sap is rising, and is easy to remove.

STRINGY-BARK, Darwin *Eucalyptus tetradonta* (Stringy-bark) Hora.

STRIP-FOR-STRIP *Cichorium intybus* (Chicory) Som. Macmillan.

STRIP-JACK-NAKED *Colchicum autumnale* (Meadow Saffron) Dev. Macmillan. The flowers appear before the leaves, hence the large number of 'naked' names.

STRIPLING, Cow *Primula veris* (Cowslip) Yks. Harland. Cf. Cow Stropple, or Cow Strupple.

STROIL; STROYL *Agropyron repens* (Couchgrass) Corn. Grigson.1959.

STROPPLE, Cow 1. *Primula veris* (Cowslip) Yks, Cumb, West. Grigson. 2. *Primula veris x vulgaris* (False Oxlip) Dur. Brockett.

STRUPPLE, Cow *Primula veris* (Cowslip) Yks, Cumb, West Grigson.

STUB-APPLE *Malus sylvestris* (Crabapple) Halliwell.

STUBBLEBERRY *Solanum nigrum* (Black Nightshade) S Africa Watt.

STUBWORT *Oxalis acetosella* (Wood Sorrel) Prior. "From its growing about the stubs of hewn trees", said Prior. But there are also Stobwort and Stabwort for this plant.

STURDY *Lolium temulentum* (Darnel) Scot, Ire. B & H. There was a disease in sheep called sturdy, or staggers. Meal was said to be sturdied when it had a lot of darnel in it.

STURDY LOWRIES *Daphne laureola* (Spurge Laurel) Dur. Grigson. A variation of 'Spurge Laurel' (Lowries was often used for laurel).

STURTION *Tropaeolum majus* (Nasturtium) Som. Elworthy.

STYPTIC WEED *Cassia occidentalis* (Coffee Senna) USA Vestal & Schultes. With a name like that, it must be a wound plant. But where are the records?

SUBALTERN'S BUTTER; MIDSHIPMAN'S BUTTER *Persea gratissima* (Avocado Pear) Grigson.1974.

SUBCLOVER *Trifolium subterraneum* (Burrowing Clover) USA Kingsbury.

SUCCAMORE *Acer pseudo-platanus* (Sycamore) Shrop. B & H. This is but one of many variations on 'sycamore'. They include Sacymore, Segumber and Seggy.

SUCCORY *Cichoriuum intybus* (Chicory) Turner. 'Succory' is a variant of 'chicory', itself from Latin cichoreum.

SUCCORY, Blue *Cichorium intybus* (Chicory) Curtis.

SUCCORY, Garden *Cichorium endivia* (Endive) Turner.

SUCCORY, Lamb's *Arnoseris minima* (Swine's Succory) McClintock.

SUCCORY, Swine's *Arnoseris minima*.

SUCCORY, Wild *Cichorium intybus* (Chicory) Thornton.

SUCCORY, Yellow *Picris hieracioides* (Hawkweed Ox-tongue) Turner.

SUCK-BOTTLE 1. *Lamium album* (White Deadnettle) Som. Macmillan; Northants. Sternberg. 2. *Trifolium pratense* (Red Clover) Northants. Grigson. It is the bees that do the sucking, in all the names in which the word occurs. Cf. Honeysuckle, etc.

SUCKERS; SUCKERS, Sweet *Symphytum officinale* (Comfrey) both Som. Macmillan.

SUCKERY *Cichorium intybus* (Chicory) A S Palmer. The result of misunderstanding Succory.

SUCKIES *Pedicularis palustris* (Red Rattle) Ayr. Grigson.

SUCKLE, Sweet; SUCKLE-BUSH *Lonicera periclymenum* (Honeysuckle) Som. Macmillan (Sweet Suckle); Norf. Grigson (Suckle-bush).

SUCKLERS 1. *Trifolium pratense* (Red Clover) N'thum. B & H. 2. *Trifolium repens* (White Clover) N'thum. B & H; Berw. Grigson.

SUCKLES 1. *Lonicera periclymenum* (Honeysuckle) Som. Grigson. 2. *Trifolium pratense* (Red Clover) Gerard.

SUCKLEYA, Poison *Suckleya suckleyana*.

SUCKLING CLOVER *Trifolium dubium* (Lesser Yellow Trefoil) Polunin.

SUCKLINGS 1. *Lonicera periclymenum* (Honeysuckle) E Ang. Forby. 2. *Trifolium pratense* (Red Clover) Shrop, Norf, Suff. Grigson. 3. *Trifolium repens* (White Clover) E Ang. Halliwell.

SUCKLINGS, Lamb's 1. *Lotus corniculatus* (Bird's-foot Trefoil) Yks. B & H. 2. *Trifolium pratense* (Red Clover) Cumb. B & H. 3. *Trifolium repens* (Wgite Clover) Yks, Cumb. B & H.

SUCKS *Trifolium pratense* (Red Clover) Tynan & Maitland.

SUCKY CALVES *Arum maculatum* (Cuckoo-pint) Som. Grigson. In this case, it is the calf that is doing the sucking. The name is part of the sequence that includes Cows-and-calves, though a little out of the category.

SUCKY SUE *Lamium album* (White Deadnettle) Som. Macmillan; Berw. B & H.

SUDAN EBONY *Dalbergia melanoxylon* (African Ebony) Dalziel. Cf. African Ebony, and Senegal Ebony. This is probably the most valuable wood from Africa.

SUDAN GRASS *Sorghum sudanense*.

SUDAN TEAK *Cordia abyssinica* (Mukumari) Dalziel.

SUDANESE TEA *Hibiscus sabdariffa* (Rozelle) Schauenberg.

SUDS *Saponaria officinalis* (Soapwort) Kent Tynan & Maitland. Soap suds, of course.

SUDS, Sour 1. *Oxalis acetosella* (Wood Sorrel) Dev. Grigson. 2. *Rumex acetosa* (Wild Sorrel). 'Suds' in this case is something like 'sops' or 'sabs' - "sour as a sab" is a Cornish saying.

SUE, Sucky *Lamium album* (White Deadnettle) Som. Macmillan.

SUFFOCATED CLOVER *Trifolium suffocatum*.

SUGAR, Brown *Rumex acetosa* (Wild Sorrel) Som. Mabey.

SUGAR, Butter-and- *Linaria vulgaris* (Yellow Toadflax) Wilts. Macmillan. Unusual - they are more often likened to Butter-and-eggs, or some similar combination.

SUGAR, Horse *Symplocos tinctoria* (Sweetleaf) Wit.

SUGAR APPLE *Annona squamosa*. Cf. Custard Apple, Sweetsop, etc.

SUGAR BASINS *Stellaria holostea* (Greater Stitchwort) Som. Macmillan. A typical West country plant name that requires some imagination in interpreting.

SUGAR BEAN *Phaseolus lunatus* (Lima Bean) Candolle.

SUGAR BEET *Beta vulgaris var. altissima*.

SUGAR-BOSSES; SUGAR-BUSSES *Trifolium pratense* (Red Clover) Som. Grigson (Boss); Macmillan (Buss). Cf. Sugar-plums, and all the 'honeysuckle' names for the clover.

SUGAR BUSH 1. *Protea angolensis*. 2. *Rhus ovata* (Sugar Sumach) USA T H Kearney. Presumably from the nectar produced at the base of the Protea flowers.

SUGAR CANDY *Rosa canina* shoots (Dog Rose) Wilts. Macmillan. There are differences of taste about the country. They may be Sugar Candy in Wiltshire, but they are Bread-and-cheese in Yorkshire, Bull Beef in Cheshire and Bacon in Somerset.

SUGAR CANE *Saccharum officinarum*.

SUGAR CODLINS *Epilobium hirsutum* (Great Willowherb) Som, Wilts. Grigson. Cf. Codlins-and-cream, Coddled Apples, etc. It is said that the plant smells like apples when it is crushed, but it is probably Gerard's Codded Willow-herbe that is the original, and not codlins.

SUGAR GRASS *Glyceria fluitans* (Flote Grass) Douglas. Cf. Floating Sweet Grass.

SUGAR HACKBERRY *Celtis laevigata* (Sugarberry) Barber & Phillips.

SUGAR HAWTHORN *Crataegus spathulata*.

SUGAR-LEAVES *Ulmus procera* (English Elm) Som. Macmillan. Apparently reserved for the young leaves, but do children eat them like those of hawthorn?

SUGAR MAPLE *Acer saccharum*. Many maples yield a sugary sap, but only this one does so in sufficient quantity to justify exploitation. The American Indians extracted the sap before the arrival of Europeans.

SUGAR MAPLE, Black *Acer nigrum*.

SUGAR MAPLE, Chinese *Sorghum saccharum* (Sorgo) Grieve.1931. Cultivated extensively for syrup production, but of course nothing at all to do with maples.

SUGAR MAPLE, Florida *Acer floridanum*.

SUGAR PINE *Pinus lambertiana*. The coagulated sap is the "sugar", much used by the American Indians in the past.

SUGAR PLUM 1. *Trifolium pratense* (Red Clover) Bucks. Grigson. 2. *Uapaca guineensis* fruit Liberia Dalziel.

SUGAR SUMACH *Rhus ovata*.

SUGARBERRY 1. *Celtis laevigata*. 2. *Celtis occidentalis* (Hackberry) USA Vestal & Schultes. 3. *Celtis reticulata* (Western Hackberry) USA Elmore.

SULLEN LADY *Fritillaria meleagris* (Snake's-head Lily) Coats. Given as a result of the sombre colouring of the flowers. Cf. Weeping Widow, and a number of similar names.

SULPHUR CINQUEFOIL *Potentilla recta*. Sulphur, to describe the colour of the flowers.

SULPHUR CLOVER *Trifolium ochroleucum*.

SULPHUR LILY *Lilium sulphureum*.

SULPHUR PLANT *Eriogonum umbellatum*.

SULPHUR ROOT *Peucedanum officinale* (Hog's Fennel). Cf. Sulphurweed, Brimstonewort, etc., from its yielding, as Coles says, "a yellow sap which waxeth quickly hard, and dry, and smelleth not unlike to brimstone".

SULPHUR ROSE *Rosa hemisphaerica*.

SULPHURWEED *Peucedanum officinale* (Hog's Fennel) Grigson. See Sulphur Root.

SULPHURWORT 1. *Oenanthe silaifolia*. 2. *Peucedanum officinale* (Hog's Fennel) Gerard. 3. *Silaum silaus* (Pepper Saxifrage) D Grose. See Sulphur Root for *Peucedanum officinale*. Pepper Saxifrage has yellow flowers, presumably the reason for the name.

SULPHURWORT, Sea *Peucedanum officinale* (Hog's Fennel) Barton & Castle.

SULTAN, Sweet *Centaurea imperialis*.

SULTAN('S) FLOWER 1. *Centaurea imperialis* (Sweet Sultan) Parkinson. 2. *Impatiens walleriana* (Busy Lizzie) Howes.

SUMACH 1. *Primula vulgaris* (Primrose) IOM Moore. 2. *Rhus sp*. An Arabic word originally, at any rate for the genus *Rhus*. It came from another word meaning red, a reference to dye. The Manx word for primrose is either rendered like this, or as Sumark.

SUMACH, Black *Rhus copallina*.

SUMACH, Chinese *Ailanthus altissima* Chopra. Probably because of the varnish obtained from it, the mark of a number of true sumachs.

SUMACH, Coral *Metopium toxiferum* (Poison-wood) Usher. Related to the sumachs.

SUMACH, Dwarf *Rhus copallina* (Black Sumach) USA Upham.

SUMACH, Elm-leaved *Rhus coriaria* (European Sumach) Hora.

SUMACH, European *Rhus coriaria*.

SUMACH, Fragrant *Rhus aromatica* (Sweet Sumach) Yarnell.

SUMACH, Ill-scented *Rhus trilobata* (Three-leaf Sumach) Le Strange. Cf. Skunk Bush and Polecat Bush, both American names for it.

SUMACH, Lemon *Rhus aromatica* (Sweet Sumach) Howes.

SUMACH, Lemonade *Rhus trilobata* (Three-leaf Sumach) USA Yanovsky. The American Indians used to crush the berries to make into a cooling drink.

SUMACH, Mountain *Rhus copallina* (Black Sumach) Le Strange.

SUMACH, Pennsylvania *Rhus glabra* (Smooth Sumach) Grieve.1931.

SUMACH, Poison *Rhus vernix* (Swamp Sumach) USA Zenkert. It is indeed very poisonous, as its botanical synonym, *Toxicodendron vernix*, implies. Cf. Poison Dogwood, Poison Elder and Poison Ash.

SUMACH, Scarlet *Rhus glabra* (Smooth Sumach) Le Strange.

SUMACH, Shining *Rhus copallina* (Black Sumach) Schery.
SUMACH, Sicilian *Rhus coriaria* (European Sumach) Grieve.1931.
SUMACH, Smooth *Rhus glabra*.
SUMACH, Stag's-horn *Rhus typhina*. "...the young branches are covered with a soft, velvet-like down, resembling greatly that of a young stag's horn, from whence the common people have given it the name of stag's horn" (Taylor).
SUMACH, Sugar *Rhus ovata*.
SUMACH, Swamp *Rhus vernix*.
SUMACH, Sweet *Rhus aromatica*.
SUMACH, Tanning *Rhus coriaria* (European Sumach) R T Gunther.1934. The leaves are computed to contain up to 35% tannin, and the shrub was actually cultivated in Spain and Italy for the tanning process. Originally the word sumach was applied only to these dried leaves.
SUMACH, Three-leaf *Rhus trilobata*.
SUMACH, Upland *Rhus glabra* (Smooth Sumach) Grieve.1931.
SUMACH, Velvet *Rhus typhina* (Stag's-horn Sumach) Grieve.1931. The "velvet" on the young branches, likened to that on a young stag's horn.
SUMACH, Venetian *Cotinus coggyria*. It used to be classed with the genus *Rhus*, as *Rhus cotinus*, hence the use of Sumach. 'Venetian', presumably as a reference to the red dye obtained from the root.
SUMACH, Venus's *Cotinus coggyria* (Venetian Sumach) W Miller. Surely it should be 'Venetian', rather than 'Venus's'?
SUMACH, Virginia *Rhus typhina* (Stag's-horn Sumach) Parkinson.
SUMACH, White *Rhus aromatica* (Sweet Sumach) Youngken.
SUMACH, Wild *Myrica gale* (Sweet Cicely) B & H.
SUMARK 1. *Primula vulgaris* (Primrose) IOM Paton. 2. *Trifolium pratense* (Red Clover) IOM Gill.1929. A Manx word, probably related to shamrock.
SUMMER, Farewell 1. *Aster paniculatus* (Michaelmas Daisy) Wilts. D & G. 2. *Saponaria officinalis* (Soapwort) Mon. Grigson. Because they bloom late in the season, when summer is almost over.
SUMMER CHRYSANTHEMUM *Chrysanthemum coronarium* (Annual Chrysanthemum) USA T H Kearney.
SUMMER DAISY *Leucanthemum vulgare* (Ox-eye Daisy) N'thants. Clkare. Cf. Midsummer Daisy for this.
SUMMER HATS *Viola tricolor* (Pansy) Som. Macmillan.
SUMMER LILAC *Buddleia davidii* (Buddleia) Grigson.
SUMMER PHEASANT'S EYE *Adonis aestivalis*.
SUMMER POPPY 1. *Kallstroemeria grandiflora* (Arizona Poppy) USA T H Kearney. 2. *Papaver rhoeas* (Red Poppies) Som. Macmillan.
SUMMER ROSE *Kerria japonica* (Jew's Mallow) Dev. Friend.1882.
SUMMER SAVORY *Satureia hortensis*.
SUMMER SQUASH *Cucurbita pepo var. condensa*.
SUMMER TREE *Acer pseudo-platanus* (Sycamore) Cork Danaher. Because in co Cork sycamore was the favourite May bough, so "bringing in the summer" meant getting sycamore boughs, rather than hawthorn, as elsewhere. But May, or May-tree, is recorded for sycamore in Cornwall.
SUMMER VIOLET *Viola riviniana* (Dog Violet) War. Grigson.
SUMMER'S BRIDE *Calendula officinalis* (Marigold) Coats. Probably connected with the plant's sun-following habit. Cf. Sun's Bride, Sunflower, and Husbandman's Dial.
SUMMER'S DARLING *Godetia amoena* USA Barrett & Gifford.
SUMMER'S FAREWELL 1. *Aster tripolium* (Sea Aster) Dev, Som. Wright. 2. *Filipendula ulmaria* (Meadowsweet) Dev, Dor. Macmillan. 3. *Senecio jacobaea* (Ragwort) Dev, Glos. Grigson. All because of the lateness of flowering.
SUMMER'S GOODBYE *Parnassia palustris*

(Grass of Parnassus) Scot. Tynan & Maitland. It blooms from July to September.
SUMMERLOCKS *Primula veris x vulgaris* (False Oxlip) Yks. Grigson.
SUMMIT CEDAR *Athrotaxus laxifolia*.
SUN *Crotalaria juncea* (Sunn Hemp) Schery.
SUN-Centre-of-the- *Centaureum erythraea* (Centaury) Worcs. Fernie. Just a corruption of Centaureum, which was also corrupted to centrum aureos, giving rise to the French Herbe à mille florins, and the German Tausendguldenkraut.
SUN, Flower-of-the- *Myrtus communis* (Myrtle) Som. Macmillan.
SUN, Golden *Taraxacum officinale* (Dandelion) Som. Grigson. Cf. the Somerset Burning Fire. Is there a connection with a sun dedication? It was once called Sonnewirbel in Germany, and there is a legend that the plant was born of the dust raised by the chariot of the sun, which is why it opens at dawn and closes at dusk.
SUN, Rising *Leucanthemum vulgare* (Ox-eye Daisy) Som. Macmillan. Cf. Sun Daisy.
SUN DAISY 1. *Helianthemum chamaecistus* (Rock Rose) Lincs. B & H. 2. *Leucanthemum vulgare* (Ox-eye Daisy) Dor. Dacombe. But Cf. Moon Daisy for Ox-eye.
SUN-FOLLOWING SPURGE *Euphorbia helioscopia* (Sun Spurge) Turner. This, as well as the common specific names, are awarded because of its habit "of turning or keeping time with the Sunne" (Gerard).
SUN PLANT *Portulaca grandiflora* Hay & Synge.
SUN ROSE *Helianthemum chamaecistus*.
SUN ROSE, Alpine *Helianthemum alpestre*.
SUN ROSE, Showy *Helianthemum venustum*.
SUN SPURGE *Euphorbia helioscopia*. See Sun-following Spurge.
SUN'S BRIDE; SUN'S HERB *Calendula officinalis* (Marigold) both T Hill. Cf. Sunflower among others, all given from the plant's supposed sun-following habit.
SUN'S EYE; SUN'S RAYS *Helianthus annuus* (Sunflower) both Som. Macmillan.

SUNBERRY *Physalis minima*.
SUNBONNET *Narcissus pseudo-narcissus* (Daffodil) Som. Macmillan.
SUNBONNET, Lilac *Gilia punctata* (Spotted Gilia) USA Jaeger.
SUNDAY, Whit 1. *Narcissus x biflorus* (Primrose Peerless) Dev. Friend.1882. 2. *Stellaria holostea* (Greater Stitchwort) Dev. Friend.1882. Cf. White Sunday for Stitchwort, and also Easter Flower or Easter Bell. One has to assume that the white flowers are the origin of the names.
SUNDAY, White 1. *Stellaria graminea* (Lesser Stitchwort) Fernie. 2. *Stellaria holostea* (Greater Stitchwort) Dev. Friend.1882. A variant of Whit Sunday.
SUNDAY WHITES *Stellaria holostea* (Greater Stitchwort) Dev. Greenoak. See Whit Sunday.
SUNDCORNS *Saxifraga granulata* (Meadow Saxifrage) Cockayne.
SUNDEW *Drosera rotundifolia*. The leaves are covered with hair-like tentacles, which secrete a drop of viscous liquid at their tip. It is these drops glistening in the sunlight that give the plant its name. Insects are attracted to the liquid, probably by scent, which is said to be fungus-like, and the fluid traps them. The generic name, *Drosera*, also means dewy (Greek droseros).
SUNDEW, Forked-leaved *Drosera binata*.
SUNDEW, Great; SUNDEW, Great-leaved *Drosera anglica*. Great Sundew is the usual name. Great-leaved Sundew is used by J A MacCulloch.1905.
SUNDEW, Long-leaved *Drosera intermedia*.
SUNDEW, Portuguese *Drosophyllum lusitanicum* (Dewy Pine) Simmons.
SUNDEW, Round-leaved *Drosera rotundifolia* (Sundew) J A MacCulloch.1905.
SUNDIAL, Shepherd's *Anagallis arvensis* (Scarlet Pimpernel) Suff. Moor. Cf. Shepherd's Clock, and also such names as John-go-to-bed-at-noon, which explains why it was a countryman's timepiece.
SUNDROPS *Oenothera fruticosa*.

SUNDROPS, Ozark *Oenothera missouriensis*.
SUNDROPS, Prairie *Oenothera pratensis*.
SUNDROPS, Small(er) *Oenothera pumila*.
SUNFLOWER 1. *Anagallis arvensis* (Scarlet Pimpernel) Cumb. Grigson. 2. *Anagallis tenella* (Bog Pimpernel) Cumb. B & H. 3. *Calendula officinalis* (Marigold) B & H. 4. *Chrysanthemum segetum* (Corn Marigold) N'thants. A E Baker. 5. *Helianthemum chamecistus* (Rock Rose) Som. Macmillan. 6. *Helianthus annuus*. 7. *Inula helenium* (Elecampane) Stir. Grigson. 8. *Ornithogalum umbellatum* (Star-of-Bethlehem) Dev. Friend.1882. Pimpernel and Star-of-Bethlehem only open their flowers when the sun is out, hence the name. The same must apply to Rock Rose, or perhaps it is the typical yellow colour that is the reason. The rest are all yellow flowered, and have the reputation of being sun-followers, either by turning their flower heads to face the sun, or by only opening during the hours of daylight.
SUNFLOWER, False 1. *Helenium autumnale* (Sneezeweed) USA House. 2. *Heliopsis helianthoides*. Plants that look like small sunflowers (*Helianthus*), but are not.
SUNFLOWER, Indian *Helianthus annuus* (Sunflower) Gerard.
SUNFLOWER, Mexican *Tithonia rotundifolia*.
SUNFLOWER, Narrow-leaved *Helianthus angustifolius*.
SUNFLOWER, Rough *Helianthus divaricatus*.
SUNFLOWER, Swamp 1. *Helenium autumnale* (Sneezeweed) USA Tolstead. 2. *Helianthus angustifolius* (Narrow-leaved Sunflower) USA. House.
SUNFLOWER, Thin-leaved *Helianthus decapitalus*.
SUNFLOWER, Tick *Bidens coronata* (Swamp Beggar-ticks) Howes.
SUNFLOWER, Tickseed *Bidens trichosperma*. 'Tickseed', from the bur-like seeds that all members of the genus have.
SUNFLOWER, Weed *Helianthus annuus var. annuus*.
SUNFLOWER, Wild 1. *Helianthus annuus var. annuus*. 2. *Helianthus annuus var. lenticularis*. 3. *Inula helenium* (Elecampane) IOW Grigson.
SUNFLOWER, Wild, Hairy *Helianthus mollis*.
SUNFLOWER, Wild, Willow-leaved *Helianthus salicifolius*.
SUNFLOWER, Woodland *Helianthus divaricatus* (Rough Sunflower) USA House.
SUNFLOWER, Woolly *Eriophyllum lanatum*.
SUNFLOWER ARTICHOKE *Helianthus tuberosus* (Jerusalem Artichoke) Grieve.1931. 'Jerusalem' is Girasole, a reference to the imaginary habit of sunflowers turning their heads to follow the sun.
SUNGREEN *Sempervivum tectorum* (Houseleek) Wilts. Dartnell & Goddard; Suss. Latham. It means evergreen, and is one of a series that starts with Singreen, and varies through Sengreen to Silgreen.
SUNN HEMP *Crotalaria juncea*. Also known as San Hemp, or just as Sun.
SUNNAN-CORN *Lithospermum officinale* (Gromwell) Cockayne. The Saxon name, according to Cockayne.
SUNSET *Lavatera arborea* (Tree Mallow) Som. Macmillan.
SUNSHADE 1. *Convolvulus arvensis* (Field Bindweed) Som. Macmillan. 2. *Helianthus annuus* (Sunflower) Dor. Macmillan. 3. *Umbilicus rupestris* (Wall Pennywort) Som. Macmillan. All descriptive in their way, either by shape of flower, as with the bindweed, the leaf, as with pennywort, or sheer size, as with sunflower. Cf. Parasol for bindweed.
SUNSHADE, Lady's *Convolvulus arvensis* (Field Bindweed) Som. Macmillan.
SUNSHADE, Larger *Calystegia sepium* (Greater Bindweed) Som. Macmillan.
SUPPLEJACK; SUPPLEJACK VINE *Rhipogonum parviflorum*.
SURGEONS, Wisdom of *Descurainia sophia*

(Flixweed) Salisbury. A translation of its old name, *Sophia chirurgorum*.
SURINAM CHERRY *Eugenia uniflora*. But it is also Brazilian, Barbados, Pitanja, or Cayenne Cherry.
SURRY *Sorbus torminalis* (Wild Service Tree) N'thants. Clare. Odd-looking, but probably only a version of 'service' originally.
SUSAN, Black-eyed 1. *Aster sp* Som. Macmillan. 2. *Hibiscus syriacus* (Syrian Katmia) Som. Macmillan. 3. *Rudbeckia hirta*. 4. *Thunbergia alata* Gannon.
SUSAN, Brown-eyed 1. *Rudbeckia hirta* (Black-eyed Susan) Howes. 2. *Rudbeckia triloba* (Three-lobed Coneflower) S M Robertson.
SUSSEX, Pride of *Phyteuma tenorum* (Round-headed Rampion) Suss. Parish. For these grow particularly on the chalk downs in Sussex.
SUSSEX WEED *Quercus robur* (Oak) Suss. Parish. For the same reason that elms used to be called Wiltshire Weed there.
SUTERBERRY *Xanthoxylum americanum* (Prickly Ash) Grieve.1931.
SUTHYWOOD *Artemisia abrotanum* (Southernwood) Suss. Friend.
SWALLOW, Grinning; SWALLOW, Groundie; SWALLOW, Grundy; SWALLOW GRUNDY *Senecio vulgaris* (Groundsel) N Eng, Scot. Grigson (Grinning, Grundy Swallow, Swallow Grundy) Jamieson (Groundie Swallow). All these very odd names are the result of OE grundeswelge, which apparently means 'ground-swallower', though some would have the OE to be gundeswelge, with the consequent meaning of 'pus-swallower'.
SWALLOW-HERB; SWALLOW-GRASS *Chelidonium majus* (Greater Celandine) Tynan & Maitland. See Swallow-wort.
SWALLOW-WORT 1. *Asclepias curassavica* (Blood-flower) Dalziel. 2. *Asclepias syriaca* (Common Milkweed) McDonald. 3. *Asclepias tuberosa* (Orange Milkweed) Grieve.1931. 4. *Calotropis procera*. 5. *Chelidonium majus* (Greater Celandine) Gerard. 6. *Cynanchum caudatum*. 7. *Cynanchum vincetoxicum*. 8. *Ranunculus ficaria* (Lesser Celandine) Som. Macmillan. Pliny said that the flowering of Greater Celandine coincided with the swallow's stay in the country, but this would not agree with conditions in England. Theophrastus linked the blooming with the time the "Swallow-wind" blew. Celandine is Greek Khelidonion, from khelidon, the swallow. The birds used the plant, Pliny said, to restore their sight.
SWALLOW-WORT GENTIAN *Gentiana asclepiadea* (Willow Gentian) Parkinson. The specific name, asclepiadea, comes from a Greek name meaning swallow-wort, which is why Parkinson used it here.
SWALLOW-PEAR *Sorbus torminalis* (Wild Service) Prior. Cf. Choke-pear, though that and Swallow-pear seem to imply opposites.
SWAMP'S COMPANION *Cardamine pratensis* (Lady's Smock) Som. Macmillan.
SWAN, Golden *Crocosmia masonorum*.
SWAN-AMONG-THE-FLOWERS *Nymphaea alba* (White Waterlily) Dor, Wilts. Macmillan.
SWAN-BILL *Iris pseudo-acarus* (Yellow Flag) Som. Macmillan. It is known as Duck's-bill, too, in the same county. Why?
SWAN FLOWER *Aristolochia gigas* (Pelican Flower) RHS.
SWAN RIVER DAISY *Brachycome iberidifolia*. Swan River is in Canada.
SWARMS *Alliaria petiolata* (Jack-by-the-hedge) Yks. B & H.
SWEATING PLANT *Eupatorium perfoliatum* (Thoroughwort) USA Henkel. Cf. the names Agueweed and Feverwort, which explain this name as well. It is still used as a remedy for colds and fever.
SWEATROOT *Polemonium reptans* (Abscess-root) USA Grieve.1931. As with *P caeruleum*, it was at one time used for promoting sweating in fevers.
SWEDE 1. *Brassica napus* (Rape). 2. *Brassica napus var. napobrassica*. One finds Swedish Turnip and Russian Turnip, too. But

the country of origin was neither Sweden nor Russia, but Bohemia.

SWEDE, Indian *Terminalia catappa* (Myrobalan) Douglas. Cf. Indian Turnip for this.

SWEDE TURNIP *Brassica napus var. napobrassica* (Swede) USA Kingsbury.

SWEDISH CLOVER *Trifolium hybridum* (Alsike Clover) S Africa Watt.

SWEDISH HEMP *Urtica dioica* (Nettle) S Africa Watt. A reference to the fibres obtainable from nettles, much in use into recent times in the Scandinavian countries.

SWEDISH JUNIPER *Juniperus communis var. suecica*.

SWEDISH TURNIP 1. *Brassica campestris var. rapa* (Turnip) USA Elmore. 2. *Brassica napus* (Rape) Clapham. 3. *Brassica napus var. napobrassica* (Swede) Bianchini. See Swede.

SWEDISH WHITEBEAM *Sorbus intermedia* (Cut-leaved Whitebeam) Perry. It grows in Sweden, and also in the other Baltic states as well as in northeast Germany, although it was introduced into Britain long ago, and is plentiful in northeast Scotland.

SWEEP 1. *Bellis perennis* (Daisy) Yks. B & H. 2. *Centaurea nigra* (Knapweed) Derb. Grigson. 3. *Hypericum calycinum* (Rose of Sharon) Wilts. Dartnell & Goddard. 4. *Luzula campestre* (Good Friday Grass) Shrop. B & H. 5. *Nigella damascena* (Love-in-a-mist) Som. Macmillan. 6. *Typha latifolia* (False Bulrush) Som. Macmillan. It is the dark red garden variety of daisy that gets the name in Yorkshire, but it still seems out of place in this company. Some of the others are vaguely brush-shaped, hence the allusion, while the blackness of the grass and the bulrush make the comparison obvious. The bunch of stamens of Rose of Sharon is the reference point.

SWEEP, Chimney 1. *Centaurea nigra* (Knapweed) Som. Macmillan. 2. *Luzula campestris* (Good Friday Grass) Ches, Lancs. B & H. 3. *Plantago lanceolata* (Ribwort Plantain) Som. Macmillan. 4. *Plantago media* (Lamb's-tongue Plantain) Som. Macmillan.

SWEEP'S BRUSH 1. *Dipsacus fullonum* (Teasel) Dev. Macmillan. 2. *Luzula campestris* (Good Friday Grass) McClintock & Fitter. 3. *Plantago lanceolata* (Ribwort Plantain) Som. Macmillan. 4. *Plantago media* (Lamb's-tongue Plantain) Som. Macmillan. 5. *Tussilago farfara* (Coltsfoot) Som. Macmillan. All descriptive in their way.

SWEEPERS, Chimney *Luzula campestris* (Good Friday Grass) Wilts. Dartnell & Goddard; Ches, Lancs. B & H.

SWEET 1. *Myrica gale* (Bog Myrtle) Yks, N'thum. Grigson. 2. *Myrrhis odorata* (Sweet Cicely) N Eng. Grigson. 'Sweet' in a plant name virtually always means sweet-smelling.

SWEET ACACIA *Acacia farnesiana* (Cassie) Usher. The flowers are the source of the perfume called Cassie.

SWEET ALICE 1. *Arabis alpina* (Alpine Rock-cress) Dev. B & H. 2. *Lobularia maritima* (Sweet Alison). 3. *Pimpinella anisum* (Anise) A W Smith. Alice for Alison and therefore Alyssum is a well-known sequence, and accounts for Lobularia (previously classified as Alyssum), and Arabis (by mistake). Anise to Alice is a common corruption.

SWEET ALISON *Lobularia maritima*. Earlier classified as Alyssum, hence Alison.

SWEET AMBER *Hypericum androsaemum* (Tutsan) Suss. B & H. People used to gather the leaves and press them in books. When dry, they have a very sweet smell, likened to ambergris, hence 'amber' in this name. Cf. Sweet Leaf and Bible Leaf, etc.

SWEET ANGELICA *Myrrhis odorata* (Sweet Cicely) B & H.

SWEET ANISE *Foeniculum vulgare* (Fennel) A W Smith.

SWEET ASH *Anthriscus sylvestris* (Cow Parsley) Glos. B & H. 'Ash' here is the old word ache, a name applied to parsley, or indeed to umbellifers in general.

SWEET BALLOCKS *Spiranthes spiralis* (Lady's Tresses) Leyel.1948.
SWEET BASIL *Ocymum basilicum*.
SWEET BAY; SWEET BAY, Southern *Magnolia virginiana*.
SWEET BAY WILLOW *Salix pentandra* (Bay-leaved Willow) Brimble.1948.
SWEET BENT *Luzula campestris* (Good Friday Grass) Scot. B & H.
SWEET BETSY 1. *Centranthus ruber* (Spur Valerian) Som, Kent Grigson. 2. *Dicentra spectabilis* (Love-lies-bleeding) Som. Macmillan. There are many Bet, Betsy or Betty names for Valerian, and Love-lies-bleeding carries many affectionate cottage garden names too.
SWEET BETTY 1. *Centranthus ruber* (Spur Valerian) Som, Dor. Macmillan. 2. *Saponaria officinalis* (Soapwort) Grieve.1931. Cf. Bouncing Bet or Betty for Soapwort.
SWEET BILLERS *Heracleum sphondyllium* (Hogweed) Dev. Grigson.
SWEET BIRCH 1. *Betula fontinalis* (Streeamside Birch) USA Elmore. 2. *Betula lenta* (Cherry Birch) USA Zenkert.
SWEET BRACKEN *Myrrhis odorata* Cumb, West. Grigson. Cf. Sweet Fern, used by Coles.
SWEET BRIAR 1. *Acacia farnesiana* (Cassie) Barbados Gooding. 2. *Rosa micrantha* (Small-leaf Sweet Briar) Phillips. 3. *Rosa rubiginosa*.
SWEET BRIAR, Narrow-leaved *Rosa agrestis*.
SWEET BRIAR, Small-leaf *Rosa micrantha*.
SWEET BROOM *Ruscus aculeatus* (Butchers' Broom) Grieve.1931.
SWEET BUCKEYE *Aesculus octandra* (Yellow Buckeye) USA Kingsbury.
SWEET BUGLE *Lycopus virginicus* (Virginia Bugleweed) Grieve.1931.
SWEET BUSH *Calycanthus floridus* (Allspice) Perry.
SWEET CALABASH *Passiflora maliformis*.
SWEET CALAMUS *Cymbopogon maritima* (Ginger Grass) Zohary.

SWEET CALOMEL *Acorus calamus* (Sweet Flag) Watt.
SWEET CANE *Acorus calamus* (Sweet Flag) Grieve.1931.
SWEET CASSAVA *Manihot dulcis*.
SWEET CHERVIL; SWEET CHERVIL, Giant *Myrrhis odorata* (Sweet Cicely) B & H (Sweet Chervil); Clair (Giant Sweet Chervil).
SWEET CHESTNUT *Castanea sativa*.
SWEET CICELY *Myrrhis odorata*. 'Cicely' is the result of confusion with the girl's name; in fact it comes through Latin from Greek seselis, and Seseli is the name of another genus of umbellifers.
SWEET CIS *Myrrhis odorata* (Sweet Cicely) Yks. Grigson.
SWEET CLOVER 1. *Melilotus alba* (White Melilot) S Africa Watt. 2. *Melilotus indica* (Small-flowered Melilot) S Africa. Watt.
SWEET CLOVER, White *Melilotus alba* (White Melilot) USA Gates.
SWEET CLOVER, Yellow *Melilotus officinale* (Melilot) USA Gates.
SWEET CODS; SWEET CULLINS *Spiranthes spiralis* (Lady's Tresses) Notes & Queries;1853. Cods and Cullins are typical orchid names for the testicle-like tubers.
SWEET COLTSFOOT 1. *Petasites fragrans* (Winter Heliotrope) Geldart. 2. *Petasites palmatus*.
SWEET CORN *Zea mays var. saccharata*.
SWEET COVEY *Erodium moschatum* (Musk Storksbill) B & H. 'Covey' is from 'Muscovy', a pun on 'musk'.
SWEET CROWFOOT *Ranunculus auricomus* (Wood Goldilocks) Flower.1859. Our only non-acrid buttercup, hence 'sweet'.
SWEET CYPRESS *Cyperus longus* (Galingale) A S Palmer. Cypress, i.e. Cyperus.
SWEET DOCK *Polygonum bistorta* (Bistort) N Eng. Schofield.
SWEET ELDER *Sambucus canadensis* (American Elder) Sanecki.
SWEET ELM *Ulmus fulva* (Slippery Elm) USA Vestal & Schultes.

SWEET EVERLASTING *Gnaphalium polycephalum* (Mouse-ear Everlasting) USA Upham.

SWEET FALSE CAMOMILE *Matricaria recutita* (Scented Mayweed) H & P.

SWEET FENNEL 1. *Anethum graveolens* (Dill) Thornton. 2. *Foeniculum dulce* (Finocchio) E Hayes.

SWEET FERN 1. *Myrica asplenifolia*. 2. *Myrrhis odorata* (Sweet Cicely)) Coles. Cf. Sweet Bracken for the latter.

SWEET FLAG *Acorus calamus*. 'Flag', from the iris-like leaves.

SWEET GALE 1. *Myrica asplenifolia* (Sweet Fern) USA Zenkert. 2. *Myrica gale* (Bog Myrtle) Lincs, Yks, Renf. Grigson. Gale is the old English name for the plant.

SWEET GRASS 1. *Asperula odorata* (Woodruff) Berw. Grigson. 2. *Glyceria plicata*. 3. *Hedeoma drummondii* (Drummond Pennyroyal) USA Elmore. 4. *Hierochloe odorata*.

SWEET GRASS, Floating *Glyceria fluitans* (Flote Grass) C P Johnson. Cf. Sugar Grass.

SWEET GREEN BASIL *Ocymum basilicum* (Sweet Basil) Webster.

SWEET GUM *Liquidambar styracifolia*.

SWEET HAIRHOOF *Asperula odorata* (Woodruff) Yks. Grigson.

SWEET HAY *Filipendula ulmaria* (Meadowsweet) Dor. Macmillan; Suss. Grigson.

SWEET HERB *Satureia hortensis* (Summer Savory) New Engl. Sanford.

SWEET HUMLICK *Myrrhis odorata* (Sweet Cicely) N Eng. Grigson.

SWEET HURTS *Vaccinium angustifolium* (Low Sweet Blueberry) USA Yarnell.

SWEET JOHN *Dianthus barbatus* (Sweet William) Gerard. Perhaps because the feast day of St John (24 June) is only the day before the celebration of St William (25 June).

SWEET LEAF *Hypericum androsaemum* (Tutsan) Dev. Grigson. A reference to the practice of gathering the leaves and pressing them in books. When dry, they have a very sweet smell. Cf. Book Leaf, Bible Leaf etc.

SWEET LIME; SWEET LEMON *Citrus limetta* Douglas (Lemon).

SWEET LOCUST *Gleditsia triacanthus* (Honey Locust) USA Elmore.

SWEET LUCERNE *Melilotus officinalis* (Melilot) Leyel.1937.

SWEET LUPIN *Lupinus luteus* (Yellow Lupin) Le Strange.

SWEET MAGNOLIA *Magnolia virginiana* (Sweet Bay) Taylor.

SWEET MARJORAM 1. *Origanum heraclioticum* (Winter Marjoram) G M Taylor. 2. *Origanum marjorana* (Knotted Marjoram) Webster.

SWEET MARY *Centranthus ruber* (Spur Valerian) Bucks. Grigson. Cf. Sweet Betty, or Betsy.

SWEET MELON *Cucumis melo* (Melon) Watt. To distinguish it in name from the Bitter melon.

SWEET MYRTLE *Acorus calamus* (Sweet Flag) Grieve.1931. Cf. Myrtle Flag, Myrtle Grass and Myrtle Sedge. The scent is usually said to be like that of violets, but cinnamon is mentioned too, and so, by these names, is myrtle.

SWEET NANCE *Stellaria holostea* (Greater Stitchwort) Som. Grigson.

SWEET NANCY 1. *Narcissus maialis* (Pheasant's Eye) Norf. B & H; Ches. Leigh; War. Bloom. 2. *Narcissus pseudonarcissus* (Daffodil) Tynan & Maitland. 3. *Stellaria holostea* (Greater Stitchwort) Som. Macmillan.

SWEET NUTS *Achillea millefolium* (Yarrow) Dor. Macmillan.

SWEET ORANGE *Citrus aurantium var. sinensis* (Malta Orange) Willis.

SWEET PEA *Lathyrus odoratus*.

SWEET PEA, Prairie *Lathyrus venosus*.

SWEET PEA, Swamp *Lathyrus palustris* (Marsh Pea) USA Tolstead.

SWEET PEA, White *Lathyrus ochroleucus*.

SWEET PEA, Wild 1. *Hedysarum mackenzii*.

2. *Ononis repens* (Rest Harrow) Som. Macmillan.
SWEET PITCHER PLANT *Sarracenia rubra*.
SWEET PIGNUT *Carya ovalis* (Red Hickory) Wilkinson.1981.
SWEET PIGWEED *Chenopodium ambrosioides* (American Wormseed) Dalziel.
SWEET POTATO *Ipomaea batatas*.
SWEET POTATO, Wild *Phytolacca heptandra* (Umbra Tree) South Africa Watt. This in spite of the fact that the roots are supposed to be toxic.
SWEET ROOT; SWEET RUSH *Acarus calamus* (Sweet Flag) Grieve.1931 (Root); C P Johnson (Rush).
SWEET ROOT-CORN *Calathea allovia*. A reference to the edible bulbs, from which Guinea arrowroot is obtained.
SWEET SAGE *Poliomentha incana*.
SWEET SCABIOUS 1. *Erigeron annuus* (White Top) USA Polunin. 2. *Scabiosa atropurpurea*.
SWEET SEA GRASS *Zostera marina* (Eel Grass) Moore.1924.
SWEET SEDGE; SWEET SEG *Acorus calamus* (Sweet Flag) B & H E Angl. B & H (Seg).
SWEET SLUMBER *Sanguinaria canadensis* (Bloodroot) Grieve.1931. Why? There is no mention in the record of any use as a sedative, though perhaps medicines made of it for such troubles as asthma and croup may have influenced the name.
SWEET SORGHUM *Sorghum saccharatum* (Sorgo) Schery.
SWEET SUCKERS *Symphytum officinale* (Comfrey) Som. Grigson. Or Suckers, without the adjective.
SWEET SULTAN *Centaurea imperialis*.
SWEET SUMACH *Rhus aromatica*.
SWEET THORN *Acacia karroo* (Mimosa Thorn) Palgrave.
SWEET VERNAL GRASS *Anthoxanthum odoratum*.
SWEET VETCH *Hedysarum boreale*.
SWEET VIOLET *Viola odorata*.

SWEET WHITE VIOLET *Viola pallens*.
SWEET WILLIAM 1. *Cheiranthus cheiri* (Wallflower) Lincs. B & H. 2. *Dianthus barbata*. 3. *Saxifraga x urbinum* (London Pride) A E Baker. 'William' is possibly the French oeillet, little eye, descriptive enough of the plant, corrupted into Willy, and then connected in some way with, and changed into, William. Archdeacon Hare said that it is dedicated to St William, whose feast day is 25 June, and that 'sweet' is a substitute for 'saint'. As far as wallflower is concerned, the name may have started with gilliflower, 'gilli-' to 'Willy' being easily negotiated. In both cases 'sweet' is a recognition of the scent.
SWEET WILLIAM, Childing *Kohlrauschia prolifer* (Childing Pink) Gerard. 'Childing', in the sense of producing young directly from the parent child, by runners.
SWEET WILLIAM, Wild 1. *Lychnis flos-cuculi* (Ragged Robin) Prior. 2. *Phlox maculata*. 3. *Saponaria officinalis* (Soapwort) Grieve.1931.
SWEET WILLIAM CATCHFLY *Silene armeria* (Catchfly) Clapham.
SWEET WILLIE 1. *Dactylorchis incarnata* (Early Marsh Orchid) Donegal. Grigson. 2. *Melandrium rubrum* (Red Campion) Shet. Grigson. 3. *Salix pentandra* (Bay-leaved Willow) Cumb. B & H. 'Willie', at least in the last example, is of course 'willow'.
SWEET WILLOW 1. *Myrica gale* (Bog Myrtle) Suss. Grigson. 2. *Salix pentandra* (Bay-leaved Willow) Cumb. Grigson.
SWEET WINTER GRAPE *Vitis cinerea* (Downy Grape) Vestal & Schultes.
SWEET WITHY 1. *Myrica gale* (Bog Myrtle) IOW Grigson. 2. *Myrrhis odorata* (Sweet Cicely) IOW Grigson.
SWEET WIVELSFIELD *Dianthus barbatus x allwoodii*.
SWEET-WOOD *Croton eleutheria*. Presumably it has a fragrant timber - certainly the bark is sweet-smelling, and is used to scent clothes and bedding.
SWEET WOODRUFF *Asperula odorata* (Woodruff) Gilmour & Walters.

SWEETBREAD *Potentilla anserina* (Silverweed) Freethy. A reference to the use of the roots as a marginal food in the Highlands. They can be ground into meal to make porridge, or a kind of bread.

SWEETBREAD, St John's *Ceratonia siliqua* fruit (Carob) Kourennoff. Are they the "locusts" eaten by St John the Baptist in the wilderness?

SWEETHEART PLANT *Philodendron scandens*.

SWEETHEARTS 1. *Agrimonia eupatoria* (Agrimony) Som. Macmillan. 2. *Arctium lappa* burrs (Burrs) Dev. Friend.1882; War. F Savage. 3. *Arum maculatum* (Cuckoo-pint) Som. Macmillan. 4. *Asperula odorata* (Woodruff) Som. Macmillan. 5. *Bidens bipinnata* (Spanish Needles) S Africa Watt. 6. *Bidens pilosa* (Bur Marigold) S Africa Watt. 7. *Galium aparine* (Goose-grass) Dev. Friend; Som. Tongue; Dor. Dacombe; Wilts. Powell; Oxf, Yks. Grigson. 8. *Listera ovata* (Twayblade) Som. Macmillan. 9. *Stellaria holostea* (Greater Stitchwort) Som. Macmillan. 10. *Verbascum thapsus* (Mullein) Som. Macmillan. 'Sweethearts' for most of these commemorates the sticking power of the burrs. Of the rest, Twayblade owes the name to the single pair of leaves, and Cuckoo-pint to the eroticism inherent in its method of growth. But why Mullein?

SWEETHEARTS, Clinging *Galium aparine* (Goose-grass) Wilts. Macmillan.

SWEETLEAF *Symplocos tinctoria*.

SWEETSIES *Myrrhis odorata* (Sweet Cicely) Yks. Carr. This is actually 'Sweet Cis'.

SWEETSOP *Annona squamosa* (Sugar Apple) USA Chopra.

SWEETWOOD *Nectandra membranacea*.

SWETH *Allium schoernoprasum* (Chives) Gerard. A relic of the medieval form siethes, which is the origin of chives.

SWICHEN *Senecio vulgaris* (Groundsel) B & H.

SWIFT *Armeria maritima* (Thrift) Suss. Friend. Obviously an original mis-hearing of 'thrift'.

SWIMMING PLANTAIN *Potamogeton natans* (Broad-leaved Pondweed) Turner.

SWINE ARNUT *Stachys palustris* (Marsh Woundwort) Banff. Grigson. 'Arnut' is 'earthnut', so given in this case from the tubers on the rhizome. Cf. Swine's Beads and Swine's Murrills.

SWINE CARSE; SWINE GRASS *Polygonum aviculare* (Knotweed) B & H (Carse); Cockayne (Grass). Cf. Swine's Cress, or Swine's Grass.

SWINE CRESS *Coronopus squamatus* (Wart-cress) Perring. Cf. Swine's Cress. They mean a cress fit only for pigs.

SWINE THISTLE 1. *Sonchus arvensis* (Corn Sow Thistle) Cumb. B & H; Yks. Carr. 2. *Sonchus oleraceus* (Sow Thistle) N Eng, Scot, Donegal. Grigson.

SWINE WEED *Heracleum sphondyllium* (Hogweed) N'thants. B & H.

SWINE'S BEADS 1. *Potentilla anserina* (Silverweed) Ork. Grigson. 2. *Scilla verna* (Spring Squill) Ork. Grigson. 3. *Stachys palustris* (Marsh Woundwort) Ork. Grigson. The 'beads' in each case are the roots, or the tubers that grow on them. Silverweed roots were not just food for pigs, though, for they have been a marginal food in the Highlands and Ireland for a very long time. They are food for fairies, too.

SWINE'S CRESS 1. *Apium nodiflorum* (Fool's Watercress) Yks, Ork. Grigson. 2. *Coronopus squamatus* (Wart-cress) Tradescant. 3. *Lapsana communis* (Nipplewort) Som. Macmillan. 4. *Polygonum aviculare* (Knotweed) Dawson. 5. *Senecio jacobaea* (Ragwort) B & H.

SWINE'S CRESS, Lesser 1. *Coronopus didymus* (Slender Wart-cress) McClintock. 2. *Coronopus squamatus* (Wart-cress) Salisbury.

SWINE'S GRASS 1. *Polygonum aviculare* (Knotweed) Som. Macmillan. 2. *Senecio jacobaea* (Ragwort) B & H.

SWINE'S GREASE *Polygonum aviculare*

(Knotweed) R T Gunther. It is glossed as such on a 12th century MS of Apuleius Barbarus. 'Grease' is presumably 'grass', or even 'cress', but it is just possible that it means just what it says, in other words pig fat, much in use in medieval ointments.

SWINE'S MASKERT; SWINE'S MOSSCORTS *Stachys palustris* (Marsh Woundwort) Scot. Jamieson. Surely Jamieson's explanation, that it means maskwort, cannot be true?

SWINE'S MURRILLS 1. *Scilla verna* (Spring Squill) Shet. Grigson. 2. *Stachys palustris* (Marsh Woundwort) Shet. Grigson. 'Murrills' are roots.

SWINE'S SKIR *Polygonum aviculare* (Knotweed) Gerard.

SWINE'S SNOUT *Taraxacum officinale* (Dandelion) B & H.

SWINE'S SUCCORY *Arnoseris minima*. Succory is a variant of chicory.

SWINE'S TUSKER *Polygonum aviculare* (Knotweed) Tynan & Maitland.

SWINEBREAD 1. *Conopodium majus* (Earthnut) Inv. B & H. 2. *Cyclamen europaeum* (Sowbread) Coles. For Earthnut, Cf. Pignut.

SWINESNAP *Symphytum officinale* (Comfrey) R T Gunther. It is glossed as such on a 12th century ms of Apuleius Barbarus. Cf. the Wiltshire Pigweed, and the fact that Norfolk pigkeepers always gave them comfrey leaves in their feed, to keep them healthy, and also to make sure they could not be bewitched.

SWINIES *Sonchus oleraceus* (Sow Thistle) Border F M T Palgrave.

SWISS CHARD *Beta vulgaris var. cicla*. Why Swiss. It is also known as Sicilian, or Chilean, Beet, and Roman Kale. Chard is French carde, and then Latin carduus, thistle.

SWISS CHEESE PLANT *Monstera deliciosa*. The name is given because the leaves have cut-like indentations and iregular holes, possibly a device to protect against high winds, or perhaps the holes are there to allow light to reach the leaves below, according to Anthony Huxley.

SWISS MELILOT *Melilotus caerulea*.

SWISS MOUNTAIN PINE *Pinus mugo* (Mountain Pine) RHS.

SWISS SALLOW *Salix helvetica*.

SWISS STONE PINE *Pinus cembra*.

SWISS TREACLE MUSTARD *Erysimum helveticum*.

SWITCH *Myrrhis odorata* (Sweet Cicely) Lancs. Grigson. Possibly the last element in the series Sweet Cis, Sweetsies, Sweets, Switch.

SWITCH, Stone *Lithospermum officinale* (Gromwell) Dyer. Cf. Stone-stitch.

SWITCH ELM *Ulmus glabra* (Wych Elm) Yks. Grigson. In this case, 'switch' means the same as 'wych' - pliant.

SWITCH GRASS *Panicum virgatum*. 'Switch' here must mean a rod, for that is what the specific name means.

SWITCH SORREL *Dodonea viscosa*. A switch-plant is usually one with long, pliant shoots. This New Zealand evergreen must come into this category. "Sorrel", because it has bitter leaves. For the same reason, in Australia it is called Hop-bush.

SWITIKS *Angelica sylvestris* (Wild Angelica) Shet. Grigson.

SWIZZLESTICK TREE *Rauwolfia vomitoria*. The twigs are a convenient implement for mixing drinks. Larger branches are used by African dyers for stirring indigo.

SWORD, Flaming *Kniphofia uvaria* (Red Hot Poker) Dor. Macmillan. Not quite so good a descriptive name as Red Hot Poker, but good enough.

SWORD, Jacob's *Iris pseudo-acarus* (Yellow Flag) Aber. B & H. 'Sword', from the shape of the leaves.

SWORD BEAN 1. *Canavalia ensiformis* (Jack Bean) Willis. 2. *Canavalia gladiata* (Parang Bean) Schery. The specific name in the latter case means 'like a sword'.

SWORD BEAN, Seaside *Canavalia obtusifolia*.

SWORD FERN *Polystichum munitum*.

SWORD FLAG 1. *Gladiolus sp* Gerard. 2. *Iris pseudo-acarus* (Yellow Flag) Grieve.1931. The shape of the leaves suggests a sword; in fact, gladiolus means 'little sword'.

SWORD GRASS *Iris pseudo-acarus* (Yellow Flag) Halliwell.

SWORD-LEAVED HELLEBORINE *Cephalanthera longifolia* (Narrow Helleborine) Turner Ettlinger.

SWORD LILY 1. *Gladiolus triphyllus*. 2. *Iris pseudo-acarus* (Yellow Flag) Som. Macmillan.

SWORDS *Iris pseudo-acarus* (Yellow Flag) Wilts. Dartnell & Goddard.1899; Yks. Nicholson.

SWORDS-AND-SPEARS *Plantago lanceolata* (Ribwort Plantain) Dor. Macmillan. A reference to the Soldiers, or Fighting Cocks, game played by children. One child holds out a flowering stalk, while the other child tries to decapitate it with his stalk. The game proceeds very like Conkers.

SWY *Salicornia europaea* (Glasswort) B & H.

SYBIE; SYBOW *Allium fistulosum* (Welsh Onion) both Scot. B & H. These are from Ciboule, which is Provencal cebula, the name for onion that originated in the Latin diminutive for cepa - cepulla.

SYCAMORE 1. *Acer pseudo-platanus*. 2. *Platanus occidentalis* (Western Plane) USA Tolstead. The name derives from a mistake, from the fig *Ficus sycomorus*. The shape of the leaves is a little similar.

SYCAMORE, American *Platanus occidentalis* (Western Plane) USA Barber & Phillips.

SYCAMORE FIG *Ficus sycomorus*. *Sycomorus* would mean 'fig-mulberry'. The leaves and wood closely resemble those of mulberry. Cf. Mulberry-fig, or Fig-mulberry.

SYCAMORE MAPLE *Acer pseudo-platanus* (Sycamore) A W Smith.

SYCOMORE *Ficus sycomorus* (Sycamore Fig) Hutchinson & Melville.

SYDNEY BLUE GUM *Eucalyptus saligna*.

SYDNEY GOLDEN WATTLE *Acacia longifolia*.

SYMYL *Primula veris* (Cowslip) Wales Roberts.

SYNDOW *Alchemilla vulgaris* (Lady's Mantle) Turner.

SYPHELT *Sempervivum tectorum* (Houseleek) Cumb. Grigson. Cf. Cyphel. They both come from the Greek kuphella, which means the hollows of the ear. In the Middle Ages the plant was often called Erewort, and used against deafness, and to this day it is in demand for ear drops.

SYRIAN BINDWEED *Convolvulus scammonia* (Scammony) Grieve.1931.

SYRIAN CHRIST-THORN *Zizyphus spina-christi* (Lotus Tree) Moldenke.

SYRIAN HYSSOP *Origanum maru* (Syrian Marjoram) Zohary. This is probably the hyssop of the Old Testament, in spite of the fact that it is not a hyssop at all.

SYRIAN JUNIPER *Juniperus drupacea*.

SYRIAN KATMIA *Hibiscus syriacus*. Actually, it is a native of China and Japan, so why 'Syrian'?

SYRIAN MARJORAM *Origanum maru*.

SYRIAN RUE *Peganum harmala* (African Rue) Emboden.

SYRIAN SCABIOUS *Cephalaria syriaca*.

SYRINGA *Philadelphus coronaria*. Misnamed, of course, for Syringa is the lilac, but the mistake is venerable. It was so named, too late, by a contemporary of Linnaeus, who had already called it *Philadelphus*.

SYRINGA, Wild *Burkea africana*.

SYRINGA, Yellow *Cestrum parqui* (Green Jessamine) Howes. Why. The facts do not support such a name.

SYVES *Allium schoernoprasum* (Chives) Dev. Whitcombe; Scot. B & H.

SZECHUAN JUNIPER *Juniperus detans*.

T

TAB-MAWN *Armeria maritima* (Thrift) Corn. Grigson. From a Cornish original.

TABASCO MAHOGANY *Swietenia macrophylla* (Honduras Mahogany) Ackermann. Tabasco is the name of a Mexican province.

TABASCO PEPPER *Capsicum annuum* (Chili) Watt.

TABLE, Fairy *Hydrocotyle vulgaris* (Pennywort) Ches. Holland.

TABLE DOGWOOD *Cornus contraversa* (Wedding-cake Tree) Hora. 'Table', because of the same "tiered arrangement" that gave Wedding-cake Tree.

TABLE FERN *Pteris cretica*.

TABLES, Chairs-and-; TABLES-AND-CHAIRS *Buxus sempervirens* seed (Box) both Som. Macmillan. There are a number of paired names for the seed, all from the West Country. Cf. Crocks-and-kettles, for instance.

TABOR BARLEY *Hordeum spontaneum*.

TACAMAHAC(A) 1. *Calophyllum inophyllum* (Alexandrian Laurel) Howes. 2. *Populus balsamifera* (American Balsam Poplar) A W Smith. Presumably an Indian name.

TACKER-GRASS; TUCKER-GRASS *Polygonum aviculare* (Knotgrass) Dev, Som. Macmillan. "From its likeness to a "tacker", or shoemaker's wax-end", according to Elworthy.

TACKER-WEED *Capsella bursa-pastoris* (Shepherd's Purse) Som. Macmillan.

TAE-GIRSE *Thymus vulgaris* (Wild Thyme) Shet. B & H. i.e. tea-grass. Thyme tea is taken for headaches.

TAG ALDER 1. *Alnus rubra* (Red Alder) O P Brown. 2. *Alnus rugosa* (Smooth Alder) Grieve.1931.

TAHL GUM *Acacia seyal*.

TAHOTTA DAISY *Machaeranthera tanacitifolia*.

TAIL PLANT *Anthurium scherzerianum* (Flamingo Plant) Harvey.

TAILDERS *Scandix pecten-veneris* (Shepherd's Needle) Corn. Quiller-Couch. Tailors, presumably, by association with the 'needle' part of the plant.

TAILOR, Gentleman *Viola tricolor* (Pansy) Dor. Macmillan. Cf. Gentleman John for the pansy.

TAILOR TREE *Decaspermum fruticosum*.

TAILOR'S NEEDLE 1. *Scandix pecten-veneris* (Shepherd's Needle) Corn. Jago. Dev. Grigson. 2. *Spergula arvensis* (Corn Spurrey) Corn. Grigson.

TAILS, Baa Lamb's *Corylus avellana* catkins (Hazel catkins) Son. Macmillan.

TAILS, Burro's; TAILS, Burrow's *Sedum morganianum*. Burro's Tails is the usual name. Burrow's Tails is in Cunningham, and is for those who have no Spanish.

TAILS, Capon's 1. *Aquilegia vulgaris* (Columbine) B & H; USA Hutchinson. 2. *Centranthus ruber* (Spur Valerian) Friend. 3. *Valeriana officinalis* (Valerian) Turner. 4. *Valeriana pyrenaica* (Giant Valerian) B & H.

TAILS, Cat, Broadleaf; TAILS, Cat, Common *Typha latifolia* (False Bulrush) E Gunther (Broadleaf); USA Kingsbury (Common).

TAILS, Cat's 1. *Aconitum napellus* (Monkshood) Shrop. B & H. 2. *Amaranthus caudatus* (Love-lies-bleeding) Dev. Friend. 3. *Corylus avellana* catkins (Hazel catkins) Dev. Friend.1882; Som. Tongue; Wilts. Dartnell & Goddard; Suss. Parish; N'thants. A E Baker. 4. *Echium vulgare* (Viper's Bugloss) Ess, Norf, Cambs. B & H. 5. *Equisetum arvense* (Common Horsetail) Som. Macmillan. Wilts. Dartnell & Goddard; Ches. Holland. 6. *Eriophorum angustifolium* (Cotton-grass) J G McKay. 7. *Hippuris vulgaris* (Mare's Tail) Dev, Som, Oxf. Grigson. Hants. Cope; E Ang. Forby. 8. *Juglans regia* catkins (Walnut catkins) Lyte. 9. *Phleum nodosum*. 10. *Salix fragilis* catkins (Crack Willow catkins) Som. Macmillan. 11. *Typha latifolia* (False Bulrush) Som. Macmillan.

TAILS, Cat's, Blue *Echium vulgare* (Viper's Bugloss) Herts. Grigson.

TAILS, Cat's, Greater *Typha latifolia* (False Bulrush) Curtis.

TAILS, Cat's, Redhot *Acalypha hispida* (Chenille Plant) Hay & Synge. The reference here is to the crimson catkin-like spikes.

TAILS, Cat's, Smaller *Typha angustifolia* (Lesser Bulrush) Curtis.

TAILS, Colt's 1. *Equisetum arvense* (Common Horsetail) Dev. Macmillan. 2. *Erigeron canadensis* (Canadian Fleabane) USA Henkel. 3. *Hippuris vulgaris* (Mare's Tail) Dev. Macmillan.

TAILS, Cow's *Erigeron canadensis* (Canadian Flebane) USA Henkel.

TAILS, Donkey's *Sedum morganianum* (Burro's Tails) Goold-Adams.

TAILS, Hare's *Eriophorum angustifolium* (Cotton-grass) Som. Macmillan.

TAILS, Kitten's *Corylus avellana* catkins (Hazel catkins) Dor. Macmillan.

TAILS, Lamb's 1. *Alnus glutinosa* catkins (Alder catkins) Som. Macmillan. 2. *Anthyllis vulneraria* (Kidney Vetch) Som. Macmillan. 3. *Corylus avellana* catkins (Hazel catkins) Dev. Friend.1882; Wilts. Dartnell & Goddard. 4. *Cotyledon oppositifolia*. 5. *Plantago lanceolata* (Ribwort Plantain) Som. Macmillan. 6. *Plantago major* (Great Plantain) Norf, Cumb, Scot. Grigson. 7. *Salix caprea* catkins (Goat Willow catkins) Grigson. 8. *Salix fragilis* catkins (Crack Willow catkins) Corn. Jago.

TAILS, Lion's 1. *Agastache mexicanum* Brownlow. 2. *Leonotis leonurus*. 3. *Leonurus cardiaca* (Motherwort) Brownlow.

TAILS, Mare's 1. *Equisetum arvense* (Common Horsetail) Som. Elworthy. 2. *Erigeron canadensis* (Canadian Fleabane) S Africa Watt. 3. *Euphorbia amygdaloides* (Wood Spurge) Ire. B & H. 4. *Hippuris vulgaris*.

TAILS, Pony's *Plantago major* (Great Plantain) Dev. Macmillan.

TAILS, Pussy's *Typha latifolia* (False Bulrush) Som. Macmillan.

TAILS, Pussy Cat's *Corylus avellana* catkins (Hazel. catkins) Dev. B & H.

TAILS, Rat's 1. *Agrimonia eupatoria* (Agrimony) Wilts. Macmillan. 2. *Plantago lanceolata* (Ribwort Plantain) Cumb. Grigson. 3. *Plantago major* (Great Plantain) Norf, Cumb, Scot. Grigson.

TAILS, Ratten *Plantago major* (Great Plantain) N'thum. Grigson.

TAILS, Scorpion's *Heliotropum europaeum* (Cherry Pie) Turner.

TAILS, Sheep's *Corylus avellana* catkins (Hazel catkins) Som. Macmillan.

TAILWORT *Borago officinalis* (Borage) H & P.

TAIWAN SPRUCE *Picea morrisonicola* (Mount Morrison Spruce) RHS.

TAINTWORT See Tentwort.

TALEWORT *Borago officinalis* (Borage) W Miller.

TALI LILY *Lilium taliense*. A reference to the Tali Range in northeast Yunnan, in western China, where this lily was discovered by Delavay in 1883.

TALIESIN'S CRESS *Sedum telephium* (Orpine) Wales Grigson. Welsh Cerwr Taliesin.

TALLOW, Japanese *Rhus succedanea* (Japanese Wax Tree) H & P.

TALLOW-TREE 1. *Pentadesma butyracea* Dalziel. 2. *Sapium sebiferum*. The seeds of *Sapium sebiferum* yield a wax suitable for lighting, and used as such in China. It is the seeds of *Pentadesma butyracea* also that yield the fat, extracted for cooking, ointment and soap. Cf. Butter-tree and Candle Tree for this.

TALLOW-WOOD *Ximenia americana* (Wild Oliver) Sofowora.

TAM, Stinking *Ononis repens* (Rest Harrow) N'thum. Grigson. Cf. Stinking Tommy. Is it so evil-smelling? The fact that it is called Ramsons in Devonshire might suggest it.

TAM FURZE; TAM FUZZ *Ulex minor*, or *Ulex gallii* (Dwarf Furze) Corn. Quiller-Couch. 'Tam' means dwarf.

TAMARACK *Larix laricina* (American

Larch) USA H Smith. This comes from the American Indian name for the tree.

TAMARIND *Tamarindus indica*. From an Arabic word, tamr-Hindi, Date of India.

TAMARIND, Wild *Cassia glandulosa var. swartzii* Barbados Gooding.

TAMARISK *Tamarix anglica*. Although tamarisk obviously comes from the Latin, it is said that they both come from a Hebrew word meaning 'sweeping broom'.

TAMARISK, False; TAMARISK, German *Myricaria germanica* False Tamarisk is the usual name; German Tamarisk is in Polunin.

TAMARISK, French *Tamarix gallica* Brimble.1948.

TAMARISK, Leafless *Tamarix aphylla*.

TAMARISK, Manna *Tamarix gallica var. manifera*. The 'manna' is caused by the puncturing of the branches by the insect Coccus manniparus, so that little honey-like drops are exuded, and solidify. This "manna" was collected by the Arabs and sold to the monks of St Katharine, who disposed of it to the pilgrims visiting the convent. This is in fact the Biblical Manna. The Bedouins still gather it, either keeping it like honey, or making it into cakes.

TAMARISK, Nile *Tamarix nilotica*.

TAMBOOTIE *Spirostachys africana*.

TAME *Ulex gallii* (Dwarf Furze) Corn. Carew. This is the Cornish word 'tam', dwarf, as in Tam Fuzz for the same plant.

TAME CORNEL *Cornus mas* (Cornel Cherry) Gerard.

TAN OAK *Quercus densiflora*.

TANA RIVER POPLAR *Populus ilicifolia*. An East African tree,growing on riversides (of, presumably, the Tana River, which runs through eastern Kenya).

TANEKAHA *Phyllocladus trichomanoides*. This is the Maori name.

TANGERINE *Citrus reticulata*. Tangerine means the orange imported from Tangier. But the plant itself seems to be a native of the southern part of Vietnam, and it has been cultivated for a very long time in China and Japan.

TANGIER IRIS *Iris tingitana*.

TANGIER PEA *Lathyrus tangitanus*. It is used as a forage crop in North Africa.

TANGLE, Love-and- *Trifolium campestre* (Hop Trefoil) Dor. Dacombe.

TANGLE, Red *Cuscuta epithymum* (Common Dodder) Norf. Grigson.

TANGLE-GRASS *Ranunculus repens* (Creeping Buttercup) Yks. Grigson. Preumably a reference to the creeping root system.

TANGLE-LEGS *Viburnum alnifolium* (Hobble-bush) USA Perry. The lower branches of this tree droop to the ground and root at the tips, hence this name, and Down-you-go, Devil's Shoestring, and the common name, Hobble-bush.

TANGLEFOOT BEECH *Nothofagus gunnii*.

TANGING NETTLE *Urtica dioica* (Nettle) Yks. Grigson. 'Tang' is an old Yorkshire word meaning to sting.

TANIAS *Xanthosoma sagittifolium*.

TANKARD, Cool *Borago officinalis* (Borage) N'thants. A E Baker. This is a local reference, but the name certainly is not confined to that county. Borage forms one of the ingredients, with water, wine, lemon and sugar, of a drink called Cool Tankard, from which the name was transferred to the plant.

TANKET *Cordyline fruticosa*.

TANNER'S APRON *Primula auricula* (Auricula) Glos. B & H. Apparently reserved only for the yellow kind.

TANNER'S BARBERRY *Berberis aristata* (Nepal Barberry) Chopra.

TANNER'S CASSIA *Cassia auriculata*. The bark is valued in tanning processes.

TANNER'S DOCK *Rumex hymenosepalus*. The roots are 25/35% tannin, and are still cultivated for this, and of course it is used for any complaint needing an astringent to cure it.

TANNING SUMACH *Rhus coriaria* (European Sumach) R T Gunther.1934. The leaves are up to 35% tannin. Originally the word 'sumach' was applied only to these dried leaves. It is Arabic summaq, from a word meaning red, a reference to the use as a dyestuff.

TANSY 1. *Achillea millefolium* (Yarrow) Ches. Holland. 2. *Chrysanthemum segetum* (Corn Marigold) Glos. J D Robertson. 3. *Potentilla anserina* (Silverweed) Cumb, N'thum, Yks. Grigson. 4. *Senecio jacobaea* (Ragwort) Aber. B & H. 5. *Tanacetum vulgare*. Ragwort and yarrow have certain similarities in the leaves to tansy, and the name probably arose by wrong identification. The other plants on the list have yellow flowers, but otherwise are in no way similar. Tansy is from medieval Latin athanasia, itself fromn Greek athanatos, undying. Athanasia became in old French tanesil, so to tansy in English. The immortality name might come from the fact that the flowers take a long time to wither, but more likely because it was used for preserving dead bodies from corruption.

TANSY, Creeping; TANSY, Trailing *Potentilla anserina* (Silverweed) Tynan & Maitland (Creeping); Grieve.1931 (Trailing).

TANSY, Curled *Tanacetum vulgare var. crispum*.

TANSY, Dog; TANSY, Goose *Potentilla anserina* (Silverweed) Scot. Jamieson (Dog); N'thants. A E Baker; Norf, Lincs, Cumb. Grigson (Goose). 'Goose' is a result of the specific name,*anserina*, and there are a number of similar 'goose' names, presumably because geese like to eat them.

TANSY, White 1. *Achillea ptarmica* (Sneezewort) B & H. 2. *Agrimonia eupatoria* (Agrimony) B & H.

TANSY, Wild 1. *Agrimonia eupatoria* (Agrimony) B & H. 2. *Potentilla anserina* (Silverweed) Turner.

TANSY ASTER *Machaeranthera sp*. Similarity in the leaves engendered the name.

TANSY-LEAVED ROCKET *Sisymbrium tanacetifolium*.

TANSY MUSTARD 1. *Descurainia pinnata*. 2. *Sisymbrium incisum*.

TANSY MUSTARD, Northern *Sisymbrium brachycarpum*.

TANSY MUSTARD, Western *Sisymbrium incisum*.

TANSY PHACELIA *Phacelia tanacetifolia* (Lacy Phacelia) Usher.

TANSY RAGWORT *Senecio jacobaea* (Ragwort) USA Allan.

TANYA *Colocasia antiquorum* (Taro) W Indies Beckwith.

TAOTOA *Phyllocladus glaucus*. The Maori name, in general use.

TAPER; TAPER, Hag; TAPER, Hag's; TAPER, Hedge; TAPER, Hig; TAPER, Higgis; TAPER, High *Verbascum thapsus* (Mullein). Cockayne (Taper); Bucks, Herts, N Ire. B & H (Hag); C P Johnson (Hag's); Som. Macmillan; Lupton (Hedge); Som, N Ire. Grigson (Hig); Turner (Higgis); Parkinson (High). All these mean Hedge Taper - OE haga, hedge. 'Taper', because the stems and leaves were once dipped in tallow or suet and burnt to give light at outdoor country gatherings, or even in the home, They were used at funerals, too.

TAPER, King's *Verbascum thapsus* (Mullein) Som. Macmillan. Fit for a king, perhaps?

TAPER, Lady's *Verbascum thapsus* (Mullein) Som. Macmillan. The 'Lady' of the name would of course be the Virgin Mary. See the next entry.

TAPER, Mary's *Galanthus nivalis* (Snowdrop) Coats. It is not clear why snowdrop bears the name. The ascription to the Virgin, though, is consistent with the flower's dedication to Candlemas.

TAPIOCA *Manihot esculenta* (Manioc) USA Kingsbury. Tapioca is a word from some south or central American language, in which it would be pronounced tipyoca.

TAPPIE, Soldier's *Plantago lanceolata* (Ribwort Plantain) Ayr. Grigson.

TAR *Vicia sativa* (Common Vetch) Lincs, Yks. Grigson. 'Tar' is 'tare', used for any vetch.

TAR-FITCH, Blue *Vicia cracca* (Tufted Vetch) Ches. Holland. 'Tar-fitch' is 'tare-vetch'.

TAR-FITCH, Yellow *Lathyrus pratensis* (Yellow Pea) Ches. Grigson.
TAR GRASS *Vicia cracca* (Tufted Vetch) B & H.
TAR-VETCH *Vicia sativa* (Common Vetch) Som. Macmillan.
TARATA *Pittosporum eugenioides* (New Zealand Lemonwood) Leathart. This is the Maori name for the tree.
TARBOTTLE *Centaurea nigra* (Knapweed) Oxf. Grigson.
TARBRUSH *Florensia cernua*. Also known as Blackbrush in America.
TARE 1. *Calystegia sepium* (Great Bindweed) Wilts. Dartnell & Goddard. 2. *Convolvulus arvensis* (Field Bindweed) Wilts. Dartnell. & Goddard. 3. *Lathyrus aphaca* (Yellow Vetchling) B & H. 4. *Vicia hirsuta* (Hairy Tare) Middx, Ess, Suff, Lincs, Notts, Ches, Yks, Scot. Grigson. 5. *Vicia sativa* (Common Vetch) Middx, Ess, Suff, Lincs, Notts, Ches, Yks, Scot. Grigson. 6. *Vicia sepium* (Bush Vetch) Suff. Grigson. Tare, of obscure origin, according to the dictionary, is a common name for the vetches. The tares of the Bible are probably darnel. Only in Wiltshire was the word used for the bindweeds, but the two names convey the same meaning.
TARE, American *Vicia americana* (American Vetch) USA Elmore.
TARE, Hairy *Vicia hirsuta*.
TARE, Slender *Vicia tenuissima*.
TARE, Smooth *Vicia tetrasperma*.
TARE, Strangle 1. *Cuscuta epithymum* (Common Dodder) Grieve.1931. 2. *Cuscuta europaea* (Greater Dodder) B & H. 3. *Vicia hirsuta* (Hairy Tare) Salisbury.
TARE, Tine; TARE, Tine, Rough-podded *Vicia hirsuta* (Hairy Tare) Kent Grigson (Tine-tare); Curtis (Rough-podded...). 'Tine' is itself a dialect word for vetch, so the combination with 'tare' is pleonastic.
TARE, Tine, Smooth-podded *Vicia tetrasperma* (Smooth Tare) Curtis.
TARE, Wild 1. *Vicia cracca* (Tufted Vetch) Scot. Grigson. 2. *Vicia sepium* (Bush Vetch) Bucks. B & H.
TARE, Wild, Great *Lathyrus sylvestris* (Everlasting Pea) Gerard.
TARE-FITCH 1. *Vicia cracca* (Tufted Vetch) W Eng, Shrop, Ches. Grigson. 2. *Vicia hirsuta* (Hairy Tare) W Eng, Shrop, Ches. Grigson. 'Fitch' is 'vetch'.
TARE-GRASS 1. *Vicia cracca* (Tufted Vetch) Kent, Staffs. Grigson. 2. *Vicia hirsuta* (Hairy Tare) Kent, Staffs. Grigson.
TARE-TINE, Yellow *Lathyrus pratensis* (Yellow Pea) Middx. Grigson.
TARE-VETCH 1. *Vicia cracca* (Tufted Vetch) Dor, Som, IOW Grigson. 2. *Vicia hirsuta* (Hairy Tare) Dor, IOW Grigson.
TARO *Colocasia antiquorum*.
TARRAGON *Artemisia dracunculus*. Tarragon is a corruption of dracunculus, derived through French estragon. Dracunculus is the diminutive of draco, dragon, the twisted root and stem being compared to a dragon's tail.
TARRAGON, False *Artemisia dracunculoides* (Russian Tarragom) USA T H Kearney.
TARRAGON, French *Artemisia dracunculus* (Tarragon) Rohde. 'French', not because it grows in France (its home is central Asia), but probably because it has always been used in French cooking, and only rarely in English.
TARRAGON, Russian *Artemisia dracunculoides*.
TARRIFY *Erisymum cheiranthoides* (Treacle Mustard) Cambs. B & H.
TARRY-TONGUE *Nasturtium officinale* (Watercress) Yks. Grigson. Cf. Teng-tongues and Tongue Grass. The names depend on the derivation of Nasturtium, which is Nasi-tortium, nose-twisting, from the tangy taste of the darker leaves.
TART RHUBARB *Rheum rhaponticum* (Rhubarb) USA Sanecki.
TARTAR-ROOT *Panax quinquefolium* (Ginseng) Leyel.1948.
TARTARIAN HONEYSUCKLE *Lonicera tatarica*.

TARTARY BUCKWHEAT *Fagopyrum tataricum* (Green Buckwheat) Candolle.
TARTS, Jam 1. *Fumaria officinalis* (Fumitory) Som. Macmillan. 2. *Geranium molle* (Dove's-foot Cranesbill) Som. Macmillan. 3. *Geranium robertianum* (Herb Robert) Som. Macmillan.
TARWEED 1. *Amsinckia intermedia*. 2. *Grindelia squarrosa* (Gumweed) USA Gates. 3. *Hemizonia fasciculata*. 4. *Madia sativa*.
TARWOOD *Dacrydium bidwillii*.
TASMANIAN BLUE GUM *Eucalyptus globosus* (Blue Gum) RHS.
TASMANIAN CEDAR *Athrotaxus selaginoides* (King William Pine) Everett.
TASMANIAN CEDAR, Smooth *Athrotaxus cupressoides*.
TASMANIAN CIDER TREE *Eucalyptus gunnii* (Cider Gum) W Miller. There is also Cedar Gum for this. But 'cider' is correct; the sap used to be a local substitute for honey, but it was called cider.
TASMANIAN CYPRESS PINE *Callitris oblonga*.
TASMANIAN DAISY BUSH *Olearia stellulata*.
TASMANIAN MYRTLE *Nothofagus cunninghamii* (Myrtle Tree) N Davey.
TASMANIAN NGAIO *Myoporum insulare*.
TASMANIAN OAK *Eucalyptus obliqua* (Stringy-bark) N Davey.
TASMANIAN PEPPERMINT *Eucalyptus amygdalana*.
TASMANIAN PODOCARP *Podocarpus alpinus*.
TASMANIAN SASSAFRAS *Atherosperma moschatum* (Australian Sassafras) RHS.
TASMANIAN SNOWGUM *Eucalyptus coccifera* (Mount Wellington Peppermint) Wilkinson.1981.
TASMANIAN TEAK *Dacrydium franklinii* (Huon Pine) Everett.
TASSEL 1. *Centaurea nigra* (Knapweed) Berw. Grigson. 2. *Dipsacus fullonum* (Teasel) Stow. 3. *Fraxinus excelsior* flowers (Ash) Som. Macmillan. Not descriptive as far as knapweed and teasel are concerned. The derivation there is from thistle.
TASSEL, Dog's *Arum maculatum* (Cuckoo-pint) Som. Grigson. Cf. Dog Cocks, Dog's Dibble, etc.
TASSEL, May *Viburnum opulus* (Guelder Rose) Dev. Macmillan. Cf. May Rose, Maypole, etc, and more especially May Tosty.
TASSEL, Milky *Sonchus oleraceus* (Sow Thistle) Corn. B & H. In this case, 'tassel' means 'thistle'.
TASSEL, Silk, Yellow-leaf *Garrya flavescens* (Quinine-bush) Jaeger.
TASSEL, Siller *Briza media* (Quaking-grass) Berw. Grigson.1959.
TASSEL, Whitsun *Viburnum opulus* (Guelder Rose) Som. Macmillan. Guelder Rose is the ecclesiastical symbol for Whitsuntide. Cf. Whitsun Balls, Boss, Flower, and Rose. It is also the May Rose, and May Tassel.
TASSEL BUSH, Silk *Garrya elliptica*.
TASSEL FLOWER 1. *Amaranthus caudatus* (Love-lies-bleeding) Hay & Synge. 2. *Cacalia sp*. 3. *Thalictrum flavum* (Meadow Rue) N'thants. Clare.
TASSEL FLOWER, Alpine *Soldanella montana*.
TASSEL PONDWEED *Ruppia maritima*.
TASSEL-RAGS *Salix caprea* catkins (Goat Willow catkins) Ches. B & H.
TATARIAN DOGWOOD *Cornus alba* (Red-barked Dogwood) Hora.
TATARIAN MAPLE *Acer tataricum*.
TATERS; TATEYS *Solanum tuberosum* (Potato) Wales Salaman (Taters); Yks. F K Robinson (Tateys).
TAURIAN BELLFLOWER *Campanula rapunculoides* (Creeping Bellflower) Allan. It was found in the 18th century in Tauria, and sent to the Chelsea Physic Garden.
TAUSLE, Dog's *Arum maculatum* (Cuckoo-pint) Som. B & H. 'Tausle' is just another spelling of 'tassel'.
TAWHAI *Nothofagus fusca* (Red Beech) Brimble.1948.
TAZZLE 1. *Cirsium heterophyllum*

(Melancholy Thistle) Yks. B & H. 2. *Dipsacus fullonum* (Teasel) B & H. Not 'tassel', but thereafter it is not easy to determine whether it is 'teasel' or 'thistle'.

TEA *Rumex acetosa* (Wild Sorrel) Som. Macmillan. Reserved for the flowers.

TEA, Abyssinian *Catha edulis* (Khat) Douglas. Also called Arabian, or Somali, Tea. Khat, in various forms, is the name usually given to a "tea" made from the buds, twigs and fresh leaves of this tree, predating coffee in Arabia by over a hundred years.

TEA, Appalachian 1. *Ilex glabra* (Gallberry) Le Strange. 2. *Viburnum cassinoides* (Witherod) New Engl. Sanford.

TEA, Arabian *Catha edulis* (Khat) Douglas. See Abyssinian Tea.

TEA, Bahama *Lantana camara* (Jamaica Mountain Sage) Dalziel.

TEA, Blue Mountain *Solidago gigantea* (Early Golden Rod) USA H Smith.

TEA, Brazil; TEA, Brazil, Jesuits' *Ilex paraguayensis* (Maté) Schery (Brazil); Le Strange (Jesuits'). Cf. Paraguay Tree.

TEA, Brigham; TEA, Brigham Young 1. *Ephedra trifurca* (Long-leaved Joint-Fir) USA Elmore. 2. *Ephedra viridis* (Mountain Joint-Fir) USA Youngken. Cf. Mormon Tea. An infusion of the branches is a popular local beverage.

TEA, Bush *Ocymum sanctum* (Holy Basil) Australia Howes.

TEA, Bush, Gambian *Lippia multiflora*.

TEA, Bushman's *Catha edulis* (Khat) Howes.

TEA, Canadian *Gaultheria procumbens* (Wintergreen) USA Sanford. Cf. Teaberry, Mountain Tea, etc. The tea in this case is a medicinal one.

TEA, China, Wild *Sapindus drummondii* (Drummond's Soapberry) Vestal & Schultes.

TEA, Desert *Ephedra trifurca* (Long-leaved Joint-Fir) USA Elmore.

TEA, European *Veronica officinalis* (Common Speedwell) Mitton.

TEA, Ground *Gaultheria procumbens* (Wintergreen) USA Sanecki.

TEA, Irish *Prunus spinosa* (Blackthorn) Ire. O Suilleabhain. Blackthorn leaves have been used as an adulterant of tea, notoriously so in Victorian times.

TEA, Java *Orthosiphon stamineus*.

TEA, Jersey *Gaultheria procumbens* (Wintergreen) USA Sanecki.

TEA, Jerusalem *Chenopodium ambrosioides* (American Wormseed) USA Henkel.

TEA, Jesuit 1. *Chenopodium ambrosioides* (American Wormseed) USA. Henkel. 2. *Ilex paraguayensis* (Mate) Schery.

TEA, Jesuit's *Psoralea glandulosa*.

TEA, Kidney, Indian *Orthosiphon stamineus* Thomson.1978. The leaves are used medicinally in a number of teas, primarily for its diuretic properties.

TEA, Labrador 1. *Ledum groendlandicum*. 2. *Ledum palustre* (Marsh Rosemary) Gilmour & Walters. A medicinal tea is made from the leaves, but it is drunk as a beverage, too.

TEA, Labrador, Pacific *Ledum columbianum*. A real confusion of a name!

TEA, Labrador, Western *Ledum glandulosum*.

TEA, Marsh *Ledum palustre* (Marsh Rosemary) Fernie.

TEA, Matura *Cassia auriculata* (Tanner's Cassia) Howes.

TEA, Mexican 1. *Chenopodium ambrosioides* (American Wormseed) USA Henkel. 2. *Ephedra trifurca* (Long-leaved Joint-Fir) USA Elmore.

TEA, Miner's *Ephedra nevadensis* (Nevada Joint-Fir) USA Bean. Cf. Teamster's Tea for this.

TEA, Moleery 1. *Achillea millefolium* (Yarrow) Caith. Grigson. 2. *Achillea ptarmica* (Sneezewort) Caith. Grigson.

TEA, Mormon *Ephedra nevadensis* (Nevada Joint-Fir) USA Bean.

TEA, Mormon, Green *Ephedra viridis* (Mountain Joint-Fir) USA Elmore. Cf. Brigham Tea - Brigham Young,that is.

TEA, Mormon, Torrey *Ephedra torreyana*.

TEA, Mountain 1. *Gaultheria procumbens* (Wintergreen) USA Sanford. 2. *Mentha* *arvensis var. canadensis* (Canadian Mint) Usa. Chamberlin.

TEA, Mountain, Blue *Solidago gigantea* (Early Golden Rod) USA H Smith.

TEA, Navajo *Heuchera bracteata* USA Elmore. The Navajo make some kind of tea with this to treat toothache.

TEA, New Jersey *Ceanothus americanus*. During both the War of Independence and the Civil War, the leaves were used as a tea substitute, though there is no caffeine in them.

TEA, Oswego *Monarda didyma* (Red Bergamot) Brownlow. This plant is common in the Oswego area, by Lake Ontario.

TEA, Paraguay *Ilex paraguayensis* (Maté) Schery. Cf. Brazil Tree, and there are also Jesuit Tea and Jesuit's Brazil Tea. Maté is made from the dried leaves, just as we make tea, though in more primitive areas the leaves are carried in a horn or gourd (the maté is actually the gourd container of the beverage), and cold water is poured over them, and allowed to steep. This is one of the important stimulants or restoratives of ancient America, for there is high caffeine content in the leaves.

TEA, Pennsylvania *Monarda didyma* (Red Bergamot) Lewin.

TEA, Poor Man's *Veronica chamaedrys* (Germander Speedwell) Cumb. Grigson. Curtis said at the end of the eighteenth century that "the leaves have been recommended by some writers as a substitute for Tea".

TEA, St James's *Ledum groendlandicum* (Labrador Tea) Grieve.1931.

TEA, Somali *Catha edulis* (Khat) Dale & Greenway. Cf. Arabian Tea and Abyssinian Tea. Khat is the name of the tea made from the twigs, buds and fresh leaves.

TEA, South Sea *Ilex vomitoria* (Cassine) J Smith. Cf. Carolina Tea. The concentrated infusion makes the well-known "black drink".

TEA, Spanish *Chenopodium ambrosioides* (American Wormseed) USA Henkel. Cf. Mexican, Jerusalem and Jesuit Tea. The "tea" made is a medicinal one.

TEA, Squaw *Ephedra trifurca* (Long-leaved Joint-Fir) USA Elmore. Cf. Mormon Tea, Brigham or Brigham Young Tea, Mexican Tea, etc. An infusion of the branches is a popular local beverage where it grows in America, but people like the Navajo used it medicinally, too.

TEA, Sudanese *Hibiscus sabdariffa* (Rozelle) Schauenberg & Paris. A slightly laxative tisane was made from it, and introduced into Europe at the end of the 19th century. However, it was regarded as offensive because of its blood-red colour, and is now used to give flavour in various herbal tea mixtures.

TEA, Teamster's 1. *Ephedra nevadensis* (Nevada Joint-Fir) USA Stevenson. 2. *Ephedra trifurca* (Long-leaved Joint-Fir) USA Elmore.

TEA, Violet *Viola odorata* (Sweet Violet) S Africa Watt. In South Africa, they may use a tea made from the flowers, but apparently the plants themselves are called Violet Tea.

TEA BALM *Melissa officinalis* (Bee Balm) Howes.

TEA BUSH *Ocymum viride* (Fever-plant) Dalziel. The leaf decoction is drunk as a tea for fever over most of West Africa.

TEA BUSH, Blue *Ceanothus americanus* (New Jersey Tea) Leyel.1937.

TEA FLOWER 1. *Epilobium montanum* (Broad-leaved Willowherb) Dor. Macmillan. 2. *Filipendula ulmaria* (Meadowsweet) Som. Macmillan. 3. *Sambucus nigra* (Elder) Som. Macmillan. Cf. Tea-tree for elder - the usage is medicinal.

TEA GRASS *Thymus vulgaris* (Wild Thyme) Shet. Grigson. Thyme tea is good for headaches.

TEA-LEAVED WILLOW *Salix phylicifolia*.

TEA OLIVE *Osmanthus fragrans* (Fragrant Olive) Howes.

TEA PLANT 1. *Agrimonia eupatoria* (Agrimony) Som. Macmillan. 2. *Lycium halimifolium* (Duke of Argyll's Tea Plant) Willis. 3. *Thea sinensis*. Agrimony tea used to be widely used as a gargle, for jaundice, and as a blood purifier or to rectify digestive disorders. And French peasants still drink it as an ordinary beverage tea. *Thea sinensis* is of course the source of genuine tea.

TEA PLANT, Duke of Argyll's *Lycium halimifolium*. It is said that the 3rd Duke of Argyll received a tea plant and one of these. Their labels, and so their identities, got mixed. He just referred to the Lycium as his tea plant. It is often cultivated under the name Tea plant, though we should be ill-advised to make tea of it.

TEA PLANT, Poisonous *Solanum dulcamara* (Woody Nightshade) Oxf. B & H.

TEA SENNA *Cassia mimosoides*. The young shoots are used as a tea, like senna.

TEA TREE 1. *Leptospermum scoparium* (Manuka) New Zealand Goldie. 2. *Lycium chinense*. 3. *Sambucus nigra* (Elder) Som. Macmillan. 4. *Symphoricarpos rivularis* (Snowberry) Ess. Gepp. Why should Snowberry get the name? There is no record of tea being made of it in this country, though the American Indians made a medicinal tea from the roots.

TEA TREE, Ceylon *Elaeodendron glaucum*.

TEA TREE, White *Melaleuca leucadendron* (Cajuput) Mitton.

TEABERRY 1. *Gaultheria humifusa*. 2. *Gaultheria procumbens* (Wintergreen) A W Smith. It is the leaves that are usually used to make a tea, but one assumes from this name that the berries can also be used.

TEACUPS 1. *Ranunculus acris* (Meadow Buttercup) Som. Grigson. 2. *Ranunculus bulbosus* (Bulbous Buttercup) Som. Grigson. 3. *Ranunculus repens* (Creeping Buttercup) Som. Grigson.

TEAK *Tectona grandis*. The name comes from the Dravidian dialects of South India and the East Indies. It is tekka in Malayalam.

TEAK, African 1. *Chlorophora excelsa* (African Oak) Wit. 2. *Oldfieldia africana* (African Oak) Sierra Leone Dalziel. 3. *Pterocarpus erinaceus* Dalziel.

TEAK, Bastard 1. *Butea monosperma*. 2. *Pterocarpus marsupium*.

TEAK, Rhodesian *Baikiaea plurijuga*.

TEAK, Sudan *Cordia abyssinica* (Mukumari) Dalziel.

TEAK, Tasmanian *Dacrydium franklinii* (Huon Pine) Everett.

TEAMSTER'S TEA 1. *Ephedra navadensis* (Nevada Joint-Fir) USA Stevenson. 2. *Ephedra trifurca* (Long-leaved Joint-Fir) USA Elmore. Cf. Mormon's Tea, Miner's Tea, etc for both of these. The tea is medicinal, used by the Indians for all sorts of complaints.

TEAR-BLANKET *Acacia greggii* (Cat's Claws) USA Jaeger. A fiercely thorny shrub, hence the name - and Cf. Wait-a-minute and Devil's Claws.

TEAR-THUMB *Polygonum sagittatum*.

TEAR-THUMB, Halberd-leaved *Polygonum arifolium*.

TEAR-YOUR-MOTHER'S-EYES-OUT; PICK-YOUR-MOTHER'S-EYES-OUT; HAWK- YOUR MOTHER'S-EYES-OUT *Veronica chamaedrys* (Germander Speedwell) Dev. Macmillan (Tear); Grigson (Pick); Dor. Dacombe (Hawk). These are all West country names, but the belief is found all over England, e.g. Yorkshire, where it is said that if a child gathers germander speedwell its mother will die during the year, or Lincolnshire - if anyone picks the flower, his eyes will be eaten. Another belief is that if you look steadily at the flower for an hour, you will become blind. Of course, the speedwell has an "eye" itself.

TEARS, Angel's 1. *Billbergia x windii*. 2. *Datura sanguinea* USA Cunningham. 3. *Helxine soleirolii* (Mind-your-own-business) USA. Cunningham. 4. *Narcissus triandrus*. 5. *Ornithogalum umbellatum* (Star-of-Bethlehem) Som. Macmillan. 6. *Veronica chamaedrys* (Germander Speedwell) Dev. Friend.

TEARS, Baby's 1. *Helxine soleirolii* (Mind-your-own-business) McClintock. 2. *Peperomia rotundifolia* USA Cunningham.

TEARS, Eve's *Galanthus nivalis* (Snowdrop) Som. Macmillan.

TEARS, Granny's *Campanula rotundifolia* (Harebell) Som. Macmillan.

TEARS, Jacob's *Convallaria maialis* (Lily-of-the-valley) Le Strange.

TEARS, Job's 1. *Polygonatum multiflorum* (Solomon's Seal) Corn. Grigson. 2. *Pulicaria dysenterica* (Yellow Fleabane) Som. Macmillan. There is a tradition that Job applied Yellow Fleabane to his boils, and obtained relief. There seems to be a lot of Biblical association with Solomon's Seal. As well as Solomon, there are references in the names to David and Jacob, the Virgin Mary, and in this instance, to Job.

TEARS, Juno's *Verbena officinalis* (Vervain) Halliwell.

TEARS, Lady's; TEARS, Virgin's *Convallaria maialis* (Lily-of-the-valley) Som. Dyer (Lady's); Tynan & Maitland (Virgin's). The "Lady" of the name, as in almost all cases, is the Virgin Mary. The legend is that the Virgin's tears at the foot of the cross were changed into this plant.

TEARS, Lady Mary's; TEARS, Mary's; TEARS, Virgin Mary's *Pulmonaria officinalis* (Lungwort) all Dor. Macmillan (Lady Mary's); J J Foster (Mary's); Grigson (Virgin Mary's). The white spots on the leaves are accounted for variously as the Virgin's tears, for the plant grew on Calvary, or her milk, which fell on the leaves during the flight into Egypt.

TEARS, Maiden's *Silene cucubalis* (Bladder Campion) Howes.

TEARS, Our Lady's *Lithospermum officinale* (Gromwell). These are the seeds, which are white.

TEARS, Queen's *Billbergia nutans*.

TEARS, Soldier's *Verbascum thapsus* (Mullein) Dor. Dacombe.

TEARS, Widow's *Tradescantia virginica* (Spiderwort) Coats. The pigmented sap colours the dewdrops that hang from the closed-up flowers in the early morning. The name, it is said cynically, was given because the drops dry in a day. Children make yesterday's flowers "weep" by pressing them with thumb and forefingers, and stain their fingers with the purple "ink".

TEARWEED, Arrow-leaved *Polygonum sagittatum* (Tear-thumb) USA Tolstead.

TEASEL 1. *Dipsacus fullonum*. 2. *Eupatorium perfoliatum* (Thoroughwort) USA Henkel. OE taesel. The verb is to tease, in the sense of opening out the fibres of fabric. So teasels are used to raise the nap on woollen cloth.

TEASEL, Burler's; TEASEL, Card; TEASEL, Draper's *Dipsacus fullonum* (Teasel) Glos. Brill (Burler's); Gerard (Card); B & H (Draper's). A burl is a knot in wool or cloth, so to burl is to remove these knots. To card has a similar meaning, and a card is an instrument for combing wool or flax. Teasels are set in a wooden frame, usually known as a handle, but sometimes called a card, which is from Latin carduus, a thistle.

TEASEL, Fuller's *Dipsacus fullonum var. fullonum*. Apparently there is a difference in the shape of the bracts of Fuller's Teasel and those of Wild Teasel.

TEASEL, Small *Dipsacus pilosum*.

TEASEL, Wild *Dipsacus fullonum var. sylvestris*.

TEASEL, Wolf's *Dipsacus fullonum* (Teasel) Cockayne. Cf. Wolf's Comb for this.

TEASER *Carduus nutans* (Musk Thistle) Som. Macmillan. Presumably traceable from thistle through teasel.

TEAZLE, Wild *Dipsacus fullonum* (Teasel) Dawson. Teazle is an alternative spelling, perhaps archaic.

TECATE CYPRESS *Cupressus forbesii*. A Californian cypress.

TEDDY BUTTONS *Knautia arvensis* (Field Scabious) Som. Macmillan. Descriptive, of course.

TEESDALE SANDWORT *Minuartia stricta*. Very rare in this country, and confined to Widdybank Fell in Upper Teesdale.

TEESDALE VIOLET *Viola rupestris*. Also very rare in this country, and thought at one time to be confined to Widdybank Fell in Upper Teesdale, but there are in fact two other limestone habitats in the north of England.

TEETH, Lion's 1. *Mycelis muralis* (Wall Lettuce) Som. Macmillan. 2. *Taraxacum officinale* (Dandelion) B & H. The reference in each case is to the jagged leaves, more obvious in dandelion than in wall lettuce.

TEIL *Tilia europaea* (Lime) Jamieson. From Tilia. It appears in various spellings, as Tile and Teyl.

TELEGRAPH PLANT *Desmodium gyrans*. At temperatures above 72F, the leaves continually gyrate round and round in an elliptical orbit - they adopt sleep movement at night.

TELL-TIME *Taraxacum officinale* (Dandelion) Mabey. Children tell the time by puffing at the seed head, a game that is celebrated in a large number of local names, like Clock-flower, Schoolboys' Clock, One-o'clock, and many others.

TEMPLE JUNIPER *Juniperus rigida*.

TEMPLE TREE *Plumeria alba* (Frangipani) Leyel.1937.

TEN COMMANDMENTS *Passiflora caerulea* (Passion Flower) Dor. Macmillan. There are a few different names for the Passion Flower, apart, that is, from the 'Passion' symbolism itself. As well as the Ten Commandments, we have the Twelve Disciples and Christ-and-the Apostles. The petals and sepals are the Apostles.

TEN-MONTHS YAM *Dioscorea alata* (Greater Yam) Dalziel.

TEN O'CLOCK 1. *Ornithogalum umbellatum* (Star-of-Bethlehem) Som. Macmillan. 2. *Portulaca quadrifida* West Africa Dalziel. In both cases, the name is an indication of the late opening of the flowers. In the case of Star-of-Bethlehem, the names show confusion in pinpointing the exact time. As well as ten, they claim eleven, twelve, one and four o'clock.

TENASSERIM PINE *Pinus merkusii*. An Indonesian species.

TENBY DAFFODIL *Narcissus obvallaris*. Known only in a few grass fields near Tenby, and long established there.

TENCHWEED *Potamogeton natans* (Broad-leaved Pondweed) E Ang. Forby.

TENDERGREENS *Brassica perviridis* (Spanish Mustard) USA Usher.

TENERIFE BROOM *Cytisus supranubens*.

TENG-TONGUE *Nasturtium officinale* (Watercress) Yks. B & H. Cf. Tarry-tongue and Tongue Grass. The derivation of *Nasturtium* is enlightening. It means nose-twisting (nasi-tortium), from the tangy taste of the darker leaves.

TENS-O'-THOUSANDS *Malcomia maritima* (Virginian Stock) Wilts. Macmillan. And from the West country, there are also Hundreds-and-thousands and Mother-of-thousands.

TENTES *Juglans regia* catkins (Walnut catkins) Lyte.

TENTWORT; TAINTWORT *Asplenium ruta-muraria* (Wall Rue) Coats.1975. 'Tent' is an old name for rickets, so this plant must have been used as a cure once.

TEOSINTE *Euchlaena mexicana*. This close relative of maize was called teocintl by the Aztecs. Teosinte is the modern Mexican.

TEPARY BEAN *Phaseolus acutifolius*.

TERE *Vicia sativa* (Common Vetch) B & H. A variation on 'tare'.

TEREBINTH *Pistacia terebinthus*.

TEREBINTH, Large *Pistacia atlantica*.

TEREBINTH, Palestine *Pistacia terebinthus var. palaestina*.

TERMINALIA, Yellow *Terminalia ivorensis*.

TERRYDIDDLE; TERRYDEVIL 1. *Polygonum convolvulus* (Black Bindweed) Ches. Holland. 2. *Solanum dulcamara* (Woody Nightshade) Ches. Holland. They are variations on 'Tether-devil', 'Devil's Tether', and a comment in the plants' binding properties.

TETHER, Devil's *Polygonum convolvulus* (Black Bindweed) Ches, Yks. Grigson. It conveys the same information as does Bindweed.

TETHER, Toad; TETHER-TOAD *Ranunculus repens* (Creeping Buttercup) both Yks. Grigson.

TETHER-DEVIL 1. *Polygonum convolvulus* (Black Bindweed) Ches. Grigson. 2. *Solanum dulcamara* (Woody Nightshade) Ches. Holland.
TETHER-GRASS *Galium aparine* (Goosegrass) N'thum. Grigson. A comment on the clinging properties of the burrs.
TETRA MAD *Atropa belladonna* (Deadly Nightshade) C J S Thompson.1897. Cf. Daftberries.
TETSAN *Hypericum androsaemme* (Tutsan) Dev. Choape. The name 'Tutsan' varies considerably. As well as this, there are Titsum, Tipsen, Titzen, Totsan and Stitson. The word is apparently from the French, as with modern French toutesaine, wholly sound, or healing all.
TETTER 1. *Chelidonium majus* (Greater Celandine) M Baker.1974. 2. *Ranunculus ficaria* (Lesser Celandine) Drury.1992. Tetter is short for Tetterwort - the juice of these plants has always been used for warts, tetters or felons, a tetter being some kind of skin eruption, ringworm possibly. Lesser Celandine is identified as tetter by Drury, and taken from Thomas Browne, *The garden of Cyrus*, 1658.
TETTER-BERRIES 1. *Bryonia dioica* (White Bryony) Som. Macmillan; Hants. Grigson. 2. *Tamus communis* (Black Bryony) Fernie. They must have been used in some way for skin eruptions, though memory of the cure seems to have passed.
TETTERWORT 1. *Chelidonium majus* (Greater Celandine) Gerard. 2. *Sanguinaria canadensis* (Bloodroot) Grieve.1931. 3. *Tamus communis* (Black Bryony) Fernie. Bloodroot must have been used for skin complaints, although there is no record of such usage.
TETTY *Solanum tuberosum* (Potato) Dev. Friend.1882. Not Tetter in this case, but the Devonshire equivalent of Taters.
TEXAN PRIDE *Phlox drummondii* W Miller.
TEXAN RHATANY *Krameria lanceolata*.
TEXAS MILLET *Panicum texanum*. Or Colorado Millet.
TEXAS MOONSEED *Menispermum canadense* (Moonseed) Le Strange. Why Texas? This is a plant of the eastern side of America.
TEXAS MULBERRY *Morus microphylla* (Mexican Mulberry) USA T H Kearney.
TEXAS REDBUD *Cercis reniformis*.
TEXAS SAGE *Salvia coccinea* (Red Sage) Watt.
TEXAS SNAKEROOT *Aristolochia reticulata*.
TEXAS THISTLE *Solanum rostratum* (Prickly Nightshade) USA Kingsbury. But it is also known as Kansas Thistle.
TEXAS UMBRELLA TREE *Melia azedarach var. umbraculiformis*.
TEXAS WALNUT *Juglans rupestris*.
TEYL *Tilia europaea* (Common Lime) W Miller. Close to the Scottish word Teil, and a variation of Tile, once the common name for the tree.
THALE CRESS *Arabidopsis thaliana*. Named for a Dr Thalius, "who published a catalogue of the plants of the Hartz mountains", according to Prior.
THAMES, Pride of the *Butomus umbellatus* (Flowering Rush) Som, Dor. Macmillan.
THANET CRESS; THANET WEED *Cardaria draba* (Hoary Cress) Salisbury (Cress); McClintock (Weed). The story is that it was accidentally introduced in 1809, the year of the ill-fated expedition to Walcheren. The fever-stricken soldiers were brought back to Ramsgate on mattresses stuffed with hay, and this was disposed of to a Thanet farmer, who ploughed it in for manure. Then the cress appeared, presumably from seeds in the hay.
THAPE *Ribes uva-crispa* (Gooseberry) Norf. Moor. Better known as Fape, or Feabe. But this is closer to the original, which is OE thefe, probably meaning a prickly bush.
THATCH 1. *Onobrychis sativa* (Sainfoin) Glos. Grigson. 2. *Vicia cracca* (Tufted Vetch) Glos. Grigson. 3. *Vicia sativa*

(Common Vetch) Wilts. D & G. i.e. vetch, of which there are several local variants. Besides this, Fitch or Fatch are the commonest.
THATCH, Yellow *Lathyrus pratensis* (Yellow Pea) Wilts. D & G.
THATCH-LEAF PALM *Howeia forsteriana* (Kentia Palm) RHS.
THEABBERRY; THEABE *Ribes uva-crispa* (Gooseberry) E Angl. Fernie (Theabberry); Moor (Theabe). Cf. Thape. They are from OE thefe, probably meaning a prickly bush.
THETCH 1. *Vicia sativa* (Common Vetch) Wilts. D & G. 2. *Vicia sepium* (Bush Vetch) Som, Wilts, Dor, Hants, Bucks. Grigson. Cf. Thatch.
THETCH, Gore *Vicia sativa* (Common Vetch) B & H.
THETCH-GRASS, Wild 1. *Vicia cracca* (Tufted Vetch) B & H. 2. *Vicia hirsuta* (Hairy Tare) B & H.
THICKENING GRASS *Pinguicula vulgaris* (Butterwort) Ayr. B & H. Because the leaves act on cow's milk like rennet. Cf. Earninggrass (to earn in the North country and Scotland is to curdle milk). Steep is another word for rennet, so one finds variations on Steepweed in Northern Ireland.
THICKET HAWTHORN *Crataegus coccinea*.
THIEF *Rubus fruticosus* (Bramble) Leics. Grigson. Cf. Lawyers - once you have been caught, they never let you go. Follower, from Sussex, conveys the same idea, and so indeed does Thief.
THIEF, Wheat *Buglossoides arvensis* (Corn Gromwell) USA Zenkert. This is a cornfield weed, or perhaps pest, rather.
THIMBLE-BERRY 1. *Rubus occidentalis*. 2. *Rubus odoratus* (Purple-flowering Raspberry) USA Zenkert. 3. *Rubus parviflorus*.
THIMBLE-FLOWER *Digitalis purpurea* (Foxglove) Dor, Som. Macmillan. For the flowers can be 'thimbles' as well as 'gloves'.
THIMBLE-FLOWER, Queen Anne's *Gilia capitata*.

THIMBLE LILY *Lilium bolanderi*.
THIMBLES 1. *Aquilegia vulgaris* (Columbine) Som. Macmillan. 2. *Campanula rotundifolia* (Harebell) Som. Grigson; Wilts. Dartnell & Goddard; Glos. J D Robertson. 3. *Digitalis purpurea* (Foxglove) Cumb, Ire. B & H; USA. Henkel. 4. *Lotus corniculatus* (Bird's-foot Trefoil) Som. Macmillan. 5. *Silene maritima* (Sea Campion) E Ang. Grigson. Foxgloves are thimbles rather than gloves in the Gaelic languages too.
THIMBLES, Dead Man's *Digitalis purpurea* (Foxglove) Som. Macmillan; Ire. Skinner. They are Dead Man's Fingers, too - probably something to do with the colour.
THIMBLES, Fairy 1. *Campanula pusilla*. 2. *Campanula rotundifolia* (Harebell) Dev, Som. Macmillan. 3. *Digitalis purpurea* (Foxglove) Som. Macmillan; USA. Henkel.
THIMBLES, Goblin's *Digitalis purpurea* (Foxglove) Hants. Boase.
THIMBLES, Granny's *Aquilegia vulgaris* (Columbine) Som. Macmillan.
THIMBLES, Lady's 1. *Campanula medium* (Canterbury Bell) Som. Macmillan. 2. *Campanula rotundifolia* (Harebell) Som. Elworthy; E Ang, N Scot, Ire. Grigson. 3. *Digitalis purpurea* (Foxglove) Som. Macmillan; Norf, N'thum. Grigson. 4. *Endymion nonscriptus* (Bluebell) Ire. Tynan & Maitland. 5. *Stellaria holostea* (Greater Stitchwort) Dor. Macmillan. 6. *Veronica chamaedrys* (Germander Speedwell) Lancs. Grigson. The bell-shaped flowers of the campanulas, foxglove and bluebell make then likely candidates for the name. Not so likely are Greater Stitchwort and the speedwell, but the stitchwort does have 'needle' names, and 'stitch' in the common name could very well be misunderstood, thus suggesting both thimble and needle.
THIMBLES, Mother's *Stellaria holostea* (Greater Stitchwort) Wilts. Grigson.
THIMBLES, Our Lady's *Campanula rotundifolia* (Harebell) Som. Macmillan. Cf. Lady's Thimbles, etc.

THIMBLES, Queen Anne's *Gilia capitata*.

THIMBLES, Sheegie; THIMBLES, Shilly *Digitalis purpurea* (Foxglove) Donegal Grigson. i.e. fairy thimbles, for both 'Sheegie' and 'Shilly' are from the Irish sidhe. Cf. Fairies', or Goblin's, Thimbles.

THIMBLES, Witches' 1. *Campanula rotundifolia* (Harebell) Som. Macmillan. 2. *Centaurea cyanus* (Cornflower) N Eng. B & H. 3. *Digitalis purpurea* (Foxglove) N Eng. Henderson. 4. *Silene maritima* (Sea Campion) N'thum. Prior. 5. *Wahlenbergia hederacea* (Ivy-leaved Bellflower) Som. Macmillan. What is cornflower doing in such company?

THIMBLEWEED 1. *Anemone cylindrica* (Slender-fruited Anemone) USA. Tolstead. 2. *Anemone virginiana* (Tall Anemone) USA House. 3. *Rudbeckia laciniata* (Rall Coneflower) USA Kingsbury.

THISTLE 1. *Arctium lappa* (Burdock) Dev. B & H. 2. *Sonchus oleraceus* (Sow Thistle) S Africa Watt. OE thistel, apparently from thydan, to stab, appropriately enough.

THISTLE, Alpine *Carlina acaulis* (Stemless Carline Thistle) RHS.

THISTLE, Ball *Echinops sp* (Globe Thistle) B & H.

THISTLE, Bank 1. *Carduus nutans* (Musk Thistle) B & H. 2. *Cirsium vulgare* (Spear Plume Thistle) B & H.

THISTLE, Bell *Cirsium vulgare* (Spear Plume Thistle) War, Yks. Grigson. Probably the same as Bull Thistle.

THISTLE, Bird *Cirsium vulgare* (Spear Plume Thistle) Worcs. Grigson. A corruption of Bur Thistle.

THISTLE, Bitter *Carduus benedictus* (Holy Thistle) USA Henkel.

THISTLE, Black 1. *Cirsium palustre* (Marsh Thistle) Som. Macmillan. 2. *Cirsium vulgare* (Spear Plume Thistle) S Africa Watt.

THISTLE, Blessed 1. *Carduus benedictus* (Holy Thistle) Ellacombe. 2. *Silybum marianum* (Milk Thistle) War. Grigson. Milk Thistle gets the name because of its white-veined leaves, the result of the Virgin Mary spilling milk from her breast. Cf. Holy Thistle and the various names associating it with the Virgin. But when a plant is called 'holy', or 'blessed', it means that it has the power of counteracting poison, or at least it was supposed to have the power. *Carduus benedictus* had that reputation, and was known as a heal-all.

THISTLE, Blessed Mary's *Silybum marianum* (Milk Thistle) RHS.

THISTLE, Blue 1. *Cirsium vulgare* (Spear Plume Thistle) Worcs. Grigson. 2. *Echium vulgare* (Viper's Bugloss) Som. Macmillan.

THISTLE, Bo; THISTLE, Bow *Cirsium vulgare* (Spear Plume Thistle) Ches. Holland (Bo); Ches. B & H (Bow). In each case, a corruption of Bur Thistle.

THISTLE, Boar 1. *Cirsium arvense* (Creeping Thistle) B & H. 2. *Cirsium vulgare* (Spear Plume Thistle) Grigson. As with the previous entry, these are both corruptions of Bur Thistle.

THISTLE, Bog *Cirsium palustre* (Marsh Thistle) B & H.

THISTLE, Boyton *Cirsium tuberosum* (Tuberous Thistle) Wilts. Flower. It used to grow in a spot in Wiltshire between Boyton House and Fonthill.

THISTLE, Buck 1. *Carduus nutans* (Musk Thistle) Yks. Grigson. 2. *Cirsium vulgare* (Spear Plume Thistle) Yks. B & H.

THISTLE, Bull 1. *Centaurea nigra* (Knapweed) Som. Grigson. 2. *Cirsium vulgare* (Spear Plume Thistle) Som. Macmillan; Dor, N Ire Grigson; USA Gates. 3. *Silybum marianum* (Milk Thistle) USA Kingsbury. At least as far as knapweed is concerned, 'bull' is OE bol, a round object, quite descriptive.

THISTLE, Bur 1. *Arctium lappa* (Burdock) Scot. Guthrie. 2. *Cirsium vulgare* (Spear Plume Thistle) B & H.

THISTLE, Cabbage *Cirsium oleraceum*. Sometimes eaten as a vegetable.

THISTLE, Canada *Cirsium arvense* (Creeping Thistle) USA H Smith.1945.

THISTLE, Card *Dipsacus fullonum* (Teasel)

B & H. 'Card' - the process of carding cloth, for which teasels are used.

THISTLE, Carline *Carlina vulgaris*. 'Carline' is French, from medieval Latin carolina (herba) or carlina (herba), from Latin carduus, altered by association with Carolus to take in the tradition that Charlemagne (or the Emperor Charles V?) dreamt that the decoction of the root would cure the plague then rife in his army.

THISTLE, Carline, Dwarf *Cirsium acaulon* (Picnic Thistle) Gerard.

THISTLE, Carline, Stemless *Carlina acaulis*.

THISTLE, Carmine *Cirsium rothrockii*.

THISTLE, Chalk *Cirsium acaulon* (Picnic Thistle) H C Long.1910. Surely it is not confined to chalk soils?

THISTLE, Cock *Onopordum acanthium* (Scotch Thistle) Dor. Macmillan.

THISTLE, Corn *Cirsium arvense* (Creeping Thistle) Cumb, N Ire B & H.

THISTLE, Cotton 1. *Cirsium eriophorum* (Woolly Thistle) Prior. 2. *Onopordum acanthium* (Scotch Thistle) Curtis.

THISTLE, Cow's *Cirsium arvense* (Creeping Thistle) Som. Macmillan.

THISTLE, Cramp *Cirsium arvense* (Creeping Thistle) War. Bloom. There is a gall insect which causes swellings on this thistle. The gall used to be carried about in the belief that it prevented cramp.

THISTLE, Creeping *Cirsium arvense*.

THISTLE, Cursed 1. *Carduus benedictus* (Holy Thistle) USA Henkel. 2. *Cirsium arvense* (Creeping Thistle) Curtis. It is difficult to understand the name when applied to Holy Thistle, especially in view of the common name and the regard in which it was held. But it is another matter for Creeping Thistle, which is a plant universally detested.

THISTLE, Distaff *Carthamus tinctorius* (Safflower) A W Smith. Presumably connected with its use as a dyeplant.

THISTLE, Dog *Cirsium arvense* (Creeping Thistle) B & H. Cf. Cursed Thistle. 'Dog' is nearly always a derogatory appellation.

THISTLE, Dog's *Sonchus oleraceus* (Sow Thistle) Som. Macmillan; Surr. Grigson. Presumably a way of saying that this is not a real thistle.

THISTLE, Donkey's *Dipsacus fullonum* (Teasel) Som. Macmillan.

THISTLE, Down 1. *Cirsium eriophorum* (Woolly Thistle) Gerard. 2. *Onopordum acanthium* (Scotch Thistle) W Miller. 'Down', in the textile sense.

THISTLE, Dwarf *Cirsium acaulon* (Picnic Thistle) Salisbury. Cf. Ground Thistle.

THISTLE, Fair Weather (Stemless Carline Thistle) *Carlina acaulis* D Parish. Because the bracts are outspread in fine weather, and folded over the flowers to protect them in damp conditions.

THISTLE, Field 1. *Cirsium arvense* (Creeping Thistle) Perring. 2. *Cirsium discolor*.

THISTLE, Fuller's *Dipsacus fullonum* (Teasel) Loudon(1844).

THISTLE, Gentle *Cirsium dissectum* (Meadow Thistle) Hill. A name coined by Hill in 1769.

THISTLE, Globe *Echinops sp*.

THISTLE, Golden 1. *Scolymus hispanicus* (Spanish Salsify) Schery. 2. *Scolymus maculatus*.

THISTLE, Ground *Cirsium acaulon* (Picnic Thistle) War, Worcs. Grigson. Cf. Dwarf, or Stemless, Thistle.

THISTLE, Gum 1. *Euphorbia resinifera* (Gum Euphorbia) Le Strange. 2. *Onopordum acanthium* (Scotch Thistle) Turner.

THISTLE, Hard *Cirsium arvense* (Creeping Thistle) E Ang. Halliwell. Cf. Sharp Thistle.

THISTLE, Hare's *Sonchus oleraceus* (Sow Thistle) Dyer. Another tradition associates the plant with hares rather than with pigs. Cf. Hare's Lettuce, Hare's Palace, etc. The legend tells how this plant gave strength to hares when they were overcome with the heat.

THISTLE, Holy 1. *Carduus benedictus*. 2. *Silybum marianum* (Milk Thistle) Bronwlow. Milk Thistle is 'holy' because of the white veins on its leaves, made, so it was said, by the spilling of some of the Virgin Mary's milk upon them. Another tradition applies to *Carduus benedictus*. When a plant is called 'holy' or 'blessed', it means it has the power of counteracting poison, so it was supposed. In addition, this thistle used to be well-known as a heal-all.

THISTLE, Horse 1. *Cirsium vulgare* (Spear Plume Thistle) Dev, Som. Grigson. 2. *Lactuca virosa* (Great Prickly Lettuce) B & H.

THISTLE, Hundred; THISTLE, Hundred-leaved *Eryngium campestre* (Field Eryngo) N'thants. B & H (Hundreed); Lyte (Hundred-leaved).

THISTLE, Jersey *Centaurea calcitrapa* (Star Thistle) Pratt.

THISTLE, Kansas *Solanum rostratum* (Prickly Nightshade) USA Kingsbury. The common name tells why this is called a thistle. Cf. Texas Thistle.

THISTLE, Ladies' *Carduus benedictus* (Holy Thistle) Halliwell. A plural form makes nonsense of the attribution to the Virgin.

THISTLE, Lady's *Silybum marianum* (Milk Thistle) Som. Macmillan; Lincs. Grigson. As in all other cases, the 'Lady' of the common name is Our Lady, the Virgin Mary. The reference in this case is to the white veins of the leaves, caused by milk spilt by the Virgin.

THISTLE, Malta; THISTLE, Maltese; THISTLE, Maltese Star *Centaurea melitensis* (Napa Thistle) Watt (Malta, Maltese); Clapham (Maltese Star).

THISTLE, Marsh *Cirsium palustre*. Also known as Bog, Moss, or Water Thistle.

THISTLE, Marsh, Great *Carduus personata*.

THISTLE, Mary's *Silybum marianum* (Milk Thistle) S Africa Watt. See Lady's Thistle.

THISTLE, Meadow *Cirsium dissectum*.

THISTLE, Melancholy *Cirsium heterophyllum*. The reference is to the hanging flower heads, and this became the "signature" of the plant, hence the use for melancholy. "The decoction of the Thistle in wine being drunk, expels superfluous Melancholy out of the Body, and makes a man as merry as a Cricket" (Gerard).

THISTLE, Mexican *Argemone mexicana* (Mexican Poppy) Barbados Gooding. The fact that it is also called Prickly Poppy is enough to explain why a poppy is called a thistle.

THISTLE, Midsummer *Centaurea solstitialis* (St Barnaby's Thistle) Parkinson. St Barnabas's Day is 11 June, when it was expected to be in bloom. It seems a bit early, so perhaps Midsummer is nearer the mark.

THISTLE, Milk 1. *Silybum marianum*. 2. *Sonchus oleraceus* (Sow Thistle) Som, Dor, War, Lincs. Grigson. For *Silybum marianum* it is the white veins on the leaves that give it the name, with consequent legends about the Virgin having spilled some of her milk on them. But Sow Thistle gets the name for a different reason - the milky sap.

THISTLE, Milk, Field *Sonchus arvensis* (Corn Sow Thistle) Clapham.

THISTLE, Milk, Our Lady's *Silybum marianum* (Milk Thistle) Bronwlow.

THISTLE, Milk, Spiny *Sonchus asper* (Prickly Sow Thistle) Clapham.

THISTLE, Milk, Striped *Silybum marianum* (Milk Thistle) Coles. The "stripes" would be the white leaf veins.

THISTLE, Moss *Cirsium palustre* (Marsh Thistle) Scot. B & H.

THISTLE, Musk *Carduus nutans*.

THISTLE, Napa *Centaurea melitensis*.

THISTLE, New Mexico *Cirsium neomexicanum*.

THISTLE, Nodding *Carduus nutans* (Musk Thistle) Salisbury.

THISTLE, Oat *Onopordum acanthium* (Scotch Thistle) Turner.

THISTLE, Our Lady's *Carduus benedictus* (Holy Thistle) USA Henkel. Cf. Virgin Mary's Thistle.

THISTLE, Pasture *Cirsium pumilum*.

THISTLE, Picnic *Cirsium acaulon*. A good name for this thistle. Being *acaulon*, without a stem, one does not notice them until one sits on them.

THISTLE, Plume, Marsh *Cirsium dissectum* (Meadow Thistle) Clapham.

THISTLE, Plume, Spear *Cirsium vulgare.*

THISTLE, Plume, Tuberous *Cirsium tuberosum* (Tuberous Thistle) Flower.

THISTLE, Pod *Cirsium acaulon* (Picnic Thistle) N'thants. N'thants. A E Baker. 'Pad' seems to make more sense than 'Pod', because it is *acaulon*, stemless. But the latter was repeated in Grigson. Was it an original misreading, or is 'pod' the local variant?

THISTLE, Pricky *Cirsium arvense* (Creeping Thistle) Yks. B & H.

THISTLE, Pyrenean *Carduus carlinoides.*

THISTLE, Red *Cirsium palustre* (Marsh Thistle) Yks. B & H.

THISTLE, Queen Anne's *Carduus nutans* (Musk Thistle) Berw. B & H.

THISTLE, Queen Mary's *Onopordium acanthium* (Scotch Thistle) N'thants. Grigson. The Scottish queen, presumably.

THISTLE, Russian 1. *Salsola pestifera.* 2. *Salsola kali* (Saltwort) S Africa Watt. Why Russian? Salsola pestifera is an American plant, and Saltwort is a native British plant.

THISTLE, Saffron *Carthamus tinctorius* (Safflower) Leyel.1948.

THISTLE, St Barnaby's *Centaurea solstitialis.* It is dedicated to St Barnabas, and should be in flower round about his feast day, which is 11 June. But that sounds a little early.

THISTLE, St Benedict's *Carduus benedictus* (Holy Thistle) USA Henkel. A mistake in confusing 'benedictus', holy, or blessed, with the saint's name.

THISTLE, Scotch 1. *Cirsium acaulon* (Picnic Thistle) Som. Macmillan. N Ire. Vickery.1995. 2. *Cirsium vulgare* (Spear Plume Thistle) Worcs. Grigson. 3. *Onopordum acanthium.* An odd name for *Onopordum acanthium*, for it is probably introduced, and uncommon in Scotland, anyway. The tradition is that when a Scottish garrison was in danger of being surprised by a Danish foe, a bare-footed Dane, creeping along in the darkness of night, trod on the sharp prickles of a thistle, and yelled with the shock of sudden pain. So the sentinel was aroused, and the garrison, thus saved, adopted the emblem and motto, which afterwards became that of the whole nation.

THISTLE, Seaside *Cirsium tenuiflorus.*

THISTLE, Sharp *Cirsium arvense* (Creeping Thistle) Cumb. B & H. Cf. Hard Thistle.

THISTLE, Sheep's *Cirsium arvense* (Creeping Thistle) Som. Macmillan.

THISTLE, Siberian *Cirsium oleraceum* (Cabbage Thistle) Douglas.

THISTLE, Slender; THISTLE, Slender-flowered *Cirsium tenuiflorus* (Seaside Thistle) Clapham (Slender); Curtis (Slender-flowered).

THISTLE, Soe *Sonchus oleraceus* (Sow Thistle) S Africa Watt. That is how it appears. Is it a misprint?

THISTLE, Sow *Sonchus oleraceus.*

THISTLE, Sow, Blue *Cicerbita macrophylla.*

THISTLE, Sow, Corn *Sonchus arvensis.*

THISTLE, Sow, Creeping *Sonchus arvensis* (Corn Sow Thistle) Salisbury.

THISTLE, Sow, Marsh *Sonchus palustris.*

THISTLE, Sow, Prickly; THISTLE, Sow, Spiny *Sonchus asper* The former is the usual name. Spiny Sow Thistle is in Clapham.

THISTLE, Sow, Smooth *Sonchus oleraceus* (Sow Thistle) McClintock.

THISTLE, Sow, Tree 1. *Sonchus arvensis* (Corn Sow Thistle) B & H. 2. *Sonchus palustris* (Marsh Sow Thistle) Curtis. 'Tree', because these are the tallest members of the genus.

THISTLE, Spanish *Centaurea iberica.*

THISTLE, Spear; THISTLE, Spear-plume *Cirsium vulgare.* 'Spear', without the 'plume', is in Perring.

THISTLE, Spotted *Carduus benedictus* (Holy Thistle) USA Henkel.

THISTLE, Star *Centaurea calcitrapa.*

THISTLE, Star, Maltese *Centaurea melitensis* (Napa Thistle) Clapham.

THISTLE, Star, Red *Centaurea calcitrapa* (Star Thistle) Blamey.

THISTLE, Star, Rough *Centaurea aspera*.

THISTLE, Star, Yellow *Centaurea solstitialis* (St Barnaby's Thistle) Blamey.

THISTLE, Stemless *Cirsium acaulon* (Picnic Thistle) Clapham.

THISTLE, Swamp *Cirsium muticum*.

THISTLE, Swine 1. *Sonchus arvensis* (Corn Sow Thistle) Cumb. B & H. 2. *Sonchus oleraceus* (Sow Thistle) N Eng, Scot, Donegal. Grigson.

THISTLE, Texas *Solanum rostratum* (Prickly Nightshade) USA Kingsbury. Cf. Kansas Thistle.

THISTLE, Tuberous *Cirsium tuberosum*.

THISTLE, Variegated *Silybum marianum* (Milk Thistle) Australia Watt; USA Kingsbury. The variegation lies in the white veins of the leaves.

THISTLE, Virgin's *Silybum marianum* (Milk Thistle) Norf. Grigson. Cf. Virgin Mary's Thistle, Virgin's Milk, etc.

THISTLE, Virgin Mary's 1. *Carduus benedictus* (Holy Thistle) Suff. Moor. 2. *Silybum marianum* (Milk Thistle) Hants. Cope. Cf. Our Lady's Thistle.

THISTLE, Water *Cirsium palustre* (Marsh Thistle) Cumb. B & H.

THISTLE, Watling Street *Eryngium campestre* (Field Eryngo) N'thants. A E Baker.

THISTLE, Way *Cirsium arvense* (Creeping Thistle) Grieve.1931. Presumably in the sense of wayside, or roadside.

THISTLE, Weather *Carlina vulgaris* (Carline Thistle) Howes.

THISTLE, Welted *Carduus crispus*.

THISTLE, White *Atriplex lentiformis* (Quailbrush) USA T H Kearney.

THISTLE, Wild *Sonchus oleraceus* (Sow Thistle) S Africa Watt.

THISTLE, Wool; THISTLE, Woolly; THISTLE, Woolly-headed *Cirsium eriophorum* Woolly Thistle is the usual name. The other two are in B & H (Wool) and Perring (Woolly-headed).

THISTLE, Yellow *Cirsium neomexicanum* (New Mexico Thistle) USA Elmore.

THISTLE, Yellow-spined *Cirsium ochrocentrum*.

THISTLE BROOMRAPE *Orobanche reticulata*.

THISTLE HEMP *Cannabis sativa* (Hemp) B & H.

THISTLE SAGE *Salvia carduacea*.

THISTLE-UPON-THISTLE *Onopordum acanthium* (Scotch Thistle) Prior. From its numerous prickles, according to Prior.

THOMPSON'S CURSE *Cardaria draba* (Hoary Cress) PLNN 10. The plant is reputed to have been introduced here accidentally after the ill-fated expedition to Walcheren in 1809. The fever-stricken soldiers were brought back to Ramsgate on mattresses stuffed with hay, and this was disposed of to a Thanet farmer who ploughed it in for manure. Then the cress appeared, presumably from seeds contained in the hay. Evidently the farmer's name was Thompson.

THONGS, Sea *Salsola kali* (Saltwort) Pomet.

THOR'S BEARD *Sempervivum tectorum* (Houseleek) Grieve.1931. The Romans called it iovis barba, which translated into English Jove's Beard, with a number of variations. Thor's Beard brings us into a different mythology, but signals 'thunder' names like Thunder Plant or Thunderwort in English, and similar names in German and Dutch.

THOR'S MANTLE 1. *Arctium lappa* (Burdock) Friend. 2. *Digitalis purpurea*. 3. *Potentilla erecta* (Tormentil) Wright. Apparently, burdock is tordenkrappe in Danish, so the link with the thunder god is established there. But there seems to be no reason why these very different plants should bear such a name in English. Certainly, the large leaves of foxglove and burdock may account for 'mantle', but tormentil does not even have this tenuous association - perhaps the first syllable of the common name, 'tor-', may have been misunderstood at some time.

THORA BUTTERCUP; THORE'S BUTTERCUP *Ranunculus thora*.
THORN, Branch *Erinacea anthyllis* (Hedgehog Broom) Whittle & Cook.
THORN, Buffalo *Zizyphus mucronata*.
THORN, Camel *Acacia giraffae*.
THORN, Cape *Zizyphus mucronata* (Buffalo Thorn) Howes.
THORN, Christ's 1. *Ilex aquifolium* (Holly) Yks. Grigson. 2. *Zizyphus spina-christi* (Lotus Tree) J Smith. For Lotus Tree, this is a translation of the specific name, spina-christi. As for holly, it earns the name by being the Christmas plant par excellence, often being known simply as Christmas.
THORN, Cockspur *Crataegus crus-galli*. Cockspur is a translation of the specific name.
THORN, Cotton *Tetradymia spinosa var. longispina*.
THORN, Desert *Lycium europaeum* (European Box-thorn) Moldenke.
THORN, Devil's *Tribulus terrestris* (Puncture Vine) S Africa Dalziel. Ascribed to the devil, for this is a pest, both for man and animals. The spiny fruits are vicious enough to puncture a bicycle tyre if wheeled over land where it is prevalent.
THORN, Dog's *Rosa canina* (Dog Rose) Gerard.
THORN, Dotted *Crataegus punctata*.
THORN, Egyptian 1. *Acacia arabica* (Babul Tree) Pomet. 2. *Acacia nilotica* (Throny Acacia) Hora. 3. *Acacia vera*. 4. *Pyracantha angustifolia* (Orange Fire Thorn) Ches. B & H.
THORN, Felt, Bald-leaved *Tetradymia glabrata* (Littleleaf Horsebrush) USA Jaeger.
THORN, Felt, Grey *Tetradymia canescens* (Spineless Horsebrush) USA Jaeger.
THORN, Felt, Narrow-scaled *Tetradymia stenolepis*.
THORN, Felt, White *Tetradymia comosa*.
THORN, Fire *Pyracantha sp*. The translation of *Pyracantha*.
THORN, Fire, Orange *Pyracantha angustifolia*.
THORN, Fire, Red *Pyracantha coccinea*.
THORN, Garland *Paliurus spina-christi* (Christ-thorn) Howes.
THORN, Glastonbury *Crataegus monogyna var. praecox*. Cf. Holy Thorn. The well-known legend tells how Joseph of Arimathaea, on his way to Glastonbury, arrived at Wearyall Hill, to the south of the town, and rested there after having pushed his staff into the ground. The stick took root, and blossomed each year on the anniversary of the birth of Christ (old style).
THORN, Goat's *Astragalus tragacantha* Thornton.
THORN, Hedge *Crataegus monogyna* (Hawthorn) Tynan & Maitland. Hawthorn is the hedging plant par excellence, and the name means exactly the same as hawthorn.
THORN, Highway *Rhamnus cathartica* (Buckthorn). Cf. Waythorn, with the same meaning.
THORN, Holy 1. *Berberis vulgaris* (Barberry) Vesey-Fitzgerald. 2. *Crataegus monogyna var. praecox* (Glastonbury Thorn) Hole. Barberry is called Holy Thorn because of traditions, especially in Italy, that they formed the Crown of Thorns. Glastonbury Thorn gets the name from its connection with Christmas, old style.
THORN, Holy, of Christmas *Salix discolor* (American Black Willow) USA Bergen. Another of the shrubs said to blossom out at old Christmas, after which the flowers "go in again".
THORN, Hook *Acacia campylacantha*. '*Campylacantha*' means having hooked thorns, which describes the prickles perfectly.
THORN, Hungarian *Crataegus nigra*.
THORN, Jerusalem *Paliurus spina-christi* (Christ-thorn) Polunin.1976.
THORN, Kaffir *Lycium afrum*.
THORN, Kangaroo *Acacia armata*.
THORN, Knob *Acacia nigrescens*.
THORN, May *Crataegus monogyna* (Hawthorn) Tynan & Maitland.

THORN, Mimosa *Acacia karroo*.
THORN, Mysore *Caesalpina decepetala*.
THORN, Naivansha *Acacia xanthophloea* (Fever Tree) Dale & Greenway.
THORN, Oriental *Crataegus laciniata*. It is not clear why this is called an oriental thorn. Certainly, it grows in eastern Europe, but it is also a native of Spain.
THORN, Paperbark *Acacia woodii*.
THORN, Peach *Lycium cooperi*.
THORN, Pear *Crataegus tomentosa*. 'Pear', from the shape of the haws, presumably.
THORN, Purging *Rhamnus cathartica* (Buckthorn) Parkinson. The berries have been known as a purgative since before the Norman Conquest. "It is a rough purge", said Hill, "but a very good one".
THORN, Rabbit *Lycium pallidum*.
THORN, Rainberry *Rhamnus cathartica* (Buckthorn) B & H. The berries are sometimes known as Rhineberries, apparently a translation from the German name, and then corrupted to Rainberry.
THORN, Red *Acacia lahai*.
THORN, St Francis's *Eryngium maritimum* (Sea Holly).
THORN, Sallow *Hippophae rhamnoides* (Sea Buckthorn) Hutchinson. Cf. Willow-thorn for this.
THORN, Slaa *Prunus spinosa* (Blackthorn) Scot. B & H.
THORN, Squaw *Lycium torreyi*.
THORN, Sweet *Acacia karroo* (Mimosa Thorn) Palgrave.
THORN, Umbrella *Acacia sieberiana*.
THORN, Wait-a-bit 1. *Acacia brevispica*. 2. *Acacia melliflora*. A tribute to the acacia's thorns.
THORN, Whistling *Acacia drepanolobium* (Black-galled Acacia) Kenya Dale & Greenway.
THORN, Whistling, Coast *Acacia zanzibarica*.
THORN, White *Acacia hockii*.
THORN, Willow *Hippophae rhamnoides* (Sea Buckthorn) W Miller. Cf. Sallow Thorn for this.

THORN, Wire *Taxus baccata* (Yew) Lincs. B & H; Yks. Addy.
THORN, Woolly *Crataegus mollis*. Cf. Downy Hawthorn.
THORN-APPLE *Datura stramonium*. Descriptive of the spiny capsules, which are the "apples".
THORN-APPLE, Downy *Datura meteloides*.
THORN-APPLE, Green *Datura stramonium* (Thorn-apple) Dalziel.
THORN-APPLE, Hairy *Datura metel*.
THORN-APPLE, Purple 1. *Datura stramonium* (Thorn-apple) S Africa Watt. 2. *Datura stramonium var. tatula*.
THORN-APPLE-LEAVED GOOSEFOOT *Chenopodium hybridum* (Sowbane) Curtis.
THORN BROOM 1. *Genista anglica* (Needle Whin) Turner. 2. *Ulex europaeus* (Gorse) Gerard.
THORN LILY *Catesbaea spinosa*.
THORN PLUM *Prunus americana* (Canada Plum) USA Elmore.
THORN TREE, Hook *Acacia campylacantha*.
THORNBERRIES *Crataegus monogyna* fruit (Haws) Ches. Holland.
THORNBUSH, Frémont *Lycium fremontii*.
THORNS, Crown of 1. *Euphorbia milii* USA Kingsbury. 2. *Nigella damascena* (Love-in-a-mist) Dor. Macmillan. 3. *Passiflora caerulea* (Passion Flower) Dor. Macmillan. The only one of these that has thorns is the Spurge. For Love-in-a-mist, the name is only one of many on a similar theme. Cf. Rose-among-the-thorns, Devil-in-a-hedge, etc. Passion flower, of course, has symbolic connections with Christ's Passion. It is said that the crown of the flower, located above the petals, stands for the crown of thorns.
THORNY ACACIA *Acacia nilotica*. Why pick on this one to call 'thorny'? Aren't they all thorny?
THORNY BURR *Arctium lappa* (Burdock) Grieve.1931.
THORNY BURWEED *Xanthium spinosum* (Spiny Cocklebur) S Africa Watt.

THORNY MALLOW *Hibiscus sabdariffa* (Rozelle) Schery.

THORNY REST HARROW *Ononis spinosa* (Spiny Rest Harrow) Flower.

THOROUGH-STEM *Eupatorium perfoliatum* (Thorough-wort) USA Henkel.

THOROUGH-WAX 1. *Bupleurum rotundifolium* (Hare's Ear) Macmillan. 2. *Eupatorium perfoliatum* (Thorough-wort) USA Henkel. 'Wax' here means to grow. The stem seems to grow through ('thorough') the leaves.

THOROUGH-WORT 1. *Eupatorium altissimum* USA Tolstead. 2. *Eupatorium cannabinum* (Hemp Agrimony) Fernie. 3. *Eupatorium perfoliatum*.

THOROUGH-WORT, Hyssop-leaved *Eupatorium hyssopifolium*.

THOROUGH-WORT, Purple *Eupatorium occidentale purpureum*.

THOROUGH-WORT, Rough *Eupatorium verbenaefolium*.

THOROUGH-WORT, Vervain *Eupatorium verbenaefolium* (Rough Thorough-wort) USA House.

THOROUGH-WORT, Western *Eupatorium occidentale*.

THOROW-WAX *Bupleurum rotundifolium* (Hare's Ear) Gerard.

THOUGHTS, Lovers' *Viola tricolor* (Pansy) Som. Macmillan. A lot of names for the pansy show erotic intent. Indeed, at one time the plant was even thought to be an aphrodisiac.

THOUSAND-FLOWER *Cymbalaria muralis* (Ivy-leaved Toadflax) Ches. Holland. Cf. Hundreds-and-thousands, Mother-of-thousands, etc., the latter explaining this name.

THOUSAND-HOLES *Hypericum hirsutum* (Hairy St John's Wort) Yks. Grigson. The "holes" of a St John's Wort are in fact the glandular dots that are visible when the leaf is held to the light.

THOUSAND-JACKET *Gaya lyalli* (Mountain Ribbonwood) Perry. Presumably for the same reason as the name Lacebark, for the appearance of the inner bark has a typically lacy appearance.

THOUSAND-LEAF; THOUSAND-LEAF GRASS *Achillea millefolium* (Yarrow) Dor, Som, Lancs. Grigson; Ches. Holland; USA Henkel (Thousand-leaf); Staffs. Hackwood (Thousand-leaf Grass). A translation of the specific name, *millefolium*, the result of observation of the very finely divided leaves.

THOUSAND-LEAVED CLOVER *Achillea millefolium* (Yarrow) Border B & H; USA Henkel.

THOUSAND-SEAL *Achillea millefolium* (Yarrow) W Miller.

THOUSAND-WEED *Achillea millefolium* (Yarrow) Grieve.1931.

THOUSANDS, Hundreds-and- 1. *Pulmonaria officinalis* (Lungwort) Brownlow. 2. *Saxifraga x urbinum* (London Pride) Som. Macmillan. 3. *Sedum acre* (Biting Stonecrop) Som. Grigson.

THREAD, Needle-and- *Stipa comata*.

THREAD FLOWER *Eupatorium cannabinum* (Hemp Agrimony) Som. Macmillan.

THREAD-LEAF GROUNDSEL *Senecio longilobus* (Woolly Groundsel) USA Kingsbury.

THREAD-LEAVED CROWFOOT *Ranunculus trichophyllus*.

THREAD-OF-LIFE 1. *Cymbalaria muralis* (Ivy-leaved Toadflax) Bucks. Harman. 2. *Saxifraga sarmentosa* (Mother-of-thousands) N'thants. B & H.

THREADS, Devil's 1. *Clematis vitalba* (Old Man's Beard) Dyer. 2. *Cuscuta epithymum* (Common Dodder) Kent Grigson. Graphically descriptive, especially in the case of dodder.

THREADS, Granny; THREADS, Old Wife's *Ranunculus repens* (Creeping Buttercup) B & H (Granny); Yks. Grigson (Old Wife's).

THREE-COLOURED VIOLET *Viola tricolor* (Pansy) Thornton.

THREE-CORNERED LEEK *Allium tri-*

quetrum. *Triquetrum*, the specific name, means triangular, hence 'three-cornered'.

THREE-FACES-IN-A-HOOD; THREE-FACES-UNDER-A-HOOD *Viola tricolor* (Pansy) Gerard (in-a-hood); Suss. Parish; Herts. Jones-Baker.1974; N'thants. A E Baker (under-a-hood). Cf. Herb Trinity, or Trinity Flower.

THREE-FINGERED JACK *Saxifraga tridactylites* (Rue-leaved Saxifrage) Som. Macmillan.

THREE-LEAVED GRASS *Trifolium pratense* (Red Clover) Gerard.

THREE-LEAVED NIGHTSHADE *Trillium erectum* (Bethroot) Le Strange.

THREE-LOBED WATER CROWFOOT *Ranunculus tripartitus* (Mud Crowfoot) Clapham.

THREE-MEN-IN-A-BOAT *Rhoeo spathacea* (Boat Lily) Goold-Adams. Cf. Moses-in-the-cradle. Both of these arise from the example of the flowers in their bracts.

THREE-THREADED WILLOW *Salix triandra* (French Willow) Curtis. Note the specific name *triandra* in this connection.

THREEFOLD; TREFOLD *Menyanthes trifoliata* (Buckbean) Yks, Kirk, Wigt. (Threefold); Shet. (Trefold) both Grigson. i.e. three-leaf, for this is a trefoil.

THREEPENNY-BIT HERB *Juniperus sabina* (Savin) E Ang. G E Evans. The dosage in horse medicine should never be more than would cover a silver threepenny-bit.

THREEPENNY-BIT ROSE *Rosa farreri var. persetosa*. From the tiny, single blooms.

THRIFT 1. *Armeria maritima*. 2. *Sedum reflexum* (Yellow Stonecrop) Turner. 'Thrift' apparently means that which thrives, or is evergreen. But that did not stop the plant figuring on the back of the old threepenny bit, as a pun.

THRIFT, Broad-leaved *Armeria arenaria*.

THRIFT, Jersey *Armeria arenaria* (Broad-leaved Thrift) Polunin.

THRIFT, Lavender *Limonium vulgare* (Sea Lavender) B & H. Cf. Sea Thrift.

THRIFT, Plantain-leaved *Armeria arenaria* (Broad-leaved Thrift) Polunin.

THRIFT, Prickly *Acantholimon glumaceum*.

THRIFT, Sea 1. *Armeria maritima* (Thrift) Leigh. 2. *Limonium vulgare* (Sea Lavender) J Smith.

THRIFT, Spiny *Armeria fasciculata*.

THRISSEL, Milky *Silybum marianum* (Milk Thistle) N'thum. Grigson. The variants on 'thistle' are spelt in a number of different ways - see below, Thrissil, and Thristle.

THRISSIL, Bur *Cirsium vulgare* (Spear Plume Thistle) Scot. Jamieson.

THRISTLE, Sow *Sonchus oleraceus* (Sow Thistle) Berw. Grigson.

THROAT, Yellow *Linaria vulgaris* (Toadflax) Dor. Northover.

THROAT ROOT *Geum virginianum* (White Avens) Le Strange. Astringent, so used for sore throats.

THROATWORT 1. *Campanula trachelium* (Nettle-leaved Bellflower) Gerard. 2. *Digitalis purpurea* (Foxglove) Som. Macmillan; USA. Henkel. 3. *Scrophularia nodosa* (Figwort) Culpepper. Figwort was a country cure for sore throat and inflamed tonsils. The name may possibly be an error as far as foxglove is concerned, but there is also Throttlewort, and an infusion has been used to cure a cold. But it is the bellflower that really earns the name. It secretes a yellow latex that was regarded as the signature of its value against sore throat and tonsilitis, and Gerard, in quoting the name, says it is given "of the vertue it hath against the paine and swelling thereof" Haskwort (Cf. modern huskiness), Uvula-wort and Neckwort are other names given to this plant.

THROATWORT, Giant *Campanula latifolia* (Giant Bellflower) Yks. Grigson. This too was apparently used for sore throats.

THROTTLEWORT *Digitalis purpurea* (Foxglove) Fernie. Cf. Throatwort for this.

THROW-WAX *Bupleurum rotundifolium* (Hare's Ear) Prior. Better known as Thorough-wax, or Thorow-wax. The plant has perfoliate leaves, hence these names, which all mean "through-grow".

THRUMWORT 1. *Amaranthus caudatus* (Love-lies-bleeding) B & H. 2. *Damasonium alisma* (Star-fruit) Godwin. 'Thrum', according to Halliwell, means green and vigorous, applied to herbage.

THRUMWORT, Greater *Alisma plantago-aquatica* (Water Plantain) Barton & Castle.

THUMB, Dead Man's 1. *Gentianella amarella* (Autumn Gentian) Shet. Grigson. 2. *Orchis mascula* (Early Purple Orchis) N Eng Dyer. The half-opened buds of Autumn Gentian "are like livid finger-nails protruding from the turf" (Grigson, quoting from Shetland Folk Book, 1947). As for the orchid, children used to say that the root was once the thumb of some unburied murderer. Cf. Dead Man's Hand.

THUMB, Gowk's *Campanula rotundifolia* (Harebell) N Scot. Gregor. In this case, 'thumb' is probably 'thimble' - bellflowers make very good 'thimble' flowers.

THUMB, Lady's *Polygonum persicaria* (Persicaria) USA Tolstead. Cf. the English names Pinchweed, or Virgin's Pinch. The legend is that the Virgin Mary plucked a root, left her mark on the leaf, and threw it aside, saying "This is useless", and useless it has remained ever since.

THUMB, Lady's, Bigroot *Polygonum muhlenbergii* (Swamp Persicaria) USA Yanovsky.

THUMB, Tear *Polygonum sagitattum*.

THUMB, Tear, Halberd-leaved *Polygonum arifolium*.

THUMB, Tom 1. *Lathyrus pratensis* (Yellow Pea) Suss, Berw. Grigson. 2. *Lotus corniculatus* (Bird's-foot Trefoil) Dev, Som, Oxf, Yks. Grigson; Wilts. Dartnell & Goddard; Bucks. Harman. 3. *Orchis mascula* (Early Purple Orchis) Glos. Vickery.1995. 4. *Trifolium campestre* (Hop Trefoil) Som. Macmillan.

THUMB, Tom, Wild *Tetragonolobus maritimum* (Dragon's Teeth) Hutchinson.

THUMBLE 1. *Campanula rotundifolia* (Harebell) Scot. Jamieson. 2. *Centaurea cyanus* (Cornflower) Scot. Jamieson. 3. *Centaurea nigra* (Knapweed) Scot. B & H. Why cornflower and knapweed should be called thimble is not apparent.

THUMBS, Bloody *Briza media* (Quaking grass) Worcs. J Salisbury.

THUMBS, Fingers-and- 1. *Anthyllis vulneraria* (Kidney Vetch) Som, Dor. Macmillan. 2. *Corydalis lutea* (Yellow Corydalis) Dor. Macmillan. 3. *Cypripedium calceolus* (Lady's Slipper) Dev. Friend.1882. 4. *Digitalis purpurea* (Foxglove) Som. Macmillan. 5. *Hippocrepis comosa* (Horseshoe Vetch) Som. Macmillan. 6. *Lathyrus pratensis* (Yellow Pea) Som. Macmillan. 7. *Linaria vulgaris* (Yellow Toadflax) Som. Macmillan. 8. *Lotus corniculatus* (Bird's-foot Trefoil) Dev, Som. Macmillan; Hants. Cope. 9. *Medicago lupulina* (Black Medick) Dor. Dacombe. 10. *Melandrium rubrum* (Red Campion) Som. Macmillan. 11. *Ulex europaeus* (Gorse) Wilts. Dartnell & Goddard. 12. *Vicia cracca* (Tufted Vetch) Som. Macmillan. 13. *Vicia sepium* (Bush Vetch) Som. Macmillan. Red Campion seems the odd one out in this sequence. The rest of them have the kind of flowers that need pressure of finger and thumb to pop open.

THUMBS, Fingers-and-, Lady's, Double *Lotus corniculatus* (Bird's-foot Trefoil) Wilts. Dartnell & Goddard.

THUMBS-AND-FINGERS 1. *Lotus corniculatus* (Bird's-foot Trefoil) Dor. Macmillan. 2. *Ulex europaeus* (Gorse) Som. Macmillan.

THUMBS-AND-FINGERS, God Almighty's *Lotus corniculatus* (Bird's-foot Trefoil) Som. Grigson; Hants. Cope.

THUMBS-AND-FINGERS, Lady's *Lotus corniculatus* (Bird's-foot Trefoil) Som. Macmillan.

THUNDER-AND-LIGHTNING 1. *Ajuga reptans* (Bugle) Glos. Grigson. 2. *Pulmonaria officinalis* (Lungwort) Banff. Grigson. Bugle growing on a building protected it from lightning, so it was said in Wiltshire. There is no special record of lungwort's enjoying a similar reputation. Perhaps the name is a convenient doublet, like Hundreds-and-thousands, to commemorate the two, or even three-, coloured flowers.

THUNDER-BALL *Papaver rhoeas* (Red Poppy) War. Grigson. Glos. PLNN35. Right across northern Europe, it was said that picking poppies would cause a thunderstorm. So they said in Wiltshire, too, but on the other hand, if you had some growing on your roof, they protected against lightning. Of course, being red, this is a typical lightning plant.

THUNDER CLOVER *Ajuga reptans* (Bugle) Cockayne. See Thunder-and-lightning.

THUNDER-CUP *Papaver rhoeas* (Red Poppy) Berw. Grigson.

THUNDER DAISY *Leucanthemum vulgare* (Ox-eye Daisy) Som. Macmillan.

THUNDER FLOWER 1. *Convolvulus arvensis* (Field Bindweed) Shrop. Vickery.1985. 2. *Leucanthemum vulgare* (Ox-eye Daisy) Dor. Macmillan. 3. *Lychnis flos-cuculi* (Ragged Robin) Yks. Grigson. 4. *Melandrium album* (White Campion) Cumb. Grigson. 5. *Papaver rhoeas* (Red Poppy) Wilts. Dartnell & Goddard; Berw. Grigson. 6. *Stellaria holostea* (Greater Stitchwort) Cumb. Grigson. If you pick bindweed it will be sure to thunder before the day is out. As red poppy also has this name, it is possible that it was given to stop children damaging crops by picking them. But would they want to pick bindweed?

THUNDER PLANT; THUNDERWORT *Sempervivum tectorum* (Houseleek) C J S Thompson.1897 (Thunder Plant); Suss. Parish (Thunderwort). Houseleek "preserves what it grows upon from Fire and Lightning", as Culpepper said; "as good as a fire insurance", they say in Wiltshire.

THUNDER VINE *Glechoma hederacea* (Ground Ivy) Vesey-Fitzgerald.

THUNDERBOLT IRIS *Iris xiphium* (Spanish Iris) W Miller. Another name, Clouded Iris, is the probable explanation of this one.

THUNDERBOLT PLANT *Sesamum indicum* (Sesame) Leyel.1937.

THUNDERBOLTS 1. *Melandrium album* (White Campion) Rut. B & H. 2. *Papaver rhoeas* (Red Poppy) Dev, W Eng, Shrop. Grigson. Ches. Holland. 3. *Silene cucubalis* (Bladder Campion) Kent Grigson. 4. *Stellaria holostea* (Greater Stitchwort) Dor. Dacombe. 5. *Veronica chamaedrys* (Germander Speedwell) Ches. Hole.1937. Picking speedwells will cause a storm, and so will picking campions, with even direr results, for it can cause a death, too. And picking poppies would have similar results.

THUNDERBOLTS, Great *Alisma plantago-aquatica* (Water Plantain) Som. Macmillan.

THURBER LOCO *Astragalus thurberi*.

THYME *Thymus vulgaris*. Thyme is derived from French thym, Latin thymum, Greek thunion, which comes from thein, to burn or sacrifice, or perhaps to fumigate, possibly from its use in religious ceremonies.

THYME, Bank *Thymus vulgaris* (Thyme) Berks. Grigson.

THYME, Basil 1. *Acinos arvense*. 2. *Calamintha ascendens* (Calamint) Grieve.1931. 3. *Calamintha nepeta* (Lesser Catmint) Webster. 4. *Clinopodium vulgare* (Wild Basil) Fernie.

THYME, Black *Thymus vulgaris* (Thyme) Hatfield.

THYME, Blue, Shepherd's *Polygala calcarea* (Chalk Milkwort) Wilts. Dartnell & Goddard.1899.

THYME, Carraway *Thymus herba-baroni*. It has got a scent like carraway seed, and the leaves were at one time rubbed on the baron of beef, to give it this distinctive carraway flavour.

THYME, Cat 1. *Teucrium marum*. 2. *Teucrium polium* (Poly) J Smith.

THYME, Corsican *Thymus herba-baroni* (Carraway Thyme) Rohde.

THYME, Creeping *Thymus vulgaris* (Thyme) Gerard.
THYME, Crimson *Thymus serpyllum var. coccineum.*
THYME, Downy *Thymus serpyllum var. villosus.*
THYME, English *Thymus serpyllum* (Breckland Wild Thyme) Sanecki.
THYME, French *Thymus zygis* Brouk.
THYME, German *Thymus serpyllum var. citriodorus* (Lemon Thyme) Sanecki.
THYME, Garden, Common *Thymus vulgaris* (Thyme) Thornton.
THYME, Horse 1. *Clinopodium vulgare* (Wild Basil) Turner. 2. *Thymus vulgaris* (Thyme) Som. Grigson.
THYME, Lemon *Thymus serpyllum var. citriodorus.*
THYME, Lemon, Golden *Thymus serpyllum var. citriodorus "aureus".*
THYME, Mastick *Thymus mastichenus.*
THYME, Mother; THYME, Mother-of- 1. *Acinos arvense* (Basil Thyme) Som. B & H. 2. *Thymus vulgaris* (Thyme) Som. Grigson. Meaning womb-thyme, according to Prior, from the supposed effect of the plant on the womb. Cf. Motherwort.
THYME, Mountain *Thymus serpyllum* (Breckland Wild Thyme) Webster.
THYME, Orange *Thymus fragrantissimus.*
THYME, Pennyroyal *Thymus pulegioides* (Large Wild Thyme) Painter.
THYME, Running *Thymus vulgaris* (Thyme) Turner.
THYME, Seed-cake *Thymus herba-baroni* (Carraway Thyme) Grant White. For the typical flavour is of caraway, and of course caraway seeds are the seeds par excellence of seed cake.
THYME, Sheep's *Thymus serpyllum* (Breckland Wild Thyme) Dor. Macmillan. Cf. Shepherd's Thyme for this.
THYME, Shepherd's 1. *Polygala calcarea* (Chalk Milkwort) Wilts. Dartnell & Goddard. 2. *Polygala vulgaris* (Milkwort) Wilts. Dartnell & Goddard. 3. *Thymus serpyllum.*

THYME, Spanish *Thymus zygis* Brouk. It is also known as French Thyme.
THYME, Spurge *Euphorbia peplus* (Petty Spurge) Turner.
THYME, Water *Elodea canadensis* (Canadian Pondweed) Salisbury.
THYME, Wild 1. *Lotus corniculatus* (Bird's-foot Trefoil) Som. Macmillan. 2. *Thymus serpyllum* (Breckland Wild Thyme) Rohde. 3. *Thymus vulgaris.*
THYME, Wild, Breckland *Thymus serpyllum.* Common enough elsewhere, but in this country it is confined in its wild state to the brecks of East Anglia.
THYME, Wild, Large *Thymus pulegioides.*
THYME, Winter *Thymus serpyllum var. citriodorus* (Lemon Thyme) Sanecki.
THYME, Woolly *Thymus lanuginosus.*
THYME BROOMRAPE *Orobanche alba.*
THYME-LEAVED FLAXSEED *Radiola linoides* (Allseed) Macmillan.
THYME-LEAVED SPEEDWELL *Veronica serpyllifolia.*
THYME-LEAVED SPURGE *Euphorbia serpyllifolia.*
THYME PENNYROYAL *Hedeoma thymoides.*
TI PLANT *Cordyline terminalis.*
TIBETAN CHERRY *Prunus serrula.*
TIBINAGUA *Eriogonum nudum.*
TICK SUNFLOWER *Bidens coronata* (Swamp Beggar-ticks) Howes.
TICKBERRY *Lantana camara* (Jamaica Mountain Sage) S Africa Watt. Whatever 'tickberry' means, they are certainly poisonous.
TICKLERS; TICKLING TOMMIES *Rosa canina* seeds (Dog Rose seeds) both Dev. Macmillan. Ticklers seems rather mild - for these seeds make the original itching powder.
TICKLERS, Nose *Tropaeolum majus* (Nasturtium) Som. Macmillan. Cf. Nose-smart and Nose-twitcher. They must all be references to the peppery cress quality of the plant, as is Nasturtium itself, which is Latin nasus torsus, twisted nose.
TICKSEED 1. *Bidens bipinnata* (Spanish

Needles) USA Elmore. 2. *Bidens involucrata* USA Gates. 3. *Coreopsis tinctoria*. 4. *Ricinus communis* (Castor Oil Plant) Turner. 'Coreopsis' means 'like a bug', and 'Ricinus' means a dog tick - the Romans thought they saw a resemblance between the seeds and the ticks.

TICKSEED, Lance-leaved *Coreopsis lanceolata*.

TICKSEED, Pink *Coreopsis rosea*.

TICKSEED, Stiff *Coreopsis palmata*.

TICKSEED, Western *Bidens aristosa*.

TICKSEED SUNFLOWER *Bidens trichosperma*.

TIDELAND SPRUCE *Picea sitchensis* (Sitka Spruce) USA Brimble.1948.

TIDDY *Solanum tuberosum* (Potato) Dor. Dacombe. Hence Tiddy-drills for the trench in which to plant them.

TIDY TIPS *Layia platyglossa*.

TIE PALM *Dracaena indivisa* (Fountain Tree) Howes.

TIEN SHAN SPRUCE *Picea schrenkiana*.

TIGER ALOE *Aloe variegata* (Partidge-breasted Aloe) USA Cunningham. A lot of these 'tiger' names simply mean having striped flowers or leaves.

TIGER FLOWER *Tigridia pavonia* (Shellflower) Hornell.

TIGER LILY 1. *Lilium catesbaei* (Catesby's Lily) USA Woodcock. 2. *Lilium columbianum* (Columbia Lily) USA E Gunther. 3. *Lilium pardalinum* (Panther Lily) USA Schenk & Gifford. 4. *Lilium tigrinum*.

TIGER LILY, Wild *Lilium superbum* (Turk's Cap Lily) USA Woodcock.

TIGER PLANT *Aphelandra squarrosa* (Zebra Plant) Nicolaisen.

TIGER'S CLAWS *Erythrina indica* (Indian Coral Tree) Malaya Gilliland.

TIGER'S MOUTH 1. *Antirrhinum majus* (Snapdragon) Suss, Suff. B & H. 2. *Digitalis purpurea* (Foxglove) Suss. Grigson. Descriptive - both of these plants have a large number of 'mouth' names.

TIGERNUT *Cyperus esculentus*.

TIGERTAIL SPRUCE *Picea polita*.

TILE; TILET-TREE; TILLET; TILLET-TREE *Tilia europaea* (Common Lime) all B & H. All these and also the generic name are from Latin tilia, the name of the tree. The name has also appeared as Teil or Teyl.

TILLS *Lens esculenta* (Lentil) A S Palmer. A contraction of Lentils.

TIME FLOWER; TIME TABLE; TIME TELLER *Taraxacum officinale* (Dandelion) Mabey (Time Flower); Fernie (Time Table); Som. Grigson (Time Teller). Dandelion flowers open with the sun, but it is a time-teller in another way; everyone knows the children's way of puffing at the seed-globe to find out the hour. There are very many names to commemorate the fact.

TIMOTHY; TIMOTHYGRASS *Phleum pratense*. Celebrating Timothy Hanson, who promoted its cultivation for cattle feed in America about 1720, and then brought it over here, under the impression it was a purely American species.

TIMOTHY, Rhodesian *Setaria sphacellata* S Africa Dalziel.

TIMOTHY, Sand *Phleum arenarium* (Sand Catstail) Hepburn.

TIMOTHY GRASS, Golden *Setaria sphacellata* S Africa Dalziel.

TINE; TINE-GRASS 1. *Vicia cracca* (Tufted Vetch) Herts. Grigson. 2. *Vicia hirsuta* (Hairy Tare) Herts. Grigson.

TINE, Tare-, Yellow *Lathyrus pratensis* (Yellow Pea) Middx. Grigson.

TINE-TARE *Vicia hirsuta* (Hairy Tare) Kent Grigson.

TINE-TARE, Rough-podded *Vicia hirsuta* (Hairy Tare) Curtis.

TINE-TARE, Smooth-podded *Vicia tetraspermum* (Smooth Tare) Curtis.

TINE-WEED 1. *Vicia cracca* (Tufted Vetch) Herts. Grigson. 2. *Vicia hirsuta* (Hairy Tare) B & H.

TINGIRINGI GUM *Eucalyptus glaucescens*.

TINKER, Wandering *Lysimachia nummularia* (Creeping Jenny) N Ire. Tynan & Maitland. Cf. Roving Sailor, Wandering Jenny etc.

TINKER-TAILOR GRASS 1. *Lolium perenne* (Rye Grass) Corn, Dev, Som. Grigson.1959. 2. *Plantago lanceolata* (Ribwort Plantain) Som. Elworthy. The game Tinker, tailor, soldier, sailor - the blow that knocks the head off ribwort marks the profession of the future husband, and the grass is treated in exactly the same way. Cf. What's-your-sweetheart, a Sussex name for the rye-grass.

TINTERN SPURGE *Euphorbia stricta*. A very rare native to Britain, confined to clearings in limestone woods in the Wye valley, and one hedgebank near Bath.

TIPSEN *Hypericum androsaemum* (Tutsan) Dev, Bucks. Grigson. One of a number of variations on tutsan, which comes from Old French, but modern French still has Toutesaine. That either means wholly sound, or healing all.

TIPSY-LEAVES *Hypericum androsaemum* (Tutsan) Som. Macmillan. Cf. Titsy-leaves, a Cornish variant.

TIPSY-WOOD *Galega frutescens* W Miller.

TIPTON-WEED *Hypericum perforatum* (St John's Wort) USA Watt. There are several places called Tipton in the USA. Exactly which one is not necessarily relevant, for the point is that St John's Wort was introduced into America and has become a pest, just as it has in Australia and New Zealand.

TISTY-TOSTY 1. *Kerria japonica* (Jew's Mallow) Dev. B & H. 2. *Primula veris* (Cowslip) Dev, Dor, Som, Glos, Heref. Grigson; Wilts. Dartnell & Goddard. 3. *Viburnum opulus* (Guelder Rose) Dev. Friend.1882; Som. Elworthy; Wilts. Tennant. Cowslips are the real tisty-tosties. Children used to play a kind of divination game with cowslips, whose blossoms were tied in a ball. Strictly, the balls themselves were the tisty-tosties, though the growing flowers got the name, too. The cowslip ball is tossed about while the names of various boys or girls are called, till it drops. The name called at that moment is taken to be the "one indicated by the oracle": Tisty-tosty tell me true. Who shall I be married to?

TITAN ARUM *Amorphophallus titanum*.

TITOKI *Alectryon excelsus*. A New Zealand tree, and this is the Maori name for it.

TITS, Sow's *Polygonatum multiflorum* (Solomon's Seal) Dor. Grigson. Graphically descriptive.

TITSUM; TITZEN *Hypericum androsaemum* (Tutsan) Dev. Friend.1882 (Titsum); Corn. Grigson (Titzen). A variation of Tutsan - see Tipsen.

TITSY-LEAF *Hypericum androsaemum* (Tutsan) Corn. B & H.

TITTERS *Vicia hirsuta* (Hairy Tare) B & H. Used by Tusser, and presumably having some affinity with 'tine'.

TITTLE-MY-FANCY *Viola tricolor* (Pansy) E Ang. Gurdon. One of the large group of the 'Jump-up-and-kiss-me' type.

TITTY-BOTTLES *Rosa canina* hips (Hips) Som. Macmillan. Descriptive of the shape - Cf. Brandy-bottles. Or the shape can be compared to pears, with names like Pixy Pears, etc.

TIVERS *Galium aparine* (Goose-grass) Bucks. B & H. A corruption of 'clivers', itself a variant of 'cleavers', "because it cleveth upon mennes clothes", according to Turner.

TOAD, Tether; TOAD TETHER *Ranunculus repens* (Creeping Buttercup) Yks. Grigson. One of the references to its method of growth, and the creeping runners it sends out.

TOAD FLOWER 1. *Stapelia variegata* (Starfish Flower) Oplt. 2. *Stachys sylvatica* (Hedge Woundwort) Yks. B & H. *Stapelia variegata* has star-shaped flowers, hence the common name; brown with darker spots on them, hence the reference to toads.

TOAD LILY *Trillium sessile* (Toadshade) Perry.

TOAD-PIPE 1. *Equisetum arvense* (Common Horsetail) H C Long.1910. 2. *Equisetum limosum* (Water Horsetail) Lincs. Peacock; Yks. Carr.

TOAD RUSH *Juncus bufonius*.

TOAD TREE *Conopharyngia elegans*.

TOAD'S BRASS *Spergula arvensis* (Corn Spurrey) Ches. Grigson. Was 'brass' a misprint for 'grass' in Grigson?

TOAD'S HEAD *Fritillaria meleagris* (Snake's-head Lily) Wilts. Dartnell & Goddard. Cf. Toad's Mouth. They are both suggested by the colouring.

TOAD'S MEAT *Arum maculatum* (Cuckoo-pint) Corn. Grigson. Cf. Frog's Meat, from Dorset.

TOAD'S MOUTH 1. *Antirrhinum majus* (Snapdragon) B & H. 2. *Fritillaria meleagris* (Snake's-head Lily) Wilts. Dartnell & Goddard.1899. 3. *Misopates orontium* (Lesser Snapdragon).

TOADFLAX 1. *Linaria sp.* 2. *Spergula arvensis* (Corn Spurrey) Ches. Grigson. One explanatioon is that 'toad' means 'dead', i.e. it is the German word tot, conveying the idea that this is a flax that is "dead", that is, useless for the purpose to which proper flax is applied. Grigson came to the same conclusion, but said the name is a simple translation of Krottenflachs (Krötenflachs), a wild, useless flax, a flax for toads. It was, of course, at one time a destructive weed of the flax fields.

TOADFLAX, Alpine *Linaria alpina*.

TOADFLAX, Bastard 1. *Comandra pallida* USA Kingsbury. 2. *Thesium humifusum*.

TOADFLAX, Blue *Linaria repens* (Pale Toadflax) Salisbury.

TOADFLAX, Daisy-leaved *Annarhinum bellidifolium*.

TOADFLAX, Field *Linaria arvensis*.

TOADFLAX, French *Linaria arenaria* (Sand Toadflax) McClintock. A native British plant, but its main range is on coastal dunes in France.

TOADFLAX, Italian *Linaria angustissima*.

TOADFLAX, Ivy-leaved *Cymbalaria muralis*.

TOADFLAX, Jersey *Linaria pelisseriana*.

TOADFLAX, Pale *Linaria repens*.

TOADFLAX, Prostrate *Linaria supina*.

TOADFLAX, Purple *Linaria purpurea*.

TOADFLAX, Pyrenean *Linaria pyrenaica*.

TOADFLAX, Sand *Linaria arenaria*.

TOADFLAX, Small *Chaenorhinum minus*.

TOADFLAX, Striped *Linaria repens* (Pale Toadflax) Howes.

TOADFLAX, Wall *Cymbalaria muralis* (Ivy-leaved Toadflax) Le Strange.

TOADFLAX, Yellow *Linaria vulgaris*.

TOADROOT 1. *Actaea alba* (White Baneberry) W Miller. 2. *Actaea spicata* (Herb Christopher) Grieve.1931. Mrs Grieve said that toads are attracted by the smell of Herb Christopher.

TOADSHADE *Trillium sessile*. Cf. Toad Lily for this, presumably from the strange colouring of the flowers.

TOADWORT *Aster tripolium* (Sea Aster) Hulme.1895. The "Ortus Sanitatis" has "when a spider stings a toad and the toad is becoming vanquished, and the spider stings it thickly and frequently, and the toad cannot avenge itself, it bursts asunder. If such a burst toad be near a toad-wort, it chews it and becomes sound again: but if it happens that the wounded toad cannot get to the plant, another toad fetches it and gives it to the wounded one". Topsell vouches for this having actually been witnessed!

TOBACCO *Nicotiana tabacum*. The name derives from Spanish tabaco, which in turn derived directly from the Arawak term for cigar. More accurately, it comes from an implement used by the Carib Indians, called a tabaco. They strewed dry tobacco leaves on the embers of a fire, and inhaled the smoke through a hollow forked reed, the two ends of which were put in the nostrils. This reed was the tabaco. By a misunderstanding, the name became transferred to the herb, and so gave tobacco.

TOBACCO, Aztec *Nicotiana rustica* (Turkish Tobacco) Howes.

TOBACCO, Coyote *Nicotiana attenuata*.

TOBACCO, Desert *Nicotiana trigonophylla*.

TOBACCO, Devil's *Heracleum sphondyllium* (Hogweed) Staffs. PLNN17. Cf. the Devonshire Boy's Bacca. Apparently, the stems were actually smoked as a tobacco substitute. Gypsies smoked them, as well as the boys.

TOBACCO, Gypsy's 1. *Clematis vitalba* (Old Man's Beard) Dor. Macmillan. 2. *Rumex acetosa* (Sorrel) Som. Macmillan. There is no record of sorrel ever having been smoked, but why else should it get this name? Old Man's Beard, though, is another matter, for boys certainly used to smoke its porous stalks. Cf. Smoke-wood, Tom Bacca, Boy's Bacca and Devil's Cut.

TOBACCO, Indian 1. *Antennaria plantaginifolia* (Plantain-leaved Everlasting). Grieve.1931. 2. *Lobelia inflata*. 3. *Nicotiana bigelovii var. exaltata* Schenk & Gifford. The Lobelia was used by the Indians as a substitute for real tobacco, and presumably so was the everlasting, which also carries the name Ladies' Tobacco.

TOBACCO, Lady's 1. *Antenarria dioica* (Cat's Foot) USA Leighton. 2. *Antennaria plantaginifolia* (Plantain-leaved Everlasting). New Engl. Sanford. 3. *Gnaphalium polycephalum* (Mouse-ear Everlasting). They must all have been tried as tobacco substitutes and found to be very mild, hence "Lady's".

TOBACCO, Mountain 1. *Arnica montana* (Arnica) Grieve.1931. 2. *Nicotiana attenuata* (Coyote Tobacco) USA Elmore. 3. *Nicotiana sylvestris*. Arnica leaves are used to make a tobacco, known in France as tabac des savoyards, tabac des Vosges, or herbe aux prêcheurs. All parts can be used for the tobacco.

TOBACCO, Poor Man's *Tussilago farfara* (Coltsfoot) Som. Macmillan. Coltsfoot leaves are still smoked in all herbal tobaccos, as it is also in Chinese medicine, for asthma and bronchitis.

TOBACCO, Sailor's *Artemisia vulgaris* (Mugwort) Hants. Grigson. The leaves often used to be smoked. There are records of the usage from as far from Hampshire as the Hebrides.

TOBACCO, Tree *Nicotiana glauca*. It can be anything up to about 18 ft tall.

TOBACCO, Turkish *Nicotiana rustica*. Like the rest of the Nicotianas, this is a South American species. 'Turkish', presumably, because the particular flavour was popular there.

TOBACCO, Wild 1. *Lobelia inflata* (Indian Tobacco) USA House. 2. *Lobelia nicotianifolia*. The latter has leaves like tobacco, and the former was often used as a subsitute for tobacco.

TOBACCO-BRUSH *Ceanothus velutinus* USA Schenk & Gifford.

TOBACCO-ROOT 1. *Lewisia rediviva* (Bitter-root) Northcote. 2. *Valeriana edulis* (Edible Valerian) USA Tolstead. 3. *Valeriana septentrionalis* Johnston. Bitter-root has the smell of tobacco when cooked.

TOBIN, Mogue *Chrysanthemum segetum* (Corn Marigold) Ire. (co Carlow) Vickery.1995.

TOCALOTE *Centaurea melitensis* (Napa Thistle) USA Schenk & Gifford.

TODAY-AND-TOMORROW 1. *Echium vulgare* (Viper's Bugloss) Leyel.1937. 2. *Pulmonaria officinalis* (Lungwort) Fernie. Merely an indication of the two-coloured flowers, particularly in the case of Lungwort, which has attracted many of these double names, which, out of context, would just be meaningless - such as Thunder and lightning, William and Mary, Hundreds-and-thousands, etc.

TODDLING GRASS *Briza media* (Quaking-grass) H C Long.1910. A 'shaking' name. Cf. Tottering, or Totty, Grass.

TOE, Crow 1. *Endymion nonscriptus* (Bluebell) Turner. 2. *Lotus corniculatus* (Bird's-foot Trefoil) Som, Scot. Prior. 3. *Orchis mascula* (Early Purple Orchis) Cumb. B & H. 4. *Ranunculus repens* (Creeping Buttercup) Greenoak.1979.

TOE, Dog's *Geranium robertianum* (Herb Robert) Donegal Grigson.

TOE, Lamb's 1. *Anthyllis vulneraria* (Kidney Vetch) Som. Macmillan; N'thants.

Sternberg; Rut. Grigson. 2. *Lotus corniculatus* (Bird's-foot Trefoil) Midl. Grigson. 3. *Medicago lupulina* (Black Medick) Staffs. Grigson.

TOE, Lark's *Delphinium ajacis* (Larkspur) Gerard. The lark has a long hind claw, and larkspur has a long calyx-spur.

TOE, Nigger *Bertholettia excelsa* (Brazil Nut) Schery.

TOE, Pussy('s) 1. *Antennaria dioica* (Cat's Foot) USA Leighton. 2. *Antennaria neglecta* USA Zenkert.

TOE, Pussy, Plantain-leaved *Antennaria plantaginifolia* (Plantain-leaved Everlasting) USA Youngken.

TOENAILS, Grandmother's; TOENAILS, Old Woman's *Lotus corniculatus* (Bird's-foot Trefoil) Dev, Som. Macmillan (Grandmother); Dev. Grigson (Old Woman).

TOES, Fingers-and- *Anthyllis vulneraria* (Kidney Vetch) Lincs. Tynan & Maitland.

TOKEN BLACKBERRY *Rubus caesius* (Dewberry) Wilts. Grigson. Grigson explains the name by the habit the plant has of producing quite a lot of fruit one year, and then another year they will be "mean, scanty, and not worth the picking or eating".

TOLMEINER; TOLMENEER *Dianthus barbatus* (Sweet William) Gerard (Tolmeiner); Prior (Tolmeneer). It appears in both Lyte and Parkinson as Toll-me-neer. Colmenier is another version. Prior suggested that they may have arisen from d'Almagne, or d'Allemagne, in other words a pink from Germany.

TOLOACHE *Datura meteloides* (Downy Thorn-apple) C Grant. Toloache is the Mexican name for the plant, and it serves too as the name of the narcotic cult associated with it.

TOLOACHE, Purple-stained *Datura discolor*.

TOM, Creeping *Sedum acre* (Biting Stonecrop) Grieve.1931.

TOM BACCA *Clematis vitalba* (Old Man's Beard) Suss. Parish. Boys used to smoke the porous stalks of Old Man's Beard - Cf. Smoking Cane, Boy's Bacca, Gypsy's Tobacco, etc.

TOM PAINE *Quercus robur* (Oak) Yks. Grigson. Perhaps it is their solid worth that merits the comparison with Tom Paine.

TOM PIMPERNEL; MALE PIMPERNEL *Anagallis arvensis* (Scarlet Pimpernel) Gerard (Male); Yks. Grigson (Tom). Blue Pimpernel is the female, but in neither case is the reason clear.

TOM THUMB 1. *Lathyrus pratensis* (Yellow Pea) Suss, Berw. Grigson. 2. *Lotus corniculatus* (Bird's-foot Trefoil) Dev, Som, Oxf, Yks. Grigson; Wilts. Dartnell & Goddard; Bucks. Harman. 3. *Orchis mascula* (Early Purple Orchis) Glos. Vickery.1995. 4. *Trifolium pratense* (Red Clover) Som. Macmillan.

TOM THUMB, Wild *Tetragonolobus maritimus* (Dragon's Teeth) Hutchinson.

TOM THUMB'S HONEYSUCKLE *Lotus corniculatus* (Bird's-foot Trefoil) Wilts. Dartnell & Goddard.

TOM THUMB'S THOUSAND FINGERS *Rumex acetosa* (Wild Sorrel) Kent B & H.

TOMATILLO 1. *Lycium pallidum* (Rabbit Thorn). 2. *Physalis ixiocarpa* Coats. The diminutive of tomato.

TOMATO *Lycopersicon esculentum*. Spanish tomate, in turn from a Mexican word, tomatl.

TOMATO, Bitter *Solanum incanum* (Hoary Nightshade) Howes.

TOMATO, Cherry *Lycopersicon cerasiforme*.

TOMATO, Children's *Solanum anomalum*.

TOMATO, Devil's *Solanum carolinense* (Carolina Nightshade) Allan. Cf. Devil's Potato for this. The two names show that the roots and fruit are not what they seem to be.

TOMATO, Gooseberry *Physalis bunyardii* (Cape Gooseberry) Willis.

TOMATO, Husk 1. *Physalis aequata* USA Driver. 2. *Physalis pubescens*. 'Husk', because the berries are covered with a papery calyx, or husk.

TOMATO, Strawberry 1. *Physalis alkekengi* (Winter Cherry) Bianchini. 2. *Physalis bun-*

yardii (Cape Gooseberry) USA Watt. 3. *Physalis pubescens* (Husk Tomato) Bianchini.

TOMATO, Tree *Cyphomandra betacea*.

TOMATO, Wild 1. *Solanum nigrum* (Black Nightshade) Som. Macmillan. 2. *Solanum trifolium* (Three-flowered Nightshade) Johnston. Extremely optimistic for plants as poisonous as these!

TOMCAT CLOVER *Trifolium tridentatum*.

TOMMIES, Tickling *Rosa canina* seeds (Dog Rose seeds) Dev. Grigson. Tickling is putting it mildly - these are the original itching powder.

TOMMY, Stinking *Ononis repens* (Rest Harrow) Fernie. Cf. Stinking Tam, from Northumberland. Do they smell so badly?

TOMMY TOTTLES *Lotus corniculatus* (Bird's-foot Trefoil) Yks. Grigson.

TONGUE *Aster macrophyllus* (Large-leaved Aster) New Engl. Sanford. New England farmers used to eat the young leaves as greens, under this local name.

TONGUE, Adder's 1. *Achillea ptarmica* (Sneezewort) Aber. Grigson. 2. *Arum maculatum* (Cuckoo-pint) Corn. B & H; Som. Elworthy. 3. *Geranium robertianum* (Herb Robert) Ess. B & H. 4. *Listera ovata* (Twayblade) Wilts. Dartnell & Goddard. 5. *Ophioglossum vulgatum*. 6. *Orchis mascula* (Early Purple Orchis) Ches. B & H. 7. *Phyllitis scolopendrium* (Hartstongue) Dev. Friend.1882. 8. *Sagittaria sagittifolia* (Water Archer) B & H. The shape of cuckoo-pint's leaf would suggest a snake's tongue. The ferns would do likewise, and the name as applied to Water Archer is obviously descriptive. Herb Robert has a distinct association with snakes - it is, for example, unlucky to pick it, for snakes would come from the stems, though surely there cannot be any descriptive associations. Sneezewort probably gets the name from the fact that the powdered leaves were once used for snuff. The common name says it all.

TONGUE, Adder's, American *Erythronium americanum* (Yellow Adder's Tongue) Grieve.1931.

TONGUE, Adder's, Early *Ophioglossum lusitanicum*.

TONGUE, Adder's, Giant *Erythronium oreganum*.

TONGUE, Adder's, White *Erythronium albidum*.

TONGUE, Adder's, Yellow *Erythronium americanum*.

TONGUE, Bird's 1. *Acer campestre* fruit (Field Maple keys) Evelyn. 2. *Anagallis arvensis* (Scarlet Pimpernel) Norf. Grigson. 3. *Fraxinus excelsior* fruit (Ash keys) B & H. 4. *Pastinaca sativa* (Wild Parsnip) B & H. 5. *Polygonum aviculare* (Knotgrass) Som, Norf. Prior; N Eng. Gerard. 6. *Senecio paludosus* (Great Fen Ragwort) B & H. 7. *Stellaria holostea* (Greater Stitchwort) B & H. 8. *Strelitzia reginae* (Bird of Paradise Flower) Whittle & Cook.

TONGUE, Bity *Polygonum hydropiper* (Water Pepper) Cumb. Grigson. The common name explains this well enough.

TONGUE, Blind; TONGUE, Blood *Galium aparine* (Goose-grass) Dyer (Blind); Ches, N'thum, Scot. Grigson (Blood). Blindtongue is apparently a children's game, but Blood Tongue conveys the same idea - Cf. Tongue-bleed and Whip-tongue.

TONGUE, Brook *Cicuta virosa* (Cowbane) Cockayne.

TONGUE, Clove *Helleborus niger* (Christmas Rose) B & H. 'Cloven', rather then 'clove', makes sense, given the shape of the leaves.

TONGUE, Dead; TONGUE, Dead Man's *Oenanthe crocata* (Hemlock Water Dropwort) Barton & Castle (Dead); Salisbury (Dead Man). "...from its paralysing effect on the organs of voice" (Prior).

TONGUE, Deer's *Liatris odoratissima*.

TONGUE, Devil's *Papaver rhoeas* (Red Poppy) Corn. Grigson.

TONGUE, Dog's *Cynoglossum officinale* (Hound's Tongue) Turner.

TONGUE, Drop, Painted *Aglaonema crispum*.

TONGUE, Goose 1. *Achillea millefolium* (Yarrow) Som. Macmillan. 2. *Achillea ptarmica* (Sneezewort) Shrop, War. B & H; N'thants. A E Baker; Yks. Carr. 3. *Galium aparine* (Goose-grass) Som. Macmillan; Ches. Grigson. 4. *Ranunculus flammula* (Lesser Spearwort) Scot. Grigson. 5. *Tanacetum balsamita* (Balsamint) Conway. Descriptive, in their various ways. For spearwort and balsamint, it is the shape of the leaves, but with the others, it is their roughness.

TONGUE, Hound's 1. *Cynoglossum officinale*. 2. *Stachys palustris* (Marsh Woundwort) Mor. Grigson. A translation of Cynoglossum, Greek kunoglossum, and explained variously from the texture of the leaf, from its "doggy" smell, and by association from the fact that it cures the bites of dogs. There is a superstition, mentioned by Coles, but taken from Albertus Magnus, that it will tie a dog's tongue, so that it will not bark, if it is laid under the soles of its feet.

TONGUE, Hound's, Green *Cynoglossum germanicum*.

TONGUE, Hound's, Himalayan *Cynoglossum nervosum*.

TONGUE, Lamb 1. *Chenopodium urbicum* (Upright Goosefoot) Som. Elworthy. 2. *Phyllitis scolopendrium* (Hartstongue) Som. Elworthy.

TONGUE, Lamb's 1. *Chenopodium album* (Fat Hen) Dev, Ches. B & H. 2. *Mentha arvensis* (Corn Mint) Scot. Grigson. 3. *Plantago lanceolata* (Ribwort Plantain) Som. Macmillan. Hants, Suss, Shrop, N'thum. Grigson. 4. *Plantago media* (Lamb's-tongue Plantain) Som. Macmillan. Suss. Grigson. 5. *Polygonum persicaria* (Persicaria) Dev. Grigson. 6. *Rhinanthus crista-galli* (Yellow Rattle) Som. Macmillan. 7. *Stachys lanata* (Lamb's Ears) Dev. Friend.1882.

TONGUE, Lion's *Linaria vulgaris* (Toadflax) Dev. Macmillan.

TONGUE, Mother-in-law's *Sansevieria trifasciata*.

TONGUE, Old Wives' *Populus tremula* (Aspen) Roxb. Grigson. Because the leaves are never still.

TONGUE, Ox 1. *Anchusa officinalis* (Alkanet) Lyte. 2. *Borago officinalis* (Borage) Lyte. 3. *Lycopsis arvensis* (Small Bugloss) Gunther. The reference in all these cases is to the shape and texture of the leaves, and incidentally to the name Bugloss, which means Ox-tongue.

TONGUE, Painted *Salpiglossis sinuata* (Velvet Trumpet Flower) Perry.

TONGUE, Serpent's 1. *Erythronium americanum* (Yellow Adder's Tongue) O P Brown. 2. *Ophioglossum vulgatum* (Adder's Tongue) Tynan & Maitland.

TONGUE, Snake's 1. *Erythronium americanum* (Yellow Adder's Tongue) Mitton. 2. *Ranunculus flammula* (Lesser Spearwort) Berw. Grigson.

TONGUE, Sparrow's *Polygonum aviculare* (Knotgrass) Som, Norf. Cockayne. Bird's Tongue is more usual.

TONGUE, Tarry; TONGUE, Teng *Nasturtium officinale* (Watercress) both Yks. Grigson (Tarry); B & H (Teng). Cf. the Irish Tongue-grass. They are explained by reference to the derivation of Nasturtium, the generic name - it is nasi-tortium, nose-twisting, from the tangy taste of the darker leaves.

TONGUE, Whip 1. *Galium aparine* (Goose-grass) Fernie. 2. *Galium mollugo* (White Bedstraw) B & H. Cf. Blind-tongue, or Blood-tongue for Goose-grass. The reference is to the bristly leaves and stems.

TONGUE, Witch's *Clerodendron serratum* Malaya Gilliland.

TONGUE, Woman's 1. *Albizia lebbeck* (Siris Tree) Everett. 2. *Briza media* (Quaking-grass) Berks. Grigson.1959. 3. *Hypericum androsaemum* (Tutsan) Fernie. 4. *Populus tremula* (Aspen) Berks. Lowsley. The allusion as far as Quaking-grass and aspen are concerned is to the fact that they are never still.

TONGUE-BLEED; TONGUE-BLUIDERS
Galium aparine (Goose-grass) Fernie (Bleed); Yks. Nicholson; Berw. Grigson (Bluiders).
TONGUE FLOWER, Spiny-stemmed *Glossopetalon spinescens*.
TONGUE FLOWER, Spiny-tipped *Glossopetalon pungens*.
TONGUE-GRASS 1. *Lepidium sativum* (Garden Cress) Ire. B & H. 2. *Nasturtium officinale* (Watercress) Ire. Grigson. 3. *Stellaria media* (Chickweed) Ire. Grigson. For the first two at least, the reason for the name must be their pungency.
TONGUE-LEAVED CROWFOOT *Ranunculus lingua* (Greater Spearwort) Flower.
TONGUE-UNDER-TONGUE *Astragalus danicus* (Purple Milk Vetch) Flower.
TONKA BEAN *Dipteryx odorata*.
TOON *Cedrela sinensis*.
TOOTH, Dog's *Erythronium dens-canis* (Dog-tooth Violet) Gerard.
TOOTH, Horse's *Melilotus officinalis* (Melilot) Henslow.
TOOTH, Hound's *Agropyron repens* (Couchgrass) Grigson.1959. Cf. Dog-grass, perhaps because it is a dog's favourite tonic grass, or perhaps it simply infers whiteness (one of the Wiltshire names is White Couch).
TOOTH CRESS *Dentaria bulbifera* (Coral-root) Prior.
TOOTH CRESS, Large *Dentaria maxima* (Large Toothwort) USA House.
TOOTH CRESS, Violet *Dentaria bulbifera* (Coral-root) Prior. Cf. Toothed Violet and Dog-toothed Violet.
TOOTH-LEAF *Stillingia paucidentata*.
TOOTH(ED) VIOLET *Dentaria bulbifera* (Coral-root) Prior. (Tooth); Gerard (Toothed).
TOOTHACHE PLANT *Spilanthes acmella* Malaya Gilliland.
TOOTHACHE TREE 1. *Aralia spinosa* (Hercules' Club) Grieve.1931. 2. *Xanthoxylum americanum* (Prickly Ash) Grieve.1931. 3. *Xanthoxylum clava-herculis* (Southern Prickly Ash) USA. Weiner. In the case of the Aralia, it is the berries that are used, in tincture, to relieve toothache, while with Prickly Ash, it is the root and bark that provide the cure.
TOOTHBRUSH TREE *Salvadora persica*. Because the twigs are used as such in Africa.
TOOTHED DODDER *Cuscuta denticulata*.
TOOTHED MEDICK *Medicago polymorpha* (Hairy Medick) McClintock.1974.
TOOTHED ORCHID *Orchis tridentata*.
TOOTHED WINTERGREEN *Ramischia secunda* (Yavering Bells) Blamey.
TOOTHWORT 1. *Capsella bursa-pastoris* (Shepherd's Purse) B & H. 2. *Dentaria laciniata* (Cut-leaved Toothwort) USA Tolstead. 3. *Lathraea squamaria*.
TOOTHWORT, Cut-leaved *Dentaria laciniata*.
TOOTHWORT, Large *Dentaria maxima*.
TOOTHWORT, Purple *Lathraea clandestina* (Willow Toothwort) Clapham.
TOOTHWORT, Two-leaved *Dentaria diphylla*.
TOOTHWORT, Willow *Lathraea clandestina*. It is a parasite on the roots of willows, and poplars.
TOOWOOMBA CANARY GRASS *Phalaris aquatica* (Harding Grass) Howes.
TOP, Black 1. *Centaura scabiosa* (Greater Knapweed) B & H. 2. *Ligustrum vulgare* (Privet) Som. Grigson.
TOP, Blackie *Plantago lanceolata* (Ribwort Plantain) Som. Grigson.
TOP, Blue 1. *Centaurea nigra* (Knapweed) Worcs. Grigson. 2. *Succisa pratensis* (Devil's-bit Scabious) Worcs. B & H.
TOP, White *Erigeron annuus*.
TOP, Yellow *Sinapis arvensis* (Charlock) N'thum. Grigson.
TOP-KNOT *Centaurea nigra* (Knapweed) Som. Macmillan.
TOPER'S PLANT *Poterium sanguisorba* (Salad Burnet) Perry. "thought to make the heart merry and glad", Gerard said. Poterion means drinking cup, because the leaves were used in the preparation of many beverages.

TOPPERS, Blackie *Typha latifolia* (False Bulrush) Som. Macmillan. Descriptive.

TORCH; TORCH PLANT; TORCHWORT *Verbascum thapsus* (Mullein) Gerard (Torch); Genders (Torch Plant); USA Henkel (Torchwort). The stems and leaves used to be dipped in tallow or suet and burnt to give light at outdoor country gatherings, or even in the home. They were used at funerals, too, and also in the French Fête de Brandons, on the first Sunday in Lent.

TORCH, Blue-flowered *Tillandsia lindenii*.

TORCH, Devil's; TORCH LILY *Kniphofia uvaria* (Red Hot Poker) Som. Macmillan (Devil's Torch); Dev. Friend (Torch Lily). Descriptive, of course.

TORMENTIL 1. *Lamium purpureum* (Red Deadnettle) Shrop. B & H. 2. *Potentilla erecta*. It is from the French tormentille, Latin tormentum, the rack, or tormine - colic, in other words. It survived as a medicine for colic until quite recently in isolated parts, particularly Northumberland and the Hebrides.

TORMENTIL, Creeping; TORMENTIL, Trailing *Potentilla anglica* Trailing Tormentil is the usual name; Creeping Tormentil is in Phillips.

TORMENTING ROOT *Potentilla erecta* (Tormentil) N Ire Grigson. One of a number of variations on Tormentil.

TORMENTERS, Gentleman's *Galium aparine* (Goose-grass) Suff. Grigson. The reference, of course, is to the clinging powers of the burrs.

TORMERIK *Potentilla erecta* (Tormentil) Turner.

TORNSOLE *Euphorbia helioscopia* (Sun Spurge) Prior. i.e. Turnsole, the idea being that Sun Spurge always turns its flowers towards the sun.

TORREY MORMON TEA *Ephedra torreyana*. The Indians used to make a medicinal tea from its branches.

TORREY PINE *Pinus torreyana*.

TORREY SALTBUSH *Atriplex torreyi*.

TORREY SEABLITE *Suaeda torreyana* 'ramosissima'.

TORY-TOP *Pinus sylvestris* cones (Pine cones) Ire. B & H.

TOSS, Snow; TOSSEL, May; TOSSY-BALLS *Viburnum opulus* (Guelder Rose) Som. Macmillan (Snow Toss, Tossy-balls); Dev. Macmillan (May Tossel). Variants on May-tosty, or Tisty-Tosty, which is usually a ball of cowslips or primroses for the May garland.

TOSTY *Primula veris* (Cowslip) Wilts. Dartnell & Goddard Som, Glos, Worcs, Pemb. Grigson. Or Tisty-tosty, the cowslip ball that children used to toss about in a kind of divination game.

TOSTY, May *Viburnum opulus* (Guelder Rose) Som. Macmillan.

TOTARA *Podocarpus totara*. A New Zealand tree, and Totara is the Maori name.

TOTSAN *Hypericum androsaemum* (Tutsan) Turner. One of a number of variations of Tutsan - see, for instance, Titsum, or Titzen.

TOTTER GRASS; TOTTERING GRASS; TOTTY GRASS *Briza media* (Quaking-grass) Som, Wilts, N'thants. (Totter); Hants, Cambs, Yks. (Tottering); Lincs. (Totty) all Grigson.1959. Totter, in the sense of tremble.

TOUCH-AND-HEAL 1. *Hypericum androsaemum* (Tutsan) Bucks. Grigson. 2. *Hypericum perforatum* (St John's Wort) N Ire. Grigson. 3. *Prunella vulgaris* (Self-heal) N Ire. Grigson. They are all, of course, great healing plants, but 'touch' in the case of Tutsan is probably a corruption of the common name.

TOUCH-LEAF; TOUCHED-LEAF; TOUCHEN-LEAF *Hypericum androsaemum* (Tutsan) Wales Grigson (Touch); Hants. Cope (Touched); Hants Boase; Wales Grigson (Touchen). The leaves have antiseptic properties, and they were certainly used to cover open flesh wounds before bandaging became common, but 'touch', nevertheless, is probably a corruption of Tutsan.

TOUCH-ME-NOT 1. *Arabidopsis thaliana* (Thale Cress) Som. Macmillan. 2. *Arctium* *lappa* (Burdock) Som. Mabey. 3. *Cardamine hirsuta* (Hairy Bittercress) Ches. B & H. 4. *Impatiens noli-me-tangere*. For the *Impatiens* and Hairy Bittercress, the reference is to the explosive method of seed distribution. The words themselves are borrowed from the words Christ spoke to Mary Magdalene after the resurrection, in the Vulgate. They have a different meaning in parts of France. In the Gironde, for instance, they used to make a girl touch this flower. If she was not a virgin, the flower would recoil and fade at once.

TOUCH-ME-NOT, Pale *Impatiens pallida*.

TOUCH-ME-NOT, Spotted *Impatiens capensis* (Orange Balsam) USA H Smith.

TOURPIN *Sempervivum tectorum* (Houseleek) Ire. Logan. Or Turpeen. Presumably from a Gaelic word, for it appears again in the Scottish Hockerie-Topner.

TOW TREE *Aesculus hippocastanum* (Horse Chestnut) Heref. Havergal.

TOWEL GOURD *Luffa cylindrica* (Loofah) Candolle.

TOWER *Dactylorchis maculata* (Heath Spotted Orchid) Som. Macmillan. One can see the point, but they make fairly diminutive towers.

TOWER, Round *Lythrum salicaria* (Purple Loosestrife) Dev. Macmillan. Descriptive.

TOWER CRESS 1. *Arabis turritis* (Tower Rock-cress) Clapham. 2. *Turritis glabra* (Tower Mustard) USA Zenkert.

TOWER MUSTARD *Turritis glabra*. Descriptive of the tall, straight stem.

TOWER MUSTARD, Smooth *Turritis glabra* (Tower Mustard) Curtis.

TOWER OF BABEL *Verbascum thapsus* (Mullein) Som. Notes & Queries, vol 168, 1935. Babel has no significance here; the allusion is solely to the height of the plant.

TOWER ROCK-CRESS *Arabis turritis*.

TOWER'S TREACLE *Turritis glabra* (Tower Mustard) Flower.1859.

TOWN-CRESS *Lepidium sativum* (Garden Cress) Dawson.

TOWN-WEED *Mercurialis perennis* (Dog's Mercury) B & H.

TOY-WORT *Capsella bursa-pastoris* (Shepherd's Purse) Gerard. Cf. the Warwickshire name Naughty Man's Plaything.

TOYON *Photinia arbutifolia* (California Holly).

TRACES, Lady *Spiranthes autumnalis* (Lady's Tresses) Turner. 'Traces' is 'tresses', of course.

TRAGACANTH, African *Sterculia tragacantha*. The bark of this tree exudes a pinkish gum like tragacanth, used by African blacksmiths in smelting iron, and also to mend broken earthenware and calabashes. The real tragacanth, which comes from two Greek words meaning 'goat thorn', comes from various members of the genus *Astragalus*.

TRAIL, Cat *Valeriana officinalis* (Valerian) Yks. Grigson. Cf. Cat's Love for this plant. Cats are apparently inordinately fond of it.

TRAMMAN *Sambucus nigra* (Elder) IOM Train. The usual name for elder on the Isle of Man. It comes from a Manx original.

TRANSVAAL DAISY *Gerbera jamesonii* (Barbaton Daisy) A W Smith.

TRAVELLER'S COMFORT *Galium aparine* (Goose-grass) Wilts. Dartnell & Goddard.1899. Perhaps because it was traditionally used to soothe wounds and ulcers.

TRAVELLER'S EASE 1. *Achillea millefolium* (Yarrow) Wilts. Dartnell & Goddard. 2. *Galium aparine* (Goose-grass) Wilts. Goddard. 3. *Potentilla anserina* (Silverweed) War. Grigson. Put a sprig of silverweed in each shoe to prevent blistering when walking long distances. It sounds as if the name as given to yarrow is misapplied from some other plant, probably tansy. But there is a Somerset belief that a yarrow leaf in the sock will prevent cramp.

TRAVELLER'S FOOT *Plantago major* (Great Plantain) War. Grigson. Acording to legend, plantain persistently follows the tracks of man. One superstition is more specific, and says that it follows Englishmen, and springs up in whatever part of the world he makes his home. In this case, White Man's Foot, which is what the North American Indians call the plant, becomes Englishman's Foot.

TRAVELLER'S JOY 1. *Clematis vitalba* (Old Man's Beard) Gerard. 2. *Lycopodium clavatum* (Stag's-horn Clubmoss) Yks. Atkinson. Old Man's Beard, "commonly called the Viorna, quasi viam ornans, of decking and adorning waies and hedges where people travel; and thereupon I have named it the Traveller's Joy" No matter that he got his derivation wrong. For clubmoss, the name is listed by Atkinson, without comment.

TRAVELLER'S JOY, Indian *Clematis gouriana*.

TRAVELLER'S LEAF *Potentilla anserina* (Silverweed) Wilts. D Grose. See Traveller's Ease.

TRAVELLER'S REST *Tanacetum vulgare* (Tansy) Wilts. D & G. There must have been at some time a belief in putting tansy leaves inside a boot to stop tiredness on a journey, but the only belief that has come down is from New Forest gypsies, who put a sprig of tansy inside the boot to prevent the onset of ague.

TRAVELLER'S TREE *Ravenala madagascariensis*. The leaf bases of this tree fill with water, and they are often picked to provide drinking water, hence this name, and also Pilgrim's Tree.

TREACLE, Churl's; TREACLE, Clown's *Allium sativum* (Garlic) Prior (Churl's); B & H (Clown's). 'Treacle' is an interesting word. It is from triacle, an antidote to the bite of venomous animals. In origin, it is Greek theriake, from therione, a name given to the viper. So what we have is an example of homeopathic magic, viper's flesh curing viper's bite. Philips, *World of Words*, defines treacle as a "physical compound made of vipers and other ingredients", and this was a favourite against all poisons. The word eventually became applied to any confection of sweet syrup, and finally and solely to the syrup of molasses. 'Churl' and 'Clown' are synonymous - they both simply mean 'countryman' Garlic, of course, has always been regarded as a great heal-all.

TREACLE, Countryman's 1. *Allium sativum* (Garlic) B & H. 2. *Ruta graveolens* (Rue) B & H. 3. *Valeriana officinalis* (Valerian) B & H.

TREACLE, English 1. *Alliaria petiolata* (Jack-by-the-hedge) B & H. 2. *Teucrium chamaedrys* (Germander) Turner. 3. *Teucrium scordium* (Water Germander) Coles. 4. *Teucrium scorodonia* (Wood Sage). Jack-by-the-hedge has a garlic-like flavour, and is often called Garlic Mustard, or some such name. Hence the inclusion as a 'treacle'. The Germanders were actually used as poison antidotes - see Gerard, talking of Water Germander - "the decoction made in wine and drunke, is good against the bitings of serpents, and deadly poisons...".

TREACLE, Gypsy's *Sambucus nigra* (Elder) Ire. Jean Philpot.

TREACLE, Poor Man's 1. *Alliaria petiolata* (Jack-by-the-hedge) B & H. 2. *Allium sativum* (Garlic) B & H. 3. *Allium vineale* (Crow Garlic) B & H. 4. *Teucrium scorodonia* (Wood Sage) Tynan & Maitland. See Churl's Treacle.

TREACLE, Tower's *Turritis glabra* (Tower Mustard) Flower.1859.

TREACLE-BERRIES *Smilacina racemosa* (False Solomon's Seal) USA Leighton.

TREACLE-LEAF *Hypericum androsaemum* (Tutsan) Cumb. Grigson. It must be given because of the curative powers of the leaves, for treacle, in its older meaning, is a sovereign remedy. Cf. Touch-and-heal.

TREACLE MUSTARD 1. *Erysimum cheiranthoides*. 2. *Erysimum repandum* USA Gates. 3. *Lepidium campestre* (Pepperwort) B & H. 4. *Thlaspi arvense* (Field Pennycress) Turner.

TREACLE MUSTARD, Garlic *Alliaria petiolata* (Jack-by-the-hedge) Barton & Castle.

TREACLE MUSTARD, Swiss *Erysimum helveticum*.

TREACLE WORMSEED *Erysimum cheiranthoides* (Treacle Mustard) B & H. A vermifuge, but actually the seeds can be quite dangerous, especially to animals.

TREBIZOND DATES *Elaeagnus angustifolia* fruit (Narrow-leaved Oleaster) Moldenke. The fruits are dried and pounded to be made into a kind of bread.

TREES *Malva sylvestris* (Common Mallow) Som. Macmillan. A much-branching plant, but all the same the name is probably a mishearing of 'Cheese', given over a wide area to the seed cases.

TREFOIL, Bean 1. *Anagyris foetida*. 2. *Laburnum anagyroides* (Laburnum) Parkinson. 3. *Menyanthes trifoliata* (Buckbean) B & H.

TREFOIL, Bird's-foot 1. *Lotus corniculatus*. 2. *Trigonella ornithopodioides* (Fenugreek) Prior.

TREFOIL, Black *Medicago lupulina* (Black Medick) Norf. Grigson.

TREFOIL, Bog *Menyanthes trifoliata* (Buckbean) Yks. Grigson.

TREFOIL, Codded *Tetragonolobus maritimus* (Dragon's Teeth) Gerard.

TREFOIL, Creeping *Trifolium repens* (White Clover) Flower.1859.

TREFOIL, Golden *Hepatica caerulea* (Hepatica) Gerard. 'Trefoil', because Hepatica has three leaflets combined in one leaf.

TREFOIL, Great *Medicago sativa var. sativa* (Lucerne) B & H.

TREFOIL, Hare's-foot *Trifolium arvense*.

TREFOIL, Heart *Medicago arabica* (Spotted Medick) B & H. 'Heart', "not only because the leaf is triangular like the heart of a man, but also because each leafe doth contain the perfection (or image) of an heart, and that in its proper colour, viz a flesh colour. It defendeth the heart against the noisome vapour of the spleen" (Coles). That is pure doctrine of signatures.

TREFOIL, Honeysuckle *Trifolium pratense* (Red Clover) Flower.1859. Cf. Honeysuckle, Meadow Honeysuckle, Honeysuck and a lot of others.

TREFOIL, Hop *Trifolium campestre*.

TREFOIL, Hop, Large *Trifolium aureum*.

TREFOIL, Hop, Lesser *Trifolium dubium* (Lesser Yellow Trefoil) Salisbury.

TREFOIL, Marsh *Menyanthes trifoliata* (Buckbean) Gerard.

TREFOIL, Meadow *Trifolium pratense* (Red Clover) Gerard.

TREFOIL, Melilot *Melilotus officinalis* (Melilot) Barton & Castle. Melilot is from a Greek word meaning a clover rich in honey.

TREFOIL, Moon *Medicago arborea* (Tree Medick) RHS.

TREFOIL, Rough 1. *Trifolium arvense* (Hare's-foot Clover) Turner. 2. *Trifolium scabrum* (Rough Clover) Curtis.

TREFOIL, Round-headed *Trifolium glomeratum* (Clustered Clover) Curtis.

TREFOIL, Sour *Oxalis acetosella* (Wood Sorrel) Gerard.

TREFOIL, Subterranean *Trifolium subterraneum* (Burrowing Clover) Curtis.

TREFOIL, Tree *Cytisus scoparius* (Broom) Parkinson.

TREFOIL, Water *Menyanthes trifoliata* (Buckbean) War. Grigson.

TREFOIL, White *Menyanthes trifoliata* (Buckbean) War. Grigson.

TREFOIL, Yellow *Medicago lupulina* (Black Medick) USA Gates. Yellow flowers, black pods.

TREFOIL, Yellow, Common *Trifolium dubium* (Lesser Yellow Trefoil) McClintock.1974.

TREFOIL, Yellow, Least *Trifolium micranthum* (Slender Yellow Trefoil) D Grose.

TREFOIL, Yellow, Lesser *Trifolium dubium*.

TREFOIL, Yellow, Slender *Trifolium micranthum*.

TREFOIL, Yellow, Small *Trifolium dubium* (Lesser Yellow Trefoil) Murdoch McNeill.

TREFOIL MILKMAID *Cardamine trifolia* Salisbury.

TREFOLD *Menyanthes trifoliata* (Buckbean) Shet. Grigson.

TREFOY *Trifolium repens* (White Clover) Som. Elworthy.
TREMBLE *Populus tremula* (Aspen) Gerard.
TREMBLING ASPEN *Populus tremuloides* (American Aspen) USA H Smith.
TREMBLING GRASS *Briza media* (Quaking Grass) Suff, Ches, Lancs, Yks. Grigson.1959.
TREMBLING JOCKIES; TREMBLING JOCKS *Briza media* (Quaking Grass) Yks. F K Robinson (Jockies); Yks. Wright (Jocks). 'Trembling' would be pronounced as 'trimmling'.
TREMBLING POPLAR 1. *Populus tremula* (Aspen) Prior. 2. *Populus tremuloides* (American Aspen) USA Zenkert.
TREMBLING SHADOW; SHAKING SHADOW *Briza media* (Quaking Grass) both Tynan & Maitland.
TRESSES, Lady's 1. *Arum maculatum* (Cuckoo-pint) Som. Macmillan. 2. *Briza media* (Quaking Grass) L Gordon. 3. *Galium verum* (Lady's Bedstraw) Som. Macmillan. 4. *Listera ovata* (Twayblade) Tynan & Maitland. 5. *Spiranthes spiralis*. The true Lady's Tresses gets the name from the peculiar method of growth, apparently spiralling like plaited hair. The others are also all descriptive, in their different ways.
TRESSES, Vicar's *Calystegia sepium* (Great Bindweed). According to a letter to The Times dated 8 August 1989, this name was given in Victorian times because the vicar was always "far too occupied on the sabbath to clear his own garden... "Poorer parishes, unable to support proper gardening staff for the vicarage, used to have a date in the calendar known as "Widubundae Saturday", when the poor of the parish would gather in the vicarage garden to clear the vines and receive a blessing and a small celebratory gift of vegetables or fruit in return".
TRICK-MADAM; TRIP-MADAM; PRICK-MADAM *Sedum reflexum* (Yellow Stonecrop) Prior (Trick); B & H (Trip); Cumb. Grigson (Prick). All from a French original - triacque madame, i.e. Lady's Treacle.
TRIDENT MAPLE *Acer buergerianum*.
TRIFOLY, Garden 1. *Melilotus caerulea* (Swiss Melilot) Turner. 2. *Oxalis corniculata* (Sleeping Beauty) Turner.
TRIFOLY, Mount- *Hepatica caerulea* (Hepatica) Turner. Cf. Herb Trinity, Golden Trefoil, etc., all because it has three leaflets combined in one leaf.
TRIGONA CUCUMBER *Cucumis trigonus*.
TRIGONEL *Trigonella ornithopodioides* (Fenugreek) Candolle. This shortened form of the name of the genus comes from the Greek meaning 'three-angled'. It is the shape of the corolla and of course the subsequent pods, that suggests this.
TRILLIUM, Nodding *Trillium cernuum*.
TRILLIUM, Painted *Trillium undulata*.
TRILLIUM, Snow *Trillium nivale*.
TRILLIUM, White *Trillium grandiflorum* (Wood Lily) Weiner.
TRINITY *Tradescantia virginica* (Spiderwort) B & H. For the flower is formed of three petals.
TRINITY, Herb 1. *Hepatica caerulea* (Hepatica) Browne. 2. *Viola tricolor* (Pansy) Hants. Jones-Baker.1974. Hepatica has three leaflets combined in one leaf, while pansy has three petals in the flower.
TRINITY, Herb of the *Viola tricolor* (Pansy) Culpepper.
TRINITY FLOWER 1. *Trillium erectum* (Bethroot) Le Strange. 2. *Trillium grandiflorum* (Wood Lily) Perry. 3. *Viola tricolor* (Pansy) Greenoak.
TRINITY VIOLET *Viola tricolor* (Pansy) Yks. Grigson.
TRIP-MADAM *Sedum reflexum* (Yellow Stonecrop) B & H. See Trick-madam.
TRIPE, Rock *Umbilicaria pustularia* (Pustulous Moss) S M Robertson. This is the French tripe de roche.
TRIPOLI SENNA *Cassia obovata* (Italian Senna) Dalziel. 'Italian', because it was once cultivated there, but this is ascribed not only to Tripoli, but to Senegal and Jamaica, too.

TRIQUETROUS GARLIC *Allium triquetrum* (Three-cornered Leek) Grigson. 'Triquetrous' means 'triangular' - the 'three-cornered' of the common name. It is the stem that is three-angled.

TRISTRAM'S KNOT *Cannabis sativa* (Hemp) B & H. The name occurs in Bullein's *Book of Simples*. Perhaps the 'knot' is the one in a gallows rope, for which hemp was always used. The use is commemorated in such names as Gallow-grass, and the Somerset Neckweed.

TRIUMPH, Bloody *Trifolium incarnatum* (Crimson Clover) Dor. Macmillan. The local tradition is that the name commemorated a battle, in which the winners decorated themselves with these flowers.

TROLL-FLOWER *Trollius europaeus* (Globe-flower) Gerard. Gerard was right according to the thinking of his day in ascribing Trollius to trolls, but the word is more likely to come from German Trollblume, probably a contraction of die rolle Blume, rolled, or closed in, petals, as is implied in such names as Locken Gowan.

TROPILLO *Solanum eleagnifolium* (Silverleaf Nightshade) USA Kinsgbury.

TROSSACHS DOCK *Rumex aquaticus*. Known only in this country on the eastern shores of Loch Lomdon.

TROTH, Love *Paris quadrifolia* (Herb Paris) Tynan & Maitland. Cf. Herb Truelove and True-lover's Knot. All probably because of the derivation of 'Paris', i.e. pair. So herba paris is pair herb, that is,the herb of equality, "from the numerical harmony of its parts" (Grigson.1974).

TROUT LILY *Erythronium americanum* (Yellow Adder's Tongue) Le Strange.

TROUT LILY, White *Erythronium albidum* (White Adder's Tongue) USA Yanivsky.

TROWIE GIRSE; TROWIE GLIV *Digitalis purpurea* (Foxglove) Ork. Leask. 'Gliv' is 'glove', and the 'trowie' epithet follows the usual fairy line that characterizes foxglove.

TRUBBA, Red, Small *Solanum ficifolium* Barbados Gooding.

TRUCKLEBERRY *Vaccinium myrtillus* (Whortleberry) Fernie.

TRUCKLES OF CHEESE *Malva sylvestris* (Common Mallow) Som. Macmillan. It is the ripe seed capsules that are referred to here. There are numerous 'cheese' names for them, all derived from the shape, but not the taste, which is claimed to be like peanuts.

TRUE ANISE *Illicium verum*.

TRUE LOVE *Lunaria annua* (Honesty) Briggs.

TRUELOVE, Four-leaved; TRUELOVE, Herb; TRUE LOVERS' KNOT *Paris quadrifolia* (Herb Paris) Midl. Tynan & Maitland (Four-leaved); Friend (Herb); Tynan & Maitland (Knot). Friend claimed that the leaves have the appearance of a true-love knot, but the real reason probably lies in the fact that this is herba paris, pair herb.

TRUFFLE *Conopodium majus* (Earth-nut) Inv. Grigson. Just to emphasise the underground factor.

TRUMPET 1. *Calystegia sepium* (Great Bindweed) Som. Macmillan. 2. *Sarracenia purpurea* (Pitcher Plant) Perry. 3. *Tropaeolum majus* (Nasturtium) Som. Macmillan. All 'trumpet' names are descriptive.

TRUMPET, Angel's 1. *Datura arborea* Australia Watt. 2. *Datura metel* (Hairy Thorn-apple). 3. *Datura sanguinea* Cunningham. 4. *Datura stramonium* (Thorn-apple) Som. Grigson. 5. *Datura suaveolens*.

TRUMPET, Blue *Endymion nonscriptus* (Bluebell) Som. Macmillan.

TRUMPET, Desert *Eriogonum inflatum*.

TRUMPET, Devil's *Datura stramonium* (Thorn-apple) USA Henkel. It is a notorious narcotic, hence the ascription to the devil.

TRUMPET, Fairy 1. *Calystegia sepium* (Great Bindweed) Som. Macmillan. 2. *Lonicera periclymenum* (Honeysuckle) Som. Macmillan.

TRUMPET, Fiddler's *Sarracenia drummondii*. A very odd pairing of epithets.

TRUMPET, Gabriel's *Solandra longiflora*.
TRUMPET, Golden *Narcissus pseudo-narcissus* (Daffodil) Som. Macmillan.
TRUMPET, Humming Bird's *Zauschneria californica* (Californian Fuchsia) S Clapham.
TRUMPET, Huntsman's *Sarracenia flava* (Trumpet-leaf) Howes.
TRUMPET, Little *Eriogonum trichopes* USA Jaeger.
TRUMPET, Long *Eriogonum nudum* (Tibinagua) Jaeger.
TRUMPET, Scarlet *Gilia attenuata* (Scarlet Gilia) USA Elmore.
TRUMPET, Yellow 1. *Narcissus pseudo-narcissus* (Daffodil) Som. Macmillan. 2. *Sarracenia sledgei*. 3. *Tussilago farfara* (Coltsfoot) Som. Grigson. Coltsfoot is not as easily identifiable by this name as the daffodil.
TRUMPET CREEPER 1. *Bignonia sp*. 2. *Campsis radicans*.
TRUMPET CUP *Mimulus guttatus* (Monkey Flower) Som. Macmillan.
TRUMPET FLOWER 1. *Calystegia sepium* (Great Bindweed) Som. Macmillan. 2. *Incarvillea delavayi*. 3. *Lonicera periclymenum* (Honeysuckle) Yks. Grigson. 4. *Tecoma stans* (Yellow Elder) Roys.
TRUMPET FLOWER, Chinese *Campsis grandiflora*.
TRUMPET FLOWER, Desert *Datura meteloides* (Downy Thorn-apple) Safford.1920.
TRUMPET FLOWER, Golden *Allemanda cathartica* (Yellow Allemande) Simmons.
TRUMPET FLOWER, Evening *Gelsemium sempervirens* (Carolina Jessamine) USA Kingsbury.
TRUMPET FLOWER, Velvet *Salpiglossis sinuata*.
TRUMPET GENTIAN *Gentiana acaulis*.
TRUMPET GENTIAN, Stemless *Gentiana clusii*.
TRUMPET GOURD *Lagenaria siceraria* (Calabash) Lindley.
TRUMPET HONEYSUCKLE 1. *Campsis radicans* (Trumpet Creeper) RHS. 2. *Lonicera sempervirens*.
TRUMPET KECK *Angelica sylvestris* (Wild Angelica) B & H. This is out of the usual mould of 'trumpet' names, and has nothing descriptive about it. It is just that boys used to employ the hollow stems like trumpets.
TRUMPET LEAF 1. *Sarracenia flava*. 2. *Sarracenia purpurea* (Pitcher Plant) Perry.
TRUMPET LILY *Zantedeschia aethiopica* (Arum Lily) J Smith.
TRUMPET TREE *Cecropia peltata*. Because the hollow stems are used for this purpose by some South American Indians.
TRUMPET VINE *Campsis radicans* (Trumpet Creeper) RHS.
TRUMPET VINE, Blue *Thunbergia grandiflora* (Sky Vine) RHS.
TRUMPET WEED *Eupatorium purpureum* (Purple Boneset) O P Brown.
TRUXILLO LEAF *Erythroxylon truxillense* Thomson.
TSAMA MELON *Citrullus vulgaris* (Water Melon) Lee. The name is given to the Kalahari variety.
TSIN BEAN *Bauhinia esculenta*.
TSINGTAU LILY *Lilium tsingtauense*.
TUBA ROOT *Derris elliptica* (Derris) Chopra.
TUBE-ROOT *Colchicum autumnale* (Meadow Saffron) B & H. i.e. tuber-root.
TUBER-ROOT *Asclepias tuberosa* (Orange Milkweed) O P Brown.
TUBEROSE *Polyanthes tuberosa*. Not tube-rose, which is false association. The word is the same as tuberous.
TUBEROSE, Yellow *Hemerocallis flava* (Day Lily) McDonald.
TUBEROUS THISTLE; TUBEROUS PLUME THISTLE *Cirsium tuberosum*.
TUCKER-GRASS; TACKER-GRASS *Polygonum aviculare* (Knotgrass) both Som. Elworthy (Tacker); Macmillan (Tucker). "From its likeness to a 'tacker', or shoemaker's wax-end", according to Elworthy. But surely a tacker is a hobnail?

TUDNOORE *Glechoma hederacea* (Ground Ivy) Langham.

TUFT, London 1. *Dianthus barbatus* (Sweet William) Gerard. 2. *Saxifraga spathularia x umbrosa* (London Pride) N'thants. A E Baker. A 'tuft' in this sense is a cluster of flowers. Cf. London Pride and London Bobs for the Sweet William. But, at least as far as the true London Pride is concerned, the reference is to Mr London, a royal gardener in the 18th century.

TUFT, Spanish *Thalictrum aquilegifolium* (Greater Meadow Rue) Parkinson. Cf. Parkinson's Tufted Columbine.

TUFTED FORGET-ME-NOT *Myosotis caespitosa*.

TUFTED SAXIFRAGE *Saxifraga caespitosa*.

TULE MINT *Mentha arvensis var. canadensis* (Canadian Mint) USA Elmore. 'Tule' is actually the name of one of the American bulrushes, and is from Nahuatl tollin.

TULE POTATO *Sagittaria latifolia* (Broadleaved Arrowhead) USA Bean.

TULE ROOT *Sagittaria arifolia* (Arumleaved Arrowhead) USA H Smith.

TULIP, Cape 1. *Haemanthus coccineus*. 2. *Homeria collina*.

TULIP, Chequered *Fritillaria meleagris* (Snake's-head Lily) B & H. Cf. Chequered Daffodil and Chequered Lily.

TULIP, Clarimond *Tulipa praecox* Geldart. 'Tulip' is derived from a Persian word, thuliban, or dulband, meaning turban. This evolved into Turkish dulbend and tulbend, Old French tulipan, Italian tulipano, Latin tulipa, and the modern Spanish tulipan, French tulipe, German tulpe and English tulip.

TULIP, Crocus *Tulipa pulchella*.

TULIP, Drooping *Fritllaria meleagris* (Snake's-head Lily) Ches. Holland. 'Drooping' is perfectly descriptive. The plant is likened to a tulip in other names, Wild Tulip and Chequered Tulip, for instance.

TULIP, Globe, Golden *Calochortus amabilis* (Golden Fairy Lantern) RHS.

TULIP, Horned *Tulipa acuminata*.

TULIP, Lady *Tulipa clusiana*.

TULIP, Mountain *Tulipa montana*.

TULIP, Sharon *Tulipa sharonensis*. In Moldenke's view, probably the "rose of Sharon" of the Song of Solomon.

TULIP, Waterlily *Tulipa kaufmanniana*.

TULIP, Wild 1. *Fritillaria meleagris* (Snake's-head Lily) Berks, N'thants, War Grigson. 2. *Tulipa sylvestris*.

TULIP, Yellow *Meconopsis cambrica* (Welsh Poppy) Som. B & H.

TULIP POPLAR *Liriodendron tulipifera* (Tulip Tree) USA Harper. Cf. Yellow Poplar and White Poplar for this tree, which does not even belong to the same family as the poplars.

TULIP POPPY *Papaver glaucum*.

TULIP TREE 1. *Acer pseudo-platanus* (Sycamore) Dev, Yks. Grigson; Wilts. Dartnell & Goddard. 2. *Liriodendron tulipifera*. 3. *Thespesia populnea* (Portia Tree) India Dalziel. Tulip Tree gets the name from a similarity in the flowers, but sycamore is different - it can only be the shape of the leaves that is compared, though Dartnell & Goddard said that children believed that the smell and taste of the young shoots resembled that of tulips.

TULIP TREE, African *Spathodea campanulata*.

TULIP TREE, Chinese *Liriodendron chinense*.

TULIP TREE, Pink *Magnolia campbelli*.

TULIP-WOOD *Physocalymna sp*.

TUMBLE MUSTARD 1. *Sisymbrium altissimum* (Tumbling Mustard) W Miller. 2. *Sisymbrium officinale* (Hedge Mustard) USA Upham.

TUMBLE PIGWEED *Amaranthus albus* (White Pigweed) USA Allan.

TUMBLEWEED 1. *Amaranthus albus* (White Pigweed) USA Gates. 2. *Amaranthus blitoides*. 3. *Salsola kali* (Saltwort) Watt. The pigweeds have the habit of breaking off at ground level once the fruits are ripe. The wind blows them away, and thus the seeds are scattered.

TUMBLEWEED, Palestinian *Anastatica hierochuntica* (Rose of Jericho) Moldenke. As the seeds ripen during the dry season, the leaves fall off and the branches curve inwards to make a round, lattice-like ball, which is blown out of the soil and which rolls about the desert until it either reaches a moist spot, or the rainy season begins.

TUMBLEWEED, Russian *Salsola kali* (Saltwort) Watt.

TUMBLING MUSTARD *Sisymbrium altissimum*.

TUMMIT *Brassica campestris var. rapa* (Turnip) Leics. A B Evans. There are all sorts of versions of 'turnip'. Besides this one, there are Turmit, Turmet and Turmop, besides various shortened forms.

TUN-FOOT *Glechoma hederacea* (Ground Ivy) Som. Macmillan. From its former use in brewing. This is a bitter herb, and was much used before the general advent of the employment of hops. 'Foot', and also 'hoof' in the next entry, is from the shape of the leaves, whether those of ground ivy or of coltsfoot.

TUN-HOOF 1. *Glechoma hederacea* (Ground Ivy) Som, E Ang. B & H. 2. *Tussilago farfara* (Coltsfoot) Grieve.1931.

TUNG OIL TREE *Aleurites fordii*. From the Chinese name, yu-t'ung.

TUNIC FLOWER *Petrorhagia saxifraga*. The older botanical classification was *Tunica saxifraga*.

TUNSING-WORT *Veratrum album* (White Hellebore) Cockayne.

TUPELO *Nyssa sylvatica*. An American tree, with the name adapted from the Indian name for it.

TUPELO, Water *Nyssa aquatica*.

TURBAN, Yellow *Eriogonum pusillum* USA Jaeger.

TURBAN BELL *Nigella damascena* (Love-in-a-mist) Dor. Macmillan.

TURBAN SQUASH *Cucurbita maxima* (Giant Pumpkin) Whitaker & Davis.

Picturesquely descriptive, given that turbans that size were actually worn.

TURBITH, Quacksalver's *Euphorbia esula* (Hungarian Spurge) W Miller. Cf. Quacksalver's Spurge. 'Quacksalver' means the same as the modern 'quack', while 'turbith' seems to convey the general idea of a medicinal root. That name was originally applied to one of the Ipomaeas from the East. The whole implies a rough-and-ready purge, given by someone who did not really know what he was doing.

TURBITH, Serapia's *Aster tripolium* (Sea Aster) Gerard. Serapia is presumably Serapis, or Sarapis, a god of the Greeks of Egypt, identified with Apis and Osiris.

TURF, Sea *Armeria maritima* (Thrift) Dev. Grigson.

TURFING DAISY *Matricaria tchihatchensis*.

TURK'S CAP 1. *Aconitum anglicum* (Wild Monkshood) N'thants. B & H. 2. *Aconitum napellus* (Monkshood) N'thants. A E Baker. 3. *Tulipa sp* (Tulip) Parkinson.

TURK'S CAP LILY 1. *Lilium martagon*. 2. *Lilium superbum*. Cf. Turncap, which conveys quite well the imagery of the reverted petals that gave Turk's Cap as well.

TURK'S CAP LILY, American *Lilium superbum* (Turk's Cap Lily) Yanovsky.

TURK'S CAP LILY, Scarlet *Lilium chalcedonicum*.

TURK'S CAP LILY, Western *Lilium michiganense*.

TURK'S CAP LILY, Yellow *Lilium pyrenaicum*.

TURK'S HEAD 1. *Fritillaria meleagris* (Snake's-head Lily) War. Grigson. 2. *Lilium tigrinum* (Tiger Lily) Som. Macmillan. For Snake's-head Lily, this name follows Turkey-hen Flower, which is another name for Guinea-hen.

TURKEY ALMOND *Prunus amygdalus* (Almond) Mitton. The almond is probably indigenous to Asia Minor and Persia, hence 'Turkey'.

TURKEY-BERRY *Mitchella repens* (Partridge-berry) Howes.

TURKEY BOX *Buxus sempervirens* the timber (Box) Brimble.1948. Boxwood has been sold under a variety of names. As well as this one, there are Abassian, European and Persian Box.

TURKEY-CAP; TURKEY-EGGS; TURKEY-HEN FLOWER *Fritillaria meleagris* (Snake's-head Lily) Hudson (cap); Berks. B & H (eggs); Gerard (Hen). A guinea hen is more appropriate (that is what *meleagris* means). The spotted petals are the point of reference. Cf. too Turk's Head.

TURKEY-CORN 1. *Dicentra canadensis*. 2. *Dicentra formosa* (Western Bleeding Heart) USA Berdoe. Cf. Turkey Pea for both of these. The reference in this case is to the bird, not the country.

TURKEY-DISH *Mentha pulegium* (Pennyroyal) Ches. B & H. Probably a variation on Lurk-in-ditch, which exists in a number of forms - see Lurgadish, Lurgeydish, etc.

TURKEY FIG *Ficus carica* (Common Fig) Dev. Friend.1882.

TURKEY-GARLIC, Great *Allium scordoprasum* (Sand Leek) Tradescant.

TURKEY GILLIFLOWER *Tagetes erecta* (African Marigold) Gerard.

TURKEY-HEN FLOWER *Fritillaria meleagris* (Snake's-head Lily) Gerard. Another name for Guinea-hen, which is what the specific name means.

TURKEY OAK *Quercus cerris*. Reasonably accurate, for it is a native of southwest Asia, as well as Europe.

TURKEY MULLEIN *Eremocarpus setigerus*.

TURKEY PEA 1. *Dicentra canadensis* (Turkey Corn) Grieve.1931. 2. *Dicentra formosa* (Western Bleeding Heart) USA Berdoe. 3. *Tephrosia virginiana* (Devil's Shoestring) USA Youngken. See Turkey-Corn.

TURKEY-POD *Sisymbrium officinale* (Hedge Mustard) B & H.

TURKEY RHUBARB 1. *Arctium lappa* (Burdock) Som. Macmillan. 2. *Petasites hybridus* (Butterbur) Som. Macmillan. 3. *Rheum officinale*. Rhubarb tends to be called after the country through which it reached the market from its native land (Tibet). Hence we find such names as Russian Rhubarb or Chinese Rhubarb, as well as the one quoted here.

TURKEY'S FOOD *Galium aparine* (Goosegrass) Som. Macmillan. Better known as food for geese.

TURKEY'S SNOUT *Amaranthus caudatus* (Love-lies-bleeding) Som. Macmillan. Very imaginative, but one can see the connection.

TURKISH COMFREY *Symphytum orientale* (Soft Comfrey) Salisbury.

TURKISH CORN; TURKISH MILLET *Zea mays* (Maize) Lehner (Corn); Turner (Millet). For the early herbalists of the 16th century believed that the plant had been brought by the Turks from Asia. The Turks invaded Europe about this time, and brought many new items into the west. Anything unusual was labelled "Turkish". Perhaps that just meant 'foreign', anyway, just as 'French' still does.

TURKISH HAZEL *Corylus colurna*. A native of Asia Minor, but it also grows in southern Europe. Cf. Constantinople Hazel.

TURKISH OAK *Quercus aegilops*.

TURKISH TOBACCO *Nicotiana rustica*. Nothing to do with Turkey - this is a South American species. Did the Turks develop a taste for its particularly pungent fumes?

TURMENTILL; TURMENTYNE *Potentilla erecta* (Tormentil) both Dawson.

TURMERIC *Curcuma longa*. 'Turmeric' is a garbled version of the French terre-merite, itself from medieval Latin terra merita, i.e. "proper earth".

TURMERIC, Long-rooted *Curcuma longa* (Turmeric) Thornton.

TURMET; TURMIT *Brassica campestris var. rapa* (Turnip) Wessex Rogers (Turmet); Som. Jennings; Hants. Cope; Lancs. Grose (Turmit).

TURMOP *Brassica campestris var. rapa* (Turnip) Yks. Robinson.

TURN-AGAIN-GENTLEMAN *Lilium martagon* (Turk's-cap Lily) Heref, Glos, Worcs, Bucks, N'thants. B & H. Because of the reverted petals, a description that also accounts for Turk's-cap and Turn-cap.

TURN-AGAIN-GENTLEMAN, Scarlet *Lilium chalcedonicum* (Scarlet Turk's-cap Lily) Coats.

TURN-MERICK *Curcuma longa* (Turmeric) A S Palmer. If Turmeric is a garbled version of terre-merite, this is an equally garbled version of the result.

TURNCAP *Lilium martagon* (Turk's-cap Lily) B & H. A more approachable description than Turk's-cap itself.

TURNHOOF *Glechoma hederacea* (Ground Ivy) Culpepper. 'Turn' is more likely to be 'tun' in this name. Cf. Tun-hoof and Tun-foot, which are the same as 'alehoof', that which will cause ale to heave, or work.

TURNIP 1. *Brassica campestris var. rapa*. 2. *Sinapis arvensis* (Charlock) Bucks. B & H. The second syllable is the operative one - it is OE naep, eventually from Latin napus. The first syllable, which may be 'turn', or from the French tour, implying roundness, is still often omitted.

TURNIP, Angel's *Apocynum androsaemifolium* (Spreading Dogbane) S USA Puckett.

TURNIP, Cabbage *Brassica oleracea var. caulo-rapa* (Kohlrabi) Schery. Probably just a translation of Kohlrabi, the German name, from Italian cavoli-rape, cabbage turnips (Latin caulis, rapa).

TURNIP, Devil's *Bryonia dioica* (White Bryony) North. 'Turnip', because of the large root, and 'devil's' because the plant is poisonous. It is navet du diable in French, too.

TURNIP, Indian 1. *Arisaema triphyllum* (Jack-in-the-pulpit) USA Kingsbury. 2. *Psoralea tenuiflora* USA Kingsbury. 3. *Terminalia catappa* (Myrobalan) Douglas. North American Indian for the first two, but Myrobalan is an Asiatic Indian plant.

TURNIP, Pepper *Arisaema triphyllum* (Jack-in-the-pulpit) Le Strange. 'Pepper', presumably, because of the acrid corms, which are perfectly edible after boiling.

TURNIP, Prairie *Psoralea esculenta* (Breadroot) USA Havard. They have turnip-shaped roots, rich in starch, and are also known as Prairie Potato.

TURNIP, Russian *Brassica napus var. napobrassica* (Swede) Bianchini. Cf. Swede Turnip and Russian Turnip. Actually it is neither Swedish nor Russian. The point of origin was Bohemia, in the 17th century.

TURNIP, St Anthony's *Ranunculus bulbosus* (Bulbous Buttercup) Gerard. This is the Sancti Antonii napus quoted by Gubernatis. 'Turnip', from the bulb-like swollen base of the stem.

TURNIP, Swede *Brassica napus var. napobrassica* (Swede) USA Kingsbury. See Russian Turnip.

TURNIP, Swedish 1. *Brassica campestris var. rapa* (Turnip) USA Elmore. 2. *Brassica napus* (Rape) Clapham. 3. *Brassica napus var. napobrassica* (Swede) Bianchini.

TURNIP, Wild 1. *Arisaema triphyllum* (Jack-in-the-pulpit) USA Le Strange. 2. *Sinapis arvensis* (Charlock) Yks. Grigson.

TURNIP, Yellow *Brassica napus var. napobrassica* (Swede) USA Kingsbury.

TURNIP BROCCOLI, Italian *Brassica ruvo*.

TURNIP-ROOTED CELERY *Apium graveolens var. rapaceum* (Celeriac) Bianchini.

TURNIP-ROOTED CHERVIL *Chaerophyllum bulbosum*.

TURNIP-ROOTED PARSLEY *Petroselinum fusiformis* (Hamburg Parsley) Clair.

TURNIP-WEED *Rapistrum rugosum* (Bastard Cabbage) New Zealand Connor.

TURNSOLE 1. *Cichorium intybus* (Chicory) Fernie. 2. *Croton tinctoria* Cennini. 3. *Euphorbia helioscopia* (Sun Spurge) Prior. 4. *Heliotropum europaeum* (Cherry Pie) Grieve.1931. 'Turnsole' usually means turning the flowers towards the sun, and that applies, at least in common superstition, to the Sun Spurge and Cherry Pie. But the word also means a deep purple coloured dye.

Cennini used the Croton to achieve this for tinting paper, while Chicory seems to have a foot in both camps. It was known in the 15th century as Sponsa Solis. But the blue colour caused it to be thought of as the favourite, not of the sun only, but of the sky as well.

TURNSOLE, Indian *Heliotropum indicum* (Indian Heliotrope) Dalziel.

TURNSOLE, Peruvian *Heliotropum europaeum* (Cherry Pie) J Smith. In spite of the specific name, this is a Peruvian plant, introduced here via France in 1757.

TURPEEN *Sempervivum tectorum* (Houseleek) Ire. Logan. Cf. Tourpin. Evidently the same name occurs in the Scottish Hockerie-Topner.

TURPENTINE, Cyprus *Pistacia terebinthus* (Terebinth) RHS.

TURPENTINE-BRUSH *Aplopappus laricifolia*.

TURPENTINE TREE 1. *Copaifera mopane* (Mopane) Palgrave. 2. *Pistacia terebinthus* (Terebinth) Polunin.1976. 3. *Pistacia terebinthus var. palaestina* (Palestine Terebinth). Moldenke. Turpentine is the product obtained from incisions in the terebinth tree. This is the name given to the Palestine Terebinth in Ecclesiasticus 24 : 16.

TURPENTINE WEED *Gutierrezia microcephala* (Broomweed) USA Kingsbury.

TURR *Ulex europaeus* (Furze) B & H. Probably the same as Fur and Furra, which are East Anglian names.

TURRAIE *Grevillea striata*.

TURTLE DOVES 1. *Aconitum napellus* (Monkshood) Som. Macmillan. 2. *Digitalis purpurea* (Foxglove) Som. Macmillan. Cf., for Monkshood, Lavinia's Dove-carriage and its variants. It is claimed that the plant's nectaries look like doves.

TURTLEBLOOM; TURTLEHEAD *Chelone glabra* Turtlehead is the usual name. Turtlebloom appears in O P Brown.

TUSHALAN; TUSHY-LUCKY GOWAN *Tussilago farfara* (Coltsfoot) N'thum. B & H (Tushalan); Dumf. Grigson (Tushy-lucky Gowan). These are attempts at Tussilago, as are Dishalaga, etc., all from the same Northcountry and Border areas.

TUSKER, Swine's *Polygonum aviculare* (Knotgrass) Tynan & Maitland. There are many 'pig' names for knotgrass. 'Tusker' may either be the same as 'skir', as in Gerard's Swine's skir, or it may be 'tacker', or 'tucker' - "from its likeness to a "tacker", or shoemaker's wax end".

TUSSAC GRASS; TUSSOCK GRASS *Deschampsia cespitosa* (Tufted Hair-grass) both H C Long.1910.

TUTSAN 1. *Hypericum androsaemum*. 2. *Hypericum perforatum* (St John's Wort) Folkard. 3. *Vinca minor* (Lesser Periwinkle) Suff. Moor. Tutsan is apparently from the French toute saine, wholly sound, or healing all. Another derivation is from tout-sang, since the flowers leave a red, blood-like stain on the fingers if rubbed. This ties in with the specific name androsaemum - andros, man and aima, blood. There was a Hampshire superstition that tutsan berries originated by germination in the blood of slaughtered Danes. But by the doctrine of signatures, the plant was applied to all wounds.

TUTSAN, Goat-scented *Hypericum hircinum* (Stinking Tutsan) McClintock.1974.

TUTSAN, Stinking *Hypericum hircinum*.

TUTSAN, Tall *Hypericum elatum* (Tall St John's Wort) McClintock.1974.

TUTTIES *Prunus cerasus* flowers (Cherry blossom) Dor. Macmillan. A tutty is a posy.

TUTU *Coriaria sarmentosa*. A small New Zealand shrub, so presumably this is the Maori name.

TUTU, Tree *Coriaria arborea*.

TUTTY PEA *Lathyrus odoratus* (Sweet Pea) Som. Macmillan. A tutty is a posy.

TUZZY-MUZZY 1. *Arctium lappa* (Burdock) Dev, Som. Mabey; Wilts. Dartnell & Goddard.1899. 2. *Clematis vitalba* (Old Man's Beard) Glos. J D Robertson. The name is probably only used to describe the burrs of burdock, so the seed heads of Old Man's Beard should also be included in all probability.

TWADGER 1. *Vicia sativa* (Common Vetch) Yks. Grigson. 2. *Vicia sepium* (Bush Vetch) Yks. Grigson.
TWAYBLADE *Listera ovata*. 'Twayblade' is literally, two leaves. So there have been other similar book names given to it - Bifoil, and Double-leaf, for example.
TWAYBLADE, Broad-leaved *Listera convallarioides*.
TWAYBLADE, Heart-leaved *Listera cordata*.
TWAYBLADE, Lesser *Listera cordata* (Heart-leaved Twayblade) Clapham.
TWAYBLADE, Southern *Listera australis*. 'Southern', but this is a North American plant.
TWEENY-LEGS; TWINY-LEGS *Parentucellia viscosa* (Yellow Bartsia) both Dev. Grigson.
TWELVE APOSTLES *Pulmonaria officinalis* (Lungwort) Som. Tongue.1968.

The Twelve Apostles in the garden plot do grow,
Some be blue, some be red and others
                      white as snow,
They cure the ill of every man, whatever ill it be,
But Judas he was hanged on an elder tree.

That is from an old song called the Twelve Apostles.
TWELVE DISCIPLES 1. *Bellis perennis* (Daisy) Som. Macmillan. 2. *Passiflora caerulea* (Blue Passion Flower) Som. Macmillan. Cf. Christ-and-the-Apostles from Devon, and The Ten Commandments from Dorset for the Passion Flower. All these are part of the complex that saw in the flowers the symbols of Christ's Passion.
TWELVE-MONTH YAM *Dioscorea cayennensis* (Yellow Yam) Perry.1979. 'Twelve-month', because they take a year to mature.
TWELVE-O'CLOCK 1. *Abutilon americanum* W Miller. 2. *Anagallis arvensis* (Scarlet Pimpernel) Som. Macmillan. 3. *Ornithogalum umbellatum* (Star-of-Bethlehem) Som. Elworthy; Dor. Dacombe; Surr. B & H. 4. *Taraxacum officinale* (Dandelion) Mabey. 5. *Tragopogon pratensis* (Goat's Beard) Som. Macmillan. Usually, names like this refer to the early opening, or closing, of the flowers, and so it is with Pimpernel and Star-of-Bethlehem. Dandelion and Goat's Beard, though, get the name from their 'clocks', the seed heads that children puff at to tell the time.
TWICE-WRITHEN *Polygonum bistorta* (Bistort) Prior. i.e. literally, bistort, which is Latin bis, twice, and torta, twisted. The word is a reference to the writhed roots, hence too the various 'snake' names like Snakeweed or Adderwort.
TWICKBAND; TWICKBINE *Sorbus aucuparia* (Rowan) Hants. Cope (Twickband); Dev. Grigson (Twickbine). 'Twick' is a south country variation of 'Quick', which is from OE cuic, alive.
TWIG-WITHY *Salix viminalis* (Osier) Ches. B & H. This is a basket-making willow, hence the name.
TWIG-BEAN *Sorbus aucuparia* (Rowan) Suss. Parish. i.e. Twickband or Twickbine, obviously misunderstood at the time of recording.
TWIGGY GLASSWORT *Salicornia ramosissina*.
TWIGGY MULLEIN *Verbascum virgatum* (Large-flowered Mullein) Clapham.
TWIKE *Agropyron repens* (Couch-grass) Grigson.1959. O E cwice, cuic - alive, a very good name for such an indestructible plant. Cf. Quitch, or Squitch, and Twitch-grass.
TWILIGHT, Evening 1. *Anemone nemorosa* (Wood Anemone) Som, Dor. Macmillan. 2. *Oxalis acetosella* (Wood Sorrel) Dor. Macmillan.
TWIN-LEAF *Jeffersonia diphylla*.
TWINBERRY 1. *Lonicera involucrata* (Bearberry Honeysuckle) USA E Gunther. 2. *Mitchella repens* (Partridge-berry) Howes.
TWINE, Devil's 1. *Clematis vitalba* (Old Man's Beard) Skinner. 2. *Convolvulus arvensis* (Field Bindweed) N Eng. Wakelin. Obviously descriptive, and assigned to the

devil with enthusiasm as far as bindweed is concerned.

TWINE-GRASS *Vicia cracca* (Tufted Vetch) Herts. Grigson. 'Tine' is a common name for the vetches, and this is merely a variation on Tine-grass, which is also recorded in Herts.

TWINFLOWER 1. *Chimaphila maculata* (Umbellate Wintergreen) Maryland. Whitney & Bullock. 2. *Linnaea americana* USA House. 3. *Linnaea borealis*.

TWINKLE-STAR *Stellaria holostea* (Greater Stitchwort) Som. Macmillan. Cf. Star-flower, Starwort, and a good deal more such descriptive names.

TWINS *Anthyllis vulneraria* (Kidney Vetch) Yks. Grigson.

TWINSPUR *Diascea barberiae*.

TWINY-LEGS 1. *Odontites verna* (Red Bartsia) Dev. Grigson. 2. *Parentucellia viscosa* (Yellow Bartsia) Dev. Grigson.

TWISTWOOD *Viburnum lantana* (Wayfaring Tree) Hants. Cope. Explained with reference to whips - to be twisted into whips. Cf. also Whipcrop and Whiptop, from the same general area - "the branches are long, tough and easie to be bowed, and hard to be broken" (Gerard).

TWITCH, Don's *Agropyron donianum*.

TWITCH, Surface *Polygonum aviculare* (Knotgrass) H C Long.1910. 'Twitch' is couchgrass, so this describes the feeling about knotgrass very well.

TWITCH-GRASS *Agropyron repens* (Couch-grass) Som. Elworthy; N'thants. Clare; E Ang. Forby; Ess. Gepp. 'Twitch' is from OE cwice, alive, which gave various vernacular names, like Quitch or Squitch, Quick, and Wick and so on. "Couch" has the same derivation.

TWO-EYED BERRY *Mitchella repens* (Partridge-berry) USA Cunningham.

TWO-FACES-IN-A-HOOD *Viola tricolor* (Pansy) Turner.

TWO-FACES-UNDER-A-HAT *Aquilegia vulgaris* (Columbine) Suss. Wright.

TWO-FACES-UNDER-ONE-HAT *Viola tricolor* (Pansy) Heref. Leather.

TWO-FACES-UNDER-THE-SUN *Viola tricolor* (Pansy) Shrop. B & H.

TWO-FLOWERED NARCISSUS *Narcissus x biflorus* (Primrose Peerless) Hulme.

TWO-LEAF PIÑON *Pinus edulis* (Arizona Nut Pine) Elmore.

TWOPENCE, Herb 1. *Lysimachia nummularia* (Creeping Jenny) Turner. 2. *Lythrum salicaria* (Purple Loosestrife) Dev. Macmillan. Creeping Jenny has a number of similar names. Cf. Gerard's Moneywort, and the Irish Strings of Sovereigns. Of course, it is the leaves on the creeping stems that provide the imagery.

TWOPENNY GRASS *Lysimachia nummularia* (Creeping Jenny) Turner. See previous entry.

TWYBLADE *Listera ovata* (Twayblade) Hill. Is this genuine?

TYLE-BERRY *Jatropha multifida* (Coral Plant) West Indies Howes.

U

UGH; UGHE *Taxus baccata* (Yew) Lowe (Ugh); Turner (Ughe). Two of the many attempts at rendering the peculiar sound of 'yew".

UGLI *Citrus reticulata x paradisii*.

UGLY, Madam *Fritillaria meleagris*. Presumably from the sombre colouring of the flowers, but that is a matter of taste. Cf. Shy Widows and Sullen Ladies.

UINTJE, Water *Aponogeton distachyum* (Cape Asparagus) S Africa Hutchinson & Melville.

ULLIMER *Artemisia absinthium* (Wormwood) IOM Moore.1924. Manx ullymar.

ULLUM *Ulmus procera* (English Elm) Som. Macmillan. The Somerset version of Ellum.

ULMO *Eucryphia cordifolia*.

UMBEL *Cypripedium parviflorum var. pubescens* (Larger Yellow Lady's Slipper) O P Brown.

UMBRA TREE *Phytolacca heptandra*.

UMBRELLA 1. *Alisma plantago-aquatica* (Water Plantain) Som. Macmillan. 2. *Calystegia sepium* (Great Bindweed) Som. Macmillan. 3. *Petasites hybridus* (Butterbur) Som. Macmillan. 4. *Sambucus nigra* flowers (Elder flowers) Som. Macmillan. 5. *Umbilicus rupestris* (Wall Pennywort) Som. Macmillan. 6. *Vinca minor* (Lesser Periwinkle) Som. Macmillan. Either from the size of the leaves (Butterbur) or their shape (Wall Pennywort), or the shape of the flowers (Bindweed, or the clusters of Elder). Lesser Periwinkle and Water Plantain are less obvious.

UMBRELLA, Fairy's *Convolvulus arvensis* (Field Bindweed) Som. Macmillan.

UMBRELLA, Gypsy's *Anthriscus sylvestris* (Cow Parsley) Som. Grigson.

UMBRELLA, Lady's 1. *Calystegia sepium* (Great Bindweed) Som. Macmillan. 2. *Convolvulus arvensis* (Field Bindweed) Dor. Macmillan. 3. *Solanum dulcamara* (Woody Nightshade) Som. Macmillan. Woody Nightshade is the newcomer here. It is the flowers, of course, that provide the name.

UMBRELLA-LEAVES *Petasites hybridus* (Butterbur) Yks. B & H.

UMBRELLA PINE 1. *Pinus pinea* (Stone Pine) Watt. 2. *Sciadophys verticillata*. Stone Pines look like umbrellas - the French think so, too - Cf. their name Pin Parasol. So, of course, do *Sciadophys*, which comes from Greek skiados, umbel and pitys, pine tree.

UMBRELLA PLANT 1. *Cyperus involucratus*. 2. *Eriogonum fasciculata* (Wild Buckwheat) USA Youngken. 3. *Petasites hybridus* (Butterbur) Som. Macmillan. 4. *Saxifraga peltata* (Indian Rhubarb) Chaplin. Probably the size of the leaves provides the imagery.

UMBRELLA THORN 1. *Acacia sieberiana*. 2. *Acacia tortilis*.

UMBRELLA TREE 1. *Cussonia kirkii* (Cabbage Tree) Palgrave. 2. *Magnolia tripetala* Wit. 3. *Musanga cecropioides*.

UMBRELLA TREE, Queensland *Brassaia actinophylla*.

UMPLESCRUMP *Heracleum sphondyllium* (Hogweed) Som. Macmillan. The hollow stems of Hogweed had their uses. You could, for instance, drink your cider through them. Hence this name, or Lumper-scrump, and quite a few more, equally picturesque.

UNDERGROUND IVY 1. *Cymbalaria muralis* (Ivy-leaved Toadflax) Som. Macmillan. 2. *Glechoma hederacea* (Ground Ivy) Som. Macmillan.

UNDERGROUND NUT *Conopodium majus* (Earth-nut) Corn. Grigson; Dev. Choape; Som. Macmillan.

UNDERGROUND SHEPHERD *Orchis mascula* (Early Purple Orchid) Wilts. Jones & Dillon.

UNGLE-PIGLE *Stellaria holostea* (Greater Stitchwort) Dawson. 'Pigle' is 'paigle', a common name for the cowslip, but both Paigle and Pigle are recorded for this stitchwort.

UNICORN PLANT *Proboscidea louisiana*. The generic name says it all, but it has been likened to a claw, as well.

UNICORN ROOT *Aletris farinosa*.

UNSHOE-THE-HORSE 1. *Hippocrepis comosa* (Horseshoe Vetch) Prior. 2. *Lunaria annua* (Honesty) Culpepper. Horseshoe Vetch is so named because of the shape of the pods. It had the reputation of being able to unshoe any horse that trod on it. Very possibly, the real reason may be that the plant grows in the sort of stony ground that is apt to lead to accidents. Honesty earns the name because of its ancient magical fame as a picklock. So great was its power that it could draw the nails from horses' hoofs.

UNTRODDEN-TO-DEATH; UNTRODDEN-TO-PIECES *Polygonum aviculare* (Knotgrass) both Cockayne.

UPAS CLIMBER *Strychnos ovalifolia*.

UPAS TREE *Antiaris toxicaria*. Upas is a Malayan word for poison. Legend has it that the Upas Tree blights all other vegetation around it, and is so poisonous that birds that fly over it drop dead, and any animal or man venturing into its shade is killed, so that the ground beneath the tree is covered with bones. The truth is not quite so dramatic, but it is actually extremely poisonous, or at least its latex is.

UPLAND BURNET *Poterium sanguisorba* (Salad Burnet) Tynan & Maitland.

UPLAND COFFEE *Coffea stenophylla* (Narrow-leaved Coffee) Dalziel.

UPLAND COTTON *Gossypium punctatum*.

UPLAND CRANBERRY *Arctostaphylos uva-ursi* (Bearberry) O P Brown.

UPLAND SUMACH *Rhus glabra* (Smooth Sumach) Grieve.1931.

UPSTART *Colchicum autumnale* (Meadow Saffron) Prior. Because the flowers shoot up before the leaves appear.

URCHIN, Ragged *Lychnis flos-cuculi* (Ragged Robin) Dev. Macmillan.

URCHIN CROWFOOT *Ranunculus arvensis* (Corn Buttercup) B & H. The reference here is to the bristly seed capsules.

URCHINS *Fraxinus excelsior* seeds (Ash keys) Halliwell.

URD BEAN *Phaseolus mungo* (Black Gram) Schery.

URLES *Vicia sativa* (Common Vetch) B & H.

URN GUM *Eucalyptus urnigera*. 'Urn', from the shape of the fruits.

URN PLANT *Aechmea fasciata*. Descriptive of the tall flower head.

URTS *Vaccinium myrtillus* (Whortleberry) Dev. T Brown.1961. Whorts is a common name in the south of England, and this has a number of variants, including Hurts, or Urts, as it is more likely to be pronounced.

URUGUAY, Ivy of *Cissus striata*.

USELESS *Polygonum persicaria* (Persicaria) Scot. Grigson. The legend is that the Virgin pulled a root, left her mark on the leaf, and threw it aside, saying, "This is useless", and it has been useless ever since. Her "mark" accounts for the names Virgin's Pinch, Pinchweed, and Lady's Thumb.

UTAH BREADROOT *Psoralea mephitica*.

UTAH FIRECRACKER *Penstemon utahensis*.

UTAH JUNIPER 1. *Juniperus californica*. 2. *Juniperus osteosperma*.

UVULA-WORT *Campanula trachelium* (Nettle-leaved Bellflower) Gerard. "of the vertue it hath against the paine and swelling thereof". He also called it Throatwort and Haskwort for the same reason, and much earlier than that, it was known as Neckwort.

V

VAGABOND'S FRIEND *Polygonatum multiflorum* (Solomon's Seal) Cumb, West. Wright. Because gypsies and others made an ointment from the leaves to apply to a bruise or black eye.

VALAERIA; VALARA *Valeriana officinalis* (Valerian) IOM Moore.1924 (Valaeria); Cumb, Donegal Grigson (Valara).

VALAIS CATCHFLY *Silene vallesia*.

VALERIAN 1. *Cypripedium acaule* (Stemless Lady's Slipper) USA Bergen. 2. *Polemonium caeruleum* (Jacob's Ladder) Turner. 3. *Valeriana officinalis*. The name valerian derives from Latin valere, to be well, or to be in good health, but the plant's main medicinal use has been as a nervine in one form or another. The fact that the Lady's Slipper has also been used as a nervine in America earns it the name there. Cats are very fond of valerian, and so they are of Jacob's Ladder, which may explain the use of the name in the latter case.

VALERIAN, American 1. *Cypripedium parviflorum* (Small Yellow Lady's Slipper) USA Sanford. 2. *Cypripedium parviflorum var. pubescens* (Larger Yellow. Lady's Slipper) USA Lloyd. Both used in American domestic medicine as nervines, hence the name.

VALERIAN, Cat's *Valeriana officinalis* (Valerian) Howes.

VALERIAN, Edible *Valeriana edulis*. The American Indians found them to be edible. They cooked the roots, to make them into soup or bread, and the seeds too were used as food. Nowadays people are not so sure.

VALERIAN, False *Senecio aureus* (Swamp Squaw-weed) O P Brown. Evidently given for medicinal reasons.

VALERIAN, Giant *Valeriana pyrenaica*.

VALERIAN, Greek 1. *Polemonium caeruleum* (Jacob's Ladder) Gerard. 2. *Polemonium reptans* (Abscess Root) USA House.

VALERIAN, Greek, American *Polemonium reptans* (Abscess Root) Hatfield. This must be one of the best examples of nonsense names!

VALERIAN, Indian *Nardostachys jatamansi* (Spikenard) Leyel.1937. Oil of spikenard is produced by distillation of the root. Oil from valerian is sometimes substituted for the true spikenard.

VALERIAN, Little *Valerianella locusta* (Cornsalad) Leyel.1937. Valerianella = little valerian.

VALERIAN, Marsh *Valeriana dioica*.

VALERIAN, Pyrenean *Valeriana pyrenaica* (Giant Valerian) Clapham.

VALERIAN, Red *Centranthus ruber* (Spur Valerian) Hudson.

VALERIAN, Spur *Centranthus ruber*. The specific name, *Centranthus*, comes from Greek kentron, spur and tanthos, flower. The corolla has a spur half as long as itself.

VALERIAN, Swamp *Valeriana uliginosa*.

VALLEY, Lily-of-the- *Convallaria maialis*.

VALLEY LILY; VALLEYS *Convallaria maialis* (Lily-of-the-valley) Gerard; Sanecki (Valleys).

VALONIA OAK *Quercus macrolepis*. Valonia, or Valona, is an Albanian port, known these days as Vlóra, from which the large acorns were exported, to be used for their tannin in the preparation of leather. Both the port and the acorn seem to be derived from the same Greek word for acorn - balanos.

VANILLA *Vanilla planifolia*.

VANILLA, Wild, American *Liatris odoratissima* (Deer's Tongue) Mitton.

VANILLA ORCHID, Black *Nigritella nigra*.

VAPOUR *Fumaria officinalis* (Fumitory) Hatfield. The reference is to the derivation of Fumitory, which is Old French fumeterre, medieval Latin fumus terrae, i.e. smoke of the earth. One reason is the belief that it did not spring up from seeds, but from the vapours of the earth. The root when fresh pulled gives a strong gaseous smell like nitric acid, and this is probably the origin of the belief in its gaseous origins.

VARDEN, Dolly *Tephrosia virginiana* (Devil's Shoestring) Howes.

VARENUT; FARENUT *Conopodium majus* (Earth-nut) Corn. Quiller-Couch (Varenut); Jago (Farenut). Cf. Hognut, another west country name. Fare is a young pig (OE fearh).

VARGES *Malus sylvestris* fruit (Crabapple) War. Palmer. Bad beer was said to be "sour as varges". No wonder - varges is actually verjuice,the notoriously sour liquid extracted from crabs. Cf. the Cheshire Wharre - "sour as wharre".

VARNISH TREE 1. *Aleurites molucanna* (Candle-nut Tree) A W Smith. 2. *Rhus verniciflua* (Lacquer-Tree) Le Strange. The oily kernels of the Candle-nut Tree are put to various uses, one being the making of varnish. Another is for making candles, hence the common name. *Rhus verniciflua* provides the raw material from which Chinese and Japanese lacquer is made.

VARNISH TREE, Japan *Ailanthus altissima* Chopra.

VARVINE *Verbena officinalis* (Vervain) IOM Gill.1963.

VASES *Geranium molle* (Dove's-foot Cranesbill) Som. Macmillan.

VATCH *Vicia sativa* (Common Vetch) Glos. Grigson.

VAUDOIS SALLOW *Salix vaudensis*.

VEGETABLE CALOMEL *Acorus calamus* (Sweet Flag) Watt. Calomel is mercurous chloride, so to liken the medicinal effects with those of Sweet Flag would necessitate the addition of 'vegetable'. However, Sweet Calomel is also used for the plant.

VEGETABLE GOLD *Coptis trifoliata* (Goldthread) Howes.

VEGETABLE MARROW *Cucurbita pepo*.

VEGETABLE PEAR *Sechium edule* (Chayote) Howes.

VEGETABLE SPONGE *Luffa cylindrica* (Loofah) Chopra.

VEIL, Widow *Fritillaria meleagris* (Snake's-head Lily) Coats. Cf. Widow-wail. The sombre colouring of the flowers accounts for the name, and for other "sorrowful" ones.

VEINED VERVAIN *Verbena venosa*.

VELANIDI OAK *Quercus macrolepis* (Valonia Oak) Argenti & Rose.

VELDT, Monarch of the *Venidium fastuosum* (Namaqualand Daisy) Hay & Synge.

VELDT, Star of the *Dimorphoteca aurantiaca*.

VELURE, Flower *Amaranthus caudatus* (Love-lies-bleeding). Cf. Velvet-flower, etc.

VELURE, Flower, Golden *Amaranthus luteus*.

VELVET, Vining *Ruellia makoyana* USA Cunningham.

VELVET ASH *Fraxinus velutina*.

VELVET BEAN *Stizolobium deeringianum*.

VELVET BEAN, Florida 1. *Mucuna pruriens* (Cowage) Willis. 2. *Stizolobium deeringianum* (Velvet Bean) Schery.

VELVET BELLS *Bartsia alpina* (Alpine Bartsia) H & P.

VELVET BENT *Agrostis canina* (Brown Bent) Howes.

VELVET DOCK 1. *Inula helenium* (Elecampane) IOW Grigson. 2. *Verbascum thapsus* (Mullein) Dev. Choape; Som. Prior; USA Henkel. 'Dock' for the large leaves, and 'velvet' for their texture, though in the case of mullein they are more usually referred to as flannel.

VELVET FLOWER 1. *Amaranthus caudatus* (Love-lies-bleeding). 2. *Amaranthus hypondriachus* (Prince's Feathers) Mitton. 3. *Sparaxis tricolor* S Africa Perry. 4. *Tagetes patula* (French Marigold) Turner.

VELVET FLOWER, Purple *Amaranthus caudatus* (Love-lies-bleeding) Turner.

VELVET FLOWER-DE-LUCE *Hermodactylus tuberosum* (Snakeshead Iris) Gerard.

VELVET GRASS *Holcus lanatus* (Yorkshire Fog) USA T H Kearney.

VELVET GRASS, German *Holcus mollis* (Creeping Soft-grass) USA Allan. 'German', because it is a fairly recent introduction into America from Europe.

VELVET LEAF 1. *Abutilon avicennae* (China Jute) USA Gates. 2. *Lavatera arborea* (Tree Mallow) Prior. 3. *Verbascum thapsus* (Mullein) USA Henkel.

VELVET MULLEIN; VELVET PLANT *Verbascum thapsus* (Mullein) both USA House (Mullein); Henkel (Plant).

VELVET POPPY *Verbascum thapsus* (Mullein) Corn. Grigson.

VELVET SUMACH *Rhus typhina* (Stag's Horn Sumach) Grieve.1931. "...the young branches are covered with a soft, velvet-like down, resembling greatly that of a young stag's horn, from whence the common people have given it the name of stag's horn".

VELVET TRUMPET FLOWER *Salpiglossis sinuata*.

VELVET WEED *Abutilon avicennae* (China Jute) Usher.

VELVET WILLIAM *Dianthus barbatus* (Sweet William) Coats.

VELVET WILLOW *Salix sitchensis*.

VENETIAN SUMACH *Cotinus coggyria*. Cf. Venus's Sumach, which is probably this name corrupted.

VENICE MALLOW *Hibiscus trionum* (Bladder Katmia) Coats. But its home is Central Africa, or at least the Old World tropics.

VENUS, Mirror of *Ophrys speculum* (Mirror Orchid) Polunin.

VENUS FLYTRAP *Dionaea muscifera*. One of the best known of the carnivorous plants.

VENUS HAIR *Adiantum capillis-veneris* (Maidenhair Fern) Turner. A direct translation of the specific name.

VENUS-IN-HER-CAR *Scrophularia aquatica* (Water Betony) Som. Macmillan. Cf. Babes-in-the-cradle.

VENUS TREE *Achillea millefolium* (Yarrow) McDonald.

> Thou pretty herb of Venus's Tree,
> Thy true name it is yarrow;
> Now who my bosom friend may be.
> Pray tell thou me tomorrow.

There are many other love divinations played with yarrow, enough to justify a name like this.

VENUS'S BASIN; VENUS'S BATH *Dipsacus fullonum* (Teasel) Gerard (Basin); B & H (Bath). Cf. Our Lady's Basin. The reference is to the cups formed by the fusing together of the plant's opposite leaves, "so fastened that they hold dew and raine water in manner of a little bason" (Gerard). The water was much prized for cosmetic use, hence the ascription to Venus.

VENUS'S COMB *Scandix pecten-veneris* (Shepherd's Needle) Gerard. A direct translation of the specific name. In both Greek and Latin (pecten), comb signifies the sexual organ, usually the female one. Hence, presumably, the connection with Venus.

VENUS'S CHARIOT; VENUS'S CHARIOT DRAWN BY DOVES *Aconitum napellus* (Monkshood) S Africa Watt (the former); Dev, Berks, Ess. Grigson (the latter). These names and a number of others derive, it is claimed, from a comparison of the nectaries with doves. Cf. Doves-in-the-ark, Lady Lavinia's Dove Carriage, etc.

VENUS'S LOOKING GLASS 1. *Legousia hybrida*. 2. *Lunaria annua* (Honesty) Som. Macmillan. *Legousia hybrida* gets the name from the shining seeds, "like brilliantly polished brass mirrors", in Richard Mabey's words. Honesty obviously gets the name from the general appearance of the capsules.

VENUS'S NAVELWORT 1. *Omphalodes linifolia*. 2. *Umbilicus rupestris* (Wall Pennywort) B & H. The leaves, with their central dimple, provide the origin of the name, and a number of variations all stem from the medieval 'umbilicus veneris'.

VENUS'S NEEDLE *Scandix pecten-veneris* (Shepherd's Needle) B & H. The 'needle' is the spiky looking seed capsule. Cf. Venus's Comb.

VENUS'S PRIDE, Blue *Houstonia caerulea* (Bluets) W Miller.

VENUS'S SUMACH *Cotinus coggyria* (Venetian Sumach) W Miller. Probably, 'Venetian' is meant here instead of 'Venus's'.

VERBENA, Lemon 1. *Hedeoma drummondii* (Drummond Pennyroyal) USA Elmore. 2. *Lippia citriodora*.

VERLOT'S MUGWORT *Artemisia verlotorum* (Chinese Mugwort) Genders.

VERNAL GRASS, Sweet *Anthoxanthum odoratum*.

VERNAL SANDWORT *Minuartia verna* (Spring Sandwort) Condry.

VERONICA, Hedge *Hebe x franciscana*.

VERONICA, Hedge, Narrow-leaved *Hebe salicifolia*.

VERSAILLES LAUREL *Prunus laurocerasus* (Cherry Laurel) Howes.

VERVAIN *Verbena officinalis*. 'Vervain' is clearly from Old French verveine, and thence from Latin verbena, which is also adopted as the generic name. In spite of the clarity of this derivation, attempts have been made to connect it with two Celtic words, fer, to drive away, and faen, a stone (it has long been used for stone in the bladder).

VERVAIN, American *Verbena hastata* (Blue Vervain) Howes.

VERVAIN, Base *Veronica chamaedrys* (Germander Speedwell) B & H. 'Base' here means low-growing. Cf. Flat Vervain.

VERVAIN, Blue *Verbena hastata*.

VERVAIN, Burry *Verbena lappulacea*.

VERVAIN, Canada *Verbena canadensis*.

VERVAIN, False *Verbena hastata* (Blue Vervain) USA House.

VERVAIN, Flat *Veronica chamaedrys* (Germander Speedwell) B & H. Cf. Base Vervain.

VERVAIN, Hoary *Verbena stricta*.

VERVAIN, Holy *Verbena officinalis* (Vervain) Northall. The Romans gave the name verbena, or more frequently the plural, verbenae, to the foliage or branches of shrubs and herbs which for their religious association, had acquired a sacred character. Hence this name, or Holy Herb. Cf. the French herbe sacrée.

VERVAIN, Jamaica *Verbena jamaicensis*.

VERVAIN, Prostrate *Verbena bracteata*.

VERVAIN, South American *Verbena bonariensis*.

VERVAIN, Upright *Verbena stricta* (Hoary Vervain) USA Elmore.

VERVAIN, Veined *Verbena venosa*.

VERVAIN, White *Verbena urticaefolia*.

VERVAIN, Wild *Verbena hastata* (Blue Vervain) USA Youngken.

VERVAIN SAGE *Salvia horminoides* (Wild Sage) Clair. An earlier specific name for this was *verbenacea*.

VERVAIN THOROUGHWORT *Eupatorium verbenaefolium* (Rough Thoroughwort) USA House.

VESPER FLOWER *Hesperis matronalis* (Sweet Rocket) Grieve.1931. Another name, Night-smelling Rocket, explains this as well as Hill's invention, Eveweed.

VESPER IRIS *Iris dichotoma* (Afternoon Iris).

VETCH, American *Vicia americana*. 'Vetch' comes through Norman French from the Latin, vicia, still used as the generic name.

VETCH, Bithynian *Vicia bithynica*. Bithynia is the part of Asia Minor opposite Istanbul. The plant is said to be native there, but it is to be found, albeit rarely, in this country.

VETCH, Bitter *Vicia orobus*.

VETCH, Black *Vicia melanops*.

VETCH, Bush 1. *Vicia cracca* (Tufted Vetch) Salisbury. 2. *Vicia sepium*.

VETCH, Chichling *Vicia lathyroides* (Spring Vetch) Camden/Gibson.

VETCH, Common *Vicia sativa*.

VETCH, Cow *Vicia cracca* (Tufted Vetch) Glos. J D Robertson.

VETCH, Creamy *Lathyrus ochroleucus* (White Sweet Pea) USA H Smith.

VETCH, Crown *Coronilla varia*. The generic name, *Coronilla*, means 'little crown'.

VETCH, Fodder *Vicia villosa* (Lesser Tufted Vetch) Blamey.

VETCH, Four-seeded *Vicia tetrasperma* (Smooth Tare) Hutchinson.

VETCH, Giant *Vicia gigantea*.

VETCH, Grass *Lathyrus nissolia* (Grass Pea) Curtis.
VETCH, Hairy 1. *Vicia hirsuta* (Hairy Tare) Hutchinson. 2. *Vicia villosa* (Lesser Tufted Vetch) USA Kingsbury.
VETCH, Hedge *Vicia sepium* (Bush Vetch) Hutchinson.
VETCH, Heath *Lathyrus montanus* (Bittervetch) Murdoch McNeill.
VETCH, Horseshoe *Hippocrepis comosa*. So called from the shape of the pods. Besides giving the name, the shape also gave rise to its reputation of being able to unshoe any horse that trod on it. But the reason may very well be its habitat - the sort of stony ground that leads to accidents.
VETCH, Italian *Galega officinalis* (Goat's Rue) Parkinson.
VETCH, Kidney *Anthyllis vulneraria*. Because it was used for kidney troubles.
VETCH, Liquorice *Astragalus glycyphyllus* (Milk Vetch) B & H. The specific name tells that this has leaves like those of the liquorice plant.
VETCH, Milk *Astragalus glycyphyllus*. "The seeds", said Gerard, "...ingender store of milke".
VETCH, Milk, Alpine *Astragalus alpinus*.
VETCH, Milk, Canadian *Astragalus carolinianus*.
VETCH, Milk, Mountain, Purple *Oxytropus halleri*.
VETCH, Milk, Purple *Astragalus danicus*.
VETCH, Milk, Sweet *Astragalus glycyphyllus* (Milk Vetch) D Grose.
VETCH, Milk, Yellow *Oxytropus campestris*.
VETCH, Narrow-leaved *Vicia angustifolia*.
VETCH, Pebble *Vicia sativa* (Common Vetch) B & H. An old cultivated variety. Lisle's Husbandry, 1757, has "The pebble-vetch is a summer-vetch, different from the goar-vetch, and not so big".
VETCH, Russian; VETCH, Siberian *Vicia villosa* (Lesser Tufted Vetch) Howes.
VETCH, Spring 1. *Vicia lathyroides*. 2. *Vicia sativa* (Common Vetch) USA E Gunther.
VETCH, Strangler *Vicia hirsuta* (Hairy Tare) Blamey.1980. Cf. Strangler Tare.
VETCH, Sweet *Hedysarum boreale*.
VETCH, Tar *Vicia sativa* (Common Vetch) Som. Grigson. 'Tar' is 'tare', an equally common name for the vetches.
VETCH, Tare 1. *Vicia cracca* (Tufted Vetch) Som, Dor, IOW Grigson. 2. *Vicia hirsuta* (Hairy Tare) Dor, IOW, S Eng. Grigson.
VETCH, Tufted *Vicia cracca*.
VETCH, Tufted, Lesser *Vicia villosa*.
VETCH, Upright *Vicia orobus* (Bitter Vetch) McClintock.
VETCH, Wild *Vicia cracca* (Tufted Vetch) Oxf, Cumb. Grigson.
VETCH, Winter *Vicia villosa* (Lesser Tufted Vetch) USA Kingsbury.
VETCH, Wood *Vicia sylvatica*.
VETCH, Yellow *Vicia lutea*.
VETCH, Yellow, Rough-podded *Vicia lutea* (Yellow Vetch) Flower.1851.
VETCH GRASS *Lathyrus nissolia* (Grass Pea) B & H.
VETCHLING, Bitter *Lathyrus montanus* (Bittervetch) Blamey.
VETCHLING, Creamy *Lathyrus ochroleucus* (White Sweet Pea) USA Yarnell.
VETCHLING, Crimson *Lathyrus nissolia* (Grass Pea) Flower.1851.
VETCHLING, Grass *Lathyrus nissolia* (Grass Pea) McClintock.
VETCHLING, Hairy *Lathyrus hirsutus* (Hairy Pea) Clapham.
VETCHLING, Marsh *Lathyrus palustris* (Marsh Pea) USA H Smith.
VETCHLING, Meadow *Lathyrus pratensis* (Yellow Pea) Curtis.
VETCHLING, Medick *Onobrychis sativa* (Sainfoin) Camden/Gibson.
VETCHLING, Narrow-leaved *Lathyrus sylvestris* (Everlasting Pea) Curtis.
VETCHLING, Nissol's *Lathyrus nissolia* (Grass Pea) Flower.1851.
VETCHLING, Wild; VETCHLING, Wood *Lathyrus sylvestris* (Everlasting Pea) both Flower.1851.
VETCHLING, Yellow *Lathyrus aphaca*.

VETHER-VO; VETHERVOW; VETHER-VAW *Tanacetum parthenium* (Feverfew) First two Som. Macmillan (Vether-vo); Elworthy (Vethervow); Dev. Hewett.1892 (Vethervaw). A variation of feverfew, which is pronounced in all sorts of ways through the west country. The immediate forebear of these is probably Featherfoe.

VEVINE *Verbena officinalis* (Vervain) IOM Moore.1924.

VEW; VEWE *Taxus baccata* (Yew) Derb, Lancs. Grigson; Ches. B & H; Yks. Hunter (Vew): Ches. Lowe (Vewe). Written to accomodate local pronunciation of 'yew'. It even appears as 'View'.

VICAR'S TRESSES *Calystegia sepium* (Great Bindweed). Reported in a letter to The Times in August 1989, and so called, the correspondent claimed, "because the poor man was always far too occupied on the sabbath to clear his own garden".

VICTOR-NUT *Corylus avellana* nut (Hazel nut) Corn. Jago. Cob-nuts used to be a game played with these nuts, much like conkers. The name cob-nut was reserved for the winner. Hence this name, and also the Devonshire Crack-nut.

VICTORIA BLOODWOOD *Eucalyptus corymbosa*.

VICTORIAN BOX *Pittosporum undulatum*.

VICTUALS, Cuckoo's 1. *Geranium robertianum* (Herb Robert) Bucks. Grigson. 2. *Oxalis acetosella* (Wood Sorrel) Dor, Bucks. Grigson; Glos. J D Robertson. 3. *Stellaria holostea* (Greater Stitchwort) Bucks. Grigson. Cf. Cuckoo's Meat for these.

VICTUALS, Rabbit's *Sonchus oleraceus* (Sow Thistle) Som. Grigson. Cf. Rabbit's Meat. Do rabbits eat them? Cowper wrote that his pet hares were very fond of them.

VICTUALS, Snake's 1. *Arum maculatum* fruit (Cuckoo-pint) Wilts. Jefferies. Glos. B & H. 2. *Mercurialis perennis* (Dog's Mercury) Suss. Grigson. Both of these are poisonous, hence the reference to snakes.

VICTUALS, Viper's *Arum maculatum* (Cuckoo-pint) Hants. Hants FWI.

VIEW *Taxus baccata* (Yew) Ches. B & H; Yks. Hunter. Cf. Vew, or Vewe.

VILIP *Viola odorata* (Sweet Violet) Dor. B & H. Very occasionally written as Vylip. They are both, of course, dialectal variations of Violet.

VILLERA *Valeriana officinalis* (Valerian) Ire. B & H.

VINE *Bryonia dioica* (White Bryony) IOW Grigson. Cf. Hedge, or Wild, Vine.

VINE, Allegheny *Adluaria fungosa*.

VINE, Apricot *Passiflora incarnata* (Maypops) USA Cunningham.

VINE, Arrow *Polygonum sagitattum* (Tearthumb) USA Cunningham. 'Arrow' - the shape of the leaves. Cf. Arrow-leaved Tearweed.

VINE, Balloon *Cardiospermum halicacabun* (Blister Creeper) USA Cunningham. Presumably a reference, as with the common name, to the highly irritant sap.

VINE, Bead *Abrus precatorius* (Precatory Bean) Roys. The "beads" are the seeds, sometimes in use as components of a rosary.

VINE, Black *Tamus communis* (Black Bryony) Pomet.

VINE, Bleeding Heart *Clerodendron thomsonae*. The flowers are cream and blood-red, hence this name.

VINE, Blood *Chamaenerion angustifolium* (Rosebay) Hants. Cope.

VINE, Blue *Clitoria ternata* (Butterfly Pea) Howes.

VINE, Bower *Pandorea jasminoides*.

VINE, Chalice *Solandra maxima*.

VINE, Coral *Antigonon leptopus*.

VINE, Condor *Marsdenia condurango* (Eagle Vine) Le Strange.

VINE, Cross *Bignonia capreolata*.

VINE, Cypress *Ipomaea quamoclit*.

VINE, Deer *Linnaea borealis* var. *americana* USA House.

VINE, Devil's *Calystegia sepium* (Great Bindweed) Salisbury. It is consigned to the devil on a number of counts - Cf. Devil's Garter, Guts, or Nightcaps.

VINE, Eagle *Marsdenia condurango*. Cf. Condor Vine for this.

VINE, Flame *Pyrostegia venusta* (Chinese Cracker Flower) Simmons. Orange flowers.

VINE, Glory, Chilean *Eccremocarpus scaber* (Glory Flower) Phillips.

VINE, Grape *Vitis vinifera*.

VINE, Heart *Ceropegia woodii* (Rosary Vine) RHS.

VINE, Hearts-and-honey *Ipomaea x multifida* (Cardinal Climber) RHS.

VINE, Hedge 1. *Bryonia dioica* (White Bryony) Hulme. 2. *Clematis vitalba* (Old Man's Beard) Turner.

VINE, Herb of *Asperula cynanchica* (Squinancywort) Gerard.

VINE, Hupeh *Actinidia chinensis* (Chinese Gooseberry) Douglas.

VINE, Indigo *Lonchocarpus cyanescens*. This is the Yoruba wild indigo, an indican-bearing plant, used in Nigeria for a blue dye.

VINE, Ink *Passiflora suberosa* Barbados Gooding.

VINE, Irish *Lonicera periclymenum* (Honeysuckle) Ire. Grigson. Presumably for the same reason as that elms used to be known as Wiltshire Weeds in Wiltshire.

VINE, Isle of Wight 1. *Bryonia dioica* (White Bryony) IOW B & H. 2. *Tamus communis* (Black Bryony) W Miller. Are they noticeably abundant there?

VINE, Jade *Strongylodon macrobotrys*. It has wistaria-like flowers of jade-green colour.

VINE, Junction *Aristolochia odoratissima* (Sweet-scented Birthwort) Barbados Gooding.

VINE, Kangaroo *Cissus antartica*. Australian, obviously.

VINE, Love 1. *Cuscuta americana* Barbados Gooding. 2. *Cuscuta compacta* USA Bergen. 3. *Cuscuta gronovii* (Gronovius's Dodder) USA Tolstead. 4. *Cuscuta indecora* Barbados Gooding. 5. *Cuscuta paradoxa* USA Gilmore. These American dodders are used for a love charm, of a similar kind to the well-known apple paring divination. One from Tennessee requires you to break off a piece of a love vine, trail it round the head three times, and drop it on a bush behind you. If it grows, the lover is true; if not, false.

VINE, Madeira *Anredera cordifolia*.

VINE, Marbles *Dioclea reflexa*. It grows in West Africa on sandbanks and seashore, and the seeds are used by children all over the area in a game played like marbles.

VINE, Matrimony 1. *Lycium halimifolium* (Duke of Argyll's Tea Plant) USA. Kingsbury. 2. *Lycium pallidum* (Rabbit Thorn) USA Elmore.

VINE, Matrimony, Chinese *Lycium chinense* (Tea Tree) Howes.

VINE, Mignonette *Anredera cordifolia* (Madeira Vine) RHS.

VINE, Monkey *Ipomaea nil*.

VINE, Moon-flower *Ipomaea bona-nox*. The white flowers open late in the evening, hence the reference to the moon in this and other names.

VINE, Natal *Cissus rhombifolia* Nicolaisen.

VINE, Orange *Eustrephus latifolius*.

VINE, Our Lady's *Tamus communis* (Black Bryony) Tynan & Maitland. Cf. Lady's Seal and Lady's Signet for this.

VINE, Pipe *Aristolochia sipho* (Dutchman's Pipe) W Miller. From the shape of the tube of the flower, like the stem of a meerschaum pipe.

VINE, Poison *Rhus radicans* (Poison Ivy) USA Kingsbury. This is the earlier name.

VINE, Pop; POPS *Physalis pubescens* (Husk Tomato) Barbados Gooding.

VINE, Potato *Solanum jasminoides*.

VINE, Potato, Wild *Ipomaea pandurata*.

VINE, Puncture *Tribulus terrestris*. 'Puncture' because of the spiny fruits, which can puncture bicycle tyres if wheeled over land where the plant is prevalent. It is in fact a pest for both men and animals.

VINE, Rosary *Ceropegia woodi*. It has hanging strings of tiny, heart-shaped leaves, hence this name and also String-of-hearts.

VINE, Rubber *Cryptostegia grandiflora*. It is a source of rubber, though probably not very important.

VINE, Russian *Polygonum baldschuanicum*.
VINE, Sandpaper *Petraea volubilis* (Purple Wreath Vine) USA Perry.
VINE, Silk *Periploca graeca*.
VINE, Silver. 1, *Actinidia polygama*. 2. *Scindapsus pictus 'argyraeus'* (Devil's Ivy) Perry.1979.
VINE, Snake *Hibbertia scandens*.
VINE, Squaw *Mitchella repens* (Partridgeberry) USA Mitton. American Indian women used it for several weeks before childbirth to ensure a safe and easy delivery.
VINE, Staff 1. *Celastrus orbiculatus* (Oriental Bittersweet) RHS. 2. *Celastrus scandens* (American Bittersweet) W Miller. Staff Tree is also recorded.
VINE, Starflower *Petraea volubilis* (Purple Wreath Vine) Simmons.
VINE, Strainer *Luffa cylindrica* (Loofah) Dalziel.
VINE, Thunder *Glechoma hederacea* (Ground Ivy) Vesey-Fitzgerald.
VINE, Trumpet *Campsis radicans* (Trumpet Creeper) RHS.
VINE, Trumpet, Blue *Thunbergia grandiflora* (Sky Vine) RHS.
VINE, Welsh *Sambucus nigra* (Elder) S Wales G E Evans.1969. An ironically-given name, probably pointing out that elderberry wind is the only wine that poor Welshmen can afford.
VINE, White *Bryonia dioica* (White Bryony) Gerard.
VINE, Wild 1. *Bryonia dioica* (White Bryony) Aubrey. 2. *Clematis vitalba* (Old Man's Beard) Skinner. 3. *Tamus communis* (Black Bryony) Gerard.
VINE, Wild, White *Bryonia dioica* (White Bryony) Parkinson.
VINE, Wonga Wonga *Pandorea pandorana*. An Australian climber.
VINE, Wood *Bryonia dioica* (White Bryony) Prior. Cf. Hedge Vine.
VINE, Wreath, Purple *Petraea volubilis*.
VINE LILAC *Hardenbergia violacea* (Blue Coral Pea) RHS.
VINE MAPLE 1. *Acer circinatum*. 2. *Menispermum canadense* (Moonseed) Le Strange.
VINE MESQUITE GRASS *Panicum obtusum*.
VINEGAR PLANT *Rhus typhina* (Stag's-horn Sumach) Taylor. "It has got the name of Vinegar Plant, from the double reason of the young germin of its fruit, when fermented, producing either new or adding to the strength of old weak vinegar, whilst its ripe berries afford an agreeable acid..." (Taylor).
VINEGAR TREE *Rhus glabra* (Smooth Sumach) H & P.
VINEYARD STINKWEED *Diplotaxis viminea*.
VINING VELVET *Ruellia makoyana* USA Cunningham.
VIOLA, Butterfly *Viola cornuta*.
VIOLET, African *Saintpaulia ionantha*.
VIOLET, Arabian *Exacum affine* (Persian Violet) Bonar.
VIOLET, Australian *Viola hederacea*.
VIOLET, Autumn *Gentiana pneumonanthe* (Marsh Gentian) B & H.
VIOLET, Bird's-foot *Viola pedata*. 'Bird's-foot', from the palmately divided leaves, looking exactly like the marks of birds' feet in mud or snow.
VIOLET, Blue 1. *Gentiana verna* (Spring Gentian) Dur. B & H. 2. *Viola riviniana* (Dog Violet) Dev, Ches. Grigson.
VIOLET, Blue, Early *Viola palmata*.
VIOLET, Blue, Marsh *Viola cucullata*.
VIOLET, Blue, Northern *Viola septentrionalis*.
VIOLET, Blue, Woolly *Viola sororia*.
VIOLET, Bog 1. *Pinguicula vulgaris* (Butterwort) N Eng. Grigson. 2. *Viola palustris* (Marsh Violet) Curtis.
VIOLET, Bulbous 1. *Galanthus nivalis* (Snowdrop) Gerard. 2. *Leucojum vernum* (Spring Snowflake) McDonald.
VIOLET, Bush *Browallia speciosa*.
VIOLET, Calathian *Gentiana pneumonanthe* (Marsh Gentian) Gerard. 'Calathian' is Latin calathus, a basket; the name was given by

Pliny to another plant, and misappropriated to this one.

VIOLET, Canada *Viola canadensis*.

VIOLET, Coast *Viola brittoniana*.

VIOLET, Confederate *Viola papilionacea* (Meadow Violet) Howes.

VIOLET, Corn *Legousia hybrida* (Venus's Looking-glass) Prior.

VIOLET, Crowfoot *Viola pedata* (Bird's-foot Violet) Whittle & Cook. 'Crowfoot', because the leaves look just like the marks of bird's feet in the mud or snow.

VIOLET, Cut-leaved *Viola pedata* (Bird's-foot Violet) Whittle & Cook. Palmately divided leaves, hence this name and Bird's-foot Violet.

VIOLET, Damask; VIOLET, Dame's *Hesperis matronalis* (Sweet Rocket) Gerard (Damask). *Viola damascena* was the old botanical name for the plant, hence 'Damask'. 'Dame's', of course, is a corruption of 'Damask'. But in French *Viola damascena* became Violette de Damas, misspelt to Violette des Dames, and so into English Dame's Violet.

VIOLET, Dog *Viola riviniana*. 'Dog' in a deprecatory sense - it has no smell.

VIOLET, Dog, American *Viola conspersa*.

VIOLET, Dog, Heath *Viola canina*.

VIOLET, Dog, Wood *Viola reichenbachiana*.

VIOLET, Dog-tooth *Erythronium dens-canis*.

VIOLET, Dog-tooth, American *Erythronium americanum* (Yellow Adder's Tongue) Grieve.1931.

VIOLET, Dog-toothed *Dentaria bulbifera* (Coral-root) Gerard. Cf. Tooth or Toothed Violet, & Tooth Cress.

VIOLET, English *Viola odorata* (Violet) Scot. B & H. Presumably because they don't grow in Scotland. But can this be true?

VIOLET, Fen *Viola stagnina*.

VIOLET, Flame *Episcia cupreata*.

VIOLET, Garnesie *Matthiola incana* (Stock) Lyte.

VIOLET, Gypsy *Viola riviniana* (Dog Violet) Som. Macmillan.

VIOLET, Hairy *Viola hirta*.

VIOLET, Heath *Viola canina* (Heath Dog Violet) Clapham.

VIOLET, Heath, Pale *Viola lactea*.

VIOLET, Hedge *Viola riviniana* (Dog Violet) Dev. Grigson.

VIOLET, Hooded, Blue *Viola cucullata* (Marsh Blue Violet). 'Cucullatus' means covered with a hood.

VIOLET, Horned *Viola cornuta* (Butterfly Viola) RHS.

VIOLET, Horse 1. *Viola riviniana* (Dog Violet) Dev. Friend.1882. 2. *Viola tricolor* (Pansy) Dev. Friend.1882. 'Horse' has the same significance as 'dog' in a plant name. They both mean coarse, or in some way inferior. In this case the inferiority lies in the fact that they have no scent.

VIOLET, Ivy-leaved *Viola hederacea* (Australian Violet) RHS.

VIOLET, Karroo *Aptosimum indivisum*.

VIOLET, March *Viola odorata* (Violet) B & H. From the time of their flowering.

VIOLET, Marsh 1. *Pinguicula vulgaris* (Butterwort) Yks. Grigson. 2. *Viola palustris*.

VIOLET, Meadow 1. *Viola papilionacea*. 2. *Viola pumila*.

VIOLET, Mercury's 1. *Campanula medium* (Canterbury Bell) Gerard. 2. *Campanula trachelium* (Nettle-leaved Bellflower) Gerard.

VIOLET, Night *Platanthera chlorantha* (Butterfly Orchid) Wilts. Dartnell & Goddard.

VIOLET, Northern *Viola selkirkii*.

VIOLET, Pansy, Yellow *Viola pedunculata*.

VIOLET, Persian *Exacum affine*. Cf. Arabian Violet.

VIOLET, Philippine *Barleria cristata*.

VIOLET, Pig *Viola riviniana* (Dog Violet) Ches. Grigson. 'Pig', a derogatory adjective in this sense, to equate with 'dog' or 'horse' - the Dog Violet has no smell.

VIOLET, Rape *Cyclamen europaeus* (Sowbread) Turner. 'Rape', to compare it with a turnip. In other words, a violet with a bulb.

VIOLET, Rhodes's *Securidaca longipedunculata* (Wild Violet Tree) S Africa Dalziel.

VIOLET, Shrubby *Viola arborescens*.

VIOLET, Sister *Viola sororia* (Woolly Blue Violet) USA House. A result of the specific name, *sororia*.

VIOLET, Snake 1. *Viola reichenbachiana* (Wood Dog Violet) Dor. Macmillan. 2. *Viola riviniana* (Dog Violet) Dor. Macmillan.

VIOLET, Sorcerer's *Vinca minor* (Lesser Periwinkle) Friend. Because of its use in charms against the evil eye. Worn in the buttonhole, or carried dried in a sachet, it was a great protection against any witch not carrying one herself.

VIOLET, Spanish *Lupinus luteus* (Yellow Lupin) Parkinson.

VIOLET, Spring *Gentiana verna* (Spring Gentian) Dur. Grigson.

VIOLET, Summer *Viola riviniana* (Dog Violet) War. Grigson.

VIOLET, Sweet *Viola odorata*.

VIOLET, Teesdale *Viola rupestris*. It was thought that the only station for this plant was on Widdybank Fell, in Teesdale, but other limestone habitats in the north of England have been found.

VIOLET, Three-coloured *Viola tricolor* (Pansy) Thornton.

VIOLET, Tooth(ed) *Dentaria bulbifera* (Coral-root) Prior (Tooth); Gerard (Toothed). Gerard also called it Dog-toothed Violet.

VIOLET, Trinity *Viola tricolor* (Pansy) Yks. Grigson. Culpepper used the name Herb of the Trinity - "this is the Herb which such Physicians as are licensed to blaspheme by Authority, without Danger of having their Tongues burned through with an hot Iron, called an Herb of the Trinity". The specific name, *tricolor*, translates as Three-coloured Violet.

VIOLET, Water *Hottonia palustris*.

VIOLET, White, Sweet *Viola pallens*.

VIOLET, Wild *Viola decumbens* S Africa Watt.

VIOLET, Wood 1. *Viola cucullata* (Marsh Blue Violet) USA Sanford. 2. *Viola reichenbachiana* (Wood Dog Violet) Hutchinson. 3. *Viola riviniana* (Dog Violet) Hutchinson. 4. *Viola sarmentosa* USA Schenk & Gifford.

VIOLET, Wood, Pale *Viola reichenbachiana* (Wood Dog Violet) Clapham.

VIOLET, Wood, Yellow *Viola biflora*.

VIOLET, Yellow 1. *Cheiranthus cheiri* (Wallflower) Coats. 2. *Viola lutea* (Mountain Pansy) Cumb. Grigson.

VIOLET, Yellow, Downy *Viola pubescens*.

VIOLET, Yellow, Early *Viola rotundifolia*.

VIOLET CRESS *Ionopsidium acaule*.

VIOLET TEA *Viola odorata* (Violet) S Africa Watt.

VIOLET TREE, Wild *Securidaca longepedunculata*. The flowers have a violet-like scent.

VIOLIN STRINGS *Plantago lanceolata* (Ribwort Plantain) Som. Macmillan.

VIPER GOURD *Trichosanthes anguina* (Snake Gourd) Whittle & Cook.

VIPER'S BUGLOSS *Echium vulgare*. So called because the seeds are like the head of a viper. Echium is from a Greek plant name ekhion, from ekhis, a viper. Or perhaps it has a stem that is spotted like a snake. So from the doctrine of signatures, it was used for snake bite nnd scorpion stings.

VIPER'S BUGLOSS, Purple *Echium plantagineum*.

VIPER'S BUGLOSS, Red *Echium wildpretii*.

VIPER'S GRASS 1. *Echium vulgare* (Viper's Bugloss) IOW B & H. 2. *Scorzonera humilis*.

VIPER'S GRASS, Purple *Scorzonera purpurea*.

VIPER'S HERB *Echium vulgare* (Viper's Bugloss) Gerard.

VIPER'S VICTUALS *Arum maculatum* (Cuckoo-pint) Hants. Hants FWI. Cf. Snake's Victualls, etc. The red berries actually are poisonous, hence the allusion to serpents.

VIPPE *Pinus sylvestris* (Scots Pine) B & H.

VIRGIN, Naked *Colchicum autumnale* (Meadow Saffron) Ches. B & H. All the 'naked' names arise from the plant's habit of flowering before the leaves appear. It also accounts for such names as Upstart, or Son-before-the-father.

VIRGIN, Rose of the *Anastatica hierochuntica* (Rose of Jericho) Ackermann. Cf. Mary's Flower. The legend tells that all the plants of this species expanded, became green and blossomed again at the birth of Jesus, and still do so in commemoration.

VIRGIN, Spotted *Pulmonaria officinalis* (Lungwort) Grigson. The white spots on the leaves are accounted for in the legend that, during the flight into Egypt, some of the Virgin's milk fell on them while she was nursing the infant Jesus. Either that, or it was her tears that spotted the leaves, for the plant grew on Calvary.

VIRGIN MARY 1. *Eupatorium cannabinum* (Hemp Agrimony) Corn. Grigson. 2. *Pulmonaria officinalis* (Lungwort) Fernie.

VIRGIN MARY'S CANDLE *Verbascum thapsus* (Mullein) Ire. Grigson. The stems and leaves of mullein used to be dipped in tallow or suet and burnt to give light at outdoor country gatherings, or in the home. They were used at funerals, too. Cf. Lady's Taper, or Lady's Candle, and also the Welsh tapr Mair.

VIRGIN MARY'S COWSLIP; VIRGIN MARY'S HONEYSUCKLE *Pulmonaria officinalis* (Lungwort) Glos, Worcs. Grigson; Shrop. Burne (Cowslip); Ches, Shrop. Burne (Honeysuckle). The connection with the Virgin lies in the spotted leaves, stained with her milk while she was nursing the infant Jesus during the flight into Egypt. Cowslip was quite commonly applied to lungwort, and we have examples such as Jerusalem, Bethlehem or Bedlam, Cowslip.

VIRGIN MARY'S MILKDROPS *Pulmonaria officinalis* (Lungwort) Wilts. Macmillan; Mon. Grigson. See Spotted Virgin, etc.

VIRGIN MARY'S NIPPLES 1. *Euphorbia amygdaloides* (Wood Spurge) Som. Grigson. 2. *Euphorbia helioscopia* (Sun Spurge) Dev. Friend. The point of reference is the milky sap common to all the spurges.

VIRGIN MARY'S TEARS *Pulmonaria officinalis* (Lungwort) Dor. Macmillan. It is usually the Virgin's milk rather than her tears that are described as being responsible for the spots on the leaves of lungwort.

VIRGIN MARY'S THISTLE 1. *Carduus benedictus* (Holy Thistle) Suff. Moor. 2. *Silybum marianum* (Milk Thistle) Hants. Cope.

VIRGIN'S BOWER 1. *Clematis virginiana* USA House. 2. *Clematis vitalba* (Old Man's Beard) Gerard. One writer says that it is a name coined as a tribute to Queen Elizabeth, but, at any rate in German legend, Mary and the Child sheltered under the Virgin's Bower on the flight into Egypt.

VIRGIN'S BOWER, Sweet *Clematis flammula* (Upright Virgin's Bower) Lindley. Cf. Fragrant Clematis.

VIRGIN'S BOWER, Upright *Clematis flammula*.

VIRGIN'S FINGERS *Digitalis purpurea* (Foxglove) Som. Macmillan. 'Fingers' are a favourite appellation for the flowers, as common as 'gloves'.

VIRGIN'S HAIR *Briza media* (Quaking-grass) Tynan & Maitland. Cf. Maiden's Hair.

VIRGIN'S MILK 1. *Silybum marianum* (Milk Thistle) Page.1978. 2. *Sonchus oleraceus* (Sow Thistle) Som. Macmillan. The 'milk' is the white sap, in each case.

VIRGIN'S PINCH *Polygonum persicaria* (Persicaria) Berks. Grigson. The legend is that the Virgin Mary pulled a root, left her mark on the leaf, and threw it aside, saying, "This is useless", and it has been useless ever since. Besides this name, and on the same theme, there are Pinchweed, Devil's Pinch, Lady's Thumb, and, in Scotland, Useless.

VIRGIN'S ROBE *Borago officinalis* (Borage) Canbs. Porter.

VIRGIN'S TEARS; LADY'S TEARS *Convallaria maialis* (Lily-of-the-valley) Tynan & Maitland (Virgin's); Som. Grigson (Lady's). The legend is that the Virgin's tears

at the foot of the Cross, or those of Mary Magdalene at the tomb of Christ, were changed into the flower.
VIRGIN'S THISTLE *Silybum marianum* (Milk Thistle) Norf. Grigson.
VIRGINIA BLUEBELL *Mertensia virginica* (Virginia Cowslip) Howes.
VIRGINIA BUGLEWEED *Lycopus virginicus*.
VIRGINIA CREEPER *Parthenocissus quinquefolia*.
VIRGINIA GROUND CHERRY *Physalis virginiana*.
VIRGINIA HOREHOUND *Lycopus virginicus* (Virginia Bugleweed) USA Zenkert.
VIRGINIA PEPPERGRASS; VIRGINIA PEPPERWEED *Lepidium virginicum*. USA Allan (Pepperweed).
VIRGINIA PINE *Pinus virginiana*.
VIRGINIA SAVIN *Juniperus virginiana* (Virginian Juniper) USA Elmore.
VIRGINIA SUMACH *Rhus typhina* (Stag's-horn Sumach) Parkinson.
VIRGINIA WATER HOREHOUND *Lycopus virginicus* (Virginia Bugleweed) Grieve.1931.
VIRGINIAN ANGELICA TREE *Aralia spinosa* (Hercules' Club) W Miller.
VIRGINIAN BIRD CHERRY *Prunus virginiana* (Choke Cherry) Ablett.
VIRGINIAN BOXWOOD *Cornus florida* (Flowering Dogwood) Grieve.1931. Cf. American, or New England, Boxwood.
VIRGINIAN CEDAR *Juniperus virginiana* (Virginian Juniper) J Smith.
VIRGINIAN CHESTNUT *Castanea pumila* (Chinquapin) Howes.
VIRGINIAN COWSLIP *Pulmonaria officinalis* (Lungwort) Grigson. Cowslip is a favourite name for lungwort - Cf. Bugloss Cowslip, Virgin Mary's Cowslip (hence 'Virginian' in this name), Jerusalem or Bethlehem Cowslip, etc.
VIRGINIAN DAFFODIL *Zephyranthes atamasco* (Atamasco Lily) Parkinson.
VIRGINIAN JUNIPER *Juniperus virginiana*.

VIRGINIAN PENCIL CEDAR *Juniperus virginiana* (Virginian Juniper) Davey.
VIRGINIAN POKE *Phytolacca decandra* (Poke-root) Brownlow.
VIRGINIAN RASPBERRY *Rubus occidentalis* (Thimble-berry) Grieve.1931.
VIRGINIAN SNAKEROOT *Aristolochia serpentaria*.
VIRGINIAN STRAWBERRY *Fragaria virginica* (Scarlet Strawberry) Grigson.
VIRGINIAN WATERLEAF *Hydrophyllum virginianum*.
VIRGINIAN WINTERBERRY *Ilex verticillata* (Winterberry) Le Strange.
VITAMIN C PLANT *Malpighia glabra* (Barbados Cherry) USA Cunningham. Apparently, the unripe fruit is the richest known source of Vitamin C, containing at least 20 times as much as an orange.
VIVVERVAW; VIVVYVEW *Tanacetum parthenium* (Feverfew) both Friend.1882. Corruptions of 'feverfew' that exist in a number of forms - Cf. for instance Vethervow, or Vether-vo, and such as Fetterfoe, all of them from the West country.
VLIX *Linum usitatissimum* (Flax) Dor. B & H. An accomodation, from Flix, to local pronunciation.
VOMIT NUT *Jatropha curcas* (Barbados Nut) Howes.
VOMITWORT *Lobelia inflata* (Indian Tobacco) USA Henkel. Cf. Gagroot, Pukeweed, etc, for this has long been used by the American Indians as an emetic or expectorant.
VOODOO LILY *Sauromatum guttatum*.
VUZ; VUZZ; VUZZEN *Ulex europaeus* (Furze) Dor. Udal (Vuz, Vuzzen); Dev. Fernie (Vuzz).

W

WABRAN-LEAF; WABRET-LEAF
Plantago major (Great Plantain) Scot. Jamieson (Wabran); Tynan & Maitland (Wabret). Evidently an offshoot of Waybread, which is OE weg-breade, way breadth. It appears as Waveran-leaf in Orkney, and has quite a lot of alternatives in northern England and Scotland.

WAD 1. *Genista tinctoria* (Dyer's Greenweed) Yks. Brockett. 2. *Isatis tinctoria* (Woad) E Ang, N'thants, Yks, Cumb. Grigson. 'Woad' for Dyer's Greenweed was quite common, particularly in the northern counties. Brockett said that 'woad' was reserved for *Isatis tinctoria*, while 'wad' was used for *Genista tinctoria*.

WADD; WADE *Isatis tinctoria* (Woad) Scot. Jamieson (Wadd); Gerard (Wade). Gerard's orthography does not necessarily imply a pronunciation normally connected with that spelling. But Cf. Wud with the Scottish version; there the pronunciation would be the same.

WADE, Nan *Verbena officinalis* (Vervain) IOM Gill in Notes & Queries.1941. She was the Isle of Man's most famous white witch, and vervain was her "trump card".

WAFER ASH *Ptelea trifoliata* (Hop Tree) USA Yarnell. Presumably because of its disc-like fruits.

WAFERS, Wag *Briza media* (Quaking-grass) Dor. Grigson.1959.

WAG-WAFERS; WAG-WAMS; WAG-WANTS; WAG-WANTONS; WAG-WANTING *Briza media* (Quaking-Grass) Dor. Grigson.1959 (Wafers); Som. Grigson.1959 (Wams); Som, Bucks. Grigson.1959; Wilts. Dartnell & Goddard; Dor. Dacombe; Hants. Cope (Wants); Dor, Bucks. Grigson.1959 (Wantons); Yks. Tynan & Maitland (Wanting).

WAGGERING GRASS *Briza media* (Quaking-grass) Yks. Grigson.1959.

WAGGLES, Wiggle *Briza media* (Quaking-grass) Som, Suss. Grigson.1959.

WAGONS, Wig *Briza media* (Quaking-grass) Wilts. Grigson.1959.

WAGTAILS *Briza media* (Quaking-grass) Wilts. Dartnell & Goddard.

WAHOO 1. *Euonymus americanus* (Strawberry Bush) USA Lloyd. 2. *Euonymus atropurpureus*. 3. *Ulmus alata* (Winged Elm) USA Harper. This is the Dakota Indian word wanhu.

WAIL, Widow 1. *Daphne mezereum* (Lady Laurel) Gerard. 2. *Fritillaria meleagris* (Snake's-head Lily) Shrop. B & H. One of the many names that reflect the fritillary's sombre colouring. Cf. Weeping Widow and Widow-veil, obviously the same as Widow-wail. The name for Lady Laurel is probably given because it is so poisonous, although the purple flowers (mourning colour) may have something to do with it.

WAINSCOT OAK *Quercus cerris* (Turkey Oak) J Smith.

WAIT-A-BIT THORN 1. *Acacia brevispica*. 2. *Acacia mellifera*. The reference is to the vicious thorns.

WAIT-A-MINUTE *Acacia greggii* (Cat's Claw) USA Jaeger. Extremely thorny.

WAITEWEED *Taraxacum officinale* (Dandelion) Wilts. Macmillan. i.e. Wetweed, one of the many local names that acknowledge the plant's diuretic qualities.

WAKE-AT-NOON *Ornithogalum umbellatum* (Star-of-Bethlehem) Wilts. Macmillan; IOW B & H. To commemorate the late opening of the flowers - Cf. Eleven-o'clock Flower, Twelve o'clock and many other forecasts of the time of opening.

WAKE-PINTEL *Arum maculatum* (Cuckoo-pint) Prior. 'Wake' is an Anglo-Saxon word that probably meant 'alive'. So the whole name means erect penis, in keeping with the commoner name, Cuckoo-pint.

WAKE-ROBIN 1. *Arum maculatum* (Cuckoo-pint) Turner. 2. *Melandrium rubrum* (Red Campion) Yks. Grigson. 3. *Orchis mascula* (Early Purple Orchid) Ches.

Holland. 4. *Trillium erectum* (Bethroot) Grieve.1931. 5. *Trillium grandiflorum* (Wood Lily) USA Densmore. 'Wake' as in Wake-pintel, and 'robin', as Prior claimed, is French robinet, meaning penis (quite possible, as the word in modern French means 'tap'). Cuckoo-pint is the main claimant to the name, but Early Purple Orchid has a number of similar names, all connected with the way the flower stem thrusts out of the ground.

WAKE-ROBIN, American *Arisaema triphyllum* (Jack-in-the-pulpit) Le Strange.

WALEWORT *Sambucus ebulus* (Dwarf Elder) Parkinson. Cf. Wallwort, in Turner. Neither means 'growing on walls'. The OE name was wealhwyrt, and it means a foreign plant (the root appears again in 'walnut', and for various plants described as 'Welsh'.

WALKING-STICK, Devil's *Aralia spinosa* (Hercules' Club) USA Kingsbury.

WALKING-STICK, Jacob's; WALKING-STICK, Joseph's *Polemonium caeruleum* (Jacob's Ladder) Hants. Wright (Jacob's); Hants. Cope (Joseph's).

WALKING-STICK CAMWOOD *Baphia polygalacea*. Walking sticks are made from the stem with the attached root.

WALKING-STICK EBONY *Diospyros monbuttensis* (Yoruba Ebony) Dalziel. A very hard wood, used in Nigeria for making walking sticks.

WALKING-STICK PLANT *Buglossoides purpuro-caerulea* (Purple Gromwell) Perry. Because of the tendency of the sprays to bend down and root at the tips.

WALL, Blood; WALL, Bloody *Cheiranthus cheiri* (Wallflower) Wilts. Macmillan; N'thants. A E Baker (Blood); McDonald (Bloody). Cf. Bloody Wallier, as well as aberrant forms like Bloody Warrior.

WALL BARLEY 1. *Lolium perenne* (Rye Grass) Lyte. 2. *Lolium temulentum* (Darnel) Coles.

WALL CRESS 1. *Aubrietia purpurea*. 2. *Diplotaxis muralis* (Stinkweed) Salisbury.

WALL GERMANDER *Teucrium chamaedrys* (Germander) C P Johnson.

WALL GILLIFLOWER; WALL JULY FLOWER *Cheiranthus cheiri* (Wallflower) Turner (Gilliflower); Wesley (July Flower). Gilliflower for wallflower used once to be quite common, but it served for a number of plants, perhaps modified for convenient identification, as in Clove Gilliflower for carnation, or Stock Gilliflower for stock, as well as in this example. July Flower, of course, is simply a version of Gilliflower.

WALL GINGER; WALL GRASS *Sedum acre* (Biting Stonecrop) Grieve.1931 (Ginger); Dev. Friend.1882 (Grass). 'Ginger' is explained by the use of 'pepper' in other names, all complemented by the common name, which describes this stonecrop as 'biting'.

WALL INK *Veronica beccabunga* (Brooklime) B & H. Nothing to do with walls - this is 'well'. Cf. Horse Well-cress, or Well-grass. More to the point is a quite common Cumbrian name, Wellink, obviously the forerunner of our Wall Ink.

WALL IRIS *Iris tectorum*.

WALL LILAC *Centranthus ruber* (Spur Valerian) Dev. Macmillan.

WALL MOSS *Sedum acre* (Biting Stonecrop) Yks. B & H.

WALL PENNYROYAL *Umbilicus rupestris* (Wall Pennywort) Grieve.1931. Not a pennyroyal, of course - this is a mishearing of pennywort.

WALL PENNYWORT *Umbilicus rupestris*. 'Pennywort', because of the round leaves, likened to coins. Besides lots of 'penny' names, we find Money-penny, Halfpennies-and-pennies, etc.

WALL PEPPER *Sedum acre* (Biting Stonecrop) Gerard. Cf. Country Pepper, also used by Gerard, and indeed the common name itself - 'biting' speaks for itself.

WALL SPEEDWELL *Veronica arvensis*. Its habitat is not only on old walls, but also in gravelly fields, as the specific name, *arvensis*, implies.

WALL STONECROP *Sedum acre* (Biting Stonecrop) Thornton.

WALLFLOWER 1. *Cheiranthus cheiri*. 2. *Erysimum asperum* (Desert Wallflower) USA Gates. 3. *Helianthus annuus* (Sunflower) Shrop. B & H. Sunflower looks very odd in this company. Surely, the name can only mean that the plant is best grown by a wall. The familiar wallflower proper will, like Spur Valerian, grow on old walls.

WALLFLOWER, Desert *Erysimum asperum*.

WALLFLOWER, False; WALLFLOWER, Wild *Erysimum asperum* (Desert Wallflower) both USA Elmore.

WALLFLOWER, Field *Raphanus raphanistrum* (Wild Radish) S Africa Watt.

WALLFLOWER, Siberian *Erysimum allionii*.

WALLFLOWER, Western 1. *Erysimum asperum* (Desert Wallflower) USA Elmore. 2. *Erysimum capitatum*.

WALLFLOWER, White *Matthiola incana* (Stock) B & H. The two plants are linked together by both having been given the name Gilliflower, usually with some distinguishing adjective.

WALLFLOWER CABBAGE *Rhyncosinapis erucastrum*.

WALLFLOWER MUSTARD *Erysimum cheiranthoides* (Treacle Mustard) Hutchinson.

WALLICH'S LILY *Lilium wallichianum*. Nathaniel Wallich, 1786-1854, a Danish botanist who first described this lily as *L longiflorum*. But that name being already occupied, it was renamed in his honour.

WALLICH'S OAK *Quercus wallichiana*.

WALLIER, Bloody *Cheiranthus cheiri* (Wallflower) Halliwell. Bloody Warrior, a West country name, is probably better known than this. But 'warrior' is not warrior at all, but 'wallyer'. So the name means the red flower that grows on walls.

WALLOW; WULLOW *Alnus glutinosa* (Alder) both Shrop. Grigson. Both extreme forms of Aul, the single syllable variation of Alder.

WALLSPRAY *Cotoneaster microphylla* (Rose Box Tree) McClintock.1978. Cf. Rockspray for this.

WALLWORT 1. *Cheiranthus cheiri* (Wallflower) T Hill. 2. *Parietaria diffusa* (Pellitory-of-the-wall) Gerard. 3. *Sambucus ebulus* (Dwarf Elder) Turner. 4. *Sedum acre* (Biting Stonecrop) Yks. B & H. 5. *Umbilicus rupestris* (Wall Pennywort) B & H.

WALLY, Water *Baccharis glutinosa* (Seep Willow) USA T H Kearney.

WALNUT *Juglans regia*. 'Walnut' means a 'foreign nut', for the first syllable is OE wealh, and has nothing to do with walls.

WALNUT, American *Juglans nigra*.

WALNUT, Black *Juglans nigra*.

WALNUT, Chinese *Juglans cathayensis*.

WALNUT, East Indian *Albizia lebbeck* (Siris Tree) Hora.

WALNUT, English *Juglans regia* (Walnut) Schery. A strange name, as it is not an English native tree, and in any case the very name Walnut proclaims its foreign-ness.

WALNUT, Indian *Aleurites moluccana* (Candlenut Tree) Leyel.1948.

WALNUT, Japanese *Juglans ailanthifolia*.

WALNUT, Little *Juglans rupestris* (Texas Walnut) RHS.

WALNUT, Manchurian *Juglans mandshurica*.

WALNUT, Persian *Juglans regia* (Walnut) Schery. Asia Minor is its original home, so 'Persian' is not a bad stab at naming its provenance.

WALNUT, Royal *Juglans regia* (Walnut) Ablett. Simply a translation of the botanical name.

WALNUT, Satin *Liquidambar styraciflua* timber (Sweet Gum) Wilkinson.1981.

WALNUT, Texas *Juglans rupestris*.

WALNUT, White *Juglans cinerea*.

WALSH NUT; WELSH NUT *Juglans regia* fruit (Walnut) Gerard (Walsh); Evelyn (Welsh). In both cases, the meaning is 'foreign nut', OE wealh. There are a number of examples of this use of 'Welsh' - Welsh Onion, for instance.

WALTER, Herb *Asperula odorata* (Woodruff) Dawson. It is said to have been named after Walter de Elvesdon, a 14th century Bishop of Norwich. But why?

WAND, Devil's *Aethusa cynapium* (Fool's Parsley) Dor. Macmillan. A very poisonous plant, hence the ascription to the devil.

WAND, Fairy's 1. *Agrimonia eupatoria* (Agrimony) Dor. Macmillan. 2. *Verbascum thapsus* (Mullein) Dor. Macmillan.

WAND-FLOWER *Sparaxis pulcherrima*. Cf. Angel's Fishing-rods for this - even more expressive.

WAND SAGE *Salvia vaseyi*.

WANDERING JACK *Cymbalaria muralis* (Ivy-leaved Toadflax) Som. Macmillan.

WANDERING JEW 1. *Cymbalaria muralis* (Ivy-leaved Toadflax) Suss. B & H. 2. *Glechoma hederacea* (Ground Ivy) Donegal Grigson. 3. *Saxifraga sarmentosa* (Mother-of-thousands) W Miller. 4. *Tradescantia crassifolia* USA Bergen. 5. *Tradescantia fluminensis* Gannon. 6. *Tradescantia fluviatilis* Rochford. 7. *Tradescantia virginica* (Spiderwort) Harvey. 8. *Zebrina pendula* Newall.1973. "Wandering" because of their method of reproducing themselves, usually achieved by means of new plants growing at the end of runners.

WANDERING SAILOR 1. *Cymbalaria muralis* (Ivy-leaved Toadflax) Dev. Friend.1882. 2. *Lysimachia nummularia* (Creeping Jenny) Dev. Macmillan. 3. *Saxifraga sarmentosa* (Mother-of-thousands) Coats.

WANDERING TINKER *Lysimachia nummularia* (Creeping Jenny) N Ire. Tynan & Maitland.

WANDERING WILLY 1. *Convolvulus arvensis* (Field Bindweed) N Eng. Wakelin. 2. *Geranium robertianum* (Herb Robert) Som. Macmillan.

WANTONS, Wag; WANTONS, Wigwag; WANTONS, Wiggle-waggle *Briza media* (Quaking-grass) Dor, Bucks (Wag); Bucks. (Wigwag); Berks. (Wiggle-waggle) all Grigson.1959.

WAORIKI *Ranunculus rivularis*. A New Zealand species, hence the Maori name.

WAPATOO 1. *Sagittaria latifolia* (Broad-leaved Arrowhead) USA Weiner. 2. *Sagitarria sagittifolia* (Water Archer) Grieve.1931. Wapatoo is the Indian name for the tuber, which used to be an important food source for them. The Chippewa, for instance, called it the wild potato, and they were cooked much as potatoes are.

WARBA-LEAVES *Plantago major* (Great Plantain) Mor. B & H. A variation on the English Waybread, OE weg breade, which means way breadth.

WARD LILY *Lilium wardii*. Ward is Frank Kingdon-Ward, who discovered it in 1924.

WARDSEED *Capsella bursa-pastoris* (Shepherd's Purse) Dev. B & H.

WARE-MOTH; WARMOT *Artemisia absinthium* (Wormwood) Coats. (Ware-moth); B & H (Warmot). Wormwood is very good at keeping clothes-moths at bay, but Ware-moth is without doubt Warmot, a variation of 'wormwood'.

WARENCE *Rubia tinctoria* (Madder) B & H.

WARLOCK *Sinapis arvensis* (Charlock) IOW Grigson. A variation of 'Charlock', which has many such.

WARMINSTER BROOM *Cytisus x praecox*.

WARNING, Shepherd's *Anagallis arvensis* (Scarlet Pimpernel) Som. Macmillan. A reference to pimpernel's reputation as a weather forecaster.

WARNUT *Juglans regia* nut (Walnut) Berks. Lowsley; IOW W Long; Worcs. J Salisbury.

WARRIGAL CABBAGE *Tetragonia tetragonioides* (New Zealand Spinach) C Macdonald.

WARRIOR, Bleeding; WARRIOR, Bleedy *Cheiranthus cheiri* (Wallflower) Som. (Bleeding); Dev. (Bleedy) both Macmillan. It is suggested that 'warrior' is in fact 'wallyer'.

Halliwell has Bloody Wallier as an alternative name, and Blood Wall used to be quite common in both Wiltshire and Northamptonshire.

WARRIOR, Bloody 1. *Cheiranthus cheiri* (Wallflower) Corn. Courtney.1880; Dev. Friend.1882; Som. Jennings; Wilts. Akerman; Dor. Dacombe; Hants. Cope; Berks. Lowsley. 2. *Fritillaria meleagris* (Snake's-head Lily) Berks. Grigson. See previous entry for wallflower.

WARRIOR, Blue *Geranium pratense* (Meadow Crane's-bill) Som. Macmillan.

WARRIOR, Old *Artemisia pontica* (Roman Wormwood) Brownlow. An odd name, possibly stemming somehow from Old Woman, given as a counterpart to Old Man, which is Southernwood.

WART, Kill *Chelidonium majus* (Greater Celandine) Dev. Hay & Synge. The juice is used as an application on warts. In fact, the plant is famous for it - see Wart-curer, Wart-flower, Wart Plant, Wartwort etc - all west country names.

WART-CRESS *Coronopus squamatus*.

WART-CRESS, Slender *Coronopus didymus*.

WART-CURER; WART-PLANT *Chelidonium majus* (Greater Celandine) both Som. Macmillan. See above under Kill-wart.

WART-FLOWER 1. *Chelidonium majus* (Greater Celandine) Dev. Grigson. 2. *Papaver rhoeas* (Red Poppy) Corn. Grigson.

WART-GRASS 1. *Euphorbia helioscopia* (Sun Spurge) Derb, Yks, Cumb. Grigson. 2. *Euphorbia peplus* (Petty Spurge) Lincs. Drury.1991. Generations of children have picked Sun Spurge, to put the milky juice on to the warts on their fingers. Petty Spurge is equally efficacious.

WART-SPURGE *Euphorbia helioscopia* (Sun Spurge) Prior. Cf. Wartweed, etc.

WARTED BUNIAS *Bunias orientalis* (Warty Cabbage) Polunin.

WARTED SPURGE, Broad-leaved *Euphorbia platyphyllos* (Broad Spurge) Flower.

WARTWEED 1. *Chelidonium majus* (Greater Celandine) E Ang. Grigson. 2. *Euphorbia helioscopia* (Sun Spurge) Som, Cambs, Suff, Ess, Norf, Yks, Cumb. Grigson; Glos. J D Robertson. 3. *Euphorbia peplus* (Petty Spurge) Lincs. Drury.1991; Yks. B & H. All well-known for application to warts.

WARTWORT 1. *Chelidonium majus* (Greater Celandine) Som. Grigson; Wilts. Dartnell & Goddard; Glos. J D Robertson. 2. *Euphorbia helioscopia* (Sun Spurge) Turner. 3. *Euphorbia peplus* (Petty Spurge) Gomme.1884. 4. *Gnaphalium uliginosum* (Marsh Cudweed) Ches. B & H.

WARTWORT, Sea *Euphorbia peplus* (Petty Spurge) Turner.

WARTY BIRCH *Betula verrucosa* (Birch) Hart. *Verrucosa* means 'warty'.

WARTY CARRION FLOWER *Stapelia verrucosa var. robusta*.

WARTY-GIRSE *Euphorbia helioscopia* (Sun Spurge) Ork. Leask. Cf. Wart-grass.

WARWICKSHIRE WEED *Ulmus procera* (English Elm) War. Ablett. Cf. Wiltshire Weed. These names must be an acknowledgment of the elm's commonness once in these two areas.

WASHING BOWL, My Lady's *Saponaria officinalis* (Soapwort) Mabey.1977. Presumably only because this is soapwort, for a lather can be got by rubbing the leaves in water, and is still used for cleaning old tapestries without damaging the fabric.

WASHINGTON HAWTHORN *Crataegus phaenopyrum*. But it is a native of the southern states of America.

WASHINGTON LILY *Lilium washingtonianum*.

WASHINGTONIA *Sequoia gigantea* (Wellingtonia) USA Potter and Sargent.

WASHRAG GOURD; WASHRAG SPONGE *Luffa cylindrica* (Loofah) Chopra.

WASP ORCHID *Ophrys apifera var. trollii*.

WASTER LEDGES *Polygonum bistorta* (Bistort) Cumb. B & H. A local variation on

Easter Ledgers, the reference being to the Herb Pudding, or Ledger Pudding, that used to be made with the leaves of bistort at Easter or Passion-tide.

WATCH, Shepherd's *Anagallis arvensis* (Scarlet Pimpernel) E Ang. Grigson. As well as being a weather forecaster, pimpernel is a time teller. It will open its flowers only when the sun is up, and will close them again at dusk.

WATCH-AND-CHAIN, Gold *Laburnum anagyroides* (Laburnum) Som. Macmillan.

WATCH-AND-CHAIN, White *Robinia pseudacacia* (Common Acacia) Som. Macmillan.

WATCH GUARDS *Laburnum anagyroides* (Laburnum) Ches. B & H.

WATCH WHEELS *Ranunculus arvensis* (Corn Buttercup) H C Long.1910. A description of the flat, spiny seed vessels, which also account for such names as Pricklebacks, and Hedgehogs.

WATCHCHAIN; WATCHCHAIN, Gold *Laburnum anagyroides* (Laburnum) both Som. Macmillan.

WATCHES 1. *Sarracenia purpurea* (Pitcher Plant) Perry. 2. *Stellaria holostea* (Greater Stitchwort) Dor. Macmillan.

WATCHES, Gold 1. *Glaucium flavum* (Yellow Horn Poppy) Dor. Grigson. 2. *Hypericum calycinum* (Rose of Sharon) Som. Macmillan. Round, yellow flowers in each case.

WATCHES-AND-CLOCKS; WATCHES, Clocks-and- *Taraxacum officinale* (Dandelion) Som. Macmillan (Watches-and-clocks); Mabey (Clocks-and-watches). A mixture of a name, for the yellow flowers are the (gold) watches, while the seed heads are the famous 'clocks' that children tell the time by.

WATER BERRY 1. *Syzygium cordatum*. 2. *Syzygium guineense*.

WATER FLOWER *Geum rivale* (Water Avens) eastern USA Grieve.1931.

WATER GRASS *Nasturtium officinale* (Watercress) N Ire. Grigson. 'Grass' and 'cress' often get misplaced.

WATER NYMPH FLOWER (Polyanthus Narcissus) *Narcissus canaliculatus orientalis*.

WATER-ROOT *Fockea monroi*. Literally correct,for these roots are water-bearing, and used by the Bushmen to survive during the dry season.

WATER TREE; WATER PEAR *Ilex mitis* (Cape Holly) S Africa Watt. Cf. Waterwood.

WATER TREE, Red *Erythrophloeum guienense* (Ordeal Tree) Verdcourt.

WATERBLEB *Caltha palustris* (Marsh Marigold) Lincs. Grigson. 'Blob', which means blister, becomes 'bleb' on the eastern side of England. Cf. Butterbleb.

WATERBUTTONS *Cotula coronopifolia* (Buttonweed) Australia Watt.

WATERCRESS 1. *Cardamine amara* (Large Bittercres) Cumb. B & H. 2. *Lepidium sativum* (Garden Cress) S Africa Watt. 3. *Nasturtium officinale*.

WATERCRESS, Brown-leaved *Nasturtium microphyllum*.

WATERCRESS, Fool's *Apium nodiflorum*. It looks vaguely like watercress, but only when there are no flowers, and sometimes grows with it. But they belong to quite different families, and one would have to be a fool indeed to mistake it.

WATERCRESS, Iceland *Rorippa palustris* (Marsh Yellow Cress) Hutchinson.

WATERCRESS, One-rowed *Nasturtium microphyllum* (Brown-leaved Watercress) Clapham.

WATERCRESS, Yellow *Rorippa amphibia* (Greater Yellow Cress) Turner.

WATERLEAF *Phacelia glandulosa*.

WATERLEAF, Virginian *Hydrophyllum virginianum*.

WATERLILY, Blue, Cape *Nymphaea capensis*.

WATERLILY, Fragrant; WATERLILY,Sweet-scented *Nymphaea odorata* (White Pond Lily) USA Yarnell (Fragrant); Whittle & Cook (Sweet-scented).

WATERLILY, Yellow *Nymphaea advena* (Large Yellow Pond Lily) USA Youngken.

WATERMELON PEPEROMIA *Peperomia argyrea*.

WATERMOTIE *Baccharis glutinosa* (Seep Willow) USA T H Kearney.

WATERSAL *Salix caprea* (Goat Willow) Ess. Gepp. 'Sal' is sallow, of course, or at least Sally.

WATERWEED, American *Elodea canadensis* (Canadian Pondweed) Howes.

WATERWHEEL PLANT *Aldrovanda vesiculosa*. Given because the plant, an aquatic, produces whorls of six to eight leaves arranged like the spokes of a wheel.

WATERWOOD *Ilex mitis* (Cape Holly) S Africa Watt. Cf. Water Tree and Water Pear.

WATERWORT 1. *Callitriche verna* (Water Starwort) Cockayne. 2. *Elatine hexandra*.

WATLING STREET THISTLE *Eryngium campestre* (Field Eryngo) N'thants. A E Baker.

WATTERY DRUMS *Senecio vulgaris* (Groundsel) Shet. Grigson.

WATTLE *Acacia sp.* Applied to several of the Australian acacias, given because the early settlers used them for makeshift wattle-and-daub dwellings, and also for wattle fences.

WATTLE, Black *Acacia decurrens*.

WATTLE, Black, African *Peltophorum africanum* (Rhodesian Wattle) Palgrave.

WATTLE, Blue-leaved *Acacia cyanophylla*.

WATTLE, Cootamundra *Acacia baileyana*.

WATTLE, Golden *Acacia pycnantha*.

WATTLE, Golden, Sydney *Acacia longifolia*.

WATTLE, Green *Acacia decurrens* (Black Wattle) Usher.

WATTLE, Knife-leaf *Acacia cultriformis*.

WATTLE, Mount Morgan *Acacia podalyrifolia*.

WATTLE, Rhodesian *Peltophorum africanum*.

WATTLE, Silver *Acacia dealbata*. 'Silver', because of the down that coats the foliage.

WATTLE, Silver, Queensland *Acacia podalyrifolia* (Mount Morgan Wattle) H & P.

WAVERAN-LEAF *Plantago major* (Great Plantain) Ork. Leask. Cf. other similar Scottish names, as Wabran-lead, Wabret-leaves, or Warba-leaves. They all seem to come originally from OE weg-breade, way breadth. Hence, too the English Waybread, or Waybroad.

WAVES, Green *Mercurialis perennis* (Dog's Mercury) Som. Macmillan.

WAVEWIND; WAVEWINE *Calystegia sepium* (Great Bindweed) Glos. Brill; Worcs. J Salisbury (Wavewind); B & H (Wavewine). Cf. rather Waywind and its many derivatives.

WAVY BITTERCRESS *Cardamine flexuosa*.

WAVYLEAF OAK *Quercus undulata*.

WAX CURRANT *Ribes cereum*.

WAX DOLLS *Fumaria officinalis* (Fumitory) Som. Macmillan; Hants, War, Yks, N'thum. Grigson; Kent Leyel.1937.

WAX FLOWER, Clustered *Stephanotis floribunda* (Madagascar Jasmine) Perry. Waxy flowers.

WAX GOURD, Chinese *Benincasa hispida*.

WAX MYRTLE *Myrica pennsylvanica* (Bayberry) USA Zenkert. Cf. Waxberry, and also Candleberry - the early settlers in New England made night lights from the white wax crust of the berries.

WAX TREE *Ligustrum lucidum*. It is cultivated for the white wax.

WAX TREE, Japanese *Rhus succedanea*. The seeds provide a wax suitable for lighting.

WAXBERRY 1. *Myrica pennsylvanica* (Bayberry) Mitton. 2. *Myrica quercifolia* S Africa Watt. 3. *Symphoricarpos albus* (Snowberry) USA Turner & Bell. 4. *Symphoricarpos orbiculatus* (Red Wolfberry) USA Elmore.

WAXFLOWER 1. *Asperula cynanchica* (Squinancywort) Dor. Grigson. 2. *Cammaphila umbellata* USA Yarnell. 3. *Hoya carnosa*.

WAXWORKS 1. *Celastrus scandens* (American Bittersweet) USA Harper. 2. *Polygala vulgaris* (Milkwort) Wilts. Macmillan.

WAXY MEADOW RUE *Thalictrum revolutum*.

WAY BENNETT *Lolium temulentum* (Darnel) Coles. Coles may have very well meant *Hordium murinum*, especially as the latter is also sometimes called Waybent. That suggests that the 'bennett' of the name is not 'benedicta' at all, but bent, a name for grass.

WAY THISTLE *Cirsium arvensis* (Creeping Thistle) Grieve.1931.

WAYAKA YAM BEAN *Pacchyrrhizus angulatus*.

WAYBENT *Hordeum murinum* (Wall Barley) H C Long.1910. An annual grass, common by roadsides, hence this name.

WAYBERRY *Plantago major* (Great Plantain) Ches. Grigson. 'Berry' is coincidental in this name, which probably started as Waybread, a common north country name for the plant - OE weg breade, way breadth.

WAYBORN; WAYBURN-LEAF *Plantago major* (Great Plantain) Ellacombe (Wayborn); Lanark. Grigson (Wayburn).

WAYBREAD; WAYBROAD; WAYBROAD-LEAF *Plantago major* (Great Plantain) N Eng, Scot. (Waybread); Worcs, Lancs. (Waybroad); Worcs. (Waybroad-leaf) all Grigson. See Wayberry.

WAYBREAD, Crowfoot *Plantago coronopus* (Buck's-horn Plantain) Turner. Cf. Crowfoot, or Crow's Foot Plantain.

WAYBREAD, Great *Plantago major* (Great Plantain) Turner.

WAYBROW *Plantago major* (Great Plantain) Ches. Grigson.

WAYFARING TREE *Viburnum lantana*. So called because of the tree's characteristically dusty appearance, like a traveller.

WAYFARING TREE, American *Viburnum alnifolium* (Hobble-bush) USA Zenkert.

WAYFORN; WAYFRON *Plantago major* (Great Plantain) Ellacombe (Wayform); Border B & H (Wayfron).

WAYGRASS *Polygonum aviculare* (Knotgrass) Kent Grigson. The meaning is clear - it is the plant that grows by roadsides.

WAYSIDE BEAUTY *Prunus spinosa* (Blackthorn) Som. Macmillan. A characteristically effusive, if accurate, Somerset name.

WAYSIDE BREAD *Plantago major* (Great Plantain) Wilts. B & H. 'Bread' here is OE breade, breadth, and Waybroad, still recorded in Worcestershire and Lancashire, is the most reasonable translation of the OE weg-breade.

WAYSIDE CUDWEED *Gnaphalium uliginosum* (Marsh Cudweed) McClintock.

WAYSIDE MOUSE-EAR CHICKWEED *Cerastium triviale* (Narrow-leaved Mouse-ear) J A MacCulloch.1905.

WAYTHORN *Rhamnus cathartica* (Buckthorn) Shrop. B & H. Cf. Highway Thorn, with the same meaning.

WAYWIND 1. *Calystegia sepium* (Great Bindweed) N'thants. A E Baker. Notts. Grigson. 2. *Convolvulus arvensis* (Field Bindweed) War, Oxf. Grigson. See Withywind. Presumably the "way" here is the usual "highway".

WAYWORT *Anagallis arvensis* (Scarlet Pimpernel) B & H. As Gerard said, "they grow in plowed fields neere path waies".

WEASEL PHACELIA *Phacelia mustelina*. 'Weasel', because of the smell.

WEASEL'S NOSE *Galeobdalon luteum* (Yellow Archangel) Dor. Macmillan. *Galeobdalon* means weasel stench, so calling it weasel's nose or weasel's snout seems to add an unnecessary further dimension to the name.

WEASEL'S SNOUT 1. *Galeobdalon luteum* (Yellow Archangel) Som. Elworthy; Glos. J D Robertson; Yks. Grigson. 2. *Linaria vulgaris* (Yellow Toadflax) Kent Grigson. 3. *Misopates orontium* (Lesser Snapdragon). See Weasel's Nose for Yellow Archangel. The other two plants. are regular recipients of 'mouth' or 'snout' names.

WEATHER, Change-of-the- *Anagallis arvensis* (Scarlet Pimpernel) Som. Macmillan. Pimpernel is the best known of all the plant weather forecasters. Gerard certainly knew its properties - he said that it will be fine or wet weather according to whether the flowers of the pimpernel in the morning are open or closed. In fact, it is any countryman's weather-forecaster. It can predict rain 24 hours ahead, so it is said in Wales.

WEATHER CLOCK *Taraxacum officinale*

(Dandelion) Som. Macmillan. When the down of the seed heads is fluffy, then there will be fine weather; when it is limp and contracted, there will be rain.

WEATHER FLOWER; WEATHER TELLER *Anagallis arvensis* (Scarlet Pimpernel) Dor. (Flower); Dor. (Teller) both Macmillan.

WEATHER PLANT *Abrus precatorius* (Precatory Bean) Howes.

WEATHER THISTLE *Carlina vulgaris* (Carline Thistle) Howes.

WEATHER WARNER, Husbandman's *Anagallis arvensis* (Scarlet Pimpoernel) Tynan & Maitland.

WEATHERGLASS *Anagallis arvensis* (Scarlet Pimpernel) Wilts. Dartnell & Goddard; Bucks. Grigson.

WEATHERGLASS, Countryman's; WEATHERGLASS, Farmer's; WEATHERGLASS, Husbandman's; WEATHERGLASS, Ploughman's *Anagallis arvensis* (Scarlet Pimpernel) Swainson.1873 (Countryman's); Som. Macmillan (Farmer's); Tynan & Maitland (Husbandman's); Wilts. Dartnell & Goddard (Ploughman's).

WEATHERGLASS, Grandfather's; WEATHERGLASS, Old Man's *Anagallis arvensis* (Scarlet Pimpernal) Dev. Macmillan (Grandfather's); Som. Grigson (Old Man's).

WEATHERGLASS, Poor Man's *Anagallis arvensis* (Scarlet Pimpernel) Corn. Deane & Shaw; Som. Macmillan; Hants. Cope; Ches. Holland; USA Whitney & Bullock.

WEATHERGLASS, Shepherd's 1. *Anagallis arvensis* (Scarlet Pimpernel) Dev. Friend.1882; Som. Macmillan; N'thants. A E Baker. 2. *Stellaria holostea* (Greater Stitchwort) Lancs. Grigson. Why is stitchwort in this company?

WEATHERWIND *Calystegia sepium* (Great Bindweed) B & H. 'Weather' is accidental in this name. It is really Withywind, or one of its derivatives.

WEATHERWIND, Cow's *Stachys sylvatica* (Hedge Betony) B & H. There is Cow's Withywind for this as well. They both sound like misinterpretations.

WEAVER'S BROOM *Spartium junceum* (Spanish Broom) Howes.

WEB, Spider's *Nigella damascena* (Love-in-a-mist) Dor. Macmillan.

WEDDING-CAKE TREE *Cornus controversa*. Because of what Huxley described as its "tiered arrangment".

WEDDING-FLOWER *Stellaria holostea* (Greater Stitchwort) Glos. Grigson.

WEDE-WIXEN *Genista tinctoria* (Dyer's Greenweed) Grieve.1931. 'Wede', or 'weed' means the dyeplant. The whole name varies very considerably, but Woadwaxen is probably the nearest to the present name. This dyeplant was mixed with woad to give a green dye.

WEDGE-LEAVED SAXIFRAGE *Saxifraga cuneifolia*.

WEDGESCALE *Atriplex truncata*.

WEEBOW 1. *Senecio jacobaea* (Ragwort) Scot. Jamieson. 2. *Tanacetum vulgare* (Tansy) Scot. Grigson.

WEEBY *Senecio jacobaea* (Ragwort) Scot. B & H.

WEEDBIND *Calystegia sepium* (Great Bindweed) B & H. Cf. Woodbind, but both of these are part of the 'withywind' sequence, and of course are just a reversal of the common name.

WEEDWIND 1. *Convolvulus arvensis* (Field Bindweed) Howes. 2. *Polygonum convolvulus* (Black Bindweed) B & H. Cf. Widwind for Field Bindweed. It is probably the origin of this name.

WEEPING BEECH, Green *Fagus sylvatica* var. *pendula*.

WEEPING BEECH, Purple *Fagus sylvatica* var. *purpurea* 'Pendula'.

WEEPING CYPRESS *Cupressus funebris* (White Fir) Wit.

WEEPING FIG *Ficus benjamina*.

WEEPING FORSYTHIA *Forsythia suspensa*.

WEEPING JUNIPER, Himalayan *Juniperus recurva* (Drooping Juniper) RHS.
WEEPING SALLY *Eucalyptus mitchelliana*.
WEEPING WIDOW *Fritillaria meleagris* (Snake's-head Lily) N'thants, Staffs. Grigson. The combination of sombre colouring and hanging head have given a number of similar names.
WEEPING WILLOW 1. *Fritillaria meleagris* (Snake's-head Lily) Coats. 2. *Laburnum anagyroides* (Laburnum) Dev. B & H. 3. *Salix alba var. tristis*. Any downward hanging plant gets the epithet 'weeping', but the willow deserves the name on another account, for it is the symbol of sorrow and forsaken love. The use of 'willow' for the fritillary is just an extension of the real name, which is 'widow'.
WEEPING WILLOW, Wisconsin *Salix blanda*.
WEGWANTS *Briza media* (Quaking-grass) Wilts. Dartnell & Goddard. The first syllable is better understood as 'wag'. Wagwants and a number of variations are also recorded.
WELCOME-HOME-HUSBAND *Euphorbia cyparissias* (Cypress Spurge) Yks. Carr.
WELCOME-HOME-HUSBAND-THOUGH-NEVER-SO-DRUNK 1. *Sedum acre* (Biting Stonecrop) Dor. Macmillan; Suff. Wright. 2. *Sempervivum tectorum* (Houseleek) Som. Tongue.
WELCOME-HOME-HUSBAND-THOUGH-NEVER-SO-LATE *Sempervivum tectorum* (Houseleek) Dor. Dacombe.
WELCOME-TO-OUR-HOUSE *Euphorbia cyparissias* (Cypress Spurge) Gerard.
WELD *Reseda luteola* (Dyer's Rocket) Gerard. 'Weld' is used both for the plant and for the yellow dye. It seems to be a Germanic word, possibly Wald, wood or forest, an earlier interpretation of which was uncultivated waste land.
WELD, Buck's-horn *Catananche caerulea* (Cupid's Darts) Gerard.
WELD, Dyer's *Reseda luteola* (Dyer's Rocket) Howes.

WELD, Greening, Dyer's *Genista tinctoria* (Dyer's Greenweed) B & H.
WELD, Mustard *Sinapis arvensis* (Charlock) J Smith.
WELGERS; WILGERS *Salix viminalis* (Osier) Dev. Choape, Grigson. They are probably, with the aid of the typically west country intrusive 'w', the same as Augers, also recorded, even though from Northamptonshire.
WELL-CRESS, Horse; WELL-GRASS, Horse *Veronica beccabunga* (Brooklime) Scot. Grigson (Cres); Scot. Jamieson (Grass).
WELL-GIRSE; WELL-KARSE; WELL-KERSE *Nasturtium officinale* (Watercress) all N'thum, Scot. Grigson (Girse); Jamieson (Karse, Kerse). All of these are variations of 'cress'.
WELLINGTONIA *Sequoia gigantea*. The name was given in honour of the Duke of Wellington. But it is known as Washingtonia in America.
WELLINK *Veronica beccabunga* (Brooklime) Cumb, Scot, Ire. Grigson. Also recorded, erroneously, as Wall Ink.
WELSH CORN *Zea mays* (Maize) Lehner. In this case, and in most of those that follow, 'Welsh' simply means 'foreign' - OE walch, or something like it.
WELSH GENTIAN *Gentianella uliginosa*. A very rare native, in dunes in South Wales.
WELSH NUT *Juglans regia* nut (Walnut) Evelyn. 'Foreign' again - in fact, with exactly the same meaning as 'walnut' itself.
WELSH ONION *Allium fistulosum*. Introduced here through Germany from Siberia.
WELSH PARSLEY *Cannabis sativa* (Hemp) Coles.
WELSH POPPY *Meconopsis cambrica*. As the specific name shows, 'Welsh' has geographical significance in this case.
WELSH RAGWORT *Senecio cambrensis*.
WELSH STONECROP, Small *Sedum forsterianum* (Rock Stonecrop) Phillips.
WELSH VINE *Sambucus nigra* (Elder) S

Wales G E Evans.1969. An ironic name - it implies that the only wine poor Welsh miners could afford was made from elderberries.

WERMOUT; WERMUD *Artemisia absinthium* (Wormwood) Pemb. (Wermout); N'thum. (Wermud) both Grigson. Variations on 'wormwood', of course, but reflecting not the modern word, but the OE weremod, or wermod.

WEST AFRICAN BLACK PEPPER *Piper guineense*.

WEST AFRICAN MULBERRY *Chlorophora excelsa* (African Oak) Dalziel. It is a member of the same family as the mulberries.

WEST AFRICAN NUTMEG *Monodora myristica* (Calabash Nutmeg) Howes.

WEST AFRICAN RUBBER TREE *Funtumia elastica* (Lagos Silk-rubber Tree) Dalziel. The latex of this tree yields about one third of its weight in pure rubber.

WEST HIMALAYAN FIR *Abies pindrow*.

WEST INDIAN ALOE *Aloe vera* (Mediterranean Aloe) Mitton. Evidently a far-flung plant, for it is called Cape, or Zanzibar Aloe, as well.

WEST INDIAN BIRCH *Bursera simaruba* Howes.

WEST INDIAN CEDAR *Cedrella odorata*.

WEST INDIAN CHERRY *Malpighia glabra* (Barbados Cherry) Brouk.

WEST INDIAN ELDER *Sambucus simpsonii*.

WEST INDIAN GHERKIN *Cucumis anguria* (Gherkin) Barbados Gooding.

WEST INDIAN IPECACUANHA *Asclepias curassivica* (Blood Flower) Chopra. Ipecacuanha apparently means "wayside-plant-emetic", and that is exactly what it is used for in the West Indies. It is known too as Bastard, or Wild, Ipecacuanha.

WEST INDIAN JASMINE *Plumeria rubra* (Frangipani) Bonar.

WEST INDIAN LOCUST *Hymenaea courbaril*.

WEST INDIAN MIGNONETTE *Lawsonia inermis* (Henna) Howes.

WEST INDIAN WILD CINNAMON *Canella alba* (White Cinnamon) Le Strange.

WESTERN BALSAM POPLAR *Populus trichocarpa*. Most of these 'western' names refer to the west side of the North American continent.

WESTERN BLEEDING HEART *Dicentra formosa*.

WESTERN BLUE FLAG *Iris missouriensis* (Missouri Iris) USA Elmore.

WESTERN BUTTERWORT *Pinguicula lusitanica*. 'Western' in the sense that it is a native of the Atlantic seaboard of Europe.

WESTERN CHOKE CHERRY *Prunus demissa*.

WESTERN CONEFLOWER *Rudbeckia occidentalis*.

WESTERN CORAL BEAN *Erythrina flabelliformis*.

WESTERN FIGWORT *Scrophularia umbrosa*.

WESTERN GLASSWORT *Salicornia rubra*.

WESTERN GORSE *Ulex gallii* (Dwarf Furze) McClintock.

WESTERN HACKBERRY *Celtis reticulata*.

WESTERN HEMLOCK *Tsuga heterophylla*.

WESTERN JIMSON *Datura meteloides* (Downy Thorn-apple) Jaeger.

WESTERN JUNIPER *Juniperus occidentalis* (Sierra Juniper) USA Elmore.

WESTERN LABRADOR TEA *Ledum glandulosum*.

WESTERN LARCH *Larix occidentalis*.

WESTERN MONKSHOOD *Aconitum columbianum*.

WESTERN PINK *Dianthus gallicus*. 'Western', referring to the Atlantic coast of Europe.

WESTERN PLANE *Platanus occidentalis*.

WESTERN POISON OAK *Rhus diversiloba* (Poison Oak) Le Strange.

WESTERN RAGWEED *Ambrosia psilostachya* (Black Sage) USA Vestal & Schultes.

WESTERN RASPBERRY *Rubus leuocodermis*.

WESTERN RED CEDAR 1. *Juniperus scop-

ulorum. 2. *Thuja plicata* (Giant Cedar) Clapham.
WESTERN SPIDERWORT *Tradescantia occidentalis.*
WESTERN SPRING BEAUTY *Claytonia lanceolata.*
WESTERN SQUAW LETTUCE *Hydrophyllum occidentale.*
WESTERN TANSY MUSTARD *Sisymbrium incisum.*
WESTERN THOROUGHWORT *Eupatorium occidentale.*
WESTERN TURK'S CAP LILY *Lilium michiganense.*
WESTERN WALLFLOWER 1. *Erysimum asperum* (Desert Wallflower) USA Elmore. 2. *Erysimum capitatum.*
WESTERN WATER HEMLOCK *Cicuta douglasi.*
WESTERN WHEAT GRASS *Agropyron smithii.*
WESTERN WHITE CEDAR *Thuja plicata* (Giant Cedar) Hora. Even though it is better known as the Western Red Cedar.
WESTERN WHITE PINE *Pinus monticola* (Mountain White Pine) USA Schery.
WESTERN WILD DAISY *Aster integrifolia.*
WESTERN WILD IRIS *Iris missouriensis* (Missouri Iris) USA Elmore.
WESTERN YARROW *Achillea lanulosa* (Woolly Yarrow) USA Gates.
WESTERN YELLOW PINE 1. *Pinus pinea* (Stone Pine) Watt. 2. *Pinus ponderosa* (Yellow Pine) USA Kluckhohn.
WESTERN YEW *Taxus brevifolia.*
WESTFELTON YEW *Taxus baccata var. dovastoniana.* A variety found by John Dovaston, of Westfelton, Shropshire, in 1771.
WET-A-BED; WET-THE-BED *Taraxacum officinale* (Dandelion) Som. Grigson. Dandelion's diuretic properties are well known, and Pissabed is probably its most famous name. "Children that eat it in the evening experience its diuretic effects, which is the reason that other European nations, as well as ourselves, vulgarly call it Pissabed".

It is said that if they even gather the flowers, they will experience the symptoms.
WETS *Avena sativa* (Oats) Dev. Hewett.1892. A form that accomodates local pronunciation of oats.
WETWEED 1. *Euphorbia helioscopia* (Sun Spurge) Norf. Grigson. 2. *Taraxacum officinale* (Dandelion) Mabey. For Sun Spurge, not 'Wetweed', but 'Wartweed', via a typical dialectal misplacement, 'Wretweed', from the same area. Children have always used the milky juice to put on the warts on their fingers.
WEYBRED *Plantago major* (Great Plantain) Gerard. OE weg-breade, literally way-breadth. Waybroad is a more obvious name.
WEYMOUTH PINE *Pinus strobus* (Western White Pine) Geldart. It was introduced into Britain by Lord Weymouth at Longleat, although it was first cultivated at Badminton by the Duchess of Beaufort. But large quantities of it were grown at Longleat by 1710.
WHARRE *Malus sylvestris* fruit (Crabapple) Ches. Holland. "Sour as wharre" is a local saying. The word is obviously akin to the Warwickshire Varges - bad beer was said to be "sour as varges".
WHAT O'CLOCK *Taraxacum officinale* (Dandelion) Putnam. Children tell the time by the number of puffs it takes to scatter all the seeds from the head.
WHAT'S-YOUR-SWEETHEART *Lolium perenne* (Rye-grass) Suss. B & H. A children's game, played either by pulling off the alternating spikelets, hence Yes-or-no, or Aye-no Bent, or by striking the heads together, and at each blow saying Tinker, tailor, soldier, sailor, etc. The blow that knocks the head off marks the profession of the future husband.
WHEAT, Bread *Triticum aestivum.* As might be suspected, 'wheat' is a word allied to 'white' - OE hwaete in the one case, and hwit in the other.
WHEAT, Club *Triticum compactum.*
WHEAT, Durum; WHEAT, Hard *Triticum durum.*

WHEAT, French *Fagopyrum esculentum* (Buckwheat) Gerard. 'French', because it is a foreign plant, not because it grows in France. It is in fact Asian.

WHEAT, Inca *Amaranthus caudatus* (Love-lies-bleeding) Perry. It is still grown in the Andean region of Bolivia, Peru and northern Argentina.

WHEAT, India *Fagopyrum tataricum* (Green Buckwheat) Douglas.

WHEAT, Indian 1. *Fagopyrum esculentum* (Buckwheat) USA Grieve.1931. 2. *Zea mays* (Maize) Cobbett.

WHEAT, Indian, Woolly *Plantago purshii* (Pursh Plantain) USA Elmore.

WHEAT, Mother of *Veronica hederifolia* (Ivy-leaved Speedwell) Border B & H. Perhaps because it seems to precede wheat in disturbed ground.

WHEAT, Rivet *Triticum turgidum*.

WHEAT BARLEY *Hordeum sativum* (Barley) Turner. Turner differentiated what he called Dutch Barley, and Wheat Barley, "because it hath no mo Huskes on it than wheat".

WHEAT-GRASS, Creeping *Agropyron repens* (Couch-grass) Barnes.

WHEAT-GRASS, Slender *Agropyron trachycaulum*.

WHEAT-GRASS, Western *Agropyron smithii*.

WHEATBINE *Convolvulus arvensis* (Field Bindweed) Herts. B & H. Cf. Cornbine. These are two of the very many "bine" or "bind" names for the plant.

WHEAT-THIEF *Buglossoides arvensis* (Corn Gromwell) USA Zenkert. For this is a weed of arable fields.

WHEATLEY ELM *Ulmus stricta var. sarniensis* (Guernsey Elm) Barber & Phillips.

WHEEL LILY *Lilium medeoloides*. A translation of the Japanese name, Kuruma-yuri, given from the whorled arrangement of the leaves.

WHEELS, Watch *Ranunculus arvensis* (Corn Buttercup) H C Long.1910. Descriptive of the flat, spiny seed vessels.

WHEELSCALE *Atriplex elegans var. fasciculata*.

WHEYS, Bulls-and- *Arum maculatum* (Cuckoo-pint) Yks, Cumb. B & H. A whey is a heifer, so this is one of the series of male plus female names like Bulls-and-cows, or Lords-and ladies. The imagery is provided by the spadix in the spathe, a symbol for copulation.

WHICKEN 1. *Agropyron repens* (Couch-grass) H C Long.1910. 2. *Sorbus aucuparia* (Rowan) Yks. B & H; N'thum. Denham/Hardy. In both cases, the word is a variation of 'Quicken', which is OE cuic, alive.

WHICKS 1. *Agropyron repens* (Couch-grass) Yks. Carr. 2. *Crataegus monogyna* (Hawthorn) Ches. B & H. See previous entry.

WHICKY *Sorbus aucuparia* (Rowan) Blount. See Whicken.

WHIMBERRY *Vaccinium vitis-idaea* (Cowberry) Howes.

WHIN 1. *Ononis repens* (Rest-harrow) N'thants. A E Baker; Cambs. Grigson. 2. *Ulex europaeus* (Furze) Gerard. 3. *Ulex minor* (Small Furze) B & H. Whin is O Norse hvin, still found in place-names like Whinburgh, Norfolk. It is the name for *Ulex europaeus* in the Scandinavian areas of England.

WHIN, Cat 1. *Genista anglica* (Needle Whin) Cumb, Yks, Kirk, Wigt. Grigson. 2. *Ononis repens* (Rest-harrow) Som, Yks. Grigson. 3. *Rosa canina* (Dog Rose) Yks. Morris. 4. *Rosa pimpinellifolia* (Burnet Rose) Yks, N'thum. Grigson. 5. *Ulex gallii* (Dwarf Furze) Cumb. Grigson. 6. *Ulex minor* (Small Furze) Cumb. B & H.

WHIN, Galloway *Genista anglica* (Needle Whin) Scot. Grigson.

WHIN, Heather 1. *Genista anglica* (Needle Whin) Berw. Grigson. 2. *Genista tinctoria* (Dyer's Greenweed) Border B & H.

WHIN, Lady *Onosis repens* (Rest-harrow) Som. Macmillan; Scot. Grigson.

WHIN, Land *Onosis repens* (Rest-harrow) E Ang. Halliwell.

WHIN, Needle *Genista anglica*.

WHIN, Moor; WHIN, Moss *Genista anglica* (Needle Whin) both Berw. Grigson.

WIIIN, Petty 1. *Genista anglica* (Needle Whin) Gerard. 2. *Ononis repens* (Rest-harrow) Gerard. 3. *Ulex minor* (Small Furze) Jones & Dillon.

WHIN-CAMMOCK *Ononis repens* (Rest-harrow) B & H. 'Cammock' is the proper English name for the plant.

WHINBERRY *Vaccinium myrtillus* (Whortleberry) Fernie.

WHINSHAG *Fraxinus excelsior* (Ash) Donegal Grigson. Presumably a Gaelic name.

WHIP-POOR-WILL'S SHOE *Cypripedium reginae* (Showy Lady's Slipper) USA House.

WHIPBEAM *Sorbus aria* (Whitebeam) Herts. B & H. Probably 'whit', i.e. white, rather than 'whip', although there are instances of whips being made from the young branches.

WHIPCROP 1. *Sorbus aria* (Whitebeam) IOW Long. 2. *Viburnum lantana* (Wayfaring Tree) IOW Grigson. 3. *Viburnum opulus* (Guelder Rose) IOW Grigson. Farmers used to make whips from the wood of Wayfaring Tree, which is very tough. Cf. such names as Whiptop and Twistwood, i.e to be twisted into whips. For whitebeam, see the previous entry.

WHIPTONGUE 1. *Galium aparine* (Goose-grass) Fernie. 2. *Galium mollugo* (White Bedstraw) B & H. Cf. Blind-tongue or Blood-tongue for Goose-grass. Blind-tongue is a children's game. Goose-tongue is also recorded, a mark of the roughness of the back of the leaves.

WHIPTOP *Viburnum lantana* (Wayfaring Tree) Dor. Macmillan. See Whipcrop.

WHIRL-MINT *Mentha pulegium* (Pennyroyal) Hants. Grigson.

WHISKERS *Adoxa moschatellina* (Moschatel) Som. Macmillan.

WHISKERS, Cat's 1. *Gynandropsis gynandra* West Indies Dalziel. 2. *Tacca chantrieri* (Bat Flower) RHS.

WHISKERS, Daddy's; WHISKERS, Grandfather's *Clematis vitalba* (Old Man's Beard) Som. Macmillan; Wilts. Dartnell & Goddard (Daddy's); Corn, Som. Grigson (Grandfather's).

WHISKERS, Old Man's *Geum triflorum* (Prairie Smoke) USA T H Kearney. Silvery feathered tails to the fruit, hence this name and others like it.

WHISTLE, Wode *Conium maculatum* (Hemlock) B & H. This is a medieval book name for the plant. In the eastern parts of Yorkshire, boys used to make whistles of the hollow stems.

WHISTLE TREE *Acer pseudo-platanus* (Sycamore) Corn. Grigson. Cf. Whistlewood from the northern counties. Presumably children used some part of the tree from which to make whistles, but what? Apparently, Lancashire children made their whistles from the bark of rowan (see next entry).

WHISTLEWOOD 1. *Acer campestre* (Field Maple) Scot. Grigson. 2. *Acer pseudo-platanus* (Sycamore) N'thum. Grigson. 3. *Alnus glutinosa* (Alder) N'thum. Grigson. 4. *Sorbus aucuparia* (Rowan) Yks. Grigson. See Whistle Tree.

WHISTLING PINE *Casuarina equisetifolia* (Beefwood) Kenya Dale & Greenway.

WHISTLING THORN *Acacia drepanolobium* (Black-galled Acacia) Kenya Dale & Greenway.

WHISTLING THORN, Coast *Acacia zanzibarica*.

WHIT-ALLER *Sambucus nigra* (Elder) Som. Elworthy. 'White elder', in spite of the 'nigra' of the botanical name.

WHITBIN PEAR *Sorbus aria* (Whitebeam) Chaplin. Cf. Widbin Pear. They must both be renderings of 'whitebeam'.

WHITE-AND-RED *Arum maculatum*

(Cuckoo-pint) Dor. Macmillan. Descriptive,and quite in keeping with the many double names for the plant, as for instance Lords-and-ladies.

WHITE FLOWER OF HELL *Silene cucubalis* (Bladder Campion) Dor. Macmillan. From a superstition that the leaves and "bladders" were poisonous. Actually, children often eat the young leaves, which are supposed to taste like green peas. They have even been used as a substitute for asparagus.

WHITE-FLOWERED GRASS *Stellaria holostea* (Greater Sitchwort) Wilts. Macmillan.

WHITE MAN'S FOOTPRINTS *Plantago major* (Great Plantain) USA Watt. There is a well-known legend of its persistently following the tracks of man. More specifically, one superstition says that it follows Englishmen, and springs up in whatever part of the world he makes his home. In this case, White Man's Footprints is what the North American Indians call the plant. All this comes from the shape of the leaf - Plantago is from Latin planta, the sole of the foot.

WHITE TREE *Melaleuca leucadendron* (Cajuput) Chopra. Cf. Paperbark for this.

WHITEBACK *Populus alba* (White Poplar) E Ang. Forby.

WHITEBARK PINE *Pinus albicaulis*.

WHITEBEAM *Sorbus aria*. 'Beam' is the OE word for a tree, which makes one of its names, Beam Tree, a nonsense.

WHITEBEAM, Cornish *Sorbus x latifolia* (French Hales) Brimble.1948.

WHITEBEAM, Cut-leaved *Sorbus intermedia*.

WHITEBEAM, Finnish *Sorbus x hybrida*.

WHITEBEAM, Himalayan *Sorbus cuspidata*.

WHITEBEAM, Scots *Sorbus intermedia* (Cut-leaved Whitebeam) Brimble.1948.

WHITEBEAM, Swedish *Sorbus intermedia* (Cut-leaved Whitebeam) Perry.

WHITEBLOW GRASS *Erophila verna* (Whitlow Grass) Gerard. It sounds almost like a pun of Gerard's on 'whitlow'. But 'whiteblow' means white flowers, 'blow' being the old usage meaning to blossom.

WHITEBRUSH *Lippia ligustrina*.

WHITEFLOWER *Stellaria holostea* (Greater Stitchwort) Wilts. Dartnell & Goddard. Cf. White-flowered Grass, from the same area.

WHITEHEAD 1. *Parthenium hysterophorus* Barbados Gooding. 2. *Typha latifolia* (False Bulrush) Dev. Friend.1882. When False Bulrush has lost its 'Blackhead' colour.

WHITELEAF *Spiraea tomentosa* (Hardhack) Grieve.1931.

WHITELEAF TREE *Sorbus aria* (Whitebeam) Evelyn.

WHITEROOT 1. *Pinguicula vulgaris* (Butterwort) Gerard. 2. *Polygonatum multiflorum* (Solomon's Seal) Gerard. In the first instance, is 'Whiteroot' rather 'White Rot'? There is a similar example in 'Sheep-rot' and 'Sheep-root'. The plant has always had the reputation of causing rot in sheep.

WHITETHORN 1. *Acacia constricta* (Mescat Acacia) T H Kearney. 2. *Crataegus monogyna* (Hawthorn) Gerard.

WHITETOP 1. *Cardaria draba* (Hoary Cress) USA Gates. 2. *Erigeron ramosus* USA Yarnell.

WHITEWEED 1. *Achillea ptarmica* (Sneezewort) N Ire. Grigson. 2. *Anthriscus sylvestris* (Cow Parsley) Yks. B & H. 3. *Leucanthemum vulgare* (Ox-eye Daisy) Le Strange. 4. *Viburnum lantana* (Wayfaring Tree) Wilts. Macmillan.

WHITEWOOD 1. *Canella alba* (White Cinnamon) Le Strange. 2. *Ilex mitis* (Cape Holly) S Africa Watt. 3. *Liriodendron tulipifera* (Tulip Tree) Harper. 4. *Melicytus ramiflorus*. 5. *Picea excelsa* (Spruce Fir) N Davey. 6. *Populus alba* (White Poplar) Herts. Grigson. 7. *Tilia europaea* (Lime) IOW Long; Worcs. B & H. 8. *Viburnum lantana* (Wayfaring Tree) Dor. B & H; Wilts. Dartnell & Goddard.

WHITEWOOD, African *Enantia chlorantha* (African Yellow-wood). The timber is yellow throughout, and as soft as any other whitewood.

WHITEWOOD, Bahama *Canella alba* (White Cinnamon) Howes.

WHITEWOOD, Canary *Liriodendron tulipifera* (Tulip Tree) Willis.

WHITEWORT 1. *Anthemis arvensis* (Corn Camomile) Hants. Cope. 2. *Polygonatum multiflorum* (Solomon's Seal) Turner. 3. *Tanacetum parthenium* (Feverfew) IOW B & H.

WHITLOW GRASS 1. *Draba sp.* 2. *Erophila verna*. 3. *Euphorbia helioscopia* (Sun Spurge) Lincs. B & H. 4. *Saxifraga tridactylites* (Rue-leaved Saxifrage) Gerard. The true Whitlow Grasses have the name from their use - "as touching the qualitie hereof, we have nothing to set downe; only it hath been taken to heale the disease of the nailes called a Whitlow, whereof it tooke his name" (Gerard). Sun Spurge is more famous for its use against warts, but presumably whitlows can have the same treatment, though it sounds unwise.

WHITLOW GRASS, Spring *Erophila verna* (Whitlow Grass) Clapham.

WHITLOW GRASS, Twisted; WHITLOW GRASS, Twisted-podded *Draba incana* (Hoary Whitlow Grass) McClintock (Twisted); Murdoch McNeill (Twisted-podded).

WHITLOW PEPPERWORT *Cardaria draba* (Hoary Cress) PLNN10. Evidently used as a whitlow cure, but no evidence of it remains.

WHITLOW WORT *Paronychia sp.*

WHIT SUNDAY 1. *Narcissus pseudo-narcissus* (Daffodil) Dev. Greenoak. 2. *Stellaria holostea* (Greater Stitchwort) Dev. Friend.1882. The name applied to daffodil seems entirely misplaced. Whitsun would have to very early indeed to find daffodils still in bloom. Much more appropriate are such names as Easter Lily or Easter Rose, and the various 'Lent' names. In any case, 'whit' means 'white' (see the next entry), so stitchwort qualifies very well. Even in that case, there is confusion between Easter and Whitsun, for both appear in the names.

WHITE SUNDAY *Stellaria holostea* (Greater Stitchwort) Dev. Friend.1882.

WHITNEY *Viburnum lantana* (Wayfaring Tree) Dev. Friend.1882. Cf. Whitten-tree, of which this a variation, They are probably from the white branches, but there has been a suggestion that they are connected with withe. Certainly there used to be 'whip' uses, and the names mirror this - see Whipcrop or Whiptop.

WHITSUN BALLS 1. *Peonia mascula* (Peony) Som. Macmillan. 2. *Viburnum opulus* (Guelder Rose) Som. Macmillan. Guelder Rose has been used as the ecclesiastical symbol for Whitsuntide, and there are a number of other names using the name of the festival. 'Balls', of course, because of the globular flower heads.

WHITSUN BOSS; WHITSUN ROSE *Viburnum opulus* (Guelder Rose) Glos. J D Robertson; Heref. Leather (Boss); Dev. Macmillan (Rose).

WHITSUN FLOWER 1. *Oxalis acetosella* (Wood Sorrel) Dor. Macmillan. 2. *Viburnum opulus* (Guelder Rose) Som. Macmillan. Whitsun would have to be very early to coincide with the flowering of Wood Sorrel.

WHITSUN GILLIES; WHITSUN GILLIFLOWER *Hesperis matronalis* (Sweet Rocket) War. Broom (Gillies); Som. Elworthy; Dor. Dacombe (Gilliflower).

WHITSUN TASSELS *Viburnum opulus* (Guelder Rose) Som. Macmillan.

WHITSUNTIDE *Syringa vulgaris* (Lilac) Som. Macmillan.

WHITTEN *Sorbus aria* (Whitebeam) Hants. Grigson.

WHITTEN-TREE 1. *Sorbus aucuparia* (Rowan) Shrop, Ire. Grigson. 2. *Viburnum lantana* (Wayfaring Tree) Dev. Friend.1882. 3. *Viburnum opulus* (Guelder Rose) Gerard. The word may mean 'white' for *Sorbus aria* in the previous entry, and also for the Viburnums, but surely it cannot be so for the rowan. It is part of a long series that starts with Gerard's Quickbeam, i.e. it is OE cuic, alive, and that becomes 'twick', in the southern counties, and Whick from the Midlands northwards.

WHITTENBEAM 1. *Sorbus aria* (Whitebeam) Hants. Grigson. 2. *Viburnum lantana* (Wayfaring Tree) common in Midl and S. Eng. Grigson.

WHITTY *Sorbus aucuparia* (Rowan) Shrop,Rad. Denham/Hardy. See Whitten-tree.

WHITTY-BUSH 1. *Acer campestre* (Field Maple) Shrop. Grigson. 2. *Sorbus torminalis* (Wild Service) Worcs. Grigson.

WHITTY-PEAR 1. *Sorbus domestica* (True Service) Rackham.1986. 2. *Sorbus torminalis* (Wild Service) Worcs. Grigson. Cf. Witten Pear. They are both mirrors of rowan names.

WHITTY-TREE 1. *Sorbus aucuparia* (Rowan) Heref. G C Lewis. 2. *Viburnum lantana* (Wayfaring Tree) Aubrey. See Whitten-tree.

WHO-STOLE-THE-DONKEY? *Galium aparine* (Goose-grass) Som. Macmillan. The plant, or rather the burrs, are called Donkeys in Somerset.

WHOREMAN'S PERMACETTY *Capsella bursa-pastoris* (Shepherd's Purse) Culpepper. Is this Culpepper's joke, or was it a misrepresentation? The original is obviously Poor Man's Parmacetty, as used by Gerard. According to Dyer, the whole thing is a joke on the Latin bursa, a purse,which to a poor man is always the best remedy for his bruises. On the other hand, Shepherd's Purse is a well known domestic remedy for a lot of ills, especially bleeding, internal as well as external.

WHORLED CARAWAY *Carum verticillatum*.

WHORLED MALLOW *Malva verticillata*.

WHORLED MILKWEED *Asclepias verticillata*.

WHORLED MINT *Mentha x verticillata*.

WHORLED SAGE *Salvia verticillata*.

WHORLED SILKWEED *Asclepias verticillata* (Whorled Milkweed) USA Elmore. Perhaps a misprint.

WHORLED SOLOMON'S SEAL *Polygonatum verticillatum*.

WHORTS *Vaccinium myrtillus* (Whortleberry) S Eng. Grigson. Whorts for whortleberry is a very common contraction through the south of England, sometimes rendered as Hurts, but there are many other variations as well.

WHORT, Black; WHORTLE, Black *Vaccinium myrtillus* (Whortleberry) both B & H. 'Black', for 'Red Whortleberry' is a name for Cowberry.

WHORTLE-BILBERRY *Vaccinium myrtillus* (Whortleberry) B & H.

WHORTLE-LEAVED WILLOW *Salix myrsinites*.

WHORTLEBERRY *Vaccinium myrtillus*.

WHORTLEBERRY, Bear *Arctostaphylos uva-ursi* (Bearberry) Gerard.

WHORTLEBERRY, Blue *Vaccinium ovalifolium*.

WHORTLEBERRY, Bog *Vaccinium uliginosum* (Northern Bilberry) Clapham.

WHORTLEBERRY, Caucasian *Vaccinium arctostaphylos*.

WHORTLEBERRY, Grouse *Vaccinium scoparium* (Low Huckleberry) Yanovsky.

WHORTLEBERRY, Marsh *Vaccinium oxycoccus* (Cranberry) Fernie.

WHORTLEBERRY, Red 1. *Vaccinium parvifolium* (Red Bilberry) Yanovsky. 2. *Vaccinium vitis-idaea* (Cowberry) Gerard.

WHORTLEBERRY, Red, Tiny *Vaccinium scoparium* Yanovsky.

WHORTLEBERRY, Rocky Mountain *Vaccinium oreophibium*.

WHUN *Ulex europaeus* (Furze) Ire. B & H. Merely the Irish pronunciation of whin.

WHURT *Vaccinium myrtillus* (Whortleberry) Corn. Carew. Dev. Hewett.1892.

WHUTTLE-GRASS *Melilotus officinalis* (Melilot) Roxb. B & H.

WHYA-TREE *Robinia pseudacacia* (Common Acacia) Yks. B & H.

WIBROW; WIBROW-WOBROW *Plantago*

major (Great Plantain) both Ches. Grigson. Evidently forms of Waybread, which is OE weg breaed, literally way breadth. That name varies into a number of forms, of which these are the most extreme, but which include Warba-leaves, Wayberry, and Wayside Bread.

WICK 1. *Agropyron repens* (Couch-grass) Grigson.1959. 2. *Crataegus monogyna* (Hawthorn) Yks. B & H. In both cases, this is 'quick' - OE cuic, alive, obvious as far as couch-grass in concerned, and for hawthorn a reference to the habit of using live thorns as a hedge in contrast to the dead material used in early fences and hedges. The name appears in various guises, as Whick, Wicken and Quickset or Quickthorn.

WICKED HERB *Filago germanica* (Cudweed) Notes & Queries. vol 7. i.e. the herba impia of the old botanists - wicked because of its way of growth. The main stem has its flowers at the top, but underneath this grow two or three more flowering shoots, all rising above the main stem. Therein lies the wickedness, for it conveys the idea that children were "undutifully disposed to exalt themselves above the parent flower".

WICKED TREE *Cuscuta epithymum* (Dodder) Dor. Macmillan. Dodder was often associated with the devil, both here and in France. This name must be a continuation of the theme that includes such names as Devil's Guts and Hellweed.

WICKEN *Agropyron repens* (Couch-grass) Yks. F K Robinson. See Wick.

WICKEN, Witch *Sorbus aucuparia* (Rowan) Lincs. Grigson. Cf. the Derbyshire Witch-wiggin. They are both examples of forgotten origins, for 'wicken', as with previous examples, means 'alive'. But 'wicken' has many variants, of which Witchen-tree and Witchwood are examples, so that a name like Witch-wicken is entirely meaningless.

WICKEN TREE *Sorbus aucuparia* (Rowan) Som. Tongue; Shrop, Ches, Cumb, Dur. Denham/Hardy; Lincs. Rudkin.

WICKENWOOD; WICKY *Sorbus aucuparia* (Rowan) Yks. Morris (Wickenwood); Shrop, Ches, Cumb, Dur. Grigson (Wicky).

WICKLOW MARSH ORCHID *Dactylorchis traunsteineri* (Pugley's Marsh Orchid) Turner Ettlinger.

WICKUP; WICOPY *Chamaenerion angustifolium* (Rosebay) USA Sanecki.

WICKY 1. *Kalmia angustifolia* (Sheep Laurel) USA House. 2. *Kalmia hirsuta* (Hairy Mountain Laurel) USA Harper.

WIDBIN 1. *Cornus sanguinea* (Dogwood) Bucks. B & H. 2. *Lonicera periclymenum* (Honeysuckle) Scot. B & H. 'Woodbine', of course, at least as far as honeysuckle is concerned. With dogwood, the likelihood is that that the first syllable means 'white', which would make the original of this name 'whitbin'.

WIDBIN PEAR TREE *Sorbus aria* (Whitebeam) Bucks. Harman. 'Widbin', as 'whitbin', is a variation on 'whitebeam'.

WIDDY *Salix viminalis* (Withy) Yks. Atkinson. A local (Yorkshire) version of 'withy'. But Cf. the next entry, which is from Somerset.

WIDDY-WINE *Convolvulus arvensis* (Field Bindweed) Som. Macmillan. 'Withywind' is the point of origin.

WIDEWIND *Lonicera periclymenum* (Honeysuckle) War. Bloom. A relative either of 'withywind', usually associated with the bindweeds, or, more likely, of 'woodwind', i.e. woodbine.

WIDOW, Grass *Sisyrinchium douglassi*. The grass-like leaves must have suggested the name, for the colour of the flowers does not seem to be very mournful.

WIDOW, Mournful 1. *Geranium phaeum* (Dusky Cranesbill) Lancs, Yks. Grigson. 2. *Knautia arvensis* (Field Scabious) Dev. Friend.1882. 3. *Scabiosa atropurpurea* (Sweet Scabious) Wright. It is the colour of the flowers of Dusky Cranesbill and Sweet Scabious that suggest mourning shades. *Knautia arvensis* gets the name by association only.

WIDOW, Mourning 1. *Geranium phaeum*

(Dusky Cranesbill) Coats. 2. *Scabiosa atropurpurea* (Sweet Scabious) Wright. Cf. Mournful Widow for both of these, and Poor Widow and Widow's Flower for the scabious.

WIDOW, Poor *Scabiosa atropurpurea* (Sweet Scabious) Wright.

WIDOW, Ranting *Chamaenerion angustifolium* (Rosebay) Ches. Wright. 'Widow' in this case is 'widdy', or willow, for this is Rosebay Willowherb. There is a resemblance of its leaves to those of the willow, hence the large number of 'willow' names.

WIDOW, Shy; WIDOW, Weeping *Fritillaria meleagris* (Snake's-head Lily) War. Grigson (Shy); N'thants, Staffs Grigson (Weeping). 'Widow', because of the sombre colouring, and 'shy' (or weeping), because of the way the plant hangs its head.

WIDOW IRIS *Hermodactylus tuberosus* (Snakeshead Iris) A W Smith. Again, from the colour of the flowers.

WIDOW-VEIL *Fritillaria meleagris* (Snake's-head Lily) Coats.

WIDOW-WAIL 1. *Fritillaria mleagris* (Snake's-head Lily) Shrop. B & H. 2. *Daphne mezereum* (Lady Laurel) Gerard. Explained by the sombre colouring in the case of the Fritillary, and the poisonous nature of Lady Laurel.

WIDOW-WAIL, Mountain *Daphne cneorum* (Flax-leaved Daphne) Gerard.

WIDOW-WISSE *Genista tinctoria* (Dyer's Greenweed) A S Palmer. It must be another version of Woadwax or one of its variants. Dyer's Greenweed was mixed with woad to give a green dye.

WIDOW'S FLOWER *Scabiosa atropurpurea* (Sweet Scabious) Folkard.

WIDOW'S TEARS *Tradescantia virginica* (Spiderwort) Coats. The purple pigmented sap colours the dewdrops that hang from the closed-up flowers in the early morning. The name, it is said cynically, was given because the drops dry in a day.

WIDOW'S WEEDS *Aquilegia vulgaris* (Columbine) Wilts. Macmillan.

WIDOW'S WILLOW *Salix fragilis* (Crack Willow) Suss. PLNN 31.

WIDWIND *Convolvulus arvensis* (Field Bindweed) Som. Macmillan. Cf. Withywind, etc.

WIG, Bishop's *Arabis alpina* (Alpine Rock-cress) Ches. B & H. From the resemblance of the tufts in full flower to the wigs once worn by bishops.

WIG TREE *Cotinus coggyria* (Venetian Sumach) Lindley. From the wig-like flower heads.

WIGEON-GRASS *Zostera marina* (Eel-grass) Greenoak.1979. Wigeon feed on it. So do Brent Geese, even more exclusively than wigeon.

WIGWAG-WANTONS; WIG-WAGONS *Briza media* (Quaking-grass) Bucks. Grigson.1959 (Wigwag-wantons); Wilts. Grigson.1959 (Wig-wagons).

WIGGAN; WIGGEN; WIGGIN; WIGGY *Sorbus aucuparia* (Rowan) N Eng. Denham/ Hardy (Wiggan); Dev, Som, Wales, Ches, Derb, Lancs, Cumb Grigson; Yks. Hunter (Wiggen); Wales, Cumb. Grigson (Wiggin); Dur. Denham/Hardy (Wiggy).

WIGGERS *Taraxacum officinale* (Dandelion) Fernie.

WIGGIN, Witch *Sorbus aucuparia* (Rowan) Derb. Addy. Cf. Witch-wicken in Lincolnshire. Both elements of the name are from OE cuic, alive, and the number of variants are legion.

WIGGLE-WAGGLE GRASS; WIGGLE-WAGGLES; WIGGLE-WAGGLE-WANTONS; WIGGLE-WANTS *Briza media* (Quaking-grass) Hants. Grigson.1959 (Wiggle-waggle Grass); Som, Suss. Grigson.1959 Hants. Boase (Wiggle-waggles); Berks. Grigson.1959 (Wiggle-waggle-wantons); Wilts. Dartnell & Goddard.1899 (Wiggle-wants).

WIGWAMS; WIGWANTS *Briza media* (Quaking-grass) Dor. Dacombe (Wigwams); Wilts. Dartnell & Goddard (Wigwants).

WILD-FIRE 1. *Achillea ptarmica* (Sneezewort) Donegal Grigson. 2. *Caltha* *palustris* (Marsh Marigold) Kirk. Grigson. As far as Marsh Marigold is concerned, the imagery probably concerns the golden-yellow flowers. Cf. The Bucks. Fire-o'-gold.

WILDING TREE *Malus sylvestris* (Crabapple) Gerard.

WILF *Salix caprea* (Goat Willow) Yks. Nicholson.

WILFIRE *Ranunculus flammula* (Lesser Spearwort) Scot. Grigson. This must be 'wild-fire', so the book name Flame-leaved Crowfoot is presumably connected.

WILGERS; WELGERS *Salix viminalis* (Osier) both Dev. Grigson (Wilgers); Choape (Welgers). Presumably the Northamptonshire name Augers is related.

WILL, Nimble *Muhlenbergia schreberi*.

WILLIAM, Bloody *Lychnis coronaria* (Rose Campion) Coats. The colour is not rose, but crimson, which accounts very well for 'bloody'.

WILLIAM, Herb *Ammi visnaga* (Bishop's Weed) Coles.

WILLIAM, Sweet 1. *Cheiranthus cheiri* (Wallflower) Lincs. B & H. 2. *Dianthus barbatus*. 3. *Saxifraga x urbinum* (London Pride) A E Baker. 'William' (in Sweet William) is probably the French name oeillet, which may very well have been corruped into 'Willy', and then connected in some way with a saint, and changed into William. Commentators have had a field day with this name. Archdeacon Hare reckoned it was dedicated to St William, whose feast is 25 June, and that 'sweet' is a substitute for 'saint'. Others talk of 'Saint Sweet William'. The 'saint' has only been dropped since the demolition of St William's shrine in Rochester Cathedral.

WILLIAM, Sweet, Wild 1. *Lychnis flos-cuculi* (Ragged Robin) Gerard. 2. *Phlox maculata*. 3. *Saponaria officinalis* (Soapwort) Grieve.1931.

WILLIAM, Velvet *Dianthus barbatus* (Sweet William) Coats.

WILLIAM, Wild *Taraxacum officinale* (Dandelion) Som. Macmillan.

WILLIAM-AND-MARY 1. *Chimaphila maculata* (Umbellate Wintergreen) Maryland. Whitney & Bullock. 2. *Malcomia maritima* (Virginian Stock) Som. Macmillan. 3. *Pulmonaria officinalis* (Lungwort) Brownlow. Double names like this are often given to two-coloured flowers.

WILLIE, Pretty *Dianthus barbatus* (Sweet William) Som. Macmillan.

WILLIE, Ragged *Lychnis flos-cuculi* (Ragged Robin) Shet. Grigson.

WILLIE, Sticky *Galium aparine* (Goose-grass) Scot. Fairweather. Very descriptive of the burrs. Cf. Sticky-weed, Sticky-balls, and a lot of others. There are many more describing the habit of the burrs very well - Cf. for instance Sweethearts, and Cleavers.

WILLIE, Stinking *Senecio jacobaea* (Ragwort) Hants. Hants FWI; Mor. Grigson; USA Kingsbury. Cf. Stinking Billy, from Lincolnshire, or another Scottish name, Stinking Davies. Stinking Willie is said to be named after "Butcher" Cumberland.

WILLIE, Sweet *Dactylorchis incarnata* (Early Marsh Orchid) Donegal Grigson.

WILLOW, African *Salix safsaf*. Willow is OE welig, from which OE wilige means a wicker basket. The Latin salix is said to come from the verb salire, to leap, given to it because of the extraordinarily quick growth of the tree. Not for nothing is there a saying, "The willow will buy a horse, before the oak will pay for a saddle".

WILLOW, Almond(-leaved) *Salix triandra* (French Willow) Grigson, Perring. The alternative specific name is '*amygdalina*', which accounts for 'Almond' in this name.

WILLOW, Apple-leaved *Salix malifolia*.

WILLOW, Apuan *Salix crataegifolia*. The reference is to the Apuan Alps, where it grows.

WILLOW, Babylon *Salix alba var. vitellina "pendula"* (Weeping Willow) Barber & Phillips. Often known as *Salix babylonica*, hence this name.

WILLOW, Basford *Salix basfordiana*.

WILLOW, Basket *Salix viminalis* (Osier) A W Smith.
WILLOW, Bay 1. *Chamaenerion angustifolium* (Rosebay Willowherb) Gerard. 2. *Salix pentandra* (Bay-leaved Willow) Clapham.
WILLOW, Bay-leaved *Salix pentandra*.
WILLOW, Beaked *Salix bebbiana*.
WILLOW, Bedford *Salix viridis*.
WILLOW, Bitter *Salix purpurea* (Purple Osier) J Smith.
WILLOW, Black 1. *Salix nigra*. 2. *Salix pentandra* (Bay-leaved Willow) Ire. Grigson.
WILLOW, Black, American *Salix discolor*.
WILLOW, Bleeding *Orchis morio* (Green-winged Orchid) Bucks. B & H. Why? It is listed without comment in B & H.
WILLOW, Blooming *Chamaenerion angustifolium* (Rosebay Willowherb) Ire. B & H. Cf. Blooming Sally, also from Ireland, and the English Flowering Withy. There is a resemblance between the leaves and those of willow.
WILLOW, Blunt-leaved *Salix retusa*.
WILLOW, Brittle *Salix fragilis* (Crack Willow) Warren-Wren. Cf. Snap Willow. All these names commemorate the fact that even a slight pressure at the base of a twig will be enough to separate it from the branch with quite a loud cracking sound.
WILLOW, Brown *Salix muhlenbergiana*.
WILLOW, Calaf of Persia *Salix aegyptica*.
WILLOW, Cape *Salix mucrinata*.
WILLOW, Caspian *Salix acutifolia*.
WILLOW, Ceylon *Ficus benjamina* (Weeping Fig) Howes.
WILLOW, Chaste *Vitex agnus-castus* (Chaste Tree) Pomet. 'Willow', because of the pliant branches. 'Chaste' is the Latin castus, also responsible for the name castor-oil.
WILLOW, Contorted *Salix matsudana var. tortuosa* (Corkscrew Willow) Mitchell.1974.
WILLOW, Coral-bark *Salix alba var. vitellina "Britzensis"* (Scarlet Willow) Mitchell.1974.
WILLOW, Corkscrew *Salix matsudana var. tortuosa*.
WILLOW, Coyote *Salix exigua* (Slender Willow) USA T H Kearney.
WILLOW, Crack *Salix fragilis*. See Brittle Willow above.
WILLOW, Creeping *Salix repens*.
WILLOW, Cricket Bat *Salix alba var. caerulea*. Cricket bats have been made commercially from the wood of this willow since about 1820, but bats of various kinds have always been cut from willows.
WILLOW, Dark-leaved *Salix nigricans*.
WILLOW, Diamond *Salix mackenzieana*.
WILLOW, Downy *Salix lapponicum* (Lapland Willow) McClintock.
WILLOW, Dragon's-claw *Salix matsudana var. tortuosa* (Corkscrew Willow) RHS. More fanciful than the other names describing the contorted branches of this tree.
WILLOW, Drooping 1. *Laburnum anagyroides* (Laburnum) Dev. Macmillan. 2. *Salix alba var. vitellina "Pendula"* (Weeping Willow) Dev. Macmillan.
WILLOW, Duck *Salix alba* (White Willow) B & H. Ducks do sometimes nest in these trees, but is that the reason for the name?
WILLOW, Dwarf 1. *Salix herbacea*. 2. *Salix humilis* (Prairie Willow) USA Harper. 3. *Salix repens* (Creeping Willow) Hutchinson.
WILLOW, European *Salix alba* (White Willow) Hatfield.
WILLOW, Fire *Salix scouleriana* (Nuttall Willow) USA T H Kearney. Because of the rapidity with which it colonises areas burnt out by forest fires.
WILLOW, French 1. *Chamaenerion angustifolium* (Rosebay) Som. Grigson. 2. *Salix triandra*. 3. *Thevetia peruviana* (Yellow Oleander) Howes. Cf. French Saugh for Rosebay. 'Willow', because of the resemblance of the leaves to those of willow - that is why the group is called Willowherb. But why 'French' for *Salix triandra*? This is a native British willow.
WILLOW, Goat *Salix caprea*. Apparently because goats are quite happy browsing on it.
WILLOW, Golden *Salix alba var. vitellina*.

WILLOW, Great *Salix alba* (White Willow) Turner.
WILLOW, Gray *Salix atrocinerea* (Gray Sallow) Hutchinson.
WILLOW, Ground *Polygonum amphibium* (Amphibious Persicaria) Ches. B & H.
WILLOW, Hairless *Salix glabra*.
WILLOW, Heart-leaved *Salix cordata*.
WILLOW, Herb *Lysimachia vulgaris* (Yellow Loosestrife) Turner. There are a number of willow names for this - Cf. Willow-herb, Willow-wort, etc.
WILLOW, Hoary 1. *Salix alba* (White Willow) Hatfield. 2. *Salix candida* (Sage Willow) USA Youngken.
WILLOW, Hooded *Scutellaria galericulata* (Skullcap) B & H. Curtis called it Hooded Willow-herb, which makes more sense.
WILLOW, Huntingdon *Salix alba* (White Willow) Brimble.1948.
WILLOW, Kilmarnock *Salix caprea var. pendula*.
WILLOW, Kit *Salix triandra* (French Willow) N'thants. Sternberg.
WILLOW, Lapland *Salix lapponicum*.
WILLOW, Least *Salix herbacea* (Dwarf Willow) McClintock.
WILLOW, Mountain *Salix arbuscula* (Small Tree Willow) Blamey.
WILLOW, Mulberry *Salix cramacile*.
WILLOW, Musk *Salix aegyptiaca* (Calaf of Persia's Willow) RHS.
WILLOW, Native *Salix mucronata* (Cape Willow) S Africa Watt. Native to South Africa, that is.
WILLOW, Net-leaved; WILLOW, Netted; WILLOW, Net-veined *Salix reticulata* Netted Willow is preferred by Warren-Wren; RHS (Net-veined).
WILLOW, Norway, Golden-flowered *Salix chrysanthos*.
WILLOW, Nuttall *Salix scouleriana*.
WILLOW, Orange *Lippia citriodora* (Lemon Verbena) Dev. Friend.1882.
WILLOW, Pacific *Salix lasiandra* (Yellow Willow) USA T H Kearney. 'Pacific', because it is a native of the western, Pacific, side of North America.
WILLOW, Palestine *Salix safsaf*.
WILLOW, Palm *Salix caprea* (Goat Willow) Leics. Grigson. Goat Willow gives the substitute "palm" for Palm Sunday. Flora Thompson tells how sprays of sallow catkins were worn in buttonholes for church-going in her day, and how they were brought indoors to decorate the house. Names include Palm itself, English Palm or Palm Tree.
WILLOW, Pekin *Salix matsudana*.
WILLOW, Persian *Chamaenerion angustifolium* (Rosebay Willowherb) Prior. Presumably 'Persian' just means foreign, as does 'French'. It is actually an American plant.
WILLOW, Plum-leaved *Salix arbuscula* (Small Tree Willow) McClintock.
WILLOW, Polished *Salix laevigata* (Red Willow) Everett.
WILLOW, Prairie *Salix humilis*.
WILLOW, Prick *Hippophae rhamnoides* (Sea Buckthorn) Turner. 'Willow' because of the resemblance of the leaves to those of willow (Cf. Willow-thorn, Sallow-thorn or Sea Willow), and 'Prick' because is has thorns.
WILLOW, Purple *Salix purpurea* (Purple Osier) Clapham.
WILLOW, Pussy 1. *Salix caprea* (Goat Willow). 2. *Salix discolor* (American Black Willow) USA Upham. The "pussies" are the catkins.
WILLOW, Red 1. *Cornus amomum* (Silky Cornel) Grieve.1931. 2. *Cornus stolonifera* (Red-osier Dogwood) USA Youngken. 3. *Salix laevigata*. Red for the dogwoods refers to the tinted colouring of the twigs.
WILLOW, Red, Belgian *Salix alba var. cardinalis*.
WILLOW, Red-wood *Salix fragilis* (Crack Willow) C P Johnson.
WILLOW, Reticulate *Salix reticulata* (Net-laved Willow) Darling & Boyd.
WILLOW, River 1. *Salix mucrinata* (Cape Willow) S Africa Watt. 2. *Salix safsaf* (Palestine Willow) S Africa Watt.

WILLOW, Roman *Syringa vulgaris* (Lilac) Lincs. B & H. Lilac is not a native British shrub, and this name probably is the result of a belief that it, along with a lot of other plants, was introduced by the Romans. Actually, it was not until medieval times that it was introduced from eastern Europe.

WILLOW, Rose *Cornus amomum* (Silky Cornel) Grieve.1931.

WILLOW, Round-leaved, Great *Salix caprea* (Goat Willow) Curtis.

WILLOW, Sadler's *Salix sadleri*.

WILLOW, Sage *Salix candida*.

WILLOW, Sallow, Great *Salix caprea* (Goat Willow) Hutchinson. Pleonastic, for sallow is OE sealh, and meant willow.

WILLOW, Sand *Salix arenaria*.

WILLOW, Sandbar 1. *Salix longifolia* USA Zenkert. 2. *Salix sessifolia var. hindsiana*.

WILLOW, Scarlet *Salix alba var. vitellina 'britzensis'*.

WILLOW, Sea *Hippophae rhamnoides* (Sea Buckthorn) Turner. See Prick Willow.

WILLOW, Seep *Baccharis glutinosa*.

WILLOW, Shining *Salix lucida*.

WILLOW, Silky *Salix alba var. sericea*.

WILLOW, Silky, Dwarf *Salix repens* (Creeping Willow) Brimble.1948.

WILLOW, Silver *Salix alba var. sericea* (Silky Willow) Barber & Phillips.

WILLOW, Slender *Salix exigua*.

WILLOW, Snake's-skin *Salix triandra* (French Willow) Wilts. Dartnell & Goddard. For this is a tree that sheds its bark.

WILLOW, Snap *Salix fragilis* (Crack Willow) Wilts. Dartnell & Goddard; Kent Grigson. See Crack Willow.

WILLOW, Spear-leaved *Salix hastata* Blamey.

WILLOW, Sticky *Arctium lappa* (Burdock) Argyll PLNN29. 'Willow' is not what it seems to be here. For 'willow', read 'Willie'. 'Sticky' is obvious when the burrs are considered.

WILLOW, Stinking *Amorpha californica* USA T H Kearney.

WILLOW, Sweet 1. *Myrica gale* (Bog Myrtle) Suss. Grigson. 2. *Salix pentandra* (Bay Willow) Cumb. Grigson.

WILLOW, Sweet Bay *Salix pentandra* (Bay Willow) Brimble.1948.

WILLOW, Tea-leaved *Salix phylicifolia*.

WILLOW, Three-threaded *Salix triandra* (French Willow) Curtis. 'Triandra' means with three anthers, or stamens.

WILLOW, Tree, Small *Salix arbuscula*.

WILLOW, Velvet *Salix sitchensis*.

WILLOW, Violet *Salix daphnoides*.

WILLOW, Water *Baccharis glutinosa* (Seep Willow) USA T H Kearney.

WILLOW, Weeping 1. *Fritillaria meleagris* (Snake's-head Lily) Coats. 2. *Laburnum anagyroides* (Laburnum) Dev. B & H. 3. *Salix alba var. pendula*. Weeping Willow for Snake's-head Lily is part of a clutch of names occasioned by the drooping heads and sombre colouring.

WILLOW, Weeping, Wisconsin *Salix blanda*.

WILLOW, White *Salix alba*. Silvery leaves, hence "white" willow.

WILLOW, Whortle-leaved *Salix myrsinites*.

WILLOW, Widow's *Salix fragilis* (Crack Willow) Suss. PLNN31. It seems more appropriate for Weeping Willow.

WILLOW, Wild 1. *Epilobium hirsutum* (Great Willowherb) Wilts. Dartnell. & Goddard. 2. *Salix mucrinata* (Cape Willow) S Africa Watt. 3. *Salix subserrata* Palgrave. 4. *Salix woodii* (Wood's Willow) S Africa Watt.

WILLOW, Wolf *Elaeagnus commutata* (Silverberry) Johnston.

WILLOW, Wood's *Salix woodii*.

WILLOW, Woolly *Salix lanata*.

WILLOW, Wrinkle-leaved *Salix reticulata* (Net-leaved Willow) Warren-Wren.

WILLOW, Yellow 1. *Salix alba var. vitellina* (Golden Yellow) Ablett. 2. *Salix lasiandra*. The reference is to the winter colour of the twigs.

WILLOW, Yew-leaf *Salix taxifolia*.

WILLOW, Yolk-of-egg *Salix alba var. vitellina*.

WILLOW-BAY *Salix pentandra* (Bay Willow) Staffs. Grigson.
WILLOW-BLOSSOM *Phlox decussata* Friend.1882.
WILLOW FLOWER *Epilobium hirsutum* (Great Willowherb) Parkinson.
WILLOW GENTIAN *Gentiana asclepiadea.*
WILLOW-GRASS 1. *Polygonum amphibium* (Amphibious Persicaria) Yks. B & H. 2. *Polygonum aviculare* (Knotgrass) Yks. Grigson. There is some resemblance between the leaves of these and willow.
WILLOW-LEAVED DOCK *Rumex mexicanus* (Pale Dock) USA Zenkert.
WILLOW-LEAVED JESSAMINE *Cestrum parqui* (Green Jessamine) USA Kingsbury.
WILLOW-LEAVED PEAR *Pyrus salicifolia.*
WILLOW-LEAVED WILD SUNFLOWER *Helianthus salicifolius.*
WILLOW LETTUCE *Lactuca saligna* (Least Lettuce) W Miller.
WILLOW OAK *Quercus phellos.*
WILLOW PATTERN TREE, Chinese *Koelreuteria paniculata* (Golden Rain Tree).
WILLOW PERSICARIA *Polygonum lapathifolium* (Pale Persicaria) USA Upham. Cf. Pale Willow-weed for this. A lot of these Polygonums have leaves somewhat similar to those of willow.
WILLOW PODOCARP *Podocarpus saligna.*
WILLOW POPLAR *Populus nigra* (Black Poplar) Cambs. Grigson. 'Willow', because this poplar grows in the same sort of places as willows, particularly riverside meadows.
WILLOW SPIRAEA; WILLOW-LEAVED SPIRAEA *Spiraea salicifolia* (Bridewort) McClintock, Hutchinson respectively.
WILLOW-STRIFE *Lythrum salicaria* (Purple Loosestrife) Som. Macmillan. A hybrid name, for this is half Loosestrife and half Willowherb. The leaves are similar to those of willow, hence this name and various "sally" names.
WILLOW-THORN *Hippophae rhamnoides* (Sea Buckthorn) W Miller. The leaves are more like those of a willow than a buckthorn, hence all the 'willow' names, the closest to this particular one being Turner's Prick-willow.
WILLOW TOOTHWORT *Lathraea clandestina.* A parasite on the roots of willow and poplars.
WILLOW-WEED 1. *Polygonum amphibium* (Amphibious Persicaria) Yks. B & H. 2. *Polygonum persicaria* (Persicaria) Allan. The leaves of both these bear some resemblance to those of willow, enough to account for a number of 'willow' references in the local names.
WILLOW-WEED, Pale *Polygonum lapathifolium* (Pale Persicaria) New Zealand Connor.
WILLOW-WEED, Swamp *Polygonum decipiens.*
WILLOW-WEED, Yellow *Lysimachia vulgaris* (Yellow Loosestrife) Tradescant. Another plant whose leaves are compared with those of willow. Cf. Turner's Herb Willow, and various 'willow' names from the English counties.
WILLOW-WIND 1. *Clematis vitalba* (Old Man's Beard) Glos. Grigson. 2. *Convolvulus arvensis* (Field Bindweed) Wilts. Rogers. 3. *Fagopyrum esculentum* (Buckwheat) Wilts. Dartnell & Goddard. A natural extension of the more usual name, Withywind.
WILLOW-WORT *Lysimachia vulgaris* (Yellow Loosestrife) Som. Macmillan. Willow-like leaves, hence a variety of 'willow' names for this. Cf. Willowherb, Herb Willow, etc.
WILLOWHERB 1. *Epilobium sp.* 2. *Lysimachia vulgaris* (Yellow Loosestrife) Som. Macmillan. All the willowherbs get the name because of their willow-like leaves.
WILLOWHERB, Alpine *Epilobium anagallidifolium.*
WILLOWHERB, American *Epilobium adenocaulon.*
WILLOWHERB, Broad-leaved *Epilobium montanum.*

WILLOWHERB, Chickweed *Epilobium alsinifolium*.

WILLOWHERB, Codded *Epilobium hirsutum* (Great Willowherb) Gerard. "the flower groweth at the top of the stalke, comming out of the end of a small longe codde". This may very well be the origin of the various 'coddled' and 'codlin' names applied to this willow-herb, in spite of explanations of such names as Coddled Apples as being the result of the fact that the plant smells like apples when it is crushed.

WILLOWHERB, Great 1. *Chamaenerion angustifolium* (Rosebay) USA House. 2. *Epilobium hirsutum*.

WILLOWHERB, Hoary *Epilobium parviflorum* (Small-flowered Willowherb) McClintock.

WILLOWHERB, Hooded *Scutellaria galericulata* (Skullcap) Curtis.

WILLOWHERB, Hooded, Small *Scutellaria minor* (Lesser Skullcap) Curtis.

WILLOWHERB, Large-flowered *Epilobium hirsutum* (Great Willowherb) Curtis.

WILLOWHERB, Marsh; WILLOWHERB, Marsh, Narrow-leaved *Epilobium palustre* (Narrow-leaved... is used by J A MacCulloch.1905).

WILLOWHERB, Narrow-leaf *Chamaenerion angustifolium* (Rosebay) USA Elmore.

WILLOWHERB, New Zealand 1. *Epilobium nerterioides* (Prostrate Willowherb) McClintock.1975. 2. *Epilobium pedunculare*.

WILLOWHERB, Northern *Epilobium adenocaulon* (American Willowherb) USA House.

WILLOWHERB, Pale *Epilobium roseum*.

WILLOWHERB, Prostrate *Epilobium nerterioides*.

WILLOWHERB, Purple *Lythrum salicaria* (Purple Loosestrife) Grieve.1931. There are a number of 'willow' names for Purple Loosestrife, presumably because of a resemblance in the leaves. Cf. Red Sally or Flowering Sally, and Willow-strife quoted above.

WILLOWHERB, Rosebay *Chamaenerion angustifolium*.

WILLOWHERB, Short-fruited *Epilobium obscurum*.

WILLOWHERB, Southern *Epilobium lamyi*.

WILLOWHERB, Spear-leaved *Epilobium lanceolatum*.

WILLOWHERB, Spiked *Chamaenerion angustifolium* (Rosebay Willowherb) USA House.

WILLOWHERB, Small-flowered *Epilobium parviflorum*.

WILLOWHERB, Smooth-leaved,Broad *Epilobium montanum* (Broad-leaved Willowherb) Phillips.

WILLOWHERB, Square-stalked *Epilobium adnatum*.

WILLOWHERB, Wood *Epilobium montanum* (Broad-leaved Willowherb) Curtis.

WILLOWMORE CEDAR *Widdringtonia schwarzii*.

WILLY, Drunken *Centranthus ruber* (Spur Valerian) Dev, Som. Elworthy.

WILLY, Sweet *Salix pentandra* (Bay-leaved Willow) Cumb. B & H. This is the only example of 'Willy' actually meaning 'willow'. Cf. Sweet Willow, and Sweet Withy, for this tree.

WILLY, Wandering 1. *Convolvulus arvensis* (Field Bindweed) N Eng. Wakelin. 2. *Geranium robertianum* (Herb Robert) Som. Macmillan. There has always been a tendency to regard Herb Robert as a kind of pink, and that may be the reason for the name 'Willy'. Sweet William may very well have been given the name from the French oeillet, meaning little eye.

WILLY-RUN-THE-HEDGE *Galium aparine* (Goose-grass) Ire. Grigson. 'Willy' is better known as 'Jack' running the hedge. But 'Robin' does it too, and so does 'Lizzie'.

WILLYWIND *Convolvulus arvensis* (Field Bindweed) S Eng. Wakelin. A variation of 'Withywind', through Willow-wind.

WILSON SPRUCE *Picea wilsonii*.
WILTSHIRE WEED *Ulmus procera* (English Elm) Wilts. Macmillan. Because of the vast numbers of elms that used to grow in the county, alas, no more. But was this ever a genuine vernacular name?
WIMBERRY *Vaccinium myrtillus* (Whortleberry) E Ang, Midl. Grigson. Presumably it means wine berry. Nodal suggested that the berries look like miniature grapes, but this name has produced Winberry, or Windberry, and even Whinberry.
WIMBLE-STRAW *Cynosurus cristatus* (Crested Dogstail) N'thants. Sternberg.
WIMMER'S OSIER *Salix calodendron*.
WIMOTE; WYMOTE *Althaea officinalis* (Marsh Mallow) Shrop. Grigson (Wimote); USA House (Wymote).
WINBERRY *Vaccinium myrtillus* (Whortleberry) Shrop, Ches. B & H; Lancs. Nodal. See Wimberry above.
WINCOPIPE *Anagallis arvensis* (Scarlet Pimpernel) Wilts. D Grose. Presumably Winkapeep. There is a Somerset name, Windpipe, which may be from the same original. Winkapeep, though, is a North Country name, of the John-go-to-bed-at-noon type, and there are a number of variants on it, all describing the plant's habit of early opening and early closing.
WIND FLOWER *Anemone sp* Gerard. Anemone is from Greek anemos, the wind, and the name of the flower means literally daughter of the wind. Friend tried to explain it by saying that some species flourish in open exposed places, or that they would not open till the March winds began to blow. That is from Pliny, but all this is a sort of folk etymology. The true origin is the Semitic na'aman, which meant the one "who was pleasant", or "lovely", and actually refers to the Poppy Anemone, *A coronaria*.
WIND FLOWER, Desert *Anemone tuberosa*.
WIND FLOWER, Fairy's *Anemone nemorosa* (Wood Anemone) Dor. Macmillan.

WIND FLOWER, Snowdrop *Anemone sylvestris*.
WIND FLOWER, Storksbill *Anemone apennina* (Blue Anemone) Gerard.
WIND PLANT *Anemone nemorosa* (Wood Anemone) Lincs. B & H.
WIND ROOT *Asclepias tuberosa* (Orange Milkweed) Grieve.1931. 'Wind' in this instance is explained by reference to another of this plant's names - Colic Root.
WIND ROSE *Papaver argemone* (Rough Long-headed Poppy) Gerard.
WINDBERRY *Vaccinium myrtillus* (Whortleberry) N Eng. B & H. Probably 'Wineberry'. Cf. Wimberry, Winberry, and Whinberry.
WINDING LILY *Calystegia sepium* (Great Bindweed) Hants. Hants FWI.
WINDLES *Plantago lanceolata* (Ribwort Plantain) N Eng. B & H. It used to be a general name for the dried stalks of many grasses.
WINDMILL PALM *Trachycarpus excelsus* (Chusan Palm). Descriptive.
WINDPIPE *Anagallis arvensis* (Scarlet Pimpernel) Som. Macmillan. Cf. the Wiltshire Wincopipe. They both probably are from something like Wink-a-peep, which is a north country name, and a comment on the plant's short opening time - see John-go-to-bed-at-noon.
WINDSOR BEAN *Vicia faba* (Broad Bean) F P Smith.
WINE, Cup-of- *Taxus baccata* (Yew) Som. Macmillan.
WINEBERRY 1. *Ribes nigrum* (Blackcurrant) Scot. Grigson. 2. *Ribes rubrum* (Redcurrant) Yks, Cumb, Scot. Grigson. 3. *Ribes uva-crispa* (Gooseberry) Yks. Hunter. Perhaps it means exactly what it says. Certainly, redcurrant wine was very popular at one time in Scotland and northern England.
WINEBERRY, Chilean *Aristotelia macqui*.
WINEBERRY, Japanese *Rubus phoenocolasium*.

WINECUP, Fairy's *Convolvulus arvensis* (Field Bindweed) Som. Macmillan.
WINEGLASS, King's *Tulipa sp* (Tulip) Som. Macmillan.
WINEGLASSES *Campanula medium* (Canterbury Bells) Som. Macmillan.
WING-WANGS *Briza media* (Quaking-grass) Wilts. Dartnell & Goddard.
WINGNUT, Caucasian *Pterocarya fraxinifolia*. *Pterocarya* is from Greek pteron, wing and koryon, nut.
WINGNUT, Japanese *Pterocarya rhoifolia*.
WINGED ELM *Ulmus alata*.
WINGED SPINDLE *Euonymus alatus*.
WINGS 1. *Acer pseudo-platanus* keys (Sycamore keys) Som. Macmillan. 2. *Fraxinus excelsior* keys (Ash keys) Som. Macmillan.
WINGS, Angel *Caladium bicolor 'Splendens'* (Fancy-leaved Caladium) USA Cunningham. From the colourful and paper-thin leaves.
WINGS, Birds' *Acer pseudo-platanus* keys (Sycamore keys) Som. Macmillan.
WINGS, Golden *Solidago virgaurea* (Golden Rod) Som. Macmillan.
WINGSCALE *Atriplex canescens* (Chamizo) USA Jaeger. Cf. Shadscale for this.
WINGSEED *Ptelea trifoliata* (Hop Tree) Leyel.1948.
WINK-A-PEEP; WINK-AND-PEEP *Anagallis arvensis* (Scarlet Pimpernel) Ches. Holland; Lancs. Wright (Wink-a-peep); Shrop, Ches, Staffs, Lancs. Grigson (Wink-and-peep). Because of the flower's short opening time. That is why it is also called John-go-to-bed-at-noon, and names of a similar import.
WINTER ACONITE *Eranthis hyemalis*.
WINTER BEAN *Lathyrus latifolius* (Garden Everlasting Pea) Suss. Parish.
WINTER CHERRY 1. *Cardiospermum halicacabum* (Blister Creeper) Chopra. 2. *Physalis alkekengi*. 3. *Physalis pubescens* (Husk Tomato) Bianchini. 4. *Solanum capsicastrum*.

WINTER CLOVER *Mitchella repens* (Partridgeberry) Le Strange.
WINTER CRESS *Barbarea vulgaris*.
WINTER FAT *Eurotia lanata* (Winter Sage) Jaeger.
WINTER FORGET-ME-NOT *Omphalodes verna*.
WINTER GILLIFLOWER 1. *Cheiranthus cheiri* (Wallflower) Gerard. 2. *Galanthus nivalis* (Snowdrop) Tynan & Maitland.
WINTER GRAPE, Sweet *Vitis cinerea* (Downy Grape) Vestal & Schultes.
WINTER GREENS *Brassica fimbricata* (Curled Kale) Som. Elworthy.
WINTER HEATH 1. *Erica carnea* (Alpine Heath) RHS. 2. *Erica erigena* (Irish Heath) Hay & Synge.
WINTER HEDGE MUSTARD *Barbarea vulgaris* (Winter Cress) Flower.1859.
WINTER HELIOTROPE *Petasites fragrans*.
WINTER HELLEBORE 1. *Eranthis hyemalis* (Winter Aconite) Prior. 2. *Helleborus niger* (Christmas Rose) Jones-Baker.1974.
WINTER HYACINTH *Scilla autumnalis* (Autumnal Squill) W Miller.
WINTER JASMINE *Jasminum nudiflorum*.
WINTER IRIS 1. *Iris stylosa* (Algerian Iris) RHS. 2. *Marica northiana* (Apostle Plant) Gannon.
WINTER KECKSIES *Prunus spinosa* fruit (Sloes) IOW Long. Kex is recorded for sloes in Hampshire. Both this and Kecksies seem quite out of place in this context, and it may be that the actual word should be something like Picks, which is a Wiltshire name for sloes; in fact Winter Picks is also recorded.
WINTER MARJORAM 1. *Origanum heracleoticum*. 2. *Origanum onites* (French Marjoram) Sanecki.
WINTER MELON *Cucumis melo var. inodorus*.
WINTER MINT *Mentha cordifolia*. This is a kind that does not die down in winter.
WINTER OAK *Quercus ilex* (Holm Oak)

Rodd. 'Winter', presumably because it is evergreen.

WINTER ONION *Allium cepa* (Onion) Turner.

WINTER PEA, Wild *Lathyrus hirsutus* (Hairy Pea) USA Kingsbury.

WINTER ROSE *Helleborus niger* (Christmas Rose) Dev. Friend.1882. Cf. Winter Hellebore.

WINTER SAGE *Eurotia lanata*.

WINTER SAVORY *Satureia montana*.

WINTER SQUASH 1. *Cucurbita maxima* (Giant Pumpkin) USA Elmore. 2. *Cucurbita moschata* (Crook-neck Squash) Whitaker & Davis. The pumpkin is also known as Autumn Squash.

WINTER STRAWBERRY *Arbutus unedo* (Strawberry Tree) Dev. Choape; Som. Macmillan.

WINTER-SWEET *Chimonanthus fragrans*.

WINTER THYME *Thymus serpyllum var. citriodorus* (Lemon Thyme) Sanecki.

WINTER VETCH *Vicia villosa* (Lesser Tufted Vetch) USA Kingsbury.

WINTER WOLF'S BANE *Eranthis hyemalis* (Winter Aconite) Gerard. 'Wolf's Bane', because this is a name for the real Aconite, Monkshood in our terms.

WINTER'S BARK; WINTER'S CINNAMON *Drimys winteri* Sanecki (Cinnamon). Named after Captain Winter, the commander of one of Drake's ship, who used the bark as a spice, and, more important, to relieve his crew of scurvy.

WINTER'S BARK, False *Cinnamodendron corticosum*. As the real Winter's Bark became more difficult to get, it was gradually replaced in medicine by the bark of this Caribbean tree.

WINTERBERRY 1. *Gaultheria procumbens* (Wintergreen) USA Perry. 2. *Ilex verticillata*.

WINTERBERRY, Evergreen *Ilex glabra* (Gallberry) Le Strange.

WINTERBERRY, Smooth *Ilex laevigata*.

WINTERBERRY, Virginian *Ilex verticillata* (Winterberry) Le Strange.

WINTERBLOOM *Hamamelis virginica* (Witch Hazel) Grieve.1931.

WINTERGREEN 1. *Gaultheria procumbens*. 2. *Pyrola minor*.

WINTERGREEN, Aromatic *Gaultheria humifusa* (Teaberry) USA Elmore.

WINTERGREEN, Bog *Pyrola uliginosa*.

WINTERGREEN, Chickweed 1. *Trientalis borealis* USA House. 2. *Trientalis europaea*.

WINTERGREEN, Creeping 1. *Gaultheria humifusa* (Teaberry) USA Elmore. 2. *Gaultheria procumbens* (Wintergreen) USA House.

WINTERGREEN, False *Pyrola rotundifolia* (Round-leaved Wintergreen) O P Brown. *Pyrola minor* being the real wintergreen of herbalists.

WINTERGREEN, Flowering *Polygala paucifolia* (Fringed Milkwort) USA House.

WINTERGREEN, Greater; WINTERGREEN, Intermediate *Pyrola media*. 'Intermediate', in keeping with the specific name, is used by Darling & Boyd.

WINTERGREEN, Larger *Pyrola rotundifolia* (Round-leaved Wintergreen) Clapham.

WINTERGREEN, Nodding *Ramischia secunda* (Yavering Bells) Polunin. This was included in the genus *Pyrola* at one time.

WINTERGREEN, Norwegian *Pyrola norvegica*.

WINTERGREEN, One-flowered *Moneses uniflora* (St Olaf's Candlesticks) Clapham. Wintergreen, because this was once classified as *Pyrola uniflora*.

WINTERGREEN, One-sided *Ramischia secunda* (Yavering Bells) USA House.

WINTERGREEN, Pear-leaf *Pyrola rotundifolia* (Round-leaved Wintergreen) O P Brown.

WINTERGREEN, Round-leaved *Pyrola rotundifolia*.

WINTERGREEN, Round-leaved, American *Pyrola americana*.

WINTERGREEN, Serrated; WINTERGREEN, Toothed *Ramischia secunda* (Yavering Bells) Clapham (Serrated); Blamey (Toothed).

WINTERGREEN, Spicy *Gaultheria procumbens* (Wintergreen) USA House.
WINTERGREEN, Spotted *Chimaphila maculata*.
WINTERGREEN, Umbellate *Chimaphila umbellata*.
WINTERGREEN, Yellow *Pyrola chlorantha*.
WINTERPICKS 1. *Prunus spinosa* fruit (Sloes) Hulme. 2. *Rubus fruticosus* fruit (Blackberry) Fernie. 'Picks' is a common Wiltshire name for sloes, and it exists in one form or another right throughout the West country. For example, it varies into Hedge-picks, and so to Hedge-speaks, or specks. 'Picks' becomes 'pegs' - Heg-pegs, for instance, from Gloucestershire, and then to 'pigs', as in Hedge Pigs, also from Gloucestershire. Winterpicks for blackberries is probably an error, for the fruit ripens much too early to qualify.
WINTERSWEET 1. *Acocanthera oblingifolium*. 2. *Origanum heracleoticum* (Winter Marjoram) W Miller. The name implies a winter-flowering plant, with a strong smell.
WINTERWEED 1. *Stellaria media* (Chickweed) B & H. 2. *Veronica agrestis* (Green Field Speedwell) Shrop. B & H. 3. *Veronica hederifolia* (Ivy-leaved Speedwell) Beds, Norf. Prior. Because they all seem to spread most in winter time.
WIPPUL-SQUIP *Heracleum sphondyllium* (Hogweed) Grigson. The hollow stems of hogweed have been used for a number of purposes, including drinking cider! Hence Wippul-squip, according to Grigson.
WIRE, Bleeding *Cheiranthus cheiri* (Wallflower) Som. Macmillan. A variant on Bleeding Warrior, where 'warrior' is not warrior at all, but 'Wallyer', in keeping with the common name.
WIRE, Cranberry 1. *Vaccinium oxycoccus* (Cranberry) Cumb. Grigson. 2. *Vaccinium vitis-idaea* (Cowberry) Scot. Grigson.
WIRE, Craneberry *Arctostapylos uva-ursi* (Bearberry) Aber. Grigson.
WIRE, Gold *Hypericum concinnum*.
WIRE, Golden *Sedum acre* (Biting Stonecrop) N'thants. Clare.
WIRE GRASS *Eleusine indica* (Goosegrass) USA Allan.
WIRE LING *Erica tetralix* (Cross-leaved Heath) Yks. Atkinson.
WIRE PLANT *Muehlenbergia complexa*.
WIRE RUSH 1. *Juncus balticus*. 2, *Juncus inflexus* (Hard Rush) F K Robinson.
WIRE THORN *Taxus baccata* (Yew) Lincs. B & H; Yks. Addy.
WIREGRASS *Polygonum aviculare* (Knotweed) Glos. J D Robertson. Cf. Irongrass for this. Perhaps the reason lies not in the toughness of the stalks but in the difficulty of pulling the plant up. It is said that the name Armstrong was also awarded because of this virtue.
WIREWEED 1. *Calystegia sepium* (Great Bindweed) Surr. Grigson. 2. *Filipendula ulmaria* (Meadowsweet) Hants. Grigson. 3. *Polygonum aviculare* (Knotweed) Wilts. D Grose; IOW. Long; Suss, Kent, Norf. Grigson.
WIRRAL; WORRAL; WURRAL *Ballota nigra* (Black Horehound) all Wilts. Dartnell & Goddard.
WIRY JACK *Sisymbrium officinale* (Hedge Mustard) Palaiseul.
WISCONSIN WEEPING WILLOW *Salix blanda*.
WISDOM OF SURGEONS *Descurainia sophia* (Flixweed) Salisbury. A translation of its old name *Sophia chirurgorum*, though it is not clear why. Flixweed signals its use for the flix, or dysentery, but that is the only medicinal use recorded.
WISE TREE; WISDOM TREE *Morus nigra* (Black Mulberry) Hants. Saunders (Wise); F Savage (Wisdom). Because, so it is said, unlike other trees, it never puts forth buds until all the frosts have ceased.
WISH-ME-WELL *Veronica chamaedrys* (Germander Speedwell) Ches. B & H. Cf. the widely used Forget-me-not, or the Yorkshire Remember-me. Speedwell means goodbye. We are asked to believe that the

plants got this widely used name because the blossoms when full fall and fly away. But Andrew Young suggests that if 'speed' means the same as in 'Speed the plough', then perhaps the name means good luck, an explanation that ties in very well with Wish-me-well.

WISHBONE BUSH *Mirabilis bigelovii var. retrorsa.*

WISHBONE FLOWER *Torenia fournieri.*

WISHES *Taraxacum officinale* (Dandelion) Wilts. Macmillan. A reminder that the seed globes can be used to find out more than the time. Cf. the American Fortune-teller.

WISHING-FLOWER *Leucanthemum vulgare* (Ox-eye Daisy) Scot. Cumming. Presumably a translation from the Gaelic, but it must connect with the daisy divination of pulling off the petals one by one (He loves me...he loves me not).

WISTARIA, Chinese *Wistaria sinensis.* The original spelling was Wisteria, but it is named after Dr Caspar Wistar, 1761-1818, professor of anatomy and surgery in Pennsylvania University.

WISTARIA, Japanese *Wistaria floribunda.*

WISTARIA, Native *Hardenbergia comptoniana/violacea* Simmons. Violet-blue flowers that hang in racemes like a wistaria.

WISTARIA, Rhodesian *Bolusanthus speciosus.*

WISTARIA, Silky *Wistaria venusta.*

WISTARIA, Wild 1. *Bolusanthus speciosus* (Rhodesian Wistaria) Palgrave. 2. *Securidaca longipedunculata* (Wild Violet Tree) S Africa. Dalziel.

WITAN ELM *Ulmus glabra* (Wych Elm) Shrop. Grigson. It looks odd, but presumably it can only be the same as 'wych'.

WITCH *Ulmus glabra* (Wych Elm) B & H. 'Wych' is from a Germanic base meaning pliant, but it all too easily becomes 'witch', simply by consonance.

WITCH ALDER *Fothergilla gardeni.* It has alder-like leaves, as an alternative specific name, alnifolia, proclaims.

WITCH BELLS 1. *Campanula rotundifolia* (Harebell) Scot. Jamieson. 2. *Centaurea cyanus* (Cornflower) N Eng. B & H; Scot. Jamieson. 3. *Digitalis purpurea* (Foxglove) N Eng. Henderson. There is nothing very bell-like about a cornflower, and there is no great connection with witches, though Witch's Thimbles is recorded, also in the northern counties. But the others are fairy flowers, and that is the reason for their ascription to witches, for the two are usually connected in folklore.

WITCH ELDER *Fothergilla sp.* See Witch Alder - alder and elder are frequently confounded.

WITCH ELM 1. *Sorbus aucuparia* (Rowan) Kelly. 2. *Ulmus glabra* (Wych Elm) Som. Elworthy; Yks. Nicholson. For rowan, 'witch' originates in OE cuic, alive. Hence Quickbeam, and other names like it. But 'quick' also becomes 'wick', with such names as Wicken-tree, which in turn descends into Witchen-tree, and thence into a totally uncalled-for, but quite understandable, connection with witches. But the elm has a different origin story. It is a straightforward confusion between two words that are pronounced the same, 'witch' and 'wych'.

WITCH GOWAN 1. *Taraxacum officinale* (Dandelion) Scot. B & H. 2. *Trollius europaeus* (Globe Flower) Grigson. 'Gowan' signifies a yellow flower, and *Trollius* proclaims a connection with trolls, even though a more likely etymology is from German Trollblume, which seems to be a contraction of die rolle Blume, rolled, or closed-in-petals. That is Grigson's explanation, but the milky sap was known in Scotland as Witch's Milk, and had a sinister reputation, for it was believed to cause blindness if put in the eye. Evidently the milky sap of the dandelion was looked on with some suspicion, too, for Devil's Milk-plant was a Scottish name, and as far away as Somerset it was called Devil's Milk-pail.

WITCH-GRASS *Agropyron repens* (Couchgrass) Allan. "Twitch" is the original of this (OE cwice, alive); it easily becomes "witch".

WITCH HALSE 1. *Corylus avellana* (Hazel) Corn. Grigson. 2. *Ulmus glabra* (Wych Elm) Som. Elworthy. 'Halse', by means of a familiar consonantal misplacement, is the same word as 'hazel', and both of the trees mentioned here seem to have been confused into interchangeablility.

WITCH HAZEL 1. *Carpinus betula* (Hornbeam) Gerard. 2. *Hamamelis virginica*. 3. *Sorbus aucuparia* (Rowan) Yks. Denham/Hardy. 4. *Ulmus glabra* (Wych Elm) B & H. Twigs of *Hamamelis* were used as divining rods in America, just as hazel twigs were used in England. As this was looked on in America as the result of occult power, the name Witch Hazel was given.

WITCH HAZEL, Japanese *Hamamelis japonica*.

WITCH HOBBLE *Viburnum alnifolia* (Hobble-bush) A W Smith. This, and a number of other names of similar import, stem from the tree's habit of letting its lower branches droop to the ground, with consequent rooting at the tips.

WITCH TREE 1. *Sambucus nigra* (Elder) War. F Savage. 2. *Ulmus glabra* (Wych Elm) Som. Elworthy. See Witch Elm.

WITCH WICKEN; WITCH WIGGIN *Sorbus aucuparia* (Rowan) Lincs. Grigson (Wicken); Derb. Addy (Wiggin). The inclusion of both elements (see Witch Elm) shows that the etymology had been forgotten by the time that these names were coined.

WITCHBANE *Sorbus aucuparia* (Rowan) N Eng. Denham/Hardy. Even though it is obviously a confusion with Witchbeam, the next entry, this is the only 'witch' name for rowan that matches its folklore. Rowan, as is well-known, is the prime protector from witchcraft in the country.

WITCHBEAM *Sorbus aucuparia* (Rowan) Dev. Grigson. 'Quickbeam', actually.

WITCHEN; WITCHEN-TREE; WITCHIN-TREE *Sorbus aucuparia* (Rowan) N'thants. Sternberg (Witchen); Kelly (Witchen-tree); Lincs, Rudkin (Witchin-tree). See Witch Elm, above.

WITCHES' ARMS *Galeopsis tetrahit* (Hemp Nettle) Dor. Macmillan.

WITCHES' BLOOD *Genista tinctoria* (Dyer's Greenweed) Mass. Leighton. Apparently given because it grew on Gallows Hill, Salem.

WITCHES' CAP *Helianthus annuus* (Sunflower) Dev. Macmillan.

WITCHES' FLOWER *Chelidonium majus* (Greater Celandine) Som. Grigson. Why? There is no evidence of any mistrust of the plant, though the juice, being extremely acrid, is toxic. On the other hand, in Ulster it was hung up to save both the cattle and the people from harm on May Eve.

WITCHES' GLOVES *Digitalis purpurea* (Foxglove) Skinner. The "gloves" are usually given to the fairies, or to the Virgin. But it was said that witches enjoyed the flowers, using them to decorate their fingers.

WITCHES' MILK *Hippuris vulgaris* (Mare's Tail) Lancs, Scot. Grigson. It sounds as if this has been mistaken for one of the spurges.

WITCHES' NEEDLE *Scandix pecten-veneris* (Shepherd's Needle) Berw. Grigson. It is the fairies' needle, as well.

WITCHES' THIMBLES 1. *Campanula rotundifolia* (Harebell) Som. Macmillan. 2. *Centaurea cyanus* (Cornflower) N Eng. Grigson. 3. *Digitalis purpurea* (Foxglove) N Eng. Henderson. 4. *Silene maritima* (Sea Campion) N'thum. Prior. 5. *Wahlenbergia hederacea* (Ivy-leaved Bellflower) Som. Macmillan. As with Witch Bells, cornflower is the odd one out. The rest could be described as thimble-like, but surely not cornflower.

WITCHES' TONGUE *Clerodendron serratum* Malaya Gilliland.

WITCHETTY BUSH *Acacia kempeana*. For the trunk harbours witchetty grubs, but so do several other acacias.

WITCHFLOWER 1. *Circaea lutetiana* (Enchanter's Nightshade) Som. Macmillan. 2. *Solanum dulcamara* (Woody Nightshade) Som. Grigson. Cf. Witchwort for the

Enchanter's Nightshade. The plant was certainly linked with witchcraft in this country. Its very name ensured that. And Witchflower for Woody Nightshade is understandable in the light of the magical associations of the plant.

WITCHWOOD 1. *Euonymus europaeus* (Spindle Tree) Suff. Grigson. 2. *Pyrus americana* USA Bergen. 3. *Sorbus aucuparia* (Rowan) N Eng. Denham/Hardy. 4. *Ulmus glabra* (Wych Elm) Cumb, Yks. B & H. It is most probable that Spindle will join Wych Elm in claiming 'wych', i.e. pliant, as the true meaning of 'witch'. The other two are more sinister.

WITCHWORT *Circaea lutetiana* (Enchanter's Nightshade) Storms. See Witchflower.

WITH-VINE 1. *Calystegia sepium* (Great Bindweed) Som. Macmillan. 2. *Convolvulus arvensis* (Field Bindweed) Som. Macmillan. See the various names like Withywind that are common in the west country.

WITHE-ROD *Viburnum cassinoides*. Because of its flexible shoots, used in America for binding purposes, and in basket-making.

WITHE-TREE *Salix aurita* (Eared Willow) Cumb. B & H.

WITHEN *Sorbus aucuparia* (Rowan) Lancs. Grigson. A variation on Witchen.

WITHERIPS *Asperula odorata* (Woodruff) Banff. Grigson. There are variations that give a clue to the meaning - see Woodrip or Woodrep and Woodrow, which is what Gerard called it. The last named is short for Wood-rowell, for as he said, it "has its leaves set about like a star or the rowell of a spurre". In other words, it means exactly the same as Woodruff.

WITHERS *Glyceria maxima* (Great Water Grass) Dev. Friend.1882.

WITHERSPAIL *Galium aparine* (Goose-grass) Roxb. B & H.

WITHEWIND *Calystegia sepium* (Great Bindweed) Barton & Castle. One of a series that describes the plant's habit. Perhaps the best known of them is Withywind.

WITHERWINE *Calystegia sepium* (Great Bindweed) Wilts, Glos. A S Palmer.

WITHWIND 1. *Calystegia sepium* (Great Bindweed) Wilts. Dartnell & Goddard; Dor, Glos. Grigson; Hants. Cope. 2. *Convolvulus arvensis* (Field Bindweed) Gerard. See Withywind.

WITHWIND, Cow's *Stachys sylvatica* (Hedge Woundwort) B & H. Cf. Cow's Weather-wind for this.

WITHWIND, Sea *Calystegia soldanella* (Sea Bindweed) Parkinson.

WITHWINE *Convolvulus arvensis* (Field Bindweed) Wessex Rogers (inc. Wilts. Dartnell & Goddard. Cf. Withwind.

WITHY 1. *Salix alba* (White Willow) Hants. Cope. 2. *Salix viminalis*. 3. *Sorbus aucuparia* (Rowan) Heref, Shrop. Grigson. Withy is OE withig, a name for the willow. From there it came to mean any flexible twig of branch, or indeed any climbing plant (hence the use in various forms for the bindweeds). But for the rowan the meaning is different, and is simply a variant of witchen, which goes back to OE cuic, alive.

WITHY, Blind *Polygonum persicaria* (Persicaria) Wilts. D Grose. A hybrid name - 'withy' because of the willow-like leaves (Cf. Willow-weed for this plant), and 'blind' to distinguish it from the Water Pepper (*P hydropiper*), which has also been called the Biting Persicaria, from the acrid taste, and Arsmart for obvious reasons. Gerard called Persicaria Dead Arsmart - "it doth not bite as the other doeth".

WITHY, Flowering *Chamenerion angustifolium* (Rosebay) Berks. Grigson. Once again, it is the resemblance of the leaves to those of willow that gave all these 'willow' names.

WITHY, Goose *Salix caprea* (Goat Willow) Hants. Grigson. Cf. Gosling-tree. These names are better understood in relation to the catkins, called descriptively Goslings, Goose-and-goslings, etc.

WITHY, Gold; WITHY, Golden *Myrica*

gale (Bog Myrtle) Hants, IOW Cope. There are a number of 'withy' or 'willow' names for Bog Myrtle, recognition again of the leaf resemblance.

WITHY, Hoar *Sorbus aria* (Whitebeam) Hants. Boase. This is 'withy' in its sense of 'witchen', as for rowan.

WITHY, Sally *Salix caprea* (Goat Willow) Wilts, Heref, Shrop. Grigson. Pleonastic, for they both mean 'willow', 'Sally' being OE sealh, willow.

WITHY, Sweet 1. *Myrica gale* (Bog Myrtle) IOW Grigson. 2. *Myrrhis odorata* (Sweet Cicely) IOW Grigson.

WITHY, Tame *Chamaenerion angustifolium* (Rosebay) IOW Grigson. Cf. Flowering Withy.

WITHY, Twig *Salix viminalis* (Osier) Ches. B & H. Acknowledgement that these are never allowed to grow into trees, the twigs being cut annually. They were often known simply as Twigs, and an osier bed was a twiggery.

WITHY PEAR *Sorbus domestica* (True Service Tree) Worcs. Gent. Mag.: PS.

WITHYBIND *Convolvulus arvensis* (Field Bindweed) Som. Macmillan. See Withywind.

WITHYHERB, Red *Epilobium hirsutum* (Great Willowherb) Lyte.

WITHYVINE 1. *Clematis vitalba* (Old Man's Beard) Wilts. Macmillan. 2. *Convolvulus arvensis* (Field Bindweed) Som. Macmillan. See Withywind.

WITHYWEED *Convolvulus arvensis* (Field Bindweed) Dev. Friend.1882; Som. Macmillan. See Withywind.

WITHYWIND 1. *Calystegia sepium* (Great Bindweed) Dev, Som, Wilts, Dor, Hants, Glos, Oxf, Berks, Bucks Grigson. 2. *Clematis vitalba* (Old Man's Beard) Som, Wilts, Glos. Grigson. 3. *Convolvulus arvensis* (Field Bindweed) Dev, Som, Wilts, Glos, Dor, Hants, Oxf, Berks. Rogers. 4. *Lonicera periclymenum* (Honeysuckle) Dev. Grigson. 5. *Myrica gale* (Bog Myrtle) Hants. Cope. Bog Myrtle is the odd one out in this list - the rest are climbing plants. The reason Bog Myrtle gets the name is its association with willows. Cf. Gold Withy or Sweet Withy, etc. A withe is a flexible twig, usually of willow, used for binding or tying, hence the string-like, or rope-like, characteristics of the climbing plants mentioned, particularly the bindweeds.

WITHYWINE; WITHYWING *Convolvulus arvensis* (Field Bindweed) Som. Jennings; Berks. Lowsley (Withywine); Dev. Friend.1882 (Withywing). See Withywind.

WITHYWINNY; WITHYWINY *Tamus communis* (Black Bryony) Dev. Macmillan. Variations on Withywind, for this is another climbing plant.

WITLOOF *Cichorium intybus* (Chicory) Schery.

WITTEN PEAR TREE *Sorbus domestica* (True Service Tree) Fernie. Cf. Whitty Pear. Both names are taken from the rowan name Witchen.

WITTLES, Chick *Stellaria media* (Chickweed) Suff. Mabey. i.e. chicken's food. Cf. Chicken's Meat, from the same area. Chickweed provides an iron rich tonic long given to cage birds.

WITTY *Sorbus aucuparia* (Rowan) N Eng. Denham/Hardy. It is more usual to find this with an 'h', as whitty, or something like it. All these names are variations of 'quicken', which is OE cuic, alive.

WIVELSFIELD, Sweet *Dianthus barbatus x allwoodii*.

WIZZARD *Agropyron repens* (Couch-grass) Grigson.1959.

WOAD 1. *Genista tinctoria* (Dyer's Greenweed) Yks. B & H. 2. *Isatis tinctoria*. 3. *Reseda luteola* (Dyer's Rocket) B & H. Woad is *Isatis tinctoria*. The other two get the name by association, for they are both dye plants. The word is OE wad, which accounts for the pronunciation change in versions like Wad, pronounced to rhyme with 'mad'.

WOAD, Dyer's *Isatis tinctoria* (Woad) Buhler.

WOAD, Wild 1. *Genista tinctoria* (Dyer's

Greenweed) Tynan & Maitland. 2. *Reseda luetola* (Dyer's Rocket) Prior.

WOADMESH *Genista tinctoria* (Dyer's Greenweed) Yks, N Eng. Grigson. This and the 'woad' names that follow probably arise from the fact that it was mixed with woad to give a green dye.

WOADWAX 1. *Genista anglica* (Needle Whin) Wilts. Grigson. 2. *Genista tinctoria* (Dyer's Greenweed) Som, Wilts, Dor. Macmillan.

WOADWAXEN; WOADWEX; WOADWISE *Genista tinctoria* (Dyer's Greenweed) Som. Macmillan; New Engl. Sanford (Woadwaxen) Dor. Grigson (Woadwex); Kirk, Wigt. Grigson (Woadwise).

WODE-WHISTLE *Conium maculatum* (Hemlock) B & H. This is a medieval book name, but small boys used to make whistles of the hollow stems right up to the end of the 19th century.

WOLD; WOULD *Reseda luteola* (Dyer's Rocket) B & H. A hybrid - obviously it is 'woad' x 'weld'.

WOLEWORT *Epilobium parviflorum* (Small-flowered Willowherb) Som. Macmillan.

WOLF, Red *Melandrium rubrum* (Red Campion) Som. Macmillan. Can there be a connection with another west country name, Red Riding Hood?

WOLF'S BANE 1. *Aconitum anglicum* (Wild Monkshood) Lyte. 2. *Aconitum napellus* (Monkshood) Krappe. 3. *Aconitum vulparia*. 4. *Eranthis hyemalis* (Winter Aconite) B & H.

WOLF'S BANE, Winter *Eranthis hyemalis* (Winter Aconite) Gerard.

WOLF'S COMB *Dipsacus fullonum* (Teasel) Blunt. Cf. Wolf's Teasel for this.

WOLF'S EYE *Lycopsis arvensis* (Small Bugloss) Som. Macmillan. The generic name, *Lycopsis*, is from Greek meaning wolf's face.

WOLF'S MILK 1. *Euphorbia cyparissias* (Cypress Spurge) Kourennoff. 2. *Euphorbia helioscopia* (Sun Spurge) Prior. 3. *Euphorbia paralias* (Sea Spurge) Gerard. All these have a juice fierce enough to raise blisters on the skin, so Wolf's Milk is a good enough name for them.

WOLF'S TEASEL *Dipsacus fullonum* (Teasel) Cockayne. Cf. Wolf's Comb.

WOLF WILLOW *Elaeagnus commutata* (Silverberry) Johnston.

WOLFBERRY, Arabian *Lycium arabicum*.

WOLFBERRY, Chinese *Lycium chinense* (Tea Tree) Douglas.

WOLFBERRY, Pale *Lycium pallidum* (Rabbit Thorn) Yanovsky.

WOLFBERRY, Red *Symphoricarpos orbiculatus*.

WOLFBERRY, Western *Symphoricarpos occidentalis*.

WOMAN-DRAKE *Bryonia dioica* (White Bryony) Lincs. Rudkin. It seems that *Tamus communis*, Black Bryony, was the mandrake in Lincolnshire. Woman-drake is a strange name anyway, for the 'man' of 'mandrake' has nothing to do with gender - the whole word is Latin mandragora, and before that, Greek. The confusion probably arose because of the mandrake root's anthropomorphism. Bryony is a local substitute for the real thing, which of course does not grow in Britain. Granted, bryony is dioecious, so the male and female are different plants, but that is hardly a satisfactory explanation of the name, as the differentiation is between plants of a different genus.

WOMAN'S NIGHTCAP *Oxalis acetosella* (Wood Sorrel) Som. Macmillan. Old Woman's Nightcap is more likely, but it is merely a descriptive name.

WOMAN'S TONGUE 1. *Albizia lebbeck* (Siris Tree) Everett. 2. *Briza media* (Quaking-grass) Berks. Grigson.1959. 3. *Hypericum androsaemum* (Tutsan) Fernie. 4. *Populus tremula* (Aspen) Berks. Lowsley. Aspen and Quaking-grass are obvious candidates for such a chauvinistic name - they are never still. Perhaps Siris Tree has a similar characteristic. But such a name for Tutsan remains mysterious.

WOMBAT-BERRY *Eustrephus latifolius* (Orange Vine) Australia Perry.

WOMEN, Men-and- *Arum maculatum* (Cuckoo-pint) Som. Macmillan. Better known as Lords-and-ladies, but there are actually a number of these male + female names, all overtly sexual and descriptive (spadix in the spathe).

WONDER, Golden *Cassia didymobotrya*.

WONDERBERRY *Solanum intrusum* (Garden Huckleberry) USA Kingsbury.

WONGA WONGA VINE *Pandorea pandorana*.

WOOD *Isatis tinctoria* (Woad) Dor, Wilts, Yks, Kirk, Wigt Grigson.

WOOD-ALONE *Adoxa moschatellina* (Moschatel) Dor. Macmillan.

WOOD-BRONEY *Fraxinus excelsior* (Ash) B & H.

WOOD-NUT *Corylus avellana* (Hazel) Yks. B & H.

WOOD-ROWE; WOOD-ROWELL *Asperula odorata* (Woodruff) both Gerard. The reference is to the rowel of a spur. As Gerard said, it "has its leaves set about like a star or the rowell of a spurre". Woodruff itself conveys the same idea.

WOOD'S WILLOW *Salix woodii*.

WOODAS *Genista tinctoria* (Dyer's Greenweed) W Eng. Grigson. Probably a variation of 'woad', which was often given in one form or another to this plant, for it was mixed with woad to give a green dye.

WOODBERRY *Veratrum album* (White Hellebore) Cockayne.

WOODBIND 1. *Hedera helix* (Ivy) Scot. B & H. 2. *Lonicera periclymenum* (Honeysuckle) Grigson.

WOODBIND, Bell *Calystegia sepium* (Great Bindweed) B & H.

WOODBINE 1. *Ampelopsis hederacea* (Virginia Creeper) O P Brown. 2. *Calystegia sepium* (Great Bindweed) Suss. Grigson; Kent Notes & Queries, vol 18, 3. *Clematis virginiana* USA House. 4. *Gelsenium sempervirens* (Carolina Jessamine) O P Brown. 5. *Lonicera periclymenum*. 6. *Solanum dulcamara* (Woody Nightshade) Palaiseul. Woodbine for honeysuckle is OE wudubind, wudu being a tree, and binden, to bind. The name is as common as honeysuckle for that plant.

WOODBINE, Cherry *Lonicera alpigena* (Alpine Honeysuckle) Polunin. Probably a reference to the size of the berries.

WOODCOCK ORCHID *Ophrys scolopax*.

WOODLAND MINT *Mentha longifolia* (Horse Mint) Genders.

WOODREEVE; WOODREP; WOODRIP; WOODROOF; WOODROSE;

WOODROW *Asperula odorata* (Woodruff) Wilts. Tanner (Woodreeve); Scot. Grigson (Woodrep,Woodrip); A S Palmer (Woodroof); Northcote (Woodrose); Barton & Castle (Woodrow).

WOODRUFF 1. *Asphodelus ramosus* (White Asphodel) Cockayne. 2. *Asperula odorata*. Woodruff is OE wudurofe, the meaning of the second element being unclear. It has been suggested that it may mean fragrant, but others (including Le Strange) think it may have evolved through French roue, wheel, to the present 'ruff', and so would be descriptive of the way the leaves grow in whorls round the stem. Gerard's Wood-rowell rather supports this view.

WOODRUFF, Blue *Asperula azurea*.

WOODRUFF, Blue-flowered *Asperula arvensis*.

WOODRUFF, Dyer's *Asperula tinctoria*. It is the roots that are used - they give a red dye.

WOODRUFF, Pink *Asperula taurina*.

WOODRUFF, Squinancy *Asperula cynanchica* (Squinancywort) Barton & Castle. 'Squinancy' is a corruption of 'quinsey'. The plant provides an astringent gargle that was used to treat the complaint.

WOODRUFF, Sweet *Asperula odorata* (Woodruff) Gilmour & Walters.

WOODRUSH, Arctic *Luzula arcuata*.

WOODRUSH, Fen *Luzula pallescens*.

WOODRUSH, Field *Luzula campestris*.

WOODRUSH, Great *Luzula sylvatica*.
WOODRUSH, Hairy *Luzula pilosa*.
WOODRUSH, Heath *Luzula multiflora*.
WOODRUSH, Narrow-leaved *Luzula forsteri*.
WOODRUSH, Spiked *Luzula spicata*.
WOODSORE *Berberis vulgaris* (Barberry) Oxf. Grigson. 'Sore' here is the same as 'sour' - see next entry.
WOODSOUR 1. *Berberis vulgaris* (Barberry) Oxf. Grigson. 2. *Oxalis acetosella* (Wood Sorrel) Gerard. 'Sour' has always been a synonym for 'sorrel', but that does not explain why barberry should get the name. Perhaps it is because of the malic acid the berries contain. Anyway, the plant is still called Sauerdorn in German.
WOODSOW; WOODSOWER *Oxalis acetosella* (Wood Sorrel) Pollock B & H.
WOODVINE *Lonicera periclymenum* (Honeysuckle) Glos. PLNN35. 'Vine' for 'bine' is an obvious variant, especially considering the plant's habit.
WOODWARD *Asperula odorata* (Woodruff) Schauenberg. An outlandish version of woodruff, probably stemming from Woodreeve, and confusing 'reeve' and 'ward' with manorial officials of earlier times.
WOODWAX 1. *Cytisus scoparius* (Broom) Dev. Macmillan. 2. *Genista anglica* (Needle Whin) Wilts. Dartnell & Goddard. 3. *Genista tinctoria* (Dyer's Greenweed) Dor. Macmillan; Wilts. Dartnell & Goddard. See Woadwax, rather - this is a dyeplant name. 'Woad' has become 'wood' in a number of cases.
WOODWAXA; WOODWAXEN *Genista tinctoria* (Dyer's Greenweed) Glos. Grigson (Woodwaxa); Gerard (Woodwaxen).
WOODWESH *Genista tinctoria* (Dyer's Greenweed) Yks. F K Robinson.
WOODWEX 1. *Genista anglica* (Needle Whin) Wilts. Grigson. 2. *Genista tinctoria* (Dyer's Greenweed) Dor. Macmillan. Cf. Woodwax.
WOODWIND *Lonicera periclymenum* (Honeysuckle) Glos, Shrop. Grigson.

WOODY-RUFFEE *Asperula odorata* (Woodruff) War, Yks. Grigson.
WOOL, Loki's *Eriophorum angustifolium* (Cotton-grass) Shet. Marwick. Cotton usually, but wool in this commemoration of the disruptive god of Scandinavian mythology.
WOOL THISTLE *Cirsium eriophorum* (Woolly Thistle) B & H.
WOOLD *Reseda luteola* (Dyer's Rocket) Prior. It seems to be a mixture of woad and weld, a common name for the plant.
WOOLLEN *Verbascum thapsus* (Mullein) Prior. The reference is to the leaves, whose texture has been likened to duffle, flannel and blankets, as well as to woollens generally.
WOOLLY BLUE VIOLET *Viola sororia*.
WOOLLY BUR-SAGE *Franseria eriocentra*.
WOOLLY BUTTERCUP *Ranunculus lanuginosus*.
WOOLLY CLARY *Salvia aethiopis*.
WOOLLY FOXGLOVE *Digitalis lanata*.
WOOLLY GROUNDSEL *Senecio longifolius*.
WOOLLY-HEADED CLOVER *Trifolium eriocephalum*.
WOOLLY-HEADED THISTLE *Cirsium eriophorum* (Woolly Thistle) Perring.
WOOLLY HEADS *Anemone nemorosa* (Wood Anemone) Som. Macmillan. An unusual plant to be described as 'woolly' - perhaps because they are white?
WOOLLY MINT, White *Mentha longifolia* (Horse Mint) Bardswell.
WOOLLY-PODDED BROOM *Cytisus grandiflorus*.
WOOLLY PYRUL *Phaseolus mungo* (Black Gram) Brouk.
WOOLLY THISTLE *Cirsium eriophorum*.
WOOLLY THORN *Crataegus mollis*.
WOOLLY YARROW 1. *Achillea lanulosa*. 2. *Achillea tomentosa* (Yellow Milfoil) Webster.
WOOTON LOCO *Astragalus wootonii*.
WOOZARD, Old Man's *Clematis vitalba* (Old Man's Beard) Bucks. Grigson.
WORLDWISE *Samolus valerandi* (Brookweed) S Africa Watt

WORM-GRASS *Sedum album* (White Stonecrop) Prior. Were any of the stonecrops used for worming? There is no record of any such practice.

WORMIT; WORMOD *Artemisia absinthium* (Wormwood) Dev. Grigson; N'thum. Brockett (Wormit); N'thum. Grigson (Wormod). Both of these are variations on the common name, as were Wermout and Warmot, which then became Ware-moth.

WORMSEED 1. *Artemisia maritima* (Sea Wormwood) Chopra. 2. *Erysimum cheiranthoides* (Treacle Mustard) Prior. 3. *Sisymbrium officinale* (Hedge Mustard) Barton & Castle. Treacle Mustard is certainly a vermifuge, though it can be quite a dangerous one. Sea Wormwood is also a known anthelmintic. Presumably Hedge Mustard was once used in the same way.

WORMSEED, American *Chenopodium ambrosioides*. A tropical American native, long used as a vermifuge by the South American Indians, and later introduced into the pharmacopeia particularly to deal with children's round-worms.

WORMSEED, Indian *Chenopodium ambrosioides* (American Wormseed) Dalziel.

WORMSEED, Jerusalem *Chenopodium botrys* (Jerusalem Oak) Watt.

WORMSEED, Levant 1. *Artemisia cina*. 2. *Artemisia maritima* Chopra.

WORMSEED, Treacle *Erysimum cheiranthoides* (Treacle Mustard) B & H.

WORMSEED MUSTARD *Erysimum cheiranthoides* (Trewacle Mustard) USA Zenkert.

WORMSEED WEED; WORMWEED *Chenopodium ambrosioides* (American Wormseed) Barbados Gooding (Wormseed Weed); Jamaica Beckwith; S Africa Watt (Wormweed).

WORMWOOD 1. *Artemisia absinthium*. 2. *Artemisia ludoviciana* (Lobed Cudweed) Calif. Merrill. 3. *Artemisia vulgaris* (Mugwort) Bucks. Grigson; USA. Barrett & Gifford. 4. *Chenopodium ambrosioides* (American Wormseed) Barbados. Gooding. 5. *Lapsana communis* (Nipplewort) Som. Macmillan. 6. *Parthenium hysterophorus* Barbados Gooding. A fanciful derivation of the name is that the plant sprang up in the track of the serpent as it writhed its way along the ground when it was driven out of Eden. Actually, the original word is Germanic, and it has never been worm + wood. It became weremod or wermod in OE, Wermut in German, and vermouth in French (wormwood is used in making the drink).

WORMWOOD, American *Ambrosia artemisiaefolia*.

WORMWOOD, Beach *Artemisia stelleriana*.

WORMWOOD, Camphor *Artemisia camphorata*. Camphor-scented, as the name implies.

WORMWOOD, Canadian *Artemisia canadensis*.

WORMWOOD, Cypress *Artemisia pontica* (Roman Wormwood) Gerard.

WORMWOOD, French 1. *Artemisia maritima* (Sea Wormwood) Turner. 2. *Artemisia pontica* (Roman Wormwood) Gerard.

WORMWOOD, Fringed *Artemisia frigida*.

WORMWOOD, Garden *Artemisia pontica* (Roman Wormwood) Gerard.

WORMWOOD, Louisiana *Artemisia ludoviciana* USA Yanovsky.

WORMWOOD, Pontic *Artemisia pontica* (Roman Wormwood) Turner.

WORMWOOD, Roman 1. *Ambrosia artemisiaefolia* (American Wormwood) Allan. 2. *Artemisia pontica*. *Artemisia pontica* is a native of southern Europe, but why should the *Ambrosia* get the name? It is an American plant, but the epithet is carried over to another of its names, Roman Ragweed.

WORMWOOD, Russian *Artemisia sacrorum*.

WORMWOOD, Scottish *Artemisia norvegica*. The specific name is definite enough, but it does grow in Scotland, too.

WORMWOOD, Sea 1. *Artemisia cina* (Levant Wormseed) Grieve.1931. 2. *Artemisia maritima*.

WORMWOOD, Shrubby *Artemisia arborescens*.
WORMWOOD, Silvery *Artemisia filifolia* (Sand Sagebrush) USA Elmore.
WORMWOOD, Sweet *Artemisia annua*. 'Sweet', because it lacks the bitterness of wormwood.
WORMWOOD, Three-lobed *Artemisia trifida* (Three-lobed Sagebrush) USA Elmore.
WORMWOOD, Tree *Artemisia arborescens* (Shrubby Wormwood) Painter.
WORMWOOD, White *Artemisia lactiflora*.
WORMWOOD, Wild *Artemisia afra* S Africa Watt.
WORMWOOD CASSIA *Cassia artemisoides* (Silver Cassia) RHS.
WORMWORT *Senecio fluviatilis* (Saracen's Woundwort) Salisbury. The implication of this name is that there is a usage for worms, but it is possible that it is a misreading of 'woundwort'.
WORRAL *Ballota nigra* (Black Horehound) Wilts. Dartnell & Goddard. Also Wirral and Wurral.
WORTS *Vaccinium myrtillus* (Whortleberry) Dev. Friend.1882; Som. Elworthy. Merely a version of Whorts, a common name throughout the south of England, itself a variant of Hurts.
WORTS, Red *Vaccinium vitis-idaea* (Cowberry) Gerard.
WOUDWIX *Genista tinctoria* (Dyer's Greenweed) Grieve.1931. See Woadwax, rather.
WOULD *Reseda luteola* (Dyer's Rocket) B & H. Cf. Wolds and Woolds, and presumably Weld.
WOUND ROCKET *Barbarea vulgaris* (Winter Cress) Turner. "For it is good for a wound", Turner said.
WOUND ROOT *Tamus communis* (Black Bryony) Pomet. "The root...apply'd fresh upon Contusions or Wounds, stops the Bleeding, and heals the Part; so that it has obtain'd the Name of the Wound-root" (Pomet). As for contusions, note the French name Herbe aux femmes battues.

WOUNDWEED; WOUNDWORT *Solidago virgaurea* (Golden Rod) Folkard (Woundweed); Som. Macmillan (Woundwort). Gerard said that Golden Rod is "extolled above all other herbes for the stopping of blood in bleeding wounds...".
WOUNDWORT 1. *Achillea millefolium* (Yarrow) Som. B & H. 2. *Anthyllis vulneraria* (Kidney Vetch) Grigson. 3. *Senecio fluviatilis* (Saracen's Woundwort) Turner. 4. *Solidago virgaurea* (Golden Rod) Som. Macmillan. All of these are wound herbs, Kidney Vetch proclaiming the fact by its specific name *vulneraria*.
WOUNDWORT, Alpine *Stachys alpina* (Limestone Woundwort) Polunin.
WOUNDWORT, Clown's *Stachys palustris* (Marsh Woundwort) Gerard. Cf. Clown's All-heal. 'Clown' here simply means countryman, and the herb is a countryman's remedy for what Gerard called "greene wounds".
WOUNDWORT, Corn *Stachys arvensis* (Field Woundwort) Murdoch McNeill.
WOUNDWORT, Downy *Stachys germanica*.
WOUNDWORT, Field *Stachys arvensis*.
WOUNDWORT, Hedge *Stachys sylvatica*.
WOUNDWORT, Hercules' *Prunella vulgaris* (Self-heal) O P Brown. The corolla is shaped like a billhook, and, from the doctrine of signatures, it was supposed to heal wounds from edged tools. But the reference to Hercules remains unexplained.
WOUNDWORT, Husbandman's 1. *Stachys palustris* (Marsh Woundwort) Gerard. 2. *Stachys sylvatica* (Hedge Woundwort) Fernie.
WOUNDWORT, Limestone *Stachys alpina*.
WOUNDWORT, Marsh *Stachys palustris*.
WOUNDWORT, Saracen's *Senecio fluviatilis*. 'Saracen' places the plant in the Near East, and particularly with the Crusaders, who brought it back to England. It was much esteemed here, and Gerard reckoned it "not inferiour to any of the wound-herbes whatsoever...".
WOUNDWORT, Soldier's *Achillea millefolium* (Yarrow) Friend USA Henkel. Yarrow is

a styptic, hence names like Sanguinary and Staunch-grass. As Gerard wrote, "this plant Achillea is thought to be the very same wherewith Achilles cured the wounds of his souldiers...". Whether it really was the same plant is doubtful, but it is still said that yarrow was always carried by the Greek and Roman armies, and this name is a witness to the legend.

WOUNDWORT, Yellow, Annual *Stachys annua*.

WOUNDWORT, Yellow, Perennial *Stachys recta*.

WRACK, Grass *Zostera marina* (Eel-grass) Hutchinson & Melville.

WRACK, Sea *Salsola kali* (Saltwort) Pomet.

WRAPPERS; WROPPERS *Digitalis purpurea* (Foxglove) Wilts. Dartnell & Goddard.1899. 'Wrapper' and 'Wropper' here are the same as 'popper'. Children have always amused themselves by "popping" the unopened buds of foxglove, and the noise they make is subject to local interpretation. Here it is 'rap' or 'rop', easily converted to 'wrap'.

WRAY *Lolium temulentum* (Darnel) Culpepper. Culpepper's spelling of Ray(-grass), which is derived from the French ivraie, drunk. That is a recognition of the toxicity of the grass, or rather of the ergot fungus that preys on it.

WREATH, Bridal 1. *Campanula pyramidalis* (Chimney Bellflower) Som. Macmillan. 2. *Spiraea arguta* RHS. The Spiraea may have been used as such, but there is no evidence that Chimney Bellflower was ever so used.

WREATH, Queen's *Petraea volubilis* (Purple Wreath Vine) W Indies Perry.

WREATH FLOWER *Vinca major* (Greater Periwinkle) Nash.

WREATH GOLDEN ROD *Solidago caesia* (Blue-stemmed Golden Rod) USA House.

WREATH VINE, Purple *Petraea volubilis*.

WREN; WREN-FLOWER; WREN, Jenny *Geranium robertianum* (Herb Robert) Som. Grigson (Wren); Som. Macmillan (Wrenflower, Jenny Wren). Herb Robert has been associated in the local names with a lot of birds - cranes, of course, for all the Geraniums are Cranesbills, and storks, as well as sparrows (Sparrow-birds is a name recorded in the west country), even Nightingales in Buckinghamshire, where they are also Cuckoo's Meat.

WREST-HARROW *Ononis repens* (Restharrow) Murdoch McNeill. It looks as if he misunderstood the meaning of "rest" in the common name. It is the same as "arrest", to stop, and certainly has no meaning as implied by the spelling of "wrest".

WRETWEED 1. *Chelidonium majus* (Greater Celandine) E Ang. Grigson. 2. *Euphorbia helioscopia* (Sun Spurge) Norf, Suff, Scot. Grigson. "Wartweed", of course, by a familiar process of transpositioning of consonants. Both of these have juice with a great reputation for curing warts. Cf. Wetweed.

WRIGHT LOTUS *Lotus wrightii*.

WRINKLED-LEAVED WILLOW *Salix reticulata* (Net-leaved Willow) WarrenWren.

WUD *Isatis tinctoria* (Woad) Kirk, Wigt. Hurry. A spelling that takes local pronunciation into account. Cf. Wadd, also from Scotland.

WUDWISE *Genista tinctoria* (Dyer's Greenweed) S Scot. B & H.

WUK *Quercus robur* (Oak) Som. Macmillan; Glos. B & H. The form in 'w' is typical of West country pronunciation.

WULLOW *Alnus glutinosa* (Alder) Shrop. B & H. Cf. Wallow. They are both fairly extreme forms of Aul, a single- syllable name for alder from roughly the same area.

WULLY-WUSS *Reseda luteola* (Dyer's Rocket) IOM Moore.1924.

WURRAL *Ballota nigra* (Black Horehound) Wilts. Dartnell & Goddard. See Wirral.

WUSHLEEN *Umbilicus rupestris* (Wall Pennywort) IOM Moore.1924. Manx woishleeyn.

WYCH ELM; WYCH, Elm *Ulmus glabra*. The aberrant form, Elm Wych, is recorded in

Northumberland, according to Grigson. 'Wych' may get transformed in popular usage into 'witch', but it bears no relation to the latter word. It is OE wice, and means pliant (switchy, in other words).

WYCH ELM, Japanese *Ulmus laciniata*.

WYCH HALSE *Ulmus glabra* (Wych Elm) Corn, Som. Grigson. 'Halse' is the same as hazel.

WYCH HAZEL 1. *Carpinus betula* (Hornbeam) Ess. Grigson. 2. *Ulmus glabra* (Wych Elm) Dev, Som, Wilts, Worcs, Ches. Grigson.

WYCH TREE *Ulmus glabra* (Wych Elm) Cumb. B & H.

WYCHEN *Sorbus aucuparia* (Rowan) Ches. B & H. Cf. Witchen, and a lot more deriving from OE cuic, alive.

WYCHWOOD *Ulmus glabra* (Wych Elm) Cumb. B & H.

WYMOTE; WIMOTE *Althaea officinalis* (Marsh Mallow) USA House (Wymote); Shrop. Grigson (Wimote).

WYTHAM, Moss *Myrica gale* (Bog Myrtle) Cumb. Grigson.

Y

YACKROD; YARKROD; YACKYAR
Senecio jacobaea (Ragwort) all Lincs. Peacock (Yackrod, Yarkrod); Grigson (Yackyar).

YAK *Quercus robur* (Oak) Yks. Nicholson. The intrusive 'y' is typical of north country pronunciation of oak, just as west country speech puts a 'w' in front, producing 'Wuk'. Yak may be pronounced that way, or as Yek or Yik, both of which have been recorded from Cumbria, but surely the spelling would be Ak or Aik?

YAKKRON *Quercus robur* acorn (Acorn) Yks. Morris. Two points here - as in the previous example, 'y' before the first vowel is typical north country pronunciation. And there is displacement of consonants, too, often found in dialect speech - 'acorn' becomes 'acron', or something like it.

YALLER *Senecio jacobaea* (Ragwort) Som. Macmillan. Colour descriptive, of course.

YALLOW *Achillea millefolium* (Yarrow) Lancs. B & H. One of a number of local variations of 'yarrow'.

YALLUC *Symphytum officinale* (Comfrey) Cockayne. It appears also as Galla and Galluc. Halliwell spells it Galloc.

YAM 1. *Dioscorea sp*. 2. *Ipomaea batatas* (Sweet Potato). Sweet Potato is often referred to in America, quite erroneously, as yam, which appears to be from a Portuguese original.

YAM, Aerial *Dioscorea bulbifera* (Potato Yam) Alexander & Coursey. Cf. Air Potato. The names are from the fact that aerial tubers are produced in this species in the leaf axils.

YAM, Bitter *Dioscorea dumetorum* (Cluster Yam) Alexander & Coursey.

YAM, Bush *Dioscorea praehensilis*.

YAM, Carib *Rajania cordata*.

YAM, Chinese 1. *Dioscorea batatas*. 2. *Dioscorea esculenta* (Lesser Yam) Alexander & Coursey. 3. *Dioscorea japonica*. 4. *Dioscorea opposita*.

YAM, Cluster *Dioscorea bulbifera*. For aerial tubers are produced in this species in the leaf axils (in a cluster).

YAM, Coco, Old *Colocasia antiquorum* (Taro) Schery. 'Old', for the specific name is *antiquorum*, though that can only mean 'of the ancients'.

YAM, Elephant *Amorphophallus campanulatus*.

YAM, Forest *Dioscorea praehensilis* (Bush Yam) Dalziel.

YAM, Greater *Dioscorea alata*.

YAM, Guinea, White *Dioscorea rotundata* (White Yam) Alexander & Coursey.

YAM, Guinea, Yellow *Dioscorea cayennensis* (Yellow Yam) Alexander & Coursey.

YAM, Lesser *Dioscorea esculenta*.

YAM, Negro *Dioscorea cayennensis* (Yellow Yam) Willis. 'Negro', because the roots are black.

YAM, Potato *Dioscorea bulbifera*.

YAM, Seaside *Ipomaea pes-caprae var. brasiliensis* Barbados Gooding.

YAM, Striped *Dioscorea vittata*. It is the decorative leaves that are striped.

YAM, Ten-months *Dioscorea alata* (Greater Yam) Dalziel.

YAM, Three-leaved *Dioscorea dumetorum* (Cluster Yam) Schery.

YAM, Water *Dioscorea alata* (Greater Yam) Alexander & Coursey.

YAM, White 1. *Dioscorea alata* (Greater Yam) Willis. 2. *Dioscorea praehensilis* (Bush Yam) Dalziel. 3. *Dioscorea rotundata*.

YAM, Winged *Dioscorea alata* (Greater Yam) Alexander & Coursey.

YAM, Yellow *Dioscorea cayennensis*.

YAM BEAN 1. *Pachyrrhizus erosus*. 2. *Pachyrrhizus tuberosus*. 3. *Sphenostylis stenocarpa*. *P tuberosus* is differentiated from *P erosus* - it has larger tubers.

YAM BEAN, Wayaka *Pachyrrhizus angulatus*.

YAMP; YAMPA *Carum gardneri*. Yampa is the usual name. Yamp is in Havard. It is an

American plant, particularly abundant apparently on the Little Snake, or Yampa, River.

YAMPI *Dioscorea trifida* (Cush-cush) Alexander & Coursey.

YAP-MOUTH *Antirrhinum majus* (Snapdragon) Som. Macmillan. More recognizable as Gap-mouth, one of the large number of descriptive "mouth" names given to the plant.

YARD, Shepherd's *Dipsacus fullonum* (Teasel) Dawson. Cf. Shepherd's Rod and Shepherd's Staff, both of which stem from the old name of the plant, virga pastoris.

YARD DAISY *Tanacetum parthenium* (Feverfew) Som. Macmillan. An acknowledgement of the fact that once one is introduced, it will always be with you.

YARD GRASS *Eleusine indica* (Goosegrass) USA Allan.

YARD-LONG BEAN *Vigna sesquipedalis* (Asparagus Cowpea) Bianchini. A yard is, if anything, a conservative estimate. The pods can easily be four feet in length.

YARN, Devil's *Clematis vitalba* (Old Man's Beard) Trevelyan. It is the Devil's Thread, too, and his twine, but perhaps the most expressive name of this class is Devil's Guts, also a west country appellation.

YARNUT *Conopodium majus* (Earth-nut) Lincs. Peacock; Yks. Carr. One of the many interpretations of local pronunciations of the word 'earthnut'. Cf. too Yernut, Yornut, etc.

YARR *Spergula arvensis* (Corn Spurrey) Lancs, Cumb, N'thum, Donegal Grigson; Scot. Jamieson.

YARRA-GRASS; YARREL *Achillea millefolium* (Yarrow) Ess. (Yarra-grass); Suff. (Yarrel) both Grigson.

YARROW *Achillea millefolium*. OE gearwe. It has been said that the word is really Greek hiera, meaning holy herb, but the dictionaries prefer a Germanic source. The modern German Garbe, and the Dutch gerwe, are close to the OE.

YARROW, Alpine *Artemisia rupestris*.

YARROW, Fernleaf *Achillea filipendula* (Yellow Yarrow) Painter.

YARROW, Golden *Eriophyllum caespitosum*.

YARROW, Greek *Achillea ageratifolia*.

YARROW, Soldier's *Stratiotes aloides* (Water Soldier) W Miller. Why is it called a yarrow? There is not much resemblance.

YARROW, Water 1. *Hottonia palustre* (Water Violet) Yks. Grigson. 2. *Myriophyllum spicatum* (Spiked Water Milfoil) Gerard.

YARROW, Western *Achillea lanulosa* (Woolly Yarrow) USA Gates.

YARROW, Woolly 1. *Achillea lanulosa*. 2. *Achillea tomentosa* (Yellow Milfoil) Webster.

YARROW, Yellow 1. *Achillea filipendula*. 2. *Achillea tomentosa* (Yellow Milfoil) Gerard.

YARROW BROOMRAPE *Orobanche purpurea*. As the name implies, this is parasitic on yarrows, though it is very rare in this country.

YARROWAY *Achillea millefolium* (Yarrow) Norf. B & H. One of a number of variations on 'yarrow'. There are also Yellow, Yarrel, Yarra-grass, Arrow-root among others.

YAUPON 1. *Ilex myrtifolia*. 2. *Ilex vomitoria* (Cassine) Schery. From an American Indian word.

YAVERING BELLS *Ramischia secunda*.

YAWL *Agropyron repens* (Couchgrass) Scilly Grigson.1959. This is Old Cornish dyawl, devil - an understandable reaction to any infestation.

YAWROOT *Stillingia treculeana* (Queen's Delight) O P Brown.

YEAKER *Quercus robur* acorn (Acorn) Hants. Cope. The original is 'acorn', and this is but one of a number of dialectal variations on the word. In this case, the nearest is the Dorset Eacor.

YEDDO SPRUCE *Picea jezoensis*.

YEDWARK *Papaver rhoeas* (Red Poppy) Derb. B & H. 'Headache', which has produced in local pronunciation Headwarke and Headwork. Poppies were used to cure headache, but the underlying folklore shows that they were believed to cause it. In Wiltshire they said that if you picked poppies from the corn, you would either have a bad headache or there would be thunder and lightning.

YEK *Quercus robur* (Oak) Cumb. B & H. Merely a local pronunciation of 'oak', which can become Yak or Yik as well. The intrusive 'y' in the north country becomes 'w' in the west country.

YELLOW 1. *Brassica campestris* (Field Mustard) Notts. B & H. 2. *Genista tinctoria* (Dyer's Greenweed) Midl. Grigson. 3. *Reseda luteola* (Dyer's Rocket) N'thants. Grigson. 4. *Sinapis arvensis* (Charlock) N'thum. Grigson. Yellow flowers, but also in the case of Dyer's Greenweed and Dyer's Rocket, a yellow dye plant.

YELLOW-BARK 1. *Cinchona calisaya* (Calisaya) Chopra. 2. *Phellodendron amurense* (Amur Cork Tree) Hyatt.

YELLOW-BLOSSOM *Ranunculus californica* (Californian Buttercup) USA Powers.

YELLOW-BOY *Senecio jacobaea* (Ragwort) Donegal Grigson.

YELLOW-EYED GRASS *Sisyrinchium californicum*.

YELLOW-FLOWER 1. *Brassica napus* (Rape) Ches. B & H. 2. *Sinapis arvensis* (Charlock) Ches. Holland.

YELLOW-GRASS *Narthecium ossifragum* (Yellow Asphodel) Shet. Freethy.

YELLOW-ROOT *Hydrastis canadensis* (Golden Seal) Mitton.

YELLOW-WEED 1. *Gutierrhezia microcephala* (Broomweed) USA Elmore. 2. *Monotes glaber*. 3. *Reseda luteola* (Dyer's Rocket) Gerard. 4. *Senecio jacobaea* (Ragwort) Hants. Hants FWI; Berw. Grigson. 5. *Sinapis arvensis* (Charlock) N Eng, Scot. Grigson. Either the colour of the flowers, or, as with Dyer's Rocket, to give a yellow dye colour.

YELLOW-WEED, Dyer's *Genista tinctoria* (Dyer's Greenweed) Grigson.

YELLOW-WOOD 1. *Berberis fremontii* (Desert Barberry) USA Jaeger. 2. *Cladrastis lutea*. 3. *Podocarpus gracilier* (Podo) N Davey. 4. *Xanthoxylum americanum* (Prickly Ash) Grieve.1931. 5. *Terminalia sericea* (Mangwe) Palgrave. Referring either to the colour of the timber, or to a usage as a dye, as is the case with Desert Barberry, which the Navajo used to get a yellow dye. Mangwe is used in Africa, or an extract of its bark is, in tanning skins, to give a yellowish colour to the product.

YELLOW-WOOD, African *Enantia chlorantha*.

YELLOW-WOOD, Chinese *Cladrastis chinensis*.

YELLOW-WOOD, East African *Podocarpus gracilier* (Podo) Dale & Greenway.

YELLOW-WOOD, Japanese *Cladastris platycarpa*.

YELLOW-WOOD, Outeniqua *Podocarpus falcata*. The tallest of South African forest trees.

YELLOW-WOOD, Real *Podocarpus latifolia* (Upright Yellow-wood) Leathart.

YELLOW-WOOD, Sharp-leaved *Podocarpus acutifolia*.

YELLOW-WOOD, Upright *Podocarpus latifolia*.

YELLOW-WORT *Blackstonia perfoliata*.

YELLOWBY *Chrysanthemum segetum* (Corn Marigold) W Miller. This is probably a shortened version of some such name as Yellow Bozzom, or Botham, both of which are recorded for Corn Marigold, as is Yellow Bottle.

YELLOWTOP 1. *Senecio jacobaea* (Ragwort) Fernie. 2. *Sinapis arvensis* (Charlock) N'thum. Grigson.

YERB-A-GRASS *Ruta graveolens* (Rue) B & H. This is the pronunciation of Herb-a-grass (Herb of Grace) in some parts of England.

YENNUT, YERNUT *Conopodium majus* (Earthnut) both Prior Versions of 'earth-nut' - Cf. Yarnut above.

YERROW *Achillea millefolium* (Yarrow) Lyte.

YES-OR-NO *Lolium perenne* (Rye-grass) Som. Grigson.1959. Children play divination games with this grass. They are on the basis of calling "He loves me, he loves me not" or some such form of words, reduced to essentials in this case, while pulling off each spikelet.

YES SMART 1. *Polygonum hydropiper* (Water Pepper) Wessex Rogers. 2. *Polygonum persicaria* (Persicaria) Wessex Rogers. This is only a local version of a common name, Arsmart, for Water Pepper. It gets the name from the extremely irritant effect of the leaves. Cockayne said "it derives its name from its use in that practical education of simple Cimons, which village jokers enjoy to impart". Persicaria looks like its vicious relative, but as Gerard wrote "it doth not bite as the other doeth", hence it is Dead, or Mild, Arsmart among other names.

YESTERDAY, TODAY AND TOMORROW *Brunfelsia calycina*. The blooms go from purple to mauve and to white as they fade, hence this name.

YETH *Calluna vulgaris* (Heather) Dev. Choape. Local pronunciation of 'heath'.

YETHNUT *Conopodium majus* (Earthnut) Ches. Holland. Cf. Yennut and Yernut above. They all mean "earthnut".

YEUGH *Taxus baccata* (Yew) Lowe. One of the many variations on 'yew'.

YEW *Taxus sp.* OE iw, eow, which, Freethy suggested, simply meant greenery. The word has appeared in print in many forms, ranging from Vew to Ugh, or even You.

YEW, American 1. *Taxus canadensis* (Canadian Yew) USA Tolstead. 2. *Taxus brevifolia* (Western Yew) Leathart.

YEW, Californian *Taxus brevifolia* (Western Yew) Dallimore.

YEW, Canadian *Taxus canadensis*.

YEW, Chinese *Taxus celebica*.

YEW, Dovaston *Taxus baccata var. dovastoniana* (Westfelton Yew) Dallimore. The variety was found by John Dovaston, of Westfelton, in Shropshire.

YEW, English *Taxus baccata*.

YEW, Florence Court *Taxus baccata var. fastigiata* (Irish Yew) Brimble.1948. The variety was first noticed growing on the mountains of co Fermanagh, near Florence Court.

YEW, He- *Juniperus sabina* (Savin) E Angl. G E Evans.

YEW, Himalayan *Taxus wallichiana*.

YEW, Irish *Taxus baccata var. fastigiata*. A chance variety, luckily noticed by a farmer named Willis in co Fermanagh in 1779.

YEW, Japanese *Taxus cuspidata*.

YEW, Mountain *Juniperus communis* (Juniper) Campbell. A translation from a Gaelic original.

YEW, Neidpath *Taxus baccata var. erecta* Lowe.

YEW, Pacific *Taxus brevifolia* (Western Yew) W Miller.

YEW, Plum-, Chinese *Cephalotaxus fortunei*. 'Plum-yew', from the plum-like appearance of the fruit.

YEW, Plum-, Japanese *Cephalotaxus drupacea*.

YEW, Plum-fruited *Podocarpus andinus*. The fruit looks like a damson.

YEW, Prince Albert's *Saxegothaea conspicua*.

YEW, Shooter *Taxus baccata* (Yew) Dallimore. Cf. Shoter. The reference must be to the use of yew for bow-making.

YEW, Western *Taxus brevifolia*.

YEW, Westfelton *Taxus baccata var. dovastoniana*. A variety found by John Dovaston, of Westfelton, in Shropshire, in 1771.

YEW-BRIMMLE *Rosa canina* (Dog Rose) Som. Macmillan. Cf. Yoe-brimmle. They mean 'ewe bramble'.

YEW-LEAF WILLOW *Salix taxifolia*.

YIK *Quercus robur* (Oak) Cumb. B & H. Another example of a stab at the local pronunciation of oak.

YIRNIN-GIRSE *Pinguicula vulgaris* (Butterwort). Shet. B & H. Further south, this is 'earning-grass'. Earning is a North-country word for cheese rennet, and to earn is to curdle milk.

YLANG-YLANG *Canangium odoratum*. The name comes from alang-ilang, the name of the tree and of the perfume obtained by distillation of the flowers, in Tagalog, the Filipino national language.

YOE-BRIMMLE 1. *Rosa canina* (Dog Rose) Dev, Som. Grigson. 2. *Rubus fruticosus* (Bramble) Som. Elworthy. Cf. Yew-brimmle, above, for Dog Rose. It is probably He-brimmle, the East Anglian equivalent of which is Cock Bramble, in one form or another, but it could appear as Ewe-bramble as well. Halliwell defined He-brimmle as a bramble or more than one year's growth.

YOKE ELM *Carpinus betula* (Hornbeam) Gerard. Obviously, yokes were once made of the wood.

YOLK-OF-EGG WILLOW *Salix alba. var. vitellina* (Golden Willow) Warren-Wren. The reference here is to the colour of the twigs in winter.

YORK AND LANCASTER ROSE *Rosa damascena 'versicolor'*. The red and white colours of the flowers symbolize the reconciliation of the houses of York and Lancaster after the Wars of the Roses.

YORKSHIRE FOG *Holcus lanatus*. The variable colour of the panicles, blending so well in pastures, has given rise to the name 'fog'. When there is a lot of this grass in the fields, they are referred to as being fogged, or foggy.

YORKSHIRE MILKWORT *Polygala amara*.

YORKSHIRE SANICLE *Pinguicula vulgaris* (Butterwort) Gerard. 'Sanicle' means a healing herb.

YORLIN, Yowie; YORNUT; YORNUT, Yowe *Conopodium majus* (Earthnut) Prior (Yornut); Cumb. Grigson (Yowe Yorlin, Yowe Yornut). Cf. Jacky Jurnals, or Jocky Jurnals, where Jurnals, or in this case Yorlins or Yornuts are the same word as earth-nut.

YORUBA EBONY *Diospyros monbuttensis*. A West African species, much used in Nigeria.

YORUBA WILD INDIGO *Lonchocarpus cyanescens* (Indigo Vine) Picton & Mack. An indican-bearing plant, used in Nigeria for a blue dye.

YOSHINO CHERRY *Prunus x yedoensis*.

YOU; YOUE; YUGH *Taxus baccata* (Yew) all Lowe.

YOUNG FUSTIC *Cotinus coggyria* (Venetian Sumach) Watt. The wood is the source of a yellow dye best known as young fustic. 'Young', to distinguish it from 'old' fustic - *Chlorophora tinctoria*.

YOUNG MAN'S DEATH *Convolvulus arvensis* (Field Bindweed) Perth Vickery.1985. Bindweed is an unlucky plant. In this case, the result of picking the flowers, if a girl did it, would be the death of her boyfriend. Perhaps the superstition alluded to the fact that the flowers fade so rapidly.

YOUNGBERRY *Rubus mirus*.

YOUTH-AND-OLD-AGE *Zinnia elegans* A W Smith.

YOUTH-ON-AGE *Tolmiea menziesii* (Pick-a-back Plant) RHS. Consideration of the common name, Pick-a-back Plant, will be enough to explain this name - the new shoots are carried on the old.

YOUTHWORT *Drosera rotundifolia* (Sundew) Grigson. Cf. Lustwort. Both are medieval book names, and probably stem from the plant's supposed aphrodisiac powers. It was even said to affect cows, whose sexual instincts were reckoned to be aroused by eating even a small quantity of the plant.

YOWE-YORNUT; YOWIE-YORLIN *Conopodium majus* (Earthnut) both Cumb. Grigson. From 'earthnut' originally, through some form like Jur-nut, which is recorded elsewhere. The 'j' of this last name may have been pronounced as 'y', hence Jocky Jurnals into Yowie-yorlins.

YUCATAN PLUM *Spondias lutea*.

YUCCA, Aloe *Yucca aloifolia*.

YUCCA, Blue *Yucca baccata* (Datil) USA T H Kearney.

YUCCA, Broad-leaved *Yucca baccata* (Datil) Whiting.

YUCCA, Mohave *Yucca schidigera*.

YUCCA, Narrow-leaved *Yucca glauca* (Soapweed) Whiting.
YUCCA, Tree *Yucca brevifolia* (Joshua Tree) Jaeger.
YULAN *Magnolia denudata*.
YULE, White *Bellis perennis* (Daisy) Som. Grigson. Cf. White Frills.
YULE GIRSE *Filipendula ulmaria* (Meadowsweet) Shet. Nicolson.
YUNNAN CYPRESS Cupressus duclouxiana.
YUNNAN HEMLOCK Tsuga yunnanensis.

Z

ZAMBAK *Jasminum sambac* (Arabian Jasmine) Grieve.1931.

ZAMBESI COFFEE *Bauhinia petersiana* (Orchid Bauhinia) Palgrave. The plant was known to all early hunters and explorers by this name. The seeds were ground to powder and used as a coffee substitute.

ZAMBESI REDWOOD *Baikaea plurijuga* (Rhodesian Teak) Palgrave.

ZANTE FUSTIC; ZANTE WOOD *Cotinus coggyria* (Venetian Sumach) Hutchinson & Melville (Fustic); W Miller (Wood). The use of 'fustic' shows this to be a dye plant, and the colour produced is yellow, although, according to the mordant used, it could give various colours - green, orange to brown or dark olive, and even black. Zante is one of the Ionian islands.

ZANZIBAR ALOE *Aloe vera* (Mediterranean Aloe) Mitton. The Mediterranean area is its region of origin, but evidently it is a far-flung plant, Zanzibar being only one of its habitats, according to the names given to it.

ZANZIBAR EBONY *Diospyros mespiliformis* timber (African Ebony) Dalziel.

ZEBRA PLANT 1. *Aphelandra squarrosa*. 2. *Calathea zebrina*. Presumably because the leaves are striped - certainly so with *Calathea zebrina*.

ZEBRA WOOD *Connarus guianensis*. The timber is a light yellowish-brown in colour, with dark vertical lines, hence the name.

ZEDOARY *Curcuma zedoaria*. The name came to us through medieval Latin, but it is originally an Arabic word, zedwar.

ZEDOARY, Black *Curcuma casia*.

ZEDOARY, Round *Curcuma zedoaria* (Zedoary) Lindley.

ZENRY *Sinapis arvensis* (Charlock) W Eng. Freethy. See the next entry, rather.

ZENVY *Sinapis arvensis* (Charlock) Som. Jennings. Cf. Senvie, or Sinvey, also from Somerset.

ZEODARY *Curcuma zedoaria* (Zedoary) Lupton. Is this genuine? Or just a misprint in Lupton?

ZEPHYR-FLOWER *Anemone nemorosa* (Wood Anemone) Som. Macmillan. Surely not a vernacular name. Anemone comes from the Greek anemos, the wind, and Wind-Flower used to be quite a common book name for the plant.

ZEPHYR LILY *Zephyranthes candida*.

ZIGZAG 1. *Acer campestre* (Field Maple) Som. Macmillan. 2. *Acer pseudo-platanus* (Sycamore) Som. Macmillan. A reference to the way the winged 'keys' fall to the ground.

ZIGZAG CLOVER *Trifolium medium*. 'Zigzag' from the way the stems grow.

ZIGZAG GOLDEN ROD *Solidago flexicaulis*. '*Flexicaulis*' in the specific name is enough to explain 'zigzag'.

ZINEGAR; SINNEGAR *Matthiola incana* (Stock) Som. Macmillan. Odd names, probably owing their existence to some local pronunciation of Jilloffer, which is Gilliflower.

ZINNIA, Creeping *Sanvitalia procumbens*.